Geological Evolution of Antarctica

Symposium sponsors

Scientific Committee on Antarctic Research
International Union of Geodesy and Geophysics
International Union of Geological Sciences
The Royal Society of London
British Antarctic Survey
Scott Polar Research Institute
Geological Society of London

SCAR Steering Committee

J.D. Bradshaw (New Zealand)
H. Charnock (IUGG)
D.H. Elliot (USA)
J.B. Jago (Australia)
Y. Kristoffersen (Norway)
J. Lameyre (France)
A.C. Rocha-Campos (Brazil)
M.R.A. Thomson (UK)
F. Thyssen (FRG)

Local Organizing Committee

M.R.A. Thomson (Chairman)
R.J. Adie
P.F. Barker
D.J. Drewry
A.G. Smith
C.W.M. Swithinbank

Supporting Organizations

Donations from the following are gratefully acknowledged:
The Trans-Antarctic Association
The Royal Society of London
Robertson Research International Ltd
Shell International Petroleum Company Ltd
British Petroleum Company plc
Enterprise Oil plc
North Sea Sun Oil Company Ltd
Occidental International Oil Inc

Geological Evolution of Antarctica

EDITED BY

M.R.A. THOMSON, J.A. CRAME & J.W. THOMSON

British Antarctic Survey,
Natural Environment Research Council, Cambridge, England

Proceedings of the Fifth International Symposium on Antarctic Earth Sciences, held at Robinson College, Cambridge, 23–28 August 1987

CAMBRIDGE UNIVERSITY PRESS

Cambridge

New York   Port Chester   Melbourne   Sydney

CAMBRIDGE UNIVERSITY PRESS
Cambridge, New York, Melbourne, Madrid, Cape Town,
Singapore, São Paulo, Delhi, Tokyo, Mexico City

Cambridge University Press
The Edinburgh Building, Cambridge CB2 8RU, UK

Published in the United States of America by Cambridge University Press, New York

www.cambridge.org
Information on this title: www.cambridge.org/9780521188906

First published 1991
First paperback edition 2011

A catalogue record for this publication is available from the British Library

Library of Congress Cataloguing in Publication data

International Symposium on Antarctic Earth Sciences
(5th : 1987 : Cambridge, England)
Geological evolution of Antarctica/edited by M. R. A. Thomson,
J. A. Crame & J. W. Thomson.
p. cm.
'Proceedings of the Fifth International Symposium on Antarctic
Earth Sciences, held at Robinson College, Cambridge,
23–28 August 1987.'
ISBN 0 521 37266 6
1. Geology – Antarctic regions – Congresses.
I. Thomson, M. R. A. (Michael Robert Alexander)
II. Crame, J. A. (J. Alistair)
III. Thomson, Janet W. IV. Title.
QE350.I54 1991
559.8′9 – dc20 89-22314 CIP

ISBN 978-0-521-37266-4 Hardback
ISBN 978-0-521-18890-6 Paperback

Additional resources for this publication at www.cambridge.org/9780521188906

Contents

Crustal development: Weddell Sea–Ross Sea region

Crustal development: the Pacific margin

Contents

Crustal development: Gondwana break-up

Evolution of Cenozoic palaeoenvironments

Preface

Scientific research in Antarctica has seen many changes and developments since the discovery of the Antarctic Peninsula in 1820 but perhaps none so rapid as those experienced in the last ten years. With the exception of a few areas where access is extremely difficult, all the exposed rock on the continent has been mapped geologically, at least at a reconnaissance level, and an increasing amount of marine geophysical work is building up a broad general picture of the geology of the Antarctic continental shelf and its surrounding oceans. Thus, although many scientific discoveries have yet to be made in Antarctica and much greater effort is needed to investigate the 98% of rock beneath the ice, a good overall picture of the continent's exposed geology now exists.

Not many years ago a good geological description of a hitherto unexplored area was sufficient to justify publication. In recent years, however, Antarctic geologists have recognized a need to place greater emphasis on interpreting their findings within a broader framework and they have been using their results to help test and constrain such important global theories as that of plate tectonics. Furthermore, the importance of Antarctica in providing a primary input into the understanding of some of the wider geological issues of our planet, e.g. the significance of the region as a centre of biological evolution in Mesozoic and Cenozoic times, and of its Cenozoic glacial record stretching back perhaps more than 30 million years, are becoming more apparent.

The number of nations actively engaged in the scientific study of Antarctica has been increasing steadily and by July 1985, when the Fifth International Symposium on Antarctic Earth Sciences was first advertized, 32 nations had signed the Antarctic Treaty. When the symposium was held in August 1987 that figure stood at 36 and, at the time of writing, it has risen further to 38.

Against such a background, the British National Committee on Antarctic Research felt that, if Britain were to host the next Antarctic Earth Sciences Symposium, then the scientific objectives of the meeting had to be more focused than they had been in the past. Such a decision did not meet entirely with international approval. A number of scientists felt that one of the primary roles of Antarctic symposia is to bring scientists of many nations together, and focusing a meeting too closely could have the effect of excluding some nations, purely by some geographical accident of their area of operation. Furthermore, it would be difficult for those nations new to Antarctica and who had only just begun their programmes, to participate in and contribute to the meeting. There was therefore a problem of finding a focus which would have some global relevance and yet would enable the widest possible participation by Antarctic earth scientists. In the event the answer was provided by the 19th Meeting of the Scientific Committee on Antarctic Research (SCAR) in San Diego, July 1986, at which the Working Groups on Geology and Solid Earth Geophysics recognized the need for research initiatives in the 'Structure and Evolution of the Antarctic Lithosphere', and in the 'Evolution of Cenozoic Palaeoenvironments of the Southern High Latitudes', and convened two Groups of Specialists to investigate these problems further.

The Fifth International Symposium on Antarctic Earth Sciences now had two major foci, albeit under the slightly different titles of 'Tectonic Evolution of the Antarctic Crust' and 'Palaeoenvironmental Evolution of Antarctica since the Late Mesozoic'. How successful the meeting was in addressing these problems must be left for the reader to judge, but a glance at the list of contents will show that evidence from a wide range of areas and rock units was brought to bear. Inevitably there were some significant gaps, none more noticeable than the dearth of papers on the Ellsworth Mountains and Marie Byrd Land, but to some extent these may be taken as pointers to those areas where further research is badly needed. In addition, the small number of papers offered on the geophysical investigation of Antarctica's continental shelves was disappointing, particularly since these are areas which have come under increasing scrutiny by several nations in recent years.

On the day, the symposium attracted nearly 200 participants from 19 countries. During the course of the five-day meeting, 138 papers were presented orally and a particularly gratifying aspect was a well-prepared and well-attended poster session with a further 34 contributions. It provided authors with the opportunity for a longer display of their work than 20 minutes on the rostrum and helped to keep the number of parallel sessions to the absolute minimum. By the nature of their work, Antarctic geologists have somewhat catholic interests and too many parallel sessions can be particularly irksome. Nevertheless, it is an inevitable consequence of Antarctica's growing scientific community and the rate at which new scientific information is being generated that parallel sessions, often more than two, are here to stay. However, the need to hold meetings more frequently than every five years or so is also clearly indicated.

As with many earth science meetings, field excursions have always been an integral part of Antarctic symposia. Two excursions, each demonstrating a classic area of British geology, were run to great acclaim from their participants. These covered the crustal evolution of north-west Scotland and the evolution of the Wessex Basin. The north–south divide was much in evidence with the southerners enjoying their geology against a backdrop of cream teas, fine weather and rural scenery, and the northerners claiming that studying the Moine Thrust in torrential rain made it all the more meaningful.

Notes

1. Place names. The use of a place name in this volume does not necessarily imply its acceptance by the editors or that it has received formal approval by an Antarctic place-names authority.

2. Stratigraphical terminology. Geochronological and chronostratigraphical terminology follows that in Owen, D. E. (1987). Commentary: usage of stratigraphic terminology in papers, illustrations, and talks. *Journal of Sedimentary Petrology*, **57(2)**, 363–72, except that formal subdivision of the Palaeozoic into two units has also been allowed.

Acknowledgements

Symposia do not just happen. They are the result of a great deal of hard work by large numbers of people, many of whom may not be particularly conspicuous during the proceedings. It is therefore my pleasure to record their names here and to express my gratitude to them. Without them the Fifth International Symposium on Antarctic Earth Sciences would never have been possible.

First there was the Local Organizing Committee which, together with George Hemmen (Executive Secretary SCAR), guided the meeting through its initial stages and gave me much needed support during the difficult early months. With an outline for the symposium established it was then the task of a small but dedicated band of helpers to make the whole thing happen. My special thanks go to Gill Clarke (cheerful ace typist, compiler of endless lists and minder) and to Peter Clarkson (treasurer) who, with Alistair Crame and Janet Thomson, helped me to knock the whole programme into shape. The excellent field excursions were ably organized and led by David Macdonald and Jane Francis (south coast, England) and by John Smellie, Bryan Storey, Philip Nell and John Mendum (north-west Scotland). Candy Smellie made sure that accompanying persons were not left out. During the symposium Gill Clarke, Peter Clarkson and Hanni Willan were always on hand to 'man' the desk and to help anyone in need. Janet Thomson organized the poster displays and many BAS geologists beavered away behind the scenes to ensure that participants arrived safely from the station and that all the rooms were set up and ready for use: Peter Butterworth, Peter Doyle, Steve Harrison, Malcolm Hole, Philip Howlett, Philip Nell, Phil Marsh, Duncan Pirrie, Tim Tranter and Andy Whitham. George and Vi Wood actually enjoyed making up all the document cases, or so they said. Dick Laws and David Drewry (previous and present Directors of the British Antarctic Survey) kindly put at my disposal all the office facilities so necessary for organizing an international meeting of this kind.

Those who provided for our creature comforts are also not forgotten and, on behalf of all symposium participants, I wish to extend my thanks to the Warden and staff of Robinson College for the use of accommodation and facilities, to the Scott Polar Research Institute for a very pleasant evening, and to the Master of Darwin College for allowing us use of the grounds of the college for the punting party.

Finally, there is this book, a truly mammoth task for the editors. There is no doubt that it would never have come to fruition if I had not had the assistance and support of Alistair Crame and my wife Janet who really did all the work. As senior editor, I take any criticism that may be due; any credit belongs to Alistair and, especially, Janet.

Michael Thomson

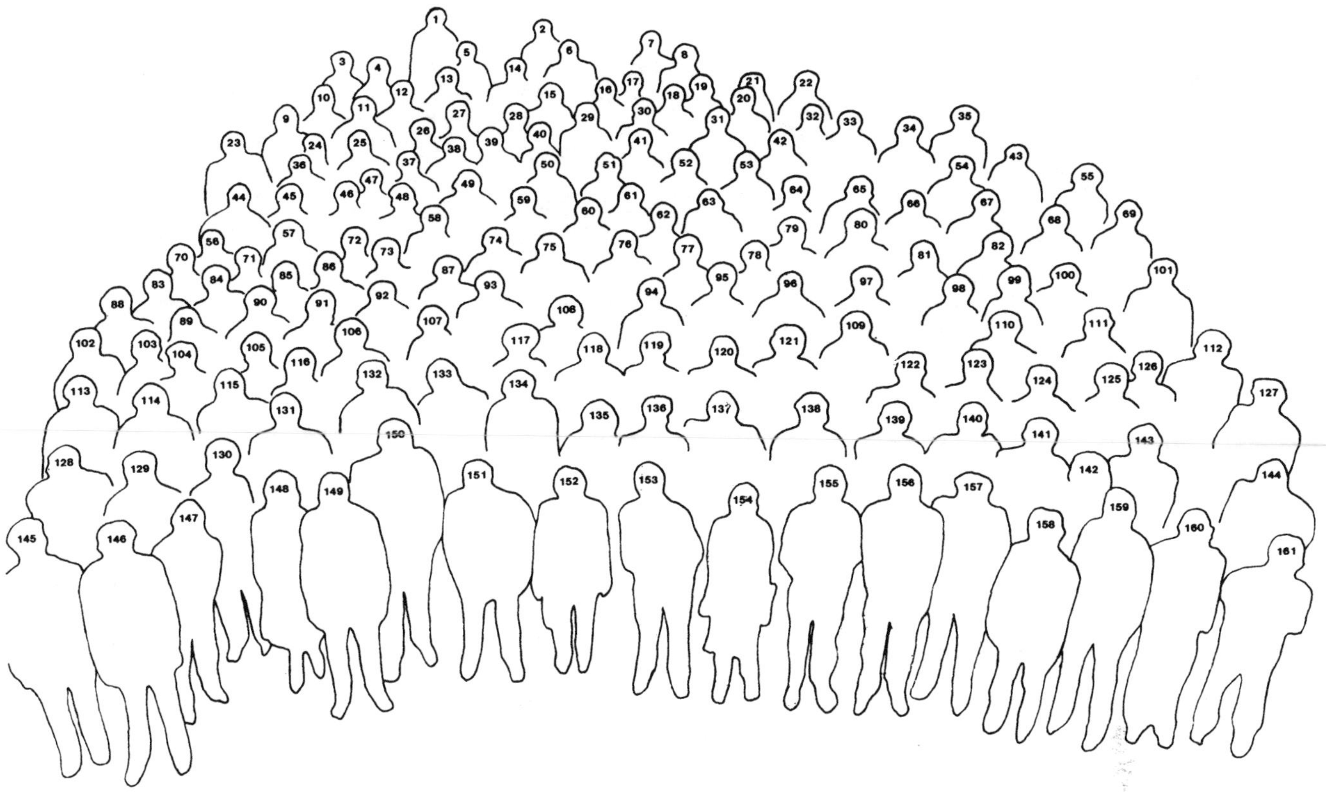

5th International symposium on Antarctic Earth Sciences, Cambridge, UK.
Photograph of participants, 28 August 1987

1 S.L. Klemperer
2 J.L. Smellie
3 E.C. King
4 B.C. McKelvey
5 P.R. Kyle
6 K. Rankama
7 —
8 T.S. Laudon
9 D.D. Blankenship
10 M.F. Osmaston
11 J.W.A. van Enst
12 J.W. Goodge
13 R.J. Pankhurst
14 D.K. Futterer
15 G. Kleinschmidt
16 Hubert Miller
17 J.W. Thomson
18 J.A. Jones
19 M.P. Maslanyj
20 S.W. Garrett
21 C.F.H. Willan
22 A.J. Milne
23 J.B. Anderson
24 W.R. Vennum
25 S. Olaussen
26 I.B. Evans
27 J.L. LaBrecque
28 K. Weber
29 Heinz Miller
30 W. Buggisch
31 A.G. Whitham
32 D. Damaske
33 I.O. Norton
34 J.P. Henriet
35 Liu Xiaohan
36 G. Bylund
37 I.J. Haapala
38 B.A. Piercy
39 H.E. Wever
40 K.J. McGibbon
41 Li Zhaonai
42 M. Richter
43 A. Gaździcki
44 P. Fitzgerald
45 R.D. Larter
46 L.A. Lawver
47 M.J. Hambrey
48 I.W. Hamilton
49 W.J. Zinsmeister
50 N.W. Roland
51 Y. Hiroi
52 T.H. Tranter
53 L.D.B. Herrod
54 R. Förster
55 J.A. Crame
56 M.N. Rees
57 S.L. Harley
58 A. Conway
59 E.B. Olivero
60 E.A. Colhoun
61 H. Zimmerman
62 D.R.H. O'Connell
63 A.L. Inderbitzen
64 S.M. Harrison
65 J. Hofman
66 K. Hahne
67 H.-J. Paech
68 M.J. Ringe
69 T. Horner
70 S.T. Rooney
71 G. Clarke
72 M.J. Lonsdale
73 J.R. Krynauw
74 C.A. Rinaldi
75 H. Decleir
76 B.K. Kamenov
77 P.A.R. Nell
78 S.E. Ishman
79 T.A. Stern
80 S.D. Weaver
81 C.J.D. Adams
82 T.S. Brewer
83 A.M. Grunow
84 J.D. Jeffers
85 C. Pudsey
86 P.F. Barker
87 B.R. Watters
88 I.W.D. Dalziel
89 C. Raymond
90 K.S. Kellogg
91 A. Hungeling
92 F. Thyssen
93 G.F. Webers
94 D.I.M. Macdonald
95 B.T. Huber
96 S.W. Wise
97 P.D. Marsh
98 F.M. McGibbon
99 M.A. Sykes
100 D. Pirrie
101 D.R. Hunter
102 P.D. Rowley
103 E.W. Domack
104 A.J. Rowell
105 J.C. Parra
106 D.N.B. Skinner
107 B.K. Lucchitta
108 A.-M. Moons
109 C. Harris
110 P. Ding
111 M.K. Wells
112 P.G. Quilty
113 D.J. Ellis
114 F. Hervé
115 R.A.J. Trouw
116 G. Wörner
117 J.M.G. Miller
118 J.W. Collinson
119 D.H. Elliot
120 D.J. Drewry
121 B. Corner
122 P. Doyle
123 P.J. Howlett
124 T. Gjelsvik
125 W. Fielitz
126 Y. Kristoffersen
127 A. Guterch
128 M.K. Kaul
129 O. Rösler
130 L. van der Maesen
131 A.K. Cooper
132 A.B. Ford
133 P.J. Barrett
134 P.-N. Webb
135 R.S. Frisch
136 M. Manzoni
137 B.A. Lombardo
138 E. Stump
139 W.E. LeMasurier
140 M.J. Hole
141 K. Birkenmajer
142 M.C. Delitala
143 Y. Yoshida
144 J.H. Berg
145 M.G. Laird
146 P.D. Clarkson
147 J.E. Francis
148 G.M. Clarke
149 M.R.A. Thomson
150 J.C. Behrendt
151 B.F. Molnia
152 E. Barrera
153 R.J. Tingey
154 S. Sengupta
155 D.N. Young
156 F.J. Davey
157 O. González-Ferrán
158 K. Kaminuma
159 S.M. Kuehner
160 G.H. Grantham
161 K. Shiraishi

Crustal development: the craton

Uplift history of the East Antarctic shield: constraints imposed by high-pressure experimental studies of Proterozoic mafic dykes

S.M. KUEHNER[1,2] & D.H. GREEN[1]

1 Geology Department, GPO Box 252C, University of Tasmania, Hobart, 7001 Australia
2 Department of Geophysical Sciences, 5743 South Ellis Avenue, The University of Chicago, Chicago, Illinois 60637, USA

Abstract

The depths at which two suites of Proterozoic mafic dykes were emplaced within the Vestfold Hills were determined by using high-pressure experimental studies to reproduce the equilibrium phenocryst assemblage of their chilled margins. The results of these studies were then applied to chemically similar dykes exposed in the Napier Complex. As the dyke suites were emplaced during the ~ 1400 m.y. interval separating major metamorphic events at ~ 2500 Ma and ~ 1100 Ma, determination of their emplacement depth constrains possible uplift interpretations of the East Antarctic shield. This study indicates that, following the peak metamorphic event at ~ 3100 Ma, the Napier Complex and the Vestfold Hills experienced ~ 2000 m.y. of crustal stability. Both terranes followed identical isobaric cooling paths to ~ 2500 Ma, after which the Napier Complex remained at deep crustal levels (28 km) until ~ 1000 Ma, while the Vestfold Hills was exhumed at a net rate of 1 cm/1000 y until at least ~ 1360 Ma. This long period of crustal stability was terminated by a Himalayan-style tectonic event which resulted in isothermal decompression at minimum net rate of 1 cm/1000 y in the Napier Complex, while crustal loading depressed the Vestfold Hills from depths of ~ 16–22 km. In the absence of more recent deformational events, it is assumed that these terranes have experienced slow, erosion-controlled uplift since ~ 1000 Ma.

Introduction

The conventional method for determining the pressure–temperature–time (*P–T–t*) path that a polymetamorphic terrane has followed is to evaluate the conditions under which metamorphic assemblages formed by applying experimentally based phase equilibria and geothermometric–barometric techniques. The application of these techniques towards resolving the *P–T–t* path relies almost entirely on deformation producing penetrative fabrics throughout the terrane that resulted in the development of successive, though not pervasive, stages of recrystallization/re-equilibration within the phase assemblages. When deformational events are separated by long periods of time, the *P–T–t* path over this period cannot be determined directly, but can only be inferred from the phase assemblage of the younger event. In the East Antarctic shield, well characterized tectonothermal events occurred at ~ 2500 Ma and ~ 1100 Ma. During the intervening tectonically quiescent 1400 m.y. several suites of chemically distinct mafic dykes of known age were emplaced throughout the shield. Thus, if the load pressure at the time of dyke emplacement could be established, the information would further characterize the uplift trajectory (*P–t*) of the shield, and constitute a deformation-independent method of estimating *P–t* paths.

We report here the results of high-pressure experimental studies in which equilibrium phenocryst–liquid assemblages in the chilled margins of selected mafic dykes were reproduced, thus establishing the depth of dyke solidification. As the studied material was collected from the Vestfold Hills, the experimental results can be applied directly to the uplift history of this region. The results can also be extended to chemically similar dykes cropping out in the Napier Complex (see Kuehner, 1986).

Geological background

The geology of the East Antarctic shield has been studied in some detail. In particular, combined structural, geochemical and petrological studies of the Archaean Napier and Proterozoic Rayner Complexes of Enderby Land (Fig. 1*A*; see reviews by James & Tingey (1983); Harley & Black (1987)), have yielded evidence for an extensive history of repeated ductile deformation in conjunction with unusually high metamorphic temperatures. The earliest recognized metamorphic episode in the Napier Complex is dated at 3072^{+35}_{-33} Ma (James & Black, 1981). This was a pervasive tectonothermal event, which produced the rare assemblages of sapphirine + quartz, orthopyroxene + sillimanite and osumilite which, when

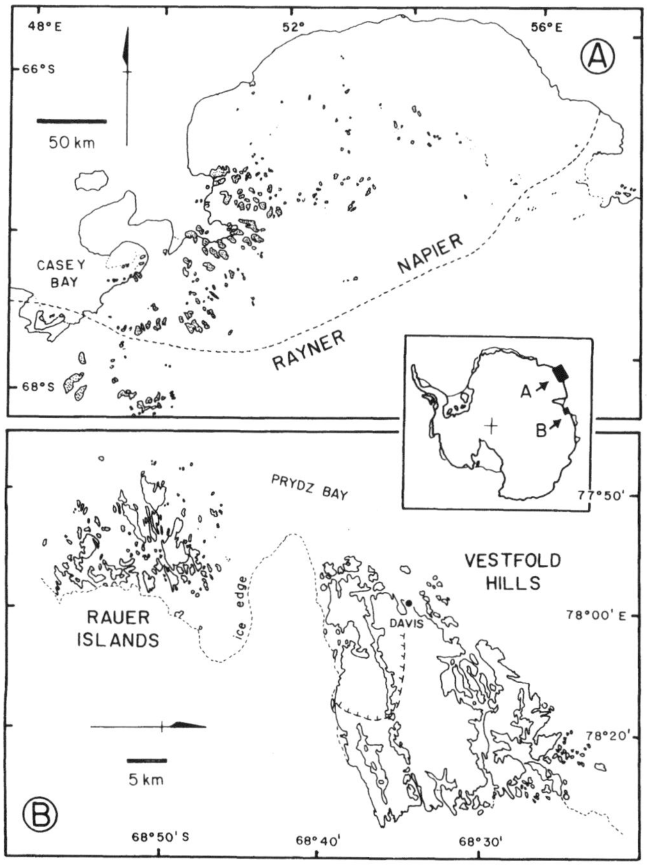

Fig. 1. ***A.*** **Map of Enderby and western Kemp lands divided into the Archaean Napier Complex and Proterozoic Rayner Complex.** ***B.*** **The Archaean Vestfold Hills and Proterozoic Rauer Islands; hatched line shows northernmost development of garnet in mafic dykes. Note differences in scale between** ***A*** **and** ***B.***

considered with other mineralogical associations, indicate unusually dry metamorphic conditions at pressures of 8–10 kbar and temperatures of at least 900–1000 °C. A second localized and less well defined deformation occurred at ~ 2900 Ma under continuing high-grade granulite facies conditions. At ~ 2500 Ma, the Napier Complex again experienced a widespread, though only locally intense, tectonothermal metamorphism. Mineral assemblages attributed to this event record a significantly lower temperature than earlier events (650–750 °C) but only marginally lower pressures (5–9 kbar). Taken together, these studies indicate that the Napier Complex, which was buried at depths of 25–30 km, followed an essentially isobaric cooling path from 900–1000 °C at ~ 3100 Ma to 630–700 °C at ~ 2500 Ma.

In marked contrast to the isobaric cooling path deduced from the Archaean Napier Complex mineral assemblages, the adjacent Late Proterozoic Rayner Complex records an isothermal decompression path. Napier Complex metamorphic conditions (8 kbar, 720 °C) are preserved in rocks variably affected by the ~ 1000 Ma metamorphism, whereas corona assemblages and chemical zoning within phases attributed to the Rayner metamorphism indicate conditions of 3–5 kbar pressure and temperatures of 570–650 °C (see Harley, 1985).

Similarly detailed metamorphic–geochronological studies of the Archaean Vestfold Hills and adjacent Proterozoic Rauer Islands terranes (Fig. 1*B*) have yet to be reported. However, previous reconnaissance work indicated a similar crustal history to that described from Enderby Land. Isotopic studies of Vestfold Hills granulites have defined Archaean metamorphic events at ~ 3000–2800 Ma and ~ 2500–2400 Ma, and a Proterozoic event in the adjacent Rauer Islands (~ 1100 Ma; Collerson, Reid, Millar & McCulloch, 1983). Peak metamorphic conditions of ~ 1000 °C and 8–10 kbar can be correlated with the oldest metamorphic episode (Collerson, Sheraton & Arriens, 1983).

Hundreds of mostly unmetamorphosed mafic dykes crop out in the Vestfold Hills (~ 450 km^2) and are common throughout the much larger area of Enderby land (100000 km^2). Detailed geochemical and geochronological studies of these dyke suites are discussed in Sheraton & Black (1981), Collerson & Sheraton (1986) and Kuehner (1986; 1989). Chemically similar, mostly quartz-normative, opx ± oliv high-Mg tholeiites (Mg-number = 100 Mg/(Mg + Fe) = 70–54) were emplaced in both the Napier Complex and the Vestfold Hills at ~ 2400 Ma. In the Napier Complex, high-Mg tholeiites (HMG) postdate the emplacement of undeformed, Fe-rich tholeiites with granoblastic textures, but are contemporaneous with an Fe-rich tholeiite suite in the Vestfold Hills. The most voluminous Fe-rich tholeiite suites were emplaced in the Vestfold Hills and Napier Complex over the period ~ 1200–1400 Ma. The 1190 ± 200 Ma Group I Amundsen dykes of the Napier Complex are cpx + plag ± opx–phyric, whereas the plag + cpx ± oliv–phyric Group II suite does not define an isochron. The Vestfold Hills Fe-rich tholeiites form a generally coherent, chemical group dated at 1362 ± 200 Ma.

Sample descriptions

In order to evaluate the high pressure liquid phase assemblage of samples selected to represent the ~ 2420 Ma and ~ 1360 Ma Vestfold Hills dyke suites with regard to their emplacement depth, it is crucial that the chosen samples are not phenocryst-enriched but instead represent liquid compositions. To minimize the possibility of collecting crystal-enriched material, fieldwork emphasized the collection of samples from chilled (quenched) dyke margins. The samples were then assessed using published crystal/liquid $K_D^{Fe/Mg}$ values. In addition to showing crystal/liquid equilibrium, or near equilibrium, the samples selected for study also contain mineral assemblages from which independent estimates of emplacement pressures and temperatures could be made using conventional geothermometric–barometric techniques (see Kuehner, 1986).

Chilled margins of HMG dykes typically contain a quench-textured groundmass and consequently do not contain an equilibrium assemblage from which independent *P*–*T* estimates can be made. HMG sample 65060 was collected 1 m from the margin of a 30 m wide dyke, and was chosen for experimental study specifically because the somewhat slower cooling resulted in a groundmass assemblage applicable to *P*–*T* estimates, while still retaining crystal/liquid equilibrium. Sample 65060 (Mg-number = 64.0, Table 1) contains phenocrysts of olivine (4%) and colourless orthopyroxene (6%) set in

Table 1. *Dyke compositions referred to in this study*

Location number	Vestfold 65206[a]	Napier 49590[a]	Vestfold 65060[a]	Napier 3793[b]
SiO_2	49.45	50.18	52.59	51.91
TiO_2	2.25	2.35	0.62	0.43
Al_2O_3	13.91	13.43	11.95	12.05
FeO	14.89	14.08	10.87	10.04
MnO	0.26	0.19	0.18	0.20
MgO	6.55	6.45	11.28	14.51
CaO	10.03	10.10	10.02	9.51
Na_2O	1.95	2.14	1.79	0.90
K_2O	0.45	0.81	0.60	0.40
P_2O_5	0.26	0.27	0.10	0.05
Total	100.00	100.00	100.00	100.00
Mg-number	44.0	45.0	64.9	72.0

[a]University of Tasmania sample numbers.
[b]Australian Bureau of Mineral Resources sample number: prefixed by 7728- (Sheraton, 1981).

a meshwork of columnar clinopyroxene crystals and granular to intersertal, reddish-brown orthopyroxene with minor quartz. Plagioclase laths (An_{65-50}) are restricted to the groundmass. The orthoyroxene phenocrysts are zoned ($Mg_{86.5-73.0}$) and are generally ~ 1 mm long. Most olivine phenocrysts have experienced partial secondary oxidation and are now composed of orthopyroxene (Wo < ~ 1%)-magnetite symplectites.

Sample 65206 is an Fe-rich ($Mg_{42.6}$) member of the ~ 1360 Ma tholeiite suite (Table 1). The chilled margin phenocryst assemblage is predominantly (27%) seriate plagioclase laths < 1 mm long (An_{70-60}) with clinopyroxene (10%; $Mg_{79.2-69.9}$) and olivine (5%). These phases occur individually, and in glomeroporphyritic clots, whereas idiomorphic, pinkish grains of orthopyroxene ($\ll$ 1%, $Mg_{72.2-64.5}$) occur only as single crystals. Olivine is completely replaced by orthopyroxene–magnetite–symplectite in this sample. The groundmass consists of a very fine grained two-pyroxene, plagioclase, quartz assemblage with abundant opaque oxides.

Experimental techniques for high-pressure studies

A 10–15 mg mix, prepared from analytical grade reagents was used in each run, with run times ranging from 0.75–40 h, depending on pressure and proximity to the liquidus. As there are no primary hydrous phases in the studied samples, all experiments were run under 'dry' conditions using Spec-pure Fe containers. The effects of Fe addition on the phase topology were evaluted by relocating the liquidus multiple saturation point of the more refractory sample (65060) using Pt-capsules (Fe-loss). The resulting differences in topology were found to be within the calibration errors of the $\frac{1}{2}$ inch piston cylinder (0.5 kbar, ± 5 °C). Detailed discussion of the experimental techniques can be found in Kuehner (1986).

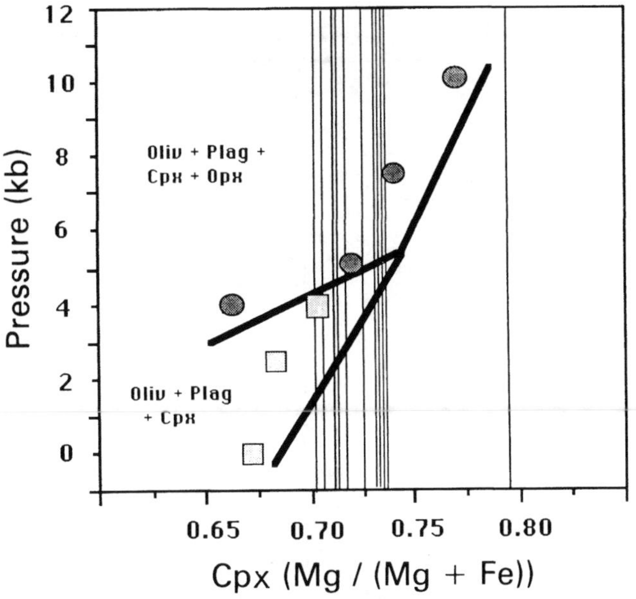

Fig. 2. Natural clinopyroxene phenocryst compositions (narrow vertical lines) compared with the pressure-dependent compositions of experimental clinopyroxene. Heavy lines mark phase field boundaries and the composition of the first formed clinopyroxene to crystallize from the studied composition. The $Mg_{79.2}$ clinopyroxene composition indicates evidence for more primitive liquid composition retained in the cores of some phenocrysts.

Experimental results of Vestfold Hills dykes

HMG sample 65060

A liquidus *P–T* diagram for sample 65060 was constructed from 20 experimental points over a pressure range of 1 atm–12 kbar. At 1 atm, olivine is the sole crystallizing phase from the 1280 °C liquidus to 1180 °C, and is joined by pigeonite, Ca-clinopyroxene and plagioclase between 1180 °C and 1150 °C. Pigeonite remains the first pyroxene to crystallize after olivine until 7.5 kbar/1280 °C, when orthopyroxene and olivine coprecipitate on the liquidus. At pressures > 7.5 kbar, pigeonitic clinopyroxene is the liquidus phase. Thus, only at 7.5 kbar is the phenocryst assemblage of sample 65060 reproduced. Furthermore, the composition of the liquidus orthopyroxene ($Mg_{85.0}$, $Wo_{5.3}$) agrees well with the most Mg-rich phenocryst composition ($Mg_{86.5}$, $Wo_{4.5}$), and substantiates the premise that the composition of sample 65060 is that of a liquid. Although the experiments have shown that olivine is the liquidus phase in sample 65060 over an extensive range of *P* and *T*, petrographically olivine comprises only 6% of the sample, and does not occur in the groundmass. These features preclude the possibility that the phenocrysts crystallized at 7.5 kbar and were subsequently transported into an expanded olivine stability field at significantly lower pressure. From these data, it is concluded that the dyke from which sample 65060 was collected, and thus logically the HMG suite, was emplaced under a load pressure of 7–8 kbar.

Fe-rich tholeiite sample 65206

The phase relations of sample 65206 are based on 19 experiments between 1 atm and 10 kbar. Olivine is the liquidus

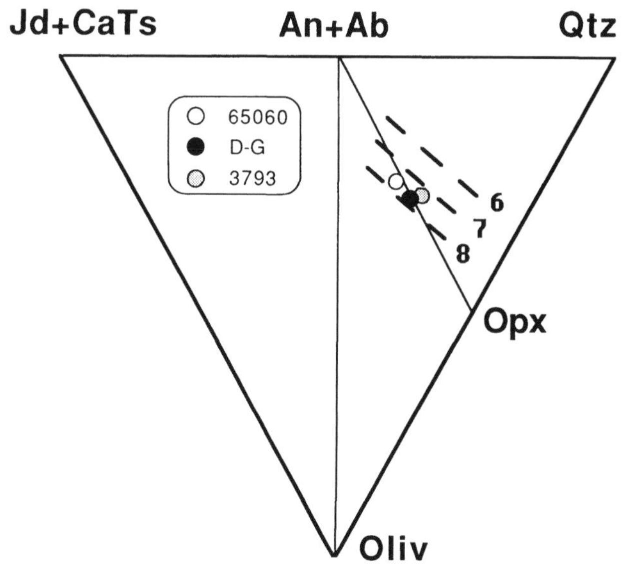

Fig. 3. High Mg basalt compositions projected from Diop onto the plane (Jd + CaTs)–Oliv–Qtz, following Green (1970). Oliv + Opx cotectics (kbar) are from Kuehner (1986). D–G, Duncan & Green (1987).

phase at 1 atm/1165 °C–8 kbar/1185 °C, above which orthopyroxene + clinopyroxene coprecipitate at the liquidus. Consecutive stability fields of oliv, oliv + opx + cpx, and oliv + opx + cpx + plag exist with decreasing temperature between 5.5–8 kbar. Below 5.5 kbar/1160 °C, and extending to 1 atm/1150–1100 °C, there exists an oliv + plag stability field, followed by oliv + plag + cpx and oliv + plag + cpx + opx at lower temperatures. Considering the glomerophyric oliv + plag + cpx in sample 65206, and the extreme rarity of opx phenocrysts, crystallization of this sample is restricted to pressures < 5.5 kbar. A comparison of the natural clinopyroxene phenocryst compositions (Fig. 2) with the pressure-dependent compositional variation determined experimentally, indicates crystallization of the ~ 1360 Ma tholeiite suite at pressures of 4–5 kbar.

Emplacement pressures of Napier Complex dykes

HMG dykes cropping out in the Napier Complex display slight–moderate degrees of recrystallization, and consequently do not meet the requirements for high-pressure studies. Thus, in order to estimate their emplacement pressure, the experimental results of 65060 and studies of several Mg-rich basaltic compositions obtained from the literature were used to construct contours within the basalt tetrahedron that display the pressure dependency on the positioning of oliv + opx cotectics with respect to major element variation. This is shown in Fig. 3, a projection from diopside onto the plane (Jd + CaTs)–Oliv–Qtz, following Green (1970), with oliv + opx cotectics drawn parallel to the cotectics defined experimentally by Jaques & Green (1980). From this construction, an estimate can be made of the pressure at which an Mg-rich liquid will be saturated in olivine and orthopyroxene.

Re-equilibration of the phenocryst phases in the Napier HMG dykes prohibits the confident application of crystal/liquid K_D values to establish the bulk rock–liquid criteria. However, the petrography of sample 3793 suggests it is a quenched liquid (Table 1) as recrystallization has only produced a very fine, granular mosaic and has not disturbed the original igneous texture. The sample is interpreted as being dominated by a meshwork of 'quench' pyroxene crystals (< 1 mm long), as is evident from preserved swallowtails and hollow cross-sections. Stocky, orthopyroxene phenocrysts (< 2% of mode; 0.5 mm long) are also recrystallized, though a few grains retain castellated terminations. The projected position of 3793 (Fig. 3) overlaps the Mg-rich basalt composition studied by Duncan & Green (1987), which, like 65060, is saturated in oliv + opx at 7–8 kbars. As there is no petrographic evidence for olivine in 3793, crystallization must have taken place at pressures greater than that of olivine saturation. Thus, 7–8 kbar is the minimum emplacement pressure of the Napier HMG suite.

Lack of suitable experimental studies on Fe-rich basaltic compositions prohibits an evaluation of the emplacement pressure of the ~ 1190 Ma dykes, following the approach used to estimate emplacement depths of the Napier HMG suite. However, a Napier Group I tholeiite sample, 49590, is nearly identical in composition to 65206 (Table 1) such that these samples overlap one another on the (Jd + CaTs)–Oliv–Qtz diagram. Judging from these similarities, the high-pressure phase relations of 49590 are not expected to vary significantly from those determined experimentally in 65206. Petrographically, however, there is a dramatic difference between these samples. Unlike the oliv + cpx + plag–phyric 65206, sample 49590 contains glomerocrysts of opx + cpx + plag, and does not contain olivine. Compared to the phase relations of the chemically similar 65206, the phenocryst assemblage of 49590 is consistent with coprecipitation at pressures above that of olivine saturation, implying a minimum emplacement pressure of the Group I Napier tholeiites of ≈ 8 kbar.

P–T estimates from garnet-bearing dykes

A detailed assessment of the *P–T* conditions which prevailed in the Vestfold Hills during the ~ 1100 Ma metamorphism has previously not been available. Although this metamorphism is characteristic of the geology in the Rauer Islands (Fig. 1), its effects are evident only in the south-west part of the Vestfold Hills where hydration/recrystallization produced garnet-bearing assemblages in Fe-rich dykes (Mg-number < 45).

The extent of hydration/recrystallization is highly variable in the mafic dykes, and is strongly influenced by the degree of shearing. In strongly foliated margin samples, a granular amph + plag + qtz ± cpx ± gar assemblage prevails. However, within the dyke interiors, where shearing is less intense or absent, igneous textures may be preserved and the metamorphic reactions identified. In these examples, corroded clinopyroxene phenocrysts have dark green, tabular hornblende coronas. This assemblage is commonly set in a relict subophitic matrix of randomly orientated plagioclase crystals possessing irregular margins, with the primary subophitic pyroxene replaced by hornblende, granular clinopyroxene,

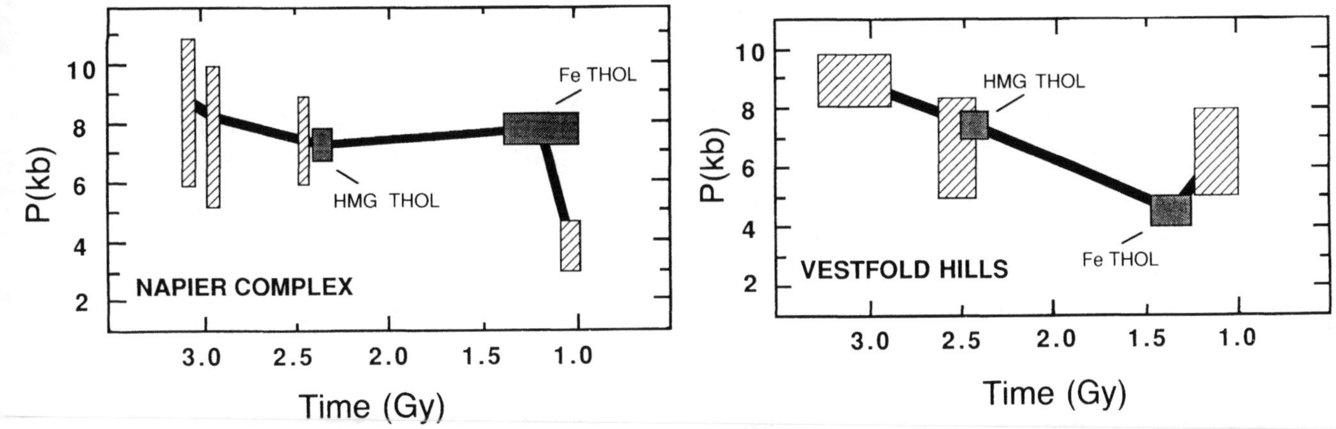

Fig. 4. Pressure (*P*) vs time (*t*) plots depicting the uplift history of the Napier Complex and the Vestfold Hills by incorporating previous metamorphic–geochronological studies with the results discussed here. Diagonal pattern, metamorphic studies; stippled boxes, emplacement *P–t* of mafic dykes.

opaque oxides and quartz. The textures indicate the following idealized reaction:

$$\text{Cpx} + \text{Plag} + H_2O \rightarrow \text{Hbd} + \text{Qtz} \qquad (1)$$

In the undeformed samples, aggregates up to 1 cm in diameter of granular, poikiloblastic garnet, typically contain numerous idioblasts of clinopyroxene, whereas hornblende is absent. The relict igneous texture, as defined by unorientated plagioclase laths, is continuous through the garnet aggregates, although plagioclase grains within the aggregates may be completely pseudomorphed by garnet. Hornblende coronas do not occur on relict clinopyroxene phenocrysts that are enclosed by garnet aggregates. These textures indicate garnet formed via the reaction

$$\text{Hbd} + \text{Plag}_1 + \text{Qtz} \rightarrow \text{Gar} + \text{Cpx} + \text{Plag}_2 + H_2O \qquad (2)$$

A comparison of these petrographically-deduced, common hydration/dehydration reactions with a Schreinemaker's evaluation of the simple system Na-CMASH (Glassley & Sørensen, 1980) indicates that a pressure increase with, or without, an increase in temperature, is required to produce garnet at the expense of amphibole. The *P–T* of garnet equilibrium was estimated by applying the Ellis & Green (1979) gar + cpx thermometer, and the Newton & Perkins (1982) gar + cpx + plag + qtz barometer to samples collected from eight garnet-bearing dykes. The resulting temperatures range from 602–660 °C (average 635 °C) at pressures of 5.6–7.5 kbar (average 6.4 kbar). Logically, hydration of the dykes took place at pressures < 6.4 kbar.

Discussion

The results of the mafic dyke studies are combined with previous metamorphic–geochronological data on pressure versus time plots (Fig. 4) which graphically summarize the uplift paths of the Vestfold Hills and the Napier Complex, from the Late Archaean through the Late Proterozoic. The emplacement of the ~ 1190 Ma Group I dykes in the Napier Complex at pressures ≥ 8 kbar confirms the previous inference, based on the isothermal decompression path recorded in ~ 1000 Ma mineral assemblages (see Harley, 1985), that the Napier Complex existed at deep crustal levels (> 25 km) for at least 2000 m.y. This long period of crustal stability was terminated by the Rayner metamorphism and, based on the low pressure phenocryst assemblage of the widespread Group II Amundsen dykes (see Sheraton & Black, 1981; Kuehner, 1986) coincided with the uplift of the entire Napier Complex.

Isothermal decompression paths are characteristic of the upper part of a doubly thickened crust produced in collisional tectonic regimes (see Thompson & England, 1984). As such, it is worth noting the similarity between the net uplift rate calculated for the Napier Complex during the Rayner metamorphism (11 cm/1000 y), and that determined for the Cretaceous granulites exposed in the north-west Himalayas (9 cm/1000 y; Zeitler, 1985). The similarity in net uplift rate over comparable periods of time suggests that the crustal thickness of the Napier Complex doubled during the Rayner metamorphism. The minimum crustal thickness prior to uplift is defined by the emplacement depth of the Group I dykes (~ 28 km). Thus, the Rayner metamorphism, which produced mineral assemblages that record minimum pressures of 3–5 kbar, may have resulted in a crustal thickness ≥ 56 km. Removal of the upper 10–18 km of crust to expose this terrane suggests a present crustal thickness of 38–46 km. Considering crustal thinning may have occurred during rifting of Gondwana, this rough calculation compares well with the 35 km estimate derived from the magnetic studies of Groushinsky & Sazhina (1982).

The unusually stable crustal conditions in the Napier Complex during the 600 m.y. following peak metamorphism at ~ 3100 Ma has been emphasized in a number of studies (see Ellis, 1987). An identical Archaean history has now been identified in the Vestfold Hills following the recognition that the HMG suite was emplaced under load pressures only slightly less than that of peak metamorphism. In addition, the absence of significant uplift following the ~ 2500 Ma metamorphism does not support the model of DePaolo *et al.* (1982), who advocated a Himalayan-style continental collision to explain the metamorphic features of this event in the Napier Complex. Furthermore, Kuehner (1989) has shown that the parental liquids to the ~ 2400 Ma HMG suite were derived from mantle peridotite at pressures of ~ 10 kbar (~ 35 km depth). This severely limits the amount of possible crustal thickening during any tectonic event in the Napier Complex and the Vestfold Hills between ~3100Ma and 2400 Ma, and also places a maximum value on the thickness of the crust in these regions during the Late Archaean.

Following the emplacement of the HMG suite, the Vestfold Hills experienced ~10 km of uplift at a net rate of ~1 cm/1000 y until intrusion of the ~1360 Ma dykes at 4–5 kbar. As in the Napier Complex, this slow rate of exhumation signifies very stable tectonic conditions during the Early–Middle Proterozoic, suggesting both terranes were part of a stable continental mass. The development of garnet (~6.5 kbar) through the dehydration of secondary amphibole as a response to crustal loading at ~1100 Ma is consistent with the experimentally-determined, higher crustal emplacement of the ~1360 Ma dykes. Evidence for the timing, and process responsible for the crustal depression and subsequent uplift has not been recognized in the mafic dykes.

Acknowledgements

This paper summarizes part of the Ph.D. work of S.M.K., which was supported by a University of Tasmania postgraduate award. Logistic support for sample collection in the Vestfold Hills during the summer of 1982–83 was provided by the Antarctic Division of the Australian Department of Science. S.M.K. also wishes to thank T. Chacko for comments and discussion.

References

Collerson, K.D., Reid, E., Millar, D. & McCulloch, M.T. (1983). Lithological and Sr–Nd isotopic relationships in the Vestfold Block: implications for Archaean and Proterozoic crustal evolution in the East Antarctic. In *Antarctic Earth Science*, ed. R.L. Oliver, P.R. James & J.B. Jago, pp. 77–84. Canberra; Australian Academy of Science and Cambridge; Cambridge University Press.

Collerson, K.D. & Sheraton, J.W. (1986). Age and geochemical characteristics of a mafic dyke swarm in the Archaean Vestfold Block, Antarctica: inferences about Proterozoic dyke emplacement in Gondwanaland. *Journal of Petrology*, **27(4)**, 853–86.

Collerson, K.D., Sheraton, J.W. & Arriens, P.A. (1983). Granulite facies metamorphic conditions during the Archaean evolution and late Proterozoic reworking of the Vestfold Block, eastern Antarctica. In *6th Australian Geological Convention,Canberra* (abstract), 53–4.

DePaolo, D.J., Manton, W.I., Grew, E.S. & Halpern, M. (1982). Sm–Nd, Rb–Sr and U–Th–Pb systematics of granulite facies rocks from Fyfe Hills, Enderby Land, Antarctica. *Nature, London*, **298(5875)**, 614–18.

Duncan, R.A. & Green, D.H. (1987). The genesis of refractory melts in the formation of oceanic crust. *Contributions to Mineralogy and Petrology*, **96(3)**, 326–42.

Ellis, D.J. (1987). Origin and evolution of granulites in normal and thickened crusts. *Geology*, **15(2)**, 167–70.

Ellis, D.J. & Green, D.H. (1979). An experimental study of the effect of Ca upon garnet–clinopyroxene Fe–Mg exchange equilibria. *Contributions to Mineralogy and Petrology*, **71(1)**, 13–22.

Glassley, W.E. & Sørensen, K. (1980). Constant P_s–T amphibolite to granulite facies transition in Agto (West Greenland) metadolerites: implications and applications. *Journal of Petrology*, **21(1)**, 69–105.

Green, D.H. (1970). The origin of basaltic and nephelinitic magmas. *Transactions of the Leicester Literary and Philosophical Society*, **64(1)**, 28–54.

Groushinsky, N.P. & Sazhina, N.B. (1982). Some features of Antarctic crustal structure. In *Antarctic Geoscience*, ed. C. Craddock, pp. 907–11. Madison; University of Wisconsin Press.

Harley, S.L. (1985). Paragenetic and mineral–chemical relationships in orthoamphibole-bearing gneisses from Enderby Land, East Antarctica: a record of Proterozoic uplift. *Journal of Metamorphic Geology*, **3(2)**, 179–200.

Harley, S.L. & Black, L.P. (1987). The Archaean geological evolution of Enderby Land, Antarctica. In *Evolution of the Lewisian and Comparable Precambrian High Grade Terraines*, ed. R.G. Park & J. Tarney, *Geological Society of London, Special Publication*, **27**, 285–96.

James, P.R. & Black, L.P. (1981). A review of the structural evolution and geochronology of the Archaean Napier Complex of Enderby Land, Australian Antarctic Territory. In *Archaean Geology*, ed. J.E. Glover & D.I. Groves, *Special Publication of the Geological Society of Australia*, **7**, 71–83.

James, P.R. & Tingey, R.J. (1983). The Precambrian geological evolution of the East Antarctic metamorphic shield – a review. In *Antarctic Earth Science*, ed. R.L. Oliver, P.R. James & J.B. Jago, pp. 5–10. Canberra; Australian Academy of Science and Cambridge; Cambridge University Press.

Jaques, A.L. & Green, D.H. (1980). Anhydrous melting of peridotite at 0–15 kb pressure and the genesis of tholeiitic basalts. *Contributions to Mineralogy and Petrology*, **73(3)**, 287–310.

Kuehner, S.M. (1986). Mafic dykes of the East Antarctic Shield: experimental, geochemical and petrological studies focusing on the Proterozoic evolution of the crust and mantle. Ph.D. thesis, University of Tasmania, 300 p. (unpublished).

Kuehner, S.M. (1989). Petrology and geochemistry of Early Proterozoic high-Mg dikes from the Vestfold Hills, Antarctica. In *Boninites and Related Rocks*, ed. A.J. Crawford, pp. 208–31. London; Unwin-Hymen Press.

Newton, R.C. & Perkins, D. (1982). Thermodynamic calibration of geobarometers based on the assemblages garnet–plagioclase–orthopyroxene (clinopyroxene)–quartz. *American Mineralogist*, **67(3)**, 203–22.

Sheraton, J.W. (1981). Chemical analyses of rocks from East Antarctica. *Bureau of Mineral Resources, Australia, Record*, **1981/14**.

Sheraton, J.W. & Black, L.P. (1981). Geochemistry and geochronology of Proterozoic tholeiite dykes of East Antarctica: evidence for mantle metasomatism. *Contributions to Mineralogy and Petrology*, **78(3)**, 305–17.

Thompson, A.B. & England, P.C. (1984). Pressure–temperature–time paths of regional metamorphism II. Their inference and interpretation using mineral assemblages in metamorphic rocks. *Journal of Petrology*, **25(4)**, 929–55.

Zeitler, P.K. (1985). Cooling history of the NW Himalayas, Pakistan. *Tectonics*, **4(1)**, 127–51.

The crustal evolution of some East Antarctic granulites

S.L. HARLEY

Department of Earth Sciences, University of Oxford, Parks Road, Oxford, OX1 3PR, UK

Abstract

Important metamorphic pressure–temperature–time (*P–T–t*) data and related features of granulites from the Archaean Napier Complex and Late Proterozoic metamorphic belt of East Antarctica are reviewed and their bearing on probable tectonic settings of metamorphism assessed. The high temperature metamorphism, near-isobaric cooling history and spatial distribution of *P–T* conditions in the Archaean Napier Complex are interpreted in terms of diffuse collisional tectonics, where an important late phase of extension of thickened crust has produced striking metamorphic features. This collision–extension model also involved extensive magmatism, induced in part by syn-extensional decompression and perhaps partial lithospheric detachment. The contrasting medium-temperature granulites of the Late Proterozoic complex, which show near-isothermal decompression *P–T* paths at mid-crustal levels, result from a collisional tectonic setting probably involving previously thinned crust, and with a heat source augmented locally by the addition of new magmas. The Archaean Napier Complex was underplated by 10–15 km of mainly younger crust in the Late Proterozoic, leading to its partial *en masse* uplift without significant retrogression or pervasive deformation.

Introduction

Pioneering research in the delineation of the pressure–temperature–time (*PTt*) paths of high-grade metamorphism has followed from studies of East Antarctic granulites. In particular, the near-isobaric cooling history of the Archaean Napier Complex (Ellis, 1980; Sheraton *et al.*, 1980; Ellis & Green, 1985; Harley, 1985a; Sandiford, 1985) remains one of the best documented granulite *P–T* paths. This *P–T* history contrasts markedly with the near-isothermal decompression *P–T* paths which appear to characterize most of the Late Proterozoic granulites from near Davis Station to Lützow-Holm Bay (Ellis, 1983; Harley, 1985b, this volume, p. 99). The purposes of this paper are to review briefly the principal lines of evidence for the contrasted *P–T* paths of the Archaean and Late Proterozoic granulites, incorporate other geological evidence bearing on the possible tectonic settings of metamorphism, and speculate upon the tectonic processes responsible for the production of these terranes.

The Napier Complex

The geological history of the Napier Complex described by Harley & Black (1987 & see references therein) is adopted here as a framework for the *P–T* paths recorded after the main tectono-thermal episodes at 3070–2900 Ma. The evidence for the stability of the Napier Complex as lower crust through 2000 m.y. of earth evolution is reviewed below.

Diversity of protoliths

Layered gneiss sequences of mainly supracrustal origin are important components of the Napier Complex. Protoliths include quartz-rich sediments, chemical sediments and evaporites, ironstones, pelites, magnesian pelites, and volcanogenic rocks.

Timing and nature of magmatism

Felsic orthogneisses constitute a major proportion of the Napier Complex. The occurrence of both Y,HREE-depleted tonalitic and Y,HREE-undepleted granitic I-type orthogneisses, interpreted as representing new continental crust and remelted felsic crustal precursors respectively (Sheraton, Ellis & Kuehner, 1985), is not diagnostic of any tectonic setting in view of the spectrum of ages for both types. However, magmatism of both types was important prior to and synchronous with the major deformations, and post-orogenic felsic magmatism was essentially absent in the highest-*T* part of the complex.

High metamorphic temperatures (950–1020 °C)

Several lines of evidence show conclusively that the early highest-grade events progressed at temperatures in excess of 950 °C in the south-western part of the complex:

(a) The FMAS mineral assemblages sapphirine (sa) + quartz (qz), hypersthene (hy) + sillimanite (sill) + qz, and spinel (sp) + qz are only stable at *P* or *T* conditions beyond those of garnet (gt) + cordierite (cd) assemblages (Ellis *et al.*, 1980). Recent experiments (D.J. Ellis & P. Bertrand, pers. comm.) confirm the view (Hensen & Green, 1973) that sa + qz is stable only above 1000 °C, whilst hy + sill + qz occurs at pressures of over 10–11 kb.
(b) The stability of osumilite + gt and of calcic mesoperthites, which also imply very low water activities (Ellis, 1980, 1983).
(c) Meta-ironstones contain exsolved and inverted metamorphic pigeonite with Mg-numbers [(Mg/Mg + Fe) × 100] in the range 42–47. These imply temperatures > 970 °C at Tonagh Island (Harley, 1987) and 1000–1020 °C at Fyfe Hills (Sandiford & Powell, 1986a).
(d) Coexisting gt–hy pairs yield minimum Fe–Mg K_D's (2.0) giving calculated temperatures > 900 °C. The initial alumina contents (10–13 wt % Al_2O_3) of hypersthene coexisting with garnet or sillimanite are also very high.

Near-isobaric cooling paths

The post-deformational, post-peak *P–T* paths in the south-western Napier Complex involved cooling through ~ 300–350 °C from 950–1020 °C, with only a minor decrease in pressures (see Fig. 1*a*; Harley & Black, 1987). Principal evidence for this near-isobaric cooling includes:

(a) Production of hy + sill (Scott Mountains; Sheraton *et al.*, 1980) or gt + cd + sill (Tula Mountains) coronas between earlier sa + qz, and secondary gt + hy + sill on earlier hy (Harley, 1985a).
(b) Formation of secondary gt in mafic (Ellis & Green, 1985) and felsic (Harley, 1985a) rock types.
(c) Absence of any late melting.
(d) A decrease in the Al contents of hy coexisting with late gt or occurring as neoblasts, coupled with a marked increase in the gt–hy Fe–Mg K_D's for secondary assemblages (Harley, 1985a).
(e) Gt rims or neoblasts are more calcic than cores in gt + sill + plag + qz and gt + hy + plag + qz assemblages.

The Late Proterozoic complex

The Late Proterozoic complex considered here includes the Rayner Complex south and south-east of the Napier Complex (Sheraton *et al.*, 1980; Ellis, 1983; Black *et al.*, 1987), and probable correlatives extending eastwards to the Rauer Group near Davis Station and westwards to Lützow–Holm Bay (Harley, this volume).

Diversity of protoliths

Layered gneiss sequences are more abundant than in the Napier Complex, and are typified by gneisses derived from

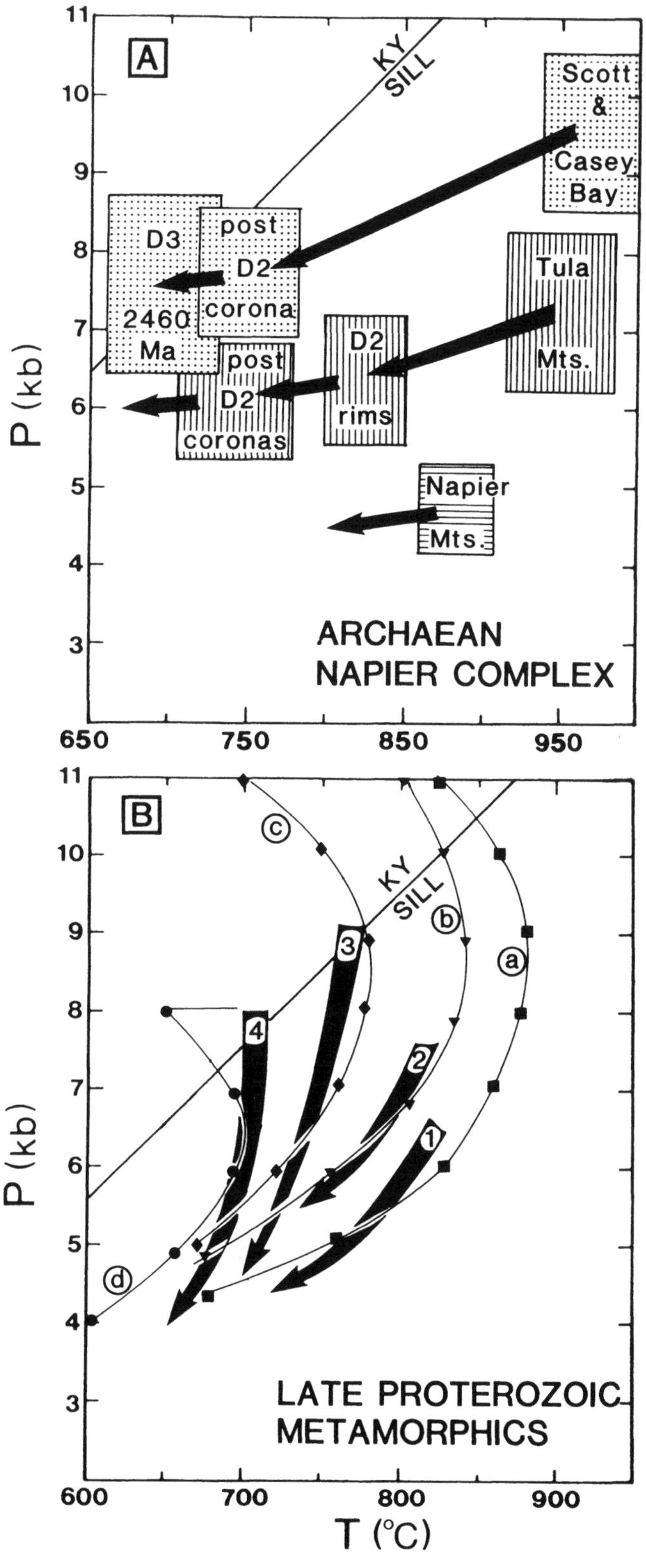

Fig. 1. Pressure–temperature paths of selected Antarctic granulites: (A) Archaean Napier Complex. Paths for granulites from different geographical areas: localities and definitions of deformation phases are given in Harley (1985a) and Harley & Black (1987). (B) Late Proterozoic metamorphic rocks: 1, Rauer Group, northern parts (Harley, this volume); 2, Rauer Group, southern parts (Harley, this volume); 3, Nye Mountains and Rayner Complex (Ellis, 1983; Black et al., 1987); 4, shear zones within the Napier Complex (Harley, 1985b). Thin lines are calculated *P–T–t* paths adapted from England & Thompson (1984). Solid symbols on each path denote 10 m.y. intervals: (a) homogeneous

meta-quartzites, Fe-rich to intermediate pelites, impure carbonates, rare ironstones, and basic volcanic and layered intrusive rocks (Harley, this volume, p. 99). Metasedimentary rocks show derivation from mid- to Late Proterozoic crustal precursors (Black *et al.*, 1987) and were deposited up to the time of the major ~ 1100–960 Ma tectonism.

Nature and timing of magmatism

Felsic orthogneisses include tonalitic I-types and distinctive anorthosites, but they are dominated by K-rich S- and I-type granites (Sheraton, Black & McCulloch, 1984). Migmatitic granulites are also common. A spectrum of derivation and emplacement ages (1900–960 Ma) is apparent (Black *et al.*, 1987), and some magmatism occurred late in and after the most intense deformation and peak of metamorphism. Magmatism continued until a relatively later stage in this *P–T* history than in the Napier situation.

Deformational features

The late deformation phases produced structures which vary in intensity, style, and orientation. Strain is often strongly partitioned so that relatively little-deformed granulites are found adjacent to narrow zones of strong late fabric development and transposition. These features result from either varied fluid access or the presence of partial melts during deformation.

P–T conditions and paths

Extensive decompression from 7–9 kb to 3–5 kb, with minor cooling through 20–100 °C, is a characteristic feature of the Late Proterozoic *P–T* paths from several regions (Fig. 1*b*; Harley, this volume, p. 99). Constraints on maximum temperatures are the stability of gt + cd + qz assemblages in metapelites; 800–900 °C temperatures calculated from gt–hy and gt–cpx Fe-Mg thermometry; and ~ 820 °C scapolite + - wollastonite assemblages in calc-silicates. Decompression with some cooling to 650–750 °C is inferred from textural indicators and geothermobarometry (Ellis, 1983; Harley, this volume, p. 99). The 720–600 °C decompressional *P–T* paths of Proterozoic shear zones within the Napier Complex (Fig. 1*b*) are well constrained by the early stability of kyanite, and phase relations between gt, cd, sill, staurolite and ortho-amphibole (Harley, 1985b).

Figure 2-1 (*cont.*)
thickening model for crust initially 30 km thick with conductivity (*k*) 1.5 W/m/K and heat source distribution III (surface heat flux 75 mW/m², basal *Q* = 40 mW/m²); corresponds to line 2, Fig. 3 of Harley (this volume, p. 99); (b) path for rocks initially 10 km below a thrust at 30 km, with *k* = 1.5 W/m/K, heat source distribution II (surface heat flux 60 mW/m², basal flux *Q* = 30 mW/m²); corresponds to line 1, Fig. 3. Harley (this volume, p. 99); (c) path for rocks initially 6 km below a thrust at 30 km, with parameters as in (b); (d) path for rocks initially 4 km *above* a thrust at 30 km, with *k* = 1.5 W/m/K, heat source distribution III. Upper plate path.

Tectonic models and implications

The Napier Complex

P–T paths (Fig. 1*a*) demonstrate a spatial variation in metamorphic pressures even in the highest-*T* areas. Granulites from the Scott Mountains–Khmara Bay region and Amundsen Bay–Tula Mountains area evolved along parallel *P–T* paths and through a similar temperature interval, but at different pressures, implying that a substantial thickness of the lower crust (maximum 16 km) experienced broadly similar but extreme metamorphic temperatures. No net thinning or thickening of the exposed lower crustal section occurred following the recorded peak conditions, apart from that caused by an unrelated deformation even at ~ 2460 Ma.

Near-isobaric cooling *P–T* paths may result from magmatic accretion (e.g. Wells, 1980) or extension of the continental crust and lithosphere, which itself often induces magmatism (e.g. Sandiford & Powell, 1986b). The presence of orthogneisses with a spectrum of ages back to 3750 Ma suggests that a significant crustal volume existed well before the events which caused the main *P–T* history. Therefore simple magmatic accretion models requiring large additions to the crust (40–70% if felsic melts) over a short time scale (50 m.y.) are not pursued further here.

Many workers (e.g. Sandiford & Powell, 1986b) have considered the *P–T* histories of extensional terranes, described by either symmetrical thinning of the lithosphere or asymmetric extension on a low-angle normal shear zone through the lithosphere. In the symmetrical extension case and for the region of the greatest crustal thinning (and hence deformation) in the simple shear models, near-isobaric cooling follows extension (paths 1 and 3, Fig. 2). As the maximum temperatures and cooling paths occur at only 5–7 kb for rocks which were initially at the base of 'normal' thickness (35 km) crust, these models do not explain isobaric cooling in granulites at pressures of up to 11 kb. Isobaric heating and cooling paths are generated near and beyond the 'discrepant zones' (Wernicke, 1985) in asymmetric simple shear models, because most of the extension is accommodated in the mantle. Although cooling in the appropriate depth range of 7–11 kb would occur (path 3, Fig. 2; Sandiford & Powell, 1986b), such non-extended crustal regions will not be intensely deformed.

An alternative model, preferred here, involves extension of crust and lithosphere previously thickened in a collisional event distributed across a broad zone of diffuse deformation (e.g. England, 1987) (paths 4 and 5a, 5b, Fig. 2). This type of collision regime is characterized by a stage where a high plateau overlying crust of doubled thickness (e.g. the Tibetan Plateau) is developed. England (1987) showed that extension of the plateau region would be expected to occur at some time (10–30 m.y.) following the cessation of convergence if temperatures near the Moho exceeded a critical value (e.g. above 750 °C) related to the temperature-dependent rheology of the lower crust. Rapid extension (for 10–20 m.y.) would then cause essentially isothermal or increasing-*T* decompression, and would be followed by near-isobaric cooling in crust of near-normal thickness.

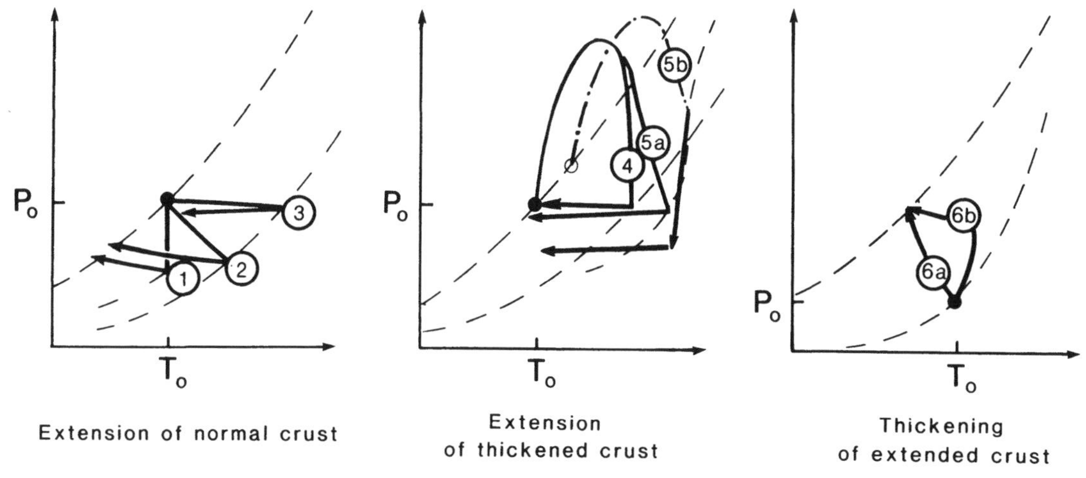

Fig. 2. Schematic *P–T* paths related to extension and magmatism. P_0 and T_0 are initial conditions: (1) symmetric extension of normal crust, without significant subcrustal thinning, or zone of maximum crustal extension in asymetric models; (2) extension of crust and whole lithosphere (intromogeneous symmetric model) (Sandiford & Powell, 1986b); (3) path for crust in 'discrepant zones' (Wernicke, 1985) where most of the extension is accommodated in the mantle (asymmetric model); (4) extension of thickened crust and remaining lithosphere (England & Thompson, 1986); (5) extension of thickened crust; whole lithosphere extended following detachment of lower portion of earlier thickened lithosphere. (5a) temperatures enhanced by asthenospheric upwelling and emplacement of magmas from (5b) or deeper levels; (6) thickening of previously thinned crust, through tectonic–sedimentation processes (6a) or magmatic additions (6b).

Magmatism will be an important feature throughout this history. England & Thompson (1986) demonstrated that melting is likely in doubly thickened crust if internal heat generation is high, conductivity low, or mantle heat flux is enhanced. Terranes where extension followed thickening will therefore contain a range of magmatic rocks which predate the main deformation (extensional) phase. During syn-extensional decompression, further felsic magmatism caused by melting of both pre-existing crustal rocks and newly accreted mafic rocks (see below) would be expected.

Extension is already proceeding in the Tibetan Plateau, contemporaneous with Indo-Asian convergence. One mechanism which may trigger such early extension (England & Thompson, 1986; P.C. England, pers. comm.) is delamination or detachment of the thermal boundary layer which comprises the lowermost 50–60 km of the thickened subcontinental lithosphere. This induces an elevation contrast which cannot be sustained by convergence. The conductive heat-flux characterizing the collision zone in this case will be greater than that in a 'whole lithosphere thickening' model, causing higher temperatures (900–1100 °C) in the lower crust prior to and during extension. Asthenospheric upwelling prior to crustal thinning will lead to extensive mantle melting and hence basaltic magmatism. These magmas, accreted within or at the base of the thickened crust, may be remelted to produce HREE-depleted felsic magmas when that crust decompresses during extension.

A collision–extension model will therefore produce many of the features observed in the Napier Complex. The role of magmatism, and maximum temperatures, will be enhanced if partial lithospheric detachment occurs. The similarity of peak metamorphic conditions in the deeper levels (20–30 km) of the Napier Complex suggests an active role for syn-extensional magmatism (path 5b, Fig. 2) in buffering maximum temperatures attained by the end of extension, at least in the south-western areas. The Napier Mountains area was probably above the emplacement zone, and thus records somewhat lower-grade granulite conditions. In this context, it is possible that the ~ 3070 Ma Proclamation Island orthogneiss (Sheraton & Black, 1983) represents an early pre-extensional melt and therefore provides a maximum age constraint on the main metamorphism.

If the pressure conditions at the low-*T* ends of the paths represent isostatic equilibrium in continental crust of typical thickness (35–40 km), then the crustal thickness just prior to the onset of cooling must have been 40–50 km (e.g. Sandiford, 1985) and the highest-*P* granulites (11 kb, 36 km) now exposed and underlain by 30–35 km of crust were within 4–14 km of the base of the crust in the late Archaean. This implies (Harley, 1985b) that the crust underlying the Napier Complex has been thickened since the Archaean without penetrative deformation. The preferred thickening mechanism is underplating by Proterozoic crust of the Rayner Complex at 1100–1000 Ma.

The Late Proterozoic Complex

The decompressional *P–T* paths determined for Late Proterozoic granulites are qualitatively similar to the later portions of *P–T* loops calculated for collision zones where erosion terminates the heating of the crust. In detail (Fig. 1*b*; Harley, this volume, p. 99) decompression paths at 850–750 °C and from 9–4 kb require rather unusual model parameter values. In particular, low and probably unrealistic conductivities ($K = 1.5–1.8$) are characteristic of the best-fit paths whether simple thrusting or homogeneous continental thickening models are used. High basal heat flux (40 mW/m^2) and crustal heat production are also features of the fitted *P–T–t* paths, which therefore probably imply thickening of crust with a high initial heat flux (Q = 75 mW/m^2) and a heat supply augmented by externally-derived magmatic additions.

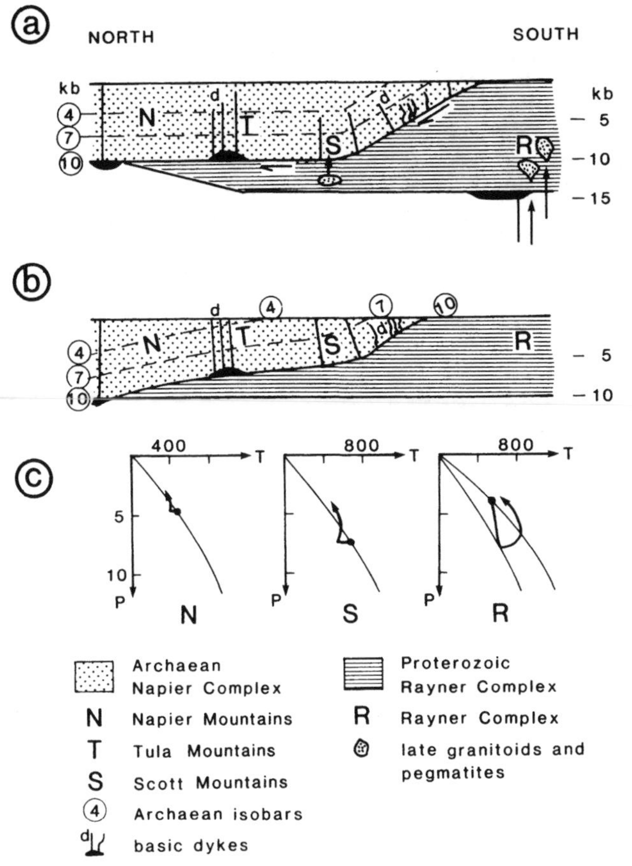

Fig. 3. Schematic cross-sections illustrating the relationship of the Rayner Complex to the Napier Complex in an underplating model: (a) note tilting of pre-Rayner Complex dykes and Napier Complex isotherms (b) subsequent to the erosion of thickened crust. Present erosion level is at 3–4 kb on right-hand pressure scale. Width of crust represented is approximately 900 km. Note distinctive late *P–T* paths (c) across the region; compare with Fig. 1.

It is suggested here that the initial crust prior to late Proterozoic collision was a relatively thinned (< 25 km) continental margin. This is consistent with the emplacement of abundant mid–Late Proterozoic dyke swarms in the surviving Archaean blocks, and with the presence of sedimentary rocks deposited after dyke emplacement but metamorphosed subsequently. Thickening of such an extended terrane might produce an orogenic belt with crustal thicknesses of only 50 km, relatively high post-collisional temperatures, common sediment-derived gneisses and magmas, and negligible late extension.

It is believed that the Late Proterozoic collision involved underplating of the Napier Complex by younger crust. The extent of underplating can be assessed from the decompressional *P–T* paths recorded in shear zones cutting the Archaean gneisses, which imply that at least a 10–14 km thickness of allochthonous crust was added under the Khmara Bay region (Harley, 1985b) and also Amundsen Bay (Harley, unpublished data). The differences between the shear zone *P–T* paths and the Late Proterozoic granulite paths (Fig. 1*b*) reflect the upper-plate setting, the absence of early to syn-decompressional magmatism, and the probable lower initial heat flux (45–55 mW/m^2) in the older (Archaean) crust in the shear zone case.

The effects of Late Proterozoic underplating of the Archaean Napier Complex are schematically depicted in Fig. 3. It is concluded, from the metamorphic studies, that continental underplating is a very effective means for the whole scale uplift and exhumation of old lower crust without significant reworking. The Proterozoic tectonism in East Antarctica played a key role in the preservation of the exposed Archaean areas which have such an important bearing on the nature of high-grade metamorphism in the continental crust.

References

Black, L.P., Harley, S.L., Sun, S.S. & McCulloch, M.T. (1987). The Rayner Complex of East Antarctica: complex isotopic systematics within a Proerozoic mobile belt. *Journal of Metamorphic Geology*, **5(1)**, 1–26.

Ellis, D.J. (1980). Osmulite–sapphirine–quartz granulites from Enderby Land, Antarctica: *P–T* conditions of metamorphism, implications for garnet–cordierite equilibria and the evolution of the deep crust. *Contributions to Mineralogy and Petrology*, **74(2)**, 201–10.

Ellis, D.J. (1983). The Napier and Rayner Complexes of Enderby Land, Antarctica: contrasting styles of metamorphism and tectonism. In *Antarctic Earth Science*, ed. R.L. Oliver, P.R. James & J.B. Jago, pp. 20–4. Canberra; Australian Academy of Science and Cambridge; Cambridge University Press.

Ellis, D.G. & Green, D.H. (1985). Garnet-forming reactions in mafic granulites from Enderby Land, Antarctica – implications for geothermometry and geobarometry. *Journal of Petrology*, **26(3)**, 633–622.

Ellis, D.J., Sheraton, J.W., England, R.N. & Dallwitz, W.B. (1980). Osmulite–sapphirine–quartz granulites from Enderby land, Antarctica – mineral assemblages and reactions. *Contributions to Mineralogy and Petrology*, **72(2)**, 123–43.

England, P.C. (1987). Diffuse continental deformation: length scales, rates and metamorphic evolution. In *Tectonic Settings of Regional Metamorphism*, ed. E.R. Oxburgh, B.W. Yardley & P.C. England. *Philosophical Transactions of the Royal Society of London*, **A-321(1557)**, 3–22.

England, P.C. & Thompson, A.B. (1984). Pressure–temperature–time paths of regional metamorphism. I. Heat transfer during the evolution of regions of thickened continental crust. *Journal of Petrology*, **25(4)**, 894–982.

England, P.C. & Thompson, A.B. (1986). Some thermal and tectonic models for crustal melting in continental collision zones. In *Collision Tectonics*, ed. M. P. Coward & A. C. Ries, *Geological Society of London Special Publication*, **19**, 83–94.

Harley, S.L. (1985a). Garnet–orthopyroxene bearing granulites from Enderby Land, Antarctica: metamorphic pressure–temperature–time evolution of the Archaean Napier Complex. *Journal of Petrology*, **26(4)**, 819–56.

Harley, S.L. (1985b). Paragenetic and mineral–chemical relationships in orthoamphibole-bearing gneisses from Enderby Land, east Antarctica: a record of Proterozoic uplift. *Journal of Metamorphic Geology*, **3(2)**, 179–200.

Harley, S.L. (1987). A pyroxene-bearing meta-ironstone and other pyroxene-granulites from Tonagh Island, Enderby Land, Antarctica: further evidence for very high temperature (> 980 °C) Archaean regional metamorphism in the Napier Complex. *Journal of Metamorphic Geology*, **5(3)**, 341–56.

Harley, S.L. & Black, L.P. (1987). The Archaean geological evolution of Enderby Land, Antarctica. In *Evolution of the Lewisian and Comparable High-Grade Terranes*, ed. R.G. Park & J. Tarney, *Geology Society of London Special Publication*, **27**, 285–96.

Hensen, B.J. & Green, D.H. (1973). Experimental study of the stability of cordierite and garnet in pelitic compositions at high pressures and temperatures. III. Synthesis of experimental data and geological applications. *Contributions to Mineralogy and Petrology*, **38(2)**, 151–66.

Sandiford, M.A. (1985). The metamorphic evolution of granulites at Fyfe Hills: implications for Archaean crustal thickness in Enderby Land, Antarctica. *Journal of Metamorphic Geology*, **3(2)**, 155–78.

Sandiford, M.A. & Powell, R. (1986a). Pyroxene exsolution in granulites from Fyfe Hills, Enderby Land, Antarctica: evidence for 100 °C metamorphic temperatures in Archaean continental crust. *American Mineralogist*, **71(7–8)**, 946–54.

Sandiford, M.A. & Powell, R. (1986b). Deep crustal metamorphism during continental extension: ancient and modern examples. *Earth and Planetary Science Letters*, **79(2)**, 151–8.

Sheraton, J.W. & Black, L.P. (1983). Geochemistry of Precambrian gneisses: relevance for the evolution of the East Antarctic shield. *Lithos*, **16(4)**, 273–96.

Sheraton, J.W., Black, L.P. & McCulloch, M.T. (1984). Regional geochemical and isotopic characteristics of high-grade metamorphism of the Prydz Bay area: the extent of Proterozoic reworking of Archaean continental crust in East Antarctica. *Precambrian Research*, **26(2)**, 169–98.

Sheraton, J.W., Ellis, D.J. & Kuehner, S.M. (1985). Rare-earth element geochemistry of Archaean orthogneisses and evolution of the East Antarctic shield. *BMR Journal of Australian Geology and Geophysics*, **9(2)**, 207–18.

Sheraton, J.W., Offe, L.O., Tingey, R.J. & Ellis, D.J. (1980). Enderby Land, Antarctica – an unusual Precambrian high-grade metamorphic terrain. *Journal of the Geological Society of Australia*, **27(1)**, 1–18.

Wells, P.R.A. (1980). Thermal models for the magmatic accretion and subsequent metamorphism of continental crust. *Earth and Planetary Science Letters*, **46(2)**, 253–65.

Wernicke, B. (1985). Uniform-sense normal simple shear of the continental lithosphere. *Canadian Journal of Earth Sciences*, **22(2)**, 108–25.

Structural evolution of the Bunger Hills area of East Antarctica

P. DING & P.R. JAMES

Department of Geology and Geophysics, University of Adelaide, S.A. 5001, Australia

Abstract

Several lithostratigraphic and tectonic units have been mapped in the Bunger Hills. A layer-parallel foliation formed during an early granulite-facies metamorphism. Four phases of ductile deformation include small F_1 isoclinal or sheath folds, and regional F_2 recumbent isoclines refolded to luniforms by regional tight upright F_3 folds; a large shear-zone, bounded by a major thrust, cuts the F_3 folds to produce a regional discordance, and it is the locus of intense amphibolite-facies retrogression. F_4 regional ductile folding followed the intrusion of major charnockite bodies. Final cratonization of the Bunger Hills is indicated by reverse faults and dykes.

Introduction

The Bunger Hills 'oasis' (Fig. 1) was first studied by Soviet geologists during 1956–57 (Ravich, Klimov & Solov'ev, 1968) and by a Polish expedition in 1979–80. Ravich *et al.* (1968) described charnockitic and gneissic lithologies in detail but included little structural geology in their report. The predominant structures of their map (Ravich *et al.*, 1968) are faults inferred along topographic lineaments and two major upright folds. No geochronological data have yet been published.

The area described here (Fig. 1) represents the main outcrops of the 'oasis', excluding the northern outlying islands. It was visited by PRJ for one day in 1977 and PD spent a six-week field season there, from January to March 1986.

Description and distribution of lithologies

Migmatized pelitic and semi-pelitic gneisses form an arcuate belt up to 1 km wide in the north-west and centre of the area (Fig. 1); they also occur as thin layers or lenticular bodies within psammitic metasedimentary rocks and charnockitic gneisses. They comprise essential biotite (bi), garnet (gt), quartz (qz) and perthite (perth) with some sillimanite (sill), cordierite (cd) and orthopyroxene (opx). Metamorphosed psammitic rocks (qz + gt + perth ± sill ± cd) show a fine compositional layering, which may be original bedding. They occur predominantly in the west, with the metapelites, and are interlayered with minor charnockitic gneiss; quartz-rich varieties form impure quartzites. There are also rare small lenticular bodies of marble.

Charnockitic gneiss (perth + qz + opx ± clinopyroxene (cpx) ± hornblende (hb) ± bi) has a well-developed compositional layering (Fig. 2). It occurs as either the dominant phase (in the east) or the minor phase interlayered with supracrustals and mafic granulites in the west.

Mafic granulites (opx + cpx + plag ± bi ± hb) form about 15% of the area. There are at least two types: an early phase which intruded the metasedimentary rocks, locally truncating gneissic layering and also occurring as xenoliths or very thin layers within the charnockitic gneisses (Figs. 2 & 3), and a later, more biotite-rich phase, which cross-cut both the early granulites and the layering of the charnockitic gneisses (Fig. 4).

A large body of weakly foliated homogeneous charnockite occurs in the central northern part of the area. The body, 6–7 km in diameter, is sheet-like and regionally concordant with the belt of supracrustal rocks. However, local discordant relations and xenoliths show it to have been intrusive.

A massive intrusive charnockite forms a lens-shaped outcrop in the south-west (Fig. 1). It is discordant at its north-eastern end and it has a similar mineral composition to the weakly foliated charnockite.

Retrograde rocks, including biotite- and hornblende-rich mylonites and plagiogneisses, formed in a major thrust and shear zone cutting across the south-eastern part of the area (Fig. 1).

Structure

Origin of the layering

No primary sedimentary features have been observed. The layering in the mafic granulites and charnockitic gneisses is always associated with a layer-parallel schistosity (Fig. 2) and mineral lineation, and probably formed during early tectonothermal activity. The pelites are migmatized throughout with fine interlayering of neo- and palaeosomes. The lithological boundaries, fine compositional layering and layer-parallel foliation are all assigned to S_0.

Deformation ($D_1 - D_4$, $T_1 - T_2$)

Mafic layers (S_0) within charnockitic gneiss form rootless F_1 isoclines transected by axial planar fabrics (S_1) (Fig. 3). The

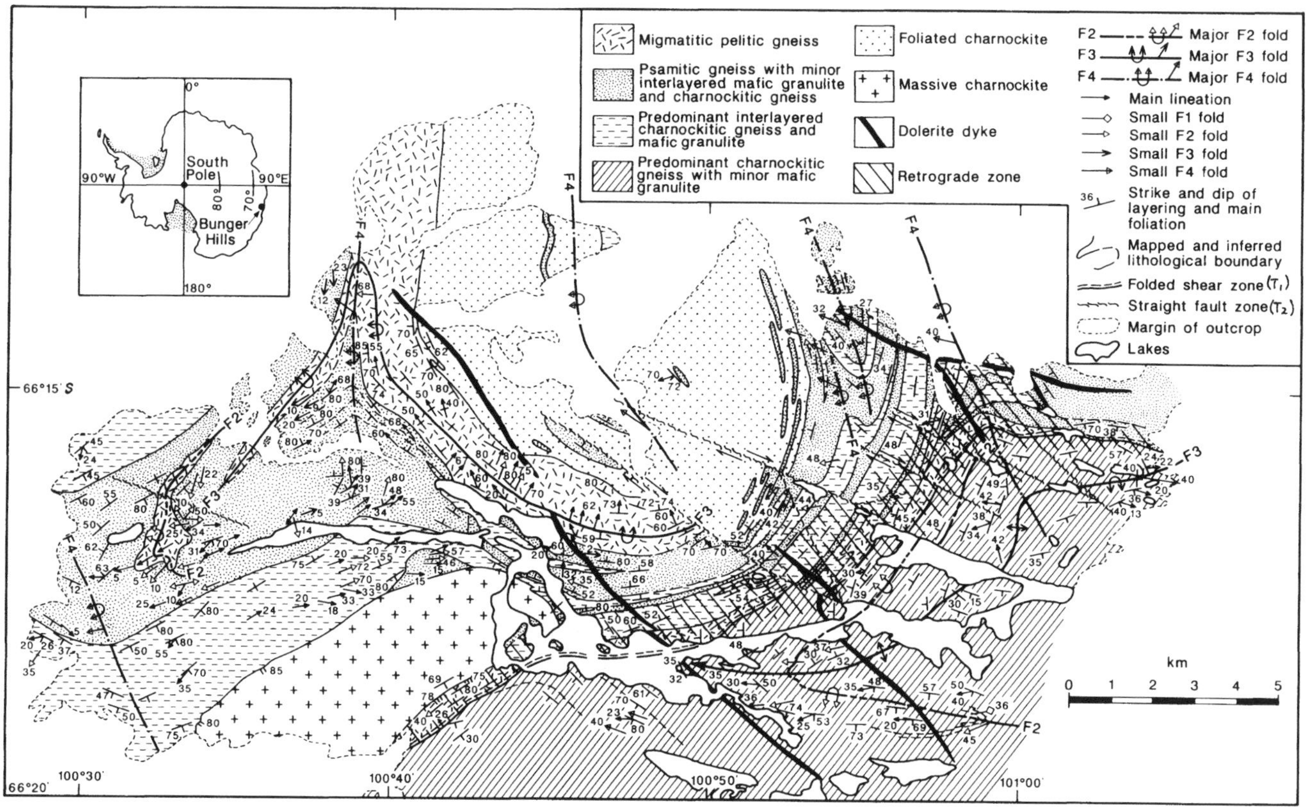

Fig. 1. Geological map of the Bunger Hills area.

folds resemble non-cylindrical sheath folds (Cobbald & Quinquis, 1980) with mineral lineations (L_1) parallel to their hinges.

Two major and many minor F_2 folds have been recognized. They have wide rounded closures and are isoclinal. The large F_2 fold in the south-east of the area (Fig. 1) plunges at 45° towards 250° and is refolded by later folds. The other major F_2 fold occurs in the west, where migmatized pelitic gneiss forms an enclosed subhorizontal isoclinal F_2 fold core. The fold plunges at 10° towards 230°, and at its SW hinge zone, there are many tight recumbent parasitic minor F_2 folds. These have shallow to subhorizontal axial planes and SW–NE-trending hinges. Refolded isoclinal F_1 folds are also common in the hinge zone.

Two major F_3 antiforms have been mapped (Fig. 1) and these are tight upright E–W-trending and plunging structures. The one in the south-eastern part of the area refolds layered gneisses and a major isoclinal F_2 fold. Its western closure, ~ 5 km across, plunges at about 30° W and SW but its eastern closure is only 1.5 km across and plunges 20–40° E. Parasitic minor F_3 folds are common around the closure of this fold, and earlier small F_2 folds are refolded. The second major F_3 fold is ~ 20 km long. It has a steep (70–85°) axial surface and shallow (0–30°) hinge, whilst the main lineations and fold axes of small F_2 folds plunge steeply on its limbs. The F_2 and F_3 fold axes vary in orientation (60–90°) and their superposition forms luniform interference folds (Ramsay, 1967). Upright parasitic F_3 folds are well developed (Fig. 5) and a spaced cleavage (S_3) has been observed locally.

A major thrust (T_1 of Fig. 1), and a 2 km wide shear zone, extend SW–NE for 20 km in the south-eastern part of the study area. These separate charnockitic gneiss from interlayered supracrustal rocks and charnockite, and represent a zone of significant displacement. F_3 folds on both sides of the shear zone are highly discordant, and none of the lithological layers can be traced across it. Rocks within this zone show intensely recrystallized mylonitic and retrograde microstructures. Some mafic layers are amphibolitized and the zone also contains small granitic gneiss bodies.

Two F_4 antiforms and one synform are the most prominent folds in the area (Figs 1 & 6). Their axial planes trend N–NNW and dip 35–85° W–WSW, and the fold axes plunge 32–80° W–WNW. The fold profiles are disharmonic due to interference by earlier folds and charnockite intrusions. The north-eastern limb of the eastern antiform trends E–SE, dipping 30–50° S–SW, while its south-western limb trends SW and dips at 30–50° NW; its axial plane trends 330°, dipping 45° SW and it has a narrow dihedral angle of ~ 50–60°. The major shear zone is folded by this antiform, and major F_4 and F_3 fold interference forms a large triangular domal structure.

The major F_4 synform in the centre of the area has the large (~ 10 km long) weakly foliated charnockite at its core. This probably intruded during D_4 and acted as a more competent body during folding. The north-eastern limb of the synform dips 20–45° SW but its south-western limb is nearly vertical, dipping steeply SW or NE; its axial plane dips 55–67° W–SW. The fold closure is rounded and up to 5 km wide, with the hinge plunging 30–45° NW. Major F_2 and F_3 folds and early lineations are folded by this synform, which is the only major syncline described by Ravich *et al.* (1968); they considered the other mesoscopic folds to represent parasitic folds.

Fig. 2. F_2 fold closure outlined by the thin layers of hornblende two-pyroxene mafic granulite and compositional layering plus possible layer-parallel fabric (all termed S_0). No axial planar fabric is developed.

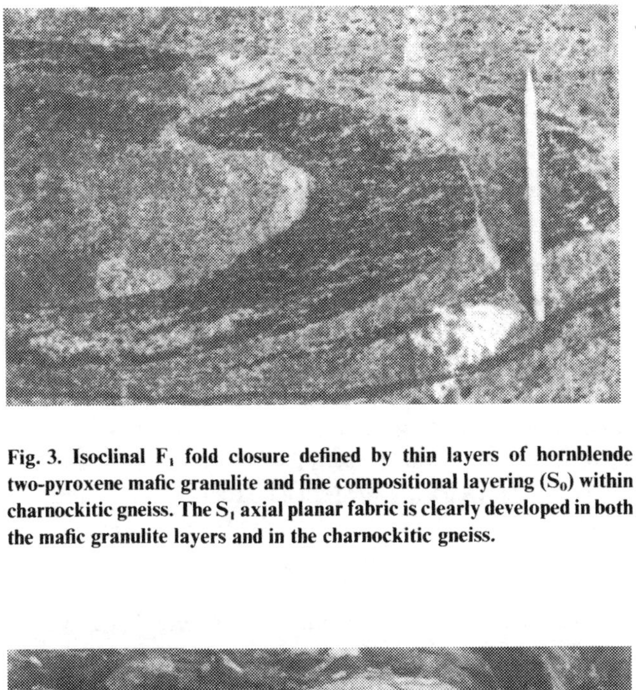

Fig. 3. Isoclinal F_1 fold closure defined by thin layers of hornblende two-pyroxene mafic granulite and fine compositional layering (S_0) within charnockitic gneiss. The S_1 axial planar fabric is clearly developed in both the mafic granulite layers and in the charnockitic gneiss.

Fig. 4. Cross-cutting relationship between two types of mafic granulite. The earlier subhorizontal one is a layered inclusion within the dominant charnockitic gneiss; the fabrics within both these rocks are parallel to each other. The later mafic granulite intruded the charnockitic gneiss and earlier mafic granulite in a direction almost perpendicular to the lithological contact and metamorphic layering.

Fig. 5. Parasitic upright F_3 folds in well-layered metasedimentary rocks and mafic granulites on the north-west side of the northern major F_3 fold (see Fig. 1). They contrast markedly with the recumbent F_1 and F_2 folds. The outcrop is about 20 m high.

Fig. 6. Large-scale tight parasitic F_4 fold in well-layered metasedimentary rocks. The closure of an early isoclinal fold (F_2 or F_1?) can be seen on its lower limb.

The western limb of the western F_4 antiform trends NE–SW, whereas its eastern limb trends NW–SE. These limbs dip 70–90° NW and SW, respectively, defining a steep (80–85°) westward-dipping axial plane and a steep (70–80°) WNW-plunging hinge. The charnockite intrusion in the south-western part of the area acted as a competent body, creating the overall triangular shape of this F_4 fold.

Cratonization of the area was essentially post-D_4. Two sets of steep reverse faults with straight traces (T_2) developed (Fig. 1), the shorter, less intense and earlier set trending at 295–310° and the later set at 315–330°. Both sets developed narrow mylonite zones (~ 1–10 m wide) which dip steeply at 60–85° NE; crenulations within the fault zones reveal a dextral and southwards uplift movement. Displacements on the faults, which were overestimated by Ravich *et al.* (1968), are usually small but two examples have estimated offsets of 500–1000 m.

Undeformed 10–20 m wide dolerite dykes intruded along both fault sets. One of the NW-trending dykes appears to be offset by a major fault (Fig. 1) but the dyke is undeformed and the fracture which it follows could have been previously offset by the W–NW-trending fault. Conjugate joints and steep normal faults represent the latest stage of deformation, followed by the intrusions of minor pegmatite and aplite veins along some of the joints.

Metamorphism, migmatization and plutonic activity

Most of the rocks belong to the granulite facies of Turner (1981), though a gradation to retrograde amphibolite-facies metamorphism occurs along the major early shear zone (Fig. 1), where the mafic granulite is altered to foliated amphibolite. Partial melting is evident in pelitic gneiss, which contains migmatitic veins involved in the earliest deformation phases. Later granitic veins were emplaced along both the layering and the axial fabric of F_4 folds in the metasedimentary rocks.

The origin of Precambrian charnockitic gneiss has been much debated (Newton & Hansen, 1983). Many Proterozoic charnockites undoubtedly had igneous precursors (e.g. the massive Stark Complex of the Adirondacks; Brock, 1980), but metasedimentary charnockites are also recognized in the Adirondacks (DeWaard, 1969) and in southern India and Sri Lanka (Cooray, 1962). The Bunger Hills charnockites probably had plutonic precursors since both massive and weakly foliated charnockite bodies clearly cross-cut metamorphic layering in the gneisses.

The origin of the well-layered charnockitic gneisses is equivocal though much evidence supports an intrusive origin:

(i) they contain major-scale xenoliths
(ii) thin layers of charnockitic gneiss intrude into the metasedimentary sequence
(iii) the composition of the charnockitic gneiss indicates 'dry', or low H_2O, high CO_2 granulite-facies conditions, while the metamorphosed supracrustal rocks represent 'hydrous' or high water content granulite-facies conditions

Reaction zones occur at the contacts between the two rock types, with hornblende-bearing mafic granulite xenoliths having hornblende-free margins. This indicates that dehydration processes were facilitated by the intrusion of 'dry' charnockitic magmas.

Summary of geological evolution and conclusions

The original shallow-water sedimentary rocks were intruded by minor mafic sills and metamorphosed to the hornblende–granulite subfacies (Turner, 1981). A metamorphic foliation and compositional layering (S_0) and mineral lineation formed, while partial melting occurred during, or soon after the peak of metamorphism. This sequence of progressive basin evolution, subsidence and syntectonic metamorphism may reflect a period of early crustal extension (Sandiford & Powell, 1986), and was followed by subconcordant intrusions of acidic to intermediate 'dry' magma which were metamorphosed to charnockitic gneiss.

The original orientation of F_1 isoclinal sheath folds is uncertain because of complex interference by later folds. Recumbent F_2 isoclines may have had NE–SW-trending hinges but they are also reorientated by later folds. F_3 folds with original E–W-trending axial planes and subhorizontal hinges, probably formed due to a subhorizontal maximum principal compression from the NNW. The WSW–ENE-trending subconcordant major thrust (T_1) which occurred after D_3, moved the metamorphosed supracrustal northern block over the mainly orthogneissic southern block, probably as part of a progressive deformation during continued subhorizontal and NNW-directed stress. Retrograde amphibolite-facies metamorphism occurred along the major T_1 shear zone, followed by mafic dyke intrusions. Although undeformed these dykes have undergone granulite-facies metamorphism. Two acidic to intermediate charnockite intrusive bodies then intruded the northern block. The final ductile deformation produced tight angular F_4 folds formed under a still subhorizontal but now WSW–ENE compression. The more competent massive southern charnockite greatly influenced the disharmonic geometry of the western F_4 antiform, whereas the more layered and weakly foliated northern charnockite formed the core of the central F_4 synform.

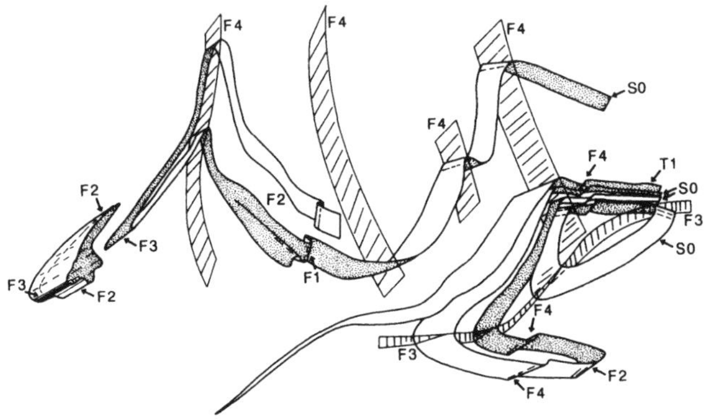

Fig. 7. Three-dimensional sketch illustrating the overall structural geometry of the Bunger Hills area. The sketch covers the area of Fig. 1 and is 27.5 km in width.

The preceding sequence of mainly ductile overprinted major folding events produced a high-grade terrane with an extremely complicated three-dimensional morphology. Fig. 7 is a simplified 3-D cartoon showing these structures. Similarly complex geometries of superposed major refolded high-grade gneisses are common in both Archaean and Early Proterozoic terranes. The characteristic feature of an early recumbent fabric (Park, 1981) and a complexly shortened later history may be a product of repeated coaxial folding and refolding by gravitational collapse (Wynne-Edwards, 1963) or it could be due to progressive unidirectional non-coaxial shear (Cobbald & Quinquis, 1980).

Acknowledgements

The Australian Antarctic Division are thanked for logistic support; the Adelaide University Department of Geology and Geophysics for financial assistance. Sherry Proferes drafted the map figures and Rick Barrett reproduced the photographic plates.

References

Brock, B.S. (1980). Stark complex (Dexter Lake area): petrology, chemistry, structure and relation to other green rock complexes and layered gneisses, north Adirondacks, New York. *Geological Society of American Bulletin*, **91(1)**, 93–7.

Cobbald, P.R. & Quinquis, H. (1980). Development of sheath folds in shear regimes. *Journal of Structural Geology*, **2(1/2)**, 119–26.

Cooray, P.G. (1962). Charnockites and their associated gneisses in the Precambrian of Ceylon. *Quarterly Journal of the Geological Society of London*, **118(3)**, 239–66.

DeWaard, D. (1969). The occurrence of charnockite in the Adirondacks: a note on the origin and definition of charnockite. *American Journal of Science*, **267(8)**, 983–7.

Newton, R.C. & Hansen, E.C. (1983). The origin of Proterozoic and Late Archean charnockites – evidence from field relations and experimental petrology. In *Proterozoic Geology*, ed. L.G. Medaris, pp. 167–78. Madison; University of Wisconsin Press.

Park, R.G. (1981). Origin of horizontal structure in Archaean high-grade terrains. In *Archaean Geology*, ed. J.E. Glover & D.I. Groves, pp. 481–90. Perth; Geological Society of Australia Inc.

Ramsay, J.G. (1967). *Folding and Fracturing of Rocks*. New York; McGraw-Hill, 568 pp.

Ravich, M.G., Klimov, L.V. & Solov'ev, D.S. (1968). *The Pre-cambrian of East Antarctica*, ed. N.G. Sudovikov. Jerusalem; Israel Program for Scientific Translations (translated from Russian). 475 pp.

Sandiford, M. & Powell, R. (1986). Deep crustal metamorphism during continental extension: modern and ancient examples. *Earth and Planetary Science Letters*, **79(1/2)**, 151–8.

Turner, F.J. (1981). *Metamorphic Petrology*. New York; McGraw-Hill, 524 pp.

Wynne-Edwards, H.R. (1963). Flow folding. *American Journal of Science*, **261**, 793–814.

Structural geology of the early Precambrian gneisses of northern Fold Island, Mawson Coast, East Antarctica

P.R. JAMES[1], P. DING[1] & L. RANKIN[2]

1 Department of Geology and Geophysics, University of Adelaide, South Australia, 5001, Australia
2 South Australia Department of Mines and Energy, PO Box 151, Eastwood, South Australia, 5063, Australia

Abstract

The relative proportions of reworked Archaean Napier Complex and new Proterozoic crust within the ~ 1000 Ma Rayner Complex of East Antarctica is controversial. The high-grade granulite-facies gneisses of northern Fold Island, within the Rayner Complex, comprise an early supracrustal sequence of quartzose paragneisses and mafic granulites interleaved with an intrusive series of layered pyroxene charnockitic gneisses. Mafic sheets within all the gneisses represent remnants of at least two suites of highly distorted mafic dykes, plus a suite of undeformed mafic dykes which retains a granulite-facies mineralogy. Five main generations of superposed fold events affected the gneisses; the three earlier ones were mostly isoclinal minor fold events. The earliest dykes may predate the earliest recognized deformation D_1, while a major suite intruded post-D_2 and pre-D_3. The later two fold events produced major folds, with D_4 forming a plunging inclined E–W overfold, and D_5 refolding it. Late mylonites and pseudotachylite-generating faults record the excavation of the gneissic terrane. Pre-D_3 activity is related to Napier Complex deformation, signifying that most of this area is Napier Complex-equivalent reworked during D_{3-5} and by semi-brittle deformation during the Rayner event.

Introduction

The East Antarctic craton is composed of Archaean high-grade (granulite-facies) stable cratons surrounded by predominantly Middle–Late Proterozoic mobile belts (James & Tingey, 1983). These younger belts contain significant supracrustal sequences, some of which are also Proterozoic (Tingey, 1982). However, the relative proportions of reworked or reactivated intracratonic continental Archaean crust, as opposed to juvenile Proterozoic crust newly accreted to continental margins or formed in orogenic collision zones, is not known.

Detailed structural, metamorphic and geochronological studies of parts of the coastal outcrops, between Mawson and Molodheznaya bases in East Antarctica (Sheraton *et al.*, 1980; Black, James & Harley, 1983a, b; Black, Sheraton & James, 1986; Black *et al.*, 1987; Ellis, 1983), have shown that the Archaean Napier Complex of Enderby Land is quite different from the surrounding Middle Proterozoic Rayner Complex. The structural evolution of the Napier Complex has been described in detail (James & Black, 1981; Black & James, 1983) and its Archaean geological history is well documented (Black *et al.*, 1986). On the other hand, the origin and evolution of the Rayner Complex are much less well understood. Sheraton *et al.* (1980), Ellis (1983) and Sheraton & Black (1983) considered that much of the Rayner Complex represents reworked, and specifically rehydrated, Napier Complex granulites, whereas Black *et al.* (1987) presented Sm–Nd isotopic evidence to show that much of the belt is a product of Middle–Late Proterozoic reworking of Early–Middle Proterozoic juvenile crust. Evidence of reworked Archaean crust in the Rayner Complex of Fold Island has been provided by Grew *et al.* (1988). Although no other geological or detailed structural studies have been reported from the Rayner Complex, Black *et al.* (1987) described mesoscopic folds which predated the prominent E–W folding characteristic of, and parallel to the regional strike of the Rayner Belt (Sheraton *et al.*, 1980).

Fold Island (Fig. 1), off the Mawson Coast of Kemp Land, lies within the broad E–W trending belt of the ~ 1000 Ma Rayner Complex (Sheraton *et al.*, 1980; Black *et al.*, 1987). The Napier Complex lies 100 km west of Fold Island, the boundary between the two in the King Edward VIII Gulf area being a transitional one over tens of kilometres (Black *et al.*, 1987). One hundred kilometres to the east lies the ~ 1100 Ma Mawson Charnockite pluton (Sheraton, 1982).

After a reconnaissance traverse of the Mac. Robertson Land coastline, Trail *et al.* (1967) described the geology of northern and central Stillwell Hills and southern Fold Island. Their geological map (Trail *et al.*, 1967; fig. 5) shows gneissic layering, and they described a layer-parallel fabric containing early isoclinal folds on the limbs of major NE–SW- and N–S-trending later folds.

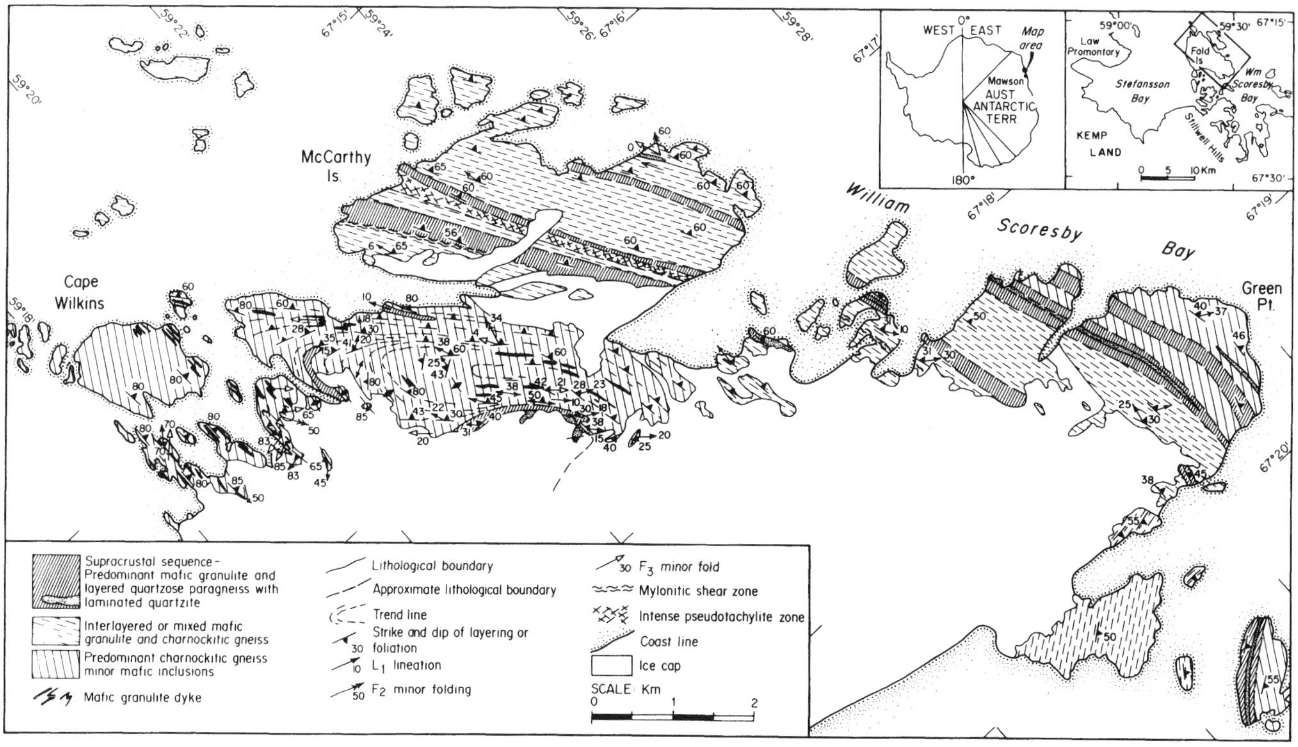

Fig. 1. Geological map of northern Fold Island.

Description of lithologies

Northern Fold Island consists of high-grade (granulite-facies), complexly deformed metamorphic tectonites which can be spatially separated into three main lithostratigraphic subdivisions (Fig. 1). The oldest lithologies are a supracrustal sequence of mixed mafic granulites and quartz-rich paragneisses which form continuous layers across the southern outcrops and the western side of McCarthy Island. Felsic granulites or pyroxene-charnockitic gneisses are dominant in the north-western outcrops south of Cape Wilkins. The charnockitic gneisses contain a variety of mafic, ultramafic and paragneiss inclusions, which are considered in part to represent the remnants of the earlier supracrustal sequence. Associated with the supracrustal sequence in the south and east are areas characterized by alternations of mafic granulite, garnetiferous paragneiss, quartzite and charnockitic gneiss; these probably represent the more deformed supracrustal sequence which had been partially intruded by charnockitic gneiss.

The supracrustal sequence

The dominant lithology is a homogenous two-pyroxene plagioclase gneiss or mafic granulite, occurring as massive continuous layers within layered paragneisses. Mafic granulite layers vary from 2–5 cm to 2 m in width, though mappable units up to 10 m wide occur (Fig. 1). Mafic granulite layers are homogeneous except for migmatitic veins and schlieren. Low-angle discordances between the granulite layers and the host gneisses indicate that at least some of the mafic layers may represent originally intrusive sheets or sills, whilst others may represent basaltic flows. Minor ultramafic units are interlayered with the mafic layers.

Paragneisses form another distinct but minor proportion of the supracrustal sequences. These are mostly quartz-rich, ranging from red to pale pink garnet-rich quartzites and gneisses, to alternating grey and white orthoquartzites. The layering in the quartzites is markedly continuous and may reflect an original, if highly transposed, sedimentary lamination. The supracrustal sequence may represent the intensely interleaved and distorted remnants of a former submarine ophiolite-like mixed sedimentary–igneous suite.

Pyroxene gneiss (charnockitic granulite)

The outcrops south of Cape Wilkins are composed of nebulitic, mixed homogeneous to layered, predominantly acidic pyroxene (charnockitic) gneisses (Fig. 1). The gneisses contain few marker horizons but are characteristically layered; this layering varies considerably in thickness and extent and is defined both by composition and texture. There are three principal components to the gneissic layering.

(a) *Finely layered nebulitic charnockite.* This is the predominant lithology of the subarea, making up 60–70% of the gneiss. Its layering (on a mm–cm scale) is defined by varying proportions of mafic (pyroxenes) v. felsic (quartz–feldspar) phases (Fig. 2). The layers are complexly folded and cross-cut by mafic pods and lenses or discordant acid migmatitic neosomes.

(b) *Charnockitic migmatite.* The pyroxene gneisses are infiltrated by coarse homogeneous acidic migmatite veins and lenses, which vary from intensely contorted ptygmatic veinlets to major discordant pegmatites, and display a uniform orthopyroxene–quartz–feldspar mineralogy. The charnockitic mig-

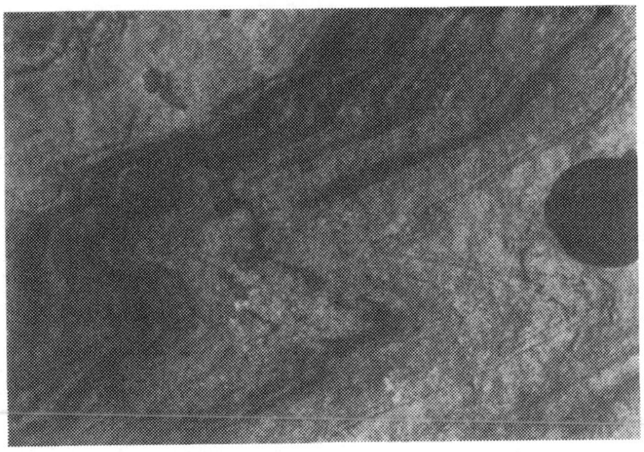

Fig. 2. Minor (F_3) fold of fine compositional layering in charnockitic gneiss.

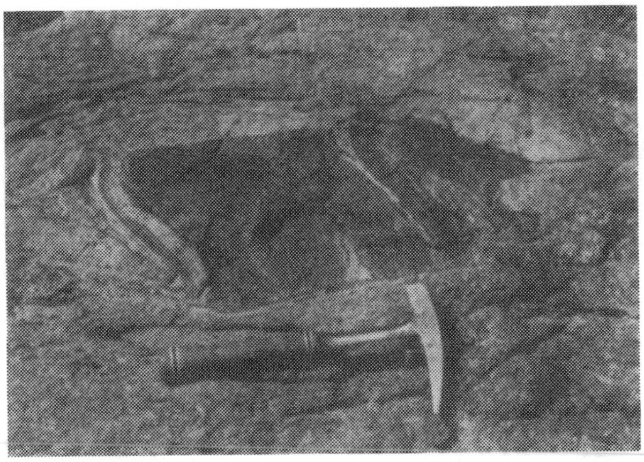

Fig. 4. Mafic xenolith with internal compositional and migmatitic layering highly oblique to, and cut by external S_0/S_1 fabric in charnockitic gneiss.

matite often grades into homogeneous charnockitic gneiss host, indicating a likely intrusive origin for both lithologies.

(c) *Mafic and ultramafic bodies*. The finely layered–homogeneous charnockitic gneiss has many inclusions of dark-coloured, compositionally homogeneous mafic and ultramafic gneiss (Fig. 3). All are broadly parallel to the layering of the host gneiss. The bodies vary in morphology from tabular sheets to near-equidimensional pods with sharply tapered terminations. Size varies from centimetres to metres in width and up to 200–400 m in length. Small and large bodies lie in trains generally parallel to the local layering orientation (Fig. 3). Some of these layers are relics of formerly continuous supracrustal units and some small pods display an internal migmatitic layering and strong layer-parallel foliation lying at a high angle to, and truncated by the first foliation developed in the host gneiss (Fig. 4). Some of the bodies cross-cut the layering showing very low angular discordances. These are boudinaged layers which were originally continuous possibly representing a highly deformed major dyke suite. The dyke remnants indicate at least two separate periods of dyke intrusion into the Fold Island orthogneisses, which support the likely Archaean origins and Proterozoic reworking of these orthogneisses (Sheraton & Black, 1981).

Fig. 3. Cliff exposure of mafic dyke trains.

Interlayered or mixed supracrustal rocks and charnockite

On McCarthy Island and to the south (Fig. 1) the supracrustal rocks are represented by broadly concordant sheets of mafic granulite and minor ultramafic rocks and quartzose paragneisses, interleaved with homogeneous and finely layered charnockitic gneiss.

Structure

Five superposed fold events occur in the area, with three earlier minor-scale fold phases (F_1, F_2 and F_3) refolded by two generations of regional-scale folds (F_4 and F_5). The structural data are shown in Fig. 5 and the structural geometry in Fig 6.

First deformation (D_1)

F_1 minor folds in the paragneisses are isoclinal, with grossly thickened hinge areas and thinned parallel limbs. They occur also as small rootless folds outlined by mafic or ultramafic lenses within charnockitic gneiss. F_1 folds are common in complex minor superimposed fold patterns, where they are overprinted by the later fold phases.

S_1 is the dominant fabric in all gneisses and involved considerable flattening. It is a shape fabric defined by orthopyroxene and quartz aggregates, and quartz–feldspar ribbons. The fabric also defines a stretching lineation (L_1), and some mafic inclusions form rods reflecting the transposed hinges of F_1 folds, which always are parallel to L_1.

Second deformation (D_2)

D_2 minor structures are recognized by interference relationships with both earlier and later fold structures. F_2 minor folds affect both the layering and S_1 foliation without the development of a new axial planar fabric (Fig. 7). The folds are isoclinal with thickened but more rounded hinges than F_1 folds.

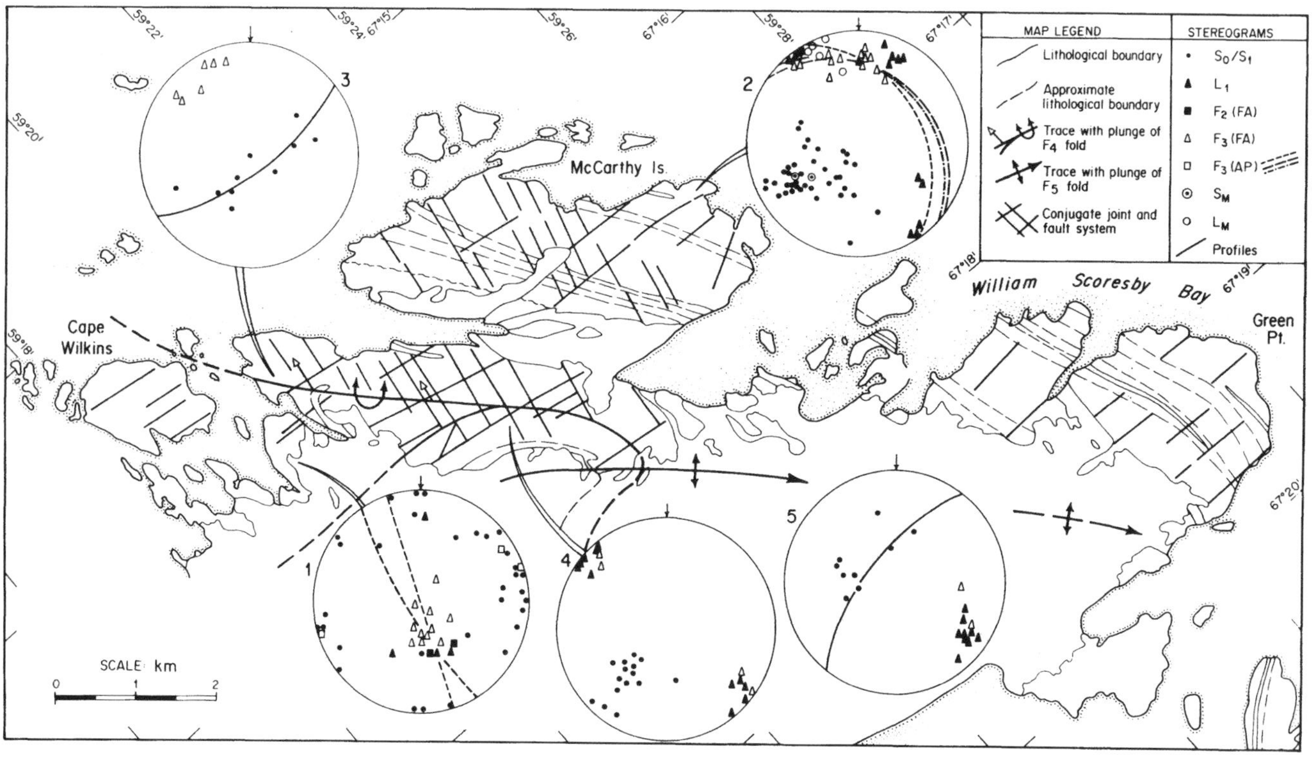

Fig. 5. Structural map and stereograms (for explanation see text) of northern Fold Island. FA, fold axes; AP, axial plane.

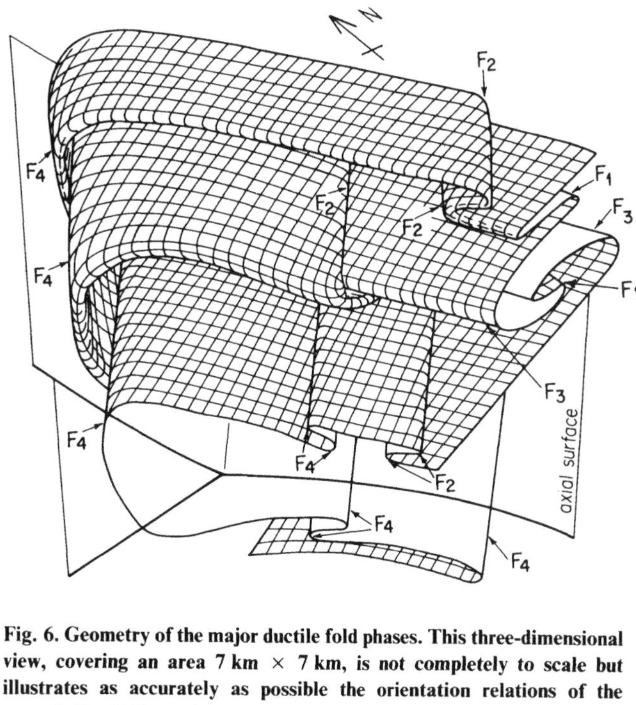

Fig. 6. Geometry of the major ductile fold phases. This three-dimensional view, covering an area 7 km × 7 km, is not completely to scale but illustrates as accurately as possible the orientation relations of the overprinting fold events.

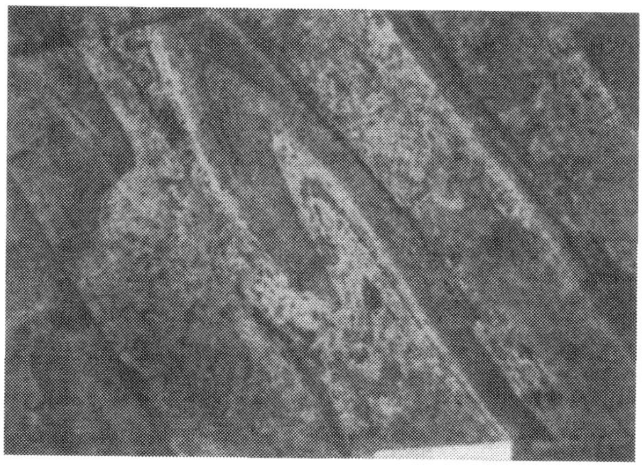

Fig. 7. F_1 isocline refolded by F_2. Scale object is 10 cm long.

Third deformation (D_3)

F_3 folds vary from tight to isoclinal and exhibit mappable asymmetry and vergence variation. However, due to the lack of marker horizons, no major F_3 folds have been recognized. F_2 and F_3 folds show wide angular discordance of plunge in minor superposed fold patterns to form luniform fold interference patterns.

The major suite of mafic dyke inclusions in the charnockitic gneisses intruded subparallel to the layering (Fig. 3). They have been folded by the F_3 folds but occasionally are discordant to the layering and F_2 axial surfaces. This suggests that the dyke suite may have been intruded post-F_2 and pre-F_3.

Fourth deformation (D_4)

The most significant regional structure is a major plunging to inclined fold which affected the whole area (Figs. 1 & 5). Only the north-eastern limb and part of the hinge and south-western limb are exposed, the remainder being covered by an ice-cap. The fold has a tight to isoclinal core and mostly parallel limbs but it has an unusual rectangular closure (Fig. 6). Fig. 5 shows that the major F_4 axial trace trends mainly NW but becomes double hinged, with a branch trending E–W south of Cape Wilkins.

The exposed hinge zone of the F_4 fold is about 3 km wide and 2 km long. In this steep zone (Fig. 5, stereogram 1), the layering and early folds define mostly steeply dipping and plunging structures. F_3, the most common minor folds, have plunges varying from 45–85° toward the S, SE and SW. Thus there is a variation of between 40–90° difference in orientation of the major F_4 and minor F_3 fold plunges. In the north-east and south-west limbs of the F_4 fold (stereograms 2 and 4, Fig. 5), layering and early planar structures have uniform trends but variable inclinations.

F_4 minor folds with close–tight and rounded profiles are common, with correct vergence variations on the shallowly inclined limbs of the major structure (Fig. 8).

Fifth deformation (D_5)

A regional major upright antiform refolded the F_4 axial trace at the south-western extremity of Fold Island (Fig. 5). This F_5 fold has a closure plunging 15–20° towards 130° (Fig. 5, stereogram 5).

Mylonite and pseudotachylite

Narrow layer-parallel mylonite zones developed after folding. These zones (2–50 mm wide) are normally bordered by thin (~ 30 cm) bands of mylonite–ultramylonite and grade inwards to protomylonitic gneiss, with anastomosing networks of small-scale faults and ultramylonite bands (Figs 1 and 5). The thickest development of pseudotachylite occurs in breccia zones (0.5–3.0 m wide) which are typically layer-parallel. These zones of intense brittle deformation are commonly bounded by thin (5 mm) ultramylonite layers.

Mylonite and pseudotachylite generation along fracture and shear zones is also associated with the emplacement of mafic dykes and coarse pegmatites. Some mafic dykes retain a granulite-facies assemblage indicating that such *P–T* conditions still acted during the initiation of pseudotachylite-generating fractures and uplift.

With continued deformation and uplift, pseudotachylite generation was replaced by the production of an intersecting conjugate set of steeply inclined joints (Fig. 5).

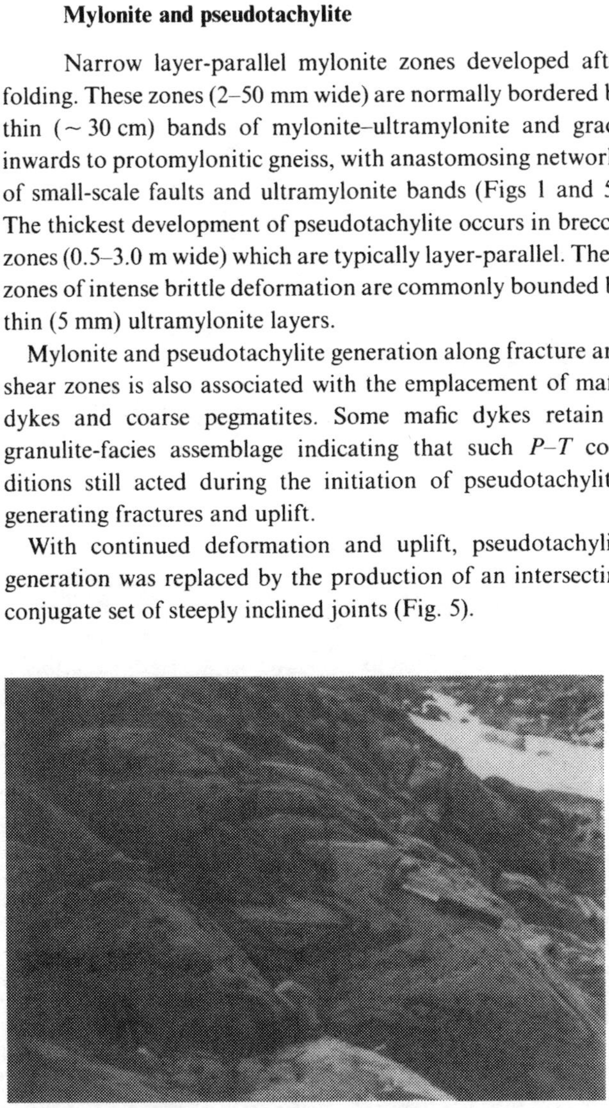

Fig. 8. Overturned minor F_4 folds in charnockitic gneiss.

References

Black, L.P., Harley, S.L., Sun, S.S. & McCulloch, M.T. (1987). The Rayner Complex of East Antarctica: complex isotopic systematics within a Proterozoic mobile belt. *Journal of Metamorphic Geology*, **5(1)**, 1–26.

Black, L.P. & James, P.R. (1983). Geological history of the Napier Complex, Enderby Land, East Antarctica. In *Antarctic Earth Science*, ed. R.L. Oliver, P.R. James & J.B. Jago, pp. 11–15. Canberra; Australian Academy of Science and Cambridge; Cambridge University Press.

Black, L.P., James, P.R. & Harley, S.L. (1983a). The geology and geochronology of reputedly ancient rocks of Fyfe Hills, Enderby Land, Antarctica. *Precambrian Research*, **21(3/4)**, 197–222.

Black, L.P., James, P.R. & Harley, S.L. (1983b). The correlation of geochronology with geological history from high-grade metamorphic tectonites of the Field Islands, Enderby Land, East Antarctica. *Journal of Metamorphic Geology*, **1(4)**, 277–303.

Black, L.P., Sheraton, J.W. & James, P.R. (1986). Late Archaean granites of the Napier Complex, Enderby Land, Antarctica: a comparison of Rb–Sr, Sm–Nd and U–Pb isotopic systematics in a complex terrain. *Precambrian Research*, **32(4)**, 343–68.

Ellis, D.J. (1983). The Napier and Rayner Complexes of Enderby Land, Antarctica: contrasting styles of metamorphism and tectonism. In *Antarctic Earth Science*, ed. R.L. Oliver, P.R. James & J.B. Jago, pp. 20–4. Canberra; Australian Academy of Science and Cambridge; Cambridge University Press.

Grew, E.S., Manton, W.I. & James, P.R. (1988). Archaean and Proterozoic U–Pb ages on granulite facies rocks from Fold Island, Kemp Coast, East Antarctica. *Precambrian Research*, **42**, 63–75.

James, P.R. & Black, L.P. (1981). A review of the structural evolution and geochronology of the Archaean Napier Complex of Enderby Land, Australian Antarctic Territory. In *Archaean Geology*, ed. J.E. Glover & D.I. Groves. *Geological Society of Australia Special Publication*, **No. 7**, 71–83.

James, P.R. & Tingey, R. (1983). The Precambrian geology of East Antarctica, a review. In *Antarctic Earth Science*, ed. R.L. Oliver, P.R. James & J.B. Jago, pp. 5–10. Canberra; Australian Academy of Science and Cambridge; Cambridge University Press.

Sheraton, J.W. (1982). Origin of charnockitic rocks of Mac. Robertson Land. In *Antarctic Geoscience*, ed. C. Craddock, pp. 489–97. Madison; University of Wisconsin Press.

Sheraton, J.W. & Black, L.P. (1981). Geochemistry and geochronology of Proterozoic tholeiitic dykes of East Antarctica: evidence for mantle metasomatism. *Contributions to Mineralogy and Petrology*, **78(3)**, 305–17.

Sheraton, J.W. & Black, L.P. (1983). Geochemistry of Precambrian gneisses: relevance for the evolution of the East Antarctic shield. *Lithos*, **16(4)**, 273–96.

Sheraton, J.W., Offe, L.A., Tingey, R.J. & Ellis, D.J. (1980). Enderby Land, Antarctica – an unusual Precambrian high grade metamorphic terrain. *Journal of the Geological Society of Australia*, **27(1)**, 1–18.

Tingey, R.J. (1982). The geologic evolution of the Prince Charles Mountains – an Antarctic Archaean Cratonic Block. In *Antarctic Geoscience*, ed. C. Craddock, pp. 455–63. Madison; University of Wisconsin Press.

Trail, D.S., McLeod, I.R., Cook, P.J. & Wallis, G.R. (1967). Geological investigations by the Australian National Antarctic Research Expeditions, 1965. *Australian National Antarctic Research Expeditions Scientific Reports, Series A (III) Geology*, **No. 100**, 48 pp.

The intrusive Mawson charnockites: evidence for a compressional plate margin setting of the Proterozoic mobile belt of East Antarctica

D.N. YOUNG & D.J. ELLIS

Department of Geology, Australian National University, PO Box 4, Canberra, Australian Capital Territory, 2601

Abstract

The deformed and metamorphosed charnockites of the Mawson Coast have relict igneous features. High magmatic temperatures are indicated by partially melted granulite facies gneissic xenoliths, relict antiperthites with high Or, and rare coarsely exsolved pyroxenes. Estimates of the age and level of emplacement are 970 Ma and 6 kb. The charnockites are intermediate in composition, unlike most Proterozoic granitoids which are markedly felsic or bimodal. A high- and a low-Ti group of charnockites are recognized. Mafic tonalitic compositions are inferred for the parent magmas of both groups, although the high-Ti rocks are notably higher in K, Ti, P and light rare earth elements (LREE) than modern or Cordilleran tonalites. Derivation by melting of mainly mafic to intermediate crust is the preferred petrogenetic model, with an old sedimentary component to account for highly evolved initial Sr- and Nd-isotopic compositions. Many rocks have very high K/Na ratios for their high CaO and low SiO_2 contents, probably caused in part by K-feldspar accumulation. Depletions of heavy rare earth elements (HREE) in the low-Ti charnockites suggest that garnet was a residual phase in partial melting, which requires high pressures and a thickened crust. The tectonic setting is thus tentatively likened to an Andean or Himalayan type plate margin, with the inference that subduction was active during the Proterozoic.

Introduction

The complex Proterozoic mobile belt of East Antarctica, which separates Archaean cratonic blocks (Ravich & Kamenev, 1975; Sheraton *et al.*, 1987), contains large bodies of massive charnockite: that is, felsic to intermediate, orthopyroxene-bearing intrusive rocks. This paper summarizes the petrology of orthopyroxene granitoids from the Mawson Coast and the Framnes Mountains of Mac. Robertson Land. There is widespread debate whether charnockites are the result of igneous or metamorphic processes (Newton & Hansen, 1983; Wickham, 1988). The Mawson charnockites serve not only as a case study of charnockite genesis but as an indicator of the palaeotectonic setting of the Proterozoic mobile belt.

Geological setting

Area of exposure

The term Mawson Charnockite has become entrenched in Antarctic literature as a term for the orthopyroxene-bearing felsic rocks around Mawson station in Mac. Robertson Land (Trail, 1970; Sheraton, 1982). It is among the largest intrusive massifs in East Antarctica with a minimum exposed area, estimated from the discontinuous exposures, of 3130 km^2 (Fig. 1, Table 1). The western limit of outcrop is about 90 km west of Mawson where intrusive contacts with banded gneiss can be seen. The northern, southern and eastern limits of the charnockites are unknown because of sea and ice cover. A central belt of scattered outcrops of basement gneiss separates the charnockite into two groups of outcrops (Fig. 1) and there are some distinctions between charnockite types on either side.

Structure and metamorphism

Three main ductile deformations are recognized in the study area and the structures produced are described in detail by Clarke (1988). All of these affected the basement gneiss sequence but the charnockites only preserve evidence for D_3, hence they intruded between D_2 and D_3. The earliest recognized deformation (D_1) produced isoclinal folds in the gneiss and the D_2 and D_3 events refolded the D_1 structures about nearly upright fold axes striking approximately easterly and northerly, respectively. Cognate enclaves and xenoliths are

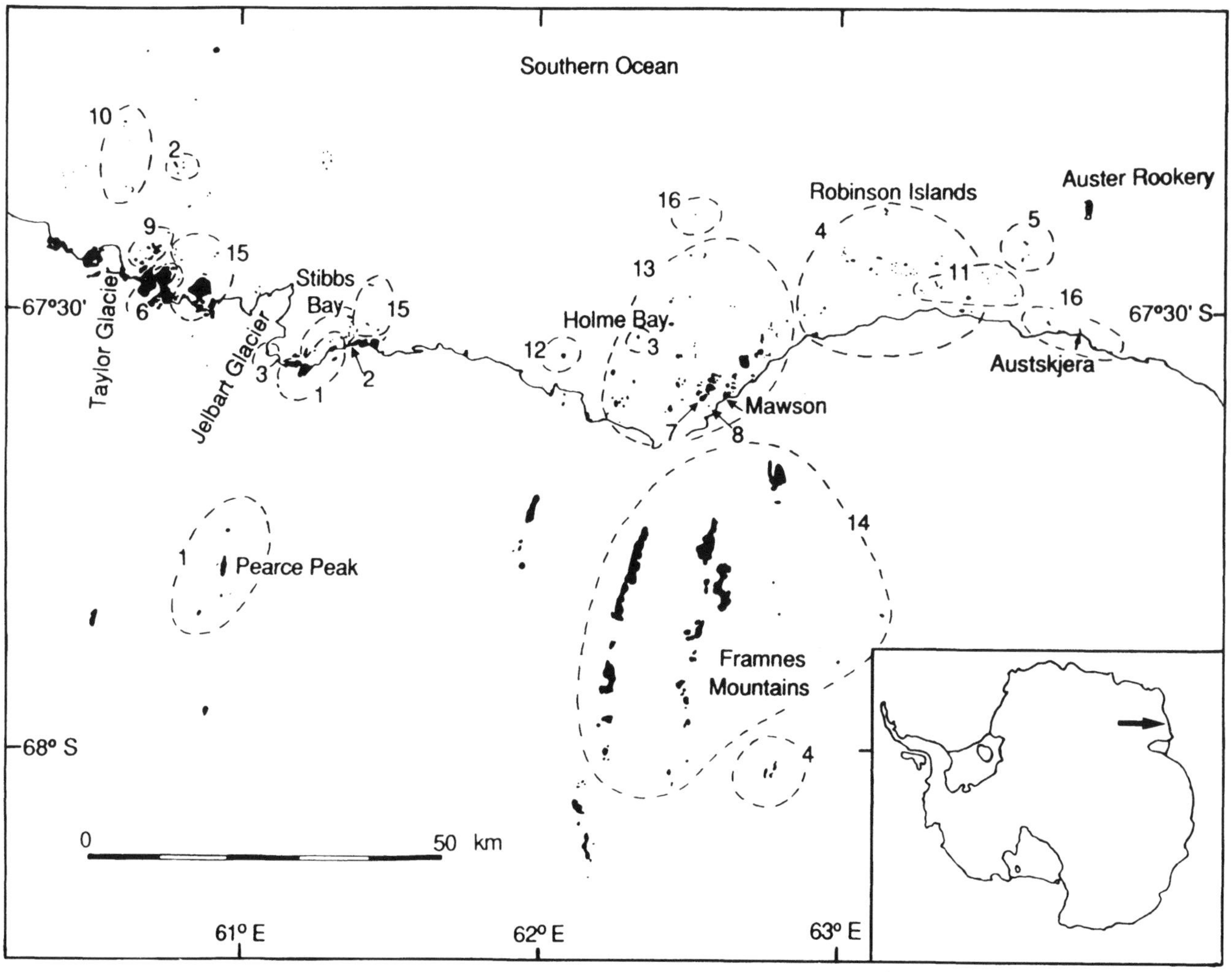

Fig. 1. Map of the Mawson Coast, Mac. Robertson Land, with mountain, nunatak and island outcrops marked in black. The location of different charnockite suites (generalized in some cases) is noted by dashed lines and numbers corresponding to those in Table 1. Outcrops of basement gneiss are those which lie outside the dashed lines. Smaller outcrops and xenolithic blocks of gneiss occur within several of the areas marked as charnockite suites but are too small to be shown here. Arrow on inset points to location of Mawson station.

flattened and rotated into near parallelism with the northerly foliation, and most dykes are also folded by D_3. Subsequent deformation is shown by a series of retrograde shear zones, cut by later mylonite–pseudotachylite zones (Clarke, 1988).

The prevailing metamorphic grade in the basement gneiss is of granulite facies, as seen by the following common mineral assemblages: orthopyroxene plus K-feldspar in felsic gneiss; orthopyroxene plus clinopyroxene in mafic gneiss; clinopyroxene, plagioclase, scapolite plus quartz in calcsilicates; and garnet, cordierite, sillimanite plus biotite in pelitic gneiss. These assemblages, and hence the metamorphic peak, were thought by Clarke, Powell & Guiraud (1989) to predate the first deformation recognized.

Pressure estimates from garnet–orthopyroxene–plagioclase–quartz assemblages in the charnockites ranges from 5.5 to 7.0 kb using the geobarometers of Perkins & Chipera (1985) and provide a minimum estimate of the depth of emplacement by recording D_3 metamorphic resetting. Temperature estimates from the charnockites using two-pyroxene (Lindsley, 1983) and garnet–orthopyroxene (Harley, 1984) thermometry range from 600 to 700 °C and almost certainly represent metamorphic resetting. Direct evidence for high temperatures is provided by relict crystals of antiperthitic plagioclase in the charnockites. A typical reintegrated bulk composition is An_{44}–Ab_{45}–Or_{11}, which suggests temperatures greater than 1000 °C (Fuhrman & Lindsley, 1988). Coexisting high temperature K-feldspar is lacking: even megacrysts have low An and Ab contents suggesting metamorphic re-equilibration.

Geochronology

The crystallization age of the charnockites is best estimated by ion-probe U–Pb dates of 954 ± 12 Ma (2 sigma error) and 985 ± 29 Ma (Young & Black, unpublished data), obtained from single zircon crystals with apparently igneous zoning and morphology. Young & Black (unplublished data) estimate the age of D_3 as 921 ± 19 Ma, using U–Pb analysis of metamorphic zircon rims. The age of the initial two deformations and the basement gneiss is not well constrained. Rb–Sr isochron dates, obtained by P. A. Arriens (pers. comm.) of 1254 ± 31 and 1153 ± 47 Ma (recalculated using the decay constant of Steiger & Jager, 1977) come from two banded

Table 1. *Summary of distinctive features and names of the charnockite suites shown geographically in Fig. 1*
Unless otherwise stated the suites are even-grained for the Stibbs Bay Group, K-feldspar megacrystic for the Holme Bay Group and intermediate in composition. Any differences from the basic mineral assemblage of orthopyroxene, biotite, feldspars and quartz are noted in parenthesis.

Name of suite	Area (km^2)	Distinguishing features
Stibbs Bay Group (low-Ti)		
1. Falla Bluff	215	Mafic, low K/Na, HREE-depleted, high Cr, Ni
2. Alfons Island	40	Felsic, HREE strongly depleted, variable K/Na
3. Jelbart Glacier	20	Enriched in HREE (garnet)
4. Robinson Islands	465	High K/Na, similar to high-Ti rocks except for high Cr, Ni, Mg, low Ti (garnet, kf-megacrysts)
5. Auster Island	35	Some felsic samples, depleted HREE
6. Byrd Head	25	High Na_2O, depleted HREE, high LREE
7. Evans Island	5	High K_2O, La, Ce, Th (kf-megacrysts in some)
8. Departure Rocks	5	Coarse-grained, biotite-poor, undepleted HREE (garnet)
9. Colbeck Archipelago	25	Mafic, higher Ti, P and low Cr (amphibole)
10. Thorfinn Islands	75	Low K/Na, undepleted HREE (garnet)
11. Macklin Island	65	Mafic gabbroic rocks (clinopyroxene)
12. Gibbney Island	15	Felsic (garnet)
Total	990	
Holme Bay Group (high-Ti)		
13. Mawson Islands	540	Some felsic, some even-grained rocks (clinopyroxene in mafic rocks, garnet in felsic rocks)
14. Framnes Mountains	1385	Wide range of composition (clinopyroxene in mafic rocks, garnet in felsic rocks)
15. Ufs Island	130	Mainly even-grained mafic rocks (some clinopyroxene, amphibole)
16. Austskjera	85	Lower K_2O, peraluminous (garnet)
Total	2140	
Overall total	3130	(minimum estimate)

gneiss outcrops in the Framnes Mountains (Fig. 1). However, it is uncertain whether these represent partial or total metamorphic overprints of the protolith age.

Charnockitic intrusives

Field geology

The Mawson coast charnockites are foliated orthogneisses with intrusive contacts into gneiss and other charnockites, and containing xenolithic blocks of banded gneiss and mafic granulite (Crohn, 1959). They presumably take the form of diapiric plutons or sheets up to a few kilometres wide, although incomplete outcrop does not enable detailed observations. Charnockite dykes 1–5 m wide also occur. For the most part the field appearance of the charnockites is homogeneous.

The basement gneisses are mainly quartzo-feldspathic, pelitic, calcareous and quartzose lithologies together with minor amounts of conformable mafic and Fe-rich horizons and discordant metadolerites. The basement gneiss in contact with the intrusions is more migmatitic and contains more felsic dykes of local anatectic origin than outcrops more distant from charnockite. The contact metamorphic effects of the charnockite appear to be partial melting of felsic, mica-bearing gneiss rather than subsolidus production of a metamorphic aureole.

Xenoliths

A wide range of gneissic and massive xenoliths occur within the charnockites but differ from typical basement gneiss in being more mafic in bulk composition and/or more anhydrous in mineralogy. No felsic, hydrate-bearing xenoliths were observed in equilibrium with the charnockites, which suggests that some assimilation has taken place by the digestion of typical felsic basement gneiss rock types. Although some large xenolithic blocks of pelitic and quartzo-feldspathic gneiss occur, these are invariably partially melted to a greater degree than the migmatitic basement gneiss. They have either an envelope of migmatite or are host to felsic dykes which sometimes emanate out into the surrounding charnockite. Partly assimilated gneissic xenoliths in some cases provide fluid and melt which can affect the nearby charnockite to produce localized zones of hydrated and/or 'bleached' rock. The zone of affected charnockite and the migmatitic coronas on xenoliths were described as 'hybrid' rocks by McCarthy & Trail (1964). A variety of cognate enclaves also occurs in the charnockites and these have been chemically analysed. They are usually even-grained and have similar mineralogy to their host charnockite but are more mafic.

Regional petrological variation

On the basis of field mapping and petrographic and geochemical studies, a number of charnockite suites are recognized. They form two first-order groups: low-Ti (Stibbs Bay Group) and high-Ti (Holme Bay Group), on the basis of different abundances of Y, TiO_2, P_2O_5, Cr and other elements for a given SiO_2 content. Rocks having consistent petrological properties tend to be located close to each other areally which aids in the recognition of suites. They are characterized by the ubiquitous presence of orthopyroxene and K-feldspar over almost the entire range of compositions. In the field and in thin section the most noticeable characteristics are the presence or absence of perthitic K-feldspar megacrysts (in part rounded by deformation into porphyroblasts), garnet, clinopyroxene and amphibole. Pyroxenes in both the high- and low-Ti groups are small (usually < 1 mm), almost unzoned chemically and free of exsolution (apart from the two-pyroxene rocks). Orthopyroxenes have similar Mg/(Mg + Fe) to the bulk rock over the whole range of bulk-rock compositions, suggesting metamorphic re-equilibration during D_3. Coarsely exsolved pyroxenes are found only in the least recrystallized rocks. Strongly antiperthitic plagioclase is also found in these rocks and is progressively obliterated by deformation, but still occurs as relict phenocrysts in many foliated rocks. A summary of important characteristics of the various charnockite suites is given in Table 1.

Low-Ti charnockites: the Stibbs Bay Group

This group of intrusive rocks features high Mg/Mg + Fe ratios and Ni, Cr contents relative to the Holme Bay Group (Table 1, Fig. 2). Low-Ti charnockites were not recognized as a separate group in a previous geochemical study of the Mawson charnockites (Sheraton, 1982). There is a broad spread of compositions from quartz diorites to granodiorites and adamellites (56–70% SiO_2) and some gabbros also occur. For the most part these rocks are mineralogically distinct from the more Fe-rich Holme Bay Group in having no metamorphic garnet and clinopyroxene only in the gabbroic rocks. K-feldspar megacrysts are found only in two suites.

High-Ti charnockites: the Holme Bay Group

This group has less compositional diversity than the Stibbs Bay Group, but greater areal extent (Table 1). Four suites, with SiO_2 mainly between 55 and 68%, are recognized and described briefly in Table 2. They typically have abundant perthitic K-feldspar megacrysts which preserve simple twins that almost certainly indicate igneous crystallization. A few even-grained, megacryst-free charnockites occur. Mafic compositions (< 58% SiO_2) contain two pyroxenes and many felsic samples have metamorphic garnet plus biotite. Hornblende, which is rare in the low-Ti charnockites, occurs in some high-Ti samples and appears to be retrograde.

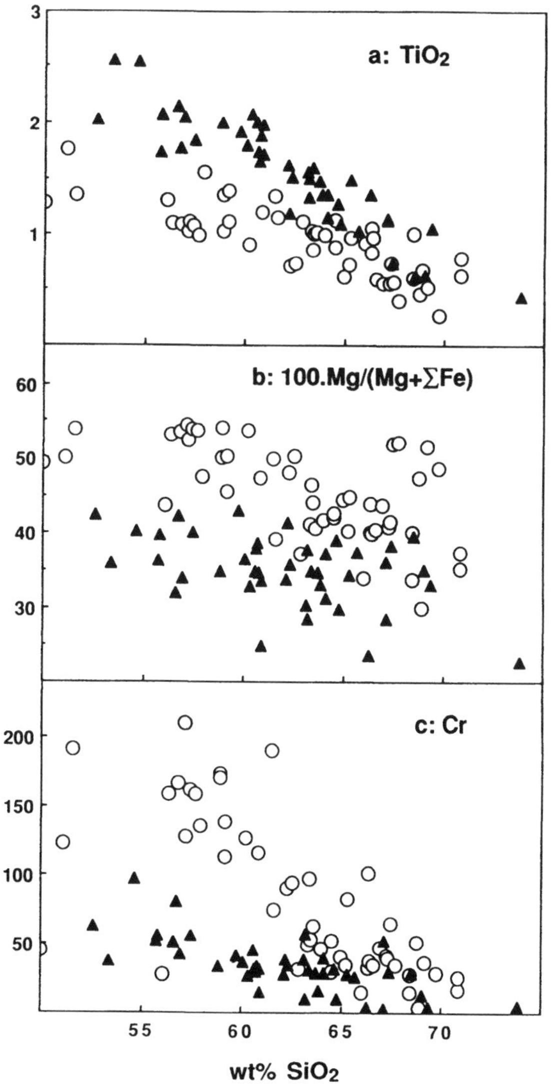

Fig. 2. Harker variation diagrams of charnockites from the Holme Bay Group (triangles) and Stibbs Bay Group (circles), including some data from Sheraton (1982, 1985). The parameters plotted are: (*a*) TiO_2; (*b*) Mg-number for total Fe; and (*c*) Cr. Three mafic Stibbs Bay Group samples with low Cr (Macklin Island Suite at 50% and 51% SiO_2, Colbeck Archipelago Suite at 56% SiO_2) probably have a different petrogenesis from other low-Ti suites. Felsic Stibbs Bay Group samples with high Mg-number are the Alfons Island Suite.

Chemical variation and petrogenesis

The charnockites are quartz-normative and vary from diopside- to slightly corundum-normative. Many samples are notable for their high Ca and K but low Na at intermediate silica contents (Table 2), a feature typical of ancient lavas which have been altered or weathered (e.g. Hildebrand, Hoffman & Bowring, 1987). The Mawson rocks, however, are unaltered and unweathered and we consider the high K/Na to be an igneous feature. Both groups of charnockite have highly radiogenic initial Sr and Nd isotopic compositions (Young & McCulloch, unpublished data) which require a component of ancient, felsic crust in their source region.

The Holme Bay Group is characterized by quite high K/Na ratios and by the most mafic samples being highest in P_2O_5 and

Table 2. *Representative XRF analyses of Mawson charnockites*
Major elements are in wt % oxide and selected trace elements in ppm. Note the high CO_2 compared to H_2O. H_2O- is equivalent to surface moisture and H_2O+ to structurally-bound water. The suite affinities of the samples are: Maw16, Falla Buff Suite; Maw31, Robinson Islands Suite; Maw11, Alfons Island Suite; Maw56, Ufs Island Suite; Maw24, Mawson Islands Suite.

Sample no.	Maw16	Maw31	Maw11	Maw56	Maw24
SiO_2	57.40	60.83	64.92	54.63	60.74
TiO_2	1.07	1.19	0.61	2.54	1.88
Al_2O_3	17.09	15.50	16.76	14.71	14.78
Fe_2O_3	1.92	1.48	1.07	1.73	0.89
FeO	5.60	5.43	3.51	7.90	7.42
MnO	0.12	0.10	0.07	0.15	0.19
MgO	4.79	3.41	2.00	3.58	2.44
CaO	6.47	4.57	4.74	6.75	5.39
Na_2O	2.47	1.56	2.81	2.29	1.86
K_2O	1.77	4.26	2.85	2.80	3.41
P_2O_5	0.25	0.51	0.15	1.03	0.47
S	0.08	0.08	0.00	0.08	0.08
H_2O-	0.19	0.20	0.15	0.24	0.17
H_2O+	0.49	0.43	0.29	0.57	0.28
CO_2	0.52	0.41	0.22	0.49	0.49
Subtotal	100.23	99.96	100.15	99.49	100.49
O=S	0.04	0.04	0.00	0.04	0.04
Total	100.19	99.92	100.15	99.45	100.45
$Mg/Mg + Fe^{2+}$	60.4	52.8	50.4	44.7	37.0
$Mg/Mg + Fe^{TOTAL}$	53.8	47.3	44.4	40.3	34.6
Nb	8.5	18.0	9.0	31.0	24.5
Y	10	42	3	54	30
La	26	37	36	83	52
Cr	162	116	40	97	33
Ni	68	27	18	37	9

TiO_2. The genesis of the mafic rocks and their connection to the intermediate samples is uncertain, but for most rocks a parent magma of around 60% SiO_2 is inferred. Enclaves provide clues to parent magma compositions because they show a continuum from rocks with low K/Na, free of K-feldspar megacrysts, to sparingly megacrystic rocks with higher K/Na. It is likely from this and other evidence that some concentration of magmatic K-feldspar occurred, perhaps due to flotation of megacrysts during magma ascent. However, a reasonably high K/Na parent magma would be needed to crystallize the K-feldspar megacrysts because the lowest K/Na tonalitic enclaves would have plagioclase as a liquidus phase (e.g. Wyllie, 1977).

The Stibbs Bay Group has widely occurring mafic quartz diorites (Falla Bluff Suite: Table 2) which have low K/Na and strongly resemble modern andesites (Ewart, 1982), except for high Ni and low HREE. However, they are distinctly lower in K/Rb, Sr and Mg than modern high-Ni andesites, or 'bajaites' (Saunders *et al.*, 1987). Fractionation from basaltic parent magma may be possible but the model preferred at this stage is one of high-degree, high-temperature melting of mafic crust. In order to explain the deleted HREE, garnet is needed in the partial-melting residue because it preferentially retains heavy rather than light REE (Schnetzler & Philpotts, 1970). To have garnet on the solidus of basalt or andesite for melting, pressures of 13 kb are needed (Green, 1982) and this strongly points to thick crust (Ellis, 1987; Gromet & Silver, 1987).

No one petrogenetic model applies to all the low-Ti suites, although high-temperature crustal melting at great depth is probably common to most. Some felsic Stibbs Bay Group suites have high K/Na and have probably had some K-feldspar accumulation. A suite of felsic samples (Alfons Island Suite; Table 2) which crop out near to the quartz diorites are very strongly depleted in HREE, and a suite of garnet-bearing rocks (Jelbart Glacier Suite) show a contrasting enrichment in HREE relative to the middle REE and provide further evidence for garnet involvement and high pressures of magma generation. For the Jelbart Glacier suite it is likely that garnet crystallized at depth, or represents restite, and was brought up to the final level of emplacement in a partly solid magma.

Conclusions

Unlike the metamorphic charnockites of India and Sri Lanka (Hansen *et al.*, 1987), these Antarctic charnockites involved igneous processes. A large composite batholith of intermediate composition was intruded into the supracrustal basement gneiss of the mobile belt during orogeny at about 1000 Ma. Two main groupings are recognized amongst the great diversity of granitoid types. Partial melting of mainly mafic to intermediate crust, accompanied by magmatic concentration of K-feldspar crystals, is the petrogenetic model favoured to explain the unusually high K/Na of some rocks. The high-Ti rocks have high REE, Ti, K and P compared to typical tonalites. This is a problem as yet unresolved, although Sheraton (1982) appealed to unusual, previously melted source rocks as a possible solution. We suggested earlier a comparison with highly evolved tholeiitic ferro-basalt systems in which these incompatible elements are enriched by fractionation (Young & Ellis, 1987).

High-Ti magmas are likely to have come from moderate depths where plagioclase rather than garnet was stable in the source. Low-Ti magmas, for the most part, are depleted in HREE and have other features consistent with garnet being residual during anatexis. This requires high pressures and an overthickened crust. Ellis (1987) proposed that I-type, HREE-depleted granitoids in ancient fold belts were 'fingerprints' for thickened crust during magmatism. Looking at the modern earth, this is consistent with the frequent association of HREE-depleted magmas with areas of thick crust (e.g. Petterson & Windley, 1985; Gromet & Silver, 1987; Hildreth & Moorbath, 1988) and their rarity in oceanic island arcs and absence from extensional terrains.

In contrast to many Northern Hemisphere Grenvillian charnockitic orthogneisses (McLelland, 1986), the Mawson Coast magmas are orogenic in nature, being intruded into a high grade gneiss terrain between the second and third deformations. They cannot represent an accreted arc batholith because they intrude gneissic basement after the peak of deformation. It is not inconceivable that within-craton compressional tectonics could produce the high pressures and temperatures suitable for the genesis of these rock types. However, active subduction and a plate-margin tectonic setting for at least this part of the Proterozoic mobile belt of East Antarctica is more consistent with the intermediate, calc-alkaline nature of the proposed parent magmas and the evidence for very thick crust.

Acknowledgements

The Australian Antarctic Division and the expeditioners at Mawson in 1984–5 are thanked for their invaluable help with fieldwork. Prof. D.H. Green initiated this study and provided funds from the University of Tasmania to support fieldwork. Geoff Clarke, Steve Foley and Scott Kuehner assisted with fieldwork and B.W. Chappell provided excellent analytical facilities at the ANU. We acknowledge useful discussions with W. Cameron, G. Corlett, J. Kilpatrick, M. Obata, E. Reid and S.R. Taylor and the careful reviews of John Sheraton, Alastair Allen and Doone Wyborn.

References

Clarke, G.L. (1988). Structural constraints on the Proterozoic reworking of Archaean crust in the Rayner Complex, MacRobertson and Kemp Land coast, East Antarctica. *Precambrian Research* **40/41**, 137–56.

Clarke, G.L., Powell, R. & Guiraud, M. (1989). Low-pressure granulite facies metapelitic assemblages and corona textures from MacRobertson Land, East Antarctica: the importance of Fe_2O_3 and TiO_2 in accounting for spinel-bearing assemblages. *Journal of Metamorphic Geology* **7**, 323–35.

Crohn, P.W. (1959). A contribution to the geology and glaciology of the western part of Australian Antarctic Territory. *Bulletin Bureau of Mineral Resources, Geology and Geophysics, Australia*, **52**.

Ellis, D.J. (1987). Origin and evolution of granulites in normal and thickened crust. *Geology*, **15**, 167–70.

Ewart, A. (1982). The mineralogy and petrology of Tertiary–Recent orogenic volcanic rocks: with special reference to the andesitic-basaltic compositional range. In *Andesites*, ed. R.S. Thorpe, pp. 25–95. New York; John Wiley & Sons.

Fuhrman, M.L. & Lindsley, D.H. (1988). Ternary feldspar modelling and thermometry. *American Mineralogist*, **73**, 201–15.

Green, T.H. (1982). Anatexis of mafic crust and high pressure crystallisation of andesite. In *Andesites*, ed. R.S. Thorpe, pp. 465–87. New York; John Wiley & Sons.

Gromet, L.P. & Silver, L.T. (1987). REE variations across the Peninsular Ranges Batholith: implications for batholithic petrogenesis and crustal growth in magmatic arcs. *Journal of Petrology*, **28**, 75–125.

Hansen, E.C., Janardhan, A.S., Newton, R.C., Prame, W.K.B.N. & Ravindra Kumar, G.R. (1987). Arrested charnockite formation in southern India and Sri Lanka. *Contributions to Mineralogy and Petrology*, **96**, 225–44.

Harley, S.L. (1984). An experimental study of the partitioning of Fe and Mg between garnet and orthopyroxene. *Contributions to Mineralogy and Petrology*, **86**, 359–73.

Hildebrand, R.S., Hoffman, P.F. & Bowring, S.A. (1987). Tectonomagmatic evolution of the 1.9-Ga Great Bear magmatic zone, Wopmay orogen, northwestern Canada. *Journal of Volcanology and Geothermal Research*, **32**, 99–118.

Hildreth, W. & Moorbath, S. (1988). Crustal contributions to arc magmatism in the Andes of Central Chile. *Contributions to Mineralogy and Petrology*, **98**, 455–89.

Lindsley, D.H. (1983). Pyroxene thermometry. *American Mineralogist*, **68**, 477–93.

McCarthy, W.R. & Trail, D.S. (1964). The high-grade metamorphic rocks of the Mac.Robertson Land and Kemp Land coast. In *Antarctic Geology*, ed. R.J. Adie, pp. 473–81. Amsterdam; North-Holland Publishing Co.

McLelland, J.M. (1986). Pre-Grenvillian history of the Adirondacks as an anorogenic, bimodal caldera complex of mid-Proterozoic age. *Geology*, **14**, 229–33.

Newton, R.C. & Hansen, E.C. (1983). The origin of Proterozoic and late Archaean charnockites – evidence from field relations and experimental petrology. *Geological Society of America Memoir*, **161**, 167–78.

Perkins, D. & Chipera, S.J. (1985). Garnet–orthopyroxene–plagioclase–quartz barometry: refinement and application to the English River subprovince and the Minnesota River valley. *Contributions to Mineralogy and Petrology*, **89**, 69–80.

Petterson, M.G. & Windley, B.F. (1985). Rb–Sr dating of the Kohistan arc batholith in the Trans-Himalaya of north Pakistan, and tectonic implications. *Earth and Planetary Science Letters*, **74**, 45–57.

Ravich, M.G. & Kamenev, E.N. (1975). *Crystalline Basement of the Antarctic Platform*. New York; John Wiley & Sons.

Saunders, A.D., Rogers, G., Marriner, G.F., Terrell, D.J. & Verma, S.P. (1987). Geochemistry of Cenozoic volcanic rocks, Baja California, Mexico: implications for the petrogenesis of post-subduction magmas. *Journal of Volcanology and Geothermal Research*, **32**, 223–45.

Schnetzler, C.C. & Philpotts, J.A. (1970). Partition coefficients of rare-earth elements and barium between igneous matrix material and rock-forming mineral phenocrysts, II. *Geochimica et Cosmochimica Acta*, **34**, 331–40.

Sheraton, J.W. (1982). Origin of charnockitic rocks of MacRobertson land. In *Antarctic Geoscience*, ed. C. Craddock, pp. 489–97. Madison; University of Wisconsin Press.

Sheraton, J.W. (1985). Chemical analyses of rocks from East Antarctica: Part 2. *Bureau of Mineral Resources, Australia, Record*, **1985/12**.

Sheraton, J.W., Tingey, R.J., Black, L.P., Offe, L.A. & Ellis, D.J. (1987). Geology of Enderby Land and western Kemp Land, Antarctica. *Bulletin, Bureau of Mineral Resources, Australia* **223**, 51 pp.

Steiger, R.H. & Jaeger, E. (1977). Subcommission on geochronology: convention on the use of decay constants in geo- and cosmochronology. *Earth and Planetary Science Letters*, **36**, 359–62.

Trail, D.S. (1970). ANARE 1961 geological traverses on the Mac.Robertson Land and Kemp Land Coast. *Bureau of Mineral Resources, Australia, Report*, **135**.

Wickham, S.M. (1988). Evolution of the lower crust. *Nature*, **333**, 119–20.

Wyllie, P.J. (1977). Crustal anatexis: an experimental review. *Tectonophysics*, **43**, 41–71.

Young, D.N. & Ellis, D.J. (1987). Plutonic charnockites from a Proterozoic mobile belt in Antarctica: evidence for fractionation of tholeiitic parent magma in thickened, orogenic crust. In *Precambrian Events in the Gondwana Fragments* (abstract volume). *Geological Society of Sri Lanka Special Publication*, **3**, 10.

A review of the field relations, petrology and geochemistry of the Borgmassivet intrusions in the Grunehogna province, western Dronning Maud Land, Antarctica

J.R. KRYNAUW[1], B.R. WATTERS[2], D.R. HUNTER[1] & A.H. WILSON[1]

1 Department of Geology, University of Natal, Pietermaritzburg, 3200 South Africa
2 Department of Geology, University of Regina, Regina, Saskatchewan, Canada

Abstract

The tholeiitic, mid-Proterozoic Borgmassivet intrusions and Straumsnutane basalts of western Dronning Maud Land vary in composition from ultramafic to mafic. The intrusions and lavas in different nunatak areas display distinctive mineralogical, textural and intrusive characteristics, but the coherence of major and trace element data suggests that the magmas may have been consanguineous. Hydrothermal alteration and crustal contamination may have generated the wide range (~ 800–1700 Ma) in isotopic age determinations. A model for extensive igneous activity during the closing stages of deposition in the Ritscherflya sedimentary basin is proposed. The present distribution of the Borgmassivet intrusions and Straumsnutane basalts reflects stratigraphically higher levels from west to east. It is suggested that western Dronning Maud Land is broken up into large tectonic blocks by the Pencksökket–Jutulstraumen-type faults, within which a series of smaller blocks exist which represent tectonic readjustment following, or synchronous with, the major faulting event.

Introduction

The mid-Proterozoic Borgmassivet intrusions of western Dronning Maud Land (Fig. 1) intrude Archaean granites and a sedimentary–volcanic sequence, the Ritscherflya Supergroup (Table 1). The intrusions are tholeiitic and vary in composition from ultramafic to mafic. They occur as sills up to 400 m thick and layered bodies of unknown lateral and vertical dimensions. These rocks constitute approximately 80% of the exposures in the Ahlmannryggen. The Straumsnutane Formation in the eastern Ahlmannryggen is the uppermost exposed unit of the Ritscherflya Supergroup and comprises a succession of basaltic lavas, with minor interlayers of volcaniclastic and chemical sediments. Previous investigations indicated these intrusions and lavas to be approximately 1700–1800 Ma old (Allsopp & Neethling, 1970; T.P. Elworthy, in Wolmarans & Kent, 1982; Barton & Copperthwaite, 1983), but recent studies have shown that the isotopic data are poorly understood and need to be re-evaluated (Allsopp & Barton, pers. comm., 1985; A.B. Moyes, pers. comm., 1987). Isotopic age determinations are summarized in Table 1.

The intrusions in different nunatak areas display distinctive mineralogical, textural and intrusive features. In order to obtain a preliminary understanding of the regional features of the igneous rocks in the Ahlmannryggen–Giaeverryggen area, four wide-spaced localities within an area of approximately 20000 km^2 were selected for detailed study (Fig. 1). These are: (i) Annandagstoppane–Juletoppane–Förstefjell; (ii) Robertskollen–Krylen; (iii) Grunehogna–Jekselen; and (iv) Straumsnutane.

The Annandagstoppane–Juletoppane–Förstefjell area

The intrusive rocks at these nunataks form part of a layered body or bodies, of which only a small part is exposed (Krynauw, Hunter & Wilson, 1984). Two groups were recognized, namely a 'main suite' comprising medium-grained gabbronorites with minor anorthosites and a 'younger suite' composed of quartz diorite pegmatite, basaltic dykes, fine- to medium-grained doleritic sills and very restricted volumes of albitites. The main suite rocks are orthocumulates in which plagioclase, orthopyroxene and clinopyroxene are primocrysts. The dykes and sills of the younger suite show textures typical of basalts and dolerites. Krynauw *et al.* (1984) suggested that the rocks of the main and younger suites may be consanguineous on the basis of major and trace element data.

Robertskollen–Krylen area

The layered complex at Robertskollen comprises a lower, cryptically layered ultramafic sequence, overlain by a mafic unit. The latter shows no evidence of rhythmic layering.

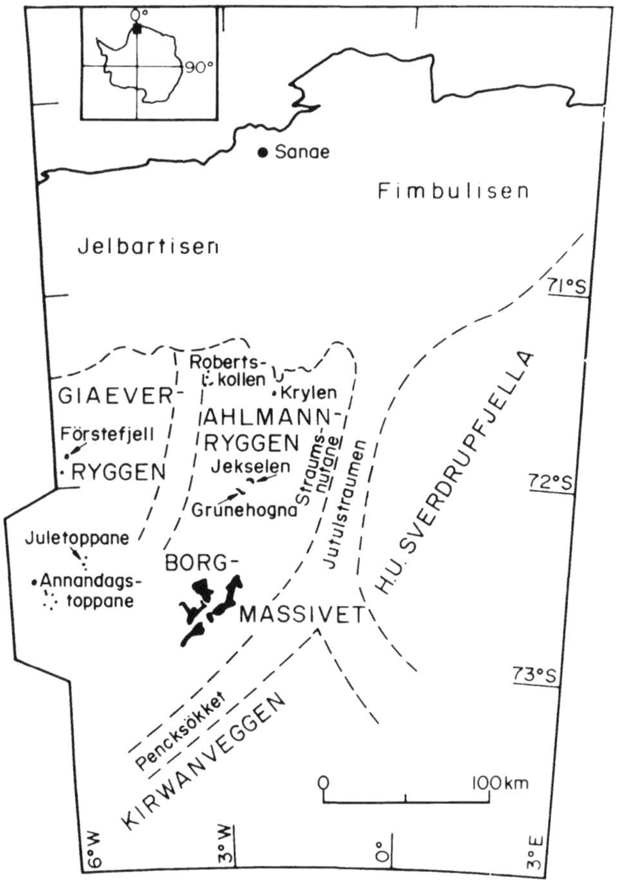

Fig. 1. Locality map of the Ahlmannryggen and Giaeverryggen.

Olivine and orthopyroxene are the dominant cumulus phases in the ultramafic rocks, whereas orthopyroxene, plagioclase and, possibly, clinopyroxene are the primary phases of the mafic rocks. Clinopyroxene and orthopyroxene are primocrysts in a 6 m thick olivine gabbronorite boundary layer at the interface between the mafic and ultramafic sequences. Quartz xenocrysts occur in metastable coexistence with olivine in the olivine gabbronorite. HFSE and incompatible elements such as Rb increase upwards in the ultramafic unit and decrease upwards in the mafic unit, whereas Mg-number and compatible elements such as Ni show the reverse trends. However, within the olivine–gabbronorite boundary layer, K, P and Rb increase sympathetically with increasing Mg-number and Ni. This apparently anomalous trace element behaviour of these incompatible elements, together with the co-existence of quartz and olivine, suggest crustal contamination of the magma(s) by partial digestion and/or incomplete mixing. The mafic rocks at Krylen have petrographic and geochemical features very similar to the Annandagstoppane and Juletoppane main suites. These features include reversed pyroxene–pigeonite zoning which has also been described in samples from the last two localities (Krynauw *et al.*, 1984; Krynauw, 1986).

Grunehogna–Jekselen

Two mafic sills intrude sedimentary rocks of the Schumacherfjellet and Högfonna Formations at Grunehogna (Table 1). The lower Grunehogna sill comprises a medium-grained diorite, of unknown thickness, overlain by quartz diorite pegmatite. Numerous vugs, partially filled with quartz, carbonate, chlorite and/or epidote are present in the sedimentary rocks up to 3 m from the contact with the sill. Partial destruction of sedimentary structures has occurred within this contact zone. However, primary sedimentary structures are preserved in tabular enclaves with oval cross-sections up to 5 × 50 cm in diameter, that are present throughout the contact zone. The sedimentary rocks in the contact zone consist mainly of meta-arenites. They are deformed into large (15 m amplitude), S-type folds where the Grunehogna sill is transgressive. Xenoliths within the sill show extensive soft sediment deformation, that has not been seen in adjacent country rocks. The contact zone and xenoliths show petrographic characteristics consistent with the formation of small amounts of partial melts that collected into granitic veins and irregular bodies less than 3 m^3 in volume. These petrographic features include development of micro-aplites along quartz grain boundaries and of granophyric, pseudogranophyric and sieve textures in quartz grains. The latter are extensively corroded. The nature of the contact zone and the presence of the vugs, deformation of the meta-arenites and xenoliths, formation of partial melts and the presence of large volumes of pegmatites have been interpreted by Krynauw (1986) and Krynauw, Hunter & Wilson (1988) as evidence for intrusion of the sills into wet, unconsolidated or partially lithified sediments.

The overlying Kullen sill varies in thickness from 120 m at Grunehogna to over 400 m at Kullen, 4 km south-east, and Preikestolen, 10 m south of Grunehogna (G.B. Snow, pers. comm., 1986). The sill varies in composition from gabbronorite and gabbro to diorite. Plagioclase, orthopyroxene and clinopyroxene are cumulus phases, with ilmenite and/or magnetite appearing in the more evolved diorites. Both the Grunehogna and Kullen sills contain xenoliths of sedimentary rocks. The Kullen sill locally contains highly corroded quartz grains and, in places, strained quartz grains occur; these grade into granophyric intergrowths of unstrained quartz with albite and K-feldspar. The corroded and strained quartz grains are interpreted as xenocrysts, probably derived from country rocks.

The Jekselen subvolcanic complex consists typically of quartz-diorites. These are coarse-grained in the lower outcrops and grade through a medium- to fine-grained variety to a fine-grained quartz-diorite at their upper contact with the Ritscherflya sediments. These latter rocks contain quartz and calcite amygdales which are locally up to 5 cm in diameter. The fine-grained quartz diorites contain small amounts of quartz amygdales and hollow, acicular, hopper pyroxene needles with chlorite and/or serpentine cores.

Straumsnutane

Outcrops of the Straumsnutane basalts occur in a number of small nunataks in the eastern Ahlmannryggen (Fig. 1). The sequence has been folded into a series of NNE-trending anticlines and synclines in the easternmost outcrops. A prominent shear zone, dipping 65° ESE, affects all of the eastern outcrops along the margin of Jutulstraumen. The rocks

Table 1. *Summary of the stratigraphy of the Ritscherflya Supergroup, suggested levels of intrusion and isotopic age determinations (IC – Isochron, EC – Errorchron, M – Model age)*

Formation	Spatial relationship of intrusions	Isotopic age determinations (Ma) (Rb–Sr unless indicated) and Reference
Overlying rocks unknown		
Straumsnutane Fm. – Fasettfjellet Fm.		821 ± 58 (IC, 1); 1644 ± 35 (M, 1); 1169 ± 35 (M, 7)
Istind Fm. Raudberget Fm.		
Tindeklypa Fm.		
Jekselen Fm.	Jekselen complex	1672 ± 79 (IC, 2); 988 ± 63 (EC, 3); 1339 ± 55; 1140 ± 56; 806 ± 28 ($^{40}Ar/^{39}Ar$, 6); 1079 ± 87; 948 ± 120 (IC, 5)
Högfonna, Fm.	Kullen sill	832 ± 4; 716 ± 10 ($^{40}Ar/^{39}Ar$, 6); 1426 ± 87 (IC, 5)
	Grunehogna sill	838 ± 112 (IC, 8)
Schumacherfjellet Fm.		
Framryggen Fm.		
Pyramiden Fm.	Krylen intrusions	1701 (K–Ar, 4); 767 ± 49; 806 ± 122; 1241 ± 61 (IC, 5)
?	Robertskollen complex	1181 ± 53; 1215 ± 128 (IC, 8); 1576 ± 177 (EC, 9); 1573 ± 92 (EC, 9); 1327 ± 364 (IC, 9); 1022 ± 172 (EC, 9)
?	Annandagstoppane intrusions	1802 ± 100;1800 ± 158; 1791 ± 164; 2518 ± 406 (IC, 5)
Archaean Granitoid basement		

(1) Eastin *et al.* (1970), quoted by Elworthy in Wolmarans & Kent (1982); (2) Allsopp & Neethling (1970); (3) Allsopp (1971), quoted by Elworthy in Wolmarans & Kent (1982); (4) McDougall (1969) quoted by Elworthy in Wolmarans & Kent (1982); (5) Barton & Copperthwaite (1983); (6) Bredell (1976) quoted by Elworthy in Wolmarans & Kent (1982); (7) Bowman (1971) quoted by Elworthy in Wolmarans & Kent (1982); (8) J.M. Barton (pers. comm., 1983, 1984); (9) A.B. Moyes (pers. comm., 1987).

are less deformed in the southern and western Straumsnutane where they have shallow to moderate dips, mainly SE.

Watters (1972) originally considered the lavas to comprise a sequence of altered andesites, but new geochemical data show them to be basalts to basaltic-andesites. The physical characteristics of the lavas are discussed by Watters *et al.* (this volume, p. 41).

Geochemistry

A total of 405 samples of the mafic intrusions (229) and Straumsnutane basalts (176) was analysed by XRF spectroscopy. Geochemical data are summarized in Krynauw (1986). The rocks show considerable variation in both major and trace element chemistry; e.g. SiO_2 (42.49–59.39 wt %) MgO (1.66–32.31%), Nb (0.8–22.3 ppm), Zr (22–477 ppm), Y (1–59 ppm), Cr (1 ppm–0.44%) and Ni (1–868 ppm). The Mg-number (100 Mg/(Mg + Fe^{2+})) varies from 20.79 to 83.92.

Whereas the most Mg-rich compositions are from the Robertskollen ultramafic unit, the most evolved occur in the Straumsnutane basalts at Snökallen nunataks and eastern Trollkjellpiggen. These last basalts (named here the Snökallen lavas) are characterized by high TiO_2, P_2O_5, Nb, Zr, Y and Cu and low MgO contents compared to the majority of the Straumsnutane basalts.

Interelement variations between high field strength (HFS, e.g. Fig. 2*a*), large ion lithophile (LIL) and compatible elements give high correlation coefficients for linear regressions. However, the Grunehogna sill and the Snökallen lavas are enriched and the sediment-contaminated Jekselen samples are depleted in Ti relative to the rest of the Borgmassivet intrusions and Straumsnutane basalts (Fig. 2*b*). Variation of Ti in the Grunehogna sill is attributed to primary ilmenite fractionation.

Samples for rare earth element (REE) analysis were selected from different nunataks on the basis of their Mg numbers (i.e. where possible, samples with high and low Mg numbers were analysed). Chondrite-normalized REE patterns of the Borgmassivet intrusions and Straumsnutane basalts are identical (Fig. 3), resembling certain continental tholeiites, e.g. lavas from the Karoo Central area (Marsh & Eales, 1984), Kirkpatrick basalts from Southern Victoria Land (Kyle, Pankhurst & Bowman, 1983) and the Sudbury Complex (Cullers & Graf, 1984). The flat heavy REE slopes and the presence of small negative Eu anomalies in the most evolved samples are consistent with shallow level fractionation of both the Borgmassivet intrusions and Straumsnutane basalts.

The chondrite-normalized trace element patterns or 'spidergrams' (Fig. 4) of the Borgmassivet intrusions and the Straumsnutane basalts resemble one another closely and are characterized by LILE enrichment that is typical of many continental tholeiites (Thompson *et al.*, 1983; Dupuy & Dostal, 1984). Troughs at Ba, Nb, Sr, P and Ti are prominent. Whereas

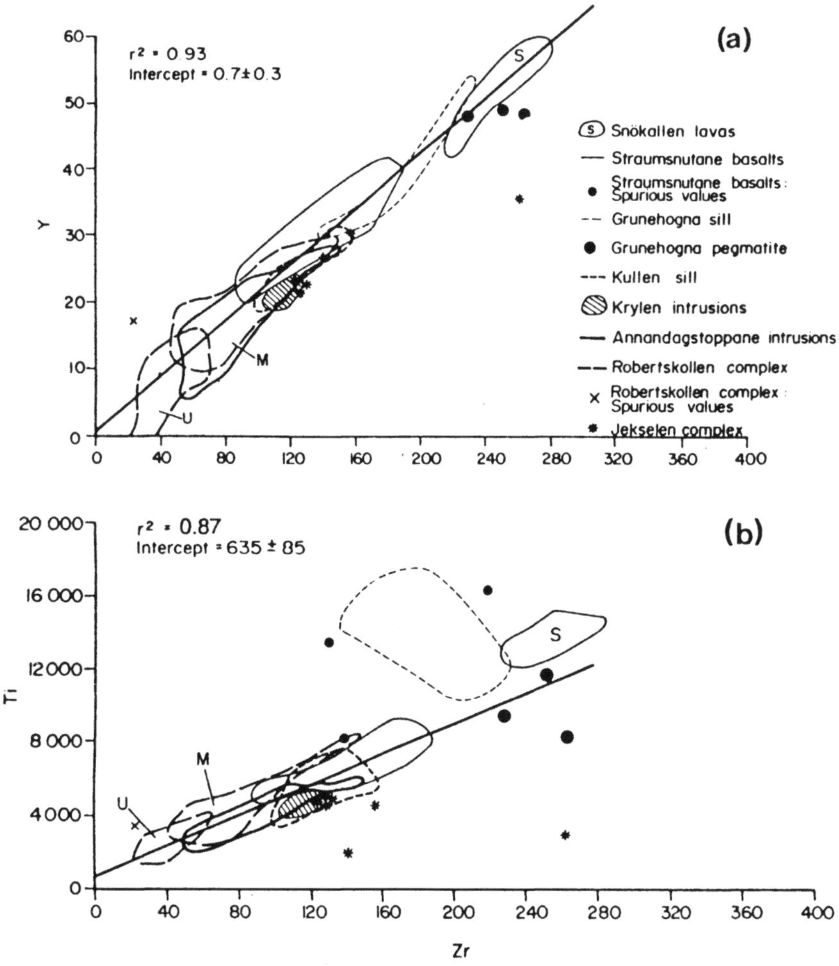

Fig. 2. Y and Ti variation with Zr (in ppm) of Borgmassivet intrusions and Straumsnutane basalts. The 's' field defines geochemical variation of Snökallen lavas. Linear regression and the correlation coefficient r_2 were calculated on the Ti–Zr diagram by excluding all values in excess of 10000 ppm Ti, the Grunehogna pegmatite samples and three sediment-contaminated samples from Jekselen (stars) plotting outside limits of the fields defined by the Borgmassivet intrusions).

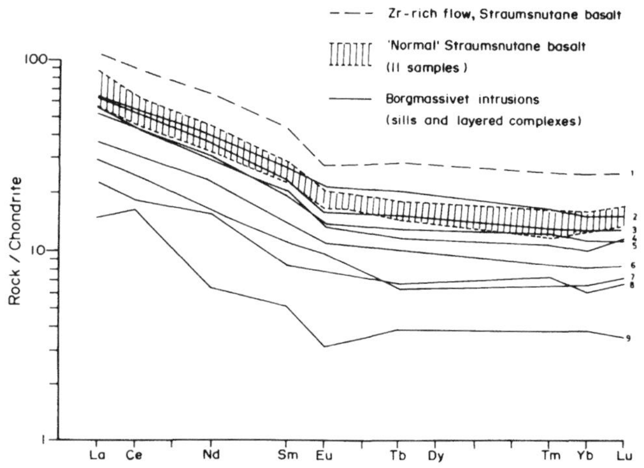

Fig. 3. The range in chondrite-normalized REE abundances in the Borgmassivet intrusions and Straumsnutane basalts. Normalizing factors are from Taylor & McLennan (1981). 1: GT-11, Snökallen lava; 2: G9/82, Grunehogna sill; 3: G18/82, Kullen sill; 4: J19/81, Jekselen complex; 5: JT10/82, Juletoppane main suite; 6: A27/82, Annandagstoppane main suite; 7: JT23/82, Juletoppane main suite; 8: A14/82, Annandagstoppane main suite; 9: R8/82, Ultramafic unit, Robertskollen.

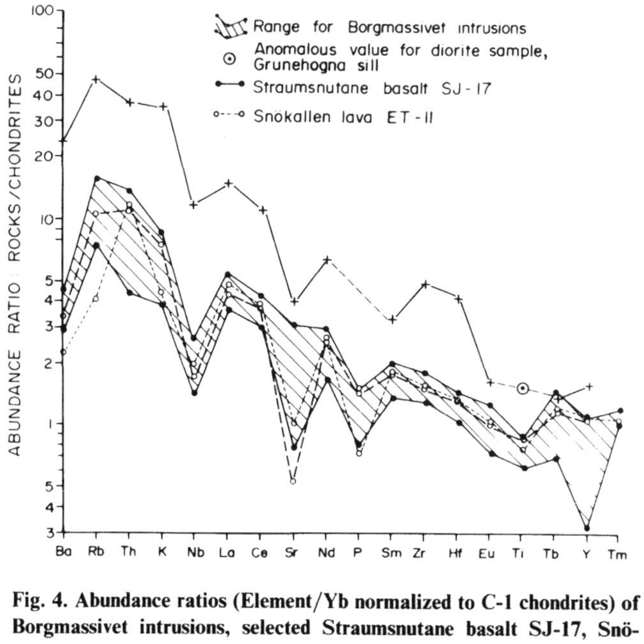

Fig. 4. Abundance ratios (Element/Yb normalized to C-1 chondrites) of Borgmassivet intrusions, selected Straumsnutane basalt SJ-17, Snökallen lava E-T 11 and 'average' Canadian Shield (Shaw, Dostal & Keays, 1976). Normalizing factors are from Thompson (1982).

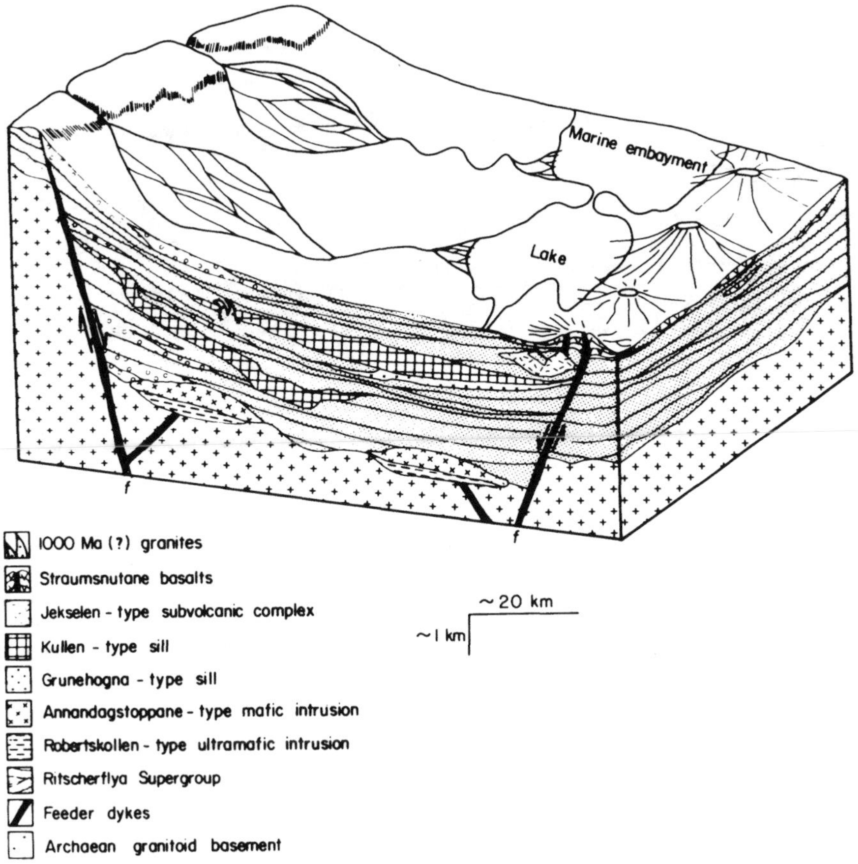

Fig. 5. Proposed model of depositional and intrusive environment in the Ritscherflya Supergroup during late stages of basin development.

LILE show a wide range of variation for a specific Mg-number, the abundances of high field strength elements display coherent increases with decreasing Mg-number, which is consistent with crystal fractionation. The scatter of LILE probably results from processes such as alteration, contamination and metamorphism superimposed on crystal fractionation. Watters (1972) stated that the Straumsnutane basalts had been affected by low grade metamorphism, epidotization and, probably, some degree of weathering.

Discussion

One of the problems in the assessment of the geochemistry of the Ahlmannrygen igneous rocks is that the available Rb–Sr-isotope data are difficult to interpret (T.P. Elworthy, in Wolmarans & Kent, 1982; Barton & Copperthwaite, 1983; J.M. Barton, pers. comm., 1985; A.B. Moyes, pers. comm., 1987). Lutz & Srogi (1986) showed on theoretical grounds that subsolidus hydrothermal alteration of rocks can cause systematic errors in age estimates if Sr addition is involved. The extensive hydrothermal alteration of the Kullen and Grunehogna sills and Straumsnutane basalts may have resulted in the generation of pseudo-isochrons. DePaolo (1981) discussed the generation of pseudo-isochrons as being the combined results of assimilation and fractionation, for which abundant field and petrographic evidence exists in the Borgmassivet intrusions (Krynauw, 1986). Lutz & Srogi (1986) showed that only small proportional changes in Sr concentration are required to produce false ages and meaningless initial ratios, and Juteau *et al.* (1984) proved that a heterogeneous Sr-isotope composition at the time of emplacement may result in considerable bias in both the age and initial ratios. This may also be the case for the Borgmassivet intrusions and Straumsnutane basalts (J.M. Barton, pers. comm., 1985).

The Rb–Sr-isotope data are therefore insufficient to determine whether or not the Borgmassivet intrusions and Straumsnutane basalts are contemporaneous. However, the coherence of geochemical data reported here suggests that the magmas, from which the intrusions and basalts solidified, may have been consanguineous. A simple model for extensive igneous activity during the closing stages of deposition in the Ritscherflya basin can thus be proposed. The model is summarized in Fig. 5 and the sequence of events may have been as follows:

(i) The depositional environment in the Ritscherflya basin was largely fluvial, dominated by braided and meandering river and local lacustrine systems (Bredell, 1982; Wolmarans & Kent, 1982; J.J.P. Swanepoel, pers. comm., 1984; Ferreira, 1986). The local presence of marine embayments was recognized by Ferreira (1986). Sediments were derived from south-western, western and north-western provenances that were possibly reactivated by block faulting (Wolmarans, 1982; J.J.P. Swanepoel, pers. comm., 1984; Ferreira, 1986).

(ii) Volcanic activity commenced during upper Högfonna times (Table 1), as shown by the presence of volcaniclas-

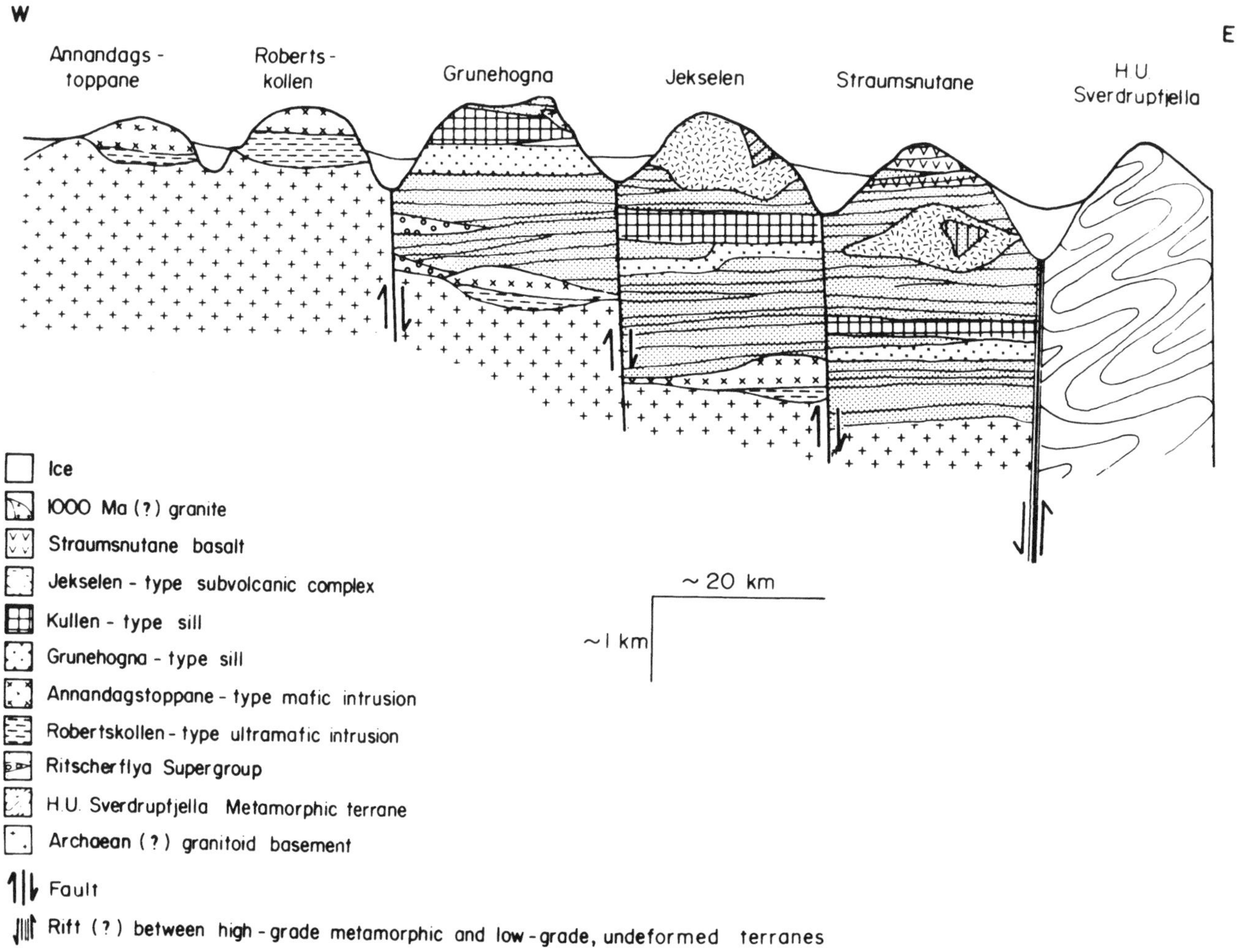

Fig. 6. Diagrammatic east–west section of the Ritscherflya Supergroup and Borgmassivet intrusions from Annandagstoppane to the western H.U. Sverdrupfjella. Ice cover prevents accurate determination of the style of faulting.

tic sediments (Bredell, 1982; Wolmarans & Kent, 1982; J.J.P. Swanepoel, pers. comm., 1984; Ferreira, 1986).

(iii) Increased volcanic activity culminated in the outpouring of the Straumsnutane basalts, which were largely subaerial flows. Local intercalations of pillowed basalts indicate that some subaqueous extrusion, marine and/or lacustrine, occurred (Watters, 1972).

(iv) Major faults at or near the margin(s) of the Ritscherflya basin acted as conduits for the Borgmassivet magmas which were generated in the upper mantle.

(v) The presence of layered intrusions and of limited outcrops of Archaean granites in western Annandagstoppane suggest that these intrusions were emplaced at or near the base of the Ritscherflya Supergroup. In contrast the Grunehogna and Kullen sills and Jekselen Complex were emplaced at shallow crustal levels. They show evidence for reaction with wet wall rocks and *in situ* contamination. The Straumsnutane basalts in the northeastern Ahlmannryggen were largely subaerial extrusions that terminated Ritscherflya sedimentation. If the intrusions and extrusions are consanguineous, as suggested by the geochemistry, the present distribution of the Borgmassivet intrusions and Straumsnutane basalts reflects stratigraphically higher levels from west to east (Fig. 6). This conclusion is consistent with the concept that the present disposition of the Ritscherflya Supergroup (with its oldest formations in the west and the youngest in the east) is a consequence of the presence of faults with downthrows to the E (Wolmarans, 1982; Ferreira, 1986).

The Pencksökket–Jutulstraumen dislocation represents a major tectonic break with its downthrown side to the W. A number of large-scale faults parallel to Pencksökket–Jutulstraumen have been recognized to the west of the Ahlmannryggen in the area bounded by 66–75° S and 0–30° W. These faults produced a series of blocks which are stepped down to the W and NW (Hinz & Krause, 1982; K. Weber & G. Spaeth, pers. comm., 1986). At the present stage of investigations in western Dronning Maud Land it is difficult to establish whether there is a temporal relationship between the major faults represented by the Pencksökket–Jutulstraumen-type, with downthrows to the W, and the smaller-scale faults in the Ahlmannryggen, with their downthrows to the E. If such a relationship exists, it is tempting to speculate that western Dronning Maud Land is broken up into large tectonic blocks by the Pencksökket–Jutulstraumen-type faults, within which a series of smaller blocks exist that represent tectonic readjustment following, or synchronous with, the major faulting event.

Acknowledgements

The Antarctic Division, Department of Environment Affairs, is thanked for logistical and financial aid. We were given permission to quote unpublished reports by J.M. Barton, A.B. Moyes, E.P. Ferreira, G.B. Snow, J.J.P. Swanepoel and the late H.L. Allsopp. K. Weber and G. Spaeth shared unpublished information with us aboard the *R.V. Polarstern*. Chris Harris and the referees gave valuable criticism of an earlier version of this paper.

References

Allsopp, H.L. & Neethling, D.C. (1970). Rb–Sr isotopic ages of Precambrian intrusives from Queen Maud Land, Antarctica. *Earth and Planetary Science Letters*, **8(1)**, 66–70.

Barton, J.M., Jr. & Copperthwaite, Y.E. (1983). Sr-isotopic studies of some intrusive rocks in the Ahlmann Ridge and Annandagstoppane, western Queen Maud Land, Antarctica. In *Antarctic Earth Science*, ed. R.L. Oliver, P.R. James & J.B. Jago, pp. 59–62. Canberra; Australian Academy of Science and Cambridge; Cambridge University Press.

Bredell, J.H. (1982). The Precambrian sedimentary–volcanic sequence and associated intrusive rocks of the Ahlmannryggen, western Dronning Maud Land: a new interpretation. In *Antarctic Geoscience*, ed. C. Craddock, pp. 591–7. Madison; The University of Wisconsin Press.

Cullers, R.L. & Graf, J.L. (1984). Rare earth elements in igneous rocks of the continental crust: predominantly basic and ultrabasic rocks. In *Rare Earth Element Geochemistry*, ed. P. Henderson, pp. 237–74. Amsterdam; Elsevier.

DePaolo, D.J. (1981). Trace element and isotopic effects of combined wallrock assimilation and fractional crystallization. *Earth and Planetary Science Letters*, **53(2)**, 189–202.

Dupuy, C. & Dostal, K. (1984). Trace element geochemistry of some continental tholeiites. *Earth and Planetary Science Letters*, **67(1)**, 61–9.

Ferreira, E.P. (1986). A sedimentological stratigraphic investigation of the sedimentary rocks in the Ahlmannryggen, Antarctica. M.Sc. thesis, University of Stellenbosch, 193 pp. (in Afrikaans) (unpublished).

Hinz, K. & Krause, W. (1982). The continental margin of Queen Maud Land/Antarctica: seismic sequences, structural elements and geological development. *Geologisches Jahrbuch*, **E23**, 17–41.

Juteau, M., Michard, A., Zimmermann, J.L. & Albarede, F. (1984). Isotopic heterogeneities in the granitic intrusion of Monte Capanne (Elba Island, Italy) and dating concepts. *Journal of Petrology*, **25(2)**, 532–45.

Krynauw, J.R. (1986). The petrology and geochemistry of intrusions at selected nunataks in the Ahlmannryggen and Giaeverryggen, western Dronning Maud land, Antarctica. Ph.D. thesis, University of Natal, Pietermaritzburg, 370 pp. (unpublished).

Krynauw, J.R., Hunter, D.R. & Wilson, A.H. (1984). A note on the layered intrusions at Annandagstoppane and Juletoppane, western Dronning Maud Land. *South African Journal of Antarctic Research*, **14**, 2–10.

Krynauw, J.R., Hunter, D.R. & Wilson, A.H. (1988). Emplacement of sills into wet sediments at Grunehogna, western Dronning Maud Land, Antarctica. *Journal of the Geological Society, London*, **145 (6)**, 1019–32.

Kyle, P.R., Pankhurst, R.J. & Bowman, J.R. (1983). Isotopic and chemical variations in Kirkpatrick Basalt Group rocks from southern Victoria Land. In *Antarctic Earth Science*, ed. R.L. Oliver, P.R. James & J.B. Jago, pp. 234–7. Canberra; Australian Academy of Science and Cambridge; Cambridge University Press.

Lutz, T.M. & Srogi, L.A. (1986). Biased isochron ages resulting from subsolidus isotope exchange. A theoretical model and results. *Chemical Geology*, **56(1)**, 63–71.

Marsh, J.S. & Eales, H.V. (1984). The chemistry and petrogenesis of igneous rocks of the Karoo central area, southern Africa. *Special Publication of the Geological Society of South Africa*, **13**, 27–67.

Shaw, D.M., Dostal, J. & Keays, R.R. (1976). Additional estimates of continental surface Precambrian shield composition in Canada. *Geochimica et Cosmochimica Acta*, **40(1)**, 73–83.

Taylor, S.R. & McLennan, S.M. (1981). The composition and evolution of the continental crust: rare earth element evidence from sedimentary rocks. *Philosophical Transactions of the Royal Society of London*, **A301**, 381–99.

Thompson, R.N. (1982). Magmatism of the British Tertiary volcanic province. *Scottish Journal of Geology*, **18(1)**, 49–107.

Thompson, R.N., Morrison, M.A., Dickin, A.P. & Hendry, G.L. (1983). Continental flood basalts . . . arachnids rule OK? In *Continental Flood Basalts and Mantle Xenoliths*, ed. C.J. Hawkesworth & M.J. Norry, pp. 158–85. Nantwich, Shiva Publishing Ltd.

Watters, B.R. (1972). The Straumsnutane volcanics, western Dronning Maud Land, *South African Journal of Antarctic Research*, **2**, 23–31.

Wolmarans, L.G. (1982). Subglacial morphology of the Ahlmannryggen and Borgmassivet, western Dronning Maud Land. In *Antarctic Geoscience*, ed. C. Craddock, pp. 963–8. Madison; University of Wisconsin Press.

Wolmarans, L.G. & Kent, L.E. (1982). Geological investigations in western Dronning Maud Land, Antarctica – a synthesis. *South African Journal of Antarctic Research*, Supplement 2, 93 pp.

Volcanic rocks of the Proterozoic Jutulstraumen Group in western Dronning Maud Land, Antarctica

B.R. WATTERS[1], J.R. KRYNAUW[2] & D.R. HUNTER[2]

1 Department of Geology, University of Regina, Regina, Canada
2 Department of Geology, University of Natal, Pietermaritzburg, South Africa

Abstract

The Jutulstraumen Group comprises a sequence of Proterozoic basaltic to basaltic-andesite lava flows with minor intercalated volcaniclastic and siliciclastic beds. The minimum thickness of the group is about 850 m. The volcanic units conformably overlie a sedimentary platform sequence (Ahlmannryggen Group) and they display field characteristics typical of a subaerial succession. The presence of local shallow bodies of water at the time of emplacement is indicated by ripple marking and desiccation features in thin, discontinuous, tuffaceous sedimentary intercalations overlain by pillowed flows. The latter may grade laterally or vertically into hyaloclastite deposits which include isolated pillows. Similar units comprising relatively large pillow- or pod-like forms (up to 4 m in diameter) may represent accumulations of pahoehoe flow 'toes'. Major and trace element geochemistry indicates the presence of two distinct tholeiitic basalt types: a dominant variety having relatively low Ti and high Mg and Ca, and a subordinate series of flows, occurring near the top of the exposed succession, characterized by relatively high Ti, total Fe, Al, Zr, Y, P, V, Rb, Cs, Hf, Ta, Th, U and rare earth elements (REE). Both types display a similar restricted range of SiO_2 content (mean 52.6%). Chondrite-normalized spidergram patterns and abundances of trace and minor elements are consistent with their interpretation as continental tholeiites.

Introduction

The mid-Proterozoic Jutulstraumen Group (Wolmarans & Kent, 1982) comprises predominantly volcanogenic rock units and occurs in the Ahlmannryggen and Borgmassivet areas of western Dronning Maud Land (Fig. 1). Four formations have been assigned to the Group (Wolmarans & Krynauw, 1981; Wolmarans & Kent, 1982), which constitutes the uppermost exposed part of the Ritscherflya Supergroup (Fig. 2):

(i) The Straumsnutane Formation (Neethling, 1970; Watters, 1972) occurring in the Straumsnutane area of the Ahlmannryggen (Fig. 1)
(ii) The Fasettfjellet Formation (Watters, 1972; Wolmarans & Kent, 1982) forming an isolated exposure at Fasettfjellet in the Borgmassivet (Fig. 1)
(iii) and (iv) The Tindeklypa and Istind formations (Neethling 1970; Bredell, 1982; Wolmarans & Kent, 1982) together form the Viddalen Subgroup occurring in the east-central part of the Ahlmannryggen (Fig. 1). The Tindeklypa Formation consists of a sequence of sedimentary breccia deposits and tuff, and is overlain conformably by the Istind Formation. The latter comprises sedimentary units, dominantly quartz arenite, with interbedded agglomerate and tuff. The stratigraphical position of the Viddalen Subgroup within the Ritscherflya Supergroup is uncertain but the presence of volcaniclastic beds and amygdaloidal sills (Van Zyl, 1974) suggests a position close to, or at, the base of the Jutulstraumen Group.

The extensive mid-Proterozoic Borgmassivet intrusions that have invaded the Ahlmannryggen Group exhibit many compositional characteristics in common with the lavas of the Jutulstraumen Group and are considered to represent the same episode of basic igneous activity (Krynauw, 1986; Krynauw *et al.*, this volume, p. 33). The present study is concerned primarily with the Straumsnutane and Fasettfjellet formations. It was undertaken because previous investigations of the volcanic rock units of the Jutulstraumen Group were not sufficiently detailed for comprehensive geochemical studies, nor did they permit stratigraphic relationships to be satisfactorily resolved.

Age relationships

The age of emplacement of the Jutulstraumen basalts has yet to be established. Rb–Sr whole-rock ages of 821 ± 58 Ma ($R_0 = 0.71051 \pm 0.00194$) (Eastin, Faure & Neethling, 1970) and 876 ± 86 Ma ($R_0 = 0.71265 \pm 0.0085$; MSWD = 79),

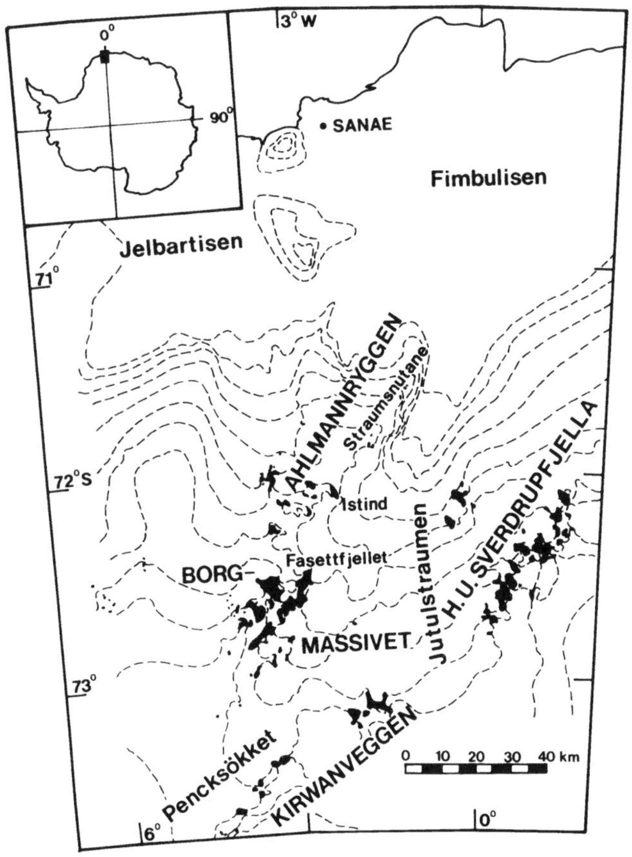

Fig. 1. Map of the Ahlmannryggen and Borgmassivet area. Contours (on ice surface) are at 200 m intervals.

obtained from different suites of lava by the present authors, are unlikely to represent the emplacement age. Rather, they probably record the timing of a thermotectonic event associated with a major phase of movement along the Jutulstraumen–Penksökket dislocation zone (Fig. 1). A model Rb–Sr age of 1664 ± 35 Ma for highly sheared lava from the north-eastern Straumsnutane area has also been reported by Eastin *et al.* (1970). A whole-rock Sm–Nd-isotopic study of the Jutulstraumen lavas has been undertaken by the present authors but the spread of Sm/Nd ratios was too small to provide a meaningful isochron. Model T^{Nd}_{chur} ages range from 1300 to 1620 Ma (mean of 1420 Ma) and would seem to represent the best estimate of emplacement age.

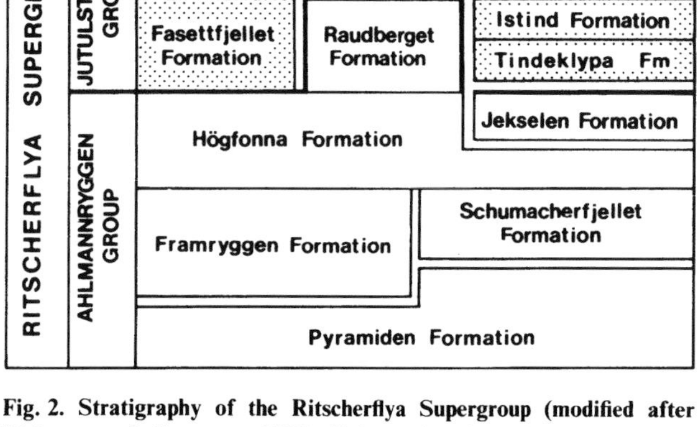

Fig. 2. Stratigraphy of the Ritscherflya Supergroup (modified after Wolmarans & Krynauw, 1981). Units assigned to the Jutulstraumen Group are stippled.

The proposed comagmatic relationship between the Jutulstraumen lavas and the Borgmassivet intrusions implies similar emplacement ages. Isotopic-age determinations for the intrusions, summarized by Krynauw *et al.* (this volume, p. 33), are mainly in the range 800–1800 Ma but are poorly understood.

Straumsnutane Formation

The isolated and relatively sparse exposures of this formation inhibit determination of stratigraphical relationships and the full extent of the succession is unknown. An estimated 850 m of the sequence is exposed but, since the base of the sequence is not seen and the top is not preserved, this must be regarded as a minimum thickness.

Lava flows

The bulk of the succession comprises basaltic to basaltic-andesite lava flows which range in thickness from about 1 m to > 50 m. Most of the lava flows typically have thin, moderately-amygdaloidal basal zones, massive- to slightly-amygdaloidal central zones and highly-amygdaloidal upper zones. Where seen, flow tops commonly have a 'wavey' or undulating character typical of pahoehoe flows.

Lava flows composed of pillow- or pod-like forms make up about 10% of the exposed succession; they do not appear to be more common at, or confined to, any particular stratigraphical level. They probably represent both pillow lava and pahoehoe 'toes' and, whereas in some outcrops a clear distinction exists between the two forms, in many cases characteristic features of both types are displayed. Pahohoe toes usually develop subaerially but can also form in a subaqueous environment (MacDonald, 1972). It would appear, therefore, that while many of the Straumsnutane flows were emplaced subaerially, subaqueous emplacement was also fairly common, with some occurrences indicating a mixed environment.

The pillow forms are generally < 1 m in diameter and display the characteristic convex upper surface with a 'moulded' lower surface. Both radial and concentric internal structures, reflected primarily in the arrangement and shape of amygdales, are present. Chilled margins are ubiquitous and spalled-off pillow rinds form a major component of the interpillow matrix breccia. Many pillow forms appear to be discrete ellipsoidal bodies but sinuous forms are also common. Pahoehoe toes have pod-like forms very similar to the pillows but are generally larger (up to 4 m in diameter), have a concentric internal structure and lack a significant development of matrix breccia. Rare examples exhibit clear evidence of having functioned as lava tubes.

With few exceptions, accumulations of lava pillows and pahoehoe flow toes grade upward into structurally homogeneous lava representing the mid- to upper parts of the same flow unit. It is suggested that, during flow progradation, pillows and toes developed at flow fronts and were overridden by massive lava. The vertical transition to massive lava is commonly marked by the presence of vague ellipsoidal forms and radiating amygdales, representing incipient pillow development.

Major mineral components of the lava are plagioclase feldspar and clinopyroxene. The latter is augite and generally completely unaltered. Plagioclase compositions are in the calcic-andesine–sodic labradorite range and the mineral invariably displays some degree of saussuritization ranging from mild to complete. Serpentinized olivine comprises a minor mineral phase in the 'Snökallen Lavas' (see Geochemistry section) forming a distinct group near the top of the succession. Much of the lava is porphyritic or vitrophyric, with phenocrysts of plagioclase feldspar. A glassy mesostasis or groundmass is commonly present and appears as a dark, cloudy material crowded with feldspar microlites. Alteration of the lava is generally limited to the feldspar component. However, in highly-sheared occurrences close to the Jutulstraumen (Fig. 1) breakdown to a foliated mass of mainly sericite, chlorite, epidote and minor quartz, is complete.

Hyaloclastite beds

Some pillowed flows grade laterally into hyaloclastite beds which usually include isolated pillows. Resulting from the disintegration or granulation of lava on coming into contact with water, these deposits apparently developed mainly along the fronts of subaqueous flows. Vertical gradations from hyaloclastite through pillow lava into non-pillowed lava are present and could also be a result of flows prograding in shallow water.

Intercalated sedimentary beds

Thin, lensoid sedimentary beds occur as intercalations throughout the succession but are most commonly found immediately beneath pillowed flow units. In many occurrences these beds are only a few centimetres in thickness and comprise red and/or green, highly siliceous, fine-grained deposits that probably represent chemical precipitates. Rarely, these beds include thin lenses rich in calcite. The deposits also fill hollows and fissures in the underlying flow surfaces and are locally deformed by, and squeezed up between, overlying lava pillows.

Locally, these sediments grade upward into red tuffaceous mudstones and siltstones, which display current ripple structures and desiccation cracks, overlain by quartz arenite. The maximum observed thickness for such sedimentary intercalations is about 20 m. Rarely, thin beds of volcanic breccia are present in these sedimentary sequences.

Fasettfjellet Formation

Conformably overlying the Hogfonna Formation, this formation comprises a sequence of volcanic breccia and tuff beds (50–80 m thick), overlain by about 50 m of basaltic lava (the lower 10 m of which is pillowed). Locally, 7 m of quartz arenite is developed above the volcaniclastic units and lies immediately below the lava. Above the arenite, the lower parts of the flow locally grade into hyaloclastite deposits including isolated pillows. The upper limits of the formation are no longer preserved.

The physical, geochemical and lithological characteristics of these volcanic units are identical to those of the Straumsnutane Formation and the Fasettfjellet occurrence is regarded as representing the basal levels of the more extensive volcanic sequence as exposed in the Straumsnutane.

Structure and metamorphism

The dominant structural feature of the Ahlmannryggen–Borgmassivet area is the Pencksökket–Jutulstraumen 'rift', a major crustal dislocation that has placed the basement gneiss complex of the Sverdrupfjella (Fig. 1) at the same level as the relatively undeformed platform succession of the Ritscherflya Supergroup. The Ahlmannryggen–Borgmassivet area is characterized by a series of faults parallel to the Pencksökket–Jutulstraumen structure, dividing the Ritscherflya succession into a series of blocks (Krynauw *et al.*, this volume p. 33).

Exposures along the eastern Straumsnutane area, adjacent to the Jutulstraumen, have been affected by intense shearing that generally dips at 65–70° ESE. This shearing, related to the Pencksökket–Jutulstraumen dislocation feature, diminishes in intensity westwards and is barely discernible in nunataks along the western side of the Straumsnutane. Spaeth (1987) has discussed the significance of these and other minor structures in the Ahlmannryggen area. In the south-western and central Straumsnutane the volcanic units have been folded into a very broad NE-trending synclinal structure with low-angle dips (up to 25°). Folding is more intense in the north-eastern Straumsnutane where the sequence has been deformed into fairly tight NE-trending folds which are overturned towards the south-east.

The volcanic rocks have been affected only by low-grade metamorphism marked by the development of chlorite, epidote and sericite. This effect, while relatively mild in western and south-western exposures, reaches a maximum in the highly-sheared rocks of north-eastern Straumsnutane, where no trace of the original mineralogy is preserved.

Geochemistry

A total of 176 samples of lava has been anlaysed for major and trace elements (Rb, Sr, Y, Zr, Nb, Pb, Th, Ni, Cu, Zn, Ga, Ba, Co, Cr, V) by XRF techniques, and a suite of 17 samples was also analysed for REE, Hf, Ta, U, Sc and Cs by Instrumental Neutron Activation. Analyses were carried out at the Open University, Milton Keynes, England. A tabulation of average compositions is available from the senior author.

Geochemical characterization

Characterization of Proterozoic volcanic suites using major element and derived normative data may be unreliable due to the ready mobilization of these elements. Although the Jutulstraumen lavas are only slightly metamorphosed and altered, it can be clearly demonstrated that the alkalis, at least, have been mobile. Therefore, characterization of the volcanics

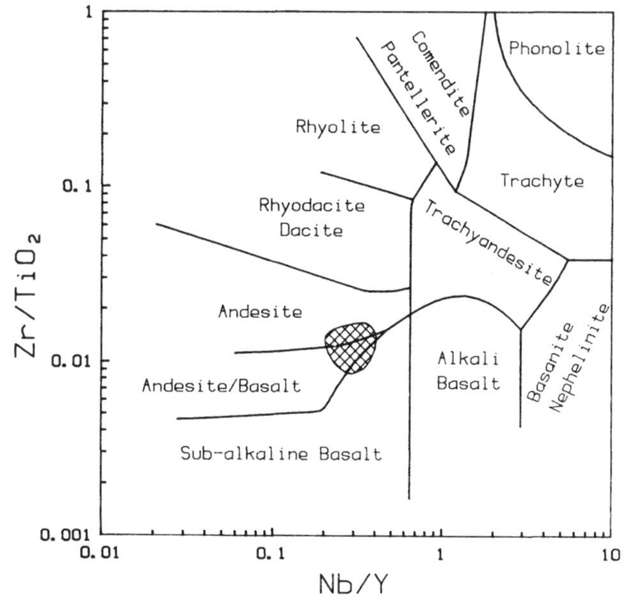

Fig. 3. Zr/TiO_2–Nb/Y diagram showing the compositional distribution (shaded areas) for lava flows of the Jutulstraumen Group. Field boundaries are from Winchester & Floyd (1977).

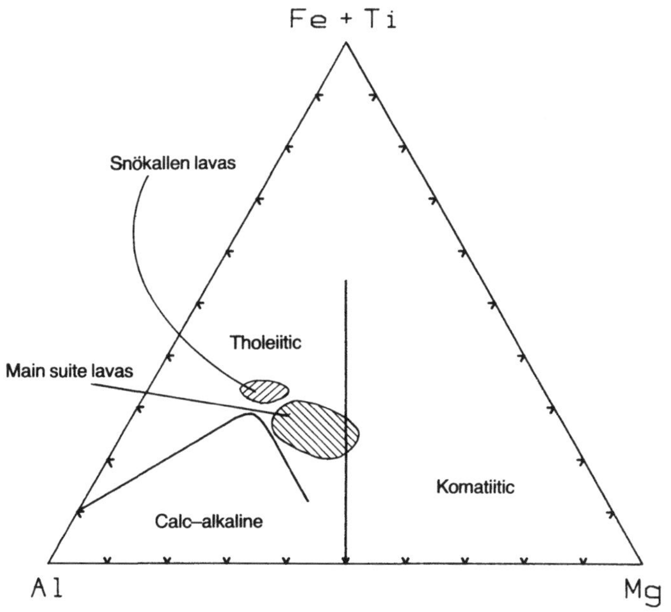

Fig. 4. Jensen cation plot for lava flows of the Jutulstraumen Group.

has been based, as far as possible, on those minor and trace elements which are generally considered to be relatively immobile during alteration or metamorphic processes (e.g. Ti, Y, Zr, Nb and REE).

Ratios between Y and Nb provide a means of identifying the petrological character of volcanic rock suites. Pearce & Cann (1973) and Winchester & Floyd (1977) have demonstrated the use of covariation between the immobile element ratios Zr/TiO_2 and Nb/Y in the characterization of metamorphosed volcanic suites. The former ratio can serve as a differentiation index (instead of SiO_2, for example), whereas Nb/Y ratios provide a reliable measure of alkalinity.

The Jutulstraumen lavas exhibit consistent Y/Nb and Zr/TiO_2 ratios, with a restricted distribution that is characteristic of subalkaline basalts and basaltic-andesites (Fig. 3).

Relationships between Al, Mg and Fe + Ti (Fig. 4) indicate a tholeiitic character for the lavas and it is apparent that the sequence comprises two compositional groupings, each with a different range of Fe + Ti/Mg ratios. A distinction between the two groups can be more clearly demonstrated by consideration of certain minor and trace element contents. For example, Zr shows a well defined bimodal distribution, as does Ti (Fig. 5).

The lesser group of lava flows are characterized by relatively low Mg and Ca, and relative enrichment in Al, total Fe, and the incompatible trace and minor elements Ti, Zr, Y,Cu, Zn, Rb, Th, P, Y, V and REE. This group of flows occurs mainly in the area of Snökallen nunatak in Straumsnutane, in what is regarded as the mid- to upper levels of the succession; they have been informally named the 'Snökallen lavas' (Krynauw *et al.*, this volume, p. 33). All other flows, comprising the bulk of the succession, will be referred to here as the 'Main Suite lavas'. Lava from the Fasettfjellet Formation in the Borgmassivet is compositionally indistinguishable from the Main Suite lavas of Straumsnutane and has not been separately identified in the diagrams presented here.

Chondrite-normalized trace and minor element patterns

Chondrite-normalized incompatible element distribution patterns (spidergrams) provide a convenient means of comparing certain geochemical characteristics of basaltic rocks, and of identifying their petrological character. Abundance ratios for incompatible elements in the Straumsnutane lavas (Fig. 6) form patterns which are unique to continental tholeiites (Thompson *et al.*, 1983; Dupuy & Dostal, 1984). Typically, the large-ion lithophile elements are relatively enriched and normalized ratios increase generally from Tb to Rb, but with troughs and peaks in the overall pattern. The elements Ba, Nb, Ta, Sr, P and Ti are relatively depleted and from Sr to Ba the patterns have characteristic M-shape configurations.

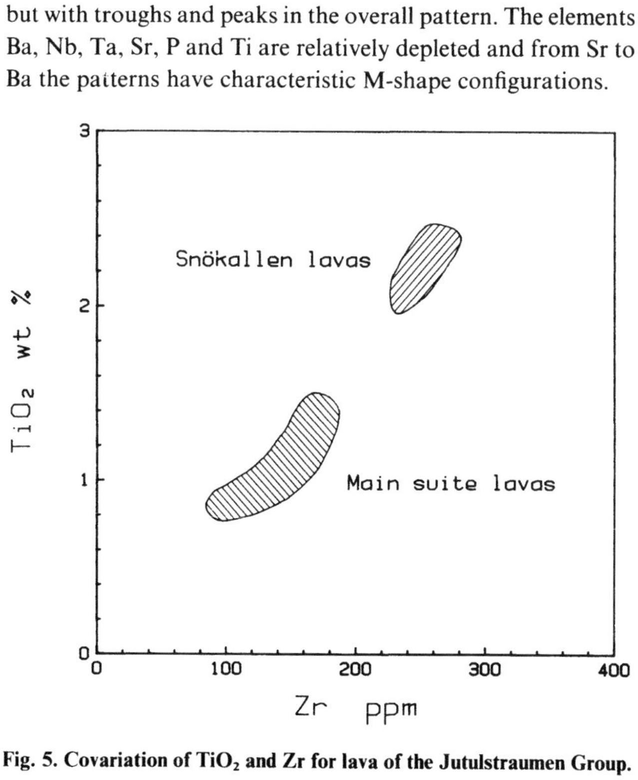

Fig. 5. Covariation of TiO_2 and Zr for lava of the Jutulstraumen Group.

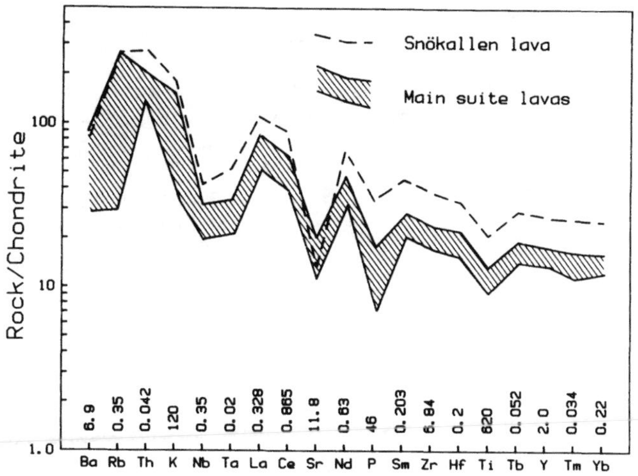

Fig. 6. Chondrite-normalized (except for K and Rb) trace and minor element distribution patterns for lava of the Jutulstraumen Group. Normalizing values as indicated are from Thompson *et al.* (1983).

Dupuy & Dostal (1984) and Thompson *et al.* (1983) have shown that many of the trace and minor element characteristics of continental tholeiites are also common to continental crust. These authors concluded that the unique incompatible element patterns displayed by continental tholeiites can be attributed largely to crustal contamination.

Chondrite-normalized REE distribution patterns for basalts of the Jutulstraumen Group (Fig. 7) show moderate total REE contents (90–190 ppm), low LREE/HREE enrichment ((La/Lu)cn = 3.9–6.2) and only very small negative Eu anomalies (Eu/Sm = 0.24–0.28); these are all features characteristic of continental tholeiites (Cullers & Graf, 1984).

The relative enrichment in incompatible elements of the Snökallen lavas over the Main Suite lavas is clearly shown by these patterns and suggests that the magmas giving rise to the Snökallen lavas evolved from Main Suite magma by an orthopyroxene dominated fractionation process. Further, the range of compositions exhibited by the Main Suite basalts (Figs 5, 6 & 7) can also be attributed to this process. Such a suggestion is consistent with the observation by Krynauw (1986) that the Borgmassivet intrusions, considered to be consanguineous with the Jutulstraumen lavas, provide evidence of dominantly orthopyroxene and plagioclase fractionation.

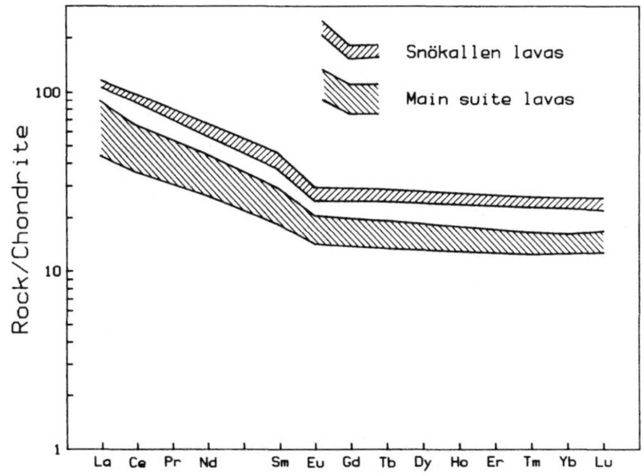

Fig. 7. Chondrite-normalized rare earth element distribution patterns for lava of the Jutulstraumen Group. Normalizing values are from Nakamura (1974).

Conclusions

The Jutulstraumen Group comprises a sequence of basaltic to basaltic-andesite lava flows and minor associated volcaniclastic and clastic units. Lava flows were probably emplaced in a dominantly subaerial environment but the presence of local, shallow bodies of water is indicated by the development of pillow lava, hyaloclastite beds, and intercalated tuffaceous sedimentary units showing current ripple structures and desiccation features.

The lavas display geochemical characteristics of continental tholeiites and two distinct, but probably cogenetic, compositional groupings make up the succession: the 'Main group lavas' and the subordinate 'Snökallen lavas'. The latter are characterized mainly by relatively low Mg and Ca, and enrichment in incompatible trace and minor elements.

The lithological and chemical characteristics of the Fasettfjellet Formation are identical to those of the Straumsnutane Formation, and the former are regarded as representing the basal parts of the volcanic sequence exposed in the Straumsnutane.

The physical and compositional characteristics of the volcanic rock units suggest that the Jutulstraumen Group developed in an intracratonic setting which was relatively stable and affected only by vertical tectonics. Major faults bounding the Ritscherflya basin provided conduits for the basic magmas (Krynauw *et al.*, this volume, p. 33). The apparently local development of the Tindeklypa and Istind formations in the east-central part of the Ahlmannryggen, close to the Penck-sökket–Jutulstraumen dislocation zone, may provide evidence in support of the suggestion that these major faults predate, or were consanguineous with, emplacement of the Jutulstraumen Group (Krynauw *et al.*, this volume, p. 33). These two formations possess all the characteristics of alluvial fan deposits and are likely to have developed in association with active fault scarps.

In a Gondwana context, correlation of the Ahlmannryggen and Jutulstraumen groups with sequences of similar character in southern Africa is difficult because of the uncertainty concerning the age of the groups. However, the most likely correlatives are the intracratonic Waterberg–Soutpansberg successions. Furthermore, in a Gondwana reconstruction such as that proposed by Norton & Sclater (1979), the Ahlmannryggen area occurs in juxtaposition with, and along an extension of, the axis of basin development for these Proterozoic stable platform successions.

References

Bredell, J.H. (1982). The Precambrian sedimentary–volcanic sequence and associated intrusive rocks of the Ahlmannryggen, western Dronnin Maud Land: a new interpretation. In *Antarctic Geoscience*, ed. C. Craddock, pp. 591–7. Madison; University of Wisconsin Press.

Cullers, R.L. & Graf, J.L. (1984). Rare earth elements in igneous rocks of the continental crust: predominantly basic and ultrabasic rocks. In *Rare Earth Element Geochemistry*, ed. P. Henderson, pp. 237–74. Amsterdam; Elsevier.

Dupuy, C. & Dostal, K. (1984). Trace element geochemistry of some continental tholeiites. *Earth and Planetary Science Letters*, **67**, 61–9.

Eastin, R., Faure, G. & Neethling, D.C. (1970). The age of the Trollkjellrygg Volcanics of western Queen Maud Land. *Antarctic Journal of the United States*, **5**, 157–8.

Krynauw, J.R. (1986). The petrology and geochemistry of intrusions at selected nunataks in the Ahlmannryggen and Gaeverryggen, western Dronning Maud Land, Antarctica. Ph.D. thesis, University of Natal, Pietermaritzburg, 370 pp. (unpublished).

MacDonald, G.A. (1972). *Volcanoes*. New Jersey; Prentice-Hall, 510 pp.

Nakamura, N. (1974). Determination of REE, Ba, Fe, Mg, Na and K in carbonaceous and ordinary chondrites. *Geochimica et Cosmochimica Acta*, **38**, 757–75.

Neethling, D.C. (1970). South African earth science exploration of western Queen Maud Land, Antarctica. Ph.D. thesis, University of Natal, Durban, 139 pp. (unpublished).

Norton, I.O. & Sclater, J.G. (1979). A model for the evolution of the Indian Ocean and the breakup of Gondwanaland. *Journal of Geophysical Research*, **84**, 6803–30.

Pearce, J.A. & Cann, J.R. (1973). Tectonic setting of basic volcanic rocks determined using trace element analysis. *Earth and Planetary Science Letters*, **19**, 290–300.

Spaeth, G. (1987). Aspects of the structural evolution and magmatism in western New Schwabenland, Antarctica. In *Gondwana Six*, ed. G.D. McKenzie, pp. 295–307. Washington, DC; American Geophysical Union, Monograph No. 40.

Thompson, R.N., Morrison, M.A., Dickin, A.P. & Hendry, G.L. (1983). Continental flood basalts . . . arachnids rule OK? In *Continental Flood Basalts and Mantle Xenoliths*, ed. C.J. Hawkesworth & M.J. Norry, pp. 158–85. Nantwich; Shiva Publishing Ltd.

Van Zyl, C.Z. (1974). The geology of Istind, western Dronning Maud Land, and the relationships between Istind and Tindeklypa Formations. *South African Journal of Antarctic Research*, **4**, 2–5.

Watters, B.R. (1972). The Straumsnutane Volcanics, western Dronning Maud Land. *South African Journal of Antarctic Research*, **2**, 22–31.

Winchester, J.A. & Floyd, P.A. (1977). Geochemical discrimination of different magma series and their differentiation products using immobile elements. *Chemical Geology*, **20**, 325–43.

Wolmarans, L.G. & Kent, L.E. (1982). Geological investigations in western Dronning Maud Land, Antarctica – a synthesis. *South African Journal of Antarctic Research*, Supplement **2**, 93 pp.

Wolmarans, L.G. & Krynauw, J.R. (1981). Reconnaissence geological map of the Ahlmannryggen area, western Dronning Maud Land, Antarctica; Sheet 1 of 3. Pretoria; SASCAR.

The timing and nature of faulting and jointing adjacent to the Pencksökket, western Dronning Maud Land, Antarctica

G.H. GRANTHAM* & D.R. HUNTER

Department of Geology, University of Natal, PO Box 375, Pietermaritzburg, South Africa

*Present address: *Department of Geology, University of Pretoria, Hillcrest, Pretoria 0002, South Africa*

Abstract

The incidence of faults and joints in H.U. Sverdrupfjella increases from the east towards the Pencksökket, which separates the high-grade Sverdrupfjella terrane from the relatively undeformed Precambrian Ritscherflya sedimentary and volcanic rocks to the west. Ages between 500 and 1000 Ma have been reported from gneisses of the Sverdrupfjella Group in Kirwanveggen, whereas the oldest ages yielded by mafic intrusions into the Ritscherflya sedimentary rocks are ± 1700 Ma. The faults and joints show similar orientations and display dominantly ENE strikes. This orientation parallels that of the Pencksökket and supports previous suggestions that the Pencksökket represents a major fracture system, probably associated with rifting. The jointing postdates the early Jurassic Tvora and Straumsvola alkaline complexes. The faults are normal, with downthrows to the NW. Dolerite dyke orientations in areas near Pencksökket differ from those of the faults and joints. Fault, joint and dyke trends from other areas nearby support the conclusions from Sverdrupfjella and show that faulting and jointing postdated the Jurassic volcanism. Recent reconstructions of Gondwana place Dronning Maud Land adjacent to Mozambique and it is suggested that the Pencksökket faulting and jointing could represent a continuation of the western limb of the East African rift system.

Introduction

The Pencksökket (trough) separates the high-grade metamorphic H.U. Sverdrupfjella terrane from the relatively undeformed Ritscherflya Supergroup exposed in Ahlmannryggen, Straumsnutane and Borgmassivet (Fig. 1). The Ritscherflya Supergroup comprises clastic sedimentary and volcanic rocks which have been intruded by ultrabasic to basic sheets and rare granitic dykes. Radiometric data yielded by these intrusions are difficult to interpret due to the spread of apparent ages (Wolmarans & Kent, 1982). However, a recent review of isotopic data suggests that emplacement occurred between 1400 and 1000 Ma (Barton *et al.*, 1986).

The terrane east of the Pencksökket comprises high-grade polydeformed quartzo-feldspathic gneisses, amphibolites and tabular granitoid intrusions. Post-tectonic, early Jurassic alkaline intrusions occur at Straumsvola and Tvora (Fig. 1). These early Jurassic intrusions have yielded ages of 170 ± 4 Ma (Rb–Sr whole rock) and 180 ± 2.8 Ma ($^{40}Ar/^{39}Ar$, amphibole) (A.R. Allen & I. Evans, pers. comms.). Only limited geochronological data are available from H.U. Sverdrupfjella. Preliminary results give Rb–Sr whole-rock ages of ± 1000 Ma and ± 500 Ma (A.B. Moyes, pers. comm.) which are consistent with ages reported from Kirwanveggen, ~ 150 km south-west of Sverdrupfjella (Elworthy, in Wolmarans & Kent, 1982).

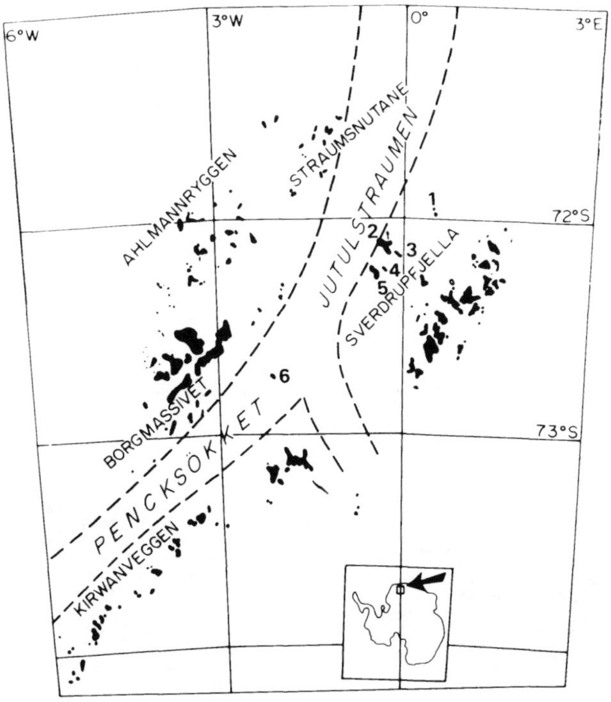

Fig. 1. Location of nunataks in western Dronning Maud Land. Nunataks mentioned in the text are: 1, Holane; 2, Straumsvola; 3, Tvora; 4, Brekkerista; 5, Jutulrora; 6, Midbresrabben.

The Pencksökket (trough) has been described as a 'classical graben downfaulted through at least 600–1500 m' (Ravich & Solov'ev, 1966). The fault and joint patterns described in this paper represent part of a regional study of the structural, metamorphic and intrusive history of H.U. Sverdrupfjella being undertaken by the South African National Antarctic Programme.

Faults

Mapping at a scale of 1: 10 000 has revealed numerous faults in north-western H.U. Sverdrupfjella. Whereas it was not possible to determine the type of fault (i.e. normal or reverse) in all examples, those faults which have identifiable displacements are normal, with downthrows to the W. The fault with the largest measurable vertical displacement (60 m) occurs at Jutulrora. Fault planes are generally vertical or have steep NW or NE dips. The faults are characterized in the field by highly fractured country rock and by pink alteration zones associated with the development of epidote. Two orientations of faults have been recognized (Fig. 2). The dominant fault set trends ENE whereas the less common set is orientated WNW. Faults are common on the nunataks located closest to the Pencksökket, i.e. Jutulrora, Brekkerista and Holane (Fig. 1) but fault development decreases towards the east.

Joints

Joints are common at Jutulrora but decrease in abundance towards the east. Jointing is particularly prominent and intense in an early Jurassic, quartz-syenite intrusion at Tvora (Fig. 3), located approximately 8 km to the east-noth-east, along-strike from the major fault at Jutulrora. The jointing at Tvora may therefore be related to this fault. Joint planes are characterized by a pink discolouration resulting from the breakdown of plagioclase and hornblende to epidote and chlorite. The dominant joint orientations are similar to the faults, namely ENE, but with a small proportion of WNW strikes (Fig. 2). Jointing is most prevalent on the nunataks located closest to the Pencksökket, i.e. Jutulrora, Brekkerista, Straumsvola and Tvora.

Mylonites

The nunatak at Midbresrabben has the only rock outcrops within the Pencksökket. Its eastern part is underlain by

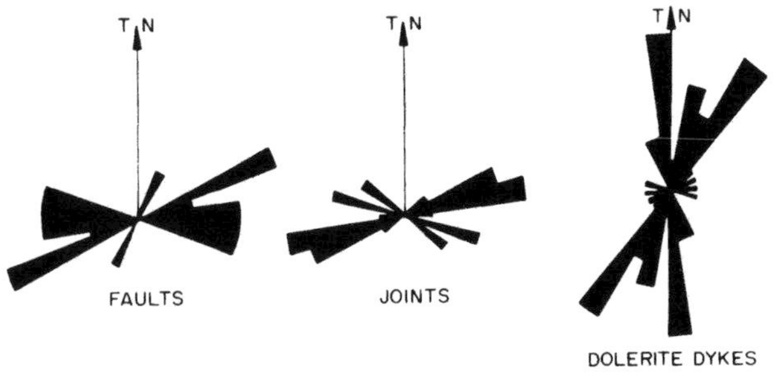

Fig. 2. Orientations of faults, joints and basic dykes from north-western H.U. Sverdrupfjella.

Fig. 3. Intense jointing developed in quartz–syenite at Tvora.

highly sheared quartzo-feldspathic gneisses and there is sheared quartz-diorite in the west. In general, the quartz-diorite becomes progressively more sheared from west to east. Mylonites are exposed over a width of ~400 m, reflecting the existence of a major structure. North–south orientated mafic dykes (up to 40 m wide) which intruded the quartz-diorite are also extremely sheared.

The westernmost outcrops of quartz-diorite are devoid of a fabric and contain weakly strained xenoliths of gneiss (Fig. 4). In thin section, these rocks show an equigranular texture and recrystallization of quartz. Patches of epidote and biotite apparently replaced an earlier mineral, probably amphibole. Plagioclase is unstrained and is partially replaced by epidote.

A planar mylonitic fabric becomes progressively more intense towards the east and it is associated with a prominent mineral lineation, defined by stretched patches of epidote and biotite developed on the foliation planes. The planar fabric strikes N–NNE and dips steeply E (70–80°); the mineral lineation plunges 70–80° SE.

Dyke orientation

Vertical to near-vertical mafic dykes, presumably of Jurassic age, trend N–S and NE–SW in western H.U. Sverdrupfjella. Mafic dykes at Straumsnutane, adjacent to the north-western margin of the Pencksökket, display a similar

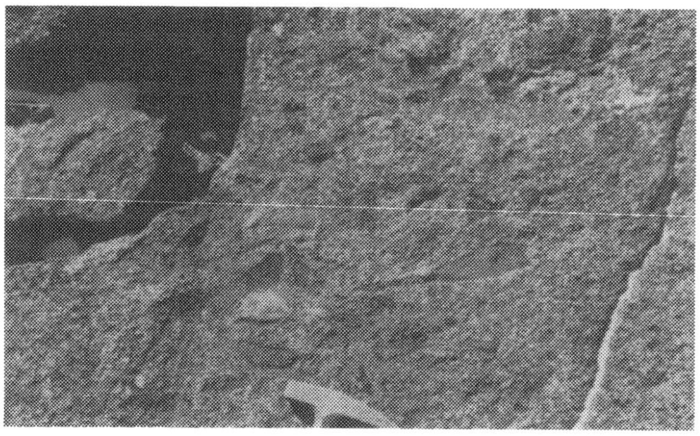

Fig. 4. Weakly strained xenoliths in diorite at Midbresrabben.

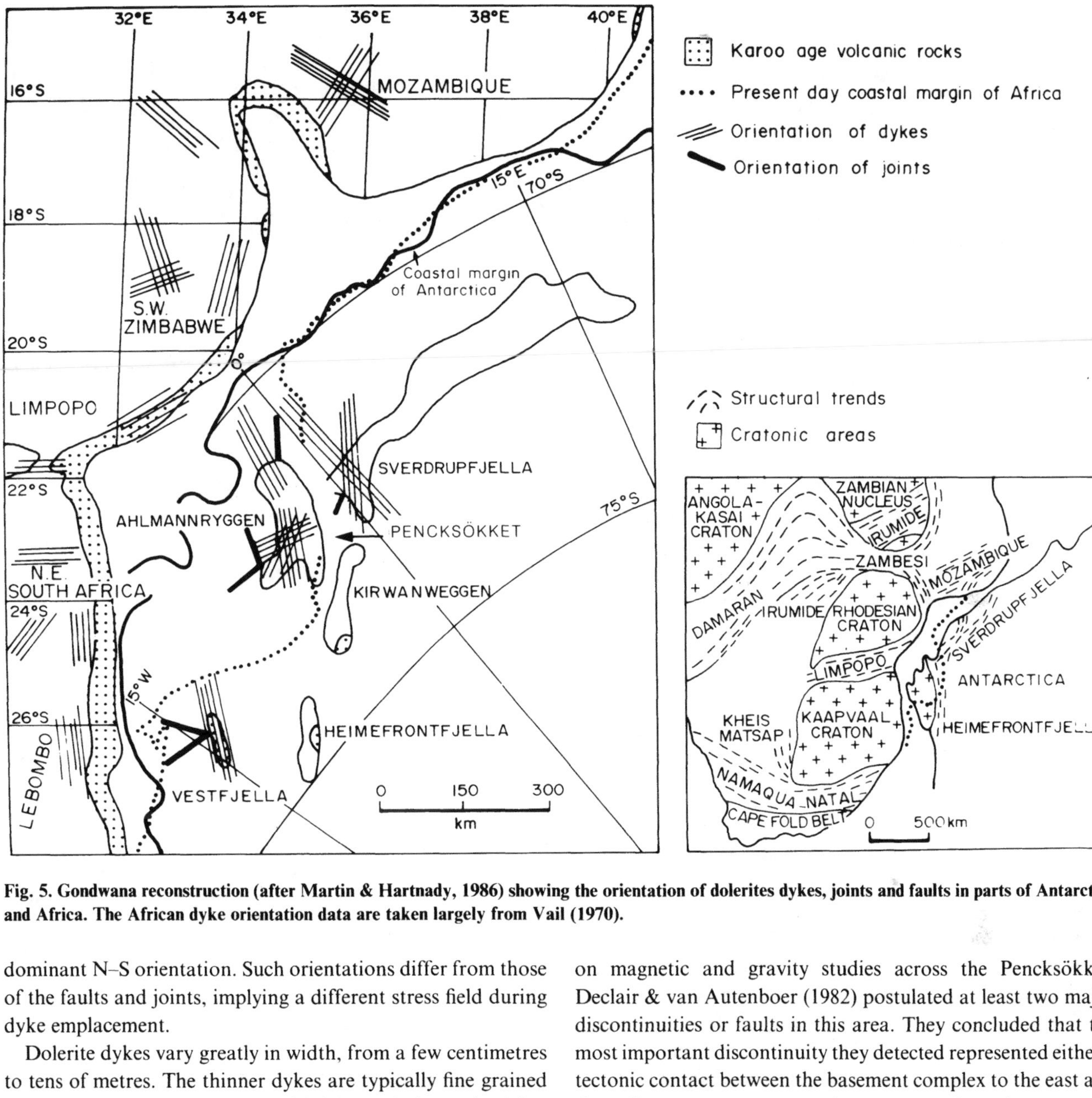

Fig. 5. Gondwana reconstruction (after Martin & Hartnady, 1986) showing the orientation of dolerites dykes, joints and faults in parts of Antarctica and Africa. The African dyke orientation data are taken largely from Vail (1970).

dominant N–S orientation. Such orientations differ from those of the faults and joints, implying a different stress field during dyke emplacement.

Dolerite dykes vary greatly in width, from a few centimetres to tens of metres. The thinner dykes are typically fine grained and commonly contain vesicles filled by prehnite and calcite. The wider dykes have fine-grained margins that grade into coarse-grained cores. Two generations of mafic dyke may be present because some dykes are fresh whereas others are extensively altered; this difference can only be seen in thin section. The unaltered dykes are composed of euhedral to subhedral olivine, a high content of opaque minerals and strongly zoned titanaugite; plagioclase is also strongly zoned. Locally fox-red biotite encloses opaque ore grains. The altered dykes are characterized by saussuritized plagioclase, partial replacement of pyroxene by chlorite, and by the absence of olivine.

Faults, joints and dykes in other areas adjacent to the Pencksökket

The faults and joints described above are spatially related to the Pencksökket which has been described as a graben structure (Ravich & Solov'ev, 1966). Fault and joint studies from other areas support this conclusion (Fig. 5). Based on magnetic and gravity studies across the Pencksökket, Declair & van Autenboer (1982) postulated at least two major discontinuities or faults in this area. They concluded that the most important discontinuity they detected represented either a tectonic contact between the basement complex to the east and the sedimentary sequence to the west or an intrusive contact.

A strong planar fabric has been reported from Straumsnutane, along the north-western margin of the Pencksökket (Watters, 1972). The planar fabric here has a NE strike (mean azimuth N 26°E) and dips steeply E, which is in the opposite direction to H.U. Sverdrupfjella, where the joints dip steeply W. The intensity of the planar fabric in Straumsnutane decreases away from the Pencksökket (Watters, 1972). The opposing dip directions of the planar fabric at Straumsnutane and joints at H.U. Sverdrupfjella, as well as the reduction in joint intensity away from the Pencksökket, support the interpretation of a graben structure as suggested by Ravich & Solov'ev (1966).

Wolmarans' (1982) joint data from the Ahlmannryggen and Borgmassivet areas identified four main joint orientations, namely N 30°E, N 95°E, N 125°E and N 165°E; the N 30°E and N 95°E trends are predominant. The N 30°E direction is approximately parallel to the NE trend of the Pencksökket. Dominant joint trends in Borgmassivet are NE and NW

(Wolmarans & Kent, 1982). Faults in Kirwanveggen strike NE and, locally, are seen to postdate dolerite dykes (Gavshon & Erasmus, 1975). Two orientations of joints have been recorded in the Jurassic lavas in Vestfjella (Fig. 5), namely, WNW and NNW (Hjelle & Winsnes, 1972). The WNW-striking joints have a similar orientation to the ESE-striking joints in Ahlmannryggen (Wolmarans, 1982). No joints with orientations similar to the NNW-striking joints of Vestfjella have been recorded in other areas of western Dronning Maud Land. Block faulting has been identified from seismic profiles along the continental shelf, in areas located along strike from the Pencksökket (Hinz & Krause, 1982). The orientations of these faults are similar to those identified in western Dronning Maud Land.

Dykes in Straumsnutane have orientations similar to the N-striking dykes in H.U. Sverdrupfjella and have a mean azimuth of N 13°E (Watters, 1972); significantly, they postdate the planar fabric (Watters, 1972). Dyke orientations in Kirwanveggen vary from N and SW, with the SW orientation being most common (Gavshon & Erasmus, 1975). Dykes in Vestfjella have a dominantly NE trend (Hjelle & Winsnes, 1972) and near Grunehogna in central Ahlmannryggen, they have WNW, NE and ENE trends (Aucamp, 1972). Aucamp (1972) stated that, whereas the WNW-trending dykes are olivine dolerites and considered to be of Jurassic age, the time of emplacement of the other dykes in the Ahlmannryggen is uncertain.

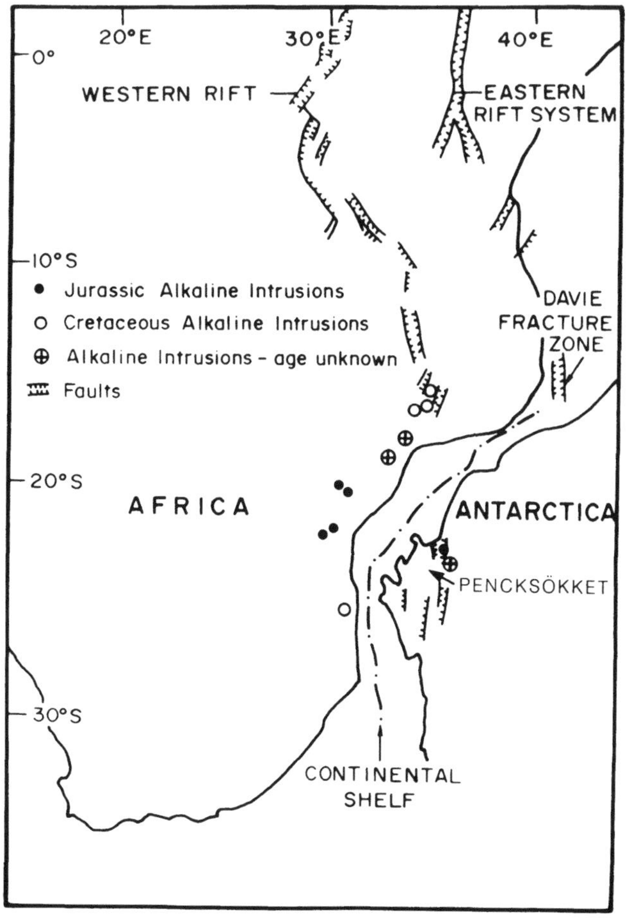

Fig. 6. Gondwana reconstruction (after Martin & Hartnady, 1986) showing a possible continuation of the East African rift system in Antarctica. The fault distributions in Africa are from Mougenet *et al.* (1986) whereas those from Antarctica are from an unpublished compilation by K. Weber (pers. comm., 1986).

The relation of faults and dykes to Gondwana

A recent reconstruction of Gondwana places Dronning Maud Land (Fig. 5) adjacent to the Mozambique coast (Martin & Hartnady, 1986). This reconstruction is supported by broad geological similarities between the H.U. Sverdrupfjella metamorphic terrane and the Mozambique Belt, and between the Ritscherflya Supergroup and the Umkondo Group in Zimbabwe and Mozambique (Martin & Hartnady, 1986; Grantham, Groenewald & Hunter, 1988). It would be reasonable therefore to expect that Jurassic dykes in Antarctica and Africa would have similar pre-drift orientations in this reconstruction. Although similarities are seen in juxtaposed areas, this is not always the case (Fig. 5). Dyke swarms in Vestfjella, H.U. Sverdrupfjella, northern Ahlmannryggen (Straumsnutane), Lebombo, south-eastern Zimbabwe and Mozambique have parallel N to NE–SW trends and some dykes in central Ahlmannryggen have trends similar to those in the north-eastern part of South Africa. However ESE-trending dykes in Ahlmannryggen are the only dykes which appear to share a similar trend to the strong E–W-trending dyke swarm in the Limpopo region.

In the context of Gondwana reconstructions (Fig. 6), studies of fault patterns in Antarctica and Africa show that the Pencksökket represents a fault system which may be a continuation of the western limb of the East African rift system. Whereas much of the volcanism and faulting in the East African rift system is of Cretaceous age and younger, Woolley & Garson (1970) stated that the southern end of the East African rift system developed in Jurassic time. Significantly the alkaline complexes intruded in southern Africa (Nuanetsi area) are Jurassic in age whereas those farther north, in Malawi, are Cretaceous (Fig. 6) (Woolley & Garson, 1970). This would suggest that the faulting and rifting associated with the East African rift proceeded from the south in Jurassic time northwards. Mougenot *et al.* (1986) have shown that the eastern limb of the East African rift may be traced across the continental margin of Tanzania and that it appears to link up with the Davie fracture zone in the Mozambique Channel. The western limb of the East African rift is obscured in southern Malawi, where rocks of Karoo age and older are overlain by Cretaceous and Tertiary sedimentary rocks constituting the Mozambique coastal plain. The distribution of faults in the East African rift system presented by Mougenot *et al.* (1986) indicates that faulting affected the Cretaceous sedimentary rocks on the Mozambique coastal plain. This faulting appears to be a continuation of structures associated with the western limb of the East African rift system. As in the alkaline complexes associated with the Pencksökket, abundant alkaline volcanism is also associated with the western limb of the East African rift. It is therefore proposed that the faulting and related structures in the Pencksökket area are of similar age to those associated with the southern part of the East African rift system, and that the faulting in both these areas is related to the fragmentation of this part of Gondwana.

References

Aucamp, A.P.H. (1972). The geology of Grunehogna, Ahlmannryggen, western Dronning Maud Land. *South African Journal of Antarctic Research*, **2**, 16–22.

Barton, J.M., Allsop, H.L., Copperthwaite, Y.E., Retief, E., Elworthy, T.P. & Wallace, R.C. (1986). A geochronological review of the tectonic evolution of western Dronning Maud Land, Antarctica. In *Geocongress '86, University of the Witwatersrand, Johannesburg, Extended Abstracts*, 711–14.

Declair, H. & van Autenboer, T. (1982). Gravity and magnetic anomalies across Jutulstraumen, a major geological feature in western Dronning Maud Land. In *Antarctic Geoscience*, ed. C. Craddock, pp. 941–8. Madison; University of Wisconsin Press.

Gavshon, R.D.J. & Erasmus, J.M. (1975). Precambrian rocks of the Neumayerskarvet area, Kirwannveggen, western Dronning Maud Land. *South African Journal of Antarctic Research*, **5**, 2–9.

Grantham, G.H., Groenewald, P.B. & Hunter, D.R. (1988). Geology of the northern H.U. Sverdrupfjella, western Dronning Maud Land and implications for Gondwana reconstructions. *South African Journal of Antarctic Research*, **18**, 2–10.

Hinz, K. & Krause, W. (1982). The continental margin of Queen Maud Land, Antarctica: seismic sequences, structural elements and geological development. *Geologisches Jahrbuch*, **E23**, 17–41.

Hjelle, A. & Winsnes, T. (1972). The sedimentary and volcanic sequences of Vestfjella, Dronning Maud Land. In *Antarctic Geology and Geophysics*, ed. R.J. Adie, pp. 539–46. Oslo; Universitetsforlaget.

Martin, A.K. & Hartnady, C.J.H. (1986). Plate tectonic development of the southwest Indian Ocean: a revised reconstruction of East Antarctica and Africa. *Journal of Geophysical Research*, **91(B5)**, 4767–86.

Mougenot, D., Recq, M., Virlogeux, P. & Lepvrier, C. (1986). Seaward extension of the East African Rift. *Nature, London*, **321**, 599–603.

Ravich, M.G. & Solov'ev, D.S. (1966). Geology and petrology of the mountains of central Queen Maud Land (Eastern Antarctica). *Transactions of the Science Research Institute of Arctic Geology, Ministry of Geology of the USSR*, **141**, 348 pp. (Translated from Russian by the Israel Program for Scientific Translations, Jerusalem, 1969.)

Vail, J.R. (1970). Tectonic control of dykes and related irruptive rocks in eastern Africa. In *African Magmatism and Tectonics*, ed. T.N. Clifford & I.G. Gass, pp. 337–54. London; Oliver & Boyd.

Watters, B.R. (1972). The Straumsnutane Volcanics, western Dronning Maud Land, Antarctica. *South African Journal of Antarctic Research*, **2**, 23–31.

Wolmarans, L.G. (1982). Subglacial morphology of the Ahlmannryggen and Borgmassivet, western Dronning Maud Land. In *Antarctic Geoscience*, ed. C. Craddock, pp. 963–8. Madison; University of Wisconsin Press.

Wolmarans, L.G. & Kent, L.E. (1982). Geological investigations in western Dronning Maud Land, Antarctica – a synthesis. *South African Journal of Antarctic Research*, **Supplement 2**, 93 pp.

Woolley, A.R. & Garson, M.S. (1970). Petrochemical and tectonic relationship of the Malawi carbonatite-alkaline province and the Lupata–Lebombo volcanics. In *African Magmatism and Tectonics*, ed. T.N. Clifford & I.G. Gass, pp. 237–62. London; Oliver & Boyd.

The tectonic and metamorphic evolution of H.U. Sverdrupfjella, western Dronning Maud Land, Antarctica

A.R. ALLEN

Department of Geology, University College, Cork, Ireland

Abstract

H.U. Sverdrupfjella, a polydeformed high-grade metamorphic terrane in western Dronning Maud Land, East Antarctica, is separated from undeformed terranes to the west by the Jutulstraumen–Pencksökket (J–P) rift, a major crustal discontinuity displaying an early strike-slip and subsequent extensional history. H.U. Sverdrupfjella consists of basement and cover sequences in tectonic juxtaposition. The basement, composed of isoclinally interfolded orthogneiss and paragneiss complexes, with abundant meta-intrusive rocks, has undergone five phases of folding and later extensional deformation, whereas the cover is layered, almost devoid of meta-intrusive rocks, and displays only three fold episodes and later extension. A Jurassic alkaline intrusive suite dominates the western margin of the terrane adjacent to the J–P rift. F_1 folds in the cover are correlated with F_3 folds in the basement, and are interpreted as reflecting north-westerly directed thrusting which resulted in interleaving of the basement and cover. The late brittle extension is related to rifting on the J–P structure during the break-up of Gondwana. Five regional metamorphic episodes are recognized, of which only the latter three affected the cover. An early high pressure metamorphism (M_1) immediately followed by isothermal decompression (M_2), reflects rapid uplift of tectonically thickened continental crust. M_3 accompanied thrusting whilst M_4 was a widespread metasomatic episode which affected both basement and cover. Greenschist-facies retrogression (M_5) was associated with late shearing.

Introduction

H.U. Sverdrupfjella is a 2000 km^2 polydeformed, polymetamorphosed terrane situated between 72 and 73° S and 2° E and 1° W, in the Precambrian East Antarctic craton (Fig. 1). The adjacent Gjelsvikfjella and Kirwanveggen terranes to the east and south-west display a similar polymetamorphic character. To the west, however, lies the relatively undeformed and unmetamorphosed Ahlmannryggen terrane, composed of a flat-lying platform of sedimentary, mafic volcanic, and shallow-level intrusive rocks, mainly of mafic–ultramafic composition. The tectonic boundary between these two terranes is a major crustal discontinuity, the Jutulstraumen-Pencksökket (J–P) rift.

H.U. Sverdrupfjella consists of a NNE-trending main range, and a number of outlying nunataks strung out westwards toward the J–P rift. Previous work in H.U. Sverdrupfjella (e.g. Roots, 1953; Ravich & Solov'ev, 1966; Van Autenboer & Loy, 1972; Hjelle, 1974) has been mainly of a reconnaissance nature, although Hjelle attempted the first stratigraphic subdivision of the terrane. This contribution presents the results of two field seasons plus an initial brief reconnaissance visit. Only about half of the terrane was visited due to lack of time and difficulty of access.

Lithostratigraphy

A preliminary lithostratigraphic subdivision of H.U. Sverdrupfjella is presented in Table 1. The major lithostratigraphic feature of the area is the existence of basement and cover sequences with distinctive lithological, structural and metamorphic characteristics. The two sequences are exposed on adjacent nunataks, but have not as yet been observed in contact. The basement forms the Rootshorga–Skarsnuten massif in the southern part of the main range, all of the nunataks from Vendeho northwards in the northern part of the main range, and all of the outlying western nunataks from Roerkulten westwards (Fig. 1). The cover sequence occupies the Nupskapa–Alanpiggen ridge in the extreme south of the main range and possibly most of the central part of the main range from Dalsnatten to Fuglefjellet. No cover rocks are exposed on the outlying western nunataks. Basement and cover sequences appear to be separated by faults, possibly occupying some of the NW–SE-trending tributary glaciers which

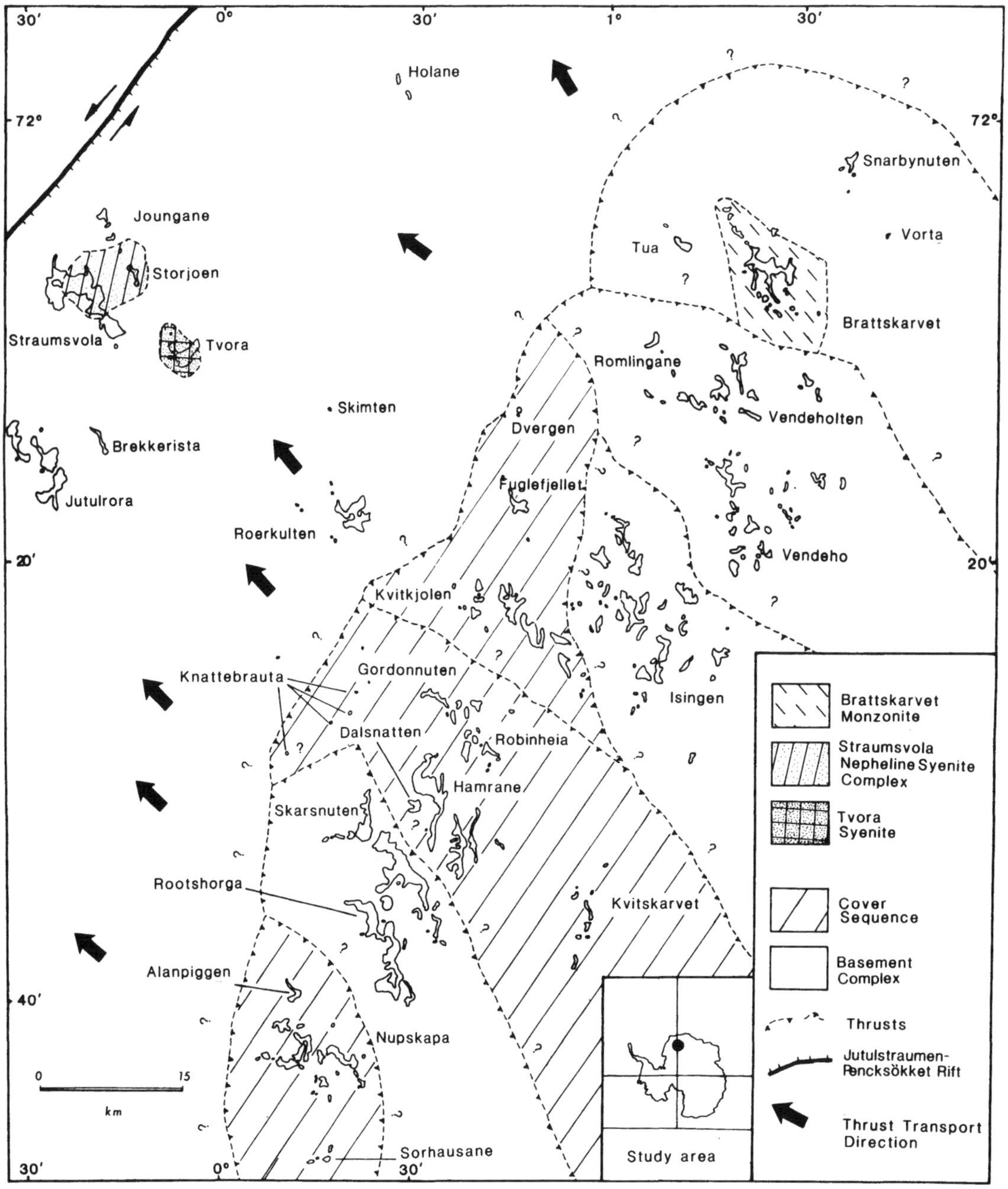

Fig. 1. Tentative tectonic map of H.U. Sverdrupfjella.

subdivide the main range. The western group of basement nunataks appears to be separated from cover rocks to the east by a NE–SW-trending fault parallel to the main range.

The basement complex is intensely deformed and extremely heterogeneous with both orthogneiss and paragneiss sequences isoclinally interfolded. A significant proportion of the basement consists of meta-intrusive rocks of various generations and compositions. Particularly important volumetrically are highly porphyritic sheet granitoids (now augen gneisses) which occupy large areas of the northern basement terrane but are also present in the other basement domains. In rare cases, relict bedding can be recognized in paragneiss sequences by the presence of isoclinally folded and boudinaged chert or calc-silicate layers, but these lithologies are rare and in general the layering in the basement is of tectonic origin.

The cover sequence is characterized by subhorizontal layering of sedimentary origin, a significant proportion of thick marble and calc-silicate units, and abundant para-amphibolites. Meta-intrusive rocks are rare and the few igneous rocks within the cover are mainly post-tectonic intrusives.

Basement sequences differ in the three separate basement terranes. The western basement consists of a coarse, banded grey orthogneiss complex of granodiorite composition, isoclinally interfolded with banded paragneiss which possibly has a large metavolcanic component. No major orthogneiss component is present in northern H.U. Sverdrupfjella, and this basement consists of homogeneous grey semipelitic paragneiss, and subordinate banded paragneiss with boudinaged chert layers and rare pyroxenitic and eclogitic lenses. Intrusive augen gneisses constitute nearly 50% of the outcrop and the terrane is

Table 1. *Provisional stratigraphic table for H.U. Sverdrupfjella (modified after Grantham & Groenewald, 1986)*

Lithostratigraphic group	Lithostratigraphic unit	Deformation episode	Tentative tectonic correlation
Younger post-tectonic intrusive suites	Younger dolerites and alkaline dyke swarms	D_7	Gondwana rifting episode
	Straumsvola & Tvora Alkaline complexes		
	Older gabbro–dolerite suite	D_6	
Older post-tectonic intrusive suites	Tourmaline pegmatites		
	Pink aplitic dykes		
	Grey aplitic dykes (Dalmation granites)		
	Ultramafic dykes	D_5	
Younger meta-intrusive suites	Brattskarvet Monzonite	D_4	Pan-African orogenic episode
		D_3	
	A_3 mafic dykes		
Cover sequence	Fuglefjellet Formation		
Older meta-intrusive suites	Magnetite–epidote pegmatites (western H.U. Sverdrupfjella)	D_2	Kibaran orogenic episode
	Roerkulten Granite		
	A_2 mafic dykes and sheets		
	Migmatitic neosome	D_1	
	A_1 mafic dykes and sheets		
	Sheet granitoids (augen gneiss)		
Basement gneiss sequences	Rootshorga Formation (southern H.U. Sverdrupfjella)		
	Romlingane Formation (northern H.U. Sverdrupfjella)		
	Jutulrora Formation (western H.U. Sverdrupfjella)		

intruded by the relatively undeformed Brattskarvet Monzonite, an extremely inhomogeneous sheet-like intrusion covering 100 km^2 and dipping eastwards. The basement of southern H.U. Sverdrupfjella is composed of large granitoid orthogneiss complexes and a paragneiss sequence containing rare dismembered calc-silicate layers and a metaconglomerate (Hjelle, 1974).

Two alkaline intrusive complexes are exposed in the extreme west of the terrane, adjacent to the J–P structure, and these are associated with mafic and alkaline dyke swarms which form 10–20% of the outcrop of Jutulrora and Straumsvola nunataks. The Straumsvola Nepheline Syenite Complex is a layered complex with a subhorizontal, discontinuously layered core surrounded by a coarse homogeneous unit, whereas the smaller Tvora Syenite is more homogeneous with dark brown and subordinate pale-coloured phases.

Alkaline dyke rocks are restricted to the vicinity of these bodies, whereas the mafic dyke rocks occur throughout H.U. Sverdrupfjella but diminish rapidly in abundance eastwards from the J–P rift. There are at least two generations of mafic dykes: an older suite which predates the intrusion of the alkaline complexes, and a younger suite which cuts the alkaline complexes. The dykes typically display en-echelon or dilational intrusive characteristics and can rarely be traced for long distances. A Rb–Sr isochron age of 170 ± 4 Ma has been obtained from the two alkaline complexes and their related alkaline dyke swarms (Allen & McNutt, unpublished data). This agrees with the K–Ar age of Ravich & Solov'ev (1966) and indicates a Jurassic age for the suite.

Structure

The main tectonic element in western Dronning Maud Land is the J–P structure separating H.U. Sverdrupfjella from the relatively undeformed and unmetamorphosed Ahlmannryggen to the west. This discontinuity is unexposed, except on Midbresrabben in Pencksökket. Here a network of anastomosing mylonitic shear zones and cataclastic zones cut banded gneisses, similar to those of the basement of H.U. Sverdrupfjella, which are intruded by relatively undeformed granite and later mafic sheets (G.H. Grantham, pers. comm.). Thus the J–P structure has undergone both brittle and ductile compressional or strike-slip movements.

The westernmost nunataks of H.U. Sverdrupfjella are dominated by the Jurassic alkaline intrusive suite, indicative of major crustal extension, probably related to crustal rifting associated with the break-up of Gondwana. Furthermore, the presence of Jurassic volcanic and intercalated sedimentary rocks in south-west Kirwanveggen (Wolmarans & Kent, 1982) suggests that the J–P structure has been reactivated as a normal fault with downthrow to the SE. Thus it has undergone a complex history of early compressional or strike-slip movements and subsequent Jurassic extensional movements.

Seven phases of deformation are recognized in the basement; five phases of folding and two subsequent episodes of brittle extension associated with dyke intrusion (Table 2). The cover sequence, however, displays only three fold episodes and the two brittle episodes (Table 2). Foliations of at least three generations are identified in the basement, whereas in the cover

Table 2. *Correlation of structural, metamorphic and tectonic evolution of H.U. Sverdrupfjella*

Deformation episode	Structural characteristics: Basement complex	Structural characteristics: Cover sequence	Metamorphic episode	Metamorphic characteristics: Basement complex	Metamorphic characteristics: Cover sequence	Associated tectonic activity	Tentative tectonic correlation
D_7 (D_5 in cover)	En échelon and dilational extension fractures – shearing of D_6 extension fractures and older dolerite dyke swarms		$M_{6?}$	Localized thermal and contact metasomatic effects associated with intrusion of the alkaline plutonic complexes and younger dolerite dyke swarm		Further rifting on J–P structure – intrusion of alkaline plutonic complexes and younger dolerite dyke swarm	Gondwana rifting episode
D_6 (D_4 in cover)	En échelon and dilational extension fractures			Localized thermal and contact metasomatic effects associated with intrusion of the older dolerite dyke swarm		Reactivation of J–P structure as an extensional normal fault with a downthrow to the NW – intrusion of older dolerite dyke swarm	
D_5 (D_3 in cover)	Occasional minor F_5 (F_3 in cover) kink folds with axes plunging moderately NE or SE – fold L_3 and L_4 lineations into small circle girdles – retrograde semi-brittle shear zones		M_5 (M_3 in cover)	Greenschist-facies retrogression – Bio replaced by Ch–Plag and Hb replaced by Ep – associated with late cross-cutting semi-brittle shear zones		Semi-brittle reactivation of J–P structure	
D_4 (D_2 in cover)	Gentle to open upright macroscopic F_4 folds – fold axes plunge NNE – L_4 stretching lineation parallel to fold axes	Gentle to open upright macroscopic F_2 folds, coaxial with F_1 folds – fold S_1 cleavage around E–W axial surfaces	M_4 (M_2 in cover)	Regional metasomatic episode – Bio cross-cutting pre-existing fabrics – K-feld replacing Plag		Intrusion of Brattskarvet Monzonite	Pan-African orogenic episode
D_3 (D_1 in cover)	Gentle to open subhorizontal upright macroscopic F_3 folds – fold axes trend NW–SE – L_3 stretching lineation trends NW–SE	Tight to isoclinal recumbent F_1 folds – dominant macroscopic folds – fold axes plunge gently E or SE – folds verge to NW – strong S_1 axial planar cleavage – L_1 stretching lineation plunging E or SE	M_3 (M_1 in cover)	Amphibolite-facies – recrystallization of post-D_2–pre-D_3 mafic dykes – Hb–Plag assemblages	Amphibolite-facies – recrystallization – Hb–Gnt–Cpx–Plag–Qtz–Diop–Zois – Bio-bearing assemblages – Hb–Plag assemblages in post-D_2–pre-D_3 mafic dykes	Oblique continental convergence – NW-directed thrusting in H.U. Sverdrupfjella – imbrication of basement and cover – initiation of J–P structure as a sinistral strike-slip fault	

D_2	Tight upright F_2 folds – dominant macroscopic folds – subhorizontal fold axes trending W–NW and E–SE – S_2 axial planar foliation locally developed – local transposition of S_1	M_2	Upper amphibolite-facies – $M_1 \rightarrow M_2$ decompression reactions e.g. Ky → Sill; Ky → Musc; Ky → Staur; Cord → Musc; Diop → Trem; Opx → Cumm; also coronitic reactions Gnt + Cpx + Qtz → Opx + Plag; Ol + Plag → Opx + Cpx + Gnt	Uplift by deep crustal thrusting	Kibaran orogenic episode
D_1	Isoclinal F_1 folds – strong S_1 axial planar foliation – transposition of S_0	M_1	Amphibolite – granulite-facies high-P assemblages – Gnt–Opx–Cpx–Ky–Sill–Cord–Diop–Hb–Anthoph–Bio-bearing assemblages	Continental collision and crustal thickening – anatexis and intrusion of sheet granitoids	

Gnt, garnet; Opx, orthopyroxene; Cpx, clinopyroxene; Ky, kyanite; Sill, sillimanite; Cord, cordierite; Staur, staurolite; Diop, diopside; Hb, hornblende; Anthoph, anthophylite; Cumm, cummingtonite; Ch, chlorite; Ep, epidote; Bio, biotite; Plag, plagioclase; K-feld, K-feldspar; Trem, tremolite; Musc, muscovite; Ol, olivine; Qtz, quartz; Zois, zoisite.

sequence the main structural element is a single prominent cleavage which contains a well defined stretching lineation consistently plunging E or SE.

Basement

In the basement, rare relict bedding identified in the paragneiss is isoclinally folded and dismembered by F_1 folds. The S_1 foliation, axial planar to these folds, is the dominant foliation in the basement and pre-F_1 meta-intrusive rocks, and it defines a layering which is thus transposed. However, in the paragneiss, layering generally parallels original bedding (S_0), except in the hinges of the F_1 folds. A similar layering in the orthogneiss, also defined by S_1, is axial planar to isoclinal F_1 folds which affected early cross-cutting veins and dykes. Migmatitic neosome in both paragneiss and orthogneiss sequences is typically concordant with S_1, and is regarded as partial melt derived from the host rock during F_1.

Second generation mesoscopic-scale folds (F_2) which fold the S_1 layering, are the most conspicuous folds in the basement. They are typically tight to isoclinal, commonly with sheared-out limbs. An S_2 axial planar cleavage is generally not developed in the strongly layered paragneiss, except in the more micaceous layers, but crenulation of S_1 is common. In the more compositionally homogeneous rock types, domainal development of a penetrative S_2 foliation occurs in zones of high strain, where tightening of F_2 folds leads to transposition of S_1. S_2 is also penetrative in post-F_1–pre-F_2 mafic sheets, which cut the S_1 foliation but are folded by F_2.

F_2 folds vary in orientation due to subsequent refolding, but they are generally upright, and within km-scale domains display relatively constant attitudes (Table 2). Sheath folds are common and axes within individual folds may vary in trend by up to 90° or more. Refolding of F_1 folds during F_2 has given rise to spectacular interference patterns, particularly prominent within the banded gneiss sequences of northern H.U. Sverdrupfjella where Type II and Type III interference patterns (Ramsay, 1967) are common.

F_3 folds in the basement are upright, open to tight mesoscopic and macroscopic folds, parasitic on major structures verging westwards. F_3 folds commonly display a strong stretching lineation parallel to subhorizontal fold axes trending NW–SE (Table 2). Some of the F_3 folds appear to shear-out the limbs of F_2 folds rather than refold them. No S_3 foliation is developed except in post-F_2–pre-F_3 mafic sheets, where it sometimes has the form of a strong penetrative cleavage.

F_4 folds are rare and as yet have only been identified in the northern basement, where they fold cleaved post-F_2 mafic dykes. They are open to tight folds with steep axial surfaces, fold axes which plunge steeply NNE, and a stretching lineation parallel to the fold axes. F_5 folds are sporadic small NW-inclined kink folds which plunge moderately NE and fold both the L_3 and L_4 stretching lineations.

Cover

The cover sequence has undergone a significantly less complex and less intense deformation history than the basement. Bedding is the dominant S-surface present, with thick marble units acting as local marker horizons.

F_1 folds in the cover sequence are recumbent tight to isoclinal folds which verge westwards and plunge moderately E in northern Fuglefjellet and to the SE in southern H.U. Sverdrupfjella. In contrast to F_1 folds in the basement, they are the dominant folds in the cover. Competent layers in the sequence are commonly boudinaged, and thin para-amphibolite horizons within thick marble layers are isoclinally folded and dismembered. S_1 is widely developed in the cover rocks as an axial planar cleavage subparallel to layering on the limbs of F_1 folds. It contains an intense stretching lineation plunging E in Fuglefjellet or SE in the southern H.U. Sverdrupfjella.

F_2 folds in the cover are upright folds coaxial with F_1, which fold the S_1 foliation round E–W axial surfaces. F_3 folds in the cover are rare late kink folds.

Late brittle deformation of both the basement and cover is indicated by dyke-filled extensional fractures. Two episodes of extension are implied by the presence of dykes predating the Straumsvola and Tvora Alkaline complexes, and later dykes which cut these two bodies. The dykes show a wide variety of orientations and cross-cutting relationships, and many of the dykes are sheared indicating reactivation of the fractures they occupy. Two episodes of late brittle extension are also indicated by kinking of fractures cutting the Brattskarvet Monzonite in northern H.U. Sverdrupfjella.

A tentative correlation is made between F_1 folds in the cover and F_3 folds in the basement, based on their general E–SE trend, their association with a SE-trending stretching lineation, and their consistent westerly vergence. However, significant differences exist between the two sets of folds. F_1 folds in the cover are tight to isoclinal, whereas F_3 folds in the basement are more open and parasitic on major structures. Furthermore, F_3 folds have steep axial surfaces which vary in orientation from the northern to the western basement outcrops, whereas the F_1 folds in the cover are recumbent. However, these differences may reflect different mechanical states of the basement and cover at the time of deformation. The former was crystalline, anhydrous and intensely deformed, and the latter layered, undeformed and unmetamorphosed. Such mechanical differences would favour decoupling of basement and cover during the initial deformation of the latter, and they would then deform as distinct entities. Deformation of the basement could be expected to have been concentrated into steep zones of high strain whereas that of the cover would be more penetrative, but with a high strain zone at the basement–cover interface. Thus initial deformation of the cover sequence resulted in nappe structures accompanied by thrusting at the basement–cover interface. The recumbent nature of the F_1 folds in the cover, and the absence of basement and cover in direct contact on the same nunatak is consistent with this, and suggests that such thrusts may underlie the tributary glaciers separating the different ridges and massifs of the main range (Fig. 1). Another possible thrust boundary may separate the cover sequence in the main range from the basement of the western nunataks, which differs from the basement in the main range both lithologically and in metamorphic characteristics (below).

Thus the main range may consist of a series of basement and

cover slices which have been imbricated by thrusting (Fig.1). Subsequent movement of this imbricated thrust sheet over the western basement terrane may account for the geological relations seen in H.U. Sverdrupfjella. The direction of such thrusting, suggested by the stretching lineations and fold vergence associated with F_1 folds in the cover and F_3 folds in the basement, is north-westwards. It is further suggested that initial movements on the J–P structure occurred during this episode, possibly in the form of northwards sinistral transpression of H.U. Sverdrupfjella relative to the undeformed terranes to the west.

Metamorphism

Only a general summary of the main metamorphic characteristics of H.U. Sverdrupfjella is presented here. More detailed analyses are presented by Groenewald, Allen & Grantham (1986) and Groenewald & Hunter (this volume, p. 61). Five phases of regional metamorphism are postulated for the terrane, only three of which affected the cover. The two earliest phases, which affected only the basement, indicate a complex metamorphic evolution for the basement prior to emplacement of the cover. The metamorphic evolution of the cover sequence is relatively uncomplicated in comparison, with little or no evidence of retrogression of early assemblages, apart from replacement of diopside by hornblende in mafic assemblages.

In the basement, metamorphic grade increases from west to east, ranging from epidote–amphibolite- to amphibolite-facies assemblages in the western basement, to upper amphibolite- or granulite-facies assemblages in the main range. In the main range, metamorphic grade also varies, with kyanite-bearing metapelites developed in the southern basement domain and sillmanite- and hypersthene-bearing metapelites in northern H.U. Sverdrupfjella. Furthermore, high-pressure M_1 assemblages are characteristic of both basement terranes in the main range (garnetiferous mafic bodies occurring in both southern and northern terranes) but they are absent from the western basement terrane. However, an abundance of garnet and rare kyanite combined with a concomitant rarity of cordierite in metapelites of the western basement suggest that this terrane also experienced moderately high pressures during M_1 in relation to the lower thermal peak it attained.

These early high-pressure assemblages have reacted to form lower-pressure assemblages during M_2 with evidence for isothermal decompression reactions. In southern H.U. Sverdrupfjella kyanite is partially replaced by either M_2 sillimanite or muscovite, suggesting that reaction took place close to the univariant boundary curve for the reaction muscovite + quartz = sillimanite + K-feldspar + water, indicating pressures of the order of 6 kb and temperatures close to the amphibolite–granulite-facies transition for these reactions. Complex coronitic reactions in pre-F_1 mafic bodies of northern H.U. Sverdrupfjella also indicate isothermal decompression from initial M_1 pressures of the order of 8–10 kb to pressures of around 6 kb during M_2 (Groenewald *et al.*, 1986). No equivalent M_2 decompression reactions have been observed in the western basement terrane, apart from the breakdown of M_1 kyanite to M_2 staurolite, suggesting the possibility of a metamorphic discontinuity between this basement terrane and the basement of the main range. Although a gradual decrease in structural level from east to west could explain all these metamorphic features, such a discontinuity would be consistent with lithological differences between the western basement and basement of the main range.

High-pressure asssemblages are not found in the cover sequence, nor do cover assemblages display complex reaction relationships equivalent to those found in the basement rocks. Assemblages are of mid-amphibolite-facies grade and no regional variation in grade is apparent. Similar mid–lower-amphibolite-facies assemblages are observed in post-F_2 dykes which cut both basement and cover, and thus an M_3 metamorphic event in the basement is correlated with M_1 in the cover. The M_4 metamorphic event is a widespread late metasomatic event affecting both basement and cover throughout the whole terrane. It is identified principally in mafic rocks where all early fabrics are cut by undeformed, randomly orientated, coarse, elongate biotite with no sign of reaction. Also, in mafic rocks, plagioclase is partially replaced by microcline, both in the form of patch perthite and as marginal alteration. These reactions suggest an influx of potassic fluids derived from an external source.

The M_5 metamorphic event is retrogressive, associated with late cross-cutting shear zones and gives rise to greenschist-facies chlorite and epidote, mainly after biotite and plagioclase. Still later localized thermal and metasomatic effects are associated with the Jurassic alkaline intrusive rocks.

The high-pressure M_1 assemblages in the H.U. Sverdrupfjella basement suggests a regime of continental collision and crustal thickening at this time (England & Thompson, 1984; Ellis, 1987). Furthermore the M_2 isothermal decompression reactions imply rapid uplift of the basement between M_1 and M_2, whilst the basement was still close to its thermal peak. This suggests a short time gap between M_1 and M_2, and that M_1 and M_2 reflect a single orogenic cycle of deformation and uplift (see England & Richardson, 1977). Both England & Richardson (1977) and Ellis (1987) consider such an uplift phase to reflect rapid erosion of thickened crust attendant upon continental collision, but an equally plausible uplift mechanism could be deep crustal thrusting associated with such collision processes (Coward, 1980, 1983; M.K. Watkeys, pers. comm., 1987). The *P–T* path defined by Groenewald & Hunter (this volume, p. 61) suggests that the basement complex of the main range of H.U. Sverdrupfjella has undergone rapid uplift associated with tectonically thickened crust. Thus it is suggested that the basement of H.U. Sverdrupfjella represents thickened crust formed during continental collision. The absence of decompression reactions in the western basement may reflect a much higher level, lower-temperature thrust sheet which did not develop high-pressure assemblages to become unstable during uplift.

Tectonic evolution of H.U. Sverdrupfjella

A tentative tectonic evolution for H.U. Sverdrupfjella is proposed on the basis of the structural and metamorphic characteristics discussed above and is presented in Table 2. The H.U. Sverdrupfjella basement developed in a continental collision environment and represents a deep-seated portion of a thickened crust. Perturbation of the geotherm arising from such thickening led to the formation of high-pressure M_1 assemblages, as temperatures increased. Subsequent rapid uplift of this thickened crust led to the instability of the high-pressure assemblages and reaction to lower-pressure assemblages under isothermal conditions. Uplift and decompression may have been brought about by rapid erosion accompanying isostatic rebound, or may be due to thick-skinned thrusting associated with F_2.

A cover sequence of carbonate-rich sediments was then deposited on this uplifted basement possibly during an extensional evolutionary phase, and both basement and cover were subsequently deformed during a later continental convergence. Oblique convergence of H.U. Sverdrupfjella and the undeformed Ahlmannryggen terrane to the west gave rise to sinistral transpression resulting in initiation of the J–P structure as a strike-slip fault which transported H.U. Sverdrupfjella northwards relative to Ahlmannryggen. Associated NNW–SSE compression brought about decoupling of basement and cover of H.U. Sverdrupfjella and thrusting of the cover northwestwards over basement. Interleaving of basement and cover led to the development of an imbricated thrust sheet corresponding to the main range of H.U. Sverdrupfjella, with this main range thrust sheet in turn thrust NW over the lower-grade basement of H.U. Sverdrupfjella. Attendant with the thrust movements, recumbent F_1 and subsequent coaxial F_2 folds developed in the cover rocks and F_3 and F_4 folds developed in the basement. As thrusting waned, K-rich metasomatic fluids penetrated the terrane, possibly migrating along the various thrust surfaces. The Brattskarvet Monzonite may also have been emplaced at this time, intruding as a sheet along one of the imbricate thrusts in the main range. Finally late-stage shearing with accompanying greenschist-facies retrogression and minor F_5 kink folds brought this episode to a close.

The final stage of evolution of H.U. Sverdrupfjella is related to the break-up of Gondwana, when the J–P thrust was reactivated as an extensional normal fault with a downthrow to the SE. Extension of the western margin of H.U. Sverdrupfjella led to intrusion of swarms of mafic and alkaline dykes and the intrusion of two plutonic alkaline complexes. The J–P structure is thus a failed Jurassic rift, formed by reactivation of a pre-existing strike-slip fault forming the suture between the polydeformed high-grade terranes of H.U. Sverdrupfjella and Kirwanveggen, and the undeformed terranes of Ahlmannryggen and Borgmassivet to the west.

Acknowledgements

The present study is the result of work done in H.U. Sverdrupfjella from 1982–84 under the auspices of the South African Antarctic Earth Sciences Programme. The other members of the University of Natal H.U. Sverdrupfjella Research Project, G.H. Grantham and P.B. Groenewald are thanked for stimulating discussions both in the field and subsequently. However, they bear no responsibility for any errors in the interpretations presented above. Thanks are also due to M.K. Watkeys for an extremely thorough and perceptive review which has greatly improved the overall interpretation. Professor D.R. Hunter and Dr. J.R. Krynauw are thanked for their administration and organization of the South African Antarctic Earth Sciences Programme. Funding and logistics for fieldwork were provided by the South African Department of Transport and CSIR.

References

Coward, M.P. (1980). Shear zones in the Precambrian crust of Southern Africa. *Journal of Structural Geology*, **2(1–2)**, 19–28.

Coward, M.P. (1983). Thrust tectonics, thin skinned or thick skinned, and the continuation of thrusts to deep in the crust. *Journal of Structural Geology*, **5(2)**, 113–23.

Ellis, D.J. (1987). Origin and evolution of granulites in normal and thickened crust. *Geology*, **15(2)**, 167–70.

England, P.C. & Richardson, S.W. (1977). The influence of erosion upon the mineral facies of rocks from different metamorphic environments. *Journal of the Geological Society, London*, **134(2)**, 201–13.

England, P.C. & Thompson, A.B. (1984). Pressure–temperature–time paths of regional metamorphism. 1. Heat transfer during the evolution of regions of thickened continental crust. *Journal of Petrology*, **25(4)**, 894–928.

Grantham, G.H. & Groenewald, P.B. (1986). Sheeted granulites in the H.U. Sverdrupfjella, western Dronning Maud Land, Antarctica. In *Geocongress '86, Johannesburg, Abstracts Volume*, pp. 723–6. Geological Society of South Africa.

Groenewald, P.B., Allen, A.R. & Grantham, G.H. (1986). Retrogressive evolution of granulite facies gneisses, Northern H.U. Sverdrupfjella, western Dronning Maud Land, Antarctica: preliminary results. In *Geocongress '86, Johannesburg, Abstracts Volume*, pp. 727–30. Geological Society of South Africa.

Hjelle, A. (1974). Some observations on the geology of H.U. Sverdrupfjella, Dronning Maud Land. *Norsk Polarinstitutt Årbok*, **1972**, 7–22.

Ramsay, J.G. (1967). *Folding and Fracturing of Rocks*. New York; McGraw-Hill, 568 pp.

Ravich, M.G. & Solov'ev, D.S. (1966). Geology and petrology of the mountains of central Queen Maud Land (Eastern Antarctica). *Transactions of the Science Research Institute of Arctic Geology, Ministry of Geology of the USSR*, **141**, 348 pp. (Translated from Russian by the Israel Program for Scientific Translations, Jerusalem, 1969).

Roots, E.F. (1953). Preliminary note on the geology of western Dronning Maud Land. *Norsk Geologiske Tidsskrifft*, **32(1)**, 17–33.

Van Autenboer, T. & Loy, H. (1972). Recent geological investigations in the Sør-Rondane Mountains, Belgicafjella and Sverdrupfjella, Dronning Maud Land. In *Antarctic Geology and Geophysics*, ed. R.J. Adie, pp. 563–71. Oslo; Universitetsforlaget.

Wolmarans, L.G. & Kent, L.R. (1982). Geological investigations in western Dronning Maud Land. *South African Journal of Antarctic Research*, **Supplement 2**, 93 pp.

Granulites of northern H.U. Sverdrupfjella, western Dronning Maud Land: metamorphic history from garnet–pyroxene assemblages, coronas and hydration reactions

P.B. GROENEWALD & D.R. HUNTER

Department of Geology, University of Natal, Pietermaritzburg, South Africa 3200

Abstract

The Proterozoic metamorphic terrane of H.U. Sverdrupfjella comprises a sequence of para- and orthogneisses. The amphibolite-facies mineralogy of these gneisses results from widespread regressive rehydration of granulite-facies assemblages, now preserved in mafic lenses and boudins. Thermobarometry using core, rim and corona assemblages of garnet, pyroxene, plagioclase and quartz indicates initial conditions of 9–11 kb at 850 °C, and retrogression to 6–7 kb at 650 °C. The prograde path was in the sillimanite stability field which requires relatively high heat flow during crustal thickening. After an initial decompressive stage, retrogression by nearly isobaric cooling was followed by decompression and cooling. The data suggest that collision tectonics were involved in the evolution of H.U. Sverdrupfjella.

Introduction

H.U. Sverdrupfjella comprises some 6000 km^2 of medium- to high-grade para- and orthogneisses which lie near the boundary between the high-grade East Antarctic structural province and the low-grade, gently deformed, volcanic and sedimentary Ritscherflya Supergroup (Wolmarans & Kent, 1982).

The stratigraphy of the H.U. Sverdrupfjella Group has been described by Hjelle (1974) and Grantham, Groenewald & Hunter (1988). The south-western extension of the H.U. Sverdrupfjella Group in Kirwanveggen has yielded ages of 1050–1173 Ma (Rb–Sr; Wolmarans & Kent, 1982) which probably reflect an early metamorphic event. Biotite Rb–Sr ages from the same area are tightly constrained at ~ 475 Ma. Preliminary Rb–Sr data from H.U. Sverdrupfjella and the area immediately to the west are consistent with these ages (Moyes, Grantham & Groenewald, unpublished data).

Layering and foliation in the area are predominantly subparallel and generally dip at 30–50° E and SE. Two initial phases of isoclinal folding (F_1 and F_2) are locally distinguishable although in many areas intense refoliation and the coplanar nature of the fold events obscures this distinction. A later episode of close, locally isoclinal folding (F_3) was accompanied by limited refoliation. At least two subsequent generations of more open folding are recognized. F_2 and F_3 were accompanied by the development of shear zones. Large-scale brittle deformation postdates the main folding events.

Several periods of magmatic activity were interspersed between the main deformational events and are represented by tabular granitoid intrusions, mafic dykes (now amphibolites) and alkaline complexes.

Metamorphism, initiated before and accompanying D_1 and D_2, reached granulite-facies conditions but there is little evidence of the prograde path preceding the attainment of this metamorphic grade. The granulite-facies event is preserved in boudins of mafic composition and in some pelitic units, but most assemblages in the area are more typical of the amphibolite-facies. More hydrous conditions prior to F_3 are reflected by the presence of biotite and hornblende within folded fabrics in post-F_1, pre-F_3 mafic intrusions. Amphibolite-facies assemblages are also present in post-F_3 dykes. Late post-tectonic intrusive bodies associated with block faulting have metamorphic assemblages typical of the greenschist-facies.

This paper presents preliminary quantitative constraints on the early metamorphic pressure–temperature (*P–T*) changes based on various mineral assemblages and published thermobarometric calibrations. The path followed during the subsequent evolution of northern H.U. Sverdrupfjella is poorly constrained by phase relations and garnet–biotite thermometry.

Petrography

Boudins

Salient details of mineral assemblages in the four main lithologies present as boudins are given in Table 1. An example of the first lithological group is a pod-like intrusion of medium-grained olivine gabbro–norite into quartzo-feldspathic

Table 1. *Mineral assemblages and P–T estimates for northern H.U. Sverdrupfjella*

Sample type & number	Assemblages	X_{Mg}^{opx}	X_{Fe}^{opx}	X_{Ca}^{opx}	Al^{opx*}	X_{Mg}^{cpx}	X_{Fe}^{cpx}	X_{Ca}^{cpx}	Al^{cpx}	X_{Py}	X_{Gr}	X_{Al}	X_{Sp}	X_{Ca}^{plag}	X_{Na}^{plag}	T (°C)	P (kb)	Relative age
Metagabbro ol–cpx–opx–plag (igneous)																		
SF8565	opx–cpx–gnt (coronas)	.807	.187	.006	.066	.550	.104	.346	.192	.494	.140	.353	.013	.548	.447	824[1]	9.7–12[3]	M_1
SF8568		.789	.178	.033	.136	.467	.061	.472	.161	.487	.170	.334	.009	.490	.499	753[1]	10–12.5[3]	M_1
Px–granulite																		
SF8521	opx–cpx–plag(–bbt–qz)	.638	.353	.009	.037	.425	.113	.462	.104					.893	.106	650[2]	10[6]	?
Gnt–px–granulite																		
SF8525	(i) gnt–opx–cpx–plag–qz–mt(–bhbl–bbt)	.485	.504	.011	.040	.371	.172	.457	.089	.165	.188	.618	.030	.949	.050	685[1]	7.9[4]	M_1
	(ii) (gnt)–plag2–px2–sp	.485	.504	.011	.040	.350	.185	.465	.077	.165	.188	.618	.030	.949	.050	653[1]	5.8[4,5]	M_2
Hi–P granulites																		
SF8528	(i) gnt–(opx)–cpx–qz–mt	.503	.443	.054	.036	.345	.170	.485	.072	.230	.218	.541	.010			822[1]	9.6[3]	M_1
	(ii) (gnt)–opx–cpx2–plag	.542	.449	.009	.050					.159	.199	.590	.052	.939	.061	650[2]	6.5[4]–7.8[3]	M_2
SF8438	(i) gnt–cpx–qz(–mt)					.340	.188	.472	.060	.146	.232	.607	.015			701[1]	7.1[6]–8.7[4]	M_1?
	(ii) opx–plag lamellae	.490	.502	.009	.040	.346	.176	.478	.056	.128	.214	.613	.046	.926	.072	640[1]	5.6[5]–6.9[3]	M_2
Metapelites																		
TUA1	opx–gnt–plag	.529	.465	.006	.041					.129	.094	.741	.036			(700)	10–12[3]	M_1

Thermobarometric calibrations used: *T*–1, Ellis & Green (1979); 2, Lindsley (1983); *P*–3, Harley & Green (1982); 4, Newton & Perkins (1982); 5, Perkins & Chipera (1985); 6, Ellis (1980). *, Al in 6 oxygen opx, cpx.
bbt, brown biotite; bhbl, brown hornblende; mt, magnetite; ol, olivine.

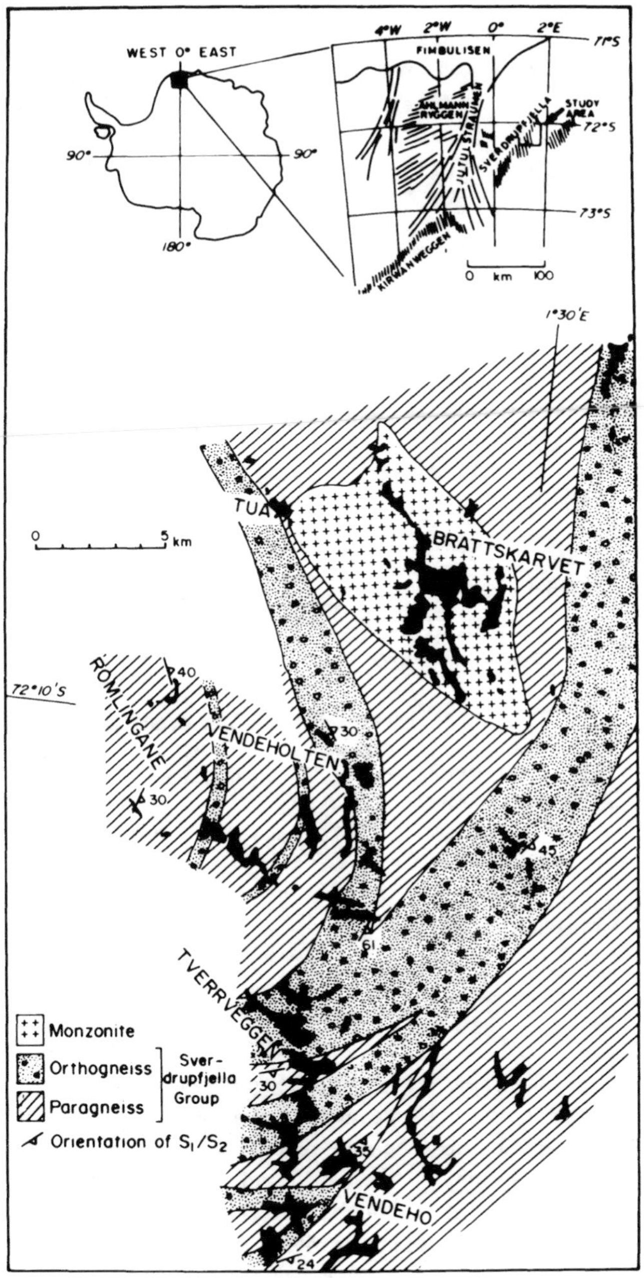

Fig. 1. Location and simplified geological maps of northern H.U. Sverdrupfjella.

gneisses in the Romlingane area (Fig. 1). This disrupted body is ~ 300 m in length and 5–30 m wide. Its igneous parentage is demonstrated by the preservation of orthocumulate textures. Olivine and plagioclase show evidence of subsolidus reaction which produced coronas between relict olivine and plagioclase: the inner shell consists of radiating enstatite lamellae with a narrow rim of randomly orientated clinopyroxene, commonly intergrown with garnet; the outer shell is an irregular zone of garnet which replaces most of the plagioclase. Small amounts of pale yellow phlogopite are present within the inner shells of some coronas. Spinel is not confined to the coronas and seems unrelated to the corona-forming reaction. The cores of all igneous minerals, with the exception of olivine, are packed with unidentified, minute bleb-like and acicular inclusions, possibly of spinel. Progressive amphibolitization occurs toward the margins of this intrusion. In the core, limited amounts of biotite and hornblende are present, both of which are pale brown in colour and may represent original granulite-facies minerals. Near the contact fine lamellae of green amphibole become common along pyroxene–plagioclase boundaries, and nearer to the contact biotite and hornblende are abundant. Small remnant pyroxenes have undergone recrystallization to fine granular subgrains in this zone. At the contact are bimineralic mixtures of plagioclase (20%) and amphibole (80%).

Boudins of the second group (up to 1.5 m in diameter) are widely distributed throughout the Romlingane area but are sporadic elsewhere. No relict igneous textures are apparent but metamorphic parageneses of at least two generations are present. These rocks are mineralogically and chemically distinct from the larger metagabbroic intrusions and they can be divided into two distinct types:

(i) Mafic rocks, devoid of garnet, in which the dominant minerals are orthopyroxene and plagioclase, with two generations of biotite (an earlier reddish-brown variety and a later pale brown to green biotite), and magnetite and quartz (qz) as minor constituents
(ii) Mafic to ultramafic lithologies in which the dominant minerals are garnet (gnt), clinopyroxene (cpx), orthopyroxene (opx), plagioclase (plag), magnetite (mt) and green hornblende (hbl).

The most common M_1 assemblage is gnt–cpx–opx–plag(-qz). This has been modified by reactions between gnt, cpx and qz to produce radial lamellae of opx and plag around remnants of garnet. Later metamorphic reactions produced green hbl at the expense of opx, cpx and plag. Ilmeno-magnetite breakdown also occurred during the formation of amphibole and sphene.

A third type of mafic boudin occurs at localities in the extreme north and west of H.U. Sverdrupfjella. They are characterized by an early high pressure mineral assemblage: gnt–cpx–qz–ilmenite(ilm). Local to pervasive reaction of this assemblage has produced opx–plag (An_{93}) kelyphitic rims on relict garnet, at the outer edge of which is a narrow zone of fine, granular, second-generation clinopyroxene.

Metapelites

Metapelites containing granulite-facies assemblages are sparse but widely distributed. Unfoliated pods in the north-west of the area have the assemblage gnt–hyp–plag–biotite (biot). Nearby are rocks containing almandine (alm)–cordierite (cd)–sillimanite (sil)–plag–biot. Both of these assemblages were modified by growth of amphibole and/or biotite during retrogression. Another characteristic is the abundance of large zoned zircons, and coarse subhedral apatite. On south Vendeholten layered and foliated metapelites also have alm–cd–sil–plag–biot assemblages, but these, apart from the introduction of pale second-generation biotite, show little evidence for retrogression. Most of the metapelites are characterized by pre- to syn-kinematic garnet, and coarse sillimanite which predates fibrolite.

Thermobarometry

The boudins contain parageneses for which reaction equilibria have been experimentally calibrated or thermodynamically calculated and thus provide a means of quantifying the early metamorhic P–T conditions. Garnet–clinopyroxene relations produce consistent temperatures for several mineral pairs in each rock, and allow discrimination between granulite-facies and subsequent amphibolite-facies events. Pressure estimates are more problematical in that the different geobarometers applied to the same mineral data yield variable pressure estimates. In the present situation, many of the assemblages under consideration are mixtures of products and reactants and equilibrium can only be expected within small domains.

Estimated P–T conditions for M_1 (Table 1) are 9–11 kb at 750–880 °C. Some mineral pairs considered to represent M_1 on petrographic grounds yielded lower temperatures than others. These may have undergone re-equilibration to the blocking temperatures during slow cooling. There is also a spatial control on the distribution of these lower temperature estimates in that they are restricted to the northern part of the area, in the vicinity of a major late-tectonic monzonite intrusion.

M_1 temperature estimates rely on a boudin from west Romlingane and the corona assemblages from the metagabbros. The latter require further comment. Whereas the pressure estimate provided by the gnt–opx assemblage is likely to represent metamorphic pressure, the corona-forming reaction may have occurred at a temperature between that of igneous crystallization and the ambient metamorphic temperature. The temperature calculated on the basis of gnt–cpx pairs in the coronas, however, agrees closely with that from the boudin.

Retrogression of M_1 assemblages in the boudins under relatively anhydrous conditions produced pyroxene and plagioclase, commonly at the expense of garnet and early clinopyroxene. Although garnet core compositions are unaffected, sharply increased Mn contents towards the rims of many corroded crystals, reflect re-equilibration of garnet. These minerals yielded temperature estimates of 600–720 °C at pressures of 6.5–8 kb, with a trend of decreasing temperature and pressure from core to rim. This trend for individual rocks is significant in that it is parallel to the inferred path for the whole data set (Fig. 2).

Post-M_2 conditions are imprecisely known because of a lack of suitable assemblages. Lower amphibolite-facies conditions were attained as indicated by the presence of amphibolitized dykes which postdate F_2/M_2. Biotite, which crystallized in F_3 fabrics, provides an indication of F_3/M_3 temperatures. Garnet-rim to biotite-core compositions provide a temperature estimate of 580 °C (equation of Indares & Martignole, 1985). The M_3 assemblage of K-feldspar–sillimanite–muscovite–qz requires relatively low pressure (~ 3.5 kb) during M_3 at this temperature.

An important aspect of the retrogressive evolution was extensive rehydration, demonstrated by the total amphibolitization of most of the metabasites and ubiquitous biotite in late-tectonic cleavage planes. The amount of water required for rehydration is difficult to estimate because the extent of prograde dehydration is unknown. The source of this water is also obscure.

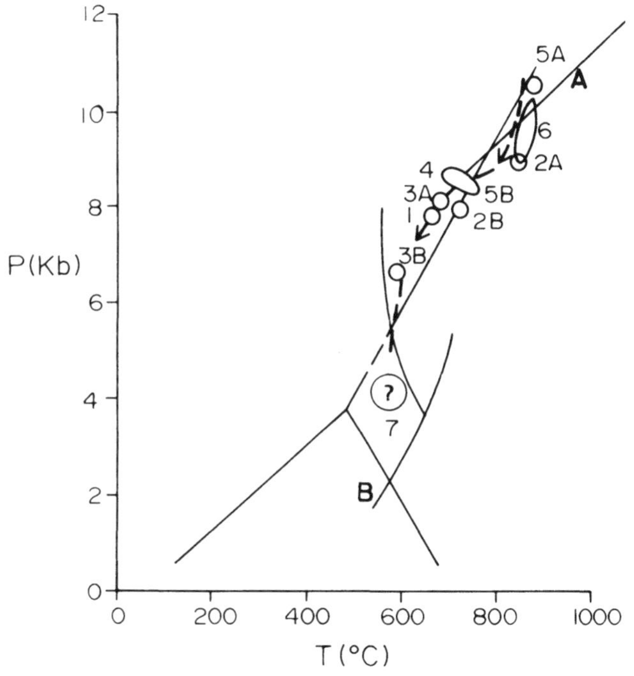

Fig. 2. P–T diagram for data in Table 1, with inferred thermobarometric path indicated for the retrogressive history. 1, SF8525; 2, SF8528; 3, SF8438 rims, 4, SF8438 cores; 5, TUA1; 6, SF8565 & SF8568; 7, inferred from gnt–biot thermometry and stability of muscovite. Phase boundaries: A, opx + an > gnt + cpx + qz (Green & Ringwood, 1966); B, muscovite stability field from Winkler (1974).

Discussion

P–T–t paths provide important insights into tectonic evolution of orogenic belts, and recent advances in the field of thermobarometry have made it practical to use P–T variation to generate tectonic models which are consistent with other lines of evidence. Thermobarometry for northern H.U. Sverdrupfjella indicates that retrogression may have been characterized initially by decompression accompanied by limited cooling, followed by two periods of cooling and decompression of different dP/dT slopes (Fig. 2).

The M_1 conditions were at the limit of cordierite stability which is in agreement with the paucity of cordierite-bearing rocks in the area. The reaction gnt + cpx + qz→opx + plag observed in the boudins indicates high pressure granulite-facies conditions at an early stage, and the transition to medium pressure conditions. Abundant coarse sillimanite in some rocks requires P–T in the sillimanite field for this stage, and constrains M_1 to the lower pressure part of the high pressure granulite field. The subsequent P–T path passes just into the kyanite field (Fig. 2). The period following the first decompressive episode is discontinuous with respect to the early decompression, and is characterized by apparently progressively steeper dP/dT after an initial stage of nearly isobaric cooling.

Interpretation of the P–T path is hampered by the lack of information about the prograde path because the retrograde path alone is not fully diagnostic of the tectonic regime. There is no direct evidence for a heating phase after the attainment of maximum pressure as predicted by calculated models of England & Thompson (1984) and others. Nor is there evidence for a long period of isobaric cooling as documented for the East Antarctic metamorphic subprovinces in Enderby Land (Harley, 1983).

The processes of crustal thickening during metamorphism in northern H.U. Sverdrupfjella can be inferred from other lines of evidence. First, the lithostratigraphy comprises a sequence of paragneisses (carbonates, pelites and impure quartzose rocks) and calc–alkaline orthogneisses. This sequence is not typically cratonic and may represent a marginal or extensional basin sequence. Where such successions are present in other orogenic belts they are commonly accompanied by evidence for emplacement by thrust mechanisms (cf. Windley, 1984, chapters 6 & 10). Second, the first recognized deformation is characterized by flat-lying isoclinal folds. Major nappe structures have not been identified in the terrane despite abundant evidence for small-scale thrusting, but there are some indications that the northern main range of H.U. Sverdrupfjella is allochthonous relative to nunataks immediately to the west. This inference relies on the presence of higher-grade mineral assemblages at higher structural levels in the eastern than in the western area, although similar rock types are present in both areas.

The main retrogressive reactions after M_1 result from decompression as recorded by incomplete breakdown of early cpx–gnt–qz assemblages to form opx–plag. This initial steep $-\mathrm{d}P/\mathrm{d}T$ suggests that the *P–T* path may have followed a clockwise loop. The abundance of coarse sillimanite in the pelites argues for conditions well within the sillimanite field, and implies an anticlockwise *P–T* path. The regional tectonic character of the gneisses in central Dronning Maud Land is, however, one of polyphase isoclinal folding accompanied by intense refoliation (Ravich & Solov'ev, 1966), which is consistent, by analogy with several Proterozoic and Archaean high-grade terranes, with an interpretation of crustal thickening through overthrusting. It is possible that accretion of a marginal basin sequence occurred as a result of NW-directed overthrusting, probably accompanied by some form of delamination and lithospheric subduction. Similar processes have been invoked for the Proterozoic metamorphic complexes of the Lützow–Holm Bay region (Shiraishi, Motoyoshi & Yanai, 1986), and for several 1000 Ma old tectonic provinces in Africa (Daly, 1986; Shackleton, 1986).

Post-M_1/F_1 deformation has considerably modified the geometry of the terrane, and is reminiscent of Pan-African overprinting on the Kibaran and Irumide tectonic provinces in eastern central Africa. Significantly, recent research provides evidence for a Gondwana reconstruction in which western Dronning Maud Land is adjacent to the south-eastern extension of the Mozambique sector of the Kibaran orogeny (Martin & Hartnady, 1986; Grantham *et al.*, 1988). The Mozambique belt, as described by Daly (1986; and references therein), contains tectonic elements similar to those presented here for the western part of Dronning Maud Land. A correlation of the gneiss complexes of the Lützow–Holm Bay area farther to the east with the northern Irumide and Kibaran structural provinces (Yoshida & Kizaki, 1983) supports a correlation between H.U. Sverdrupfjella and the southern part of these provinces. The implications of this correlation are important because a considerable body of evidence is accumulating for major and repeated accretion of continental and oceanic crust onto Africa during the Proterozoic (Daly, 1986; Shackleton, 1986). It seems likely that collision orogeny was equally important in the evolution of the East Antarctic tectonic province.

The last part of the *P–T* path represents tectonic reactivation involving uplift. If this event occurred at the time represented by development of biotite in F_3 tectonic fabrics, then the 475 Ma age obtained by Elworthy (in Wolmarans & Kent, 1982) is consistent with a correlation with the Ross or Pan-African orogenies.

Acknowledgements

The sponsorship and logistic support provided by the Departments of Transport and Environmental Affairs are gratefully acknowledged. We are indebted to Professor Hugh Eales for the use of microprobe facilities at Rhodes University, Grahamstown.

References

Daly, M.C. (1986). The intracratonic Irumide belt of Zambia and its bearing on collision orogeny during the Proterozoic of Africa. In *Collision Tectonics*, ed. M.P. Coward & A.C. Ries, *Geological Society of London, Special Publication*, **19**, 321–8.

Ellis, D.J. (1980). Osumilite–sapphirine–quartz granulites from Enderby Land, Antarctica: *P–T* conditions of metamorphism, implications for garnet–cordierite equilibria and the evolution of the deep crust. *Contributions to Mineralogy and Petrology*, **74(2)**, 201–10.

Ellis, D.J. & Green, D.H. (1979). An experimental study of the effect of Ca upon garnet–clinopyroxene Fe–Mg exchange equilibria. *Contributions to Mineralogy and Petrology*, **71(1)**, 13–22.

England, P.C. & Thompson, A.B. (1984). Pressure-temperature-time paths of regional metamorphism. I. Heat transfer during the evolution of regions of thickened continental crust. *Journal of Petrology*, **25(3)**, 894–928.

Grantham, G.H., Groenewald, P.B. & Hunter, D.R. (1988). Geology of the northern H.U. Sverdrupfjella, western Dronning Maud Land, and implications for Gondwana reconstructions. *South African Journal of Antarctic Research*, **18**, 2–10.

Green, D.H. & Ringwood, A.E. (1966). An experimental investigation of the gabbro-eclogite transformation and some petrological applications. *Geochimica et Cosmochimica Acta*, **31(6)**, 767–833.

Harley, S.L. (1983). Regional geobarometry–geothermometry and metamorphic evolution of Enderby Land, Antarctica. In *Antarctic Earth Science*, ed. R.L. Oliver, P.R. James & J.B. Jago, pp. 25–30. Canberra; Australian Academy of Science and Cambridge; Cambridge University Press.

Harley, S.L. & Green, D.H. (1982). Garnet–orthopyroxene barometry for granulites and peridotites. *Nature, London*, **300**, 697–701.

Hjelle, A. (1974). Some observations on the geology of the H.U. Sverdrupfjella, Dronning Maud Land, Antarctica. *Norsk Polarinstitutt Årbok*, **1972**, 7–22.

Indares, A. & Martignole, J. (1985). Biotite–garnet geothermometry in the granulite facies: the influence of Ti and Al in biotite. *American Mineralogist*, **70(2)**, 272–8.

Lindsley, D.H. (1983). Pyroxene thermometry. *American Mineralogist*, **68(5)**, 477–93.
Martin, P.K. & Hartnady, C. (1986). Plate tectonic development of the southwest Indian Ocean: a revised reconstruction of East Antarctica and Africa. *Journal of Geophysical Research*, **91(B5)**, 4767–85.
Newton, R.C. & Perkins, D., III (1982). Thermodynamic calibration of geobarometers based on the assemblages garnet–plagioclase–orthopyroxene (clinopyroxene)–quartz. *American Mineralogist*, **67(1)**, 203–22.
Perkins, D., III & Chipera, S.J. (1985). Garnet–orthopyroxene–plagioclase–quartz barometry: refinement and application to the English River subprovince and the Minnesota River valley. *Contributions to Mineralogy and Petrology*, **89(1)**, 69–80.
Ravich, M.G. & Solov'ev, D.S. (1966). Geology and petrology of the mountains of central Queen Maud Land (Eastern Antarctica). *Transactions of the Science Research Institute of Arctic Geology, Ministry of Geology of the USSR*, **141**, 348 pp. (Translation from Russian by the Israel Program for Scientific Translations, Jerusalem, 1969).
Shackleton, R.M. (1986). Precambrian collision tectonics in Africa. In *Collision Tectonics*, ed. M.P. Coward & A.C. Ries. *Geological Society of London, Special Publication*, **19**, 329–49.
Shiraishi, K., Motoyoshi, Y. & Yanai, K. (1986). Plate tectonic development of the upper Proterozoic metamorphic complexes in eastern Queen Maud Land, East Antarctica. *6th Gondwana Symposium, Abstracts*, p. 84. Columbus, Ohio; Institute for Polar Studies.
Yoshida, M. & Kizaki, K. (1983). Tectonic situation of the Lützow-Holm Bay in East Antarctica and its significance in Gondwanaland. In *Antarctic Earth Science*, ed. R.L. Oliver, P.R. James & J.B. Jago, pp. 36–9. Canberra; Australian Academy of Science and Cambridge; Cambridge University Press.
Windley, B.F. (1984). *The Evolving Continents*, 2nd edn, 399 pp. Chichester; Wiley.
Winkler, H.G. (1974). *Petrogenesis of the Metamorphic Rocks*, 3rd edn, 320 pp. Berlin; Springer-Verlag.
Wolmarans, L.G. & Kent, K.E. (1982). Geological investigations in western Dronning Maud Land, Antarctica – a synthesis. *South African Journal of Antarctic Research*, **Supplement 2**, 93 pp.

A structural survey of Precambrian rocks, Heimefrontfjella, western Neuschwabenland, with special reference to the basic dykes

W. FIELITZ[1] & G. SPAETH[2]

1. *Geologisches Institut, Universität Karlsruhe, Kaiserstrasse 12, D-7500 Karlsruhe 1, Federal Republic of Germany*
2. *Lehr- und Forschungsgebiet für Geologie–Endogene Dynamik der RWTH Aachen, Lochnerstrasse 4–20, D-5100 Aachen, Federal Republic of Germany*

Abstract

Heimefrontfjella consists of metavolcanic and metasedimentary rocks intruded by several phases of granitic rocks; most were metamorphosed to an amphibolite-facies grade and a few to a granulite-facies grade. A major shear zone, several kilometres wide, transects at least part of the mountain range. Although the metamorphic complex at Heimefrontfjella underwent polyphase deformation, only two phases of folding are clearly recognizable. The older set of folds trends NW–SE, the younger ones trend NE–SW. Numerous overthrusts in the north-eastern part of Heimefrontfjella developed during the later phase of folding and this deformation also affected the shear zone. A number of predominantly N–S-trending subalkaline basalt dykes cut the metamorphic rocks. These range from weakly metamorphosed and undeformed dykes to amphibolitic and/or strongly sheared rocks. Although a few are probably of Mesozoic age, most belong to an older intrusive phase. Tectonic events affecting the Precambrian rocks of Heimefrontfjella and adjacent areas of the East Antarctic shield may be correlated with the Ross Orogeny and older orogenies in western Neuschwabenland, and with geological structures in eastern South Africa.

Introduction

The metamorphic rocks of Heimefrontfjella form part of the East Antarctic shield, near its western boundary against the younger fold belts of West Antarctica. Reconnaissance mapping by Worsfold (1967) and Juckes (1972) has been followed by more detailed geological, especially structural, investigations by the authors. The distribution of the six major rock units found in Heimefrontfjella is shown in Fig. 1. The main outcrops are composed of Precambrian metamorphic rocks (including metamorphosed sedimentary–volcanogenic rocks, amphibolite-facies orthogneisses, granulite-facies rocks and strongly sheared units), but Permian sedimentary rocks and Jurassic basic rocks are also present. All the Precambrian metamorphic rocks have been affected to some degree by retrograde metamorphism.

Precambrian rock units

Lithologically varied sedimentary–volcanogenic metamorphic rocks are common throughout the whole of Heimefrontfjella. The metasedimentary rocks are mostly paragneisses and mica-schists with subordinate amphibolites, metaquartzites, marbles and calc-silicate rocks. Bimodal metavolcanic rocks form amphibolite and leucocratic quartz–feldspar gneiss sequences; the latter occur in part as augen gneisses.

Granitic–dioritic orthogneisses are also common. They are mostly coarse-grained rocks with widespread augen structures, and commonly they contain xenoliths of ortho- and paragneisses; at a few localities there is no gneissose fabric.

Granulite-facies rocks (charnockites, anorthosites and granulites) crop out in south-western Tottanfjella (Fig. 1). Migmatitic granulitic gneisses are of partly sedimentary origin and they occur sometimes as large xenoliths in the non-schistose, dark charnockites. Most of these rock types (especially the augen gneisses mentioned above) have been mylonitized where they occur in the shear zone, which extends from north-eastern Sivorgfjella to southern Tottanfjella.

Structures

S-planes

Bedding (mostly corresponding to a bedding-parallel schistosity) and schistosity have been measured in all the Precambrian rock units; there is a well-developed foliation in the shear zone. The stereograms (Fig. 2) show that the S-planes strike NE–SW and dip mainly SE in Sivorgfjella and Tottanfjella. In the Kottasberge (Milorgfjella) region they strike more E–W and dip S. The plot of Sivorgfjella (Fig. 2) shows a distinct girdle, with a pi-pole in the north-east which corresponds to the

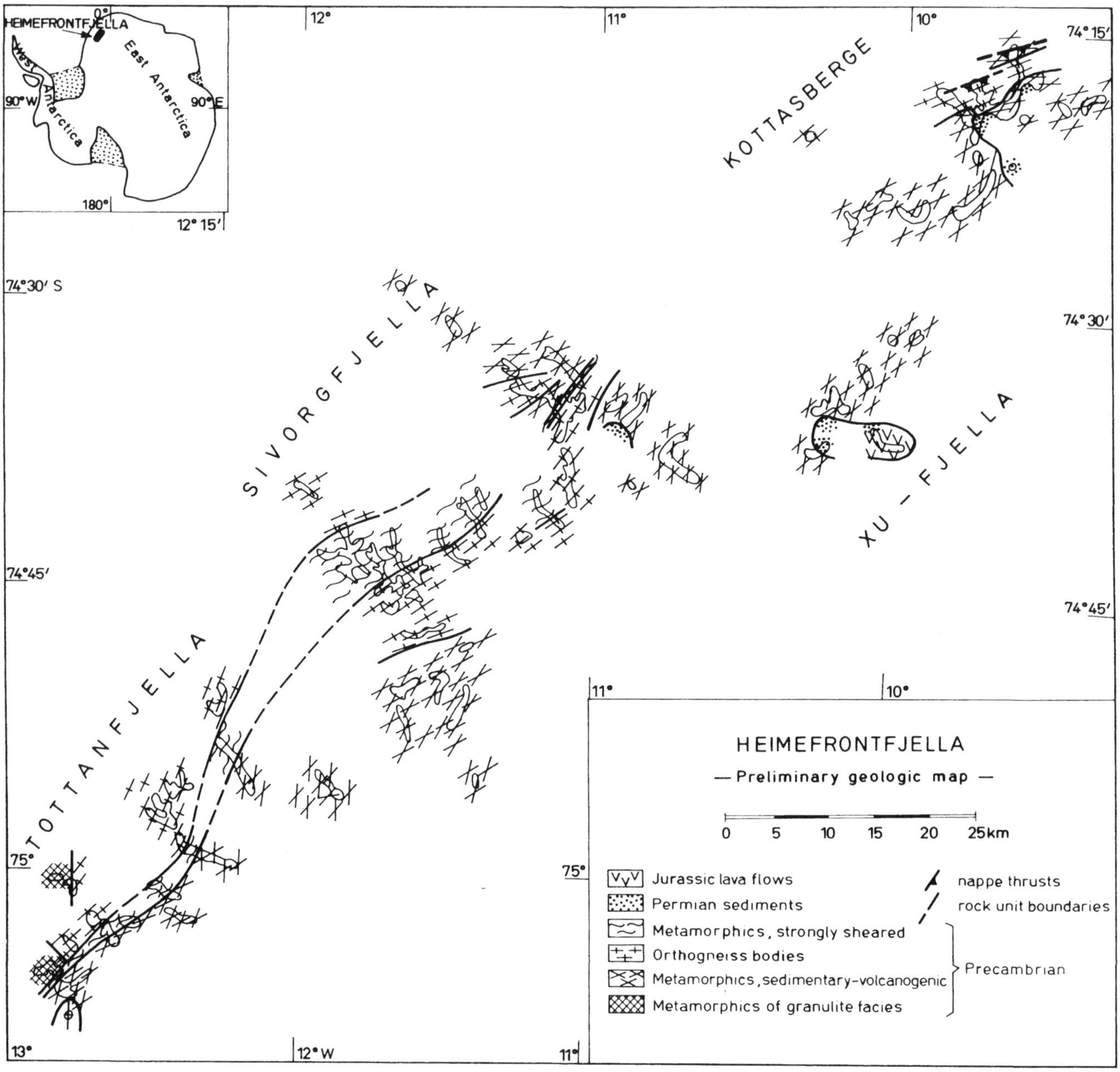

Fig. 1. Preliminary geological map of Heimefrontfjella, western Neuschwabenland (partly after Worsfold (1967) and Juckes (1972)).

B_2 folds. The Tottanfjella plot has a second, weaker maximum, which corresponds to ESE–WNW-striking, SSW-dipping S-planes; these measurements are from granulite-facies metamorphic rocks and their different S-plane orientations indicate a different structural unit.

Folds

Folds are common, especially in the sedimentary–volcanogenic rocks and strongly sheared metamorphic rocks. Two superimposed fold generations have been recognized, although most outcrops show only one phase of folding. There is evidence for a third, older phase of deformation producing intrafolial folds, but it was not possible to determine their direction and plunge. The plots in Fig. 2, therefore, show all undifferentiated B-axes combined with two clearly recognizable fold generations, referred to as B_1 and B_2 by Spaeth & Fielitz (1987).

B_1 folds are mostly similar, in places non-cylindrical, with a NE-vergence. They are often combined with imbricate structures. They have a NW–SE trend and occur particularly in Kottasberge and northern Sivorgfjella, and to a lesser extent in the rest of Sivorgfjella and Tottanfjella (Fig. 2), possibly because shearing and mylonitization connected with the shear zone obliterated the older folds. The direction of the B_1 fold axes shows a distinct scatter as a result of refolding by B_2 folds.

B_2 folds are similar or parallel cylindrical folds and they are mostly more open than the B_1 folds. They have a NE–SW trend and commonly a NW-vergence; some folds have a SE-vergence possibly due to backfolding. Homoaxial refolding has been recorded at some outcrops. B_2 folds occur mainly in the Sivorgfjella and Tottanfjella area (Fig. 2) where some of the

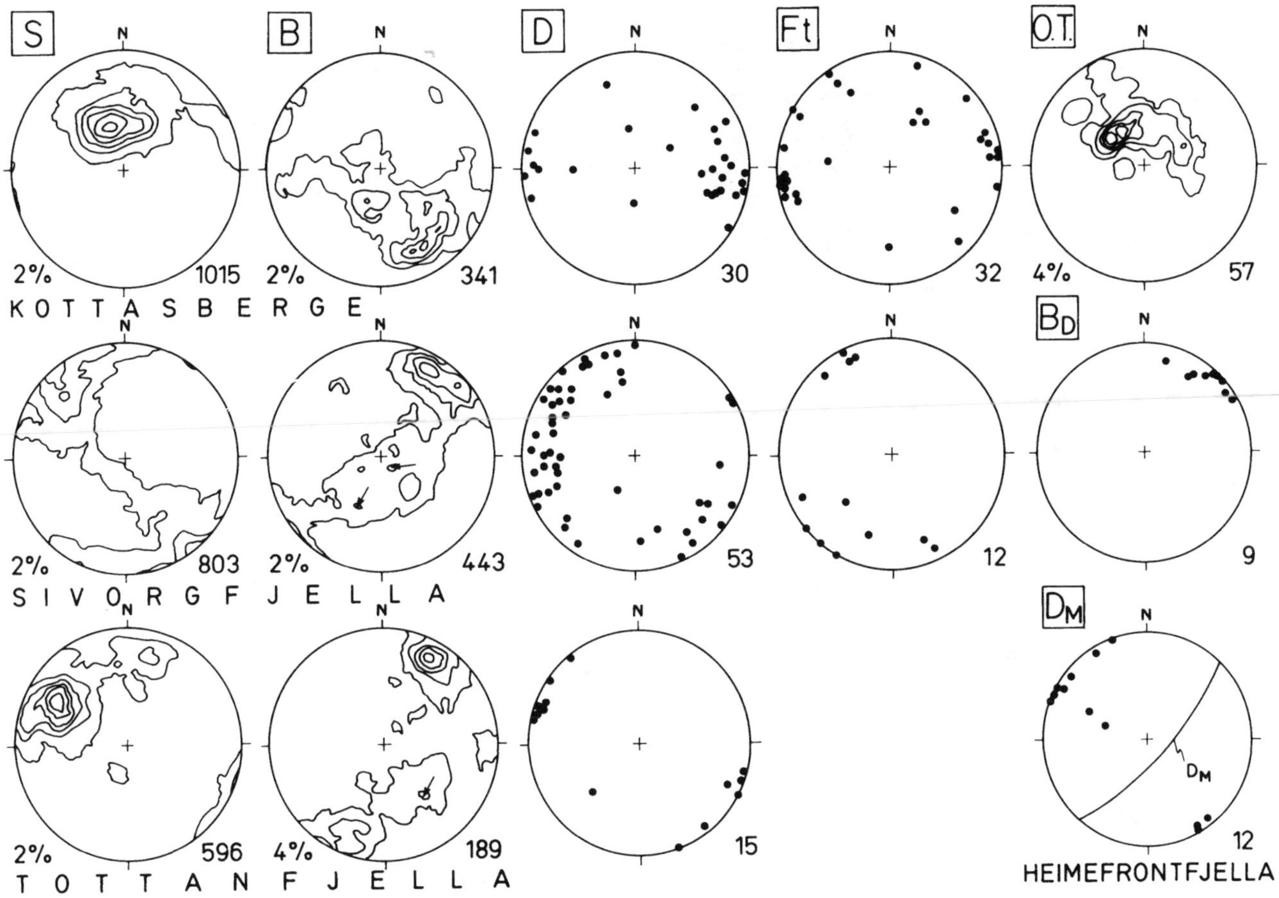

Fig. 2. Stereograms with structural data for Heimefrontfjella: S, schistosity and shear planes; B, fold axes; D, metamorphosed basic dykes; Ft, high-angle faults; O.T., overthrusts; B_D, fold axes of folded metamorphosed basic dykes; D_M, Mesozoic basic dykes. First contour-line at 1%, subsequently at intervals of 2 or 4%. Number of poles are indicated beside each diagram. Schmidt net, lower hemisphere.

sheared gneisses are strongly folded by the B_2 deformation. Because the B_2 folds have a consistent direction, showing less scatter than the B_1 folds, and their plunge is mostly gentle, they probably represent the youngest major phase of deformation in the area. B_2 folding is often associated with a strong stretching lineation or deformation parallel to its axes. The ENE–WSW trend of the S-planes at Kottasberge changes to a more NNE–SSW trend at Tottanfjella (Fig. 2). This trend is repeated less distinctly by the B_2 folds, showing that the B_2 deformation has not a linear but an arcuate structural trend.

Recognizable folds in the granulite-facies metamorphic rocks occur only in the Vardeklettane area of western Tottanfjella. These have an E–W trend, which is different from the rest of Heimefrontfjella (see the weak maximum which they form in the plot for Tottanfjella (in Fig. 2). These folds provide further support for a different structural position for the granulite-facies rocks, which cannot be correlated with the main part of the remaining metamorphic massif.

Overthrusts

Overthrusts are common in the main Kottasberge massif. They represent the youngest tectonic elements in the region, cutting all previous structures (Fig. 3), and they are associated with thin ductile deformed bands of mylonite and ultramylonite. They strike E–W to NE–SW (Fig. 2) and dip gently S or SE; the sense of movement is from S to N and SE to NW, respectively. Displacement due to overthrusting is only a few metres to a few tens of metres but the overthrusts are regularly spaced about 20–50 m apart. Younger, nappe-like overthrusts (Fig. 1) are associated with thick zones of cataclastic rocks, indicating a higher structural level with more brittle deformation, and they seem to have larger displacements. The overthrusts apparently replaced the B_2 folds since areas with abundant B_2 folds (Sivorgfjella and Tottanfjella) have little or no overthrusts and the area with few B_2 folds (Kottasberge) has the most overthrusts.

High-angle faults

There are a limited number of high-angle, probably normal faults in the area (Fig. 2). The preferred trend at Kottasberge is N–S but elsewhere the measurements are too scattered or too few to give a clear structural trend. At Sivorgfjella NW–SE and NE–SW fault trends are apparent. Air photographs and satellite images revealed major NW–SE-striking fractures, which are occupied now by glacier-filled valleys.

Shear zone

The north-western parts of Sivorgfjella and Tottanfjella are cut by a steeply dipping, NE–SW-trending shear zone, with

Fig. 3. South-easterly dipping biotite gneisses and amphibolites belonging to the sedimentary–volcanogenic metamorphic rocks of south-western Leabotnen, northern Kottasberge. The rocks are cut by numerous pegmatite dykes, which are themselves crossed by weakly metamorphosed basic dykes; a south-easterly dipping overthrust cuts all previous structures. Height of outcrop is ~ 120 m.

a maximum width of 2–3 km. In south-western Tottanfjella this shear zone separates granulite-facies rocks (to the north-west) from amphibolite-facies rocks (in the south-east; see Fig. 1). The shear zone is composed of strongly sheared gneisses in which lenses of less-deformed granulite- or amphibolite-facies gneisses are preserved; the lenses vary from a few metres to several hundred metres in size. The strongly sheared gneisses are in places intensely deformed by B_2 folds.

Basic dykes

Apart from a few unmetamorphosed, probably Mesozoic dolerite dykes (discussed in detail by Spaeth & Schüll (1987)), a large number of metamorphosed basic dykes occur throughout Heimefrontfjella, particularly in its north-eastern part. Although most of them have been sampled (with the exception of those at XU–Fjella), details of their frequency of occurrence and regional distribution are incomplete. A total of 98 dykes was studied (29 from northern Kottasberge, 36 from Boyesennuten–Scharffenbergbotnen (north-eastern Sivorgfjella), 8 from Sanengenrusta (western central Sivorgfjella), 11 from Ardusberget (south-western Tottanfjella), and the remaining 14 from localities scattered throughout the rest of Heimefrontfjella).

Although the thickness of the dykes ranges from 8 cm to 20 m, most are 0.5–10 m wide. The metamorphosed dykes have a preferred N–S strike in the Kottasberge region and Scharffenbergbotnen area, and a more NW–SE to WNW–ESE trend elsewhere in Sivorgfjella and Tottanfjella (Fig. 2); the latter trend is slightly different from the NW–SE strike of the unmetamorphosed Mesozoic dolerite dykes in Heimefrontfjella (see Fig. 2). Some of the older basic dykes occupy faults and there is a good correlation between the dyke and high-angle fault trends (Fig. 2) for the Kottasberge area and, for the NE–SW-striking faults, at Sivorgfjella. The dykes are mostly nearly vertical or steeply dipping (Fig. 3) but gently dipping dykes occur locally. Although the dykes have undergone low-grade metamorphism, sometimes with the development of garnet porphyroblasts, they have retained their primary macroscopic texture (plagioclase and/or pyroxene phenocrysts, chilled margins and columnar jointing). True amphibolites with plagioclase + hornblende ± biotite are also common. All the dykes are deformed to varying degrees, from weakly sheared and/or boudinaged to strongly mylonitized rocks.

In thin section the fresh-looking rocks have a well-preserved ophitic texture with partially altered titanaugite, strongly sericitized plagioclase and a chlorite + epidote + sericite-rich groundmass. Other, hornblende-rich rocks still preserve remnants of their original ophitic texture. Higher-grade metamorphic rocks, composed largely of green hornblende and plagioclase (oligoclase–andesine) ± biotite and a granoblastic–nematoblastic texture, can be subdivided into less-metamorphosed and highly metamorphosed amphibolites on the texture of the hornblende crystals (irregular, poikiloblastic or fresh, subidiomorphic crystals, respectively). The few amphibolitic dykes cutting the granulite-facies units have a granular, non-foliated texture, and additional pyroxene in one case. Shearing resulted in a higher chlorite content in the less metamorphosed dykes and in a higher biotite content in the amphibolite dykes.

The gradation from nearly fresh basic rocks to highly metamorphosed amphibolites shows a distinct regional distribution:

(i) the freshest dykes occur in central Kottasberge and the northernmost nunataks of Sivorgfjella;
(ii) the more altered dykes are concentrated in other parts of Kottasberge and in north-eastern Sivorgfjella;
(iii) the amphibolite dykes are predominant in central and southern Sivorgfjella and in Tottanfjella, with the exception of the granulite-facies rocks which are cut by the altered as well as the non-foliated amphibolite dykes.

Major and trace element analyses of 35 samples from 34 basic dykes were determined by X-ray fluorescence spectroscopy and these complement the few analyses published by Juckes (1972). A preliminary interpretation of the data, using a SiO_2 versus $Zr–TiO_2$ diagram after Winchester & Floyd (1977), shows that most of the samples plot in the field of subalkaline basalts, irrespective of their metamorphic grade. Samples with a lower SiO_2 content than these basalts for the most part represent strongly sheared dykes, indicating element migration. Nevertheless a Zr–TiO_2 versus Nb–Y diagram (after Winchester & Floyd, 1977) also demonstrates that most samples lie in the subalkaline basalt field, supporting Winchester & Floyd's (1977) claim that their diagrams should be applicable to extensively altered and metamorphosed volcanic rocks.

Apart from the amphibolite dykes, which intruded granulite-facies rocks and have different textures and slightly different geochemistry, all the other dykes seem to fit into one group of basic rocks. Changes in their metamorphic grade seem to be a result of regional rather than petrological differences. This implies an increase in the metamorphic grade of the dykes from north/north-east to south/south-west. A difference in structural levels between Kottasberge and the rest of Heimefrontfjella, which could be related to differences in metamorphic grade, has already been discussed with respect to overthrusts and B_2 folds (see above and Spaeth & Fielitz, 1987). These differences are exemplified by the weakly altered dykes cut by an overthrust in northern Kottasberge (Fig. 3) and by the sheared and folded amphibolite dykes in central Sivorgfjella (Fig. 4). The structural relationships between dykes and the overthrust/shear zone and B_2 folds give some indication of the relative ages of the dykes.

Although the majority of the metamorphosed basic dykes belong to one magmatic event, there may have been an older phase of dyke intrusion indicated, for example, by the two generations of dykes shown in Fig. 4. The lower-grade metamorphic dykes in the granulite-facies rocks may indicate a relatively high structural position of the country rock after their intrusion and thus during the pre-B_2 formation of the shear zone.

Fig. 4. Strongly sheared augen gneisses in the wide shear zone at northern Sanengenrusta, central Sivorgfjella. A strongly sheared amphibolite dyke (outlined in white) and a younger, less-sheared amphibolite dyke cut the gneisses. Both dykes are folded by B_2 folds and together they are approximately 1 m thick.

Discussion

The rocks at Heimefrontfjella have been affected by at least two distinct fold generations (B_1 and B_2); Kirwanveggen, situated to the north-east, has comparable NW–SE-trending folds refolded by a second set of folds with NE–SW trends (Wolmarans & Kent, 1982). The Shackleton Range, situated much farther south, has crystalline Precambrian rocks affected by at least two interfering phases of folding (Hofmann & Paech, 1980; Clarkson, 1982). However, the relationship of these folds to the folding at Heimefrontfjella is unknown.

Age determinations on the gneissic rocks from Heimefrontfjella have yielded dates of 1032 and 1158 Ma (Arndt, Tapfer & Weber, 1986), which can be correlated with B_1 folding. These ages agree well with the ~ 1100 Ma dates obtained from Ahlmannryggen and Kirwanveggen (Wolmarans & Kent, 1982), and all of them may correlate with the Nimrod event. The metamorphosed basic dykes of Heimefrontfjella are younger than this event; one dyke has yielded an age of 455 Ma (Rex, 1972). Comparable ages are known from Ahlmannryggen where numerous small, NE–SW-trending overthrusts (Spaeth & Fielitz, 1987) cut a steep shear zone with an age of 526 Ma (Peters *et al.*, 1986). Ages of 460 and 485 Ma have been obtained from Kirwanveggen (Wolmarans & Kent, 1982). These younger ages, combined with the structural trends of the adjacent areas of western Neuschwabenland, may indicate a belt of crystalline rock units affected by Ross orogenic movements and rejuvenation. Such movements could correlate with the NE–SW-trending structures (B_2 folds) in Heimefrontfjella and Kirwanveggen. A less-deformed foreland may lie to the west and north-west (at Borgmassivet and Ahlmannryggen, where there are NE–SW-trending overthrusts; Spaeth & Fielitz, 1987). In this structural setting the metamorphosed basic dykes of Heimefrontfjella could represent an early extensional event of the Ross Orogeny.

The structural trends of western Neuschwabenland are important in the context of Gondwana reconstructions. Thus the presumed Archaean basement of Annandagstoppane may be correlated with parts of the Kapvaal craton of South Africa (Wolmarans & Kent, 1982; Paech, 1985; Spaeth & Fielitz, 1987), and the NW–SE- and younger NE–SW-trends of the metamorphic complexes of Heimefrontfjella and Kirwanveggen could correspond to the Namaqua–Natal Mobile Belt in eastern South Africa; the latter also is affected by two phases of deformation, with comparable ages (Kibaran and Pan-African events).

Acknowledgements

We thank the Alfred-Wegener-Institut, Bremerhaven, for logistic support of the expedition and the Deutsche Forschungsgemeinschaft for financial support. We are grateful to the other 1985–86 expedition members N. Arndt, H. Miller, G. Patzelt, M. Tapfer and K. Weber for discussions and assistance during the fieldwork. We also owe our gratitude to G. Friedrich, W. Plüger and H. Siemes, Institut für Mineralogie und Lagerstättenlehre, RWTH Aachen, for geochemical analyses and for making available computer facilities for the evaluation of the field data. Many thanks also to J.W. Thomson for critically reading the first draft of the typescript.

References

Arndt, T.N., Tapfer, M. & Weber, K. (1986). Die Kottasberge Neuschwabenlands (Antarktika) – Teil eines polyphas geformten präkambrischen Orogens? *Nachrichten der Deutschen Geologischen Gesellschaft*, **35**, 9–10.

Clarkson, P.D. (1982). Tectonic significance of the Shackleton Range. In *Antarctic Geoscience*, ed. C. Craddock, pp. 835–9. Madison; University of Wisconsin Press.

Hofmann, J. & Paech, H.-J. (1980). Zum strukturgeologischen Bau am Westrand der Ostantarktischen Tafel. *Zeitschrift für Geologische Wissenschaften*, **8(4)**, 425–37.

Juckes, L.M. (1972). The geology of north-eastern Heimefrontfjella, *Dronning Maud Land. British Antarctic Survey Scientific Reports*, **No. 65**, 44 pp.

Paech, H.-J. (1985). Comparison of the geologic development of Southern Africa and Antarctica. *Zeitschrift für Geologische Wissenschaften*, **13(3)**, 399–415.

Peters, M., Behr, H.J., Emmermann, R., Kohnen, H. & Weber, K. (1986). Alter und geodynamische Stellung der Vulkanitserien im westlichen Neuschwabenland (Antarktika). *Nachrichten der Deutschen Geologischen Gesellschaft*, **35**, 64–5.

Rex, D.C. (1972). K–Ar age determinations on volcanic and associated rocks from the Antarctic Peninsula and Dronning Maud Land. In *Antarctic Geology and Geophysics*, ed. R.J. Adie, pp. 133–6. Oslo; Universitetsforlaget.

Spaeth, G. & Fielitz, W. (1987). Structural investigations in the Precambrian of western Neuschwabenland, Antarctica. *Polarforschung*, **57(1/2)**, 71–92.

Spaeth, G. & Schüll, P. (1987). A survey of Mesozoic dolerite dikes from western Neuschwabenland, Antarctica, and their geotectonic significance. *Polarforschung*, **57(1/2)**, 93–113.

Winchester, J.A. & Floyd, P.A. (1977). Geochemical discrimination of different magma series and their differentiation products using immobile elements. *Chemical Geology*, **20**, 325–43.

Wolmarans, L.G. & Kent, L.E. (1982). Geological investigations in western Dronning Maud Land – a synthesis. *South African Journal of Antarctic Research*, **Supplement 2**, 93 pp.

Worsfold, R.J. (1967). The geology of southern Heimefrontfjella, Dronning Maud Land. Ph.D. thesis, University of Birmingham, 176 pp. (unpublished).

Reflection seismic measurements in western Neuschwabenland

A. HUNGELING & F. THYSSEN

Department of Geophysics, University of Münster, Corrensstrasse 24, D–4400 Münster, Federal Republic of Germany

Abstract

During the 1985–86 season reflection seismic measurements were carried out at 18 stations north of Heimefrontfjella (Kottas Mountains), to study the structure of the earth's crust. The differing ice thicknesses defined a graben-like structure extending NE–SW, parallel to Heimefrontfjella, with the mountains forming the southern boundary. This structure is more than 40 km wide and its total length has not yet been explored. The surface of the graben lies up to 1090 m below sea level and here the graben is filled with 2550 m thick ice. At most stations the ice is underlain by crystalline basement. Some seismograms show reflections from the 600 m thick Permian sedimentary rocks which partly overlie the crystalline basement. The calculated velocity of these sedimentary rocks is 4.7 km/s. The graben-like structure and the different depths of the sedimentary rock/basement boundary can be explained by large fractures in western Neuschwabenland. The crystalline rocks show reflection signals with typical anomalous and/or reverse move-out. At depths of 7–12 km and 28 km two boundaries can be recognized within the crust. The total thickness of the crust is estimated to be nearly 38 km.

Seismic reflection recordings with 12 channels were carried out at 18 stations along three profiles (Fig. 1). Six shots at 50 m intervals were recorded for a six-fold coverage at each station. The amount of charge ranged from the usual 3 kg up to an occasional 14 kg in bore holes of 4–45 m depth; most shots were fired at depths of 10 m. In addition to the usual reflection seismic soundings, 12 shots with 40–375 kg charges were recorded. The distance of the first geophone to the shot varied between 100 m and 350 m. Geophones, with a resonant frequency of either 4.5 cps or 7.5 cps, were located at 50 m intervals and were connected to 15 m-long linear arrays, with four geophones per channel, for some shots. Geo Space GSC 111 G seismic equipment and an additional PCM recording system were used. A doppler satellite navigation system for the estimation of the ice surface elevation and the coordinates of the shot points, and a hot-water drill for the shot holes were made available by H. Miller/AWI, Bremerhaven. On comparable profiles the surface elevations correspond to the results from radar- and baro-altimeter measurements (Hoppe & Thyssen, 1988).

For faster measurements, we have developed an ice streamer which was tested during the 1985–86 field season. 12 small gimbal-mounted geophones with a resonant frequency of 4.5 cps were built into special cases and connected to the geophone cable at 18 m intervals. A wire hawser was fixed to the cases and cable, and attached to a sledge; a snow vehicle pulled the whole system along the ice. The comparison between

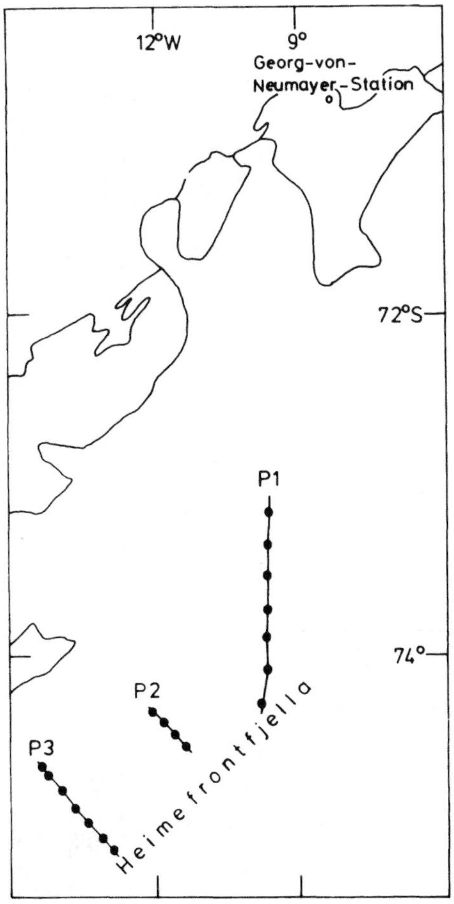

Fig. 1. Sketch map of western Neuschwabenland showing the location of reflection seismic measurement stations along three profiles.

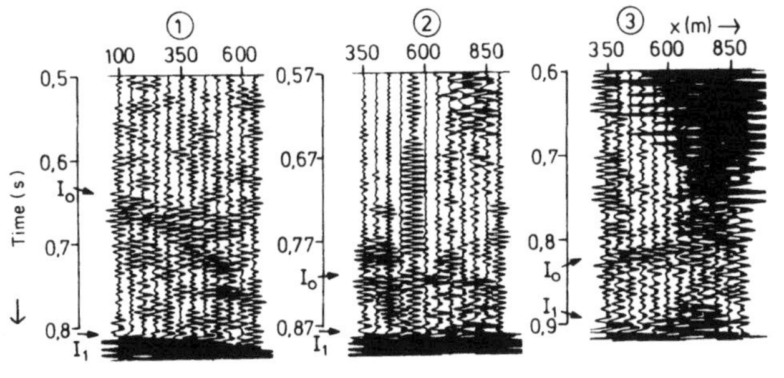

Fig. 2. Reflection signals from a basal layer (I_0) and from the ice base (I_1) at three different stations (1, P1; 2 and 3, P2).

seismograms recorded with normally planted geophones and those with the streamer showed an adequate correspondence. Owing to vibrations which it was only possible to suppress at a later date using a modified streamer, only nine of the 132 shots were recorded with the streamer.

The seismograms were processed at the computer centre, University of Münster. The processing included notch filtering, muting and tapering, time-dependent frequency filtering and gain control. The seismograms were corrected to a reference elevation of 1000 m above sea level (a.s.l.) and are presented as six-fold stacked sections or as single processed seisomograms. The latter was necessary because several seismograms recorded at the same station showed differences in quality. This was due to different shot conditions, including varying bore-hole depths and charges, and repeated shooting in the same bore-hole, which resulted in differences in amplitude and frequency content of the reflection signals. Under such conditions, stacking does not improve the quality significantly.

Earlier reflection signals than the ice-base signals were recognized at three stations (profile 1, third station in the north; profile 2, both middle stations; Fig. 2). The reflection signals from the ice base (I_1) begin at the end of the ~ 0.3 s long time windows and are attenuated in amplitude. The earlier signals (I_0) are well known from reflection seismic measurements in West Antarctica during the early 1960s (Bentley, 1971). The layer between these signals and the ice base is defined as the basal layer, which probably consists of ice and moraine. Assuming a velocity range of 3.6–4.3 km/s, the calculated thickness of the basal layer is 380 ± 50 m for the station on profile 1 (P1) and 210 ± 20 m for both stations on profile 2 (P2).

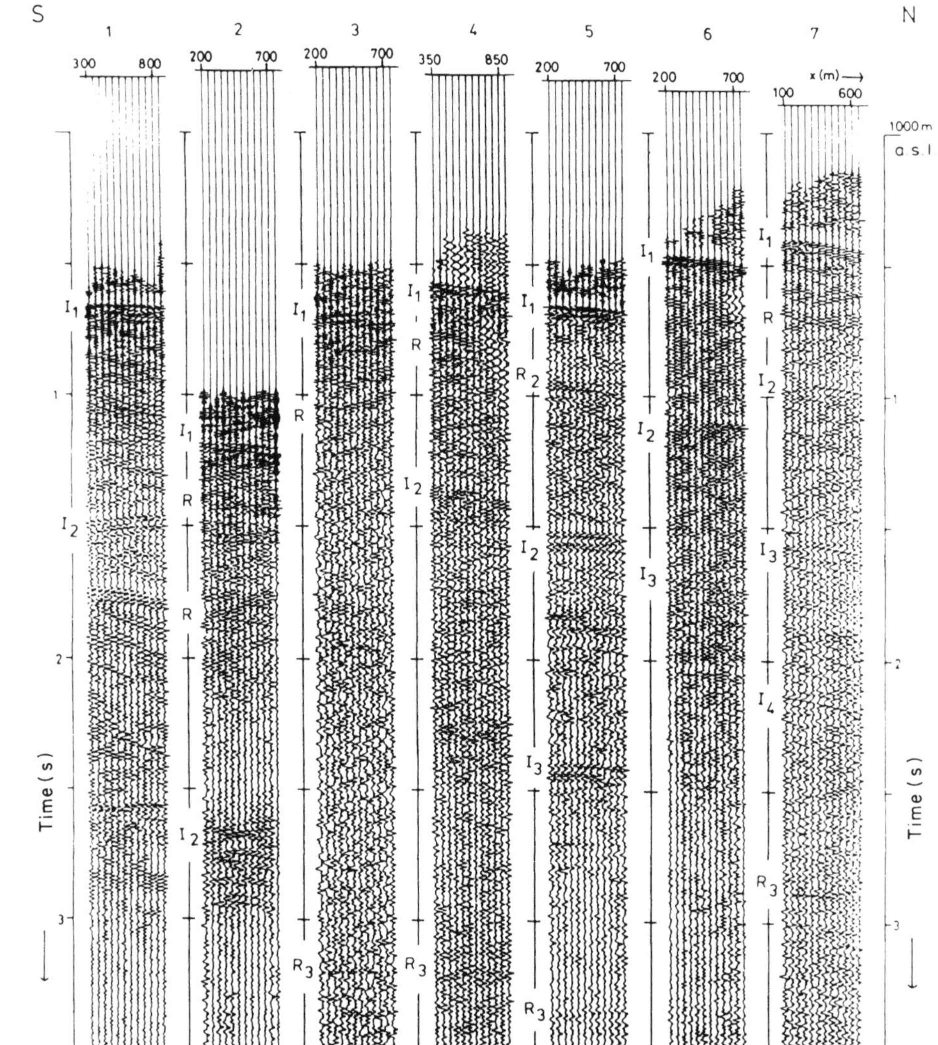

Fig. 3. Profile 1: seismograms from all stations, spaced at nearly 20 km intervals. Data processing: muting, notch filter, time varying frequency filter, gain control.

Fig. 3 shows seven seismograms from profile 1. The southernmost was recorded in north-eastern Heimefrontfjella. All seismograms show reflection signals from the ice base (I_1). The conditions at the ice–rock interface differ from station to station. Seismograms 2 and 3 show exceptionally weak reflection signals and partly weak multiples (I_2, I_3, I_4), which may represent a transition zone consisting of ice, moraine, rocks, permafrost, etc. at the ice base.

The thickest ice (2550 ± 30 m) occurs at station 2, where the ice base lies 1090 m below sea level (sea level not adjusted). To the north and south, the ice base rises to 250 m above–below sea level. Together with radio-echo soundings (Hoppe & Thyssen, 1988), the seismic data define a graben-like structure parallel to Heimefrontfjella, with the mountains forming the southern boundary. This structure is more than 40 km wide and its total length has not yet been explored. One supposition is that large fractures were formed here during the break-up of Gondwana. Glacier movement may have excavated this structure to a depth of 1090 m below sea level (b.s.l.).

Several strong reflection signals (R) can be correlated beneath the ice. Most of these show anomalous and/or reverse move-out, indicating high dips. This is typical for reflection seismic measurements in crystalline rocks (Smithson, Shive & Brown, 1977). Thus we assume that the base of the ice is underlain by crystalline basement. The southernmost recorded seismogram may contain side reflections caused by the mountains and it is therefore excluded from the interpretation. Clear reflection signals are marked R_2 and R_3. The nature of R_2 is not known and there is no correlation with neighbouring seismograms. However, assuming a velocity of 5.5 km/s, the calculated depth is 1 km b.s.l. R_3 is identified as a boundary within the crystalline rocks. Taking into consideration the velocity–depth function derived for the area near the Soviet station Novolazarevskaya (Kogan, 1972; Kurinin & Grikurov, 1982), we assume a velocity increase from 6.0 km/s to 6.2 km/s at R_3. From this a mean depth of 8.5 ± 1 km b.s.l. is calculated.

Fig. 4 shows four six-fold stacked sections from P2. In the south-east the ice base lies 350 m a.s.l., descending to 620–690 m b.s.l. to the north-west. Only section 1 is characterized by signals with anomalous travel times indicating side reflections. Up to three horizons (R_1, S and R_2) are identified within this profile. The stacking velocities for R_1 show no significant differences to those of the ice base, so a velocity of 3.75 km/s is assumed up to the horizon R_1. It may be a part of the transition zone at the ice base.

In section 4 signals (S) were obtained from the base of Permian sedimentary rocks. From the stacked sections a velocity of 4.7 ± 0.3 km/s was calculated for the layer R_1–S. This corresponds to velocity determinations on Permian sandstones from the Beaver Lake area (Kurinin & Grikurov, 1982) and rocks of the Beacon Supergroup (Crary, 1963). We estimate that the boundary between the nearly 600 m thick Permian sedimentary rocks and the crystalline basement is at a depth of ~ 1.3 km b.s.l. The pattern of reflection signals between R_1 and S are a result of internal layering, consisting probably of basaltic intrusions which are common elsewhere in this area. In Milorgfjella, north-eastern Heimefrontfjella, more than 60 km to the east of this seismic station, the base of the sedimentary

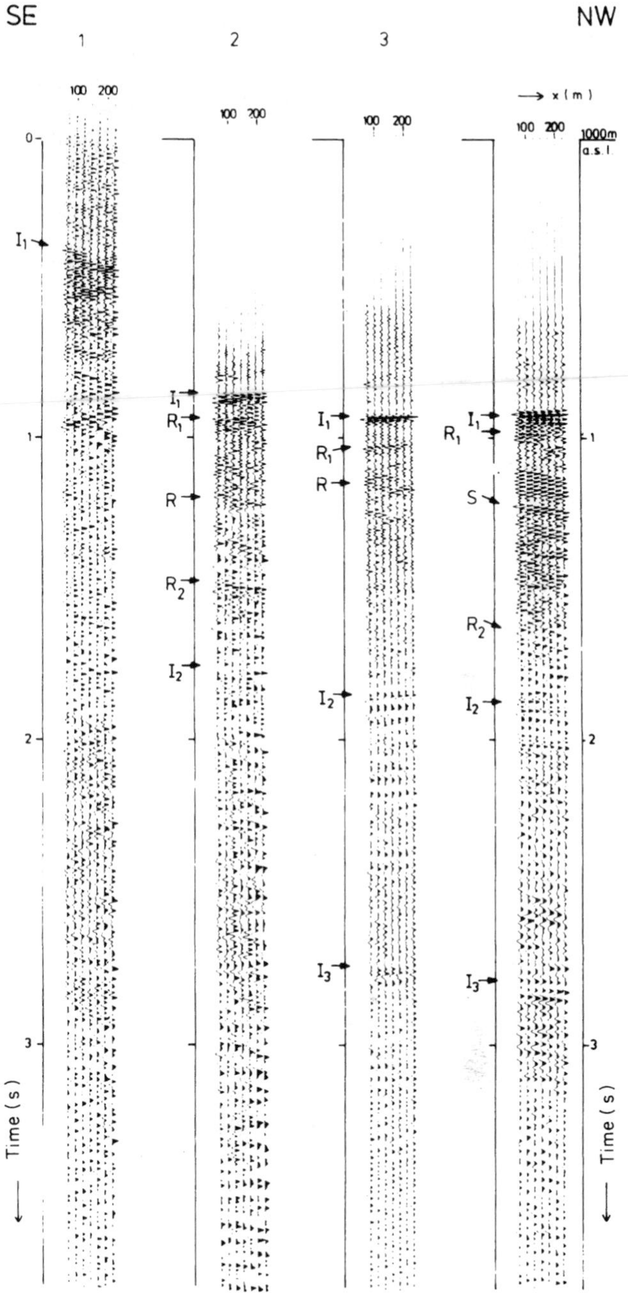

Fig. 4. Profile 2: six-fold stacked sections, spaced at nearly 10 km intervals.

rocks is exposed at nearly 2.0 km a.s.l. The 3.3 km difference in height is a second indication of large fractures in western Neuschwabenland. We cannot decide whether all R signals are identical with S because they are very weak at stations 2 and 3. R_2 is thought to be a boundary within crystalline rocks of the upper crust. Assuming a velocity of 5.5 km/s, the depth is calculated to be 2.0 km (section 2) and 2.7 km (section 4) b.s.l. Owing to energy losses in the overlying sedimentary rocks, the reflection signals in section 4 are weak.

Profile 3 is more or less comparable with P1 and is not described here.

Several seismograms with large charges were processed for a 16 s long time window. Fig. 5 shows one such seismogram recorded with a charge of 375 kg at the northernmost station on profile P2. Although many sufficiently strong reflection

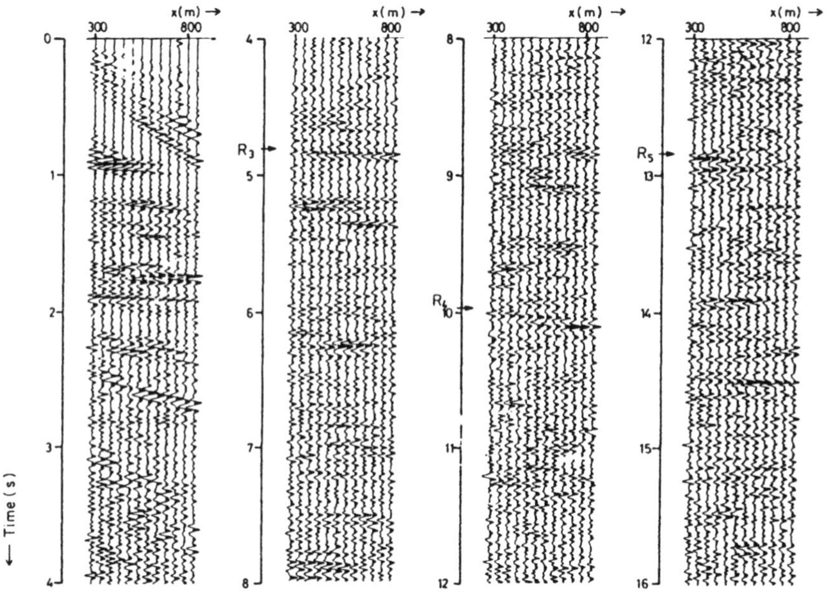

Fig. 5. Seismogram from profile 2: charge 375 kg. Data processing: frequency filter (5–40 Hz), gain control.

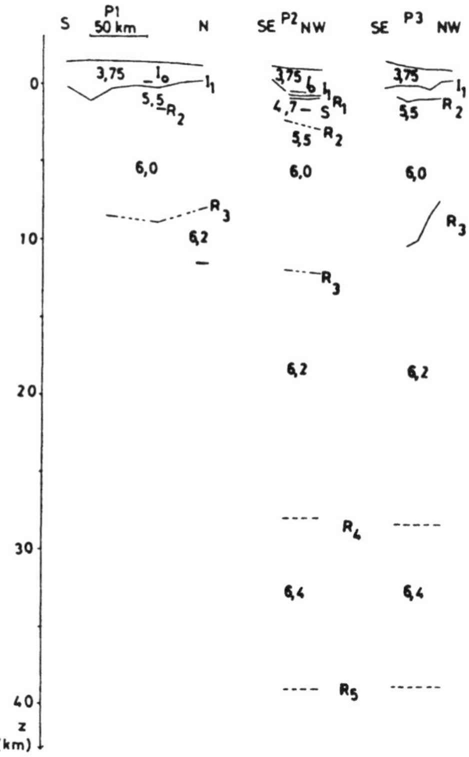

Fig. 6. Interpreted structures at three profiles in western Neuschwabenland; velocities in km/s, partly assumed.

signals can be correlated, we can only interpret those signals which are recognizable at comparable travel times in neighbouring seismograms. Reflection signals were identified at 4.8 s (R_3), 9.95 s (R_4) and 12.8 s (R_5). Assuming a velocity increase from 6.0 to 6.2 km/s at R_3 and from 6.2 to 6.4 km/s at R_4, the depths are estimated to be 12.1 km (R_3), 28.1 km (R_4) and 37.2 km (R_5). These signals were recognized in several seismograms.

A compilation of these reflection seismic measurements across western Neuschwabenland is shown in Fig. 6. The velocity values are partly assumed. Three boundaries are recognized within the crust: R_3 at 7–12 km, R_4 at 28 km and R_5 at 39 km. The values for R_4 and R_5 are mean values calculated from all estimated depths. The deepest boundary may be the Mohorovicic discontinuity. The depth of 39 km is an expected value for this area. Just at or beneath the ice base several horizons can be interpreted: the basal layer (I_0), the ice base (I_1), a transition zone (R_1), the Permian sedimentary rocks (S) and a 1–2 km thick layer (R_2) in the upper crust.

In conclusion, reflection seismic soundings with standard techniques and data processing offer the capability of exploring the earth's crust in Antarctica. For future measurements longer profiles with multiple coverage will be more successful. By using the ice streamer, faster measurements with multiple coverage (24-fold and more) can be achieved.

References

Bentley, C.R. (1971). Seismic evidence for moraine within the basal Antarctic ice sheet. In *Antarctic Snow and Ice Studies – II*, ed. A.P. Crary, pp. 89–129, Antarctic Research Series, 16, Washington, DC; American Geophysical Union.

Crary, A.P. (1963). *Results of the United States Traverses in East Antarctica, 1958–61*, IGY World Data Center A: Glaciology. New York; International Geophysical Year Glaciological Report No. 7, 144 pp.

Hoppe, H. & Thyssen, F. (1988). Ice thicknesses and bedrock elevation in Western Neuschwabenland and Berkner Island, Antarctica. *Annals of Glaciology*, **11**, 42–5.

Kogan, A.L. (1972). Results of deep seismic sounding of the earth's crust in East Antarctica. In *Antarctic Geology and Geophysics*, ed. R.J. Adie, pp. 485–9. Oslo; Universitetsforlaget.

Kurinin, R.G. & Grikurov, G.E. (1982). Crustal structure of part of East Antarctica from geophysical data. In *Antarctic Geoscience*, ed. C. Craddock, pp. 895–901. Madison; University of Wisconsin Press.

Smithson, S.B., Shive, P.N. & Brown, S.K. (1977). Seismic velocity, reflections, and structure of the crystalline crust. In *The Earth's Crust*, ed. J.G. Heacock, pp. 254–70. Geophysical Monograph 20. Washington, DC; American Geophysical Union.

Geology and metamorphism of the Sør Rondane Mountains, East Antarctica

K. SHIRAISHI[1], M. ASAMI[2], H. ISHIZUKA[3], H. KOJIMA[1], S. KOJIMA[4], Y. OSANAI[5], T. SAKIYAMA[6], Y. TAKAHASHI[7], M. YAMAZAKI[3] & S. YOSHIKURA[3]

1 National Institute of Polar Research, Itabashi-ku, Tokyo 173, Japan
2 Department of Geological Sciences, College of Liberal Arts, Okayama University, Okayama 700, Japan
3 Department of Geology, Kochi University, Kochi 780, Japan
4 Department of Earth Sciences, School of Science, Nagoya University, Chikusa-ku, Nagoya 464, Japan
5 Department of Geology and Mineralogy, Hokkaido Univerity, Sapporo 060, Japan
6 Institute of Geology and Mineralogy, Hiroshima University, Hiroshima 730, Japan
7 Geological Survey of Japan, Tsukuba 305, Japan

Abstract

The Sør Rondane Mountains are underlain by metamorphic rocks of probable Late Proterozoic age, and plutonic rocks and minor dykes of latest Proterozoic to Early Palaeozoic age. The majority of the rocks are pelitic, psammitic and intermediate gneisses, associated with subordinate basic and calc-silicate gneisses. Regional metamorphism was of medium-pressure type grading up to granulite-facies, and the peak temperature–pressure (*T–P*) conditions are estimated to have been 750–830°C and 7–8.5 kb. There has been extensive retrogression during mylonitization and plutonic intrusion, resulting in the restricted occurrence of kyanite in biotite that embays garnet in a sillimanite-bearing rock, and in the local development of granoblastic andalusite in some other sillimanite-bearing rocks. In the south-western part of the mountains, a pronounced shear belt at least 10 km in width trends E–W. The shear belt consists mainly of gneissose tonalite containing abundant fragments of basic schist, and of minor amounts of psammitic schist. All these rocks have undergone intense mylonitization and cataclasis under conditions of the greenschist- to epidote–amphibolite-facies. The discrepancy between the compositions of the rocks and their metamorphic grade in the two regions suggest tectonic movements such as rifting in the latest Proterozoic to Early Palaeozoic.

Introduction

The Sør Rondane Mountains are one of the largest inland mountain ranges in East Antarctica, occupying an area of 25000 km^2. The mountains were studied by Belgian geologists, who published a geological map at a scale of 1:500,000 (Van Autenboer, 1969). Two groups of metamorphic rocks were distinguished: the Teltet–Vengen Group, composed of high-grade gneisses of various compositions, in the northern and eastern parts of the mountains, and the Nils Larsen Group, in the south-western part of the area, composed mainly of mylonitic 'gneisses' of basic to intermediate compositions. Geochronological studies indicate intense thermal activity during the Early Palaeozoic (Van Autenboer, 1969). The oldest reported age is 2700 Ma, determined by U–Pb analysis of detrital zircon from gneisses, but the age of the main phase of regional metamorphism remains uncertain (Van Autenboer & Loy, 1972).

Since 1984, we have been carrying out fieldwork and mapped more than 60% of all the outcrops in the mountains (Kojima & Shiraishi, 1986; Ishizuka & Kojima, 1987). This paper summarizes the results of the field surveys over the past three seasons, and discusses the metamorphism of the Teltet–Vengen Group.

Outline of geology

Outcrops of various kinds of metamorphic and plutonic rocks, and minor dykes occur in the area (Fig. 1). Pelitic to psammitic gneisses, and intermediate gneiss which possibly represents mixtures of metasedimentary and metavolcanic rocks, are dominant among the metamorphic rocks; thin layers and lenses of calcareous and basic rocks also occur in many places. The plutonic rocks contain rare ultramafic xenoliths.

Foliation of the gneisses strikes generally E–W and dips monoclinally S in the western part, whereas it is more complicated in the central and eastern parts of the area. The most remarkable major structural feature in the area is a pronounced shear belt, at least 10 km in width. This trends E–W in

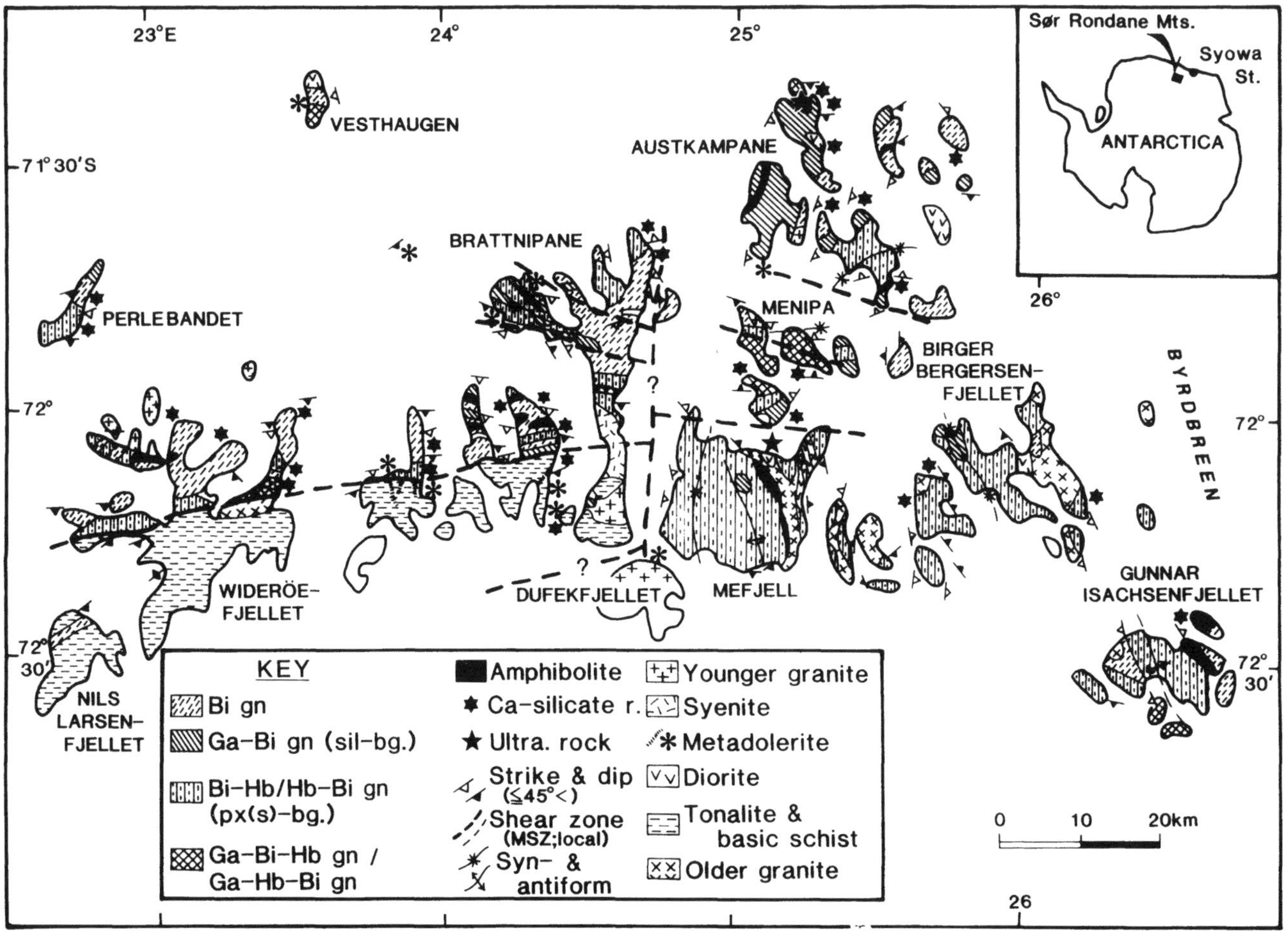

Fig. 1. Geological map of the main part of the Sør Rondane Mountains; for mineral abbreviations see Table 1. Other abbreviations: px(s), pyroxene(s); Ultra, Ultramafic; r, rock; gn, gneiss; bg, bearing.

the south-western part of the area, its southern limit is not exposed, its eastern end (in the central part of the mountains) is cut by syenite and granite plutons and its western end is obscured by ice. The northern limit of the shear belt, referred to as the Main Shear Zone (MSZ), is composed of intensely mylonitized metamorphic and plutonic rocks (ultramylonite and cataclasite) (Kojima & Shiraishi, 1986). Following the Belgian terminology, the metamorphic rocks in the shear belt are assigned to the Nils Larsen Group (NLG), whereas other metamorphic rocks underlying the major part of the area are called the Teltet–Vengen Group (TVG). Large xenoliths of rock types similar to the NLG occur in the granite mass in the eastern extension of the shear belt. Many local shear zones (mylonite zones) subparallel to the MSZ are present in the TVG.

Plutonic rocks of granitic, syenitic and dioritic compositions are widespread, and some have had a contact metamorphic effect on the surrounding rocks. Mafic dykes of basaltic to andesitic composition (metadolerite) have been thermally metamorphosed by the plutonic rocks.

Metamorphic rocks

Teltet–Vengen Group

The Teltet–Vengen Group (TVG) was named after one of the northern ridges of the western part of the area (Van Autenboer, 1969). Although the metamorphic rocks of this group are largely composed of psammitic and intermediate rocks throughout the area, typical pelitic rocks predominate in the central-northern part. In addition, thin layers and lenses (cm to m in size) of basic and calcareous rocks (e.g. marble and skarn) occur in many places.

Characteristic mineral assemblages in this group (Table 1) represent a primary, upper amphibolite–granulite-facies metamorphic event and a retrograde stage, probably the result of plutonic intrusions and mylonitization. For example, a peculiar rock containing sapphirine, gedrite, corundum, cordierite, spinel, orthopyroxene, chlorite and biotite occurs as a xenolithic block in a younger granite stock. This mineral assemblage is not apparently in equilibrium and may represent two or more stages of re-equilibration. Fig. 2 shows the distribution of the characteristic mineral assemblages of the major rock types, together with retrograde minerals. Regional metamorphism up to granulite-facies is indicated by the occurrence of mineral assemblages such as sillimanite (Sil)–garnet (Ga)–biotite (Bi)–K-feldspar (Kf)–plagioclase (Pl)–quartz (Qu) ± cordierite (Cd) ($X_{Fe} = 0.15$) and orthopyroxene (Op)–Ga–Bi–Kf–Pl–Qu in the pelitic to psammitic rocks, and two-pyroxene–hornblende–Ga–Pl–Qu in the basic to intermediate rocks. Among the primary minerals, spinel (ZnO–9.35 wt %) occurs only as inclusions in sillimanite, garnet and plagioclase,

Table 1. *Characteristic mineral assemblages in the metamorphic rocks from the Sør Rondane Mountains*

Rocks	Primary assemblages	Retrograde assemblages
Pelitic–psammitic	$P_1$1 Sil + Ga + Bi + Ru (± Sp)	$P_2$1 Ky + Ga + Bi + Mu + Qu
	$P_1$2 Sil + Ga + Cd + Bi (± Sp)	$P_2$2 Ad + Sil + Bi + Pl + Qu ± Ga
	$P_1$3 Op + Ga + Bi	
	$P_1$4 Op + Bi	
	$P_1$5 Ga + Bi	
	[+ Kf + Pl + Qu + Il]	
Basic–intermediate	$B_1$1 Op + Cp + Ga + Hb + Il ± Bi	
	$B_1$2 Op + Cp + Hb + Il ± Bi	
	$B_1$3 Cp + Ga + Hb ± Bi	
	$B_1$4 Cp + Hb ± Bi	
	$B_1$5 Ga + Hb ± Bi	
	[+ Pl ± Qu + Mt]	
Ultrabasic	$U_1$1 Ol + Op + Sp + Hb	
Calcareous	$C_1$1 Ol + Hu + Sp + Ph + Do + Cc	
	$C_1$2 Ga + Cp + Ep + Sc + Cc + Qu	

Mineral abbreviations: Ad, andalusite; Bi, biotite; Cc, calcite; Cd, cordierite; Cp, clinopyroxene; Do, dolomite; Ep, epidote; Ga, garnet; Hb, hornblende; Hu, clinohumite; Il, ilmenite; Kf, K-feldspar; Ky, kyanite; Mt, magnetite; Mu, muscovite; Ol, olivine; Op, orthopyroxene; Ph, phlogopite; Pl, plagioclase; Qu, quartz; Ru, rutile; Sc, scapolite; Sil, sillimanite; Sp, spinel.

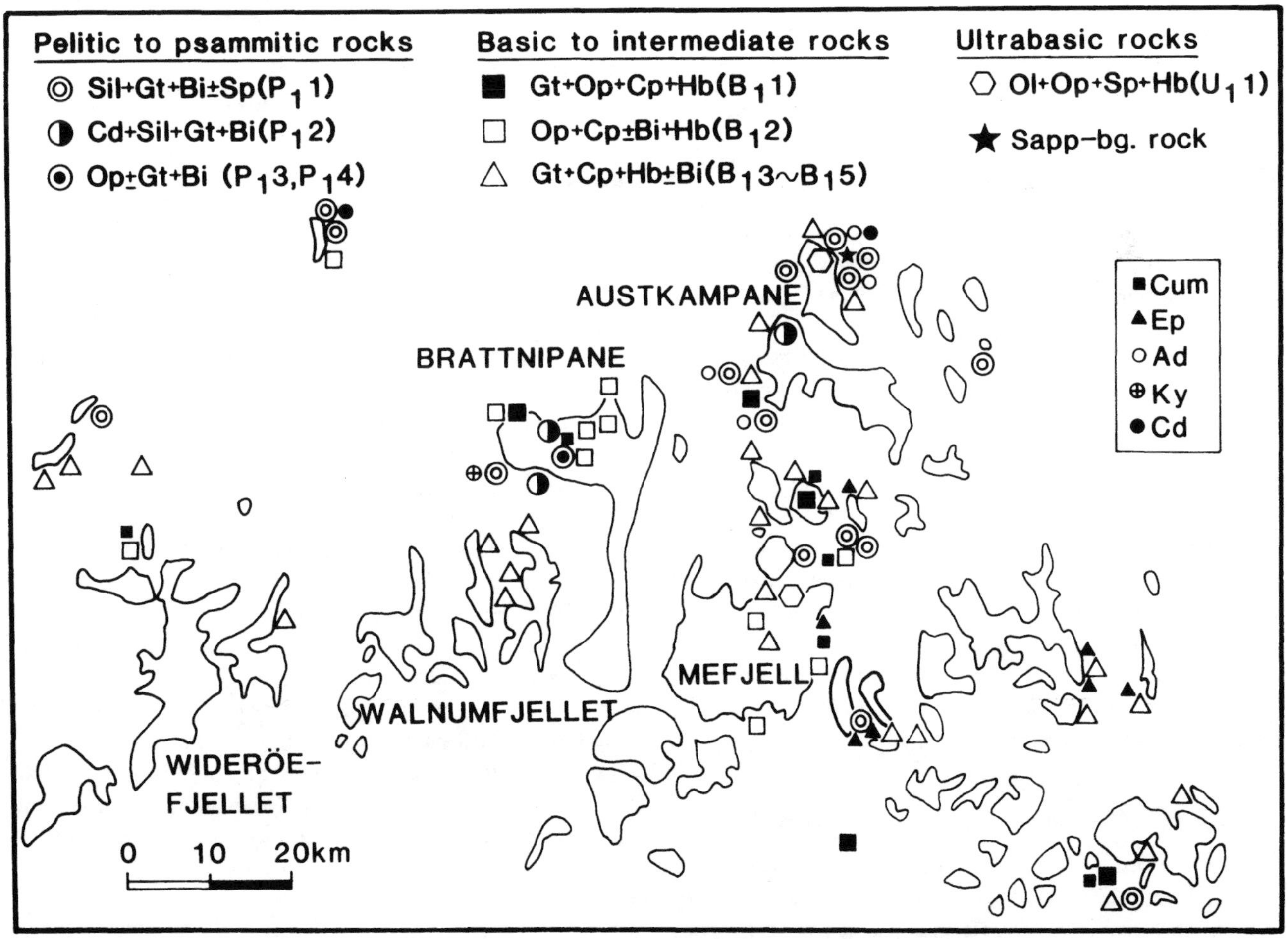

Fig. 2. Map showing the distribution of characteristic metamorphic minerals in the TVG. The mineral assemblages show the highest metamorphic grade at the sample location; probable secondary minerals are shown as small symbols. For mineral abbreviations see Table 1, except for Gt, garnet and Sapp, sapphirine.

suggesting the prograde breakdown product of staurolite. Garnet in the pelitic rocks characteristically has a homogeneous, relatively magnesian interior and a rim enriched in Fe (e.g. $X_{Fe} = 0.64$ and 0.79, respectively). The Fe-rich rim is of retrograde origin because it is replaced by secondary biotite. Orthopyroxene in basic and intermediate gneisses is also replaced by cummingtonite and by tschermakitic hornblende (Shiraishi & Kojima, 1987). Although upper amphibolite-facies rocks occur in the south-western part of the TVG area, a metamorphic zonation has not been established.

The kyanite-bearing assemblage (P_21 in Table 1) is recognized in one pelitic rock of the P_11 assemblage. Kyanite, accompanied by some quartz and muscovite, occurs exclusively in biotite aggregates which embay many garnet porphyroblasts, and thus kyanite is isolated from the matrix. The andalusite-bearing assemblage (P_22) occurs in other pelitic rocks of the P_11 assemblage. Andalusite either forms very fine-grained granoblastic domains together with quartz, plagioclase, biotite and sillimanite, or it forms poikiloblasts some of which, intergrown with sillimanite, obliquely cut tabular biotite in the coarser-grained gneissose matrix. Such kyanite-bearing and andalusite-bearing assemblages suggest two kinds of retrograde metamorphic episodes.

T–P conditions of the granulite-facies regional metamorphism were estimated to be 750–830 °C based on garnet (core)–biotite (matrix), garnet (core)–pyroxenes (core) and pyroxenes (core) compositions (Wood & Banno, 1973; Thompson, 1976; Ellis & Green, 1979; Sen & Bhattacharya, 1984), and 7–8.5 kb based on garnet (core)–plagioclase (matrix) and garnet (core)–pyroxene (core)–plagioclase compositions (Newton & Haselton, 1981; Perkins & Newton, 1981; Bohlen, Wall & Boettcher, 1983). As shown in Fig. 3, these estimates are consistent with the stability of the sillimanite–K-feldspar association, which is included in the assemblages P_11 and P_12 in Table 1, and the stability of the mineral association Sil–Ga–Cd ($X_{Fe} = 0.15$)–Bi–Kf–Qu (I_1 and $I_{0.5}$), which corresponds to the assemblage P_12. Such conditions indicate that the metamorphism is of medium-pressure type, and this is supported by the type of spinel present.

On the other hand, the pairs of garnet rim and biotite in direct contact with garnet, garnet rim and plagioclase in direct contact with garnet, and kyanite stability (Holdaway, 1971) indicate *T–P* conditions around 560 °C and more than 4.5 kb for the retrograde kyanite-forming episode, and around 550 °C and 3 kb for the retrograde andalusite-forming episode. The former conditions are compatible with the stability of the muscovite–quartz association, found in the P_21 assemblage, and the latter with the stability of the andalusite–sillimanite association in the P_22 assemblage (Fig. 3).

Nils Larsen Group

In contrast with the TVG rock types, the NLG has a characteristically simple lithology and is composed mainly of mylonitized tonalitic rocks with intercalated layers, ribbons and blocks of basic schists; psammitic and calcareous rocks are rare. The distribution of the basic schist fragments suggests that they represent a mafic dyke swarm in the tonalite. The

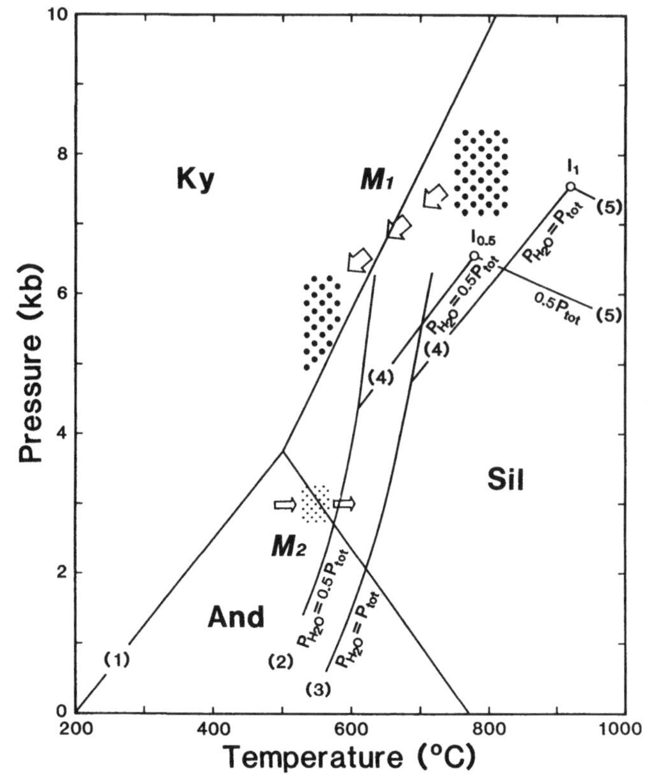

Fig. 3. *T–P* diagram showing metamorphic conditions and history. (1) phase boundaries of Al_2SiO_5 polymorphs (Holdaway, 1971); (2) and (3) muscovite + quartz = Al_2SiO_5 + K-feldspar + H_2O (after Kerrick, 1972 and Chatterjee, 1974, respectively); (4) biotite + sillimanite + quartz = cordierite ($X_{Fe} = 0.15$) + K-feldspar + H_2O (Holdaway & Lee, 1977); (5) cordierite ($X_{Fe} = 0.15$) = garnet + sillimanite + quartz + H_2O (Holdaway & Lee, 1977). The six mineral phases, sillimanite–garnet–cordierite ($X_{Fe} = 0.15$)–biotite–K-feldspar–quartz coexist at I_1 under $P_{H_2O} = P_{total}$ and at $I_{0.5}$ under $P_{H_2O} = 0.5\ P_{total}$. M_1, main regional metamorphism (Late Proterozoic); M_2, later thermal metamorphism (500 Ma).

schistosity trends and elongation of the basic schist fragments are parallel to the mylonitic foliation.

The mineral assemblages of the NLG are rather uniform. The typical tonalitic rock is composed of bluish-green calcic amphibole, biotite, epidote, plagioclase and quartz; chlorite and saussuritized plagioclase are common and garnet is found in some places. As mylonitization increases, the proportions of chlorite and epidote increase and plagioclase is more saussuritized. The basic schists are mineralogically identical to the tonalite. The psamitic schists are composed of biotite, muscovite, plagioclase, and quartz with or without garnet. It is suggested that all these rocks have undergone intense mylonitization and cataclasis under conditions of the greenschist- to epidote–amphibolite-facies. Pre-existing higher-grade metamorphic assemblages are not identified in all of the constituent rocks, whereas igneous plagioclase retaining oscillatory zoning is commonly observed in the tonalite.

Intrusive rocks

Plutonic rocks and dykes are widespread in the area and they have been subdivided into older and younger groups based on their relative ages to each other and to the regional mylonitization.

The older intrusive rocks consist of tonalite and mafic dykes (I; basic schist at present) in the NLG, and older granites and diorites. The younger intrusions are mafic dyke rocks (II; metadolerite at present), syenites, younger granite and the latest pegmatite and aplite. The older intrusions were affected by the regional mylonitization, and intruded by the mafic dykes (II).

The older granites, medium- to coarse-grained gneissose biotite–hornblende-granodiorite and biotite-granite, occur mainly in the central to eastern part of the area. Their contact with the host gneisses commonly has an agmatitic appearance and they include many xenoliths derived from the neighbouring gneisses. The younger granite, on the other hand, is generally a coarse-grained massive hornblende–biotite-granite which has a sharp contact with the host gneisses. The relationship between the older granite and the NLG tonalite is uncertain. Field relationships suggest that the mafic dykes (I) intruded the tonalite and that they were metamorphosed under greenschist- to epidote–amphibolite-facies conditions coeval with shear deformation. The older granites are intruded by dioritic rocks, and xenoliths of the older granites are present in the dioritic rocks.

Mafic dykes (II), ranging from a few tens of centimetres to a few metres wide, are especially predominant in the western part of the area. They are in sharp contact with the host gneisses and distinct chilled margins are commonly observed. The latest aplite and pegmatite dykes cut the mafic dykes in the northern nunatak (Vesthaugen) but the younger granite in the southernmost massif (Dufekfjellet) is intruded by the mafic dykes. These dykes, however, are slightly metamorphosed, with randomly orientated fine-grained biotite throughout and a typical subophitic texture. Since a whole-rock K–Ar age (536 ± 27 Ma) is consistent with previously reported radiometric ages of the granite, this age probably indicates the age of the time of reheating (Kojima & Shiraishi, 1986). Major element chemistry of this rock is characterized by high-alkalis ($Na_2O + K_2O$ = 5–6 wt %), high-TiO_2 (2.5–3.5 wt %) and high-P_2O_5 (0.9–1.3 wt %) with basic to intermediate compositions (SiO_2 = 51–54 wt %) (K. Shiraishi, unpub. data; S. Kanisawa, pers. comm.).

Syenitic rocks exposed in the southern ridge of the central part of the area represent multiple intrusions: an early heterogeneous syenite, characterized by rhythmic layering of dark mafic bands and leucocratic felsic bands on a cm–m-scale, and a later homogeneous leucocratic quartz-syenite which intrudes the heterogeneous syenite. Although the syenites are emplaced at the eastern end of the shear belt, no distinct shear zone is found in the syenite masses.

Geological structure

Teltet–Vengen Group

The main foliation is parallel to the compositional banding of the constituent rocks. Mesoscopic isoclinal folds, some of which are of an intrafolial type, are commonly observed; these formed during the earliest deformation, probably under the peak metamorphic conditions. Open folds of various scales are found throughout the group. Major open folds, which range up to a few kilometres in amplitude, occur in the central and eastern parts of the area.

Local shear zones, up to a few metres wide, are found in many places in the TVG and pegmatite and aplite dykes were intruded along the shear planes (mylonitic foliation). Shear folds are common in the surrounding gneisses; in some places, the shear planes developed parallel to the axial plane of the tight to isoclinal folds. Preferred orientation, defined by elongated quartz aggregates and mafic mineral lineation (such as amphibole and chlorite), are found on the shear planes; these plunge gently SSW–SW, suggesting the slip direction of shearing.

Nils Larsen Group

The foliation observed in the NLG is a mylonitic foliation in the tonalite and in the basic schists. It is concordant with the elongation of the basic schist fragments and the shear planes. Although the mylonitic foliation strikes roughly parallel to the gneissosity of the TVG, it dips steeply southwards, occasionally northwards or it is vertical.

Discussion

Metamorphism in the TVG

According to Hammarstrom & Zen (1986), crystallization pressures of 2.7–3 kb were inferred from hornblende compositions in a younger granite intruding the TVG. The pressure conditions are in good agreement with those of the andalusite-forming retrograde episode. Therefore, it is probable that the low-pressure retrograde episode (M_2) is genetically related to the acid intrusions at a shallow crustal level. Taking the medium-pressure character of the granulite-facies metamorphism into consideration, the higher-pressure retrograde episode would have taken place during the later stage of the main metamorphism (M_1). It is possible that the shear deformation was related to this episode because kyanite occurs near the shear zone. A possible metamorphic T–P history in the TVG is also depicted in Fig. 3.

Tectonic interpretation

Van Autenboer (1969) made a distinction between the gneisses forming the northern ridges (TVG) and those forming the southern massifs of the western part of the Sør Rondane Mountains (NLG). However, structural and geochronological differences between them are obscure. In the present study, the discrepancy between the compositions of the major rock types and the metamorphic grade of both groups suggests the following interpretations.

The peak metamorphic grade reached by the TVG was granulite-facies and there is no evidence to show that metamorphism occurred below lower amphibolite-facies conditions throughout the TVG area, although there are extensive retrograde effects (Ishizuka & Kojima, 1987; Shiraishi & Kojima, 1987). On the other hand, it is difficult to estimate the early metamorphic history of the NLG, because intense mylonitization associated with the greenschist- to epidote–amphibolite-facies metamorphism obscured the pre-existing nature of the rocks. We cannot recognize any evidence of higher-grade minerals, even in the least mylonitized rocks.

There are two possible interpretations of the relationships between the TVG and NLG. One is that the tectonic positions of the two groups differed greatly during the regional metamorphism, even if the NLG existed at the time of the regional metamorphism, i.e. the NLG was at a much shallower crustal level than the TVG. Depression of the NLG related to the formation of the wide shear belt caused the present tectonic situation. Another possibility is that the tonalite and mafic dykes (I) of the NLG were emplaced immediately after the peak regional metamorphism and that they were metamorphosed during a retrograde phase. We prefer the former possibility because of the existence of rare low-grade metasedimentary rocks, and abundant tonalite and mafic dyke (I) intrusions suggesting extensive magmatic activity before the shear movements in the NLG. Finally, emplacement of the NLG in a wide shear belt may be related to the rifting which occurred before the intrusion of the younger granite and mafic dykes (II), i.e. in the Early Palaeozoic at the latest.

The role of the mafic dykes (II; metadolerite) is important, not only as a marker for the age relationships of the plutonic rocks and the development of mylonitization, but also to compare with similar rock types occurring throughout Dronning Maud Land. The whole-rock K–Ar age of 536 ± 27 Ma indicates that the metadolerite dykes are not equivalent to the Ferrar and Karoo dolerites reported elsewhere in Dronning Maud Land and southern Africa (e.g. Kaiser & Wand, 1985). Moreover, the dykes have been thermally metamorphosed by granite of a similar age. The fact that the metadolerite dykes show distinct chilled margins indicates that the gneiss terrane had been considerably uplifted before granite emplacement.

In spite of the large number of age determinations for Mesozoic dolerites, only a few dolerites in Dronning Maud Land have yielded Early Palaeozoic ages (see compilation by Kaiser & Wand, 1985). However, many mafic dykes which may be of similar ages have been described from elsewhere in East Antarctica: Prince Olav Coast (Hiroi, Shiraishi & Sasaki, 1986), Yamato Mountains (K. Shiraishi, unpub. data), central Dronning Maud Land (Wolmarans & Kent, 1982; Kaiser & Wand, 1985) and northern Prince Charles Mountains (Sheraton, 1983). Although alkalic-mafic rocks seem to be predominant among them, geochemical studies are not sufficient. It is noteworthy that both acid and basic magmatism of latest Proterozoic to Early Palaeozoic age may be widespread in Dronning Maud Land.

References

Bohlen, S.R., Wall, V.J. & Boettcher, A.L. (1983). Geobarometry in granulites. In *Kinetics and Equilibrium in Mineral Reactions*, ed. S.K. Saxena, pp. 141–71. New York; Springer-Verlag.

Chatterjee, N.D. (1974). Thermal stability and standard thermodynamic properties of synthetic $2M_1$-muscovite, $KAl_2[AlSi_3O_{10}(OH)_2]$. *Contributions to Mineralogy and Petrology*, **48**, 89–114.

Ellis, D.J. & Green, D.H. (1979). An experimental study of the effect of Ca upon garnet–clinopyroxene exchange equilibria. *Contributions to Mineralogy and Petrology*, **71**, 13–22.

Hammarstrom, J.M. & Zen, E-An (1986). Aluminium in hornblende: an empirical igneous geobarometer. *American Mineralogist*, **71(11)**, 1297–313.

Hiroi, Y., Shiraishi, K. & Sasaki, K. (1986). Geological map of Akebono Rock, East Antarctica. *Antarctic Geological Map Series*, sheet 16. Tokyo; National Institute of Polar Research.

Holdaway, M.J. (1971). Stability of andalusite and the aluminum silicate phase diagram. *American Journal of Science*, **271(2)**, 97–131.

Holdaway, M.J. & Lee, S.M. (1977). Fe–Mg cordierite stability in high-grade pelitic rocks based on experimental, theoretical, and natural observations. *Contributions to Mineralogy and Petrology*, **63**, 175–98.

Ishizuka, H. & Kojima, H. (1987). A preliminary report on the geology of the central part of the Sør Rondane Mountains, East Antarctica. *Proceedings of the National Institute of Polar Research Symposium on Antarctic Geosciences, No. 1*, pp. 113–28.

Kaiser, G. & Wand, U. (1985). K–Ar dating of basalt dykes in the Schirmacher Oasis area, Dronning Maud Land, East Antarctica. *Zeitschrift für Geologische Wissenschaften*, **13(3)**, 299–307.

Kerrick, D.M. (1972). Experimental determination of muscovite + quartz stability with $P_{H_2O} < P_{total}$. *American Journal of Science*, **272(10)**, 946–58.

Kojima, S. & Shiraishi, K. (1986). Note on the geology of the western part of the Sør Rondane Mountains, East Antarctica. *Memoirs of the National Institute of Polar Research, Special Issue*, **43**, 116–31.

Newton, R.C. & Haselton, H.J. (1981). Thermodynamics of the garnet–plagioclase–Al_2SiO_5–quartz geobarometer. In *Thermodynamics of Minerals and Melts*, ed. R.C. Newton, A. Navrotsky & B.J. Wood, pp. 125–45. New York; Springer-Verlag.

Perkins, D., III & Newton, R.C. (1981). Charnockite geobarometers on coexisting garnet–pyroxene–plagioclase–quartz. *Nature, London*, **292(5819)**, 144–6.

Sen, S.K. & Bhattacharya, A. (1984). An orthopyroxene–garnet thermometer and its application to the Madras charnockites. *Contributions to Mineralogy and Petrology*, **88**, 64–71.

Sheraton, J.W. (1983). Geochemistry of mafic igneous rocks of the northern Prince Charles Mountains, Antarctica. *Journal of the Geological Society of Australia*, **30**, 295–304.

Shiraishi, K. & Kojima, S. (1987). Basic and intermediate gneisses from the western part of the Sør Rondane Mountains, East Antarctica. *Proceedings of the National Institute of Polar Research Symposium on Antarctic Geosciences, No. 1*, pp. 129–49.

Thompson, A.B. (1976). Mineral reactions in pelitic rocks; II. Calculation of some P–T–X (Fe–Mg) phase relations. *American Journal of Science*, **276(4)**, 425–54.

Van Autenboer, T. (1969). Geology of the Sør Rondane Mountains. In *Geologic Maps of Antarctica*, ed. V.C. Bushnell & C. Craddock, Pl. VIII. Antarctic map folio series, Folio 12. Washington, DC; American Geographical Society.

Van Autenboer, T. & Loy, W. (1972). Recent geological investigations in the Sør Rondane Mountains, Belgicafjella and Sverdrupfjella, Dronning Maud Land. In *Antarctic Geology and Geophysics*, ed. R.J. Adie, pp. 563–71. Oslo; Universitetsforlaget.

Wolmarans, L.G. & Kent, L.E. (1982). Geological investigations in western Dronning Maud Land, Antarctica – a synthesis. *South African Journal for Antarctic Research*, **Supplement 2**, 93 pp.

Wood, B.J. & Banno, S. (1973). Garnet–orthopyroxene and orthopyroxene–clinopyroxene relationships in simple and complex systems. *Contributions to Mineralogy and Petrology*, **42**, 109–24.

Late Proterozoic paired metamorphic complexes in East Antarctica, with special reference to the tectonic significance of ultramafic rocks

Y. HIROI[1], K. SHIRAISHI[2] & Y. MOTOYOSHI[3]

1 Department of Earth Sciences, Faculty of Science, Chiba University, 1–33, Yayoi-cho, Chiba 260, Japan
2 National Institute of Polar Research, 9–10, Kaga 1-chome, Itabashi-ku, Tokyo 173, Japan
3 Department of Applied Geology, University of New South Wales, PO Box 1, Kensington, NSW 2033, Australia

Abstract

The general geological and petrological characteristics of the Late Proterozoic Lützow–Holm and Yamato–Belgica complexes of eastern Queen Maud Land, East Antarctica, are summarized. The Lützow–Holm Complex is characterized by a continuous south-westward increase in metamorphic grade (progressive metamorphism of the medium-pressure type from amphibolite- to granulite-facies), prograde P–T (pressure–temperature) paths of rocks from earlier relatively high-pressure/low-temperature conditions to later lower-pressure/higher-temperature conditions, and low initial $^{87}Sr/^{86}Sr$ ratios for the metasedimentary rocks. The Yamato–Belgica Complex is characterized by widespread igneous activity and low-pressure type metamorphism. Thus, these complexes may have formed Late Proterozoic paired metamorphic belts. The presence of compositionally variable ultramafic rocks in isolated blocks that are unique to the Lützow–Holm Complex suggests that this complex formed in a suture zone between the older Yamato–Belgica Complex to the west and the Rayner Complex in the east. Continent–continent collision tectonics may therefore have been important in the development of this part of the East Antarctic shield.

Introduction

Recent geological, petrological and geochronological studies have revealed that the East Antarctic shield is not uniform but is composed of various geological units. To the west of the well-known Archaean Napier and Proterozoic Rayner complexes in Enderby Land (45–60° E), two Late Proterozoic complexes, Lützow–Holm and Yamato–Belgica, have been recognized in eastern Queen Maud Land, between 30 and 45° E longitude (Hiroi & Shiraishi, 1986; Shiraishi *et al.*, 1987; Fig. 1). This paper summarizes the general characteristics of these two complexes and presents new data on the metamorphosed ultramafic rocks peculiar to the Lützow–Holm Complex.

Two Late Proterozoic complexes in eastern Queen Maud Land

Eastern Queen Maud Land includes the Prince Olav Coast, Lützow–Holm Bay area, Yamato Mountains, and Belgica Mountains (Fig. 1). In addition to high-grade metamorphic rocks, Early Palaeozoic granites and pegmatites are widely exposed throughout the region. About 50 mineral and whole-rock K–Ar, Rb–Sr and U–Pb ages of ~ 500 Ma have been reported for the granites, pegmatites and metamorphic rocks (Yanai & Ueda, 1974; Kojima, Yanai & Nishida, 1982), and Shibata, Yanai & Shiraishi (1985) determined Rb–Sr mineral isochron ages of 469–493 Ma, biotite K–Ar ages of 469–483 Ma, and a hornblende K–Ar age of 502 Ma for the granulite-facies rocks in this area. Most radiometric ages of the metamorphic rocks in the region were reset probably during emplacement of the Early Palaeozoic granites and pegmatites. The geological and petrological features of the two complexes are summarized below.

Lützow–Holm Complex

This complex is composed mainly of well-layered pelitic–psammatic and intermediate (between basic and pelitic–psammitic) gneisses with some migmatitic rocks of granitic to granodioritic compositions. Lesser amounts of metamorphosed calcareous, basic, and ultramafic rocks are also present. The occurrence of ultramafic rocks is peculiar to this complex. All the rocks have been deformed at least twice; the

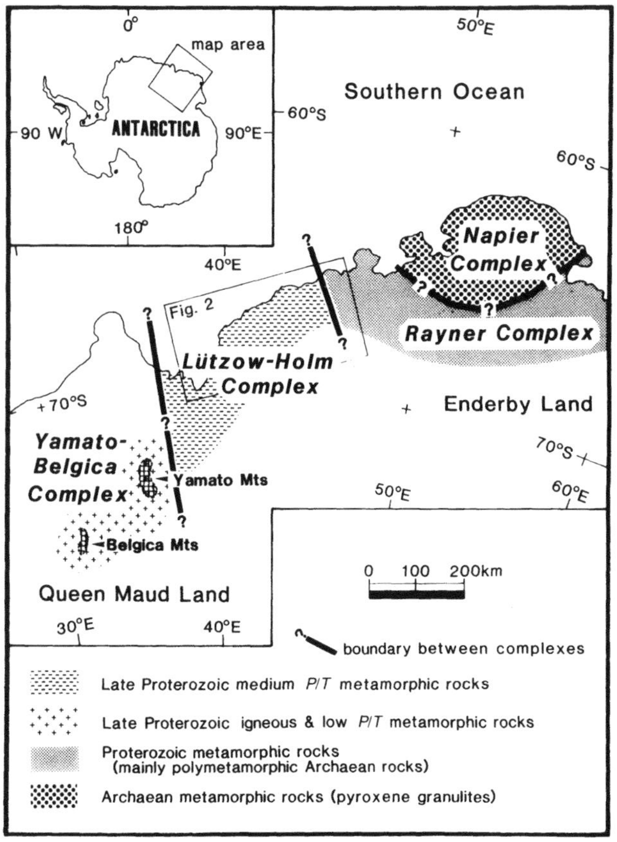

Fig. 1. Map of eastern Queen Maud Land and Enderby Land, East Antarctica, showing the distribution of four distinct Precambrian complexes.

axial planes of earlier isoclinal folds trend N–S to NW–SE and those of later open to close folds NE–SW. Besides the reset K–Ar and Rb–Sr ages of about 500 Ma, Rb–Sr whole-rock ages ranging from 680 to 1200 Ma have been obtained from the metamorphic rocks (e.g. Shibata, Yanai & Shiraishi, 1986). These ages probably date an earlier regional metamorphism of the complex. Initial $^{87}Sr/^{86}Sr$ ratios for the metasedimentary rocks are relatively low (0.705–0.706) (Shibata *et al.*, 1986), suggesting that most, if not all metasedimentary rocks did not have a long pre-metamorphic history. The most characteristic feature of the Lützow–Holm Complex is the south-westward increase in medium-pressure type metamorphism, from amphibolite to granulite-facies, over a distance of about 300 km. In this connection, it is significant and peculiar to this complex that small to trace amounts of metastable kyanite occur as inclusions within garnet and plagioclase grains in many sillimanite-bearing rocks, regardless of metamorphic grade (Fig. 2). Andalusite occurs locally as a contact metamorphic effect in rocks intruded by Early Palaeozoic granites and pegmatites. These isotopic and metamorphic characteristics of the Lützow–Holm Complex are clearly different from those of the adjacent Rayner Complex in Enderby Land. The boundary between the Lützow–Holm and Rayner complexes is not exposed, but it is most probably tectonic.

Yamato–Belgica Complex

This complex is characterized by widespread igneous intrusions, mostly syenites; metamorphic rocks have a sporadic distribution in the Yamato Mountains. Because the contact between rocks of amphibolite- and granulite-facies grade is tectonic, their original relationship is unknown. The granulite-facies rocks, exposed only in the Yamato Mountains, are closely associated with two-pyroxene syenite intrusions. Metamorphic temperatures of about 750 °C were evaluated using two-pyroxene geothermometers. The wollastonite + anorthite assemblage occurring in the granulite-facies calc-silicate rocks suggests pressures below 5 kb at 750 °C (Fig. 3). Thus, the metamorphic rocks of this complex belong to the low-pressure facies series. A Rb–Sr whole-rock isochron age of about 720 Ma, with an initial $^{87}Sr/^{86}Sr$ ratio of 0.709, was obtained for the granulite-facies rocks (Shibata *et al.*, 1986). The age probably dates the granulite-facies metamorphism and is similar to the age of regional metamorphism of the Lützow–Holm Complex. The relatively high initial $^{87}Sr/^{86}Sr$ ratio is similar to that of the Rayner Complex at Molodezhnaya Station (Grew, 1978). The boundary between the Yamato–Belgica and Lützow–Holm complexes is covered by thick continental ice.

Progressive and prograde metamorphism of Lützow–Holm Complex

South-westward progressive metamorphism of the medium-pressure type, from amphibolite- to granulite-facies, has been confirmed by mapping six isograds in pelitic, calc-silicate, and basic–intermediate rocks on the Prince Olav and Sōya coasts (Fig. 2). These isograds are based on the following continuous and discontinuous reactions.

Pelitic rocks (Hiroi *et al.*, 1983a, b; Motoyoshi *et al.*, 1985)

Staurolite = aluminium silicate(s) + garnet + hercynite + H_2O (1)

Sillimanite + garnet = spinel + quartz (2)

Calc-silicate rocks (Hiroi *et al.*, 1987)

Epidote + quartz = ugrandite garnet + plagioclase + Fe-oxide(s) + H_2O ± scapolite (3)

Grossular + quartz = anorthite + wollastonite (4)

Calcite + quartz = wollastonite + CO_2 (5)

Basic-intermediate rocks (Hiroi *et al.*, 1983a, b; Shiraishi, Hiroi & Onuki, 1984)

Ca-poor amphibole(s) = orthopyroxene + quartz + H_2O (6)

Tremolite = orthopyroxene + clinopyroxene + quartz + H_2O (7)

Tschermakite + quartz = orthopyroxene + plagioclase + H_2O (8)

The occurrence of grossular ± quartz in the granulite-facies terrane indicates that the metamorphic facies series of the complex is of the medium-pressure type (Fig. 3). In addition to the mineral changes mentioned above, gradual changes in partition coefficients of elements between minerals, and in chemical compositions of solid-solution minerals such as hornblende, indicate a progressive increase in temperature from

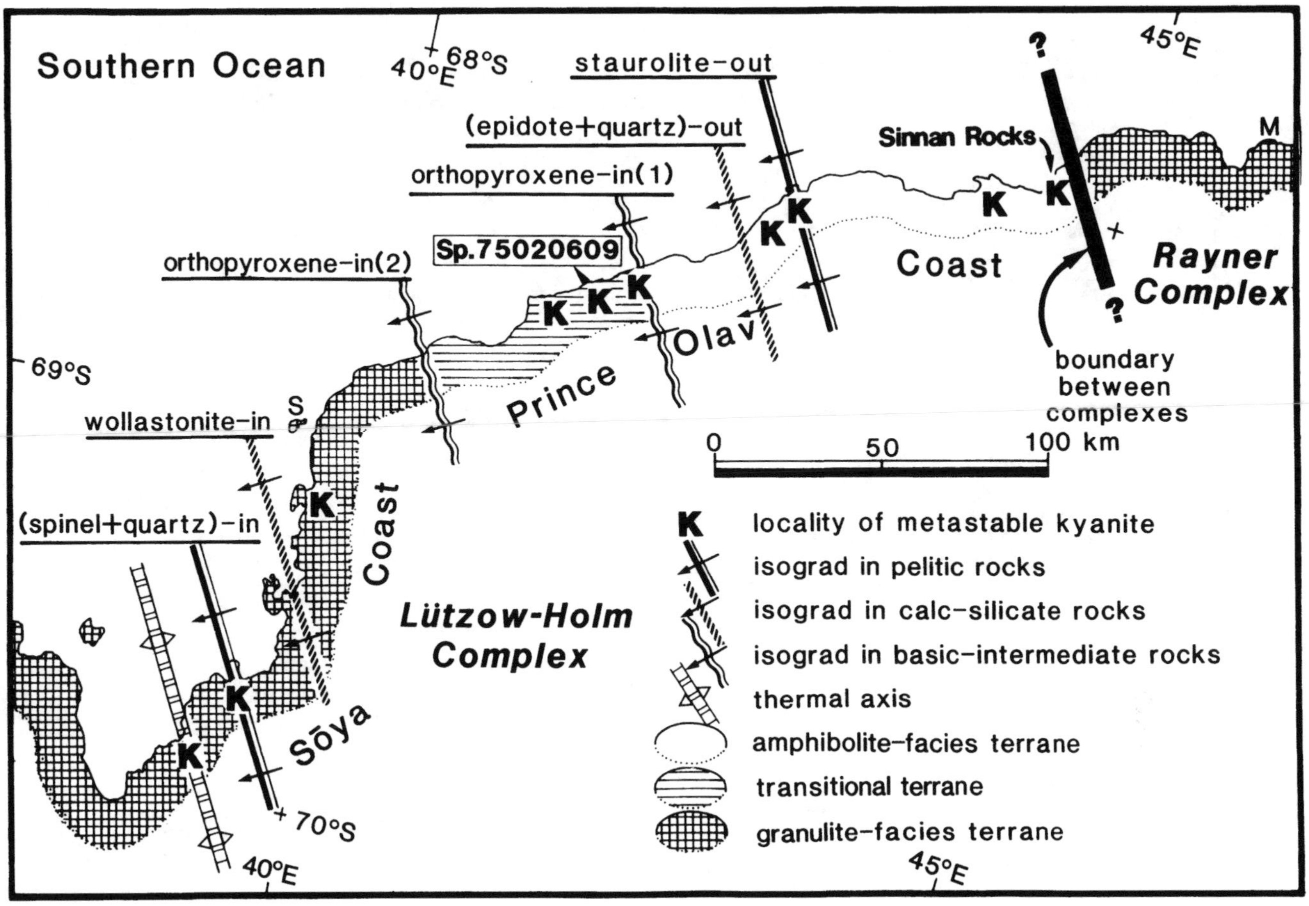

Fig. 2. Map of Prince Olav and Sōya coasts, showing the boundary between the two complexes, their metamorphic facies and the location of metastable kyanite, Sp. 75020609, isograds, and the thermal axis. The thermal axis is after Motoyoshi (1986). S, Syowa Station; M, Molodezhnaya Station.

north-east to south-west. Estimated peak metamorphic conditions for the progressive metamorphic sequence are collectively shown as the 'metamorphic geotherm' of the Lützow–Holm Complex in Fig. 3.

The occurrence of metastable kyanite in many sillimanite-bearing rocks provides evidence of prograde metamorphism. Staurolite and its breakdown products by reaction (1) occur always as inclusions in garnet and plagioclase grains. Such staurolite may be a relic of other prograde staurolite-consuming reactions because reaction (1) is a terminal reaction. It is significant that the aluminium silicate in the staurolite-breakdown products is sillimanite and kyanite, respectively, in the lower-grade and higher-grade parts of the Lützow–Holm Complex (Hiroi *et al.*, 1983b). This suggests that the *P–T* paths followed by the metasedimentary rocks are not identical to the medium-pressure type metamorphic geotherm based on progressive mineral zones, because, in the higher-grade rocks, reaction (1) may have taken place in the kyanite stability field (Fig. 3). Similar evidence is also found in the ultramafic rocks.

Tectonic significance of ultramafic rocks

Metamorphosed ultramafic rocks occur mainly in the south-western part of the Lützow–Holm Complex, where they form small lenticular masses and thin concordant layers within the metasedimentary gneisses. The spatial distribution, mode of field occurrence, bulk chemical compositions, mineral assemblages and textures of the ultramafic rocks have been summarized by Hiroi *et al.* (1986). Their bulk chemical compositions are highly variable. Some are poor in SiO_2 (~ 40 wt %) and extremely rich in MgO (> 30 wt %) or Al_2O_3 (> 20 wt %), and are similar to olivine-rich or plagioclase-rich troctolites. Such troctolitic rocks, along with other rocks such as pyroxenites, were produced most probably by fractional crystallization of basic magmas under plagioclase–lherzolite facies conditions. Some rocks have MgO/(MgO + FeO) ratios as high as 0.85, suggesting that fractionating magmas were close to primitive magmas generated in the upper mantle. These chemical features suggest that the metamorphosed ultramafic rocks were derived from various parts of layered gabbros probably constituting an ophiolitic complex. They now have spinel–amphibole–lherzolite-facies mineral asssemblages. Although neither olivine + plagioclase nor olivine + garnet assemblages have been observed, olivine-free rocks of troctolitic compositions usually contain garnet. It is significant that orthopyroxene never occurs as inclusions in garnet porphyroblasts which contain abundant inclusions of hornblende, plagioclase and spinel ± sapphirine. On the other hand, garnet commonly breaks down partially to a symplektite of orthopyroxene + plagioclase ± spinel. The partial breakdown of garnet is described by the continuous reaction garnet ± hornblende = orthopyroxene + plagioclase ± spinel ± hornblende.

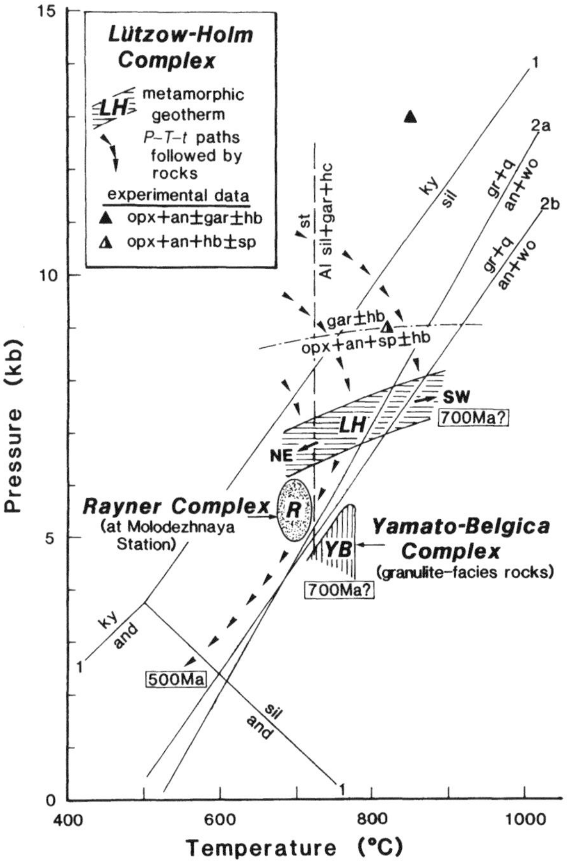

Fig. 3. *P–T* diagram showing a medium-pressure type metamorphic geotherm based on the progressive mineral zones and schematic *P–T* paths followed by rocks of the Lützow–Holm Complex. Estimated *P–T* conditions for the granulite-facies metamorphic rocks of the Yamato–Belgica and Rayner complexes, and preliminary mineral synthesis data are also shown. Data sources: Lützow–Holm Complex (Hiroi *et al.*, 1987); Yamato–Belgica Complex (Asami & Shiraishi, 1985); Rayner Complex (Grew, 1981). 1, Holdaway (1971); 2a, Newton (1966), Boettcher (1970) and Windom & Boettcher (1976); 2b, Huckenholz, Lindhuber & Fehr (1981). Staurolite-breakdown reaction is for reference only. an, anorthite; and, andalusite; gar, garnet; gr, grossular; hb, hornblende; hc, hercynite; ky, kyanite; opx, orthopyroxene; q, quartz; sil, sillimanite; sp, spinel; st, staurolite; wo, wollastonite.

These mineral textures suggest changing *P–T* conditions during prograde metamorphism of the ultramafic rocks, from relatively high-pressure and low-temperature conditions (to form garnet free of orthopyroxene inclusions) to later lower-pressure and higher-temperature conditions resulting in the formation of the symplektite (Fig. 3).

Preliminary experimental work, carried out by Dr Y. Tatsumi at Kyoto University, on a symplektitic intergrowth of orthopyroxene, plagioclase and spinel in a sapphirine-bearing troctolitic rock (Sp. 75020609), suggests that the symplektite may have recrystallized at pressures > 9 kb (Fig. 3).

The *P–T* paths followed by the ultramafic and metasedimentary rocks are not identical to the medium-pressure type metamorphic geotherm based on progressive mineral zones (Fig. 3), and they suggest a higher-pressure and lower-temperature geothermal regime similar to that in, or near a subduction zone at an early stage of metamorphism. The relatively high-pressure and later medium-pressure character of the Lützow–Holm Complex contrasts with the low-pressure character of the Yamato–Belgica Complex (Fig. 3). These two complexes may have formed Late Proterozoic paired metamorphic belts. As discussed by Miyashiro (1973) and others, a plate tectonic model would be most suitable for the interpretation of the development of paired metamorphic belts. Likewise, the spatially limited distribution, mode of field occurrence as small isolated blocks, and wide range of bulk chemical compositions of the metamorphosed ultramafic rocks of the Lützow–Holm Complex are accounted for most satisfactorily by a plate tectonic model, as follows.

Stage 1. Fractional crystallization of basic magmas (some were primitive), under plagioclase–lherzolite-facies conditions, formed layered gabbros as part of an ophiolitic complex which constituted the missing oceanic crust between older continents (represented by the Yamato–Belgica and Rayner complexes).

Stage 2. Tectonic emplacement of fractured blocks of the layered gabbros in the original sedimentary piles of the Lützow–Holm Complex, and metamorphism under relatively high-pressure and low-temperature conditions of spinel–amphibole–lherzolite-facies in or near a subduction zone.

Stage 3. Later metamorphism under lower-pressure and higher-temperature conditions of spinel–amphibole–lherzolite-facies as a result of erosion and recovery of the geothermal gradient after the collision of the older continents (Yamato–Belgica and Rayner complexes).

Thus, it is probable that the Lützow–Holm Complex formed in a suture zone between the older Yamato–Belgica and Rayner Complexes.

Conclusions

The Late Proterozoic Lützow–Holm and Yamato–Belgica complexes in eastern Queen Maud Land may have formed paired metamorphic belts. In contrast with the low-pressure metamorphic character of the Yamato–Belgica Complex, the relatively high- to medium-pressure metamorphic character of the Lützow–Holm Complex is confirmed by petrological studies of pelitic, calc-silicate and ultramafic rocks and by preliminary mineral synthesis experiments. The development of the paired metamorphic belts, in conjunction with the spatially limited distribution, mode of field occurrence as isolated small blocks and wide range of bulk chemical compositions of the metamorphosed ultramafic rocks in the Lützow–Holm Complex, is best explained by a Late Proterozoic continent–continent collision tectonic model.

Acknowledgements

We would like to express our sincere thanks to Dr Y. Tatsumi of Kyoto University for the high *P–T* mineral synthesis experiments. We also would like to thank the many geologists who carried out geological surveys in eastern Queen Maud Land for the use of their field data and specimens.

References

Asami, M. & Shiraishi, K. (1985). Retrograde metamorphism in the Yamato Mountains, East Antarctica. *Memoirs of the National Institute of Polar Research, Special Issue*, **37**, 147–63.

Boettcher, A.L. (1970). The system $CaO–Al_2O_3–SiO_2–H_2O$ at high pressures and temperatures. *Journal of Petrology*, **11(2)**, 337–79.

Grew, E.S. (1978). Precambrian basement at Molodezhnaya Station, East Antarctica. *Geological Society of America Bulletin*, **89(6)**, 801–13.

Grew, E.S. (1981). Granulite-facies metamorphism at Molodezhnaya Station, East Antarctica. *Journal of Petrology*, **22(3)**, 297–336.

Hiroi, Y. & Shiraishi, K. (1986). Geology and petrology of the area around Syowa Station. In *Science in Antarctica, 5, Earth Sciences*, ed. National Institute of Polar Research, pp. 45–84. Tokyo; Kokon Shoin (in Japanese).

Hiroi, Y., Shiraishi, K., Motoyoshi, Y. & Katsushima, T. (1987). Progressive metamorphism of calc-silicate rocks from the Prince Olav and Soya Coasts, East Antarctica. *Proceedings of the National Institute of Polar Research Symposium on Antarctic Geosciences, No. 1*, pp. 73–97. Tokyo; NIPR.

Hiroi, Y. Shiraishi, K., Motoyoshi, Y., Kanisawa, S., Yanai, K. & Kizaki, K. (1986). Mode of occurrence, bulk chemical compositions, and mineral textures of ultramafic rocks in the Lützow–Holm Complex, East Antarctica. *Memoirs of the National Institute of Polar Research, Special Issue*, **43**, 62–84.

Hiroi, Y., Shiraishi, K., Nakai, Y., Kano, T. & Yoshikura, S. (1983a). Geology and petrology of Prince Olav Coast, East Antarctica. In *Antarctic Earth Science*, ed. R.L. Oliver, P.R. James & J.B. Jago, pp. 32–5. Canberra; Australian Academy of Science and Cambridge; Cambridge University Press.

Hiroi, Y., Shiraishi, K., Yanai, K. & Kizaki, K. (1983b). Aluminum silicates in the Prince Olav and Sōya Coasts, East Antarctica. *Memoirs of the National Institute of Polar Research, Special Issue*, **28**, 115–31.

Holdaway, M.J. (1971). Stability of andalusite and the aluminium silicate phase diagram. *American Journal of Science*, **271(2)**, 97–131.

Huckenholz, H.G., Lindhuber, W. & Fehr, K.T. (1981). Stability relationships of grossular + quartz + wollastonite + anorthite. I. The effect of andradite and albite. *Neues Jahrbuch für Mineralogie; Abhandlungen*, **142(3)**, 223–47.

Kojima, H., Yanai, K. & Nishida, T. (1982). Geology of the Belgica Mountains. *Memoirs of the National Institute of Polar Research, Special Issue*, **21**, 32–64.

Miyashiro, A. (1973). *Metamorphism and Metamorphic Belts*. London; George Allen & Unwin, 492 pp.

Motoyoshi, Y. (1986). Prograde and progressive metamorphism of the granulite-facies Lützow–Holm Bay region, East Antarctica. D.Sc. thesis, Hokkaido University, 238 pp. (unpublished).

Motoyoshi, Y., Matsubara, S., Matsueda, H. & Matsumoto, Y. (1985). Garnet–sillimanite gneisses from the Lützow–Holm Bay region, East Antarctica. *Memoirs of the National Institute of Polar Research, Special Issue*, **37**, 82–94.

Newton, R.C. (1966). Some calc-silicate equilibrium relations. *American Journal of Science*, **264(3)**, 204–22.

Shibata, K., Yanai, K. & Shiraishi, K. (1985). Rb–Sr mineral isochron ages of metamorphic rocks around Syowa Station and the Yamato Mountains, East Antarctica. *Memoirs of the National Institute of Polar Research, Special Issue*, **37**, 164–71.

Shibata, K., Yanai, K. & Shiraishi, K. (1986). Rb–Sr whole rock ages of metamorphic rocks from eastern Queen Maud Land, East Antarctica. *Memoirs of the National Institute of Polar Research, Special Issue*, **43**, 133–48.

Shiraishi, K., Hiroi, Y., Motoyoshi, Y. & Yanai, K. (1987). Plate tectonic development of Late Proterozoic paired metamorphic complexes in eastern Queen Maud Land, East Antarctica. In *Gondwana Six: Structure, Tectonics, and Geophysics*, Geophysical Monograph 40, ed. G.D. McKenzie, pp. 309–18. Washington, D.C.; American Geophysical Union.

Shiraishi, K., Hiroi, Y. & Onuki, H. (1984). Orthopyroxene-bearing rocks from the Tenmondai and Naga-iwa Rocks in the Prince Olav Coast, East Antarctica; first appearance of orthopyroxene in progressive metamorphic sequence. *Memoirs of the National Institute of Polar Research, Special Issue*, **33**, 126–44.

Windom, K.E. & Boettcher, R.L. (1976). The effect of reduced activity of anorthite on the reaction grossular + quartz = anorthite + wollastonite: a model for plagioclase in earth's crust and upper mantle. *American Mineralogist*, **61(5)**, 889–96.

Yanai, K. & Ueda, Y. (1974). Absolute ages and geological investigations on the rocks in the area of around Syowa Station, East Antarctica. *Nankyoku Shiryo*, **48**, 70–81. (in Japanese, with English abstract).

Petrographic and structural characteristics of a part of the East Antarctic craton, Queen Maud Land, Antarctica

M.K. KAUL, R.K. SINGH, D. SRIVASTAVA, S. JAYARAM & S. MUKERJI

Geological Survey of India, NH-5P, NIT, Faridabad, India

Abstract

Geological investigations recorded here were carried out in central Queen Maud Land during the fifth Indian expedition to Antarctica. The rock types encountered were gneisses, amphibolites, basic granulites/charnockites and anorthosites. Petrological studies have revealed that the metamorphic grade increases from sillimanite–almandine sub-facies of almandine-amphibolite facies in the Schirmacher Hills to pyroxene-granulite facies in the eastern part of the Wohlthat Mountains. Anorthosites of the Gruber Mountains are suggested to be Proterozoic massif type. A tectonic control for the anorthosite emplacement is suggested on the basis of their spatial disposition and megastructures in the area.

Introduction

In the Schirmacher Hills and the eastern Wohlthat Mountain range of East Antarctica, Precambrian gneisses, anorthosites, charnockites, amphibolites and their variant rock types are exposed between latitudes 70°40′ S and 72°00′ S, and longitudes 11°00′ E and 14°00′ E. The area studied forms part of central Queen Maud Land of the East Antarctic craton (Fig. 1).

Distribution of rock types

Gneisses of various types are exposed along the entire stretch of the Schirmacher Hills, in the nunataks between the Schirmacher Hills and the Wohlthat Mountains, in the northernmost parts of the Gruber Mountains, and in the Petermann Ranges of the Wohlthat Mountains (Fig. 1).

Leucocratic, augen- and biotite-bearing streaky gneisses are the main varieties of rock type exposed in the Schirmacher Hills. In the Gruber Mountains and Petermann Ranges, augen gneiss predominates with alternate compositional banding of biotite–amphibole/pyroxene-rich layers and quartzo-feldspathic layers. Minute grains of garnet are evenly disseminated in the rock. Syn-tectonic to post-tectonic growth of feldspar porphyroblasts is seen megascopically.

In the Baalsrudfjellet, Sonstebynuten, Pevikhornet and Starheimtind nunataks between the Schirmacher Hills and the Wohlthat Mountains, leucocratic gneiss alternates regularly with melanocratic gneiss. Whereas the leucocratic types are essentially quartz–feldspar gneisses, the melanocratic ones are composed of coarse-grained quartz, feldspar, biotite, amphibole, garnet and, in some cases, sillimanite needles.

Anorthosites, varying in grain size, texture and mineral composition, comprise a major part of the Gruber Mountains. Totally monomineralic anorthosites are found, but some varieties contain dark grey or brownish-grey-coloured pyroxenes and rare quartz. Secondary biotite occurs in very small amounts, particularly in pyroxene-bearing anorthosites, where small mica flakes are disseminated through the rock body. Occasional instances of ilmenite/magnetite mineralization, in the form of stringers or isolated grains, also occur. Within some anorthosites there are 10–20 cm wide bands of garnet–pyroxene rock in which garnet occurs as elongated grains.

Very coarse-grained quartz–syenite with large (> 2 cm diameter) grains of feldspars are found in the Petermann Ranges. They are typically light brown to grey-coloured with < 10% mafic mineral component, and are markedly different in their physical appearance from the rocks of Gruber Mountains.

Pyroxene–garnet–quartz–feldspar-bearing basic granulites/charnockites/enderbites are well exposed in the Gruber Mountains area between gneisses and anorthosites. These rocks are medium- to coarse-grained, and occasionally contain porphyroblasts of dark grey feldspars. Garnet is disseminated in the form of small grains with concentrations around pyroxene crystals. In general, these rocks are devoid of biotite; some pyroxene-bearing and garnet-free anorthosites are associated with them. They are of dark grey colour, in contrast to the dark brownish-grey colour of the above-mentioned granulites.

Amphibolites are present in a very few exposures within gneisses of the Schirmacher Hills. These occur as dykes, veins, small pockets and lenses that are well foliated and lineated. Lineation is marked by parallel alignment of prismatic hornblende. In Sonstebynuten, Pevikhornet, Starheimtind, Tallaksenvarden, and Andersensata nunataks lying between the Schirmacher Hills and the Wohlthat Mountains, basic intrusive bodies are found within the gneisses. They mostly occur in the form of sills.

Garnet–sillimanite–graphite-bearing gneisses resembling the

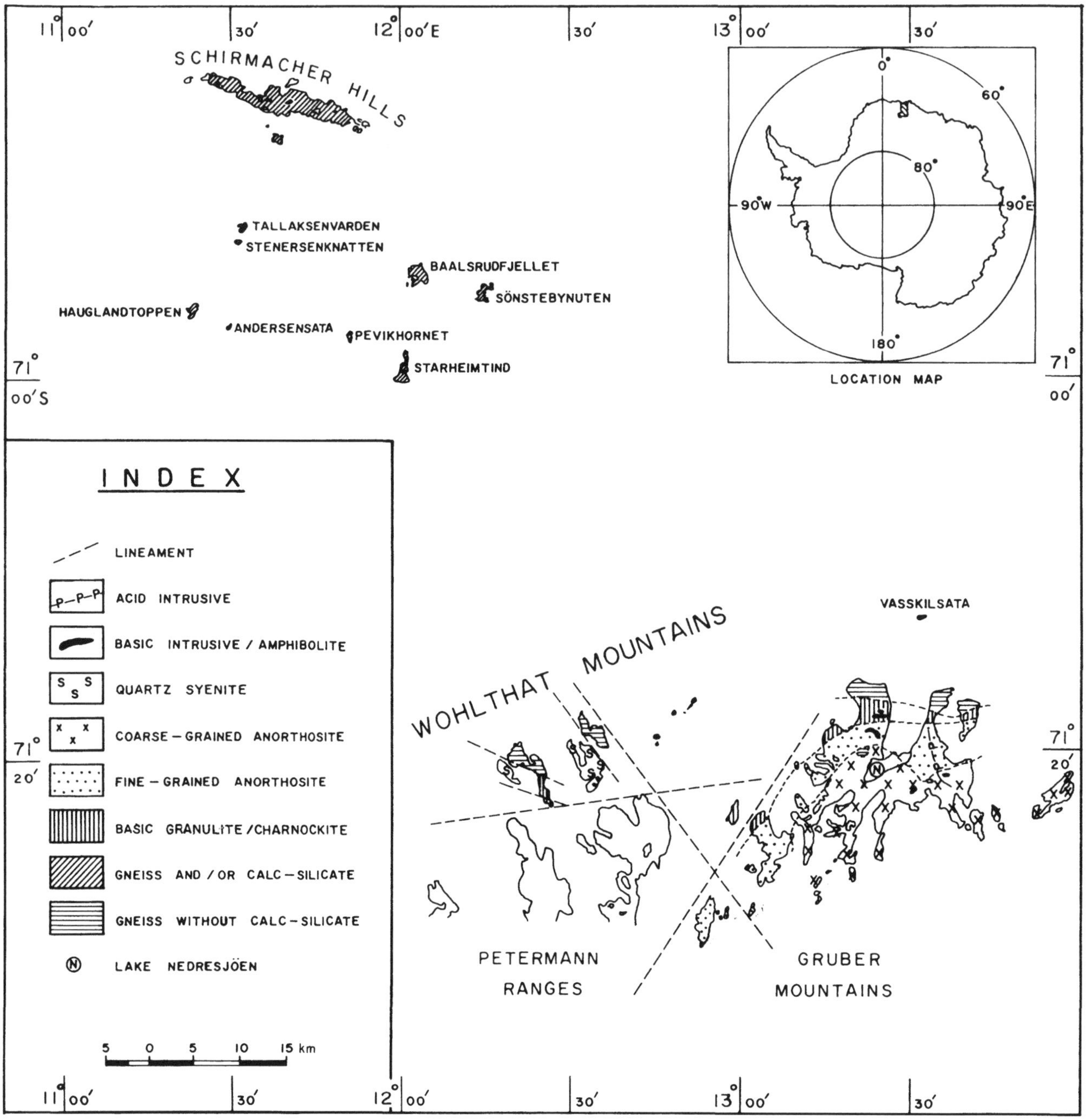

Fig. 1. Geological map of part of central Queen Maud Land, East Antarctica.

khondalites (Walker, 1902) of South India are exposed in conformity with other gneisses of the Schirmacher Hills and are also in contact with calc-silicate rocks. The latter are coarse-grained rocks comprising diopside–calcite–tremolite–quartz as mineral constituents. Bands of this rock are also present in the central-southern part of the Sonstebynuten nunatak (Fig. 1). Quartzo-feldspathic layers in these rocks give them a gneissic appearance.

All rock types are profusely inundated by basic and acid intrusives; these include basalt–dolerite dykes and aplite–quartz–pegmatite veins, respectively.

Petrographic characters

Gneisses

Kaul, Chakraborty & Raina (1985a) have discussed in detail the petrography of the gneisses of the Schirmacher Hills. They described three pyroxene-free, quartzo-feldspathic assemblages. In the Gruber Mountains, the following parageneses have been noticed in the gneisses: (1) garnet–orthopyroxene–clinopyroxene–perthite–plagioclase–quartz, with apatite as a minor constituent; (2) garnet–biotite–hornblende–perthite–plagioclase–quartz, with apatite, zircon and ilmenite as minor constituents; and (3) quartz–plagioclase–garnet–perthite–orthopyroxene, with apatite and ilmenite as minor constituents.

The gneissic texture in the gneisses of Gruber Mountains is defined by the banded form of quartzo-feldspathic portions alternating with garnet–pyroxene layers. The former show granoblastic texture while the latter have a preferred orientation of prismatic pyroxene, elongated garnet (in some cases) and inequant quartz (Fig. 2). Garnet is present, both as discrete equant grains, and as reaction rims round orthopyroxene and clinopyroxene, and also around biotite.

Anorthosites

These are medium- to coarse-grained rocks with a plagioclase component of ≥ 90%. The anorthite content of the plagioclases varies from 28 to 55%. These anorthosites present a conspicuously granoblastic texture (Fig. 3), although stretched and deformed grains of plagioclase are very common. The plagioclase grains are occasionally rimmed by clusters of smaller, fresh grains of recrystallized plagioclase (Fig. 4). In such cases coarser grains are strained and show undulatory extinction as well as deformed twin lamellae; grain boundaries are frequently crenulated. Quartz is rarely present, while pyroxenes and biotite are more common. Opaque minerals present include ilmenite and magnetite, and chlorite, white mica and carbonates are associated with fracture zones in anorthosites.

Basic granulites

The granulites of the Gruber Mountains show a mineral paragenesis of orthopyroxene–clinopyroxene–garnet–plagioclase–potash feldspar (mostly in microperthite) ± quartz ± biotite ± hornblende. These rocks present a characteristic granoblastic texture with the presence of equant and more or less equigranular grains of plagioclase, pyroxene and garnet. Whereas quartz is present in some samples in very minor amounts, perthite is ubiquitous. Plagioclase composition varies between An_{30} and An_{60}. In basic granulites there is textural evidence for the formation of garnet + quartz from clinopyroxene and plagioclase. In addition, there is evidence of orthopyroxene and plagioclase reacting to form garnet and clinopyroxene. Brownish-green recrystallized hornblende is occasionally found and ilmenite, magnetite and apatite comprise minor constituents. In a few samples of enderbite, biotite–quartz symplectite is present where it occurs at the grain boundary of a coarse biotite crystal which contains inclusions of orthopyroxene. Adjacent to, and in contact with, this coarse grain of biotite is a coarse grain of orthopyroxene which encloses a corroded crystal of biotite. The two textures indicate that the rock has been subjected to both prograde and retrograde metamorphic reactions.

Amphibolites/basic intrusives

Kaul, Chakraborty & Raina (1985b) have identified three zones of amphibolite occurrence in the Schirmacher Hills, on the basis of distinct mineral paragenesis. Although no amphibolite occurrence was noticed in the area south of the Schirmacher Hills, dolerites and lamprophyres are exposed in the Gruber Mountains, and the Pevikhornet and Sonstebynuten nunataks, respectively. Dolerites are also found in the Starheimtind, Tallaksenvarden and Andersensata nunataks. The dolerites, which are fine- to medium-grained showing a

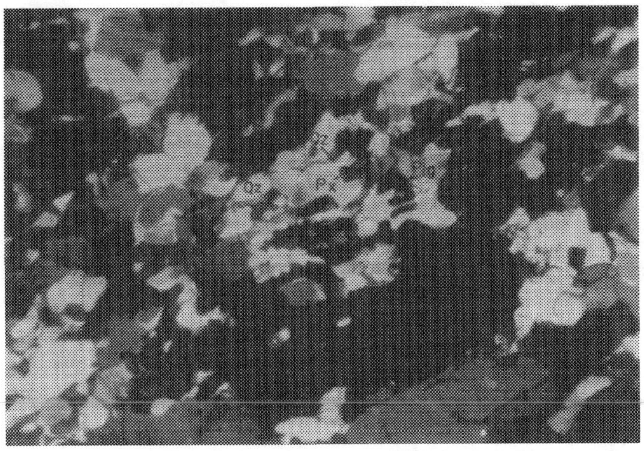

Fig. 2. Inequant quartz and prismatic pyroxenes showing preferred orientation in the gneisses, Gruber Mountains, East Antarctica. Crossed-polarisers (C.P.) × 32. Qz, quartz; Px, pyroxene; Plg, plagioclase.

Fig. 3. Plagioclase crystals in anorthosites of the Gruber Mountains, East Antarctica, showing granoblasic texture. C.P. × 32.

Fig. 4. Recrystallized grains of plagioclase in the vicinity of a coarse plagioclase crystal. Anorthosite. Gruber Mountains, East Antarctica. C.P. × 32.

Table 1. *Whole-rock analyses by XRF of anorthosites from the Gruber Massif, Wohlthat Mountains, East Antarctica (values in wt %)*

Oxides	Sample no. G-1	G-7	G-8	G-13	G-33	G-38	G-43	G-47	G-52	G-62	G-63	G-65	G-66	G-69
SiO_2	52.93	53.00	52.10	53.08	52.90	53.18	53.34	51.90	54.02	51.61	54.03	54.10	53.47	50.80
TiO_2	1.17	0.27	0.27	0.17	0.41	1.37	1.17	0.89	0.63	2.13	0.13	0.16	0.19	2.47
Al_2O_3	25.70	26.65	27.49	28.11	26.89	26.57	28.11	25.40	28.29	27.27	28.64	29.03	28.16	25.43
Fe_2O_3 +	2.67	1.06	0.86	0.80	1.69	3.29	2.36	5.75	1.07	2.59	0.33	0.31	0.55	3.67
MnO	0.03	0.03	0.02	0.02	0.03	0.04	0.03	0.07	0.02	0.03	0.01	0.01	0.01	0.04
MgO	bld	bld	bld	bld	bld	0.68	bld	bld	bld	0.06	bld	bld	bld	bld
CaO	11.30	11.87	11.56	10.86	11.43	10.48	11.76	10.41	10.93	10.90	11.17	11.51	10.97	11.14
Na_2O	3.22	3.71	3.40	3.79	3.75	3.15	2.75	2.63	3.81	2.35	3.47	4.26	4.19	3.55
K_2O	0.82	0.88	0.54	0.84	0.81	0.78	0.81	0.77	0.89	0.95	0.96	0.93	1.00	0.84
P_2O_5	0.06	0.05	0.05	0.06	0.05	0.05	0.04	0.25	0.05	0.15	0.04	0.05	0.06	0.46
LOI	0.17	0.27	1.68	1.79	0.90	0.35	0.36	1.59	0.57	0.83	0.36	0.20	0.20	0.11
Total	98.07	97.79	97.97	99.52	98.86	99.94	100.73	99.66	100.28	98.87	99.14	100.56	98.80	98.51

+ : Total Fe as Fe_2O_3
bld: below level of detection. Detection limit for MgO is 0.05.
LOI: Loss on ignition.

holocrystalline texture, are essentially composed of plagioclase, clinopyroxene, orthopyroxene, amphibole and olivine. Subophitic texture is a common feature of these dolerites. Lamprophyres show a porphyritic texture with phenocrysts of hornblende and biotite lying in a matrix of fine-grained feldspars.

Chemistry of the rocks

Chemical analyses of 14 anorthosite samples are given in Table 1. The chemical data of all these anorthosites have been plotted in an Alk–F–M diagram (Fig. 5*a*). The positions of the Gruber Mountains, Adirondacks, Sittampundi, Perinthatta, Bushveld, Fiskenaesset and Sakeny anorthosites are shown in Fig. 5*b*, along with plots of two samples of lunar anorthosite for comparison. Table 2 shows chemical analyses of Gruber Mountains anorthosites as compared to those from several other areas of the world.

The chemical analyses of Gruber Mountains anorthosites are also plotted on a variation diagram ($CaO + Al_2O_3$)–SiO_2–($Na_2O + K_2O$) (Fig. 6). It is interesting to note that all the plots cluster in such a manner as to suggest silica enrichment of the anorthosites. This is a unique feature of Gruber Mountains anorthosites compared with the plots of other terrestrial anorthosites.

Indian geologists first found an unusual occurrence of loose pebbles of basalt at the bottom of a small gully (Raina, Kaul & Chakraborty, 1985) in the western-central part of the Schirmacher Hills. *In situ* outcrops were subsequently traced to a thin dyke of a basaltic rock (average width approximately 50 cm) cutting across augen gneiss. Although it is yet to be discovered how these perfectly rounded pebbles or nodules were formed, a preliminary investigation revealed that the rock is olivine-rich. Chemical analyses of two samples of the same basalt, when plotted in a TiO_2 versus FeO/MgO diagram (Fig. 7), have affinity with the 'oceanic island basalt' field. This observation gains support when the samples are plotted in a TiO_2–K_2O–P_2O_5 diagram (Fig. 8).

Structure

The disposition of outcrops between the shelf ice–continental ice contact in the north and the Wohlthat Mountains in the south is such that they are restricted to the south-western side of a presumed line which is parallel to the NW–SE-trending major lineament in Wohlthat Mountains (identified with the help of satellite imagery and passing along the eastern flank of the Petermann Ranges – Fig. 1). Schirmacher Hills, and Tallaksenvarden, Stenersenknatten, Hauglandtoppen, Andersensata, Baalsrudfjellet, Sonstebynuten, Pevikhornet and Starheimtind nunataks, all lie west of this presumed line which, if extended, will touch the easternmost extremity of the Schirmacher Hills and the northern tip of the Sonstebynuten nunatak. The authors agree with Ravich & Kamenev (1975) that major tectonic dislocations may be due to large-scale block faulting in the region. The NW–SE fracture, coupled with major ENE–WSW and NE–SW lineaments, might have been genetically related to the emplacement of anorthosites within the rocks of granulite facies in Queen Maud Land. The most common direction of shear planes marked by mylonitization in the Schirmacher Hills and the Wohlthat Mountains, is ENE–WSW, indicating their affinity with the longest traced lineament passing through the northern limits of the Petermann Ranges and abutting against a NE–SW lineament, the latter demarcating the western edge of the Gruber Mountains.

On the south-eastern margins of Lake Nedresjoen there are deposits of lacustrine clay which show warping along a NW–SE axis. This is probably a manifestation of neotectonic activity in the area.

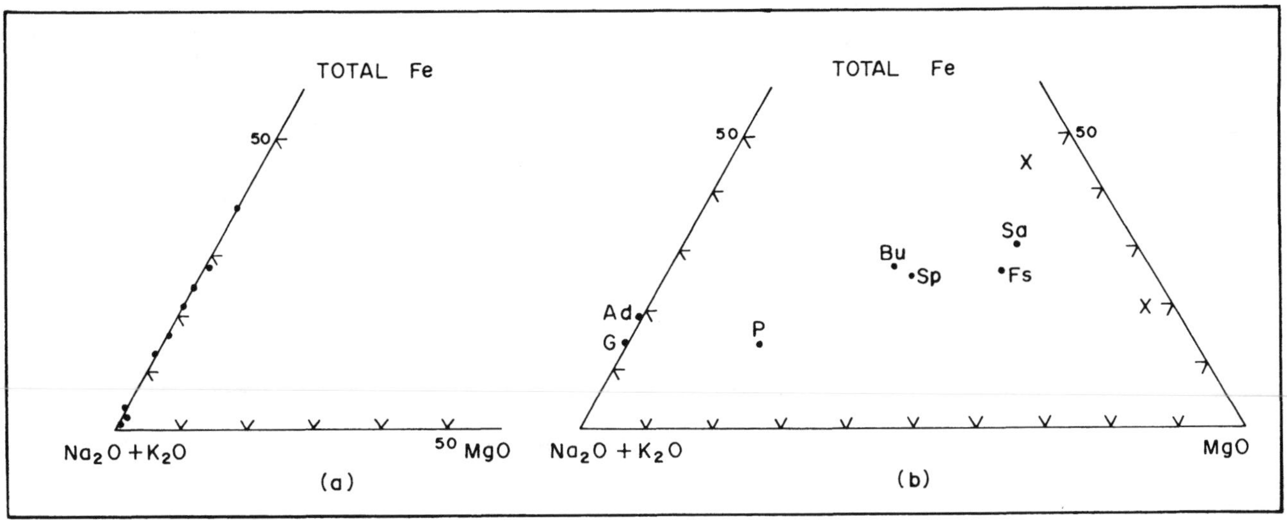

Fig. 5. (*a*) Total Fe–(Na_2O + K_2O)–MgO variation diagram showing plots of anorthosite samples from the Gruber Mountains, Antarctica: (*b*) Total Fe–(Na_2O + K_2O)–MgO variation diagram showing plots of anorthosites from G, Gruber Mountains, Antarctica; Ad, Adirondacks, USA; P, Perinthatta, India; Sp, Sittampundi, India; Bu, Bushveld, South Africa; Fs, Fiskenaesset, Greenland; Sa, Sakeny, Malagasy; X, lunar surface. G, Ad and P are plotted from chemical analyses given in Table 2.

Table 2. *Chemical composition of anorthosites from various parts of the world*

Oxides	Sample no. 1	2	3	4	5
SiO_2	52.89	50.28	44.35	55.90	52.40
TiO_2	0.82	0.64	0.05	0.24	0.40
Al_2O_3	27.41	25.86	31.34	27.20	27.11
Fe_2O_3	1.93[a]	0.96	1.06	—	0.49
FeO	—	2.07	0.82	1.28[b]	1.16
MnO	0.03	0.05	0.02	0.02	0.01
MgO	0.05[c]	2.12	1.12	0.21	1.05
CaO	11.15	12.48	18.18	9.65	8.77
Na_2O	3.43	3.15	1.22	5.23	5.78
K_2O	0.84	0.65	0.05	1.10	0.96
P_2O_5	0.10	0.09	0.01	—	0.27
BaO	0.02	—	—	—	—
S	0.05	—	—	—	—
CO_2	—	0.14	0.20	—	—
H_2O^+	—	1.17	0.62	—	—
H_2O^-	—	0.14	0.10	—	—
LOI	0.63	—	—	—	—
Total	99.35	99.80	99.14	100.83	98.40

Key to sample numbers:
1 Mean chemical composition of 14 anorthosites from the Gruber Mountains, East Antarctica.
2 Mean chemical composition of 104 anorthosites (after Le Maitre, 1976).
3 Mean chemical composition of three Archaean anorthosites from Sittampundi Complex, Madras, India (after Subramaniam, 1956); also contains 0.01% Cr_2O_3.
4 Chemical composition of massif-type anorthosite from Tahawas Marcy Massif, Adirondack Mts, USA (after Simmons & Hanson, 1978).
5 Chemical composition of two massif-type anorthosites from Perinthatta, Kerala, India (after Nambiar *et al.*, 1982).

[a]Total Fe as Fe_2O_3.
[b]Total Fe as FeO.
[c]MgO detected in two samples only out of 14.

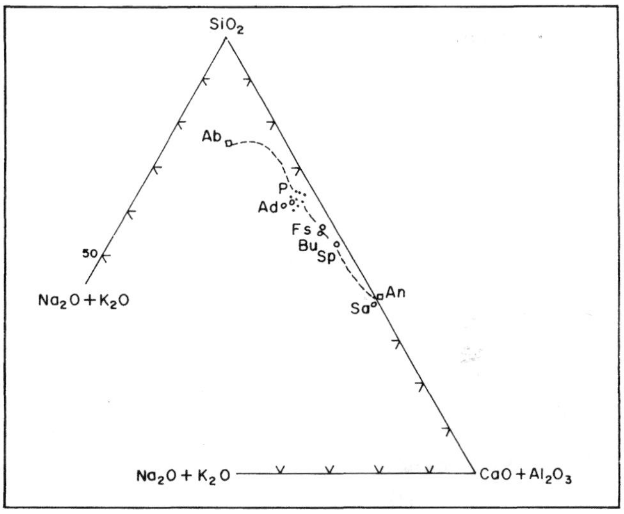

Fig. 6. SiO_2–(Na_2O + K_2O) – (CaO + Al_2O_3) variation diagram showing plots of Gruber Mountains anorthosites (solid circles), and anorthosites from other areas (open circles). Locality abbreviations as given in Fig. 5(*b*). Ab, albite; An, anorthite.

Conclusions

Gneisses from the Wohlthat Mountains have textural evidence of prograde metamorphism under granulite facies conditions. The model reactions involved are likely to be:

$$\text{Orthopyroxene} + \text{plagioclase} \rightleftharpoons \text{clinopyroxene} + \text{garnet} + \text{quartz} \quad (1)$$

$$\text{Clinopyroxene} + \text{plagioclase} \rightleftharpoons \text{garnet} + \text{quartz} \quad (2)$$

$$\text{Biotite} + \text{quartz} \rightleftharpoons \text{orthopyroxene} + \text{K-feldspar} + \text{vapour} \quad (3)$$

Reaction (3) explains the biotite–quartz symplectite found in enderbite.

Textures indicative of a prograde character of granulite facies metamorphism have not been found in the rocks of the

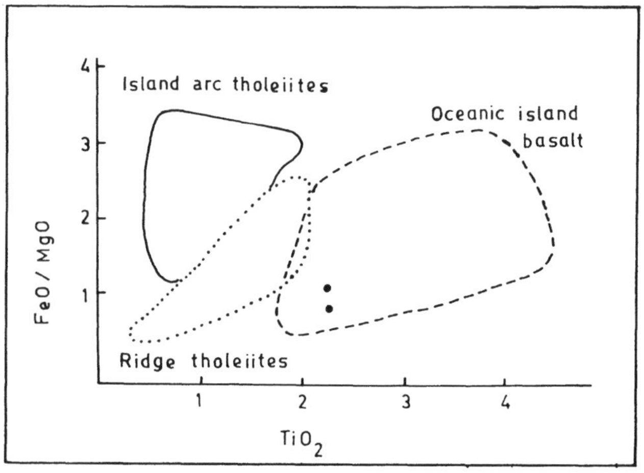

Fig. 7. FeO/MgO versus TiO_2 diagram showing two basalt samples from the Schirmacher Hills, Antarctica, plotting in the field of 'oceanic island basalt' (after Glassley, 1974).

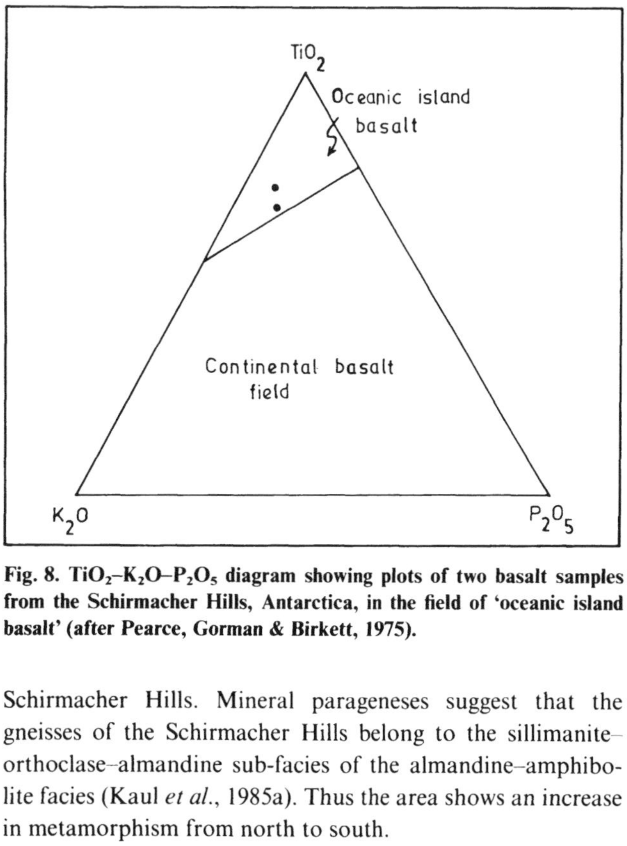

Fig. 8. TiO_2–K_2O–P_2O_5 diagram showing plots of two basalt samples from the Schirmacher Hills, Antarctica, in the field of 'oceanic island basalt' (after Pearce, Gorman & Birkett, 1975).

Schirmacher Hills. Mineral parageneses suggest that the gneisses of the Schirmacher Hills belong to the sillimanite–orthoclase–almandine sub-facies of the almandine–amphibolite facies (Kaul *et al.*, 1985a). Thus the area shows an increase in metamorphism from north to south.

A comparative study of anorthosites of the Gruber Mountains with those of other areas of the world reveals that the Perinthatta anorthosites from Kerala State of South India (Nambiar, Vidyadharan & Sukumaran, 1982) are petrographically similar to anorthosites of the Gruber Mountains. Chemical analyses show that both have iron-poor, alkali-rich environs and both have a K_2O/CaO ratio of 0.1, corresponding to that of the Adirondack anorthosites. It is, therefore, tentatively suggested that the anorthosites of the Gruber Mountains of East Antarctica are the Proterozoic massif type. Their total absence in the areas north of the Gruber Mountains links their spatial distribution with a tectonic boundary defined by the ENE–WSW-trending lineament.

The basalt dyke found in the Schirmacher Hills and containing olivine-rich nodules is of the 'oceanic island basalt' type. Such an oceanic character of continental rocks may be related to a major flexure or tensional feature approximately parallel to an adjacent mid-oceanic ridge (Gibson, 1966).

Acknowledgements

This work was carried out during the fifth Indian expedition to Antarctica organized by the Department of Ocean Development (DOD), New Delhi. Thanks are due to the Secretary, DOD, for providing all the necessary support to carry out these studies in Antarctica. M.K.K. is grateful to the Director General, Geological Survey of India, for allowing him to present this paper at the 5th International Symposium on Antarctic Earth Sciences held at Cambridge.

References

Gibson, I.L. (1966). Crustal flexures and flood basalts. *Tectonophysics*, 3, 447–56.

Glassley, W. (1974). Geochemistry and tectonics of Crescent volcanic rocks, Olympia Peninsula, Washington. *Geological Society of America Bulletin*, **85**, 785–94.

Kaul, M.K., Chakraborty, S.K. & Raina, V.K. (1985a). Petrography of the Biotite–Hornblende Bearing Quartzofeldspathic Gneisses from Dakshin Gangotri, Schirmacher Hill, Antarctica. *In Scientific Report of the Second Indian Expedition*. Department of Ocean Development Technical Publication, New Delhi, No. 2, pp. 29–38.

Kaul, M.K., Chakraborty, S.K. & Raina, V.K. (1985b). On the Amphibolites from the Indian Research Station at Antarctica. *In Scientific Report of the Second Indian Expedition*. Department of Ocean Development Technical Publication, New Delhi, No. 2, pp. 15–22.

Le Maitre, R.W. (1976). The chemical variability of some common igneous rocks. *Journal of Petrology*, **17(4)**, 589–637.

Nambiar, A.R., Vidyadharan, K.T. & Sukumaran, P.V. (1982). Petrology of anorthosites of Kerala. *Record Geological Survey of India*, **144(5)**, 60–70.

Pearce, T.H., Gorman, B.E. & Birkett, T.C. (1975). The TiO_2–K_2O–P_2O_5 diagram. A method for discriminating between oceanic and non-oceanic basalts. *Earth and Planetary Sciences Letters*, **24**, 419–26.

Raina, V.K., Kaul, M.K. & Chakraborty, S.K. (1985). On the Nature of the Basic Rock Erratics from the Dakshin Gangotri Landmass, Antarctica. *In Scientific Report of the Second Indian Expedition*. Department of Ocean Development Technical Publication, New Delhi, No. 2, pp. 11–14.

Ravich, M.G. & Kamenev, E.N. (1975). *Crystalline Basement of the Antarctic Platform*. New York; John Wiley & Sons. 582 pp.

Simmons, E.C. & Hanson, G.N. (1978). Geochemistry and origin of massif type anorthosites. *Contributions to Mineralogy and Petrology*, **66**, 119–35.

Subramaniam, A.P. (1956). Mineralogy and Petrology of Sittampundi complex, Salem District, Madras State, India. *Geological Society of America Bulletin*, **67**, 317–90.

Walker, T.L. (1902). Geology of Kalahandi State, Central Provinces. *Memoir Geological Survey of India*, **33(3)**.

Structural and petrological evolution of basement rocks in the Schirmacher Hills, Queen Maud Land, East Antarctica (Extended abstract)

S. SENGUPTA

Department of Geological Sciences, Jadavpur University, Calcutta 700032, India

Introduction

The Precambrian rocks of the Schirmacher Hills area, belonging to the East Antarctic Charnockitic Province (Ravich & Kamenev, 1975), record a complex history of repeated metamorphism, migmatization, folding and shearing. Compositional and textural differences can be used to subdivide the gneissic complex into six units in this area: (i) streaky gneiss; (ii) augen gneiss; (iii) calc-gneiss, khondalite and associated migmatites; (iv) garnet–biotite gneiss; (v) garnetiferous alaskite; and (vi) banded gneiss with intercalated pyroxene-granulite, charnockite and quartzo-feldspathic gneiss (Fig. 1). The present paper outlines the interrelationships between the structural and metamorphic events in the Schirmacher Hills area.

Metamorphism and migmatization

The rocks have been affected by at least two episodes of metamorphism and migmatization: an early granulite-facies metamorphism and charnockitization, followed by an amphibolite-facies metamorphism and granitization. The initial granulite-facies metamorphism gave rise to pyroxene-granulite, calc-gneiss and khondalite, and was closely followed by a period of charnockitization. The pyroxene-granulites, with an orthopyroxene–clinopyroxene–plagioclase assemblage, have a xenoblastic granular texture and a crude foliation. Fe–Mg partitioning of coexisting ortho- and clinopyroxenes (Wood & Banno, 1973; Wells, 1977) indicate an average crystallization temperature of about 850 °C. The temperature of the same samples calculated from Ca solubility in clinopyroxene coexisting with orthopyroxene (Kretz, 1982) would be around 750 °C (cf. Grew, 1983; Sengupta, 1988). Calc-gneiss (plagioclase–garnet–diopside–quartz with scapolite, calcite and sphene) and khondalite (K-feldspar–quartz–sillimanite–garnet with minor plagioclase and graphite) are closely associated. The initial compositional layering of these rocks has been strongly modified by transposition, lit-par-lit migmatization and ductile shearing on a mm- and cm-scale.

The charnockites, associated mostly with the banded gneisses, are typical migmatites and show gradational contacts with both the mafic granulites and quartzo-feldspathic gneisses. The rocks are commonly traversed by concordant and discordant pegmatites which have diffuse borders and contain coarse crystals of hypersthene and garnet. Local variations in the composition of the charnockites are considerable. Common assemblages are hypersthene (hy)–garnet (gt)–perthite (perth)–plagioclase (plag)–quartz (qtz), hy–gt–perth–plag–qtz, hy–diopside–gt–plag–qtz–perth. The majority of the rocks are garnetiferous and enderbitic.

The superposed amphibolite-facies metamorphism produced polymetamorphic assemblages. Complete retrogression occurred mostly within zones of ductile shearing, e.g. pyroxene has been replaced by biotite–amphibole, garnet by biotite, and sillimanite by kyanite–muscovite.

Amphibolite-facies metamorphism broadly coincided with extensive granitization and the development of a variety of

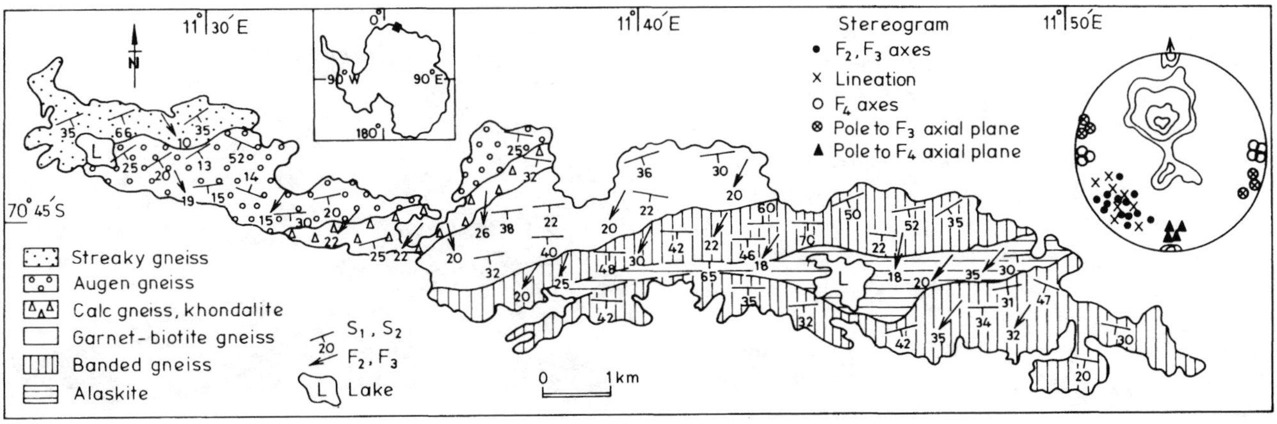

Fig. 1. Geological map of the Schirmacher Hills. Insets show location, and lower-hemisphere equal-area projection of structural data; contours of 227 S_2 foliation poles at 2, 5, 10 and 20% intervals.

Fig. 2. (a) Veined gneiss with isoclinal F_2 folds partly replaced by a more homogeneous gneiss. (b) Agmatite with fragments of pyroxene-granulites and migmatitic banded gneiss in a later granitic orthogneiss, central Schirmacher Hills.

migmatitic fabrics. Several phases of migmatization are indicated by: veined gneiss invaded by a more homogeneous granitic neosome (Fig. 2*a*), augen gneiss cross-cut by leucocratic gneiss and later pegmatite, and the neosome of an agmatite containing not only paleosome of pyroxene-granulite, calc-gneiss or other metamorphic rocks but also fragments of banded gneiss with an earlier migmatitic banding (Fig. 2*b*; Sengupta, 1988).

Folding

At least four phases of deformation have affected the rocks of the Schirmacher Hills. The earliest deformation (D_1) is represented by a foliation marked by parallel orientation of sillimanite, hypersthene and other granulite-facies minerals, indicating that D_1 was contemporaneous with granulite-facies metamorphism. Relict early isoclinal crenulations marked by sillimanite needles and overgrown by foliation-parallel elongated garnet porphyroblasts have been recognized in some rocks. Since both these generations of S-surfaces are marked by sillimanite needles, and since garnet grains in the khondalites have overgrown the crenulations, both S-surfaces must have developed during the granulite-facies metamorphism. Both deformations are grouped together as D_1 and all pre-D_2 fabrics are designated as S_1.

D_2, the dominant deformation of the area, produced the earliest recognizable mesoscopic folds (F_2) and a prominent fabric (S_2). F_2 folds are isoclinal or very tight, with axial surfaces striking roughly E–W and dipping moderately S, and the fold axes plunge SW (Fig. 1). Two generations of coaxial F_2 folds have been identified. Since both sets of folds have similar style and orientation, they can only be identified where one is overprinted by the other. Hence these folds are grouped together as F_2 folds. Screens of pegmatite with diffuse borders have been emplaced parallel to S_2, indicating that the post-charnockitic granitization is contemporaneous with D_2 (Fig. 3). Newly formed biotite and amphibole define a foliation parallel to S_2, with a mineral lineation parallel to the F_2 fold hinges.

F_2 folds are deformed by two sets of open folds, F_3 and F_4. F_3 folds have steep N–S-striking axial surfaces and axes which are essentially parallel to F_2 hinges. F_4 folds have steep N-dipping axial surfaces with subhorizontal fold axes (Fig. 1). The contoured equal area projection (see Fig. 1) shows S_2 poles clustering around a single maximum as expected in areas of isoclinal folding; the north–south spread of the contours is the result of F_4 folding.

Successive development of shear zones

Shear zones ranging in width from a few millimetres to a few metres occur throughout the area. Almost all outcrops contain narrow discontinuous shear zones enclosing lenses of less sheared rocks. Three categories of shear zones have been recognized:

(i) ductile shear zones parallel to the gneissic banding, with mylonitic foliation parallel to S_2
(ii) subvertical ductile shear zones cutting across S_2 at high angles
(iii) single or rarely conjugate sets of discrete shear fractures whose sense of shear is consistent with an overall extension parallel to S_2

Ductile shearing took place during amphibolite-facies metamorphism, as indicated by recrystallization of small equant grains of diopside in the sheared groundmass of mylonitized calc-gneiss, and recrystallization of fine-grained crystals of garnet in the mylonitized garnet–biotite gneiss. Fe–Mg par-

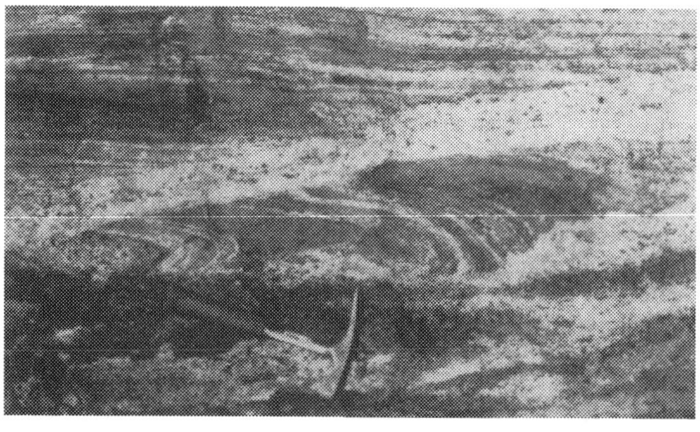

Fig. 3. Emplacement of diffuse bands of coarse-grained quartzo-feldspathic gneiss along F_2 axial planes, western Schirmacher Hills.

Table 1. *Metamorphic–migmatitic–tectonic history of the Schirmacher Hills*

Events	Metamorphism–migmatization	Tectonics
D_1	Granulite-facies, development of charnockites	Initial gneissic foliation and its transposition (F_1)
D_2	Upper amphibolite-facies, extensive migmatization	Isoclinal folding (F_2), ductile shearing
$D_3 + D_4$	Amphibolite-facies, local migmatization	Open, upright folds ($F_3 + F_4$), steep ductile shear zones
	Local pegmatites	Discrete shear fractures

titioning of coexisting recrystallized garnet and biotite in the groundmass (Goldman & Albee, 1977) indicates that the temperature of mylonitization was about 550 °C. Emplacement of pegmatites along the ductile shear zones suggests that the shearing was broadly contemporaneous with a period of granitic activity. Late pegmatitic veins sometimes cut across these ductile shear zones and a small number of pegmatites was emplaced along the discrete shear fractures.

Summary and conclusions

The Precambrian rocks of Schirmacher Hills have been affected by at least two episodes of metamorphism–migmatization and four phases of deformation (Table 1). The initial granulite-facies metamorphism and charnockitization was associated with an early deformation D_1. The temperature of this metamorphism, calculated from two-pyroxene geothermometry, is ~ 800 °C. An amphibolite-facies metamorphism and contemporaneous regional granitization were superimposed on these granulites. The dominant deformation of the area, D_2, was broadly synchronous with the amphibolite-facies metamorphism and included two sets of coaxial isoclinal folds with E–W-striking, southerly dipping axial-planar foliations (S_2). Concurrent ductile shearing took place parallel to S_2. The temperature of mylonitization estimated from the neocrystallized garnet and biotite in the shear zones is around 550 °C. There are also two generations of upright folds, with northerly- and easterly-striking steep axial surfaces (F_3 and F_4), and a set of steep ductile shear zones at high angles to S_2. Discrete shear surfaces cut across all the earlier structures.

References

Goldman, D.S. & Albee, A.L. (1977). Correlation of Mg/Fe partitioning between garnet and biotite with O^{18}/O^{16} partitioning between garnet and magnetite. *American Journal of Science*, **277**, 750–67.

Grew, E.S. (1983). Sapphirine–garnet and associated parageneses in Antarctica. In *Antarctic Earth Science*, ed. R.L. Oliver, P.R. James & J.B. Jago, pp. 40–3. Canberra; Australian Academy of Science and Cambridge; Cambridge University Press.

Kretz, R. (1982). Transfer and exchange equilibria in a portion of the pyroxene quadrilateral as deduced from natural and experimental data. *Geochimica et Cosmochimica Acta*, **46**, 411–21.

Ravich, M.G. & Kamenev, E.N. (1975). *Crystalline Basement of the Antarctic Platform*. New York; John Wiley & Sons. 582 pp.

Sengupta, S. (1988). History of successive deformation in relation to metamorphic–migmatitic events in the Schirmacher Hills, Queen Maud Land, East Antarctica. *Journal of the Geological Society of India*, **32**, 295–319.

Wells, P.R.A. (1977). Pyroxene thermometry in simple and complex systems. *Contributions to Mineralogy and Petrology*, **62**, 129–39.

Wood, B.J. & Banno, S. (1973). Garnet–orthopyroxene and orthopyroxene–clinopyroxene relationships in simple and complex systems. *Contributions to Mineralogy and Petrology*, **42(2)**, 109–24.

Metamorphic evolution of granulites from the Rauer Group, East Antarctica: evidence for decompression following Proterozoic collision

S.L. HARLEY

Department of Earth Sciences, University of Oxford, Parks Road, Oxford, OX1 3PR, UK

Abstract

Granulites from the Rauer Group, Prydz Bay, were metamorphosed at 840 ± 30 °C during an intense deformation event (D_3), which produced flat-lying isoclinal folds in felsic orthogneisses intruded into and interleaved with earlier-formed basic, pelitic and migmatitic composite gneisses. A spatial variation in pressures of metamorphism, from 6–7.5 kb in the northern part of the area to 7–8.5 kb in southern regions, is calculated. All rock types were intruded by basic dykes prior to folding about S-plunging upright folds in D_4, a deformation which increased in intensity southwards and is thought to correlate with the ~ 960 Ma event documented for the Rayner Complex. Post-deformational symplektitic reaction textures in a variety of rock types indicate substantial decompression to calculated pressures of 3–5 kb (northern outcrops) and 5–6 kb (southern outcrops) at 750 °C. This *P–T* evolution, which indicates a uniform unroofing of some 6–9 km while lower–mid-crustal temperatures only decreased by ~ 100 °C, is interpreted to represent the later stages of a prolonged collision-related thermal evolution.

Introduction

The Rauer Group (Fig. 1) comprises a granulite-facies basement region (Sheraton & Collerson, 1983; Harley, 1987) of Late Proterozoic age (1106 ± 110 Ma; Tingey, 1981). Despite its proximity to the Archaean Vestfold Hills, Sm–Nd model age calculations for orthogneisses (Sheraton, Black & McCulloch, 1984) mostly demonstrate Early–Middle Proterozoic crustal accretion and suggest that reworking of the nearby Archaean granulites was of only minor significance. The presence of deformed and metamorphosed mafic dykes (Sheraton & Collerson, 1983; Harley, 1987) demonstrates that the Rauer Group was profoundly affected by Late Proterozoic tectonism, in contrast to rocks from the Vestfold Hills where comparable dykes are undeformed and only locally metamorphosed.

Geological setting and history

The regional setting and geological history of the Rauer Group are discussed in detail in Harley (1987). Three generalized lithological groups have been recognized:

(1) well-layered gneisses dominated by pyroxene-granulites, migmatitic metapelitic paragneisses and quartzitic garnet-gneisses, with less common lenses and boudins of aluminous paragneiss, calc-silicates, and ironstones. These sequences and relicts preserve early (D_1, D_2) foliations and folds
(2) composite layered mafic–felsic-gneisses comprised of migmatitic mafic pyroxene granulite intimately interleaved with, and veined by, gneisses of intermediate and felsic composition
(3) felsic–intermediate orthogeneisses which preserve intrusive contacts with (1) and (2) in areas of low (D_3–D_4) strain and show occasional relict igneous structures, which are usually obscured by structures produced in an intense deformation, D_3.

These major groups were interleaved and folded about major (0.5–2 km), NW–SE-trending, reclined–recumbent isoclines during D_3. This deformation is inferred to be Late Proterozoic in age from the isotopic constraints on orthogneiss provenance (Sheraton *et al.*, 1984), but predated the emplacement of:

(4) local leucogranite stocks and lenses
(5) metamorphosed mafic dykes which intrude lithologies (1)–(4), and therefore are the youngest pervasively deformed rock types in the area. If they correspond to the ~ 1400–1200 Ma undeformed dyke swarms in the Vestfold Hills, the D_3 event in the Rauer Group must significantly predate the 960 Ma metamorphism documented by Black *et al.* (1987) for the Rayner Complex. The dykes and all other lithologies were folded about

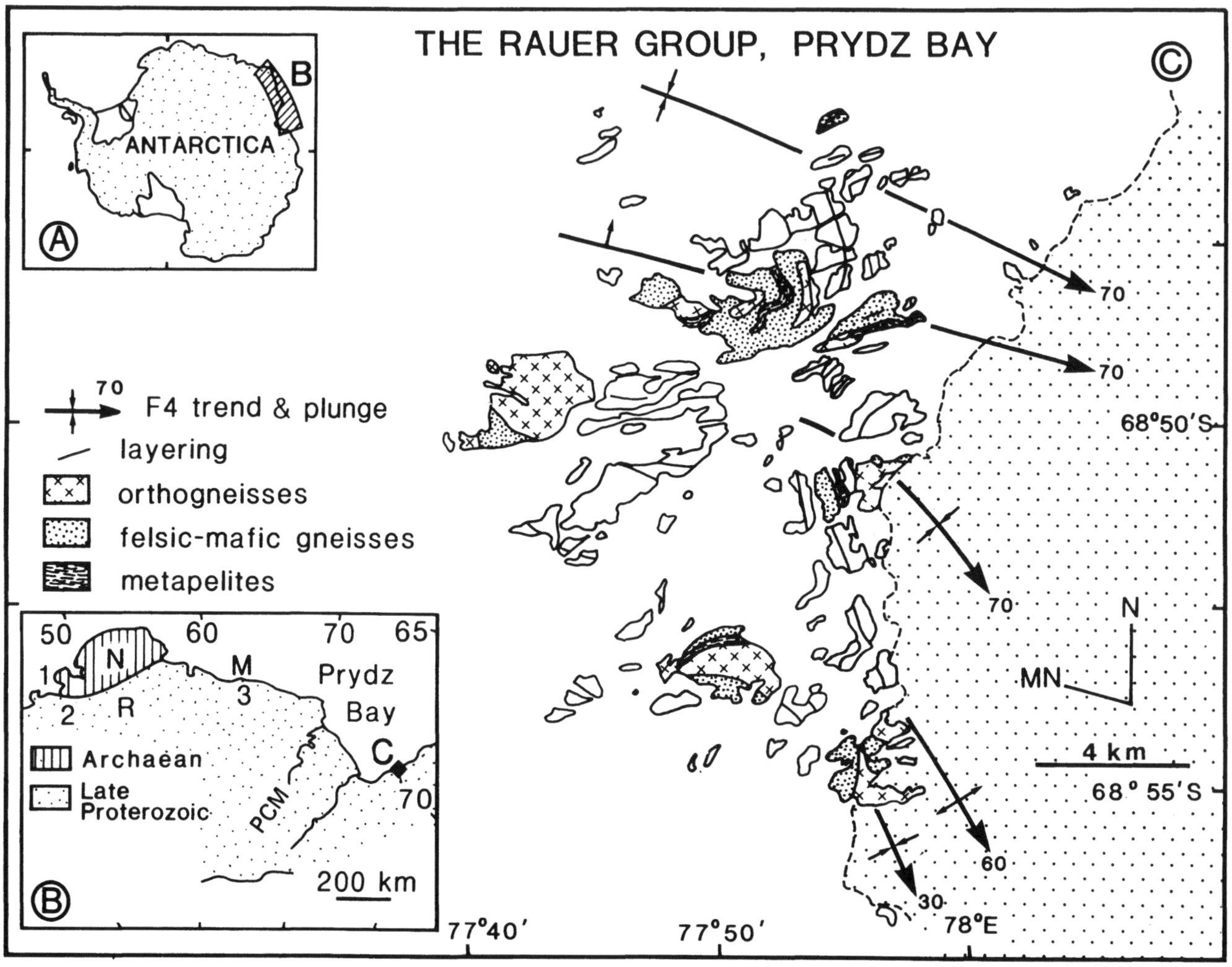

Fig. 1. Sketch maps to show the general geology and distribution of the Rauer Group rocks. (A) Antarctica showing the locations of (B). (B) distribution of Archaean and Proterozoic rocks in East Antarctica: N, Napier Complex; R, Rayner Complex; M, Mawson Station; PCM, Prince Charles Mountains; C is position of main map. (1) Napier shear zones; (2) Nye Mountains; (3) Mawson Station. (C) Geological sketch map of the Rauer Group (open stipple, ice sheet).

open–tight, SE-plunging, upright folds in D_4. This deformation increased in intensity southwards and probably defines the major structures farther south-west in the Prydz Bay region.

Mineral assemblages, chemistry, and reaction features

Mafic pyroxene granulites

Quartz-absent types may contain garnet (gt) and pargasitic amphibole (amp) in addition to pyroxenes (cpx, opx), plagioclase (plag), and ilmenite (ilm), and therefore belong to the medium-pressure granulite-facies of Green & Ringwood (1967). Garnet occurs as a porphyroblast phase only in maficgranulites of intermediate or low X_{Mg} (= Mg/Mg + Fe) (Fig. 2*a*). Important chemical and textural features are:

(a) consistency in the distribution of Fe and Mg between garnet and other ferromagnesian phases
(b) garnets range up to $X_{Mg} = 0.42$, but typically show rimward zoning to lower X_{Mg} at near-constant grossular content (Ca/Ca + Mg + Fe = X_{gr}) (Fig. 2*b*)
(c) Al contents of initial pyroxenes are low, and plagioclase cores are typically An_{40-70} (An = (Ca/(Ca + Na + K) × 100)
(d) symplektites of optically continuous plagioclase and orthopyroxene, often with additional opaque minerals, are developed between garnet and amphibole or clinopyroxene. Resorbed garnets show the zoning features outlined above whereas the symplektite phases are notably homogeneous. Late orthopyroxene is marginally more aluminous than primary grains (Fig. 2*c*), and the secondary plagioclase is more calcic than early cores (An_{74-95}) (Fig. 2*b*). These textural and chemical features are consistent with the production of opx + Ca-plag from gt + cpx + amp) on decompression and with minor oxidation

Garnet occurs in Fe-rich quartz-bearing mafic pyroxene granulites (Fig. 2*a*), but is always more Fe-rich than $X_{Mg} = 0.20$. Zoning and textural relationships are similar to

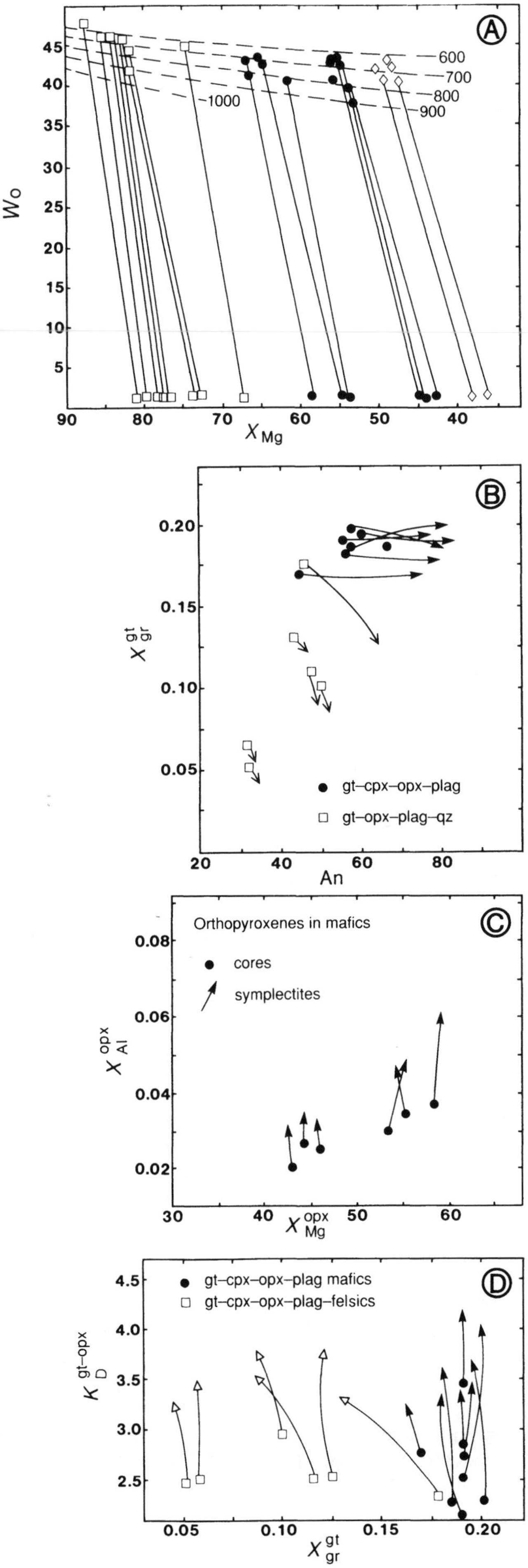

Fig. 2. (A) Compositions of coexisting pyroxenes in mafic granulites from the Rauer Group (Wo – corrected $Ca_2Si_2O_6$ component). Isotherms (in °C) relevant to clinopyroxene compositions are from Lindsley (1983). Open squares, garnet-absent; solid circles, quartz-absent, garnet-bearing; open diamonds, garnet + quartz-bearing. (B) Zoning relationships between garnet and plagioclase in mafic (solid circles) and felsic (open squares) granulites; X^{gt}_{gr} = Ca/(Ca + Mg + Fe + Mn) in garnet; An = (Ca/(Ca + Na)) × 100 in plagioclase. (C) Compositional trends in orthopyroxenes from garnet-bearing (quartz-absent) mafic granulites. (D) K^{gt-opx}_{D} data for mafic and felsic granulites. Arrowed lines show the rimward increase in K_D for each sample, often associated with corona or symplektite textures.

those in the quartz-absent granulites; however, late orthopyroxene also commonly occurs as rims and mantles on clinopyroxene, consistent with the generalized reactions below occurring as a result of decompression:

$$\text{gt + quartz (qz) = plag + opx} \qquad \text{(GOPS)}$$

and,

$$gr_{33}py_{67} + \text{diopside} = gr_{67}py_{23} + \text{enstatite}, \qquad \text{(CAMG)}$$

where gr and py are the Ca and Mg components of garnet.

Felsic orthogneisses

Charnockites are subordinate to enderbites containing the assemblage gt + opx + plag + qz (+ biotite (bi) + ilm + K-feldspar (ksp)) which show the following features:

(a) the more calcic garnets (up to $X_{gr} = 0.18$) coexist with higher An plagioclase (cores up to An_{50}), consistent with buffering by the (GOPS) reaction
(b) the Al contents of orthopyroxenes vary inversely with X_{gr} of coexisting garnet
(c) plagioclase moats and rims occur between calcic garnet and orthopyroxene, reflecting the progress of (GOPS) to the right with decompression. Chemical effects associated with the moats include: (i) rimward zoning of garnet to lower X_{Mg} and X_{gr} (Fig. 2); (ii) rimward zoning of orthopyroxene to higher X_{Mg}, often with a minor increase in Al content; and (iii) the formation of calcic moat plagioclase (An_{50-70}). K_D (gt–opx) is higher for rim compositions (3.3–3.8) (Fig. 2*d*) than for cores (2.4–2.9), suggesting cooling either accompanying moat formation or post-dating it. Maximum K_D values obtained from gt–opx rim pairs still in local contact suggest that moat formation occurred prior to much of the inferred cooling.

Metapelites

Garnet–sillimanite (sill) gneisses usually also contain biotite (early and secondary), K-feldspar, plagioclase, quartz, rutile (rut), and ilmenite, so that the continuous reactions:

$$\text{gt + rut = ilm + sill + qz}, \qquad \text{(GRAIL)}$$

and

$$\text{gt (gr) + sill + qz = plag}, \qquad \text{(GASP)}$$

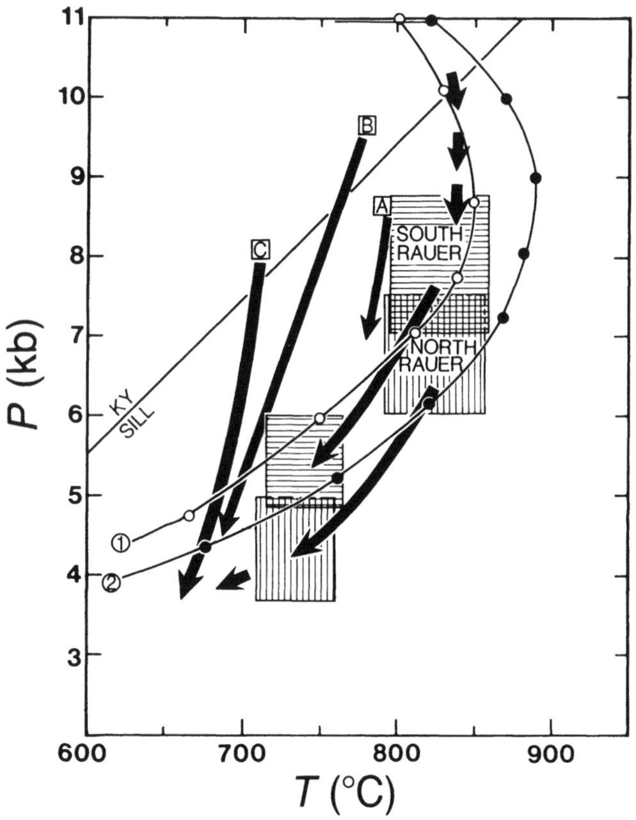

Fig. 3. Pressure (*P*)–temperature (*T*) paths of granulites from the Rauer Group (boxes and unlabelled arrowed lines) and other Proterozoic granulites. (A) Lützow–Holm Bay granulites (Hiroi *et al.*, 1987). (B) Nye Mountains and Turbulence Bluffs (Ellis, 1983; Black *et al.*, 1987). (C) schists in retrograde shear zones within the Napier Complex (Harley, 1985).

Lines (1) and (2) are calculated erosion-controlled *PTt* paths (after England & Thompson, 1984) with 10 Ma time intervals indicated by circles: 1, path for rocks initially 10 km below a thrust at 30 km depth, with conductivities, *k*, of 1.5 W/m//K, and heat source distribution II; 2, path for the case of homogeneous thickening (doubling) of 30 km crust alone, with *k* = 1.5 W/m/K and heat source distribution III.

control the main phase relations. Garnet porphyroblasts (X_{Mg} = 0.26–0.31) show only minor rimwards zoning in X_{Mg}, but may be zoned to lower X_{gr} adjacent to plagioclase, consistent with the progress of the (GASP) reaction to the right with decompression. Ilmenite is also mostly a late phase, consistent with the progress of the (GRAIL) reaction to the right on decompression.

Gt + cordierite (cd) + opx + qz, K-feldspar-absent assemblages contain relatively magnesian garnets (X_{Mg} = 0.42–0.47 in cores). The occurrence of symplektites of cordierite, orthopyroxene and ilmenite on ragged, embayed garnets showing rimward zoning to lower X_{Mg} (0.4–0.37) (Fig. 3) is consistent with the progress of the continuous reaction:

gt + qz = opx + cd, (COGS)

from left to right with decreasing pressure.

The most magnesian garnets yet recorded from Proterozoic granulite metaplites in East Antarctica (X_{Mg} = 0.53) occur as porphyroblasts in a quartz-absent gneiss from Long Point (Harley, unpublished data). These blasts are partly wrapped by a foliation involving biotite, sillimanite, cordierite, and minor green spinel. The garnets are strongly zoned to lower X_{Mg} (0.42) and X_{gr} adjacent to coronas of cordierite and spinel, consistent with the progress of the continuous reaction:

gt + sill = cd + sp,

from left to right in response to decompression.

Calc-silicates

Quartz-bearing calc-silicates contain the high-temperature assemblages (780–830 °C; Harley, 1987) wollastonite + calcite + scapolite (90–94% meionite) + diopside and wollastonite + plag (An_{97}) + scapolite (92% meionite). Scapolite is partially pseudomorphed by symplektites of calcite + plag + qz, consistent with crossing of the reaction:

meionite = anorthite + calcite,

with cooling below 780 °C. Grossular garnet occurs as secondary grains formed in the symplektites, as rims between scapolite and diopside, or associated with late zoisite in pseudomorphs after scapolite. These features indicate the operation of internally buffered conditions during cooling, and constrain X_{CO_2} to be < 0.2 (Harley, 1987 & unpublished data).

Pressure–temperature estimates and post-D_3 *P–T* paths

Textural, chemical and phase equilibria cited in the previous sections qualitatively indicate that metamorphic temperatures of at least 800 °C were attained in D_3, and that the post-deformational metamorphic *P–T* history in the Rauer Group was characterized by decompression at temperatures > 700 °C. Mineral compositional data on cores, rims, and homogeneous symplektites or coronas have been used in the estimation of the pressures and temperatures relevant to early (pre- or syn-D_3) and late (post-$D_{3/4}$) assemblages (Table 1 & Fig. 3). As local near-rim Fe–Mg exchange may continue to lower temperatures than those relevant to the formation of coronas and symplektites, average compositions of the new phases (e.g. opx) have been coupled with near-rim compositions of reactant phases (e.g. gt) adjacent to phases which do not participate in Fe–Mg exchange (e.g. plag) in the calculation of probable late *P–T* conditions. Maximum K_D data can be shown to reflect temperatures somewhat lower than those at which decompression proceeded.

Temperature estimates

Metabasites yield the best estimates for the maximum temperatures of metamorphism. Whilst pyroxene miscibility thermometry (Lindsley, 1983) gives scattered temperature estimates in the range 900–700 °C, gt–cpx Fe–Mg exchange thermometry (Ellis & Green, 1979) indicates that most cores were in equilibrium at 860 ± 30 °C (K_D = 3.5–4.3). Estimates based on gt–opx (Harley, 1984a) and gt–amp (Graham & Powell, 1984) Fe–Mg partitioning are some 20–60 °C below the gt–cpx estimates.

Symplektites in metabasites developed at 750 ± 40 °C,

Table 1. *Summary of P–T estimates*

Method	Rock type	Cores	Rims	Coronas/symplektites
Temperatures (°C)				
gt–cpx FeMg	Mafic	920–810	790–730	790–730
gt–opx FeMg	Mafic	900–770	750–650	780–680
	Felsic	820–750	680–620	790–700
	Pelite	820–750	650	700–650
gt–bi FeMg	Felsic + pelite	840–750	730–620	—
gt–cd FeMg	Pelite	—	660–600	660–600
Pressures (kb) – southern Rauer Group				
GOPS	Mafic	< 9–10	—	< 6.5–7.5
	Felsic	6.5–8	—	5–6
gt–opx Al	Mafic	9.5–11	—	6.3–7.5
	Felsic	7–10	—	6–7.5
GRAIL	Pelite	8.0	< 7.4	—
GASP	Pelite	< 9.5	< 7.0	< 6.5
Pressures (kb) – northern Rauer Group				
GOPS	Mafic	< 7.5–9	—	< 5–5.7
	Felsic	5.2–6.5	—	3.2–5
gt–opx Al	Mafic	> 8.0	—	6–7.5
	Felsic	6.5–8	—	3–7
	Pelite	6.8	—	3.3–4.5
COGS	Pelite	7.8	—	< 6.5

Methods: gt–cpx . . . Ellis & Green (1979); gt–opx FeMg . . . Harley (1984a); gt–bi . . . Thompson (1976); gt–cd . . . Thompson (1976); GOPS . . . Perkins & Newton (1981); gt–opx Al . . . Harley (1984b); GRAIL . . . Bohlen, Wall & Boettcher (1983); GASP . . . Newton & Haselton (1981).

according to gt–cpx rim data and the higher range of gt–opx temperatures derived using average symplektite data. Plagioclase moats in the felsic granulites may have formed at 710 ± 40 °C; however, this estimate is based on gt–opx thermometry of phases no longer in contact. Gt–opx grains still in mutual contact equilibrated to lower temperatures (620–680 °C).

Metapelites yield gt–opx temperatures of 750–820 °C for core data. Estimates of 650–700 °C for opx–cd coronas on relict garnet rims probably reflect continued Fe–Mg exchange postdating corona formation. Gt–bi thermometry (Thompson, 1976) (Table 1), in the range 840–750 °C for core garnet compositions, may not be realistic because of re-equilibration of matrix biotite. Gt–bi temperatures based on rim data (730–640 °C) provide minimum constraints on post-decompressional metamorphic conditions.

Pressure estimates

Despite containing gt + opx + plag, the metabasites are relatively poor for geobarometry because: (a) gt–opx barometry requires major extrapolation from its experimental base (Harley, 1984b) and involves large uncertainties because of the low Al contents in opx; and (b) only maximum pressure estimates can be obtained from gt–pyroxene–plag–qz barometers (GOPS) in the absence of quartz. Initial maximum pressures defined by GOPS barometers were higher in the southern Rauer Group (8–10 kb) than in the northern areas (6.8–9 kb). The maximum GOPS pressures calculated for symplektites at 750 °C are uniformly lower than these initial maximum pressures by 1.5–2.2 kb. Although gt–opx barometry often yields somewhat scattered absolute pressure estimates, it also indicates decompression through at least 2 kb (or up to 6 kb).

Gt–opx barometry of felsic orthogneisses yields imprecise results which suggest decompression through 1–2 kb from initial pressures of 8 ± 2 kb (southern Rauer Group) and 7 ± 1.5 kb (northern areas). GOPS barometry yields precise pressure estimates which are generally some 1.7–2.2 kb less than the maximum pressures derived from metabasites. Thus, the southern Rauer Group orthogneisses indicate decompression from 8–6 kb to 4.5–6 kb, while decompression from 7–5 kb to 3.5–5 kb occurred in the northern areas.

Barometry of metapelites supports the results outlined above. GRAIL and GASP barometry (southern Rauer Group) yield pressures of 7.5–8.5 kb. Gt–opx barometry of gt + cd + opx + qz assemblages (northern Rauer Group) yields initial pressures of 6.4–7 kb followed by corona growth at 4.5–3.3 kb.

The barometry of a diverse range of granulites therefore shows that post-D_3/D_4 decompression through at least 2 kb occurred throughout the Rauer Group. Initial pressures of at least 7–8.5 kb in the southern areas, and 6–7.5 kb in the northern outcrops, imply a small regional depth gradient in the syn-D_3 metamorphism.

Pressure–temperature paths (P–T paths)

Late Proterozoic *P–T* paths for the Rauer Group (Fig. 3) are characterized by decompression from lower- to mid-crustal levels accompanied by minor cooling through 50–100 °C to 750 °C. As most of the recognized decompression features are post-deformational, an important question is whether significant decompression from the late D_3 conditions occurred prior to D_4. Post D_3 leucogranites (group 4) crystallized at 700–760 °C and 4–6 kb (Harley, unpublished data), whilst metabasite dykes (group 5) of intermediate X_{Mg} suggest maximum pressures of 5 kb at 750–800 °C. Both these observations qualitatively imply that some, if not most, of the decompression occurred prior to D_4. There is some indication of higher pressure conditions ($\simeq$ 10 kb) prior to D_3 in the core compositions of garnets from pre-migmatite metabasites in the composite gneisses (group 2 above). Thus, the *P–T* paths derived here from studies of reaction textures are thought to reflect the later segments of a more extensive and lengthy D_3–D_4 decompressive *P–T* history, denoted by the dashed arrows in Fig. 3.

Discussion and conclusions

P–T paths derived for the Rayner Complex in Enderby and Kemp lands (Ellis, 1983; Black *et al.*, 1987), the Lützow–Holm Bay granulites (Hiroi *et al.*, 1987), and ~ 1000 Ma shear zones within the Napier Complex (Harley, 1985) are qualitatively similar to those determined for the Rauer Group granulites (Fig. 3). *P–T* paths involving near-isothermal decompression are therefore typical of the Late Proterozoic granulite complexes of a large part of East Antarctica. Variations occur in the temperatures at which decompression progressed in different regions. This is believed to reflect variations in the extent of magmatism, basal heat flux or deep heat source distribution, and the age and composition of pre-existing lower crust remobilized in the Late Proterozoic events. The age and timing of the metamorphism and decompression may also vary. For example, older ages in shear zones within the Napier Complex (1070 Ma; Black, Fitzgerald & Harley, 1984) compared with ages obtained from the Rayner Complex (Black *et al.*, 1987) might reflect the formation of the shears in an upper plate offset from, and little affected by, late magmatic events occurring mostly within the region of maximum crustal thickening.

The observed and inferred *P–T* paths are compared with specific *PTt* paths calculated for continental collisional zones in Fig. 3. The Rauer Group and other Late Proterozoic *P–T* paths are consistent with the later stages of the thermal evolution of crust tectonically thickened to at least 50 km and allowed to relax, mainly through erosion modified by the internal and upward redistribution of heat through magmatism. The initial high heat flux (*Q*) boundary conditions required in the models (Harley, this volume, p. 7) might be provided if the collision involved previously thinned and young crust–lithosphere. The final phase of decompression would post-date collision by 40–70 m.y., implying a short D_3–D_4 time interval. This basic tectonic model can therefore be tested through isotope geochronology of rock groups (3), (4) and (5).

References

Black, L.P., Fitzgerald, J.D. & Harley, S.L. (1984). Pb isotopic composition, colour, and microstructure of monazites from a polymetamorphic rock in Antarctica. *Contributions to Mineralogy and Petrology*, **85(2)**, 141–8.

Black, L.P., Harley, S.L., Sun, S.S. & McCulloch, M.T. (1987). The Rayner Complex of East Antarctica: complex isotopic systematics within a Proterozoic mobile belt. *Journal of Metamorphic Geology*, **5(1)**, 1–26.

Bohlen, S.R., Wall, V.J. & Boettcher, A.L. (1983). Geobarometry in granulites. In *Kinetics and Equilibrium in Mineral Reactions, Advances in physical geochemistry, Vol. 3*, ed. S.K. Saxena, pp. 141–72. New York: Springer-Verlag.

Ellis, D.J. (1983). The Napier and Rayner Complexes of Enderby Land, Antarctica: contrasting styles of metamorphism and tectonism. In *Antarctic Earth Science*, ed. R.L. Oliver, P.R. James & J.B. Jago, pp. 20–4. Canberra; Australian Academy of Science and Cambridge; Cambridge University Press.

Ellis, D.J. & Green, D.H. (1979). An experimental study of the effect of Ca upon garnet–clinopyroxene Fe–Mg exchange equilibria. *Contributions to Mineralogy and Petrology*, **71(1)**, 13–22.

England, P.C. & Thompson, A.B. (1984). Pressure–temperature–time paths of regional metamorphism, I. Heat transfer during the evolution of regions of thickened continental crust. *Journal of Petrology*, **25(4)**, 894–982.

Graham, C.M. & Powell, R. (1984). A garnet–hornblende geothermometer: calibration, testing and application to the Pelona schist, Southern California. *Journal of Metamorphic Geology*, **2(1)**, 13–31.

Green, D.H. & Ringwood, A.E. (1967). An experimental investigation of the gabbro to eclogite transformation and its petrological applications. *Geochimica et Cosmochimica Acta*, **31(5)**, 767–833.

Harley, S.L. (1984a). An experimental study of the partitioning of Fe and Mg between garnet and orthopyroxene. *Contributions to Mineralogy and Petrology*, **86(4)**, 359–73.

Harley, S.L. (1984b). The solubility of alumina in orthopyroxene coexisting with garnet in FeO–MgO–Al_2O_3–SiO_2 and CaO–FeO–MgO–Al_2O_3–SiO_2. *Journal of Petrology*, **25(3)**, 665–96.

Harley, S.L. (1985). Paragenetic and mineral–chemical relationships in orthoamphibole-bearing gneisses from Enderby Land, East Antarctica: a record of Proterozoic uplift. *Journal of Metamorphic Geology*, **3(2)**, 179–200.

Harley, S.L. (1987). Precambrian geological relationships in high-grade gneisses of the Rauer Islands, East Antarctica. *Australian Journal of Earth Sciences*, **34(2)**, 175–207.

Hiroi, Y., Shiraishi, K., Motoyoshi, Y. & Katsushima, T. (1987). Progressive metamorphism of calc-silicate rocks from the Prince Olav and Sôya coasts, East Antarctica. *Proceedings of the Japanese National Institute of Polar Research Symposium on Antarctic Geoscience, Vol. 1*, pp. 73–97.

Lindsley, D.H. (1983). Pyroxene thermometry. *American Mineralogist*, **68(5–6)**, 477–93.

Newton, R.C. & Haselton, H.T. (1981). Thermodynamics of the garnet–plagioclase–Al_2SiO_5–quartz geobarometer. In *Thermodynamics of Minerals and Melts*, ed. R.C. Newton, A. Navrotsky & B.J. Wood, pp. 131–47. New York; Springer-Verlag.

Perkins, D., III & Newton, R.C. (1981). Charnockite geobarometers based on coexisting garnet–pyroxene–plagioclase–quartz. *Nature, London*, **292(5819)**, 144–6.

Sheraton, J.W., Black, L.P. & McCulloch, M.T. (1984). Regional geochemical and isotopic characteristics of high-grade metamorphism of the Prydz Bay area: the extent of Proterozoic reworking of Archaean continental crust in East Antarctica. *Precambrian Research*, **26(2)**, 169–98.

Sheraton, J.W. & Collerson, K.D. (1983). Archaean and Proterozoic geological relationships in the Vestfold Hills – Prydz Bay area, Antarctica. *BMR Journal of Australian Geology and Geophysics*, **8(2)**, 119–28.

Thompson, A.B. (1976). Mineral reactions in pelitic rocks. II. Calcula-

tion of some P–T–X (Fe–Mg) phase relations. *American Journal of Science*, **276(4)**, 425–54.
Tingey, R.J. (1981). Geological investigations in Antarctica 1968–1969: the Prydz Bay–Amery Ice Shelf–Prince Charles Mountains area. *Bureau of Mineral Resources, Australia, Record*, **1981/34**, (unpublished).

Fault tectonics and magmatic ages in the Jetty Oasis area, Mac. Robertson Land: a contribution to the Lambert rift development

J. HOFMANN

Sektion Geowissenschaften, Bergakademie Freiberg, 9200 Freiberg, German Democratic Republic

Abstract

Structural and geochronological investigations (K–Ar data) of Archaean–Proterozoic basement rocks and Permian sedimentary rocks exposed in the Platforma Kamenistaja and the Else Platform, Mac. Robertson Land, East Antarctica are discussed. Both platforms show patterns of meridional and latitudinal faults connected with igneous intrusions. The meridional system consists of N–S to NE–SW-trending normal and oblique-slip faults dipping 55–40° W or NW. Late Carboniferous mafic dykes and Early Triassic porphyries and quartz veins followed this fault system, which was re-activated repeatedly during Mesozoic and Cenozic times. It is regarded as an antithetic system of the western main fault zone of the Lambert rift. The latitudinal system is represented by strike-slip faults, dykes and small pipes of Early Cretaceous brecciated and massive monchiquites. Magmatic and hydrothermal activity at the western rift margin, between 70 and 74° S, are discussed with reference to published data.

Introduction

The graben nature of the Lambert Glacier area was recognized during the first Australian expeditions to Mac. Robertson Land, when Permian sedimentary rocks (Amery Group) disconformably underlain by Precambrian metamorphic rocks were discovered in the Jetty Oasis area (Crohn, 1959), and erratics of Permian rocks were found 300 km to the south in moraines of Mount Rymill (Trail, 1964). The rift character of the Lambert Glacier and Prydz Bay areas was established by Soviet authors using seismic data, and also by the discovery of an alkali–ultramafic explosive breccia in the Jetty Oasis area (Ravich, Soloviev & Fedorov, 1978; Grikurov, Orlenko & Fedorov, 1980; Masolov, Kurinin & Grikurov, 1980). In the light of seismic data from Prydz Bay (Stagg, 1985), the Lambert rift is now considered to be a failed rift at a Late Palaeozoic–Mesozoic triple or four-armed junction. Its neotectonic development was discussed by Tingey (1985).

Little is known, however, about the structure and development of the Lambert rift. The Jetty Oasis, situated between two main faults of the western rift margin (Kurinin & Grikurov, 1980, fig. 3), represents a key area for understanding the development of the Lambert rift (Figs 1 & 2). Parts of the western marginal rift fault system and its associated dykes and veins are exposed on the Platforma Kamenistaja and Else Platform, two flat-topped outcrops of some 20 and 120 km^2 located between Lambert and Stagnant glaciers and forming the northern oasis. Fault patterns and the cross-cutting relationships of the dykes enable the reconstruction of the structural and chronological (Palaeozoic–Cenozoic) development of the rift.

Basement structure

Both platforms are underlain by granulite-facies rock sequences intruded by granites and pegmatites. The most important rock types are charnockite, granulite, garnet-bearing biotite–plagioclase gneiss and partially migmatized sillimanite–cordierite–garnet gneiss; except for the massive charnockite these rocks have a well-developed foliation or gneissosity and a mineral or mineral aggregate lineation (Ravich *et al.*, 1978; Grew, 1985). Ravich *et al.* (1978) regarded these rocks as members of the Reinbolt and Larsemann series, parts of the Archaean Humboldt Complex. Isotope studies in the 'Prydz Bay metasediments', to which these rocks belong (Sheraton, Black & McCulloch, 1984), indicate a multistage tectono-metamorphic history of Early Archaean precursors subjected to granulite-facies events at 3000, 2500 and 1100 Ma. Local high-pressure metamorphism accompanied by intensive shearing and late-stage granite emplacement was connected with the last event.

The tectonics of the Platforma Kamenistaja and south-eastern Else Platform are characterized by steeply dipping commonly vertical planar elements. On the Platforma Kamenistaja, charnockite and granulite sequences interlayered with garnet–biotite–plagioclase gneiss trend SE–NW and dip steeply NE or SW (Fig. 3A). The south-eastern Else Platform has a synformal structure whose axis plunges 50° ESE (Figs 3E & 4). Possibly, these structures are comparable to F_2 folds in the Bunger Hills area (Ding & James, this volume, p. 13).

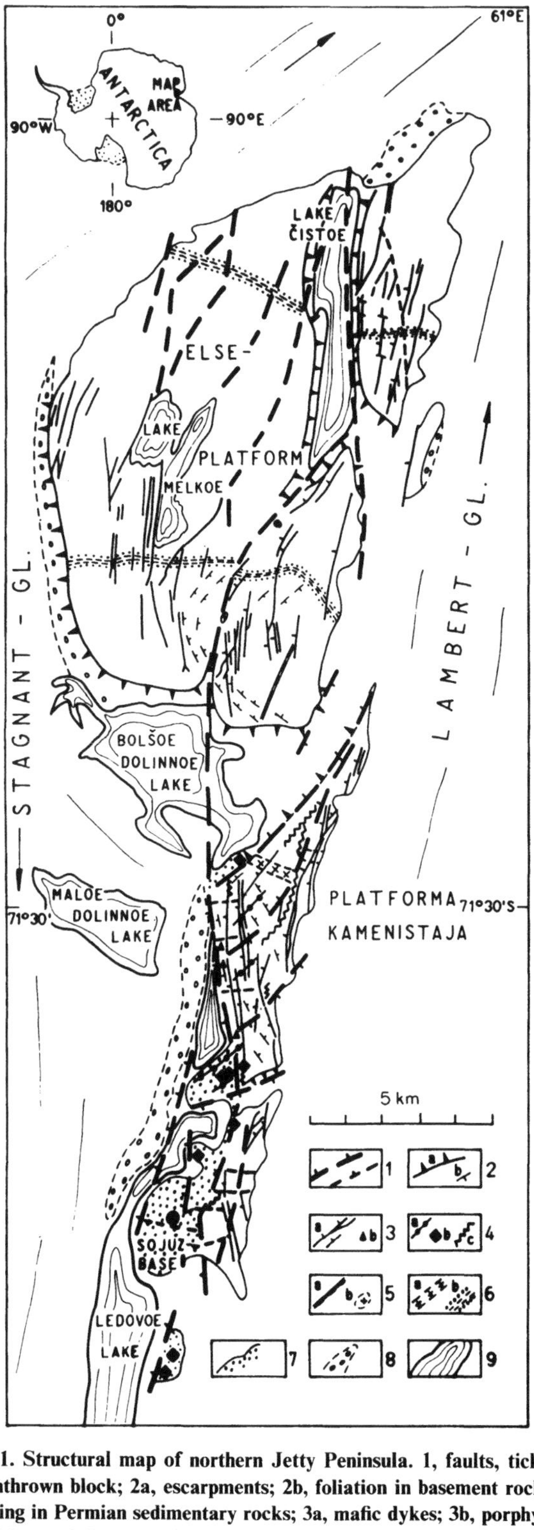

Fig. 1. Structural map of northern Jetty Peninsula. 1, faults, ticks on downthrown block; 2a, escarpments; 2b, foliation in basement rocks or bedding in Permian sedimentary rocks; 3a, mafic dykes; 3b, porphyries, 4a, dykes and 4b, pipes of ultramafic alkaline rocks; 4c, quartz veins and breccia zones, > 2 m wide; 5a, granite porphyry; 5b, biotite-granite; 6a, foliated garnet-bearing granite; 6b, basement rock sequences, partially identified by air photography; 7, Amery Group (Permian); 8, moraine; 9, meltwater lake. The map is based on the Soviet topographic map of Prince Charles Mountains (1:100,000), Lake Bolšoe Dolinnoe sheet. Monchiquite pipes from Ravich *et al.* (1978), Grew (1985) and author's field data.

The synformal structure on the Else Platform is bounded in the north by an E–W-trending, steeply S-dipping zone of blastomylonites ('augen-granulites'; Figs 1 & 2) which resembles the retrograde ductile shear zones of the Napier Complex, Enderby Land (Sandiford, 1985). Considering the results of Sheraton *et al.* (1984), this zone could be contemporaneous with the youngest granulite-facies event.

Small bodies (0.05 km^2) and dykes of fine-grained biotite-granite and a granite porphyry dyke (Fig. 1) are exposed on the south-eastern Else Platform. Their ages are Pan-African (Grew, Hofmann & Manton, in press). Metamorphic rocks and granites are cut by pegmatite veins (garnet, amphibole, biotite, plagioclase, K-feldspar, quartz).

Fault structures and igneous rocks

The meridional system

This system (Fig. 1) has a horse-tail pattern comprising N–S to NE–SW-trending elements, dipping at 40–55° W or NW. Its most obvious element is the fault bordering the Platforma Kamenistaja in the west and dividing the eastern and western Else Platform (Figs. 1 & 2); this fault is represented by a zone of diaphthorites, mylonites and breccias, 250–600 m thick. Permian sedimentary rocks at the western side of the Platforma Kamenistaja and at the fault, which strike E–W and dip 15–25° N are rotated to a N–S strike at the fault and dip vertically, indicating north-westward, oblique downthrow of the western block. A N–S-trending foliation in basement rocks at the eastern side of the fault on the Platforma Kamenistaja supports this picture (Fig. 3A). Considering the thickness of the Permian rocks given by Ravich *et al.* (1978), the total vertical displacement at the fault is between 500 and 1000 m; the horizontal displacement is 400–600 m according to offset along the shear zone on the Else Platform (Fig. 1).

The orientation of minor faults in this fault system is shown in Figs 3B, C & F. Slickensides along minor faults within the above-mentioned fault, and from minor faults occurring in basement rocks on both platforms and in the mafic dykes, mainly indicate normal downthrow of the western block to the W or oblique downthrow to the NW (Fig. 3C). Moderate western dips are controlled by *ac*-jointing in the basement rocks (compare Figs 3B, C & F with Figs 3A & E). These faults represent an antithetic oblique-slip system between the three marginal faults of the Lambert rift, located between Lambert Glacier and the eastern escarpment of the northern Prince Charles Mountains (Fig. 2).

Mafic dykes, porphyries and quartz veins are associated with the meridional fault system (Figs 1 & 5). North–south to NE–SW-trending mafic dykes are widely distributed in the basement rocks. They do not occur in the Amery Group sedimentary rocks (Figs 3D & G) and are deformed and brecciated within the fault zone at the western margin of the Platforma Kamenistaja. A K–Ar isochron of 308 ± 10 Ma indicates a Late Carboniferous age for the mafic dykes (Table 1). Their pre-Permian age is supported by the highly volatile bituminous rank of the Amery Group coals, which indicates a lack of magmatic influence on them (Rose & McElroy, 1987). Two brecciated porphyry dykes were observed within this fault zone and two samples yielded whole-rock ages of 245 and 238 Ma (Table 1), indicating a period of Early Triassic magmatic activity. The age of the quartz veins and breccia zones is

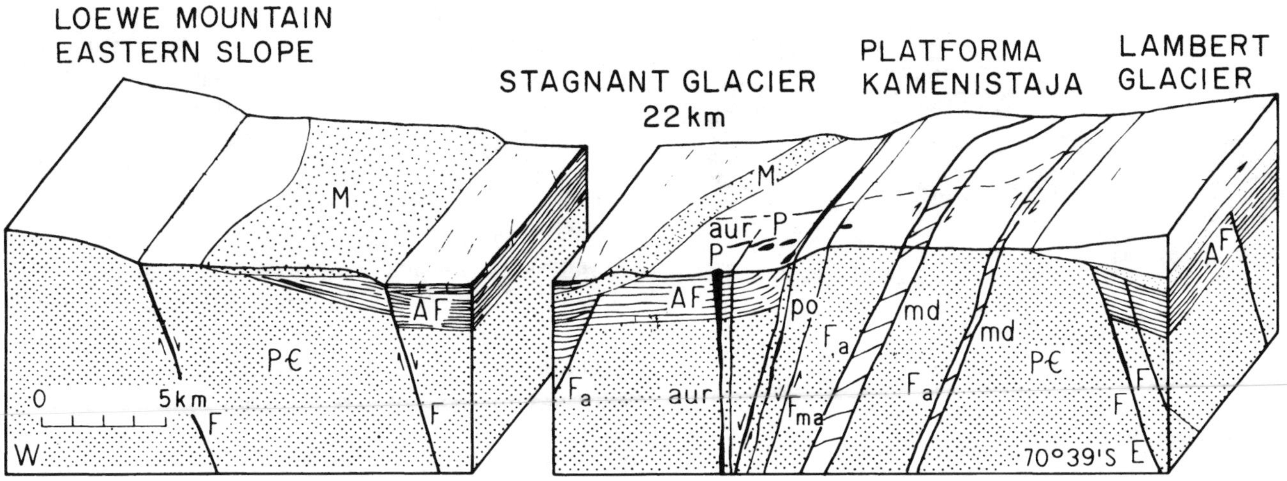

Fig. 2. Schematic block diagram of the western Lambert rift margin in the Jetty Oasis area. PC, metamorphic Precambrian basement; AG, Amery Group (Permian); M, moraines; md, mafic dykes; po, porphyries; aur, monchiquite dykes; p, monchiquite pipes; f, main faults; fma, main faults of the antithetic system; fa, antithetic faults.

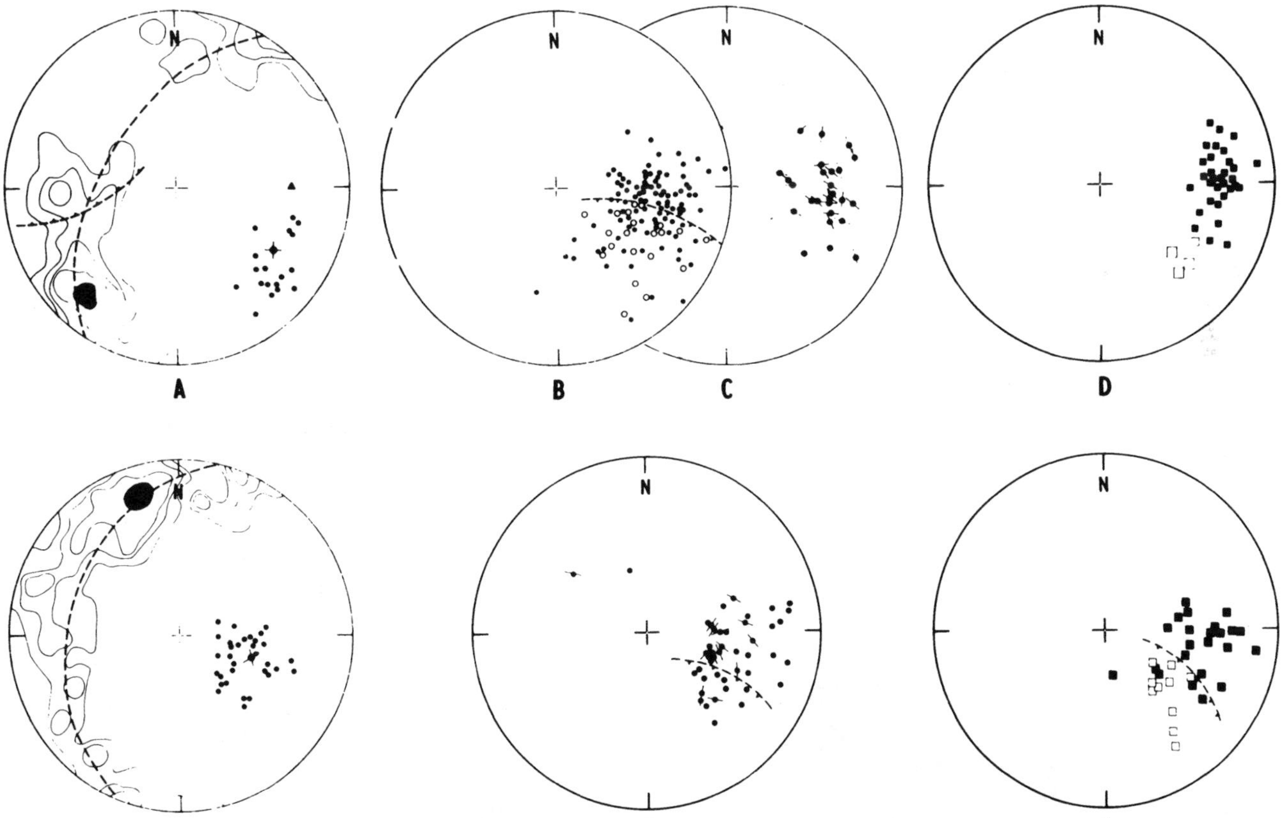

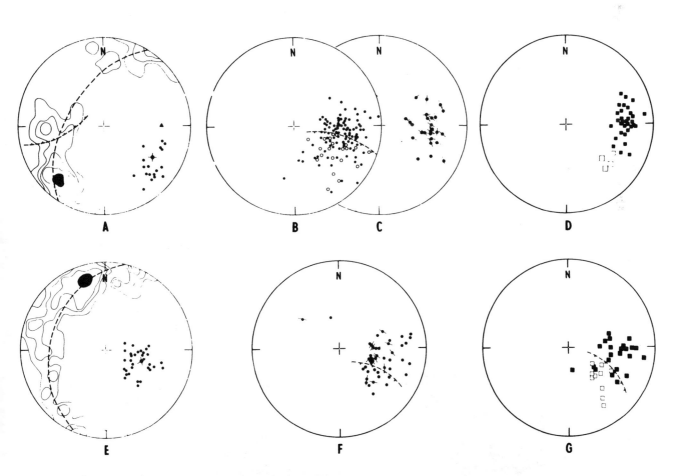

Fig. 3. Equal area projections of structural data for the metamorphic basement, central and northern Platforma Kamenistaja (PK) (A–D) and south-eastern Else Platform (E–G). A, foliation (81 poles), lineation (●) and one fold axes (▲) in granulites, biotite–plagioclase-gneisses and charnockites (contours at 0, 1, 3, 5, 10 and 15%), lineation poles averaged from 18 outcrops, maximum above the dashed line, foliation at the fault at the western margin of the PK; B, 93 faults in metamorphic rocks and mafic dykes, (●) < 500 m from the fault at the western margin of the PK, (○) > 500 m from the fault; C, 27 faults with slickensides in metamorphic rocks and mafic dykes; D, 32 mafic dykes (■) and 4 mafic dykes (□) from the north-western margin of the Platforma Kamenistaja; E, foliation (103 poles) and lineation (●) in granulite, biotite–plagioclase-gneiss and augen-granulite (contours at 0, 1, 3, and 5%), lineation poles averaged from 32 outcrops; F, 45 fault planes, partially with slickensides, in metamorphic rocks and mafic dykes, poles beneath the dashed line are faults south and south-west of Čistoe Lake; G, 23 mafic dykes (■) from the south-eastern Else Platform and 10 mafic dykes (□) south and south-west of Čistoe Lake.

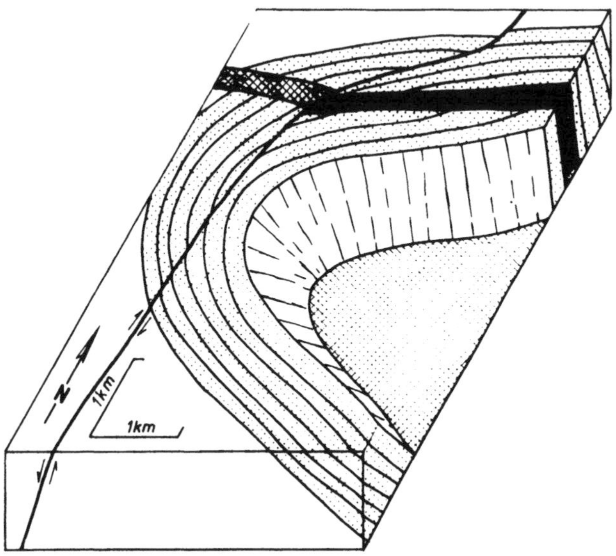

Fig. 4. Schematic representation of basement structures in the south-eastern part of Else Platform (cf. Fig. 3E).

uncertain. Although they cut the mafic dykes and are themselves cut by the Early Cretaceous monchiquites (see below), their age relationships to the Permian sedimentary rocks and the porphyries are not known. Quartz veinlets occur in both rocks but they may be younger than the quartz veins and breccia zones.

The latitudinal system

East–west to WSW–ENE-trending strike-slip faults cut the meridional elements and in north-eastern Platforma Kamenistaja they grade into the NE–SW-trending elements of the

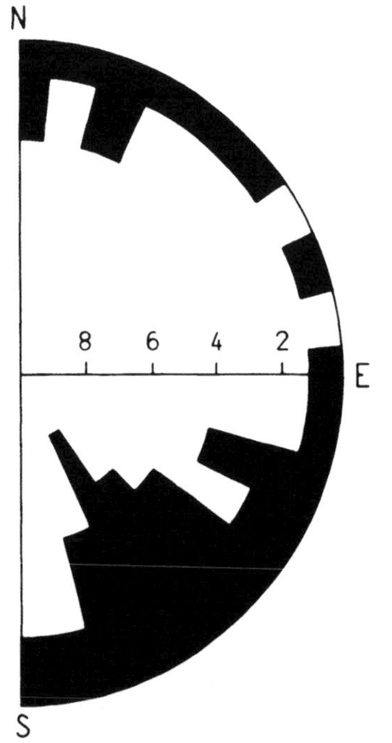

Fig. 5. Plot of 56 trends from single quartz veins and quartz breccia zones (> 2 m wide). Central and northern Platforma Kamenistaja and south-eastern Else Platform.

meridional system. At the western margin of the Platforma Kamenistaja, small monchiquite veins and pipes of breccia (25–45 m in diameter) containing metamorphic rocks and fritted Permian sedimentary rocks clasts embedded in a monchiquite matrix, occur in a zone 1–1.5 km east and west of the main fault (Figs 1 & 2). The K–Ar date of 110 Ma (Table 1) indicates an Early Cretaceous age for these rocks. The dykes are displaced by N–S-trending dextral faults belonging to the meridional system, as is seen for example where the southern escarpment of the Else Platform is displaced by the main fault (Fig. 1).

The latitudinal system is regarded here to be of Early Cretaceous age and to have no genetic connection with the meridional system. The possibility must be mentioned, however, that the two systems are part of one dextral–normal oblique fault system. More detailed field data are needed to determine the true relationship of the two fault systems.

History of tectono-magmatic activity

Structural data, together with K–Ar age determinations on igneous rocks associated with both fault systems, indicate the following history for the development of the western Lambert rift margin:

1. crustal fracturing during the formation of the meridional fault system and emplacement of Late Carboniferous mafic dykes
2. the probable formation of quartz veins and zones controlled by the meridional system
3. initial graben subsidence and sedimentation of the Upper Permian Amery Group
4. re-activation of the meridional system and main period of graben subsidence, accompanied by cataclasis and mylonitization of basement and Permian rocks; emplacement of Early Triassic porphyries
5. formation of the latitudinal system, explosive volcanism and emplacement of Early Cretaceous monchiquites, possibly connected with the formation of the East Antarctic continental margin (Stagg, 1985)
6. Cretaceous–Quaternary multiphase block movements and general elevation due to the re-activation of the meridional system.

These data support the scheme for the development of the Lambert rift proposed by Stagg (1985), which was derived mainly from seismic data from Prydz Bay.

Tectono-magmatic and hydrothermal activity at the western margin of the Lambert Rift – summary

Published age data from the dykes and volcanic rocks indicate a long history (Ordovician–Palaeogene) of magmatic activity in the mountains at the western edge of the Lambert rift (Table 2). Radiometric ages and structural data indicate that rift development possibly began after the Pan-African Orogeny by meridional crustal fracturing accompanied by the intrusion of Late Ordovician–Silurian alkaline to mafic dykes; this predated subsidence and the formation of the Amery Group in

Table 1. *K–Ar ages from mafic dykes, porphyries and monchiquites from the Platforma Kamenistaja*

Sample	Rock type	K (%)	^{40}Ar rad. ($\times 10^{-9}$/g)	^{40}Ar rad. (%)	Age (Ma)	
1854	Mafic dyke	1.90 ± 0.08	1.449 ± 0.049	96.1	394 ± 19	
1855	Mafic dyke	2.31 ± 0.18	1.559 ± 0.053	97.9	353 ± 27	308 ± 10[1]
1856	Mafic dyke	1.97 ± 0.08	1.406 ± 0.048	90.3	371 ± 18	
1857	Mafic dyke	1.64 ± 0.08	1.902 ± 0.065	96.5	596 ± 29	
1871	Porphyry	3.33 ± 0.13	1.473 ± 0.051	93.5	239 ± 12	
1872	Porphyry	3.96 ± 0.14	1.798 ± 0.063	93.6	245 ± 11	
1867	Monchiquite	1.51 ± 0.05	0.305 ± 0.026	95.8	113 ± 10	
754-1	Monchiquite	1.33 ± 0.10	0.244 ± 0.008	48.0	103 ± 5	
754-2	Monchiquite	1.81 ± 0.10	0.367 ± 0.012	53.0	113 ± 7	119[1] ± 6
754-3	Monchiquite	1.46 ± 0.10	0.286 ± 0.010	46.0	110 ± 8	
754-4	Monchiquite	1.89 ± 0.10	0.374 ± 0.013	52.0	111 ± 7	
174-5	Monchiquite	4.29 ± 0.10	0.902 ± 0.026	82.0	117 ± 4	

Samples 754-1–754-5 from G.E. Grikurov.
[1]K–Ar isochron age.

Table 2. *K–Ar ages from the western margin of the Lambert rift*

Rock type	Locality	Age (Ma)	
Leucite tristanite[1]	Manning Massif*	49.8 ± 2	(a)
		51.8 ± 2	(a)
Alnoite[1]	Radok Lake*	108 ± 3	(b)
		110 ± 3	(b)
Monchiquite[3]	Platforma Kamenistaja*	119 ± 6	(a)†
Monchiquite[2]	Platforma Kamenistaja*	130	(a)
Porphyry[3]	Platforma Kamenistaja*	239 ± 12	(a)
		245 ± 11	(a)
Calc-alkaline basalt[1]	Taylor Platform*	246 ± 6	(b)
Mafic dykes[3]	Platforma Kamenistaja*	308 ± 10	(a)†
Alkali melasyenite[1]	Mount Bayliss‡	414 ± 10	(b)
		430 ± 10	(b)
Alkali olivine basalt[1]	Fox Ridge*	504 ± 20	(b)

(a) Whole-rock ages; (b) mineral ages.
†K–Ar isochron ages.
*Northern Prince Charles Mountains; ‡Southern Prince Charles Mountains.
[1]Sheraton (1983); [2]Halpern & Grikurov (1975); [3]this paper.

Permian times. In general the granitic rocks on the Else Platform follow the pattern of the meridional system, indicating an even earlier, Pan-African age for the initial crustal fracturing. Post-Permian to Recent oblique-slip movements at the western rift margin were interrupted by Early Cretaceous latitudinal transform movements connected with crustal break-up and the formation of the continental margin. A leucite–tristanite lava flow in northern Prince Charles Mountains (Sheraton, 1983) represents the youngest magmatic event.

Hydrothermal activity has occurred at a number of localities along the western rift margin. In southern Prince Charles Mountains, a sphalerite- and galena-bearing fluorite–barytes vein and chalcopyrite mineralization have been recorded (Hofmann, 1978). Calcite- and pyrite-bearing quartz veins occur in the northern part of the mountains as well as on the Fisher Massif (L.F. Fedorov, pers. comm.), on Mount Izabelle (author's observations) and on the Platforma Kamenistaja and Else Platform. The genetic character and age of this mineralization, however, requires further investigation.

Acknowledgements

Fieldwork for this contribution was carried out during the 1984–85 austral summer with the geological party of the 30th Soviet Antarctic Expedition (SAE), under the cooperative program of the Arctic and Antarctic Research Institute, Leningrad and the Central Institute for Physics of the Earth, Academy of Sciences, Potsdam, GDR. I thank the Bergakademie Freiberg for academic leave which enabled my participation in this expedition. I express my gratitude for active support in all stages of the expedition to the staff of the 30th SAE, Dr G.E. Grikurov, L.F. Fedorov, R.G. Kurinin and many other Antarctic friends from the 'Sojuz' base, among whom I include Dr E.S. Grew, US exchange scientist from the University of Maine at Orono.

The K–Ar age determinations were carried out in the Isotope Laboratory of the Sektion Geowissenschaften, Bergakademie Freiberg, for which I thank Dr G. Kaiser and Mrs Braun.

References

Crohn, P.W. (1959). A contribution to the geology and glaciology of the western part of Australian Antarctic Territory. *Bulletin of the Bureau of Mineral Resources, Geology & Geophysics*, **52**, 103 pp.

Grew, E.S. (1985). Field studies on the Jetty Peninsula (Amery Ice Shelf area) with the 30th Soviet Antarctic Expedition. *Antarctic Journal of the United States*, **20(5)**, 52–53.

Grew, E.S., Hofmann, J. & Manton, W. (in press). Radiometric ages of granites from the Jetty Oasis, Mac. Robertson Land (East Antarctica). *Zeitschrift für Geologische Wissenschaften.*

Grikurov, G.E., Orlenko, E.M. & Fedorov, L.V. (1980). Sčeločno-ul'traosnovnye porody rajona Ozera Biver, Vostočnaja Antarktida (Alkali–ultramafic rocks of the Beaver Lake Area, East Antarctica). *Trudy Sovetskoj Antarktičeskoj Ekspedicii*, **70**, 87–99.

Halpern, M. & Grikurov, G.E. (1975). Rubidium/strontium data from the southern Prince Charles Mountains. *Antarctic Journal of the United States*, **10(1)**, 9–15.

Hofmann, J. (1978). Tektonische Beobachtungen im hoch- und schwachmetamorphen Präkambrium der Gebirgsumrandung des Lambert-Gletschers (Ostantarktis). *Freiberger Forschungshefte*, **Reihe C 335**, 7–112.

Kurinin, R.G. & Grikurov, G.E. (1980). Stroenie riftovoj zony lednika Lamberta. (Structure of the Lambert Glacier rift zone). *Trudy Sovetskoj Antarktičeskoj Ekspedicii*, **70**, 76–86.

Masolov, V.N., Kurinin, R.G. & Grikurov, G.E. (1980). Crustal structure and tectonic significance of Antarctic rift zones. In *Gondwana Five*, ed. M.M. Cresswell & P. Vella, pp. 303–9. Rotterdam; A.A. Balkema Publishers.

Ravich, G.M. (1974). Razrez permskich uglenosnych otloženij oz. Biver (A profile of Permian coal-bearing sediments of the Beaver Lake). *Antarktika, Doklady Kommissii*, **13**, 19–35.

Ravich, M.G., Soloviev, D.S. & Fedorov, L.F. (1978). *Geologičeskoe stroenie Zemli Mak-Robertsona, Vostočnaja Antarktida* (Geology of Mac. Robertson Land, East Antarctica), Leningrad; Gidrometeoizdat. 230 pp.

Rose, G. & McElroy, C.T. (1987). *Coal Potential of Antarctica.* Bureau of Mineral Resources, Geology & Geophysics, Research Report 2, 24 pp.

Sandiford, M. (1985). The origin of retrograde shear zones in the Napier Complex: implications for the tectonic evolution of Enderby Land, Antarctica. *Journal of Structural Geology*, **7(3/4)**, 477–88.

Sheraton, J.W. (1983). Geochemistry of mafic igneous rocks of the northern Prince Charles Mountains. *Journal of the Geological Society of Australia*, **30**, 295–304.

Sheraton, J.W., Black, L.P. & McCulloch, M.T. (1984). Regional geochemical and isotopic characteristics of high-grade metamorphics of the Prydz Bay area. *Precambrian Research*, **26**, 169–98.

Stagg, H.M.J. (1985). The structure and origin of Prydz Bay and Mac. Robertson Shelf, East Antarctica. *Tectonophysics*, **114**, 315–40.

Tingey, R.J. (1985). Uplift in Antarctica. *Zeitschrift für Geomorphologie, NF*, **Supplement-Bd. 54**, 85–99.

Trail, D.S. (1964). The glacial geology of the Prince Charles Mountains. In *Antarctic Geology*, ed. R.J. Adie, pp. 143–51. Amsterdam; North-Holland Publishing Company.

Major fracture trends near the western margin of East Antarctica

P.D. MARSH

British Antarctic Survey, Natural Environment Research Council, High Cross, Madingley Road, Cambridge CB3 0ET, UK

Abstract

Comparison of the trends of ice-surface lineaments seen in satellite imagery of Coats Land and western Dronning Maud Land, with fracture directions and structural trends known from outcrop studies, supports the conclusion that the lineaments provide a summary of major fault trends in the region. Some are parallel to pre-Mesozoic trends but their present expression probably relates mainly to structures associated with Mesozoic rifting. The three most persistent trends intersect at approximately 120° and are parallel to adjacent rifted margins of East Antarctica. It is suggested that they are associated with a tri-radial rift system formed during the early stage of break-up between east and west Gondwana. The trends revealed by the imagery may aid in defining the shape of major structures identified by geophysical studies and the continental fragments in the adjacent part of West Antarctica.

Introduction

Satellite images of Coats Land and western Dronning Maud Land reveal discontinuities in the topography of the ice surface which can be inferred to indicate changes in bedrock morphology or elevation (Marsh, 1985). The recognition and interpretation of surface features produced by ice movement, atmospheric effects and bedrock topography have been discussed by Bud'ko (1985) and by Marsh (1985), who gave a summary of the literature treating the relationship between ice-surface and bedrock topography.

In this study the only inference drawn from the discontinuities is that they indicate the location of some change at the bedrock surface. Attention is given mainly to features at and beyond the margins of the exposed mountain ranges, in order to utilize the filtering effect of the ice sheet in emphasizing the larger-scale bedrock changes. This approach differs from that of Apostolescu (1977) who mapped lineaments mainly within the areas of outcrop, and hence found a larger number of features with less well defined trends.

The images used were paper prints from Landsat MSS Band 7 (0.8–1.1 μm, nominal resolution 80 m) and US National Oceanic and Atmospheric Administration (NOAA) weather satellite channel 2 (0.725–1.1 μm, nominal resolution 1.1 km). As noted by Bud'ko (1985), the lower resolution of the latter provides useful generalization in addition to the filtering effect of the ice sheet.

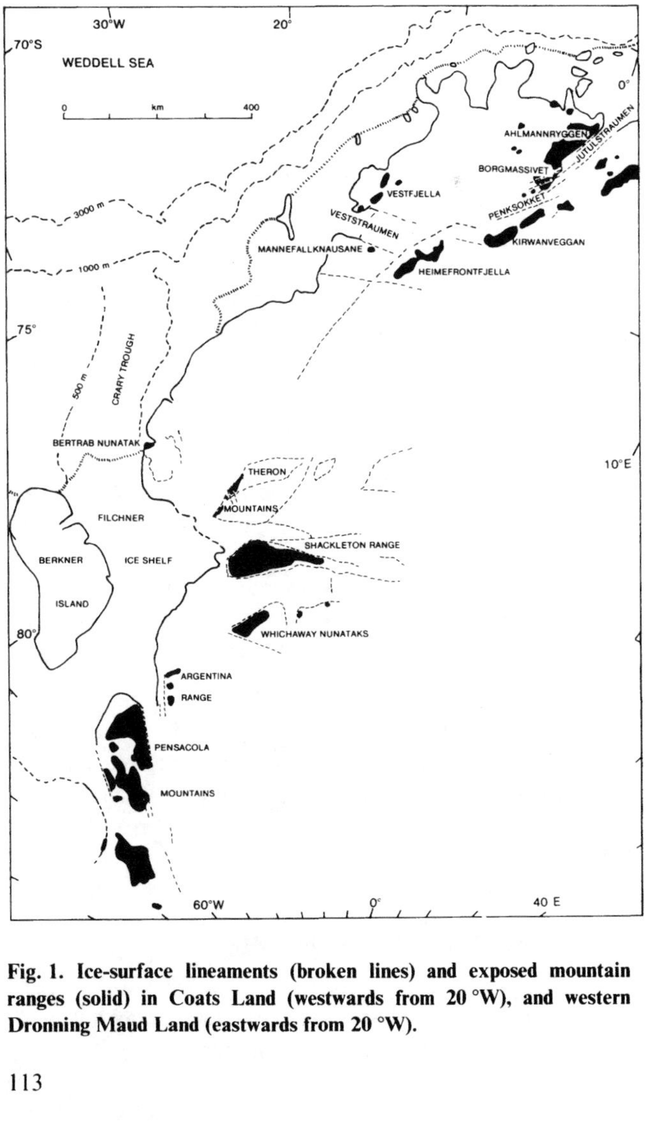

Fig. 1. Ice-surface lineaments (broken lines) and exposed mountain ranges (solid) in Coats Land (westwards from 20 °W), and western Dronning Maud Land (eastwards from 20 °W).

Inferred bedrock features

The features and feature trends recognized in Fig. 1 are based on Marsh (1985, fig. 4) between Heimefrontfjella and the Argentina Range, and on new work outside this sector. Only

NOAA imagery is available south of 81° S. The majority of the inferred bedrock features are approximately linear and some can be traced for over 100 km. They are most striking near outcrops of Beacon Supergroup rocks in the Theron Mountains (see Marsh, 1985, fig. 5).

The main lineaments are commonly associated with areas which have numereous shorter (5–10 km long) features parallel to the main trends. Local consistency in the orientation of these minor features suggests that the main trends are accurate indications of some trend in the local bedrock, and hence that regional variations in the orientation of the main sets are genuine, rather than the result of imprecise delineation at the surface.

The minor features also suggest that in some areas a prominent lineament is represented by two sets of features: for example, at the change in trend of the Penksökket–Jutulstraumen trough where two sets of features (present to varying extents throughout western Dronning Maud Land) are seen to instersect at approximately 15°. No example of a gradual change in orientation was observed, suggesting that the main features are rectilinear; infrequent curved outlines are probably the result of smoothing by the ice of changes in orientation at the intersection of two sets of structures.

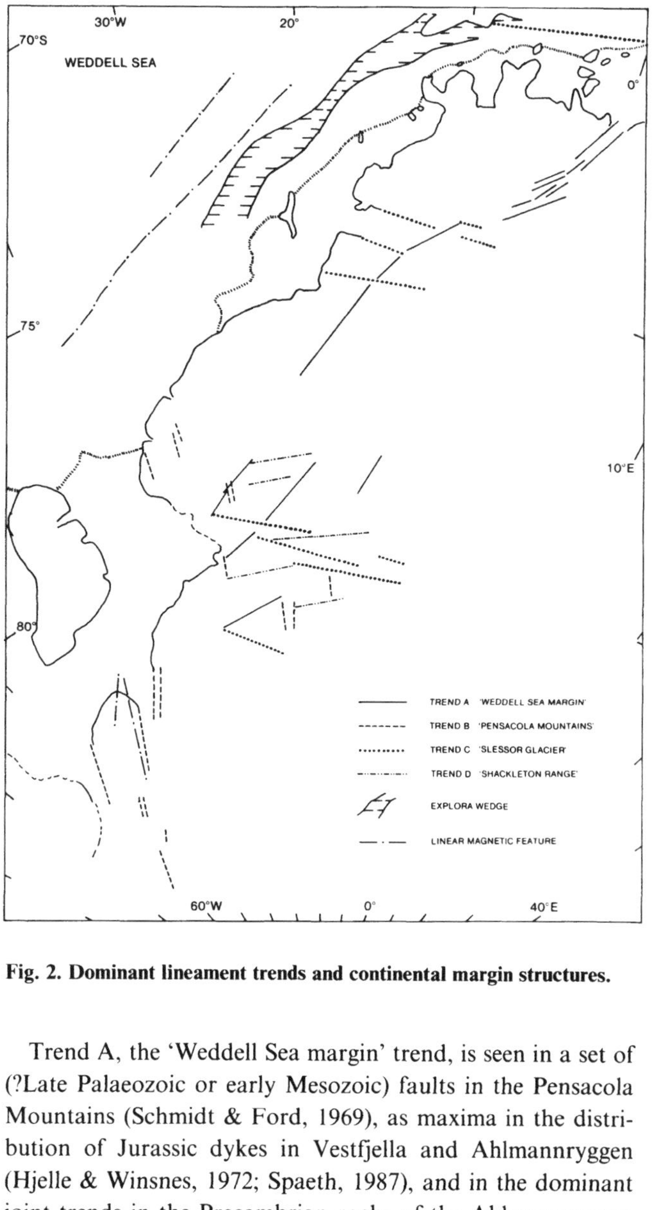

Fig. 2. Dominant lineament trends and continental margin structures.

Major trends

Features throughout the region can be grouped into three major and one minor trend (Fig. 2). Local representatives of the three main trends intersect at approximately 120°.

Trend A. Subparallel to the Weddell Sea coast. Examples: the Penksökket–Jutulstraumen trough, the Theron Mountains escarpment, the south-eastern margin of the Filchner Ice Shelf.

Trend B. Subparallel to the glaciers bounding the Pensacola Mountains. Example: the main valleys dividing the Theron Mountains (see Marsh, 1985, fig. 5).

Trend C. Features approximately bisecting the obtuse angle between A and B. Examples: the eastern margins of the Shackleton Range, the margins of Veststraumen. This trend is prominent in eastern Dronning Maud Land (Apostolescu, 1977; NOAA imagery of this study) where it is subparallel to the continental margin.

Trend D. Subparallel to the southern margin of the Shackleton Range. Examples of this orientation are restricted to the Shackleton Range–Theron Mountains area.

The range of orientations within each group is less than the difference between them and no major features lie outside the groups. In all areas the main trends correspond to peaks in the polymodal distribution of lineaments mapped by Apostolescu (1977).

Relationship to geological structures

The orientation of the majority of features seen in the imagery can be matched to structures observed at outcrop and/or those inferred from geophysical evidence.

Trend A, the 'Weddell Sea margin' trend, is seen in a set of (?Late Palaeozoic or early Mesozoic) faults in the Pensacola Mountains (Schmidt & Ford, 1969), as maxima in the distribution of Jurassic dykes in Vestfjella and Ahlmannryggen (Hjelle & Winsnes, 1972; Spaeth, 1987), and in the dominant joint trends in the Precambrian rocks of the Ahlmannryggen and Borgmassivet (Wolmarans & Kent, 1982). The orientation of the continental margin itself can be taken from the parallel trends of the central portion of the continental slope, the main section of the Explora wedge (Hinz, 1981) and magnetic anomalies mapped by Barker & Jahn (1980, fig. 1).

Trend B, the 'Pensacola Mountains' trend, is that of Early Palaeozoic and early Mesozoic folding in the Pensacola Mountains (Schmidt & Ford, 1969; Ford, 1972) and western Shackleton Range (Clarkson, 1982; Marsh, 1983), and of magnetic lineaments interpreted by Behrendt, Drewry & Jankowski (1981) to be due to faulting of the Early Jurassic Dufek intrusion.

Trend C occurs as maxima in the distribution of joints in Vestfjella (Hjelle & Winsnes, 1972), Ahlmannryggen and Borgmassivet (Wolmarans & Kent, 1982), and in the strike of faults cutting the ?Lower Palaeozoic Urfjell Group in Kirwanveggan (Aucamp, Wolmarans & Neethling, 1972). An independent indication of this trend in the west of the region is provided by Grikurov *et al.* (1980), who inferred major faults with this trend at the margins of Academy Glacier (Pensacola Mountains).

The local 'Shackleton Range', trend D, is that of a major fault and the late Precambrian–Early Palaeozoic metamorphic foliation in the Shackleton Range (Clarkson, 1982; Marsh, 1983). It is similar to the trend of the foliation in Kirwanveggan, but Roots (1969) specifically pointed out that the escarpment there (a 'Weddell Sea margin' lineament) is oblique to the foliation.

Structural significance

The prominence of the lineaments at the margins of the main outcrop areas, the observed relationship to mapped fractures, the consistency of orientation throughout the region, and the large scale of individual features combine to suggest that the lineaments represent the main fault trends in the region.

Some features are parallel to Early Palaeozoic and older basement structures but representatives of the three main trends are parallel to Late Palaeozoic or younger fractures. The coincidence of trends A and C with the rifted oceanic margins and the prominence of trend B close to the poorly delineated 'Filchner graben' (Masolov, Kuranin & Grikurov, 1980) at the margin with West Antarctica, suggest that the lineaments indicate structures associated with Mesozoic rifting.

Significance as Gondwana break-up structures

It is possible to speculate that the lineaments represent either pre-rifting fractures that influenced the orientation of Gondwana break-up structures or structures formed as a response to the stress pattern associated with rifting. They could also represent late structures active during the equilibration of the new rifted margins.

The rectilinear pattern and dominance of the 'Weddell Sea margin' trend suggest that, as in the east coast of southern Africa, which is adjacent to this part of Antarctica in most reconstructions, the structures associated with Gondwana break-up were extensional rifts (as envisaged by Hinz (1981) and Kristoffersen & Hinz (this volume, p. 225)) rather than basins formed by transform faulting (as for example, in the south coast of Africa (Dingle & Scrutton, 1974)). This model implies that the overall lateral movement of Africa relative to the Weddell Sea coast of Dronning Maud Land during the Mesozoic (Norton & Sclater, 1979) did not begin until considerable decoupling had taken place.

The lineament pattern is consistent with the suggestion of Hofmann & Weber (1983) that initial rifting in the region involved a tri-radial rift system (Burke & Dewey, 1973), with a 'triple junction' near the Dufek intrusion. The series of lineaments extending along the southern margin of the Filchner Ice Shelf and across Coats Land towards the Penksökket–Jutulstraumen rift would represent a failed arm of this junction.

However, separation from Africa did not take place near the Dufek intrusion. It is suggested here (Fig. 3) that a more important junction was situated at the intersection of the 'Filchner graben' and the rift along the site of the present continental margin. A rift with the 'Weddell Sea margin' trend intersecting the southern part of the Filchner graben near the

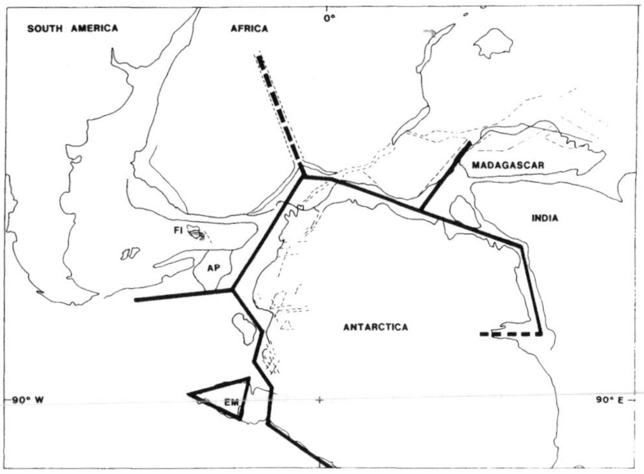

Fig. 3. Possible rift system at the margin of East Antarctica. Continental reconstruction of Lawver & Scotese (1987); reconstructed position of Alghulas Plateau (AP) after Allen & Tucholke (1981), Falkland Islands (FI) after Mitchell *et al.* (1986). Geological boundaries and fracture trends in Africa from Förster (1975), Reeves (1978), Cannon, Simiyu Siambi & Karanja (1981), Martin, Hartnady & Goodland (1981). The Ellsworth Mountain block (EM) is shown in its present day position in Antarctica; the remainder of West Antarctica is omitted.

Dufek intrusion could form the deep bedrock depressions at the northern termination of the Pensacola Mountains (Behrendt *et al.*, 1981) and the Thiel Mountains–Stewart Hills bedrock high (Jankowski & Drewry, 1981, fig. 3). In this model, 'trend C' lineaments would be parallel to a failed third arm of the southern junction, extending into the continent, but could also be related to a rift junction at the 'African' end of the Weddell Sea rift. Reeves (1978) has presented evidence for the latter.

The dominance of the three main lineament trends throughout the region suggests that structures with these orientations may also have defined the margins of crustal fragments in the adjacent complex boundary with West Antarctica (see reviews by Dalziel & Elliot (1982) and Hofmann & Weber (1983)). If the Ellsworth Mountains block has been rotated clockwise from a position in which the dominant fold trend was parallel to that in the Pensacola Mountains, then the deep trough underlying the Rutford Ice Stream (Doake, Crabtree & Dalziel, 1983) and the almost linear margin with the Byrd subglacial basin (Jankowski & Drewry, 1981) would have had trends close to the Filchner graben and the Weddell Sea margin, respectively. However, the tri-radial geometry would also support two other orientations. On a larger scale the rift junction north-west of Coats Land would be ideally situated in most Gondwana reconstruction for a third arm to follow the southern margin of the Falkland Plateau and be continuous with the Jurassic 'back-arc' basin of South America (Dalziel, 1981) and South Georgia (Tanner, Storey & Macdonald, 1981) which it has been suggested (Dalziel, 1974; Suárez, 1976) opened into the Weddell Sea.

References

Allen, R.B. & Tucholke, B.E. (1981). Petrography and implications of continental rocks from the Alghulas Plateau, southwest Indian Ocean. *Geology*, **9**, 463–8.

Apostolescu, V. (1977). *Carte Géologique de l'Antartique, 1: 2500000*. 7 Sheets and explanatory notes (23 pp.). Paris; Éditions Technip.

Aucamp, A.P.H., Wolmarans, L.G. & Neethling, D.C. (1972). The Urfjell Group, a deformed (?)early Palaeozoic sedimentary sequence, Kirwanveggen, western Dronning Maud Land. In *Antarctic Geology and Geophysics*, ed. R.J. Adie, pp. 557–62. Oslo; Universitetsforlaget.

Barker, P.F. & Jahn, R.A. (1980). A marine geophysical reconnaissance of the Weddell Sea. *Geophysical Journal of the Royal Astronomical Society*, **63**, 271–83.

Behrendt, J.C., Drewry, D.J. & Jankowski, E. (1981). Aeromagnetic and radio echo ice-sounding measurements over the Dufek intrusion, Antarctica. *Journal of Geophysical Research*, **86(B4)**, 3014–20.

Bud'ko, V.M. (1985). Geologic interpretations of Antarctica's mountainous regions with space imagery. *Mapping Sciences and Remote Sensing*, **22**, 27–33.

Burke, K. & Dewey, J.F. (1973). Plume generated triple junctions: key indicators in applying plate tectonics to old rocks. *Journal of Geology*, **81**, 406–33.

Cannon, R.T., Simiyu Siambi, W.M.N. & Karanja, F.M. (1981). The proto-Indian Ocean and a probable Palaeozoic/Mesozoic triradial rift system in East Africa. *Earth and Planetary Science Letters*, **52**, 419–26.

Clarkson, P.D. (1982). Tectonic significance of the Shackleton Range. In *Antarctic Geoscience*, ed. C. Craddock, pp. 835–9. Madison; University of Wisconsin Press.

Dalziel, I.W.D. (1974). Margins of the Scotia Sea. In *The Geology of Continental Margins*, ed. C.A. Burke & C.L. Drake, pp. 567–80. New York; Springer-Verlag.

Dalziel, I.W.D. (1981). Back-arc extension in the southern Andes: a review and critical appraisal. *Philosophical Transactions of the Royal Society of London*, **A.300**, 319–35.

Dalziel, I.W.D. & Elliot, D.H. (1982). West Antarctica: problem child of Gondwanaland. *Tectonics*, **1**, 3–19.

Dingle, I.W. & Scrutton, R.A. (1974). Continental breakup and the development of post-Paleozoic sedimentary basins around southern Africa. *Geological Society of America Bulletin* **85**, 1467–74.

Doake, C.S.M., Crabtree, R.D. & Dalziel, I.W.D. (1983). Subglacial morphology between Ellsworth Mountains and Antarctic Peninsula: new data and tectonic significance. In *Antarctic Earth Science*, ed. R.L. Oliver, P.R. James & J.B. Jago, pp. 270–3. Canberra; Australian Academy of Science and Cambridge; Cambridge University Press.

Ford, A.B. (1972). Weddell Orogeny: latest Permian to early Mesozoic deformation at the Weddell Sea margin of the Transantarctic Mountains. In *Antarctic Geology and Geophysics*, ed. R.J. Adie, pp. 419–25. Oslo; Universitetsforlaget.

Förster, R. (1975). The geological history of the sedimentary basin of southern Mozambique and some aspects of the origin of the Mozambique Channel. *Palaeogeography, Palaeoclimatology, Palaeoecology*, **17**, 267–87.

Grikurov, G.E., Kadima, I.N., Kamenev, E.N., Kurinin, R.G., Masolov, V.N. & Shulyatin, O.G. (1980). Tektonicheskoe stroenie basseyna morya Ueddella. In *Geofizicheskie Issledovaniya v Antarktide* (Geophysical investigations in Antarctica), ed. G.I. Gaponenko, G.E. Grikurov & V.N. Masolov, pp. 29–43. Leningrad: Research Institute of the Geology of the Arctic (in Russian).

Hinz, K. (1981). A hypothesis on terrestrial catastrophies. Wedges of very thick oceanward dipping layers beneath passive continental margins. Their origin and palaeo-environmental significance. *Geologisches Jahrbuch*, **E22**, 28 pp.

Hjelle, A. & Winsnes, T. (1972). The sedimentary and volcanic sequence of Vestfjella, Dronning Maud Land. In *Antarctic Geology and Geophysics*, ed. R.J. Adie, pp. 539–46. Oslo; Universitetsforlaget.

Hofmann, J. & Weber, W. (1983). A Gondwana reconstruction between Antarctica and South Africa. In *Antarctic Earth Science*, ed. R.L. Oliver, P.R. James & J.B. Jago, pp. 584–9. Canberra; Australian Academy of Science and Cambridge; Cambridge University Press.

Jankowski, E. & Drewry, D.J. (1981). The structure of west Antarctica from geophysical studies. *Nature*, **291**, 17–21.

Lawver, L.A. & Scotese, C.R. (1987). A revised reconstruction of Gondwanaland. In *Gondwana Six: Structure, Tectonics and Geophysics*, ed. G.M. Mckenzie, pp. 17–24. Washington DC; American Geophysical Union.

Marsh, P.D. (1983). The late Precambrian and early Palaeozoic history of the Shackleton Range, Coats Land. In *Antarctic Earth Science*, ed. R.L. Oliver, P.R. James & J.B. Jago, pp. 190–3. Canberra; Australian Academy of Science and Cambridge; Cambridge University Press.

Marsh, P.D. (1985). Ice surface and bedrock topography in Coats Land and part of Dronning Maud Land, Antarctica, from satellite imagery. *British Antarctic Survey Bulletin*, **No. 68**, 19–36.

Martin, A.K., Hartnady, C.J.H. & Goodland, S.W. (1981). A revised fit of South America and south central Africa. *Earth and Planetary Science Letters*, **54**, 293–305.

Masolov, V.N., Kuranin, R.G. & Grikurov, G.E. (1980). Crustal structures and tectonic significance of Antarctic rift zones (from geophysical evidence). In *Gondwana Five*, ed. M.M. Cresswell & P. Vella, pp. 303–10. Rotterdam; A.A. Balkema.

Mitchell, C., Taylor, G.K., Cox, K.G. & Shaw, J. (1986). Are the Falkland Islands a rotated microplate? *Nature*, **319**, 131–3.

Norton, I.O. & Sclater, J.G. (1979). A model for the evolution of the Indian Ocean and the breakup of Gondwanaland. *Journal of Geophysical Research*, **84(B12)**, 6802–30.

Reeves, C.V. (1978). A failed Gondwana spreading axis in southern Africa. *Nature*, **273**, 222–3.

Roots, E.F. (1969). Geology of Western Queen Maud Land. In *Geologic Maps of Antarctica*, ed. V.C. Bushnell & C. Craddock. Antarctic Map Folio Series, Folio 12, Pl. 6.

Schmidt, D.L. & Ford, A.B. (1969). Geology of the Pensacola and Thiel Mountains. In *Geologic Maps of Antarctica*, ed. V.C. Bushnell & C. Craddock. Antarctic Map Folio Series, Folio 12, Pl. 5.

Spaeth, G. (1987). Aspects of the structural evolution and magmatism in western New Schwabenland, Antarctica. In *Gondwana Six: Structure, Tectonics and Geophysics*, ed. G.M. Mckenzie, pp. 295–307. Washington DC; American Geophysical Union.

Suárez, M. (1976). Plate-tectonic model for southern Antarctic Peninsula and its relation to southern Andes. *Geology*, **4**, 211–14.

Tanner, P.W.G., Storey, B.C. & Macdonald, D.I.M. (1981). Geology of a Jurassic–lower Cretaceous island-arc assemblage in Hauge Reef, the Pickersgill Islands and adjoining areas of South Georgia. *British Antarctic Survey Bulletin*, **53**, 77–117.

Wolmarans, L.G. & Kent, L.E. (1982). Geological investigations in western Dronning Maud Land, Antarctica – a synthesis. *South African Journal for Antarctic Research*, Supplement 2, 93 pp.

Mesozoic magmatism in Greater Antarctica: implications for Precambrian plate tectonics

T.S. BREWER[1] & P.D. CLARKSON[2]

1 Department of Mining, University Park, Nottingham NG7 2RD, UK
2 British Antarctic Survey, Natural Environment Research Council, High Cross, Madingley Road, Cambridge CB3 0ET, UK
Present address: *Scientific Committee on Antarctic Research, Scott Polar Research Institute, Lensfield Road, Cambridge, CB2 1ER, UK*

Abstract

Mesozoic tholeiites from Coats Land and western Dronning Maud Land can be divided into two suites, characterized by their Ti content. These tholeiites have subduction-related trace element signatures which contrast with their known intracontinental setting. Similar geochemical signatures have been noted for the Ferrar Supergroup in the Transantarctic Mountains. Sm–Nd-isotope studies suggest that the high-Ti group have an age of about 1000 Ma which contrasts with the known emplacement age of about 160 Ma. The geochemical data suggest that the tholeiites were derived in part from an enriched source region, probably continental lithosphere. This lithosphere was probably produced during a Precambrian subduction event. Due to their proximity to the Transantarctic Mountains, the Cenozoic alkali basalts of Marie Byrd Land and the McMurdo Volcanic Group might be expected to show a similar trace element signature to the Mesozoic tholeiites. However, the available data suggest a depleted source giving rise to ocean–island type magmas. It is suggested that the Mesozoic tholeiites were derived, in part, from an old enriched lithospheric source distinct from the source of the Cenozoic magmas. This implies that the lithosphere beneath the Transantarctic Mountains is very different to that beneath Marie Byrd Land, which represents an exotic accreted fragment.

Introduction

Lesser Antarctica formed part of the Pacific margin of Gondwana prior to the break-up of the supercontinent during the Mesozoic. However, reconstructions of this region are complicated by the recognition that Lesser Antarctica comprises a number of crustal blocks which have moved independently of each other and the craton of Greater Antarctica. The variety of reconstructions which have been proposed, e.g. Dalziel & Elliot (1982) and Grindley & Davey (1982), merely emphasizes the complexity of the problem. There is good evidence that the northern tip of the Antarctic Peninsula was located close to the southern tip of South America (Hill & Barker, 1980) although the orientation of the peninsula is far from certain; there is controversy over the position of the Ellsworth Mountains region; there is a general consensus that Thurston Island and Marie Byrd Land were situated formerly much closer to the Transantarctic Mountains. The Palaeozoic and earlier history of these crustal blocks is not known. However, in north-eastern Victoria Land there is now evidence that the Bowers and Robertson Bay terranes were accreted during the Ross Orogeny in Cambro-Ordovician times (Gibson & Wright, 1985).

One of the earliest manifestations of continental rifting was the widespread intrusion along the Transantarctic Mountains of tholeiitic sills of the Ferrar Supergroup (Kyle, Elliot & Sutter, 1981). These were emplaced into the Beacon Supergroup terrestrial sedimentary rocks about 160 m.y. ago. Despite their obvious continental position, and that of their equivalents in Brazil, Africa and Tasmania, they possess trace element enrichment patterns reminiscent of a subduction event. Recent Sm–Nd-isotope results (Brewer & Brook, this volume, p. 569) approximate to a 1000 Ma reference isochron implying that this was probably the age of the enrichment event, at least in the Theron Mountains and, by analogy, along the length of the Transantarctic Mountains. Therefore, it is inferred that the Transantarctic Mountains formed part of the Gondwana margin during Late Proterozoic times, conflicting with the assumption that Marie Byrd Land formed the continental margin in this region. A model is presented here for the subduction history of this part of the Gondwana margin from about 1000 m.y. ago, to account for the observed magmatic and tectonic events during this time span.

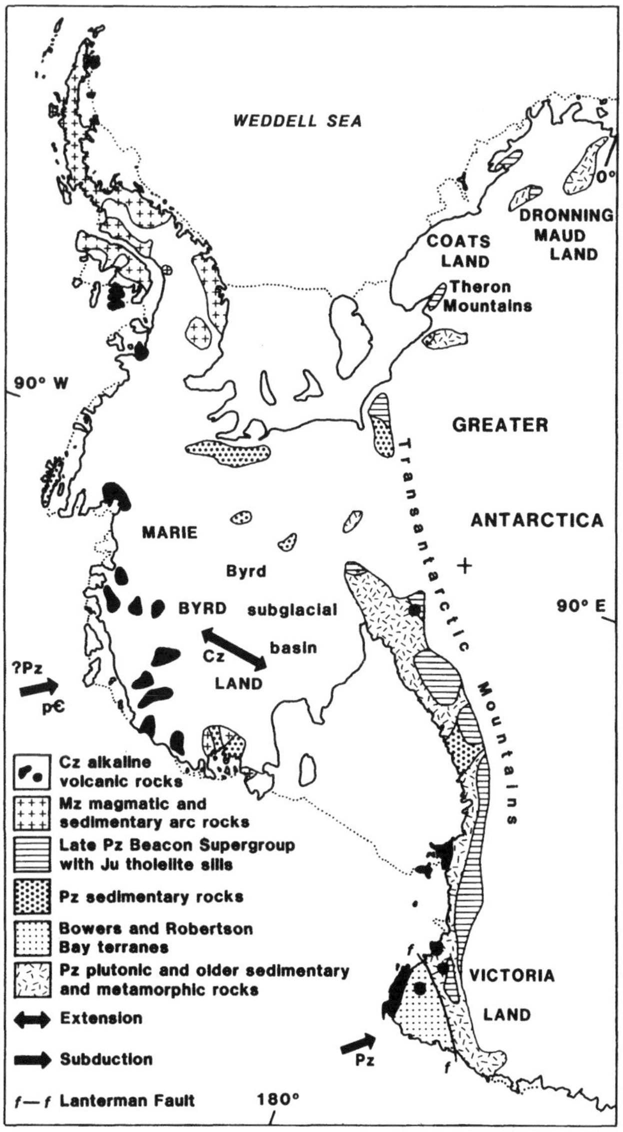

Fig. 1. Diagrammatic geological map of Lesser Antarctica and the Transantarctic Mountains showing possible subduction and extension regimes. Cz, Cenozoic; Ju, Jurassic; Mz, Mesozoic; p€, Precambrian; Pz, Palaeozoic.

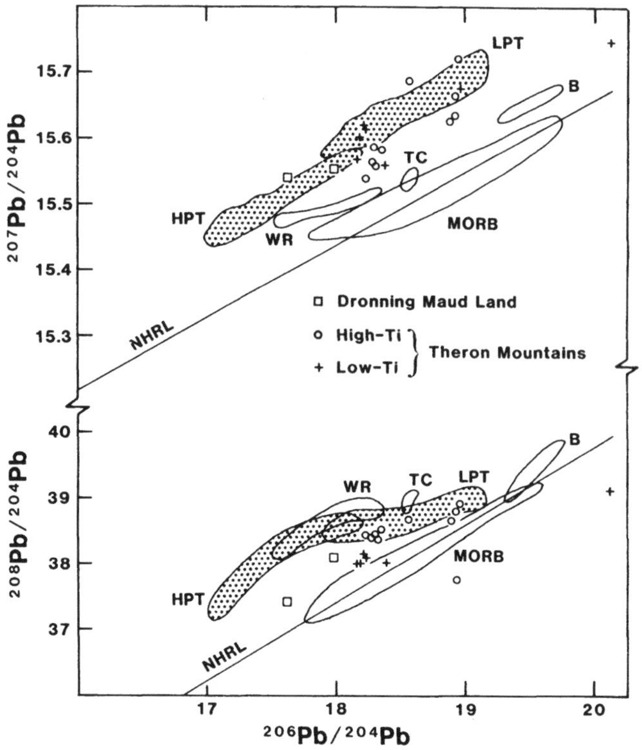

Fig. 2. Pb-isotope data for tholeiites from the Theron Mountains and western Dronning Maud Land. Additional fields after Hawkesworth *et al.* (1986): stippled, Paraná, low (LPT) and high (HPT) Ti and P; WR, Walvis Ridge; TC, Tristan da Cunha; B, Bouvet Island; MORB, mid-ocean ridge basalt; NHRL, Northern Hemisphere Reference Line after Hart (1984).

Geochemical evidence for Proterozoic subduction

Basic Mesozoic magmatism associated with the break-up of Gondwana forms a large province extending from Dronning Maud Land along the Transantarctic Mountains to Victoria Land (Fig. 1). Within this province the magmatism is represented predominantly by two geochemically distinct groups of dolerite sills and basaltic lavas. In Dronning Maud Land these rocks are characterized by low initial $^{87}Sr/^{86}Sr$ (Sr_0) ratios, whereas the Ferrar Supergroup of the Transantarctic Mountains is characterized by high Sr_0 ratios (i.e. > 0.7090). In the Theron Mountains of Coats Land, the basic rocks can be further separated into high- and low-Ti compositions (Brewer & Brook, this volume, p. 569) and they appear to be of both Dronning Maud Land and Ferrar Supergroup types. All of these rocks have trace element signatures more diagnostic of subduction-related magmatism than of within-plate magmatism (e.g. Ba/Nb > 15, Ce/Nb > 1). Such ratios show no significant correlation with SiO_2, MgO or Sr_0, suggesting that they may not be related to AFC (Assimilation–Fractional Crystallization) models of crustal contamination but may reflect source-region characteristics (see Devey & Cox, 1987).

The range in Sr_0 of the tholeiites is extremely large (0.7052–0.7129), with values > 0.7090 being more similar to those of crustal-derived magmas. The relationship between SiO_2 and Sr_0 is somewhat complicated by the Ti-grouping but, in both the high- and low-Ti groups, there is some positive correlation (Brewer & Brook, this volume), suggesting that the more evolved compositions in both groups record some degree of crustal contamination. A lead-isotope study of the Coats Land and Dronning Maud Land specimens (Fig. 2) shows that they plot above the Northern Hemisphere Reference Line (NHRL; Hart, 1984), along an array similar to that defined by the Paraná and Karoo data. These arrays give poor ages (~ 1000 Ma) which have been interpreted by Hawkesworth *et al.* (1986) as reflecting the age of lithosphere formation. A similar age (~ 1000 Ma) can be obtained from the Sm–Nd diagram for the Coats Land data (Fig. 3) and it may also reflect a lithospheric event; when coupled with the trace element signatures, the Sm–Nd data suggest a subduction component, implying that these Mesozoic magmas are derived in part from an ancient enriched lithosphere. A similar, more constrained age of 1050 Ma has also been obtained from Karoo picrite basalts (R.M. Ellam & K.G. Cox, pers. comm., 1988).

The data set from the Cenozoic volcanic rocks of Marie Byrd Land is extremely limited but it indicates that the basalts have

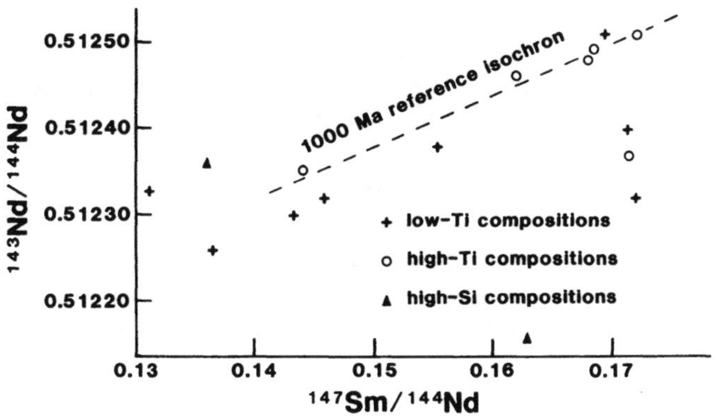

Fig. 3. Sm–Nd diagram for the Theron Mountains tholeiites showing the 1000 Ma reference isochron.

alkaline compositions. Their Sr and Nd contents plot in the depleted mantle field (Fig. 4), in contrast to the Mesozoic magmas and much closer to a typical ocean–island basalt source. From these data and more recent Sr and Nd studies on Alexander Island (Hole, Smellie & Marriner, this volume, p. 521), the Cenozoic basalts seem to be derived from an ocean–island basalt source and do not show any significant lithospheric signatures.

Evolution of the Gondwana margin

The geology of the region is shown in Fig. 1, emphasizing the principal magmatic and tectonic features central to the proposed model. The extent of the alkali volcanism in Marie Byrd Land contrasts markedly with that of the McMurdo Volcanic Group in Victoria Land. As indicators of crustal extension, these rocks imply that Marie Byrd Land has moved a considerable distance oceanward from the Transantarctic Mountains whereas there has been virtually no movement in Victoria Land. The pre-Mesozoic coastal fringe of Marie Byrd Land may have moved as much as 500 km (Grindley & Oliver,

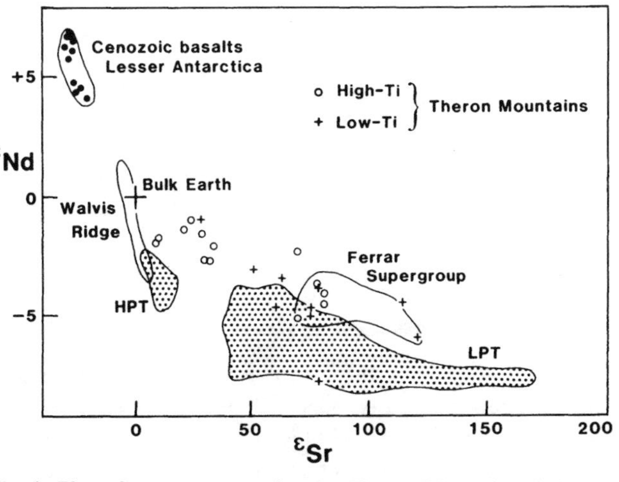

Fig. 4. Plot of ϵ_{Nd} versus ϵ_{Sr} for the Theron Mountains tholeiites. Cenozoic basalt field after Futa & LeMasurier (1983); bulk Earth, Walvis Ridge and Paraná high (HPT) and low (LPT) titanium and phosphorus fields after Hawkesworth *et al.* (1986), Ferrar Supergroup field after Pankhurst *et al.* (1986).

1983) if the entire width of the Byrd Subglacial Basin is extensional in origin.

Proposed sequence of events

Trace element enrichment of the Mesozoic tholeiites of the Ferrar Supergroup is ascribed to a subduction event along the proto-Pacific margin of Gondwana about 1000 m.y. ago. During this period the crust forming the coastal fringe of Marie Byrd Land was accreted, and subduction ceased or continued by an oceanward step beyond the accreted fragment. Subduction probably also occurred in the Victoria Land region but may not necessarily have had an identical history. According to Gibson & Wright (1985) subduction certainly took place in Victoria Land during the Early Palaeozoic, when the Bowers and Robertson Bay terranes were accreted, and probably continued into the Late Palaeozoic when underplating and back-thrusting compressed these accreted terranes.

During the Mesozoic one of the first indications of the pending fragmentation of Gondwana was the widespread intrusion of tholeiitic sills throughout a belt extending from Brazil, through Africa and Antarctica to Tasmania. The source of these was the subcontinental mantle wedge. Mesozoic to Recent subduction along the margin of Lesser Antarctica ceased progressively northward, at least along the Antarctic Peninsula (Barker, 1982) and presumably ceased first in the Marie Byrd Land region.

The first indications of extension in Marie Byrd Land were the Late Cretaceous dykes; the oldest alkaline volcanic rocks yet known have Eocene ages (LeMasurier, 1972) and the youngest are < 0.5 Ma. The rocks of the McMurdo Volcanic Group are younger and date only from the mid-Miocene (Kyle & Muncy, 1983). Crustal extension in Victoria Land is minimal but this may be reflected in the relative youth of the volcanic group as a whole.

The proximity of the alkaline eruptive centres to the outcrop of the Ferrar Supergroup tholeiites would suggest that their source regions might be related, at least horizontally if not vertically within the mantle. However, the available chemical data show that whereas the tholeiitic magmas are trace element-enriched, the alkali magmas are unrelated, depleted ocean–island basalt types.

The model proposed here for Marie Byrd Land (Fig. 5) shows the foundered down-going slab from the Late Proterozoic subduction event forming an effective barrier within the mantle, between the subcontinental mantle wedge (source of the enriched Ferrar tholeiites) and the suboceanic mantle (source of the alkali magmas). It has been suggested (Olson, Schubert & Anderson, 1987) that ocean–island basalt-type magmas originate from a deep mantle source, at the base of the upper mantle, and the ascent of these magmas could be triggered by the disturbance created by a foundered subducted slab. Thus, there is a possible connection between the two magmas in this case.

The situation in Victoria Land is more complex because of the juxtaposition of alkali basalt and Ferrar Supergroup tholeiitic outcrops. It is suggested that the complex accretion and underplating of the Bowers and Robertson Bay terranes on to

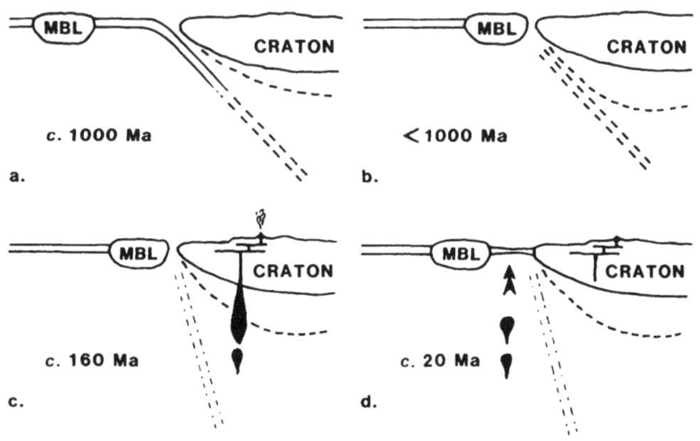

Fig. 5. A cartoon sequence to illustrate the main features of the proposed model: (*a*) active subduction and enrichment of the subcontinental lithosphere during the Precambrian; (*b*) accretion of Marie Byrd Land and cessation of subduction; (*c*) Mesozoic magmatism with a component of subcontinental lithosphere; and (*d*) Cenozoic extension and alkaline volcanism.

the craton margin during the Early Palaeozoic, coupled with major back-thrusting along the Leap Year and Lanterman faults, would have created additional pathways within the crust which could be exploited by a rising magma. This contrasts with a simple accretionary suture being formed between Marie Byrd Land and the Transantarctic Mountains which could have been reopened with the advent of extension; even this model is complicated by known, and possibly related, Miocene alkali-basalts in the central Transantarctic Mountains (Treves, 1967; Stump *et al.*, 1980).

Discussion

The continental reconstruction for the Cambrian (Smith, Hurley & Briden, 1981) shows that Lesser Antarctica was a central part of a continental mass facing a major ocean. Therefore, it seems likely that the region has been a potential subduction site for at least the last 600 m.y. and probably for much longer. It is also likely that exotic crustal fragments will have been accreted at times so that the situations of northern Victoria Land and Marie Byrd Land need not be considered exceptional.

The geology of Marie Byrd Land bears a strong resemblance to that of northern Victoria Land when the Swanson Formation and the Robertson Bay Group are compared. Wade & Couch (1982) considered that, despite their similarities, Marie Byrd Land is totally unrelated to Greater Antarctica. In this, they agreed with Elliot's (1975) suggestion, that Marie Byrd Land was accreted to the craton during late Precambrian or Early Palaeozoic time. Grindley & Oliver (1983) concluded from their palaeomagnetic study that Marie Byrd Land has been part of Lesser Antarctica at least since the Early Cretaceous and probably since the Early Palaeozoic. Hence there is an underlying assumption that Marie Byrd Land is an exotic crustal fragment.

Many of the tholeiites of the Mesozoic Ferrar Supergroup of the Transantarctic Mountains and its equivalents from Brazil, Africa and Tasmania show anomalous $^{87}Sr/^{86}Sr$ initial ratios (Mantovani *et al.*, 1985) which have been attributed to crustal contamination (e.g. Faure *et al.*, 1972; Erlank *et al.*, 1984; Petrini *et al.*, 1987). However, the pattern of trace element enrichment could reflect an ancient subduction event and this seems a better model to explain similar effects over several thousand kilometres of almost linear outcrop. Kyle, Pankhurst & Bowman (1983) noted that many trace element ratios for some Kirkpatrick basalts were more akin to calc-alkaline andesites than to mid-ocean ridge tholeiites. The problem of enrichment by subduction well within a continent may be resolved by considering the position of the continental margin during the Late Proterozoic and not the position of the modern margin. Bartholomew (1984) has interpreted the belts of late Precambrian plutons across South America from northernmost Chile south-eastward to Argentina, south of Buenos Aires, as the late Precambrian continental margin. This effectively places the Mesozoic Paraná magmatic province within reach of a Late Proterozoic subduction event. Therefore, enrichment of the Mesozoic magmas by a common subduction event along a continental margin from Brazil to Tasmania seems entirely plausible. This subduction event probably created an enriched lithosphere mantle keel to the adjacent craton.

Conclusions

Trace element and isotope studies of the Mesozoic tholeiites of the Theron Mountains, Coats Land, and basaltic lavas of Dronning Maud Land indicate that they were derived from a subcontinental mantle source which was enriched, probably by a subduction event, about 1000 m.y. ago. These rocks are correlated with the petrogenetically similar Ferrar Supergroup magmas of the Transantarctic Mountains. The Cenozoic alkali basalts of Marie Byrd Land and the McMurdo Volcanic Group are adjacent to the Transantarctic Mountains but they are depleted ocean-island basalt types of a contrasting source region, despite their proximity. The proposed model shows that the foundered slab from the 1000 Ma event provided an effective barrier within the mantle between the two adjacent source regions, and that it possibly also triggered the rise of the alkali magmas. Thus it is deduced that the present coastal fringe of Marie Byrd Land was an exotic crustal fragment accreted to the margin of the Antarctic craton about 1000 m.y. ago and that it was subsequently separated from the craton during the Cenozoic along the line of its original accretionary suture.

Acknowledgements

We acknowledge the use of British Antarctic Survey rock collections originally made by Drs L.M. Juckes and D. Brook, the use of analytical facilities and expertise at the Department of Geology, University of Nottingham and the Department of Earth Sciences, The Open University, and discussions with colleagues at these departments and the British Antarctic Survey. We are grateful to Dr K.G. Cox for a critical review of the manuscript.

References

Barker, P.F. (1982). The Cenozoic subduction history of the Pacific margin of the Antarctic Peninsula: ridge crest–trench interactions. *Journal of the Geological Society, London*, **139(6)**, 787–801.

Bartholomew, D.S. (1984). Geology and geochemistry of the Patagonian batholith (45°–46° S), Chile. PhD thesis, University of Leicester. (unpublished).

Dalziel, I.W.D. & Elliot, D.H. (1982). West Antarctica: problem child of Gondwanaland. *Tectonics*, **1(1)**, 3–19.

Devey, C.W. & Cox, K.G. (1987). Relationships between crustal contamination and crystallisation in continental flood basalt magmas with special reference to the Deccan Traps of the Western Ghats, India. *Earth and Planetary Science Letters*, **84(1)**, 59–68.

Elliot, D.H. (1975). Tectonics of Antarctica: a review. *American Journal of Science*, **275-A**, 45–106.

Erlank, A.J., Marsh, J.S., Duncan, A.R., Miller, R. McG., Hawkesworth, C.J., Betton, P.J. & Rex, D.C. (1984). Geochemistry and petrogenesis of the Etendeka volcanic rocks from SWA/Namibia. *Geological Society of South Africa, Special Publication*, **13**, 195–245.

Faure, G., Hill, R.L., Jones, L.M. & Elliot, D.H. (1972). Isotope composition of strontium and silica content of Mesozoic basalt and dolerite from Antarctica. In *Antarctic Geology and Geophysics*, ed. R.J. Adie, pp. 617–24. Oslo; Universitetsforlaget.

Futa, K. & LeMasurier, W.E. (1983). Nd and Sr isotopic studies on Cenozoic mafic lavas from West Antarctica: another source for continental alkali basalts. *Contributions to Mineralogy and Petrology*, **83(1/2)**, 38–44.

Gibson, G.M. & Wright, T.O. (1985). Importance of thrust faulting in the tectonic development of northern Victoria Land, Antarctica. *Nature, London*, **315(6019)**, 480–3.

Grindley, G.W. & Davey, F.J. (1982). The reconstruction of New Zealand, Australia and Antarctica. In *Antarctic Geoscience*, ed. C. Craddock, pp. 15–29. Madison; University of Wisconsin Press.

Grindley, G.W. & Oliver, P.J. (1983). Palaeomagnetism of Cretaceous volcanic rocks from Marie Byrd Land, Antarctica. In *Antarctic Earth Science*, ed. R.L. Oliver, P.R. James & J.B. Jago, pp. 573–8. Canberra; Australian Academy of Science and Cambridge; Cambridge University Press.

Hart, S.R. (1984). A large isotopic anomaly in the Southern Hemisphere mantle. *Nature, London*, **309(5971)**, 753–7.

Hawkesworth, C.J., Mantovani, M.S.M., Taylor, P.N. & Palacz, Z. (1986). Evidence from the Paraná of south Brazil for a continental contribution to Dupal basalts. *Nature, London*, **322(6077)**, 356–9.

Hill, I.A. & Barker, P.F. (1980). Evidence for Miocene back-arc spreading in the central Scotia Sea. *Geophysical Journal of the Royal Astronomical Society*, **63(2)**, 427–40.

Kyle, P.R., Elliot, D.H. & Sutter, J.F. (1981). Jurassic Ferrar Supergroup tholeiites from the Transantarctic Mountains, Antarctica, and their relationship to the initial fragmentation of Gondwana. In *Gondwana Five*, ed. V.M. Cresswell & P.J. Vella, pp. 283–7. Rotterdam; A.A. Balkema.

Kyle, P.R. & Muncy, H.L. (1983). The geology of the mid-Miocene McMurdo Volcanic Group at Mount Morning, McMurdo Sound, Antarctica. In *Antarctic Earth Science*, ed. R.L. Oliver, P.R. James & J.B. Jago, p. 675. Canberra; Australian Academy of Science and Cambridge; Cambridge University Press.

Kyle, P.R., Pankhurst, R.J. & Bowman, J.R. (1983). Isotopic and chemical variations in Kirkpatrick Basalt Group rocks from southern Victoria Land. In *Antarctic Earth Science*, ed. R.L. Oliver, P.R. James & J.B. Jago, pp. 234–7. Canberra; Australian Academy of Science and Cambridge; Cambridge University Press.

LeMasurier, W.E. (1972). Volcanic record of Cenozoic glacial history of Marie Byrd Land. In *Antarctic Geology and Geophysics*, ed. R.J. Adie, pp. 251–9. Oslo; Universitetsforlaget.

Mantovani, M.S.M., Marques, L.S., Sousa, M.A. de, Civetta, L., Atalla, L. & Innocenti, F. (1985). Trace element and strontium isotope constraints on the origin and evolution of Paraná continental flood basalts of Santa Catarina State (southern Brazil). *Journal of Petrology*, **26(1)**, 187–209.

Olson, P., Schubert, G. & Anderson, C. (1987). Plume formation in the D″-layer and the roughness of the core–mantle boundary. *Nature, London*, **327(6121)**, 409–13.

Pankhurst, R.J., Millar, I.L., McGibbon, F., Menzies, M., Hawkesworth, C.J. & Kyle, P.R. (1986). Evidence for the enriched source of the Jurassic Ferrar magmas of Antarctica. *Terra Cognita*, **6**, 204 (abstract only).

Petrini, R., Civetta, L., Piccirillo, E.M., Bellieni, G., Comin-Chiaramonti, P., Marques, L.S. & Melfi, A.J. (1987). Mantle heterogeneity and crustal contamination in the genesis of low-Ti continental flood basalts from the Paraná Plateau (Brazil): Sr–Nd isotope and geochemical evidence. *Journal of Petrology*, **28(4)**, 701–26.

Smith, A.G., Hurley, A.M. & Briden, J.C. (1981). *Phanerozoic Paleocontinental World Maps*. Cambridge; Cambridge University Press.

Stump, E., Sheridan, M.F., Borg, S.G. & Sutter, J.F. (1980). Early Miocene subglacial basalts, the East Antarctic ice sheet, and uplift of the Transantarctic Mountains. *Science*, **207(4432)**, 757–9.

Treves, S.B. (1967). Volcanic rocks from the Ross Island, Marguerite Bay and Mt Weaver areas, Antarctica. In *Proceedings of the Symposium on Pacific–Antarctic Sciences*, ed. T. Nagate, pp. 136–49. Tokyo; National Science Museum. (Japanese Antarctic Research Expedition Scientific Reports, Special issue, No. 1.)

Wade, F.A. & Couch, D.R. (1982). The Swanson Formation, Ford Ranges, Marie Byrd Land – evidence for and against a direct relationship with the Robertson Bay Group, northern Victoria Land. In *Antarctic Geoscience*, ed. C. Craddock, pp. 609–16. Madison; University of Wisconsin Press.

Crustal development: the Transantarctic Mountains

Sedimentary palaeoenvironments of the Riphaean Turnpike Bluff Group, Shackleton Range

H.-J. PAECH, K. HAHNE & P. VOGLER

Zentralinstitut für Physik der Erde, Akademie der Wissenschaften der DDR, Telegrafenberg, Potsdam, 1561, GDR

Abstract

A sedimentological characterization of the Riphaean Turnpike Bluff Group is given on the basis of field observations compiled during the 22nd Soviet Antarctic Expedition, microscope studies, and whole-rock mineral, geochemical and isotopic analyses. An incipient weathering crust on Turnpike Bluff Group sediments is indicated by depletion of Zn, Co, Mn, Cu, Ba, enrichment of the Al_2O_3 content and oxidation of the Fe compounds. The overlying quartzite sequence (attaining 10 m in thickness) is considered to be of aeolian origin. In the following carbonate-bearing formation (about 70 m in thickness), sedimentation occurred under shallow marine conditions. Evidently, the subsequent history was controlled by continuing subsidence which was almost completely compensated by arenite–pelite alternations of > 1 km thickness. There are no indications of deep-water turbidity currents within this succession. The source area of the Turnpike Bluff Group was to the north-east of the present outcrops. The Turnpike Bluff Group was evidently deposited in an epicratonic basin despite the subsequent 'Pan-African' weak deformations which are associated with upthrusting of rigid basement blocks.

Geological setting

The Turnpike Bluff Group (TBG), described by Stephenson (1966), Clarkson (1983), Marsh (1983) and others, is a sedimentary sequence of > 1 km thickness, exposed on the southern flank of the Shackleton Range (Fig. 1). On the basis of the discovery of stromatolite colonies occurring in the lower part of the TBG, it is assigned to a Riphaean age (Golovanov, Mikhailov & Shulyatin, 1980). This is in general accordance with microfossil assemblages identified by Weber (1990). These comprise locally abundant occurrences of spheromorphic microfossils of the genera *Nucellosphaeridium*, *Trachysphaeridium* and *Palaeoanacystis*, and of *Bavinella*-like specimens. The age is confirmed by a Rb–Sr radiometric date (526 ± 6 Ma) which, in contrast to the opinion of Pankhurst, Marsh & Clarkson (1983), is considered to indicate the age of metamorphism.

Contrary to the lithostratigraphic terminology of Clarkson (1983) and Marsh (1983), the following description is based on the probable correlation of the lowermost units (i–iii, Table 1). The thickest sequence (iv) shows a lateral transition from an arenite-dominated series on the southern limb of the Read Mountains anticlinorium (and its western prolongation) to the psephite–arenite succession in the easternmost outcrops and the pelite-dominated TBG on the northern limb of the anticlinorium.

With respect to the tectonic structure being attributed to the Beardmore tectogenesis, the TBG is associated with the limbs of the Read Mountains anticlinorium and its prolongation. The deformation intensity varies considerably from flat-lying almost undeformed deposits (Hofmann & Paech, 1983) to phyllite grade in places.

In view of the general trend, the intensity decreases in a southerly direction due to the diminishing influence of the thrust of Shackleton Range crystalline blocks. The corresponding grade of metamorphism associated with the tectonic

Table 1. *Lithostratigraphical subdivision of the Turnpike Bluff Group*

(iv)	Arenite–pelite alternation, > 1 km thick, containing thin carbonate and psephite intercalations; dominant in easternmost outcrops.
(iii)	Carbonate-bearing clastics (in association with stromatolites); about 70 m thick.
(ii)	Basal quartzite; up to 10 m thick.
(i)	Basal red-coloured clastics, including weathering debris; < 3 m thick.
Unconformity	~~~~~~~~~~
	Basement represented by tourmaline-bearing granitoid of about 1400 Ma (Rex, 1972; Hofmann *et al.*, 1980) and by high-grade metamorphics of possible Archaean origin.

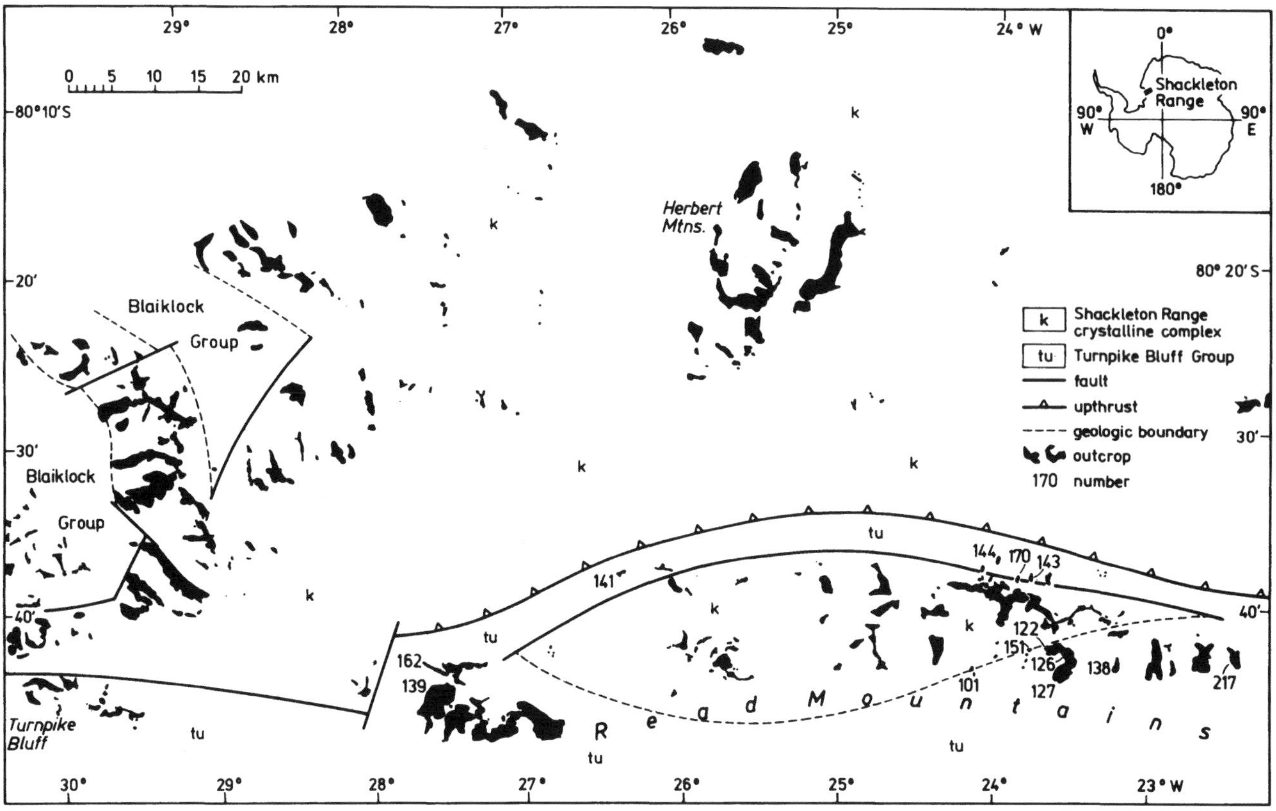

Fig. 1. Geological sketch map of the Shackleton Range.

deformation is reflected in mineral assemblages (Table 2). Whereas the pervasively affected series contain biotite, in weakly deformed deposits remnants of kaolinite may be present. Besides the occurrence of biotite formed during metamorphism, transverse biotite is also present on the northern limbs of the Read Mountains anticlinorium.

Sedimentological characterization

In the two outcrops (101-O, 122-O) exhibiting the unconformity beneath the TBG at Nicol Crags and the northwesternmost ridge of Mount Wegener, the basement is represented by a 1400 Ma old granitoid with weakly discernible gneissic structure. It is characterized by upwardly increasing red colouration and rock disintegration caused by Riphaean weathering.

Basal red-coloured clastics

The basal red-coloured clastics are composed of arenite and siltstone containing pebbles (maximum size 6 cm) and conglomerate intercalations. Evidently, the arenitic and coarser fraction represents debris of the directly underlying basement granitoid. Besides the basement pebbles, this is confirmed by the occurrence of clastic tourmaline. The fine-grained material can be referred to Riphaean weathering processes. The stratification is well developed but other primary sedimentary structures are lacking. They locally contain microfossils in abundance; these are frequently well preserved and indicate a shallow marine environment.

Basal quartzite

In the outcrops studied the basal quartzite begins abruptly. It is represented by medium- to coarse-grained (0.2–0.8 mm) pale grey-coloured arenite, predominantly (> 95%) of quartz composition. The roundness of the quartz grains in the coarsest fraction is well preserved. The quartzite sequence is subdivided into strata whose mean thickness increases upwards from metres to decametres (in the uppermost part). It is characterized by the occurrence of cross-bedding with laminae inclined at 25° to the general stratification; in association with current striation this indicates palaeotransport from a southern direction. The problem of dark-coloured lamination bands (composed of chlorite and other dark accessories) which simulate cross-bedding has not been solved yet. Pseudo-cross-bedding is dragged to the top of the strata with axes oriented perpendicularly to the primary lamina inclination. Locally, the quartzite contains concentrations of carbonate-cemented concretions.

Carbonate-bearing clastics

The carbonate-bearing clastics pass gradationally upwards into pelites, greywackes and carbonates. Of particular interest are the intercalated grey carbonates which comprise about one-third of the sequence, and have individual beds up to 4 m thick. These have a biogenic origin, being formed by bioherms of *in-situ* stromatolites plus clastics. The colour of the siliclastic strata is predominantly pale to dark grey. Fine-

Table 2. *Mineralogical composition and geochemical analyses of the TBG (mean values)*

	Phyllite $\bar{X}$	Slate $\bar{X}$	Arenite $\bar{X}$	Basal clastics $\bar{X}$	Quartzite $\bar{X}$
X-ray diffractometry (%)					
Quartz	22	23	36	54	
Feldspar (K-orthoclase)	16	7	20	0	
Muscovite	29	53	24	40	
Chlorite	13	17	15	2	
Biotite	20				
Calcite			5		
Haematite				4	
Major elements (%)					
SiO_2	57.0	58.62	68.7		
TiO_2	0.68	0.81	0.61		
Al_2O_3	18.15	19.34	12.1		
Fe_2O_3 (total)	7.50	6.48	4.20		
FeO	4.78	4.09	3.14		
Fe_2O_3	2.20	1.93	0.71		
MnO	0.18	0.048	0.053		
MgO	3.70	2.74	1.90		
CaO	1.05	0.69	2.30		
Na_2O	1.41	1.04	2.50		
K_2O	4.00	4.62	2.60		
P_2O_5	0.16	0.122	0.17		
Trace elements (ppm)					
Li	58	44	26	23	5
Be	3.33	4.0	1.6	2.25	0.17
Sc	16.9	16.6	6.6	9.5	0.2
Ti	4486	5117	2710	3348	78
V	159	115	46	57	4
Cr	90	89	28	26	5
Mn	1434	323	389	19	47
Co	26.5	25.6	15.9	4	4
Ni	49	42	19	10	4
Cu	33	15	20	3	5
Zn	129	71	37	14	5.5
Sr	103	59	101	55	3
Y	27.4	23.3	22.1	30.1	3.2
Ba	589	580	457	141	4
La	34.5	47.3	29.7	42.8	2.0

grained clastics containing a small amount of carbonate matrix predominate.

Arenite–pelite alternation

The majority of TBG outcrops belong to the arenite–pelite succession. This is represented by greywacke and quartzitic arenite, in alternation with pelite containing a varying amount of carbonate matrix or thin carbonate bands. At Mount Wegener (122-O) the transition from the underlying carbonate-bearing sequence is gradational over 10 m, as shown by the disappearance of carbonate banks and the occurrence of light green-coloured deposits. Additionally, the transition is reflected in the increasing deformation intensity, which varies from the uncleaved lowermost part of the TBG to the phyllite at the top of Mount Wegener (revealing two cleavages; Hofmann & Paech, 1983).

The thickness of conglomerate intercalations only rarely exceeds 1 m (Clayton Ramparts, 162-O; Flett Crags; easternmost Read Mountains, 217-O; Fig. 1). Generally, the clast size ranges from 2–10 mm. The provenance of the TBG debris can be deduced from the composition of clastic material. The clasts are mostly represented by carbonates which are commonly deformed pervasively. They can be referred either to the marble of the Skidmore Group (Paech, 1985) or to the lowermost part of the TBG. Due to the consequent deformation in small pebbles, the carbonate types cannot be identified exactly. Seemingly, the occurrence of gneiss and granite is extremely restricted. Only Clarkson (1983) has described corresponding rock debris from the eastern part of the Read Mountains,

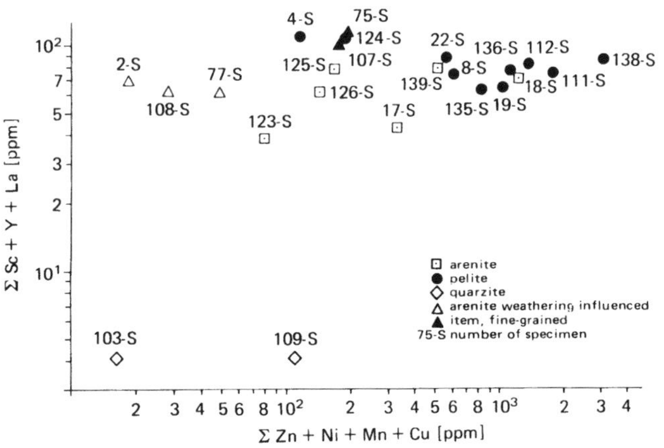

Fig. 2. Correlation of redox-sensitive with redox-insensitive elements.

where carbonate pebbles attain maximum diameters (10 cm). Other clasts are represented by quartzite, sericitized slate, quartz and subordinate ophitic mafics.

The arenite fraction of the pelite–arenite alternation differs in composition from that of the pebbles. The arenaceous grains are predominantly quartz and feldspar with varying amounts of orthoclase, microcline and plagioclase; within the study area, pure quartzite is missing. Myrmekitic extinction and tourmaline grains indicate, at least in part, a magmatic composition of the source area. Furthermore, the presence of high- to medium-grade metamorphic rocks is proved by the occurrence of garnet-enriched arenite. However, the relative proportions of magmatic and metamorphic rocks in the source area is not clear.

The arenites within the succession show only a few primary sedimentary structures. There are occasional large-scale sets of cross-bedding, but more common is horizontal lamination in fine-grained lithologies. Sedimentary structures characterizing flysch deposits are almost completely absent; only fine groove casts (indicating slow palaeocurrents) have been found.

Geochemical characterization

The limited quantity of geochemical data (Table 2) is complemented by X-ray diffractometry analyses indicating the presence of quartz, plagioclase, orthoclase, sericite, chlorite, biotite, calcite, haematite and, rarely, pyrophyllite. Together these are sufficient to characterize the TBG sediments.

The deposits formed from debris weathered from the Riphaean are marked by depletion in Zn, Ni, Co, Mn, and Ba, and the oxidization of the Fe compounds (as indicated by the red colouration). The relatively high content of the Al_2O_3 in sample O8-S can probably be related to the weathering processes in Riphaean times.

A plot of elements sensitive to redox conditions versus insensitive elements (Fig. 2) shows the geochemical depletion trend inferred to be due to Riphaean weathering processes. The content of Zn, Ni, Mn and Cu clearly increases from the fine-grained weathered debris, through coarse-grained weathering-influenced material, to arenite and pelite; however, in contrast, the amount of Sc + Y + La shows an insignificant increase. Due to the overwhelming proportion of quartz, the

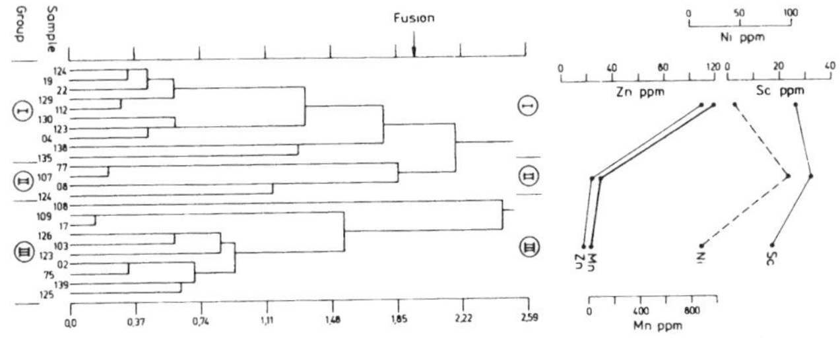

Fig. 3. Cluster analysis of trace elements in TBG sedimentary rocks.

quartzite samples are separately located from the above trend. The depleted content of Ba in weathered debris indirectly indicates the breakdown of mica and feldspar into kaolinite.

This typical depletion of single elements due to weathering processes is not restricted to deposits in the lowermost part of the TBG. Thus samples 124-S and 125-S from Mount Greenfield (139-O), of unknown vertical distance above the basal unconformity, show depletion in Cu, Mn, Zn. The weathering-induced composition of the slate (045) sampled directly above the basal quartzite ((ii) in Table 1) is also evident. These phenomena are further reflected in a cluster analysis (Fig. 3), with the samples of group I being characterized by a high content of redox-sensitive elements (Zn, Mn, Cu) and mostly represented by biotite-bearing fine-grained clastics. Group II contains rocks with muscovite and is characterized by slow enrichment of lithophile elements (Sc, Y, La and Ni). Group III comprises arenites depleted in redox-sensitive elements. The differences here are caused by the varying mineralogical composition. The correlation matrix (Table 3) shows a strong correlation of the lithophile elements (V, Sc, Be, Cr) with redox-sensitive elements (Mn, Cu and Zn).

The rare earth elements (REE) pattern is known only from sample 112-S and 2-S (Table 4). The ratio of light (LREE) versus heavy (HREE) elements of 27.6 shows enrichment of LREE. This is also reflected in the curve normalized to the North American slate (NASC) (Fig. 4). With respect to the facies, the HREE content, the low Hf/Ta ratio and the high Rb/Hf ratio (in 112-S versus sample 2-S), all suggest enrichment of LREE due to sedimentary accumulation some way from a coastline.

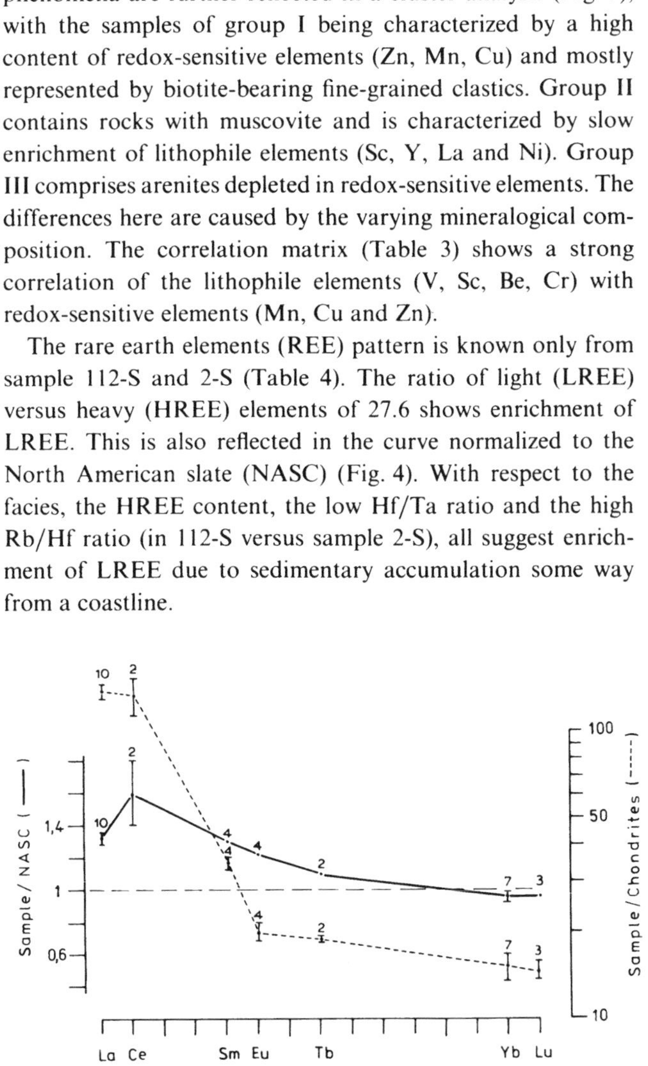

Fig. 4. REE pattern of TBG sample 112.

Table 3. *Correlation matrix of the trace elements determined in the Turnpike Bluff Group* ($n = 24$; $P = 99\%$; $v^L > 0.538$)

	Zn	Ni	Co	Mn	Ti	Cr	V	Be	Sc	Y	La	Cu
Zn	—											
Ni		—										
Co		0.529	—									
Mn	0.739			—								
Ti												
Cr				0.611	0.807	—						
V	0.687				0.864	0.844	—					
Be					0.819	0.715	0.832	—				
Sc	0.683				0.868	0.830	0.961	0.908	—			
Y					0.647				0.536	—		
La					0.762			0.690	0.597	0.694	—	
Cu	0.860			0.763		0.647	0.580		0.569			—
Sr										0.628		
Ba					0.639	0.586	0.650	0.709	0.672			
Li	0.865			0.557		0.573	0.679	0.674	0.731	0.597		0.724

Table 4. *INAA results of TBG samples 112-S (phyllite) and 2-S (basal clastics) in ppm*

	112-S	2-S
Sc	18.3	19.9
Co	30.5	1.3
Rb	169	104
Cs	11.0	5.8
La	43.8	34.4
Ce	112	74
Sm	7.4	5.5
Eu	1.51	1.14
Tb	2.93	1.4
Yb	3.0	3.2
Lu	0.46	0.65
Hf	3.7	19.8
Ta	1.27	1.2
Th	14.4	10.5
U	2.85	3.8

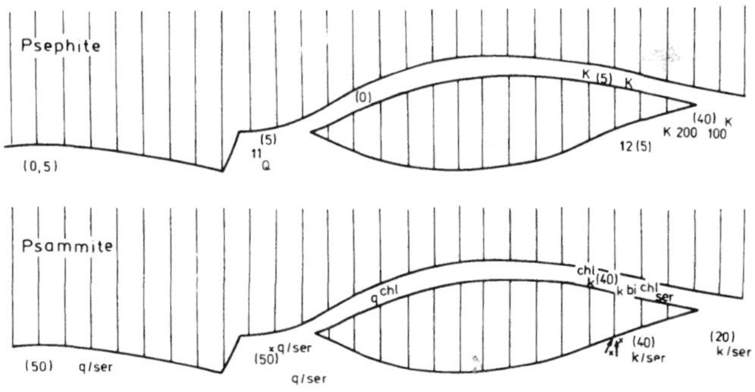

Fig. 5. Palaeogeographical reconstruction of the basin environment (for geological position see Fig. 1). Key: (40), portion of the given fraction in the sequence. Pebble composition: Q, quartz-dominated; K, carbonate-dominated; 200, maximum pebble size in mm. Matrix composition: q, quartzose; ser, sericitic; bi, biotitic; chl, chloritic; x, geochemical indication of weathered debris; arrow, palaeowind current.

Conclusions

The results outlined here allow some conclusions to be drawn on the palaeoenvironments during TBG sedimentation. In the TBG source area Riphaean weathering activity was dominated by oxidizing processes. This is shown by the occurrence of Fe^{3+} compounds, the depletion of Zn, Co, Mn, Cu, and, indirectly, by poor Ba content (as the result of kaolinite formation). The weathered debris is concentrated in the basal red-coloured clastics ((i) in Table 1), but occurs in intercalations within the carbonate-bearing formation (iii). Evidently, in the TBG sequence (iv) cropping-out at Mount Greenfield (139-O), the depletion of Zn, Mn and Cu (Table 3) indicates the presence of weathered debris in the clastic material. The basal quartzite (ii) is considered to be a dune formation. This is indicated by the maturity of the quartzite, the roundness and sorting of the quartz grains and the planar cross-bedding (characterized by mean laminae inclination of 25°, which reveals wind-controlled transport from a southerly direction).

The history of the TBG basin began with continental weathering and marine redeposition or aeolian accumulation, and then continued with the shallow marine environment of the carbonate-bearing sequence. Subsequently, the subsidence of the basin was largely compensated by the accumulation of the thick arenite–pelite succession ((iv) in Table 1); this allowed only a restricted precipitation of carbonate. Due to the absence of turbidity currents, it can be concluded that the sedimentary environment did not attain deep water conditions. Thus there exists a significant difference from the Riphaean Patuxent Formation (Frischbutter & Vogler, 1985) of the Pensacola Mountains. Whereas the TBG was an epicratonic formation, the Patuxent Formation comprises geosynclinal deposits.

The composition of the source area can be deduced from the clastic fragments in the TBG. The arenaceous fraction is inferred to be derived from an acid crystalline complex of either magmatic or metamorphic origin. On the other hand, the gravel and pebble component indicates a provenance from a sedimentary rock complex which was probably part of the

Table 5. *List of principal outcrops and sample numbers* (for location see Fig. 1)

Outcrop	Geographical position	Samples
101-O	Nicol Crags	(75, 77, 103, 107)-S
122-O	Northern corner, Mount Wegener	(2, 4, 8, 108, 109)-S
126-O	Southern corner, Mount Wegener	(17, 18)-S
127-O	Western slope, Mount Wegener	(19, 22)-S
139-O	Mount Greenfield	(123, 124, 125, 126)-S
141-O	Nunatak at Crossover Pass	129-S
143-O	Nunatak north of Eskola Cirque	(135, 136)-S
144-O	Nunatak north of Eskola Cirque	(138, 139)-S
151-O	Northern Bubnoff Nunatak	(111, 112)-S
162-O	Clayton Ramparts	
170-O	Nunatak north of Eskola Cirque	

Skidmore Group (Paech, 1985). However it could also represent, in part, redeposited rock debris of the TBG.

With respect to the palaeogeographical configuration of the TBG basin, the sedimentological data allow only a very simplified model to be constructed (Fig. 5). The occurrence of fragments greater than 5 cm in only the eastern part of the Read Mountains reveals the proximal position of this area. This is confirmed by the increasing portion of the psephite fraction. Due to the occurrence of probable weathering debris in the TBG sequences at Mount Greenfield (139-O), the position of the source area is assumed to be north-east of the Read Mountains.

Acknowledgements

The authors thank the staff of the 22nd Soviet Antarctic Expedition (especially Dr G. Shulyatin and V. Masolov) for fruitful cooperation and logistic support in the Shackleton Range. Dr G. Grikurov and Dr V. Ivanov are thanked for discussion of the results.

References

Clarkson, P.D. (1983). Geology of the Shackleton Range: II. The Turnpike Bluff Group. *British Antarctic Survey Bulletin*, **52**, 109–24.

Frischbutter, A. & Vogler, P. (1985). Contributions to the geochemistry of magmatic rocks in the Upper Precambrian–Lower Palaeozoic profile of the Neptune Range, Transantarctic Mountains, Antarctica. *Zeitschrift für Geologische Wissenschaften*, **13(3)**, 345–57.

Golovanov, N.P., Mikhailov, V.M. & Shulyatin, O.G. (1980). Pervye diagnostiruemye stromatolity Antarktidy i ich biostratigrafichekoe znachenie (First determinable stromatolites of Antarctica and their biostratigraphic bearing). *Antarktika*, **19**, 152–9.

Hofmann, J., Kaiser, G., Klemm, W. & Paech, H.-J. (1980). K/Ar-Alter von Doleriten und Metamorphiten der Shackleton Range und der Whichaway Nunataks, Ost- und Südostumrandung des Filchner-Eisschelfs (Antarktis). *Zeitschrift für Geologische Wissenschaften*, **8(9)**, 1227–32.

Hofmann, J. & Paech, H.-J. (1983). Tectonics and relationships between structural stages in the Precambrian of the Shackleton Range, western margin of the East Antarctic Craton. In *Antarctic Earth Science*, R.L. Oliver, P.R. James & J.G. Jago, pp. 183–9. Canberra; Australian Academy of Science and Cambridge; Cambridge University Press.

Marsh, P.D. (1983). Late Proterozoic and early Palaeozoic history of the Shackleton Range, Coats Land. In *Antarctic Earth Science*, R.L. Oliver, P.R. James & J.G. Jago, pp. 190–3. Canberra; Australian Academy of Science and Cambridge; Cambridge University Press.

Paech, H.-J. (1985). Tectònic structures of the crystalline basement in the Shackleton Range, Antarctica. *Zeitschrift für Geologische Wissenschaften*, **13(3)**, 309–19.

Pankhurst, R.J., Marsh, P.D. & Clarkson, P.D. (1983). A geochronological investigation of the Shackleton Range. In *Antarctic Earth Science*, R.L. Oliver, P.R. James & J.G. Jago, pp. 176–82. Canberra; Australian Academy of Science and Cambridge; Cambridge University Press.

Rex, D. (1972). K–Ar age determinations on volcanic and associated rocks from the Antarctic Peninsula and Dronning Maud Land. In *Antarctic Geology and Geophysics*, ed. R.J. Adie, pp. 133–6. Oslo; Universitetsvorlaget.

Stephenson, P.J. (1966). Geology 1. Theron Mountains, Shackleton Range and Whichaway Nunataks. *Scientific Reports of the Trans-Antarctic Expedition*, **8**, 1–79.

Weber, B. (1990). Microfossils from the Turnpike Bluff Group, Shackleton Range. *Zeitschrift für Geologische Wissenschaften*, **18** (in press).

Precambrian ancestry of the Asgard Formation (Skelton Group): Rb–Sr age of basement metamorphic rocks in the Dry Valley region, Antarctica

C.J. ADAMS & P.F. WHITLA

Institute of Nuclear Sciences, DSIR, Lower Hutt, New Zealand

Abstract

Rb–Sr whole-rock isochron ages are reported for greenschist–amphibolite-facies quartzo-feldspathic schists, pelitic schists and marbles from the Asgard Formation (Skelton Group) in the lower Wright and Victoria valleys, Antarctica. Mica-schists (pelitic protolithology) give evidence of regional metamorphism at least 615 ± 15 Ma ago, whilst quartzo-feldspathic schists (psammite protolithology) give evidence of incomplete strontium isotopic homogenization at ~ 670 Ma. Rb–Sr ages of impure marbles are 840 ± 30 Ma, suggesting a late Precambrian age of sedimentation and correlation with the Koettlitz Group of southern Victoria Land rather than the Byrd Group (Cambrian) in the central Transantarctic Mountains.

Introduction

In the Dry Valley region of southern Victoria Land (Fig. 1) the local basement comprises medium–high-grade metamorphic rocks (Skelton Group) and granitoids (Wright Intrusives) (McKelvey & Webb, 1962). Rb–Sr and K–Ar dating show that most granitoids are 460–510 Ma old and were emplaced in Late Cambrian–Early Ordovician times during the Ross Orogeny (data summarized by Faure & Jones, 1974). The Wright Intrusives are a complex suite of granitoid plutons (some gneissic) and dykes intruded into the metasedimentary Skelton Group. In the Wright and Victoria valleys (Fig. 1), the Skelton Group is represented by a single unit (Asgard Formation) of predominantly medium–high-grade marble, calc-schist, some hornfels and very subordinate pelitic and quartzo-feldspathic schist (McKelvey & Webb, 1962; Haskell *et al.*, 1965). All are strongly deformed and regionally metamorphosed to upper greenschist–amphibolite-facies and grade locally into paragneiss, migmatite and orthogneiss (Olympus Granite-gneiss). On all scales, the metamorphic rocks appear as large enclaves within Olympus Granite-gneiss, which is probably an early syn-metamorphic Wright Intrusive (i.e. part of the Ross orogenic cycle) or possibly an older, (?) Precambrian unit (Faure & Jones, 1974; Vocke & Hanson, 1981). In detail the gneiss is heterogeneous and complex; U–Pb, Rb–Sr and K–Ar dating have yielded inconclusive ages, there is evidence of early Precambrian (Archaean) and late Precambrian (638 Ma) ancestries for zircons (U–Pb evidence, Deutsch & Grögler, 1966; Vocke & Hanson, 1981), for Late Cambrian–Early Ordovician emplacement and/or metamorphism (Rb–Sr whole-rock dating, Faure & Jones, 1974), and for further reheating (K–Ar and fission-track age evidence, Vocke & Hanson, 1981) during intrusion of the Ferrar dolerites. To examine the Precambrian ancestry of Dry Valley basement further we have used Rb–Sr whole-rock methods to date the metamorphism of metapelitic and metapsammatic horizons in the Asgard Formation.

The Asgard Formation (Skelton Group)

The Asgard Formation is a thick sequence of complexly and strongly deformed medium–high-grade metamorphic rocks. It is dominated by calcareous rocks (marbles and calc-schists), with only minor intercalations of psammitic and pelitic protolithologies (quartzo-feldspathic and mica-schists) and even rarer metabasic horizons (amphibolitic schists). All are strongly recrystallized schists which show no relict sedimentary features apart from major large-scale (> 10 m) bedding. Amphibolitic schists are small bodies a few metres thick and 25–50 m long. At the time of metamorphism, the sequence would have been dominated by limestones (low Rb/Sr ratio) with $^{87}Sr/^{86}Sr$ ratios close to contemporary seawater (~ 0.708; Veizer & Compston, 1976). Since this is the principal Sr reservoir, isotopic homogenization during metamorphism should have converged on the marble $^{87}Sr/^{86}Sr$ ratio.

Rb–Sr dating

Detailed collections of 1–2 kg schist samples were made from the Asgard Formation in the lower Wright and lower Victoria valleys (Fig. 1), in particular between Meserve and Hart (type section) and Hart and Goodspeed glaciers, and south of Mount Theseus and on Sponsors Peak. Rb–Sr analytical data are listed in Tables 1 & 2, and sample locations are

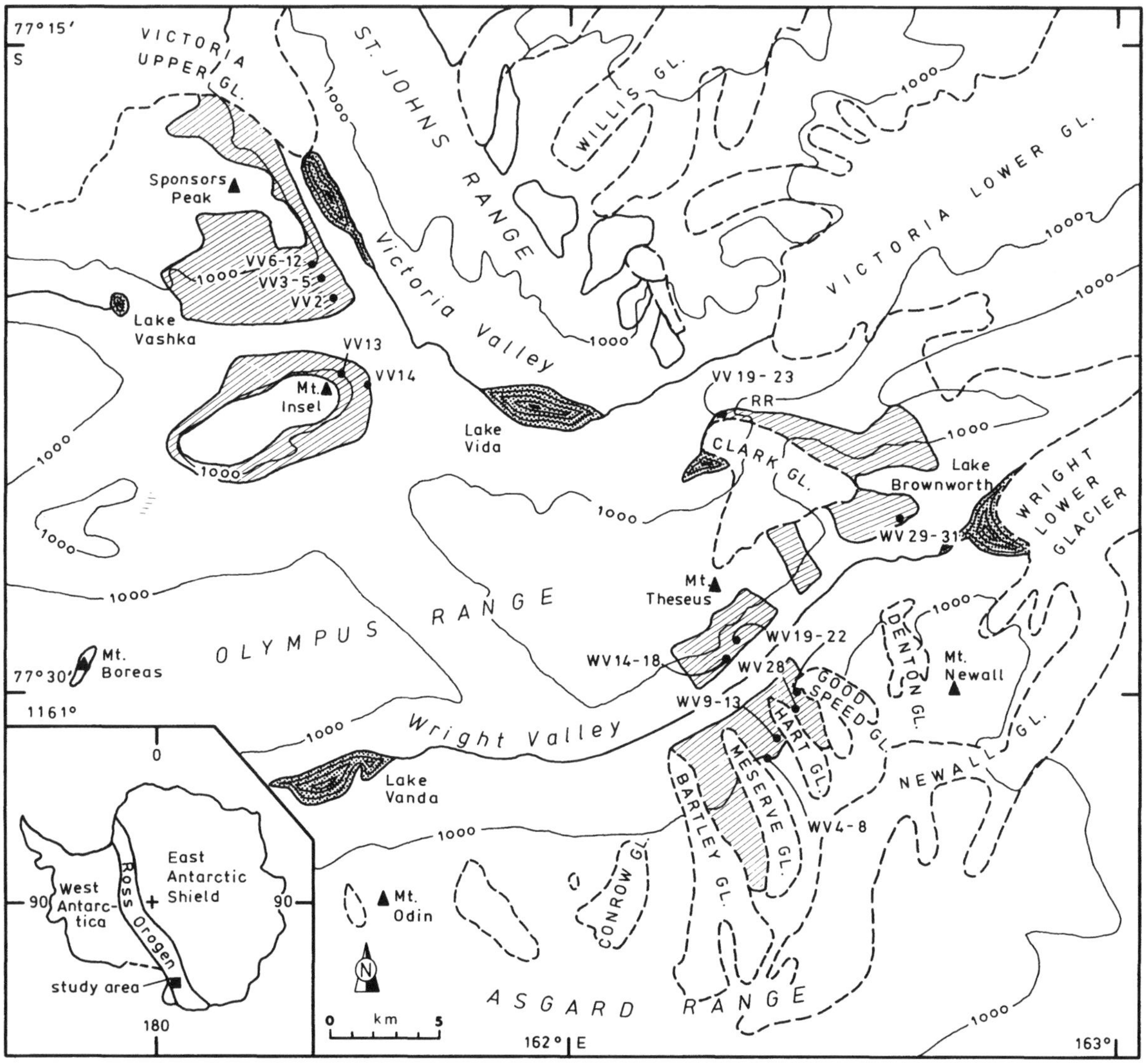

Fig. 1. Geological outcrop map of the Asgard Formation (diagonal ruling) in the lower Wright and Victoria valleys, showing the location of samples dated. RR = Robertson Ridge.

shown in Fig. 1. Representative XRF chemical analyses are given in Table 3. Analytical techniques were similar to those described by Adams (1986); isochron ages were calculated using a method similar to York (1966) and modified at the Institute of Nuclear Sciences (INS) by I.J. Graham, with decay constants recommended by Steiger and Jäger (1977).

Rb–Sr dating of schists assumes that inherently variable $^{87}Sr/^{86}Sr$ ratios of the precursors are homogenized during metamorphic recrystallization, not only amongst component minerals at the hand-specimen scale but also between several sedimentary horizons at the outcrop scale. Rb–Sr isochrons are often obtained for lower-grade metapelites (cf. Adams, 1986) but Sr isotopes are perhaps only incompletely homogenized during recrystallization of coarser-grained protolithologies. Accordingly we have sampled only finer-grained protolithologies (pelites and siltstones) which show strong metamorphic foliation and/or segregation. During the total recrystallization of proto-minerals there must have been substantial *in situ* exchange of K, Ca, Rb, Sr with consequent Sr isotopic homogenization.

Wright Valley Rb–Sr data

Rb–Sr data for 18 quartzo-feldspathic schist samples are shown in Fig. 2; data for marbles and mica-schist samples are discussed separately (see below). The data do not define a single isochron, rather only an errorchron age (709 ± 57 Ma) and initial $^{87}Sr/^{86}Sr$ ratio of 0.7080 ± 0.0007 (MSWD 4.8) can be calculated. The data may be subdivided further into three local groups: (1) Meserve–Hart glaciers area; (2) Hart–Goodspeed glaciers area; and (3) Olympus Range (Mount Theseus–Lake Brownworth). In each case meaningful isochron ages have not been obtained, the data falling within elongate envelopes which partly overlap and are subparallel. Since each set comes from a small area of demonstrably similar metamorphic age, it is clear that the significant geological data scatter is caused by incomplete strontium isotope homogenization between whole-rock samples at the time of metamorphism. It is striking that the data do appear to be confined to a narrow band; the dotted lines in Fig. 2 are drawn to confine all the data with minimum separation. This could suggest that the initial ratio did indeed vary at

Table 1. *Rb–Sr age data: Asgard Formation, Wright Valley*

INS R. No.[a] (field no.)	Sample details Location	Lat. °S Long. °E	Rb (ppm) Sr (ppm)	$^{87}Rb/^{86}Sr$ ($\pm$ 1.0%)	$^{87}Sr/^{86}Sr$ ($\pm$ 0.00010)[b]
12293 (WV4)	Biotite-schist; between Meserve and Hart glaciers, alt. 950 m	77°32′ 166°22′	98.9 66.5	4.329	0.76547 ± 6
12294 (WV5)	Pelitic quartz–biotite-schist; between Meserve and Hart glaciers, alt. 800 m	77°31′ 162°22′	61 445	0.394	0.71108 ± 9
12295 (WV6)	Pelitic quartz–biotite-schist; between Meserve and Hart glaciers, alt. 800 m	77°31′ 162°22′	68 436	0.446	0.71180 ± 7
12296 (WV7)	Quartzo-feldspathic-schist; psammite unit as 20 m screen; loc. as for 12293 above	77°31′ 162°22′	53.5 284.6	0.546	0.71612 ± 7
12297 (WV8)	Quartz–biotite-schist, unit in biotite, amphibolitic and calc-schists; loc. as for 12294	77°31′ 162°22′	142 170	2.374	0.73485 ± 7) 0.73499 ± 6)
12298 (WV9)	Amphibolitic-schist; 1 m lens in gneiss; Hart Glacier, west margin, alt. 900 m.	77°31′ 162°23′	17.0 331.8	0.148	0.70484 ± 8
12300 (WV11)	Marble; in schist lens in gneiss; Hart Glacier, west margin, alt. 900 m	77°31′ 162°24′	9 76	0.356	0.71236 ± 12
12301 (WV12)	Biotite-schist; in psammite unit; loc. as 12300 above	77°31′ 162°24′	145 35	11.88	0.83172 ± 21
12302 (WV13)	Quartzo-feldspathic–biotite-schist; Hart Glacier, west margin alt. 700 m	77°31′ 162°23′	114 350	0.933	0.71634 ± 8) 0.71647 ± 9)
12304 (WV15)	Quartzo-feldspathic–biotite-schist; enclave in foliated granitoid; Olympus Range 400 m	77°29′ 162°19′	106.9 627.8	0.493	0.71171 ± 9
12305 (WV16)	Biotite-schist; loc. as 12305, alt. 450 m	77°29′ 162°19′	86 845	0.298	0.70920 ± 9
12306 (WV17)	Biotite-schist; calc-schist–marble unit; Olympus Range, south side, alt. 500 m	77°29′ 162°20′	73 664	0.321	0.70977 ± 5
12307 (WV18)	Marble; in calc-schist–mica-schist unit; loc. as 12306	77°29′ 162°20′	2.3 77.8	0.086	0.70921 ± 10
12308 (WV19)	Biotite–hornblende-schist; enclave in gneiss; Olympus Range, south side, alt. 540 m	77°28′ 162°20′	75 599	0.362	0.71248 ± 6) 0.71218 ± 6)
12309 (WV20)	Calc-schist; in thick calc-unit (100 m); loc. as 12308 but alt. 600 m	77°28′ 162°20′	67.5 149.4	1.310	0.72492 ± 10
12310 (WV21)	Quartzo-feldspathic–biotite-schist; in calc-schist unit; loc. as 12309	77°28′ 162°20′	104 424	0.695	0.71456 ± 8
12311 (WV22)	Quartzo-feldspathic–biotite-schist; Olympus Range, south side alt. 580 m	77°28′ 162°21′	70.2 160.0	1.271	0.72077 ± 9
12312 (WV23)	Quartzo-feldspathic–biotite-schist; Goodspeed Glacier, west margin, alt. 660 m	77°30′ 162°25′	99.1 64.5	4.465	0.75036 ± 7
12313 (WV24)	Quartzo-feldspathic–biotite-schist; similar to 12312; same loc.	77°30′ 162°25′	82.5 71.1	3.368	0.73855 ± 5
12315 (WV26)	Quartz–biotite-schist; thick (> 100 m) metapelite–metapsammite unit; Goodspeed Glacier, west margin, alt. 680 m	77°30′ 162°25′	102 66	4.395	0.75214 ± 12
12317 (WV28)	Quartzo-feldspathic schist; in thick metasediment unit; between Hart and Goodspeed glaciers, alt. 720 m	77°30′ 162°31′	83 85	2.737	0.73574 ± 5
12318 (WV29)	Quartzo-feldspathic–biotite-schist; (possibly not *in situ*); Olympus Range, east spur, 3 km from Lake Brownworth Hut, alt. 530 m	77°26′ 162°31′	169.9 285.3	1.727	0.72779 ± 10
12320 (WV31)	Quartzo-feldspathic–biotite-schist; in calc-schist lens in gneiss; Olympus Range, east spur, south slopes, at 580 m	77°26 162°30′	204.2 640.3	0.924	0.71780 ± 10

[a]All samples are total-rock powders.
[b]Reproducibility used in isochron age calculations.

the time of metamorphism, in the range 0.707–0.713, with the lower value representing that of relatively unradiogenic strontium in plagioclase-rich schists and/or adjacent calcium-rich marbles, and the higher value that of relatively radiogenic strontium in mica or K-feldspar-rich schists. With such a small data set, this interpretation can only be tentative but, if the range of initial strontium ratio is accepted as realistic, then the slope of the data band can be related to their metamorphic age, in this case ~ 670 Ma. This late Precambrian age is important since it is quite clearly older than the metamorphic–intrusive

Table 2. *Rb–Sr age data: Asgard Formation, Victoria Valley*

INS R. No.[a] (field no.)	Sample details Location	Lat. °S Long. °E	Rb (ppm) Sr (ppm)	$^{87}Rb/^{86}Sr$ (± 1.0%)	$^{87}Sr/^{86}Sr$ (± 0.00010)[b]
12322	Biotite–calc-schist; in thick marble sequence; Sponsors	77°21′	88	1.293	0.72224 ± 6
(VV2)	Peak, southeast spur, alt. 730 m	161°35′	195		
12323	Biotite–hornblende–calc-schist; thin schist unit in thick	77°21′	143	2.195	0.72938 ± 5
(VV3)	marble sequence; Sponsors Peak, south-east spur, alt. 830 m	161°34′	186		
12324	Quartz–biotite-schist (± calc); loc. as 12323	77°21′	134	1.011	0.72178 ± 5
(VV4)		161°34′	380		
12325	Quartzo-feldspathic–biotite-schist; in pelitic sequence;	77°21′	183	3.467	0.74270 ± 6)
(VV5)	Sponsors Peak, southeast spur, alt. 840 m	161°33′	150		0.74251 ± 11)
12326	Hornblende-schist; in thick schist sequence; Sponsors Peak,	77°21′	18	0.217	0.70701 ± 4
(VV6)	southeast spur, alt. 960 m	161°32′	251		
12327	Quartzo-feldspathic–biotite-schist; loc. as 12326	77°20′	122	1.708	0.72840 ± 10
(VV7)		161°32′	204		
12328	Biotite–quartz-schist; Sponsors Peak, south-east spur, alt.	77°20′	86	2.050	0.72942 ± 7
(VV8)	960 m	161°32′	120		
12330	Biotite–quartz-schist; in thick calc-schist sequence; loc. as	77°20′	77	1.154	0.72422 ± 11
(VV10)	12328, alt. 980 m	161°32′	193		
12331	Schist as 12330; Sponsors Peak, south-east spur, alt. 990 m	77°20′	124	1.641	0.72729 ± 7
(VV11)		161°31′	215		
12332	Marble; in thick marble unit; Sponsors Peak, south-east	77°21′	108	0.794	0.71780 ± 14
(VV12)	spur, alt. 970 m	161°32′	385		
12339	Biotite-schist; (possibly not *in situ*), Robertson Ridge, centre	77°24′	124	1.443	0.72092 ± 14
(VV19)	section, alt. 870 m	162°15′	245		
12341	Biotite–quartz-schist; Robertson Ridge, centre section, alt.	77°24′	181.6	4.317	0.74386 ± 10
(VV21)	930 m	162°15′	122.2		
12342	Biotite–quartz-schist; loc. as 12341	77°24′	84.8	1.165	0.72391 ± 6
(VV22)		162°15′	210.4		
12344	Quartz–mica-schist; loc. as 12341, alt. 850 m	77°24′	21	0.307	0.71500 ± 4
(VV24)		162°15′	194		

[a]All samples are total-rock powders.
[b]Reproducibility used in isochron age calculations.

igneous events of the Late Cambrian–Early Ordovician Ross Orogeny in the Dry Valleys.

Rb–Sr data for marble samples (including one from Victoria Valley) are shown in the upper inset of Fig. 2. These could have initial Sr-isotope ratios quite different from the quartzo-feldspathic schists since the original limestones would have acquired homogeneous Sr-isotopic compositions over large areas during sedimentation, by precipitation of carbonates (and clays) in contemporary seawater; coarse detrital minerals are negligible in this case. The three samples define an isochron age of 840 ± 30 Ma (MSWD 1.4), with an initial ratio of 0.7081 ± 0.0001 that is similar to calculated late Precambrian seawater $^{87}Sr/^{86}Sr$ values (Veizer & Compston, 1976). It is argued that this is a sedimentary rather than a metamorphic age because, within the overwhelmingly large reservoir of 'common' Sr in the limestones, it is far more likely that the tiny amounts of radiogenic Sr would only be homogenized on a small (hand-specimen) scale and not on a large (kilometre) scale.

Two biotite-schist samples analysed (12283, 12301, Table 1) seem unrelated to either quartzo-feldspathic schist or marble data. Both are mica-rich and consist almost entirely of quartz and mica. Since there is no Sr-rich phase present to provide a 'sink' for radiogenic Sr diffusing out of micas, it could be lost from such a sample during any subsequent thermal event. Although similar to a biotite mineral age, the calculated isochron age for the biotite-schists (615 ± 15 Ma; initial ratio, 0.727 ± 0.001) could only be regarded as a minimum metamorphic age.

Two amphibolites analysed (point A, Fig. 2) are hornblende-rich metabasic rocks (Table 3) which represent intercalated volcanic horizons. They parallel marble and/or metapsammite units and do not cut across them. They are therefore not older than the limestones (< 840 Ma) and not younger than the metamorphism (> 670 Ma). With these age constraints we can deduce that their intial Sr ratios are 0.7030–0.7034 (sample 12298, Wright Valley) and 0.7045–0.7050 (sample 12326, Victoria Valley).

Victoria Valley Rb–Sr data

Eleven Asgard Formation schists from the Victoria Valley area (Fig. 3) represents two sites: (1) an extensive outcrop at Sponsors Peak; and (2) an enclave at Robertson Ridge. Their Rb/Sr ratios have a larger range than the Wright Valley suites. As with the Wright Valley quartzo-feldspathic

Table 3. *Representative chemical analyses of the Asgard Formation*

INS R. No. field no./area[b] rock type[a]	12293 WV4/M bi-sch	12296 WV7/M q-f-sch	12298 WV9/M hb-sch	12304 WV15/T q-f-sch	12307 WV18/T marble	12309 WV20/T calc-sch	12311 WV22/T q-f-sch	12312 WV23/G q-f-sch	12313 WV24/G q-f-sch	12318 WV29/B bi-sch	12320 WV31/B q-f-sch	12342 VV22/R q-f-sch	12343 VV23/R q-f-sch
SiO_2	80.68	72.47	37.21	48.26	17.05	70.07	80.75	83.78	85.86	55.64	52.46	55.30	56.73
TiO_2	0.52	0.64	8.04	1.28	—	0.61	0.58	0.44	0.39	0.72	0.71	0.59	0.58
Al_2O_3	7.39	8.31	10.71	20.13	0.12	8.71	7.61	6.53	5.24	17.85	18.01	9.79	9.50
Fe_2O_3	4.08	4.50	23.94	10.56	0.26	4.62	3.88	3.38	3.23	7.25	7.35	4.85	4.94
MnO	0.06	0.06	0.38	0.16	0.02	0.06	0.04	0.04	0.04	0.09	0.07	0.08	0.09
MgO	1.57	3.39	6.13	4.01	20.98	3.59	1.55	1.16	1.16	5.13	5.17	11.44	10.84
CaO	0.83	6.67	9.48	8.44	32.23	8.70	1.73	0.47	0.50	6.06	7.63	13.02	11.02
Na_2O	1.32	1.51	2.47	3.50	0.18	0.82	1.40	1.24	0.63	1.92	3.42	1.55	1.79
K_2O	2.29	1.31	0.73	2.36	0.01	1.86	1.57	2.55	2.06	4.10	4.21	2.15	2.78
P_2O_5	0.04	0.12	0.95	0.33	—	0.12	0.12	0.11	0.07	0.17	0.16	0.13	0.13
Y ppm	21.5	31.1	38.6	46.4	—	36.5	9.9	18.1	18.1	42.1	41.8	30.2	30.9
Sr	67.1	283.9	333.6	621.6	78.7	150.4	159.6	64.9	71.9	284.5	634.1	211.2	199.8
Th	13.0	7.7	—	5.4	0.8	9.4	7.6	6.4	7.2	14.9	14.9	9.4	8.9
Pb	13.6	8.8	9.0	10.1	7.0	13.8	6.5	11.5	8.7	22.1	23.8	10.1	11.3
Ga	9.3	12.5	28.2	26.2	0.3	13.0	11.2	9.7	6.9	23.7	23.6	13.0	12.8
U	1.1	1.4	0.4	1.5	1.0	1.8	1.0	0.9	0.5	2.8	2.3	1.8	1.5
Rb	102.8	54.9	17.6	109.2	2.6	69.5	72.4	102.6	85.2	175.1	210.5	85.1	120.3
LOI	0.48	0.45	0.00	1.19	29.90	0.61	0.56	0.43	0.46	1.10	0.67	1.27	1.72
Total	99.25	99.42	100.03	100.22	100.73	99.78	99.78	100.13	99.63	100.02	99.88	100.18	100.12

[a]Areas: G, Hart–Goodspeed glaciers; M, Meserve–Hart glaciers; R, Robertson Ridge; T, Mt Theseus (Olympus Range).
[b]Rocks types: bi-sch, biotite-schist; q-f-sch, quartzo-feldspathic-schist; hb-sch, hornblende-schist; calc-sch, calc-schist.

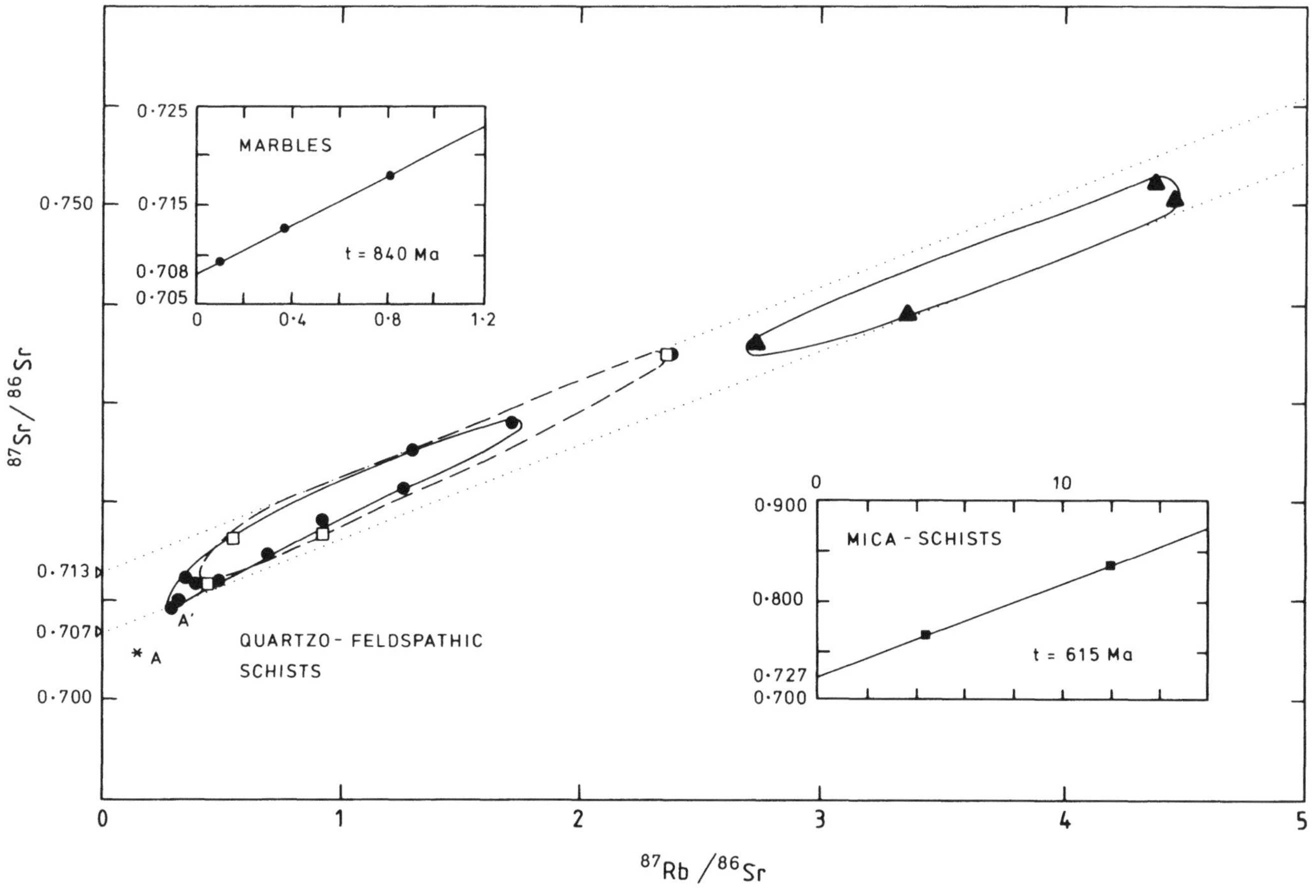

Fig. 2. Rb–Sr isochron diagram of Asgard Formation metamorphic rocks in lower Wright Valley. Main diagram: quartzo-feldpathic-schist data; lower inset: mica-schist data; upper inset: marble data (including one sample from Victoria Valley). Open squares, Meserve–Hart glaciers area; solid circles, Olympus Range area; solid triangles, Hart–Goodspeed glacier areas; A, amphibolitic schist; A′, hornblendic quartzo-feldspathic schist.

schists, neither data set falls on an isochron. However, all but one sample fall in the same data band (defined by dotted boundary lines on Fig. 3) as the Wright Valley quartzo-feldspathic schists, which suggests that they had a similar metamorphic history. However, the Robertson Ridge envelope may be skewed with respect to the others. This may indicate that they have been affected on a whole-rock scale by a younger isotopic event during the formation of the surrounding Olympus Granite-gneiss.

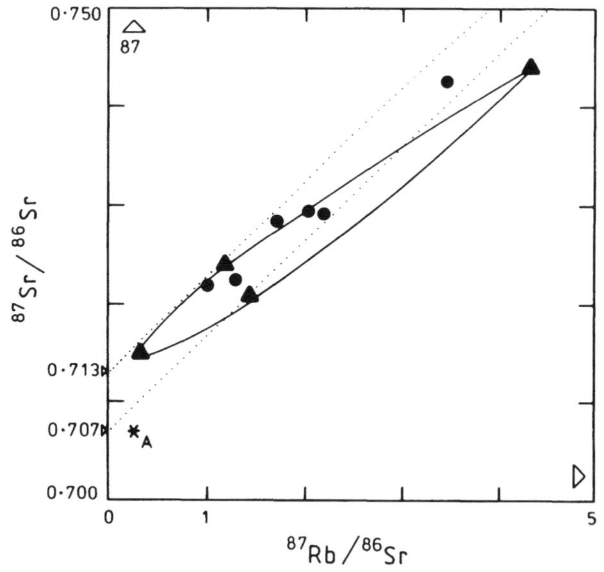

Fig. 3. Rb–Sr isochron diagram of Asgard Formation metamorphic rocks in Victoria Valley. Circles, Sponsors Peak area; triangles, Robertson Ridge area; A, amphibolitic schist. Dotted lines confining data points are transferred from Fig. 2.

Precambrian ancestry of the Asgard Formation (Skelton Group)

The limited Rb–Sr data for marble samples suggest that Asgard Formation sedimentation occurred 840 ± 30 Ma ago. Regional metamorphism to greenschist–amphibolite-facies then accomplished only incomplete Sr-isotopic homogenization (initial ratio 0.707–0.713) in late Precambrian times (~ 670 Ma ago). This event is unrelated to the later Ross Orogeny and emplacement of the Wright Intrusives.

Evidence for late Precambrian metamorphism has been suggested elsewhere in the Transantarctic Mountains (see review by Grindley, 1982). Age data are imprecise but a Beardmore Orogeny has been defined on the basis of: (1) K–Ar and Rb–Sr ages in the Nimrod Group, central Transantarctic Mountains (Gunner & Faure, 1972; Adams, Gabites & Grindley, 1982); and (2) U–Pb zircon ages in the Olympus Granite-gneiss, Dry Valley region, (Deutsch & Grögler, 1966). Slightly younger associated plutonism is suggested by latest Precambrian ages of ~ 600 Ma (mostly Rb–Sr) from the Scott Glacier, central Transantarctic Mountains and Carlyon Glacier areas, southern Victoria Land (Faure, Murtaugh & Montigny, 1968; Felder & Faure, 1980). These data provide evidence for high-grade regional metamorphism, similar to that of the Asgard Formation (Skelton Group), occurring between the two major sedimentary cycles of the Beardmore Group (late Precambrian) in the central Transantarctic Mountains, and the Bowers Supergroup (Cambrian) in northern Victoria Land. Similarly the present age data would suggest that the Skelton Group in the Dry Valley region is entirely Precambrian and a correlative of the Koettlitz Group (Precambrian) of southern Victoria Land, rather than the Byrd Group and Bowers Supergroup.

Acknowledgements

Sampling was done during the 1985–86 NZARP field season and the Antarctic Division, DSIR, Christchurch are thanked for their encouragement, logistic advice and support. Dr S. Weaver and Mr K. Palmer are thanked for XRF analyses at University of Canterbury and Victoria University, Wellington, respectively. Dr I. Graham (INS) is thanked for use of his computer programme for age calculations. Margaret MacDonald prepared the diagrams and Anne Hepenstall typed the manuscript. CJA thanks the Symposium Organizing Committee and the Trans-Antarctic Association Committee for financial support to present papers at the symposium in Cambridge.

References

Adams, C.J. (1986). Geochronological studies of the Swanson Formation of Marie Byrd Land, West Antarctica, and correlation with northern Victoria Land, East Antarctica and South Island, New Zealand. *New Zealand Journal of Geology and Geophysics*, **29**, 345–58.

Adams, C.J., Gabites, J.E. & Grindley, G.W. (1982). Orogenic history of the central Transantarctic Mountains: new K–Ar age data on the Precambrian–Lower Paleozoic basement. In *Antarctic Geoscience*, ed. C. Craddock, pp. 817–26. Madison; University of Wisconsin Press.

Deutsch, S. & Grögler, N. (1966). Isotopic age of Olympus Granite–Gneiss (Victoria Land–Antarctica). *Earth and Planetary Science Letters*, **1**, 82–4.

Faure, G. & Jones, L.M. (1974). Isotopic composition of strontium and geologic history of the basement rocks of Wright Valley, southern Victoria Land, Antarctica. *New Zealand Journal of Geology and Geophysics*, **17**, 611–27.

Faure, G., Murtaugh, J.G. & Montigny, R. (1968). Geology and geochronology of the basement complex, Wisconsin Range, Transantarctic Mountains. *Canadian Journal of Earth Sciences*, **5**, 555–60.

Felder, R.P. & Faure, G. (1980). Rubidium–strontium age determination of part of the basement complex of the Brown Hills, central Transantarctic Mountains. *Antarctic Journal of the United States*, **15(5)**, 16–17.

Grindley, G.N. (1982). Precambrian rocks of the Ross Sea region. *Journal of the Royal Society of New Zealand*, **11**, 411–23.

Gunner, J. & Faure, G. (1972). Rubidium–strontium geochronology of the Nimrod Group, central Transantarctic Mountains. In *Antarctic Geology*, ed. R.J. Adie, pp. 305–11. Amsterdam; North-Holland Publishing Co.

Haskell, T.R., Kennett, J.P., Prebble, W.M., Smith, G. & Willis, I.A.G. (1965). The geology of the middle and lower Taylor Valley of south Victoria Land, Antarctica. *Transactions of the Royal Society of New Zealand*, **2**, 169–86.

McKelvey, B.C. & Webb, P.N. (1962). Geological investigations in southern Victoria Land, Antarctica. *New Zealand Journal of Geology and Geophysics*, **5**, 143–62.

Steiger, R.H. & Jäger, E. (1977). Subcommission on geochronology: convention on the use of decay constants in geochronology and cosmochronology. *Earth and Planetary Science Letters*, **36**, 359–62.

Veizer, J. & Compston, W. (1976). $^{87}Sr/^{86}Sr$ in Precambrian carbonates as an index of crustal evolution. *Geochimica et Cosmochimica Acta*, **40**, 905–14.

Vocke, R.D., Jr & Hanson, G.N. (1981). U–Pb zircon ages and petrogenetic implications for two basement units from Victoria Valley, Antarctica. In *Dry Valley Drilling Project*, Antarctic Research Series, 33, ed. L.D. McGinnis, pp. 248–55. Washington, DC; American Geophysical Union.

York, D. (1966). Least-squares fitting of a straight line. *Canadian Journal of Physics*, **44**, 1079–86.

The Priestley Formation, Terra Nova Bay, and its regional significance

D.N.B. SKINNER

New Zealand Geological Survey, PO Box 30368, Lower Hutt, New Zealand

Abstract

In spite of isoclinal, two-fold deformation and low- to medium-grade metamorphism, preserved sedimentary features in the Late Proterozoic Priestley Formation suggest a range of depositional environments. The lower 880 m is a fining-upwards sequence of quartzose, pelitic sandstone to siltstone interbedded with thin shale, suggestive of deposition in an upper mid-fan environment. A sudden change to a 480 m thick pyritic-ferruginous and mud cracked quartzite–carbonate sequence implies a rapid shallowing to a restricted anoxic basin. The following 50 m of thinly interbedded calcareous quartz sandstone and sandy shale, and 300 m of ferruginous interbedded sandstone and pelitic siltstone have undergone large-scale soft sediment deformation and mark a further change to a more unstable environment. The overlying 700 m are still pyritic and ferruginous but show an increasing influx of quartzose sandy detritus, with graphite and amphibolitic tuffaceous sandstone. Above this quartzose and volcaniclastic sequence, the section ends with 60 m of amphibolitized pillow lava interlayered with volcaniclastic sandstone and quartzite. A suitable environment for the anoxic–sandy–volcanic sequence is a starved lagoon behind an encroaching prograding sand barrier, perhaps bounding the outer shelf edge adjacent to the East Antarctic Proterozoic craton, and within a volcanic back-arc basin.

Introduction

Probable Precambrian metasedimentary units in north Victoria Land extend north and north-west from the Ross Sea margin to the Oates Coast west of the Matusevich Glacier. They include: the Priestley Formation at Terra Nova Bay (Skinner, 1983a), the Dessent Formation in the Mountaineer Range (Mountaineer Metamorphics of Kleinschmidt, Roland & Schubert, 1984), the Rennick Schist of the Campbell Glacier–Sequence Hills (Gair, 1967), the Wilson Group of the Daniels Range and USARP Mountains (Kleinschmidt & Skinner, 1981), the Lanterman Metamorphics (Roland *et al.*, 1984), unnamed metasedimentary rocks cropping out at scattered localities in the Wilson Hills west of the lower Rennick Glacier (field notes and samples of Sturm & Carryer, 1970), and the Berg Group of the Lev Berg Mountains (Ravich, Klimov & Solov'ev, 1968). The rocks are predominantly of pelitic and greywacke facies, but include carbonates, calc-silicates, and quartzites. In addition, pyritic and jarositic shales and ferruginous quartzites are common to the Priestley Formation and the Wilson Group. Minor igneous rocks are present in the Dessent Formation, the Priestley Formation and the Lanterman Metamorphics, and both the Dessent and Priestley formations contain graphite.

Priestley Formation

Background

The Priestley Formation at Terra Nova Bay (Fig. 1) is intruded by both the late Vendian Larsen Granodiorite (Skinner, 1983b) and the Terra Nova Batholith, some younger granites of which have yielded a Rb–Sr isochron of 535 ± 26 Ma (Vetter *et al.*, 1984). At least two phases of deformation took place prior to granite intrusion, the most obvious folds being attenuated, large-scale isoclines at the ends of each of two measured sections (one on either side of the upper O'Kane Canyon). The gross structure is not clear, but appears to be a large upright antiform with a near horizontal axis trending NW, more or less parallel to the canyon. In contrast, synestral microfolds and folded early granite dikes in the Lowry Bluff section (Fig. 2*a*) suggest that it could be the inverted limb of a NE-verging overturned fold. As it is separated from the O'Kane Canyon sections by a large area of granites and icefields, no direct connection can as yet be demonstrated between the two localities. However, lithological similarities suggest that a possible correlation, based on the order of facies changes from pelitic quartzose sandstone and siltstone to carbonate and pyritic shale, can be made between O'Kane Canyon and Lowry Bluff.

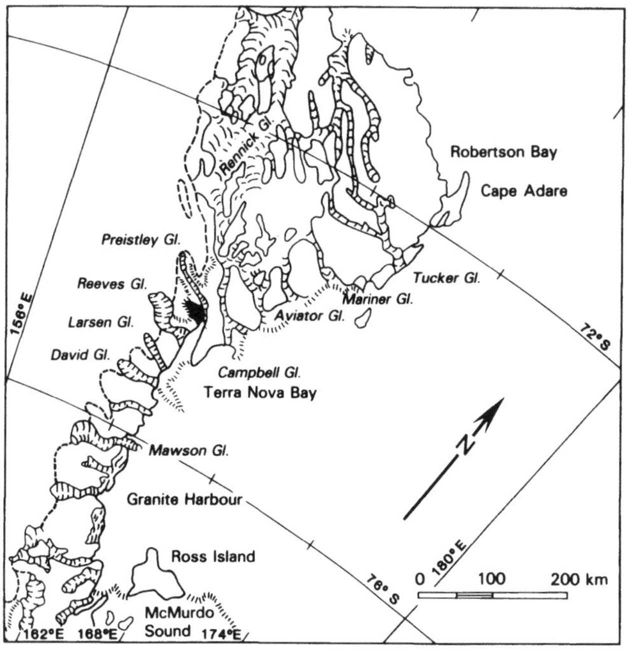

Fig. 1. Locality map of Terra Nova Bay; O'Kane Canyon is blacked-in.

It must be emphasized that all the rocks have been metamorphosed, and rise in grade from albite–epidote hornfels (greenschist) facies at the type locality in O'Kane Canyon, to amphibolite facies with migmatites along strike to the southeast (Priestley schist; Skinner, 1983a). Contact metamorphism is dominant in O'Kane Canyon, but biotite, deformed during the isoclinal folding, defines an overprinted earlier metamorphism. Nonetheless, the following sedimentary features are still preserved: grading, load casts, flame structures, possible mud crack casts, laminations, current ripples and cross-bedding. On the whole, the rocks may still be treated as sedimentary, albeit recrystallized and with the more pelitic and originally shaly constituents now being essentially biotite schists.

Measured Sections

In the upper O'Kane Canyon (Fig. 2*a*), sections were measured on: (1) the spur to Mt Baxter above 162°36′ E, 74°20.5′ S; (2) further west above 162°26′ E, 74°19′ S; and (3) on the south spur off Eskimo Point below 162°38.5′ E, 74°18′ S. Sections (1) and (2) can be directly correlated along strike (Fig. 2*b*). Because of avalanche danger, the Lowry Bluff section (163°20′ E, 74°21′ S to 163°20′ E, 74°24′ S) could only be approximated from oblique aerial photo (ANT 15 F33 052) interpolation of toboggan odometer readings (Skinner, 1983c); apparent overturning indicated this direction to be in inverted stratigraphic order.

Lithologies

In the O'Kane Canyon sections (Fig. 2*b*), the lower rocks constitute a fining-upwards sequence of laminated quartzose fine pelitic sandstone to siltstone interbedded with thin shale and medium sandstone. Bed thickness changes upwards from 5 to 50 mm, in 20 m cyclic packages, the thicker beds being more weakly laminated to almost massive and structureless. Grain size also decreases upwards and more pelitic beds become prominent. Cross-bedding, usually defined by calcareous laminae, is not uncommon in cm-thick layers, and grading was observed locally. Load casts, ripples, mud crack-like features, and linear branching trail-like casts are concentrated at a few upper horizons only.

At about 800 m above the base of the section, there is a sudden increase in finer, more pelitic lithologies. Brown pelitic and black pyritic shales commonly with yellow jarosite, laminated quartzite, and carbonate beds, are interspersed with hard layers of limestone (< 1 m thick), pyritic dense ferruginous quartzite, and dark grey laminated quartzose siltstone. At the top of the sequence, thinly interbedded calcereous sandstone and shale contain stylolitic limestone layers from 1–5 m thick.

The lower part of the sequence at Lowry Bluff (Fig. 2*a*) lacks the fining-upwards succession seen in O'Kane Canyon; before metamorphism it was a more evenly interbedded sandy siltstone and fine sandstone unit. However, just as at the canyon, there is an abrupt change to thinly bedded more pelitic lithologies, and then to pyritic calcareous brown and black shales (mm–cm layered into 1–3 m bands) with interbedded limestone and pyritic quartzite–shale (Fig. 2*b*). This is overlain by 50 m of varve-like, mm–cm-thick bedded red–brown calcareous quartz sandstone and blue–grey calcareous sandy shale, both of which have suffered considerable soft-sediment deformation, including large-scale slumping and convolute bedding.

Cascade folding, slump balls, and dislocated bedding are features of many of the upper rock units, although carbonate virtually disappears above the varve-like zone. Instead, there is an increasing amount of quartz, in a sequence ranging up from silty sandstone and pelitic siltstone to cross-bedded and laminated sandstone and quartzite. Within this sequence there are pyritic and jarositic sandstone and minor shale, and ferruginous quartzite with scattered graphite lenses. About 400 m from the top of the section there is a brief influx of volcaniclastic detritus. However, the last 60 m above the Priestley Glacier surface is predominantly volcanic, consisting of pillow lava-like bodies interlayered with amphibolite, amphibolitic (tuffaceous) quartzose sandstone, and quartzite. Graphite streaks and lenses are particularly prominent, along with pyrite and jarosite.

Interpretation

The Priestley Formation and its regionally metamorphosed equivalent, the Priestley schist (Skinner, 1983a),

Fig. 2. (*a*) Priestley Glacier–O'Kane Canyon locality map showing the lines of measured section (marked 1, 2, & 3, and Lowry Bluff), and generalized strikes and dips in the Priestley Formation. (*b*) Representational stratigraphic columns for the measured sections: width of box for interbedded lithologies is proportional to the relative % of each lithology; double vertical lines indicate alternating sequences of the lithologies on either side of the lines. Key: f, fine; m, medium; mm, cm, dm, mt, modal bedding thicknesses; lam, laminated; mssv, massive; q, quartzose; Q, quartzite; br, brown; gy, grey; y, yellow; C, graphite lenses; Fe, ferruginous; Jar, jarosite; pyr, pyritic; Su, sulphurous.

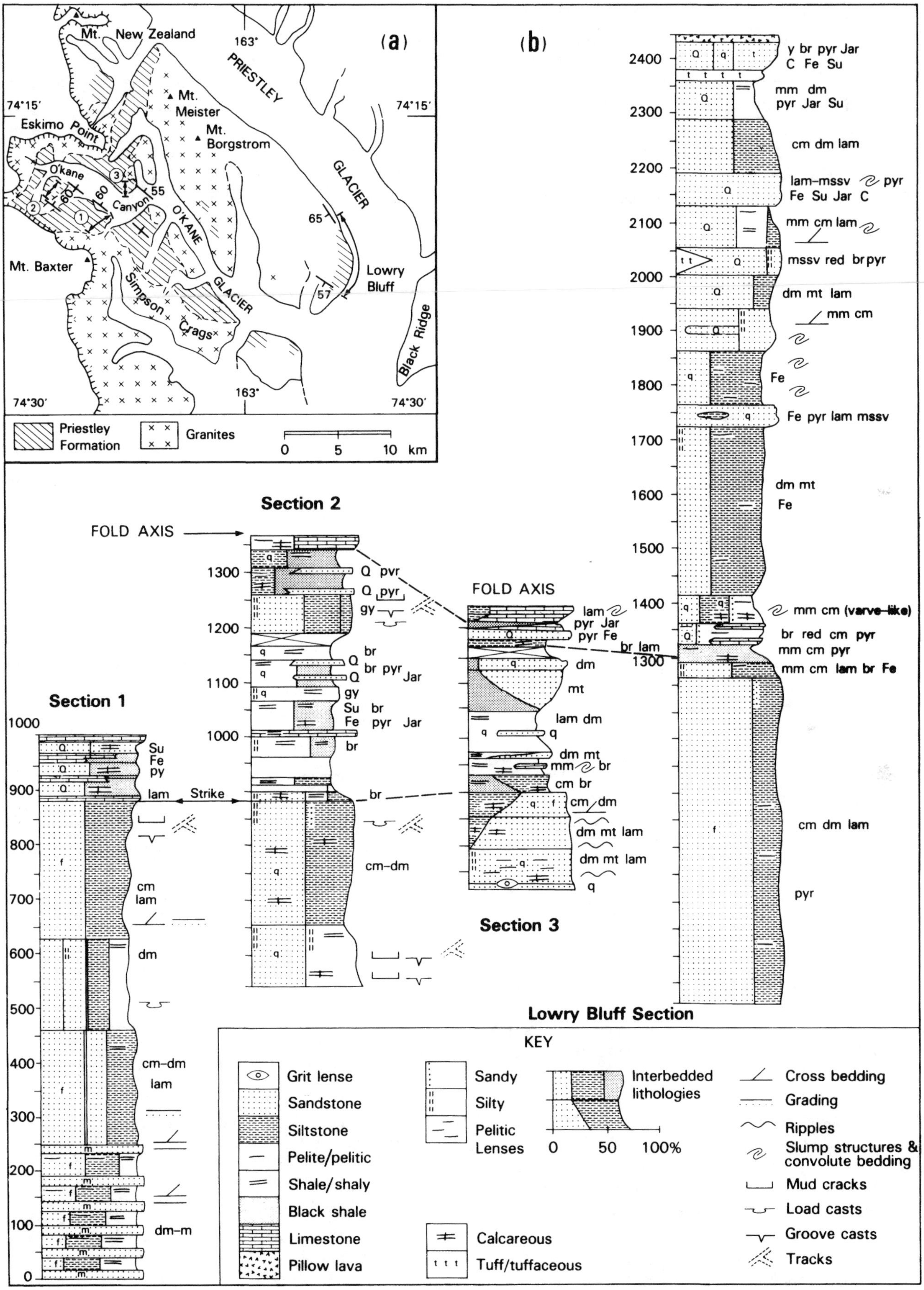
(a)
Mt. New Zealand
163°
PRIESTLEY
74°15'
Mt. Meister
Mt. Borgstrom
74°15'
Eskimo Point
O'kane
Canyon
O'KANE
GLACIER
Mt. Baxter
Simpson Crags
GLACIER
Lowry Bluff
Black Ridge
74°30'
163°
74°30'
Priestley Formation
Granites
0 5 10 km
(b)
Section 1
Section 2
Section 3
Lowry Bluff Section
FOLD AXIS
FOLD AXIS
Strike
KEY
Grit lense
Sandstone
Siltstone
Pelite/pelitic
Shale/shaly
Black shale
Limestone
Pillow lava
Sandy
Silty
Pelitic Lenses
Interbedded lithologies
0 50 100%
Calcareous
Tuff/tuffaceous
Cross bedding
Grading
Ripples
Slump structures & convolute bedding
Mud cracks
Load casts
Groove casts
Tracks

include rock facies of widely different depositional environments; together they comprise a composite sequence totalling over 2.4 km (Fig. 2*b*). This is not a clear stratigraphic succession because of uncertainties in the structural relationships of the few horizons with unequivocal geopetal indicators. Nevertheless, when the metamorphic overprints are ignored, several conclusions may be drawn from the lithologies and the facies that they represent. Not the least important are the distinct differences between the detailed lithological successions at O'Kane Canyon (especially within the carbonate–shale sequences; Fig. 2*b*), in the short distance along strike between sections (1) and (2), and across the canyon to section (3). There is thus no problem in accepting the correlation with the relatively thin carbonate–shale unit at Lowry Bluff. Indeed, it serves to emphasize the variability of facies changes that are possible, even during deposition within a single basin system.

Although the lower sandy part of the total section may be of turbidite origin, there are few observed features to confirm this unequivocally. No flysch-like sequences or typical Bouma cycles were seen, and the rocks are relatively massive and structureless (except for the laminations and minor calcareous cross-bedding). In the terminology of Mutti & Ricci Lucchi (1978), they form a positive megasequence (i.e. thinning and fining upwards) of perhaps a shallow middle fan environment.

The rather rapid change to the carbonate–black shale–quartzite–pyrite association implies either rapid shallowing to a restricted anoxic environment, or a time of oxygen-deficient oceans. In either case, the lithologies may be compared to those formed during Cretaceous anoxic events (e.g. marine black shale–limestone–marl–green grey pelites; Arthur & Natland, 1979). These in turn have been compared to sediments being deposited at the present day in oxygen minimum–anoxic areas of the slope and outer shelf off south-west Africa (Demaison & Moore, 1980). Although the upper, increasingly quartzose sequence of Lowry Bluff does not show any certain facies-distinctive features, a more vigorous environment (albeit still rather anoxic) is suggested by the increased supply of quartz, the continued production of pyrite, and the depletion of pelitic lithologies (especially black shale). Taken together, the carbonate and quartzose sequences could well have been the deposits of restricted nearshore shallow-water conditions. This possibility is supported by the mud crack-like load casts, the coarse current bedding of some of the quartzites, the ripple casts, and the possible linear branching trail-like casts. An example of a shallow environment with both a restricted basin and a more vigorous domain is the enclosure of a starved lagoon behind an encroaching, prograding quartz sand barrier (Reinson, 1984).

However, the extreme soft-sediment slump deformation of the upper rocks also indicates a relatively unstable environment, such as might be present on a sediment-laden outer shelf. It may, perhaps, be taken to indicate a sudden deepening at the sand barrier's outer edge. The origin of the peculiar varve-like facies remains obscure. Their structures are very similar to syn-sedimentary creep-distorted, very thinly bedded lime-mudstones figured in McIlreath & James (1984); however, in the Priestley Formation, the rocks also have a high quartz sand component. Because of the regional and contact metamorphism, thin sections no longer show any original microstructures that might help indicate their origin.

Finally, the tuffaceous sediments and pillow lavas indicate the nearby presence of both subaerial and subaqueous volcanic activity. Petrographically, the high proportion of porphyroblastic K-feldspar and muscovite (the former enclosing original detrital quartz) suggests that both the tuffaceous material and the pillow lavas were rather acidic, with a high K_2O content. Unfortunately, no relict igneous textures were observed that might indicate an ignimbritic rather than rhyolitic origin for the tuffs. However, such volcanic rocks would be indicative of a continental crustal source rather than oceanic, and would tend to confirm the proximity of the sedimentary basin to the Late Proterozoic continental margin.

Regional speculations

The lithological facies succession for the Priestley Formation suggests a possible change in depositional environments from deeper water slope-fan to perhaps a near shore sediment-starved anoxic basin (such as a lagoon behind a quartz sand barrier). Ultimately, the sand barrier may have encroached over the lagoon at the same time as a regional deepening occurred to form an unstable shelf–slope edge environment. Concurrently, acidic volcanic activity commenced along the continental margin, with both subaerial eruptions and the effusion of lava into the soft sediments. Such a scenario can be envisaged as occurring in a back-arc basin with the volcanic arc on the outer, oceanic side of the Late Proterozoic East Antarctic shield. The similarities of the Priestley Formation lithologies to those of the other probable Late Proterozoic rock units across north Victoria Land leads to the speculation that these too were deposited in a similar range of environments. However, the detailed differences between the rock sequences suggest that there were several distinct basins around the East Antarctic shield, each with its own particular variations of littoral to shallow water to slope-fan facies deposition.

References

Arthur, M.A. & Natland, J.H. (1979). Carbonaceous sediments in the North and South Atlantic: the role of salinity in stable stratification of Early Cretaceous basins. *Maurice Ewing Series, American Geophysical Union*, **3**, 375–401.

Demaison, G.J. & Moore, G.T. (1980). Anoxic environments and oil source bed genesis. *Bulletin of the American Association of Petroleum Geologists*, **64(8)**, 1179–209.

Gair, H.S. (1967). The geology from the upper Rennick Glacier to the coast, northern Victoria Land, Antarctica. *New Zealand Journal of Geology and Geophysics*, **10(2)**, 309–44.

Kleinschmidt, G., Roland, N.W. & Schubert, W. (1984). The metamorphic basement complex in the Mountaineer Range north Victoria Land, Antarctica. *Geologisches Jahrbuch*, **B60**, 213–51.

Kleinschmidt, G. & Skinner, D.N.B. (1981). Deformation styles in the basement rocks of north Victoria Land, Antarctica. *Geologisches Jahrbuch*, **B41**, 155–99.

McIlreath, I.A. & James, N.P. (1984). Carbonate slopes. In *Facies Models*, ed. R.G. Walker, pp. 245–58. St John's, Newfoundland; Geoscience Canada Reprint Series, 1.

Mutti, E. & Ricci-Lucchi, F. (1978). Turbidites of the northern Appenines. *International Geology Review*, **20**, 125–66.

Ravich, M.G., Klimov, L.V. & Solov'ev, D.S. (1968). *The Precambrian of East Antarctica*, ed. N. G. Sudovikov. Jerusalem; Israel program for Scientific Translations (translated from Russian). 475 pp.

Reinson, G.E. (1984). Barrier island and associated strand-plain systems. In *Facies Models*, ed. R.G. Walker, pp. 119–40. St John's, Newfoundland; Geoscience Canada Reprint Series, 1.

Roland, N.W., Gibson, G.M., Kleinschmidt, G. & Schubert, W. (1984). Metamorphism and structural relations of the Lanterman Metamorphics north Victoria Land, Antarctica. *Geologisches Jahrbuch*, **B60**, 319–61.

Skinner, D.N.B. (1983a). The Geology of Terra Nova Bay. In *Antarctic Earth Science*, ed. R.L. Oliver, P.R. James & J.B. Jago, pp. 150–5. Canberra; Australian Academy of Science and Cambridge; Cambridge University Press.

Skinner, D.N.B. (1983b). The granites and two orogenies of south Victoria Land. In *Antarctic Earth Science*, ed. R.L. Oliver, P.R. James & J.B. Jago, pp. 160–3. Canberra; Australian Academy of Science and Cambridge; Cambridge University Press.

Skinner, D.N.B. (1983c). Terra Nova Bay Geological Expedition III NZARP 1982–83. *New Zealand Geological Survey Report*, **G72**, 1–40.

Sturm, A. & Carryer, S. (1970). Geology of the region between the Matusevich and Tucker Glaciers, north Victoria Land, Antarctica. *New Zealand Journal of Geology and Geophysics*, **13(2)**, 408–35.

Vetter, U., Lenz, H., Kreuzer, H. & Besang, C. (1984). Pre-Ross granites at the Pacific margin of the Robertson Bay terrane, north Victoria Land, Antarctica. *Geologisches Jahrbuch*, **B60**, 363–9.

The myth of the Nimrod and Beardmore orogenies

E. STUMP[1], R.J. KORSCH[2] & D.G. EDGERTON[1]

1 Department of Geology, Arizona State University, Tempe, Arizona 85287, USA
2 Bureau of Mineral Resources, Geology & Geophysics, Canberra, ACT 2601, Australia

Abstract

The Beardmore Orogeny was designated to encompass Late Proterozoic deformation and magmatism in the Transantarctic Mountains. An angular unconformity in the Nimrod Glacier area, with Lower Cambrian Shackleton Limestone overlying folded Goldie Formation, indicates that the deformation was Precambrian. In the Cobham Range the Goldie Formation conformably overlies the Cobham Formation. Our work indicates that the Goldie and Cobham formations were both deformed during two separate events. The first (Precambrian) produced an axial-plane cleavage orientated N–S and mesofolds with an eastward vergence. The second, associated with folding of the Shackleton Limestone (Cambro-Ordovician Ross Orogeny), produced a NW–SE-orientated axial-plane cleavage and mesofolds with a westward vergence. Outcrops of the Nimrod Group in the Miller Range were deformed and metamorphosed during the Nimrod Orogeny. Five episodes of deformation have been identified in this group, the first three being assigned to the Nimrod Orogeny. Both mesofolds and a shear zone associated with the Nimrod Orogeny have eastward vergence. We propose that the Cobham Formation is a lower-grade correlative of a part of the Nimrod Group and that the Precambrian, E-vergent deformation of the Beardmore and Nimrod groups is equivalent. This implies that the Nimrod and Beardmore orogenies were the same event, with more intense deformation and metamorphism of the Miller Range decreasing eastwards.

Introduction

The upper Nimrod Glacier area of Antarctica (Fig. 1) has been critical in the formulation of the tectonic history of the Transantarctic Mountains, for within a radius of 20 km field evidence has been cited for three separate orogenies during the Proterozoic and Early Palaeozoic. The first parties to work in the area recognized archaeocyathid-bearing, Lower–Middle Cambrian Shackleton Limestone, and unfossiliferous turbidites of the Goldie Formation, assumed to be late Precambrian in age (Gunn & Walcott, 1962; Laird & Waterhouse, 1962; Laird, 1963). Both formations were assigned to the Ross Series (now Supergroup) and initially were thought to be in conformable contact (Grindley, 1963; Grindley, McGregor & Walcott, 1964; Laird, 1964). Both are of low metamorphic grade and contain upright folds.

The first reconnaissance survey of the Miller Range recorded an assemblage of complexly deformed, amphibolite-grade metamorphic rocks that were named the Nimrod Group (Grindley *et al.*, 1964). Because of differences in structural style and metamorphic grade, the Nimrod Group was considered to be older than the Ross Supergroup. The name Nimrod Orogeny was coined for the main episode of deformation and metamorphism that affected these rocks (Grindley & Laird, 1969).

At several localities, notably Cotton Plateau, Laird, Mansergh & Chappell (1971) discovered that Shackleton Limestone unconformably overlies folded Goldie Formation. The Shackleton Limestone itself was folded during the Cambro-Ordovician Ross Orogeny. Consequently, the folding of the Goldie Formation was placed as an event between the Nimrod and Ross orogenies. A second notable discovery by Laird *et al.* (1971) was that in the Cobham Range the Goldie Formation conformably overlies a suite of schists, calc-schists, marbles, and quartzites that they named the Cobham Formation.

K–Ar isotopic studies were undertaken to determine the age of the Nimrod Group (McDougall & Grindley, 1965; Grindley & McDougall, 1969). Whereas the majority of the dates fell in the range of 450–520 Ma, reflecting cooling during the Ross Orogeny, several dates on hornblende separates were approximately 1000 Ma. These dates were interpreted to indicate the time of cooling following the Nimrod Orogeny (Grindley & McDougall, 1969). That the Nimrod Group was of greater antiquity appeared to be indicated by a Rb–Sr study which concluded, from a very filtered data set, that isotopic equilibration had 'occurred during sedimentation and diagenesis or

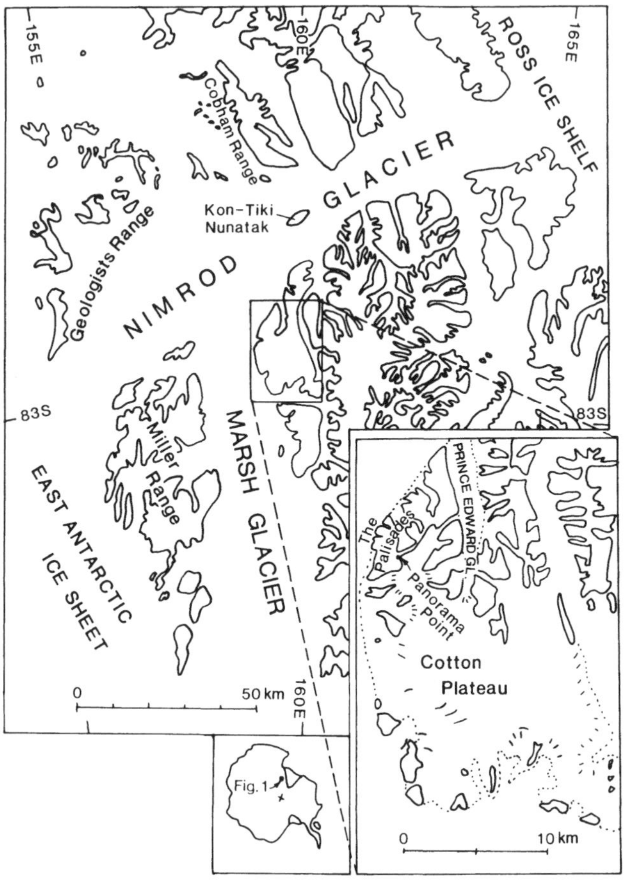

Fig. 1. Location map of the Nimrod Glacier area.

during metamorphism' at approximately 1980 Ma (Gunner & Faure, 1972, p. 305).

Grindley & McDougall (1969) introduced the term Beardmore Orogeny for the pre-Lower Cambrian deformation of the Beardmore Group (Goldie and Cobham formations). In seeking to date this event they were forced to look beyond the Nimrod Glacier area. Magmatic activity 600 km to the southeast, in the area between Scott Glacier and the Thiel Mountains, had been dated at 620–680 Ma (Ford, Hubbard & Stern, 1963; Ford, 1964; Faure, Murtaugh & Montigny, 1968). This magmatism fell in the time range that they sought, between the Nimrod and Ross orogenies, and by associating it with the deformation in the Nimrod Glacier area, Grindley & McDougall (1969) were able to date their Beardmore Orogeny.

Investigation in Nimrod Glacier area

During the 1985–86 field season, working from tent camps on Cotton Plateau and at the Cobham Range, we undertook a structural study in the Nimrod Glacier area in order to characterize the effects of the Beardmore and Ross orogenies on the Beardmore Group. We also had the opportunity, with helicopter support, to make a reconnaissance examination of the Nimrod Group in the Miller Range. The most far-reaching conclusion of the investigation is that the Nimrod and Beardmore orogenies were the same event. The evidence that has been used to state this hypothesis is summarized in the sections that follow.

Structural geology

Cotton Plateau

The northern end of Cotton Plateau displays some remarkable geology. As described and pictured by Laird *et al.* (1971), the eastern limb of a syncline of Shackleton Limestone lies with marked angular unconformity on the Goldie Formation, as seen in the west-facing cliff of The Palisades (Fig. 1). The Shackleton Limestone changes from horizontal in the lower portions of its outcrop to vertical and slightly overturned at its highest. The beta-axis of the Shackleton syncline is orientated 04 to 140° (Edgerton, 1987).

A structural boundary appears to occur in the Goldie Formation along The Palisades. In outcrops to the east of Panorama Point, the Goldie Formation consistently dips steeply westward and contains upright mesofolds. To the west, on the lower slopes of The Palisades, it occurs in tight recumbent folds. Along the crest of The Palisades the Goldie Formation contains a thick unit of pillow basalt (see below) that forms the core of a pre-unconformity syncline in the Goldie Formation. A broad shear zone occurs in parts of the basalt along the hinge of this syncline.

Two episodes of deformation are apparent at many outcrops of the Goldie Formation throughout the northern part of Cotton Plateau. The earlier deformation (pre-Shackleton Limestone) produced a cleavage striking N–S (Fig. 2), axial-planar to mesofolds with a consistent *eastward* vergence. The older cleavage is crosscut by a younger cleavage orientated NW–SE, coincident with the orientation of the Shackleton syncline and axial-planar to mesofolds with a consistent *westward* vergence.

Cobham Range

The Cobham Formation is exposed along the southwestern part of the Cobham Range and it is conformably overlain by the Goldie Formation, which crops out on the eastern spurs of the range. The formations occur as a homoclinal sequence dipping moderately or steeply to the E, striking NW parallel to the elongation of the range.

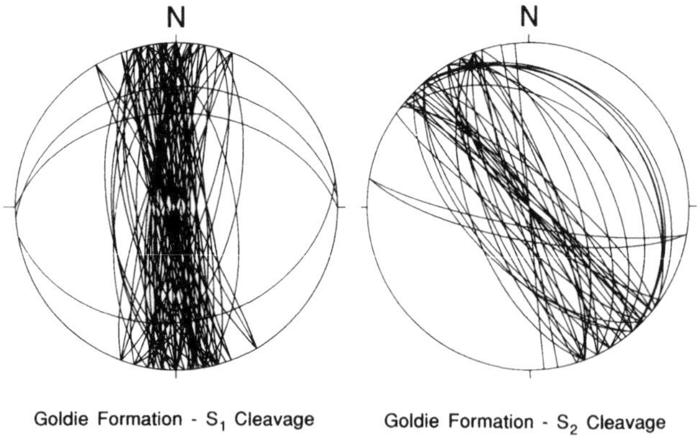

Fig. 2. Orientations of cleavage in the Goldie Formation, northern Cotton Plateau. S_1 cleavage produced during Precambrian deformation; S_2 cleavage produced during the Cambro-Ordovician Ross Orogeny.

Fig. 3. Oblique aerial view of the northern face of Kon-Tiki Nunatak showing recumbent structure at east (right) side of exposure deformed to upright folds on the west (left) side.

Within the Cobham Formation and the lower part of the Goldie Formation, the later deformation (Ross Orogeny) has produced mesofolds with a N–S trend and shallow, northerly plunge. The mesofolds have a westward vergence, but unlike the Cotton Plateau structures, an axial-planar cleavage is not widely developed.

Structures associated with the earlier deformation have a more varible orientation than is found on Cotton Plateau. Mesofolds plunge shallowly to moderately in both the NE and SE quadrants but in all cases they have an easterly vergence. A prominent recumbent fold occurs in the Cobham Formation (Laird *et al.*, 1971, fig. 19) but most of the older mesofolds are smaller (wavelengths of < 1 m) and have axial planes with moderate to steep dips.

Kon-Tiki Nunatak

The two generations of folding in the Goldie Formation are exposed on a spectacular cliff face along the north side of Kon-Tiki Nunatak. Isoclinal, recumbent folds occurring on the western half of the face are refolded into tight, upright structures to the east (Fig. 3). The orientations of structures are similar to those found at the northern end of Cotton Plateau.

Lithologies of Goldie and Cobham formations

Goldie Formation

As described by Laird *et al.* (1971) the Goldie Formation is predominantly metagreywacke and argillite. Graded bedding is the most common sedimentary structure. Calcite is a conspicuous fraction of the groundmass in many of the samples from the area, although the Goldie Formation appears to be less calcareous in exposures to the east.

A previously undescribed occurrence of mafic pillow lavas crops out along the crest of The Palisades. The lavas conformably overlie metasedimentary rocks and form the core of a major syncline in the Goldie Formation. Mafic lavas crop out at the head of Prince Edward Glacier, though at that locality pillows were not observed. The lavas contain an alteration assemblage of sodic-plagioclase, calc-amphibole, zoisite, and biotite. Inclusions of coarse-grained gabbro are present in parts of the lava at Panorama Point.

Cobham Formation

Laird *et al.* (1971) described a measured section of the Cobham Formation. Essentially, the formation is a sequence of interbedded schist, calc-schists, marbles and quartzites. We affirm its conformable contact with the overlying Goldie Formation in the Cobham Range.

Correlation of Beardmore and Nimrod groups

Correlation of the Nimrod and Beardmore groups has always been precluded by the assertion that they were separated in time by the Nimrod Orogeny (Grindley & Laird, 1969; Grindley & McDougall, 1969). However, except for metamorphic grade, an almost one-to-one comparison can be made between rock types of the Nimrod and Beardmore groups. The schists, calc-schists, marbles and quartzites of the Cobham Formation, and the schists, metagreywackes, and argillites of the Goldie Formation all have higher-grade equivalents in the Nimrod Group (Gunner, 1969). The presence of amphibolites (originally mafic volcanic rocks) in the Nimrod Group could weigh against correlation with the Beardmore Group. However, their occurrence in the Nimrod Group is minor and pillow lavas have now been found in the Goldie Formation on Cotton Plateau.

At present it is not possible to make precise correlations between the two groups. The stratigraphy of the Cobham and lowermost Goldie formations is well established (Laird *et al.*, 1971). A transition from a shallow-water carbonate–quartzite association up into a deeper-water greywacke–shale association is apparent. Similar details from measured sections in the Nimrod Group await future work. However, the broad stratigraphy of the Nimrod Group has been established (Grindley *et al.*, 1964). The tectonically highest Miller Formation is allochthonous, and its stratigraphical relationship to the other formations is uncertain. Three conformable formations in the Nimrod Group display the same general transition of lithologies as the Beardmore Group. The lowermost Worsley Formation with its marbles and schists, gives way to the Argosy Formation with schists, marbles and quartzites. This is overlain by the Aurora Formation, characterized by biotite gneiss and the lack of marbles and quartzites.

On the basis of both similar lithologies and gross stratigraphical trend we suggest the correlation of the Nimrod and Beardmore groups. Certainly more work is needed on the stratigraphy of the Nimrod Group and geochemistry of the rocks in both groups before such a correlation can be accepted or rejected.

The Geologists Range, a key area to this problem, has been visited only once (New Zealand party, 1964–65) and descriptions of the work have never been published. Grindley & McDougall (1969) included the basement rocks of the Geologists Range in the Nimrod Group, as did Grindley & Laird (1969). However, Laird *et al.* (1971, p. 432) stated 'Similar lithologies to those found in the Cobham Formation are found in the Geologists Range 40 km to the west (R.G. Adamson, pers. comm.), and the sequences may be correlatives'. And further, 'thus it does not necessarily follow that the Geologists Range rocks should correlate with the Nimrod Group. They could be equally correlated with the Cobham Formation'.

Structural correlations

Grindley (1972) reported five phases of folding in the Nimrod Group in the Miller Range. F_1–F_3 were associated with the Nimrod Orogeny. F_1 involved emplacement of the Miller Formation allochthon along the Endurance thrust and the formation of recumbent mesoscopic folds. F_2 produced tightly compressed mesofolds. Both F_1 and F_2 trend NW–SE. F_3 produced tightly compressed folds orientated NE–SW. In all three phases the vergence of mesostructures is to the east. F_4 folds are limited in occurrence, have an open style, and have been associated with the Ross Orogeny. F_5 folds are local flexures associated with emplacement of the Cambro-Ordovician Hope Granite.

From our structural observations we suggest that the Precambrian deformation of the Beardmore Group coincided with Grindley's (1972) F_{1-3} deformation of the Nimrod Group. Most importantly, the eastward vergence direction is the same in both groups. The recumbent style of folding in the Nimrod Group is seen in the Goldie Formation on the western parts of Cotton Plateau and Kon-Tiki Nunatak, and to limited extent in the Cobham Formation on the western edge of the Cobham Range. East of these areas, folding of the Goldie Formation becomes more upright.

The differences in metamorphic grade, structural style, and number of recorded phases of deformation between the Nimrod and Beardmore groups, rather than arguing for separate orogenic episodes, are better interpreted as reflecting changes across the same orogenic belt from a more mobile core to outer, less-deformed and less-metamorphosed parts.

Geochronology

Application of geochronological techniques to the problem of the Nimrod and Beardmore orogenies remains inconclusive. Most dates (recalculated where appropriate using the accepted decay constants of Steiger and Jäger (1977)) from the Nimrod Glacier area (K–Ar, Rb–Sr and U–Pb) record the effects of the Ross Orogeny (500 ± 50 Ma) (McDougall & Grindley, 1965; Grindley & McDougall, 1969; Gunner & Faure, 1972; Gunner & Mattinson, 1975; Adams, Gabites & Grindley, 1982). As mentioned above, the initial dating of the Beardmore Orogeny was based on 620–680 Ma dates from magmatic rocks 600 km south-east of the Nimrod Glacier area (Grindley & McDougall, 1969). Since then any isotopic dates in the 600 Ma range on the continent have been discussed in terms of the Beardmore Orogeny (Gunner & Faure, 1972; Faure *et al.*, 1979; Adams *et al.*, 1982; Vetter *et al.*, 1984).

However, based on mapping in the La Gorce Mountains, (86°45′ S, 146° W), Stump, Smit & Self (1986) demonstrated that the La Gorce Formation (Goldie Formation correlative) was folded and regionally metamorphosed to biotite grade prior to intrusion of the Wyatt Formation, one of the magmatic units used originally to date the Beardmore Orogeny. The conclusion was that Beardmore deformation must be appreciably older than the isotopically dated late Precambrian magmatism. Rb–Sr dating by Pankhurst *et al.* (1988) has shown with reasonable certainty that magmatism in the Thiel Mountains (Wyatt Formation correlative) occurred at ~ 490 Ma, rather than in the Late Proterozoic as previously thought. This then loosens the constraint on the age of deformation in the La Gorce Mountains (Stump *et al.*, 1986).

Adams *et al.* (1982) added considerably to the K–Ar data of the Nimrod Glacier area. In a compilation of all existing K–Ar dates from the Nimrod Group, a clear clustering of concordant biotite and hornblende ages between 500–560 Ma were interpreted as representing the cooling time of the dynamic phase of the Ross Orogeny, and ages between 450–500 Ma as representing the time of emplacement of the Hope Granite.

The remainder of the K–Ar dates (9 hornblende, 1 biotite) are spread rather evenly in time from 633–1153 Ma. Using arguments for excess argon, Adams *et al.* (1982) interpreted all the older dates as resulting from the effects of the Beardmore Orogeny, though they present no *a priori* evidence as to why the excess argon should have migrated during the Beardmore and not the Ross Orogeny. Adams *et al.* (1982, p. 825) were correct when they stated 'Little direct evidence now exists for a separate Nimrod Orogeny'; however, that they picked the 600 Ma (Beardmore) timeframe as responsible for the spread in ages in the Nimrod Group is, likewise, not warranted from the data.

Conclusions

1. Parts of the Nimrod Group correlate with the Cobham Formation and lower Goldie Formation.

2. Precambrian, E-vergent deformation in the Nimrod Group coincided with Precambrian, E-vergent deformation in the Cobham and Goldie formations.

3. The Nimrod and Beardmore orogenies were the same event.

Acknowledgements

Thanks to John Sutter for discussion of an early version of the manuscript. Support was provided by NSF Grant DPP-8418088.

References

Adams, C.J.D., Gabites, J. & Grindley, G.W. (1982). Orogenic history of the central Transantarctic Mountains: new K–Ar age data on the Precambrian–Early Paleozoic basement. In *Antarctic Geoscience*, ed. C. Craddock, pp. 817–26. Madison; University of Wisconsin Press.

Edgerton, D.G. (1987). Kinematic orientations of Precambrian Beardmore and Cambro-Ordovician Ross Orogenies, Goldie Formation, Nimrod Glacier area, Antarctica. MS Thesis, Tempe; Arizona State University (unpublished).

Faure, G., Eastin, R., Ray, P.T., McLelland, D. & Schultz, C.H. (1979). Geochronology of igneous and metamorphic rocks, central Transantarctic Mountains. In *Fourth International Gondwana Symposium Calcutta, India, 1977–: Papers* vol. II, ed. B. Laskar & C.S.. Raja Rao, pp. 805–13. Calcutta; Hindustan Publishing Corporation.

Faure, G., Murtaugh, J.G. & Montigny, R.J.E. (1968). The geology and geochronology of the basement complex of the central Transantarctic Mountains. *Canadian Journal of Earth Sciences*, **5(3)**, 555–60.

Ford, A.B. (1964). Cordierite-bearing, hypersthene–quartz–monzonite porphyry in the Thiel Mountains, and its regional importance. In *Antarctic Geology*, ed. R.J. Adie, pp. 429–41. Amsterdam; North-Holland Publishing Company.

Ford, A.B., Hubbard, H.A. & Stern, T.W. (1963). Lead-alpha ages of zircon in quartz monzonite porphyry, Thiel Mountains, Antarctica. A preliminary report. *US Geological Survey Professional Paper*, **450-E**, 105–7.

Grindley, G.W. (1963). The geology of the Queen Alexandra Range, Beardmore Glacier, Ross Dependency, Antarctica; with notes on the correlation of Gondwana sequences. *New Zealand Journal of Geology and Geophysics*, **6(3)**, 307–47.

Grindley, G.W. (1972). Polyphase deformation of the Precambrian Nimrod Group, central Transantarctic Mountains. In *Antarctic Geology and Geophysics*, ed. R.J. Adie, pp. 313–18, Oslo; Universitetsforlaget.

Grindley, G.W. & Laird, M.G. (1969). Geology of the Shackleton Coast. In *Geologic Maps of Antarctica*, ed. V.C. Bushnell & C. Craddock, Pl. XIV: Antarctic map folio series, Folio 12. Washington, DC; American Geographical Society.

Grindley, G.W. & McDougall, I. (1969). Age and correlation of the Nimrod Group and other Precambrian rock units in the central Transantarctic Mountains, Antarctica. *New Zealand Journal of Geology and Geophysics*, **12(2 & 3)**, 391–411.

Grindley, G.W., McGregor, V.R. & Walcott, R.I. (1964). Outline of the geology of the Nimrod–Beardmore–Axel Heiberg Glaciers region, Ross Dependency. In *Antarctic Geology*, ed. R.J. Adie, pp. 206–19. Amsterdam; North-Holland Publishing Company.

Gunn, B.M. & Walcott, R.I. (1962). The geology of the Mt Markham region, Ross Dependency, Antarctica. *New Zealand Journal of Geology and Geophysics*, **5(3)**, 407–26.

Gunner, J.D. (1969). *Petrography of Metamorphic Rocks from the Miller Range, Antarctica.* Columbus; Ohio State University, Institute of Polar Studies, Report no. 32, 44 pp.

Gunner, J.D. & Faure, G. (1972). Rubidium–strontium geochronology of the Nimrod Group, central Transantarctic Mountains. In *Antarctic Geology and Geophysics*, ed. R.J. Adie, pp. 305–11, Oslo; Universitetsforlaget.

Gunner, J.D. & Mattinson, J.M. (1975). Rb–Sr and U–Pb isotopic ages of granites in the central Transantarctic Mountains. *Geological Magazine*, **112(1)**, 25–31.

Laird, M.G. (1963). Geomorphology and stratigraphy of the Nimrod Glacier–Beaumont Bay region, southern Victoria Land, Antarctica. *New Zealand Journal of Geology and Geophysics*, **6(3)**, 465–84.

Laird, M.G. (1964). Petrology of rocks from the Nimrod Glacier–Starshot Glacier region, Ross Dependency. In *Antarctic Geology*, ed. R.J. Adie, pp. 463–72. Amsterdam; North-Holland Publishing Company.

Laird, M.G., Mansergh, G.D. & Chappell, J.M.A. (1971). Geology of the central Nimrod Glacier area, Antarctica. *New Zealand Journal of Geology and Geophysics*, **14(3)**, 427–68.

Laird, M.G. & Waterhouse, J.B. (1962). Archeocyathine limestones of Antarctica. *Nature, London*, **194(4831)**, 861.

McDougall, I. & Grindley, G.W. (1965). Potassium–argon dates on micas from the Nimrod–Beardmore–Axel Heiberg region, Ross Dependency, Antarctica. *New Zealand Journal of Geology and Geophysics*, **8(2)**, 304–13.

Pankhurst, R.J., Storey, B.C., Millar, I.L., Macdonald, D.I.M. & Vennum, W. (1988). Cambrian–Ordovician magmatism in the Thiel Mountains, Transantarctic Mountains, and implications for the Beardmore orogeny. *Geology*, **16(3)**, 246–9.

Steiger, R.H. & Jäger, E. (1977). Subcommission on geochronology: convention on use of decay constants in geo- and cosmochronology. *Earth and Planetary Science Letters*, **36(3)**, 359–62.

Stump, E., Smit, J.H. & Self, S. (1986). Timing of events during the late Proterozoic Beardmore Orogeny, Antarctica: geological evidence from the La Gorce Mountains. *Geological Society of America Bulletin*, **97(8)**, 953–65.

Vetter, U., Lenz, H., Kreuzer, H. & Besand, C. (1984). Pre-Ross granites at the Pacific margin of the Robertson Bay terrane, north Victoria Land, Antarctica. *Geologisches Jahrbuch*, **B60**, 363–9.

Age of the metamorphic basement of the Salamander and Lanterman ranges, northern Victoria Land, Antarctica

C.J. ADAMS[1] & A. HÖHNDORF[2]

1 Institute of Nuclear Sciences, DSIR, Lower Hutt, New Zealand
2 Bundesanstalt für Geowissenschaften und Rohstoffe, Hannover, West Germany

Abstract

Rb–Sr whole-rock isochron ages are reported for greenschist-facies metasedimentary rocks in the Salamander and Lanterman ranges, northern Victoria Land; these rocks are probable correlatives of the Wilson Group basement metamorphic rocks in the Usarp Mountains. The ages, 558 ± 17 Ma and 550 ± 20 Ma, respectively (initial $^{87}Sr/^{86}Sr$ ratios of 0.7156 ± 2 and 0.7155 ± 3), indicate an (?)Early Cambrian metamorphism of the Wilson Group metamorphic rocks and correlate with similar metamorphic ages for the Rennick schists in the Usarp Mountains. This event probably predates mid-Cambrian–Early Ordovician sedimentation (Bowers Supergroup) in northern Victoria Land.

Introduction

In the western part of northern Victoria Land, high-grade metamorphic rocks form a regional basement, possibly part of the Precambrian Shield of East Antarctica (Fig. 1, inset A). They occur principally in the Usarp Mountains (Daniels Range) and Wilson Hills, at the western margin of Rennick Glacier (see review by Tessensohn *et al.*, 1981). Ravich & Krylov (1964) defined the Wilson Group as high-grade gneisses of igneous and sedimentary parentage but Sturm & Carryer (1970) restricted the term to paragneiss and migmatite (Wilson gneisses) and schists (Rennick schists). Later work (Vetter *et al.*, 1983; Adams, 1987) has shown that most of the heterogeneous orthogneisses and migmatite are probably part of an intrusive granitoid complex (Granite Harbour Intrusives) emplaced during the Cambro-Ordovician Ross Orogeny. At this time there was widespread regional metamorphism which is reflected by numerous cooling ages (mostly K–Ar) of 465–500 Ma (Kreuzer *et al.*, 1981; Adams & Kreuzer, 1984) throughout the Wilson Terrane (I in Fig. 1A). Rb–Sr whole-rock isochron ages for the Rennick schists are somewhat older, 520–550 Ma (Adams, 1987) and may date an early Ross metamorphic event or an older orogenic phase.

East of Rennick Glacier is the Bowers Terrane (II in Fig. 1A) of Middle Cambrian–Early Ordovician Bowers Supergroup sedimentary rocks, which were folded and metamorphosed during the Ross Orogeny. Mineral age patterns in the Bowers Terrane are similar both to those in the Wilson Terrane to the west and the Robertson Bay Terrane (III in Fig. 1A) to the east.

At the eastern edge of the Wilson Terrane (Rennick and Lanterman faults) and directly adjacent to the Bowers Terrane there are gneisses in the Lanterman and Salamander ranges (Fig. 1) which are correlative with the Wilson Group in the Daniels Range. In this report we present new Rb–Sr whole-rock ages of schists and gneisses from these two areas to investigate their age relationship to the Wilson Group (to the west) and the Bowers Supergroup (to the east).

Lanterman and Salamander metamorphic rocks

In the Usarp Mountains the Wilson Group (*sensu lato*) is predominantly granodiorite orthogneiss with minor quartzo-feldspathic schist, whilst in the Lanterman Range (Bradshaw & Laird, 1983) it is dominantly paragneiss. Schists and gneisses are locally variable, including garnet–mica schist, quartzo-feldspathic schist, chloritic schist, amphibolitic schist, calc-schist and marble and small bodies of gneissic diorite and/or granitoid. A few small granitoid plutons have yielded K–Ar ages of 480–490 Ma (Granite Harbour Intrusives) (Kreuzer *et al.*, 1981; Vetter *et al.*, 1983).

In the Salamander Range to the south, there is a simpler metamorphic terrane of predominantly quartzo-feldspathic greenschist (eastern side) and minor calc-schist and marble (western side). Granitoid dykes and diorite pods occur in the central sector but larger plutons are absent. The geological history is simplest in the eastern Salamander Range and is the most appropriate area for Rb–Sr dating of the metamorphic history.

Rb–Sr dating

Representative whole-rock samples were prepared from those medium–coarsely crystalline schistose rocks from the Lanterman Range which had been used previously for K–Ar

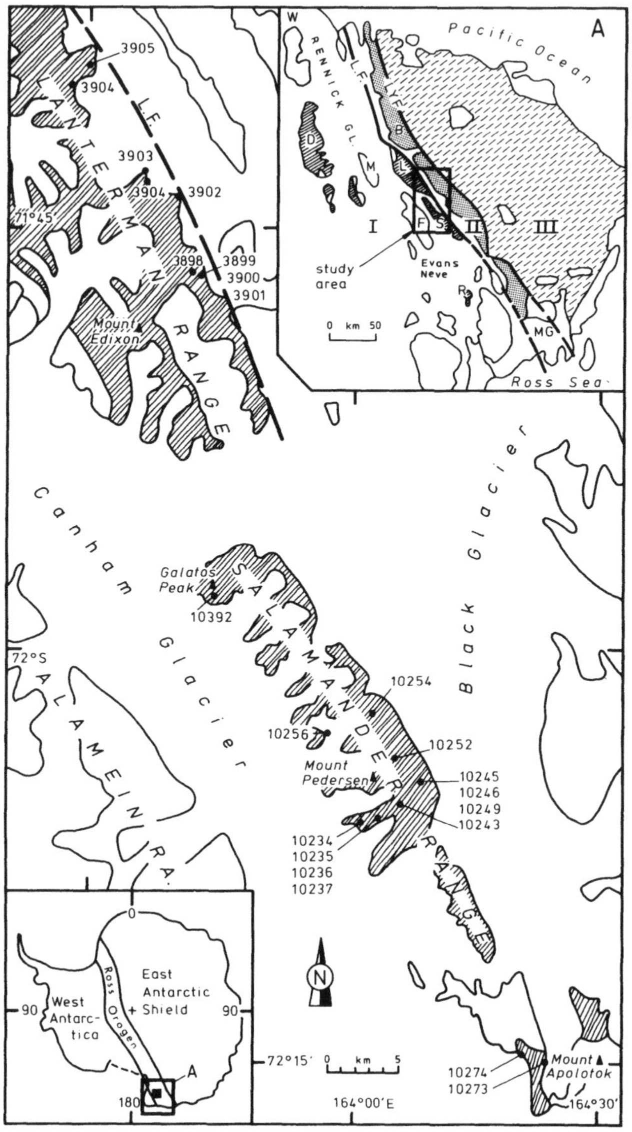

Fig. 1. Location map of samples in medium–high-grade metasedimentary rocks (Wilson Group) in the Lanterman and Salamander ranges, northern Victoria land (diagonal ruling). Inset A (N.B. north to top of page), terrane map of northern Victoria land: I, Wilson Terrane and Wilson Group (diagonal ruling); II, Bowers Terrane (stipple); III, Robertson Bay Terrane (dashed diagonal ruling); B, Bowers Mountains; D, Daniels Range; F, Freyberg Mountains; L, Rennick Glacier–Lanterman Range; LF, Lanterman Fault; LYF, Leap Year Fault; M, Morozumi Range; MG, Mariner Glacier; R, Retreat Hills; S, Salamander Range; W, Wilson Hills.

dating (Adams *et al.*, 1982b). The samples are mostly quartzo-feldspathic schists with minor garnet and/or amphibole. Amphibolitic schists (3900, 3903, Table 1) are metabasic horizons whose initial $^{87}Sr/^{86}Sr$ ratio could differ from the surrounding metasedimentary rocks.

More suitable quartzo-feldspathic schists were collected from the eastern margin of the Salamander Range, from a uniform sequence of psammitic metasedimentary rocks (Fig. 1). They are fully recrystallized to greenschist-facies, having lost all sedimentary features except large-scale (10 m) bedding. A small unfoliated intrusive diorite pod (~ 10 m in diameter) was sampled (RG259) for K–Ar dating.

Brief petrographic details, sample locations and analytical data are given in Table 1. Experimental methods used are essentially similar to those described by Adams (1986). Decay constants used are those recommended by Steiger & Jäger (1977).

Salamander Range

Rb–Sr data for 13 samples of quartzo-feldspathic-schists from the Salamander Range are shown in Fig. 2 (lower section). They define an isochron, yielding an age of 555 ± 35 Ma (95% confidence level errors) and an initial ratio of 0.7158 ± 0.0004. The MSWD of 4.6 indicates that the data scatter outside the $^{87}Sr/^{86}Sr$ and $^{87}Rb/^{86}Sr$ error limits and that there is geological variation in the age and/or initial ratio. In this case, the metamorphic age is the same over the outcrop scale involved and the geological variation must reflect incomplete strontium isotope homogenization at the time of metamorphism. Although the dominant proto-lithology was a uniform psammite (quartz + feldspar; low Rb/Sr ratio ~ 1–2), there could be: (i) minor intercalated pelitic laminae (i.e. micas; high Rb/Sr) with high $^{87}Sr/^{86}Sr$ ratios immediately prior to metamorphism; or (ii) calcareous laminae ($CaCO_3$; low Rb/Sr ratio < 0.1) with contemporary $^{87}Sr/^{86}Sr$ seawater ratio (< 0.710). Although minor, both could have been responsible for a small variability of $^{87}Sr/^{86}Sr$ ratios in the whole-rock samples immediately prior to metamorphism. These variable $^{87}Sr/^{86}Sr$ ratios have not been completely homogenized during metamorphism and hence there is geological scatter of the isochron data. The scatter is almost removed if samples 10236, 10249 and 10274 are withdrawn from the isochron calculation, although the age and initial Sr ratio remains essentially similar, i.e. 558 ± 17 Ma, 0.7156 ± 0.0002 (MSWD 2.1).

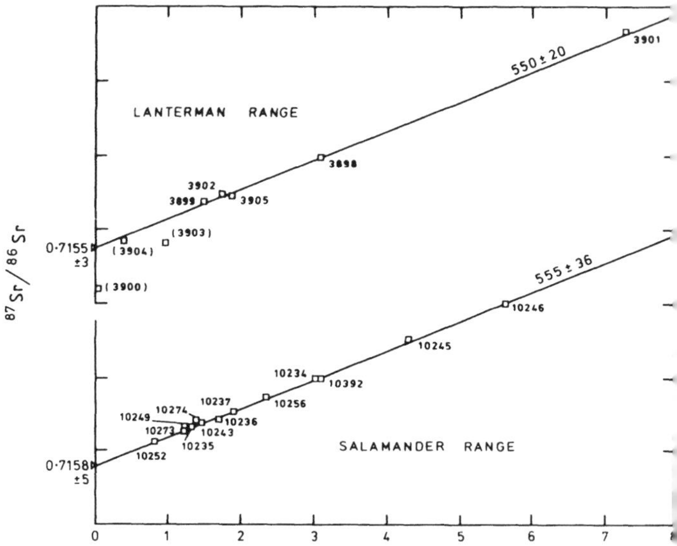

Fig. 2. Rb–Sr whole-rock isochron diagrams for medium–high-grade metasedimentary rocks from Lanterman (upper section) and Salamander ranges (lower section). Sample numbers in brackets are hornblende-bearing quartzo-feldspathic schists or amphibolitic schists of probable volcanic parentage which are excluded from isochron age calculations. $^{87}Sr/^{86}Sr$ axis graduations are 0.02 intervals. In each case the graduation below the initial ratio is 0.700.

Table 1. *Rb–Sr age data from Wilson Group metamorphic rocks*

INS R. No.[a] (field no.)	Sample details Location	Lat. °S Long. °E	Rb (ppm) Sr (ppm)	$^{87}Rb/^{86}Sr$ (± 1.0%)[b]	$^{87}Sr/^{86}Sr$ (± 0.00010)[b]
Salamander Range					
10234 (RG230)	Biotite–quartz-schist; in thick metasediments; Mt Pedersen, SSW of summit	72°06′ 163°59′	235 214	3.065	0.73990
10235 (RG231)	Quartzo-feldspathic biotite-schist; in gneissic metasediments; Mt Pedersen; south of summit	72°06′ 164°00′	127 264	1.350	0.72659
10236 (RG232)	Quartzo-feldspathic biotite-schist; location as for 10235	72°06′ 164°00′	163 266	1.716	0.72826
10237 (RG233)	Biotite–quartzo-feldspathic schist; location as for 10235	72°06′ 164°00′	136 197	1.925	0.73068
10243 (RG239)	Quartzo-feldspathic biotite-schist; in coarse psammitic sequence; Mt Pedersen, SE of summit	72°06′ 164°07′	144 270	1.489	0.72776
10245 (RG241)	Quartzo-feldspathic biotite-schist; in coarse psammitic sequence; Mt Pedersen, ESE of summit	72°05′ 164°09′	224 145	4.315	0.75070
10246 (RG242)	Quartzo-feldspathic biotite–muscovite-schist; location as for 10245	72°05′ 164°09′	254 125	5.663	0.75988
10249 (RG245)	Quartzo-feldspathic schist; in calc-silicate sequence; Mt Pedersen, east of summit	72°05′ 164°09′	112 247	1.261	0.72654
10252 (RG248)	Quartz–biotite-schist; in psammitic sillimanite-schists; Mt Pedersen, NE of summit	72°04′ 164°06′	95 311	0.851	0.72216
10256 (RG252)	Quartz-schist; in calc-schist sequence; Mt Pedersen, NW of summit	72°03′ 163°57′	153 188	2.362	0.73461
10273 (RG269)	Quartz–biotite–calc-schist; in calc-schist sequence; Mt Apolotok, west of summit	72°15′ 164°23′	101.1 238.3	1.230	0.72544
10274 (RG270)	Quartz–biotite-schist; in metapelite sequence; Mt Apolotok, NW of summit	72°13′ 164°17′	101 209	1.401	0.72799
10392 (SR5)	Biotite-schist; in psammitic and calc-gneiss sequence; Galatos Peak	71°58′ 163°43′	204 183	3.113	0.73965
Lanterman Range					
3898 (W1)	Quartzo-feldspathic garnet–biotite–muscovite–granodiorite-gneiss; Zenith Glacier, upper east side	71°47′ 163°41′	171 159	3.118	0.7394
3899 (W2)	Quartzo-feldspathic biotite–muscovite–chlorite–granodiorite-gneiss; Zenith Glacier, upper east side	71°47′ 163°42′	114 221	1.502	0.72734
3900 (W3)	Hornblende–albite–epidote–amphibolitic gneiss; Zenith Glacier, upper east side	71°47′ 163°42′	1.6 115	0.041	0.70371
3901 (W4a)	Quartzo-feldspathic biotite–muscovite–chlorite–granodiorite-gneiss; head of Sledgers Glacier, ridge on SW side	71°44′ 163°40′	186 74	7.358	0.77312
3902 (W7)	Quartzo-feldspathic biotite–muscovite–garnet gneiss; head of Sledgers Glacier, ridge on SW side	71°44′ 163°40′	119 196	1.757	0.72902
3903 (W11)	Hornblende–amphibolitic gneiss; head of Sledgers Glacier, ridge on SW side	71°43′ 163°36	21 61	0.991	0.71637
3904 (W13)	Quartzo-feldspathic biotite–muscovite–oligoclase–amphibolitic gneiss; head of Sledgers Glacier, ridge on SW side	71°40′ 163°27′	14 106	0.388	0.71618
3905 (W15)	Quartzo-feldspathic biotite–muscovite–hornblende gneiss; head of Sledgers Glacier, ridge on SW side	71°39′ 163°29′	89 137	1.885	0.72832

[a]All samples are whole-rock powders.
[b]Reproducibility, used in isochron age calculations.

Lanterman Range

Five metasedimentary samples from the Lanterman Range metamorphic rocks yielded an age of 558 ± 77 Ma and an initial ratio of 0.7148 ± 12 (Fig. 2, upper portion). The higher MSWD value of 6.0 reflects the considerable geological contribution to the age error. Almost certainly this reflects greater lithological heterogeneity of the Lanterman Range metasedimentary rocks. When compared with the Salamander metamorphic rocks, there is an additional component of basic (volcanic or volcaniclastic) composition with $^{87}Sr/^{86}Sr$ ratios of 0.703–0.705. Chlorite and amphibole in the Lanterman meta-

Table 2. *K–Ar age data, diorite pegmatite, Salamander Range*

INS R. No. (field no.)	Location	K (wt %)	^{40}Ar radiogenic (nl/g)	(% total)	Age (Ma)
10254 (RG250) hornblende	4 km north of Mount Pedersen (72°11′ S, 164°00′ E)	1.077	23.52	98	498 ± 6

Decay constants ^{40}K: $\lambda_\beta = 0.4962 \times 10^{-9}/y$; $\lambda_e = 0.581 \times 10^{-10}/y$.
Isotopic abundance: $^{40}K/K = 0.01167\%$. Errors are two standard deviations.

morphic rocks reflect this, e.g. samples 3903, 3904, and the metabasite (sample 3900) with $^{87}Sr/^{86}Sr = 0.7037$ falls well below the isochron line. If sample 3905, a hornblende–quartzofeldspathic schist, is excluded, then the isochron age becomes 550 ± 20 Ma with an initial ratio of 0.7155 ± 3 and MSWD 1.6. These improved values are similar to those obtained from the Salamander metamorphic rocks and confirm their common metamorphic history.

Metamorphic history of the Wilson Group

The Rb–Sr ages of the Salamander Range and Lanterman Range metamorphic rocks, 558 ± 12 and 550 ± 20 Ma, respectively, may be compared to those of the Rennick schists in the Usarp Mountains (527 ± 24 and 536 ± 28 Ma; Adams, 1987). All are significantly older than the maximum mineral ages (490–500 Ma) of the Wilson Group gneisses and Granite Harbour Intrusives in the Wilson Terrane. Coarse hornblende from a diorite–pegmatite cutting schists at Mount Pedersen (sample 10254, Table 2) has yielded a K–Ar age of 489 ± 6 Ma which confirms this mineral age maximum in the Salamander Range.

The Rb–Sr ages suggest that the metamorphism is entirely Cambrian (505–590 Ma; Harland *et al.*, 1982) but within their error limits it cannot be decided whether metamorphism predates earliest Cambrian sedimentation in the adjacent Bowers Terrane. The oldest fossils in the Bowers Supergroup (Molar Formation) are Middle Cambrian, i.e. 523–540 Ma (Cooper & Shergold, in press). The contact between the high-grade gneisses and the Middle Cambrian sedimentary rocks in the Lanterman Range is controversial; it is highly tectonized and a boundary fault could separate entirely different exotic terranes (Bradshaw & Laird, 1983; Tessensohn, 1984). On the other hand, if they were contiguous in Cambrian times it is unlikely that high-grade metamorphism could have occurred so close to low-grade Cambrian sedimentary rocks. The Ross metamorphic mineral age pattern (470–500 Ma) crosses over this boundary, suggesting that large vertical or lateral displacements are unlikely. The most precise Rb–Sr age (558 ± 17 Ma) for the Salamander metamorphic rocks suggests that early metamorphism of the Wilson Group occurred before mid-Cambrian sedimentation commenced about 540 m.y. ago. Bowers Supergroup sedimentation ended in latest Cambrian or Early Ordovician times (495–505 Ma) with widespread regional deformation and metamorphism during the Ross Orogeny.

The new Rb–Sr ages imply a pre-Ross orogeny in the Wilson Group. However, throughout the Ross Orogeny there are only scattered age data for pre-Ross events. Grindley & McDougall (1969) defined a Beardmore Orogeny in the Transantarctic Mountains on the basis of pre-Cambrian gneisses in the Nimrod Glacier area, and possible correlatives in the southern Victoria Land and Dry Valley region. The data are imprecise, with age patterns being much disturbed by the pervasive overprint of the Ross Orogeny. Some old U–Pb ages (638 ± 92 Ma) for zircons in the Olympus Granite-gneiss in the Dry Valley region (Deutsch & Grögler, 1966) may derive from Precambrian Skelton Group schists nearby. Hornblende ages (600–1100 Ma) from the Nimrod Group, central Transantarctic Mountains, although complicated by excess argon concentrations, were interpreted by Adams, Gabites & Grindley (1982a) as dating a late Precambrian metamorphism, 580–650 m.y. ago.

In northern Victoria Land, Vetter *et al.* (1983) reported rather anomalous Rb–Sr isochron ages for granites: (i) in the Robertson Bay Terrane, 530 ± 30 Ma; (ii) at Surgeon Island, 599 ± 21 Ma; and (iii) at Mariner Glacier, 535 ± 26 Ma. In southern Victoria Land, a Rb–Sr whole-rock age of 568 ± 9 Ma for the Carlyon Granodiorite (Felder & Faure, 1980) is significantly older than the majority of ages from the Granite Harbour Intrusives (465–500 Ma) and suggests that the structural–metamorphic history of the metasedimentary basement in this region must be older. These age data point to increasing evidence for a metamorphic event extending from the Transantarctic Mountains into northern Victoria Land, with regional metamorphism and some plutonism occurring in latest Precambrian–earliest Cambrian times (550–650 Ma), and separated from the Ross Orogeny by the Cambrian Bowers Supergroup–Byrd Group sedimentary cycle.

Acknowledgements

The major part of this study was completed whilst CJA was an Alexander von Humboldt Fellow at the Bundesanstalt für Geowissenschaften, West Germany in 1983–84. The assistance of all those in the Geochronology laboratory at BGR, in particular Dr H. Kreuzer and Mr H. Lenz, is gratefully acknowledged. Preliminary XRF analyses were made at the Department of Geology, Victoria University, Wellington, New Zealand. Dr I. Graham and Pamela Whitla are thanked for their assistance and for permission to use Rb–Sr computer programs. Samples from the Salamander Range were collected by CJA and Dr G. Grindley (sample 10392) during the 1981–82 International Northern Victoria Land Expedition (NZARP–USARP–ANARE) and numerous colleagues are thanked for their logistic help and enthusiasm in the field. Margaret MacDonald prepared the diagrams and Anne Hepenstall typed the manuscript. CJA thanks the Symposium Organizing Committee and the Trans-Antarctic Association Committee for financial support to present papers at the symposium in Cambridge.

References

Adams, C.J. (1986). Geochronological studies of the Swanson Formation of Marie Byrd Land, West Antarctica, and correlation with northern Victoria Land, East Antarctica and South Island, New Zealand. *New Zealand Journal of Geology and Geophysics*, **29**, 345–58.

Adams, C.J. (1987). Age and ancestry of the metamorphic rocks of the Daniels Range, USARP Mountains, Antarctica. In *Geological Investigations in Victoria Land*, Antarctic Research Series, 46, ed. E. Stump, pp. 25–38. Washington, DC; American Geophysical Union.

Adams, C.J., Gabites, J.E. & Grindley, G.W. (1982a). Orogenic history of the central Transantarctic Mountains: new K–Ar age data on the Precambrian–Lower Paleozoic basement. In *Antarctic Geoscience*, ed. C. Craddock, pp. 817–26. Madison; University of Wisconsin Press.

Adams, C.J., Gabites, J.E., Wodzicki, A., Laird, M.G. & Bradshaw, J.D. (1982b). Potassium–argon geochronology of the Precambrian–Cambrian Wilson and Robertson Bay Groups and Bowers Supergroup, northern Victoria Land, Antarctica. In *Antarctic Geoscience*, ed. C. Craddock, pp. 543–8. Madison; University of Wisconsin Press.

Adams, C.J. & Kreuzer, H. (1984). Potassium–argon age studies of slates and phyllites from the Bowers and Robertson Bay Terranes, North Victoria Land, Antarctica. *Geologisches Jahrbuch*, **B60**, 265–88.

Bradshaw, J.D. & Laird, M.G. (1983). The pre-Beacon geology of northern Victoria Land: a review. In *Antarctic Earth Science*, ed. R.L. Oliver, P.R. James & J.B. Jago, pp. 98–101. Canberra; Australian Academy of Science and Cambridge; Cambridge University Press.

Cooper, R.A. & Shergold, J.H. (in press). Paleozoic invertebrates of Antarctica. In *The Geology of Antarctica*, ed. R.J. Tingey. Oxford; Oxford University Press.

Deutsch, S. & Grögler, N. (1966). Isotopic age of Olympus Granite–Gneiss (Victoria Land–Antarctica). *Earth and Planetary Science Letters*, **1**, 82–4.

Felder, R.P. & Faure, G. (1980). Rubidium–strontium age determination of part of the basement complex of the Brown Hills, central Transantarctic Mountains. *Antarctic Journal of the United States*, **15(5)**, 16–17.

Grindley, G.W. & McDougall, I. (1969). Age and correlation of the Nimrod Group and other Precambrian rocks units in the central Transantarctic Mountains, Antarctica *New Zealand Journal of Geology and Geophysics*, **12**, 391–411.

Harland, W.B., Cox, A.V., Llewellyn, P.G., Picton, C.A.G., Smith, A.G. & Walters, R. (1982). *A Geologic Time Scale*. Cambridge; Cambridge University Press. 131 pp.

Kreuzer, H., Höhndorf, A., Lenz, H., Vetter, U., Tessensohn, F., Müller, P., Jordan, H., Harre, W. & Besang, C. (1981). K/Ar and Rb/Sr dating of igneous rocks from North Victoria Land, Antarctica. *Geologisches Jahrbuch*, **B41**, 267–73.

Ravich, M.G. & Krylov, A.J. (1964). Absolute ages from East Antarctica. In *Antarctic Geology*, ed. R.J. Adie, pp. 579–89. Amsterdam; North-Holland Publishing Co.

Steiger, R.H. & Jäger, E. (1977). Subcommission on geochronology: convention on the use of decay constants in geochronology and cosmochronology. *Earth and Planetary Science Letters*, **36**, 359–62.

Sturm, A. & Carryer, S.J. (1970). Geology of the region between the Matusevich and Tucker Glaciers, north Victoria Land, Antarctica. *New Zealand Journal of Geology and Geophysics*, **13**, 408–35.

Tessensohn, F. (1984). Geological and tectonic history of the Bowers Structural Zone, North Victoria Land, Antarctica. *Geologisches Jahrbuch*, **Reihe B**, 371–96.

Tessensohn, F., Duphorn, K., Jordan, H., Kleinschmidt, G., Skinner, D.N.B., Vetter, U., Wright, T.O. & Wyborn, D. (1981). Geological comparison of basement units in north Victoria Land, Antarctica. *Geologisches Jahrbuch*, **B41**, 31–88.

Vetter, U., Roland, N.W., Kreuzer, H., Höhndorf, A., Lenz, H. & Besang, C. (1983). Geochemistry, petrography and geochronology of the Cambro-Ordovician and Devonian–Carboniferous granitoids of northern Victoria Land, Antarctica. In *Antarctic Earth Science*, ed. R.L. Oliver, P.R. James & J.B. Jago, pp. 140–3. Canberra; Australian Academy of Science and Cambridge; Cambridge University Press.

Recovery and recrystallization of quartz and 'crystallinity' of illite in the Bowers and Robertson Bay terranes, northern Victoria Land, Antarctica

W. BUGGISCH[1] & G. KLEINSCHMIDT[2]

1 Institut für Geologie und Mineralogie, Universität Erlangen–Nürnberg, Schlossgarten 5, D-8520 Erlangen, Federal Republic of Germany
2 Geologisch–Paläontologisches Institut, Universität Frankfurt, Senckenberganlage 32–34, D-6000 Frankfurt, Federal Republic of Germany

Abstract

Metamorphism of the rocks of the Bowers and Robertson Bay terranes (BT and RBT) in northern Victoria Land has been studied by means of illite 'crystallinity' and the behaviour of quartz in deformed veins. Low-grade metamorphism, especially within RBT, is uniform over an area of 150×300 km^2. Such metamorphism required a uniform overburden of at least 10–15 km thickness above the RBT rocks. This cover could not have been entirely depositional but must also have involved tectonic processes. Two feasible mechanisms for this crustal thickening are: (a) BT was thrust eastwards over RBT for more than 100 km; (b) thickening was caused by multiple thrusting, mainly within RBT. The first mechanism could be proven if a klippe of BT was shown to exist in eastern RBT and the second by the existence of more than the two thrust faults known at present (BT/RBT and 30 km east of RBT's western margin).

Introduction

Northern Victoria Land is composed of three fault-bounded terranes: the Wilson Terrane (WT), the Bowers Terrane (BT) and the Robertson Bay Terrane (RBT). BT and RBT consist predominantly of very low-grade to low-grade rocks and WT is composed of medium- to high-grade metamorphic rocks. The RBT exhibits a thick sequence of flysch-like sediments of (?)Vendian–Cambrian age – the Robertson Bay Group (Tessensohn *et al.*, 1981). Bouldery and pebbly mudstones and conglomerates of the Handler Formation occur in the uppermost part of the Robertson Bay Group (Bradshaw *et al.*, 1985; Wright & Brodie, 1987). A rich conodont fauna from the reworked carbonates within the Handler Formation indicate a Tremadocian age for the primary deposition of the shallow-water limestones (Burrett & Findlay, 1984; Wright, Ross & Repetski, 1984; Buggisch & Repetski, 1988). The Bowers Supergroup of the BT mainly consists of shales, volcanic rocks, conglomerates and reworked carbonate clasts and slabs (Reilly Ridge); a Middle–early Late Cambrian age is indicated by a trilobite fauna. The Bowers Supergroup is overlain by cross-bedded red sandstones of the Leap Year Group. Both the Bowers and Robertson Bay terranes are intruded by Lower Devonian Admiralty Intrusives.

Approximately 275 samples of shale and quartz veins were collected from the Robertson Bay Group and the Bowers Supergroup throughout these terranes. Investigations of the 'crystallinity' of illite and the behaviour of quartz in deformed veins (recovery and recrystallization, respectively) form the basis of our study of the metamorphism in the RBT and BT.

Methods

Different parameters are used in order to describe the 'crystallinity' of illite (Weaver, 1960; Kubler, 1967; Weber, 1972; Kisch, 1983). Illite becomes more 'crystalline' with increasing burial and very low-grade metamorphism. Initially broad (001) peaks become narrower and $\Delta 2\theta$ decreases. In this paper the half peak width of (001) is measured as $\Delta 2\theta$. The value obtained is very dependent on the instrumental conditions, the grain size of the sample and on the method of sample preparation (e.g. grinding and thickness; see Buggisch, 1986). Samples from northern Victoria Land were ground in a porcelain mortar mill in a 0.01 N ammonia solution. This method enables the fine-grained material to float and the illite crystals are not affected by further grinding. The results were reproducible when a very dilute suspension of the fraction $< 2\ \mu$m was sedimented on glass (0.1 mg of sediment/cm^2). The values are not compatible with other data from the literature and are therefore only used as a relative parameter in this paper (conditions: Philips PW 1729, 20 KV, 30 mA, diffractometer PW 1840: speed 0.005 2°/min; chart 50 mm/2°; range 1×10^3–1×10^4; TC 5, slit 0.2 mm).

The characteristics of quartz have been studied in deformed quartz veins of the BT and RBT. Deformed quartz recovers or recrystallizes progressively with an increasing strain rate (e.g.

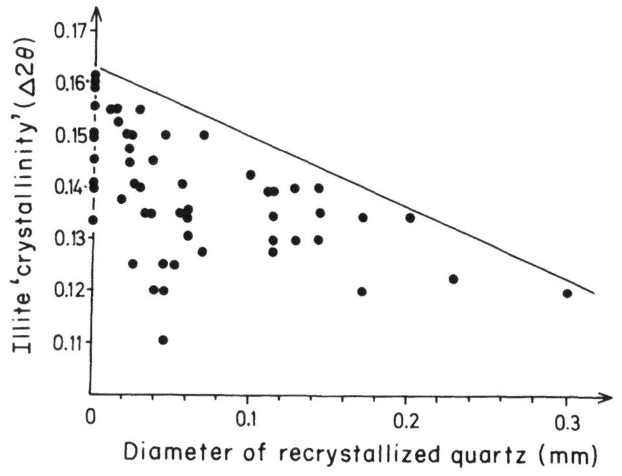

Fig. 1. Illite 'crystallinity' versus grain size of recrystallized quartz from northern Victoria Land.

Courrioux, 1987) and/or with increasing temperature (e.g. Voll, 1976) and/or depending on stress (e.g. Behrmann, 1984). The recovery starts with subgrain formation and strain-induced boundary migration. Quartz begins to recrystallize at about 290° C. The grain size and the percentage of recrystallized quartz increases simultaneously during prograding metamorphism.

Both methods, illite 'crystallinity' and quartz-behaviour, can be applied in the range from burial diagenesis to low-grade metamorphism when common critical minerals are rare or lacking. All of our data for illite 'crystallinity', and percentage and grain size of recrystallized quartz represent mean values of a large number of readings.

A comparison of quartz behaviour and the 'crystallinity' of illite is shown in two plots of our data from northern Victoria Land (Figs 1 & 2). Both plots show a poor negative correlation but the upper limit of the dots is quite distinct. Therefore, both the behaviour of quartz and illite 'crystallinity' result from the same process.

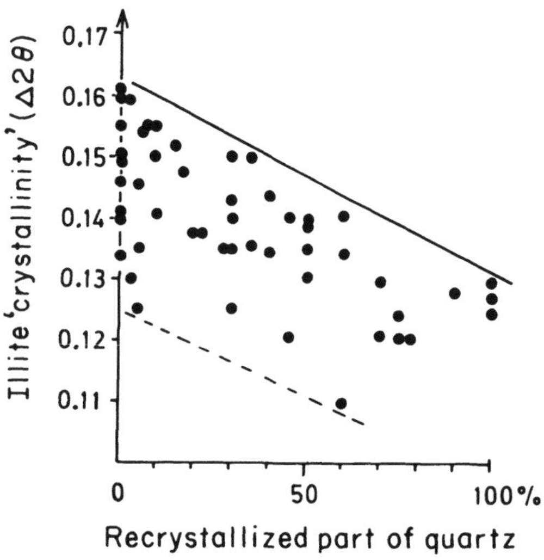

Fig. 2. Illite 'crystallinity' versus amount of recrystallized quartz from northern Victoria Land.

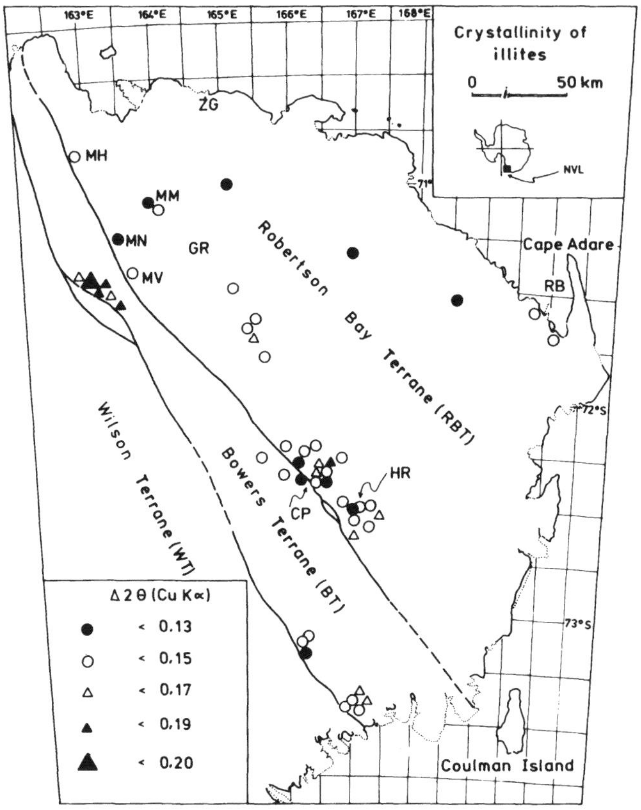

Fig. 3. Illite 'crystallinity' in northern Victoria Land. CP, Crosscut Peak; GR, Griffith Ridge; MH, Mount Hager; MM, Mount Mulach; MN, McKenzie Nunatak; MV, Mount Verhage; HR, Handler Ridge; RB, Robertson Bay and ZG, Zykov Glacier.

Results

Our results modify the assumptions made by Kleinschmidt (1983). The 'crystallinity' of illite (Figs 3 & 5) ranges from 0.12 to 0.20 $\Delta 2\theta$ in the BT and RBT. Broad peaks (solid triangles in Fig. 3) are restricted to the BT where a mean of 0.15 $\Delta 2\theta$ was calculated from 124 readings. The illite 'crystallinity' in RBT is somewhat better and more constant, with a mean value of 0.13 $\Delta 2\theta$ calculated from 148 readings.

Only quartz recovery occurs within and near the eastern margin of the BT (Fig. 4). Toward its western margin, the recrystallization of quartz increases rapidly in both grain size and percentage. The behaviour of quartz in the RBT seems to be more complicated: there are hardly any areas without recrystallization and other areas (Fig. 4, circles in brackets, mouth of Zykov Glacier and Griffith Ridge) show strong recrystallization which is due to contact metamorphism by the Admiralty Intrusives. In the Mount Mulach region, different degrees of recrystallization and recovery of quartz only occur together. This pattern is partly due also to contact metamorphism, especially since the nearest granite contact is observed only about 10 km away. Consequently, quartz recrystallization behaves rather uniformly throughout the RBT when it is not affected by contact metamorphism. There is a slight increase in grain size and percentage of recrystallized quartz grains in some places near the western boundary of the RBT (McKenzie Nunatak and the western end of Handler Ridge). In other areas (Mount Hager, Mount Verhage and south of

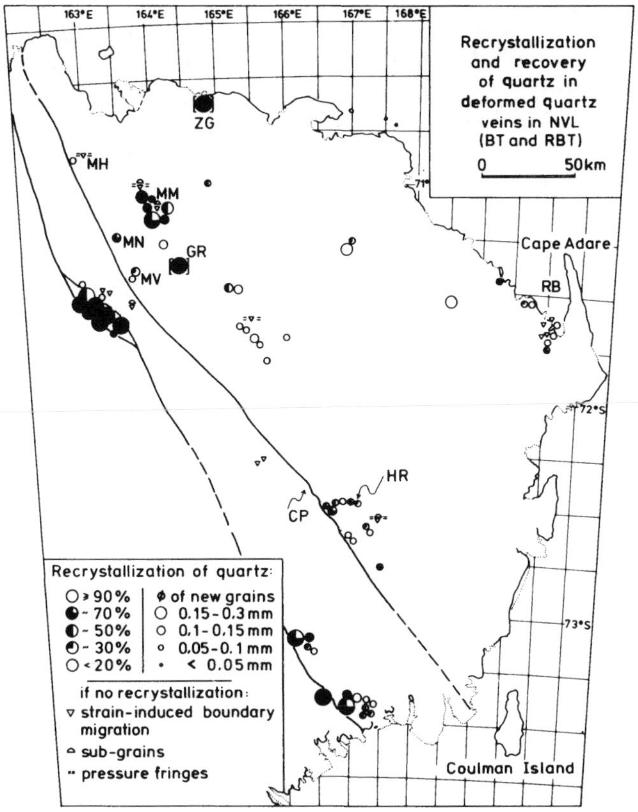

Fig. 4. Recrystallization and recovery of quartz in deformed quartz veins, northern Victoria Land.

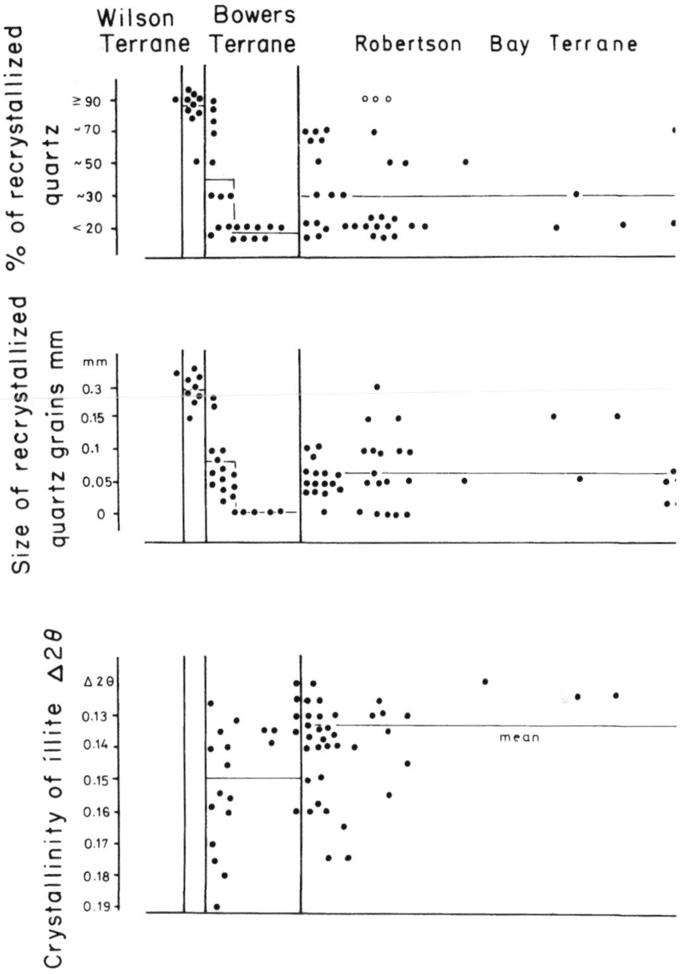

Fig. 5. Mean values of illite 'crystallinity', grain size, and the amount of recrystallized quartz projected onto a SW–NE-trending line across northern Victoria Land.

Handler Ridge), even the very schistose rocks do not exhibit increased quartz recrystallization.

In addition to our data, the conodont 'colour alteration index' (CAI 4: Buggisch & Repetski, 1988) points to temperatures between 190 and 300 °C, and $^{40}Ar/^{39}Ar$ results (Wright & Dallmeyer, 1986) require temperatures of 300 to 350 °C.

A projection of all mean values onto a SW–NE-trending line is shown in Fig. 5. The mean value of all the readings for illite 'crystallinity' is slightly better in the RBT than in the BT. The size and amount of recrystallized quartz grains is greater in the RBT than in the BT. Therefore, each terrane has its specific metamorphic character. The metamorphism of the RBT is extremely uniform within an area of 150×300 km^2 (Fig. 6). Metamorphism in the BT is of a slightly lower level but also appears to be uniform, even if the BT seems to be too narrow for this statement to have much validity. The increase of metamorphism at the western border is more pronounced in the BT, reaching greenschist-facies.

Metamorphism and deformation of the BT and RBT occurred during the Ross Orogeny. K–Ar whole-rock age determinations (Adams & Kreuzer, 1984) yielded an age of ~ 470–500 Ma for the RBT slates and a similar age but more scattered data for shales of the BT. $^{40}Ar/^{39}Ar$ determinations (Wright & Dallmeyer, 1986) indicate an age of 500–505 Ma for the phyllites at Handler Ridge (RBT). Therefore, the metamorphism is scarcely younger than the Tremadocian age of the conodont faunas from the Handler Formation.

Conclusions

The uniform, almost low-grade metamorphism of the RBT requires a uniform burial of the Robertson Bay Group. Assuming a 'normal' geothermal gradient of about 25 °C/km, a cover of 10–15 km above the RBT is necessary. A cover of as much as 20 km should be considered according to Winkler's (1979) gradient. These data agree well with the results of Engel (1985), who postulated a pressure of 2 kb during contact metamorphism of the Devonian Admiralty Intrusives.

The age of metamorphism is barely younger than that of the youngest preserved sedimentary rocks of the Robertson Bay Group. Therefore, burial cannot have been due solely to a cover by sedimentary deposition but involved additional thickening by tectonic processes. Two possibilities have to be considered:

(i) RBT sedimentary rocks were covered by rocks of the Bowers Supergroup. Hence, the thrust between the BT and RBT, which can be observed at Crosscut Peak (Crowder, 1968; Findlay & Field, 1983; Wright & Findlay, 1984), would have extended more than 100 km eastward. Proof of this model would be the presence of a klippe of the Bowers Supergroup within the RBT!

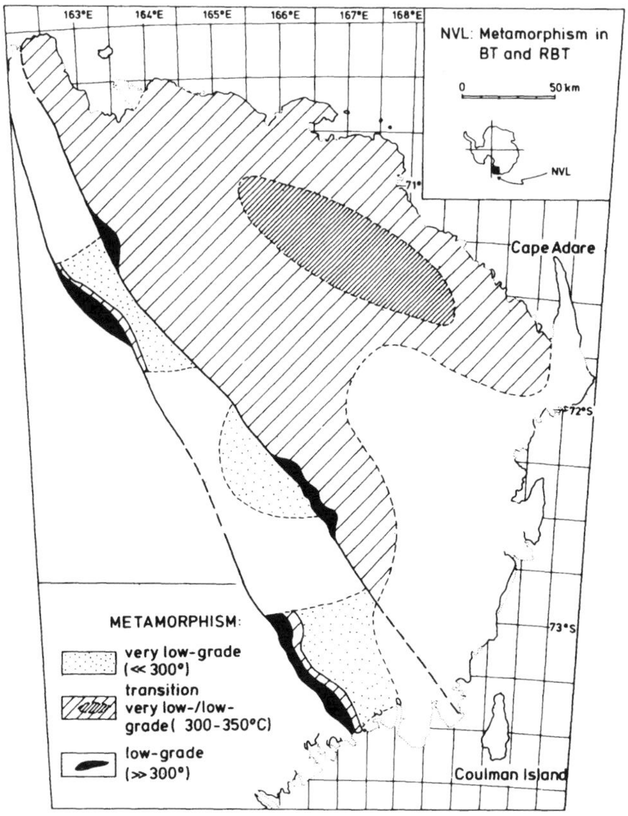

Fig. 6. Simplified map showing the distribution of metamorphic grades in northern Victoria Land (BT and RBT).

(ii) The required tectonic thickening was caused by thin-skinned tectonics, e.g. multiple thrusting involving tectonic ramps and abrading at the BT–RBT boundary and within the RBT. The boundary between the BT and RBT would be only one of several thrust-ramps. A second one could be easily assumed 30 km east of this boundary. Additional thrust ramps would have to be substantiated farther to the east.

The westward increase of metamorphism may be due to frictional heating at the BT–RBT boundary and to the overriding of the (hot) WT above the BT at their boundary. Thus the BT obtained progressive imprint from above and the WT a retrogressive imprint from below (Gibson & Wright, 1985).

The metamorphic pattern of northern Victoria Land and the two models suggested above are only partly consistent with any strike–slip-accretionary model. However, our results agree well with the subduction models of Gibson & Wright (1985) and Kleinschmidt & Tessensohn (1987). The model of Gibson & Wright is the more convenient one: the youngest sedimentary rocks (Handler Formation) occur at the western border of the Robertson Bay basin, indicating the proximity of and/or docking with a continental margin having a carbonate platform. Thus, these sedimentary rocks could be easily explained as wildflysch of westward-directed subduction or by eastward thrusting of BT over RBT.

Acknowledgements

The field work was carried out during GANOVEX I–IV (1979–85) and the NZARP expedition K 051 (1984–85). Additional samples were collected by H. Jordan, M. Schmidt-Thomé and T. Wright. We are also indebted to the leaders of GANOVEX and K 051 expeditions (F. Tessensohn and J. Bradshaw). This work was supported by grants of the 'Deutsche Forschungsgemeinschaft' (Bu 312/12 and Kl 429/2).

References

Adams, C.J. & Kreuzer, H. (1984). Potassium–argon age studies of slates and phyllites from the Bowers and Robertson Bay Terranes, North Victoria Land, Antarctica. *Geologisches Jahrbuch*, **B60**, 265–88.

Behrmann, J.H. (1984). Patterns of paleostress and strain beneath the Aguilon Nappe, Betic Cordilleras (Spain). *Zeitschrift der Deutschen Geologischen Gesellschaft*, **135(1)**, 293–305.

Bradshaw, J.D., Begg, J.C., Buggisch, W., Brodie, C., Tessensohn, F. & Wright, T.O. (1985). New data on Paleozoic stratigraphy and structure in North Victoria Land. *New Zealand Antarctic Record*, **6(3)**, 1–6.

Buggisch, W. (1986). Diagenese und Anchimetamorphose aufgrund von Conodontenfarbe (CAI) und 'Illit-Kristallinität' (CI), Methodische Untersuchungen und Daten zum Oberdevon und Unterkarbon der Dillmulde (Rheinisches Schiefergebirge). *Geologisches Jahrbuch*, **114**, 181–200.

Buggisch, W. & Repetski, J.E. (1988). Uppermost Cambrian (?) and Tremadocian conodonts from Handler Ridge (Robertson Bay terrane, northern Victoria Land, Antarctica). *Geologisches Jahrbuch*, **B66**, 145–85.

Burrett, C.F. & Findlay, R.H. (1984). Cambrian and Ordovician conodonts from the Robertson Bay Group, Antarctica and their tectonic significance. *Nature, London*, **307(5953)**, 723–5.

Courrioux, G. (1987). Oblique diapirism: the Criffle granodiorite/granite zoned pluton (southwest Scotland). *Journal of Structural Geology*, **9(3)**, 313–30.

Crowder, D.F. (1968). Geology of a part of North Victoria Land, Antarctica. *US Geological Survey Professional Paper*, **600-D**, D95–107.

Engel, S. (1985). Untersuchungen zur Kontaktmetamorphose von Flysch–Sedimenten durch granitische und gabbroide Intrusive im Nord-Victoria-Land/Antarktis. Dissertation (Dr.-Thesis), University of Darmstadt (THD), W. Germany, 141 pp. (unpublished).

Findlay, R.H. & Field, B.D. (1983). Tectonic significance of deformations affecting the Robertson Bay Group and associated rocks, northern Victoria Land, Antarctica. In *Antarctic Earth Science*, ed. R.L. Oliver, P.R. James & J.B. Jago, pp. 107–12. Canberra; Australian Academy of Science and Cambridge; Cambridge University Press.

Gibson, G.M. & Wright, T.O. (1985). Importance of thrust faulting in the tectonic development of northern Victoria Land, Antarctica. *Nature, London*, **315(6019)**, 480–3.

Kisch, H.J. (1983). Mineralogy and petrology of burial diagenesis (burial metamorphism) and incipient metamorphism in clastic rocks. In *Developments in Sedimentology*, 25B, ed. G. Larsen & G.V. Chilingar, pp. 513–41. Amsterdam; Elsevier.

Kleinschmidt, G. (1983). Trends in regional metamorphism and deformation in northern Victoria Land, Antarctica. In *Antarctic Earth Science*, ed. R.L. Oliver, P.R. James & J.B. Jago, pp. 119–22. Canberra; Australian Academy of Science and Cambridge; Cambridge University Press.

Kleinschmidt, G. & Tessensohn, F. (1987). Early Paleozoic westward directed subduction at the Pacific margin of Antarctica. In *Gondwana Six: Structure, Tectonics, and Geophysics*, Geophysical Monograph, 40, ed. G.D. McKenzie, pp. 89–105. Washington, DC; American Geophysical Union.

Kubler, B. (1967). La cristallinité d'illite et les zones tout à fait supérieures du métamorphisme. In *Colloque Etages Tectoniques, Baconnière*, pp. 105–22. Neuchâtel (Suisse).

Tessensohn, F., Duphorn, K., Jordan, H., Kleinschmidt, G., Skinner, D.N.B., Vetter, U., Wright, T.O. & Wyborn, D. (1981). Geological comparison of basement units in North Victoria Land, Antarctica. *Geologisches Jahrbuch*, **B41**, 31–88.

Voll, G. (1976). Recrystallization of quartz, biotite and feldspars from Erstfeld to the Leventina Nappe, Swiss Alps, and its geological significance. *Schweizerische mineralogisch–petrographische Mitteilungen*, **56**, 641–7.

Weaver, C.E. (1960). Possible use of clay minerals in the search for oil. *Bulletin of the American Association of Petroleum Geologists*, **44**, 1505–18.

Weber, K. (1972). Kristallinität des Illits in Tonschiefern und andere Kriterien schwacher Metamorphose im nördlichen Rheinischen Schiefergebirge. *Neues Jahrbuch für Geologie und Paläontologie, Abhandlungen*, **141(3)**, 332–63.

Winkler, H.G.F. (1979). *Petrogenesis of Metamorphic Rocks*, 5th edn. Berlin, Heidelberg; Springer-Verlag. 348 pp.

Wright, T.O. & Brodie, C. (1987). The Handler Formation, a new unit of the Robertson Bay Group, northern Victoria Land, Antarctica. In *Gondwana Six: Structure, Tectonics, and Geophysics*, Geophysical Monograph, 40, ed. G.D. McKenzie, pp. 25–9. Washington, DC; American Geophysical Union.

Wright, T.O. & Dallmeyer, R.D. (1986). $^{40}Ar/^{39}Ar$ whole rock slate ages: timing of tectonothermal events in northern Victoria Land, Antarctica. *Geological Society of America, Abstracts with Programs*, **18**, 795.

Wright, T.O. & Findlay, R.H. (1984). Relationships between the Robertson Bay Group and the Bowers Supergroup – new progress and complications from the Victory Mountains, North Victoria Land. *Geologisches Jahrbuch*, **B60**, 105–16.

Wright, T.O., Ross, R.J. & Repetski, J.E. (1984). Newly discovered youngest Cambrian or oldest Ordovician fossils from the Robertson Bay Terrane (formerly Precambrian), northern Victoria Land, Antarctica. *Geology*, **12(4)**, 301–12.

The boundary of the East Antarctic craton on the Pacific margin

N.W. ROLAND

Bundesanstalt für Geowissenschaften und Rohstoffe, Postfach 51 01 53, D-3000 Hannover 51, Federal Republic of Germany

Abstract

Along the Pacific coast of northern Victoria Land (NVL), three different terranes are encountered; from east to west these are: the Robertson Bay, the Bowers, and the Wilson terranes. The high-grade metamorphic rocks of the Wilson Terrane comply with the requirements of a platform and have to be assigned to the East Antarctic craton. Though radiometric age determinations mainly provide ages related to the Ross Orogeny, Late Precambrian ages also appear, which may point to the existence of an older orogeny in NVL. There are indications that a rejuvenation took place during the Ross Orogeny. New evidence for the positioning of the boundary between shield and Precambrian platform was obtained by the aeromagnetic survey of GANOVEX IV in 1984–85. The shield is inferred to begin west of the Priestley Glacier, whilst the metamorphics of the Precambrian platform extend at least to the Bowers Terrane.

Introduction

The morphological and geological subdivision of the Antarctic continent into the East Antarctic shield and the West Antarctic fold belts is obvious. The Transantarctic Mountains, folded during the Ross Orogeny, were supposed to delineate the boundary between these two complexes (e.g. Hamilton, 1964). However, the exact position of this boundary has always been problematical and a number of possibilities exist: (i) in the vicinity of the Adelie coast 'somewhat west of Cape Gunther' at 142°20′ E (Klimov & Solov'ev, 1958); (ii) at 145° E (Klimov, 1967), respectively at the western edge of the Wilkes Basin (Steed, 1983); (iii) between 150 and 155° E (e.g. Ushakov, 1960); and (iv) west of the Rennick Glacier (Adams, 1986), i.e. west of 162° E. As new geological and geophysical data are available from northern Victoria Land NVL, the boundary question will be broached again.

Definitions

Definitions of the terms 'shield', 'craton', and 'platform' are given first as the same expression is often used in a different sense by different authors. A 'shield' is a large area of exposed basement rocks in a craton, commonly with a gently convex surface, surrounded by sediment-covered platforms. The rocks of virtually all shield areas are Precambrian. A 'platform' is that part of a continent that is covered by flat-lying or gently tilted strata, mainly sedimentary, which are underlain at varying depths by a basement of rocks that were consolidated during earlier deformations. Both platforms and shields are part of the 'craton' (Bates & Jackson, 1980).

A more detailed definition is given by Ravich & Kamenev (1975, p. 534), although their term 'platform' is used in the sense of 'craton'. Areas 'classified as shields of the Antarctic platform' possess the following characteristics:

- prolonged mild uplifts and a high degree of erosion
- the ubiquitous development of regional metamorphic rocks of the granulite facies (charnockites and enderbites)
- the absence of polymetamorphic formations but with complex intrusions of charnockitoids
- the possible presence of rocks with the oldest absolute ages (3000–3300 Ma).

The shields can be considered as Archaean nuclei (> 2500 Ma) which are surrounded by Proterozoic (2500–600 Ma) polymetamorphic mobile zones in which retrograde greenschist facies rocks may occur (e.g. Queen Maud Land). In these mobile belts the crystalline basement was subjected to polyphase metamorphism in the Proterozoic, with reworking of granulite facies rocks under high-grade amphibolite facies conditions and the intrusion of magmatic masses (Ravich & Kamenev, 1975).

As there is not a single 'East Antarctic shield', but probably several separated shields, and also Precambrian platforms, the term 'East Antarctic craton' is recommended and will be applied in the following for the Precambrian metamorphic basement of greater Antarctica.

Main features of northern Victoria Land geology

There are three main complexes in NVL, which differ in metamorphic grade and tectonic development, and which are discussed as terranes in a plate tectonic sense; from east to west these are: the Robertson Bay, Bowers and Wilson terranes. They are separated by NW–SE-striking lineaments which cut

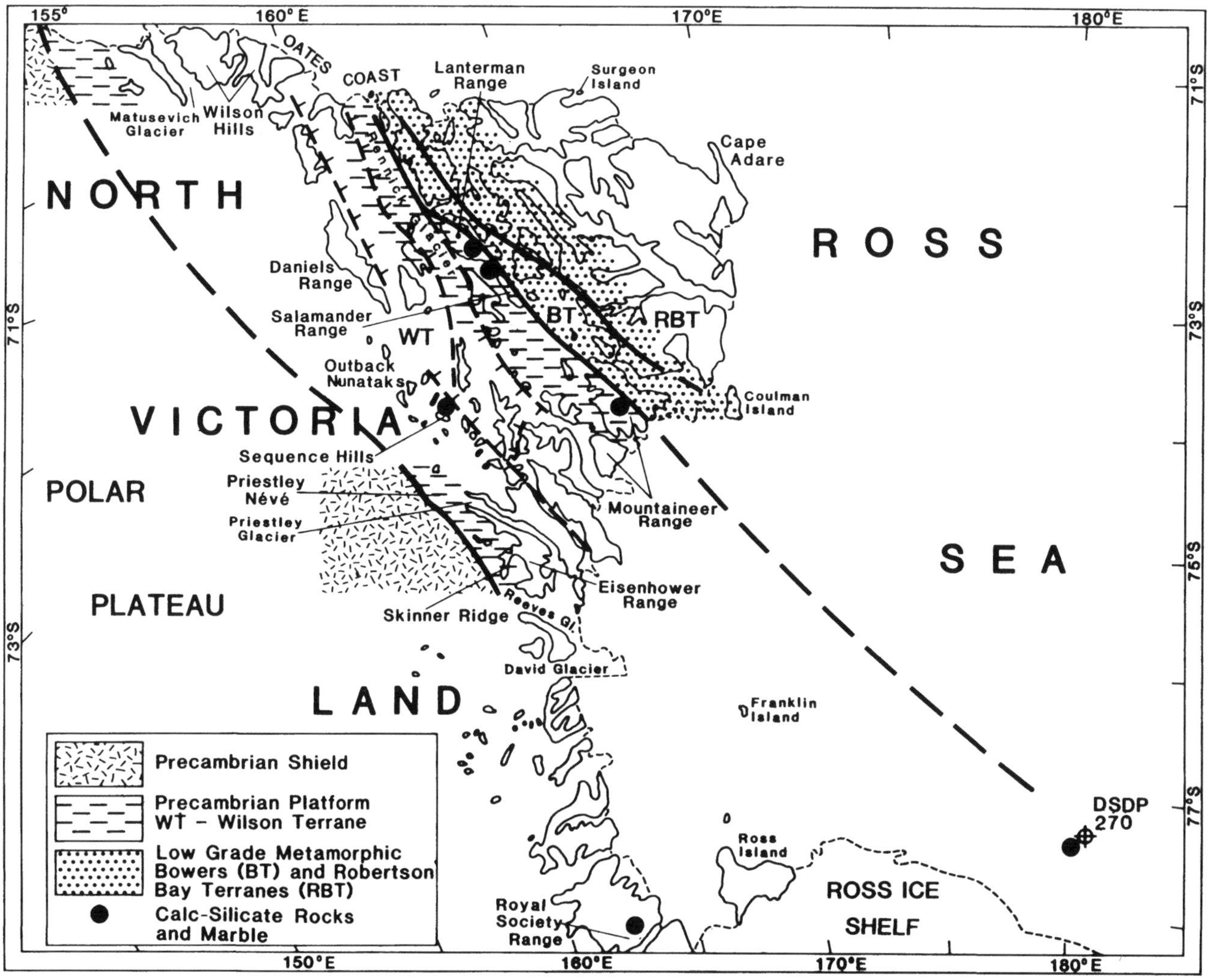

Fig. 1. The main structural units in northern Victoria Land.

through NVL and can be traced from the Pacific to the Ross Sea coasts (Fig. 1). Both lineaments are accompanied by mylonite and/or retrograde greenschist facies zones which have to be seen in context with an Early Palaeozoic subduction (Kleinschmidt & Tessensohn, 1987).

Metamorphic grade

The Robertson Bay Terrane comprises turbiditic greywackes and slates of the Robertson Bay Group. The regional metamorphism is of very low–low grade. Low greenschist facies occurs in the boundary zone adjacent to the Bowers Terrane. Kleinschmidt (1983) proved that an increase in metamorphic grade from east to west can be demonstrated by the investigation of the illite crystallinity, the appearance of biotite near the Bowers Terrane, and the recrystallization of quartz in syn- or pretectonic quartz veins.

In the Bowers Terrane the metamorphic grade of the Bowers Supergroup rocks also varies between very low-grade and greenschist facies. The latter is mainly developed near the boundary faults of the Bowers Terrane.

The metamorphic rocks in the Wilson Terrane range from the medium-grade Rennick schists to the high-grade Wilson gneisses which occasionally include migmatites. Polyphase metamorphism is developed at least in parts of the Daniels Range where a progressive medium–high-grade (garnet, fibrolite) metamorphism is overprinted by a retrogressive greenschist facies (biotite, chlorite). Also, in the Lanterman Range, polyphase metamorphism was observed: the greenschist facies overprint was preceded by a metamorphism under 5–8 kb P_{H_2O} and 570–700 °C P–T conditions based on the presence of relict kyanite and staurolite and one sample that was placed in the garnet granulite field (Roland *et al.*, 1984). The Rennick schists were metamorphosed under medium pressure conditions (4 kb) and are postulated to have been deposited in an intracratonic trough (aulacogen?) (Roland, Olesch & Schubert, 1989).

Deformation

Across NVL, there is also an increase in the intensity of deformation from east to west. Regular open folds are typical for the eastern Robertson Bay Terrane, but towards its western part, isoclinal, tightly folded rocks prevail, and adjacent to the Bowers Terrane the folds are more complicated.

In the Bowers Terrane the sediments and volcanics of the

Bowers Supergroup are essentially concordant, but folded by Late Cambrian–Early Ordovician tectonic activity.

The Wilson Terrane shows the highest degree of deformation, the rocks having been intensely folded by polyphase deformation in at least four episodes. In the Lanterman Range, the last deformation was closely related to the retrograde metamorphism (Roland *et al.*, 1984).

Radiometric age determinations

The radiometric ages from the Robertson Bay Terrane are concentrated in the range from 460 to 500 Ma, according to K–Ar whole-rock age determinations (Adams & Kreuzer, 1984). They are in the same age range as the higher grade metamorphic rocks of the Wilson Terrane and the associated Granite Harbour intrusives farther west.

The same upper limit of radiometric dates was found for slates of the Bowers Terrane, but the age range is much broader and covers dates between 275 and 505 Ma (Adams & Kreuzer, 1984). There is no grouping of the age data according to stratigraphy but there is a significant relation between age and altitude at which the samples were taken. The younger ages are related to the present day lower altitude, the older rocks are found at higher altitudes. Most likely the age pattern was caused 'by prolonged deep burial of the lower part of the Bowers graben sequence at sufficient depth for argon loss to occur continuously from the Bowers Supergroup slates, irrespective of their metamorphic (Ross) grade, stratigraphic age, or proximity to the boundary faults' (Adams & Kreuzer, 1984, p. 276).

The Rennick schists yielded Rb–Sr WR-isochron ages of 534 ± 24 Ma and 531 ± 17 Ma (IR 0.7187 ± 0.0009 and 0.7243 ± 0.0009, respectively). Model ages, based on averaged isochron data, suggest sedimentary ages of the Rennick schists not older than 1150 Ma (Adams, 1986). For the Wilson gneisses Adams (1986) mentioned a Rb/Sr age of 490 ± 30 Ma and U–Pb ages of 484–488 Ma.

The Granite Harbour intrusives revealed maximum ages of 500 Ma. The initial Rb–Sr isotope ratios are in the range of 0.709–0.715. As these ratios are significantly lower than those of the Wilson gneisses (0.7205 ± 16), Adams (1986) argued that the Granite Harbour intrusives cannot be derived by anatexis of the Rennick schists, the precursors of the Wilson gneisses.

A Precambrian age was mentioned by Faure & Gair (1970) for the Rennick schists from Bleak Peak, Sequence Hills. Quartz–biotite schist from this locality had a Rb–Sr whole-rock age of 770 ± 20 Ma. The authors suggest the possibility that the date is an underestimate of the time elapsed since deposition of the original sediment, but Adams (1981) stated that the Rb–Sr rock age has a much larger age error when calculated with a realistic range of initial $^{87}Sr/^{86}Sr$ ratios.

'Beardmore' (i.e. late Precambrian) ages were also reported from the upper Zenith Glacier and from Husky Pass, Lanterman Range; Rb–Sr scatterchrons yielded 610 ± 26 Ma (IR = 0.7132 ± 0.0008, MSWD = 4.0). The data were determined on four granodioritic and one amphibolitic gneiss (Adams & Höhndorf, this vol. p. 149). An indication for a Beardmore age is also reported from the north-eastern margin of the Robertson Bay Terrane, where a foliated granodiorite of Surgeon Island revealed 599 ± 21 Ma (IR = 0.7590 ± 0.0014, MSWD = 0.25) (Vetter *et al.*, 1984; Kreuzer *et al.*, 1987). Single Precambrian ages occur, though Ross ages prevail even in the high-grade metamorphic Wilson Terrane.

For comparison it might be helpful to consider the dating problems in the Antarctic Shield areas. From Enderby Land, the classical shield area from which ages older than even 4000 Ma are known, Early Palaeozoic ages (400–600 Ma) were reported from a granulite complex and only U–Pb determinations on chevkinites gave a greater age of 1500–1550 Ma for this complex: 'Apparently the entire zone extending from the coast of the Alasheyev Bight (western Enderby Land) to the head of the Lambert Glacier can be regarded as a Ross intracratonal trough with a reactivated pre-Riphean basement at the bottom' (Klimov, 1967, p. 475).

Queen Maud Land, the classical mobile belt area, may be another example of the dominance of Ross ages. But the apparent dates of 450–550 Ma for the metamorphic basement rocks are put in correct perspective by the overlying volcanic platform rocks which yield an age of 1050 Ma. Folding cannot be the reason for a rejuvenation of the basement in Queen Maud Land. Although a distinct predominance of Early Palaeozoic absolute age data for the various basement rocks is obvious, there is still no reliable evidence for the development of Ross geosynclinal fold systems in this region.

The migmatization of the oldest strata of East Antarctica falls within the limits of the Proterozoic. However, there are exceptions; e.g. the block of gneisses in the region of Ainsworth Bay and the Wilson Hills, where the weakly migmatized biotite gneisses have absolute ages of 425–485 Ma. They are intruded by porphyroblastic granites of 430–525 Ma (Starik *et al.*, 1960). An intense contact-metamorphic influence was thought to be the reason for the rejuvenation of the gneisses.

A similar rejuvenation has to be postulated not only for the Wilson Hills but for NVL in general. Some features cannot otherwise be explained:

- similar ages for granites and surrounding schists or even older granite and younger schist ages
- similar ages of foliated granodiorites and non-foliated, post-tectonic granites in the Lanterman Range
- younger ages in the more highly deformed and high-grade metamorphic Wilson gneisses than in the less deformed, medium-grade Rennick schists
- the inverse vertical age distribution in the Bowers Terrane.

In the Wilson Terrane the 460–500 Ma radiometric ages represent the effect of the last metamorphic episode, i.e. the effects of the Granite Harbour intrusives and the tectonic processes of the Ross Orogeny which were superimposed on the pre-existing metamorphics. It can also be demonstrated that the country rock was metamorphosed and folded prior to the emplacement of the Granite Harbour intrusives (see below).

In the Robertson Bay Terrane the radiometric age determinations of the low grade schists reflect the time of deformation of the sediments.

Geological and petrological results

Geological and petrological observations can provide further arguments for a cratonization older than the Ross Orogeny. At Miller Butte, Outback Nunataks, totally sheared leucocratic dykes were found. As the Granite Harbour aplites and pegmatites generally cut the folded Rennick schists without any sign of tectonic overprint, an older dyke generation can be assumed.

The xenoliths in the Granite Harbour intrusives may give some information on the basement at the time of intrusion. In the tonalite intrusions at the western border of the Lanterman Range, xenoliths of biotite schist were found which showed folding, probably F_2 folds (Roland *et al.*, 1984). This constitutes evidence that the Wilson metamorphics had already suffered polyphase deformation when the Granite Harbour granites were intruded.

DSDP results, Ross Sea region

Four of the DSDP Leg 28 drill sites were located in the Ross Sea. At Site 270 in the southern Ross Sea (77°26.48′ S, 178°30.19′ W) a foliated, fine-grained marble and calc-silicate gneiss, cut by calcite veins, was encountered beneath 413.2 m of Recent–Oligocene sediments. It is overlain by 30 m of a coarse lithified breccia with clasts of leucogranite, biotite schist and garnet-bearing leucogneiss. The marble was compared with the Koettlitz Group (redefinition of the Koettlitz marble), situated in the Royal Society Range, 400 km west of Site 270, and therefore an Early Palaeozoic age was assumed (Glomar Challenger Scientific Party, 1973).

The Koettlitz marble was defined as a steeply dipping chondrodite marble and sillimanite schist. However, considering the general strike direction in NVL, the Site 270 marble could more easily be the continuation of the calc-silicate gneisses and marble of the NW–SE-striking metamorphic belt which can be traced from the Lanterman and Salamander ranges to the Mountaineer Range at the Ross Sea coast (Fig. 1).

Geophysical results

Gravity and magnetic observations near the Oates Coast indicated a major structural break near longitude 155° E, about 80 km west of the Wilson Hills (Ushakov, 1960). This break was supposed to mark 'the transition from the shield area of East Antarctica to the (?)Caledonian fold zone bordering the Ross Sea' (McLeod, 1964), i.e. the transition from the shield to the Ross Orogen.

During the GANOVEX IV aeromagnetic survey, four areas differing in their anomaly pattern were distinguished. The most westerly one, which is located west of the Priestley Névé and the Eisenhower Range, showed the typical shield pattern (Fig. 2). Its boundary strikes NNW–NW and probably has to be drawn less than 50 km west of the upper Priestley Glacier, about 10 km west of Shephard Cliff and 10–20 km west of Skinner Ridge (Bosum *et al.*, 1989). No geological information is available from this area as it is totally covered by the ice of the polar plateau. The area east of the possible shield has a different anomaly pattern and geological information is available. It can be interpreted as a platform, i.e. a consolidated, block-faulted and partially rifted area.

The Transantarctic Mountains are not reflected in the anomaly pattern. The same is true in regard to the Ross Orogen; a pattern typical of fold belts is not developed. This is understandable as the present mountain chain was created by a relatively recent and quick uplift in NVL in connection with rift-forming processes of the Rennick graben and the Ross Sea Rift (Roland & Tessensohn, 1987). They do not represent a consistent geological unit but are intersected at an acute angle by different terranes in NVL.

Conclusions

According to the aeromagnetic results, shield rocks and structures can be presumed in the area west of the upper Priestly Glacier and Skinner Ridge. The question is, can this break be correlated with the one mentioned from 155° E at the Oates Coast by Ushakov (1960)?

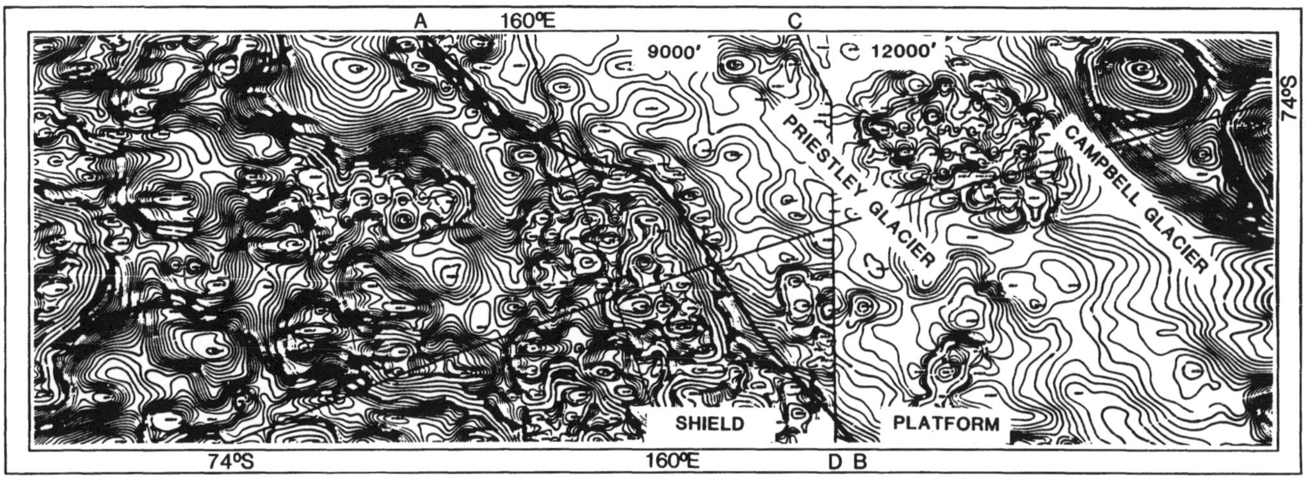

Fig. 2. Section of the aeromagnetic map of GANOVEX IV, with the probable shield–platform boundary (A–B). The line C–D presents the boundary between the 9000 and 12000 ft flight level.

The probable shield is bordered by a Precambrian platform and the Wilson Terrane has all the prerequisites for such a platform: the metamorphics show multiple deformation, polymetamorphism, intrusion by granite, uplift and denudation followed by the sedimentation of typical flat-lying cover rocks, the intrusion of dolerites, effusion of tholeiitic basalts and, finally, germanotype tectonics including graben formation. The dominance of Ross ages is explained by the rejuvenation of the Precambrian basement during the Early Palaeozoic, induced mainly by the intrusion of the Granite Harbour intrusives as a consequence of subduction processes. Data from the DSDP 270 drill site in the Ross Sea show that the metamorphic basement continues beneath the present southwestern Ross Sea and it is likely that this area belongs to the Precambrian platform as well.

In NVL the western boundary of the NW–SE-striking Bowers Terrane can be considered as the former edge of the craton (Fig. 2) where a major structural break is located, and where isotopic compositions as well as major and trace elements of the NVL granites indicate that the boundary also coincides with a major discontinuity in the lower crust (Borg *et al.*, 1987). Obviously, the East Antarctic craton extends farther towards West Antarctica than was originally thought.

References

Adams, C.J. (1981). Geochronological correlation of Precambrian and Paleozoic orogens in New Zealand, Marie Byrd Land (West Antarctica), northern Victoria Land (East Antarctica) and Tasmania. In *Gondwana Five*, ed. M.M. Cresswell & P. Vella, pp. 191–208, Rotterdam; A.A. Balkema.

Adams, C.J. (1986). Age and ancestry of metamorphic rocks of the Daniels Range, USARP Mountains, Antarctica. In *Geological Investigation in Northern Victoria Land*, Antarctic Research Series, 46, ed. E. Stump, pp. 25–38. Washington, DC; American Geophysical Union.

Adams, C.J. & Kreuzer, H. (1984). Potassium–argon age studies of slates and phyllites from the Bowers and Robertson Bay terranes, north Victoria Land, Antarctica. In *German Antarctic North Victoria Land Expedition 1982/83, GANOVEX III*, vol. I, ed. N.W. Roland, *Geologisches Jahrbuch*, **B60**, 265–88.

Bates, R.L. & Jackson, J.A. (ed.) (1980). *Glossary of Geology*, 2nd edn. Falls Church; American Geological Institute. 751 pp.

Borg, S.G., Stump, E., Chappell, B.W., McCulloch, M.T., Wyborn, D., Armstrong, R.L. & Holloway, J.R. (1987). Granitoids of northern Victoria Land, Antarctica: implications of chemical and isotopic variations to regional crustal structure and tectonics. *American Journal of Science*, **287**, 127–69.

Bosum, W., Damaske, D., Roland, N.W., Behrendt, J. & Saltus, R. (1989). Results of the aeromagnetic investigations of the GANOVEX IV Ross Sea/Victoria Land survey, Antarctica. *Geologisches Jahrbuch*, **E38**, 153–230.

Faure, G. & Gair, H.S. (1970). Age terminations of rocks from northern Victoria Land, Antarctica. *New Zealand Journal of Geology and Geophysics*, **13(4)**, 1024–6.

Glomar Challenger Scientific Party (1973). Leg 28 deep-sea drilling in the southern ocean. *Geotimes*, **18(6)**, 19–24.

Hamilton, W. (1964). Tectonic map of Antarctica – a progress report. In *Antarctic Geology*, ed. R.J. Adie, pp. 676–9. Amsterdam; North-Holland Publishing Co.

Kleinschmidt, G. (1983). Trends in regional metamorphism and deformation in northern Victoria Land, Antarctica. In *Antarctic Earth Science*, eds. R.L. Oliver, P.R. James & J.B. Jago, pp. 119–22. Canberra; Australian Academy of Science and Cambridge; Cambridge University Press.

Kleinschmidt, G. & Tessensohn, F. (1987). Early Paleozoic westward directed subduction at the Pacific margin of Antarctica. In *Gondwana Six: Structure, Tectonics and Geophysics*. Geophysical Monograph Series, 40, ed. G.D. McKenzie, pp. 89–105. Washington, DC; American Geophysical Union.

Klimov, L.V. (1967). Main features of the geologic structure of Antarctica. *Information Bulletin of the Soviet Antarctic Expedition, VI*, **65**, 473–9.

Klimov, L.V. & Solov'ev, D.S. (1958). Preliminary report on geological observations in East Antarctica. *Information Bulletin of the Soviet Antarctic Expedition* (Elsevier Translation), **1(1)**, 15–18.

Kreuzer, H., Höhndorf, A., Lenz, H., Müller, P. & Vetter, U. (1987). Radiometric ages of pre-Mesozoic rocks from northern Victoria Land, Antarctica. In *Gondwana Six: Structure, Tectonics and Geophysics*. Geophysical Monograph Series, 40, ed. G.D. McKenzie, pp. 31–47. Washington, DC; American Geophysical Union.

McLeod, I.R. (1964). Geological observations in Oates Land. In *Antarctic Geology*, ed. R.J. Adie, pp. 482–6. Amsterdam; North-Holland Publishing Co.

Ravich, M.G. & Kamenev, E.N. (1975). *Crystalline Basement of the Antarctic Platform*, New York; John Wiley & Sons. 582 pp.

Roland, N.W., Gibson, G.M., Kleinschmidt, G. & Schubert, W. (1984). Metamorphism and structural relations of the Lanterman Metamorphics, north Victoria Land, Antarctica. In *German Antarctic North Victoria Land Expedition 1982/83, GANOVEX III*, vol. I, ed. N.W. Roland, *Geologisches Jahrbuch*, **B60**, 319–61.

Roland, N.W., Olesch, M. & Schubert, W. (1989). Geology and petrology of the western border of the Transantarctic Mountains between the Outback Nunataks and the Reeves Glacier, north Victoria Land, Antarctica. *Geologisches Jahrbuch*, **E38**, 119–41.

Roland, N.W. & Tessensohn, F. (1987). Rennick faulting, distinct post-Jurassic tectonics in north Victoria Land, Antarctica. *Geologisches Jahrbuch*, **B66**, 131–57.

Starik, I.E., Krylov, A.J., Ravich, M.G. & Silin, J.I. (1960). The absolute age of the rocks of the East-Antarctic platform. *Doklady (Proceedings) of the Academy of Sciences of the USSR*, **126(1–6)**, 429–31.

Steed, R.H.N. (1983). Structural interpretation of Wilkes Land, Antarctica. In *Antarctic Earth Science*, eds. R.L. OLiver, P.R. James & J.B. Jago, pp. 567–72. Canberra; Australian Academy of Science and Cambridge; Cambridge University Press.

Ushakov, S.A. (1960). Some features of the structure of George V Coast and Oates Coast according to geophysical data. *Information Bulletin of the Soviet Antarctic Expedition*, **18**, 11.

Vetter, U., Lenz, H., Kreuzer, H. & Besang, C. (1984). Pre-Ross granites at the Pacific margin of the Robertson Bay Terrane, north Victoria Land, Antarctica. In *German Antarctic North Victoria Land Expedition 1982/83, GANOVEX III*, vol. I, ed. N.W. Roland, *Geologisches Jahrbuch*, **B60**, 363–9.

Northern Victoria Land, Antarctica: hybrid geological, aeromagnetic and Landsat-physiographic maps

B.K. LUCCHITTA[1], J.A. BOWELL[1], F. TESSENSOHN[2] & J.C. BEHRENDT[3]

1 US Geological Survey, 2255 N. Gemini Drive, Flagstaff, Arizona 86001, USA
2 Bundesanstalt für Geowissenschaften und Rohstoffe, Postfach 51 01 53, 3000 Hannover 51, Federal Republic of Germany
3 US Geological Survey, Box 25046, Denver Federal Center, Denver, Colorado 80225, USA

Abstract

Correlations between geological units and aeromagnetic anomalies were facilitated by superposing a generalized geological map of an area in northern Victoria Land on: (1) Landsat physiography; and (2) aeromagnetic anomalies displayed in shaded relief. The hybrid maps show that basement rocks of the Wilson and Bowers groups and the Admiralty Intrusives have low magnetic relief, that Jurassic and Cenozoic mafic rocks correlate well with positive anomalies, and that small outcrops of Cenozoic alkalic intrusions coincide with conspicuous positive anomalies that may reflect more mafic igneous differentiates at depth.

Hybrid maps of superposed geological, aeromagnetic, and physiographic data of northern Victoria Land (Fig. 1) have proved to be very useful for correlating geological and aeromagnetic information and thereby aiding geological interpretations. Fig. 2*a* shows a hybrid map in which the geology is superposed on physiography, Fig. 2*b* a map in which the geology is superposed on a representation of aeromagnetic anomalies.

Geological data were acquired under the auspices of the Bundesanstalt für Geowissenschaften und Rohstoffe (BGR) during the German Antarctic North Victoria Land Expeditions (GANOVEX) of the austral summers of 1983–84 and 1984–85. Field work during the expedition yielded geological maps of rock exposures near the Mariner, Aviator, Campbell, and Priestley glaciers (Figs 1 & 2*a*, *b*). For this report, the geological maps were generalized and digitized.

The aeromagnetic survey covers the same region as the geological map and an additional area on the polar plateau of East Antarctica, west of upper Priestley Glacier (Fig. 2*b*). The aeromagnetic data were acquired by contract during the austral summer of 1984–85 by a cooperative programme of the BGR and the US Geological Survey (USGS). The land survey was flown by fixed-wing aircraft at a flight-line spacing of 4.4 by 20 km. The raw field data were reduced by the contractor in the Federal Republic of Germany, who removed the background magnetic field and placed the data on a grid of 440-m spacing. A number of interpretive maps were produced by the BGR and USGS from the aeromagnetic data; in this report we show one example, a shaded-relief rendition of the aeromagnetic anomalies that shows positive anomalies as topographic highs (Fig. 2*b*).

The Landsat mosaic is composed of five images of MSS band 7 (wavelength 0.8–1.1 μm). The scenes were computer-

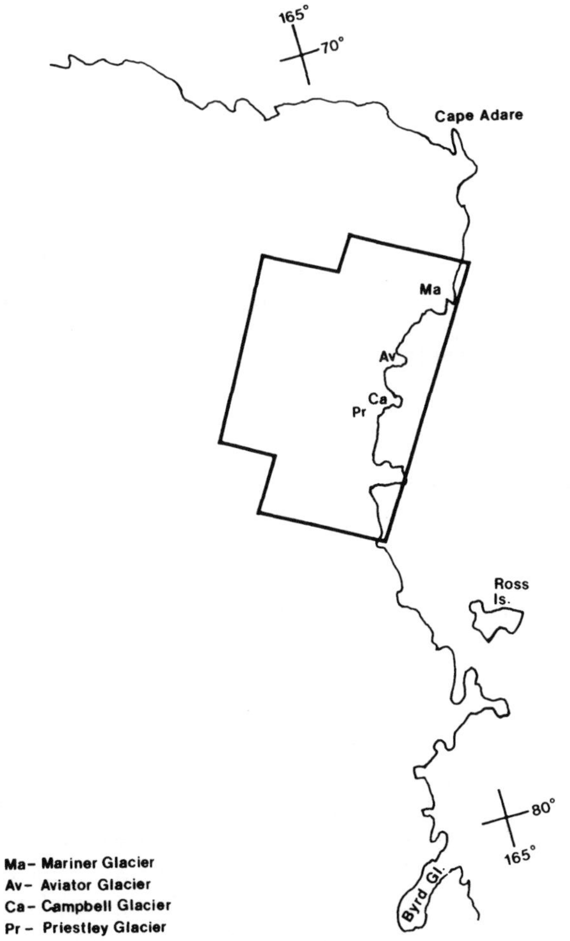

Fig. 1. Map of northern Victoria Land showing the outline of the Landsat mosaic.

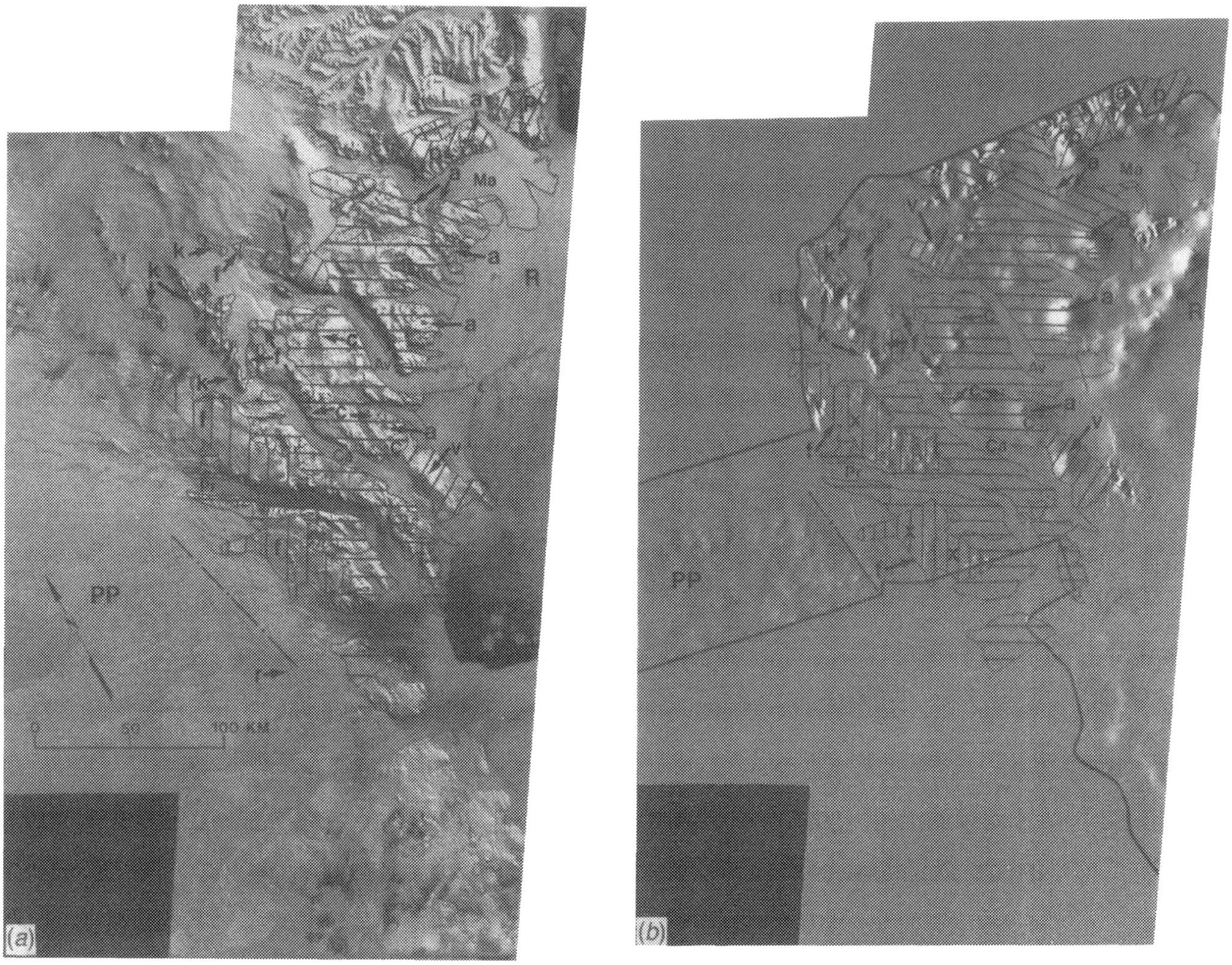

Fig. 2. Hybrid maps of northern Victoria Land. (*a*) Geology superposed on Landsat mosaic (illumination from right; Landsat MSS images 1128-20281, 1128-20284, 1128-20290, 1131-20453, 1131-20455.) (*b*) Geology superposed on aeromagnetic anomalies in shaded relief version (illumination from left). Heavy line delimits area of aeromagnetic survey. Geologic units: horizontal hachures, Wilson Group; right-sloping hachures, Bowers Group; cross hachures, Admiralty Intrusives; vertical hachures, Ferrar Dolerite (f) and Kirkpatrick Basalt (k); left-sloping hachures, Cenozoic volcanic rocks. Features of interest: v, volcanoes; p, high volcanic plateaux; c, cones and plugs; a, alkalic intrusions; r, ridge (in (*a*) only); x, areas of snow lacking dolerite exposures (in (*b*) only). Dash-dot line marks discontinuities in physiographic and aeromagnetic patterns. Physiographic features: Pr, Priestley Glacier; Ca, Campbell Glacier; Av, Aviator Glacier; Ma, Mariner Glacier; PP, polar plateau; R, Ross Sea.

enhanced to emphasize detail in the snow- and ice-covered mountains (Lucchitta *et al.*, 1987) and were placed in the same projection (Lambert conformal conic) as the aeromagnetic and geological maps. The image mosaic (Fig. 2*a*) shows topographic features exceptionally well because of the uniform albedo of ice-covered scenes and the low sun angle on Landsat images. Thus, the mountain configuration is seen not only in outcrop areas but also where the ice cover is thin enough to reflect the subjacent topography.

The generalized geological map (Fig. 2*a*, *b*) shows three major subdivisions: (1) basement composed of crystalline rocks of the Wilson Group, sedimentary and volcanic rocks of the Bowers Group, and granites of the Admiralty Intrusives; (2) cover composed of the Beacon Supergroup intruded by Ferrar Dolerite and overlain by Kirkpatrick Basalt; and (3) Cenozoic volcanic rocks.

The gneisses, metasedimentary rocks, and granites of the late Precambrian–Early Palaeozoic Wilson Group (Grindley & Warren, 1964) (Fig. 2*a*, *b*, horizontal hachures) have broad negative aeromagnetic anomalies which show on Fig. 2*b* as bland areas with little magnetic relief. The relief that does occur in places in this area is probably attributable to other units, as shown below. The lack of pronounced magnetic signature is surprising for an exposed fold belt. Also surprising is the absence of a reflection of the marked NW–SE strike of the basement rocks in the magnetic anomalies. This strike is well reflected in the trend of the middle section of several of the major glaciers.

The fossiliferous sedimentary and submarine volcanic rocks of the Cambrian and Ordovician Bowers Group (Sturm & Carryer, 1970) (Fig. 2*a*, *b*, right-sloping hachures) form a narrow tectonic zone that crosses northern Victoria Land in a north-westerly direction. The zone is characterized by low magnetic relief (Fig. 2*b*), that again is surprising, considering the presence of mafic rocks that generally have high magnetic susceptibilities.

The granite plutons of the Devonian–Carboniferous Admiralty Intrusives (Grindley & Warren, 1964) are exposed along the northern map boundary (Fig. 2*a*, *b*, cross hachures). They are characterized by broad, shallow, positive anomalies.

A thin Permian and Carboniferous sandstone sequence of the Beacon Supergroup (Gunn & Warren, 1962) rests unconformably on the basement rocks, is intruded by flat-lying, tholeiitic sills of the Jurassic Ferrar Dolerite (**f** in Fig. 2*a*, *b*) and is overlain by Jurassic Kirkpatrick Basalt (Nathan & Schulte, 1968) (**k** in Fig. 2*a*, *b*). These intrusive and extrusive rocks show as clusters of small aeromagnetic highs that match the outcrop areas; the lavas display slightly stronger anomalies than the sills. Where correlation is poor between anomalies and inferred extent of intrusive units, generally in areas of snow cover lacking rock exposures (**x**'s in Fig. 2*b*), the units may not be as extensive as shown on this generalized geological map.

Cenozoic volcanic rocks (Hamilton, 1972) (Fig. 2*a*, *b*, left-sloping hachures) form stratovolcanoes (**v**), high volcanic plateaus (**p**), and scattered small plugs and cones (**c**). Small intrusions of alkalic granites and syenites (**a**) occur along the coast.

The stratovolcanoes (Mount Melbourne and Mount Overlord) coincide with distinct positive anomalies. The high volcanic plateaus of mafic and alkalic rocks near the north-eastern corner of the map area show an irregular aeromagnetic pattern. However, outcrops of quartzites of the Bowers Group (small area of right-sloping hachures north of Mariner Glacier), which are magnetically quiet elsewhere, in places coincide with a line of broad, domal positive anomalies along the southern margin of this lava plateau. Unexposed mafic bodies at depth are assumed to account for this line of anomalies.

Positive anomalies similar to those of the stratovolcanoes occur in areas of scattered outcrops of plugs and cones and may indicate the presence of larger igneous bodies at depth. Also, small outcrops of alkalic intrusions lie within very strong magnetic anomalies. The exposed intrusive rocks are much too small and have magnetic susceptibilities much too low to account for these strong anomalies. However, some of these intrusions are associated with mafic dykes or, in one case, a mafic layered intrusion, suggesting that the outcrops of alkalic rocks may be considered differentiates of more mafic igneous bodies at depth, perhaps even differentiates of large ultramafic bodies.

Other magnetic anomalies occur in areas of ice or ocean with no rock exposure. Under the polar-plateau ice, a sharp magnetic boundary (Fig. 2*a*, *b*, dash-dot line) coincides with aligned NW-trending ridges (**r** in Fig. 2*a*) and depressions on the ice surface, which reflect subsurface bedrock topography. The boundary may reflect a fault or suture. The pattern of positive anomalies on the west side of this boundary has been interpreted as typical for shield areas (Bosum *et al.*, 1989) but no Precambrian shield rocks are exposed along-strike where this belt of rocks reaches the coast farther south. The physiographic pattern on the ice, as well as the aeromagnetic pattern, is similar to that associated with the Jurassic intrusive and extrusive rocks exposed along strike farther south (Warren, 1969). The relationships suggest that dolerite sills or associated extrusive rocks may be present beneath the ice in this area of the polar plateau.

The conspicuous NE-trending line of anomalies in the Ross Sea (**R** in Fig. 2*a*, *b*) consists of broad domes in a pattern similar to that associated with the Cenozoic volcanic rocks. The alignment of the Ross Sea anomalies cuts across basement-rock trends, suggesting that the anomalies may reflect a relatively young feature associated with Cenozoic magmatism. Perhaps this line of anomalies is associated with Cenozoic ultramafic bodies at depth, similar to the alkalic intrusions. The cause of such magmatism is unknown, but it may be due to Cenozoic rifting of the Ross Sea depression.

This discussion is a brief synopsis of some of the correlations between geology, aeromagnetic anomalies, and physiography deduced from the hybrid maps. Much more detailed relations can be recognized when larger-scale colour versions and other interpretive aeromagnetic maps are inspected. Overall, such hybrid maps are very useful in Antarctica, where the terrain is partly obscured and of difficult access, and where geological mapping at present is at a reconnaissance level.

References

Bosum, W., Damaske, D., Roland, N. & Saltus, R. (1989). Interpretation of the aeromagnetic anomalies of the Ross Sea/Victoria Land survey of GANOVEX IV, Antarctica. *Geologisches Jahrbuch*, **E38**, 153–230.

Grindley, G.W. & Warren, G. (1964). Stratigraphic nomenclature and correlation in the Ross Sea region. In *Antarctic Geology*, ed. R.J. Adie, pp. 314–33. Amsterdam; North-Holland Publishing Company.

Gunn, B.M. & Warren, G. (1962). Geology of Victoria Land between the Mawson and Mulock Glaciers, Antarctica. *Bulletin of the New Zealand Geological Survey*, **71**, 157 pp.

Hamilton, W.H. (1972). The Hallett volcanic province, Antarctica. *US Geological Survey Professional Paper*, **456-C**, 62 pp.

Lucchitta, B.K., Bowell, J.A., Edwards, K., Eliason, E.M. & Ferguson, H.M. (1987). Multispectral Landsat images of Antarctica. *US Geological Survey Bulletin*, **1696**, 21 pp.

Nathan, S. & Schulte, F.J. (1968). Geology and petrology of the Campbell–Aviator Divide, northern Victoria Land, Antarctica. Part 1, Post-Paleozoic rocks. *New Zealand Journal of Geology and Geophysics*, **11(4)**, 940–75.

Sturm, A. & Carryer, S.J. (1970). Geology of the region between Matusevich and Tucker Glaciers, north Victoria Land, Antarctica. *New Zealand Journal of Geology and Geophysics*, **13(2)**, 408–35.

Warren, G. (1969). Geology of Terra Nova Bay–McMurdo Sound area, Victoria Land (Sheet 14). In *Geologic Maps of Antarctica*, ed. V.C. Bushnell & C. Craddock, Plate XIII, Antarctic Map Folio Series, Folio 12. Washington, DC; American Geographical Society.

Setting and significance of the Shackleton Limestone, central Transantarctic Mountains

A.J. ROWELL[1] & M.N. REES[2]

1 Museum of Invertebrate Paleontology and Department of Geology, University of Kansas, Lawrence, Kansas 66045, USA
2 Department of Geosciences, University of Nevada, Las Vegas, Las Vegas, Nevada 89154, USA

Abstract

During the Cambrian, a carbonate platform was developed intermittently along the palaeo-Pacific margin of Gondwana. The Shackleton Limestone provides an isolated record of its Early Cambrian history in Antarctica. Although this formation is strongly deformed, its lithofacies associations show that most of it accumulated in shallow subtidal conditions and that oolite bars and sand sheets, together with *Epiphyton*–archaeocyathan build-ups, developed in higher energy regimes. Similarity of rock type and fauna are consistent with the Shackleton Limestone being a continuation of the South Australian platform and suggest that they developed in broadly comparable settings. Outboard of the carbonate platform, shallow-water and terrestrial volcanic and volcaniclastic rocks are associated with shallow-water carbonate lithofacies. Deep-water lithologies are unknown. Platform development along the margin was diachronous and in the Weddell Sea area was not extensive until the Middle Cambrian. In this region, the platform is truncated by Mesozoic faulting and location of its displaced remnants are unknown. Conceivably, the Ellsworth Mountains are a former outboard part of it. Faunal and structural data suggest that a fault with major lateral displacement separates the Pensacola Mountains and Shackleton Range.

Introduction

The configuration of many cratonic regions changed radically towards the close of the Proterozoic (Bond, Nicheson & Kominz, 1984). Extensive rift faulting followed by Early Cambrian continental break-up was a feature of the palaeo-Pacific facing margin of Gondwana along its Australian segment (von der Borch, 1980; Cook, 1982; Veevers & Powell, 1984). Consideration of Cambrian sequences in Antarctica is important in any interpretation of the lateral extent of such movements.

A turbulent tectonic history veils events that occurred during the Cambrian in Antarctica. Interpretation of the sequences is complicated by tectonic deformation and metamorphism associated with the Cambro-Ordovician Ross Orogeny; the consequences of Early Palaeozoic terrane accretion; uncertainties regarding the number, disposition and trajectories of Late Palaeozoic microplates; and the unknown extent and consequences of probable Mesozoic rifting and transform faulting associated with the break-up of Gondwana.

Along extensive stretches of the Antarctic palaeo-Pacific margin, no record of these events nor the Cambrian is retained: the Shackleton Limestone crops out in a critical segment. The formation occurs widely in the Churchill Mountains between Byrd and Nimrod glaciers, and small exposures are present at

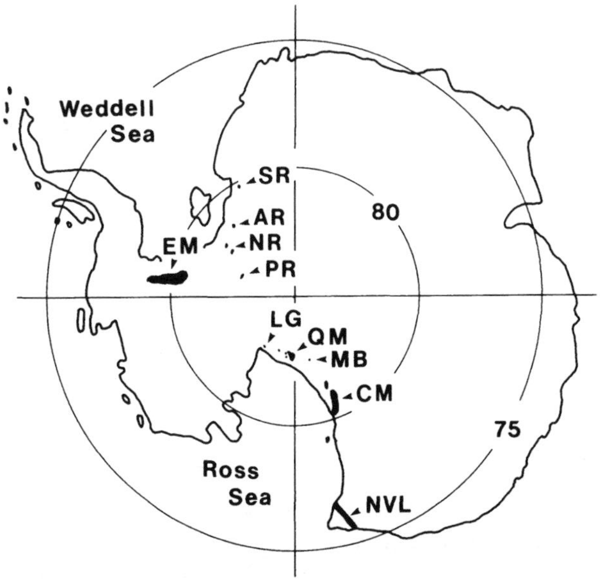

Fig. 1. Principal Cambrian outcrops of Antarctica (modified from Laird (1981)); Cambrian outcrops shown in black. Abbreviations: SR, Shackleton Range (Cambrian known only from morainic material); AR, Argentina Range; CM, Churchill Mountains; EM, Ellsworth Mountains; LG, Leverett Glacier; MB, Mount Bowers; NR, Neptune Range; NVL, Northern Victoria Land; PR, Patuxent Range; QM, Queen Maud Mountains. The Pensacola Mountains (shown as PM in Fig. 2) include the Argentina Range (AR), the Neptune Range (NR) and the Patuxent Range (PR).

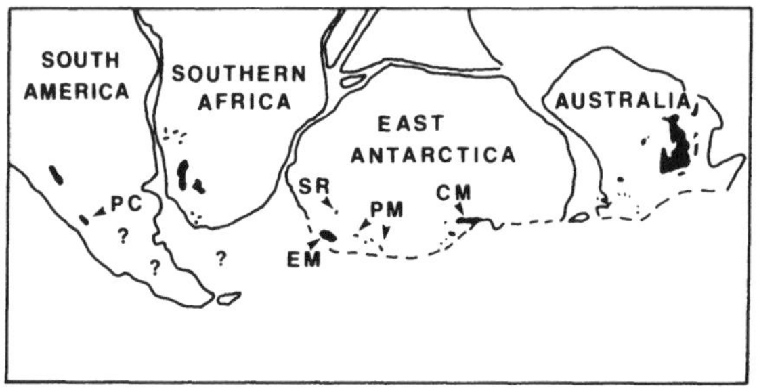

Fig. 2. Diagrammatic reconstruction of some of the principal components of the palaeo-Pacific margin of Gondwana during the Early Cambrian; Cambrian outcrops in black. Principal offshore terranes that were accreted subsequently are not shown. Ellsworth Mountains are plotted as a continuation of the Transantarctic Mountains. Abbreviations: SR, Shackleton Range (Cambrian known only from morainic material); EM, Ellsworth Mountains; CM, Churchill Mountains; NVL, Northern Victoria Land; PC, Precordillera; PM, Pensacola Mountains.

the head of Beardmore Glacier adjacent to Mount Bowers (Fig. 1). Its outcrops are almost equidistant from the nearest unaltered Cambrian platform deposits in South Australia and the Pensacola Mountains (Fig. 2). In any acceptable reconstruction, both are > 1000 km distant. The Shackleton Limestone thus affords an isolated but significant glimpse of the Cambrian platform.

Implications of our recent work on the depositional setting and faunal affinities of this formation, lead us to conclude that Cambrian platformal deposits were developed at least intermittently between Australia and East Antarctica and that their basal beds vary in age. The carbonate platform is truncated by Mesozoic faults near the Weddell Sea and its former extent is unknown: its lithologies are not exposed in Africa nor in South America. Stratigraphic considerations suggest that the Ellsworth Mountains may be a rotated distal slice of the cratonic margin but the evidence is equivocal. We feel greater confidence in asserting that there has been significant lateral movement between the Cambrian rocks in the Shackleton Range and those of the Pensacola Mountains (consisting of the Argentina, Neptune and Patuxent ranges) (Fig. 1). In this region, one or more suspect terranes may be present that were brought into position largely by transform movement.

Shackleton Limestone

Depositional setting and age

Laird (1981) reviewed the Cambrian of the continent in his pioneering studies on the Shackleton Limestone. We have had the opportunity to clarify some aspects of the stratigraphic succession with two seasons of additional field work. Structural complexity is characteristic of much of the Shackleton Limestone and it precludes any precise attempt to determine the thickness of the formation, which probably is in the order of 1–2 km.

The formation typically is faulted against the underlying Late Proterozoic Goldie Formation, an argillaceous turbidite-rich sequence. The unconformable contact between the two formations is rarely exposed but the basal beds of the Shackleton Limestone are metamorphosed quartz-rich sandstones and interbedded sandstone and dolomite, whose relict sedimentary structures suggest that they accumulated as subtidal bars on a shallow shelf. Earlier reconnaissance studies of the Shackleton Limestone were misled by thick sequences of conglomerate and breccia that were interpreted to occur in its lower part. Our field work, which included detailed mapping around the type locality (Rees & Rowell, this volume, p. 187), showed that these coarse rudstones locally rest unconformably on the Shackleton Limestone and elsewhere are fault-bounded blocks within the formation: in both settings, they are part of the younger Douglas Conglomerate.

The majority of the Shackleton Limestone consists of carbonate rocks. Lithofacies characteristic of intertidal settings are relatively rare in the southern Churchill Mountains and although fenestral cryptalgal laminates, fenestral lime mudstones and rudstones composed of flat pebbles of dolomitic mudstone occur, they constitute but a minor part of the sequence (Rees, Rowell & Pratt, 1985). Only towards the Byrd Glacier are peritidal deposits more common (Burgess & Lammerink, 1979). The most characteristic rock types are burrow-mottled lime mudstone and wackestone that include very thin layers of bioclastic grainstone. These lithologies form thick sequences of recessively weathered rocks. They are interbedded with grainstone successions that are characteristically ooidal and between 10 and 50 m thick. Commonly, the oolites are planar or cross-bedded; more rarely bimodal cross-stratified sets are developed. Bioherms are also significant components in the Shackleton Limestone. Small, isolated cyanobacterial mounds occur in the burrow-mottled sequences. The boundstones that form them are relatively muddy and *Renalcis* is their dominant microorganism. The majority of the bioherms, however, are *Epiphyton* boundstones whose composition may include *Girvanella*, with or without *Renalcis*. Archaeocyathans are commonly present and may comprise up to 30% of the mound. Such *Epiphyton* bioherms are typically small, usually < 1 m in height and some 2–3 m in diameter. They occur as isolated mounds, but also coalesced to form build-ups up to 50 m thick. Synoptic relief on even the largest structures was low, only in the order of 1 m, but the open cavities characteristic of *Epiphyton* bioherms, together with their common association with ooidal grainstones, suggest that they developed in relatively high-energy settings.

We consider that the majority of the formation was deposited on a shallow-water platform that was protected by a belt of archaeocyathan–cyanobacterial bioherms and build-ups together with migrating carbonate sand bars and sheets. The burrow-mottled lithologies accumulated below normal wave base, but storms episodically washed the bottom producing thin, bioclastic grainstones. The presence of *Epiphyton* in the build ups suggests proximity to a carbonate margin, but deep-water laminated rocks are unknown. Previous comments on mass movement (Burgess & Lammerink, 1979) apply to beds that we regard as part of the overlying Douglas Conglomerate, not the Shackleton Limestone (Rees & Rowell, this volume, p. 187), and we conclude that the mounds formed in a high-energy setting on the shelf (Rees *et al.*, 1985).

The basal siliciclastic rocks have yielded no fauna but

seemingly all the limestones are Early Cambrian. Archaeocyathans suggest that much of the formation is Botomian and that it may extend into the overlying Toyonian (Debrenne & Kruse, 1986). Some trilobites indicate that the older part of the limestone is mid-Early Cambrian, Atdabanian in age (Rowell *et al.*, 1988). An episode of deformation, uplift and erosion, expressed as an angular unconformity, separates the Douglas Conglomerate from the Shackleton Limestone (Rowell *et al.*, 1988).

Comparison with South Australia

The Cambrian sequences and settings seen in the central Transantarctic Mountains are comparable to those of southern and south-eastern Australia. There are, however, distinctions in the timing of initial subsidence and the complete lack of autochthonous deep-water lithofacies in Antarctica.

Although significant differences of opinion remain regarding correlation of individual tectono-stratigraphic terranes between northern Victoria Land, Tasmania, and south-east Australia, most reconstructions agree that the autochthonous platform of South Australia continued into East Antarctica (Fig. 2). There are strong similarities between the shallow-water Lower Cambrian sequences of South Australia, as found in the Flinders Ranges and Yorke Peninsula (Cook, 1982) and rocks of the Shackleton Limestone. As in Antarctica, the Australian sequences commence with basal siliciclastic rocks that are overlain by middle and late Early Cambrian rocks, predominantly limestones. In the Flinders Ranges, the limestones include *Epiphyton*–archaeocyathan bioherms (James & Gravestock, 1986) broadly comparable to those in the Shackleton Limestone. Strong faunal resemblances exist also. Debrenne & Kruse (1986) have shown that of the 31 archaeocyathan species presently known from the Shackleton Limestone, 16 are common to Australia. We have recently found phosphatic sclerites of the problematic *Dailyatia* in the Shackleton Limestone: previously, the organism was known only from Australia.

Volcanic rocks occur outboard of the Australian carbonate platform that flanked the cratonic nucleus. There is not universal agreement on their setting and origin. Veevers & Powell (1984) regarded them as having formed in rifts and grabens under terrestrial and marginal marine conditions. Scheibner (1978) seemingly perceived them as having originated in an arc setting, a view supported by Cook (1982). Many of the sequences include andesites; they occur as approximately contemporaneous volcanic successions in the present-day Bancannia Trough, Warburton Basin and Gnalta Shelf (Cook, 1982). The volcanic rocks typically are overlain by upper Lower or Middle Cambrian rocks that include relatively shallow-water limestones. In Antarctica, metavolcanic rocks and associated metamorphosed siliciclastics, volcaniclastic and shallow-water carbonate lithofacies accumulated in a comparable position in the Queen Maud Mountains (Fig. 1), where they form the Liv Group (Stump, 1982). Unidentifiable trilobites and (?)*Cloudina* (Yochelson & Stump, 1977) provide a rather insecure Early Cambrian age. The volcanic rocks include basalts, together with porphyritic rhyolite lavas and ashfall and welded tuffs, a bimodal succession commonly associated with extensional settings. The rhyolites are calc-alkaline and Stump (1976) considered them to be arc-related. As in Australia, the relationship of these volcanic-rich rocks to the carbonate platform towards the craton is not clear. The similarity of basement types suggests that the Liv Group is autochthonous or para-autochthonous relative to the Shackleton Limestone in the Churchill Mountains. It would be difficult, however, to demonstrate convincingly that it had not been moved by transform faults trending subparallel to the Transantarctic Mountains.

Although the setting of Early Cambrian rocks is comparable in South Australia and the central Transantarctic Mountains, there are significant differences. The carbonate rocks of South Australia probably were deposited in a rimmed basin (Veevers & Powell, 1984) not on an open shelf. Associated with this difference but perhaps of greater significance, rifting commenced in Australia in the Late Proterozoic with movement along the intraplate Adelaide Rift axis (von der Borch, 1980). In latest Precambrian and early Early Cambrian time, the spreading axis shifted outboard of the rift (Scheibner, 1985) and may have involved extensive transform faulting as the Tasman Line was established (Veevers & Powell, 1984). In the central Transantarctic Mountains, although there was some Late Proterozoic volcanic activity, there are no data to locate the rift axis. The continental margin may have been defined earlier, for the late Precambrian is predominantly a continental rise sequence that was deformed by the Beardmore Orogeny towards the close of the Proterozoic. Wherever they are known, autochthonous Cambrian platform rocks rest unconformably on this deformed rise succession. It is conceivable that an Early Cambrian extensional event occurred outboard of the carbonate platform, but no autochthonous deep-water Lower Cambrian lithofacies have been recognized in Antarctica, and there is no substantiated record even of slope deposits. There are no rocks comparable to those of the Kanmantoo Trough of South Australia but the extent to which the Kanmantoo sequence is allochthonous is not completely resolved (Scheibner, 1985). The considerable thickness of the Shackleton Limestone does, however, suggest that there was a prior episode of crustal thinning.

Continuation of the carbonate platform towards the Weddell Sea

The Pensacola Mountains and Leverett Glacier sequence

Widespread development of Lower Cambrian carbonate platform rocks is unknown from Mount Bowers to the Weddell Sea. In the Patuxent and Neptune ranges of the Pensacola Mountains (Fig. 1), platformal rocks rest unconformably on Precambrian argillites and turbidites, but, where known, the lower part of the succession is largely or entirely of Middle Cambrian age (Schmidt *et al.*, 1965). Archaeocyathans have been reported from talus material in the Neptune Range but no Lower Cambrian fauna has been found *in situ*. If Lower Cambrian rocks are present in these two ranges, they are thin and primarily sandstones and conglomerates. Limestone outcrops with archaeocyathans occur in the southern Argentina Range (Fig. 1). These rocks are of late Early Cambrian age (Konyushkov & Shulyatin, 1980), but nothing more is known from this region.

A metamorphosed but very similar sequence to that in the Pensacola Mountains occurs in the Leverett Glacier region (Fig. 1), and this too has yielded a Middle Cambrian fauna from limestones (Laird, 1981). The lower strata are cross-bedded sandstones, and all of the 1000 m thick sequence appears to have been deposited in shallow-water settings in spite of the outboard location of the exposures.

Lack of information hinders interpretation of this segment of the Transantarctic Mountains. Seemingly, development of a carbonate platform was areally restricted in Early Cambrian time, for carbonate accumulation was not well established across the region until the Middle Cambrian.

Termination of the carbonate platform

The original areal extent of the carbonate platform is unknown: its outcrops terminate at the fault-bounded margin of Antarctica. The Cambrian outcrops of the Neptune and Argentina ranges are the nearest exposures of the platform rocks to the present-day margins of the continent. The structures in these ranges strike N–S, nearly normal to the continental margin. Aeromagnetic data (Masolov, 1980; Fig. 3) show narrow N–S anomalies over the Neptune Range aligned parallel with axes of deformation. These anomalies continue northward towards the Filchner Ice Shelf and terminate against broad negative anomalies that trend SW from the Andenes Escarpment (Kristoffersen & Haugland, 1986). The latter, together with its western extension and the Explora Escarpment, probably marks the faulted margin of the continent developed during the break-up of Gondwana (Kristoffersen & Haugland, 1986). If this supposition is correct, outcrops of the Cambrian carbonate platform will not lie beyond it and the present whereabouts of the former continuation of the platform is uncertain. Seemingly, it did not extend to Africa nor to South America. Cambrian rocks in southern Africa consist entirely of siliciclastic strata (Rust, 1981) comprising the Fish River Subgroup and its correlatives. Trace fossils occur in them and elements of the late Precambrian Ediacaran fauna (Glaessner, 1984) are present in older strata, but no Cambrian body fossils are known from the region.

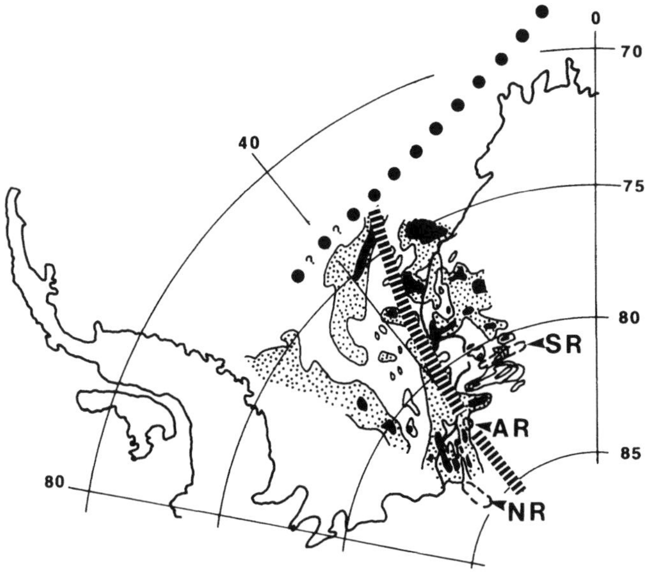

Fig. 3. Simplified version of compilation of aeromagnetic data from part of the Weddell Sea area (based on Masolov (1980) and Kristoffersen & Haugland (1986)). Areas of negative anomaly between 0 and −1 with dotted ornament, < −1 shown in black; areas of positive anomaly > 2 bounded by solid lines and unornamented. Abbreviations: SR, Shackleton Range; AR, Argentina Range; NR, Neptune Range. Heavy broken line marks possible transform boundary; heavy dotted line is the position of the Andenes–Explora Escarpment.

In South America, Cambrian carbonates occur in the Precordillera (Fig. 2), but their trilobite fauna is North American and bears little resemblance to that of Antarctica and Australia. Ramos *et al.* (1986) have offered the suggestion that the Precordillera is an accreted terrane: no data from our study negate the hypothesis. Autochthonous Upper Precambrian rocks occur in the northern Sierras of Buenos Aires and in north-west Argentina. In the latter region, siliciclastic rocks of the upper Puncoviscana Formation may be of Early Cambrian age but have yielded only trace fossils (Acenolaza & Durand, 1986). Thus, there are broad resemblances between the South American and southern African Cambrian successions, but neither region retains any indication of autochthonous carbonate platform deposits of this age.

Relationship of the Ellsworth Mountains

It is conceivable that much of the former extension of the Cambrian platform now resides in the Ellsworth Mountains, whose original position remains enigmatic. Recent palaeomagnetic data from the Falkland Islands (Mitchell *et al.*, 1986) suggest that they may be a far-travelled microplate. In addition, these data support earlier models that postulated a large anticlockwise rotation of the Ellsworth Mountains block from an original Permo-Triassic position in line with, and an extension of, the Transantarctic Mountains (Martin, 1986).

The Cambrian of the Shackleton Range

Cambrian rocks of the Shackleton Range are anomalous and probably allochthonous relative to sequences in the Neptune and Argentina ranges. They occur only as erratic blocks and slabs on Mount Provender (Thomson, 1972) but carry important biostratigraphic and palaeogeographic information.

Solov'ev & Grikurov (1978) documented the Middle Cambrian age of the blocks and suggested that they had not been transported far, but their provenance is still uncertain. It is significant, however, that the fauna includes ptychagnostids, for these organisms are characteristic of open-ocean facing settings, typically on the seaward side of carbonate shoals. It is difficult to reconcile such a palaeogeographic position with the present-day location of the range relative to the platformal sequence of the Transantarctic Mountains without invoking extensive lateral tectonic displacement. There is independent evidence for such displacement.

It has long been known that the structural trend of the Shackleton Range is almost normal to that of the Transantarctic Mountains. Aeromagnetic data reveal that there is an abrupt boundary between E–W anomalies through and to the south of the Shackleton Range and N–S-trending anomalies to the west. The boundary runs through the Filchner Ice Shelf to the coast of Berkner Island (Fig. 3); its form suggests that it marks a major structural discontinuity, and its position is consistent with it having extensive dextral movement of the type postulated by Schmidt & Rowley (1986). This suggestion does not solve the problem of the origin of Cambrian rocks in the Shackleton Range. By invoking large-scale lateral displacement, however, it does remove the incongruity of a deeper-water fauna laying craton-ward of a shallow-water carbonate platform.

Acknowledgements

We are grateful to our colleagues, B.R. Pratt, Ray Waters and Peter Braddock for their help and support in the field. We are indebted to Margery Rowell for translation of the relevant Russian literature. Rowell acknowledges the support of NSF grants DPP 8317966 and 8519722, Rees is supported by grant DPP 8518157.

References

Acenolaza, F.G. & Durand, F.R. (1986). Upper Precambrian–Lower Cambrian biota from the northwest of Argentina. *Geological Magazine*, **123**, 367–75.

Bond, G.C., Nickeson, P.A. & Kominz, M.A. (1984). Breakup of a supercontinent between 625 Ma and 555 Ma: new evidence and implications for continental histories. *Earth and Planetary Science Letters*, **70**, 325–45.

Burgess, C.J. & Lammerink, W. (1979). Geology of the Shackleton Limestone (Cambrian) in the Byrd Glacier area. *New Zealand Antarctic Record*, **2**, 12–16.

Cook, P.J. (1982). The Cambrian palaeogeograhy of Australia and opportunities for petroleum exploration. *Journal of Australian Petroleum Exploration Association*, **22(1)**, 42–64.

Debrenne, F. & Kruse, P.D. (1986). Shackleton Limestone archaeocyaths. *Alcheringa*, **10**, 235–78.

Glaessner, F.G. (1984). *The Dawn of Animal Life*. Cambridge; Cambridge University Press. 244 pp.

James, N.P. & Gravestock, D.I. (1986). Lower Cambrian carbonate shelf and shelf margin buildups, South Australia. *12th International Sedimentological Congress, Abstracts, Canberra*, p. 154. (unpublished).

Konyushkov, K.N. & Shulyatin, O.G. (1980). Ob arkheotsiatakh Antarktidy i ikh sopostavlenii s arkheotsiatami Sibiri. (On the archaeocyaths of Antarctica and their comparison with the archaeocyaths of Siberia.) In *Kembriy Altae–Sayanskoy Skladchatoy Oblasti*, ed. I.T. Zhuravleva, pp. 143–50. Moscow; Nauka.

Kristoffersen, Y. & Haugland, K. (1986). Geophysical evidence for the East Antarctic plate boundary in the Weddell Sea. *Nature, London*, **322**, 538–41.

Laird, M.G. (1981). Lower Palaeozoic rocks of Antarctica. In *Lower Palaeozoic of the Middle East, Eastern and Southern Africa, and Antarctica*, ed. C.H. Holland, pp. 257–314. Chichester, New York, Brisbane & Toronto; John Wiley and Sons.

Martin, A.K. (1986). Microplates in Antarctica. *Nature, London*, **319**, 100–1.

Masolov, V.I. (1980). Geofizicheskie issledovaniya v Antarktide. (Geophysical investigations in Antarctica.) *Nauchno-Issledovatel'skij Institut Geologii Arktiki*, pp. 14–28. Leningrad; Ministerstva Geologii SSSR.

Mitchell, C., Taylor, G.K., Cox, K.G. & Shaw, J. (1986). Are the Falkland Islands a rotated microplate? *Nature, London*, **319**, 131–4.

Ramos, V.A., Jordan, T.E., Allmendinger, R.W., Mpodozis, C., Kay, S.M., Corte's, J.M. & Palma, M. (1986). Paleozoic terranes of the central Argentine–Chilean Andes. *Tectonics*, **5**, 855–80.

Rees, M.N., Rowell, A.J. & Pratt, B.R. (1985). The Byrd Group of the Holyoake Range, central Transantarctic Mountains. *Antarctic Journal of the United States*, **20(5)**, 3–5.

Rowell, A.J., Rees, M.N., Cooper, R.A. & Pratt, B.R. (1988). Early Paleozoic history of the Central Transantarctic Mountains: evidence from the Holyoake Range, Antarctica. *New Zealand Journal of Geology and Geophysics* **31**, 397–404.

Rust, I.C. (1981). Lower Palaeozoic rocks of Southern Africa. In *Lower Palaeozoic of the Middle East, Eastern and Southern Africa, and Antarctica*, ed. C.H. Holland, pp. 165–87. Chichester, New York, Brisbane & Toronto; John Wiley and Sons.

Scheibner, E. (1978). Tasman Fold Belt System in New South Wales – general description. *Tectonophysics*, **48**, 207–16.

Scheibner, E. (1985). Suspect terranes in the Tasman Fold Belt System, Eastern Australia. In *Tectonostratigraphic Terranes of the Circum-Pacific Region*, ed. D.G. Howell, pp. 493–514. Houston; Circum-Pacific Council for Energy and Mineral Resources.

Schmidt, D.L. & Rowley, P.D. (1986). Continental rifting and transform faulting along the Jurassic Transantarctic Rift, Antarctica. *Tectonics*, **5**, 279–91.

Schmidt, D.L., Williams, P.L., Nelson, W.H. & Ege, J.R. (1965). Upper Precambrian and Paleozoic stratigraphy and structure of the Neptune Range, Antarctica. *United States Geological Survey Professional Paper*, **525-D**, D112–19.

Solov'ev, I.A. & Grikurov, G.E. (1978). Pervye nakhodki Srednekembrijskikh trilobitov v khrebte Shekltov (Antarktida). (First records of Middle Cambrian trilobites in the Shackleton Range (Antarctica).) *Antarktika*, **17**, 187–98.

Stump, E. (1976). *On the Late Precambrian–Early Paleozoic metavolcanic and metasedimentary rocks of the Queen Maud Mountains, Antarctica, and a comparison with rocks of similar age in Southern Africa.* Columbus; Ohio State University, Institute of Polar Studies, Report no. 62, 1–212 pp.

Stump, E. (1982). The Ross Supergroup in the Queen Maud Mountains. In *Antarctic Geoscience*, ed. C. Craddock, pp. 565–9. Madison; University of Wisconsin Press.

Thomson, M.R.A. (1972). Inarticulate Brachiopoda from the Shackleton Range and their stratigraphical significance. *British Antarctic Survey Bulletin*, **31**, 17–20.

Veevers, J.J. & Powell, C.Mc.A. (1984). Cambrian: plate divergence on the east and northwest followed by plate convergence on the east. In *Phanerozoic Earth History of Australia*, ed. J.J. Veevers, pp. 282–9. Oxford; Clarendon Press.

Von der Borch, C.C. (1980). Evolution of Late Proterozoic to Early Paleozoic Adelaide Fold belt, Australia: comparisons with post-Permian rifts and passive margins. *Tectonophysics*, **70**, 115–34.

Yochelson, E.L. & Stump, E. (1977). Discovery of Early Cambrian fossils at Taylor Nunatak, Antarctica. *Journal of Paleontology*, **51**, 872–5.

Lower–mid-Palaeozoic sedimentation and tectonic patterns on the palaeo-Pacific margin of Antarctica

M.G. LAIRD

NZ Geological Survey, University of Canterbury, Private Bag, Christchurch, New Zealand

Abstract

Regional patterns of sedimentation, tectonics, and magmatism in the Lower–mid-Palaeozoic pre-Beacon basement rocks of Antarctica suggest that they can be grouped into two major sedimentary–tectonic provinces.

In the Transantarctic Mountains Province, which includes the Ellsworth Mountains and most of the Transantarctic Mountains, Lower Palaeozoic rocks constitute two contrasting sequences bounded by unconformities or by major lithofacies disjunctions. Cambrian rocks consist largely of shallow-marine shelf deposits, while strata of Ordovician–(?)Early Devonian age comprise coarse mainly non-marine post-orogenic clastic sediments. An important Early Devonian(?) tectonic event (Shackleton Event) caused uplift, erosion, and, locally, folding before deposition of the Beacon Supergroup.

By contrast, in the Ross Sea Province, which includes Marie Byrd Land and much of northern Victoria Land, Lower Palaeozoic sedimentation, restricted to the Cambrian–Early Ordovician, occurred in submarine fan or island arc environments. Rocks of the province were affected by Middle Devonian–Early Carboniferous igneous activity and deformation (Borchgrevink Event). The Ross Sea Province represents a collage of terranes which, until Gondwana break-up time, probably included the then contiguous Campbell Plateau and parts of New Zealand and south-eastern Australia. This continent-sized allochthonous block probably accreted to East Antarctica in the mid-Devonian–Carboniferous.

Introduction

Outcrops of lower–mid-Palaeozoic rocks in Antarctica are restricted, with minor exceptions, to the present sites of the Transantarctic Mountains, Marie Byrd Land, and the Ellsworth Mountains (Fig. 1). Although these regions are extensive and widely separated, a pattern is emerging to suggest that the Ellsworth Mountains and Transantarctic Mountains from the Shackleton Range to Byrd Glacier were part of a single major sedimentary–tectonic province, encompassing similar sedimentary environments and tectonic events. Marie Byrd Land and much of northern Victoria Land formed part of another, contrasting, probably allochthonous province.

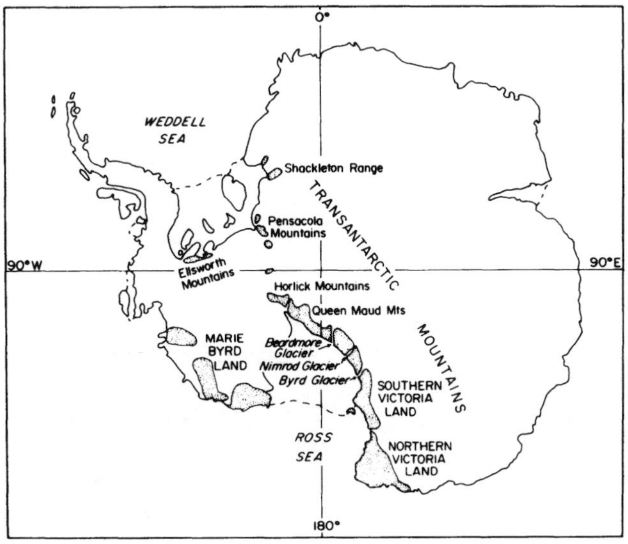

Fig. 1. Locality map of Antarctica. Stippled areas show outcrops of rocks discussed in the text.

Transantarctic Mountains

In the Transantarctic Mountains, from Byrd Glacier to the Shackleton Range, Lower Palaeozoic rocks form two unconformity-bounded sequences. The older one consists dominantly of Cambrian shallow-marine sediments and silicic volcanics resting on a Proterozoic basement and the younger, of probably Ordovocian–Early Devonian age, comprises

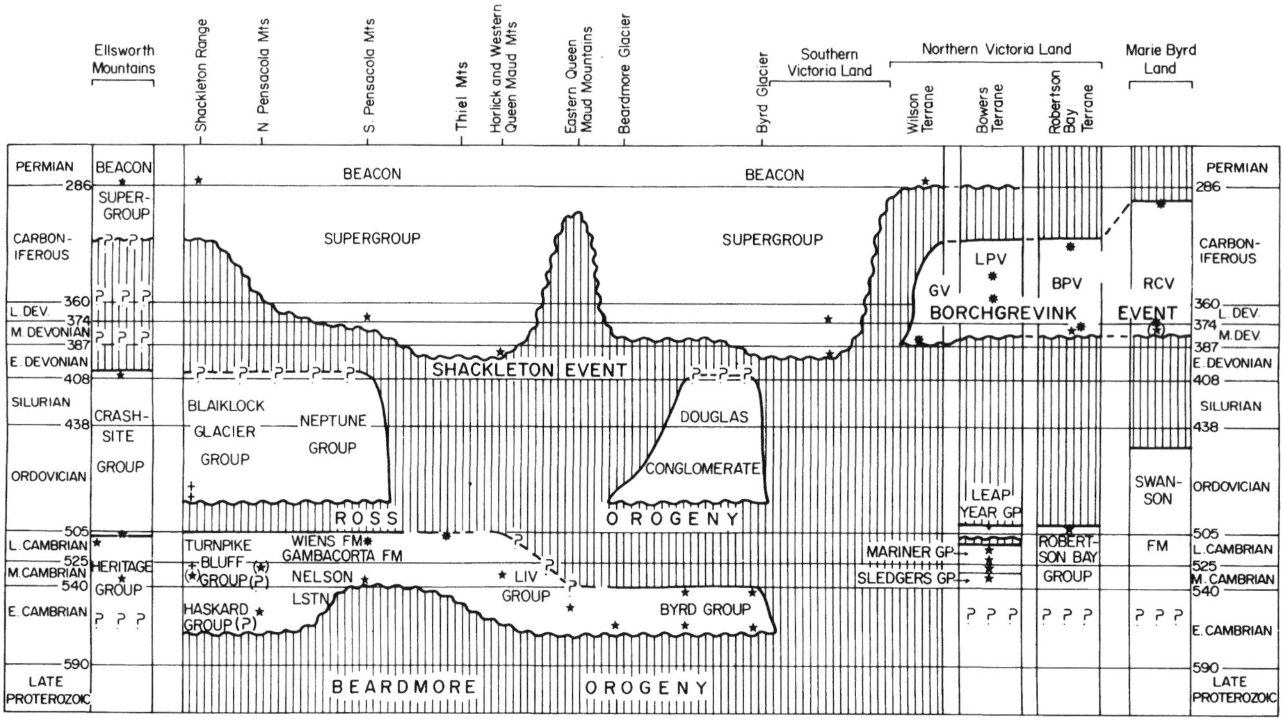

Fig. 2. Chronostratigraphic chart showing correlations between Lower and mid-Palaeozoic rock units in the Transantarctic Mountains, Ellsworth Mountains, and Marie Byrd Land. Vertical line pattern shows periods of non-deposition or erosion: wavy lines depict unconformable relationships. Unconformities within the Beacon Supergroup are not shown. GV, Gallipoli Volcanics; LPV, Lawrence Peaks Volcanics; BPV, Black Prince Volcanics; and RCV, Ruppert Coast Metavolcanics. *, fossil horizons; (*), fossils not *in situ*; +, isotopic ages (diagenesis) on sediments; ★, isotopic ages on volcanics. Time scale after Harland *et al.*, 1982.

coarse-grained, mainly non-marine, clastic post-orogenic deposits (Fig. 2). These latter are unconformably overlain by Devonian or younger rocks of the Beacon Supergroup.

Cambrian sequences

In the Shackleton Range, two units, the Haskard and Turnpike Bluff groups, are of either Late Proterozoic or Cambrian age (Pankhurst, Marsh & Clarkson, 1983; Marsh, 1984). The two groups are separated geographically, but both rest unconformably on the Archaean–Middle Proterozoic Shackleton Range Metamorphic Complex. The Haskard Group, which is metamorphosed to amphibolite-facies, consists of sandy metalimestone, calcareous metasandstone, mica-schist, and quartzite. The Turnpike Bluff Group consists mainly of quartzite and slate with minor conglomerate, sandstone, and limestone horizons. Isotopic studies of the Turnpike Bluff Group give an isochron age of 526 ± 6 Ma, which suggests a late Middle or early Late Cambrian age for diagenesis of the sediments (Pankhurst *et al.*, 1983). No fossils have been located *in situ* in either group, although erratics in a Quaternary moraine flanking the Blaiklock Glacier contain a Middle Cambrian fauna (Clarkson, Hughes & Thomson, 1979). These have previously been considered to originate from a concealed unit of the Blaiklock Glacier Group (now thought to be of Ordovician age – see later). However, a more appropriate source may be the Turnpike Bluff Group, outcrops of which occur at the head of the Blaiklock Glacier.

In the central and southern Pensacola Mountains three conformable formations, the Nelson Limestone, the Gambacorta Formation, and the Wiens Formation overlie the Upper Proterozoic Patuxent Formation with marked angular unconformity (Schmidt & Ford, 1969). The Nelson Limestone consists of a thin basal red-bed clastic succession overlain by thin- and thick-bedded grey limestone, commonly oolitic and pisolitic. Trilobites from the limestone in the Neptune Range indicate a medial or late Middle Cambrian age for the succession. At the northern end of the Pensacola Mountains (Argentina Range), Early Cambrian archaeocyathan limestone (Konyushkov & Shulyatin, 1980) crops out extensively, and limestone erratics containing faunas ranging in age from Early Cambrian to late Middle Cambrian occur nearby. The conformably overlying Gambacorta Formation consists of interlayered rhyolite flows, volcanic breccias and conglomerate, pyroclastic deposits, and detrital volcanic-derived sandstone and conglomerate. It has not yielded any fossils, but an isotopic date of 510 ± 35 Ma (Schmidt, 1983) from associated rhyolitic ignimbrite suggests a latest Cambrian age. The unfossiliferous Wiens Formation, which both conformably overlies and intertongues with the upper part of the Gambacorta Formation, consists of interlayered shale, siltstone, and fine sandstone, with rare thin-bedded grey oolitic limestone units. These three Cambrian formations represent shallow-marine deposits.

In the Thiel, Horlick, and Queen Maud Mountains, the sequences contain prominent silicic volcanics, giving an Rb–Sr whole-rock isochron age of 502 ± 5 Ma in the Thiel Mountains (Pankhurst *et al.*, 1988) and a whole-rock Rb–Sr age of 483 ± 9 Ma (recalculated) in the western Queen Maud Mountains (Faure *et al.*, 1979), dating the volcanism as close to the Cambro-Ordovician boundary. In the Horlick and Queen Maud mountains, Cambrian – earliest Ordovocian successions are referred to the Liv Group (Stump, 1982). The Leverett Formation, which crops out in the western Queen Maud Mountains, consists of interbedded coarse-grained conglomeratic sandstone and rhyolite in its upper half, which conformably overlies 1000 m of interbedded clastic rocks and limestone containing trilobites of Middle Cambrian age (Stump, 1982).

The Taylor and Fairweather formations of the eastern Queen Maud Mountains are dominated by silicic volcanics, particularly in their lower part, but also contain quartzite and conglomerate. Both sequences pass gradationally upwards into white, coarsely crystalline marble, and may be equivalent. The Taylor Formation contains the Early Cambrian fossil *Cloudina*? (Stump, 1982). The Greenlee Formation, which lies stratigraphically below the Taylor Formation, is a sequence of fine-grained quartzite and phyllite. Contacts between the Cambrian rocks and the underlying sediments of the Late Proterozoic Beardmore Group are not exposed, but an unconformity is inferred.

In the region between the Beardmore and Byrd glaciers, a very extensive sequence of fossiliferous Lower Cambrian sediments (Byrd Group), perhaps up to 5000 m thick, occurs; it rests with marked unconformity on the Late Proterozoic Beardmore Group (Laird, 1981). The strata consists dominantly of shallow water (in some instances peritidal), pure limestone, with subordinate quartzite and conglomerate (Shackleton Limestone – Laird, 1981; Rowell *et al.*, 1988), and minor siliceous argillite. Common archaeocyathans and other scattered fossils indicate a middle–late Early Cambrian age for the Shackleton Limestone (Rowell *et al.*, 1988).

Cambrian rocks have not been recognized in southern Victoria Land, and occur only east of the Rennick Glacier in northern Victoria Land. Here, however, they form part of terranes considered to be allochthonous, and will be discussed separately.

Ross Orogeny

During Late Cambrian–Early Ordovician times, the entire length of the Transantarctic Mountains from the Shackleton Range to northern Victoria Land experienced uplift, folding, and metamorphism (Ross Orogeny) and, over most of the region, intrusion by calc-alkaline granitoid plutons (Granite Harbour Intrusives). The Ross Orogeny has been interpreted as a cratonization even associated with an east-dipping subduction zone along the palaeo-Pacific margin of Antarctica (Stump, 1982; Borg, 1983; Schmidt, 1983; Borg *et al.*, 1987). Following the Ross Orogeny, a new cycle of erosion and deposition occurred, resulting in the formation of coarse clastics of mainly continental origin over much of the region.

Ordovician–(?)Early Devonian sequences

In the Shackleton Range, the youngest of the folded sedimentary and metasedimentary sequences, the Blaiklock Glacier Group, unconformably overlies the Proterozoic Shackleton Range Metamorphic Complex. The group consists dominantly of feldspathic sandstone with subordinate grit and conglomerate beds inferred to have been deposited under terrestrial–deltaic conditions (Clarkson *et al.*, 1979). An Rb–Sr isochron age of 475 ± 40 Ma (Early Ordovician) is presumed to reflect diagenesis of the sediments soon after deposition (Pankhurst *et al.*, 1983). This is consistent with their unconformable relationship upon schists recording a 500 ± 5 Ma metamorphism probably associated with the Ross Orogeny (Pankhurst *et al.*, 1983).

In the Pensacola Mountains, the Neptune Group overlies the Cambrian succession with marked unconformity. This group has been divided into four formations (Brown Ridge Conglomerate, Elliot Sandstone, Elbow Formation, and Heiser Sandstone) which consist largely of conglomerate and quartzose sandstone (Schmidt & Ford, 1969). The two oldest units are considered to represent continental orogenic and alluvial fan deposits, whereas the Elbow Formation and Heiser Sandstone indicate more quiescent conditions of sedimentation, perhaps at shallow depths in an epineritic environment on a slowly subsiding continental shelf (Williams, 1969). The Neptune Group contains no diagnostic fossils and is of uncertain age. However, since the group rests unconformably on rocks dated as Late Cambrian and its upper formation is disconformably overlain by the Dover Sandstone (Beacon Supergroup) with plant fossils of probable Late Devonian age (Schmidt & Ford, 1969), an Ordovician–Devonian age is inferred.

No post-Cambrian Lower Palaeozoic rocks have been recognized in the Horlick or Queen Maud Mountains, but, between the Nimrod and Byrd Glaciers, the Lower Cambrian Byrd Group is overlain unconformably by the Douglas Conglomerate (Rowell *et al.*, 1988). This is a polymictic conglomerate, up to 1000 m thick, consisting largely of clasts from the underlying Shackleton Limestone, plus quartzite, mudstone, and various igneous rocks. The beds are considered to have accumulated largely in the middle portion of an alluvial fan. Although outcrops in the vicinity have not yet been recorded, morainic material in the Beardmore Glacier indicates that the Douglas Conglomerate extends at least this far south (Rowell, Rees & Braddock, 1986). The Douglas Conglomerate is unfossiliferous and its age is poorly constrained. However, its deposition postdated folding of the Lower Cambrian Shackleton Limestone, an event which was probably associated with the Late Cambrian–Early Ordovician Ross Orogeny (Rowell *et al.*, 1988). It was deformed and eroded prior to deposition of the unconformably overlying Middle or Upper Devonian Beacon Supergroup (Rowell *et al.*, 1988). Thus, the age falls within the range Early Ordovician–Middle Devonian. No post-Ross Orogeny Lower Palaeozoic strata are known from either southern or northern Victoria Land.

Mid-Palaeozoic tectonism – the Shackleton Event

It is clear that in at least the three regions of the Transantarctic Mountains where post-Cambrian Lower Palaeozoic strata are preserved (Shackleton Range, Pensacola Mountains, and Beardmore–Byrd glaciers area) a tectonic event has caused uplift, erosion and, with the possible exception of the Pensacola Mountains, folding, before deposition of the overlying Beacon Supergroup. This important and apparently geographically widespread period of tectonism is here termed the Shackleton Event, after the Shackleton Coast, which stretches between the Byrd and Beardmore glaciers. It was in this region that mid-Palaeozoic deformation was first clearly documented (Rowell *et al.*, 1988).

The Shackleton Event can be dated only indirectly. In the Beardmore–Byrd glaciers region deformation and erosion must have occurred not later than the Middle Devonian. In the Pensacola Mountains, the age of the Beacon Supergroup (Dover Sandstone) disconformably overlying the Lower Palaeozoic rocks is Late Devonian: a tectonic event, possibly associated with the disconformity, may be reflected in a Rb–Sr whole-rock isochron age of 394 ± 5 Ma (recalculated – Early Devonian) from the Proterozoic Patuxent Formation (Eastin, 1970). This date is interpreted as indicating isotopic resetting by the last metamorphic event which affected the unit. Although in the region separating the Pensacola Mountains from the Beardmore Glacier no post-Cambrian pre-Beacon sediments are preserved, in the Horlick Mountains the fossiliferous flat-lying basal Beacon Horlick Formation of late Early Devonian age (McCartan & Bradshaw, 1987) rests directly on Early Ordovician granite. This indicates that any mid-Palaeozoic tectonic event in this region can be no younger than Early Devonian. In the Shackleton Range the age of the event which folded the Ordovician (and younger?) Blaiklock Glacier Group is less well constrained. Clearly, however, it must have occurred before deposition of flat-lying Beacon Supergroup sediments of Permian age exposed 100 km to the north and south. On the balance of the evidence, an Early Devonian or perhaps Silurian age for the Shackleton Event seems probable, although the limited constraints do not rule out dischroneity.

Ellsworth Mountains

The region that has most stratigraphic similarity to the Transantarctic Mountains is the Ellsworth Mountains, which contain the only Lower Palaeozoic sediments with shelly fossils known from West Antarctica (Figs 1 & 2). Here more than 10000 m of Lower Palaeozoic sediments are exposed, overlain by more than 2000 m of Upper Palaeozoic rocks.

The lower part of the succession consists of the Middle–Upper Cambrian Heritage Group (> 7400 m) (Webers & Spörli, 1983; Craddock *et al.*, 1986). The lowest third of the Heritage Group, whose base is not exposed, consists mainly of dark green volcanic diamictite composed of basalt clasts and highly-deformed pumice fragments in a tuffaceous matrix, interbedded with subordinate conglomerate and tuffaceous argillite horizons. The volcanic diamictite is considered to be a terrestrial deposit of lahar and/or ashflow tuff origin. This unit is overlain conformably by a heterogeneous clastic unit consisting of green and maroon tuffaceous argillite, greywacke, polymict conglomerate and quartzite. These rock types are thought to have been deposited under deltaic conditions.

An immediately overlying unit of black shale and interbedded limestone, with fossiliferous horizons, indicates a transition to shallow marine conditions. This sequence passes upwards into a thick clastic unit consisting of argillaceous quartzite and polymictic conglomerate, interpreted as fluvial and/or shallow marine in origin. The upper part of the Heritage Group includes a range of lithologies showing rapid lateral and vertical facies changes. In the south, it consists of basaltic flows, volcanic breccia, polymict conglomerate, quartzite, and green argillite which were probably deposited in fluvial and shallow-marine environments. These facies pass laterally northwards into sequences dominated by argillite and quartzite, with subordinate limestone, deposited under marine and deltaic conditions. The youngest unit of the Heritage Group is composed almost entirely of white–grey marble and is considered to represent an open marine, shallow-water environment (Buggisch, 1983). Thus, like the Cambrian deposits of the Transantarctic Mountains, those of the Ellsworth Mountains represent shallow-marine shelf to marginal and non-marine environments.

Ordovician–(?)Early Devonian sequence

Unlike the situation in the Transantarctic Mountains, the Ordovician and younger rocks of the Ellsworth Mountains rest without apparent regional break on the Cambrian strata. There is however a marked facies change in latest Cambrian times from the marble or argillite of the uppermost beds of the Heritage Group to the highly quartzose sediments of the overlying Crashsite Group (Spörli, in press). This 3000 m thick sequence of quartzite and minor argillite is considered to have been emplaced under dominantly shallow-marine conditions, but with periodic emergence suggested by horizons of desiccation cracks. The age of the Crashsite Group is fairly well constrained. Basal units contain Late Cambrian fossils while the upper part has yielded a diverse Early Devonian fauna (Boucot *et al.*, 1967; Webers & Spörli, 1983). It is overlain by the glaciomarine Whiteout Conglomerate which is correlated with the (?)Carboniferous–Permian glacigene deposits of the Transantarctic Mountains (Frakes, Matthews & Crowell, 1971). The contact was considered to be unconformable by Hjelle, Ohta & Winsnes (1982), and conformable by Spörli (in press). The pronounced increase in angular plagioclase and rock fragments, coupled with poor sorting of the uppermost (Early Devonian) part of the Crashsite Group (Spörli, in press) suggests at least uplift of the source area. Intrusion into the Heritage Group of massive dolerite sills yielding whole-rock K–Ar dates of 381 and 396 Ma (Early and Middle Devonian) (Yoshida, 1982) may be associated with a related tectonic event.

Northern Victoria Land

In northern Victoria Land, there is a marked contrast in the depositional and tectonic environments of Lower Palaeozoic deposits with those of the rest of the Transantarctic Mountains. Sedimentary rocks crop out in three tectono-stratigraphic terranes, the Wilson, Bowers, and Robertson Bay terranes, which show contrasting stratigraphy and which are separated from each other by major tectonic zones (Fig. 2). The Bowers and Robertson Bay terranes, which contain rocks of Cambro-Ordovician age, are considered to be allochthonous with respect to each other and to the westernmost Wilson Terrane, which contains late Precambrian sedimentary rocks and is regarded to be an autochthonous part of the Transantarctic Mountains belt (Weaver, Bradshaw & Laird, 1984a; Borg *et al.*, 1987; Bradshaw, 1987).

Bowers Terrane: Cambro–Ordovician sequence

The Bowers Terrane comprises mainly the Early Palaeozoic Bowers Supergroup (Laird & Bradshaw, 1983). This consists of three groups, the Sledgers (oldest), Mariner, and Leap Year (youngest) groups, totalling over 10000 m, as well as fault-involved conglomerates of uncertain affiliation. The Middle Cambrian Sledgers Group consists of a volcanic association (Glasgow Formation) and an interfingering clastic sedimentary sequence (Molar Formation). The volcanic association consists of porphyritic vesicular basaltic and andesitic (and also subordinate rhyolitic) pillow lavas and breccias which are interpreted as primitive island arc tholeiites (Weaver *et al.*, 1984a). The interfingering Molar Formation consists dominantly of dark mudstone and thin-bedded fine sandstone: however, it also includes coarse-grained fining-upward mass-flow dominated sequences interpreted as infills of channels cut into submarine slopes. Sedimentological analysis indicates that the Molar Formation sediments represent infill of a tectonically-controlled trough-shaped basin.

The mainly conformably overlying Mariner Group of late Middle Cambrian–late Late Cambrian age represents a slowing of subsidence and infilling of the basin. Most of the group is represented by fissile mudstone with limestone lenses up to 30 m thick, although in the northern part of the region quartzite, tuff, conglomerate, and grey and red mudstone compose the basal part of the sequence. Locally, fissile mudstone passes gradationally upwards into a regressive sandy sequence culminating in grey quartzose sandstone and minor mudstone with numerous trace-fossil horizons and mudcracks. These are interpreted as intertidal deposits (Laird & Bradshaw, 1983).

The overlying Leap Year Group is separated from the older units by a marked erosion surface. Most of the succession consists of red–brown- or buff-coloured quartzose sandstone (commonly trough cross-bedded), quartzose conglomerate, and minor mudstone at least 4000 m thick, with local development of basal polymictic conglomerate. The group is considered to be dominantly fluvial in origin, although abundant trace fossils at some horizons near the base of the succession suggest localized marine influence.

Although there is a marked erosion surface separating the Leap Year Group from older units in the Bowers Supergroup, angular discordance can only be detected on a regional scale, and tectonic activity appears to have been restricted to uplift and tilting, without folding, prior to deposition of the Leap Year Group. No shelly fossils are known from the group, but an upper age limit is given by a K–Ar date of 482 ± 4 Ma; this probably represents the time of cleavage formation during Early Ordovician folding (Adams & Kreuzer, 1984).

Robertson Bay Terrane: Cambro-Ordovician sequence

The Bowers Terrane is separated from the Robertson Bay Terrane by a complex tectonic contact zone in which major overthrusting of one sequence over the other has occurred, and in which local schistosity is present (Bradshaw, 1987). The Robertson Bay Group, which occupies most of the terrane, consists almost entirely of alternating quartzose sandstone and mudstone several thousands of metres thick. These beds are interpreted to represent turbidites deposited on a submarine fan (Field & Findlay, 1983). Locally, in the south, turbidites are overlain apparently conformably by pebbly sandstone and conglomerate-bearing horizons containing fossiliferous limestone blocks of latest Cambrian or earliest Ordovician age (Wright & Brodie, 1987). The K–Ar age pattern for the Robertson Bay Group shows an age range from 455–505 Ma, probably reflecting the stages of post-orogenic cooling after low-grade metamorphism (Adams & Kreuzer, 1984). Thus the minimum age of metamorphism of the group is also close to the Cambro-Ordovician boundary. The maximum age of the Robertson Bay Group is uncertain, but it probably ranges from Cambrian to earliest Ordovician.

Ross Orogeny

Late Proterozoic rocks of the Wilson Terrane are intruded by granitoids of the Granite Harbour Intrusives, as in most of the Transantarctic Mountains. Rb–Sr whole-rock isochrons calculated on granitoid samples from different areas yielded dates mainly between 515 and 478 Ma (Vetter *et al.*, 1983); similarly, isochrons calculated on gneisses give an age of 490 ± 33 Ma (Adams, 1986a), indicating a Late Cambrian–Early Ordovician age of high-grade regional metamorphism and associated syn-metamorphic plutonism. No Lower Palaeozoic granitoids intrude either the Bowers Terrane or the Robertson Bay Terrane. However, the K–Ar age pattern for the Robertson Bay Terrane is similar to that of the Wilson Terrane, and a limited age range from 455 to 505 Ma is apparent. In the Bowers Terrane, there is a much more even spread of ages from 510 to 275 Ma; however, there is a minor but significant grouping of ages at the older end, in the range 510–470 Ma (Adams & Kreuzer, 1984). This suggests that although the three terranes may not have been immediately adjacent at the time of the Ross Orogeny, they were subjected to a common tectonic and metamorphic (but not igneous) event during the Late Cambrian–Early Ordovician.

Middle Devonian–Early Carboniferous magmatism

Sedimentation was brought to a halt in northern Victoria Land by Early Ordovician times, as indicated in the preceding section, and there is no further record of geological events over this entire region until Devonian times. During this period high level, strongly discordant I-type granitoids, the Admiralty Intrusives, were intruded into at least the Robertson Bay and Bowers terranes, and perhaps also the Wilson Terrane, although this point is still debated. Emplacement and cooling of the Admiralty Intrusives probably occurred between

~ 390 and 360 Ma (Borg *et al.*, 1987). Mid-Palaeozoic plutonism in northern Victoria Land was accompanied or immediately succeeded in all three terranes by calc-alkaline volcanism. In the Wilson Terrane the Gallipoli Volcanics, which are dominated by porphyritic rhyolites with some ignimbrites, yielded a Rb–Sr age of 382 Ma (Middle Devonian) (Grindley & Oliver, 1983). In the Bowers Terrane, the Lawrence Peak Volcanics, which are a moderate–high-K, calc-alkaline series of andesite, dacite, and rhyolite (Weaver, Bradshaw & Laird, 1984b), gave K–Ar total-rock dates of between 345 Ma and 175 Ma (C.J. Adams, unpublished preliminary data). At Mount Black Prince in the Robertson Bay Terrane, interbedded lava flows and volcaniclastic sediments are exposed (Findlay & Jordan, 1984). The volcanics are mostly altered andesite and basaltic-andesite with minor olivine basalt together with shallow intrusive and probably extrusive quartz and feldspar porphyry. Plant fragments of Middle Devonian–Early Carboniferous age from sediments associated with the volcanics are compatible with K–Ar whole-rock isotopic dates of 374 and 323 Ma for two samples of the volcanics (Findlay & Jordan, 1984).

Steep dips in outcrops of Middle Devonian–Early Carboniferous volcanics show that deformation must have occurred in post-Middle Devonian, pre-Beacon (i.e. pre-Permian) times, as, in the Bowers and Wilson terranes, flat-lying or shallow-dipping Beacon rocks are present close by.

Marie Byrd Land

Cambro–Ordovician sequence

In Marie Byrd Land, the Swanson Formation consists of grey, interbedded quartz-rich sandstone and mudstone (Bradshaw, Andrews & Field, 1983), representing a turbidite sequence very similar in appearance to the Robertson Bay Group (Fig. 2). It is also interpreted as representing a submarine fan setting. Shelly fossils have not been found but simple burrows have been seen at several localities and a single occurrence of *Palaeodictyon* high in one of the sections has been recorded. K–Ar whole-rock dating of slates at ~ 450 Ma, and Rb–Sr whole-rock isochron ages of 444 Ma (Adams, 1986b), suggest a Late Ordovician tectonic event. If *Palaeodictyon* was restricted to Ordovician and younger rocks, then a Cambro-Ordovician age of sedimentation for the Swanson Formation, followed by Late Ordovician deformation and uplift, is indicated (Bradshaw *et al.*, 1983).

Middle Devonian–Early Carboniferous magmatism

As in northern Victoria Land, the turbidites were intruded in the mid-Palaeozoic by I-type granitoids, collectively termed the Ford Granodiorite. Isotopic data suggest that it was emplaced between 380–345 Ma (Middle Devonian–Early Carboniferous) (Adams, 1987). Calc-alkaline metavolcanics (Ruppert Coast Metavolcanics; Grindley, Mildenhall & Schopf, 1980), consisting mainly of grey feldsparphyric andesite but including a range of lithologies from spilitic basalt through andesite and dacite to rhyolite, also occur. Locally, they conformably overlie thin-bedded sediments with scattered carbonaceous shaly layers. Nearby carbonaceous erratics, probably derived from this sequence, contain plant fossils indicating a late Middle–early Late Devonian age (Grindley *et al.*, 1980); thus the conformably overlying volcanics are inferred to be of similar or slightly younger age. Probably correlative volcanics in eastern Marie Byrd Land give isotopic dates of 370–295 Ma and confirm the Late Devonian–Carboniferous ages for the extrusives (Grindley *et al.*, 1980). Like the volcanic sequences in northern Victoria Land, the volcanic and associated sedimentary sequences are strongly deformed.

Tectono-stratigraphic provinces

From the preceding discussion it appears that a Middle Devonian–Early Carboniferous magmatic event or events occurred in both northern Victoria Land and Marie Byrd Land. This resulted in the widespread intrusion of I-type granitoids and of approximtely coeval volcanism at a time of tectonic calm in the remainder of the Transantarctic Mountains when Beacon Supergroup deposition was proceeding. This was accompanied or followed by deformation. The magmatic event and a possible earlier metamorphic event (now discounted) were termed the Borchgrevink Orogeny by Craddock (1972). However, until the tectonic environment of the Devonian–Carboniferous magmatism and the nature of the associated deformation are better understood, Borchgrevink Event may be a more objective term. In the Transantarctic Mountains, at least between the Byrd Glacier and the Shackleton Range, there is strong evidence for a separate and earlier uplift and/or folding event (Shackleton Event). This was not associated with magmatism and probably occurred in the Early Devonian, prior to deposition of Beacon sequences or their equivalents. Scattered evidence for a tectono-magmatic event, approximately coeval with the Shackleton Event, also occurs in the Ellsworth Mountains succession.

The geographical distribution of these contrasting mid-Palaeozoic tectono-magmatic events is closely related to the distribution of contrasting Early Palaeozoic sedimentary environments. The Transantarctic Mountains and Ellsworth Mountains sequences are characterized by: shallow-water non-marine and marine shelf environments for the Cambrian strata; an unconformity or abrupt facies change close to the Cambro-Ordovician boundary; and non-marine to very shallow-marine coarse-grained clastic sediments, either sandstones or conglomerates, during Ordovician–(?)Early Devonian times. The Marie Byrd Land and Robertson Bay Terrane sediments, on the other hand, were deposited in a deeper water submarine fan setting. The Cambrian–Early Ordovician sediments and volcanics of the Bowers Terrane contrast with both, having been deposited in a rapidly subsiding basin in an active island arc setting: however, the Terrane shows the same mid-Palaeozoic tectono-magmatic imprint as the remainder of eastern northern Victoria Land and Marie Byrd Land. Thus two tectono-stratigraphic provinces can be defined, here termed the Transantarctic Mountains Province and the Ross Sea Province, respectively (Fig. 3). The latter province is only a remnant of what was probably originally a larger continent-

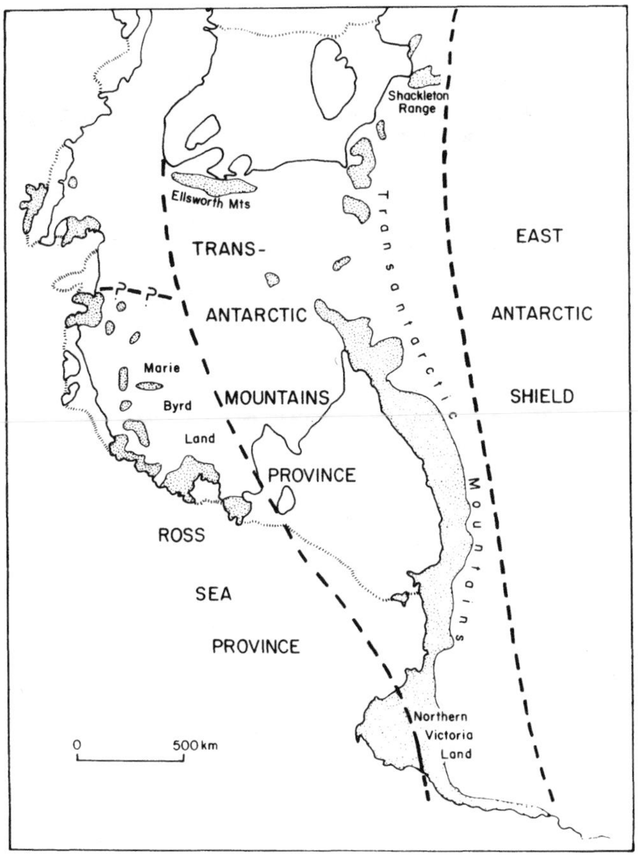

Fig. 3. Sketch map showing Early–mid-Palaezoic tectono-stratigraphic provinces of Antarctica as defined in this paper.

sized allochthonous block which is likely to have included the Campbell Plateau, the west coast of the South Island of New Zealand and perhaps also parts of southern Australia and Tasmania (Adams, 1986b, 1987).

Although it is clear that the Bowers and Robertson Bay terranes were amalgamated by the time the Devonian Admiralty Intrusives were emplaced (Borg *et al.*, 1987; Bradshaw, 1987), the timing of accretion of these terranes to the Wilson Terrane, and therefore of the Ross Sea Province to the Transantarctic Province, is still speculative. An upper age limit is provided by the presence of nearly flat-lying Permian Beacon sediments in both the Wilson and Bowers terranes. A lower limit is less certain. Although the occurrence of Admiralty Intrusives of Devonian age in the Wilson Terrane is uncertain (Borg *et al.*, 1987), it is clear that approximately coeval Devonian–Carboniferous calc-alkaline volcanics were common to all three terranes, and, perhaps significantly in the Wilson Terrane, are known only from its eastern margin. Although there is as yet no proof that these volcanics are genetically related, their petrographic similarity and their apparently coeval eruption suggests that their extrusion and/or subsequent deformation could have been related to amalgamation of the two provinces in the Mid-Devonian–Carboniferous.

The stratigraphy and structure of the Ellsworth Mountains are closely similar in many respects to those of the Transantarctic Mountains, in particular to the Pensacola Mountains which do not have an angular discordance separating Ordovician–Early Devonian(?) strata from overlying Beacon Supergroup. It has been suggested by various writers that the Ellsworth Mountains originally formed part of the Transantarctic Mountain block, having been rotated and translated into its present position during the break-up of Gondwana in post-mid-Jurassic times (Watts & Bramall, 1981; Grunow, Dalziel & Kent, 1987). Although there is continuing debate about the original specific position and orientation of the Ellsworth Mountains, there seems to be agreement that they probably lay adjacent to the Pensacola Mountains–Shackleton Range region. The conclusions of the present study reinforce this impression, and further suggest that the Ellsworth Mountain block originally lay close to, and at least in post-Cambrian times, seaward of the Pensacola Mountains.

References

Adams, C.J. (1986a). Age and ancestry of the metamorphic rocks of the Wilson Group of the Daniels Range, USARP Mountains, Antarctica. In *Geological Investigations in northern Victoria Land, Antarctica*, Antarctic Research Series, ed. E. Stump, pp. 25–38. Washington, DC; American Geophysical Union.

Adams, C.J. (1986b). Geochronological studies of the Swanson Formation of Marie Byrd Land, West Antarctica, and correlation with northern Victoria Land, East Antarctica, and South Island, New Zealand. *New Zealand Journal of Geology and Geophysics*, **29(3)**, 345–58.

Adams, C.J. (1987). Geochronology of granitic terranes in the Ford Ranges, Marie Byrd Land, West Antarctica. *New Zealand Journal of Geology and Geophysics*, **30(1)**, 51–72.

Adams, C.H. & Kreuzer, H. (1984). Potessium–argon age studies of slates and phyllites from the Bowers and Robertson Bay terranes, northern Victoria Land, Antarctica. *Geologisches Jahrbuch*, **B60**, 265–88.

Borg, S.G. (1983). Petrology and geochemistry of the Queen Maud Batholith, central Transantarctic Mountains, with implications for the Ross Orogeny. In *Antarctic Earth Science*, eds. R.L. Oliver, P.R. James & J.B. Jago, pp. 165–9. Canberra; Australian Academy of Science and Cambridge; Cambridge University Press.

Borg, S.G., Stump, E., Chappell, B.W., McCulloch, M.T., Wyborn, D. & Holloway, J.R. (1987). Granitoids of northern Victoria Land, Antarctica: implications of chemical and isotopic variations to regional crustal structure and tectonics. *American Journal of Science*, **287(2)**, 127–69.

Boucot, A.J., Doumani, G.A., Johnson, J.G. & Webers, G.F. (1967). Devonian of Antarctica. In *International Symposium on the Devonian System, Papers*, vol. I, ed. D.H. Oswald, pp. 639–48. Calgary; Alberta Society of Petroleum Geologists.

Bradshaw, J.D. (1987). Terrane boundaries and terrane displacement in northern Victoria Land, Antarctica: some problems and constraints. In *Terrane Accretion and Orogenic Belts*, ed. E. Leitch & E. Scheibner, pp. 199–205. Washington, DC; American Geophysical Union.

Bradshaw, J.D., Andrews, P.B. & Field, B.D. (1983). Swanson Formation and related rocks of Marie Byrd Land and a comparison with the Robertson Bay Group of northern Victoria Land. In *Antarctic Earth Science*, eds. R.L. Oliver, P.R. James & J.B. Jago, pp. 274–9. Canberra; Australian Academy of Science and Cambridge; Cambridge University Press.

Buggisch, W. (1983). Palaeozoic rocks of the Ellsworth Mountains, west Antarctica (abstract). In *Antarctic Earth Science*, eds. R.L. Oliver, P.R. James & J.B. Jago, pp. 265. Canberra; Australian Academy of Science and Cambridge; Cambridge University Press.

Clarkson, P.D., Hughes, C.P. & Thomson, M.R.A. (1979). Geological significance of a Middle Cambrian fauna from Antarctica. *Nature, London*, **279(5716)**, 791–2.

Craddock, C. (1972). Antarctic tectonics. In *Antarctic Geology and Geophysics*, ed. R.J. Adie, pp. 449–55. Oslo; Universitetsforlaget.

Craddock, C., Webers, G.F., Rutford, R.H., Spörli, K.B. & Anderson, J.J. (1986). *Geological Map of the Ellsworth Mountains, Antarctica*. Washington, DC; Geological Society of America, Map and Chart Series MC-57.

Eastin, R. (1970). Geochronology of the basement rocks of the central Transantarctic Mountains. In *Studies in the Geochronology and Geochemistry of the Transantarctic Mountains*, ed. G. Faure, R. Eastin, J. Gunner, R.L. Hill, L.M. Jones & D.H. Elliot, pp. 67–180 Columbus, Ohio; Report No. 5, Laboratory for Isotope Geology and Geochemistry, Institute of Polar Studies and Department of Geology, The Ohio State University.

Faure, G., Eastin, R., Ray, P.T., McLelland, D. & Schulz, C.H. (1979). Geochronology of igneous and metamorphic rocks, central Transantarctic Mountains. In *Fourth International Gondwana Symposium Calcutta, India, 1977– Papers, vol. II*, ed. B. Laskar & C.S. Raja Rao, pp. 805–13. Delhi; Hindustan Publishing Corporation.

Field, B.D. & Findlay, R.H. (1983). The sedimentology of the Robertson Bay Group, northern Victoria Land. In *Antarctic Earth Science*, eds. R.L. Oliver, P.R. James & J.B. Jago, pp. 102–6. Canberra; Australian Academy of Science and Cambridge; Cambridge University Press.

Findlay, R.H. & Jordan, H. (1984). The volcanic rocks of Mt Black Prince and Lawrence Peaks. *Geologisches Jahrbuch*, **B60**, 143–51.

Frakes, L.A., Matthews, J.L. & Crowell, J.C. (1971). Late Paleozoic Glaciation: Part III, Antarctica. *Geological Society of America Bulletin*, **82(6)**, 1581–604.

Grindley, G.W., Mildenhall, D.C. & Schopf, J.M. (1980). A mid-Late Devonian flora from the Ruppert Coast, Marie Byrd Land, Antarctica. *Journal of the Royal Society of New Zealand*, **10(3)**, 271–85.

Grindley, G.W. & Oliver, P.J. (1983). Post-Ross Orogeny cratonisation of northern Victoria Land. In *Antarctic Earth Science*, eds. R.L. Oliver, P.R. James & J.B. Jago, pp. 133–9. Canberra; Australian Academy of Science and Cambridge; Cambridge University Press.

Grunow, A.M., Dalziel, I.W.D. & Kent, D.V. (1987). Ellsworth Mountains–Whitmore Mountains Crustal Block, Western Antarctica: new paleomagnetic results and their tectonic significance. In *Gondwana Six: Structure, Tectonics, and Geophysics*, ed. G.D. McKenzie, pp. 161–72. Washington, DC; American Geophysical Union.

Harland, W.B., Cox, A.V., Llewellyn, P.G., Pickton, C.A.G., Smith, A.G. & Walters, R. (1982). *A Geologic Time Scale*. Cambridge; Cambridge University Press. 131 pp.

Hjelle, A., Ohta, Y. & Winsnes, T.S. (1982). Geology and petrology of the southern Heritage Range, Ellsworth Mountains. In *Antarctic Geoscience*, ed. C. Craddock, pp. 599–608. Madison; University of Wisconsin Press.

Konyuschkov, K.N. & Shulyatin, O.G. (1980). Ob arkheotsiatakh Antarktidy i ikh sopostavlenii s arkheotsintami Sibiri. (On the archaeocyathids of Antarctica and their comparison with archaeocyathids of Siberia.) In *Kembrii Altae–Sayanskoi Skladchatoi Oblasti* (Cambrian of the Altay–Sayan folded region), ed. I.T. Zhuravleva, pp. 143–50. Nauka, Moskva; Akademiya Nauk SSSR, Sibirskoe Otdelenie, Instituta Geologii i Geofiziki.

Laird, M.G. (1981). Lower Paleozoic rocks of the Ross Sea area and their significance in the Gondwana context. *Journal of the Royal Society of New Zealand*, **11(4)**, 425–38.

Laird, M.G. & Bradshaw, J.D. (1983). New data on the Lower Palaeozoic Bowers Supergroup, northern Victoria Land. In *Antarctic Earth Science*, eds. R.L. Oliver, P.R. James & J.B. Jago, pp. 123–6. Canberra; Australian Academy of Science and Cambridge; Cambridge University Press.

McCartan, L. & Bradshaw, M.A. (1987). Petrology and sedimentology of the Horlick Formation (Lower Devonian), Ohio Range, Transantarctic Mountains. *US Geological Survey Bulletin*, **1780**, 31 pp.

Marsh, P.D. (1984). The stratigraphy and structure of the Lagrange Nunataks, northern Fuchs Dome and Herbert Mountains of the Shackleton Range. *British Antarctic Survey Bulletin*, **63**, 19–40.

Pankhurst, R.J., Marsh, P.D. & Clarkson, P.D. (1983). A geochronological investigation of the Shackleton Range. In *Antarctic Earth Science*, eds. R.L. Oliver, P.R. James & J.B. Jago, pp. 176–82. Canberra; Australian Academy of Science and Cambridge; Cambridge University Press.

Pankhurst, R.J., Storey, B.C., Millar, I.L., Macdonald, D.I.M. & Vennum, W.R. (1988). Cambro-Ordovician magmatism in the Thiel Mountains, Transantarctic Mountains, and implications for the Beardmore Orogeny. *Geology*, **16**, 246–9.

Rowell, A.J., Rees, M.N. & Braddock, P. (1986). Pre-Devonian Paleozoic rocks of the central Transantarctic Mountains. *Antarctic Journal of the United States*, **21(5)**, 48–50.

Rowell, A.J., Rees, M.N., Cooper, R.A. & Pratt, B.R. (1988). Early Paleozoic history of the central Transantarctic Mountains: evidence from the Holyoake Range, Antarctica. *New Zealand Journal of Geology and Geophysics*, **31(4)**, 397–404.

Schmidt, D.L. (1983). Proterozoic to Mesozoic mobile-belt geology, Pensacola Mountains, Antarctica. *United States Geological Survey Polar Research Symposium, Abstract with Program.* Geological Survey Circular 911, pp. 12–13.

Schmidt, D.L. & Ford, A.B. (1969). *Geology of the Pensacola and Thiel Mountains*, Antarctic Map Folio Series, Folio 12-Geology, Sheet 5, ed. C. Craddock & V. Bushnell. Washington, DC; American Geographical Society.

Spörli, K.B. (in press). Stratigraphy of the Crashsite Group, Ellsworth Mountains, west Antarctica. In *Geology of the Ellsworth Mountains, Antarctica*, ed. G.F. Webers, C. Craddock & J.F. Splettstoesser. Geological Society of America, Memoir.

Stump, E. (1982). The Ross Supergroup in the Queen Maud Mountains. In *Antarctic Geoscience*, ed. C. Craddock, pp. 565–9. Madison; University of Wisconsin Press.

Vetter, U., Roland, N.W., Kreuzer, H., Hohndorf, A., Lenz, H. & Besang, C. (1983). Geochemistry, petrography and geochronology of the Cambro-Ordovician and Devonian–Carboniferous granitoids of northern Victoria Land, Antarctica. In *Antarctic Earth Science*, eds. R.L. Oliver, P.R. James & J.B. Jago, pp. 140–3. Canberra; Australian Academy of Science and Cambridge; Cambridge University Press.

Watts, D.R. & Bramall, A.M. (1981). Palaeomagnetic evidence for a displaced terrain in western Antarctica. *Nature, London*, **293(5834)**, 638–41.

Weaver, S.D., Bradshaw, J.D. & Laird, M.G. (1984a). Geochemistry of Cambrian volcanics of the Bowers Supergroup and implications for the Early Palaeozoic tectonic evolution of northern Victoria Land, Antarctica. *Earth and Planetary Science Letters*, **68(1)**, 128–40.

Weaver, S.D., Bradshaw, J.D. & Laird, M.G. (1984b). Lawrence Peaks Volcanics, north Victoria Land. *New Zealand Antarctic Record*, **5(3)**, 18–22.

Webers, G.F. & Spörli, K.B. (1983). Paleontological and stratigraphic investigations in the Ellsworth Mountains, West Antarctica. In *Antarctic Earth Science*, eds. R.L. Oliver, P.R. James & J.B. Jago, pp. 261–4. Canberra; Australian Academy of Science and Cambridge; Cambridge University Press.

Williams, P.L. (1969). Petrology of Upper Precambrian and Paleozoic sandstones in the Pensacola Mountains, Antarctica. *Journal of Sedimentary Petrology*, **39(4)**, 1455–65.

Wright, T.O. & Brodie, C. (1987). The Handler Formation, a new unit of the Robertson Bay Group, northern Victoria Land, Ant-

arctica. In *Gondwana Six: Structure, Tectonics, and Geophysics*, ed. G.D. McKenzie, pp. 25–9. Washington, DC; American Geophysical Union.
Yoshida, M. (1982). Supposed deformation and its implication to the geologic history of the Ellsworth Mountains, West Antarctica. *Proceedings of the Second Symposium on Antarctic Geosciences, 1980*, Memoirs of National Institute of Polar Research Special Issue, 21, pp. 120–71.

The pre-Devonian Palaeozoic clastics of the central Transantarctic Mountains: stratigraphy and depositional settings

M.N. REES & A.J. ROWELL

Department of Geoscience, University of Nevada, Las Vegas, Las Vegas, Nevada 89154, USA
Museum of Invertebrate Paleontology and Department of Geology, University of Kansas, Lawrence, Kansas 66045, USA

Abstract

Pre-Devonian clastic beds forming the Douglas Conglomerate crop out in the Churchill Mountains where locally they rest unconformably on the Lower Cambrian Shackleton Limestone. Following at least one episode of deformation, the Shackleton Limestone was uplifted and eroded and contributed extensively to the deposition of the Douglas Conglomerate. Clasts derived from it are a significant component of the polymictic terrigenous-rich sequences that had a variety of sources and constitute most of the material in the limestone breccia beds of the Douglas Conglomerate. Following deposition and prior to the Devonian, the Douglas Conglomerate was subjected to at least three periods of deformation. The sediment was deposited by a variety of mechanisms in alluvial fan environments and in marine and possibly lacustrine environments of fan or braid deltas. Pebble–boulder clast-supported conglomerates with minor amounts of sandstone were deposited by: density-modified mass flows in proximal alluvial fans; deep high-discharge streams in main channels; shallow braided stream and sheet floods in interchannel areas; and by flood discharge in distal settings. Coarse-grained turbidites and bioturbated silty limestone represent submarine deposition associated with fan or braid deltas. Fine-grained terrigenous rocks with thin Bouma sequences and abundant ripples, but lacking trace or body fossils, may signify lacustrine depositional environments.

Introduction

Numerous isolated coarse clastic sequences crop out in the Churchill Mountains between the Byrd and the Nimrod glaciers (Fig. 1) and we tentatively assign them to the Douglas Conglomerate. Their age is poorly constrained, but they contain clasts of the deformed underlying Lower Cambrian Shackleton Limestone and everywhere their deposition and complex deformation predate the initial beds of the Beacon Supergroup, which are Devonian in age.

This paper focusses on the lithofacies, depositional environments and stratigraphic relationships of these clastic units. They vary in lithology from limestone breccia to terrigenous conglomerate, sandstone and shale, and they were deposited in a diversity of environments ranging from shallow marine to alluvial fan. We speculate that this diversity is the result of accumulation at different times in response to changing tectonic patterns, sporadic basin development, and laterally and temporally variable source areas. Locally, the basal beds of the succession rest with angular unconformity on the Shackleton Limestone: elsewhere, they are in fault contact with it. The coarse clastic sequences thus are the only sedimentary record of a 150 m.y. period during which the region was affected by at least two, and probably more than three, deformational events (Rees *et al.*, 1988). In a complex fashion, they document the history of an Early Palaeozoic orogen.

Relationship of clastic sequences to the Shackleton Limestone

We consider that none of the coarse clastic sequences formerly included within the Shackleton Limestone are part of that formation. All of them are younger, are of pre-Devonian age, and are best regarded as belonging to the Douglas Conglomerate. We arrive at this conclusion because coarse breccia and conglomerates are never seen in conformable depositional contact with beds that undoubtedly belong to the Shackleton Limestone, and commonly their clasts include those that demonstrably are deformed Shackleton Limestone.

Limestone breccia, quartzite and limestone conglomerate and shale units were regarded as part of the Shackleton Limestone in the initial reconnaissance of the southern Churchill Mountains (Laird, 1963; Laird, Mansergh & Chappell, 1971). At its type locality on the east side of the Holyoake Range (Fig. 2), Laird (1963) recognized that many beds had sharp contacts, the dip of strata was not everywhere similar,

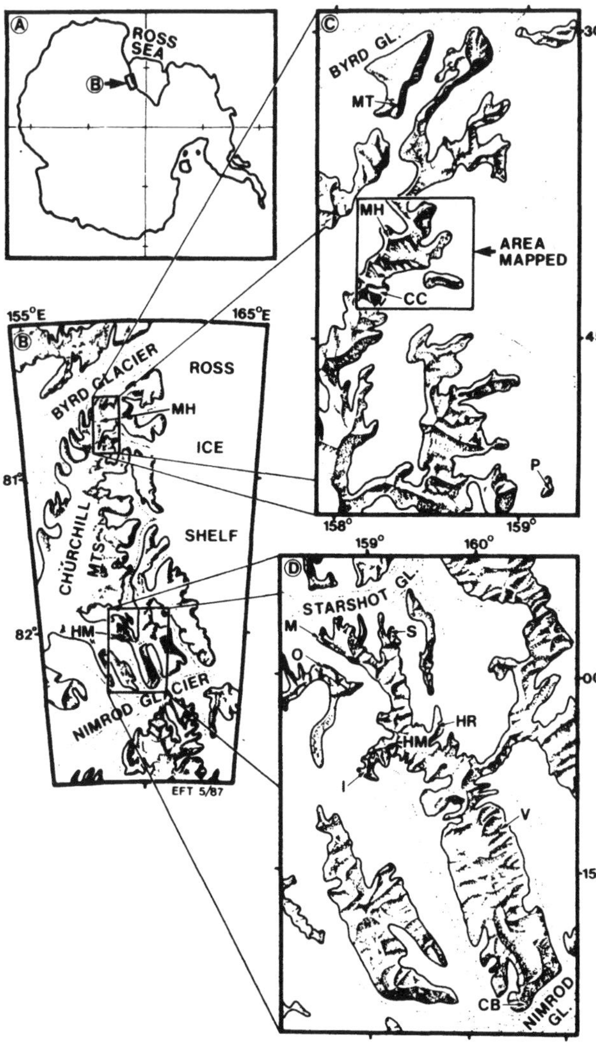

Fig. 1. ***A & B.*** **Location map of Churchill Mountains. Enlarged map *C* shows area mapped by Girty & Rees (unpublished data) and the location of Mt Tuatara (MT), Mt Hamilton (MH), Crackling Cwm (CC), and type locality of the Douglas Conglomerate (P). Enlarged map *D* of the Holyoake Range indicates the location of Hunt Mountain (HM) and Cambrian Bluff (CB) and measured sections or outcrops of Douglas Conglomerate discussed in the text (O, M, S, HR, I & V).**

topping directions were inconsistent, and some 'bedding faults' were present. Our re-examination of the type ridge (Rowell, Rees & Braddock, 1988a) indicated that the area is highly faulted and folded, but four general rock units could be mapped (Fig. 2): two limestone units that are clearly part of the Shackleton Limestone and two coarse clastic units that we assign to the Douglas Conglomerate. The complexity of large-scale structures was most easily observed from the air near Cambrian Bluff about 50 km to the south (Rowell *et al.*, 1988a), but similar deformation is pervasive throughout the range. Even though the structure is complex, depositional contacts commonly were observed within fine–medium-grained limestone sequences and within clastic successions, but we never observed conformable depositional contact between the Shackleton Limestone and any coarse clastic beds.

The coarse clastic sequences in the Holyoake Range typically are either in fault contact with Shackleton Limestone or the contacts are not exposed. Locally at spur M and possibly at section V (Fig. 1), however, the units are unconformable. At spur M, the underlying Shackleton Limestone is strongly deformed and the overlying 500 m thick conglomerate and sandstone sequence is steeply dipping. The unconformity is exposed across the spur for approximately 130 m, and it represents an irregular erosional surface along which there was at least 30 m of relief prior to deposition of the conglomerate (Rowell *et al.*, 1988b).

The relationship between the Shackleton Limestone and a 200 m thick limestone breccia sequence at section V (Fig. 2) is less clear, but probably it also is an unconformity. The critical exposures are largely covered by snow and scree, but the boundary can be located within a 3 m wide zone. This zone does trend parallel to faults in the area, but shearing and veining are minimal in rocks on either side of it. Furthermore, the thin-bedded Shackleton Limestone is isoclinally folded with limbs striking N 30°W and having a steep SE dip. In contrast, the basal breccia beds are massive but, about 10 m up section, the beds strike N 10°W and dip steeply W.

Mapping in the northern Churchill Mountains (Fig. 1) indicates that quartz–limestone conglomerates and sandstone sequences are more widespread and much more deformed (Rees *et al.*, 1988) than initial maps of the region suggested (Skinner, 1964). Others have also noted additional localities of terrigenous sequences near Mt Hamilton (Burgess & Lammerink, 1979) and Mt Tuatara (Stump *et al.*, 1979) and have recognized multiple phases of deformation in the Shackleton Limestone (Stump *et al.*, 1979). The terrigenous sequences have suffered at least three deformational events and, within the map area (Fig. 1), these units are everywhere in fault contact and commonly in thrust contact with either the Shackleton Limestone or a very coarse limestone breccia unit (Rees *et al.*, 1988). In turn, the limestone breccia is everywhere in fault contact with the Shackleton Limestone and terrigenous sequences as at Crackling Cwm (Fig. 1). In contrast to earlier views (Burgess & Lammerink, 1979; Debrenne & Kruse, 1986), we consider the evidence suggests that the limestone breccia and terrigenous sequence are fault-bounded units whose deposition significantly post-dates that of the Shackleton Limestone.

Such a view is supported by examination of clast composition. All clastic sequences examined contain varying amounts of limestone clasts whose lithologies are identical to those of the Shackleton Limestone. More importantly, many of the clasts were deformed prior to their incorporation into the conglomerate or breccia. No fossils have been found in folded clasts, but their lithology and style of tectonic deformation are closely comparable to those in nearby Shackleton Limestone outcrops, and they were probably derived from that formation. Some brittlely deformed clasts have lithologies identical to those found in the Shackleton Limestone and furthermore contain archaeocyathans. They have fractures filled with blocky calcite spar that do not extend into the matrix or from clast to clast. Such fractures in different clasts appear to be haphazardly oriented, and they are unlike later-formed tension gashes that have consistent orientations and clearly cross-cut numerous clasts and intervening matrix. These deformed clasts demonstrate that the Lower Palaeozoic coarse clastic

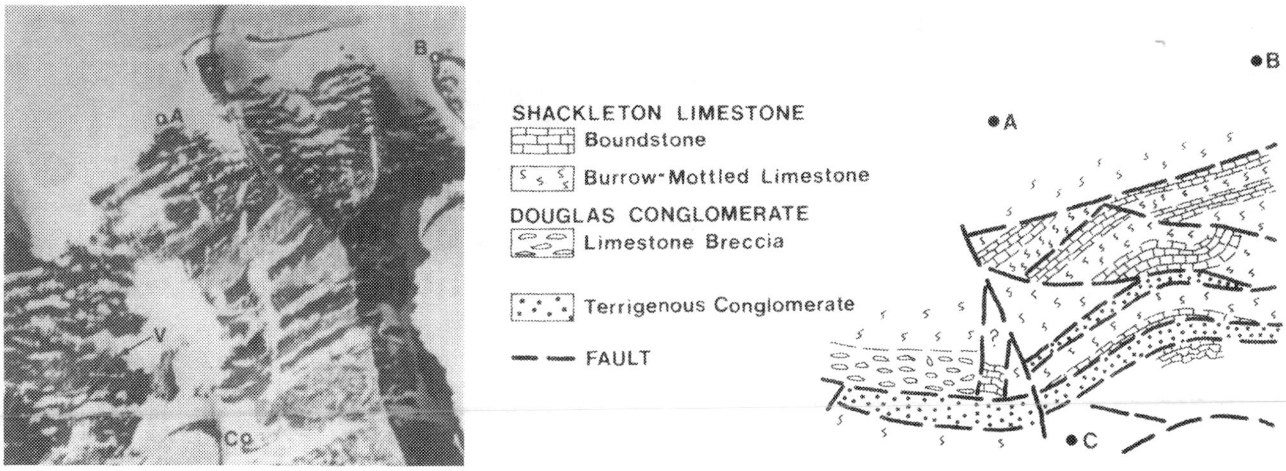

Fig. 2. Oblique aerial photograph covering the type locality of the Shackleton Limestone and measured section V of the Douglas Conglomerate (US Navy Photograph Flight TMA 769, Photo number 237 F33). Right: points A, B and C are arbitrary corresponding points on both photograph and sketch map. Points A and B are approximately 1.7 km apart and a line from B to A points approximately N35 °W.

sequences in the Churchill Mountains were derived in part from the Shackleton Limestone after it was lithified, deformed at least once, uplifted and exposed to erosion. In addition to limestone clasts, other sedimentary rock fragments and igneous and metamorphic clasts commonly are present in the clastic units and locally produce terrigenous-rich sequences.

Nomenclature and correlation of coarse clastic sequences

We tentatively use the term Douglas Conglomerate for all pre-Beacon clastic sequences in the Churchill Mountains that contain Shackleton Limestone clasts. Such usage avoids proliferation of names, but we recognize numerous stratigraphic and nomenclatural difficulties inherent in it. Some of the problems arise because no complete sequence of the Douglas Conglomerate is known and because we are unable to lithologically or chronologically correlate coarse clastic sequences with confidence from fault-bounded unit to fault-bounded unit within one locality or between localities. We do not suggest that all these coarse clastic units are time or lithologic equivalents. We know only that they post-date at least one period of deformation of the Shackleton Limestone and predate the Devonian beds of the Beacon Supergroup.

The Douglas Conglomerate was defined by Skinner (1964), who designated exposures of conglomerate on an isolated nunatak near Byrd Glacier (Fig. 1) as the type locality. There, only the lower 140 m of the formation crop out and overlie an argillite and sandstone sequence, the Dick Formation, whose age is unknown (Skinner, 1964). At that locality, the clasts in the conglomerate consist mainly of quartz and Shackleton Limestone fragments together with minor amounts of argillite, siltstone, metaquartzite, dark arkosic sandstone, diorite, two-mica granite, gneissic granite and haematite-gneiss (Skinner, 1964, 1965). Laird (1981) recognized that the conglomerates at the localities we designated as spur M and locality O (Fig. 1) should be included in the Douglas Conglomerate rather than the Shackleton Limestone because of lithologic similarities with the Douglas Conglomerate as described from its type locality. We concur with this assignment because, both at the type locality (Skinner, 1964) and at spur M, the units are predominantly coarse terrigenous sequences and, even though their composition differs in detail, they both contain clasts of Shackleton Limestone. However, at spur M the 500 m thick coarse clastic succession rests unconformably on the Shackleton Limestone instead of the Dick Formation, and the stratigraphic sequences are different. Indeed, variability is a characteristic feature of these clastic sequences. At its type locality, the Douglas Conglomerate is a pebble–boulder conglomerate with minor sandstone lenses, whereas at locality O it consits predominantly of sandstone and shales. Other coarse clastic sequences in the Douglas Conglomerate show yet greater diversity because the clasts in some of the fault-bounded sequences, such as those at section V, HR and possibly I, are dominantly limestone. Nevertheless, we do not believe separate formal formations or members can be recognized consistently at this time. We speculate that the limestone-dominated breccia and conglomerates maybe somewhat older than the terrigenous sequences, but this has not been demonstrated conclusively.

Depositional settings of the Douglas Conglomerate

The sequences of Douglas Conglomerate that we have examined in the Churchill Mountains are characterized by thick units of pebble–boulder conglomerate and breccia with thinner and less abundant successions of sandstones and shales. These rocks are characterized by lithofacies that can conveniently be designated using codes developed by Miall (1978) and Rust (1978), and those of the Bouma sequence. The lithofacies and their vertical associations, as illustrated by four measured sections from the Holyoake Range (Figs 3 & 4), are the product of processes active in alluvial fan environments and marine and lacustrine environments of fan or braid deltas.

The development of a coherent palaeogeographic model or basin evolution scenario based on our environmental interpretations is yet severely limited for numerous reasons. First, most

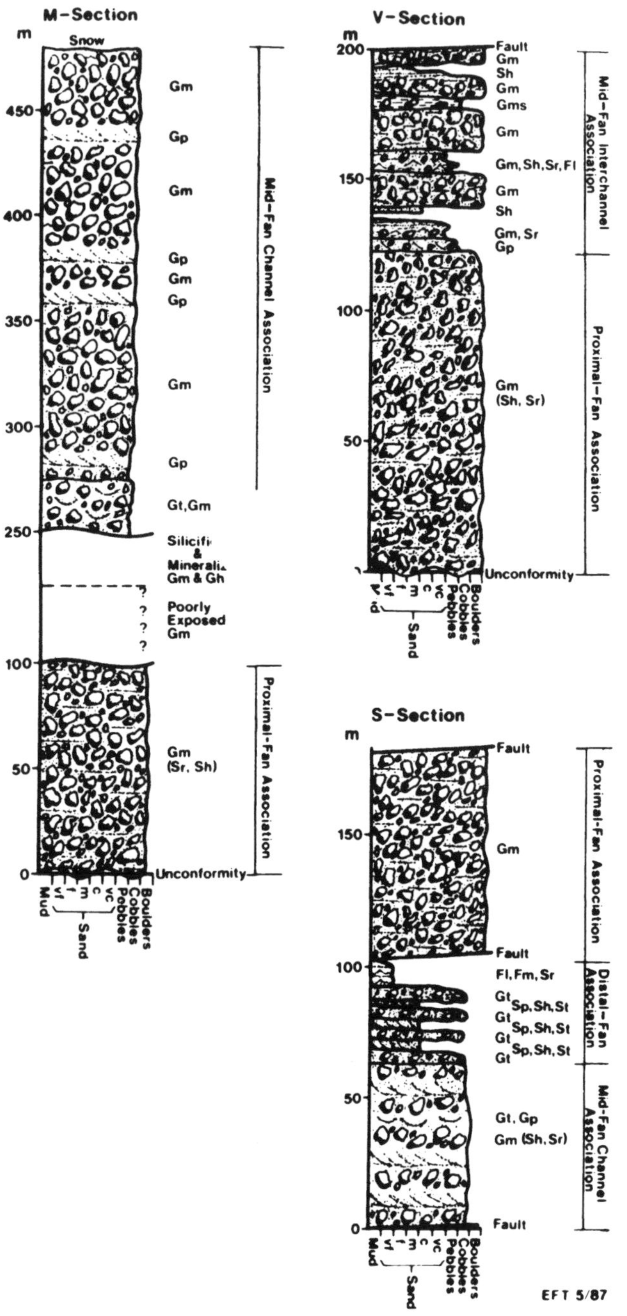

Fig. 3. Measured sections of the Douglas Conglomerate in the Holyoake Range indicating grain size, lithofacies and interpretation of depositional environments. There is no evidence that these sections are temporally equivalent. Lithofacies codes from Miall (1978) and Rust (1978). Gravels: G_{ms}, massive, matrix supported; G_m, massive or crudely bedded; G_t, trough crossbedded; G_p planar crossbedded. Sands: S_t, trough crossbedded; S_p, planar crossbedded; S_r, ripple cross-stratified; S_h, horizontally laminated; S_l, low angle crossbeds. Muds and silts: F_l, horizontally laminated and small-scale ripple laminated; F_m, massive, desiccation cracks.

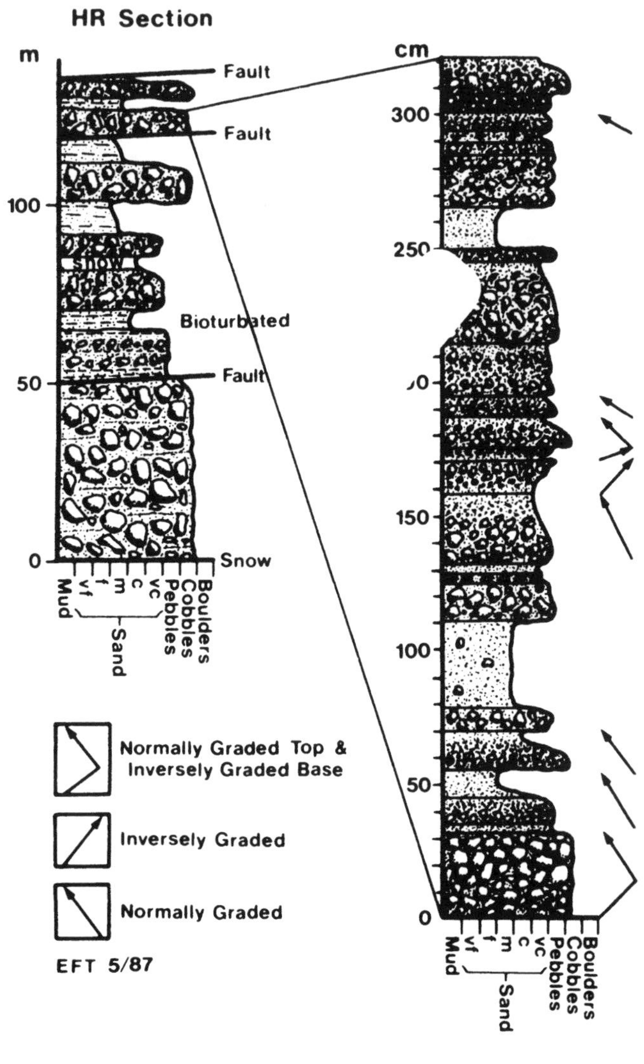

Fig. 4. Measured section at Hunt Ridge (HR) in the Holyoake Range illustrating part of the Douglas Conglomerate that was deposited in a marine environment.

of the coarse clastic sequences occur in geographically isolated or fault-bounded outcrops of limited lateral and vertical extent. Second, the lack of biostratigraphic and geochronologic data limits our ability to correlate from locality to locality. There is little evidence to suggest that successions at different localities were temporal equivalents or even that they formed in diverse parts of the same depositional system. Third, deformation locally has been so intense that stratigraphic information is lost beyond recovery and original depositional features are obliterated. Provenance and structural studies and further radiometric dating projects, that are presently under way, may help to resolve some of these problems and allow us to develop reasonable tectonic and basin models. Nevertheless, within some of the less deformed successions, lithofacies analysis has contributed significantly to our understanding of the depositional regimes of the Douglas Conglomerate.

Alluvial fan deposits

Lithofacies within measured sections M, V, and S are grouped into four associations (Fig. 3) that were produced by various fluvial and subaerial mass flow processes and are representative of deposition on different parts of alluvial fans. The four associations are proximal fan, mid-fan channel, mid-fan interchannel and distal fan.

The Proximal Fan Association is dominated by massive–crudely stratified, clast-supported, pebble–boulder conglomerate and breccia (G_m). Bedding is difficult to discern in this lithofacies because many contacts or grain size changes are gradational, beds of similar grain size are superimposed, and beds are amalgamated. Where recognizable, beds range from 0.5 to 3 m in thickness, but some may be much thicker. They have flat–undulatory bases and the upper surfaces of some are marked by the projection of large clasts out of the bed. Locally, massive beds and protruding clasts are buried by thin lenticular sandstones. The sandstone lenses are typically < 0.2 m thick and display horizontal lamination and cross-stratification (S_h, S_l).

The coarse-grained units have a polymodal–bimodal sorting with a matrix composed of coarse- to medium-sized sand and only a few percent of finer grained material. Matrix-free open framework breccias occluded by carbonate cement are rare. The average maximum clast size ranges from 25 to 70 cm and the largest isolated megaclast observed was 2.7 m across. Angular–subangular limestone clasts dominate in section V and polymictic angular–subrounded clasts compose the lower part of section M. Size distribution within beds is typically random with rare graded or size-segregated units. Although the majority of clasts are irregularly shaped and equant, elongated clasts occur: these are typically unoriented or show only crude horizontal imbrication.

We interpret these coarse-grained breccias from conglomerates (G_m) as representing deposition primarily from subaerial mass flows because of their general lack of stratification, their lack of obvious erosion surfaces, and their generally poor clast organization. The clast-supported texture and low percentage of fine matrix suggest emplacement by density-modified grain-flow processes (Middleton & Southard, 1977) or by cohesionless debris flows (Lowe, 1982). The few large isolated megaclasts and large clasts projecting from some beds do suggest that some flows had substantial matrix strength. The rare open-framework breccias are interpreted as sieve deposits that are common in modern alluvial fans but have a low preservation potential. The thin lenticular sandstone interbeds may have been deposited by either rill wash or sheet flow across the surface of the fan or by the waning traction current subsequent to mass flow. All of these processes, the coarse grain size and the abundance of mass flow deposits, suggest deposition in the proximal reaches of an alluvial fan system.

The distinguishing feature of the Mid-Fan Channel Association is the presence and local abundance of cross-stratified conglomerate and breccia (G_t, G_p) interbedded with clast-supported, massive–moderately well stratified, coarse clastics (G_m) (Fig. 3). In section M, some trough and more common planar–tabular crossbeds develop 15–30 cm thick sets enclosed in, and associated with, thin horizontally stratified conglomerates that display horizontal and inclined imbrication. Clasts in other horizontally stratified units show no preferred orientation. Some beds have channelized bases with up to 35 cm of relief. The cross-stratified and horizontally stratified units typically are composed of poorly sorted pebble conglomerates, but locally clasts up to 45 cm lie isolated from one another along erosive bedding surfaces. Much of the upper part of section M is composed of massive conglomerate similar to that in its lower part.

At section S, this association is composed, in part, of the gravel massive lithofacies (G_m), but is dominated by multiple sets of trough cross-stratified, clast-supported, pebble–boulder conglomerates. The trough sets are up to 1.5 m thick.

The Mid-Fan Channel Association represents deposition from both subaerial mass flow and channelized high discharge stream flow. Stream flow is responsible for the erosive channels, isolated boulder lags, cross-stratification resulting from bar migration, and imbricated, horizontally stratified conglomerates. Some of the horizontally–poorly stratified, poorly imbricated conglomerates may have resulted from hyperconcentrated flood flows (Smith, 1986) or other fluidized sediment flows.

The Mid-Fan Interchannel Association is characterized by approximately 15 m thick, massive–crudely stratified, amalgamated, pebble–boulder conglomerate alternating with finer grained, well stratified conglomerate and sandstone sequences up to 7 m thick (Fig. 3). The massive conglomerate units have attributes similar to those in the other two associations (G_m), but some grade laterally into tabular cross-stratified conglomerates with sets up to 45 cm thick. Thinner sets, up to 15 cm thick, grade laterally and vertically into, or are overlain by, finer-grained conglomerate or coarse–fine-grained sandstone. Other thin conglomerate units (< 40 cm) typically have erosive bases and flat tops and are overlain by sandstone beds, which are up to 10 cm thick, display horizontal stratification (S_h) and tabular (S_p) and ripple cross-stratification (S_r). Rare matrix-supported pebble conglomerates (G_{ms}) are interbedded with coarser, more massive, clast-supported units.

We interpret this association to represent interchannel areas where massive gravels were periodically emplaced by mass flows that escaped their channels. These events were followed by development of shallow braided streams and sheet floods that deposited thin layers of imbricated gravels and cross-stratified, fine-grained sediment.

The Distal Fan Association is comprised of thinning-upward and fining-upward packages whose lower beds consist of cross-stratified, pebble–cobble conglomerates that pass upwards into horizontally stratified and cross-stratified sandstones (Fig. 3). These packages are overlain by a 10 m thick sequence dominated by fine-grained sandstone, siltstone and mudrock with very minor amounts of granule–pebble conglomerate. The fine-grained rocks are commonly carbonate-rich and display horizontal lamination and ripple cross-lamination, small scour channels, large desiccation cracks, and wrinkle mark-like features.

The thinning- and fining-upward packages may represent either flood events across an alluvial fan or in a braided channel complex, or alternatively, the gradual abandonment of segments of a braided stream system. The overlying fine-grained sequences are unusual in alluvial fan environments and strongly suggest that they were deposited in distal reaches. The desiccation cracks signify alternating periods of wet and dry conditions and the wrinkle mark-like features indicate episodic inundation of very shallow water. No conclusive wave-induced features were observed that might signify lacustrine or marine environments, nor were any fossils found.

Marine deposits

Marine rocks are often recognized in these Lower Palaeozoic sequences by the presence of abundant trace fossils. At one level in section HR (Fig. 4), silty limestones with abundant horizontal trace fossils (signifying marine deposition) are interbedded with coarse conglomerates. The conglomerates are composed of a wide variety of well-rounded limestone clasts requiring at least a modest distance of transport prior to deposition. However, the conglomerate units carry a signature of subaqueous deposition. They typically are normally graded and commonly display full or partial Bouma sequences whose grain sizes range from pebbles to fine-grained sand and rare shales. Conglomerate units 3–7 m thick were formed by the amalgamation of abundant graded beds 10–20 cm thick and rare ungraded and inversely graded beds of about the same thickness. Interbedded fine-grained sandstone and shale successions are up to 7 m thick and consist of thin graded beds (3 cm) and complete Bouma sequences. The basal beds of section HR are comprised of 50 m of massive–crudely stratified, pebble–cobble conglomerate (G_m), which may also contain features of subaqueous mass flow deposition as described by Nemec & Steel (1984), but the types of exposures limited observations. Collectively, the sequence at section HR may represent the submarine part of a deltaic complex fed by coarse-grained fluvial system.

Lacustrine deposits

Rocks representative of lacustrine deposition form thick sequences of strikingly finer grain size than those previously discussed. These sequences, however, typically are in small fault-bounded outcrops or are strongly deformed. They are characterized by: coarse sandstones–siltstones that display abundant ripple cross-laminations, many of which are wave-formed; thin graded beds; flute casts; and rare ball-and-pillow and dewatering structures. Highly cleaved shale is also common. These rocks characteristically lack any body or trace fossils and typically are not calcareous. Although interbedded with the fine-grained sequences are thin clast-supported (and less common matrix-supported) granule–pebble conglomerates, they do not characterize the sequences. For example, at location O, in a 250 m long exposure, no conglomerates are present. Thus, these lacustrine deposits are typified by their fine grain size, lack of fauna, and abundant ripples suggesting shallow rather than deep water. Nevertheless, the thin interbedded conglomerates indicate nearby fluvial discharge.

Acknowledgements

We are grateful to Peter Braddock, Ray Waters, Gary Girty, Susan Panttaja, and Brian Pratt for their assistance. Edward Thomas drafted the figures. Rowell acknowledges the support of NSF grant DPP-8317966 and DPP-8519722. Rees acknowledges the support of NSF grant DPP-8518157.

References

Burgess, C.J. & Lammerink, W. (1979). Geology of the Shackleton Limestone (Cambrian) in the Byrd Glacier area. *New Zealand Antarctic Record*, **2(1)**, 12–16.

Debrenne, F. & Kruse, P.D. (1986). Shackleton Limestone archaeocyaths. *Alcheringa*, **10**, 235–78.

Laird, M.G. (1963). Geomorphology and stratigraphy of the Nimrod Glacier–Beaumont Bay region, southern Victoria Land, Antarctica. *New Zealand Journal of Geology and Geophysics*, **6**, 465–84.

Laird, M.G. (1981). Lower Palaeozoic rocks of Antarctica. In *Lower Palaeozoic of the Middle East, Eastern and Southern Africa, and Antarctica*, ed. C.H. Holland, pp. 257–314. Chichester, New York, Brisbane & Toronto; John Wiley and Sons.

Laird, M.G., Mansergh, G.D. & Chappell, J.M. (1971). Geology of the central Nimrod Glacier area, Antarctica. *New Zealand Journal of Geology and Geophysics*, **14(3)**, 427–68.

Lowe, D.R. (1982). Sediment gravity flows: II. Depositional models with special reference to the deposits of high-density turbidity currents. *Journal of Sedimentary Petrology*, **52(1)**, 279–97.

Miall, A.D. (1978). Lithofacies types and vertical profile models in braided river deposits: a summary. In *Fluvial Sedimentology*, ed. A.D. Miall, pp. 597–604. Calgary; Canadian Society of Petroleum Geologists, Memoir 5.

Middleton, G.V. & Southard, J.B. (1977). *Mechanics of Sediment Movement*. Binghamton, New York; Society of Economic Paleontologists and Mineralogists, Short Course No. 3, 242 pp.

Nemec, W. & Steel, R.J. (1984). Alluvial and coastal conglomerates: their significant features and some comments on gravelly mass-flow deposits. In *Sedimentology of Gravels and Conglomerates*, ed. E.H. Koster & R.J. Steel, pp. 1–31. Calgary; Canadian Society of Petroleum Geologists, Memoir 10.

Rees, M.N., Girty, G.H., Panttaja, S.D. & Braddock, P. (1988). Multiple phases of early Paleozoic deformation in the central Transantarctic Mountains. *Antarctica Journal of the United States*, 1987 Review **22(5)**, 33–5.

Rowell, A.J., Rees, M.N. & Braddock, P. (1988a). Pre-Devonian Paleozoic rocks of the central Transantarctic Mountains. *Antarctic Journal of the United States*, 1986 Review, **21(5)**, 48–50.

Rowell, A.J., Rees, M.N., Cooper, R.A. & Pratt, B.R. (1988b). Early Paleozoic history of the central Transantarctic Mountains: evidence from the Holyoake Range, Antarctica. *New Zealand Journal of Geology and Geophysics*, **31**, 397–404.

Rust, B.R. (1978). Depositional models for braided alluvium. In *Fluvial Sedimentology*, ed. A.D. Miall, pp. 605–25. Calgary; Canadian Society of Petroleum Geologists, Memoir 5.

Skinner, D.N.B. (1964). A summary of the geology of the region between Byrd and Starshot Glaciers, south Victoria Land. In *Antarctic Geology*, ed. R.J. Adie, pp. 284–92. Amsterdam; North-Holland Publishing Co.

Skinner, D.N.B. (1965). Petrographic criteria of the rock units between the Byrd and Starshot Glaciers, south Victoria Land, Antarctica. *New Zealand Journal of Geology and Geophysics*, **8**, 292–303.

Smith, G.A. (1986). Coarse-grained nonmarine volcaniclastic sediment: terminology and depositional process. *Geological Society of America Bulletin*, **97(1)**, 1–10.

Stump, E., Sheridan, M.R., Borg, S.G., Lowry, P.H. & Colbert, P.V. (1979). Geological investigations in the Scott Glacier and Byrd Glacier areas. *Antarctic Journal of the United States*, **14(5)**, 39–40.

The Devonian Pacific margin of Antarctica

M.A. BRADSHAW

Canterbury Museum, Rolleston Avenue, Christchurch, New Zealand

Abstract

Three stratigraphic patterns are recognizable in the Devonian of Antarctica: (1) Early Devonian shallow-marine basins which developed on the consolidated continental crust of East Antarctica are represented in most parts of the Transantarctic Mountains by flat-lying Devonian Beacon Supergroup sedimentary rocks. Trimming of a low-relief landscape during an Early Devonian marine transgression probably produced the Kukri Erosion Surface, upon which the clastic sequence was deposited unconformably; (2) Middle–Late Devonian magmatic arc rocks occur in Marie Byrd Land and northern Victoria Land. Both regions are characterized by late Devonian I-type granitoids which cut Cambrian–Ordovician siliceous flysch. Comparisons with Australasia suggest that the Antarctic magmatic arc lay in the same orogenic belt as New Zealand and eastern Australia; (3) the Ellsworth Mountains represent a region of strong crustal subsidence during the Early Palaeozoic, with the accumulation of a thick, apparently conformable sequence of shallow-water sedimentary rocks in a marginal basin setting. A shelly Devonian fauna from the Crashsite Quartzite suggests a connection with the Ohio–Pensacola shallow sea to the south.

Introduction

Three areas of Devonian rocks in Antarctica that have contrasting stratigraphic histories reflect different positions along the Pacific margin of Antarctica during the mid-Palaeozoic (Fig. 1). Detailed descriptions of these sequences are given by Bradshaw & Webers (1988), and only brief overviews are presented here.

Transantarctic Mountains

Throughout most of the Transantarctic Mountains, excluding northern Victoria Land, a flat-lying and relatively thin sequence of clastic sediments (the Beacon Supergroup) blankets the level or gently undulating Kukri Erosion Surface, which cuts across Ordovician granitoids and older metasedimentary rocks. A Devonian age for the lower Beacon Supergroup, or Taylor Group, is based on fossils obtained from a 1400 m thick sequence of predominantly mature quartzarenites in southern Victoria Land (Fig. 2). Upper Devonian fossil fish from the highest formation (Aztec Siltstone) provide an upper age limit (Ritchie, 1971), and poorly preserved Early Devonian palynomorphs from the second lowest formation (Terra Cotta Siltstone) give an approximate lower age (Kyle, 1977). Beyond the Dry Valley region, unfossiliferous sedimentary rocks occurring above the Kukri Erosion Surface, and below the unconformably overlying Permian glacigene sediments, are assumed to be Devonian in age.

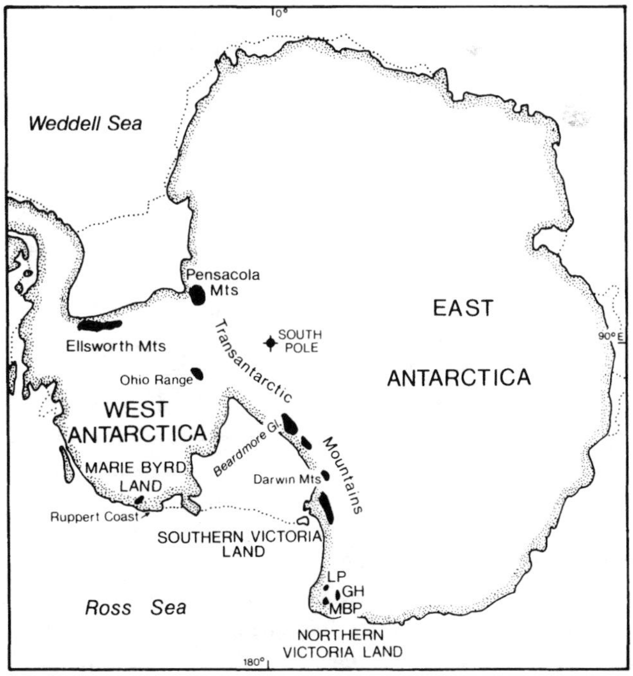

Fig. 1. Devonian sequences of Antarctica. GH, Gallipoli Heights; LP, Lawrence Peaks; MBP, Mount Black Prince.

The Aztec Siltstone Formation shows many features consistent with an alluvial plain environment (McPherson, 1979). However, below this formation in southern Victoria Land, the Taylor Group contains no body fossils and there has been

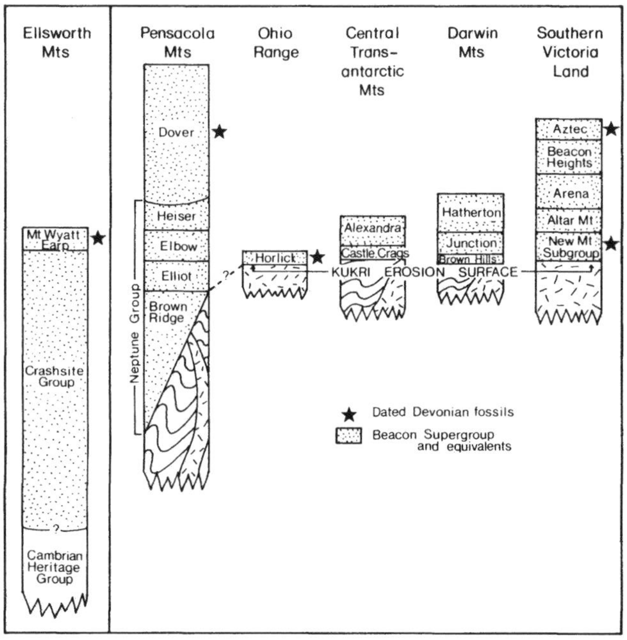

Fig. 2. Devonian, and probable Devonian sedimentary rocks of the Ellsworth and Transantarctic mountains. Unless otherwise indicated, all names are abbreviated formation names.

much discussion as to whether these, and similar rocks elsewhere in the Transantarctic Mountains, were deposited in a marine or non-marine environment (see Bradshaw, 1981). Widespread bioturbation in the lower four formations of the Taylor Group in southern Victoria Land, including several trace fossils generally associated with marine sequences (*Tigillites*, *Diplocraterion*, *Skolithos*), tends to favour a shallow-water marine origin for the sandstones, and a possible lagoonal origin for an interbedded argillaceous formation (Terra Cotta Siltstone) (Bradshaw, 1981). There is a pronounced reduction in bioturbation above the Altar Mountain Formation, with the disappearance of earlier trace fossils. This, together with the appearance of *Beaconites*, a trace fossil associated with Devonian–Carboniferous fluvio-deltaic sequences elsewhere (e.g. Allen & Williams, 1981), suggests that marine conditions were progressively replaced by coastal and non-marine environments coincident with the development of a prograding sequence (represented by the Arena Sandstone, Beacon Heights Orthoquartzite and Aztec Siltstone formations).

The only known marine sedimentary rocks are those of the Horlick Formation in the Ohio Range (0–50 m). Early Devonian shelly fossils occur in poorly sorted arkosic arenites, and are rare in the orthoquartzite lithology typical of the Taylor Group. Eight marine and one fluvial lithofacies were established by Bradshaw & McCartan (1983), and a storm-influenced inshore environment of changing coarse sand bars with finer troughs has been suggested (McCartan & Bradshaw, 1987). Invertebrate fossils in particular show strong affinities with the cold-water Malvinokaffric Realm fauna that was present during the Early Devonian in epicontinental seas which covered the then conjoined South American and South African parts of Gondwana.

Erosion associated with the Permo-Carboniferous glaciation has resulted in various thicknesses of Taylor Group rocks throughout the Transantarctic Mountains, the thickest undeformed succession being in southern Victoria Land. The Kukri Erosion Surface maintains a relatively even character from southern Victoria Land to the Pensacola Mountains, a distance of over 1600 km, although an irregular pre-Devonian karstic surface is recorded in the Churchill Mountains, south of the Darwin Mountains (Anderson, 1979).

The depth of weathering of the basement below the Kukri Erosion Surface varies from unweathered basement to weathered zones up to 9 m deep. In the Ohio Range, where marine invertebrates occur in the base of the lowest sandstone unit, the Kukri Erosion Surface clearly truncates deep weathering zones that mirrored a low-lying topography; the Horlick Formation rests on both fresh and weathered granitoid in different parts of the range (Fig. 3). At the same locality, spherical domes of unweathered granitoid (maximum diameter 7.5 m) are onlapped and buried by immature sediment at the bottom of the marine sequence. The domes appear to be cores of spheroidally weathered granitoid that were stripped of loose material by wave action as the sea encroached. The consistent nature of the Kukri Erosion Surface in the Transantarctic Mountains, below sedimentary rocks that are: (a) definitely marine; (b) suspected to be marine on the basis of trace fossils; and (c) deposited in an uncertain environment, suggests that this surface may have been progressively developed during an Early Devonian marine transgression.

A thin basal conglomerate usually overlies the erosion surface, and is followed by crossbedded quartz-arenites. One interesting exception is the Castle Crags Formation, in the central Transantarctic Mountains. Here 123 m of predominantly argillaceous sedimentary rock with minor sandstone, succeeds a 5 m thick basal conglomerate, and is overlain by more typical Taylor Group sandstones of the Alexandra Formation (Laird, Mansergh & Chappell, 1971).

In the Pensacola Mountains a thick sequence of predominantly arenaceous clastic sedimentary rocks (3100 m), consisting of the Neptune Group and Dover Sandstone Formation, overlies an erosion surface developed on folded Cambrian

Fig. 3. Pale-coloured basal sandstones of the Horlick Formation are clearly visible in the centre of this outcrop in the Ohio Range (Darling Ridge). Weathering zones in the quartz–monzonite below (9 m deep to the right and tapering to 2 m at left) have been trimmed during the Early Devonian transgression that produced the Kukri Erosion Surface.

metasedimentary and volcanic rocks which were intruded by minor granitoids during the Ordovician (Schmidt & Ford, 1969). The sequence was extensively folded during post-Permian times. In the extreme south of the area, Middle–Upper Devonian plant fossils occur in undeformed sandstone that has been correlated with the Dover Sandstone Formation of the folded sequence, suggesting that the underlying Neptune Group may contain Silurian as well as Devonian sedimentary rocks.

The Neptune Group begins with a lithic basal conglomerate (Brown Ridge Conglomerate Formation) of variable thickness (0–1000 m). The overlying cross-bedded quartzose sandstone (Elliot Sandstone Formation) has an inverse relationship to the succeeding Elbow Formation which thins northwards. The lower two formations are regarded as alluvial fan and coastal plain deposits (Williams, 1969). The Elbow Formation contains several argillaceous horizons and, where it thins southwards, sandstones become coarser and contain both primary phosphorites and reworked phosphatic pebbles (Cathcart & Schmidt, 1977). Lithologically this formation shows similarities to the inshore Horlick Formation of the Ohio Range, particularly in the presence of phosphatic pebble horizons. The Elbow Formation probably represents a marine transgression into a large fluvial or coastal basin. The establishment of fully marine conditions is also suggested by abundant bioturbation in the overlying Heiser Sandstone Formation. The Dover Sandstone Formation rests disconformably on the Neptune Group, and contains phosphatic pebble layers in its lowest beds. Cathcart & Schmidt (1977) suggested that the pebbles may have been derived by uplift and erosion of the older Elbow Formation, but they could equally well have been derived from primary phosphorites forming close to the shoreline during a hiatus in sedimentation.

In summary, the sequences throughout the Transantarctic Mountains may represent the progressive encroachment of shallow seas onto the East Antarctic continent following tectonism, magmatism, and erosion associated with the Ross Orogeny. Sedimentation following the erosive interval may have begun earlier in the Pensacola Mountains, coinciding with the unroofing of large Ordovician intrusions elsewhere. Sedimentation in the Pensacola Mountains initially may have been fluviatile, but was followed by a marine incursion southwards.

Northern Victoria Land and Marie Byrd Land

Research over the last decade in northern Victoria Land and Marie Byrd Land suggests that these two areas were once part of the same terrane (Harrington, 1987). Subaerial lavas were erupted in both regions during the Late Devonian–Early Carboniferous, onto a land surface underlain by Late Cambrian–Ordovician siliceous flysch (Robertson Bay and Swanson groups). These lavas are today represented by scattered outcrops in northern Victoria Land, and limited exposures on the Ruppert Coast of Marie Byrd Land.

In northern Victoria Land thick sequences of Late Devonian ignimbrite, rhyolite and andesite occur at Gallipoli Heights (Dow & Neall, 1974), Black Stump (Grindley & Oliver, 1983), and Mount Black Prince (Findlay & Jordan, 1984). Associated volcaniclastic rocks contain plant material at Gallipoli Heights and Mount Black Prince. The lavas may have been contemporaneous with Late Devonian I-type granitoids (Admiralty Intrusives) exposed mainly in the Admiralty Range (Vetter *et al.*, 1983; Borg *et al.*, 1986). A slightly younger sequence of volcanic rocks at Lawrence Peaks, radiometrically dated as Late Devonian–Early Carboniferous, has a calc-alkaline geochemistry that is consistent with a subduction-related origin (Weaver, Bradshaw & Laird, 1984). In Marie Byrd Land similar lavas are represented by the Ruppert Coast Metavolcanics, although reliable radiometric dates have yet to be obtained. The metavolcanic rocks succeed, and are folded with, shallow-water clastic sedimentary rocks (Wilkins Formation) containing identifiable Middle–Late Devonian plant stems (Grindley, Mildenhall & Schopf, 1980; Grindley & Mildenhall, 1981). The lavas may have been contemporaneous with I-type granitoids (Ford Granodiorite) that have yielded Late Devonian–Early Carboniferous isochron ages (Adams, 1987); the granitoids are widespread in Marie Byrd Land and geochemically compare well with the Admiralty Intrusives of northern Victoria Land.

Ellsworth Mountains

The Ellsworth Mountains region was an area of persistent subsidence for most of the Palaeozoic, during which 8000 + m of predominantly shallow-water sediment accumulated. The sequence has been extensively folded, probably in the Jurassic. In the Late Cambrian there was a change from shallow-water carbonate sedimentation (upper Heritage Group) to siliciclastic, predominantly arenaceous deposition (Crashsite Group). The change of lithology coincides with a reversal in sediment-derivation direction, from a southerly direction for the Heritage Group, to a northerly one for the Crashsite Group (Spörli, in press). Lower Devonian fossils occur in the Mount Wyatt Earp Formation, the youngest part of this clastic wedge; the lower part of the wedge may therefore be older than Devonian.

The contact between the Heritage and Crashsite groups is regarded as gradational, with Late Cambrian faunas occurring in the junction beds (Webers & Spörli, 1983). Spörli (in press) noted substantial variation in the rocks above and below the boundary as well as local erosional contacts, and the boundary may represent a hiatus of unknown duration. Crashsite Group rocks which locally rest on a karstic surface of Heritage Group marble suggest that this hiatus was significant.

The Crashsite Group comprises 3000 m of crossbedded quartzite deposited under shallow-water marine conditions (Spörli, in press). Wedges of sandstone containing lithic fragments indicate derivation from the north-west; better-sorted orthoquartzites and quartzose conglomerates suggest derivation from the north-east. Mudstone horizons of variable thickness are also present.

An Early Devonian shelly fauna has been obtained from the Mount Wyatt Earp Formation at only one locality (Fig. 2). Spörli (in press) noted that this formation contained a higher proportion of lith-arenite than earlier formations. Sediment source appears to have been a high-grade metamorphic terrane that also contained basic volcanic rocks and associated plutons. No sense of source direction is indicated, however. The fauna of the Mount Wyatt Earp Formation shows strong affinities to that of the Horlick Formation in the Ohio Range, particularly in the presence of the brachiopod *Orbiculoidea falklandensis* and the fish spine *Machaeracanthus* (Bradshaw & Webers, 1988).

Palaeogeographic summary

Generally uninterrupted shallow-water sedimentation occurred in a subsidising basin off the margin of East Antarctica during the Early–mid Palaeozoic (Ellsworth Basin). Along the edge of the continent, however, there was a pronounced break in sedimentation following tectonism and magmatism associated with the Ross Orogeny (Transantarctic Mountains). In the Early Devonian a marine transgression onto the continent locally trimmed weathered subaerial surfaces to form the Kukri Erosion Surface, on which predominantly quartzose sandstones were deposited in a large epicontinental marine basin. An Early Devonian eustatic rise in sea level is recorded in many Gondwana sequences, including those of South America, Australia, possibly New Zealand, and South Africa where there was deepening of an existing marine basin. In Antarctica sedimentation probably began earlier in the Pensacola Mountains region than elsewhere, with the Ohio Range sequence representing onlap onto the Antarctic continent.

If restored to a probable Palaeozoic position north of the Pensacola Mountains, the Ellsworth Mountains may represent a marginal basin comparable to that which occupied South Africa at the same time. The remoteness of this basin from tectonic events occurring along the continental margin allowed relatively uninterrupted sedimentation. In South Africa, as in the Ellsworth and Pensacola mountains, Devonian rocks form a relatively small part of a thick shallow-water, predominantly arenaceous, clastic marine sequence that accumulated in a subsiding basin, with northward onlap onto the African continent. Pre-Devonian crossbedded quartz-arenites are considered to be shallow-water deposits, but usually contain only trace fossils. The Early Devonian Bokkeveld Beds, which are comparable in age to the Horlick and Mount Wyatt Earp formations, are more argillaceous; they contain a similar but more varied Malvinokaffric fauna. Cyclic sedimentation in the Bokkeveld Beds has been linked to progradation of major lobate deltas into the shallow depositional basin (Theron, 1987).

The Malvinokaffric elements of both the Ellsworth Mountains and Ohio Range faunas indicate a clear marine connection with the basins that extended across the conjoined South African and South American blocks. The exact route of intercommunication is uncertain. It may not have been via the more obvious route of the Falkland Islands, whose faunas show closer affinities with the South African Bokkeveld Beds than to the Brazilian sequences of the Paraná Basin, to which the Horlick Formation faunas are particularly close. A widespread Early Devonian unconformity in many parts of South America records a transgression onto the continent and the establishment of narrow, interconnecting intracratonic basins, with marginal basins along Gondwana's western edge. There is some evidence of a compressive margin south of central Chile (Barrett & Isaacson, 1988). The critical southern part of South America was probably accreted after the Devonian and information from this area is, at present, incomplete and confusing (Barrett, 1987).

Farther east along the Gondwana margin, shallow-water quartzose marine sandstones and fluvial sediments were deposited in epicontinental basins following an Early Devonian transgression onto the Australian continent. Such sediments in the Darling, Drummond and Ardvale basins show many similarities to the Taylor Group. Those of the west Darling Basin contain thick marine sandstones with abundant trace fossils (Amphitheatre Group) overlain by fluvial sediments of the Mulga Downs Group, thus repeating a pattern of transgression–regression already suggested for the Taylor Group. Eastwards the Darling epicontinental basin passed into a marine-shelf environment represented by the Cobar Supergroup (Veevers, 1984).

The presence of the Malvinokaffric brachiopod *Pleurothyrella* in Lower Devonian beds at Reefton, New Zealand, and the Reefton rhynchonellid *Tanerhynchia* in the Ohio Range, indicates that a connection must have existed between the two basins during the early Devonian via shelf seas marginal to the Antarctic continent. Although showing some faunal affinities with south-eastern Australia, the Reefton sequence of the Buller Terrane contains at least three endemic genera, and this area must have experienced some degree of isolation from its Gondwana neighbours to allow the establishment of a distinctive faunal subprovince. The Reefton sequence is markedly different to Devonian rocks of comparable age a short distance away at Baton River in the Takaka Terrane, which faunally show strong affinities with south-eastern Australia. The Devonian Baton Formation forms the upper part of a thick lower Palaeozoic volcanic, clastic and carbonate sequence that probably formed in a comparable marginal basin to that represented by the Ellsworth Mountains succession. The Anatoki Thrust which separates the two New Zealand terranes may mark a zone of substantial post-Palaeozoic strike-slip movement.

Strike-slip movement in the New Zealand–eastern Australian part of Gondwana makes the Devonian Pacific margin difficult to reconstruct. Anomalies between adjacent Devonian sequences, like those in New Zealand, also occur in eastern Australia, and have been explained by post-Devonian strike-slip movement of up to 1500 km (Talent, 1987).

The relationship of a Late Devonian–Early Carboniferous magmatic arc (represented by sequences in northern Victoria Land and Marie Byrd Land) to the margin of Antarctica is also difficult to assess, again due to post-Devonian faulting. Siliceous flysch similar to that underlying the Antarctic magmatic arc is found in eastern Tasmania (Mathinna Beds) and in the Buller Terrane of New Zealand (Greenland Group). In addition, Late Devonian–Early Carboniferous granitoids occur in eastern Australia, in Tasmania (I-type), and New Zealand (S-type), and it is likely that these areas, together with Marie

Byrd Land and northern Victoria Land, lay within the same orogenic belt during the Devonian (Weaver, Bradshaw & Adams, this volume, p. 345).

Volcanic material in the Mount Wyatt Earp Formation may indicate the proximity of a Devonian magmatic arc to the Ellsworth Mountains. However, the detritus could have come from an older source, particularly as the Mount Wyatt Earp Formation appears to be older than the northern Victoria Land and Marie Byrd Land lavas. A Late Devonian–Early Carboniferous arc near Antarctica may have provided material for ash beds occurring in the Early Carboniferous Witteburg Formation of South Africa, the source of which is presently unknown (J.N. Theron, pers. comm.).

Acknowledgements

Financial assistance from the R.S. Allan Trust Fund in the preparation of this paper is gratefully acknowledged.

References

Adams, C.J. (1987). Geochronology of granite terranes in the Ford Ranges, Marie Byrd Land, West Antarctica. *New Zealand Journal of Geology and Geophysics*, **30(1)**, 51–72.

Allen, J.R.L. & Williams, B.P.J. (1981). *Beaconites antarcticus*: a giant channel-associated trace fossil from the lower Old Red Sandstone of South Wales and the Welsh Borders. *Geological Journal*, **16(4)**, 255–69.

Anderson, J.M. (1979). The geology of the Taylor Group, Beacon Supergroup, Byrd Glacier area, Antarctica. *New Zealand Antarctic Record*, **2(1)**, 6–11.

Barrett, S.F. & Isaacson, C.E. (1988). Devonian paleogeography of South America. In *Devonian of the World*, ed. N.J. McMillan, A.S. Embray & D.J. Glass, Vol. I, pp. 655–67. Calgary; Canadian Association of Petroleum Geologists.

Borg, S.G., Stump, E., Chappell, B.W., McCulloch, M.T., Wyborn, D., Armstrong, R.L. & Holloway, J.R. (1986). Granitoids of northern Victoria Land, Antarctica: implications of chemical and isotopic variations to regional crustal structure and tectonics. *American Journal of Science*, **287(1)**, 127–69.

Bradshaw, M.A. (1981). Paleoenvironmental interpretations and systematics of Devonian trace fossils from the Taylor Group (lower Beacon Supergroup), Antarctica. *New Zealand Journal of Geology and Geophysics*, **24(5 & 6)**, 615–52.

Bradshaw, M.A. & McCartan, L. (1983). The depositional environment of the Lower Devonian Horlick Formation, Ohio Range. In *Antarctic Earth Science*, eds. R.L. Oliver, P.R. James & J.B. Jago, pp. 238–41. Canberra; Australian Academy of Science and Cambridge; Cambridge University Press.

Bradshaw, M.A. & Webers, G.F. (1988). The Devonian rocks of Antarctica. In *Devonian of the World*, ed. N.J. McMillan, A.S. Embray & D.J. Glass, Vol. I, pp. 783–95. Calgary; Canadian Association of Petroleum Geologists.

Cathcart, J.B. & Schmist, D.L. (1977). Middle Paleozoic sedimentary phosphate in the Pensacola Mountains, Antarctica. *US Geological Survey Professional Paper*, **456-E**, 1–18.

Dow, J.A.S. & Neall, V.E. (1974). Geology of the lower Rennick Glacier, northern Victoria Land, Antarctica. *New Zealand Journal of Geology and Geophysics*, **17(3)**, 659–714.

Findlay, R.H. & Jordan, H. (1984). The volcanic rocks of Mount Black Prince and Lawrence Peaks, northern Victoria Land, Antarctica. *Geologisches Jahrbuch*, **B60**, 143–51.

Grindley, G.W. & Mildenhall, D.C. (1981). Geological background to a Devonian plant fossil discovery, Ruppert Coast, Marie Byrd Land, West Antarctica. In *Gondwana Five*, ed. M.A. Cresswell & P. Vella, pp. 23–30. Rotterdam; A.A. Balkema.

Grindley, G.W., Mildenhall, D.C. & Schopf, J.M. (1980). A mid-Late Devonian flora from the Ruppert Coast, Marie Byrd Land, West Antarctica. *Journal of the Royal Society of New Zealand*, **10(3)**, 271–85.

Grindley, G.W. & Oliver, P.J. (1983). Post-Ross Orogeny cratonisation of northern Victoria Land. In *Antarctic Earth Science*, eds. R.L. Oliver, P.R. James & J.B. Jago, pp. 133–9. Canberra; Australian Academy of Science and Cambridge; Cambridge University Press.

Harrington, H.J. (1987). Geological units common to Eastern Australia and New Zealand. *Proceedings of the Pacific Rim Congress 87*, pp. 801–4. Parkville, Victoria; Australian Institute of Mining & Metallurgy.

Kyle, R.A. (1977). Devonian palynomorphs from the basal Beacon Supergroup of south Victoria Land, Antarctica (note). *New Zealand Journal of Geology and Geophysics*, **20(6)**, 1147–50.

Laird, M.G., Mansergh, G.D. & Chappell, J.M.A. (1971). Geology of the central Nimrod Glacier area, Antarctica. *New Zealand Journal of Geology and Geophysics*, **14(3)**, 427–68.

McCartan, L. & Bradshaw, M.A. (1987). Petrology and sedimentology of the Horlick Formation (Lower Devonian), Ohio Range, Transantarctic Mountains. *US Geological Survey Bulletin*, **1780**, 31 pp.

McPherson, J.G. (1979). Calcrete (caliche) palaeosols in fluvial redbeds of the Aztec Siltstone (Upper Devonian), southern Victoria Land, Antarctica. *Sedimentary Geology*, **22(3/4)**, 267–85.

Ritchie, A. (1971). Fossil fish discoveries in Antarctica. *Australian Natural History*, **17**, 65–71.

Schmidt, D.L. & Ford, A.R. (1969). Geology of the Pensacola and Thiel Mountains. In *Geologic Maps of Antarctica*, Antarctica Map Folio Series, Folio 12, ed. V.C. Bushnell & C. Craddock, Pl. V. Washington, DC; American Geographical Society.

Spörli, K.B. (in press). Stratigraphy of the Crashsite Group, Ellsworth Mountains, West Antarctica. In *The Geology and Palaeontology of the Ellsworth Mountains, West Antarctica*, ed. G.F. Webers, C. Craddock & J.F. Splettstoesser. Washington, DC; *Geological Society of America Memoir*.

Talent, J.A. (1988). Dating diastrophism in the Devonian of eastern Australia: biostratigraphic constraints. In *Devonian of the World*, ed. N.J. McMillan, A.S. Embray & D.J. Glass, Vol. II, pp. 313–20. Calgary; Canadian Association of Petroleum Geologists.

Theron, J.N. & Loock, J.C. (1988). Devonian deltas of the Cape Supergroup, South Africa. In *Devonian of the World*, ed. N.J. McMillan, A.S. Embray & D.J. Glass, Vol. I, pp. 729–40. Calgary; Canadian Association of Petroleum Geologists.

Veevers, J.J. (1984). *Phanerozoic Earth History of Australia*. Oxford; Clarendon Press. (418 pp.)

Vetter, U., Roland, N.W., Kreuzer, H., Hohndorf, A., Lenz, H. & Besang, C. (1983). Geochemistry, petrography, and geochronology of the Cambro-Ordovidian and Devonian–Carboniferous granitoids of northern Victoria Land. In *Antarctic Earth Science*, ed. R.L. Oliver, P.R. James & J.B. Jago, pp. 140–3. Canberra; Australian Academy of Science and Cambridge; Cambridge University Press.

Weaver, S.D., Bradshaw, J.D. & Laird, M.G. (1984). Lawrence Peaks Volcanics, North Victoria Land. *New Zealand Antarctic Record*, **5(3)**, 18–22.

Webers, G.F. & Spörli, K.B. (1983). Palaeontological and stratigraphic investigations in the Ellsworth Mountains, West Antarctica. In *Antarctic Earth Science*, eds. R.L. Oliver, P.R. James & J.B. Jago, pp. 261–4. Canberra; Australian Academy of Science and Cambridge; Cambridge University Press.

Williams, P.L. (1969). Petrology of Upper Precambrian and Paleozoic sandstones in the Pensacola Mountains, Antarctica. *Journal of Sedimentary Petrology*, **39(4)**, 1455–65.

The palaeo-Pacific margin as seen from East Antarctica

J.W. COLLINSON

Byrd Polar Research Center & Department of Geology & Mineralogy, The Ohio State University, Columbus, Ohio 43210, USA

Abstract

Indirect evidence of the nature of source terranes and tectonism along the palaeo-Pacific margin of Antarctica is found in the Upper Carboniferous–Triassic sedimentary sequence of the Transantarctic and Ellsworth mountains. Four stratigraphic basins are recognized: Ellsworth Mountains, central Transantarctic Mountains, southern Victoria Land, and northern Victoria Land. Several lines of evidence suggest that these basins were components of a larger foreland basin that bordered the East Antarctic craton. The oldest volcaniclastic sediment in the foreland basin suggests that subduction and associated calc-alkaline volcanism began by Early Permian time. The earliest evidence of foreland folding and thrusting is mid-Permian, when a large prism of volcaniclastic sediment from West Antarctica reached the central Transantarctic Mountains basin.

Introduction

Upper Palaeozoic–lower Mesozoic sedimentary rocks in the Transantarctic Mountains (TM) and Ellsworth Mountains (EM) can be grouped into at least four distinct stratigraphic basins (Fig. 1): (1) Ellsworth Mountains, (2) central Transantarctic Mountains (CTM), (3) southern Victoria Land (SVL), and (4) northern Victoria Land (NVL). These basins are exposed obliquely along the trend of TM and EM. Patterns of sedimentary facies and palaeocurrent directions suggest that these were elongate, trough-shaped basins that developed parallel to the margin of the East Antarctic craton (EAC). The more cratonward basins (CTM, SVL, NVL) may have been separated by palaeohighs during the Late Carboniferous–Early Permian. An important step in reconstructing the geological history of the palaeo-Pacific margin (PPM) of Antarctica is to classify these basins into a plate tectonic framework. In West Antarctica, where little of the plate margin is exposed, interior sedimentary basins contain the most direct evidence related to the character of the history of PPM.

Until means are devised to look under the ice, any attempt to reconstruct PPM is highly speculative. The approach here will be to make the following assumptions that are both reasonable and supported by the existing, albeit sparse, evidence: (1) sedimentation and subsidence in basins along the margin of EAC were partly controlled by events along PPM; (2) the highly oblique cross-section exposed in TM and EM is generally representative of EAC margin; (3) each stratigraphic basin is a segment of TM foreland basin (Fig. 2) that developed between PPM and EAC in the following order: EM > CTM > SVL > NLV; and (4) the hypothetical TM foreland basin was similar to foreland basins that developed along PPM of eastern Australia and southern Africa.

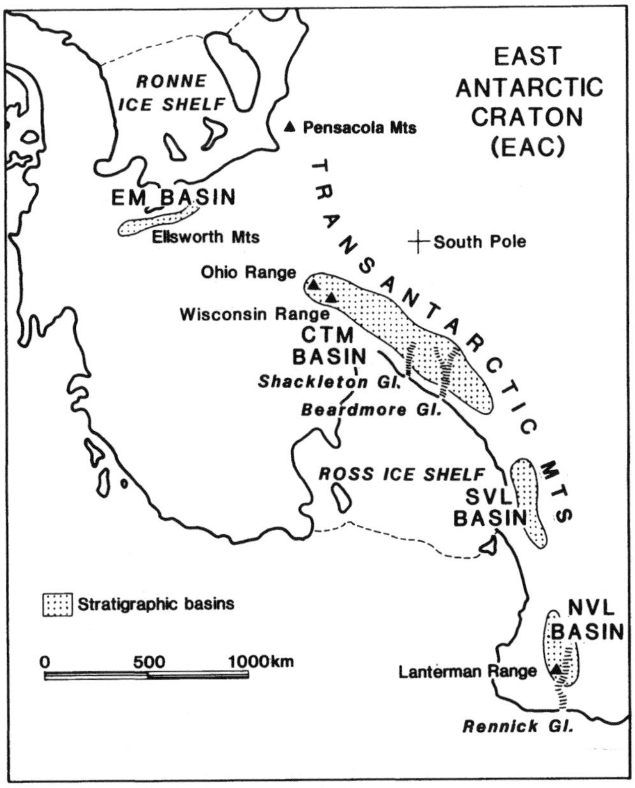

Fig. 1. Location map showing extent of stratigraphic basins: EM, Ellsworth Mountains; CTM, central Transantarctic Mountains; SVL, southern Victoria Land; NVL, northern Victoria Land.

Stratigraphic basins

The stratigraphy of these Antarctic basins has been summarized at length by Elliot (1975) and Barrett (1982). An updated summary is shown for each basin in Tables 1–4. Each basin is identified by the continuity of its stratigraphic units.

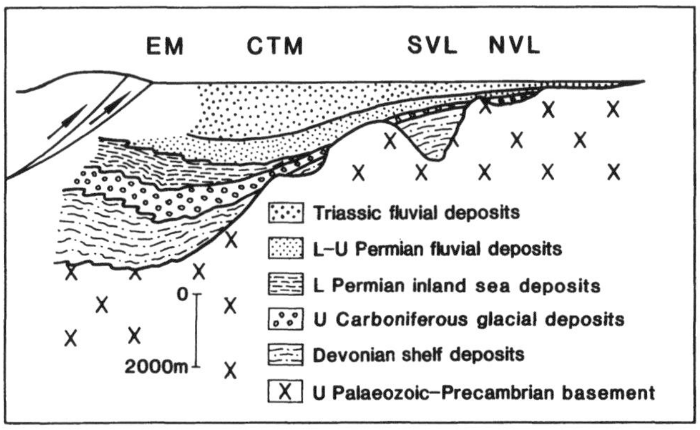

Fig. 2. Hypothetical cross-section across the Transantarctic Mountains foreland basin.

Palaeocurrent and provenance data suggest that these basins may have been isolated from each other for at least part of their history, but there is little evidence to suggest that they were separate structural basins. Strata in EM and part of the Pensacola Mountains have undergone folding. Little deformation has occurred in CTM, SVL and NVL basins. The major structural discontinuity along the front of TM, which separates East from West Antarctica, is a relatively new feature, which cuts obliquely across the strike of CTM, SVL and NLV basins and separates the similar and probably related EM and CTM basins.

The position of EM block has been the subject of much speculation (Dalziel & Elliot, 1982). The orientation of EM and its folds is approximately at 90° to the trend of folding in the Pensacola Mountains. Schopf (1969) suggested that EM block was a microplate that had been rotated and translated from a position between and in line with the Cape Fold Belt in South Africa and the Pensacola Mountains. This alignment also accounted for the stratigraphic similarities in the upper Palaeozoic of the Karoo Basin in South African and EM and CTM basins. Preliminary results from a palaeomagnetic study of Cambrian rocks in EM appeared to confirm a 90° anticlockwise rotation (Watts & Bramall, 1981). More recent studies suggest that, since the Middle Jurassic, EM has been rotated no more than 15°, with the possibility of some translation along the trend of TM since Middle Jurassic time (Grunow, Dalziel & Kent, 1987). Whether EM should be adjacent to the Pensacola Mountains or CTM is not critical to the cross-section (Fig. 2), because the relative position of EM basin would still be outermost in the foreland basin.

Evidence for TM foreland basin

Stratigraphic

Equivalent time–stratigraphic intervals thin progressively from EM basin toward NVL basin. In fact Permian and Lower Triassic strata wedge out against EAC in NVL where very thin Middle–Upper Triassic rests directly on basement (Table 4). The 1000 m thick section that Steed & Drewry (1982) project under the ice of EAC from NVL is probably Jurassic basalt. Further evidence that the Beacon Supergroup wedges out under the ice behind the Transantarctic Mountains is the abundance of basement rocks in Neogene and recent moraines everywhere in TM.

The thickness of Upper Carboniferous diamictites greatly varies in the various basins, probably because of pre-glacial topography. However, the change in facies from non-marine tillite to glaciomarine diamictite, and increase in thickness from south to north in EM basin (Table 1) (Ojakangas & Matsch, 1981), indicate a shoreline and possible hinge line within EM plate. All glacial strata in TM were deposited in terrestrial or lacustrine environments.

Lower Permian strata are represented in all basins. Post glacial black shale to deltaic deposits that occur in EM and CTM basins may have been deposited in a vast inland sea similar to the present Baltic (Elliot, 1975). Trace fossil assemblages suggest marine connections somewhere along PPM and a decreasing salinity gradient from EM to CTM basins. Collinson, Vavra & Zawiskie (in press) reported a marine trace fossil assemblage in EM; Minshew (1967) illustrated an example of *Palaeodictyon* (typically a marine trace fossil) from the Wisconsin Range, and Frisch & Miller (this volume, p. 219)

Table 1. *Stratigraphy of EM basin (from Collinson et al. (in press); Craddock et al., 1986; Ojakangas & Matsch, 1981)*

Age	Name and thickness	Lithologic description	Depositional setting and provenance
Permian	Polarstar Formation 1000 m	*Upper:* fining-upward cycles of medium sandstone, siltstone, argillite and coal. *Glossopteris* flora. *Middle:* coarsening-upward cycles of argillite–ripple-laminated, bioturbated fine sandstone–cross-bedded medium sandstone. *Lower:* laminated argillite.	*Upper:* fluvially-dominated coastal plain. *Middle:* prograding delta. *Lower:* anaerobic–dysaerobic basin. PPM source.
	Whiteout Conglomerate 1000 m	Diamictite with dropstones to north; with striated boulder pavements to south.	Glaciomarine to north and terrestial tillite to south. EAC source.
----------	unconformity ----------		
Devonian	Mt. Wyatt Earp Formation	Quartzite	Marine–fluvial. EAC source.

Table 2. *Stratigraphic summary of CTM basin (from Barrett, Elliot & Lindsay, 1986; Collinson & Elliot, 1984; William R. Hammer, pers. comm.; Kyle & Schopf, 1982)*

Age	Name and thickness	Lithologic description	Depositional setting and provenance
Jurassic	Prebble Formation	Volcanic conglomerate, tuff and tuffaceous sandstone.	Lahars and pyroclastic deposits. Local source
---------- unconformity ----------			
Upper–Middle Triassic	Falla Formation 160–530 m	*Upper:* volcanic tuffs & tuffaceous sandstone. *Lower:* fining-upward cycles of medium quartzose–volcaniclastic sandstone, siltstone, and carbonaceous shale. *Dicroidium* flora.	*Upper:* air-fall and water-worked tuffs from local source. *Lower:* braided streams on large low-gradient fan system. EAC and PPM source.
Middle–Lower Triassic	Fremouw Formation 620–750 m	*Upper:* fine–medium volcaniclastic sandstone. *Dicroidium* flora near top. *Cynognathus* fauna near base. *Middle:* mudstone and fining-upward cycles of medium–fine volcaniclastic sandstone. *Lower:* fining-upward cycles of medium quartzose–volcaniclastic sandstone–mudstone. Quartzose near Beardmore Glacier. *Lystrosaurus* fauna.	Braided streams on large high–low-gradient fan system. Trunk stream on west side of Beardmore glacier. EAC and PPM source.
---------- disconformity ----------			
Upper Permian	Buckley Formation 750 m	Fining-upward cycles of coarse sandstone–carbonaceous shale and coal. Sand/shale ratio decreases upward. Change from arkosic to volcaniclastic sandstone in middle of section. *Glossopteris* flora.	Sandy braided streams on large low-gradient fan system. Change from EAC to PPM source with palaeocurrent reversal in middle part.
Lower Permian	Fairchild Formation 130–220 m	Massive fine–medium arkosic sandstone.	Braided streams marginal–large inland sea. EAC source.
	Mackellar Formation 60–140 m	*Upper:* coarsening upward cycles of shale to ripple-laminated fine sandstone. *Lower:* shale.	Prograding delta into shallow inland sea. EAC source.
	Pagoda Formation 126–200 m	Diamictite.	Terrestrial and lacustrine glacial deposits EAC source.
---------- unconformity ----------			
Devonian	Alexandra Formation or basement	Quartzose sandstone. Brachiopod fauna in equivalent Horlick Formation in Ohio Range.	Marine–fluvial. EAC source.

suggested a non-marine setting in the Beardmore Glacier area based on sedimentological features including the trace fossil assemblage. If these assumptions are correct, the post glacial rise in sea level may have been responsible for the deposition of this unit.

The Triassic, which is preserved in the northern CTM basin and SVL and NLV basins, is similar throughout its extent. Collinson, Kemp & Eggert (1987) suggested that Triassic streams flowed into a major trunk stream that occupied the axis of TM foreland basin.

Provenance

The most compelling evidence for TM foreland basin is the bimodal provenance of quartzitic and arkosic sands from EAC, and volcaniclastic sands from a calc-alkaline source which was most likely PPM. The occurrence of volcanic tuffs in the Lower Permian in EM and throughout the Triassic in CTM indicates contemporaneous volcanism. The introduction of volcanic detritus was diachronous toward EAC. The oldest volcaniclastic sediments are in EM basin in the Lower Permian; they appear in CTM basin in the Upper Permian, and in SVL and NVL basins in the Triassic (see Tables 1–4). Major drainage was longitudinal to the axis of TM foreland basin and composition of sandstone was determined by location with respect to the axis of deposition. A very large low-gradient alluvial fan, comparable to the present Kosi Fan that drains the Himalayas (Wells & Dorr, 1987), prograded from PPM gradually shifting volcaniclastic deposition in the foreland basin cratonward (Isbell, this volume, p. 215).

Table 3. *Stratigraphy of SVL basin (from Collinson, Pennington & Kemp, 1983; McElroy & Rose, 1987; Kyle, 1977)*

Age	Name and thickness	Lithologic description	Depositional setting and provenance
Jurassic	Mawson Formation	Volcanic breccia and basaltic lava.	Local source.
	unconformity		
Triassic	Lashly Formation 325 + m	*Member D:* fining-upward cycles of coarse, quartzose sandstone–carbonaceous shale. *Dicroidium* flora. *Member C:* fining-upward cycles of medium–fine volcaniclastic sandstone, carbonaceous shale and coal. *Dicroidium* flora. *Member B*: massive, medium–fine volcaniclastic sandstone. *Member A:* fining-upward cycles of medium–fine volcaniclastic sandstone–mudstone.	*Members D and C:* meandering stream. *Members B and A:* braided stream. EAC and PPM source.
	Feather Formation 215 m	Coarse–medium quartzose sandstone–siltstone. Coarsens to pebbly sandstone to south.	Braided stream. EAC source.
	disconformity		
Lower Permian	Weller Coal Measures 222 m	Fining-upward cycles of coarse quartzose–arkosic sandstone–carbonaceous shale and coal.	Meandering stream. EAC source.
	disconformity		
	Metschel Tillite 0–85 m	Diamictite, minor sandstone, siltstone, conglomerate.	Terrestrial and lacustrine glacial deposits. EAC source.
	unconformity		
Devonian	Taylor Group	Quartzose–arkosic sandstone and siltstone.	Marine–fluvial. EAC source.

Barrett & Kohn (1975) identified a Permian drainage divide between CTM and SVL basins across which palaeocurrent data indicated dispersal in opposite directions. Drainage reversal supposedly occurred with uplift in CTM basin at the transition from Permian into Triassic, after which time all drainage flowed toward NVL. However, new palaeocurrent data in CTM indicate that the drainage reversal occurred in mid-Permian time concurrently with the introduction of a wedge of volcaniclastic sand from PPM into the basin (Table 2) (Isbell, this volume, p. 215). Uplift in CTM basin seems unlikely as a cause for drainage reversal, because no evidence of major erosion or rejuvenation exists for that time. The progradation of the large Kosi-type fan into the area may have caused the drainage reversal.

Evidence for the fan model is found in the Lower Triassic between the Shackleton and Beardmore glaciers. In Early Triassic time a large trunk stream that flowed along the axis of the foreland basin in CTM was fed quartzose sands from EAC on one side and PPM volcaniclastic sands on the other (Collinson *et al.*, 1987). Palaeocurrent vectors west of the Beardmore Glacier are generally north-westward toward SVL, but to the east, particularly around the Shackleton Glacier, they appear to shift south-westward from the direction of PPM (Collinson & Elliot, 1984). Concomitantly, Lower Triassic sandstones are increasingly volcaniclastic eastward toward the Shackleton glacier area and the PPM calc-alkaline volcanic source (Vavra, 1984).

Comparisons with basins on other Gondwana continents

The foreland basin model described here is similar to the better exposed Karoo Basin in South Africa and the Sydney Basin in eastern Australia. The folded rocks in EM are the same age as folded rocks in the Cape Fold Belt and folding in the two areas may have been contemporaneous (Elliot, 1975). The Upper Carboniferous to Triassic sedimentary record is virtually identical in the Antarctic and Karoo foreland basins (see Tankard *et al.*, 1982). It is possible that the black shale sequence in EM–CTM basins was deposited in the same inland sea as the similar sequences in the Karoo Basin and Paraná Basin of South America.

Triassic drainage patterns in the Sydney Basin were very similar to those in CTM and SVL basins; tributaries contributed volcanic detritus from PPM on one side and the Australian craton on the other to a major trunk stream that flowed down the axis of the foreland basin (Conaghan *et al.*, 1982).

Timing of events

The oldest volcaniclastic suite in the outermost preserved part of the TM foreland basin (EM) suggests that subduction and associated calc-alkaline volcanism had begun at PPM by the Early Permian. The earliest evidence of major uplift and possible foreland folding and thrusting is mid-Permian, when a large prism of volcaniclastic sediments was

Table 4. *Stratigraphy of NVL basin (from Laird & Bradshaw, 1981; Collinson, Pennington & Kemp, 1986; Elliot, Haban & Siders, 1986)*

	Name and maximum thickness				
Age	Lanterman Range	Lower Rennick Glacier	Upper Rennick Glacier	Lithologic description	Depositional setting and provenance
Jurassic			Exposure Hill Formation	Volcanic breccia, tuff and tuffaceous sandstone.	Pyroclastic deposits. Local source.
			unconformity		
Upper Triassic			Section Peak Formation 160 m	Medium arkosic and tuffaceous sandstone. *Dicroidium* flora.	Braided streams. Air fall tuff. EAC and PPM source.
			unconformity Basement		
Lower Permian		Takrouna Formation 280+ m		Coarse arkosic sandstone on margin of basin; medium sandstone and carbonaceous shale in centre of basin.	Braided streams controlled by moderate relief on basement. EAC source.
		unconformity Basement		*Glossopteris* flora.	
Upper Carboniferous	Unnamed unit 350 m			Diamictite.	Terrestrial and lacustrine glacial deposits. EAC source.
	unconformity Basement				

introduced into CTM. Continued tectonic activity on PPM is indicated by the increasing supply of volcanic detritus throughout the Triassic to CTM, SVL and NVL. Changes in depositional facies in the various parts of TM foreland basin may eventually be linked to uplift events on PPM and increased subsidence in TM foreland basin.

References

Barrett, P.J. (1982). History of the Ross Sea region during the deposition of the Beacon Supergroup 400–180 million years ago. *Journal of the Royal Society of New Zealand*, **11(4)**, 447–58.

Barrett, P.J., Elliot, D.H. & Lindsay, J.F. (1986). The Beacon Supergroup (Devonian–Triassic) and Ferrar Group (Jurassic) in the Beardmore Glacier area, Antarctica. In *Geology of the Central Transantarctic Mountains*, Antarctic Research Series, 36, ed. M.D. Turner & J.F. Splettstoesser, pp. 339–428. Washington, DC; American Geophysical Union.

Barrett, P.J. & Kohn, B.P. (1975). Changing sediment transport directions from Devonian to Triassic in the Beacon Supergroup of south Victoria Land, Antarctica. In *Gondwana Geology*, ed. K.S.W. Campbell, pp. 329–32. Canberra; Australian National University Press.

Collinson, J.W. & Elliot, D.H. (1984). Triassic stratigraphy of the Shackleton Glacier area. In *Geology of the Central Transantarctic Mountains*, Antarctic Research Series, 36, ed. M.D. Turner & J.F. Splettstoesser, pp. 103–17. Washington, DC; American Geophysical Union.

Collinson, J.W., Kemp, N.R. & Eggert, J.T. (1987). Comparison of the Triassic Gondwana sequences in the Transantarctic Mountains and Tasmania. In *Gondwana Six: Stratigraphy, Sedimentology, and Paleontology*, Geophysical Monograph Series, 41. ed. G.D. McKenzie, pp. 51–61. Washington, DC; American Geophysical Union.

Collinson, J.W., Pennington, D.C. & Kemp, N.R. (1983). Sedimentary petrology of Permian–Triassic fluvial rocks in Allan Hills, central Victoria Land. *Antarctic Journal of the United States*, **17(5)**, 15–17.

Collinson, J.W., Pennington, D.C. & Kemp, N.R. (1986). Stratigraphy and petrology of Permian and Triassic fluvial deposits in northern Victoria Land, Antarctica. In *Geological Investigations in Northern Victoria Land*, Antarctica Research Series, 46. ed. E. Stump, pp. 211–42. Washington, DC; American Geophysical Union.

Collinson, J.W., Vavra, C.L. & Zawiskie, J.M. (in press). Sedimentology of the Polarstar Formation, Permian, Ellsworth Mountains, Antarctica. In *Geology and Paleontology of the Ellsworth Mountains, Antarctica*, ed. G.F. Webers, C. Craddock & J.F. Splettstoesser, Geological Society of American Memoir.

Conaghan, P.J., Jones, J.G., McDonnell, K.L. & Royce, K. (1982). A dynamic fluvial model for the Sydney Basin. *Journal of the Geological Society of Australia*, **29**, 55–70.

Craddock, C., Webers, G.F., Rutford, R.H., Spörli, K.B. & Anderson, J.J. (1986). *Geologic Map of the Ellsworth Mountains, Antarctica.* Boulder, Colorado; Geological Society of America, Map and Chart Series MC-57.

Dalziel, I.W.D. & Elliot, D.H. (1982). West Antarctica: problem child of Gondwanaland. *Tectonics*, **1(1)**, 3–19.

Elliot, D.H. (1975). Gondwana basins in Antarctica. In *Gondwana Geology*, ed. K.S.W. Campbell, pp. 439–536. Canberra; Australian National University Press.

Elliot, D.H., Haban, M.A. & Siders, M.A. (1986). The Exposure Hill Formation, Mesa Range. In *Geological Investigations in Northern Victoria Land*, Antarctica Research Series, 46. ed. E. Stump, pp. 267–78. Washington, DC; American Geophysical Union.

Grunow, A.M., Dalziel, I.W.D. & Kent, D.V. (1987). Ellsworth–Whitmore Mountains crustal block, Western Antarctica: new paleomagnetic results and their tectonic significance. In *Gondwana Six: Structure, Tectonics, and Geophysics*, Geological Monograph Series, 40, ed. G.D. McKenzie, pp. 161–71. Washington, DC; American Geophysical Union.
Kyle, R.A. (1977). Palynostratigraphy of the Victoria Group of South Victoria Land, Antarctica. *New Zealand Journal of Geology and Geophysics*, **20(6)**, 1081–102.
Kyle, R.A. & Schopf, J.M. (1982). Permian and Triassic palynostratigraphy of the Victoria Group, Transantarctic Mountains. In *Antarctic Geoscience*, ed. C. Craddock, pp. 649–59. Madison; University of Wisconsin Press.
Laird, M.G. & Bradshaw, J.D. (1981). Permian tillites of north Victoria Land, Antarctica. In *Earth's Pre-Pleistocene Glacial Record*, ed. M.J. Hambrey & W.B. Harland, pp. 237–40. Cambridge; Cambridge University Press.
McElroy, C.T. & Rose, G. (1987). *Geology of the Beacon Heights area, southern Victoria Land, Antarctica*, 1:50000, New Zealand Geological Survey miscellaneous series map 15 (1 sheet) and notes. Wellington, New Zealand; Department of Scientific and Industrial Research.
Minshew, V.H. (1967). Geology of the Scott Glacier and Wisconsin Range areas, central Transantarctic Mountains, Antarctica. PhD thesis, Ohio State University, 268 pp. (unpublished).
Ojakangas, R.W. & Matsch, C.L. (1981). The Lower Palaeozoic Whiteout Conglomerate: a glacial and glaciomarine sequence in the Ellsworth Mountains, West Antarctica. In *Earth's Pre-Pleistocene Glacial Record*, ed. M.J. Hambrey & W.B. Harland, pp. 241–4. Cambridge; Cambridge University Press.
Schopf, J.M. (1969). Ellsworth Mountains: position in West Antarctica due to sea-floor spreading. *Science*, **164**, 63–6.
Steed, R.H.N. & Drewry, D.J. (1982). Radio-echo sounding investigations of Wilkes Land, Antarctica. In *Antarctic Geoscience*, ed. C. Craddock, pp. 969–83. Madison; University of Wisconsin Press.
Tankard, A.J., Jackson, M.P.A., Eriksson, K.A., Hobday, D.K., Hunter, D.R. & Minter, W.E.L. (1982). *Crustal Evolution of Southern Africa: 3.8 Billion Years of Earth History*. New York, Heidelberg, Berlin; Springer-Verlag. 523 pp.
Vavra, C.L. (1984). *Triassic Fremouw and Falla Formations, Central Transantarctic Mountains, Antarctica*. Columbus; Institute of Polar Studies Report 87, Ohio State University. 98 pp.
Watts, D.R. & Bramall, A.M. (1981). Palaeomagnetic evidence for a displaced terrain in Western Antarctica. *Nature, London*, **293**, 639–41.
Wells, N.A. & Dorr, J.A. Jr. (1987). Shifting of the Kosi River, northern India. *Geology*, **15(3)**, 204–7.

Permo-Carboniferous glacial sedimentation in the central Transantarctic Mountains and its palaeotectonic implications (Extended abstract)

J.M.G. MILLER & B.J. WAUGH

Department of Geology, Vanderbilt University, Nashville, TN 37235, USA

The Pagoda Formation records terrestrial glacial deposition in the Beardmore Glacier area of the Transantarctic Mountains during Permo-Carboniferous time. The formation is part of an elongate outcrop belt of coeval glaciogenic rocks which trends subparallel to the inferred palaeo-Pacific margin of Antarctica (Fig. 1). This paper outlines sedimentological data from the Pagoda Formation which are relevant to the palaeotectonic setting of the central Transantarctic Mountains during Permo-Carboniferous time. In particular it pertains to the proximity of this region to the palaeo-Pacific margin of the East Antarctic sector of Gondwana. Glaciogenic beds form the base of the Victoria Group (see Fig. 1 in Frisch & Miller, this volume, p. 219). In the central Transantarctic Mountains, Permo-Triassic units of the upper Victoria Group were deposited on the cratonic side of a back-arc or foreland basin which lay behind a calc-alkaline volcanic arc (Collinson, this volume, p. 199).

The thickest Permo-Carboniferous glaciogenic rocks in Antarctica occur in the Pensacola and northern Ellsworth mountains (Fig. 1). In the northern Ellsworth Mountains the facies are glaciomarine, deposited on a subsiding continental margin (Ojakangas & Matsch, 1981). Elsewhere the facies are dominantly terrestrial (Frakes, Matthews & Crowell, 1971; Nelson, 1981). Between the Ohio Range and northern Victoria Land, thicknesses of the glacial sequence range from 0 to 375 m (Fig. 1). Within this region the Pagoda Formation is unusually thick (100–375 m, Figs 1 and 2), particularly considering that it records terrestrial deposition and preservation of several erosional and depositional episodes (glacial advances and retreats; see below). The thickness may be due to either greater subsidence in the Beardmore Glacier area as compared to neighbouring areas or infilling of an irregular palaeotopography.

Palaeo-ice flow and palaeocurrent transport in the Beardmore Glacier area were dominantly towards the south-east (Fig. 1; Lindsay, 1968, 1970; Miller & Waugh, 1988). Regionally in the Transantarctic Mountains, the palaeo-ice flow directions approximately parallel the trend of the outcrop belt with some divergence in the vicinity of the Queen Maud Mountains (Fig. 1; Coates, 1985). Ice centres may have existed in southern Victoria Land (Lindsay, 1970), close to the southern end of the present Ross Ice Shelf (Coates, 1985), possibly between the Pensacola Mountains and Ohio Range, and in the Weddell Sea region (Frakes *et al.*, 1971; Fig. 1). The distribution of ice centres and the directions of ice movement indicate that the Permo-Carboniferous continental margin did not lie close to the outcrop belt between the Pensacola Mountains and northern Victoria Land.

Lithofacies within the Pagoda Formation include various types of diamictite, sandstone and shale (Table 1). Massive diamictite, with a sandy or silty matrix, predominates (Fig. 2). Within diamictite units, deformed sandstone inclusions, striated stones, striated surfaces and boulder beds are common. The diamictites represent lodgement, melt-out and redeposited tills (Lindsay, 1968, 1970; Miller, 1989). Sandstones range from coarse-grained and pebbly to fine-grained. They form both tabular and lenticular units and may be massive or show cross-bedding, ripple marks or parallel laminae. They record englacial meltwater deposits and proglacial fluvial and glaciolacustrine outwash. Shales are commonly massive, but parallel laminae are locally present. Soft-sediment deformation of sandstone beds and lenses within shales is common. Shales are interpreted as glaciolacustrine deposits; they are not present in all Pagoda Formation sections (Lindsay, 1968, 1970; Coates, 1972; Miller, 1989).

The facies association of the Pagoda Formation is interpreted to record deposition in a terrestrial region close to the terminus of a temperate glacier (Lindsay, 1970; Miller, 1989). A very high C/S ratio in two Pagoda shale samples indicates a non-marine depositional environment (Berner, R.A., written communication to M.F. Miller, 1986). The absence of marine body or trace fossils and local abundance of plant debris in the formation, the formation's erosional base and its gradational top contact with the non-marine Mackellar Formation (Miller & Frisch, 1988) all corroborate a terrestrial depositional setting. Thus, there is no sedimentological evidence which suggests proximity to the continental margin.

Within the formation episodes of glacial advance and retreat can be recognized, principally through analysis of facies sequences. Massive diamictite overlain by stratified or sheared diamictite, capped by sandstone or shale, constitutes the simplest glacial retreat sequence (Fig. 2; Miller, 1989). On the basis of the preserved sedimentary record, major advance–retreat cycles can be distinguished from more minor pauses in sedimentation, fluctuations of the ice margin, or changes in basal-ice dynamics, which are represented by individual sedimentary features (e.g. sand-filled wedges, striated surfaces, boulder beds) within diamictite units. Advance and retreat sequences are from 5 to 50 m thick. Three or four major glacial cycles are commonly present, and up to six may be visible in thicker, northern Pagoda Formation sections (Fig. 2).

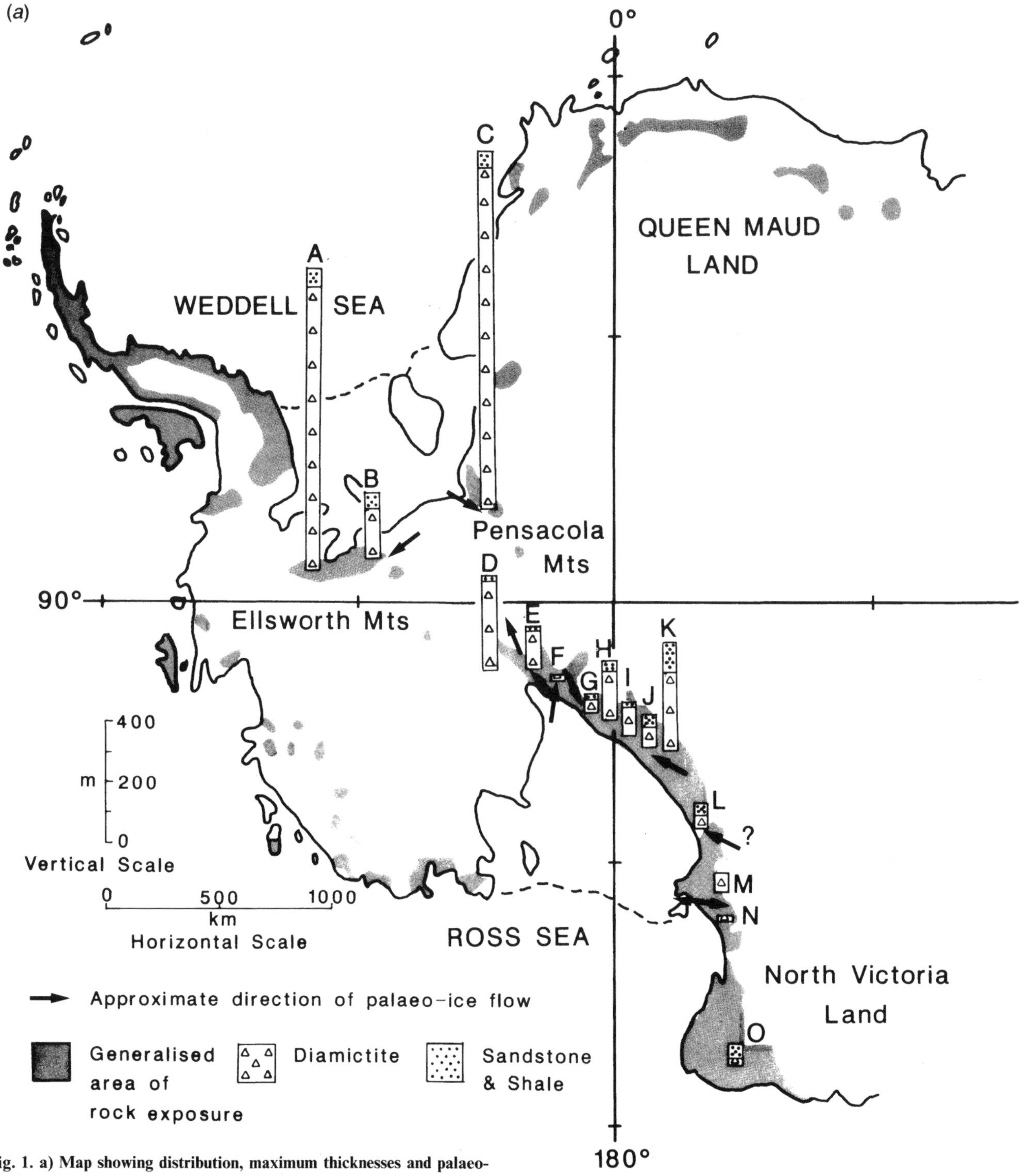

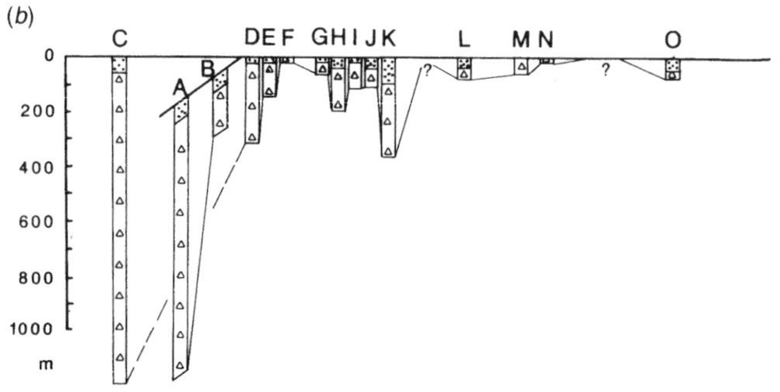

Fig. 1. a) Map showing distribution, maximum thicknesses and palaeo-ice flow directions for Permo-Carboniferous glaciogenic rocks in Antarctica. A: Sentinel Range (Ojakangas & Matsch, 1981); B: Meyer Hills (Ojakangas & Matsch, 1981); C: Gale Ridge (Nelson, 1981; Frakes *et al.*, 1966); D: Discovery Ridge, Ohio Range (Long, 1964; Bradshaw, Newman & Aitchison, 1984); E: Wisconsin Range (Minshew, 1967); F: Watson Escarpment (Coates, 1985); G: Shackleton Glacier (La Prade, 1972); H: Mt. Butters (Coates, 1985); I: Mt. Hermanson (this paper); J: Mt. Elizabeth (this paper); K: Mt. Counts (Lindsay, 1968); L: Darwin Mountains (Barrett & Kyle, 1975); M: South Victoria Land (Barrett & Kyle, 1975); N: Dry Valleys (Pinet, Matz & Hayes, 1971); O: North Victoria Land (Laird & Bradshaw, 1981).

b) Profile showing distribution and maximum thicknesses of glaciogenic rocks from map (a). Stratigraphical datum is top of Pagoda Formation. Note that definite palinspastic position of Ellsworth Mountains block, columns A and B, is unknown.

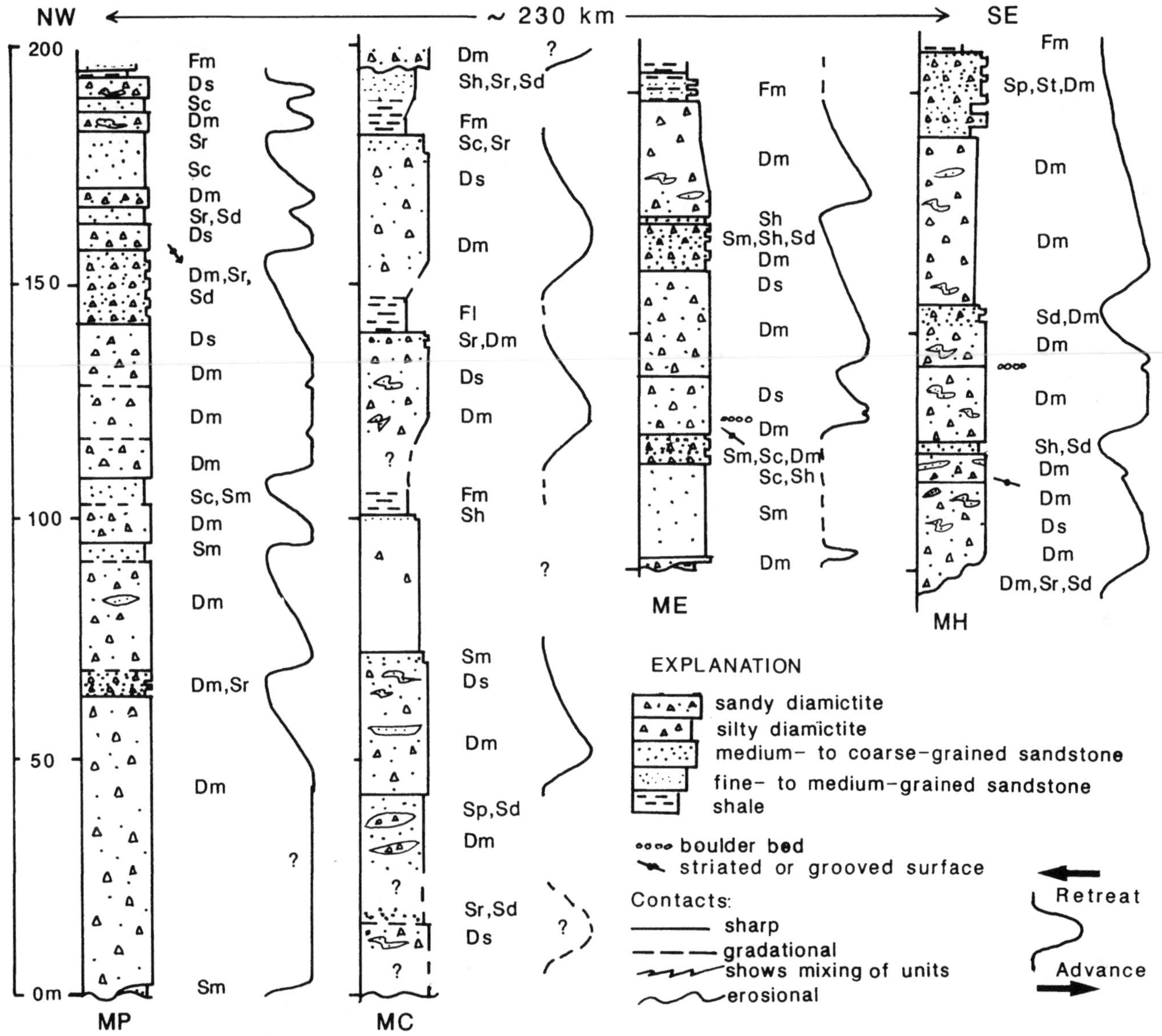

Fig. 2. Selected stratigraphical columns through the Pagoda Formation in the Beardmore Glacier area showing interpretation of glacial advance and retreat. MP = Markham Plateau; MC = Mt. Counts (north face); ME = Mt. Elizabeth; MH = Mt. Hermanson. Geographical distance between MP and MH is about 230 km. For more details see Miller (1989).

Table 1. *Lithofacies code used in Fig. 2*

Diamictite	Massive	Dm
	Stratified	Ds
	Sheared	Dms
Sandstone	Massive	Sm
	Planar cross-bedded	Sp
	Trough cross-bedded	St
	Cross-bedded, cross-bed type not specified	Sc
	Horizontally-laminated	Sh
	Ripple cross-laminated	Sr
	Normally-graded	Sg
	With soft-sediment deformation	Sd
Shale	Massive	Fm
	Parallel laminated	Fl
	With soft-sediment deformation	Fd
	With dropstones	F d

Rare occurrence of structure denoted by parentheses.

Clasts in Pagoda diamictites are dominantly granitic rocks and quartzite. Other high- and low-grade metamorphic rocks are also common and some sedimentary rocks occur (see also Lindsay, 1968). Petrographic data for Pagoda sandstones and selected diamictites are shown in Fig. 3. Together the clast and petrographic data show that Pagoda sediments were derived from a stable continental block region with some contribution from a quartz-rich recycled orogen (Dickinson *et al.*, 1983). Precambrian and Palaeozoic rocks of the East Antarctic craton provided most of the debris.

In summary, the terrestrial glacial facies assemblage, palaeo-ice flow directions and provenance data for the Pagoda Formation indicate deposition in a stable cratonic region distant from the palaeo-Pacific margin of East Antarctica. There is no evidence for an active continental margin in the vicinity of the present central Transantarctic Mountains during Permo-Carboniferous time. These data therefore lend support to reconstructions of the Pacific margin of Gondwana which

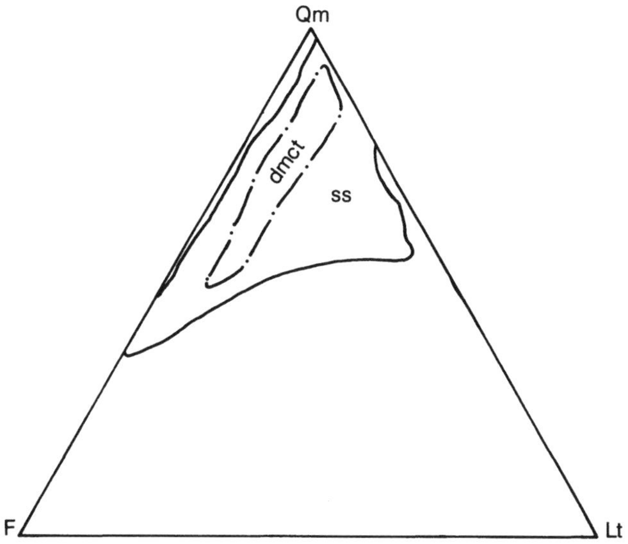

Fig. 3. Triangular diagram showing petrographic composition of Pagoda Formation sandstones (ss) and diamictites (dmct). Qm = monocrystalline quartz; F = feldspar; Lt = lithic fragments, including polycrystalline quartz.

place microcontinental blocks, now in West Antarctica and New Zealand, adjacent to the present continental outline of East Antarctica (Dalziel & Elliot, 1982).

Acknowledgements

This research was funded by National Science Foundation grant DPP-8418445. We thank J.A. Crame and D.I.M. Macdonald for suggestions which improved the manuscript.

References

Barrett, P.J. & Kyle, R.A. (1975). The early Permian glacial beds of South Victoria Land and the Darwin Mountains, Antarctica. In *Gondwana Geology*, ed. K.S.W. Campbell, pp. 333–46. Canberra; Australia National University Press.

Bradshaw, M.A., Newman, J. & Aitchison, J.C. (1984). Preliminary geological results of the 1983–84 Ohio Range expedition. *New Zealand Antarctic Record*, **5(3)**, 1–17.

Coates, D.A. (1972). Pagoda Formation: evidence of Permian glaciation in the central Transantarctic Mountains. In *Antarctic Geology and Geophysics*, ed. R.J. Adie, pp. 359–64. Oslo; Universitetsforlaget.

Coates, D.A. (1985). Late Paleozoic glacial patterns in the central Transantarctic Mountains, Antarctica. In *Geology of the Central Transantarctic Mountains*, Antarctic Research Series, vol. 36, ed. M.D. Turner & J.F. Splettstoesser, pp. 275–324. Washington, D.C.; American Geophysical Union.

Dalziel, I.W.D. & Elliot, D.H. (1982). West Antarctica: problem child of Gondwanaland. *Tectonics*, **1(1)**, 3–19.

Dickinson, W.R., Beard, L.S., Brakenridge, G.R., Ergavec, J.C., Ferguson, R.C., Inman, K.F., Knapp, R.A., Lindberg, F.A. & Ryberg, P.T. (1983). Provenance of North American Phanerozoic sandstones in relation to tectonic setting. *Geological Society of American Bulletin*, **94(2)**, 222–35.

Frakes, L.A., Matthews, J.L. & Crowell, J.C. (1971). Late Paleozoic Glaciation: Part III, Antarctica. *Geological Society of America Bulletin*, **82(6)**, 1581–4.

Frakes, L.A., Matthew, J.C., Neder, I.R. & Crowell, J.C. (1966). Movement directions in Late Paleozoic glacial rocks of the Horlick and Pensacola Mountains, Antarctica. *Science*, **153(3737)**, 746–9.

La Prade, K.E. (1972). Permian–Triassic supergroup of the Shackleton Glacier area, Queen Maud Range, Transantarctic Mountains. In *Antarctic Geology and Geophysics*, ed. R.J. Adie, pp. 373–8. Oslo; Universitetsforlaget.

Laird, M.G. & Bradshaw, J.D. (1981). Permian tillites of north Victoria Land, Antarctica. In *Earth's Pre-Pleistocene Glacial Record*, ed. M.J. Hambrey & W.B. Harland, pp. 237–40. Cambridge; Cambridge University Press.

Lindsay, J.F. (1968). Stratigraphy and sedimentation of the lower Beacon rocks of the Queen Alexandra, Queen Elizabeth, and Holland Ranges, Antarctica, with emphasis on Paleozoic glaciation. PhD thesis, Ohio State University, 300 pp. (unpublished).

Lindsay, J.F. (1970). Depositional environment of Paleozoic glacial rocks in the central Transantarctic Mountains. *Geological Society of America Bulletin*, **83(4)**, 1149–72.

Long, W.E. (1964). The stratigraphy of the Ohio Range, Antarctica. PhD thesis, Ohio State University, 340 pp. (unpublished).

Miller, J.M.G. (1989). Glacial advance and retreat sequences in a Permo-Carboniferous section, central Transantarctic Mountains. *Sedimentology*, **36**, 419–30.

Miller, J.M.G. & Waugh, B.J. (1988). Paleotectonic implications of the Permo-Carboniferous Pagoda Formation, Beardmore Glacier area. *Antarctic Journal of the United States*, 1987 Review, **23(5)**, 19–21.

Miller, M.F. & Frisch, R.S. (1988). Early Permian paleogeography and tectonics of the central Transantarctic Mountains: inferences from the Mackellar Formation. *Antarctic Journal of the United States*, 1987 Review, **23(5)**, 24–5.

Minshew, V.H. (1967). Geology of the Scott Glacier and Wisconsin Range areas, central Transantarctic Mountains. PhD thesis, Ohio State University, 268 pp. (unpublished).

Nelson, W.H. (1981). The Gale Mudstone: a Permian (?) tillite in the Pensacola Mountains and neighbouring parts of the Transantarctic Mountains. In *Earth's Pre-Pleistocene Glacial Record*, ed. M.J. Hambrey & W.B. Harland, pp. 227–9. Cambridge; Cambridge University Press.

Ojakangas, R.W. & Matsch, C.L. (1981). The Late Palaeozoic Whiteout Conglomerate: a glacial and glaciomarine sequence in the Ellsworth Mountains, West Antarctica. In *Earth's Pre-Pleistocene Glacial Record*, ed. M.J. Hambrey & W.B. Harland, pp. 241–7. Cambridge; Cambridge University Press.

Pinet, P.R., Matz, D.B. & Hayes, M.O. (1971). An Upper Paleozoic tillite in the Dry Valleys, south Victoria Land, Antarctica. *Journal of Sedimentary Petrology*, **41(3)**, 835–83.

Clay mineralogy and provenance of fine-grained Permian clastics, central Transantarctic Mountains

L.A. KRISSEK & T.C. HORNER

Byrd Polar Research Center and Department of Geology and Mineralogy, The Ohio State University, Columbus, Ohio 43210, USA

Abstract

On the basis of field analyses, the (?)Upper Carboniferous–Permian sequence of the central Transantarctic Mountains has been interpreted to record a transition from glacial to subaqueous to fluvial environments, with associated climatic moderation. In this study, the palaeoclimatic record of this sequence is evaluated using clay mineral and bulk geochemical data from its fine-grained facies, and a similar palaeoenvironmental transition is recognized. Because the clay mineral composition has been subjected to regional burial diagenesis and contact thermal metamorphism, least altered samples were selected using criteria based on clay mineral presence/absence and relative abundance, and on illite order/disorder parameters. The regional compositional patterns of least-altered samples indicate that provenance evolved as follows: 1) Pagoda Formation sediments throughout the study area were derived from sources dominated by physical weathering, 2) Mackellar Formation sediments in the northern portion of the study area were derived from chemically-weathered sources, while the southern portion of the Mackellar was dominated by physically-weathered detritus, 3) Fairchild Formation sources provided a mixture of physically- and chemically-weathered material, but the exact nature and geographic position of the sources is poorly constrained by these data, and 4) Buckley Formation sediments throughout the study area were derived from sources dominated by chemical weathering, and the clay minerals may also record the onset of volcanic input to the region. Limited geochemical data support this pattern of provenance evolution in space and time. The results of this study indicate that careful consideration of the mineral and chemical composition of fine-grained sediments can provide valuable provenance and palaeoclimatic data to a basin analysis study.

Introduction

In modern marine sediments, the global distribution of clay minerals has been related to the lithologies and the weathering regimes in the continental source areas (Griffin, Windom & Goldberg, 1968), and numerous other studies have demonstrated that similar relationships exist on local and regional scales. De Segonzac (1970), Keller (1970), and Dixon & Weed (1977), among others, have discussed the source lithology and weathering regime controls on clay mineral formation. As a result of these studies, provenance interpretations based on clay mineral occurrences are common in the geological literature, and can be summarized as follows:

(1) Smectites are developed by continental or submarine alteration of basic igneous rocks
(2) Chlorites are developed by physical weathering of basic source rocks, and are especially common in glacial environments
(3) Illites are developed by moderate weathering of acidic source rocks in temperate climates
(4) Kaolinites are developed by intense chemical weathering in warm, humid climates.

In older sedimentary sequences, the occurrences of these clay minerals can be modified by diagenesis, so that provenance interpretations are not necessarily straightforward. A diagenetic sequence for clay-rich sediments undergoing burial and increased temperatures was described in detail by Hower *et al.* (1976), and has subsequently been widely applied in other studies. This sequence is characterized by the conversion of smectite to successively better ordered forms of illite in the temperature range of 70–190 °C. The degree of illite ordering, which indicates either the abundance of physically-weathered detrital illite or the extent of thermal diagenesis (Weaver, 1960; de Segonzac, 1970; Guthrie, Houseknecht & Johns, 1986), can be determined by X-ray diffractometry. The abundance of various illite polytypes carries similar information, with the

Table 1. *Upper Carboniferous(?)–Permian stratigraphy, central Transantarctic Mountains*

Formation	Age	Thickness	Lithologies	Environment
Buckley	Permian	< 750 m	Sandstones, coals, mudstones	Fluvial (braided and anastomosing)
Fairchild	Permian	120–220 m	Sandstones	Fluvial (braided)
Mackellar	Permian	45–140 m	Sandstones, siltstones, black shales	Delta(?) and 'deep' basinal shales
Pagoda	Permian–Late Carboniferous(?)	126–395 m	Diamictites, sandstones, shales	Glacial and glaciofluvial

disordered 1Md illite polytype dominating most sedimentary clays, while the ordered 2M illite polytype dominates physically-weathered detrital illites and illites that have experienced high temperature diagenesis (Maxwell & Hower, 1967; de Segonzac, 1970; Velde, 1983). Other workers have demonstrated that a smectite to chlorite transition can be the dominant reaction (Potter, Maynard & Pryor, 1980), but smectite to illite and smectite to chlorite transitions do not occur simultaneously in the same stratigraphic horizon. The composition of the source material determines which transition will occur. Kaolinite is rarely developed by diagenesis in clay-rich deposits, and does not form under the conditions that favour illite or chlorite formation.

Provenance and palaeoclimatic information can be extracted, therefore, by carefully determining the composition of a clay-rich sequence. Because the Permian sequence in the central Transantarctic Mountains (Table 1) records the transition from a glacial regime (Pagoda Formation), through subaqueous clastic (deltaic) deposits (Mackellar Formation), to fluvial sequences (Fairchild Formation) with coals (Buckley Formation), such an examination promises to provide valuable insight into the timing and nature of this palaeoenvironmental change. To provide such insight, this study has two objectives: 1) to identify samples that have experienced minimal amounts of thermal alteration and, therefore, provide the least altered provenance signal, and 2) to determine and interpret the stratigraphic and geographic record of source lithology and palaeoclimatic change as recorded by least-altered Permian samples from the central Transantarctic Mountains.

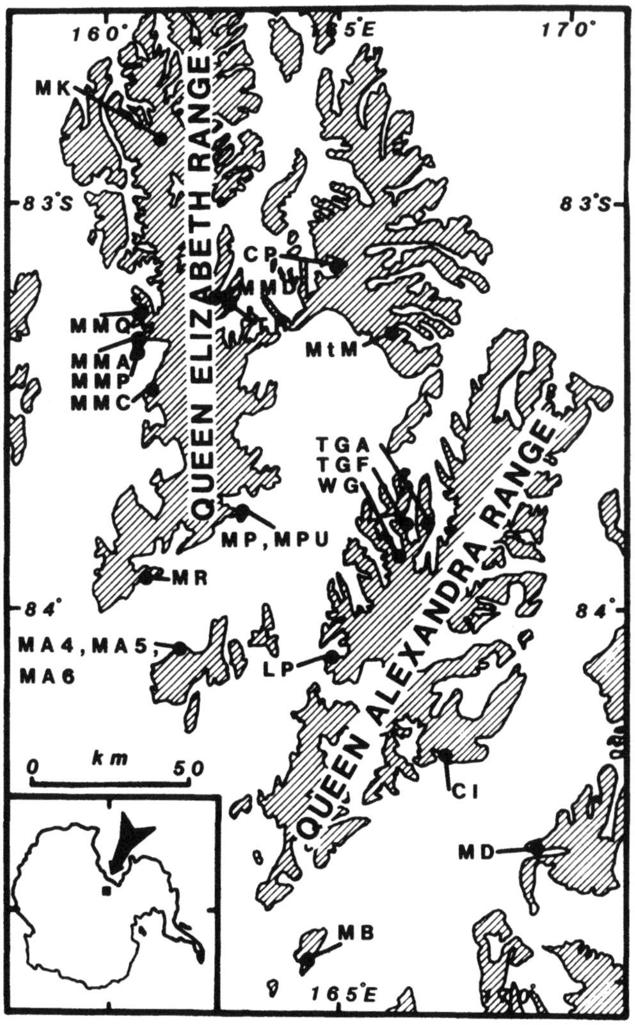

Fig. 1. Geographical distribution of measured sections, Beardmore Glacier area, central Transantarctic Mountains.

Samples and analytical techniques

During the austral summer of 1985–86, approximately 310 samples of fine-grained sediments were collected from 24 measured sections in the Permian sequence of the central Transantarctic Mountains (Fig. 1; Krissek & Horner, 1988). A representative subset of 53 samples was examined for this study. A split of each bulk sample was powdered in a ball mill, mounted as a pressed powder, and analyzed by X-ray diffractometry on a Philips XRD system. Each sample was scanned from 2–60° 2θ with Cu K-alpha radiation. The resulting strip chart records were then analyzed for clay mineral presence (using basal diffraction peaks for identification; Carroll, 1970), illite crystallinity index (Weaver, 1960; Guthrie *et al.*, 1986; ratio of signal intensity at 10 Å to intensity at 10.5 Å), illite polytype parameter (Maxwell & Hower, 1967; ratio of signal intensity at 2.80 Å, due to the 2M illite polytype, to signal intensity at 2.58 Å, due to all illite polytypes), and relative abundance of the clay phases present (based on basal diffraction peak areas). Organic carbon contents of most of the samples were measured using a Coulometrics carbon analyzer. Vitrinite reflectance values for approximately 15 samples and major/minor/trace element abundances of 10 samples were measured by commercial laboratories.

Identification of 'least-altered' sample

Because of the complications imposed by diagenesis, significant effort was expended to identify those samples that have experienced minimal amounts of post-depositional alteration. Such samples have been recognized on the basis of five criteria, which are justified on the basis of the studies discussed previously. These five criteria are listed below in decreasing order of significance:

(1) Presence of kaolinite. Detrital kaolinite is stable to temperatures between 200 and 550 °C (Carroll, 1970; de Segonzac, 1970), so its presence indicates relatively low levels of thermal alteration
(2) Low illite crystallinity index, which indicates that the less-ordered (i.e. low temperature) illites are relatively abundant (Weaver, 1960; Guthrie *et al.*, 1986)
(3) Low illite polytype parameter, which indicates that the less-ordered (i.e. low temperature) 1Md illite polytype is relatively abundant (Maxwell & Hower, 1967).
(4) Low vitrinite reflectance value, which indicates that the organic components have experienced relatively limited heating (Dow, 1977; Guthrie *et al.*, 1986)
(5) High organic carbon content. Although the organic carbon content of a sediment is affected by several controls, the presence of abundant organic carbon supports an interpretation that thermal alteration has been limited. Low organic carbon values, however, do not necessarily imply significant thermal alteration.

While each criterion is individually subject to a range of controls and interpretations, samples with favourable characteristics for several criteria are strongly indicated as least-altered samples. It should be noted, however, that the input of detrital well-ordered illites, dominated by the 2M polytype, can produce relatively high illite crystallinity index and illite polytype parameter values for samples that experienced little thermal alteration. Under such input conditions, the unaltered nature of the sediment must be interpreted from the other criteria provided.

Data

Of the 53 samples examined in this study, 19 meet two or more of the criteria for least-altered samples outlined above. The data for these samples are listed in Table 2, where the samples are grouped stratigraphically and geographically (arranged from north to south for each formation). The stratigraphic groupings in Table 2 are tentative, as the level listed for each sample is only determined relative to the base of its stratigraphic section. Correlations between sections are unavailable at this time. In general, the samples that are best constrained as minimally altered contain kaolinite, have an illite crystallinity index of ≤ 15 (and most are < 5), have an illite polytype parameter of < 0.3, have a vitrinite reflectance value of < 3, and contain $> 1\%$ organic carbon.

Provenance interpretations

Four samples from the Pagoda Formation appear to meet the criteria for minimal alteration, especially as defined by the illite crystallinity index. These samples have low organic carbon contents, which probably reflect the low productivity and high clastic dilution rates that characterize glacial and pro-glacial environments. The range in illite polytype parameter values suggests that variable amounts of detrital 1Md and 2M illite were mixed during Pagoda deposition; the detrital 2M illite input probably reflects the dominance of physical weathering processes under glacial conditions, while the detrital 1Md illites may have been eroded from pre-existing soils or sedimentary rocks. Petrographic data from Pagoda sandstones (B. Waugh, pers. comm., 1987) show that illite/muscovite is the predominant authigenic phyllosilicate, so that the chlorite present can be considered detrital. As a result, the high relative chlorite/illite abundance ratios also indicate the importance of physical weathering processes in the Pagoda Formation source areas.

Four samples from the Mackellar Formation meet the criteria for minimal alteration, but these samples exhibit a marked geographical variability in composition. The two northern samples, from sections TGA and MMC, have low illite crystallinity index values and contain very little 2M illite. The sample from TGA contains kaolinite, while the sample from MMC has a relative chlorite/illite abundance ratio approximately equal to that of the Pagoda Formation samples. The two southern samples, both from section MB, lack kaolinite, have moderate to high illite crystallinity index values, contain common to abundant 2M illite, and have high relative chlorite/illite abundance ratios. Petrographic data from Mackellar sandstones (Frisch & Miller, this volume, p. 219) indicate that authigenic chlorite is as abundant as detrital chlorite in some samples, suggesting that the smectite–chlorite conversion dominated clay mineral diagenesis in the Mackellar Formation. If this hypothesis is correct, then the relative chlorite/illite abundance ratios in the Mackellar Formation decrease significantly from those in the Pagoda. The criteria examined here suggest that a northern source to the Mackellar Formation experienced relatively humid weathering conditions during this time, thereby providing kaolinites, less-ordered illite forms, and a predominance of illite over chlorite to the basin. A southern or western source to the Mackellar Formation, however, was still dominated by physical weathering, as recorded by detrital 2M illites and abundant chlorite at section MB.

Four samples from the Fairchild Formation have been identified as minimally-altered; although none of these samples contain kaolinite, their characteristics do suggest that mixing of physically- and chemically-weathered clastics occurred throughout the basin. All four samples have low to moderate illite crystallinity index values, indicating that these samples experienced limited post-depositional thermal alteration. The range in illite polytype parameter values and chlorite/illite relative abundance ratios, therefore, is best explained by mixing variable amounts of physically-weathered 2M illite and chlorite from one source with chemically-weathered 1Md illites derived from another source. The locations of such sources are not well constrained by these four samples, but future study of other Fairchild Formation samples should provide better geographical control, especially if kaolinite is identified in Fairchild Formation samples.

Three samples from the lower halves of the Buckley For-

Table 2. *Compositional data for least-altered Permian samples from the central Transantarctic Mountains*

Formation	Section	Level	% C	ICI	IPP	% R_0	RMA
'Upper' Buckley	MPU	223 m	9.20	2.0	0.11		I > C
	WG	#10	1.10	1.0	0.29		I ≫ C
	WG	#15	1.10	2.0	0.40		I ≫ C
	LP	40 m	51.00	1.0	0.18		I
'Lower' Buckley	MMD	284 m	5.30	2.0	0.24	0.93	K > I > C
	MA6	45 m	1.80	8.0	0.10		I > C
	MBb	90 m	12.30	1.5	0.00	1.40	K > I = C
Fairchild	MK	92 m	0.02	4.5	0.30		I
	CP	202 m	5.70	3.5	0.00		C > I
	MD	91 m	0.10	2.0	0.14	5.40	C > I
	MB		2.80	2.0	0.28		C > I
Mackellar	MMC	38 m	0.05	2.5	0.00		C > I
	TGA	160 m	1.05	3.0	0.00	3.10	K > I
	MB	38 m	0.96	10.0	0.25	3.90	C > I
	MB	104 m	1.02	4.0	0.40	2.23	C > I
Pagoda	MMZ	129 m	0.32	3.0	0.25		C = I
	MMP	24 m	0.11	2.5	0.50		C = I
	TGA	41 m	0.19	1.7	0.21		C > I
	CI	54 m		3.0	0.18		C > I

% C = organic carbon content (weight %); ICI = illite crystallinity index; IPP = illite polytype parameter; % R_0 = vitrinite reflectance; RMA = relative mineral abundance, based on basal diffraction peak area (C = chlorite, I = illite, K = kaolinite)

mation stratigraphic sections have been identified as minimally altered. These exhibit a marked compositional similarity across the study area, especially as shown at sections MMD and MB(B). These two samples have low 2M illite abundances and contain abundant kaolinite, while all three samples have low relative chlorite/illite abundance ratios. These three compositional characteristics indicate that chemical weathering processes were dominant in all source areas during deposition of the 'lower' Buckley Formation. Because chemical weathering processes are favoured in warm, humid climates, these compositional characteristics point directly to the regional onset of a warm, humid palaeoclimate during deposition of the Buckley Formation; similar conditions are also suggested by the abundance of coals in the Buckley Formation.

Four samples from the upper halves of the Buckley Formation stratigraphic sections have been identified as minimally altered. These four samples are restricted to the central portion of the study area, consistently contain illite as the dominant phyllosilicate, and have low illite crystallinity index values. The range of illite polytype parameter values shown by these samples, however, suggests variable mixing of ordered and disordered illites across the study area. Although the exact sources of the illite are unknown at this time, one hypothesis is that the disordered illites are detrital (chemically-weathered), while the ordered illite polytypes are the products of low temperature diagenesis of smectites, which were initially formed by alteration of volcanic glass. This hypothesis is supported by the observation of significant amounts of volcanic detritus in sandstones of the 'upper' Buckley Formation (Collinson & Isbell, 1988; Isbell, this volume, p. 215).

To summarize, the data presented here outline the transition from source areas dominated by physical weathering to source areas dominated by chemical weathering during Permian deposition in the central Transantarctic Mountains. In detail, however, this transition first appears in the northern portion of the study area in the Mackellar Formation, and may not be observed in the southern portion of the study area until the 'lower' Buckley Formation. These interpretations of temporal and lateral variations in provenance are supported by the limited number of geochemical analyses available at this time. In order to determine the geochemical signature of palaeoweathering conditions, the chemical alteration index (CIA = $(Al_2O_3/(Al_2O_3 + CaO + Na_2O + K_2O) \times 100)$ of Nesbitt & Young (1982) has been calculated for each of 10 bulk rock analyses. As chemical weathering becomes more important, the CIA of the solid weathering residue increases; Pleistocene glacially-weathered clays have a CIA of 65, while modern tropically-weathered muds have a CIA of 82 (Nesbitt & Young, 1982). CIA values for these samples are tabulated in stratigraphic order in Table 3, and convincingly demonstrate the up-section increase in chemical weathering recognized previously in the mineral data. Samples from section MB, the southernmost section examined, indicate that physical weathering dominated southern sources through Mackellar deposition, while the shift in CIA from MMD-11 m to MMD-115 m records the input of chemically-weathered components during Mackellar deposition to the north. Because time correlations between sections are unvailable, the stratigraphic position of CI-132 m is unclear; its high CIA value, however, may indicate that chemically-weathered sediments

Table 3. *Chemical index of alteration (CIA; Nesbitt & Young, 1982) values of Permian samples from the central Transantarctic Mountains*

Formation	Section	Level	CIA
Buckley	MMD	417 m	83.7
	MMD	270 m	83.1
	MBb	80 m	79.6
Fairchild	MMD	115 m	78.0
Mackellar	MK	79 m	78.6
	MMD	11 m	73.8
	CI	132 m	77.2
	MB	125 m	71.8
	MB	38 m	69.8
Pagoda	TGA	41 m	73.0

reached the southern end of the study area near the end of Mackellar deposition.

Detailed examinations of lateral and temporal provenance variability will require the development of a much larger data base, as well as geochronological correlations between stratigraphic sections. A consistent general sequence has already been inferred from several independent lines of evidence, however, and traces the transition from predominantly physical to predominantly chemical weathering at the sources of these Permian clastics. This sequence also supports regional-scale palaeoenvironmental interpretations based on facies analyses (Miller & Waugh, 1988; Miller & Frisch, 1988; Collinson & Isbell, 1988).

Acknowledgements

This project was funded by US National Science Foundation grant DPP-8418354. This support is gratefully acknowledged.

References

Carroll, D. (1970). *Clay Minerals: A Guide to Their X-Ray Identification.* Boulder, Colorado; Geological Society of America Special Paper 126, 80 pp.

Collinson, J.W. & Isbell, J.L. (1988). Permian–Triassic sedimentology of the Beardmore Glacier region. *Antarctic Journal of the United States*; 1986 Review, **21(5)**, 29–30.

De Segonzac, G.D. (1970). The transformation of clay minerals during diagenesis and low-grade metamorphism: a review. *Sedimentology*, **15(3/4)**, 281–346.

Dixon, J.B. & Weed, S.B. (ed.) (1977). *Minerals in Soils Environments.* Madison, Wisconsin; Soils Science Society of America. 974 pp.

Dow, W.G. (1977). Kerogen studies and geological interpretations. *Journal of Geochemical Exploration*, **7(2)**, 79–99.

Griffin, J., Windom, H. & Goldberg, E.D. (1968). The distribution of clay minerals in the world ocean. *Deep-Sea Research*, **15(4)**, 433–59.

Guthrie, J.M., Houseknecht, D.W. & Johns, W.D. (1986). Relationships among virtinite reflectance, illite crystallinity, and organic geochemistry in Carboniferous strata, Ouachita Mountains, Oklahoma and Arkansas. *Bulletin of the American Association of Petroleum Geologists*, **70(1)**, 26–33.

Hower, J., Eslinger, E.V., Hower, M.E. & Perry, E.A. (1976). Mechanism of burial metamorphism of argillaceous sediment: 1. Mineralogical and chemical evidence. *Geological Society of America Bulletin*, **87(5)**, 725–37.

Keller, W.D. (1970). Environmental aspects of clay minerals. *Journal of Sedimentary Petrology*, **40(3)**, 788–854.

Krissek, L.A. & Horner, T.C. (1988). Sedimentology of fine-grained Permian clastics, central Transantarctic Mountains. *Antarctic Journal of the United States*, 1986 Review, **21(5)**, 30–32.

Maxwell, D.T. & Hower, J. (1967). High-grade diagenesis and low-grade metamorphism of illite in the Precambrian Belt series. *American Mineralogist*, **52(5/6)**, 843–57.

Miller, J.M.G. & Waugh, B.J. (1988). Sedimentology of the Pagoda Formation (Permian), Beardmore Glacier area. *Antarctic Journal of the United States*, 1986 Review, **21(5)**, 45–6.

Miller, M.F. & Frisch, R.S. (1988). Depositional setting of the (Permian) Mackellar Formation, Beardmore Glacier area. *Antarctic Journal of the United States*, 1986 Review, **21(5)**, 37–8.

Nesbitt, H.W. & Young, G.M. (1982). Early Proterozoic climates and plate motions inferred from major element chemistry of lutites. *Nature, London*, **299(5885)**, 715–17.

Potter, P.E., Maynard, J.B. & Pryor, W.A. (1980). *Sedimentology of Shale.* New York; Springer-Verlag. 303 pp.

Velde, B. (1983). Diagenetic reactions in clays. In *Sediment Diagenesis*, ed. A. Parker & B.W. Sellwood, pp. 215–68. Boston; D. Reidel Publishing Company.

Weaver, C.E. (1960). Possible uses of clay minerals in search for oil. *Bulletin of the American Association of Petroleum Geologists*, **44(9)**, 1505–18.

Evidence for a low-gradient alluvial fan from the palaeo-Pacific margin in the Upper Permian Buckley Formation, Beardmore Glacier area, Antarctica

J.L. ISBELL

Byrd Polar Research Center and Department of Geology and Mineralogy, The Ohio State University, 125 South Oval Mall, Columbus, Ohio 43210, USA

Abstract

In the central Transantarctic Mountains, fluvial sediments of the Permian Fairchild (230 m thick) and Buckley (750 m thick) formations were deposited in an elongate basin that paralleled the margin of the East Antarctic craton. Lower Permian palaeocurrents indicate southerly flow with deposition occurring along the axis of the basin. During the Late Permian, a thick prism of volcaniclastic sand prograded as a humid, low-gradient, alluvial fan into the basin from the palaeo-Pacific margin. Fan progradation resulted in a reversal of regional palaeoslope.

Introduction

In the Transantarctic Mountains, Late Palaeozoic–Early Mesozoic fluvial sediments were deposited in a foreland basin which formed between the East Antarctic craton (EAC) and the palaeo-Pacific margin (PPM) (Collinson, this volume, p. 199). Changes in fluvial architecture, provenance, and palaeoslope, within the Fairchild and Buckley formations, imply that tectonic activity along PPM controlled sedimentation and dispersal patterns in the central Transantarctic Mountains (CTM) as early as Late Permian time.

In the Transantarctic Mountains, previous investigators recognized a regional palaeoslope reversal. This reversal was believed to have occurred during the Permian–Triassic lacuna (Barrett & Kohn, 1975; Barrett, Elliot & Lindsay, 1986). The Buckley Formation was interpreted as a meandering stream deposit (Barrett *et al.*, 1986). New data presented here indicate that Buckley sediments were deposited within a braided fluvial system and that palaeoslope reversal occurred well before the Triassic, in Late Permian time.

Fluvial architecture

Fluvial architecture was determined by detailed study of lithofacies and facies geometries along extensive cliff faces in CTM. The Permian fluvial sequence is subdivided into two formations, Fairchild (Lower Permian) and Buckley (Lower–Upper Permian). The Fairchild Formation conformably overlies the Mackellar Formation (Lower Permian). The contact with the Mackellar Formation (lowest thick bed of massive medium-grained sandstone) marks a change from deltaic to fluvial facies. The Fairchild Formation (230 m thick) consists of fine–medium-grained sandstone composed of multistoried sand-filled channel bodies. Channels are tens to hundreds of metres wide and 5–20 m thick. This unit is interpreted as a sandy braided stream system that fed into the Mackellar inland sea.

The base of the overlying Buckley Formation is gradational and is identified by the lowest occurrence of quartz–pebble lenses. The Buckley Formation (750 m thick) consists of interbedded sandstone, shale and coal. In the lower part of the formation, sandstone predominates. At higher stratigraphic levels, channels are subordinate to fine-grained overbank deposits; however, channel sandstones may dominate locally.

In the Buckley Formation, channel sandstones are 10–20 m thick, sheet-like, and laterally continuous for several kilometres. Within channel sand bodies, laterally continuous, scour-bounded, fining-upward cycles 1–2 m thick occur. Cycles consist of basal medium–coarse-grained, large-scale trough crossbeds overlain by fine-grained sandstone containing small-scale crossbeds, planar beds, and/or sandy organic layers. The final cycle grades into overlying shales. Multistoried channel sandstones are common; abandoned channels are sand-filled and lateral accretion surfaces are rare. Interchannel deposits are 5–50 + m thick and include sandstone, siltstone, shale and coal. Interchannel sands are 0.1–5 m thick, sheet-like, and in gradational, sharp or erosional contact with underlying units. Interchannel sandstones increase in abundance below channels. These sandstones are less common in strata overlying channels.

Channel sandstones in the Buckley Formation are interpreted as sandy braided stream deposits based on the following criteria: (1) the presence of laterally continuous sandstone bodies; (2) the presence of sand-filled abandoned channels; (3) the abundance of scour-bounded cycles; (4) the absence of fine-grained abandoned channels; and (5) the rarity of lateral

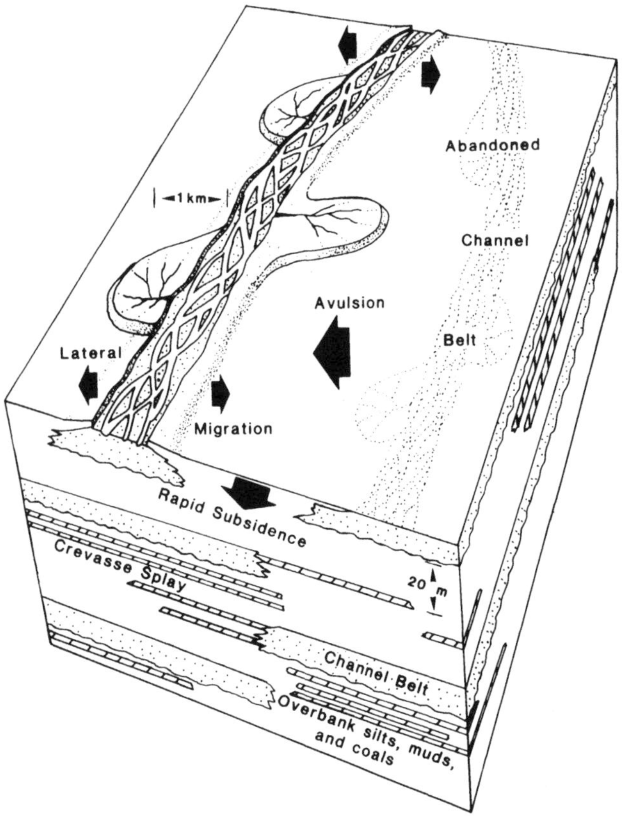

Fig. 1. Generalized block diagram of the Buckley Formation showing vertical and lateral facies relationships. Fluvial architecture in the Buckley Formation resulted from the interaction between: (1) braided stream deposition on a low-gradient, humid, alluvial fan; (2) channel belt avulsion; and (3) rapid basin subsidence.

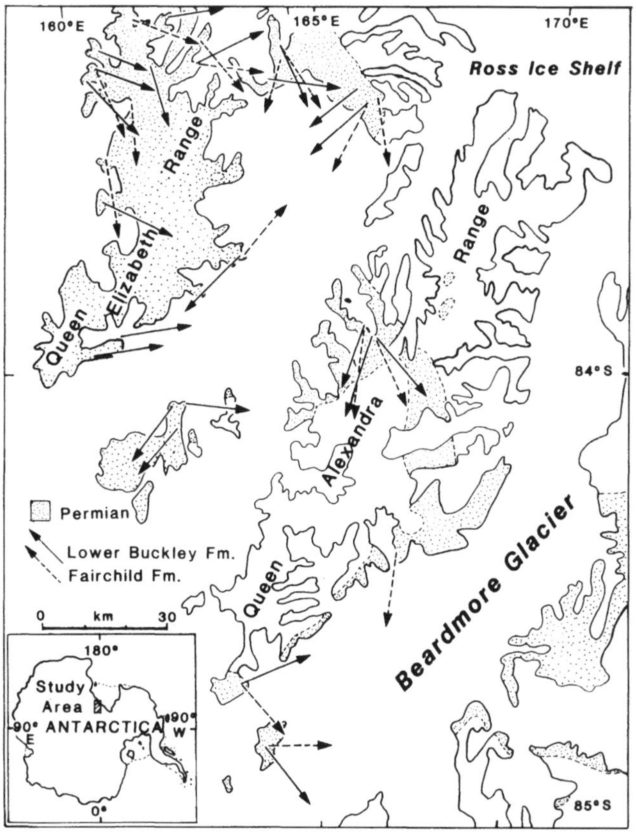

Fig. 2. Palaeocurrent orientations from the Fairchild and lower Buckley formations. Orientations are vector means from data collected from the 1985–86 field season and those reported by Barrett (1968). Measurements are taken from cross-stratifications contained within channels; sand bodies > 5 m thick which have erosional bases and contain medium–coarse-grained sandstone are recognized as channels.

accretion surfaces. Geometries and sedimentary structures attributed to meandering streams are absent. Fine-grained sediments were deposited in overbank environments and inter-channel sandstones are interpreted as crevasse splays. Channel migration by avulsion is indicated by: (1) the occurrence of laterally extensive channels interstratified with overbank deposits; (2) the increase in abundance of crevasse splays below channels; and (3) the absence of splays overlying channels (Fig. 1).

The Buckley Formation is unusual for a braided stream deposit in that a relatively high percentage of flood-plain sediment is preserved. The concepts of flood-plain preservation and channel avulsion in braided systems are not well documented. Emphasis in the literature is on confined non-aggrading systems. Unconfined braided streams have been described from humid, low-gradient, alluvial fans in the Himalayan foreland (Brahmaputra River, Coleman, 1969; Kosi River, Wells & Dorr, 1987). These fans have radii of approximately 150 km, and large-scale stream migration across them is by avulsion. During floods, enormous areas of the flood-plain are covered (Coleman, 1969; Wells & Dorr, 1987). For low-gradient, humid, alluvial fans, thick flood-plain deposits can be preserved under conditions of rapid subsidence relative to avulsion periodicity (cf. Allen, 1978; Bridge & Leeder, 1979). In rapidly subsiding basins, sheet sands are produced by bank erosion and small-scale migrations of the braided system (km). Large-scale channel migration by avulsion (tens of km) then isolates these sheets within fine-grained deposits. In the Buckley Formation, flood-plain preservation is consistent with deposition in an unconfined braided fluvial system on a humid, low-gradient alluvial fan.

Palaeoslope-controlled dispersal

Palaeoslope was identified from palaeocurrents taken from cross-stratification contained within channels; non-channel structures (i.e. crevasse splay crossbeds) were not used in reconstructing palaeoslope.

Palaeocurrents from the Fairchild Formation (Fig. 2) indicate that flow from the craton was directed into the basin with regional flow towards the present Queen Maud Mountains (southward). Measurements from the lower Buckley Formation (Fig. 2), collected from basin margin areas, display convergent palaeoflow towards the axis of the basin; regional drainage was longitudinal down the basin to the south (present true north). A palaeocurrent reversal occurred during upper Buckley sedimentation (Fig. 3), causing streams to flow north toward the Victoria Land sector.

Late Permian palaeocurrent reversal in CTM is contained within a conformable fluvial sequence. This conformity suggests the following: (1) that sedimentation within the basin was continuous; (2) that the reversal was produced by extra-basinal

Fig. 3. Palaeocurrent orientations from the upper part of the Buckley Formation and the lower portion of the Triassic Fremouw Formation. Criteria used for orientations and measurements are the same as in Fig. 2. Palaeocurrents from the Lower Triassic (lower–middle) Fremouw Formation are from Vavra (1984).

controls; and (3) that clastic influx from a 'new' source overwhelmed Early Permian dispersal trends within the basin.

Provenance of Permian sandstones

Sandstones in the Fairchild and lower Buckley Formations are quartzitic–arkosic with an average composition of $Q_{68\%}$ $F_{29\%}$ $R_{3\%}$ (Barrett *et al.*, 1986). Lithic fragments include granitic and pelitic grains. In the upper Buckley, a change in composition occurs. These sands are represented by sandstone compositions of $Q_{13.9\%}$ $F_{36.8\%}$ $R_{49.3\%}$, with lithic grains consisting almost entirely of volcaniclastic fragments (Barrett *et al.*, 1986).

Lower Permian detritus was shed into CTM from exposed basement terranes located on EAC. A change in provenance during deposition of the upper part of the Buckley Formation resulted in the introduction of PPM volcaniclastic detritus into CTM in Late Permian time.

Conclusions

Late Palaeozoic sediments were deposited in an elongate basin parallel to the margin of EAC (Elliot, 1975; Collinson, this volume, p. 199) and palaeocurrents suggest deposition along the axis of the basin. The Buckley Formation was deposited by braided streams. Preservation of its flood-plain sediments suggests lateral channel migration by avulsion coupled with rapid basin subsidence. Regional palaeoslope reversal and concurrent volcaniclastic influx in CTM occurred during Late Permian time and is contained within a conformable fluvial sequence. Fluvial architecture, depositional conformity, and palaeoslope reversal all suggest that volcaniclastics were introduced into CTM from a prograding, low-gradient, humid, alluvial fan which inundated the basin with detritus. Previous dispersal trends within the basin were overwhelmed and regional gradients were reversed.

Acknowledgements

This research was supported by National Science Foundation Grant DPP-841835A01 to James W. Collinson, and I acknowledge financial support from the Ohio State University and Shell Oil Company.

References

Allen, J.R.L. (1978). Studies in fluvial sedimentation: an exploratory quantitative model for the architecture of avulsion-controlled alluvial suites. *Sedimentary Geology*, **21(2)**, 129–47.

Barrett, P.J. (1968). The post-glacial Permian and Triassic Beacon rocks in the Beardmore Glacier area, central Transantarctic Mountains, Antarctica. PhD thesis, Ohio State University, (unpublished).

Barrett, P.J., Elliot, D.H. & Lindsay, J.F. (1986). The Beacon Supergroup (Devonian–Triassic) and Ferrar Group (Jurassic) in the Beardmore Glacier area, Antarctica. In *Geology of the Central Transantarctic Mountains*, Antarctic Research Series, vol. 36, ed. M.D. Turner & J.F. Splettstoesser, pp. 339–428. Washington, DC; American Geophysical Union.

Barrett, P.J. & Kohn, B.B. (1975). Changing sediment transport directions from Devonian to Triassic in the Beacon Supergroup of south Victoria Land, Antarctica. In *Gondwana Geology*, ed. K.S.W. Campbell, pp. 15–35. Canberra; Australian National University Press.

Bridge, J.S. & Leeder, M.R. (1979). A simulation model of alluvial stratigraphy. *Sedimentology*, **26(5)**, 617–44.

Coleman, J.M. (1969). Brahmaputra River: channel processes and sedimentation. *Sedimentary Geology*, **3(2/3)**, 129–239.

Elliot, D.H. (1975). Gondwana basins in Antarctica. In *Gondwana Geology*, ed. K.S.W. Campbell, pp. 493–536. Canberra; Australian National University Press.

Vavra, C.L. (1984). *Provenance and Alteration of the Triassic Fremouw and Falla Formations, central Transantarctic Mountains, Antarctica*, Institute of Polar Studies Report, 87. Columbus, Ohio; Ohio State University. 98 pp.

Wells, N.A. & Dorr, J.A. (1987). Shifting of the Kosi River, northern India. *Geology*, **15(3)**, 204–7.

Provenance and tectonic implications of sandstones within the Permian Mackellar Formation, Beacon Supergroup of East Antarctica

R.S. FRISCH & M.F. MILLER

Department of Geology, Vanderbilt University, Nashville, TN 37235, USA

Abstract

The Mackellar Formation of the Beardmore Glacier region is comprised of interbedded black shales and fine sandstones and is lithostratigraphically equivalent to other post-glacial units exposed elsewhere in the Transantarctic and Ellsworth mountains. It was deposited by southward prograding deltaic lobes during the initial infill of a post-glacial starved basin. The inferred tectonic setting is an elongate foreland basin behind a volcanic arc on the palaeo-Pacific margin of East Antarctica. Framework mineralogy of sandstones from the Mackellar Formation indicates a primarily continental block provenance. The probable sources are the granites and metamorphic rocks exposed to the north of the Beardmore area; this is consistent with both petrological and sedimentological data. The Polarstar Formation of the Ellsworth Mountains is stratigraphically equivalent and lithologically similar to the Mackellar Formation. However, it is thicker and has marine trace fossils and volcanic detritus; both of these features are absent from the Mackellar Formation. Closer proximity to the volcanic arc, as suggested by palaeogeographical reconstructions, would account for these differences.

Introduction

The Mackellar Formation is part of the Upper Carboniferous–Triassic sedimentary sequence found throughout the Central Transantarctic Mountains (CTM). Like Gondwana sequences exposed on other continents, it contains sedimentological evidence of the geological and tectonic history of Antarctica. This comprises a vertical sequence of glacial diamictite conformably overlain by interbedded black shale and sandstone, and followed by *Glossopteris*-bearing fluvial sandstone.

In the Beardmore Glacier area the black shale unit, which contains subequal amounts of interbedded sandstone, is called the Mackellar Formation. Lithologically and stratigraphically similar units with different names crop out southward through the Transantarctic Mountains (from the Nimrod Glacier area) and in the Ellsworth and Pensacola mountains (Collinson, this volume, p. 199). According to recent interpretations of Permian palaeogeography of this sector of Gondwana, the Permo–Triassic siliciclastics accumulated in a foreland basin behind a volcanic arc on the palaeo-Pacific margin of East Antarctica (Elliot, 1975; Collinson, this volume, p. 199). The primary purpose for studying the sandstones of the Mackellar Formation was to determine the provenance of the Mackellar Formation and thus to help constrain the presence and/or timing of the volcanic arc postulated for the palaeo-Pacific margin.

CENTRAL TRANSANTARCTIC MOUNTAINS					ELLSWORTH MTNS	
GROUP	PERIOD	FORMATION	THICKNESS (m)	REFERENCES	FORMATION	THICKNESS (m)
VICTORIA	JURASSIC	PREBBLE	460	Barrett et al., 1986		
	TRIASSIC	FALLA	530	Collinson and Elliot, 1984		
		FREMOUW	750	Barrett et al., 1986		
	PERMIAN	BUCKLEY	750	Barrett et al., 1986	POLARSTAR FORMATION (Collinson, this vol.)	1000
		FAIRCHILD	250	Barrett et al., 1986		
		MACKELLAR	140			
		PAGODA	200	Lindsay, 1969	WHITEOUT CONGLOMERATE	1000
TAYLOR	DEVONIAN	ALEXANDRA		Barrett et al., 1986	WYATT EARP FM.	

Fig. 1. Stratigraphic units of Central Transantarctic and Ellsworth mountains.

Stratigraphical and depositional setting

Lower Permian stratigraphy of the Queen Elizabeth and Queen Alexandra ranges is shown in Fig. 1. The Mackellar

Formation is typically 100 m thick, ranging from 60 to 140 m. Occurrences of Mackellar-like rock types in both the underlying Pagoda Formation and the overlying Fairchild Formation indicate continuous or nearly continuous deposition of the Pagoda–Mackellar–Fairchild sequence. The Mackellar Formation pinches out in the Nimrod Glacier area (Laird, Mansergh & Chappel, 1971) and is more sandy in the Mt Markham region than farther south in the Queen Elizabeth Range or in the Queen Alexandra Range.

The Mackellar Formation consists primarily of interbedded shale and fine–very fine-grained sandstone, with uninterrupted thicknesses of either rock type commonly < 2 m. Sandstone interbeds contain divisions T_{a-d} of the Bouma sequence and were deposited by episodic, waning, sediment-laden density currents. Sandstone beds < 0.25 m thick are most common and are characterized by a vertical change from ripple cross-lamination, including climbing-ripple lamination, to parallel lamination (divisions T_{c-e} or T_{c-d} of Bouma sequence; Howell & Normark, 1982). Neither individual sandstone beds, nor the 2–5 m coarsening-upward sequences into which they are grouped, are continuous on the scale of the outcrop. These coarsening-upward sequences are grouped into larger coarsening-upward sequences 25–30 m thick, capped by large-scale channel-form sand bodies.

We interpret the Mackellar Formation to record progradation of deltaic lobes (represented by coarsening-upward sequences) which spawned turbidity currents. Palaeocurrent directions indicate flow from north to south, suggesting delta progradation from the north, possibly with a component from the east (Fig. 2). This is consistent with the observed coarsening of the formation and pinching out to the north in the Nimrod Glacier area.

The Mackellar Formation resembles post-glacial black shale units found to the south in the Transantarctic Mountains and in the Ellsworth Mountains (Collinson, this volume, p. 199), and is broadly similar to equivalent units in the Karoo Basin of South Africa (Tankard *et al.*, 1982). However, it differs significantly from lithologically correlative units in the Ellsworth Mountains, the Ohio Range and Mt Weaver in lacking marine trace fossils. The Mackellar Formation contains only simple burrows and endostratal trails similar to those of the slightly younger fluvial deposits of the Buckley Formation. Deposition in fresh–slightly brackish water is corroborated by very low C/S ratios (R.A. Berner, pers. comm., 1986; see Berner & Raiswell, 1984 for explanation of method.)

In summary, the Mackellar Formation represents the initial infilling of a post-glacial starved basin. Deltaic lobes grew as turbidity currents flowed south through and over channels cut through gentle slopes. The Beardmore area was near the northern margin of the basin and did not have free water exchange with the palaeo-Pacific Ocean.

Tectonic setting

The Mackellar Formation is inferred to have been deposited in a back-arc or foreland basin behind a volcanic arc associated with the subduction of the Pacific oceanic crust beneath the East Antarctic craton (Collinson, this volume, p. 199). Permian palaeogeographic reconstructions based on palaeomagnetic data (Dalziel & Elliot, 1982) place the present CTM adjacent to the Karoo Basin of South Africa. Although alternative hypotheses have been suggested (Grunow & Dalziel, 1987), Dalziel & Elliot (1982) postulated a 90° clockwise rotation of the Ellsworth block to a position contiguous with the East Antarctic craton, and position several continental fragments on the seaward margin of East Antarctica. Such reconstructions place the Mackellar basin a considerable distance from the Permian continental margin (Fig. 3).

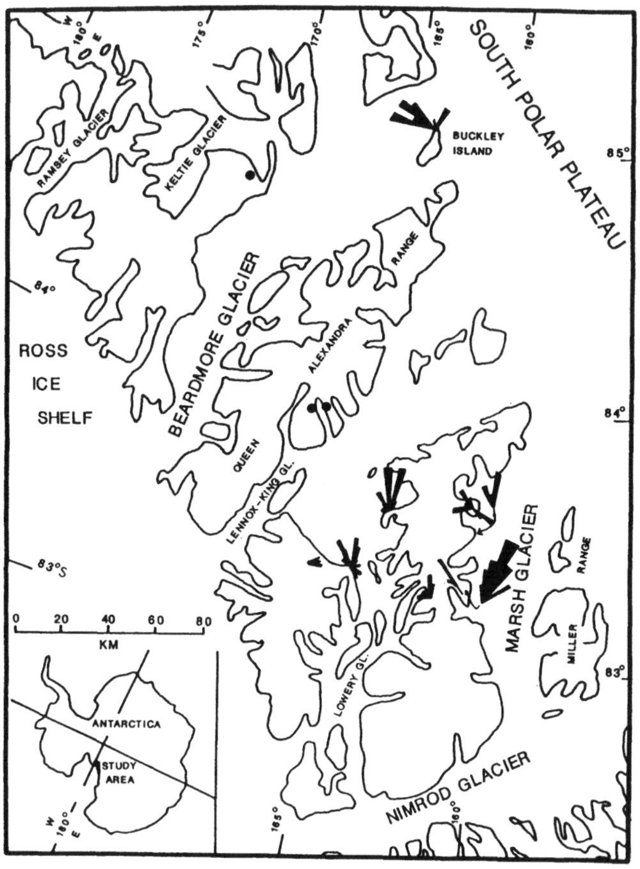

Fig. 2. Map of Beardmore Glacier region showing localities and summarized palaeocurrent data. Breakdown of palaeocurrent data available in Miller & Frisch (1986).

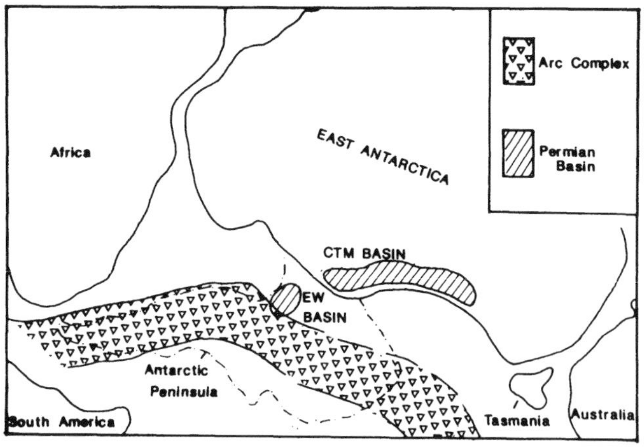

Fig. 3. Palaeogeographic reconstruction of the Permian Pacific margin of Gondwana, with inferred volcanic arc and sedimentary basins. Note position of Ellsworth Mountains (EW). Modified after Collinson *et al.*, in press.

A volcanic arc may have been present during Mackellar deposition. The (Permian) Buckley Formation in the CTM contains volcanic rock fragments (Isbell, this volume, p. 215), as does the Polarstar Formation, the Mackellar/Fairchild lithological equivalent in the Ellsworth Mountains. Polarstar sediments probably were deposited cratonward of the arc, close enough for abundant volcanic lithic fragments and plagioclase microlites, and trace amounts of pyroclastic material (including volcanic tuff), to accumulate. The Polarstar source terrane was dominated by silicic–andesitic volcanic rocks with lesser amounts of mafic volcanic, plutonic and low-grade metasedimentary rocks (Collinson, Vavra & Zawiskie, in press).

Methods

13 stratigraphical sections were measured and described as part of a larger study of the Mackellar depositional environment and processes (see Fig. 2) (Miller & Frisch, 1986). Most accessible sand bodies thicker than 15 cm were sampled. Petrographic thin sections were stained with sodium cobaltinitrate to facilitate potassium feldspar identification. 40 thin sections of Mackellar Formation sandstones were point-counted using the Gazzi–Dickinson method (Dickinson, 1970; Ingersoll *et al.*, 1984); 300 points/section were counted.

Results

Texture

Sandstones from the Mackellar Formation are immature and fine–very fine-grained in texture. Matrix comprises between 5 and 42% of the rock. It consists of detrital clay and authigenic clay, quartz, feldspar and calcite. Pseudomatrix, an interstitial powder derived from compaction and disaggregation of less stable grains in pore space, is probably also contributing to the matrix fraction. There is a continuum between grains with minor alteration and those which are entirely disaggregated. Most grains are angular–subrounded, although a few are very angular or well rounded.

Grain types

Detrital constituents in Mackellar sandstones include quartz, feldspar, biotite, muscovite, chlorite, and minor amounts of lithic fragments, garnet, zircon, sphene, rutile, apatite, tourmaline and opaque minerals.

Monocrystalline quartz dominates the quartz population and is present in both straight and undulose forms. Although polycrystalline quartz may have been abundant in source rocks, the fine-grained nature of Mackellar detritus implies that quartz grains have been disaggregated into their constituent monocrystalline segments; only minor polycrystalline quartz remains. Many grains exhibit multiple quartz overgrowths, suggesting multicyclic quartz from a sedimentary source.

Total feldspar content ranges from 8 to 62% of Mackellar Formation sandstones. Detrital plagioclase ranges in composition from andesine to oligoclase and constitutes from 3 to 35% of the framework grains. K-feldspars constitute from 3 to 34% of the framework grains. They include both microcline and orthoclase. Other K-feldspars may have originally been present, but identification was precluded by extensive alteration and by etching during the staining process. Biotite, muscovite and chlorite are all present in varying amounts, collectively constituting as much as 20% of detrital grains. Biotite is commonly partially or entirely replaced by chlorite. Authigenic chlorite constitutes more than half of the chlorite present. Muscovite and biotite are commonly deformed around stable framework grains, indicating a detrital rather than authigenic origin.

Lithic fragments are uncommon and very difficult to identify in sandstones from the Mackellar Formation. A few grains of sedimentary, metamorphic and igneous lithics were all tentatively identified, but post-depositional alteration has masked original internal structure.

Heavy minerals are present in trace amounts. Zircon and garnet are the most common types and rarely constitute as much as 2% of framework grains. Additionally, trace amounts of tourmaline, rutile, apatite and opaques are present. Although many of the heavy minerals are significantly smaller than other grains used in these modal analyses, they are included because of their importance in provenance determinations.

Composition

Modal analyses of framework grains of the Mackellar Formation sandstones plotted on Q: F: L ternary diagrams (Fig. 4) fall within the Folk (1974) classification of arkoses–subarkoses. Textural classification including matrix places Mackellar sands in the greywacke classification of Pettijohn (1954) and the arkosic wacke classification of Dott (1964).

Post-depositional alteration

Mackellar Formation sandstones have undergone post-depositional alteration caused by contact metamorphism near dolerite intrusions. Chemical modifications include grain dissolution and replacement, cementation, and destruction of framework grains. Authigenic minerals up to greenschist-facies metamorphism are present. Mechanical alterations consist of the disaggregation of grains and accumulation in interstitial pore space. Although features indicative of these processes are ubiquitous, alteration has not been sufficiently extensive to mask the original sandstone texture.

Discussion

The Mackellar Formation sandstones fall within or very close to the continental block provenance field of Dickinson & Suczek (1979). These fields were determined by analysing the composition of medium-grained modern sands from known tectonic settings. No vertical or lateral compositional trends are present, indicating a single source for Mackellar sandstones in the Beardmore Glacier region.

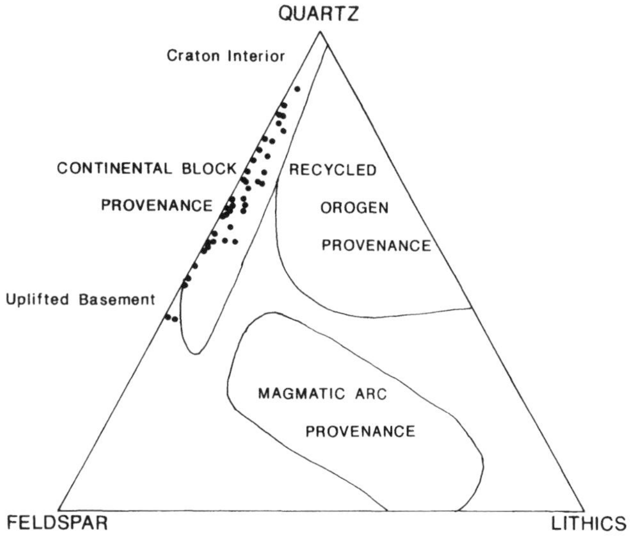

Fig. 4. Composition of sandstones from the Mackellar Formation. Provenance fields from Dickinson & Suczek, 1979.

The association of heavy minerals gives corroborating evidence of a cratonic source. The co-occurrence of zircon, tourmaline, rutile, apatite and opaques as in the Mackellar Formation has been proposed to be characteristic of a granitic source (Stattegger, 1987). Abundance of garnets (> 2%) may indicate a metamorphic source (Stattegger, 1987). However, the euhedral–subhedral nature of some of the garnets may imply a post-depositional provenance, with garnets forming during contact metamorphism; they cannot be considered reliable provenance indicators until their origin is interpreted by determination of their chemical composition (M. Cornerford, pers. comm., 1987).

Composition of Mackellar Formation sands spans the range of cratonic interior to uplifted basement, within the continental block provenance field (Fig. 4). Further sedimentological and stratigraphical evidence for a cratonic source is: (1) north–south palaeocurrents, suggesting a source to the north toward Victoria Land (Barrett, 1970; Miller & Frisch, 1986); and (2) northward increase in the amount of sandstone within the Mackellar interval, and the absence of the Mackellar Formation farther north in the Nimrod Glacier area (Grindley, 1963). Rocks exposed to the north of the Beardmore Glacier region would provide a source consistent with the provenance inferred from these sandstone compositions. They include: the Goldie Formation (Precambrian sediments and metasediments); the Nimrod Group (Precambrian metamorphic rocks); and the Hope Granites (crystalline two-mica granites of Ordovician age) (Barrett, Lindsay & Gunner, 1970; Lindsay, Gunner & Barrett, 1973; S. Borg, pers. comm., 1987). Detrital quartz grains with multiple overgrowths indicate a component of sedimentary source.

Great care must be taken in extrapolating provenance determination based on studies of medium-grained modern sands from known tectonic settings to finer-grained ancient sandstones. On a worldwide basis sandstone composition has been shown to vary systematically with grain size (Blatt, Middleton & Murray, 1980). Recent studies have attempted to assess the complex relationships between depositional environment (including transport distance, relief, energy and climate), grain size and mineralogical composition (Pittman, 1969; Davies & Ethridge, 1975; Odum, Doe & Dott, 1976; Suttner, Basu & Mack, 1981). Given a single source, finer-grained sands, which typically have undergone greater transport, are more quartz-rich than their medium-grained counterparts. If this were true for Mackellar sands, one would predict that, if they existed, medium-grained Mackellar Formation sandstones would contain more feldspar and lithic fragments. Therefore, although the abundance of feldspars in these rocks suggests that they have not undergone considerable compositional change due to depositional processes, a conservative interpretation must include the possibility that source rocks contained more feldspar and lithic fragments.

Comparison of Mackellar Formation with Polarstar Formation

The Mackellar Formation strongly resembles the Polarstar Formation of the Ellsworth Mountains both in stratigraphy and in inferred depositional environment (Collinson *et al.*, in press). The major differences are that the Polarstar Formation is thicker, and contains both marine trace fossile and volcanic rock fragments.

Differences between the Polarstar and Mackellar formations may be accounted for by differences in position relative to the palaeo-Pacific margin. The Polarstar Formation accumulated in a part of the basin near the palaeo-Pacific margin (Elliot, 1975), which was located sufficiently close to the arc to receive volcanic detritus, be affected by subduction–collision-associated subsidence, and have relatively free exchange with the open ocean. Conversely, the Mackellar Formation was deposited farther cratonward, and thus cut off from the supply of volcanic-rich sediment, removed from open circulation with the palaeo-Pacific Ocean, and relatively unaffected by collision tectonics.

Volcaniclastic sediments did not reach the Beardmore Glacier area until the Middle Permian, during deposition of the Buckley Formation. Introduction of volcanic fragments was accompanied in the Buckley Formation by addition of a south to north component in palaeocurrent directions (Collinson, this volume, p. 199; Isbell, this volume, p. 215). At least two explanations for the apparently diachronous introduction of volcaniclastic sediments are possible. (1) The Ellsworth Mountains were close enough to the palaeo-Pacific margin to receive volcanic sediments soon after initiation of volcanic activity. Thrusting and folding to close the gap between the margin and the more cratonward position of the CTM was then necessary before volcanic debris reached the Beardmore area. (2) Subduction and formation of the volcanic arc proceeded from south to north. The arc may not have been present as far north as the CTM during Mackellar deposition.

Conclusions

1. Composition of sandstones from the Mackellar Formation indicates a primarily granitic source with a subordinate

metamorphic and sedimentary component. The source did not change during deposition of the Mackellar Formation, as shown by absence of lateral and vertical trends.

2. South-directed palaeocurrents within the Mackellar Formation suggest a northerly source, probably the compositionally appropriate granites and metamorphic rocks exposed to the north of the Beardmore Glacier area. Northward coarsening of the Mackellar Formation and its absence north of the Nimrod Glacier suggests a shoreline in the Nimrod Glacier region. The Mackellar Formation was probably deposited as a series of southward-prograding deltaic lobes.

3. The Mackellar Formation is similar to the Polarstar Formation of the Ellsworth Mountains; however, it is thinner, and lacks marine trace fossils and volcanic detritus. These differences may be explained by the location of the Ellsworth Mountains adjacent to the active palaeo-Pacific margin and of the Beardmore Glacier area farther towards the centre of the craton (and/or by northward migration of volcanism through the Permian).

4. Contact metamorphism caused by intrusion of dolerites associated with Jurassic rifting (marking the break-up of Gondwana) has chemically and mechanically altered sandstones of the Mackellar Formation. Although authigenic minerals up to greenschist-facies metamorphism are present, they generally constitute < 20% of the sandstones and alteration is not complete.

Acknowledgements

We appreciate the constructive suggestions of J.W. Collinson, J.A. Crame and D. Macdonald. This research was funded by National Science Foundation grant DP-8418445.

References

Barrett, P.J. (1970). Paleocurrent analysis of the mainly fluviatile Permian and Triassic Beacon rocks, Beardmore Glacier area, Antarctica. *Journal of Sedimentary Petrology*, **40(1)**, 395–411.

Barrett, P.J., Elliot, D.H. & Lindsay, J.F. (1986). The Beacon Supergroup (Devonian–Triassic) and Ferrar Group (Jurassic) in the Beardmore Glacier area, Antarctica. In *Geology of the Central Transantarctic Mountains*, Antarctic Research Series, 36, ed. M.D. Turner & J.F. Splettstoesser, pp. 339–429. Washington, DC; American Geophysical Union.

Barrett, P.J., Lindsay, J.F. & Gunner, J. (1970). *Reconnaissance Geologic Map of the Mount Rabot Quadrangle, Transantarctic Mountains, Antarctica.* Washington, DC; US Antarctic Research Program, US Geological Survey.

Berner, R.A. & Raiswell, R. (1984). C/S method for distinguishing freshwater from marine rocks. *Geology*, **12(6)**, 365–8.

Blatt, H., Middleton, G.V. & Murray, R.C. (1980). *Origin of Sedimentary Rocks.* New Jersey; Prentice-Hall Inc. 728 pp.

Collinson, J.W. & Elliot, D.H. (1984). Triassic stratigraphy of the Shackleton Glacier area. In *Geology of the Central Transantarctic Mountains*, Antarctic Research Series, 36, ed. M.D. Turner & J.F. Splettstoesser, pp. 103–17. Washington, DC; American Geophysical Union.

Collinson, J.W., Vavra, C.L. & Zawiskie, J.M. (in press). Sedimentology of the Polarstar Formation, Permian, Ellsworth Mountains, Antarctica. In *Geology and Paleontology of the Ellsworth Mountains, Antarctica*, ed. G.F. Webers, C. Craddock & J.F. Splettstoesser, Washington, DC; Geological Society of America Memoir.

Dalziel, I.W.D. & Elliot, D.H. (1982). West Antarctica: problem child of Gondwanaland. *Tectonics*, **1(1)**, 3–19.

Davies, D.K. & Ethridge, F.G. (1975). Sandstone composition and depositional environment. *Bulletin of the American Association of Petroleum Geologists*, **59(3)**, 239–64.

Dickinson, W.R. (1970). Interpreting detrital modes of graywacke and arkose. *Journal of Sedimetary Petrology*, **40(2)**, 695–707.

Dickinson, W.R. & Suczek, C.A. (1979). Plate tectonics and sandstone compositions. *Bulletin of the American Association of Petroleum Geologists*, **63(12)**, 2164–82.

Dott, R.H., Jr. (1964). Wacke, greywacke and matrix – what approach to immature sandstone classification? *Journal of Sedimentary Petrology*, **34(3)**, 625–32.

Elliot, D.H. (1975). Gondwana basins in Antarctica. In *Gondwana Geology*, ed. K.S.W. Campbell, pp. 493–536. Canberra; Australian National University Press.

Folk, R.L. (1974). *Petrology of Sedimentary Rocks.* Austin, Texas; Hemphill Publishing Co. 184 pp.

Grindley, G.W. (1963). The geology of the Queen Alexandra Range, Beardmore Glacier, Ross Dependency, Antarctica. *New Zealand Journal of Geology and Geophysics*, **6(3)**, 307–47.

Grunow, A.M., Dalziel, I.W.D. & Kent, D.V. (1987). Ellsworth–Whitmore Mountains Crustal Block, Western Antarctica: new paleomagnetic results and their tectonic significance. In *Gondwana Six: Structure, Tectonics, and Geophysics*, Geophysical Monograph 40, ed. G.D. McKenzie, pp. 161–72. Washington, DC; American Geophysical Union.

Howell, D.G. & Normark, W.R. (1982). Sedimentology of submarine fans. In *Sandstone Depositional Environments*, ed. P.A. Scholle & D. Spearing, pp. 365–404. Tulsa, Oklahoma; American Association of Petroleum Geologists.

Ingersoll, R.V., Bullard, T.F., Ford, R.L., Grimm, J.P., Pickle, J.D. & Sares, S.W. (1984). The effect of grain size on detrital modes: a test of the Gazzi–Dickinson point-counting method. *Journal of Sedimentary Petrology*, **54(1)**, 0103–16.

Laird, M.G., Mansergh, G.D. & Chappell, J.M.A. (1971). Geology of the central Nimrod Glacier area, Antarctica. *New Zealand Journal of Geology and Geophysics*, **14(3)**, 427–68.

Lindsay, J.F. (1969). *Stratigraphy and Sedimentation of Lower Beacon Rocks in the Central Transantarctic Mountains, Antarctica.* Columbus; Ohio State University, Institute of Polar Studies, Report 33, 58 pp.

Lindsay, J.F., Gunner, J. & Barrett, P.J. (1973). *Reconnaissance Geologic Map of the Mount Elizabeth and Mount Kathleen Quadrangles, Antarctica.* Washington, DC; US Antarctic Research Program, US Geological Survey.

Miller, M.F. & Frisch, R.S. (1986). Depositional setting of the (Permian) Mackellar Formation, Beardmore Glacier area. *Antarctic Journal of the United States*, **21(5)**, 37–8.

Odum, I.E., Doe, T.W. & Dott, R.H., Jr. (1976). Nature of feldspar–grainsize relations in some quartz-rich sandstones. *Journal of Sedimentary Petrology*, **46(4)**, 862–70.

Pettijohn, F.J. (1954). Classification of sandstones. *Journal of Geology*, **62(4)**, 360–5.

Pittman, E.D. (1969). Destruction of plagioclase twins by stream transport. *Journal of Sedimentary Petrology*, **39(4)**, 1432–7.

Stattegger, K. (1987). Heavy minerals and provenance of sands: modeling of lithologic end members from river sands of northern Austria of the Austroalpine Gosau Formation (Late Cretaceous). *Journal of Sedimentary Petrology*, **57(2)**, 301–10.

Suttner, L.J., Basu, A. & Mack, G.H. (1981). Climate and the origin of quartz arenites. *Journal of Sedimentary Petrology*, **51(4)**, 1235–46.

Tankard, A.J., Jackson, M.P.A., Eriksson, K.A., Hobdy, D.K., Hunter, D.R. & Minter, W.E.L. (1982). *Crustal Evolution of Southern Africa: 3.8 Billion Years of Earth History.* New York, Heidelberg, Berlin; Springer-Verlag. 523 pp.

Crustal development: Weddell Sea – Ross Sea region

Evolution of the Gondwana plate boundary in the Weddell Sea area

Y. KRISTOFFERSEN[1] & K. HINZ[2]

1 Seismological Observatory, University of Bergen, Norway
2 Bundesanstalt für Geowissenschaften und Rohstoffe (BGR), Hannover, Federal Republic of Germany

Abstract

Multichannel seismic surveys have outlined two sets of basement structures below the East Antarctic continental margin of the Weddell Sea, between 0° and 40° W. One set is formed by the Explora–Andenes Escarpment (EAE), trending N60–80° E and bounded on the seaward side by oceanic crust. The other set lies landward of the EAE and represents a failed rift system trending N50° E. The presence of a failed rift system (named Weddell Rift) with symmetrical volcanic wedges demonstrates that the initial motion was rifting accompanied by prolific volcanism, and that an area of oceanic crust > 40 km wide may have been generated. A subsequent change in the regional stress field initiated transtensional movements between South Africa and Antarctica, resulting in the formation of the Explora–Andenes Escarpment as a new plate boundary, and the opening of the Weddell Sea by seafloor spreading.

Introduction

Jurassic volcanism has been recorded along the entire length of the present Transantarctic Mountains (Elliot, 1970; Behrendt *et al.*, 1980) and it manifests a rift regime of continent-wide proportions which initiated the fragmentation of Gondwana. In the Weddell Sea sector of the East Antarctic craton (Fig. 1), amygdaloidal plagioclase basalts were emplaced over continental crust in Vestfjella, Heimefrontfjella, Kirwanveggen and Ahlmannryggen (Aucamp, Wolmarans & Neethling, 1972; Furnes & Mitchell, 1978). However, the main volcanic activity took place farther north and west, with the emplacement of the elongated Explora Wedge (Hinz & Krause, 1982). A recent German multichannel seismic survey has defined a failed rift system (Hinz & Kristoffersen, 1987) paralleling the trend of the Transantarctic Mountains and terminating against the Explora–Andenes Escarpment (Hinz & Krause, 1982; Kristoffersen & Haugland, 1986). This paper discusses the tectonic evolution of the East Antarctic continental margin of the Weddell Sea on the basis of new geophysical data.

Structural elements of the East Antarctic continental margin between 0° and 40° W

The basement, mapped from multichannel seismic data (Figs 1 & 2), is often characterized by a very distinct unconformity, named 'unconformity U9' by Hinz & Kristoffersen (1987) and 'Weddell Sea continental margin unconformity' by Hinz & Krause (1982). A late Middle Jurassic age for this unconformity has been assumed. The basement is overlain by a thick section (up to 6000 m) of flat-lying sediments subdivided by several unconformities.

A prominent escarpment, called the Explora–Andenes Escarpment (EAE), has been recognized between 72° S, 40° W and 69°20′ S, 10° W (Hinz & Krause, 1982; Kristoffersen & Haugland, 1986). The EAE is ~ 1000 km long and consists of several segments that trend about ENE (60°–80°). Between 24° W and 28° W the EAE is overprinted by volcanic intrusions, and between 34° W and 40° W the exact location of the EAE is difficult to define because the recognized escarpment may represent the seaward edge of a succession of lava flows in this area.

Despite the local uncertainties concerning the position of the EAE, which is associated with a linear magnetic anomaly high, named 'Orion anomaly' by J.L. LaBrecque (pers. comm.), it is certain that the EAE represents the major structural boundary in the southern Weddell Sea area. It probably represents the ocean–continent transition.

The basement seaward of the EAE often has a hummocky and/or undulating relief and exhibits seismic characteristics typical of oceanic crust. Landward of the EAE, the top of the basement is generally the pronounced U9 unconformity. The difference in the level of the basement on the two sides of the EAE varies between 1 and 2 seconds of two-way traveltime (twt) along the length of the EAE.

On the landward side of the EAE, our recent seismic data reveal the presence of an elongated and approximately N50° E-trending depression in the basement west of 25° W, flanked by some remarkably symmetrical basement structures (Fig. 2). These units are characterized by a pattern of divergent

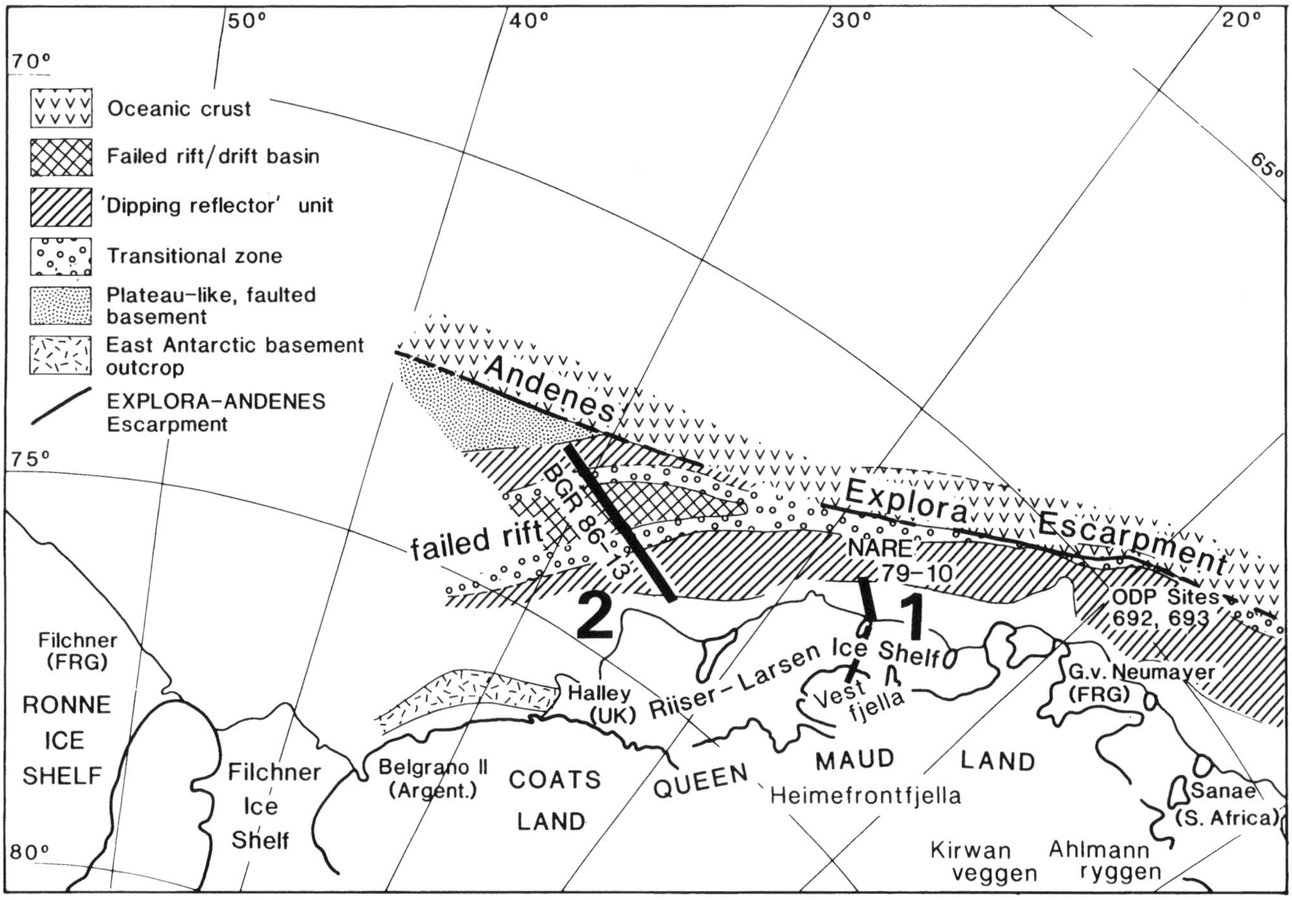

Fig. 1. Major geological structures of the East Antarctic continental margin between 0 ° and 40 °W. Bold lines indicate location of seismic line BGR 86-13, NARE 79-10 and the composite crustal transect shown in Fig. 4.

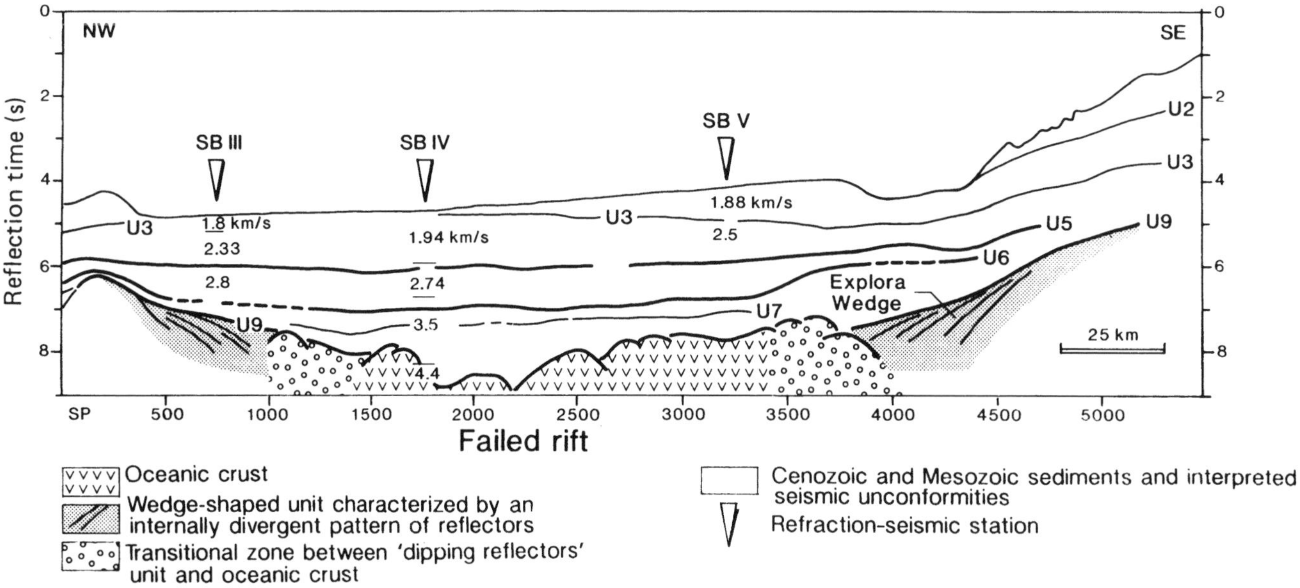

Fig. 2. Geoseismic section of Line BGR 86-13. For location see Fig. 1.

reflectors which, without exception, dip toward the basement depression, where the basement relief is relatively rough and similar to that of oceanic basement (Fig. 3). The northern wedge terminates against the EAE between 25° W and 30° W longitude. The southern wedge, known as the 'Explora Wedge', has been mapped as a continuous unit of northward-dipping sub-basement reflectors from 20° E through 35° W, where the unit is no longer detectable due to repeated multiple reflections from the seafloor and of depositional boundaries superposed with peg-leg multiples.

The characteristic reflection pattern associated with the Explora Wedge emerges below the continental shelf and

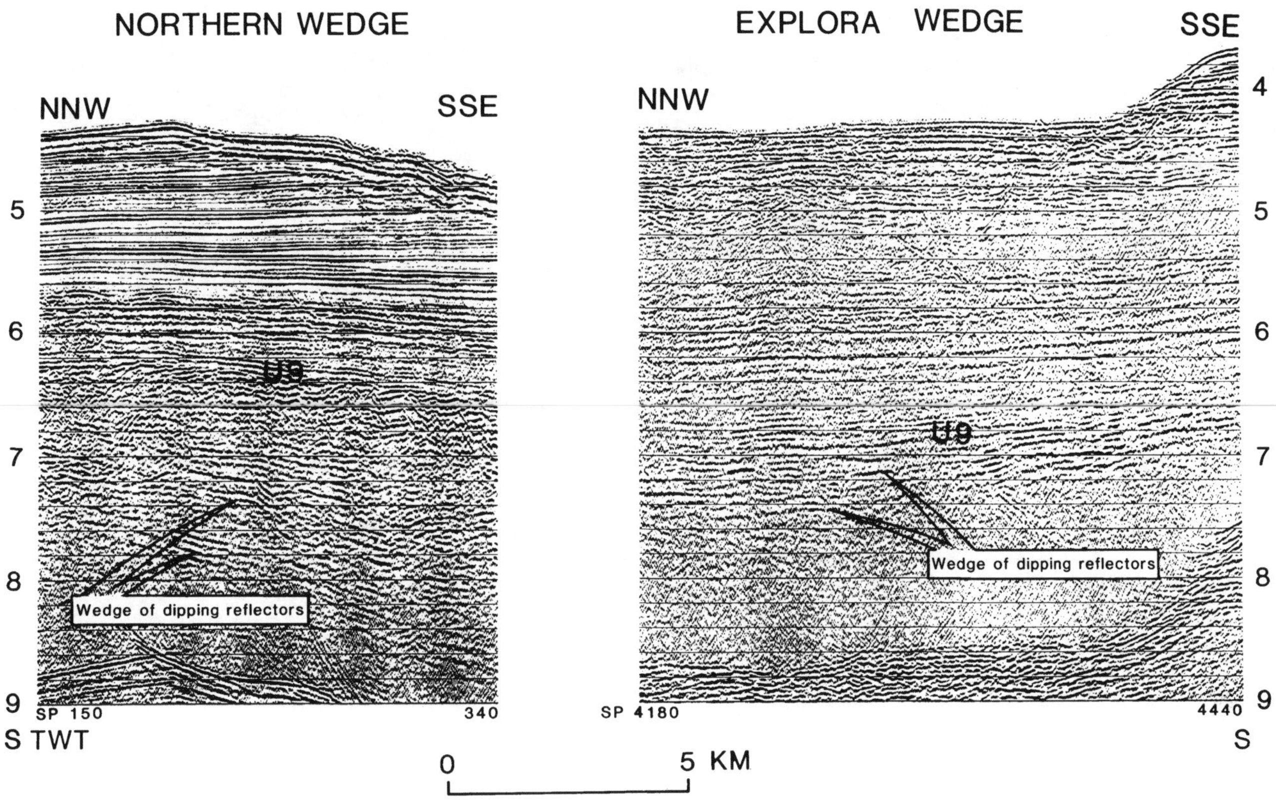

Fig. 3. Sections of multichannel seismic data from Line BGR 86-13 (shot points 150–340 and 4180–4440) demonstrating characteristic dipping reflectors beneath unconformity U9, i.e. Explora Wedge (right) and its symmetrical northern wedge (left).

extends northward to the EAE east of 15° W. West of this longitude its northern limit diverges south-westward and becomes the symmetrical element to the northern wedge (Figs 1 & 2).

The dipping sub-basement reflections generally exhibit a good lateral continuity in both the northern and southern wedge, but strong variation in amplitude and also apparent roughness may occur (Fig. 3). Invariably, sonobuoy measurements and seismic interval velocities derived from NMO velocities show velocities from 3.5 km/s to 4.5 km/s at the top of a wedge. The Explora Wedge was the target of ODP Sites 691 and 692, but operational difficulties prevented sufficient penetration. However, sub-basement units with similar seismic characteristics have been observed on several continental margins around the world (Hinz, 1981) and have been drilled during ODP Leg 104 on the Vøring Plateau in the Norwegian Sea. Here, ODP Site 642 passed through more than 900 m of a sequence of dipping reflectors and encountered a volcanic sequence comprising 132 lava flows, 60 volcaniclastic units and 7 units that may be dykes (Eldholm *et al.*, 1987). These drilling results, as well as the seismic characteristics (a divergent pattern and the seismic velocities), suggest that both wedges on the western Weddell Sea margin consist of volcanic rocks. We favour a mode of emplacement for these symmetrical wedge-shaped units by eruption of massive sequences of volcanic rocks onto highly distended continental crust in a subaerial to shallow-marine environment (Hinz, 1981).

A failed rift below the western Weddell Sea continental margin

The basement depression trending N50° E west of 23° W longitude is associated with a positive free-air gravity anomaly of about 50 mgal relative to the field over its flanking stratified basement wedges, and reaching over 100 mgal where the trend crosses the shelf edge at about 74°35′ S, 33°00′ W (Hinz & Kristoffersen, 1987). Gravity modelling along Line BGR 86-13, assuming Airy-type local compensation, indicates substantial thinning below the basement depression (Fig. 4, upper left). We interpret the conjugate basement wedges characterized by a divergent reflection pattern, the crustal structure and the distinct basement depression as elements of a failed rift for which we propose the name 'Weddell Rift'. This rift system is truncated by the EAE between 19° W and 27° W longitude, in a segment of the EAE characterized by an elevated region of the volcanic rock unit (Kristoffersen & Haugland, 1986). West of 35° W, like the Explora Wedge, the failed rift is no longer recognizable on the seismic records due to very strong multiples and a thick sedimentary cover. However, the reflection pattern and the seismic velocities of the basement on Line NARE 77-7 along the Filchner Ice Shelf suggest the presence of the southern wedge onlapping the East Antarctic craton (Haugland, Kristoffersen & Velde, 1985).

Soviet aeromagnetic data show a magnetic total intensity low associated with the general location of the Weddell Rift, continuing south-westward toward the northern end of

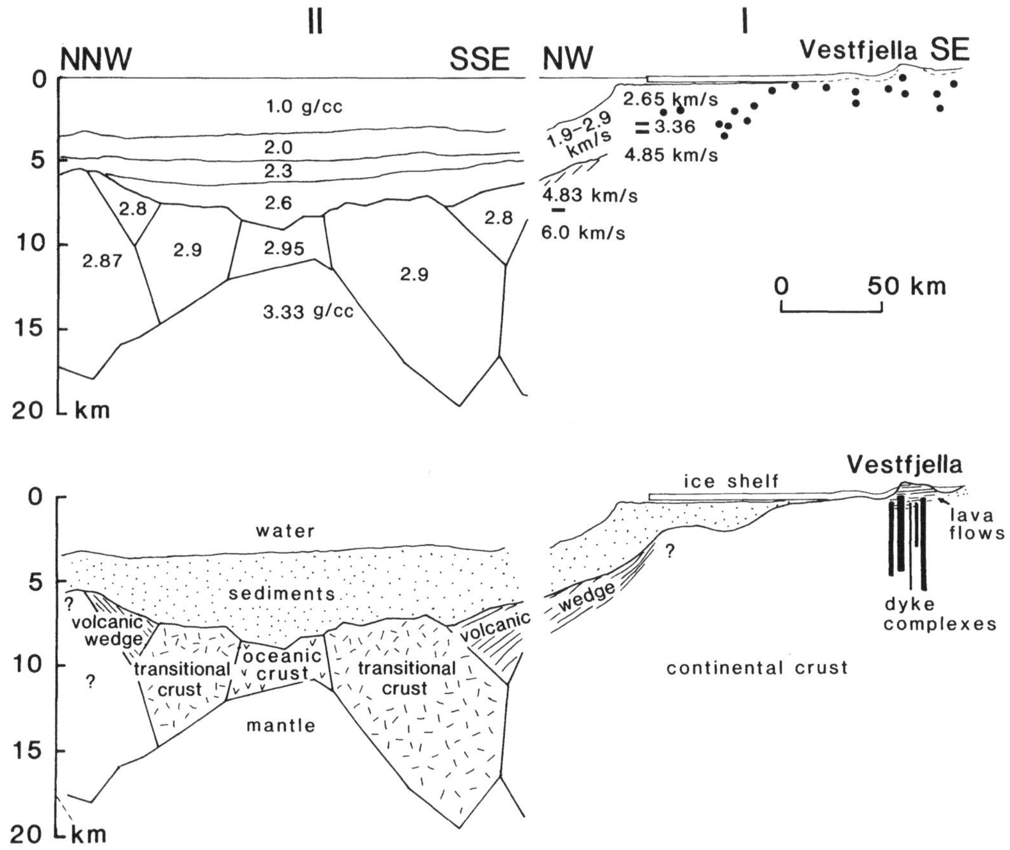

Fig. 4. Upper: Compilation of geophysical data bearing on the crustal structure along a composite transect across the Gondwana plate boundary. For location see Fig. 1. Section I from Vestfjella across Riiser–Larsen Ice Shelf and including line NARE 79-10 after Kristoffersen & Aalerud (1988). Heavy dots indicate estimated depth to magnetic source. Section II along Line BRG 86-13 derived from modelling of the free-air gravity field. Basement topography and layers above constrained by reflection seismic data. Lower: Interpretive crustal structure along the composite transect.

Berkner Island. Estimates of depth to magnetic sources from the same data set suggest that a 10–15 km deep sedimentary basin is present between the Antarctic Peninsula and the East Antarctic craton, divided by a basement horst situated at depth below the position of Berkner Island (Masolov, 1980). This is also supported by the results of deep seismic soundings (Kadmina *et al.*, 1983). We therefore believe that the rift may continue for more than 500 km farther SSW than the position where it is presently established by MCS data.

Crustal transect across the Gondwana plate boundary

The trend of the failed rift system parallels the strike of the adjacent segment of the Transantarctic Mountains. A compilation of available geophysical and geological data along a composite transect across the continental margin of the western Weddell Sea is shown in Fig. 4.

Basement is not exposed at Vestfjella, but biotite–amphibolites are exposed in Mannefallknausane, and amphibolite-facies paragneisses in Heimefrontfjella (Juckes, 1972), 110 and 150 km to the south-east of Vestfjella (see Figs 1 & 4). Middle Jurassic basalts are present at Vestfjella, Heimefrontfjella and Kirwanveggen (Aucamp *et al.*, 1972; Juckes, 1972; Furnes & Mitchell, 1978). The estimated thickness of the gently westward-dipping volcanic section at Vestfjella ranges from 800 to 2000 m and the pile is cut by numerous dykes of predominantly NE–SW strike (Hjelle & Winsnes, 1972; Spaeth, 1987). Dyke swarms may be more abundant at depth and bound the Vestfjella fault block complex (Kristoffersen & Aalerud, 1988).

Depth to magnetic sources continues to be shallow for about 50 km seaward of the last geological exposure and then a deepening occurs (Fig. 4). The Explora Wedge has been recognized from the seismic data beneath the upper slope and shelf off the Riiser–Larsen Ice Shelf. The landward extent of the wedge is not known but can be followed by sonobuoy measurements from its velocity characteristics to within 20 km of the Riiser–Larsen Ice Shelf. The Explora Wedge and its recognized northern conjugate part, with maximum thicknesses of the order of 5 km, presumably rest on highly distended continental crust of the East Antarctic craton and terminate at a transition zone flanking the central rift.

The thinning of the crust is remarkable below the rift axis, and we assume that at least a 40 km wide strip of oceanic crust has been generated, along Line BGR 86-13 (Fig. 4, upper left), before drifting ceased.

The transition zone, > 25 km wide and characterized by a relatively smooth basement relief, represents the transition of

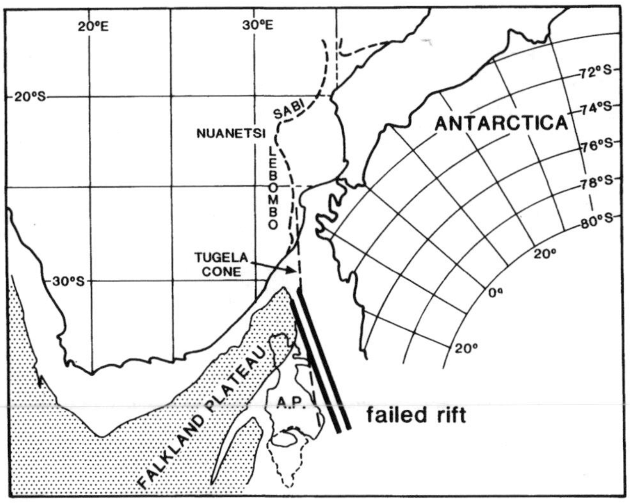

Fig. 5. Reconstruction for magnetic anomaly M21 (145 Ma) from Martin & Hartnady (1986), with the assumed position of the Weddell Rift. A.P., Antarctic Peninsula.

thick igneous crust, associated with the basinward-dipping reflectors, to oceanic crust.

The spectrum of structural elements and the symmetry observed in the Weddell Rift appear to offer a remarkable case study for understanding the evolution of a passive continental margin of the volcanic type.

Tectonic evolution of the continental margin of the western Weddell Sea: a discussion

The abundance of NE-trending dykes in Vestfjella, the presence of undeformed volcanic rocks in linear chains of nunataks along the Transantarctic Mountains, and the recognized NE-trending Weddell Rift with symmetrical structural elements suggest that the East Antarctica sector between 0° and 40° W longitude was under NW–SE-directed regional tension during Middle Jurassic time. Although crustal extension of the order of 5% has been observed locally in Vestfjella (Spaeth, 1987), it appears that the area between East Antarctica and the proto-Antarctic Peninsula was mainly affected by tension, which resulted in high strain within relatively narrow zones of low mechanical strength in the continental crust. The Weddell Rift was originally such a zone, and there are some indications for the presence of another NE-trending zone around 73° S, 35° W. Rift zones associated with wedges of dipping reflectors are apparently characterized by a large reduction in the thickness of continental crust within a zone only a few tens of kilometres wide (Hinz & Krause, 1982; Hinz *et al.*, 1987).

Voluminous melts intruded the highly distended continental crust and erupted onto it in a subaerial to shallow-marine environment. Gradual sagging under the load of the intrusive and extrusive rocks caused reversal of dip of the volcanic rocks toward the rift axis. The intra-continental rift axis later became an initial seafloor-spreading axis. When this occurred the thinned and contaminated crust, with its overlying volcanic pile, split and subsided.

Unconformity U9 (Hinz & Kristoffersen, 1987) marks the end of this initial Jurassic fragmentation of Gondwana that was accompanied by prolific volcanism and followed by deposition of Mesozoic sedimentary rocks. The Lower Cretaceous claystones recovered at ODP Site 692 were deposited in water depths of 500–1000 m, in anoxic or weakly aerobic conditions (Barker *et al.*, 1987).

After the formation of unconformity U9, for which a late Middle Jurassic age has been assumed (Hinz & Krause, 1982), a change in the regional stress field occurred, resulting in transtensional movements between South Africa and Antarctica. This phase was accompanied by volcanism and was heralded by the formation of the Explora–Andenes Escarpment as a new plate boundary, and the opening of the Weddell Sea by seafloor spreading.

The Gondwana plate geometry between South Africa and East Antarctica requires a significant component of Late Jurassic–Early Cretaceous shear (Fig. 5) in current reconstructions (Lawver, Sclater & Meinke, 1985; Martin & Hartnady, 1986). However, the discovery of the failed Weddell Rift, with its remarkable symmetrical elements, clearly demonstrates that the initial motion was rifting accompanied by prolific volcanism. The EAE cuts across the early rift structures, and therefore one can expect that a counterpart of the distinct Explora Wedge is present on the conjugate South African margin.

The separation between South Africa and Antarctica commenced by transtensional movements initially accompanied by volcanism, presumably in Late Jurassic–Early Cretaceous time, and followed by oblique spreading.

Conclusions

Our geophysical surveys on this part of the East Antarctic continental margin of the Weddell Sea suggest that the area was affected by two rift phases. The evidence for the first phase is the presence of wedges of basinward-dipping acoustic sub-basement reflectors that are developed to a high degree of symmetry north and south of a basement depression trending N50° E. The latter exhibits seismic characteristics and a crustal structure, derived from gravity modelling, typical of oceanic crust. The northern wedge and the basement depression terminated against the 1000 km-long Explora–Andenes Escarpment, trending N60°–80° E. The southern wedge (Explora Wedge) has been mapped as a continuous unit from 20° E to 35° W. ODP drilling in the Norwegian Sea (Eldholm *et al.*, 1987) of a similar wedge, encountered subaerially erupted volcanic successions.

We interpret the basement depression and its associated thin crust flanked by volcanic wedges to be the elements of a failed rift related to the initial fragmentation of Gondwana. The second phase of rifting is related to a change in the regional stress field during the Late Jurassic–Early Cretaceous. This resulted in transtensional movements between South Africa and Antarctica, in the formation of the N60°–80° E-trending Explora–Andenes Escarpment as a new plate boundary, and the opening of the Weddell Sea by seafloor spreading.

Acknowledgements

The results described herein would not have been obtained without the expertise of the officers and crew of the S/V Explora, K.V. Andenes, M/S Polarsirkel and F/S Polarstern, and the dedicated efforts of the BGR's and the University of Bergen's scientists and technicians on board. We acknowledge the assistance of the Alfred-Wegener-Institute for Polar Research, the Norwegian Polar Research Institute, the Federal Ministry for Research and Technology, the Norwegian Petroleum Directorate, Statoil and Norsk Hydro.

References

Aucamp, A.P.H., Wolmarans, L.G. & Neethling, D.C. (1972). The Urfjell Group, a deformed Early Palaeozoic sedimentary sequence, Kirwanveggen, western Dronning Maud Land. In *Antarctic Geology and Geophysics*, ed. R.J. Adie, pp. 557–61. Oslo; Universitetsforlaget.

Barker, P.F., Kennett, J.P. & Leg 113 Shipboard Party. (1987). Glacial history of Antarctica. *Nature, London*, **328(6126)**, 115–16.

Behrendt, J.C., Drewry, D.J., Jankowsky, E. & Grim, M.S. (1980). Aeromagnetic and radio echo sounding measurements show much greater area of the Dufek Intrusion, Antarctica. *Science*, **209(4460)**, 1014–17.

Eldholm, O., Thiede, J., Taylor, E. & Leg 104 Shipboard Scientific Party. (1987). *Proceedings ODP, Initial Report* No. 104, 783 pp. College Station, Texas (Ocean Drilling Program).

Elliot, D.H. (1970). Jurassic tholeiites of the central Transantarctic Mountains, Antarctica. In *Proceedings of the Second Columbia River Basalt Symposium*, ed. E.H. Gilmour & D. Stradling, pp. 301–25. Cheney, Washington; Eastern Washington State College Press.

Furnes, H. & Mitchell, J.C. (1978). Age relationships of Mesozoic basalt lavas and dykes in Vestfjella, Dronning Maud Land, Antarctica. *Norsk Polarinstitutt Skrifter*, **169**, 45–68.

Haugland, K., Kristoffersen, Y. & Velde, A. (1985). Seismic investigations in the Weddell Sea Embayment. *Tectonophysics*, **114(1–4)**, 293–313.

Hinz, K. (1981). A hypothesis on terrestrial catastrophes – wedges of very thick oceanward dipping layers beneath passive continental margins – their origin and paleoenvironmental significance. *Geologisches Jahrbuch*, **E22**, 3–28.

Hinz, K. & Krause, W. (1982). The continental margin of Queen Maud Land/Antarctica: seismic sequences, structural elements and geological development. *Geologisches Jahrbuch*, **E23**, 17–41.

Hinz, K. & Kristoffersen, Y. (1987). Antarctica – recent advances in the understanding of the continental margin. *Geologisches Jahrbuch*, **E37**, 3–54.

Hinz, K., Mutter, J.C., Zehnder, C.M. & NGT Study Group. (1987). Symmetric conjugation of the continent–ocean boundary structures off the Norwegian and East Greenland margins. *Marine and Petroleum Geology*, **4**, 166–87.

Hjelle, A. & Winsnes, T. (1972). The sedimentary and volcanic sequence of Vestfjella, Dronning Maud Land. In *Antarctic Geology and Geophysics*, ed. R.J. Adie, pp. 539–47. Oslo; Universitetsforlaget.

Juckes, L.M. (1972). *The Geology of North-eastern Heimefrontfjella, Dronning Maud Land.* London; British Antarctic Survey Scientific Reports, No. 65, 44 pp.

Kadmina, I.N., Kurinin, R.G., Masolov, V.N. & Grikurov, E.G. (1983). Antarctic crustal structure from geophysical evidence: a review. In *Antarctic Earth Science*, ed. R.L. Oliver, P.R. James & J.B. Jago, pp. 498–502. Canberra; Australian Academy of Science and Cambridge; Cambridge University Press.

Kristoffersen, Y. & Aalerud, J. (1988). Aeromagnetic reconnaissance over the Riiser–Larsen Ice Shelf, East Antarctica. *Polar Research*, **6**, 123–8.

Kristoffersen, Y. & Haugland, K. (1986). Evidence for the East Antarctic plate boundary in the Weddell Sea. *Nature, London*, **322(6079)**, 538–41.

Lawver, L.A., Sclater, J.G. & Meinke, L. (1985). Mesozoic reconstructions of the South Atlantic. In *Geophysics of the Polar Regions*, ed. E.S. Husebye, G.L. Johnson & Y. Kristoffersen. *Tectonophysics*, **114(1–4)**, 233–55.

Martin, A.K. & Hartnady, C.J.H. (1986). Plate tectonic development of the South Indian Ocean: a revised reconstruction of East Antarctica and Africa. *Journal of Geophysical Research*, **91(B5)**, 4767–80.

Masolov, V.N. (1980). Structure of the magnetically active basement of the southeastern part of the Weddell Sea basin. In *Geophysical Investigations in Antarctica*, pp. 16–28. Leningrad; Nauchno-Issledovatel'skij Institut Geologii Arktiki. (translated from Russian).

Spaeth, G. (1987). Aspects of the structural evolution and magmatism in western Neuschwabenland, Antarctica. In *Gondwana Six: Structure, Tectonics, and Geophysics*, Geophysical Monograph, 40, ed. G.D. McKenzie, pp. 295–305. Washington, DC; American Geophysical Union.

Petrology and palynology of Weddell Sea glacial sediments: implications for subglacial geology

J.B. ANDERSON[1], B.A. ANDREWS[1], L.R. BARTEK[1] & E.M. TRUSWELL[2]

1 Department of Geology and Geophysics, Rice University, Houston, Texas 77251, USA
2 Bureau of Mineral Resources, Geology and Geophysics, Canberra City, ACT 2601, Australia

Abstract

Petrological analyses and examination of recycled palynomorphs (spores, pollen and dinoflagellates) were conducted on basal tills and glacial marine sediments (subice shelf deposits) of the eastern Weddell Sea shelf. Five petrological provinces whose boundaries parallel glacial flow lines of both the East and West Antarctic ice sheets are recognized. To some degree, the pebbles within glacial deposits of the continental shelf mimic the exposed geology of regions situated upstream (along palaeoflow lines) of these deposits. However, a proportion of pebbles, sand-sized minerals, and recycled palynomorphs suggest that sedimentary basins containing Late Jurassic–Late Cretaceous sedimentary deposits are situated beneath the East Antarctic ice sheet and on the adjacent continental shelf. These same data indicate that, during that time, a seaway extended along what is now the eastern shelf south to approximately 75° S, and that Tertiary sedimentary deposits are either absent or are of limited extent.

Introduction

Knowledge of Antarctica's geology is restricted due to limited outcrop exposure. Evidence for the geology of ice-covered areas can be obtained from investigations of terrigenous sediments deposited on the continental shelf by glacial ice, and sediment gravity flow deposits whose source areas can be reasonably well constrained. Petrological analyses of pebbles and coarse sands were conducted on basal tills and transitional glacial marine deposits (sediments deposited close to the grounding line of ice shelves), and on selected sediment gravity flow deposits (debris flows and turbidites) acquired in piston cores from the eastern Weddell Sea shelf. Selected cores were also studied for recycled palynomorphs. The combined results provide important information about the subglacial geology and continental shelf sequences of the region.

Results from petrological analyses

The 22 piston cores used for this study were collected during cruises of the USCGC *Glacier* in 1969–70 and by the *Islas Orcadas* in 1978 (Fig. 1). Basal tills and transitional glacial marine sediments were distinguished using a variety of sedimentological criteria (Anderson *et al.*, 1980). Both sediment types are compositionally homogeneous downcore, and each depositional unit is compositionally unique, presumably representing the character of underlying bedrock along particular ice flow lines.

Over 700 pebbles were identified from these cores, and more

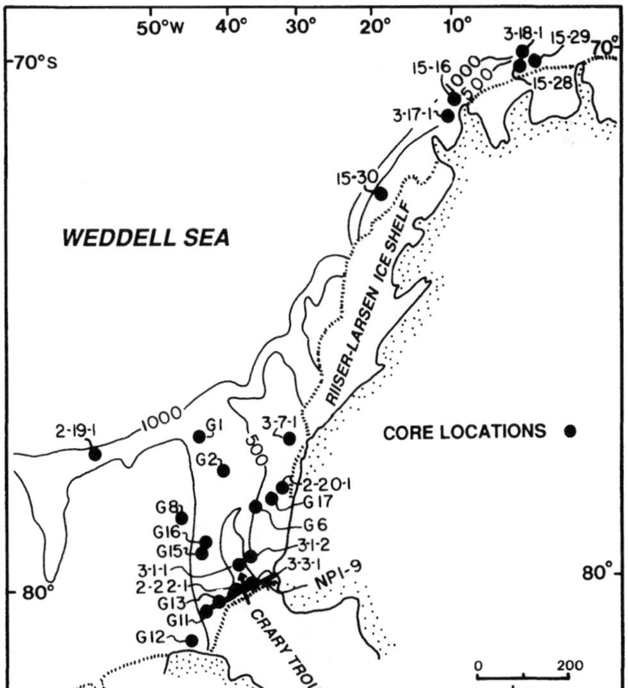

Fig. 1. Bathymetric sketch map (depths in m) of the south-eastern Weddell Sea showing the locations of piston cores used in this study and the location of seismic line NPI-9.

Table 1. *Characteristics of the petrological and palynological provinces of the eastern Weddell Sea*

	Province I	Province II	Province III	Province IV	Province V
Cores	G15, 2-22-1, 2-19-1	G1, 3-1-1	G6, G2, 3-7-1	G17, 2-20-1, 3-1-2, 3-3-1	16, 28, 29, 30, 3-18-1, 3-17-1
Location	West of Crary trough	Central Crary trough	Eastern edge of Crary trough and basement high	East of Crary trough	Princess Martha Coast
Pebble content	(Qtz.-arenite, > qtz. wacke)^, & (basalt/dolerite/ andesite)	Greywacke, metaqtz., slate/phyllite, shale, ≫ arkose, & ≫ high grade met.	Diorite, > granite, ≫ metaqtz., & ≫ qtz. wacke	Basalt/dolerite/ andesite	Basalt/dolerite andesite
Sand content	Qtz./feldspar, greywacke, ≫ sed. and met. rock frags.	Qtz./feldspar, greywacke, ≫ sed. and met. rock frags.	Qtz./feldspar, greywacke, ≫ sed. and met. rock frags.	(Qtz./feldspar, greywacke, ≫ sed. and met. rock fracs.)^, volcanic, > basic, > intermediate	volcanic, > basic, > intermediate
Palynomorph content	Terrestrial	Terrestrial	Terrestrial	Terrestrial	Marine

^, most prominent component of mixture; >, components on the right are less prolific than components on the left; ≫ components on the right are much less prolific than components on the left, but they are still significant

than 100 coarse sand (– 1.0–1.0 phi) samples were extracted by sieving samples at 50 cm intervals. Grain counts were made on 200–300 grains from each sample in order to obtain a statistical basis for identifying petrological suites. Factor analysis was used to define the major petrological suites and provinces using pebble and coarse sand data. The results are summarized in Table 1, and the distribution of petrological provinces shown in Fig. 2.

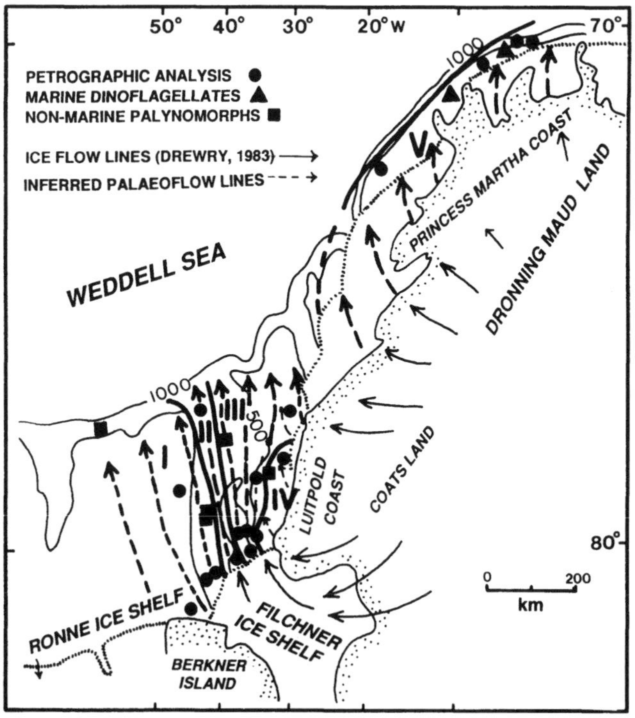

Fig. 2. The distribution of petrological and palaeontological provinces (I–V).

Province I

Province I is situated west of Crary trough (Fig. 2) and includes cores G-15, 2-22-1, and 2-19-1 (Fig. 1). These cores contain a high percentage of sedimentary rocks, notably quartz-arenites and quartz wackes with lesser amounts of volcanic pebbles. Coarse sands in these cores are dominated by quartz. Two cores (G8 and G12) penetrated outcrops of unconsolidated aeolian (Core G8) and beach (Core G12) quartz sand, which are presumed to underlie glacial sediments. This province is situated within the drainage area of ice streams flowing from the Pensacola Mountains. Two other cores from the area, and collected from the flanks of Berkner Island (G11 and G13, Fig. 1), penetrated sediment gravity flow deposits (debris flows and turbidites). The pebbles from these deposits consist mainly of basalts, implying that Berkner Island has a volcanic basement.

Province II

Although Province II is recognized on the basis of only two cores (Cores G1 and 3-1-1) from the western flank of Crary trough (Fig. 2), the constituent pebbles and coarse sands are quite distinct from those of adjacent provinces. The glacial deposits of this province contain a greater variety of rock types than any of the other provinces, including greywacke, metaquartzite, black and grey phyllites, biotite-schists, biotite gneisses, slate, shale, arkose, diorite, granite, basalt, andesite, and rhyolite. Arkose does not occur in any other province. Approximately 50% of the coarse sands are lithic fragments; the remainder are quartz and feldspar grains. Province II is associated with ice streams which drain an extensive basin that is partly bounded by the Shackleton Range, and with

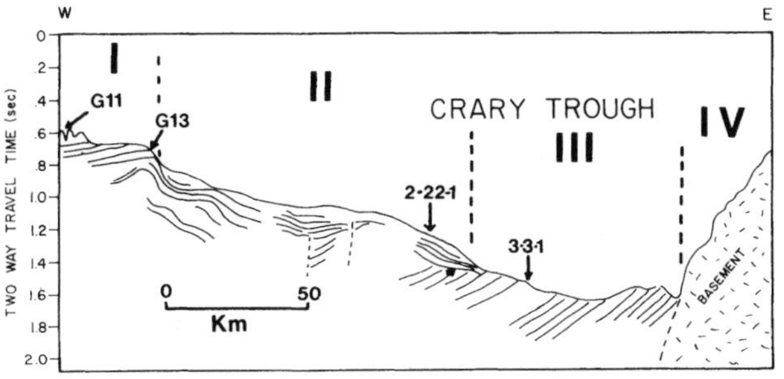

Fig. 3. Interpreted single channel (sparker) seismic-reflection profile (NPI-9) across Crary trough showing exposed acoustic basement in the eastern flank of the trough and exposed westward-dipping sequences in the western flank of the trough (from Elverhøi & Maisey, 1983). See Fig. 1 for profile location. Also shown are the locations of piston cores acquired along this line and corresponding petrological province boundaries.

westward-dipping strata exposed in the western flank of Crary trough (Fig. 3).

Province III

Piston cores taken along the north-eastern flank of Crary trough (Cores G6 and G2) and on the inner shelf near the Brunt Ice Shelf (Core 3-7-1) have similar pebble and coarse sand components and constitute Province III (Fig. 2). This province is characterized by abundant igneous rock fragments, chiefly basalts, andesites, and diorites with notable occurrences of white granite, and subordinate amounts of tan and grey quartzites, slate, amphibolite, schist, and medium-grained quartz wackes.

Province IV

A seismic profile across Crary trough (Elverhøi & Maisey, 1983) shows exposed acoustic basement on the eastern flank of this feature (Fig. 3), the exposed roots of the Transantarctic Mountains. Glacial sediments in piston cores G17, 2-20-1, 3-1-2, and 3-3-1 from this area contain predominantly volcanic pebbles (basalt, dolerite and andesite), and subordinate amounts of metaquartzites and amphibolites. Sedimentary rock types are virtually absent. The coarse sands of these cores are dominated by lithic grains, consisting of approximately equal proportions of igneous, metamorphic and sedimentary rock fragments. The nearest outcrops in the area of Province IV, at Bertrab and Littlewood nunataks, consist of Precambrian volcanic rocks (Marsh & Thomson, 1984).

Province V

Province V is situated off Princess Martha Coast (Fig. 2) and is based on similarities in the sand and pebble contents of glacial sediments in cores 15-16, 15-28, 15-29, 15-30 and 3-17-1, and 3-18-1. Lithologies are mostly basalt, andesite, and grey quartzite, although moderately sorted, medium-grained quartz arenites and poorly sorted quartz wackes occur in limited amounts. White granites were found in cores 15-16 and 15-30 only. The component of quartz + feldspar grains in the coarse sands is generally < 25%, and lithic grains are dominated by igneous rocks, chiefly volcanic. Quartzites and sandstones occur in smaller proportions.

Palynologic results

Truswell & Drewry (1984) have demonstrated the value of using recycled palynomorphs to infer the geology of ice-covered regions. In the Weddell Sea, recycled palynomorphs were reported previously (Kemp, 1972; Truswell, 1983) but sampling was too sparse for these occurrences to be related to source areas. In this study palynomorphs were examined in selected cores (G2, G15, G16, G17, 2-19-1, 3-1-2, 3-17-1 and 15-28) and the results provide a basis for distinguishing two palaeontological provinces.

All of the cores studied have palynomorph assemblages dominated by taxa which elsewhere have Late Jurassic–Late Cretaceous stratigraphical ranges. Tertiary forms are present but in limited numbers. Most of the material suggests derivation from source beds of Neocomian–Cenomanian age. Cores from the south-eastern shelf (G2, G15, G16, G17, 2-19-1 and 3-1-2) contain abundant, well-preserved spores and pollen. Most common are fern spores referable to *Cicatricosisporites*, *Contignisporites* and *Leptolepidites verrucatus*. Common too is pollen referable to the podocarpaceous taxa *Podocarpidites* and *Microcachryidites*. A few species have restricted ranges, including *Appendicisporites distocarinatus*, which in Australia ranges from Albian to Turonian. Also present are *Coptospora paradoxa* (an Albian zone fossil in Australia) and *Balmeisporites glenelgensis* (Cenomanian–Santonian). Thus, the total microflora in these cores reflects a source which ranges from Early to Late Cretaceous (Neocomian–(?)Santonian). The presence of fine woody debris and plant tissue, plus the absence of dinoflagellates, suggest derivation from terrestrial strata.

Glacial sediments from the north-eastern Weddell Sea shelf (cores 3-17-1 and 15-28) contain palynomorph assemblages dominated by marine dinoflagellates and subordinate spores and pollen. *Cribroperidinium muderongense* (Aptian–Cenomanian in Australia) is the dominant dinoflagellate. Also present in all samples is *Disconodinium davidii* which, in Australia, ranges from mid-Aptian to Albian (Morgan, 1979). Other less common species include *Cyclonephelium membraniphorum*, (late Albian–Cenomanian) and *Pareodinia ceratophora* (late Neocomian–Albian). Spores and pollen recovered from these cores are mainly long-ranging within the Cretaceous. This collective assemblage suggests derivation from Aptian–Cenomanian source beds of marine origin. Eocene dinoflagellates (*Areosphaeridium diktyoplokus*, *Vozzhenikovia rotundum*, *Deflandrea macmurdoense* and *Deflandrea antarctica*) occur in small numbers in these cores.

Discussion

Outcrops in western Dronning Maud and Coats lands are dominated by Precambrian gneisses, schists, and metasedimentary rocks. Farther south, outcrops occur in the Shackleton Range and Pensacola Mountains, and at Littlewood and Bertrab nunataks. High-grade metamorphic rocks of the Turnpike Bluff Formation are exposed in the Shackleton Range (Clarkson, 1971) and granites and acidic intrusives are exposed in the Pensacola Mountains (Elliot, 1975). The dominant lithologies of Bertrab and Littlewood nunataks are 'massive, brick red, oligoclase–phyric rocks of granitic composition' (Marsh & Thomson, 1984). At Bertrab Nunatak these granitic rocks are intruded by fine-grained basic and acidic dykes (Toubes Spinelli, 1983).

When considered in relation to principal outcrops of the study area, the abundance of clastic and basic volcanic rocks and the relatively small proportion of metamorphic rocks in most Weddell Sea glacial sediments implies that the mountains that project above the 3000 m thick Antarctic ice sheet are not representative of the overall subglacial geology of the region. Glacial sediments contain much higher proportions of volcanic rocks and quartz-arenites (possibly Beacon rocks) than are described for surface outcrops. Accordingly, the clastic sedimentary and associated volcanic rocks must be derived from areas entirely beneath the ice sheet. The relatively low percentages of metamorphic clasts in glacial sediments is inconsistent with an origin from a region dominated by Precambrian gneisses, schists, and metasedimentary rocks as suggested by outcrops in western Dronning Maud and Coats lands.

Only glacial sediments of Provinces II and III have rock and mineral compositions that are consistent with their having been derived from the roots of mountains. These include high-grade metamorphic rocks, similar in lithology to the Turnpike Bluff Formation in the Shackleton Range (Clarkson, 1971), and granitic pebbles and lithic fragments similar to Cambrian acidic intrusive rocks of the Pensacola Mountains (Elliot, 1975). The various other sedimentary (including arkose) and igneous rock types in Province II and III glacial sediments are probably derived from Gondwanian rocks and overlying Jurassic volcanics and associated Ferrar dolerites. The recycled palynomorphs in these glacial sediments are from a source that is predominantly Cretaceous in age and non-marine. This implies that marine invasions of the southern Weddell Sea did not occur until latest Cretaceous time, and possibly well into the Tertiary.

Province IV and V glacial sediments were derived from a sedimentary basin, analogous to the Karoo Basin of southern Africa, situated somewhere beneath the East Antarctic ice sheet in Dronning Maud Land. Beacon-type rocks of this basin are capped by volcanic rocks of probable Jurassic age. Recycled palynomorphs within glacial sediments of the shelf indicate that Cretaceous marine deposits occur, either within continental basins or on the continental shelf.

Our results indicate that glaciation of Antarctica apparently has not resulted everywhere in a stripping of its sedimentary cover down to basement. Rather, a number of subglacial sedimentary basins must still exist in the hinterland of the Weddell Sea. This suggestion is supported by depths to magnetic basement which indicate the presence of up to 4 km of sedimentary cover in areas east of the Pensacola Mountains (Jankowski, 1983), and by the presence of abundant recycled Cretaceous palynomorphs.

Conclusions

1. Independently conducted petrographic and palynological examinations of glacial sediments from the eastern Weddell Sea continental shelf show that discrete sediment provinces are present. The boundaries of these provinces are aligned with modern glacial flow lines of the East Antarctic ice sheet and provide a basis for reconstructing palaeoflow lines of a much expanded ice sheet.

2. Glacial sediments of the continental shelf are composed chiefly of basic volcanic and quartzose sedimentary rocks whereas outcrops of western Dronning Maud and Coats lands are dominated by metamorphic lithologies. This suggests that the mountainous outcrops are not representative of the overall subglacial geology of the region, and that a number of sedimentary basins must exist beneath the Antarctic ice sheet.

3. Glacial sediments of the eastern Weddell Sea shelf are derived from three general geological settings: (1) Cretaceous–(?)Tertiary sequences of a basin situated west of the Transantarctic Mountains; (2) the recently uplifted Transantarctic Mountains; and (3) a Palaeozoic–Mesozoic Beacon sedimentary basin situated somewhere in Dronning Maud Land.

4. Recycled palynomorphs in glacial sediments of the shelf consist mainly of Cretaceous forms. This corroborates our petrographic results which imply the existence of sedimentary basins beneath the East Antarctic ice sheet.

5. The absence of marine Cretaceous palynomorphs in the southern Weddell Sea (south of approximately 76° S) could indicate that marine incursion of this region did not occur until some time in the Tertiary. Tertiary palynomorphs are extremely rare in glacial sediments of this region.

References

Anderson, J.B., Kurtz, D.D., Domack, E.W. & Balshaw, K.M. (1980). Glacial and glacial marine sediments of the Antarctic continental shelf. *Journal of Geology*, **88(4)**, 399–414.

Clarkson, P.D. (1971). Shackleton Range geological survey 1970–1971: *Antarctic Journal of the United States*, **6(4)**, 121–2.

Elliot, D.H. (1975). Tectonics of Antarctica: a review. *American Journal of Science*, **275A**, 45–106.

Elverhøi, A. & Maisey, G. (1983). Glacial erosion and morphology of the eastern and southeastern Weddell Sea Shelf. In *Antarctic Earth Science*, ed. R.L. Oliver, P.R. James & J.B. Jago, pp. 483–7. Canberra; Australian Academy of Science and Cambridge; Cambridge University Press.

Jankowski, E.J. (1983). Depth to magnetic basement in West Antarctica. In *Antarctica: Glaciological and Geophysical Folio*; Sheet 8. ed. D.J. Drewry. Cambridge; Scott Polar Research Institute.

Kemp, E.M. (1972). Recycled palynomorphs in continental shelf sediments from Antarctica. *Antarctic Journal of the United States*, **7(5)**, 190–1.

Marsh, P.D. & Thomson, J.W. (1984). Location and geology of nunataks in north-western Coats Land. *British Antarctic Survey Bulletin*, **65**, 33–9.

Morgan, R. (1979). Palynostratigraphy of the Australian Early and Middle Cretaceous. *Memoirs of the Geological Survey, New South Wales, Palaeontology*, **18**, 1–153.

Toubes Spinelli, R.O. (1983). *Geología del Nunatak Bertrab, Sector Antártico Argentino.* Buenos Aires; Contribución Instituto Antártico Argentino, **296**, 9 pp.

Truswell, E.M. (1983). Recycled Cretaceous and Tertiary pollen and spores in Antarctic marine sediments: a catalogue. *Paleontographica B*, **186(4–6)**, 121–74.

Truswell, E.M. & Drewry, D.J. (1984). Distribution and provenance of recycled palynomorphs in surficial sediments of the Ross Sea, Antarctica. *Marine Geology*, **59(1–4)**, 187–214.

A multichannel seismic profile across the Weddell Sea margin of the Antarctic Peninsula: regional tectonic implications

P.F. BARKER[1,2] & M.J. LONSDALE[2]

1 British Antarctic Survey, Madingley Road, Cambridge CB3 0ET, UK
2 Department of Geological Sciences, University of Birmingham, Birmingham B15 2TT, UK

Abstract

Sediments beneath the continental slope and rise of the Weddell Sea margin of the Antarctic Peninsula are very thick. Basement (presumed oceanic) is seen only beneath the rise, where the minimum sediment thickness is 4.7 s TWT. After isostatic correction for sediment loading, basement depths clearly predict a late Mesozoic (Early Cretaceous or Late Jurassic) age for oceanic basement. Although there are no controls on sediment age, this basement age estimate and the great thickness of overlying sediment argue against the existence of an active Cenozoic plate boundary along the eastern margin of the Antarctic Peninsula, and thus against Cenozoic motion between the peninsula and East Antarctica. The age estimate also constrains models of the early evolution of the Scotia arc: east-directed subduction originated at a southward extension of the Magallanes Basin fold–thrust belt.

Introduction

The western Weddell Sea and its margin with the Antarctic Peninsula are poorly known, because of almost continuous pack-ice cover. This has left some major problems unresolved. For example, some models of Cenozoic global plate motion postulate relative motion of the Antarctic Peninsula and East Antarctica as recently as 40 Ma: the line of such motion cannot be detected in the better known eastern Weddell Sea, but in the absence of relevant data it could extend south along the Antarctic Peninsula margin. A related problem concerns the sequence of ridge crest–trench collisions, younging eastward, which helped form the South Scotia Ridge. The westward extent of the collisions is unknown, and *could* include part of the peninsula margin. The onshore geology of the margin suggests a Late Jurassic onset of Weddell Sea opening, but the sedimentary record is so discontinuous that subsequent tectonic events of this kind cannot be ruled out.

Unusually favourable sea-ice conditions in February 1985 allowed a multichannel seismic-reflection line across the north-eastern end of the Antarctic Peninsula (Fig. 1) during *Discovery* cruise 154. At the south-eastern end of this line, in the north-western Weddell Sea, acoustic penetration of the thick sediment pile was sufficient to provide an estimated minimum age for the underlying oceanic basement. Here we consider the implications of that age determination for the history of relative motion of the Antarctic Peninsula and East Antarctica and for the early stages of Scotia arc development.

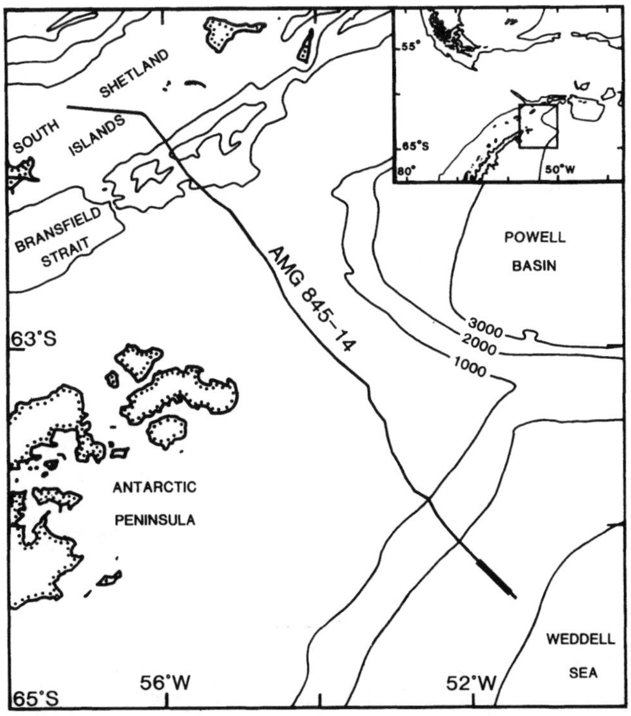

Fig. 1. Location of multichannel seismic line AMG845-14, from the South Shetland shelf across Bransfield Strait and the Antarctic Peninsula shelf to the north-western Weddell Sea. That part of the line shown in Fig. 2 and discussed in the text is drawn thicker.

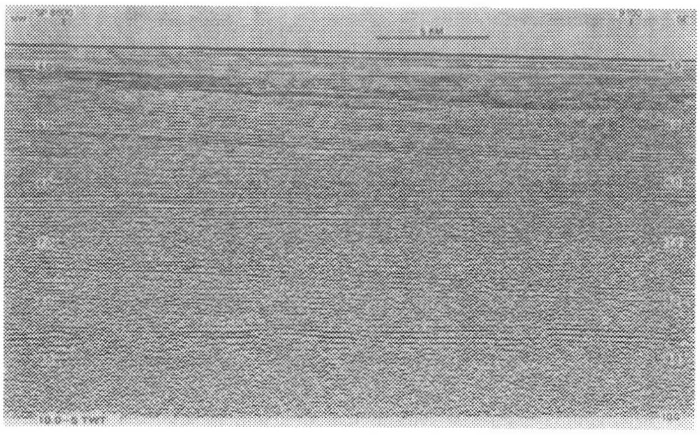

Fig. 2. Part of profile AMG845-14, located in Fig. 1. Strong reflections near 8.6 s TWT are either from oceanic basement or from sediment directly overlying basement.

Seismic-reflection profile

Profile AMG845-14 was acquired using a 48 channel, 2.4 km streamer and a small (8.5 litre) four-airgun source fired at 50 m intervals, recorded to 10 s and processed routinely by GECO UK to 24-fold CMP stack.

The section in Fig. 2 crosses the lower continental slope; the seabed lies at between 3.6 and 3.9 s of 2-way time (TWT), and primary reflections extend down to 8.6 s. The profile shows generally continuous, down-dip thinning and parallel-layered planar reflectors, suggesting a succession largely of lower-slope and basin turbidites. A change to higher-energy deposition and stronger bottom currents suggested by the uppermost 0.5–1.0 s of sediment is supported by the presence of a sediment drift higher up the slope. Beneath about 7 s TWT the only coherent primary reflector is the prominent but rather discontinuous reflector at 8.6 s. This is either oceanic basement or a basal sedimentary sequence ponded between basement highs. If the former, then basement age may be estimated using empirical age–depth relationships for oceanic basement, after isostatically unloading the overlying sediments. If the latter, then the same procedure provides a minimum basement age.

Sediment unloading and age–depth estimates

A range of empirical velocity–depth and density–depth curves is available to determine the isostatic correction for sediment load (e.g. Horowitz, 1976; Sclater & Christie, 1980; LeDouaran & Parsons, 1982; Crough, 1983). Most are derived from measurements on sediments from shallow depths of burial (principally from DSDP holes), and have to be extrapolated to unload the 4.7–5.0 s of sediment recognized here. However, both density and velocity vary with burial depth in a non-linear fashion, and the form of many curves is inappropriate for such extrapolation, because (for example) higher-order terms dominate unjustifiably at greater depth. Crough (1983) combined velocity–depth and density–depth data from several DSDP holes to eliminate the depth variable, demonstrating a linear relationship between TWT to the base of a sediment pile, and the correction to seafloor depth resulting from its removal. Substantiated down to 1.8 s TWT, this relationship yields a 'correction velocity' of 600 m/s. This is used to convert TWT to a number which is added to the present seabed depth to give unloaded depth. The linearity of this relationship suggests it can be extrapolated without becoming grossly inappropriate. Comparing data from COST wells and empirical curves (Nafe & Drake, 1957), Crough (1983) concluded that a 'correction velocity' of 680 m/s is more appropriate below 3 s. Using these two 'velocities' to unload the sediments in Fig. 2 gives a depth of about 5820 m at either end.

The general relation of ocean-floor subsidence to the square root of age has been demonstrated empirically back to about 80 Ma (e.g. Trehu, 1975; Parsons & Sclater, 1977), and justified theoretically in terms of a thermal boundary-layer model of lithospheric thickening (Parker & Oldenburg, 1973; Davis & Lister, 1974). Such empirical curves yield ages of 90 Ma (Parsons & Sclater, 1977) and 96 Ma (Trehu, 1975) for a depth of 5820 m. However, ocean floor older than about 80 Ma is shallower than these curves predict, possibly because the data used to construct them are 'contaminated' by data from ocean floor re-elevated by passage over a hotspot (Heestand & Crough, 1981). Alternatively, the thermal boundary-layer model may need modification to include the more general possibility of heat supply to the base of the plate (Parsons & Sclater, 1977). Empirical curves revised to take these factors into account provide ages of 107 Ma (Parsons & Sclater, 1977) and 112 Ma (Heestand & Crough, 1981). Allowing for more recent revision of the Magnetic Reversal Time Scale (Harland *et al.*, 1982), these ages become 117 and 122 Ma (early Aptian or Barremian). Since the reflector may lie above the mean depth of oceanic basement, these are minimum ages. Taking into consideration the various sources of uncertainty, which are many, oceanic basement in this part of the Weddell Sea is therefore most likely to be Early Cretaceous or older.

Antarctic Peninsula–East Antarctic motion

An Early Cretaceous age for the north-western Weddell Sea bears on the question of the age of any motion between the Antarctic Peninsula and East Antarctica, and the location of the plate boundary between them. The unacceptable overlap of the Antarctic Peninsula and Falkland Plateau is a feature of most otherwise satisfactory Gondwana reconstructions (Smith & Hallam, 1970; Norton & Sclater, 1979), if no relative motion between the peninsula and East Antarctica is allowed. During dispersal of the fragments after Gondwana break-up, the overlap disappears by about M0 time (~ 118 Ma), when the northern tip of the peninsula lay close to southernmost Tierra del Fuego (Lawver, Sclater & Meinke, 1985). Many have taken this to suggest that continuity of the Tierra del Fuego–Antarctic Peninsula connection was older, and was preserved through an earlier phase of Antarctic Peninsula–East Antarctic movement which had ceased by M0 time. However, suggestions of Early Tertiary motion between blocks now all within the Antarctic plate have been made on the basis of misfits around the East Antarctic–Australia–New Zealand–Pacific–West Antarctic plate motion circuit (Molnar *et al.*, 1975; Jurdy, 1978; Gordon & Cox, 1980; Stock & Molnar, 1982). Most recently Stock & Molnar (1987) postulated a separate Bellingshausen plate, which may have included the Antarctic Peninsula, until anomaly 18 time (~ 43 Ma). Antarctic–Bellingshausen motion in the Weddell Sea region could have been extensional.

The age and structure of the northern and eastern Weddell Sea floor is now reasonably well known (Barker & Jahn, 1980; LaBrecque & Barker, 1981; Hinz & Krause, 1982; Haugland, Kristoffersen & Velde, 1985). Over the last 50 Ma at least, the Weddell Sea has grown as a result of South American–East Antarctic (SAM–ANT) plate motion (Barker & Lawver, 1988). Weddell Sea floor becomes younger northward: a progressive change in SAM–ANT motion, from an initial or early N–S direction, through NW–SE to the present W–E, was accommodated through the creation firstly of small-offset and then of large-offset fracture zones, but the gross orientation of magnetic anomalies remained almost E–W.

The near permanent sea-ice cover in the western Weddell Sea has hindered exploration, and the westward extent of magnetic anomalies identified farther east was unknown (until recently: see the aeromagnetic data of LaBrecque *et al.*, 1986). It remained possible that a Tertiary plate boundary extended along the eastern margin of the Antarctic Peninsula. Motion there could have been extensional (Stock & Molnar, 1987) or convergent as suggested by earlier work (e.g. Molnar *et al.*, 1975). If the latter, the boundary could have been a westerly extension of the ridge crest–trench collision zones along the South Scotia Ridge (Barker, Barber & King, 1984).

It is clear from the new aeromagnetic data (LaBrecque *et al.*, 1986) that the SAM–East Antarctic boundary has extended no farther west than the north-east tip of the Antarctic Peninsula since the mid- or Late Cretaceous. Any Weddell Sea floor produced by hypothetical extensional ANT–Bellingshausen motion before anomaly 18 time (Stock & Molnar, 1987) must lie within a narrow strip directly east of the Antarctic Peninsula. The Early Cretaceous age determined here for ocean floor along a length of profile AMG845-14 suggests that ocean floor at the peninsula margin was formed during earlier (Early Cretaceous or Late Jurassic) motion, probably between a separate Antarctic Peninsula plate and South America. No evidence supports the presence of Tertiary-age ocean floor in the western Weddell Sea.

Scotia arc evolution

Our present understanding of Scotia arc evolution has been reviewed recently by Barker *et al.* (in press). Essentially, progress is being made in understanding the geometry of a reconstructed Gondwana and the early stages of break-up, mainly from onshore geological, geochemical, isotopic and palaeomagnetic studies, and in working back from the present-day using marine geophysical methods. The least well known period lies between about 50 and 100 Ma, for which as yet there are no reliable marine data constraining SAM–ANT plate motion, and very little pertinent onshore geology. Using the additional constraint of the Early Cretaceous age for the north-west Weddell Sea on line AMG845-14, we offer a crude, speculative model of regional tectonic evolution through this period, considering firstly the more recent evolution of the Scotia Sea.

The Scotia Sea and surrounding Scotia Ridge have evolved through a series of extensional episodes behind an east-migrating island arc and trench ancestral to the present South Sandwich subduction zone, over the past 40 m.y. Changes in the mode of back-arc extension may have been triggered by ridge crest–trench collisions along the South Scotia Ridge, at progressively earlier times westward (Barker *et al.*, 1984). The main obstacles to detailed reconstruction of this evolution have been ignorance of motion of the bounding major (SAM–ANT) plates, of the coupling between short-lived spreading regimes in the back-arc, and of the nature and age of elevated blocks within the Scotia Ridge and Scotia Sea. Recently, Barker & Lawver (1988) have estimated SAM–ANT motion over the past 50 m.y., and new data bear on the age of Powell Basin opening and the nature of the South Orkney microcontinent (King & Barker, 1988; Barber, Barker & Pankhurst, this volume, p. 361), and the age of Bruce Bank (Toker, Barker & Wise, this volume, p. 639). The 20 Ma reconstruction of Barker *et al.* (1984) has been extended to 35 Ma (King & Barker, 1988) and more speculatively to 50 Ma (Toker *et al.*, this volume, p. 639). Many uncertainties remain, but these reconstructions have two persistent features in common: (a) the east-facing subduction zone remains continuous in the north with the foreland fold-thrust belt of the Magallanes Basin; and (b) southward, the subduction zone terminates at the SAM–ANT plate boundary, migrating eastward with successive ridge crest–trench collisions. The existence of these anomalous features at the *start* of Scotia Sea growth, about 50 Ma ago, provides a clue to earlier evolution.

Our knowledge of the tectonic evolution of the Scotia arc region, between latest Jurassic and Palaeogene, may be summarized as:

(a) Global plate motions are known approximately (e.g. Lawver *et al.*, 1985). Cretaceous SAM–ANT motion was about a south-east Pacific pole, but the early Cenozoic pole was closer to the South Geographic Pole. SAM–AFR motion remained east–west, after South Atlantic opening started in the Early Cretaceous, but was much faster in the Late Cretaceous than before or since.

(b) Palaeomagnetic constraints on the position of the Antarctic Peninsula are tightening (e.g. Grunow, Dalziel & Kent, 1987), but an initial position subparallel to the Falkland Plateau (Grunow *et al.*, 1987) or near-perpendicular (Longshaw & Griffiths, 1983) both remain possible. Independent motion of the peninsula appears essential, but its time extent remains uncertain.

(c) The timing of Rocas Verdes Basin opening and closure (Dalziel, 1981, 1985) is known, but not its original relation to other intra-Gondwana spreading systems, because the Scotia Sea has overridden the region south of the Falkland Plateau where the evidence lay. The Rocas Verdes Basin opened in the Late Jurassic, before the South Atlantic, so probably it is related to the earlier East–West Gondwana separation.

(d) Subduction at the Pacific margin of South America and the Antarctic Peninsula continued throughout.

(e) When Scotia Sea opening started, the microcontinental fragments (now dispersed) formed a compact, cuspate connection between South America and the Antarctic Peninsula. Directly to the east, a continuous convergent boundary extended from the Magallanes foreland fold-thrust belt in the north to the extensional SAM–ANT boundary in the Weddell Sea to the south.

(f) The SAM–ANT boundary did not continue down the eastern margin of the Antarctic Peninsula in the early Tertiary.

Three stages in our speculative model of the development of the SAM–ANT boundary are shown in Fig. 3, representing earliest Cretaceous, Late Cretaceous and Eocene. Major plate positions are based on Lawver *et al.* (1985). Besides the evidence listed above, the model incorporates the following assumptions.

(1) A strong causal connection between westward motion of South America and Rocas Verdes Basin evolution. The basin opened (subparallel to the Cordillera; Dalziel, de Wit & Palmer (1974)) before South Atlantic opening started. When South Atlantic opening started, basin opening stopped. When South Atlantic opening accelerated at about M0 time, the basin started to close.

(2) The Pacific margin of the Rocas Verdes Basin remained flexibly attached to the northern Antarctic Peninsula, from the Late Jurassic, when Gondwana break-up started, to earliest Tertiary.

(3) The Antarctic Peninsula's motion with respect to East Antarctica was probably confined to the period before South Atlantic opening (pre-M12). Fig. 3*a* reflects this early phase; it assumes the initial peninsula position given by Grunow *et al.* (1987), although the alternative (Longshaw & Griffiths, 1983) position is not ruled out. As the peninsula moved with respect to South America, first independently, then with East Antarctica about a south-east Pacific near pole, the narrow link with the Rocas Verdes outer arc flexed progressively and by the early Tertiary formed the broken cusp of Fig. 3*c*.

(4) The connection between the Rocas Verdes spreading ridge and the extensional SAM–ANT plate boundary may be oversimplified in Fig. 3*a*, but the combined boundary would have lain approximately as shown. As the Antarctic plate rotated about its near pole, the direction of motion along the boundary would have slowly changed, and its north-western part would inevitably have become convergent. Fig. 3*b* shows convergence at an early stage, extending south from the foreland fold–thrust belt into the region of oceanic lithosphere south-west of the Falkland Plateau. Fig. 3*c* shows the situation further advanced. Whereas in the narrow northern (Rocas Verdes) Basin the convergence zone soon meets buoyant continental lithosphere of the foreland, farther south it encounters SAM oceanic lithosphere. After this had been forced to some threshold depth it would continue sinking under its own weight, whereupon subduction proper would begin. This was the origin of east-facing subduction, featured in the 50 Ma (Toker *et al.*, this volume, p. 000 fig. 6) and subsequent reconstructions.

(5) The new subduction could only proceed by tearing the sinking SAM oceanic lithosphere along an east–west line near its northern margin (the Falkland Plateau), and by eastward roll-back of the hinge of subduction. As Pacific ocean floor was being subducted at the western margin at the same time, roll-back led inevitably to back-arc extension, creating the Scotia Sea.

This story suggests how the SAM–ANT boundary evolved since the Early Cretaceous, and how east-directed subduction began. The model in Fig. 3 is approximate, because of uncertainties in, for example, the age of older Weddell Sea

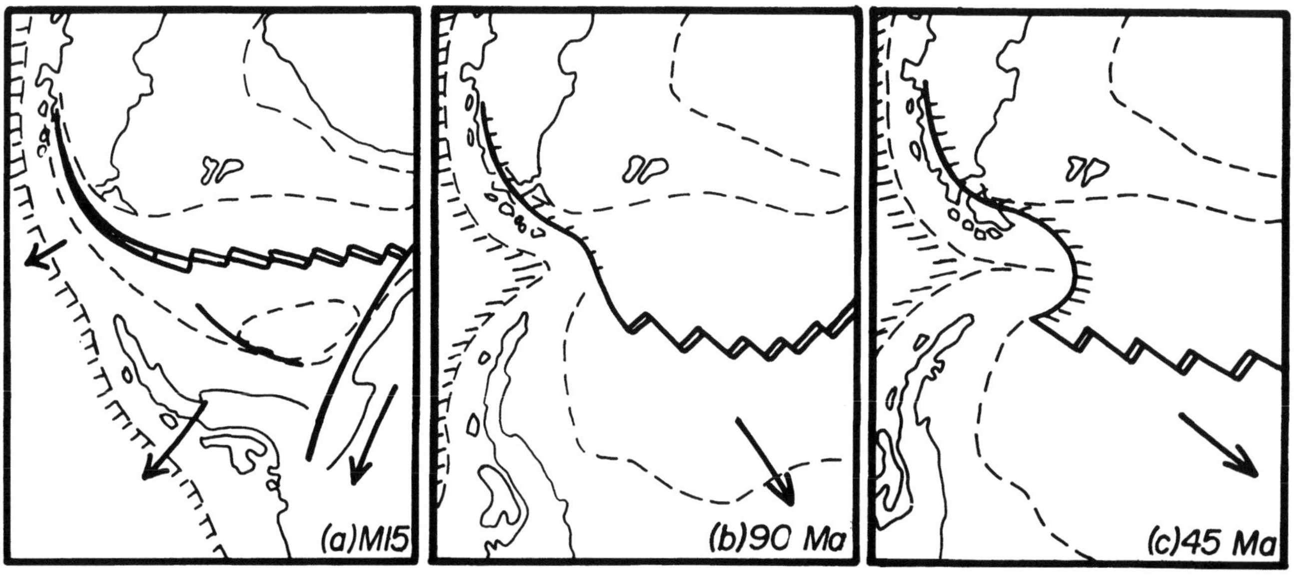

Fig. 3. Model of Scotia arc evolution at (*a*) M15 time (~ 140 Ma); (*b*) 90 Ma; and (*c*) 45 Ma, suggesting how changes in South American–Antarctic Plate motion led to closure of the Rocas Verdes Basin in the north, and to subduction of South Atlantic ocean floor with complementary back-arc extension in the south.

anomalies and the original position of the Antarctic Peninsula. More precise data may reveal the conditions necessary for the onset of subduction, and provide better estimates of the palaeomotion of the Antarctic Peninsula.

Acknowledgements

The work on which this paper is based was carried out under a British Antarctic Survey research contract to the Department of Geological Sciences, University of Birmingham.

References

Barker, P.F., Barber, P.L. & King, E.C. (1984). An Early Miocene ridge crest–trench collision on the South Scotia Ridge near 36W. *Tectonophysics*, **102**, 315–32.

Barker, P.F., Dalziel, I.W.D. & Storey, B.C. (in press). Tectonic development of the Scotia arc region. In *Antarctic Geology*, ed. R.J. Tingey. Oxford; Oxford University Press.

Barker, P.F. & Jahn, R.A. (1980). A marine geophysical reconnaissance of the Weddell Sea. *Geophysical Journal of the Royal Astronomical Society*, **63**, 271–83.

Barker, P.F. & Lawver, L.A. (1988). South American–Antarctic plate motion over the past 50 Ma, and the evolution of the South American–Antarctic Ridge. *Geophysical Journal of the Royal Astronomical Society*, **94**, 377–86.

Crough, S.T. (1983). The correction for sediment loading on the seafloor. *Journal of Geophysical Research*, **88**, 6449–54.

Dalziel, I.W.D. (1981). Back-arc extension in the Southern Andes: a review and critical reappraisal. *Philosophical Transactions of the Royal Society of London*, **A300**, 300–35.

Dalziel, I.W.D. (1985). Collision and Cordilleran orogenesis: an Andean perspective. In *Collision Tectonics*, ed. M.P. Coward & A.C. Ries, pp. 389–404. Oxford; Blackwells Scientific Publications, Geological Society of London, Special Publication **19**.

Dalziel, I.W.D., de Wit, M.S. & Palmer, K.F. (1974). A fossil marginal basin in the southern Andes. *Nature, London*, **250**, 291–4.

Davis, E.E. & Lister, C.R.B. (1974). Fundamentals of ridge crest topography. *Earth and Planetary Science Letters* **21**, 405–13.

Gordon, R.G. & Cox, A. (1980). Paleomagnetic test of the Early Tertiary plate circuit between the Pacific basin plates and the Indian plate. *Journal of Geophysical Research*, **85**, 6534–46.

Grunow, A.M., Dalziel, I.W.D. & Kent, D.V. (1987). Ellsworth Mountains–Whitmore Mountains crustal block, western Antarctica: new paleomagnetic results and their tectonic significance. In *Gondwana Six: Structure, Tectonics, and Geophysics*, ed. G.D. McKenzie, pp. 161–71. Washington, DC; American Geophysical Union, Geophysical Monograph 40.

Harland, W.B., Cox, A.V., Llewellyn, P.G., Pickton, C.A.G., Smith, A.G. & Walters, R. (1982). *A Geologic Time Scale*. Cambridge; Cambridge University Press. 131 pp.

Haugland, K., Kristoffersen, Y. & Velde, A. (1985). Seismic investigations in the Weddell Sea embayment. *Tectonophysics*, **114**, 293–313.

Heestand, R.L. & Crough, S.T. (1981). The effects of hot spots on the oceanic age-depth relation. *Journal of Geophysical Research*, **86**, 6107–14.

Hinz, K. & Krause, W. (1982). The continental margin of Queen Maud Land, Antarctica: seismic sequences, structural elements and geological development. *Geologische Jahrbuch*, **E23**, 17–41.

Horowitz, D.H. (1976). Mathematical modelling of sediment accumulations in prograding deltaic systems. In *Quantitative Techniques for the Analysis of Sediments*, ed. D.F. Merriam, pp. 105–19. London; Pergamon Press.

Jurdy, D.M. (1978). An alternative model for Early Tertiary absolute plate motions. *Geology*, **6**, 469–72.

King, E.C. & Barker, P.F. (1988). The margins of the South Orkney microcontinent. *Journal of the Geological Society, London*, **145(2)**, 317–31.

LaBrecque, J.L. & Barker, P.F. (1981). The age of the Weddell Basin. *Nature, London*, **290**, 489–92.

LaBrecque, J.L., Cande, S.C., Bell, R., Raymond, C., Brozena, J., Keller, M., Parra, J.C. & Yanez, G. (1986). Aerogeophysical survey yields new data in the Weddell Sea. *Antarctic Journal of the United States*, **21(5)**, 69–71.

Lawver, L.A., Sclater, J.G. & Meinke, L. (1985). Mesozoic and Cenozoic reconstructions of the South Atlantic. *Tectonophysics*, **114**, 233–54.

LeDouaran, S. & Parsons, B. (1982). A note on the correction of ocean floor depths for sediment loading. *Journal of Geophysical Research*, **87**, 4715–22.

Longshaw, S.K. & Griffiths, D.H. (1983). A palaeomagnetic study of Jurassic rocks from the Antarctic Peninsula and its implications. *Journal of the Geological Society, London*, **140**, 945–54.

Molnar, P., Atwater, T., Mammerickx, J. & Smith, S.M. (1975). Magnetic anomalies, bathymetry and the tectonic evolution of the South Pacific since the Late Cretaceous. *Geophysical Journal of the Royal Astronomical Society*, **40**, 383–420.

Nafe, J.E. & Drake, C.L. (1957). Variation with depth in shallow and deep water marine sediments of porosity, density and the velocities of compressional and shear waves. *Geophysics*, **22**, 523–52.

Norton, I.O. & Sclater, J.G. (1979). A model for the evolution of the Indian Ocean and the break-up of Gondwanaland. *Journal of Geophysical Research*, **84**, 6803–30.

Parker, R.L. & Oldenburg, D.W. (1973). Thermal model of ocean ridges. *Nature, Physical Sciences*, **242**, 137–9.

Parsons, B.L. & Sclater, J.G. (1977). An analysis of the variation of ocean floor bathymetry and heat flow with age. *Journal of Geophysical Research*, **82**, 803–27.

Sclater, J.G. & Christie, P.A.F. (1980). Continental stretching: an explanation of the post-mid-Cretaceous subsidence of the central North Sea basin. *Journal of Geophysical Research*, **85**, 3711–39.

Smith, A.G. & Hallam, A. (1970). The fit of the southern continents. *Nature, London*, **225**, 139–44.

Stock, J.M. & Molnar, P. (1982). Uncertainties in the relative positions of the Australia, Antarctica, Lord Howe and Pacific plates since the Late Cretaceous. *Journal of Geophysical Research*, **87**, 4697–714.

Stock, J.M. & Molnar, P. (1987). Revised history of early Tertiary plate motion in the south-west Pacific. *Nature, London*, **325**, 495–9.

Trehu, A.M. (1975). Depth versus (age)$^{1/2}$: a perspective on mid-ocean rises. *Earth and Planetary Science Letters*, **27**, 287–304.

Verification of crustal sources for satellite elevation magnetic anomalies in West Antarctica and the Weddell Sea and their regional tectonic implications

M.E. GHIDELLA[1], C.A. RAYMOND[2] & J.L. LABRECQUE[2]

1 Direccion Nacional del Antarctico, Cerrito 1248, Buenos Aires, Argentina and Comision Nacional de Investigaciones Espaciales, San Miguel, Buenos Aires, Argentina
2 Lamont–Doherty Geological Observatory of Columbia University, Palisades, New York, USA

Abstract

We have investigated the correlation between dense aeromagnetic data and the Magsat satellite magnetic anomaly field for the region of the northern Antarctic Peninsula, South Shetland Islands and Bransfield Strait. We demonstrate, via two- and three-dimensional models, that crustal structure variations arising from volcanic and plutonic activity related to palaeosubduction at the Pacific margin of the Antarctic Peninsula are the source of these satellite elevation anomalies. Our ability to assign crustal sources to the satellite anomalies in this region has encouraged us to extrapolate our modelling effort, on a regional scale, to the margin of West Antarctica and the Weddell Sea. Several interesting inferences can be drawn from this modelling, including the continuation of the Andean Intrusive Suite from the base of the Antarctic Peninsula to Thurston Island, and the existence of a very broad region of anomalous oceanic crust at the Dronning Maud Land margin, adjacent to previously mapped seaward dipping reflectors.

Introduction

Despite the interference of external field noise at high geomagnetic latitudes, the Magsat satellite magnetic anomaly field over Antarctica appears to contain valuable information which can be used to infer regional scale variations in crustal structure. The Magsat mission, flown by NASA in 1979, measured both the scalar and vector components of Earth's magnetic field. Various techniques have been used by investigators to isolate the crustal signal of the field from the main field and external field contributions. This task is made more difficult at high latitude because of the presence of the auroral electrojet and field-aligned currents in the ionosphere. As the mission was flown during the austral summer, the worst contamination occurred in the southern polar region because of nearly perpetual illumination. Of the existing versions of the Magsat gridded scalar anomaly field which have been compiled for high latitudes, we have considered two versions: those of Ritzwoller & Bentley (1982, 1983) and Arkani–Hamed & Strangway (1985). Figs 1*a* and *b* show the two fields. We have included only the sector corresponding to our present region of interest; West Antarctica and the Weddell Sea. The techniques used in deriving each map differ. In the map of Ritzwoller & Bentley (R&B), the authors have carefully eliminated all satellite passes which appeared to be contaminated by external fields, using cut-off values for K_p (the planetary magnetic storm

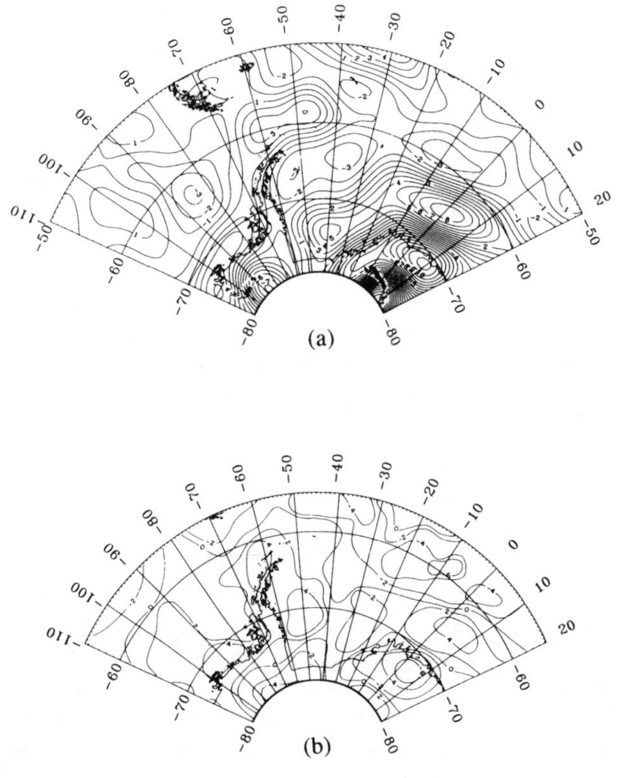

Fig. 1. **(*a*) Magsat field of Arkani–Hamed & Strangway (1985); and (*b*) Ritzwoller & Bentley (1983). Contours in nT. Negative numbers denote west longitudes and south latitudes. See text for explanation of the derivation of the two fields.**

activity index), and anomaly amplitude, and subsequent removal of a least-squares polynomial fit to the profiles. In the map of Arkani–Hamed & Strangway (A&S), the authors have taken into account the asymmetry between dawn and dusk satellite passes, eliminating the antivariant and contaminated covariant harmonics of the two data sets, in order to remove quasi-stable components of the external magnetic field, and isolate the crustal field.

Comparison of the two maps reveals similarities but also important differences. Both maps show positive anomalies that extend along the west coast of the Antarctic Peninsula, from Thurston Island (73° S, 97° W) to the South Orkney Islands (62° S, 45° W), but their amplitudes and shapes differ. The strong positive anomaly of A&S smoothly follows the margin and extends northeastwards along the South Scotia Ridge. The map of R&B shows isolated positive anomalies in the region of Thurston Island and the South Shetland Islands, and a negative gradient towards the South Orkney Islands. Both maps display negative anomalies over the Dronning Maud Land margin and Ellsworth Mountains region. A negative correlation between the two maps exists within the Weddell Sea, R&B showing a positive anomaly which reaches 4 nT, A&S a negative anomaly of 2 nT. This is, however, the region of maximum auroral current activity, so the differences may be an artifact of the processing techniques used to eliminate the external field noise.

Despite the aforementioned differences, the anomalies in general seem to correlate fairly well with sources of crustal origin, as was recognized by Ritzwoller & Bentley (1982). However, in order to arrive at a more definitive conclusion regarding their origin, and to evaluate the differences between the two maps, it is necessary to compare them with an independently derived anomaly field, as has been done for Magsat anomalies in other regions of the world. A positive correlation of the satellite field with crustal composition and low level magnetic observations enhances our confidence in the Magsat field for other unsurveyed regions of the Antarctic. The USAC (United States–Argentina–Chile) aerogeophysical survey of the basins surrounding the Antarctic Peninsula has added an independent geomagnetic data set for comparison with the Magsat field, providing a dense pattern of flights extending from Tierra del Fuego to the southern Weddell Basin. We have taken advantage of this extensive new data source to identify the crustal component present in the Magsat field in this region of the Antarctic, and to characterize the sources through two- and three-dimensional models which incorporate geological and geophysical information.

The USAC aerosurvey has added 150 000 km of aeromagnetic data in previously unsurveyed regions of the Weddell Basin, with a track spacing generally on the order of 30 km, far exceeding our required sampling density. Transits to and from the survey area in the Weddell Sea resulted in dense two-dimensional coverage of the southern Drake Passage, South Shetland Islands (SSI), Bransfield Strait (BS) and northern Antarctic Peninsula (Trinity Peninsula), as shown in Fig. 3*c*. Following detailed modelling of this area, we will expand our crustal modelling on a regional scale to the Weddell and Bellingshausen margins.

The Bransfield Strait region

Two-dimensional modelling

Selected aeromagnetic tracks, collected during the 1985–86 and 1986–87 field seasons of the USAC aerosurvey, were used in a two-dimensional modelling study of this region. The magnetic profiles were projected along a trend of 150° (normal to the strike of Bransfield Strait) and averaged to create a composite profile. Averaging enhances the two dimensionality of the profile by reducing contributions from local anomalies of the Drake Passage and Weddell Sea and measurement noise. The average profile at 7 km altitude (Fig. 2)

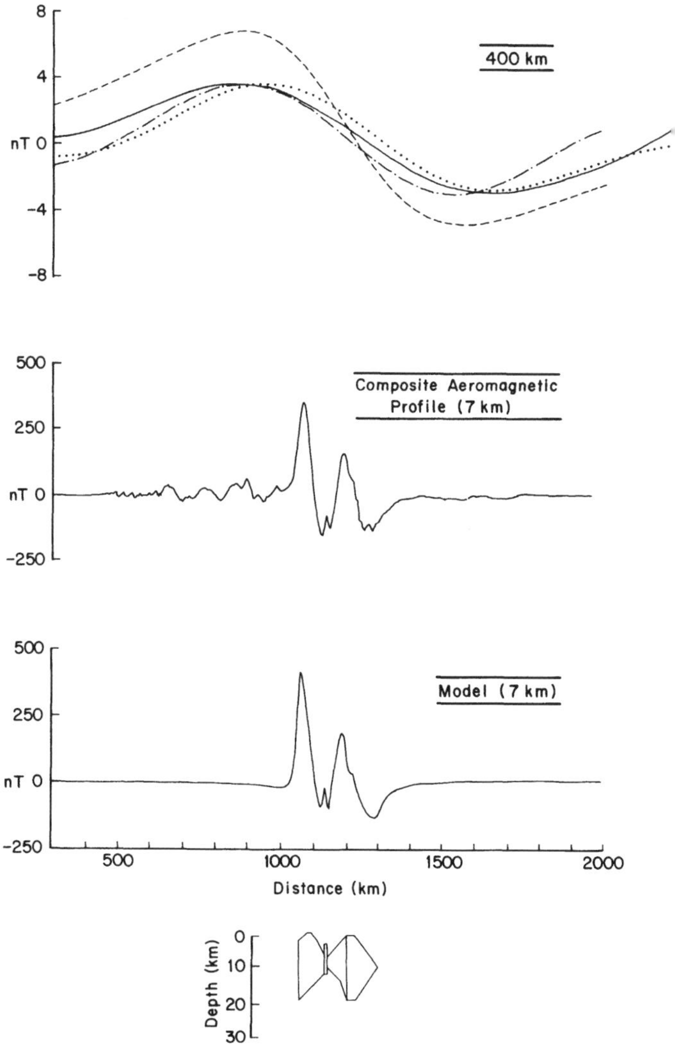

Fig. 2. Two-dimensional model of the Bransfield Strait region. At bottom are the model bodies used to calculate the anomaly profile shown immediately above, which was calculated at 7 km altitude. The middle profile is the average composite aeromagnetic profile across the region, collected at 7 km altitude. Shown at the top of the figure are satellite elevation profiles as follows: solid line, Magsat field of Arkani–Hamed & Strangway (1985); dashed line, upward continued average aeromagnetic profile; dot–dash line, filtered upward continued average aeromagnetic profile; dotted line, filtered upward continued model profile. The filtering, which is done in the wavenumber domain allows a proper comparison to be made between the model and aeromagnetic data and the Magsat field. We have tried to match both the aeromagnetic profile and the satellite profile with our model of structural variations beneath Bransfield Strait.

exhibits an average maximum of 350 nT over the South Shetland Islands which is attributed to the highly magnetized Andean Intrusive Suite by Parra, Gonzalez-Ferran & Bannister (1984). A sediment-filled trough in Bransfield Strait results in a negative anomaly, while the small positive anomaly within the trough defines the axis of recent volcanism in this extensional envirionment. The Andean Intrusive Suite also crops out on Trinity Peninsula, resulting in the 150 nT high over the western Antarctic Peninsula margin (Parra *et al.*, 1984). The magnetized zone on the Antarctic Peninsula ends abruptly at the contact of the Andean Intrusive Suite with the metasedimentary rock of the Trinity Peninsula Group, which was shown by Parra *et al.* (1984) to be essentially non-magnetic. The margin of the Weddell Sea exhibits a low amplitude, long wavelength negative anomaly.

In Fig. 2 we show the composite aeromagnetic profile at 7 km altitude (middle) and upward continued to satellite elevation (top, broken line). When the aeromagnetic profile is upward continued (via a continuation operator in the Fourier domain) to satellite elevation (400 km), a smoothly varying field is observed, with a high of 6.2 nT over the Drake Passage and a low of −4.8 nT in the Weddell Sea. A strong gradient in the field occurs over the South Shetland Islands, Bransfield Strait and Antarctic Peninsula. To compare the aeromagnetic data with the satellite field, we have sampled the gridded field of A&S along the composite profile (Fig. 2, top, solid line). Since the aeromagnetic data support the field of A&S, we will compare our models to this field. The field of R&B is similar outside the Weddell Basin, so the correlations will generally apply to the field of R&B as well.

A comparison between the upward continued average aeromagnetic profile and the Magsat profile shows a good agreement in phase but a noticeable difference in amplitude and wavelength. These differences may be attributed to the mathematical treatment of the Magsat data: in order to suppress the dawn–dusk asymmetry, A&S have bandpass filtered the spherical harmonic components of the data (see Arkani-Hamed & Strangway, 1986). To make a proper comparison, our profile must also be filtered, to attenuate the high and very low order components present in the broadband aeromagnetic data. As can be seen from the filtered profile (Fig. 2, top, dot-dash line), bandpass filtering does account for most of the difference.

To quantify this comparison, and to determine its implications for the magnetic composition of the crust, we have created a two-dimensional model which expands upon the modelling of Parra *et al.* (1984) to include deep structure under the South Shetland Islands and the Antarctic Peninsula. The model considers magnetizations aligned along the present field direction and utilizes susceptibilities measured in the laboratory on rock samples (Allen, 1966), as reported in Parra *et al.* (1984). The model is shown in Fig. 2 (bottom), directly above is the anomaly profile calculated at 7 km, and at the top the anomaly profile calculated at 400 km (dotted line). The model exhibits a good match with both the low altitude composite aeromagnetic profile and the Magsat profile (after the model has been bandpass filtered equivalently to the Magsat data). We conclude that the source bodies used to model the aeromagnetic profile at low altitude can also explain the satellite elevation anomaly profile. Although the model lacks fine detail, it illustrates that the source of the satellite anomalies resides in the highly magnetized Andean Intrusive Suite, which we estimate extends to a depth of nearly 20 km beneath the SSI and north-western coast of Trinity Peninsula. The upper surface of the model bodies corresponds to the known topography, which has been compiled in atlas form by LaBrecque (1986), and we have used the seismic result of Ashcroft (1972) as a guide in constructing the lower bounds of the source bodies.

In our modelling, we have considered that the oceanic crust of the Drake Passage will not contribute to the satellite elevation anomaly pattern (because the very short wavelengths of Cenozoic seafloor-spreading anomalies tend to average to zero at longer wavelength). We have tried another version of the model which takes into account the induced magnetization of the slab of oceanic crust dipping beneath the South Shetland Islands and Bransfield Strait. The contribution of this hypothetical slab is extremely small and, considering the lack of seismic evidence for such a slab, we conclude that it is unnecessary to include it when modelling the region.

Three-dimensional modelling

Aeromagnetic data density in the Bransfield Strait region is adequate to allow a gridded anomaly field to be constructed with high resolution. Therefore, although the structure of the Bransfield Strait region is approximately two dimensional, we have investigated three-dimensional models at low altitudes as well for this region. The three-dimensional model employs simple blocks of constant susceptibility representing the South Shetland Islands (0.006 cgs), Bransfield Strait (0.008 cgs), the intrusive suite bordering the western margin of Trinity Peninsula (also 0.006 cgs), and South Scotia Ridge (0.003 cgs).

Fig. 3*d* shows the areal extension of the magnetized bodies employed in the modelling, line A–A′ indicates where the cross-section of the bodies is similar to that in the two-dimensional model (Fig. 2, bottom), although without so much detail. These large crustal blocks can account for the main features of the gridded aeromagnetic data. Fig. 3*a* shows the gridded aeromagnetic anomaly field and Fig. 3*b* the modelled anomaly field. Examination of the fields in Figs 3*a* & *b* reveals several interesting features: the South Shetland Islands block terminates abruptly in the north-east as does the magnetized portion of the Bransfield Strait extensional zone; the negative anomaly associated with the bathymetric trough in Bransfield Strait appears to continue farther to the north-east; the deep intrusive body that follows the north-western margin of the Antarctic Peninsula continues north-eastwards about 200 km past the tip of the peninsula towards Elephant Island.

We find that a body which corresponds to the South Scotia Ridge is necessary in order to account for the linear positive anomaly extending north-east from the Bransfield Trough. This rectangular body shows fair agreement with the findings of Watters (1971); however, we employed a susceptibility of 0.003 cgs with a top at 3 km and a bottom at 12 km in our

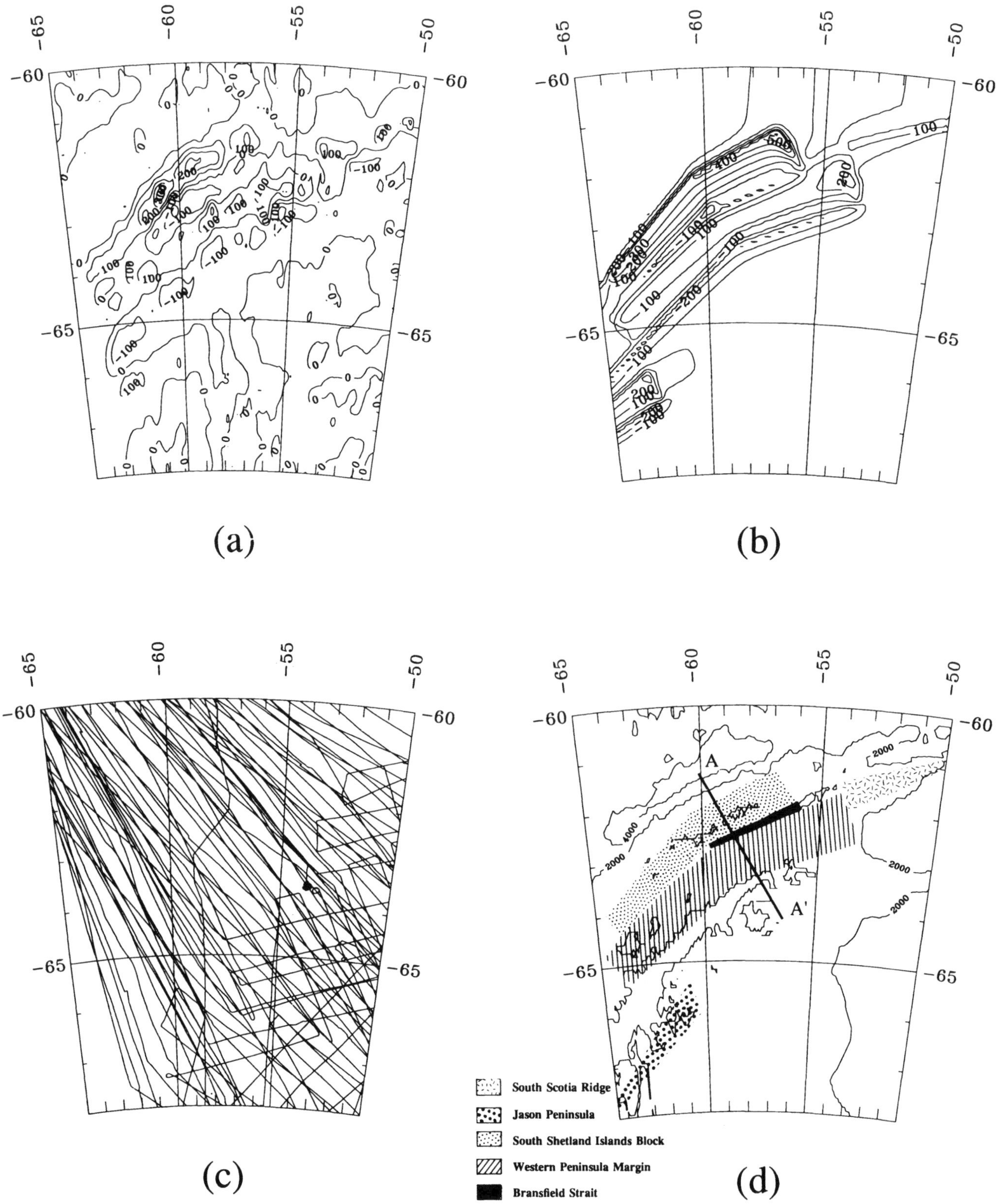

Fig. 3. ***(a)*** **Gridded USAC aeromagnetic data in the Bransfield Strait region;** ***(b)*** **anomaly field calculated at 7 km altitude for the model bodies shown in** ***(d)*****;** ***(c)*** **aeromagnetic tracks used to derive the gridded field shown in** ***(a)*****;** ***(d)*** **plan view of the model bodies incorporated into a low altitude, three-dimensional model of the region. Line A–A′ marks the position where the model corresponds to the 2-D model in cross-section. Susceptibility contrast employed in the model was 0.006 cgs for the Antarctic Peninsula margin, Jason Peninsula and South Shetland Islands, 0.008 cgs for Bransfield Strait and 0.003 cgs for the South Scotia Ridge.**

model, whereas Watters (1971) used 0.006 cgs and 16 km thickness for a flat bottomed prism-shaped body. Watters' objective was to model the high frequency anomalies found on two sea-surface profiles crossing the South Scotia Ridge. At 7 km altitude the high frequencies are attenuated, so that we are modelling the bulk magnetization contrast, which is more a function of broad structural variations or changes in petrology than shallow intrusions.

At the south-western end, the magnetized blocks continue past the border shown in Fig. 3*d*, but this is where the aero-

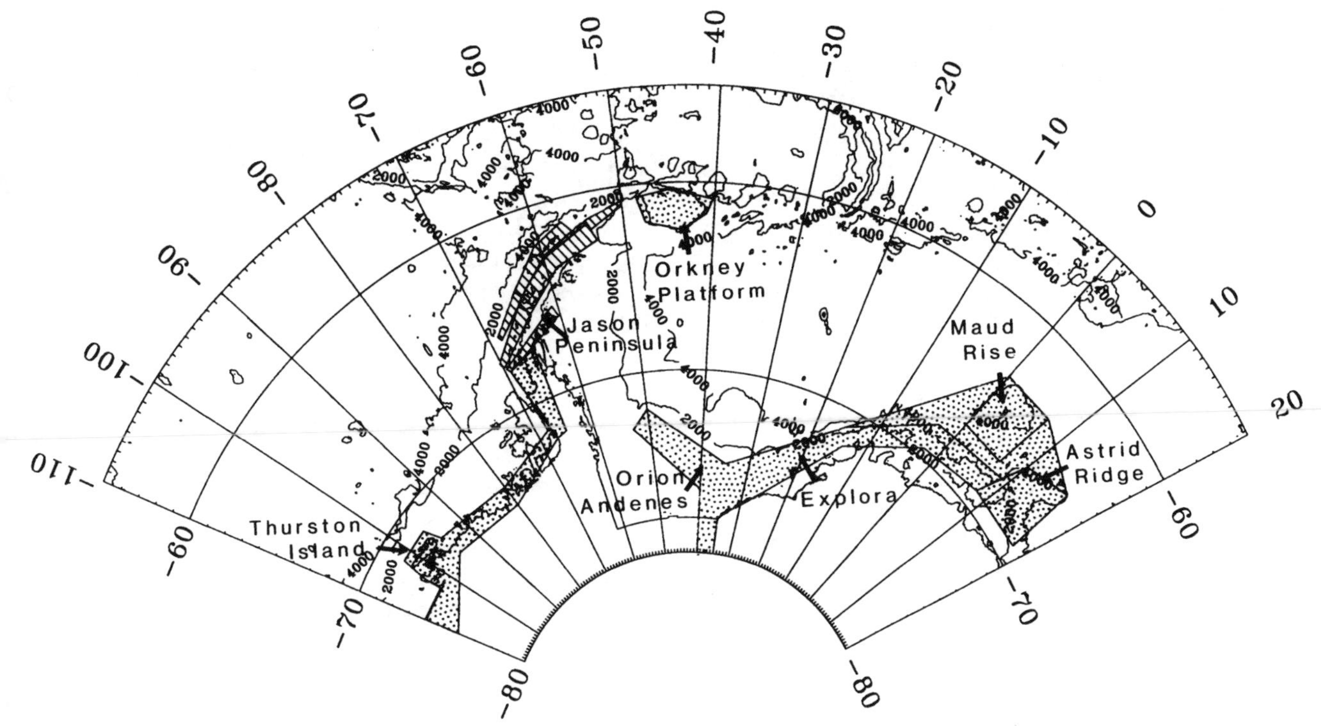

Fig. 4. Plan view of the model bodies employed in the satellite elevation regional three-dimensional model. The lined pattern denotes bodies discussed in the low-altitude section; other bodies are highlighted by the dotted pattern. The thickness, shape and susceptibility of each body is given in the text. GEBCO bathymetry, contoured in metres, is shown for reference.

magnetic data set ends. We will discuss their extension in the next section. We have included a body representing the Jason Peninsula, whose anomaly field was mapped by the aerosurvey. Just the north-eastern tip of this body is apparent in the model field (lower left). We find the Jason Peninsula anomaly to be lineated NE–SW, contrary to Renner, Sturgeon & Garrett (1985), who show it as two discreet bullseyes. The Jason Peninsula area may be comprised of discreet shallow intrusions upon a more continuous buried layer.

Regional modelling of the satellite field

Many intriguing correlations with structural features can be seen in the Magsat map of A&S. Although dense aeromagnetic data coverage such as we used to constrain the low altitude modelling is absent, we have utilized available geological and geophysical data in the sector extending from Thurston Island to the Astrid Ridge and southern Weddell Basin to the South Orkney Islands to infer broad source distributions for the satellite elevation anomalies. In Fig. 5*a* a three-dimensional modelled anomaly field is shown, calculated at 400 km altitude, for which the source body configuration is given in Fig. 4. For comparison, the field of A&S is shown in Fig. 5*b*. The following discussion is split into three geographical areas: the northern Antarctic Peninsula, the western Antarctic Peninsula margin and the Weddell Sea–East Antarctic margin. Throughout the three-dimensional modelling, with the exception of the bodies already discussed in the low altitude modelling, we have employed a susceptibility of 0.006 cgs for the blocks, which range in thickness from 6 to

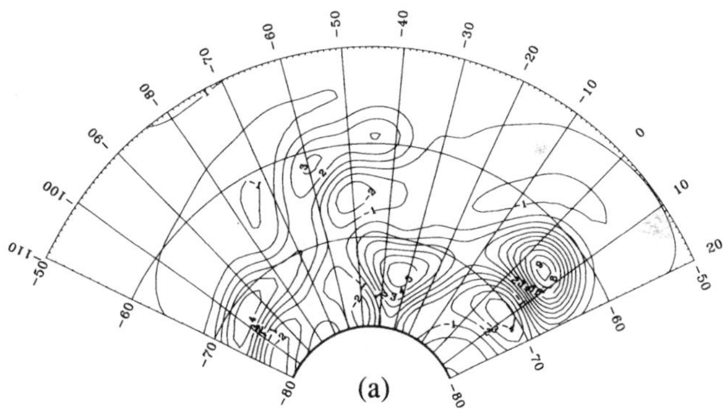

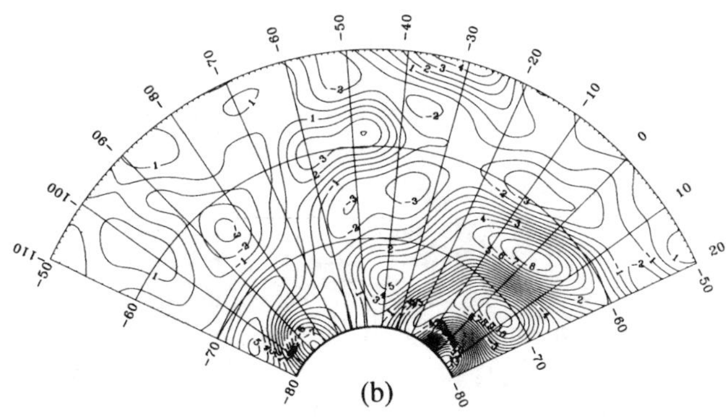

Fig. 5. (*a*) Model anomaly field generated by the bodies shown in Fig. 4; (*b*) Magsat field of Arkani–Hamed & Strangway (1985). Contours in nT.

15 km. This susceptibility value is generally higher than the values which appear in the susceptibility map of Arkani–Hamed & Strangway (1985) which they derived from inverting the anomaly field assuming induced magnetization. However, we chose to place the source bodies in the upper crust, whereas Arkani–Hamed & Strangway (1985) assumed the magnetized crust extended to a depth of 50 km. Our value of susceptibility is reasonable for calc-alkaline island arc plutonic assemblages, and it also agrees with values derived from forward modelling of oceanic plateaus by Frey (1983) and Raymond & LaBrecque (1984), which applies to the case of Maud Rise. Continental margins whose seismic sequences display seaward-dipping reflectors also have large amplitude magnetic anomalies associated with them (Kristoffersen & Haugland, 1986; Hinz, Mutter & Zehnder, 1987). Therefore, we feel our susceptibility estimate is justified; nevertheless, more detailed modelling at low altitude will be required to refine the estimate of susceptibility contrast.

Northern Antarctic Peninsula

The high altitude field calculated for the source bodies of Fig. 3*d* correlates well with the field of A&S, reproducing the western portion of the 3 nT elongate anomaly which extends from the area of the SSI to the Orkney Platform. At this geomagnetic latitude, the anomalies are displaced to the north of the source bodies. We have investigated the source of the eastern half of this anomaly, and find it to be associated with the Orkney Platform, a crustal block which has an affinity to the Antarctic Peninsula (Harrington, Barker & Griffiths, 1972; Garrett *et al.*, 1987). Seafloor-spreading anomalies identified by the USAC aerosurvey in the Powell Basin confirm the restoration of the South Orkney Block as a continuation of the Antarctic Peninsula. Our source body for the South Orkney Block has a lower bound of 18 km; its upper surface deepens from a depth of 2 km in the south to 7 km in the north, after seismic results of Harrington *et al.* (1972), and its outline follows the bathymetry. The elongate satellite anomaly is truncated to the east of the South Orkney Block by a negative anomaly. The elongate anomaly is attenuated to the south-west; however positive energy continues in the field along the entire west coast of the peninsula. Therefore, we have continued the SSI and peninsula margin bodies (Andean Intrusive Suite) south-west; they terminate near Adelaide Island. Their top is flat, at 1 km depth, and their cross-section is rectangular, the lower bound being at a depth of 10 km. Extension of these bodies is necessary to avoid a negative edge effect anomaly near Anvers Island. Continuation of the bodies which represent the calc-alkaline plutons of the Andean Intrusive Suite along the margin is consistent with the aeromagnetic data of the British Antarctic Survey (BAS) over the Antarctic Peninsula, which displays a pronounced ridge of positive energy in the anomaly field, following the west coast of the Antarctic Peninsula (Renner *et al.*, 1985) and also agrees with geological information (for review, see Saunders & Tarney, 1982).

The USAC aeromagnetic data were used to define the Jason Peninsula body. We constructed two blocks, one on top of the other, with flat tops and bottoms; one of them extends from sea level to 6 km depth, the other from 6 to 10 km. The eastern limit of the shallow body is the east coast of the Peninsula, while the deeper body extends farther out into the Weddell Sea, following the outline of the Jason Peninsula, as can be seen in Fig. 4. The western termination of the bodies was based on a Project Magnet aeromagnetic profile, which shows a large positive anomaly over the Antarctic Peninsula aligned with the Jason Peninsula trend. The contoured anomaly map of Renner *et al.* (1985), shows a low over Marguerite Bay, south of Adelaide Island, which could be an edge effect anomaly resulting from the termination of this magnetized zone.

Western Antarctic Peninsula margin

This region was modelled using the constraints of the BAS aeromagnetic survey and geological information. According to the satellite anomaly map of A&S, the magnetized crust continues south along the western Peninsula margin, and the anomaly intensifies as Thurston Island is approached. In the aeromagnetic contour map of Renner *et al.* (1985), Alexander Island appears magnetically quiet, while the crust of western Palmer Land produces anomalies that reach 600 nT. Therefore we have added a body which roughly outlines the high intensity zone of western Palmer Land as seen in the BAS map. It is also rectangular in cross-section, with a top of 1 km depth and a lower bound of 12 km. It is met at its southern termination by a body which extends along the Bellingshausen margin from south of Alexander Island to Thurston Island and into the Amundsen Sea. This body is largely unconstrained by supporting geological and geophysical data. However, geological mapping of Thurston Island and the Eights Coast has revealed volcanic and gabbroic sequences which are very intensely magnetized, as well as calc-alkaline intrusive rocks (Storey *et al.*, this volume, p. 399). Our modelling of the satellite anomaly field suggests that the West Coast Magnetic Anomaly (WCMA) of Renner *et al.* (1985) continues along the Bellingshausen margin and beyond, and is supplemented by a local positive anomaly over Thurston Island. The positive anomaly terminates in the Amundsen Sea, which is consistent with the plate tectonic history of the area deduced by Weissel, Hayes & Herron (1977) and Watts *et al.* (1988). The Amundsen Sea marks the transition from the formerly subducting West Antarctic margin to a passive margin formed during separation of New Zealand from West Antarctica.

Weddell Sea–East Antarctic margin

The satellite field also shows a positive anomaly over the southern margin of the Weddell Sea which is in clear correspondence with the margin parallel Orion anomaly detected by the USAC aerosurvey. This anomaly is interpreted as a delineation of the ocean–continent boundary at a passive margin. We have modelled its gross structure with a body at 4 km depth and 11 km thick, to simulate thick igneous crust similar to that found at other volcanic passive margins. The Orion anomaly merges with the Andenes magnetic anomaly (Kristoffersen & Haugland, 1986) indicating its association with the Andenes and Explora Escarpments which have been mapped seismically farther to the east (Haugland, Kristoffersen & Velde, 1985; Kristoffersen & Haugland, 1986; Hinz & Kristoffersen, 1987). We have extended the Orion body along the Andenes–Explora Escarpment trend up to the Maud Rise. Another branch of this body extends south to 85° S, following the contours of the satellite field. The satellite magnetic field data are considered to be contaminated by unresolved external field components below 75° S. However, the satellite anomaly extending into the Filchner Ice Shelf seems to be associated with the Orion anomaly so that we tentatively conclude that it may have a crustal source. A large positive anomaly occurs over the Maud Rise, which we have modelled as a rectangular body 14 km thick. The intense positive which is associated with the Maud Rise continues farther to the east. Therefore we have also included a block representing the thickened oceanic crust of Astrid Ridge, 16 km thick, and an intervening thickened oceanic crustal layer, following the results of Hinz & Krause (1982), LaBrecque & Keller (1982) and Bergh (1987). This results in a good agreement with the satellite field.

Discussion

We have been able to show, through two- and three-dimensional models at low and high altitude, and comparison to an extensive new aeromagnetic data set, that the bulk of the sources for satellite elevation magnetic anomalies in West Antarctica and the Weddell Sea may reside in the mid–upper crust. Furthermore, we can relate the positive satellite anomaly following the western margin of the Antarctic Peninsula to the calc-alkaline magmatic arc rocks exposed on the peninsula and South Shetland Islands and also occurring on Thurston Island and Eights Coast. Our modelling suggests continuity of this structure, at least on a broad scale, along the entire western Antarctic Peninsula margin and the margin of the Bellingshausen Sea, as far as the Amundsen Sea. We can also relate another intense satellite anomaly which follows the southern Weddell Sea margin and Dronning Maud Land margin to buried volcanic sequences which have been mapped seismically east of the Filchner Ice Shelf (Hinz & Kristoffersen, 1987). It appears that the Orion anomaly mapped by the USAC aerosurvey is associated with a structure which is continuous (on a regional basis) with the Andenes Escarpment, and there is a possibility that a third arm of this structure enters the continent, beneath the Filchner Ice Shelf, in the vicinity of the Transantarctic Mountains. The continuous nature of this anomaly in the satellite field may belie important structural breaks. It cannot be ascertained from examining the satellite field whether the Andenes and Orion anomalies arise from structures which are continuous, or if they are separated by a boundary such as a shear zone.

Our three-dimensional modelling of the satellite magnetic anomaly field is not meant to resolve specific information about the source bodies of the high altitude anomalies. It is meant instead to be used as a reconnaissance tool for predicting the structure and composition of the crust on a regional basis in poorly surveyed areas. Our assumption of purely induced magnetization is one which we ourselves would call into question. Although remanent magnetism can be effectively ignored in modelling continental crustal sources for satellite anomalies, because the source regions are generally of small dimensions (< 200 km) and would tend to cancel each other out when averaged over a broad area, there is still a possibility that remanent magnetism may be important in the satellite magnetic anomaly field. In particular, plutonic rocks of the Andean Intrusive Suite as mapped on the Antarctic Peninsula show ages which cluster in the mid-Cretaceous, so that their magnetization would have been acquired during the Cretaceous Long Normal interval of constant polarity of the geomagnetic field. This would result in a much more coherent source of remanent magnetism than if the rocks were intruded during periods of rapid fluctuations in the geomagnetic field, such as during the Cenozoic. Furthermore, subsequent remagnetization or viscous magnetizations would be coeval with the primary magnetization direction, enhancing the importance of remanent components of magnetization as sources of the satellite anomalies. It has been recognized in the Andes of Central Chile that mid-Cretaceous plutonic rocks appear to be more intensely magnetized than plutons of other ages (Parra & Yanez, in press). However, for this region of the world, the remanent direction is scarcely different from the present field direction so that we may offer the possibility that the sources could be combined remanent and induced. In the case of the South Scotia Ridge, which is presumed to be composed of several fragments which have rotated independently since emplacement of the plutonic suite (Garrett *et al.*, 1987), the observed decrease in magnetization for this block relative to the peninsula intrusive suite and the South Orkney block could be conveniently explained by a randomization of the remanent magnetization on the scale of the block, with an accompanying decrease in the bulk magnetization.

Several new hypotheses arise from the results of this modelling. Is the belt of intrusive rocks on the peninsula as continuous (at depth) as it appears to be in the anomaly field of A&S, and does this belt continue uninterrupted along the Bellingshausen margin up to Thurston Island and beyond? The anomaly map of R&B would indicate that there is a break in the linear belt at the western Palmer Land margin; however, the continuous WCMA of the BAS aeromagnetic anomaly map supports the satellite field of A&S. Therefore, we suggest from our modelling that the source rocks are continuous and we estimate they extend to a depth of 10–12 km at the western Peninsula margin, in agreement with Renner *et al.* (1985). The intensely magnetized rocks on Thurston Island are good evidence for a continuation of this zone along the Bellingshausen margin, although we have no low level aeromagnetic data to support this conjecture. Another intriguing question concerns Jason Peninsula. How does it fit in with the tectonic trends of the Antarctic Peninsula which are clearly dominated by previous subduction along the Pacific margin? We observe that only the later stage basic intrusive rocks emplaced closest to the palaeotrench axis seem to contribute to the satellite and aeromagnetic anomaly fields. Could the Jason Peninsula be controlled by an underlying zone of weakness, such as a Pacific fracture zone intersecting the Antarctic Peninsula, which allowed the later stage magmas to migrate farther away from the trench axis and be erupted on the eastern peninsula margin? Or, alternately, is the anomaly caused by the presence of lower crustal rocks residing at shallow–mid-crustal levels?

An important observation made during this modelling is that the positive magnetic anomaly which enters the Antarctic continent in the vicinity of the Filchner Ice Shelf (seen in both the satellite field and independently in the Russian aeromagnetic data, shown by Kristoffersen & Haugland, 1986), may represent the West Antarctica–East Antarctica Plate boundary. Its source is almost certainly related to the Dufek massif and Ferrar group of layered intrusions (Behrendt *et al.*, 1980). It also appears to be continuous with the Explora anomaly, and trends at an angle to the Orion–Andenes trend. This group of intense anomalies is associated with seaward-dipping reflectors on the East Antarctic margin. We have also identified a zone of anomalous oceanic crust offshore of the seaward-dipping reflectors and surrounding Maud Rise, which may have important implications for the tectonic evolution of this region.

Acknowledgements

This research was supported by the Division of Polar Programs of the National Science Foundation grant DPP-85-17635. This work was also made possible by a grant from the Tinker Foundation of New York for international scientific cooperation. We are indebted to our colleagues at the Naval Research Laboratory, particularly John Brozena, and the Flight Support Detachment of NRL for dedication and professionalism in carrying out long hours of surveying to acquire the aeromagnetic data on which this study relied.

References

Allen, A. (1966). A magnetic survey of north-east Trinity Peninsula, Graham Land: II. Mount Bransfield and Duse Bay to Victory Glacier. *British Antarctic Survey Scientific Reports*, **49**, 32 pp.

Arkani-Hamed, J. & Strangway, D.W. (1985). Lateral variations of apparent susceptibility of lithosphere deduced from Magsat. *Journal of Geophysical Research*, **90**, 2655–64.

Arkani-Hamed, J. & Strangway, D.W. (1986). Band-limited global scalar magnetic anomaly map of the Earth derived from Magsat data. *Journal of Geophysical Research*, **91**, 8193–203.

Ashcroft, W.A. (1972). Crustal structure of the South Shetland Islands and Bransfield Strait. *British Antarctic Survey Scientific Reports*, **66**, 43 pp.

Barker, P.F. (1982). The Cenozoic subduction history of the Pacific margin of the Antarctic Peninsula: ridge-crest interactions. *Journal of the Geological Society, London*, **139**, 787–801.

Behrendt, J.C., Drewry, D.J., Jankowsky, E. & Grim, M.S. (1980). Aeromagnetic and radio echo sounding ice measurements show much greater area of the Dufek Intrusion, Antarctica. *Science*, **209**, 1014–17.

Bergh, H.W. (1987). Underlying fracture zone nature of Astrid Ridge off Antarctica's Queen Maud Land. *Journal of Geophysical Research*, **92**, 475–84.

Frey, H. (1983). Satellite elevation magnetic anomalies over oceanic plateaus. *EOS Transactions of the American Geophysical Union*, **64**, 214.

Garrett, S.W., Renner, R.G.B., Jones, J.A. & McGibbon, K.J. (1987). Continental magnetic anomalies and the evolution of the Scotia arc. *Earth and Planetary Science Letters*, **81**, 273–81.

Griffiths, D.H. & Barker, P.F. (1971). Review of marine geophysical investigations in the Scotia Sea. In *Antarctic Geology and Geophysics*, ed. R.J. Adie, pp. 3–11. Oslo; Universitetsforlaget.

Harrington, P.K., Barker, P.F. & Griffiths, D.H. (1972). Crustal structure of the South Orkney Islands area from seismic refraction and magnetic measurements. In *Antarctic Geology and Geophysics*, ed. R.J. Adie, pp. 27–32. Oslo; Universitetsforlaget.

Haugland, K., Kristoffersen, Y. & Velde, A. (1985). Seismic investigations in the Weddell Sea embayment. *Tectonophysics*, **114**, 293–313.

Hinz, K. & Krause, W. (1982). The continental margin of Queen Maud Land/Antarctica: seismic sequences, structural elements and geological development. *Geologishes Jahrbuch*, **E23**, 17–41.

Hinz, K. & Kristoffersen, Y. (1987). Antarctica: recent advances in the understanding of the continental shelf. *Geologishes Jahrbuch*, **E37**, 3–54.

Hinz, K., Mutter, J.C. & Zehnder, C.M. (1987). Symmetric conjugation of continent–ocean boundary structures along the Norwegian and East Greenland margins. *Marine and Petroleum Geology*, **4**, 166–87.

Kristoffersen, Y. & Haugland, K. (1986). Geophysical evidence for the East Antarctic plate boundary in the Weddell Sea. *Nature, London*, **322**, 538–41.

LaBrecque, J.L. (1986). *South Atlantic Ocean and Adjacent Antarctic Continental Margin, Atlas 13.* Ocean Marine Drilling Program, Regional Atlas Series. Woods Hole, Massachusetts; Marine Science International. 21 sheets.

LaBrecque, J.L. & Keller, M. (1982). A geophysical study of the Indo-Atlantic basin. In *Antarctic Geoscience*, ed. C. Craddock, pp. 331–7. Madison; University of Wisconsin Press.

Pankhurst, R.J. (1982). Rb–Sr geochronology of Graham Land, Antarctica. *Journal of the Geological Society, London*, **139**, 701–11.

Parra, J-C., Gonzalez-Ferran, O. & Bannister, J. (1984). Aeromagnetic survey over the South Shetland Islands, Bransfield Strait and part of the Antarctic Peninsula. *Revista Geológica de Chile*, **23**, 3–20.

Parra, J-C. & Janez, G. (in press). Provincias magneticas en Chile central. *Revista Geologica de Chile.*

Renner, R.G.B., Sturgeon, L.J.S. & Garrett, S.W. (1985). Reconnaissance gravity and aeromagnetic surveys of the Antarctic Peninsula. *British Antarctic Survey Scientific Reports*, **110**, 50 pp.

Raymond, C.A. & LaBrecque, J.L. (1984). Constraints on the regional magnetization of the oceanic crust. *EOS, Transactions of the American Geophysical Union*, **65**, 201.

Ritzwoller, M.H. & Bentley, C.R. (1982). Magsat magnetic anomalies over Antarctica and the surrounding oceans. *Geophysical Research Letters*, **9**, 285–8.

Ritzwoller, M.H. & Bentley, C.R. (1983e). Magnetic anomalies over Antarctica measured by Magsat. In *Antarctic Earth Science*, ed. R.L. Oliver, P.R. James & J.B. Jago, pp. 504–7. Canberra; Australian Academy of Science and Cambridge; Cambridge University Press.

Saunders, S. & Tarney, J. (1982). Igneous activity in the Southern Andes and northern Antarctic Peninsula, a review. *Journal of the Geological Society, London*, **139**, 691–700.

Watters, D.G. (1971). Geophysical investigation of a section of the South Scotia Ridge. In *Antarctic Geology and Geophysics*, ed. R.J. Adie, pp. 125–31. Oslo; Universitetsforlaget.

Watts, A.B., Weissel, J.K., Duncan, R.A. & Larson, R.L. (1988). Origin of the Louisville Ridge and its relationship to the Eltanin fracture zone system. *Journal of Geophysical Research*, **93**, 3051–77.

Weissel, J.K., Hayes, D.E. & Herron, E.M. (1977). Plate tectonic synthesis: the displacements between Australia, New Zealand and Antarctica since the Late Cretaceous. *Marine Geology*, **25**, 231–77.

Aeromagnetic studies of crustal blocks and basins in West Antarctica: a review

S.W. GARRETT

British Antarctic Survey, Natural Environment Research Council, High Cross, Madingley Road, Cambridge CB3 0ET, UK
Present address: Chevron UK Ltd, 2 Portman Street, London W1H 0AN, UK.

Abstract

Topographical and geological data have led to the conceptual division of West Antarctica into discrete crustal blocks which are separated in places by large basins. Regional aeromagnetic data from the Weddell Sea sector of Antarctica support this approach. In general, large magnetic anomalies appear to be caused by Precambrian–Lower Palaeozoic crystalline basement and by Mesozoic–Cenozoic mafic intrusions. Disruptions of the magnetic pattern indicate major vertical and transcurrent displacements of the crustal fabric within, and at the margins of, the crustal blocks.

Introduction

Aeromagnetic surveys completed in the last 10 years by the British Antarctic Survey (BAS), Scott Polar Research Institute (SPRI) and United States Antarctic Program (USAP), together with the work of Chilean and Soviet Antarctic expeditions, provide a valuable and extensive regional geophysical database covering the Weddell Sea sector of Antarctica (Fig. 1). Although the line spacings vary between 10 and 100 km, the area covered amounts to several million square kilometres and the data reveal several major features of crustal structure which were previously unknown. Available contour maps (Masolov, 1980; Parra, Gonzalez-Ferran & Bannister, 1984; Renner, Sturgeon & Garrett, 1985; Jones & Maslanyj, this volume, p. 405) and stacked profiles (Jankowski, 1983a; Garrett, Herrod & Mantripp, 1987a) have been used to compile a map showing areas containing large (> 200 nT) anomalies of regional significance (Fig. 2). Representative aeromagnetic profiles (Fig. 3) have been smoothed to remove noise and have had a spherical harmonic model of Earth's magnetic field removed (e.g. Barraclough, 1981). A specific problem of aeromagnetic surveying at high latitudes is the rapid, large-amplitude temporal variations in the ambient magnetic field caused by auroral activity (Herrod & Garrett, 1985; Maslanyj & Damaske, 1987). Diurnal control and strong correlations between different datasets indicate that the profiles presented here adequately reflect the crustal magnetic signature.

This brief paper attempts some broad generalizations as to the origin of the bodies imaged by the aeromagnetic data. In particular, it is generally acknowledged that Antarctica comprises several major crustal blocks of contrasting geology (Dalziel & Elliot, 1982) and the associated boundaries may coincide with changes in the pattern of crustal magnetization.

Characteristic aeromagnetic signatures of crustal blocks and basins

Pacific margin

The dominant feature of the residual magnetic field along the Pacific margin (Fig. 2) is the West Coast Magnetic Anomaly (WCMA) of the Antarctic Peninsula (Renner *et al.*, 1985) (Fig. 3, profile 1). This belt of 500–1500 nT anomalies extends for over 1500 km along-strike from the South Shetland Islands (Parra *et al.*, 1984) to southern Palmer Land (Jones & Maslanyj, this volume, p. 405). The source of the anomalies is a plutonic complex of batholithic proportions which may be modelled using magnetizations of approximately 2 A/m (Parra *et al.*, 1984; Garrett *et al.*, 1987b; Jones & Maslanyj, this volume, p. 405). Whilst the batholith is undoubtedly a result of Mesozoic–Cenozoic subduction-related processes, it is not clearly expressed as a distinctive plutonic complex at outcrop. Associated positive isostatic residual gravity anomalies indicate that the batholith contains a large proportion of mafic material (Garrett, in press). Disruptions of the pattern of the WCMA coincide with major physiographic features which may reveal major fractures within the Antarctic Peninsula (AP) (Fig. 2). For example, the transition zone between Palmer Land and Graham Land coincides with a zone of sinistral displacements of the WCMA totalling 75 km (Garrett, in press). The western shore of northern Palmer Land coincides with the steep magnetic gradients along the western edge of the WCMA suggesting the presence of a major faulted boundary between the fore-arc rocks exposed on Alexander Island and the arc rocks of Palmer Land. The western edge of the WCMA shows a dramatic right-angled bend offshore of southern Palmer Land (Jones & Maslanyj, this volume, p. 405).

Another characteristic magnetic feature of the Antarctic

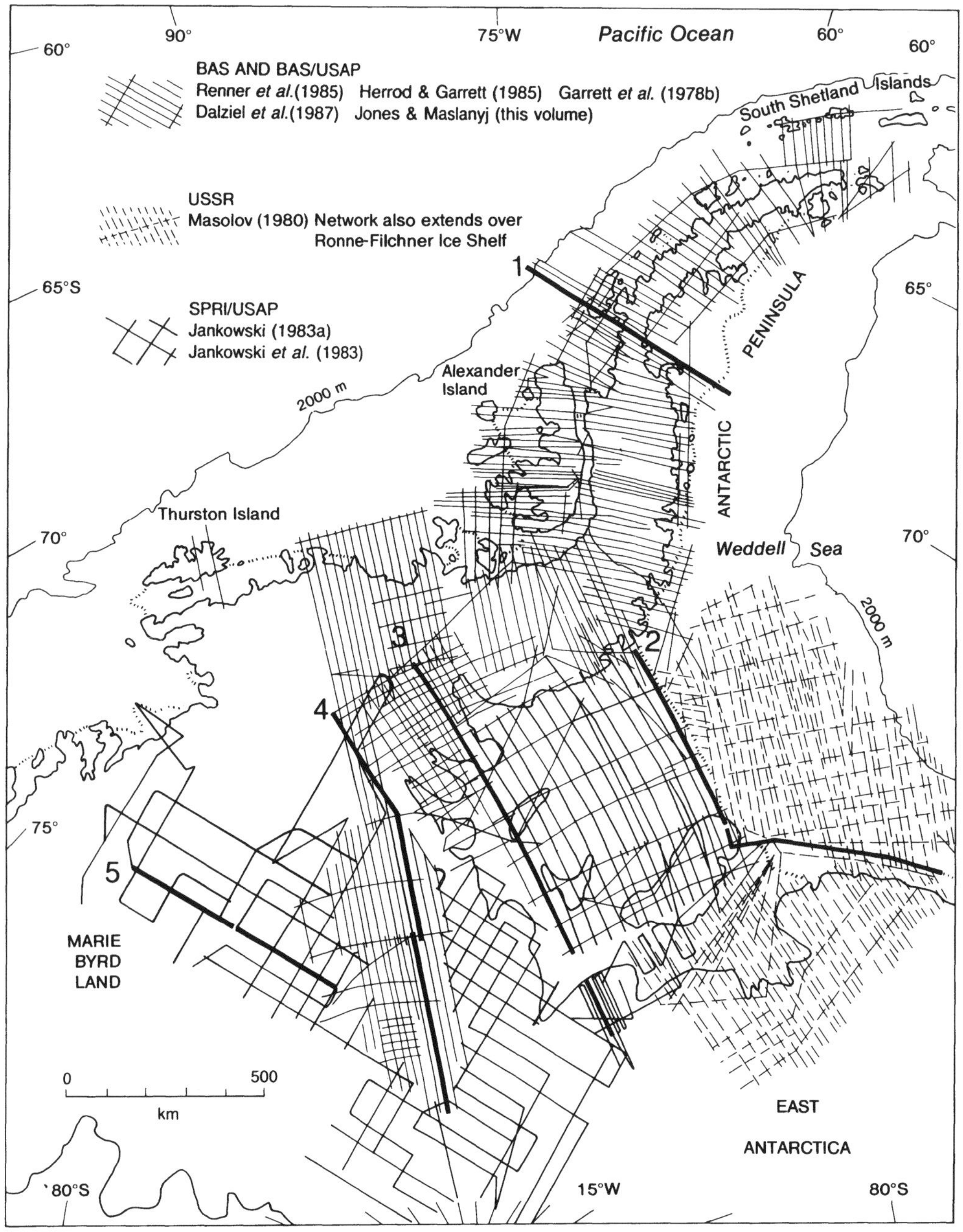

Fig. 1. Aeromagnetic flightlines over West Antarctica, 1973–87. The detailed Chilean survey over northern Antarctic Peninsula (Parra *et al.*, 1984) and Soviet coverage of the Ronne–Filchner Ice Shelf are excluded. Selected profiles (numbered 1–5) are shown in Fig. 3.

Peninsula is a belt of moderate amplitude (100 nT) magnetic anomalies over the mainland plateau (e.g. Fig. 3, profile 1 over Graham Land) which reflects igneous rocks of the Mesozoic arc. Other anomalies correspond to the Cenozoic arc on Alexander Island (Care, 1983; Crawford, Girdler & Renner, 1986). Larger magnetic anomalies in the eastern half of the mainland peninsula (Fig. 2) correspond to mafic plutons (Jones & Maslanyj, this volume, p. 405; McGibbon & Wever, this volume, p. 395). The topographic escarpment at the southern end of the Antarctic Peninsula crustal block is the limit of these isolated anomalies (Behrendt, 1964; Garrett *et al.*, 1987a).

The magnetic field is quiet along the east coast (Fig. 3, profile 1 southeast of Graham Land) over the 5 km thick back-arc sedimentary basin believed to exist along the western Weddell Sea (Farquharson, Hamer & Ineson, 1983; Macdonald *et al.*, 1988). In contrast, profiles recovered over the continental shelf on the Pacific margin show short wavelength, low-amplitude anomalies which suggest only a thin (< 2 km) sedimentary sequence (Fig. 3, profile 1 north-west of WCMA). These terminate at a 100 nT anomaly (Fig. 3, north-western end of profile 1) over the shelf edge which probably marks the transition from continental to oceanic crust.

A coastal 300 nT anomaly between the Antarctic Peninsula and Thurston Island is probably related to the WCMA (M.P. Maslanyj & J.A. Jones, pers. comm.). The large anomaly

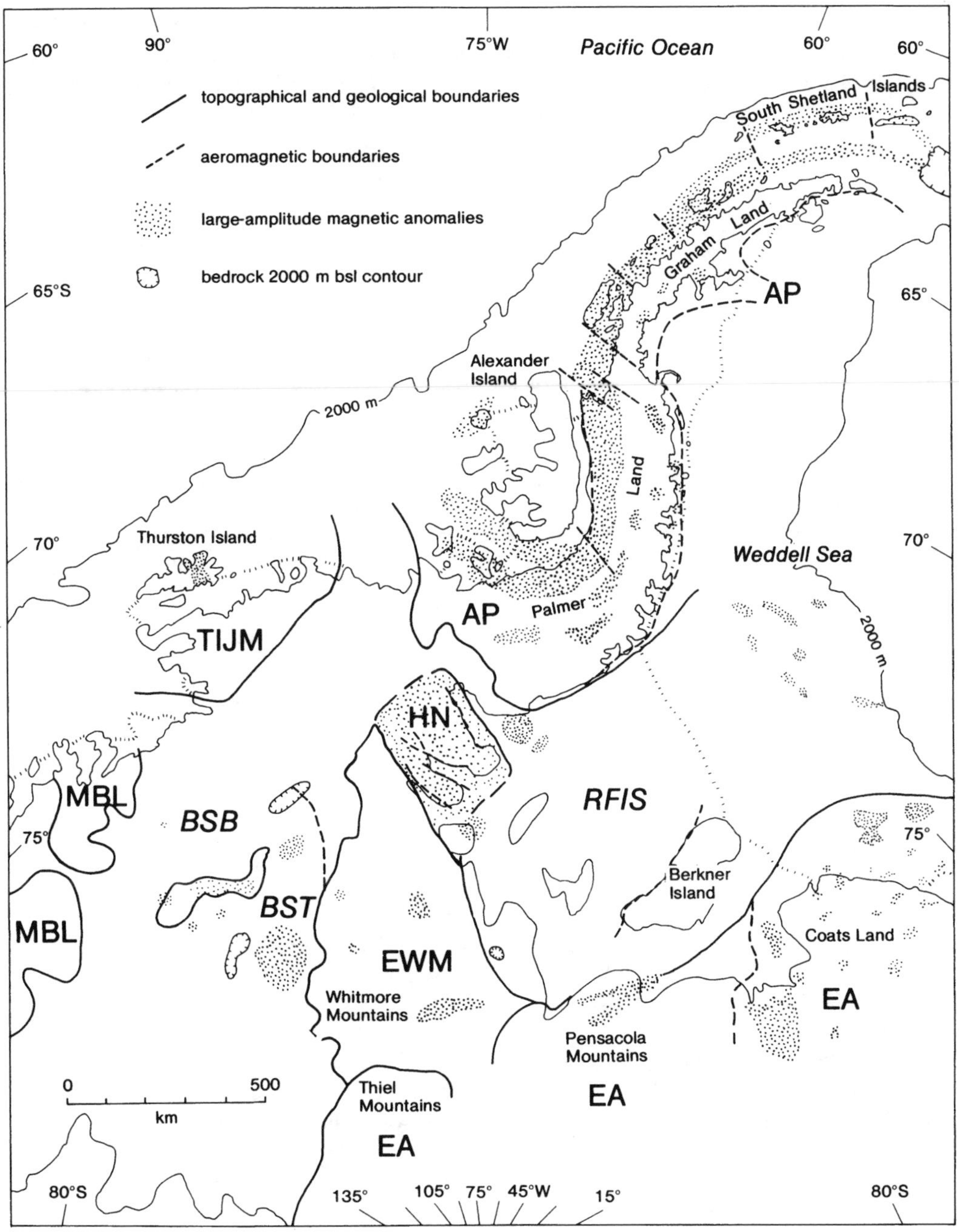

Fig. 2. Crustal blocks of West Antarctica: AP, Antarctic Peninsula; TIJM, Thurston Island–Jones Mountains; MBL, Marie Byrd Land; HN, Haag Nunataks; EWM, Ellsworth–Whitmore Mountains; BSB, Byrd Subglacial Basin; BST, Bentley Subglacial Trench; RFIS, Ronne–Filchner Ice Shelf; EA, East Antarctica. Large (> 200 nT) regional positive magnetic anomalies are shaded. See text for sources of information.

recorded over Thurston Island (Fig. 2) (Storey *et al.*, this volume, p. 399) may reflect mafic igneous rocks and be genetically related to the WCMA. Such a comparison is supported by the large MAGSAT anomalies over Thurston Island and the west coast of the Antarctic Peninsula (Ritzwoller & Bentley, 1983).

Late Cenozoic alkaline basalts on the Antarctic Peninsula are associated with short wavelength magnetic anomalies which may also show reversed polarities. This suggests the absence of large underlying plutonic bodies, the basalts probably having arisen through narrow fissures (Garrett & Storey, 1987). Profiles gathered over the Cenozoic volcanic province of Marie Byrd Land indicate similar short wavelength anomalies (Behrendt & Wold, 1963; Behrendt, 1964) superimposed on a broader framework of long wavelength anomalies which continues into the Byrd Subglacial Basin (Figs 2 & 3, BSB) (Jankowski, 1983a; Jankowski, Drewry & Behrendt, 1983).

Haag Nunataks

The isolated exposure of Precambrian gneiss at Haag Nunataks (Figs 1 & 2) lies within a region of distinctive high amplitude, high frequency magnetic anomalies (Fig. 3, profile 3) (Garrett *et al.*, 1987a). Faults suggested by linear physiogra-

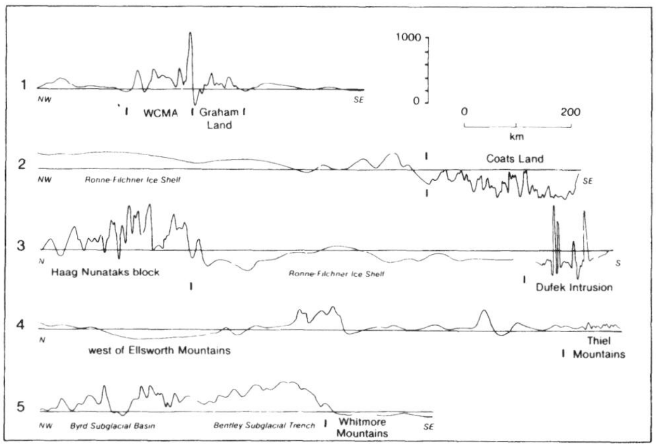

Fig. 3. Aeromagnetic profiles across West Antarctica. See Fig. 1 for locations. Flight heights vary between 0.15 and 3.0 km above the bedrock surface. Based on BAS data, with the exception of profile 3 over the Dufek Intrusion (Behrendt *et al.*, 1981) and profile 5 over the Byrd Subglacial Basin/Bentley Subglacial Trench (Jankowski, 1983a).

phic escarpments (Doake, Crabtree & Dalziel, 1983) are also marked by changes in anomaly wavelength. The anomalies are probably caused by the metamorphic rocks. The moderate wavelength (50 km), large-amplitude (> 200 nT) anomalies in the surrounding region testify to the continuation of this basement at several kilometres depth. The characteristic anomalies of the Haag Nunataks crustal block (Storey *et al.*, 1988) (Fig. 2, HN) are terminated at the eastern edge of the Ellsworth Mountains.

Ellsworth–Whitemore Mountains

The Ellsworth–Whitemore Mountains crustal block (EWM) shows a generally flat magnetic signature (Behrendt, 1964; Jankowski & Drewry, 1981; Jankowski *et al.*, 1983). Small amplitude (< 100 nT) anomalies of > 10 km wavelength are attributable to shallow sources within the thick Palaeozoic sedimentary and igneous succession. The quiet magnetic signature is punctuated by 100–500 nT anomalies (Fig. 3, profile 4) associated with large exposures of granite of Middle Jurassic age (Dalziel *et al.*, 1987). The measured magnetizations of these rocks (< 0.01 A/m) is insufficient to cause the observed anomalies and so it is necessary to postulate the existence of deeper, magnetite-rich intrusions related to the granites (Garrett, Maslanyj & Damaske, 1989). The most likely candidate for the magnetic sources are mafic rocks of the Ferrar Supergroup which are exposed in the Transantarctic Mountains, and are the same age as the granites.

Byrd Subglacial Basin/Bentley Subglacial Trench

The Byrd Subglacial Basin and Bentley Subglacial Trench (Fig. 2, BSB/BST) are characterized by a pattern of 500 nT anomalies with moderate wavelengths (30–50 km) (Fig. 3, profile 5) (Jankowski, 1983a). The largest of these anomalies is associated with a sinuous bedrock ridge (Fig. 2) (Jankowski & Drewry, 1981). The source of these predominantly positive anomalies is probably igneous rocks, as suggested by high crustal seismic velocities (Bentley & Clough, 1972). The boundary between the BSB/BST and EWM to the west of longitude 100° W is clearly defined by an increase in magnetic activity (Fig. 3, profile 5) accompanying the decrease in bedrock elevation (Jankowski & Drewry, 1981). To the east of this longitude, the quiet magnetic signature of the EWM continues northward towards the Pacific margin (Behrendt, 1964) (Fig. 3, profile 4).

Ronne–Filchner Ice Shelf and southern Weddell Sea

The magnetic signature over the Ronne–Filchner Ice Shelf (RFIS) (Herrod & Garrett, 1985) is characterized by 200 km wavelength, 100–300 nT anomalies (Fig. 3, profiles 2 & 3) which represent magnetic sources at 10–20 km depth (Masolov, 1980). The data are consistent with a thick (> 10 km) sedimentary sequence overlying crystalline basement (Kadmina & Kurinin, 1983; Kamenev & Ivanov, 1983). This basement is probably affected by deep-seated faults, such as that indicated by linear magnetic gradients along the west coast of Berkner Island (Fig. 3). The anomalies of the Ronne–Filchner Ice Shelf continue northwards over the continental shelf of the Weddell Sea (Fig. 2) but change trend towards the shelf edge where they have shorter wavelengths, higher amplitudes and are associated with the Explora–Andenes Escarpment (Kristoffersen & Haugland, 1986).

The boundary between East and West Antarctica

The boundary between the Mesozoic–Cenozoic mobile belts of West Antarctica and the shield region of East Antarctica (Fig. 2, EA) shows three characteristic magnetic signatures in this region. The magnetic signature of the RFIS shows an eastward decrease in anomaly wavelength accompanied by an increase in anomaly amplitude towards Coats Land (Fig. 3, profile 2). This indicates a gradual shallowing of crystalline magnetic basement towards the shield region, an interpretation consistent with seismic-reflection data (Haugland, 1982). Farther to the south-west the boundary is marked by the 1000 nT anomalies associated with the layered gabbro Dufek Intrusion (Behrendt *et al.*, 1981) (Fig. 2, profile 3). These may only be traced eastward over the Filchner Ice Shelf for approximately 100 km (Fig. 3). Still farther to the south-west, the boundary is marked by an abrupt change in magnetic signature between the Thiel Mountains and EWM (Fig. 2, profile 4) (Dalziel *et al.*, 1987). This separates the Lower Palaeozoic igneous rocks and Ferrar dolerite sills of East Antarctica, which have high magnetizations (approximately 1 A/m), from the metasediments of the EWM, which have low magnetizations (< 0.1 A/m).

Discussion

The geological interpretation of the magnetic anomalies in ice-covered regions can rely heavily upon correlation with sparse exposures. Thus, maps of depth to 'magnetic basement' (e.g. Jankowski, 1983b), although useful in the preliminary stages of interpretation, can only be meaningful if the magnetic sources represented are of the same age and tectonic significance. The regional aeromagnetic anomalies appear to image two types of source. First, anomalies over Haag Nunataks and Coats Land are associated with Precambrian crystalline basement. It is probable that long wavelength anomalies over the intervening area of the RFIS and southern Weddell Sea are caused largely by subsided basement of this type. Second, anomalies along the Pacific Margin, over the Dufek Intrusion and on the EWM are caused by large Mesozoic–Cenozoic mafic plutons. Those along the Pacific margin (e.g. WCMA) originated by subduction-related processes such as intra-arc extension (Storey & Garrett, 1985). Those farther inland (e.g. Dufek Intrusion, EWM) are related to the early stages of continental rifting during Middle Jurassic times. The anomalies of the BSB/BST may be caused by mafic igneous rocks whose formation was influenced by regional transtension (Storey, this volume, p. 587). The anomalies close to the continent–ocean transition in the Weddell Sea are probably caused by igneous rocks, imaged on seismic-reflection data, which formed during the early stages of rifting (Kristoffersen & Haugland, 1986).

The aeromagnetic anomalies show a consistent pattern within each of the crustal blocks defined on the basis of physiographical and geological data. However, some of the accepted block boundaries do not coincide with marked changes in anomaly pattern, especially in areas of flat magnetic signature. The boundary between the AP and RFIS cannot be defined magnetically. Neither can the northern boundary of the EWM nor the divide between the EWM and Pensacola Mountains. This situation arises when thick sediments or metasedimentary rocks are present on both sides of a boundary.

Storey (this volume, p. 587) discusses the movement between the various crustal blocks of West Antarctica throughout the Mesozoic and Cenozoic. Future studies involving aeromagnetic data may help to add further constraints to this type of tectonic analysis. Suggested topics are as follows: (i) mapping and interpretation of the WCMA between the Antarctic Peninsula, Thurston Island and Marie Byrd Land could contribute to a reconstruction of the Pacific margin (such as that recently attempted for the Scotia arc; Garrett *et al.*, 1987b); (ii) completion of a more detailed flight network over the BSB/BST would contribute to confident mapping of anomaly patterns which may in turn indicate a sense of opening for the basin; (iii) interpretation of deep fault patterns beneath the RFIS and southern Weddell Sea may suggest the tectonic controls on the subsidence of this area; (iv) an extension of the regional network over the Ross Ice Shelf and Marie Byrd Land is needed; (v) compilation of data from different surveys in a coherent gridded database will allow integration of aeromagnetic and MAGSAT data, the application of image processing techniques and filtering to enhance fault trends. These further studies are an essential prelude to the planning and execution of overland crustal seismic traverses which may form the next stage of the geophysical exploration of Antarctica.

Conclusions

The most pervasive feature of the regional magnetic field in the Weddell Sea sector of Antarctica is the West Coast Magnetic Anomaly of the Antarctic Peninsula crustal block. This reflects a Mesozoic–Cenozoic subduction-related batholith. The Haag Nunataks block is also defined by a region of large magnetic anomalies but these are probably caused by Proterozoic basement. The flat magnetic signature of the Ellsworth–Whitmore Mountains block is interrupted by anomalies which represent deep-seated mafic intrusions of Middle Jurassic age related to the Dufek Intrusion of the Pensacola Mountains. The contrasting magnetic signatures of the Byrd Subglacial Basin/Bentley Subglacial Trench and Ronne–Filchner Ice Shelf basins reflect their igneous and sedimentary floors, respectively. The magnetic data indicate complex boundaries between East and West Antarctica in this region.

Acknowledgements

The BAS regional aeromagnetic program was managed by R.G.B. Renner, who provided support and encouragement throughout the course of the surveys. I am indebted to all my other colleagues in the BAS Field Geophysics section for advice and help.

References

Barraclough, D.R. (1981). The 1980 geomagnetic reference field. *Nature, London*, **294**, 14–15.

Behrendt, J.C. (1964). Distribution of narrow-width magnetic anomalies in Antarctica. *Science*, **144**, 993–9.

Behrendt, J.C., Drewry, D.J., Jankowski, E.J. & Grim, M.S. (1981). Aeromagnetic and radio-echo ice sounding measurements over the Dufek Intrusion, Antarctica. *Journal of Geophysical Research*, **86**, 3014–20.

Behrendt, J.C. & Wold, R.J. (1963). Depth to magnetic 'basement' in West Antarctica. *Journal of Geophysical Research*, **68**, 1145–53.

Bentley, C.R. & Clough, J.W. (1972). Antarctic subglacial structure from seismic refraction measurements. In *Antarctic Geology and Geophysics*, ed. R.J. Adie, pp. 683–91. Oslo; Universitetsforlaget.

Care, B.W. (1983). The petrology of the Rouen Moutains, northern Alexander Island. *British Antarctic Survey Bulletin*, **52**, 63–86.

Crawford, I.A., Girdler, R.W. & Renner, R.G.B. (1986). Interpretation of aeromagnetic anomalies over the Staccato Peaks area of Alexander Island. *British Antarctic Survey Bulletin*, **70**, 41–53.

Dalziel, I.W.D. & Elliot, D.H. (1982). West Antarctica: problem child of Gondwanaland. *Tectonics*, **1**, 3–19.

Dalziel, I.W.D., Garrett, S.W., Grunow, A.M., Pankhurst, R.J., Storey, B.C. & Vennum, W.R. (1987). The Ellsworth–Whitmore Mountains crustal block: its role in the tectonic evolution of West Antarctica. In *Gondwana Six: Structure, Tectonics, and Geophysics*, Geophysical Monograph 40, ed. G.D. McKenzie, pp. 173–82. Washington, DC; American Geophysical Union.

Doake, C.S.M., Crabtree, R.D. & Dalziel, I.W.D. 1983. Bedrock morphology between the Antarctic Peninsula and Ellsworth Mountains: new data and tectonic significance. In *Antarctic Earth Science*, ed. R.L. Oliver, P.R. James & J.B. Jago, pp. 270–3. Canberra; Australian Academy of Science and Cambridge; Cambridge University Press.

Farquharson, G.W., Hamer, R.D. & Ineson, J.R. (1983). Proximal volcaniclastic sedimentation in a Cretaceous back-arc basin, northern Antarctic Peninsula. In *Marginal Basin Geology*, ed.

B.P. Kokelaar & M.F. Howells, pp. 219–29. Geological Society of London Special Publication 16. Oxford; Blackwell Scientific Publications

Garrett, S.W. (in press). Interpretation of regional gravity and aeromagnetic surveys of the Antarctic Peninsula. *Journal of Geophysical Research.*

Garrett, S.W., Herrod, L.D.B. & Mantripp, D.R. (1987a). Crustal structure of the area around Haag Nunataks, West Antarctica: new aeromagnetic and bedrock elevation data. In *Gondwana Six: Structure, Tectonics, and Geophysics*, Geophysical Monograph 40, ed. G.D. McKenzie, pp. 109–16. Washington, DC; American Geophysical Union.

Garrett, S.W., Renner, R.G.B., Jones, J.A. & McGibbon, K.J. (1987b). Continental magnetic anomalies and the evolution of the Scotia arc. *Earth and Planetary Science Letters*, **81**, 273–81.

Garrett, S.W. & Storey, B.C. (1987). Lithospheric extension on the Antarctic Peninsula during Cenozoic subduction. In *Continental Extensional Tectonics*, ed. M.P. Coward, J.F. Dewey & P.L. Hancock, pp. 419–31. Geological Society of London, Special Publication 28, Oxford; Blackwell Scientific Publications.

Garrett, S.W., Maslanyj, M.P. & Damaske, D.J. (1989). Interpretation of aeromagnetic data from the Ellsworth Mountains–Thiel Mountains Ridge, West Antarctica. *Journal of the Geological Society, London*, **145**, 1009–17.

Haugland, K. (1982). Seismic reconnaissance survey in the Weddell Sea. In *Antarctic Geoscience*, ed. C. Craddock, pp. 405–13. Madison; University of Wisconsin Press.

Herrod, L.D.B. & Garrett, S.W. (1985). Geophysical fieldwork on the Ronne Ice Shelf, Antarctica. *First Break*, **4**, 9–14.

Jankowski, E.J. (1983a). Residual magnetic field in West Antarctica. In *Antarctica: Glaciological and Geophysical Folio*, ed. D.J. Drewry, Sheet 7. Cambridge; Scott Polar Research Institute.

Jankowski, E.J. (1983b). Depth to magnetic basement in West Antarctica. In *Antarctica: Glaciological and Geophysical Folio*, ed. D.J. Drewry, Sheet 8. Cambridge; Scott Polar Research Institute.

Jankowski, E.J. & Drewry, D.J. (1981). The structure of West Antarctica from geophysical studies. *Nature, London*, **291**, 17–21.

Jankowski, E.J., Drewry, D.J. & Behrendt, J.C. (1983). Magnetic studies of upper crustal structure in West Antarctica and the boundary with East Antarctica. In *Antarctic Earth Science*, ed. R.L. Oliver, P.R. James & J.B. Jago, pp. 197–203. Canberra; Australian Academy of Science and Cambridge; Cambridge University Press.

Kadmina, I.N. & Kurinin, R.G. (1983). Antarctic crustal structure from geophysical evidence: a review. In *Antarctic Earth Science*, ed. R.L. Oliver, P.R. James & J.B. Jago, pp. 498–502. Canberra; Australian Academy of Science and Cambridge; Cambridge University Press.

Kamenev, E.N. & Ivanov, V.L. (1983). Structure and outline of the history of the southern Weddell Sea basin. In *Antarctic Earth Science*, ed. R.L. Oliver, P.R. James & J.B. Jago, pp. 194–7. Canberra; Australian Academy of Science and Cambridge; Cambridge University Press.

Kristoffersen, Y. & Haugland, K. (1986) Geophysical evidence for the East Antarctic plate boundary in the Weddell Sea. *Nature, London*, **322**, 538–41.

Macdonald, D.I.M., Barker, P.F., Garrett, S.W., Ineson, J.R., Kinghorn, R.R.F., Marshall, J.E.A., Pirrie, D., Storey, B.C. & Whitham, A.G. (1988). A preliminary assessment of the hydrocarbon potential of the Larsen Basin, Antarctica. *Marine and Petroleum Geology*, **5**, 34–53.

Maslanyj, M.P. & Damaske, D.J. (1987). Lessons regarding aeromagnetic surveys during magnetic disturbances in polar regions. *British Antarctic Survey Bulletin*, **73**, 9–17.

Masolov, V.N. (1980). *Geophysical Investigations in Antarctica*, pp. 16–28. Leningrad; NIIGA (in Russian).

Parra, J.C., Gonźalez-Ferrán, O. & Bannister, J.P. (1984). Aeromagnetic survey over the South Shetland Islands, Bransfield Strait and part of the Antarctic Peninsula. *Revista Geológica de Chile*, **23**, 3–20.

Renner, R.G.B., Sturgeon, L.J.S. & Garrett, S.W. (1985). Reconnaissance gravity and aeromagnetic surveys of the Antarctic Peninsula. *British Antarctic Survey Scientific Reports*, **110**, 50 pp.

Ritzwoller, M.H. & Bentley, C.R. (1983). Magnetic anomalies over Antarctica measured by MAGSAT. In *Antarctic Earth Science*, ed. R.L. Oliver, P.R. James & J.B. Jago, pp. 504–7. Canberra; Australian Academy of Science and Cambridge; Cambridge University Press.

Storey, B.C., Dalziel, I.W.D., Garrett, S.W., Grunow, A.M., Pankhurst, R.J. & Vennum, R.W. (1988). West Antarctica in Gondwanaland: crustal blocks, reconstruction and breakup processes. *Tectonophysics*, **155**, 381–90.

Storey, B.C. & Garrett, S.W. (1985). Crustal growth of the Antarctic Peninsula by accretion, magmatism and extension. *Geological Magazine*, **122**, 5–14.

Palaeomagnetic studies of Palaeozoic rocks from the Ellsworth Mountains, West Antarctica

M. FUNAKI[1], M. YOSHIDA[2] & H. MATSUEDA[3]

1 National Institute of Polar Research, 9–10 Kaga 1-chome, Itabashi-ku, Tokyo 173, Japan
2 Department of Geosciences, Faculty of Science, Osaka City University, 3–138 Sugimoto 3-chome, Sumiyoshi-ku, Osaka 558, Japan
3 Department of Geology and Mineralogy, Faculty of Science, Hokkaido University, N10 W8, Sapporo, 060, Japan

Abstract

Natural remanent magnetizations (NRMs) of 37 samples from the Ellsworth Mountains were investigated palaeomagnetically. The NRM directions of sedimentary rocks form two low-latitude clusters, after tilt corrections of the strata; these samples were magnetized early (pre-folding) in their Cambrian history. The NRM directions from dykes of Cambrian and Ordovician age form a cluster without tilt corrections. The inclinations of magnetite-bearing dykes changed from steep to flat during thermal demagnetization but samples containing haematite grains show widely scattered NRM directions. The relatively high inclinations were probably acquired during the Mesozoic. The virtual geomagnetic pole (VGP) positions obtained from sedimentary rocks are displaced 21° eastward and 42° westward from the Cambro-Ordovician VGP position for East Antarctica. The Mesozoic VGP position obtained from the dykes is located 42° westward from the Jurassic VGP position for East Antarctica. The results indicate a possibility that in post-Jurassic times the Ellsworth Mountains were rotated 42° counterclockwise with respect to East Antarctica. The sense of rotation is in agreement with a previous perpendicular counterclockwise rotation model for the Ellsworth Mountains.

Introduction

The Ellsworth Mountains of West Antarctica consist of the Sentinel Range in the north and Heritage Range in the south. Although their dominant structural trends are perpendicular to those of the Pensacola Mountains, in the Transantarctic Mountains, the two mountain ranges are geologically similar (Schopf, 1969).

The geology of the Ellsworth Mountains has been investigated by a number of workers, including Hjelle, Ohta & Winsnes (1982) and Yoshida (1982). They described a total thickness of 13 000 m of late Precambrian(?), Cambrian, Devonian, Permo-Carboniferous and Permian sedimentary rocks intruded by Cambrian and Ordovician dykes. At least four phases of low-grade metamorphism (under 500 °C maximum temperature) affected these rocks during the Cambrian–Ordovician (first and second phases), Devonian–Carboniferous (third phase) and the Mesozoic (fourth phase).

Palaeomagnetic studies on argillites collected from the upper Heritage Group (Upper Cambrian) of Pipe Peak were made by Watts & Bramall (1980, 1981). After thermal demagnetization they obtained significant natural remanent magnetization (NRM) from the highest blocking temperature component (650–670 °C) caused by haematite. The mean NRM direction of + 13° inclination and 22° E declination was obtained after tilt corrections and thermal demagnetization. This NRM direction and the virtual geomagnetic pole (VGP) position (4° N latitude, 269° E longitude) differ by approximately 90° counterclockwise with respect to those of the early Palaeozoic of the East Antarctic craton. They concluded therefore that the Ellsworth Mountains had been rotated 90° counterclockwise with respect to East Antarctica.

For the present palaeomagnetic study a total of 43 orientated hand samples were selected from a number of rock samples collected over a wide area of the Heritage Range and at Polarstar Peak, in the Sentinel Range. Although small in number, the samples were collected from a wide area, and the palaeomagnetic data obtained appear to give some meaningful results. Samples of sandstone, argillite and limestone, together with tuff and volcanic breccia were assumed to be of Cambrian and Permian age, and the dolerite dykes and lavas of Cambrian–Ordovician age.

Experiments

Every sample was AF demagnetized to 50 mT to remove the soft magnetic components. Based on preliminary demagnetization results, a total of 37 samples with stable NRM com-

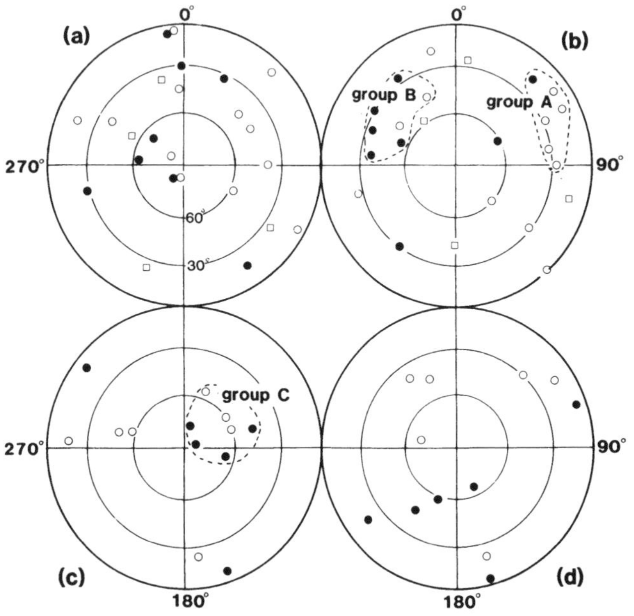

Fig. 1. Distribution of NRM directions for sedimentary rocks (a, b) and dykes (c, d) from the Ellsworth Mountains; a and c without tilt corrections; b and d after tilt corrections.

ponents were selected for thermal demagnetization and thermomagnetic analyses. The results of thermal demagnetization from 30 to 580 °C in steps of 50 °C, suggest that the original steep inclination of several dykes shifted to a lower inclination above 430 °C, although the declination shift is random. No such tendency was observed in the sedimentary rocks. The most reliable NRM components after thermal demagnetization from 280 to 380 °C were adopted for this study, because of possible reduction and oxidation of magnetic grains during high-temperature thermal demagnetization. As fractured samples were likely to be destroyed during thermal demagnetization, the NRMs demagnetized to 15 mT were used for these samples.

The NRM distributions before and after tilt corrections of the sedimentary rocks and dykes are shown in Fig. 1. The Cambrian sedimentary rocks form two clusters (Fig. 1*b*, groups A and B) with moderate inclinations after tilt corrections; no significant cluster is observed without such corrections (Fig. 1*a*). However, the stable NRM directions of four Permian sedimentary rocks do not make a cluster, even though tilt corrections were applied. Stable NRM directions from seven dyke samples (Fig. 1*c*, group C) cluster at high inclinations without tilt corrections but they are dispersed by tilt corrections referred to strikes and dips of the surrounding sedimentary formations (Fig. 1*d*). The mean NRM directions and the VGP positions obtained from these clusters are listed in Table 1, where the mean NRM inclinations were assumed to be normal polarity for group B and to be reversed for groups A and C, based on the number of polarities in the groups.

Thermomagnetic curves (I_S–T curves) were obtained from room temperature to 750 °C under a 0.6 T external magnetic field and 10^{-2} Pa atmospheric pressure. The curves obtained were divided into five types, based on the reversibility of the first run heating cycle (Fig. 2). Type 1 is a reversible I_S–T curve with clearly defined Curie point at 580 °C; type 2 is an almost reversible one with a Curie point at 580 °C and small thermomagnetic humps between 180 and 350 °C; type 3 is an irreversible one with a small hump around 400 °C, characterized by increasingly spontaneous magnetization after heat treatment and a clearly defined Curie point at 580 °C; type 4 is a reversible one with weak spontaneous magnetization and a poorly defined Curie point between 300 and 580 °C; type 5 is an extremely irreversible one characterized by a Curie point at 580 °C and at 675 °C and a thermomagnetic hump at 400 °C.

Magnetic minerals in the representative samples of I_S–T types 1–5 were identified under the microscope and chemically analyzed by electron probe microanalyzer. Type 1 samples contain only primary magnetite with a minor amount of maghemite; type 2 contain maghemite and titanomaghemite associated with a small amount of goethite but apparently no iron sulphide minerals. In type 3 samples secondary haematite or goethite replace iron sulphides and oxides but some small magnetite grains are also present. Type 4 samples contain magnetite, titanomagnetite, haematite, ilmenite and iron sulphide but those of type 5 have only secondary origin haematite. Every sample has undergone some degree of low-temperature retrogressive metamorphism, the intensity of which increases with the order of types, from 1, 2, and 3–5. However, the intensity of type 4 can not be estimated because of the small size of the magnetic grains.

Discussion

Magnetic minerals derived from the thermomagnetic analyses can be related to the results of mineralogical studies. For example, magnetite for type 1, maghemite for type 2, haematite and goethite for type 3, titanomagnetite, magnetite and haematite for type 4, and haematite for type 5. Thermomagnetic humps at 180–380 °C in type 2 probably result from the break-up of titanomaghemite or the mineralogical change from goethite to haematite, rather than reflecting the phase transition temperature of pyrrhotite since no iron sulphide grains have been detected in samples of this type. Observed Curie points at 580 °C in types 3 and 5 represent reduction of haematite to magnetite; haematite changes to magnetite during heat treatment at more than 400 °C under 10^{-2} Pa atmospheric

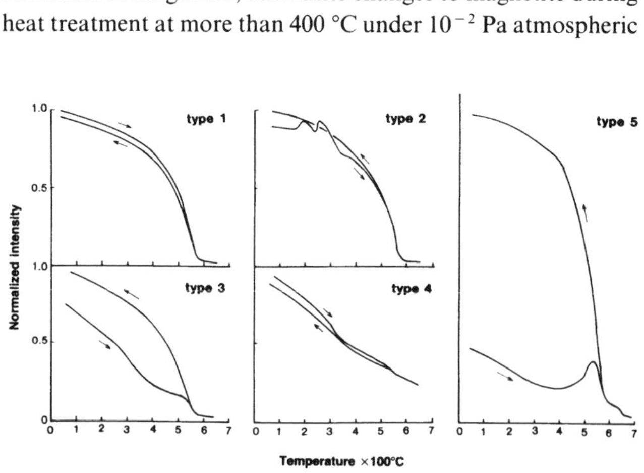

Fig. 2. Five representative types of thermomagnetic (I_S–T) curves for samples from the Ellsworth Mountains.

Table 1. *Mean NRM directions and VGP positions from the Ellsworth Mountains*

Group	Tilt correction	N	Inclination	Declination	K	α_{95}	Latitude	Longitude
A	After	6	− 23.3°	64.7°	21	15.0°	16.2° N	18.0° W
B	After	7	37.9	307.3°	12	18.5°	15.0° N	44.9° E
C	Before	7	− 65.9	60.5°	23	12.8°	52.3° S	166.5° E

pressure. Therefore the origin of the representative NRM may be presumed to be thermal remanent magnetization (TRM) for type 1, chemical remanent magnetization (CRM) for types 3 and 5, TRM with partial CRM for type 2 and detrital remanent magnetization (DRM) or TRM with partial CRM for type 4.

As the NRM distribution clusters are observed without tilt corrections for dykes and with corrections for sedimentary rocks (Fig. 1), it may be concluded that the dykes intruded sedimentary strata which were already folded. The significant NRM directions were obtained by tilt corrections for the sedimentary rocks and without such corrections for the dykes. Fig. 3 shows the relationship between NRM inclination variations against thermal demagnetization and I_S–T types. The NRM inclinations from Cambrian (Ca) and Ordovician (Or) dykes of types 1 and 2 changed from steep to flat during thermal demagnetization. However, type 5 samples show a random distribution of inclinations with small variability during demagnetization. Type 3 inclination variations represent a mixture of types 1 and 5; Cambrian sedimentary rocks belonging to type 4 show variable distributions with small inclination changes. These results suggest that magnetite grains recorded at least two kinds of TRM which decomposed into a steep inclination at low temperatures and a flat one at high temperatures, whereas haematite grains recorded variable directions. Ordovician dykes and Cambrian dykes and lavas are classified as types 3 and 5, respectively (probably reflecting differences in mineralogy or metamorphic history between the two types); types 1 and 2 are a mixture of these rocks (Fig. 3).

If the Ellsworth Mountains were a fragment of Gondwana, the expected NRM inclinations estimated from the apparent polar wander path of Gondwana (McElhinny, 1973) would

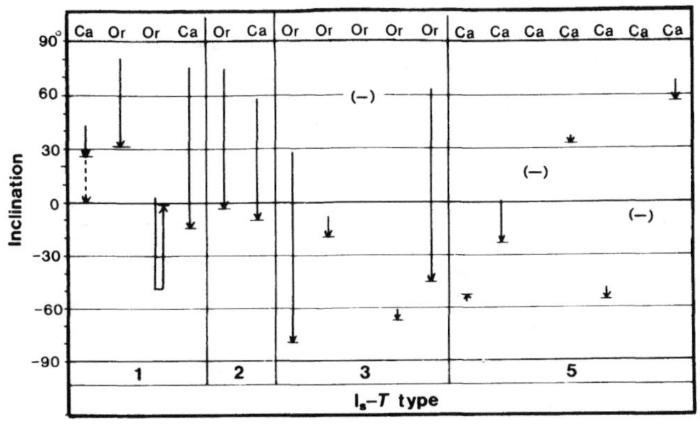

Fig. 3. NRM inclination variations of dykes and lavas during thermal demagnetization to 580 °C, and AF demagnetization for dykes from the Ellsworth Mountains. (−), inclination level by AF demagnetization; Ca, Cambrian; Or, Ordovician; Nos. 1, 2, 3 and 5, types of thermomagnetic (I_S–T) curves.

tend to be relatively low in the Early Palaeozoic, moderate in the mid-Palaeozoic, high in the Late Palaeozoic and moderate again during the Mesozoic. Temperatures did not exceed 500 °C throughout the four periods of metamorphism (Yoshida, 1982) which affected the rocks of the Ellsworth Mountains. The inclination variations based on the above estimation can be explained as follows. The relatively flat NRM components were acquired during the formation of Cambrian or Ordovician rocks or their metamorphism and the steep ones were acquired as CRM or TRM during the last (Mesozoic) metamorphic episode. The diverse distribution of NRM inclinations for samples of types 3 and 4 shows that the Ellsworth Mountains experienced several phases of metamorphism, characterized by haematite formation, throughout their geological history.

After tilt corrections, the NRM directions of six and seven sedimentary rock samples cluster at two sites (groups A and B). The clusters are the representative NRM directions resulting from the DRM when the Heritage Group was formed in Cambrian times (Yoshida, 1982). The calculated VGP positions (Table 1, A and B) from mean NRM directions are illustrated in Fig. 4. The VGP positions of A and B are located about 42° westward and 21° eastward, respectively, compared with the standard VGP position at 3° S, 24° E, with $\alpha_{95} = 5.2°$ which Funaki (1984) obtained from Cambro-Ordovician dykes from Wright Valley in southern Victoria Land, East Antarctica. These positions are completely different from each other considering their α_{95} values.

The NRM directions for seven dyke samples cluster to form group C whereas the six other samples show a scattered distribution without tilt corrections. A VGP position of the cluster is obtained from the mean NRM direction as shown in Table 1 and Fig. 4. The VGP positions of a sample of Permo-Triassic Beacon Supergroup from Allan Hills, in southern Victoria Land (Funaki, 1983b) and a Jurassic Ferrar dolerite from the McMurdo Sound region (Funaki, 1983a) are at 62.3° S, 151.4° W, with $\alpha_{95} = 2.4°$ and 45.3° S, 152.0° W, with $\alpha_{95} = 5.3°$, respectively. The VGP position of the dykes is completely isolated from the Jurassic VGP position, considering the α_{95} value and 42° eastward rotation from that VGP position, but the α_{95} value does not separate them from the Permo-Triassic VGP position.

Although the NRM inclinations of types 1 and 2 dykes are changed from high to flat by thermal demagnetization, the declinations differ among the samples. Furthermore the VGP positions obtained by this study (groups A and B) are not consistent with those obtained by Watts & Bramall (1981). These experimental disagreements may be caused by the rotation of Gondwana during Cambrian–Ordovician times, as

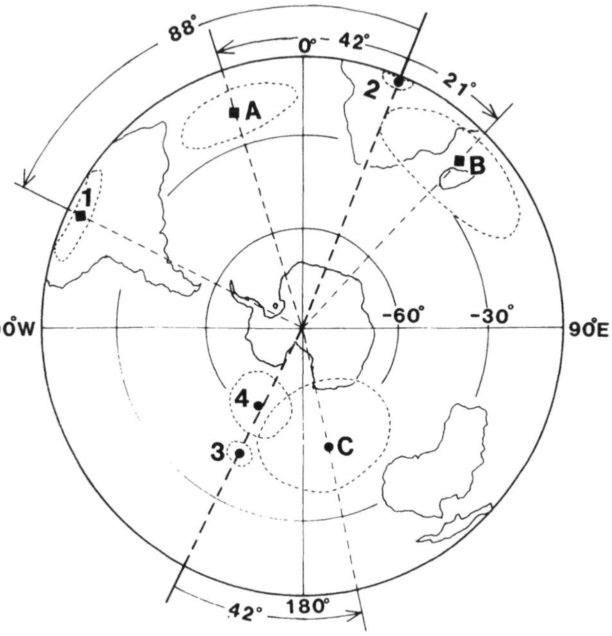

Fig. 4. VGP positions with α_{95} values from the Ellsworth Mountains (*A–C* and 1) and East Antarctica (2–4). *A* and *B*, this study (Cambrian); *C*, this study (Mesozoic); 1, Watts & Bramall (1981; Cambrian); 2, Funaki (1984; Cambro–Ordovician); 3, Funaki (1983a: Jurassic); 4, Funaki (1983b; Permo–Triassic); solid circle (square), VGP position of the southern (northern) hemisphere.

discussed by Klootwijk (1980) and Watts & Bramall (1981). Since the variation of the VGP positions for Gondwana is probably small, at least from the Triassic to the Jurassic (McElhinny, 1973), the VGP position of group C (located 42° westward from that of Jurassic rocks in East Antarctica) suggests significant counterclockwise rotation of the Ellsworth Mountains with respect to East Antarctica after the Jurassic. The sense of rotation is in agreement with the 90 ° counterclockwise rotation model of the Ellsworth Mountains proposed by Watts & Bramall (1981).

Conclusions

The rocks of the Ellsworth Mountains have undergone several phases of retrogressive low temperature metamorphism. Consequently many samples were remagnetized before the Mesozoic, although primary NRMs have survived in the sedimentary rocks and some of the dykes. The resulting Cambrian VGP positions for the primary NRM directions are located at sites A and B in Fig. 4. These positions are inconsistent with the VGP position obtained from Watts & Bramall (1981), taking the α_{95} values into consideration. The angular deviations of the longitude for these VGP positions are 42° westward and 21° eastward, compared with the standard VGP position for the Cambro-Ordovician period as reported from East Antarctica (Funaki, 1984). The VGP position resulting from NRM directions of secondary magnetization is located at site C (Fig. 4), at 42° westward compared with the Jurassic VGP position of East Antarctica. This position suggests a post-Jurassic 42° counterclockwise rotation of the Ellsworth Mountains with respect to East Antarctica, a result which supports the sense of 90 ° counterclockwise rotation proposed by Watts & Bramall (1981), although the rotation angle is 42°.

References

Funaki, M. (1983a). Paleomagnetic investigation of Ferrar dolerite in the McMurdo Sound region, Antarctica. *Antarctic Record*, **77**, 20–32.

Funaki, M. (1983b). Paleomagnetic investigation of the Beacon Group in the McMurdo Sound region, Antarctica. *Antarctic Record*, **78**, 1–14.

Funaki, M. (1984). Investigation of the paleomagnetism of the basement complex of Wright Valley, southern Victoria Land, Antarctica. *Journal of Geomagnetism and Geoelectricity*, **36(11)**, 529–63.

Hjelle, A., Ohta, Y. & Winsnes, T.S. (1982). Geology and petrology of southern Heritage Range, Ellsworth Mountains. In *Antarctic Geoscience*, ed. C. Craddock, pp. 559–608. Madison; University of Wisconsin Press.

Klootwijk, C.T. (1980). Early Palaeozoic paleomagnetism in Australia. *Tectonophysics*, **64(3/4)**, 249–332.

McElhinny, M.W. (1973). *Paleomagnetism and Plate Tectonics*. Cambridge; Cambridge University Press. 260 pp.

Schopf, J.M. (1969). Ellsworth Mountains: position in West Antarctica due to sea-floor spreading. *Science*, **164(3875)**, 63–6.

Watts, D.R. & Bramall, A.M. (1980). Paleomagnetic investigation in the Ellsworth Mountains. *Antarctic Journal of the United States*, **15(5)**, 34–6.

Watts, D.R. & Bramall, A.M. (1981). Palaeomagnetic evidence for a displaced terrain in Western Antarctica. *Nature, London*, **293(5834)**, 638–41.

Yoshida, M. (1982). Superposed Deformation and its Implication to the Geologic History of the Ellsworth Mountains, West Antarctica, *Memoirs of the National Institute of Polar Research, Special Issue* **21**, 120–71.

Seismic reflection profiling of a sediment-filled graben beneath ice stream B, West Antarctica

S.T. ROONEY[1], D.D. BLANKENSHIP[2], R.B. ALLEY[3] & C.R. BENTLEY[4]

1 Mobil Research and Development Corp., PO Box 819047, Dallas, TX 75381–9047, USA
2 Byrd Polar Research Center, Ohio State University, 125 South Oval Mall, Columbus, OH 43210, USA
3 Pennsylvania State University, 248 Deike Building, University Park, PA 16802, USA
4 Geophysical and Polar Research Center, University of Wisconsin, 1215 W. Dayton Street, Madison, WI 53706-1692, USA

Abstract

Multifold seismic-reflection data from ice stream B, on the Siple Coast of West Antarctica, show lithified sedimentary beds beneath the ice that dip grid north-east about 0.5° and are truncated in an angular unconformity near the base of the ice. The sediments have compressional-wave velocities ranging from 1.90 to 2.25 km/s and are probably late Oligocene or younger in age. The sedimentary basin beneath ice stream B is at least 1 km thick, and possibly bounded by a large normal fault. We speculate that these sediments fill a graben formed by rifting in the Siple Coast region. The sedimentary sequence observed beneath ice stream B is essentially identical to that observed beneath the Ross Sea, where a few metres of till rest unconformably (the Ross Sea unconformity) on a sedimentary sequence several kilometres thick. The late Oligocene and younger rocks beneath the Ross Sea have seismic velocities of about 1.7–2.5 km/s. This similarity suggests that the Ross Sea unconformity probably was created by erosion beneath a grounded ice sheet by a deforming till, and that the sediment beneath ice stream B is probably of glacial–marine origin similar to that filling the grabens of the Ross embayment.

Introduction

Geophysical information regarding the rock beneath central West Antarctica is scarce. Isolated seismic-refraction experiments (Bentley, 1973) have provided information on acoustic-basement velocities and depths, and regional aeromagnetic profiling has indicated depth to magnetic basement over some areas of West Antarctica (Jankowski, 1983). Recently, a glaciological research effort on ice stream B has shown a subglacial till layer several metres thick (Blankenship *et al.*, 1986, 1987) and a substantial sedimentary section (Rooney, Blankenship & Bentley, 1987b) in the vicinity of the upstream B camp (Fig. 1). As part of this effort the University of Wisconsin Geophysical and Polar Research Center also collected 4 km of multifold seismic-reflection data on ice stream B during the 1984–85 Antarctic field season. These data show reflections from a thickness of about 600 m of sedimentary rock. In conjunction with the previously reported seismic-refraction results (Rooney *et al.*, 1987b), we now are able to provide a seismic cross-section of the upper crust for one location in West Antarctica.

Data collection and processing

Compressional-wave seismic-reflection data were collected along a 4.3 km line which extended from near the grid north-eastern (true southern) edge of ice stream B toward the centre of the ice stream near upstream B camp (Fig. 1). A 690 m long spread with 24 detectors spaced at 30 m intervals made up the receiving array. Each detector consisted of a single geophone with 28 Hz natural frequency.

Five pound explosive charges detonated in boreholes spaced 360 m apart and averaging 16 m deep were used to obtain three-fold common depth point coverage for one-half of the line and four-fold coverage for the other half. Data collection occupied two people for the period 7–10 January, 1985. A data logger designed and built at the Geophysical and Polar Research Center was used for all recording. Before processing, all data were bandpass filtered using a passband of 50–300 Hz.

During data processing it was our goal to determine both a stacking velocity function (necessary to produce a seismic section) and interval velocities of the sediments beneath the ice. First, examination of a large number of stacked sections, each

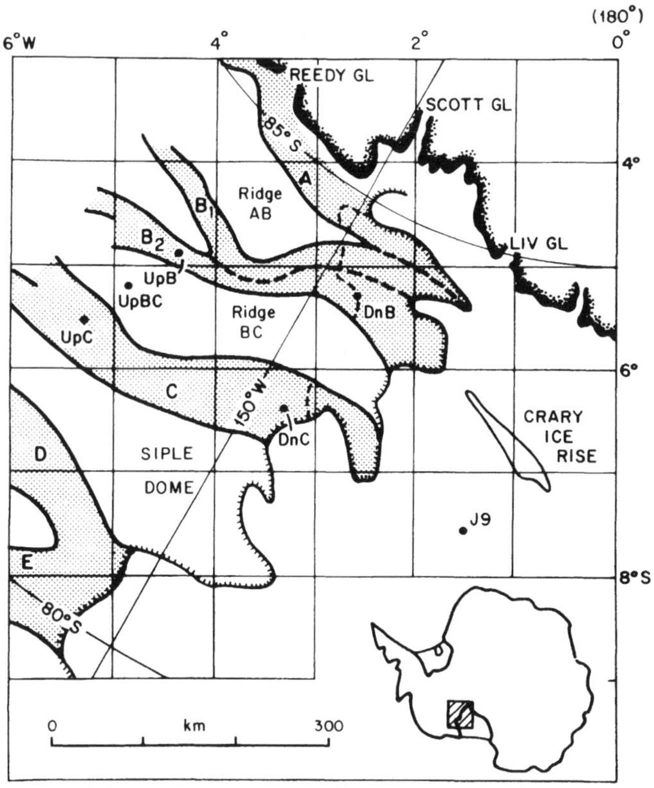

Fig. 1. Map showing the locations of ice stream B and Upstream B camp (UpB); stippled areas are the ice streams. The seismic profile was located near UpB. Other camp locations indicated are Downstream B camp (DnB), Upstream C (UpC), Downstream C (DnC), Ridge BC camp (UpBC), and the Ross Ice Shelf Project drillhole site J9. In the rectangular grid coordinate system, 0 ° grid longitude lies along the Greenwich and 180 ° meridians with grid north toward Greenwich, and the grid equator passes through the geographic South Pole. This map is a modified version of that presented by Shabtaie & Bentley (1987).

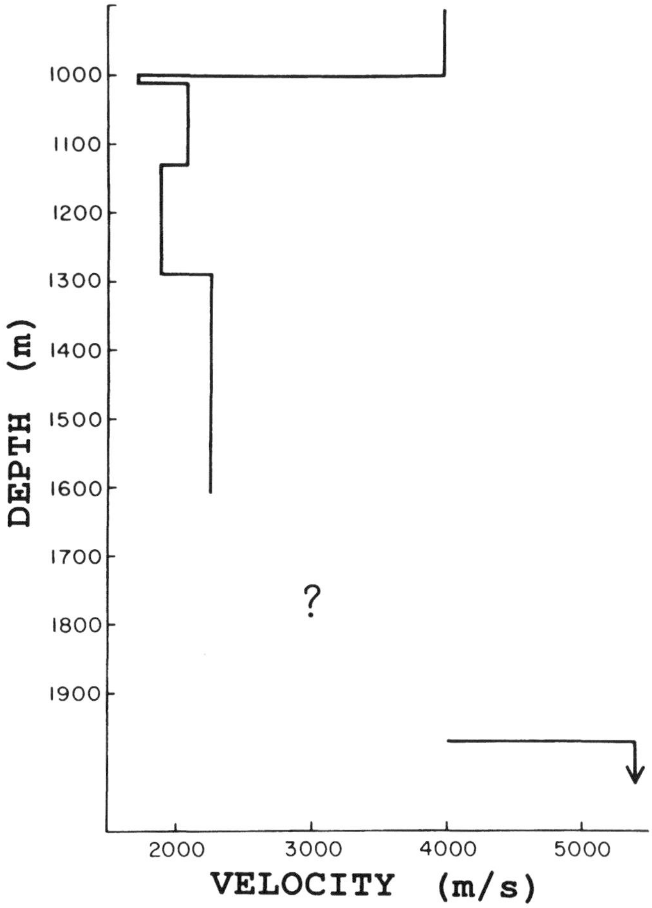

Fig. 2. Line drawing of the interval velocity function of depth derived from our data. The uppermost layer (3.8 km/s) represents ice and the second layer (1.7 km/s) represents unconsolidated subglacial sediment (from Blankenship *et al.*, 1986, 1987). The velocities and thicknesses of the three layers beneath the subglacial sediment were derived in this study. The lowermost velocity (5.4 km/s) is taken from Rooney *et al.* (1987a) with the depth recalculated in this study. The question mark denotes a region of unknown velocity ($\leq$ 5.4 km/s), as discussed in the text.

prepared with a different (constant) stacking velocity, allowed us to determine an approximate stacking velocity versus time function. This stacking velocity function then was used to compute a preliminary stacked section. Examination of this section indicated two important features: a negligibly small dip and apparent lateral homogeneity of the sub-ice reflectors across the 4 km section.

Assuming negligible dip and lateral homogeneity of the sediments, we then computed a velocity semblance plot (Taner & Koehler, 1969) for each 360 m portion of our 4.3 km line. Each portion provided a semblance plot showing coherent reflection energy after the reflection from the ice bottom. This band of reflections starts at about 0.5 s of vertical travel time and often ends abruptly at travel times of about 1.1 s. Three strong reflectors that appeared on both the preliminary stacked section and the original record sections were chosen for interval velocity analysis. The strongest peak on the velocity semblance plots corresponding to each of the chosen reflections was then picked. The result was a stacking velocity function for each 360 m portion of the line.

Interval velocities and layer thicknesses of the sediments beneath the 1 km thick ice and 6 m thick subglacial till (Fig. 2) were then determined from the stacking velocity function using the model simulation method of Al-Chalabi (1974) and the standard Dix (1955) equations. From top to bottom, the interval velocities (with errors at the 95% confidence level) and layer thicknesses are 2.08 ± 0.12 km/s, 120 m; 1.90 ± 0.15 km/s, 160 m; and 2.25 ± 0.13 km/s, 320 m. The stacking velocity function corresponding to this earth model was then used to create the final stacked seismic section (Fig. 3).

Discussion

The main objectives of this study were to determine the likely type, age, and tectonic history of the rock beneath ice stream B. Available data include the interval velocity function beneath the ice and the observed orientations of the reflectors (seismic structure). These data are compared below with similar data collected around West Antarctica.

Interval velocity results

Upstream B camp lies near the grid northern end of a physiographic province characterized by linear topography

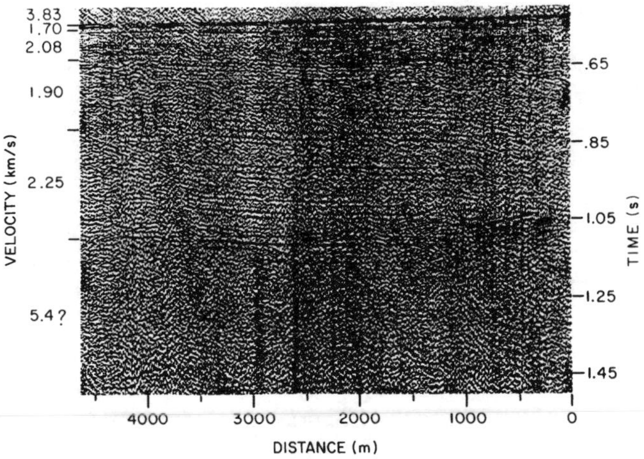

Fig. 3. Stacked seismogram produced from our data. The layers for which velocities were determined are shown along the left edge. The high frequency arrivals occurring at about 1.05 s, especially at a lateral position of about 0–1.0 km, are due to the first ice multiple.

(Hayes & Davey, 1975; Robertson *et al.*, 1982). In the Ross Sea such topography is believed to be the result of extensional tectonic activity between Marie Byrd Land and the Transantarctic Mountains (Davey, 1981; Cooper, Davey & Behrendt, 1987). Interpretation of gravity data beneath the Ross Ice Shelf has indicated deep structural control of the sea-floor topography rather than a glacial origin (Robertson *et al.*, 1982).

The sedimentary basins beneath the Ross Sea are among the best studied in Antarctica (e.g. Hayes & Frakes, 1975; Davey, Hinz & Schroeder, 1983; Cooper, *et al.*, 1987). Because Upstream B is in the same physiographic province as these basins, they are obvious analogues to use for interpreting the type and age of the sediments beneath ice stream B.

A 1300 m sequence of mostly late Oligocene and younger sedimentary rock was sampled at three Deep Sea Drilling Project sites (270–272) in the grid north-central Ross Sea (Hayes *et al.*, 1975). Laboratory sonic velocity measurements showed that only the early Miocene and younger rocks (from the upper 1100 m of the sequence) were characterized by sonic velocities < 2.55 km/s (Barrett & Froggatt, 1978). The samples tested were all of glacial–marine origin. Barrett & Froggatt (1978) also determined sonic velocities of a large number of rocks from Victoria Land; only one sample (of Arena sandstone) provided a velocity within the 1.90–2.25 km/s range observed for ice stream B.

A 230 m section of glacial–marine sediments of late Oligocene and younger age was sampled by the MSSTS-1 drillhole in McMurdo Sound. This sequence yielded widely scattered sonic velocity measurements, several within the range observed beneath ice stream B (Davey & Christoffel, 1984; Harwood, 1986).

Reflection and refraction experiments in the Victoria Land basin (Cooper *et al.*, 1987) and in the Weddell Sea embayment (Haugland, Kristoffersen & Velde, 1985) indicate that only sediments interpreted to be late Oligocene and younger are characterized by velocities of < 2.7 km/s.

One piece of geological evidence for the age of the sediments beneath ice stream B comes from a 1 m sediment sample collected beneath the J9 drillhole, which was located on the Ross Ice Shelf directly downstream (with respect to ice flow) from ice stream B (Fig. 1). Sedimentary clasts of Miocene age were found in these sediments as well as clasts of granite, gneiss and schist (Webb *et al.*, 1979). Assuming that recent models for sediment transport beneath ice stream B are correct (Alley *et al.*, 1987b), the upper 1 m of sediment found beneath J9 is likely to have originated beneath ice streams B or C or their catchments. The presence of Miocene clasts then indicates that Miocene-age sediments occur somewhere in the drainage basins of these ice streams.

Rooney *et al.* (1987b) interpreted seismic-refraction experiments performed near upstream B as indicating a thickness of 2 km of low-velocity rock on a basement surface that dips 4° to the grid south-east. The measured wave velocity in the basement was 5.4 km/s. For the analysis a velocity for the low-velocity rock overlying basement of 4.0 km/s was assumed; in the light of the present reflection results that velocity was much too high. Recalculation of the sedimentary thickness and basement dip using our observed average sedimentary velocity of 2.1 km/s gives an apparent dip on the basement surface of 1° to the grid south-east. The thickness of the sediments overlying the basement then would be 970 m at a location about 4 km to the grid south-west of the reflection profile. The deepest reflector to which a velocity determination could be made in this reflection experiment lies beneath about 600 m of sediment. Whether this represents acoustic basement and the overlying sediments thicken to the grid south-west, or whether about 360 m of rock of unknown velocity overlies the acoustic basement at this location, is uncertain.

The type and age of the rock interpreted here as acoustic basement are very difficult to determine. A large number of West Antarctic rocks have measured velocities around 5.4 km/s (Barrett & Froggatt, 1978). Cooper *et al.* (1987) have shown convincingly that what had been interpreted as basement in the Victoria Land basin, on the basis of seismic velocities > 5 km/s, actually represents deeper sedimentary strata. They interpreted these high-velocity sedimentary strata as being of marine origin. Because of this we must leave open the question of whether the 5.4 km/s rock is crystalline basement or high-velocity sedimentary strata. Regional aeromagnetic studies around the region of our reflection profile have been interpreted as indicating a thickness of up to 8 km of non-magnetic, probably sedimentary rock (Jankowski, 1983). Gravity studies currently in progress at the Geophysical and Polar Research Center may help resolve this question.

We believe that a consistent interpretation of the low velocities beneath ice stream B is that they represent late Oligocene or younger glacial–marine strata. We know of no West Antarctic rock older than late Oligocene that has acoustic velocities consistently as low as those observed beneath ice stream B. Likewise, the glacial history of West Antarctica is such that any sediments deposited after the late Oligocene in the area now covered by ice stream B were probably glacial–marine deposits (Hayes & Frakes, 1975).

Seismic structure

The 600 m of presumed late Oligocene or younger strata beneath ice stream B are essentially undeformed (Fig. 3). The strata dip < 0.5° to the grid north-east, whereas the ice bottom dips about 0.5° to the grid south-west. An angular unconformity exists within a few metres of the ice bottom where the consolidated sedimentary layers are truncated by what is believed to be an unconsolidated layer of saturated till lying roughly parallel to the ice bottom (see Rooney *et al.*, 1987a, fig. 6). The till is thought to be deforming throughout its thickness, eroding its base, and generating the unconformity (Alley *et al.*, 1986, 1987a). If the ice were removed the till, which averages about 6 m in thickness, might then resemble the diamicton (a few metres to a few tens of metres thick) observed on the Ross Sea floor (Hayes *et al.*, 1975; Karl, Reimnitz & Edwards, 1987). The Ross Sea diamicton overlies a widespread unconformity which has been mapped over much of the Ross Sea (Karl *et al.*, 1987). We suggest that the unconformity beneath ice stream B represents a modern example of the Ross Sea unconformity (Alley *et al.*, 1987a).

Previous refraction studies in the upstream B region gave some indication that a large, high-angle fault or fault zone may trend roughly parallel to ice stream B (Rooney *et al.*, 1987b). The inferred location of the fault zone where it passes the reflection profile is 2–4 km to the grid north-east of the north-eastern end of the profile. The total throw across the fault zone (recalculated from Rooney *et al.* (1987b) using the new sedimentary section velocities) is about 1 km, with the reflection profile located over the downthrown block. If our lowest reflector (at 1.6 km depth) represents the 5.4 km/s acoustic basement a second fault farther to the grid south-west could lower the basement to the 2.0 km depth observed in the refraction experiment. Recent gravity results from ice stream B indicate that a steep gravity gradient exists over the inferred fault zone, with the reflection profile located on the low gravity side (S. Atre, pers. comm., 1988). This is consistent with the seismic-refraction interpretation.

On the basis of the thickness, lack of deformation and presumed age of the sedimentary rock, plus the evidence for large-scale, high-angle faulting, we believe that the depositional environment of these rocks is similar to that proposed by Cooper *et al.* (1987) for late Oligocene and younger rocks within the Victoria Land basin (VLB). The VLB is one of a number of grabens beneath the Ross Sea adjacent to, and roughly colinear with, the Transantarctic Mountains. The origin of these grabens is believed to involve two phases of rifting. The first phase began during rifting of Gondwana in Jurassic time; the VLB was created by down-faulting during this extension and filled by thick Cretaceous–Paleocene–Eocene sedimentary rocks. The late phase of rifting was coeval with uplift of the Transantarctic Mountains which began in the Eocene. Late Oligocene and younger sediments were deposited during the late rifting phase in the VLB. The late phase of rifting probably continues today.

We propose that the late rifting phase in the Ross Sea was contemporaneous with a similar extensional event beneath ice stream B. A graben created by this extension was filled by Late Oligocene and/or younger sediments. A recent estimate of the geothermal flow at a site about 50 km to the grid south-west of Upstream B was about 2.0 HFU (Alley, 1987). This relatively high geothermal flow is consistent with active or recent rifting. The thickness of the presumed late Oligocene or younger rocks beneath ice stream B is similar to that within the VLB and thin relative to the grid-western Ross Sea. As interpreted in the VLB (Cooper *et al.*, 1987) this may indicate that the subsidence beneath ice stream B was slower and/or the sedimentation rate was less than in the grid-western Ross Sea. We suspect that initiation of the late rifting phase beneath ice stream B was later than Eocene, possibly as late as late Oligocene, because we observe no rock with velocities similar to the Palaeocene–Eocene rock in the VLB (2.7–4.9 km/s (Cooper *et al.*, 1987)). However, it is possible that our 5.4 km/s acoustic basement represents Palaeocene or Eocene sediments with higher velocities than those observed in the VLB or that the region of unknown velocity between 1.6 and 2.0 km depth beneath upstream B camp is Palaeocene–Eocene rock. If so, then the initiation of extension beneath the grid-eastern Ross Sea and ice stream B may have been coeval. We cannot determine from our data whether the area around ice stream B participated in the early (Jurassic) phase of rifting as well.

Conclusions

Our reflection data from ice stream B indicate that a 600 m sequence of sedimentary rock with seismic velocities ranging from 1.90 to 2.25 km/s exists beneath the ice stream. The upper interface of the sedimentary section is an angular unconformity truncated by an unconsolidated till layer which is several metres thick. The acoustic basement has a velocity of 5.4 km/s and occurs about 1 km beneath the ice–rock interface. The low-velocity sedimentary rock is interpreted as late Oligocene or younger glacial–marine strata deposited in a graben formed during extension which began in Palaeocene–Oligocene time. The graben is probably bounded on the grid north-eastern side by a high-angle fault with about 1 km of offset. The 5.4 km/s acoustic basement may represent either crystalline or sedimentary rock.

Acknowledgements

We wish to thank B.R. Weertman for help collecting these data, K. Kuivinen, B. Boller, and J. Litwak of the Polar Ice Coring Office for shot hole drilling, A.N. Mares for manuscript preparation, and P. Dombrowski for figure preparation. Support for this research was provided under National Science Foundation grant DPP-8412404. This is contribution number 480 of the Geophysical and Polar Research Center, University of Wisconsin–Madison.

References

Al-Chalabi, M. (1974). An analysis of stacking, rms, average, and interval velocities over a horizontally layered ground. *Geophysical Prospecting*, **22(3)**, 458–75.

Alley, R.B. (1987). Transformations in Polar Firn. PhD thesis, University of Wisconsin-Madison, 413 pp. (unpublished).

Alley, R.B., Blankenship, D.D., Bentley, C.R. & Rooney, S.T. (1987a). Till beneath ice stream B, 3, till deformation: evidence and implications. *Journal of Geophysical Research*, **92(B9)**, 8921–9.

Alley, R.B., Blankenship, D.D., Rooney, S.T. & Bentley, C.R. (1986). Deformation of till beneath ice stream B, West Antarctica. *Nature, London*, **322(6074)**, 57–9.

Alley, R.B., Blankenship, D.D., Rooney, S.T. & Bentley, C.R. (1987b). Continuous till deformation beneath ice sheets. *International Assocation of Hydrological Sciences Publication*, **170**, 81–91.

Barrett, P.J. & Froggatt, P.C. (1978). Densities, porosities, and seismic velocities of some rocks from Victoria Land, Antarctica. *New Zealand Journal of Geology and Geophysics*, **21(2)**, 175–87.

Bentley, C.R. (1973). Crustal structure of Antarctica. *Tectonophysics*, **20**, 229–40.

Blankenship, D.D., Bentley, C.R., Rooney, S.T. & Alley, R.B. (1986). Seismic measurements reveal a saturated, porous layer beneath an active Antarctic ice stream. *Nature, London*, **322(6074)**, 54–7.

Blankenship, D.D., Bentley, C.R., Rooney, S.T. & Alley, R.B. (1987). Till beneath ice stream B, 1, properties derived from seismic travel times. *Journal of Geophysical Research*, **92(B9)**, 8903–11.

Cooper, A.K., Davey, F.J. & Behrendt, J.C. (1987). Seismic stratigraphy and structure of the Victoria Land basin, western Ross Sea, Antarctica. In *The Antarctic Continental Margin: Geology and Geophysics of the Western Ross Sea*, ed. A.K. Cooper & F.J. Davey, pp. 27–65. Houston, Texas; Circum-Pacific Council for Energy and Mineral Resources, Earth Science Series, 5B.

Davey, F.J. (1981). Geophysical studies in the Ross sea region. *Journal of the Royal Society of New Zealand*, **11(4)**, 465–79.

Davey, F.J. & Christoffel, D.A. (1984). The correlation of the MSSTS-1 drillhole results with seismic reflection data from McMurdo Sound, Antarctica. *New Zealand Journal of Geology and Geophysics*, **27(4)**, 405–12.

Davey, F.J., Hinz, K. & Schroeder, H. (1983). Sedimentary basins of the Ross Sea, Antarctica. In *Antarctic Earth Science*, ed. R.L. Oliver, P.R. James & J.B. Jago, pp. 533–8. Canberra; Australian Academy of Science and Cambridge; Cambridge University Press.

Dix, C.H. (1955). Seismic velocities from surface measurements. *Geophysics*, **20(1)**, 68–86.

Harwood, D.M. (1986). Diatom biostratigraphy and paleoecology with a Cenozoic history of Antarctic ice sheets. PhD thesis, Ohio State University, Columbus, 2 vols, 592 pp. (unpublished).

Haugland, K., Kristoffersen, Y. & Velde, A. (1985). Seismic investigations in the Weddell Sea embayment. *Tectonophysics*, **114**, 293–313.

Hayes, D.E. & Davey, F.J. (1975). A geophysical study of the Ross Sea, Antarctica. In *Initial Reports of the DSDP 28*, ed. D.E. Hayes *et al.*, pp. 887–907. Washington, DC; US Government Printing Office.

Hayes, D.E. & Frakes, L.A. (1975). General synthesis, deep sea drilling project leg 28. In *Initial Reports of the DSDP 28*, ed. D.E. Hayes *et al.*, pp. 919–42. Washington, DC; US Government Printing Office.

Hayes, D.E., Frakes, L.A. *et al.* (1975). Sites 270, 271, 272. In *Initial Reports of the DSDP 28*, ed. D.E. Hayes *et al.*, pp. 211–334. Washington, DC; US Government Printing Office.

Jankowski, E. (1983). Depth to magnetic basement in West Antarctica. In *Antarctica: Glaciological and Geophysical Folio*, ed. D.J. Drewry, Sheet 8. Cambridge; Scott Polar Research Institute, University of Cambridge.

Karl, H.A., Reimnitz, E. & Edwards, B.D. (1987). Extent and nature of Ross Sea unconformity in the western Ross Sea, Antarctica. In *The Antarctic Continental Margin: Geology and Geophysics of the Western Ross Sea*, ed. A.K. Cooper & F.J. Davey, pp. 77–92. Houston, Texas; Circum-Pacific Council for Energy and Mineral Resources, Earth Science Series, 5B.

Robertson, J.D., Bentley, C.R., Clough, J.W. & Greischar, L.L. (1982). Sea-bottom topography and crustal structure below the Ross ice shelf, Antarctica. In *Antarctic Geoscience*, ed. C. Craddock, pp. 1083–90. Madison; University of Wisconsin Press.

Rooney, S.T., Blankenship, D.D., Alley, R.B. & Bentley, C.R. (1987a). Till beneath ice stream B, 2, structure and continuity. *Journal of Geophysical Research*, **92(B9)**, 8913–20.

Rooney, S.T., Blankenship, D.D. & Bentley, C.R. (1987b). Seismic refraction measurements of crustal structure in West Antarctica. In *Gondwana Six: Structure, Tectonics, and Geophysics*, Geophysical Monograph Series, 40, ed. G.D. McKenzie, pp. 1–7. Washington, DC; American Geophysical Union.

Shabtaie, S. & Bentley, C.R. (1987). West Antarctic ice streams draining into the Ross Ice Shelf: configuration and mass balance. *Journal of Geophysical Research*, **92(B9)**, 1311–36.

Taner, M.T. & Koehler, F. (1969). Velocity spectra-digital computer derivation and applications of velocity functions. *Geophysics*, **36(6)**, 859–81.

Webb, P.N., Ronan, T.E., Jr., Lipps, J.H. & DeLaca, T.E. (1979). Miocene glaciomarine sediments from beneath the southern Ross Ice Shelf, Antarctica. *Science*, **203(4379)**, 435–7.

The aeromagnetic survey of northern Victoria Land and the western Ross Sea during GANOVEX IV and a geophysical–geological interpretation

W. BOSUM[1], D. DAMASKE[1], J.C. BEHRENDT[2] & R. SALTUS[2]

1 Bundesanstalt für Geowissenschaften und Rohstoffe, Postfach 51 01 53, D-3000 Hannover 51, Federal Republic of Germany
2 US Geological Survey, Box 25046, Denver Federal Center, Denver, CO 80225, USA

Abstract

In 1984–85 aeromagnetic investigations were carried out over an area extending from the East Antarctic shield, through the Transantarctic Mountains to the Ross Sea. The survey was planned like a conventional aeromagnetic survey, with relatively small distances (4.4 km) between profile lines and correspondingly low flight levels over the estimated depths of magnetic sources. This method enabled identification of large-scale anomalies and provided information on the smaller-scale magnetic patterns. Exact flight positioning was obtained using a special ground-based navigation system. As the area is located within the auroral zone, special attention was given to recording magnetic time variations. Profiles totalling 50000 line kilometres were flown. The Victoria Land basin shows a distinct magnetic character, strongly supporting a graben-type structure within continental crust. To the north, an abrupt change in magnetic pattern is marked by large circular anomalies. These are connected to a magnetic unit, the most impressive anomaly of which is the 'Polar 3 anomaly', extending about 150 km WSW–ENE, with amplitudes of up to 1600 nT. Farther west, over the polar plateau, the survey revealed features typical of cratonic areas. Some magnetic features over the partly exposed mountain areas can be correlated with known geological structures.

Introduction

In 1984–85 the Bundesanstalt für Geowissenschaften und Rohstoffe (BGR) of the Federal Republic of Germany, in cooperation with the US Geological Survey, carried out an aeromagnetic survey over parts of northern Victoria Land (NVL) and the western Ross Sea. The aim was to cover an area extending from the East Antarctic shield, through the younger formations of the Transantarctic Mountains, to the Ross Sea which, according to seismic and gravity data, consists of a number of N–S-aligned basins (Hinz & Block, 1984; Cooper, Davey & Behrendt, 1985).

The investigations should be viewed in the context of the general framework of tectonic evolution in this sector of Antarctica. NVL forms a key area for three major problems:

(i) its relationship to Australia, Tasmania, and New Zealand
(ii) the significance of mobile belts at the fringe of the East Antarctic shield
(iii) the Ross Sea and its development in connection with recent plate tectonics.

The specific tasks of GANOVEX IV were to determine the boundaries between different tectonic provinces and to follow known structures underneath the ice sheet and the sea, and thus place them in the regional framework. One of the most important questions to be answered was the position of the boundary between the Antarctic shield and the younger mobile belts. According to all available information this boundary should lie to the west of the most remote outcrops of the Transantarctic Mountains, in an area covered totally by the ice sheet of the polar plateau. The coastal area on the other side of the Transantarctic Mountains has not been investigated widely. Here the land drops from heights of 4000 m to the down-faulted block of the Ross Sea. The western Ross Sea itself consists of a number of N–S-aligned morphological rises and troughs which are not well understood.

The study area comprises the ice-covered area of the polar plateau west of the Transantarctic Mountains, a central section of the Transantarctic Mountains, and the western Ross Sea, from Coulman Island to Ross Island (Fig. 1). A number of ground-based field parties worked in this area at the same time as, but independently of the airborne programme, to provide additional data for interpretation of the aeromagnetic survey.

Organization of the survey

The spacing of flight lines is dependent on the presumed depth to source bodies. Ideally, distance to the magnetic source

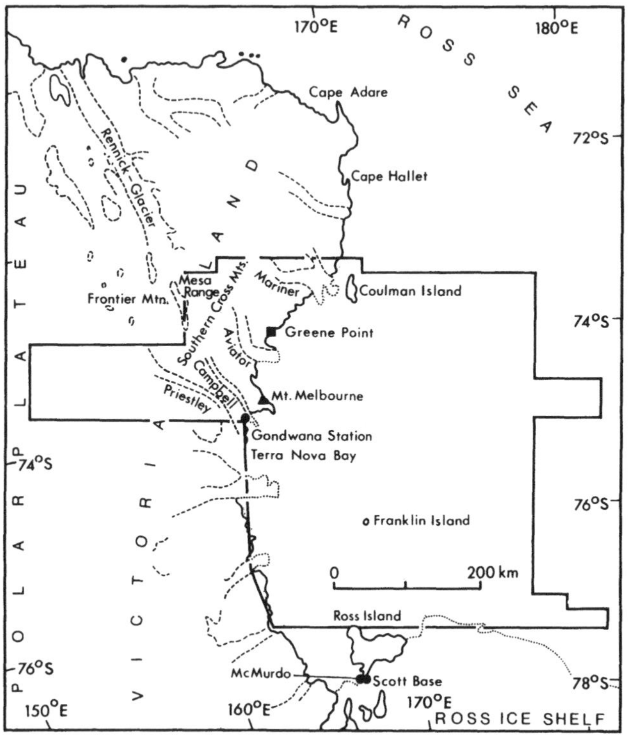

Fig. 1. Sketch map showing the location of the aeromagnetic survey area (outlined) during Ganovex IV, 1984–85.

and the spacing should be equal, yielding an unambiguous picture of the magnetic anomaly. However, even ratios of 1:2–1:4 between distance to source and spacing can provide good information on the source body. For a mean minimum distance to magnetic source below flight levels of around 1.5 km therefore, a profile line distance of 4.4 km was chosen. This was a compromise between an optimum line distance and the time available to cover the given survey area.

Because of the morphology of the area, the survey was carried out at three altitude levels. The topography of the Transantarctic Mountains, with peaks of more than 3600 m, necessitated a minimum flight level of 1200 ft (3660 m). Flights over the polar ice sheet were carried out at an altitude of 900 ft (2740 m), a compromise in recognition of the fact that the ice thickness measurements (carried out simultaneously with the aeromagnetic survey) needed a low flying level. For the survey over the Ross Sea, geophysical considerations and the envisaged profile line distance of 4.4 km, led to the choice of a maximum flight level of 2000 ft (610 m).

The survey

General

The airborne survey was carried out with two Dornier Do-228 aircraft leased from the German Alfred-Wegener-Institute of Polar Research and the Dornier Company. The United States airlifted three German Ecureuil AS-350 helicopters from New Zealand to McMurdo. These were used to install the navigation beacons, to support ground fieldwork, to maintain the link with McMurdo and Scott Base over a distance of 360 km, and to shuttle between Gondwana station and the landing site for the aircraft at Browning Pass.

At first, both aircrafts operated simultaneously. Polar 2, equipped for radio-echo sounding, was used predominantly for flights over the plateau ice until the front ski of the plane was badly damaged during a hard landing on the polar plateau. Ice thickness measurements then came to an end. After the accident, both crews operated the second aircraft and more than 50000 km of survey lines were flown during the season.

Navigation

Special attention had to be given to precise navigation along survey lines. The narrow line spacing demanded flight positioning which was not possible with the standard equipment in the aircraft. The VLF/Omega System was inadequate for flights over the interior land masses whilst the TANS-Doppler System could not be used over the sea, due to drift of the reflecting surface. In geomagnetically high latitudes, the TANS-Doppler System, supported by a magnetic compass, has additional problems: the horizontal component of the magnetic field is small, the variation changes markedly over a short distance and strong time variations influence the system. Optical navigation is only possible in the mountainous area, and crevassed areas over the polar plateau could only be used in a few cases for later flight-path reconstruction.

To solve these problems, a special ground-based navigation system was set up in the survey area, operating in the so-called 'range-to-range mode'. The transceiver installed in the aircraft activated ground beacons which were within the signal range and received a coded signal from which the distance to the beacon was computed. With only two such distances, it was possible to calculate the aircraft's position. A monitor indicated the aircraft deviation from the set course, enabling the pilot to correct the flight. With ground-beacon control it was possible to guide the aircraft to within 100 m. Altogether six beacons were available per survey covering most of the survey area. For areas beyond the range of the ground transmitter, e.g. those parts of a survey line far out over the sea, the aircraft were navigated by standard systems, their position being checked as soon as the aircraft returned into the range of the beacon system. In this way it was possible to locate flight lines within an accuracy of $< 5\%$ of the line spacing (i.e. < 250 m) for most of the area.

Magnetic time variations

As the survey area is located within the auroral zone, special attention was given to the magnetic time variations. A magnetic base station monitored changes in the magnetic field throughout the season to select the most favourable times of day for the survey flights.

At Gondwana station the values for the magnetic total intensity were recorded at two-minute intervals on 75 consecutive days. They were used to define a magnetic quiet period favourable for the daily survey flights and to follow the general behaviour of the geomagnetic pattern throughout the survey. The time variations on each day were characterized by a high level of activity in the early morning hours until noon local time. Relatively quiet conditions occurred from early afternoon until after local midnight. This behaviour was observed on all days and showed no detectable shift throughout the season.

Base stations were set up parallel to these continuous recordings, for a comparison of magnetic activity at other locations in the survey area. For the magnetically quiet period of a day, the recordings at stations even more than 100 km away from Gondwana station show a remarkably similar trend. However, the observed differences in amplitude and phase of individual variations make it impossible to extrapolate from one station to another in the survey area. This was particularly obvious as soon as the activity level was increased.

Control of magnetic activity during a flight was given by base station sampling at 10 s intervals. Because these recordings could not be extrapolated over the whole survey area, control lines played an even more vital role in the survey. Their distance was chosen in such a way that the flying time between intersections with profile lines allowed a linear interpolation of the magnetic variation. This was achieved by a 22 km spacing of the control lines, corresponding to 5 min flying time.

Results

The aeromagnetic survey produced data for more than 50000 flight line kilometres, covering an area of about 250000 km^2. The resulting magnetic anomaly map of northern Victoria Land and the western Ross Sea is shown in Fig. 2. Due to the near-pole position of the survey area, with a mean inclination of 84°, the magnetic anomaly map permits a direct interpretation of symmetric positive and negative anomalies: high levels of the magnetic crystalline basement or nearly vertical magnetic bodies show up as maxima while minima can be interpreted as depressions in the crystalline basement.

The south-western Ross Sea

The south-western Ross Sea area (Figs 1 & 2) generally shows anomalies of moderate amplitude (~ 50–100 nT) which strike predominantly N–S. The most prominent structure is the nearly N–S-trending minimum flanked by maxima to the west and east. It is a direct response to a depression in the magnetic crystalline basement. The form of the anomaly, especially its nearly linear course over more than 250 km, suggests a graben structure for this area which coincides with the Victoria Land basin (VLB), known from marine seismic work. An interpretation by two-dimensional models (Fig. 3) yielded simple step-like bodies, the upper face about 2–4 km below sea level, the lower face extending to a depth of 20 km. The flanks of the models dip E on the western and W on the eastern part. The magnetization ranges from 0.3 to 0.8 A/m. Based on this interpretation a three-dimensional model has been calculated, the anomalies of which compare well with the measured data. Magnetic profiles along cross-sections through the basin revealed a large graben structure which, in the central part, is divided by an intermediate high and bordered by another graben to the east. A detailed interpretation of single magnetic profiles gives an average depth of the magnetic basement of 15 km in the centre of the graben.

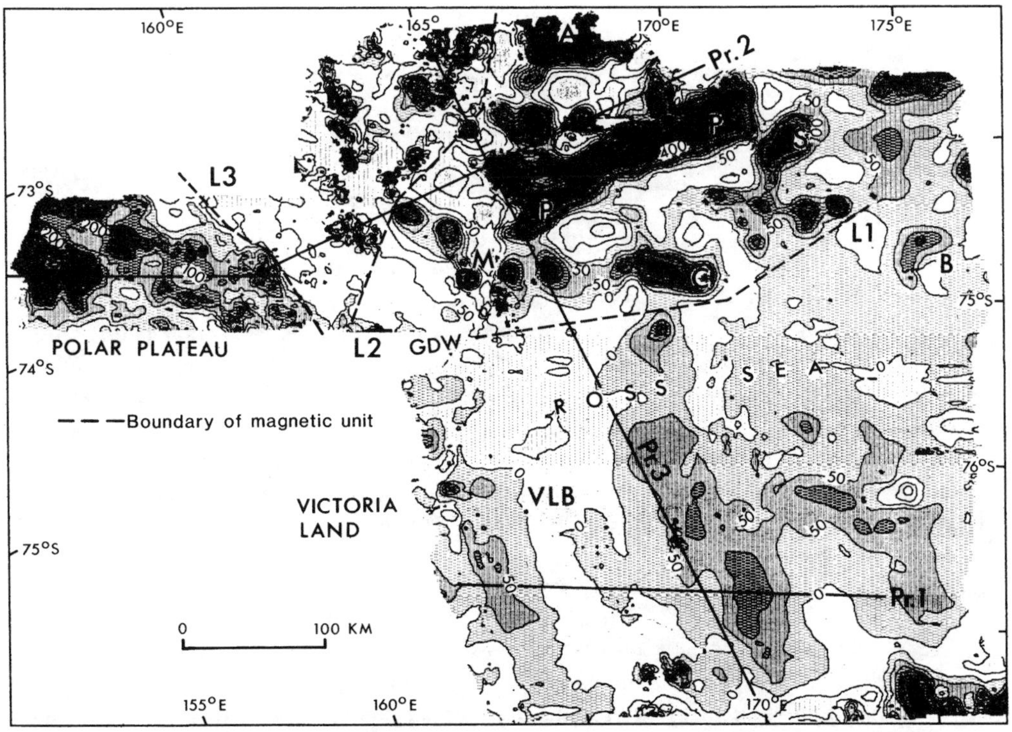

Fig. 2. Anomalies of the total magnetic field in the northern Victoria Land and western Ross Sea areas, as derived from airbone measurements during GANOVEX IV, 1984–85. The positions of the profiles Pr. 1 and 2 (Figs 3 and 4) and Pr. 3 (Fig. 5) are shown; P, Polar 3 anomaly; GDW, Gondwana station; GP, Greene Point; M, Mount Melbourne; A, B and C, other anomalies.

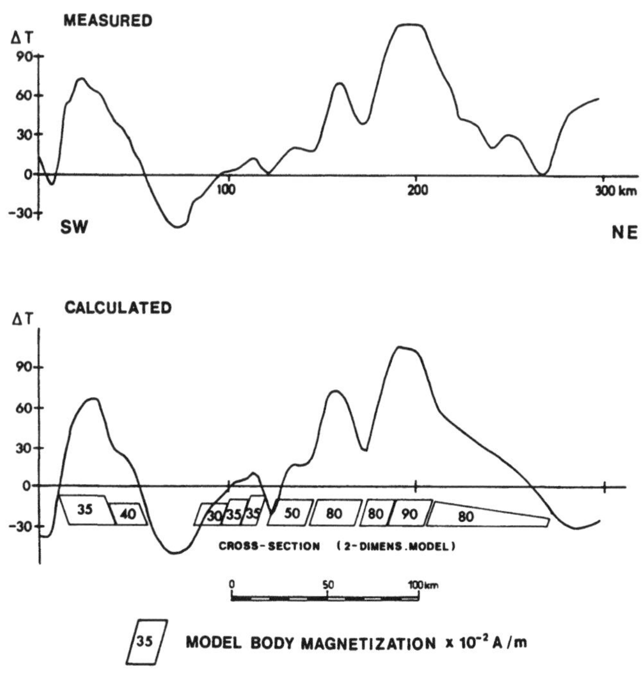

Fig. 3. Interpretation by two-dimensional models for profile Pr. 1 across the Victoria Land basin.

Smaller-scale anomalies are scattered over parts of this area. Near Franklin Island they are undoubtedly of volcanic origin and it is feasible to assume that the others have a similar origin. However, as they occur only occasionally it can be concluded that here volcanic activity was relatively limited.

Farther east the central basin area, deduced from seismic results, displayed a magnetic pattern which differed totally from the Victoria Land basin's graben-type structure, i.e. irregular shapes of long wavelengths with no distinct character. Calculating magnetizations for the crystalline basement yielded typical continental crust values of 0.3 A/m.

The Southern Cross unit and the Transantarctic Mountains

The Southern Cross unit is situated between the central Ross Sea area in the south and the transition zone in the west and north-west, and is separated from both areas by lineaments. One lineament (L1) trends nearly E–W, starting in Terra Nova Bay, south of Gondwana station (GDW). Other lineaments which are situated farther east might be interpreted as the continuation of L1. It should be emphasized that no Southern Cross unit anomalies extend into the central Ross Sea area. The boundary is marked by a strong gradient over long distances but there is no indication of a strike-slip fault along this boundary.

In the north-west the Southern Cross unit is bordered by further lineaments (L2). Here, the separation of the adjacent units is not as well defined. NW–SE-trending negative anomalies continue from the transition zone into the Southern Cross unit, especially in the Aviator Glacier, Campbell Glacier and Priestley Glacier minima.

The Southern Cross unit is unique because of its strong anomalies (up to 1700 nT). These show either well-defined circular or elongated forms. The former indicate magmatic complexes, e.g. volcanoes such as the Mount Melbourne shield volcano (M), or alkaline complexes similar to those known in, for example, the Brazilian Craton (Bosum, 1981). They are partly arranged on NW–SE- to WNW–ESE-trending lines. Anomalies A and C (Fig. 2) are included in this system as well as the conspicuous 'Polar 3 anomaly' (P) and its satellite anomaly to the east, both of which show a WSW–ENE strike.

The largest anomaly in the area, located around Greene Point (GP), Lady Newness Bay, cannot readily be associated with any known magnetized body. Granite outcrops there do not show any significant magnetization. However, the occurrence of dark veins in many places within these granites and the presence of basalts at other outcrops nearby, suggests that the source body for the strong magnetic anomaly might be a basic intrusion beneath the granitic complex. Model calculations indicate that the upper boundary of such an intrusion would be at about sea level.

More difficult to interpret are the elongated anomalies, especially the 'Polar 3 anomaly' which extends approximately parallel to the coast from just north of Mount Melbourne to east of Coulman Island. The calculated magnetization of about 5.0 A/m points to a highly basic or serpentinized ultrabasic body. The depth of the upper surface of such a source body is estimated to be 2 km, comparing well with the 2 km depths calculated for the more circular anomalies. The nearly E–W strike of the 'Polar 3 anomaly' cuts off other NW–SE-trending anomalies and anomaly alignments which are dominant farther north-west, in the Transantarctic Mountains.

Apart from the round and broad, elongated anomalies, the Southern Cross unit is characterized by weak anomalies which probably reflect a continental crust penetrated by alkali-intrusions and large basic–ultrabasic 'dykes'. These intrusions may represent upper mantle material emplaced in stable continental crust during tectonism. This areas contrasts with the western Ross Sea where the continental crust is less stable and volcanism is of comparatively minor extent.

Outcrops in the Transantarctic Mountains, to the north-west of this area, provided some information on the surface geology. Magnetic measurements showed that the exposed granites and gneisses are almost non-magnetic. Magnetizations were observed for some granodiorites, gabbros and younger volcanic rocks. These results can be seen in the aeromagnetic map (Fig. 2), where a quiet pattern of slightly negative field values and some complexes of small-scale anomalies reflect the Kirkpatrick basalts of the Mesa Range, and the Ferrar dolerites of the Deep Freeze Range. This pattern is comparable to the typical 'basalt anomaly pattern' (Bosum, 1981). The latter anomalies are of known near-surface origin, thus indicating that the deeper structures have only a weak magnetic field (continental crust). The alignment of broad minima trending roughly NW–SE (possibly representing lows in the magnetic crystalline basement at depths of 4–5 km) follows the major valleys of Mariner, Aviator, Campbell and Priestley glaciers, suggesting that these valleys are a result of old structural trends already present in the old continental crust.

It is obvious that the Transantarctic Mountains do not form

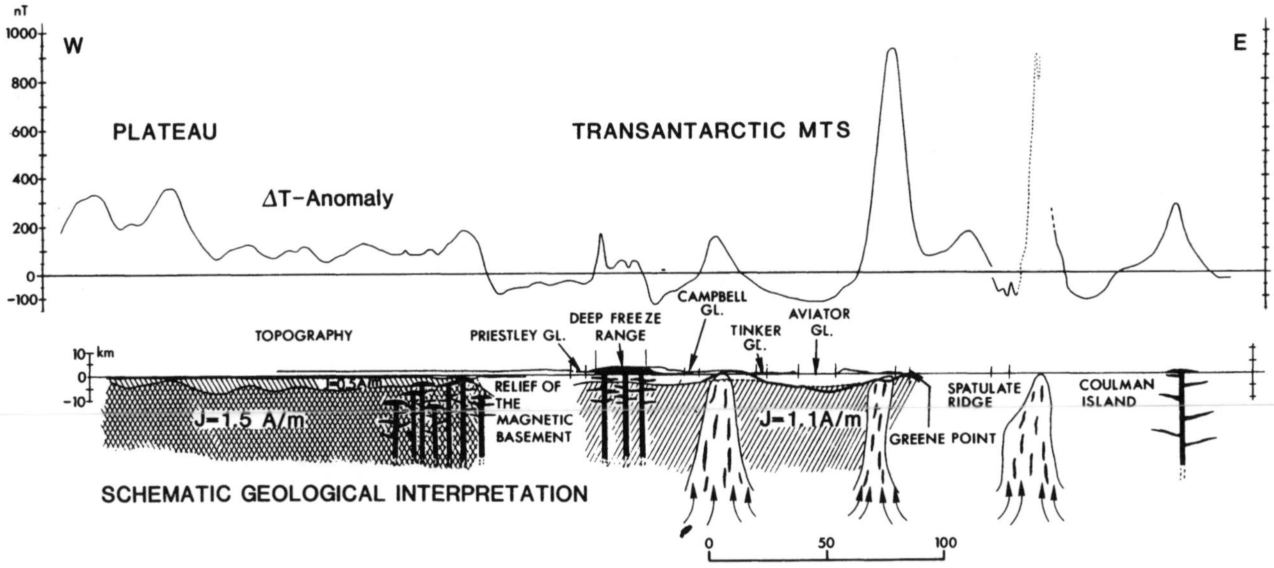

Fig. 4. West to east profile (Pr. 2) from the polar plateau across the Transantarctic Mountains to Coulman Island.

a well defined unit on the aeromagnetic map. Only anomalies of volcanic origin (e.g. Kirkpatrick basalts and Ferrar dolerites) point to magmatism accompanying the formation of this mountain range.

The polar plateau

To the west of the Transantarctic Mountains there is a prominent boundary (L3) in the anomaly pattern. The magnetic field values change from a broad minimum to a large maximum. Thus this area consists of more strongly magnetized material similar to typical 'high-grade' shield areas. A two-dimensional harmonic analysis shows a subdivision of the magnetic basement into two layers: an upper part with a surface at ~ 500 m below sea level (i.e. about 2000–2500 m below the ice surface) extending to a depth of 5–6 km. This higher layer is only weakly magnetized whereas the deeper one shows a higher magnetization, responsible for the general maximum area.

Conclusion

A schematic geological interpretation is given in Fig. 4. In the area of the polar plateau the relatively strong anomalies of the craton dominate, reflecting the higher magnetized, lower magnetic layer. The upper magnetic layer shows a relatively smooth surface at about 500 m below sea level. At its eastern side the short wavelength pattern similar to that of the Deep Freeze Range suggests that here also dolerites might be the source. Known Ferrar dolerites are responsible for the magnetic pattern on the eastern side of the Priestley Glacier minimum. This minimum separates the shield area in the west from an area of typical continental crust in the east which underlies the Transantarctic Mountains. Known magmatic bodies, e.g. the Mount Melbourne volcano or unknown ones, as around the anomaly at Greene Point, penetrate the otherwise moderately magnetized crust. Large glacier valleys follow the morphology of the magnetic basement. Fig. 5 shows the corresponding north to south profile, starting in the Southern Cross unit and crossing the western Ross Sea at about Franklin Island.

References

Bosum, W. (1981). Anlage und Interpretation aeromagnetischer Vermessungen im Rahmen der Erzprospektion. *Geologisches Jahrbuch*, **E20**, 3–63.

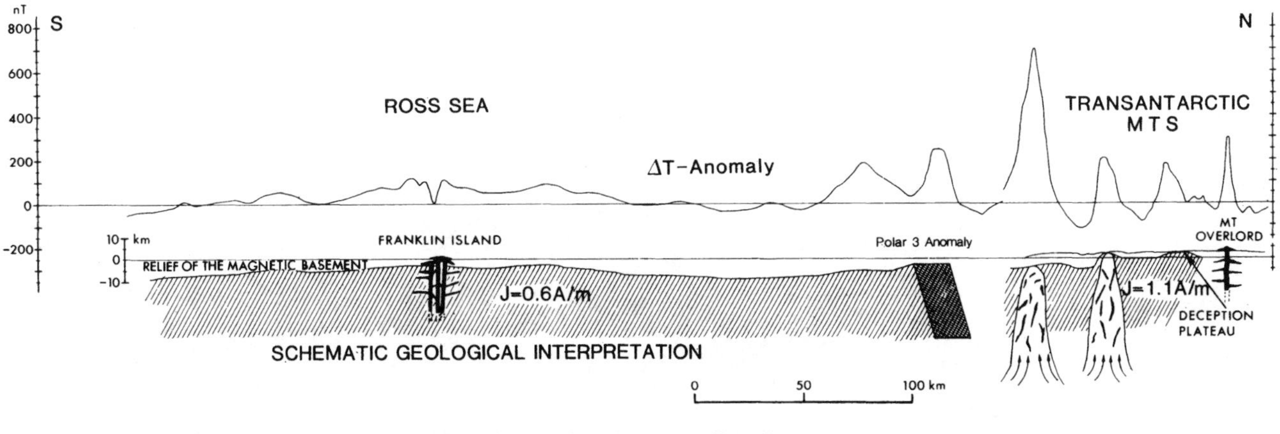

Fig. 5. North to south profile (Pr. 3) from the Southern Cross unit to the western Ross Sea area.

Cooper, A.K., Davey, F.J. & Behrendt, J.C. (1985). Structure and evolution of the Victoria Land basin, western Ross Sea, Antarctica. *Sixth Gondwana Symposium, Abstracts*, p. 23. Columbus, Ohio; Institute of Polar Studies, Miscellaneous Publication, 231.

Hinz, K. & Block, M. (1984). Results of geophysical investigations in the Weddell Sea and in the Ross Sea, Antarctica. In *Proceedings, 11th World Petroleum Congress, London, 1983*, vol. 2, pp. 79–92. Chichester, New York; Wiley & Sons Ltd.

The Ross Sea rift system, Antarctica: structure, evolution and analogues

F. TESSENSOHN[1] & G. WÖRNER[2]

1 Bundesanstalt für Geowissenschaften und Rohstoffe, Stilleweg 2, 3000 Hannover 51, FRG
2 Institut für Mineralogie, Ruhr Universität Bochum, 4630 Bochum, FRG

Abstract

Based on the combination of available geological and geophysical information it is concluded that the Ross Sea depression is a major continental rift system comparable to other important rift provinces in size, morphology, structure and magmatism. It shows a two-phase evolution: a first phase related to the Australia–Antarctica separation at around 95 Ma was amagmatic and characterized by diffuse crustal attenuation (Basin & Range-type crustal thinning). A second phase commenced at about 50 Ma and caused more focussed basin subsidence, shoulder uplift and rift-type magmatism from 25 Ma to the present. This second phase resembles the development of the East African rift system.

Introduction

Similarities of Mount Erebus kenytes with East African volcanics and block faulting of the 'horst of the Transantarctic Mountains' were noted by the geologists of Scott's expeditions. Extensional tectonism was postulated later for the Ross depression from observations in the following different fields:

(i) Bathymetric and seismic surveys carried out in the Ross Sea (Hayes *et al.*, 1975; Davey, Hinz & Schroeder, 1983; Cooper & Davey, 1987) indicated the existence of sedimentary basins and thin crust in the Ross Sea area
(ii) Gravity and aeromagnetic surveys in coastal regions (McGinnis *et al.*, 1983; Bosum *et al.*, 1989) further established the elongated basin structure and a shallow Moho
(iii) Volcanism in the Transantarctic Mountains and Marie Byrd Land (Kyle & Cole, 1974; LeMasurier & Rex, 1982; Wörner *et al.*, 1989) is of typical 'rift' character
(iv) The uplift history of the Transantarctic Mountains was derived from fission-track data by Fitzgerald *et al.* (1987)
(v) Stratigraphic information from drillholes in the Ross Sea area (Hayes *et al.*, 1975; Barrett, 1986) indicates the presence of thick Cenozoic sediments in the basin.

In this paper, we treat the Ross Sea depression and the Transantarctic Mountains as a tectonic entity and propose that both are parts of one large crustal feature, the Ross Sea rift system (RSR). We compare it to its closest analogues: the Basin & Range Province (B&R) and the East African rift system (EAR).

The Ross Sea rift system

Setting and structure

Continental break-up between Australia and Antarctica commenced at about 95 ± 5 Ma (Veevers, 1986). Whereas the larger western part of the two continents split fairly regularly, the break-up was more complicated in the east, where several microplates developed: Lord Howe Rise, Tasmania and South Tasman Rise, New Zealand and Chatham Plateau, and Marie Byrd Land (Grindley & Davey, 1982).

The Ross Sea depression (RSD) is located in this area of former microplates. It is about 1000 km wide and extends inland from the continental margin as a curved depression for more than 3000 km, between the Transantarctic Mountains (TAM) and Marie Byrd Land (MBL) (Fig. 1). The Transantarctic Mountains form a fault-scarp range up to 4000 m high parallel to the western border of the depression and display features of an uplifted and tilted rift shoulder. The lack of a comparable scarp in Marie Byrd Land indicates asymmetry of the depression. Along-strike it comprises three segments: (a) the sedimentary basins of the Ross Sea proper; (b) the morphological low under the Ross Ice Shelf; and (c) the Byrd Subglacial Basin.

In the Ross Sea, the depression is divided into three subparallel sedimentary basins (Fig. 2). The Eastern basin and the Central trough are 5–6 k m deep (Davey *et al.*, 1983), whereas the Victoria Land basin (VLB) immediately in front of the Transantarctic Mountains is 10–14 km deep (Cooper & Davey, 1987). The latter terminates in the north against an offshore basement block (Coulman high), but possibly has a northward

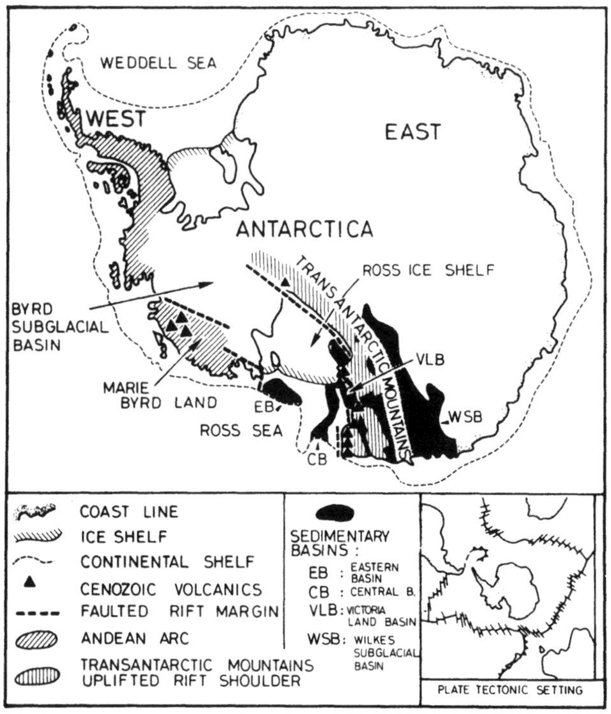

Fig. 1. Setting of the Ross Sea rift in relation to Antarctic continental geology and regional geotectonic framework (inset).

extension onshore in northern Victoria Land (Fig. 2), represented by the Rennick graben (Roland & Tessensohn, 1987; Wörner *et al.*, 1989), separated today by uplifted granitic basement.

Bedrock morphology, and magnetic and gravity data indicate a continuation of the sedimentary basins from the Ross Sea under the Ross Ice Shelf (Robertson *et al.*, 1982), although data are poor and there is little information about the deeper subsurface. However, one can still deduce the existence of several basins, particularly a deep one in front of the TAM which has a maximum depth south of the Erebus volcanic province. In the region of the Ross Ice Shelf the bordering TAM swing around parallel to the Marie Byrd Land continental margin (Fig. 1).

This direction continues in the Byrd Subglacial Basin where the floor of the Bentley trough extends down to 2000 m below sea level. Little is known about thickness and age of sediments in this depression (Jankowski & Drewry, 1981) but from its position it can be regarded as a back-arc basin to the Andean arc (Fig. 1).

Scarce crustal data (McGinnis *et al.*, 1983; Cooper & Davey, 1987) show a dramatic change in thickness from about 20–28 km in the Ross Sea depression to > 40 km under the TAM and about 32 km under Marie Byrd Land. Fitzgerald *et al.* (1987) synthesized geological, geophysical and fission-track data on the TAM uplift history and emphasized the close correlation between crustal thinning in the basin and TAM uplift. They further called upon structural similarities to the Basin & Range province.

Magmatism

Rift-related magmatism (Fig. 2) in the RSD area (Kyle, 1981) is concentrated at the western margin of the Ross Sea. Linear elements in the north alternate with central volcanic complexes farther south. Seismic records indicate magmatic activity also in the VLB (Cooper & Davey, 1987) whereas the Central trough and Eastern basin appear to be amagmatic (Davey *et al.*, 1983). Apart from a few small centres no magmatism is represented in the rift shoulder of the Ross Ice Shelf segment of the RSD. Cenozoic alkaline volcanism also occurs in Marie Byrd Land, beyond the eastern margin of the Ross Sea depression (LeMasurier & Rex, 1982). Several large stratovolcanoes are arranged there on a N–S- and E–W-orientated rectilinear grid.

Magmatic activity in the RSD area commenced at about 25 Ma with the intrusion of alkali granites and syenites (Schmidt-Thomé, Müller & Tessensohn, 1990), followed by the onset of volcanism at around 19 Ma (Armstrong, 1978). In Marie Byrd Land, ages of 27–28 Ma for the oldest volcanic rocks above the basement and for a subvolcanic alkali intrusion are remarkably similar to the ages from the McMurdo Volcanic Group.

Common features of the whole Cenozoic magmatic province are its alkaline character and the contemporaneous eruption of two contrasting, critically *saturated* potassic and strongly *undersaturated* sodic suites (e.g. Kyle, 1981). The two geochemical trends originate from distinct mantle magmas – basanite and alkali basalt.

We conclude that there are at least *two* major centres of magmatic activity forming the Erebus and Melbourne volcanic provinces (Fig. 2). The latter includes abundant intrusive complexes seen in the magnetic data (Bosum *et al.*, 1989). These two areas are located at the northern and southern termination of the VLB where we suggest major offsets of this structure. These offsets may represent 'leaky' transfer faults allowing an easier magma ascent and higher accumulation rates in these critical areas.

On the basis of all the features discussed it seems appropriate to use the genetic term 'rift' instead of the descriptive 'depression'. The suggested term *Ross Sea rift system* (RSR) takes into account that the depression consists of several basins and troughs, and that the term rift has already been used for single features (e.g. Terror rift; Cooper & Davey, 1987).

Evolution

Two distinct phases of tectonic activity are documented in the RSR (Cooper & Davey, 1987; Roland & Tessensohn, 1987):

(1) An early phase is characterized by the lack of volcanism and diffuse crustal attenuation. Onshore, a series of tilted blocks is exposed along the flanks of the Rennick graben (Roland & Tessensohn, 1987). Offshore, numerous small half-grabens developed on Coulman high. On Iselin Bank, normal faults of an early tectonic phase are now covered by about 2 km of undisturbed oceanic sediments. Deeper basins developed in the present Ross Sea, where the VLB was filled by about 5–8 km of sediments (Cooper & Davey, 1987), whereas the Central trough and Eastern basin did not receive much sediment input during this period. Due to the lack of calibration of the lower seismic horizons the first phase is difficult

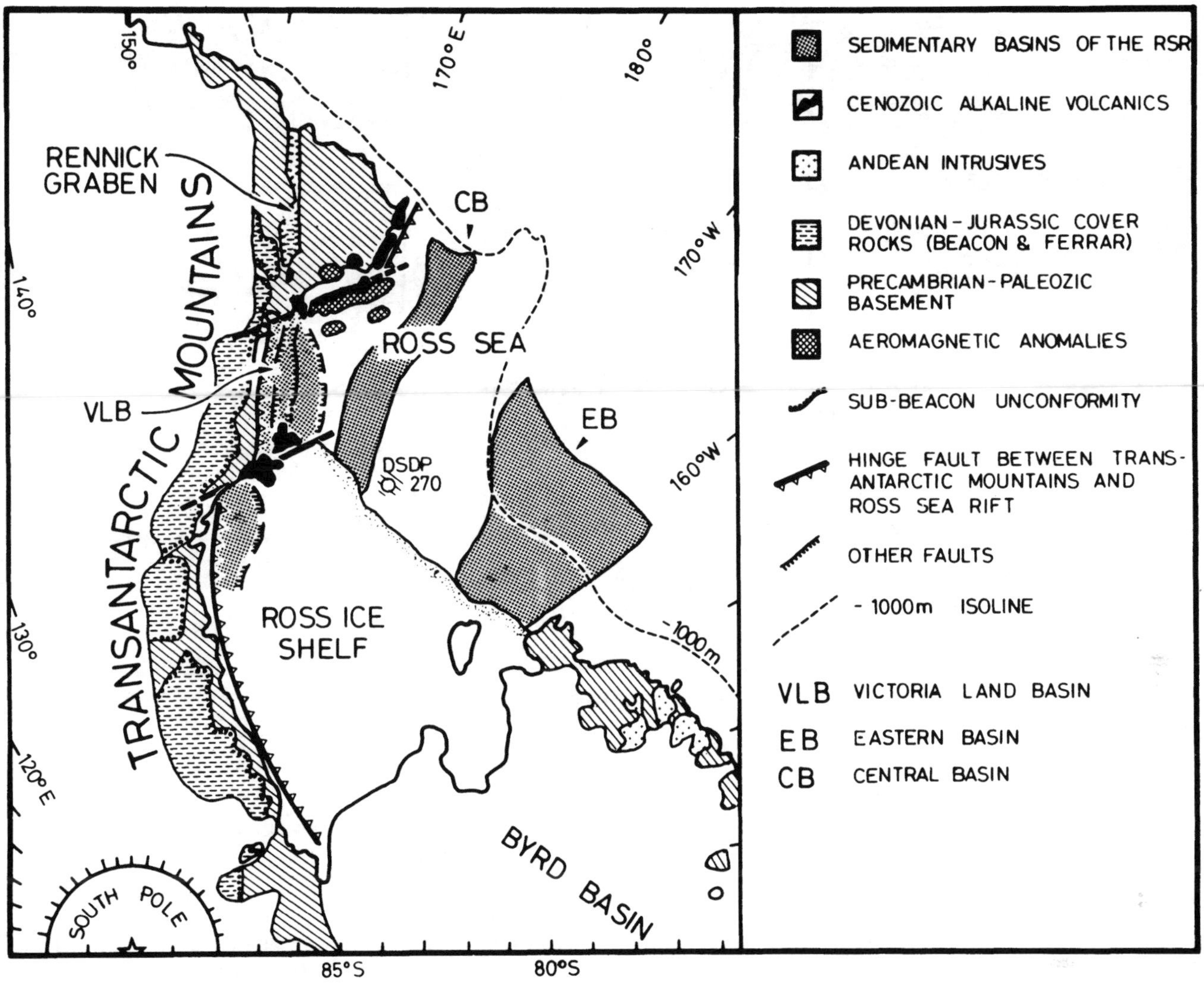

Fig. 2. The Ross Sea rift system RSR based on geophysical and geological evidence. See text for discussion.

to date at present. It may be related to back-arc tectonism associated with the Andean arc (Fig. 2) as well as to the continental break-up between Antarctica and Australia at around 90 Ma (Veevers, 1986);

(2) The second rift phase is characterized by uplift and tilting of the TAM from 50 Ma to the present (Fitzgerald *et al.*, 1987), asymmetric subsidence of the RSR with the particularly deep VLB, and the onset of alkali magmatism at about 25 Ma. Tectonic and magmatic activity are now focussed along the main fault between TAM and VLB where the maximum vertical displacement reaches 15–20 km.

Related continental rifts

Continental extensional provinces share many common features (Bosworth, Lambiase & Keisler, 1986). However, every single system has its own characteristic evolution and tectonic setting. The best analogues of the Ross Sea rift system appear to be the East African rift (EAR), with its deep and very pronounced rift valleys, and the Basin & Range province (B&R) in the western USA, with its contrasting tectonic style of tilted blocks and asymmetrical basins.

Both the EAR and the B&R rift provinces are of similar size to the RSR (Fig. 3). However, the former are related to broad regional *uplift* on the order of thousands of metres (Baker & Wohlenberg, 1971; Eaton *et al.*, 1978) whereas the RSR has *subsided* on a similar scale. The regional separation into three individual subparallel basins resembles the EAR whereas the overall topography, with a wide central depression and two uplifted shoulders (although at different elevations), relates to the B&R province.

Sengör & Burke (1978) defined *active* rifts as characterized by a rising mantle plume which causes the initial updoming through lithospheric heating, magmatic intrusion and volatile metasomatism (Bailey, 1974; McKenzie, 1978). Large volumes of magma are produced prior to the main phase of lithospheric extension. In a *passive* rift, thinning and asthenospheric uprise predates magmatism or may involve no magmatism at all.

The EAR represents an example of an active rift. As Bailey (1974) pointed out the determining factor for its long lifetime, extreme alkalinity of magmas and unsuccessful continental separation, may be that the African plate has remained stationary relative to the asthenosphere. In contrast, the B&R is considered a passive rift. The RSR is also a passive rift in the

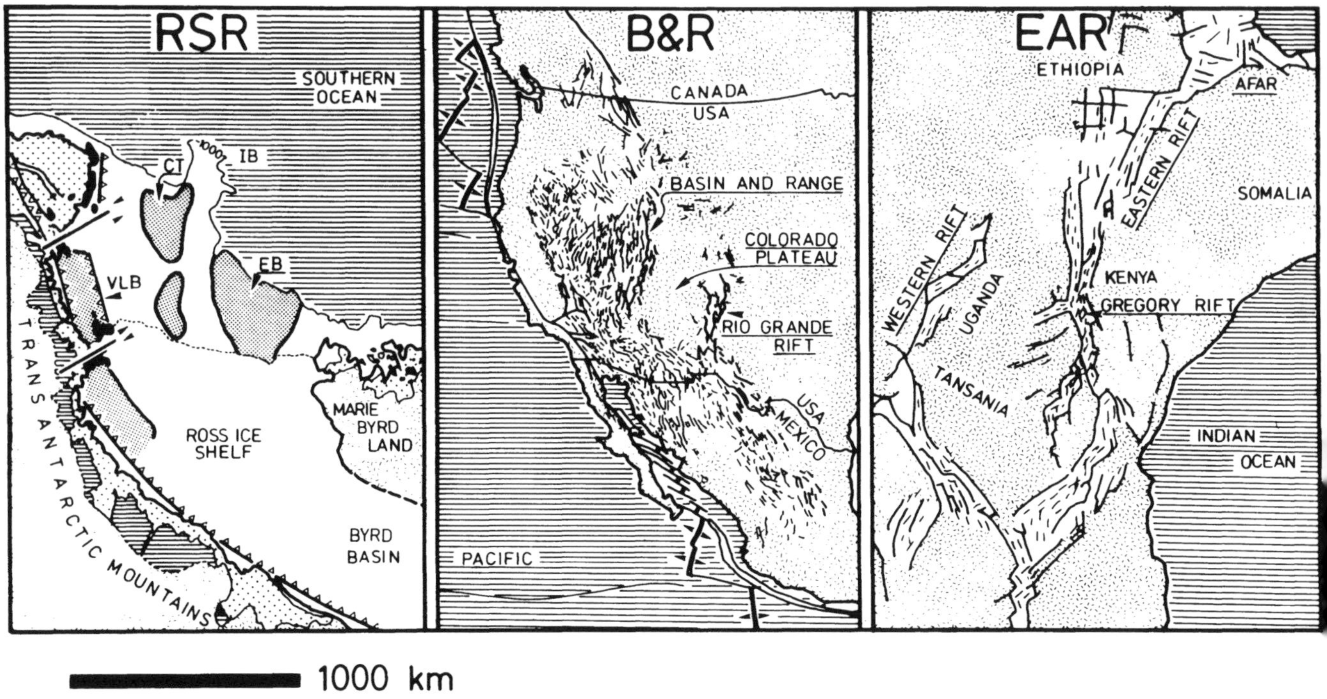

Fig. 3. Structural comparison between the Ross Sea rift system (RSR), the Basin & Range province (B&R) and the East African rift system (EAR). B&R structures after Stewart (1978), EAR structures from various sources including Baker & Wohlenberg (1971). CT, Central trough; IB, Iselin Bank; EB, Eastern basin; VLB, Victoria Land basin.

sense of Sengör & Burke (1978), involving stretching of the lithosphere during an early phase and delayed asthenospheric upflow in a second phase.

The B&R province is a typical example of bimodal (alkali)–basaltic and rhyolitic activity (Lipman, 1980), whereas in the EAR there is a complete range from extremely alkaline (carbonatitic) and oversaturated highly evolved (rhyolitic) lavas to undersaturated nephelinites and hypersthene–normative tholeiitic basalts (Baker *et al.*, 1977). High magma alkalinity and production rate have been taken as evidence for plume-controlled rifting of a stationary plate (Bailey, 1974).

Over- and undersaturated lava suites are also typical for the RSR. Magma production and alkalinity appear to be lower in the RSR compared to the EAR.

The EAR evolved over more than 20 m.y., but tectonic and magmatic activity is concentrated at the rift axes. The B&R province formed after subduction of the Farallon plate ceased 17 Ma ago. Most of the extension took place within the first few million years; tectonic and magmatic activity is now concentrated near the margins of the basin. For the RSR a two-phase history has been postulated with accumulated relative vertical motion of 15–20 km (maximum). Thus, extent and duration of rifting and uplift are larger in the RSR and it probably had a more complex history. The first phase relates in style to the B&R province. This may be due to the fact that the whole depression at that time formed in a back-arc position with regard to the Andean subduction zone. The second phase resembles the EAR style of continental rifting. This similarity may be caused by a similar stress field in the stationary Antarctic and African plates.

Models of crustal extension developed at the B&R and EAR (e.g. Bosworth *et al.*, 1986; Allmendinger *et al.*, 1987) involve a shallow (15 km) horizontal detachment zone at the brittle–ductile transition in the crust. They are also applicable to the RSR (Wörner *et al.*, 1989).

Unlike the B&R and EAR, which trend parallel to an active and passive margin, respectively, the RSR runs perpendicular to the Antarctic margin. This setting, the longer lifetime and the two-phase evolution of the RSR are the main differences to the EAR and B&R. In this respect, the Mesozoic–Tertiary North Sea rift system may be the closest analogue to the RSR.

Stratigraphic calibration of the seismic horizons in the basins, crustal geophysical investigations under the TAM, and seismic traverses across the TAM–RSR transition are particularly needed to date the proposed evolution more precisely and to gain an understanding of the prominent crustal flexure between the RSR and the TAM.

References

Allmendinger, R.W., Hauge, T.A., Hauser, E.C., Potter, C.J., Klemperer, S.L., Nelson, K.D., Knuepfer, P. & Oliver, J. (1987). Overview of the COCORP 40° N transect, western United States: the fabric of an orogenic belt. *Geological Society of America Bulletin*, **98(3)**, 308–19.

Armstrong, R.L. (1978). K–Ar dating: Late Cenozoic McMurdo Volcanic Group and Dry Valley glacial history, Victoria Land, Antarctica. *New Zealand Journal of Geology and Geophysics*, **21(6)**, 685–9.

Bailey, D.K. (1974). Continental rifting and alkaline magmatism. In *The Alkaline Rocks*, ed. H. Sørensen, pp. 148–59. New York; John Wiley and Sons.

Baker, B.H. & Wohlenberg, J. (1971). Structure and evolution of the Kenya rift valley. *Nature, London*, **229**, 538–42.

Baker, B.H., Goles, G.G., Leeman, W.P. & Lindstrom, M.M. (1977). Geochemistry and petrogenesis of a basalt–benmoreite–trachyte suite from the southern part of the Gregory Rift, Kenya. *Contributions to Mineralogy and Petrology*, **64**, 303–32.

Barrett, P.J., (ed.) (1986). *Antarctic Cenozoic History from the MSSTS-1 Drillhole, McMurdo Sound.* Department of Scientific and Industrial Research, Bulletin 237, 174 pp. Wellington; Science Information Publishing Centre.

Bosum, W., Damaske, D., Roland, N.W., Behrendt, J.C. & Saltus, R. (1989). The GANOVEX IV Ross Sea aeromagnetic survey: interpretation of anomalies *Geologisches Jahrbuch*, **E38**, 153–220.

Bosworth, W., Lambiase, J. & Keisler, R. (1986). A new look at Gregory's rift: the structural style of continental rifting. *Eos, Transactions of the American Geophysical Union*, **July 22**, 577.

Cooper, A.K. & Davey, F.J. (ed.) (1987). *The Antarctic Continental Margin: Geology and Geophysics of the Western Ross Sea*, Earth Science Series, vol. 5B, pp. 93–118. Houston, Texas; Circum-Pacific Council for Energy and Mineral Resources.

Davey, F.J., Hinz, K. & Schroeder, H. (1983). Sedimentary basins in the Ross Sea, Antarctica. In *Antarctic Earth Science*, eds. R.L. Oliver, P.R. James & J.B. Jago, pp. 533–8. Canberra; Australian Academy of Science and Cambridge; Cambridge University Press.

Eaton, G.P., Wahl, R.R., Prostka, H.J., Mabey, D.R. & Kleinkopf, R.D. (1978). Regional gravity and tectonic patterns: their relation to late Cenozoic epeirogeny and lateral spreading in the western Cordillera. *Tectonophysics*, **15**, 51–92.

Fitzgerald, P.G., Sandiford, M., Barrett, P.J. & Gleadow, A.J.W. (1987). Asymmetric extension associated with the uplift and subsidence in the Transantarctic Mountains and Ross Sea Embayment. *Earth and Planetary Science Letters*, **81**, 67–78.

Grindley, G.W. & Davey, F.J. (1982). The reconstruction of New Zealand, Australia and Antarctica. In *Antarctic Geoscience*, ed. C. Craddock, pp. 15–29. Madison; University of Wisconsin Press.

Hayes, D.E., Frakes, L.A. *et al.*, (ed.) (1975). *Initial Reports of the Deep Sea Drilling Project*, *28*, 1017 pp. Washington, DC; US Government Printing Office.

Jankowski, E.J. & Drewry, D.J. (1981). The structure of West Antarctica from geophysical studies. *Nature, London*, **291**, 17–21.

Kyle, P.R. (1981). Mineralogy and geochemistry of a basanite to phonolite sequence at Hut Point Peninsula, Antarctica, based on core from Dry Valley Drilling Project drillholes 1, 2, and 3. *Journal of Petrology*, **22**, 451–500.

Kyle, P.R. & Cole, J.W. (1974). Structural control of volcanism in the McMurdo Volcanic Group, Antarctica. *Bulletin Volcanologique*, **38**, 16–25.

LeMasurier, W.E. & Rex, D.C. (1982). Volcanic record of Cenozoic glacial history in Marie Byrd Land and western Ellsworth Land: revised chronology and evaluation of tectonic factors. In *Antarctic Geoscience*, ed. C. Craddock, pp. 725–34. Madison; University of Wisconsin Press.

Lipman, P.W. (1980). Cenozoic volcanism in the western United States: implications for continental tectonics. In *Studies in Geophysics: Continental Tectonics*, pp. 161–74. Washington, DC; National Academy of Science.

McGinnis, L.D., Wilson, D.D., Burdelik, W.J. & Larson, T.H. (1983). Crust and upper mantle study of McMurdo Sound. In *Antarctic Earth Science*, eds. R.L. Oliver, P.R. James & J.B. Jago, pp. 204–8. Canberra; Australian Academy of Science and Cambridge; Cambridge University Press.

McKenzie, D.P. (1978). Some remarks on the development of sedimentary basins. *Earth and Planetary Science Letters*, **40**, 25–32.

Robertson, J.D., Bentley, C.R., Clough, J.W. & Greischar, L.L. (1982). Sea-bottom topography and crustal structure below the Ross Ice Shelf, Antarctica. In *Antarctic Geoscience*, ed. C. Craddock, pp. 1083–90. Madison; University of Wisconsin Press.

Roland, N.W. & Tessensohn, F. (1987). Rennick faulting – an early phase of Ross Sea rifting. *Geologisches Jahrbuch*, **B66**, 203–29.

Schmidt-Thomé, M., Müller, P. & Tessensohn, F. (1990). Individual Volcano Description A.6 Malta Plateau. In *Volcanoes of the Antarctic Plate and Southern Oceans*, Antarctic Research Series, 48, ed. W.E. LeMasurier & J.W. Thomson. Washington, DC; American Geophysical Union.

Sengör, A.M.C. & Burke, K. (1978). Relative timing of rifting and volcanism on earth and its tectonic implications. *Geophysical Research Letters*, **5(6)**, 419–21.

Stewart, J.H. (1978). Basin-range structure in western North America: a review. *Tectonophysics*, **15**, 1–32.

Veevers, J.J. (1986). Breakup of Australia and Antarctica estimated as mid-Cretaceous (95 ± 5 Ma) from magnetic and seismic data at the continental margin. *Earth and Planetary Science Letters*, **77**, 91–9.

Wörner, G., Viereck, L., Hertogen, J. & Niephaus, H. (1989). The Mount Melbourne volcanic field (Victoria Land, Antarctica). II. Geochemistry and magma genesis. *Geologisches Jahrbuch*, **E38**, 395–433.

Structural and depositional controls on Cenozoic and (?)Mesozoic strata beneath the western Ross Sea

A.K. COOPER[1], F.J. DAVEY[2] & J.C. BEHRENDT[3]

1 US Geological Survey, 345 Middlefield Road, Menlo Park, CA 94025, USA
2 Geophysics Division (DSIR), PO Box 1320, Wellington, New Zealand
3 US Geological Survey, Box 25046 MS 903, Denver Federal Center, Denver, CO 80225, USA

Abstract

The western Ross Sea (WRS) is underlain by up to 14 km of mostly flat-lying sedimentary strata, deformed along the axis and edges of the Victoria Land basin (VLB). A deep layered section (V5) is confined to the basement graben beneath the VLB. Shallower units (V1–V4) unconformably cover the basin, are intruded by volcanic rocks (V6) and extend elsewhere into the Ross Sea. CIROS and MSSTS drilling indicates early Oligocene ages for glacial-marine units V1–V3. Glacial erratics suggest a Late Cretaceous–Palaeogene age for V4 and the upper part of V5. Volcanic rocks (V6) may be of Jurassic–late Cenozoic age. Acoustic basement (V7) is probably pre-Jurassic sedimentary and igneous units. In most WRS areas, bathymetric and subsurface features are located where crustal-rifting uplifts occur, yet bathymetric orientations reflect glacial erosion. Greater subsidence and lesser sedimentation rates in the WRS than in the eastern Ross Sea (ERS) probably explain the structural trough beneath the VLB (WRS) versus the prograding sediment wedge beneath the ERS.

Introduction

Multichannel seismic-reflection surveys of the western Ross Sea (WRS) show up to 14 km of sedimentary strata filling the Victoria Land basin (VLB; Cooper, Davey & Behrendt, 1987; Hinz & Kristoffersen, 1987; Fig. 1). These strata are locally deformed and glacially eroded into NE–SW-trending features. The VLB has the thickest strata of the Ross Sea basins, and, unlike the other basins, is the site of the now-active axial rift zone, the Terror rift. Here, we review the acoustic stratigraphy and deformation of the VLB, emphasizing possible ages and origins of its strata.

Geological data

Onshore geology gives little evidence for the age and composition of strata filling the VLB (Davey, 1987). Onshore, Precambrian–Early Palaeozoic basement is unconformably overlain by Devonian–Triassic sedimentary strata. Jurassic dolerite and volcanic rocks intrude and cover the older section. Post-Jurassic rocks include limited exposures of post-Oligocene volcanic and sedimentary rocks.

Offshore drilling has recovered early Oligocene and younger glacial marine rocks (Hayes & Frakes, 1975; Barrett *et al.*, 1987; Barrett, Hambrey & Robinson, this volume, p. 651; Fig. 1). Glacial erratics of Late Cretaceous, Eocene and younger marine and non-marine strata (Webb, 1981) suggest deeper VLB units. The oldest Cenozoic volcanic debris is from the CIROS-1 drill site (early Oligocene; Barrett, *et al.*, this volume, p. 651) yet most surface volcanic exposures yield ages of < 19 Ma (González-Ferrán, 1982).

Seismic data

Onshore and offshore multichannel seismic-reflection data in the WRS (Kim, McGinnis & Bower, 1986; Cooper *et al.*, 1987) provide the first image of deep strata filling the VLB, whereas previous surveys indicated only a shallow basin (Davey, Bennett & Houtz, 1982).

Acoustic stratigraphy

For the VLB, Cooper *et al.* (1987) defined seven acoustic units, V1–V7 (Fig. 2), which are separated by angular unconformities that become disconformities near the basin centre (except V7). Units V1–V5 comprise mostly flat-lying or gently dipping continuous reflections, except near the axis and edges of the basin, where they are faulted or pierced by an acoustically-chaotic unit (V6) which occurs principally in the southern half of the basin. Reflections are commonly truncated near the seafloor (V1), and weak or discontinuous at great depth (V5). Acoustic basement (V7), the deepest unit, contains hyperbolic and discontinuous reflections.

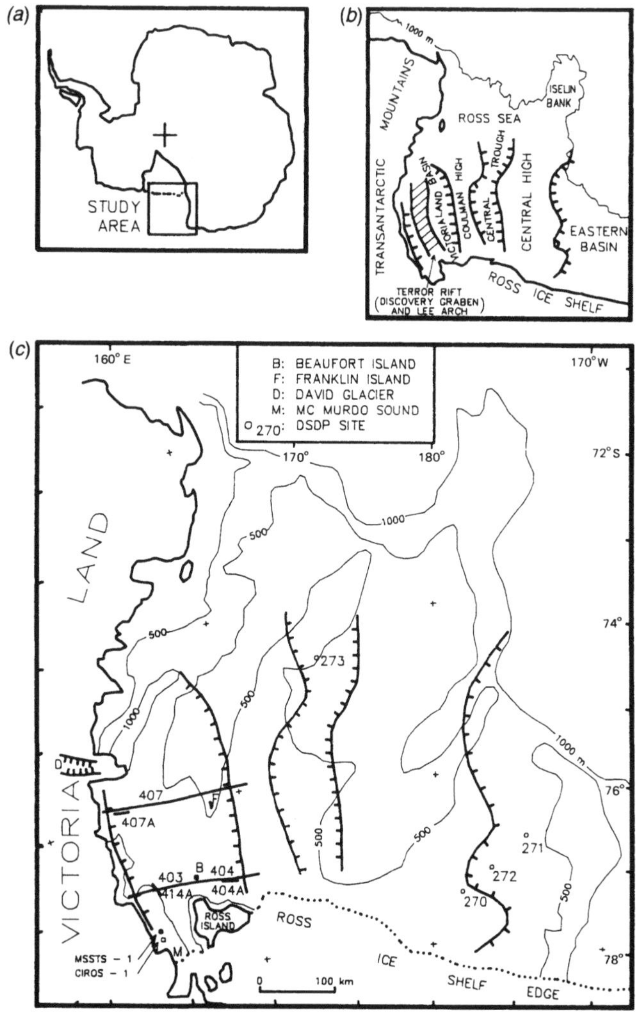

Fig. 1. **(*a*) Index map of the study area. (*b*) Location of major structures in the Ross Sea. (*c*) Location of seismic lines in Figs 2 & 3 and of the Ross Sea drilling sites.**

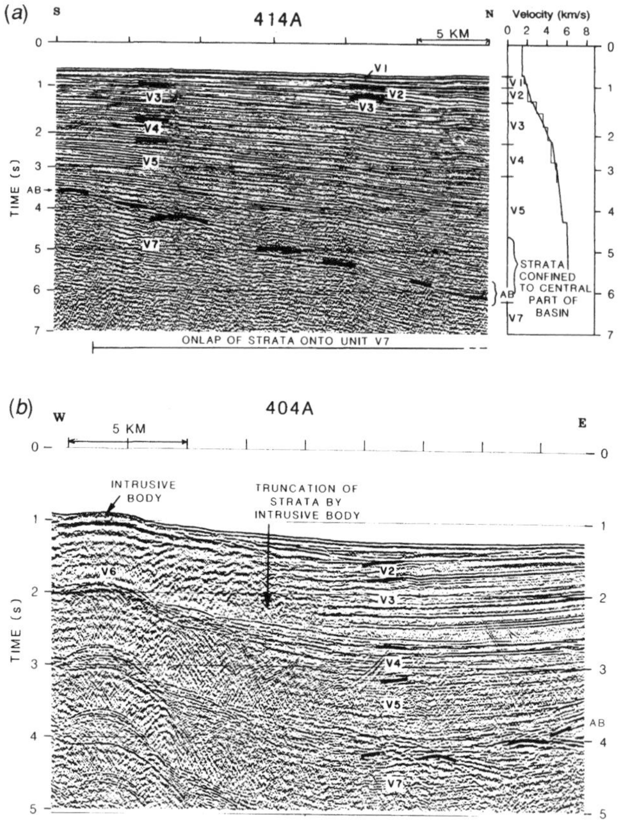

Fig. 2. Unmigrated multichannel seismic-reflection (MCS) data from the VLB showing seismic units V1–V7 (Cooper *et al.*, 1987); see Figs 1 & 3 for location. (*a*) Line 414A showing well-layered, high-velocity strata beneath the western flank of the VLB. Velocities are from a sonobuoy deployed at the northern end of the profile (see Cooper *et al.*, this volume, p. 285). (*b*) Line 404A showing abrupt truncation of basin strata (V2–V5) by subvolcanic intrusive rocks (V6) near Beaufort Island.

Seismic line 407 (Fig. 3*a*) illustrates the important characteristics of VLB acoustic units:

(i) Units V1–V4 can be traced across the basin to other parts of the Ross Sea. Unit V5, however, is confined to the basement graben underlying the VLB. Acoustic units like V5 are buried beneath other Ross Sea basins (Hinz & Block, 1984; Cooper, Davey & Hinz this volume, p. 285).

(ii) The unconformities separating V1–V5 are easily resolved only along the uplifted (and shallow) western edge of the basin, the Terror rift, and the Coulman high, showing the strong influence of tectonism (and water depth) on the location of unconformities

(iii) Major slump bodies or debris fans originating from the Transantarctic Mountains are not evident beneath the western flank of the VLB; instead, subparallel strata suggest longitudinal deposition within an existing depression

(iv) Beneath the VLB eastern flank, flat-lying strata cover basement relief, suggesting sediment deposition within an existing, rather than a rapidly subsiding, depression

(v) Piercement structures, such as those near the volcanic Beaufort and Franklin islands (Figs 2*b* & 3*b*), occur in the southern half of the VLB; these structures (unit V6) penetrate to the seafloor and are probably late Cenozoic subvolcanic intrusives, whereas those of V6 buried at great depth may be Mesozoic in age.

Acoustic correlations

The shallowest units (V1 and V2) have been correlated with early Miocene and younger glacial marine rocks at the MSSTS-1 and DSDP (273) drill sites (Fig. 1; Cooper *et al.*, 1987). Drilling at CIROS-1 recovered glacial-marine strata of early Oligocene age (Barrett *et al.*, this volume, p. 651). The upper part of unit V3 is probably coeval with the CIROS-1 Palaeogene section, and deeper strata of V3, V4, and V5 are early Palaeogene and older.

The reflection from the major unconformity at the top of unit V5 in the VLB, is considered to be equivalent with the one from unconformity U6 in the central and eastern Ross Sea (Cooper *et al.*, 1987; K. Hinz, pers. comm.), yet these reflections cannot be continuously traced between basins. The pre-early Oligocene age (based on CIROS-1) at the top of unit

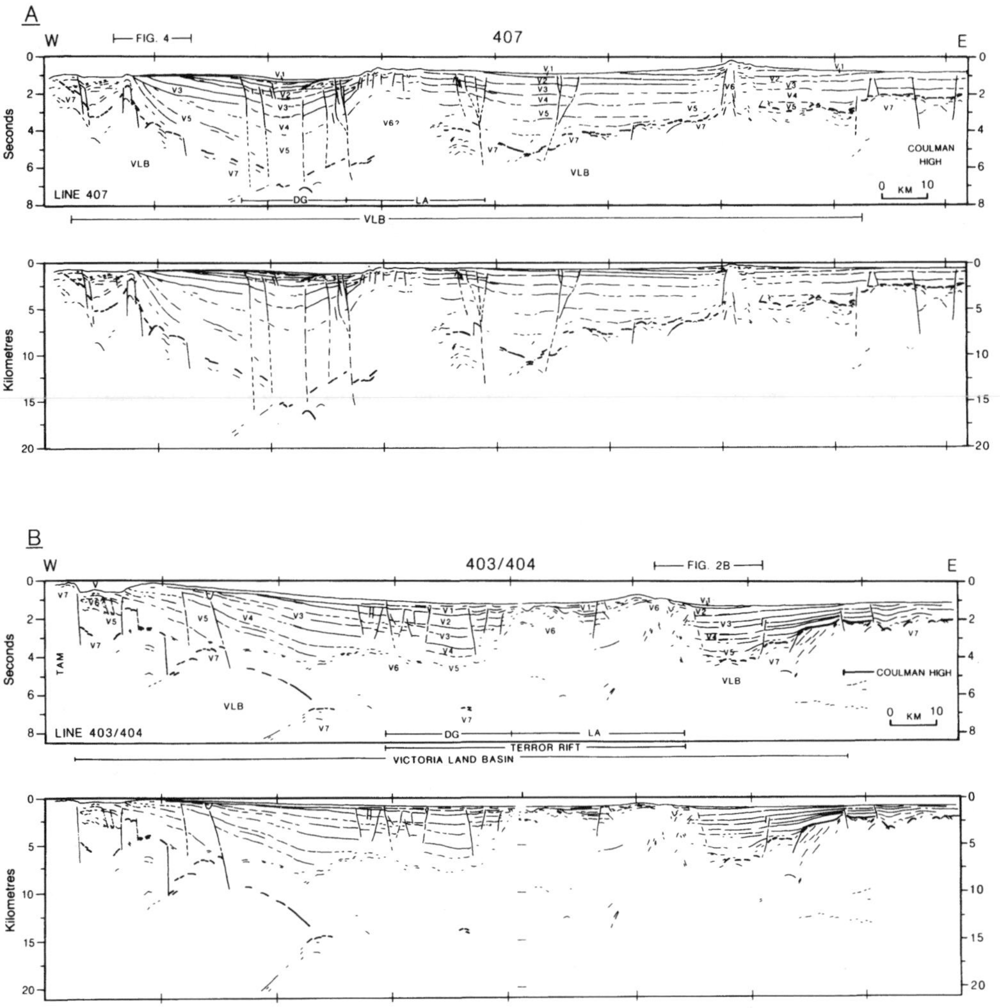

Fig. 3. Line drawings of unmigrated MCS profiles showing acoustic units beneath the VLB. The upper profile for each set is a 'time-section' and the lower is a 'depth-section'. Basin-filling strata (V1–V5) are deformed at faults and intrusions (V6) along the Discovery graben (DG) and Lee arch (LA) in the Palaeogene and younger Terror rift.

V5 conflicts with the late Oligocene age of U6 (from DSDP site 270; Hinz & Block, 1984). Resolution of the apparent age conflict for this important regional unconformity requires further drilling.

Discussion

Structural control of basin strata

Absolute ages for the VLB sedimentary section are mostly unknown. However, seismic-reflection data provide a relative sequence of depositional and structural events. Initially, now deeply buried sedimentary units were probably deposited in local rift-grabens underlying the centre of the basin and locally the Coulman high (Cooper *et al.*, this volume, p. 285; Fig. 5). Major graben subsidence apparently was completed prior to deposition of unit V5, based on flat-lying strata (V5) covering the basement. The western edge of the VLB graben was probably near the present coastline, even prior to the initial Eocene uplift of the Transantarctic Mountains, because post-Jurassic sedimentary rocks are rarely found onshore.

Unit V5 was then uplifted and eroded along the western flank of the VLB, as evidenced by steep dips and the trunction of V5 and younger strata (Figs 3–5). At least two cycles of relative subsidence–sedimentation and uplift–erosion are recorded by the prominent angular unconformities beneath the western flank of the basin (Fig. 4). Similar uplift is recorded at the MSSTS-1 hole where Oligocene rocks, previously more deeply (1 km) buried, now lie unconformably at shallow depth beneath Pliocene strata (Barrett & McKelvey, 1986).

In the Terror rift, downfaulting of the Discovery graben and uplift of the Lee arch (Fig. 3) mostly followed deposition of unit V3 (post-mid-Oligocene time; Cooper *et al.*, 1987), as suggested by thinning of unit V2 against the Lee arch. Units V1 and V2 show the greatest internal disruption, caused by late Cenozoic tectonism in the Terror rift (Fig. 3) and by likely glacial erosion in shallow-water areas (Hayes & Frakes, 1975; Karl, Reimnitz & Edwards, 1987). In general where the seafloor is shallow, due in large part to rift-related tectonism (e.g. Transantarctic Mountains, Lee arch, Franklin Island, Coulman high, Figs 3 & 4), erosion is most severe.

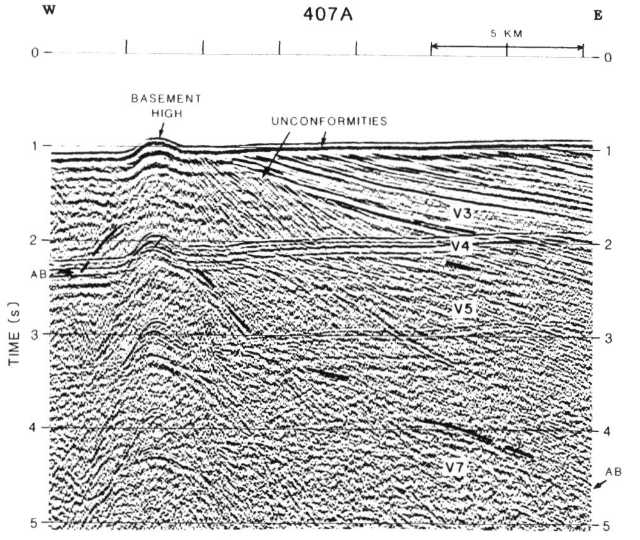

Fig. 4. Unmigrated MCS profile 407A across the western edge of the VLB. Uplift at the Transantarctic Mountain front has resulted in at least two angular unconformities; see Fig. 1 for location.

Age and origin of strata

Uplifted strata of units V4 and V5 along the western edge of the VLB (Figs *3a* & *b*) are likely sources for the Late Cretaceous and Palaeogene glacial erratics found near McMurdo Sound.

The oldest rocks of unit V5 occur only in the centre of the VLB (Fig. *2a*), and would not be exposed along the uplifted basin flank. Cooper *et al.* (1987) suggested a Cretaceous and Palaeogene age for unit V5, although the high velocity (5.0–6.0 km/s) of the unit could indicate older rocks. They considered Early Palaeozoic and older metasedimentary rocks of Victoria Land too highly deformed to comprise the well-layered sequence. Additionally, they suggested that well-stratified Beacon Supergroup rocks probably lie within acoustic basement under the VLB.

Rocks filling the VLB rift graben may be coeval with those in rift basins of the formerly nearby areas of Tasmania and New Zealand. Several hundred kilometres of crustal extension are postulated between Australia–Tasmania–Antarctica and probably in the Ross embayment prior to Southern Ocean seafloor spreading (Lawver & Scotese, 1987; Veevers & Eittreim, 1988). Pre-break-up extension between Australia–Tasmania–Antarctica probably occurred from 160 to 96 Ma ago (Veevers & Eittreim, 1988) based on geomorphic, structural and drilling information. In the Ross Sea, crustal extension is estimated as Late Cretaceous (71–63 Ma) from ages of seafloor magnetic lineations (Molnar *et al.*, 1975). An earlier rift history, such as proposed by Veevers & Eittreim (1988), is likely for the Ross Sea based on multichannel seismic-reflection evidence, but confirmation requires further Ross Sea drilling.

Deep strata (V5) were most likely deposited along the axis of the VLB under principally marine, and also non-marine, conditions. These strata thicken northwards to where a trough crosses the Transantarctic Mountains beneath David Glacier. Sediment may have been transported along this trough between the Wilkes Land and Ross Sea basins during late Mesozoic time (Cooper *et al.*, 1987), prior to Eocene uplift of the Transantarctic Mountains (Fitzgerald, Sandiford & Barrett, 1987). Sediments may also have been deposited at times within a broad inland sea or floodplain across the basin. The principal sediment-transport direction, based on seismic evidence, was along the axis rather than down the flanks of the grabens.

Shallow strata (V1–V4) are considered to be of marine origin, from shallow drilling, glacial erratics and seismic-velocity studies (Cooper *et al.*, 1987). They were probably derived in large part from pre- and interglacial sources in the Transantarctic Mountains and from marine biogenic deposition (Barrett *et al.*, this volume, p. 651).

Unconformities of late Oligocene and younger age, defined by seismic data in the eastern Ross Sea (Hinz & Block, 1984) cannot be reliably correlated with those in the western Ross Sea (Cooper *et al.*, 1987) due to local structural deformation and widespread erosion of intervening basement highs. Apparent sedimentation rates for post–mid-Oligocene time are about 2–3 times greater in the eastern than western Ross Sea, based on the present seismic stratigraphy. Glacial sediment has apparently been supplied to the eastern Ross Sea faster than the basin has subsided (by flexural down-warping), resulting in a seaward prograding sequence beneath the outer continental shelf. In contrast, glacial-sediment supply to the western Ross Sea has been slower than basin subsidence (by active downfaulting), yielding a structural-bathymetric trough.

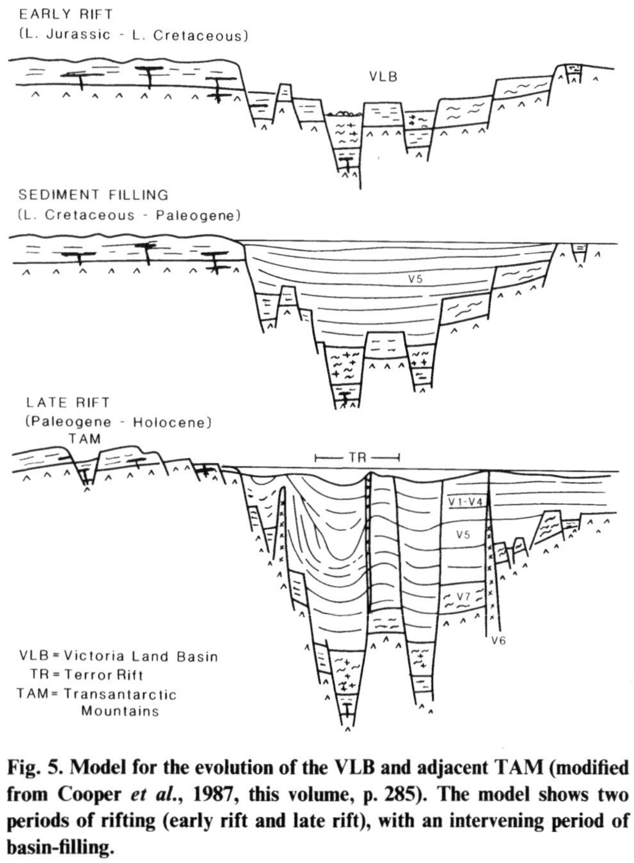

Fig. 5. Model for the evolution of the VLB and adjacent TAM (modified from Cooper *et al.*, 1987, this volume, p. 285). The model shows two periods of rifting (early rift and late rift), with an intervening period of basin-filling.

Summary

The western Ross Sea is underlain by a major basement graben, the VLB, filled with up to 14 km of (?)late Mesozoic and younger strata (units V1–V5) pierced by (?)Mesozoic–Cenozoic subvolcanic units (unit V6), and overlying (?)Mesozoic and older acoustic basement (unit V7). The VLB graben probably formed initially during Gondwana break-up, with the principal infilling of marine and non-marine strata (V5) occurring during Cretaceous and early Palaeogene time. Resumption of rifting since late Palaeogene time, coeval with uplift of the Transantarctic Mountains, has greatly affected basinal deposition (by subsidence), sediment-deformation (by faulting and intrusion) and sediment erosion (by uplift). Glacial erosion since Palaeogene time is most evident in areas uplifted largely by rifting. The oldest basin-filling rocks beneath the Ross Sea region lie near the seafloor, and are accessible to drilling only along the uplifted western edge of the VLB, at the Transantarctic Mountains front.

Acknowledgements

We thank M. Marlow, H. Karl, T. Bruns, and an unidentified reviewer for helpful reviews of the manuscript.

References

Barrett, P.J., Elston, D.P., Harwood, D.M., McKelvey, B.C. & Webb, P.N. (1987). Mid-Cenozoic record of glaciation and sea-level change on the margin of the Victoria Land basin, Antarctica. *Geology*, **15**, 634–7.

Barrett, P.J. & McKelvey, B.C. (1986). Stratigraphy. In *Antarctic Cenozoic History from the MSSTS-1 Drill Hole, McMurdo Sound*, ed. P.J. Barrett, pp. 9–52. Wellington; New Zealand Department of Scientific and Industrial Research Miscellaneous Bulletin No. 237.

Cooper, A.K., Davey, F.J. & Behrendt, J.C. (1987). Seismic stratigraphy and structure of the Victoria Land basin, western Ross Sea, Antarctica. In *The Antarctic Continental Margin: Geology and Geophysics of the Western Ross Sea*, Earth Science Series, vol. 5B, ed. A.K. Cooper & F.J. Davey, pp. 27–76. Houston; Circum-Pacific Council for Energy and Mineral Resources.

Davey, F.J. (1987). Geology and structure of the Ross Sea region. In *The Antarctic Continental Margin: Geology and Geophysics of the Western Ross Sea*, Earth Science Series, vol. 5B, ed. A.K. Cooper & F.J. Davey, pp. 1–16. Houston; Circum-Pacific Council for Energy and Mineral Resources.

Davey, F.J., Bennett, D.J. & Houtz, R.E. (1982). Sedimentary basins of the Ross Sea, Antarctica. *New Zealand Journal of Geology and Geophysics*, **25**, 245–55.

Fitzgerald, P.G., Sandiford, T., Barrett, P.J. & Gleadow, A.J.W. (1987). Asymmetric extension associated with uplift and subsidence in the Transantarctic Mountains and Ross Embayment. *Earth and Planetary Science Letters*, **81**, 67–78.

González-Ferrán, O. (1982). The Antarctic Cenozoic volcanic provinces and their implications in plate tectonic provinces. In *Antarctic Geoscience*, ed. C. Craddock, pp. 687–94. Madison; University of Wisconsin Press.

Hayes, D.E. & Frakes, L.A. (1975). General Synthesis, Deep Sea Drilling Project Leg 28. In *Initial Reports of the Deep Sea Drilling Project, 28*, D.E. Hayes & L.A. Frakes *et al.*, pp. 919–41. Washington, DC; US Government Printing Office.

Hinz, K. & Block, M. (1984). Results of geophysical investigations in the Weddell Sea and in the Ross Sea, Antarctica. *Proceedings of 11th World Petroleum Congress, (London) 1983*, vol. 2, Geology Exploration Reserves, pp. 79–81. Chichester, New York, John Wiley & Sons.

Hinz, K. & Kristoffersen, Y. (1987). Antarctica – recent advances in the understanding of the continental shelf. *Geologisches Jahrbuch*, **E37**, 1–54.

Karl, H.A., Reimnitz, E. & Edwards, B.D. (1987). Extent and nature of Ross Sea unconformity in the western Ross Sea, Antarctica. In *The Antarctic Continental Margin: Geology and Geophysics of the Western Ross Sea*, Earth Science Series, vol. 5B, ed. A.K. Cooper & F.J. Davey, pp. 77–92. Houston; Circum-Pacific Council for Energy and Mineral Resources.

Kim, Y., McGinnis, L.D. & Bower, R.H. (1986). The Victoria Land basin: part of an extended crustal complex between East and West Antarctica. In *Reflection Seismology: A Global Perspective*, Geodynamics Series, ed. M.Barazangi & L. Brown, pp. 323–30. Washington, DC; American Geophysical Union.

Lawver, L.A. & Scotese, C.R. (1987). A revised reconstruction of Gondwanaland. In *Gondwana Six: Structure, Tectonics, and Geophysics*, Geophysical Monograph 40, ed. G.D. McKenzie, pp. 17–24. Washington, DC; American Geophysical Union.

Molnar, P., Atwater, T., Mammerickx, J. & Smith, S.M. (1975). Magnetic anomalies, bathymetry, and the tectonic evolution of the South Pacific since Late Cretaceous. *Geophysical Journal of the Royal Astronomical Society*, **40**, 383–420.

Veevers, J.J. & Eittreim, S.L. (1988). Reconstruction of Antarctica and Australia at breakup (95 ± 5 Ma) and before rifting (160 Ma). *Australian Journal of Earth Science*, **35**, 355–62.

Webb, P.N. (1981). Late Mesozoic–Cenozoic geology of the Ross Sector, Antarctica. *Journal of the Royal Society of New Zealand*, **11(4)**, 439–46.

Crustal extension and origin of sedimentary basins beneath the Ross Sea and Ross Ice Shelf, Antarctica

A.K. COOPER[1], F.J. DAVEY[2], & K. HINZ[3]

1 US Geological Survey, 345 Middlefield Road, Menlo Park, CA 94025, USA

2 Geophysics Division (DSIR), PO Box 1320, Wellington, New Zealand

3 Bundesanstalt für Geowissenschaften und Rohstoffe, PO Box 51053, D-3000 Hannover 51, Federal Republic of Germany

Abstract

The Ross Sea is underlain by basement grabens filled with up to 8 km of high-velocity, (?)Mesozoic strata that are unconformably overlain by up to 6 km of flat-lying Cenozoic rocks. The 2–14 km thick sedimentary section is strongly deformed only in the Terror rift of the western Ross Sea. S–SE-trending positive gravity anomalies, probably marking locally thinner crust, coincide with the three major Ross Sea basement grabens, and continue over 1000 km beneath Ross embayment. Offsets in gravity anomalies, structures and physiographic features of the Ross embayment region suggest that major transverse basement faults have controlled locations of horizontal and vertical displacements due to rifting. Crustal extension in the Ross embayment includes: (1) an early rift period ((?)late Mesozoic) of widespread graben downfaulting, crustal thinning, and later sediment infilling; and (2) a late rift period (Cenozoic) of more localized deformation, principally along the Transantarctic Mountains, Terror rift and in Marie Byrd Land. The change from widespread to more localized deformation may coincide with the documented Eocene change in oceanic plate motions.

Introduction

The Ross embayment (Ross Sea and Ross Ice Shelf) has been cited as a major zone of crustal extension between East and West Antarctica, based on regional mapping (Elliot, 1975; Hayes & Davey, 1975; Davey, 1981; Bentley, 1983; Kadmina *et al.*, 1983; McGinnis *et al.*, 1985; Schmidt & Rowley, 1986; Tessensohn & Wörner, this volume, p. 273) and global plate-motion studies (Molnar *et al.*, 1975; Jurdy, 1978; Gordon & Cox, 1980; Lawver & Scotese, 1987; Stock & Molnar, 1987). However, the nature and extent of subsurface rift structures has only been clearly imaged by post-1980 multichannel seismic-reflection (MCS) surveys of the Ross Sea (Hinz & Block, 1984; Sato *et al.*, 1984; Kim, McGinnis & Bower, 1986; Cooper, Davey & Behrendt, 1987a; Hinz & Kristoffersen, 1987). The MCS data reveal up to 14 km of sedimentary strata lying within the three major depositional centres of the Ross Sea: the Victoria Land basin (VLB), the Central trough and the Eastern basin (Fig. 1).

Herein, we describe the major subcrustal features of the Ross Sea region, and relate them to episodes of crustal extension since late Mesozoic time.

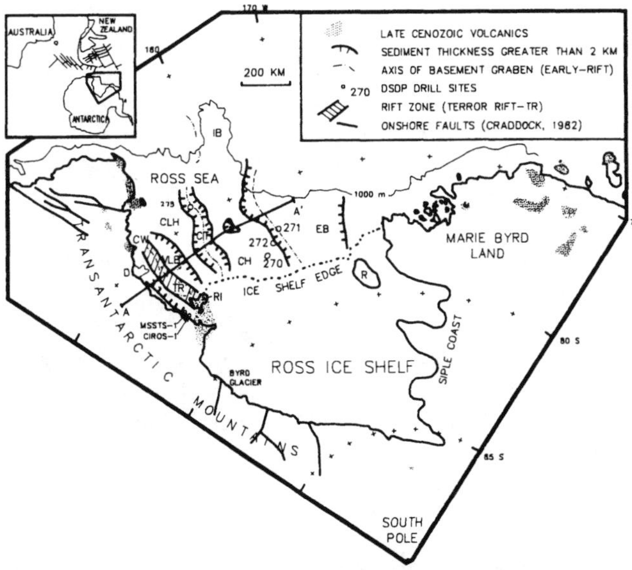

Fig. 1. Index map of the Ross Sea showing the major sedimentary basins, early rift basement grabens and the active Terror rift; basin outlines from Hinz & Block (1984) and Cooper *et al.* (1987a). CP, Campbell Plateau; CH, Central high; CLH, Coulman high; CT, Central trough; CW, Cape Washington; D, Drygalski Ice Tonge; EB, Eastern basin; IB, Iselin Bank; M, McMurdo Sound; TR, Terror rift; R, Roosevelt Island; RI, Ross Island; VLB, Victoria Land basin.

Regional setting

The Ross embayment is bordered by the Transantarctic Mountains (TAM) on the west and Marie Byrd Land on the

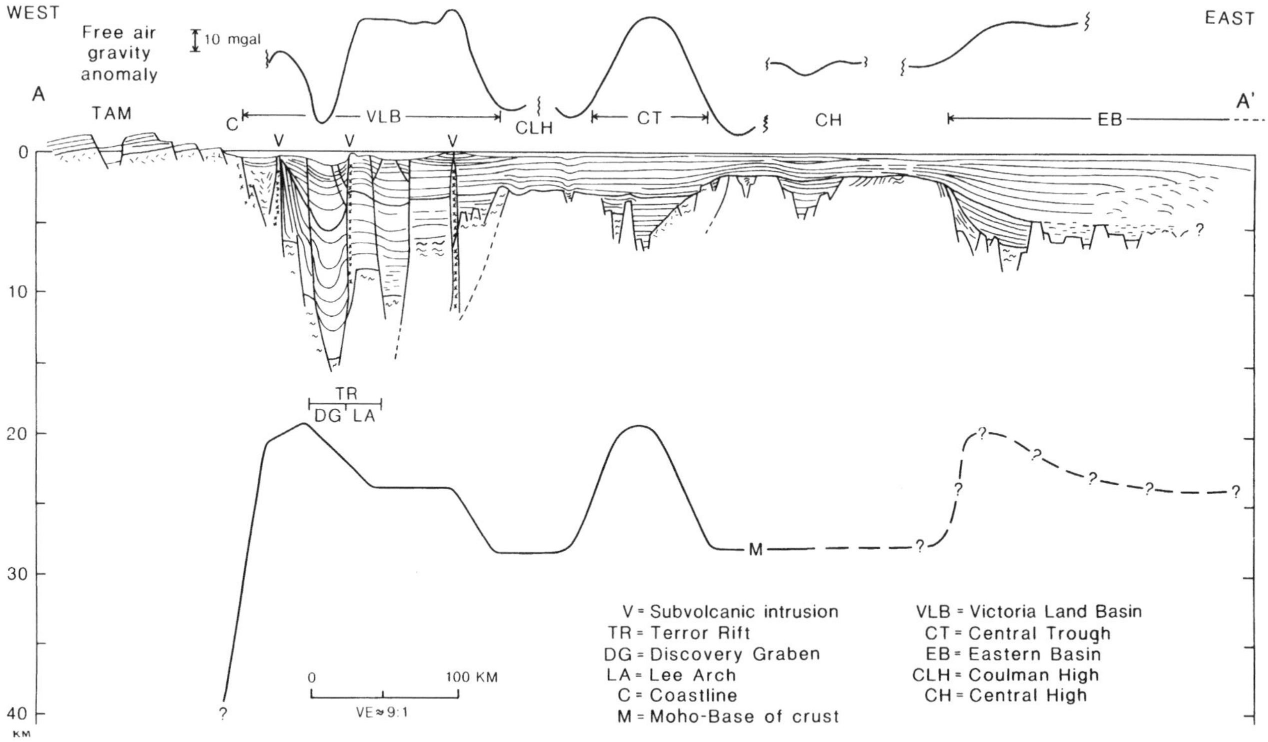

Fig. 2. Generalized profile across the Ross Sea, based in part on offshore multichannel seismic-reflection data (Hinz & Block, 1984; Cooper *et al.*, 1987a) and gravity-model studies (Davey & Cooper, 1987). Early rift grabens, delineated by positive free-air gravity anomalies and thin crust, lie beneath a buried regional unconformity (heavy line). Late-rift faults and intrusive rocks deform the Victoria Land basin and some small basement grabens.

east (Fig. 1). In the TAM, Precambrian and Early Palaeozoic basement rocks are unconformably overlain by gently dipping Devonian–Jurassic strata of the Beacon Supergroup (Davey, 1987). Rocks younger than Jurassic are limited to late Cenozoic volcanic rocks and minor Pliocene and younger glacial deposits. Asymmetric uplift of the TAM occurred principally since Eocene time (Fitzgerald *et al.*, 1987). In Marie Byrd Land, Palaeozoic basement rocks are intruded by mid–late Cretaceous plutons. Late Cenozoic volcanic rocks rest on an uplifted erosional platform of probable early Tertiary or Late Cretaceous age (LeMasurier & Rex, 1983).

Drilling in the Ross Sea (DSDP, Hayes & Frakes, 1975; MSSTS-1, Barrett & McKelvey, 1986; CIROS-1, Barrett, Hambrey & Robinson, this volume, p. 651) has recovered early Oligocene and younger glacial-marine strata, and Early Palaeozoic basement. Rocks of Late Cretaceous and Palaeogene age occur only as glacial erratics, and are thought to underline the offshore basins (Davey, Bennett & Houtz, 1982; Cooper, Davey & Behrendt, this volume, p. 279).

Ross Sea sedimentary basins

Crustal structure deduced from seismic data

Herein, we refer to basement as the acoustic basement in offshore seismic-reflection records. Acoustic basement may, in places, be younger than onshore Precambrian and Early Palaeozoic basement rocks (Hinz & Block, 1984; Cooper *et al.*, 1987a, b).

The Ross Sea is underlain by three asymmetrical basement grabens, up to 175 km wide, that contain up to 8 km of layered high-velocity (4.2–6.0 km/s) strata (Fig. 2). A buried regional unconformity cuts the high-velocity strata and intervening basement platforms. The unconformity is covered by up to 6 km of nearly flat-lying rocks that generally are thickest in the seaward prograding deposits at the continental shelf edge of the eastern Ross Sea. The total sedimentary thickness in the Ross Sea is up to 14 km in the VLB, decreasing to the south.

Small grabens lie beneath basement platforms and the flanks of the major grabens. Commonly, some small grabens contain deformed strata, and their bounding faults do not displace the eroded basement or overlying sediment. Other grabens, however, downwarp the eroded basement and overlying sedimentary section. The two graben types suggest two periods of deformation.

The major basement grabens of the Ross Sea trend N–S (Fig. 1). In the west and central Ross Sea, the 2-km isopach delineates the grabens. In the east, however, a major basement graben lies beneath only the west side of the Eastern basin (Fig. 2), which principally comprises thick seaward-prograding shelf deposits burying the major graben (Hinz & Block, 1984).

The VLB, unlike other Ross Sea basins, is underlain by an active axial rift zone, 50–60 km wide, extending from active volcanoes at Cape Washington to those at Ross Island (Terror rift; Cooper *et al.*, 1987a). The rift is defined by an axial graben (Discovery graben, coincident with the axis of the basement graben), and an adjacent subvolcanically-intruded basement horst (Lee arch, Fig. 2; Cooper *et al.*, 1987a).

Basement grabens can be traced seismically more than 450 km from seaward of the continental shelf edge to the Ross

Ice Shelf edge (Fig. 1; Hinz & Block, 1984; Davey & Cooper, 1987). The VLB ends abruptly at Cape Washington, about 300 km south of the continental shelf edge (Cooper *et al.*, 1987a). All three basins may extend several hundred kilometres S–SE beneath the Ross Ice Shelf.

Most normal faults parallel the basin axes. The basement faults, along which the grabens have been downfaulted, commonly do not disrupt overlying strata, with the exception of the Terror rift where faults extend upward through the entire sedimentary section. Along the front of the TAM, basement faults are steep and planar to depths of 1–2 km, but may flatten at depth. Seismic evidence for low-angle basement faulting, such as is seen in other highly extended regions (Allmendinger *et al.*, 1983) is poor, but weak intrabasement reflections at the eastern edge of the VLB may be from a fault that extends beneath the entire basin (Fig. 2).

Volcanic rocks and shallow subvolcanic intrusions are apparently associated with faulting and rifting in the VLB, as these rocks occur in the southern half of the basin, along the Terror rift and near the eastern and western edges of the basin (Fig. 2; Cooper *et al.*, 1987a). Jurassic and Cenozoic volcanic deposits, which are common onshore, are likely offshore but may not comprise a major part of the sedimentary section, except near Ross Island (late Cenozoic). (?)Cenozoic magmatic rocks may also locally intrude the shallow basement platform directly north of the VLB (Behrendt, Cooper & Yuan, 1987).

Crustal thickness has been measured seismically only beneath the VLB in McMurdo Sound, where it is 21 km (McGinnis *et al.*, 1985). Layered reflections from 16 to 22 km depths are seen in sonobuoy and multichannel seismic-reflection data from the Terror rift and Coulman basement high (McGinnis & Kim, 1986; Cooper *et al.*, 1987b), and are interpreted as coming from a magma chamber or layered intrusion. High heat flow values support this interpretation.

Aeromagnetic data over the VLB show a distinct long-wavelength magnetic low that contrasts with the irregular pattern of anomalies over the Central trough. This broad magnetic low can be explained either by a 10–14 km thick sedimentary section, consistent with seismic data (Behrendt *et al.*, this volume, p. 299), or by a thinner sedimentary section and a rise in the Curie isotherm, consistent with higher heat flow (Behrendt *et al.*, 1987).

Crustal structure deduced from gravity and bathymetric data

Free-air gravity anomalies in the Ross Sea (Davey & Cooper, 1987) and over the Ross Ice Shelf (Davey, 1981; Robertson *et al.*, 1982) are nearly linear and continuous over large distances from seaward of the Ross Sea continental shelf edge to the southern limit of the data at about 86° S (Fig. 3). Gravity anomalies trend N–S in the Ross Sea and NW–SE beneath the Ross Ice Shelf, south of a line connecting Byrd Glacier and Roosevelt Island (Fig. 1). In the Ross Sea, gravity trends commonly cut across bathymetric trends. Beneath the Ross Ice Shelf, the two trends converge to the south-east and finally coincide (bathymetric ridges and positive gravity anomalies) at the head of the ice shelf (Fig. 3; Robertson *et al.*, 1982).

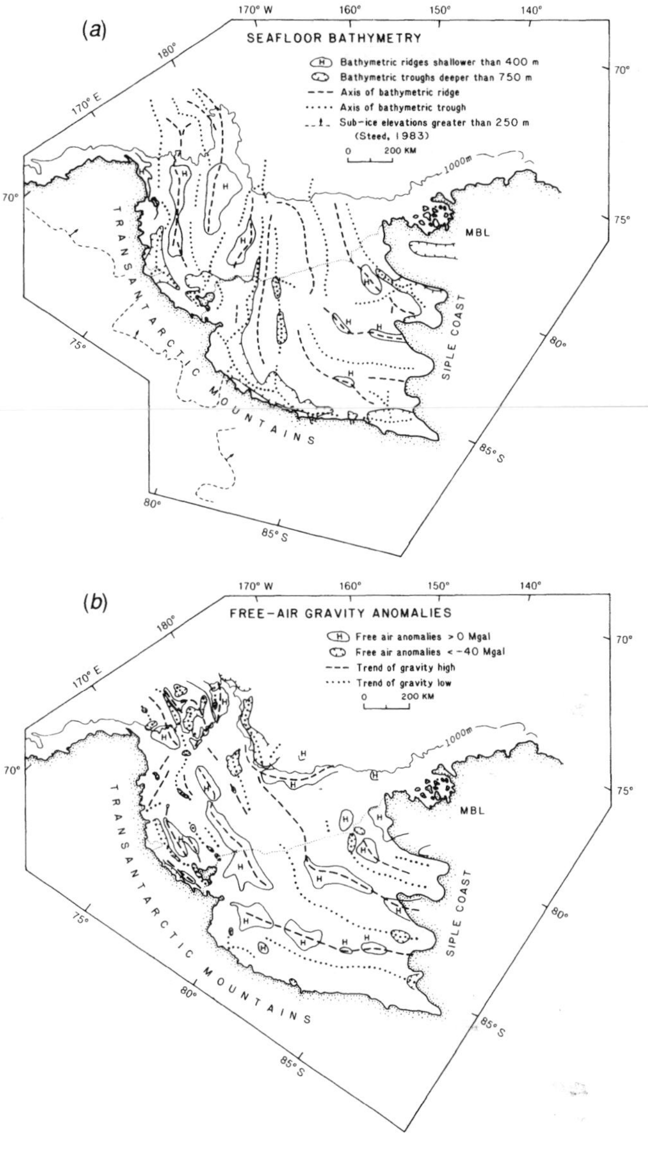

Fig. 3. Geophysical maps of the Ross embayment: (*a*) Seafloor bathymetry (Robertson *et al.*, 1982; Davey & Cooper, 1987) and subice elevation under the Transantarctic Mountains; (*b*) free-air gravity anomalies (Robertson *et al.*, 1982; Davey & Cooper, 1987). Gravity and bathymetry trends differ in most areas indicating tectonic (rather than ice erosion) control on gravity anomalies, with the exception of ice-stream erosion near the Siple Coast.

The three major basement grabens beneath the Ross Sea are marked by relatively positive gravity anomalies (Figs 2 & 3). This uncommon inverse relation can be explained by either crustal thinning or possible deep crustal intrusions, bringing high density rocks to shallow depths beneath the Central trough (Hayes & Davey, 1975), the VLB (Davey & Cooper, 1987), and the Eastern basin (Fig. 2). Small basement grabens, however, apparently are not compensated by thinner crust, as they are marked by negative anomalies (Fig. 2).

A narrow, large-amplitude, negative gravity anomaly coincides with the Discovery graben (Fig. 2), and is superimposed on the broad positive anomaly covering most of the VLB. The gravity low results from the deep bathymetric depression and low density strata at the top of the 12–14 km thick sedimentary section (Davey & Cooper, 1987). Gravity models indicate that the crust ranges in thickness from 19 to 23 km under the VLB, to about 27 km under the Ross Sea basement highs, and about 40 km under the TAM (Davey & Cooper, 1987). These thicknesses are consistent with previous interpretations (Smithson, 1972; Bentley, 1973; Robinson & Splettstoesser, 1984; McGinnis *et al.*, 1985).

Discussion

Episodic crustal extension

Seismic-reflection data (Fig. 2) provide evidence for episodic crustal extension in the Ross Sea and probably under the Ross Ice Shelf:

- (i) younger fault and intrusive structures (Terror rift) deform the older sediment-filled graben underlying the VLB
- (ii) two types (and ages) of small basement grabens are seen – grabens with bounding faults that displace the eroded basement surface and overlying sediments, and grabens that do not displace this basement surface
- (iii) some basement faults of the VLB displace overlying strata and are associated with shallow intrusions whereas other basement faults do not show this deformation
- (iv) more than one basement unconformity exists indicating at least two periods of uplift and erosion (and rifting).

Cooper *et al.* (1987a) recognized early rift and late rift periods in the VLB. We suggest similar rift periods for the Ross embayment. During the early rift period, the basement was downfaulted in a series of rift grabens. Crustal thinning by extension was accompanied by elevation of dense mantle and by rapid subsidence of the basin, resulting in positive gravity anomalies. Sediment infilling following subsidence formed flat-lying graben strata. The late rift period was characterized by widespread uplift, initial erosion of the regional unconformity that cuts across basement beneath the Coulman and Central highs, development of the Terror rift and uplift of the TAM.

Age of crustal extension

The age of the two rift periods is ill-defined because the layered sedimentary sections filling the basement grabens have not been sampled. Evidence for late Mesozoic and younger crustal extension from the regional geologic record, includes:

- (i) emplacement of Jurassic rift-related tholeiitic dolerite dykes, sills and volcanic rocks throughout the TAM (Elliot, 1975)
- (ii) downfaulting of the Rennick graben of northern Victoria Land in post-Early Cretaceous time (Grindley & Oliver, 1983a)
- (iii) emplacement of mid-Cretaceous mafic dykes in Marie Byrd Land (90–110 Ma; Grindley & Oliver, 1983b)
- (iv) uplift and (?)Late Cretaceous erosion of a regional peneplain (now at 500–2700 m elevation) in Marie Byrd Land (LeMasurier & Rex, 1983)
- (v) initial Eocene uplift of the TAM (50 Ma; Fitzgerald *et al.*, 1987), and downfaulting of axial grabens along the central TAM in Tertiary time (Katz, 1982)
- (vi) eruption of alkalic volcanic rocks in Victoria Land since probably early Oligocene time (38 Ma; Barrett *et al.*, this volume, p. 651) and Marie Byrd Land since about 27 Ma (LeMasurier & Rex, 1983)
- (vii) uplift and erosion of a now-buried regional basement unconformity beneath the Ross Sea before the late Oligocene (26 Ma at DSSDP site 270; Hayes & Frakes, 1975; Fig. 2).

Our proposed ages for the two rift periods are based on: (1) the above-listed events; (2) the relative ages from seismic stratigraphy for the Ross Sea (Hinz & Block, 1984; Cooper *et al.*, 1987a, this volume, p. 279); and (3) the Jurassic and younger ages for rocks drilled from formerly nearby rift basin areas of Australia–Tasmania (Veevers, 1982; Williamson *et al.*, 1987) and New Zealand (Cooper *et al.*, 1982; Shirley, 1983). We suggest that early rifting was mostly Cretaceous (rapid graben subsidence), but commenced in the Jurassic. Infilling of the three grabens may have been principally during the Late Cretaceous and Palaeogene. Late rifting probably began in the Eocene (initial uplift of TAM and development of Terror rift) but was more intense in the late Cenozoic (areally extensive alkalic volcanism).

We tentatively relate the early rifting and late rifting periods to widely recognized periods of major plate motions and plate re-organizations. The early rift period may correspond with a pre-80 Ma episode of slow continental rifting between Australia and Tasmania, and with the post-80 Ma opening of the Tasman Basin and the initial separation of New Zealand and Antarctica. The late rift period probably commenced during the Eocene re-organization of plate motions, marked by a 40° change in magnetic anomaly trend south of New Zealand (50–42 Ma; Stock & Molnar, 1987) and by increased spreading rates in the Indian Ocean (43 Ma; Cande & Mutter, 1982).

Crustal extension beneath the Ross Ice Shelf

Crustal extension beneath ice-covered areas south-east of the Ross Sea, such as the Ross Ice Shelf, Byrd Subglacial Basin, Marie Byrd Land and areas farther south-east, has been postulated principally on geophysical evidence for thin crust (25–30 km thick) and block-faulted bedrock ridges (Jankowski & Drewry, 1981; Bentley, 1983; Jankowski, Drewry & Behrendt, 1983; LeMasurier & Rex, 1983; Robinson & Splettstoesser, 1984; McGinnis *et al.*, 1985). We suspect that the linear positive-gravity anomalies of the Ross embayment (Fig. 3), which correspond with deeply buried rift grabens in the Ross Sea, may delineate similar early rift grabens beneath the Ross Ice Shelf.

Aeromagnetic data (Behrendt, 1983) indicate that 8 km of non-magnetic strata lie beneath the head (south-eastern end) of the Ross Ice Shelf. The bathymetric troughs in this areas are aligned with the major ice streams (Jankowski & Drewry, 1981), but Robertson *et al.* (1982) postulated, from gravity data, that the troughs and ridges have 'deep structural control'.

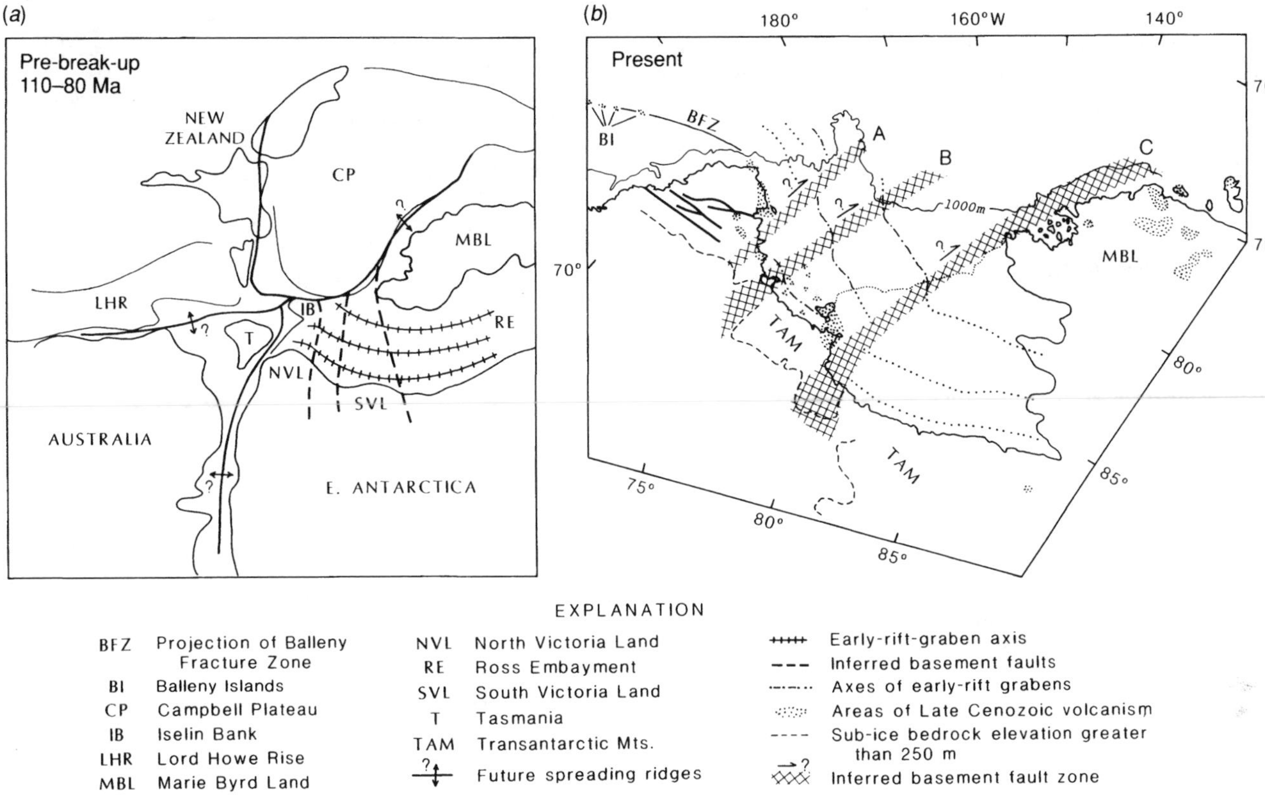

Fig. 4. Models for the Ross embayment: (*a*) 110–80 Ma (modified from Lawver & Scotese, 1987), showing early rift development of the three major graben systems beneath the Ross embayment, which is one-half its present width; (*b*) present-day, showing offset and (?) rotated position of the early rift grabens. Horizontal and vertical displacements on transverse basement faults is likely to have occurred in Mesozoic and Cenozoic times.

The poor correlation between bathymetric highs and positive free-air gravity anomalies, except near the head of the ice shelf, suggests to us that parts of the Ross Ice Shelf may be underlain by high-density sedimentary strata. The positive gravity anomalies may also be due, in part, to thinner crust.

Limited seismic-refraction data near the head of the Ross Ice Shelf (Robertson *et al.*, 1982; Rooney, Blackenship & Bentley, 1987) show < 1.5 km of rocks above a 5.7 km/s basement. This basement velocity falls in the 4.8–6.0 km/s range of velocities measured for early rift strata beneath the Ross Sea (Cooper *et al.*, 1987a, this volume, p. 279), thus leading to an uncertain interpretation for the 5.7 km/s rocks. Seismic-reflection data are needed to resolve the nature of these basement rocks (see Rooney *et al.*, this volume, p. 261). Similar basement velocities, and interpretational uncertainties, occur farther south-east beneath the Byrd Subglacial Basin, which is an area attributed to dissection by initial rifting in middle–late Mesozoic time and by further Cenozoic rifting (Jankowski *et al.*, 1983).

Transverse fault zones beneath the Ross embayment

At several locations in the Ross embayment region, major structural and physiographic features exhibit large horizontal offsets. These offsets are aligned along three subparallel zones (A, B, C, Fig. 4*b*), which we interpret as major transverse basement faults. Zone C, possibly the most significant, trends SW–NE, along the major topographic trough beneath Byrd Glacier, along the points where gravity and bathymetric trends change direction by 40° beneath the Ross Ice Shelf, and along the 400 km long bathymetric scarp in Marie Byrd Land. Horizontal offsets along zone C occur in the subice elevation contours for the TAM, the coastline of southern Victoria Land, and the linear gravity anomalies (and early rift grabens?) of the Ross embayment. Additionally, zone C marks the southernmost termination of major Cenozoic volcanism along the Transantarctic Mountains (with the exception of two localities in the central TAM) and the westernmost location in Marie Byrd Land (González-Ferrán, 1982; LeMasurier & Rex, 1982; Figs 1 & 4*b*).

Two basement fault zones (A & B, Fig. 4*b*) lie beneath the Ross Sea. Zone B lies along the deep trough beneath the Drygalski Ice Tongue and along points where horizontal offsets occur in elevation contours of the TAM and in the axes of the three Ross Sea early rift grabens. Additionally, zone B marks the northernmost extents of offshore (?)Neogene subvolcanic intrusions in the VLB and of short-wavelength linear magnetic anomalies that characterize the south-western Ross Sea (see Behrendt *et al.*, this volume, p. 299). The northernmost zone (A, Fig. 4*b*) is more tentative. Zone A lies, W–E, along a small horizontal offset in TAM elevation contours, along the major Polar 3 magnetic anomaly (Behrendt *et al.*, this volume, p. 299), and near the abrupt terminations of the VLB structure (Fig. 1) and the Central trough gravity anomaly (Fig. 3*b*).

The transverse fault zones are basement structures with a Cenozoic history of movement as they offset late Mesozoic rift-graben axes and post-Eocene mountain topography and coastlines, and they apparently control the location of Oligocene and younger volcanic centres. Based on limited seismic data, the fault zones do not appear to have extensively deformed the (?)late Mesozoic and younger sedimentary sections under (through) which they pass. The fault zones could have initiated either in Palaeozoic time as structural boundaries between different basement terranes, or in Mesozoic time as fractures delineating the break-up of Gondwana. In places, these faults appear to determine the location of: (1) horizontal continental break-up (zone C: Marie Byrd Land and the Campbell Plateau); (2) differential, vertical displacement (zones A, B and C: across the TAM); and (3) possible large-scale crustal rotation (zone C: rotation between rift grabens under the Ross Sea and Ross Ice Shelf).

The orientation of the transverse faults may reflect an early rift stress regime that controlled the initial trend and (?)offsets of the late Mesozoic rift grabens beneath the Ross embayment. With the Eocene reorganization of plate motions, the stress regime changed, leading to late rift crustal deformation and volcanism located principally along the western Ross Sea and in Marie Byrd Land (Fig. 4*b*).

Summary

In the Ross embayment, significant crustal extension (100%, Lawver & Scotese, 1987) has probably occurred episodically in response to motions of surrounding major plates. We think that most regional extension occurred during a late Mesozoic, early rift period marked by downfaulting of the major and minor graben systems. A late rift period, commencing with a major Eocene change in plate motions, resulted in crustal deformation localized principally in the western Ross Sea and Marie Byrd Land. Major transverse basement faults, newly mapped at three sites across the Ross embayment region, appear to be long-active (?post-Palaeozoic) features controlling horizontal (possibly strike-slip and rotation) and vertical displacements associated with the early rift and late rift periods.

Although we propose two distinct rift periods, crustal extension beneath the Ross embayment may have been more or less continuous since the initial break-up of Gondwana.

Acknowledgements

We thank A.B. Ford, R. Bohannon, S. Eittreim, and J.C. Behrendt for thoughtful discussions and reviews of early versions of the manuscript.

References

Allmendinger, R.A., Sharp, J.W., Tish, D. Von, Serpa, L., Brown, L., Kaufman, S., Oliver, J. & Smith, R.B. (1983). Cenozoic and Mesozoic structure of the eastern Basin and Range province, Utah, from COCOR seismic-reflection data. *Geology*, **11**, 532–6.

Barrett, P.J. & McKelvey, B.C. (1986). Stratigraphy. In *Antarctic Cenozoic History from the MSSTS-1 Drill Hole, McMurdo Sound*, Miscellaneous Bulletin No. 237, ed. P.J. Barrett, pp. 9–52. Wellington; New Zealand Department of Scientific and Industrial Research.

Behrendt, J.C. (1983). Geophysical and geological studies relevant to assessment of the petroleum resources of Antarctica. In *Antarctic Earth Science*, ed. R.L. Oliver, P.R. James & J.B. Jago, pp. 423–8. Canberra; Australian Academy of Science and Cambridge; Cambridge University Press.

Behrendt, J.C., Cooper, A.K. & Yuan, A. (1987). Interpretation of marine magnetic gradiometer and multichannel seismic-reflection observations over the western Ross Sea shelf, Antarctica. In *The Antarctic Continental Margin: Geology and Geophysics of the Western Ross Sea*, Earth Science Series, vol. 5B, ed. A.K. Cooper & F.J. Davey, pp. 155–78. Houston; Circum-Pacific Council for Energy and Mineral Resources.

Bentley, C.R. (1973). Crustal structure of Antarctica. *Tectonophysics*, **20**, 229–40.

Bentley, C.R. (1983). Crustal structure of Antarctica from geophysical evidence – a review. In *Antarctic Earth Science*, ed. R.L. Oliver, P.R. James & J.B. Jago, pp. 491–7. Canberra; Australian Academy of Science and Cambridge; Cambridge University Press.

Cande, S.C. & Mutter, J.C. (1982). A revised identification of the oldest sea-floor spreading anomalies between Australia and Antarctica. *Earth and Planetary Science Letters*, **58**, 151–60.

Cooper, A.K., Davey, F.J. & Behrendt, J.C. (1987a). Seismic stratigraphy and structure of the Victoria Land basin, western Ross Sea, Antarctica. In *The Antarctic Continental Margin: Geology and Geophysics of the Western Ross Sea*, Earth Science Series, vol. 5B, ed. A.K. Cooper & F.J. Davey, pp. 27–76. Houston; Circum-Pacific Council for Energy and Mineral Resources.

Cooper, A.K., Davey, F.J. & Cochrane, G.R. (1987b). Structure of extensionally rifted crust beneath the western Ross Sea and Iselin Bank, Antarctica, from sonobuoy seismic data. In *The Antarctic Continental Margin: Geology and Geophysics of the Western Ross Sea*, Earth Science Series, vol. 5B, ed. A.K. Cooper & F.J. Davey, pp. 93–118. Houston; Circum-Pacific Council for Energy and Mineral Resources.

Cooper, R.A., Landis, C.A., LeMasurier, W.E. & Speden, I.G. (1982). Geologic history and regional patterns in New Zealand and Antarctica – their paleotectonic and paleogeographic significance. In *Antarctic Geoscience*, ed. C. Craddock, pp. 43–54. Madison; University of Wisconsin Press.

Davey, F.J. (1981). Geophysical studies in the Ross Sea region. *Journal of the Royal Society of New Zealand*, **11**, 465–79.

Davey, F.J. (1987). Geology and structure of the Ross Sea region. In *Antarctic Continental Margin: Geology and Geophysics of the Western Ross Sea*, Earth Science Series, vol. 5B, ed. A.K. Cooper & F.J. Davey, pp. 1–16. Houston; Circum-Pacific Council for Energy and Mineral Resources.

Davey, F.J., Bennett, D.J. & Houtz, R.E. (1982). Sedimentary basins of the Ross Sea, Antarctica. *New Zealand Journal of Geology and Geophysics*, **25**, 245–55.

Davey, F.J. & Cooper, A.K. (1987). Gravity studies of the Victoria Land basin and Iselin Bank. In *The Antarctic Continental Margin: Geology and Geophysics of the Western Ross Sea*, Earth Science Series, vol. 5B, ed. A.K. Cooper & F.J. Davey, pp. 119–38. Houston; Circum-Pacific Council for Energy and Mineral Resources.

Elliot, D.H. (1975). Tectonics of Antarctica: a review. *American Journal of Science*, **275-A**, 45–106.

Fitzgerald, P.G., Sandiford, T., Barrett, P.J. & Gleadow, A.J.W. (1987). Asymmetric extension associated with uplift and subsidence in the Transantarctic Mountains and Ross Embayment. *Earth and Planetary Science Letters*, **81**, 67–78.

González-Ferrán, O. (1982). The Antarctic Cenozoic volcanic provinces and their implications in plate tectonic provinces. In *Antarctic Geoscience*, ed. C. Craddock, pp. 687–94. Madison; University of Wisconsin Press.

Gordon, R.G. & Cox, A. (1980). Paleomagnetic test of the early Tertiary plate circuit between the Pacific Basin plates and the Indian plate. *Journal of Geophysical Research*, **85**, 6534–46.

Grindley, G.W. & Oliver, P.J. (1983a). Post-Ross Orogeny cratonisation of northern Victoria Land. In *Antarctic Earth Science*, ed. R.L. Oliver, P.R. James & J.B. Jago, pp. 133–9. Canberra; Australian Academy of Science and Cambridge; Cambridge University Press.

Grindley, G.W. & Oliver, P.J. (1983b). Palaeomagnetism of Cretaceous volcanic rocks from Marie Byrd Land, Antarctica. In *Antarctic Earth Science*, ed. R.L. Oliver, P.R. James & J.B. Jago, pp. 573–8. Canberra; Australian Academy of Science and Cambridge; Cambridge University Press.

Hayes, D.E. & Davey, F.J. (1975). A geophysical study of the Ross Sea, Antarctica. In *Initial Reports of the Deep Sea Drilling Project 28*, D.E. Hayes, L.A. Frakes *et al.*, pp. 887–907. Washington, DC; US Government Printing Office.

Hayes, D.E. & Frakes, L.A. (1975). General Synthesis, Deep Sea Drilling Project Leg 28. In *Initial Reports of the Deep Sea Drilling Project, 28*, D.E. Hayes, L.A. Frakes *et al.*, pp. 919–42. Washington, DC; US Government Printing Office.

Hinz, K. & Block, M. (1984). Results of geophysical investigations in the Weddell Sea and in the Ross Sea, Antarctica. *Proceedings of 11th World Petroleum Congress, (London) 1983*, vol. 2, Geology Exploration Reserves, pp. 79–81. Chichester, New York, John Wiley & Sons.

Hinz, K. & Kristoffersen, Y. (1987). Antarctica – recent advances in the understanding of the continental shelf. *Geologisches Jahrbuch*, **E37**, 1–54.

Jankowski, E.J. & Drewry, D.J. (1981). The structure of West Antarctica from geophysical studies. *Nature, London*, **291**, 17–21.

Jankowski, E.J., Drewry, D.J. & Behrendt, J.C. (1983). Magnetic studies of upper crustal structure in West Antarctica and the boundary with East Antarctica. In *Antarctic Earth Science*, ed. R.L. Oliver, P.R. James & J.B. Jago, pp. 197–203. Canberra; Australian Academy of Science and Cambridge; Cambridge University Press.

Jurdy, D.M. (1978). An alternative model for early Tertiary plate motions. *Geology*, **6**, 469–88.

Kadmina, I.N., Kurinin, R.G., Masolov, V.N. & Grikurov, G.E. (1983). Antarctic crustal structure from geophysical evidence: a review. In *Antarctic Earth Science*, ed. R.L. Oliver, P.R. James & J.B. Jago, pp. 498–502. Canberra; Australian Academy of Science and Cambridge; Cambridge University Press.

Katz, H.R. (1982). Post-Beacon tectonics in the region of Amundsen and Scott Glaciers, Queen Maud Range, Transantarctic Mountains. In *Antarctic Geoscience*, ed. C. Craddock, pp. 827–34. Madison; University of Wisconsin Press.

Kim, Y., McGinnis, L.D. & Bower, R.H. (1986). The Victoria Land basin: part of an extended crustal complex between East and West Antarctica. In *Reflection Seismology: A Global Perspective*, Geodynamics Series, ed. M. Barazangi & L. Brown, pp. 323–30. Washington, DC; American Geophysical Union.

Lawver, L.A. & Scotese, C.R. (1987). A revised reconstruction of Gondwanaland. In *Gondwana Six: Structure, Tectonics, and Geophysics*, Geophysical Monograph 40, ed. G.D. McKenzie, pp. 17–24. Washington, DC; American Geophysical Union.

LeMasurier, W.E. & Rex, D.C. (1982). Volcanic record of Cenozoic glacial history in Marie Byrd Land and western Ellsworth Land: revised chronology and evaluation of tectonic factors. In *Antarctic Geoscience*, ed. C. Craddock, pp. 725–34. Madison; University of Wisconsin Press.

LeMasurier, W.E. & Rex, D.C. (1983). Rates of uplift and the scale of ice level instabilities recorded by volcanic rocks in Marie Byrd Land, West Antarctica. In *Antarctic Earth Science*, eds. R.L. Oliver, P.R. James & J.B. Jago, pp. 663–70. Canberra; Australian Academy of Science and Cambridge; Cambridge University Press.

McGinnis, L.D., Bowen, R.H., Erickson, J.M., Allred, B.J. & Kreamer, J.L. (1985). East–west antarctic boundary in McMurdo Sound. *Tectonophysics*, **114**, 341–56.

McGinnis, L.D. & Kim, Y. (1986). Deep seismic soundings along the boundary between East and West Antarctica. *Antarctic Journal of the United States*, **20(5)**, 21–2.

Molnar, P., Atwater, T., Mammerickx, J. & Smith, S.M. (1975). Magnetic anomalies, bathymetry, and the tectonic evolution of the South Pacific since Late Cretaceous. *Geophysical Journal of the Royal Astronomical Society*, **40**, 383–420.

Robertson, J.D., Bentley, C.R., Clough, J.W. & Greischar, L.L. (1982). Sea-bottom topography and crustal structure below the Ross Ice Shelf, Antarctica. In *Antarctic Geoscience*, ed. C. Craddock, pp. 1083–90. Madison; University of Wisconsin Press.

Robinson, E.S. & Splettstoesser, J.F. (1984). Structure of the Transantarctic Mountains determined from geophysical surveys. In *Geology of the Central Transantarctic Mountains*, Antarctic Research Series, 36, ed. M.D. Turner & J.F. Splettstoesser, pp. 119–62. Washington, DC; American Geophysical Union.

Rooney, S.T., Blankenship, D.D. & Bentley, C.R. (1987). Seismic refraction measurements of crustal structure in West Antarctica. In *Gondwana Six: Structure, Tectonics, and Geophysics*, Geophysical Monograph 40, ed. G.D. McKenzie, pp. 1–8. Washington, DC; American Geophysical Union.

Sato, S., Asakura, N., Saki, T., Oikawa, N. & Kaneda, Y. (1984). Preliminary Results of Geological and Geophysical Surveys in the Ross Sea and in the Dumont D'Urville Sea, off Antarctica, *Memoirs of the National Institute of Polar Research, Special Issue* **33**, 66–92.

Schmidt, D.L. & Rowley, P.D. (1986). Continental rifting and transform faulting along the Jurassic Transantarctic rift, Antarctica. *Tectonics*, **5**, 279–91.

Shirley, J.E. (1983). Oil exploration, prospects along northwest flank of Great South Basin, New Zealand. *Oil and Gas Journal*, **March 21**, 199–206.

Smithson, S.B. (1972). Gravity interpretation in the Transantarctic Mountains near McMurdo Sound, Antarctica. *Geological Society of America Bulletin*, **83**, 3437–42.

Stock, J. & Molnar, P. (1987). Revised history of early Tertiary plate motion in the south-west Pacific. *Nature, London*, **325**, 495–9.

Veevers, J.J. (1982). Australian rifted margins. In *Dynamics of Passive Margins*, Geodynamics Series, **6**, ed. R.A. Scrutton, pp. 72–90. Washington, DC; American Geophysical Union.

Williamson, P.E., Pigram, C.J., Colwell, J.B., Scherl, A.S., Lockwood, K.L. & Branson, J.C. (1987). Review of stratigraphy, structure, and hydrocarbon potential of Bass Basin, Australia. *Bulletin of the American Association of Petroleum Geologists*, **71**, 253–80.

Chemical characteristics of greywacke and palaeosol of early Oligocene or older sedimentary breccia, Ross Sea DSDP Site 270

A.B. FORD

US Geological Survey, 345 Middlefield Road, Menlo Park, California 94025, USA

Abstract

A 25 m thick greywacke-bearing sedimentary breccia unconformably overlies early (?)Palaeozoic metamorphic rocks and underlies Oligocene (26 Ma) greensand at DSDP Site 270, Ross Sea. Calcareous greywacke forms the breccia matrix and sandy intervals. A 3 m thick highly altered zone at the top of the breccia is an apparent palaeosol. Greywacke below the zone has an SiO_2 content (57%, 20 analyses) of typical quartz-poor greywacke, and an unusually high CaO content (7%) reflecting a local marble-bearing source. Chemical concentration ratios show strong (Co) or minor (Ni, B, Al_2O_3) apparent gain in the palaeosol and strong (CaO, CO_2, MnO, FeO) or minor loss (P_2O_5, MgO, Rb, Sr). Variations suggest non-arid climate weathering and leaching. The palaeosol, which largely represents the parent greywacke minus original carbonate, has an SiO_2 content (70%, volatile-free basis, 7 samples) typical of quartz-rich greywacke of passive margin and rift settings and similar to greywacke of some Transantarctic Mountains units (e.g. Robertson Bay Group). The breccia was possibly deposited during Cretaceous or Palaeogene Gondwana rifting or early (50 Ma) Transantarctic Mountains uplift and Ross Sea graben development, but it may be older.

Introduction

In February 1973 the Deep Sea Drilling Project (DSDP) vessel *Glomar Challenger* drilled a 25 m thick unit of sedimentary breccia beneath mid-Oligocene greensand (26 Ma; McDougall, 1976) at Site 270, beneath the western flank of the Eastern basin, southern Ross Sea (Fig. 1). The regional geology and structure of the Ross Sea have been summarized by Davey (1987).

The sedimentary breccia is unfossiliferous and its age may be older than the Oligocene age reported by The Shipboard Scientific Party (1975). It consists of poorly sorted, angular to subrounded pebble to cobble-size (or larger) lithic clasts embedded in a calcareous greywacke matrix and it contains rare thin sandy intervals. Lithic clasts include abundant marble and gneiss similar to rocks cored beneath the breccia; these possibly correlate with Early Palaeozoic units near McMurdo Sound (Ford & Barrett, 1975). The term 'greywacke' is used in this report in the broadly defined sense (Pettijohn, 1963) as being a primarily feldspathic or lithic sandstone with a prominent silt- and clay-sized detrital matrix.

The breccia is pervasively altered and its matrix contains many patches of diagenetic chlorite, altered mica, clay and calcite and common veinlets of calcite. An unusually strongly altered zone, about 3 m thick, at the top of the breccia was considered possibly to represent a palaeosol (Ford & Barrett, 1975) which, if formed subaerially, has important tectonic

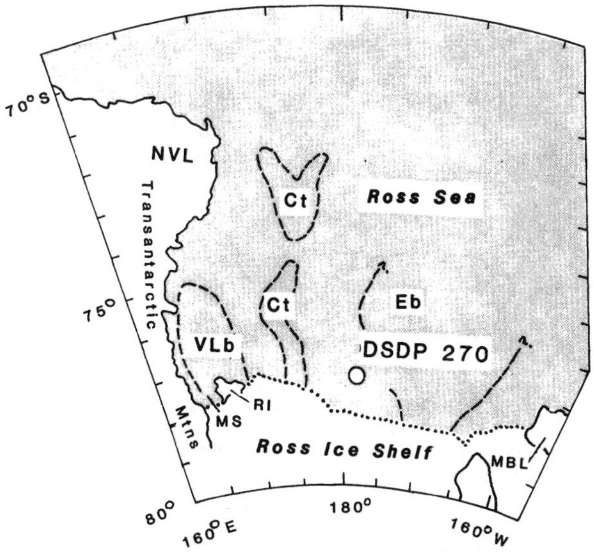

Fig. 1. Index map showing location of DSDP Site 270 and sedimentary basins of the Ross Sea (after Davey (1987) with modifications by A.K. Cooper (pers. comm., 1987)). Eb, Eastern basin; Ct, Central trough; VLb, Victoria Land basin; MBL, Marie Byrd Land; MS, McMurdo Sound; NVL, northern Victoria Land; RI, Ross Island.

Table 1. *Average composition of greywacke and palaeosol, DSDP Site 270 sedimentary breccia unit, and comparisons with other greywackes*

Location code and number of samples	A (20)	B (7)	C (3)	D (1)	E (3)	F (78)	G (6)
SiO_2	56.7	62.8	72.0	68.93	73.8	58.04	72.14
TiO_2	0.60	0.64	0.52	0.69	0.62	1.13	0.67
Al_2O_3	12.0	15.6	11.5	14.15	10.8	14.42	12.20
Fe_2O_3	3.1	3.2	1.6	0.26	1.0	6.38	0.88
FeO	1.3	0.4	2.6	4.33	3.1	3.60	3.68
MnO	0.06	0.01	0.1	—	0.06	0.17	0.06
MgO	1.8	1.5	2.0	2.49	1.7	3.32	2.32
CaO	7.3	1.1	2.4	0.83	1.9	3.64	0.87
Na_2O	1.3	1.3	1.3	1.30	1.7	3.35	1.47
K_2O	2.9	3.0	2.7	3.87	2.1	4.15	2.68
P_2O_5	0.20	0.09	0.13	0.15	0.19	0.17	0.19
H_2O^+	2.3	2.7	1.8	—	1.8	—	—
H_2O^-	3.6	5.0	0.1	0.95	0.18	—	—
CO_2	5.3	0.36	1.4	—	0.19	2.01	—
LOI	0.9	1.05	—	2.05	—	—	2.84
Total	99.36	98.75	100.15	100.00	99.14	100.38	100.00
Rb	123	84	—	165	—	41	136
Sr	956	744	—	75	—	224	64
B	105	137	—	—	33	—	—
Co	12	32	—	13.3	—	—	10.4
Ni	39	64	—	—	25	13	41
Zr	166	166	—	112	—	163	—

Notes: (3, etc.), number of samples averaged; major elements in wt. %, trace elements as ppm; —, not reported; LOI, loss on ignition less ($H_2O + CO_2$).
A. DSDP Site 270 greywacke below palaeosol (see text and Fig. 2).
B. DSDP Site 270 palaeosol (Fig. 2).
C. Metagreywacke, Robertson Bay Group (Harrington *et al.*, 1967).
D. Greywacke, Robertson Bay Group (Nathan, 1976).
E. Greywacke, Patuxent Formation, Pensacola Mountains (D.L. Schmidt, pers. comm.).
F. Greywacke, Cumberland Bay Formation, South Georgia (Clayton, 1982).
G. Greywacke, Greenland Group, New Zealand (Nathan, 1976).

implications for the evolution of the Ross Sea (Fitzgerald *et al.*, 1986). The present study investigates geochemical characteristics of this inferred palaeosol and of its parent greywacke of the breccia deposit.

The shipboard log of core involved in the study is given by The Shipboard Scientific Party (1975). Photographs of recovered core show: (1) the sharp unconformable upper contact of the sedimentary breccia with a thin carbonaceous sandstone, below the mid-Oligocene greensand; (2) the highly altered palaeosol and its gradation downward into less altered breccia; and (3) the basement gneiss and marble, for which the contact with the overlying breccia was not recovered (The Shipboard Scientific Party, 1975; but note that labels for Cores 44-3 and 44-4 are interchanged).

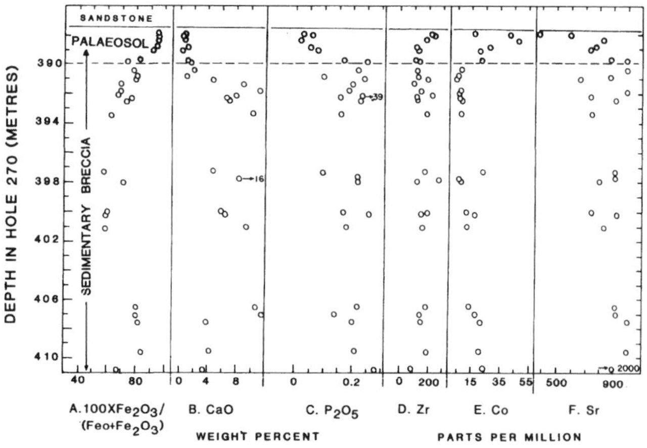

Fig. 2. Chemical variation with depth of selected elements in greywacke and palaeosol of the sedimentary breccia unit, DSDP Site 270.

Geochemistry

Major elements were analysed by the method of Shapiro (1975), and Rb, Sr, B, Co, Ni and Zr were analysed by quantitative emission spectroscopy.

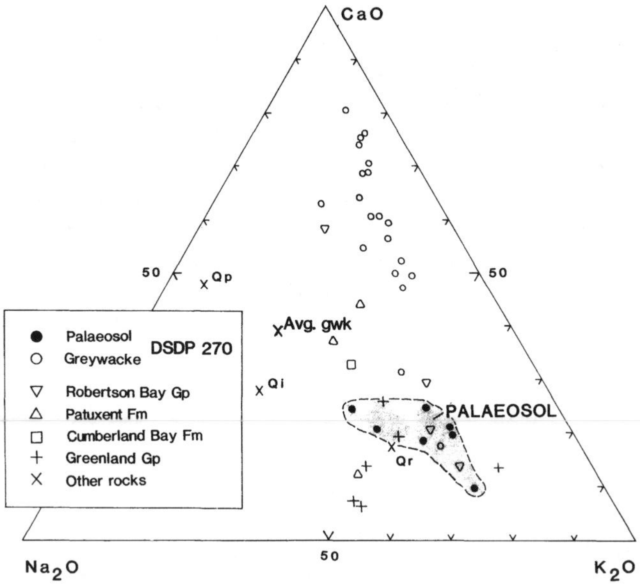

Fig. 3. CaO–Na_2O–K_2O diagram showing comparisons of DSDP Site 270 greywacke and palaeosol compositions. Data sources: Table 1; average greywacke (gwk) (Pettijohn, 1963); average quartz-rich (Qr), quartz-intermediate (Qi) and quartz-poor (Qp) greywacke (Crook, 1974).

Greywacke composition

The average composition of 20 greywacke samples from the sandy matrix between larger clasts, and from sandy intervals (e.g. Ford & Barrett, 1975, Figs. 7 & 8) of the breccia below the inferred palaeosol is given in Table 1, A. Abundances of elements vary greatly but non-systematically with depth (Fig. 2). Compared with average greywacke, the Site 270 greywacke is unusually calcareous (Fig. 3) and has unusually high K_2O/Na_2O ratios approximating that of average arkose (Fig. 4).

Palaeosol composition

The average composition of seven samples of highly altered and poorly lithified clay-rich greywacke from the upper

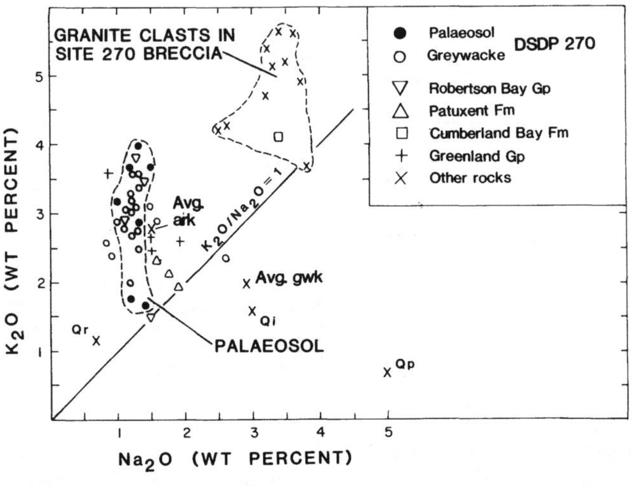

Fig. 4. K_2O v. Na_2O variation diagram showing comparisons of DSDP Site 270 greywacke and palaeosol compositions. Data sources and symbols as in Fig. 3; average arkose (ark) (Crook, 1974).

3 m of the breccia is given in Table 1, B. Abundances of some elements have narrower ranges and more regular variation with depth than in the underlying greywacke (Fig. 2*B*, *C* & *D*), in contrast to others (e.g. Fig. 2*E* & *F*). Compositional variations across the 390 m hole depth (Fig. 2) geochemically mark the approximate base of the inferred palaeosol of Ford & Barrett (1975). The palaeosol is much less calcareous than the underlying greywacke (Table 1, A & B; Figs 2 & 3), but K_2O/Na_2O ratios are similar (Fig. 4).

Discussion

Sedimentary breccia

A multilithological provenance of the sedimentary breccia is indicated by the great variety of pebble-size and larger clasts, among which marble, gneiss and granitic rock are abundant (Ford & Barrett, 1975). Much of the sand-size and smaller material was derived either by breakdown of such coarse clasts or by erosion from a similar source, which probably accounts for the greywacke's high K_2O/Na_2O ratio and similarity with granitoid clast ratios (Fig. 4), as well as its unusually high CaO and CO_2 contents (Table 1; Fig. 3).

Palaeosol

The general similarity between lithologies of relict coarse clasts in the altered zone (though marble is rare), and of clasts in the underlying sedimentary breccia, indicates a probable in-place derivation from parent breccia. The zone is clay- and chlorite-rich, has a mostly light greenish-grey colour, and shows strong subhorizontal fissility, but lacks bedding. Diagnostic criteria, such as root traces, burrows and soil horizons, that identify palaeosols (Harden, 1987; Retallack, 1988) have not been found in the Site 270 core, but biological material may not be preserved in ancient soils (Wright, 1986) and core recovery was incomplete. The altered zone does, however, show some features reported for Cretaceous palaeosols of Canada (Leckie & Foscolos, 1986), such as drab colouration, lack of primary sedimentary structures, cracking possibly due to dehydration shrinkage, and gradual downward decrease in extent of alteration.

In the absence of more diagnostic features, comparisons of chemical variations with known soils can identify an ancient soil, even if metamorphosed (Barrientos & Selverstone, 1987). Chemical variations (Fig. 2) reflect the gradual downward decrease in alteration in the Site 270 altered zone. Different elements show changes in trends or breaks in trends at slightly different depths (Fig. 2). Using relatively immobile TiO_2 as a normalizing basis, concentration ratios (CR) show apparent gains and losses in soils according to the relation CR = $(element_w/element_p)/(TiO_{2_w}/TiO_{2_p})$, where w is the constituent of weathered rock (palaeosol) and p is the constituent of parent (Feakes & Retallack, 1988). Element gain or loss shown by the CR value is apparent because it may reflect loss or gain of other elements. The average CR value based on Table 1 is highest for Co (CR = 2.5, suggesting strong gain), followed by Ni (CR = 1.5) and Al_2O_3 and B (CR = 1.2) and is lowest for CaO

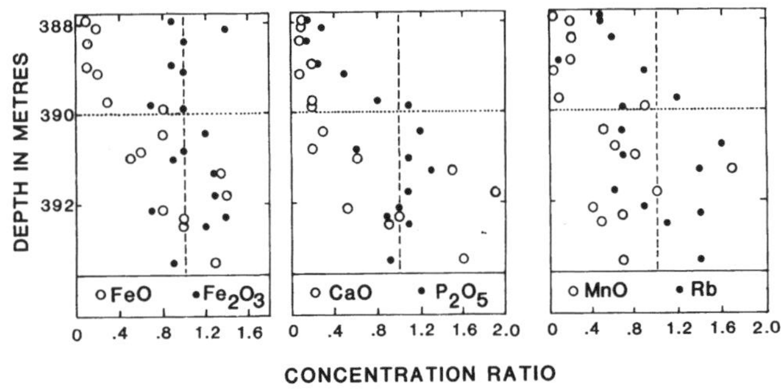

Fig. 5. Variation with depth of chemical concentration ratios (CR) in upper part of DSDP Site 270 sedimentary breccia unit. Dotted line, approximate base of palaeosol layer; dashed vertical line, CR = 1, for reference.

and CO_2 (CR = 0.1, suggesting strong loss), followed closely by MnO (CR = 0.2), FeO (CR = 0.3), P_2O_5 (CR = 0.4), and Rb, Sr and MgO (CR = 0.6–0.8). CR variations with depth shown in Fig. 5 mark apparent losses of some elements (except Fe_2O_3) across the altered zone and show the downward chemical gradation into less altered rocks in the upper part of the sedimentary breccia deposit.

Interpretation and conclusions

Palaeosol

The geochemical trends of Figs 2 & 5 and comparisons with known soils (Harden, 1987) strongly support the palaeosol nature of the Site 270 altered zone proposed by Ford & Barrett (1975). The apparent gains or losses of elements (Fig. 5; and CR averages) approximate to variations found in soils formed by humid subaerial weathering and leaching by downward percolating water (Retallack, 1986; Harden, 1987). CaO was leached to greater depths than other elements (Fig. 5). The apparent Al > Si gain reflects relative gain or loss rates in clay- and silt-size fractions (Harden, 1987). In lack of reddening, the apparent Fe oxidation of the palaeosol (Fig. 2*A*) reflects $Fe \gg Fe_2O_3$ leaching (Fig. 5). The palaeosol does not show overall accumulation of Zr, as expected in leached soils (Harden, 1987), but Zr appears to have accumulated near its top (Fig. 2*D*). The chemical variations in the Site 270 palaeosol are much more like those of soils produced under non-arid than arid conditions in which Ca, Mg, K and Na accumulate (Dregne, 1976).

Age and origin of the sedimentary breccia

The age of the sedimentary breccia and time of formation of its palaeosol, which may have been much later, are known only to be older than mid-Oligocene. The source of the deposit must have been nearby and deposition rapid, with little water transport, as shown by the generally poor sorting and coarseness and angularity of the clasts. The source was similar to the Site 270 basement, although this lacks known granitoids (only 4 m core recovery), and to Early Palaeozoic terrane (e.g. Skelton Group; Skinner, 1982), including granitoids, of the Transantarctic Mountains near McMurdo Sound (Ford & Barrett, 1975). The breccia possibly records local uplift, as a talus deposit, associated with an early stage of Transantarctic Mountains uplift beginning about 50 m.y. ago (Gleadow & Fitzgerald, 1987). However, a conceivably older age makes correlations possible with units such as the calcareous (?)Cambrian 'Teall Greywacke' of Skinner (1982) or other greywackes of Table 1, though lacking their common low-grade metamorphic features. The greywacke of the Site 270 breccia differs greatly in composition from other greywackes of Table 1, owing probably to a more calcareous source.

The subaerial nature of the Site 270 palaeosol and probably of its parent breccia indicates that this part of the Ross Sea was above sea level before mid-Oligocene time. The palaeosol thickness suggests lengthy emergence of the breccia, and indirect evidence points to much longer emergence. If the Devonian–Jurassic Beacon Supergroup and the Jurassic Ferrar Supergroup of the Transantarctic Mountains were originally present not far from the Site 270 area, as might be expected, the absence of their lithologies (e.g. quartzite and sandstone of the Beacon Supergroup and basalt and dolerite of the Ferrar Supergroup) as clasts in the breccia indicates removal by long subaerial erosion prior to breccia deposition, unless the breccia is pre-Devonian.

Tectonic setting

The composition of greywacke portions of the sedimentary breccia provides inconclusive evidence of tectonic setting (Fig. 6). Whereas the high K_2O/Na_2O ratios (> 1; Fig. 4) are characteristic of quartz-rich greywacke of Atlantic-type (rift) margins, the low SiO_2 content (57%; Table 1, A) is similar to that (58%) of quartz-poor greywacke of magmatic arcs (Crook, 1974; and Table 1, F).

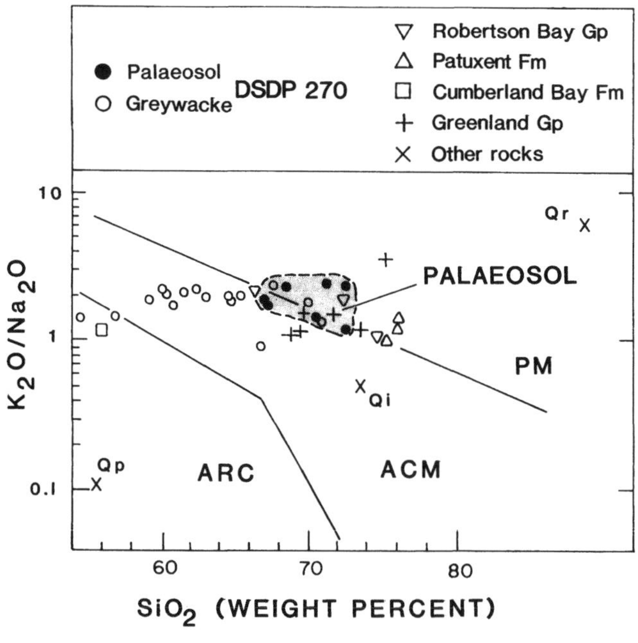

Fig. 6. Tectonic discrimination diagram (Roser & Korsch, 1986) showing comparisons of DSDP Site 270 greywacke and palaeosol compositions. From data normalized volatile-free. Data sources and symbols as in Fig. 3. PM, passive margin; ACM, active continental margin; ARC, oceanic island arc margin.

The highly calcareous nature of the greywacke (Table 1, A) strongly influences relative abundances of non-calcareous elements, particularly SiO_2. The palaeosol, from which the calcareous material was largely removed (Fig. 5), is probably a better basis than the original greywacke for making comparisons. It would have an average SiO_2 content of about 70%, approximating to quartz-rich greywacke (Crook, 1974), if abundances (Table 1, B) are normalized volatile-free and adjusted for apparent changes (e.g. minor Al_2O_3 gain and large FeO loss). The Ca-depleted greywacke represented by the palaeosol is of a quartz-rich type in Fig. 3 and shows a passive margin setting in Roser & Korsch's (1986) discrimination diagram, though slightly overlapping the field of active continental margin greywacke (Fig. 6). A similar setting is indicated using Bhatia's (1983) discrimination diagrams. Quartz-rich greywacke is characteristic of basin deposits along rifted edges of continents (Crook, 1974; Bhatia, 1983), such as in the Ross Sea (Cooper, Davey & Hinz this volume, p. 285). The greywacke composition represented by the palaeosol is thus compatible with an origin of the sedimentary breccia related to Late Cretaceous or Palaeogene rift activity (Fitzgerald *et al.*, 1986); this is more likely than the magmatic arc origin suggested by the composition of the little-weathered greywacke.

Acknowledgements

Chemical analysis of major elements by Hezekiah Smith and trace elements by Karen Duvall (B, Co, Ni, Sr and Zr) and Judith Kent (Rb) are appreciated. Core samples were obtained through the US National Science Foundation's Deep Sea Drilling Project. D.L. Schmidt provided use of unpublished analyses of greywacke of the Patuxent Formation (Table 1, E). Discussions and comments on various manuscript stages by J.W. Harden, J.C. Dohrenwend, T.E. Moore, A.K. Cooper, K.S. Kellogg, P.J. Barrett and J.W. Thomson provided much help.

References

Barrientos, X. & Selverstone, J. (1987). Metamorphosed soils as stratigraphic indicators in deformed terranes: an example from the eastern Alps. *Geology*, **15(9)**, 841–4.

Bhatia, M.R. (1983). Plate tectonics and geochemical composition of sandstones. *Journal of Geology*, **91(6)**, 611–27.

Clayton, R.A.S. (1982). A preliminary investigation of the geochemistry of greywackes from South Georgia. *British Antarctic Survey Bulletin*, **51**, 89–109.

Crook, K.A.W. (1974). *Lithogenesis and Geotectonics: the Significance of Compositional Variation in Flysch Arenites (graywackes)*, pp. 304–10. Tulsa, Oklahoma; Society of Economic Paleontologists and Mineralogists, Special Publication No. 19.

Davey, F.J. (1987). Geology and structure of the Ross Sea region. In *The Antarctic Continental Margin: Geology and Geophysics of the Western Ross Sea*, Earth Science Series, vol. 5B, ed. A.K. Cooper & F.J. Davey, pp. 1–15. Houston; Circum-Pacific Council for Energy and Mineral Resources.

Dregne, H.E. (1976). *Soils of Arid Regions*. Amsterdam; Elsevier Scientific Publishing Company. 237 pp.

Feakes, C.R. & Retallack, G.J. (1988). Recognition and chemical characterization of fossil soils developed on alluvium; a Late Ordovician example. *Geological Society of America Special Paper*, **216**, 35–48.

Fitzgerald, P.G., Sandiford, M., Barrett, P.J. & Gleadow, A.J.W. (1986). Asymetric extension associated with uplift and subsidence in the Transantarctic Mountains and Ross Embayment. *Earth and Planetary Science Letters*, **81(1)**, 67–78.

Ford, A.B. & Barrett, P.J. (1975). Basement rocks of the south-central Ross Sea, Site 270, DSDP Leg 28. In *Initial Reports of the Deep Sea Drilling Project*, *28*, ed. D.E. Hayes, L.A. Frakes *et al.*, pp. 861–8. Washington, DC; US Government Printing Office.

Gleadow, A.J.W. & Fitzgerland, P.G. (1987). Uplift history and structure of the Transantarctic Mountains: new evidence from fission track dating of basement apatites in the Dry Valleys area, southern Victoria Land. *Earth and Planetary Science Letters*, **82(1)**, 1–14.

Harden, J.W. (1987). Soils developed in granitic alluvium near Merced, California. *US Geological Survey Bulletin*, **1590A**, 65 pp.

Harrington, H.J., Wood, B.L., McKellar, I.C. & Lenssen, G.J. (1967). Topography and geology of the Cape Hallett District, Victoria Land, Antarctica. *New Zealand Geological Survey Bulletin*, N.S., **80**, 100 pp.

Leckie, D.A. & Foscolos, A.E. (1986). Palaeosols and Late Albian sea level fluctuations: preliminary observations from the north-eastern British Columbia foothills. *Geological Survey of Canada, Paper* **86-1B**, 429–41.

McDougall, I. (1976). Potassium–argon dating of glauconite from a greensand drilled at Site 270 in the Ross Sea, DSDP Leg 28. In *Initial Reports of the Deep Sea Drilling Project*, *36*, ed. P.F. Barker, I.W.D. Dalziel *et al.*, pp. 1071–2. Washington, DC; US Government Printing Office.

Nathan, S. (1976). Geochemistry of the Greenland Group (Early Ordovician), New Zealand. *New Zealand Journal of Geology and Geophysics*, **19(5)**, 683–706.

Pettijohn, F.J. (1963). Chemical composition of sandstone – excluding carbonate and volcanic sands. In *Data of Geochemistry*, 6th edn, ed. M. Fleischer. *US Geological Survey Professional Paper*, **440-S**, 21 pp.

Retallack, G.J. (1986). The fossil record of soils. In *Paleosols, their Recognition and Interpretation*, ed. V.P. Wright, pp. 1–57. Princeton; Princeton University Press.

Retallack, G.J. (1988). Field recognition of paleosols. *Geological Society of America Special Paper*, **216**, 1–20.

Roser, B.P. & Korsch, R.J. (1986). Determination of tectonic setting of sandstone–mudstone suites using SiO_2 content and K_2O/Na_2O ratio. *Journal of Geology*, **94(5)**, 635–50.

Shapiro, L. (1975). Rapid analysis of silicate, carbonate, and phosphate rocks. *US Geological Survey Bulletin*, **1401**, 76 pp.

Skinner, D.N.B. (1982). Stratigraphy and structure of low-grade metasedimentary rocks of the Skelton Group, southern Victoria Land – does Teall Greywacke really exist? In *Antarctic Geoscience*, ed. C. Craddock, pp. 555–63. Madison; University of Wisconsin Press.

The Shipboard Scientific Party. (1975). Sites 270, 271, 272. In *Initial Reports of the Deep Sea Drilling Project*, *28*, ed. D.E. Hayes & L.A. Frakes *et al.*, pp. 211–334. Washington, DC; US Government Printing Office.

Wright, V.P. (1986). Introduction. In *Paleosols, their Recognition and Interpretation*, ed. V.P. Wright, pp. x–xiv. Princeton; Princeton University Press.

Extensive volcanism and related tectonism beneath the western Ross Sea continental shelf, Antarctica: interpretation of an aeromagnetic survey

J.C. BEHRENDT[1], H.J. DUERBAUM[3], D. DAMASKE[3], R. SALTUS[1], W. BOSUM[3] & A.K. COOPER[2]

1 US Geological Survey, Denver, CO 80225, USA
2 US Geological Survey, Menlo Park, CA 94025, USA
3 Federal Institute for Geosciences and Natural Resources, Hannover, 51, Federal Republic of Germany

Abstract

A state of the art 50000 km aeromagnetic survey (the first in Antarctica) was flown over northern Victoria Land and the western Ross Sea continental shelf in 1984–85. Our interpretations over the continental shelf complement previous interpretations based on multichannel seismic-reflection, refraction, gravity and magnetic profiles and allowed us to extend those results over a broad area away from the ship tracks. The aeromagnetic map indicates three discontinuous N-striking zones along the Terror rift in the Victoria Land basin (VLB) of about 100 short-wavelength (1–10 km), 20–500 nT anomalies that appear to be caused by submarine volcanoes and subvolcanic intrusions. The late Mesozoic(?)–Cenozoic VLB has an 80–100 km wide, 80–100 nT negative anomaly centered over the Terror rift that extends north from the Ross Ice Shelf for 300 km, probably caused by relief on the magnetic basement underlying the basin combined with the demagnetizing effect of an upwarped Curie isotherm associated with this active rift zone. The basement beneath the northern Coulman high (or the Southern Cross Terrane) bordering the VLB on the north is characterized by a number of 15–40 km wide anomalies having amplitudes > 200 nT including the 200 km long ENE-striking 1700 nT Polar 3 anomaly. These anomalies are probably caused by (late Cenozoic?) magnetic magmatic rocks buried in at least one case beneath the 1–2 km thick Cenozoic sedimentary section defined by seismic-reflection data. The Polar 3 anomaly may represent an offset in the Terror rift. The Central trough, a 6 km deep sedimentary basin, does not appear to have a significant long-wavelength magnetic anomaly.

Aeromagnetic survey

Widely spaced aeromagnetic profiles, flown over the Ross Sea area in 1960–61 and 1963–64 (Behrendt, 1964; Behrendt, Cooper & Yuan, 1987), indicated short-wavelength, high-amplitude anomalies that were interpreted to be caused by volcanic rocks; interpretations indicated sedimentary rocks several kilometres thick in places. More recently, marine magnetic gradiometer observations and multichannel seismic-reflection data over the western Ross Sea were combined to study the Cenozoic volcanic and subvolcanic structures, the Coulman high, and the Victoria Land basin along the ship (*S.P. Lee*) track lines surveyed in February, 1984 (Figs 1 & 2).

During the 1984–85 austral summer, the Federal Republic of Germany (FRG), Federal Institute of Geosciences and Natural Resources (BGR) and the US Geological Survey (USGS) conducted a cooperative aeromagnetic survey over North Victoria Land and the western Ross Sea as part of the German Antarctic North Victoria Land Expedition IV (GANOVEX IV) and the US Antarctic Research Program (USARP). This was the first, and so far only, aeromagnetic survey made in Antarctica using high precision navigation, and closely spaced lines and tie lines (for adjustment of field level time variations), methods standard over land in the US (and elsewhere) for 40 years. (In 1946, USGS commenced comparable routine aeromagnetic surveys in the US, with closely spaced, accurately positioned track lines (e.g. 2 × 20 km).) The survey was flown by two ski-equipped Dornier 228-100 aircraft. The E–W survey lines were nominally spaced 4.4 and 8.8 km apart with N–S tie lines at about 22 km intervals. A total of 50000 km of survey lines was completed.

Radio transponder beacons (located by transit satellite), installed at nine locations along the Transantarctic Mountains and on Franklin and Ross islands (Fig. 1) in the Ross Sea,

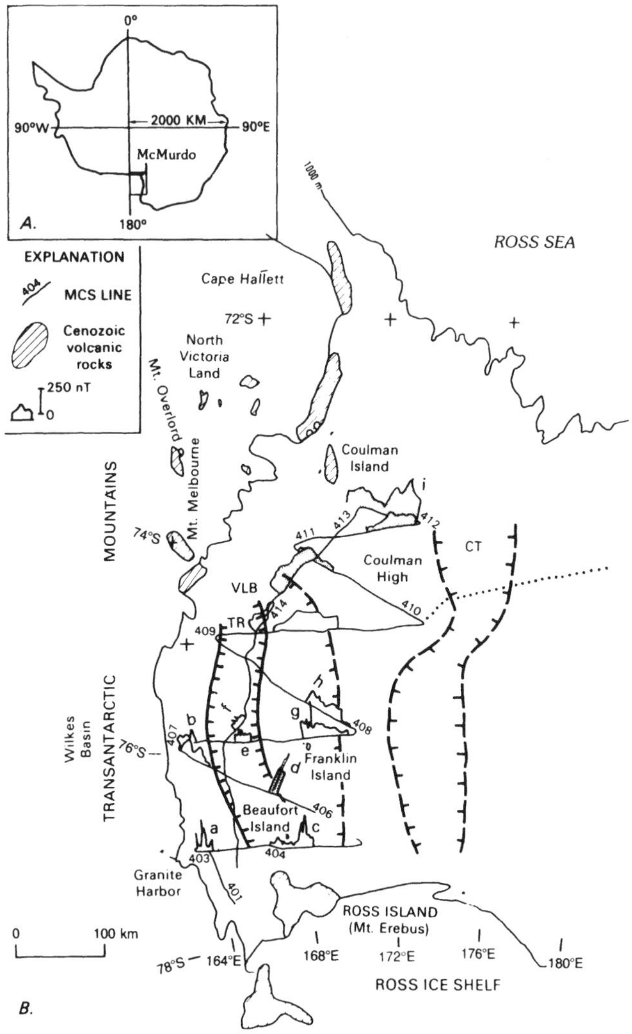

Fig. 1. (A) Index map of Antarctica showing area covered by offshore part of aeromagnetic survey; (B) Map of the western Ross Sea continental shelf and adjacent land showing marine magnetic anomalies, along multichannel seismic-reflection profiles (MCS) (collected by *S.P. Lee* in (1984). Geographic and geological structural features are modified from (Cooper *et al.*, 1987). The Victoria Land basin (VLB), Terror rift (TR), and Central trough (CT) are indicated. The Coulman high extends west and south of its labelled position. Mount Melbourne and Mount Erebus are active volcanoes. The 1000 m bathymetric contour marks the approximate edge of the continental shelf bordering the Ross Sea near line 419 at the north.

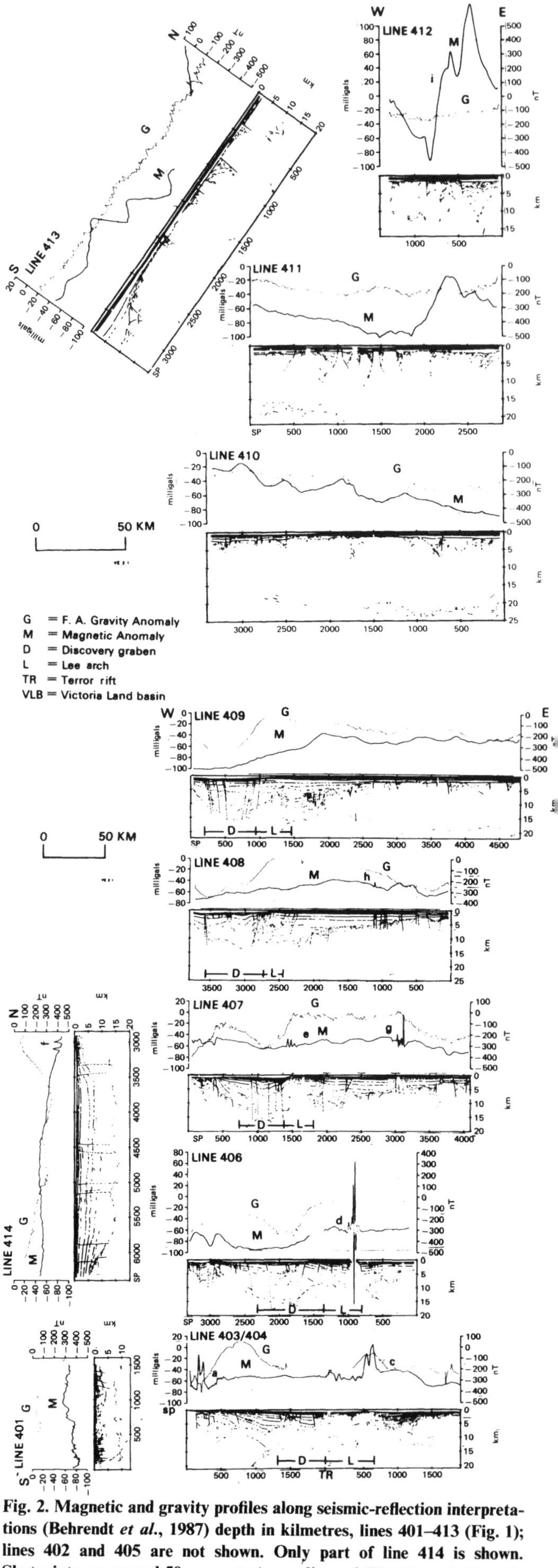

Fig. 2. Magnetic and gravity profiles along seismic-reflection interpretations (Behrendt *et al.*, 1987) depth in kilmetres, lines 401–413 (Fig. 1); lines 402 and 405 are not shown. Only part of line 414 is shown. Shotpoints are spaced 50 m apart. Anomalies *a–i* (Fig. 1) are indicated.

supplemented by doppler radar and continuous photography over the mountains, provided navigation control supplemented by VLF–OMEGA beyond 300 km range. The overall absolute error for the two sources as estimated by the contractor (Bachem & Boie, 1985) is: (1) radio transponder–Transit satellite, ± 50 – ± 200 m; (2) VLF–OMEGA, ± 500–1000 m at the eastern edge of the survey.

The International Geomagnetic Reference Field (IGRF) 1985.0 for 660 m (the flight elevation over the Ross Sea) was used to define the regional field over the Ross Sea (Bachem & Boie, 1985) that was removed from the survey data. The adjusted profile data were gridded at a 440 m spacing and contoured at 5 nT intervals at 1:250000 scale (see Fig. 3). The small-scale colour contour map (Fig. 4) shows the data at a 10 nT interval. Bosum *et al.*, (this volume, p. 267) discuss the

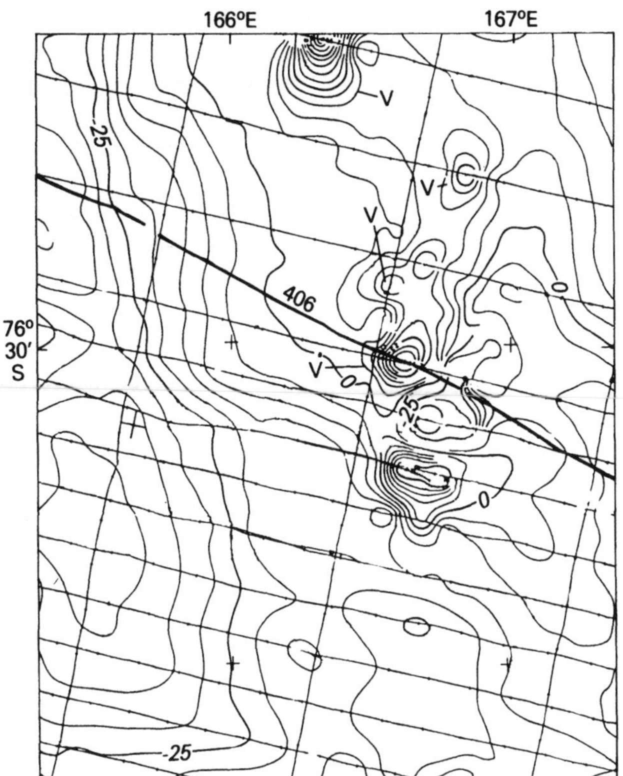

Fig. 3. Large-scale example of part of aeromagnetic survey (Fig. 4), showing flight lines. Note the 8.8 km line spacing (the result of time limitation) in the north contrasted with 4.4 km in the south. Contour interval is 5 nT. Location of *S.P. Lee* multichannel seismic-reflection line 406 is indicated. The short-wavelength semicircular anomalies (of the type labelled V) are interpreted as caused by submarine volcanoes and subvolcanic intrusions in the Terror rift zone. The north-trending 40 nT negative anomaly in the west is part of the 300 km long regional negative VLB anomaly. Compare with Fig. 4.

data reduction in more detail. In this report we present results and interpretations for the Ross Sea part of the survey (Fig. 1), concentrating on the western area covered by the *S.P. Lee* data (Behrendt *et al.*, 1987; Cooper, Davey & Behrendt, 1987). Bosum *et al.* (in press and this volume, p. 267) interpreted the aeromagnetic results over the mountains and ice-covered area to the west of these data.

Structure of the Victoria Land basin and adjacent terrane

The results of the aeromagnetic survey are in general accordance with those of the seismic surveys and supplement them over broad areas where the seismic grid lines are widely spaced (Figs 1 & 4) (Hinz & Block, 1984; Cooper *et al.*, 1987). Three N-trending basins, the Victoria Land basin (VLB), the Central trough (Fig. 1) and Eastern basin (not shown) were defined by the seismic surveys and are all part of the Cenozoic West Antarctic rift system extending throughout the Ross embayment and Marie Byrd Land. The complex, 150 km wide VLB is deepest (10 + km) along its western margin, in a 70 km wide zone named the Terror rift (Cooper *et al.*, 1987). The VLB is bounded on the west by the Transantarctic Mountains and in the north and east by the Coulman high defined from the reflection survey. The northern part of the Coulman high is underlain by the Southern Cross terrane (Bosum *et al.*, in press), which is very clearly defined by the aeromagnetic survey and magnetically characterized by conspicuous circular and extended anomalies. From seismic-reflection data we know this area is a basement platform cut by very narrow grabens and half-grabens, possibly of equivalent age to the VLB (Cooper *et al.*, 1987) and covered by 1–2 km of Cenozoic sediments.

Two episodes of extensional deformation have been proposed for the VLB region, based on seismic-reflection interpretations (Cooper *et al.*, 1987). The first episode commenced in Late Jurassic(?)–Early Cretaceous(?) time and formed the VLB half-graben. A second, more intense, extensional episode started during Palaeogene time, coeval with uplift of the Transantarctic Mountains (about 50 Ma BP) and produced the still active Terror rift and its associated volcanic activity and low-level seismicity. Short-wavelength, 200–1300 nT marine magnetic anomalies reported along the *S.P. Lee* track (Figs 1 & 2) along the west side of VLB, in the centre of the Terror rift, and in the eastern part of VLB, mark volcanic extrusions and associated intrusions which occurred from 50 Ma to the present (interpreted from comparisons with the multichannel seismic-reflection data; Behrendt *et al.*, 1987).

The adjacent 4 km high Transantarctic Mountains mark the boundary between East and West Antarctica and extend more than 3500 km across Antarctica. The active onshore uplift and associated volcanism of the Transantarctic Mountains (Fig. 1) and the offshore tectonism along the Terror rift system are all related, although there is no general agreement as to their cause. The general geological and tectonic setting is described more extensively in other reports (e.g., Cooper *et al.*, 1987), but the major surface features are clearly indicative of an active volcanic–tectonic zone. The aeromagnetic survey (Fig. 4) is interpreted in this context.

The rocks that are known to have high magnetization and to produce measurable magnetic anomalies in the adjacent land area to our survey have been discussed by several authors (Behrendt, 1964; Robinson, 1964; Pederson *et al.*, 1981; Behrendt *et al.*, 1987; Bosum *et al.*, in press). Table 1 shows a summary from Pederson *et al.* (1981) and Behrendt *et al.* (1987). The Jurassic Ferrar Dolerite, exposed in the Transantarctic Mountains, is quite magnetic and capable of producing significant anomalies (Pederson *et al.*, 1981). These diabases and their extrusive equivalents may be present in the deeper sedimentary section in VLB and may make a significant contribution to the mean susceptibility of the crust above the Curie isotherm, which produces the regional magnetic anomaly over the VLB.

Discussion

McMurdo Volcanic Group

The Cenozoic McMurdo Volcanic Group rocks (19 Ma–present; Kyle & Muncy, 1983), associated with tectonic activity along the Terror rift and the Transantarctic Mountains, are

Table 1. *Magnetic susceptibilities determined for rocks cropping out adjacent to the Victoria Land basin, as summarized from various sources by Pederson et al. (1981) and Behrendt et al. (1987)*

Location	Susceptibilities (cgs units)	Number of samples	Range (cgs units)
McMurdo Volcanic Group			
Ross Island lavas and flows	0.003	12	0.0007–0.018
Ross Island pyroclastic rocks	0.0002	4	0.00008–0.0038
Dry valley cones	0.001	44	0.00008–0.003
Ferrar dolerite	0.001	5	0.0008–0.012
Vida granite	0.00006	2	0.00004–0.00006
Olympus granite gneiss	0.00005	2	0.00003–0.00006
Beacon Supergroup sedimentary rocks	0.00002	1	

Susceptibility: 1 cgs unit $= \frac{1}{4\pi}$ SI unit.
Magnetization = susceptibility × magnetic field strength of about 0.64 Oe in the Ross Sea area.

very magnetic (Table 1). They produce anomalies of several thousand nT amplitude over Ross Island, at the south-west edge of the survey in Fig. 4, and adjacent areas of outcrop (Behrendt, 1964; Robinson, 1964; Pederson *et al.*, 1981; Behrendt *et al.*, 1987). A 350 nT negative anomaly is visible over the tip of Ross Island (Fig. 4). McMurdo volcanic rocks are inferred to be the source of most of the short-wavelength anomalies in the VLB area discussed in this report. Granites and granite gneisses in the Transantarctic Mountains west of VLB have very low susceptibilities, as inferred from only four samples (Table 1), and we would not expect these rocks to contribute significantly to magnetic anomalies in this area. On the other hand, gabbros, some granites, and diorites from the northern part of the survey over North Victoria Land (Fig. 1) have significant magnetizations (Bosum *et al.*, in press).

Geologically significant anomalies observed in the aeromagnetic data (Fig. 4) over the eastern Ross Sea range in amplitude and horizontal dimensions over three orders of magnitude (10–1000 nT and 3–300 km, respectively). Therefore, a single contour map (Fig. 4) is inadequate to display properly the complete range of anomalies but serves as a first approximation for the results presented here. Note that the contour interval is 5 nT between −50 and +50 nT, and is stretched beyond this range to show the highest amplitude anomalies. Fig. 3 illustrates the data for a small area of Fig. 4 in detail. About 100 short-wavelength anomalies, 20–500 nT in amplitude (of the type labelled V in Fig. 3) and caused by submarine volcanoes and subvolcanic intrusions (Figs 1 & 2) along the edges of the VLB and along the centre of the Terror rift (Behrendt *et al.*, 1987; Cooper *et al.*, 1987), can be seen to extend discontinuously through the length of the VLB. This is an order of magnitude greater number than those observed along the *S.P. Lee* track (Fig. 1). The associated volcanic and subvolcanic structures penetrate the sedimentary section (Fig. 2) along these zones, as was demonstrated (Behrendt *et al.*, 1987) using the seismic-reflection and magnetic data on the *S.P. Lee* profiles in Figs 1 & 2. Several marine magnetic anomalies crossed by the *S.P. Lee* have been analysed in detail (Behrendt *et al.*, 1987) and the interpretations of these anomalies can be extended using the aeromagnetic map.

Several of the short-wavelength anomalies observed in the *S.P. Lee* data and correlated with a submarine volcano (e.g. anomaly *d*, Figs 1 & 2) along the Lee arch within the Terror rift were missed by the aeromagnetic survey because of the 4.4–8.8 km line spacing. However, the approximately 20-fold coverage by aeromagnetic profiles, compared with the marine data, greatly extends the magnetic coverage. Although the lines are discretely spaced, continuous data long the lines allowed us to measure anomaly widths. Examination of 36 of these (assumed circular) short-wavelength volcanic anomalies along flight lines over the VLB, made using the large-scale map (e.g. Fig. 2), indicated ranges of widths from about 1.2 to 9.5 km, with a mean of 3.7 ± 1.9 km calculated. Therefore, considering the line spacing, flight elevation (660 m) and mean depth to the seafloor (500–900 m), which is about the shallowest possible magnetic source depth, we estimate that this survey has missed half or less of these volcanic anomalies.

Terror rift

A broad (80–100 km wide) 80–100 nT negative anomaly (crossed by A–A′, Fig. 4) extends over the Terror rift zone (Figs 1 & 2) of the VLB from the outer edge of the survey to the Coulman high (about 74°30′ S), where it is terminated abruptly by a significant change in anomaly patterns. Although three of the marine profiles partly crossed this anomaly (Behrendt *et al.*, 1987), which we here call the Victoria Land basin (VLB) anomaly, the aeromagnetic survey first defined its extent. One possible explanation for the source of this anomaly, is relief on the magnetic basement beneath the thick sedimentary section. A theoretical model fit (Fig. 5) to a profile across the VLB, constrained by the seismic-reflection data from line 414 (Figs 1 & 2), illustrates this interpretation. The steep boundaries in the

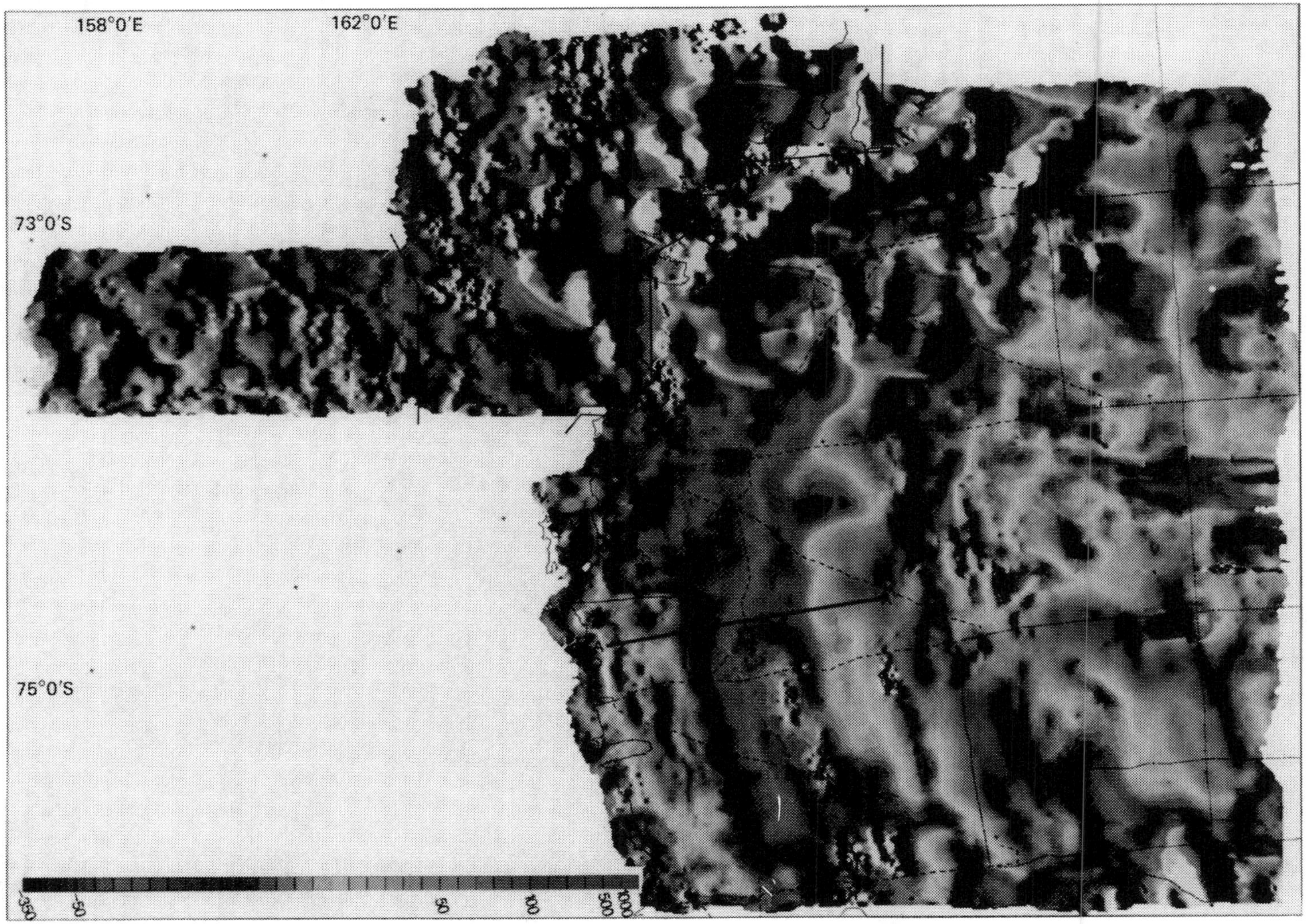

Fig. 4. Colour contour and shaded relief map of the aeromagnetic survey over the western Ross Sea continental shelf (see Fig. 1) and part of north Victoria Land. Contour interval is 5 nT from – 50 nT to + 50 nT, and is stretched beyond this range as indicated on colour bar on side to show highest amplitude anomalies. Multichannel seismic-reflection lines from *S.P. Lee* (west) and *Explora* (Hinz & Block, 1984) (east) are indicated. The coastline is shown at the west as well as on several islands; compare Fig. 1. Three flight elevations are separated by heavy lines; 2740 m, 3660 m and 610 m west–east, respectively. Location of profile AA′ modelled in Figs 5 & 6 is indicated.

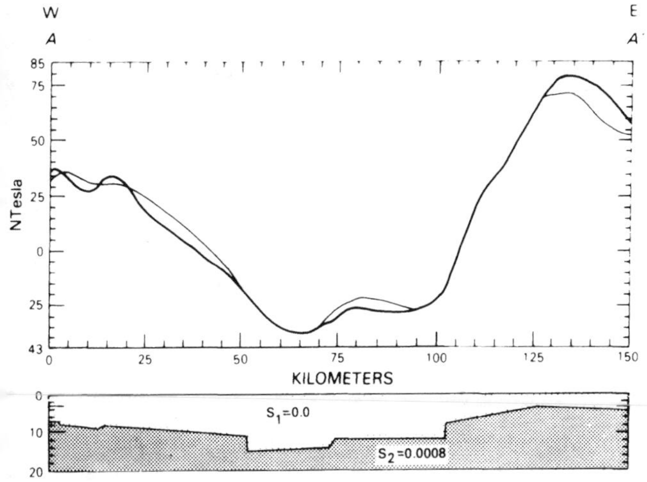

Fig. 5. Profile AA′ (Fig. 4) showing theoretical profile (thin line) fit to observed data (heavy line) for the indicated model of the Terror rift zone. The susceptibility of the basement rock is 0.0008 cgs. Induced magnetization is assumed but normal remanent magnetization in the present field direction is equally plausible and no doubt contributes as well. The model is constrained by MCS line 414, (Figs 1 & 2) which crosses AA′, and is interpreted as non-magnetic sedimentary rock, which fills the basin, contrasted with basement relief. The steep boundaries in the deepest part of the rift probably indicate faults. No vertical exaggeration.

model possibly indicate faults. This is consistent with the seismic-reflection interpretation (Fig. 2).

There is an alternative interpretation for the VLB anomaly which should be considered. This anomaly overlies an active rift zone with a large number of young volcanoes having associated subvolcanic intrusions penetrating the sedimentary section as discussed previously. Regionally high heat flow is indicated by this magmatic activity as well as interpretations of sparse thermal data from the *S.P. Lee* cruise discussed by Blackman, Von Herzen & Lawver (1987). Therefore we computed an alternate model (Fig. 6) illustrating the idea that thermal demagnetization beneath the rift may be a partial cause of the VLB anomaly. As in Fig. 5, a thick non-magnetic sedimentary section is part of the model but an upwarped Curie isotherm also contributes significantly. We infer that the more magnetic body at the base of the sedimentary section

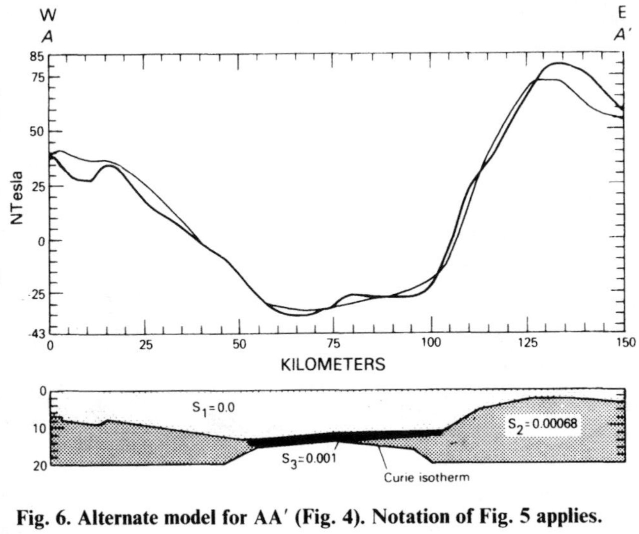

Fig. 6. Alternate model for AA′ (Fig. 4). Notation of Fig. 5 applies.

indicates the possible existence of magmatic rock, either Jurassic (i.e. early rift stage) volcanic flows, or sills, or late Cenozoic intrusions (i.e. associated with the McMurdo volcanic rocks and the active rifting). Either case is supported by the magnetic properties of rocks illustrated in Table 1.

Considering that other rifts contain thick (20–30 km) sequences of high-velocity reflectors interpreted as volcanic flows (Behrendt *et al.*, 1988), a significant part of the Terror rift fill (Fig. 2 and Cooper *et al.*, 1987) interpreted as comprising sedimentary rock may be volcanic. If this were the case, a model somewhere between those of Figs 5 & 6 (i.e. a thinner section of non-magnetic sedimentary rock and a shallower Curie isotherm) would be more appropriate.

Coulman high and Polar 3 anomaly

The basement underlying the northern Coulman high (also called the Southern Cross magnetic terrane) is characterized by a complex pattern of high-amplitude anomalies (200–1700 nT), 15–40 km in width. The highest amplitude anomaly observed in the survey is the 200 km long ENE-trending Polar 3 anomaly in the northern part of Fig. 4 (the striking red feature). The maximum amplitude on its several peaks is 1700 nT. We infer highly magnetic magmatic sources, probably similar to volcanic rocks of the McMurdo–Hallet volcanic group on Coulman Island (Hamilton, 1972) (Figs 1 & 4), are the underlying cause of this anomaly. The linearity of the Polar 3 anomaly suggests that it is associated with rifting; the anomaly may be the result of an offset in the Terror rift zone.

Multichannel seismic profile 412 at the north-east extreme of the *S.P. Lee* track lines (Figs 1, 2 & 4) shows no evidence of any volcano or buried volcanic flows or intrusion beneath the 800 nT anomaly crossed there just east of the Polar 3 anomaly. This anomaly, labelled *i* in the marine survey (Fig. 1), was 1200 nT where crossed at sea level. Two circular 15 km wide anomalies similar to that over Mount Melbourne (Figs 1 & 4 and Bosum *et al.*, this vol., p. 267) occur just east of that active volcano and probably have a similar source, as do other comparable width anomalies over the Coulman high. There is obviously a significant magnetic difference between these inferred volcanic rocks and those in the VLB area to the south. Possibly the greater areal extent of the anomalies in the northern Coulman high area is the result of the more competent basement rock through which the rocks causing the anomalies intruded, compared with the less competent thick sedimentary rock in the VLB (i.e. a greater amount of an intrusion within the sedimentary rock is in the form of low anomaly-producing sills surrounding a high anomaly-producing vertical intrusion).

The area to the east of the *S.P. Lee* track (Figs 1 & 4), which includes the Central trough crossed by the *Explora* profiles (Hinz & Block, 1984), is characterized by anomalies mostly of 100 nT or less amplitude and a few tens of kilometres extent. There is no apparent regional negative anomaly associated with the Central trough comparable with that over the VLB; this supports the hypothesis of an upwarped Curie isotherm contributing to the VLB anomaly. A prominent exception to the generally smooth magnetic field, in the south-eastern quarter of Fig. 4, occurs at the south-east corner of the survey where a 500 nT NNW-trending anomaly suggestive of buried volcanic rock can be seen (Fig. 3). Other anomalies of similar appearance (Behrendt *et al.*, 1987) exist south and east of our survey area (in aeromagnetic profiles over the Ross Ice Shelf and Marie Byrd Land) and may also be caused by buried volcanic rocks. All are probably associated with late Cenozoic tectonic activity in the West Antarctic rift system.

The results of the aeromagnetic survey clearly illustrate the existence of many important tectonic, volcanic, and magmatic features associated with the complex VLB and the Coulman high. The aeromagnetic data significantly complement and extend the results of the marine geophysical surveys (Hinz & Block, 1984; Behrendt *et al.*, 1987; Cooper *et al.*, 1987).

Acknowledgements

We thank the US Navy, Air Force, and Coast Guard for providing extensive flights of people and cargo from New Zealand to Antarctica, and within Antarctica. The US National Science Foundation (NSF) supplied these logistics by a grant to the US Geological Survey. US base facilities at McMurdo and South Pole were used as well as the FRG Gondwana camp, the staging area for GANOVEX IV. Alfred Wegener Institute (FRG) provided the Polar 2 aircraft and other support. Prakla Seismos made the aeromagnetic survey and compiled the data working with BGR, USGS, and a number of subcontractors. We thank K. Moulton, D. Breshnahan, and E. Chang (NSF) USARP for their support in Antarctica; without their enthusiastic help and understanding, the aeromagnetic survey would not have been possible. New Zealand Antarctic Division at Scott Base and Christchurch provided invaluable assistance. The continuing interpretation is a joint effort of BGR and USGS, using facilities and people in Hannover, FRG and Denver, USA. Annette Yuan computed the theoretical magnetic profiles. One of the aircraft, Polar 2, was damaged in the field so that the bulk of the survey was flown by Polar 3. The crew of Polar 3, Herbert Hampel, Richard Moebius, and Joseph Schmidt, and the aircraft were tragically lost while flying across Morocco on the return from Antarctica.

References

Bachem, H.C. & Boie, D.C. (1985). Report (Prakla-Seismos GMBH, Hannover, FRG, 1985).

Behrendt, J.C. (1964). Distribution of narrow width magnetic anomalies in Antarctica. *Science*, **144(3641)**, 995–9.

Behrendt, J.C., Cooper, A.K. & Yuan, A. (1987). Interpretation of marine magnetic gradiometer and multichannel seismic reflection observations over the western Ross Sea Shelf, Antarctica. In *The Antarctic Continental Margin: Geology and Geophysics of the Western Ross Sea*, Earth Science Series, vol. 5B, ed. A.K. Cooper & F.J. Davey, pp. 155–77. Houston; Circum-Pacific Council for Energy and Mineral Resources.

Behrendt, J.C., Green, A.G., Cannon, W.F., Hutchinson, D.R., Lee, M.W., Milkereit, B., Agena, W.F. & Spencer, C. (1988). Crustal structure of the mid-Continent Rift system – result from GLIMPCE deep seismic reflection profiles. *Geology*, **16**, 81–5.

Blackman, D.K., von Herzen, R.P. & Lawver, L.H. (1987). Heat flow and tectonics in the Western Ross Sea, Antarctica. In *The Antarctic Continental Margin: Geology and Geophysics of the Western Ross Sea*, Earth Science Series, vol. 5B, ed. A.K. Cooper & F.J. Davey, pp. 179–90. Houston; Circum-Pacific Council for Energy and Mineral Resources.

Bosum, W., Damaske, D., Roland, R.N., Behrendt, J.C. & Saltus, R. (in press). Interpretation and results of the aeromagnetic anomalies of the Ross Sea/Victoria Land survey of GANOVEX IV, *Geologisches Jahrbuch*, **E38**.

Cooper, A.K., Davey, F.J. & Behrendt, J.C. (1987). Seismic stratigraphy and structure of the Victoria Land Basin, western Ross Sea, Antarctica. In *The Antarctic Continental Margin: Geology and Geophysics of the Western Ross Sea*, Earth Science Series, vol. 5B, ed. A.K. Cooper & F.J. Davey, pp. 27–76. Houston; Circum-Pacific Council for Energy and Mineral Resources.

Hamilton, W. (1972). The Hallett volcanic province, Antarctica. *US Geological Survey Professional Paper*, **456-C**, 1–62.

Hinz, K. & Block, M. (1984). Results of geophysical investigations in the Weddell Sea and in the Ross Sea, Antarctica. *Proceedings of 11th World Petroleum Congress, (London) 1983*, vol. 2, Geology Exploration Reserves, pp. 79–81. Chichester; John Wiley & Sons.

Kyle, P.R. & Muncy, H.L. (1983). The geology of the mid-Miocene McMurdo Volcanic Group at Mt Morning, McMurdo Sound, Antarctica. In *Antarctic Earth Science*, eds. R.L. Oliver, P.R. James & J.B. Jago, pp. 675. Canberra; Australian Academy of Science and Cambridge; Cambridge University Press.

Pederson, D.M., Montgomery, G.E., McGinnis, L.P., Ervin, C.P. & Wong, H.K. (1981). Aeromagnetic survey of Ross Island, McMurdo Sound, and the Dry Valleys. In *Dry Valley Drilling Project*, Antarctic Research Series, vol. 33, ed. L.D. McGinnis, pp. 7–25. Washington, DC; American Geophysical Union.

Robinson, E.S. (1964). Correlation of magnetic anomalies with bedrock geology in McMurdo Sound area, Antarctica. *Journal of Geophysical Research*, **69**, 4319–25.

Geochemistry and tectonic implications of lower-crustal granulites included in Cenozoic volcanic rocks of southern Victoria Land

R.I. KALAMARIDES & J.H. BERG

Department of Geology, Northern Illinois University, DeKalb, Illinois, 60115, USA

Abstract

Inclusions of lower-crustal granulites have been collected from Cenozoic volcanic rocks that have perforated the crust on both sides of the boundary between the uplifted Transantarctic Mountains (TM)) and the lowlying Ross embayment (RE). Major and trace element analyses of these granulites indicate that the RE lower crust is dominated by cumulates from basic melts of transitional or alkaline affinity. In contrast, the TM lower crust is dominated by basic-melt (or near-melt) compositions having a clearly calc-alkaline affinity. The TM and RE granulites plot in well separated fields on a $\delta^{18}O$–I_{Sr} diagram and are also dissimilar to the Jurassic Ferrar magmatism. Correlations between isotopic values and element concentrations as well as inter-element correlations are clearly evident in the TM granulite suite but are weak or absent within the RE granulite suite. These results strongly indicate that the composition of the lower crust under the Transantarctic Mountains is quite different from that under the Ross embayment. Because of the striking differences in the nature of the lower crusts, the current faulting and rifting associated with this topographic boundary apparently coincide with an ancient crustal suture.

Introduction

In the McMurdo Sound region of Antarctica (Fig. 1), inclusions of basic lower-crustal granulites have been ejected as bombs or erupted in lava flows of the Cenozoic Erebus volcanic province. We have collected inclusions from volcanic centres distributed across an area of 12000 km^2, spanning the boundary between East Antarctica (Transantarctic Mountains – TM) and West Antarctica (Ross embayment – RE). The abundance and wide areal distribution of these inclusions give them a vital role in delineating horizontal lithospheric variation across the uplifted Transantarctic Mountains and the lowlying Ross embayment. This information offers critical constraints on the origin of the Transantarctic Mountains and the tectonic evolution of this enigmatic part of Antarctica. We have recently reported profound isotopic differences in the lower crust across the boundary between East and West Antarctica (Kalamarides, Berg & Hank, 1987). Here we review those isotopic differences and report additional major and trace element data which reinforce the conclusion that there is a major chemical discontinuity in the lower crust across this topographic boundary.

Most interpretations for the origin of the Transantarctic Mountains and the boundary between East and West Antarctica may be explained in terms of a convergent-margin model, employing subduction, collision, and suturing, or a divergent-margin model, employing continental rifting and block faulting. We present isotopic and chemical evidence which suggests that both of these types of plate motions may have played important roles at different times in the tectonic evolution of this region of Antarctica.

Description of the lower-crustal inclusions

The most abundant inclusions brought to the surface are lower-crustal two-pyroxene granulites. The xenoliths range from massive to layered to gneissic. Although these rocks have undergone granulite-facies metamorphism (see Berg, this volume, p. 311), many of the RE inclusions display relict cumulate textures. Such relict cumulate textures are rare, however, in the TM inclusions and most display a distinctive mortar texture. Mortar texture is virtually absent in the RE suite of inclusions.

The samples used in this study cover as much of the range of two-pyroxene-granulite compositions as possible in order to examine the degree of compositional diversity. Samples were chosen with a wide range in mineralogy. In fact the two RE inclusions with the highest $^{87}Sr/^{86}Sr$ are not at all representative of RE granulites but were chosen for analysis because of their extreme compositional character and rarity within the RE suite. Indeed, the selection of samples for analysis has been skewed in a way that would over-emphasize the compositional similarities between the two suites. Major element, trace element and isotopic analyses of seven representative two-pyroxene lower-crustal granulites are listed in Table 1.

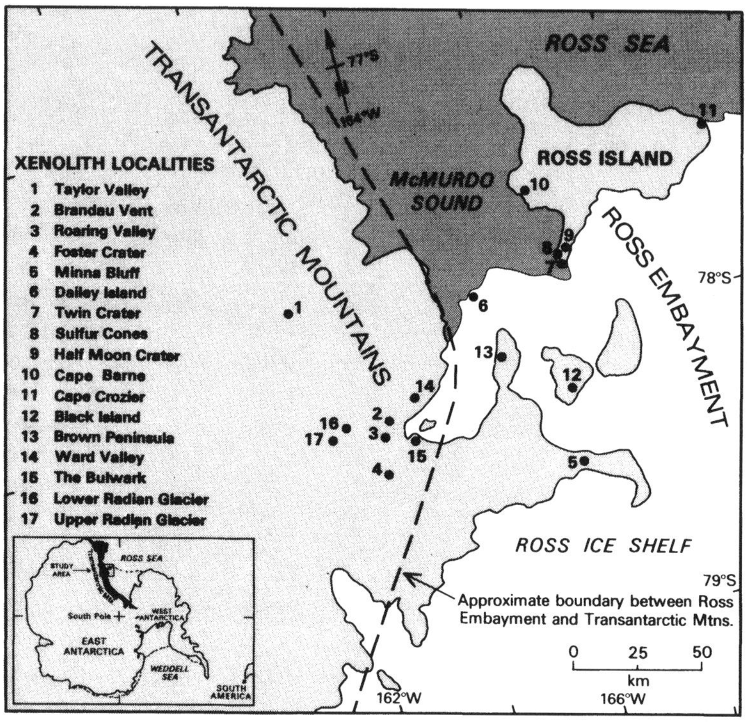

Fig. 1. Generalized map of the McMurdo Sound–Royal Society Range region showing the locations of the xenolith-bearing volcanic centres located on either side of the Ross embayment–Transantarctic Mountains boundary. Inset shows the location of study areas.

Discussion of isotope, major element and trace element systematics

The Rb–Sr systematics do not result in an isochron for either the RE or the TM xenoliths. This is not unreasonable because the samples have a wide areal distribution and are derived from varying depths within the crust (Kalamarides *et al.*, 1987). An 'age' of 900 Ma has been assumed in order to derive initial $^{87}Sr/^{86}Sr$ ratios. This age is probably close to the minimum age for the TM xenoliths, which we assume to be represented by the maximum age (675–950 Ma; Skinner, 1983) of the overlying metamorphic rocks (Kalamarides *et al.*, 1987). Because most of the samples have extremely low $^{87}Rb/^{86}Sr$ ratios essentially no error is introduced by errors in age assumptions. A $\delta^{18}O$–initial $^{87}Sr/^{86}Sr$ (I_{Sr}) diagram for the crustal xenoliths (Fig. 2) indicates that the RE and TM suites are distinctly different isotopically except for one RE inclusion. All of the TM inclusions have high $\delta^{18}O$ and some have unusually high values $\approx +13$). Moreover, the TM suite also exhibits an unusual negative correlation between $\delta^{18}O$ and I_{Sr}. In addition, the combination of $\delta^{18}O$ and I_{Sr} clearly differentiates the suites of xenoliths from any relationship to the Jurassic magmatic event manifested in the Ferrar dolerites and the Kirkpatrick basalts. The Sr-isotope data seem to preclude a relationship between the RE granulites and the Ordovician Granite Harbour Intrusives ($I_{Sr} = 0.707$–0.719; Vetter *et al.*, 1983; Borg *et al.*, 1987), but a possible relationship between the TM granulites and the Granite Harbour Intrusives cannot be ruled out solely on the basis of Sr isotopes. $\delta^{18}O$ values are not yet available for the latter. The elevated $\delta^{18}O$ of some of the RE granulites and the wide range in I_{Sr} for these samples, along with the predominant metamorphic textures, appear to deny any obvious cumulate-magma relationship between the granulites and their Cenozoic alkaline hosts.

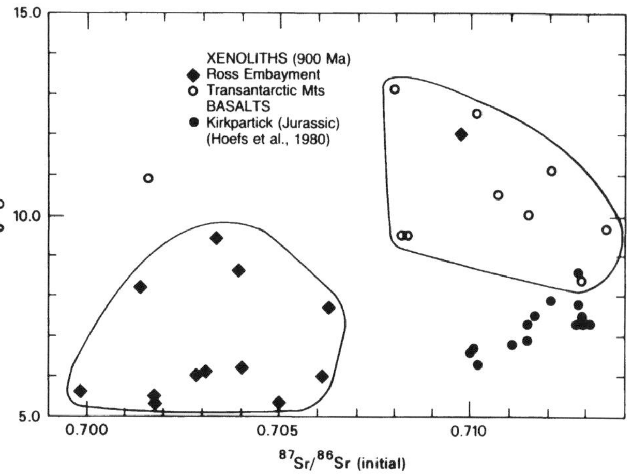

Fig. 2. A combined strontium initial ratio (see text for explanation of assumed age) and $\delta^{18}O$ plot. Note that the host lavas from the Ross embayment possess restricted isotopic values of $^{87}Sr/^{86}Sr = 0.7031$–0.7034 (Stuckless & Ericksen, 1976), and $\delta^{18}O$ values on either side of the RE–TM boundary = + 6.1 (from Kalamarides *et al.*, 1987).

More than 65 samples have been analysed for major elements and 15 trace elements. In terms of major elements, the bulk of the RE granulites appear to represent basic igneous cumulates (Table 1). The Fe/Mg ratios are almost uniformly low, although MgO + FeO varies widely, reflecting varying propor-

Table 1. *Chemical and isotopic analyses of lower-crustal xenoliths from the Ross embayment and Transantarctic Mountains*

Sample[a]	82-119	81-7	83-15	83-82	82-160	82-215	82-220a
SiO_2	43.52	47.82	48.81	48.18	51.55	53.53	62.91
TiO_2	0.23	0.09	0.26	1.55	0.87	0.80	1.74
Al_2O_3	10.04	31.27	13.02	19.92	16.69	15.72	15.23
FeO	11.71	0.61	5.26	10.54	11.00	8.44	3.58
MnO	0.19	0.01	0.10	0.19	0.16	0.14	0.05
MgO	24.44	1.54	15.06	6.05	5.99	4.98	1.83
CaO	9.04	15.97	16.32	10.46	9.89	11.14	10.98
Na_2O	0.77	2.54	1.06	2.59	3.23	3.42	1.08
K_2O	0.06	0.15	0.11	0.29	0.52	1.56	2.36
P_2O_5	0.01	0.00	0.00	0.25	0.13	0.28	0.25
Total[b]	100.01	100.00	100.00	100.02	100.03	100.01	100.01
Sc	26.0	12.8	41.0	27.9	18.3	21.0	8.8
Ba	14.8	35.0	30.0	106.0	357.0	745.0	3050.0
Co	160.0	6.3	60.0	30.0	25.3	29.0	7.1
V	91.0	23.0	109.0	177.0	146.3	123.0	66.0
Cr	1571.0	81.0	939.0	61.0	125.0	87.0	32.0
Ni	765.0	31.0	223.0	20.5	36.7	25.0	11.6
Cu	57.0	2.7	37.0	18.2	29.3	17.5	5.6
Zn	71.0	6.8	34.0	94.9	111.6	107.0	37.0
Rb	1.4	1.4	5.3	0.6	4.0	15.3	127.0
Sr	126.0	397.0	164.0	761.0	727.0	524.0	1112.0
Y	5.2	2.7	4.8	29.5	27.8	32.0	17.8
Zr	13.5	9.8	12.3	65.2	0.0	147.0	39.0
Nb	5.6	n.d.	4.9	9.2	10.1	9.4	13.9
Pb	2.2	0.8	1.6	2.1	1.9	6.7	4.2
Ga	7.4	n.d.	7.4	n.d.	n.d.	21.0	n.d.
$\delta^{18}O$[c]	6.1	6.0	6.2	9.5	11.1	10.0	13.1
$^{87}Rb/^{86}Sr$[d]	0.0227	0.0080	0.0533	0.0009	0.0176	0.1076	0.2600
$^{87}Sr/^{86}Sr$[e]	0.7033	0.7062	0.7047	0.7083	0.7123	0.7129	0.7113

[a]The first three are analyses of xenoliths from the RE; the last four are from the TM.
[b]Normalized to a water-free basis.
[c] $\pm$ 0.2‰, average mean deviation for all analyses, NBS-28 = 9.6‰.
[d]$2\sigma = 0.0014$, reported at the 95% confidence level.
[e]$2\sigma = 0.0009$, reported at the 95% confidence level.

tions of mafics and plagioclase. Likewise, the Al_2O_3 content is highly variable. Low TiO_2 and very low P_sO_5 contents of the RE samples further support the cumulate hypothesis. In contrast, the TM granulites as a group tend to resemble melt or near-melt compositions. TiO_2 and P_2O_5 are quite high, Al_2O_3 is not highly variable, and Fe/Mg is not high and can be correlated with variation in MgO + FeO. The cumulate versus melt characteristics of the two suites are perhaps best illustrated in a P_2O_5–Al_2O_3 diagram (Fig. 3). The highly variable Al_2O_3 and uniformly low P_2O_5 contents of the RE granulites are characteristic of basic cumulates, reflecting consistently small amounts of trapped liquid but variable proportions of cumulate phases. This contrasts strikingly with the variable but high P_2O_5 and relatively uniform Al_2O_3 contents of the TM granulites. Most of the TM granulites have Al_2O_3 contents in the range 14–20 wt %, which is similar to the range for most melt compositions normally encountered in the Earth's crust.

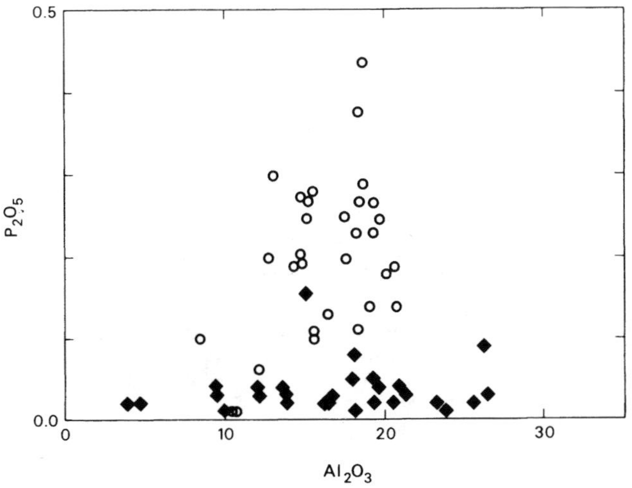

Fig. 3. P_2O_5–Al_2O_3 plot for the lower crustal inclusions. Open circles, Transantarctic Mountains; closed diamonds, Ross embayment.

Not only are the two suites distinguished on a cumulate versus melt basis, they also appear to have affiliations with

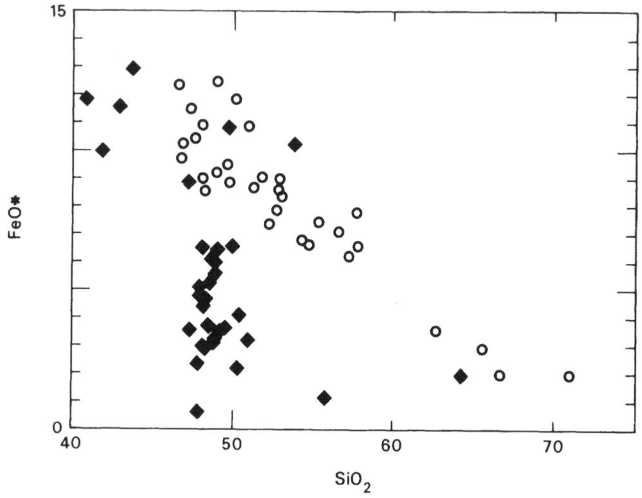

Fig. 4. FeO*–SiO_2 plot for the two-pyroxene lower crustal inclusions (symbols as in Fig. 3).

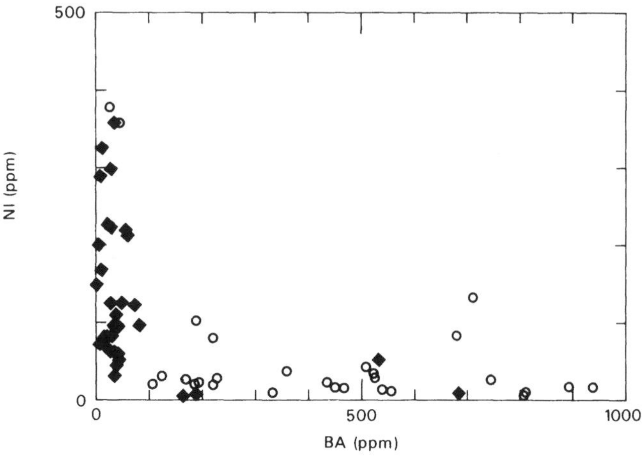

Fig. 6. Ni–Ba plot for the RE and TM xenoliths illustrating the striking differences in incompatible- and compatible-element concentrations between the two groups (symbols as in Fig. 3).

different magmatic series or associations. In Fig. 4, the strong negative correlation of total iron with silica for the TM granulites is characteristic of an andesitic or calc-alkaline series (Gill, 1981). The RE granulites show no obvious trend but instead maintain a relatively constant, and generally lower, SiO_2 content of around 48–49 wt %. In an AFM diagram (Fig. 5), the TM granulites again show the characteristics of a calc-alkaline fractionation trend (Gill, 1981; Ewart, 1982). The RE granulites show considerable scatter on the AFM diagram, largely a result of extremely low concentrations of MgO and FeO due to very high abundances of plagioclase in some granulites. Nevertheless, they do not appear to be similar to either analysed or modelled gabbroic cumulates associated with calc-alkaline magmatism (Snoke, Quick & Bowman, 1981; Kay & Kay, 1985) in that, for the most part, their Mg/Fe ratios are too high, especially for the amounts of total alkalies.

Trace element data are broadly consistent with the above conclusions. The RE granulites have generally low concentrations of incompatible elements, e.g. Ba < 50 ppm, Rb < 6 ppm, Y < 10 ppm, Zr < 15 ppm, and Nb < 6 ppm. Coupled with the high concentrations of compatible elements, such as Ni and Cr, in the more melanocratic samples, these data support a cumulate origin for the RE granulites. Assuming that the bulk of the Ti and Zr in these rocks originally resulted from trapped liquid, and that the trapped liquid accurately reflected the composition of the magma, the Ti/Zr ratios are apparently too high to be derived from calc-alkaline liquids (Pearce & Cann, 1973).

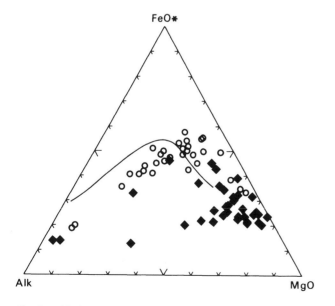

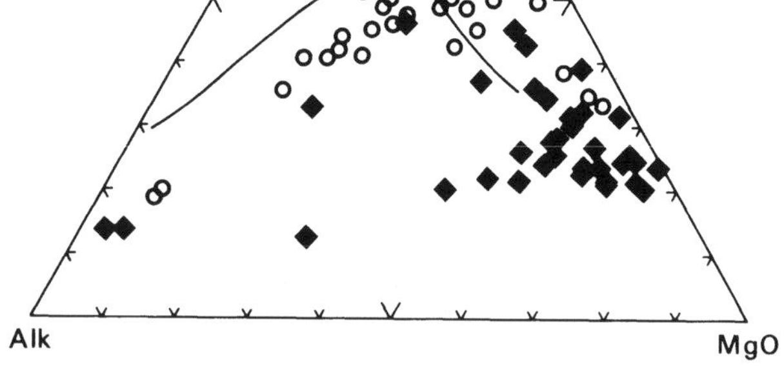

Fig. 5. AFM diagram for the lower-crustal two-pyroxene inclusion suites from either side of the RE–TM boundary (symbols as in Fig. 3).

The TM granulites have much higher concentrations of incompatible elements and generally lower concentrations of compatible elements relative to the RE granulites, consistent with the hypothesis that they represent melt compositions. Striking differences in compatible and incompatible elements between the two suites are exemplified in a Ni–Ba plot (Fig. 6). The low Ni contents (typically < 80 ppm) of the TM granulites further established their calc-alkaline character. Some of the TM granulites appear to be depleted in Rb (e.g. sample 83-82 in Table 1). It is conceivable that some of these granulites have at a late stage undergone a very small amount of partial melting, or the Rb has been scavenged by some other process. Because many of the granulites do not have low Rb contents and most have relatively high concentrations of other incompatible elements, any such depletion process clearly could not have been pervasive.

Distinctions between the two suites are also evident in plots of O and Sr isotopes versus major or trace elements (Kalamarides *et al.*, 1987). Fig. 7 is a typical example of the relevant relationships. For the most part, each suite tends to plot separately. The RE granulites typically show no, or at best weak, correlations between the isotopes and major or trace elements. In contrast, the TM granulites show moderate to good correlations, implying either a more homogeneous suite of samples (in terms of common origins) or a simpler history than the RE granulites. These correlations indicate that the high $\delta^{18}O$ for the TM samples are primary, rather than values elevated by some form of $\delta^{18}O$ enrichment due to hydrothermal processes. Furthermore, the correlations are suggestive of a mixing process being involved in the origin of the TM granu-

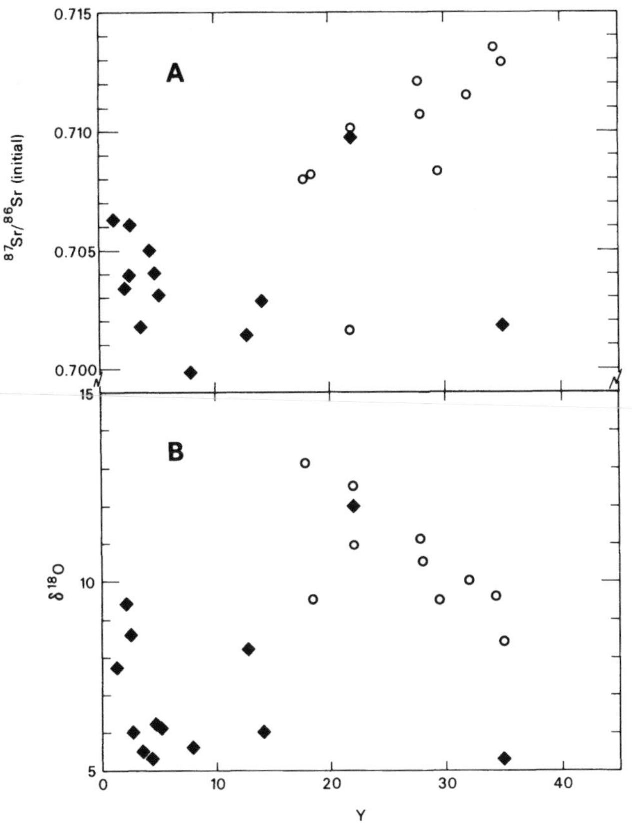

Fig. 7. (A) Y–I_{Sr} plot for the lower crustal xenoliths from either side of the RE–TM boundary (symbols as in Fig. 3). (B) Y–$\delta^{18}O$ plot for the same samples as in (A).

lites. The possible mixing trends are between an end-member having low SiO_2, low Fe/Mg, low incompatible elements (except Y), high compatible elements, *relatively* low $\delta^{18}O$, and high I_{Sr} and another end-member having high SiO_2, high Fe/Mg, high incompatible elements (except Y), low compatible elements, very high $\delta^{18}O$, and low I_{Sr}.

Conclusions

Data presented here and in Kalamarides *et al.* (1987) reveal a major chemical discontinuity in the lower crust across the boundary between East and West Antarctica in the vicinity of McMurdo Sound. The granulites in the lower crust of the Transantarctic Mountains (East Antarctica) have higher $\delta^{18}O$, $^{87}Sr/^{86}Sr$, and SiO_2 content, and despite granulite-facies metamorphism, retain geochemical characteristics of calc-alkaline liquids. This is in striking contrast to the more tholeiitic character of most lower crustal basic inclusions (Dostal, Dupuy & Leyreloup, 1980; Kay & Kay, 1981; Taylor & McLennan, 1985). The granulites in the lower crust of the Ross embayment (West Antarctica) have lower $\delta^{18}O$, $^{87}Sr/^{86}Sr$, and SiO_2 content, and have other geochemical signatures indicative of basic cumulates originating not from calc-alkaline magmas but perhaps from tholeiitic or alkaline parents.

Our data reveal that the lower crust of the Ross embayment is lower in silica and more basic than the lower crust of the Transantarctic Mountains, thus confirming the hypothesis proposed by Smithson (1972), based on gravity studies. Although continental rifting appears to adequately explain the topographic, structural, and geothermal aspects of the region (Kyle & Cole, 1974; Cooper & Davey, 1985; Berg & Herz, 1986), simple block faulting cannot explain the lateral discontinuity in the composition and nature of the lower crust across the RE–TM boundary. This compositional discontinuity is perhaps best explained if the Transantarctic Mountains front in this region marks the position of an ancient crustal suture which has been reactivated to form a zone of weakness during the present period of extensional tectonics. The suture must be as old as the Cambro-Ordovician Ross Orogeny, but could be Precambrian in age (Kalamarides *et al.*, 1987).

Acknowledgements

This research has been supported by National Science Foundation grants DPP-8213943, DPP-8317712, and DPP-8614071. Oxygen-isotope work was partially supported by a grant from the American Philosophical Society. Acknowledgement is made to the Donors of the Petroleum Research Fund administered by the American Chemical Society for partial support for RIK.

References

Berg, J.H. & Herz, D.L. (1986). Thermobarometry of two-pyroxene-granulite inclusions in Cenozoic volcanic rocks of the McMurdo Sound region. *Antarctic Journal of the United States*, **21(5)**, 19–200.

Borg, S.G., Stump, E., Chappell, B.W., McCulloch, M.T., Wyborn, D., Armstrong, R.L. & Holloway, J.R. (1987). Granitoids of northern Victoria Land, Antarctica: implications of chemical and isotopic variations to regional crustal structure and tectonics. *American Journal of Science*, **287**, 127–69.

Cooper, A.K. & Davey, F.J. (1985). Episodic rifting of Phanerozoic rocks in the Victoria Land basin, western Ross Sea, Antarctica. *Science*, **229**, 1085–7.

Dostal, J., Dupuy, C. & Leyreloup, A. (1980). Geochemistry and petrology of meta-igneous granulitic xenoliths in Neogene volcanic rocks of the Massif Central, France – implications for the lower crust. *Early and Planetary Science Letters*, **50**, 31–40.

Ewart, A. (1982). The mineralogy and petrology of Tertiary–Recent orogenic volcanic rocks: with special reference to the andesite-basaltic compositional range. In *Andesites*, ed. R.S. Thorpe, pp. 25–95. Chichester; John Wiley and Sons.

Gill, J.B. (1981). *Orogenic Andesites and Plate Tectonics*. Berlin; Springer-Verlag. 390 pp.

Hoefs, J., Faure, G. & Elliot, D.H. (1980). Correlation of $\delta^{18}O$ and initial $^{87}Sr/^{86}Sr$ ratios in Kirkpatrick Basalt on Mt Falla, Transantarctic Mountains. *Contributions to Mineralogy and Petrology*, **75**, 199–203.

Kalamarides, R.I., Berg, J.H. & Hank, R.A. (1987). Lateral isotopic discontinuity in the lower crust: an example from Antarctica. *Science*, **237**, 1192–5.

Kay, R.W. & Kay, S.M. (1981). The nature of the lower continental crust: inferences from geophysics, surface geology, and crustal xenoliths. *Reviews of Geophysics and Space Physics*, **19**, 271–97.

Kay, S.M. & Kay, R.W. (1985). Role of crystal cumulates and the oceanic crust in the formation of the lower crust of the Aleutian arc. *Geology*, **13**, 461–4.

Kyle, P.R. & Cole, J.W. (1974). Structural control of volcanism in the McMurdo Volcanic Group, Antarctica. *Bulletin Volcanologique*, **38**, 16–25.

Pearce, J.A. & Cann, J.R. (1973). Tectonic setting of basic volcanic rocks determined using trace element analyses. *Earth and Planetary Science Letters*, **19(2)**, 290–300.

Skinner, D.N.B. (1983). The granites and two orogenies of southern Victoria Land. In *Antarctic Earth Science*, ed. R.L. Oliver, P.R. James & J.B. Jago, pp. 160–3. Canberra; Australian Academy of Science and Cambridge; Cambridge University Press.

Smithson, S.B. (1972). Gravity interpretation in the Transantarctic Mountains near McMurdo. *Geological Society of America Bulletin*, **83**, 3437–42.

Snoke, A.W., Quick, J.E. & Bowman, H.R. (1981). Bear Mountain igneous complex, Klamath Mountains, California: an ultrabasic to silicic calc-alkaline suite. *Journal of Petrology*, **22**, 501–52.

Stuckless, J.S. & Ericksen, R.L. (1976). Strontium isotopic geochemistry of the volcanic rocks and associated megacrysts and inclusions from Ross Island and vicinity, Antarctica. *Contributions to Mineralogy and Petrology*, **58**, 111–26.

Taylor, S.R. & McLennan, S.M. (1985). *The Continental Crust: Its Composition and Evolution.* Oxford; Blackwell Scientific Publications Ltd. 312 pp.

Vetter, U., Roland, N.W., Kreuzer, H., Hohndorf, A., Lenz, H. & Besang, C. (1983). Geochemistry, petrography, and geochronology of the Cambro-Ordovician and Devonian–Carboniferous Granitoids of northern Victoria Land, Antarctica. In *Antarctic Earth Science*, ed. R.L. Oliver, P.R. James & J.B. Jago, pp. 140–3. Canberra; Australian Academy of Science and Cambridge; Cambridge University Press.

Geology, petrology and tectonic implications of crustal xenoliths in Cenozoic volcanic rocks of southern Victoria Land

J.H. BERG

Department of Geology, Northern Illinois University, DeKalb, IL 60115, USA

Abstract

The Cenozoic volcanic rocks that were erupted in both the Ross embayment (RE) and adjacent Transantarctic Mountains (TM) of the McMurdo Sound region apparently brought to the surface inclusions representing essentially all levels of the crust, namely supracrustal (sandstone or dolerite), upper-crustal (granite, quartzite, schist, or marble) and lower-crustal (two-pyroxene, clinopyroxene, and ultramafic granulite) inclusions. The lower-crustal garnet- and spinel-bearing granulites yielded pressures and temperatures of equilibration that define an extremely high geothermal gradient for the Transantarctic Mountains and perhaps an even higher one for the Ross embayment. The pressures indicate a thicker TM crust (45–47 km) and a thinner RE crust (20–30 km). The high geothermal gradient is indicative of an active continental-rifting tectonic environment for the origin of the Ross embayment and Transantarctic Mountains. The absence of Beacon/Ferrar inclusions from most of the RE volcanic rocks is consistent with the hypothesis that these rocks are missing at least locally from the RE crust. This conclusion implies that uplift and erosion may have occurred in the Ross embayment region after the Jurassic and prior to the rifting and subsidence of the RE crust.

Introduction

Because of the lack of orogenic deformation and volcanism associated with the uplift of the Transantarctic Mountains (TM) and the lack of exposed rock in the adjacent Ross embayment (RE), no consensus has developed concerning hypotheses for the origin of these geological features. Convergence, divergence, no relative motion between East and West Antarctica, and various sequential combinations of the above have been used to explain the uplift of the mountains and the subsidence of the embayment. Smithson (1972) seemed to imply that the uplift was related to rebound of an underplated section of crust, although a delay of several hundred million years between hypothesized collision and uplift makes this unlikely. On the other hand, Kyle & Cole (1974), Davey (1981), Cooper & Davey (1985), Fitzgerald *et al.* (1986) and others have suggested that some or all of the RE subsidence has resulted from crustal thinning due to extension, and that uplift of the TM resulted from associated block faulting (Katz, 1982). Herron & Tucholke (1975) and Davey, Bennett & Houtz (1982) explained the formation of the RE as being largely due to crustal extension in the late Cretaceous and early Tertiary, but called on compression due to convergence between East and West Antarctica to explain further development of the embayment and most of the uplift of the mountains in the late Cenozoic, continuing up to the present. Smith & Drewry (1984) also suggested that the RE formed by crustal extension in the late Cretaceous and early Tertiary, but they gave no indication of any relative movement between East and West Antarctica since then.

Berg & Herz (1986) presented *P–T* data derived from xenoliths of lower-crustal granulites which essentially require a current rifting environment for the TM and adjacent RE. Together with these data, the lithological and petrographical details of inclusions in the McMurdo Volcanic Group provide evidence for additional working hypotheses about the post-Jurassic tectonic history of the region.

Field occurrence of inclusions

Inclusions have been collected from more than 20 volcanic sites (Fig. 1), typically small to medium-sized cinder or spatter cones. Most of the crustal inclusions have been ejected as bombs and have lava mantles ranging in thickness from 0–10 cm, but many also occur in lava flows. Identifiable inclusions range in size from ~ 1 cm to 50 cm. Most of the unequivocal mantle inclusions (lherzolite) occur in flows rather than as bombs; the largest lherzolite inclusion observed is 30 cm in diameter. Most of the inclusions are subrounded to rounded, but some are very well rounded. Because the more refractory inclusions (e.g. lherzolite, dunite, pyroxenite, etc.) typically display the highest degree of rounding, the roundness is probably a result of milling and abrasion rather than melting.

At most of the sites, inclusions represent < 1% of the total

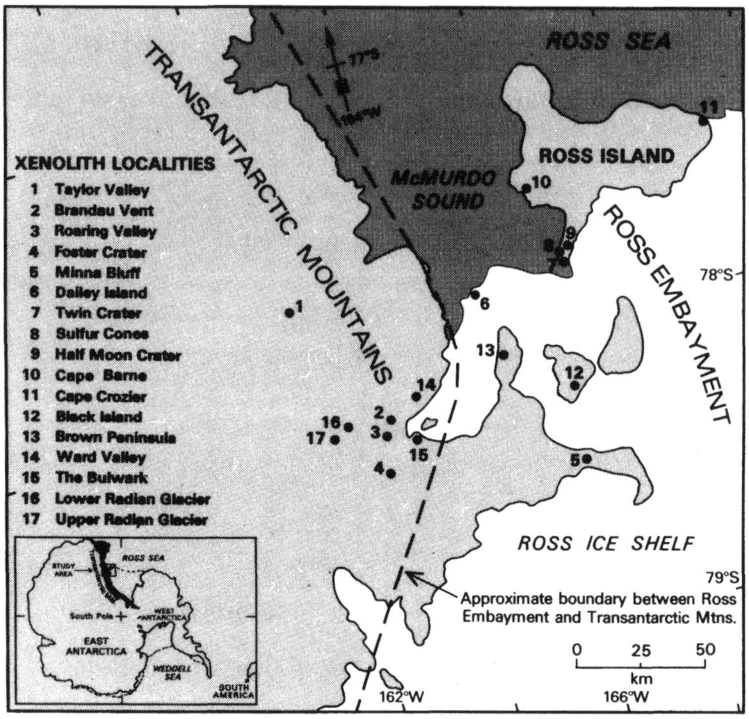

Fig. 1. Map of the McMurdo Sound region showing the primary sites where inclusions were collected.

volume of erupted material. However, at the sites where inclusions are most abundant (e.g. Black Island for crustal inclusions and Howchin Glacier for lherzolite inclusions) the inclusions represent 5–15% of the total volume of the erupted unit. The Black Island site is even more impressive because locally the tephra have been preferentially eroded, leaving a spectacular lag deposit of crustal-inclusion boulders.

Descriptive petrography and petrology

Ignoring cognate xenoliths, the inclusions can be broadly classified as: (1) supracrustal; (2) upper-crustal basement; (3) lower-crustal mafic granulite; or (4) ultramafic. Although relative abundance of the above types varies from site to site, the overall order from most to least abundant is as follows: (3) > (4) > (1) = (2).

Supracrustal inclusions

Supracrustal inclusions include Ferrar dolerite and partially fused or recrystallized rocks that probably were undeformed sandstones, calcareous sandstones, and sandy carbonate rocks. Ferrar dolerite incusions are relatively common at several volcanic centres in the TM, especially in Taylor Valley and Roaring Valley (Fig. 1). However, only one inclusion of Ferrar dolerite has been found in the RE, and the near absence of Ferrar inclusions is based not on random sampling but on a comprehensive search for Ferrar inclusions at every site. Also, the single Ferrar inclusion was found at the Dailey Islands site, which is the RE volcanic site closest to the TM. Thomson (1916) reported dolerite inclusions from near Cape Barne, but his description of them being partially recrystallized and melted is unlike the dolerite inclusions found in this study and suggests that, even if they are related to the Ferrar dolerites, they probably have a deep-seated origin.

Although inclusions of sandstone have previously been referred to as Beacon sandstone (Thomson, 1916), recrystallization or extensive partial fusion of the matrix precludes positive identification. Many of them have a fine-grained calcareous or calc-magnesian matrix now consisting of diopside, carbonate, or wollastonite. They are probably derived from Cenozoic sedimentary rocks restricted to the RE although they are unexposed, except as erratics (Vella, 1969). The fact that other sandstone inclusions were extensively melted clearly shows that they were matrix-rich; this indicates a greater similarity to the matrix-rich Cenozoic sandstones than to the matrix-poor Beacon sandstones (Barrett, Grindley & Webb, 1972). Some of the sandstone inclusions contain grains of what appears to be Ferrar dolerite, which would preclude their belonging to the Beacon Supergroup.

Upper-crustal inclusions

Upper-crustal inclusions consist of granite and metamorphic rocks. The granites range from undeformed post-tectonic to gneissic syn- or pre-tectonic types; most are extensively fused. The metamorphic rocks are predominantly quartzites (some of which have suffered strong deformation and may contain hypersthene), but calc-silicate gneisses and a few pelitic schists (some containing sillimanite) are present also; both minerals are indicative of somewhat higher grades of metamorphism than is generally evident in the exposed rocks of this part of the TM. Upper-crustal inclusions are present at nearly every inclusion-bearing site, but they are more abundant at sites in the TM.

Lower-crustal inclusions

Lower-crustal inclusions are typically pyroxene granulites and ultramafic rocks. The pyroxene granulites can be subdivided into two-pyroxene granulites (the dominant type) and clinopyroxene granulites (subordinate). Because ultramafic rocks are considered separately below, all pyroxene granulites discussed here are plagioclase-bearing by definition.

In addition to plagioclase, the two-pyroxene granulites contain orthopyroxene, clinopyroxene, and typically either olivine, spinel, or garnet. Hornblende, quartz, ilmenite, apatite, biotite, pigeonite, sanidine, or scapolite are present in some of the inclusions, and in some cases one of the two pyroxenes is absent. Garnet is rare, occurring only as corroded cores surrounded by fine-grained symplektites of olivine, plagioclase, spinel and orthopyroxene. However, symplektitic pseudomorphs of garnet are relatively common. Locally spinel shows evidence of having reacted with pyroxenes to form olivine–plagioclase symplektites. Not uncommonly the mafic minerals occur in lenses or clusters; where present, red–brown to orange–red hornblende is typically concentrated near the edges of these clusters but may occur throughout the rock.

Several minerals in the two-pyroxene granulites have regionally restricted distribution. Garnet and fine-grained garnet pseudomorphs occur only in inclusions of two-pyroxene granulites erupted through the TM, whereas primary olivine in two-pyroxene granulite inclusions is restricted to volcanic rocks erupted in the RE. Primary spinel is not common in two-pyroxene granulites from most of the RE sites; however, at the Black Island site it is very common. The presence of non-primary spinel or olivine in coronas, symplektites and intergranular glass is not regionally restricted. Hornblende occurs in inclusions from both the TM and RE, but is much more common in the RE inclusions. Quartz occurs in a few RE inclusions but is far more common in those from the TM.

Inclusions of clinopyroxene granulite contain plagioclase and clinopyroxene, and apatite, scapolite, sphene or quartz. A subgroup of these granulites consists of the assemblage clinopyroxene–sanidine–nepheline–scapolite–sphene. The scapolite in these rocks is invariably surrounded by a corona of plagioclase and fine-grained carbonate, fluid inclusions and other unidentifiable material. The clinopyroxene granulites are absent from the RE but are very common in the TM.

Inclusions of ultramafic rocks occur at virtually every site where inclusions are found. However, typical mantle lherzolites are rare at all but the Howchin Glacier site (Fig. 1). Most of the ultramafic inclusions are clinopyroxenites or dunites, with or without spinel or orthopyroxene. These rock types appear to be gradational with the more mafic two-pyroxene granulites, and in some of the larger inclusions they occur as layers alternating with two-pyroxene granulite. Therefore, it is unlikely that most of the clinopyroxenites and dunites are from the mantle; instead they probably represent lower crustal cumulates. However, their isotopic compositions and variability (Kalamarides, Berg & Hank, 1987; McGibbon, this volume, p. 317) do not permit a simple relationship to either the Jurassic Ferrar magmatism or the Cenozoic McMurdo magmatism.

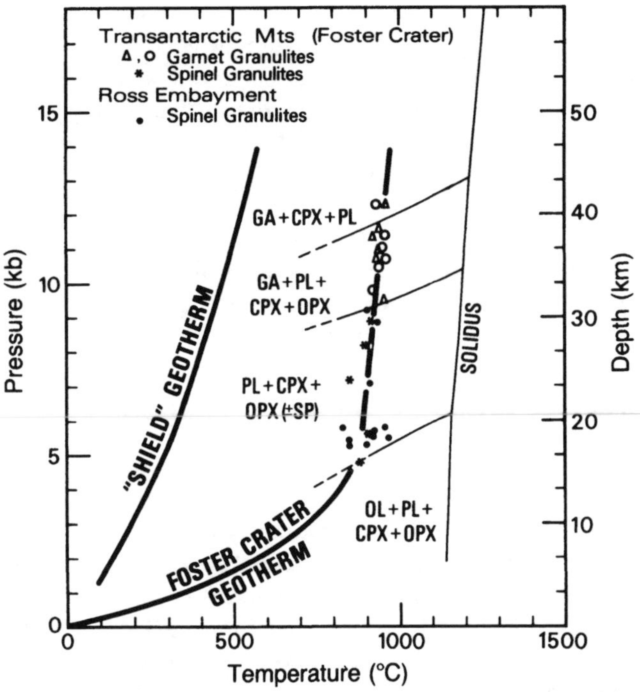

Fig. 2. Thermobarometry of garnet- and spinel-granulite inclusions. Garnet-granulite data are based on two-pyroxene thermometry (Wells, 1977) and garnet–pyroxene (open circles; Harley, 1984) and garnet–pyroxene–feldspar–quartz (open triangles; Bohlen *et al.*, 1983) barometry. The spinel-granulite data are based on two-pyroxene thermometry (Wells, 1977) and Al content of clinopyroxene (Gasparik, 1984); phase equilibria are from Irving (1974). (After Berg & Herz, 1986.)

Although relict cumulate textures are present in some of the lower-crustal inclusions, especially in the RE inclusions, most show textural evidence of moderate to strong shearing and deformation. Augen structures, granulation, bent plagioclase twin lamellae, bent exsolution lamellae and cleavages in pyroxene, and undulatory extinction in these minerals are relatively common. Mortar texture, while not as obvious in the RE inclusions, is ubiquitous in the TM granulites.

Thermobarometry

Mineral chemistry from garnet and spinel granulites in the TM and spinel granulites in the RE has been used to determine pressures and temperatures of equilibration for these lower crustal inclusions (Fig. 2; Berg & Herz, 1986). The lack of compositional zoning in the minerals, the presence of pigeonite and sanidine in some of the inclusions, and a mineral isochron that is indistinguishable from the eruption age (Berg & Herz, 1986 Kalamarides *et al.*, 1987) provide convincing evidence that the derived pressures and temperatures reflect the present-day environment of the lower crust and not some ancient 'fossilized' environment (Harte, Jackson & Macintyre, 1981). Although symplektitic coronas and pseudomorphs demonstrate disequilibria, the lack of zoning in the primary minerals suggests that the symplektites are due more to decomposition of primary phases during incorporation into the magma and eruption than to reaction between primary minerals. Furthermore, many of the symplektities are so fine grained that microprobe analyses of the multi-phase symplektites yield compositions that are *identical* to the nearby primary phase. There can be little doubt that the compositions of the primary minerals record the pre-eruption equilibria in the lower crust.

The garnet granulite data are based on compositions of coexisting garnet, clinopyroxene, orthopyroxene, and plagioclase. The close agreement between the independent barometers of Bohlen, Wall & Boettcher (1983) and Harley (1984) indicates that the accuracy of these pressures is quite good. The spinel granulite data are based on the compositions of coexisting clinopyroxene and orthopyroxene coexisting with spinel and plagioclase. The pressures were determined using the Al^{VI} content of clinopyroxene (Gasparik, 1984). Al^{VI} content of orthopyroxene was not used because it consistently yielded unreasonably low pressures, presumably due to Al^{VI}-substitution mechanisms in the natural orthopyroxene that are quite different from the mechanisms in the $CaO–MgO–Al_2O_3–SiO_2$ system studied by Gasparik (1984). There is considerably greater uncertainty for the pressures derived from the spinel granulites than for those derived from the garnet granulites. Also, the presence of a substantial albite component in some of the plagioclases, and hercynite component in some spinels, would probably mean that the actual pressures were somewhat higher than the calculated pressures for some of the spinel granulites (Gasparik, 1984).

The *P–T* data are reasonably consistent with the phase relations shown in Fig. 2 and determined experimentally on a broadly similar two-pyroxene granulite from Australia (Irving, 1974). The *P–T* data and phase relations indicate a thicker TM crust (45–47 km) and a thinner RE crust (locally up to 30 km, but perhaps typically not much over 20 km). These conclusions are consistent with thicknesses inferred from geophysical studies such as Smithson (1972) and McGinnis *et al.* (1983). The agreement of the *P–T* data with the phase relations and the geophysical data offers further support for the relative accuracy of these *P–T* results.

All of the TM data were determined on inclusions from 'Foster Crater'. A geotherm drawn through the 'Foster Crater' data and extrapolated from the top of the lower crust to the surface is shown in Fig. 2. The RE data fall on or close to this geotherm. These data require a very high geothermal gradient in the upper crust, ranging from 60 to 100 °C/km, which should result in very high surface heat flow. Consistent with this, high surface heat flow ranging up to more than 160 mW/m^2 has been measured in the TM and RE (Risk & Hochstein, 1974; Decker & Bucher, 1982; Blackman, von Herzen & Lawver, 1984).

Tectonic implications

Kalamarides *et al.* (1987) and Kalamarides & Berg (this volume, p. 305) have presented geochemical evidence from the two-pyroxene granulites that suggests a discontinuity and probable suture in the lower crust between the TM and RE. The abundance of distinctive scapolite–apatite–sphene-bearing clinopyroxene granulites in the TM suite, yet their complete absence from the RE suite is consistent with and supports the suture interpretation. As discussed by them, however, this suture is probably Early Palaeozoic or Precambrian in age.

The extremely high geothermal gradient discussed above requires an active continental-rifting environment (Wickham & Oxburgh, 1985; Chapman, 1986). Simply drifting the lithosphere over passive, previously heated asthenosphere (Smith & Drewry, 1985) would not elevate lower crustal temperatures to the extent shown in Fig. 2 and would not explain the very high heat flow. Models that call on compression or subduction for the uplift of the TM (Davey *et al.*, 1982) are clearly at odds with the high geothermal gradient and high heat flow. In addition to the extremely high geothermal gradient, the active-rifting model is consistent with the high heat flow, block faulting along normal faults (Katz, 1982), and graben formation in the Ross Sea (Cooper & Davey, 1985).

The absence of Beacon/Ferrar inclusions in the RE cannot easily be dismissed. Although one might argue that it is a negative sort of evidence, there are reasons to conclude that their absence from the RE inclusion suite indicates their absence from the RE crust. Although Ferrar dykes and sills in the sub-Beacon basement are not very common, they are moderately well represented as inclusions in the TM suite. If they, along with Beacon rocks, underlie the adjacent RE, they should be even better represented in the RE suite, not absent from it. If typical matrix-poor Beacon sandstones were not brought to the surface, why were supposedly underlying upper-crustal quartzites and overlying matrix-rich sandstones?

If the Beacon/Ferrar rocks are missing (at least locally) from the RE crust, either they were not deposited or emplaced there or they have been subsequently eroded. Although the deposition/emplacement of these rocks may have been broadly confined to the TM region (Barrett, 1981), their great thicknesses in the adjacent TM make it seem unlikely that they would pinch out abruptly. If they have been eroded, differential uplift of the RE crust prior to its subsidence, and a place to dispose of the sediment are needed. Solutions to the latter are dependent on models for the former. A plausible model for uplift of the RE crust prior to its subsidence can be taken from the known evolution of many continental rifts. In many of these cases there is evidence of broad uplift in the early stages of the rifting process (Neugebauer, 1978). It is therefore conceivable that post-Jurassic uplift and erosion may have occurred in the RE region, probably just prior to the rifting and subsidence of the RE crust. If this uplift was punctuated early by horst- and graben-formation, the eroded material could have been deposited in graben basins. If the uplift involved broad domal uplift, eroded sediment that would have been shed to the west could have been deposited on or to the west of the TM. Any sediment deposited on the site of the present-day TM would probably have been removed by erosion during the uplift of the TM.

The timing for the onset of rifting in the RE region is problematical. Many authors suggest Late Cretaceous rifting, followed either by no plate interaction (Smith & Drewry, 1984) or by convergence (Herron & Tucholke, 1975; Molnar *et al.*, 1975; Davey *et al.*, 1982). Although the data presented here offer no temporal constraints on the initiation of continental rifting, they do suggest that part of the RE may have undergone broad uplift before its eventual subsidence, and that the region is currently undergoing active rifting. Scientific drilling down to the pre-Cretaceous basement, perhaps on the eastern flank of the Victoria Land basin (Cooper & Davey, 1985), would provide an excellent evaluation of the early-uplift model proposed herein, and would provide additional information critical to a comprehensive understanding of the tectonic evolution of the Ross embayment and Transantarctic Mountains.

Acknowledgements

This research has been supported by National Science Foundation grants DPP-8213943, DPP-8317712, and DPP-8614071.

References

Barrett, P.J. (1981). History of the Ross Sea region during the deposition of the Beacon Supergroup 400–180 million years ago. *Journal of the Royal Society of New Zealand*, **11(4)**, 447–58.

Barrett, P.J., Grindley, G.W. & Webb, P.N. (1972). The Beacon Supergroup of East Antarctica (review). In *Antarctic Geology and Geophysics*, ed. R.J. Adie, pp. 319–22. Oslo; Universitetsforlaget.

Berg, J.H. & Herz, D.L. (1986). Thermobarometry of two-pyroxene granulite inclusions in Cenozoic volcanic rocks of the McMurdo Sound region. *Antarctic Journal of the United States*, **21(5)**, 19–20.

Blackman, D.K., von Herzen, R.P. & Lawver, L.A. (1984). Heat flow and tectonics in the Ross Sea. *Eos, Transactions of the American Geophysical Union*, **65(45)**, 1120.

Bohlen, S.R., Wall, V.J. & Boettcher, A.L. (1983). Experimental investigation and application of garnet granulite equilibria. *Contributions to Mineralogy and Petrology*, **83(1–2)**, 52–61.

Chapman, D.S. (1986). Thermal gradients in the continental crust. In *The Nature of the Lower Continental Crust*, ed. J.B. Dawson, D.A. Carswell, J. Hall & K.H. Wedepohl, pp. 63–70. Oxford; Blackwell Scientific Publications; Special Publication of the Geological Society of London No. 24.

Cooper, A.K. & Davey, F.J. (1985). Episodic rifting of Phanerozoic rocks in the Victoria Land Basin, western Ross Sea, Antarctica. *Science*, **229(4718)**, 1085–7.

Davey, F.J. (1981). Geophysical studies in the Ross Sea region. *Journal of the Royal Society of New Zealand*, **11(4)**, 465–79.

Davey, F.J., Bennett, D.J. & Houtz, R.E. (1982). Sedimentary basins of the Ross Sea, Antarctica. *New Zealand Journal of Geology and Geophysics*, **25(2)**, 245–55.

Decker, E.R. & Bucher, G.J. (1982). Geothermal studies in the Ross Island – Dry Valley region. In *Antarctic Geoscience*, ed. C. Craddock, pp. 887–94. Madison; University of Wisconsin Press.

Fitzgerald, P.G., Sandiford, M., Barrett, P.J. & Gleadow, A.J.W. (1986). Asymmetric extension associated with uplift and subsidence in the Transantarctic Mountains and Ross Embayment. *Earth and Planetary Science Letters*, **81**, 67–78.

Gasparik, T. (1984). Two-pyroxene thermobarometry with new experimental data in the system $CaO–MgO–Al_2O_3–SiO_2$. *Contributions to Mineralogy and Petrology*, **87(1)**, 87–97.

Harley, S.L. (1984). The solubility of alumina in orthopyroxene coexisting with garnet in $FeO–MgO–Al_2O_3–SiO_2$ and $CaO–FeO–MgO–Al_2O_3–SiO_2$. *Journal of Petrology*, **25(3)**, 665–96.

Harte, B., Jackson, P.M. & Macintyre, R.M. (1981). Age of mineral equilibria in granulite facies nodules from kimberlites. *Nature, London*, **291(5811)**, 147–8.

Herron, E.M. & Tucholke, B.E. (1975). Seafloor magnetic pattern and basement structure in the southeastern Pacific. In *Initial Reports of the Deep-Sea Drilling Project*, *35*, ed. C.D. Hollister *et al.*, pp. 263–78. Washington, DC; US Government Printing Office.

Irving, A.J. (1974). Geochemical and high pressure experimental studies of garnet pyroxenite and pyroxene granulite xenoliths from the Delegate basaltic pipes, Australia. *Journal of Petrology*, **15(1)**, 1–40.

Kalamarides, R.I., Berg, J.H. & Hank, R.A. (1987). Lateral isotopic discontinuity in the lower crust: an example from Antarctica. *Science*, **237(4819)**, 1192–5.

Katz, H.R. (1982). Post-Beacon tectonics in the region of Amundsen and Scott Glaciers, Queen Maud Range, Transantarctic Mountains. In *Antarctic Geoscience*, ed. C. Craddock, pp. 827–34. Madison; University of Wisconsin Press.

Kyle, P.R. & Cole, J.W. (1974). Structural control of volcanism in the McMurdo Volcanic Group, Antarctica. *Bulletin Volcanologique*, **38(1)**, 16–25.

McGinnis, L.D., Wilson, D.D., Burdelik, W.J. & Larson, T.L. (1983). Crust and upper mantle study of McMurdo Sound. In *Antarctic Earth Science*, ed. R.L. Oliver, P.R. James & J.B. Jago, pp. 204–8. Canberra; Australian Academy of Science and Cambridge; Cambridge University Press.

Molnar, P., Atwater, T., Mammerickx, J. & Smith, S.W. (1975). Magnetic anomalies, bathymetry and the tectonic evolutions of the South Pacific since the Late Cretaceous. *Geophysical Journal of the Royal Astronomical Society*, **40(3)**, 383–420.

Neugebauer, H.J. (1978). Crustal doming and the mechanism of rifting. Part I: rift formation. *Tectonophysics*, **45(2–3)**, 159–86.

Risk, G.F. & Hochstein, M.P. (1974). Heatflow at Arrival Heights, Ross Island, Antarctica. *New Zealand Journal of Geology and Geophysics*, **17(3)**, 629–44.

Smith, A.G. & Drewry, D.J. (1984). Delayed phase change due to hot asthenosphere causes Transantarctic uplift? *Nature, London*, **309(5968)**, 536–8.

Smithson, S.B. (1972). Gravity interpretation in the Transantarctic Mountains near McMurdo. *Geological Society of America Bulletin*, **83(11)**, 3437–42.

Thomson, J.A. (1916). Report on the inclusions of the volcanic rocks of the Ross archipelago. *Report of the British Antarctic Expedition 1907–9, Geology*, **2(8)**, 129–51.

Vella, P. (1969). Surficial geological sequence, Black Island and Brown Peninsula, McMurdo Sound, Antarctica. *New Zealand Journal of Geology and Geophysics*, **12(4)**, 761–70.

Wells, P.R.A. (1977). Pyroxene thermometry in simple and complex systems. *Contributions to Mineralogy and Petrology*, **62(2)**, 129–39.

Wickham, S.M. & Oxburgh, E.R. (1985). Continental rifts as a setting for regional metamorphism. *Nature, London*, **318(6044)**, 330–3.

Geochemistry and petrology of ultramafic xenoliths of the Erebus volcanic province

F.M. McGIBBON

Department of Earth Sciences, The Open University, Milton Keynes, MK7 6AA, UK

Abstract

The ultramafic xenoliths of the Erebus volcanic province can be subdivided into three main groups on the basis of mineral chemistry: (1) spinel lherzolite with Mg- and Cr-rich diopside; (2) augitic clinopyroxenites with Fe- and Ti-rich clinopyroxenes; and (3) fassaitic clinopyroxenites with calcic clinopyroxenes. Whole-rock major, trace, and isotopic chemistry demonstrates that spinel lherzolites are mantle residua whereas the augitic clinopyroxenites represent cumulates from asthenospheric, silica-undersaturated liquids similar to their host basanites. The fassaitic clinopyroxenites (found only at 'Foster Crater'), however, are in marked chemical contrast to both their present host lavas and the tholeiitic magmas of the Ferrar Supergroup, to which they have been previously linked; they are interpreted as cumulates from alkalic LREE-enriched liquids which had a high initial $^{87}Sr/^{86}Sr$ ratio of at least 0.7080. The abundant low-Ti phlogopite in these samples is texturally secondary but was isotopically indistinguishable from the hosting clinopyroxenite at the time of its formation. These fassaitic clinopyroxenites retain a Siluro-Ordovician isochron age. Whether this represents a formation age or a metamorphic overprint (during the Ross Orogeny) cannot be resolved at present.

Erebus volcanic province xenoliths

The Cenozoic volcanic province lies in a 2000 km long belt parallel to the trend of the Transantarctic Mountains. One subdivision of this belt is the Erebus volcanic province (Kyle & Cole, 1974), with volcanism concentrated around the two main central volcanoes of Mount Erebus and Mount Discovery but also occurring as abundant small cinder cones in the Royal Society Range. It is at these Recent basanitic cinder cones that abundant upper crustal, lower crustal and upper mantle xenoliths occur (Kyle, Wright & Kirsch, 1987). The age of volcanism ranges from 18 Ma to the present with an average age of 4 Ma (Armstrong, 1978). A marked gravity gradient has been interpreted as evidence of crustal thinning from 40 km in the Royal Society Range to 27 km in the Ross embayment (Smithson, 1972). Recent rifting and volcanism are probably similarly related to this extensional tectonic regime.

Xenolith and host basanite samples for this study were collected at several localities in the Erebus volcanic province (EVP), in particular at Cape Crozier, on the east coast of Ross Island, several sites on Hut Point Peninsula, and on Black Island in western McMurdo Sound. Many localities in the Royal Society Range, including 'Foster Crater', were also sampled (Fig. 1).

Mafic, ultramafic, granulitic and crustal xenoliths are common in the Erebus volcanic province although they occur only in basanitic scoria cones. Xenoliths range from a few centimetres to 0.5 m in size and are usually rounded, but can be blocky. Xenolith types include spinel lherzolite, wehrlite, dunite, clinopyroxenite, glimmerite and felsic granulites. The wehrlites and dunites are volumetrically insignificant at the localities discussed here and will not be considered. The granulite xenoliths are the focus of a separate study (Kalamarides,

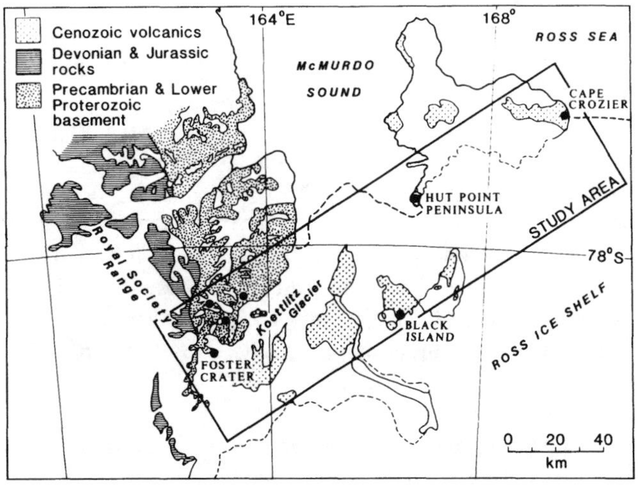

Fig. 1. Geological map of the McMurdo Sound area. Main sites of sample collection indicated by dots.

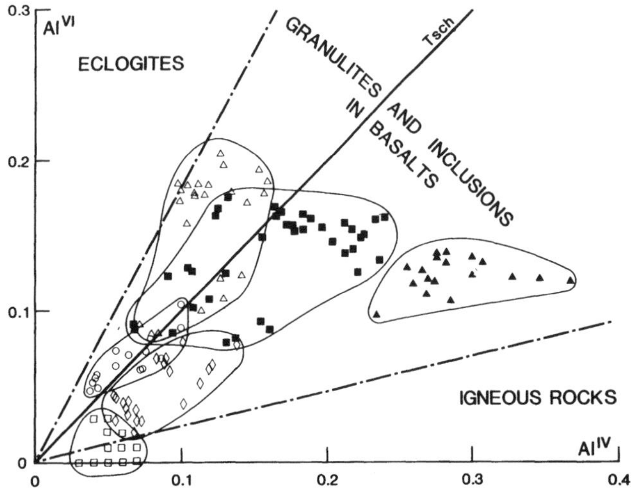

Fig. 2. Relative proportions of Al^{VI} and Al^{IV} in clinopyroxenes from various lithologies using the field boundaries of Aoki & Shiba (1973). Solid squares, 'Foster Crater' fassaitic clinopyroxenites; solid triangles, host basanites; open triangles, Type I spinel lherzolites; open diamonds, Type II augitic clinopyroxenites; open circles, granulites; open squares, clinopyroxenes of the Dufek layered intrusion (data of Himmelberg & Ford, 1976).

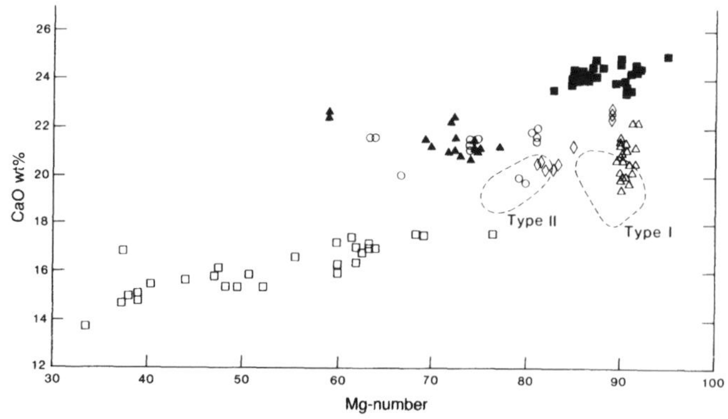

Fig. 3. Correlation of wt % CaO with Mg-number in clinopyroxenes, symbols as in Fig. 2; Type I and Type II fields from Frey & Prinz (1978).

Berg & Hank, 1987) and are shown for reference only. EVP spinel lherzolites are characterized by coarse, protogranular textures with large anhedral chrome-spinel and chrome-rich diopside. Clinopyroxenites are the most abundant xenolith type within the EVP and are usually coarse grained, consisting mainly of Fe-rich augite with minor spinel.

Xenoliths from the 'Foster Crater' locality are unique in that phlogopite-rich clinopyroxenites and glimmerites are predominant. Phlogopite-free samples consist almost entirely of calcic clinopyroxene (up to 98 modal %), pale greyish-green in colour, with minor aluminous spinel (hercynite). Low-Ti phlogopite constitutes as much as 80% of some samples and can occur disseminated throughout the xenolith or as discrete veinlets. In the glimmerites, anorthitic plagioclase is quite common as small inclusions in clinopyroxene (possibly an annealed symplektite). Some samples have been mildly sheared, resulting in granulation at the boundaries of large grains, and it is along these zones that phlogopite is most common, generally replacing clinopyroxene. Some deformation post-dates phlogopite introduction as evidenced by kink-banding and granulation of the phlogopite.

Geothermometric calculations for clinopyroxene–orthopyroxene pairs in spinel lherzolites provide a temperature range of 880–1100 °C, depending on the thermometer used. The temperature range suggests an approximate depth of origin of 35–70 km, which implies the upper mantle in this region. The clinopyroxenites and glimmerites are not mineralogically suited to thermobarometry but their depth of origin is restricted to the stability range of the included plagioclase (9–10 kb, i.e. < 30 km; Wyllie, 1979). The ratios of Al^{VI}/Al^{IV} for clinopyroxenes in the various lithologies described are shown in Fig. 2. Such diagrams have been used as qualitative indicators of P–T conditions during crystallization (Aoki & Shiba, 1973) since higher P–T regimes favour increased octahedral–tetrahedral alumina. Thus spinel lherzolites typically have higher Al^{VI}/Al^{IV} close to the eclogite field. Fe-rich pyroxenites from this region have lower octahedral alumina plotting near the field for igneous reactions, suggesting a relatively shallow depth of origin. 'Foster Crater' clinopyroxenites, however, suggest a range of conditions falling between the fields for spinel lherzolites and host phenocrysts.

Geochemical results

Ultramafic xenoliths from the EVP thus fall into three mineralogical and petrological groups. Two of these, the spinel lherzolite and the Fe-augite clinopyroxenite, are analogous to Type I and Type II xenoliths, respectively, of the Frey & Prinz (1978) classification observed worldwide. In this scheme spinel lherzolites, characterized by Mg- and Cr-rich spinel and diopside, are interpreted as mantle residua with no genetic relationship to their host lava. Type II pyroxenites and wehrlites, which typically have Fe-rich augite and igneous textures, are interpreted as cumulates from silica-undersaturated magmas similar to the basalts which entrain them. Clinopyroxenes from Fe-rich pyroxenites have similar CaO contents to clinopyroxene phenocrysts of their basanite hosts, but are more magnesian (Fig. 3), suggesting that they may be related to the parental magmas of their present host. Clinopyroxenes from the 'Foster Crater' clinopyroxenites, however, have up to 25 wt % CaO, resulting in an extremely high Ca-Tschermak component. These clinopyroxenes are specifically termed fassaites, part of the diopside solid–solution series (Gamble *et al.*, 1988). On the basis of texture and their monomineralic nature these xenoliths may also be cumulates, but the parental liquids were clearly different from those of other Type II xenoliths (i.e. unlike their present host), as well as distinct from the Ferrar tholeiites. Also shown in Fig. 3 are clinopyroxenes from the Dufek layered intrusion, the plutonic equivalent of the Ferrar rocks. Although these pyroxenes exhibit the large variation in Mg-number (Mg/Mg + Fe) typical of a fractionating sequence, they have significantly lower CaO contents than the fassaites.

EVP ultramafic xenoliths can also be subdivided on the basis of their whole-rock trace element chemistry (see Fig. 4). Type I spinel lherzolites display slightly LREE-depleted patterns, at close to chondritic abundances. EVP Type II clinopyroxenites have concave downward patterns typical of Type II from

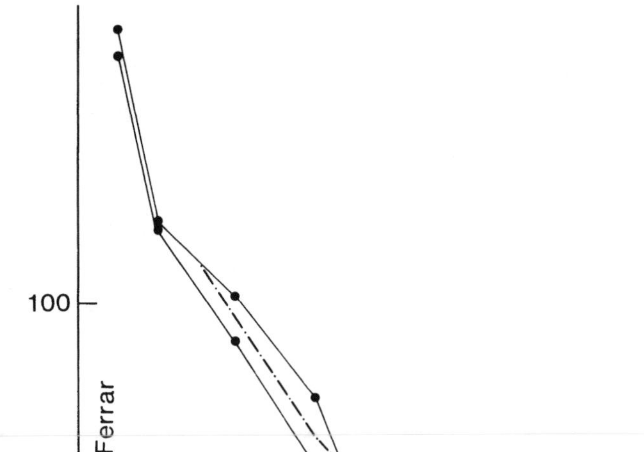

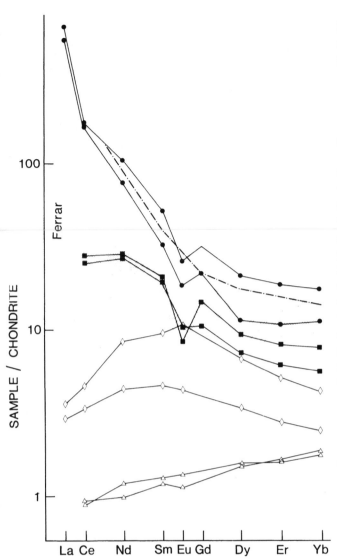

Fig. 4. Chondrite-normalized whole-rock rare-earth element concentrations as determined by isotope dilution. Symbols as in Fig. 2, with solid circles representing the calculated coexisting liquids in equilibrium with the 'Foster Crater' clinopyroxenes (distribution coefficients as in Henderson, 1982). Stippled field represents the range of concentrations of the Ferrar Supergroup Kirkpatrick Basalts (data from Kyle, 1980). Typical 'Foster Crater' basanite; dashed line.

elsewhere in the world (e.g., Frey & Prinz, 1978; Kempton, Menzies & Dungan, 1984). In contrast, clinopyroxenite and glimmerite whole-rock samples and clinopyroxene separates from 'Foster Crater' show extreme LREE enrichment and have high overall abundances. These clinopyroxenes also show marked negative Eu anomalies suggesting that plagioclase has played an important role in their genesis.

Possible parental liquid compositions calculated to be in equilibrium with the 'Foster Crater' clinopyroxenes are shown in Fig. 4. Even with the high partition coefficients used here, the concentrations of LREE in such a liquid are significantly higher than in tholeiitic magmas. These calculated liquids more closely resemble an alkali basalt such as the 'Foster Crater' basanite shown for comparison. The 'Foster Crater' clinopyroxenites can therefore be interpreted as cumulates from a magma at least as enriched in rare earth elements as their own host alkali basalt, but they cannot be cognate cumulates due to the extreme isotopic heterogeneity which exists between these xenoliths and their host.

Isotopic results

Present-day Nd–Sr isotopic values for EVP host basanites display a narrow range ($^{87}Sr/^{86}Sr = 0.70302$–$0.70396$; $^{143}Nd/^{144}Nd = 0.51285$–$0.51293$), plotting within the mantle array in the deleted quadrant of Fig. 5. Type II pyroxenites and wehrlites plot close to the basanite values, as would be expected of recent cumulates from such melts. Type I spinel lherzolite whole-rock samples also plot in this field, but diopside separates from spinel lherzolites have slightly more depleted compositions of 0.70220 and 0.51340, suggesting that contamination by the host of the respective whole-rock samples has occurred. In contrast 'Foster Crater' whole-rock samples and clinopyroxene separates have more radiogenic compositions that are in sharp istopic contrast to their host basanites. Both clinopyroxene separates and whole-rock samples exhibit a similar range in $^{143}Nd/^{144}Nd$ (0.51215–0.51233) but clinopyroxene separates display a smaller range in $^{87}Sr/^{86}Sr$ (0.70851–0.71216). Mica separates from 'Foster Crater' clinopyroxenites have extremely radiogenic Sr values (up to 0.77130) and plot outside the limits of Fig. 5, suggesting that isotopic variations in the 'Foster Crater' rocks can be thought of simply as a result of an increasing modal proportion of phlogopite relative to clinopyroxene in the whole-rock samples.

The range of present-day Sr and Nd isotopic values for the 'Foster Crater' clinopyroxenes is like that of the nearby Ferrar tholeiites (R.J. Pankhurst, unpublished data). However, if the age of the 'Foster Crater' clinopyroxenites is corrected to the Jurassic age of the Ferrar tholeiites (~ 175 Ma), the 'Foster Crater' data fall at ϵ_{Nd} values of -8, significantly lower than the Ferrar value of -5 (Fig. 5). Therefore, from the previous consideration of mineral chemistry, and also in terms of isotopic data, the 'Foster Crater' clinopyroxenites cannot repre-

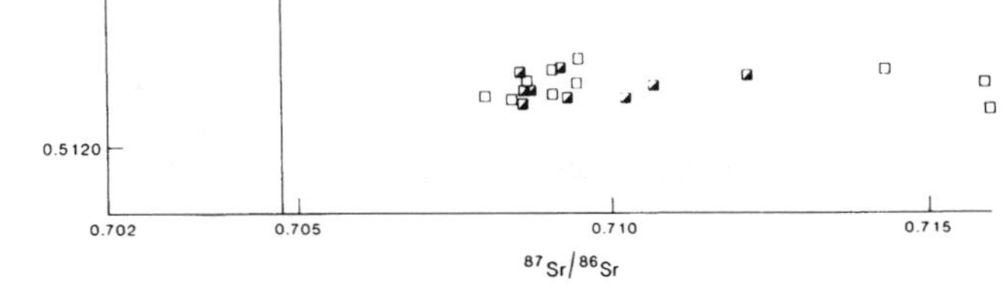

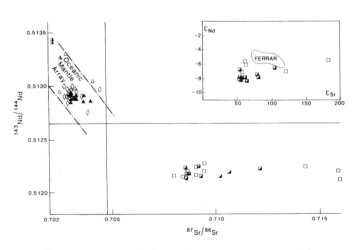

Fig. 5. Present-day $^{143}Nd/^{144}Nd$ versus $^{87}Sr/^{86}Sr$ ratios. Symbols as in Fig. 2 with half-filled squares representing 'Foster Crater' clinopyroxene separates. Inset diagram of Jurassic ϵ_{Nd} versus ϵ_{Sr} for 'Foster Crater' data only, with Ferrar tholeiite field for comparison (R.J. Pankhurst, unpublished data).

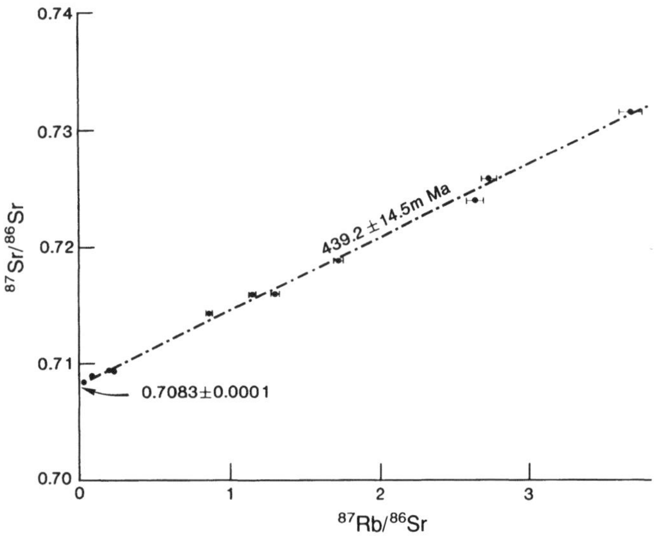

Fig. 6. Rb–Sr whole-rock isochron for 'Foster Crater' clinopyroxenites and mica-clinopyroxenites. Errors on the age and intercept value are to one standard deviation; error bars are shown on the data points.

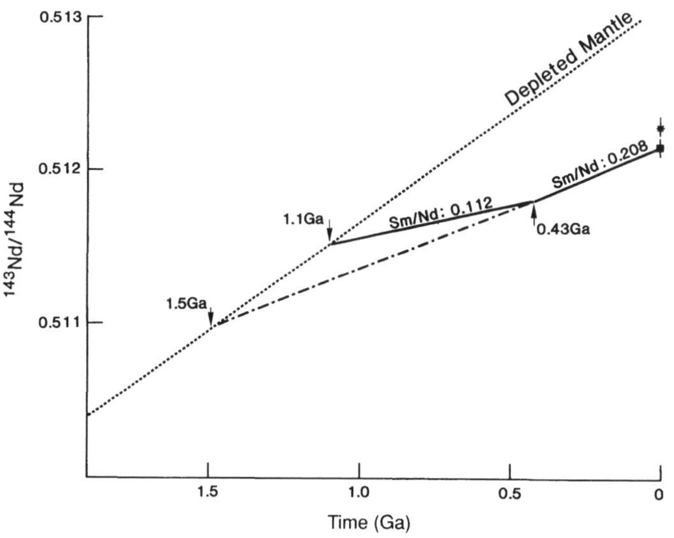

Fig. 7. Example neodymium evolution diagram with the present-day range of 'Foster Crater' clinopyroxenes (solid square) and Ferrar tholeiites (asterisk), (R.J. Pankhurst, unpublished data).

sent the source, source residue, or cumulates from Ferrar magmas.

Timing of events

Neodymium model ages, calculated using a depleted mantle composition, range from 1.4 to 2.3 Ga, with an average of 1.7 Ga for 'Foster Crater' clinopyroxene separates. This range is a result of variable Sm/Nd ratios created at the time of fractionation of the clinopyroxene, which must have been sufficiently recent that the $^{143}Nd/^{144}Nd$ ratios do not reflect the variation. New Pb data on clinopyroxene and mica separates each show a large and completely overlapping range of $^{206}Pb/^{204}Pb$ (18.0–24.0), and produce a well-constrained Pb–Pb isochron corresponding to an age of 1.095 ± 0.005 Ga, which is confirmed by the intersection of this trend with the lead ore growth curve at 1.1 Ga. It is interesting to note that both the clinopyroxenes and micas plot on this line. Also, in Fig. 6 the 'Foster Crater' clinopyroxenites and glimmerites define a Rb–Sr whole-rock isochron of 440.1 ± 9.1 Ma, with an initial $^{87}Sr/^{86}Sr$ ratio of 0.70826. On this diagram clinopyroxene separates plot near the initial value (0.70851–0.71216) while mica separates with much higher $^{87}Rb/^{86}Sr$ ratios plot on a similar trend at higher $^{87}Sr/^{86}Sr$ ratios (0.72622–0.77130). Although the linear trend exhibited by these data appears to be a function of phlogopite modal abundance, suggesting mixing between phlogopite and clinopyroxene, its slope would appear to be age-significant. The introduction of phlogopite is probably not recent as mantle-derived melts with such radiogenic signatures are not known. Furthermore, a crustal melt cannot be invoked as it could not crystallize mica of phlogopitic composition. Mica separates independently display an isochron relationship corresponding to an age of 324.5 ± 23.6 Ma, with an intercept of 0.70865. This suggests that the mica could post-date the clinopyroxene by 130 m.y. but the Pb data suggest identical source ages for the two phases. It is also possible that, since the temperature of the host basanite far exceeded the mica-blocking temperature, partial resetting has occurred in some of the smaller samples. It is interesting that the whole-rock isochron age of 440 ± 9.1 Ma is within the age range of metamorphism and plutonic intrusion during the Ross Orogeny (409–570 Ma; Elliot, 1975).

In preliminary isotopic studies Menzies, Kyle & Gamble (1985) and McGibbon *et al.* (1987) argued that the 'Foster Crater' clinopyroxenites and Ferrar tholeiites were cogenetic because they had similar model Nd ages. However, modelling based on several isotopic systems, summarized in Fig. 7, offers a different interpretation. Because the partition coefficients between clinopyroxene and liquid are higher for Sm than for Nd, clinopyroxene cumulates have higher Sm/Nd ratios than the liquids from which they crystallized. Therefore, when a model age is calculated using the measured Sm/Nd and $^{143}Nd/^{144}Nd$ ratios in a clinopyroxene separate, it may yield an unrealistically old age. Using the available partition coefficients for Sm and Nd in clinopyroxene, the calculated Sm/Nd ratio of the parental liquids to the 'Foster Crater' clinopyroxenites is 0.112. If the clinopyroxenes crystallized at the age suggested by the Rb–Sr whole-rock isochron (440 Ma), and if the parental liquid composition is projected back in time, it intersects the depleted mantle evolution line at 1.1 Ga, which is the observed Pb age. Thus it would appear that the source of the magmas from which the 'Foster Crater' clinopyroxenites crystallized is younger than that of the Ferrar tholeiites, and that in this area lithospheric source regions of different age and isotopic composition were closely juxtaposed.

Conclusions

High precision Sr-, Nd- and Pb-isotope, together with minor- and trace-element data on both whole-rocks and separated minerals has elucidated a complex history within the subcontinental mantle of Antarctica.

(1) The 'Foster Crater' calcic clinopyroxenites represent a rare clinopyroxene chemistry, not included in existing xenolith classification schemes.
(2) Other mafic clinopyroxenites from the McMurdo Sound region bear similarities to Type II xenoliths from elsewhere in the world, and are interpreted as cumulates from liquids of asthenospheric origin.
(3) 'Foster Crater' clinopyroxenites do not represent the source, source residue, or cumulates from Ferrar melts, and calculated liquids in equilibrium with 'Foster Crater' clinopyroxenes are similar to alkali basalts.
(4) Clinopyroxenite whole-rock samples and mica separates define a 439.2 ± 14.5 Ma Rb–Sr isochron with an initial $^{87}Sr/^{86}Sr$ ratio of 0.708. The mica, although texturally secondary was not isotopically distinguishable at the time of its formation.
(5) The clinopyroxene T_{DM} ages of 1.4–2.3 Ga are attributed to clinopyroxene–liquid fractionation during the Phanerozoic.

Acknowledgements

I wish to thank my colleagues P.D. Kempton, C.J. Hawkesworth and P.W. Van Calsteren for much helpful discussion; M. Kunka and A. Tindle for assistance in the laboratory; J. Taylor for drafting the diagrams; N. Dunbar and J. Gamble for assistance in the field; and especially P.R. Kyle and M. Menzies for setting up this project. I am also grateful to the National Science Foundation (grant to P.R. Kyle) for the rare opportunity to work in Antarctica, and to the Open University for partial field work support and continued salary support.

References

Aoki, K. & Shiba, I. (1973). Pyroxenes from lherzolite inclusions of Itonome-gata, Japan. *Lithos*, **6**, 41–51.

Armstrong, R.L. (1978). K–Ar dating: Late Cenozoic McMurdo Volcanic Group and dry valley glacial history, Victoria Land, Antarctica. *New Zealand Journal of Geology and Geophysics*, **21(6)**, 685–98.

Elliot, D.H. (1975). Tectonics of Antarctica: a review. *American Journal of Science*, **275A**, 45–106.

Frey, F.A. & Prinz, M. (1978). Ultramafic inclusions from San Carlos, Arizona: petrologic and geochemical data bearing on their petrogenesis. *Earth and Planetary Science Letters*, **38**, 129–76.

Gamble, J.A., McGibbon, F.M., Kyle, P.R., Menzies, M.A. & Kirsch, I. (1988). Metasomatised xenoliths from Foster Crater, Antarctica. Implications for lithospheric structure and processes beneath the Transantarctic Mountains front. *Journal of Petrology*, Special Lithosphere Issue, 109–38.

Henderson, P. (1982). *Inorganic Geochemistry*. Oxford; Pergamon Press. 353 pp.

Himmelberg, G.R. & Ford, A.B. (1976). Pyroxenes of the Dufek Intrusion, Antarctica. *Journal of Petrology*, **17(2)**, 219–43.

Kalamarides, R.I., Berg, J.H. & Hank, R.A. (1987). Lateral isotopic discontinuity in the lower crust: an example from Antarctica. *Science*, **237**, 1192–5.

Kempton, P.D., Menzies, M.A. & Dungan, M.A. (1984). Petrography, petrology, and geochemistry of xenoliths and megacrysts from the Geronimo volcanic field, southeastern Arizona. In *Kimberlites – II. The Mantle and Crust/Mantle Relationship*, ed. J. Kornprobst, pp. 71–83. Amsterdam; Elsevier Press.

Kyle, P.R. (1980). Development of heterogeneities in the subcontinental mantle: evidence from the Ferrar Group, Antarctica. *Contributions to Mineralogy and Petrology*, **73**, 89–104.

Kyle, P.R. & Cole, J.W. (1974). Structural control of volcanism in the McMurdo Volcanic Group, Antarctica. *Bulletin Volcanologique*, **38**, 16–25.

Kyle, P.R., Wright, A. & Kirsch, I. (1987). Ultramafic xenoliths in the Late Cenozoic McMurdo Volcanic Group, western Ross Sea embayment, Antarctica. In *Mantle Xenoliths*, ed. P.H. Nixon, pp. 287–93. Chichester; John Wiley & Sons.

McGibbon, F.M., Hawkesworth, C.J., Menzies, M.A. & Kyle, P.R. (1987). Geochemistry of the ultramafic xenoliths from McMurdo Sound, Antarctica: implications on Jurassic Ferrar Magmatism. *Terra Cognita*, **7(2)**, 396.

Menzies, M.A., Kyle, P.R. & Gamble, J. (1985). Fragments of aged enriched sub-continental lithosphere entrained in asthenospheric alkaline magmas erupted at Foster Crater, Antarctica (abstract). *Eos Transactions of the American Geophysical Union*, **66**, 409.

Smithson, S.B. (1972). Gravity interpretation in the Transantarctic Mountains near McMurdo Sound, Antarctica. *Geological Society of America Bulletin*, **83**, 3437–42.

Wyllie, P.J. (1979). Petrogenesis and the physics of the Earth. In *The Evolution of the Igneous Rocks*, ed. H.S. Yoder, pp. 483–520. Princeton; Princeton University Press.

Lithospheric flexure induced by the load of Ross Archipelago, southern Victoria Land, Antarctica

T.A. STERN[1], F.J. DAVEY[1] & G. DELISLE[2]

1 Geophysics Division, DSIR, PO Box 1320, Wellington, New Zealand
2 Bundesanstalt für Geowissenschaften und Rohstoffe, Postfach 51 01 53, D-3000 Hannover 51, Federal Republic of Germany

Abstract

Observations of gravity, radio echo sounding and multichannel seismic reflection are used to deduce bathymetry, ice-shelf thickness and sediment structure to the south-east of Ross Island. The data are applied to a flexural analysis of the deformation induced by the superposed load of the Ross Archipelago. The best-fit model is that of a discontinuous elastic plate (the lithosphere) overlying a weak fluid (the asthenosphere). A maximum deflection of about 1800 m is inferred to occur beneath the Ross Archipelago and a compensating outer flexural bulge of about 100 m, at a distance of about 200 km from the centre of the Ross Archipelago, is also predicted. A flexural rigidity of about 10^{23} N.m is estimated for the lithosphere underlying the Ross Archipelago. This value for rigidity is low for 500 Ma old lithosphere and it suggests that the lithosphere within the Ross embayment has been thermally reactivated in the Cenozoic.

Introduction

The Transantarctic Mountains extend for nearly 3000 km in a gentle arc across Antarctica, and divide the continent into east and west cratons (Robinson & Splettstoesser, 1984). The mountains are breached at regular intervals by outflowing glaciers of the East Antarctic ice sheet. At one 150-km wide locality along the mountain chain the discharge appears to be disrupted, and an area of deglaciated terrain has evolved. This is the Dry Valleys area, or the 'McMurdo Oasis' (Clark, 1965), of southern Victoria Land (Fig. 1). The origin of these valleys, the largest ice-free oasis of East Antarctica, has remained an enigma since the Dry Valleys were first explored by members of Captain Scott's *Discovery* expedition at the beginning of the century (e.g. Bull, McKelvey & Webb, 1962; Armstrong, 1978).

The main purpose of this study was to test for a possible mechanical link between induced flexure associated with the load of the Ross Archipelago (Fig. 1), and the origin of the Dry Valleys. Beneath superposed loads, like volcanoes or seamounts, the lithosphere is depressed but at distances of typically 150–300 km from the load, elastically supported upwarps develop; these are called outer flexural bulges. Such bulges are, for example, especially well developed adjacent to the Hawaiian volcanoes (Walcott, 1970a; Watts, Cochran & Selzer, 1985). The wavelength and amplitude of these outer bulges are a function of both the magnitude of the superposed load, and the strength, or flexural rigidity (defined later), of the lithosphere on which the load is emplaced. Shown on Fig. 1 is part of a 200 km radius arc centred on Mount Erebus (the main volcanic edifice of Ross Island). This might be a typical position for a flexural bulge to be found. The apparent coincidence of the 200 km radius arc (Fig. 1) with the catchment area of the Dry Valleys initially aroused suspicion that there is a causal relationship between the Ross Archipelago and the Dry Valleys. Moreover, a temporal link is also suggested by the approximate 5 Ma ages for both the main edifice of Ross Island and the Dry Valleys development (Armstrong, 1978). Hence the specific proposal to test is that an outer flexural high, induced by the loading of the Ross Archipelago, inhibited ice flow into the Dry Valleys and therefore initiated their development.

The Ross Archipelago

The Ross Archipelago (Thomson, 1916) is taken to include the basaltic–phonolitic volcanic cones of Ross, White and Black islands, Mount Discovery and Mount Morning (Fig. 1). The archipelago extends up to nearly 4 km above, and 1 km below, sea level. Taking 2600 and 1700 kg/m^3 as the average effective density of rocks above and below sea level, respectively, then the total load of the archipelago is estimated, by integration of the topography, to be about 2.3×10^{16} kg.

Further Cenozoic volcanic intrusions and small volcanic islands occur within the rift structure of the Victoria Land basin (Fig. 1), to the north-west of the Ross Archipelago (Cooper & Davey, 1985), including another active, 3 km high volcano (Mount Melbourne) 300 km distant from the archipel-

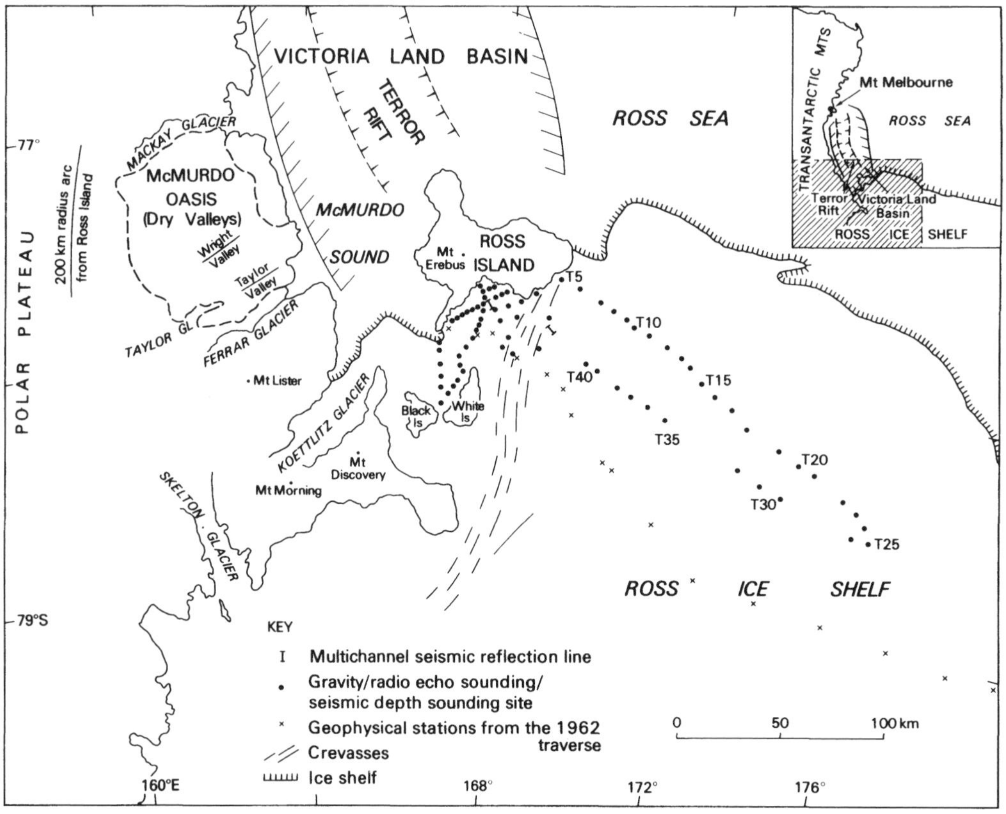

Fig. 1. Locality map for the geophysical measurements made in the vicinity of the Ross Archipelago (Ross Archipelago is taken to include the volcanoes of Mount Morning and Mount Discovery, White and Black islands and the volcanoes of Ross Island). Terror rift and the Victoria Land basin after Cooper & Davey (1985).

ago. Thus the Ross Archipelago appears to be the largest extrusion of volcanic rock at the southern end of a 500 km long rift structure; accordingly the load will be modelled as an equivalent two-dimensional line load. It should be noted that this is a first-order approximation to the load distribution as the mass of the load is not distributed evenly along the rift axis, but is concentrated within the vicinity of Ross Island.

The amplitude of an equivalent line load along the 140 km long Mount Morning–Ross Island axis is calculated to be 1.6×10^{11} kg/m.

Flexural loading models

Mathematical treatments of flexure for various types of geological problems have previously been described by Walcott (1970a, b), Lambeck (1981) and Turcotte (1979). The underlying assumption of these flexural models is that when the lithosphere is subjected to external or internal loading, with the time scale of the loading being greater than about 10^5 years, it can be modelled as an elastic sheet overlying an inviscid fluid (the asthenosphere).

Fig. 2 shows two simple cases of volcanic loading that demonstrate the manner in which a lithosphere with finite strength will distribute deformation laterally about the load. The upper model is that of a two-dimensional load on a continuous lithosphere, whereas the lower model is that of a similar load on a lithosphere that has a discontinuity, or fracture, beneath the axis of the load. In both cases the

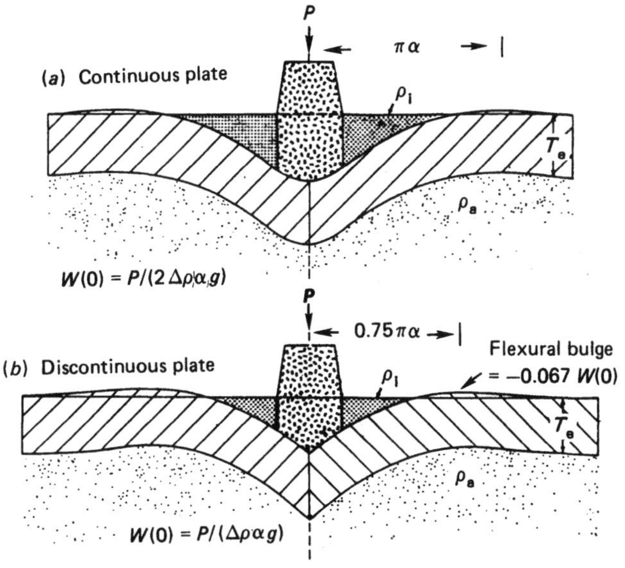

Fig. 2. (*a*) Schematic model for the flexure of a continuous elastic plate due to a 2-D volcanic load. ρ_a and ρ_l represent densities of the asthenosphere, and infilling material on top of the plate, respectively; other symbols are explained in the text. The distance between the centre of the load and outer flexural bulge is given by $X_b = \pi\alpha$ where α is the flexural parameter. (*b*) As for (*a*) but the plate has a discontinuity, or a fracture, beneath the axis of the load. Shown on the figure are analytic expressions derived from equation (1) for the maximum depression beneath the axis of the load ($W(0)$), and the distance (X_b) and amplitude ($W(x_b)$) of the flexural bulge, i.e. for this case $X_b = 0.75\,\pi\alpha$ and $W(x_b) = -0.067\,W(0)$.

lithosphere is displaced downward beneath the load, and a smaller amplitude upwarp, or flexural bulge, occurs far from the load axis. For identical loads placed on otherwise equivalent elastic plates, the fractured plate will deform with a shorter wavelength but with greater amplitude.

For some studies of seamount chains like the Hawaiian chain, the fractured plate model (Fig. 2*b*) is found to be appropriate (Walcott, 1970a; Watts *et al.*, 1985). We also find that the fractured plate model provides the best fit to our geophysical observations from the Ross Ice Shelf. We note, however, that the two models of Fig. 2 represent end-members for a continuum of possible models. Physically, the fractured plate model may be interpreted to represent the case where the lithosphere becomes strongly heated and weak beneath the axis of the volcanic load, such that the transmission of shear stresses beneath the load is effectively ceased.

A solution to the thin-plate equation (Turcotte, 1979) appropriate for the fractured plate model of Fig. 2*b* is in the form of a damped sinusoid given by:

$$W(x) = (P/\Delta\rho\alpha g)\exp(-x/\alpha)\cos(x/\alpha) \qquad (1)$$

where $W(x)$ is the displacement profile, P = superposed load, $\Delta\rho$ is the density contrast (ρ_a–ρ_i) (Fig. 2), $g = 9.8$ m s^{-2} and α is the flexural parameter defined by:

$$\alpha^4 = 4D/\Delta\rho g \qquad (2)$$

The flexural parameter (in units of length) is a measure of the wavelength and amplitude of deformation for a particular loading situation. D, the flexural rigidity, is a measure of the 'stiffness', or resistance to bending, of the lithosphere (Walcott, 1970b).

Other useful relationships, readily derived from equation (1), are shown on Fig. 2 for the distance to the flexural bulge (X_b), the maximum deflection beneath the load ($W(0)$) and the amplitude of the flexural bulge ($W(X_b)$). These relationships apply for a vanishingly thin line load. In reality a volcanic chain is a distributed load and Menke (1981) gave formulae for various load geometries that correct for this difference.

Geophysical data from the Ross Island region

Seismic profiling by Wong & Christoffel (1981) in McMurdo Sound (Fig. 1) first raised the possibility that the weight of the young (~ 5 Ma) Ross Island had downwardly flexed the 500 Ma (Armstrong, 1978) basement. Subsequent multichannel seismic-reflection work in McMurdo Sound (Kim, McGinnis & Bowen, 1986) also indicates crustal reflectors appearing to be bent down beneath Ross Island. These areas to the west and north of Ross Island (Fig. 1) have, however, proved to be areas of rifting, normal faulting, thinned crust and in general much crustal complexity (Cooper & Davey, 1985; McGinnis *et al.*, 1985). These are not favourable conditions for deciphering the long wavelength and subtle deformation associated with flexure. Accordingly, under the assumption that flexural deformation will be distributed symmetrically about the Ross Archipelago, we made our seismic and gravity observations to the south-east of Ross Island (Figs 1 & 3), where gravity and bathymetry gradients appear to be more subdued.

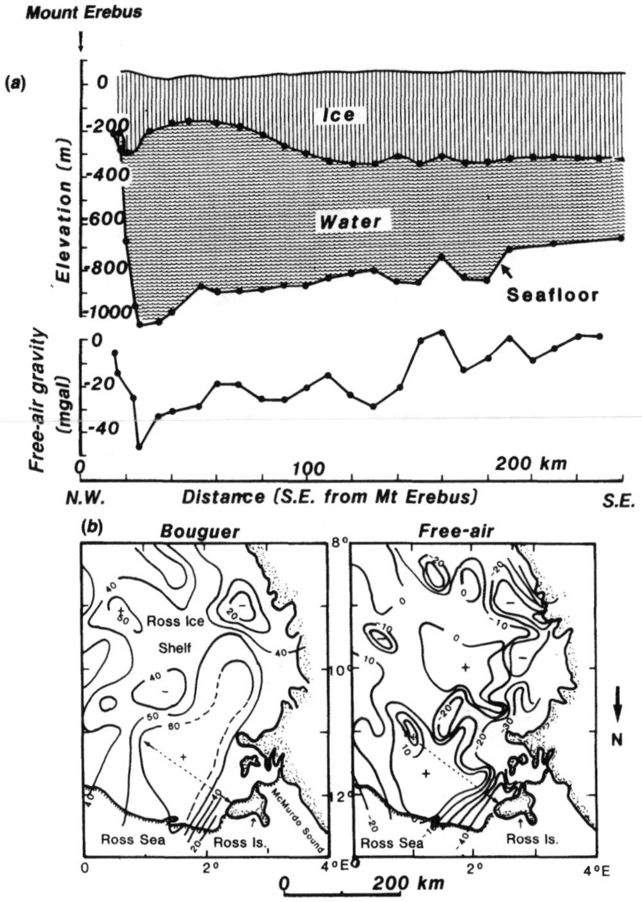

Fig. 3. (*a*) Ice shelf thickness, bathymetry, and free-air gravity along the 220 km profile south-east of Mount Erebus (Fig. 1). (*b*) Maps of free-air and Bouguer gravity fields for the area to the south-east of Ross Island. Contour interval = 10 mgal. Note that south is to the top of the page (after Robertson *et al.*, 1982). The route of the traverse shown in Fig. 1 is represented by the dashed lines with arrow heads.

Fig. 3*a* shows the data from our gravity, radio echo sounding and seismic depth sounding profile south-east of Ross Island. A portable radio echo sounder gave ice thickness values along the profile, and from these data elevations above sea level were calculated for each station. Seismic reflections from the seafloor, coupled with the ice thickness determinations, gave accurate bathymetric data. The principal features to note from Fig. 3*a* are the deepening sea bottom, culminating in a flexural 'moat' adjacent to Ross Island, and a similar pattern in the free-air gravity anomalies. These are typical patterns associated with lithospheric flexure adjacent to volcanic loads (e.g. Walcott, 1970a).

Also of note in Fig. 3*a* is that the bathymetry and gravity profiles appear to be approaching a maximum at their eastern extremity. Our profile evidently did not quite go far enough to unequivocally establish this maximum. Inspection of the 50 km-grid maps of free-air and Bouguer gravity anomalies given by Robertson *et al.* (1982), and partly reproduced in Fig. 3*b*, indeed show broad gravity highs at a distance of some 200–220 km east and south-east of Ross Island. The expression for distance to the flexural bulge is given by $X_b = 0.75\,\pi\alpha$ (Fig. 2*b*). For $X_b = 200$ km, a provisional estimate for the flexural parameter (α) of about 90 km is obtained.

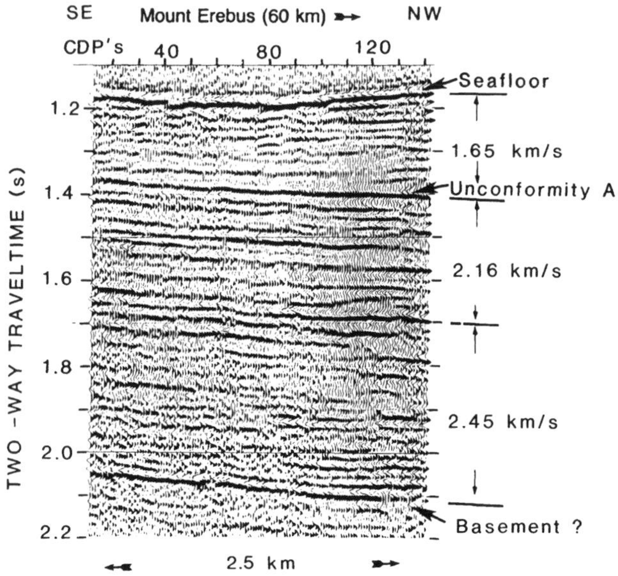

Fig. 4. Multichannel seismic-reflection data from line 602 (Fig. 1) showing an approximately 1 km thick sedimentary section from the seafloor down to an apparent seismic basement. Shots consisted of 2 × 0.5 kg charges placed in 2 m deep drill holes in the ice shelf. Shot point interval = 35 m, CDP interval = 17.5 m. The data represent a six-fold stack that has been deconvoluted both before and after stack. Velocities at the margin represent interval velocities as calculated from the observed stacking velocities. Unconformity A is estimated to be 5 Ma and to represent the 'birth' of Ross Island, as discussed in the text.

Multi-channel seismic-reflection data

Dips and unconformities

Two common depth point (CDP) seismic-reflection profiles were carried out south-east of Ross Island as shown in Fig. 1. Each profile was about 5 km long, giving a six-fold coverage of about 2.5 km. Initial brute-stacks revealed severe ringing from ice-shelf multiples. Deconvolution both before and after stack proved an effective means of diminishing these short-period multiples. A processed section of line 602, situated 60 km south-east of Mount Erebus (Fig. 1), is shown in Fig. 4.

A principal feature of the 1 km thick section shown in Fig. 4 is the unconformity (unconformity A), at about 200 m beneath the seafloor, which separates upper near-horizontal layers from lower layers that dip towards Mount Erebus at about 1°. Alongside the section are interval velocities calculated from the stacking velocities using the Dix equation (Badley, 1985); the range of velocities from 1.65 to 2.5 km/s is interpreted to indicate sedimentary rocks.

A depth section was created from the time section of Fig. 4 and the true dips of the various layers below unconformity A were measured. Dips varied from 0.9° to 1.3°, with the highest dips being on the deepest recorded reflector at a depth of about 2 km below sea level. Using the (wavelength/8) criterion of seismic resolution (Badley, 1985), an uncertainty of about ± 0.2° is estimated for these dip measurements. Accordingly, an average dip of about 1° is proposed for the sediments beneath unconformity A.

From recent drilling (Fig. 1) in McMurdo Sound a Cenozoic sedimentation rate of about 20 m/Ma has been calculated (P.J. Barrett, pers. comm., 1986). If this rate is also appropriate to the line 602 site, it would imply an age of about 7.5–10 Ma for unconformity A. This age estimate should be treated with caution for at least two reasons: (i) the flexural moat created would tend to pond sediments preferentially; and (ii) glacial erosion (Davey & Christoffel, 1984) has probably had less influence at site 602 than in McMurdo Sound.

It is thus proposed that unconformity A has an age somewhat < 7.5 Ma, and therefore probably marks the principal loading episode of Ross Island about 5 m.y. ago. Sediments below unconformity A are interpreted as being flexed down by the load of Ross Island, whereas sediments above A are considered to represent infilling of the flexural moat.

Single channel marine seismic-reflection data

Marine single-channel seismic data to the north and north-east of Ross Island were collected by the SP *Lee* cruise in 1984 (Cooper & Davey, 1985). Locations for the two unpublished lines and line drawing representations of the data are shown in Fig. 5. As in line 602, the data of Fig. 5 show a

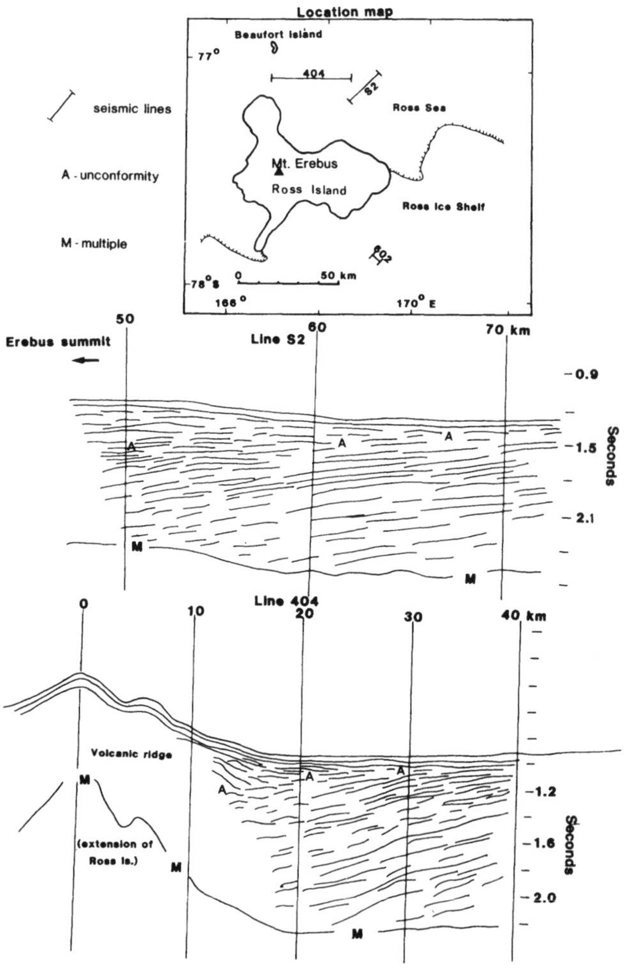

Fig. 5. The inset map of the Ross Island region shows the location of the multichannel seismic line 602 (Fig. 4) and two single-channel marine surveys, lines 404 and S2. Line drawings beneath represent the data from lines 404 and S2. Note unconformity A dipping towards the axis of the volcanic load in a similar manner to the data from line 602, as shown in Fig. 4.

dipping sequence capped by an unconformity that separates the dipping sequence below from near-horizontal reflectors above. The dip direction of the dipping sequence in all cases is toward the volcanic load, and the angle of dip is $1 \pm 0.2°$.

Flexural parameter estimates from dip angles

The first spatial derivative of equation (2) gives a simple expression for the gradient of the flexed lithosphere as a function of distance from the centre of the load. Lambeck (1981), among others, demonstrated the usefulness of using this gradient criterion in solving for the flexural parameters (α). We find that in order to produce a $1 \pm 0.2°$ dip on a discontinuous elastic plate, as measured at a distance of 60 km from the load axis, and with a line load of 1.6×10^{11} kg/m, a flexural parameter of about $\alpha = 80 \pm 10$ km is required. This value is not significantly different to the $\alpha = 90$ km estimated in a previous section from the distance to what is interpreted to be the outer flexural bulge. For an equivalent continuous elastic plate, subjected to the same load, the dip of the flexed horizon as measured at 60 km from the load axis would be about 0.5°. Thus in order to satisfy both the observed dip on the pre-Ross Island sedimentary layers, and the distance to the outer flexural bulge, it is necessary to propose a discontinuous lithospheric plate with a flexural parameter of $\alpha = 80 \pm 10$ km.

Flexural bulge and maximum deflection estimates

Expressions for the maximum deflection and the amplitude and position of the flexural bulge associated with line loads are shown in Fig. 2. These require modification for distributed loads on a broken lithosphere (Menke, 1981). For example, given a triangular load of base length, d, the distance to the flexural bulge becomes:

$$X_b = 0.75\ \pi\alpha + d/3 + d^2/9\alpha \qquad (3)$$

and the maximum deflection beneath the load is:

$$W(0) = (P/\Delta\rho g\alpha)(1 - d/3\alpha + 2d^2/9\alpha^2) \qquad (4)$$

Letting $\alpha = 80$ km and $d = 60$ km in equation (3) gives $X_b = 213 \pm 15$ km, which is in good agreement with the observed 200–220 km distance of the broad gravity maxima from the centre of the load (Fig. 3). Taking $P = 1.6 \times 10^{11}$ kg/m, $\Delta\rho = 1000$ kg/m^3, and values of α and d as above in equation (4), gives a value for the maximum deflection beneath the load of about 1800 m. Furthermore, the amplitude of the flexural bulge is about 7% of $W(0)$ (Fig. 2) and is thus predicted to be about 120 m. Finally, from equation (2) we can calculate with the above parameters a value for lithospheric flexural rigidity of about 1×10^{23} N.m; this in turn implies an 'effective' elastic thickness for the lithosphere of about 24 km (see Walcott, 1970a).

Discussion

Flexural models, rigidity and heat flow

The principal finding of this study is that isostatic loading of the Ross Archipelago can, to a first approximation, be modelled as the effect of a distributed load on a thin but discontinuous, or fractured, elastic plate that overlies an inviscid fluid. A line-load model on a continuous elastic plate does not provide a consistent fit to the observations. With the acquisition of further seismic data, more complex models will be warranted. For example, the Ross Archipelago is of limited strike length, rather than being a linear, uniformly distributed two-dimensional load as tacitly assumed in our model. Therefore a finite element approach, similar to that used by Watts *et al.* (1975) for three-dimensional seamounts, is probably required.

It may be that flexural rigidity is a function of distance from the load axis in a volcanic setting. A hint of this is given by the manner in which the stratigraphic layers increase their dip with depth, and hence tend to diverge towards Ross Island on line 602. This phenomenon is expected if the rigidity decreases, in accordance with increasing heat flux, as one approaches the volcanic line (ten Brink & Watts, 1985).

A value of 1×10^{23} N.m appears to be anomalously low for 500 Ma lithosphere, based on the compilation of Karner, Steckler & Thorne (1983). The Ross Island–McMurdo Sound–Dry Valleys area is, nevertheless, a high heat flow area where observed values of conductive heat flow are about 60–90 mW/m^2 (Decker & Bucher, 1982); Decker & Bucher further argue that the heat flow regime here is similar to that of the Basin & Range province of the western USA. As flexural rigidity, and hence elastic thickness, appear to be dependent on the geotherm, in addition to the age of the lithosphere at the time of loading (Karner *et al.*, 1983), a later heating event will effectively reset the apparent thermal age of the lithosphere (Willet, Chapman & Neugebauer, 1985). Thus the anomalously low value of flexural rigidity is ascribed to a general heating of the lithosphere in the western Ross embayment, associated with Cenozoic volcanism and rifting.

Origin of the Dry Valleys

A primary motivation for this study was to investigate a possible link between flexure and the deglaciation of the Dry Valleys region of southern Victoria Land. Our model predicts a maximum negative deflection of about 1800 m beneath the load and a flexural bulge of about 120 m at a distance of 213 ± 15 km from the centre of the load. Such a location and amplitude for the bulge maximum would appear to broadly match those for the 'dome' that Drewry (1980) identified inland from the head of Taylor Glacier. He showed the elongate dome-like feature to stand about 100 m above the adjacent portions of the ice sheet, with the dome being situated some 200–250 km from the centre of Ross Island.

There is, therefore, an apparent coincidence, in both position and amplitude, of a predicted flexural bulge within the catchment area of the Dry Valleys, and an observed ice bulge at the edge of the East Antarctic ice sheet. Between the Ross Archipelago and the Dry Valleys there is, however, a series of steeply dipping normal faults associated with the Transantarctic Mountains front (Fitzgerald *et al.*, 1986). Seismic evidence from just offshore indicates that these faults have been active in the Cenozoic and in some cases faults displace the seafloor

(Cooper & Davey, 1985). The question thus arises: can flexure 'see' through the Transantarctic Mountains front, or do these normal faults effectively decouple the transmission of shear stresses between the Ross embayment and East Antarctica? If the latter, then the coincidence of the predicted position for the flexural outer bulge and the ice dome observed by Drewry (1980), must be regarded as fortuitous.

Clearly we are not yet in a position to give an unequivocal answer to the question of whether or not induced flexure is responsible for initiation of the Dry Valleys system. What is first required is an appreciation of what sort of tectonic boundary the Transantarctic Mountains represents, and in particular, an understanding of the faulting process within the Transantarctic Mountains front.

Acknowledgements

We thank the various officers of Antarctic Division, DSIR for logistic support. In particular we are indebted to our Antarctic Division field assistant, Brian Smith, for the energy and enthusiasm he put into the project, and for designing and implementing an effective navigation system for travel on the Ross Ice Shelf. We thank Dick Walcott, Uri ten Brink and Peter Barrett for comments on the initial version of this manuscript.

References

Armstrong, R.L. (1978). K–Ar dating: Late Cenozoic McMurdo Volcanic Group and dry valley glacial history, Victoria Land, Antarctica. *New Zealand Journal of Geology and Geophysics*, **21(6)**, 685–98.

Badley, M.E. (1985). *Practical Seismic Interpretation.* Boston; International Human Resources Development Corporation.

Bull, C., McKelvey, B.C. & Webb, P.N. (1962). Quaternary glaciation in southern Victoria Land, Antarctica. *Journal of Glaciology*, **4**, 63–78.

Clark, R.H. (1965). The oases in the ice. In *Antarctica*, ed. T. Hatherton, pp. 321–30. Wellington; Reed & Reed.

Cooper, A.K. & Davey, F.J. (1985). Episodic rifting of Phanerozoic rocks in the Victoria Land Basin, western Ross Sea, Antarctica. *Science*, **229**, 1085–7.

Davey, F.J. & Christoffel, D.A. (1984). The correlation of MSSTS-1 drill hole results with seismic reflection data from McMurdo Sound, Antarctica. *New Zealand Journal of Geology and Geophysics*, **27**, 405–12.

Decker, E.R. & Bucher, G.J. (1982). Geothermal studies in the Ross Island–Dry Valley region. In *Antarctic Geoscience*, ed. C. Craddock, pp. 887–94. Madison; University of Wisconsin Press.

Drewry, D.J. (1980). Pleistocene bimodal response of Antarctic ice. *Nature, London*, **287**, 214–16.

Fitzgerald, P.G., Sandiford, M., Barrett, P.J. & Gleadow, A.J.W. (1986). Asymmetric extension associated with uplift and subsidence in the Transantarctic Mountains and the Ross Embayment. *Earth and Planetary Science Letters*, **81**, 67–78.

Karner, G.D., Steckler, M.S. & Thorne, J.A. (1983). Long-term thermo-mechanical properties of the continental lithosphere. *Nature, London*, **304**, 250–3.

Kim, Y., McGinnis, L.D. & Bowen, R.H. (1986). The Victoria Land Basin; part of an extended crustal complex between East and West Antarctica. In *Reflection Seismology: The Continental Crust*, Geodynamics Series, 14, ed. M. Barazangi & L. Brown, pp. 323–30. Washington, DC; American Geophysical Union.

Lambeck, K. (1981). Flexure of the oceanic lithosphere from island uplift, bathymetry and geoid height observations: the Society Island. *Geophysical Journal of the Royal Astronomical Society*, **67**, 91–114.

McGinnis, L.D., Bowen, R.H., Erickson, J.M., Allred, B.J. & Kreamer, J.L. (1985). East–West Antarctic boundary in McMurdo Sound. *Tectonophysics*, **114**, 341–56.

Menke, W. (1981). The effect of load shape on the deflection of thin elastic plates. *Geophysical Journal of the Royal Astronomical Society*, **65**, 571–7.

Robertson, J.D., Bentley, C.R., Clough, J.W. & Greischar, L.L. (1982). Sea bottom topography and crustal structure below the Ross Ice Shelf, Antarctica. In *Antarctic Geoscience*, ed. C. Craddock, pp. 1083–90. Madison; University of Wisconsin Press.

Robinson, E.S. & Splettstoesser, J.F. (1984). Structure of the Transantarctic Mountains determined from geophysical surveys. In *Geology of the Central Transantarctic Mountains*, Antarctic Research Series, 36, ed. M.D. Turner & J.F. Splettstoesser, pp. 119–62. Washington, DC; American Geophysical Union.

ten Brink, U.S. & Watts, A.B. (1985). Seismic stratigraphy of the flexural moat flanking the Hawaiian Islands. *Nature, London*, **317**, 421–4.

Thomson, J.A. (1916). Report on inclusions of the volcanic rocks of the Ross Archipelago. British Antarctic Expedition, 1907–1909. Report of scientific investigations. *Geology*, **2**, 129–48.

Turcotte, D.L. (1979). Flexure. *Advances in Geophysics*, **21**, 51–86.

Walcott, R.I. (1970a). Flexure of the lithosphere at Hawaii. *Tectonophysics*, **9**, 435–46.

Walcott, R.I. (1970b). Flexural rigidity, thickness and viscosity of the lithosphere. *Journal of Geophysical Research*, **75**, 3941–54.

Watts, A.B., Cochran, J.R. & Selzer, G. (1975). Gravity anomalies and flexure of the lithosphere: a three dimensional study of the Great Meteor seamount, northeast Atlantic. *Journal of Geophysical Research*, **80**, 1391–8.

Watts, A.B., ten Brink, U.S., Buhl, P. & Brocher, T.M. (1985). A multichannel seismic study of lithospheric flexure across the Hawaiian–Emperor seamount chain. *Nature, London*, **315**, 105–11.

Willett, S.D., Chapman, D.S. & Neugebauer, D. (1985). A thermo-mechanical model of continental lithosphere. *Nature, London*, **314**, 520–3.

Wong, H.K. & Christoffel, D.A. (1981). A reconnaissance seismic survey of McMurdo Sound and Terra Nova Bay, Ross Sea. In *Results of the Dry Valley Drilling Project*, Antarctic Research Series, 33, ed. L.D. McGinnis, pp. 37–62. Washington, DC; American Geophysical Union.

The structure and seismic activity of Mount Erebus, Ross Island

K. KAMINUMA & K. SHIBUYA

National Institute of Polar Research, 9–10 Kaga-1, Itabashi-ku, Tokyo 173, Japan

Abstract

Mount Erebus, an active volcano on Ross Island, began its present continuously eruptive phase in 1972; it was monitored seismically from December 1980 until December 1986. From December 1980 to August 1984 the number of seismic events around Mount Erebus averaged 20–150/d, including several earthquake swarms. A change in the volcanic activity (between 13 September and 31 December 1984) was accompanied by a distinct decrease in the intensity of background seismicity, and only a few earthquakes were recorded subsequently. The number of seismic events during 1985 and 1986 averaged 10–20/d, with one and two earthquake swarms, respectively, for those years. Seven shots were detonated at four sites on Mount Erebus during November and December 1984. The P-wave velocity of the top layer in the Mount Erebus area is 3.07 km/s with a thickness of 2.2–2.4 km; those of the second and third layers are 4.67 and 6.27 km/s, respectively. The depth of the boundary between the second and third layers is 6.6 km from the summit.

Introduction

Mount Erebus (77°37′ S, 167°09′ E, 3794 m), on Ross Island, is the only active volcano in Antarctica currently showing sustained eruptive activity. A programme to monitor continuously the seismic activity of Mount Erebus and to identify the mechanism of eruptions began in December 1980 as an international cooperative project between Japan, the United States and New Zealand, the 'International Mount Erebus Seismic Study' (IMESS).

The seismic activity observed on and around Mount Erebus during the period 1980–84 included: (1) volcanic earthquakes at a rate of 20–150 events/d; (2) widely distributed earthquakes around Mount Erebus; and (3) some earthquake swarms on the flanks of Mount Erebus which had more than several hundred events per day and continued for several days after initiation (Shibuya *et al.*, 1983; Kaminuma, Ueki & Kienle, 1985).

New volcanic activity commenced on September 1984 and lasted until the end of December 1984. A distinct change in the background seismicity was recognized before and after the new eruptive activity (Kaminuma, 1987).

To improve our knowledge of the crustal structure, IMESS carried out explosion seismic experiments during November and December 1984 (Kaminuma *et al.*, 1985; Kienle *et al.*, 1985). This paper is a report of the seismic activities from 1982 to 1986, and a preliminary report of the explosion experiments.

Seismic observations

IMESS began seismic observations in December 1980, using three seismic stations. By late 1983, 10 stations had been installed at the localities shown in Fig. 1. These stations were linked by radiotelemetry to Scott Base (77°51′03″ S, 166°45′45″ E), about 38 km south of Mount Erebus summit. The data were recorded on a 14-channel magnetic tape and an analogue chart recorder with a quartz clock timer (Kienle *et al.*, 1981; Takanami *et al.*, 1983a, b).

Seven temporary seismic stations operated around Mount Erebus during November and December 1984 (Fig. 1). Each

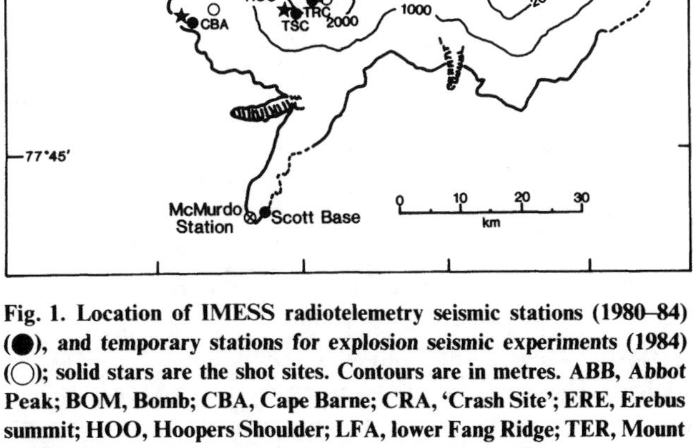

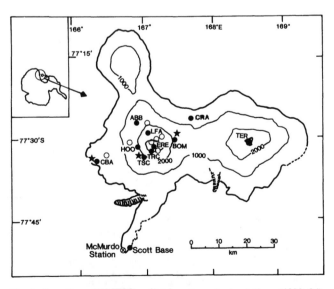

Fig. 1. Location of IMESS radiotelemetry seismic stations (1980–84) (●), and temporary stations for explosion seismic experiments (1984) (○); solid stars are the shot sites. Contours are in metres. ABB, Abbot Peak; BOM, Bomb; CBA, Cape Barne; CRA, 'Crash Site'; ERE, Erebus summit; HOO, Hoopers Shoulder; LFA, lower Fang Ridge; TER, Mount Terror; TRC, 'Truncated Cones'; TSC, Three Sisters Cones.

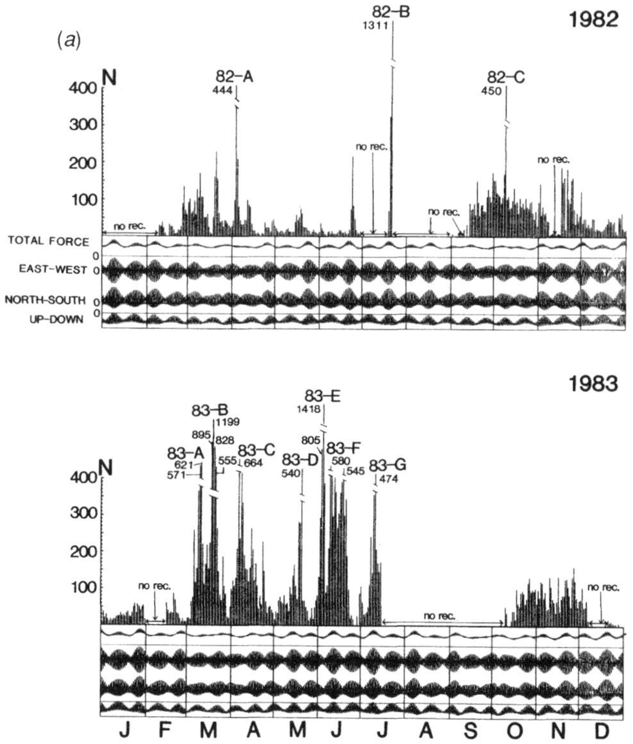

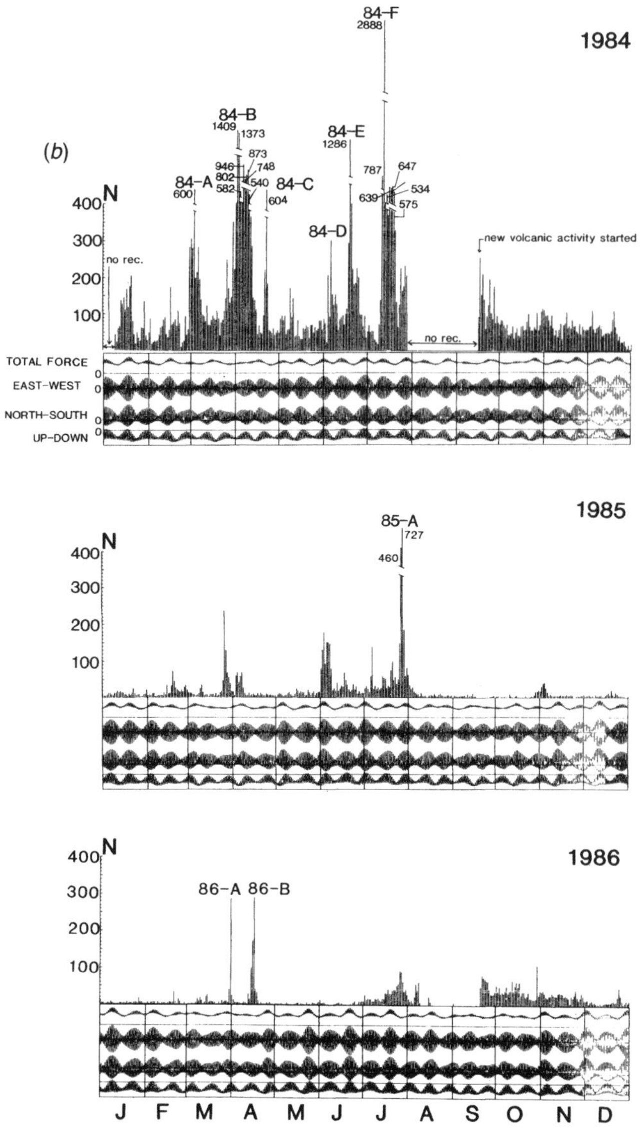

Fig. 2. (*a*) Daily number of earthquakes counted at the Hoopers Shoulder station, 1982–83. Earthquake swarms are shown as 82-A, 82-B, etc. (*b*) Daily number of earthquakes counted at the Hoopers Shoulder station, 1984–86.

station was equipped with a 2-Hz natural frequency vertical geophone and a slow-speed tape recorder (Ueki *et al.*, 1984). Each station clock was synchronized with a base station master crystal clock, which was also synchronized with the radio-telemetry network quartz clock.

Daily Frequency

The daily number of earthquakes counted at the Hoopers Shoulder station in 1982–86 is shown in Figs 2*a* & *b*. Although the magnitude of the earthquakes was not determined, the magnitude of all the events shown in Fig. 2 was estimated to be < 2.0, from the amplitude on seismograms. Only events with (S – P)-times < 10 s and amplitudes > 2 mm on a chart recorder operated at a gain of 4 V/cm, were counted. Teleseisms and local earthquakes which occurred outside Ross Island, and icequakes were not counted.

Between January 1982 and June 1984 the daily number of earthquakes varied from 20 to over 250, with averages of 64, 134 and 146 in 1982, 1983 and January–June 1984, respectively. The four components of earth tide (Total, E–W, N–S, vertical) at 77°30′ S are shown in Figs 2*a* & *b*. No clear relationship between the earth tides and the daily count has been found.

New volcanic eruptive activity began on 13 September 1984 with a number of large explosions (Kienle *et al.*, 1985; Kaminuma, 1987) and lasted until the end of December 1984, when earthquake activity decreased to < 20/d. The low level of seismicity continued during 1985 and 1986, with 23 and 17 events/d, respectively. From August 1985 to June 1986, only 9 events/d were recorded but from July until 16 August 1986, the number of earthquakes increased to 27/d. However, between 17 August 1986 and 19 September 1986, a total of only 11 earthquakes was recorded. After this period of extreme low seismicity, earthquakes increased again to 16 events/d.

Earthquake swarms occur frequently on Mount Erebus. An earthquake swarm is defined as > 250 events/24 h, with a relatively large number of earthquakes on either side of the 250-event day. 16 swarms were recorded during a 30-month period in 1982–84 (Kaminuma, Baba & Ueki, 1986; Fig. 2).

Only three earthquake swarms occurred during 1985 and 1986 (Table 1), the duration and total number of earthquakes in these swarms being smaller than the average values of the 16

Table 1. *Earthquake swarms recorded at Mount Erebus in 1985 and 1986*

Swarm	Period	Duration (h)	Total number of earthquakes
85-A	2200; July 26–0800; July 29 1985	59	2195
86-A	1700; March 30–2200; March 31 1986	30	1165
86-B	2300; April 13–0200; April 18 1986	100	3606

swarms reported by Kaminuma *et al.* (1986). It is clear that a distinct decrease in seismicity, as marked by daily numbers of earthquakes and the frequency of earthquake swarms, occurred after the 1984 volcanic activity (Fig. 2).

Explosion seismic experiments

Seven shots were detonated at four different sites on Mount Erebus during November and December 1984 (Fig. 1). Shots were located adjacent to seismic stations so that explosion times could be recorded.

Fig. 3 is a reduced travel time graph, with a reduction velocity of 2.9 km/s, for the shot site at lower Fang Ridge. The charge size of each shot was ~ 10 kg, and the elevation of the shot point was about 2570 m. The shots were recorded at stations 0–10 km away. An apparent P-wave velocity of 3.07 km/s is the velocity of the top layer shown in Fig. 5.

Fig. 4 shows the reduced travel time graphs, with a reduction velocity of 2.9 km/s, for the shot sites at Three Sisters Cones (2240 m), Bomb (2230 m) and Cape Barne (at sea level). The

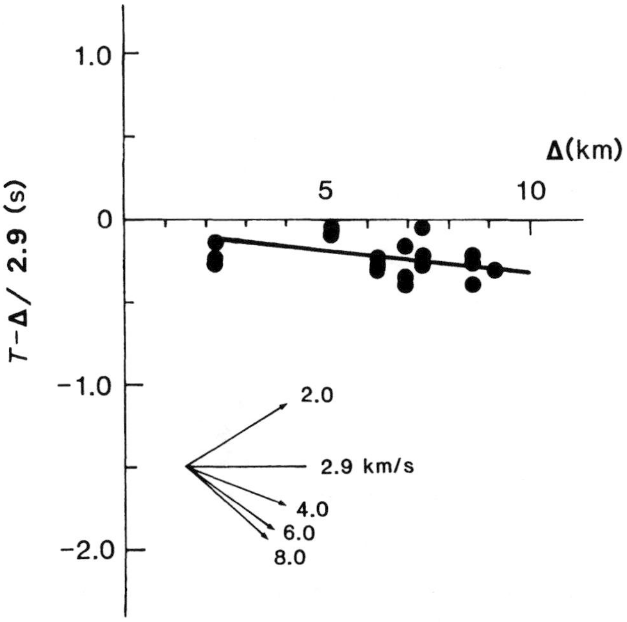

Fig. 3. Reduced travel-time graph of the three explosions at the site on lower Fang Ridge.

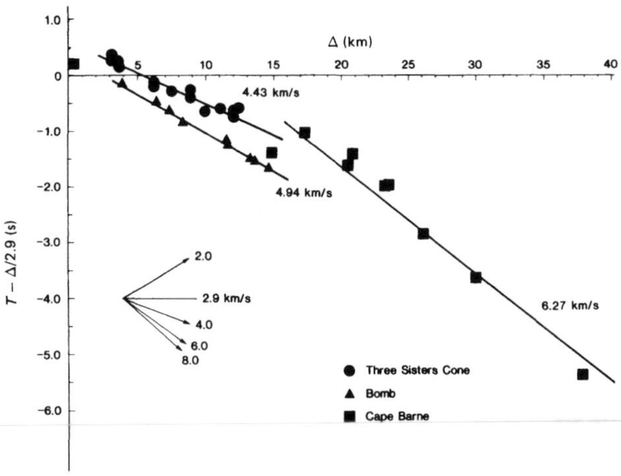

Fig. 4. Reduced travel-time graphs of the explosions at Three Sisters Cones, Cape Barne and Bomb sites.

epicentral distances from the shots at Three Sisters Cones and Bomb varied from 0 to 15 km. The measurement lines of the Three Sisters Cones and Bomb explosions are consistently reversed from each other. The apparent velocity of 4.43 km/s is obtained from two explosions at Three Sisters Cones whereas the 4.94 km/s came from the Bomb explosion. From these two apparent velocities, a velocity of 4.67 km/s was obtained for the P-wave velocity of the second layer. The shot at Cape Barne was recorded at stations 0–40 km away. An apparent velocity of 6.27 km/s (Fig. 4) was obtained using recordings made at stations 15 or more kilometres distant.

A subsurface structure obtained from the explosion experiments is shown in Fig. 5. The thickness of the first layer is 2.2–2.4 km with an apparent velocity of 3.07 km/s. The boundary between the first and second layers is slightly inclined, the thickness of the second layer changing from 4.2 km beneath the Three Sisters Cones site to 4.4 km beneath the Bomb site. The apparent velocity of the third layer is 6.27 km/s and the depth of the boundary between the second and third layers is 6.6 km from the summit.

Discussion

Earthquakes were relocated on the basis of the structural model reduced by the explosion experiments. Figs 6 & 7 show the hypocentre distributions during the two periods February

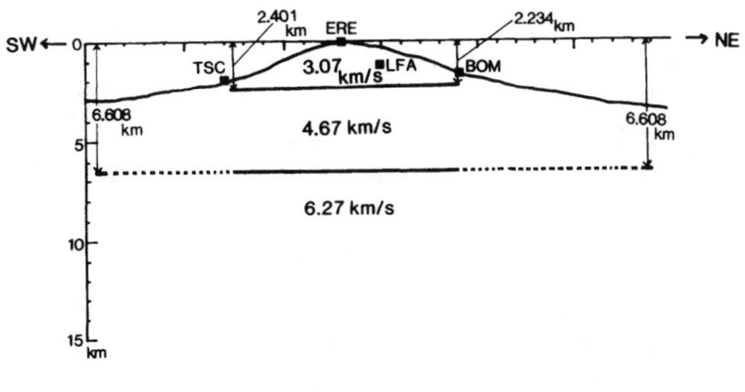

Fig. 5. The P-wave velocity structure of Mount Erebus reduced from the explosion seismic experiments. For BOM, ERE, LFA and TSC locations, see Fig. 1.

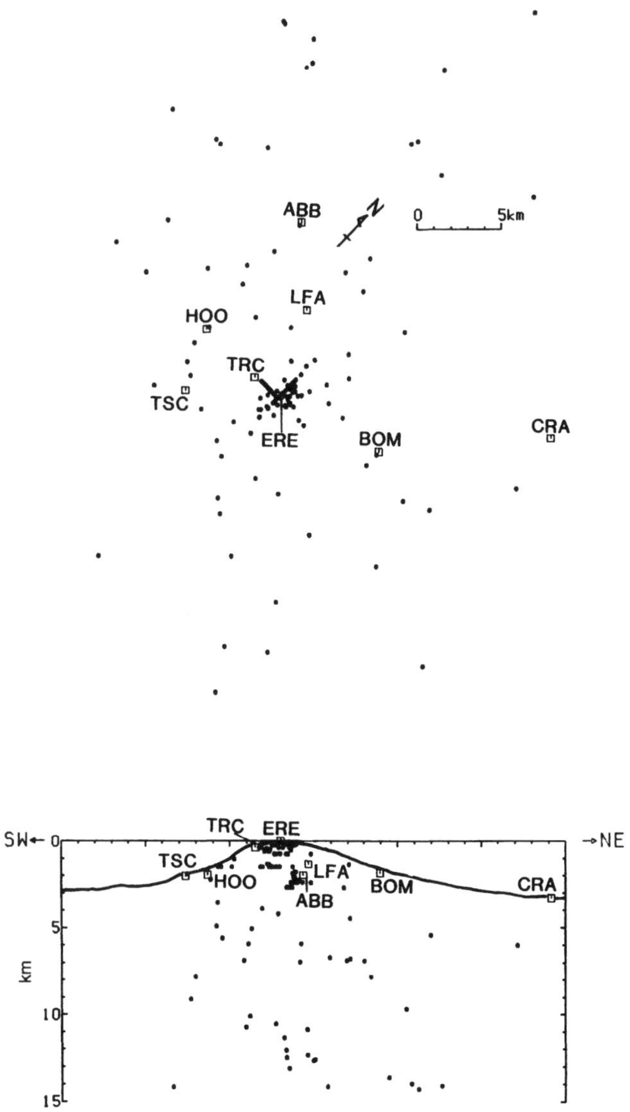

Fig. 6. Locations of earthquake hypocentres, February 1982–May 1984. See Fig. 5 for the structural setting of these earthquake locations and Fig. 1 for place-name abbreviations.

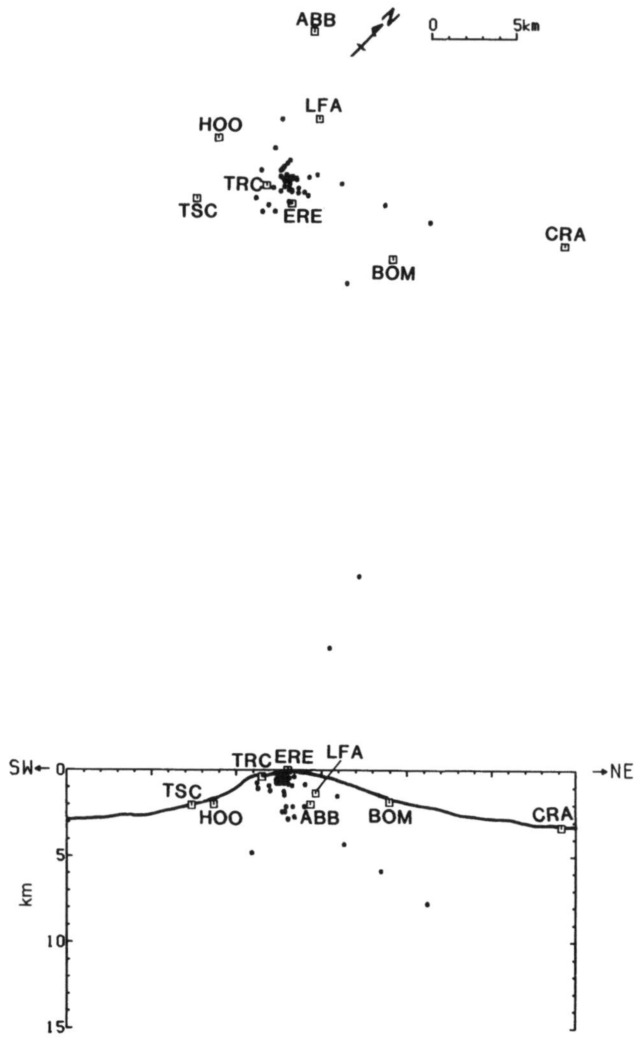

Fig. 7. Locations of earthquake hypocentres, September 1984–December 1986. See Fig. 5 for the structural setting of these earthquake locations and Fig. 1 for place-name abbrivations.

1982–May 1984 and September 1984–December 1986; vertical and horizontal location errors are < 0.5 and 1 km, respectively.

The P-wave velocity structure used in earlier papers (Kaminuma *et al.*, 1985) is an empirical model obtained from the structures of other volcanoes (Table 2) (e.g. Minakami *et al.*, 1970). The P-wave velocities at every depth of this model are smaller than those of the new model obtained from the explosion experiments. Rowe & Kienle (1986) also used a model in which P-wave velocities near the surface are from 1.25 to 2.0 km/s. The velocity of the top layer of both these earlier models is 1.2–1.3 km/s. This layer was not found during the explosion experiments described here.

Table 2. *P-wave velocity model for Mount Erebus*

Thickness (km)	P-wave velocity (km/s)
1.0	1.25
1.0	1.62
1.0	1.99
1.0	2.36
1.0	2.73
2.75	3.16
0.25	4.64
2.0	5.66
	6.61

Acknowledgements

The authors wish to thank all IMESS personnel, especially Prof. J. Kienle of the University of Alaska and Dr R.R. Dibble of Victoria University, Wellington, for their cooperation in IMESS. The maintenance of the seismic recording system was performed by the wintering parties of Scott Base, New Zealand. Ms Y. Shudo and A. Takahashi are acknowledged for typing the manuscript.

References

Kaminuma, K. (1987). Seismic activity of Erebus volcano, Antarctica. *Pure and Applied Geophysics*, **125(6)**, 993–1008.

Kaminuma, K., Baba, M. & Ueki, S. (1986). Earthquake swarms on Mount Erebus, Antarctica. *Journal of Geodynamics*, **6(1–4)**, 391–404.

Kaminuma, K., Ueki, S. & Kienle, J. (1985). Volcanic earthquake swarms at Mt Erebus, Antarctica. *Tectonophysics*, **114(1–4)**, 357–69.

Kienle, J., Kyle, P.R., Estes, S., Takanami, T. & Dibble, R.R. (1981). Seismicity of Mount Erebus 1980–1981. *Antarctic Journal of the United States*, **16(5)**, 35–6.

Kienle, J., Rowe, C.A., Kyle, P.R., McIntosh, W.C. & Dibble, R.R. (1985). Eruption of Mount Erebus and Ross Island seismicity, 1984–1985. *Antarctic Journal of the United States*, **20(5)**, 25–7.

Minakami, T., Utibori, S., Miyazaki, T., Hiraga, S., Terao, H. & Hirai, K. (1970). Seismometrical studies of Volcano Asama, Part 2. *Bulletin of the Earthquake Research Institute*, **48(3)**, 431–89.

Rowe, C.A. & Kienle, J. (1986). Seismicity in the vicinity of Ross Island, Antarctica. *Journal of Geodynamics*, **6(1–4)**, 375–85.

Shibuya, K., Baba, M., Kienle, J., Dibble, R.R. & Kyle, P.R. (1983). A study of the seismic and volcanic activity of Mount Erebus, Antarctica, 1981–1982, pp.54–66. Tokyo; Memoirs of National Institute of Polar Research, Special Issue, No. 28.

Takanami, T., Kaminuma, K., Terai, K. & Osada, N. (1983a). Seismological observations on Mount Erebus, Ross Island, Antarctica, 1980–1981, pp. 46–53. Tokyo; Memoirs of National Institute of Research, Special Issue, No. 28.

Takanami, T., Kienle, J., Kyle, P.R., Dibble, R.R., Kaminuma, K. & Shibuya, K. (1983b). Seismological observations on Mount Erebus, 1980–1981. In *Antarctic Earth Science*, ed. R.L. Oliver, P.R. James & J.B. Jago, pp. 671–4. Canberra; Australian Academy of Science and Cambridge; Cambridge University Press.

Ueki, S., Kaminuma, K., Baba, M., Koyama, E. & Kienle, J. (1984). Seismic activity of Mount Erebus, Antarctica in 1982–1983, pp. 29–40. Tokyo; Memoirs of National Institute of Polar Research, Special Issue, No. 33.

Crustal development: the Pacific margin

Mid-Palaeozoic basement in eastern Graham Land and its relation to the Pacific margin of Gondwana

A.J. MILNE & I.L. MILLAR

British Antarctic Survey, Natural Environment Research Council, High Cross, Madingley Road, Cambridge CB3 0ET, UK

Abstract

Geochronological study of the orthogneissic country rock to the Mesozoic magmatic arc of Graham Land confirms for the first time the presence of a mid-Palaeozoic basement. Rb–Sr data on whole-rock samples of the orthogneiss indicate a Silurian origin involving little crustal contribution. Sm–Nd studies on garnet–whole-rock pairs indicate that subsequent amphibolite-facies metamorphism occurred during the Late Carboniferous. The extent of this basement is unknown but it may well underlie much of the Antarctic Peninsula. In mid-Palaeozoic time Graham Land may have been attached to the Pacific margin of Gondwana and this plutonic event may be correlated with subduction-related intrusive activity elsewhere along Gondwana's Pacific margin.

Introduction

Although the exposed geology of the Antarctic Peninsula is almost entirely Mesozoic and Cenozoic, it has been clear for some time now that the Palaeozoic history of this region is important with respect to the Pacific margin of Gondwana (Dalziel & Elliot, 1982). Much of what had initially been thought of as pre-Mesozoic basement (Adie, 1954) was subsequently shown to be of Mesozoic age (e.g. Gledhill, Rex & Tanner, 1982). Petrological and isotopic studies, however, suggested the presence of a substantial amount of pre-Mesozoic crustal material at depth beneath Graham Land (Hamer & Moyes, 1982; Pankhurst, 1982, 1983), although it is only now that the first reliable radiometric age determination on basement rocks from Graham Land has been made.

The problems which have hindered the isotopic search for the basement are the extensive Mesozoic magmatism and, in eastern Graham Land, the Triassic amphibolite-grade metamorphism (Rex, 1976; Pankhurst, 1983). The result of these events is the resetting of most pre-existing isotopic systems. However, migmatitic gneisses from east Graham Land analysed by Pankhurst (1983) were found to contain much more radiogenic Sr ($\epsilon_{Sr}^{200} > +150$) than could be explained by an exclusively post-Palaeozoic history involving no older crustal material. Three whole-rock samples from granitic sheets at Target Hill in central-eastern Graham Land (Fig. 1), also analysed by Pankhurst, suggested a middle–late Palaeozoic age, although more data were obviously needed. This area, from which the strongest indications of a pre-Mesozoic history had been obtained, was revisited to study the history of possible basement terranes.

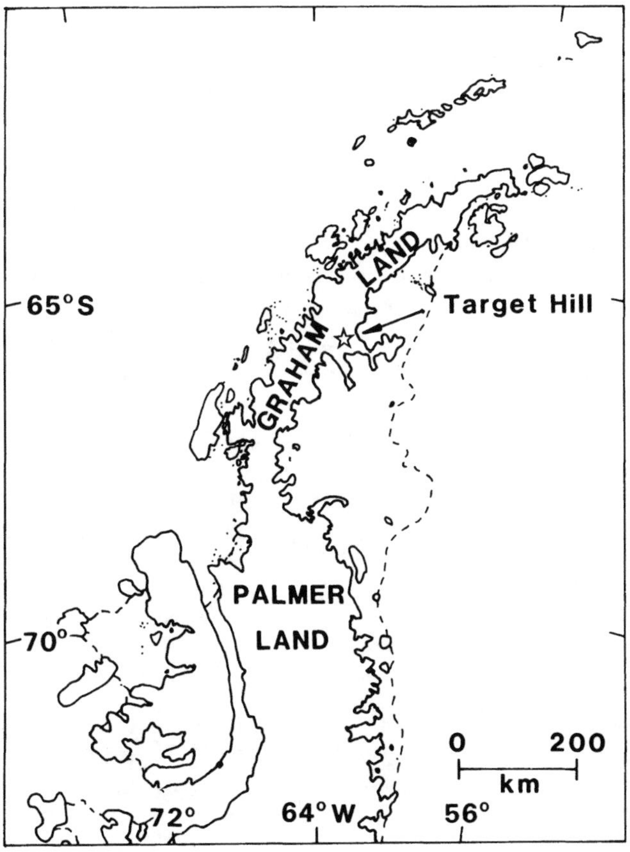

Fig. 1. Map of the Antarctic Peninsula showing the location of Target Hill.

The geology of central-eastern Graham Land

The geology of central-eastern Graham Land, in common with most of the Antarctic Peninsula, is dominated by magmatic arc rocks, which have been shown in this area to be of Middle Jurassic age (eight isochrons reported in Pankhurst (1982)). Field relations indicate that three lithologies are of pre-Middle Jurassic age: granitic orthogneiss, migmatitic banded gneiss and metasedimentary rocks (Milne, 1987). Further age constraints have not yet been placed on the migmatitic banded gneiss nor the metasedimentary rocks but the granitic orthogneiss has been determined to be mid-Palaeozoic in age.

The largest exposure of granitic orthogneiss occurs at Target Hill (Fig. 1). Here the granitic orthogneiss contains a large proportion (up to 30%) of concordant amphibolite sheets. This banded complex is cross-cut by many dykes and veins, which are from oldest to youngest: Jurassic pegmatite and aplite veins; large medium-grained amphibole-rich mafic dykes; porphyritic dykes and faults; and thin fine-grained mafic dykes.

Petrography

The granitic orthogneiss at Target Hill has a mineralogy of quartz + plagioclase + K-feldspar + biotite ± garnet ± muscovite. The foliation is developed to varying degrees and the samples may be subdivided on this basis. Those samples with the poorest fabric tend to be fine grained or pegmatitic and are of tonalitic or granodioritic composition. In the fine-grained samples the feldspar crystals are generally preserved as porphyroclasts whereas the quartz has been completely recrystallized and reduced in grain size. The pegmatitic samples often contain large K-feldspar phenocrysts. The more strongly foliated orthogneisses contain more K-feldspar than plagioclase and consist of a fine-grained granoblastic aggregate of quartz and feldspar surrounding some perthitic K-feldspar augen. The foliation is accentuated by biotite and coarser quartz segregations which show a strong preferred orientation. The textures of all the orthogneiss samples indicate that a tectonothermal event has affected these rocks. Garnet, where present, is clearly of metamorphic origin and contains many inclusions of quartz and feldspar.

The concordant amphibolite sheets have a pronounced foliation parallel to that in the granitic orthogneiss, though the amphibolite sometimes shows a crenulation not present in the orthogneiss. The mineralogy of hornblende + quartz + plagioclase + biotite + sphene ± garnet indicates that these sheets have undergone an amphibolite-facies metamorphism.

Isotope geochemistry

Analytical techniques

Sr and Nd isotopic compositions were analysed on a VG354 multicollector mass spectrometer. $^{143}Nd/^{144}Nd$ isotopic ratios are normalized to $^{146}Nd/^{144}Nd = 0.7219$. Rb/Sr atomic ratios were determined by X-ray fluorescence spectrometry, and Sm and Nd concentrations were determined by isotope dilution, using ^{149}Sm and ^{150}Nd isotopic tracers, respectively. All analyses are presented in Figs 2 & 3; full data are available from the authors upon request.

Rb–Sr isotopes

Rb–Sr analyses were made of 17 whole-rock samples from two localities at Target Hill. Six samples of weakly foliated orthogneiss from western Target Hill define an isochron (MSWD = 2.8), yielding an age of 410 ± 15 Ma with an initial $^{87}Sr/^{86}Sr$ ratio of 0.7030 ± 0.0001 (Fig. 2*a*). Four samples of strongly foliated orthogneiss, collected from the same locality, define an isochron (MSWD = 0.3) giving an age of 426 ± 12 Ma with an initial ratio of 0.7040 ± 0.0002 (Fig. 2*b*).

While the two ages obtained agree within analytical error, and the low initial ratios suggest the ages are close to the date of emplacement of the orthogneiss protolith, the initial ratio of the more strongly foliated gneisses is significantly elevated, perhaps indicating interaction with a more radiogenic fluid phase during deformation. The low initial ratio of the weakly foliated gneisses with ϵ_{Sr}^{410} around −15, precludes any significant involvement of older crustal material in their genesis.

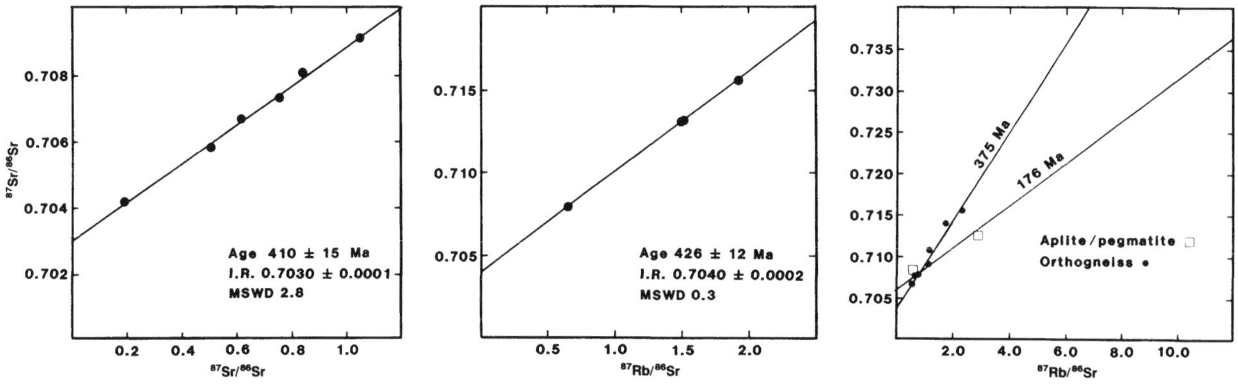

Fig. 2. Rb–Sr isochrons for (*a*) weakly foliated and (*b*) strongly foliated orthogneiss from western Target Hill; and (*c*) orthogneisses and cross-cutting pegmatites from central Target Hill. The best-fit line for the orthogneiss is shown, together with a reference isochron obtained from a nearby Jurassic pluton. Errors are 0.5% on Rb–Sr, 0.01% on $^{87}Sr/^{86}Sr$ (1-sigma); errors on ages and initial ratios (IR) are quoted at the 2-sigma level.

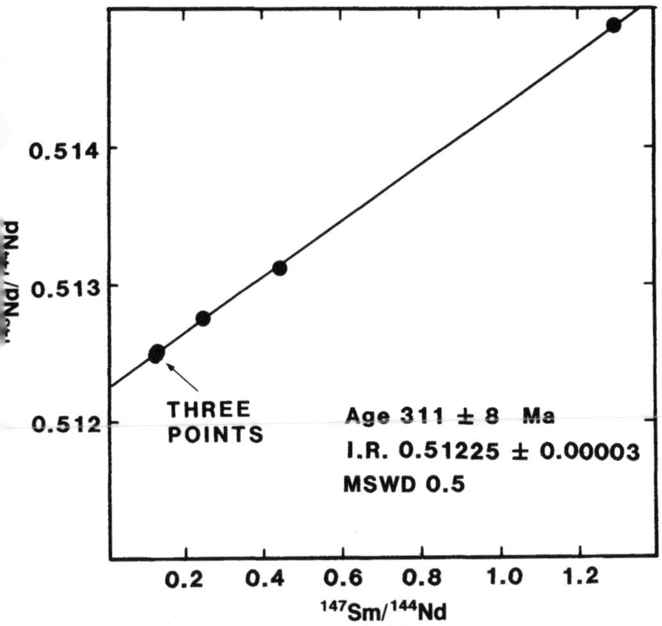

Fig. 3. Sm–Nd isochron for the garnet–whole-rock pairs from the granitic-orthogneiss and migmatitic banded gneiss of Target Hill. Errors are 0.2% on Sm–Nd and 0.005% on ^{143}Nd/^{144}Nd (1-sigma); errors on age and initial ratio (IR) are quoted at the 2-sigma level.

Seven samples of orthogneiss from central Target Hill form a linear array with a MSWD of 73.4, which defines an age of 375 ± 68 Ma with an initial ratio of 0.7039 ± 0.001 (Fig. 2*c*). Three samples of cross-cutting aplite and pegmatite from this locality lie close to a reference isochron obtained from the nearby Jurassic Moider Peak granite (Milne, unpublished data).

Sm–Nd isotopes

In an attempt to constrain the age of the amphibolite-facies metamorphism which affected the orthogneiss, Sm–Nd isotopic analyses were attempted on three garnet–whole-rock pairs. Garnets from a pegmatitic segregation within the granitic orthogneiss have a considerably higher Sm/Nd ratio than their host whole-rock and define an age of 309 ± 11 Ma. Garnets from a sample of the orthogneiss itself show moderate enrichment in Sm relative to Nd and define an age of 292 ± 35 Ma. Garnets analysed from the melanosome of a banded migmatitic gneiss showed no significant fractionation of Sm and Nd.

Data from all three garnet–whole-rock pairs plot on a single isochron (MSWD = 0.5), defining an age of 311 ± 8 Ma with an initial ^{143}Nd/^{144}Nd ratio of 0.51225 ± 0.00003 (Fig. 3). This must represent the age of the amphibolite-facies metamorphism. Pankhurst (1983) obtained an age of 336 ± 34 Ma by Rb–Sr whole-rock analysis of three samples from nearby garnet-bearing granite sheets which may have been emplaced during the same event. The Target Hill gneisses have ϵ_{Nd}^{410} from − 0.2 to + 1.2, confirming the lack of older crustal involvement in their genesis.

Mid-Palaeozoic plutonism on the Pacific margin of Gondwana

The recognition of a mid-Palaeozoic plutonic event in the Antarctic Peninsula is important when considering the pre-Mesozoic development of the Pacific margin of Gondwana (Fig. 4), and may be correlated with plutonism of comparable age on the Pacific sides of central South America, northern Victoria Land (NVL), Marie Byrd Land (MBL), New Zealand, Tasmania and Australia. The fact that Palaeozoic plutonism on Thurston Island is significantly younger is important, as is its apparent absence in Patagonia.

South America

In northern and central Chile and north-western Argentina, Early–Late Palaeozoic ages have been obtained from granitoids which have been interpreted as subduction-related calc-alkaline plutons (Davidson, Mpodozis & Rivano, 1981; Rapela, Heaman & McNutt, 1982; Mpodozis *et al.*, 1983; Brook *et al.*, 1987). This magmatic belt may be traced southwards through South America (Ramos & Ramos, 1979) but rather than being parallel to the present day Pacific coast, it swings eastward to the Atlantic coast at 42° S (Fig. 4). A Late Palaeozoic–early Mesozoic batholith follows a similar trend to the south and west of this (Bartholomew & Tarney, 1984). Despite the absence of any known exposures of Palaeozoic plutonic rocks south of 48° S, isotopic studies of the upper Palaeozoic metasedimentary basement in this region do not preclude a mid-Palaeozoic crustal component (Brook *et al.*, 1987).

Northern Victoria Land

A detailed study of the granitoids from NVL has been made by Borg *et al.* (1987). Trend surface analysis on 112 chemical analyses from Cambro-Ordovician Granite Harbour Intrusive (GHI) and Devonian Admiralty Intrusive (AI) granitoids indicate that there is a strong compositional polarity for both these suites. Assuming the I-type GHI and AI granitoids to be subduction-related, this compositional polarity may be related to the polarity of subduction. Somewhat surprisingly, there is a marked difference in the azimuth of this polarity between the GHI and AI suites. Borg *et al.* (1987) suggested that the terrane into which the AI granitoids were emplaced is allochthonous with respect to the rest of NVL and that the juxtaposing of these two crustal blocks occurred some time in the Devonian or Carboniferous (i.e. after AI emplacement).

One interpretation of these data (which is supported by the Sr and Nd isotopic data presented by Borg *et al.* (1987)) is that the subduction zone, to which the AI granitoids were related, dipped towards the Pacific, and consumed oceanic crust which was present between the two allochthonous components of NVL. The microcontinent which hosted AI emplacement was then accreted onto NVL during the Devonian or Carboniferous. After this, subduction of proto-Pacific plate resumed farther out on the Pacific side of the then complete NVL.

Tasmania

A considerable amount of isotope geochronology has been done on the granitoids of Tasmania and their Proterozoic–Palaeozoic country rocks (e.g. Cocker, 1982; Adams

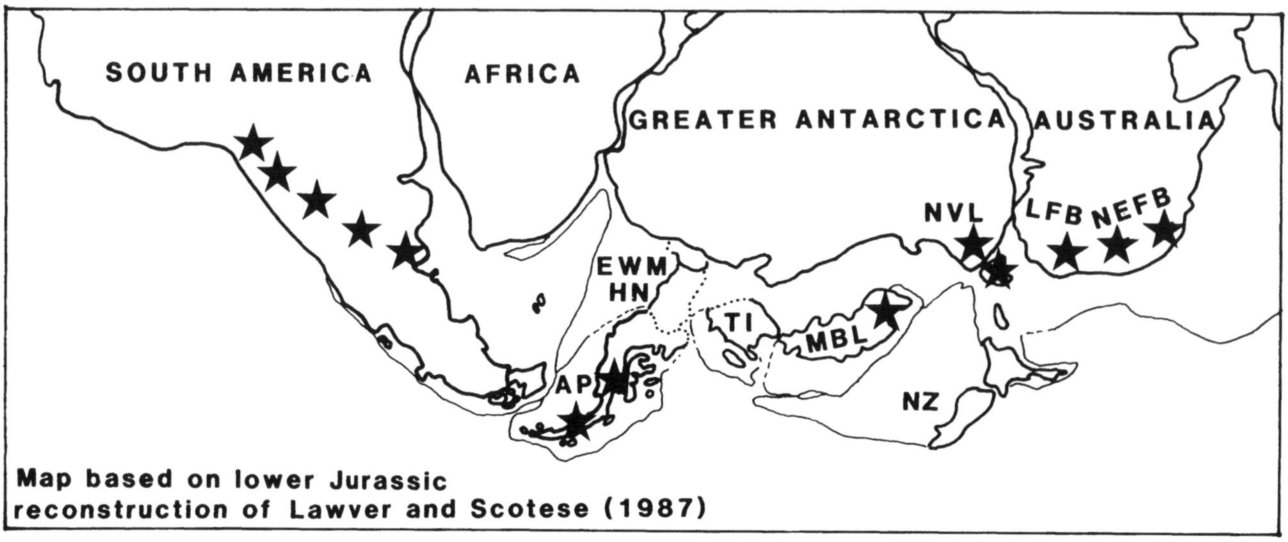

Fig. 4. Map of the Pacific margin of Gondwana (from Lawver & Scotese, 1987) showing the location of mid-Palaeozoic subduction-related plutonism (stars) in South America, the Antarctic Peninsula (AP), Marie Byrd Land (MBL), northern Victoria Land (NVL), Tasmania, the Lachlan Ford Belt (LFB) and the New England Fold Belt (NEFB). Also indicated are the Ellsworth and Whitmore mountains (EWM), Haag Nunataks (HN), Thurston Island (TI) and New Zealand (NZ).

et al., 1985). The predominantly Devonian S- and I-type granitoids (Rb–Sr biotite ages are 395–370 Ma in eastern, and 375–340 Ma in western, Tasmania) are supposed to have been emplaced after the Tabberabberan Orogeny, for which a biostratigraphical age range is 395–380 Ma. However, Adams *et al.* (1985) noticed a great preponderance of K–Ar and Rb–Sr ages of pre-Tabberabberan slates and schists from western Tasmania in the range 420–400 Ma, which could not be entirely explained by partial resetting during the Tabberabberan Orogeny. There may well have been a tectonothermal event during the period 420–400 Ma which is not apparent from the stratigraphic record.

Eastern Australia

Mid-Palaeozoic granitoids are extensively exposed along the entire length of the Pacific margin of Australia in the New England Fold Belt (NEFB) and the Lachlan Fold Belt (LFB) (Fig. 4). The LFB granitoids in the south are generally older (Late Silurian–Early Devonian; Brooks & Leggo, 1972; Richards & Singleton, 1981) than the NEFB granitoids to the north (mainly Devonian; Day, Murray & Whitaker, 1978; Harrington & Korsch, 1985a).

Subduction of proto-Pacific oceanic crust beneath the Pacific margin of Australia throughout most of the Palaeozoic caused considerable crustal anatexis, which resulted in the emplacement of diverse suites of granitoids in the LFB and NEFB.

It has been proposed that a major island arc (~ 1000 km long) existed between the Late Silurian and Middle Devonian to the Pacific side of the NEFB (Day *et al.*, 1978). This arc was supposed by Harrington & Korsch (1985a, b) to be related to a subduction zone which dipped towards the Pacific and consumed a marginal sea during the middle Palaeozoic, and ended in the Carboniferous when the arc was accreted onto the Australian craton, and westward-dipping subduction of the Pacific plate was resumed.

New Zealand

The plutonic rocks of the Karamea suite, South Island, New Zealand are middle–late Palaeozoic (430–280 Ma) S-type granitoids and may be split into two groups, the older of which is thought to be Siluro-Devonian (Tulloch, 1983) and related to an early phase of the Tuhuan Orogeny. The main tectonothermal event of the Tuhuan Orogeny was Devonian and was associated with post-tectonic granitic plutonism (370–280 Ma; Adams, 1981). There is no evidence for a mid-Palaeozoic active margin in the vicinity of New Zealand (Crook & Feary, 1982).

Marie Byrd Land

Mid-Palaeozoic plutonism in MBL is represented by the I-type Ford granodiorite whose intrusion is thought to have occurred in the range 380–350 Ma (Adams, 1986) and caused Ar loss from Lower Palaeozoic metasedimentary rocks whose metamorphism occurred in the Late Ordovician.

Thurston Island, Haag Nunataks and Ellsworth–Whitmore mountains

In the remaining three components of the Pacific margin of Gondwana there is no evidence for any mid-Palaeozoic plutonism. The basement to Thurston Island is no older than ~ 300 Ma (Storey *et al.*, this volume, p. 399). The youngest rocks found at Haag Nunataks are Precambrian and, with the exception of a minimum 400 Ma age on altered dolerites (Yoshida, 1983), the onset of magmatism in the Ellsworth and Whitmore mountains did not occur until the Mesozoic (Storey *et al.*, 1988).

Discussion

The presence of mid-Palaeozoic orthogneissic basement to the Antarctic Peninsula is demonstrated in central eastern Graham Land using both Rb–Sr and Sm–Nd isotopic methods. The intrusion of the granitoid protolith is shown to have occurred at around 410 Ma and the subsequent amphibolite-facies metamorphism, which resulted in garnet growth, occurred at 311 ± 8 Ma. Elsewhere along the Antarctic Peninsula evidence of mid-Palaeozoic plutonism is sparse. Pankhurst (1983) described a suite of granophyre cobbles from within the Trinity Peninsula Group metasedimentary rocks which yielded a 386 ± 39 Ma errorchron (MSWD = 80.3). Harrison & Piercy (this volume, p. 341) described orthogneiss from north-western Palmer Land which yielded ages of 399 ± 35 and 440 ± 57 Ma. Thus it is possible that this basement underlies much of the Antarctic Peninsula.

Mid-Palaeozoic subduction-related plutonism occurred along much of the Pacific margin of Gondwana. It is possible that the subducting slab was proto-Pacific oceanic crust throughout, but there have been speculations by several independent workers that periods of subduction of the opposite polarity were involved (Crook & Feary, 1982; Dalziel & Forsythe, 1985; Harrington & Korsch, 1985b; Ramos *et al.*, 1986). Regardless of the subduction polarity, it is clear that accretion of large allochthonous terranes played an important role in the mid-Palaeozoic history of the Pacific margin of Gondwana (e.g. 'Chilenia' (Ramos *et al.*, 1986), part of NVL (Borg *et al.*, 1987) and NEFB (Harrington & Korsch, 1985b)).

The boundary between Gondwana and some of the allochthonous terranes was most likely a zone of subduction but the direction in which the oceanic crust (which lay between the two crustal masses) was consumed is contentious, as is the degree of allochthoneity of some of the accreting blocks. The presence of supra-subduction zone magmatic rocks of pre-accretion age in the block which was destined to accrete necessitates a slab dipping beneath that allochthonous terrane. This may be achieved in several different ways, however, not all of which require a slab which dips towards the Pacific.

Acknowledgements

The authors would like to thank all their colleagues at the British Antarctic Survey (BAS) for useful discussions and constructive comments during the preparation of this paper, in particular Bryan Storey. The fieldwork on which this paper is based would not have been possible without the assistance of Alasdair Cain and Chris Griffiths and the support staff at the BAS Rothera Base.

References

Adams, C.J. (1981). Geochronological correlations of Precambrian and Palaeozoic orogens in New Zealand, Marie Byrd Land (West Antarctica), northern Victoria Land (East Antarctica) and Tasmania. In *Gondwana Five*, ed. M.M. Cresswell & P. Vella, pp. 191–8. Rotterdam; A.A. Balkema.

Adams, C.J. (1986). Geochronological studies of the Swanson Formation of Marie Byrd Land, West Antarctica, and correlation with northern Victoria Land, East Antarctica, and South Island, New Zealand. *New Zealand Journal of Geology and Geophysics*, **29(3)**, 345–58.

Adams, C.J., Black, L.P., Corbett, K.D. & Green, G.R. (1985). Reconnaissance isotopic studies bearing on the tectonothermal history of early Palaeozoic and late Proterozoic sequences in western Tasmania. *Australian Journal of Earth Sciences*, **32(1)**, 7–36.

Adie, R.J. (1954). *The Petrology of Graham Land: I. The Basement Complex; Early Palaeozoic Plutonic and Volcanic Rocks.* London; Falkland Islands Dependencies Survey Scientific Reports, No. 11, 22 pp.

Bartholomew, D.S. & Tarney, J. (1984). Crustal extension in the southern Andes (45°–46° S). In *Marginal Basin Geology*, ed. B.P. Kokelaar & M.F. Howells, pp. 195–206. Geological Society of London Special Publication 16, Oxford; Blackwell Scientific Publications.

Borg, S.G., Stump, E., Chappell, B.W., McCulloch, M.T., Wyborn, D., Armstrong, R.L. & Holloway, J.R. (1987). Granitoids of northern Victoria Land, Antarctica: implications of chemical and isotopic variations to regional crustal structure and tectonics. *American Journal of Science*, **287(2)**, 127–69.

Brook, M., Pankhurst, R.J., Shepherd, T.J. & Spiro, B. (1987). *ANDCHRON, Andean geochronology and metallogenesis.* Report on BGS/ODA research and development project 1983–1986. 137 pp. (unpublished open file report) British Geological Survey.

Brooks, C. & Leggo, M.D. (1972). The local chronology and regional implications of a Rb–Sr investigation of granitic rocks from the Corryong district, southeastern Australia. *Journal of the Geological Society of Australia*, **19(1)**, 1–19.

Cocker, J.D. (1982). Rb–Sr geochronology and Sr isotopic composition of Devonian granitoids, eastern Tasmania. *Journal of the Geological Society of Australia*, **29(2)**, 139–58.

Crook, K.A.W. & Feary, D.A. (1982). Development of New Zealand according to the fore-arc model of crustal evolution. *Tectonophysics* **87(1)**, 65–107.

Dalziel, I.W.D. & Elliot, D.H. (1982). West Antarctica: problem child of Gondwanaland. *Tectonics*, **1(1)**, 3–19.

Dalziel, I.W.D. & Forsythe, R.D. (1985). Andean evolution and the terrane concept. In *Tectonostratigraphic Terranes of the Circum-Pacific Region*, ed. D.G. Howell, pp. 565–81. Houston; Circum-Pacific Council for Energy & Mineral Resources.

Davidson, J., Mpodozis, C. & Rivano, S. (1981). El Paleozoico de Sierra de Almeida, al oeste de Monturaqui, Alta Cordillera de Antofagasta, Chile. *Revista Geológica de Chile*, **12(1)**, 3–23.

Day, R.W., Murray, C.G. & Whitaker, W.G. (1978). The eastern part of the Tasman orogenic zone. *Tectonophysics*, **48(1–2)**, 327–64.

Gledhill, A., Rex, D.C. & Tanner, P.W.G. (1982). Rb–Sr and K–Ar geochronology of rocks from the Antarctic Peninsula between Anvers Island and Marguerite Bay. In *Antarctic Geoscience*, ed. C. Craddock, pp. 315–23. Madison; University of Wisconsin Press.

Hamer, R.D. & Moyes, A.B. (1982). Composition and origin of garnet from the Antarctic Peninsula Volcanic Group of Trinity Peninsula. *Journal of the Geological Society, London*, **139(6)**, 713–20.

Harrington, H.J. & Korsch, R.J. (1985a). Tectonic model for the Devonian to middle Permian of the New England Orogen. *Australian Journal of Earth Sciences*, **32(2)**, 163–79.

Harrington, H.J. & Korsch, R.J. (1985b). Late Permian to Cainozoic tectonics of the New England Orogen. *Australian Journal of Earth Sciences*, **32(2)**, 181–203.

Lawver, L.A. & Scotese, C.R. (1987). A revised reconstruction of Gondwanaland. In *Gondwana Six: Structure, Tectonics, and Geophysics*, ed. G.D. McKenzie, Geophysical Monograph 40, pp. 17–24. Washington, DC; American Geophysical Union.

Milne, A.J. (1987). The geology of southern Oscar II Coast, Graham Land. *British Antarctic Survey Bulletin*, **75**, 73–81.

Mpodozis, C., Hervé, F., Davidson, J. & Rivano, S. (1983). Los granitoides de Cerros de Lila, manifestaciones de un episodo intusivo y termal del Paleozoico inferior en los Andes del norte de Chile. *Revista Geológica de Chile*, **18(1)**, 3–14.

Pankhurst, R.J. (1982). Rb–Sr geochronology of Graham Land, Antarctica. *Journal of the Geological Society, London*, **139(6)**, 701–11.

Pankhurst, R.J. (1983). Rb–Sr constraints on the ages of basement rocks of the Antarctic Peninsula. In *Antarctic Earth Science*, ed. R.L. Oliver, P.R. James & J.B. Jago, pp. 367–71. Canberra; Australian Academy of Science and Cambridge; Cambridge University Press.

Ramos, E.D. & Ramos, V.A. (1979). Los ciclos magmaticos de la Republica Argentina. *Actas VII Congresso Geologico Argentino*, vol. 1, pp. 771–86. Neuquén.

Ramos, V.A., Jordan, T.E., Allmendinger, R.W., Mpodozis, C., Kay, S.M., Cortes, J.M. & Palma, M. (1986). Paleozoic terranes of the central Argentine–Chilean Andes. *Tectonics*, **5(6)**, 855–80.

Rapela, C.W., Heaman, L.M. & McNutt, R.H. (1982). Rb–Sr geochronology of granitoid rocks from the Pampean Ranges, Argentina. *Journal of Geology*, **90(5)**, 574–82.

Rex, D.C. (1976). Geochronology in relation to the stratigraphy of the Antarctic Peninsula. *British Antarctic Survey Bulletin*, **43**, 49–58.

Richards, J.R. & Singleton, O.P. (1981). Palaeozoic Victoria, Australia: igneous rocks, ages and their interpretation. *Journal of the Geological Society of Australia*, **28(4)**, 395–421.

Storey, B.C., Dalziel, I.W.D., Garrett, S.W., Grunow, A.M., Pankhurst, R.J. & Vennum, W.R. (1988). West Antarctica in Gondwanaland: crustal blocks, reconstruction and breakup processes. *Tectonophysics*, **155(1–4)**, 381–90.

Tulloch, A.J. (1983). Granitoid rocks of New Zealand – a brief review. *Geological Society of America Memoir*, **159**, 5–20.

Yoshida, M. (1983). Structural and metamorphic history of the Ellsworth Mountains, West Antarctica. In *Antarctic Earth Science*, ed. R.L. Oliver, P.R. James & J.B. Jago, pp. 266–9. Canberra; Australian Academy of Science and Cambridge: Cambridge University Press.

Basement gneisses in north-western Palmer Land: further evidence for pre-Mesozoic rocks in Lesser Antarctica

S.M. HARRISON & B.A. PIERCY

British Antarctic Survey, Natural Environment Research Council, High Cross, Madingley Road, Cambridge CB3 0ET, UK

Abstract

Within north-western Palmer Land, orthogneiss, metapelite, marble and associated calc-silicate gneiss and granulite form the basement to the rocks of the Mesozoic magmatic arc. The orthogneiss represents I-type calc-alkaline granitoids and is chemically very similar to the Mesozoic granitoids. Two Rb–Sr whole-rock isochrons indicate that the age of the original igneous crystallization of the orthogneiss was approximately 400–440 Ma. The calc-alkaline nature of the rock suggests that it is a product of subduction-related plutonism, and it is likely that north-western Palmer Land represented a part of the Pacific margin of Gondwana during the mid-Palaeozoic. Age constraints on the paragneiss and granulite are limited, but they must be of Triassic or pre-Mesozoic age since they were metamorphosed, together with the orthogneiss, to high-grade amphibolite-facies at approximately 200 Ma. These rocks could represent either subduction–accretion material similar to some components of the Scotia metamorphic complex, or more likely, correlatives of the Lower Palaeozoic sedimentary sequence that is exposed intermittently along the Transantarctic Mountains on the western edge of Greater Antarctica.

Introduction

North-western Palmer Land is geologically situated on the western side of the Mesozoic magmatic arc of the Antarctic Peninsula. The arc formed as a result of essentially continuous eastward-directed subduction of proto-Pacific oceanic crust beneath the Antarctic Peninsula from at least the Late Triassic until the Tertiary. Storey & Garrett (1985) provided the most up-to-date model describing the evolution of the Antarctic Peninsula.

The exposed geology of north-western Palmer Land can be divided into: (a) plutonic, hypabyssal, and volcanic rocks of the Mesozoic magmatic arc; and (b) pre-Mesozoic rocks which form the basement to (a). The regional metamorphism of the basement during the early Mesozoic was contemporaneous with the earliest recorded activity of the magmatic arc. The Mesozoic geological history of north-western Palmer Land is summarized by Piercy & Harrison (this volume, p. 381). This paper describes the evidence for a pre-Mesozoic basement in this area and discusses the possible origins of the component rocks.

Basement rocks

The main rock types forming the basement are orthogneiss, metapelite, granulite, marble and associated calc-silicate gneiss. All these rocks were metamorphosed and migmatized during the early Mesozoic (~200 Ma) to estimated temperatures and pressures of about 800–900 °C and 5–6 kb, respectively (Harrison, unpublished data). The main textures and fabrics of that metamorphism are described by Piercy & Harrison (this volume, p. 381). The field relationships in some areas (e.g. Campbell Ridges, Fig. 1) provide direct field evidence, supported by radiometric dating, of early–late Mesozoic plutons being emplaced into a gneissic basement. However, many of the early Mesozoic plutons were emplaced synkinematically and in areas where this occurs no clear intrusive relationships are seen.

The main outcrops of orthogneiss occur at Goettel Escarpment, Orion Massif, Sirius Cliffs and Campbell Ridges (Fig. 1). These are leucocratic quartz–plagioclase–K-feldspar–biotite gneiss. Only at Campbell Ridges can it be clearly shown that it is cut by Early Jurassic plutons. The chemistry of the orthogneiss indicates typical I-type, calc-alkaline characteristics, very similar to those of the Mesozoic plutons. This is shown in Fig. 2, where pre-Mesozoic orthogneiss and Mesozoic granitoids are compared using a multi-element variation diagram. The evidence suggests that the orthogneiss bodies were emplaced as granitoids in the same tectonic environment, and under similar processes of fractionation and crystallization, as those of the Mesozoic granitoids.

The remainder of the basement gneisses are restricted mainly

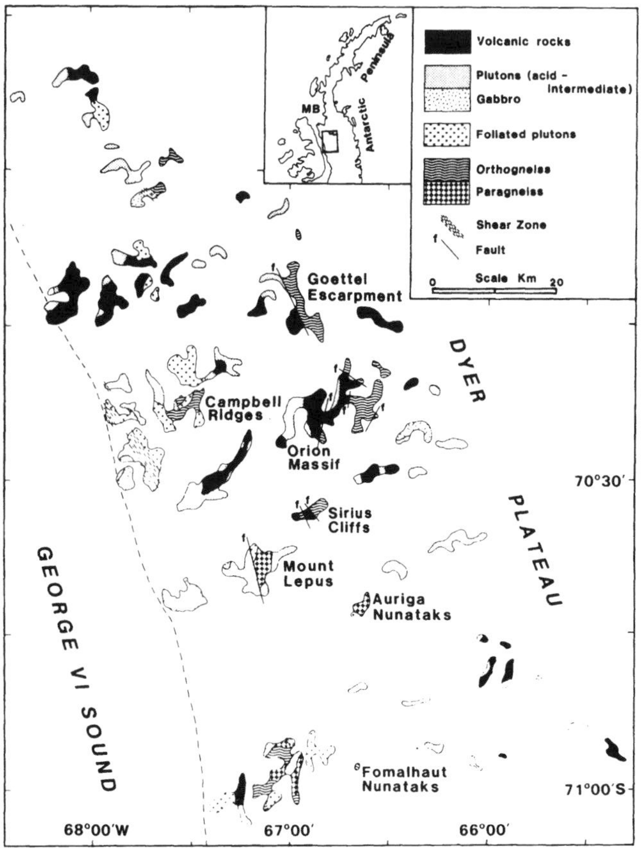

Fig. 1. Location and geological map of north-western Palmer Land.

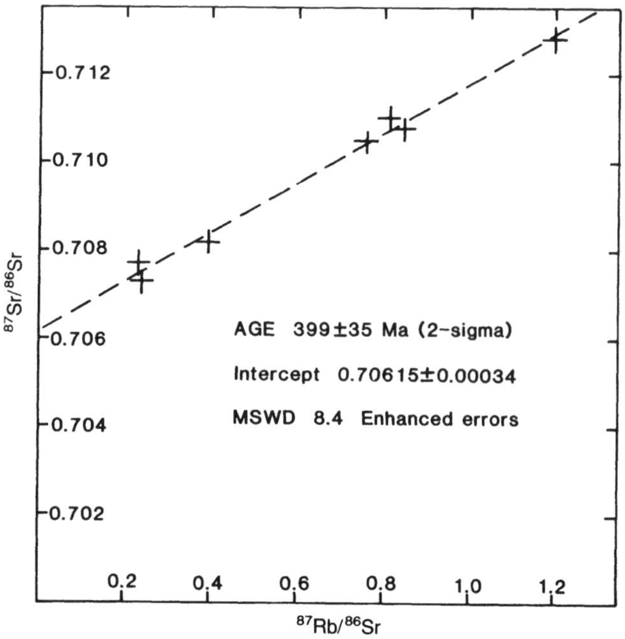

Fig. 3. Rb–Sr isochron of orthogneisses from Goettel Escarpment and Sirius Cliffs.

to Mount Lepus and Auriga Nunataks and form sub-parallel banded gneissic units. The metapelite, comprising biotite–garnet gneiss, is the main rock type. Infrequently, this is interspersed with bands, ≤ 30 cm thick, of quartzo-feldspathic-gneiss and quartzite. Most of the quartzo-feldspathic-gneiss represents partial melts but some may represent a previous sedimentary lithology. The two main mineral assemblages of the metapelite are: (a) biotite–garnet–andalusite–cordierite–K-feldspar–quartz–hercynite–plagioclase; and (b) biotite–garnet–plagioclase quartz.

At Auriga Nunataks the metapelite sandwiches a 250 m thick band of white calcite marble which can be traced for 2.5 km along-strike. Within the marble are numerous calc-silicate enclaves up to 2 m in diameter, possibly representing sweating out of partial melts from the surrounding metapelite. These enclaves are dominated by calcic-plagioclase but contain abundant diopside, scapolite and sphene.

The granulite gneiss is of much more restricted occurrence. One band approximately 25 m thick is exposed at Auriga Nunataks (apparently concordant with the metapelite and marble), and at Fomalhaut Nunataks it occurs as relict patches within migmatite. The rock is granular plagioclase–hypersthene–biotite gneiss with hypersthene showing partial replacement by the biotite.

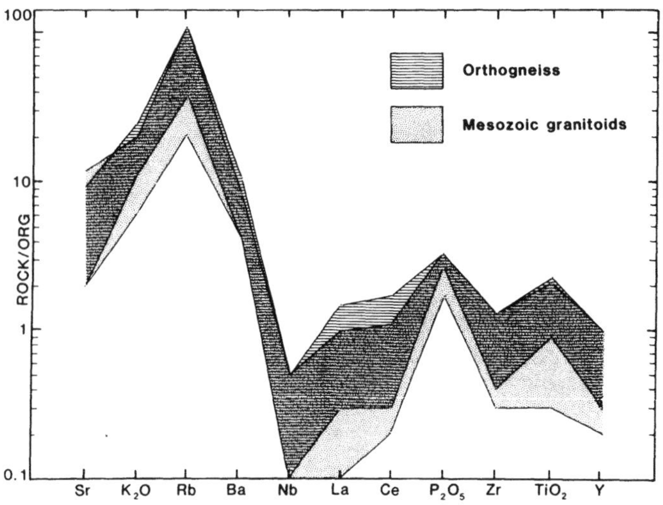

Fig. 2. Multi-element variation diagram comparing the chemical patterns of the orthogneisses with the Mesozoic granitoids. Analyses from 12 orthogneisses and 13 Mesozoic granitoids were used, all of which have between 65–70% SiO_2 values. Ranges represent minimum–maximum values. All analyses were carried out by X-ray fluorescence spectrometry.

Age constraints

Orthogneiss

Two separate Rb–Sr whole-rock isochrons have been obtained from the orthogneiss (Figs 3 & 4 and Table 1). Fig. 3 shows an isochron obtained from the combined samples of Goettel Escarpment and Sirius Cliffs, and indicates an age of 399 ± 35 Ma with an initial $^{87}Sr/^{86}Sr$ ratio of 0.7062. Fig. 4 shows an isochron obtained from Campbell Ridges. Two separate lines (shown inset) were in fact produced from 11 points. The steeper line, though showing a poor range of Rb–Sr ratios, gives an age of 440 ± 57 Ma with an initial $^{87}Sr/^{86}Sr$ ratio of 0.7053. The shallower line represents an age of 153 ± 10 Ma. The older ages are within error of each other and it is assumed that they represent the original age of

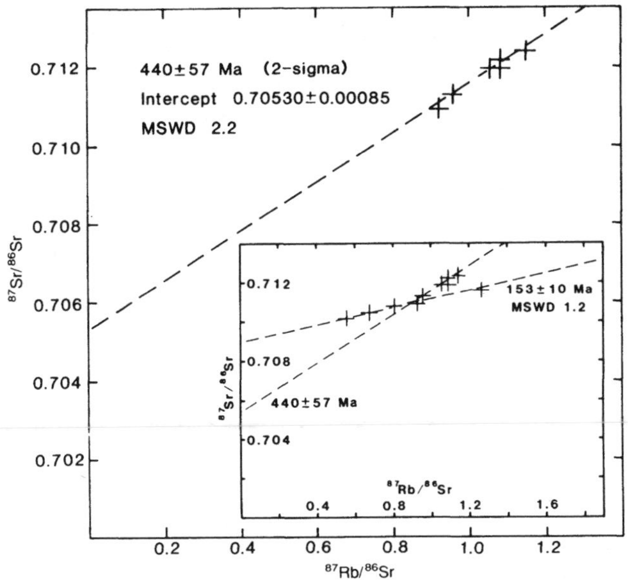

Fig. 4. Rb–Sr isochron of orthogneisses from Cambell Ridges. The inset shows two separate lines obtained from 11 samples.

Table 1. *New Rb–Sr data for orthogneiss suites of north-western Palmer Land*

Sample number	Rb (ppm)	Sr (ppm)	$^{87}Rb/^{86}Sr$	$^{87}Sr/^{86}Sr$
Data for Fig. 3: Goettel Escarpment and Sirius Cliffs				
R2559.1	148	527	0.8118	0.711020
R2559.4	122	465	0.7563	0.710504
R2562.1	112	840	0.3842	0.708200
R3305.2	143	346	1.1967	0.712874
R3349.1	91	1102	0.2382	0.707300
R3349.2	72	886	0.2345	0.707740
R3350.1	147	506	0.8428	0.710770
Data for Fig. 4: Campbell Ridges				
R2723.1	149	640	0.6733	0.71048
R2724.2(a)	174	626	0.8043	0.71081
R2724.2(b)	122	638	0.1957	0.71022
R2727.4	143	451	0.3243	0.71097
R2731.2	148	448	0.3384	0.71133
R2732.2	221	559	0.4055	0.71240
R3239.1	205	327	1.8158	0.71302
R3239.2	193	440	1.2654	0.71163
R3239.3	182	485	1.0848	0.71197
R3239.4	180	482	1.0827	0.71222
R3239.5	176	484	1.0515	0.71198

All analyses were carried out at the British Geological Survey Isotope Unit, Grays Inn Road, London. Sr isotopes were determined on a VG354 multicollector mass spectrometer. Rb–Sr analyses were determined by X-ray fluorescence spectrometry. All ages are calculated with $Rb = 1.42 \times 10^{-11}$ y. Analytical errors were taken as ± 0.5% on Rb–Sr ratios and 0.01% on $^{87}Sr/^{86}Sr$ ratios.

plutonic crystallization of the orthogneiss. The apparent large errors are to be expected considering the degree of metamorphism that these rocks have undergone. The Jurassic age obtained from the Campbell Ridges orthogneiss is interpreted as a resetting or a metamorphic cooling age due to a thermal event associated with basic magmatism (see Piercy & Harrison, this volume, p. 381).

Paragneiss and granulite

The age of these rocks is much less clear. Obviously the paragneiss (metapelite and marble) must represent sediments of Triassic or pre-Mesozoic age, since they underwent metamorphism at approximately 200 Ma. This is based on Rb–Sr whole-rock ages obtained from migmatites and Sm–Nd garnet–whole-rock ages from paragneiss (Piercy & Harrison, this volume, p. 000). As yet there are no other age constraints on these rocks. The granulite is more enigmatic. The hypersthene in this rock shows partial replacement by biotite indicating that it is a relict phase. However, there is no other available evidence to suggest that the granulite represents an even older metamorphism and it may well be that this rock is just a reflection of varying H_2O–CO_2 conditions during the early Mesozoic metamorphism.

No contact relationships are seen between orthogneiss that has been dated as mid-Palaeozoic and the paragneiss and granulite. At one locality (Mount Lepus), orthogneiss is seen to contain xenoliths of metapelite, but as the former is not yet dated they cannot be correlated with certainty with the mid-Palaeozoic orthogneiss.

Discussion and summary

The evidence presented indicates that both continental crustal material (comprising a sedimentary sequence of unknown age) and calc-alkaline granitoids of mid-Palaeozoic age were present in the region of north-western Palmer Land prior to the development of the Mesozoic magmatic arc.

The origin of the sedimentary rocks is obviously speculative. One possibility is that they represent subduction–accretion material similar to parts of the Scotia metamorphic complex exposed in the South Orkney Islands (Tanner, Pankhurst & Hyden, 1982; Storey & Meneilly, 1985). Storey & Meneilly (1985) described a sequence of polydeformed metamorphic rocks from Signy Island consisting of mica-schist, quartzite, marble and amphibolite. They interpreted these as representing metamorphosed ocean floor and oceanic island basalts, arenaceous and argillaceous sediments, limestone and chert which have all been subducted and accreted to the Scotia arc. Pressures of 8 kb and temperatures of approximately 550 °C were estimated for conditions of metamorphism, with a minimum age for this event set at 180–200 Ma. Although these features appear very similar to those described for north-western Palmer Land, there are significant differences. In Palmer Land these include: (a) no amphibolites of pre-Mesozoic age have been distinguished; (b) the marble forms one large outcrop (2.56 km × 250 m) rather than thin, predominantly centimetre-scale layers (rarely up to 10 m); and (c) the estimated conditions of metamorphism are for lower pressure (~ 5–6 kb) but higher temperature (~ 800–900 °C). The apparent differences in lithologies may not be that important but, superficially, the paragneiss of north-western Palmer Land is more readily correlated with the autochthonous sequence of Early Palaeozoic sediments that crop out along the length of the Transantarctic Mountains (Laird, 1981). In particular comparison can be made with the extensive limestone formations of Cambrian age, such as the Nelson Limestone of the Pensacola Mountains (Schmidt & Ford, 1969) and the Shackleton Limestone of the Holyoake Range (Laird, Mansergh & Chappell, 1971).

The petrogenesis of the orthogneiss is much clearer since reliable mid-Palaeozoic ages have been obtained and the chemistry clearly shows that it represents I-type calc-alkaline granitoids. This indicates that north-western Palmer Land lay on an active margin during the mid-Palaeozoic and this was most likely the Pacific edge of Gondwana. Milne & Millar (this volume, p. 335) have also described evidence for mid-Palaeozoic plutonism in eastern Graham Land and this may well be the same event as seen in Palmer Land. These authors make regional correlations with other areas of Gondwana where mid-Palaeozoic plutonism has been described. The two most important of these are northern Victoria Land and Marie Byrd Land.

Acknowledgements

We wish to thank all our colleagues at the BAS who helped with discussions and criticisms. In particular we would like to acknowledge Ian Millar who supervized and helped with the isotope dating.

References

Laird, M.G. (1981). Lower Palaeozoic rocks of Antarctica. In *Lower Palaeozoic Rocks of the World*, vol. 3, ed. C.H. Holland, pp. 257–314. New York; John Wiley and Sons.

Laird, M.G., Mansergh, G.D. & Chappell, J.M.. (1971). Geology of the Central Nimrod Glacier area, Antarctica. *New Zealand Journal of Geology and Geophysics*, **14(3)**, 427–33.

Schmidt, D.L. & Ford, A.B. (1969). Geology of the Pensacola and Thiel Mountains. (Sheet 5). In *Geologic Maps of Antarctica*, Antarctic Map Folio Series, ed. V.C. Bushnell & C. Craddock, Folio 12, Plate 5. New York; American Geographical Society.

Storey, B.C. & Garrett, S.W. (1985). Crustal growth of the Antarctic Peninsula by accretion, magmatism and extension. *Geological Magazine*, **122(1)**, 5–14.

Storey, B.C. & Meneilly, A.W. (1985). Petrogenesis of metamorphic rocks within a subduction–accretion terrane, Signy Island, South Orkney Islands. *Journal of Metamorphic Geology*, **3(3)**, 21–42.

Tanner, P.W.G., Pankhurst, R.J. & Hyden, G. (1982). Radiometric evidence for the age of the subduction complex in the South Orkney and South Shetland Islands, West Antarctica. *Journal of the Geological Society, London*, **139(6)**, 683–90.

Granitoids of the Ford Ranges, Marie Byrd Land, Antarctica

S.D. WEAVER[1], J.D. BRADSHAW[1] & C.J. ADAMS[2]

1 Geology Department, University of Canterbury, Christchurch, NZ
2 Institute of Nuclear Sciences, DSIR, Lower Hutt, NZ

Abstract

In the Ford Ranges, Swanson Formation sedimentary rocks are cut by the Late Palaeozoic Ford Granodiorite and Mesozoic Byrd Coast Granite. The Ford Granodiorite is a granodiorite–monzogranite suite of biotite-rich, hornblende-bearing rocks of 350–380 Ma age. The suite is calc-alkaline, metaluminous to peraluminous, with initial $^{87}Sr/^{86}Sr$ of 0.704–0.706, features characteristic of I-type granitoids. Trace-element tectonic discriminants suggest that the Ford Granodiorite represents magmas emplaced in either an active, continental margin or post-tectonic environment. Petrologically, the Ford Granodiorite matches the southern Admiralty Intrusives of northern Victoria Land. Marie Byrd Land, the Robertson Bay and Bowers terranes of northern Victoria Land, the Campbell Plateau and the Western Province of New Zealand are part of an extensive continental mass which accreted to the Gondwana margin during the Late Palaeozoic. The Byrd Coast Granite is typically biotite-bearing leucogranite and may have been emplaced as two pulses at about 140 Ma and 110 Ma. The suite is calc-alkaline, mostly peraluminous and has initial $^{87}Sr/^{86}Sr$ of 0.704–0.708. Trace element geochemistry designates granitoids of the Clark Mountains and Mount McClung as anorogenic A-types. At least these components of the Byrd Coast Granite may not be strictly part of the subduction-related Mesozoic magmatic arc of West Antarctica.

Introduction

Calc-alkaline granitoids of Upper Palaeozoic and Mesozoic age crop out extensively in the Ford Ranges of western Marie Byrd Land, West Antarctica (Fig. 1). The first geochemical account of these granitoids, sampled during the 1978–79 NZARP Marie Byrd Land Expedition, is presented here and their tectonic settings during emplacement are evaluated. Brief comparisons are made between the Ford Ranges granitoids and those of other sectors of the formerly contiguous Pacific margin of Gondwana.

Geological outline

Regional geology

The oldest rocks of the region are quartzose metaflysch of the Swanson Formation (Bradshaw, Andrews & Field, 1983) of late Precambrian–Early Ordovician age (Adams, 1986). These sedimentary rocks suffered low-grade regional metamorphism and deformation during the Late Ordovician. High-grade rocks of the Fosdick metamorphic complex were interpreted by Bradshaw *et al.* (1983) as Swanson Formation sedimentary rocks migmatized during the injection of granitoids, considered by Adams (1986) to be Cretaceous in age.

The plutonic rocks have been described by Wade & Wilbanks (1972) and references therein. Geological maps by Wade, Cathey & Oldham (1977a, b, 1978) divided the plutonic rocks into the Ford Granodiorite of Devonian–Early Carboniferous age and the Byrd Coast Granite of mid-Cretaceous age.

Geochronological summary

In a detailed study by Adams (1987), Rb–Sr whole-rock isochrons for the Ford Granodiorite have been obtained for Saunders Mountain (~ 360 Ma), Gutenko Nunataks–Post Ridge (371 ± 10 Ma), Sarnoff Mountains (353 ± 7 Ma), central Denfield Mountains (379 ± 8 Ma) and Mounts Little and Swan (361 ± 6 Ma) (all errors are 1-sigma). The Ford Granodiorite was therefore emplaced as a number of plutons within about a 30 Ma period during the Late Devonian–Early Carboniferous.

Rb–Sr isochrons for the Byrd Coast Granite indicate two phases: (1) Early Cretaceous intrusions of the Clark Mountains (142 ± 4 Ma) and Mount Corey (131 ± 4 Ma); (2) mid-Cretaceous intrusions in the Allegheny Mountains (111 ± 1 Ma) and at Mount McClung (110 ± 2 Ma).

The present study is based on the geochronological framework established by Adams (1987) and most of the samples chemically analysed have been dated. The distribution of the granitoids and the main geographical features are shown in Fig. 1.

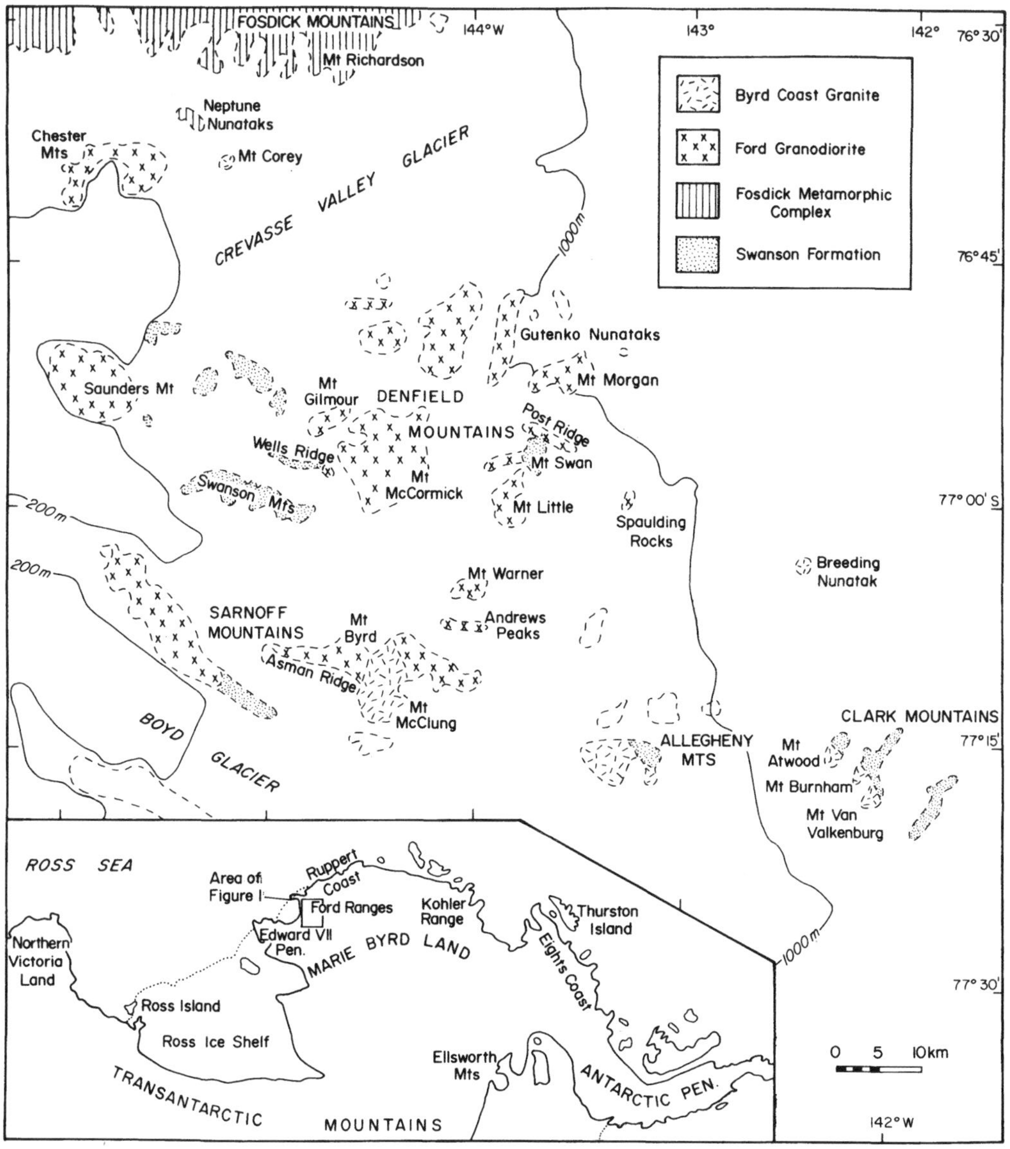

Fig. 1. Geological map of the Ford Ranges, Marie Byrd Land, showing the distribution of the Ford Granodiorite and Byrd Coast Granite. Modified from Adams (1987).

Field and petrographic characteristics

Ford Granodiorite

This occurs as plutons, a few kilometres in diameter, having sharp discordant contacts, narrow thermal aureoles and rare marginal foliations, and which are tectonically undeformed. The granitoids are grey, or black and white, have coarse equigranular or less commonly porphyritic textures and comprise a tonalite (rare)–granodiorite (abundant)–monzogranite (common) suite. In the granodiorites, brown biotite predominates over hornblende, and titanite, magnetite and ilmenite are generally present. Monzogranites contain brown biotite, rarely muscovite and usually no hornblende or titanite. Accessory minerals include apatite, zircon, allanite and rare tourmaline.

Byrd Coast Granite

This occurs as small plutons which are associated with aplites, pegmatites and acidic dykes, sills and volcanic rocks. The granitoids are pinkish, coarse equigranular or porphyritic rocks, predominantly leucogranite and monzogranite. Brown biotite is always present and accessories include ilmenite, apatite, zircon, allanite and rarely muscovite. Both hornblende and titanite are absent except in the black and white tonalite–granodiorite rocks of Neptune Nunataks which closely resemble the Ford Granodiorite.

Geochemistry and tectonic settings

Ford Granodiorite

Representative data from a group of 40 analysed rocks are given in Table 1 (sample locality coordinates are listed in Adams (1987)). The suite is calc-alkaline and both peraluminous (24 samples with ~ < 1.1 and averaging 0.4) and metaluminous (15 samples) compositions are present. Initial $^{87}Sr/^{86}Sr$ ratios, derived from the isochrons of Adams (1987), are as follows: Gutenko Nunataks–Post Ridge, 0.7051 ± 0.0002;

Table 1. *Geochemistry of the Ford Granodiorite (Devonian–Carboniferous), Ford Ranges, Marie Byrd Land*

Location	Saunders Mountain	Gutenko Nunataks–Post Ridge			Sarnoff Mountains			Denfield Mountains			Mount Little–Mount Swan	
Sample number	D1/3	R7211	R7355	R7218	R7458	R7460	R7443	R7213	R7212	R7307	R7200	R7205
SiO_2	64.08	66.47	69.74	71.13	64.72	70.06	73.95	64.79	68.66	74.58	72.90	72.73
TiO_2	0.80	0.62	0.53	0.38	0.84	0.44	0.26	0.76	0.52	0.16	0.30	0.28
Al_2O_3	15.35	15.11	13.96	13.97	15.01	14.16	13.01	15.42	14.32	13.33	13.35	13.69
Total Fe_2O_3	4.80	3.71	3.23	2.44	4.45	2.88	1.73	4.71	2.99	1.49	2.12	1.85
MnO	0.04	0.06	0.09	0.05	0.07	0.05	0.03	0.08	0.05	0.04	0.06	0.04
MgO	2.85	1.88	0.91	0.91	2.08	1.13	0.34	2.17	1.13	0.26	0.59	0.31
CaO	4.46	3.55	2.36	2.36	4.01	2.58	1.30	4.10	3.01	1.09	1.98	1.35
Na_2O	3.46	3.80	3.33	3.45	3.70	3.92	3.37	3.78	3.64	3.70	3.87	3.37
K_2O	2.84	2.92	3.76	3.64	2.65	3.30	4.85	2.53	3.36	4.63	4.47	5.11
P_2O_5	0.20	0.18	0.18	0.14	0.27	0.14	0.08	0.26	0.16	0.04	0.09	0.10
LOI	0.43	0.80	0.93	1.05	1.00	0.56	0.73	0.86	1.26	0.33	1.06	0.36
Total	99.30	99.10	99.01	99.51	100.07	99.22	99.66	99.46	99.09	99.64	99.79	99.19
Norm												
q	19.9	23.3	30.1	31.2	23.0	27.8	32.8	21.5	27.3	32.6	32.3	30.5
or	17.0	17.6	22.7	21.9	15.9	19.8	29.0	15.2	20.3	27.6	20.8	30.6
ab	29.7	32.8	28.8	29.7	31.8	33.7	28.9	32.5	31.5	31.6	33.2	28.9
an	18.2	15.9	10.8	11.0	16.8	11.5	6.0	17.9	13.1	5.2	8.9	6.1
c	—	—	0.6	0.4	—	—	0.0	—	—	0.4	—	0.4
Ba	433	413	562	358	544	533	330	464	544	315	495	395
Ga	18	18	18	18	18	17	15	19	16	15	16	19
La	37	24	60	29	38	44	27	29	30	29	25	46
Nb	8	15	20	16	17	15	25	17	14	10	16	20
Rb	122	177	166	165	101	129	216	114	119	154	138	304
Sr	366	370	276	257	379	306	134	431	315	103	245	142
Th	16	12	22	21	13	19	22	11	10	21	15	28
Y	28	23	38	19	31	22	41	20	22	24	23	31
Zr	153	147	193	135	206	145	147	162	156	96	104	158

D1/3; granodiorite, Saunders Mt: R7211; granodiorite, Post Ridge: R7355; granodiorite, Spaulding Rocks: R7218; granodiorite, Mt Morgan: R7458; granodiorite, Andrews Peaks: R7460; monzogranite, Mt Warner: R7443; monzogranite, Asman Ridge: R7213; granodiorite, Mt McCormick: R7212; granodiorite, Mt McCormick: R7307; monzogranite, Mt Gilmour: R7200; granodiorite, Mt Little: R7205; monzogranite, Mt Swan.
All elements by XRF analysis of fusion beads (majors) and pellets (traces). Loss on ignition (LOI) at 1000 °C. Fe as total Fe_2O_3 and CIPW norm calculated with $Fe_2O_3/FeO = 0.75$. Samples with prefix R have been dated by Adams (1987).

Sarnoff Mountains, 0.7060 ± 0.0003; Denfield Mountains, 0.7057 ± 0.0001 and Mount Little–Mount Swan, 0.7042 ± 0.0002 (1-sigma errors). No compositional polarity can be detected within the area studied.

Following the terminology of White & Chappell (1983), the Ford Granodiorite is an I-type suite (Figs 2 & 3). Modal and normative mineralogy and low initial $^{87}Sr/^{86}Sr$ ratios confirm this designation.

Pearce, Harris & Tindle (1984) and Harris, Pearce & Tindle (1986) have proposed trace element discriminants for the tectonic interpretation of granitoids. On a Rb versus (Y + Nb) diagram (Fig. 4), the Ford Granodiorite samples plot predominantly in the field representing volcanic arc granites (VAG) and scatter into the syn-collision granites (syn-COLG) and within-plate granites (WPG) fields. On a Rb–Zr versus SiO_2 plot (Fig. 5), the Ford Granodiorite samples are virtually excluded from the syn-COLG field. Most within-plate granites are A-types, excepting those metaluminous I-types intruded into strongly attenuated continental crust such as the Tertiary granites of Scotland (Pearce *et al.*, 1984); thus a WPG interpretation for the Ford Ranges granitoids can be confidently rejected. It should be noted that post-collision granites also plot in the VAG fields of Figs 4 & 5. The geochemistry of the Ford Granodiorite is therefore consistent with an I-type suite intruded into an active continental margin (VAG) or a late/post-collision tectonic setting. These two environments correspond to Pitcher's (1983) division of I-type granites into Cordilleran and Caledonian.

Byrd Coast Granite

Analyses of representative samples from the Clark

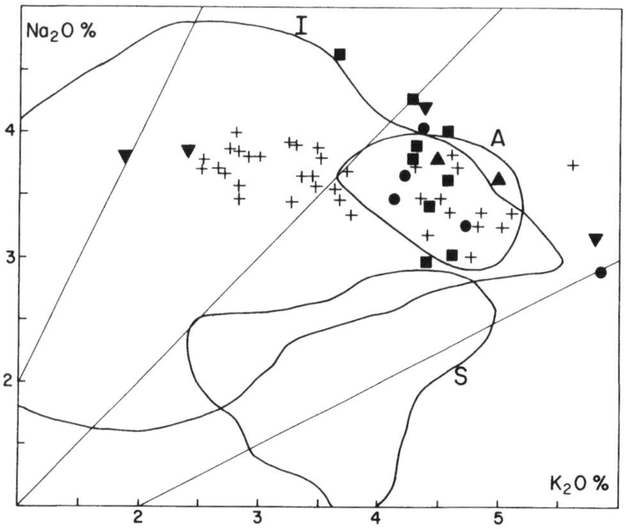

Fig. 2. Plot of Na_2O versus K_2O for granitoids of the Ford Ranges (+ , Ford Granodiorite; solid symbols, Byrd Coast Granite: ■, Clark Mountains; ●, Mount McClung; ▲, Mount Corey; ▼, Neptune Nunataks and Mount Richardson). Fields for I- (igneous source), S- (sedimentary source) and A-type (anorogenic) granitoids of the Lachlan Fold Belt after White & Chappell (1983).

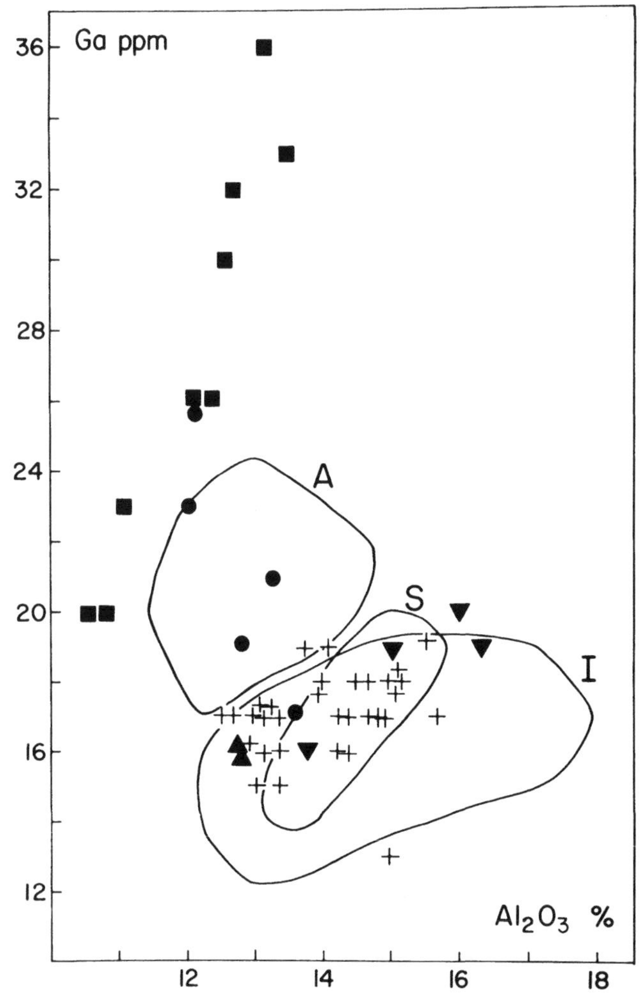

Fig. 3. Plot of Ga versus Al_2O_3 for granitoids of the Ford Ranges (symbols as for Fig. 2). Fields for I-, S- and A-type granitoids after White & Chappell (1983).

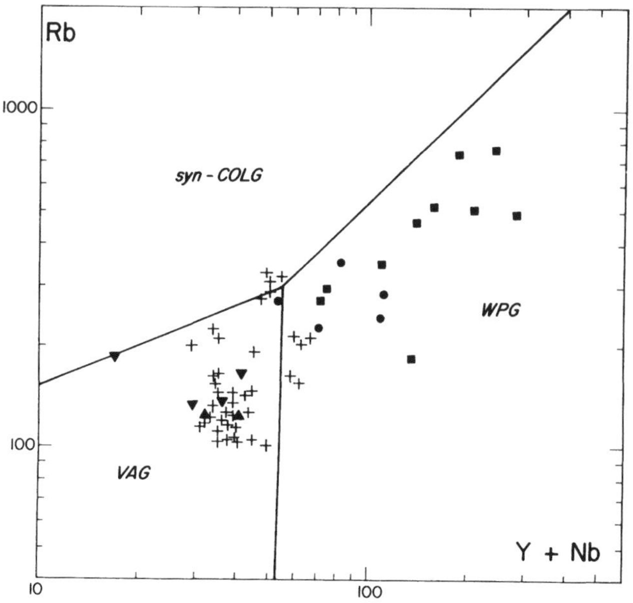

Fig. 4. Granitoids of the Ford Ranges plotted in the Rb versus (Y + Nb) discriminant diagram for syn-collision (syn-COLG), volcanic arc (VAG) and within-plate (WPG) granitoids of Pearce *et al.* (1984). Note that post-collision calc-alkaline granitoids (post-COLG) can plot in each of the three fields outlined. Symbols as for Fig. 2.

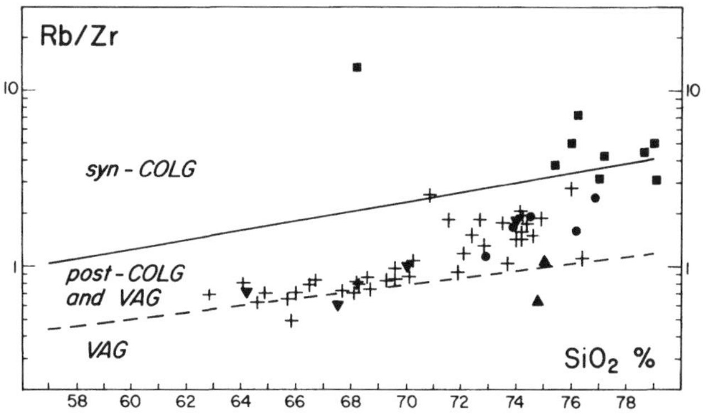

Fig. 5. Rb/Zr versus SiO_2 plot for Ford Ranges granitoids (symbols as for Fig. 2). Fields for syn-collision peraluminous granitoids (syn-COLG), calc-alkaline volcanic arc granitoids (VAG) and late or post-collision calc-alkaline granitoids (post-COLG) after Harris *et al.* (1986). Note that within-plate alkaline granitoids (WPG) can plot in each field.

Mountains, the Mount McClung area and the Mount Corey region are given in Table 2.

The Clark Mountains and Mount McClung samples are leucogranites and most are peraluminous with ~ < 0.8; one sample from Mount McClung is weakly peralkaline (~ = 0.6). Compared with the Ford Granodiorite, the Clark Mountains samples have notably higher Ga, Nb, Rb and Y, and lower Ba and Sr concentrations. The compositions of the Mount McClung samples are intermediate between these two groups. Adams (1987) obtained initial $^{87}Sr/^{86}Sr$ ratios of 0.7079 ± 0.0090 (Clark Mountains), 0.7056 ± 0.0003 (Mount McClung) and 0.7061 ± 0.0002 (Allegheny Mountains). Though scattered on Figs 2 & 3, the Clark Mountains and Mount McClung samples are best interpreted as A-type granitoids. Rocks from mounts Van Valkenburg and Atwood have

Table 2. *Geochemistry of the Byrd Coast Granite (Jurassic-Cretaceous), Ford Ranges, Marie Byrd Land*

Location	Clark Mountains			Mount McClung		Mount Corey–Mount Richardson–Neptune Nunataks			
Sample number	R7388	D19/3	R7384	D31/1	R7456	R7328	D11/1	D12/2	D12/3
SiO_2	77.05	77.17	78.70	73.87	76.20	75.04	73.99	64.20	70.15
TiO_2	0.07	0.04	0.07	0.32	0.12	0.24	0.13	0.80	0.47
Al_2O_3	11.88	12.69	11.09	12.81	12.00	12.80	13.72	16.34	14.98
Total Fe_2O_3	1.21	0.33	0.93	2.43	1.02	1.38	0.98	4.73	3.35
MnO	0.04	0.00	0.02	0.08	0.16	0.06	0.02	0.06	0.04
MgO	0.01	0.00	0.06	0.54	0.04	0.27	0.26	2.60	1.50
CaO	0.48	0.42	0.45	0.67	0.34	0.74	0.99	4.15	2.91
Na_2O	3.81	4.03	3.43	2.90	4.47	3.81	3.16	3.86	3.82
K_2O	4.30	4.53	4.40	5.83	4.14	4.48	5.79	2.40	1.85
P_2O_5	0.02	0.01	0.02	0.06	0.02	0.05	0.08	0.26	0.04
LOI	0.46	1.00	0.33	0.57	0.80	0.40	0.87	0.77	1.07
Total	99.33	100.19	99.48	100.07	99.29	99.27	99.97	100.20	100.18
Norm									
q	37.6	35.7	41.1	32.3	34.0	34.0	31.3	19.9	32.2
or	25.7	27.0	26.2	34.7	24.9	26.8	34.5	14.3	11.1
ab	32.6	34.4	29.3	24.7	38.4	32.6	27.0	33.0	32.7
an	2.3	2.0	1.9	2.9	0.5	3.4	4.4	19.1	14.3
c	0.1	0.4	—	0.7	—	0.5	0.7	0.5	1.5
Ba	7	9	32	60	10	547	454	515	76
Ga	26	32	23	19	23	17	16	19	19
La	41	45	36	13	35	53	26	42	15
Nb	63	70	38	39	54	13	6	16	15
Rb	509	518	353	244	286	122	186	137	133
Sr	5	7	7	30	1	108	164	401	222
Th	74	63	37	32	53	14	16	13	9
Y	142	86	68	68	57	19	11	20	14
Zr	164	119	78	140	176	114	100	192	132

R7388; leucogranite, Mt Burnham: D19/3; leucogranite, Mt Van Valkenburg: R7384; leucogranite, Mt Atwood: D31/1; granite, Asman Ridge: R7456; leucogranite, Mt McClung: R7328; monzogranite, Mt Corey: D11/1; monzogranite, Mt Richardson: D12/2; tonalite, Neptune Ntks: D12/3; granodiorite, Neptune Ntks.
Analytical details – see Table 1. Samples with prefix R have been dated by Adams (1986).

remarkably high Ga/Al ratios, interpreted as the products of extreme differentiation in the roof zone of a single small pluton. On Fig. 4, Clark–McClung samples plot in the WPG field and this setting is also consistent with their distribution in Fig. 5, as WPG plots in both fields. These granitoids may therefore be interpreted as being of anorogenic type, unrelated to subduction.

Samples from Mount Corey, Neptune Nunataks and Mount Richardson are geochemically (Figs 2–5) and mineralogically similar to the Ford Granodiorite. Although a 131 Ma age has been obtained for Mount Corey (Adams, 1987), the initial $^{87}Sr/^{86}Sr$ ratio of 0.7036 is suspiciously low for a subalkaline leucogranite. Moreover Halpern's (1968) age for Mount Corey of 100 ± 3 Ma (recalculated) and an initial ratio of 0.7060 are inconsistent with the above data. At Neptune Nunataks and at Mount Richardson, high-grade rocks of the Fosdick metamorphic complex are invaded by several generations of granitoids. The possibility arises that the Ford Granodiorite may be present at Neptune Nunataks and Mount Richardson and may have been involved in the migmatization of the Fosdick metamorphic complex. The ages obtained date the subsequent Cretaceous plutonism and thermal overprinting (Adams, 1986).

Discussion: regional tectonic implications

Palaeozoic

In Marie Byrd Land, granitoids of Devonian age occur only in the Ford Ranges. They are unknown in the Antarctic Peninsula but are extensive in northern Victoria Land, New Zealand and eastern Australia.

The Admiralty Intrusives of northern Victoria Land are a granodiorite-dominated, I-type suite of Late Devonian age.

They show strong compositional polarity consistent with increasing involvement of older crustal material toward the north-north-east, implying subduction which dipped in that direction (Borg *et al.*, 1987). Such an interpretation provides strong evidence for an allochthonous relationship between the Robertson Bay–Bowers terranes, into which the Admiralty Intrusives are emplaced, and East Antarctica. The geochemistry and mineralogy of the Ford Granodiorite, described here, closely match the southern and central Admiralty Intrusives (Borg, Stump & Holloway, 1986).

In the South Island of New Zealand, granitoids of the Karamea Batholith of the Western Province are mainly Devonian–Carboniferous in age but all are S-types (Tulloch, 1983). They invade quartzose sedimentary rocks (Greenland Group) which, together with similar lithologies on Campbell Island (Campbell Plateau), have been correlated with the Swanson Formation of Marie Byrd Land. The juxtaposition of the Western Province of New Zealand, including the Campbell Plateau, and Marie Byrd Land is therefore strongly indicated (Cooper *et al.*, 1982). Moreover, the sedimentology and structure of the Swanson Formation and the Robertson Bay Group, either side of the Ross Sea embayment, are closely comparable (Bradshaw *et al.*, 1983). Like the Robertson Bay and Bowers terranes of northern Victoria Land, Marie Byrd Land and New Zealand must have had (in their pre-Gondwana dispersal configuration) an allochthonous relationship to East Antarctica.

Devonian granitoids in Marie Byrd Land and northern Victoria Land could represent Caledonian-type post-tectonic magmatism generated during uplift and adiabatic melting of an igneous crustal underplate, as suggested by Vetter & Tessensohn (1987). If so, docking of the allochthon was pre-Late Devonian and may be recorded in Early Palaeozoic deformations. However, although regional metamorphism in both Marie Byrd Land and New Zealand took place in Late Ordovician times (440–450 Ma), deformation in northern Victoria Land was probably Late Cambrian–Early Ordovician (500–510 Ma), some 50 Ma earlier (Adams, 1986).

Alternatively, Devonian granitoids may represent subduction-related Cordilleran-type magmatism recording oceanic closure and the somewhat passive accretion of exotic terranes onto the Gondwana nucleus. In this case assembly was probably post-Devonian and pre-Beacon (Permian) (Borg *et al.*, 1987). The Devonian S-type granitoids of New Zealand cannot easily be accommodated in the first model but may represent the continental 'inboard' side of the active margin of an allochthonous terrane in the second, and as such, could be regarded as the logical extension of the compositional trend established for the Admiralty Intrusives.

Devonian granitoids are well represented in Tasmania, south-eastern Australia and northern Queensland, and clearly represent a major phase of crustal accretion onto the Gondwana margin, along what appears to have been a continuous orogenic belt, characterized by complex tectonic variation along its length (Fig. 6*a*).

In Marie Byrd Land, the period from Carboniferous to Cretaceous times appears to have been primarily one of erosion, and it is likely that the Devonian magmatic rocks were

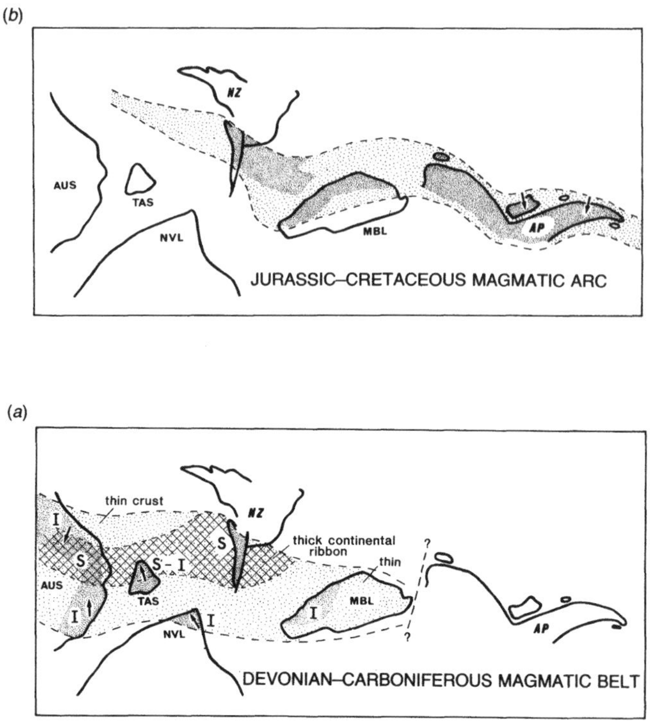

Fig. 6. (*a*) Distribution of Devonian–Carboniferous I- and S-type granitoids along the pre-dispersal, palaeo-Pacific margin of Gondwana. Arrows indicate inferred subduction polarity. The distribution of S-types delimits a ribbon of thick, old continental crust within the south-eastern Australia–northern Victoria Land–western New Zealand–Marie Byrd Land allochthon. (*b*) Distribution of granitoids which define the Jurassic–Cretaceous magmatic arc of the palaeo-Pacific sector of the Gondwana margin.

the primary source for quartzo-feldspathic sediments which were later accreted onto the New Zealand Palaeozoic foreland as the Torlesse Terrane (Bradshaw, Andrews & Adams 1981).

Mesozoic

There is general agreement that the Mesozoic Andean arc extends along the Antarctic Peninsula, across the Marie Byrd Land coast and Campbell Plateau to New Zealand. Mesozoic granitoids, similar to those of the Clark Mountains of the Ford Ranges, occur on Edward VII Peninsula (Adams, 1987) but are absent from northern Victoria Land, Tasmania and south-eastern Australia (Fig. 6*b*). It remains to be seen how widespread are A-type granitoids like those of the Clark Mountains and Mount McClung. For example, Mesozoic–Cenozoic plutonic rocks of the Antarctic Peninsula (Saunders, Tarney & Weaver, 1980) and the Cretaceous Separation Point Batholith of New Zealand (Tulloch, 1983) are all I-type, volcanic arc granites. Present evidence suggests that at least the Clark–McClung components of the Byrd Coast Granite should be considered anorogenic.

Acknowledgements

We thank Drs M.G. Laird and A.J. Tulloch of the New Zealand Geological Survey for their helpful critical comments on the manuscript.

References

Adams, C.J. (1986). Geochronological studies of the Swanson Formation of Marie Byrd Land, West Antarctica, and correlation with northern Victoria Land, East Antarctica and the South Island, New Zealand. *New Zealand Journal of Geology and Geophysics* **29**, 345–58.

Adams, C.J. (1987). Geochronology of granite terranes in the Ford Ranges, Marie Byrd Land, West Antarctica. *New Zealand Journal of Geology and Geophysics*, **30**, 51–72.

Borg, S.G., Stump, E., Chappell, B.W., McCulloch, M.T., Wyborn, D., Armstrong, R.L. & Holloway, J.R. (1987). Granitoids of northern Victoria Land, Antarctica: implications of chemical and isotopic variations to regional crustal structure and tectonics. *American Journal of Science*, **287**, 127–69.

Borg, S.G., Stump, E. & Holloway, J.R. (1986). Granitoids of northern Victoria Land, Antarctica: a reconnaissance study of field relations, petrography and geochemistry. In *Geological Investigations in Northern Victoria Land,* Antarctic Research Series, 46, ed. E. Stump, pp. 115–88. Washington, DC; American Geophysical Union.

Bradshaw, J.D., Andrews, P.B. & Adams, C.J. (1981). Carboniferous to Cretaceous on the Pacific margin of Gondwana. In *Gondwana Five*, ed. M.M. Cresswell & P. Vella, pp. 217–21. Rotterdam; A.A. Balkema.

Bradshaw, J.D., Andrews, P.B. & Field, B.D. (1983). Swanson Formation and related rocks of Marie Byrd Land and a comparison with the Robertson Bay Group of northern Victoria Land. In *Antarctic Earth Science*, ed. R.L. Oliver, P.R. James & J.B. Jago, pp. 274–9. Canberra; Australian Academy of Science and Cambridge; Cambridge University Press.

Cooper, R.A., Landis, C.A., LeMasurier, W.E. & Speden, I.G. (1982). Geologic history and regional patterns in New Zealand and West Antarctica – their palaeotectonic and palaeogeographic significance. In *Antarctic Geoscience*, ed. C. Craddock, pp. 43–53. Madison; University of Wisconsin Press.

Halpern, M. (1968). Ages of Antarctic and Argentine rocks bearing on continental drift. *Earth and Planetary Science Letters*, **5**, 159–67.

Harris, N.B.W., Pearce, J.A. & Tindle, A.G. (1986). Geochemical characteristics of collision-zone magmatism. In *Collision Tectonics*, ed. M.P. Coward & A.C. Ries, pp. 67–81. Geological Society Special Publication, No. 19. Oxford; Blackwell Scientific Publications.

Pearce, J.A., Harris, N.B.W. & Tindle, A.G. (1984). Trace element discrimination diagrams for the tectonic interpretation of granitic rocks. *Journal of Petrology*, **25**, 956–83.

Pitcher, W.S. (1983). Granite type and tectonic environment. In *Mountain Building Processes*, ed. K.J. Hsu, pp. 19–40. London & New York; Academic Press.

Saunders, A.D., Tarney, J. & Weaver, S.D. (1980). Transverse geochemical variations across the Antarctic Peninsula: implications for the genesis of calc-alkaline magmas. *Earth and Planetary Science Letters*, **46**, 344–60.

Tulloch, A.J. (1983). Granitoid rocks of New Zealand – a brief review. *Geological Society of America Memoir*, **159**, 5–19.

Vetter, U. & Tessensohn, F. (1987). S- and I-type granitoids of north Victoria Land, Antarctica, and their inferred geotectonic setting. *Geologische Rundschau*, **76(1)**, 233–43.

Wade, F.A., Cathey, C.A. & Oldham, J.B. (1977a). *Reconnaissance Geologic Map of the Boyd Glacier Quadrangle, Marie Byrd Land, Antarctica, 1:250 000*, US Antarctic Research Program Map A-6. Washington, DC; US Geological Survey.

Wade, F.A., Cathey, C.A. & Oldham, J.B. (1977b). *Reconnaissance Geologic Map of the Guest Peninsula Quadrangle, Marie Byrd Land, Antarctica, 1:250 000*, US Antarctic Research Program Map A-7. Washington, DC; US Geological Survey.

Wade, F.A., Cathey, C.A. & Oldham, J.B. (1978). *Reconnnaissance Geologic Map of the Gutenko Nunataks Quadrangle, Marie Byrd Land, Antarctica, 1:250 000*, US Antarctic Research Program Map A-11. Washington, DC; US Geological Survey.

Wade, F.A. & Wilbanks, J.R. (1972). Geology of Marie Byrd and Ellsworth Lands. In *Antarctic Geology and Geophysics*, ed. R.J. Adie, pp. 207–14. Oslo; Universitetsforlaget.

White, A.J.R. & Chappell, B.W. (1983). Granitoid types and their distribution in the Lachlan Fold Belt, southeastern Australia. *Geological Society of America Memoir*, **159**, 21–34.

Turbidite sequences on South Georgia, South Atlantic: their structural relationship and provenance

P.W.G. TANNER

Department of Geology, University of Glasgow, Glasgow G12 8QQ, UK

Abstract

Two major turbidite sequences, the Cumberland Bay (CBF) and Sandebugten Formations (SF), occupy the South Georgia portion of the southern Andes back-arc basin. Structural mapping around Cumberland East Bay shows that the two formations are separated by the Dartmouth Point Thrust. Contrary to some previous interpretations, the early (D_1) structures on either side of the thrust cannot be correlated. True-scale down-plunge profiles drawn using computed means of structural data show that the folds above the thrust verge NE and the thrust cuts up-section in that direction in the CBF. These folds probably developed broadly contemporaneously with the thrust movement, whereas tight chevron folds in the footwall (SF) are cut at a high angle by the thrust plane and relate to an earlier phase of deformation. 90 whole-rock analyses of coarse-grained sandstones from the CBF (including the Barff Point Member – BPM) and SF, when plotted on geochemical discrimination diagrams, substantiate previous conclusions from petrographic studies that the CBF was derived from an andesitic island-arc source, the SF from a continental magmatic arc, and that the BPM is of mixed provenance. One model consistent with both geochemical and structural data is that the basin underwent an early closure event during which the SF was deformed, and a final closure during which the CBF was deformed and thrust over the SF. Alternatively, the SF may be of Palaeozoic or early Mesozoic age and was deformed prior to the opening of the back-arc basin.

Introduction

Two major turbidite sequences occur on South Georgia: the Cumberland Bay and the Sandebugten formations (Fig. 1). The Cumberland Bay Formation (CBF) is of Early Cretaceous age (Thomson, Tanner & Rex, 1982) and consists mainly of andesitic volcaniclastic greywackes. This material was derived from an active island-arc terrain to the south-west of present day South Georgia and deposited in a back-arc basin trending NW–SE which opened during the Late Jurassic–mid-Cretaceous (Dalziel *et al.*, 1975). The Sandebugten Formation (SF) is a siliciclastic turbidite sequence of probable, but not proven, Early Cretaceous age which is thought to have been derived from the continental (South American) side of the same basin.

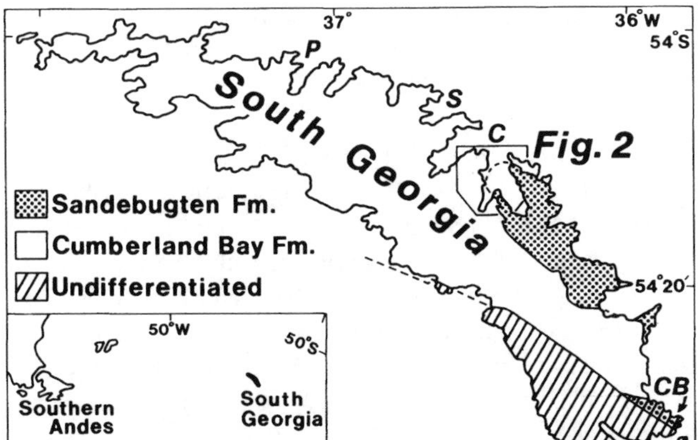

Fig. 1. Outline geological map of South Georgia showing the outcrops of the Cumberland Bay and Sandebugten formations. C, Cumberland Bay; CB, Cooper Bay Formation (ruled ornament with dots); P, Prince Olav Harbour; S, Stromness Bay. The undifferentiated rocks include elements of the floor to the back-arc basin and of the island-arc assemblage etc. (see Tanner, Storey & Macdonald, 1981). Inset shows the location of South Georgia in the South Atlantic.

Both formations have been strongly folded and affected by low-grade regional metamorphism and the contact between them is seen at three places around Cumberland Bay East (Fig. 2, localities 1–3). Following the pioneering work of Trendall (1953, 1959; see Stone (1980) for a review of earlier work), who considered all of the rocks on the Barff Peninsula to be of 'Sandebugten type', Aitkenhead & Nelson (1962)

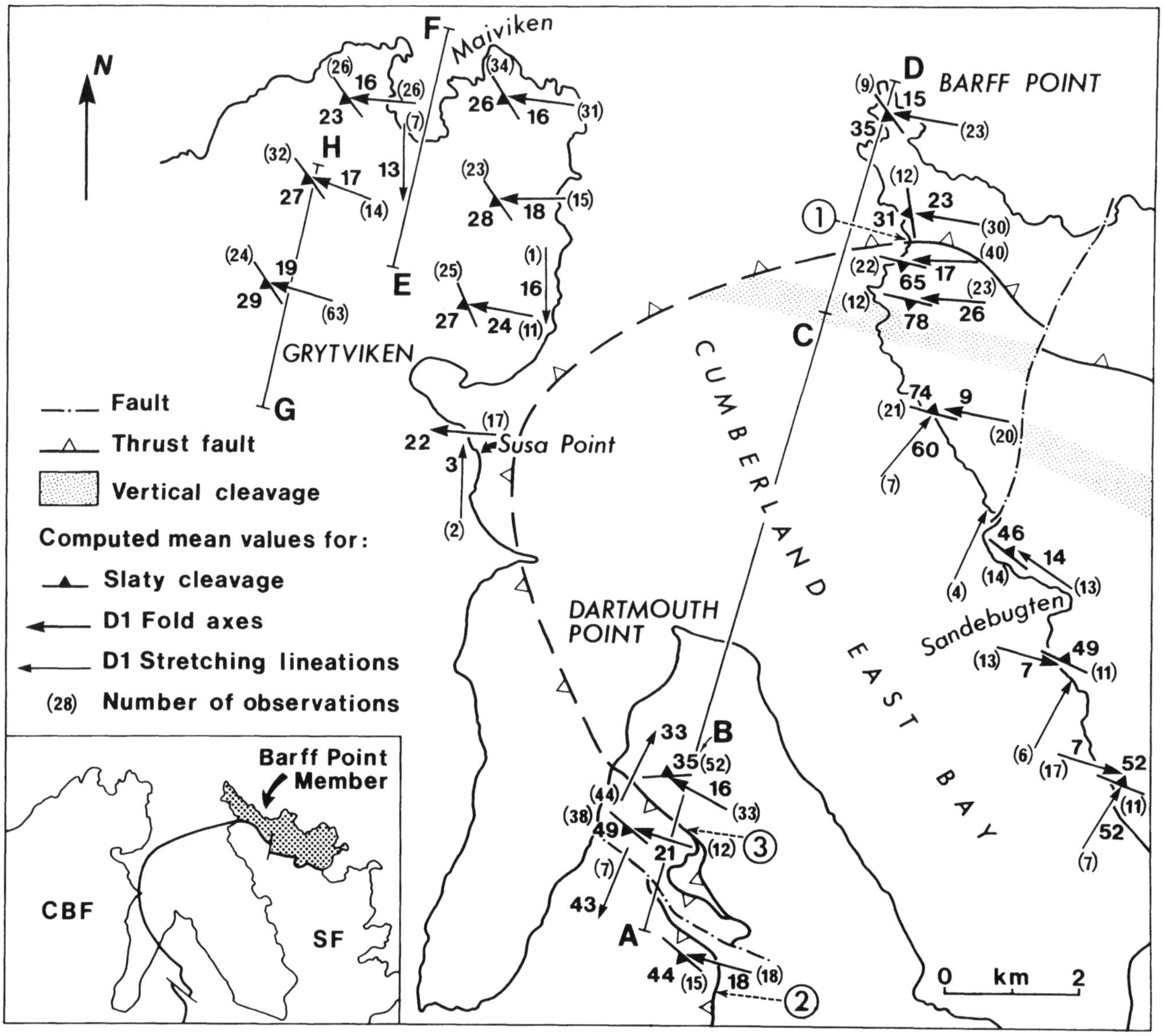

Fig. 2. Structural map of the area around Cumberland East Bay. For location, see Fig. 1. AB, CD etc. are the lines of the down-plunge profiles shown in Fig. 3. The inset shows the outcrop of the Barff Point Member with respect to the trace of the Dartmouth Point Thrust and the outcrops of the Cumberland Bay (CBF) and Sandebugten (SF) formations.

reported an unconformable contact, modified by shearing or thrusting, between the two formations at locality 1. Although Dalziel *et al.* (1975) supported Trendall's interpretation, Stone (1980) relocated the contact, considered it to be a thrust, and mapped the boundary between the CBF and SF as shown on Fig. 2. In view of the controversy over the identification of Cumberland Bay rocks on the Barff Peninsula, where they are atypical and more siliceous than normal, he named the unit the 'Barff Point Member' (Fig. 2, inset).

At locality 2 (Fig. 2) near Dartmouth Point, following reconnaissance work by Trendall (1959, fig. 12), Dalziel *et al.* (1975) recognized a thrust contact between the two formations. Both this contact and that on the Barff Point peninsula were confirmed as thrusts by Tanner (1982), and a third exposed contact (locality 3) is reported here.

The aims of this paper are as follows: to describe from the area around Cumberland Bay East the precise geometry of the major structures on either side of the thrust plane which separates the CBF and SF; to analyse the structural relationships between the two formations and, to test possible tectonic models using geochemical discrimination diagrams prepared from both existing data (Clayton, 1982) and from new XRF analyses.

Structure of the Cumberland Bay Formation

A structural synthesis of the CBF on South Georgia is given by Tanner & Macdonald (1982). The rocks are everywhere affected by D_1 chevron-style folds which vary from upward- to NE-facing. They show a systematic decrease in interlimb angle (ILA) from > 150° to 30°, and in SSW dip of their axial surfaces from vertical to < 30°, as they are traced across South Georgia from SW to NE. This geometry is consistent with the deformation of the CBF above a gently-inclined shear zone with a top-to-NE sense of shear.

Structures within three main areas, the Barff Point, Dart-

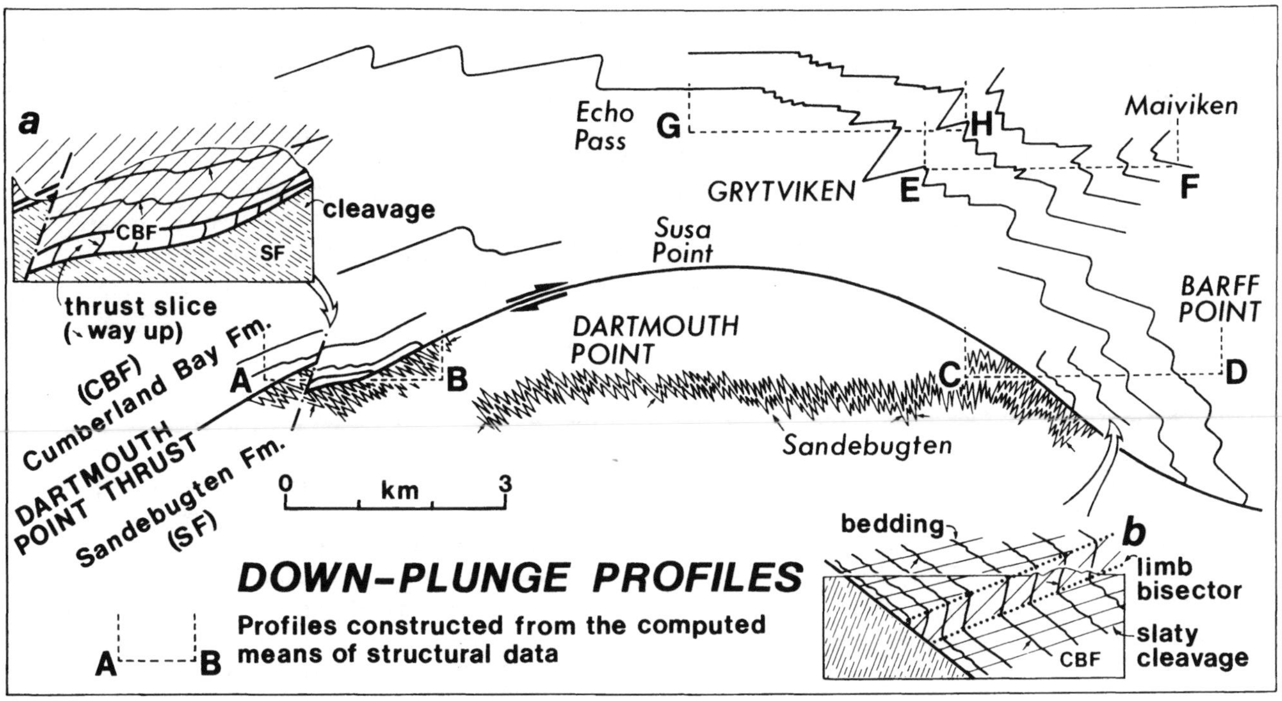

Fig. 3. Down-plunge profiles drawn along AB, CD etc. on Fig. 2. The section boxes are 2000 ft (610 m) high. Profiles AB, CD are constructed parallel to an axis plunging at 20° to 295°, and 23° to 295°, respectively, and EF, GH parallel to an axis plunging at 19° to 292°; see text for further explanation. Insets (*a*) and (*b*) show portions of these down-plunge profiles, drawn at true-scale and including the topographic profile.

mouth Point and Grytviken peninsulas, are described here. The mean values for structural measurements from subareas on each peninsula (Fig. 2) were computed using STATIS. Fold axis orientations were derived from measured cleavage–bedding intersection lineations in the Grytviken and Dartmouth Point areas, and from π-pole data at Barff Point. The major fold axes vary only slightly in trend and plunge across the area (Fig. 2), and the structures approximate to cylindroidal folds over distances of a few kilometres. Fold axes have been computed separately for each major fold from bedding, cleavage and intersection lineation data, and a gross mean value was obtained for each peninsula (see Fig. 3, caption) for use in the construction of down-plunge profiles.

The down-plunge profiles AB, CD, etc. on Fig. 3 were constructed as follows. Firstly, the mean attitude of each fold limb was computed from bedding measurements, and the orientations of the limb bisectors calculated. Then each fold profile was constructed (at 1:10 000) using the apparent dips of the limb bisector and of each fold limb in the profile plane. Finally the fold profiles were joined up by drawing bedding form-lines at an arbitrary spacing. The composite down-plunge profile on Fig. 3 has been constructed by using ABCD as a base-line at sea level and projecting the profiles EF, GH up-plunge on to the ABCD profile plane.

Slaty cleavage (D_1) is generally constant in orientation on each fold limb but varies from limb-to-limb with local cleavage fanning in the hinge zone. The divergence between computed cleavage means for any two fold limbs on the Grytviken peninsula is < 25° and the bisector is parallel, or at a small angle, to the limb bisector. For the purpose of regional comparison, only slaty cleavage measurements from the right-way-up limbs of folds have been used to derive the computed means on Fig. 2. This cleavage is fairly constant in orientation across the area, being most steeply dipping in the south-west. A stretching lineation is rarely seen on the cleavage surfaces and is at 85–90° to the local fold axis.

On the south-west side of the Dartmouth Point peninsula (Fig. 2), bedding is right-way-up, dips at about 30° SW, and is cut by a consistently steeper slaty cleavage (Fig. 3, inset *a*). In the north-east of this area a thrust sheet of near vertical Cumberland Bay rocks, strongly mylonitized and much brecciated and quartz-veined, lies above the main thrust plane. All of the rocks are strongly affected by minor D_2 folds and by the local development of a D_2 crenulation cleavage.

The Barff Point Member exposed along the south-west coastline of the Barff Point peninsula has been affected by two major folds with ILA's of 111° and 114°. In contrast to the Grytviken folds, the slaty cleavage is strongly fanned about the limb bisectors of these folds (Fig. 3, inset *b*) and this variation in the attitude of the D_1 cleavage led Stone (1980) to identify them incorrectly as major D_2 folds. Small-scale D_2 deformation is largely restricted to the fold limb closest to the thrust contact with the SF.

The Grytviken peninsula has been mapped at 1:10 000 and 32 major folds identified in a cross-strike distance of some 6 km have an average wavelength (λ) of 0.4 km. They are classical chevron folds with straight limbs, very narrow hinge zones and ILA's of 32–82° (mean 55°).

Structure of the Sandebugten Formation

A comprehensive account of the structure was given by Stone (1980). During the present work the coastal section on the north-east side of Cumberland East Bay was mapped, together with an area at the northern end of the Dartmouth Point peninsula. The rocks are affected by tight, upward-facing chevron folds associated with a penetrative slaty cleavage in mudrocks. The folds are more numerous than in the CBF and in a measured section 5 km north-west of Sandebugten, 22 folds occur within a cross-strike distance of 400 m: an average $\lambda = 38$ m. If drawn to scale the schematic cross-section through the SF (Fig. 3) would show 2–3 times more folds/km than it does. The folds are also tighter than those in the CBF, 15 folds from the measured section having a mean ILA = 25°.

As noted by previous workers (cf. Aitkenhead & Nelson, 1962; Stone, 1980) the axial surfaces of the folds in the SF change progressively in dip direction from north to south across the outcrop and define a crude anticlinorium. This feature is shown by the mean values for cleavage orientation (Fig. 2) (taken from both right-way-up and inverted fold limbs as cleavage fanning is negligible) which have a southward dip of 65° in the north, pass through the vertical and decrease to a northward dip of 35° near Dartmouth Point.

Fold axis orientations on Fig. 2 are computed from π-pole data, and agree with the attitude of local bedding–cleavage intersection lineations. The fold axes on the Barff Point peninsula change from WNW- to ESE-plunging as they are traced southwards past Sandebugten, and individual fold hinges are occasionally seen to be curvilinear. An equal area projection of these data shows that the D_1 axes are dispersed within the cleavage plane, suggesting that a partial rotation of the axes towards the X-direction has occurred. The latter, which is seen as a strongly developed stretching fibre on the cleavage planes, maintains a constant NNE trend across the area. The angle between the computed means for stretching lineations and fold axes varies from 73–84°.

Relationship between the two formations

The CBF and SF are always seen in tectonic contact (Stone, 1980) and the fault separating them is here named the 'Dartmouth Point Thrust'.

The thrust contact is exposed at three places around Cumberland Bay (Fig. 2, locs 1–3). The contact zone has the same character at each locality and consists of a unit, up to 50 cm thick, of schistose, mylonitized, and sometimes brecciated rock with thick quartz lenses, which separates recognizably bedded and cleaved members of the two formations. At one place at locality 2, en-echelon, sigmoidal quartz veins within the zone indicate a top-to-NE shear sense, but no other unambiguous small-scale kinematic indicators were found. As noted by Dalziel *et al.* (1975) and Stone (1980), there is no evidence to suggest that the contact zone is a modified unconformity. The hanging wall rocks are cut by many thin quartz veins parallel to bedding and/or cleavage, or cross-cutting both, and at locality 1 the contact zone cuts cleanly through three chevron folds in the footwall. The maximum angle between the slaty cleavages found in rocks on either side of the thrust varies from 58° (locality 1) to 78° (localities 2 & 3).

Within the thrust slice on the Dartmouth Point peninsula (Figs 2 & 3) steeply-dipping bedding and mylonitic layering are folded by minor D_2 folds which have a similar geometry to the D_2 folds which affect the gently-dipping beds in the overlying unit. This relationship suggests that the beds in the thrust slice were detached and rotated *clockwise* some 120° by NE-directed movement after considerable thrust displacement had taken place, but before the D_2 deformation.

A sharply defined break occurs within the contact zone and is oriented as follows: dip 47° N, strike 253° at locality 1; dip 33° SW, strike 160° at locality 2; dip 24–28°SW at locality 3. The smoothed profile which results from the projection of the measurements at localities 1 and 3 on to the profile plane ABCD is shown on Fig. 3. Projection of points on the thrust profile WNW parallel to the axis of this structure gives the outcrop pattern of the thrust at sea level (Fig. 2). This construction explains why the SF does not crop out on the Grytviken Peninsula; indeed use of the lower dip values at locality 2 would have resulted in the outcrop closing farther out to sea. The postulated outcrop of the thrust is in agreement with the observation that the CBF at Susa Point shows features such as: strong D_2 deformation; abundant quartz veining, especially parallel to the D_1 cleavage; and extreme, small-scale cleavage diffraction. These features have only been seen elsewhere in the hanging wall to the Dartmouth Point Thrust within < 1 km of the thrust plane.

Regional relationships

The down-plunge profiles (Fig. 3) show that folds in the CBF have a consistent easterly vergence, with long, flat-lying right-way-up limbs and shorter inverted limbs, as seen on true-scale cross-sections drawn elsewhere across South Georgia (Tanner & Macdonald, 1982). Folds such as those at Barff Point are probably thrust-related, and on a regional scale the thrust appears to climb up sequence and bring older parts of the CBF into contact with the SF as the thrust is traced westwards.

The antiformal warp of the Dartmouth Point Thrust (Fig. 3) may be an original feature of the thrusting, be due to a blind thrust(s) at depth, or result from later deformation. However, if the two limbs of the antiform are untilted by rotating the thrust planes at localities 1 and 3 about their calculated lines of intersection until the mean values for cleavage in the hanging wall at these two places coincide, it is seen that the resultant cleavage orientation, dip 38° SW, strike 156°, is in quite good agreement with the 'regional' cleavage attitude on the Grytviken peninsula. This suggests that the antiform originated as an open warp which was then accentuated by later deformation. No minor structures or fabrics have been seen which are geometrically related to the antiform: the upright D_3 structures seen in the SF farther south (Stone, 1980) trend NNE–SSW at right angles to the trace of the antiform. The continuation of the thrust on the Barff peninsula makes it unlikely that a system of conjugate shear zones or overthrusts (Dalziel *et al.*, 1975) was responsible for the present geometry.

A comparison of the structures affecting the CBF and SF shows that there are marked differences between them: (i) the folds in the CBF are more than an order of magnitude larger than those affecting the SF and have an opposing or unrelated vergence to the latter; (ii) the axial surfaces and slaty cleavages in the two groups of rocks are consistently at a very high angle to one another; and (iii), most importantly, the Dartmouth Point Thrust cuts across tight chevron folds in the SF yet appears to be genetically linked with the development of folds in the overlying CBF. Dalziel *et al.* (1975) used the coincidence in the trend of the stretching lineations in the two formations on the Dartmouth Point Peninsula (Fig. 2) to argue that the two sequences were deformed at the same time. This conclusion cannot be supported, for two reasons: (i) over a wider area, the two sets of stretching lineations do not have the same trend (Fig. 2) and each set is related to fold structures which are at different stages of development, and (ii) the mean orientations of stretching lineations from the CBF and SF are at 76° to one another, and lie within cleavages which subtend an angle of 78° between them, and thus can have no possible common origin. The sole factor common to both sequences is that the D_1 fold axes have approximately the same trend and this in isolation may be coincidental.

It is concluded that the Dartmouth Point Thrust and the D_1 folds in the CBF post-date the main folding and cleavage development in the SF. Later shortening led to a slight arching of the thrust plane, the development of minor D_2 structures locally in the CBF, and conjugate kink bands and late folds in the SF. This is contrary to the conclusions of Dalziel *et al.* (1975) that both sequences were deformed at the same time, and of Stone (1980) that the thrusting post-dated the D_1 deformation in both sequences and probably took place late in the D_2 deformation.

Geochemistry

Petrographic comparisons of the CBF and SF made by several authors show that they occupy separate fields on QFR and QFL (quartz–feldspar rock fragment or lithic fragment) diagrams, but with some overlap. This conclusion is somewhat blurred in the cases of Dalziel *et al.* (1975) and Winn (1978) as they identified the Barff Point Member as part of the SF, and not the CBF as presently accepted. From a new petrographic study, Stone (1980) found that the Barff Point Member (BPM) plotted in an intermediate position between rocks of the two main formations. This work led to the first major geochemical study (Clayton, 1982) which showed that, at the 99% significance level, there is no difference in major or trace element chemistry between the BPM and the CBF, but that these two groups show significant differences from the SF.

The present work is based upon 26 new whole-rock analyses of greywackes from the BPM and SF from Cumberland East Bay, together with 64 unpublished analyses referred to by Clayton (1982) of greywackes from the BPM and SF, and from the CBF from the north-east coast of South Georgia between Prince Olav Harbour and Stromness Bay (Fig. 1). Medium–coarse-grained greywackes consisting largely of lithic fragments have been selected for analysis from all three groups of rocks as these are most likely to reflect the geochemical nature of the source terrain. Choice of a particular rock type also limits possible variations in geochemistry due to grain size effects.

The motivation for this work comes from the recent progress that has been made in the use of both major element (Roser & Korsch, 1986) and trace element (Bhatia & Crook, 1986) data from greywacke suites to discriminate between different depositional environments. It seems logical to conclude that, if these diagrams provide the best geochemical means of identifying the tectonic setting, then they will also provide the most sensitive means of searching for similarities and differences between greywacke suites.

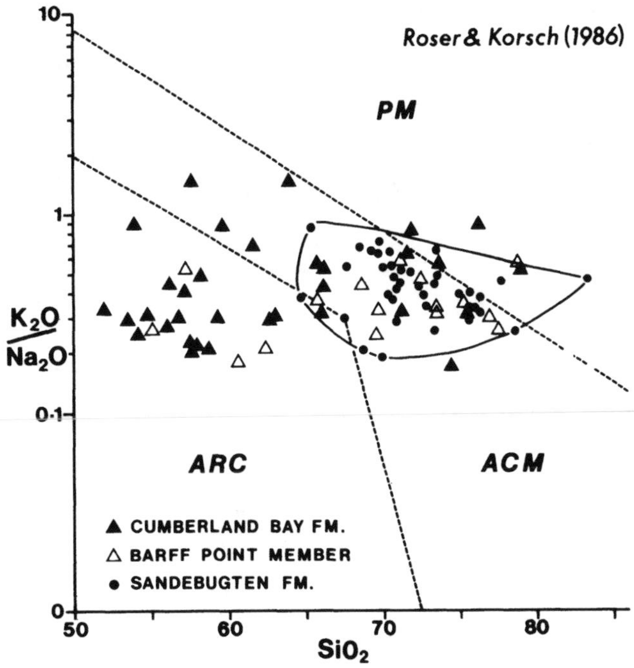

Fig. 4. Greywackes from South Georgia plotted on the SiO_2 versus K_2O Na_2O diagram of Roser & Korsch (1986). ACM, active continental margin; ARC, oceanic island margin; PM, passive margin.

SiO_2 versus K_2O Na_2O diagram

This plot was proposed by Roser & Korsch (1986) to distinguish between greywackes from three main tectonic settings (Fig. 4). SiO_2 values are from analyses calculated to 100%, volatile-free. The ACM field includes back-arc settings where the greywackes are derived from a continental margin magmatic arc and the ARC field includes greywackes from an andesitic island-arc source. The SF samples occupy a small area in the ACM field, whereas the CBF samples show a wider spread and many lie within the ARC field. The two groups overlap and it is significant that the samples from the BPM lie mainly within the SF field, but include several samples with 'CBF' characteristics.

Based on a study of Palaeozoic turbidite suites from Australia, Bhatia & Crook (1986) have selected a number of trace element plots which achieve an optimum discrimination of tectonic environments such as the Zr–Th and La–Th plots used here.

Zr–Th diagram

The SF analyses define a discrete field on this plot (Fig. 5*a*). Those of the CBF overlap part of the SF field but many of the points lie to the low-Th side of it, within the area

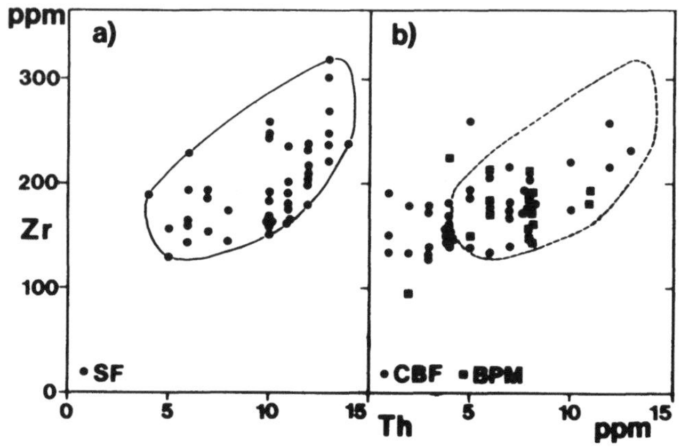

Fig. 5. Greywackes from South Georgia plotted on the Zr–Th diagram of Bhatia & Crook (1986) with the Sandebugten field from (*a*) shown pecked on (*b*). BPM, Barff Point Member; CBF, Cumberland Bay Formation; SF, Sandebugten Formation.

occupied by oceanic island arc greywackes (Bhatia & Crook, 1986). Samples of the BPM again lie mainly within the overlap between these two fields.

La–Th diagram

The Sandebugten samples occupy a narrow field which lies to the high-Th side of the CBF data points. The CBF samples have La–Th ratios characteristic of oceanic island-arcs, those of the SF are closer to continental island-arc values (Roser & Korsch, 1986). The BPM lies once more in the overlap area between the major groups, but away from the main concentration of SF points.

These preliminary geochemical data agree with the conclusions from previous petrographic work that the CBF has a largely andesitic island-arc provenance, and the SF a continental margin magmatic-arc provenance. The close relationship between the fields occupied by samples from the two formations on both major and trace element plots, the degree of overlap of these fields, and the consistent position of the BPM

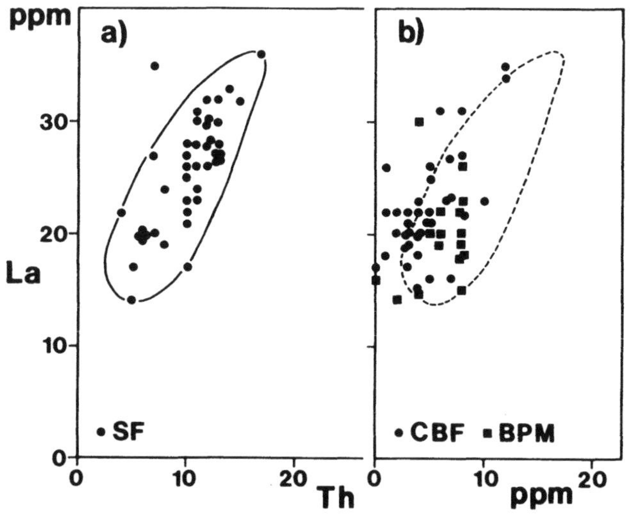

Fig. 6. Greywackes from South Georgia plotted on the La–Th diagram of Bhatia & Crook (1986). Key as for Fig. 5.

in the overlap area, also suggest that the two formations sampled the same source terrains, but to a different degree.

Discussion

From the structural and geochemical data presented in this paper, together with results of earlier work, the following conclusions may be drawn:

(i) The CBF and SF were derived largely from different provenance areas having andesitic island-arc and continental margin magmatic-arc characteristics, respectively.
(ii) The geochemistry of the BPM shows that the north-east part of the outcrop of the CBF had in part the same provenance as the SF.
(iii) The main (D_1) deformation of the SF occurred *before* the two formations were brought into contact by the Dartmouth Point Thrust.
(iv) The main (D_1) deformation of the CBF probably accompanied the development of the thrust.
(v) Later deformation (D_2) in the Cumberland East Bay area only locally affected both formations and possibly caused warping of the thrust plane which separates them.

These conclusions apply considerable constraints to possible models for the development of the South Georgia portion of the Southern Andes back-arc basin. Stone (1980) proposed a modification of Trendall's (1959) concept of the two formations being deposited contemporaneously in the same basin and then being deformed together, in which the SF at the continental margin was deformed before deposition of the CBF had been completed. The CBF was later folded and thrust over the SF. Stone invoked A-subduction of the basin contents beneath the continental edge, but work from the Southern Andes has shown that the back-arc basin has been partially obducted over the continent (Bruhn & Dalziel, 1977). Regardless of whether A- or B-type subduction is involved it is difficult to see how the CBF can be brought into thrust contact with the more highly deformed and metamorphosed (deeper level) SF during a single progressive deformation event.

An alternative model is that the SF was deposited during the early development of the back-arc basin, being derived from the so-called remnant arc (Bruhn & Dalziel, 1977) at the continental side of it and from rocks of the Tobifera Formation, or from proto back-arc silicic volcanism. A basin closure event then followed, resulting in strong deformation of the SF. Renewed Pacific-ward migration of the arc caused extensional faulting, the formation of a new back-arc basin closer to the island arc, and deposition of the CBF. The latter was of andesitic composition close to the arc but farther north-east contained detritus derived from the same provenance as the SF (cf. BPM). Final closure of the basin then occurred, with the CBF being deformed and thrust over the SF by reversal of movement (inversion) on one of the major basinal faults. Originally upright, open chevron folds were rotated and tightened during this north-eastward translation of the Cumberland Bay rocks (Tanner & Macdonald, 1982).

A further possibility which cannot be rejected is that the SF is of Palaeozoic (Trendall, 1959) or early Mesozoic age and was deposited in a fore-arc setting and deformed prior to the initiation of the back-arc basin.

Acknowledgements

This paper is based on fieldwork carried out in March–April, 1976, whilst the author was employed by the British Antarctic Survey. John McClure and David Orchard are thanked for their assistance in the field and for plane-table mapping.

References

Aitkenhead, N. & Nelson, P.H.H. (1962). *The Geology of the Area Between Cumberland West Bay and Cape George, South Georgia.* British Antarctic Survey Preliminary Geological Report, 15, 13 pp.

Bhatia, M.R. & Crook, K.A.W. (1986). Trace element characteristics of graywackes and tectonic setting discrimination of sedimentary basins. *Contributions to Mineralogy and Petrology*, **92**, 181–93.

Bruhn, R.L. & Dalziel, I.W.D. (1977). Destruction of the Early Cretaceous Marginal Basin in the Andes of Tierra del Fuego. In *Island Arcs, Deep Sea TRenches and Back-Arc Basins*, ed. M. Talwani, & W.C. Pittman, pp. 395–405. Washington, DC; American Geophysical Union, Maurice Ewing Series, 1.

Clayton, R.A.S. (1982). A preliminary investigation of the geochemistry of greywackes from South Georgia. *Bulletin of the British Antarctic Survey*, **51**, 89–109.

Dalziel, I.W.D., Dott, R.H., Winn, R.D. & Bruhn, R.L. (1975). Tectonic relations of South Georgia Island to the southernmost Andes. *Geological Society of America Bulletin*, **86**, 1034–40.

Roser, B.P. & Korsch, R.J. (1986). Determination of tectonic setting of sandstone–mudstone suites using SiO_2 content and K_2O/Na_2O ratio. *Journal of Geology*, **94**, 635–50.

Stone, P. (1980). *The Geology of South Georgia: IV. Barff Peninsula and Royal Bay Areas.* Cambridge; British Antarctic Survey Scientific Reports, No. 96, 45pp.

Tanner, P.W.G. (1982). Geologic evolution of South Georgia. In *Antarctic Geoscience*, ed. C. Craddock, pp. 167–76. Madison; University of Wisconsin Press.

Tanner, P.W.G. & Macdonald, D.I.M. (1982). Models for the deposition and simple shear deformation of a turbidite sequence in the South Georgia portion of the southern Andes back-arc basin. *Journal of the Geological Society, London*, **139**, 739–54.

Tanner, P.W.G., Storey, B.C. & Macdonald, D.I.M. (1981). Geology of an Upper Jurassic–Lower Cretaceous island-arc assemblage in Hauge Reef, the Pickersgill Islands and adjoining areas of South Georgia. *Bulletin of the British Antarctic Survey*, **53**, 77–117.

Thomson, M.R.A., Tanner, P.W.G. & Rex, D.C. (1982). Fossil and radiometric evidence for ages of deposition and metamorphism of sedimentary sequences on South Georgia. In *Antarctic Geoscience*, ed. C. Craddock, pp. 177–84. Madison; University of Wisconsin Press.

Trendall, A.F. (1953). *The Geology of South Georgia: I.* London; Falkland Islands Dependencies Survey Scientific Reports, No. 7, 26pp.

Trendall, A.F. (1959). *The Geology of South Georgia: II.* London; Falkland Islands Dependencies Survey Scientific Reports, No. 19, 48pp.

Winn, R.D. (1978). Upper flysch of Tierra del Fuego and South Georgia island: a sedimentological approach to lithosphere plate reconstruction. *Geological Society of America Bulletin*, **89**, 533–47.

Dredged rocks from Powell Basin and the South Orkney microcontinent

P.L. BARBER[1], P.F. BARKER[1,2] & R.J. PANKHURST[2]

1 Department of Geological Sciences, University of Birmingham, Birmingham B15 2TT, UK
2 British Antarctic Survey, Madingley Road, Cambridge CB3 0ET, UK

Abstract

The results of geochemical analyses and isotopic determinations on dredged rocks from around Powell Basin and on the South Orkney microcontinental block are reported. Among the *in situ* rocks are two main groups: a calc-alkaline suite of Late Cretaceous age, and an alkali-basaltic suite of Pliocene–Recent age with an additional, possibly related Eocene occurrence. The calc-alkaline rocks reinforce earlier interpretations that the main magnetic province on the South Orkney block is an extension of the Antarctic Peninsula magmatic arc, produced by subduction of Pacific oceanic lithosphere. The alkaline rocks are very similar to those exposed elsewhere along the Antarctic Peninsula (e.g. Alexander Island, Seal Nunataks, James Ross Island, etc.). The extension of the geographical range of this suite emphasizes the enigmatic nature of its relationship (if any) with subduction at the Pacific margin. The Eocene alkali basalts may be precursors to the Powell Basin opening.

Introduction

Powell Basin was created by the separation of the South Orkney microcontinent (SOM) from the Antarctic Peninsula (e.g. Hawkes, 1962; Barker & Griffiths, 1972; Dalziel, 1982). The most detailed interpretation of its origin and age (King & Barker, 1988) is based on geophysical investigation of the basin margins and subsidence history. Here we describe rocks dredged from Powell Basin, with the main aim of further constraining both the pre-drift reconstruction and the timing of break-up.

The rocks were dredged from six sites during the 1980–81 cruise of the RRS *Shackleton*, using combined rock and pipe dredges. Sites were selected on steep scarps and rough seabed topography to maximize the chance of recovering *in situ* rather than ice-rafted material. Sites were dredged upslope and the dredge tracks located using satellite navigation. Rocks were considered either *in situ* or locally-derived scree if they formed a significant proportion of a dredge haul. Striated, faceted, well-rounded and polished rocks were regarded as erratics. Other dredge hauls from the region, not reported here, contained a wide range of lithologies. In these hauls no *in situ* component was recognized, but they serve to characterize the regional compositional range of ice-rafted material.

Major- and trace-element abundances were determined using a Philips 1450 XRF spectrometer at the Department of Geological Sciences, University of Birmingham. K–Ar ages were determined on washed 80–120 mesh fractions of clean whole-rock material using standard techniques (Pankhurst & Smellie, 1983), at the NERC Isotope Geology Centre, London.

Dredge descriptions

Site 70

Dredge site 70 (61°10·3′S, 50°54·8′W) lies 160 km east of Clarence Island, on the southern margin of the South Scotia Ridge (see Fig. 1). This was probably a strike-slip margin during the (?)Oligocene opening of Powell Basin (King & Barker, 1988). The middle section of the continental slope adjoining Powell Basin was dredged over 1.5 km, between 2030 and 1810 m depths. This part of the South Scotia Ridge is strongly magnetized (Watters, 1972), but the dredge traversed an area of more rugged seabed topography south of the main slope. A haul of 46 kg was recovered of which one-third was very fresh alkali olivine basalt, including subangular fragments of sparsely phyric basaltic-glass and two radially jointed pillow lava segments.

Site 93

Dredge 93 (61°39·6′S, 41°45·0′W) lies on the south-eastern flank of the SOM, within the East Margin Magnetic High (King & Barker, 1988). The site, a minor north-easterly trending scarp on the upper continental slope, was dredged over 0.75 km between depths of 1250 and 1040 m. Of 40 kg of rock recovered, approximately 75% is locally-derived submarine scree consisting of hydrothermally altered, subangular fragments of subalkaline basalt and basaltic-andesite.

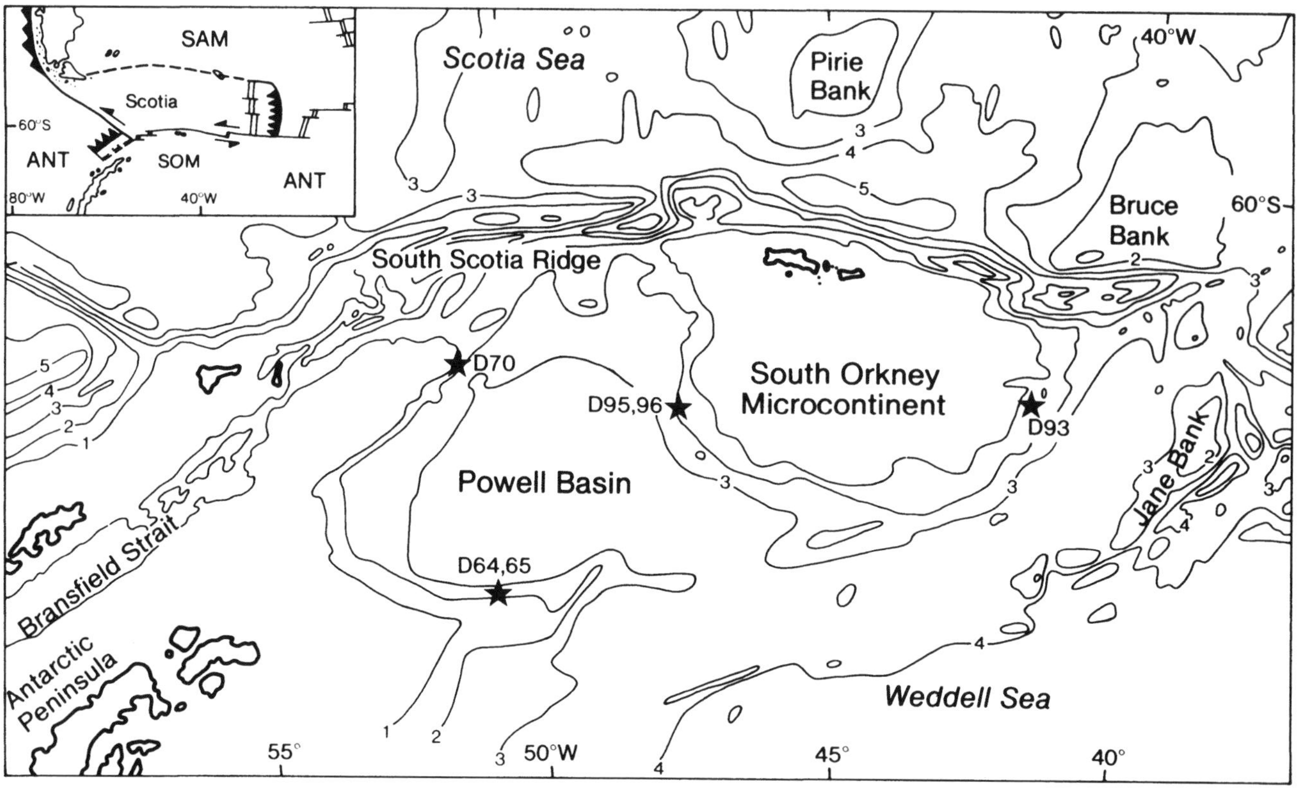

Fig. 1. Location of dredge sites from the South Orkney microcontinent and the margins of Powell Basin. Bathymetry at 1000 m intervals is taken from British Antarctic Survey (1985). The inset shows the wider plate-tectonic setting of the region.

Sites 95 and 96

The south-western margin of the SOM was sampled along two independent tracks close to 61°43·8′S, 47°34·7′W. This was an extensional margin associated with Powell Basin opening; the dredge site lies within the West Margin Magnetic High (King & Barker, 1988). At site 95 the dredge traversed the continental slope between 2470 and 2000 m depth, and recovered 22 kg of rock. About 75% of the dredge haul was hydrothermally-altered subalkaline basalts, with less common fresh alkali olivine basalts and related volcanic glass. In view of the restricted extent of the dredge haul, a longer section (3210–1940 m) was dredged approximately 1 km north of site 95. Dredge 96 yielded 41 kg, of which 45% was the olivine alkali basalt previously identified at site 95, and subordinate altered subalkaline basalt.

Sites 64 and 65

Both sites lie on the southern margin of Powell Basin, on an easterly-trending salient of presumed continental crust of the Antarctic Peninsula. This was a strike-slip margin during the (?)Oligocene opening of Powell Basin (King & Barker, 1988). Dredge site 64 lies at 63°02·80′S, 50°35·40′W, between depths of 2450 and 1430 m. Of 85 kg dredged, approximately 30% is a local assemblage of two types: 9 kg of buff-coloured siltstones and fine-grained sandstones and 19 kg of hydrothermally-modified alkali basalts displaying limonitic weathering crusts. Complementary lithologies were recognized from site 65, at 63°03·2′S, 50°38·6′W. This deeper dredge (2450–2060 m) yielded 33 kg, of which 70% is locally-derived scree (21 kg of sediment and 3 kg of altered alkali basalt). The stratigraphic relationship between sedimentary rocks and lavas is unknown, as seismic reflection profiling has been prevented by persistent sea ice. Alkali basalts from the other sites reported here are not hydrothermally altered. As discussed below the young age, fresh character and, for dredge 70, pillow structures, are consistent with submarine eruptions in the slope environment. In contrast, deeper levels of a lava pile are accessible at sites 64 and 65, suggesting tectonic exposure.

Dredge geochronology and geochemistry

On the basis of 26 analyses and 7 whole-rock K–Ar age determinations, two broad geochemical and age groupings have been identified. Group 1 rocks (from sites 93, 95 and 96 on the SOM) form a late Mesozoic subduction-related extrusive suite. Group 2 rocks (sites 64, 65, 70 and 95/96, distributed around the margins of Powell Basin) are Cenozoic alkali basalts.

Group 1

The four analysed dredge 93 samples are hydrothermally-modified basalts and basaltic-andesites, with a restricted range of silica levels (49.58–54.06%). Their feldspar-phyric character is reflected in high alumina values (17.97–19.89%), and their extensively albitized plagioclase in relatively high Na_2O. On the basis of Na_2O/CaO ratios, the lavas are strongly spilitic (Graham, 1976) and, as is often the case with

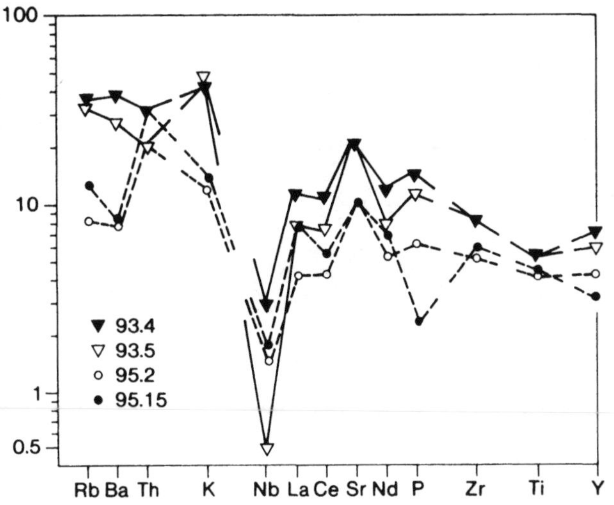

Fig. 2. Normalized incompatible element abundance diagram for representative subduction-related Mesozoic lavas recovered from dredge sites; sample locations are shown in Fig. 1. Normalizing values after Wood *et al.* (1979a). Geochemistry of the 26 analysed dredge samples is available in tabular form from Dr P.F. Barker, British Antarctic Survey.

Table 1. *Radiometric age determinations on dredged rocks from the South Orkney microcontinent and margins of Powell Basin*

Sample	K_2 (%)	^{40}Ar (rad nl/g)	% Ar (atmos.)	Age (Ma)
Group 1				
93.5	1.272	3.8506	34.3	76.0 ± 2.0
95.1	0.311	1.0466	59.7	84.6 ± 2.5
95.15	0.180	0.4864	43.9	68.3 ± 4.1
Group 2				
70.9	1.130	0.1704	76.5	3.9 ± 0.2
96.5	0.773	0.1355	72.2	4.5 ± 0.2
96.10	1.026	0.0041	99.5	0.1 ± 0.5
64.1	2.403	4.6448	19.6	49.1 ± 1.6
		4.5126	28.7	47.7 ± 1.5

Decay constants: $\lambda_B = 4.962 \times 10^{-10}$ y; $\lambda_C = 0.581 \times 10^{-10}$ y; ^{40}K/K atomic = 1.167×10^{-4}. Errors on K content assumed to be ±1%, 1 sigma.

such rocks, they plot as alkaline basalts on the basis of total alkali–silica distribution (Herrman, Potts & Knake, 1974; Vallance, 1974). This alkaline character is considered to be secondary; the samples have low Ce/Y, Nb/Zr and Nb/Y ratios, noted by Weaver, Sceal & Gibson (1972) and Floyd & Winchester (1975) as features of subalkaline basalts. Despite the limited number of analysed samples and their hydrothermal overprint, some indication of petrotectonic environment is provided by the relative enrichment of large-ion lithophile (LIL) components (Rb, K, Ba and Sr) with respect to high field strength (HFS) elements (notably immobile Ti and Zr) when plotted on the 'primordial mantle' normalized diagram of Wood *et al.* (1979a). Fig. 2 shows this decoupling and also relative Nb depletion, which together suggest a subduction-related origin (Wood, Joron & Treuil, 1979b; Saunders, Tarney & Weaver, 1980). Sample 93.5 has yielded a Late Cretaceous K–Ar age of 76.0 ± 2 Ma (Table 1). Dredge 93 lavas are of similar age and geochemistry to the Late Cretaceous onshore dyke swarm of the western South Orkney Islands (E.C. King, pers. comm., 1987).

Group 1 lavas from dredge sites 95 and 96 are represented by four analysed samples (95.1, 95.2, 95.15 and 96.12). These are subalkaline basalts with restricted silica range (49.44–50.39%). Some hydrothermally-induced element mobility is probable, particularly in sample 95.2 which registers 5% normative nepheline. Irrespective of variable alteration patterns, the dredge 95 samples show broadly concordant major and trace element chemistry, and are petrogenetically distinct from dredge 93 basalts (Fig. 2). They display lower levels of LIL and HFS elements than dredge 93 lavas, but share a decoupling of the two element groups and a subduction-related depletion in Nb. They additionally have comparable low Ce/Y ratios of ~0.5. Samples 95.1 and 95.15 yielded Late Cretaceous K–Ar ages of 84.6 ± 2.5 and 68.3 ± 4.1 Ma, respectively (Table 1).

Group 2

Four samples from the South Scotia Ridge (dredge 70) pillow basalts and associated glass were analysed. Based on total alkali–silica relationships these are alkali basalts (MacDonald & Katsura, 1964). The samples contain olivine phenocrysts and are strongly olivine-normative (14.52–19.70%). Except for sample 70.1, the rocks are also strongly nepheline-normative (12.67–13.94%) and have higher total alkalies than reported from eruptive centres at Paulet Island (Baker, Buckley & Rex, 1977), James Ross Island (Nelson, 1975) and Jason Peninsula (Saunders, 1982). This distinctive undersaturation may reflect a north-eastward increase in undersaturation for onshore alkali basalts as suggested by González-Ferrán (1982). Otherwise the dredge 70 rocks are chemically similar to the Cenozoic volcanic centres of the Antarctic Peninsula noted above. In major element geochemistry, sample 70.9 most closely resembles sample 27788 from Paulet Island (Baker *et al.*, 1977), shown in Fig. 3. Trace element distributions additionally support an affinity with the onshore alkali basalts. Sample 70.9 has been dated at 3.9 ± 0.2 Ma, which falls within the age range established for the onshore eruptive province by Rex (1976), Baker *et al.* (1977) and Smellie *et al.* (1988).

Group 2 lavas from the western margin of the SOM (sites 95 and 96) are also alkali olivine basalts, but slightly less undersaturated than dredge 70 samples, with lower levels of normative nepheline (0.0–10.0%). The specimens also vary more than those of site 70 and may represent several flow units. Of the more undersaturated samples, 96.1 and 96.5 are the most similar geochemically, but differ in texture. Interflow variation is also suggested by the unfractionated basalt, sample 96.20. This is virtually a picrite containing high Mg (11.63%), Ni and Cr; normative olivine registers 24.94%. Sample 96.5 has an early Pliocene K–Ar date of 4.5 ± 0.2 Ma. Two remaining samples, 96.10 (microporphyritic-olivine basalt) and 95.17 (vesicular palagonite), are more evolved, containing higher Si and Al and lower Ca and Mg. Given the restricted nature of the dredges it is unclear if these samples represent a separate

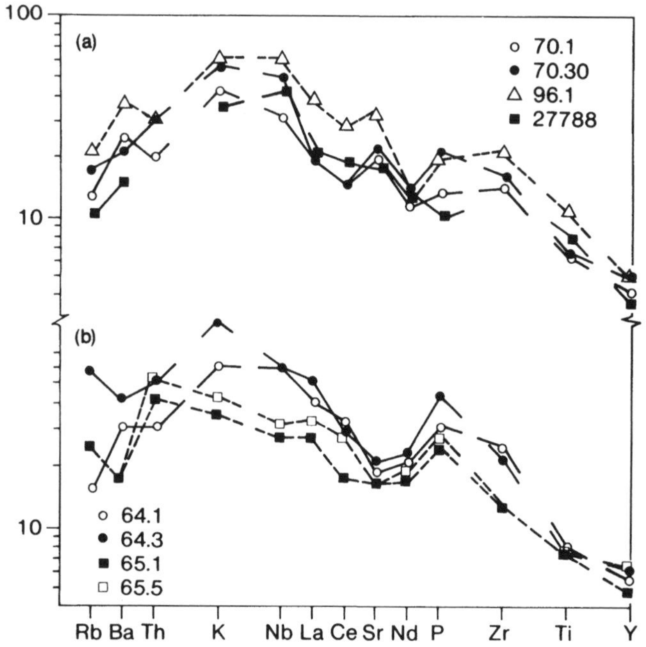

Fig. 3. (*a*) **Normalized incompatible element abundance diagram for representative Pliocene and Quaternary alkali basalts from circum-Powell Basin dredge sites. Sample 27788 (hawaiite from Paulet Island) is plotted for comparison; data are taken from analyses by Baker *et al.* (1977) and Pankhurst (1982). (b) Normalized incompatible element abundance diagram for representative Eocene alkali basalts dredged from the eastern continental slope of the Antarctic Peninsula.**

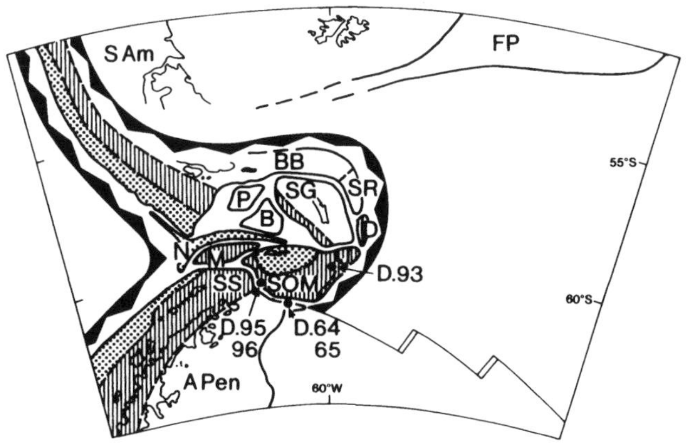

Fig. 4. Reconstruction of the Scotia arc region at 35 Ma (King & Barker, 1988) showing positions of dredge sites 64/65, 93 and 95/96. Stippled area, non-magnetic fore-arc; lined area, magnetic arc; FP, Falkland Plateau. Fragments are: BB, Burdwood Bank; SG, South Georgia; SR, Shag Rocks; D, Discovery Bank; P, Pirie Bank; B, Bruce Bank; J, Jane Bank; N, non-magnetic and M, magnetic part of South Scotia Ridge; SS, South Shetland Islands.

seafloor assemblage. Sample 96.10 has yielded a notably young date of 0.1 ± 0.5 Ma.

Group 2 alkali olivine basalts from the southern margin of Powell Basin (sites 64 and 65) have a variable hydrothermal overprint, and some show seafloor weathering. Loss on ignition values are higher (1.08–3.47%) than other group 2 rocks and most samples fall outside one or both of the standard chemical screens: the 'igneous spectrum' of Hughes (1974), based on Na_2O, K_2O values, or Stillman & Williams' (1978) criterion of CaO + MgO between 12 and 20%. However, their high incompatible/compatible ratios (e.g. Ce/Y ~ 2), high Ni and Cr and immobile element ratios such as Zr/Ti and Zr/Nb (Winchester & Floyd, 1977), make the basalts clearly alkaline. Their incompatible element abundances are compared with other group 2 dredge rocks in Fig. 3. Although the number of analyses was limited by fragment size, rocks from sites 64 and 65 appear to form discrete assemblages on this diagram. The least altered specimen (64.1) was radiometrically dated. Although less evolved than other dredge 64 samples, it has a clear affinity to other analysed samples in Fig. 3. The K–Ar age of 48.4 ± 1.5 Ma is the oldest for both onshore and offshore group 2 rocks within the Antarctic Peninsula and Scotia Sea, and has been closely checked by duplicate Ar analysis.

Discussion

Group 1: Mesozoic subduction-related volcanism

Group 1 lavas from sites 93 and 95/96 are the only Mesozoic extrusive rocks from the SOM; they complement basic dykes exposed onshore (Thomson, 1974; King & Barker, 1988) and a major offshore batholith inferred on geophysical grounds (Harrington, Barker & Griffiths, 1972). They pre-date the estimated Oligocene separation of the SOM from the Antarctic Peninsula (King & Barker, 1988). In view of their subduction-related geochemical signature and Late Cretaceous age, the dredged rocks are considered to be an easterly extension of the Antarctic Peninsula Volcanic Group (Thomson, 1982).

Late Cretaceous and early Tertiary volcanic sequences occur throughout the Antarctic Peninsula. Lavas of comparable age to the youngest dredged subduction-related rocks occur on the western side of the Antarctic Peninsula, on Byers Peninsula, Livingston Island (Pankhurst *et al.*, 1979), Tower Island (Rex, 1976) and Alexander Island (Burn, 1981). These centres are co-linear with dredge site 95/96 when Powell Basin is closed in King & Barker's (1988) 35 Ma reconstruction (Fig. 4). Site 95/96 lies within the hachured area of magnetic anomalies roughly coincident with the main batholith. In contrast, the more easterly position of site 93, more remote from the Pacific margin, makes the tectonic setting of these lavas uncertain. However, dredge 93 lavas are similar in age and chemistry to easterly members of the South Orkney Islands basic dyke swarm (E.C. King, pers. comm., 1987) and may reflect a broadly extensional continental back-arc environment.

Group 2: Cenozoic alkaline volcanism

Cenozoic alkali basaltic volcanism along the Antarctic Peninsula is sparsely distributed and episodic. Dated onshore exposures range over the last 8 m.y., with the exception of 15 Ma old dykes on Alexander Island (Smellie, 1987; Smellie *et al.*, 1988). Group 2 rocks from dredge sites 70, 95 and 96 represent a considerable north-eastward extension of this volcanic activity (Fig. 5). However, the Eocene age of sample 64.1 is much older than any previously determined. The extensive Cenozoic alkali volcanism of Marie Byrd Land (González-

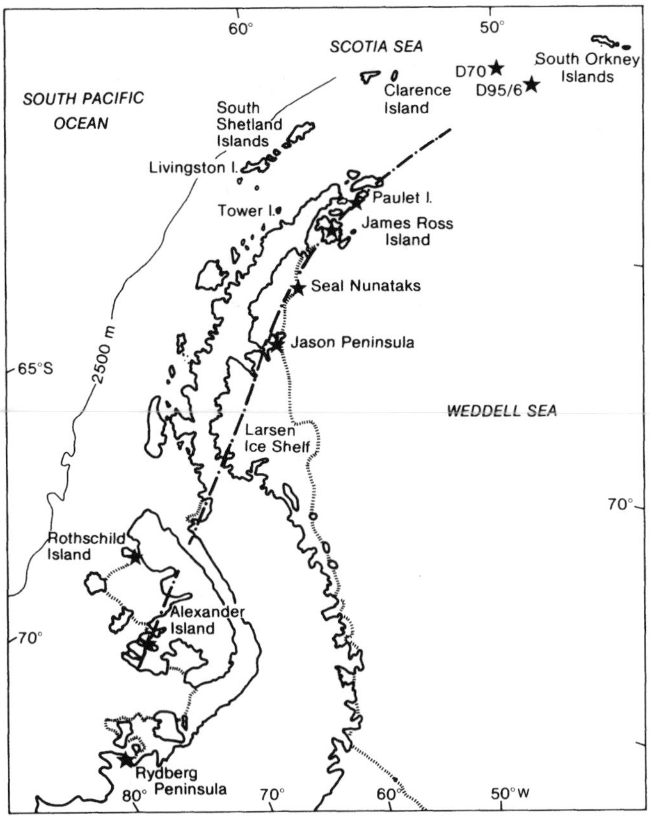

Fig. 5. Location of the younger Powell Basin alkaline volcanic centres (sites 70, 95 and 96) with respect to late Miocene–Recent onshore centres of the Antarctic Peninsula (from Smellie *et al.*, 1988). The 2500 m isobath is shown to the west of Graham Land to illustrate the apparently constant distance from the centres of alkali basalt eruption to the Pacific margin.

Ferrán & Vergara, 1972) is mainly of Pliocene–Recent age, but extends back to the late Oligocene (LeMasurier & Rex, 1982). The Patagonian plateau basalts of southern South America, dominantly alkaline since the late Oligocene, have been similarly episodic in production (57–42 Ma, and 4–0.2 Ma; Baker *et al.*, 1981).

The tectonic significance of this activity has not been explained. Antarctic Peninsula alkali volcanic rocks have been related to the inception of extension following the cessation of subduction at the Pacific margin (Barker, 1976, 1982; Baker *et al.*, 1977) and, with the clear exception of the back-arc extensional Bransfield Strait, are genetically divorced from earlier pervasive calc-alkaline magmatism. Pankhurst (1982) and Saunders (1982) noted that alkali-rich, Sr-rich basalts, with low initial $^{87}Sr/^{86}Sr$ ratios (~ 0.7030) and highly-fractionated REE patterns, typify the intraplate volcanism of oceanic islands. These may have originated by small degrees of asthenospheric melting (Kay & Gast, 1973; Sun & Hanson, 1975). The possibility of LIL enrichment of the source region by addition of ancient subducted lithosphere has also been considered (Hofmann & White, 1982). However, most authors regard alkali basaltic magmatism as originating at much greater depths than those of subduction zones beneath calc-alkaline volcanic arcs, and forming independently of the chemical influence of the subducted plate.

Nevertheless, the distribution of alkali volcanism parallel to the Pacific margin (Fig. 5), the recognition that extension can occur at subducting margins, and the evolution of the Patagonian plateau basalts away from a subduction-related, back-arc initial chemistry (Baker *et al.*, 1981) suggest that the trigger for this regional alkali volcanism could be associated indirectly with subduction. For example Skewes & Stern (1979) suggested that subduction-related thermal or mechanical perturbation of the subcontinental mantle had influenced production of the Patagonian Plateau basalts. The involvement of subduction, directly or indirectly, in producing the Powell Basin alkali volcanic rocks therefore needs to be assessed.

Subduction at the Pacific margin of the Antarctic Peninsula ceased progressively later north-eastwards, after a series of ridge crest–trench collisions (Barker, 1982; Cande, Herron & Hall, 1982). Only off the South Shetland Islands might very slow subduction be continuing at present, enabled by back-arc extension in Bransfield Strait (Barker, 1976, 1982). The younger Powell Basin alkali basalt sites (70, 95 and 96) are more remote from these subduction events than other onshore centres. They have been further isolated from any Pacific margin subduction by the opening of Drake Passage, which commenced before 28 Ma ago (Barker & Burrell, 1977).

The separation of the SOM from the Antarctic Peninsula (King & Barker, 1988), which created Powell Basin, is itself a back-arc extension linked to the subduction of South American oceanic lithosphere beneath an easterly migrating arc (now Jane Bank, Fig. 1), analogous to the present South Sandwich Islands arc (Barker, Barber & King, 1984). Powell Basin probably opened between 30 and 20 Ma ago, with possible precursory volcanism extending back 10 Ma or more earlier (King & Barker, 1988). Because subduction beneath Jane Bank ceased about 20 Ma ago, the Pliocene–Recent alkaline activity at sites 70, 95 and 96 is possibly not directly subduction-linked. However, the Eocene alkali volcanism represented at sites 64 and 65 could be precursory to Powell Basin opening, as it is otherwise regionally anomalous.

The younger Powell Basin rocks and the onshore late Miocene–Recent alkali volcanism of the Antarctic Peninsula share an intraplate signature and are synchronous over much of the peninsula. The oldest ages were obtained from Alexander Island (Smellie *et al.*, 1988) but there is no clear migratory trend related to the staggered subduction history of the peninsula. Another striking feature is the near-uniform distance (about 300 km) of nearly all known occurrences (Bransfield Strait area excepted) from the present Pacific margin (Fig. 5). This relationship exists from Rydberg Peninsula and Alexander Island in the south to Powell Basin in the north. The distance from the margin is systematically greater than that of the Cenozoic calc-alkaline magmatism, as in the South American alkali volcanic occurrences. These systematic features seem inevitably to be linked indirectly to subduction, but remain to be explained.

Acknowledgements

The authors thank the ship's company of RRS *Shackleton*, and technical officers of Research Vessel Services, Barry, for unstinting support during a three month cruise. We also thank Dr E.C. King and Andrew Wilson for their work during dredging and for subsequent helpful discussion. Dr G.L. Hendry of the University of Birmingham gave much advice and assistance with geochemical analyses. The work was supported by a NERC research grant to the University of Birmingham.

References

Baker, P.E., Buckley, F. & Rex, D.C. (1977). Cenozoic volcanism in the Antarctic. *Philosophical Transactions of the Royal Society, London*, **B279**, 131–42.

Baker, P.E., Rea, W.J., Skarmeta, J., Carmenos, R. & Rex, D.C. (1981). Igneous history of the Andean Cordillera and Patagonian Plateau around Latitude 46° S. *Philosophical Transactions of the Royal Society, London*, **A303**, 105–49.

Barker, P.F. (1976). The tectonic framework of Cenozoic volcanism in the Scotia Sea region – a review. In *Symposium on Andean and Antarctic Volcanology Problems*, ed. O. González-Ferrán, pp. 330–46. Rome; International Association of Volcanology and Chemistry of the Earth's Interior.

Barker, P.F. (1982). The Cenozoic subduction history of the Pacific margin of the Antarctic Peninsula: ridge crest–trench interactions. *Journal of the Geological Society, London*, **139(6)**, 787–801.

Barker, P.F., Barber, P.L. & King, E.C. (1984). An Early Miocene ridge crest–trench collision on the South Scotia Ridge. *Tectonophysics*, **102**, 315–32.

Barker, P.F. & Burrell, J. (1977). The opening of Drake Passage. *Marine Geology*, **25(1)**, 15–34.

Barker, P.F. & Griffiths, D.H. (1972). The evolution of the Scotia Ridge and Scotia Sea. *Philosophical Transactions of the Royal Society, London*, **A271**, 151–83.

British Antarctic Survey. (1985). *Tectonic map of the Scotia Arc, 1: 3 000 000*, BAS (Misc) 3. Cambridge; British Antarctic Survey.

Burn, R.W. (1981). Early Tertiary calc-alkaline volcanism on Alexander Island. *British Antarctic Survey Bulletin*, **53**, 175–93.

Cande, S.C., Herron, E.M. & Hall, B.R. (1982). The Early Cenozoic tectonic history of the southeast Pacific. *Earth and Planetary Science Letters*, **57(1)**, 63–74.

Dalziel, I.W.D. (1982). The early (pre-Middle Jurassic) history of the Scotia arc region: a review and progress report. In *Antarctic Geoscience*, ed. C. Craddock, pp. 111–26. Madison; University of Wisconsin Press.

Floyd, P.A. & Winchester, J.A. (1975). Magma type and tectonic setting discrimination using immobile elements. *Earth and Planetary Science Letters*, **27(3)**, 211–18.

González-Ferrán, O. (1982). The Antarctic Cenozoic volcanic provinces and their implications in plate tectonic processes. In *Antarctic Geoscience*, ed. C. Craddock, pp. 687–94. Madison; University of Wisconsin Press.

González-Ferrán, O. & Vergara, M. (1972). Post-Miocene volcanic petrographic provinces of West Antarctica and their relation to the southern Andes of South America. In *Antarctic Geology and Geophysics*, ed. R.J. Adie, pp. 187–95. Oslo; Universitetsforlaget.

Graham, C.M. (1976). Petrochemistry and tectonic significance of Dalradian metabasaltic rocks of the southwestern Scottish Highlands. *Journal of the Geological Society, London*, **132(1)**, 61–84.

Harrington, P.K., Barker, P.F. & Griffiths, D.H. (1972). Crustal structure of the South Orkney Islands area from seismic refraction and magnetic measurements. In *Antarctic Geology and Geophysics*, ed. R.J. Adie, pp. 27–32. Oslo; Universitetsforlaget.

Hawkes, D.D. (1962). The structure of the Scotia Arc. *Geological Magazine*, **99(1)**, 85–91.

Herrman, A.G., Potts, M.J. & Knake, D. (1974). Geochemistry of the rare earth elements in spilites from oceanic and continental crust. *Contributions to Mineralogy and Petrology*, **44(1)**, 1–16.

Hofmann, A.W. & White, W.M. (1982). Mantle plumes from ancient ocean crust. *Earth and Planetary Science Letters*, **57(5)**, 421–36.

Hughes, C.J. (1974). Spilites, keratophyres, and the igneous spectrum. *Geological Magazine*, **109(6)**, 513–27.

Kay, R.W. & Gast, P.W. (1973). The rare-earth content and origin of alkali-rich basalts. *Journal of Geology*, **81(6)**, 653–82.

King, E.C. & Barker, P.F. (1988). The margins of the South Orkney microcontinent. *Journal of the Geological Society, London*, **145(2)**, 317–31.

LeMasurier, W.E. & Rex, D.C. (1982). Volcanic record of Cenozoic glacial history in Marie Byrd Land and western Ellsworth Land: revised chronology and evaluation of tectonic factors. In *Antarctic Geoscience*, ed. C. Craddock, pp. 725–34. Madison; University of Wisconsin Press.

MacDonald, R. & Katsura, T. (1964). Geochemical composition of Hawaiian lavas. *Journal of Petrology*, **5(2)**, 82–133.

Nelson, P.H.H. (1975). *The James Ross Island Volcanic Group of North-east Graham Land*. London; British Antarctic Survey Scientific Reports, No. 54, 62 pp.

Pankhurst, R.J. (1982). Sr-isotope and trace-element geochemistry of Cenozoic volcanic rocks from the Scotia arc and the northern Antarctic Peninsula. In *Antarctic Geoscience*, ed. C. Craddock, pp. 229–34. Madison; University of Wisconsin Press.

Pankhurst, R.J. & Smellie, J.L. (1983). K–Ar geochronology of the South Shetland Islands, lesser Antarctica: apparent lateral migration of Jurassic to Quaternary island-arc volcanism. *Earth and Planetary Science Letters*, **66(2)**, 214–22.

Pankhurst, R.J., Weaver, S.D., Brook, M. & Saunders, A.D. (1979). K–Ar chronology of Byers Peninsula, Livingston Island, South Shetland Islands. *British Antarctic Survey Bulletin*, **49**, 277–82.

Rex, D.C. (1976). Geochronology in relation to the stratigraphy of the Antarctic Peninsula. *British Antarctic Survey Bulletin*, **43**, 49–58.

Saunders, A.D. (1982). Petrology and geochemistry of alkali basalts from the Jason Peninsula, Oscar II Coast, Graham Land. *British Antarctic Survey Bulletin*, **55**, 1–9.

Saunders, A.D., Tarney, J. & Weaver, S.D. (1980). Transverse geochemical variations across the Antarctic Peninsula: implications for the genesis of calc-alkaline magmas. *Earth and Planetary Science Letters*, **46(4)**, 344–60.

Skewes, M.A. & Stern, C.R. (1979). Petrology and geochemistry of alkali basalts and ultramafic inclusions from the Palei-Aike volcanic field in southern Chile and the origin of the Patagonian plateau lavas. *Journal of Volcanology and Geothermal Research*, **6(1)**, 3–25.

Smellie, J.L. (1987). Geochemistry and tectonic setting of alkaline volcanic rocks in the Antarctic Peninsula. *Journal of Volcanology and Geothermal Research*, **32(1–3)**, 269–85.

Smellie, J.L., Pankhurst, R.J., Hole, M.J. & Thomson, J.W. (1988). Age, distribution and eruptive conditions of Late Cenozoic alkaline volcanism in the Antarctic Peninsula and eastern Ellsworth Land: review. *British Antarctic Survey Bulletin*, **80**, 21–49.

Stillman, C.J. & Williams, C.T. (1978). Geochemistry and tectonic setting of some Upper Ordovician volcanic rocks in east and southeast Ireland. *Earth and Planetary Science Letters*, **41(4)**, 288–310.

Sun, S.S. & Hanson, G.N. (1975). Evolution of the Mantle: geochemical evidence from alkali basalts. *Geology*, **3(7)**, 297–302.

Thomson, J.W. (1974). *The geology of the South Orkney Islands: III. Coronation Island*. London; British Antarctic Survey Scientific Reports, No. 86, 39 pp.

Thomson, M.R.A. (1982). Mesozoic paleogeography of West Antarctica. In *Antarctic Geoscience*, ed. C. Craddock, pp. 331–7. Madison; University of Wisconsin Press.

Vallance, T.G. (1974). Spilitic degradation of a tholeiitic basalt. *Journal of Petrology*, **15(1)**, 79–96.

Watters, D.G. (1972). Geophysical investigations of a section of the South Scotia Ridge. In *Antarctic Geology and Geophysics*, ed. R.J. Adie, pp. 33–8. Oslo; Universitetsforlaget.

Weaver, S.D., Sceal, J.S.C. & Gibson, I.L. (1972). Trace element data relevant to the origin of trachytic and pantelleritic lavas in the East African Rift system. *Contributions to Mineralogy and Petrology*, **36(2)**, 181–94.

Winchester, J.A. & Floyd, P.A. (1977). Geochemical discrimination of different magma series and the differentiation products using immobile trace elements. *Chemical Geology*, **20(4)**, 325–43.

Wood, D.A., Joron, J.L. & Treuil, M. (1979b). A re-appraisal of the use of trace elements to classify and discriminate between magma series erupted in different tectonic settings. *Earth and Planetary Science Letters*, **45(2)**, 326–36.

Wood, D.A., Tarney, J., Varet, J., Saunders, A.D., Bougault, H., Joron, J.L., Treuil, M. & Cann, J.R. (1979a). Geochemistry of basalts drilled in the North Atlantic by IPOD Leg 49: implications for mantle heterogeneity. *Earth and Planetary Science Letters*, **42(1)**, 77–97.

Geochemical evolution of the Antarctic Peninsula magmatic arc: the importance of mantle–crust interactions during granitoid genesis

M.J. HOLE[1], R.J. PANKHURST[1] & A.D. SAUNDERS[2]

1 British Antarctic Survey, Natural Environment Research Council, High Cross, Madingley Road, Cambridge CB3 0ET, UK
2 Department of Geology, Bennett Building, University of Leicester, University Road, Leicester LE1 7RH, UK

Abstract

Four distinct groups of subduction-related intrusive rocks have been identified within Graham Land, northern Antarctic Peninsula. Group I plutons (40–80 Ma) are geographically restricted to the west coast archipelago and exhibit relatively low initial $^{87}Sr/^{86}Sr$ ratios (0.7038–0.7045) and high initial $^{143}Nd/^{144}Nd$ ratios (0.51277–0.51288). Group II plutons (~ 100 Ma) are geographically widespread with initial $^{87}Sr/^{86}Sr$ and initial $^{143}Nd/^{144}Nd$ ratios of ~ 0.7055 and ~ 0.5125, respectively. Groups III and IV (210–160 Ma) are restricted to eastern Graham Land and are separated on the basis of Sr- and Nd-isotope characteristics: Group III plutons (central-eastern Graham Land) exhibit a restricted range of initial $^{87}Sr/^{86}Sr$ and $^{143}Nd/^{144}Nd$ ratios (0.7060–0.7067 and 0.51224–0.51228, respectively) whilst Group IV plutons (south-eastern Graham Land) are characterized by higher initial $^{87}Sr/^{86}Sr$ ratios (0.7090–0.7162) and lower initial $^{143}Nd/^{144}Nd$ ratios (0.51220–0.51205) than Group III. The temporal and spatial variations in isotopic and trace element characteristics of these plutons are best explained by variable degrees of magma–crust interaction within the arc. The proportion of crustal material involved in magmagenesis increases from Group I to Group IV. Along-arc variations in crustal thickness have resulted in variable degrees of mantle–crust interaction both in space and time. However, variations in Nb/Y ratios throughout the arc may reflect a degree of heterogeneity within the subcontinental lithosphere beneath Graham Land.

Introduction

The relative importance of subducted oceanic crust, mantle-derived magmas and assimilated continental crust in the formation of subduction-related granitoids has recently received much attention. Studies in areas such as South America and the western USA have shown that in some cases, the ascending magmas do indeed contain significant proportions of crustal material, possibly incorporated by assimilation of the country rocks at depth. This is most easily recognized where compositionally distinctive crust is present (e.g. Precambrian basement with characteristic isotopic signatures in the Sierra Nevada Batholith; DePaolo, 1981). The Antarctic Peninsula potentially affords an important opportunity to assess mechanisms of crustal involvement. Calc-alkaline magmatism is both widespread and long-lived and the peninsula appears to be longitudinally segmented, so that the magmas were intruded through crust of varying thickness and composition. Comparison of different igneous suites may thus be used to monitor the relative contributions from different crustal rock-types.

Subduction-related magmatism within the Antarctic Peninsula was essentially continuous from Late Triassic–early Tertiary times (Saunders, Weaver & Tarney, 1980; Pankhurst, 1982). During this period there was an overall westward migration of the locus of the magmatic arc. Consequently, the outcrop pattern is now such that a major belt of Late Triassic–middle Jurassic (219–160 Ma) plutons are located in central and south-eastern Graham Land, whereas the majority of plutons in north-eastern Graham Land are < 120 Ma (Pankhurst, 1982). A belt of younger (< 95 Ma) plutons form the archipelago of the west coast of Graham Land (Fig. 1).

In addition to this space–time relationship for the granitoids of Graham Land, Pankhurst (1982) noted that their initial $^{87}Sr/^{86}Sr$ ratios decrease with decreasing age (Fig. 2). Pankhurst (1982) related this observed decrease in initial $^{87}Sr/^{86}Sr$ ratios with age to the waning influence of the continental crust on subduction-related granitoids with time.

A major crustal feature of Graham Land is its tectonically-segmented nature. The northernmost segment (Fig. 1) is characterized by exposures of predominantly low-grade metase-

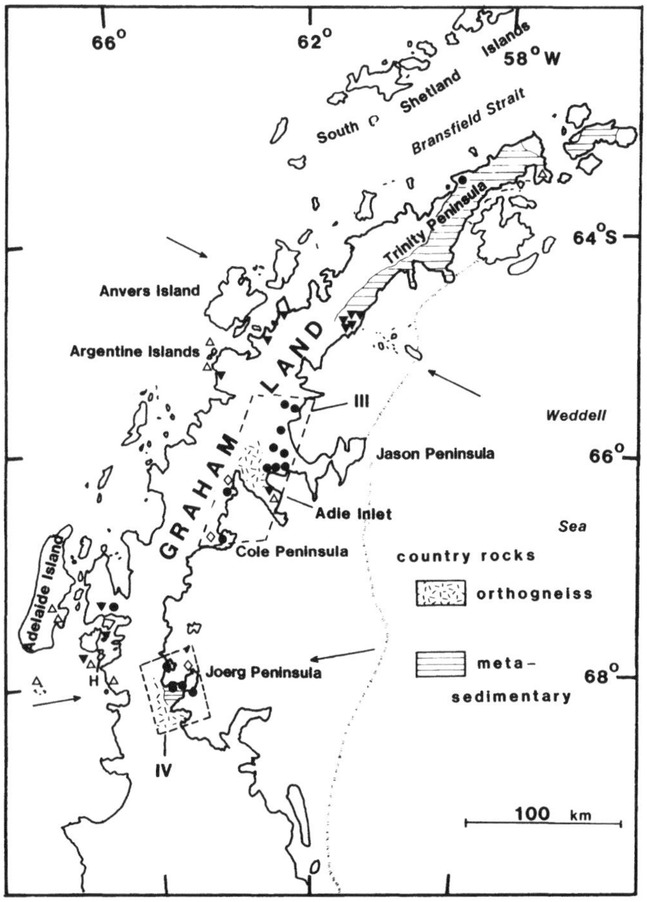

Fig. 1. Map of Graham Land showing the distribution of plutons and major country rock-types. The arrows indicate the position of probable tectonic segment boundaries. Roman numerals and pecked lines indicate the distribution of Group III and IV plutons (see text for details). Key for ages of plutons: open diamonds, > 200 Ma; solid circles, 180–160 Ma; solid inverted triangles, 125–100 Ma; open triangles, < 90 Ma; H, Horseshoe Island. Data from Pankhurst (1982), Moyes (1985) and this study.

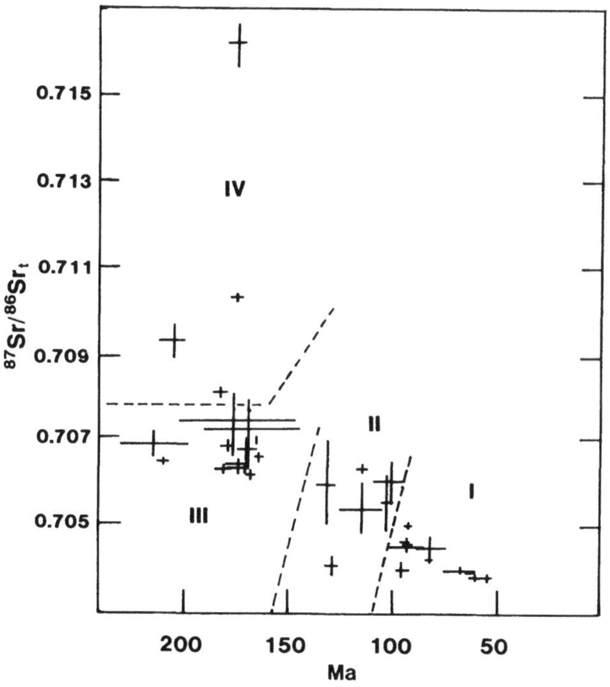

Fig. 2. Initial $^{87}Sr/^{86}Sr$ ratios versus age (Ma) for Graham Land plutons. Roman numerals and pecked lines refer to the four groups of plutons outlined in the text. Data from Pankhurst (1982), Moyes (1985) and this study.

dimentary country rock (Trinity Peninsula), whereas orthogneissose country rocks predominate in the central portion (64–68° S). Thomson, Pankhurst & Clarkson (1983) related such segmentation to the uplift and erosion of central-eastern Graham Land. A southern boundary to this central segment may occur at ~ 68° S in south-eastern Graham Land, near Joerg Peninsula, where there are further major exposures of metasedimentary country rocks.

In this paper we summarize new Rb–Sr geochronological results from this southern segment boundary, which was outside the study area of Pankhurst (1982). In addition, the first Nd-isotope data from the Antarctic Peninsula are reported and reviewed, together with new detailed trace element analyses of representative rocks from the entire Mesozoic–Cenozoic igneous suite. On the basis of these data, four chronologically- and geochemically-distinct groups of plutons are recognized and examined in the light of the models for mantle–crust interaction.

Geochronology of the area around Joerg Peninsula

Rb–Sr whole-rock and mineral isochron data for four largely unfoliated granitoids and orthogneisses from Joerg Peninsula are summarized in Table 1. These Triassic–Jurassic ages (204–175 Ma) are comparable with those of granitoids farther north on the east coast of Graham Land. However, all the plutons of the Joerg Peninsula exhibit significantly higher initial $^{87}Sr/^{86}Sr$ ratios (0.7081–0.7162) than those reported for middle Jurassic plutons from central-eastern Graham Land (initial $^{87}Sr/^{86}Sr$ 0.7060–0.7070; Fig. 2).

Clearly these new Rb–Sr data demonstrate that initial $^{87}Sr/^{86}Sr$ ratios do not simply vary with the age of individual plutons, but show a significant correlation with spatial distribution of metasedimentary and orthogneissose country rocks. However, they do add support to the general pattern previously established and enable us to propose four main chronological and geochemical groupings of the plutonic rocks of Graham Land (Table 2, Fig. 1).

Correlated Sr- and Nd-isotope variations

In order to assist identification of the source of the various plutons of Graham Land, a representative set of Nd-isotope analyses have been carried out on the different country rock-types and plutons of different age and spatial distribution.

Mesozoic and Tertiary plutons

Interplutonic Nd-isotope variations generally mirror those for Sr isotopes (Fig. 3). Initial $^{143}Nd/^{144}Nd$ ratios decrease with increasing age and thus increasing initial $^{87}Sr/^{86}Sr$ ratios, from ~ 0.5128 for Group I through ~ 0.5125

Table 1. *Summary of Rb–Sr data for Joerg Peninsula plutons*

Location	Rock type	Country rocks	Points	Age (Ma)	$^{87}Sr/^{86}Sr$	MSWD
Central Joerg Peninsula	Orthogneisses	—	4	236 ± 7	0.7095 ± 3	1.3
Pylon Point	Foliated diorite and sheared granite	Orthogneiss	5	204 ± 6	0.7093 ± 4	9.6
R.2606–R.2607	Adamellite–granite (*sensu stricto*)	Metasedimentary Rocks	5	175 ± 3	0.7103 ± 3	2.0
R.2620	Adamellite	Orthogneiss	4	181 ± 2	0.7081 ± 1	2.9
R.2614	Two-mica granite	Metasedimentary rocks	4[a]	175 ± 3	0.7162 ± 5	4.6

[a]whole-rock + muscovite + K-feldspar.

Table 2. *Summary of the isotopic characteristics and spatial distribution of the four groups of plutons from Graham Land*

Group	I	II	III	IV
Age (Ma)	48–90	~ 100	160–209	160–204
$^{87}Sr/^{86}Sri$	< 0.7045	~ 0.7055	0.7060–0.7070	0.7071–0.7062
$^{143}Nd/^{144}Ndi$	0.51277–0.51288	~ 0.51240	0.51224–0.51228	0.51207–0.51220
Distribution	West coast archipelago	Scattered	Central-eastern Graham Land	South-eastern Graham Land

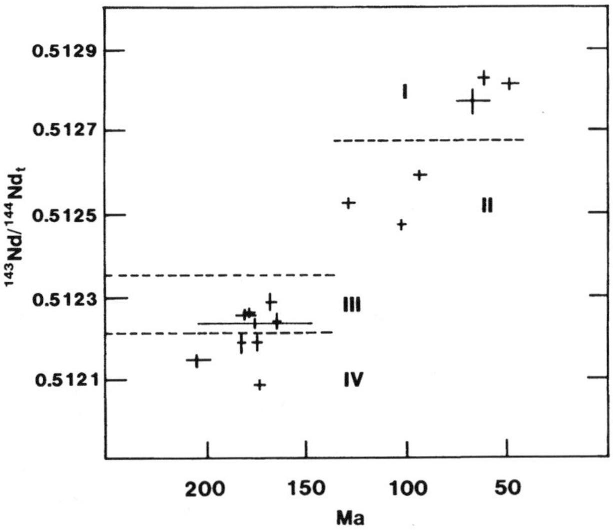

Fig. 3. Initial $^{143}Nd/^{144}Nd$ ratios versus age for Graham Land plutons. Key as for Fig. 2.

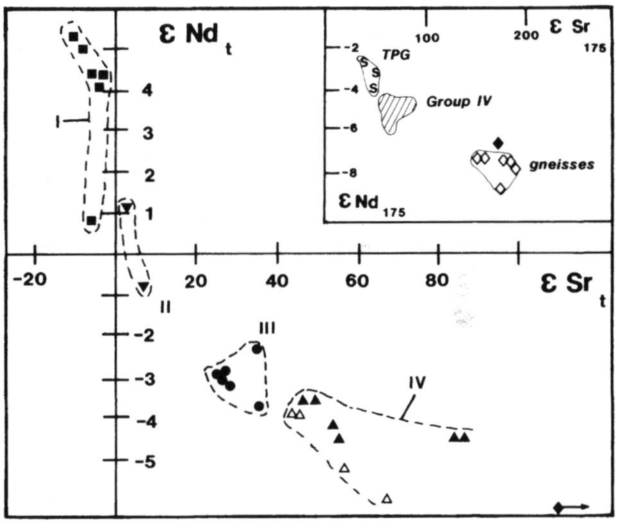

Fig. 4. ϵ_{Nd_t} versus ϵ_{Sr_t} for the plutons of Graham Land. Key: solid squares, Group I; solid inverted triangles, Group II; open triangles, Group IV monzogabbros; solid circles, Group III; solid triangles, Group IV acid plutonic rocks; solid diamond, two-mica granite. Inset: $\epsilon_{Sr_{175}}$ versus $\epsilon_{Nd_{175}}$ for the Adie Inlet acid gneisses (open diamonds), Trinity Peninsula Group (TPG) metasedimentary rocks (S), and two-mica granite (solid diamond; $\epsilon_{Sr_{175}}$ + 180); the field for Group IV plutons is also shown. Data from Pankhurst (1983) and this study.

(Group II) to ~ 0.5124 for Group III. Group IV granitoids exhibit significantly lower initial $^{143}Nd/^{144}Nd$ ratios than Group III.

All Group I plutons fall in the depleted source quadrant of an ϵ_{Nd}–ϵ_{Sr} plot (Fig. 4), with ϵ_{Nd} and ϵ_{Sr} values similar to those for magmas derived directly from a depleted mantle reservoir. Groups II, III and IV plot in $-\epsilon_{Nd}$, $+\epsilon_{Sr}$ space, with a progressive trend towards more negative ϵ_{Nd} and more positive ϵ_{Sr} values in the order Group II–III–IV, forming a hyperbolic curve asymptotic to $\epsilon_{Nd} \sim -6.5$. This suggests that the isotope geochemistry of the plutons may have been influenced by the continental crust in a way related to local crustal structure. Before discussing models for the mechanism of crustal incorporation, it is instructive to compare the most extreme Group IV compositions, where the crustal influence is strongest, with observed characteristics of the different types of exposed country rock.

Country rocks

Epsilon values for representative analyses of metasedimentary and orthogneissose country rocks have been calculated at 175 Ma, the mid-point in the major intrusive episode in eastern Graham Land. Trinity Peninsula Group (TPG) metasedimentary rocks have significantly lower $\epsilon_{Sr_{175}}$ (+ 25 to + 30) and less negative $\epsilon_{Nd_{175}}$ values (− 2.5 to − 3.0) than Group IV plutons (Fig. 4, inset) and consequently assimilation of metasedimentary rocks cannot have resulted in the high initial $^{87}Sr/^{86}Sr$ and low initial $^{143}Nd/^{144}Nd$ ratios exhibited by the plutons of south-eastern Graham Land.

Granitic orthogneisses from Adie Inlet, central-eastern Graham Land, which were considered by Pankhurst (1983) to have been ultimately derived from an Early Palaeozoic parent, exhibit very similar $\epsilon_{Nd_{175}}$ (− 6.5 to − 7.2) and $\epsilon_{Sr_{175}}$ values (+ 150 to + 180) to the Jurassic (175 Ma) two-mica granite from Joerg Peninsula (ϵ_{Nd_t} − 6.2, ϵ_{Sr_t} + 175; Table 1). The crystallization age of the Adie Inlet gneisses is uncertain. The only well defined pre-Mesozoic event within Graham Land is Silurian (~ 420 Ma) emplacement of orthogneissose precursors at Target Hill, central-eastern Graham Land (Milne & Millar, this volume, p. 335). The Target Hill gneisses yielded $\epsilon_{Sr_{420}}$ and $\epsilon_{Nd_{420}}$ values which plot in the depleted mantle field on an epsilon t plot. However, the Adie Inlet acid gneisses have $\epsilon_{Sr_{420}}$ (+ 80 to + 100) and $\epsilon_{Nd_{420}}$ (− 5.2 to − 6.8) values which are suggestive of a pre-Silurian crustal component within them. The Adie Inlet acid gneisses may thus represent an example of crustal material of pre-Silurian parentage within Graham Land.

The major- and trace-element characteristics and extremely high initial $^{87}Sr/^{86}Sr$ ratio (0.7169 ± 0.0001) of the Joerg Peninsula two-mica granite suggests that it may represent a crustal melt, i.e. an S-type granite; it is highly aluminous (~ 5% normative *cc*) and has a strongly fractionated rare earth element (REE) profile (La/Yb_n ~ 40, Yb 1.0 ppm) suggestive of derivation by partial melting of a crustal source leaving a HREE-enriched residuum, probably the result of garnet or amphibole as a restite phase. The Adie Inlet acid gneisses (La_n/Yb_n ~ 10, YB ~ 1.5 ppm) could thus provide an ideal crustal source for this two-mica granite.

Assimilation of continental crust

The Joerg Peninsula two-mica granite and Adie Inlet gneisses may represent derivatives of easily melted basement rock which would have the ability to produce the elevated initial $^{87}Sr/^{86}Sr$ and low initial $^{143}Nd/^{144}Nd$ ratios of the plutons of Group III and IV. It seems possible that the along-arc variations in Sr- and Nd-isotope characteristics seen in eastern Graham Land could thus be regarded as the result of mixing between a basic, mantle-derived magma and crustal material similar in composition to the Adie Inlet gneisses and/or the Joerg Peninsula two-mica granite. However, temporal and/or spatial variations in the isotopic and trace element characteristics of the subcontinental lithosphere beneath Graham Land may also have contributed to the along- and across-arc geochemical variations described above.

The positive identification of crustal and mantle components within the granitoids of Graham Land is not possible at this stage. Initially it is our intention to try and assess to what extent the geochemical variations seen within Graham Land could be explained solely by interaction of a basic, mantle-derived magma and continental crust similar in isotopic and trace element composition to the Adie Inlet gneisses or Joerg Peninsula two-mica granite. This type of model is necessarily a gross over-simplification, and takes no account of variations in the composition of the subcontinental mantle beneath Graham Land. An assessment of crust–mantle interactions is, however, a useful baseline for further modelling of arc evolution.

Mixing models

The isotopic composition of Group I plutons (initial $^{87}Sr/^{86}Sr$ ~ 0.7040, initial $^{143}Nd/^{144}Nd$ ~ 0.5129) may be used as an estimate of the composition of a depleted mantle-derived magma which has undergone little or no interaction with the continental crust. The average $^{87}Sr/^{86}Sr$ and $^{143}Nd/^{144}Nd$ of the two-mica granite and Adie Inlet gneisses at 175 Ma may be used as an estimate of the isotopic composition of the crustal component. Results of mixing models are summarized in Table 3. Applying a simple mixing model to these data requires more than 50% of the crustal material to be added to a mafic, mantle-derived magma to produce the isotopic characteristics of Group IV plutons. Clearly this is geologically implausible.

Applying the assimilation with fractional crystallization (AFC) model of DePaolo (1981) and James (1981), and assuming contamination occurred with concomitant crystallization of plagioclase-rich cumulate, reduces the degree of contamination required to produce the isotopic characteristics of Group III intrusions to < 10% and to ~ 15% for Group IV. This is almost entirely due to the high distribution coefficients (K_D) for Sr in plagioclase (~ 1.8; Arth, 1976); a decrease in DSr would require greater degrees of crustal assimilation (Table 3). The AFC model produces a best fit for an assimilated rock/crystal cumulate ratio (AR/CC) of 0.9. AFC mixing seems to produce geologically feasible degrees of contamination for Groups III and IV plutons.

The consistency of ϵ_{Sr} and ϵ_{Nd} values across a broad range of compositions for Groups III and IV (i.e. 60–75% SiO_2; Pankhurst, 1982; Hole, 1986) demonstrates that the contamination event must have occurred before differentiation of individual plutons. After the initial mafic magma–crust interaction, differentiation must thus have occurred in a closed system.

By applying a simplistic homogeneous mantle–crust interaction model, the interplutonic isotopic variations within Graham Land could be explained purely by varying degrees of magma–crust interaction along the arc. Similar models have been applied to problems of crustal contamination in a number of locations, e.g. western USA and Chile (DePaolo, 1981; James, 1981). It is now important to consider if interplutonic trace element variations can be explained in the same manner.

Correlated trace element and isotope variations

Saunders *et al.* (1980) demonstrated that there are marked transverse geochemical variations across Graham Land, and attributed increases in large ion lithophile element (LILE; Rb, Th, K, La), but consistency of high field strength element (HFSE; Nb, Ta, P) and heavy rare earth element (HREE) abundances, to a combination of variable LILE-

Table 3. *Summary of parameters used in AFC models*

Src/Srs	Ndc/Nds	DSr	DNd	AR/CC	% crystallization for Group IV
0.30	5.00	1.8	0.08	0.90	< 15
0.30	5.00	1.0	0.08	0.90	< 30

These two models are calculated for source $^{87}Sr/^{86}Sr$ and $^{143}Nd/^{144}Nd$ ratios of 0.7040 and 0.5129, respectively, and contaminant $^{87}Sr/^{86}Sr$ and $^{143}Nd/^{144}Nd$ ratios of 0.7165 and 0.5120, respectively. DNd and DSr, bulk distribution coefficients for Nd and Sr in the cumulate; Src/Srs and Ndc/Nds, contaminant/source ppm ratios; AR/CC, assimilated rock/crystal cumulate ratio.

enrichment of the subcontinental mantle and partial melting effects. When Saunders *et al.* (1980) published their model, virtually no Rb–Sr data existed and no Sm–Nd isotope analyses had been carried out on the plutons of Graham Land. New trace element analyses on plutons for which both Nd- and Sr-isotope data are now available, show trace element trends which are generally consistent with those reported by Saunders *et al.* (1980). In detail, however, few trace elements behave perfectly incompatibly during granitoid differentiation. Sr and Eu decrease with increasing SiO_2 reflecting plagioclase + K-feldspar fractionation. The LREE, Zr, Hf and to a certain extent Th, are rendered compatible in most Group III and IV plutons as a result of the early crystallization of minor phases (e.g. allanite, monazite, zircon). In general, however, minor phase crystallization appears to have had a less pronounced effect on the distribution of the REE and HFSE in Group I intrusions. As hornblende is an important crystallizing phase in most of the plutons of Graham Land, HREE and Y abundances of most of the granitoids have been modified during magmatic differentiation.

For the granitic rocks of Graham Land (i.e. SiO_2 ~ 70%), Rb and K abundances and La_n/Yb_n ratios increase with increasing initial $^{87}Sr/^{86}Sr$ ratios. This general increase in the degree of REE fractionation is the result of both LREE enrichment of Groups III and IV relative to Group I, and higher absolute abundances of the HREE in Group I. Nb abundances, although they may have been slightly modified by magnetite crystallization (cf. Pearce, Harris & Tindle, 1984), are significantly lower in Group I than in any other group. Nb/Y and Rb/Y ratios exhibit a positive linear correlation (Fig. 5), Group I intrusions having the lowest Nb/Y and Rb/Y ratios, with a progressive increase in both ratios in the order Group II–III–IV. Clearly both Rb/Y and Nb/Y ratios also increase with increasing initial $^{87}Sr/^{86}Sr$ ratios. An exception to this general trend is a 67 Ma alkalic Group I pluton from south-western Graham Land, which has high absolute Nb abundances (~ 35 ppm at 69% SiO_2), high Nb/Y ratios (~ 0.6–0.8) but low Rb/Y and initial $^{87}Sr/^{86}Sr$ ratios (~ 2 and 0.7040, respectively).

Absolute Rb, K and LREE abundances and La_n/Yb_n ratio of the two-mica granite are significantly higher than those for Group IV plutons. The mafic, uncontaminated precursor to Group III and IV plutons is assumed to have similar trace element and isotopic characteristics to Group I intrusions. Consequently, the addition of a two-mica granite-type crustal melt to such a mafic precursor would be capable of producing both the observed increases in LILE and LREE abundances,

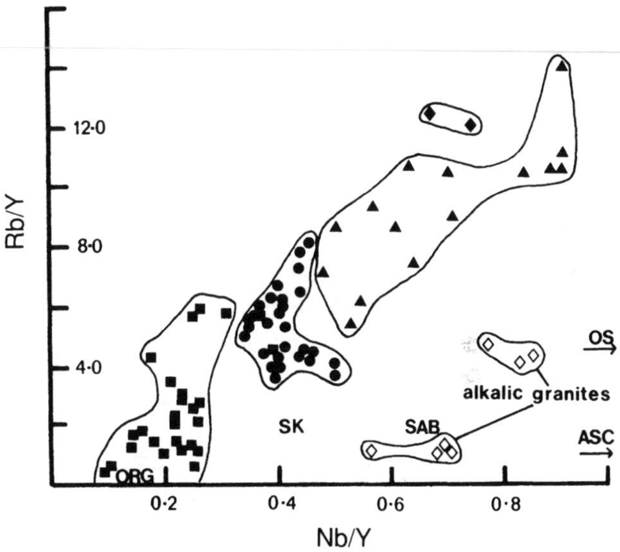

Fig. 5. Rb/Y versus Nb/Y for the four groups of plutons from Graham Land. Solid squares, Group I; solid circles, Group III; solid triangles, Group IV; open diamonds, alkalic plutons from western and eastern Graham Land; solid diamonds, two-mica granite. Data from Hole (1986) and Saunders *et al.* (1980). Other data for 'within plate' granites from Pearce *et al.* (1984); SK, Skye; SAB, Sabaloka (Sudan); OS, Oslo rift; ASC, Ascension Island; ORG, ocean ridge granite.

and the elevated initial $^{87}Sr/^{86}Sr$ and lower initial $^{143}Nd/^{144}Nd$ ratios of Groups III and IV. Whether the observed increases in Nb/Y ratios could be produced in the same manner is open to question. Indeed the Joerg Peninsula two-mica granite does exhibit a high Nb/Y ratio (0.8 of Group IV), but the low $^{87}Sr/^{86}Sr$ Group I pluton from Horseshoe Island exhibits a similarly high Nb/Y ratio which cannot be a result of crustal contamination or crustal melting. It is possible that varying Nb/Y ratios reflect source heterogeneities in the subcontinental lithosphere beneath Graham Land. The high Nb/Y but low Rb/Y ratio Group I intrusions demonstrate that an HFSE-enriched but relatively LILE-depleted source region was tapped during their generation. The mixing of a crustal melt with a high Nb/Y ratio magma derived from an enriched mantle source region may provide a convenient method of producing the observed variations in HFSE abundances within the arc. However, derivation of Group IV intrusions solely from such an enriched mantle source, with no later addition of crustal component, could probably not explain the high abundances of Rb and K or the observed Sr- and Nd-isotope characteristics of Group IV plutons.

Magma–crust interactions and crustal thickness

Wyllie (1983) demonstrated experimentally that granitoids produced above subduction zones are not primary magmas derived directly from the mantle or subducted ocean crust, and he emphasized the importance of the hybridization of mantle-derived magmas and crustal anatectic melts in the production of subduction-related granitoids. He showed that mantle–crust hybridization would occur in the lower crust or at the mantle–crust interface, the dominant mixing process being simple mixing or deep crustal AFC (i.e. high AR/CC ratio mixing). The degree of mantle–crust hybridization is thus dependent on: (1) the thermal regime within the crust and mantle wedge; and (2) the total thickness of the continental crust above the mantle wedge. In broad terms, an increase in crustal thickness would result in a greater potential for mantle–crust hybridization, and granitoids with isotopic characteristics tending towards those for the crustal component in areas of thickest continental crust.

The variable Sr- and Nd-isotope characteristics of the plutons of Groups III and IV can satisfactorily be accommodated by a mantle–crust hybridization model involving along-arc variations in crustal thickness. The tectonic segmentation of Graham Land and the presence of an uplifted central segment (Thomson *et al.*, 1983) suggest that crustal thicknesses are greatest in south-east Graham Land, where Group IV plutons and metasedimentary rocks are exposed.

However, as illustrated by combined trace element–isotope correlations, the presence of subcontinental lithosphere variably enriched in the HFSE (? and LILE) beneath Graham Land, may also have contributed to the generation of the distinct isotopic and trace element characteristics of the plutons of Graham Land.

Conclusions

We conclude that, in general, along- and across-arc variations in the isotopic and trace element characteristics of the plutons of Graham Land can be explained by variable mantle–crust interaction within the arc. Crustal contamination at deep crustal levels was greatest in south-eastern Graham Land, whereas the Tertiary plutons of Group I on the west coast of the peninsula have little or no isotopic contribution from the continental crust. Variations in the degree of mantle–crust interaction are likely to be directly linked to variable crustal thickness within the arc. Furthermore, differences in crustal thickness may be influenced by time-controlled variations in tectonic regimes within the arc, i.e. tectonic segmentation of Graham Land and Late Cretaceous–early Tertiary crustal extension in western Graham Land (Storey & Garrett, 1985). However, a contribution from enriched subcontinental lithosphere to some of the plutons of Graham Land may also have been important during the magmatic evolution of the arc.

References

Arth, J.G. (1976). The behaviour of trace elements during magmatic processes – a summary of theoretical models and their application. *Journal of Research, US Geological Survey*, **4**, 41–7.

DePaolo, D.J. (1981). A neodymium and strontium isotopic study of the Mesozoic calc-alkaline granitic batholiths of the Sierra Nevada and Peninsular Ranges, California. *Journal of Geophysical Research*, **86**, 10470–88.

Hole, M.J. (1986). Time controlled geochemistry of igneous rocks of the Antarctic Peninsula. PhD thesis, University of London, 396 pp. (unpublished).

James, D.E. (1981). The combined use of oxygen and radiogenic isotopes as indicators of crustal contamination. *Annual Review in Earth and Planetary Science*, **9**, 311–44.

Moyes, A.B. (1985). The petrology and geochemistry of plutonic rocks across southern Graham Land, Antarctica. PhD thesis, University of London, 184 pp. (unpublished).

Pankhurst, R.J. (1982). Rb–Sr geochronology of Graham Land, Antarctica. *Journal of the Geological Society, London*, **139(6)**, 701–11.

Pankhurst, R.J. (1983). Rb–Sr constraints on the ages of basement rocks within the Antarctic Peninsula. In *Antarctic Earth Science*, ed. R.L. Oliver, P.R. James & J.B. Jago, pp. 367–71. Canberra; Australian Academy of Science and Cambridge; Cambridge University Press.

Pearce, J.A., Harris, N.B.W. & Tindle, A.G. (1984). Trace element discrimination diagrams for the tectonic interpretation of granitic rocks. *Journal of Petrology*, **25(4)**, 956–83.

Saunders, A.D., Weaver, S.D. & Tarney, J. (1980). Transverse geochemical variations across the Antarctic Peninsula: implications for the genesis of calc-alkaline magmas. *Earth and Planetary Science Letters*, **46**, 344–60.

Storey, B.C. & Garrett, S.W. (1985). Crustal growth of the Antarctic Peninsula by accretion, magmatism and extension. *Geological Magazine*, **122(1)**, 5–14.

Thomson, M.R.A., Pankhurst, R.J. & Clarkson, P.D. (1983). The Antarctic Pensinsula – a Mesozoic–Cenozoic arc (review). In *Antarctic Earth Science*, ed. R.L. Oliver, P.R. James & J.B. Jago, pp. 289–94. Canberra; Australian Academy of Science and Cambridge; Cambridge University Press.

Wyllie, P.J. (1983). Experimental and thermal constraints on the deep-seated parentage of some granitoid magmas at subduction zones. In *Migmatites, Melting and Metamorphism*, Shiva Geology Series, ed. M.P. Atherton & C. Gribble, pp. 37–51. Nantwich; Shiva Publishing Ltd.

Variation in amphibole composition from the Andean Intrusive Suite across the Antarctic Peninsula

A.B. MOYES*

British Antarctic Survey, Natural Environment Research Council, High Cross, Madingley Road, Cambridge CB3 0ET, UK
**Present address: Isotope Geochemistry Unit, BPI Geophysics, University of the Witwatersrand, PO WITS, Johannesburg 2050, South Africa*

Abstract

Amphibole is a common constituent throughout the entire suite of Mesozoic–Cenozoic subduction-related calc-alkaline plutonic rocks of the Antarctic Peninsula, referred to as the Andean Intrusive Suite. Microprobe data on 51 samples, covering the range from gabbronorite to monzogranite, indicate that all the amphibole is calcic ($(Ca + Na)_B > 1.34$ and $Na_B > 0.67$), and that the compositional variation is dominated by the Ti-tschermakite ($Ti + 2Al^{IV} \leftrightarrow Mg + Si$) and edenite ($(Na + K)_A + Al^{IV} \leftrightarrow [\ ]_A + Si$) substitutions. Intercumulus, poikilitic brown or greenish-brown amphibole in gabbroic rocks is tschermakite (Al- and Ti-rich), whereas euhedral or subhedral olive-green or greenish-blue amphibole in intermediate and granitic rocks is more Si-rich magnesio- or ferro-hornblende. This variation results from higher temperatures and pressures during amphibole crystallization in the more mafic rocks. The $Mg/(Mg + Fe^{2+})$ ratio (Mg*) varies regionally in that ferro-hornblende (Mg*0.5) does not occur in samples closest to the trench, but becomes common farther from the subduction site. This variation in Mg* is independent of host rock compositional control and correlates with a change in the opaque phase from magnetite to ilmenite, resulting from lower fO_2 conditions during crystallization with increasing distance from the trench.

Introduction

The geology of the Antarctic Peninsula is dominated by a calc-alkaline suite of plutonic, volcanic and volcaniclastic rocks which resulted from subduction of Pacific ocean floor under the western margin of the peninsula during most of the Mesozoic and Cenozoic. Stratigraphical correlation of these rocks is hindered by poor field control but their overall similarity to the 'Andean' rocks of southern South America suggests that this calc-alkaline suite is a remnant of the now-fragmented western margin of Gondwana, extending from South America through the Antarctic Peninsula to Marie Byrd Land and New Zealand. Radiometric dating has suggested that magmatism was essentially continuous from at least latest Triassic to Miocene times, with distinct surges in the Early Jurassic, Early Cretaceous, Late Cretaceous and early Tertiary (Pankhurst, 1982).

The plutonic components of this suite (the Andean Intrusive Suite, AIS) form a major part of the total outcrop in the peninsula. They are represented by a continuous sequence from gabbronorite to syenogranite, although intermediate members such as granodiorite and monzogranite (adamellite) often predominate. Many petrographical and whole-rock geochemical studies have been carried out on these rocks (see summary in Moyes, 1985) but no detailed mineral chemistry has yet been published. This paper attempts to assess the compositional variation of amphibole in relation to host rock composition since amphibole occurs commonly in all members of the suite, and because of its significance in the genesis of calc-alkaline magmas. In addition, consideration is given to compositional variation across the Antarctic Peninsula by comparing three distinct geographical areas at approximately 68° S: Adelaide Island (68.5° W), the Arrowsmith Peninsula (67° W) and the Foyn Coast (65.5° W). Radiometric data from these areas suggest that the rocks generally young westwards, e.g. granitic rocks from the Foyn Coast have given ages of 209–174 Ma, from the Arrowsmith Peninsula 168–100 Ma, and from Adelaide Island ~ 60 Ma (Pankhurst, 1982; Moyes, 1985).

Amphibole occurrence and petrography

Modal data indicate that amphibole is most prevalent in mafic rocks from Adelaide Island and Arrowsmith Peninsula, showing a progressive decrease with increasing acidity of the host rock (Moyes, 1985). Foyn Coast samples are consistently poorer in amphibole, except for granodiorite and monzogranite samples. In the mafic rocks of Adelaide Island and Arrowsmith Peninsula, brown or brownish-green amphibole occurs as poikilitic, intercumulus crystals enclosing plagioclase, olivine and clino- and orthopyroxene. In some samples there is a clear colour zoning from brown core to green rim. Foyn Coast gabbro samples, by contrast, are olivine-free and contain subhedral amphibole (with yellowish or olive-green pleochroism) as a reaction product of clinopyroxene. In all the more felsic rock-types the amphibole is subhedral or euhedral, with olive-green or greenish-blue pleochroism, and commonly encloses remnant crystals of both pyroxenes.

Overall amphibole compositional variation

A total of 618 amphibole analyses were obtained from 51 samples representing the complete spectrum of rock-types in each area. Analyses were performed using an energy-dispersive electron microprobe at the University of Cambridge. For comparative host-rock analyses, major elements were determined on lithium tetraborate/carbonate fusion beads using a Philips PW1400 X-ray fluorescence spectrometer at Bedford College, University of London. Amphibole analyses were recalculated on the basis of 23 anions and 13 cations, exclusive of Ca, Na and K (Leake, 1978).

Representative analyses and structural formulae are given in Table 1, and plotted on a Si versus Mg* diagram (Fig. 1), corresponding to the classification of Leake (1978). The amphibole is exclusively calcic with $(Ca + Na)_B > 1.34$ and $Na_B < 0.67$ cations per unit formula (p.u.f.); the $(Na + K)_A$ is always < 0.5 cations p.u.f., except for two samples from the Arrowsmith Peninsula (see Fig. 1). The amphibole ranges commonly from magnesio-hornblende to ferro-hornblende, with less abundant tschermakitic varieties in the more mafic rocks. The two exceptions noted above are magnesio-hastingsitic hornblende and ferro-edenite.

Compositional variation of amphibole is the result of a number of different cation substitutions. The most important of these is the diadochic substitution of Fe^{2+} and Mg, which is reflected in the variation from magnesio-hornblende ($Mg^* > 0.5$) to ferro-hornblende ($Mg^* < 0.5$). More complex substitutions were defined by Gilbert *et al.* (1982). From the data obtained here, it is apparent that there is very little variation in the Ca content, and that Na_B has no significant correlation with any other cations (see Moyes, 1985). Thus the only complex substitutions that need be considered are edenite and Al- and Ti-tschermakite (see Gilbert *et al.*, 1982). On a plot of Al^{IV} versus Total $A + Al^{VI} + Fe^{3+} + 2Ti$ (Fig. 2) the data closely follow a 1:1 slope, as would be expected if edenite and tschermakite substitutions were dominant (Czamanske & Wones, 1973). The residual component on the X-axis is due to maximum Fe^{3+} contents as a result of the analysis recalculation.

Some amphibole from gabbronorite samples is relatively Ti-poor compared to Al^{IV}, correlating with a colour zoning from a brown core to green rim (see Table 1, analyses 1 & 4). There is much evidence from both natural and experimental studies that both Al^{IV} and Ti contents have a positive correlation with temperature, but that Ti has an additional negative correlation with fO_2 and a bulk compositional control (Gilbert *et al.*, 1982; Hammarstrom & Zen, 1986). The observed variation in the intercumulus amphibole may result from either a higher fO_2 due to higher water contents in the interstitial magma, or a paucity of Ti in this magma due to fractionation of a Ti-bearing phase, both processes occurring without significant temperature decrease.

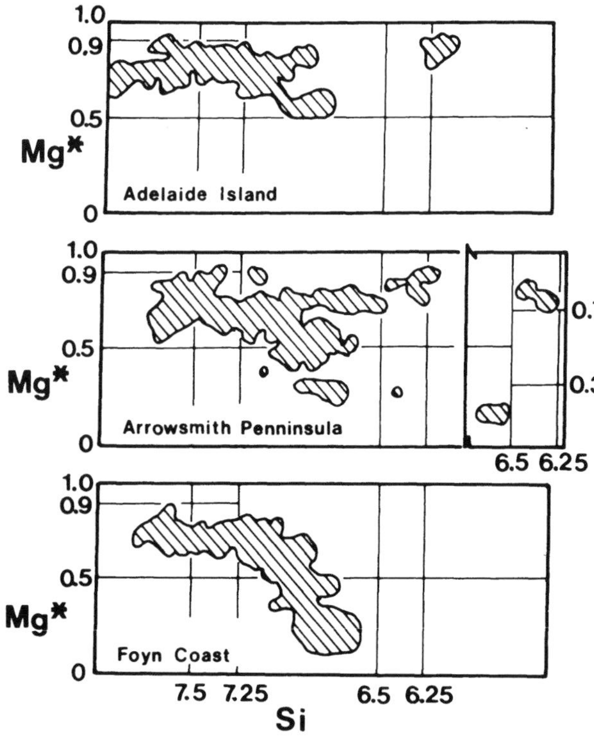

Fig. 1. Mg* ($Mg/(Mg + Fe^{2+})$) versus Si amphibole classification diagram; for nomenclature see Leake (1978). Note that all samples have $(Na + K)_A < 0.5$ cations per formula unit, with the exception of the two samples of from the Arrowsmith Peninsula given on the insert to the right of the main diagram.

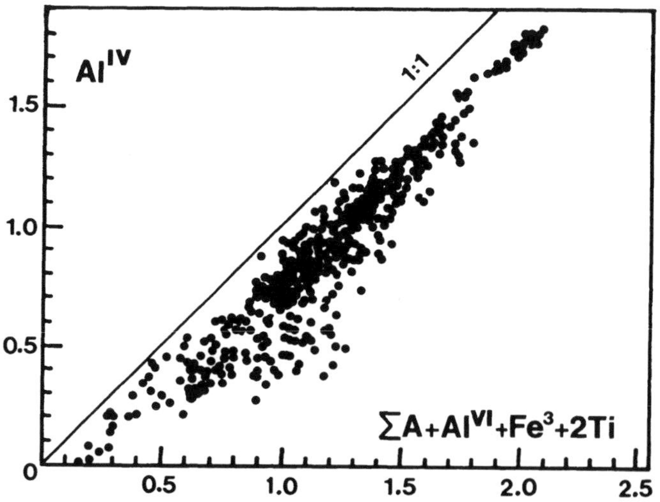

Fig. 2. Al^{IV} versus $A + Al^{VI} + Fe^{3+} + 2Ti$, as a measure of combined edenite and tschermakite substitution. Samples should follow a 1:1 slope if these are the dominant substitutions.

Table 1. *Representative amphibole analyses*

	1	2	3	4	5	6	7	8	9	10	11
SiO_2	43.76	43.61	42.83	44.44	44.42	43.41	43.98	46.72	49.99	48.64	48.48
TiO_2	0.72	3.02	3.19	0.64	1.00	1.92	0.62	1.04	1.37	0.94	0.58
Al_2O_3	12.88	11.52	11.05	12.90	11.02	8.07	10.44	6.96	6.11	5.08	6.23
FeO	11.05	10.44	10.93	10.80	11.67	24.94	23.92	21.09	11.51	16.82	17.50
MnO	—	—	0.11	—	—	0.98	0.34	0.39	0.15	0.44	0.60
MgO	15.06	15.17	14.47	14.96	14.58	5.24	4.90	8.54	15.43	12.57	10.99
CaO	11.11	11.25	11.35	11.57	11.51	10.14	11.56	11.35	11.68	10.94	10.92
Na_2O	2.73	2.19	2.35	2.09	2.34	1.88	0.70	0.96	1.09	1.37	1.23
K_2O	0.25	0.19	0.24	0.31	0.76	0.98	1.16	0.63	0.35	0.57	0.55
Cl	0.26	—	0.06	0.29	0.26	—	—	—	—	0.10	0.11
Total	97.82	97.39	96.58	98.00	97.56	97.56	97.62	97.68	97.68	97.47	97.19
Si	6.24	6.24	6.25	6.32	6.44	6.73	6.78	7.03	7.14	7.15	7.18
Al^{IV}	1.76	1.76	1.75	1.68	1.56	1.27	1.22	0.97	0.86	0.86	0.82
Al^{VI}	0.41	0.19	0.16	0.48	0.33	0.20	0.68	0.26	0.17	0.03	0.27
Ti	0.08	0.33	0.35	0.07	0.11	0.22	0.07	0.12	0.15	0.10	0.06
Fe^{3+}	1.00	0.82	0.63	0.91	0.64	0.49	0.15	0.42	0.45	0.68	0.50
Mg	3.20	3.24	3.15	3.17	3.15	1.21	1.13	1.92	3.29	2.75	2.43
Fe^{2+}	0.31	0.43	0.70	0.37	0.77	2.74	2.94	2.24	0.93	1.38	1.67
Mn	—	—	0.01	—	—	0.13	0.05	0.05	0.02	0.06	0.08
Ca	1.70	1.73	1.78	1.76	1.79	1.69	1.91	1.83	1.79	1.72	1.73
Na_B	0.30	0.27	0.23	0.24	0.21	0.31	0.09	0.17	0.21	0.28	0.27
Na_A	0.45	0.33	0.44	0.34	0.45	0.25	0.19	0.11	0.09	0.11	0.09
K	0.05	0.04	0.05	0.06	0.14	0.19	0.23	0.12	0.06	0.11	0.10
Total	15.50	15.38	15.50	15.40	15.59	15.43	15.44	15.24	15.06	15.23	15.20
Mg*	0.91	0.88	0.82	0.90	0.80	0.31	0.28	0.46	0.78	0.67	0.59

Adelaide Island, AI; Arrowsmith Peninsula, AP; Foyn Coast, FC.
Key to analysis numbers: 1–2. Al-tschermakite, gabbronorite, AI; 3. Tschermakitic hornblende, gabbronorite, AI; 4. Tschermakitic hornblende, gabbronorite, AP; 5. Magnesio-hastingsitic hornblende, gabbronorite, AP; 6. Ferro-hornblende, granodiorite, FC; 7. Ferro-hornblende, granodiorite, AP; 8. Ferro-hornblende, quartz-diorite, AP; 9. Magnesio-hornblende, gabbronorite, AI; 10. Magnesio-hornblende, granodiorite, FC; 11. Magnesio-hornblende, granodiorite, AP.

Regional variations

All the amphiboles studied show the same Ti-tschermakite and edenite substitutions regardless of their sample location. However, regional differences apparent in Mg* (Fig. 2) and in the Al^{IV} and Ti contents (Table 1) may be summarized thus: (i) no ferro-hornblende occurs in the Adelaide Island samples; and (ii) no Al- and Ti-rich amphiboles are found in the Foyn Coast mafic samples. The mafic part of the AIS varies, therefore, from olivine gabbronorite (with Al-Ti-rich amphibole) nearest the trench to amphibole gabbro (with magnesio-hornblende) farthest from the trench. Similar variations have been observed in other gabbroic samples from the Antarctic Peninsula and from many other areas (see Moyes & Storey, 1986). In the Foyn Coast olivine-free gabbronorites, the occurrence of non-cumulus magnesio-hornblende rather than tschermakitic varieties suggests that the amphibole crystallized at lower temperatures and pressures, as its reaction relationship to pyroxene indicates. Thus the conditions of mafic magma crystallization were different farther from the trench, possibly due to an increased thickness of continental crust on the east coast of the peninsula (see Moyes & Hamer, 1983); uprising magmas trapped at lower crustal levels crystallize and fractionate under different conditions to magmas nearer the trench. Moyes & Storey (1986) identified two different parental magma types in the Antarctic Peninsula: one associated with a compressional phase (typified here by the Foyn Coast samples) and the other with a post-subduction extensional phase (represented by the Adelaide Island samples). Thus the tschermakitic amphibole-bearing mafic rocks are not consanguineous with the calc-alkaline AIS.

A change from Mg-rich to Fe-rich amphibole with increasing distance from the trench is another important distinction in the felsic rocks. Ferro-hornblende occurs typically in granodioritic rocks of the Foyn Coast and Arrowsmith Pensinsula. The Mg–Fe substitution in an amphibole is not totally independent of other possible cation substitutions, but these do not vary geographically. Experimental data (Gilbert *et al.*, 1982) suggest that Mg/Fe ratios are positively correlated with temperature (at constant fO_2) and also with fO_2. Increasing pressure can result in lower Mg-number (see Hammarstrom & Zen, 1986) but there is a paucity of data to substantiate this. Although changes in temperature and pressure might result in

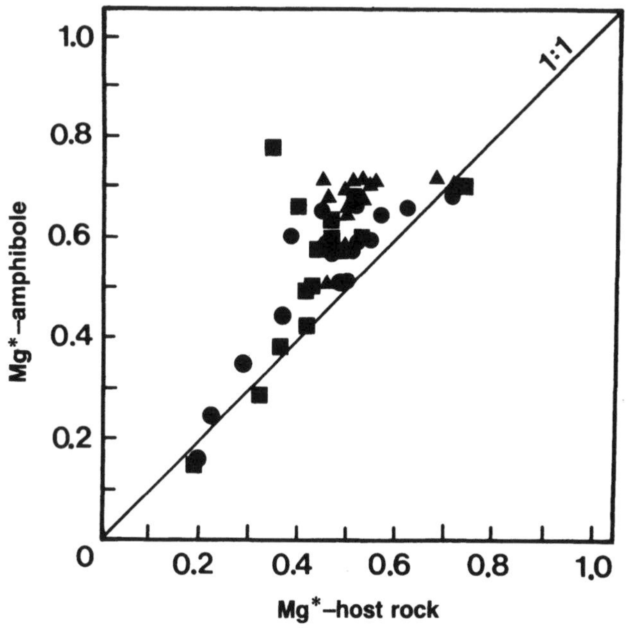

Fig. 3. Mg* in amphibole versus host rock. Solid triangles, Adelaide Island; open squares, Arrowsmith Peninsula; solid circles, Foyn Coast.

some of the variation observed here, they are unlikely to explain regional differences between similar rock types which appear to have crystallized under similar conditions.

Amphibole–host rock relationships

The amphiboles studied here become progressively more Fe-rich with the evolving magma. There is, however, considerable scatter in the granodiorite and monzogranite fields, in that some amphiboles have relatively low Fe concentrations. The Na and K contents are approximately constant, except for a slight increase in the more felsic host rocks. These data are in close agreement with calcic amphibole from elsewhere (see review by Wones & Gilbert, 1982). The decreasing modal abundance of amphibole in granitic rocks is due primarily to low total Ca contents of the host rock, but major amphibole compositional variation is not closely controlled by changes in rock composition (Gilbert *et al.*, 1982).

Fig. 3 clearly shows that most samples have higher Mg* ratios in the amphibole compared with the host rock, although some samples follow a 1:1 relationship. The Mg* of the amphibole appears to be largely independent of that of the host rock Mg*. In order to separate the cluster of data on Fig. 3, the Mg* of the amphibole has been plotted against the modified Larsen index of the host rock (Fig. 4). The general trend of the Mg* in the host rock is shown (dashed line) for comparison, and it is interesting to note a stepped decrease in Mg* with increasing acidity. In contrast, the amphibole Mg* shows the following features: (i) there is virtually no change in the amphibole Mg* of about 0.7 from all geographical areas up to a modified Larsen index of + 10: this corresponds to approximately 60 wt % SiO_2 and thus the boundary between diorite and granodiorite–tonalite (or the mafic and felsic suites); (ii) with only one exception, the Mg* of amphibole from the Adelaide Island samples remains roughly constant above a modified Larsen index of + 10; and (iii) the Mg* of amphibole from Arrowsmith Peninsula and Foyn Coast samples initially drops to approximately 0.6, then remains constant or continues to decrease until it closely corresponds to the host rock Mg*. The distinctive decrease in amphibole Mg* at the boundary between the mafic and felsic suites suggests an important change in crystallization conditions. Up to this point, amphibole Mg* is independent of host rock composition, and must therefore be a consequence of the other ferromagnesian phases, particularly pyroxene and opaque minerals. In the gabbroic rocks pyroxenes are common, whereas in the granitic rocks they are rarely present. The opaque phase changes from magnetite in the Adelaide Island samples to ilmenite in the Foyn Coast samples. Thus the distinct break in Mg* at the boundary between the mafic and felsic suites indicates either a large change in crystallization conditions, or that the mafic and felsic suites are not consanguineous.

The change from magnetite towards ilmenite crystallization with increasing distance from the trench could be accounted for chiefly by changes in fO_2 during crystallization, as the opaque phase is much more sensitive to this than to other parameters such as pressure and temperature. Samples from Adelaide Island correspond to a magnetite-series, whereas those from the Foyn Coast form in part an ilmenite-series, reflecting lower fO_2 conditions at increasing distance from the trench. Crystallization of ilmenite would allow the coexisting amphibole to be much more Fe-enriched than if magnetite were crystallizing, hence the restriction of ferro-hornblende to samples farther from the trench. A similar variation in both amphibole and opaque mineral chemistry has been described from Japanese granites (Kanisawa, 1975; Czamanske, Ishihara & Atkin, 1981), and correlates broadly also with the distinction between I- and S-type granites (Pitcher, 1983). Granitic rocks closest to the trench are I-type only (and have lower initial $^{87}Sr/^{86}Sr$ ratios) compared to granites farthest from the trench, which not only are lower in fO_2 but also may have had significant interaction with thicker continental crust. The scatter observed in Fig. 4 reflects a combination of decreasing fO_2 across the Antarctic Peninsula from west to east at any

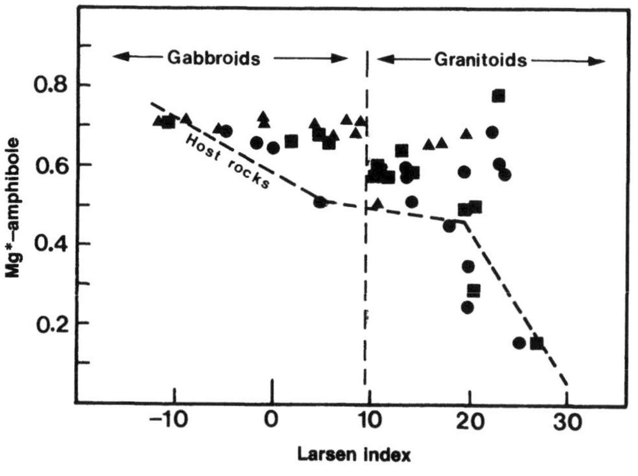

Fig. 4. Mg* in amphibole versus modified Larsen index (($\frac{1}{3}$Si + K – (Ca + Mg)) of host rock. The trend of Mg* in host rock is indicated by a dashed line; symbols as given in Fig. 3.

given time, coupled with possible time-dependent changes, at least for the Foyn Coast. Radiometric data are insufficient to separate changes with age because suitable coeval samples have not been identified as yet.

Summary and conclusions

The dominance of Ti-tschermakite and edenite substitutions in the amphibole studied here is a common feature of many other calc-alkaline suites, and also of amphibole produced experimentally. The greater modal abundance of amphibole in mafic rocks compared to felsic ones is due to bulk rock compositional control. Compositional variation in amphibole reflects variations in temperature and pressure (Al and Ti contents) and fO_2 (co-existing opaque phase, thus Mg* ratio). Amphibole was a possible liquidus phase only in the intermediate rock-types (granodiorite and tonalite), where Mg* ratios are lower farther from the trench (ferro- versus magnesio-hornblende). In the more mafic rocks, amphibole does not appear to have been an early liquidus phase and cannot, therefore, have contributed significantly to the early development of the calc-alkaline suite through fractionation. Regional differences in amphibole composition from the mafic rocks are possibly due to thicker continental crust on the eastern side of the Antarctic Peninsula (Moyes & Hamer, 1983), where magmas may have been trapped and crystallized at deeper structural levels, allowing greater interaction with crustal material than similar magmas on Adelaide Island. However, many west coast samples are related to a post-subduction extensional tectonic phase (Moyes & Storey, 1986) involving fundamentally different parental magma types. The mafic and felsic suites are not necessarily comagmatic, possibly reflected in the significant decrease of the amphibole Mg* at approximately 60 wt % SiO_2, although this may also be one result of amphibole first appearing as a liquidus phase in the felsic rocks. The lack of a distinctive compositional break for the suite as a whole on major element variation diagrams is possibly a facet of such limited diagrams, and further work is necessary to investigate this point.

Acknowledgements

The author would like to express his gratitude to all those personnel who helped organize and collect the specimens, notably Mr J.A. Jewel and Mr G.U. Summers for their guidance in the field. The microprobe data were collected with assistance from Drs P.J. Treloar and A. Buckley, the University of Cambridge, and the host rock XRF data with assistance from Drs A.D. Saunders and G.F. Marriner, Bedford College, University of London.

References

Czamanske, G.K., Ishihara, S. & Atkin, S.A. (1981). Chemistry of rock-forming minerals of the Cretaceous–Paleocene batholith in southwestern Japan and implications for magma genesis. *Journal of Geophysical Research*, **86(B11)**, 10431–69.

Czamanske, G.K. & Wones, D.R. (1973). Oxidation during magmatic differentiation, Finnmarka complex, Oslo area, Norway: Part 2, the mafic silicates. *Journal of Petrology*, **14(3)**, 349–80.

Gilbert, M.C., Helz, R.T., Popp, R.K. & Spear, F.S. (1982). Experimental studies of amphibole stability. In *Reviews in Mineralogy, 9B. Amphiboles: Petrology and Experimental Phase Relations*, ed. D.R. Veblen & P.H. Ribbe, pp. 229–353. Washington, DC; Mineralogical Society of America.

Hammarstrom, J.M. & Zen, E. (1986). Aluminium in hornblende: an empirical igneous geobarometer. *American Mineralogist*, **71(11–12)**, 1297–313.

Kanisawa, S. (1975). Chemical composition of hornblendes of some Ryoke granites, central Japan. *Journal of the Japanese Association of Mineralogists, Petrologists and Economic Geologists*, **70**, 200–11.

Leake, B.E. (1978). Nomenclature of amphiboles. *Mineralogical Magazine*, **42(324)**, 533–63.

Moyes, A.B. (1985). The petrology and geochemistry of plutonic rocks across southern Graham Land, Antarctica. PhD thesis, University of London, 184 pp. (unpublished).

Moyes, A.B. & Hamer, R.D. (1983). Contrasting origins and implications of garnet in rocks of the Antarctic Peninsula. In *Antarctic Earth Science*, ed. R.L. Oliver, P.R. James & J.B. Jago, pp. 358–62. Canberra; Australian Academy of Science and Cambridge; Cambridge University Press.

Moyes, A.B. & Storey, B.C. (1986). The geochemistry and tectonic setting of gabbroic rocks in the Scotia arc region. *British Antarctic Survey Bulletin*, **73**, 51–69.

Pankhurst, R.J. (1982). Rb–Sr geochronology of Graham Land, Antarctica. *Journal of the Geological Society, London*, **139(6)**, 701–12.

Pitcher, W.S. (1983). Granite: typology, geological environment and melting relationships. In *Migmatites, Melting and Metamorphism*, ed. M.P. Atherton & C.D. Gribble, pp. 277–85. Nantwich; Shiva Publishing Co.

Wones, D.R. & Gilbert, M.C. (1982). Amphiboles in the igneous environment. In *Reviews in Mineralogy, 9B. Amphiboles: Petrology and Experimental Phase Relations*, ed. D.R. Veblen & P.H. Ribbe, pp. 355–90. Washington, DC; Mineralogical Society of America.

Mesozoic metamorphism, deformation and plutonism in the southern Antarctic Peninsula: evidence from north-western Palmer Land

B.A. PIERCY & S.M. HARRISON

British Antarctic Survey, Natural Environment Research Council, High Cross, Madingley Road, Cambridge CB3 0ET, UK

Abstract

Studies within the magmatic arc of Palmer Land have identified several suites of plutonic rocks, of both pre-Mesozoic and Mesozoic age, in various deformation states. An undeformed plutonic suite, of Jurassic–Cretaceous age, corresponds to the Andean Intrusive Suite of Graham Land. This diorite–granite sequence intrudes all the other rock groups. More evolved rocks form small, high level, granophyric stocks. Deformed rocks range from weakly foliated plutons to orthogneisses and migmatites. These rocks, along with porphyroclastic mylonites and boudinaged amphibolite sheets, were produced by recrystallization and heterogeneous deformation in ductile shear zones. Metamorphosed sedimentary rocks form paragneisses which contain minerals whose parageneses indicate high temperature and moderate pressure metamorphic conditions. In terms of arc evolution this suggests that pre-Mesozoic crustal materials were regionally metamorphosed and deformed contemporaneously with magmatism in the Early Jurassic. Basic magmatism at that time drove the metamorphism, and produced the deformation recorded in the gneisses. These events may correspond with the break-up of the Pacific margin of Gondwana.

Introduction

The Antarctic Peninsula is believed to have been the site of essentially continuous subduction of Pacific and proto-Pacific oceanic crust from at least early Mesozoic times until the late Tertiary (Storey & Garrett, 1985). This produced an arc–trench system consisting of an accretionary prism, magmatic arc, and fore- and back-arc basins (Suárez, 1976).

The work presented here was conducted within the magmatic arc of the Antarctic Peninsula in north-western Palmer Land (Fig. 1). It details the geological history of this segment of the arc and allows correlations and comparisons to be made with other events elsewhere in Gondwana.

The strato-tectonic framework of the Mesozoic rocks is shown in Table 1 (Harrison & Piercy, 1986). The rocks are divided according to their field stratigraphical relationships, lithology and petrography. The events relating to arc evolution during the Mesozoic are considered here, but the basement rocks of this Mesozoic arc are also important as they record the effects of Mesozoic processes. These basement rocks now consist of variably deformed intermediate–acid orthogneisses, and high-grade paragneisses. The original protoliths are believed to have been calc-alkaline plutonic rocks of Devonian (Harrison & Piercy, this volume, p. 341) intrusive age, and sediments of unknown age.

Early Jurassic foliated plutonic rocks

These comprise a suite of basic–acid plutonic rocks, which now show foliated fabrics recording varying amounts of deformation and recrystallization (see below). The plutonic rocks range in composition from gabbros to diorites, quartz-monzonites and granites. They cut the ortho- and paragneisses and are themselves intruded by undeformed Cretaceous plutons (Fig. 1). Mineralogy consists of quartz, K-feldspar, plagioclase and biotite; mafic phases such as amphibole and pyroxene are restricted to the more basic members of the suite. Whole-rock chemistry suggests that they are very similar to the undeformed Cretaceous plutons (see below), forming a metaluminous–peraluminous sequence with calc-alkaline, I-type affinities. Silica contents range from about 50 to 75%, with Na and K increasing and Fe, Ca, Mg, Ti and Al decreasing with increasing silica. Trace-element discrimination diagrams (Pearce, Harris & Tindle, 1984) indicate a volcanic-arc tectonic setting. Two Rb–Sr age determinations are around 170–180 Ma for more acid members of the suite (determinations 10 and 11, Table 1).

Late Jurassic basic intrusions

Two groups of basic intrusive rocks of Late Jurassic age can be recognized. These are gabbroic plutons and mafic

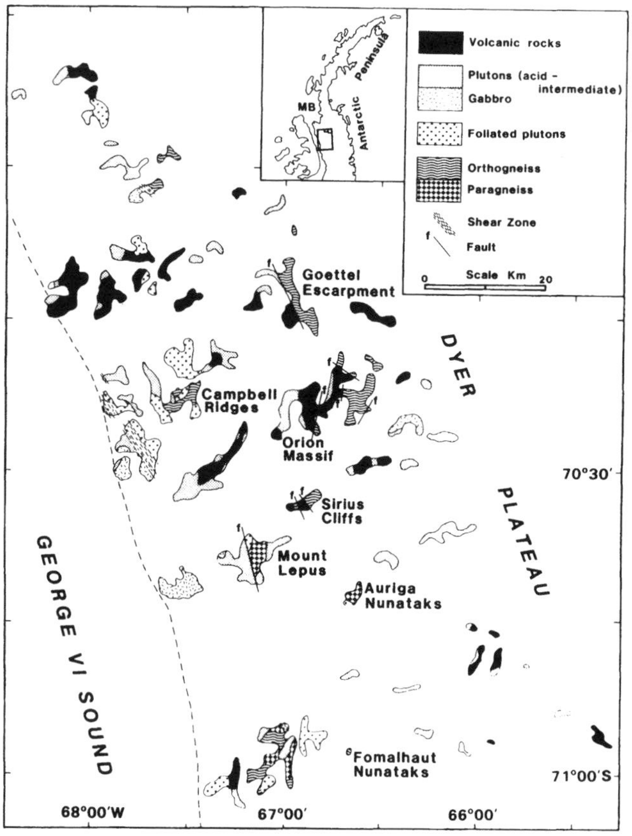

Fig. 1. Geological sketch map of north-western Palmer Land, with a location map of the Antarctic Peninsula inset. M.B., Marguerite Bay.

hypabyssal intrusions, both of which show variable recrystallization and deformation (see below). The hypabyssal intrusions form sills and dykes which intrude the ortho- and paragneisses, as well as the Early Jurassic foliated plutons. Individual intrusions vary in thickness from 0.5 to 5 m for the dykes, and up to 30 m for some sills. Their mineralogy is dominated by amphibole, plagioclase and biotite. Chemically these rocks are basaltic and show a calc-alkaline affinity on AFM triangular plots. Further trace-element discrimination diagrams such as Hf/3 versus Th versus Ta (Wood, 1980) also confirm a volcanic-arc tectonic setting. A single K–Ar whole-rock age determination 152 Ma (Rex, 1976) on one of the larger, less deformed sills must represent a minimum age (determination 8, Table 1).

The best exposures of gabbro are in the west of the area, where they form discrete intrusions. Their original mineralogy is plagioclase, pyroxene and olivine, and many outcrops display spectacular compositional layering and cumulate textures. Whole-rock Rb–Sr dating of marginal migmatites adjacent with these intrusions gives a minimum age of 140 Ma (Table 1).

Metamorphism and deformation

Deformation style

The metamorphosed rocks are variably deformed and show fabrics ranging from the foliated plutons through to porphyroclastic and ultramylonitic orthogneisses. Deformation is concentrated in narrow shear zones; in parts whole outcrops are virtually undeformed, whereas elsewhere the same rock-types have been reduced to mylonite. Three deformational phases have been identified. An early fabric forming deformation D_1 was produced by heterogeneous ductile shear, and this was followed by formation of W-verging asymmetric folds which constitute D_2. The final event, D_3, was formation of a major E-dipping shear zone (Fig. 1). Structural data from this D_3 shear, such as lineation directions and orientation of C–S planes (C parallel to shear and S parallel to the X–Y plane of the train ellipsoid; Berthé, Choukronne & Jegouzo, 1979), suggest that shear was in an oblique, normal sense, with possibly a considerable element of strike-slip movement.

Metamorphic fabrics

The pre-Mesozoic plutonic rocks now form a series of intermediate–acid orthogneisses and migmatites. These are variably deformed and range texturally from partially recrystallized granoblastic granite-gneiss through to porphyroclasticgneiss. Mineralogy is quartz, K-feldspar microperthite, plagioclase and biotite (Harrison & Piercy, this volume, p. 341).

The paragneisses form small isolated outcrops of garnet–biotite-gneiss, calc-silicate-gneiss and marble. The garnet–biotite-gneiss in particular shows complex textures representing an early garnet–hercynite assemblage, with reaction coronas formed around the hercynite and biotite.

Migmatites are generally associated with both the ortho- and paragneisses. Many of the gneisses contain millimetre-scale leucocratic segregations parallel to foliation. Leucosome mobilizates locally cutting the surrounding gneisses are medium–coarse-grained, pink or buff coloured, and consist of quartz, K-feldspar and plagioclase. The K-feldspar is microcline microperthite and plagioclase commonly contains blocky antiperthite. Intruding veins occur as metre-scale sheets to small veins and agmatitic, schlieren, phlebitic and folded migmatite forms are all present (terminology after Mehnert, 1968). Also within the migmatites are many amphibole pods as well as localities where numerous pillows of mafic material, with dark margins, are enclosed in granoblastic granite-gneiss; this suggests basic/acid magma mixing.

The foliated Early Jurassic plutonic rocks are generally less deformed and recrystallized than the gneisses. In the more acid compositions this foliation is defined by an alignment of mafic phases, and elongation of enclaves and mafic clots into schlieren. Foliated gabbros and diorites are partially recrystallized, as shown by amphibole, quartz and idiomorphically zoned plagioclase. This can give these rocks a granoblastic-polygonal texture, although a well developed foliation, with feldspar clasts, aligned biotite and quartzo-feldspathic segregations, is also common. However, some Early Jurassic rocks may be in a more deformed state, as dioritic- and granodioritic-gneisses are intimately mixed with the orthogneisses. If sufficiently deformed and recrystallized, these rocks are indistinguishable from the older orthogneisses without the aid of isotopic dating (see below).

Deformation and recrystallization fabrics in the Late Jur-

Table 1. *Stratigraphical scheme showing chronological relationships, age determinations and estimated time ranges of the different rock groups (in Ma)*

Period	Hypabyssal	Plutonic	Metamorphic	Age determination
Tertiary				
50	Basic–intermediate dykes 1		Block faulting	1. 92 ± 4 K–Ar Min DCR 2. 88 ± 3 Rb–Sr Min DCR 3. 112 ± 5 K–Ar Min DCR 119 ± 33 Rb–Sr Min DCR 4. 118 ± 1 Rb–Sr Min BAP 122 ± 2 Rb–Sr Min BAP 129 ± 1 Rb–Sr Min BAP
Cretaceous		2 3 Granitoids 4 5 Diorites		5. 134 ± 26 Rb–Sr WR SMH 6. 140 ± 5 Rb–Sr WR SMH 7. 143 ± 1 Rb–Sr Min BAP
150	8 Basic intrusions (amphibolite)	6 Gabbros (west coast)	7 Shear 9 Partial recrystallization	8. 152 ± 7 K–Ar WR DCR 9. 153 ± 10 Rb–Sr WR SMH
Jurassic		10 Granitoids 11 (foliated)		10. 170 ± 37 Rb–Sr WR SMH 11. 179 ± 41 Rb–Sr WR SMH 12. 186 ± 8 Nd–Sm Min BAP
200		Gabbroids (foliated)	12 Amphibolite facies 13 Migmatization Deformation	13. 198 ± 31 Rb–Sr WR BAP 14. 214 ± 82 Rb–Sr WR SMH
Triassic			14	

DCR, age determination from Rex (1976); BAP & SMH, age determinations by the authors; WR, whole-rock age; Min, mineral age.

assic intrusions are variable and best seen in the sills. Shearing, which was commonly parallel to the sill margins, produced a mineral alignment with minor quartzo-feldspathic segregations. Amphibole, plagioclase and biotite are recrystallized and aligned with the overall fabric; plagioclase grains often contain idiomorphic zoning and biotite forms strings along the foliation. Some sills show pinch and swell structure, or boudinage. The gabbroic rocks contain corona structures of amphibole, formed around pyroxene and olivine.

Pressure–temperature conditions

The best evidence for metamorphic conditions comes from the paragneisses which contain the most suitable parageneses for pressure–temperature (*P–T*) determinations. As already mentioned, two distinct assemblages are recorded in some of these rocks. The earlier paragenesis of garnet + hercynite + quartz is indicative of high temperature with moderate pressure (approx. 800 °C, 5–6 kb) (S.M. Harrison, unpublished data), as shown by the Mg/(Mg + Fe) ratio of garnet cores, and the presence of anatectic melts. The later paragenesis, comprising garnet (rims) + biotite + cordierite + andalusite + K-feldspar, was lower temperature with low pressure (~ 600 °C, < 2 kb) shown by biotite–garnet geothermometry and aluminosilicate geobarometry.

Both of these sets of conditions are typical of metamorphic facies of the low-pressure types as described by Myashiro (1973, chapter 11). Such conditions are produced where metamorphism is controlled by magmatic heating in an arc rather than by burial and compaction such as in a mountain belt. The higher *P* in the earlier metamorphism suggests recrystallization at a deeper level in the arc.

The more basic members of the Early Jurassic intrusive suite were particularly important in providing the heat input required to produce the high temperatures recorded in the first metamorphism. It is therefore suggested that the basic members of this suite were the initial driving force of metamorphism. Both the age data, described below, and the evidence for magma mixing, described above, reinforce this.

Age

The orthogneiss protolith is believed to be Devonian plutonic rock (~ 400 Ma intrusive age; Harrison & Piercy, this volume). However, the age of metamorphism of these rocks is considered to be Mesozoic, by correlation with dated rocks and minerals from associated migmatites believed to be products of the same metamorphic event. Thus the timing of metamorphism is deduced indirectly, with the spread of ages giving an indication of duration. Isotopic age determinations only record thermal events, so the timing of periods of deformation has to be deduced using a combination of ages and field structural criteria.

Rb–Sr whole-rock age determinations of migmatitic melt with diorite-gneiss from the Goettel Escarpment scatter around a 198 Ma reference isochron (determination 13, Table 1) and represent the initial high *T* metamorphism. The diorite-gneisses are probably the highly metamorphosed equivalents of Early Jurassic plutonic rocks (described above), in which case 200 Ma may represent an intrusive age. Nd–Sm dating of garnet–whole rock pairs from migmatites give an age of 186 Ma (determination 12, Table 1). This is the age of garnet growth in these rocks, again part of the initial high *T*, moderate *P* event.

Younger metamorphic ages relate to the lower *T*, low *P* retrogressive metamorphism and cooling. These are 153 Ma for Rb–Sr whole-rock determinations on reset orthogneisses (determination 9, Table 1), and 143 Ma (determination 7, Table 1) for Rb–Sr determinations on mineral separates taken from diorite gneiss which has a 198 Ma whole-rock isotope signature (determination 13, Table 1). These ages correspond well with the age of intrusion of the Late Jurassic basic magmas.

Post-deformational intrusions

Undeformed plutonic rocks form a distinctive group in north-western Palmer Land. They can be subdivided on structural style of intrusion into high-level and deeper-level suites. The high-level plutons form small-scale stocks with sharp margins against hornfelsed country rock. They consist of pink and grey granodiorites and quartz-monzonites. Mineralogy is dominantly quartz with both plagioclase and K-feldspars; biotite and minor amphibole textures are granophyric or porphyritic. The deeper-level plutons show a complete range of compositions from diorite to granodiorite; they form larger intrusions than the shallow stocks and have complex enclave-rich contact zones, often involving extensive veining into surrounding rocks. The rocks are medium–coarse-grained, non-porphyritic and dark grey–pale grey in colour; their mineralogy is dominantly quartz, plagioclase, biotite and amphibole, with relict clinopyroxene.

Chemically these rocks form a series from metaluminous to peraluminous and have I-type calc-alkaline affinities; the silica range is from 45 to 75%, with granophyres first appearing at 60% silica. Modelling of petrogenetic pathways using Rb, Ba, Sr and Zr suggests that chemical evolution was controlled largely by the fractionation of plagioclase, along with mafic phases such as pyroxene and amphibole. The higher-level plutons appear to have evolved from the deeper plutons by the removal of K-feldspar. This was probably by filter pressing which produced small batches of magma with high K_2O, Rb and Zr, but low Ba and Sr. Age determinations, using a variety of methods, range from about 135 to 90 Ma, peaking at 120–110 Ma (determinations 2–5, Table 1).

A series of late dykes, the youngest rocks in the area, comprise two compositional groups: mafic dykes and acid porphyries. The mafic dykes are fine-grained, steep-dipping and 1–2 m wide with chilled margins; their mineralogy is plagioclase, amphibole and clinopyroxene. Some plagioclase forms normally-zoned phenocrysts and amphibole may also be zoned. The acid porphyries are often associated with block-faulted zones and form buff-coloured bodies of medium-grained quartz, plagioclase-phyric rocks up to 100 m in width. Their textures are very characteristic comprising rounded phenocrysts in a very fine-grained matrix. Chemically these intrusions are calc-alkaline and have a compositional range from basalt and basaltic-andesite through to rhyolite. K–Ar dating (Rex, 1976) suggests an intrusive age of approximately 90 Ma.

Discussion

The sequence of geological events described (Table 1) suggests that the Mesozoic history of this segment of the Antarctic Peninsula magmatic arc can be viewed as a cyclic series of magmatic and tectonic episodes. The earliest recognized Mesozoic magmatism was in the Early Jurassic when the basic members of a basic–acid plutonic suite were intruded and produced the peak metamorphic conditions recorded in the paragneisses and migmatites. Deformation D_1 also occurred at this time. In eastern Palmer Land an Early Jurassic two-mica granite (at Mount Jackson; 71°21′S, 63°30′W) with high initial $^{87}Sr/^{86}Sr$ ratio may have formed by partial melting related to this same metamorphism (Meneilly *et al.*, 1987). By the Middle Jurassic more acid plutons were being emplaced with further deformation (D_2); metamorphosed plutons of a similar age are also present in Marguerite Bay (Gledhill, Rex & Tanner, 1982). During the Middle–Late Jurassic basic sills, dykes and plutons were emplaced, producing the retrograde metamorphic textures in the paragneisses and Early Jurassic plutons, and causing the resetting of some Early Jurassic metamorphic ages. This style of magmatism suggests that extension was taking place. This is reinforced by the evidence for extensional movements indicated by the E-dipping normal shear zone. Basic sills of similar age are also present in eastern Graham Land (Meneilly *et al.*, 1987).

By the Early Cretaceous deformation had ceased and a renewed phase of more acid magmatism commenced. Early plutons are dioritic, but by mid-Cretaceous time, shallow-level granophyres had evolved. These rocks correlate well, both temporally and chemically, with rocks of the Andean Intrusive Suite in Graham Land (Adie, 1955). A second episode of Late Cretaceous extension is indicated by the intrusion of the late mafic dykes and accompanying block faulting.

One question arising from these events is why was the metamorphism associated with the Early Jurassic plutonism so pronounced, whereas that with the Late Jurassic or Cretaceous plutons was much less so? This could be explained as a result of more intense magmatic activity at that time but erosion level also appears to be crucial. The Early Jurassic rocks were metamorphosed at a moderately deep level, as shown by pressures recorded in the paragneisses. However, by the Late Jurassic uplift had occurred, so that a shallower, less intense, metamorphic overprinting is recorded. The Early Jurassic metamorphic rocks probably represent the roots of the arc at that time.

The recognition of the Late Jurassic basic magmatic event is also important because it coincides with several other well documented events along the western margin of Gondwana, such as the opening of back-arc basins in southern South America (Dalziel, DeWitt & Palmer, 1974), and South Georgia (Storey & Macdonald, 1984). Extensive basic magmatism was also taking place around this time in Greater Antarctica in the form of the dolerites of the Ferrar Supergroup in the Transantarctic Mountains (Kyle, Elliot & Sutter, 1980). Events in Palmer Land also correlate closely with the earliest sedimentation in the fore-arc Fossil Bluff Formation, including large-scale slump zones (Macdonald & Butterworth, 1986). This segment of the Antarctic Penzinsula thus contains abundant evidence of Middle–Late Jurassic extensional and magmatic features. These probably represent a local expression of the processes known to have been operating around Gondwana at that time which relate to the break-up of its Pacific margin.

References

Adie, R.J. (1955). *The Petrology of Graham Land II. The Andean Granite-Gabbro Intrusive Suite*. London; Falkland Islands Dependencies Survey Scientific Reports, No. 12, 39 pp.

Berthé, D., Choukronne, P. & Jegouzo, P. (1979). Orthogneiss, mylonite, and noncoaxial deformation of granites: the example of the South American shear zone. *Journal of Structural Geology*, **1(1)**, 31–42.

Dalziel, I.W.D., DeWitt, M.J. & Palmer, K.F. (1974). Fossil marginal basin in the southern Andes. *Nature, London*, **250(5464)**, 291–4.

Gledhill, A., Rex, D.C. & Tanner, P.W.G. (1982). Rb–Sr and K–Ar geochronology of rocks from the Antarctic Peninsula between Anvers Island and Marguerite Bay. In *Antarctic Geoscience*, ed. C. Craddock, pp. 315–23. Madison; University of Wisconsin Press.

Harrison, S.M. & Piercy, B.A. (1986). Reports on Antarctic fieldwork: the geology of north-western Palmer Land. *British Antarctic Survey Bulletin*, **71**, 45–56.

Kyle, P.R., Elliot, D.H. & Sutter, J.F. (1980). Jurassic Ferrar Supergroup tholeiites from the Transantarctic Mountains. Antarctica, and their relationship to the initial fragmentation of Gondwana. In *Gondwana Five*, ed. M.M. Cresswell and P. Vella. pp. 283–7. Rotterdam; A.A. Balkema.

Macdonald, D.I.M. & Butterworth, P.J. (1986). Slope collapse deposits in a Mesozoic marine fore-arc basin, Antarctica. *12th International Sedimentological Congress, Abstracts, Canberra*, pp. 193–4 (unpublished).

Mehnert, K.R. (1968). *Migmatites*, pp. 7–42. Amsterdam; Elsevier.

Meneilly, A.W., Harrison, S.M., Piercy, B.A. & Storey, B.C. (1987). Major shear zones and faults in Palmer Land and their role in the development of the Antarctic Peninsula orogenic belt. In *Gondwana Six, Structure, Tectonics, and Geophysics*, ed. G.D. McKenzie, Monograph 40 pp. 209–19. Washington, DC; American Geophysical Union.

Myashiro, A. (1973). *Metamorphism and Metamorphic Belts*. London; George Allen & Unwin. 492pp.

Pearce, J.A., Harris, B.W. & Tindle, A.G. (1984). Trace element discrimination diagrams for the tectonic interpretation of granitic rocks. *Journal of Petrology*, **25(4)**, 956–83.

Rex, D.C. (1976). Geochronology in relation to the stratigraphy of the Antarctic Peninsula. *British Antarctic Survey Bulletin*, **43**, 49–58.

Storey, B.C. & Garrett, S.W. (1985). Crustal growth of the Antarctic Pensinsula by accretion, magmatism and extension. *Geological Magazine*, **122(1)**, 5–14.

Storey, B.C. & Macdonald, D.I.M. (1984). Processes of formation and filling of a Mesozoic back-arc basin on the island of South Georgia. In *Marginal Basin Geology*, ed. B.P. Kokelaar & M.F. Howells, pp. 207–16. London; Special Publication of the Geological Society of London, No. 16.

Suárez, M. (1976). Plate tectonic model for southern Antarctic Peninsula and its relation to southern Andes. *Geology*, **4**, 211–14.

Wood, D.A. (1980). The application of Th–Hf–Ta diagram to problems of tectonomagmatic classification and to establishing the nature of crustal contamination of basaltic lavas of the British Tertiary Volcanic Province. *Earth and Planetary Science Letters*, **50**, 11–30.

Rb–Sr study of Cretaceous plutons from southern Antarctic Peninsula and eastern Ellsworth Land, Antarctica

R.J. PANKHURST[1] & P.D. ROWLEY[2]

1 British Antarctic Survey, Natural Environment Research Council, High Cross, Madingley Road, Cambridge CB3 0ET, UK
2 US Geological Survey, Box 25046, Federal Center, Denver, Colorado 80225, USA

Abstract

Calc-alkaline plutons of the Early Cretaceous Lassiter Coast Intrusive Suite are abundant in the south-eastern Antarctic Peninsula and in eastern Ellsworth Land. They intrude the folded products of a Jurassic magmatic arc. The plutonic rocks, ranging in composition from gabbro to granite, are typical circum-Pacific (Andean) granitoids. New Rb–Sr mineral and whole-rock age determinations for 10 of the plutons range from 128 to 96 Ma, in excellent agreement with previous K–Ar dates. They indicate rapid cooling. The emplacement episode was relatively short, whereas plutonism in the northern Antarctic Peninsula spanned a period from Late Triassic to Tertiary time. Initial $^{87}Sr/^{86}Sr$ ratios for about 50 whole-rock samples from 14 plutons are relatively low (0.704–0.706), indicating that remelting of ancient upper crust was not an important petrogenetic factor. Unlike most comparable plutonic rocks of northern Antarctic Peninsula, those of the Lassiter Coast Intrusive Suite exhibit primary isotopic heterogeneity, interpreted as indicating significant crustal contamination during emplacement through thick but isotopically juvenile continental crust.

Introduction

The Lassiter Coast Intrusive Suite (LCIS) is a set of calc-alkaline stocks and batholiths almost entirely of Early Cretaceous age. The suite was named after exposures in the Lassiter Coast region of the southern Antarctic Peninsula (Rowley *et al.*, 1983), but includes virtually all plutonic rocks exposed in southern Palmer Land and extends westward into eastern Ellsworth Land (Fig. 1). The rocks range in composition from granite to gabbro, although most are granite or granodiorite (Vennum & Rowley, 1986). They are typical circum-Pacific ('Andean') biotite granitoids generated in response to subduction of the Pacific Ocean lithosphere and they may be correlated with similar rocks in the Triassic–Tertiary magmatic province of northern Antarctic Peninsula and southern South America.

The existing chronology of the LCIS is based on intensive dating by the conventional K–Ar technique. The present study used the Rb–Sr method to determine additional ages (mostly on the same samples used for K–Ar determinations). The results are generally concordant and support an overall range of 128 to 96 Ma. All dates are calculated using decay constants recommended by Steiger & Jäger (1977). This work also provides the first systematic study of $^{87}Sr/^{86}Sr$ variations in the plutonic rocks of this large area, supported by a few Sm–Nd analyses.

Geological setting

Reconnaissance geological mapping in this region has been conducted over more than 20 years by US Geological Survey parties (Williams *et al.*, 1972; Rowley & Williams, 1982; Rowley *et al.*, 1983, this volume, p. 467; Rowley, Kellogg & Vennum, 1985). The bedrock is dominated by the products of a long-lived subduction-related magmatic arc comparable to those of northern Antarctic Peninsula (e.g. Adie, 1955; Saunders & Tarney, 1982), although here it exhibits a shorter interval of magmatism, restricted to two main episodes. The first was Middle–Late Jurassic volcanism, resulting in a volcanic highland that roughly coincided with the present axis of southern Palmer Land and its westward extension through eastern Ellsworth Land. The basalt–rhyolite rocks of this phase (Mount Poster Formation) intertongue with a contemporaneous back-arc basin sedimentary assemblage towards the east (Latady Formation). The second igneous episode was dominantly plutonic and resulted in the LCIS: no contemporaneous volcanic rocks appear to have survived.

Several nunataks in the English Coast appear to consist of pre-Jurassic igneous and sedimentary rocks, one locality yielding a *Glossopteris* flora of probable Permian age (Laudon *et al.*, 1987; Laudon, this volume, p. 455). These rocks seem to represent the pre-Jurassic continental crust presumed to underlie the magmatic arc.

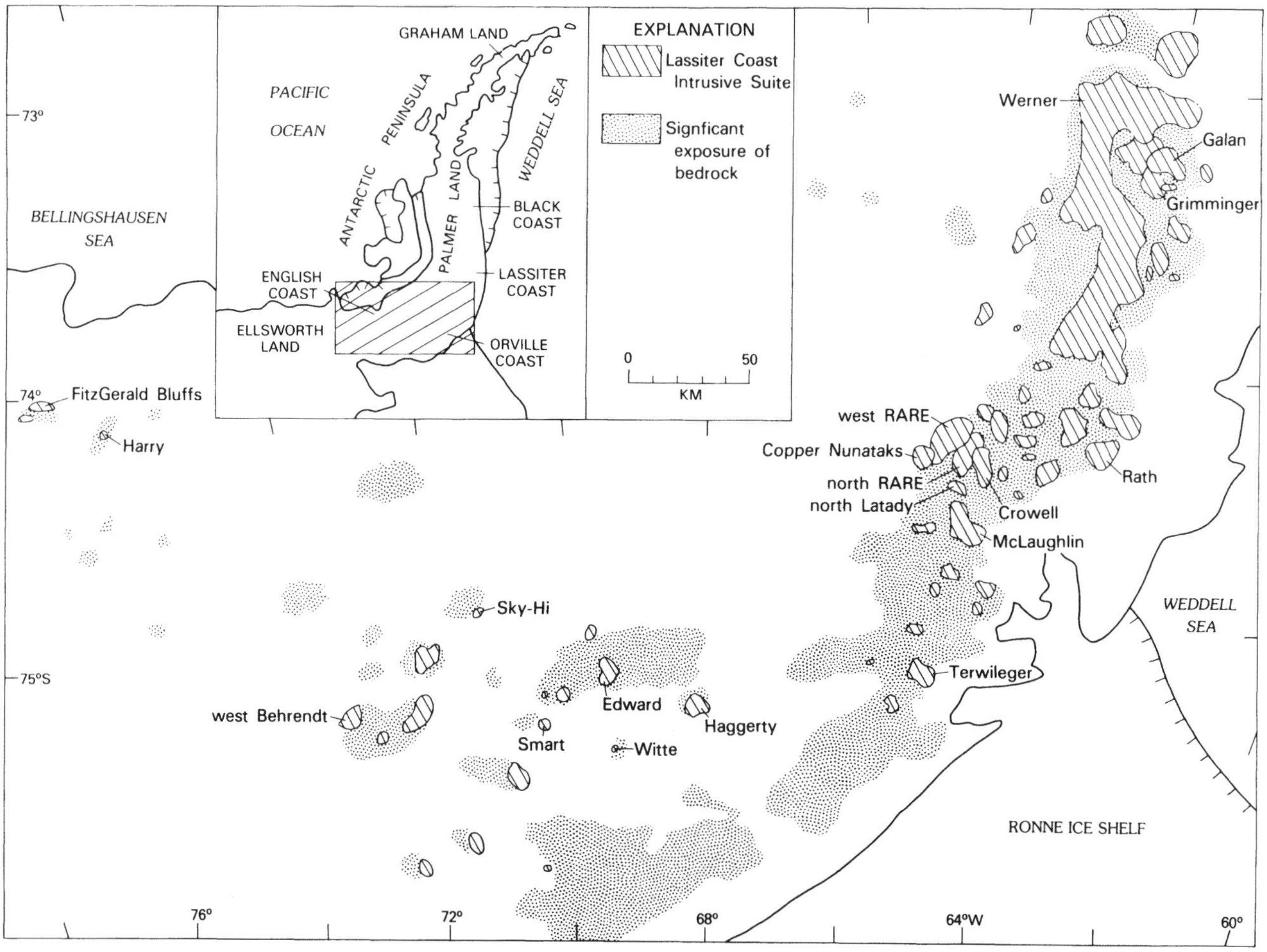

Fig. 1. Sketch map showing the distribution of the Lassiter Coast Intrusive Suite and the names of the individual plutons mentioned in the text.

Petrology and geochronology of the plutons

The new Rb–Sr data (Table 1) have been used to derive 12 new age determinations (Table 2). The data are arranged pluton-by-pluton in the same order as in the following discussion, i.e. north–south in south-eastern Palmer Land, then east–west in eastern Ellsworth Land and the English Coast area.

Werner batholith

This body (Vennum, 1978) is tentatively considered to be the largest pluton in the region, although it possibly consists of several plutons of the same general composition. It is concentrically zoned, having a thin diorite margin that passes abruptly inwards into quartz monzonite, granodiorite and granite. The mafic margin is in places intruded by the more silicic core, e.g. in the central outcrops of the south-western Dana Mountains, where there are abundant roof pendants of Latady Formation, and the main granodiorite phase has partially assimilated both wall rocks and the marginal phase (Vennum, 1978). Three samples of the core phases and one aplite from just north of the Meinardus Glacier give a Rb–Sr whole-rock isochron age of 108.8 ± 4.8 Ma. This may be compared with K–Ar ages for a nearby sample of quartz diorite (V39f) of 117.2 ± 7.0 Ma on hornblende and 103.8 ± 3.7 Ma on biotite (Farrar, McBride & Rowley, 1982). Throughout the region, hornblende generally gives older ages than coexisting biotite, often by more than the combined analytical error. This effect is due to the lower blocking temperature for Ar retention in biotite and can be explained either by slow cooling or locally, as here, by resetting of the biotite during intrusion of a later phase or nearby pluton. The new Rb–Sr isochron is dominated by the aplite data point and is more likely to record the second phase of intrusion in the batholith. A biotite separate from the silicic phase in the same area (Ke33h) gave a similar K–Ar age of 107.5 ± 4.8 Ma, and one from 25 km to the south-east (Ro241a) gave 110.6 ± 5.3 Ma. The new Rb–Sr whole-rock point for the second of these samples, and another from nearby (Ro242a), both lie on the isochron but do not affect either age or error. K–Ar mineral ages reported by Farrar *et al.* (1982) for other parts of the batholith (Table 2) are: 103.1 ± 4.4 Ma from the silicic phase 60 km to the north, and 100.7 ± 3.1 and 103.4 ± 3.1 from a simple sample of the silicic phase 60 km to the south. These ages appear to be slightly younger than those from the Dana Mountains, although individually they overlap within the errors. Allowing for sight Ar loss from the biotites during subsequent extended igneous history of the region, we suggest that the mafic phase is no older than about 115 Ma and that the main granodiorite is about 105–110 Ma.

Grimminger, Galan and Rath

These three plutons, dated by Farrar *et al.* (1982), were analysed by the Rb–Sr whole-rock method but failed to yield new ages. Of the two plutons east of the Werner batholith, the older is the Grimminger pluton, a large stock of diorite and quartz diorite with a K–Ar biotite age of 110.8 ± 2.8 Ma. It is intruded and locally hydrothermally-altered and sheared by the zoned but predominantly granodioritic Galan batholith (K–Ar biotite age = 106.9 ± 3.9 Ma). The Rath pluton, 20 km south of the Werner batholith, is also concentrically zoned, from diorite at the margin to granodiorite in the core and has given a less precise K–Ar biotite age of 111.3 ± 7.0 Ma. These three bodies are essentially synchronous with the Werner batholith.

North RARE and Crowell

These two cross-cutting plutons in the RARE Range were dated by Mehnert, Rowley & Smith (1975). The north RARE quartz diorite pluton is one of the oldest in the region, with K–Ar ages of 122.4 ± 2.6 Ma (hornblende) and 116.2 ± 2.4 Ma (biotite). The biotite was probably partially reset during emplacement of the Crowell pluton, which consists of homogeneous granite with a thin margin of granodiorite. The granite gave concordant K–Ar ages of 102.5 ± 2.2 Ma (biotite) and 101.2 ± 2.2 Ma (hornblende), indicating rapid cooling. The new Rb–Sr age, based on only two whole-rock points (granite and aplite), is 95.9 ± 1.4 Ma: broadly in agreement with the conclusion that this pluton was emplaced very close to 100 Ma ago.

West RARE and Copper Nunataks

Copper Nunataks (Fig. 1) also contain two cross-cutting plutons. The west RARE batholith of homogeneous granodiorite, with a K–Ar biotite age of 107.5 ± 3.3 Ma (Farrar *et al.*, 1982), extends into the RARE Range where it cuts the north RARE pluton and confirms the antiquity of the latter. The late stages of emplacement of the batholith were accompanied by closely spaced faulting, hydrothermal alteration and pyrite mineralization (Rowley, Williams & Schmidt, 1977). These events were followed in turn by intrusion of the Copper Nunataks pluton, a concentrically-zoned stock with a quartz-monzodiorite and granodiorite margin and a granite core. A K–Ar biotite age of 98.0 ± 3.1 Ma is confirmed by a three-point Rb–Sr whole-rock isochron age of 97.1 ± 2.2 Ma for samples from Nunatak A (Rowley *et al.*, 1977), which is also concordant with a K–Ar biotite age of 97.6 ± 3.1 Ma for a sample of the related granodiorite-porphyry dykes that intrude the pluton. This same dyke sample, together with rock samples from outcrops on other nunataks, defines a less precise but compatible isochron of 105 ± 10 Ma, with a significantly different initial $^{87}Sr/^{86}Sr$ ratio. The intrusive phases were followed by shearing, intense hydrothermal alteration and Cu–Mo mineralization of the porphyry-copper type (Rowley *et al.*, 1988).

North Latady, McLaughlin and Terwileger

These southern Lassiter Coast plutons were dated by Mehnert *et al.* (1975). The north Latady stock ranges in composition from gabbro to granodiorite and gave K–Ar ages of 119.9 ± 2.5 Ma and 111.3 ± 2.4 Ma for hornblende and biotite, respectively. The McLaughlin batholith is irregularly zoned inwards from quartz diorite to granite: hornblende and biotite gave concordant K–Ar ages of 109.8 ± 2.4 Ma and 108.6 ± 2.3 Ma, respectively, both consistent with the new Rb–Sr mineral isochron age of 107.6 ± 1.1 Ma. The Terwileger pluton is a large concentrically-zoned quartz-diorite–granodiorite stock, with K–Ar ages of 105.4 ± 2.3 Ma (hornblende) and 101.9 ± 2.2 Ma (biotite). The latter age agrees well with the new Rb–Sr mineral isochron age of 99.8 ± 1.0 Ma.

Haggerty, Witte, Smart and Edward

K–Ar ages for three of these southern Orville Coast plutons were reported by Farrar & Rowley (1980) and Rowley *et al.* (1988). The Haggerty stock, south-eastern Sweeney Mountains, is concentrically zoned inwards from monzodiorite to granite. The mean of two discordant K–Ar hornblende ages is 108.7 ± 1.6 Ma. The single biotite age of 116.0 ± 1.6 Ma seems to be more reliable as it is in reasonable agreement with the new Rb–Sr hornblende–biotite pair age of 112.6 ± 1.2 Ma (Table 2). The zoned diorite–granodiorite Witte stock to the south-west has given K–Ar hornblende and biotite ages of 111.2 ± 1.6 Ma (mean of two) and 108.8 ± 1.6 Ma, respectively, and a mineral pair Rb–Sr age of 106.5 ± 1.1 Ma. The zoned diorite–granite Smart stock gave discrepant K–Ar biotite ages of 103.4 ± 1.5 and 110.8 ± 1.5 Ma. The younger of these two ages is the more consistent with a Rb–Sr model age of 101.0 ± 1.0 Ma, assuming a probable initial $^{87}Sr/^{86}Sr$ ratio of 0.705 (see below). The Edward pluton, central Sweeney Mountains, consists mostly of gabbro–quartz monzodiorite and has not been dated previously. New Rb–Sr data yield an imprecise whole-rock age of 104 ± 11 Ma for the most silicic samples; the mafic rocks do not fit the line and may be older.

Sky-Hi Nunataks and west Behrendt

The Sky-Hi granodiorite stock in the north-western Sweeney Mountains gave consistent duplicate K–Ar hornblende ages averaging 121.8 ± 1.8 Ma (Farrar & Rowley, 1980; Rowley *et al.*, 1988). It is intruded by related porphyry dykes and affected by later hydrothermal alteration and porphyry-copper-type mineralization (Rowley *et al.*, 1988). The west Behrendt batholith of eastern Ellsworth Land is concentrically zoned, from a quartz diorite margin to a granodiorite core, and has yielded K–Ar ages of 109.9 ± 1.6 Ma on hornblende and 104.5 ± 1.5 Ma on biotite. The Rb–Sr ge of 109.6 ± 1.1 Ma (based on a mineral isochron) is more closely in agreement with the hornblende date, suggesting that the true age of emplacement is close to 109 Ma and that the biotite has lost some Ar.

Table 1. *Rb–Sr analytical results*

Sample	Type	Rb (ppm)	Sr (ppm)	$^{87}Rb/^{86}Sr$	$^{87}Sr/^{86}Sr$	Initial $^{87}Sr/^{86}Sr^{a}$
Werner batholith						
Bo68a	GD	131	480	0.789	0.70621	0.7050
V80c	QMD	67	600	0.321	0.70432	0.7038
V71	GD	143	289	1.434	0.70784	0.7056
V35a	QMD	51	455	0.321	0.70481	0.7043
V39f	QD	15	529	0.083	0.70428	0.7042
V39g	GD	28	553	0.148	0.70457	0.7043
Ke26d	A	224	9.4	69.71	0.81181	
Ro241a	D	59	466	0.364	0.70482	0.7043
Ro242a	QMD	91	473	0.555	0.70510	0.7042
Ro207a	GD	135	439	0.889	0.70671	0.7053
Bo6	GD	148	460	0.931	0.70647	0.7050
Grimminger pluton						
Wa32	D	1	668	0.006	0.70453	0.7045
Ro304a	QD	36	523	0.196	0.70492	0.7046
Galan pluton						
Ro307a	QD	139	731	0.551	0.70513	0.7043
Ke61f	GD	118	456	0.746	0.70627	0.7051
Rath pluton						
Ro160a	QMD	109	583	0.542	0.70574	0.7049
W163a	GD	120	443	0.785	0.70587	0.7046
North RARE pluton						
S54a	GD	64	525	0.350	0.70500	0.7044
W134d2	D	94	458	0.596	0.70551	0.7045
Crowell pluton						
W56	G	189	354	1.544	0.70780	0.7056
W54b	A	328	54	17.55	0.72960	
West RARE pluton						
M308a	GD	86	419	0.591	0.70596	0.7051
W144a	GD	145	304	1.377	0.70744	0.7053
Copper Nunataks pluton						
M311a	PD	136	452	0.871	0.70642	0.7052
M430	G	192	306	1.812	0.70786	0.7053
M321i	QMD	148	398	1.078	0.70680	0.7053
M429c	G	185	257	2.083	0.70824	0.7053
M315a	G	166	515	0.934	0.70615	0.7049
M318a	GD	148	548	0.785	0.70587	0.7048
M314b	A	309	120	7.440	0.71509	0.7047
North Latady pluton						
S9a	D	11	847	0.037	0.70424	0.7042
McLaughlin pluton						
S16x	GD	169	433	1.104	0.70730	0.7056
,,	Plag	19.8	510	0.113	0.70577	
,,	Hnb	17.5	51.3	0.988	0.70721	
,,	Bio	815	14.7	164.7	0.95727	
S28b	GD	178	157	3.204	0.71422	0.7093
Terwileger pluton						
S47e	G	162	460	1.024	0.70635	0.7049
,,	Plag	11.5	635	0.052	0.70491	
,,	Hnb	10.3	38.9	0.767	0.70608	
,,	Bio	740	17.7	122.9	0.89710	
S49	GD	159	499	0.922	0.70620	0.7049

Table 1. (*cont.*)

Sample	Type	Rb (ppm)	Sr (ppm)	$^{87}Rb/^{86}Sr$	$^{87}Sr/^{86}Sr$	Initial $^{87}Sr/^{86}Sr$[a]
Haggerty pluton						
V148a	G Hnb	31.1	74.4	1.173	0.70716	0.7053
,,	Bio	635	20.5	91.2	0.85118	
Witte pluton						
Ro441d	GD Hnb	16.0	42.9	1.076	0.70677	0.7051
,,	Bio	584	13.9	123.8	0.89250	
Smart pluton						
C2	G Bio	550	21.1	76.1	0.81512	
Edward pluton						
1	D	14	910	0.045	0.70388	0.7038
2	D	18	900	0.057	0.70400	0.7039
3	D	19	749	0.072	0.70396	0.7039
6	D	40	903	0.128	0.70396	0.7038
4	D	60	961	0.180	0.70471	0.7044
V91c	QMD	87	691	0.363	0.70496	0.7044
5		136	293	1.340	0.70641	0.7044
Th25	QMD	73.9	825	0.259	0.70399	0.7036
Sky-Hi pluton						
Ro498b	GD	93	546	0.494	0.70598	0.7051
Ro498j	GD	88	547	0.468	0.70594	0.7051
West Behrendt pluton						
Ke192a	GD	119	489	0.705	0.70624	0.7052
Ke193d	GD	103	412	0.724	0.70638	0.7053
,,	Hnb	11.0	39.8	0.797	0.70638	
,,	Bio	613	18.0	99.8	0.86063	
Harry pluton						
V201k	QM	83	367	0.656	0.70677	0.7057
V20111	X	86	342	0.728	0.70679	0.7056
V20112	QM	88	342	0.741	0.70674	0.7055
V201m	QM	104	355	0.850	0.70747	0.7060
South-west FitzGerald pluton						
L103a	G	101	313	0.935	0.70758	0.7061
L103b	A	120	46	7.591	0.71834	
L103e2	A	148	75	5.728	0.71523	
L103g	PD	97	155	1.808	0.70923	0.7063
L103h	PD	114	243	1.364	0.70806	0.7059
L103j	PD	96	250	1.105	0.70790	0.7061
North FitzGerald pluton						
DL1b	G	87	312	0.807	0.70762	0.7061
V200a	GD	84	324	0.747	0.70756	0.7062
V202a	G	142	136	3.029	0.71162	
V202j	G	82	380	0.623	0.70712	0.7060
V200g	P	147	123	3.447	0.71274	
V200b	A	122	58	6.091	0.71618	
V200d	PD	100	206	1.297	0.71040	0.7080

[a]Initial $^{87}Sr/^{86}Sr$ ratios are calculated for low Rb–Sr samples at the individual ages of emplacement estimated from Table 2 and are considered accurate to $\pm$ 0.0002 or better where quoted.

Rb and Sr concentrations and $^{87}Rb/^{86}Sr$ ratios for whole-rocks are derived from X-ray fluorescence data and are generally considered accurate to $\pm$ 5% and 0.5%, respectively. Data for minerals were determined by isotope dilution.

$^{87}Sr/^{86}Sr$ ratios were determined by automated mass spectrometry on a VG354 instrument in switched triple-collector mode and are accurate to better than $\pm$ 0.01%.

Key: D, diorite; QD, quartz diorite; GD, granodiorite; QMD, quartz monzodiorite; QM, quartz monzonite; G, granite; A, aplite; P, pegmatite; PD, porphyry dyke; X, inclusion. Plag, plagioclase feldspar; Hnb, hornblende; Bio, biotite; all other samples are whole-rock.

Table 2. *Rb–Sr ages for Lassiter Coast Intrusive Suite*

Location	Sample	Age (Ma)
Werner pluton	V39f, V39g, V35a, Ke26d	108.8 ± 4.8 (MSWD = 2.1)
Crowell pluton	W56, W54b	95.9 ± 1.4
Copper Nunataks pluton	M314b, M315a, M318a	97.1 ± 2.2 (MSWD = 0.6)
McLaughlin pluton	S16x WR, Plag, Hnb, Bio	107.6 ± 1.1 (MSWD = 0.2)
Terwileger pluton	S47e WR, Plag, Hnb, Bio	99.8 ± 1.0 (MSWD = 1.2)
Haggerty pluton	V148a Hnb, Bio	112.6 ± 1.2
Witte pluton	Ro441d Hnb, Bio	106.5 ± 1.1
Smart pluton	C2 Bio	101 ± 1.2 (model age with IR = 0.705 ± 0.001)
Edward pluton	4, 5, V91c	104 ± 11 (MSWD = 0.1)
West Behrendt pluton	Ke 193d WR, Hnb, Bio	109.6 ± 1.1 (MSWD = 1.3).
South-west FitzGerald pluton	L103a, b, e2, g, h, i	113.3 ± 4.2 (MSWD = 5.2)
North FitzGerald pluton	V200a, DL1b, V202a, V202j	128.3 ± 5.3 (MSWD = 2.6)

WR, whole-rock; Plag, plagioclase feldspar; Hnb, hornblende; Bio, biotite MSWD, mean-square of weighted deviates.

FitzGerald Bluffs and Mount Harry

These English Coast plutons are described by Rowley *et al.* (this volume, p. 467). At FitzGerald Bluffs, two plutons are separated by a septum of quartzite about 1 km wide. The south-west FitzGerald pluton consists of a granodiorite cut by a granite, but samples of both phases and of a later dyke suite, have yielded a reasonably well defined Rb–Sr isochron age of 113.3 ± 4.3 Ma, indicating their essential synchroneity. The north FitzGerald pluton is mostly composed of granite and granodiorite, which yielded a Rb–Sr isochron age of 128.3 ± 5.3 Ma. Two samples of later dykes lie off this line, implying resetting during a later event, perhaps the intrusion of the south-west FitzGerald pluton. The Harry pluton, a small hypabyssal stock of quartz monzonite that intrudes folded Jurassic volcanic rocks 20 km south-east of FitzGerald Bluffs, failed to give a meaningful age because the range of Rb/Sr ratios is too small. This pluton could be a feeder for the Jurassic volcanic rocks.

Summary

19 plutons in the LCIS have now been dated by K–Ar and Rb–Sr methods, with variable certainty. The total range of ages, from 128 ± 5.3 Ma (north FitzGerald Bluffs) to 97.1 ± 2.2 Ma (Copper Nunataks), is well outside the limits of analytical error. K–Ar ages as old as 123 ± 4 Ma were reported by Pankhurst (1980) for the main granodiorite phase of the central Black Coast area to the north of the Lassiter Coast and have been confirmed by unpublished Rb–Sr data. Of the other dated plutons, those clearly older than 110 Ma are Sky-Hi, north RARE, north Latady, Hagerty and perhaps Grimminger, whereas Crowell could be as young as Copper Nunataks. The remainder have ages within error of the interval 110–100 Ma. The composite plutons, especially those in the Lassiter Coast itself, have age patterns exhibiting a significant difference in the time of closure to Ar diffusion between early mafic intrusive phases and later silicic phases, which may reflect differences in emplacement age. There is no obvious systematic geographical distribution of the ages, although the FitzGerald Bluffs plutons have higher ages and $^{87}Sr/^{86}Sr$ ratios than most others in the area. It is not clear whether this reflects a progression in these parameters away from the Pacific margin.

Petrogenesis

Vennum & Rowley (1986) demonstrated that the LCIS rocks are predominantly calc-alkaline, I-type granitoids of characteristic Andean composition. Granodiorite is most common, with a transition to quartz-monzonite and granite in the late stages of some plutons. LIL element contents and LIL/HFS element ratios increase with differentiation. REE patterns are relatively unfractionated in the mafic rocks, but LREE and LREE/HREE increase markedly in intermediate rocks, whereas LREE eventually decrease in the felsic differentiates. Only small negative Eu-anomalies are observed. Such behaviour may be ascribed to primary magma formation in the upper mantle, followed by fractional crystallization in the crust of plagioclase feldspar and clinopyroxene, followed by amphibole and, finally, K-feldspar and accessory minerals. These features match those observed in northern Antarctic Peninsula (e.g. Saunders & Tarney, 1982) and are typical of magmas generated above an active subduction zone. Plutonism probably results from melting within the overlying lithospheric wedge, triggered by fluids driven off from the descending slab.

Crustal geochemical signatures may arise from the oceanic crustal slab, from within metasomatized lithosphere or by interaction with continental crust beneath the magmatic arc, but a strong mantle influence predominates in most areas. Initial $^{87}Sr/^{86}Sr$ ratios of the LCIS are generally low (Table 1, Fig. 2), ranging from 0.7038 to 0.7063 (excluding dykes and one obviously contaminated sample from the McLaughlin pluton which gives 0.7093). Although this range is well beyond the limits of analytical error (± 0.0002), the low value for all

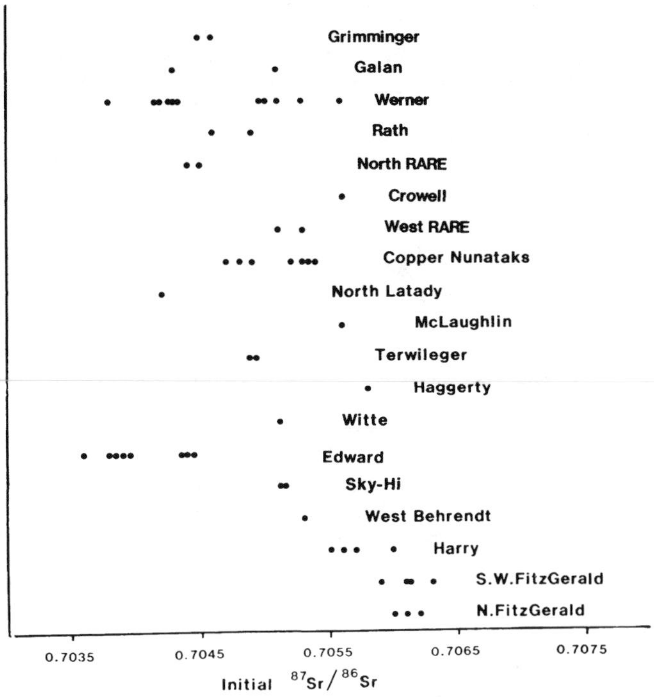

Fig. 2. Plot showing the range of initial $^{87}Sr/^{86}Sr$ ratios determined for each analysed pluton in the Lassiter Coast Intrusive Suite, arranged approximately along the axis of the peninsula from north (top) to south-west (bottom).

LCIS dioritic rocks (< 0.7050) confirms their mantle origin. The spread to higher ratios is probably best explained by interaction with the continental crust during the formation of more evolved magmas. Pankhurst (1982) ascribed the range of initial $^{87}Sr/^{86}Sr$ ratios of granitoids from Graham Land to such interaction. In Graham Land, however, the initial $^{87}Sr/^{86}Sr$ ratios clearly vary with the age of the magmas, decreasing from ~ 0.707 in Jurassic rocks to ~ 0.7035 in Tertiary rocks, due to progressive migration of the site of the arc away from the continental edge. No such relationship exists for the LCIS, although this may result from their more restricted time span. Of seven Graham Land plutons between 130 and 95 Ma in age, all but one have initial $^{87}Sr/^{86}Sr$ ratios in the range 0.7041–0.7063, comparable with the contemporaneous LCIS. The few new Sm–Nd analyses (Table 3) yield initial $^{143}Nd/^{144}Nd$ ratios that suggest a small addition of crustal Nd to magmas derived from a geochemically-depleted mantle source (ϵNd_t values of -1 to -4 for samples from the Werner batholith with initial $^{87}Sr/^{86}Sr$ ratios of 0.7038–0.7056).

These last results also emphasize the main isotopic difference between the Graham Land plutons and those of the LCIS. Whereas the former show a high degree of internal homogeneity (generally resulting in good isochrons for individual plutons), this is not so for the LCIS, where the large plutons for which adequate data are available show a significant range of initial $^{87}Sr/^{86}Sr$ ratios. Some are clearly complex (e.g. Werner batholith); others show an apparently simple bimodality suggesting two separate magmatic sequences (e.g. Copper Nunataks and Edward plutons). Although this aspect cannot be pursued further in the present reconnaissance study, it seem likely that the heterogeneity is not related to post-magmatic effects, such as hydrothermal alteration, but reflects variable crustal interaction during the generation of individual plutons in southern Antarctic Peninsula. This is most easily explained by a south-westerly increase in the thickness of pre-Cretaceous continental crust underlying the arc, as suggested by geophysical data for Graham Land and Palmer Land (Renner, Sturgeon & Garrett, 1985). The absence of an elevated radiogenic Sr signature shows that this thick crust is isotopically juvenile and, stratigraphically, probably not much older than Palaeozoic. This is consistent with emplacement of the LCIS into a magmatic arc and back-arc basin environment, where the Jurassic, or locally Palaeozoic (Laudon, this volume, p. 455), fill consists mainly of sedimentary and volcanic detritus.

Acknowledgements

The Rb–Sr analyses were carried out at the laboratories of the British Geological Survey, Geochemical Division, London. We thank T.K. Smith, Anne Robertson and Mark Ingram for their assistance with X-ray fluorescence determination of Rb and Sr contents. D.L. Schmidt and M.A. Kuntz helped to improve the manuscript.

References

Adie, R.J. (1955). *The Petrology of Graham Land – II. The Andean Granite-Gabbro Intrusive Suite*. London; Falkland Islands Dependencies Survey Scientific Reports, No. 12, 39 pp.

Table 3. *Sm–Nd data for Werner batholith*

Sample number	V39g	V71	V80c	Bo6
Sm (ppm)	0.69	3.25	6.24	4.97
Nd (ppm)	4.81	15.61	26.07	22.96
$^{147}Sm/^{144}Nd$	0.0872	0.1247	0.1447	0.1308
$^{143}Nd/^{144}Nd$	0.512502	0.512424	0.512541	0.512410
$^{143}Nd/^{144}Nd_t$	0.512440	0.512335	0.512438	0.512317
ϵNd_t	-1.2	-3.2	-1.2	-3.6
T_{DM}	662	1027	1060	1116

Analyses by mass-spectrometry at British Geological Survey, London. Sm and Nd ± 0.2%; $^{143}Nd/^{144}Nd$ ± 0.005%; T_{DM} (model age for separation from depleted mantle) ± 10% (approximately).

Farrar, E., McBride, S.L. & Rowley, P.D. (1982). Ages and tectonic implications of Andean plutonism in the southern Antarctic Peninsula. In *Antarctic Geoscience*, ed. C. Craddock, pp. 349–56. Madison; University of Wisconsin Press.

Farrar, E. & Rowley, P.D. (1980). Potassium–argon ages of Upper Cretaceous plutonic rocks of Orville Coast and eastern Ellsworth Land. *Antarctic Journal of the United States*, **15(5)**, 26–8.

Laudon, T.S., Lidke, D.J., Delevoryas, T. & Gee, C.T. (1987). Sedimentary rocks of the English Coast, eastern Ellsworth Land, Antarctica. In *Gondwana Six: Structure, Tectonics, and Geophysics*, Geophyusical Monograph 40 ed. G.D. McKenzie, pp. 183–9. Washington, DC; American Geophysical Union.

Mehnert, H.H., Rowley, P.D. & Schmidt, D.L. (1975). K–Ar ages of plutonic rocks in the Lassiter Coast area, Antarctica. *US Geological Survey Journal of Research*, **3(2)**, 233–6.

Pankhurst, R.J. (1980). Appendix: Radiometric dating. In *The Geology of the Central Black Coast, Palmer Land*, by. D.H. Singleton, pp. 49–50. Cambridge; British Antarctic Survey Scientific Reports, No. 102.

Pankhurst, R.J. (1982). Rb–Sr geochronology of Graham Land, Antarctica. *Journal of the Geological Society, London*, **139(6)**, 701–11.

Renner, R.G.B. Sturgeon, L.J.S. & Garrett, S.W. (1985). *Reconnaissance Gravity and Aeromagnetic Surveys of the Antarctic Peninsula*. Cambridge; British Antarctic Survey Scientific Reports, No. 110, 50 pp.

Rowley, P.D., Farrar, E., Carrara, P.E., Vennum, W.R. & Kellogg, K.S. (1988). Porphyry-type copper deposits and K–Ar ages of plutonic rocks of the Orville Coast and eastern Ellsworth land, Antarctica. *US Geological Survey Professional Paper*, **1351-C**, 35–49.

Rowley, P.D., Kellogg, K.S. & Vennum, W.R. (1985). Geological studies in the English Coast, eastern Ellsworth Land, Antarctica. *Antarctic Journal of the United States*, **19(5)**, 34–6.

Rowley, P.D., Vennum, W.R., Kellogg, K.S., Laudon, T.S., Carrara, P.E., Boyles, J.M. & Thomson, M.R.A. (1983). Geology and plate tectonic setting of the Orville Coast and eastern Ellsworth Land, Antarctica. In *Antarctic Earth Science*, eds. R.L. Oliver, P.R. James & J.B. Jago, pp. 245–50. Canberra; Australian Academy of Science and Cambridge; Cambridge University Press.

Rowley, P.D. & Williams, P.L. (1982). Geology of the northern Lassiter Coast and southern Black Coast, Antarctic Peninsula. In *Antarctic Geoscience*, ed. C. Craddock, pp. 339–48. Madison; University of Wisconsin Press.

Rowley, P.D., Williams, P.L. & Schmidt, D.L. (1977). Geology of an Upper Cretaceous copper deposit in the Andean province, Lassiter Coast, Antarctic Peninsula. *US Geological Survey Professional Paper*, **984**, 36 pp.

Saunders, A.D. & Tarney, J. (1982). Igneous activity in the southern Andes and northern Antarctic Peninsula – a review. *Journal of the Geological Society, London*, **139(6)**, 691–700.

Steiger, R.H. & Jäger, E. (1977). Subcommission on geochronology – a convention on the use of decay constants in geo- and cosmochronology. *Earth and Planetary Science Letters*, **36(3)**, 359–62.

Vennum, W.R. (1978). Igneous and metamorphic petrology of the southwestern Dana Mountains, Lassiter Coast, Antarctic Peninsula. *US Geological Survey Journal of Research*, **6(1)**, 95–101.

Vennum, W.R. & Rowley, P.D. (1986). Reconnaissance geochemistry of the Lassiter Coast Intrusive Suite, southern Antarctic Peninsula. *Geological Society of America Bulletin*, **97(12)**, 1521–33.

Williams, P.L., Schmidt, D.L., Plummer, C.C. & Brown, L.E. (1972). Geology of the Lassiter Coast area, Antarctic Peninsula – preliminary report. In *Antarctic Geology and Geophysics*, ed. R.J. Adie, pp. 143–8. Oslo; Universitetsforlaget.

Magnetic evidence for gabbroic plutons in the Black Coast area, Palmer Land (Extended abstract)

K.J. McGIBBON & H.E. WEVER

British Antarctic Survey, Natural Environment Research Council, High Cross, Madingley Road, Cambridge CB3 0ET, UK

Introduction

Regional aeromagnetic surveys over the Antarctic Peninsula (Renner, Sturgeon & Garrett, 1985) have revealed a N–S-aligned group of high-amplitude (800–2000 nT), short-wavelength (10–20 km) magnetic anomalies along the Black Coast of Palmer Land (Fig. 1). Detailed oversnow traverses were carried out during the 1984–85 field season by one of the authors (KJM) and the causative bodies were modelled and interpreted as a suite of highly magnetic mafic intrusions (McGibbon & Garrett, 1987).

The geology of eastern Palmer Land is dominated by calc-alkaline plutonic and volcanic rocks representing a Mesozoic arc and, further to the east, by deformed metasedimentary and metavolcanic rocks formed in a Jurassic back-arc basin (Storey & Garrett, 1985; Meneilly *et al.*, 1987).

The intrusive magmatic arc-related rocks are predominantly intermediate–acid plutons of Early Cretaceous age (Singleton, 1980). Recent geological investigations, conducted during the 1986–87 field season in the previously unmapped central Black Coast, showed the presence of large, variably deformed mafic intrusives along the eastern margin of the peninsula (Storey *et al.*, 1987). This paper describes the magnetic anomalies along the Black Coast and discusses the geological importance of the gabbroic source rocks.

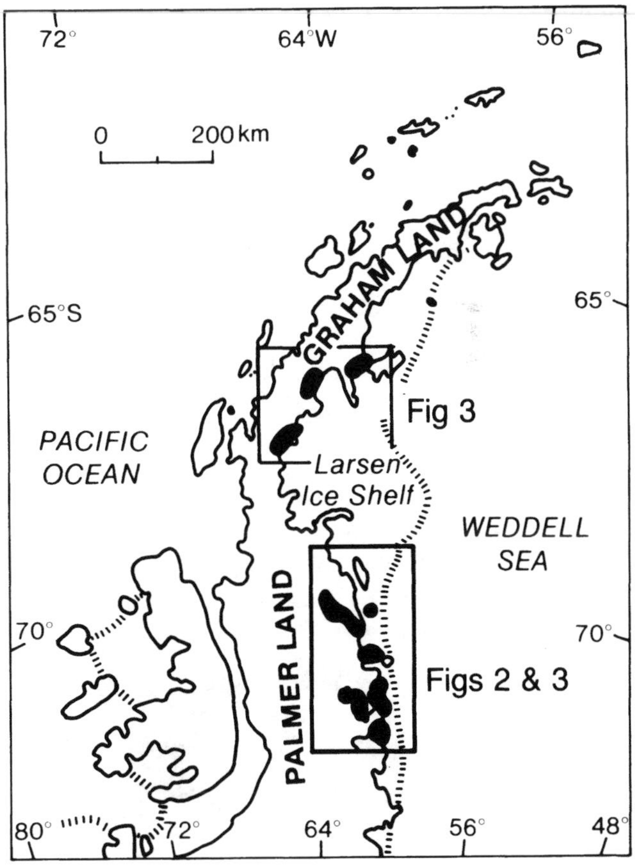

Fig. 1. The Antarctic Peninsula. Large aeromagnetic anomalies on the east coast (Renner *et al.*, 1985) are in black; enclosed areas indicate Figs 2 & 3 (after McGibbon & Garrett, 1987, fig. 1).

Magnetic surveys

Airborne survey

A de Havilland Twin Otter aircraft equipped with aeromagnetic survey equipment has been employed by the British Antarctic Survey (BAS) since 1973 on regional surveys of the Antarctic Peninsula. The procedures for data collection and processing were described by Renner *et al.* (1985). Fig. 2*a* shows the flightline network and regional aeromagnetic anomaly map for the Wilkins Coast and Black Coast areas. Selected flightlines and anomalies from other parts of the network are shown in Fig. 3.

Ground surveys

Ground magnetic traverses were conducted during the 1984–85 field season by a two-man BAS party. The procedure for data collection, reduction and interpretation was described by McGibbon & Garrett (1987). Three of the four major magnetic anomalies along the Black Coast were traversed and detailed contour maps of two of these were produced, enabling three-dimensional modelling to be carried out.

The magnetic anomalies

The regional aeromagnetic anomaly map (Fig. 2*b*) only poorly resolves the short wavelength components because of the filtering involved in its production (Renner *et al.*, 1985). The following description and interpretation therefore relies upon aeromagnetic profiles and oversnow data. The interpreted models are all from McGibbon & Garrett (1987) unless otherwise stated.

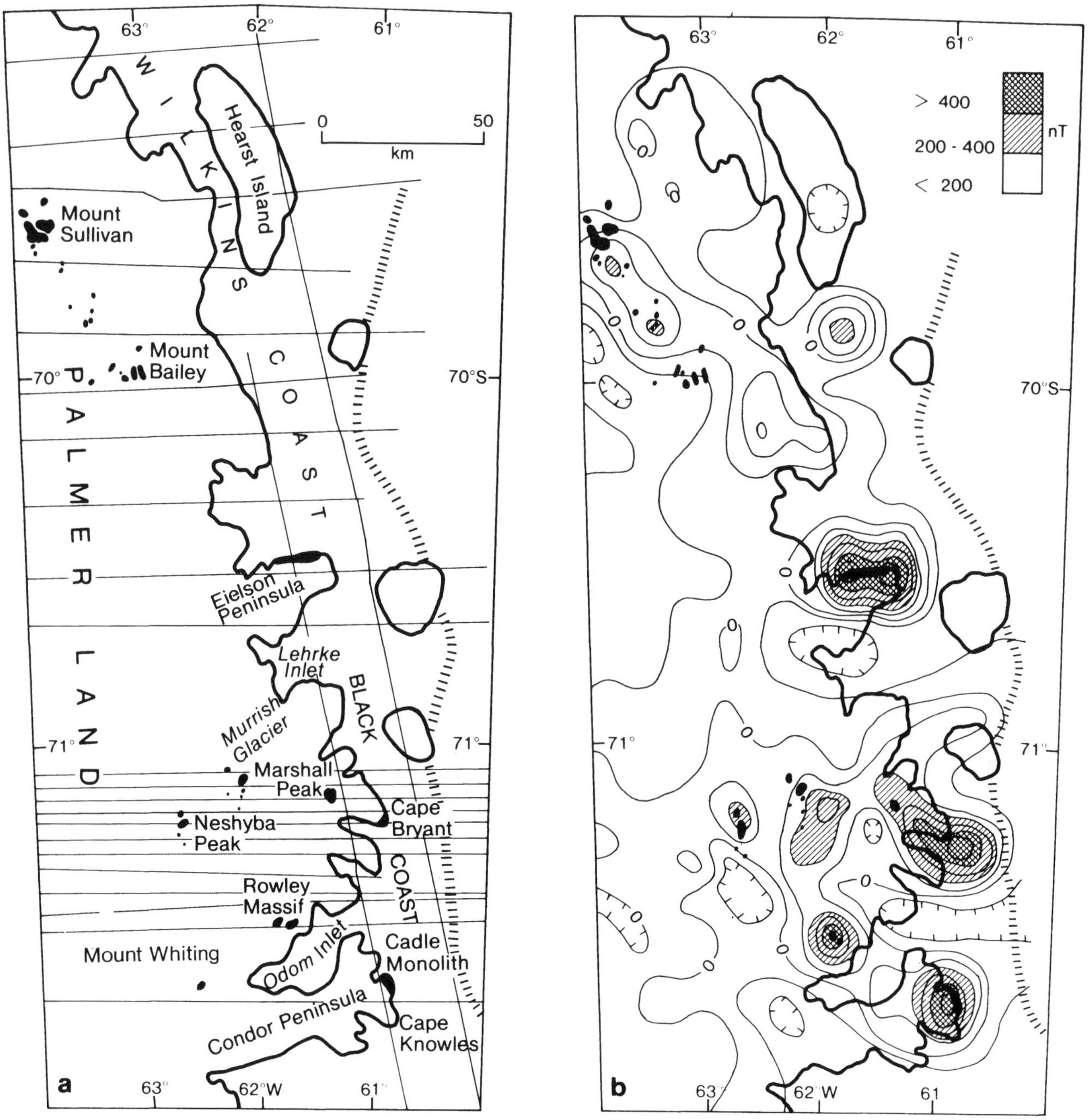

Fig. 2. (*a*) **Aeromagnetic flightline network over eastern Palmer Land. Rock outcrops referred to in the text in black.** (*b*) **Residual aeromagnetic anomaly map of eastern Palmer Land. Contour interval 100 nT. (Fig. 2 is reproduced from McGibbon & Garrett (1987, fig. 2) by kind permission of the British Antarctic Survey.)**

South of Hearst Island

Profile B089 (Fig. 3) reveals an anomaly with a wavelength of 25 km and an amplitude of 800 nT to the south of Hearst Island. This is tentatively grouped with the main Black Coast anomalies described below. The anomaly can be modelled by a composite body 30 km wide with a magnetization of 3 A/m.

Eielson Peninsula

At Elder Bluff on the Eielson Peninsula (Figs. 2*a* & 3), exposures of a basic pluton coincide with intense magnetic anomalies in excess of 2000 nT, as measured by the airborne survey, and 6000 nT as measured on the ground. Two-dimensional modelling shows that it could be caused by a 10 km wide body with a high magnetization of 8 A/m.

Marshall Peak–Cape Bryant

This NW–SE-aligned anomaly (Fig. 3, D108, B031) is approximately 50 km long and 20–30 km wide, and has two components. A 1200 nT anomaly over Marshall Peak merges with a longer wavelength 700 nT peak over Cape Bryant, where the ground data show a peak–trough amplitude of 6800 nT. A three-dimensional model of the ground data

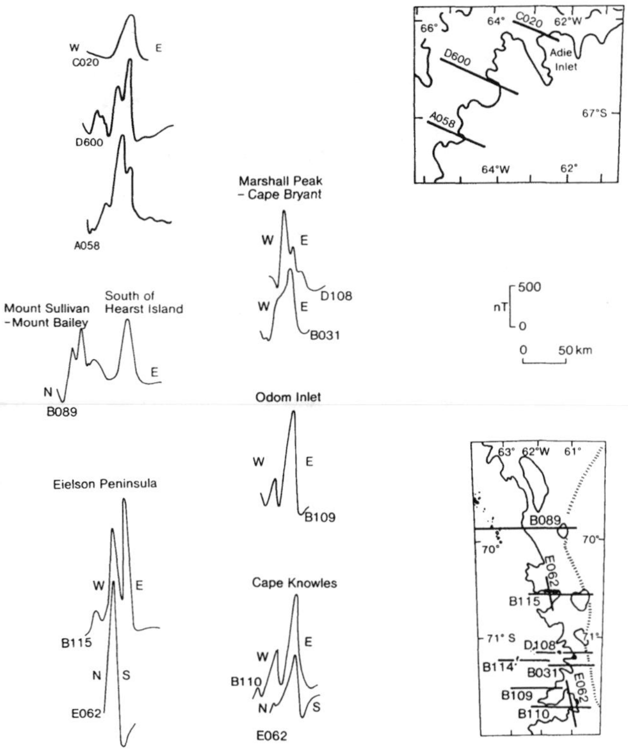

Fig. 3. Aeromagnetic profiles from the east cost of the Antarctic Peninsula. Location of inset maps is shown in Fig. 1.

reveals an intensely magnetized body (16 A/m), 6 km long, L-shaped in plan and with an upper surface at approximately 0.3 km below sea level. The depth to the base of the body is approximately 3.0 km, but owing to the limited constraints available for modelling the depth to base is ambiguous.

Rowley Massif–Odom Inlet

This anomaly, seen on the aeromagnetic profile B109 (Fig. 3), comprises two peaks of wavelength 20 km and amplitudes of 300 and 1000 nT. Modelling shows the body may consist of two parts each approximately 20 km wide, with magnetizations of 3 and 6 A/m, respectively.

Cape Knowles–Cadle Monolith

Two magnetic anomalies of wavelength 20 km were identified over Cape Knowles and Cadle Monolith by aeromagnetic profile B110 (Fig. 3). The western feature has an amplitude of 600 nT and the eastern 1300 nT.

Only the eastern part of the anomaly was covered by the oversnow survey, which revealed a series of intense positive anomalies in N–S alignment. At the southern end lies an elliptical anomaly of 3000 nT followed to the north by an intense 5000 nT peak with two smaller peaks. Two bodies with bases approximately 5 km below sea level, 25 and 20 km in width, respectively, and intensities of 3.4 and 6.0 A/m, are inferred from the aeromagnetic profile.

Possible sources of anomalies

The magnetic susceptibilities of 138 samples from eastern Palmer Land were given in McGibbon & Garrett (1987). Only gabbros, metagabbros and amygdaloidal basalts have sufficient susceptibility to produce the intense anomalies seen on the Black Coast.

A biotite–clinopyroxene-gabbro collected from Eielson Peninsula is considered a likely source rock for the adjacent intense ground and aeromagnetic anomaly, because of its high susceptibility and remanent magnetization (a resultant magnetization of 7 A/m). Samples from Elder Bluff, Eielson Peninsula collected during the 1986–87 field season by C. Griffiths show even higher values of susceptibility, indicating induced magnetizations reaching 14 A/m.

Singleton (1980) described gabbroic rocks north of Neshyba Peak which show rhythmic repetitions of magnetite- and clinopyroxene–plagioclase-rich layers. Diorites from Marshall Peak and Cape Bryant are similar to the sample from Eielson Peninsula. A recent visit to Cape Knowles and Rowley Massif demonstrated the outcrops to be large exposures of diorites and gabbros (Storey *et al.*, 1987).

A new K–Ar date of biotite from the magnetite-rich gabbro of Eielson Peninsula indicates an age of 118 ± 3 Ma (I. Millar, pers. comm.). Weak straining of biotite and plagioclase show that the gabbro is slightly deformed and therefore the date is regarded as a minimum age.

Discussion

The magnetic anomalies over eastern Palmer Land occur in a N–S-trending zone along the eastern margin of the Antarctic Peninsula (Fig. 1). Recent geological investigations confirm the presence of large gabbroic plutons along the Black Coast as suggested by McGibbon & Garrett (1987). There is a good correlation between the magnetic anomalies and the occurrence of some gabbroic rocks.

According to Moyes & Storey (1986), amphibole is the most abundant ferromagnesian mineral in the gabbros from the east coast. However it appears that high magnetite concentrations are mainly associated with a clinopyroxene-dominated gabbro. This may explain the absence of a magnetic anomaly associated with a large exposure of hornblende-gabbro at Mount Whiting (Storey *et al.*, 1987).

Other anomalies observed over eastern Graham Land resemble those of eastern Palmer Land. An anomaly over Adie Inlet (Fig. 3, C020) can be correlated with a clinopyroxene-gabbro (Marsh, 1968) which gave an age of 82 ± 1 Ma (Pankhurst, 1982).

The mafic plutonic rocks of the central Black Coast are variably deformed and at some locations intruded by the Early Cretaceous granodiorite (Singleton, 1980; Storey *et al.*, 1987). Singleton (1980) thought that the metagabbros of the northern Black Coast were part of the calc-alkaline suite but older than the main granodiorite. Concordant K–Ar ages from hornblende–biotite pairs give an age of 125 ± 3 Ma for the main granodiorite (revised values from Pankhurst, in Singleton, 1980). This is nearly similar to the minimum age of the Eielson Peninsula gabbro, which may indicate resetting by the granodiorite.

Piercy & Harrison (this volume, p. 381) describe olivine- and pyroxene-gabbros from the west coast of Palmer Land which have minimum ages of approximately 140 Ma. They suggest that these basic intrusions are related to a period of extension. Structural evidence for arc and back-arc extension in Palmer Land during the Late Jurassic–Early Cretaceous is presented by Meneilly *et al.* (1987).

Magnetic and geological evidence shows that the gabbroic bodies of the Black Coast occur mainly in a N–S-trending zone east of the main granodiorite. This distribution indicates a strong tectonic control over their emplacement. The minimum age and the distribution suggest a shift in magmatic activity along the east coast of Palmer Land during the Late Jurassic–Early Cretaceous, possibly related to a change in the tectonic regime of the arc and back-arc terranes. More intensive radiometric and geochemical studies are needed to constrain the upper age limit of these gabbros and their precise tectonic setting.

Acknowledgements

Our colleagues at the British Antarctic Survey are all thanked for useful discussions.

References

Marsh, A.F. (1968). Geology of parts of Oscar II and Foyn coasts, Graham Land. PhD thesis, University of Birmingham, 291 pp. (unpublished).

McGibbon, K.J. & Garrett, S.W. (1987). Magnetic anomalies over the Black Coast, Palmer Land. *British Antarctic Survey Bulletin*, **76**, 7–20.

Meneilly, A.W., Harrison, S.M., Piercy, B.A. & Storey, B.C. (1987). Structural evolution of the magmatic arc in northern Palmer Land, Antarctic Peninsula. In *Gondwana Six: Structure, Tectonics and Geophysics*, Geophysical Monograph 40, ed. G.D. McKenzie, pp. 209–19. Washington, DC; American Geophysical Union.

Moyes, A.B. & Storey, B.C. (1986). The geochemistry and tectonic setting of gabbroic rocks in the Scotia arc region. *British Antarctic Survey Bulletin*, **73**, 51–69.

Pankhurst, R.J. (1982). Rb–Sr geochronology of Graham Land, Antarctica. *Journal of the Geological Society, London*, **139**, 701–11.

Renner, R.G.B., Sturgeon, L.J.S. & Garrett, S.W. (1985). *Reconnaissance Gravity and Aeromagnetic Surveys of the Antarctic Peninsula.* Cambridge; British Antarctic Survey Scientific Reports, No. 110, 50 pp.

Singleton, D.G. (1980). *The Geology of Central Black Coast, Palmer Land.* Cambridge; British Antarctic Survey Scientific Reports, No. 102, 47 pp.

Storey, B.C. & Garrett, S.W. (1985). Crustal growth of the Antarctic Peninsula by accretion, magmatism and extension. *Geological Magazine*, **122**, 5–14.

Storey, B.C., Wever, H.E., Rowley, P.D. & Ford, A.B. (1987). The Geology of the Central Black Coast, Eastern Palmer Land, *British Antarctic Survey Bulletin*, **77**, 145–55.

A new look at the geology of Thurston Island

B.C. STOREY[1], R.J. PANKHURST[1], I.L. MILLAR[1], I.W.D. DALZIEL[2] & A.M. GRUNOW[3]

1 British Antarctic Survey, Natural Environment Research Council, High Cross, Madingley Road, Cambridge CB3 0ET, UK
2 Institute for Geophysics, University of Texas at Austin, 4920 North IH 35, Austin, Texas 78751, USA
3 Lamont–Doherty Geological Observatory, Palisades, New York 10964, USA

Abstract

The regional geology of the Thurston Island–Eights Coast crustal block (TI), West Antarctica, is re-assessed using recent field and laboratory data. It is clearly established as part of the Pacific active margin of Gondwana throughout Late Palaeozoic–Mesozoic times, with no direct relationship to East Antarctic cratonic evolution. The oldest rocks exposed are high-Sr granodioritic orthogneisses at Morgan Inlet, which are Carboniferous in age and may represent the Late Palaeozoic magmatic arc with which the fore-arc assemblages of the Antarctic Peninsula and New Zealand could have been associated. The younger history of Thurston Island is dominated by the development of Mesozoic igneous rocks. Upper Jurassic–Lower Cretaceous calc-alkaline granitoids with moderately low initial $^{87}Sr/^{86}Sr$ ratios (0.704–0.706) occupy most of the southern and western parts of the island although some granites have characteristics of tholeiitic differentiates. The northern and eastern parts are composed of a coarse-grained pyroxene-gabbro suite associated with a linear magnetic anomaly. An age of 122 Ma for a complex of fine-grained gabbros and aplogranite dykes near Belknap Nunatak again implies a minimum age of Early Cretaceous. The centre of the island is capped by a sequence of hydrothermally-altered volcanic rocks and associated pyroclastics. By analogy with a fuller sequence of similar rocks in the Jones Mountains to the south, these may be of Late Cretaceous age. An east–west dyke swarm intrudes the plutonic and volcanic rocks. The gabbros, dyke suite and tholeiitic granitic differentiates may represent an important extensional episode within the Mesozoic subduction-related arc. This may be related to changes in the rates of subduction, New Zealand rifting or cessation of subduction associated with collision of a spreading ridge and the trench.

Introduction

Thurston Island, the largest island off the Eights Coast of West Antarctica, is situated between the Bellingshausen and Amundsen sea coasts. It is approximately 240 km long (E–W) and 100 km (N–S) at its widest. It is separated from the Eights Coast by the Abbot Ice Shelf. An ice-cap covers much of the island, the rocks being exposed only as numerous isolated nunataks throughout the island and some cliff exposures on the coast. Smaller islands, some with outcrop, extend eastwards along the Eights Coast (Fig. 1).

Early investigations of the area were carried out by US expeditions to the Bellingshausen Sea (Craddock & Hubbard, 1961; Drake, 1962) and Craddock, White and Rutford visited all significant and accessible outcrops on Thurston Island during the Ellsworth Land Survey 1966–68 (Craddock, White & Rutford, 1969; Wade & Craddock, 1969). Published geological maps of Thurston Island show varied distributions of basement gneisses, plutonic and volcanic rocks (Craddock, 1972; Lopatin & Orlenko, 1972; Wade & Wilbanks, 1972) and a wide range of isotopic ages (Craddock, 1972). According to these workers basement gneisses and migmatites with Early Palaeozoic ages are intruded by a widespread suite of upper Palaeozoic gabbros, diorites and granitoids, and lower Mesozoic granitoids (Craddock *et al.*, 1964a, 1969; Drake, Stern & Thomas, 1964; Lopatin & Orlenko, 1972; Munigaza, 1972). Although Craddock (1972) suggested that the volcanic rocks are Mesozoic in age, Wade & Wilbanks (1972) and Lopatin & Orlenko (1972) considered them to be part of the Marie Byrd Land Palaeozoic metavolcanic complex.

Thurston, Dustin, Langhofer and McNamara Islands, as well as Lepley Nunatak and the Jones Mountains (Fig. 1), were visited in 1984–85 by the authors, as part of the joint BAS/USARP West Antarctica Tectonics Project (Dalziel & Pankhurst, 1985) and two aeromagnetic flight lines were flown by BAS in conjunction with USARP during the 1985–86 field season. The aims of this work are to reappraise the stratigraphy and field relations, and to relate the geology of

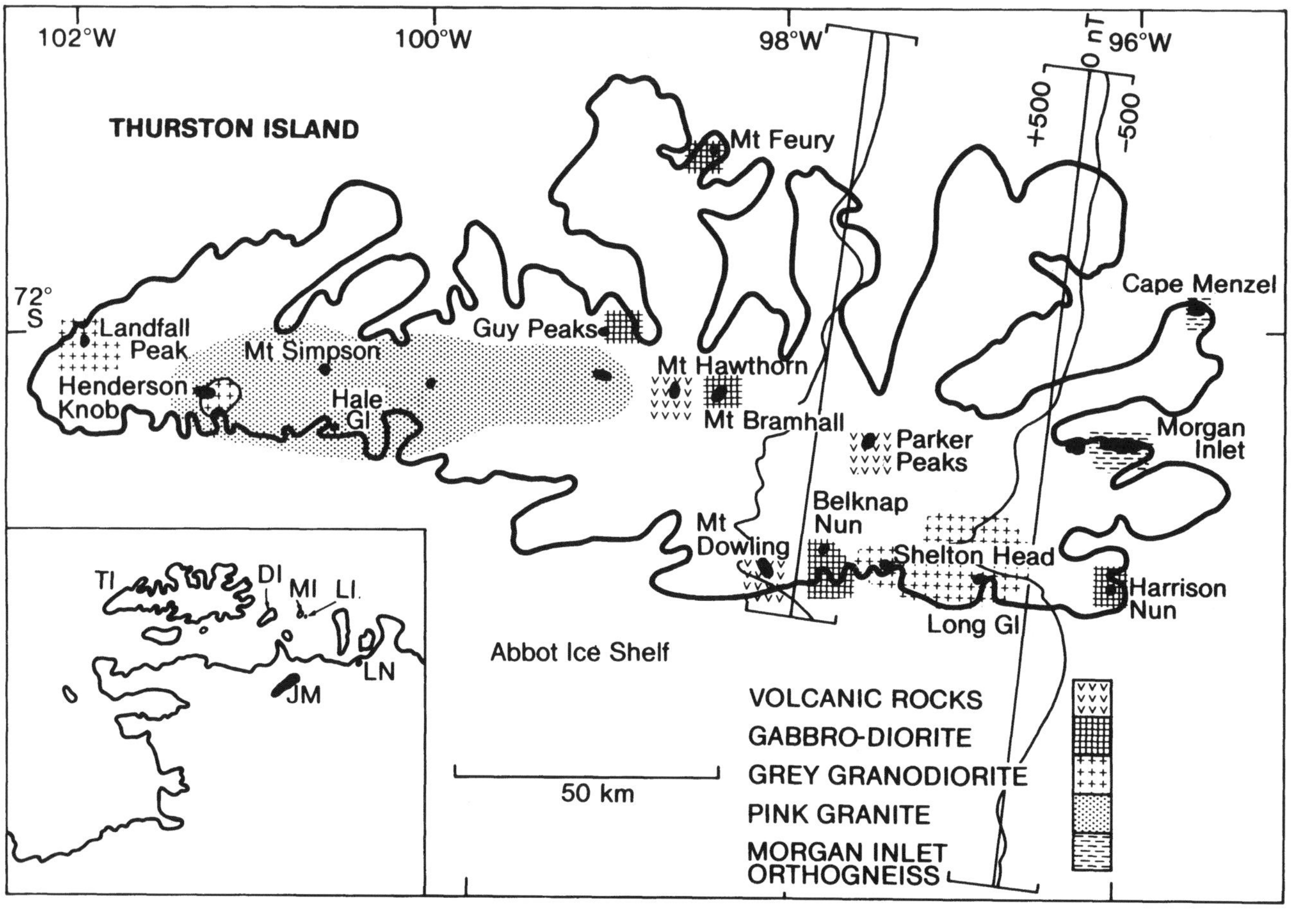

Fig. 1. Geological map of Thurston Island showing aeromagnetic profiles. TI, Thurston Island; DI, Dustin Island; MI, McNamara Island; LI, Langhofer Island; JM, Jones Mountains; and LN, Lepley Nunatak. Inset shows position of Thurston Island off the Eights Coast.

Thurston Island, one of five crustal blocks within West Antarctica (Storey *et al.*, 1988), to the neighbouring blocks by a detailed geochemical, geochronological and palaeomagnetic study. This paper is based on research in progress and is an update of our current knowledge on this important part of the Pacific margin of West Antarctica.

The geology of Thurston Island is dominated by magmatic rocks and we can clearly establish Thurston Island as having been part of the proto-Pacific margin of Gondwana throughout late Palaeozoic–Mesozoic times. The rocks are divided into four lithostratigraphic units: (1) Morgan Inlet orthogneiss; (2) Mesozoic plutonic rocks; (3) Mesozoic volcanic rocks; and (4) Cretaceous dyke rocks.

Morgan Inlet orthogneiss

Granitic orthogneiss crops out along the southern side of Morgan Inlet in the eastern part of Thurston Island (Fig. 1). The main lithology is a medium–coarse-grained biotite–hornblende-granodiorite which contains an alignment fabric, is locally compositionally layered on a 5–10 cm scale and contains some amphibolite layers. The layering strikes E–W, dips moderately northwards and is often folded. Garnet-bearing pegmatites and mafic–felsic dyke complexes, within which the mafic phase is brecciated by the felsic phase, are also common within the orthogneiss.

The orthogneiss is enriched in Sr and not ideal for Rb–Sr whole-rock dating. The work so far undertaken suggests a low initial $^{87}Sr/^{86}Sr$ ratio and a maximum possible age for the magmatic protolith of close to 300 Ma. This is the oldest age so far obtained during this investigation from Thurston Island: these are apparently the oldest rocks exposed on Thurston Island. Although Craddock *et al.* (1964a) report a scattered range of ages from this domain, a Rb–Sr biotite mineral age of 280 ± 10 Ma is in broad agreement with our new data.

At Cape Menzel, on the north side of Morgan Inlet there is a small exposure of sodium-rich hornblende-bearing quartz-diorite. Although its age is unknown it is tentatively included within the orthogneiss complex due to its foliation, a feature not present within the Mesozoic plutonic rocks.

Mesozoic plutonic rocks

Thurston Island is predominantly formed of Mesozoic plutonic rocks which are divided into three lithological groups: (1) pink granite; (2) grey granodiorite; and (3) gabbro-diorite.

Pink granite

A coarse-grained pink granite forms a large pluton in the western part of Thurston Island. It is a leucocratic hornblende–biotite-granite with up to 5% K_2O. It has moderately high Y (30 ppm) and Nb (20 ppm) contents and is similar chemically to the pink granite that forms much of the basement beneath the Cenozoic volcanic rocks in the nearby Jones Mountains (Craddock, Bastien & Rutford, 1964b). Field relations and preliminary radiometric dating both suggest this is the oldest of the Mesozoic plutons on Thurston Island with a Rb–Sr whole-rock age of about 155 Ma and a low initial $^{87}Sr/^{86}Sr$ (~ 0.7045). This is in contrast to the assumed age of the Jones Mountains granite of 199 ± 6 Ma based on K–Ar muscovite dating (Craddock *et al.*, 1964b).

Grey granodiorite

Grey, coarse-grained granitoids form individual nunataks at Long Glacier, Shelton Head, Mount Bramhall and Landfall Peak on Thurston Island and occur on Langhofer Island and at Lepley Nunatak, Eights Coast (Fig. 1). They are predominantly hornblende–biotite-granodiorite in composition, although some samples are quartz-diorite. In contrast to the pink granites they have lower SiO_2 and K_2O contents and lower Y and Nb. However, the grey granite which intrudes the pink granitoid at Henderson Knob is unusual; it is a hornblende–biotite-quartz-monzodiorite with abnormally high Zr (500 ppm), Y (40 ppm) and Nb (30 ppm) contents.

Rb–Sr radiometric dating confirms the field relations, giving a slightly younger age than the pink granite and possibly a higher initial $^{87}Sr/^{86}Sr$ ratio (~ 0.7055) for some members of this suite. The intrusions at Long Glacier, Landfall Peak and Henderson Knob are all about 145 Ma old. This is similar to K–Ar biotite and hornblende ages of 150 and 145 Ma from Landfall Peak (Drake *et al.*, 1964) and slightly older than the Rb–Sr biotite age of 132 Ma from Long Glacier (Munigaza, 1972).

Gabbro-diorite

Coarse-grained gabbro-diorite forms a large proportion of the Mesozoic plutonic suite occurring at Mount Feury, Guy Peaks, Harrison Nunatak, Belknap Nunatak and Morgan Inlet on Thurston Island and on neighbouring Dustin and McNamara islands (Fig. 1). They are predominantly gabbro in composition and contain varied proportions of olivine, spinel, orthopyroxene, clinopyroxene and plagioclase, and some contain a characteristic brown amphibole. Some are layered and most contain cross-cutting gabbro-pegmatite phases. On the nunatak 5 km south of Belknap Nunatak and on McNamara Island the gabbro is intruded by biotite–hornblende-granitoid sheets. The correlation of these with either of the above granitoid suites is uncertain and awaits further geochemical investigation. However, isotopic studies of the mafic–felsic components indicate that, at best, some of the gabbro-diorites are younger than both the pink and grey granitoids, based on a Rb–Sr whole-rock age of 122 ± 2 Ma for Belknap Nunatak.

The two aeromagnetic flight lines (unpublished data of J.A. Jones & M.P. Maslanyj) across the eastern part of Thurston Island indicate a 35 km broad long-wavelength magnetic anomaly up to 1000 nT in the vicinity of Belknap and Harrison nunataks (Fig. 1). This suggests that the southern part of Thurston Island may be underlain by a large gabbro body of high magnetization.

Mesozoic volcanic rocks

Volcanic rocks occur in the centre of Thurston Island and crop out at Parker Peak, Mount Dowling and Mount Hawthorn. At Parker Peak and Mount Dowling the rocks are well-bedded tuffs and agglomerates whereas at Mount Hawthorn volcanic breccias predominate. The volcanic rocks are variable in composition and include tuffs and breccias with SiO_2 varying from 50 to 74% and a Na_2O/K_2O ratio close to 1. The volcanic fragments contain a range of quartz, plagioclase, hornblende and clinopyroxene phenocrysts in a fine-grained groundmass.

Rb–Sr age dating suggests an age of no more than about 105 Ma, although this is not based on a perfect isochron. This Cretaceous age is consistent with the presence of some fossil wood fragments collected from similar volcanic rocks in the nearby Jones Mountains. The wood is a coniferous type and has anatomical features which suggest that it is more like late Mesozoic–Tertiary types than older more primitive forms (pers. comm., Dr J.E. Francis).

Cretaceous dyke rocks

Black, green and tan-coloured, often porphyritic, dykes intrude all the lithological groups on Thurston Island. They are generally steeply dipping and predominantly trend E–W parallel to the length of the island. The dykes are basalt, basaltic-andesite, andesite and rhyolite in composition. Clinopyroxene, hornblende and plagioclase are the common phenocryst phases and some dykes are composite with a mafic margin and felsic interior. Rb–Sr age dating of dykes from Shelton Head at 105–145 Ma is consistent with the field evidence which demonstrates that the dykes are the youngest magmatic suite in the Thurston Island area. This overlaps with a K–Ar whole-rock age (104 ± 4 Ma) of a quartz latite porphyry dyke from the Jones Mountains (Craddock *et al.*, 1964b).

Discussion

Thurston Island is dominated by Carboniferous and late Jurassic–Cretaceous calc-alkaline magmatic rocks, suggesting that this crustal block formed part of the large and complex proto-Pacific margin of Gondwana and West Antarctica. This is supported for the Cretaceous period by the palaeomagnetic data of Grunow, Kent & Dalziel (1987) from Belknap Nunatak but there are no constraints on earlier positions. However, Middle Devonian–Carboniferous mineral ages (300–380 Ma) in New Zealand, Tasmania and Marie Byrd Land (Adams, 1981, 1986) record the thermal effects of Devonian–Carboniferous plutons (the Tuhua and Karamea granites in New Zealand and the Ford Granodiorites in Marie Byrd Land) and confirm the presence of a larger magmatic province along this margin of Gondwana during Carboniferous times.

Evidence for this Carboniferous magmatic province to the east is more scarce. The only comparable ages come from plutonic rocks on the east coast of the Antarctic Peninsula but these are interpreted as the age of metamorphism (300 Ma) of a Silurian granitoid (Milne & Millar, this volume, p. 335). There is, however, evidence, as reviewed by Hervé *et al.* (this volume, p. 429), that part of the Trinity Peninsula Group and the Scotia metamorphic complex represent a fore-arc assemblage of comparable age to the Morgan Inlet orthogneiss. Two $^{40}Ar/^{39}Ar$ age spectra obtained from crossitic amphibole separates from Smith Island, South Shetland Islands also support this interpretation (Grunow, Dalziel & Harrison, 1987). Furthermore, granitic cobbles (386 ± 40 Ma) (Pankhurst, 1983) and Carboniferous detrital zircons from the Trinity Peninsula Group, Miers Bluff Formation and Scotia metamorphic complex suggest erosion of a nearby Devonian–Carboniferous magmatic terrane (Loske, Miller & Kramm, 1985, Hervé *et al.*, (this volume, p. 429). The Morgan Inlet orthogneiss may consequently represent the first known Carboniferous granitic rocks along this part of the proto-Pacific margin of Antarctica. Their occurrence close to the present-day continental margin suggests that if associated fore-arc terranes formed, they may have either rifted off this part of West Antarctica during break-up of the supercontinent or have been removed by tectonic erosion. If the former is correct then these terranes may now form part of either the Late Palaeozoic fore-arc of New Zealand or the Antarctic Peninsula.

Late Jurassic and Cretaceous magmatic activity is well known in both the Antarctic Peninsula, Marie Byrd Land and Thurston Island and formed part of a continuous Mesozoic Pacific margin arc province. Our preliminary dating suggests that at least some gabbros and volcanic rocks are part of this province and not, as previously thought, part of the Marie Byrd Land Late Palaeozoic gabbro and metavolcanic province.

The presence of Cretaceous high-Zr, Y and Nb granitoids within this magmatic province suggests that some of the granitoids may be tholeiitic gabbroic differentiates. Their occurrence together with the large gabbro plutons, the associated linear magnetic anomaly and the coast-parallel, E–W dyke swarms are characteristic of an intra-arc extensional regime. Coast-parallel dyke swarms and a marginal magnetic anomaly, termed the West Coast Magnetic Anomaly, are features of the Antarctic Peninsula magmatic arc and have been related to extensional phases within the arc (Storey & Garrett, 1985). These were caused by changes in the subduction parameters or by the collision of a ridge crest with the trench. The latter process is known to have taken place along this margin during the Late Cretaceous and Tertiary and ultimately caused the cessation of subduction (Barker, 1982). The extensional phase on the Thurston Island margin could also have been associated with the separation of New Zealand during the Late Cretaceous, although fracture zones shown on new GEOSAT data suggest New Zealand may have rifted from the Marie Byrd Land coast just west of Thurston Island (Sandwell & McAdoo, 1988).

Acknowledgements

This work is part of a joint BAS–USARP programme investigating the tectonic history of West Antarctica. We are very grateful to the British Antarctic Survey air unit, VXE6 Squadron of the US Navy, and our field assistants Pete Cleary, Chris Griffiths and Chuck Kroger for their support during the field programme. The US side was supported by the Division of Polar Programs, National Science Foundation through grant DPP82-13798 to IWDD.

References

Adams, J.A. (1981). Geochronological correlations of Precambrian and Paleozoic orogens in New Zealand, Marie Byrd Land (West Antarctica), northern Victoria Land (East Antarctica) and Tasmania. In *Gondwana Five*, ed. M.M. Cresswell & P. Vella, pp. 191–8. Rotterdam; A.A. Balkema.

Adams, J.A. (1986). Geochronological studies of the Swanson Formation of Marie Byrd Land, West Antarctica, and correlation with northern Victoria Land, East Antarctica, and South Island, New Zealand. *New Zealand Journal of Geology and Geophysics*, **29(4)**, 345–58.

Barker, P.F. (1982). The Cenozoic subduction history of the Pacific margin of the Antarctic Peninsula: ridge crest–trench interactions. *Journal of the Geological Society, London*, **139(6)**, 547–51.

Craddock, C. (1972). *Geological Map of Antarctica, 1:5 000 000*. New York; American Geographical Society.

Craddock, C., Bastien, T.W. & Rutford, R.H. (1964b). Geology of the Jones Mountains area. In *Antarctic Geology*, ed. R.J. Adie, pp. 171–87. Amsterdam; North-Holland Publishing Company.

Craddock, C., Gast, P.W., Hanson, G.N. & Linder, H. (1964a). Rubidium–strontium ages from Antarctica. *Geological Society of America Bulletin*, **75(3)**, 237–40.

Craddock, C. & Hubbard, H.A. (1961). Preliminary geologic report on the 1960 US expedition to Bellingshausen Sea, Antarctica. *Science*, **133(3456)**, 886–7.

Craddock, C., White, C.M. & Rutford, R.H. (1969). The geology of the Eights Coast. *Antarctic Journal of the United States*, **4(4)**, 93–4.

Dalziel, I.W.D. & Pankhurst, R.J. (1985). Tectonics of West Antarctica and its relation to East Antarctica: USARP-BAS geology/geophysics. *Antarctic Journal of the United States*, **19**, 40–1.

Drake, A.A. (1962). Preliminary geologic report on the 1961 US expedition to the Bellingshausen Sea, Antarctica. *Science*, **135**, 671–2.

Drake, A.A., Stern, T.W. & Thomas, H.H. (1964). Radiometric ages of zircon and biotite in quartz diorite, Eights Coast, Antarctica. *Professional Paper of the United States Geological Survey*, **501-D**, 50–3.

Grunow, A.M., Dalziel, I.W.D. & Harrison, T.M. (1987). Structural and $^{40}Ar/^{39}Ar$ constraints on the evolution of blue schists from the Scotia Arc. *Geological Society of America Abstracts with Programs*, **19**, 686–7.

Grunow, A.M., Kent, D.V. & Dalziel, I.W.D. (1987). Evolution of the Weddell Sea Basin: new paleomagnetic constraints. *Earth and Planetary Science Letters*, **86**, 16–26.

Lopatin, B.G. & Orlenko, E.M. (1972). Outline of the geology of Marie Byrd Land and the Eights Coast. In *Antarctic Geology and Geophysics*, ed. R.J. Adie, pp. 245–50. Oslo; Universitetsforlaget.

Loske, W., Miller, H. & Kramm, U. (1985). Dataciones radiométricas U–Pb en rocas de las islas del grupo Piloto Pardo, Antártica. *Serie Científica Instituto Antártico Chileno*, **32**, 39–46.

Munigaza, F. (1972). Rb–Sr isotopic ages of intrusive rocks from Thurston Island and adjacent islands. In *Antarctic Geology and Geophysics*, ed. R.J. Adie, pp. 205–6. Oslo; Universitetsforlaget.

Pankhurst, R.J. (1983). Rb–Sr constraints on the ages of basement rocks on the Antarctic Peninsula. In *Antarctic Earth Science*, ed. R.L. Oliver, P.R. James & J.B. Jago, pp. 367–71. Canberra; Australian Academy of Science and Cambridge; Cambridge University Press.

Sandwell, D.T. & McAdoo, D.C. (1988). Marine gravity of the Southern Ocean and Antarctic margin from GEOSAT: tectonic implications. *Journal of Geophysical Research*, **93(B9)**, 10389–96.

Storey, B.C., Dalziel, I.W.D., Garrett, S.W., Grunow, A.M., Pankhurst, R.J. & Vennum, W.R. (1988). West Antarctica in Gondwanaland: crustal blocks, reconstruction and breakup processes. *Tectonophysics*, **155**, 381–90.

Storey, B.C. & Garrett, S.W. (1985). Crustal growth of the Antarctic Peninsula by accretion, magmatism and extension. *Geological Magazine*, **122(1)**, 5–14.

Wade, F.A. & Craddock, C. (1969). The Ellsworth Land Survey. *Antarctic Journal of the United States*, **4(4)**, 92.

Wade, F.A. & Wilbanks, J.R. (1972). Geology of Marie Byrd and Ellsworth Lands. In *Antarctic Geology and Geophysics*, ed. R.J. Adie, pp. 207–14. Oslo; Universitetsforlaget.

An aeromagnetic study of southern Palmer Land and eastern Ellsworth Land

J.A. JONES & M.P. MASLANYJ

British Antarctic Survey, Natural Environment Research Council, High Cross, Madingley Road, Cambridge CB3 0ET, UK

Abstract

Aeromagnetic data show southern Palmer Land and eastern Ellsworth Land to be characterized by broad (> 20 km) anomalies which generally follow the arcuate structure of the area. Two-dimensional modelling studies suggest sources are large bodies buried at depths of between 5 and 12 km. High body magnetizations are required to reproduce observed anomalies, indicating that mafic plutonism may be more significant in the area than is apparent at outcrop. The West Coast Magnetic Anomaly, observed throughout Graham Land and northern Palmer Land, is seen to extend beyond 72° S, and, south of 73° S, follows the curvature of the coastline.

Introduction

Aeromagnetic surveys

The data presented here represent 12 500 line km of aeromagnetic surveying undertaken in January–February 1986, employing the equipment and technique previously described by Renner, Sturgeon & Garrett (1985). During flights a Geometrics G-866 base station magnetometer was used to monitor diurnal variations. The lines (Fig. 1) were flown with a spacing of about 20 km and a terrain clearance of 150 m. The area covered (Fig. 2) is a roughly rectangular swathe of southern Palmer Land and eastern Ellsworth Land, bounded by Smyley Island in the north-west, Butler Island in the north-east, Bowman Peninsula in the south-east and Behrendt Mountains in the south-west; it links existing British Antarctic Survey (BAS) aeromagnetic networks over the Antarctic Peninsula and Ronne Ice Shelf.

The Antarctic Peninsula magnetic anomaly field is dominated by the West Coast Magnetic Anomaly (WCMA) (Renner *et al.*, 1985), a linear feature which follows the arcuate coastline of the peninsula from the South Shetland Islands to the southern edge of the network at 72° S. This high amplitude (500–1000 nT) broad (75–120 km) anomaly is not clearly reflected in the surface geology but is believed to be related to a linear batholith of mafic – intermediate composition.

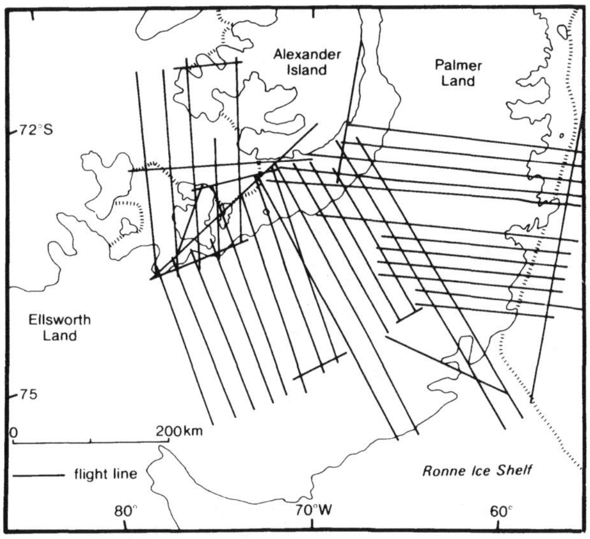

Fig. 1. Aeromagnetic flight line location map.

Geology

The surveyed area is part of the Mesozoic–Cenozoic Andean orogenic belt (Rowley & Williams, 1982). The majority of rock exposure occurs in the south and east of the region, along and inland of the Orville and Lassiter coasts. These outcrops consist of folded Middle–Upper Jurassic mudstone, siltstone and sandstone of the Latady Formation, and silicic–intermediate calc-alkaline volcanic rocks of the Mount Poster Formation (Rowley *et al.*, 1983). Cretaceous rocks belonging to the Lassiter Coast Intrusive Suite are found throughout the area (Fig. 2). These plutons, aged 90–135 Ma, vary in composition from gabbro to granite, although granite and granodiorite predominate (Vennum & Rowley, 1986).

Exposures on the English Coast and adjacent parts of north-eastern Ellsworth Land are mainly of calc-alkaline volcanic rocks deposited along the axis of the magmatic arc (Rowley, Kellogg & Vennum, 1985). However, the youngest rocks found in the region, at Henry and Snow nunataks, consist of alkaline basaltic hyaloclastites and pillow lavas of probable late Tertiary age.

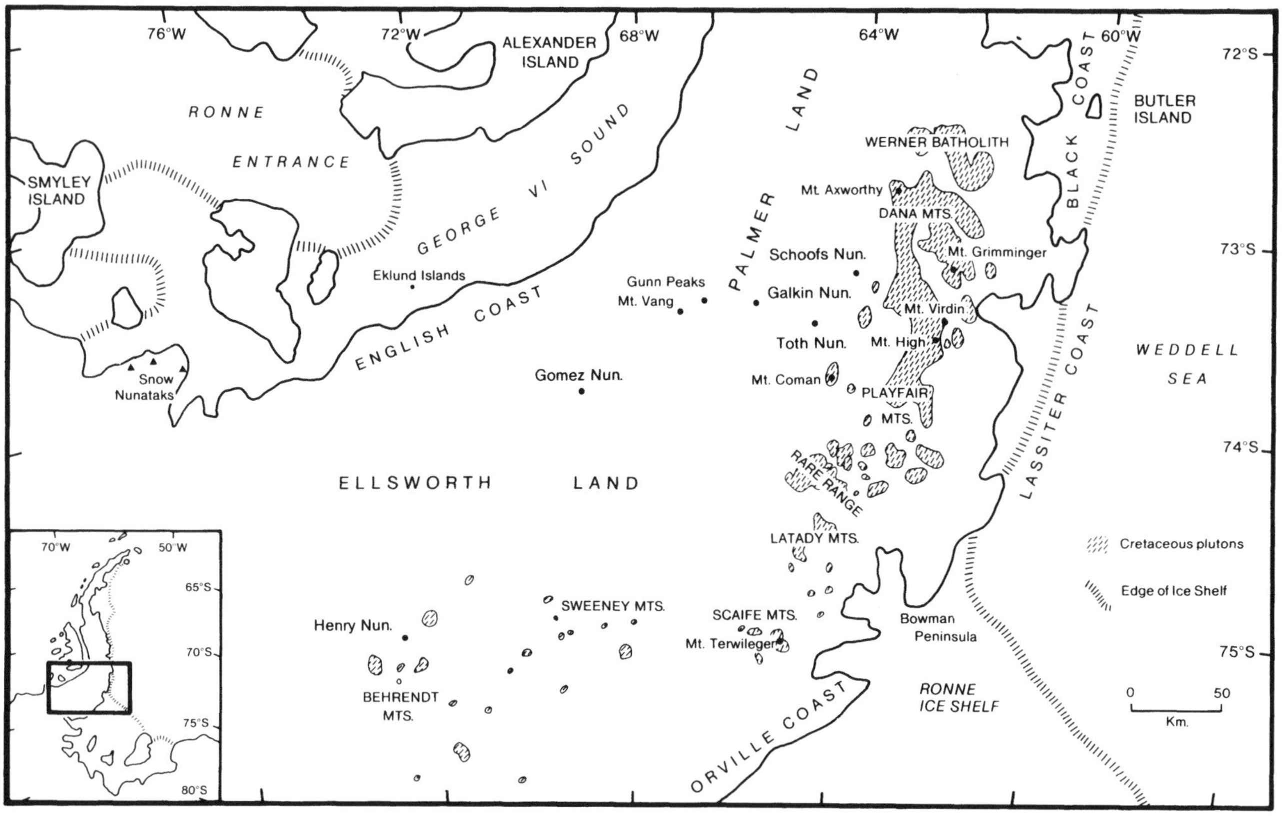

Fig. 2. Sketch map of southern Palmer Land and eastern Ellsworth Land, showing the distribution of Cretaceous plutons as outcrops.

Aeromagnetic anomaly map

The aeromagnetic anomaly map (Fig. 3) is characterized by a series of broad (> 20 km) positive anomalies, producing a magnetic grain which generally follows the arcuate trend of the Antarctic Peninsula. The 1985 IGRF (Barraclough, 1985) was removed from the data before production of the contour map. Also, due to the wide line spacing (~ 20 km), a 27 km cosine taper filter was applied to the data to remove the higher frequency anomalies, which could not be correlated between lines.

The three main positive anomalies are numbered 1, 2 and 3, in Fig. 3. Anomaly 1 is of variable width (50–100 km) and curves with, but more acutely than, the coastline of the peninsula. It trends NNE–SSW in the north of the area, where it is an obvious continuation of the WCMA. South of 73° S the anomaly widens, swings west to trend E–W along southern George VI Sound and the English Coast, and branches into two positive components separated by a belt of negative anomalies; the largest of these lie over the Eklund Islands. In the west of the area, the two positive components change trend again to lie almost NW–SE.

Anomaly 2 extends for over 150 km in southern Palmer Island, lying over Schoofs and Toth nunataks and Mount Coman. Anomaly 3 is a group of discrete anomalies that form a broadly linear belt trending parallel to and inland of the Black, Lassiter and Orville coasts. In the north of the area the anomalies are aligned approximately NNE–SSW, but in the south the trend is nearer E–W.

Magnetic anomalies and surface geology

Comparison of the regional anomaly map with the outcrop geology (Fig. 2) shows that large parts of the positive anomalies occur over areas which are either ice-covered or of unknown geology. Even after 'reduction-to-pole' (Blakely & Cox, 1972) there is no obvious correlation between magnetic anomalies and rock outcrops.

In some parts, such as at Mount Virdin, Mount High, Playfair Mountains, RARE Range and Mount Terwileger, positive anomalies lie over the outcrops of the Cretaceous plutons of the Lassiter Coast Intrusive Suite. However, over similar plutonic rocks in the northern Werner batholith, and at Mount Axworthy and Mount Grimminger, there is no corresponding positive anomaly. A similar situation can be seen in the south-west of the area, around Sweeney and Behrendt mountains.

In the vicinity of the English Coast, outcrops of predominantly silicic–andesitic volcanic rocks at Gunn Peaks, Galkin Nunatak, Mount Vang and Gomez Nunatak (B.C. Storey, pers. comm.) coincide with a positive magnetic anomaly.

Rock magnetization

Little is known about the magnetic properties of rocks in the study area, although samples of Late Cretaceous age from the Lassiter Coast (Kellogg & Reynolds, 1978) reveal normal remanent magnetization. Volume susceptibilities of rocks from eastern Palmer Land have been measured by McGibbon &

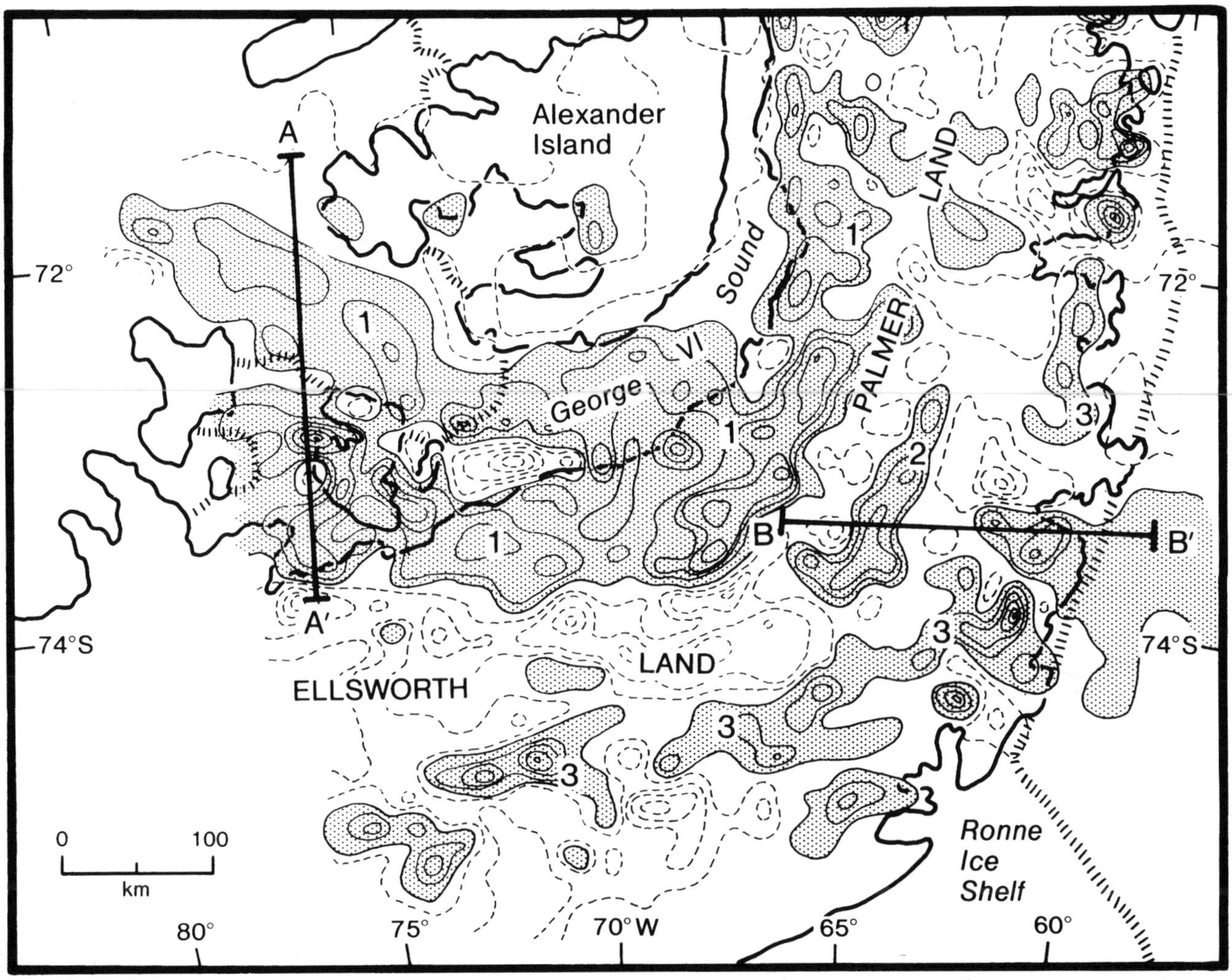

Fig. 3. Aeromagnetic anomaly map of southern Palmer Land and eastern Ellsworth Land; contour interval 100 nT. Negative contours shown as broken lines; positive anomalies are shaded. 1, 2 and 3 indicate the three main positive anomalies.

Garrett (1987). Applying their values to this area gives induced magnetizations for gabbros of about 2.5 A/m, and for silicic intrusions, < 0.5 A/m. Susceptibilities for calc-alkaline and alkaline volcanic rocks from other parts of the Antarctic Peninsula give induced magnetizations of 0.04–1.5 A/m and 0.04–5.5 A/m, respectively when applied here.

Interpretation

Depth to magnetic source

Estimates of the maximum depth to the top of source bodies were made using the automated technique described by Hartman, Tesky & Friedberg (1971). This method mainly gave estimates between 1 and 3 km, although deeper sources at 5–10 km were recognized. These deeper estimates were refined using two-dimensional power spectrum analysis (Spector & Grant, 1970) on the profiles.

Two-dimensional modelling

Two profiles, AA′ and BB′, have been modelled employing information from sub-ice topography, outcrop geology, rock magnetization and depth-to-source solutions (Fig. 4). Although AA′ and BB′ are marked on Fig. 3 it should be noted that the modelling is done on the profile data rather than the contour data. The total magnetization of the bodies is assumed coincident with the ambient field since palaeomagnetic studies in the area (Kellogg, 1980) have shown remanence inclinations to be similar to the Earth's present field.

Profile AA′ (Fig. 4*a*) extends south from Ronne Entrance on to the English Coast and, in the absence of outcrops, has been modelled with much reliance on depth-to-source solutions. The broad anomaly seen over Ronne Entrance gives depths between 8 and 13 km, whereas the higher frequency anomalies of the English Coast produce solutions at 1–3 km. Power spectrum analysis indicates this latter area is also underlain by magnetic bodies approximately 5 km below the surface. A body at a minimum depth of 8 km and a magnetization of 4.1 A/m is used to model the Ronne Entrance anomaly, whereas a body with a magnetization of 3.5 A/m is placed 5 km beneath the English Coast. The higher frequency anomalies are modelled by near surface bodies with lower magnetizations (0.6–1.3 A/m). Over Spaatz Island the 1400 nT anomaly is reproduced by a near-surface body with a high magnetization of 6.5 A/m.

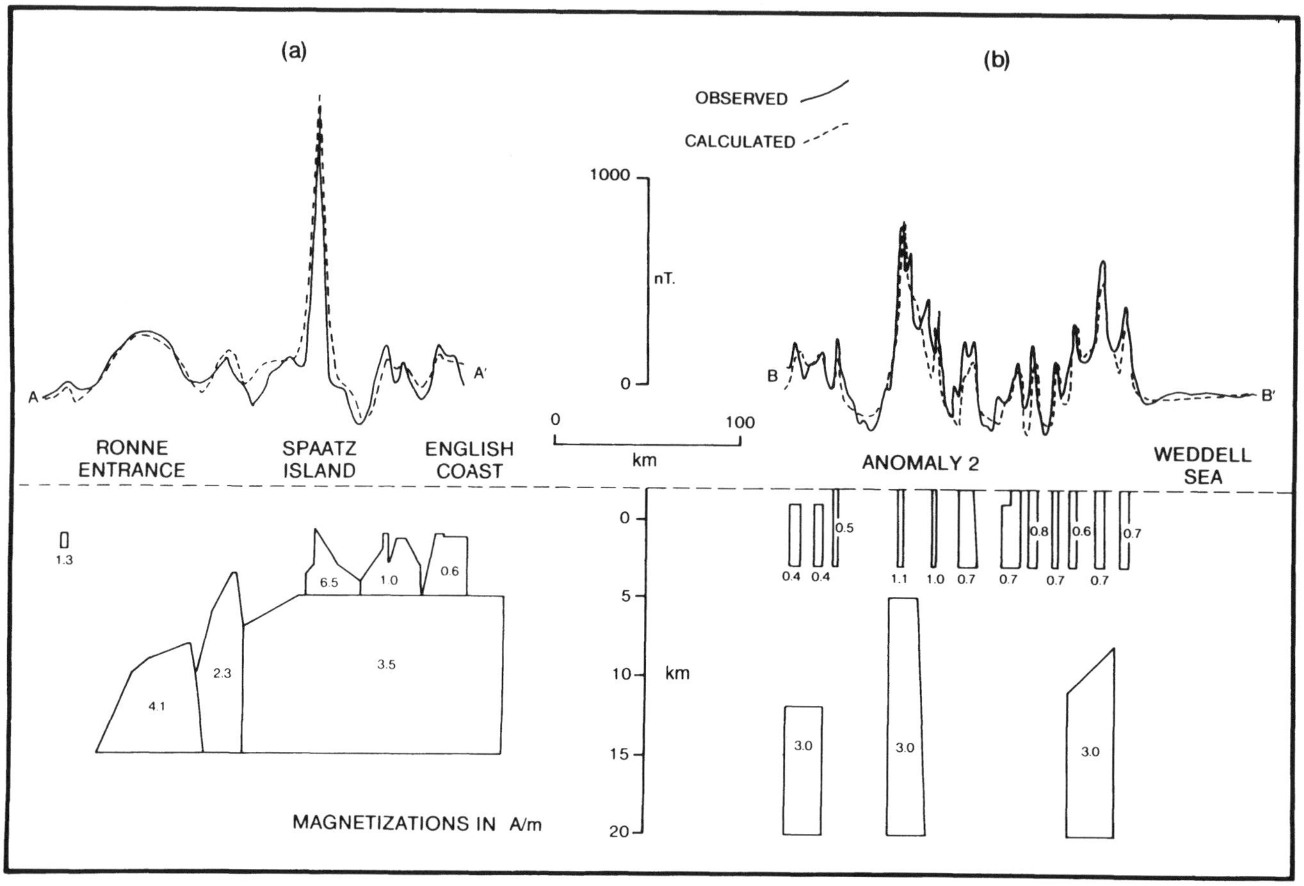

Fig. 4. Models for two profiles located in Fig. 3 as (*a*) AA′ and (*b*) BB′. The calculated profile is shown as a dashed line. Body magnetizations (in A/m) are indicated.

Profile BB′ (FIg. 4*b*) extends from the central peninsula to the Lassiter Coast, crossing the central linear anomaly (2; Fig. 3). Inspection of this and parallel profiles produces depth solutions of 2–3 km and > 5 km. These deeper solutions are taken to represent bodies responsible for the broader anomalies, and are modelled using magnetizations of 3.0 A/m. The high-frequency anomalies are modelled by near-surface bodies with susceptibilities typical of the Cretaceous Andean plutons which crop out along the profile (Behrendt, 1964).

Discussion

Mafic plutonism

The Cretaceous plutons of the Lassiter Coast Intrusive Suite, present in the east of the area, are not prominently reflected in the magnetic anomalies of regional extent. Given this lack of correlation, and the generally low susceptibilities of the plutons, it is suggested that the regional anomalies are related to more magnetic, and presumably more mafic material, buried at depths of between 5 and 12 km. However, although the models presented show this mafic material only at depth, it may reach the surface locally because gabbro crops out in the region (Vennum & Rowley, 1986). The Cretaceous plutons are thought to be responsible for the higher-frequency anomalies superimposed on the regional pattern.

The large volumes of mafic material inferred could have originated during subduction as the more mafic components of plutons observed at outcrop. Alternatively the mafic bodies could be related to arc extension, as is the case elsewhere in the Antarctic Peninsula (Storey & Garrett, 1985; Meneilly *et al.*, 1987). Whatever the origin, the results suggest mafic plutonism has been more important during the geological evolution of the area than is apparent at outcrop.

Aeromagnetic anomalies

The magnetic anomaly pattern is seen to follow the curvature of southern Antarctic Peninsula. The strike of the causal bodies is thus defined as essentially colinear with this curvature, indicating that the geological structure follows the arcuate trend. This points to the curvature having a tectonic origin, rather than being a purely glacial feature. The WCMA of northern Antarctic Peninsula is seen to continue along the west coast of southern Palmer Land to 73° S, where it swings westwards to form at least part of a broad two-limbed anomaly that extends E–W along the English Coast and southern George VI Sound. The exact relationship between the WCMA and this broad anomaly is not certain. Maslanyj (this volume, p. 527) suggests that the northern limb of this anomaly, seen in southern George VI Sound, could be separate from the WCMA and have its origins in an extensional event that moved Alexander Island north-west relative to the Antarctic Peninsula. Whatever the details, the WCMA clearly extends through the region, essentially following the arcuate coastline.

Conclusions

The anomaly pattern generally follows the arcuate trend of southern Antarctic Peninsula, thereby indicating that the trend is tectonic in origin. Modelling studies on the broader anomalies provide evidence for large magnetic bodies, suggesting mafic plutonism has been more important in the area than is apparent at the outcrop. The WCMA is seen to continue through the region, essentially following the arcuate coastline.

Acknowledgements

The authors would like to thank their colleagues at the British Antarctic Survey for their comments and advice during the preparation of this paper.

References

Barraclough, D.R. (1985). International geomagnetic reference field revision. *Nature, London*, **318(6044)**, 316.

Behrendt, J.C. (1964). Crustal geology of Ellsworth Land and the southern Antarctic Peninsula from gravity and magnetic anomalies. *Journal of Geophysical Research*, **69(10)**, 2047–62.

Blakely, R. & Cox, A. (1972). Identification of short polarity events by transforming marine magnetic profiles to the pole. *Journal of Geophysical Research*, **77(23)**, 4339–49.

Hartman, R.R., Tesky, D.J. & Friedberg, J.L. (1971). A system of rapid digital aeromagnetic interpretation. *Geophysics*, **36(5)**, 891–918.

Kellogg, K.S. (1980). Paleomagnetic evidence for oroclinal bending of the southern Antarctic Peninsula. *Geological Society of America Bulletin, Part I*, **91(7)**, 414–20.

Kellogg, K.S. & Reynolds, R.L. (1978). Paleomagnetic results from the Lassiter Coast, Antarctica, and a test for oroclinal bending of the Antarctic. *Journal of Geophysical Research*, **83(B5)**, 2293–9.

McGibbon, K.J. & Garrett, S.W. (1987). Magnetic anomalies on the Black Coast, Antarctic Peninsula. *British Antarctic Survey Bulletin*, **76**, 7–20.

Meneilly, A.M., Harrison, S.M., Piercy, B.A. & Storey, B.C. (1987). Structural evolution of the magmatic arc in northern Palmer Land, Antarctic Peninsula. In *Gondwana Six: Structure, Tectonics, and Geophysics*, ed. G.D. McKenzie, Geophysical Monograph 40, pp. 209–19. Washington, DC; American Geophysical Union.

Renner, R.G.B., Sturgeon, L.J.S. & Garrett, S.W. (1985). *Reconnaissance Gravity and Aeromagnetic Surveys of the Antarctic Peninsula*. Cambridge; British Antarctic Survey Scientific Reports, No. 110, 50 pp.

Rowley, P.D., Kellogg, K.S. & Vennum, W.R. (1985). Geologic studies in the English Coast, eastern Ellsworth Land, Antarctica. *Antarctic Journal of the United States*, **20(5)**, 34–6.

Rowley, P.D., Vennum, W.R., Kellogg, K.S., Laudon, T.S., Carrara, P.E., Boyles, J.M. & Thomson, M.R.A. (1983). Geology and plate tectonic setting of the Orville Coast and eastern Ellsworth Land, Antarctica. In *Antarctic Earth Science*, ed. R.L. Oliver, P.R. James & J.B. Jago, pp. 245–50. Canberra; Australian Academy of Science and Cambridge; Cambridge University Press.

Rowley, P.D. & Williams, P.L. (1982). Geology of the northern Lassiter Coast and southern Black Coast, Antarctic Peninsula. In *Antarctic Geoscience*, ed. C. Craddock, pp. 339–48. Madison; University of Wisconsin Press.

Spector, A. & Grant, F.S. (1970). Statistical models for interpreting aeromagnetic data. *Geophysics*, **35(2)**, 293–302.

Storey, B.C. & Garrett, S.W. (1985). Crustal growth of the Antarctic Peninsula by accretion, magmatism and extension. *Geological Magazine*, **122(1)**, 5–14.

Vennum, W.R. & Rowley, P.D. (1986). Reconnaissance geochemistry of the Lassiter Coast Intrusive Suite, southern Antarctic Peninsula. *Geological Society of American Bulletin*, **97(12)**, 1521–33.

Stratigraphy, provenance and tectonic setting of (?)Late Palaeozoic–Triassic sedimentary sequences in northern Graham Land and South Scotia Ridge

J.L. SMELLIE

British Antarctic Survey, Natural Environment Research Council, High Cross, Madingley Road, Cambridge CB3 0ET, UK

Abstract

The (?)Permo-Triassic Trinity Peninsula Group, Miers Bluff Formation and Greywacke–Shale Formation are amongst the oldest rocks known in northern Graham Land and South Scotia Ridge. They contain at least three distinctive sandstone petrofacies. Stratigraphical, structural and metamorphic considerations suggest the possibility of a crude regional stratigraphy. The main phase of deformation comprises gently plunging, large-scale isoclines with NW-dipping axial surfaces and south-easterly vergence. Regional metamorphism ranged from prehnite–pumpellyite to greenschist-facies, and was probably of medium-pressure type. Four textural zones are recognized, with a broad distribution pattern consistent with strong structural–stratigraphical control. The tectonic setting of these sequences is equivocal. The majority of the detrital modes indicate derivation from a major continental margin magmatic arc, although its location and active–inactive state remain unclear. The dominant SE-verging (i.e. 'arcward') large-scale structures and the presence of interbedded and intrusive volcanic rocks with alkaline compositions are clearly anomalous in any fore-arc model but can be accommodated in a sedimentary basin located behind an arc. There is still no unambiguous evidence for the timing of the initiation of subduction in this part of the Pacific margin of Gondwana.

Introduction

Widespread low-grade metasedimentary sequences are amongst the oldest rocks exposed in Graham Land (Trinity Peninsula Group: TPG), South Shetland (Miers Bluff Formation: MBF) and South Orkney Islands (Greywacke–Shale Formation: GSF; Fig. 1). The sediments were pervasively deformed during the Late Triassic–Early Jurassic Peninsula Orogeny (formerly Gondwanian Orogeny: Storey, Thomson & Meneilly, 1987). They form the local basement to the largely undeformed Mesozoic–Tertiary Antarctic Peninsula magmatic arc and are comparable with numerous largely Mesozoic circum-Pacific greywacke suites that are commonly interpreted as accreted subduction complexes (e.g. Forsythe & Allen, 1980; McKinnon, 1983).

The tectonic position of the TPG, MBF and GSF is contentious although most workers have favoured some association with an active margin, probably in a fore-arc setting (e.g. Smellie, 1981, 1987; Dalziel, 1984; Storey & Garrett, 1985). The sequences occupy a unique position in Antarctic Peninsula stratigraphy in providing the sole link between Palaeozoic and Mesozoic tectonics. The interpretation of their stratigraphy and tectonic setting is therefore critical to our understanding of the early history of arc development in the Antarctic Peninsula region.

Stratigraphy

The sparse age evidence supports a general Triassic age for deposition of much of the TPG and related sequences, although older (Permian?) sediments may occur in the structurally deeper levels of the sequence exposed in Graham Land (M.R.A. Thomson, 1975; Dalziel *et al.*, 1981; Hyden & Tanner, 1981; Pankhurst, 1983). Although Pankhurst (1983) attributed a Rb–Sr age of 281 ± 16 Ma to Permo-Carboniferous diagenesis of TPG mudstones from Hope Bay, the presence of abundant detrital biotite in TPG sediments (Smellie, 1987) suggests that the colinear 'errorchron' array could be a mixing line reflecting the provenance of unequilibrated radiogenic-biotite crystals. Deformation and metamorphism are also poorly constrained but probably reached a peak during latest Triassic–Early Jurassic times (Peninsula Orogeny).

In northern Graham Land, where it is least metamorphosed and deformed, the TPG has been divided stratigraphically into several lithostratigraphical units (Halpern, 1965; Alarcón *et al.*, 1976; Hyden & Tanner, 1981; Smellie, 1987). The View

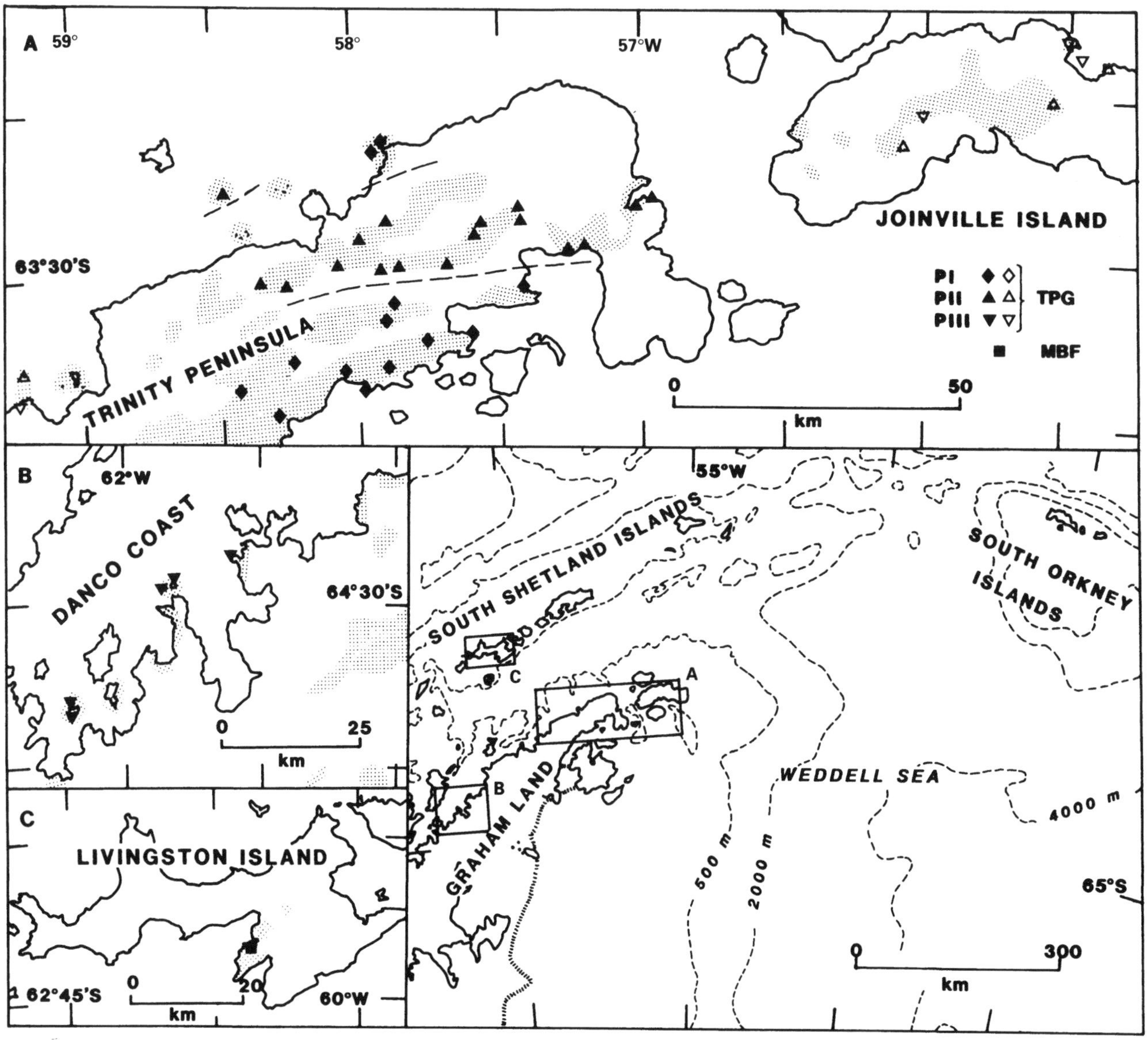

Fig. 1. Sketch maps showing TPG and MBF sample localities and petrofacies distributions in (*a* & *b*) Graham Land and (*c*) South Shetland Islands, and their location. For clarity, localities with multiple samples are shown by one symbol only. Sandstones from type areas are designated by solid symbols, other sandstones by open symbols. Stippled areas are TPG, MBF outcrops. Abbreviations are explained in the text.

Point and Legoupil formations (VPF, LF; Petrofacies I of Smellie, 1987) are quartzose arkosic-arenite sequences, in beds usually overturned and with well developed cleavage(s). They contain the only known outcrops of polymict conglomerate, chert and greenschist. Isolated outcrops of basic volcanic rocks interbedded with TPG strata were formerly thought to be restricted to the VPF but are now known to occur sparsely elsewhere. The Hope Bay Formation (HBF) and an unnamed unit (Petrofacies II of Smellie, 1987) are predominantly upright sequences of feldspathic-arkosic-arenites with a poorly developed cleavage. In the Danco Coast, the poorly known Formación Bahía Charlotte has been divided lithologically into two units (Alarcón *et al.*, 1976): Facie Bahıa Wilhelmina, consisting of fine metagreywackes, quartz breccias and pelites; and Facie Isla Gándara (sic), consisting of folded and cleaved fine sandstones and mudstones. The MBF is composed mainly of arkosic arenites and wackes, (Smellie *et al.*, 1984). Published data for the GSF are few but the descriptions by J.W. Thomson (1973) indicate that at least part of the succession is formed of quartzose arkosic wackes similar to the TPG (Petrofacies I). Deposition of the TPG may have occurred in a fan environment, in channels, levées, interchannel areas and suprafan lobes of the inner and outer mid-fan (Hyden & Tanner, 1981; Smellie, unpublished data).

Sandstone detrital modes

Detrital modes have been determined on 58 TPG and 5 MBF sandstones; 12 GSF sandstones analysed by J.W. Thomson (1973) are also included. Except for the latter, the Gazzi–Dickinson point-counting method was used (details in Smellie, 1987). The principal clasts are crystals of quartz and feldspar (mainly plagioclase, but with minor orthoclase, microcline and perthite), with intermediate–acidic volcanic clasts being the most abundant lithic fragments (Fig. 2). Accessory framework minerals include biotite, muscovite, hornblende,

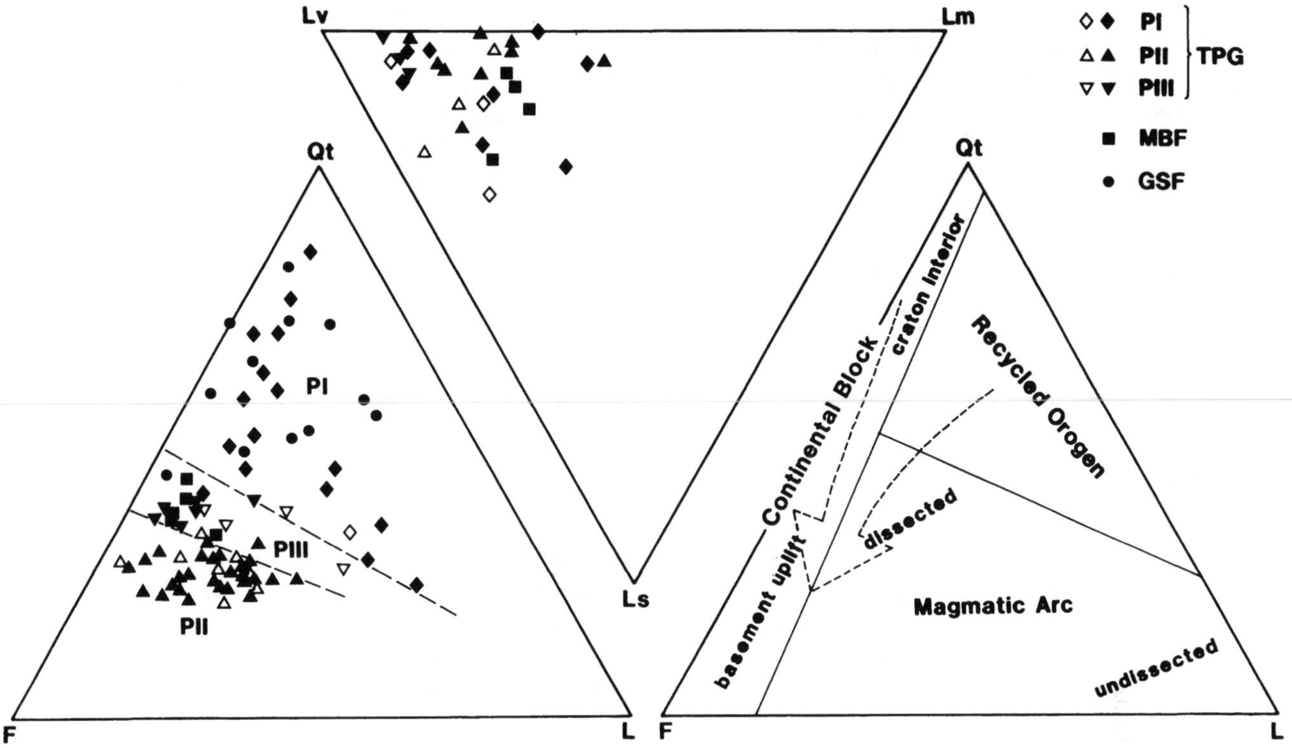

Fig. 2. Summary plots of TPG, MBF and GSF sandstone detrital modes, and suggested evolutionary trend (large dashed arrow) superimposed on fields of provenance tectonic settings (after Dickinson, 1985). (Q, total mono- and polycrystalline quartz grains; F, total feldspar grains; L, total unstable lithic grains; Lv, volcanic lithic grains; Ls, sedimentary lithic grains; Lm, metamorphic lithic grains. For other abbreviations and symbols see text and Fig. 1).

epidote, sphene (or rutile), opaque oxide, garnet, zircon, allanite, tourmaline and apatite.

Smellie (1987) recognized two major TPG petrofacies. They differ in detrital modes principally in significantly higher quartz content, quartz/feldspar and plagioclase/total feldspar ratios, and possibly a lower proportion of volcanic clasts in Petrofacies I (PI) compared with Petrofacies II (PII). The new data support these distinctions but they also clearly indicate the presence of a third major, geographically widespread sandstone suite (Petrofacies III (PIII)), with modal characteristics intermediate between PI and PII and including all five MBF sandstones analysed (Figs. 1 & 2). Despite the very large spread of sandstone compositions in the GSF, there is a clear affinity with PI sandstones.

The data for PIII are too few and from localities too widely scattered to indicate any significant variation within the petrofacies. Variation within PII is minor, with the exception of a regular south-westward decrease in the proportion of volcanic clasts within PIIa (the unnamed unit). PI displays striking cross-strike variation, which is resolvable into two principal components: variable quartz/feldspar ratios, and proportion of lithic clasts (cf. Smellie, 1987; Fig. 2). The effect of increasing the proportion of lithic clasts is simply a net dilution of the other clast types, whereas variations in Q/F ratios can be linked to apparent stratigraphical position: despite poor stratigraphical control, sandstones with lower Q/F values are restricted to the inferred upper parts of the petrofacies based on consistent younging directions and a homoclinal sequence. Although the LF is statistically separable from the VPF (Smellie, 1987), the differences are slight and it may be more realistic to regard the two formations simply as geographically separated outcrops of the same sandstone suite.

No clast types are exclusive to any petrofacies and the data set effectively demonstrates a continuum of modal sandstone compositions. The simplest interpretation of the data is of progressive, time-dependent changes in the composition of the detritus shed by an evolving provenance (Fig. 2; cf. Dickinson, Lawton & Inman, 1986). Regular cross-strike variations in sandstone compositions, apparently linked to stratigraphical position, occur within the LF and VPF, some well-sampled outcrops in the Danco Coast and possibly Joinville Island. The compositional trends observed, together with the lower degree of recrystallization, lower metamorphic grade and less intense deformation (see below) of PII and parts of PIII compared with PI, are consistent with an inferred order of stratigraphical superimposition: PII younger than PIII younger than PI.

Provenance

The TPG, MBF and GSF shared a common provenance composed principally of plutonic and high-rank metamorphic detritus together with subordinate acid–intermediate volcanic clasts. A continental provenance is suggested, composed mainly of unroofed batholiths and their metamorphic envelope, capped locally by isolated volcanoes (Smellie, 1987). A magmatic arc provenance is clearly indicated in the preponderance of volcanic clasts in the lithic populations (Fig. 2). This is well illustrated in QFL plots, in which data for PII, PIII and some of PI define a broad compositional field overlapping those of sandstones derived from dissected magmatic arcs and continental blocks (Fig. 2). However, data for PI (VPF, GSF and some LF) are ambiguous. The lithic clast population is largely indistinguishable from PII and PIII, but QFL proportions suggest recycled orogen and continental block provenances (Fig. 2). These conflicting characteristics can be accommodated in a provenance composed of a magmatic arc containing large outcrops of continental basement rocks (Carboniferous?: Hervé *et al.*, this volume, p. 429) exposed in uplifted horsts and/or within a non-volcanic frontal arc (cf. Smellie, 1987), although there is no direct evidence that the arc was active during deposition of PI.

Structure

There is a paucity of detailed structural data for many outcrops, particularly within the TPG. Moreover, because of wide geographical separation, and effects of Mesozoic continental fragmentation, it cannot be demonstrated that deformation was contemporaneous throughout the region, although the remarkable similarities in structural style suggest that it may have been so.

Strata nearly everywhere are moderate to steeply dipping and have been subjected to two principal deformations (Dalziel, 1984). The earliest recognizable structures (FM, formed during the main phase of deformation, DM) are gently-plunging asymmetrical to isoclinal mesoscopic folds and axial planar cleavage. They define a set of isoclines of nappe proportions which are locally recumbent. A second set of mesoscopic folds (FL) is also developed in most areas. These are typically more open than FM, with a less penetrative axial planar cleavage and variably plunging hinges striking approximately parallel to the first set of folds. A post-FM, brittle phase of deformation occurs locally in the VPF, and the GSF contains small, isolated outcrops of muddy mélange (Hyden & Tanner, 1981; Dalziel, 1984).

The TPG and MBF are dominated by NE-trending fold axes (FM and FL) and steep to gently NW–W-dipping FM axial surfaces and cleavage, with a consistent south-eastward sense of vergence. Structural trends are more heterogeneous in the GSF, and four fault-bounded structural domains have been identified.

Metamorphism and textural zonation

The rocks range from almost unaltered and unrecrystallized greywackes and mudstones to strongly foliated, completley recrystallized psammitic schists, pelites and greenschists. In the TPG, these changes occur progressively from north-east to south-west. The widespread occurrence of prehnite in veins and replacing plagioclase in chloritic greywackes, and coexisting prehnite and pumpellyite in associated metabasites, suggests that prehnite–pumpellyite facies conditions were reached in north-eastern areas (Table 1; Fig. 3). The striking restriction of prehnite to PII, probably reflects compositional control by the more feldspar-rich greywackes. In the Torlesse terrane (New Zealand), detrital K-feldspar breaks down low in the pumpellyite–actinolite-facies, releasing potassium possibly to form new white mica and stilpnomelane (Bishop, 1972). In the TPG, K-feldspar breaks down by replacement by albite in the prehnite–pumpellyite-facies; coexisting stilpnomelane and K-feldspar have not been observed, with stilpnomelane restricted to apparently higher-grade metasediments, suggesting a possible reaction relationship. The transition to pumpellyite–actinolite-facies is probably recorded by greenschists bearing tremolite–actinolite and Fe-epidote. It is unclear where the transition to full greenschist-facies takes place, but it probably occurs in south-western outcrops in which clinozoisite appears to be the stable epidote, in contrast to Fe-rich pistacite in north-eastern areas.

Hyden & Tanner (1981) suggested that the TPG was affected by medium to high-pressure (prehnite–pumpellyite-facies) metamorphism. This is supported by the widespread distribution of stilpnomelane, prehnite and pumpellyite. However, the observed transition from prehnite–pumpellyite to greenschist rather than blueschist-facies may be significant. Unless all traces of a blueschist-facies have been erased, the observed facies series suggests a trajectory in *P–T* space corresponding to a medium-pressure gradient.

The relationship between textural zonation (*sensu* Bishop, 1972) and metamorphism is uncertain (Table 1) and the sample density is insufficient to establish the precise positions of isograds and isotects. However, the observed south-westward increase in deformation intensity and metamorphic grade, and an interpreted pattern of interdigitating (NE-plunging(?)) 'textural culminations and troughs' (Fig. 3) are consistent with structural interpretations and suggest that deeper erosion levels are exposed in the south-west. The data also suggest that at least some deformation continued after the main prograde metamorphism.

Tectonic setting

Previous models for the tectonic setting of the TPG, MBF and GSF have argued that the strata accumulated on the Pacific side of a contemporaneous magmatic arc, although a passive margin setting for early (quartz-rich) TPG sediments was not precluded (e.g. Smellie, 1981, 1987; Askin & Elliot, 1982; Dalziel, 1984; Smellie *et al.*, 1984; Storey & Garrett, 1985). However, the precise location of the contemporaneous arc and unambiguous evidence for its activity have thus far not been forthcoming.

Several key factors support the interpretation of the TPG, MBF and GSF as tectonically-fragmented remnants of an originally continuous sequence deposited within a major sedimentary basin during (?)Permo-Triassic times: (1) similarity of clast types in lithology and (where known) isotopic ages, together with the mutually gradational modal characteristics and inferred stratigraphy of the major petrofacies, suggest a simple, comformable series of sequences linked to a common, evolving, continental provenance; (2) the considerable similarities in structural style and metamorphism suggest a shared post-depositional history prior to continental fragmentation.

The detrital modes clearly indicate the significant involvement of a major continental margin magmatic arc. The evi-

Table 1. *Summary of metamorphic and textural zonation in the Trinity Peninsula Group*

Stratigraphy		PII PIII PI			
Textural zones		1 Essentially unrecrystallized; grain boundaries distinct and unmodified	2*a* Slightly recrystallized; grains show marginal recrystallization; clastic textures but some samples show faint tectonic foliation	2*b* Moderate–extensive recrystallization; relict clastic textures modified by weak–strong tectonic foliation; mortar texture in most modified sandstones	3 Completely recrystallized strongly foliated schists
Metamorphic grade		prehnite–pumpellyite-facies		pumpellyite–actinolite-facies	greenschist-facies
Metagreywackes	Quartz				
	Calcic-plagioclase				
	Albite				
	K-feldspar				
	Epidote				
	Sericite				
	Muscovite				
	Chlorite	uniaxial	biaxial		
	Calcite				
	Stilpnomelane				
	Actinolite		?		
	Prehnite				
Metabasites	Pumpellyite				
	Chlorite				
	Epidote				
	Albite				
	Actinolite				
	Stilpnomelane				
	Sphene				

dence is unambiguous for PII and parts of PIII, but is conflicting for PI, although the high proportion of volcanic clasts in the lithic population is consistent with an arc provenance. However, although detrital modes accurately reflect the provenance of a sandstone suite (e.g. Dickinson, 1980), they need not fully constrain the tectonic setting of the depositional basin (cf. Valloni & Mezzadri, 1984; Potter, 1986). A variety of tectonic settings is possible for modern basins containing feldspatholithic and litho-volcanic sands similar to PII and PIII. Although predominantly associated with basins proximal to magmatic arcs, they are also occasionally found in strike-slip basins, adjacent to collision orogens, and (unusually) rifted continental margins. Similarly, modern quartzolithic sands (cf. PI) are commonly found in a wide variety of basin types located at both active and passive margins, although the latter are perhaps the commonest location. However, the high volcanic/total lithic clast ratios characteristic of PI are anomalous compared with modern passive margin sands and, together with the admittedly sparse occurrences of likely airfall interbeds in the LF and VPF (M.R.A. Thomson, 1975; Hyden & Tanner, 1981), indicate an association with an active magmatic arc.

The classic structural association expected in accretionary complexes consists of landward-dipping thrust planes and

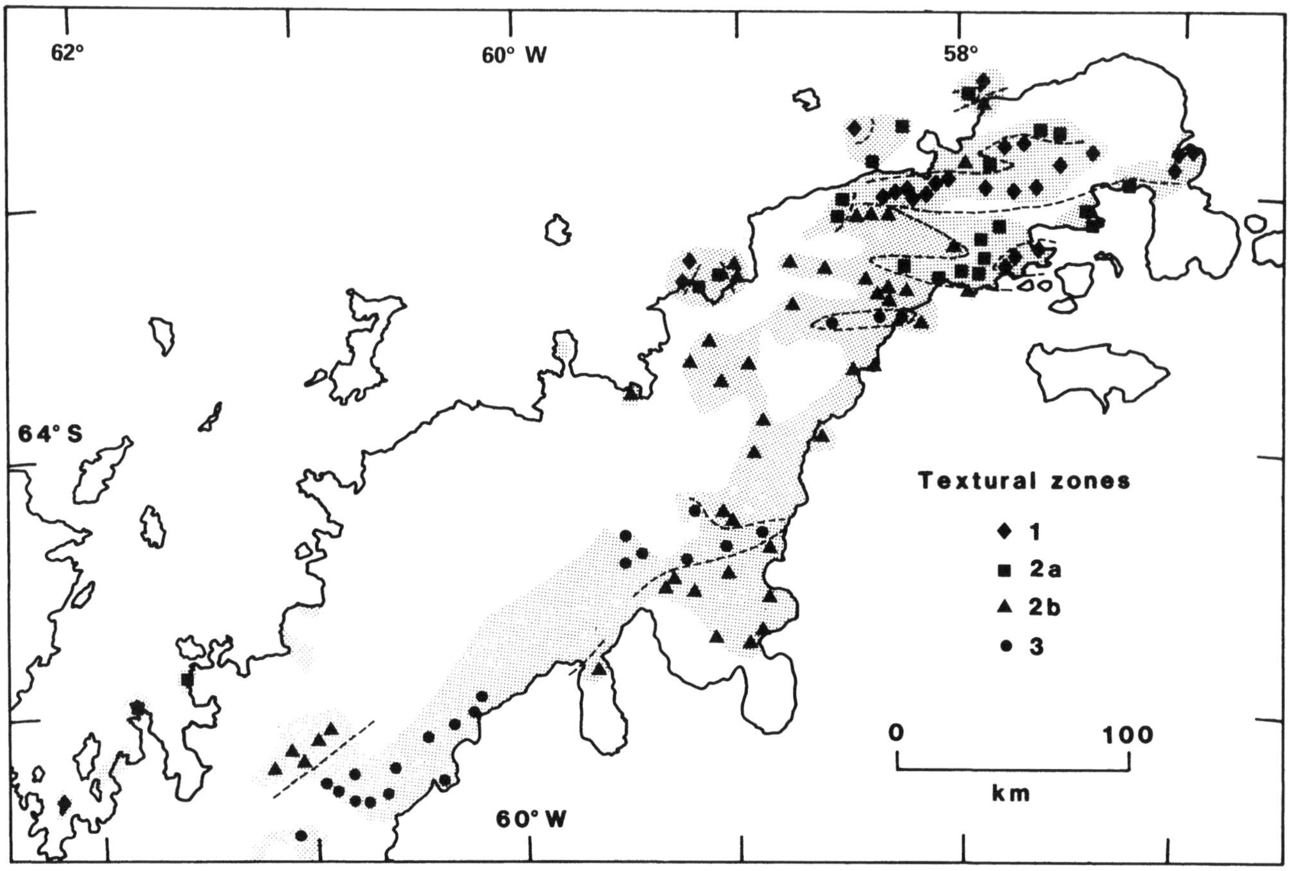

Fig. 3. Sketch map of northern Graham Land, showing TPG outcrops (stippled), sample distribution and textural zone classification (cf. Table 1).

seaward-verging structures (e.g. Seely, 1977). Dalziel (1984) acknowledged the difficulty in reconciling arcward-verging structures in the MBF by comparing them with backthrusted strata at the rear of accretionary prisms. Indeed, arcward-verging structures are not uncommon in accretionary prisms (e.g. Seely, 1977; Platt, 1986), although the predominant structures verge towards the trench. Within the MBF and TPG, at least, south-east-verging, large-scale isoclines are the characteristic structural style. This either renders unlikely any fore-arc model with a magmatic arc situated to the south-east (present coordinates), or it indicates that the polarity of subduction was to the north-west (for which there is no supporting evidence), with an arc in that direction.

Autochthonous magmatism in the TPG is represented by sparse but widespread outcrops of interbedded and intruded volcanic rocks with alkaline compositions (cf. Hyden & Tanner, 1981; Smellie, unpublished geochemical data), probably emplaced in a tensional regime in common with other occurrences worldwide. However, fore-arc volcanism is typically calc-alkaline or (less commonly) MORB-like in composition (e.g. Hussong & Uyeda, 1982; Miyake, 1985). Extension can occur in mature accretionary complexes (e.g. Platt, 1986), but is not associated with alkaline volcanism. Although regional extension could be invoked to explain the eruption of the TPG alkaline magmas, it can also be explained as a local phenomenon induced beneath a thickening sediment pile by crustal loading. Metamorphic considerations suggest that the basin was able to accommodate a minimum thickness of 9–10 km of sediment, implying substantial crustal subsidence, which was probably achieved by lateral spreading and/or normal faulting.

In summary, it is suggested that the TPG, MBF and GSF shared a common, evolving, continental margin, magmatic arc-like provenance. There is no clear evidence, in the data presently available, that the arc was active, but it probably was for at least PII and PIII. The timing of the initiation of subduction in this region remains uncertain. The weight of the evidence, in particular 'arcward'-verging large-scale structures and autochthonous alkaline volcanism, argues against a fore-arc depositional setting. However, the possibility of structural shuffling by strike-slip faults cannot be precluded and may yet render meaningless any currently envisaged reconstructions.

References

Alarcón, B., Ambrus, J., Olcay, L. & Vieira, C. (1976). Geologia del Estrecho de Gerlache entre los parallelos 64 y 65 lat. sur, Antártica chilena. *Serie Científica del Instituto Antártico Chileno*, **4**, 7–51.

Askin, R.A. & Elliot, D.H. (1982). Geologic implications of recycled Permian and Triassic palynomorphs in Tertiary rocks of Seymour Island, Antarctic Peninsula. *Geology*, **10(10)**, 547–51.

Bishop, D.G. (1972). Progressive metamorphism from prehnite–pumpellyite to greenschist-facies in the Dansey Pass area, Otago, New Zealand. *Geological Society of America Bulletin*, **83(11)**, 3177–97.

Dalziel, I.W.D. (1984). *Tectonic evolution of a Fore-arc Terrane, Southern Scotia Ridge, Antarctica.* Washington, DC; Geological Society of America, Special Paper, **200**, 32 pp.

Dalziel, I.W.D., Elliot, D.H., Jones, D.L., Thomson, J.W., Thomson, M.R.A., Wells, N.A. & Zinsmeister, W.J. (1981). The geological significance of some Triassic microfossils from the South Orkney Islands, Scotia Ridge. *Geological Magazine*, **118(1)**, 15–25.

Dickinson, W.R. (1980). Interpreting provenance relations from detrital modes of sandstones. In *Provenance of Arenites*, ed. G.G. Zuffa, pp. 333–61. Dordrecht; D. Reidel Publishing Company.

Dickinson, W.R., Lawton, T.F. & Inman, K.F. (1986). Sandstone detrital modes, central Utah foreland region: stratigraphic record of Cretaceous–Paleogene tectonic evolution. *Journal of Sedimentary Petrology*, **56(2)**, 276–93.

Forsythe, R.D. & Allen, R.B. (1980). The basement rocks of Peninsula Staines, Region XII, Province of Ultima Esperanza, Chile. *Revista Geológica de Chile*, **10**, 3–15.

Halpern, M. (1965). The geology of the Bernardo O'Higgins area, northwest Antarctic Peninsula. In *Geology and Paleontology of the Antarctic*, Antarctic Research Series Vol. 6, ed. J.B. Hadley, pp. 177–209. Washington DC; American Geophysical Union.

Hussong, D.M. & Uyeda, S. (1982). Tectonic processes and the history of the Marianas Arc: a synthesis of the results of the Deep Sea Drilling Project leg 60. In *Initial Reports of the Deep Sea Drilling Project, 60*, 909–29.

Hyden, G. & Tanner, P.W.G. (1981). Late Palaeozoic–early Mesozoic fore-arc basin sedimentary rocks at the Pacific margin in Western Antarctica. *Geologische Rundschau*, **70(2)**, 529–41.

McKinnon, T.C. (1983). Origin of the Torlesse terrane and coeval rocks, South Island, New Zealand. *Geological Society of America Bulletin*, **94(8)**, 967–85.

Miyake, Y. (1985). MORB-like tholeiites formed within the Miocene fore-arc basin, S.W. Japan. *Lithos*, **18**, 23–34.

Pankhurst, R.J. (1983). Rb–Sr constraints on the ages of basement rocks of the Antarctic Peninsula. In *Antarctic Earth Science*, ed. R.L. Oliver, P.R. James & J.B. Jago, pp. 367–71. Canberra; Australian Academy of Science and Cambridge; Cambridge University Press.

Platt, J.P. (1986). Dynamics of orogenic wedges and the uplift of high-pressure metamorphic rocks. *Geological Society of America Bulletin*, **97(9)**, 1037–53.

Potter, P.E. (1986). South America and a few grains of sand: Part 1 – beach sands. *Journal of Geology*, **94(3)**, 301–15.

Seely, D.R. (1977). The significance of landward vergence and oblique structural trends on trench inner slopes. In *Island Arcs, Deep Sea Trenches and Back-arc Basins*, ed. M. Talwani & W.C. Pitman, pp. 187–98. Washington DC; American Geophysical Union.

Smellie, J.L. (1981). A complete arc–trench system recognized in Gondwana sequences of the Antarctic Peninsula region. *Geological Magazine*, **118(2)**, 139–59.

Smellie, J.L. (1987). Sandstone detrital modes and basinal setting of the Trinity Peninsula Group, northern Graham Land, Antarctic Peninsula: a preliminary survey. In *Gondwana Six: Structure, Tectonics, and Geophysics*. Geophysical Monograph 40, ed. G.D. Mackenzie, pp. 199–207. Washington DC; American Geophysical Union.

Smellie, J.L., Pankhurst, R.J., Thomson, M.R.A. & Davies, R.E.S. (1984). *The Geology of the South Shetland Islands: VI. Stratigraphy, Geochemistry and Evolution*. Cambridge; British Antarctic Survey Scientific Reports, No. 87, 85 pp.

Storey, B.C. & Garrett, S.W. (1985). Crustal growth of the Antarctic Peninsula by accretion, magmatism and extension. *Geological Magazine*, **122(1)**, 5–14.

Storey, B.C., Thomson, M.R.A. & Meneilly, A.W. (1987). The Gondwanian orogeny within the Antarctic Peninsula: a discussion. In *Gondwana Six: Structure, Tectonics and Geophysics*. Geophysical Monograph 40, ed. G.D. Mackenzie, pp. 191–8. Washington DC; American Geophysical Union.

Thomson, J.W. (1973). The geology of Powell, Christoffersen and Michelsen Isands, South Orkney Islands. *British Antarctic Survey Bulletin*, **33/34**, 137–68.

Thomson, M.R.A. (1975). New palaeontological and lithological observations on the Legoupil Formation, north-west Antarctic Peninsula. *British Antarctic Survey Bulletin*, **41/42**, 169–85.

Valloni, R. & Mezzadri, G. (1984). Compositional suites of terrigenous deep-sea sands of the present continental margins. *Sedimentology*, **31(3)**, 353–64.

Subduction-complex rocks on Diego Ramirez Islands, Chile

T.J. WILSON[1], A.M. GRUNOW[2], R.E. HANSON[1] & K.R. SCHMITT[2]

1 Byrd Polar Research Center and Department of Geology & Mineralogy, The Ohio State University, 125 South Oval Mall, Columbus, Ohio 43210 USA
2 Lamont–Doherty Geological Observatory, Palisades, NY 10964 USA

Abstract

Recent mapping of the Diego Ramirez Islands, southernmost Chile, has shown that the islands consist of tectonic *mélange* structurally interleaved with portions of a dismembered and metamorphosed basaltic seamount. Crossite is present in the metabasalts, indicating comparatively high-*P*–low-*T* metamorphism. The rocks show evidence of a prolonged, polyphase structural history, including several generations of ductile and brittle structures that are interpreted to be the products of deformation, tectonic mixing, and *mélange* formation at different levels in a subduction complex. The Diego Ramirez Islands thus form part of the fore-arc accretionary complex developed along the Pacific margin of Gondwana, and provide a link between the Palaeozoic–Mesozoic subduction-complex rocks exposed along the coast of central and southern Chile, and the Scotia metamorphic complex of the Antarctic Peninsula region.

Introduction

The Diego Ramirez Islands, southernmost Chile, are located 100 km south-west of Cape Horn (56°30'S, 68°43'W), where the Shackleton Fracture Zone intersects the continental shelf edge of South America (Fig. 1*a*). The islands represent the southernmost exposures of fore-arc rocks along the western convergent margin of South America. Brief descriptions of the geology of the islands have been given by de la Rue (1959) and Dalziel (1983). Results of recent studies by Chilean geologists are given in Davidson *et al.* (1989). The present contribution presents the results of fieldwork carried out during two separate cruises to the islands in 1985 and 1986 on board US Antarctic research vessel R/V *Polar Duke*. It includes the first detailed geological map of the islands (Fig. 1*b*).

Rock units

The Diego Ramirez Islands consist of two principal rock units; metabasalt and tectonic *mélange* (Fig. 1*b*). Metabasalt occurs in large, fault-bounded masses and consists predominantly of pillow breccia, with subordinate amounts of closely packed pillow lava. These two facies are distinguished in Fig. 1*b*, with areas shown as undifferentiated metabasalt being poorly exposed or too highly deformed to determine the original character. The pillow breccia contains whole or broken pillows set within a hyaloclastite matrix, and in places shows a gross stratification defined by changes in clast size and content between layers, or by thin tuff horizons separating individual thick, massive breccia units. The pillows characteristically show high vesicularity ($\geq$ 50% in some pillows), which indicates extrusion in a shallow-marine environment and points to a seamount as the original site of eruption. The large volume of pillow breccia relative to pillow lava (Fig. 1*b*) makes the metabasalts similar to other described examples of ancient seamounts (e.g., Staudigel & Schmincke, 1984) and contrasts with typical ocean-floor basalts (which are usually dominated by closely packed pillow lavas and sheet flows). Additional evidence for a seamount setting is provided by limited immobile trace element data for the metabasalts given by Davidson *et al.* (1989), which indicate affinities to within plate basalts.

Tectonic *mélange* is divided into three types in Fig. 1*b*. Phyllite–metabasalt *mélange* is most abundant and consists of phacoidal masses of phyllite and metabasalt in a cataclastically disrupted matrix. Numerous thin, disrupted horizons of basaltic tuff are interlayered with the phyllites. Areas mapped as phyllite *mélange* contain tectonic inclusions of ribbon chert, greywacke, metabasalt and minor limestone in a phyllitic matrix. The inclusions occur as widely dispersed, angular–subrounded blocks from 0.50 to 10 m in width. Chert–metabasalt *mélange* is present in a few small areas, and consists of reddish ribbon-chert tectonically intermixed with metabasalt.

Structural evolution and metamorphism

Contacts between horizons of basaltic tuff and phyllite are highly irregular within the phyllite–metabasalt *mélange*. Discontinuous tuff layers terminate laterally against phyllite with complex, interdigitating contacts, and small, irregular, wispy pockets of tuff are dispersed within the phyllite. These features are cut by a ductile S_1 metamorphic fabric, and are

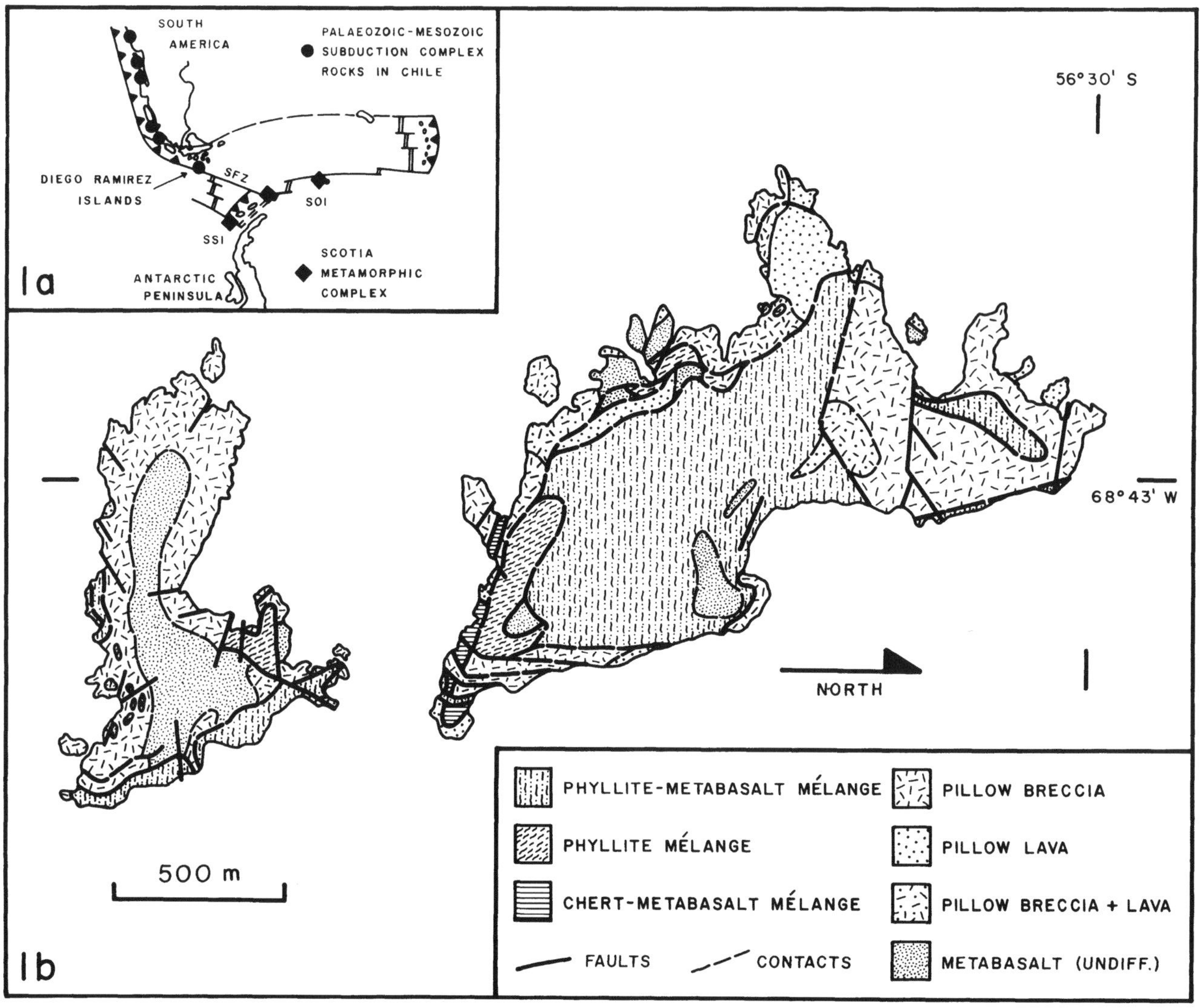

Fig. 1. Geological map and regional framework of the Diego Ramirez Islands, southern Chile. (a) The tectonic setting of the islands: comparatively high-*P*–low-*T* subduction-complex rocks are shown by dots, and occurrences of the Scotia metamorphic complex in the South Shetland Islands (SSI) and the South Orkney Islands (SOI) are shown by diamonds; after Hervé, Godoy & Davidson (1983); SFZ, Shackleton Fracture Zone. (b) Geological map of the main islands in the Diego Ramirez group, omitting smaller islands that consist primarily of metabasalt.

interpreted to represent an early stage of intermixing while the rocks were incompletely lithified.

Three phases of ductile deformation are recognized. D_1 produced the main foliation in the phyllites and imposed a penetrative flattening fabric on the metabasalts. S_1 is tightly to isoclinally folded by mesoscopic F_2 folds having a spaced, axial planar S_2 crenulation cleavage. In discrete ductile shear zones, S_2 becomes penetrative and completely transposes earlier surfaces. F_3 folds are less well developed and kink D_1 and D_2 structures.

Dynamothermal metamorphism occurring during the ductile deformation produced the assemblage quartz + albite + chlorite + stilpnomelane + white mica in the phyllites, and albite + blue amphibole + chlorite + epidote ± calcite ± white mica in the metabasalts. Microprobe analyses show the amphibole to be crossite (Grunow, Dalziel & Harrison, 1987), but lawsonite has not been detected either petrographically or by X-ray diffraction. The assemblages indicate comparatively high-*P*–low-*T* metamorphism, probably transitional between blueschist- and greenschist-facies.

Pervasive post-metamorphic cataclasis disrupted the earlier ductile fabrics and resulted in tectonic mixing of the metasedimentary rocks and metabasalt to produce *mélange*. The tectonic intermixing involved rock units that were metamorphosed and deformed together prior to *mélange* formation; exotic inclusions are absent.

Brittle fault zones cut all the rock units and are responsible for the present outcrop pattern (Fig. 1*b*).

Discussion

The lithology, structural style and comparatively high-*P*–low-*T* metamorphic assemblages of the Diego Ramirez Islands indicate that they are part of an accretionary complex that developed during subduction of oceanic lithosphere beneath the convergent margin of southern South America. The metabasalts appear to represent portions of a dismembered seamount that was tectonically incorporated into the accretionary complex during subduction, as also suggested by Davidson *et al.* (1989).

Rocks of the Diego Ramirez Islands exhibit a prolonged structural evolution, recording progressive stages of deformation at different levels in the accretionary complex. Early, soft-sediment intermixing of basaltic tuff and argillaceous sediment (now phyllite) is interpreted to represent deformation at high structural levels during initial underthrusting beneath the toe of the accretionary complex. This was followed by penetrative ductile deformation at deeper levels, presumably in the zone of underplating within the accretionary complex. Cataclastic shearing and tectonic intermixing to produce *mélange*, followed by late, discrete brittle faulting, are inferred to have occurred during structural uplift of the metamorphic rocks to high levels within the accretionary complex.

The only available geochronological data for the Diego Ramirez Islands consist of a Rb–Sr whole-rock 'errorchron' date of 169 ± 16 Ma for four phyllites given by Davidson *et al.* (1989). If this result can be taken to date the time of recrystallization of the phyllites, it implies that metamorphism at relatively deep levels in the accretionary complex occurred in the Middle Jurassic.

The Diego Ramirez Islands form part of a belt of fore-arc accretionary rocks of Late Palaeozoic–early Mesozoic age along the coast of central and southern Chile (Fig. 1a; Hervé *et al.*, 1981; Forsythe, 1982), and of Palaeozoic and Mesozoic–(?)Cenozoic age in the Scotia metamorphic complex along the western margin of the Antarctic Peninsula (Fig. 1a; Tanner, Pankhurst & Hyden, 1982; Dalziel, 1984; Smellie *et al.*, 1984). The islands show lithological affinities with the lower-grade parts of the Scotia metamorphic complex exposed in the South Shetland Islands, where similar blueschist- and greenschist-facies phyllites, metacherts, metacarbonate rocks and metavolcanic rocks are present. The Diego Ramirez rock units show a structural style similar to that of the Scotia metamorphic complex as a whole, in that both areas contain polyphase ductile fabrics inferred to have developed at relatively high pressures during shearing within the zone of underplating in the accretionary prism (Dalziel, 1984; Meneilly & Storey, 1986).

Identification of subduction complex rocks on the Diego Ramirez Islands increases the known extent of fore-arc accretionary assemblages developed along the Pacific margin of Gondwana. The Diego Ramirez Islands represent a key link between the South American and south Scotia Ridge fore-arc provinces, and must be taken into account in reconstructing the configuration of crustal blocks along the Pacific margin of Gondwana. It should be noted that comparatively high-P–low-T metamorphic rocks are now known from either end of the Shackleton Fracture Zone (Fig. 1a), both in the Diego Ramirez Islands to the north, and in the Elephant–Clarence Island group to the south. As discussed by Dalziel (1984) and Grunow *et al.* (1987), such a relation implies that Cenozoic fracture zone tectonics may have played a fundamental role in causing uplift of deep levels of the Gondwanide accretionary complex in the Scotia arc region.

Acknowledgements

This research forms part of the project entitled 'Kinematic evolution of the Scotia Ridge, Antarctica and southernmost South America: a study of orogenic processes at convergent plate margins', directed by I.W.D. Dalziel and funded by the National Science Foundation (DPP 86-43441). We are indebted to Captains Gates and Mueller and the crew of the Polar Duke for their support during the fieldwork, and to J. Davidson, Servicio Nacional de Geológia y Minería, Chile, for permission to cite data in press.

References

Dalziel, I.W.D. (1983). Geologic transect across the southernmost Chilean Andes: report of R/V HERO cruise 83–4. *Antarctic Journal of the United States*, **18(4)**, 8–12.

Dalziel, I.W.D. (1984). *Tectonic Evolution of a Forearc Terrane, Southern Scotia Ridge, Antarctica.* Washington, DC; Geological Society of America Special Paper, 200. 32 pp.

Davidson, J., Mpodozis, C., Godoy, E., Hervé, F. & Muñoz, N. (1989). Lower Mesozoic accretion of high buoyancy guyots on the Gondwanaland margin: Diego Ramirez Islands, southernmost South America. *Revista Geologica de Chile*, **16(2)**, 247–51.

Forsythe, R.D. (1982). The late Paleozoic to early Mesozoic evolution of southern South America: a plate tectonic interpretation. *Journal of the Geological Society of London*, **139(6)**, 671–82.

Grunow, A.M., Dalziel, I.W.D. & Harrison, T.M. (1987). Structural and $^{40}Ar/^{39}Ar$ constraints on the evolution of blueschists from the Scotia Arc. *Geological Society of America Abstracts with Programs*, **19(7)**, 686–7.

Hervé, F., Davidson, J., Godoy, E., Mpodozis, C. & Covaccevich, V. (1981). The Late Palaeozoic in Chile: stratigraphy, structure and possible tectonic framework. *Anais da Academia Brasileira de Ciencias*, **53(2)**, 363–73.

Hervé, F., Godoy, E. & Davidson, J. (1983). Blueschist relic clinopyroxenes of Smith Island (South Shetland Islands): their composition, origin and some tectonic implications. In *Antarctic Earth Science*, ed. R.L. Oliver, P.R. James, & J.B. Jago, pp. 363–66. Canberra; Australian Academy of Science and Cambridge, Cambridge University Press.

Meneilly, A.E. & Storey, B.C. (1986). Ductile thrusting within subduction complex rocks on Signy Island, South Orkney Islands. *Journal of Structural Geology*, **8(3/4)**, 457–72.

de la Rue, E.A. (1959). Quelques observations faites aux Iles Diego Ramirez (Chile). *Bulletin du Museum*, **31**, 387–91.

Smellie, J.L., Pankhurst, R.J., Thomson, M.R.A. & Davies, R.E.S. (1984). *The Geology of the South Shetland Islands: VI. Stratigraphy, Geochemistry and Evolution.* Cambridge; British Antarctic Survey Scientific Reports, 87, 85 pp.

Staudigel, H. & Schmincke, H.-U. (1984). The Pliocene seamount series of La Palma/Canary Islands. *Journal of Geophysical Research*, **89**, 11195–215.

Tanner, P.W.G., Pankhurst, R.J. & Hyden, G.M. (1982). Radiometric evidence for the age of the subduction complex in the South Orkney and South Shetland Islands, West Antarctica. *Journal of the Geological Society of London*, **139(6)**, 683–90.

Structural and metamorphic evolution of the Elephant Island group and Smith Island, South Shetland Islands

R.A.J. TROUW, A. RIBEIRO & F.V.P. PACIULLO

Universidade Federal do Rio de Janeiro, Departamento de Geologia, IG, Ilha do Fundão, Rio de Janeiro, RJ. CEP 21.910, Brazil

Abstract

Three distinct metamorphic zones were identified on Elephant Island: a low-grade chlorite zone with pumpellyite, a garnet zone with blue amphibole, and a biotite zone, with garnet, hornblende and albite. Contrary to the concept of a major tectonic break, separating north from south Elephant Island, the gradual changes between these zones suggest one prograde sequence of subduction-associated metamorphism of Sanbagawa-type. Clarence Island is comparable to the chlorite zone and Smith Island to the garnet zone. Three deformational phases affected the complex: D_1, D_2 and D_3. D_1 is responsible for a penetrative cleavage, an elongation lineation and tight folds. Syntectonic snowball garnets, local sheath folds and highly elongated volcanic fragments indicate strong shear movements during this phase, which is tentatively correlated with subduction, due to the presence of syntectonic blue amphibole. D_2 resulted in folding with southward vergence and was probably related to southward ductile overthrusting, accompanied by the growth of lower pressure/temperature (P/T) minerals. Post-metamorphic D_3 structures, mainly open folds and kinks with variable orientations, are attributed to isostatic uplift and denudation. Sedimentary rocks, mainly siltstones and conglomerates, have been discovered on the Seal Islands. Radiolarians and planktonic foraminifera reveal a marine environment and a maximum age of early Miocene for these rocks.

Introduction

The Elephant Island Group (65°15′ S, 55° W, at the north-eastern extreme of the South Shetland Islands) and Smith Island (63°S, 62°30′W, at the south-western end) are composed of metamorphic rocks that form part of the Scotia metamorphic complex (Tanner, Pankhurst & Hyden, 1982; Dalziel, 1984; Fig. 1). Previous work (Dalziel, 1982, 1984; Tanner *et al.*, 1982; Hervé & Pankhurst, 1984; Hervé, Godoy & Davidson, 1983; Hervé, Marambio & Pankhurst, 1984; Marsh & Thomson, 1985) established a metamorphic subdivision in a blueschist–greenschist terrain which includes Smith Island, northern Elephant Island and Clarence Island, and an albite–epidote–amphibolite facies terrain that comprises south-western Elephant Island, Gibbs Island, Aspland, Eadie and O'Brien islands (Fig. 1*c*).

The age of these rocks is still a matter of debate (Tanner *et al.*, 1982), but most radiometric data (by both K–Ar and Rb–Sr methods) are in the range 70–100 Ma for both terrains (Dalziel, 1982; Tanner *et al.*, 1982).

During the second, third and fourth Brazilian Antarctic Expeditions some 80 outcrop areas were studied. A synthesis of the main results of metamorphic and structural analyses is presented in this paper. Some preliminary results have already been published (Trouw, Ribeiro & Paciullo, 1986).

Metamorphism and protoliths

The study of many previously unvisited outcrops, especially in the central part of Elephant Island, permitted a more detailed subdivision of the metamorphism. Three metamorphic zones, separated by garnet and biotite isograds, are proposed for Elephant Island (Fig. 1*a*): a low-grade chlorite zone, a garnet zone and a biotite zone. The transitions between these zones are gradual. The distribution of the principal metamorphic minerals is indicated in Fig. 1 and in Table 1. The three transitional zones are at variance with the earlier concept of a major tectonic break (Dalziel, 1982, fig. 12.5*b*). This new interpretation casts doubt on the different ages and evolutions postulated for the rocks from north and south Elephant Island (Tanner *et al.*, 1982).

The metamorphism of Clarence Island is comparable to the chlorite zone, although no pumpellyite has been detected there. The local occurrence of green biotite in greenschist from Cape Lloyd is certainly not comparable to the biotite zone on Elephant Island. The rocks of Smith Island (Fig. 1*b*) are very

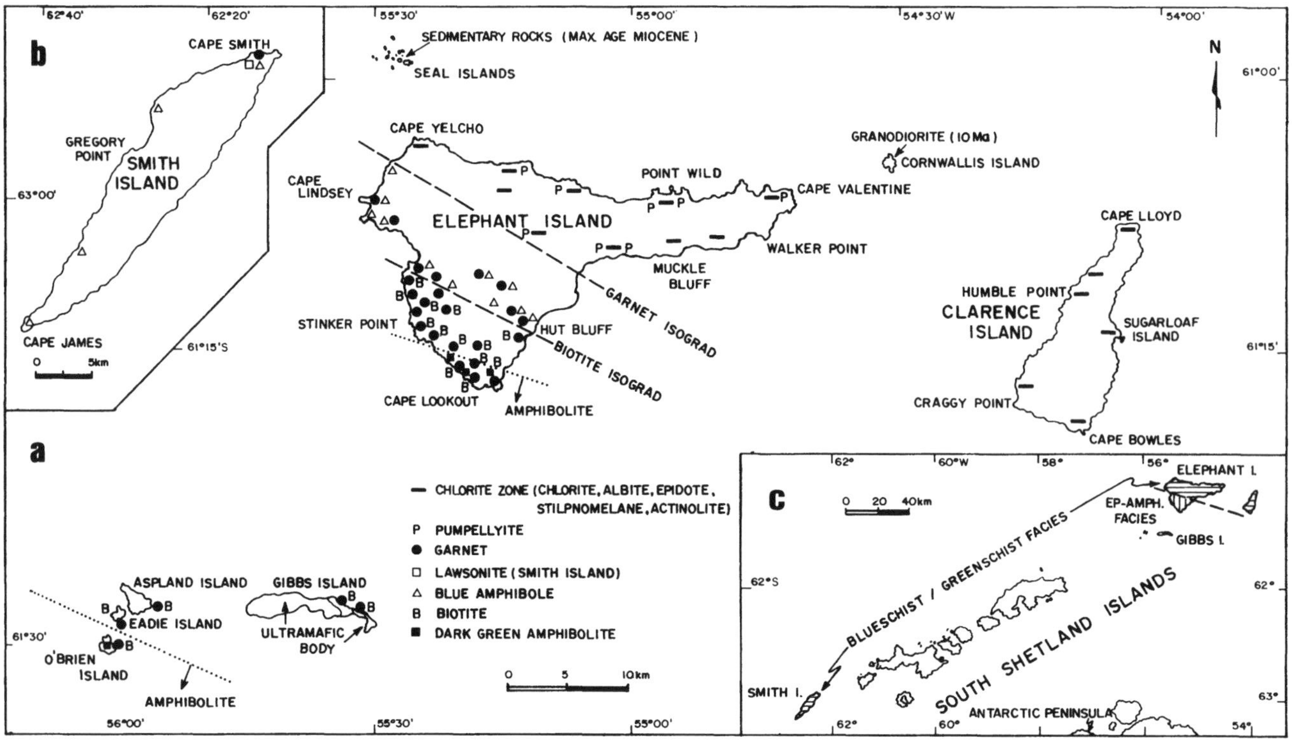

Fig. 1. ***(a)*** **Simplified locality and metamorphic map of the Elephant Island group and** ***(b)*** **Smith Island.** ***(c)*** **General setting in the South Shetland Islands, with two-fold subdivision (after Dalziel, 1982).**

Table 1

		Elephant I. and Gibbs I. group			Clarence I.	Smith I.
		Chlorite zone	Garnet zone	Biotite zone	Chlorite zone	Garnet zone
Pelite or semipelite	Chlorite	——	——	——	——	——
	Garnet		- - ———	———		—
	Biotite			- - ———		
	Hornblende		- -	———		
	Albite	——	——	——	——	——
	Quartz	——	——	——	——	——
	Epidote	——	——	——	——	——
	White Mica	——	——	——	——	——
	Stilpnomelane	——		- - -	- - -	
		—— - - -				
Metabasite	Pumpellyite	—— - -				
	Actinolite	——	—— - - -	- -	——	
	Na-Amph.		—— - -	- -		—— - -
	Hornblende		- -	———		
	Garnet			- - -		- -
	Biotite			- - -		
	White Mica	- - -	- - -	- - -	- - -	- - -
	Epidote	——	——	——	——	——
	Albite	——	——	——	——	——
	Quartz	——	——	——	——	——
	Stilpnomelane	——	—— - -	- -	- - -	- - -
	Chlorite	——	——	——	——	——
	Lawsonite					- -

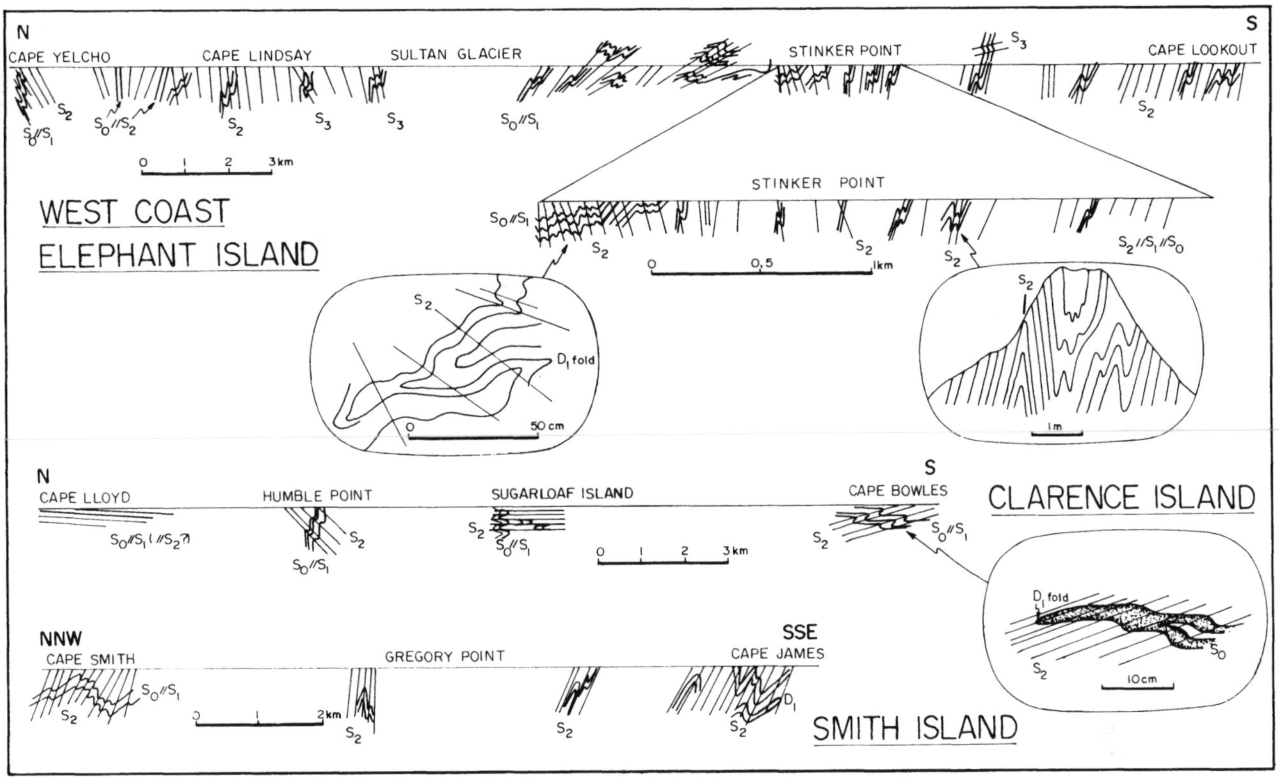

Fig. 2. Simplified structural sections, emphasizing asymmetry and vergence of D_2 structures. Minor folds are not to scale.

similar to those of the garnet zone. Both contain abundant blueschists, but the mineral garnet has only been found at Cape Smith, the same locality where lawsonite occurs (Smellie & Clarkson, 1975; Rivano & Cortés, 1976). The Gibbs Island Group fits into the biotite zone. It is remarkable that mafic rocks on O'Brien Island are amphibolites, whereas on the other islands of this group only greenschists have been found. This same transition, which is not a mineral isograd, has also been mapped at southern Elephant Island (Fig. 1*a*).

The similarity between mineral occurrences presented in Table 1 and the ones described from the Sanbagawa belt (Brothers & Yokoyama, 1982) is striking. The only major difference is that Na-amphibole occurs both in the chlorite and garnet zones in Japan, whereas at Elephant Island the blue amphiboles are mainly restricted to the garnet zone.

The main rock types in all three metamorphic zones are grey and green phyllites and schists, metachert, calc-silicate, and a few thin marble layers (for details see Dalziel, 1984; Hervé *et al.*, 1984; Marsh & Thomson, 1985; Trouw *et al.*, 1986). At northern Elephant Island, a volcanic metaconglomerate represents one of the few rock types which are restricted to a specific metamorphic zone. A dunite–serpentinite sheet on Gibbs Island has been described by de Wit *et al.* (1977).

The lithological similarity between the metamorphic zones suggests similar protoliths. Most of them are typical of an ocean-floor environment, such as chert, carbonaceous chert, grey pelitic sediment and metavolcanics. Grey schists from southern Elephant Island, rich in albite, seem to be derived from feldspathic greywacke, possibly with contributions from a continental source towards the south. Hervé *et al.* (1983) suggested an ocean-floor origin for greenschists with relict clinopyroxenes from Smith Island. Our chemical analyses support a similar origin for greenschists and amphibolites from Elephant and adjacent islands.

Structures

Dalziel (1984) described the structure of these islands in considerable detail and defined three deformational phases: D_E, D_M and D_L. Because of the limited scope of this paper only brief comments are given on these phases, labelled by us D_1, D_2 and D_3.

D_3 structures are open to tight post-metamorphic folds and kinks, with very variable orientations, intensities and distributions. Our interpretation is that this phase, or possible combination of phases, is related to the post-metamorphic rapid uplift of the area.

D_2 structures are shown in schematic composite cross sections in Fig. 2. A striking fact is that the predominant vergence of D_2 folds is towards the S or S–E, and this is difficult to match with subduction from north to south. It is also noteworthy that the form-lines of the S_0/S_1 foliation, folded by D_2 folds, make a high angle with the metamorphic isograds (Fig. 3).

D_1 created a penetrative foliation subparallel to bedding or compositional layering. Few D_1 folds have been detected with certainty. Sheath folds at Cape Lookout with E–W long axes (Fig. 4; Trouw *et al.*, 1986) are ascribed to this phase, as is a strong elongation lineation in metaconglomerates at northern Elephant Island. Intense ductile shearing during D_1 is also indicated by syn-D_1 snowball garnets that rotated over 400° (Fig. 5*a*). These occur in the hinges of fairly open D_2 folds, in

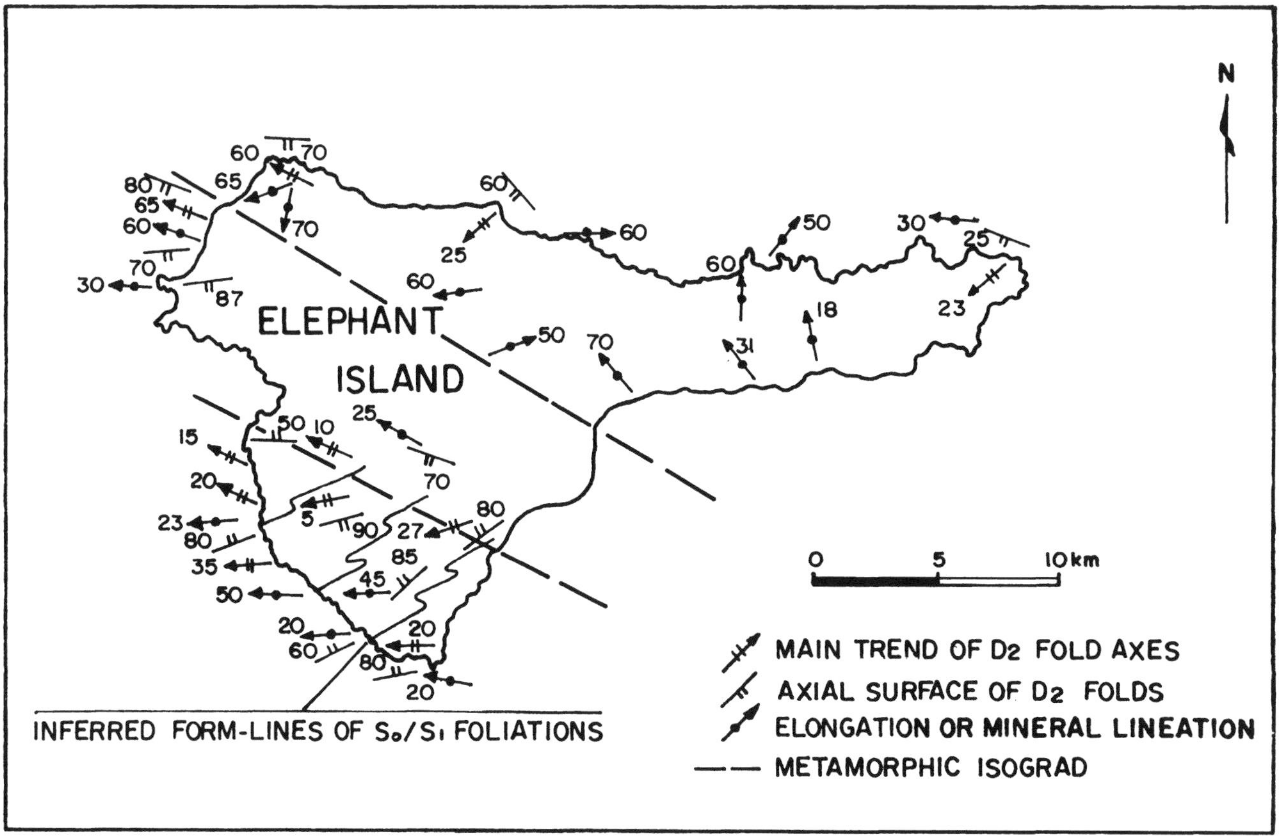

Fig. 3. Simplified structural map of Elephant Island.

such a way that the rotation cannot possibly be ascribed to D_2.

Tectono-metamorphic evolution

Microstructural observations, some of which are shown in Fig. 5 (see also Trouw *et al.*, 1986), are the basis of the evolutionary scheme presented in Table 2. In this, D_1 is interpreted as the deformational phase that accompanied subduction, because of the syn-D_1 blue amphiboles and the deformational style (characterized by intense ductile shear).

As can be observed in Fig. 3, most of the elongation and mineral lineations are oriented approximately E–W, parallel to the direction of local sheath folds. The current tectonic interpretation of this orientation (e.g. Cobbold & Quinquis, 1980) would be that D_1 was associated with a large component of

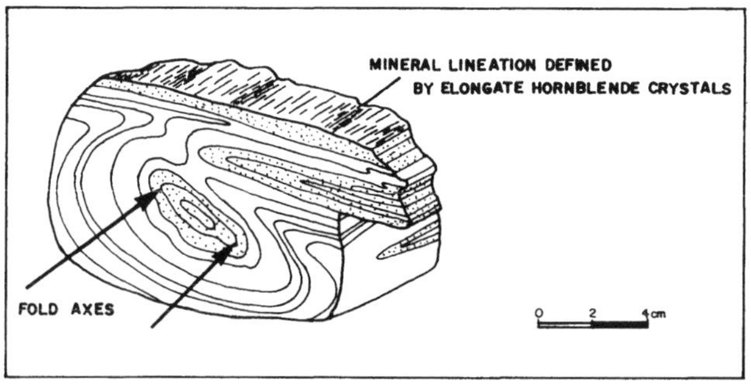

Fig. 4. Sheath fold in garnetiferous quartzite (metachert), Cape Lookout, Elephant Island. Dotted ornament: amphibolite; no ornament: garnetiferous quartzite.

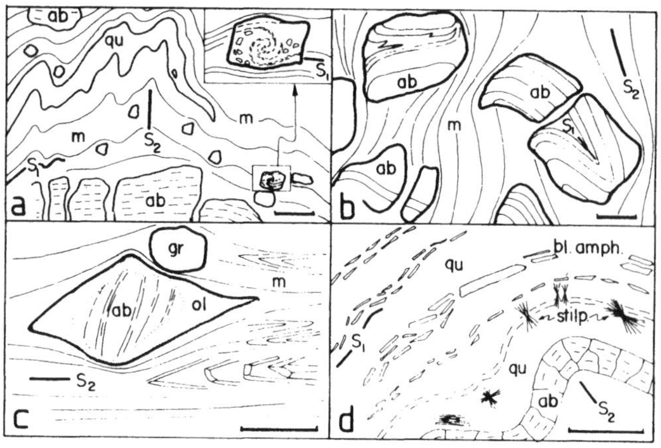

Fig. 5. Sketches of some microstructures (m = matrix, scale bar is 1 mm): (*a*) highly rotated syn-D_1 garnet in hinge of D_2 fold; (*b*) albite porphyroblasts with included D_1 folds, slightly refolded in the rims; albite growth between D_1 and D_2, and during D_2; (*c*) albite porphyroblast with oligoclase rim; albite early syn-D_2, oligoclase late syn-D_2; (*d*) syn-D_1 blue amphibole follows S_1, whereas late syn-D_2 albites overgrow D_2 fold.

E–W simple shear (E–W subduction?). However, the relatively large variation of directions and a possible rotation of D_1 structures by D_2, make this conclusion somewhat dubious. More detailed structural analyses are currently being worked out in order to clarify better this point.

After D_1 and during D_2, the growth of high *P*/*T* minerals apparently stopped. Extensive albite porphyroblast growth was possibly the result of the breakdown of blue amphibole. Upper greenschist–lower amphibolite facies parageneses were

Table 2

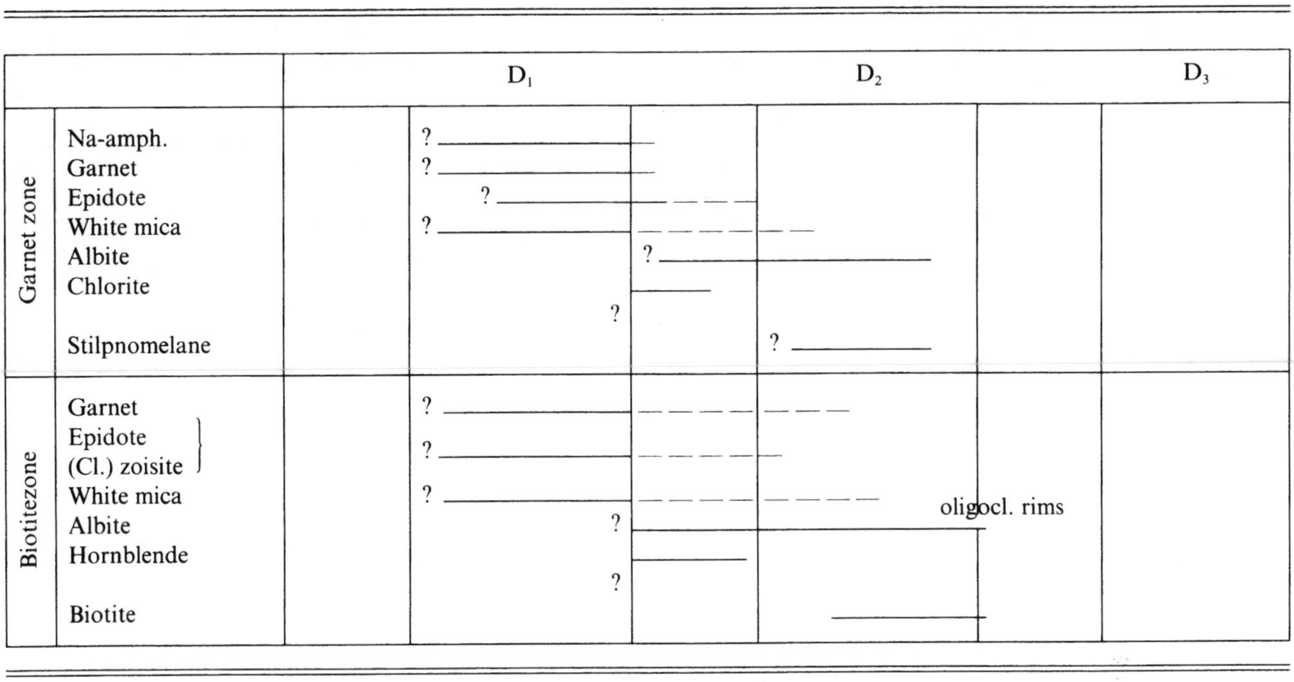
D1
D2
D3
Garnet zone
Na-amph.
Garnet
Epidote
White mica
Albite
Chlorite
Stilpnomelane
Biotitezone
Garnet
Epidote
(Cl.) zoisite
White mica
Albite
Hornblende
Biotite
oligocl. rims

created in the southern part of Elephant Island, with growth of green amphibole, biotite and local oligoclase rims around albites (Fig. 5*c*). In one thin-section from Stinker Point a blue core is preserved within a hornblende crystal, indicating progressive evolution of these rocks from blueschists to greenschists. The southward vergence and N–NW-dipping axial surfaces of D_2 folds indicate a NNW–SSE D_2 ductile overthrusting, possibly a product of subduction related backthrusting. D_2 deformation may alternatively be related to early movements along the Shackleton Fracture Zone that has a sinistral strike-slip movement. This possibility would explain overthrusting of the southern block at its abutment against the continental margin. D_2 geometry on Smith Island resembles that on Elephant Island (Fig. 2). A similar process related to the Hero Fracture Zone could have occurred there.

Cenozoic sedimentary rocks from the Seal Islands

The Seal Islands are a group of tiny islets, a few miles to the north of Elephant Island (Fig. 1). They have been described as part of the Scotia metamorphic complex, with *mélange*-type deformation (Dalziel, 1984). In fact they are sedimentary siltstones, volcanic greywackes and matrix-supported conglomerates. These conglomerates contain micritic carbonate, volcanic greywacke and quartzite pebbles ranging in size from about 0.01–1 m. Steeply-dipping bedding has a variable strike because of irregular folding, probably the result of gravity collapse. No penetrative cleavage is developed but many centimetre-scale faults and fractures are present.

The micritic carbonate pebbles and also some siltstones contain microfossils. These include the following species of planktonic foraminifera determined by R.L.M. Azevedo: *Globigerinoides ruber* (s.l.); *Globigerinoides* spp.; *Globigerina* ssp.; and *Orbulina* spp. These determinations constrain the possible age of these rocks to the period from early Miocene to Recent, and thus they are certainly younger than the Elephant Island metamorphics. It is proposed that these rocks from Seal Islands should be named the Seal Islands Formation.

Acknowledgements

This research was funded by CIRM, through a research agreement CIRM/FUJB, subproject no. 9597 of the Brazilian Antarctic Program.

We wish to thank the following participants in the project: Luis Sergio A. Simões, Claudio de M. Valeriano, Aimara Linn, Antonio Carlos Magalhães, Coriolano M.D. Neto, Ivo Karmann, Monica Heilbron, Monica Lopes, Rômulo Machado and Sheila Bittar. Ricardo L.M. Azevedo from CENPES, the research laboratory of PETROBRÁS S.A., is acknowledged for the fossil determinations.

References

Brothers, R.N. & Yokoyama, K. (1982). Comparison of the high-pressure schist belts of New Caledonia and Sanbagawa, Japan. *Contributions to Mineralogy and Petrology*, **79(2)**, 219–29.

Cobbold, P.R. & Quinquis, H. (1980). Development of sheath folds in shear regimes. *Journal of Structural Geology*, **2(2)**, 119–26.

Dalziel, I.W.D. (1982). The pre-Jurassic history of the Scotia Arc: a review and progress report. In *Antarctic Geoscience*, ed. C. Craddock, pp. 111–26. Madison; University of Wisconsin Press.

Dalziel, I.W.D. (1984). *Tectonic Evolution of a Fore-arc Terrane, Southern Scotia Ridge, Antarctica.* Washington, DC; Geological Society of America Special Paper, 200. 32 pp.

De Wit, M.J., Dutch, S., Kligfield, R., Allen, R. & Stern, C. (1977). Deformation, serpentinization and emplacement of a dunite complex, Gibbs Island, South Shetland Islands: possible fracture zone tectonics. *Journal of Geology*, **85(6)**, 745–62.

Hervé, F., Godoy, E. & Davidson, J. (1983). Blueschist relic clinopyroxenes of Smith Island (South Shetland Islands): their composition, origin and some tectonic implications. In *Antarctic Earth Science*, ed. R.L. Oliver, P.R. James & J.B. Jago, pp. 363–6. Canberra; Australian Academy of Science and Cambridge; Cambridge University Press.

Hervé, F. Marambio, F. & Pankhurst, R.J. (1984). El Complejo Metamórfico de Scotia en cabo Lookout, isla Elefante, islas Shetland del Sur, Antártica: evidencias de un metamorfismo cretácico. *Serie Cientifica INACH*, **31**, 23–37.

Hervé, F. & Pankhurst, R.J. (1984). The Scotia metamorphic complex at Cape Bowles, Clarence Island, South Shetland Islands, western Antarctica. *British Antarctic Survey Bulletin*, **62**, 15–24.

Marsh, P.D. & Thomson, J.W. (1985). Report on Antarctic Fieldwork. The Scotia metamorphic complex on Elephant Island and Clarence Island, South Shetland Islands. *British Antarctic Survey Bulletin*, **69**, 71–5.

Rivano, S. & Cortes, R. (1976). Note on the presence of the lawsonite–sodic amphibole association on Smith Island, South Shetland Islands, Antarctica. *Earth and Planetary Science Letters*, **29(1)**, 34–6.

Smellie, J.L. & Clarkson, P.D. (1975). Evidence for pre-Jurassic subduction in Western Antarctica. *Nature, London*, **258(5537)**, 701–2.

Tanner, P.W.G., Pankhurst, R.J. & Hyden, G. (1982). Radiometric evidence for the age of the subduction complex in the South Orkney and South Shetland Islands, West Antarctica. *Journal of the Geological Society, London*, **139(6)**, 683–90.

Trouw, R.A.J., Ribeiro, A. & Paciullo, F.V.P. (1986). Contribuição à geologia da Ilha Elefante, Ilhas Shetland do Sul. *Anais da Academia Brasileira de Ciências, 58, Suplemento*, 157–69.

Chronology of provenance, deposition and metamorphism of deformed fore-arc sequences, southern Scotia arc

F. HERVÉ[1], W. LOSKE[2], H. MILLER[2] & R.J. PANKHURST[3]

1 Departamento de Geología, Universidad de Chile, Casilla 13518, Correo 21, Santiago, Chile
2 Institut für Allgemeine und Angewandte Geologie, Universität München, Luisenstrasse 37/1, D-8000 München, West Germany
3 British Antarctic Survey, Natural Environment Research Council, High Cross, Madingley Road, Cambridge CB3 0ET, UK

Abstract

The Trinity Peninsula Group (TPG) is a Permo-Carboniferous–Triassic fore-arc sequence metamorphosed to very low grade in Triassic times. The euhedral form and highly discordant U–Pb ages of most detrital zircons indicate proximal provenance from juvenile Carboniferous granitoids. Scarce pink and well-rounded zircons may reflect minor source rocks with a Precambrian history. The Scotia metamorphic complex (SMC) contains rocks from a variety of tectonic settings, stacked together and metamorphosed in a subduction–accretionary wedge. Scarce idiomorphic detrital zircons suggest a component from a Late Palaeozoic granitic source terrain similar to that of the TPG. A Rb–Sr whole-rock isochron of 287 ± 48 Ma for Gibbs Island schists suggests a Late Palaeozoic metamorphism, also recorded in nearby Elephant Island epidote–amphibolite schists, where Rb–Sr whole-rock data allow a maximum protolith/provenance age of Permian. It is possible that the majority of the amphibolite-grade part of the SMC was derived from isotopically-primitive sources during Late Carboniferous–Early Permian times, with resetting of Rb–Sr, K–Ar and U–Pb sphene systems during an (?)Albian metamorphic event. By contrast, the low-grade SMC rocks of northern Elephant Island are probably no older than Mesozoic in terms of both provenance and deposition.

Introduction

The once-continuous link between Antarctica and South America, which included parts of the Scotia arc, was dismembered during Cenozoic times (Dalziel & Elliot, 1971). Among the disrupted remnants, northern Graham Land and the South Shetland Islands contain units of metamorphic basement, including the low-grade metamorphosed sandstone–shale sequence of the Trinity Peninsula Group (TPG) and its presumed equivalents, and the sub-greenschist- to epidote-amphibolite-facies Scotia metamorphic complex (SMC) (Fig. 1). This paper reviews published geochronological data for these rocks, presents new data for some localities and compiles evidence for the age and lithology of the source areas, the maximum possible ages of deposition for the non-fossiliferous SMC and the timing of the metamorphic events affecting the southern limb of the Scotia arc. The new data come from U–Pb determinations on zircons and sphenes (Table 1), and Rb–Sr whole-rock analyses (Table 2).

Trinity Peninsula Group (TPG)

The TPG (Hyden & Tanner, 1981) consists mainly of low-grade prehnite–pumpellyite-facies metamorphosed sandstones and shales, with occasional basaltic flows. Its main outcrops are in northern Graham Land, but small, lithologically similar outcrops on Livingston Island are also included in the definition. Because of structural complexity, stratigraphical relationships between the main outcrops are unknown, but there is evidence that deposition spanned Carboniferous–Triassic times; Pankhurst (1983) reported a 386 ± 39 Ma Rb–Sr whole-rock errorchron for granitic cobbles in the View Point Formation which was thought to record provenance. Deformation and metamorphism predated the unconformable deposition of the Antarctic Peninsula Volcanic Group from mid-Jurassic times onwards (Thomson & Pankhurst, 1983), and Pankhurst (1983) reported Rb–Sr whole-rock isochrons of ~ 245 Ma for TPG samples from Hurd Peninsula and from high-grade rocks in Graham Land, interpreted to be due to metamorphic re-equilibration.

Miller, Loske & Kramm (in press) and Loske, Miller & Kramm (in press) discussed the morphology and U–Pb systematics of detrital zircons from Hurd Peninsula, Livingston Island and from Gandara Island near Cape Legoupil. Zircon populations from these localities are similar, with clear and colourless euhedral grains predominating over red, well-rounded crystals. The U–Pb isotopic pattern of these morpho-

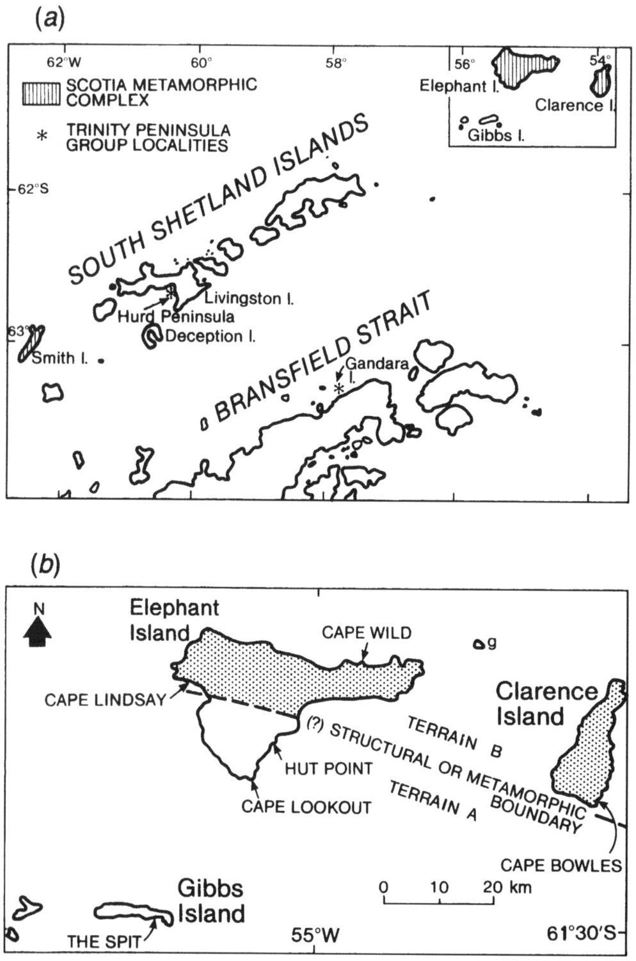

Fig. 1. (*a*) Outcrop areas of the Trinity Peninsula Group (TPG) and the Scotia metamorphic complex (SMC) with (*b*) sampled localities in the Elephant Island group.

logical types are characteristic (Fig. 2*a*). The euhedral crystals are highly discordant: those from Livingston Island define fans of discordia which converge at lower intercept dates of close to 320 Ma, whilst the Gandara Island data yield a single discordia of cogenetic zircons with a lower intercept of 322 + 7/ − 8 Ma. These are interpreted as the intrusion ages of granitoid rocks from which the zircons were last derived. The non-abraded shape of the crystals suggests that the sites of the primary host rocks and sandstone deposition were relatively close to each other. The undefined upper intercepts of the discordia suggest an ultimate Precambrian source before incorporation into the granites. The red, rounded zircons of both localities are characterized by much higher apparent $^{206}Pb/^{238}U$ ages of 676–872 Ma: a reference line through the data points intersects concordia at about 560 Ma and about 2000 Ma.

Scotia metamorphic complex (SMC)

The SMC is thought to represent a subduction-related accretionary wedge developed on the western margin of Gondwana, with metamorphism and tectonism continuing after break-up of the supercontinent. Several authors have divided the SMC into two geological units, referred to here as *terrains* (rather than *terranes*, since they are not obviously allochthonous). The higher temperature unit (Terrane B; Tanner, Hyden & Pankhurst, 1982) mainly comprises the epidote–amphibolite-facies of southern Elephant Island and Gibbs Island, whereas the other (Terrane A; Tanner *et al.*, 1982) consists of the comparatively high-P/T assemblages of northern Elephant Island, Clarence and Smith islands. These two terrains were thought to be separated by a structural or metamorphic boundary.

Rocks from a variety of tectonic settings have been identified in the complex. On the basis of relict clinopyroxene chemistry, the presence was established of ocean-floor basalts on Smith Island (Hervé, Godoy & Davidson, 1983) and of island-arc basalts in north-eastern Elephant Island (Hervé, in press). A slice of oceanic lithosphere is represented by the dunite on Gibbs Island (de Wit, 1977). The Clarence Island phyllites were interpreted as a metamorphosed sequence of arkosic metasandstones, derived from a continental granitic source (Hervé & Pankhurst, 1984). These lithological–stratigraphical elements appear to have been stacked together in the subduction complex.

Gibbs Island

Gibbs Island is mainly composed of a dunite body, in contact with metamorphic rocks at its eastern end, along the Gibbs shear zone. The metamorphic rocks are grey phyllites, with small albite and garnet microporphyroblasts in a white mica–biotite–chlorite–quartz matrix. The main foliation appears to be the first one, no relict minerals or structure having been recognized. A four-point whole-rock isochron plot for a locality just south of the shear zone yields an age of 287 ± 48 Ma (MSWD = 0.8) with an initial $^{87}Sr/^{86}Sr$ ratio of 0.7074 ± 0.0005 (Table 2, Fig. 3*a*). This probably indicates a metamorphic episode during Late Carboniferous to Early Triassic times.

One of the samples included in the above isochron contains subidiomorphic zircon characterized by clearly visible cores with colourless overgrowths. These polyphase zircons are discordant: a correlation line for five size fractions analysed by Loske, Miller & Kramm (1985) appeared to cut concordia at ~ 230 Ma and ~ 1700 Ma, suggesting a Precambrian provenance and a Triassic metamorphism equivalent to that for the TPG. However, analysis of further separates (Table 1) casts doubt on this lower intercept, as the rounded zircons, together with the original data points, now define a discordia with a very young intercept of 19 Ma (Fig. 2*b*). This could be interpreted as reflecting either a Cenozoic metamorphic overprint or Pb-loss associated with uplift, possibly correlated with the beginning of spreading along Bransfield Strait. A young age of 29 Ma was also obtained by the K–Ar method on hornblende from nearby O'Brien Island (Tanner *et al.*, 1982).

Clarence Island

The grey phyllites of Cape Bowles, south-eastern Clarence Island, have yielded a Rb–Sr errorchron of 71 ± 40 Ma (MSWD = 8.4) with a relatively high initial

Table 1. *U–Pb zircon analyses*

Size	Weight	Measured isotope ratios			U	Pb	Corrected isotope ratios			Apparent ages		
(μm)	(mg)	$^{208}Pb/^{206}Pb$	$^{207}Pb/^{206}Pb$	$^{206}Pb/^{204}Pb$	(ppm)	(ppm)	$^{206}Pb/^{238}U$	$^{207}Pb/^{235}U$	$^{207}Pb/^{206}Pb$	$^{208}Pb/^{238}U$	$^{207}Pb/^{235}U$	$^{207}Pb/^{206}Pb$
Gibbs Island[a]												
> 100	6.0	0.13493	0.08725	2761	274	21.5	0.07414	0.8397	0.08214	461	619	1249
100–80	4.1	0.13649	0.08553	2071	308	24.6	0.07570	0.8214	0.07870	470	609	1165
80–63	5.7	0.24324	0.13634	265.8	349	36.7	0.08257	0.9488	0.08334	511	677	1277
63–40	3.9	0.23715	0.13225	280.9	367	39.9	0.08651	0.9782	0.08201	535	693	1246
< 40	1.2	0.12007	0.09023	2339	380	33.9	0.08670	1.0070	0.08424	536	707	1298
Subhedral zircons												
> 100	2.9	0.14948	0.09077	1678	288	16.5	0.05380	0.6111	0.08238	338	484	1255
100–80	2.6	0.13944	0.08767	2089	337	19.9	0.05623	0.6273	0.08091	353	494	1219
80–63	2.2	0.14896	0.09327	1197	351	23.6	0.06218	0.6985	0.08147	389	538	1233
Well-rounded zircons												
> 100	1.5	0.11219	0.08019	4669	277	30.6	0.10796	1.1480	0.07713	661	776	1124
100–80	2.7	0.11011	0.08346	4411	328	33.3	0.09910	1.0968	0.08027	609	752	1204
80–63	0.6	0.10691	0.09243	2568	357	42.8	0.11663	1.3987	0.08698	711	888	1360
Elephant Island												
Sphenes												
100–63	0.8	0.26606	0.09231	477	373	12.1	0.02851	0.2440	0.06206	181	221	676
< 63	2.3	0.44669	0.16008	141	274	10.9	0.02552	0.1999	0.05682	163	185	485
63–40	0.8	0.50187	0.17325	120	333	10.4	0.01991	0.1400	0.05101	127	134	241

[a]Data used by Loske *et al.* (1985), not previously published.

Analyses carried out at the University of Münster, with the assistance of U. Kramm.

Table 2. *New Rb–Sr whole-rock analyses*

Sample	Rb (ppm)	Sr (ppm)	$^{87}Rb/^{86}Sr^a$	$^{87}Sr/^{86}Sr$	$(^{87}Sr/^{86}Sr)_o{}^b$
The Spit, Gibbs Island (phyllites)					
GIB 3	53.0	294	0.5222 (0.6)	0.70954	0.70880
GIB 7	53.1	258	0.6668 (0.5)	0.71022	0.70927
GIB 9	61.9	282	0.7001 (0.5)	0.71023	0.70924
GIB 10	70.0	248	0.8178 (0.5)	0.71076	0.70960
Point Wild, Elephant Island (Terrain A metasandstones)					
ELE 46A	11.3	103	0.3154 (2.2)	0.70493	0.70448
ELE 46B	6.2	75.9	0.2356 (3.9)	0.70485	0.70452
ELE 47A	54.1	63.3	2.474 (0.7)	0.70811	0.70459
Cape Lindsay, Elephant Island (Terrain A metapelites)					
ELE 38	33.2	179	0.5353 (0.8)	0.70582	0.70506
ELE 39	30.6	134	0.6578 (1.0)	0.70612	0.70519
ELE 40	2.5	144	0.050 (8.3)	0.70507	0.70500
ELE 40A	42.2	56.0	2.181 (0.8)	0.71014	0.70704
ELE 49	1.1	46.8	0.064 (25)	0.70607	0.70598
ELE 50A	56.2	167	0.9764 (0.6)	0.70631	0.70492
Hut Point, Elephant Island (Terrain B)					
ELE 42A	13.8	307	0.1304 (1.8)	0.70513	0.70494
ELE 43A	21.4	313	0.1978 (1.2)	0.70510	0.70482
ELE 43B	19.3	257	0.2173 (1.3)	0.70521	0.70490
ELE 43C	19.1	248	0.2238 (1.3)	0.70520	0.70488
ELE 43D	19.5	254	0.2218 (1.3)	0.70525	0.70493
ELE 45E	16.2	206	0.2277 (1.5)	0.70507	0.70475

Analyses at British Geological Survey, with the help of M. Brook. [a]Determined by X-ray fluorescence spectrometry, by courtesy of T.K. Smith and staff (% 1-sigma errors in parentheses).
[b]Measured $^{87}Sr/^{86}Sr \pm 0.01\%$; $(^{87}Sr/^{86}Sr)_o$ calculated for 100 Ma ago.

$^{87}Sr/^{86}Sr$ ratio of 0.7094 (Hervé & Pankhurst, 1984). The latter was taken as suggesting derivation of the phyllites from an arkosic sedimentary pile of early Mesozoic age, metamorphosed during Late Cretaceous times. Two episodes of metamorphism were identifed from the mineralogy, low-grade clinozoisite–actinolite-bearing assemblages replacing previous very low-grade assemblages. No typical high-*P* phases were recognized.

Elephant Island

The largest island of the group, Elephant Island, also has the most complex geology. Terrain A probably represents the rather high-*P*/*T* metamorphic outboard part of the accretionary wedge. A metamorphic zonation has recently been established within it, with feebly recrystallized pumpellyite-bearing schists around Cape Lindsay. Trouw, Ribeiro & Paciullo (this volume, p. 423) have established the approximate location of the isograds. These rocks represent the accumulation of fragmented volcanic material probably related to an oceanic island arc (Hervé, in press). Three samples of metavolcanic sandstones from Cape Wild (Table 2), which preserve relict structures and minerals and have only one foliation, yield a good Rb–Sr isochron of 103 ± 6 Ma (MSWD = 0.1) and an initial $^{87}Sr/^{86}Sr$ ratio of 0.7045) (Fig. 3*b*). From Cape Lindsay, four samples of metapelites from a set of six define a reasonable isochron of 98 ± 29 Ma (MSWD = 3.6) with an initial $^{87}Sr/^{86}Sr$ ratio of 0.7050 (Fig. 3*c*). These ages agree with previous dates for the Late Cretaceous metamorphism of Terrain A (e.g. Tanner *et al.*, 1982). One of the other samples (ELE 40A) is a stilpnomelane-rich metachert unexpectedly rich in radiogenic ^{87}Sr, perhaps due to equilibration with Cretaceous seawater.

The Cape Lookout area (Terrain B) is composed of grey schists, greenschists, minor metachert and marble. The schists are characteristically rich in albite porphyroblasts, which develop progressively from Cape Lindsay southwards, along the western coast of the island (Marsh & Thomson, 1985). Major element chemistry variation can be explained by a mixture of clay and calcite compositions, so that the greenschists appear to be paraschists rather than meta-igneous rocks. Two Rb–Sr whole-rock errorchrons for mica-schists gave ages of 237 ± 86 Ma (MSWD = 44) and 235 ± 91 Ma (MSWD = 12), with initial $^{87}Sr/^{86}Sr$ ratios of 0.7041 and 0.7046, respectively (Hervé, Marambio & Pankhurst, 1984). The mean values of $^{87}Sr/^{86}Sr$ (0.7059) and of $^{87}Rb/^{86}Sr$ (0.4580) yield a model age of either 350 Ma with an assumed initial ratio of 0.7035, or 300 Ma with 0.7040, indicating that the protolith is probably not older than Carboniferous. Rb–Sr mineral isochrons on marble (93 ± 12 Ma), amphibolite (99 ± 1 Ma) and mica-schist (99 ± 2 Ma) probably record isotopic homogenization associated with the last metamorphism. Zircons are

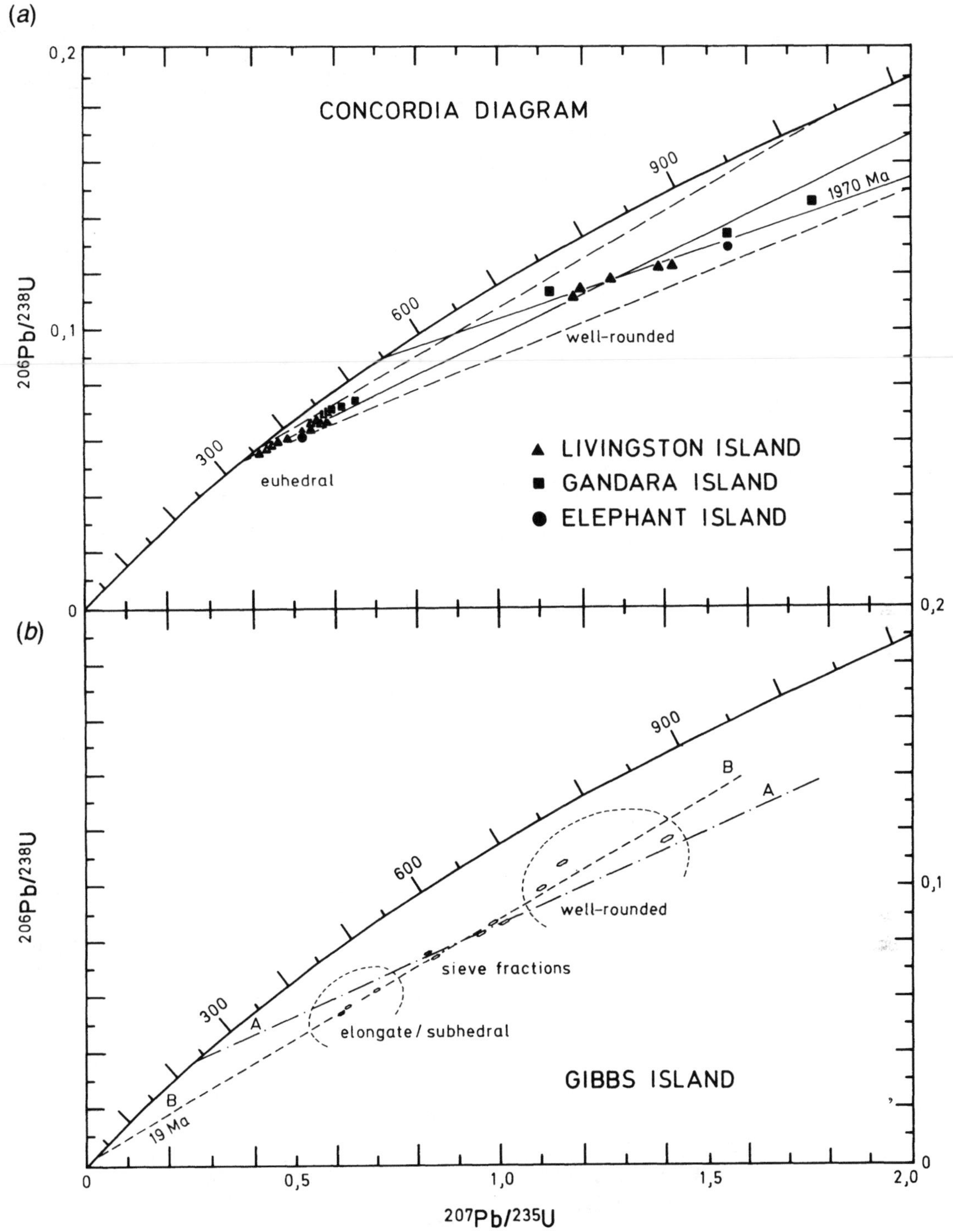

Fig. 2. U–Pb concordia diagrams for detrital zircons from (*a*) Livingston, Gandara and Elephant islands, and (*b*) Gibbs Island.

rare in the rocks sampled at Cape Lookout, but both idiomorphic and rounded grains have been identified. Both are discordant and have a two-stage history (Loske *et al.*, 1985). Their measured isotope ratios fit excellently with the isotopic pattern from Livingston and Gandara islands, and the lower intercept of 320 Ma suggests a Late Palaeozoic provenance, in accordance with the Rb–Sr whole-rock data. Additionally, idiomorphic sphene present in the rocks gives a discordia with a lower intercept close to 120 Ma, probably indicating the Cretaceous metamorphism recorded by the Rb–Sr mineral dates. This event is also seen in three K–Ar ages of 105, 108 and 110 Ma for biotites from amphibolites (Hervé & Kawashita, 1986). The rocks of the Hut Point area show too little spread in Rb/Sr ratio to yield meaningful Rb–Sr ages (Table 1), but the data are consistent with the 'normal' 100 Ma and low initial $^{87}Sr/^{86}Sr$ ratios observed from other localities on Elephant Island. However, in these rocks, there is no evidence to indicate older inherited or reset Rb–Sr systematics, as is thought to be the case for the Cape Lookout area.

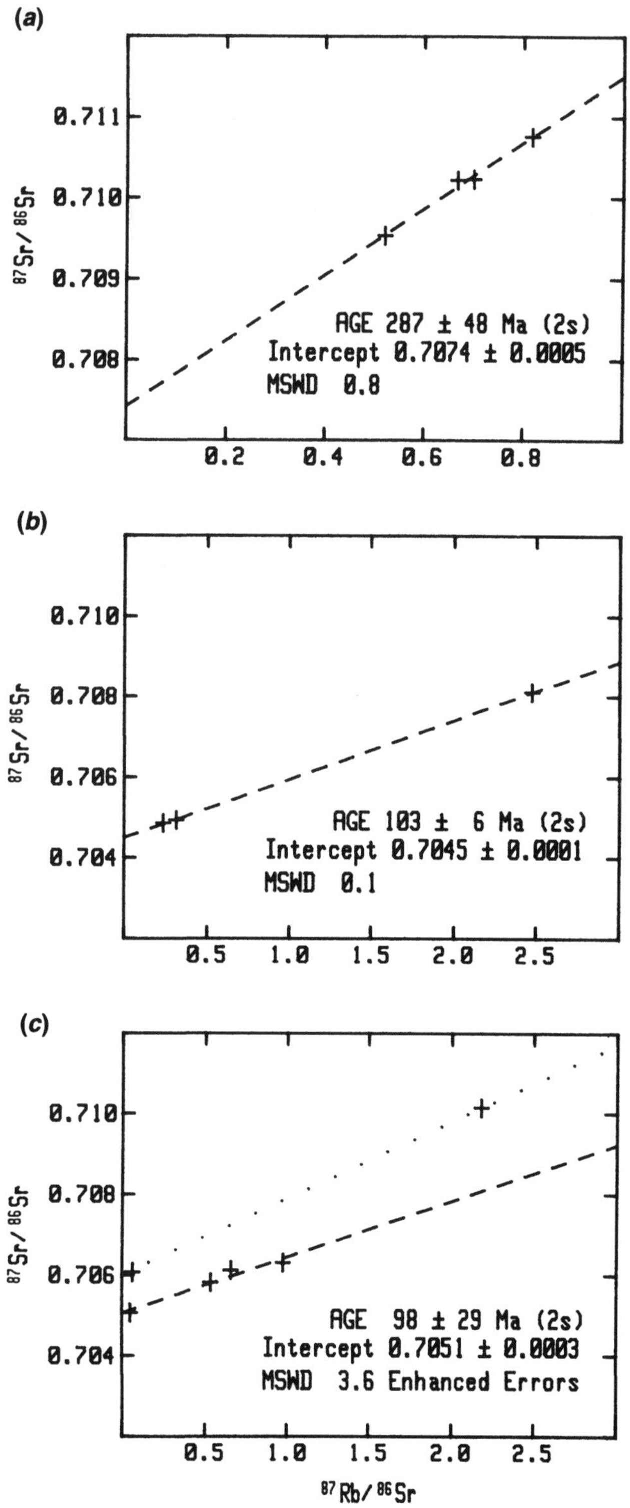

Fig. 3. Previously unpublished Rb–Sr whole-rock isochrons for *(a)* Gibbs Island, *(b)* Cape Wild and *(c)* Cape Lindsay. The upper (reference) line in *(c)* corresponds to 135 Ma, but may be meaningless since these samples have probably partially equilibrated with seawater-Sr.

Smith Island and the South Orkney Islands

Although we have no new geochronological data for Smith Island, Tanner *et al.* (1982) presented scattered Rb–Sr data compatible with a maximum age of ~ 135 Ma with an initial $^{87}Sr/^{86}Sr$ ratio of 0.703. This agrees with our data for Terrain A of Elephant Island, which has similar metamorphic assemblages, including blueschists. Terrain B rocks in the South Orkney Islands gave Rb–Sr whole-rock evidence of a Late Palaeozoic event (Rex, 1976) and abundant K–Ar data supporting a 200 Ma metamorphism (Tanner *et al.*, 1982).

Discussion

Previous conclusions about the timing of the main metamorphic events in the region are confirmed by this study. The TPG is hardly metamorphosed in northern Graham Land but has given evidence of a Triassic event farther south. This is possibly also reflected in the new Rb–Sr and U–Pb zircon data from Gibbs Island. The SMC in general, but especially Terrain A rocks, shows clear evidence for an Early–Late Cretaceous metamorphism (130–70 Ma). Except for the late Cenozoic ages from Gibbs and O'Brien Islands, of uncertain significance, this is the last event known in the SMC.

The U–Pb zircon data mostly yield ages which exceed the presumed maximum depositional ages of these rocks (~ 300 Ma), although no zircons are known from Terrain A of the SMC. All display discordant patterns with Proterozoic or older upper intercepts, indicating an ultimate source in Precambrian shield areas. The varied lower intercepts of the U–Pb discordia can be attributed to more specific events. The ~ 500 Ma lower intercept for well-rounded zircons from the TPG on Livingston Island probably also records a metamorphic event, presumably affecting the source region rocks prior to erosion and transport of the detrital zircons. However, the well defined Late Palaeozoic event (~ 320 Ma) recorded in the TPG on Livingston Island and at Cape Legoupil, and also in the SMC on southern Elephant Island, is invariably associated with idiomorphic zircons and is thus interpreted as dating the igneous crystallization of source region granitic rocks.

The Rb–Sr systematics can be used to assess maximum ages of mantle separation for the rock materials involved in the SMC. Fig. 4 is a histogram of initial $^{87}Sr/^{86}Sr$ ratios 100 Ma ago. This provides the simplest indication of the possible antiquity of the protolith, although there is some uncertainty due to possible alteration and variation in Rb/Sr ratios from one area to another. It shows no distinction between the two terrains on Elephant Island itself but there is a very clear distinction between Elephant and Smith islands on the one hand, and Gibbs and Clarence islands on the other. Only for the latter is it possible to envisage closed-system resetting of Palaeozoic rocks. The Elephant Island rocks, particularly those from the northern part, are all young, and apparently the products of mid-Cretaceous metamorphism of rocks deposited only a short time beforehand.

A major difference in protolith is thus apparent between the SMC on Smith and Elephant islands (a complex of oceanic and tholeiitic volcaniclastic arc deposits) and that on Clarence and Gibbs islands (a sandstone sequence derived from a continental margin). Apparently, the first corresponds to accreted material, and the second to sedimentation near the active front of the accretionary wedge. The sedimentary debris of the latter is distinct from the TPG material of Livingston and Gandara islands, but both share a Palaeozoic history in their source

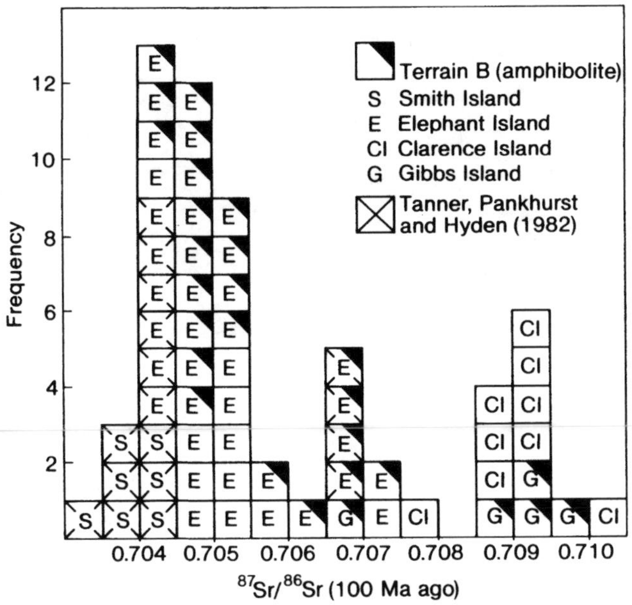

Fig. 4. Histogram of initial $^{87}Sr/^{86}Sr$ ratios at 100 Ma ago for all samples analysed from the SMC. Sources: Tanner *et al.* (1982), Hervé *et al.* (1984), this paper.

regions. This implies a palaeogeographic location of the accreted material relatively far away from any Precambrian cratonic area.

In situ Palaeozoic granitoids that could be the source material for the 320 Ma zircons are unknown from the Antarctic Peninsula and have been described up to now only from southern South America. Recent work in the Thurston Island block of West Antarctica has identified a single occurrence of 300 Ma granitic gneisses (Storey *et al.*, this volume, p. 399). New ODP data, suggesting that the Weddell Sea is an old, pre-Cretaceous basin (D. Fütterer, pers. comm.) disqualify the Transantarctic Mountains as a possible provenance area and palaeogeographic reconstructions which locate the Antarctic Peninsula near South Africa or near the East Antarctic craton must be regarded as doubtful. Our data support a pre-drift location of the Antarctic Peninsula south or south-west of southern South America (e.g. Quilty, 1982; Miller, 1983).

Acknowledgements

Samples were collected during Instituto Antártico Chileno field expeditions and 'Polarstern' cruise ANT II/2 (Alfred-Wegener Institute), with assistance from G. Marambio, I. Garrido, A. Sina, R. Toloza and students from the Universidad de Chile. Laboratory work was financially supported by a Royal Society – Comisión Nacional de Investigación Científica y Tecnológica grant to F.H., and by the Deutsche Forschungsgemeinschaft and British Antarctic Survey.

References

Dalziel, I.W.D. & Elliot, D.H. (1971). Evolution of the Scotia Arc. *Nature, London*, **233**, 246–52.

Hervé, F. (in press). Relic clinopyroxene from the metagreywackes of the Scotia Metamorphic Complex, northeastern Elephant Island, Western Antarctica: evidence for a volcanic arc province. *Revista Geológica de Chile*, **28/29**.

Hervé, F., Godoy, E. & Davidson, J. (1983). Blueschist relic clinopyroxenes of Smith Island (South Shetland Islands): their composition, origin and some tectonic implications. In *Antarctic Earth Science*, eds. R.L. Oliver, P.R. James & J.B. Jago, pp. 363–6. Canberra; Australian Academy of Science and Cambridge; Cambridge University Press.

Hervé, F. & Kawashita, K. (1986). Edades K–Ar cretácicas del Complejo Metamórfico de Scotia en Cabo Lookout, isla Elefante. *Serie Científica, Instituto Antártico Chileno*, **34**, 55–6.

Hervé, F., Marambio, F. & Pankhurst, R.J. (1984). El Complejo Metamórfico de Scotia en cabo Lookout, isla Elefante, islas Shetland del Sur, Antártica: evidencias de un metamorfismo cretácico. *Serie Científica, Instituto Antártico Chileno*, **31**, 23–37.

Hervé, F. & Pankhurst, R.J. (1984). The Scotia metamorphic complex at Cape Bowles, Clarence Island, South Shetland Islands, Western Antarctica. *British Antarctic Survey Bulletin*, **62**, 15–24.

Hyden, G. & Tanner, P.W.G. (1981). Late Palaeozoic–early Mesozoic fore-arc basin sedimentary rocks at the Pacific margin in western Antarctica. *Geologische Rundschau*, **70(1)**, 529–41.

Loske, W., Miller, H. & Kramm, U. (1985). Dataciones radiométricas U–Pb en rocas de las islas del grup Piloto Pardo, Antártica. *Serie Científica, Instituto Antártico Chileno*, **32**, 39–46.

Loske, W., Miller, H. & Kramm, U. (in press). Charakter und Alter der Liefergesteine des Trinity Peninsula Formation Metasandsteins auf Cape Legoupil, Antarktisches Halbinsel: U–Pb Isotopenuntersuchungen an detritischen Zirkonen. *Neues Jahrbuch für Geologie und Paläontologie Monatshefte*.

Marsh, P.D. & Thomson, J.W. (1985). Report on Antarctic fieldwork: the Scotia Metamorphic Complex on Elephant Island and Clarence Island, South Shetland Islands. *British Antarctic Survey Bulletin*, **69**, 71–5.

Miller, H. (1983). The position of Antarctica within Gondwana in the light of Palaeozoic orogenic development. In *Antarctic Earth Science*, eds. R.L. Oliver, P.R. James & J.B. Jago, pp. 579–81. Canberra; Australian Academy of Science and Cambridge; Cambridge University Press.

Miller, H., Loske, W. & Kramm, U. (in press). Zircon provenance and Gondwana reconstruction – U–Pb data of detrital zircons from Triassic Trinity Peninsula Formation metasandstones. *Polarforschung*.

Pankhurst, R.J. (1983). Rb–Sr age constraints on the ages of basement rocks of the Antarctic Peninsula. In *Antarctic Earth Science*, eds. R.L. Oliver, P.R. James & J.B. Jago, pp. 367–71. Canberra; Australian Academy of Science and Cambridge; Cambridge University Press.

Quilty, P.G. (1982). Tectonic and other implications of the Middle–Upper Jurassic rocks and marine faunas from Ellsworth Land, Antarctica. In *Antarctic Geoscience*, ed. C. Craddock, pp. 669–78. Madison; University of Wisconsin Press.

Rex, D.C. (1976). Geochronology in relation to the stratigraphy of the Antarctic Peninsula. *British Antarctic Survey Bulletin*, **43**, 49–58.

Tanner, P.W.G., Hyden, G. & Pankhurst, R.J. (1982). Radiometric evidence for the age of the subduction complex in the South Orkney and South Shetland Islands, West Antarctica. *Journal of the Geological Society, London*, **139(6)**, 683–90.

Thomson, M.R.A. & Pankhurst, R.J. (1983). Ages of post-Gondwanian calc-alkaline volcanism in the Antarctic Peninsula region. In *Antarctic Earth Science*, eds. R.L. Oliver, P.R. James & J.B. Jago, pp. 328–33. Canberra; Australian Academy of Science and Cambridge; Cambridge University Press.

De Wit, M.J. (1977). The evolution of the Scotia Arc as a key to the reconstruction of southwestern Gondwanaland. *Tectonophysics*, **37(2)**, 53–81.

Accretion and subduction processes along the Pacific margin of Gondwana, central Alexander Island

T.H. TRANTER

British Antarctic Survey, Natural Environment Research Council, High Cross, Madingley Road, Cambridge CB3 0ET, UK
Present address: Shell Internationale Petroleum Maatschappij, Postbus 162, 2501 AN, Den Haag, Nethelands

Abstract

Four tectonostratigraphic units are recognized in the LeMay Group of central Alexander Island, Antarctica. They are interpreted as arc-derived clastic material and allochthonous volcanic rocks of oceanic origin incorporated in a subduction–accretion complex. Deformation is variable, but can mainly be related to accretionary processes acting at different levels within the complex. Metamorphic grade is generally low, but blue amphibole-bearing metabasalts indicate locally elevated *P–T* gradients. The accreted bodies include strata of Jurassic–Cretaceous age and suggest that correlations drawn with (?)Triassic sequences elsewhere in the Antarctic Peninsula region may not be strictly valid. The LeMay Group shows similarities with the Mesozoic accretionary fore-arc of New Zealand, and these two areas may be related.

Introduction

The LeMay Group (LMG) of Alexander Island is a thick sequence of variably-deformed and metamorphosed sedimentary and igneous rocks situated on the west coast of the Antarctic Peninsula (Edwards, 1980a; Burn, 1984), the site of a Mesozoic–Cenozoic magmatic arc associated with eastward subduction of Pacific and proto-Pacific oceanic crust. In recent regional syntheses (Smellie, 1981; Storey & Garrett, 1985) it has been interpreted as a subduction–accretion complex. The age of the LMG is equivocal, although parts are unconformably overlain by the Upper Jurassic–Lower Cretaceous Fossil Bluff Formation (Edwards, 1980b). Palaeontological data indicate the presence of Early Jurassic (Thomson & Tranter, 1986) and Cretaceous units (Burn, 1984), but evidence presented below shows that these may be allochthonous. Sparse K–Ar data indicate a Late Jurassic–Early Cretaceous deformational or metamorphic event (Burn, 1984).

The LeMay Group of central Alexander Island

Although the LMG is dominated by deformed arkosic sedimentary rocks, in central Alexander Island a more heterogeneous assemblage including lavas and volcaniclastic sedimentary rocks has been identified (see Burn, 1984 for a review). Recent fieldwork in this area, between latitudes 70°41′S and 70°57′ S (Fig. 1), has allowed the delineation of four geographically, lithologically, and to an extent, structurally distinct associations.

Conglomerate–sandstone–mudstone association A

Large thicknesses of this association crop out along the eastern flanks of the LeMay Range (Fig. 1). Three facies can be recognized:

(i) *Conglomerate–sandstone facies.* Polymict, polymodal matrix-supported conglomerates, often normally and rarely inversely graded, with flame structures or scours at their bases, are interbedded with plane-laminated, normally graded sandstones with rare fluid escape structures. The conglomerates and sandstones form beds up to 10 m thick. This facies is interpreted as high-density sediment gravity flow deposits (Lowe, 1982).

(ii) *Sandstone–siltstone–mudstone facies.* This facies consists of packets of locally graded, laterally continuous sandstone beds with flute and groove casts and shale intraclasts, passing upwards into thin (< 20 cm) repetitions of siltstone and cleaved mudstone. In the siltstone beds, planar laminations consistently pass upwards into cross- and convolute-laminations. This facies is interpreted as the deposits of low-density turbidity currents, transitional to high-density flows (Lowe, 1982). In places, facies (ii) overlies (i) to form broadly fining-upwards sequences.

(iii) *Pebbly mudstone facies.* This is volumetrically insignificant, and consists of rounded polymodal clasts randomly distributed within a cleaved mudstone matrix, with rare detached slump noses of fine sandstone. It is associated with facies (i) conglomerate, and represents debris flow deposits initiated by movement of high-density sediment gravity flows.

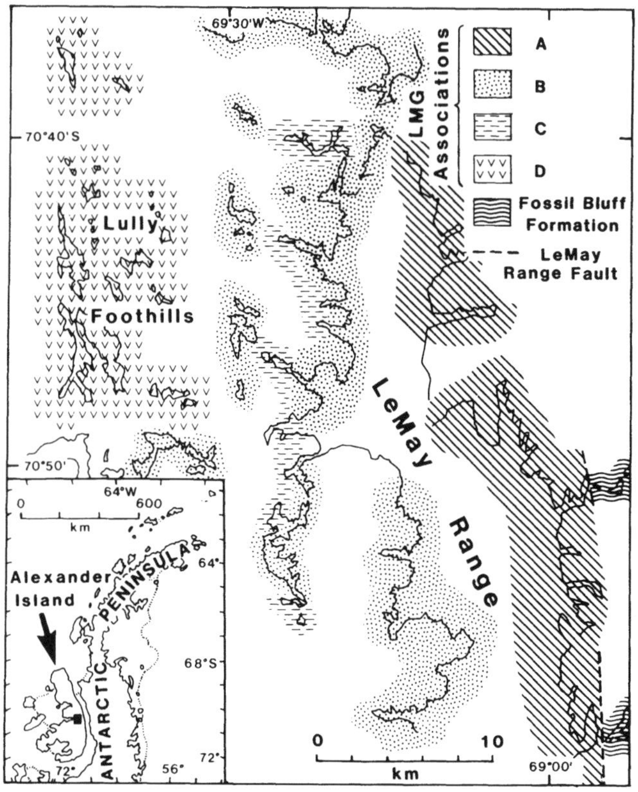

Fig. 1. Geological map of the LeMay Group of central Alexander Island. Solid box on inset shows area of main map.

Sandstone–mudstone association B

This association crops out throughout the western LeMay Range (Fig. 1) and is seen in both thrust and depositional contact with association C (below). However, exposures showing relationships between associations A and B are not seen. The association consists of a single facies, of thin- to medium-bedded sandstones and mudstones (generally < 10 cm thick) with some thicker sandstone beds, and rare conglomerates. Siltstones are almost totally absent. Much of the sequence is structureless but planar, cross- and convolute-laminations were observed together with rare syn-sedimentary faulting. Intraformational mud-flake breccias and flute casts attest to erosive current action, and sandstone load balls in mudstone indicate post-depositional instability. In terms of their generally thinly bedded and fine-grained nature, this association resembles the thinly bedded facies (ii) of association A. However, there is a higher percentage of sandstones in association B, siltstones are lacking, and the beds are generally thinner. Although good turbidite sequences were not recognized, the sedimentary structures are consistent with deposition from sand-laden erosive currents, and the rocks probably represent thin-bedded turbidites.

Chert–basalt association C

A narrow and discontinuous strip of basaltic lavas and associated sedimentary rocks is exposed along the western flank of the LeMay Range (Fig. 1). In places it passes into sandstones and mudstones of association B, but faulted (usually thrusted) contacts are also apparent. The lavas are pillowed, sheet-like or massive. Pillows range from 50–100 cm in long dimensions, with glassy or finely-crystalline margins, and vesicles and inter-pillow sedimentary material are scarce. Tabular or lenticular lava flows or sills occur above, below or within pillowed units. Interbedded with the lavas are bedded red and green radiolarian cherts, siltstones and mudstones. In most cases, the radiolarians are highly recrystallized, but a red mudstone in the central western LeMay Range yielded identifiable fragments of Jurassic–Cretaceous age (B.K. Holdsworth, pers. comm., 1987). There is evidence for some terrigeneous input with detrital angular fragments of quartz and feldspar in cherty siltstones, and a high volume of muddy material.

Basalt-tuff association D

A distinctive sequence of pillowed and massive basic lavas and coarser igneous rocks, together with tuffaceous sedimentary rocks, forms the Lully Foothills Formation, proposed by Burn (1984) as a subdivision of the LMG. The pillow lavas form units up to 30 m thick and, in contrast to those of association C, they are generally vesicular. Beds of hyaloclastic basaltic breccia, up to 10 m in thickness, occur between pillowed units. Gradation from complete pillows through broken fragments to an unsorted breccia can be traced in places. Locally-derived basic lithic and vitric tuffs, probably deposited in shallow water, are well developed in the north and west Lully Foothills, and nearby subaerial volcanism is suggested by the local occurrence of volcanic bombs and the shard-rich nature of the tuffs. Fossils collected from tuffs in the northern Lully Foothills include a diagnostic ammonite of Sinemurian (Early Jurassic) age (Thomson & Tranter, 1986).

Deformational and metamorphic history

The current study confirms the polyphase nature of deformation in the LMG summarized by Burn (1984), but has highlighted the problems of structural correlation in an environment in which deformation may have been diachronous. A summary of the structural history is given below: for a more detailed treatment see Tranter (1987).

The earliest deformation recognized throughout the area has resulted in the development of bedding-parallel or sub-parallel fabrics. In the clastic associations A and B widespread stratal disruption and local westward-directed thrusting of poorly-lithified sediment is apparent. Elsewhere in the western LeMay Range, pressure solution fabrics and metamorphic mineral growth are common. These fabrics are generally bedding-parallel and overprint earlier bedding-parallel stratal disruption in places, indicating the continued application of a similar stress regime at a variety of depths and states of lithification. In association C cherts and siltstones form *décollements* for westward-directed thrusting of lavas, with associated cleavage development. A major tectonic boundary between association D and the rocks to the east is marked by a > 150 m wide *mélange* zone consisting of inclusions of basalt and sedimentary rocks up to several metres in size enclosed in a pervasively cleaved red and green phyllitic matrix.

In the eastern LeMay Range, these early structures are deformed by eastward-(arcward-) verging folds and eastward-directed thrusts, whilst to the west a zone of upright N–S-trending downward-facing folds of equivocal origin is developed in the sandstones and shales of association B. These in turn are cut by a later phase of westward-directed thrusting.

Spaced cleavages, quartz veins and joints occur throughout the area, but with few coherent patterns. N–S normal and oblique-slip faulting are probably related to major Cenozoic structures, such as George VI Sound between Alexander Island and the Antarctic Peninsula (Crabtree, Storey & Doake, 1985), and the LeMay Range fault (Edwards, 1980b), which forms the boundary between the LMG and the Fossil Bluff Formation.

Metamorphic grade increases from east to west, and although the nature of the exposure precludes detailed interpretation, some changes in grade may be coincident with association boundaries. The clastic rocks of association A are the least metamorphosed, containing detrital K-feldspar and plagioclase. Patches and veins of prehnite, quartz and calcite are developed. To the west, associations B and C reach low greenschist-facies in places. Detrital K-feldspar is generally absent, plagioclase is largely altered, and chlorite, sericite and muscovite are developed in schistose parts. Transitional greenschist–glaucophane-schist-facies metabasalts containing blue amphibole were reported from the north-west LeMay Range by Edwards (1980a) and identification of a second locality 12 km to the south suggests that these may represent parts of a more extensive belt of high *P–T* metamorphism.

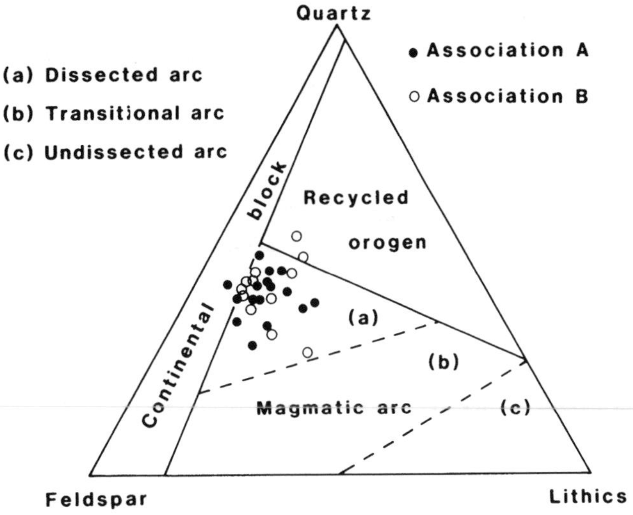

Fig. 2. Modal compositions of sandstones from the LMG. Provenance field boundaries are those of Dickinson *et al.* (1983).

Interpretation

Associations A and B

Edwards (1980a) and Burn (1984) suggested that the clastic rocks of the LMG were derived from the east and deposited in a submarine fan system at an active continental margin. The sedimentology of associations A and B is broadly consistent with such an interpretation, but few indications of overall fan geometry are evident, due to the great difficulty of establishing a stratigraphy in these deformed and unfossiliferous strata. The conglomerate–sandstone bodies interbedded with thinly bedded turbidites of association A may represent inner fan channel sediments with associated levées and interchannel deposits, whereas the thinly bedded turbidites of association B may represent suprafan or lower fan deposition (see Walker (1984) for fan terminology).

Associations A and B can also be considered in terms of a convergent margin facies model (Underwood & Bachmann, 1981), in which aspects of the submarine fan model are incorporated into a scheme more applicable to an active margin. In this, submarine topography parallel to the active margin leads to axial transport, and to the ponding of sediment in trench slope basins characterized by a high percentage of coarse sediment due to bypass of finer material. Debris flow and hemipelagic deposits also occur whilst the structural and metamorphic history is distinct from that of the underlying accretionary complex (Moore & Karig, 1976). In contrast to the LMG elsewhere in central Alexander Island, association A contains thick conglomerates and sandstones, is of consistently low metamorphic grade and has an early structural history dominated by stratal disruption and extension with only minor thrusting (Tranter, 1987). It is thus possible that associations A and B represent a ponded sediment accumulation and part of the underlying accretionary complex, respectively, although nowhere are the critical contacts necessary to test this hypothesis exposed.

Modal analysis of sandstones from associations A and B indicates derivation from a magmatic arc source (Fig. 2). Although both sandstones and conglomerates include an arc-derived igneous component, lithologies atypical of such a source, including almost pure quartzites, occur as clasts in conglomerates. Apart from some recently discovered quartzites of indeterminate age from the eastern coast of the Antarctic Peninsula (Storey *et al.*, 1987), no possible peninsula source areas for these lithologies are presently known. Unless they are derived from some currently unrecognized basement fragment on the peninsula, their presence may indicate a more distant continental source for parts of the LMG, and has important implications for palaeogeography during fore-arc development.

In summary, associations A and B both represent parts of the same suite of clastic rocks of mixed provenance, deposited in an active marginal setting and subjected to deformation at various levels within the upper parts of the subduction-accretion complex.

Associations C and D

Igneous rock associations in a subduction complex can occur in two settings: (i) as autochthonous accumulations due to fore-arc volcanism; and (ii) as allochthonous blocks accreted to the margin. The characteristic occurrence of pillow lavas and radiolarian cherts in association C indicates some type of ocean-floor environment, although a terrigenous sedimentary component, and sedimentary contacts with sandstones of association B, suggest incomplete isolation from clastic sources. The structural and metamorphic history of the association, commencing with a phase of westward-directed thrusting and including local metamorphism to transitional glaucophane-schist-facies, is consistent with the accretion and disruption of a sliver of oceanic crustal material, perhaps at a subsurface zone of underplating at some distance arcward of the accretionary toe (Tranter, 1987).

Smellie (1981) proposed an autochthonous fore-arc basin origin for the rocks of the Lully Foothills Formation (association D), suggesting that they were deposited in shallow water on the arcward side of a shoaling trench–slope break, in a basin floored by subduction complex. However, more recent evidence shows that the Lully Foothills Formation may be an allochthonous block. Modelling of the Bouguer anomaly over these basic rocks shows a thick, steeply sided body which extends to a depth of at least 4 km (Garrett & Storey, 1987). Although the geophysical boundaries of this body are interpreted by these authors as Cenozoic faults, the eastern margin correlates well with the *mélange* zone. This zone formed at an early stage in the structural evolution of the region, and could represent the original suture between an allochthonous Lully Foothills Formation block and other parts of the accretionary complex (which was later reactivated by the Cenozoic structures). There is no structural evidence that the subsurface morphology is the expression of a tectonically thickened basin fill; nor is there any evidence for marginal facies variations and unconformable contacts with other parts of the LMG, as might be expected in a ponded basin. Preliminary geochemistry of rocks of associations C and D shows a varied suite of basic rocks, some of which have within-plate affinities, with chemical variations extending across association boundaries.

Association D is therefore interpreted as a large block of volcanic material, probably representing part of an oceanic island accreted onto the margin.

In summary, the LMG in central Alexander Island contains both arc-derived clastic sedimentary rocks and allochthonous volcanogenic sequences, and includes rocks of Jurassic and Cretaceous age. Although no one aspect of the geology is diagnostic, the overall sedimentological, structural and metamorphic evolution is consistent with an accretion–subduction complex setting.

Regional considerations and correlations

The LMG forms part of a belt of Palaeozoic–Cenozoic subduction complexes along the Pacific margins of southern South America, the Scotia arc, Lesser Antarctica and New Zealand (Dalziel & Grunow, 1985). Prior to Gondwana break-up, this belt formed part of the Panthalassic margin of the supercontinent, but gradual post-mid-Jurassic continental fragmentation led to increasingly independent subduction histories in the component parts of the once continuous margin.

The accretionary fore-arc terrane of Alexander Island has been correlated with the Trinity Peninsula Group (TPG) and similar sequences in northern Antarctic Peninsula and adjacent areas (e.g. Storey & Garrett, 1985). However, the TPG is probably no younger than Triassic and is of highly debateable tectonic setting (Smellie, this volume, p. 411). Although some of the LMG could be of a similar age to the TPG, the presence of accreted Mesozoic strata in the LMG indicates that substantial parts are younger and, at present, a direct correlation between the two seems unlikely.

The Scotia metamorphic complex of the South Shetland and South Orkney islands represents the deeper levels of a subduction complex. Accretion and metamorphism range from Late Palaeozoic–Cretaceous (Tanner, Pankhurst & Hyden, 1982), suggesting that parts of the complex are contemporaneous with the LMG. However, apart from fragmentary (?)TPG equivalents, low-grade, high-level LMG-type sequences are absent in the Scotia metamorphic complex.

In southern South America, the accretionary fore-arc is mainly of Late Palaeozoic–earliest Mesozoic age prior to a Triassic–Middle Jurassic hiatus in subduction (Bruhn & Dalziel, 1977). In contrast, the post-Middle Jurassic Andean arc has only a very poorly developed accretionary complex (Dalziel & Forsythe, 1985). Subduction erosion, uplift and subaerial erosion, or Cenozoic strike-slip faulting, may have removed the fore-arc (Bruhn & Dalziel, 1977), or it may never have developed. Correlations with the LMG are therefore hard to draw.

The Marie Byrd Land–Thurston Island–Eights Coast margin has belts of Late Palaeozoic–early Mesozoic and late Mesozoic (?)subduction-related plutonism (Cooper *et al.*, 1982), but related fore-arc sequences are not preserved. Syntheses of regional geology (below) indicate that parts of the fore-arc are now incorporated into the New Zealand margin.

In New Zealand, a contemporaneous island arc, fore-arc and accretionary complex system of Permian–Cretaceous age is recognized. A major quartzo-feldspathic accretionary complex of Carboniferous–Cretaceous age (the Torlesse Terrane), is interpreted as allochthonous. It is derived from a continental-margin volcanic arc to the east of New Zealand (Marie Byrd Land) and accreted to the autochthonous fore-arc by Permian–Cretaceous sinistral strike-slip associated with oblique subduction (Bradshaw, Adams & Andrews, 1981; Crook & Feary, 1982). The Torlesse Terrane is in part contemporaneous with the LMG, and very similar in terms of sedimentology, provenance, metamorphism and deformation, and Cooper *et al.* (1982) correlated the two (together with the TPG). It is interesting to note that, as a consequence ridge crest–trench collision in the plate configuration proposed by Crook & Feary (1982), Permian–Early Cretaceous dextral transport of material from the Marie Byrd Land–Thurston Island–Eights Coast margin towards the Antarctic Peninsula is possible. This raises the possibility that parts of the LMG may be derived from this part of the margin, by a combination of tectonic and sedimentary processes. Such a source could explain the presence of quartzite clasts within the LMG, since pre-Mesozoic quartzites are known from a number of localities to the south of Alexander Island: e.g. Fitzgerald Bluffs (Laudon *et al.*, 1987), the Ellsworth Mountains (Hjelle, Yoshida & Winses, 1982) and the Transantarctic Mountains (Laird & Bradshaw, 1982). Transport of material along the continental margin is also consistent with the recent recognition of dextral strike-slip on Alexander Island (Nell & Storey, this volume, p. 443) and could perhaps explain the absence of autochthonous forearc terranes along the Marie Byrd Land–Thurston Island–Eights Coast margin.

The LMG thus forms part of a long-lived and widespread belt of subduction complexes along the Panthalassic margin of Gondwana. The evidence for post-Early Jurassic–Cretaceous accretion in the LMG suggests that the New Zealand fore-arc and parts of the Scotia arc provide the closest correlatives.

Acknowledgements

The author wishes to thank colleagues at the British Antarctic Survey for helpful discussions and comments during the preparation of this paper.

References

Bradshaw, J.D., Adams, C.J. & Andrews, P.B. (1981). Carboniferous to Cretaceous on the Pacific margin of Gondwana: the Rangitata phase of New Zealand. In *Gondwana Five*, ed. M.M. Cresswell & P. Vella, pp. 217–21. Rotterdam; A.A. Balkema.

Bruhn, R.L. & Dalziel, I.W.D. (1977). Destruction of the Early Cretaceous marginal basin in the Andes of Tierra del Fuego. In *Island Arcs, Deep-sea Trenches and Back Arc Basins*, ed. M. Talwani & W.C. Pitman III, pp. 395–406. Washington, DC; American Geophysical Union.

Burn, R.W. (1984). *The Geology of the LeMay Group, Alexander Island.* Cambridge; British Antarctic Survey Scientific Reports, No. 109. 65 pp.

Cooper, R.A., Landis, C.A., LeMasurier, W.E. & Speden, I.G. (1982). Geologic history and regional patterns in New Zealand and West Antarctica – their paleotectonic and paleogeographic significance. In *Antarctic Geoscience*, ed. C. Craddock, pp. 43–53. Madison; University of Wisconsin Press.

Crabtree, R.D., Storey, B.C. & Doake, C.S.M. (1985). The structural evolution of George VI Sound, Antarctic Peninsula. *Tectonophysics*, **114**, 431–42.

Crook, K.A.W. & Feary, D.A. (1982). Development of New Zealand according to the forearc model of crustal evolution. *Tectonophysics*, **87**, 65–107.

Dalziel, I.W.D. & Forsythe, R.D. (1985). Andean evolution and the terrane concept. In *Tectonostratigraphic Terranes of the Circum-Pacific Region*, ed. D.G. Howell, pp. 565–81. Houston; Circum-Pacific Council for Energy & Mineral Resources, Earth Science Series, No. 1.

Dalziel, I.W.D. & Grunow, A.M. (1985). The Pacific margin of Antarctica: terranes within terranes within terranes. In *Tectonostratigraphic Terranes of the Circum-Pacific Region*, ed. D.G. Howell, pp. 555–64. Houston; Circum-Pacific Council for Energy & Mineral Resources, Earth Science Series, No. 1.

Dickinson, W.R., Beard, L.S., Brakenridge, G.R., Erjavec, J.L., Ferguson, R.C., Inman, K.F., Knepp, R.A., Lindberg, F.A. & Rydberg, P.T. (1983). Provenance of North American Phanerozoic sandstones in relation to tectonic setting. *Geological Society of America Bulletin*, **94(2)**, 222–35.

Edwards, C.W. (1980a). The geology of eastern and central Alexander Island. PhD thesis, University of Birmingham, 228 pp. (unpublished).

Edwards, C.W. (1980b). New evidence of major faulting on Alexander Island. *British Antarctic Survey Bulletin*, **49**, 15–20.

Garrett, S.W. & Storey, B.C. (1987). Lithospheric extension on the Antarctic Peninsula during Cenozoic subduction. In *Continental Extensional Tectonics*, ed. M.P. Coward, J.F. Dewey & P.L. Hancock, pp. 419–31. London; Special Publication of the Geological Society of London, No. 28.

Hjelle, A., Yoshida, O. & Winses, T.S. (1982). Geology and petrology of the southern Heritage Range, Ellsworth Mountains. In *Antarctic Geoscience*, ed. C. Cradock, pp. 599–608. Madison; University of Wisconsin Press.

Laird, M.G. & Bradshaw, J.D. (1982). Uppermost Proterozoic and Lower Palaeozoic geology of the Transantarctic Mountains. In *Antarctic Geoscience*, ed. C. Craddock, pp. 525–33. Madison; University of Wisconsin Press.

Laudon, T.S., Lidke, D.J., Delevoryas, T. & Gee, C.T. (1987). Sedimentary rocks of the English Coast, eastern Ellsworth Land. In *Gondwana Six: Structure, Tectonics, and Geophysics*, ed. G.D. McKenzie, Geophysical Monograph, No. 40 pp. 183–9. Washington, DC; American Geophysical Union.

Lowe, D.R. (1982). Sediment gravity flows: II. Depositional models with special reference to the deposits of high-density turbidity currents. *Journal of Sedimentary Petrology*, **52(1)**, 279–97.

Moore, G.F. & Karig, D.E. (1976). Development of sedimentary basins on the lower trench slope. *Geology*, **4(11)**, 693–7.

Smellie, J.L. (1981). A complete arc–trench system recognised in Gondwana sequences of the Antarctic Peninsula region. *Geological Magazine*, **118(2)**, 139–59.

Storey, B.C. & Garrett, S.W. (1985). Crustal growth of the Antarctic Peninsula by accretion, magmatism and extension. *Geological Magazine*, **122(1)**, 5–14.

Storey, B.C., Wever, H.E., Rowley, P.D. & Ford, A.B. (1987). Report on Antarctic fieldwork: the geology of the central Black Coast, eastern Palmer Land. *British Antarctic Survey Bulletin*, **77**, 145–55.

Tanner, P.W.G., Pankhurst, R.J. & Hyden, G. (1982). Radiometric evidence for the age of the subduction complex in the South Orkney and South Shetland Islands, West Antarctica. *Journal of the Geological Society, London*, **139(6)**, 683–90.

Thomson, M.R.A. & Tranter, T.H. (1986). Early Jurassic fossils from central Alexander Island and their geological setting. *British Antarctic Survey Bulletin*, **70**, 23–39.

Tranter, T.H. (1987). The structural history of the LeMay Group of central Alexander Island, Antarctica. *British Antarctic Survey Bulletin*, **77**, 61–80.

Underwood, M.B. & Bachmann, S.B. (1981). Sedimentary facies associations within subduction complexes. In *Trench Forearc Geology*, ed. J.K. Leggett, pp. 537–50. London; Special Publication of the Geological Society of London, No. 10.

Walker, R.G. (1984). Turbidites and associated coarse clastic deposits. In *Facies Models*, 2nd edn, ed. R.G. Walker, pp. 171–88. St John's, Newfoundland; Geoscience Canada, Reprint Series 1.

Strike-slip tectonics within the Antarctic Peninsula fore-arc

P.A.R. NELL & B.C. STOREY

British Antarctic Survey, Natural Environment Research Council, High Cross, Madingley Road, Cambridge CB3 0ET, UK

Abstract

The Antarctic Peninsula was the site of magmatic arc activity associated with the eastward subduction of proto-Pacific oceanic lithosphere for much of the Mesozoic. It is composed of calc-alkaline igneous rocks, accretionary complexes, and extensive fore- and back-arc sedimentary basins. Structural data from the accretionary complex of Alexander Island suggest that transcurrent motion associated with accretionary processes may have been caused by oblique subduction. Later, long-lived, strike-slip faulting may have a different cause, and affected both the accretionary complex, and fore-arc basin. In the latter, its effects include NW–SE-trending, oblique folds and thrusts, faults with NW–SE extension, and contemporaneous sandstone dykes. Strike-slip motion has played an important, but previously unrecognized, control on the development of the Antarctic Peninsula fore-arc, and may have controlled the formation of sedimentary basins.

Introduction

The Antarctic Peninsula, the southward continuation of the Andes, has been the site of proto-Pacific subduction since at least the Triassic (see review by Storey & Garrett, 1985). Associated with subduction has been the growth of an accretionary prism, including accretionary complex and fore-arc basin rocks such as the LeMay Group (LMG) and Fossil Bluff Formation (FBF) of Alexander Island (Fig. 1). Intrusion of a plutonic suite has dominated calc-alkaline, subduction-related magmatic activity and extrusion of a thick volcanic sequence has occurred along the spine of the peninsula. Magmatic activity migrated trenchwards as the fore-arc grew to the west. Subduction ceased following the northward migration of a colliding spreading ridge with the trench (Barker, 1982). This marked the cessation of accretion and calc-alkaline magmatism.

The LeMay Group – an accretionary prism complex

The geology of the LeMay Group has been reviewed by Burn (1984). It is a thick sequence of unfossiliferous arc-derived, turbiditic greywackes, black mudstones and conglomerates, with basaltic volcanic rocks, cherts and pelagic shales of oceanic character (Tranter, this volume, p. 437). The LMG has undergone complex deformation, with the production of N–S-trending belts of *mélange* and broken formation. The metamorphic grade in northern Alexander Island increases north-eastwards from zeolite-facies to lower amphibolite-facies, with the local development of lawsonite and blue amphibole in restricted areas of central Alexander Island. The

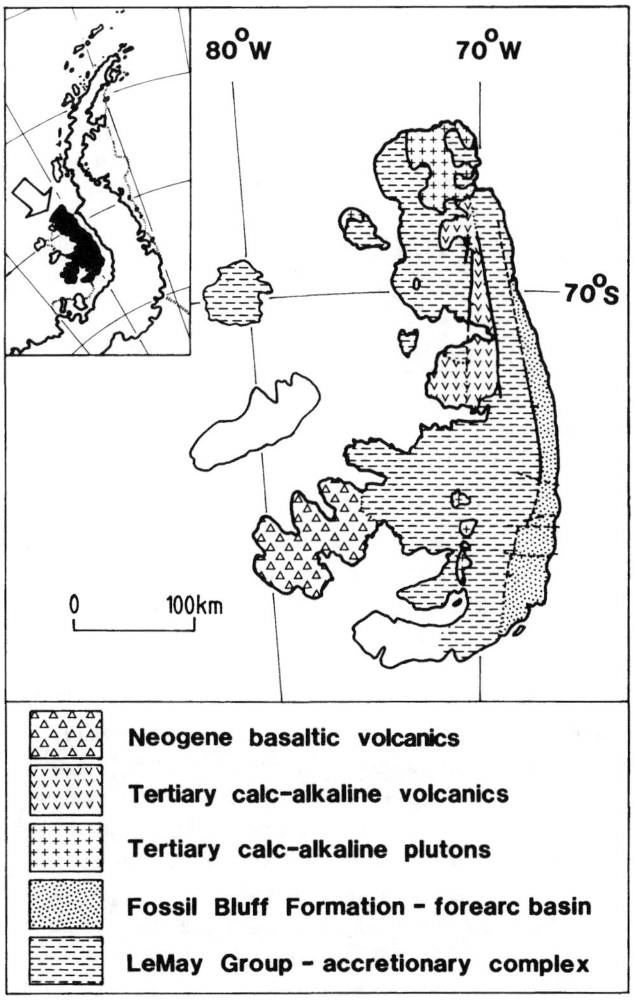

Fig. 1. Simplified geological map of Alexander Island.

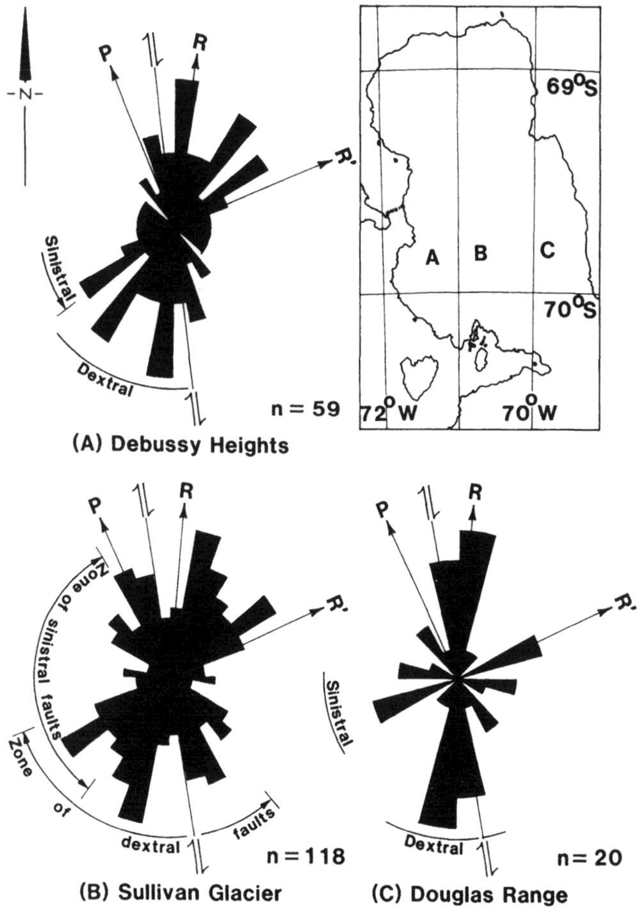

Fig. 2. Fault-strike orientations of steeply-dipping transcurrent faults in the LeMay Group, northern Alexander Island, showing angular relationships of the Riedel model.

lithologies, structures, and metamorphism suggest that these rocks are part of an accretionary prism. The age of sedimentation (fossils show Early Jurassic–mid-Cretaceous ages) and deformation (cooling ages 165–105 Ma) of the LMG is reviewed by Tranter (this volume, p. 437).

Strike-slip history of the LeMay Group

Fieldwork studying the structural evolution of the LMG in a west–east traverse across northern Alexander Island has been recently completed. Many late strike-slip faults with good movement indicators were found. These faults cross-cut, and are unrelated to, the *mélanges*, thrusts, folds, and cleavage.

The largest are sub-vertical, dextral, strike-slip faults which trend 350°, and can be traced for many kilometres. Other smaller strike-slip faults are seen in a complex association with flat-lying detachments showing northerly or NNW slickensides, positive and negative flower structures, arrays of strike-parallel compressional and extensional dip-slip faults related to tip-strains, and shear-related oblique extension. Fig. 2 shows the orientations of mesoscopic and megascopic strike-slip faults. Using the trend and sense of the major faults as indicators of the overall shear, a dextral Riedel, rather than a sinistral Riedel, or block-rotation model (McKenzie & Jackson, 1986; Nur, Ron & Scotti, 1986), explains the dominant NNE–SSW, ENE–WSW and NNW–SSE directions as R, R′, and P fractures. However, some sinistral faults are present in the NW–SE quadrants at a higher angle than P fractures, and these cannot be explained by a Riedel model. They also show some reverse motion.

The McKenzie & Jackson model (1986, fig. 1) predicts that block-rotation faults in the NW–SE quadrants should be associated with transpression. Limited block rotation may have occurred, and could explain the field-association with dip-slip fault arrays. The suppression of transtension-related NE–SW block-rotation faults may be due to the overall shear geometry.

Evidence for strike-slip deformation prior to the above faults is more circumstantial. The earliest tectonic deformation recognized in northern Alexander Island is the formation of zones of *mélange* and broken formation. These show an essentially oblate bedding-parallel extensional strain, spanning the period of dewatering and cleavage formation. Field relations suggest that these zones represent major west-directed early thrusts, but internally mesoscopic shear criteria are inconsistent or lacking. Oblate strains are difficult to achieve without along-strike extension. However, regional considerations make this unlikely in a deep tectonic setting. Strike-slip deformation may be the key, where a combination of layer-parallel, extensional, pure shear with layer-normal, simple shear (Coward & Kim, 1981) can produce an oblate strain. In an accretionary prism, spreading of material on thrusts (Platt, 1986), combined with a component of transport parallel to the prism, could produce the large, strongly-oblate strains observed. Differential movement along a thrust plane, unless internal rotations were large, could only build up slightly oblate strains.

One unusual feature of the Sullivan Glacier area is the presence of some major NNW–SSE-trending, steeply-plunging or reclined, tight, cleavage-related folds. The onset of stratal-disruption predates cleavage formation in the *mélange* zones, suggesting that they may be earlier than these folds. The folds have a wavelength of up to 5 km, and the steep plunge can be followed for 7 km with no change. They show a consistent southward facing on the associated cleavage, which is generally upwards, or locally downwards, depending on the plunge. The absence of reversals in facing direction, and the low tectonic strains present, suggest a 'sheath-fold' model is inappropriate. In addition, the absence of any major associated thrusts suggests that these folds do not represent local, lateral-ramp-related folds or lateral tip zones. It is probable that these folds initiated perpendicular to the incremental shortening direction above an oceanic slab descending to the north-east, and have been little rotated since (Fig. 3). Further accretion-related back-rotation would serve to steepen the plunge of the folds.

In association with the steeply plunging folds are more moderately plunging, cleavage-related folds with an anticlockwise transecting cleavage. This could result from a swing in the orientation of the slipped region of a fold-related detachment, producing (Fig. 3) a region of dextral transpression (Sanderson & Marchini, 1984), perhaps controlled by variations in pore pressure or lithology. Oblique subduction is clearly required by this model. Transecting cleavage has been used to infer a

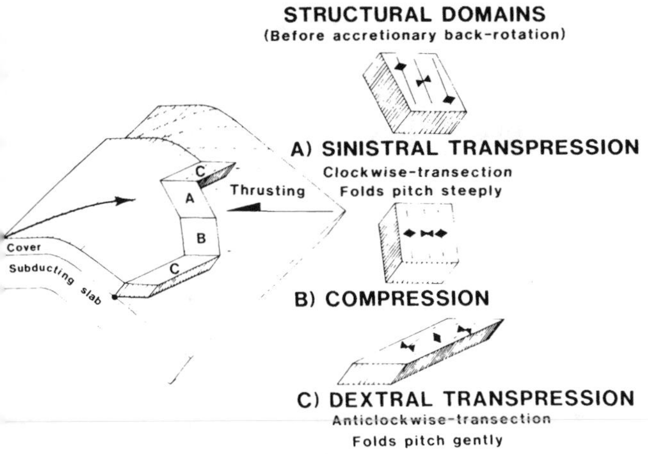

Fig. 3. The relationship between steeply-plunging and transected folds, in an oblique subduction model.

component of strike-parallel deformation in other accretionary prisms (e.g. Phillips, Flegg & Anderson, 1979).

The Fossil Bluff Formation, a fore-arc basin sequence

This succession of mostly marine mudstones and sandstones of Kimmeridgian–Albian age was reviewed by Taylor, Thomson & Willey (1979). Edwards (1980) noted that the FBF was usually downfaulted to the east against the LMG on the LeMay Range fault, a major fault zone trending N–S parallel to George VI Sound along much of Alexander Island (Fig. 4). Eastwards, the FBF is presumed to be cut-out against the magmatic arc by a major fault zone extending beneath George VI Sound.

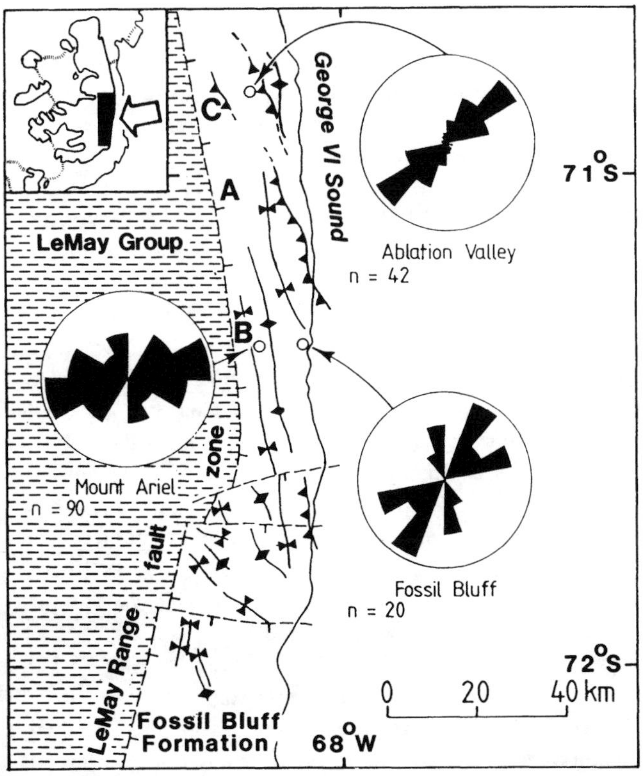

Fig. 4. Sandstone dyke, fold and thrust orientations in the FBF. Includes data from Taylor (1982), D.I.M. Macdonald and P.J. Butterworth (pers. comm.).

Recent work has confirmed the presence of an unconformity between older, strongly cleaved, crenulated, and thrusted LMG, and little-deformed overlying FBF at two localities within the LeMay Range fault zone. Conglomerates within the FBF are dominated by rounded, arc-derived, plutonic and volcanic clasts (Horne, 1968), and this provenance is supported by the palaeocurrents (Butterworth & Macdonald, this volume, p. 449). However, near the LeMay Range fault, some of the conglomerates contain numerous sandstone clasts at Nonplus Crag (Locality C, Fig. 4) (T.H. Tranter, pers. comm.), and angular phyllite clasts west of Lunar Crag (Locality A, Fig. 4), which have an atypical local provenance (Edwards, 1980) from the LMG. During deposition of the FBF, the LMG was exposed close to the present position of the LeMay Range fault, and marked the edge of the main FBF basin. This shows the FBF was deposited in an elongate basin similar in dimensions to its present outcrop. Butterworth & Macdonald (this volume, p. 449) describe the change from a deep-marine-facies in the north, to a deltaic-facies with fossil forests in the south. These changes may reflect local deepenings of the basin.

One spectacular feature of the FBF is the presence of 'disturbed zones' which have been interpreted as major slope collapse features (Taylor *et al.*, 1979), and indicate an active tectonic environment during deposition. Minor unconformities are present in the FBF and indicate local erosion. Whether the original cause is sedimentary or tectonic is unknown.

Evidence of strike-slip motion in the FBF

Work in the LeMay Range fault zone (Locality A, Fig. 4) has shown that, on the east side of the northern Milky Way, the LMG–FBF unconformity is faulted by several N–S faults dipping steeply to the east. These show slickensides indicating some history of dextral/normal oblique motion.

In the vicinity of these faults, and parallel to them, the unconformity is stepped up and down by closely spaced microfaults. Above the strongly faulted unconformity in these regions, mudstone intraclasts in FBF conglomerates have a steeply dipping NW–SE cleavage supporting a dextral sense of shear (Fig. 5b). The matrix sandstones are not cleaved, apparently having been deformed by grain-boundary sliding before lithification.

Open upright folds of the unconformity and FBF at this locality plunge gently to the SSE. This is typical of folding elsewhere in the FBF (Bell, 1975). Conjugate normal faults indicating 20% NW–SE extension were present in the FBF and LMG here (Fig. 5c). Other analogous faults at Mount Ariel similarly post-date the folding and related veining (see below). To the south of the Milky Way at Atoll Nunataks (Locality B, Fig. 4), the FBF dips steeply away from the LeMay Range fault, and contains a fold-rotated, but otherwise similar, set of conjugate extensional faults (Fig. 5a). It appears that NW–SE extension and NE–SW compression were coeval during deformation of the FBF. Many of these conjugate 'normal' faults show initial oblique slickensides, related to a dextral component of shear during extension. Cross-cutting of conjugate faults ends dip–slip motion on the earlier fault of a pair. Such faults often showed continued dextral strike-slip motion paral-

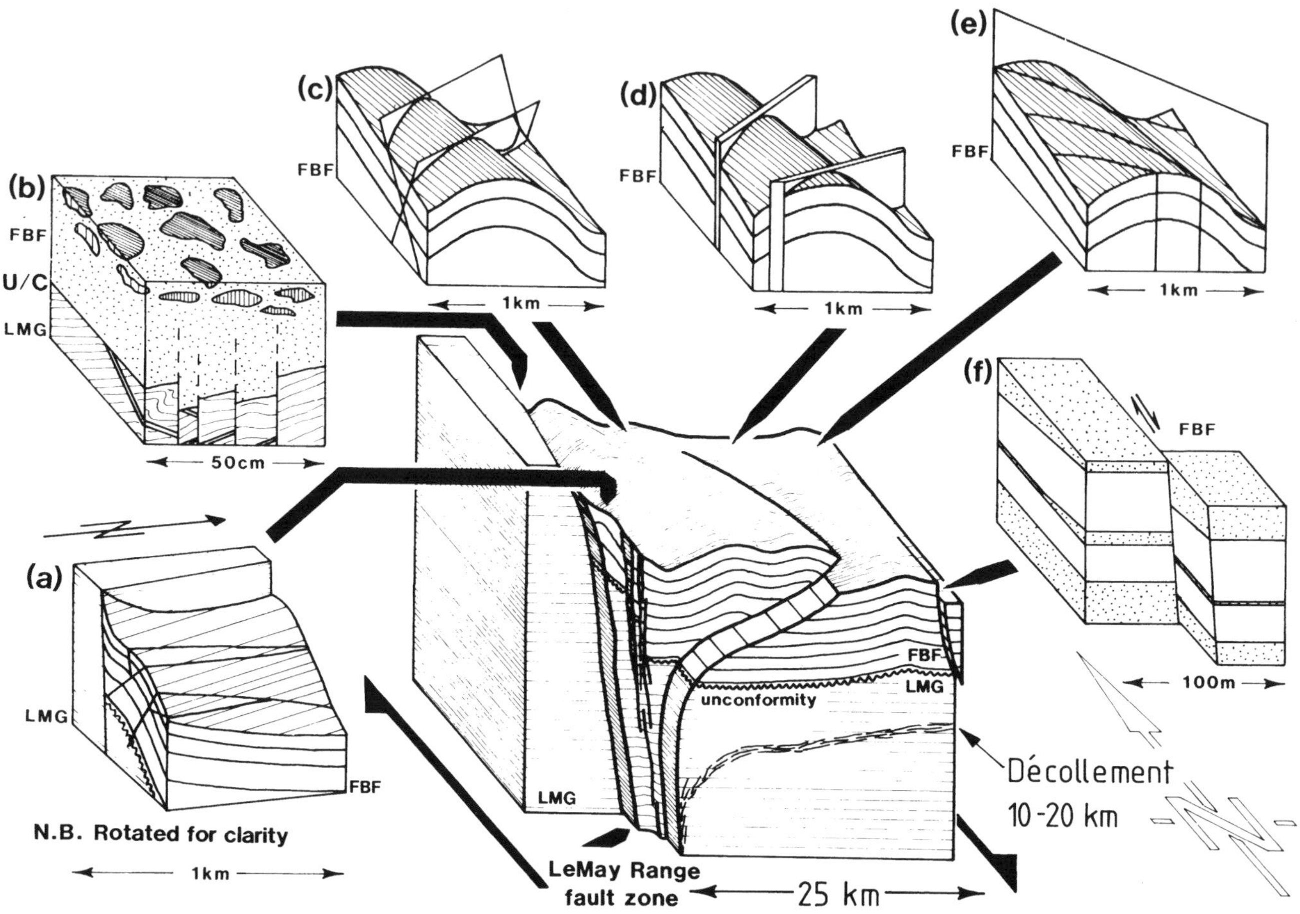

Fig. 5. The relationship of minor and major structures in the FBF. See text for explanation.

lel to the fault/fault intersection. This suggests that the simultaneous extension and compression was related to a regional shear, and not simple fold-parallel extension.

A weak, steeply dipping cleavage is recognizable in many parts of the FBF. Bell (1975) noted that the cleavage/bedding intersection, trending NW–SE, transected the gently curving, NNW–SSE fold axes, which he took as evidence of two phases of deformation (Fig. 5e). A simpler interpretation is that these folds formed in a single deformation by a non-coaxial, transpressive strain. Soper (1986) gave criteria for non-coaxial strains, including: regionally curvilinear tectonic trends, and the association of folds and cleavage with strike-slip faults. One mechanism by which a transecting cleavage may be formed (Soper, 1986) is by transpression accompanying early deformation, by dewatering or grain-boundary sliding, which would impart no initial fabric to the rock. The fold axes rotate as material lines. The cleavage can be superimposed later, parallel to the X–Y plane of the incremental strain ellipsoid as deformation mechanisms change when the rock becomes lithified. If this mechanism operated in the FBF, the anticlockwise sense of transection is evidence of dextral transpression.

Preliminary results from strain analysis on deformed ammonites and serpulids from near Fossil Bluff show low bedding-plane strains ($Rs = 1.1–1.2$), with long axes oriented close to the cleavage/bedding intersection. Strain factorization (Coward & Kim, 1981) allows these strains to be resolved into bedding-parallel shortening and bedding-normal shear, supporting dextral transpression. These ductile strains suggest the FBF basin was shortened by about half as much as the western margin was translated northwards.

Sandstone dykes are found in many parts of the FBF (Taylor, 1982). The FBF dykes are sub-vertical, and frequently occur in en-echelon or bifurcating arrays. The sediments were sufficiently cohesive to allow mudstones to be stoped off, preserve sharp, unmixed dyke margins, and permit belemnites to be broken and separated in opposing walls. However, at depth, the sandstones still contained sufficient pore fluid to allow sandstones to be fluidized, carrying bivalve shells unbroken. Some dykes can be traced for at least 200 m vertically up hillsides, which provides a minimum for the depth of burial at the time of intrusion. Dykes in mudstones show evidence of vertical shortening in the form of mullions and horizontally plunging, close- to ptygmatic-folds associated with a bedding-parallel compactional fabric. The dykes formed at depth in an actively-dewatering sedimentary pile.

As Taylor (1982) noted, they generally strike ENE–WSW, their preferred orientation suggesting some tectonic control (Fig. 4). They are kinematically analogous to large-scale, en-echelon vein arrays. Dykes in other orientations occur, but show different minor structures. Those striking in the NE–SW quadrants frequently show wall-normal extension on minor, conjugate, normal faults. Shortening also occurs along-strike on minor thrust pairs, conjugate, en-echelon vein arrays, and vertically plunging folds. Other, less common dykes with NW–SE strikes, show wall-normal shortening on minor, reverse faults, and/or faulting or boudinage, suggesting along-strike or vertical extension. This again suggests simultaneous compression and extension.

Although most dykes are parallel-sided, rare steps in the opposing walls (and broken belemnites) can occasionally be matched, showing some dextral shear on the main ENE–WSW set of dykes. This set forms some en-echelon arrays, and the dominant left-stepping arrangement of dykelets supports this (Riedel) shear sense (Nicholson & Pollard, 1985). The NW–SE-striking set most often occurs in right-stepping sets.

Fibrous, bedding-parallel, calcite–quartz veins are common in the FBF. They are related to the folding and thrusting, and accommodated bedding-parallel slip as a result of overpressuring which periodically jacked-apart bedding planes. Fold–vein, thrust–vein and vein–dyke relationships show that dyke injection mostly preceded vein formation, thrusting and folding, but did overlap in time, and all were associated with high fluid pressures.

In summary, the period of folding and thrusting, which we relate to strike-slip deformation, spanned that of dewatering and lithification. The NNW–SSE orientation of folds and thrusts, and the ENE–WSW trend of sandstone dykes in the FBF are important as they indicate the compression and extension directions in the basin. The low strains measured preclude large shear-related rotations. These suggest that deformation in the basin has been by dextral transpression (Sanderson & Marchini, 1984, fig. 5). This is compatible with observed NE–SW strike-slip faults being Riedel fractures, and ESE–WNW sinistral faults being conjugate-Riedel fractures.

Late, dextral N–S, or NE–SW strike-slip faults displace sandstone dykes, and show smaller displacements of the bedding-parallel veins. A similar set of steeply dipping, oblique-slip, dextral/extensional faults with NNW–SSE strikes is present in the region of Fossil Bluff (Fig. 5f), and cuts thrusts 10 km to the north-west. East-dipping faults are the most frequent. These faults are the most recent seen, and represent a period of dextral transtension, probably related to the formation of George VI Sound.

Conclusions

Structures in the LMG suggest that oblique subduction probably caused some transcurrent motion in the fore-arc during accretion. The age of this deformation in northern Alexander Island is poorly constrained, although elsewhere in the LMG accretion continued from at least the Kimmeridgian to about 20 Ma ago when subduction ceased (Barker, 1982). In the final stages of subduction, movement of the Phoenix (Aluk) Plate was constrained by transform faults to be at about 70° to the continental margin (Barker, 1982, fig. 4), and at a much higher angle northwards, owing to the curvature of the peninsula. Beck (1983) showed that, if movement on a shallow-dipping subduction zone (favoured by young subducting oceanic crust, and a wide accretionary prism – Karig, 1980), was sufficiently oblique, it caused oblique convergence to be partitioned into thrusts and strike-slip faults parallel to the continental margin. It is probable that fore-arc slivers have been transported dextrally along the western margin of Gondwana (this angle is sufficiently oblique, see Jarrard, 1986). Watts, Watts & Bramall (1984) suggested that oroclinal bending has not occurred in the Antarctic Peninsula since 100 Ma ago (cf. Kellogg, 1980). If so, the embayment of the Cretaceous magmatic arc would make Alexander Island an ideal location for the docking of such slivers, and occurs where there is an exceptionally wide accretionary prism. Crook & Feary (1982, fig. 8) suggested that spreading in the Pacific off Marie Byrd Land caused major dextral transcurrent motion along the West Antarctic margin from the Permian to the Late Jurassic, which is compatible with some dextral transpression in Alexander Island.

Barker (1982, fig. 5) noted that, from 100 Ma onwards (the age of the youngest sediments in the FBF), the subduction direction of the proto-Pacific beneath the peninsula was in the same orientation as the present vector. Dextral N–S faulting in the LMG and FBF could not be produced by this direction. Consequently this faulting is not subduction-related. This deformation occurred prior to lithification of the FBF, soon after, or during sedimentation. The FBF was deposited in a deep, narrow, tectonically active basin while the LMG was being accreted. Although the tectonic origin of the basin is unknown, its geometry is compatible with it being a series of linked pull-apart basins. In northern Alexander Island, late strike-slip faults affect the Tertiary volcanic rocks and probable Tertiary dykes. Dextral transcurrent motion was long-lived, possibly starting with the formation of the Fossil Bluff basin as a trough resembling the modern George VI Sound, continuing with the inversion of the basin, and probably the final extension of the sound.

Given that the late strike-slip faults were not subduction-related, the driving mechanism was probably extension between New Zealand and Marie Byrd Land (Storey, this volume, p. 587 & fig. 2). This commenced at about 85 Ma (e.g. Barker, 1982), with dextral transtension in the Ross Sea (Kamp, 1986, fig. 5). A continuation of this dextral displacement through West Antarctica could be the cause of the dextral faulting observed.

Note

Since this paper was written, the 'Fossil Bluff Formation' was assigned Group status, and should be referred to as the Fossil Bluff Group. (Butterworth, P.J., Crame, J.A., Howlett, P.J. & Macdonald, D.I.M. (1988). Lithostratigraphy of Upper Jurassic–Lower Cretaceous strata of eastern Alexander Island, Antarctica. *Cretaceous Research*, **9(3)**, 249–64).

References

Barker, P.F. (1982). The Cenozoic subduction history of the Pacific margin of the Antarctic Peninsula: ridge crest–trench interactions. *Journal of the Geological Society, London*, **139(6)**, 787–801.

Beck Jr. M.E. (1983). On the mechanism of tectonic transport in zones of oblique subduction. *Tectonophysics*, **93**, 1–11.

Bell, C.M. (1975). Structural geology of parts of Alexander Island. *British Antarctic Survey Bulletin*, **41 & 42**, 43–58.

Burn, R.W. (1984). *The Geology of the Lemay Group, Alexander Island.* Cambridge; British Antarctic Survey Scientific Reports, No. 109. 64 pp.

Coward, M.P. & Kim, J.H. (1981). Strain within thrust sheets. In *Thrust and Nappe Tectonics*, ed. K.R. McClay & N.J. Price, pp. 275–92. London; Special Publication of the Geological Society of London, No. 9.

Crook, K.A.W. & Feary, D.A. (1982). Development of New Zealand according to the fore-arc model of crustal evolution. *Tectonophysics*, **87**, 65–107.

Edwards, C.W. (1980). New evidence of major faulting on Alexander Island. *British Antarctic Survey Bulletin*, **49**, 15–20.

Horne, R.R. (1968). Petrology and provenance of the Cretaceous sediments of south-eastern Alexander Island. *British Antarctic Survey Bulletin*, **17**, 73–82.

Jarrard, R.D. (1986). Terrane motion by strike-slip faulting of forearc-slivers. *Geology*, **14(9)**, 780–3.

Kamp, P.J.J. (1986). Late Cretaceous–Cenozoic tectonic development of the southwest Pacific region. *Tectonophysics*, **121**, 225–51.

Karig, D.E. (1980). Material transport within accretionary prisms and the 'Knocker' problem. *Journal of Geology*, **88(1)**, 27–39.

Kellogg, K.S. (1980). Paleomagnetic evidence for oroclinal bending of the southern Antarctic Peninsula. *Geological Society of America Bulletin*, **91(7)**, 414–20.

McKenzie, D. & Jackson, J. (1986). A block model of distributed deformation by faulting. *Journal of the Geological Society, London*, **143(2)**, 349–54.

Nicholson, R. & Pollard, D.D. (1985). Dilation and linkage of echelon cracks. *Journal of Structural Geology*, **7(5)**, 583–90.

Nur, A., Ron, H. & Scotti, O. (1986). Fault mechanics and the kinematics of block rotations. *Geology*, **14(9)**, 746–9.

Phillips, W.E.A., Flegg, A.M. & Anderson, T.B. (1979). Strain adjacent to the Iapetus suture in Ireland. In *The Caledonides of the British Isles Reviewed*, ed. A.L. Harris, C.H. Holland & B.E. Leake, pp. 257–62. London; Special Publication of the Geological Society of London, No. 8.

Platt, J.P. (1986). Dynamics of orogenic wedges and the uplift of high-pressure metamorphic rocks. *Geological Society of America Bulletin*, **97(9)**, 1037–53.

Sanderson, D.J. & Marchini, W.R.D. (1984). Transpression. *Journal of Structural Geology*, **6(5)**, 449–58.

Soper, N.J. (1986). Geometry of transecting, anastomosing solution cleavage in transpression zones. *Journal of Structural Geology*, **8(8)**, 937–40.

Storey, B.C. & Garrett, S.W. (1985). Crustal growth of the Antarctic Peninsula by accretion, magmatism and extension. *Geological Magazine*, **122(1)**, 5–14.

Taylor, B.J. (1982). Sedimentary dykes, pipes and related structures in the Mesozoic sediments of south-eastern Alexander Island. *British Antarctic Survey Bulletin*, **51**, 1–42.

Taylor, B.J., Thomson, M.R.A. & Willey, L.E. (1979). *The geology of the Ablation Point–Keystone Cliffs Area, Alexander Island.* Cambridge; British Antarctic Survey Scientific Reports, No. 82. 65 pp.

Watts, D.R., Watts, G.C. & Bramall, A.M. (1984). Cretaceous and Early Tertiary paleomagnetic results from the Antarctic Peninsula. *Tectonics*, **3**, 333–46.

Basin shallowing from the Mesozoic Fossil Bluff Group of Alexander Island and its regional tectonic significance

P.J. BUTTERWORTH & D.I.M. MACDONALD

British Antarctic Survey, Natural Environment Research Council, High Cross, Madingley Road, Cambridge CB3 0ET, UK

Abstract

A 4000 m thick Kimmeridgian–Albian fore-arc sedimentary succession crops out as a belt 250 km long × 30 km wide on the eastern coast of Alexander Island. Three major units are recognized in the northern to central part of the outcrop belt: deep marine submarine fan conglomerates, slope mudstones and shallow-marine sandstones.

This paper describes the sedimentology of the Lower Cretaceous slope mudstone succession which defines an overall regression from deep- to shallow-marine environments. This regressive sequence implies a relative fall in sea-level from Berriasian to Albian times, which contrasts with the worldwide sea-level rise during the Early Cretaceous. This suggests that tectonism in the fore-arc region was the dominant controlling factor in basin evolution during the Early Cretaceous. Syndepositional tectonic episodes within the slope assemblage are reflected in sandstone-rich units and localized unconformities, and the occurrence of some relatively large synsedimentary *mélange* zones.

Introduction

During Mesozoic times the present day Antarctic Peninsula was the site of an active volcanic arc. The best exposed fore-arc basin sequence is the 4000 m thick Fossil Bluff Group (Butterworth *et al.*, 1988) which crops out as a belt 250 km long × 30 km wide along the eastern coast of Alexander Island (Fig. 1). The central to northernmost exposures of this Upper Jurassic (Kimmeridgian) Lower Cretaceous (Albian) succession can be broadly subdivided into three main stratigraphic intervals. These are deep-marine sediments, an overlying mudstone assemblage and shallow-marine deposits (Fig. 1). This paper concentrates on the sedimentology of the fine-grained assemblage which crops out extensively between Ganymede Heights and Fossil Bluff (Fig. 1). This assemblage is interpreted as a slope deposit which accumulated as the result of basin shallowing; an apparent direct contrast to the worldwide sea-level rise during the Early Cretaceous (Vail *et al.*, 1977; Haq, Hardenbol & Vail, 1987).

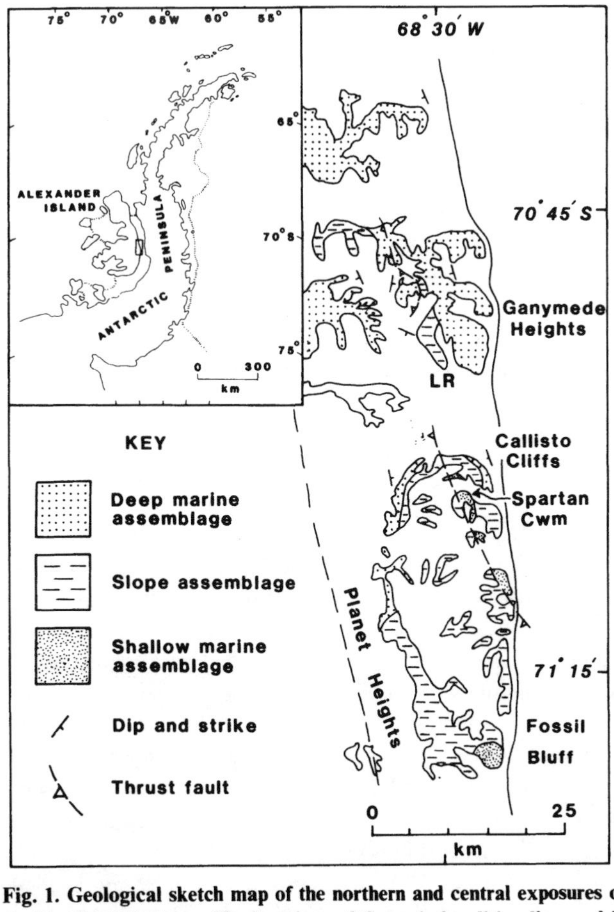

Fig. 1. Geological sketch map of the northern and central exposures of the Fossil Bluff Group. The locations of the main localities discussed in the text are shown. LR, Leda Ridge.

Deep marine assemblage

The basal 2200 m thick Kimmeridgian–Berriasian deep-marine assemblage is best exposed on Ganymede Heights (Fig. 1), and comprises four thick (70–180 m), channelled conglomerate complexes and an associated varied assemblage of mass flow sediments (Butterworth, 1985). This unit is also exposed on Callisto Cliffs and the northern end of Planet Heights (Fig. 1), where it is rather thinner than at Ganymede

Heights. Clast lithologies and imbrication data from the conglomerate facies indicate that sediment was derived from the volcanic arc to the east (Palmer Land) and was dispersed in a radial pattern towards the west. This unit is interpreted as a conglomeratic fan–channel deposit.

The transition into the overlying mud-rich assemblage is gradational, but broadly corresponds to the top of the sandy fining-upward cycle above the topmost conglomerate complex (Fig. 2*a*). This overall transition from conglomeratic mass flow-dominated sedimentation to a mudstone-rich sequence coincides with a distinctive change in the fauna. The deep-marine assemblage typically contains perisphinctid and berriaselid ammonites, *Hibolithes* belemnites and buchiid bivalves; there is also a limited occurrence of wood fragments. In contrast, the overlying mud-rich sequence has a fauna of rare ammonites, belemnopseid belemnites and inoceramid bivalves (Crame & Howlett, 1988); wood fragments (including several branches) are common. At the base of the assemblage there is a locally abundant fauna, including serpulids, *Myophorella* and *Pinna*, apparently in life position.

Slope assemblage

This predominantly fine-grained assemblage crops out over a wide area (Fig. 1), and attains a maximum thickness at Spartan Cwm of 950 m (see later, Fig. 3). This assemblage can be split conveniently into six major facies, which form a spectrum of increasing sandstone percentage.

Facies 1 comprises structureless mudstone, generally medium–thick-bedded. The lack of structure is predominantly the result of extensive bioturbation, and the facies represents quiet-water suspension sedimentation.

Facies 2 consists of laminated mudstone, which occurs in packets 1–3 m thick. Laminations are defined either by subtle grain size or colour variations, or by one–two grain thick sandstone stringers within shaley intervals; bioturbation is rare. Deposition was dominantly from suspension and dilute turbidity currents.

Facies 3 (Fig. 2*b*) comprises interbeds of graded siltstone, mudstone and sandstone. Siltstone beds are typically structureless or rarely laminated with irregular or undulating sharp basal contacts and graded tops. Beds range from 0.1–2 cm thick. Sandstone beds are typically 0.1–0.5 cm thick and often display rippled tops; this facies may be bioturbated (Fig. 2*b*). It occurs in packets 1–3 m thick and is intermediate between facies 2 and 4. The facies represents deposition from dilute low density turbidity currents and suspension sedimentation.

Facies 4 (Fig. 2*c*) comprises fine-grained, 1–3 cm thick, turbiditic sandstones (T_b, T_c, T_d) interbedded with laminated or more commonly structureless mudstone (3–10 cm). Thin sandstone beds commonly form starved ripples. Bioturbation is common, and often destroys all internal structures and bedding characteristics (Fig. 2*c*). This facies typically forms discrete packets 1–3 m thick, and is interpreted as deposition from low density turbidity currents.

Facies 5 (Figs 2*d* & *e*) comprises 2–10 cm thick interbedded sandstone and mudstone. Unbioturbated, turbiditic sandstone beds are commonly laterally continuous with sharp, undulating, rarely scoured bases and graded (rippled) tops with well developed T_a, T_{bc} and T_c divisions (Fig. 2*d*). Bioturbation often obliterates all bedding characteristics and detached slump fold noses are rare. This facies typically occurs in distinct 1–5 m thick packets (Fig. 2*e*). Deposition is envisaged from a combination of high (T_a) and low ($T_{b/c}$)-density turbidity currents.

Facies 6 beds form very distinctive thick-bedded, massive and laterally discontinuous sandstone units. Typically these units are up to 150 m wide × 2–8 m thick (Fig. 2*f*), and comprise medium–coarse-grained massive angular sandstone in beds 0.5–6 m thick; bed amalgamation is common. A few beds have either single or multiple pebble-graded basal zones. Small (1–5 cm) intraformational mudflakes are sporadically common. Characteristically this facies develops a honey comb-weathered appearance. Deposition is envisaged from turbiditic mass flows. Commonly these thick massive sandstone packets are successively overlain by facies 5–4–3–2/1, a result of active channel-fill (facies 6) followed by abandonment.

Mudstone-rich facies (facies 1, 2, 3) comprise over 80% of the sediment within this sequence (Fig. 3), reflecting prolonged intervals of quiet-water, suspension sedimentation. This mudstone-rich sequence is punctuated by low-density turbidity currents (facies 4, 5) and episodic high-density mass flows (facies 6), and is broadly assigned to a slope setting.

Although sandstone deposition (facies 4, 5, 6) is generally restricted to 1–3 m thick packets, these packets commonly form mappable units broadly confined to the middle zone of the assemblage (Fig. 3). The occurrence and lateral variability of these sandstone units is highlighted by good exposures at three localities spanning 50 km along strike. The thickest sandstone unit crops out at Spartan Cwm and comprises a 160 m thick sandstone-rich unit of fine-grained, heavily bioturbated sandstone (facies 3, 4, 5). At Leda Ridge, some 25 km to the north of Spartan Cwm, the sequence is more complex and comprises three, relatively thin (20–50 m thick) sharp-based sandstone-rich units, often with a basal channelled sandstone horizon (facies 6). Associated with these sandstone-rich units are sporadic channels and low-angle (1–2 m set thick) cross-beds and, at 1900 m (Fig. 3), a 5–10° angular unconformity with the mudstone section below. The biostratigraphical data, however, preclude a large time-gap (Crame & Howlett, 1988). Two relatively thin sandstone units of similar age crop out at Planet Heights. The extensive lateral exposure at this locality shows the lowermost sandstone unit to thin southwards from 60 m to 20 m over a distance of 4 km (Fig. 3). A feature common to each sandstone unit is the occurrence of a distinctive faunal assemblage. This generally forms as a locally abundant fauna with orientated belemnites and seams of relatively large shells (e.g. *Myophorella*, *Pinna*), some of which may be in life position (Crame & Howlett, 1988). A faunal assemblage of this sort is generally assigned to a burrowing and/or bottom-dwelling association (e.g. Fürsich, 1984), and is unlike the sparse fauna from the muds both below and above.

Ripple cross-lamination within the lowermost 550 m of this assemblage (Fig. 3) indicates a south-westerly to subordinate westerly trend, with belemnites dominantly orientated NW–SE. In contrast to this, the upper 400 m of the assemblage contains belemnites which are dominantly NE–SW-orientated.

Fig. 2. Field photographs of the slope assemblage. (*a*) View looking south-west over Ganymede Heights from Himalia Ridge, showing a 2000 m uninterrupted section through the conglomerate-dominated deep-marine assemblage to the fine-grained slope assemblage exposed on Leda Ridge (arrow defines the contact). (*b*) Facies 3 (bioturbated); pencil is 15 cm long. (*c*) Facies 4 (bioturbated); lens cap is 6 cm across. (*d*) Facies 5 (unbioturbated); lens cap is 6 cm across. (*e*) Typical packet of facies 5 beds. Note the bed amalgamation/scouring; Hammer shaft is 35 cm long. (*f*) View looking from Planet Heights south-east towards Fossil Bluff through the slope assemblage. Bedding dips gently to the south, with approximately 800 m of uninterrupted succession exposed. Note the laterally discontinuous sandstone unit in the foreground (facies 6). The sequence is capped by shallow-marine sandstones exposed at Fossil Bluff (arrow).

This abrupt change in the direction of sediment input corresponds to the cessation of sedimentation associated with the broad zone of sandstone-dominated units.

Throughout this assemblage there is evidence of local slope instability. Three synsedimentary *mélanges*, ranging from 40 to 120 m thick, crop out within mudstone-rich units (Fig. 3). Each *mélange* is similar in character, and consists of a series of rotated coherent blocks of facies 1–5, reflecting a brittle style of

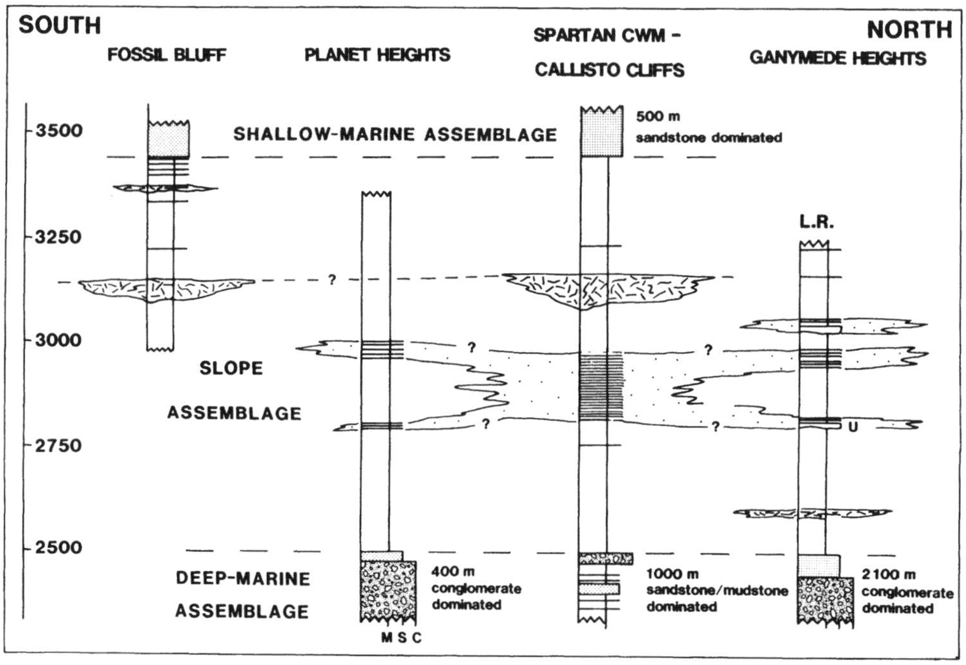

Fig. 3. Simplified geological sections highlighting the slope assemblage from Fossil Bluff in the south to Ganymede Heights in the north. Lateral correlations are based on the topmost conglomeratic unit plus biostratigraphical data (Crame & Howlett, 1988); the scale bar indicates the true thickness from the base of the Fossil Bluff Group as exposed on Ganymede Heights. M, mudstone; S, sandstone; C, conglomerate; U, unconformity; LR, Leda Ridge; random hachure indicates synsedimentary *mélange*; horizontal ruling indicates thin sandstone beds.

deformation; ductile structures are rare. The *mélange* at Fossil Bluff also includes blocks of shallow-marine sandstones, similar to the overlying assemblage, which are interpreted as a near-shore correlative of the mudstone. *Mélanges* thin along strike to both the north and south over a few kilometres. Further evidence for a depositional slope within these mud-rich units is the occurrence of rare, detached slump fold noses and synsedimentary normal faults.

Shallow marine assemblage

Sandstones of shallow-marine origin crop out at both Fossil Bluff and Spartan Cwm. The transition from the underlying slope assemblage is generally sharp and flat. At Fossil Bluff these sandstones form the cap to a 70 m thick coarsening upward sequence (Fig. 3).

This assemblage typically comprises 1–5 m thick, heavily bioturbated, mottled, well-sorted medium-grained sandstones. Pebbles are very rare. Herringbone cross-stratification is common where bioturbation is less intense. Palaeocurrent data from the herringbone cross-stratification indicates a strong southerly direction with a small but significant northerly dispersion. The assemblage contains a distinctive shallow marine benthic faunal assemblage (Crame & Howlett, 1988).

Implications

During the Early Cretaceous, the sedimentological evidence from the Fossil Bluff Group indicates an overall, if complex, regression from deep to shallow marine environments. However, during the Early Cretaceous there is a well documented global sea-level rise based on data from both seismic stratigraphy and outcrops (Vail *et al.*, 1977; Haq *et al.*, 1987). Hence there is a conflict between the idea of a global, eustatically-controlled sea-level rise during the Early Cretaceous and the relative fall in sea-level as indicated by the sediments of the Fossil Bluff Group. Simplistically, there are two models in which to accommodate this; either the unit reflects a pure 'bucket-type' fill, with sedimentation infilling a topographic low, and countering the worldwide sea-level rise, or there was some form of active uplift. Throughout the slope assemblage there is evidence of syndepositional tectonic activity, lending support to the active uplift hypothesis. The widespread occurrence of relatively thick synsedimentary *mélanges* (formed as a single *en masse* movement of sedimentary blocks) was probably as a result of a major earthquake associated with the volcanic arc to the east, or with uplift of the accretionary complex to the west. The two thickest *mélanges*, at Fossil Bluff and Spartan Cwm, are of approximately the same age and are thought to represent either the same event, or closely related episodes. A tectonic control is also proposed for the sandstone-rich units which reflect sporadic shallowing during Hauterivian–Barremian times, and are interpreted as prograded/remobilized shallow-marine sandstones. It is proposed that the easiest way to explain basin fill is with punctuated rather than gradual uplift during the Early Cretaceous.

These episodes of intermittent tectonic activity along the margins of the fore-arc basin may be related to a period of strike-slip movement (Nell & Storey, this volume, p. 443). In a wider context, it is probably related to the larger-scale extensional tectonic activity associated with the fracturing of central Gondwana (Dalziel *et al.*, 1987).

Acknowledgements

We would like to thank Drs P.J. Howlett and J.A. Crame for much useful discussion and data, Dr M.R.A. Thomson for reading an earlier draft of the manuscript, and J. Ball, C. Griffiths, I. Mills and P. McAra for their patience and enthusiasm in the field. We are grateful to an anonymous referee for a constructive review.

References

Butterworth, P.J. (1985). Sedimentology of Ablation Valley, Alexander Island. *British Antarctic Survey Bulletin*, **66**, 73–82.

Butterworth, P.J., Crame, J.A., Howlett, P.J. & Macdonald, D.I.M. (1988). Lithostratigraphy of Upper Jurassic–Lower Cretaceous strata of eastern Alexander Island, Antarctica. *Cretaceous Research* 249–64.

Crame, J.A. & Howlett, P.J. (1988). Late Jurassic and Early Cretaceous biostratigraphy of the Fossil Bluff Formation of Alexander Island. *British Antarctic Survey Bulletin*, **78**, 1–35.

Dalziel, I.W.D., Storey, B.C., Garrett, S.W., Grunow, A.M., Herrod, L.D.B. & Pankhurst, R.J. (1987). Extensional tectonics and the fragmentation of Gondwanaland. In *Continental Extensional Tectonics*, ed. M.P. Coward, J.F. Dewey & P.L. Hancock, pp. 433–44. London; Special Publication of the Geological Society of London, No. 28.

Fürsich, F.T. (1984). Benthic Macroinvertebrate Associations from the Boreal Upper Jurassic of Milne Land, Central East Greenland. *Grønlands Geologiske Undersøgelse Bulletin* No. 149, 72 pp.

Haq, B.U., Hardenbol, J. & Vail, P.R. (1987). Chronology of fluctuating sea levels since the Triassic. *Science*, **235**, 1156–67.

Vail, P.R., Mitchum, R.M., Todd, R.G., Widnier, J.M., Bubb, J.W. & Hattelid, W.G. (1977). Seismic stratigraphy and global changes of sea level. *Memoir of the American Association of Petroleum Geologists*, **26**, 49–212.

Petrology of sedimentary rocks from the English Coast, eastern Ellsworth Land

T.S. LAUDON

Department of Geology, University of Wisconsin Oshkosh, Oshkosh, Wisconsin 54901, USA

Abstract

The English Coast of West Antarctica contains the south-westernmost exposures of the Antarctic Peninsula tectonic province (APTP). Sedimentary rocks are assigned to three units on the basis of their possible ages, lithologies and structural attitudes. Metamorphosed quartz sandstone of unknown age and continental interior provenance, named the FitzGerald quartzite beds, was deposited in a passive margin environment, and probably represents pre-Upper Palaeozoic basement of the APTP. Stratigraphical linkage to the Devonian Crashsite Quartzite of the Ellsworth Mountains is suggested. Large clasts of quartzite petrographically similar to the FitzGerald quartzite beds occur in conglomerate of the Jurassic Latady and Mount Poster formations in the English Coast and Orville Coast areas, suggesting widespread distribution of quartzite in the basement of the region. Magmatic arc-derived clastic rocks of Permian age, which contain *Glossopteris* flora at Erehwon Nunatak, are named the Erehwon beds. They indicate that subduction-related magmatism had begun in the English Coast by Late Palaeozoic time. Stratigraphical linkage to the Polarstar Formation of the Ellsworth Mountains is suggested. Clastic rocks of (?)Jurassic age and recycled orogenic provenance are correlated with the Latady Formation. They were deposited in terrestrial and paralic environments, with associated coeval volcanic activity.

Introduction

The English Coast of the Bellingshausen Sea (Pacific Ocean) was first visited and mapped in reconnaissance in 1984–85. Nunataks there (Fig. 1) form the south-westernmost exposures of the Antarctic Peninsula tectonic province (Rowley, Kellogg & Vennum, 1985), which includes eastern Ellsworth Land and Alexander Island.

Snow Nunataks, located at the coast (Fig. 1), are composed of (?)upper Cenozoic mafic volcanic and volcaniclastic rocks (O'Neill & Thomson, 1985), and a group of nunataks composed predominantly of Mesozoic volcanic and plutonic rocks (Rowley *et al.*, 1985) is located approximately 70 km south of this coast. One such nunatak contains a sequence of metamorphosed quartz sandstone (Laudon *et al.*, 1987), herein named the FitzGerald quartzite beds.

Clastic sedimentary rocks occur in nunataks located approximately 100–150 km inland (Fig. 1). These are assigned to two units: Permian Erehwon beds, named herein and the Jurassic Latady Formation.

Sedimentary petrology

This paper presents the results of studies of thin sections and stained chips of representative samples from all sedimentary rock outcrops in the English Coast area. Modal analyses of sandstone slides (at least 400 points/slide) followed the Gazzi–Dickinson approach (Dickinson, 1970) as far as possible. Sandstone classification follows Dott (1964).

Sandstone modes are summarized in Table 1 and Fig. 2. Their primary value is descriptive; interpretations must be undertaken with caution because: all of the rocks have been subjected to at least greenschist-facies metamorphism; most unstable framework clasts and interstitial material have been altered; and most sandstone is fine grained. Compositions of large conglomerate-clasts (> sand-sized) in thin sections from the English and Orville coasts are summarized in Table 2.

FitzGerald quartzite beds

Near the western end of FitzGerald Bluffs (Fig. 1) ~ 300 m of quartzite beds are intruded by sheared Cretaceous granitic rocks (Rowley *et al.*, 1985; Pankhurst & Rowley, this volume, p. 387). The metasedimentary rocks are here informally named the FitzGerald quartzite beds. They consist of cross-bedded, cross-laminated, grey–green, vitreous, quartz–mica hornfels. Their age (other than pre-Cretaceous) is unknown.

Textural elements include granoblastic polygonal frameworks composed of quartz and accessory lithic grains, and

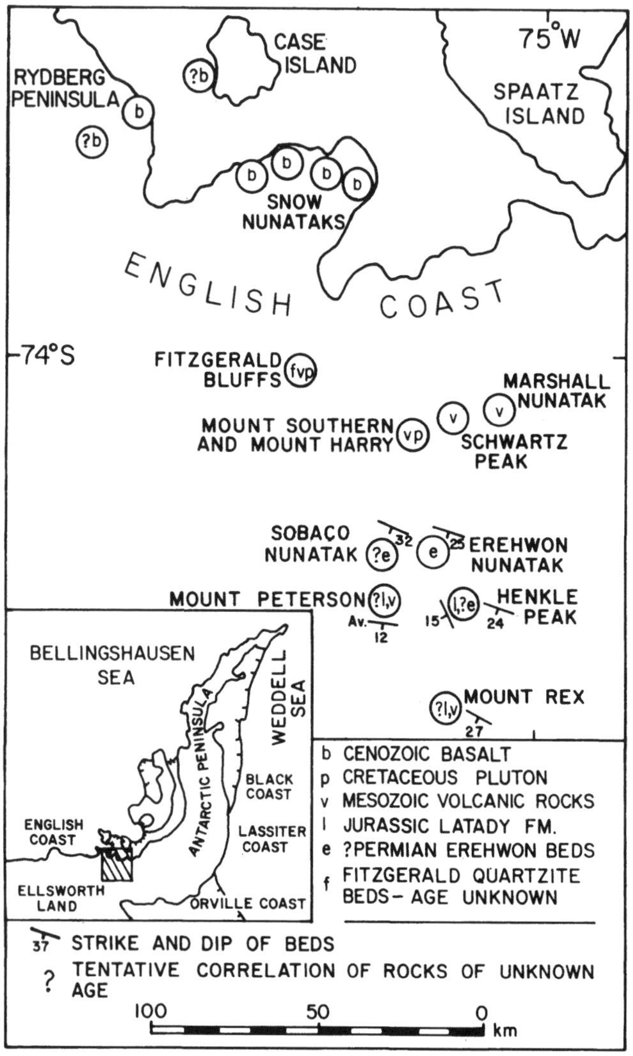

Fig. 1. Geological sketch map of the English Coast. Circles indicate outcrop locations but at this scale they are larger than the outcrops. Strike and dip at Mount Peterson is the mean for three fault blocks.

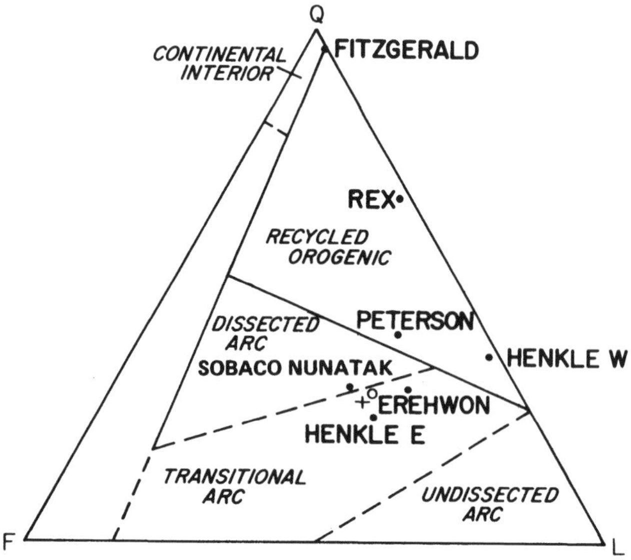

Fig. 2. Modal *QFL* framework compositions for sandstones from the English Coast. Provenance field boundaries are from Dickinson (1985). ●, mean composition of sandstones from indicated locality; ○, mean composition of sandstones from the Erehwon beds; +, mean composition of 37 sandstones from the Polarstar Formation (Collinson *et al.*, in press).

dark-coloured laminae composed of metamorphic muscovite, biotite, chlorite and epidote, and detrital magnetite and zircon. The only feldspar occurs in microfelsitic veins. Protoliths were well-sorted, medium-grained quartz-wackes of continental interior provenance (Fig. 2).

Erehwon beds

Two tiny outcrops at Erehwon Nunatak and 'Sobaco Nunatak', and small outcrops on the north-east side of Henkle Peak (Fig. 1), are composed of fine-grained clastic sedimentary rocks with strikingly similar petrological characteristics and nearly congruent structural attitudes. Thicknesses of exposed strata are approximately 2 m at Erehwon Nunatak, 28 m at 'Sobaco Nunatak', and < 10 m at Henkle Peak. These rocks are here informally named the Erehwon beds. One bed at Erehwon Nunatak contains abundant leaves of *Glossopteris erehwonensis* of Permian (probably Late Permian) age (Gee, 1989).

The Erehwon beds consist of cross-laminated, dark-coloured mudstone, siltstone and fine-grained sandstone. Sandstone textures are characterized by angularity and moderate sorting of framework grains, and by alteration of unstable framework grains and interstitial material. Siltstone textures are suggestive of tuffs, with abundant curved, splintery clasts of monocrystalline quartz, and polycrystalline quartz and microfelsite that may be devitrified shards.

Sandstone modes (Table 1) are characterized by abundant framework quartz (Q), feldspar (F) and lithic clasts (L). High plagioclase (P)/F ratios result partly from alteration of K-feldspar. Most lithic grains are altered, many beyond recognition of their original nature. Microfelsitic volcanic rock fragments (VRFs) are the most abundant recognizable lithic clasts. A few VRFs with trachytic or intergranular textures, and a few micaceous metamorphic rock fragments (MRFs) occur in every slide. Commonly occurring accessory components include plant fragments, stringers of organic carbon, detrital zircon, muscovite, chlorite and opaque minerals, authigenic pyrite and clay, and metamorphic biotite and actinolite. Sandstones are lithic arenites and wackes of magmatic arc provenance (Fig. 2).

Latady Formation

Clastic sedimentary rocks exposed on the south-west side of Henkle Peak, at Mount Rex, and at Mount Peterson (Fig. 1) are correlated with the Middle–Upper Jurassic Latady Formation which is widely exposed in the Orville and Lassiter coast areas (Fig. 3). On the south-western side of Henkle Peak 30 m of black carbonaceous conglomerate, sandstone, and siltstone contain abundant plant fossils including *Elatocladus planus* (Laudon *et al.*, 1985, fig. 3, 1987, fig. 2), which is commonly found in Jurassic rocks including the Latady Formation (Gee, 1989). The conglomerate occurs in massive units several metres thick, that have sharp, planar contacts with

Table 1. *Modal sandstone compositions; values given are means and (ranges)*

Location	Number of samples	Grain size[a]	Interstitial material (modal %)	Framework %			
				Quartz	Feldspar	Lithic fragment	Plagioclase/feldspar
Latady Formation							
Henkle SW	2	f–c	27 (± 3)	36 (± 2)	2 (± 1)	62 (± 1)	1.0
Rex	2	f–m	6 (± 2)	67 (± 1)	2 (± 2)	31 (± 4)	1.0
Peterson	4	m	18 (± 2)	40 (± 6)	16 (± 3)	44 (± 9)	> 0.9
Erehwon beds							
Erehwon	1	f	33	30	19	51	1.0
'Sobaco Nunatak'	2	f	6 (± 3)	30 (± 2)	29 (± 7)	39 (± 7)	0.57 (± 0.11)
Henkle NE	1	f	12	24	28	48	1.0
FitzGerald quartzite beds							
FitzGerald	5	m	21 (± 7)	96 (± 3)	0	4 (± 3)	

Data based on counts of at least 400 points/thin section.
[a]f, fine sand; m, medium sand; c, coarse sand.

Table 2. *Summary of pebble and granule compositions in conglomerate thin sections*

Location	N	Vv (%)	Vf (%)	Vi (%)	Vm (%)	Vp (%)	Sms (%)	Qss (%)	QLs (%)	MQz (%)	Pl (%)	Qvn (%)
Henkle	449	19	—	—	—	21	59	—	—	1	—	—
Peterson	1169	13	—	13	34	30	9	—	—	< 1	< 1	—
Marshall	355	24	21	6	3	—	< 1	11	12	23	—	—
Behrendt[a]	58	—	69	—	—	1	—	—	1	22	5	1
Hauberg[a]	653	54	9	28	0.5	3	2	< 1	1	2	—	—

[a]Located on the Orville Coast (see Fig. 2). N, total number of clasts identified; Vv, vitric volcanic rocks (vitric textures or structures without phenocrysts); Vf, felsic volcanic rocks (with quartz phenocrysts); Vi, intermediate volcanic rocks (with trachytic textures); Vm, mafic volcanic rocks (with intergranular textures); Vp; tuffaceous rocks (with pyroclastic textures); Sms, epiclastic mudstone and siltstone; Qss, quartz sandstone; QLs, quartzo-lithic sandstone; MQz, quartzite; Pl, plutonic rock fragments; Qvn, vein quartz.

overlying and underlying units of finely laminated sandstone and siltstone. Sandstone is black, and is interlaminated with finer-grained material. Thin sections show poorly sorted, extremely angular to well-rounded, very fine sand–granule-sized clasts, in a black carbonaceous matrix.

Most sand-sized quartz clasts are angular; most VRFs are subangular. Larger clasts are predominantly plants and rounded or subrounded, fine-grained sedimentary rock fragments (SRFs) that are commonly bent or squashed. Only trace amounts of detrital feldspar are present. Accessory constituents include detrital muscovite, biotite, and chlorite, patches of authigenic pyrite, and sericite and chlorite cement. Sandstones are lithic wackes of recycled orogenic provenance (Fig. 2).

Clast-supported conglomerate consists of well-rounded pebbles as long as 7 cm, in a poorly sorted sandy matrix that may be tuffaceous. Conglomerate–pebble populations (Table 2) are dominated by fine-grained SRFs thought to be of intrabasinal origin, and VRFs. Sparse quartzite pebbles are petrographically similar to the FitzGerald quartzite beds.

Mount Rex (Fig. 1) consists predominantly of porphyritic dacite correlated with the Mount Poster Formation, but two interbedded sandstone sequences, respectively 28 m and 22 m thick, resemble the Latady Formation (Laudon *et al.*, 1987). The sandstone is fine-grained and moderately sorted, with predominantly angular–subangular grains. Modes (Table 1) are characterized by significantly higher than average *Q* for the Latady Formation and very low (near zero) *F* contents. Many quartz grains show former rounded outlines with abraded overgrowths from an earlier cycle of sedimentation. Microfelsitic and microlitic VRFs are the most abundant lithic clasts; schistose MRFs are also present. Sandstones are lithic arenites of recycled orogenic provenance (Fig. 2).

At Mount Peterson (Fig. 1), as much as 54 m of red and green conglomerate, sandstone, siltstone and mudstone are tentatively correlated with the Latady Formation because they are interbedded with, and overlain by felsic volcanic rocks similar to the Mount Poster Formation. These rocks have been interpreted as alluvial fan deposits (Laudon *et al.*, 1987). The sandstone tends to be moderately sorted and the conglomerate

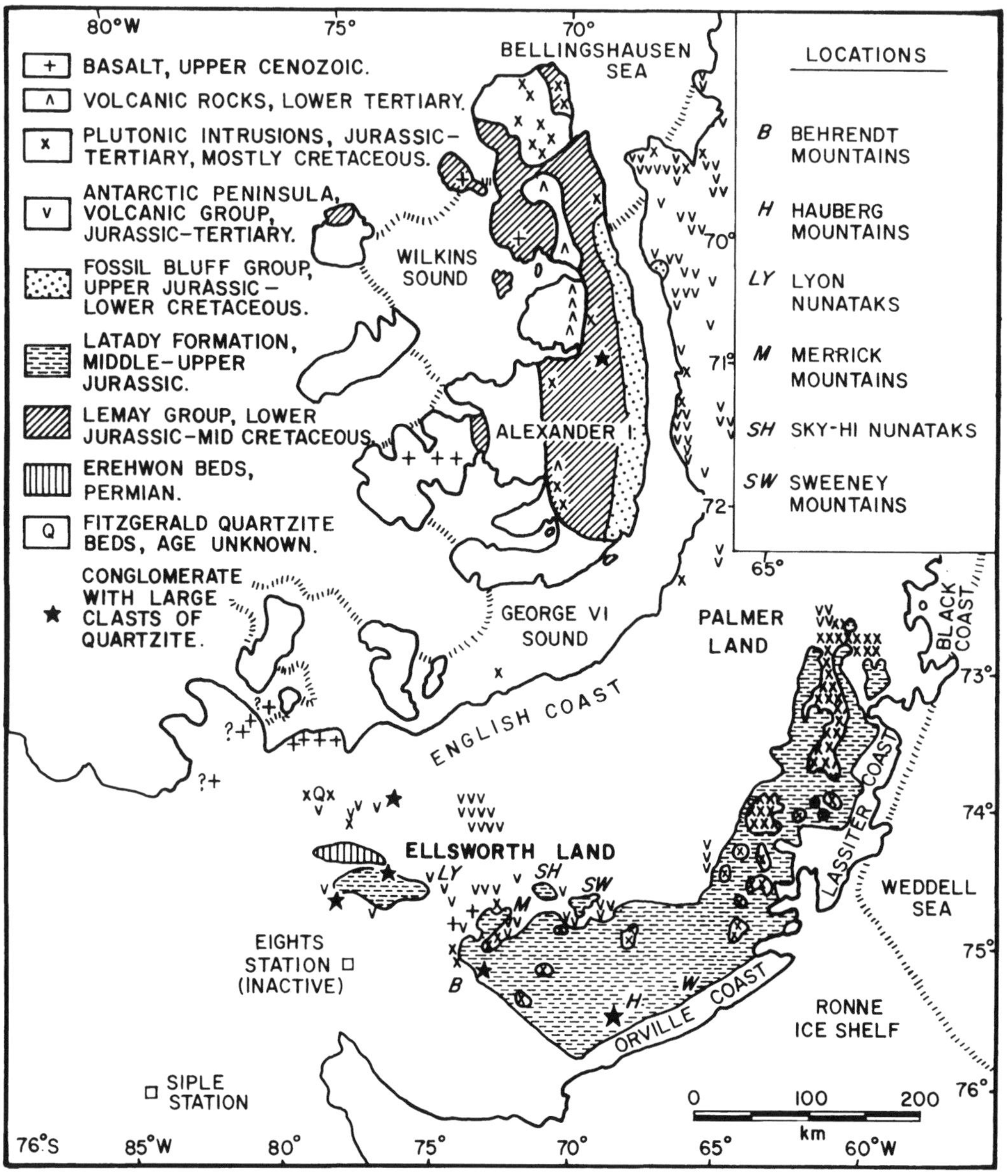

Fig. 3. Generalized geological map of the southern Antarctic Peninsula region showing the approximate distribution of the major units. In southern Palmer Land and eastern Ellsworth land, plutonic intrusions are part of the Cretaceous Lassiter Coast Intrusive Suite, and rocks of the Antarctic Peninsula Volcanic Group belong to the Jurassic Mount Poster Formation. Geology of western Palmer Land and Alexander Island from British Antarctic Survey (1981); geology of Lassiter Coast from Rowley & Williams (1982).

is poorly sorted, with a sandstone matrix. Sand-sized clasts are predominantly angular but most of the larger clasts are rounded.

Sandstone (including the conglomerate matrix) compositions are characterized by abundant *Q*, *F* and *L* (Table 1). Detrital quartz includes volcanic grains with resorbed crystal faces and bleb inclusions, reworked sedimentary grains with abraded overgrowths, and polycrystalline metamorphic grains. Detrital feldspar is almost entirely plagioclase, although a few grains of microcline were observed. Sand-sized lithic grains are predominantly VRFs, although sparse schistose MRFs are present. Accessory detrital grains include chlorite, epidote, muscovite, and prehnite. Sandstones are lithic wackes. On the *QLF* triangle (Fig. 2) sandstone modes plot near the boundary between the dissected arc and the recycled orogenic provenance fields.

Conglomerate–pebble populations (Table 2) are dominated by VRFs, of which mafic and intermediate types are the most abundant. Sparse plutonic rock fragments (PRFs), and quartzite clasts that are petrographically similar to the FitzGerald quartzite beds are present. It is interesting to note that the conglomerate matrix contains abundant sand-sized grains of volcanic quartz, but no sand-sized quartz phenocyrsts were seen in nearly 700 granule- and pebble-sized VRFs identified in thin sections of the conglomerate.

Marshall Nunatak (Fig. 1) is composed of felsic volcanic rocks and several lenses of boulder conglomerate correlated with the Mount Poster Formation. The conglomerate appears to have been deposited as mudflows (Laudon *et al.*, 1987). Conglomerate–pebble populations are composed of 54% VRFs, 11% quartz-arenite, 12% lithic-arenite (without felspar), 23% metaquartzite and < 1% epiclastic mudstone (Table 2). Conglomerate clasts of sedimentary and metasedimentary origin are of particular interest as they probably represent the basement upon which the volcanic rocks were erupted. They are dominated by quartzose sandstones and metasandstones that are petrographically similar to the FitzGerald quartzite beds.

Discussion

The Antarctic Peninsula tectonic province (APTP), which includes eastern Ellsworth Land and Alexander Island (Figs 1 & 3), is the largest of five small accreted plates of continental lithosphere that comprise West Antarctica (Dalziel & Elliot, 1982).

The geology of the APTP is dominated by calc-alkaline volcanic and plutonic rocks, and sedimentary rocks, all associated with a magmatic arc of at least Middle Jurassic–Tertiary age. This arc formed in response to subduction of Pacific Ocean lithosphere beneath continental lithosphere of the Antarctic Peninsula during Gondwana break-up (e.g. Dalziel, 1982; Elliot, 1983; Thomson & Pankhurst, 1983; Thomson, Pankhurst & Clarkson, 1983). It extended at least as far west as the English Coast, where predominantly terrestrial Jurassic sedimentary rocks are correlated with back-arc basin rocks of the Latady Formation, (?)Mesozoic volcanic rocks are correlated with the Mount Poster Formation, and Cretaceous plutonic rocks are correlated with the Lassiter Coast Intrusive Suite (Rowley *et al.*, this volume, p. 467); these units are widespread in southern Antarctic Peninsula and eastern Ellsworth Land (Fig. 3).

In the northern part of the APTP less abundant Upper Palaeozoic–Lower Jurassic rocks have been called basement (e.g. Dalziel, 1982; Pankhurst, 1983). They were affected by early Mesozoic deformational and metamorphic events, and they are separated from the overlying rocks of the upper Mesozoic magmatic arc by a Middle Jurassic angular unconformity (Smellie, 1981; Dalziel, 1982; Elliot, 1983; Thomson & Pankhurst, 1983).

More recently a tectonic model based on subduction-related processes that were active from the Late Palaeozoic onward has been proposed for the Antarctic Peninsula (Storey & Garrett, 1985). In this model, basement (known only from the east coast of Graham Land) includes only continental crustal rocks that existed prior to the accretion–subduction system (Pankhurst, 1983). Other pre-Middle Jurassic rocks are included in the accretionary prism component of the model. In the southern part of the APTP (Fig. 3) outside the English Coast, the only rocks known to be of pre-Middle Jurassic age are Lower Jurassic (Thomson & Tranter, 1986) sedimentary rocks of the LeMay Group on Alexander Island that are part of the accretionary prism.

On the English Coast, the Fitzgerald quartzite beds are of continental interior provenance (Fig. 2) and were probably deposited in a passive margin tectonic environment. Although the age and correlation of the unit are unknown, it probably predates the onset of subduction, and is part of the pre-Upper Palaeozoic basement of the APTP. This suggests pre-Upper Palaeozoic stratigraphical linkage with some cratonic block, probably East Antarctica. Petrologically similar rocks are abundant in the Devonian Crashsite Quartzite of the Ellsworth Mountains (Hjelle, Yoshida & Winsnes, 1982), and in Beacon Supergroup and older rocks throughout the Transantarctic Mountains (e.g. Laird & Bradshaw, 1982).

Glacial erratics of similar cross-bedded quartzite have been recovered from the Rydberg Peninsula (R.G.B. Renner, unpublished field notes, 1975) and from eastern FitzGerald Bluffs (Fig. 1). Large clasts of both metamorphosed and unmetamorphosed quartz-arenite occur in Mesozoic conglomerates in several places in eastern Ellsworth Land (Table 2, Fig. 3) and in the LeMay Group in central Alexander Island (Tranter, 1986). This suggests wide distribution of rocks similar to the Fitzgerald quartzite beds in the basement of southern Palmer Land and eastern Ellsworth Land.

The Erehwon beds, of Permian age and magmatic arc provenance, suggest that by Late Palaeozoic time the English Coast was part of a subducted margin of Gondwana. Upper Palaeozoic stratigraphical linkage with the *Glossopteris* floral province is indicated. The nearest known occurrence of *Glossopteris* is in the Permian Polarstar Formation of the Ellsworth Mountains, ~ 350 km south of the English Coast. The Polarstar Formation, which is petrographically similar to the Erehwon beds (e.g. see Fig. 2), was probably deposited in prodelta, delta and coastal plain environments in a back-arc basin between the Pacific margin of Gondwana and the East Antarctic craton (Collinson, Vavra & Zawiskie, in press). The Erehwon beds, which were deposited in quiet, terrestrial or paralic environments, may have been deposited on or near the magmatic arc on the Pacific margin of the Polarstar Basin.

The Latady Formation on the English Coast is of recycled orogenic provenance (Fig. 2) and was deposited in terrestrial or paralic environments. Sandstone is characterized by significantly higher than average Q content (for the Latady Formation; see Laudon *et al.*, 1983, fig. 6), and at Henkle Peak and Mount Rex by very low F content (Fig. 2). Sandstone of similar composition occurs in the Latady Formation in an E-trending belt that includes Lyon and Sky-Hi nunataks, Merrick Mountains, and Sweeney Mountains (Fig. 2), and appears to represent a distinct petrofacies within the formation.

Clastic rocks at Mount Peterson (Fig. 3) are part of an alluvial fan, and are interbedded with felsic volcanic rocks. Their source terrane was composed predominantly of volcanic rocks, but also included sedimentary, metamorphic and plutonic rocks. These features suggest deposition accompanied by contemporaneous volcanicity in a down-faulted basin adjacent to an uplifted basement block with younger volcanic cover. This contrasts with other terrestrial and paralic facies of the Latady Formation, which were deposited in quiet aqueous environments (Laudon *et al.*, 1983).

Conglomerate–pebble populations at Mount Peterson are dominated by mafic and intermediate VRFs, and lack felsic VRFs. This contrasts with other Latady conglomerate–pebble populations, which are dominated by felsic VRFs (Table 2).

Structural attitudes of sedimentary rocks on the English Coast (Fig. 1) appear to represent continuations of NW-trending Cretaceous structures of the Latady and Mount Poster formations in the Orville Coast (Fig. 3). Evidence for earlier deformation or metamorphism of the Erehwon beds has not been observed.

Acknowledgements

The other members of the 1984–85 field party to the English Coast were Party Chief Peter Rowley, and Karl Kellogg, Dave Lidke, Mike O'Neill, Janet Thomson and Walt Vennum. I thank them for many things including the use of their field notes and samples. Ted Delevoryas and Carole Gee identified the plant fossils. Cam Craddock provided access to samples and thin sections of the Crashsite Quartzite and the Polarstar Formation. This project was undertaken at the British Antarctic Survey (BAS), and supported by the University of Wisconsin Oshkosh Faculty Development Program. I thank BAS colleagues for access to samples and unpublished data, and for many worthwhile discussions related to the project. Reviews by Pete Rowley, Mike Thomson, Joe Laudon, and Andy Whitham greatly improved the manuscript.

References

British Antarctic Survey. (1981) *British Antarctic Territory Geological Map, Sheet 4, Alexander Island*, 1:500 000, BAS 500G series. Cambridge; British Antarctic Survey.

Collinson, J.R., Vavra, C.L. & Zawiskie, J.L. (in press). Sedimentology of Polarstar Formation Permian, Ellsworth Mountains, Antarctica. In *Geology and Paleontology of the Ellsworth Mountains, Antarctica*, ed. G.F. Webers, C. Craddock & J.F. Splettstoesser. Washington, DC; Geological Society of America Memoir.

Dalziel, I.W.D. (1982). The early (pre-Middle Jurassic) history of the Scotia arc region – a review and progress report (review paper). In *Antarctic Geoscience*, ed. C. Craddock, pp. 111–26. Madison; University of Wisconsin Press.

Dalziel, I.W.D. & Elliot, D.H. (1982). West Antarctica – problem child of Gondwanaland. *Tectonics*, **1(1)**, 3–19.

Dickinson, W.R. (1970). Interpreting detrital modes of graywacke and arkose. *Journal of Sedimentary Petrology*, **40(2)**, 675–707.

Dickinson, W.R. (1985). Interpreting provenance relations from detrital modes of sandstones. In *Provenance of Arenites*, ed. G.G. Zuffa, pp. 333–761. Hingham, Massachusetts; Dordrecht Reidel Publishing Company.

Dott, R.H. (1964). Wacke, graywacke and matrix – what approach to immature sandstone classification? *Journal of Sedimentary Petrology*, **34(3)**, 625–32.

Elliot, D.H. (1983). The mid-Mesozoic to mid-Cenozoic active plate margin of the Antarctic Peninsula. In *Antarctic Earth Science*, ed. R.L. Oliver, P.R. James & J.B. Jago, pp. 347–51. Canberra; Australian Academy of Science and Cambridge; Cambridge University Press.

Gee, C.T. (1989). Permian *Glossopteris* and *Elatocladus* megafossil floras from the English Coast, eastern Ellsworth Land, Antarctica. *Antarctic Science*, **1(1)**, 35–44.

Hjelle, A., Yoshida, O. & Winsnes, T.S. (1982). Geology and petrology of the southern Heritage Range, Ellsworth Mountains. In *Antarctic Geoscience*, ed. C. Craddock, pp. 599–608. Madison; University of Wisconsin Press.

Laird, M.G. & Bradshaw, J.D. (1982). Uppermost Proterozoic and lower Paleozoic geology of the Transantarctic Mountains, (review paper). In *Antarctic Geoscience*, ed. C. Craddock, pp. 525–34. Madison; University of Wisconsin Press.

Laudon, T.S., Lidke, D.J., Delevoryas, T. & Gee, C.T. (1985). Sedimentary rocks of the English Coast, eastern Ellsworth Land, Antarctica. *Antarctic Journal of the United States*, **20(5)**, 38–40.

Laudon, T.S., Lidke, D.J., Delevoryas, T. & Gee, C.T. (1987). Sedimentary rocks of the English Coast, eastern Ellsworth Land. In *Gondwana Six: Structure, Tectonics, and Geophysics*, Geophysical Monograph 40, ed. G.D. McKenzie, pp. 183–7. Washington DC; American Geophysical Union.

Laudon, T.S., Thomson, M.R.A., Williams, P.L., Milliken, K.L., Rowley, P.D. & Boyles, J.M. (1983). The Jurassic Latady Formation, southern Antarctic Peninsula. In *Antarctic Earth Science*, ed. R.L. Oliver, P.R. James & J.B. Jago, pp. 308–14. Canberra; Australian Academy of Science and Cambridge; Cambridge University Press.

O'Neill, J.M. & Thomson, J.W. (1985). Tertiary mafic volcanic and volcaniclastic rocks of the English Coast, Antarctica. *Antarctic Journal of the United States*, **20(5)**, 36–8.

Pankhurst, R.J. (1983). Rb–Sr constraints on the ages of basement rocks of the Antarctic Peninsula. In *Antarctic Earth Science*, ed. R.L. Oliver, P.R. James & J.B. Jago, pp. 367–71. Canberra; Australian Academy of Science and Cambridge; Cambridge University Press.

Rowley, P.D., Kellogg, K.S. & Vennum, W.R. (1985). Geologic studies in the English Coast, eastern Ellsworth Land, Antarctica. *Antarctic Journal of the US*, **20(5)**, 34–6.

Rowley, P.D. & Williams, P.L. (1982). Geology of the northern Lassiter Coast and southern Black Coast, Antarctic Peninsula. In *Antarctic Geoscience*, ed. C. Craddock, pp. 339–48. Madison; University of Wisconsin Press.

Smellie, J.L. (1981). A complete arc-trench system recognized in Gondwana sequences of the Antarctic Peninsula region. *Geological Magazine*, **118(2)**, 139–59.

Storey, B.C. & Garrett, S.W. (1985). Crustal growth of the Antarctic Peninsula by accretion, magmatism and extension. *Geological Magazine*, **122(1)**, 5–14.

Thomson, M.R.A. & Pankhurst, R.J. (1983). Age of post-Gondwanian calc-alkaline volcanism in the Antarctic Peninsula region. In *Antarctic Earth Science*, ed. R.L. Oliver, P.R. James & J.B. Jago, pp. 328–33. Canberra; Australian Academy of Science and Cambridge; Cambridge University Press.

Thomson, M.R.A., Pankhurst, R.J. & Clarkson, P.D. (1983). The Antarctic Peninsula – a Late Mesozoic–Cenozoic arc (review). In *Antarctic Earth Science*, ed. R.L. Oliver, P.R. James & J.B. Jago, pp. 289–94. Canberra; Australian Academy of Science and Cambridge; Cambridge University Press.

Thomson, M.R.A. & Tranter, T.H. (1986). Early Jurassic fossils from central Alexander Island and their geological setting. *British Antarctic Survey Bulletin*, **70**, 23–39.

Tranter, T.H. (1986). The LeMay Group of central Alexander Island. *British Antarctic Survey Bulletin*, **71**, 57–67.

Tectonic evolution of the south-eastern Antarctic Peninsula

K.S. KELLOGG & P.D. ROWLEY

US Geological Survey, Federal Center, Box 25046, Denver, Colorado 80225, USA

Abstract

The oldest exposed rocks of south-eastern Antarctic Peninsula have been recognized as part of a Middle–Late Jurassic volcanic arc which developed in response to subduction beneath the Pacific margin of the peninsula. Many of the calc-alkaline volcanic rocks which occupy the axial part of the arc (Mount Poster Formation), and the sedimentary sequence that occupies the back-arc basin to the east (Latady Formation), were intensely folded and thrust before the formation of an Early Cretaceous (mostly 113–97 Ma) calc-alkalic plutonic arc. The Early Cretaceous arc overlaps both the Jurassic volcanic arc and the back-arc basin to the east and south, a feature which may be related to relative flattening of the subducting slab under the southern peninsula during the Early Cretaceous. Most faults in the southern peninsula are small, orientated about N70° W, have right-lateral apparent offset, and are probably parallel to a large, right-lateral transform (the Ellsworth fault) that forms the southern tectonic boundary of the peninsula. The Ellsworth fault may be a continuation of the Tharp fracture zone in the south-east Pacific, which was active about 85–50 m.y. ago.

Introduction

The south-eastern Antarctic Peninsula (Lassiter Coast, Orville Coast, and eastern Ellsworth Land; Fig. 1) comprises part of the mostly Mesozoic Andean orogen which extends the length of the Antarctic Peninsula, bends westward in the south, and continues into western Ellsworth Land and Marie Byrd Land. The south-eastern Antarctic Peninsula lies near the juncture of several proposed crustal blocks that make up West Antarctica (e.g. Dalziel & Elliot, 1982). Thus, its tectonic history is particularly important to the overall history of West Antarctica.

Tectonic history

Middle–Late Jurassic arc magmatism and back-arc sedimentation

The oldest known rocks in the area include the volcanogenic shale, siltstone, sandstone and sparse conglomerate of the Middle–Late Jurassic Latady Formation which crops out along the south-eastern and southern margins of the peninsula (Rowley & Williams, 1982; Rowley *et al.*, 1983). The formation is at least several kilometres thick; neither the base nor top has been observed. Clasts of metamorphic and plutonic rock in the Latady Formation (Laudon *et al.*, 1983) indicate that crystalline rocks of unknown age were exposed nearby. With the exception of these clasts, gneissic (and sheared) rocks like those reported from north-eastern Palmer Land (Meneilly *et al.*, 1987) have not been reported in south-eastern Antarctic Peninsula.

The Latady Formation interfingers to the north-west and north with mostly subaerial rhyodacitic–andesitic lava flows, ash-flow tuffs, and volcanic breccias of the Mount Poster Formation (Rowley *et al.*, 1983; Fig. 1). Although it is apparent that the Mount Poster Formation is as old as Middle Jurassic, parts of it may be as young as Early Cretaceous, the age of voluminous plutons of the Lassiter Coast Intrusive Suite (Vennum & Rowley, 1986).

The Latady Formation is interpreted to have been deposited in a back-arc basin that formed in response to subduction beneath the Pacific side of the peninsula (Suárez, 1976; Rowley & Williams, 1982; Rowley *et al.*, 1983; Storey & Garrett, 1985). Active subduction ceased west of the southern part of the peninsula about 50 Ma ago, when the Aluk–West Antarctic Ridge intersected the trench (Fig. 2). Northward along the peninsula, subduction ceased by increments as successive transform fault-bounded segments of the ridge intersected the trench (Barker, 1982).

The Latady Formation and parts of the Mount Poster Formation were open to tightly folded before the onset of Early Cretaceous plutonism (Rowley *et al.*, 1983) during what we propose to call the Palmer Land deformational event. Fold axes generally are horizontal to gently plunging, except where Early Cretaceous intrusives disharmonically refold the rocks, and trend subparallel to the general geographical shape of the peninsula (Fig. 1). Most of the folds are asymmetric with S–SE vergence. Observed thrust faults are rare, dip toward the

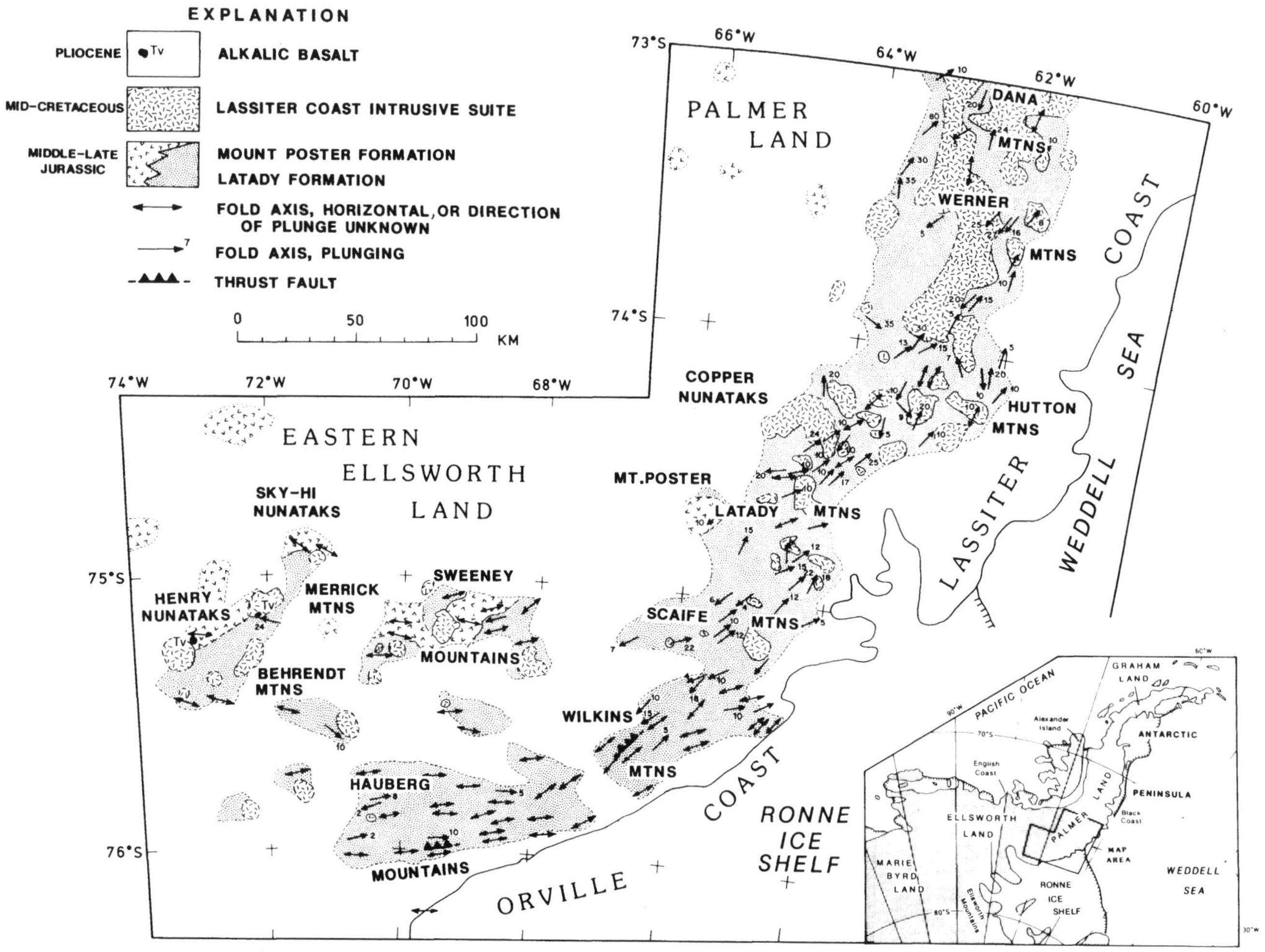

Fig. 1. Location map (inset) and geological map of south-eastern Antarctic Peninsula. Hachured lines mark the approximate edge of ice shelves; non-patterned areas on the geological map are either ice-covered or the geology is unknown.

volcanic arc, and have been mapped only in the Orville Coast area (Rowley *et al.*, 1983). However, repetition of fossil assemblages in the southern Behrendt Mountains also suggests that thrusts exist there (Laudon *et al.*, 1970).

The Palmer Land deformational event may be related to the main-phase Rangitata orogeny recognized in New Zealand (Bradshaw, Adams & Andrews, 1981), during which time New Zealand and the Campbell Plateau were attached to Marie Byrd Land (e.g. Barker, 1982). The plate collision inferred to have caused the Rangitata orogeny, whose effects (folding, metamorphism and plutonism) are apparently more intense in New Zealand than in the Antarctic Peninsula, impacted more profoundly in Palmer Land than in the more distant Graham Land; the stratified Jurassic–Early Cretaceous rocks in Graham Land are not folded (Storey & Garrett, 1985)

Lower Cretaceous plutonism

Intrusion of the calc-alkaline plutonic rocks of the Lassiter Coast Intrusive Suite appears to have occurred during a relatively short-lived but intense magmatic pulse (Vennum & Rowley, 1986). Rb–Sr and K–Ar ages on plutonic rocks are between 130–95 Ma (Farrar, McBride & Rowley, 1982; Pankhurst & Rowley, this volume, p. 387); > 90% of the radiometric ages are in the range 113–97 Ma and about 60% are between 110 and 106 Ma. No systematic geographical trend in the ages is apparent, although younger rocks are progressively more silicic (Farrar *et al.*, 1982). A geochemical trend across the peninsula in eastern Ellsworth Land is shown by the southward increase of K relative to SiO_2 away from the former subduction zone (Vennum & Rowley, 1986). No coeval volcanic rocks have been identified and exposures of intrusive rock increase northward.

The cross-peninsular extent of Cretaceous plutonism is large and includes extensive areas within the Latady back-arc basin, south and east of the Jurassic volcanic arc. This transgression of plutonism towards the back-arc suggests that a period of rapid Early Cretaceous subduction beneath the southern peninsula caused the subducting slab to flatten, effectively widening the zone of magma generation at about 100 km and displacing the locus of plutonism to the east and south.

Uplift, rifting and Neogene alkalic volcanism

Arching and rifting, following a possible period of peneplanation, occurred during or after Late Cretaceous time and may be related to the slowing of subduction as the Aluk–West Antarctic ridge approached the subduction zone

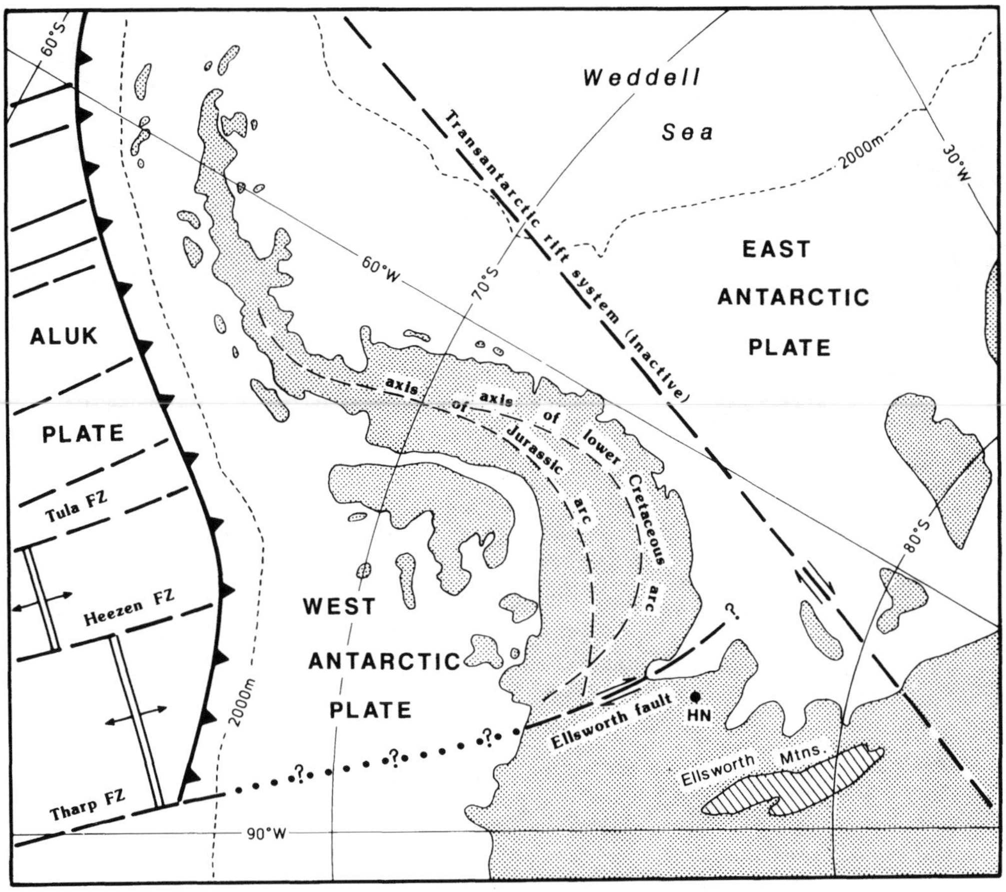

Fig. 2. Schematic map showing inferred tectonic elements of the southern Antarctic Peninsula region at the close of the Cretaceous Period. Outline of modern shoreline, 2000 m depth contour (pecked line; after LaBrecque, 1986), and positions of Haag Nunataks (HN) and Ellsworth Mountains are shown for reference. Line with teeth, trench; double line with arrows, spreading centre; FZ, fracture zone.

(Storey & Garrett, 1985). Fault-block topography imposed on an arched and tilted (?)early Tertiary erosion surface is inferred in eastern Ellsworth Land (Laudon, 1972), where seismic-reflection and gravity data define rough subice topography, including valleys nearly 2000 m below sea level (Behrendt, 1964).

The presence of Pliocene alkalic olivine basalts in the Merrick Mountains (Laudon, 1972) and Henry Nunataks (Fig. 1) indicate that rifting continued into late Tertiary time in the southern part of the peninsula. These occurrences mark the eastern end of a zone of rift-related alkalic volcanism that extends across West Antarctica, including the English Coast (Rowley *et al.*, this volume, p. 467), other parts of Ellsworth Land and Marie Byrd Land (LeMasurier & Wade, 1976).

Opening of the Weddell Sea basin

The Weddell Sea basin was opening during the formation of the Jurassic arc in the southern part of the peninsula. Marine geophysical studies in parts of the Weddell Sea basin indicate that the seafloor north of about 71° S is underlain by oceanic crust marked by W-trending, northward-younging linear magnetic anomalies that converge westward (LaBrecque, 1986); the oldest recognized anomaly (M29) is 165 Ma (Middle Jurassic). The western tectonic boundary of the opening Weddell Sea basin is inferred to be a N-striking, right-lateral transform fault (Barker & Griffiths, 1977; Fig. 2). This proposed transform fault was part of a Mesozoic fault system separating East and West Antarctica (Transantarctic rift system of Schmidt & Rowley, 1986). During the opening of the Weddell Sea basin, West Antarctica may have slid about 500–1000 km along faults parallel to the Transantarctic rift system. During all stages of the opening of the Weddell Sea basin, the Latady basin was backed by the Gondwana craton, which may have contributed sediment to the mostly volcaniclastic deposits of the Latady basin. Pre-Mesozoic crystalline rocks are also known from Graham Land (Storey & Garrett, 1985) and may also have contributed sediment to the Latady basin.

High-angle faults and their relationship to the 'Ellsworth fault' – a proposed transform

The southern tectonic boundary of the Antarctic Peninsula, which separates the peninsula from the ice-covered region to the south, is a 1 km high subice escarpment (Behrendt, 1964; Garrett, Herrod & Mantripp, 1987) which trends about N70° W. Although the morphology suggests a large component of normal faulting (or strike-slip reactivation along an older normal fault), Ford (1972) suggested that the escarp-

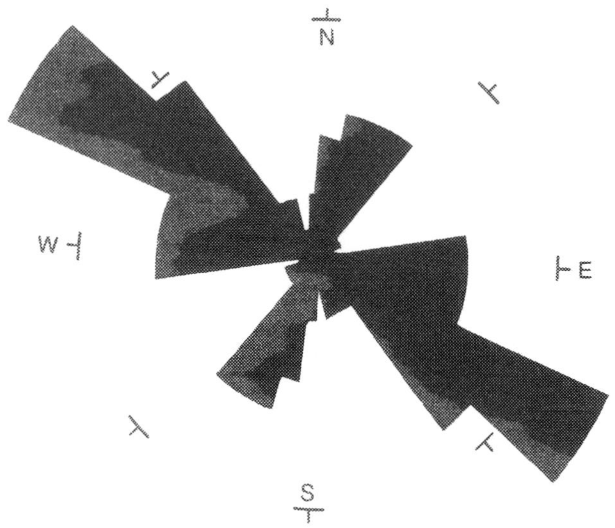

Fig. 3. Rose diagram showing the strike direction of all observed high-angle faults in the Orville Coast and eastern Ellsworth Land area; 69 observations grouped in 10° intervals.

ment marks a major right-lateral strike-slip fault, which he named the Ellsworth fault.

The orientations of observed faults in south-eastern Antarctic Peninsula support the existence of the Ellsworth fault. Along the Lassiter Coast, faults are apparently rare, although they have been noted striking perpendicular to the axis of the peninsula at Copper Nunataks (Fig. 1), parallel to a regional extension-joint direction. In the Orville Coast area and eastern Ellsworth Land, however, steep faults of apparently small displacement commonly cut both the Latady Formation and intrusive rocks. The fault zones are generally unsilicified and < 20 cm wide. Mylonitic zones, like those noted in northern Palmer Land (Meneilly *et al.*, 1987), have not been observed anywhere in south-eastern Antarctic Peninsula, indicating more shallow levels of erosion in the south.

The dominant strike direction for 69 measured faults throughout the Orville Coast and eastern Ellsworth Land is about N70° W (Fig. 3), although a subordinate group of faults is orientated perpendicular to the dominant set. Strike separation could be measured on 17 faults within the dominant set; of these, 14 show an apparent right-lateral displacement. It seems likely that the dominant set formed in the same stress field as the Ellsworth fault and marks the north edge of a zone of right-lateral shear.

It is worth considering the possibility that the Ellsworth fault is a continuation of the Tharp fracture zone in the south-east Pacific (Barker, 1982). The Tharp fracture zone, whose position adjacent to the Antarctic Peninsula is poorly constrained, apparently marked the south-west end of the Aluk–West Antarctic ridge during the Late Cretaceous–early Tertiary (Fig. 2). Movement along the Ellsworth fault may have begun after about 85 Ma ago, when the Campbell Plateau margin of the New Zealand block separated from Marie Byrd Land (Barker, 1982), and ceased about 50 Ma ago, when the Aluk–West Antarctic ridge subducted into the trench off the western side of the peninsula and subduction ceased.

Strike-slip movement along the Ellsworth fault could not have been large; palaeomagnetic data indicate that, within the limits of uncertainty, the Antarctic Peninsula has been fixed relative to the region of West Antarctica to the south of the peninsula since Middle Jurassic time (Grunow, Dalziel & Kent, 1987).

Kellogg (1980) proposed that right-lateral drag along the Ellsworth fault rotated parts of the Orville Coast, forming or at least accentuating the orocline through the region; this hypothesis is supported by clockwise rotation (by about 50°) of the palaeomagnetic declination from mid-Cretaceous plutons in the Orville Coast as compared to the declination from equivalent-age plutons from the rest of the Antarctic Peninsula. Although Watts, Watts & Bramall (1984) believed that the data instead indicated apparent polar wander rather than tectonic rotation, we find no evidence to support the suggestion that the rocks containing a rotated magnetic vector are any older than 'unrotated' plutons, a conclusion supported by new Rb–Sr dating (Pankhurst & Rowley, this volume, p. 387). We still see no problem with local oroclinal bending or block rotation adjacent to the Ellsworth fault.

Acknowledgements

We are indebted to W.W. Hamilton, J.M. O'Neill, W.E. LeMasurier, J.C. Behrendt, A.B. Ford, R.L. Reynolds and an unknown referee for critical and constructive reviews of the manuscript.

References

Barker, P.F. (1982). The Cenozoic subduction history of the Pacific margin of the Antarctic Peninsula: ridge crest–trench interactions. *Journal of the Geological Society, London*, **139(5)**, 787–801.

Barker, P.F. & Griffiths, D.H. (1977). Towards a more certain reconstruction of Gondwanaland. *Philosophical Transactions of the Royal Society of London*, **B279**, 143–59.

Behrendt, J.C. (1964). The crustal geology of Ellsworth Land and southern Antarctic Peninsula from gravity and magnetic anomalies. *Journal of Geophysical Research*, **69(10)**, 2047–63.

Bradshaw, J.D., Adams, C.J. & Andrews, P.B. (1981). Carboniferous to Cretaceous on the Pacific margin of Gondwana: the Rangitata Phase of New Zealand. In *Gondwana Five*, ed. M.M. Cresswell & P. Vella, pp. 217–22. Rotterdam; A.A. Balkema.

Dalziel, I.W.D. & Elliot, D.H. (1982). West Antarctica: problem child of Gondwanaland. *Tectonics*, **1(1)**, 3–9.

Farrar, E., McBride, S.L. & Rowley, P.D. (1982). Ages and tectonic implications of Andean plutonism in the southern Antarctic Peninsula. In *Antarctic Geoscience*, ed. C. Craddock, pp. 349–56. Madison; University of Wisconsin Press.

Ford, A.B. (1972). Fit of Gondwana continents – drift reconstructions from the Antarctic continental viewpoint. *24th International Geological Congress, Montreal, 1972*, section 3, 113–21.

Garrett, S.W., Herrod, L.D.B. & Mantripp, D.R. (1987). Crustal structure of the area around Haag Nunataks, West Antarctica: implications of rifting for Gondwana reconstructions. In *Gondwana Six: Structure, Tectonics, and Geophysics*, Geophysical Monography 40, ed. G.D. McKenzie, pp. 109–115. Washington, DC; American Geophysical Union.

Grunow, A.M., Dalziel, I.W.D. & Kent, D.V. (1987). Ellsworth

Mountains – Whitmore Mountains crustal block, Western Antarctica: new paleomagnetic results and their tectonic significance. In *Gondwana Six: Structure, Tectonics and Geophysics*, ed. G.D. McKenzie, pp. 161–71. Washington, DC; American Geophysical Union, Geophysical Monograph 40.

Kellogg, K.S. (1980). Paleomagnetic evidence for oroclinal bending of the southern Antarctic Peninsula. *Geological Society of America Bulletin*, **91(7)**, 414–20.

LaBrecque, J.L., ed. (1986). *South Atlantic Ocean and Adjacent Antarctic Continental Margin*, Atlas 13, Ocean Margin Drilling Program, Regional Atlas Series. Marine Science International.

Laudon, T.S. (1972). Stratigraphy of eastern Ellsworth Land. In *Antarctic Geology and Geophysics*, ed. R.J. Adie, pp. 215–23. Oslo; Universitetsforlaget.

Laudon, T.S., Lackey, L.L., Quilty, P.G. & Otway, P.M. (1970). Geology of eastern Ellsworth Land. In *Geologic Maps of Antarctica*, ed. V.C. Bushnell & C. Craddock. Antarctic Map Folio Series, Folio 12, Pl. III. New York; American Geographical Society.

Laudon, T.S., Thomson, M.R.A., Williams, P.L., Miliken, K.L., Rowley, P.D. & Boyles, J.M. (1983). The Jurassic Latady Formation, southern Antarctic Peninsula. In *Antarctic Earth Science*, ed. R.L. Oliver, P.R. James & J.B. Jago, pp. 308–19. Canberra; Australian Academy of Science and Cambridge; Cambridge University Press.

LeMasurier, W.E. & Wade, F.A. (1976). Volcanic history in Marie Byrd Land: implications with regard to southern hemisphere tectonic reconstructions. In *Andean and Antarctic Volcanology Problems*, ed. O. González-Ferrán, pp. 398–424. Rome; International Association of Volcanology and Chemistry of the Earth's Interior.

Meneilly, A.W., Harrison, S.M., Piercy, B.A. & Storey, B.C. (1987). Structural evolution of the magmatic arc in northern Palmer Land, Antarctic Peninsula. In *Gondwana Six: Structure, Tectonics, and Geophysics*, Geophysical Monograph 40, ed. G.D. McKenzie, pp. 209–19. Washington, DC; American Geophysical Union.

Rowley, P.D., Vennum, W.R., Kellogg, K.S., Laudon, T.S., Carrara, P.E., Boyles, J.M. & Thomson, M.R.A. (1983). Geology and plate tectonic setting of the Orville Coast and eastern Ellsworth Land, Antarctica. In *Antarctic Earth Science*, ed. R.L. Oliver, P.R. James & J.B. Jago, pp. 245–50. Canberra; Australian Academy of Science and Cambridge; Cambridge University Press.

Rowley, P.D. & Williams, P.L. (1982). Geology of the northern Lassiter Coast and southern Black Coast, Antarctic Peninsula. In *Antarctic Geoscience*, ed. C. Craddock, pp. 339–48. Madison; University of Wisconsin Press.

Schmidt, D.L. & Rowley, P.D. (1986). Continental rifting and transform faulting along the Jurassic Transantarctic rift, Antarctica. *Tectonics*, **5(2)**, 279–91.

Storey, B.C. & Garrett, S.W. (1985). Crustal growth of the Antarctic Peninsula by accretion, magmatism, and extension. *Geological Magazine*, **122(1)**, 5–14.

Suárez, M. (1976). Plate tectonic models for southern Antarctic Peninsula and its relation to southern Andes. *Geology*, **4(4)**, 211–14.

Vennum, W.R. & Rowley, P.D. (1986). Reconnaissance geochemistry of the Lassiter Coast Intrusive Suite, southern Antarctic Peninsula. *Geological Society of America Bulletin*, **97(12)**, 1521–33.

Watts, D.R., Watts, G.C. & Bramall, A.M. (1984). Cretaceous and Early Tertiary paleomagnetic results for the Antarctic Peninsula. *Tectonics*, **3(3)**, 333–46.

Tectonic setting of the English Coast, eastern Ellsworth Land, Antarctica

P.D. ROWLEY[1], K.S. KELLOGG[1], W.R. VENNUM[2], T.S. LAUDON[3], J.W. THOMSON[4], J.M. O'NEILL[1] & D.J. LIDKE[1]

1 US Geological Survey, Box 25046, Federal Center, Denver, Colorado 80225, USA
2 Department of Geology, Sonoma State University, Rohnert Park, California 94928, USA
3 Department of Geology, University of Wisconsin Oshkosh, Oshkosh, Wisconsin 54901, USA
4 British Antarctic Survey, Natural Environment Research Council, High Cross, Madingley Road, Cambridge CB3 0ET, UK

Abstract

The English Coast is the western continuation of the Antarctic Peninsula tectonic province. Most exposed rocks belong to a folded sequence of chloritized calc-alkalic volcanic rocks, correlated with the Mount Poster Formation, and intertongued subordinate sedimentary rocks, correlated with the Jurassic Latady Formation. Older rocks are locally exposed: the Erehwon beds of volcanogenic sandstone containing a (?)Permian *Glossopteris* flora, and the FitzGerald quartzite beds of cratonic provenance. The folded rocks were intruded by Lower Cretaceous granodiorite plutons, then overlain by Tertiary lava flows and hyaloclastite of alkalic basalt. The Erewhon beds probably represent an early phase of arc magmatism along the Pacific edge of Gondwana, and the FitzGerald quartzite beds may predate this magmatism. Break-up of Gondwana involved right-lateral strike-slip displacement along the Jurassic Transantarctic Rift. One result of break-up was numerous lithospheric plates that now comprise West Antarctica. Continuing Andean-type ensialic arc magmatism along the Pacific edge of these plates resulted in the intertongued volcanic and sedimentary rocks of the English Coast. All rocks were folded during the Late Jurassic–Early Cretaceous Palmer Land event, perhaps caused by plate collision. Cretaceous plutons represent renewal of arc magmatism, which terminated with Tertiary crustal thinning and extension.

Introduction

The first investigation of the previously unexplored English Coast (Fig. 1) of West Antarctica was made from December 1984 to February 1985. Preliminary reports were published by O'Neill & Thomson (1985), Rowley, Kellogg & Vennum (1985) and Laudon *et al.* (1987). These show that the English Coast is a westward continuation of the Antarctic Peninsula tectonic province (APTP) and is geologically similar to adjacent parts of the province (Laudon, 1972; Williams *et al.*, 1972; Rowley & Williams, 1982; Rowley *et al.*, 1983). However, significant differences occur, including sedimentary rocks containing a *Glossopteris* flora (Laudon *et al.*, 1987; Laudon, this volume, p. 455). These rocks are significantly older than any previously known from this part of the province.

West Antarctica appears to consist of at least four small plates of continental lithosphere, including the Antarctic Peninsula, western Ellsworth Land, Marie Byrd Land and Ellsworth–Whitmore mountains area (Dalziel & Elliot, 1982). Probably the plates were formed by relatively small plate adjustments during break-up of Gondwana that started in the Jurassic (Rowley, 1983; Schmidt & Rowley, 1986; Dalziel *et al.*, 1987). The APTP is a continuation of the Andes and similarly was formed by subduction of Pacific Ocean lithosphere. The province consists of an Andean-type ensialic magmatic arc and related fore-arc and back-arc basins that formed mostly between the Jurassic and late Cenozoic. The magmatic arc follows the shape of the Peninsula southward, then is oroclinally bent westward, parallel to the Pacific margin.

Sedimentary rocks

Folded dark, mostly continental clastic sedimentary rocks are scattered throughout the English Coast, but their relationships are poorly constrained because most nunataks are widely separated and expose only single rock types. Studies of thin sections and plant fossils suggest that the strata are of three widely different ages (Laudon, this volume, p. 455).

Probably the oldest rocks consist of the FitzGerald quartzite beds, made up of 300 m of conspicuously bedded, light–

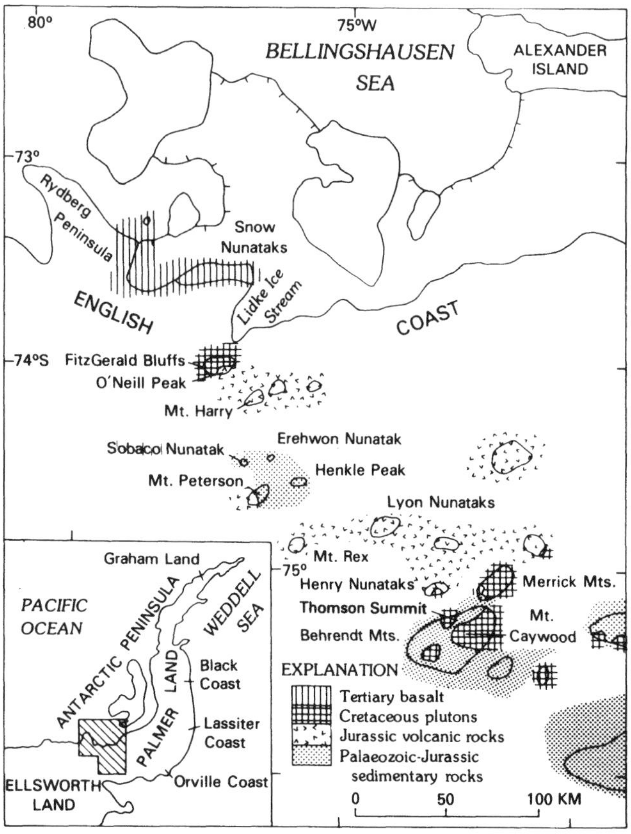

Fig. 1. Generalized geological map of the English Coast and eastern Ellsworth Land. Hachured lines mark the edges of ice shelves.

medium grey quartzite exposed only in the western FitzGerald Bluffs area. The rocks consist of well-sorted, well-rounded quartz grains suggestive of a provenance of cratonic rock that predates subduction in the area. The quartzite is petrologically similar to the Devonian Crashsite Quartzite of the Ellsworth Mountains (Laudon, this volume, p. 455). This undated quartzite is intruded by Lower Cretaceous plutons. Clasts of the unit are widely scattered in Jurassic–mid-Cretaceous sedimentary units in the southern Antarctic Peninsula and central Alexander Island (Laudon, this volume).

Old sedimentary rocks for which the age is better known are the Erewhon beds, of thin, dark-grey sandstone, siltstone and mudstone exposed at Erehwon Nunatak, 18 km NNE of Mount Peterson. There a rich collection of plant fossils contain *Glossopteris* and related species (Laudon *et al.*, 1987; Gee, 1989) of probable Permian age. Petrologically-similar rocks that lack fossils also occur at 'Sobaco Nunatak', about 15 km WNW of Erehwon Nunatak, and at Henkle Peak, 28 km east-north-east of Erehwon Nunatak. The provenance of the Erewhon beds is mostly silicic volcanic rocks, suggesting that subduction and arc magmatism had started in the area by the time they were deposited. The rocks resemble the Permian Polarstar Formation of the Ellsworth Mountains (Laudon, this volume, p. 455).

The third sequence is composed of dark-grey carbonaceous, fine-grained siltstone, shale, sandstone and conglomerate high in volcanic rock fragments of silicic to intermediate composition. The sequence is correlated with the Middle–Upper Jurassic Latady Formation of the Lassiter Coast (Williams *et al.*, 1972). On the south-western side of Henkle Peak, these rocks represent a swamp or lacustrine depositional environment because they contain leafy twigs of *Elatocladus planus* and other plant fossils known also in continental facies of the Latady Formation (Laudon *et al.*, 1983; Rowley *et al.*, 1983; Gee, 1989; J.M. Schopf and P.D. Rowley, unpublished data, 1980). At another locality, in south-western Lyon Nunataks, fossil belemnites and bivalves indicate a marine origin. At Mount Peterson and several other places, unfossiliferous fluvial sedimentary rocks are interbedded with airfall and (?)ash-flow tuff. The Latady Formation is interpreted as a back-arc basin deposit, synchronous with and lying east and south of, the magmatic arc (Suárez, 1976; Rowley & Williams, 1982; Laudon *et al.*, 1983). Basaltic rocks that appear to be related to the opening of the back-arc basin are especially abundant in the Black Coast (Rowley *et al.*, 1987; Storey *et al.*, 1987a). In inland (northern and western) parts of its basin, the Latady Formation intertongues with calc-alkalic volcanic rocks (Mount Poster Formation) of the Jurassic magmatic arc and is largely of continental origin.

Mount Poster Formation

Basaltic–rhyolitic, calc-alkalic volcanic rocks of Jurassic–Cretaceous–Tertiary age, which make up the Antarctic Peninsula Volcanic Group (Thomson, 1982), comprise the Andean magmatic arc of the APTP. In the southern part of the province, most of these rocks (Mount Poster Formation; Rowley *et al.*, 1983) were subaerially deposited, folded with the Latady Formation and then subjected to regional metamorphism of lower greenschist-facies and, adjacent to plutons, contact metamorphism of hornblende–hornfels and albite–epidote–hornfels-facies. They locally intertongue with sedimentary rocks assigned to the Latady Formation and are considered to be Middle–Late Jurassic.

Folded volcanic rocks are the most abundant rock type in the English Coast. They are lithologically and chemically similar to those of the Mount Poster Formation, with which they are correlated. Most beds originated as ash-flow tuff, lava flows, mudflow breccia and airfall tuff of mostly dacite and rhyolite compositions. Clasts in sedimentary rocks at Mount Peterson are about one-third mafic volcanic rocks (Laudon, this volume, p. 455), unusually mafic for volcanic rock fragments in the Latady Formation. At Marshall Nunatak, most rocks are silicic lava flows that are locally high in alkali metals; several lenses of (?)fluvial boulder conglomerate also are exposed. Laudon (this volume, p. 455) found that most clasts in the conglomerate are volcanic but that there are some sedimentary clasts, many from the FitzGerald quartzite beds. In the Mount Rex area, lava flows and ash-flow tuff dominate, but two thin sequences of fluvial sandstone of Latady type (Laudon, this volume, p. 000) also were mapped.

At two places in the English Coast area, volcanic rocks of Mount Poster Formation are intruded by small hypabyssal plugs that appear to be consanguineous. At O'Neill Peak, lava flows intruded by a lithologically similar plug appear to define a vent complex probably marking the roots of a volcano. In the

Mount Harry area, most rocks are ash-flow tuff and tuff breccia that are high in lithic and pumice clasts. A coarse megabreccia that resembles deposits formed by collapse of caldera walls is also exposed. These apparent intracaldera rocks are intruded by a fresh, medium-grained (1–5 mm) to fine-grained quartz–monzonite plug about 50 m in diameter. The interior of the body is porphyritic (mostly plagioclase phenocrysts as large as 5 mm but some quartz of beta morphology); fine-grained crystalline groundmass, probably originally glass, makes up about 10–20% of the rock. Toward the margin, the shallow nature of the body is even more apparent, as the groundmass becomes predominant and is finer grained than in the interior; the body has a chilled margin 5–10 cm wide. Preliminary palaeomagnetic studies of rocks collected from the plug (K.S. Kellogg, unpublished data, 1987) show magnetic poles similar to those from rocks of the Mount Poster Formation, and the plug is interpreted to be part of the caldera complex. Intrusive rocks of Mount Poster age have not been found elsewhere in the southern APTP.

Lassiter Coast Intrusive Suite (Lower Cretaceous)

Folded rocks of the field area are intruded by calc-alkalic plutons correlated with the Lower Cretaceous Lassiter Coast Intrusive Suite (Rowley *et al.*, 1983; Vennum & Rowley, 1986). In the FitzGerald Bluffs area, two coarse-grained (0.5–3 cm), equigranular–porphyritic plutons, designated the south-west and north FitzGerald Bluffs plutons, are separated by a 1 km wide septum of FitzGerald quartzite beds. The south-west FitzGerald Bluffs pluton consists of two cross-cutting phases, an older finer-grained granodiorite and a younger granite. A Rb–Sr whole-rock six-point isochron (Pankhurst & Rowley, this volume) for plutonic phases and cross-cutting dykes gives an age of 113.3 ± 4 Ma. The north FitzGerald Bluffs pluton consists of two areas of exposed plutonic rock separated by a 5 km wide area of snow and ice. The area to the east contains less K-feldspar but its samples fall within the same compositional field of the rocks to the west so it probably represents the same granodioritic–granitic batholith. The north FitzGerald Bluffs pluton has a preferred age of 128.3 ± 5.3 Ma using four points, two each from the eastern and western exposures. An alternative age of 114.7 ± 27 Ma based on these points, plus one from a porphyry dyke and another from an aplite dyke cutting the plutonic rocks, is less preferable because of the higher error in the age. This error arises because the dykes are distinctly younger; probably they were derived from the south-west FitzGerald Bluffs pluton.

Plutonic rocks in the Behrendt Mountains include the Thomson Summit pluton, which has an exposed diameter of 1 km and is one of the most mafic plutons (gabbro) known from the southern part of the province. A plutonic complex in the northern Behrendt Mountains, the Mount Caywood batholith, was largely mapped by Laudon (1972) and consists of coarse, equigranular–porphyritic granodiorite and quartz–monzodiorite. It is cut locally by gently-dipping sheets of medium-grained biotite–alaskite, as thick as 20 m, that contain abundant bodies of pegmatite and aplite. Halpern (1967) determined Rb–Sr whole-rock and mineral isochrons of 101 ± 5 Ma and 100 ± 2 Ma from granodiorite at two places in the batholith.

Tertiary basaltic-hyaloclastite

The youngest rocks in the English Coast area are sub-horizontally-bedded alkalic-hyaloclastite and pillow-basalt lava flows resembling Tertiary volcanic rocks exposed throughout West Antarctica (LeMasurier & Wade, 1976; Smellie, 1987). Most are in the W-trending Snow Nunataks, along the coast of the Bellingshausen Sea. Each nunatak may overlie or be near a small eruptive centre along a W-striking fracture (O'Neill & Thomson, 1985). The most common rock type is tan and brown, granular palagonitized, hyaloclastite tuff that contains bombs and pillows of flow rock. Mount Benkert, eastern Snow Nunataks, is about 600 m high and consists largely of huge channel fills as thick as 50 m of cross-bedded hyaloclastite probably deposited by pyroclastic eruptions (O'Neill & Thomson, 1985). Massive pillow basalts are locally abundant at Snow Nunataks; all lithologies are cut by dykes, plugs, and sills of the same composition. The rocks cluster in the nepheline-normative basalt and hawaiite compositional fields.

Other exposures of basalts near Snow Nunataks were not visited: (1) Sims Island, a dark-brown peak 20 km north-west of Snow Nunataks that appears to be a remnant of a small volcano; (2) Mount Combs, 40 km to the west that, on air photographs, appears to be another small but entirely ice-covered volcano; and (3) sparse exposures of vesicular hawaiite lava flows on Rydberg Peninsula (BAS, unpublished data, 1985) 35 km to the west-north-west. About 200–250 m of olivine-bearing scoriaceous hawaiite lava flows crop out at Henry Nunataks. These rocks resemble the only other outcrops of Tertiary rocks in the southern APTP, a basanite in south-western Merrick Mountains (Vennum & Laudon, 1988).

Structural geology

The Latady and Mount Poster Formations in the English Coast and adjacent parts of eastern Ellsworth Land are folded along W and WNW horizontal or gently-plunging axes. Folds are open and most are asymmetric or overturned. Chevron folds, containing axes 100–300 m apart and limbs dipping 45–60°, deform rocks of the Mount Poster Formation at Henry Nunataks. Axial-plane cleavage and most axial planes dip northerly. Thrust faults, indicating a sense of yielding toward the south, have been mapped in the Orville Coast (Rowley *et al.*, 1983) but were not seen in the English Coast area, probably because of lack of exposures. Sheared plutonic and volcanic rocks in the Black Coast (Meneilly *et al.*, 1987; Rowley *et al.*, 1987; Storey *et al.*, 1987a) may be the effects on massive rocks of the same deformation. Folding, thrusting, and shearing generally predate emplacement of the Lassiter Coast Intrusive Suite, as elsewhere in this part of the tectonic province, and it is referred to the Palmer Land deformational event of Kellogg & Rowley (this volume, p. 461).

Steep faults were seen in two areas. At south-westernmost

FitzGerald Bluffs, many shear zones of small apparent displacement within plutonic rocks mostly strike NNW. The few slickensides observed have horizontal to oblique rakes. Excellent exposures along the high southern side of Mount Peterson display many faults striking mostly NW to NNW and dipping west at 50–70°. Dip-slip displacement seems to dominate over strike-slip displacement. Both areas are in the western part of the English Coast, just east of a N- or NW-trending geophysical discontinuity (Behrendt, 1964; Bentley & Clough, 1972) that may be a suture between the Antarctic Peninsula and western Ellsworth Land lithospheric plates and that is part of the hypothetical Ellsworth fault (Kellogg & Rowley, this volume, p. 461). The dominant strike of the faults at both locations is parallel to the discontinuity, and faulting apparently post-dates plutonism. The strike and apparent age of these faults is similar to high-angle, mostly strike-slip faults in the Orville Coast that appear to be related to oroclinal bending of the southern Antarctic Peninsula (Kellogg, 1980).

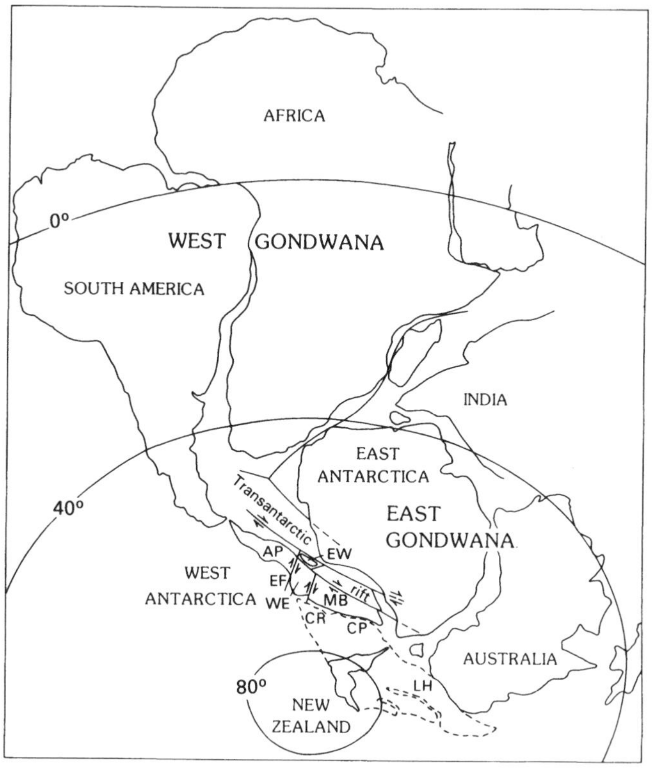

Fig. 2. Reconstruction of Gondwana during Middle Jurassic continental rifting but before Late Jurassic seafloor spreading between West and East Gondwana. The configuration is after Rowley (1983) and Schmidt & Rowley (1986), with the New Zealand reconstruction after Kamp (1986), the New Zealand–West Antarctic fit after Cooper *et al.* (1982), and the India–Sri Lanka position after Collerson & Sheraton (1986). AP, Antarctic Peninsula plate; EW, Ellsworth–Whitmore mountains plate; WE, western Ellsworth Land plate; MB, Marie Byrd Land plate; EF, Ellsworth Fault; CP, Campbell Plateau; CR, Chatham Rise; LH, Lord Howe Rise.

Discussion

Pre-Jurassic rocks were unknown in the southern part of the APTP until our discovery at Erehwon Nunatak. The exposures at this and nearby nunataks represent a probably Palaeozoic continental landmass and magmatic arc. We assume that the rocks were folded during the Late Triassic–Early Jurassic Peninsula deformation (Storey, Thomson & Meneilly, 1987b), then unconformably overlain by Jurassic volcanic rocks of the Andean magmatic arc.

A Triassic–Cenozoic age for the magmatic arc of the APTP is well documented (Smellie, 1981; Dalziel & Elliot, 1982; Pankhurst, 1982; Thomson, Pankhurst & Clarkson, 1983; Storey & Garrett, 1985; Storey *et al.*, 1987b), but evidence for pre-Triassic rocks is confined to sparse isotopic ages of old rocks in Graham Land (Pankhurst, 1983; Storey *et al.*, 1987b; Milne & Millar, this volume, p. 335) and north-western Palmer Land (Harrison & Piercy, this volume, p. 341). Initial $^{87}Sr/^{86}Sr$ ratios of the oldest dated plutons (Late Triassic and Early Jurassic) in the northern part of the arc range from 0.706 to 0.707 (Pankhurst, 1982), consistent with an interpretation based also on a study of garnet xenocrysts (Hamer & Moyes, 1982) for thick pre-Jurassic continental lithosphere beneath this part of the Antarctic Peninsula. Initial Sr-isotope ratios of younger arc plutons in the northern and central parts of the province, however, have generally lower values with decreasing age, suggesting a decreasing influence through time by this continental lithosphere, perhaps due to increasing depth of magma generation (Pankhurst, 1982; Hole, Pankhurst & Saunders, this volume, p. 369) or migration of the arc westward away from the continental lithosphere (Pankhurst & Rowley, this volume, p. 387). In the southern part of the province, only younger arc plutons (mostly Lassiter Coast Intrusive Suite) seem to be exposed and their initial Sr-isotope ratios are generally low (0.704–0.706) but show considerable local variation. Of these initial Sr-isotope ratios, the highest are in plutons of the northern English Coast, supporting the evidence from the newly discovered sedimentary units that basement rocks here are Permian or perhaps older. An influence of a mantle source is thus indicated for these plutons, with a variable degree of crustal interaction (Pankhurst & Rowley, this volume). Based on these data and the highly evolved, silicic-calc-alkalic nature of igneous rocks throughout the province, we infer that a sliver of East Gondwana (Rowley, 1983), with irregular thickness and shape, underlies much of the province, especially the eastern part. However, there is no evidence that it is older than Early Palaeozoic (Harrison & Peircy, this volume; Milne & Millar, this volume, p. 000) and it may be younger than this beneath much of the province (Pankhurst & Rowley, this volume, p. 000).

Regional tectonics

Gondwana in Palaeozoic time included cratonic parts of West Antarctica and most of East Antarctica (Fig. 2). Probably by the Late Palaeozoic, subduction of ancestral Pacific Ocean lithosphere built an Andean-type magmatic arc along the western edge of Gondwana, in the present Andes (Forsythe, 1982), Antarctic Peninsula (Smellie, 1981; Dalziel & Elliot, 1982; Thomson *et al.*, 1983; Storey & Garrett, 1985; Storey *et al.*, 1987b) and other Pacific-margin parts of West Antarctica.

Widespread compressional deformation took place in interior parts of Gondwana during what was called by Du Toit (1937) the Gondwanide orogeny, of Permian–Triassic age (Storey *et al.*, 1987b). Affected rocks extend through the Sierra de la Ventana of South America, the Cape fold belt of South Africa, and parts of the Transantarctic Mountains (Du Toit, 1937). These rocks were later hypothesized to have been deposited in back-arc basins behind a magmatic arc, which was considered to be the one along the Pacific edge of Gondwana (Dalziel & Elliot, 1982). But the classic Gondwanide fold belts of Du Toit are well inland of the leading edge of western Gondwanaland, and the evidence is tenuous for a back-arc interpretation (Storey *et al.*, 1987b). It is more likely that these fold belts represent intracratonic deformation marking the final consolidation of Gondwanaland (Schmidt & Rowley, 1986). Under our interpretation, the Late Triassic–Early Jurassic folding of the magmatic-arc rocks in the Antarctic Peninsula, formerly assigned to Gondwanide deformation (Dalziel, 1982) but recently renamed Peninsula deformation (Storey *et al.*, 1987b), represents instead a separate event related to Pacific subduction.

Folding of the pre-Middle Jurassic magmatic-arc rocks probably was due to collision. Hamilton (1989) suggested that this may have been against an as yet unidentified ensimatic island arc in the proto-Pacific Ocean, of opposite subduction polarity to the Andean subduction system. Collision resulted in accretion of the arc to Gondwana, after which the subduction zone moved Pacific-ward to the new leading edge of Gondwana and another arc formed over the western part of the older one. The post-Triassic magmatism formed most rocks now exposed along the Pacific rim of Antarctica, and in places magmatism continued into the Tertiary; this period has traditionally been called Andean magmatism. Some geologists have used 'orogeny' for parts or all of the event, but Pacific magmatism is hardly an orogeny in the classical sense and in fact deformation is not a significant feature of the overall belt. The igneous products of Andean magmatism appear to have the same general petrology as the pre-Jurassic arc, and the geometry of the Andean arc has a similar trend; thus, Storey & Garrett (1985) noted that the distinction between the two is artificial.

Break-up of Gondwana (Fig. 2) by transform faulting and rifting along the Transantarctic rift began probably in the Middle Jurassic and seafloor spreading began probably in the Late Jurassic (Schmidt & Rowley, 1986). In the Late Jurassic (about 150 Ma), West Antarctica began to be offset in a right-lateral sense with respect to East Antarctica. This sense of offset is required for separation of East and West Gondwana if one assumes that the APTP and southern South America remained in a tip-to-tip join of more or less linear trend (Rowley, 1983; Schmidt & Rowley, 1986). At or after that time, West Antarctica broke into continental lithospheric plates.

Andean magmatism continued during Gondwana break-up and the resultant arc was superimposed on the Pacific edge of the collage of small continental plates of West Antarctica. By comparison with the many rapidly moving lithospheric plates of Indonesia (Hamilton, 1989), and suggested by complex sea-floor anomalies (Barker, 1982), the APTP probably had many collisions and other complexities during its long history, but relatively few are identified. Magmatism was largely continuous, but four main pulses have been noted and some of these are confined geographically (Pankhurst, 1982). The two most widespread pulses of the Pacific margin of West Antarctica are a Middle Jurassic pulse (180–160 Ma), which coincided with the start and early part of Gondwana break-up, and an Early Cretaceous pulse (130–100 Ma), which coincided with the start and much of the opening of the Atlantic Ocean. At about 100 Ma, the Antarctic Peninsula took its approximate present position with respect to East Antarctica (Kellogg & Reynolds, 1978). Some of the oroclinal bending of the peninsula pre-dated this time (Kellogg & Reynolds, 1978, fig. 3) but other bending took place later (Kellogg, 1980). Calc-alkalic magmatism in West Antarctica ceased first in Marie Byrd Land and Ellsworth Land at about 80 Ma (LeMasurier & Wade, 1976) and later in progressively more northern parts of the Antarctic Peninsula as the Aluk Ridge was consumed (Barker, 1982). Rifting and alkalic bimodal (rhyolite and basalt) volcanism, probably the result of crustal thinning in the middle crust, began in West Antarctica at about 50 Ma (LeMasurier & Wade, 1976).

Conclusions

The English Coast area is geologically part of the APTP. Continental rocks of probable Permian age provide evidence that the Andean ensialic magmatic arc that makes up most exposures in the province is underlain in this area by older continental lithosphere of a similar type arc, formed by west–east subduction of proto-Pacific lithosphere. The (?)Permian rocks may in turn be underlain by pre-arc cratonic rocks. All these rocks were folded during the Peninsula deformation, after which Andean-type subduction and ensialic arc magmatism continued. Gondwana break-up started in the Middle–Late Jurassic (Fig. 2); the Mount Poster and Latady Formations are of the same age. The Latady Formation is generally considered to be deposited in a basin formed by extension resulting from back-arc spreading, a requirement of most subduction-related magmatic arcs (Hamilton, 1989). A seaway formed parallel to and inland from the leading edge of the supercontinent; thus it developed within continental lithosphere. In part the seaway may be the result of rifting along a segment of the Transantarctic rift (Schmidt & Rowley, 1986; Kellogg & Rowley, this volume, p. 461).

Folding, thrusting and local shearing of arc rocks between the Late Jurassic and Early Cretaceous, the Palmer Land event, took place only in the southern APTP. Unlike back-arc spreading, compression is not a natural consequence of subduction processes (Hamilton, 1989). Thus in this area it probably resulted from collision of the southern end of the province with an ensimatic island arc such as that which produced the Rangitata deformation of New Zealand (Fig. 2), which was adjacent to Marie Byrd Land at the time (Cooper *et al.*, 1982; Kamp, 1986; Kellogg & Rowley, this volume, p. 461). Alternatively, deformation in the southern part of the province may have been due to eastward gravitational spreading resulting from crustal thickening caused by emplacement of plutonic and volcanic rocks of the magmatic arc (see Hamilton, 1989).

Early Cretaceous magmatism demonstrates continuing or renewed subduction, with additional development of the magmatic arc. The bedrock geological history of the English Coast closed with Cenozoic crustal thinning, resulting in normal faulting and alkalic-basaltic volcanism.

Acknowledgements

We are grateful for isotopic age determinations by R.J. Pankhurst and for analysis of plant fossils by C.T. Gee and Theodore Delevoryas. Conversations with Warren Hamilton greatly helped us interpret regional tectonic relations, and comments by H.J. Harrington and J.D. Bradshaw led to improvements in our Gondwana reconstruction (Fig. 2). Critical reviews by W. Hamilton, G.W. Weir and M.R.A. Thomson considerably improved the manuscript. Our research was partly funded by NSF grant DPP83-18183; we especially thank M.D. Turner for logistical support and US Navy air unit VXE-6 for air support.

References

Barker, P.F. (1982). The Cenozoic subduction history of the Pacific margin of the Antarctic Peninsula: ridge crest–trench interactions. *Journal of the Geological Society, London*, **139(6)**, 787–801.

Behrendt, J.C. (1964). Crustal geology of Ellsworth Land and the southern Antarctic Peninsula from gravity and magnetic anomalies. *Journal of Geophysical Research*, **69(10)**, 2047–63.

Bentley, C.R. & Clough, J.W. (1972). Antarctic subglacial structure from seismic refraction measurements. In *Antarctic Geology and Geophysics*, ed. R.J. Adie, pp. 683–91. Oslo; Universitetsforlaget.

Collerson, K.D. & Sheraton, J.W. (1986). Age and geochemical characteristics of a mafic dyke swarm in the Archaean Vestfold block, Antarctica: inferences about Proterozoic dyke emplacement in Gondwana. *Journal of Petrology*, **27(4)**, 853–86.

Cooper, R.A., Landis, C.A., LeMasurier, W.E. & Speden, I.G. (1982). Geologic history and regional patterns in New Zealand and West Antarctica – their paleotectonic and paleogeographic significance. In *Antarctic Geoscience*, ed. C. Craddock, pp. 43–54. Madison; University of Wisconsin Press.

Dalziel, I.W.D. (1982). The early (pre-Middle Jurassic) history of the Scotia arc region: a review and progress report. In *Antarctic Geoscience*, ed. C. Craddock, pp. 111–26. Madison; University of Wisconsin Press.

Dalziel, I.W.D. & Elliot, D.H. (1982). West Antarctica: problem child of Gondwanaland. *Tectonics*, **1(1)**, 3–19.

Dalziel, I.W.D., Garrett, S.W., Grunow, A.M., Pankhurst, R.J., Storey, B.C. & Vennum, W.R. (1987). The Ellsworth–Whitmore Mountains crustal block: its role in the tectonic evolution of West Antarctica. In *Gondwana Six: Structure, Tectonics, and Geophysics*, ed. G.D. McKenzie, pp. 173–82. Washington, DC; American Geophysical Union, Geophysical Monograph 40.

Du Toit, A.L. (1937). *Our Wandering Continents*, Edinburgh; Oliver and Boyd. 366 pp.

Forsythe, R. (1982). The late Palaeozoic to early Mesozoic evolution of southern South America: a plate tectonic interpretation. *Journal of the Geological Society, London*, **139(6)**, 671–82.

Gee, C.T. (1989). Permian *Glossopteris* and *Elatocladus* megafossil floras from the English Coast, eastern Ellsworth Land, Antarctica. *Antarctic Science*, **1(1)**, 35–44.

Halpern, M. (1967). Rubidium–strontium isotopic age measurements of plutonic igneous rocks in eastern Ellsworth Land and northern Antarctic Peninsula, Antarctica. *Journal of Geophysical Research*, **72(20)**, 5138–42.

Hamer, R.D. & Moyes, A.B. (1982). Composition and origin of garnet from the Antarctic Peninsula Volcanic Group of Trinity Peninsula. *Journal of the Geological Society, London*, **139(6)**, 713–20.

Hamilton, W. (1989). Convergent-plate tectonics viewed from the Indonesian region. In *Tectonic Evolution of the Tethyan Region*, ed. A.M.C. Sengör, pp. 655–98. Dordrecht, Kluwer Academic Publishers.

Kamp, P.J.J. (1986). Late Cretaceous–Cenozoic tectonic development of the southwest Pacific region. *Tectonophysics*, **121**, 225–51.

Kellogg, K.S. (1980). Paleomagnetic evidence for oroclinal bending of the southern Antarctic Peninsula. *Geological Society of America Bulletin*, **91(7)**, 414–20.

Kellogg, K.S. & Reynolds, R.L. (1978). Paleomagnetic results from the Lassiter Coast, Antarctica, and a test for oroclinal bending of the Antarctic Peninsula. *Journal of Geophysical Research*, **83**, 2293–9.

Laudon, T.S. (1972). Stratigraphy of eastern Ellsworth Land. In *Antarctic Geology and Geophysics*, ed. R.J. Adie, pp. 215–223. Oslo; Universitetsforlaget.

Laudon, T.S., Lidke, D.J., Delevoryas, T. & Gee, C.T. (1987). Sedimentary rocks of the English Coast, eastern Ellsworth Land, Antarctica. In *Gondwana Six: Structure, Tectonics, and Geophysics*, ed. G.D. McKenzie, pp. 183–90. Washington, DC; American Geophysical Union, Geophysical Monograph 40.

Laudon, T.S., Thomson, M.R.A., Williams, P.L., Milliken, K.L., Rowley, P.D. & Boyles, J.M. (1983). The Jurassic Latady Formation, southern Antarctic Peninsula. In *Antarctic Earth Science*, ed. R.L. Oliver, P.R. James & J.B. Jago, pp. 308–14. Canberra; Australian Academy of Science and Cambridge; Cambridge University Press.

LeMasurier, W.E. & Wade, F.A. (1976). Volcanic history in Marie Byrd Land – implications with regard to southern hemisphere tectonic reconstructions. In *Andean and Antarctic Volcanology Problems*, ed. O. González-Ferrán, pp. 398–424. Rome; International Association of Volcanology and Chemistry of the Earth's Interior.

Meneilly, A.W., Harrison, S.M., Piercy, B.A. & Storey, B.C. (1987). Structural evolution of the magmatic arc in northern Palmer Land, Antarctic Peninsula. In *Gondwana Six: Structure, Tectonics, and Geophysics*, ed. G.D. McKenzie, pp. 209–220. Washington, DC; American Geophysical Union, Geophysical Monograph 40.

O'Neill, J.M. & Thomson, J.W. (1985). Tertiary mafic volcanic and volcaniclastic rocks of the English Coast, Antarctica. *Antarctic Journal of the United States*, **20(5)**, 36–8.

Pankhurst, R.J. (1982). Rb–Sr geochronology of Graham Land, Antarctica. *Journal of the Geological Society, London*, **139(6)**, 701–11.

Pankhurst, R.J. (1983). Rb–Sr constraints on the ages of basement rocks of the Antarctic Peninsula. In *Antarctic Earth Science*, ed. R.L. Oliver, P.R. James & J.B. Jago, pp. 367–71. Canberra; Australian Academy of Science and Cambridge; Cambridge University Press.

Rowley, P.D. (1983). Developments in Antarctic geology during the past half century. In *Revolution in the Earth Sciences – Advances in the Past Half-Century*, ed. S.J. Boardman, pp. 112–35. Dubuque, Iowa; Kendall/Hunt Publishing Company.

Rowley, P.D., Kellogg, K.S. & Vennum, W.R. (1985). Geologic studies in the English Coast, eastern Ellsworth Land, Antarctica. *Antarctic Journal of the United States*, **20(5)**, 34–6.

Rowley, P.D., Storey, B.C., Ford, A.B. & Wever, H.E. (1987). Geologic studies of the Black Coast, Antarctic Peninsula. *Antarctic Journal of the United States*, **22(5)**, 1–3.

Rowley, P.D., Vennum, W.R., Kellogg, K.S., Laudon, T.S., Carrara, P.E., Boyles, J.M. & Thomson, M.R.A. (1983). Geology and plate tectonic setting of the Orville Coast and eastern Ellsworth Land, Antarctica. In *Antarctic Earth Science*, ed. R.L. Oliver, P.R. James & J.B. Jago, pp. 245–50. Canberra; Australian Academy of Science and Cambridge; Cambridge University Press.

Rowley, P.D. & Williams, P.L. (1982). Geology of the northern Lassiter Coast and southern Black Coast. In *Antarctic Geoscience*, ed. C. Craddock, pp. 339–48. Madison; University of Wisconsin Press.

Schmidt, D.L. & Rowley, P.D. (1986). Continental rifting and transform faulting along the Jurassic Transantarctic rift, Antarctica. *Tectonics*, **5(2)**, 279–91.

Smellie, J.L. (1981). A complete arc–trench system recognized in Gondwana sequences of the Antarctic Peninsula region. *Geological Magazine*, **118(2)**, 139–59.

Smellie, J.L. (1987). Geochemistry and tectonic setting of alkaline volcanic rocks in the Antarctic Peninsula: a review. *Journal of Volcanology and Geothermal Research*, **32(1–3)**, 269–85.

Storey, B.C. & Garrett, S.W. (1985). Crustal growth of the Antarctic Peninsula by accretion, magmatism and extension. *Geological Magazine*, **122(1)**, 5–14.

Storey, B.C., Thomson, M.R.A. & Meneilly, A.W. (1987b). The Gondwanian orogeny within the Antarctic Peninsula: a discussion. In *Gondwana Six: Structure, Tectonics, and Geophysics*, ed. G.D. McKenzie, pp. 191–8. Washington, DC; American Geophysical Union, Geophysical Monograph 40.

Storey, B.C., Wever, H.E., Rowley, P.D. & Ford, A.B. (1987a). Report of Antarctic fieldwork. The geology of the central Black Coast, eastern Palmer Land. *British Antarctic Survey Bulletin*, **77**, 145–55.

Suárez, M. (1976). Plate-tectonic model for southern Antarctic Peninsula and its relation to southern Andes. *Geology*, **4(4)**, 211–14.

Thomson, M.R.A. (1982). Mesozoic paleogeography of West Antarctica. In *Antarctic Geoscience*, ed. C. Craddock, pp. 331–7. Madison; University of Wisconsin Press.

Thomson, M.R.A., Pankhurst, R.J. & Clarkson, P.D. (1983). The Antarctic Peninsula – a Late Mesozoic–Cenozoic arc (review). In *Antarctic Earth Science*, ed. R.L. Oliver, P.R. James & J.B. Jago, pp. 289–94. Canberra; Australian Academy of Science and Cambridge; Cambridge University Press.

Vennum, W.R. & Laudon, T.S. (1988). Igneous petrology of the Merrick Mountains, eastern Ellsworth Land, Antarctica. *US Geological Survey Professional Paper*, **1351-B**, pp. 21–34.

Vennum, W.R. & Rowley, P.D. (1986). Reconnaissance geochemistry of the Lassiter Coast Intrusive Suite, southern Antarctic Peninsula. *Geological Society of American Bulletin*, **97(12)**, 1521–33.

Williams, P.L., Schmidt, D.L., Plummer, C.C. & Brown, L.E. (1972). Geology of the Lassiter Coast area, Antarctic Peninsula: preliminary report. In *Antarctic Geology and Geophysics*, ed. R.J. Adie, pp. 143–8. Oslo; Universitetsforlaget.

New geophysical results and preliminary interpretation of crustal structure between the Antarctic Peninsula and Ellsworth Land

K.J. McGIBBON & A.M. SMITH

British Antarctic Survey, Natural Environment Research Council, High Cross, Madingley Road, Cambridge CB3 0ET, UK

Abstract

New geophysical results from eastern Ellsworth Land have been combined with existing data to produce Bouguer-anomaly and bedrock maps of the region. The bedrock topography data confirm the morphological contrast between the Antarctic Peninsula and Haag Nunataks crustal blocks and the deep bedrock around Siple Station and Evans Ice Stream. Bouguer anomalies as low as − 1300 gu characterize the Antarctic Peninsula as far south as a major bedrock scarp. To the south-west, Bouguer anomalies are dominantly positive with only gentle variations. Bedrock topography and Bouguer-anomaly data are used to draw preliminary conclusions about crustal structure in the area. They delimit the bedrock scarp at the southern end of the Antarctic Peninsula and distinguish areas of elevated bedrock from the Antarctic Peninsula crustal block. Farther south-west, previously proposed crustal fractures along deep subglacial troughs are not reflected on the Bouguer-anomaly map. Furthermore, elevated bedrock around Haag Nunataks and the Ellsworth Mountains shows little correlation with observed Bouguer anomalies. The reasons for the gently undulating positive Bouguer anomaly over this severe bedrock topography are briefly discussed, but clarification of crustal structure awaits further modelling.

A geophysical traverse over eastern Ellsworth Land undertaken during the 1986–87 field season included gravity, magnetic field, surface elevation and ice-thickness measurements at 10 km intervals. The route (Fig. 1) crossed south of the base of Antarctic Peninsula, Evans Ice Stream, Fowler Peninsula, Carlson Inlet and the head of Rutford Ice Stream. Further data were gathered on a traverse northwards from Rutford Ice Stream to Siple Station and on to Schwartz Peak on the English Coast.

Geological knowledge of southern Antarctic Peninsula and Ellsworth Land is restricted to isolated nunataks (Dalziel *et al.*, 1987; Rowley *et al.*, 1983). A major subice topographic boundary along the base of the Antarctic Peninsula was identified by Behrendt (1964a) and further delimited by Doake, Crabtree & Dalziel (1983). Aeromagnetic data have been used to identify the subglacial boundaries of the Precambrian Haag Nunataks crustal block (Storey *et al.*, 1988) and also to propose a series of fault-bounded blocks between the Ellsworth Mountains and the north-eastern edge of Evans Ice Stream (Garrett, Herrod & Mantripp, 1987).

The route was planned to cross these major boundaries between crustal blocks and to complement Behrendt's (1964b) Antarctic Peninsula traverse data. The objectives were to identify the different gravity signatures related to changes in crustal density and structure. We present the bedrock topography and gravity data. Future work will integrate this with the new magnetic field results and available aeromagnetic data.

The survey

Control for the gravity survey was achieved by gravity ties made to field base stations at Evans Ice Stream, Fowler Peninsula and Schwartz Peak (Fig. 1), which were tied by two-way flights to Sky-Hi Nunataks. The latter station was tied by several flights to Fossil Bluff (Sturgeon & Renner, 1983) and Rothera Base (Renner, Sturgeon & Garrett, 1985) gravity stations at the beginning and end of the traverse. With the one exception when a Worden gravimeter was used, all ties were made with a LaCoste and Romberg meter.

Behrendt, Wold & Laudon's (1962) value for absolute gravity was referred to the same Potsdam datum presently in use at Rothera base station (Renner *et al.*, 1985); hence their results were easily incorporated into our new data set.

The height of each station above sea level was calculated by the interval method of barometry using two pairs of Wallace and Tiernan altimeters. Simultaneous observations of height were made at successive pairs of adjacent stations to correct for local, short-term changes in sea-level pressure. Absolute height values were taken from Transit doppler satellite fixes at Sky-Hi Nunataks (D.G. Vaughan, pers. comm), Rutford Ice Stream

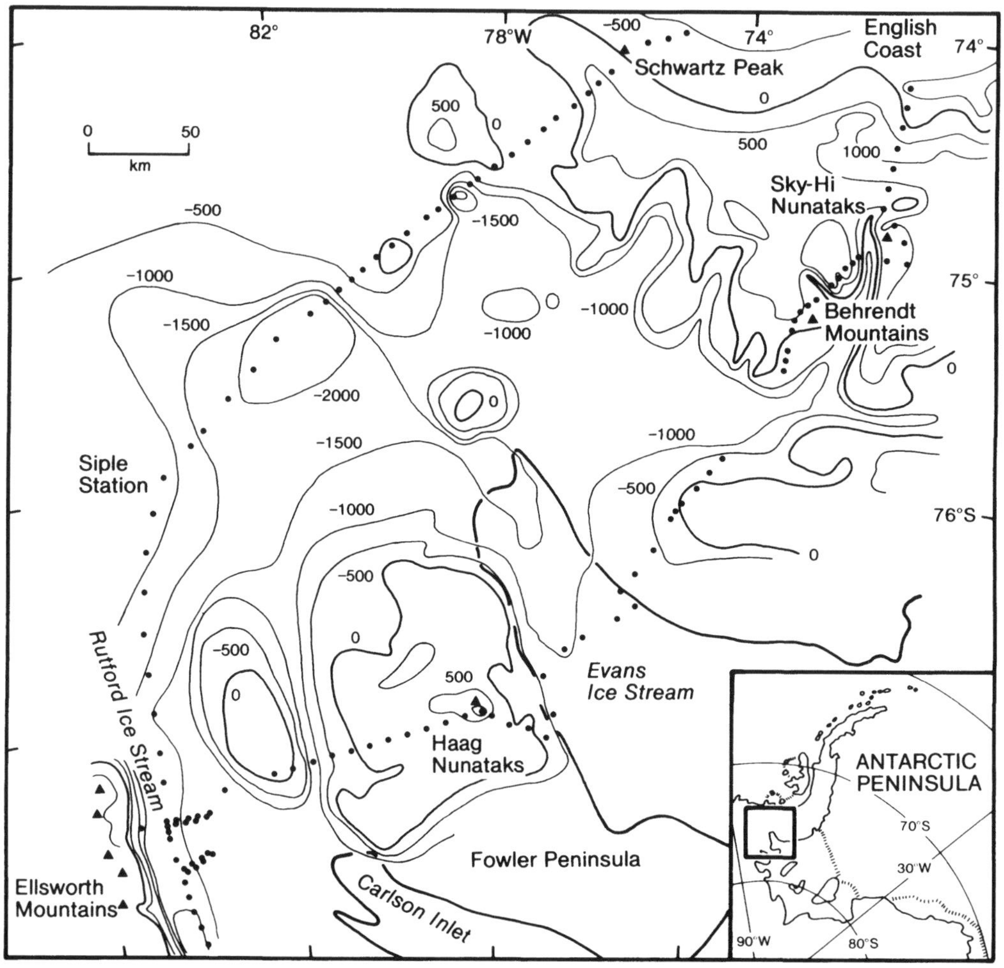

Fig. 1. Bedrock elevation map of eastern Ellsworth Land. Contour interval 500 m; points indicate stations where new ice-thickness data have been used to modify earlier maps; heavy line shows coastline.

(Doake *et al.*, 1987), and Haag Nunataks (Sturgeon & Renner, 1983). In addition, a Global Positioning System (GPS) receiver was employed to measure altitude and positions. Its use was restricted, however, by the currently incomplete satellite availability.

Behrendt (1964b) discussed the errors involved in the internal method. Based on similar considerations we estimate the maximum error in surface elevation at stations farthest from fixed sites to be ± 30 m. Errors closer to fixed sites are significantly less. The station positions are believed to be within ± 300 m or better.

Most of the field stations were occupied by two gravimeters. The LaCoste and Romberg (No. 784) and Worden (No. 886) readings were usually within 10 gu (1 mGal), although the LaCoste and Romberg reading was normally used in the data reduction. Stations on Rutford Ice Stream were measured by another British Antarctic Survey (BAS) field party using a Worden meter (No. 556). During the period of the survey the overall drift rate of the LaCoste and Romberg meter was + 0.004 gu/h. Overnight drift rates were normally better than − 0.042 gu/h, though an unexplained value of − 0.129 gu/h was recorded on one occasion. Observed values of absolute gravity are believed to be accurate to within ± 3 gu relative to the Antarctic Peninsula Geoceiver gravity base station network (Sturgeon & Renner, 1983). Reduction of gravity data was assisted by either radio-echo or seismic ice-thickness measurements. Gravity stations were mostly on grounded ice and seismic sounding was usually only necessary where the ice was thicker than 2 km. Over Evans Ice Stream both radio-echo ice thickness and seismic depth-to-bedrock were required. Ice thicknesses are believed to be accurate to ± 12 m.

The Bouguer gravity anomalies were calculated assuming rock and ice densities of 2.67 and 0.918 Mg/m^3, respectively. Behrendt *et al.*'s (1962) data were recalculated from the original absolute gravity values using the same densities and the 1967 international gravity formula (Jeffreys, 1976).

The major limitations to the accuracy of the gravity survey arise from the uncertainty of station elevations and from terrain corrections. The absolute error in surface elevation is greatest at stations farthest from fixed sites. In the worst case the error in Bouguer anomaly is ± 100 gu, but closer to fixed sites this error is significantly less, as is the error between adjacent sites. Terrain corrections are difficult to calculate due to the lack of detailed bedrock elevation data and have not been applied. However, estimates suggest that in most cases they would be < 50 gu.

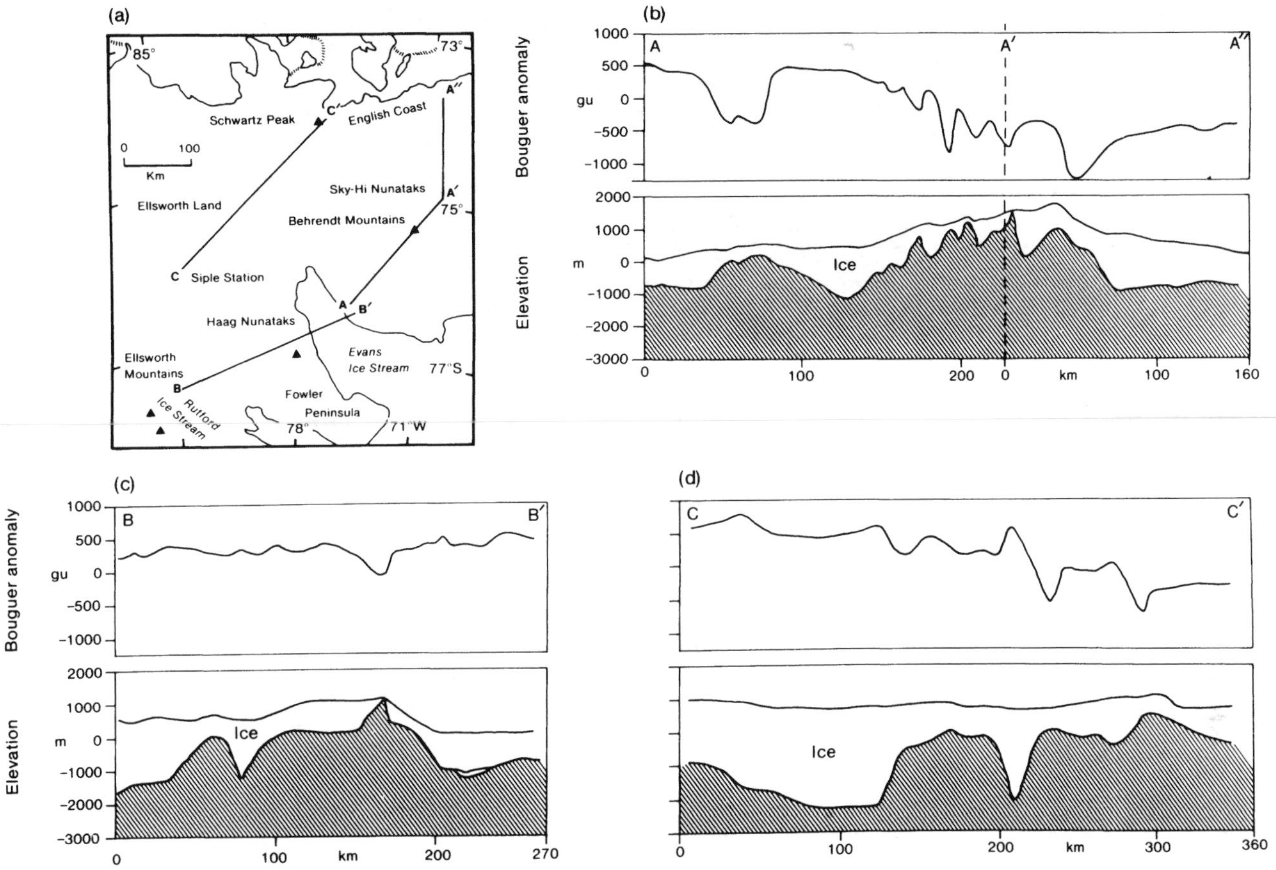

Fig. 2. (a–d) Bedrock elevation, ice thickness and Bouguer anomaly along the selected profiles shown in Fig. 2*a*; Figs 2*b–d* are all to the same scale.

Bedrock topography and free-air anomalies

Our surface-elevation and ice-thickness data have been used to modify the bedrock elevation maps of Doake *et al.* (1983) and Garrett *et al.* (1987). The main improvements in the new map (Fig. 1) are the seismic bedrock soundings over the floating section of Evans Ice Stream and along the traverse from Rutford Ice Stream to Schwartz Peak (Fig. 2*a*). The new data are in good agreement where they overlap with older airborne data.

Fig. 2*b* shows the bedrock topography and surface elevation between Evans Ice Stream and the English Coast. The traversed section of Evans Ice Stream (Fig. 2*c*) is floating over its central part; the water depth varies from 50 to 90 m, with bedrock lying nearly 1200 m below sea level.

Between Rutford Ice Stream and Siple Station the ice thickness varies from 1900 to 2550 m. At Siple Station the ice is 1920 m thick and bedrock is 850 m below sea level. Between Siple Station and Schwartz Peak (Fig. 2*d*) the ice thickness changes rapidly with the severe bedrock topography and reaches a maximum thickness of 2950 m, with bedrock at 2200 m below sea level. Two deep bedrock depressions separated by a 70 km wide subglacial high were located in this region (Fig. 1). The first is 100 km broad and extends 2200 m below sea level; the second is only 30 km wide and extends 2000 m below sea level.

The new bedrock map confirms the topographical contrast between the Antarctic Peninsula, the Haag Nunataks crustal block and the deep bedrock in the Siple Station and Evans Ice Stream regions.

The free-air anomaly generally follows the bedrock topography closely, as would be expected. It reaches a maximum of over 1400 gu at Haag Nunataks and is at a minimum of − 950 gu north-east of Siple Station.

Bouguer anomalies

The Bouguer-anomaly map in Fig. 3 was compiled from new data, supplemented by recalculated values from Behrendt (1964b). The density value of 2.67 Mg/m^3, used in the Bouguer correction, is unlikely to be correct for the whole area. However, there is little information at present on which to base more accurate estimates.

Our new data reveal that the gravity low associated with the axis of the Antarctic Peninsula continues into eastern Ellsworth Land with a minimum of − 1300 gu. To the north-east of Evans Ice Stream a local Bouguer-gravity low of − 400 gu is coincident with an elevated spur of bedrock (Fig. 1).

A large linear Bouguer-anomaly gradient is associated with the southern boundary of the Antarctic Peninsula. This major anomaly approaches 1000 gu in amplitude and extends over a distance of approximately 80 km. It is coincident with the topographic scarp seen on the bedrock map (Fig. 1). The gravity low of the Antarctic Peninsula terminates at this boundary. The anomaly gradient is greatest to the west, and highlights the distinctive gravity signature of the Antarctic

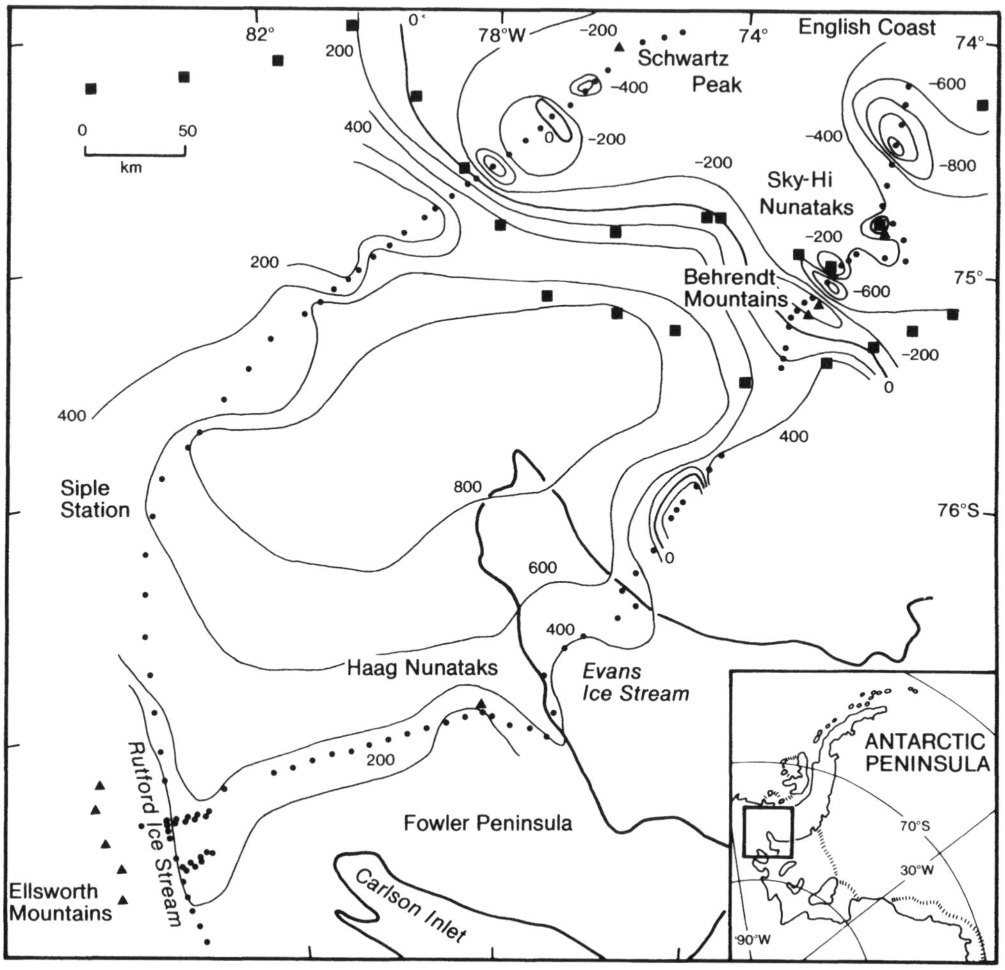

Fig. 3. Bouguer anomaly map of eastern Ellsworth Land. The contour interval is 200 gu; points indicate new Bouguer gravity sites; squares indicate Behrendt's (1964b) Bouguer gravity sites; the heavy line shows the coastline.

Peninsula compared with areas to the south. Shorter wavelength Bouguer-gravity anomalies are found near the Behrendt Mountains (Figs 2*b* & 3). The Bouguer-anomaly values on a line from Evans Ice Stream to Rutford Ice Stream (Fig. 2*c*), however, vary surprisingly little considering the dramatic topographic variations. The only exception to an overall average of + 200–400 gu is a 400 gu negative anomaly of wavelength 40 km over Haag Nunataks. The anomaly field between Rutford Ice Stream and Siple Station (Fig. 3) is fairly flat and increases slowly to 800 gu. Northwards to Schwartz Peak (Fig. 2*d*) the field over the first wide bedrock trough falls to + 500 gu then drops further over the bedrock high. The most notable feature on this traverse is a peak–trough anomaly of 1100 gu over the narrower depression in the bedrock. A second step-like feature in the gravity profile (Fig. 2*d*) is seen 80 km farther northwards, but this one is not associated with any marked bedrock feature.

Preliminary interpretation

The negative Bouguer anomaly over the central plateau region of the Antarctic Peninsula is thought to indicate a thick crust in a state of isostatic equilibrium (Renner *et al.*, 1985). Our new data show that this gravity low extends into eastern Ellsworth Land and suggest that the whole region has a similarly thickened crustal structure.

Bentley & Robertson (1982) calculated isostatic gravity anomalies from free-air anomalies measured in West Antarctica. They concluded that large areas were not detectably out of isostatic equilibrium. However, over and around the Behrendt Mountains they found a positive isostatic anomaly of 350 gu. On Fig. 3 a local increase in Bouguer anomaly is evident which could be consistent with Bentley & Robertson's interpretation of local crustal thinning. Alternatively it may be due to discrete plutonic bodies intruded into less dense sedimentary rocks, as suggested by Garrett *et al.* (1987) to explain the regional aeromagnetic anomalies.

The previous interpretation of the topographic scarp seen on the bedrock map as the southern limit of the Antarctic Peninsula crustal block (Doake *et al.*, 1983; Behrendt, 1964b), is strengthened by the coincident Bouguer anomaly which indicates a prominent structural boundary related to significant changes in crustal thickness or density.

South of the Antarctic Peninsula crustal block the Bouguer-anomaly signature is much subdued and exhibits little correlation with topography. This area includes the deep subglacial troughs of Evans and Rutford ice streams and Carlson Inlet and also the elevated topography of Fowler Peninsula and the promontory to the west ('Promontory A' in Garrett *et al.*, 1987). As indicated in Figs 2*c* and 3 the Bouguer anomaly remains relatively flat at around 300 gu between Haag Nunataks and the Ellsworth Mountains. Further work on isostatic modelling is required to try and clarify the reasons for this but possibilities include a dense upper crust or a lack of isostatic equilibrium. Interpretation is, however, severely hampered by the lack of crustal thickness measurements.

There is no marked change in gravity on approaching the Ellsworth Mountains but, without further data over the mountains themselves, it is difficult to draw any conclusions.

On the profile from Siple Station to Schwartz Peak (Fig. 2*d*) the Bouguer anomaly shows a relatively minor change over the first steep bedrock rise. Over the narrow bedrock depression 70 km farther north there is a more significant change from a positive to a negative Bouguer anomaly. The distinction between the relatively high Bouguer anomaly over the Haag Nunataks crustal block and the negative anomaly of the Antarctic Peninsula is also clear.

Conclusions

New gravity and bedrock maps of southern Palmer Land and Ellsworth Land have enabled a preliminary interpretation of crustal structure. The data confirm and clarify the distinction between the Bouguer-anomaly signatures of eastern Ellsworth Land and the regions to the west and south previously noted by Behrendt (1964a). The boundary between the two areas can be traced as a linear 1000 gu Bouguer anomaly along a bedrock scarp. A strong negative Bouguer anomaly over the central plateau region of eastern Ellsworth Land suggests a state of isostatic equilibrium. The Bouguer anomaly gradient can be used to delimit the southern margin of the Antarctic Peninsula crustal block. The elevated bedrock to the northeast of Evans Ice Stream shows a Bouguer anomaly similar to that of the Antarctic Peninsula.

The area to the south-west of the bedrock scarp shows more complex crustal structure. The observed Bouguer anomaly cannot be explained by simple variations in crustal thickness and probably has a deep-seated source. Modelling of more complex variations in both crustal thickness and density and analysis of available magnetic data will be required to further understand the crustal structure in this area.

Acknowledgements

The authors would like to thank W.K. Dark, D.A. Roberts, D.A. Waldron and N. Armstrong for assistance in the field, and D.G. Vaughan for operating a gravimeter on Rutford Ice Stream. We are grateful to R.G.B. Renner for his support and all other colleagues at the British Antarctic Survey for useful discussions of the manuscript.

References

Behrendt, J.C. (1964a). Crustal geology of Ellsworth Land and the southern Antarctic Peninsula from gravity and magnetic anomalies. *Journal of Geophysical Research*, **69(10)**, 2047–63.

Behrendt, J.C. (1964b). *Antarctic Peninsula Traverse. Geophysical Results Relating to Glaciological and Geological Studies*. Geophysical and Polar Research Center Report, 64-1. Madison; University of Wisconsin.

Behrendt, J.C., Wold, R.J. & Laudon, T.S. (1962). Gravity base stations in Antarctica. *Geophysical Journal of the Royal Astronomical Society*, **6(3)**, 400–5.

Bentley, C.R. & Robertson, J.D. (1982). Isostatic gravity anomalies in West Antarctica. In *Antarctic Geoscience*, ed. C. Craddock, pp. 949–54. Madison; University of Wisconsin Press.

Dalziel, I.W.D., Garrett, S.W., Grunow, A.M., Pankhurst, R.J., Storey, B.C. & Vennum, W.R. (1987). The Ellsworth–Whitmore crustal block: its role in the tectonic evolution of West Antarctica. In *Gondwana Six: Structure, Tectonics, and Geophysics*, Geophysical Monograph 40, ed. G.D. McKenzie, pp. 173–82. Washington, DC; American Geophysical Union.

Doake, C.S.M., Crabtree, R.C. & Dalziel, I.W.D. (1983). Subglacial morphology between Ellsworth Mountains and Antarctic Peninsula: new data and tectonic significance. In *Antarctic Earth Science*, ed. R.L. Oliver, P.R. James & J.B. Jago, pp. 270–3. Canberra; Australian Academy of Science and Cambridge; Cambridge University Press.

Doake, C.S.M., Frolich, R.M., Mantripp, D.R., Smith, A.M. & Vaughan, D.G. (1987). Glaciological studies on Rutford Ice Stream, Antarctica. *Journal of Geophysical Research*, **92, B9**, 8951–60.

Garrett, S.W., Herrod, L.D.B. & Mantripp, D.R. (1987). Crustal structure of the area around Haag Nunataks, West Antarctica: new aeromagnetic and bedrock elevation data. In *Gondwana Six: Structure, Tectonics, and Geophysics*, Geophysical Monograph 40, ed. G.D. McKenzie, pp. 109–16. Washington, DC; American Geophysical Union.

Jeffreys, H. (1976). *The Earth. Its Origin, History and Physical Construction*, 6th edn. Cambridge; Cambridge University Press. 574 pp.

Renner, R.G.B., Sturgeon, L.J.S. & Garrett, S.W. (1985). *Reconnaissance Gravity and Aeromagnetic Surveys of the Antarctic Peninsula*. Cambridge; British Antarctic Survey Scientific Reports, No. 110, 50 pp.

Rowley, P.D., Vennum, W.R., Kellogg, K.S., Laudon, T.S., Carrara, P.E., Boyles, J.M. & Thomson, M.R.A. (1983). Geology and plate tectonic setting of the Orville Coast and eastern Ellsworth Land, Antarctica. In *Antarctic Earth Science*, ed. R.L. Oliver, P.R. James & J.B. Jago, pp. 245–50. Canberra; Australian Academy of Science and Cambridge; Cambridge University Press.

Storey, B.C., Dalziel, I.W.D., Garrett, S.W., Grunow, A.M., Pankhurst, R.J. & Vennum, W.R. (1988). West Antarctica in Gondwanaland: crustal blocks, reconstruction and breakup processes. *Tectonophysics*, **155**, 381–90.

Sturgeon, L.J.S. & Renner, R.G.B. (1983). New Doppler satellite controlled gravity stations in Antarctica. *British Antarctic Survey Bulletin*, **59**, 9–14.

Evolution of the Bransfield basin, Antarctic Peninsula

J.D. JEFFERS[1], J.B. ANDERSON[1] & L.A. LAWVER[2]

1 Department of Geology and Geophysics, Rice University, Houston, Texas 77251 USA
2 Institute for Geophysics, University of Texas, Austin, Texas 78759, USA

Abstract

Marine geophysical data collected during five seasons in Bransfield Strait provide the basis for reconstructing the evolution of this complex basin. Seismic-reflection profiles and bathymetry show three main sub-basins and numerous smaller basin segments. Five stages in the evolution of Bransfield Strait are related to the progressive cessation of subduction along the margin of the northern Antarctic Peninsula during the Cenozoic. The early–mid-Neogene magmatic arc of the South Shetland Islands was later deformed by transpressional strike-slip faulting related to oblique convergence between the Aluk and Antarctic plates. Boyd Strait opened in response to the late Miocene–early Pliocene ridge–trench collision south of the Hero fracture zone, while oblique convergence continued to the north. Bransfield Strait rifting began in the early Pliocene, but rapid sedimentation and diffuse igneous activity prevented the formation of correlatable magnetic anomalies. Recent igneous activity may be progressing toward a crudely organized seafloor-spreading system since short period high-amplitude magnetic anomalies are associated with the central volcanic ridge.

Introduction

Bransfield Strait, in the south-western Scotia arc, is an extensional basin separating the South Shetland Islands from the northern Antarctic Peninsula. Marine geophysical data collected during five austral summers provide the basis for assessment of the important tectonic controls on its development as a sedimentary basin and allow construction of a model for its evolution. While previous workers (Barker, 1982; Barker & Dalziel, 1983) explained the timing and mechanism of formation of Bransfield Strait primarily from the seafloor magnetic-anomaly patterns in Drake Passage, this study examines the evolution of the basin from the marine sedimentary record. Since the basin is relatively young (< 4 Ma; Barker & Dalziel, 1983), shallow penetration, high resolution seismic-reflection techniques are ideal for studying the late Cenozoic history of Bransfield Strait.

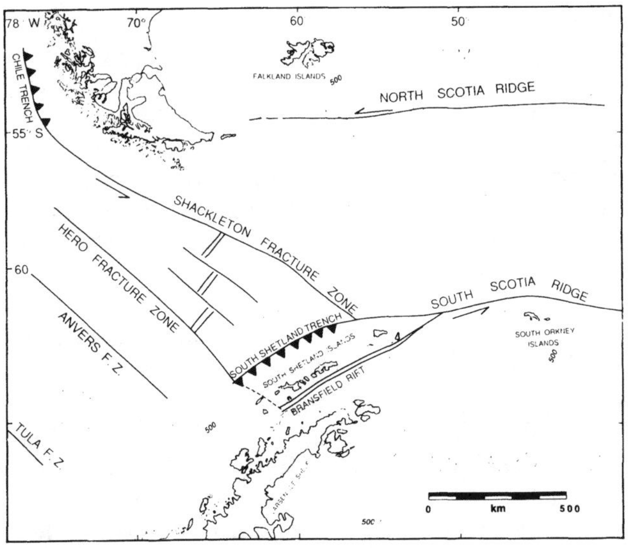

Fig. 1. Simplified tectonic map of the western Scotia arc. Compiled from Craddock (1981) and British Antarctic Survey (1985).

Present setting

The south-western Scotia arc region (Fig. 1) occupies an extremely complex plate tectonic setting (Craddock, 1981; British Antarctic Survey, 1985; Barker, Dalziel & Storey, in press). The oceanic lithosphere of the Antarctic plate is being subducted under the South American plate at the Chile trench from 46 to 52° S. Farther south, Scotia–Antarctic motion becomes left-oblique convergent along the Shackleton fracture zone. Three extinct segments of the Antarctic-Aluk ridge separate the former Drake microplate from the Antarctic plate; if this ridge is still actively spreading, the Drake microplate still exists. The Shetland microplate is defined by the South Shetland trench and the postulated back-arc spreading centre in Bransfield Strait. The Drake(?) and Shetland microplates are bounded on the south-west by the Hero fracture zone. Bransfield Strait is a young extensional basin, approximately 100 km wide, which is confined to the segment of the Antarctic Peninsula between the Hero and Shackleton fracture zones. Recent volcanic activity on Deception Island (Baker *et al.*, 1975) and seismic activity along the axis of the Bransfield basin (Forsyth, 1975; Pelayo & Wiens, 1986) indicate that extension is continuing, while the continental margin of the Antarctic Peninsula south of the Hero fracture zone is passive (Barker, 1982).

Previous work

Marine geophysical surveys during the 1960s first confirmed the extensional nature of Bransfield Strait. Ashcroft (1972) concluded from seismic-refraction data that thin crust (14 km) underlies the deep axial trough, while the crustal sections beneath the South Shetland Islands and the Trinity Peninsula are somewhat thicker and very similar to one another. He also tentatively located several normal faults on the margins of the basin and proposed that the islands and the peninsula were united until they were separated by a Tertiary rifting event.

Marine magnetic surveys in Drake Passage (Barker, 1976, 1982) identified seafloor magnetic-anomalies which were vital in reconstructing the Cenozoic history of the active margin of the Antarctic Peninsula. The youngest anomalies identified were anomaly 3 (~ 4 Ma) at three spreading ridge-segments between the Hero and Shackleton fracture zones (Fig. 1) adjacent to the South Shetland trench; symmetrical magnetic anomaly patterns flank this ridge. South of the Hero fracture zone, no ridge crest is present; the anomalies are youngest at the continental margin, with ages increasing seaward, indicating that the spreading ridge which produced this seafloor has been subducted (Barker, 1982). Between the Tula and Anvers fracture zones (Fig. 1), subduction stopped at anomaly 6 time (20 Ma), while at anomaly 3 or 4 time (4–7 Ma) subduction stopped between the Anvers and Hero fracture zones, and spreading stopped at the Aluk–Antarctic ridge adjacent to the South Shetland Islands. Barker & Dalziel (1983) postulated that the coincident cessation of subduction beneath the South Shetland trench resulted in extension in Bransfield Strait, as the subducting slab continued to cool and sink; they placed the inception of Bransfield Strait rifting at 4 Ma. Roach (1978) concluded from magnetic data that symmetrical seafloor spreading in Bransfield Strait began at 1.3 Ma.

From stratigraphic and structural relations on King George Island, Birkenmajer (1982) and colleagues have identified four phases of late Cenozoic deformation. NE–SW-trending right-lateral strike-slip faults and associated folds, mappable the entire length of King George Island, were formed in a transpressional regime during the late Miocene Ezcurra phase. Numerous shorter strike-slip faults, spaced 5–15 km apart, offset the Ezcurra faults and reflect the transition from compression to extension during the (?)early Pliocene Admiralty phase. Normal faulting and associated volcanism marked the onset of Bransfield Strait rifting during the Pliocene–early Pleistocene Bransfield phase. The Penguin phase, characterized by submarine and subaerial volcanism along the line connecting Bridgeman and Deception islands and a subparallel line through Penguin Island, continues to the present.

Results

Bathymetry

Bathymetric profiles were collected on all six cruises, totalling approximately 5500 km (Fig. 2). Important aspects of the bathymetry of the Bransfield Strait have been discussed by many earlier workers (e.g. Ashcroft, 1972). The main part of Bransfield Strait is a deep asymmetric trough, trending NE–SW, bounded by a steep slope down from the South Shetland Islands and a broader, more gently sloping shelf on the Trinity Peninsula margin. The basin is composed of three distinct sub-basins. The widest of these three, approximately 40 km wide and 1850 m deep, lies south of King George and Livingston islands. Irregular basement topography is exposed at the seafloor, mainly as a line of seaknolls between Bridgeman and Deception islands. Several deep troughs, roughly aligned with bays of the South Shetland Islands, incise the relatively flat platform of Trinity Peninsula. To the northeast of the central basin, a narrower, deeper trough extends north-eastward at least as far as Elephant Island. One seamount in this northern trough rises 900 m above the seafloor. The irregularly shaped trough south and west of Deception Island trends NE–SW toward the Gerlache Strait and also branches north-westward through Boyd Strait. These three sub-basins of the Bransfield Strait are separated by the two bathymetric divides associated with Deception and Bridgeman islands.

Seismic-reflection profiles

Over 2200 km of continuous seismic-reflection profiles were collected in and around Bransfield Strait using a variety of sparker arrays on USCGC *Glacier* and water guns and air guns on R/V *Polar Duke*. Line drawings presented in Fig. 3 are from the shipboard analogue monitor records.

Dip-orientated profiles display several common features. In all cases, Bransfield Strait is a deep, flat-bottomed trough, with thick sedimentary accumulations in the central deep. Our sparker system was not powerful enough to penetrate to basement in this area, but the water gun used on line PD86-11 (Fig. 3) penetrated a maximum of 0.75 s of sediment overlying volcanic(?) basement. Normal faults are identified on both flanks of the central trough. Irregular basement topography rises above the basin floor in many of our seismic profiles; in two lines (DF86-11, DF85-A; Fig. 3) a twin-peaked ridge, crudely resembling a mid-ocean ridge is present. Elsewhere, seafloor topography is more subdued (DF86-12; Fig. 3) or completely absent (DF86-14, Fig. 3; DF86-8, not shown).

Aside from these common features, there are many significant differences. The strike line (PD86-7 and PD87-6; not shown) shows the division of Bransfield Strait into three major sub-basins. The extensional nature of Boyd Strait also is apparent. The style of normal faulting, especially on the Trinity Peninsula platform, varies among our profiles. In lines PD86-11 and DF85-A (Fig. 3) a single major fault has down-dropped a large block, with numerous closely spaced faults have created a gentle basinward slope on line DF86-11 (Fig. 3).

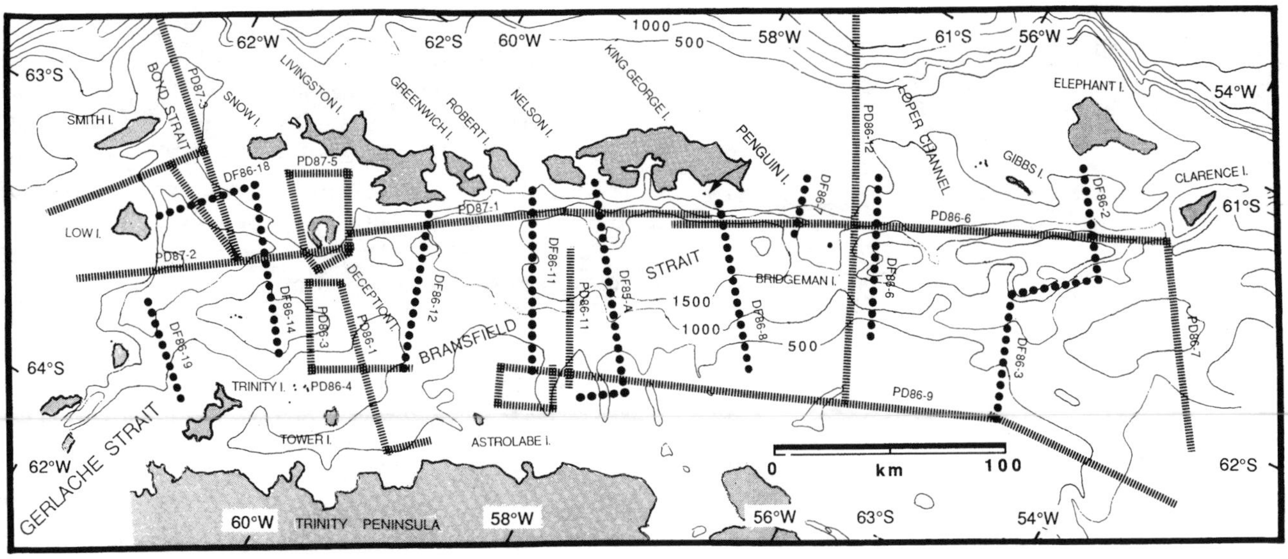

Fig. 2. Seismic-reflection profiles in the Bransfield Strait region. Dotted lines are 4.6 kJ sparker profiles; broken lines are air gun and water gun profiles; bathymetry contoured at 500 m.

Distribution of slope sediments is largely fault controlled. The style of igneous activity also varies; line DF86-11 (Fig. 3) shows an off-axis basement high on the South Shetland Islands side of the basin and several strong reflectors which may be sills or lava flows. Sills apparently abruptly restricted acoustic penetration in line PD86-11 (Fig. 3).

These differences probably reflect segmentation of the basin by structures trending normal to the extensional axis. Differences in width and depth among the three main sub-basins may result from differing rates and amounts of extension. Smaller scale transverse structures may further dissect the basin. In particular, the fault-bounded troughs on the Trinity Peninsula shelf appear to separate provinces of different structural style. The trend of one of these faults was measured as 317° ± 5°, which agrees with the 315° azimuth of least compressive stress computed by Forsyth (1975) for a recent extensional earthquake in Bransfield Strait. Line DF86-12 (Fig. 3), crosses the basin along the trend of one of these troughs (Fig. 2) and shows diffuse volcanic centres which may be related to the intersection of a transform zone with the axial rift. However, there is little direct evidence of transform faulting on as fine a scale as the shelf morphology and the spacing of Admiralty phase faults on King George Island would suggest. The thick sedimentary cover and the probable small offset and near-vertical orientation of transforms related to back-arc spreading conspire to render them invisible in most seismic-reflection profiles.

Discussion

Given the complex tectonic setting of Bransfield Strait, at least two hypotheses for its origin should be considered as alternatives to the subduction–cessation mechanism proposed by Barker & Dalziel (1983). First, with the prevalence of strike-slip (transcurrent) faulting in the region (i.e. Shackleton fracture zone, South Scotia Ridge, and the faults mapped on the South Shetland Islands by Birkenmajer (1982)), the possibility that Bransfield Strait originated as a strike-slip-dominated basin, similar to the Andaman Sea (Curray *et al.*, 1979), deserves consideration. Second, the temporal coincidence of the cessation of spreading in Drake Passage with the inception of Bransfield Strait rifting requires that the possibility of a spreading-ridge jump be addressed. Earthquake focal mechanisms (Forsyth, 1975; Pelayo & Wiens, 1986) indicate that present motion in south-western Bransfield Strait is purely extensional, and approximately normal to the trend of the basin; the possibility of a significant contribution of strike-slip faulting to the Bransfield Strait extension is remote, although data in the north-eastern part of the basic are sparse. The thermal and geodynamic implications of moving the locus of spreading through an active subduction zone make a ridge-jump an unlikely alternative. Subduction-related back-arc spreading, probably induced by the cessation of seafloor spreading in Drake Passage, seems the more plausible explanation for the origin of Bransfield Strait.

Tectonic model for evolution of Bransfield Strait

The following model integrates our observations with the work of others to describe the evolution of the Bransfield basin. Further investigations of the region may require revision of this model, particularly since no deep sampling of marine sediments has been made in order to date the basement and overlying sediments. For simplicity, the segment of Pacific oceanic crust between the Hero and Shackleton fracture zones will be named the 'Shetland' segment, the segment between the Anvers and Hero fracture zones will be called the 'Brabant' segment (after Hawkes, 1981), and the segment between the Tula and Anvers fracture zones will be referred to as the 'Biscoe' segment.

Until the mid-Neogene, subduction of the Pacific (Aluk) plate continued along the Antarctic continental margin north-east of the Tula fracture zone, and arc volcanism was active on the South Shetland Islands. At around anomaly 6 time (20 Ma)

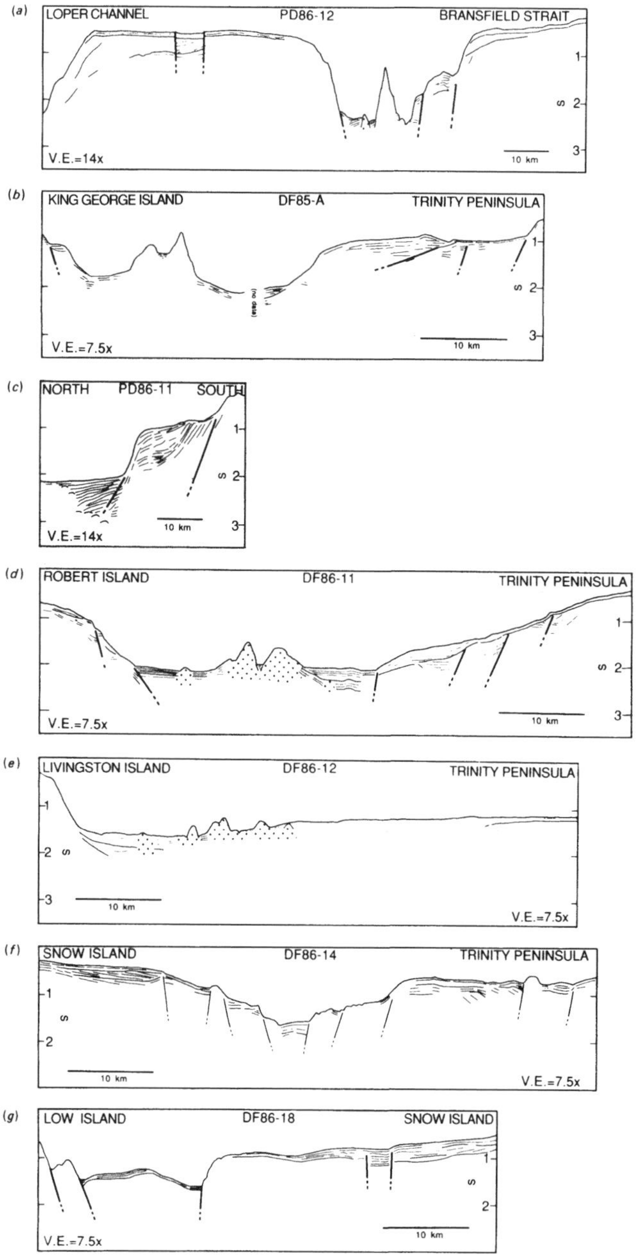

Fig. 3. Line drawings of representative seismic-reflection profiles from Bransfield Strait. Vertical exaggeration at the seafloor is indicated; please note the variation in horizontal scale. See Fig. 2 for locations.

subduction ceased in the Biscoe segment as the ridge crest collided with the trench. By this time, and perhaps even earlier, convergence in the northern Antarctic Peninsula was right-oblique (Minster & Jordan, 1978; Barker, 1982). This oblique convergence is manifested in the Ezcurra system of transpressional dextral strike-slip faults and associated folds on King George Island (Birkenmajer, 1982), which accommodated the trench-parallel component of the relative convergence vector (Walcott, 1978).

The Brabant segment of the Aluk–Antarctic ridge crest collided with the trench at anomaly 4–3 time (6–4 Ma), causing spreading and subduction to cease as this part of the Pacific plate became rigidly locked to the Antarctic plate (Fig. 4*a*). Spreading may have continued slightly longer on the Shetland segment with continued convergence at the South Shetland trench and consequent dextral shear along the arc-parallel faults during this phase. Continued strike-slip motion in the Shetland segment while the margin to the south was passive (Fig. 4*a*) would result in extension normal to the segment boundary (i.e. the Hero fracture zone). We propose this type of extension for the origin of the Boyd Strait basin (Fig. 4*a*).

Extension in Bransfield Strait began in response to the cooling and sinking of the subducted oceanic slab. On land, this transition from a compressional regime to an extensional regime probably correlates with the development of the Admiralty phase strike-slip faults in the early Pliocene (Birkenmajer, 1982). The subsequent Bransfield phase normal faults (late Pliocene–Pleistocene) reflect the early rifting of the Bransfield basin (Fig. 4*b*). Crustal failure may have occurred along previously existing arc-parallel strike-slip faults. It is attractive to relate the segmentation of the basin to differing rates of sinking of the subducted slab, with the older, denser slab below the north-eastern sub-basin sinking faster and driving extension more rapidly than farther to the south-west. As Bransfield

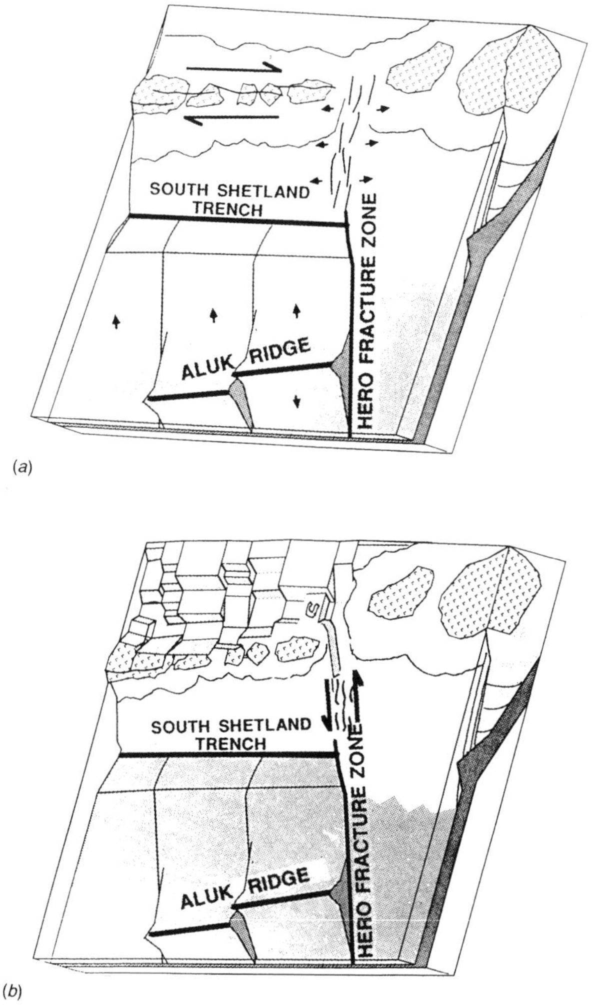

Fig. 4. Schematic block diagrams of the evolution of Bransfield Strait. (*a*) Late Miocene–early Pliocene, showing the subducted ridge crest in the Brabant segment, right-lateral strike-slip faulting in the South Shetland Islands, and the extension across Boyd Strait. (*b*) Early–middle Pliocene, showing the inception of rifting in Bransfield Strait.

Strait became extensional. Boyd Strait became a sinistral transform zone (Fig. 4*b*).

As extension and crustal thinning continued, Bransfield Strait evolved from a simple rift basin to a marginal basin displaying the characteristics of early seafloor spreading. Much of the igneous activity involved injection of dykes and sills into the sedimentary strata, creating thick intercalations of lavas and terrigenous sediment, as is common in young marginal basins with rapid sedimentation (Carey & Sigurdsson, 1984). Material formed in this fashion is likely to be much more diffusely distributed and less strongly magnetized than normal oceanic crust (Lawver & Hawkins, 1978), so well defined seafloor-spreading anomalies should not be expected. The intermediate crustal thickness, and diffuse and disorganized seafloor spreading associated with rapid sedimentation make Bransfield Strait very similar in many respects to a young rifted continental margin. The crust presently underlying Bransfield Strait may be analogous to the transitional continental–oceanic crust underlying passive continental margins.

References

Ashcroft, W.A. (1972). *Crustal structure of the South Shetland Islands and Bransfield Strait.* Cambridge; British Antarctic Survey Scientific Reports, No. 66, 43 pp.

Baker, P.E., McReath, I., Harvey, M.R., Roobol, M.J. & Davies, T.G. (1975). *The Geology of the South Shetland Islands: V. Volcanic Evolution of Deception Island.* Cambridge; British Antarctic Survey Scientific Reports, No. 78, 81 pp.

Barker, P.F. (1976). The tectonic framework of Cenozoic volcanism in the Scotia Sea region, a review. In *Andean and Antarctic Volcanology Problems*, ed. O. González-Ferrán, pp. 330–46. Rome; International Association of Volcanology and Chemistry of the Earth's Interior.

Barker, P.F. (1982). The Cenozoic subduction history of the Pacific margin of the Antarctic Peninsula: ridge crest–trench interactions. *Journal of the Geological Society, London*, **139(6)**, 787–802.

Barker, P.F. & Dalziel, I.W.D. (1983). *Progress in Geodynamics in the Scotia Arc Region.* American Geophysical Union, Geodynamic Series, **9**, 137–70.

Barker, P.F., Dalziel, I.W.D. & Storey, B.C. (in press). Tectonic development of the Scotia arc region. In *Antarctic Geology*, ed. R.J. Tingey. Oxford; Oxford University Press.

Birkenmajer, K. (1982). Late Cenozoic phases of block-faulting on King George Island (South Shetland Islands, West Antarctica). *Bulletin de l'Académie Polonaise des Sciences, Serie des Sciences de la Terre*, **30(1–2)**, 21–32.

British Antarctic Survey. (1985). *Tectonic Map of the Scotia Arc*, 1:3 000 000. BAS (Misc.) 3. Cambridge; British Antarctic Survey.

Carey, S. & Sigurdsson, H. (1984). A model of volcanogenic sedimentation in marginal basins. In *Marginal Basin Geology: Volcanic and Associated Sedimentary and Tectonic Processes in Modern and Ancient Marginal Basins*, ed. P.B. Kokelaar & M.F. Howells, pp. 37–58. Oxford; Blackwells Scientific Publications. Geological Society of London Special Publication, No. 16.

Craddock, C., ed. (1981). *Plate-Tectonic Map of the Circum-Pacific Region, Antarctica Sheet*, 1:10 000 000. Tulsa; American Association of Petroleum Geologists.

Curray, J.R., Moore, D.G., Lawver, L.A., Emmel, F.J., Raitt, R.W., Henry, M. & Kieckhefer, R. (1979). Tectonics of the Andaman Sea and Burma. In *Geological and Geophysical Investigations of Continental Margins*, ed. J.S. Watkins, L. Montadert & P.W. Dickenson, pp. 189–98. Tulsa; American Association of Petroleum Geologists. Memoir 29.

Forsyth, D.W. (1975). Fault plane solutions and tectonics of the South Atlantic and Scotia Sea. *Journal of Geophysical Research*, **80(11)**, 1429–43.

Hawkes, D.D. (1981). Tectonic segmentation of the northern Antarctic Peninsula. *Geology*, **9(5)**, 220–4.

Lawver, L.A. & Hawkins, J.W. (1978). Diffuse magnetic anomalies in marginal basins: their possible tectonic and petrologic significance. *Tectonophysics*, **45(4)**, 323–39.

Minster, J.B. & Jordan, T.H. (1978). Present-day plate motions. *Journal of Geophysical Research*, **83(B11)**, 5331–54.

Pelayo, A.M. & Wiens, D.A. (1986). Seismotectonics of the western Scotia Sea region (abstract). *Eos Transactions, American Geophysical Union*, **67(44)**, 1239.

Roach, P.J. (1978). The nature of back-arc extension in Bransfield Strait (abstract). *Geophysical Journal of the Royal Astronomical Society*, **53(1)**, 165.

Walcott, R.I. (1978). Geodetic strains and large earthquakes in the axial tectonic belt of North Island, New Zealand. *Journal of Geophysical Research*, **83(B9)**, 4419–29.

The geological and geochemical evolution of Cenozoic volcanism in central and southern Fildes Peninsula, King George Island, South Shetland Islands

LI ZHAONAI[1] & LIU XIAOHAN[2]

1 Institute of Geology, Chinese Academy of Geological Sciences, Baiwanzhuang Road, Beijing, China
2 Institute of Geology, Academia Sinica, PO Box 634, Beijing, China

Abstract

Volcanism on Fildes Peninsula occurred in three stages: I (tholeiitic rocks), II (calc-alkaline basalts and basaltic-andesites) and III (rocks with transitional characteristics); most of the volcanic rocks are of Eocene age. Isotopic studies show that these volcanic rocks have higher $^{87}Sr/^{86}Sr$ ratios than most mantle-derived volcanic rocks, and relatively more ^{207}Pb than MORBs and oceanic island basalts. $\delta^{18}O$ values for stage III lavas vary widely but are within the range for fresh mantle-derived volcanic rocks, whereas stage II lavas and some stage I rocks have $\delta^{18}O$ values close to fresh oceanic basalts. The remaining stage I lavas have slightly high $\delta^{18}O$ values due to interaction with seawater. MORB-normalized trace element patterns show an enrichment in Sr, K, Rb and Ba relative to the other trace elements (Ta–Cr), possibly due to alkali-rich fluids migrating from subducted ocean crust into the mantle source. Stage I basalts have low concentrations of high ionic elements and incompatible and compatible elements, but slightly higher LREE concentrations. Stage II lavas are enriched in alkali and alkaline-earth elements but have low concentrations of most incompatible elements; LREE concentrations are higher than in stage I rocks.

Introduction

Cenozoic volcanic rocks on Fildes Peninsula, King George Island (Fig. 1) have been studied by many geologists during the past 25 years (e.g. Barton, 1965; Birkenmajer, 1979, 1982; Pankhurst, 1982; Watts, 1982; Weaver, Saunders & Tarney, 1982; Pankhurst & Smellie, 1983; Smellie *et al*., 1984). Together with volcanic rocks elsewhere in the South Shetland Islands, they form part of a Jurassic–Quaternary magmatic island arc (Pankhurst & Smellie, 1983; Smellie *et al*., 1984), which is bounded to the south-east by Bransfield Strait, a young marginal basin (Ashcroft, 1972; Barker, 1982).

In 1985–86, the second Chinese National Antarctic Research Expedition studied the volcanic stratigraphy of a 20 km^2 area near the Chinese Great Wall station on Fildes Peninsula (Fig. 1). The results of this survey and of the subsequent analytical work are presented here.

Geology

Most outcrops in the central and southern parts of Fildes Peninsula are of basalt and basaltic-andesite lavas and volcaniclastic rocks belonging to the Fildes Formation (Smellie *et al*., 1984). The present study has subdivided these volcanic rocks into five litho-stratigraphical units which, from the base of the succession upwards, are:

(i) Jasper Hill Member (E*b*), a 60–120 m thick sequence of basalt and basaltic-andesite lavas, with basal volcanic breccias and agglomerates; it occurs mainly in the south-western part of Fildes Peninsula

(ii) Agate Beach Member (E*m*), 30–100 m of amygdaloidal basalt and basaltic-andesite lavas, with basal volcanic breccias and agglomerates; the E*m* is widespread in the area and extends northwards to Collins Glacier

(iii) Fossil Hill Member (E*h*), 6–24 m thick, which is predominantly volcaniclastic but has a few fossiliferous intercalations. The E*h* crops out mainly in the eastern and south-eastern parts of the peninsula

(iv) Block Hill Member (E*y*), 30–70 m of andesitic and basaltic-andesite lavas, and agglomeratic and brecciated lavas with variable clast size. This member is exposed mostly in the eastern and south-eastern parts of the study area

(v) Long Hill Member (E*ch*), which occurs mostly as sporadic outcrops of subvolcanic bodies.

Radiometric ages determined during the present study range from 58 to 42 Ma, and they young progressively from west to

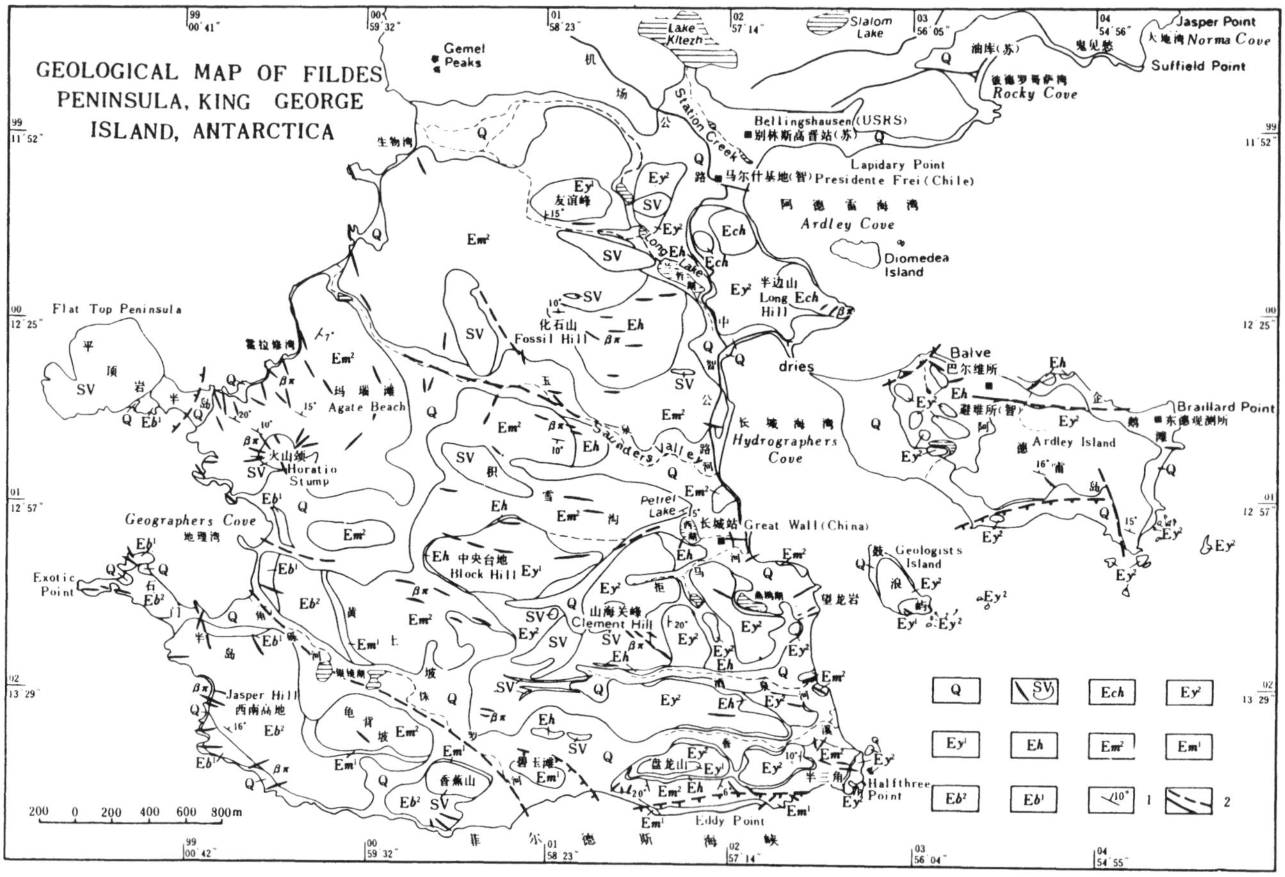

Fig. 1. Geological sketch map of Fildes Peninsula. Q, Quaternary; SV, dykes, plugs and subvolcanic intrusions; E*ch*, Long Hill Member, basaltic lavas; E*y*², Block Hill Member, lavas; E*y*¹, Block Hill Member, brecciated lavas; E*h*, Fossil Hill Member, pyroclastic and other volcaniclastic rocks; E*m*², Agate Beach Member, lavas; E*m*¹, Agate Beach Member, breccias and agglomerates; E*b*², Jasper Hill Member, lavas; E*b*¹, Jasper Hill Member, breccias and agglomerates; 1, dip and strike of bedding; 2, fault/inferred fault.

east, in general accordance with the north-eastward migration of magmatism already noted along the entire chain of the South Shetland Islands (Pankhurst & Smellie, 1983). Since at least early Tertiary times, tectonic uplift and the accumulation of large quantities of volcanic material have raised the Fildes Peninsula area above sea level, except at Halfthree Point.

Although widespread in the area, late-stage (pre-E*h*) subvolcanic bodies of porphyritic dolerite, porphyritic microgabbro and porphyritic basaltic-andesite are concentrated around the volcanic centres and fracture–fissure zones. The bodies are mostly dykes and plug-like apophyses with a few sills; their intrusive contacts are usually steep and well defined, with chilled margins. Brecciation and alteration has been observed at some margins and others have steeply dipping amygdaloidal zones and flat-lying marginal columnar joints.

Most of the pre-E*h* volcanic centres occur on the western side of Fildes Peninsula, where they are preserved as remnant volcanic necks and parasitic cones surrounded by coarse deposits of agglomerate and volcanic breccia (clast size 3–5 m); dykes radiate from the volcanic necks. Flat-lying columnar jointing is developed at the edges of some volcanic necks, and hydrothermal alteration increases towards the volcanic centres.

Post-E*h* volcanic eruptions were probably concentrated on the eastern side of the peninsula, at volcanic centres now below sea level, since coarse agglomerates occur locally on the east and south-east coasts and volcanic breccias tend to thin westwards. Arcuate normal faults trending parallel to the present coastline, well developed dyke trends perpendicular to the coast, and the increase in alteration near the coast provide further evidence for the easterly location of these centres.

The volcanic strata dip 10–15° (locally 20–30°) NNE–NE on an undulating monoclinal structure; some small-scale folds are related to faulting. Regional fractures and fissures strike mainly WNW.

Geochemistry and evolutionary mechanism

Late Jurassic–late Tertiary magmatism in the South Shetland Islands was characterized by the emplacement and eruption of primitive magmas intermediate between island-arc tholeiites and calc-alkaline types (Weaver *et al.*, 1982; Smellie *et al.*, 1984). The volcanic rocks of Fildes Peninsula display similar transitional characteristics (Fig. 2), with a range in compositions which varies in both lateral and vertical stratigraphical senses. The five lithostratigraphical units can be related to three geochemical stages: (I) tholeiitic basalts (E*b* and E*m*); (II) calc-alkaline basalts (E*h* and E*y*); and (III) intermediate compositions, with high REE content (E*ch*).

Mid-ocean ridge basalt (MORB)-normalized trace element patterns (Fig. 3) show an enrichment in Sr, K, Rb and Ba relative to the remaining trace elements Ta–Cr. The distin-

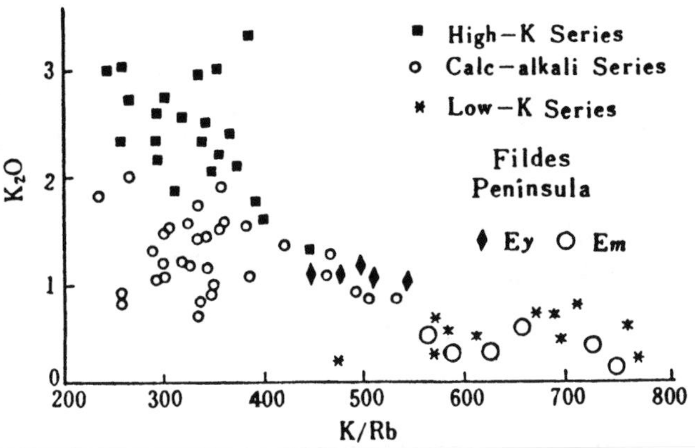

Fig. 2. Plot of K_2O versus K/Rb for the low-K to shoshonitic volcanic rocks from Filde Peninsula and other orogenic regions (after Ewart, 1982). *Em*, Agate Beach Member; *Ey*, Block Hill Member.

guishing feature of these enriched elements in their low ionic potential and hence their greater tendency to be mobilized by aqueous fluids. For this reason it is considered that such enrichment of a mantle-source region would be by aqueous fluids containing alkali and alkaline earth elements.

Stage I basalts are closest to the theoleiitic series, with low absolute abundances of elements of high ionic potential (Fig. 3) and of the incompatible elements Th–Yb (which are enriched in the melt during fractional crystallization) and the compatible element Cr (which is depleted during fractional crystallization). The chondrite-normalized REE pattern (Fig. 4) shows that concentrations of LREE are 10–25 times the chondritic values, and that HREE concentrations are 4–8 times such values. These features could be explained by a high degree of melting, stability of minor residual oxide phases, and remelting of depleted mantle. All are thought to be due to the

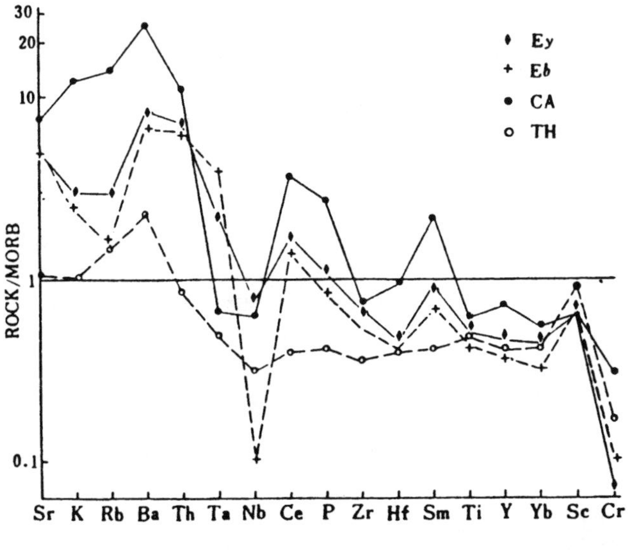

Fig. 3. MORB-normalized trace element data for volcanic rocks from Fildes Peninsula (after Pearce, 1982). *Ey*, Block Hill Member; *Eb*, Jasper Hill Member; CA and TH, after Pearce (1982).

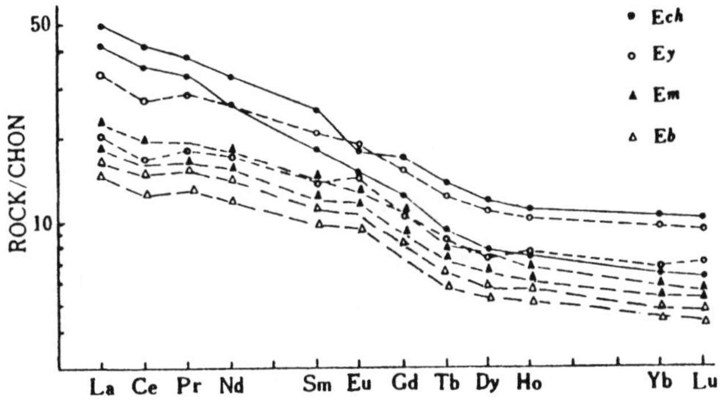

Fig. 4. Chondrite-normalized REE patterns for volcanic rocks from Fildes Peninsula. *Em*, Agate Beach Member; *Ey*, Block Hill Member; *Eb*, Jasper Hill Member; *Ech*, Long Hill Member.

input of aqueous fluids from subducted oceanic crust into the mantle-source regions.

Stage II lavas, belonging to the calc-alkaline series, are enriched not only in the alkali and alkaline earth elements but also in Th, Ce, P and Sm (Fig. 3); they are low in Ta, Nb, Zr, Hf, Ti, Y and Yb. Fig. 4 shows that LREE concentrations are 20–25 times the chondritic values and HREE concentrations are 7–12 times these values, indicating that in most cases these rocks could not have been derived by fractional crystallization of tholeiitic magmas or by simple partial melting of subducted oceanic crust. The calc-alkaline magmas contain material from both the subducted crust and the overriding mantle wedge. Such calc-alkaline rocks must include significant amounts of sedimentary material. Subduction of significant quantities of sediment would facilitate melting at lower temperatures and the melt could be expected to contain less mobile elements such as the rare-earth elements and P.

Faure (1977) calculated mean $^{87}Sr/^{86}Sr$ ratios of 0.7028 and 0.7039, respectively, for the available results on mid-ocean ridge and oceanic-island magmatic rocks. It is now known that such variations in Sr isotopes are usually accompanied by complementary changes in the isotope composition of Nd. Thus, on a $^{143}Nd/^{144}Nd$ versus $^{87}Sr/^{86}Sr$ plot, the majority of recent mantle-derived rocks fall on a single broad negative trend (Fig. 5).

The Sr-isotope compositions of the volcanic rocks from Fildes Peninsula tend to be more radiogenic than those of MORB but similar to those of many oceanic-island basalts. They appear to have higher $^{87}Sr/^{86}Sr$ ratios compared with the main trend of mantle-derived volcanic rocks and it is significant that the ratios of stage II calc-alkaline rocks are higher than those of stage I tholeiitic basalts. It is argued that subducted ocean crust contributed to the Fildes Peninsula magmas because similar $^{87}Sr/^{86}Sr$ ratios occur in both altered ocean-floor basalts and deep-sea sediments.

Pb-isotope results for volcanic rocks from the Fildes Formation are more radiogenic than MORBs and oceanic-island basalts and typically they form steeper slopes on a $^{207}Pb/^{204}Pb$ versus $^{206}Pb/^{204}Pb$ graph (Fig. 6). Similar results were obtained during early work on rocks from the Cascade Range (Church, 1973, 1976; Church & Tilton, 1973), Aleutian Islands (Kay, Sun & Lee-Hu, 1978) and Taiwan (Sun, 1980). The steep trend

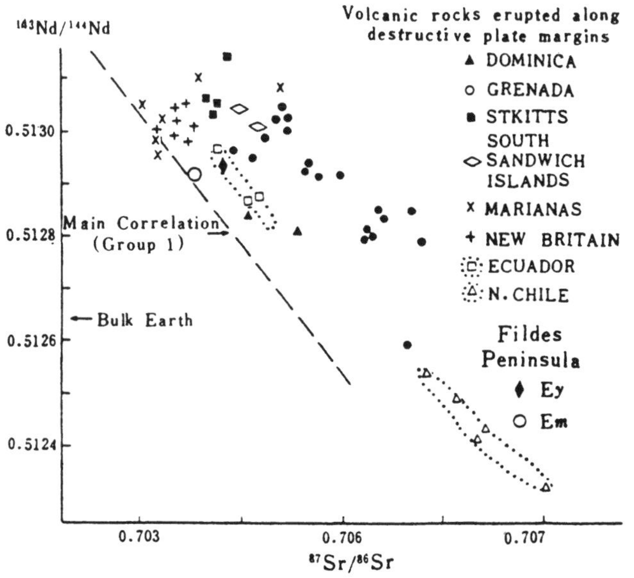

Fig. 5. $^{143}Nd/^{144}Nd$ **versus** $^{87}Sr/^{86}Sr$ **plot comparing Fildes Peninsula rocks with those from other regions (after Hawkesworth, 1982).** ***Em*, Agate Beach Member; *Ey*, Block Hill Member.**

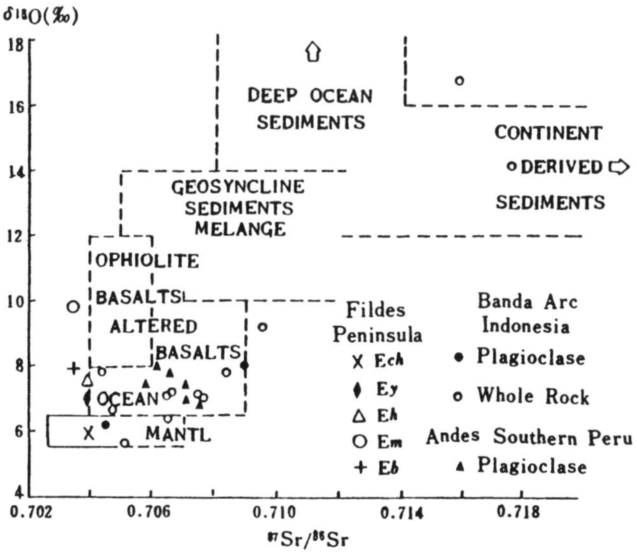

Fig. 7. Ranges of δ ^{18}**O relative to SMOW (standard mean ocean water) and** $^{87}Sr/^{86}Sr$ **in volcanic rocks from Fildes Peninsula (after Hawkesworth, 1982).** ***Em*, Agate Beach Member; *Ey*, Block Hill Member; *Eh*, Fossil Hill Member; *Eb*, Jasper Hill Member; *Ech*, Long Hill Member.**

of these data, lying between the fields for MORB and recent sediments, has led to their interpretation in terms of mixing between an upper mantle and sedimentary (or continental) component. The latter might be subducted ocean-floor basalt, material derived from ocean-floor basalts or material derived from the overriding mantle wedge; contributions from the sediments are usually small (< 5%), because of their relatively high Pb contents.

Fig. 7 summarizes $\delta^{18}O$ and $^{87}Sr/^{86}Sr$ values of common rock-types which might contribute to the composition of magmas generated above subduction zones. The $\delta^{18}O$ values are fractionated by phase changes and water–rock interaction at low temperature (e.g. Savin & Epstein, 1970), whereas $^{87}Sr/^{86}Sr$ depends on Rb/Sr ratios and the decay of ^{87}Rb over long periods of time. Thus a correlation between $\delta^{18}O$ and $^{87}Sr/^{86}Sr$ may be interpreted as evidence for some comparatively recent mixing process (see Turi & Taylor, 1976).

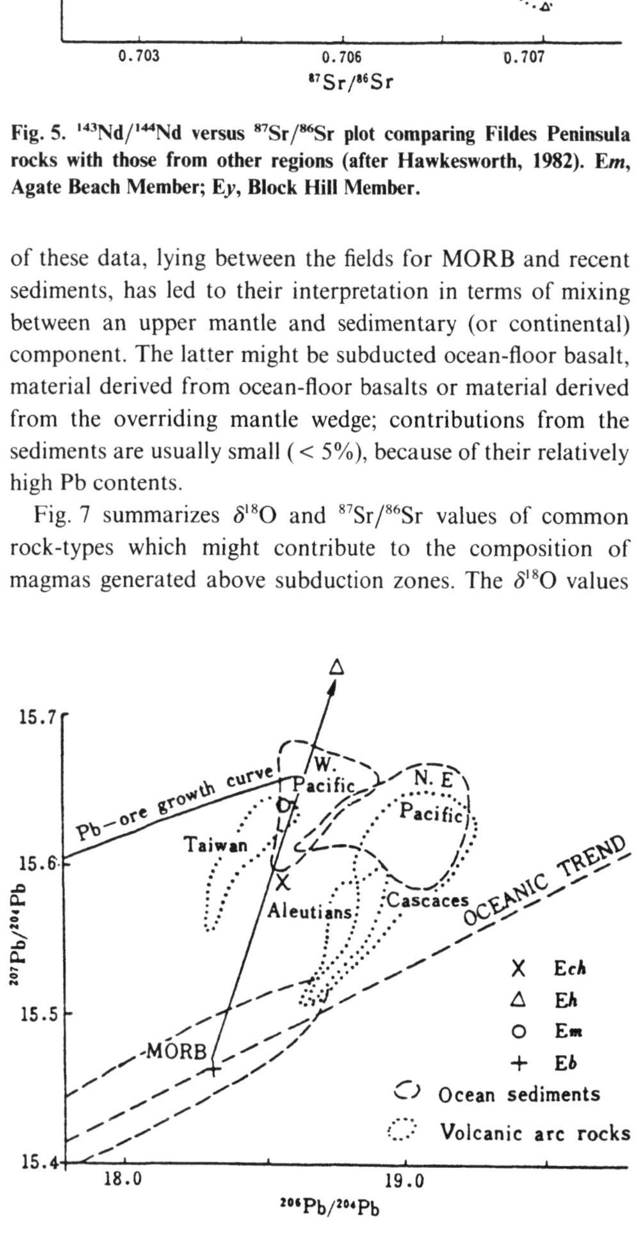

Fig. 6. Pb-isotope plot of volcanic rocks from Fildes Peninsula. The main oceanic trend and the fields for Pacific sediments and arc volcanic rocks are also shown (after Hawkesworth, 1982). ***Ech*, Long Hill Member; *Eh*, Fossil Hill Member; *Em*, Agate Beach Member; *Eb*, Jasper Hill Member.**

There is a wide variation in oxygen-isotope compositions in the different stages of volcanic rocks from Fildes Peninsula (Fig. 7). Stage III basalts have $\delta^{18}O$ values which are well within the range obtained from fresh mantle-derived volcanic rocks. Stage II lavas have values close to those obtained from fresh oceanic-basalts (Taylor, 1968; Matsuhisa, 1979) and it is unlikely that they were derived directly from subducted oceanic crust. Because oxygen is so abundant in magmatic rocks, its isotopic composition is insensitive to small additions of material with high $\delta^{18}O$; 5% of oceanic sediment or 10% of altered basalt would not significantly change the $\delta^{18}O$ value of a mantle-derived magma. Stage I basalts have higher $\delta^{18}O$ values, due to their interaction with seawater; their $\delta^{18}O$ values rise if the temperature is below 200–300 °C (Muehlenbachs & Clayton, 1976).

The isotope data demonstrate that, compared with magmas erupted at mid-ocean ridges and in intraplate environments, the volcanic rocks of Fildes Peninsula are commonly relatively enriched in ^{87}Sr and ^{207}Pb. These features are attributed to the release of material from the subducted oceanic crust. However, the available $\delta^{18}O$ data appear to preclude very large contributions from subducted material.

References

Ashcroft, W.A. (1972). *Crustal Structure of the South Shetland Islands and Bransfield Strait.* London; British Antarctic Survey Scientific Reports, No. 66, 43 pp.

Barker, P.F. (1982). The Cenozoic subduction history of the Pacific margin of the Antarctic Peninsula: ridge crest–trench interactions. *Journal of the Geological Society, London*, **139(6)**, 787–801.

Barton, C.M. (1965). *The Geology of the South Shetland Islands: II. The stratigraphy of King George Island.* London; British Antarctic Survey Scientific Reports, No. 44, 33 pp.

Birkenmajer, K. (1979). A revised lithostratigraphic standard for the Tertiary of King George Island, South Shetland Islands (West Antarctica). *Bulletin of the Polish Academy of Sciences, Earth Sciences*, **27(1 & 2)**, 49–57.

Birkenmajer, K. (1982). Report on geological investigations on King George Island and Nelson Island (South Shetland Islands, West Antarctica) in 1980–81. *Studia Geologica Polonica*, **74(3)**, 175–97.

Church, S.E. (1973). Limits of sediment involvement in the genesis of orogenic volcanic rocks. *Contributions to Mineralogy and Petrology*, **39**, 17–32.

Church, S.E. (1976). The Cascade Mountains revisited: a re-evaluation in the light of new lead isotopic data. *Earth and Planetary Science Letters*, **29**, 175–88.

Church, S.E. & Tilton, G.R. (1973). Lead and strontium isotopic studies in the Cascade Mountains: bearing on andesite genesis. *Geological Society of America Bulletin*, **84**, 431–54.

Ewart, A. (1982). The mineralogy and petrology of Tertiary–recent orogenic volcanic rocks, with special reference to the andesitic–basaltic compositional range. In *Andesites: Orogenic Andesites and Related Rocks*, ed. R.S. Thorpe, pp. 25–87. Chichester; John Wiley & Sons.

Faure, G. (1977). *Principles of Isotope Geology*. New York; John Wiley & Sons. 589 pp.

Hawkesworth, C.J. (1982). Isotope characteristics of magmas erupted along destructive plate margins. In *Andesites: Orogenic Andesites and Related Rocks*, ed. R.S. Thorpe, pp. 549–71. Chichester; John Wiley & Sons.

Kay, R.W., Sun, S.-S. & Lee-Hu, C.N. (1978). Pb and Sr isotopes in volcanic rocks from the Aleutian Islands and Pribilof Islands, Alaska. *Geochimica et Cosmochimica Acta*, **42**, 263–73.

Matsuhisa, Y. (1979). Oxygen isotopic composition of volcanic rocks from the East Japan island arcs and their bearing on petrogenesis. *Journal of Volcanology and Geothermal Research*, **5**, 271–96.

Muehlenbachs, K. & Clayton, R.N. (1976). Oxygen isotope composition of the oceanic crust and its bearing on sea water. *Journal of Geophysical Research*, **81**, 4365–9.

Pankhurst, R.J. (1982). Sr-isotope and trace-element geochemistry of Cenozoic volcanic rocks from the Scotia arc and northern Antarctic Peninsula. In *Antarctic Geoscience*, ed. C. Craddock, pp. 229–34. Madison; University of Wisconsin Press.

Pankhurst, R.J. & Smellie, J.L. (1983). K–Ar geochronology of the South Shetland Islands, Lesser Antarctica: apparent lateral migration of Jurassic to Quaternary island arc volcanism. *Earth and Planetary Science Letters*, **66**, 214–22.

Pearce, J.A. (1982). Trace element characteristics of lavas from destructive plate boundaries. In *Andesites: Orogenic Andesites and Related Rocks*, ed. R.S. Thorpe, pp. 527–45. Chichester; John Wiley & Sons.

Savin, S.M. & Epstein, S. (1970). The oxygen and hydrogen isotope geochemistry of ocean sediments and shales. *Geochimica et Cosmochimica Acta*, **34**, 43–63.

Smellie, J.L., Pankhurst, R.J., Thomson, M.R.A. & Davies, R.E.S. (1984). *The Geology of the South Shetland Islands: VI. Stratigraphy, Geochemistry and Evolution.* Cambridge; British Antarctic Survey Scientific Reports, No. 87, 85 pp.

Sun, S.-S. (1980). Lead isotope study of young volcanic rocks from mid-ocean ridge, ocean islands and island arcs. *Philosophical Transactions of the Royal Society, London*, **A297**, 409–46.

Taylor, H.P. (1968). The oxygen isotope geochemistry of igneous rocks. *Contributions to Mineralogy and Petrology*, **19**, 1–71.

Turi, B. & Taylor, H.P. (1976). Oxygen isotope studies of potassic volcanic rocks of the Roman province, central Italy. *Contributions to Mineralogy and Petrology*, **55**, 1–31.

Watts, D.R. (1982). Potassium–argon ages and paleomagnetic results from King George Island, South Shetland Islands. In *Antarctic Geoscience*, ed. C. Craddock, pp. 255–61. Madison; University of Wisconsin Press.

Weaver, S.D., Saunders, A.D. & Tarney, J. (1982). Mesozoic–Cenozoic volcanism in the South Shetland Islands and the Antarctic Peninsula: geochemical nature and plate-tectonic significance. In *Antarctic Geoscience*, ed. C. Craddock, pp. 263–73. Madison; University of Wisconsin Press.

The Late Cretaceous–Cenozoic structural history of King George Island, South Shetland Islands, and its plate-tectonic setting

A.K. TOKARSKI

Polish Academy of Sciences, Institute of Geological Sciences, 31–002 Kraków, ul. Senacka 3, Poland

Abstract

King George Island (KGI) comprises largely calc-alkaline volcanic and volcanogenic rocks, pierced by small calc-alkaline intrusions. Brittle deformation has affected most of this magmatic pile, resulting in tension gashes and strike-slip faults. Folding affected only those portions of the pile containing more clastic intercalations. Structural development of the KGI magmatic pile has been controlled largely by a strike-slip tectonic regime for the last 77 m.y. Three sets of extensional joints and dykes, parallel to the respective orientations of the main principal stresses, correspond to three deformation stages: (1) development of an ESE-orientated set of joints and dykes (set I), attributed to the eastward subduction of the ancient Pacific ocean crust (up to about 23 Ma); (2) development of another set (set II) reflecting the clockwise rotation of the main principal stress up to its recent WSW orientation, attributed to main plate reorganization which resulted in cessation of subduction and opening of the Scotia Sea, and (3) development of set III, associated with the opening of the Bransfield Strait ((?)1.3 Ma–Recent). Bransfield Strait basin is interpreted as one of the zones of young volcanism in the northern Graham Land region. These zones reflect the push of the spreading ridges surrounding the Antarctic plate.

Introduction

Plate-tectonic reconstructions in the Antarctic Peninsula region have been based on interpretations of magnetic anomalies, facies distribution and geochemistry of the magmatic rocks. The purpose of the present study has been to check these reconstructions by structural analyses. This paper summarizes the results of a structural analysis of the rocks on King George Island (KGI). For a more detailed discussion of the structural data, see Tokarski (1987) and references therein.

Geological setting

King George Island (Fig. 1) is built largely of continental volcanic and volcanogenic rocks. This volcanic pile (KGIVP) is > 2000 m thick and is composed of numerous volcanic edifices. The lithology of the KGIVP is varied, changing from parts composed mainly of lava flows to those composed mostly of detrital rocks. Glaciomarine strata are intercalated in the upper part of the sequence (Birkenmajer, 1985).

According to Birkenmajer (1982), the backbone of King George Island is formed by the Barton horst. Within the horst the volcanic rocks are commonly altered (e.g. calcite and chlorite), whereas the rocks outside the horst are mostly fresh. All rocks are cut by numerous andesitic and basaltic dykes, and those in the Barton horst are also intruded by numerous calc-alkaline plutons, ranging in composition from quartz-gabbro to granodiorite.

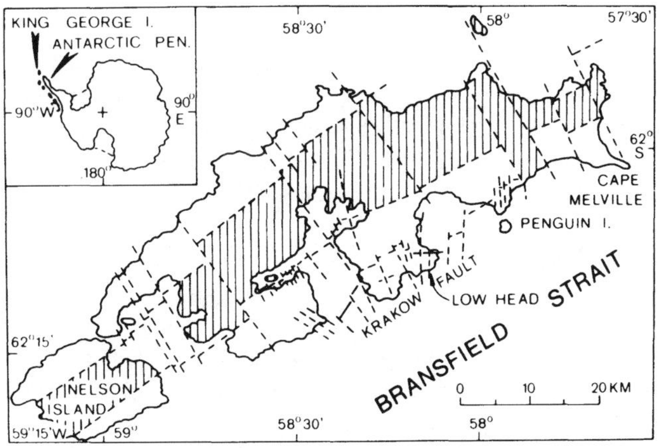

Fig. 1. Structural scheme of King George Island (after Birkenmajer, 1982); Barton horst, vertical shading.

The structure of the KGIVP is largely controlled by a system of strike-slip faults (Birkenmajer, 1982). The system comprises an ENE–WSW set, and a NNW–SSE set. A younger group of faults (mostly dip-slip) is represented by the Kraków fault system (Birkenmajer, 1982), which can be traced in the southern part of the island.

The volcanic rocks of the KGIVP, and the dykes and plutons

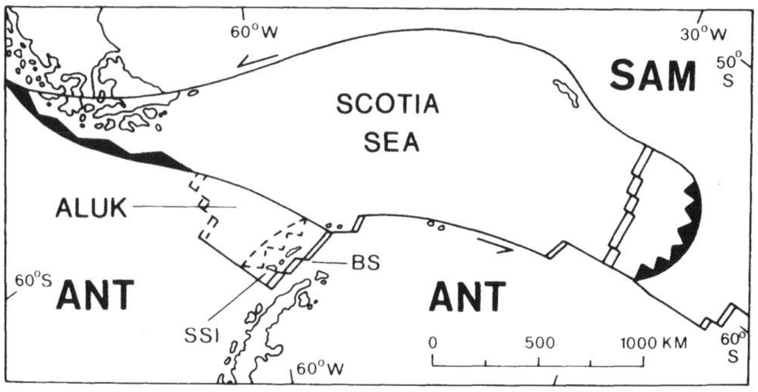

Fig. 2. Recent plate boundaries in the Scotia Sea region (after British Antarctic Survey, 1985). ANT, Antarctic plate; SAM, South American plate; ALUK, Aluk plate; BS, Bransfield Strait; SSI, South Shetland Islands.

have yielded numerous K–Ar dates from ~ 110– ~ 14 Ma (for review see Smellie *et al.*, 1984; Birkenmajer *et al.*, 1986).

The KGIVP is predominantly calc-alkaline and corresponds to the Antarctic Peninsula Volcanic Group, related to Mesozoic–Cenozoic eastward subduction of ancient Pacific crust (Aluk plate) (Fig. 2) which ceased about 4 m.y. ago (Barker, 1982). Subsequently, the Aluk and Antarctic plates became coupled. The present stress regime in the KGIVP appears to be typified by opening of the Bransfield Strait, believed to have started about 1.3 m.y. ago (Roach in Barker & Dalziel, 1983). The uppermost part of the KGIVP formed during this last period; its rocks are mildly alkaline on Penguin Island (Fig. 1).

Structural style

Brittle deformation

The mesostructural pattern of the KGIVP is dominated by extensional joints which have been identified in all the localities studied (Fig. 3). The joints are usually non-planar, are commonly arranged en échelon and are filled by mineral veins. Neptunian dykes commonly follow the extensional joints at Low Head (Fig. 1).

On a larger scale, the same brittle extensional deformation is expressed by the dykes. Their irregular shapes and courses are similar to those of the veins filling the extensional joints and they are commonly arranged en échelon. Stepped dykes occur also, following either a single set of extensional joints or offset along pre-existing joints.

Shear deformation is much less marked than extensional deformation. Shear-joints have been identified in only some of the localities studied. A paucity of minor faults is a characteristic feature of the KGIVP. Tectonic striae have been found on only 65 out of about 10 000 joint surfaces examined. These are: 38 strike-slip faults, 11 dip-slip faults, 10 oblique-slip faults and 6 reverse faults.

Extensional joints follow the direction of σ_1/σ_2 planes. In the KGIVP, extensional-joints as well as shear-joints are always normal to bedding and/or vertical. This indicates a horizontal position of σ_1. Strike-slip-dominated movement patterns on minor faults support this supposition.

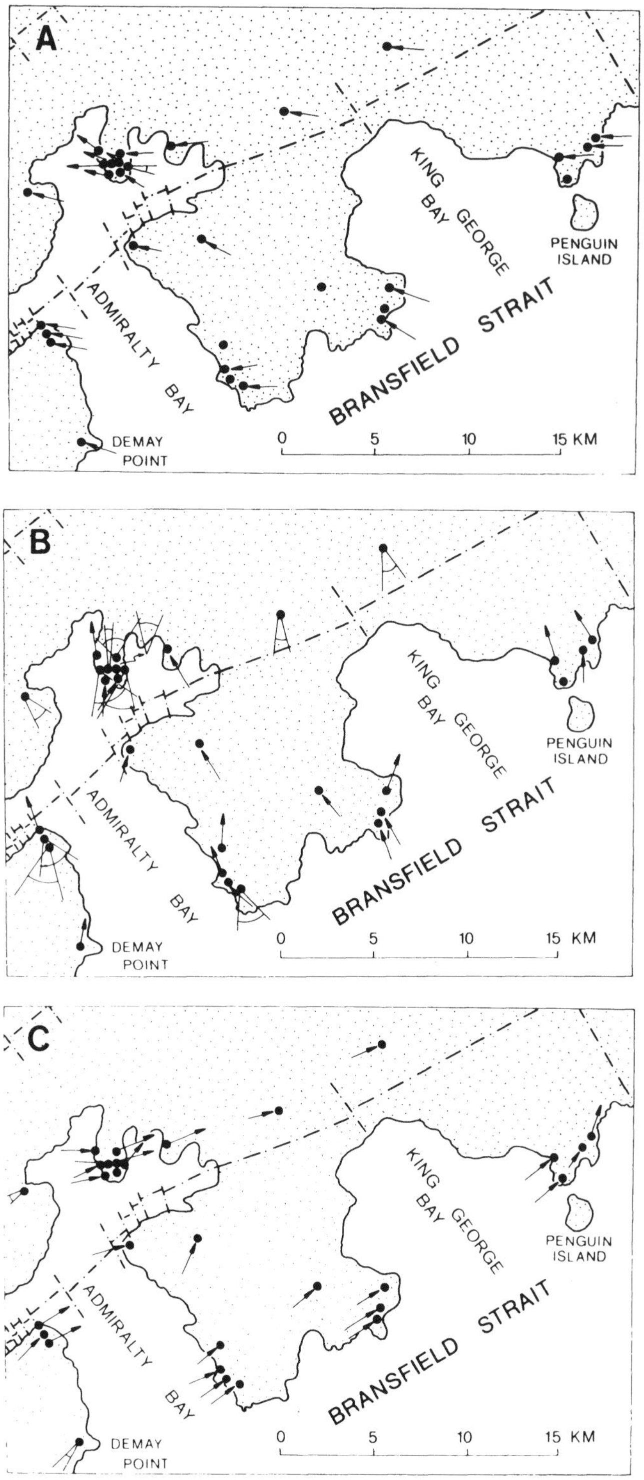

Fig. 3. Directions of main principal stress (σ_1) responsible for the origin of joints in the central part of King George Island (after Tokarski, 1987). A, set I; B, set II; C, set III.

Penetrative joints have developed uniformly throughout the KGIVP and they provide the basis for reconstructions of its structural development.

Folding

Tightly folded rocks occur in several places within the Barton horst (Tokarski, 1987), whereas other parts of the

KGIVP are virtually unfolded. Inside the horst, the style of deformation changes with lithology: the intrusive bodies (least ductile) are deformed in a brittle way whereas folds may affect the volcanic sequence, changing from more open folds in successions containing thick lava flows (more ductile) to tighter folds in predominantly detrital (most ductile) successions. Fold orientations in the more ductile bodies conform to the shapes of unfolded (less ductile) bodies. It follows that neither the position of folded rocks in the sequence, nor the directions of fold axes can be used for regional tectonic reconstructions.

Structural development

Succession of joint sets

Joints have been studied at 30 stations (Fig. 3): in the KGIVP sequence, from rocks older than 77 Ma (Demay Point) to the Recent rocks on Penguin Island; in plutonic intrusions; and in dykes. Three sets of extensional-joints cut the lower part of the pile (> 23 Ma). The higher part is cut by only two sets of extensional-joints and no tectonic-joints were found in the Recent rocks on Penguin Island.

There is a marked consistency of joint directions within sets I (~ 105°) and III (~240°) at all localities whereas the directions of set II joints are more scattered (120–220°). The cross-cutting relationships observed at numerous stations show that set I is the oldest and set III the youngest. This succession is confirmed by two independent lines of evidence. (1) The dykes which cut the KGIVP form three sets parallel to the joints of sets I, II and III, respectively. Moreover, the lower part of the pile (> 23 Ma) is cut by all three sets of dykes (Fig. 4A), whereas in the upper part of the pile (< 23 Ma) only dykes of sets II and III are present (Fig. 4B). (2) Three dykes parallel to set II joints and cutting the KGIVP outside the Barton horst have yielded K–Ar ages: one at Low Head (Fig. 1) (striking 130°) is ~ 22 Ma and two at Cape Melville (striking 135° and 150°) are ~ 20 Ma.

Pattern of movement

The time–space patterns of joints and dykes, and the analysis of minor faults, form the basis for the reconstruction of the KGIVP movement pattern (Fig. 5). This interpretation is based on the clockwise rotation of σ_1 from ~ 105° to ~ 240°.

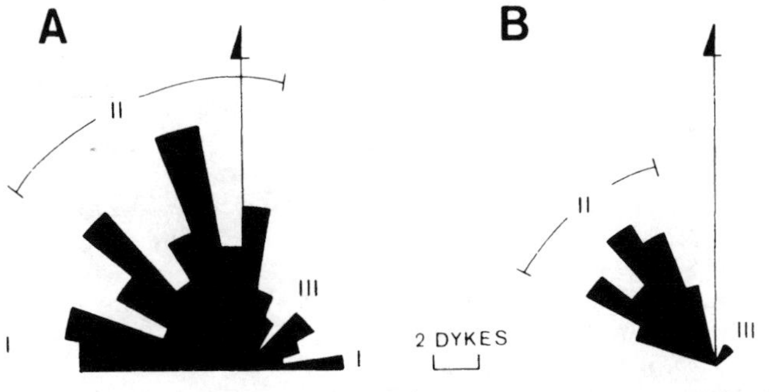

Fig. 4. Orientation of dykes cutting the King George Island volcanic pile: (A) lower part of the pile (> 23 Ma); (B) upper part of the pile (< 23 Ma).

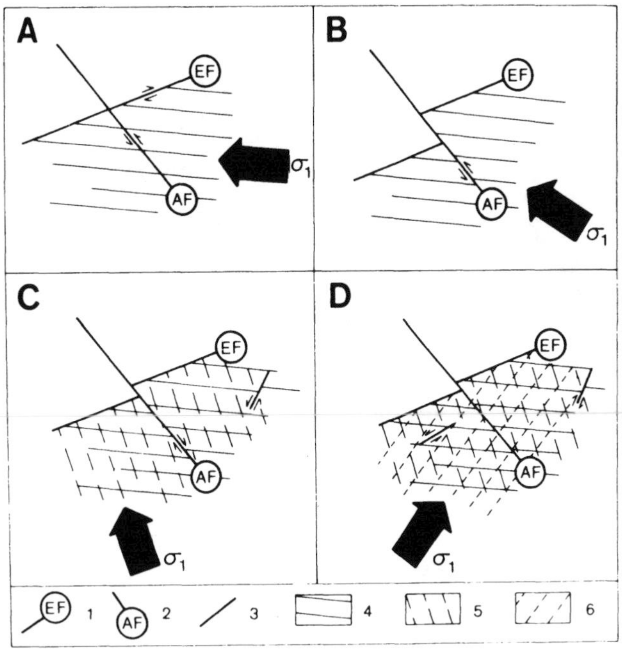

Fig. 5. Sequential model for the development of strike-slip tectonics in the King George Island volcanic pile: (1–2) two basic sets of strike-slip faults; (3) other faults; (4) joints of set I; (5) joints of set II; (6) joints of set III.

Three successive sets of extensional-joints and dykes record three stages of deformation. In the first stage, two sets of strike-slip faults were initiated as a conjugate pair when σ_1 was orientated at 105°. During rotation, the strike-slip movement on these faults at first continued in the same direction, then ceased, and finally was reorientated to the actual position of σ_1.

Plate tectonic setting

Stage I – subduction

The lower part of the KGIVP formed in a volcanic arc controlled by long-lived eastward subduction. The persistent orientation of set I joints, approximately perpendicular to the fossil spreading zone (Fig. 2), and formed from at least 77 Ma to about 23 Ma ago, shows that no major reorganization of plates took place during that period.

State II – reorganization of plates

The two stages with a stabilized position for σ_1, the subduction-controlled stage (I) and the stage of young extensional deformation (III – see below), were separated by a stage of clockwise rotation of σ_1 (II). It seems that this rotation was related to the major reorganization of plates in the region (Weissel, Hayes & Herron, 1977), which resulted in the opening of the Scotia Sea (Barker & Dalziel, 1983). The rotation (Fig. 6) was due to interference between the waning subduction-controlled regime (Barker, 1982) and the increasing influence of the left-lateral movement between the South American and Antarctic plates (Barker & Dalziel, 1983). The rotation stopped with completion of the plate reorganization

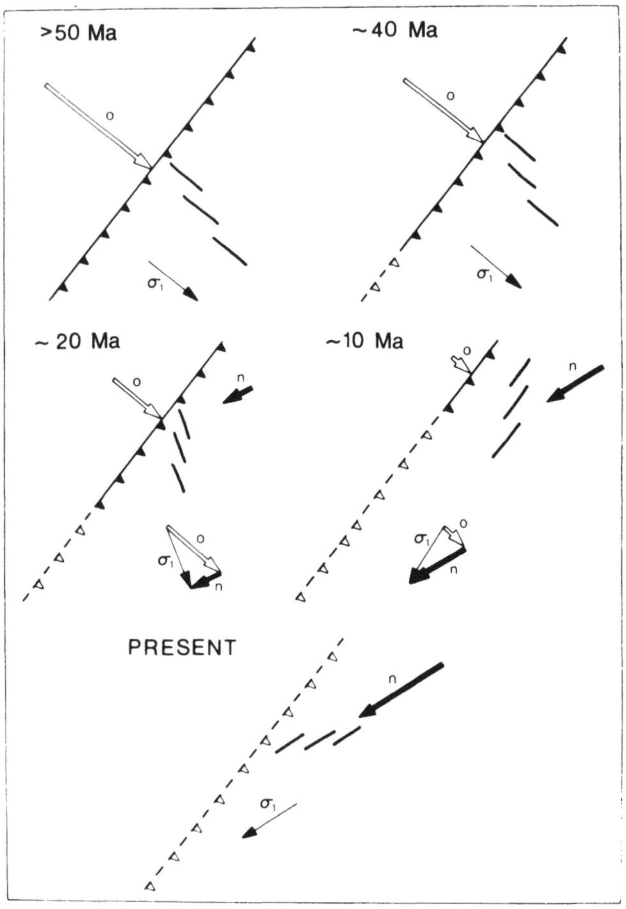

Fig. 6. Inferred rotation of the main principal stress (σ_1) due to interference of old, subduction-controlled (o) and new (n) stress regimes. Barbed line represents subduction zone (dashed, inactive), and thick short lines (parallel to σ_1) show the direction of extensional-joints and dykes.

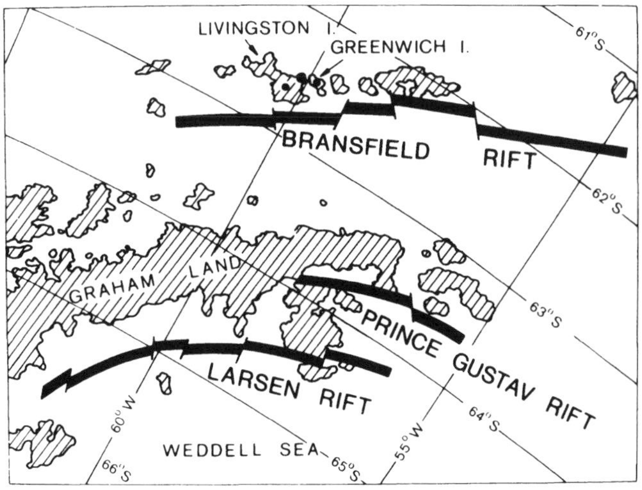

Fig. 7. Course of axial parts of the Bransfield Strait, Prince Gustav and Larsen rifts, which are loci of Quaternary volcanism (after González-Ferrán, 1985). Sites of young (< 0.5 Ma) volcanism on Livingston and Greenwich islands (dots) after Smellie *et al.* (1984).

about 4 Ma ago, when subduction finally ceased (Barker, 1982).

Stage III – origin of Bransfield Strait

It is commonly considered that Bransfield Strait is a back-arc basin related to subduction of the Aluk plate (e.g. Weaver *et al.*, 1979; Barker & Dalziel, 1983). However, this interpretation is inconsistent with three independent lines of evidence.

(1) The associated volcanic arc is extinct and subduction ceased 4 Ma ago (Barker, 1982).

(2) In northern Graham Land, Quaternary volcanism (mostly alkaline) occurs not in one (Bransfield Strait) but in four linear zones (Fig. 7). These zones are arranged roughly en-echelon and are orientated obliquely to the direction of subduction. These features hardly fit the back-arc setting. Moreover, late Cenozoic volcanism in Antarctica is not restricted to the northern Graham Land region. It occurs in numerous other places, unrelated to subduction.

(3) In the majority of back-arc areas, σ_1 acts vertically (Nakamura & Uyeda, 1980). The situation is different in the Bransfield Strait, where the tectonic regime is controlled by a horizontal σ_1, as indicated by the structural analysis of the KGIVP.

The parallelism between the zones of young volcanism in northern Graham Land and set III joints suggests their common origin. These features are interpreted here as manifestations of internal deformation of the Antarctic plate due to the stress regime set up by the surrounding spreading centres.

References

Barker, P.F. (1982). The Cenozoic subduction history of the Pacific margin of the Antarctic Peninsula: ridge crest–trench interactions. *Journal of the Geological Society, London*, **139(6)**, 787–801.

Barker, P.F. & Dalziel, I.W.D. (1983). Progress in geodynamics in the Scotia Arc region. In *Geodynamics of the Eastern Pacific Region, Caribbean and Scotia Arcs*, ed. R. Cabré, pp. 137–70. Washington, DC; American Geophysical Union. Geodynamics Series 9

Birkenmajer, K. (1982). Late Cenozoic phases of block-faulting on King George Island (South Shetland Islands, West Antarctica). *Bulletin de l'Académie Polonaise des Sciences, Série des Sciences de la Terre*, **30(1–2)**, 21–32.

Birkenmajer, K. (1985). Onset of Tertiary continental glaciation in the Antarctic Peninsula sector (West Antarctica). *Acta Geologica Polonica*, **35(1–2)**, 1–31.

Birkenmajer, K., Delitala, M.C., Narębski, W., Nicoletti, M. & Petrucciani, C. (1986). Geochronology of Tertiary island-arc volcanics and glacigenic deposits, King George Island, South Shetland Islands (West Antarctica). *Bulletin of the Polish Academy of Sciences, Earth Sciences*, **34(3)**, 257–73.

British Antarctic Survey. (1985). *Tectonic Map of the Scotia Arc*, 1:3000000, BAS (Misc) 3. Cambridge; British Antarctic Survey.

González-Ferrán, O. (1985). Volcanic and tectonic evolution of the northern Antarctic Peninsula – Late Cenozoic to Recent. *Tectonophysics*, **114(1–4)**, 389–409.

Nakamura, K. & Uyeda, S. (1980). Stress gradient in arc–backarc regions and plate subduction. *Journal of Geophysical Research*, **85(B11)**, 6419–28.

Smellie, J.L., Pankhurst, R.J., Thomson, M.R.A. & Davies, R.E.S.

(1984). *The Geology of the South Shetland Islands: VI. Stratigraphy, Geochemistry and Evolution.* Cambridge; British Antarctic Survey Scientific Reports, No. 87, 85 pp.

Tokarski, A.K. (1987). Structural events in the South Shetland Islands (Antarctica). III. Barton Horst, King George Island. *Studia Geologica Polonica*, **90**, 7–38.

Weaver, S.D., Saunders, A.D., Pankhurst, R.J. & Tarney, J. (1979). A geochemical study of magmatism associated with the initial stages of back-arc spreading. The Quaternary volcanics of Bransfield Strait, from South Shetland Islands. *Contributions to Mineralogy and Petrology*, **68**, 151–69.

Weissel, J.K., Hayes, D.E. & Herron, E.M. (1977). Plate tectonics synthesis: the displacement between Australia, New Zealand, and Antarctica since the Late Cretaceous. *Marine Geology*, **25(1–3)**, 231–77.

Tectonophysical models of the crust between the Antarctic Pensinula and the South Shetland trench

A. GUTERCH[1], M. GRAD[2], T. JANIK[1] & E. PERCHUĆ[1]

1 Institute of Geophysics, Polish Academy of Sciences, Pasteura 3, 00-973 Warsaw, Poland
2 Institute of Geophysics, University of Warsaw, Pasteura 7, 02-093 Warsaw, Poland

Abstract

Seismic measurements, including multichannel seismic-reflection and deep seismic soundings, were carried out in the region of the west coast of the Antarctic Peninsula, Bransfield Strait, South Shetland Islands and South Shetland trench. The measurements were made along several lines with a total length of about 2500 km. Crustal sections and one- and two-dimensional models of the crust for this area are discussed in detail. The thickness of the crust ranges from 30–33 km in the South Shetland Islands to 38–45 km near the coast of the Antarctic Peninsula. The crustal structure beneath the trough of Bransfield Strait is highly anomalous; a seismic discontinuity with velocities of 7.0–7.2 km/s was found at a depth of about 10 km, and a second discontinuity with velocities of 7.6–7.7 km/s was found at a depth of 20–25 km. A seismic inhomogeneity along the Deception–Penguin–Bridgeman islands volcanic line has also been found. A scheme for the geotectonic division and a geodynamic model of the area are discussed.

Introduction

During the Polish Antarctic Geodynamical Expeditions in 1979–80 and 1984–85, multi-channel seismic-reflection and deep seismic-refraction measurements were made by the Institute of Geophysics of the Polish Academy of Sciences. Seismic measurements were carried out in the region of the Antarctic Peninsula and the South Shetland Islands, between Antarctic Sound and Adelaide Island. Special attention was paid to tectonically active zones and to the contact zones between crustal blocks and lithospheric plates.

Fig. 1 shows the locations of the seismic land stations during various stages of the experiment, and of the profile lines along which crustal sections and models of the crust were determined. The distance between shot points was about 5 km. The shots, which varied in size between 25 and 120 kg, were electrically detonated from a ship at depths of 70–80 m along the profile lines. Three- and five-channel refraction stations on land recorded all the shots made along different profiles. Reflection profiles were made by a 24-channel seismic system (DFS IV) with the Bolt airgun system. The data have been processed by Geophysical Enterprise Toruń, using standard processing procedures. The total length of the reflection and refraction profiles undertaken was about 2500 km.

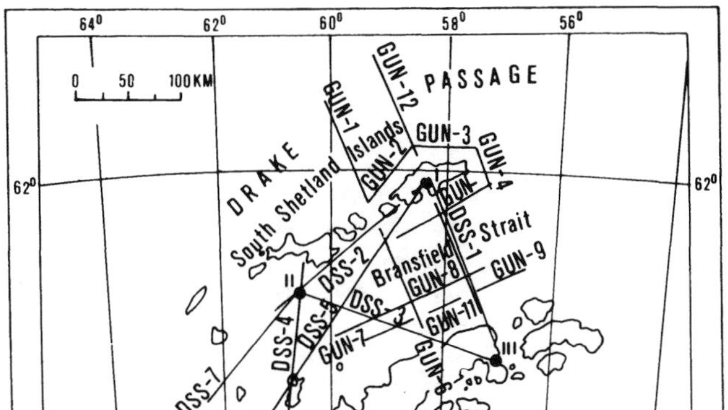

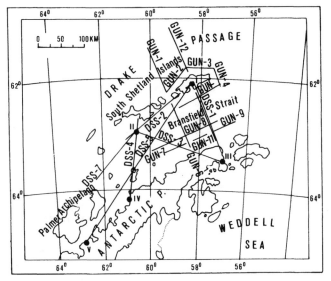

Fig. 1. Location of seismic-refraction and reflection profiles. I–V, seismic refraction stations; DSS-1–DSS-7, deep seismic sounding profiles; GUN-1, GUN-2, etc., seismic-reflection profiles.

Interpretation

The results of the crustal studies presented in this paper are based on the interpretation of travel times of refracted as well as reflected waves. The correlation of wave groups was based on their kinematic and dynamic properties, using composite records and seismic record sections. Initial interpretations were achieved using simple methods of effective and apparent velocities. Effective velocities were calculated for particular branches of the travel times of reflected waves and, by considering the relations between effective and mean veloci-

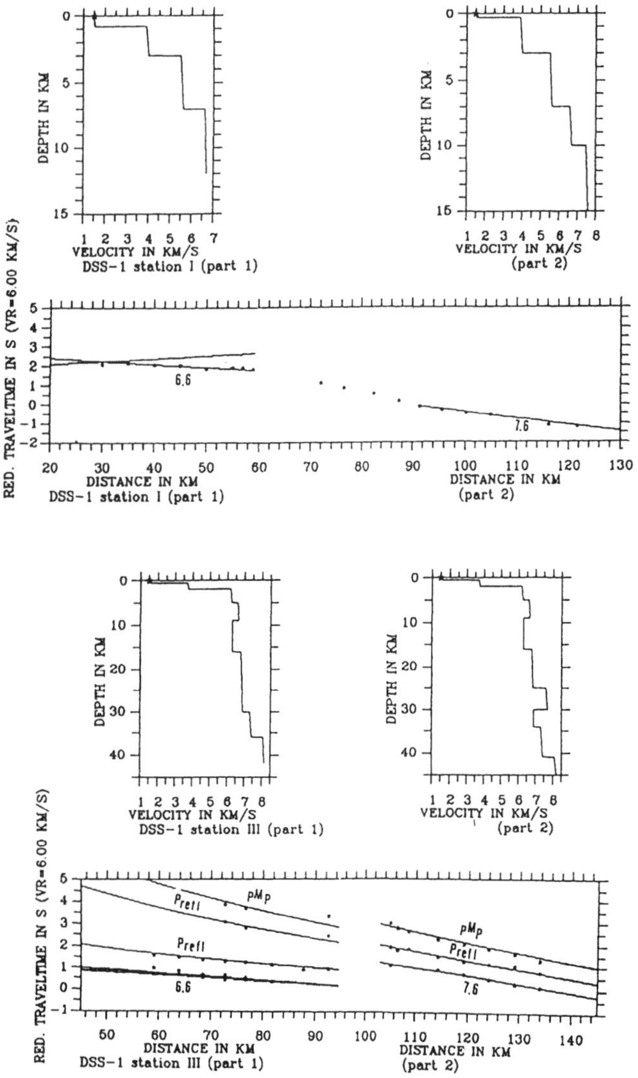

Fig. 2. **Travel times and crustal models for DSS-1 profile, stations I and III, between the Antarctic Peninsula and the South Shetland Islands. Experimental data are indicated by the dots. Each point on the travel times corresponds to a 3- or 5-channel seismic record, with 200 m distance between them. Theoretical travel times for the corresponding models are indicated by the solid curve. 6.6, 7.6, etc., apparent velocities of refracted waves in km/s; P_{refl}, reflected waves; $_PM_P$, waves reflected from the Moho discontinuity.**

ties, the depths of the reflection boundaries were determined (Grad, 1983). Depths of refraction boundaries were determined by assigning mean velocities to refracted waves. By testing one-dimensional (1D) models of the crust, it is possible to evaluate the validity of correlations and also to determine the mean velocities and depths of the boundaries. Theoretical travel times were calculated within the individual crustal blocks using the ray method for multilayer models (Alekseev & Gelchinsky, 1959; Fuchs & Müller, 1971). The 1D models were then used to develop two-dimensional (2D) models with curvilinear boundaries and complex velocity distribution $v(x, z)$ (Červený & Pšenčík, 1983). As a result of a multi-stage process of interpretation, good agreement between theoretical and experimental travel times, and also qualitative agreement between the amplitudes of main wave types, were obtained.

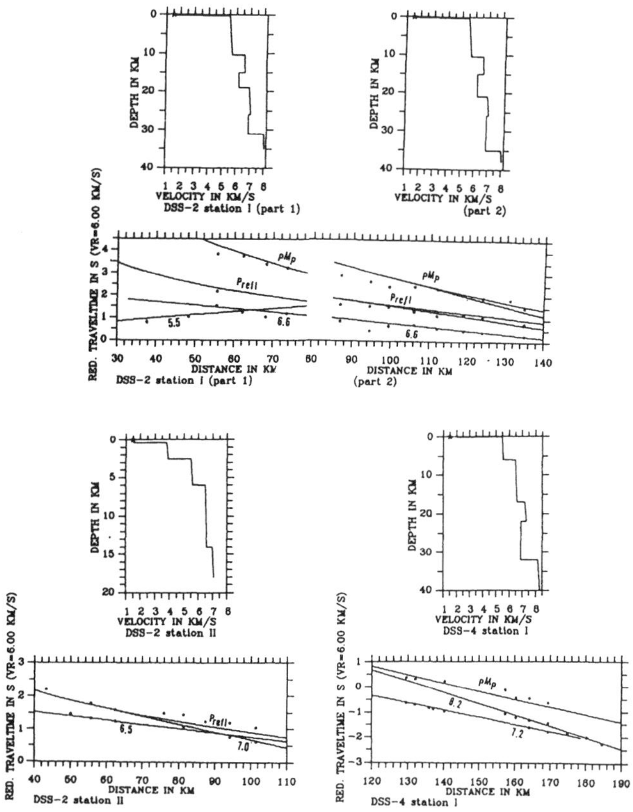

Fig. 3. **Travel times and crustal models for DSS-2 profile, stations I and II, and for DSS-4 profile, station I. For explanation see Fig. 2.**

Main results

The deep crustal structure over the area from the western margin of the Antarctic Peninsula, across Bransfield trough in Bransfield Strait, to the South Shetland Islands and the South Shetland trench was determined along profiles shown in Fig. 1. These are:

- DSS-1: King George Island–Hope Bay, between stations I and III,
- DSS-2: King George Island–Deception Island, between stations I and II,
- DSS-3: Deception Island–Hope Bay, between stations II and III,
- DSS-4: Deception Island–'Primavera' base, between stations II and IV,
- DSS-5: King George Island–'Primavera' base, between stations I and IV,
- DSS-7: Deception Island–Snow Island, one branch from station II.

Moreover, records made at station I from shots in profile DSS-4 were also used. Data from reflection profiles (Fig. 1) complemented the crustal refraction data.

The travel times of correlated refracted and reflected waves with 1D crustal models are presented in Figs 2–6. Each point on the travel time corresponds to a 3-channel seismic record. The pattern of the wave field in Bransfield Strait is very complex. For example, on the South Shetland Islands side of lines DSS-1 and DSS-3, the first arrivals involve consecutive

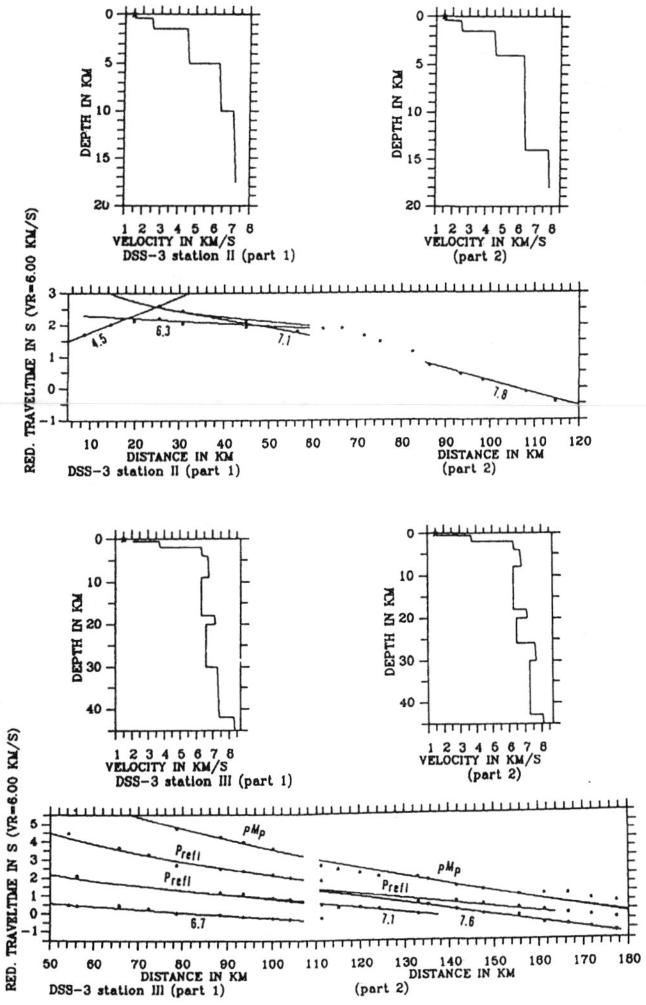

Fig. 4. Travel times and crustal models for DSS-3 profile, stations II and III. For explanation see Fig. 2.

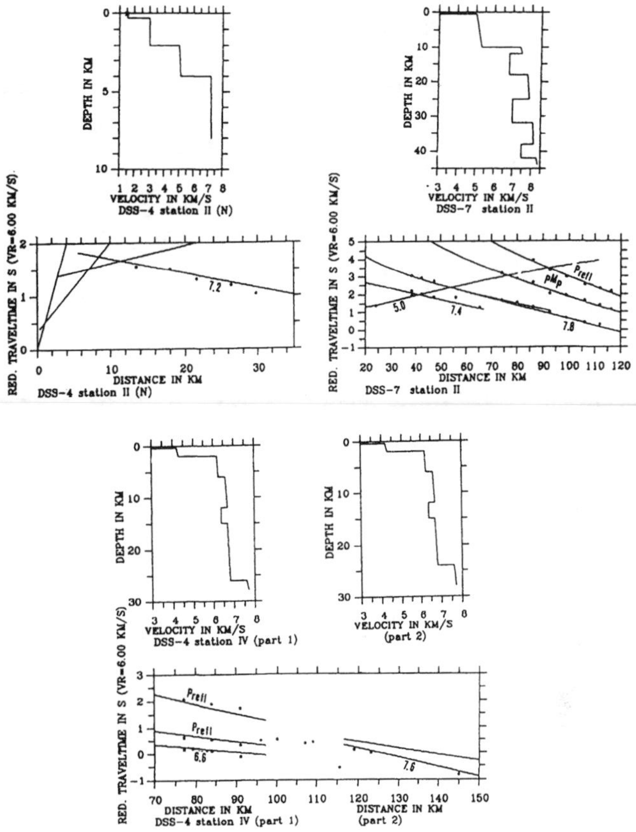

Fig. 5. Travel times and crustal models for DSS-4 profile, stations II and IV, and for DSS-7 profile, station II. For explanation see Fig. 2.

waves with velocities of 5.7 and 6.5 km/s, whereas refracted waves with anomalously high velocities of about 9.6 km/s were recorded at distances of about 80 km. Farther way, at distances exceeding 100 km, refracted waves with velocities of about 7.6 km/s were recorded. In turn, along the Antarctic Peninsula side of the same lines, 6.6 km/s refracted waves were followed by waves with anomalously low velocities of 5.7–5.9 km/s; at distances of about 90–150 km refracted waves with velocities of about 7.6 km/s were recorded. Because of this, it was necessary to compute no less than two 1D crustal models for successive travel times. These contain high- and low-velocity zones, in both the lower and upper parts of the crust.

Generalized crustal sections (Fig. 7) were developed from the interpretation of travel times of refracted and reflected waves, as well as 1D and 2D crustal modelling. The northern part of profile DSS-1 was complemented with data from profile DSS-2. The boundary velocities measured, or the probable layer velocities estimated from 1D modelling, are marked on the sections. The zones of the contact between crustal blocks marked on these sections, which may be zones of deep fractures, were determined on the basis of distinct variations in seismic wave fields, the strong attenuation of seismic waves, and discordances between the depths of the determined seismic

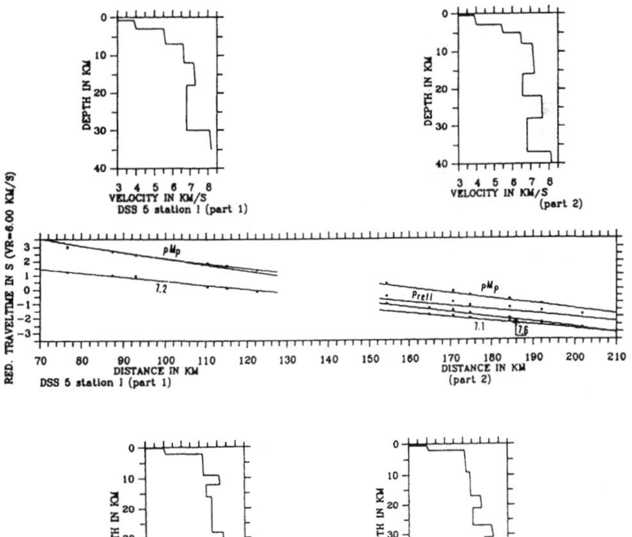

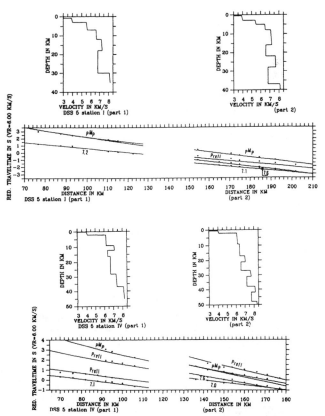

Fig. 6. Travel times and crustal models for DSS-5 profile, stations I and IV. For explanation see Fig. 2.

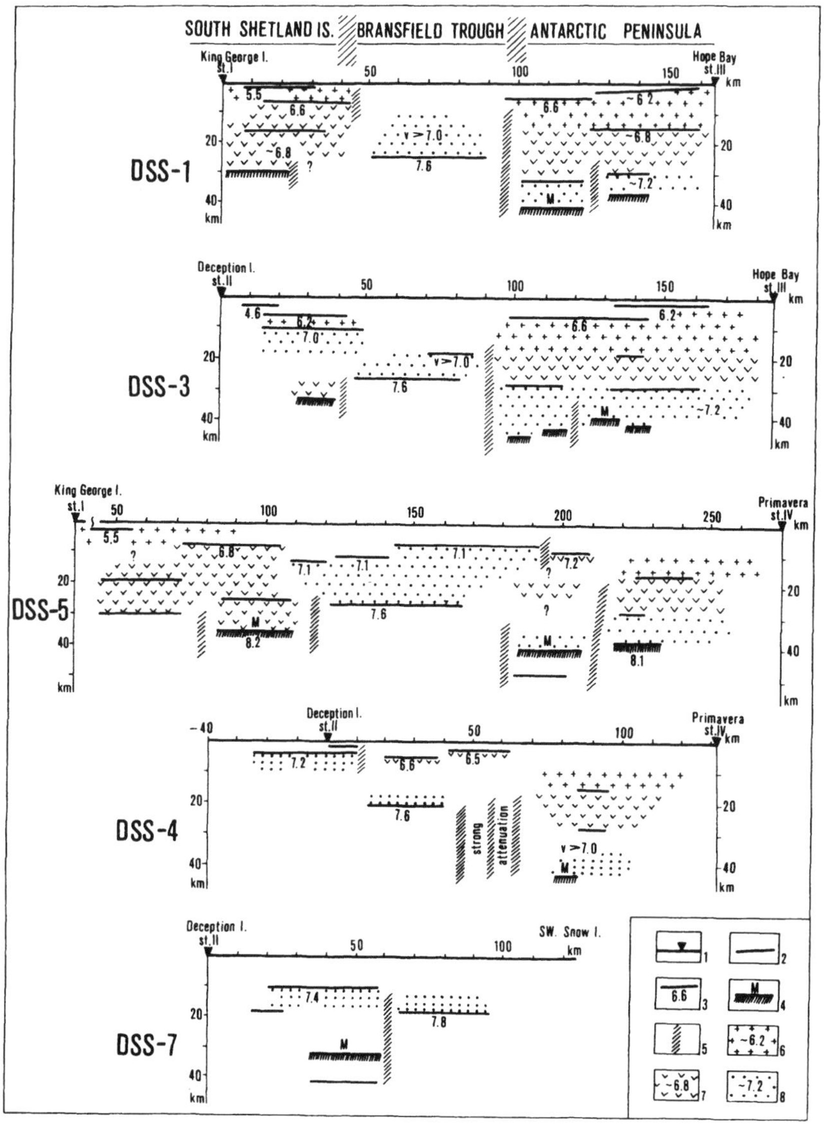

Fig. 7. Generalized crustal sections of the transition between the Antarctic Peninsula and the South Shetland Islands. 1, position of refraction stations; 2, reflection discontinuities; 3, refraction discontinuities and boundary velocities in km/s; 4, Moho discontinuity; 5, contact zones of the crustal blocks (deep fractures?) and zones of strong attenuation of seismic waves; 6, upper part of the crust and approximate layer velocities; 7, lower part of the crust and approximate layer velocities; 8, high velocity crustal layer.

boundaries. Such contact zones occur mainly in the lower part of the crust.

Seismic boundaries with anomalous seismic wave velocities were found in Bransfield trough. Seismic boundaries with velocities of 7.0–7.2 km/s occur at depths as shallow as about 10–15 km, whereas boundaries with velocities of about 7.6 km/s are present at depths of 20–25 km. A special case occurs near Deception Island, where a seismic boundary was found with velocities of 7.4 km/s at a depth of only 8 km, and of 7.8 km/s at a depth of about 18 km. In general, the Moho discontinuity in the South Shetland Islands study area occurs at depths of 30–33 km, and along the western margin of the Antarctic Peninsula it is present at depths of about 38–45 km. The seismic wave field in the crust of the Antarctic Peninsula is characterized by the presence of numerous strong reflected waves which are related mainly to discontinuities in the lower crust. The most effective seismic-reflection boundaries above the Moho discontinuity usually occur at depths of 25–35 km. In the Antarctic Peninsula area, distinct S-waves (both refracted and reflected) were also recorded. By contrast the seismic wave field in the South Shetland Islands area is poor, with no strong reflected waves and no S-waves; seismic waves are strongly attentuated in the area of the crustal blocks here. In general the crustal structure of the South Shetland Islands is quite distinct from that of the Antarctic Peninsula.

It is difficult to carry out 2D modelling of Bransfield trough because, with the complex tectonic structure of this region and the large distances between the seismic recording stations (150–180 km), the seismic data obtained are relatively poor. An example of 2D crustal modelling (Fig. 8) shows a record section for station I on profile DSS-1, a synthetic section and a crustal model of Bransfield trough.

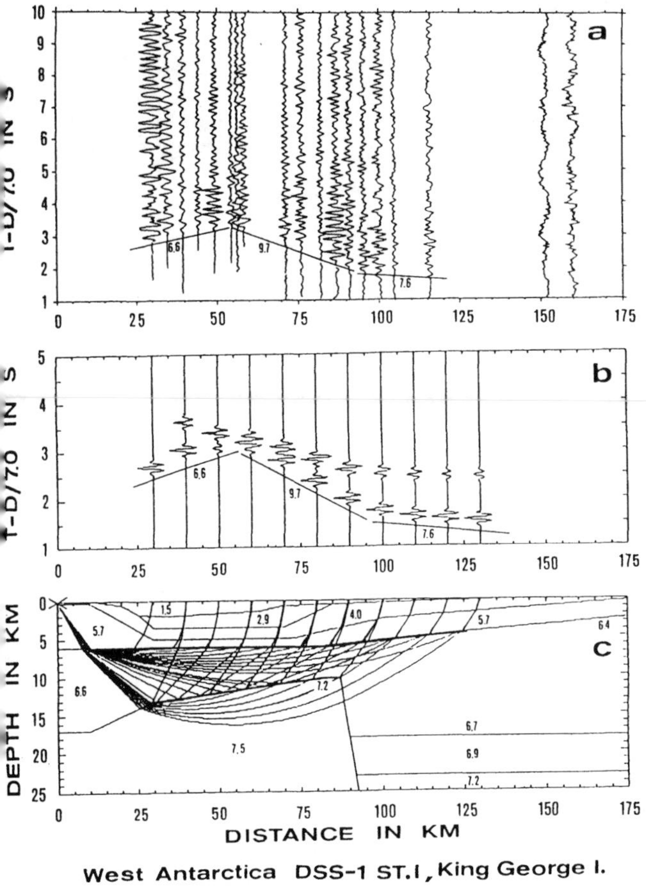

Fig. 8. Record section (a), synthetic record section (b) and two-dimensional crustal model (c) for DSS-1 profile. Velocity of seismic waves in km/s.

Contact zones between crustal blocks (Fig. 7) made it possible to determine the limits of Bransfield trough, which separates the Shetland block from that of the Antarctic Peninsula. The position of Bransfield trough, as determined in the lower part of the crust, lies to the east of that located by bathymetric and seismic-reflection measurements (Fig. 9). The position of submarine volcanoes found on the seismic-reflection sections in Bransfield Strait are also shown. These volcanic cones occur west of the zone of deep contact between crustal blocks in Bransfield Strait, along the line of the volcanoes of Deception, Penguin and Bridgeman islands.

The geotectonic situation described is illustrated by the general model for the central part of the transect, shown in Fig. 10. Seismic-reflection profiles GUN-6 and GUN-12 were used (Fig. 1) to construct this transect. On profile GUN-6, we can see the outline of a volcanic cone in Bransfield Strait (cf. Guterch *et al.*, 1985). The transect clearly shows a shift in the outline of Bransfield trough marked on seismic-reflection profiles and bathymetry with respect to the deep refraction part of the section.

Discussion

In the light of the data presented, the thickness of the crust in this part of West Antarctica is calculated to be 30–47 km. These values are considerably higher than those of 25–30 km generally accepted until now (Ashcroft, 1972; Dalziel & Elliot, 1982; Barker & Dalziel, 1983). According to Ashcroft (1972) the Bransfield Strait trough resembles that of an oceanic ridge crest with a 7.6–7.7 km/s boundary at about

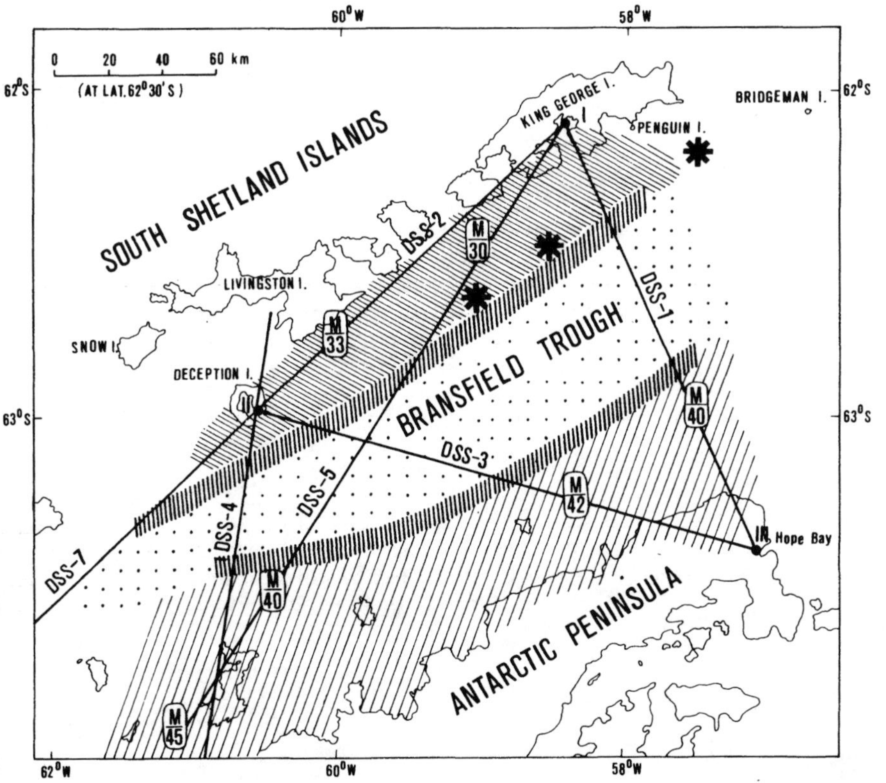

Fig. 9. Deep tectonic scheme of the transition between the Antarctic Peninsula and the South Shetland Islands. Asterisks, submarine volcanic cones in Bransfield Strait located on seismic-reflection section; M/40, etc., depth of the Moho discontinuity in km.

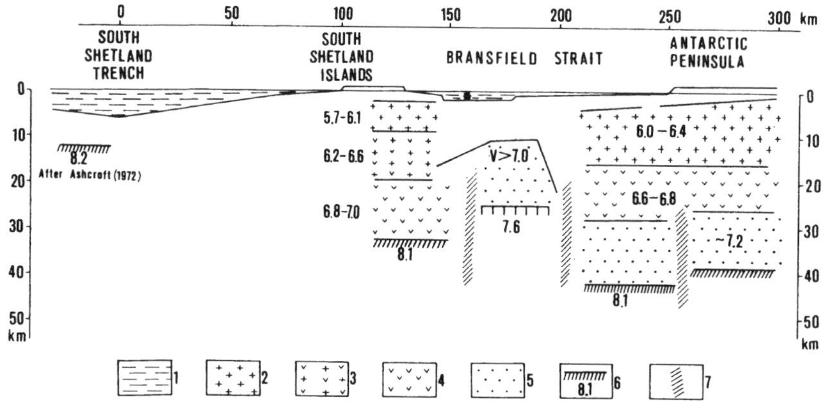

Fig. 10. General crustal model of the transition between the Antarctic Peninsula and the South Shetland trench, based on reflection profiles GUN-6 and GUN-12 (see Fig. 1). 1, sea; 2, 3, 4, upper, middle, lower part of the crust, and layer velocity in km/s; 5, high velocity crustal layer; 6, Moho discontinuity and boundary velocity in km/s; 7 contact zones of the crustal blocks.

13–16 km below sea level. Davey (1972) used marine gravity measurements made in this area to prove the existence of a semi-oceanic crust beneath Bransfield Strait.

The greater thicknesses of the Earth's crust obtained in this experiment are confirmed by interpretation of refracted and reflected waves. Lengths of travel times are 130–210 km. In the Antarctic Peninsula region refracted waves with first arrival velocities of about 8.0 km/s were not observed within distances of 180–200 km. Strong waves reflected from the Moho discontinuity were recorded initially at distances of 80–100 km. In this case seismic modelling of the wave field gives explicitly high thicknesses of the crust (30–45 km). Much shorter travel times (40–100 km) presented by Ashcroft (1972) allowed the structure of the upper crust to be determined. Our much longer travel times (130–210 km) provide data for us to determine the structure of the lower crust also.

Conclusions

The crustal structure beneath the tectonic trough of Bransfield Strait is highly anomalous. P-wave velocities of 7.0–7.2 km/s were found at depths of 10–15 km below the Bransfield trough; at depths greater than this, an increase in velocity was observed (see 2D model, Fig. 8). Moreover, in the northern part of Bransfield Strait the 7.6 km/s velocity was found at a depth of about 25 km, whereas in the southern part the same velocity was found at a depth of about 20 km. This boundary becomes distinctly shallower in the area of Deception Island. It seems that the P-wave velocity close to 7.6 km/s can be interpreted as corresponding to the top of an anomalous upper mantle below Bransfield trough. Detailed modelling of the seismic wave field shows that in the crust in Bransfield Strait there are zones with normal seismic velocities of 6.3–6.8 km/s, and zones of high velocities (7.2–7.8 km/s). Thus, only 2D modelling of the seismic waves gives correct results.

The position of Bransfield trough in the lower crust (Fig. 10) lies to the east of that located by bathymetric and seismic-reflection measurements.

References

Alekseev, A.S. & Gelchinski, B. Ya. (1959). The ray method of wave field calculation for inhomogeneous media with curvilinear boundaries. In *Voprosy Dinamicheskoy Teorii Rasprostronieniya Seismicheskikh Voln*, ed. G.I. Petrashen, vol. III, pp. 107–60. Leningrad; Nauka (in Russian).

Ashcroft, W.A. (1972). *Crustal Structure of the South Shetland Islands and Bransfield Strait*. London; British Antarctic Survey Scientific Reports, No. 66, 43 pp.

Barker, P.F. & Dalziel, I.W.D. (1983). Progress in geodynamics in the Scotia Arc region. In *Geodynamics of the Eastern Pacific Region, Caribbean and Scotia Arcs*, ed. R. Cabré, Geodynamics Series 9, pp. 137–70. Washington, DC; American Geophysical Union.

Červený, V. & Pšenčík, I. (1983). *Ray Method in Seismology*. Prague; Charles University Press.

Dalziel, I.W.D. & Elliot, D.H. (1982). West Antarctica: problem child of Gondwanaland. *Tectonics*, **1(1)**, 3–19.

Davey, F.J. (1972). Marine gravity measurements in Bransfield Strait and adjacent areas. In *Antarctic Geology and Geophysics*, ed. R.J. Adie, pp. 39–46. Oslo; Universitetsforlaget.

Fuchs, K. & Müller, G. (1971). Computation of synthetic seismograms with the reflectivity method and comparison with observations. *Geophysical Journal of the Royal Astronomical Society*, **23**, 417–33.

Grad, M. (1983). Determination of mean velocities and depths of boundaries in the Earth's crust from reflected waves. *Acta Geophysica Polonica*, **31(3)**, 231–41.

Guterch, A., Grad, M., Janik, T., Perchuć, E. & Pajchel, J. (1985). Seismic studies of the crustal structure in West Antarctica 1979–1980 – preliminary results. In *Geophysics of the Polar Regions*, ed. E.S. Husebye, G.L. Johnson & Y. Kristoffersen, *Tectonophysics*, **114(1–4)**, 411–29.

The Bransfield rift and its active volcanism

O. GONZÁLEZ-FERRÁN

Universidad de Chile, Departamento de Geología y Geofísica, Casilla 27116, Santiago 27, Chile

Abstract

During Plio-Pleistocene to Recent times, extensional processes were dominant in Bransfield Strait; these resulted in the formation of a series of NE-trending normal faults, which are subparallel to the north-western margin of the Antarctic Peninsula and the opposite coast of the South Shetland Islands. A qualitative examination of the aeromagnetic map of Bransfield Strait reveals a well defined pattern of magnetic anomalies which is closely related to that of the normal faulting. These magnetic anomalies are interpreted as representing basic intrusions into the continental crust. The axis of the rift is defined by an offset spreading centre, with which are associated the volcanoes of Deception, Penguin and Bridgeman islands, and a number of submarine volcanoes. The rift has opened between 5 and 15 km since its inception at the end of the Pliocene. Thus, the spreading rate can be estimated at 0.25–0.75 cm/y.

Introduction

This paper presents the results of aeromagnetic surveys conducted during the Antarctic summers of 1983 and 1985, over the South Shetland Islands, Bransfield Strait and the northern part of the Antarctic Peninsula (Fig. 1). The surveys were undertaken to promote a better understanding of the geological evolution of this active tectonic area and to provide the basis for a seismic network to forecast future volcanic eruptions.

Bransfield rift

The existence of a rift along Bransfield Strait has been widely accepted by various authors on the basis of both geophysical and geological data (Barker, 1976; Birkenmajer *et al.*, 1981; Thomson, Pankhurst & Clarkson, 1983; Parra, González-Ferrán & Bannister, 1984; González-Ferrán, 1985; Renner, Sturgeon & Garrett, 1985). The 1983 and 1985 aeromagnetic regional surveys permit a more precise identification of the rift and show its relationship to active volcanism in the area (Fig. 1).

Aeromagnetic data

Aeromagnetic surveys over Bransfield Strait revealed three major positive magnetic anomalies, with a widespread magnetic low representing a thick package of rocks with low magnetic susceptibility, i.e. sedimentary rocks. All these anomalies are nearly parallel to the N60°E trend of Bransfield Strait (Fig. 1). Furthermore, there are a number of magnetic lineaments corresponding to a system of normal NE-trending faults, and others with a NW strike. The latter could be related to, or represent traces of a transform fault parallel to the Hero and Shackleton fracture zones (Parra *et al.*, 1984).

The rift aeromagnetic anomaly

The aeromagnetic map (Fig. 1) shows a series of linear en échelon magnetic anomalies which are broadly parallel to the main structural trend of Bransfield Strait. This feature is asymmetrically located near the South Shetland Islands block, and is referred to as the 'Northwest Rift Zone' (Fig. 2). Two more groups of positive anomalies, probably corresponding to thin rift segments, have been detected near the Antarctic Peninsula block (Fig. 2) and have been named the 'Southeast Rift Zone' and 'Southwest Rift Zone'. The southern rift zones are separated from each other by a distance of about 140 km, and their trend is parallel to the main 'Northwest Rift Zone'.

These anomalies are thought to represent high magnetic susceptibility materials, e.g. basic hypabyssal and/or volcanic rocks covered by more than 1000 m of water, rather than magnetic sediments. They represent a zone of basic dyke injection into continental crust, the dykes also acting as feeders to the volcanic centres; thus they prove the existence of a rift along Bransfield Strait. The model (Fig. 3) suggests that in the first stage of the rift, instead of a typical oceanic crustal layer, there was a mixture of successive injections of basic material into marine sedimentary rocks (Mesozoic, Cenozoic and Recent in age), probably with a total thickness of 5–10 km.

Quantitative interpretations indicate that the magnetic susceptibility of basic material in the Bransfield rift is not as high as that of oceanic crust. This could be due partly to its chemical composition, possibly a transitional type between material related to the lithospheric continental block and asthenospheric material.

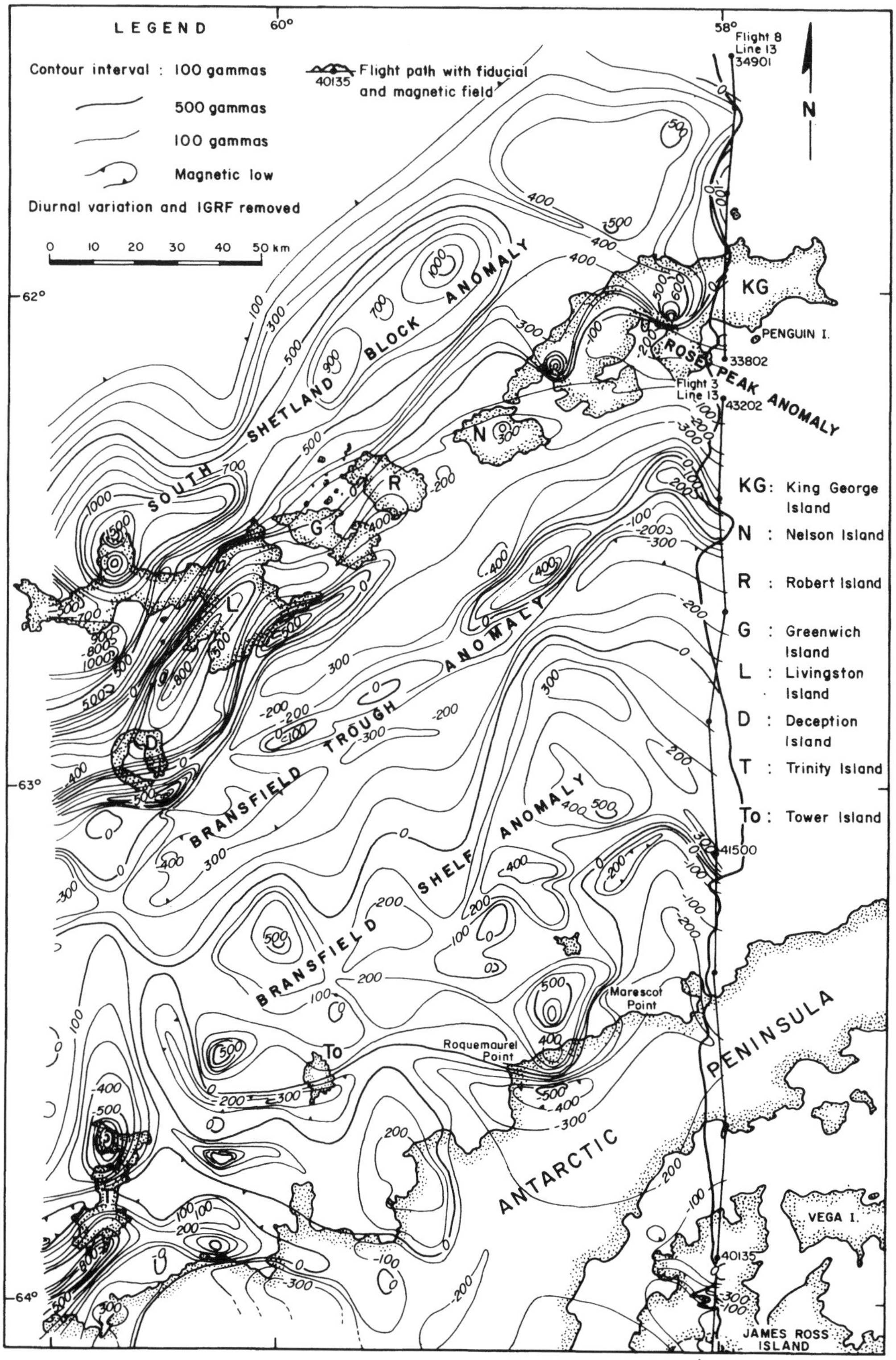

Fig. 1. Total field aeromagnetic anomaly map of the Bransfield Strait area.

Quantitative interpretation

According to Parra *et al.* (1984), a quantitative interpretation across Bransfield Strait was made to determine the depth of magnetic source, shape and magnetic susceptibility of the rock bodies creating the observed anomalies. Two-dimensional models were made using automatic interpretation with finite extension in the direction of strike. The computer programme for the model-fitting solution was based on Shuey & Pascuale (1973). In this method a theoretical anomaly is calculated for a specified two-dimensional polygon in a direction perpendicular to the strike (Bott, 1960, 1963; Heirtzler, Peter & Talwani, 1962; Bhattacharyya, 1966; Bhattacharyya & Navolio, 1975).

A good fit was obtained between the observed and computed magnetic anomalies (Fig. 3). An average susceptibility of 6000×10^{-6} cgs unit/cm^3 was estimated for the Andean Intru-

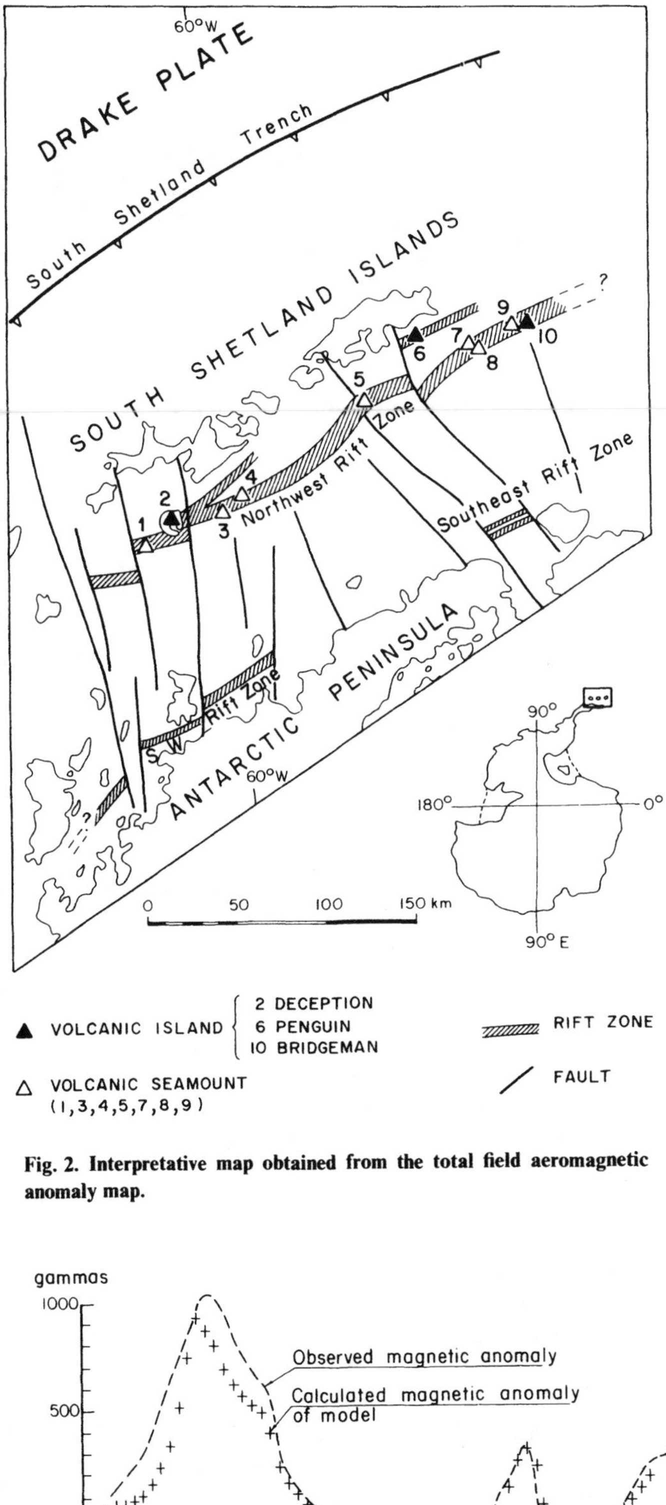

Fig. 2. Interpretative map obtained from the total field aeromagnetic anomaly map.

sive Suite and relative vertical fault displacements of about 7000 m can be inferred from the model. The model also shows a thick layer of sediments in Bransfield trough and a faulted contact between rocks of the Trinity Peninsula Group, and the Andean Intrusive Suite in the Antarctic Peninsula. The central positive anomaly of Bransfield Strait was modelled by a high susceptibility body representing the rift zone.

Evolution, age and velocity of the rift

González-Ferrán (1985) noted that in some ways the evolution of Bransfield Strait, during its initial phase, involved similar tectonic processes to those affecting the South American western margin of the Andes during late Miocene–Pliocene times. For example, the initial cycle of back-arc extension was characterized by longitudinal faulting associated with vertical movements. This favoured the development of differential block tectonics, thus causing the formation of a graben or central longitudinal valley. In Bransfield Strait, this graben would have been about 100 km wide (Ashcroft, 1972). Its development was similar to, and nearly contemporaneous with, the formation of the blocks of the Coastal Range, Central Valley and the Andean Range of Chile.

In the Bransfield Strait region en échelon block movements, probably controlled by fault systems parallel to the Hero and Shackleton fracture zones, contributed to the main morphological features of the South Shetland Islands and the Antarctic Peninsula. Late-stage block uplift would probably be due to eustatic movements generated by the loss of the ice cover, as shown by marine terraces and raised beaches.

The continuity of crustal extension, the thin continental crust and the proximity of asthenospheric material which, according to Ashcroft (1972), would be at a depth of only 14 km below the axis of the Bransfield trough, would have generated a weak zone which allowed the rupture and inception of volcanic activity along the rift. As has already been pointed out, this activity developed en échelon segments, asymmetric to the axis of Bransfield Strait (Fig. 2).

According to aeromagnetic data (Fig. 3), the estimated

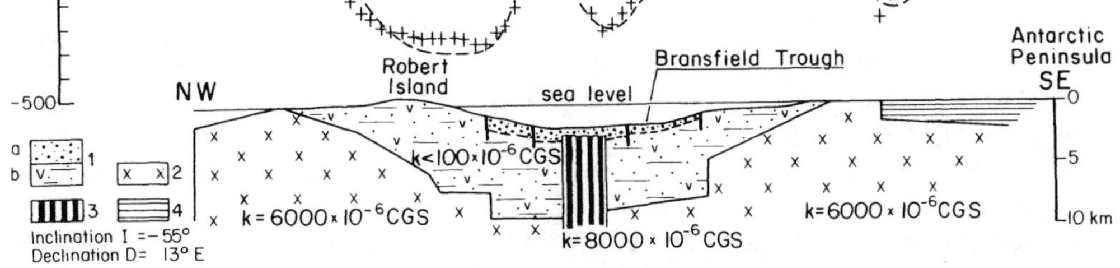

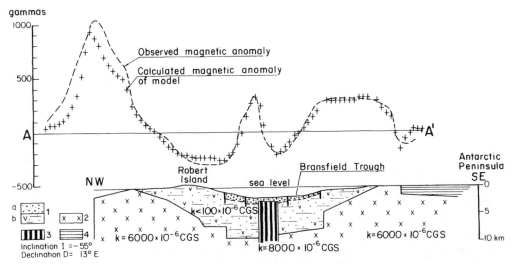

Fig. 3. Interpretative model along a line across Bransfield Strait, between Robert Island and the Antarctic Peninsula. 1, Sedimentary and volcanic rocks: (*a*) Late Cenozoic–Recent; (*b*) Mesozoic–Cenozoic; 2, Andean Intrusive Suite; 3, Zone of basic intrusive and volcanic rocks; 4, Sedimentary rocks (mainly belonging to the Trinity Peninsula Group).

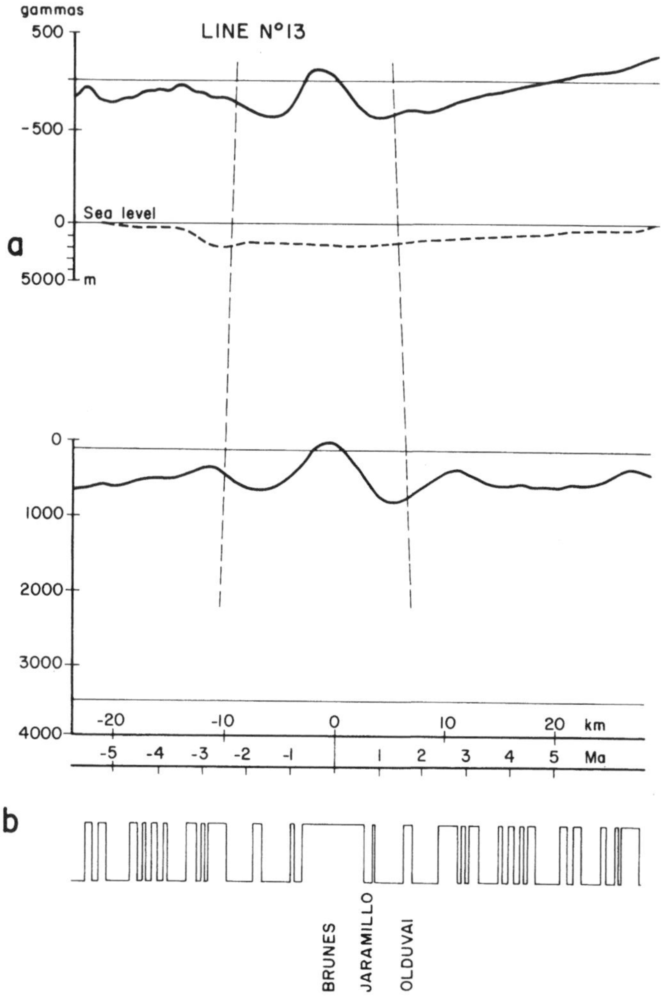

Fig. 4. Anomaly profile for flight-line 13. (*a*) Residual magnetic profile obtained after IGRF and diurnal correction. (*b*) Theoretical anomaly corresponding to an ideal spreading centre.

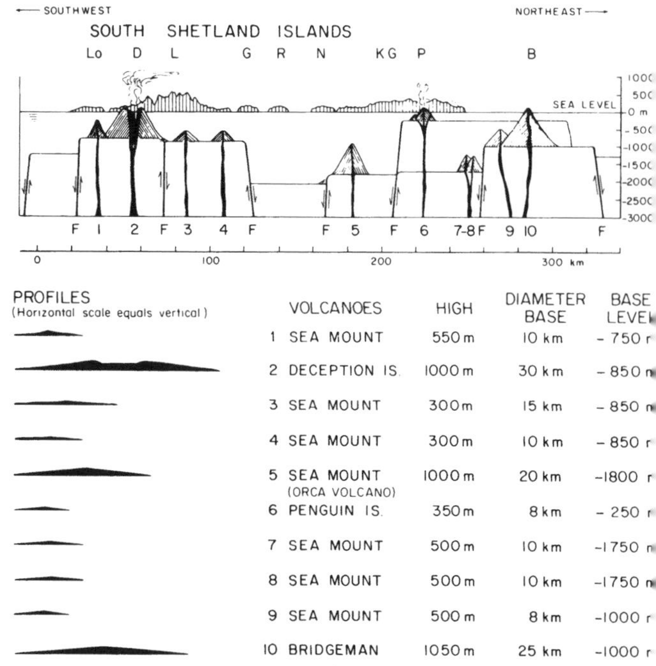

Fig. 5. The submarine volcanic ridge along the Bransfield rift. B, Bridgeman Island; D, Deception Island; G, Greenwich Island; KG, King George Island; L, Livingston Island; LO, Low Island; N, Nelson Island; P, Penguin Island; R, Robert Island.

width of the rift opening is 5–15 km; this value only represents new igneous material injected during rifting. It is substantially narrower than previous estimates which considered that the South Shetland Islands had been displaced to the north-west by about 65 km (Thomson *et al.*, 1983). The latter figure probably represents the total width of the initial Bransfield Strait graben.

The residual magnetic profile for flight-line 13 in Bransfield Strait, after IGRF and diurnal corrections (Figs 3 & 4) has the typical form of a spreading centre during the initial stages of rifting (J.C. Parra, pers. comm.). A theoretical model simulating a spreading centre is shown in Fig. 4*b*. This anomaly corresponds to an ideal spreading centre with the following characteristics: strike, N60°E; inclination of the present magnetic field, $-55°$; spreading rate, 0.40 cm/y; magnetization, 10000×10^{-6} e.m.u./cm^3; depth to the top of the magnetic layer, 3.5 km; thickness of the magnetized layer, 700 m. This model was obtained using a computer programme which referred to the revised magnetic polarity time scale for Late Cretaceous and Cenozoic time by LaBrecque, Kent & Cande (1977). Comparing Figs 4*a* & *b*, there is a close similarity in the central part of the anomalies, representing the last 2 m.y. (about 8 km either side of the spreading centre). This age is consistent with the one postulated by Weaver, Saunders & Tarney (1982), who considered that the opening of Bransfield Strait had occurred during the last 2 m.y..

If we accept this 2 Ma age, as well as the 5–15 km width of the rift and its continued activity up to the present (as demonstrated by volcanic eruptions and historically recorded earthquakes, e.g. 21 February 1971 (magnitude 7.1), we can postulate that the spreading rate of the rift would be 0.25–0.75 cm/y, as a mean, for the past 2 m.y.

Submarine volcanic ridge

A series of submarine volcanoes developed in association with the Bransfield rift, especially along the 'Northwest Rift Zone'. Some of these emerged above sea level to form Deception, Penguin and Bridgeman islands (Figs 2 & 5). Aeromagnetic and bathymetric data for Bransfield Strait confirm the presence of these volcanoes along the ridge (Figs 5 & 6); seven are submarine volcanoes. Their structures correspond to oceanic shield volcanoes (Fig. 5). Due to differential block faulting during the initial stage of rifting, the volcanoes were emplaced on a stepped surface (Fig. 5) and fed by hot spots within the system of injected dykes. At least three of these eruptive centres (Deception and Bridgeman islands and the submarine volcano 'Orca') reached significant dimensions, with an estimated lava volume of 235 km^3 each. It is postulated here that the petrographic and geochemical characteristics of these submarine volcanoes should be similar to the essentially alkaline basic lavas of Deception, Bridgeman and Penguin islands.

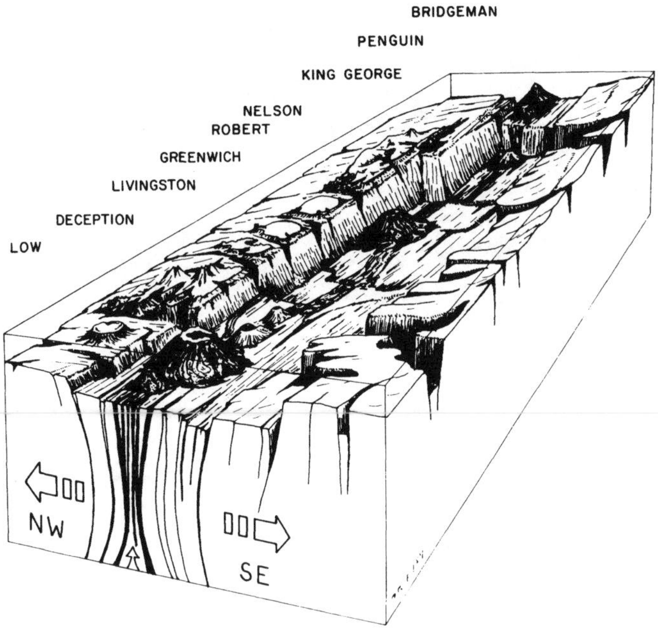

Fig. 6. Block diagram of the Bransfield rift and its volcanic ridge.

Active volcanism

Since the beginning of historical records, there have been frequent eruptions at Deception and Penguin islands. Nearby 'Orca Seamount', off the southern coast of King George Island, is the largest submarine volcano in the area. Deception, Penguin and Bridgeman islands represent the tops of stratovolcanics formed during Pleistocene–Recent times. They represent the only visible part of a large submarine ridge ('Bransfield ridge'), which trends parallel to the south-west coast of the South Shetland Islands. There are at least ten volcanoes along the ridge (Fig. 5), spanning a distance of 300 km from Bridgeman Island to the submarine volcano south-west of Deception Island (Fig. 6); this distance does not take account of overlap.

Considering that the opening of Bransfield Strait and magmatism associated with rifting occurred during the last 2 m.y. (Weaver *et al.*, 1982; González-Ferrán, 1985), and that historical activity has been recorded at Deception Island (Baker *et al.*, 1975) and possibly at Penguin Island (González-Ferrán & Katsui, 1970), it can be estimated that volcanic activity along Bransfield Strait is still in the process of development. New eruptions (both subaerial and submarine) could be recorded in the future at any of the volcanic centres shown on Fig. 5 or at other points associated with the rift.

Acknowledgements

This study was financed by Grant E2606-8712 of the Departamento de Investigación y Bibliotecas (DIB) de la Universidad de Chile, the Chilean Air Force and Corporacion de Fomento a la Produccion (CORFO). The Chilean Air Force supported the cost of the aeromagnetic operation and also gave safe and efficient logistic support. The author would like to express his sincere appreciation to those who assisted in the field, and to my colleague J.C. Parra (SERNAGEOMIN) for the aeromagnetic work during the Antarctic Project; Mrs. Oriana González re-drew all the figures.

References

Ashcroft, W.A. (1972). *Crustal Structure of the South Shetland Islands and Bransfield Strait.* London; British Antarctic Survey Scientific Reports, No. 66, 43 pp.

Baker, P.E., McReath, I., Harvey, M.R., Roobol, M.J. & Davies, T.G. (1975). *The Geology of the South Shetland Islands: V. Volcanic Evolution of Deception Island.* Cambridge; British Antarctic Survey Scientific Reports, No. 78, 81 pp.

Barker, P.F. (1976). The tectonic framework of Cenozoic volcanism in the Scotia Sea region: a review. In *Andean Antarctic Volcanology Problems*, ed. O. González-Ferrán, pp. 330–46. Rome; International Association of Volcanology and Chemistry of the Earth's Interior.

Bhattacharyya, B.K. (1966). Continuous spectrum of the total magnetic-field anomaly, due to a rectangular prismatic body. *Geophysics*, **31**, 97–121.

Bhattacharyya, B.K. & Navolio, M.E. (1975). Digital convolution for computing gravity and magnetic anomalies due to arbitrary bodies. *Geophysics*, **40**, 981–92.

Birkenmajer, K., Narebski, W., Skupinski, A. & Bakun-Czubarow, N. (1981). Geochemistry and origin of the Tertiary island-arc calc-alkaline volcanic suite at Admiralty Bay, King George Island, South Shetland Islands, Antarctica. *Studia Geologia Polonica*, **72**, 7–57.

Bott, M.H.P. (1960). The use of rapid digital computing methods for direct gravity interpretation of sedimentary basins. *Geophysical Journal of the Royal Astronomical Society*, **3(1)**, 63–7.

Bott, M.H.P. (1963). Two methods applicable to computers for evaluating magnetic anomalies due to finite three basins. *Geophysic Prospecting*, **11(3)**, 292–9.

González-Ferrán, O. (1985). Volcanic and tectonic evolution of the northern Antarctic Peninsula–Late Cenozoic to Recent. *Tectonophysics*, **114**, 389–409.

González-Ferrán, O. & Katsui, Y. (1970). Estudio integral del volcanismo cenozoico superior de las islas Shetland del Sur, Antártica. *Instituto Antártico Chileno, Serie Científica*, **1(2)**, 123–74.

Heirtzler, J.R., Peter, G. & Talwani, M. (1962). *Magnetic Anomalies Caused by Two-Dimensional Structures: Interpretation.* Lamond–Doherty Geological Observatory, Technical Report No. 6, 123 pp.

LaBrecque, J.L., Kent, D.V. & Cande, S.C. (1977). Revised magnetic polarity time scale for Late Cretaceous and Cenozoic time. *Geology*, **5**, 330–5.

Parra, J.C., González-Ferrán, O. & Bannister, J. (1984). Aeromagnetic Survey over the South Shetland Islands, Bransfield Strait and part of the Antarctic Peninsula. *Revista Geológica de Chile*, **23**, 3–20.

Renner, R.G.B., Sturgeon, L.J.S. & Garrett, S.W. (1985). *Reconnaissance Gravity and Aeromagnetic Surveys of the Antarctic Peninsula.* Cambridge; British Antarctic Survey Scientific Reports, No. 110, 50 pp.

Shuey, R.T. & Pascuale, A.S. (1973). End corrections in magnetic profile interpretation. *Geophysics*, **38(8)**, 570.

Thomson, M.R.A., Pankhurst, R.J. & Clarkson, P.D. (1983). The Antarctic Peninsula–a Late Mesozoic–Cenozoic arc (review). In *Antarctic Earth Science*, eds. R.L. Oliver, P.R. James & J.B. Jago, pp. 289–97. Canberra; Australian Academy of Science and Cambridge; Cambridge University Press.

Weaver, S.D., Saunders, A.D. & Tarney, J. (1982). Mesozoic–Cenozoic volcanism in the South Shetland Islands and the Antarctic Peninsula: geochemical nature and plate tectonic significance. In *Antarctic Geoscience*, ed. C. Craddock, pp. 263–73. Madison; University of Wisconsin Press.

^{210}Pb activities in Bransfield Strait sediments, Antarctic Peninsula: additional proof of hydrothermal activity

J.W.A. VAN ENST

Ministry of Economic Affairs, Directorate General for Industry and Regional Policy, Bezuidenhoutseweg 20, PO Box 20101, 2500 EC The Hague, The Netherlands

Abstract

^{210}Pb in marine sediments may have two sources: atmospheric ^{210}Pb as the result of ^{222}Rn decay (excess $^{210}Pb_x$) and from within the sediment as the result of ^{226}Ra decay (supported $^{210}Pb_s$). Activity profiles of ^{210}Pb and ^{226}Ra, measured on samples of core 1327-1 (sampled during ANT IV/2, of FS *Polarstern*), suggest that the major ^{210}Pb source for Bransfield Strait basin sediments is from within the sediment (supported $^{210}Pb_s$). This is inferred from average constant activity down to 2 m, and strong coherence between the profiles. The ^{210}Pb source itself can be attributed to hydrothermal fluids within the sediments which establish: a constant supply of ^{238}U-series isotopes and secular equilibria between the daughter isotopes. Furthermore, a relation between grain-size and ^{210}Pb activity exists, based on a comparison between the grain-size distribution curve and the activity profiles.

Introduction

^{210}Pb is a member of the ^{238}U-decay series (Fig. 1) with a half-life of 22.3 y. Usually ^{210}Pb is used to date sediments up to 150–200 y old (van der Wijk & Mook, 1987) and in sedimentation rate calculations (DeMaster, 1982). ^{210}Pb dating is based on the assumption that ^{222}Rn, an inert gas, escapes from the Earth's crust at an average rate of 7.0×10^3–7.5×10^3 atoms/(M^2 s) (van der Wijk & Mook, 1987). ^{222}Rn decays through a series of short-lived daughter isotopes to ^{210}Pb, which can enter the Earth's lithosphere or hydrosphere. Apart from atmospheric production (excess $^{210}Pb_x$), ^{210}Pb is also produced within the sediment by means of the decay of ^{226}Ra (supported $^{210}Pb_s$).

In sediment dating, the total ^{210}Pb ($^{210}Pb_t$) activity is measured and the excess activity ($^{210}Pb_x$) calculated by subtracting the supported activity ($^{210}Pb_s$) at each depth (z).

$$^{210}Pb_x = {}^{210}Pb_t - {}^{210}Pb_s \qquad (1)$$

Normal ^{210}Pb profiles show from top to bottom: (1) a surface mixed layer (5–15 cm thick) with constant ^{210}Pb activities as the result of particle mixing (Nittrouer *et al.*, 1979; DeMaster, 1982); (2) a zone of radioactive decay in which the ^{210}Pb activity decreases with depth as the result of the decay of ^{210}Pb; and (3) a zone of background level which shows that contribution of supported $^{210}Pb_s$ to the $^{210}Pb_t$ activity. In this zone a secular equilibrium exists between $^{210}Pb_s$ and ^{226}Ra (van der Wijk & Mook, 1987). ^{226}Ra profiles show a constant activity with increasing depth. It should be noticed here that ^{226}Ra and $^{210}Pb_t$ activities are very low compared to the $^{210}Pb_t$ activity at the top, where the contribution of excess $^{210}Pb_x$ is larger.

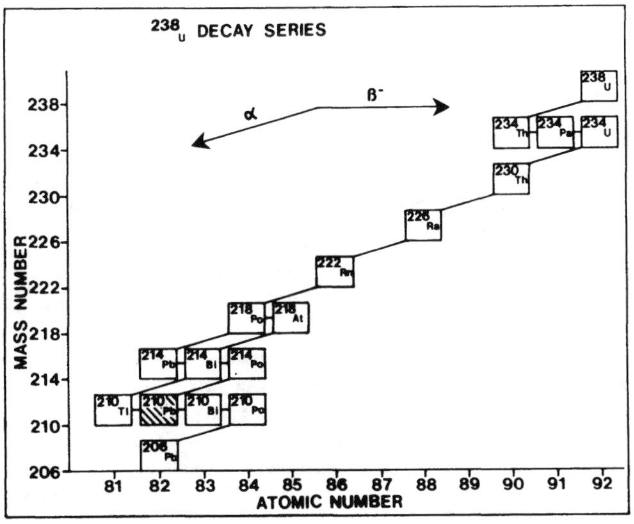

Fig. 1. ^{238}**U decay series.**

Geological and seismic data from the Bransfield Strait basin (Fig. 2*a*) show that the basin formed as the result of extensional and related strike-slip processes and associated rifting (Weaver *et al.*, 1979; Jeffers *et al.*, this volume, p. 481). It is suggested that hydrothermal fluids associated with volcanism can be active within the sediments of the Bransfield Strait basin.

^{210}Pb and ^{226}Ra activity profiles show that hydrothermal fluids are active within the basin sediment and that ^{210}Pb dating here may be an ambiguous approach because of these hydrothermal activities.

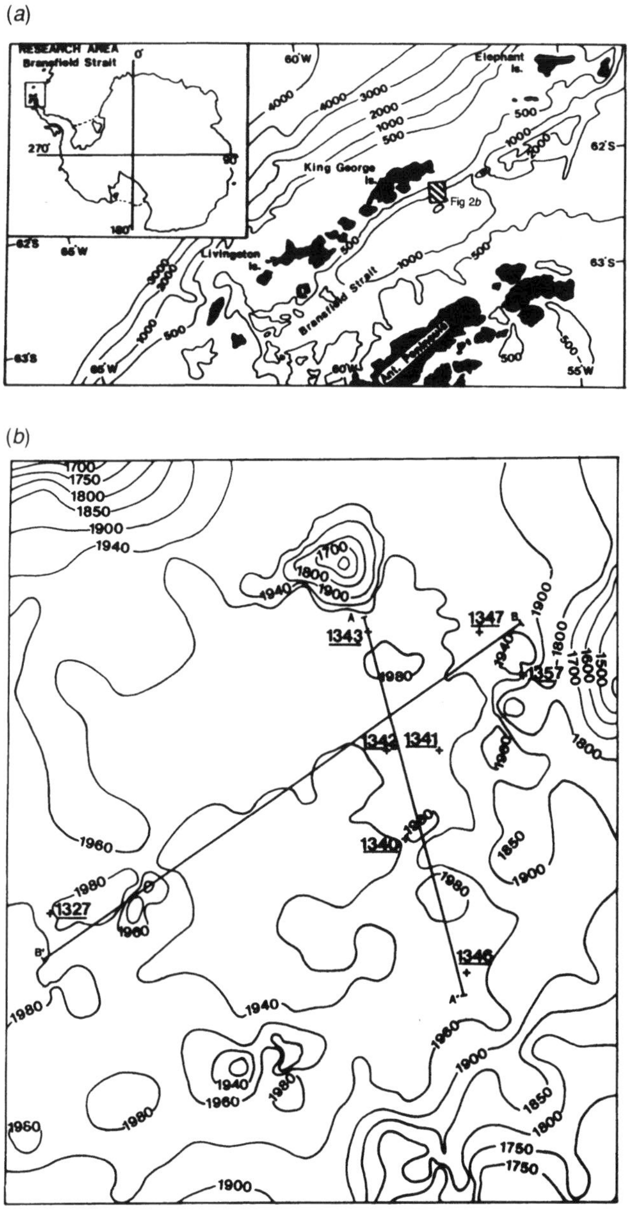

Fig. 2. (*a*) Sketch map of Bransfield Strait basin showing the location of the research area (Fig. 2*b*). Depth contours in metres. (*b*) Detail of the Bransfield Strait King George basin, showing the positions of the cores taken during ANT IV/2 (1985, FS *Polarstern*, AWI, Bremerhaven). Bathymetric detail from compiled Seabeam results obtained during ANT IV/2 and presented by E. Suess, Corvallis, Oregon State University, USA.

Sampling and analyses

In 1985 the Geological Survey of the Netherlands (RGD), Marine Geology division was invited to join Antarktis IV/2 aboard FS *Polarstern*. ANT IV/2 was organized by the Alfred Wegener Institute for Polar and Marine Research, Bremerhaven, FRG. During this cruise cores were taken in the Bransfield Strait (King George basin, Fig. 2*b*). Subsamples taken from core 1327-1 were analysed for ^{210}Pb and ^{226}Ra activities by Drs A. van der Wijk and R.J. Fransens at the Laboratory for Isotope Physics, Groningen State University, Groningen, The Netherlands. Granulometric analyses were carried out by the RGD on the same core. The samples were split into four grain-size classes (< 2 μm; 2 μm–16 μm; 16 μm–35 μm; > 35 μm) using a combined method of dry and wet sieving and the behaviour of particles to Stokes Law.

Core description

In Fig. 3 the core description of the upper 2 m of core 1327-1 is presented. This core (total length 7.73 m) shows a succession of pelagic and hemipelagic deposits with thin intercalated turbidites (Laban & de Groot, 1986). The pelagic clays locally change to silty clays with intense bioturbation and black mottling. In core 1327-1 the turbidites include several silt beds, although these are not very well defined. Granulometric analyses show that the clay–fine silt fraction (< 16 μm) is dominant in the upper 2 m of the core (Fig. 4). At 59–71 cm there is a sharp decrease in the 16–35 μm fraction, whereas there is an increase in the < 2 μm and 2–16 μm fractions. Sharp sedimentary contacts observed in the core at 62–73 cm are probably related to an intercalated turbidite.

Results and interpretations

^{210}Pb and ^{226}Ra activity profiles are presented in Fig. 5. These profiles show three important features.

(1) *Average constant activity down to 2 m depth.* Apart from the apparent $^{210}Pb_t$ activity maximum (Fig. 5), the $^{210}Pb_t$ activity stays rather constant down to 2 m depth. A measured ^{210}Pb activity profile can theoretically reflect two sources: (1) Atmospherically-produced $^{210}Pb_x$ as the dominating source (excess $^{210}Pb_x$); and (2) Supported $^{210}Pb_s$ produced within the sediment as the result of ^{226}Ra decay.

If excess $^{210}Pb_x$ is the major source, it follows from equation (1) that the $^{210}Pb_s$ must have a lower value than the total measured $^{210}Pb_t$ activity. Furthermore, a decrease in activity with depth, due to the constant decay of ^{210}Pb, should be found in the profile, unless some active source produces new ^{210}Pb isotopes.

In general, the supported $^{210}Pb_s$ activity is assumed to be in secular equilibrium with ^{226}Ra and therefore the measured activity of ^{226}Ra can be considered as equal to the $^{210}Pb_s$ activity in equation (1) (van der Wijk & Mook, 1987). However substantially higher $^{210}Pb_t$ values than ^{226}Ra (= $^{210}Pb_s$) and a decreasing ^{210}Pb activity depth relation are not apparent from the activity profiles in Fig. 5, and therefore it is suggested that atmospherically-produced excess $^{210}Pb_x$ can be excluded as the major ^{210}Pb source.

(2) *Strong correlation between ^{210}Pb and ^{226}Ra profiles.* In order to check the theory of supported $^{210}Pb_s$ as a major source, the ^{226}Ra activity was also measured. When comparing the two activity profiles the strong correlation and the larger ^{226}Ra activities (which can be considered as equal to the $^{210}Pb_s$ activity) down to 160 cm are obvious. At this stage it can be concluded that with respect to equation (1), the contribution of $^{210}Pb_x$ activities to the total measured $^{210}Pb_t$ is very low, and there is a constant supply mechanism of ^{210}Pb and ^{226}Ra isotopes. This constant supply results in a secular equilibrium between ^{226}Ra and ^{210}Pb, and a constant activity down to a depth of 2 m.

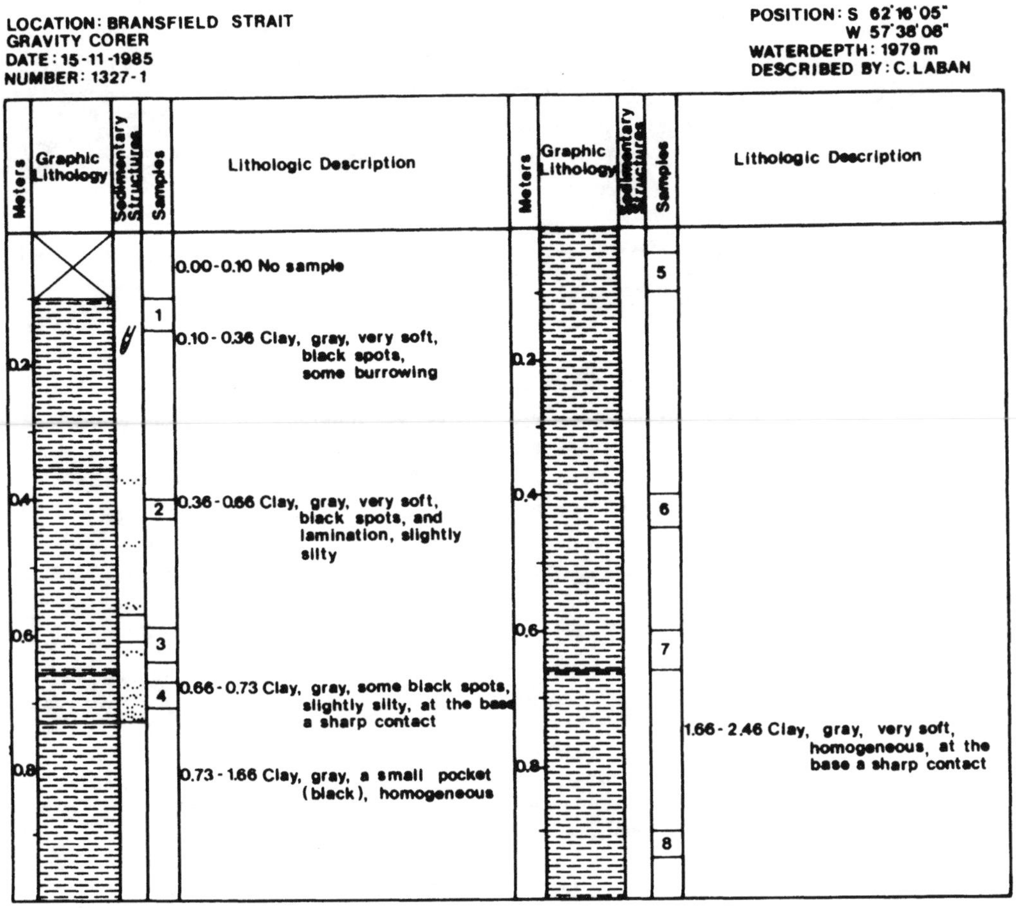

Fig. 3. Descriptions of core 1327-1 (total core length = 7.73 m).

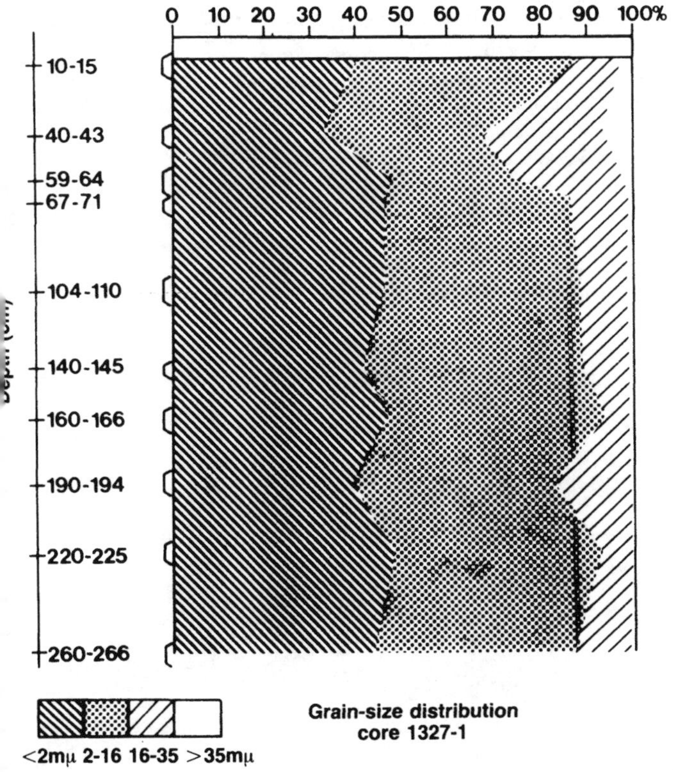

Fig. 4. Grain-size distribution in core 1327-1.

Thus secular equilibrium between ^{226}Ra and ^{210}Pb explains features (1) and (2) better than the assumption that excess atmospheric $^{210}Pb_x$ could be the major ^{210}Pb source. The reason for the dominance of supported $^{210}Pb_s$ over excess $^{210}Pb_x$ (atmospheric origin) may be the presence of hydrothermal fluids containing ^{238}U and daughter products. A constant supply of these isotopes creates equilibria between several daughter products of the ^{238}U-series, and also explains the observed constant activity–depth relation. Further evidence for the hydrothermal activity and hydrothermal alteration effects in Bransfield Strait basin sediments was given by the occurrence of thermogenic hydrocarbons (Whiticar, Suess & Wehner, 1985) and by near-surface thermal alteration of marine organic matter and slow advection of hydrothermal fluids within the sediment (Suess, 1986).

(3) *Strong activity increase at 59–64 cm.* An activity increase was detected in both the ^{210}Pb and ^{226}Ra profiles at 59–64 cm. This peak corresponds to the interval of siltier, perhaps turbiditic sediment (Fig. 4).

Conclusions

From the above results it may be concluded that the *in situ* production of ^{210}Pb (supported ^{210}Pb) in the Bransfield Strait basin sediments forms the dominant ^{210}Pb source. Furthermore, ^{238}U-series isotopes are advecting through the

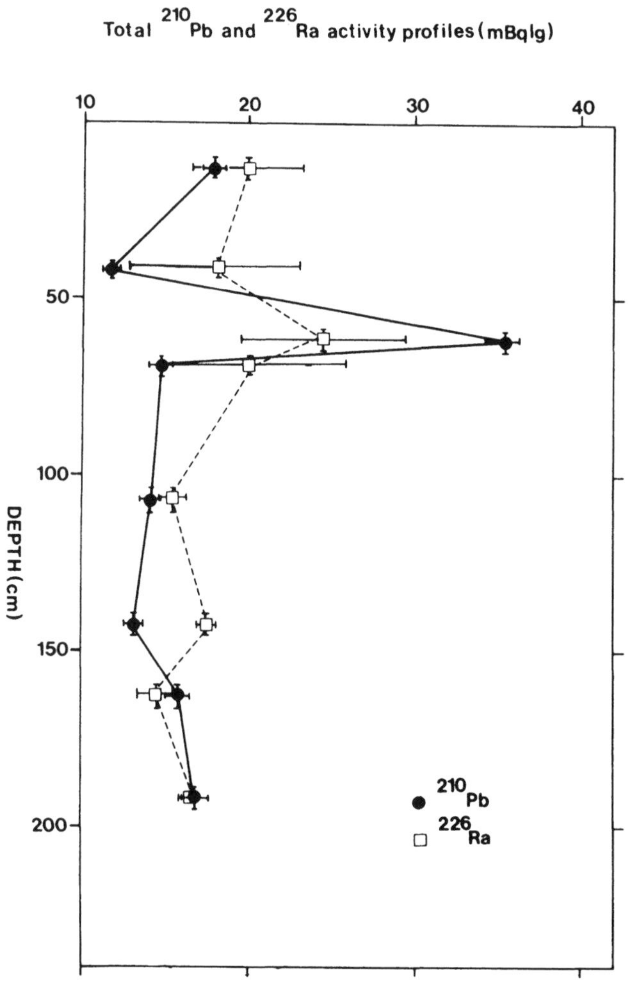

Fig. 5. ^{210}Pb and ^{226}Ra activity profiles in mBq/g from core 1327-1 (location shown in Fig. 2*b*).

sediment as part of the hydrothermal fluids. The ^{238}U isotopes maintain constant secular equilibria between each other, which is reflected by the average constant activity–depth relation for ^{210}Pb and ^{226}Ra and the coherence between the profiles.

There is a correlation between the strong activity increase at 59–64 cm and the increasing importance of smaller grain-sizes at the same depth (cf. Figs 4 & 5).

In view of hydrothermal activity reflected by the strong ^{210}Pb anomalies within the Bransfield Strait Basin sediments, ^{210}Pb dating is probably an unreliable method with respect to these sediments.

Acknowledgements

A. van der Wijk is thanked for the ^{210}Pb analyses and the details concerning ^{210}Pb dating. R.J. Fransens is acknowledged for the ^{226}Ra analyses. C. Laban made the shipboard descriptions of the sediment cores taken during ANT IV/2. Mrs S. Sendouk carried out the grain-size analyses. I am very much indebted to P. Frantsen who completed all the figures.

References

DeMaster, D.J. (1982). Rates of particle mixing in Antarctic deep-sea sediments using ^{210}Pb measurements. In *Antarctic Geoscience*, ed. C. Craddock, pp. 1039–44. Madison; University of Wisconsin Press.

Laban, C. & de Groot, Th. (1986). A preliminary interpretation of the gravity core. *Berichte zur Polarforschung*, **32**, 96–8.

Nittrouer, C.A., Sternberg, R.W., Carpenter, R. & Bennett, J.T. (1979). The use of Pb-210 geochronology as a sedimentological tool: application to the Washington continental shelf. *Marine Geology*, **31**, 297–316.

Suess, E. (1986). Thermal interaction between back-arc volcanism and basin sediments in the Bransfield Strait. *Berichte zur Polarforschung*, **32**, 98–101.

Van der Wijk, A. & Mook, W.G. (1987). ^{210}Pb dating in shallow moorland pools. *Geologie en Mijnbouw*, **66(1)**, 43–56.

Weaver, S.D., Saunders, A.D., Pankhurst, R.J. & Tarney, J. (1979). A geochemical study of magmatism associated with the initial stages of back-arc spreading: the Quaternary volcanics of Bransfield Strait, South Shetland Islands. *Contributions to Mineralogy and Petrology*, **68**, 151–69.

Whiticar, M.J., Suess, E. & Wehner, M. (1985). Thermogenic hydrocarbons in surface sediments of the Bransfield Strait, Antarctic Peninsula. *Nature, London*, **314(6006)**, 87–90.

Volcanism on Brabant Island, Antarctica

M.J. RINGE

Department of Geology, University of Nottingham, Nottingham, NG7 2RD, UK

Abstract

A brief description of the geology of Brabant Island is given. Cretaceous or older tuffs and agglomerates are predominant in the southern part of the island and these are overlain by a thick and lithologically variable buff-coloured volcaniclastic deposit. A thick lava and tuff pile which forms the main ridge on the west coast of the island represents continuing magmatism during the Tertiary. Ages of 3.07 ± 0.1 Ma (from an andesite lava) and 24.0 ± 0.7 Ma and ~ 10 Ma (for two intrusions) confirm this time span. Activity ended when subduction of the Aluk ridge ceased 4 Ma ago. There followed a period of extension in the Bransfield Strait area and it is envisaged that Recent volcanism on Brabant Island was related to this episode. A multi-phase dyke swarm in south-eastern Brabant Island is indicative of extensional tectonics related to the ridge–trench movements farther west, and subsequent rifting in Gerlache Strait. A 52 ± 2 Ma age suggests extension may have begun in the early Tertiary. Petrological and geochemical data are given for the Recent volcanic rocks.

Introduction

Brabant Island (64°S, 62°30′W) is a mountainous island situated off the west coast of the Antarctic Peninsula (Fig. 1). A major N–S ridge, reaching > 2200 m in height, extends the length of the west coast. Ice and snow cover most of the island and the few rocks exposed are commonly inaccessible. Fieldwork was undertaken by the author as a member of the Joint Services Expedition to Brabant Island, 1983–85 (Furse, 1986).

Previous geological research on the island has been limited, the only major investigation being by Alarcón *et al.* (1976). Subduction-related magmatism has persisted along the Antarctic Peninsula since Jurassic times (Thomson, Pankhurst & Clarkson, 1983). From early Tertiary time this magmatism ceased in a northwardly progressive pattern along the peninsula as the Aluk ridge was subducted to the west (Barker, 1982). In the latitude of Brabant and Anvers islands subduction ceased 4 Ma ago, before the Aluk ridge was subducted. During the last 2–3 m.y. a change to extensional tectonics took place in the northern Antarctic Peninsula region. This led to the opening of Bransfield Strait 1.3 Ma ago (Roach, 1979) and a concomitant change of magmatism, as shown by the volcanic rocks of Deception, Bridgeman and Penguin islands (Weaver *et al.*, 1979).

Geological setting

The geology of Brabant Island is dominated by volcaniclastic deposits which range in age from at least Cretaceous to Recent. Because of the limited and often inaccessible nature of the exposure it is, however, difficult to define geological boundaries with any degree of confidence.

Cretaceous or older purple tuffs and agglomerates are found in the south of Brabant Island, in the Mount Bulcke area (Fig. 1). The tuffs are fine grained, often welded and in places ignimbritic in character. The agglomerates show considerable variation in the angularity of the predominantly basaltic clasts, and locally indicate considerable reworking. A coarse pebble-conglomerate overlies and is interbedded with the purple tuffs at Mount Bulcke but it has not been found elsewhere on the island. The conglomerate clasts are well rounded, occasionally > 40 cm in size and consist predominantly of coarse granite, granodiorite and diorite, with local occurrences of gneiss. The southern section of Mount Bulcke is composed of subhorizontally bedded lavas and tuffs. Although inaccessible, it is thought they are of much younger age and that they correspond to the andesite lavas and tuffs of northern Anvers Island described by Hooper (1962).

East of Mount Bulcke and south of Buls Bay a grey–green volcanic agglomerate overlies the purple tuffs and agglomerates. Although angular to subangular basaltic and intermediate volcanic clasts are predominant, more exotic clasts (including schists) occur locally. In this area there are a large number of mainly basaltic cross-cutting dykes. These dykes, of which there are at least three phases, show a range of alteration including epidotization and silicification. The majority are aligned along the NNE–SSW trend of Gerlache and Bransfield straits, apparently indicating a period of extension. Sample 26249, which cross-cuts the agglomerate and is itself cross-cut by other dykes, gives a K–Ar age of 52 ± 2 Ma. This suggests early Tertiary extension and indicates a Cretaceous or older age for the agglomerate.

Overlying this agglomerate, possibly unconformably, is a

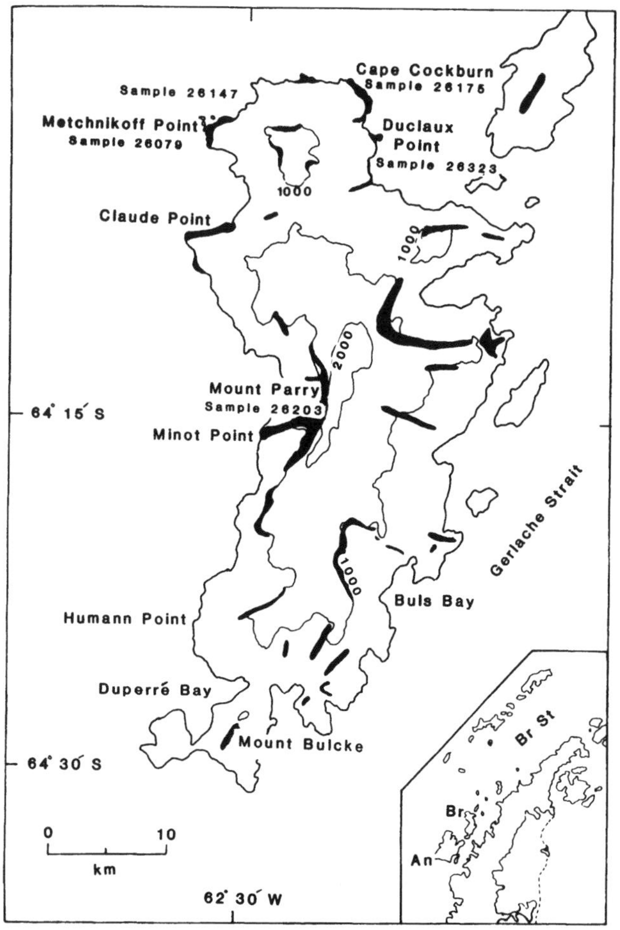

Fig. 1. Sketch map showing the distribution of volcanic rocks (black areas) on Brabant Island. Contours are at 1000 m intervals. Inset map of the Antarctic Peninsula: Br St, Bransfield Strait; Br, Brabant Island; An, Anvers Island.

thick buff-coloured deposit ranging in lithology from conglomeratic channel fill to tuffaceous beds; cross-bedding is common. This unit occurs north of the Buls Bay area and is well exposed in the central section of the island. Clasts are predominantly basaltic, sometimes vesicular and glassy and commonly set in a sandy matrix. The unit represents a period of erosion and deposition during which there was at least sporadic volcanic activity in the area.

The high ridge on the west coast of the island exhibits good but inaccessible exposures on precipitous sea-cliffs 1000 m high. The buff-coloured agglomerate seen in the lower sections is overlain by subhorizontal lavas and tuffs. An extremely fresh andesite (sample 26203) collected from near the top of this pile, at 2220 m on Mount Parry, has yielded a whole-rock K–Ar date of 3.07 ± 0.1 Ma, i.e. about 1 m.y. after subduction ceased to the west. It is probable that the lava pile represents continuing subduction during mid–late Tertiary time.

The northern section of the island is dominated by two phases of Recent volcanism. The initial extrusive phase (lower lava unit) of palagonitized lavas and tuffs indicates shallow submarine or subglacial emplacement. This unit is seen at Duclaux Point and along the northern coast, west of Metchnikoff Point. The lower unit was followed by subaerial basaltic lavas (upper lava unit) which have little or no associated volcaniclastic material. These flows, generally 3–5 m thick and horizontal, form the cap rocks of the northern coastline but are also present as far south as Buls Bay. One of the upper lava unit flows, sample 26079 from Metchnikoff Point, has yielded a whole-rock K–Ar age of < 100000 y.

Intrusive rocks form the coastal headlands, offshore islands and skerries. Although generally dioritic they show variation in composition. A diorite intrusion at Metchnikoff Point (sample 26045) has an eroded contact beneath the overlying tuffs and has given a biotite K–Ar age of 11.9 ± 0.5 Ma and a hornblende K–Ar age of 8.9 ± 0.4 Ma. These are amongst the youngest ages determined for intrusive rocks in the Antarctic Peninsula area (cf. Rex & Baker, 1973). A slightly more siliceous diorite from Minot Point (sample 26343) yielded a hornblende K–Ar age of 24.0 ± 0.7 Ma, indicating that magmatic activity was continuous throughout mid–late Tertiary times. Other intrusions are present farther south on the west coast (in Duperré Bay), on the east coast in Buls Bay, and to the east of Mount Bulcke.

Petrography of the Recent lavas

Lavas of the lower unit are all extremely fresh, commonly vesicular and basaltic in composition. In general they are fine-grained, porphyritic basalts with phenocrysts of forsteritic olivine and plagioclase set in a groundmass of flow-aligned labradorite laths, and crystals of olivine, euhedral magnetite and augitic clinopyroxene. The opaque oxides (mainly magnetite) occur both in the groundmass and as slightly larger euhedral crystals, and they are also common as inclusions in olivine phenocrysts. Extremely porphyritic basalts with large phenocrysts of strongly-zoned plagioclase and olivine also occur. Zeolite phases are present in some, but not all, of the vesicular lavas.

The upper lava unit flows in general have a coarser grain size than those of the lower unit. They are extremely fresh, often vesicular and porphyritic. Olivine and plagioclase phenocrysts are present in all these flows but commonly the dominant phenocryst phase is ophitic and subophitic pink titanaugite (as crystals up to 6 mm in length). Magnetite crystals are common, occurring both as euhedral inclusions in the olivine phenocrysts but also as smaller, anhedral crystals in the groundmass. The groundmass is composed of labradorite laths, olivine, clinopyroxene and opaque crystals, with haematite rims formed around magnetite. There is little evidence of secondary infilling of vesicles with zeolite.

Alarcón *et al.* (1976) noted iron-rich lava flows at the northern end of Brabant Island, and these have been described subsequently as magnetite lava flows analogous to those of northern Chile (Hawkes, 1982). Magnetite is present in all the youngest flows (see above) but none are geochemically iron-enriched. However, the lava flows 2 km east of Metchnikoff Point do show enrichment in modal magnetite and martite and they have a red colouration. Modal olivine in these rocks is low and there is evidence that the alteration of olivine has led to this increase in magnetite. Nevertheless, there is no evidence of rocks similar to the true magnetite lava flows of Chile.

Table 1. *Selected major and trace element analyses of the lower and upper lava units of Brabant Island*

	Upper lava unit			Lower lava unit		
	Min.	Mean	Max.	Min.	Mean	Max.
SiO_2	46.84	47.86	48.31	48.01	48.56	48.77
TiO_2	1.28	1.36	1.47	1.62	1.82	2.18
Al_2O_3	16.26	17.20	17.78	16.88	17.45	17.83
Fe_2O_3	8.80	9.15	10.53	8.92	9.34	10.68
MnO	0.14	0.16	0.17	0.13	0.16	0.18
MgO	8.42	9.03	9.43	6.87	7.52	7.86
CaO	10.04	10.49	11.03	8.64	8.97	9.62
Na_2O	3.09	3.32	3.63	3.87	4.19	4.77
K_2O	0.06	0.19	0.31	0.33	0.60	0.82
P_2O_5	0.09	0.18	0.30	0.24	0.35	0.42
Ba	14	56	102	31	70	119
Co	39	43	49	33	37	42
Cr	176	258	357	67	170	201
Cu	61	80	98	56	64	74
Nb	0	2	4	3	9	15
Ni	125	157	183	83	107	124
Rb	0	2	4	3	5	9
Sr	284	355	397	456	580	846
Zn	54	60	68	56	61	69
Zr	98	112	123	171	181	202

The lower lava suite consists of 18 samples, the upper lava suite 14 samples; all analyses by XRF.

Geochemistry

Table 1 shows the geochemical variations found in the upper and lower lava units of Brabant Island. All are basaltic in composition and all are *ne*-normative, the lower unit having 3.4% *ne* and the upper 1–2%. It is also clear that the lower lava unit is the more evolved of the two, having notably lower Cr and Ni levels than the upper unit. On a total alkali–silica plot (Fig. 2), the lower lava unit plots in the alkali field of Irvine & Baragar (1971) whereas the upper lava unit, having very low levels of K_2O (down to 0.06%), plots in the subalkali field.

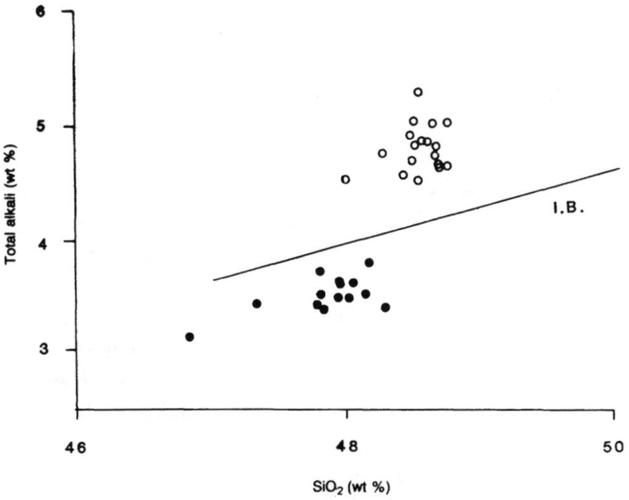

Fig. 2. Total alkali versus silica plot of the Recent volcanic rocks of Brabant Island. Open circles, lower lava unit; solid circles, upper lava unit; I.B. is the line separating the alkali from the subalkali fields of Irvine & Baragar (1971).

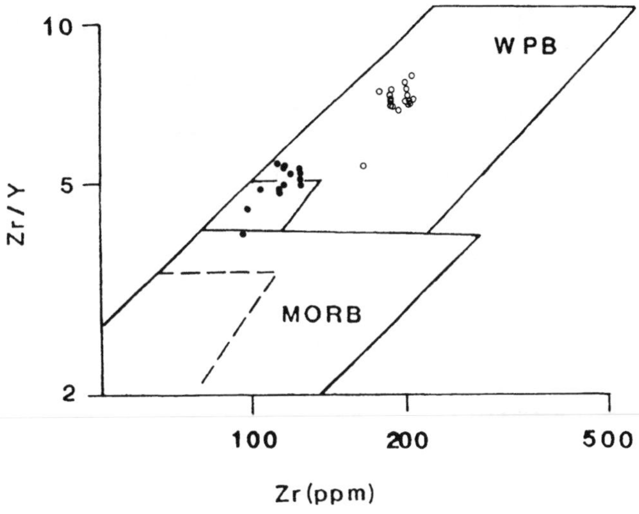

Fig. 3. Zr versus Zr/Y plot of the recent volcanic rocks of Brabant Island (Pearce & Norry, 1979). Open circles, lower lava unit; solid circles, upper lava unit; WPB, within-plate basalt field; MORB, mid-ocean ridge basalt field.

Fig. 3 is a Zr–Zr/Y plot (Pearce & Norry, 1979), showing the fields for within-plate basalts (WPB) and mid-ocean ridge basalt field (MORB). The lower lava unit plots in the WPB field whereas the upper lava unit plots towards the MORB field. It is also noteworthy that the Nb content of the lower lava unit actually shows a considerable range, from 15 ppm (which may be expected in WPB) down to 3 ppm.

A basaltic-andesite from the Mount Parry ridge (sample 26203) has been plotted on a rock/MORB spidergram (Fig. 4) only because it has a similar chemistry to basalts from this ridge, and has given a K–Ar date of 3.07 ± 0.1 Ma; sample 26079, from the upper lava unit, has yielded a K–Ar date of < 100000 y. Fig. 4 therefore represents the changing geochemistry over the last 3 m.y. on the island. Sample 26203 has a typical calc-alkaline signature, with enrichment in large ion lithophile (LIL) elements and a marked (but not large) depletion in Nb. Sample 26147 (lower lava unit) shows a distinct depletion from these levels in the LIL elements, slight depletion in the high field strength (HFS) elements and distinctly lower Nb levels. Other lower lava unit flows show that the Nb-depletion appears to coincide with the depletion of the LIL elements. Sample 26079 (upper lava unit) shows a further depletion in the LIL elements. However, there is a pronounced depletion in the K and Rb contents compared to Sr and Ba; Ba shows considerable variation but the distinct U-shape in the LIL part of the diagram remains. There is also continuing depletion in the Nb levels whilst the HFS elements remain at similar levels to the lower unit. In sample 26175 the Sr and Ba levels are still high compared to MORB but the depletion of K, Rb and Nb is extreme. Again there has been virtually no depletion of the HFS elements.

The REE contents of the same samples (Fig. 5) show a similar pattern of depletion over time. The calc-alkaline sample shows enrichment in the light rare earth elements (LREE). The lower lava unit shows a slight LREE-enrichment whereas the

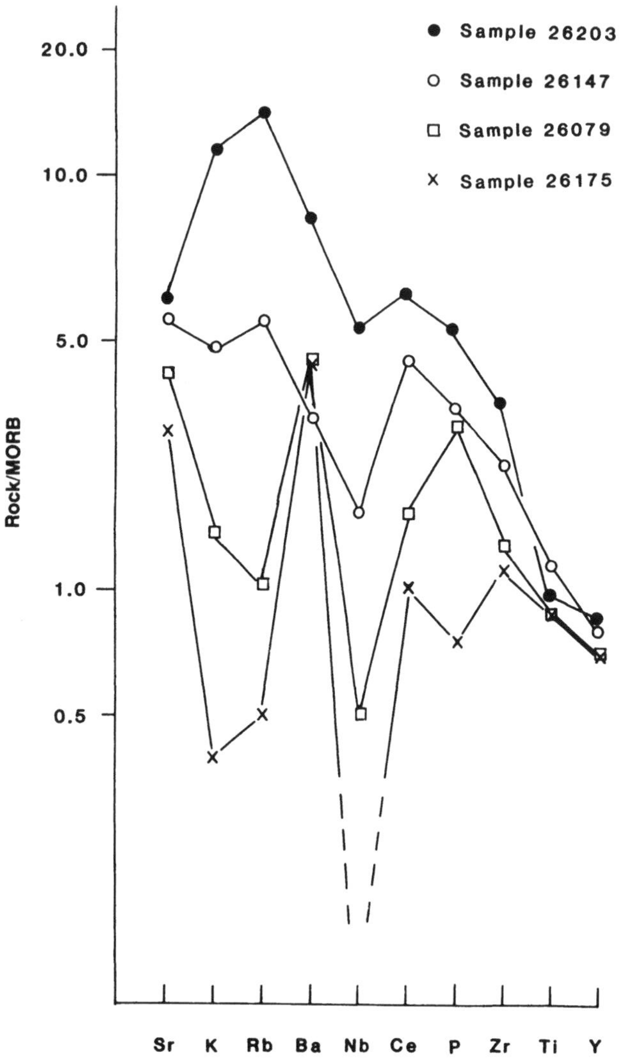

Fig. 4. Rock/MORB spidergram for the Recent volcanic rocks of Brabant Island; values normalized to Pearce (1983). Sample 26147 (lower lava unit); samples 26079 and 26175 (upper lava unit).

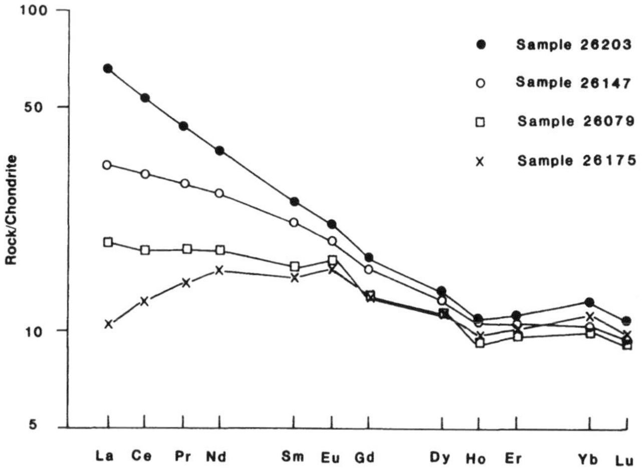

Fig. 5. Chondrite-normalized REE plot for the Recent volcanic rocks of Brabant Island; values normalized to Nakamura (1974). Analyses by ICP.

upper lava unit shows a distinct depletion of the LREE compared to the HREE. The youngest flows also show a small but distinct positive Eu-anomaly.

Discussion

As noted already subduction-related magmatism has been present in the area throughout the Tertiary, and probably for considerably longer. However, about 4 Ma ago subduction finally ceased as the ridge crest out to the west approached the trench, but was not subducted. The presence of the 3 Ma andesite suggests that activity continued for at least 1 m.y. after subduction ceased, possibly related to continuing downward motion of the plate. Since that time there has been considerable change in the type of magmatism seen on Brabant Island. The geochemistry of the basalts shows a progressive depletion history, from the enrichment of the calc-alkaline lavas to the strongly depleted upper lava flows. K, Rb, and Nb show the strongest depletion whilst Sr, Ba and the HFS elements are depleted to a lesser degree. It is logical to relate this changing geochemistry to the mantle wedge above the subduction zone. During active subduction the LIL elements were derived from the downgoing slab and the HFS elements from the mantle wedge. Cessation of magmatism would result in the supply of the LIL elements being cut off, and continued melting of the mantle wedge would result therefore in progressively lower amounts of these elements. The low levels of K and Rb compared to Sr and Ba suggest that a residual phlogopite phase may have enhanced the depletion of the LIL elements further but, if continued melting was sufficient to deplete the wedge to these levels, it may not be necessary to invoke this phase. The seemingly gradual depletion of Nb appears to be related to the same process. The relatively high Nb values found in some of the lower lava unit flows (15 ppm) may be the result of initial depletion of the LIL elements before depletion in Nb reached a significant level. As melting continued the Nb contents subsequently decreased to lower values (the average of the lower lava unit is 9 ppm, dropping to 3 ppm). It would seem, however, that the HFS elements have not been depleted to the same extent, as the levels found in both the upper and lower units are roughly similar.

To continue melting the mantle wedge to the degree suggested requires an input of energy into the system. Extension has been prevalent in the last 2 m.y. to the north of Brabant Island and has resulted in the formation of the Bransfield Strait back-arc basin. It is likely that the extensional crustal thinning has resulted in the raising of the geotherms over a considerable area (including Brabant Island) and this has allowed melting to continue in the mantle wedge. In Bransfield Strait there has been a similar switch from calc-alkaline to more depleted type of magmatism (Weaver *et al.*, 1979) and there are similarities in the geochemistry of basalts from these islands (especially Deception Island) and the upper lava unit from Brabant Island. What is uncertain is whether the upper lava unit of Brabant Island is entirely derived from the depleted mantle wedge or if there has been additional input of upwelling mantle from depth. Preliminary isotope data suggest that this may be the case.

Acknowledgements

I wish to acknowledge NERC funding for this project and am indebted to Commander C. Furse, RN and all the members of the Joint Services Expedition to Brabant Island (1983–85) without whom the fieldwork would have been impossible. I also wish to thank D.C. Rex at the University of Leeds for undertaking the K–Ar dating.

References

Alarcón, B., Ambrus, J., Olcay, L. & Vieira, C. (1976). Geología del Estrecho de Gerlache entre los paralelos 64° y 65° lat. sur, Antártica chilena. *Instituto Antártico Chileno, Serie Científica*, **4(1)**, 7–51.

Barker, P.F. (1982). The Cenozoic subduction history of the Pacific margin of the Antarctic Peninsula: ridge crest–trench interactions. *Journal of the Geological Society, London*, **139(6)**, 787–801.

Furse, C. (1986). *Antarctic Year: Brabant Island Expedition*. Beckenham, Kent; Croom Helm Publishers Ltd. 223 pp.

Hawkes, D.D. (1982). Nature and distribution of metalliferous mineralization in the northern Antarctic Peninsula. *Journal of the Geological Society, London*, **139(6)**, 803–9.

Hooper, P.R. (1962). *The Petrology of Anvers Island and adjacent islands*. London; Falkland Island Dependencies Survey Scientific Reports, No. 34, 69 pp.

Irvine, T.N. & Baragar, W.R.A. (1971). A guide to the classification of the common igneous rocks. *Canadian Journal of Earth Sciences*, **8**, 523–48.

Nakamura, N. (1974). Determination of REE, Ba, Fe, Mg, Na and K in carbonaceous and ordinary chondrites. *Geochimica et Cosmochimica Acta*, **38**, 757–75.

Pearce, J.A. (1983). Role of sub-continental lithosphere in magma genesis at active continental margins. In *Continental Basalts and Mantle Xenoliths*, Shiva Geology Series, ed. C.J. Hawkesworth & M.J. Norry, pp. 230–49. Nantwich; Shiva Publishing Ltd.

Pearce, J.A. & Norry, M.J. (1979). Petrogenetic implications of Ti, Zr, Y and Nb variations in volcanic rocks. *Contributions to Mineralogy and Petrology*, **69**, 33–47.

Rex, D.C. & Baker, P.E. (1973). Age and petrology of the Cornwallis Island granodiorite. *British Antarctic Survey Bulletin*, **32**, 55–61.

Roach, P.J. (1979). The nature of back-arc extension in the Bransfield Strait. *Geophysical Journal of the Royal Astronomical Society*, **53**, 165.

Thomson, M.R.A., Pankhurst, R.J. & Clarkson, P.D. (1983). The Antarctic Peninsula – a Late Mesozoic–Cenozoic arc (review). In *Antarctic Earth Science*, ed. R.L. Oliver, P.R. James & J.B. Jago, pp. 289–94. Canberra; Australian Academy of Science and Cambridge; Cambridge University Press.

Weaver, S.D., Saunders, A.D., Pankhurst, R.J. & Tarney, J. (1979). A geochemical study of magmatism associated with the initial stages of back-arc spreading. The Quaternary volcanics of Bransfield Strait, from South Shetland Islands. *Contributions to Mineralogy and Petrology*, **68**, 151–69.

Geochemistry and tectonic setting of Cenozoic alkaline basalts from Alexander Island, Antarctic Peninsula

M.J. HOLE[1], J.L. SMELLIE[1] & G.F. MARRINER[2]

1 British Antarctic Survey, Natural Environment Research Council, High Cross, Madingley Road, Cambridge CB3 0ET, UK
2 Department of Geology, Royal Holloway and Bedford New College, Egham Hill, Egham, Surrey TW20 0EX, UK

Abstract

Scattered outcrops of Cenozoic alkaline basalts on Alexander Island represent part of the major alkaline basalt province which extends from New Zealand to South America. All the Alexander Island basalts post-date the cessation of subduction along the western margin of the Antarctic Peninsula. Those in northern Alexander Island fall in the field of alkalic magmas in Ta/Yb, Th/Yb space, whilst samples from Beethoven Peninsula, southern Alexander Island, plot just above the field for transitional magma types. The Beethoven Peninsula samples exhibit lesser degrees of LREE-enrichment and have significantly higher Th/Ta and Yb/Ta ratios than the samples from northern Alexander Island, even though the latter exhibit the highest absolute Th abundances. $^{87}Sr/^{86}Sr$ ratios show a positive correlation with both Th/Ta and Yb/Ta ratios, the Beethoven Peninsula samples exhibiting significantly higher Sr-isotope ratios (~ 0.7034) than those from northern Alexander Island (0.7027–0.7030). The varying degrees of LREE-, LIL-element and ^{87}Sr enrichment must reflect variable source compositions and/or variable degrees of partial melting within the different areas of Alexander Island. The higher Th/Ta and $^{87}Sr/^{86}Sr$ ratios of the Beethoven Peninsula samples may reflect a minor subduction zone component inherited from the previous subduction history of the west coast of the Antarctic Peninsula.

Introduction

Cenozoic alkali basalts are widespread throughout West Antarctica, with major outcrops in Marie Byrd Land (LeMasurier, 1972a, b), Ross Island (Kyle, 1981) and the Antarctic Peninsula (Smellie, 1987). The basanite–alkali basalt sequences of Alexander Island, south-western Antarctic Peninsula (Fig. 1), are probably the least known members of this alkaline volcanic province, some occurrences being undescribed until the mid-1970s (Burn, 1976). Field work during the 1985–86 austral summer initiated a programme of detailed geochemical and volcanological studies of eight scattered outcrops of alkali basalts on Alexander Island. The new data presented here represent the preliminary results of this programme.

Tectonic setting

The geology of Alexander Island can be attributed mainly to processes associated with the subduction of proto-Pacific oceanic crust along the western margin of the Antarctic Peninsula, from latest Triassic to late Tertiary times. Metasedimentary rocks of an accretionary prism complex (LeMay Group) and calc-alkaline plutonic and associated volcanic rocks dominate the geology of western Alexander Island, whereas a narrow band of Upper Jurassic to Lower Cretaceous marginal basin sedimentary rocks (Fossil Bluff Group) is present in eastern Alexander Island. Inception of alkaline volcanism on Alexander Island followed the cessation of subduction along the west coast of the island by progressive northward ridge crest–trench collision (Barker, 1982). The alkali basalts post-date the cessation of subduction by ~ 45 m.y. in southern Alexander Island and by < 30 m.y. in the north of the island (Smellie *et al.*, 1988).

Volcanology

Three distinct modes of occurrence of alkali basalts and associated volcaniclastic rocks have been recognized within Alexander Island.

(1) The late Miocene (Table 1) sequences around Mount Pinafore, Elgar Uplands, and at Rothschild Island unconformably overlie accretionary prism metasedimentary rocks or Tertiary (40–60 Ma) calc-alkaline volcanic rocks (Burn, 1981). At two localities near Mount Pinafore,

Table 1. *K–Ar ages for Alexander Island alkaline basalts*

	Age range (Ma)	Location	Eruptive environment
Basanites	2.5	Hornpipe Heights	Subaerial
	5.5	Rothschild Island	Subglacial/subaqueous
Tephrites	5.5–7.6	Mount Pinafore	(?)Subglacial
Alkali & olivine basalts	< 1–2.5	Beethoven Peninsula	(?)Subglacial

Data from Smellie (1987) and Smellie *et al.* (1988).

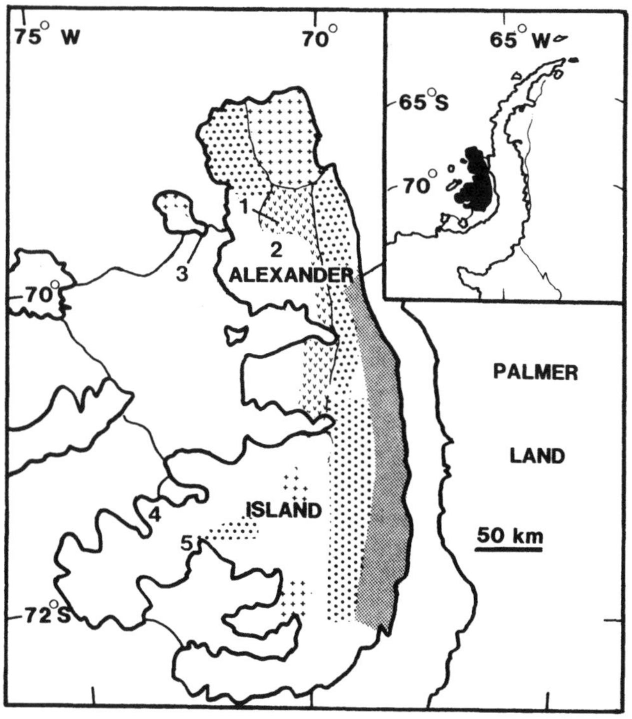

Fig. 1. Diagrammatic map showing the location and general geology of Alexander Island. Key: 1, Mount Pinafore tephrites; 2, Hornpipe Heights basanites; 3, Rothschild Island basanites; 4, Mussorgsky Peaks olivine basalts; 5, Gluck Peak alkali basalts; [+ +] Tertiary plutons; [v v] Tertiary calc-alkaline volcanic rocks; fine stipple, Jurassic–Cretaceous fore-arc sedimentary rocks (Fossil Bluff Group); coarse stipple, accretionary prism metasedimentary rocks (LeMay Group).

ortho- and paraconglomeratic breccias containing abundant glacially striated metasedimentary or volcanic clasts form the basal 1–2 m of each succession. Buff- to yellow-coloured palagonitized lapillistones and breccias form the majority of the exposures, with crystalline basalt generally occurring as thick (up to 60 m) columnar-jointed pooled flows. Burn & Thomson (1981) have compared the conglomeratic breccia at Mount Pinafore with similar outcrops in the Jones Mountains and interpreted them as tillites, implying that eruptions took place subglacially or during interglacial periods. At a newly discovered locality south-east of Mount Pinafore, paraconglomeratic breccia of uncertain thickness (> 1 m) is overlain by a 60 m thick columnar basalt flow. This breccia consists of sub- to well-rounded, glacially striated boulders and cobbles of metasedimentary and calc-alkaline volcanic rocks set in a buff-coloured sandy matrix. It is probably the best example of a tillite exposed within this area, and again suggests that some of the alkali basalts of Elgar Uplands were erupted in an interglacial or subglacial environment.

(2) At Hornpipe Heights, Elgar Uplands (Fig. 1), a relatively thin succession (~ 20 m) of lapillistones, tuffs and incipiently to completely welded agglutinates is undoubtedly of subaerial origin. The whole sequence unconformably overlies an irregular surface of LeMay Group metasedimentary rocks, the basal bed being a moderate to well sorted, poorly stratified lapillistone consisting of black, vesicular olivine–vitrophyric basalt lapilli. This bed is laterally discontinuous and infills and drapes over gulleys and irregularities in the underlying metasedimentary rocks. The majority of the exposure consists of thin (< 0.75 m) massive flows interbedded with thicker (up to 6 m) agglutinates with welded lenses. The agglutinates are almost entirely made up of brick red, aerodynamically moulded, incipiently welded bombs of vesicular olivine–phyric basalt. In places the bombs may be up to 0.75 m along their longest axis. Fragmented bombs form the majority of some of the agglutinate units.

(3) Thick sequences (up to 1000 m) of weakly palagonitized volcaniclastic rocks and associated pillow lava crop out in five isolated nunataks on Beethoven Peninsula (Fig. 1). At Mussorgsky Peaks, north-eastern Beethoven Peninsula, 200 m of interbedded palagonitized lapillistones and thin palagonite tuffs are overlain by 100 m of pillow breccia and pillow lava. The cap-rock is characterized by widely spaced pillows up to 0.75 m in diameter and lensoid bodies of olivine–phyric basalt up to 3 m long and 0.5 m thick, set in a poorly sorted pillow breccia matrix almost entirely composed of hyaloclastic shards. The majority of the lapilli in the underlying lapillistones are composed of highly vesicular, subrounded or fragmented sideromelane showing marginal alteration to palagonite. These lapilli are likely to be of pyroclastic rather than hyaloclastic origin. Sedimentary structures occur on both small and large scles; flame structures and small-scale features characteristic of syndepositional dewatering are present within the tuffaceous horizons, whereas broad, shallow channels are common within the lapillistones. Macro-faulting is confined to individual lapillistone units up to 5 m thick, and at the western end of Mussorgsky Peaks large-scale slump

folds, in which pillow breccia has slumped into the underlying lapillistone–tuff sequence, are clearly exposed.

Regional comparisons

There are considerable similarities between the Beethoven Peninsula volcanic outcrops and those described by LeMasurier (1972a) in Marie Byrd Land. The occurrence of widely spaced pillows and lensoid bodies of basalt within thick hyaloclastite sequences characterizes a number of volcanic centres in Marie Byrd Land (e.g. Mount Steere and Mount Murphy). LeMasurier (1972a) suggested that such hyaloclastite sequences are a result of the interaction between basaltic lavas and water in a subglacial environment. However, the occurrence of abundant pyroclasts in the Beethoven Peninsula lapillistones contrasts with the predominance of hyaloclasts within the Marie Byrd Land exposures. Sideromelane pyroclasts similar to those found at Mussorgsky Peaks are reported from Surtsey, Iceland, and are probably indicative of near-synchronous pyroclastic disruption and quenching in shallow water (Fisher & Schmincke, 1984), and as such may represent the expression of subglacial volcanism immediately prior to emergence of the volcano. Following emergence, the overlying hyaloclastites with isolated basalt lenses and pillows may have formed where lava flowed into a glacial lake, the water being generated by natural melt or magmatic heat. Significantly, both the subaerial volcanic sequences of Hornpipe Heights and the probably subglacial volcanic rocks of Beethoven Peninsula yield K–Ar ages within error of one another (Table 1). The occurrence of temporally restricted subaerial and subglacial volcanic sequences is also a characteristic of the Marie Byrd Land volcanic province (cf. the subaerial stratovolcano of Toney Mountain (0.5 ± 0.2 Ma) and subglacial sequences at Mount Murphy (0.8 ± 0.14 Ma: LeMasurier, 1972a)). Changes from subglacial to subaerial volcanic activity thus provide a record of the extent of glaciation in both Marie Byrd Land and Alexander Island. The Elgar Uplands and Beethoven Peninsula volcanic centres provide further evidence substantiating the existence of a major ice-sheet in West Antarctica during the Miocene.

Geochemistry

The alkalic basalts of Alexander Island cover a broad compositional spectrum, ranging from highly undersaturated basanites (up to 13% *ne*-normative) through moderately undersaturated tephrites (up to 8% *ne*-normative) to slightly *hy*-normative (< 5%) olivine basalts. The *hy*-normative olivine basalts are geographically restricted to Beethoven Peninsula whilst the basanites (Hornpipe Heights and Rothschild Island) and tephrites (around Mount Pinafore) are restricted to the north of the island. Slightly *ne*-normative (< 4%) alkali basalts occur at Gluck Peak, Beethoven Peninsula. Individual volcanic centres are compositionally restricted but there are major variations in absolute incompatible element enrichment and interelement ratios throughout the island. All samples analysed fall in the alkaline field on a total alkalis–silica plot and exhibit the high Zr/Y, Nb/Y, Zr/Nb and La_n/Yb_n ratios and a positive Nb, Ta anomaly (i.e. $La/Nb \leq 1$) on primordial mantle normalized trace element diagrams (Fig. 2) which are characteristic of anorogenic alkali basalts from continental and oceanic settings (Thompson *et al.*, 1984).

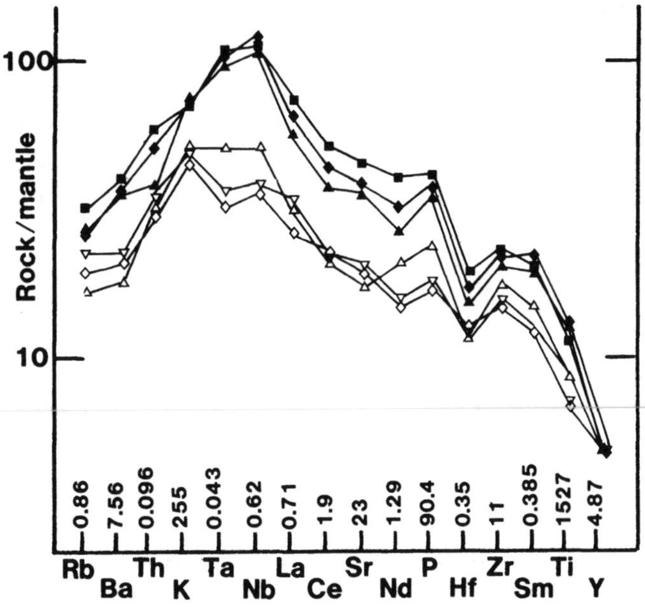

Fig. 2. Primordial mantle normalized trace element patterns for basanites (solid symbols) and olivine basalts (open symbols) of Alexander Island. For clarity the tephrites have been omitted as they plot at intermediate positions between the basanites and olivine basalts. Normalizing values from Wood *et al.* (1979).

Absolute Nb abundances increase systematically in the sequence: olivine basalts (23–32 ppm), tephrites (42–56 ppm), basanites (63–80 ppm). The degree of light rare earth element (LREE) enrichment increases in the same sequence, the olivine basalts (La_n/Yb_n 7–9) being less LREE-enriched than the basanites or tephrites (La_n/Yb_n 11–14) (Fig. 3). Absolute Yb abundances (1.8–2.2 ppm) are, however, similar for all the basalts.

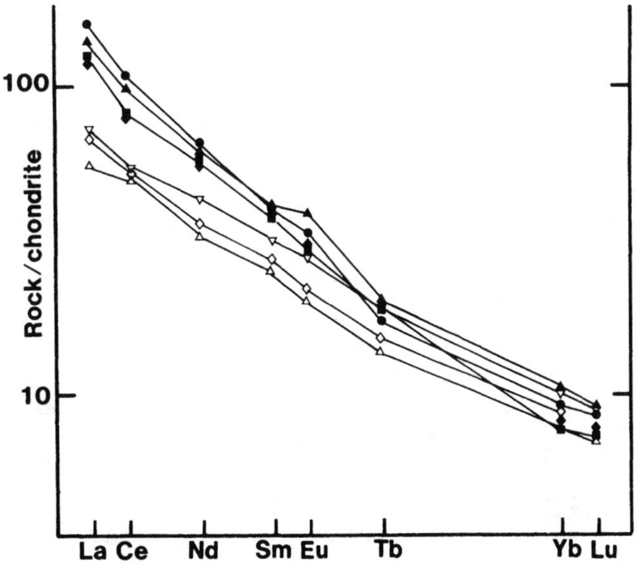

Fig. 3. Chondrite-normalized rare earth element patterns for Alexander Island basanites (solid symbols) and olivine basalts (open symbols).

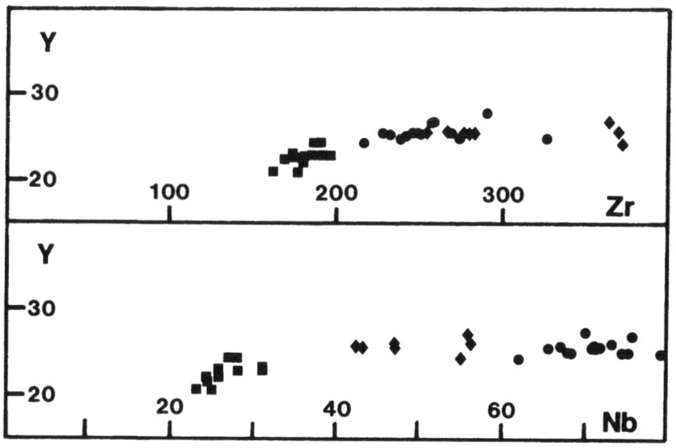

Fig. 4. Plots of Zr versus Y and Nb versus Y for Alexander Island basanites (circles), tephrites (diamonds) and olivine basalts (squares).

$^{87}Sr/^{86}Sr$ ratios for the tephrites (~ 0.7027) are slightly lower than for the basanites (~ 0.7030), whereas olivine basalts from Mussorgsky Peaks exhibit the highest $^{87}Sr/^{86}Sr$ ratios (~ 0.7034). However, a sample of alkali basalt (4% normative *ne*) from Gluck Peak has a $^{87}Sr/^{86}Sr$ ratio (0.7030) similar to those of the basanites.

Consistency of Yb abundances but variable degrees of LREE-enrichment suggests that all the basalts may have been derived by small, but variable degrees of partial melting of a mantle source containing a heavy rare earth element (HREE)-enriched phase (e.g. garnet lherzolite), the olivine basalts and basanites representing the largest and smallest degrees of partial melting respectively (cf. Smellie, 1987).

Zr and Nb versus Y plots (Fig. 4) demonstrate that the alkali basalts define linear arrays with constant Zr/Y (~ 7) and Nb/Y (~ 1) ratios whereas the basanites exhibit variable Zr/Y and Nb/Y ratios. Clearly, Nb, Zr and Y are all behaving incompatibly in the olivine basalts, but Y is behaving compatibly (bulk distribution coefficient (D) ~ 1) in the basanites, resulting in variable Nb/Y and Zr/Y ratios. Crystallization of the observed phenocryst phases in the olivine basalts (mainly olivine with rare plagioclase) would not be expected to fractionate Nb or Zr relative to Y during fractional crystallization, and it is likely that the olivine basalts were derived by relatively large degrees of partial melting of a garnet-bearing source in which garnet was not completely consumed, and subsequently differentiated mainly by olivine crystallization. The relatively low MgO (< 8%) and Ni (< 131 ppm) as well as higher SiO_2 (~ 50%) contents of the olivine basalts compared with the basanites (MgO up to 12%, Ni up to 344 ppm, SiO_2 < 45%) suggest that the olivine basalts are more evolved than the basanites.

The variable Nb/Y and Zr/Y ratios exhibited by the basanites suggest that two possible petrogenetic processes may have operated during their genesis: 1) derivation by extremely small (< 1%) but slightly variable degrees of partial melting from a garnet–lherzolite source; or 2) derivation by small degrees of partial melting of a garnet–lherzolite source followed by crystallization of an HREE-enriched phase which was capable of rendering the HREE and Y compatible during fractional crystallization. The occurrence of rare 'sieve'-textured, disequilibrium titaniferous clinopyroxene xenocrysts in the basanites indicates that clinopyroxene was an important crystallizing phase at some stage during their genesis (Smellie, 1987). Reported whole-rock/mineral distribution coefficients (k_D's) for clinopyroxenes in basaltic liquids show that, in general, clinopyroxenes are HREE-enriched ($La_n/Yb_n < 1$; Arth, 1976), and so partial melting of a garnet-bearing source, followed by fractional crystallization of clinopyroxene, may result in the fractionation of Zr and Nb relative to Y and thus variable Zr/Y and Nb/Y ratios. Further evidence for the importance of clinopyroxene crystallization during the genesis of the basanites comes from Cr abundances; although the basanites appear to be the most primitive alkalic basalts in the area and were produced by the smallest degrees of partial melting, their absolute Cr abundances cover a similar range to those for the more evolved alkali basalts. As reported k_D values for Cr in clinopyroxene are high (> 5; Henderson, 1982), clinopyroxene crystallization could be the cause of the relatively low absolute Cr abundances of the basanites. Therefore it is clear that some of the compositional differences between the alkali basalts and basanites cannot solely be attributed to low, but variable degrees of partial melting of the same source; variations in source compositions and/or variations in the crystallization histories of both basalt types must also have contributed to the generation of their observed trace element characteristics.

Conversely, although there are gross variations in the absolute abundances of trace elements between the olivine basalts and tephrites, both rock types have similar inter-element ratios which are distinct from the basanites. Zr/Nb (~ 6), Sr/Nb (17–20) and P/Nb (~ 65) ratios are significantly higher in the olivine basalts and tephrites compared to the basanites (Zr/Nb ~ 3.5, Sr/Nb ~ 12, P/Nb ~ 45). Tarney *et al.* (1978) demonstrated that variable Zr/Nb ratios are likely to reflect variations in the Zr and Nb contents of the mantle source rather than being artefacts of partial melting. The similarity in Zr/Nb ratios between the olivine basalts and tephrites thus suggests that both rock types may have been derived by low, but variable degrees of partial melting from a compositionally similar source.

Two distinct geochemical lineages can now be recognized for the alkali basalts of Alexander Island; a high Zr/Nb, Sr/Nb and P/Nb group consisting of olivine basalts and tephrites and a low Zr/Nb, Sr/Nb and P/Nb group of basanites.

Subduction components in the alkali basalts

Although the olivine basalts and tephrites are indistinguishable in terms of HFSE (high field strength elements, i.e. Nb, Ta, Zr, the middle REE, P, Sr and Ti) ratios, the olivine basalts from Mussorgsky Peaks exhibit higher LILE/HFSE (LILE; large ion lithophile elements, i.e. Ba, Rb, Th, K and La) ratios than the Gluck Peak alkali basalts and the tephrites (Table 2). Rb/Nb (~ 1), Ba/Nb (~ 7), Th/Ta (~ 2) and K/Nb (~ 510) ratios are significantly higher in the Mussorgsky Peaks olivine basalts compared with the tephrites (Rb/Nb < 0.4; Ba/Nb ~ 4; Th/Ta ~ 1.2; K/Nb ~ 280). This apparent decoupling of LIL- from HFS-elements is generally a process associ-

Table 2. *Summary of the geochemical characteristics of Alexander Island alkalic basalts*

	Zr/Nb	Th/Ta	K/Nb	Ba/Nb	Rb/Nb	$^{87}Sr/^{86}Sr$
Basanites	3.6	1.2	280	4.0	0.37	0.7030
Tephrites	6.4	1.2	337	3.0	0.30	0.7027
Alkali basalts (Gluck Peak)	6.0	1.4	420	4.3	0.43	0.7030
Olivine basalts (Mussorgsky Peaks)	6.9	2.2	510	6.9	0.77	0.7034

ated with subduction-related arc and back-arc magmatism, and is likely to be a result of the input of LILE-enriched, high $^{87}Sr/^{86}Sr$-ratio fluids derived from the downgoing slab into a mantle-derived magma. The long period of subduction (> 150 m.y.?) prior to the inception of basic anorogenic volcanism on Alexander Island, is likely to have resulted in LILE-enrichment of the subcontinental lithosphere beneath the island. Although anorogenic basalts are considered to be derived from the asthenospheric mantle, they may interact with the subcontinental lithosphere prior to eruption (Morrison, Thompson & Dickin, 1986). In this case, the interaction between LILE-enriched subcontinental lithosphere and an unmodified asthenosphere-derived magma provides an ideal mechanism for the decoupling of the LIL- from HFS-elements. The fact that only the Mussorgsky Peaks olivine basalts exhibit high LILE/HFSE ratios suggests that there may be lateral heterogeneities in the degree of LILE-enrichment of the subcontinental lithosphere beneath Alexander Island. The slight elevation in $^{87}Sr/^{86}Sr$ ratios (~0.7034) in the high LILE/HFSE ratio olivine basalts relative to the tephrites ($^{87}Sr/^{86}Sr$ ~0.7027) and low LILE/HFSE ratio alkali basalts from Gluck Peak ($^{87}Sr/^{86}Sr$ 0.7030) is consistent with a subduction-enrichment model.

Pearce (1982) has produced a discrimination diagram to allow the identification of subduction components in a variety of basalt types. The Mussorgsky Peaks high LILE/HFSE olivine basalts plot at higher Th/Yb ratios than the basanites and tephrites. The Mussorgsky Peaks olivine basalts are thus vertically displaced away from the other Alexander Island samples in Th/Yb–Ta/Yb space. Vertical displacement of data points in Fig. 5 to Th/Ta ratios greater than unity is caused by the preferential enrichment in Th relative to Yb and Ta during subduction-related magmatism. Clearly, the Mussorgsky Peaks olivine basalts contain a small but geochemically significant subduction component.

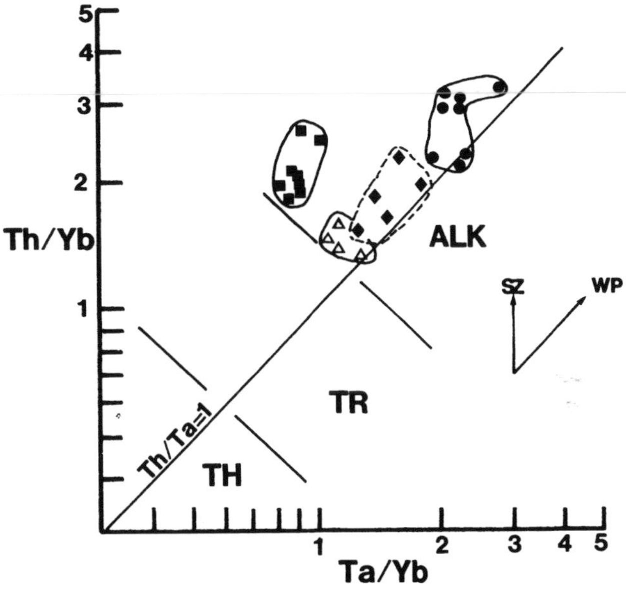

Fig. 5. Th/Yb–Ta/Yb variations in Alexander Island alkaline basalts. Symbols as for Fig. 4 except: open triangles, Gluck Peak low Th/Ta, Rb/Nb and K/Ta alkali basalt; SZ, vector representing subduction zone enrichment; WP, vector representing 'within plate' enrichment. Basalt fields: TH, tholeiitic; TR, transitional; ALK, alkalic from Pearce (1982).

References

Arth, J.G. (1976). The behaviour of trace elements during magmagenesis: a summary of theoretical models and their applications. *Journal of Research, US Geological Survey*, **4(1)**, 41–7.

Barker, P.F. (1982). The Cenozoic subduction history of the Pacific margin of the Antarctic Peninsula: ridge crest–trench interactions. *Journal of the Geological Society, London*, **139(6)**, 787–801.

Burn, R.W. (1976). *Northern Alexander Island Geology 1975/76. The Elgar Uplands and Area to the South.* Cambridge; British Antarctic Survey Geological Field Report, G2/1975/T, 10 pp. (unpublished).

Burn, R.W. (1981). Early Tertiary calc-alkaline volcanism on Alexander Island. *British Antarctic Survey Bulletin*, **53**, 175–93.

Burn, R.W. & Thomson, M.R.A. (1981). Late Cenozoic tillites with intraglacial volcanic rocks, Lesser Antarctica. In *Pre-Pleistocene Tillites: a Record of Earth's Glacial History*, ed. M.J. Hambrey & W.B. Harland, pp. 199–203. Cambridge; Cambridge University Press.

Fisher, R.V. & Schmincke, H.-U. (1984). *Pyroclastic Rocks*. Berlin; Springer Verlag. 276 pp.

Henderson, P. (1982). *Inorganic Geochemistry*. Oxford; Pergamon Press. 184 pp.

Kyle, P.R. (1981). The mineralogy and geochemistry of a basanite to phonolite sequence at Hut Point, Antarctica, based on a core from the Dry Valleys Drilling Project. *Journal of Petrology*, **22(2)**, 451–500.

LeMasurier, W.E. (1972a). Volcanic record of Cenozoic glacial history of Marie Byrd Land. In *Antarctic Geology and Geophysics*, ed. R.J. Adie, pp. 251–60. Oslo; Universitetsforlaget.

LeMasurier, W.E. (1972b). Volcanic record of Antarctic Glacial history: implications with regard to Cenozoic sea levels. In *Polar Geomorphology*, ed. R.J. Price & D.E. Sugden, pp. 59–74. London; Special Publication of the Institute of British Geographers, No. 4.

Morrison, M.A., Thompson, R.N. & Dickin, A.P. (1986). Geochemical evidence for complex magmatic plumbing during the development of a continental volcanic center. *Geology*, **13**, 581–4.

Pearce, J.A. (1982). Trace element characteristics of lavas from destructive plate boundaries. In *Andesites*, ed. R.S. Thorpe, pp. 525–47. Chichester; John Wiley & Sons.

Smellie, J.L. (1987). Geochemistry and tectonic setting of alkaline volcanic rocks in the Antarctic Peninsula: a review. *Journal of Volcanology and Geothermal Research*, **32**, 269–85.

Smellie, J.L., Pankhurst, R.J., Hole, M.J. & Thomson, J.W. (1988). Age, distribution and eruptive conditions of Late Cenozoic alkaline volcanism in the Antarctic Peninsula and eastern Ellsworth Land: review. *British Antarctic Survey Bulletin*, **80**, 21–49.

Tarney, J., Saunders, A.D., Donellan, N.C.B. & Hendry, G.L. (1978). Minor-element geochemistry of basalts from Leg 49, North Atlantic Ocean. In *Initial Reports of the Deep Sea Drilling Project, 49*, ed. B.P. Luyendyk, J.R. Cann *et al.*, pp. 657–91. Washington, DC; US Government Printing Office.

Thompson, R.N., Morrison, M.A., Hendry, G.L. & Parry, S.J. (1984). An assessment of the relative roles of crust and mantle in magmagenesis: an elemental approach. *Philosophical Transactions of the Royal Society of London*, **A310**, 549–90.

Wood, D.A., Joron, J-L, Treuil, M., Norry, M.J. & Tarney, J. (1979). Elemental and Sr-isotope variations in basic lavas from Iceland and the surrounding ocean floor. *Contributions to Mineralogy and Petrology*, **70**, 310–39.

Geophysical investigation of George VI Sound, Antarctic Peninsula

M.P. MASLANYJ

British Antarctic Survey, Natural Environment Research Council, High Cross, Madingley Road, Cambridge CB3 0ET, UK

Abstract

Geophysical data are presented with a view to understanding the structure of George VI Sound, a major curvilinear channel on the west coast of southern Antarctic Peninsula. Northern and southern George VI Sound have quite different topographic, gravity and magnetic characteristics. The northern part is a deep (> 800 m) elongated trough, which trends N–S and exhibits low Bouguer gravity and a quiet residual magnetic field. The southern part is a deeper (> 1000 m) and broader trough, which trends E–W, with higher Bouguer gravity and a long wavelength (50 km), positive (300 nT) magnetic anomaly. The results suggest that George VI Sound is floored by a thick sequence of non-magnetic rock, and in the south is underlain by a magnetic body at an estimated depth of 8–13 km. In contrast to the north, southern George VI Sound exhibits local anomalies indicating variable shallow structure, which may include significant amounts of low-density sediments. If the geophysical features are related to crustal extension, then the evidence suggests that this was more pronounced in the south than in the north. One consistent model involves NW-directed movement of Alexander Island relative to the Antarctic Peninsula, and requires predominantly strike-slip motion in the north and extension in the south, although a more complicated history cannot be excluded.

Introduction

George VI Sound is a curvilinear channel along the west coast of southern Antarctic Peninsula (Fig. 1). Geological evidence (Butterworth & Macdonald, this volume, p. 449; Piercy & Harrison, this volume, p. 381; Tranter, this volume, p. 437) indicates that it is primarily tectonic rather than glacial in origin. It has been interpreted as a rift (Crabtree, Storey & Doake, 1985) but geomorphological evidence which might support this (King, 1964) is inconclusive and geological mapping (British Antarctic Survey, 1981) is incomplete. In this paper, geophysical data acquired in the vicinity are described with a view to defining the bedrock and underlying structure.

Geophysical surveys

Bedrock topography

Between 70° S and 73° S George VI Sound is covered with shelf ice. Depths to bedrock beneath the ice shelf (Fig. 1) have been calculated from travel times of radio-echo and seismic reflections (Maslanyj, 1987). Its topography is characterized by a deep, steep-sided, elongated trough trending 170° in the north, and 270° in the south, with bedrock depths exceeding 800 m and 1000 m, respectively. It curves parallel to the arcuate topographic axis of the Antarctic Peninsula and is oblique to the Pacific continental margin. The change in trend occurs between 71°30′S and 72°30′S where George VI Sound widens from 30 to 70 km (Fig. 1).

Gravity anomalies

210 gravity stations were established on the ice shelf during the 1983–84 and 1984–85 field seasons. Radio-echo and seismic reflections provided ice thicknesses and bedrock-depth control. Absolute station heights were derived from the ice-shelf thickness assuming hydrostatic equilibrium (Maslanyj, 1987). Over grounded ice, Wallace and Tiernan barometric altimeters were used. Values from adjacent on-rock gravity stations (Renner, Sturgeon & Garrett, 1985) have been incorporated in the contour maps.

The contour map of Fig. 2 shows Bouguer anomalies calculated using a crustal density of 2.67 Mg/m^3, consistent with the map of Renner *et al.* (1985). Bouguer gravity contours tend to be parallel to topographic trends. Over northern George VI Sound, anomalies are close to zero, except in the extreme north where positive anomalies are observed over Marguerite Bay (Renner *et al.*, 1985). There is a gradual increase in Bouguer gravity from north to south, reaching almost + 500 gu (gu = gravity unit = 1 $\mu m/s^{-2}$) in the south. Short wavelength anomalies of local extent are superimposed on the higher overall Bouguer gravity over the southern part of the study area. Positive anomalies of 400–500 gu are observed over the ice shelf near Steeple Peaks and Buttress Nunataks (Fig. 1).

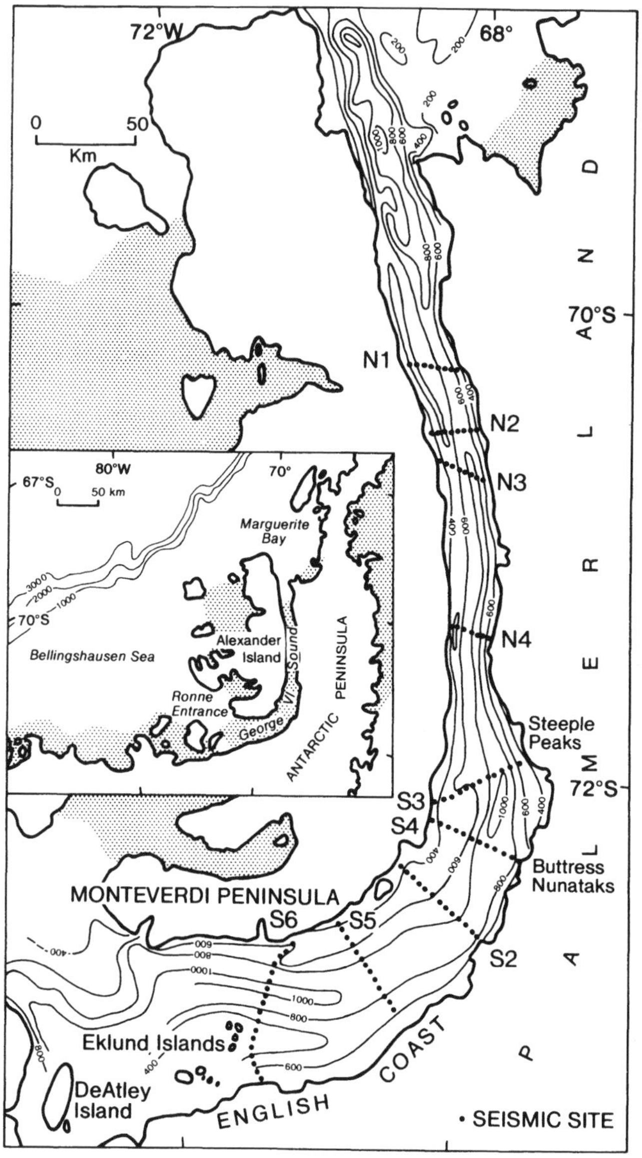

Fig. 1. Location map and contour map of the bedrock topography (contour interval 200 m) in George VI Sound. Ice shelves are shaded, except in George VI Sound. Numbered dotted lines are gravity–seismic profiles.

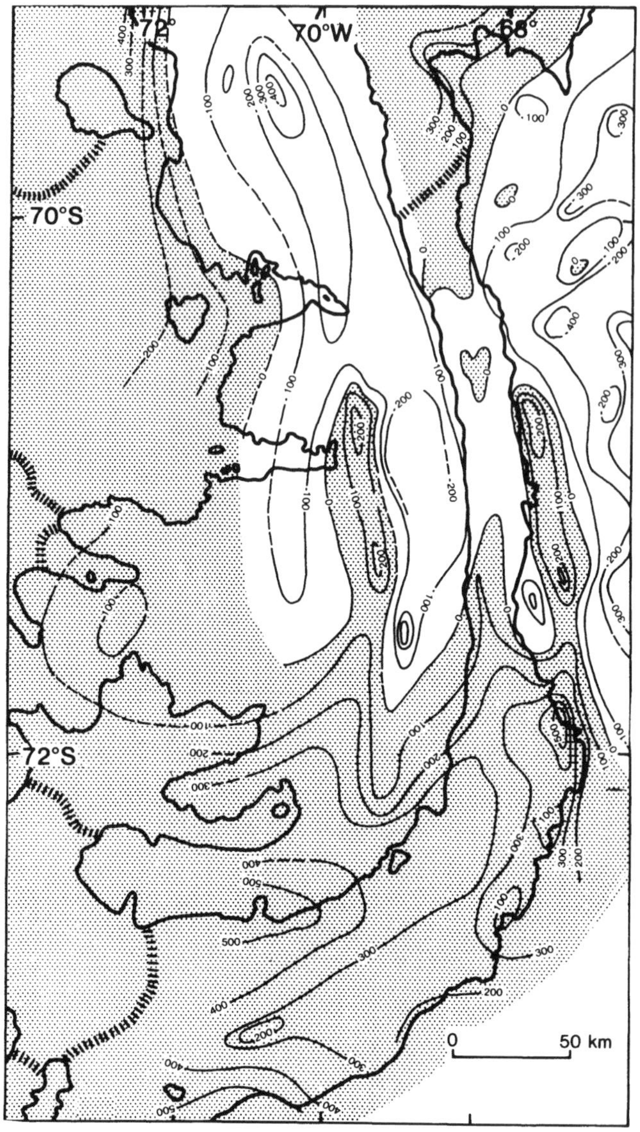

Fig. 2. Bouguer gravity anomaly map of George VI Sound area; contour interval is 100 gu; shaded areas represent positive anomalies.

Positive gravity anomalies of unknown extent occur over the southern edge of Monteverdi Peninsula and near Eklund Islands. Except over the Alexander Island coast, steep Bouguer gravity gradients generally coincide with the coastlines of George VI Sound.

Aeromagnetic anomalies

The aeromagnetic anomaly contour map (Fig. 3) is compiled from data of Renner *et al.* (1985) and Jones & Maslanyj (this volume, p. 405). Over George VI Sound north of 72°30′S, long wavelength magnetic anomalies are observed. These are similar to those over the adjacent area of Alexander Island, but contrast with the high amplitude (200–600 nT) and moderate wavelength (20 km) anomalies of Palmer Land. The boundary between these magnetic provinces is marked by a major linear magnetic discontinuity related to the West Coast Magnetic Anomaly (WCMA) (Renner *et al.*, 1985), and coincident with the west coast of Palmer Land in the north.

A smooth, broad (50 km), linear (> 300 km) magnetic anomaly of 200–300 nT amplitude, extends from E–W over southern George VI Sound. Its southern limit is marked by a linear magnetic break (less prominent than in the north). This diverges on to the ice shelf parallel to the English Coast and extends north of Eklund Islands, into Ronne Entrance. Several short-wavelength anomalies of local extent were observed over the southern ice shelf.

Geological interpretation

Shallow structure

In many broad basins, the gravity effect of low-density sediments can be cancelled by the effect of a deeper, higher-density body (Simpson *et al.*, 1986). Simple models (Maslanyj,

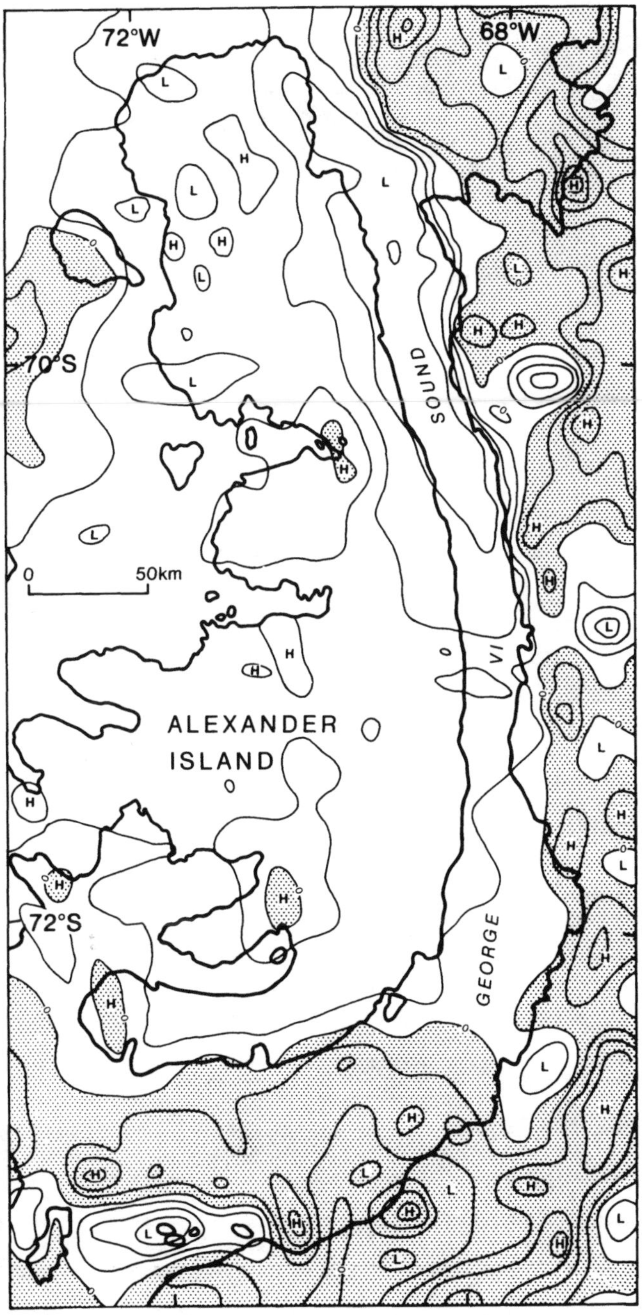

Fig. 3. Aeromagnetic anomaly map of George VI Sound area; contour interval is 100 nT; shaded areas correspond to positive anomalies.

1988) indicate that in a narrow feature such as northern George VI Sound, low-density sediment fill would produce a negative anomaly despite deep compensation. The Bouguer gravity (Fig. 2) from northern George VI Sound indicates that, unlike many 'grabens', it is not extensively floored by thick low-density sediments, unless a mass excess exists at shallow depths to mask the sediment fill. Absence of sediments may imply significant glacial erosion during a glacial episode.

In contrast, gravity and magnetic anomalies of local areal extent over southern George VI Sound indicate variable shallow structure (Maslanyj, 1988). The negative anomaly (Fig. 2) could represent up to 1 km of unconsolidated sediment (density 2.2 Mg/m^3). Consequently, the true bedrock may well be significantly deeper than seismic depth measurements indicate. Dense and magnetic near-surface bodies beneath the southern ice shelf probably indicate volcanic rocks and/or minor intrusions. Similar bodies are required on the eastern margins of profiles N1, S3 and S4 (Fig. 1) and coincide with the development of magnetic anomalies, possibly related to the WCMA.

Deep structure

The generally quiet magnetic field over Alexander Island reflects the thick, non-magnetic, mainly sedimentary sequences of the LeMay Group and Fossil Bluff Formation (Renner, Dikstra & Martin, 1982). Continuity and similar grain of the magnetic and gravity fields over northern George VI Sound suggest the presence of a thick sequence of similar material, which is cut out against a major fault, exemplified by the well-developed magnetic discontinuity along the western coastline of Palmer Land. The eastern coastline of Alexander Island, on the other hand, is not a major geophysical boundary and, if faulted, either juxtaposes rock of similar density and magnetic properties by vertical movement, or involves predominantly strike-slip movement. The aeromagnetic anomaly over southern George VI Sound indicates that it is underlain by a large magnetic body at depths of 8–13 km below sea level (Jones & Maslanyj, this volume, p. 405), with a minimum cover of 7 km of non-magnetic rock.

Assuming that they can be directly related, the long-wavelength Bouguer gravity anomalies can be used to postulate relative changes in crustal thickness. Garrett & Storey (1987) considered northern George VI Sound to be underlain by thinner crust. The higher Bouguer gravity over the southern region could be related to even thinner crust (Maslanyj, 1988), which would be consistent with simple isostatic assumptions. It is also likely to be related to the magnetic body described earlier.

Regional structure and tectonic setting

The grain of the gravity and magnetic anomalies generally mirrors the arcuate regional topography and confirms that George VI Sound is a tectonic rather than purely a glacial feature. Changes in subice topography, gravity and magnetic signature occur where George VI Sound and the Antarctic Peninsula exhibit maximum curvature. A more abrupt change in orientation of WCMA is observed in adjacent Palmer Land (Jones & Maslanyj, this volume, p. 405).

Northern and southern George VI Sound are quite different in geophysical character. The structural significance of the southern part has not been recognized previously due to the more subdued topography and less outcrop geology in adjoining areas. If the formation of George VI Sound is related to crustal extension, then the evidence suggests that extension was more pronounced in the south than in the north.

The geophysical configuration can be explained in terms of extension normal to the Pacific continental margin, moving Alexander Island in a north-westerly direction relative to the Antarctic Peninsula (Maslanyj, 1987). Although the exact movement cannot be defined, this would require predominantly strike-slip in the north and extension in the south, consistent with the evidence of late-state (?Tertiary), dextral transtensional motion which affected the accretionary complex and fore-arc basin in Alexander Island (Nell & Storey, this volume p. 443). Within this model the relative influence of convergent margin processes (Garrett & Storey, 1987) or margin-independent processes (Nell & Storey, this volume, p. 443) remains difficult to assess.

The data do not provide a unique solution and in the absence of better age and structural constraints, there is still ambiguity concerning the origin of George VI Sound. All the same, a new perspective has been placed; one which is at variance with the north–south rift hypothesis (King, 1964).

Acknowledgements

I would like to thank R.G.B. Renner for advice, support and giving me the opportunity to do this work.

References

British Antarctic Survey. (19981). *British Antarctic Territory Geological Map, Sheet 4, Alexander Island,* 1:500000, BAS 500G series. Cambridge; British Antarctic Survey.

Crabtree, R.D., Storey, B.C. & Doake, C.S.M. (1985). The structural evolution of George VI Sound, Antarctic Peninsula. In *Geophysics of the Polar Regions*, ed. E.S. Husebye, G.L. Johnson & Y. Kristoffersen, *Tectonophysics*, **114**, 431–42.

Garrett, S.W. & Storey, B.C. (1987). Lithospheric extension on the Antarctic Peninsula during Cenozoic subduction. In *Continental Extensional Tectonics*, ed. M.P. Coward, J.F. Dewey & P.L. Hancock, pp. 419–31. Oxford, London; Geological Society Special Publication No. 28.

King, L. (1964). Pre-glacial geomorphology of Alexander Island. In *Antarctic Geology*, ed. R.J. Adie, pp. 53–64. Amsterdam; North-Holland Publishing Company.

Maslanyj, M.P. (1987). Seismic bedrock depth measurements and the origin of George VI Sound, Antarctic Peninsula. *British Antarctic Survey Bulletin*, **75**, 51–65.

Maslanyj, M.P. (1988). Gravity and aeromagnetic evidence for the crustal structure of George VI Sound, Antarctic Peninsula. *British Antarctic Survey Bulletin*, **79**, 1–16.

Renner, R.G.B., Dikstra, B.J. & Martin, J.L. (1982). Aeromagnetic surveys over the Antarctic Peninsula. In *Antarctic Geoscience*, ed. C. Craddock, pp. 363–7. Madison; University of Wisconsin Press.

Renner, R.G.B., Sturgeon, L.J.S. & Garrett, S.W. (1985). *Reconnaissance Gravity and Aeromagnetic Surveys of the Antarctic Peninsula.* British Antarctic Survey Scientific Reports, No. 110, 49 pp.

Simpson, R.W., Jachens, R.C., Blakely, R.W. & Saltus, R.W. (1986). A new isostatic residual gravity map of the conterminous United States with a discussion on the significance of isostatic residual anomalies. *Journal of Geophysical Research*, **91(B8)**, 8348–72.

Tectonic significance of linear volcanic ranges in Marie Byrd Land in late Cenozoic time (Extended abstract)

W.E. LEMASURIER[1] & D.C. REX[2]

1 Department of Geology, University of Colorado at Denver, 1200 Larimer St, Denver, CO 80204, USA
2 Department of Earth Sciences, University of Leeds, Leeds, LS2 9JT, UK

Linear volcanic chains are a conspicuous feature of Marie Byrd Land (MBL) geology. 14 of the 18 major volcanoes in the MBL Cenozoic volcanic province occur in five N–S- or E–W-orientated ranges, and in three of these, chronological data and equidistant caldera spacings suggest systematic migrations of felsic activity. In the Executive Committee Range (ECR), the focus of activity moved 90 km from north to south, between about 14 Ma and 1 Ma. In the Flood Range (FR) volcanic activity migrated 90 km from east to west, between about 9 Ma and 2.5 Ma, and a 154 km northward migration appears to have taken place from the south end of the Ames Range (AR) to Shepard Island between about 12.7 Ma and 0.6 Ma (Fig. 1); details of the relevant K–Ar data are presented in LeMasurier & Rex, 1989. The chronology and volcano spacings within these three ranges suggest average rates of migration of about 1 cm/y.

In contrast to the more familiar Hawaiian chain, each volcano in the MBL chains appears to be composed mainly of the end-products of fractional crystallization sequences, and each often appears to represent a petrological evolution that was different from, and independent of, other volcanoes in the chain. For example, peralkaline phonolitic volcanoes alternate with comenditic and pantelleritic volcanoes in the ECR chain. Systematic compositional changes are rare within individual volcanoes and there is commonly an abrupt change in petrological character from one volcano to the next along the chains. In spite of the abundance of felsic rock, crustal contamination is difficult to demonstrate, suggesting that virtually the entire volume of lava is derived ultimately from the mantle, but modified by fractionation during temporary storage in the crust.

Plate motion is not a viable explanation for the migrations of volcanic activity in MBL because of the opposing and orthogonal directions of contemporaneous migration in the FR, AR, and ECR and because geophysical evidence indicates that the Antarctic plate has remained stationary in middle–late Cenozoic time (Duncan, 1981). Fracture propagation is a more likely mechanism, but geochemical and chronological data impose two important constraints:

(i) the abrupt compositional change from one volcano to the next, within distances of 18 km (and even 6 km in

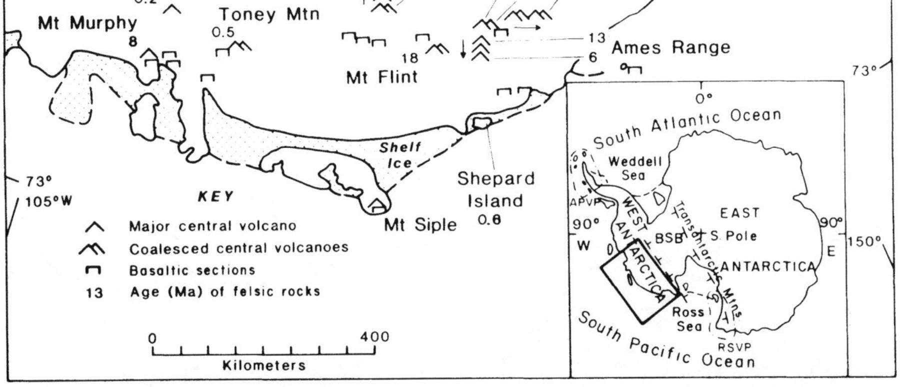

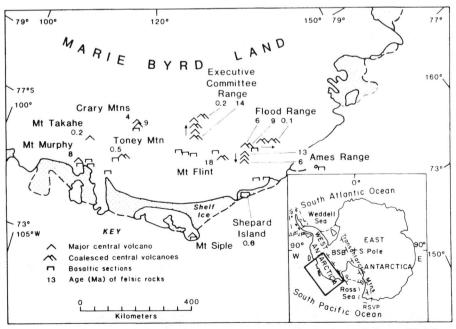

Fig. 1. Schematic map of the Marie Byrd Land late Cenozoic volcanic province, modified from LeMasurier & Rex (1989). K–Ar ages are for felsic and intermediate rocks only. Inset shows the inferred position of the West Antarctic rift system (hachures), the Byrd Subglacial Basin (BSB), and the Ross Sea and Antarctic Peninsula volcanic provinces (RSVP and APVP, respectively).

two cases), would seem to eliminate magma-driven mechanisms of fracture propagation, which might be expected to produce more homogeneous petrological patterns; and,

(ii) migrations of activity are a characteristic of felsic volcanism only; no spatial or chronological pattern in mafic activity is discernible (Le Masurier & Res, 1989). We therefore believe that the fracture propagation mechanism is tectonic, and that the rise of felsic magma is a dilatant response to extensional crack propagation.

The only indication of a province-wide pattern that can be resolved from the chronological data is an apparent migration of felsic activity toward the perimeter of the province. This can be observed not only in the linear chains, but also in the distribution of the oldest and youngest single volcanoes. The three oldest felsic volcanoes in MBL occur within 100 km of Mount Flint (Fig. 1) and the youngest occur at the province margin. Mount Murphy seems out of place with respect to this pattern but Mount Siple probably is not. Although no felsic rocks have been found yet at Mount Siple, they are likely to be found at the summit; a basalt flow at the base of the section has yielded a $^{39}Ar/^{40}Ar$ age of 2.0 Ma.

The model that we propose as an explanation of the patterns described above takes cognizance of the fact that the MBL province lies on the flank of the West Antarctic rift system, and that volcanism in this region has been accompanied by extensional block faulting throughout the past 25 m.y. The major points of the model are as follows:

1. Rise of basalt from the mantle source region has been random in time and space, but volumes have been highly variable, throughout the past 25–30 m.y. The largest volumes have been observed beneath felsic volcanoes.
2. The larger volumes of basalt have been impeded in their ascent by slow extension (1 cm/y), leading to entrapment within the crust.
3. Trapped magma evolved to felsic compositions in crustal reservoirs.
4. Continued tectonic extension eventually reopened sealed fractures in a sequence.
5. The apparent centrifugal migration pattern implies doming, centred in the vicinity of the oldest felsic volcanoes. The topographically highest horst of pre-Cenozoic basement rock occurs adjacent to Mount Flint.
6. The lack of an accompanying radial fracture pattern implies that the rectilinear pattern expressed by the linear volcanic ranges is inherited from a deformation event that preceded this latest rift episode.
7. The rectilinear pattern may be related to a mid-ocean ridge–transform fault pattern that originated when the New Zealand–Campbell Plateau block was separated from MBL in Late Mesozoic (80 Ma) time (cf. Kellogg & Rowley, this volume, p. 461).

This model implies that rifting has been recurrent in MBL. Since it does not depend on plate motion to form the linear ranges, the model also implies that the late Cenozoic rifting is a thermally driven event, but a less intense one than the late Mesozoic rifting that produced the Pacific–Antarctic ridge. The isotopic composition of the basalts is consistent with an asthenospheric source (Futa & LeMasurier, 1983), and a rise of the asthenosphere/lithosphere boundary may be the fundamental cause of crustal doming and associated rifting in this province.

References

Duncan, R.A. (1981). Hotspots in the southern oceans – an absolute frame of reference for motion of the Gondwana continents. *Tectonophysics*, **74**, 29–42.

Futa, K. & LeMasurier, W.E. (1983). Nd and Sr isotopic studies on Cenozoic mafic lavas from West Antarctica: another source for continental alkali basalts. *Contributions to Mineralogy and Petrology*, **83(1/2)**, 38–44.

LeMasurier, W.E. & Rex, D.C. (1989). Evolution of linear volcanic ranges in Marie Byrd Land, West Antarctica. *Journal of Geophysical Research*, **94(B6)**, 7223–36.

Crustal development: Gondwana break-up

Evolution of the Antarctic continental margins

L.A. LAWVER[1], J.-Y. ROYER[1], D.T. SANDWELL[1] & C.R. SCOTESE[2]

1 Institute for Geophysics, University of Texas, Austin. 8701 N. MOPAC Blvd, Austin, Texas 78759-8345 USA
2 Now at: Shell Oil Company, Bellaire Research Center, PO Box 481, Houston, Texas 77001, USA

Abstract

With the exception of the Pacific-facing margin of West Antarctica between Thurston Island and the tip of the Antarctic Peninsula, all of the continental margins of Antarctica are either rifted passive margins or sheared transform margins. The exception was a convergent margin where subduction was active from before the break-up of Gondwana until very recently. Starting in the south-western Weddell Sea, which rifted as part of a back-arc basin connected with Middle–Late Jurassic (~ 170 Ma) back-arc spreading in the Rocas Verdes Basin of southern South America, the continental margins of Antarctica seem to young clockwise. A sheared margin along the Explora Escarpment between 25° W and 10° W connected the south-western Weddell Sea rifting with contemporaneous rifting in the Mozambique Basin. This resulted in a Middle Jurassic rifted passive margin along Dronning Maud Land. East of the Gunnerus Ridge at 35° E, Sri Lanka and India rifted off of Antarctica some time between 127 and 118 Ma. Rifting between Australia and Antarctica, stretching in the Ross Sea embayment and rifting between the Campbell Plateau–Chatham Rise and Marie Byrd Land, all started at about 95 ± 5 Ma. Active subduction ceased about 4 Ma ago off the South Shetland Islands.

Introduction

The reconstruction of an early Mesozoic Gondwana (Fig. 1) indicates that most of the present-day margin of Antarctica was bounded by other continental masses. With the exception of West Antarctica, which was a subduction margin from Thurston Island to the tip of the Antarctic Peninsula during part of the period since break-up, the rest of the Antarctic margin was formed either as a rifted passive margin or as a sheared transform margin.

The identified marine magnetic anomalies that record the initial break-up of Gondwana (Fig. 2) are observed in the Mozambique Basin (Ségoufin, 1978; Simpson *et al.*, 1979) and in the western Somali Basin (Ségoufin & Patriat, 1980). Recently Cochran (1988) has extended the initial break-up to the northern Somali Basin as well. The oldest anomaly identified by Cochran in the western Somali Basin is M22 (152 Ma; Palmer, 1983). There is sufficient room between the identified anomalies and the continental margin of Madagascar in the south and Africa in the north to extend seafloor spreading as much as 300 km beyond both M22 to the north and to the south. This translates to as much as 20 m.y. of seafloor spreading prior to the first identified anomaly in the western Somali Basin if a constant spreading rate is assumed. In the Mozambique Basin, the anomalies older than M16 (142 Ma) are not well identified but upwards of 300 km of seafloor exist that are presumably older than the tentatively identified M22 anomaly. Again this might imply that seafloor spreading commenced as early as 170 Ma ago.

Dronning Maud Land/Africa

The magnetic anomalies conjugate to those in the Mozambique Basin were found by Bergh (1987) near the Astrid Ridge off Dronning Maud Land. He identified magnetic anomalies M0 (118 Ma)–M11 (134 Ma) but other anomalies are clearly visible between M11 and the continental margin of Dronning Maud Land. Even so, there is at most 100 km between what might be M22 and the margin, representing only 6–7 m.y. of additional seafloor spreading prior to 152 Ma. Bergh calculated the same spreading rate of 15 mm/y (half-rate) for anomalies M1 (122 Ma)–M5 (127 Ma) as Ségoufin (1978) found in the Mozambique Basin. Bergh (1987) presented two scenarios for the Mesozoic opening of the Mozambique Basin; one fits the Astrid Ridge against the Mozambique Escarpment whereas the other matches it with a fracture zone postulated by Simpson *et al.* (1979) about 100 km to the east of the Mozambique Escarpment, at 25–28° S. Since the Mozambique Escarpment is not only steep, > 27° in some places, but is parallel to the general direction of opening of the Mozambique Basin, it is assumed to be a sheared margin that was produced by the seafloor spreading in the Mozambique Basin.

Worldwide, the average angle of the continental slope is 4° (Kennett, 1982). Strike-slip transform margins generally have

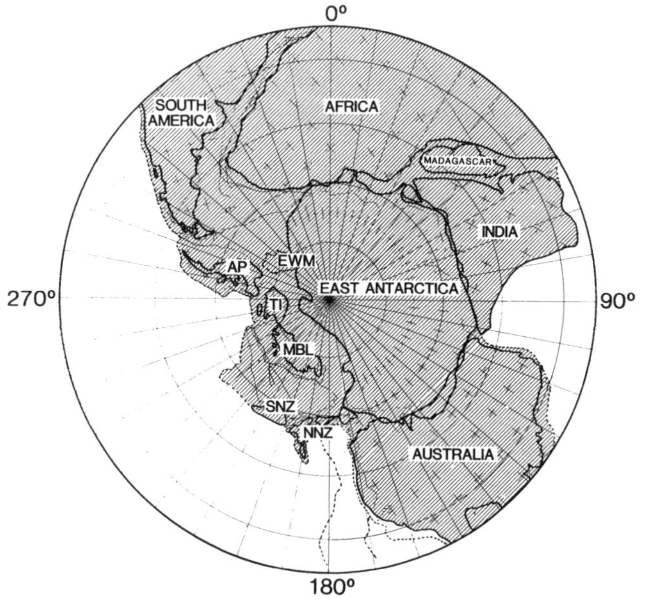

Fig. 1. Early Mesozoic reconstruction of Gondwana based on Lawver & Scotese (1987). AP, Antarctic Peninsula; MBL, Marie Byrd Land; SNZ, south New Zealand; NNZ, north New Zealand. Continental blocks are reconstructed to a fixed East Antarctica. The Antarctic Peninsula has been rotated with respect to East Antarctica.

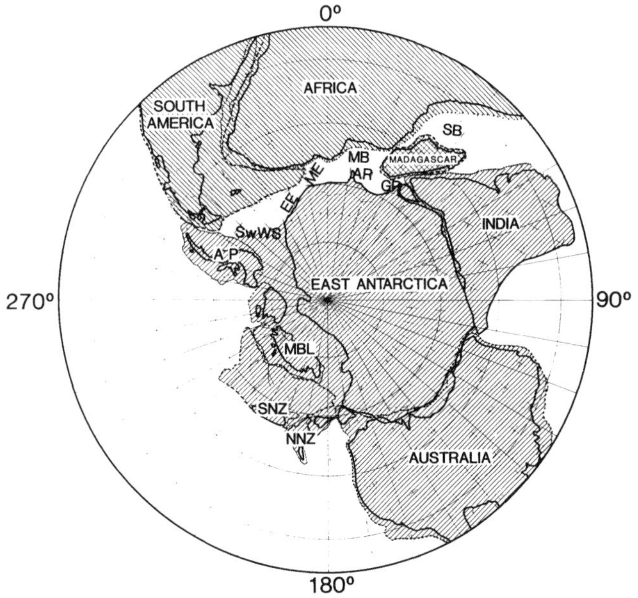

Fig. 2. Magnetic anomaly M15 time (140 Ma) reconstruction of West Gondwana (South America–Africa) and East Gondwana. Seafloor spreading has taken place in the Somali Basin, Mozambique Basin and south-western Weddell Sea. A sheared transform margin is present along the Explora Escarpment. SB, Somali Basin; MB, Mozambique Basin; SwWS, south-western Weddell Sea; ME, Mozambique Escarpment; EE, Explora Escarpment; AR, Astrid Ridge; GR, Gunnerus Ridge.

very steep slopes (up to 45°) whereas subduction margins, which may be steep in places, nearly always have obvious subduction-related features. Rifted passive margins usually have gentle slopes of 4° or less. If the Mozambique Escarpment is a sheared margin, it is reasonable to ask what was sheared off it. Part of the East Antarctic margin (Explora Escarpment) from about 10° to about 25° W also has a very steep slope. A SEABEAM survey (unpublished data, Alfred-Wegener-Institute) between 13°30′ and 15° W revealed a continental slope of 45°. The strike of this margin is roughly parallel to the direction of spreading recorded by the anomalies to the east of Astrid Ridge (Bergh, 1987). The Explora Escarpment between 10° and 25° W may have been the conjugate sheared transform margin to the Mozambique Escarpment. If that is correct, then the reconstruction of Gondwana by Lawver & Scotese (1987) is too tight for a Middle Jurassic fit, particularly considering the position of Madagascar with respect to Africa.

Reeves, Karanja & Macleod (1987) produced an identical fit for Madagascar and Africa as Lawver & Scotese (1987) but suggested that active rifting in a triple junction centred on the Kenyan coast may have begun as early as the Permo-Carboniferous, contemporaneous with the initial Karoo deposition. It is possible that this early rifting along the Mombasan coast (Reeves *et al.*, 1987) which was contemporaneous with the initial Karoo deposition, extended south to Durban. This slow early extension may have produced the Mozambique Plateau between the eastern margin of the Falkland Plateau and East Antarctica.

Kristoffersen & Haugland (1986) discussed the geophysical evidence for the East Antarctic plate boundary in the Weddell Sea. While they discussed the possibility that the Explora Escarpment section of the East Antarctic margin between 12° and 19° W represents a sheared margin overprinted by rifting, they connected it with the colinear Andenes Escarpment that extends to 40° W and suggested that both escarpments were formed by rifting. The 45° slope of the Explora Escarpment is more indicative of a sheared margin than a rifted one. As suggested by Lawver, Sclater & Meinke (1985) the sheared margin of Dronning Maud Land only operated as such from break-up (170 Ma) to approximately 130 Ma when separation between South America and Africa (Fig. 3) resulted in a triple junction that replaced the original two-plate break-up between West (South America–Africa) and East Gondwana. Consequently we propose that the Explora Wedge was first rifted during a period of pre-break-up extension similar in age to the Permo-Carboniferous rifting along the Mombasan Coast (Reeves *et al.*, 1987) and then sheared between 170 and 130 Ma. The Explora Wedge was sliced off during the shearing period and is now represented by the deep sediments landward of the Explora Escarpment and those left behind on the Mozambique Ridge. The Andenes Escarpment is undoubtedly related to the younger rifting event (~ 130 Ma to present) that produced much of the seafloor in the Weddell Sea and is only coincidentally in-line with the Explora Escarpment.

Many researchers have discussed the problem of the overlap of the Antarctic Peninsula with the Falkland Plateau in Gondwana reconstructions (Dalziel & Elliot, 1982; Lawver *et al.*, 1985; Dalziel *et al.*, 1987a). The most reasonable solution seems to be a clockwise rotation of the Antarctic Peninsula with respect to East Antarctica during the early stages of break-up since any overlap problem disappears by M0 (118 Ma) time according to Lawver *et al.* (1985). Such a scenario is supported by palaeomagnetic data from the Ellsworth Mountains, the Antarctic Peninsula and Thurston Island (Grunow, Kent & Dalziel, 1987). If the Late Jurassic–Early Cretaceous opening of the Mozambique Basin was

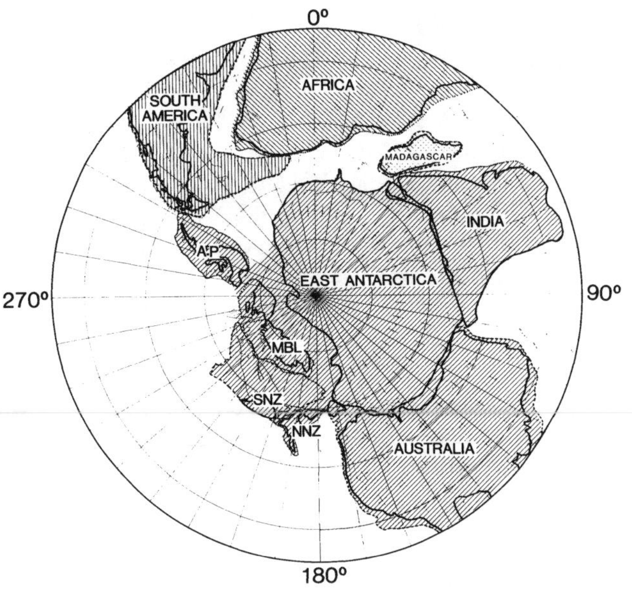

Fig. 3. Magnetic anomaly M10 time (130 Ma) reconstruction of Gondwana. Rifting has started in the South Atlantic between South America and Africa producing a reorganization of the seafloor spreading in the Weddell Sea at M12 (132 Ma) time.

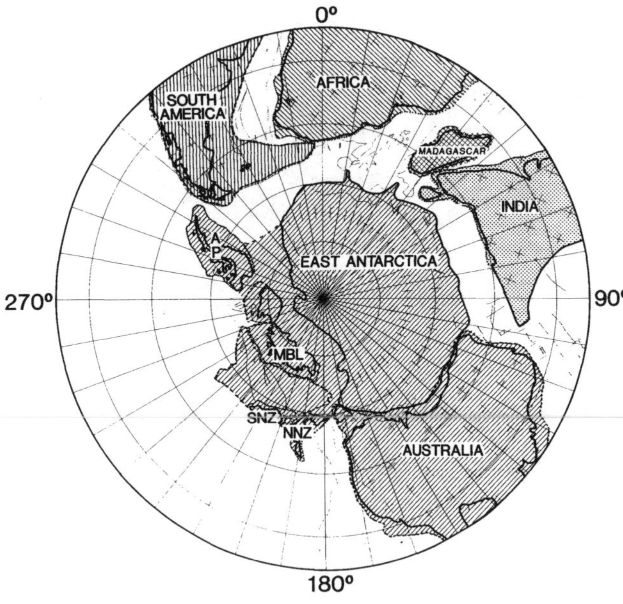

Fig. 4. Magnetic anomaly M10 time (118 Ma) reconstruction of Gondwana. Rifting between India and Antarctica has begun with the cessation of seafloor spreading in the Somali Basin. West Antarctica, with the exception of Marie Byrd Land, has assumed its present-day geographical location with respect to East Antarctica.

translated along the Explora Escarpment to the south-westernmost Weddell Sea, earliest opening of the Weddell Sea would have resulted in a clockwise rotation of the Antarctic Peninsula with respect to East Antarctica. In contrast, if the Weddell Sea began opening in a N–S direction prior to anomaly M25 (156 Ma) time (LaBrecque & Barker, 1981; LaBreceque & Cande, 1987), then it is difficult to produce clockwise rotation of the Antarctic Peninsula between break-up and M0 time. As Gondwana first broke up, the Antarctic Peninsula is assumed to have slid along the western edge of the tip of South America. Since there is no geological evidence to support such a sheared margin, it is premature to assume that that is what actually happened.

Enderby Land–Queen Mary Coast/Sri Lanka–India

The next major event to affect the continental margin of present-day Antarctica was the rifting of India/Sri Lanka from East Antarctica between 35° and 95° E (Fig. 4). The Proterozoic Shillong Plateau in north-eastern India (present location 27° N, 94° E) is considered to be the easternmost point of lesser India that must have fitted against East Antarctica. With the Australia–Wilkes Land fit constrained by GEOSAT data (Sandwell & McAdoo, 1988) and the location of Sri Lanka/lesser India constrained by the Gunnerus Ridge (Lawver *et al.*, 1985), the very western tip of the Naturaliste Plateau fits against the Shillong Plateau. This assumes that there has been no motion between the Naturaliste Plateau and Australia since break-up. Marine magnetic anomalies immediately north-west of Naturaliste Plateau have been dated as M0–M5 (118–127 Ma) by Veevers *et al.* (1985). If Naturaliste Plateau is not a continental fragment older than the initial break-up (Coleman, Michael & Mutter, 1982) then M5 (127 Ma) may not be the age of the break-up of India–Antarctica. To the north, the magnetic anomalies closest to the Continent–Ocean Boundary (COB) of western Australia are identified as M9 or M10 (129–130 Ma). In the Argo Abyssal Plain off north-western Australia, the oldest identified anomalies are older than 160 Ma (Heirtzler *et al.*, 1978). While it is clear that whatever rifted off north-western Australia was not part of greater India, whatever rifted off western Australia from south of the Argo Abyssal Plain to the Naturaliste Plateau may have been attached or partially attached to lesser India. The difference in total spreading between M9 or M10 and M5 time is only 230 km. Such an offset could have taken place on a few strike-slip faults in greater India. R.A. Livermore & F.J.Vine (pers. comm.) envisage a similar scenario for the early evolution of the Indian Ocean but argue that some other block, perhaps Tibet, rifted off western Australia rather than greater India. We suggest that the difference of only a few million years in age, makes it unlikely that whatever rifted off western Australia was totally unattached to lesser India, and that Tibet may have rifted off north-western Australia 160 m.y. ago.

Space problems with East Antarctica, Madagascar and Africa preclude India moving away from Antarctica at M10 (130 Ma) time or earlier (Lawver *et al.*, 1985; R.A. Livermore & F.J. Vine, pers. comm.). If the M5 anomalies off the Naturaliste Plateau indicate initial India–Antarctica spreading in the east then the active spreading centre off western Australia may have initiated westward propagation of the rift. Without identifiable magnetic anomalies off Enderby Land or off the east coast of India, the exact age of the seafloor spreading between India and East Antarctica can not be determined to closer than M5–M0 time (127–118 Ma). Rifting between Sri Lanka and India started in the Early Cretaceous (130 Ma) according to Katz (1978), who suggested that as much as 200 km of separation between Sri Lanka and India

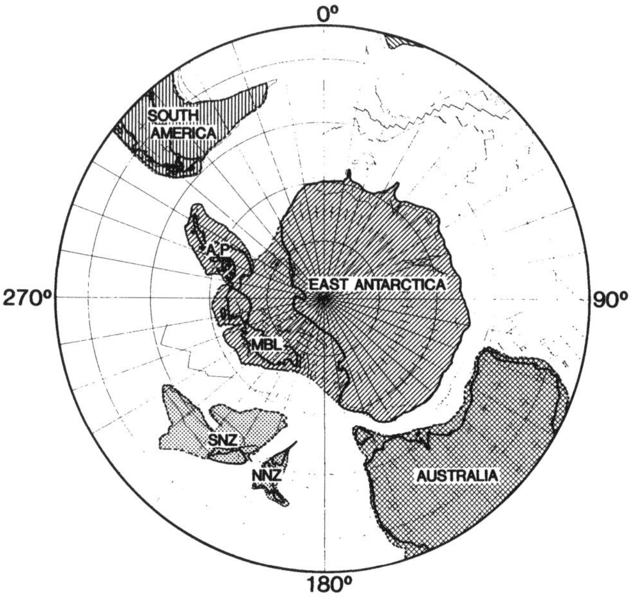

Fig. 5. Magnetic anomaly A30 time (70 Ma) reconstruction of Gondwana. Stretching between Marie Byrd Land and East Antarctica has produced the Antarctic continental mass as it is presently configured. Rifting between Australia and Antarctica has begun as well as rifting between south New Zealand and Marie Byrd Land.

resulted in the formation of the NE–SW Cauvery–Palk Basin in the Gulf of Mannar. Such rifting may have represented the initiation of eastwardly propagating rifting between India and East Antarctica–Sri Lanka. At roughly the same time, westwardly propagating rifting was initiated between India and the Naturaliste Plateau of Australia–Antarctica. The Cauvery–Palk Basin rifting, which may have started as an extension of the Mozambique Basin rifting, failed and Sri Lanka remained as part of India when the successful westwardly propagating rifting detached it from Antarctica.

Wilkes Land/Australia

Cande & Mutter (1982) revised the identification of seafloor-spreading anomalies between Australia and Antarctica. They suggested that spreading could be separated into two distinct phases: break-up to anomaly 20 (45 Ma) at 4.5 mm/y, and anomaly 20 to present at 22–38 mm/y (half-rate). Very slow spreading in regions of continental break-up frequently produces confused magnetic signatures because of high sedimentation blanketing the spreading centres. Cande & Mutter (1982) concluded that break-up and initial rifting could have occurred anytime between 110 and 90 Ma ago (Fig. 5).

Veevers (1986) put the age of the break-up as mid-Cretaceous (95 ± 5 Ma) based on seismic data from the Australian margin. There is a short mid-Cretaceous lacuna in the Otway Basin between the block-faulted Early Cretaceous rift-valley sediments of the Otway Group and the overlying relatively unfaulted Late Cretaceous Sherbrook Group (Veevers, 1986). Although Veevers (1986) was able clearly to identify the COB on the Australian margin, he had only limited data on the conjugate Antarctic margin. With the recent GEOSAT data (Sandwell & McAdoo, 1988), the COB gravity anomaly is now clearly delineated on the Antarctic margin as well. In fact, the meanderings of the COB on the Australian margin at 118°, 120°, 124° and 126° E, are mirrored in the GEOSAT gravity anomaly on the Antarctic margin at 110°, 115°, 123° and 127° E. In addition the GEOSAT-derived gravity anomalies clearly show the tracks that Tasmania and the Tasman Rise followed off the Antarctic margin. It is clear that the initial slow rifting of Australia away from Antarctica had a certain westward component as evidenced by the GEOSAT anomalies.

Ross Sea/New Zealand/Marie Byrd Land

Marine magnetic anomalies have been identified off New Zealand in the south-west Pacific as anomaly 32 (72 Ma) by Stock & Molnar (1987). In addition to the seafloor spreading that occurred to the west between Australia and Antarctica, seafloor spreading was active in the Tasman Sea from prior to anomaly 33 (76 Ma) time according to Stock & Molnar (1982). Stretching in the Ross Sea has been suggested by Cooper, Davey & Hinz (this volume, p. 285), Bradshaw (this volume, p. 581) and Fitzgerald *et al.* (unpublished data). They variously date the extension as Palaeocene (Bradshaw), late Mesozoic 'early-rift' (Cooper *et al.*), and Late Cretaceous–early Cenozoic (Fitzgerald *et al.*). Bradshaw (this volume, p. 581) discusses the convergent Gondwanide margins in north-eastern Australia and New Zealand that continued to develop in the Early Cretaceous but 'ended abruptly about 100 Ma ago, probably due to the oblique convergence and collision of the Pacific–Phoenix Ridge'. He also states that the evidence for rifting and extension between Antarctica and New Zealand dates back to 100 Ma ago, which suggests that the subduction of the Pacific–Phoenix Ridge to the north, may have been a major factor in the initiation of the extension and rifting between New Zealand and Antarctica.

In the reconstruction of Gondwana shown in Fig. 1, 40–50% extension between Marie Byrd Land and East Antarctica has been assumed (Lawver & Scotese, 1987). This extension coincides with earliest rifting between Australia–East Antarctica but seems to predate the south-west Pacific seafloor spreading (Stock & Molnar, 1987). Bradshaw (this volume, p. 581) indicates that extension seems to have started near the Antarctic–Indian–Australian triple junction and spread eastwards, being older south of Australia than in the Tasman region and youngest in the Ross Sea. It is possible that the Ross Sea embayment represents a failed rift and that the seafloor spreading then propagated between Marie Byrd Land and Campbell Plateau–Chatham Rise to join the active spreading to the east.

The GEOSAT-derived gravity anomalies shown in Fig. 6 indicate that the prominent Udintsev fracture zone (50° S, 210° E–65° S, 240° E) can be extended into the Antarctic margin at Pine Island Bay, between Thurston Island and Marie Byrd Land. The northern extension of the Udintsev fracture zone continues to the north-eastern point of Chatham Rise. This confirms that the Grindley & Davey (1982) reconstruction of New Zealand in Gondwana is essentially correct. The relatively fast spreading between the Pacific and Antarctic plates (36 mm/y (half-rate); Stock & Molnar, 1987) and the moderately fast spreading in the Tasman Sea (21 mm/y (half-

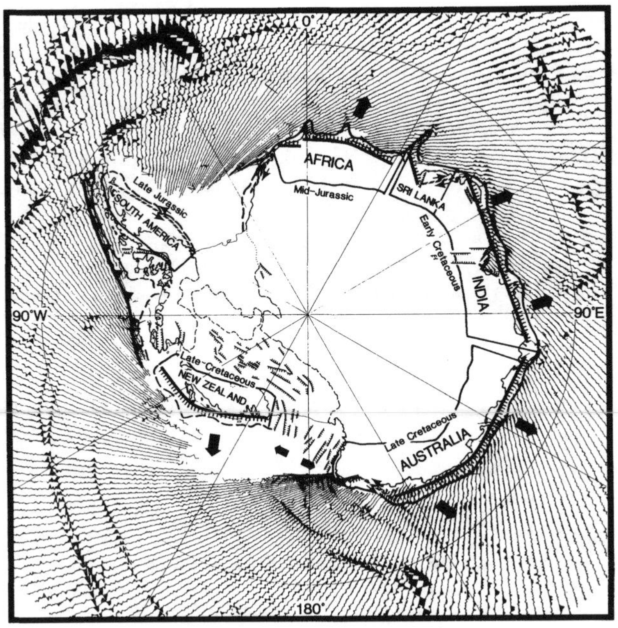

Fig. 6. Summary chart of present-day Antarctica showing the continental blocks that rifted off Antarctica and the time of the rifting. Chart shows the GEOSAT-derived gravity anomalies (Sandwell & McAdoo, 1988). The prominent Udintsev fracture zone can be seen at 60 °S, 125 °W. The tracks marking the northward path of Tasmania are easily seen at 60 °S, 150 °E. Undulations in the continental-edge gravity anomaly along the Wilkes Land margin mirror the continental slope of Australia.

rate); Hayes & Ringis, 1973) contrast sharply with the very slow spreading between Antarctica and Australia (4.5 mm/y; Veevers, 1986) and the stretching in the Ross Sea that was occurring at the same time. The four areas all suggest roughly the same time for initiation of stretching and seafloor spreading where it occurred. Since seafloor spreading died out in the Tasman Sea just before the major increase in spreading rate between Antarctica and Australia, a suggested scenario for this very complicated region is that seafloor spreading was propagating slowly eastward as suggested by Veevers (1986) between Antarctica and Australia. At the same time spreading between Antarctica and the Pacific plate was propagating westward since there appears to be slightly more room between Chatham Rise and Marie Byrd Land than there is between Campbell Plateau and western Marie Byrd Land. The Pacific–Antarctic seafloor spreading continued into the Tasman Sea with two triple junctions producing changes in ridge orientation. The eastern one resulted in stretching in the Ross Sea region whereas the other formed the intersection of the spreading centres between the East Antarctic, Australian and Pacific plates with two relatively fast spreading ridges and one very slow one (Antarctic–Australian).

Convergent margin: Pine Island Bay–Antarctic Peninsula

The convergent Gonwanide margin in north-eastern Australia and New Zealand was active in the Early Cretaceous but ceased about 100 Ma ago (Bradshaw, this volume, p. 581). Barker (1982) showed the subduction boundary between the Phoenix plate and the remnant of the Gondwana plate at 100 Ma as a continuous subduction zone from north-west of 'New Zealand' to the tip of the Antarctic Peninsula. Because there are no identified magnetic anomalies immediately off West Antarctica between the Udintsev fracture zone and the Tharp fracture zone, it is not possible to speculate as to the timing and nature of this boundary but we assume that it was probably a subduction boundary until some time between 100 and 80 Ma ago.

Anomaly 31 (70 Ma) is identified more than 1000 km offshore of Thurston Island, so it can be assumed that seafloor older than anomaly 34 exists closer inshore. Since the conjugate seafloor to that off Thurston Island extends more than 1000 km past the Chatham Rise to the Kermadec Trench region, there is no real space problem that would necessitate continued subduction. Conversely, since there is active subduction at the Kermadec Trench, it can be assumed that some seafloor was subducted beneath the Thurston Island region. Unlike the seafloor north of the Tharp fracture zone, where it is documented (Cande, Herron & Hall, 1982) that a Phoenix–Antarctic spreading centre was subducted at the plate margin resulting in a cessation of subduction (Barker, 1982), the undated older (probably Cretaceous Quiet Zone) anomalies between the Udintsev and Tharp fracture zones do not imply when subduction ceased since we can not be sure that an active spreading centre was subducted.

If the magmatic activity in the Transantarctic Mountains can be related to subduction along the pacific margin of Gondwana then subduction was occurring prior to 175 Ma (Dalziel *et al.*, 1987b). From the Tharp fracture zone northwards, Barker (1982) dated the cessation of subduction by the age of the oldest identified magnetic anomalies offshore and the age of the magmatic rocks of the Antarctic Peninsula. Subduction ceased about anomaly 21 time (49 Ma) between the Tharp and Heezen fracture zones, about anomaly 6 time (25–20 Ma) between the Heezen and Anvers fracture zones and within the last 5–8 m.y. for the remainder of the Antarctic Peninsula.

Conclusions

The present-day Antarctic plate boundary is far removed from the Antarctic continental margin except in the immediate vicinity of the northern tip of the Antarctic Peninsula. The continental margin of the Antarctic Peninsula has had the longest history of tectonic complications of any of the Antarctic margin in the last 175 m.y. Prior to the break-up of Gondwana it was a westward-facing subduction margin. Back-arc basin seafloor spreading in the Rocas Verdes region of southern South America, extension between the Antarctic Peninsula and East Antarctica coupled with the intrusion of Jurassic magmas along the Transantarctic Mountains and in southern Africa, and seafloor spreading in the northern and western Somali basins and the Mozambique Basin, all resulted in the two-plate break-up of West and East Gondwana. At about 130 Ma, South America and Africa broke apart resulting in a complete reorganization of spreading in the Weddell Sea. The previously rifted, then sheared Explora Escarpment, once again became a rifted margin that coincided with the

Andenes Escarpment to as far west as 44° W (Kristoffersen & Haugland, 1986).

Although rifting in the Mozambique Basin may have contributed to some extension between India and Sri Lanka (in the present-day Gulf of Mannar) between anomaly M10 and M0 time, major rifting between India and Antarctica probably started in the east and propagated westward between M5 (127 Ma) and M0 time (118 Ma). It is impossible to constrain the time of break-up between India and Antarctica any closer than 127–118 Ma. With the break-up of India and East Antarctica, seafloor spreading stopped in the Somali Basin. The clockwise rotation of the Antarctic Peninsula with respect to East Antarctica stopped, leaving Marie Byrd Land as perhaps the only part of continental Antarctica not in its present-day location. Marie Byrd Land reached its present position with respect to Antarctica by stretching of the Ross Sea embayment region during the Late Cretaceous and early Cenozoic.

By 100 Ma, Gondwanide subduction beneath north-eastern Australia and New Zealand stopped abruptly (Bradshaw, this volume, p. 581) and initial stretching, premonitory to seafloor spreading, was beginning between Antarctica, Australia, New Zealand and Marie Byrd Land. By anomaly 34 time (82 Ma) the last of the rifted passive margins of present-day Antarctica had formed opposite Australia and New Zealand. Active subduction continued along the Antarctic Peninsula coast until almost 4 Ma ago off the South Shetland Islands. At approximately 30 Ma seafloor spreading began in the Scotia Sea which bounds the northern tip of the Peninsula with a left-lateral transform fault (Barker & Dalziel, 1983). Although the tectonic evolution of the Antarctic continent seems to be almost complete, it may in fact be starting all over again with the recently active extension in the Ross Sea region (Cooper *et al.*, this volume, p. 285).

Acknowledgements

This paper would not have been possible without the previous work of numerous Antarctic investigators, not all of whom could be cited in the reference list. We thank Ian Dalziel, Cathy Mayes and John Sclater for reading this manuscript. We appreciate the comments of Mike Thomson and an anonymous reviewer. We thank Prof. Heinz Miller for use of his unpublished Seabeam map. This work was supported by Grant No. DPP 86-15307 and by the Paleoceanographic Mapping Project of the Institute for Geophysics. This is the University of Texas, Institute for Geophysics contribution no. 0742.

References

Barker, P.F. (1982). The Cenozoic subduction history of the Pacific margin of the Antarctic Peninsula: ridge crest–trench interaction. *Journal of the Geological Society, London*, **139(6)**, 787–801.

Barker, P.F. & Dalziel, I.W.D. (1983). Progress in geodynamics in the Scotia Arc region. In *Geodynamics of the Eastern Pacific Region, Caribbean and Scotia Arcs*, Geodynamics Series 9, ed. R. Cabré, pp. 137–70. Washington, DC; American Geophysical Union.

Bergh, H.W. (1987). Underlying fracture zone nature of Astrid Ridge off Antarctica's Queen Maud Land. *Journal of Geophysical Research*, **92(B1)**, 475–84.

Cande, S.C., Herron, E.M. & Hall, B.R. (1982). The early Cenozoic tectonic history of the southeast Pacific. *Earth and Planetary Science Letters*, **57(1)**, 63–74.

Cande, S.C. & Mutter, J.C. (1982). A revised identification of the oldest sea-floor spreading anomalies between Australia and Antarctica. *Earth and Planetary Science Letters*, **58(2)**, 151–60.

Cochran, J.R. (1988). The Somali Basin, Chain Ridge and the origin of the Northern Somali Basin gravity and geoid low. *Journal of Geophysical Research*, **93(B10)**, 11985–2008.

Coleman, P.J., Michael, P.J. & Mutter, J.C. (1982). The origin of the Naturaliste Plateau, SE Indian Ocean: implications from dredged basalts. *Journal of the Geological Society of Australia*, **29**, 457–68.

Dalziel, I.W.D. & Elliot, D.H. (1982). West Antarctica: problem child of Gondwanaland. *Tectonics* **1(1)**, 3–19.

Dalziel, I.W.D., Garrett, S.W., Grunow, A.M., Pankhurst, R.J., Storey, B.C. & Vennum, W.R. (1987a). The Ellsworth–Whitmore Mountains crustal block: its role in the tectonic evolution of West Antarctica. In *Gondwana Six: Structure, Tectonics, and Geophysics*, Geophysical Monograph 40, ed. G.D. McKenzie, pp. 173–82. Washington, DC; American Geophysical Union.

Dalziel, I.W.D., Storey, B.C., Garrett, S.W., Grunow, A.M., Herrod, L.D.B. & Pankhurst, R.J. (1987b). Extensional tectonics and the fragmentation of Gondwanaland. In *Continental Extensional Tectonics*, ed. J.F. Dewey, M.P. Coward & P. Hancock, pp. 433–41. Oxford; Special Publication of the Geological Society of London, No. 28.

Grindley, G.W. & Davey, F.J. (1982). The reconstruction of New Zealand, Australia, and Antarctica. In *Antarctic Geoscience*, ed. C. Craddock, pp. 15–29. Madison; University of Wisconsin Press.

Grunow, A.M., Kent, D.V. & Dalziel, I.W.D. (1987). Mesozoic evolution of West Antarctica and the Weddell Sea Basin: new paleomagnetic constraints. *Earth and Planetary Science Letters*, **86(1)**, 16–26.

Hayes, D.E. & Ringis, J. (1973). Seafloor spreading in the Tasman Sea. *Nature, London*, **243(5408)**, 454–8.

Heirtzler, J.R., Cameron, P., Cook, P.J., Powell, T., Roeser, H.A., Sukardi, S. & Veevers, J.J. (1978). The Argo Abyssal Plain. *Earth and Planetary Science Letters*, **41(1)**, 21–31.

Katz, M.B. (1978). Sri Lanka in Gondwanaland and the evolution of the Indian Ocean. *Geological Magazine*, **115(4)**, 237–44.

Kennett, J.P. (1982). *Marine Geology*, Englewood Cliffs, NJ; Prentice-Hall, Inc.

Kristoffersen, Y. & Haugland, K. (1986). Geophysical evidence for the East Antarctic plate boundary in the Weddell Sea. *Nature, London*, **322(6079)**, 538–41.

LaBrecque, J.L. & Barker, P.F. (1981). The age of the Weddell Basin. *Nature, London*, **290(5806)**, 489–92.

LaBrecque, J.L. & Cande, S.C. (1987). Total intensity magnetic anomaly profiles, south. In *South Atlantic Ocean and Adjacent Antarctic Continental Margin, Atlas 13*, sheet 9, ed. J.L. LaBrecque. Woods Hole, MA; Marine Science International. Ocean Margin Drilling Program, Regional Atlas Series.

Lawver, L.A. & Scotese, C.R. (1987). A revised reconstruction of Gondwanaland. In *Gondwana Six: Structure, Tectonics, and Geophysics*, Geophysical Monograph 40, ed. G.D. McKenzie, pp. 17–24. Washington, DC; American Geophysical Union.

Lawver, L.A., Sclater, J.G. & Meinke, L. (1985). Mesozoic and Cenozoic reconstruction of the South Atlantic. *Tectonophysics*, **114**, 233–54.

Palmer, A.R. (1983). The decade of North American geology 1983 geologic time scale. *Geology*, **11(11)**, 503–4.

Reeves, C.V., Karanja, F.M. & Macleod, I.N. (1987). Geophysical evidence for a failed Jurassic rift and triple junction in Kenya. *Earth and Planetary Science Letters*, **81(2/3)**, 299–311.

Sandwell, D.T. & McAdoo, D.C. (1988). Marine gravity of the Southern Ocean and Antarctic margin from GEOSAT: tectonic implications. *Journal of Geophysical Research* **93(B9)**, 10389–96.

Ségoufin, J. (1978). Anomalies magnetiques mesozoiques dans le bassin de Mozambique. *Comptes rendus hebdomadaires des séances, Académie des sciences, Paris*, **287D(2)**, 109–12.

Ségoufin, J. & Patriat, P. (1980). Existence d'anomalies mesozoiques dans le bassin de Somalie, Implications pour les relations Afrique–Antarctique–Madagascar. *Comptes rendus séances, Académie des sciences*, **291B(2)**, 85–8.

Simpson, E.S.W., Sclater, J.G., Parsons, B., Norton, I.O. & Meinke, L. (1979). Mesozoic magnetic lineations in the Mozambique Basin. *Earth and Planetary Science Letters*, **43(2)**, 260–4.

Stock, J. & Molnar, P. (1982). Uncertainties in the relative positions of the Australia, Antarctica, Lord Howe, and Pacific plates since the Late Cretaceous. *Journal of Geophysical Research*, **87(B6)**, 4697–714.

Stock, J. & Mclnar, P. (1987). Revised history of early Tertiary plate motion in the southwest Pacific. *Nature, London*, **325(6104)**, 495–9.

Veevers, J.J. (1986). Break-up of Australia and Antarctica estimated as mid-Cretaceous (95 ± 5 Ma) from magnetic and seismic data at the continental margin. *Earth and Planetary Science Letters*, **77(1)**, 91–9.

Veevers, J.J., Tayton, J.W., Johnson, B.D. & Hansen, L. (1985). Magnetic expression of the continent–ocean boundary between the western margin of Australia and the eastern Indian Ocean. *Journal of Geophysics*, **56(2)**, 106–20.

Triassic–Early Cretaceous evolution of Antarctica

D.H. ELLIOT

Byrd Polar Research Center and Department of Geology and Mineralogy, The Ohio State University, Columbus, Ohio 43210, USA

Abstract

Palaeomagnetic data suggest that Antarctica has existed in its present form from the mid-Cretaceous, and therefore all major reorganization of the original Pacific margin of Gondwana took place in late Middle Jurassic–Early Cretaceous time. Geological evidence indicates that the major translations and rotations must have occurred in the back-arc region of the Gondwana active plate margin which spanned at least the Permian–Triassic. In this region, thermal perturbations associated with break-up induced doming and erosion, and silicic and tholeiitic magmatism. This was also accompanied or followed by crustal extension. Initial movements between East and West Gondwana must have been accommodated in this extensional zone while subduction was active along the Pacific margin. Constraints on reassembly and movements of crustal blocks are given by palaeomagnetic data and Mesozoic seafloor magnetic anomaly data off Queen Maud Land, including the Weddell Sea. The Explora and Andenes escarpments may reflect early movements as well as later rifting. Critical data for understanding the early evolution of the Antarctic Plate include the Weddell Sea M anomalies, and the nature and age of the Explora–Andenes Escarpment.

Introduction

In the evolution of Antarctica, the Jurassic–Early Cretaceous periods encompass a critical time, during which Gondwana started to break up and the fragments to disperse. The original configuration of Antarctica within a reconstructed Gondwana remains uncertain. However, West Antarctica consists of a limited number of continental blocks (Dalziel & Grunow, 1985) that represent fragments of the Gondwana plate margin (Dalziel & Elliot, 1982) rearranged by translation and rotation into their present configuration (largely by 100 Ma; Watts, Watts & Bramall, 1984). The purpose of this paper is to review the relevant aspects of the Permian–Early Cretaceous geology of the continent, examine some of the constraints on understanding the early stages of the reorganization of West Antarctica, and to discuss some of the implications of those constraints.

Geological history

Gondwana plate margin

The active plate margin of Gondwana existed from the Permian to the Late Triassic–Early Jurassic. Lithotectonic terrains associated with this margin can be recognized in parts of the Transantarctic Mountains, West Antarctica and the Antarctic Peninsula (Fig. 1), as well as the formerly adjacent continents. The fore-arc region is well developed in the South Orkney Islands, the Antarctic Peninsula and New Zealand, which at that time was located adjacent to the Ross Sea and Marie Byrd Land. In the peninsula sector, fore-arc terrain rocks (Dalziel, 1984) consist of thick greywacke–shale sequences that locally contain Late Triassic–Early Jurassic fossils (Thomson & Tranter, 1986), and metamorphic rocks representing pelagic cherts, volcanic rocks, limestones and clastic sediments. These sequences include both fore-arc basin deposits and subduction complexes. Deformation in this sector can be dated as post-Triassic or, locally in Alexander Island, post-Early Jurassic and pre-Late Jurassic. In fact the Alexander Island rocks probably record a continuum of subduction complex sedimentation and episodes of deformation during the mid–late Mesozoic, because Cretaceous radiolaria have been recovered from this thick deformed sequence (Burn, 1984). Evidence for the fore-arc terrane is lacking in the rest of West Antarctica, but is abundant in New Zealand where volcaniclastic and volcanic strata of calc-alkaline affinities (Bradshaw, Adams & Andrews, 1981) span the Permian, Triassic and Jurassic. Deformation of the sequences in New Zealand occurred in the Late Permian and the Late Triassic (Rangitata I), though the main orogenic episode (Rangitata II) was late Early Cretaceous (Bradshaw *et al.*, 1981).

The magmatic arc is documented by volcanic detritus in both the fore-arc terrain and back-arc (*sensu lato*) region which includes the Transantarctic and Ellsworth mountains; however, plutonic and metamorphic rocks of the arc itself are poorly represented. Plutons crop out on the Antarctic Pen-

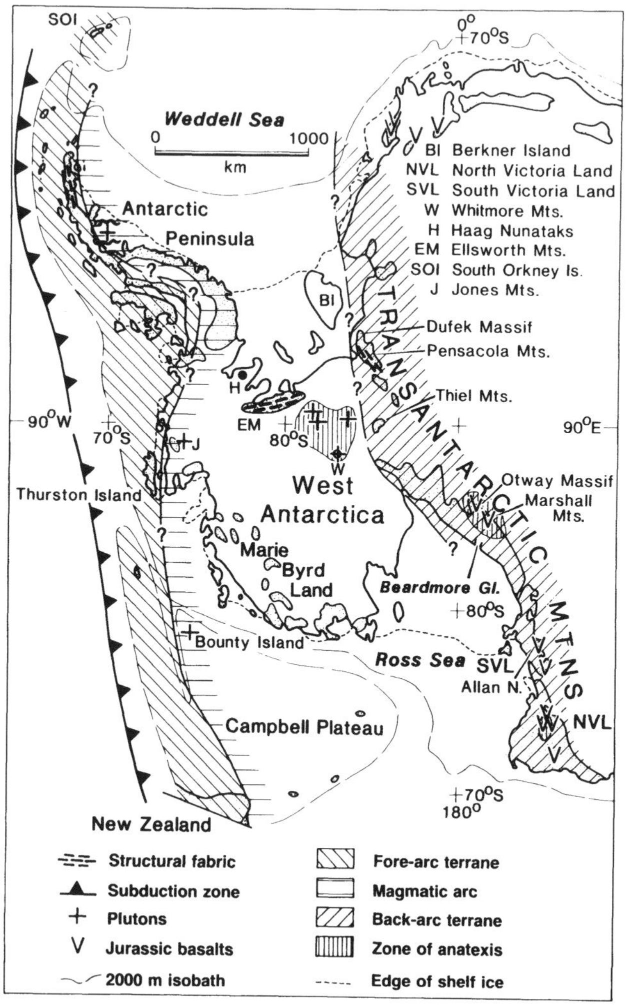

Fig. 1. Schematic illustration of the Gondwana plate margin in the Antarctica–New Zealand sector during the Permian–Early Jurassic. Because of uncertainties in reconstruction of the Gondwana margin, present geography is used. Removal of crustal extension in the Ross Embayment would close up Marie Byrd Land and the Transantarctic Mountains. Similarly, if the Campbell Plateau crust has been extended, restoration would close up New Zealand and the craton. Rotations and translations in the Weddell Sea sector and removal of crustal extension in the Weddell embayment also have to be accommodated in any reconstruction. The Middle Jurassic plutonic rocks of West Antarctica, and the Lower and Middle Jurassic silicic and basaltic rocks of the Transantarctic Mountains are superimposed. The Middle Jurassic magmatic rocks of the northern Palmer Land region are omitted in order to simplify the figure. Back-arc is used here and in Fig. 3 in a generalized sense.

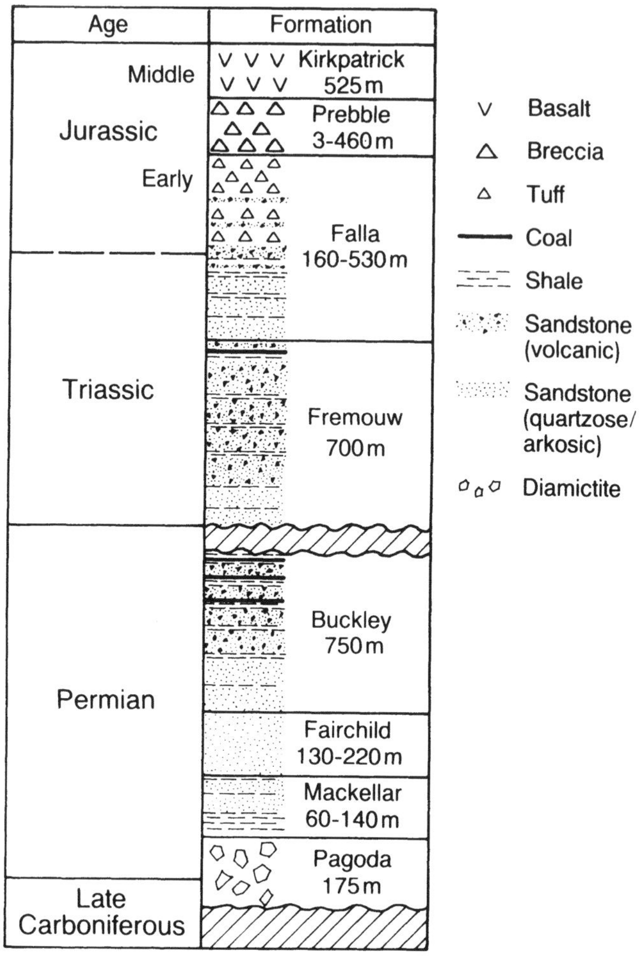

Fig. 2. Simplified stratigraphic column for the Victoria Group of the Beacon Supergroup in the central Transantarctic Mountains.

insula (Pankhurst, 1982; Meneilly *et al.*, 1987) and in the Jones Mountains, and have ages ranging between 209 and 199 Ma; a single Permian pluton is known in Marie Byrd Land. Gneisses in the central Antarctic Peninsula also have compatible radiometric ages (Pankhurst, 1983). The Jurassic plutons on the Ellsworth–Thiel Ridge are discussed later. Although Permian plutons occur in New Zealand, younger intrusive bodies are known only from Bounty Island where a granite has yielded a 193 Ma date.

The back-arc terrain in Antarctica (Dalziel & Elliot, 1982) is represented by the Permian–Lower Jurassic Victoria Group of the Beacon Supergroup (Fig. 2) and the Permian strata in the Ellsworth Mountains. Both constitute part of the depositional record of a retro-arc foreland basin, and both contain abundant volcanic detritus as well as airfall material. In the Victoria Group in the central Transantarctic Mountains (CTM) the influx of volcanic detritus from West Antarctica, recorded in Permian–Triassic sandstones and sparse tuffs, is succeeded in the upper Falla and Prebble formations by locally-derived volcaniclastic deposits (Barrett, Elliot & Lindsay, 1986). The lower part of the Falla, below the tuffaceous sequence, contains the Middle–Late Triassic *Dicroidium* flora and palynomorphs assigned a possible Late Triassic age. The upper part lacks palaeontological control, but radiometric data, which will be discussed later, suggests the sequence passes up into the Early Jurassic. In south Victoria Land, the upper Beacon rocks, the Lashly Formation, have a large volcanic component (Collinson, Pennington & Kemp, 1983), but apparently lack the airfall debris present in the Beardmore region. The Lashly Formation is dated palaeontologically as Middle–Late Triassic. Triassic strata also crop out in north Victoria Land and consist of a thin sequence of quartzose sediments which contain silicic shards (Collinson, Pennington & Kemp, 1986).

In the Antarctic sector, here considered to include New Zealand and the Campbell Plateau, the history of the lithotectonic terrains identified with the Gondwana plate margin is moderately well documented for the Permian and Triassic, but is only sparsely so for the Early Jurassic. During the Early

Jurassic, subduction-related magmatic activity diminished and its record is largely localized on the Pacific margin. However, a new phase of magmatic activity with associated rifting was initiated at this time in West Antarctica and the Transantarctic Mountains.

Jurassic rifting and magmatism

In the CTM, the Falla Formation records the waning stages of sediment infilling of the Beacon basin, and replacement of dominantly clastic sedimentation by locally-erupted silicic–volcanic deposits. These are overlain by the coarse pyroclastic rocks of the Prebble Formation and the tholeiitic lavas of the Ferrar Supergroup (Barrett *et al.*, 1986).

The base of the Prebble Formation is marked by the incoming of coarse pyroclastic deposits. Both the upper Falla tuff sequence and the Prebble Formation contain beds of accretionary lapilli, and together with the volcanic breccias argue for proximal volcanic sources. This dramatic increase in the intensity of volcanism has been attributed to the centres of volcanism encroaching from the magmatic arc in West Antarctica into the Beacon basin (Barrett *et al.*, 1986). However, it is more likely that the upper Falla and Prebble volcanism represents magmatism within the retro-arc foreland basin that is unrelated to the arc volcanism recorded in the underlying Victoria Group strata. The age of this silicic volcanism is not well constrained. The upper part of the Falla Formation lacks palaeontological age control but a minimum K–Ar age of 203 Ma was obtained for a volcanic pebble near the base of the sandstone–tuff sequence, and a Rb–Sr isochron on five tuffs from 100–200 m below the upper contact gave a minimum age of 186 ± 9 Ma (see Barrett *et al.*, 1986, for references). The overlying Prebble Formation has a minimum age of 180 ± 5 Ma, the best estimate of the age of Ferrar magmatism (Elliot, Fleck & Sutter, 1985). The upper Falla and Prebble formations were probably deposited during the Early Jurassic.

Upper Falla Formation rocks contain only silicic ash. The Prebble Formation, however, marks the onset of tholeiitic igneous activity and the accumulation of coarse basaltic beds formed by explosive eruptions, as well as continuing silicic tuff deposition. Furthermore, thin interbeds carrying silicic shards occur in the overlying flood basalt sequence. These rocks, taken as a whole, constitute a major episode of bimodal volcanism in which silicic magmas were formed by crustal anatexis.

The Jurassic granitic plutons exposed in the Ellsworth–Thiel mountains region have K–Ar ages between 190 and 163 Ma and Rb–Sr isochron ages of 182–173 Ma (Millar & Pankhurst, 1987). These plutons are reported to have characteristics of S-type granites, and therefore are probably the result of crustal anatexis. They are part of the same igneous province as the bimodal rocks of the CTM.

At Allan Nunatak in south Victoria Land, basaltic pyroclastic rocks rest on an erosion surface cut into Permian and Triassic strata, and demonstrate at least 500 m of relief prior to onset of tholeiitic magmatism (Collinson *et al.*, 1983). In the CTM, field relations and characteristics of the Prebble Formation rocks point to the presence of topographic relief which was most likely the result of extensional tectonism and vertical movements (Larsen, 1988; Elliot, unpublished data).

The evidence for uplift and erosion in south Victoria Land, the overlapping tholeiitic and silicic (bimodal) volcanism of the CTM, and the S-type plutons on the Ellsworth–Thiel ridge lead to the possibility of a common causal process. The thermal perturbation associated with the generation of tholeiitic magmas and the disruption of Gondwana may have caused doming and uplift of at least part of the Transantarctic Mountains prior to the onset of basaltic activity. The early tholeiitic magmas were emplaced in the crust, causing anatectic melting which was followed by the eruption of upper Falla and Prebble Formation volcanic rocks and the emplacement of S-type plutons in the Ellsworth–Thiel ridge. Finally, basaltic magmas reached the surface and were erupted explosively, although that activity was soon replaced by extrusion of flood basalts. Emplacement of hypabyssal and plutonic bodies accompanied eruption of lavas. Anatectic silicic volcanism persisted at a low level during the time of lava extrusion.

Direct evidence for Jurassic faulting is limited to Coalsack Bluff (Collinson & Elliot, 1984). Nevertheless the Ferrar must have been accompanied by crustal extension associated with Gondwana break-up. The thin crust of West Antarctica and the normal cratonic crustal thickness for East Antarctica suggest the major part of the extension was likely to have been in what is now West Antarctica and thus located on the magmatic arc flank of the former Beacon basin.

The Middle Jurassic–Early Cretaceous plate margin

No record exists in the Transantarctic Mountains of events between eruption of Kirkpatrick Basalt lavas and formation of the late Cenozoic glacial deposits and alkaline volcanic rocks. However, abundant evidence is present in the Antarctic Peninsula–Ellsworth Land region (Thomson & Pankhurst, 1983; Elliot, 1988) for an active magmatic arc in Bajocian time (Quilty, 1983), or possibly earlier (Meneilly *et al.*, 1987), and thus overlapping the emplacement of the Ferrar. The arc is represented by volcanic and plutonic rocks which are dominantly Late Jurassic–early Tertiary in age (Fig. 3). Palaeontologically dated Middle Jurassic volcanic strata are confined to Ellsworth Land, whereas radiometrically-dated plutonic and volcanic rocks of that age (and older) occur sparsely in the peninsula. Late Jurassic and younger volcaniclastic fore-arc basin strata and accretionary prism rocks are known from Alexander Island and the South Shetland Islands. Back-arc (*sensu lato*) strata crop out in the northern Antarctic Peninsula on or adjacent to James Ross Island, and are represented by the Upper Jurassic Nordenskjöld Formation and the Cretaceous Gustav Group rocks and their equivalents (Elliot, 1988). On the Weddell Sea flank of Palmer Land sedimentation began in Middle Jurassic time and is represented by the Latady Formation. The basin-directed folding and thrusting of the Latady Formation suggests a foreland basin depositional setting. In the general absence of such deformation in the James Ross sequence (although local structures of uncertain significance are known; see Elliot, 1988), basin subsidence was likely to have been associated with an extensio-

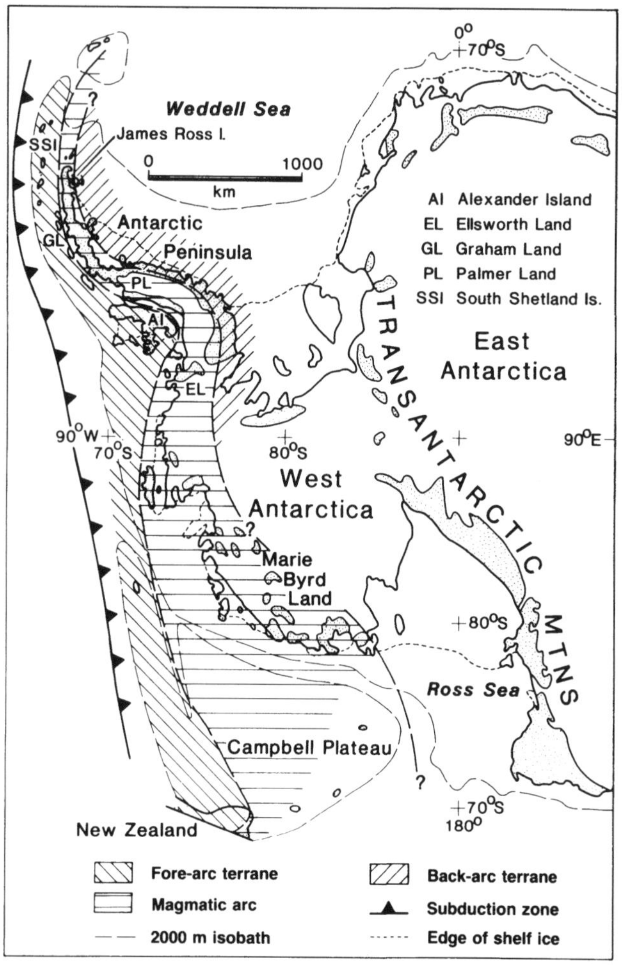

Fig. 3. Schematic illustration of the Late Jurassic–Early Cretaceous plate margin of Antarctica and the New Zealand–Campbell Plateau region.

nal regime of true back-arc type. In Marie Byrd Land, Late Jurassic and Cretaceous granitic rocks are widespread but outcrops of volcanic strata are sparse. No back-arc (*sensu lato*) deposits are known in this region; however, the fore-arc terrain is well developed in New Zealand which was adjacent to Marie Byrd Land till late in the Cretaceous (~ 85 Ma ago). In New Zealand thick volcaniclastic strata, ranging in age from Triassic to Jurassic (the Murihiku Terrane; Bradshaw *et al.*, 1981), form a fore-arc accretionary prism that was deformed and intruded by granite in the Early Cretaceous Rangitata II Orogeny.

Discussion

From the Pacific margin perspective, despite the presence of a widespread unconformity, albeit intermittently exposed, the distinction between the Gondwanian and the Andean events is becoming increasingly blurred as geochronological and other data accumulate (Storey & Garrett, 1985; Meneilly *et al.*, 1987; see also Storey, Thomson & Meneilly, 1987). The same lack of distinction is evident in the New Zealand record for the Late Palaeozoic and Mesozoic, the orogenic events being synchronous with neither the peninsula nor the classical Gondwanian. Nevertheless, there is a distinction to be made. Non-marine deposition in the Gondwana foreland basins continued to at least the end of the Triassic. Thermal perturbations associated with the break-up are reflected in the pre-Ferrar erosion in the Transantarctic Mountains, and a major episode of magmatism which reached a peak in the early Middle Jurassic at about 180 Ma. Even if subduction-related processes were virtually continuous across this time interval, there was a significant outboard shift in the location of the various elements of the active plate margin. Furthermore, the depositional environment of the back-arc region changed to a marine setting, indicating considerable subsidence of what had been part of the former arc terrain. The reason for this shift is unclear but might be the result of steepening of the subduction zone or outboard migration of the trench.

Constraints on the geological evolution

Although the reassembly of the cratonic regions of East–West Gondwana is now fairly well constrained (e.g. Lawver, Sclater & Meinke, 1985; Martin & Hartnady, 1986), the reconstruction of the Pacific margin remains a problem (see Dalziel & Elliot, 1982; Dalziel *et al.*, 1987). The problems are centred on the requirement to eliminate overlap of continental crust, and a general space problem associated with a fixed relationship between South America and Africa, and between New Zealand and Australia. Assuming the continental block nature of West Antarctica, palaeomagnetic investigations offer the possibility of constraining the reassembly. The most recent data (Grunow *et al.*, 1987a) suggest that the Antarctic Peninsula and the Ellsworth–Whitmore mountains crustal blocks rotated about 20° clockwise by the mid-Cretaceous as either one unit or two discrete units. The original location for a single unit block (Fig. 4A) is near the limit allowed by the palaeomagnetic data. This reconstruction can accommodate a Filchner block (Filchner and Ronne ice and continental shelf region). In a two unit reconstruction (Fig. 4B), with the Ellsworth–Whitmore block adjacent to the West Gondwana margin, the location of the Filchner block becomes a major problem even allowing for the considerable crustal extension that has undoubtedly occurred. No data are as yet available for the Thurston Island region which is a critical piece in the reassembly. Grindley & Oliver (1983) interpreted palaeomagnetic data for Marie Byrd Land to suggest a possible

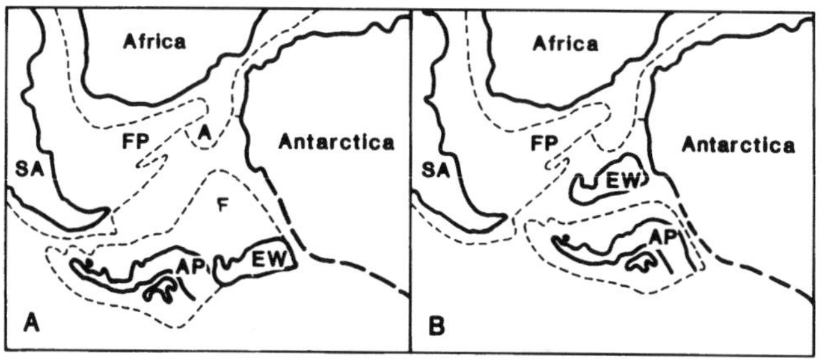

Fig. 4. Alternative reconstructions for the Antarctic Peninsula and Ellsworth–Whitmore crustal blocks, based on palaeomagnetic data (adapted from Grunow *et al.*, 1987a). SA, South America; FP, Falkland Plateau; A, Agulhas Plateau; AP, Antarctic Peninsula; F, Filchner block.

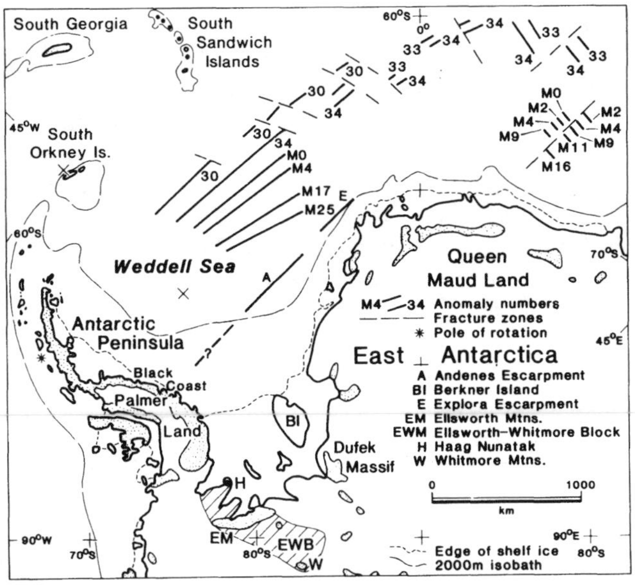

Fig. 5. Antarctica and Mesozoic seafloor anomaly data in the Weddell Sea and south-west Indian Ocean region (from LaBrecque, 1986; Martin & Hartnady, 1986). Pole of rotation for the Weddell M anomalies from LaBrecque (1986).

200–500 km of extension in the Ross embayment, and 10–45° of rotation, although a tilt correction of 10° would eliminate the need for any extension. On a simple two-dimensional model, assuming a 40 km original crustal thickness, the extension would be approximately 200 km. At present, there is insufficient palaeomagnetic information to adequately constrain the reconstruction and the sequence and timing of the Mesozoic displacements of the blocks making up West Antarctica and the Antarctic Peninsula.

Seafloor magnetic anomalies in the Weddell Sea–Queen Maud Land region (Fig. 5) provide important information on the early development of the ocean floor between East and West Gondwana and thus on possible configurations of a reconstructed Gondwana. Anomalies off Queen Maud Land can be correlated with conjugate anomalies in the Somali and Mozambique basins, and have been interpreted by Martin & Hartnady (1986) to show initial separation starting by M22 time (152 Ma), although Rabinowitz, Coffin & Falvey (1983) suggested an age of more than 160 Ma (older than M29) for the Somali Basin (ages for the anomalies are taken from Palmer, 1983). The identification of the earliest Weddell Sea anomaly remains uncertain; M29 (160 Ma) was suggested by LaBrecque & Barker (1981), M25 (156 Ma) by LaBrecque (1986), and M10 (130 Ma) by Martin & Hartnady (1986). The M10 identification is based on their interpretation of the evolution of the spreading systems in the south-west Indian Ocean, and the fact that no change in spreading direction is detected in the Weddell anomalies on initiation of South Atlantic spreading. The simplest interpretation is that the oldest anomaly is associated with formation of a triple junction on opening of the South Atlantic. Older ages require that a more complicated set of spreading ridges existed. Whatever the age span of the Weddell M anomalies, the associated pole of rotation lies on or to the west of the Antarctic Peninsula (LaBrecque, 1986). The anomalies record separation of East Antarctica from the South America plate and they must terminate in a fracture zone system parallel or subparallel to the peninsula margin.

The other constraint on the early evolution of the Weddell Sea is given by the Explora–Andenes Escarpment which is interpreted by Kristoffersen & Haugland (1986) as most likely to be a rifted or obliquely rifted margin. As they point out, if this feature extends to the peninsula and is a Gondwana plate margin (i.e. associated with Gondwana rifting), then any rotation and translation of West Antarctic crustal blocks did not involve the Weddell embayment but occurred around its periphery. The region south-east of the Andenes Escarpment could be the remnants of a rift associated with the earliest stages of break-up of Gondwana, or might be floored by oceanic crust that predates the Weddell anomalies.

Geophysical data from the continent itself clearly show the presence of grabens in the Ross Sea (Cooper, Davey & Cochrane, 1987) and Haag Nunataks (Garrett, Herrod & Mantripp, 1987) regions. Extensional features are less well documented for the Weddell embayment; although the region has been interpreted by Masolov, Kurinin & Grikurov (1981) as dominated by rift systems, the identification of an extension of the Dufek Intrusion to Berkner Island calls this into question (Behrendt *et al.*, 1981). At least some of the graben topography is late Cenozoic, but may reflect reactivated older structures. Certainly, the available data on sediment and crustal thickness suggest the Weddell embayment is a deep basin, floored by thinned continental crust, and with as much as 14 km of sediment fill that must date back to the Cretaceous. The region as a whole is an extensional zone associated with Mesozoic rifting and plate margin reorganization.

Discussion

Review of the Jurassic–early Cretaceous geology suggests that the early stages of Gondwana break-up occurred while subduction beneath the Pacific margin was continuously, or nearly continuously, active. Even though the Early Jurassic is not well represented along much of the plate margin, a pronounced break in subduction history is not evident in arc-related igneous activity. The early stages, therefore, occurred in a framework of an active plate margin and concurrent rifting plus reorganization in the back-arc terrain of the earlier Gondwana plate margin.

Reconstructions of Gondwana are constrained by seafloor anomaly data (e.g. Martin & Hartnady, 1986), and by geological information such as truncated tectonic, structural and geochronological provinces in both the cratons and plate margins. In the West Antarctica–Antarctic Peninsula sector of the margin, the lithotectonic terrains that can be identified in the plate margin provide insufficient constraints. The reconstruction favoured by Dalziel *et al.* (1987) leads to the craton margin lying much further outboard than in other reconstructions (e.g. Dalziel & Elliot, 1982), in which the Ellsworth block is placed adjacent to the Queen Maud Land margin and Falkland–Agulhas plateau region. The space problems evident in reconstructions such as that in Dalziel & Elliot (1982) could be alleviated by displacement, with left lateral sense, of the

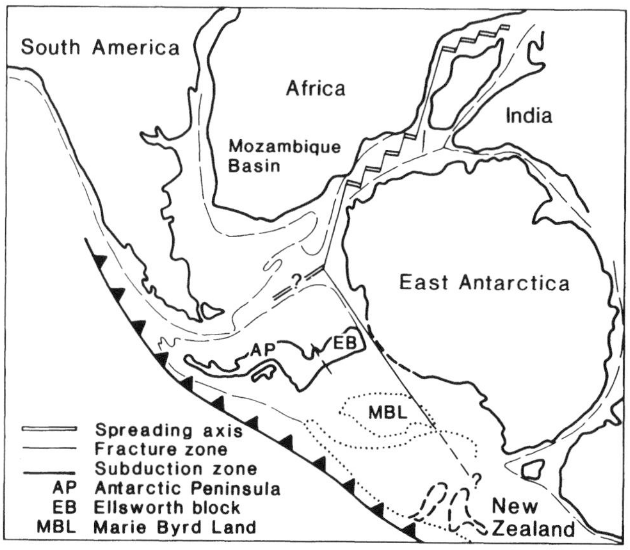

Fig. 6. Schematic reconstruction for the early stages in the evolution of the south-west Indian Ocean and Weddell Sea regions. Back-arc extension in the Marie Byrd Land and New Zealand region has been removed. The Thurston Island region, which leads to space problems, is not illustrated.

West Antarctica–New Zealand sector relative to the craton. Initial movements, with right lateral sense, between East and West Gondwana would remove that displacement, and the zone along which movements occurred would have been the previously extended back-arc terrain of West Antarctica and New Zealand.

Doming and rifting occurred during the Early–Middle Jurassic and the earliest translational movements must have accompanied the opening of the Somali and Mozambique basins at or before about 150 Ma (i.e., during the Late Jurassic). The Explora Escarpment is located along an expected shear (or fracture) associated with the early stages of opening, but the continuation of that system through West Antarctica can only be speculated upon. It could terminate in a triple junction (ridge–fault–fault), one arm (a fault) penetrating through the back-arc region (Fig. 6). It is not yet clear that the colinear Explora and Andenes escarpments have similar origins and evolutions. If the Explora–Andenes Escarpment is the Gondwana margin (Kristoffersen & Haugland, 1986), and this is not certain, then most of the extension and crustal thinning of the Weddell embayment must have occurred during, or before, the earliest stages of break-up in middle Jurassic time. Whatever the age of the Weddell anomalies, some movements parallel to the Explora Escarpment must have occurred in Late Jurassic time and necessarily were accompanied by some increase in convergence rate along parts of the Pacific margin. The model proposed by Schmidt & Rowley (1986) seems incompatible with the geophysical data from the Weddell Sea.

If the Weddell anomalies date from M25 (156 Ma) or older time, then the opening preceded or coincided with initial separation of East and West Gondwana, and right lateral motion in the Explora Escarpment region. Such an anomaly identification would require a more complicated set of M anomalies associated with the south-west Indian Ocean than has so far been recognized. If the Weddell anomalies date from M10 (130 Ma) and opening of the South Atlantic, movements along the Queen Maud Land Margin still must have occurred earlier and the Explora–Andenes Escarpment therefore has a multi-stage history of transcurrent motion followed by rifting.

Palaeomagnetic data for the Ellsworth–Antarctic Peninsula block (Grunow *et al.*, 1987a) require a 20° clockwise rotation relative to the Antarctic craton by mid-Cretaceous time. It cannot be assumed that the rotation was coincident with the opening of the Weddell Sea because the identified anomalies do not relate to the peninsula. In fact, the rotation could be associated with the Late Jurassic separation of East and West Gondwana which requires Pacific-ward movement of the plate margin. However, if the Andenes–Explora Escarpment is the plate margin, then rotation of the Antarctic Peninsula and the Ellsworth Mountains would require extension perpendicular to the length of the combined block and compression subparallel to it. The present horst and graben structure in the Haag Nunataks region may reflect reactivation of extensional structures formed in the Mesozoic. Finally, if the Andenes Escarpment does, in fact, extend to the Black Coast (Kristoffersen & Haugland, 1986), it appears to separate the southern part of Palmer Land, which has Mesozoic batholithic rocks, from the northern part, which has batholithic rocks and orthogneisses exposed at the surface. Presumably this reflects greater uplift of the magmatic arc.

Although palaeomagnetic data provide important constraints for reconstruction and dispersal of the Gondwana plate margin, the lack of sufficient precision makes the Weddell Sea region critical to understanding the Mesozoic evolution of the continent. Particularly critical are the confirmation, or revision, of the Weddell Sea M anomaly identifications, the identification of the magnetic anomaly pattern to the west and east of those anomalies, and the age and origin of the Explora and Andenes escarpments.

Acknowledgements

This research has been supported by NSF grants DPP 8519080 and DPP 8419529. Comments by David Macdonald are appreciated and have improved this paper. Note added (*in press*). Palaeomagnetic results for Thurston Island have recently been reported by Grunow, Kent & Dalziel (1987b), and suggest that region was fixed with respect to the Antarctic Peninsula from Middle Jurassic time onward.

References

Barrett, P.J., Elliot, D.H. & Lindsay, J.F. (1986). The Beacon Supergroup (Devonian–Triassic) and Ferrar Group (Jurassic) in the Beardmore Glacier area, Antarctica. In *Geology of the Central Transantarctic Mountains*, Antarctic Research Series, 36, ed. M.D. Turner & J.F. Splettstoesser, pp. 339–428, Washington, DC; American Geophysical Union.

Behrendt, J.C., Drewry, D.J., Jankowski, E. & Grim, M.S. (1981). Aeromagnetic and radio-echo sounding measurements over the Dufek Intrusion, Antarctica. *Journal of Geophysical Research*, **84(B4)**, 3014–20.

Bradshaw, J.D., Adams, C.J. & Andrews, P.B. (1981). Carboniferous to Cretaceous on the Pacific margin of Gondwana: the Rangitata phase of New Zealand. In *Gondwana Five*, ed. M.M. Cresswell & P. Vella, pp. 217–21. Rotterdam; A.A. Balkema.

Burn, R.W. (1984). *The Geology of the LeMay Group, Alexander Island.* Cambridge; British Antarctic Survey Scientific Reports, No. 109, 65 pp.

Collinson, J.W. & Elliot, D.H. (1984). Geology of Coalsack Bluff, Antarctica. In *Geology of the Central Transantarctic Mountains*, Antarctic Research Series, 36, ed. M.D. Turner & J.F. Splettstoesser, pp. 97–102. Washington, DC; American Geophysical Union.

Collinson, J.W., Pennington, D.C. & Kemp, N.R. (1983). Sedimentary petrology of Permian–Triassic fluvial rocks in the Allan Hills, central Victoria Land. *Antarctic Journal of the United States*, **18(5)**, 20–2.

Collinson, J.W., Pennington, D.C. & Kemp, N.R. (1986). Stratigraphy and petrology of Permian and Triassic fluvial deposits in northern Victoria Land, Antarctica. In *Geological Investigations in Northern Victoria Land*, Antarctic Research Series, 46, ed. E. Stump, pp. 211–42. Washington, DC; American Geophysical Union.

Cooper, A.K., Davey, F.J. & Cochrane, G.R. (1987). Structure of extensionally rifted crust beneath the western Ross Sea and Iselin Bank, Antarctica, from sonobuoy seismic data. In *The Antarctic Continental Margin: Geology and Geophysics of the Western Ross Sea*, CPCEMR Earth Science Series, 5B, pp. 93–118. Houston; Circum-Pacific Council for Energy and Mineral Resources.

Dalziel, I.W.D. (1984). *Tectonic Evolution of a Forearc Terrane, Southern Scotia Ridge, Antarctica.* Boulder; The Geological Society of America Special Paper 200, 32 pp.

Dalziel, I.W.D. & Elliot, D.H. (1982). West Antarctica: problem child of Gondwanaland. *Tectonics*, **1(1)**, 3–19.

Dalziel, I.W.D., Garrett, S.W., Grunow, A.M., Pankhurst, R.J., Storey, B.C. & Vennum, W.R. (1987). The Ellsworth–Whitmore crustal block: its role in the tectonic evolution of West Antarctica. In *Gondwana Six: Structure, Tectonics, and Geophysics*, ed. G.D. McKenzie, pp. 173–82. Washington, DC; American Geophysical Union, Geophysical Monograph No. 40.

Dalziel, I.W.D. & Grunow, A.M. (1985). The Pacific margin of Antarctica: terranes within terranes within terranes. In *Tectonostratigraphic Terranes of the Circum-Pacific Region*, vol. 1, Earth Science Series, ed. D.G. Howell, pp. 555–64. Houston, Texas; American Association of Petroleum Geologists.

Elliot, D.H. (1988). Tectonic setting and evolution of the James Ross basin, northern Antarctic Peninsula. In *Geology and Palaeontology of Seymour Island*, ed. R.M. Feldmann & M.O. Woodburne pp. 541–55. Boulder, Colorado; Geological Society of America, Memoir 169.

Elliot, D.H., Fleck, R.J. & Sutter, J.F. (1985). Potassium–argon age determinations of Ferrar Group rocks, central Transantarctic Mountains. In *Geology of the Central Transantarctic Mountains*, Antarctic Research Series, 36, ed. M.D. Turner & J.F. Splettstoesser, pp. 197–224. Washington, DC; American Geophysical Union.

Garrett, S.W., Herrod, L.D. & Mantripp, D.R. (1987). Crustal structure of the area around Haag Nunataks, West Antarctica, new aeromagnetic and bedrock elevation data. In *Gondwana Six: Structure, Tectonics, and Geophysics*, Geophysical Monograph No. 40, ed. G.D. McKenzie, pp. 109–16. Washington, DC; American Geophysical Union.

Grindley, G.W. & Oliver, P.J. (1983). Paleomagnetism of Cretaceous volcanic rocks from Marie Byrd Land. In *Antarctic Earth Science*, ed. R.L. Oliver, P.R. James & J.B. Jago, pp. 573–8. Canberra; Australian Academy of Science and Cambridge; Cambridge University Press.

Grunow, A.M., Dalziel, I.W.D. & Kent, D.V. (1987a). Ellsworth–Whitmore Mountains crustal block, Western Antarctica: new paleomagnetic results and their tectonic significance. In *Gondwana Six: Structure, Tectonics, and Geophysics*, Geophysical Monograph 40, ed. G.D. McKenzie, pp. 161–71. Washington, DC; American Geophysical Union.

Grunow, A.M., Kent, D.V. & Dalziel, I.W.D. (1987b). Mesozoic evolution of West Antarctica and the Weddell Sea Basin: new paleomagnetic constraints. *Earth and Planetary Science Letters*, **86**, 16–26.

Kristoffersen, Y. & Haugland, K. (1986). Geophysical evidence for the East Antarctic plate boundary in the Weddell Sea. *Nature, London*, **322(6079)**, 538–41.

LaBrecque, J.L. (1986). *South Atlantic Ocean and Adjacent Continental Margin*, Atlas 13, Ocean Margin Drilling Program, Regional Atlas Series, 21 sheets. Woods Hole, Massachusetts; Marine Science International.

LaBrecque, J.L. & Barker, P.F. (1981). The age of the Weddell Basin. *Nature, London*, **290**, 489–92.

Larsen, D. (1988). The petrology and geochemistry of the volcaniclastic upper part of the Falla Formation and Prebble Formation, Beardmore Glacier area, Antarctica. MS thesis, Ohio State University, Columbus (unpublished).

Lawver, L.A., Sclater, J.G. & Meinke, L. (1985). Mesozoic and Cenozoic reconstructions of the South Atlantic. *Tectonophysics*, **114(1–4)**, 233–54.

Martin, A.K. & Hartnady, C.J.H. (1986). Plate tectonic development of the southwest Indian Ocean: a revised reconstruction of East Antarctica and Africa. *Journal of Geophysical Research*, **91(B5)**, 4767–86.

Masolov, V.N., Kurinin, R.G. & Grikurov, G.E. (1981). Crustal structure and tectonic significance of Antarctic rift zones (from geophysical evidence). In *Gondwana Five*, ed. M.M. Creswell & P. Vella, pp. 303–9. Rotterdam; A.A. Balkema.

Meneilly, A.W., Harrison, S.M., Piercy, B.A. & Storey, B.C. (1987). Structural evolution of the magmatic arc in northern Palmer Land, Antarctic Peninsula. In *Gondwana Six: Structure, Tectonics, and Geophysics*, Geophysical Monograph 40, ed. G.D. McKenzie, pp. 209–19, Washington, DC; American Geophysical Union.

Millar, I.L. & Pankhurst, R.J. (1987). Rubidium–strontium geochronology of the region between the Antarctic Peninsula and the Transantarctic Mountains: Haag Nunataks and Mesozoic granitoids. In *Gondwana Six: Structure, Tectonics, and Geophysics*, Geophysical Monograph 40, ed. G.D. McKenzie, pp. 151–60. Washington, DC; American Geophysical Union.

Palmer, A.R. (1983). The decade of North American geology, geologic time scale. *Geology*, **11(9)**, 503–4.

Pankhurst, R.J. (1982). Rb–Sr geochronology of Graham Land, Antarctica. *Journal of the Geological Society, London*, **139(6)**, 701–11.

Pankhurst, R.J. (1983). Rb–Sr constraints on the ages of basement rocks of the Antarctic Peninsula. In *Antarctic Earth Science*, ed. R.L. Oliver, P.R.James & J.B. Jago, pp. 367–71. Canberra; Australian Academy of Science and Cambridge; Cambridge University Press.

Quilty, P.G. (1983). Bajocian bivalves from Ellsworth Land, Antarctica. *New Zealand Journal of Geology and Geophysics*, **26(4)**, 395–418.

Rabinowitz, P.D., Coffin, M.F. & Falvey, D. (1983). The separation of Madagascar and Africa. *Nature, London*, **220(4592)**, 67–9.

Schmidt, D.L. & Rowley, P.D. (1986). Continental rifting and transform faulting along the Jurassic Transantarctic rift, Antarctica. *Tectonics*, **5(2)**, 279–91.

Storey, B.C. & Garrett, S.W. (1985). Crustal growth of the Antarctic Peninsula by accretion, magmatism and extension. *Geological Magazine*, **122(1)**, 5–14.

Storey, B.C., Thomson, M.R.A. & Meneilly, A.W. (1987). The Gondwanian Orogeny within the Antarctic Peninsula: a discussion. In *Gondwana Six: Structure, Tectonics, and Geophysics*, Geophysical Monograph 40, ed. G.D. McKenzie, pp. 191–8. Washington, DC; American Geophysical Union.

Thomson, M.R.A. & Pankhurst, R.J. (1983). Age of post-Gondwanian calc-alkaline volcanism in the Antarctic Peninsula region. In *Antarctic Earth Science*, ed. R.L. Oliver, P.R. James & J.B. Jago, pp. 328–33. Canberra; Australian Academy of Science and Cambridge; Cambridge University Press.
Thomson, M.R.A. & Tranter, T.H. (1986). Early Jurassic fossils from central Alexander Island and their geological setting. *British Antarctic Survey Bulletin*, **70**, 23–39.
Watts, D.R., Watts, G.C. & Bramall, A.M. (1984). Cretaceous and Early Tertiary paleomagnetic results from the Antarctic Peninsula. *Tectonics*, **3(3)**, 333–46.

Palaeomagnetism, K–Ar dating and geodynamic setting of igneous rocks in western and central Neuschwabenland, Antarctica

M. PETERS[1], B. HAVERKAMP[2], R. EMMERMANN[3], H. KOHNEN[4] & K. WEBER[1]

1 Institut für Geologie und Dynamik der Lithosphäre der Universität Göttingen, Goldschmidtstr. 3, D-3400, Göttingen, FRG
2 Institut für Geophysik der Universität Münster, Corrensstr. 24, D-4400, Münster, FRG
3 Institut für Geowissenschaften und Lithosphärenforschung der Universität Giessen, Senckenbergstr. 3, D-6300, Giessen, FRG
4 Alfred-Wegener-Institut für Polarforschung, Columbusstr., D-2850, Bremerhaven, FRG

Abstract

The Proterozoic country rocks at Ahlmannryggen consist of lava flows and sedimentary rocks intruded by sills (Borgmassivet Intrusives). These suites are intruded by dykes dated at ~ 1150 Ma. Palaeomagnetic data indicate that these magmatic rocks are all of Proterozoic age. Geochemical data indicate that they resemble modern tholeiitic island-arc suites. Locally they have a slaty cleavage grading into mylonitic texture which strikes parallel to the Jutulstraumen-Pencksökket(J-P) graben. Such tectonic structures were dated at 525 Ma using syntectonic white micas. Evidence for the initial break-up of Gondwana during the Jurassic is given by the ~ 200 Ma dykes in Ahlmannryggen and the ~ 180 Ma lava flows, dykes and sills at Vestfjella. The mean pole position for the dykes in Ahlmannryggen and for the stratigraphically younger flows and cross-cutting dykes at Vestfjella is in good agreement with a mean Jurassic pole position for East Antarctica. The mean pole position for older lava flows at Vestfjella, however, has a significantly lower latitude and indicates an older age for these rocks. Whereas the Early Jurassic dykes in Ahlmannryggen have tholeiitic and subordinate alkaline affinities typical for continental rift magmatism, crustal contamination has obscured the affinities of the flows, dykes and sills at Vestfjella.

Introduction and geological setting

The area described comprises the northern Vestfjella, central and north-eastern Ahlmannryggen and Boreas and Robertskollen nunataks (Fig. 1). The most conspicuous geological structure in this region is the J-P graben, which was interpreted as a rift structure by Neethling (1970, 1972).

Ahlmannryggen

Precambrian amygdaloidal basaltic–andesitic lava flows interlayered with pillow lavas and thin beds of tuffaceous quartzites are exposed in north-eastern Ahlmannryggen (Straumsnutane), on the western margin of the graben. The total thickness of this sequence is ~ 860 m (Watters, 1972); individual, mostly horizontal, lava flows are < 10–55 m thick. Sedimentary intercalations display structures typical of shallow-water deposition (ripple marks, mudcracks and cross-bedding).

There are no lava flows in central Ahlmannryggen (Grunehogna), nor at Boreas Nunatak, and the area is composed of horizontal, Precambrian shallow- and deep-water sedimentary rocks intruded by dioritic sills (Borgmassivet Intrusives of

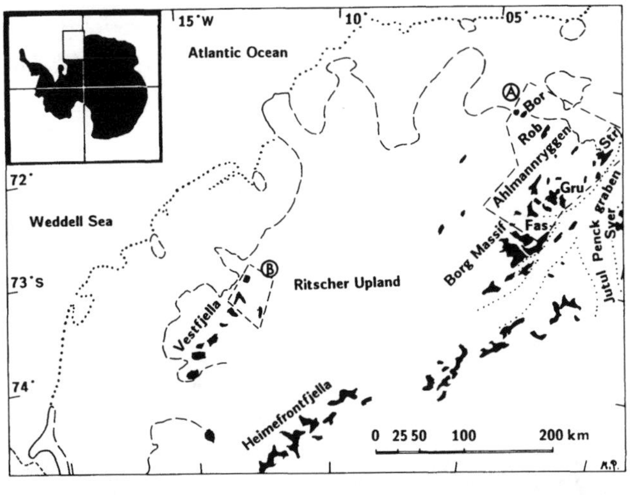

Fig. 1. Location map for Ahlmannryggen (A) and northern Vestfjella (B). Bor, Boreas Nunatak; Rob, Robertskollen Nunatak; Str, Straumsnutane; Gru, Grunehogna; Fas, Fasettfjellet; Sver, Sverdrupfjella.

Wolmarans & Kent, 1982) up to 200 m thick, the majority of which are extensively altered. It is not known whether the effusive rocks of the Straumsnutane and the Borgmassivet Intrusives are related temporally. Previous workers (see refer-

Table 1. *Palaeomagnetic and geochronological results for Ahlmannryggen and northern Vestfjella*

Site		N	D	I	*k*	α95	Latitude	Longitude	Pol.	Age (Ma)	Mineral
Ahlmannryggen: Proterozoic suites											
A1	Flow	5	247.0	−1.0	164.8	7.2	6.6	66.4	−		
A2	Flow	6	221.4	−7.5	9.8	22.5	9.9	40.5	−	1115 ± 37	p
A3	Flow	5	237.3	11.8	10.8	24.4	15.6	58.8	−	872 ± 25	p
A3	Flow									465 ± 16	g
A4	Flow	8	238.8	7.4	307.4	3.8	13.0	59.6	−		
A5	Flow	5	240.2	6.3	67.0	11.3	12.0	60.9	−		
A6	Flow	5	237.3	−7.3	48.1	13.4	6.2	56.2	−		
A7	Flow	8	236.5	−2.2	14.5	20.8	8.9	56.1	−		
A8	Flow									666 ± 22	p
A9	Flow									460 ± 16	g
A10	Flow									1013 ± 36	p
A10	Flow									699 ± 14	g
A11	BMI	8	244.5	3.3	65.1	11.5	9.2	63.2	−	666 ± 26	p
A12	BMI	8	251.2	−8.2	55.5	7.5	1.4	57.8	−		
A13	Dyke	8	252.1	−3.0	48.7	8.7	4.0	69.6	−	842 ± 30	p
										1143 ± 39	g
										1183 ± 33	b
A14	Dyke									751 ± 29	p
A14	Dyke									1143 ± 39	b
A15	M									526 ± 11	m
A15	M									522 ± 11	m
Ahlmannryggen: Mesozoic suites											
A16	Dyke	5	227.4	49.1	57.7	12.2	41.3	56.5	−		
A17	Dyke	5	47.8	−67.9	55.1	9.1	60.2	70.2	+		
A18	Dyke	6	218.4	52.1	206.1	5.3	46.2	48.0	−		
A19	Dyke	5	227.7	48.7	33.9	16.0	40.8	56.6	−		
A20	Dyke	8	51.2	−58.1	121.1	11.3	48.4	64.8	+		
A21	Dyke	7	220.3	57.7	96.4	6.9	51.0	52.2	−	281 ± 18	g
A22	Dyke	6	200.0	56.5	57.0	8.1	53.9	27.1	−		
A23	Dyke	6	49.1	−41.8	123.3	8.3	35.2	56.5	+		
A24	Dyke	5	213.7	44.6	119.1	11.3	40.9	11.3	−		
A25	Dyke	6	207.0	59.7	77.7	14.1	55.9	35.2	−	202 ± 18	g
A26	Dyke									246 ± 20	g
Northern Vestfjella: Mesozoic suites											
B1	Flow									176 ± 10	w
B2	Flow									325 ± 2	w
B3	Dyke									295 ± 19	p
B3	Dyke									197 ± 10	g
B4	Flow	8	196.6	17.9	32.3	9.9	24.8	4.6	−		
B5	Flow	8	199.3	38.0	212.8	3.8	37.0	8.9	−		
B6	Flow	8	42.0	−51.9	298.6	3.2	44.2	37.9	+	148 ± 10	g
B6	Flow									189 ± 10	p
B7	Flow	6	28.5	−56.2	270.0	4.1	50.9	23.5	+		
B8	Flow	8	201.1	32.4	107.0	5.4	33.3	10.8	−		
B9	Flow	5	37.4	−50.2	307.2	4.4	43.7	32.7	+	90 ± 9	g
B10	Flow	8	231.6	56.3	269.1	3.4	45.8	50.4	−	179 ± 13	p
B10	Flow									169 ± 13	g
B11	Dyke	5	35.4	−74.8	309.8	4.4	72.6	52.8	+		
B12	Dyke	5	46.5	−58.4	55.6	10.4	49.4	45.8	+		
B13	Dyke	8	211.2	38.0	25.3	11.2	35.5	23.0	−	180 ± 11	g
B14	Dyke	8	49.7	−30.8	396.4	2.8	26.8	42.0	+		
B15	Dyke	8	27.6	−76.6	53.1	7.7	76.9	48.7	+		
B16	Dyke	6	28.2	−59.3	121.3	1.5	54.1	25.0	+	174 ± 13	p
B16	Dyke									183 ± 20	g
B17	Dyke	7	9.4	−60.9	163.6	4.7	58.4	359.6	+	160 ± 16	p

Table 1 (*cont.*)

B18	Dyke									171 ± 15	g
B19	Sill	8	304.2	− 63.3	56.1	7.5	52.4	273.2	+ / −	174 ± 16	g

Mean Jurassic pole position from Livermore *et al.* (1984): 53.5° N, 38.5° E, $\alpha95 = 7.0°$, N = 7. Mean pole positions of igneous rocks at northern Vestfjella: All sites (except for sill B19): 47.8° N, 26.7° E, $\alpha95 = 9.7°$, N = 14; dykes and stratigraphically younger flows: 52.0° N, 31.9° E, $\alpha95 = 9.5°$, N = 10; older flows (except for flow B13): 31.8° N, 7.9° E, $\alpha95 = 10.3°$, N = 3.

Mean pole positions of igneous rocks at Ahlmannryggen: All Proterozoic suites: 8.8° N, 59.9° E, $\alpha95 = 5.5°$, N = 10; all Mesozoic suites: 48.0° N; 50.8° E, $\alpha95 = 7.0°$, N = 10.

N, number of independent samples per site (sites per mean pole). D, I, declination and inclination of mean direction calculated by Fisher-statistic from the characteristic remanent magnetizations of different samples. k, $\alpha95$, precision parameter and radius of 95% confidence circle derived from the Fisher-statistic. Lat., Long.: geographic latitude and longitude of the resulting pole position. +, −, + / −, normal, inverse and intermediate polarity. p, g, b, m, w, K–Ar ages on concentrates of plagioclase, groundmass, Ti-biotite, white mica, whole rock. BMI, Borgmassivet Intrusives. M: mylonite zone.

ences in Wolmarans & Kent, 1982, table 18) determined a Proterozoic age for the magmatic rocks in Ahlmannryggen. The volcanic rocks of north-eastern Ahlmannryggen and the intrusions exposed in its central and north-western parts have yielded ages of ~ 1700 Ma, 1000–1200 Ma and 800–900 Ma, respectively (Wolmarans & Kent, 1982; K–Ar- and Rb–Sr-techniques, mainly whole-rock analyses). Whether these episodes were characterized by three different magmatic events, or reflect one magmatic episode of 1700 Ma overprinted by two thermal events at 1000–1200 Ma and 800–900 Ma is, however, still open to question. An age of 192 Ma was determined for one Mesozoic dyke (Wolmarans & Kent, 1982).

Palaeomagnetic studies in this region had been made previously on Precambrian rocks and one Mesozoic dyke. These investigations did not provide any useful results (Wolmarans & Kent, 1982).

At Straumsnutane a slaty cleavage, which grades into a mylonitic texture, has a strike parallel to the axis of the J-P graben. Both these minor structures are marked by strongly orientated flakes of chlorite and white mica. Vesicles are elongated and their secondary mineral fillings of quartz and calcite are partly recrystallized.

Dolerite dykes which intruded these deformed Proterozoic rocks along pre-existing, near-vertical faults are undeformed and only slightly affected by alteration. Less common, slightly inclined dykes are coarser grained and more altered than the near-vertical fine-grained dykes. Those which have developed a slaty cleavage are also altered. However, albitization and sericitization of plagioclase are the only effects of alteration in the less-deformed dykes.

Northern Vestfjella

The nunataks in northern Vestfjella (Fig. 1) consist mainly of subaerial basalt flows intruded by a large number of dykes and sills. Initial K–Ar whole-rock analyses on samples from northern Vestfjella yielded ages of 220 Ma for a dolerite sill, and 400 Ma for a lava flow (Krylov in Hjelle & Winsnes, 1972). Other K–Ar whole-rock analyses on samples from dykes and lava flows in south-western Vestfjella yielded ages of 154–172 and 200–695 Ma, respectively (Furnes & Mitchell, 1978; Furnes *et al.*, 1987). The lava flows have undergone secondary mineralization centred on voids and gas vesicles, and intensive pseudomorphic replacement of magnetic mineral phases. The dykes and sills, however, show no other alteration effects except for serpentinization and saponitization of orthopyroxene and slight albitization of plagioclase.

Geochemical analyses of igneous rocks from southern Vestfjella (Furnes, Neumann & Sundvoll, 1982; Furnes *et al.*, 1987) led to the assumption that magma genesis was influenced by crustal contamination, magma mixing and crystal fractionation. The trace-element spidergrams obtained show similarities to N-Type mid-ocean ridge basalts (MORB), on the one hand, and initial rift tholeiites, on the other.

Age determinations

K–Ar age determinations were made on 33 samples using the freshest possible concentrations of plagioclase, groundmass, and Ti-rich biotite and white mica (from one of the mylonite zones). The results are summarized in Table 1.

Plagioclase and groundmass samples from northern Vestfjella yielded ages of 150–190 Ma, with a distinct maximum at ~ 180 Ma for flows, dykes and sills (Table 1). Only two older values (> 200 Ma) were obtained, for one dyke and one flow, possibly due to excess argon. A 90 Ma age for a lava flow from the Basen massif was obtained twice by a double-Ar dating method but it is meaningless in the geological context.

Fine-grained, partially porphyritic dykes from Ahlmannryggen have been dated at 202, 246 and 281 Ma (post-deformation age according to the mylonitization mentioned below). Ages of 1109 and 1183 Ma were obtained on Ti-biotites from the coarser-grained dykes. The oldest ages obtained from the plagioclase and groundmass from the lava flows at Straumsnutane and Borgmassivet Intrusives are close to these values. All younger ages yielded by the Borgmassivet Intrusives and Straumsnutane volcanic rocks are interpreted as overprinted ages. These younger ages can be attributed to a tectnothermal event which was dated at ~ 525 Ma using syntectonic white micas from a mylonite zone exposed in Straumsnutane.

Palaeomagnetic studies

To supplement the radiometric data 15 sites in northern Vestfjella and 20 in Ahlmannryggen were sampled for palaeomagnetic studies. A total of 231 orientated samples were collected, usually by taking 5–8 samples/site. Normally two specimens could be cut out of one sample.

Magnetomineralogy

At least one sample from each site was investigated by ore microscopy. These studies showed that opaque minerals in the lava flows and Borgmassivet Intrusives have nearly all undergone intensive hydrothermal alteration. Magnetite may be totally replaced by haematite and pseudobrookite. These pseudomorphs retain the original crystal shape of magnetite and may preserve ilmenite exsolution lamellae, representing relicts of titanomagnetite of oxidation classes II–III. Although not visible, small relicts of magnetite may be present still in these samples. In addition to flows with high proportions of secondary haematite, there are lavas in which titanomagnetite of oxidation classes II–IV have been affected only by maghemitization. Magnetic minerals in the dykes are mostly skeletal to euhedral titanomagnetite grains of oxidation classes I–III, which are sometimes slightly maghemitized. Isolated granulation was also observed and chromite with magnetite-rich rims occur as well.

To investigate the carriers of the remanent magnetization, IRM acquisition and hysteresis experiments were carried out in addition to stepwise and continuous thermal and AF demagnetization. Stepwise thermal demagnetization indicates a dominant blocking temperature of 580 °C which suggests Ti-poor titanomagnetite in this case. In three flows in northern Vestfjella and all flows at Straumsnutane, and in the Borgmassivet Intrusives, haematite is present as well as magnetite and contributes significantly to the NRM. Several sites also show blocking temperature components of about 300 °C and 450 °C, which probably correspond to maghemite and titanomagnetite, respectively. These results are also supported by the other magnetic experiments.

Demagnetization experiments

To determine the direction of the stable remanence, at least 5–8 specimens from each site were subjected to stepwise AF-demagnetization. An additional 2–3 specimens/site from Vestfjella were stepwise thermally demagnetized. 20 specimens from four sites in Ahlmannryggen were thermally demagnetized. To obtain a good comparability of the results for the thermal treatment, specimens were selected from the same samples as the AF-pilot-specimens. The stable directions of remanence detected by the different treatments did not show any significant differences at Vestfjella. The same is true for specimens in Ahlmannryggen.

With the exception of flow B7, magnetite is the carrier of stable magnetization at all sites in northern Vestfjella. Thermal demagnetization of specimens containing relicts of primary magnetite and secondary haematite as main carriers of remanence showed the same direction of remanence for both minerals. Similar results were obtained on the Proterozoic and Mesozoic igneous rocks from Ahlmannryggen. It can be assumed therefore that the hydrothermal alteration is of syngenetic origin instead of being caused by the later magmatic or tectonothermal events. This result implies that the stable magnetization is a primary one and not due to a later overprinting. Apart from the present direction of the Earth's magnetic field and the random directions of viscous magnetization, no secondary directions were found.

Palaeomagnetic results

The mean results of the stable remanence for the 15 sites are summarized in Table 1. According to field observations a tectonic correction has been applied only to the flows of northern Vestfjella.

AHLMANNRYGGEN

One Precambrian dyke (dated at ~ 1183 Ma) gave a palaeopole position of 4.02° S, 249.6° E, $\alpha95$ = 8.7°. The pole postions of samples from Boreas Nunatak, Borgmassivet Intrusives and lava flows at Straumsnutane (mean palaeopole position: 8.8° S, 239.9° E, $\alpha95$ = 5.5°) are grouped around the pole of this dated dyke.

The mean pole positions for the dolerite dykes in Ahlmannryggen (K–Ar age of ~ 202–281 Ma) show mean values of 48.0° S, 230.8° E, $\alpha95$ = 7°. This value is in good agreement with the mean Jurassic pole position of East Antarctica according to Livermore, Vine & Smith. (1984; 53.5° S, 218.5° E, $\alpha95$ = 7°).

VESTFJELLA

The sampled flows in northern Vestfjella can be subdivided on stratigraphical grounds into an older age group and a younger one. It is evident that the older flows all have inverse magnetizations with significantly lower pole latitudes whereas the younger flows have normal directions and relatively higher pole latitudes like the dykes, where only one out of seven sites has an inverted magnetization. This result strongly supports the assumption that the stable magnetization of the highly altered flows is really a primary one and has not been overprinted during the emplacement of the dykes. The calculation of a mean pole position for all sites, with the exception of site B19 which has an intermediate polarity, yields a pole position of 47.8° N and 26.7° E ($\alpha95$ = 9.7°). Considering the $\alpha95$ values, this result agrees with a mean Jurassic pole position of 53.5° N and 38.5° E ($\alpha95$ = 7°) (derived from seven Jurassic poles for East Antarctica regarded as reliable by Livermore *et al.* (1984)).

If a mean pole is calculated for the dykes and younger flows separately, the $\alpha95$ circles of confidence for the two poles overlap totally. A common pole for both groups was calculated which is situated at 52.0° N and 31.9° E ($\alpha95$ = 9.5°); this agrees more with the mean Jurassic pole and the pole for dykes from south-western Vestfjella (54.4° N and 27.6° E, $\alpha95$ = ~ 5.0°; Løvlie, 1979). In calculating a mean pole for the older flows, the pole for flow B10 was rejected because of its great angular distance to the remaining well-grouped three poles (B4, B5 and B8). With a position of 31.8° N and 7.9° E ($\alpha95$ = 10.3°), the resulting pole differs significantly from a Jurassic pole towards lower latitudes. These different pole positions could indicate that there might have been a longer time span between the eruption of the oldest flows and the intrusion of the dykes than would be expected from the K–Ar dates.

On the other hand it is obvious that the pole position for flow B10, which is stratigraphically the oldest flow of the group, is not only different from the pole with lower latitudes but also roughly consistent with a Jurassic pole. If this single pole position is taken to reflect the real mean pole position, corresponding to the time of the extrusion of flow B10, this could lead to a different interpretation of the data. In this case the whole volcanic sequence was extruded during Jurassic times, including a period with a palaeodirection of abnormally low inclinations, possibly due to an excursion or field reversal during which the three older flows (with low-latitude poles) were extruded. However, the available data support the first interpretation. It is concluded that the palaeomagnetic investigations support the K–Ar determinations of a Jurassic age for the dykes and younger flows. In contrast, the pole positions from the lower parts of this volcanic sequence indicate an older age for these flows. However, the number of sites investigated is too small to allow any definite conclusions to be drawn.

Geochemical investigations

The Proterozoic magmatic rocks in Ahlmannryggen, in particular, are intensely altered. Consequently, only elements which were relatively immobile during post-magmatic tectonothermal overprinting (P, Ti, Nb, Zr, Y, REE) have been used for geodynamic interpretations.

Based on their Zr–Y–Nb–TiO_2 proportions, the Proterozoic magmatic rocks can be classified as calc-alkaline andesites and subordinate calc-alkaline basalts (Pearce & Cann, 1973; Floyd & Winchester, 1978). However, since the general applicability of the triangular plot (Ti/100)/Zr/Y × 3 as a reliable indicator for the geotectonic environment has been questioned by several authors (e.g. Holm, 1982), additional geochemical criteria have had to be adopted. In Mg_v variation diagrams (Mg_v = atomic proportion of Mg/Mg + Fe) the magmatic rocks can be subdivided into a high-Ti and a low-Ti series (Fig. 2). Each of these series is characterized by typical trace-element patterns that suggest compositionally different starting melts, but in between both series there is a strong chemical consanguinity. In contrast to the high-Ti series, the low-Ti series displays very regular variations that can be explained by crystal fractionation of olivine (ol), clinopyroxene (cpx) and plagioclase (plag). Fe- and Ti-enrichment trends are typical for tholeiitic igneous suites. According to our investigations the initial features of the trace-element geochemistry of these magmas is strongly modified by fractionation of ilmenite–magnetite and apatite in addition to ol, cpx, plag fractionation and alteration. Because of this distinct chemical evolution we believe that it is more appropriate to define the geotectonic environment by Nb and rare earth element (REE) concentrations and proportions, respectively. Both series (low-Ti and high-Ti) are characterized by La/Nb = 3.36–4.1 and $(La/Lu)_{cn}$ = 4.1–7.1, respectively.

When applying the alkali/silica ratio (Irvine & Baragar, 1971) and the normative mineral composition (Yoder & Tilley, 1962) for classification, the weakly altered Mesozoic dykes in

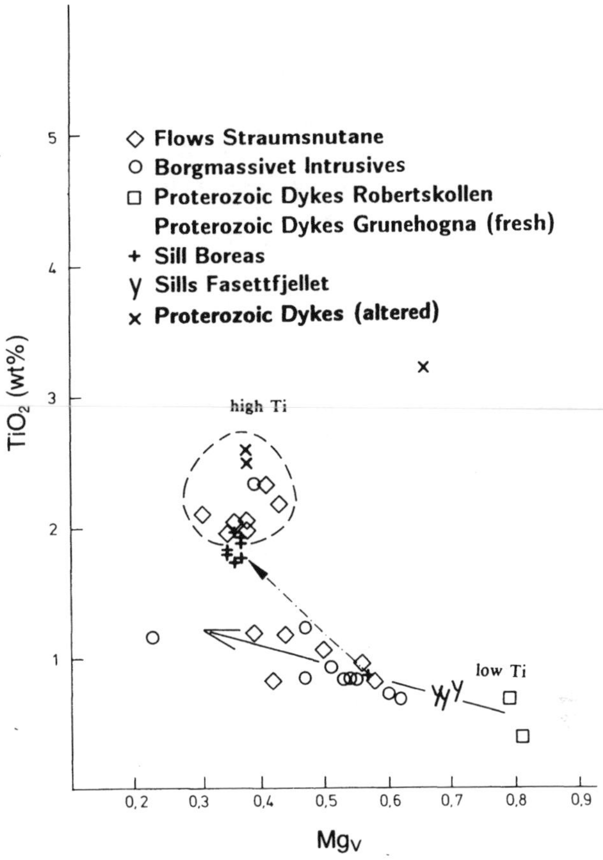

Fig. 2. TiO_2 versus Mg_v (atomic proportion of Mg/Mg + Fe) plot of the Proterozoic high-Ti and low-Ti series in Ahlmannryggen.

Table 2. *Representative element values in igneous rocks from Ahlmannryggen and northern Vestfjella*

	Proterozoic Ahlmannryggen			Triassic/Jurassic Vestfjella		
	high-Ti	low-Ti	Dyke	Flow	Dyke	Sill
TiO_2	2.33	0.73	3.28	1.0	1.42	1.56
Cr	79	171	500	131	367	267
La	41	16	13	7	7	5
Nb	10	< 5	6	< 5	< 5	< 5
Ni	54	66	232	111	109	99
Y	52	22	38	22	20	20
Zr	320	113	183	94	87	91

TiO_2 in wt %, trace elements in ppm.

Ahlmannryggen can be called quartz tholeiites, tholeiites, olivine tholeiites and alkali-olivine basalts. In these magmas Nb is slightly depleted relative to La (Table 2). Fig. 3 demonstrates that two different starting melts have to be assumed due to a garnet component in their initial peridotite. Strong variations in Cr and Ni concentrations indicate olivine, and clinopyroxene fractionation, and the low La level is typical for tholeiitic basalts (Table 2).

Considering the volcanic rocks in northern Vestfjella, only immobile elements were used due to strong post-magmatic

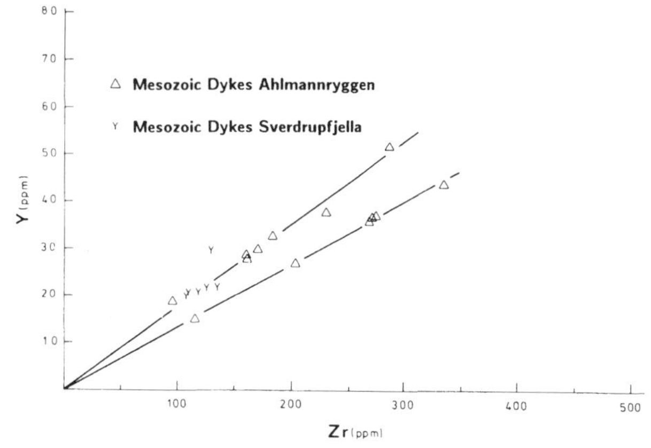

Fig. 3. Zr versus Y plot for Jurassic dykes from Ahlmannryggen.

alteration of the lava flows. The chemical evolution of the flows, dykes and sills is dominated by crystal fractionation. In addition to separation of ol, (opx), and cpx (Ni and Cr variations, Table 2) plagioclase fractionation also occurred. This evolution led to andesitic-basalts and cumulates, respectively. According to their Y/Nb proportions (≥ 2) (Pearce & Cann, 1973) these rocks are tholeiitic and, less commonly, transitional basalts. Strong variations within the whole suite, for example Zr/Y proportions, cannot be explained by crystal fractionation alone (Fig. 4).

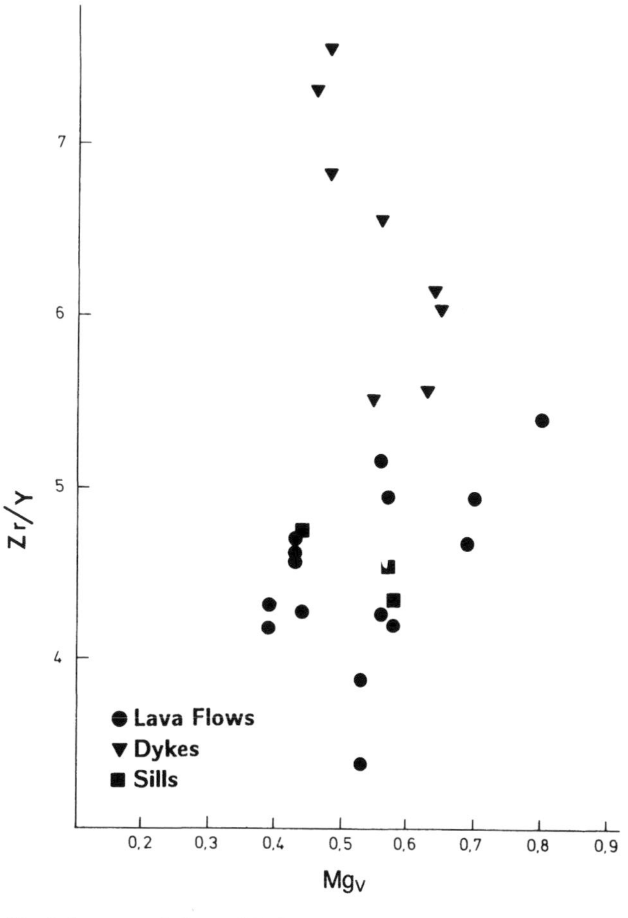

Fig. 4. Strong variations of Zr/Y ratios versus Mg_r in the flows, dykes and sills from northern Vestfjella.

Conclusions

The calculated virtual palaeomagnetic pole positions of all the Proterozoic magmatic rocks in Ahlmannryggen are grouped around the mean pole position of the dolerite dyke dated at ~ 1200 Ma. This age is interpreted as the crystallization age for all the Proterozoic rocks in Ahlmannryggen. The initial geochemistry of the lavas has been obscured by intensive fractionation and alteration processes. The enrichment of chrondrite-normalized light REE relative to heavy REE is indicative both of continental and island-arc tholeiites. According to Thompson *et al.* (1983) and Holm (1984), high La/Nb ratios may be of help in distinguishing continental tholeiites from island-arc tholeiitic suites. The Proterozoic lavas and intrusions in Ahlmannryggen are all characterized by distinctly higher La/Nb ratios (3.36–4.1) compared to those of the majority of continental tholeiites (1–2). Therefore tholeiitic island-arc suites cannot be excluded when considering the Proterozoic igneous rocks in Ahlmannryggen. The chemical features of the high-Ti series and the low-Ti series from Straumsnutane (lava flows), central Ahlmannryggen (Borgmassivet Intrusives and Proterozoic dykes), Boreas Nunatak and Fasettfjellet (sills) are nearly identical and therefore a close genetic correlation can be concluded.

The development of the observed mylonite zones was caused by a tectonothermal event dated at 525 Ma, believed to be an expression of the Ross Orogeny in this region.

The mean palaeopole position of the Jurassic dykes in Ahlmannryggen coincides with the mean Jurassic pole for East Antarctica so that a maximum age of 200 Ma has to be assumed instead of the possibly higher ages implied by geochronology. The chemical features of the tholeiitic dykes (slight depletion of Nb relative to La) is typical for continental basalts and initial rift tholeiites (Holm, 1984). Also the subordinate occurrence of alkaline magmas may indicate initial rifting (cf. Wedepohl, 1985).

In northern Vestfjella the lower latitudes for the palaeo-pole positions of the stratigraphically lower flows, compared to those of the upper flows and dykes, indicate greater ages for the lower flows. Because of their distinct variations in trace-element ratios it is concluded that their initial chemical characteristics have been obscured by intense crustal contamination, but that they have tholeiitic and subordinate transitional affinities. In addition to crustal contamination, fractionation of ol, (opx), cpx and plag produced basaltic-andesites and cumulates. Lava flows, dykes and sills cannot be subdivided on their chemistry and they show the same chemical evolution. Hence, it is concluded that they all belong to the same magmatic event.

Dykes are more numerous in the lava flows of northern Vestfjella than they are in Ahlmannryggen. Their total thickness in northern Vestfjella indicates an extension of > 4% in a NW–SE direction (Spaeth, 1987). In Ahlmannryggen much smaller amounts of extension must be assumed, based on the lower number of Jurassic dykes. This observation can be interpreted as an indication of increase in crustal extension to the west, towards the present continental margin. Thus, the dykes in Ahlmannryggen and the older flows at Vestfjella may represent the initial magmatic rocks of the Gondwana break-up at ~ 200 Ma, whereas the younger flows, dykes and sills at Vestfjella may represent a later stage of this rifting at ~ 180 Ma.

The conspicuous parallelism in strike between the Middle Jurassic dykes of Vestfjella and the Explora Escarpment, proven by Hinz & Krause (1982), and between the Early Jurassic dykes in Ahlmannryggen and the J-P graben, is conclusive evidence for their mutual geodynamic formation. However, in Ahlmannryggen older lineaments, marked by slaty cleavage and mylonitic shear zones, were established during the Ross Orogeny and then reactivated during the Mesozoic break-up of Gondwana.

Acknowledgements

We wish to thank the Deutsche Forschungsgemeinschaft for financial support during these studies. We also thank the Alfred-Wegener-Institut, Bremerhaven, and the Council for Scientific and Industrial Research, RSA, for supplying logistical requirements for the expeditions. Furthermore, we are grateful to J. Krynauw, team leader of the South African Earth Science Programme 1983/84, G. Spaeth, and B. Watters for discussion and advice during fieldwork.

Thanks are due also to H.J. Lippolt and collaborators, Laboratory for Geochronology of the University of Heidelberg, for making the K–Ar age determinations.

References

Floyd, P.A. & Winchester, J.A. (1978). Identification and discrimination of altered and metamorphosed volcanic rocks using immobile elements. *Chemical Geology*, **21**, 291–306.

Furnes, H. & Mitchell, J.G. (1978). Age relationships of Mesozoic basalt lavas and dykes in Vestfjella Dronning Maud Land, Antarctica. *Norsk Polarinstitutt Skrifter*, **169**, 45–68.

Furnes, H., Neumann, E.-R. & Sundvoll, B. (1982). Petrology and geochemistry of Jurassic basalt dykes from Vestfjella, Dronning Maud Land, Antarctica. *Lithos*, **15**, 295–304.

Furnes, H., Vad, E., Austerheim, H., Mitchell, J.G. & Garmann, L.B. (1987). Geochemistry of basalt lavas from Vestfjella and adjacent areas, Dronning Maud Land, Antarctica. *Lithos*, **20**, 337–56.

Hinz, K. & Krause, W. (1982). The continental margin of Queen Maud Land, Antarctica: seismic sequences, structural elements and geological development. *Geologisches Jahrbuch*, **E23**, 17–41.

Hjelle, A. & Winsnes, T. (1972). The sedimentary and volcanic sequence of Vestfjella, Dronning Maud Land. In *Antarctic Geology and Geophysics*, ed. R.J. Adie, pp. 539–46. Oslo; Universitetsforlaget.

Holm, P.E. (1982). Non-recognition of continental tholeiites using the Ti–Y–Zr diagram. *Contributions to Mineralogy and Petrology*, **79**, 308–10.

Holm, P.E. (1984). The geochemical fingerprints of different tectonomagmatic environments using hygromagmatophile element abundances of tholeiitic basalts and basaltic andesites. *Chemical Geology*, **51**, 303–23.

Irvine, T.N. & Baragar, W.R.A. (1971). A guide to the chemical classification of the common volcanic rocks. *Canadian Journal of Earth Sciences*, **8**, 523–48.

Livermore, R.A., Vine, F.J. & Smith, A.G. (1984). Plate motions and the geomagnetic field-II. Jurassic to Tertiary. *Geophysical Journal of the Royal Astronomical Society*, **79**, 939–61.

Løvlie, R. (1979). Mesozoic palaeomagnetism in Vestfjella, Dronning Maud Land, East Antarctica. *Geophysical Journal of the Royal Astronomical Society*, **59**, 529–37.

Neethling, D.C. (1970). South African Earth Science exploration of western Dronning Maud Land, Antarctica. PhD thesis, University of Natal, Pietermaritzburg. (unpublished).

Neethling, D.C. (1972). Comparative geochemistry of Proterozoic and Palaeo-Mesozoic tholeiites of western Dronning Maud Land. In *Antarctic Geology and Geophysics*, ed. R.J. Adie, pp. 603–16. Oslo; Universitetsforlaget.

Pearce, J.A. & Cann, J.R. (1973). Tectonic setting of basic volcanic rocks determined using trace element analysis. *Earth and Planetary Science Letters*, **20**, 290–300.

Spaeth, G. (1987). Aspects of the structural evolution and magmatism in western New Schwabenland. In *Gondwana Six: Structure, Tectonics, and Geophysics*, Geophysical Monograph, 40, ed. G.D. McKenzie, pp. 295–307. Washington, DC; American Geophysical Union.

Thompson, R.N., Morrison, M.A., Dickin, A.P. & Hendry, G.L. (1983). Continental flood basalts – arachnids rule OK? In *Continental Flood Basalts and Mantle Xenoliths*, ed. C.J. Hawkesworth & M.J. Norry, pp. 158–85. Nantwich; Shiva Publishing Ltd.

Watters, B.R. (1972). The Straumsnutane Volcanics, western Dronning Maud Land. *South African Journal of Antarctic Research*, **2**, 23–31.

Wedepohl, K.-H. (1985). Origin of the Tertiary basaltic volcanism in the northern Hessian Depression. *Contributions to Mineralogy and Petrology*, **89**, 122–43.

Wolmarans, L.G. & Kent, L.E. (1982). Geological investigations in Western Dronning Maud Land, Antarctica – a synthesis. *South African Journal of Antarctic Research*, Supplement 2, 1–93.

Yoder, H.S. & Tilley, C.E. (1962). Origin of basalt magmas: an experimental study of natural and synthetic rock systems. *Journal of Petrology*, **3**, 342–532.

Crustal evolution in the Pensacola Mountains: inferences from chemistry and petrology of the igneous rocks and nodule-bearing lamprophyre dykes

W.W. BOYD

Kaivosrinteentie 2-G-55; Vantaa, Finland

Abstract

Tholeiitic and calc-alkaline basalts are associated with late Precambrian–Mesozoic sedimentary rocks of the Pensacola Mountains, East Antarctica. The majority of these intrusive and extrusive rocks are exposed in the central and northern ranges of the mountains. Late Precambrian dolerite sills also occur in the southern Patuxent Range, where Late Permian ultramafic lamprophyre dykes, bearing a varied suite of ultramafic and mafic nodules and megacrysts, intrude late Precambrian greywackes. Reference to the relevant experimental data for these lamprophyre assemblages as well as the basalts provides some constraints on speculation as to crustal evolution in the area.

Introduction

The Pensacola Mountains lie at the Weddell Sea end of the Transantarctic Mountains, which form the border between East and West Antarctica (Fig. 1). The boundary also marks an apparent change in crustal thickness, from ~ 40–45 km in East Antarctica to ~ 25–30 km under West Antarctica (Bentley, 1983). There has been repeated but sporadic igneous activity throughout the Pensacola Mountains since at least the late Precambrian.

Precambrian pillow lavas and dolerite dykes of tholeiitic and transitional basalt lineages represent the earliest recorded igneous activity in the Pensacola Mountains. The duration of the activity is uncertain; the volume of both intrusive and extrusive rocks is in the order of 100 km^3. Deformation in the constituent feldspars and pyroxenes suggests that some, if not all, of the suite was intruded at the time of intense folding of the Patuxent Formation greywackes. This is thought to have occurred during the Beardmore Orogeny, 650 ± 50 Ma ago (Schmidt, Williams & Nelson, 1978).

Remnants of the flows and pillow lavas are best seen in the Williams Hills, whereas the intrusive rocks are concentrated in the Schmidt Hills (Fig. 1). Isolated exposures of similar rocks occur in the Patuxent Range and are presumed to be stratigraphically equivalent. Both flows and intrusive rocks have undergone greenschist-facies metamorphism, which apparently affected measured $^{87}Sr/^{86}Sr$ ratios in the flows more than in the dolerite sills, and possibly the major- and trace-element chemistry as well. Nonetheless, Eastin (1970) was able to plot an isochron, giving an initial $^{87}Sr/^{86}Sr$ ratio (R_0) of 0.7065 ± 0.0003 and an age of 778 ± 59 Ma. Additional Sr-isotope analyses for the intrusive rocks and the associated felsic flows corroborate the earlier results; a single $^{143}Nd/^{144}Nd$ measurement on a rhyolite flow gave 0.51229 (C.J. Hawkesworth, pers. comm., 1986). These data may be interpreted as indicative of derivation from an enriched mantle source, lying within the bounds of the mantle array (Hawkesworth *et al.*, 1983).

Late Cambrian (510 ± 35 Ma) rhyolitic–dacitic volcanic rocks in the Gambacorta Formation are exposed within a caldera complex in the southern Neptune Range. R_0 values of 0.7052 ± 0.0015 (Eastin, 1970) can also be interpreted as indicative of derivation from an enriched mantle source, but crustal contamination or remelting of crustal rocks would be an equally plausible explanation. The estimated volume of all the units is some 300 km^3.

The major igneous contribution to crustal development in the Pensacola Mountains occurred during a relatively brief interval of time during the Jurassic, with the emplacement of a layered intrusion in the northern part of the range. With the associated satellitic dykes there is a minimum volume of ~ 400 000 km^3 (Ford & Kistler, 1980). The intrusion is considered to be penecontemporaneous with the intrusive rocks of the Ferrar Group that occur throughout the Transantarctic Mountains province (TMP). These are usually classified as 'tholeiitic', mainly on petrological characteristics, but chemically they have calc-alkaline affinities. Coeval intrusions to the north (in the Theron Mountains) apparently have both calc-alkaline and tholeiitic lineages (Brook, 1972), whereas to the east, in the Dronning Maud Land province (DMP), the igneous rocks are tholeiitic (Furnes, Neumann & Sundvoll, 1982). The primary geochemical basis for discrimination between the TMP and DMP has been their initial Sr-isotope ratios, the TMP having ratios of ~ 0.709 and the DMP a range of 0.704–0.706 (Kyle, 1980; Faure, Pace & Elliot, 1982; Furnes *et al.*, 1982). Another diagnostic geochemical characteristic is

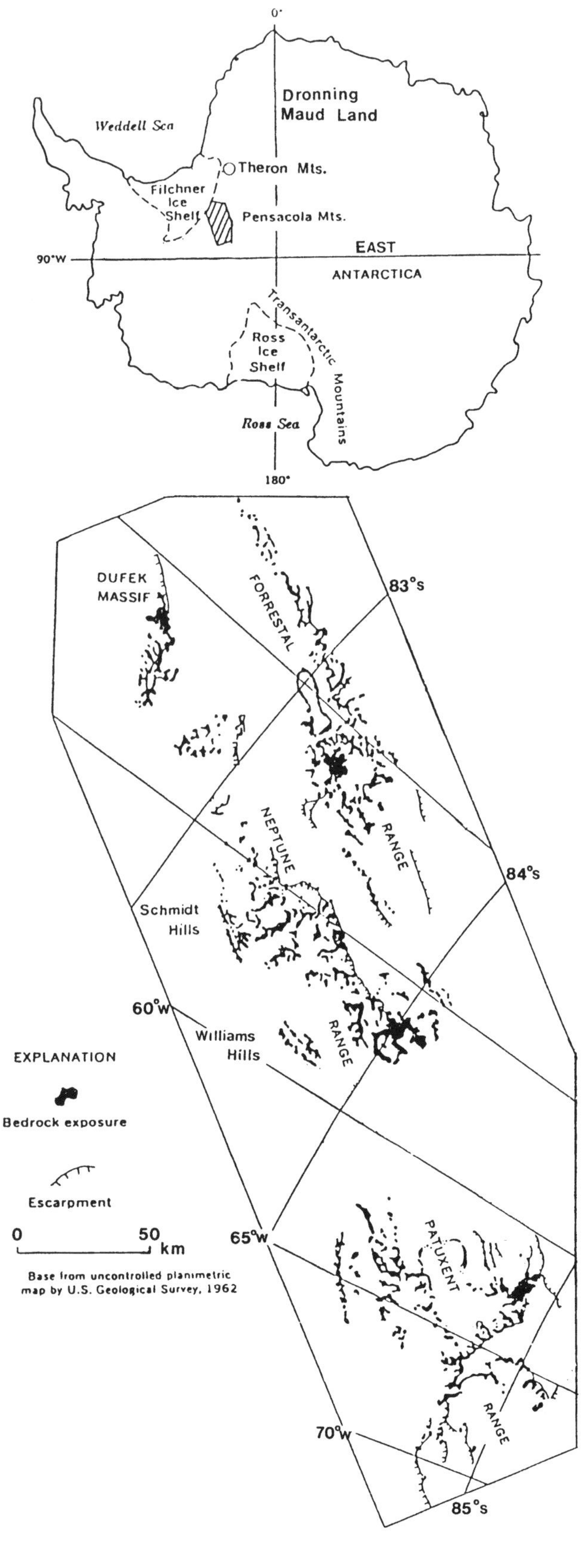

Fig. 1. Map of Antarctica with the Pensacola Mountains shown in relation to the other localities mentioned in the text. The expanded inset shows the location of individual features.

that the TMP basaltic rocks with high R_0 have low TiO_2 and P_2O_5 compared with the less radiogenic basalts of the DMP. These geochemical characteristics have also been noted in the continental flood basalts (CFB) elsewhere in Gondwana.

The (?)Late Permian ultramafic lamprophyre dykes (~ 230 Ma) intruded into late Precambrian greywackes in the Patuxent Mountains are volumetrically unimportant but they are significant in that they alone, out of the varied igneous rocks in the Pensacola Mountains, bear upper mantle–lower crustal samples. Thermobarometric data from lherzolite nodules can be fitted to theoretical (conductive) geothermal gradients but, since the dykes are a possible manifestation of rifting, it cannot be said whether this defines steady-state or transitory phenomenon. The R_0 measured on nearly aphyric rocks at four localities are equally divided between low (0.70469–0.70507) and high (0.70713–0.70831) ratios (R.J. Pankhurst, pers. comm., 1983).

The increase in R_0 that occurs between 230 and 170 Ma in the mafic igneous rocks of the region may reflect either enrichment in the continental lithosphere or contamination by more radiogenic rocks. The Ferrar dolerites throughout the TMP gave a pseudo-isochron ('mantle isochron') of 240 ± 70 Ma (Pankhurst, 1977, table 1) which, provided certain conditions are fulfilled, defines the last time at which the source region was isotopically homogenous. Two conditions, namely the absence of contamination and lack of extensive plagioclase fractionation, probably are not fulfilled but to what extent this affects the pseudo-isochron is unknown. What relation the near coincidence of lamprophyre emplacement, rise in R_0 and the Ferrar dolerite pseudo-isochron have to the possible initiation of rifting in the area is unknown and largely a matter for speculation.

The nature of igneous crustal additions in the Pensacola Mountains

Dependent upon the petrogenetic model chosen, the various mafic igneous suites exposed in the Pensacola Mountains are represented at depth within the crust by an almost equal volume of cumulates. The only direct evidence as to the nature and existence of such bodies lies in the mafic nodules of the lamprophyre dykes. Indirectly their existence can be deduced from the paths that presumed coeval magmas lying on liquid lines of descent followed; these are best shown by a relevant pseudoternary projection. The major igneous contribution to crustal development in the Pensacola Mountains is from basalts with tholeiitic and calc-alkaline affinities. Derivation of these basalts, and specifically the more evolved Jurassic basalts from more primitive basalts in the Theron Mountains (Brook, 1972), shown as Group A in Fig. 2, is discussed below.

No melting experiments on the basalts of the TMP and DMP have established the topology of the relevant phase diagrams. Melting experiments on similar suites of basalts elsewhere have, however, and it is a projection (Fig. 2) from Grove & Baker (1984) that is used as a first approximation, realizing that the configuration of the cotectics in Fig. 2 might be different for the TMP and DMP basalts. An estimate of the

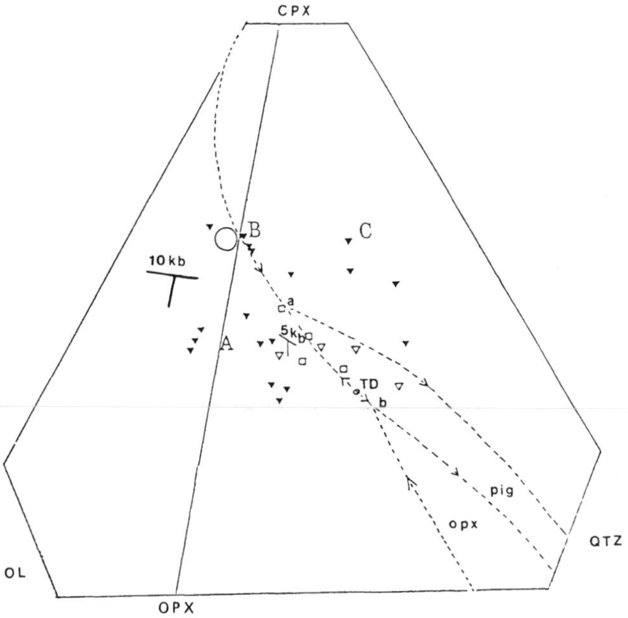

Fig. 2. Low-pressure phase relations (dashed lines) for plagioclase saturated liquids (cf. Grove & Baker, 1984); the 5 kb and 10 kb cotectics (solid lines) are labelled; TD marks a thermal divide. Both *a* and *b* are reaction points where olivine reacts with liquid to give: (a) pigeonite (pig) + augite + plag and (b) orthopyroxene (opx) + pig + plag. Arrows indicate the direction of falling temperature. Different populations of Mesozoic basaltic rocks are represented on the diagram by the following symbols: open squares, TMP intrusive rocks (Dufek dykes, olivine-bearing Tasmanian dolerites, Ferrar sills); open triangles, TMP extrusive rocks (northern Victoria Land Kirkpatrick basalts); solid triangles, the Theron Mountains intrusive rocks, divided into Groups A, B, C. The position of the least-evolved DMP basalts is shown by an open circle (near 'B').

magnitude of such differences is seen in the shift of projected analyses in Group C away from the main trend lying along the low-pressure cotectic lines between 'B' and the reaction point at 'a', caused apparently by a higher TiO_2 content in Group C basalts. Other ambiguities arise from the lack of olivine in any of the more evolved rocks lying along the low-pressure cotectics, as well as the absence of any earlier fractionates. Despite these limitations, general trends in both provinces can be shown.

The distinctive petrological and chemical features of the more evolved basalts appear to have developed at pressures of < 10 kb. The least-evolved basalts (Group A) in either province lie between the 10 kb and 5 kb cotectics; these nearly aphyric rocks contain < 10% plagioclase, clinopyroxene and olivine phenocrysts. The other TMP basalts that lie along the low-pressure cotectics lack olivine, with the notable exception of the dolerites at Mount Sedgewick in Tasmania (Edwards, 1942).

Both tholeiitic and calc-alkaline trends can be generated by gabbroic assemblage fractionation at pressures of < 10 kb. The most important difference lies in the ratio of fractionates necessary to generate tholeiitic trends, i.e. 50–60:30–20:15–20 (plag:cpx:ol) (Biggar, 1983). These proportions remain nearly constant despite considerable changes in the An content of the plagioclase or in the MgO/FeO ratio of clinopyroxene and olivine. Calc-alkaline trends are generated by all three phases being in roughly equal proportions (Grove & Baker, 1984), although there appears to be more tolerance for variation.

The two trends can be shown with reference to Fig. 2. If it is assumed that rocks in Group A represent the origin of possible parental liquid lines that terminate in the more evolved rocks lying along the low-pressure cotectics, then a tholeiitic trend leads to Group B whereas a calc-alkaline trend leads to those analyses clustered about 'a'; these are the dykes associated with the layered intrusion in the northern ranges and are representative of the widespread 'hypersthene' tholeiites of Gunn (1966) in the TMP. Further evolution at low pressures by crystal fractionation alone accounts for most of the points representing 'pigeonite' tholeiites which lie between 'a' and the thermal divide (TD). Despite the importance of olivine in generating a calc-alkaline trend, its extreme rarity in the TMP suggests that reaction point 'a' was not bypassed by magmas that constitute most of the igneous rocks in the TMP. These magmas, in general, show minimal iron enrichment concomitant with a relatively minor increase in alkalis and SiO_2, as well as a slight decrease in Al_2O_3. Such trends characterize calc-alkaline suites in general and fractionation of gabbroic cumulates in particular. Those cumulates generated at low pressures will resemble the gabbroic assemblages of the Dufek intrusion; of potentially equal volumetric significance will be the cumulates generated in deriving 'hypersthene' tholeiites from the yet more primitive olivine tholeiites, as found in the Theron Mountains or more rarely in the rest of the TMP. The presence of olivine in basalts that project near the 5 kb cotectics suggests that olivine-bearing cumulates may exist at deeper levels within the crust, with modes dependent upon the trend followed. How much the idealized modes described above differ from those actually found depends partly on whether crystal fractionation models alone are adequate as a paradigm. Combined assimilation–fractionation (AFC) models in which 10–15% 'granitic' rocks are assimilated are compatible with the trends shown in Fig. 2. The relevant melting experiments, besides providing an accurate guide to phase relations, would serve to constrain such general models.

The geochemical evidence from the TMP basalts is ambiguous and has been interpreted as evidence for the contamination of a tholeiitic magma by crustal rocks (Faure *et al.*, 1982), or genesis in a metasomatically altered continental lithosphere (Kyle, 1980).

Direct evidence as to the nature of the source region in the continental lithosphere, and the existence of mafic cumulate bodies within the crust or at the crust–mantle interface, is provided by a suite of nodules in the ultramafic lamprophyre dykes of the southern Pensacola Mountains. These dykes occur throughout the Patuxent Range, varying in width from several metres to a few centimetres; the dykes have pronounced compositional and textural layering, with up to 30% of unsorted megacrysts and nodules in the central portions. Both ultramafic and mafic nodules are present; the former occur in most dykes sampled, as fragments usually < 3 cm across with serpentinized rims 1–2 mm thick. Grains from disaggregated nodules are distinguished from megacrysts on the basis of their morphology, composition and the nature of their reaction rims.

Only two types of ultramafic nodules have been identified, both with mosaic or tabular equigranular textures. The more abundant spinel lherzolite has variable modal compositions which can have up to 91% olivine; the second type has no spinel but contains primary phlogopite ($\leq 2\%$), clinopyroxene (2–3%), orthopyroxene ($\sim 10\%$) and olivine (80–85%). It is possible that uncommon angular pyrope grains are from disaggregated nodules of this second type. Nodules in this latter group, which conform to Harte's (1983) 'modally metasomatized' peridotites, have large intra- and intergranular variation in Al_2O_3 and CaO in the orthopyroxenes. If, indeed, the pyrope grains are from this type of nodule, then it is possible to calculate provisionally both temperature and pressure at which the hypothetical reconstituted nodule last equilibrated. Pressures (25 ± 3 kb) and temperatures (1225 ± 30 °C) so calculated, using Finnerty & Boyd's (1984) preferred geobarometer and Bertrand & Mercier's (1985–86) geothermometer, should be used with caution. The values derived and fitted to one of the calculated geothermal gradients (Chapman & Pollack, 1977) imply a thickness of 100–140 km for the continental lithosphere.

The mafic (troctolite) nodules are present in only one set of dykes; they are larger (≤ 10 cm across) than the ultramafic nodules, with cumulate textures. Plagioclase (An_{80-82}) predominates, with clinopyroxene ($En_{45}Fs_8Wo_{47}$) and olivine (Fo_{69}) constituting $< 25\%$. The ratios of the three phases do not fit either of the trends described above, but variations are to be expected in such small samples of cumulate sequences. The proximity of Precambrian sills to these particular dykes suggests that the nodules may represent samples of cumulate bodies associated with this earlier period of magmatism, and emplaced within the lower crust or at the crust–mantle interface. Their scarcity in the dykes of the Patuxent Range may indicate that the magmatism had a restricted distribution in the area, confined largely to those outcrops now exposed.

The megacrysts can be divided into those that are probably cognate (most clinopyroxenes and phlogopites) and those, on the basis of disequilibrium textures, likely to be xenocrysts. One of the probable cognate megacrysts, a clinopyroxene–ilmenite intergrowth, provides a crystallization temperature (1090 ± 20 °C) based on experimental work (Bishop, 1980); in conjunction with the thermal gradient derived from the ultrabasic nodules it implies equilibration (? origin) at ~ 16 kb. The xenocrysts are sporadically distributed in the dykes and none can be shown to have coexisted. Their individual stability fields are generally too extensive (i.e. fluorapatite, K-richterite) to define a petrogenetic locus, but for some (melilite) the relevant experimental studies (Yoder, 1973) suggest pressures of ~ 10 kb. As a group, however, they constitute reasonable evidence of metasomatic event(s) in the continental lithosphere. No isotope analyses are available to specify when or what relationship either nodules or megacrysts bear to the host lamprophyres.

Conclusions

Igneous activity in the Pensacola Mountains was concentrated largely in the northern ranges during the Jurassic; indirect evidence suggests that of the total crustal thickness ($\sim$ 40–45 km) nearly 40% is comprised of gabbroic cumulates. Earlier magmatic activity appears to have added very little as cumulates but where present they may either underplate, or reside within the lower crust. The contrasted proportions and styles of igneous activity possibly relate to the extent of subsidence at the rifted margin of the Transantarctic Mountains and to the temperature of upwelling magmas in the rift zone itself (White *et al.*, 1987).

Initial $^{87}Sr/^{86}Sr$ ratios of the source region in the continental lithosphere appear to have remained relatively constant over a long period (~ 550 Ma), but underwent a considerable increase in a shorter period (50–70 Ma) during or just prior to the emplacement of large quantities of magmas with calc-alkaline affinities. Metasomatic infiltration of the continental lithosphere may have occurred at this time during the initial stages of rifting, as marked by the intrusion of ultramafic lamprophyre dykes. A provisional geothermal gradient from included ultramafic nodules is not abnormally steep.

References

Bentley, C.R. (1983). Crustal structure of Antarctica from geophysical evidence – a review. In *Antarctic Earth Science*, ed. R.L. Oliver, P.R. James & J.B. Jago, pp. 491–7. Canberra; Australian Academy of Science and Cambridge; Cambridge University Press.

Bertrand, P. & Mercier J.-C.C. (1985–86). The mutual solubility of coexisting ortho- and clinopyroxene; toward an absolute geothermometer for the natural system? *Earth and Planetary Science Letters*, **76(1)**, 109–22.

Biggar, G.M. (1983). Crystallization of plagioclase, augite and olivine in synthetic systems and in tholeiites. *Mineralogical Magazine*, **47(1)**, 161–76.

Bishop, F.C. (1980). The distribution of Fe^{2+} and Mg between coexisting ilmenite and pyroxene with applications to geothermometry. *American Journal of Science*, **280(1)**, 46–77.

Brook, D. (1972). Geology of the Theron Mountains, Antarctica. PhD thesis, University of Birmingham, 267 pages. (unpublished).

Chapman, D.S. & Pollack, H.N. (1977). Regional geotherms and lithospheric thickness. *Geology*, **5(3)**, 265–8.

Eastin, R. (1970). Geochronology of the basement rocks of the central Transantarctic Mountains, Antarctica. PhD thesis, Ohio State University, Columbus, Ohio. (unpublished).

Edwards, A.B. (1942). Differentiation of the dolerites of Tasmania. *Journal of Geology*, **50(4)**, 451–80 and **(5)**, 579–610.

Faure, G., Pace, K.K. & Elliot, D.H. (1982). Systematic variations of $^{87}Sr/^{86}Sr$ ratios and major element concentrations in the Kirkpatrick Basalt of Mt Falla, Queen Alexandria Range, Transantarctic Mountains. In *Antarctic Geoscience*, ed. C. Craddock, pp. 715–23. Madison; University of Wisconsin Press.

Finnerty, A.A. & Boyd, F.R. (1984). Evaluation of thermobarometers for garnet peridotites. *Geochimica et Cosmochimica Acta* **48(1)**, 15–27.

Ford, A.B. & Kistler, R.W. (1980). K–Ar age, composition, and origin of Mesozoic mafic rocks related to the Ferrar Group, Pensacola Mountains, Antarctica. *New Zealand Journal of Geology and Geophysics*, **23(3)**, 371–90.

Furnes, H., Neumann, E.R. & Sundvoll, B. (1982). Petrology and geochemistry of Jurassic basalt dykes from Vestfjella, Dronning Maud Land, Antarctica. *Lithos*, **15(3)**, 295–304.

Grove, T.L. & Baker, M.B. (1984). Phase equilibrium controls on the tholeiitic versus calc-alkaline differentiation trends. *Journal of Geophysical Research*, **89(B5)**, 3253–74.

Gunn, B.M. (1966). Modal and element variation in Antarctic tholeiites. *Geochimica et Cosmochimica Acta*, **30(9)**, 881–920.

Harte, B. (1983). Mantle peridotites and processes – the kimberlite sample. In *Continental Basalts and Mantle Xenoliths*, ed. C.J. Hawkesworth & M.J. Norry, pp. 46–91. Nantwich; Shiva Publishing Ltd.

Hawkesworth, C.J., Erlank, A.J., Marsh, J.S., Menzies, M.A. & van Calsteren, P. (1983). Evolution of the continental lithosphere: evidence from volcanics and xenoliths in southern Africa. In *Continental Basalts and Mantle Xenoliths*, ed. C.J. Hawkesworth & M.J. Norry, pp. 111–38. Nantwich; Shiva Publishing Ltd.

Kyle, P.R. (1980). Development of heterogeneities in the subcontinental mantle; evidence from the Ferrar Group, Antarctica. *Contributions to Mineralogy and Petrology*, **73(1)**, 89–104.

Pankhurst, R.J. (1977). Strontium isotope evidence for mantle events in the continental lithosphere. *Journal of the Geological Society, London*, **134(2)**, 255–68.

Schmidt, D.L., Williams, P.L. & Nelson, W.H. (1978). *Geologic Map of the Schmidt Hills Quadrangle and Part of the Gambacorta Peak Quadrangle, Pensacola Mountains, Antarctica.* Antarctic Geologic Map A-8, 1:250000. Washington, DC; US Geological Survey.

White, R.S., Spence, G.D., Fowler, S.R., McKenzie, D.P., Westbrook, G.K. & Bowen, A.N. (1987). Magmatism at rifted continental margins. *Nature, London*, **330(6147)**, 439–44.

Yoder, H.S., Jr. (1973). Melilite stability and paragenesis. *Fortschritte der Mineralogie*, **50(1)**, 140–73.

The geochemistry of the Kirwan and other Jurassic basalts of Dronning Maud Land, and their significance for Gondwana reconstruction

C. HARRIS[1], A.J. ERLANK[1], A.R. DUNCAN[1] & J.S. MARSH[2]

1 Department of Geochemistry, University of Cape Town, Rondebosch 7700, South Africa
2 Department of Geology, Rhodes University, Grahamstown 6140, South Africa

Abstract

The Kirwan Volcanics are all basalts with a restricted compositional range (SiO_2 50.1–52.1 wt %; MgO 5.1–6.6 wt %). They are very similar in composition to the adjacent Heimefrontfjella basalts but differ from the more varied Vestfjella basalts (MgO 5–25 wt %). The Kirwan basalts are chemically and isotopically similar to the Karoo basalts of the southern Lebombo in southern Africa, which is consistent with the juxtaposition of these areas in most accepted reconstructions of Gondwana. Contemporaneous basic dykes to the east of Kirwanveggen are chemically more variable and their higher concentration of certain incompatible elements is reminiscent of the 'enriched' northern province of the Karoo volcanics. In southern Africa, geochemical provinces are considered to be the result of large-scale mantle heterogeneity and it is suggested that a similar situation exists in Antarctica.

Introduction

The Kirwan basalts crop out along the southernmost 50 km of Kirwanveggen in Dronning Maud Land (74° S, 6° W). The basalts overlie the Permian–Triassic Amelang Plateau Formation, which rests unconformably on a variety of Precambrian schists and gneisses or Lower Palaeozoic quartzites. The Kirwan basalts are the most easterly exposed outcrops of Jurassic basalt in that part of Dronning Maud Land west of Jutulstraumen (Fig. 1). Father west are the Heimefrontfjella and Vestfjella basalts, which appear to be of similar age (Rex, 1972).

The Kirwan basalts have been dated at 172 Ma (Faure, Bowman & Elliot, 1979) and volcanic rocks of similar age occur elsewhere in Antarctica and on the other southern continents. This paper discusses the nature and petrogenesis of the Kirwan basalts and compares them with the other Jurassic basalts of Dronning Maud Land. Comparison of the Kirwan basalts with the Karoo basalts of southern Africa provides useful geological constraints on Gondwana reconstructions.

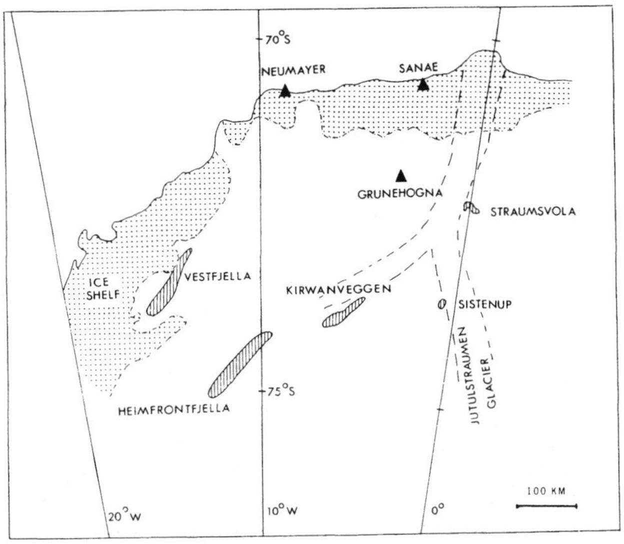

Fig. 1. Map of Dronning Maud Land showing the location of Kirwanveggen and other outcrops of Jurassic basalt lavas at Vestfjella and Heimefrontfjella. Straumsvola is a syenite–Ne–syenite complex and at Sistenup, Q-syenite intrudes basalt lavas of uncertain age (? post-Precambrian). Ahlmannryggen is the range of mountains around the Grunehogna base.

Previous work

Reconnaissance geological mapping and surveying of the Kirwan basalts were carried out by a South African expedition in 1969–70, and a limited number of samples were collected (Aucamp, Wolmarans & Neethling, 1972; Wolmarans & Kent, 1982). The samples were analysed for their Sr-isotopic composition (Faure *et al.*, 1979) and one sample of lava gave a K–Ar age of 172 ± 10 Ma; these samples were later analysed for rare earth elements (REE) by Furnes *et al.* (1987). This paper is based on data from basalts collected in 1985 and 1986 from Kirwanveggen.

Table 1. *Comparison of representative basalts from Dronning Maud Land and the Karoo*

	Kirwan		Heimefrontfjella		Vestfjella		South Lebombo		Lesotho Formation	
	Mean	Standard deviation	Mean	Standard deviation	Mean	Standard deviation	Mean	Standard deviation	Mean	Standard deviation
SiO_2	51.19	0.48	51.14	1.34	49.96	1.98	53.49	2.49	52.06	0.78
TiO_2	1.47	0.15	1.38	0.28	1.59	0.33	1.46	0.46	0.96	0.06
Al_2O_3	14.79	0.41	14.78	0.47	14.39	1.81	15.05	2.32	15.86	0.53
FeO*	12.52	0.35	11.88	1.02	11.99	1.18	11.23	1.69	9.97	0.43
MnO	0.20	0.02	0.20	0.04	0.21	0.01	0.18	0.03	0.17	0.02
MgO	5.88	0.38	6.59	0.79	8.47	3.73	5.58	1.27	7.09	0.42
CaO	10.48	0.40	10.91	0.68	9.64	1.22	9.38	1.28	10.81	0.61
Na_2O	2.74	0.22	2.42	0.16	2.47	0.70	2.58	0.56	2.19	0.32
K_2O	0.53	0.14	0.52	0.17	1.00	0.50	0.79	0.43	0.71	0.21
P_2O_5	0.19	0.02	0.19	0.03	0.28	0.07	0.26	0.10	0.17	0.02
Ba	147	26	223	47	489	299	295	169	177	46
Ce	21	3	27	8	19	6				
Cr	123	28	195	99			136	81	283	61
Nb	4	1	5	1	6	2	5	3	5	1
Ni	75	18	85	19			76	64	94	15
Rb	10	4	10	5	17	15	19	16	12	9
Sr	189	24	232	97	331	109	316	155	192	31
Y	32	3	23	5	25	7	29	8	24	2
Zr	109	12	114	23	98	26	134	49	94	12
	(N = 78)		(N = 9)		(N = 34)		(N = 49)		(N = 37)	

N = number of analyses. All major element oxide data were recalculated to 100% on volatile-free basis.
Data for Heimefrontfjella and Vestfjella from Juckes (1972) and Furnes *et al.* (1987). Data for S. Lebombo and Lesotho from Duncan, Erlank & Marsh (1984, table III Zululand SRBF and table IV).

Petrography and chemistry of the Kirwan lavas

Petrography and mineralogy

The majority of the lavas are plagioclase-phyric with plagioclase + augite being the most common phenocryst assemblage. The most porphyritic lavas contain about 20–25% modal phenocrysts but the majority have less than 15% phenocrysts. Plagioclase varies from $An_{82}Ab_{18}Or_0$ in phenocryst cores in the most primitive of the basalts, to $An_{32}Ab_{64}Or_4$ in coarse-grained pegmatoid patches in some of the evolved lavas. Augite and pigeonite phenocrysts occur in most lavas though the latter are relatively uncommon. Although pseudomorphs of olivine are present in about 30% of the basalts, fresh olivine was only found in one lava (Fo_{77}–Fo_{62}). The opaque oxide present is in most cases a Ti-rich magnetite ($Usp_{50}Mt_{50}$) but isolated grains of ilmenite ($Ilm_{96}Hm_4$) occur in some lavas.

Major and trace-element chemistry of the Kirwan lavas

The mean and standard deviation of 78 XRF analyses of Kirwan basalt are given in Table 1. The rocks show a restricted range in composition, particularly in the major elements, e.g. SiO_2 (50.1–52.1 wt %), MgO (5.1–6.6 wt %). Trace elements show rather more variation, with incompatible elements (e.g. Zr, Ba, Y) varying by factors of up to 2.2. Zr correlates well with TiO_2, P_2O_5, Ba and Y but more scattered plots are

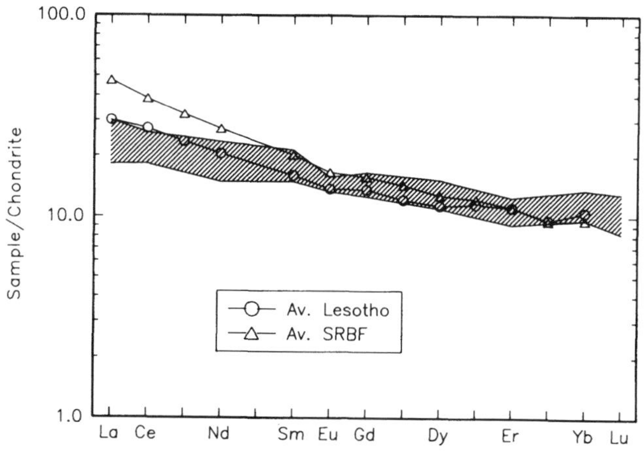

Fig. 2. Range of Kirwan REE data (shaded area) compared to average southern Lebombo SRBF and average Lesotho Formation basalts. 10 samples were selected to cover the compositional range of the Kirwan basalts. Normalizing values from Frey *et al.* (1968) and Nakamura (1974).

obtained for MgO, SiO_2 and Sr versus Zr. There is no evidence for more than one magma type.

A selection of 10 lavas have been analysed for REE (by ICP). Most notable is the small spread of the data (Fig. 2). All the individual patterns are LREE-enriched (La_N/Yb_N about 2) and most show slight Eu anomalies.

Isotopic composition of the Kirwan basalts

Thirteen samples of Kirwan basalts have been analysed for $^{87}Sr/^{86}Sr$ by Faure *et al.* (1979). The main feature of these data is the large variation in initial $^{87}Sr/^{86}Sr$ ratios (R_0) from 0.7029 to 0.7063, with a mean of 0.7044. The data from 29 basalts analysed for Sr isotopes during the present work differ from those obtained by Faure *et al.* (1979) in that all samples have greater average R_0 (corrected to 172 Ma). R_0 varies from 0.70469 to 0.70648 with one altered sample having a ratio of 0.70743. The range in initial $^{143}Nd/^{144}Nd$ ratios is small for the seven samples analysed (0.512324–0.512521, mean 0.512354) and straddles the bulk Earth value at 172 Ma (0.5124).

The samples chosen for oxygen-isotope analysis were from among the freshest of the lavas. Where possible, suitable inclusion-free plagioclase phenocrysts were separated and analysed together with the whole rock sample. A small range of values for both whole rock and plagioclase separates was obtained (6.2–6.7‰), with one altered sample having a $\delta^{18}O$ of 7.4‰. The values of $\delta^{18}O$ obtained for the samples are within the range accepted for primary mantle oxygen. The plagioclase–magma fractionation factor observed in the Kirwan basalts is small (average Δ plag–WR = − 0.26‰). The accepted value (Kyser, O'Neil & Carmichael, 1981) is about + 0.3‰ at 1000 °C. The Kirwan basalts therefore give whole-rock $\delta^{18}O$ values which are about 0.5‰ heavier than the value indicated for the original magma by the $\delta^{18}O$ for plagioclase phenocrysts. This discrepancy is probably due to slight deuteric alteration, even in the freshest of the basalts.

Petrogenesis of the Kirwan basalts

Processes responsible for the variation in chemistry of the lavas and dykes are likely to be some combination of the following: crystal fractionation, variation in degree of partial melting of the source during melt extraction, assimilation of continental material, a heterogeneous source on a small scale (either vertically or horizontally) and alteration. All these processes have been invoked in the literature as contributing to the compositional variation of the various Mesozoic basalts from Antarctica (Faure *et al.*, 1979; Hoefs, Faure & Elliot, 1980; Kyle, 1980; Furnes *et al.*, 1987).

Crystal fractionation and partial melting

Least-squares mixing models show that the most evolved basalts could have been generated from the least evolved by removal from the latter of about 30% of an assemblage of plagioclase, clinopyroxene, olivine and magnetite in the proportions 52:37:8:3. The incompatible trace elements, however, show enrichment factors ranging from about 1.5 (Y) to 2.2 (Ba and Ce) which cannot be explained by crystal fractionation (the maximum amount of enrichment possible for 30% fractionation is 1.43, i.e. with a partition coefficient $[D] = $ zero).

For the suite as a whole there is a reasonable intercorrelation of incompatible elements, e.g. Zr, Ce, Ba, Y, P and Nb, and there is no apparent decoupling of Ba, K and Rb from Zr, Y and Nb other than that attributable to alteration. This decoupling would be expected in a series of basaltic lavas contaminated by granitic crust (Cox & Hawkesworth, 1985). A model which could account for these higher than expected enrichments of trace elements is one in which the basalts were derived from liquids extracted after varying degrees of partial melting of the source. Highly incompatible elements in the source would be strongly partitioned into the liquid initially formed during melting but their concentration in the melt would decrease with small increases in the degree of melting. Major and compatible trace elements would remain buffered by the restite assemblage and show little change for degrees of melting less than about 15%. In this model then, the major element and compatible element variation is due to crystal fractionation, but the incompatible element variation is due to crystal fractionation superimposed on already varying concentrations caused by slightly different degrees of melting of the same source. This type of two-stage model would account for the rather poor correlation between the incompatible elements and MgO.

Role of crustal contamination

Contamination of the low R_0 Kirwan basalts by material with higher R_0 and Sr may have produced the high R_0 basalts. For the fractionating assemblage suggested above, the bulk D for Sr is close to unity with the result that AFC (De Paolo, 1981) is equivalent to simple mixing for the special case of R_0 versus Sr, provided that the amount of contamination is small (i.e. D_{Sr} remains close to 1). R_0 versus 1/Sr for the 10 freshest samples (with one aberrant point omitted) plot on a straight line with $r = 0.79$, which is consistent with a simple mixing model. The fresh Kirwan basalt with the highest R_0 could have been produced by 7.5% contamination of the lowest R_0 basalt by granitic material having 800 ppm Sr with $^{87}Sr/^{86}Sr$ of 0.708 (calculated using equations in Faure *et al.*, 1974). Such material would appear to exist among the gneisses of the Sverdrup Group (Wolmarans & Kent, 1982) and it is plausible that small amounts of contamination may have occurred.

Crustal contamination is, however, unlikely to have been important as a mechanism for generating the observed chemical and isotopic variations in the Kirwan lavas, for the following reasons.

1. There is no good correlation between parameters likely to have been affected by contamination (e.g. $\delta^{18}O$, R_0, Ba, Rb and K_2O) apart from the case of R_0 versus Sr discussed above.

2. The Kirwan lavas have mantle oxygen values which would make contamination by significant amounts of granite ($\delta^{18}O = 10$?) implausible. However, assimilation of about 7% granite would not be detectable in the oxygen values.

3. There is no decoupling of elements like Y, Zr, Ce from Ba, Rb (especially in the subset of the freshest basalts discussed above) which would be expected if these basalts had been contaminated by granitic continental material (Cox & Hawkesworth, 1985).

In summary, there is some evidence from Sr-isotopes that small amounts of contamination may have affected the Kirwan lavas, but the trace element data do not support this hypo-

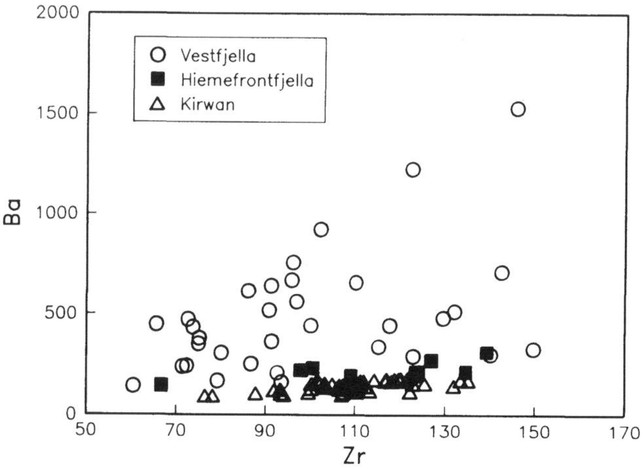

Fig. 3. Ba versus Zr for the three Mesozoic basalt 'provinces' of Dronning Maud Land. Note the similarity between the Kirwan and Heimefrontfjella basalts and the large variation shown by those from Vestfjella. Data from Juckes (1972) and Furnes *et al.* (1987).

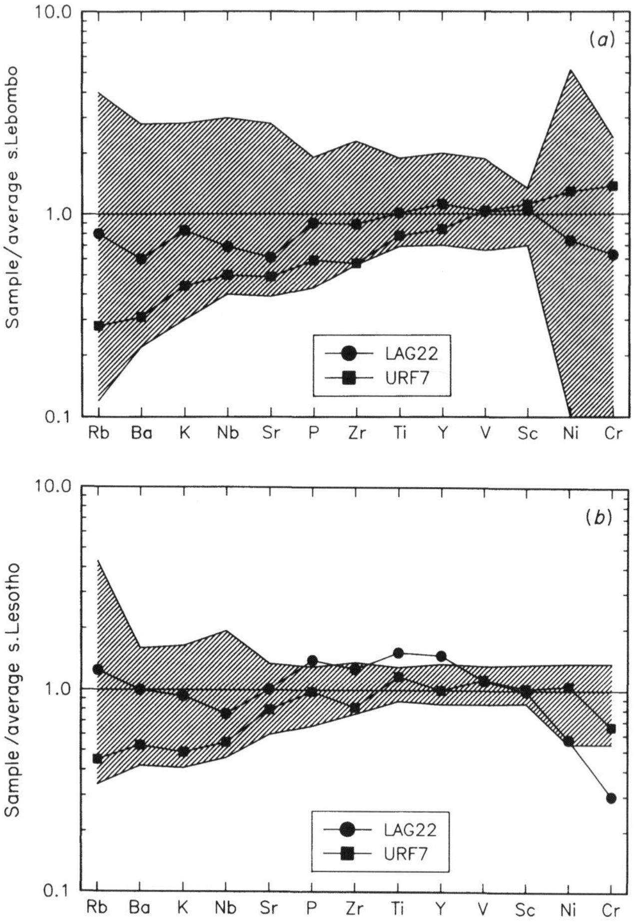

Fig. 4. Kirwan basalts (LAG22, evolved and URF7, primitive) normalized to: (*a*) average southern Lebombo SRBF; and (*b*) average Lesotho basalt. The shaded areas represent the range for the southern Lebombo SRBF (*a*) and Lesotho Formation (*b*). Data for normalizing values from Duncan *et al.* (1984).

thesis. In any event, the amount of contamination was small (< 7% in the lavas with the highest R_0).

Role of mantle heterogeneity

Mantle heterogeneity may have played a small additional role in determining the variation seen in the Kirwan lavas, but it is not required to explain it. Mantle heterogeneity has been invoked by numerous authors (e.g. papers in Erlank, 1984) to explain compositional differences between Karoo basalts from different areas in southern Africa. The relationship between the mantle sources for Karoo basalts and those of the Kirwan basalts will be discussed below.

Comparison of the Jurassic lavas in Dronning Maud Land

The Heimefrontfjella basalts (Juckes, 1972) are very similar in both major and trace element composition to the Kirwan lavas, whereas the Vestfjella lavas (Furnes *et al.*, 1987) show a much greater variation in chemistry, e.g. MgO 5–24 wt %, SiO_2 48–52 wt % (Table 1), some of which is presumably due to cumulus enrichment. Furthermore, in the Vestfjella lavas there is a strong decoupling of K, Rb and Ba from Nb and Zr, with Ba concentrations as high as 1500 ppm (Fig. 3). This possibly reflects some addition of crustal material and/or alteration to these lavas. The Heimefrontfjella and Kirwan lavas overlap the fields for the Vestfjella lavas on any variation diagram but it is not known at present whether those from Heimefrontfjella are compositionally equivalent to 'uncontaminated' Vestfjella basalt.

Comparison of Kirwan lavas with the Karoo lavas of southern Africa

The Karoo basalts belong to two distinct geochemical provinces (Cox, MacDonald & Hornung, 1967). These are the northern incompatible-rich (high-K of Cox *et al.*, 1967) and the southern incompatible-poor (low-K of Cox *et al.*, 1967). The northern province is characterized by enrichment of K, Ti, P, Ba, Sr, Zr and Nb relative to the southern province. A comparison of the Kirwan basalts with the Karoo basalts shows that they are similar in chemistry to the least evolved basalts of the southern province. Fig. 4*a* shows two basalts from either end of the Kirwan compositional range, normalized to average southern Lebombo variety of the Sabie River Basalt Formation (SRBF). In Fig. 4*b* the basalts are normalized to average Lesotho Formation basalt. It is evident that all three basalt types are broadly similar but the straight line pattern shown by the Kirwan basalts (Fig. 4*a*) is what would be expected for basalts derived from a similar source to the southern Lebombo SRBF by slightly higher degrees of partial melting. Fig. 4*b* indicates that the Kirwan lavas could not have been directly derived from the same source as the Lesotho Formation lavas. The REE data (Fig. 2) show that the Kirwan basalts are similar to average Lesotho and SRBF basalts but have less steep patterns. This is also consistent with derivation from a similar source by higher degrees of partial melting. The Sr- and Nd-isotope data are close to bulk Earth values, and on Fig. 5 the Kirwan data plot in or close to the Central Area basalts. The low-k-type southern Lebombo variant of the SRBF plots just in the top left quadrant of Fig. 5. Given the

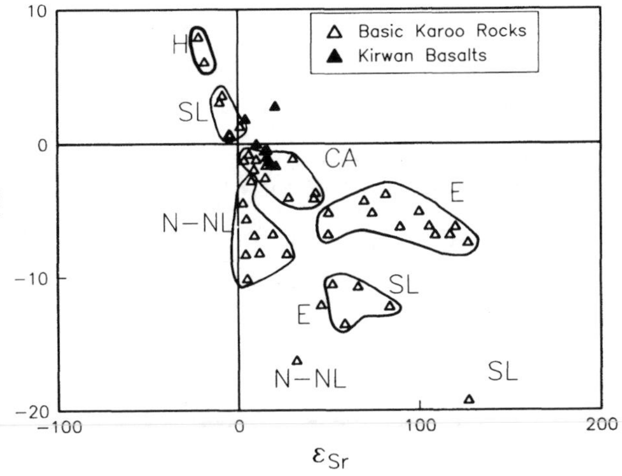

Fig. 5. ϵ_{Nd} versus ϵ_{Sr} for Karoo basalts with values for the Kirwan samples shown. Karoo data from Hawkesworth *et al.* (1984); H, Horingbaai; SL, southern Lebombo; N–NL, Nuanetsi and northern Lebombo, CA, Central Area; E, Etendeka. Kirwan age corrected to 172 Ma, Karoo to 190 Ma and Etendeka to 121 Ma.

small number of analyses available, it is possible that there is an overlap in isotopic composition between the southern Lebombo and the Central Area, which coincides with the composition of the Kirwan basalts.

The Kirwan basalts were extruded while Dronning Maud Land was adjacent to the south-east coast of Africa. Reconstructions of this part of Gondwana are numerous, the latest being Martin & Hartnady (1986). This reconstruction is favoured because of the proximity of Dronning Maud Land to Africa and the positioning of the Kirwan basalts closer to the southern part of the Lebombo Monocline than in other reconstructions. This is consistent with the similarity in geochemistry between the Kirwan basalts and the southern Lebombo variety of the SRBF.

Ahlmannryggen dykes – an 'enriched' basalt type?

A suite of dykes crops out at Ahlmannryggen, 100 km north-east of Kirwanveggen. Several of these have been dated at 170–200 Ma by the K–Ar technique (Wolmarans & Kent, 1982; B.R. Watters, pers. comm., 1985). Preliminary study of these dykes show them to be much more varied in composition than the Kirwan lavas (e.g. MgO 5–18%), with more magnesian varieties petrographically resembling the picrites of the northern Lebombo. These dykes show enrichment of incompatible elements relative to the Kirwan basalts.

It has been shown above that none of the Kirwan basalts shows the enriched chemistry associated with the northern part of the Karoo province (Cox *et al.*, 1967). Bristow (1980) and Cox & Bristow (1984) have shown that the boundary between these two provinces is located close to the Komati River in the central Lebombo and trends approximately E–W. If this boundary is projected into Antarctica using the reconstruction of Martin & Hartnady (1986), then it lies close to the Jutulstraumen trough (Fig. 6).

Fig. 7 shows that relative to the Kirwan basalts, and hence the Lesotho Formation basalts (Fig. 4), the Jurassic dykes

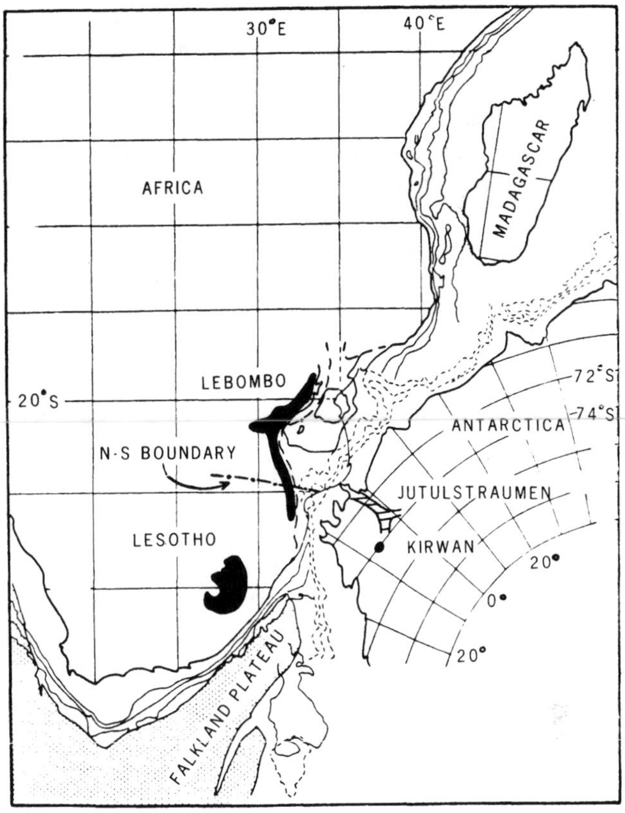

Fig. 6. Reconstruction of southern Africa and Dronning Maud Land after Martin & Hartnady (1986) for magnetic anomaly 21 (145 Ma). Note that Kirwanveggen is much closer to the southern Lebombo than it is to Lesotho. Also shown is the boundary between the northern and southern provinces of the Karoo basalts, and the position of Jutulstraumen.

from Ahlmannryggen are enriched in Ti, Sr, P, Zr and Nb but not K, Rb and Ba. It is hard to envisage any petrogenetic process by which these dykes could have been derived from the same source as the Kirwan basalts. Lower degrees of partial melting of the same source would have resulted in enrichment of K, Rb and Ba and contamination is equally unlikely to have imported the enriched character because of the unusual composition of contaminant required. It is therefore suggested that

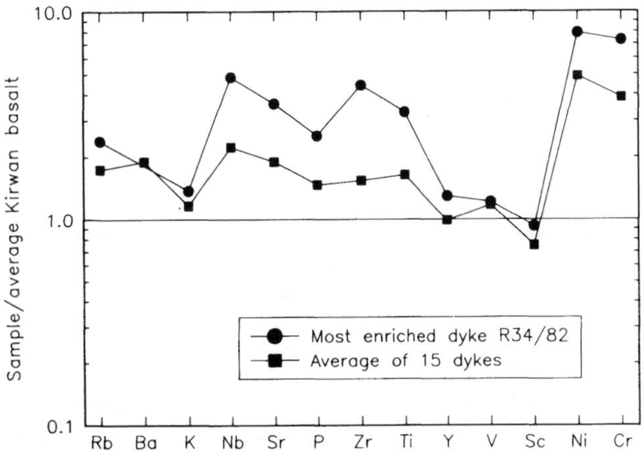

Fig. 7. Ahlmannryggen dykes normalized to average Kirwan basalt. R34/82 (Krynauw, 1986) is the most enriched dyke and the other pattern is the average for the remaining 15 dykes in the UCT collection.

the dykes were derived from 'enriched' mantle similar, but not identical, to that for the northern 'high-K' Karoo province. It is suggested that the basalt provinces recognized in the Karoo extend into Antarctica and that Jutulstraumen marks the boundary between the two provinces. Further work will be necessary to confirm this suggestion and to see if there is any connection between this major tectonic feature and the existence of these basalt provinces.

Conclusions

The Kirwan lavas show many features in common with their counterparts in the Karoo Province. In particular their chemical composition is not easily explained by a single process. The Kirwan basalts are similar to the Heimefrontfjella basalts and both correspond to southern type (incompatible-poor) Karoo basalts. The Kirwan basalts could have been derived from a similar source to the southern Lebombo and this is good supporting geological evidence that the reconstruction of Gondwana by Martin & Hartnady (1986) is correct. The Jurassic dykes to the east of the Kirwan basalts are enriched in certain incompatible elements and are similar in many respects to the northern 'high-K' Karoo basalts. It seems that a similar geochemical provinciality exists in this part of Antarctica.

Acknowledgements

We would like to thank the following people and institutions: the CSIR through its SASCAR programme for financial support; the SAAF (30 squadron) for logistical support in the field; colleagues in the Geochemistry and Geology Departments at UCT for much valuable help; BPI Geophysics and NPRL for help in obtaining the isotope data; Colin Devey of the CRPG Nancy, France for provision of REE data; Johan Krynauw and Brian Watters for sharing unpublished data; K.G. Cox for a helpful review.

References

Aucamp, A.P.H., Wolmarans, L.G. & Neethling, D.C. (1972). The Urfjell Group, a deformed (?)Early Palaeozoic sedimentary sequence, Kirwanveggen, western Dronning Maud Land. In *Antarctic Geology and Geophysics*, ed. R.J. Adie, pp. 557–62. Oslo; Universitetsforlaget.

Bristow, J.W. (1980). The geochronology and geochemistry of Karoo volcanics in the Lebombo and adjacent area. PhD thesis, University of Cape Town, 257 pp. (unpublished).

Cox, K.G. & Bristow, J.W. (1984). The Sabie River Basalt Formation of the Lebombo Monocline and south east Zimbabwe. *Geological Society of South Africa, Special Publication*, **13**, 125–48.

Cox, K.G. & Hawkesworth, C.J. (1985). Geochemical stratigraphy of the Deccan Traps at Mahabaleshwar, western Ghats, India, with implications for open system magmatic processes. *Journal of Petrology*, **26(2)**, 355–77.

Cox, K.G., Macdonald, R. & Hornung, G. (1967). Geochemical and petrographic provinces in the Karoo basalts of southern Africa. *American Mineralogist*, **52(9, 10)**, 1451–74.

De Paolo, D.J. (1981). Trace element and isotopic effects of combined wall rock assimilation and fractional crystallization. *Earth and Planetary Science Letters*, **53**, 182–202.

Duncan, A.R., Erlank, A.J. & Marsh, J.S. (1984). Regional geochemistry of the Karoo igneous province. *Geological Society of South Africa, Special Publication*, **13**, 355–88.

Erlank, A.J., (ed.) (1984). Petrogenesis of the volcanic rocks of the Karoo Province. *Geological Society of South Africa, Special Publication*, **13**, 395 pp.

Faure, G., Bowman, J.R. & Elliot, D.H. (1979). The initial $^{87}Sr/^{86}Sr$ ratios of the Kirwan volcanics of Dronning Maud Land: comparison with Kirkpatrick Basalt, Transantarctic Mountains. *Chemical Geology*, **26**, 77–90.

Faure, G., Bowman, J.R., Elliot, D.H. & Jones, L.M. (1974). Strontium isotope composition and petrogenesis of the Kirkpatrick Basalt, Queen Alexandra Range, Antarctica. *Contributions to Mineralogy and Petrology*, **48**, 153–69.

Frey, F.A., Haskin, M.A., Poetz, J.A. & Haskin, L.A. (1968). Rare earth abundances in some basic rocks. *Journal of Geophysical Research*, **73(18)**, 6085–98.

Furnes, H., Vad, E., Austrheim, H., Mitchell, J.G. & Garmann, L.B. (1987). Geochemistry of basalt lavas from Vestfjella and adjacent areas, Dronning Maud Land, Antarctica. *Lithos*, **20(3)**, 337–56.

Hawkesworth, C.J., Marsh, J.S., Duncan, A.R., Erlank, A.J. & Norry, M.J. (1984). The role of continental lithosphere in the generation of the Karoo volcanic rocks: evidence from combined Nd- and Sr-isotope studies. *Geological Society of South Africa, Special Publication*, **13**, 341–54.

Hoefs, J., Faure, G. & Elliot, D.H. (1980). Correlation of $\delta^{18}O$ and initial $^{87}Sr/^{86}Sr$ ratios in Kirkpatrick Basalt on Mt Falla, Transantarctic Mountains. *Contributions to Mineralogy and Petrology*, **75**, 199–203.

Juckes, L.M. (1972). *The Geology of North-Eastern Heimefrontfjella, Dronning Maud Land.* London; British Antarctic Survey Scientific Reports, No. 65, 44 pp.

Krynauw, J.R. (1986). The petrology and geochemistry of intrusions at selected nunataks in the Ahlmannryggen and the Giaeverryggen, western Dronning Maud Land, Antarctica. PhD thesis, University of Natal (Pietermaritzburg), 370 pp. (unpublished).

Kyle, P.R. (1980). Development of heterogeneities in subcontinental mantle: evidence from the Ferrar Group, Antarctica. *Contributions to Mineralogy and Petrology*, **73**, 89–104.

Kyser, T.K., O'Neil, J.R. & Carmichael, I.S.E. (1981). Oxygen isotope thermometry of basic lavas and mantle nodules. *Contributions to Mineralogy and Petrology*, **77(1)**, 11–23.

Martin, A.K. & Hartnady, C.J. (1986). Plate tectonic development of the southwest Indian Ocean: a revised reconstruction of east Antarctica and Africa. *Journal of Geophysical Research*, **91(B5)**, 4767–86.

Nakamura, N. (1974). Determination of REE, Ba, Fe, Mg, Na and K in carbonaceous and ordinary chondrites. *Geochimica et Cosmochimica Acta*, **38**, 757–75.

Rex, D.C. (1972). K–Ar age determinations on volcanic and associated rocks from the Antarctic Peninsula and Dronning Maud Land. In *Antarctic Geology and Geophysics*, ed. R.J. Adie, pp. 133–6. Oslo; Universitetsforlaget.

Wolmarans, L.G. & Kent, L.E. (1982). Geological investigations in western Dronning Maud Land, Antarctica – a synthesis. *Southern African of Antarctic Research*, **Supplement 2**, 93 pp.

The geochemistry of Mesozoic tholeiites from Coats Land and Dronning Maud Land

T.S. BREWER[1] & D. BROOK[2]

1 Department of Mining, University of Nottingham, Nottingham NG7 2RD, UK
2 Department of the Environment, Marsham St, London SW1P 3EB, UK

Abstract

Mesozoic tholeiites from Coats Land and Dronning Maud Land can be divided into two groups: those with high TiO_2 (1.9–3.4%) and those with low TiO_2 (0.6–1.4%). All of these compositions have trace element signatures that are more characteristic of subduction-related volcanism than within-plate volcanism. However, this interpretation is inconsistent with the major element compositions and the tectonic environment of formation of the tholeiites. Samples from the Theron Mountains have Nd model and isochron ages (~ 1000 Ma) considerably older than their stratigraphical age (~ 150 Ma). In an attempt to explain the trace element and isotopic systematics of the tholeiites, a model involving a previous mantle-enrichment event has been developed. This model is similar to those developed for the Karoo and Paraná Mesozoic provinces. Thus, it would appear that the subcontinental Gondwana mantle inherited its geochemical characteristics from earlier geological events in Africa, South America and Antarctica.

Introduction

Mesozoic magmatism associated with the break-up of Gondwana is widespread in southern Africa, Brazil and Antarctica. In Antarctica remnants of this magmatism form a large province, extending from Dronning Maud Land along the Transantarctic Mountains to Victoria Land (Fig. 1). Within this province the magmatism is represented by either basaltic flows or dolerite intrusions, which were divided into two petrological provinces by Faure, Bowman & Elliot (1979). The Dronning Maud Land Province is characterized by 'typical' continental tholeiites, whereas the Ferrar Province is distinguished by unusually high initial $^{87}Sr/^{86}Sr$ ratios and high large ionic lithophile elements (LILE)) (Kyle, Pankhurst & Bowman, 1983). The Sr-isotopic compositions of the Ferrar Supergroup (FS) are high compared to many continental tholeiites and more akin to crustal isotopic compositions than mantle-derived magmas.

In order to account for the unusual geochemistry of the FS, several models have been proposed which can be divided into two broad categories: those involving crustal contamination and those derived from an anomalous mantle source (Kyle *et al.*, 1983; Siders & Elliot, 1985). The majority of these models are based on data from Victoria Land but this paper presents preliminary results of a comprehensive geochemical study of Coats Land and Dronning Maud Land. The Coats Land area is of particular interest because it has been suggested that two petrographic provinces are present (Brook, 1972). A detailed discussion of the geochemistry and analytical data will be

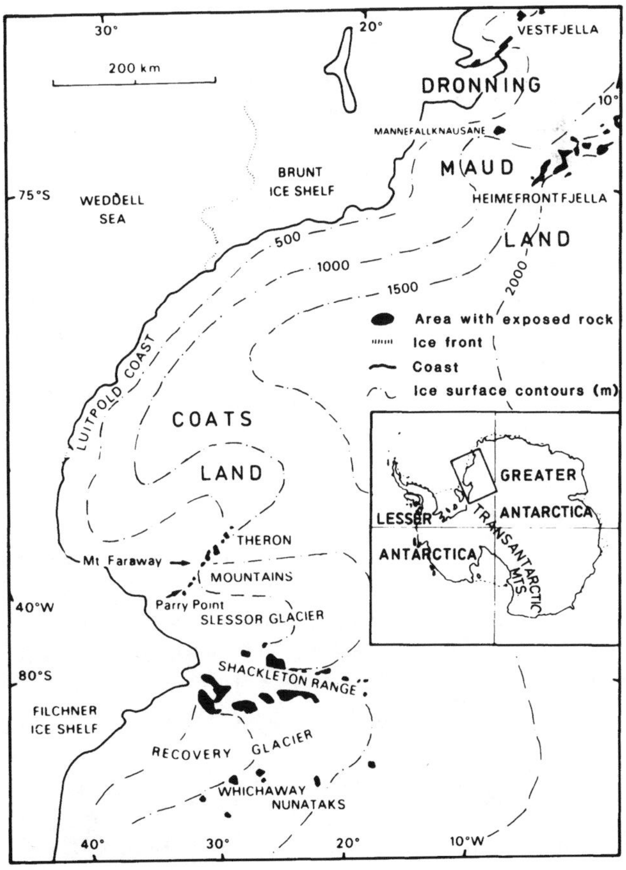

Fig.1 Location map; ice contours from Marsh (1986).

Table 1. *Petrology of Mezozoic dolerites and lavas from Coats Land and Dronning Maud Land*

Region		Petrographic type
Theron Mountains and Whichaway Nunataks	High Ti	Ti-augite + plagioclase + opaques ± granophyric quartz and K-feldspar ± olivine ± apatite ± zircon
	Low Ti	Augite + plagioclase + opaques ± orthopyroxene ± olivine ± glass ± granophyric quartz and K-feldspar
Shackleton Range		Augite + plagioclase ± glass
Heimefrontfjella	Lavas	Plagioclase + augite + opaques ± glass
	Intrusions	Plagioclase + augite + opaques ± olivine

presented elsewhere. However, the data are available from the senior author on request.

Field relationships and petrography

In Coats Land dolerite sills and/or dykes occur in the Theron Mountains, Shackleton Range and Whichaway Nunataks (Fig. 1). In the Theron Mountains and Whichaway Nunataks the dolerites are intruded into Lower Permian continental sedimentary rocks of the Beacon Supergroup (Stephenson, 1966; Brook, 1972). These intrusions have K–Ar ages ranging from 169 ± 6 to 154 ± 6 Ma (Rex, 1972); the larger sills of the Theron Mountains display olivine layering. In the Shackleton Range a single 3 m wide Jurassic dyke has been reported, which intruded the Cambro-Ordovician Blaiklock Glacier Group (Clarkson, 1981). On the basis of the clinopyroxene composition, the dolerites can be divided into two groups (Table 1). The most obvious mineralogical variations in each group are in the proportions of olivine and granophyric quartz and alkali-feldspar intergrowths.

The Dronning Maud Land samples are from Mannefallknausane and Vestfjella, in the Heimefrontfjella area (Fig. 1). The Mannefallknausane nunataks expose a basement complex of granites and charnockitic gneisses which are intruded by Mesozoic dolerites (Juckes, 1968). In contrast, lavas and intrusions are exposed at Vestfjella (Juckes, 1968). K–Ar ages from both of these nunataks are in the range 179 ± 7 to 162 ± 6 Ma (Rex, 1972). The petrology of the dolerites and lavas from the Heimefrontfjella region are summarized in Table 1.

Major and trace element geochemistry

All of the analysed samples (Table 2) have tholeiitic affinities and are subalkaline in composition. On the basis of the TiO_2 content the data can be divided into two major groups (Fig. 2). One group is characterized by high Ti values ($TiO_2 > 2.0\%$), hereafter referred to as HTG, and the second group is defined by relatively low Ti values ($TiO_2 < 1.5\%$), hereafter referred to as the LTG.

All of the LTG compositions are tholeiites with Mg-number, ranging from 44.6 to 78.1. This group shows good positive correlations between SiO_2 and TiO_2, P_2O_5, Na_2O and K_2O, and negative correlations between SiO_2 and Al_2O_3, MgO and CaO. Compositions with extremely low SiO_2(42–45%) have very

Table 2. *Representative analyses from the Theron Mountains. Iron reported as Fe_2O_3, LOI equals loss on ignition at 1000 °C; selected trace elements in parts per million*

	Low Ti	High Ti			
Sample	Z.453.1	Z.455.1	Z.471.10	Z.474.1	Z.469.3
subgroup		I	II	III	IV
SiO_2	49.93	48.69	51.76	53.46	55.25
Al_2O_3	15.76	14.29	12.95	11.33	11.82
TiO_2	0.98	2.23	2.98	2.54	1.98
Fe_2O_3	11.10	16.82	15.09	17.80	15.78
MgO	9.68	4.47	3.60	2.70	2.52
CaO	9.44	8.74	8.16	7.43	6.81
Na_2O	1.79	2.49	2.84	2.30	2.30
K_2O	0.72	1.37	1.45	1.65	1.51
MnO	0.16	0.22	0.23	0.22	0.21
P_2O_5	0.15	0.30	0.52	0.23	0.28
LOI	0.43	0.00	0.10	0.00	1.24
Total	100.14	99.62	99.68	99.66	99.70
Ba	246	429	522	297	417
Nb	7	12	20	10	13
Rb	18	63	50	43	61
Sr	179	144	216	185	145
Y	27	49	74	51	54
Zr	94	187	324	208	205

high MgO (> 12%) and are the samples which contain cumulate olivine; consequently they are excluded from the following discussion.

HTG rocks are either tholeiites or tholeiitic andesites, with Mg number ranging from 22.9 to 53.6. In contrast to other Antarctic Mesozoic high Ti tholeiites (Siders & Elliot, 1985), the HTG has a range of SiO_2 values (48–58%) and is not solely restricted to high SiO_2 compositions. From Fig. 2 it can be seen that the HTG can be further divided into four subgroups (I–IV). Of these, groups I and II have higher Fe, K, Na, P, LILE, HFSE and REE concentrations for equivalent SiO_2 values relative to LTG samples, whereas groups III and IV have similar elemental concentrations to the more evolved (i.e. high SiO_2) LTG samples.

Many of the trace element ratios of these tholeiites are unusual in that they are very similar to those of arc-related

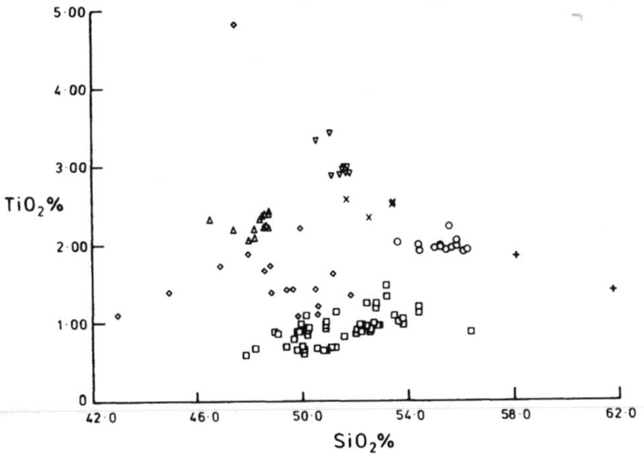

Fig. 2. TiO_2 versus SiO_2 plot showing the subdivision of the tholeiites into high Ti and low Ti groups. Dronning Maud Land (◇), Theron Mountains: Low Ti (□); High Ti, I (△), II (▽), III (×), IV (○); high Si (+).

magmatism (e.g. high Ba/Nb, Ce/Nb; cf. Gill, 1981). These ratios do not show any correlations with fractionation indices for either of the two main groups.

Strontium and neodymium isotopes

Sr and Nd analyses were performed at the Open University, Milton Keynes, England; details of the technique are given in Hawkesworth *et al.* (1986). The decay constant of ^{87}Sr was taken as $1.42 \times 10^{-11}y^{-1}$. Initial ratios for the Theron Mountains were calculated assuming an age of 160 Ma. The initial $^{87}Sr/^{86}Sr$ ratios in the Theron Mountains range from 0.7062 to 0.7130 and ϵ_{Nd} range from -2 to -9. There is no systematic variation of either initial $^{87}Sr/^{86}Sr$ (Sr_0) or $^{143}Nd/^{144}Nd$ (Nd_0) with TiO_2, although the LTG samples do have the highest Sr_0 and Nd_0. The ranges of Sr_0 and Nd_0 are similar to those from Victoria Land (Pankhurst *et al.*, 1986), except that the Theron Mountains data have an absolute lower limit less than that proposed for the minimum Sr_0 ratio of FS parental magma (0.709; Kyle *et al.*, 1983). The initial ϵ_{Sr} of $+21$ to $+121$ and ϵ_{Nd} of -2 to -9 form a well defined cluster in the enriched mantle quadrant which is similar to those trends defined by the Karoo and Paraná data (Hawkesworth *et al.*, 1983, 1986). The more restricted range of Sr_0 of the HTG relative to the LTG is such that they plot closer to the bulk earth value relative to the LTG. This is closely similar to the relationships between the Highland low Ti compositions of the Paraná Province (Hawkesworth *et al.*, 1986).

Petrogenesis

Previous workers (Hoefs, Faure & Elliot, 1980; Faure, Pace & Elliot, 1982), in explaining the anomalously high Sr_0 ratios of the FS magmas, invoked crustal contamination (bulk contamination of a basic magma by continental crust). This mechanism was supported by the linear relationship between $^{87}Sr/^{86}Sr$ and other chemical parameters (i.e. 1/Sr). The problems with such an interpretation are: (i) the volume of contamination required, (ii) the composition and origin of the contaminant, and (iii) why, over such a vast region (3500 km in length), should the contaminant be so similar in composition?

The Theron Mountain samples do not show good correlation between SiO_2 and Sr_0, Nd_0 or selected trace element ratios (e.g. Ba/Nb), which probably suggests that a simple bulk contamination model cannot be used to explain their petrogenesis. However, as Kyle *et al.* (1983) demonstrated, more complex contamination processes, such as combined assimilation and fractional crystallization (AFC; DePaolo, 1981), or assimilation–equilibrium crystallization (AEC; Devey & Cox, 1987), must be considered.

In an attempt to consider the role of AFC, a plot of Rb/Sr v. $^{87}Sr/^{86}Sr$ has been used (Fig. 3). All of the LTG and subgroups III and IV of the HTG form a curved trend, whereas subgroups I and II plot in a separate region. The trend defined by the LTG and HTG samples is closely similar to those defined by DePaolo (1981) as indicative of AFC. One possible interpretation is that these compositions are recording AFC, whereas subgroups I and II are not. If this is so then the high Sr_0 of these two groups (I & II) is probably a feature of their source regions. Two important questions arise from this conclusion: (i) What are the mechanisms of this source region enrichment? (ii) What is the age of the enrichment?

On a conventional Sm–Nd isochron diagram (Brewer & Clarkson, this volume, p. 117), subgroups I and II define an isochron giving an age of ~ 1000 Ma. This age is considerably older than the stratigraphical age of the dolerites (~ 160 Ma; Rex, 1972); it is a common age obtained from the East Antarctic craton and it is also similar to model ages for the Ferrar Group (Pankhurst *et al.*, 1986). If this age is 'real' then it may represent an ancient enrichment event, but only further work will substantiate this. It is also interesting to note that the tholeiites in the present study have calc-alkaline trace element characteristics; if these were generated by such an enrichment event, it may represent an ancient subduction episode.

Although this model is an attractive one and is similar to

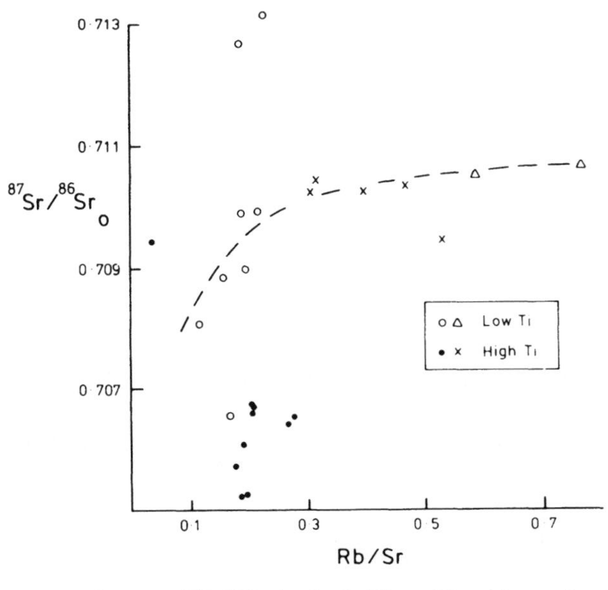

Fig. 3. Rb/Sr versus $^{87}Sr/^{86}Sr$ plot for the Theron Mountain samples; the pecked line illustrates a possible AFC evolutionary trend.

models proposed for the Paraná and Karoo (Cox, 1983; Hawkesworth *et al.*, 1986), further work is needed to substantiate it. Detailed comparisons with the new isotopic data from Victoria Land (P.R. Kyle, pers. comm., 1987) are needed also.

Acknowledgements

The samples in this study were provided by the British Antarctic Survey. This manuscript benefited greatly from discussions with Drs B.P. Atkin, P.K. Harvey (Nottingham), P.D. Clarkson (BAS), Z. Palacz (Open University) and Professors P.E. Baker (Nottingham) and C.J. Hawkesworth (Open University).

References

Brook, D. (1972). Geology of the Theron Mountains, Antarctica. PhD thesis, University of Birmingham, 267 pp. (unpublished).

Clarkson, P.D. (1981). Geology of the Shackleton Range: IV. The dolerite dykes. *British Antarctic Survey Bulletin*, **53**, 201–12.

Cox, K.G. (1983). The Karoo Province of Southern Africa: origin of trace element enrichment patterns. In *Continental Basalts and Mantle Xenoliths*, ed. C.J. Hawkesworth & M.J. Norry, pp. 139–57. Nantwich; Shiva Publishing Co.

DePaolo, D.J. (1981). Trace element and isotopic effects of combined wallrock assimilation and fractional crystallisation. *Earth and Planetary Science Letters*, **53**, 189–202.

Devey, C.W. & Cox, K.G. (1987). Relationships between crustal contamination and crystallisation in continental flood basalt magmas with special reference to the Deccan Traps of the Western Ghats, India. *Earth and Planetary Science Letters*, **84**, 59–68.

Faure, G., Bowman, J.R. & Elliot, D.H. (1979). The initial $^{87}Sr/^{86}Sr$ ratios of the Kirwan volcanics of Dronning Maud Land: comparison with the Kirkpatrick basalt, Transantarctic Mountains. *Chemical Geology*, **26**, 77–90.

Faure, G., Pace, K.K. & Elliot, D.H. (1982). Systematic variations of $^{87}Sr/^{86}Sr$ ratios and major element concentrations in Kirkpatrick Basalts of Mount Falla, Queen Alexendra Range, Transantarctic Mountains. In *Antarctic Geoscience*, ed. C. Craddock, pp. 715–23. Madison; University of Wisconsin Press.

Gill, J.B. (1981). *Orogenic Andesites and Plate Tectonics*. New York; Springer-Verlag, 390 pp.

Hawkesworth, C.J., Mantovani, M.S.M., Taylor, P.N. & Palacz, Z. (1986). Evidence from the Paraná of south Brazil for a continental contribution to Dupal Basalts. *Nature, London*, **322(6007)**, 356–9.

Hawkesworth, C.J., Marsh, J.S., Duncan, A.R., Erlank, A.J. & Norry, M.J. (1983). The role of continental lithosphere in the generation of the Karoo volcanic rocks: evidence from combined Nd- and Sr-isotope studies. *Geological Society of South Africa, Special Publication*, **13**, 341–54.

Hoefs, J., Faure, G. & Elliot, D.H. (1980). Correlation of $\delta^{18}O$ and initial $^{87}Sr/^{86}Sr$ ratios in Kirkpatrick Basalt on Mt Falla, Transantarctic Mountains. *Contributions to Mineralogy and Petrology*, **75(3)**, 199–203.

Juckes, L.M. (1968). The geology of Mannefallknausane and part of Vestfjella, Dronning Maud Land. *British Antarctic Survey Bulletin*, **18**, 65–78.

Kyle, P.R., Pankhurst, R.J. & Bowman, J.R. (1983). Isotopic and chemical variations in Kirkpatrick Basalt Group rocks from southern Victoria Land. In *Antarctic Earth Science*, eds. R.L. Oliver, P.R. James & J.B. Jago, pp. 234–7. Canberra; Australian Academy of Science and Cambridge; Cambridge University Press.

Marsh, P.D. (1986). Ice surface and bedrock topography in Coats land and part of Dronning Maud Land, Antarctica, from satellite imagery. *British Antarctic Survey, Bulletin*, **68**, 19–36.

Pankhurst, R.J., Millar, I.L., McGibbon, F., Menzies, M., Hawkesworth, C.J. & Kyle, P.R. (1986). Evidence for the enriched source of the Jurassic Ferrar magmas of Antarctica. *Terra Cognita*, **6**, 204.

Rex, D.C. (1972). K–Ar age determinations on volcanic and associated rocks from the Antarctic Peninsula and Dronning Maud Land. In *Antarctic Geology and Geophysics*, ed. R.J. Adie, pp. 133–6. Oslo; Universitetsforlaget.

Siders, M.A. & Elliot, D.H. (1985). Major trace element geochemistry of the Kirkpatrick Basalts, Mesa Range, Antarctica. *Earth and Planetary Science Letters*, **72**, 54–64.

Stephenson, P.J. (1966). Geology. 1. Theron Mountains, Shackleton Range and Whichaway Nunataks. *Transantarctic Expedition Scientific Reports*, **8**, 1–79.

Magmatism related to the break-up of Gondwana

R.J. PANKHURST, B.C. STOREY & I.L. MILLAR

British Antarctic Survey, Natural Environment Research Council, High Cross, Madingley Road, Cambridge CB3 0ET, UK

Abstract

Early–Middle Jurassic magmatism is widespread throughout West Antarctica. In the Ellsworth–Whitmore mountains (EWM) crustal block it is mostly granitic, whereas in the Transantarctic Mountains (TAM) it is largely volcanic or hypabyssal in style and basaltic in composition. New Rb–Sr data are presented for plutons from the Whitmore Mountains and the Martin and Hart hills (EWM), and Lewis Nunatak (TAM). The Whitmore Mountains granites record two separate intrusive episodes. The Linck Nunatak granite gives an errorchron of 176 ± 5 Ma, typical of the previously recognized Ellsworth–Whitmore granite suite. Its initial $^{87}Sr/^{86}Sr$ ratio of 0.7232 ± 0.0008 reflects a very high crustal contribution, probably through anatexis of underlying basement rocks. The Mount Seelig granite gives an errorchron age of 203 ± 8 Ma with an initial $^{87}Sr/^{86}Sr$ ratio of 0.7068 ± 0.0007. The Martin Hills granitic stock displays variable initial $^{87}Sr/^{86}Sr$ ratios of ~ 0.718. The gabbro outcrops at Hart Hills and Lewis Nunatak have isotopic characteristics comparable with the Ferrar Supergroup (TAM). The Middle Jurassic igneous rocks of Antarctica form a distinctive bimodal, within-plate igneous suite, connected with the initial break-up of the Gondwana supercontinent.

Introduction

It is commonly accepted that West Antarctica should be regarded as a mosaic of continental microplates, which may have moved independently of each other during and after the break-up of Gondwana (Dalziel & Elliot, 1982). This view is partly based on the distinctive geology of the different crustal blocks, whose boundaries are defined by deep subglacial rifts. However, at least two of the blocks, the Antarctic Peninsula and the Ellsworth–Whitmore mountains block (EWM), share with each other and with surrounding fragments of Gondwana, evidence of important early Mesozoic magmatism concentrated, in part at least, along large-scale linear belts. Late Triassic–Middle Jurassic calc-alkaline magmatism on the east coast of the Antarctic Peninsula (Hole, Pankhurst & Saunders, this volume, p. 369) is usually considered as an integral part of Pacific Ocean subduction-related activity which continued throughout the Cretaceous and Tertiary periods. Similarly, the Middle Jurassic Tobífera Formation volcanism of southern South America (Gust *et al.*, 1985) is seen as subduction-related in part, although both this and the Darwin suite of granites (Nelson, Dalziel & Milnes, 1980) show evidence of within-plate crustal melting, indicating extensional tectonics within the subduction-dominated regime. Jurassic magmatism in the Transantarctic Mountains (TAM) is almost entirely of an anomalous mafic type; Ferrar tholeiites, being noted for their 'continental' trace-element and isotopic characteristics, are also most easily ascribed to lithospheric melting during a major crustal rifting event. This paper presents new geochronological and isotopic data for the contemporaneous plutons of the Ellsworth–Whitmore mountains block of West Antarctica and shows that they also represent essentially within-plate magma generation.

The EWM (Fig. 1*a*) is transected by a line of nunataks which are the highest parts of a mainly subglacial topographic ridge. Spectacular exposures of granite are seen in the Pirrit and Nash hills, Pagano Nunatak and the Whitmore Mountains (Fig. 1*b*). In the Nash and Martin hills and Whitmore Mountains they have sharp roof contacts against low-grade metasedimentary rocks with a similar structural history to the Cambrian–Permian Ellsworth Mountains succession (Storey & Dalziel, 1987). In the Hart Hills there is a body of altered post-tectonic gabbro. The Thiel Mountains are composed of Cambro-Ordovician igneous and volcanogenic rocks that relate to the earlier Gondwana history of the Transantarctic Mountains rather than to West Antarctica.

Most of the EWM granites are fairly uniform in appearance, consisting predominantly of medium–coarse-grained white or pale pink leucocratic biotite (± muscovite) granite (Vennum & Storey, 1987a). There is a range of textures from porphyritic varieties, with up to 40% K-feldspar phenocrysts (often strongly perthitic), to fine-grained aplogranites and veins and small bodies of aplite and alaskite. Pegmatitic segregations containing tourmaline, beryl and muscovite occur at the Pirrit Hills, but otherwise alteration and xenoliths are rare. Previous radiometric dating of these granites has been summarized by

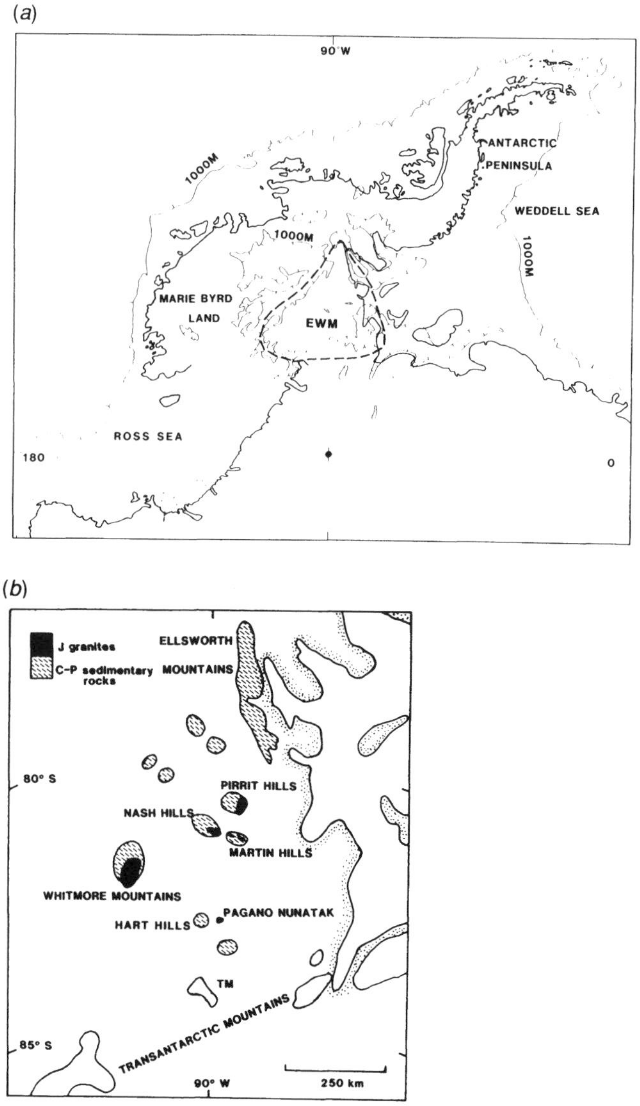

Fig. 1. (*a*) **Sketch map showing the location of the Ellsworth–Whitmore mountains crustal block (EWM).** (*b*) **Geological sketch map of the Ellsworth–Whitmore mountains crustal block: J, Jurassic; C–P, Cambrian-Permian. Geology of the Transantarctic Mountains is not differentiated; unenclosed stipple indicates the generalized edge of the continental ice cap. TM, Thiel Mountains.**

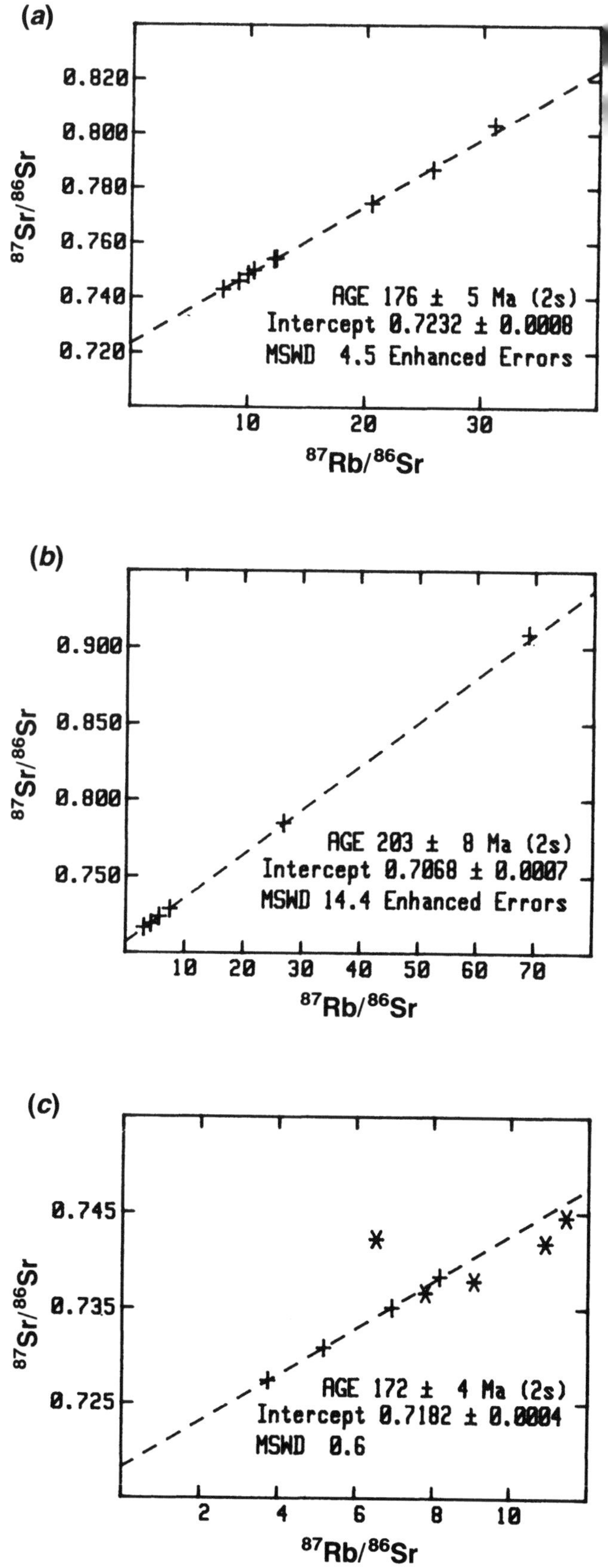

Fig. 2. Rb–Sr isochron plots for (*a*) **Linck Nunatak granite,** (*b*) **Mount Seelig granite and** (*c*) **Martin Hills granite (asterisks represent dta points not included in the fitted isochron).**

Millar & Pankhurst (1987), who presented Rb–Sr whole-rock isochron ages of 173 ± 5, 175 ± 8 and 175 ± 8 Ma, respectively for the Pirrit Hills, Nash Hills and Pagano Nunatak plutons. They proposed that these constitute a coherent and contemporaneous suite. The largest plutonic outcrop area in the EWM, the Whitmore Mountains, exhibits two principle granite types. The coarse-grained Mount Seelig granite (Webers *et al.*, 1982) differs from the rest in that it is hornblende-bearing and commonly possesses a slight mica fabric. It is cut by the fine grained leucocratic Linck Nunatak granite which is more typical of the remainder of the suite.

Geochronology and isotope geology

New Rb–Sr whole-rock data for igneous rocks of the EWM are presented in Table 1 and isochron plots are shown in Fig. 2.

Table 1. *New Rb–Sr whole-rock data for the Whitmore Mountains, Martin Hills and Hart Hills*

Locality & sample number	Description	Rb (ppm)	Sr (ppm)	$^{87}Rb/^{86}Sr$	$^{87}Sr/^{86}Sr$	$(^{87}Sr/^{86}Sr)_0$[a]
Linck Nunatak granite						
Mount Chapman						
R.2226.4	Grey microgranite	269	84.9	9.224	0.74592	
R.2226.5	Grey microgranite	289	68.7	12.20	0.75405	
R.2226.6	Pink microgranite	288	67.6	12.40	0.75429	
R.2227.1	Grey microgranite	262	96.3	7.876	0.74283	
Linck Nunatak						
R.2246.1	Fine-grained granite	248	28.1	25.78 (0.7)	0.78676	
R.2246.2	Fine-grained granite	285	40.4	20.57 (0.6)	0.77436	
R.2246.3	Pegmatitic segregation	153	14.4	31.04 (1.2)	0.80336	
R.2246.4	Foliated muscovite-granite	272	79.1	9.991	0.74833	
R.2246.5		275	76.1	10.49	0.74987	
Mount Seelig granite						
Mount Chapman						
R.2226.1	Porphyritic granodiorite	233	127	5.325	0.72754	
R.2226.2	Foliated granodiorite	223	104	6.191	0.72872	
R.2226.3	Foliated granodiorite	255	126	5.879	0.72659	
Mount Radlinski						
R.2248.1	Coarse granite	220	170	3.756	0.72219	
Mount Seelig						
R.2228.1	Porphyritic granodiorite	221	205	3.118	0.71621	
R.2228.2	Pink aplogranite	415	17.8	68.86 (1.0)	0.90915	
R.2228.4	Fine-grained grey granite	332	36.1	26.85 (0.6)	0.78479	
R.2229.1	Porphyritic granite	396	153	7.523	0.72829	
R.2229.2	Porphyritic granite	253	169	4.348	0.71886	
R.2229.3	Porphyritic granite	276	140	5.696	0.72328	
Martin Hills stock						
R.2230.7	Porphyry	251	194	3.750	0.72741	(0.71808)
R.2230.8	Porphyry	297	111	7.797	0.73661	(0.71721)
R.2230.9	Porphyry	300	168	5.177	0.73088	(0.71800)
R.2230.10	Porphyry	355	114	9.068	0.73786	(0.71530)
R.2230.11	Porphyry	384	97.6	11.44	0.74462	(0.71615)
R.2230.12	Porphyry	360	95.7	10.92	0.74189	(0.71472)
R.2232.5	Felsite dyke	290	129	6.530	0.74427	(0.72602)
R.2232.6	Felsite dyke	279	117	6.930	0.73509	(0.71785)
R.2232.7	Felsite dyke	301	107	8.169	0.73830	(0.71797)
Hart Hills Gabbro						
R.2211.2	gabbro	29.9	123	0.704	0.71316	(0.71141)
R.2212.1	Gabbro	36.2	146	0.718	0.71371	(0.71192)

$^{87}Rb/^{86}Sr$ determined from X-ray fluorescence spectrometric analysis of Rb and Sr at BGS, 1-sigma error ± 0.5% except where shown. $^{87}Sr/^{86}Sr$ by automatic mass spectrometry, 1-sigma error ± 0.01%.
[a]Initial $^{87}Sr/^{86}Sr$ ratios in parentheses for Martin and Hart hills samples are calculated for 175 Ma ago.

Millar & Pankhurst (1987) reported a tentative, four-point Rb–Sr whole-rock isochron age of 182 ± 5 Ma for the Linck Nunatak granite of the Whitmore Mountains. This is now superseded by a nine-sample suite including the original samples of the microgranite phase from Mount Chapman and five new analyses from Linck Nunatak itself (Fig. 2*a*). These yield a well defined errorchron (MSWD = 4.5), corresponding to an age of 176 ± 5 Ma and an initial $^{87}Sr/^{86}Sr$ ratio of 0.7232 ± 0.0008. This result is compatible with a single K–Ar biotite age of 180 ± 5 Ma (recalculated with current decay constants), reported by Webers *et al.* (1982). In its time of emplacement, the Linck Nunatak granite is thus indistinguishable from those of the Ellsworth–Thiel mountains ridge: i.e. the Pirrit and Nash hills and Pagano Nunatak, with which it also shares many geochemical characteristics. These four plutons are fully representative of the EWM granitic suite (Millar & Pankhurst, 1987).

Data for the Mount Seelig granite type differ from this

pattern. In an isochron diagram (Fig. 2*b*), excess scatter provides evidence for initial heterogeneity or open system behaviour, especially for the samples from Mount Chapman. The latter are fresh, unaltered granites, although they show clear signs of foliation, possibly indicating stress-induced isotopic migration. Six samples from the area of Mount Seelig itself, including one of highly evolved aplogranite, give an errorchron (MSWD = 14) corresponding to an age of 203 ± 8 Ma. This substantiates the claim of Webers *et al.* (1982), based on a single K–Ar biotite age of 194 ± 8 Ma, that this granite is significantly older than the Linck Nunatak granite. Comparison with geological time-scales shows that neither result need be as old as Triassic, 200 Ma generally being within the earliest Jurassic. The initial $^{87}Sr/^{86}Sr$ ratio of the Mount Seelig granite is 0.7068 ± 0.0007, which is again distinct from that of the Linck Nuntak granite (0.723) but indistinguishable from that of the Pirrit Hills granite (0.707) (Millar & Pankhurst, 1987).

The Martin Hills outcrop consists almost entirely of mixed argillitic sedimentary rocks, intruded and baked at the eastern end by an irregular dyke or stock of porphyritic granite 2–3 m across. It is extensively fractured and altered, with epidote staining on the fractures. At the western end of the outcrop the sedimentary rocks are cut by narrow (0.1–0.5 m) dykes of a fresher felsite. Rb–Sr results for the granitic rocks scatter on an isochron diagram (Fig. 2*c*), presumably largely due to mobility of Rb and Sr during alteration. It may be noted that four of the nine data points fit a line corresponding to an age of 172 ± 4 Ma. Initial $^{87}Sr/^{86}Sr$ ratios mostly fall in the range 0.715–0.718. Although not conclusive, these data are consistent with the idea of the Martin Hills stock being a high-level, hydrothermally-altered and contaminated correlative of the granite body exposed nearby in the Nash Hills (175 ± 8 Ma; 0.7122 ± 0.0008).

Also presented in Table 1 are Rb–Sr data for the Johnson Peak gabbro from the Hart Hills (Webers *et al.*, 1983; Vennum & Storey, 1987b). These may be compared with analyses of the Lewis Nunatak gabbro, Thiel Mountains (Millar & Pankhurst, 1987). In particular, both display high Rb/Sr ratios (for mafic igneous rocks) and high initial $^{87}Sr/^{86}Sr$ ratios (assuming a Middle Jurassic age) characteristic of the Ferrar Supergroup, i.e. 0.710–0.712.

Table 2 contains the results of Sm–Nd analyses of representative samples of the entire suite of rocks. Initial $^{143}Nd/^{144}Nd$ ratios for the 175 Ma granites are very consistent at 0.5120–0.5122. This reflects depletion in radiogenic ^{143}Nd relative to a mantle source (ϵNd_t mostly ~ -5), indicating significant incorporation of crustal Nd. The T_{DM} model ages of the granites mostly range from 1250 to 1750 Ma, probably reflecting a Precambrian age for the crustal component, but those of some of the aplitic rocks are meaningless, presumably due to severe fractionation of Sm/Nd ratios during differentiation. The data for the Lewis Nunatak gabbro, which is representative of the Ferrar basic magmas (Pankhurst *et al.*, 1986), are essentially identical in these parameters. The isotopic characteristics of the Mount Seelig granite are discordant, the initial $^{143}Nd/^{144}Nd$ ratio of one of the two analysed samples being lower than those of the Ellsworth–Whitmore granite suite, and that of the other being higher.

Geochemistry and petrogenesis

The most important feature of these data is the confirmation that the majority of granite plutons in the Ellsworth–Whitmore mountains block are members of a coherent, essentially synchronous, magmatic episode in Middle Jurassic times (~ 175 m.y. ago). The only obvious exception is the 200 Ma Mount Seelig granite of the Whitmore Mountains, which also differs in being metaluminous. The remainder are highly differentiated, peraluminous S-type granites (Vennum & Storey, 1987a), substantiated by their Sr- and Nd-isotope characteristics (above). Within each pluton, petrological variation from coarse granites to more evolved aplogranites is accompanied by sharply falling Mg, Fe, Ti and Mn, with Al and Ca falling more gradually, together with slightly rising Na and K. These trends may be explained by liquid–crystal fractionation of feldspar and ferromagnesian minerals. Trace elements show generally consistent variations: Rb and Nb remaining incompatible throughout but others, including Sr, Ba, Zr and Eu, falling in the most differentiated rocks. Thus there is a very wide overall range, for example, in Rb/Sr ratios (~ 1–1000). The rare earth elements (REE) behave as incompatible in rocks with SiO_2 less than ~ 73%, but become compatible thereafter, with a more restricted range of heavy REE (HREE) than light REE (LREE). Negative Eu-anomalies increase with differentiation throughout, reaching extreme values of < 0.1. In terms of their trace-element distributions compared to ocean-ridge granite (Pearce, Harris & Tindle, 1984), they possess 'crust-dominated profiles', characterized by high lithophile-element abundances with especially high Rb and Th compared to K and Ba, and moderate HFS-element abundances (Sm, Y and Yb although Zr is relatively low). Such patterns are exhibited by a variety of granite suites, including within-plate granites (e.g. the islands of Mull and Skye in Scotland), volcanic arc granites (e.g. Chile and the Sn-bearing granites of Malaysia) as well as collision-related granites (e.g. the Hercynides and Himalaya).

The Pirrit Hills, Martin Hills and Linck Nuntak granites are restricted to the more differentiated end of the spectrum, but for the remainder there is a clearly defined evolution from less evolved coarse–medium-grained granites to evolved aplogranites and alaskites. However, the individual plutons have trace-element and isotopic differences which require different sources and/or differing degrees of crustal contamination. In particular, there is a strong inverse correlation between the degree of REE fractionation and initial $^{87}Sr/^{86}Sr$ ratios (Fig. 3). The isotopically most primitive granite (Pirrit Hills) has the flattest REE pattern and this steepens largely by depletion in HREE with increasing initial $^{87}Sr/^{86}Sr$ ratio. This is accompanied by a regular trend in the Rb versus (Y + Nb) discrimination diagram of Pearce *et al.* (1984), from a clear within-plate assignment for the Pirrit Hills granite to the syn-collisional field for the Linck Nunatak granite (Fig. 4). This trend was modelled by Storey *et al.* (1988), who proposed mixing between mantle-derived and crust-derived material or magmas. A within-plate mantle source was postulated for the Pirrit Hills parent magma (from which the granite was derived by *extensive* fractional crystallization with little crustal conta-

Table 2. *Sm–Nd analyses of igneous rocks from the Ellsworth–Whitmore Mountains crustal block*

Locality & sample number	Sm (ppm)	Nd (ppm)	$^{147}Sm/^{144}Nd$	$^{143}Nd/^{144}Nd$	$(^{143}Nd/^{144}Nd)_0$	ϵNd_t	T_{DM} (Ma)
Mount Seelig							
R.2228.1 (G)	6.87	25.0	0.1663	0.51199	0.51177	− 11.9	2867
R.2229.1 (G)	10.2	49.7	0.1241	0.51248	0.51232	− 1.2	935
Pirrit Hills							
R.2243.4 (G)	12.5	48.3	0.1563	0.51232	0.51214	− 5.4	1736
R.2243.6 (G)	11.5	22.7	0.3066	0.51252	0.51217	− 4.8	− 1092
R.2243.9 (A)	10.0	18.8	0.3229	0.51238	0.51201	− 7.9	− 1128
R.2243.13 (G)	13.3	41.9	0.1924	0.51239	0.51217	− 4.7	3276
Nash Hills							
R.2242.1 (G)	11.4	53.0	0.1303	0.51231	0.51216	− 4.9	1268
R.2242.6 (G)	11.1	52.5	0.1275	0.51231	0.51217	− 4.9	1232
R.2242.9 (A)	2.63	8.33	0.1907	0.51238	0.51216	− 5.0	3176
Pagano Nunatak							
R.2256.1 (G)	7.28	32.4	0.1357	0.51231	0.51215	− 5.1	1352
R.2215.4 (G)	6.65	29.2	0.1375	0.51227	0.51211	− 5.9	1447
R.2215.8 (A)	7.06	30.9	0.1380	0.51229	0.51213	− 5.5	1413
Martin Hills							
R.2230.7 (G)	11.9	52.5	0.1369	0.51234	0.51219	− 4.5	1309
Linck Nunataks							
R.2226.4 (G)	7.57	31.6	0.1450	0.51223	0.51212	− 5.7	1656
R.2227.1 (G)	6.32	27.9	0.1371	0.51221	0.51212	− 7.1	1544
Lewis Nunatak							
R.2206.1 (Ga)	4.69	20.0	0.1419	0.51231	0.51214	− 5.3	1453

Key to rock types: G, granite; A, aplite, Ga, gabbro.
Initial $^{143}Nd/^{144}Nd$ and ϵNd_t calculated for the estimated time of emplacement: i.e. 203 Ma for the Mount Seelig granite, 175 Ma for the remainder.
Approximate analytical errors: 0.1% on Sm and Nd, 0.005% on isotope ratios, 0.5–1 on ϵ values and 80–200 Ma on model ages.
T_{DM} is the model age for separation from a typical depleted-mantle evolution path.

mination). Similar parent magmas combined with an increasing contribution from crustal sources to produce the trends observed in the other granites, with the Linck Nunatak granite representing almost total crustal melting (on account of its very high initial $^{87}Sr/^{86}Sr$ ratio). The Mount Seelig granite, which was not considered by Storey *et al.* (1988), has a more primitive geochemistry than the Pirrit Hills granite but similar isotopic characteristics. It also falls in the 'within-plate granite' field of Fig. 4, although its fractionated REE pattern is distinct and more reminiscent of calc-alkaline granites. At present its tectonic significance is unclear.

Relationship to Ferrar magmatism and continental break-up

The other major suite of within-plate magmatic rocks in central Antarctica is the Ferrar Supergroup of continental tholeiites. Notwithstanding the obvious differences from granites, they exhibit some remarkable similarities. These include a predominantly Middle Jurassic age (Kyle, Elliot & Sutter, 1981), low Na/K ratios, 'crust-dominated' trace-element abundance pattern and an identical range of old continental Sr and Nd isotopic compositions (see above). The similarities are so striking as to suggest a genetic link. Storey *et al.* (1988) proposed that the undifferentiated mafic parent magma of the Pirrit Hills granite could have been a Ferrar-type magma, the mafic products of which are exposed elsewhere in the Elsworth–Whitmore mountains block as the Hart Hills and Lewis Nunatak gabbros. The presence of further mafic bodies, underlying the granites at depth, may be inferred from aeromagnetic survey data (Dalziel *et al.*, 1987). The Ferrar Supergroup is well known as being representative of a Middle Jurassic (and partly later), Gondwana-wide continental flood basalts event which includes the Tasmanian dolerites, the Karoo Supergroup of southern Africa and the Serra Geral basalts of central South America. Wide contemporaneous acid magmatism is represented by Karoo rhyolites and the Tobífera series of Patagonia, although plutonic granites of this age are rare. This is the first recognition of such bimodal within-plate igneous activity in Antarctica.

Since continental flood basalts in modern times are commonly related to large-scale extensional rift systems, it is

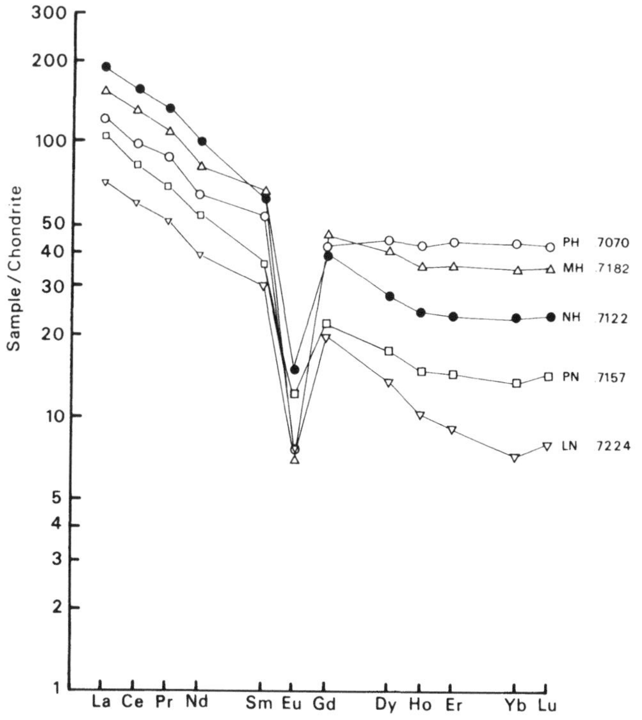

Fig. 3. Correlation of chondrite-normalized REE pattern with initial $^{87}Sr/^{86}Sr$ ratio for the Ellsworth–Whitmore granite suite. PH, Pirrit Hills; MH, Martin Hills; NH, Nash Hills; PN, Pagano Nunatak; LN, Linck Nunatak.

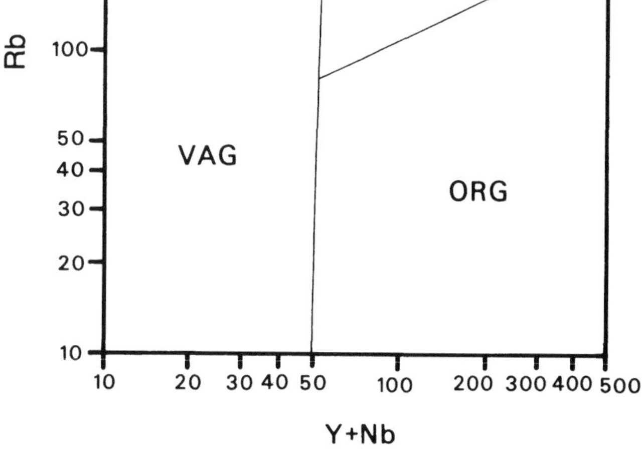

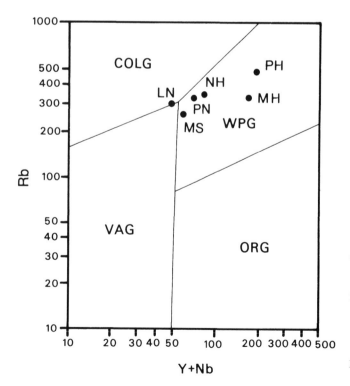

Fig. 4. Plot of Rb versus (Y + Nb) for the mean composition of the least differentiated granites of the Ellsworth-Whitmore granite suite. PH, Pirrit Hills; MH, Martin Hills; NH, Nash Hills; PN, Pagano Nunatak; MS, Mount Seelig; WPG, within-plate granite; COLG, collisional granite; VAG, volcanic-arc granite; ORG, ocean-ridge granite.

probable that such intense Middle Jurassic magmatic activity throughout the area around Antarctica and the South Atlantic was linked in some way with the subsequent break-up of Gondwana (e.g. Kyle *et al.*, 1981). This began in earnest some 25 m.y. after the magmatic episode recognized here, with the initial separation of East Antarctica from Africa and the opening of the Mozambique Basin (Bergh, 1977). We propose that the Jurassic bimodal igneous event in Antarctica represented by the EWM granites and the Ferrar Supergroup is a response to the thermal disturbance preceding, but associated with, break-up. The dominant effect was large-scale production of the mafic suite by melting in the subcontinental mantle. Most authors believe that the anomalous geochemistry of the Ferrar magmas reflects an origin within an old LIL-enriched lithosphere. These magmas reached the surface relatively unaltered except for minor crustal contamination along the rift-controlled lineament of the Transantarctic Mountains. Within the EWM crustal block the magmas were ponded within, or at the base of, the crust. This led to extensive crystal fractionation as well as the formation of large amounts of anatectic melt within the crust, with which the mafic magmas partially hybridized to produce the observed granites.

Acknowledgements

X-ray fluorescence analysis for Rb, Sr and Rb/Sr was carried out at the British Geological Survey (BGS) Geochemical Division, by courtesy of Dr T.K. Smith, and with the helpful assistance of M. Ingham and A. Robertson. Isotopic analyses were performed in the laboratories of BGS (now the NERC Isotope Geology Centre).

References

Bergh, H.W. (1977). Mesozoic sea floor off Dronning Maud Land, Antarctica. *Nature, London*, **269**, 686–7.

Dalziel, I.W.D. & Elliot, D.H. (1982). West Antarctica, problem child of Gondwanaland. *Tectonophysics*, **1(1)**, 3–19.

Dalziel, I.W.D., Garrett, S.W., Grunow, A.M., Storey, B.C., Pankhurst, R.J. & Vennum, W.R. (1987). The Ellsworth–Whitmore crustal block: its role in the tectonic evolution of West Antarctica. In *Gondwana Six: Structure, Tectonics, and Geophysics*, Monograph 40, ed. G. McKenzie, pp. 173–82. Washington, DC; American Geophysical Union.

Gust, D.A., Biddle, K.T., Phelps, D.W. & Uliana, M.A. (1985). Associated Middle to Late Jurassic volcanism and extension in southern South America. *Tectonophysics*, **116(3/4)**, 223–53.

Kyle, P.R., Elliot, D.H. & Sutter, J.F. (1981). Jurassic Ferrar Supergroup tholeiites from the Transantarctic Mountains, Antarctica, and their relationship to the initial fragmentation of Gondwana. In *Gondwana Five*, ed. M.M. Cresswell & P. Vella, pp. 283–7. Rotterdam; A.A. Balkema.

Millar, I.L. & Pankhurst, R.J. (1987). Rb–Sr geochronology of the region between the Antarctic Peninsula and the Transantarctic Mountains: Haag Nunataks and Mesozoic granitoids. In *Gondwana Six: Structure, Tectonics, and Geophysics*, Monograph 40, ed. G. McKenzie, pp. 151–60. Washington, DC; American Geophysical Union.

Nelson, E.P., Dalziel, I.W.D. & Milnes, A.C. (1980). Structural geology of the Cordillera Darwin: collision-style orogenesis in the southernmost Chilean Andes. *Eclogae Geologicae Helvetiae*, **73(3)**, 729–51.

Pankhurst, R.J., Millar, I.L., McGibbon, F., Menzies, M. & Hawkesworth, C.J. (1986). Evidence for the enriched source of the Ferrar magmas of Antarctica. *Terra Cognita*, **6(2)**, 204. (abstract).

Pearce, J.A., Harris, N.B.W. & Tindle, A.G. (1984). Trace element discrimination diagrams for the tectonic interpretation of granitic rocks. *Journal of Petrology*, **25(4)**, 956–83.

Storey, B.C. & Dalziel, I.W.D. (1987). Outline of the structural and tectonic history of the Ellsworth Mountains–Thiel Mountains ridge. In *Gondwana Six: Structure, Tectonics, and Geophysics*, Monograph 40, ed. G. McKenzie, pp. 117–28. Washington, DC; American Geophysical Union.

Storey, B.C., Hole, M.J., Pankhurst, R.J., Millar, I.L. & Vennum, W.R. (1988). Within-plate granites in Lesser Antarctica and their bearing on the break-up of Gondwanaland. *Journal of the Geological Society, London*, **145(6)**, 999–1007.

Vennum, W.R. & Storey, B.C. (1987a). Petrology, geochemistry and tectonic setting of granitic rocks from the Ellsworth–Whitmore mountains crustal block and Thiel Mountains, West Antarctica. In *Gondwana Six: Structure, Tectonics, and Geophysics*, Monograph 40, ed. G. McKenzie, pp. 139–50. Washington, DC; American Geophysical Union.

Vennum, W.R. & Storey, B.C. (1987b). Correlation of gabbroic and diabasic rocks from the Ellsworth Mountains, Hart Hills and Thiel Mountains, West Antarctica. In *Gondwana Six: Structure, Tectonics, and Geophysics*, Monograph 40, ed. G. McKenzie, pp. 129–38. Washington, DC; American Geophysical Union.

Webers, G.F., Craddock, C., Rogers, M.A. & Anderson, J.J. (1982). Geology of the Whitmore Mountains. In *Antarctic Geoscience*, ed. C. Craddock, pp. 841–7. Madison; University of Wisconsin Press.

Webers, G.F., Craddock, C., Rogers, M.A. & Anderson, J.J. (1983). Geology of Pagano Nunatak and the Hart Hills. In *Antarctic Earth Science*, eds. R.L. Oliver, P.R. James & J.B. Jago, pp. 251–5. Canberra; Australian Academy of Science and Cambridge; Cambridge University Press.

Cretaceous dispersion of Gondwana: continental and oceanic spreading in the south-west Pacific–Antarctic sector

J.D. BRADSHAW

Geology Department, University of Canterbury, Christchurch, New Zealand

Abstract

There has been a lack of unanimity on the relationship of crustal blocks in the south-west Pacific region and an appropriate reassembly of the Gondwana margin. Published reconstructions have major gaps or geologically unlikely structures. The south-west Pacific sector differs from other areas in having large embayments and extensive submarine plateaux floored by thin continental crust. Current models of passive margin development suggest that such large thin crustal remnants are the result of major crustal extension. Consequently the crustal fragments are larger now than in the Mesozoic. This is particularly important in reassembly if it can be shown that some of the extension post-dates continental rupture. Partial closing of the embayments and reduction of the plateaux in proportion to their probable extension leads to a better reassembly. The revised assembly and the disparity of spreading rates between the south Tasman and south-west Pacific sectors suggests a major transform extending along the east side of the Iselin Bank and passing west of the Campbell Plateau. Crustal dispersion around the Antarctic continental margin was accompanied by major extension within the continental crust, at least some of which clearly post-dates continental separation. Consequently reassemblies and plate-motion studies which rely wholly on geometric fit and magnetic lineation analysis are likely to be misleading.

Introduction

Although it is widely accepted that the crustal blocks of the south-west Pacific region, Antarctica, Australia and New Zealand are Gondwana fragments that have become separated by Cretaceous–Cenozoic spreading, there is little unanimity about their exact relationships. Australia and Antarctica have been assembled in various ways (compare Laird, 1981a with Stump, White & Borge, 1986), and there has been widespread disagreement about the configuration of New Zealand and its relationship to the other two. The many attempts to solve this problem (e.g. Molnar *et al.*, 1975; Weissel, Hayes & Herron, 1977; Crook & Belbin, 1978; Dalziel & Elliot, 1982; Veevers, 1984) are characterized by unacceptable overlaps, major gaps, or massive strains within New Zealand. There is no doubt that New Zealand changed shape in the Cenozoic as a result of the propagation of a new plate boundary across the continent, and several new reconstructions have taken into account the dextral offset with related shear strain, and post-Miocene shortening perpendicular to the Alpine Fault (Walcott, 1984). Kamp (1986), after reviewing all pertinent data, presented a superior Antarctica–Australia–New Zealand reconstruction but was still unable to resolve the problem of the large 'gap' in the southern Tasman Sea.

In the Pacific–Indian Ocean sector the margins of Antarctica are of 'passive' type, and face similar passive margins in Australia and New Zealand. Passive margins also border the oceanic crust of the Tasman Sea, except where modified by the new late Cenozoic plate boundary (Fig. 1). The intervening oceanic crust is Late Cretaceous and younger (Weissel *et al.*, 1977; Cande & Mutter, 1982). Unlike many passive margins, these particular continents are bordered by extensive submarine plateaux and major embayments that are underlain by thin crust. Where exposed in islands (Fig. 1) or penetrated by Deep Sea Drilling Project holes (sites 270, 281) or commercial wells (Anderton *et al.*, 1982), the submerged areas are underlain by normal continental crust that is closely comparable to onshore areas and separated by angular unconformities from Cretaceous or younger sediments. It is clear that in many places the crust was subjected to subaerial erosion in the Jurassic–Early Cretaceous, and in the Campbell Plateau and Great Australian Bight overlying cover sequences include thick non-marine sediments (Falvey & Mutter, 1981; Anderton *et al.*, 1982). Thus it seems certain that the submerged continental crust was normal prior to rifting and that the current thickness of 17–26 km (Adams, 1962; Willcox *et al.*, 1980), which locally includes over 5 km of Cretaceous–Cenozoic sediment, is a consequence of crustal thinning. Furthermore, in places such as the northern Campbell Plateau, which is underlain by Mesozoic Haast Schist (Davey & Christoffel, 1978), it is

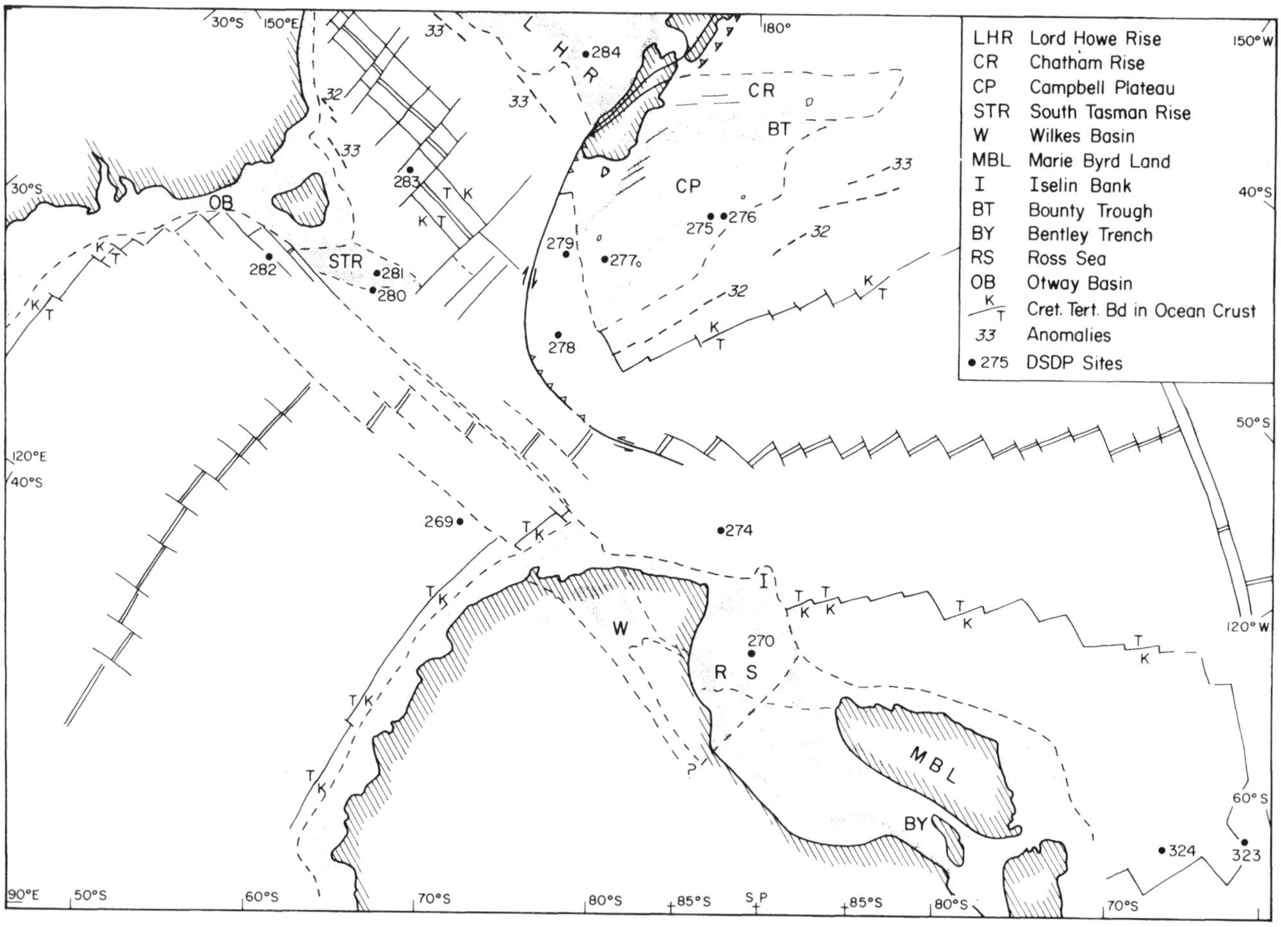

Fig. 1. Simplified map of the salient features of continental and oceanic crust in the south-west Pacific–Antarctic region (Polar projection).

probable that the pre-break-up thickness was significantly more than 40 km.

Continental thinning and rifting

Mechanisms for crustal thinning and rifting have been widely discussed but only recently have studies of western North America drawn attention to the importance of asymmetric low-angle normal faulting in continental crustal thinning. On this basis, Lister, Etheridge & Symonds (1986) proposed a model for crustal rifting that involves asymmetric subhorizontal ductile extension in the lower crust and listric normal faults in the upper layers, leading to the development of relatively abrupt 'upper plate' margins which face extensively thinned 'lower plate' margins. The model appears particularly relevant to margins in the Antarctic–Australian region where marked asymmetry and extensive thinned crust is widespread. For example, the New South Wales–eastern Tasmanian margin conforms well to the 'upper plate' type of Lister *et al.* (1986) and faces the broad continental ribbon of the Lord Howe Rise, which is itself strongly asymmetric. Many parts of the submarine plateaux, however, seem to reflect a more complex situation where multiple and intersecting faults produce thinned slabs of varying geometry (Lister *et al.*, 1986, fig. 4). In addition to the main rifted margins, subsidiary rifted basins such as the West Coast Rift (Laird, 1981b) and the Gippsland Basin, diverge from the main axis like incipient aulacogens. The Wilkes Basin and Ross Sea are similar diverging structures on a much larger scale.

Extended margins and continental reassembly

The amounts of continental extension are usually expressed as either percentages or an extension factor 'β' (McKenzie, 1978). Their significance, however, is not apparent until the size of the extended region is considered. The Campbell Plateau extends for over 750 km east of New Zealand and, although the extension factor cannot be determined exactly due to uncertainties about the Cretaceous versus present crustal thickness, it is probably between 1.4 and 2.3. The use of an intermediate value indicates that the Campbell Plateau may be 200 km wider now than it was in the Early Cretaceous and that this extension, coupled with a much smaller figure for the Marie Byrd Land shelf, represents between 10 and 12% of the total separation between reference points within each continent.

Apart from a possible influence on the interpretation of palaeomagnetic data and the possible deflection of structural trends, intracontinental crustal extension does not affect continental reassembly, provided all extension is completed prior to initial rifting. This appears to be the case in many areas, for example the eastern US margin (Sawyer *et al.*, 1982), where sedimentation is much more regular and less influenced by contemporaneous faults above the break-up unconformity. A similar situation pertains in the North Sea where active extension indicated by listric faulting declined sharply at the time of opening of the North Atlantic. Several lines of evidence suggest that a simple relationship between the timing of break-up and continental crustal extension does not hold true in the southwest Pacific–Antarctic region. Analysis of magnetic lineation patterns indicates that ocean floor generation predates anomaly 33 and that continental separation was complete before 83 Ma (Cande & Mutter, 1982; Kamp, 1986), well before the end of active faulting.

Extension and break-up in the Pacific–Antarctic sector

New Zealand

The New Zealand region was geologically active throughout the Mesozoic, and a succession of terranes (Bishop, Bradshaw & Landis, 1985) was developed in a convergent margin setting. Sedimentation and deformation continued until the mid-Cretaceous (about 100 Ma or slightly later (Laird, 1981b)), when convergence abruptly ceased, probably due to the oblique approach and collision of the Phoenix–Pacific Ridge. Permian–Mesozoic terranes underlie the northern Campbell Plateau and the Chatham Rise. South and west of the median tectonic line, older basement is cut by Early Cretaceous granitoids, some with K–Ar cooling ages that extend into the early part of the Late Cretaceous (95–100 Ma). Similar granitoids also occur in Marie Byrd Land (Adams, 1987).

The eastern margin of the Campbell Plateau cuts across both the Mesozoic terranes and older basement, and new Cretaceous basins within the Plateau are discordant to earlier structures. The oldest fault-bounded extensional basins on land date from approximately 100 Ma and are almost contemporaneous with the last events of the convergent margin. In some places (e.g. Clipper Graben; Hawkes & Mound, 1984) fault-controlled sedimentation continued through Campanian, and possibly Maastrichtian, times, while in the western part of the Great South Basin (Anderton *et al.*, 1982) new basins with at least 2.1, and probably 3.5 km, of non-marine and marginal marine sediments developed in the Campanian–early Paleocene. Although Anderton *et al.* (1982) attributed this sedimentation to compaction which post-dated Senonian faulting, this explanation is not entirely satisfactory because the younger sediments are sometimes thicker than those beneath, and some of the thickest sections rest directly on basement. Seismic sections (Anderton *et al.*, 1982) show faults which flatten downwards with thickness variation and basement depths that suggest that the basins and faults were developed in an extensional regime.

A similar history characterizes western New Zealand and its western shelf, where the basement consists of the same Palaeozoic sedimentary rocks and Palaeozoic–Mesozoic plutonic rocks. Early extensional basins of Albian–Cenomanian age are fault-bounded depressions with up to 4 km of non-marine clastics. Following a pause, a second extensional phase led to new fault-controlled basins of Campanian–Maastrichtian age in which 1–2 km of mainly non-marine rocks accumulated (Pilaar & Wakefield, 1978; Nathan, 1986). Sections of the basins suggest that they result from listric extensional failure. Some faults appear to have been reactivated in a similar mode in the Eocene.

Southern Australia

Study of the southern Australian margin has contributed significantly to general theories of passive margin development (Falvey & Mutter, 1981). In many respects these margins differ from those farther to the east in that they dissect stable crust and are characterized by a longer development history with great thicknesses of pre-rift and rift sedimentation prior to continental rupture. Reidentification of magnetic anomalies and changes to the magnetic polarity scale mean that rifting is now placed before anomaly 34 (90 + Ma) (Cande & Mutter, 1982), rather than at about 55 Ma as previously supposed. Thus considerable thicknesses of sediment (1–3 km) in several areas previously regarded as rift basins must post-date rifting and are now termed 'rim' basin sediments (Veevers, 1984). Seismic sections (e.g. Falvey & Mutter, 1981) all show listric normal faults cutting Late Cretaceous rocks. Sections of the Gippsland Basin show similar faulting cutting the Late Cretaceous and Paleocene. In general, however, margin development is much more protracted, with a far higher proportion of pre-rift and syn-rift sediment.

Antarctica

The continental shelf of Antarctica is relatively narrow, except to the north of the Wilkes Basin and in the Ross Sea sector. The Wilkes Basin is a roughly wedge-shaped depression, 500 km wide near the coast and tapering towards 82° S over a distance of 1500 km. After compensation for ice removal (Drewry, 1983), the bathymetry suggests thinned continental crust, and Steed (1983) suggested that the basin resulted from Mesozoic crustal extension. Autochthonous and allochthonous non-marine Cretaceous microfloras show strong similarities with the Otway Basin of Australia and suggest basin development during the Cretaceous (Quilty, 1986). The sedimentary fill may reach 6.5 km but generally appears somewhat thinner (Drewry, 1976).

The Ross Sea is much larger than the Wilkes Basin. It is 800 km wide in the north and extends to 85° S in a broad arc curving to the east where it merges with the Bentley Trench and the Byrd Basin (Drewry, 1983). An origin by extension has been suggested (Davey, 1981; Cooper & Davey, 1985) and subsidiary rift-type basins have been identified. Sedimentary fill in the Victoria Basin exceeds 10 km, the lower half of which is probably Late Palaeozoic–early Mesozoic in age. Cooper & Davey (1985) favoured a Paleogene age for the upper 4–5 km of sediments, consistent with the suggestion of crustal extension in the Ross Sea starting at about 50 Ma and concurrent with the uplift of the Transantarctic Mountains (Fitzgerald *et al.*, 1986). The latter authors reviewed the geophysical data and concluded that the crust in the Ross Sea Basin is 21–30 km thick, compared to 40–45 km in the Transantarctic Mountains and 35 km in Marie Byrd Land. As with the Campbell Plateau, uncertainties about past and present crustal thickness make calculation of the extension difficult, but a figure of 100–200 km seems reasonable. Once again extension seems to have post-dated the main phase of crustal separation in the Cretaceous.

Reassembly

In addition to the well recognized change in shape of some of the elements of the Gondwana margin, it appears that some components may be somewhat larger and that the periphery of Antarctica may have increased since the late Mesozoic. Incorporation of these factors, coupled with a small clockwise flexure of the Lord Howe Rise about its point of inflexion at 32° S, permits a reassembly without major gaps or overlaps (Fig. 2). The configuration proposed incorporates the small strike-slip offsets north of Tasmania (Stump *et al.*, 1986) and in the Ross Sea (Grindley, 1981), although these are not vital. Of more significance is dextral deformation in the Campbell Plateau. During petroleum exploration, Anderton *et al.* (1982) could not detect the Campbell Fault postulated by Davey & Christoffel (1978), although they did find a number of en-echelon dextral faults. Consequently I have replaced the Campbell Fault with a broad dextral transtensional zone. Kamp (1986) argued that the full 330 km dextral offset of the magnetic anomalies was closely linked to the development of the Bounty Trough (Fig. 1) and is post 'mid-Cretaceous', but, if the Bounty Trough is underlain by thinned continental crust, less than half of that amount of Late Cretaceous movement would be required.

The distribution of new Cretaceous oceanic crust poses problems (Figs 2 & 3). Spreading south of Australia was very slow, while the Tasman Sea opened rapidly as a south-easterly-widening wedge and the south-west Pacific as an easterly-widening wedge. Although originally the two ridge segments were roughly aligned, a mismatch is indicated by the termination of Cretaceous ocean crust at the southern tip of the Campbell Plateau in the north, and on the east side of the Iselin Bank in the south (Fig. 1). There is a related disparity in spreading rates, with the Cretaceous south Tasman ocean crust being wider than that of the south-west Pacific. Together these facts point to a ridge–ridge transform extending from the Iselin Bank towards the south-west tip of New Zealand with continued dextral displacement to the north-east. In some earlier models (e.g. Weissel *et al.*, 1977; Crook & Belbin, 1978) this displacement was accommodated by excessive dextral shift on the line of the Alpine Fault, producing spurious offsets of matching structures within New Zealand. Seismic profiles of the southern Lord Howe Rise (Bentz, 1974) show abundant faulting in basement and Cretaceous–Paleogene rocks in the region of fault zones 'A' and 'B', and it is suggested that these have a dextral component. The line of the transform is extended southwards from the Iselin Bank to the Bentley Trench as a fault bounding the Ross Sea sector.

Marie Byrd Land is considered to have been an integral part of the New Zealand block until the Late Cretaceous, as indicated by their strong geological similarities. Initial movement involved clockwise rotation and closure towards the Bentley Trench, and subsequently a small southward translation and anticlockwise rotation, an overall pattern closely resembling that proposed by Molnar *et al.* (1975).

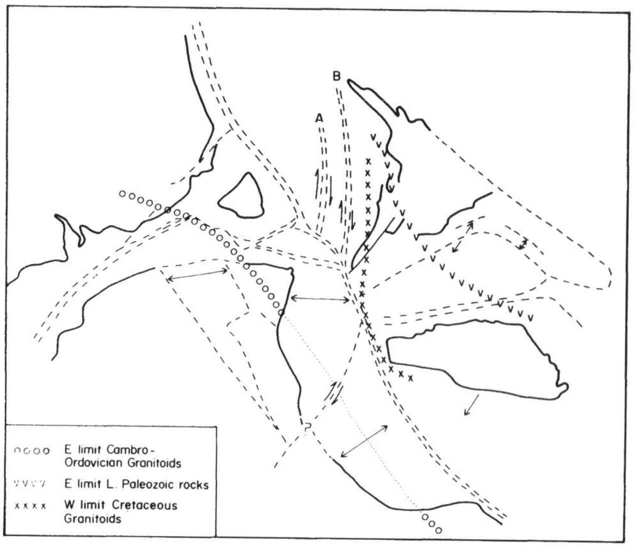

Fig. 2. Reassembly of the south-west Pacific–Antarctic region before break-up. Areas of currently thinned continental crust in the Wilkes Basin, Ross Sea, Campbell Plateau, South Tasman Rise, etc. have been reduced in area.

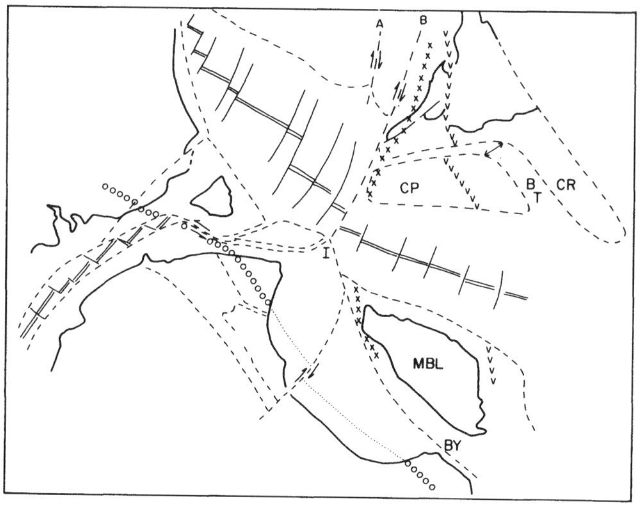

Fig. 3. Suggested configuration of the south-west Pacific–Antarctic region at the end of the Cretaceous. Symbols as in Fig. 2.

Conclusions

Dalziel & Elliot (1982) have argued that West Antarctica must be considered a collection of independent crustal blocks, and many accept that New Zealand has changed in shape. It is argued herein that there have also been small but significant changes in size due to continued crustal extension after continental break-up, a phenomenon which may be related to the rapid pace of rifting close to the Mesozoic convergent margin. Consequently, modelling of crustal dispersion requires detailed knowledge of the crustal structure and history of the adjacent continents. Plate-motion studies cannot rely wholly on magnetic lineation patterns and geometrical arguments.

References

Adams, C.J. (1987). Geochronology of granite terranes in the Ford Ranges, Marie Byrd Land, West Antarctica. *New Zealand Journal of Geology and Geophysics*, **29(1)**, 51–72.

Adams, R.D. (1962). Thickness of the earth's crust beneath the Campbell Plateau. *New Zealand Journal of Geology and Geophysics*, **5(1)**, 74–85.

Anderton, P.W., Holloway, N.H., Engerstrom, J.C., Ahmad, J.C. & Chong, B. (1982). *Evaluation of Geology and Hydrocarbon Potential of the Great South and Campbell Basins.* Phillips Petroleum Report, New Zealand Geological Survey Open File Petroleum Report, 828, 220 pp. (Unpublished.)

Bentz, F.P. (1974). Marine geology of the southern Lord Howe Rise, southwest Pacific. In *The Geology of Continental Margins*, ed. C.A. Burke & C.L. Drake, pp. 537–47. New York; Springer Verlag.

Bishop, D.G., Bradshaw, J.D. & Landis, C.A. (1985). Provisional terrane map of South Island, New Zealand. In *Tectonostratigraphic Terranes of the Circum-Pacific Region*, ed. D.G. Howell, pp. 515–21. Houston; Circum-Pacific Council for Minerals and Energy.

Cande, S.C. & Mutter, J.C. (1982). A revised identification of the oldest sea-floor spreading anomalies between Australia and Antarctica. *Earth and Planetary Science Letters*, **58(2)**, 151–60.

Cooper, A.K. & Davey, F.J. (1985). Episodic rifting of the Phanerozoic rocks of the Victoria Land Basin, Western Ross Sea, Antarctica. *Science*, **229(4718)**, 1085–7.

Crook, K.A.W. & Belbin, L. (1978). The southwest Pacific during the last 90 million years. *Journal of the Geological Society of Australia*, **25(1)**, 23–40.

Dalziel, I.W.D. & Elliot, D.H. (1982). West Antarctica: problem child of Gondwanaland. *Tectonics*, **1(1)**, 3–19.

Davey, F.J. (1981). Geophysical studies in the Ross Sea Region. *Journal of the Royal Society of New Zealand*, **11(4)**, 465–79.

Davey, F.J. & Christoffel, D.A. (1978). Magnetic anomalies across Campbell Plateau, New Zealand. *Earth and Planetary Science Letters*, **21(1)**, 61–72.

Drewry, D.J. (1976). Sedimentary basins of the East Antarctic craton from geophysical evidence. *Tectonophysics*, **36(2)**, 301–14.

Drewry, D.J. (1983). Isostatically adjusted bedrock surface of Antarctica. In *Antarctica: Glaciological and Geophysical Folio*, ed. D.J. Drewry, Sheet 6. Cambridge; Scott Polar Research Institute.

Falvey, D.A. & Mutter, J.C. (1981). Regional plate tectonics and the evolution of Australia's passive continental margins. *Journal of Australian Geology and Geophysics*, **6(1)**, 1–29.

Fitzgerald, P.G., Sandiford, M., Barrett, P.J. & Gleadow, A.J. (1986). Asymmetric extension associated with uplift and subsidence in the Transantarctic Mountains and Ross Embayment. *Earth and Planetary Science Letters*, **81(1)**, 67–78.

Grindley, G.W. (1981). Precambrian rocks of the Ross Sea region. *Journal of the Royal Society of New Zealand*, **11(4)**, 411–23.

Hawkes, P.W. & Mound, D.G. (1984). *Clipper – 1: Geological Completion Report*, BP Shell Todd. New Zealand Geological Survey Open File Petroleum Report, 1036, 79 pp. (Unpublished.)

Kamp, P.J.J. (1986). Late Cretaceous–Cenozoic tectonic development of the southwest Pacific. *Tectonophysics*, **121**, 225–51.

Laird, M.G. (1981a). Lower Paleozoic rocks of the Ross Sea area and their significance in a Gondwana context. *Journal of the Royal Society of New Zealand*, **11(4)**, 425–38.

Laird, M.G. (1981b). The Late Mesozoic fragmentation of the New Zealand segment of Gondwana. In *Gondwana Five*, ed. M.M. Cresswell & P. Vella, pp. 311–18. Rotterdam; A.A. Balkema.

Lister, G.S., Etheridge, M.A. & Symonds, P.A. (1986). Detachment faulting and the evolution of passive continental margins. *Geology*, **14(3)**, 246–50.

McKenzie, D.P. (1978). Some remarks on the development of sedimentary basins. *Earth and Planetary Science Letters*, **40(1)**, 25–32.

Molnar, P., Atwater, T., Mammerickx, J. & Smith, S.M. (1975). Magnetic anomalies, bathymetry and tectonic evolution of the South Pacific since the Late Cretaceous. *Geophysical Journal of the Royal Astronomical Society*, **40**, 383–420.

Nathan, S.J. (1986). *Cretaceous and Cenozoic Sedimentary Basins of the West Coast Region, South Island, New Zealand.* Wellington; New Zealand Geological Survey, 90 pp.

Pilaar, W.F.H. & Wakefield, L.L. (1978). Structural and stratigraphic evolution of the Taranaki Basin, offshore North Island, New Zealand. *Australia Petroleum Exploration Association Journal*, **18(1)**, 93–101.

Quilty, P.G. (1986). Australia's Antarctic basins: identification and evoluton. *Petroleum Exploration Society of Australia Journal*, **8(1)**, 20–46.

Sawyer, D.S., Swift, B.A., Sclater, J.G. & Toksoz, M.N. (1982). Extension model for the subsidence of the northern United States Atlantic continental margin. *Geology*, **10(3)**, 134–40.

Steed, R.H.N. (1983). Structural interpretation of Wilkes Land, Antarctica. In *Antarctic Earth Science*, ed. R.L. Oliver, P.R. James & J.B. Jago, pp. 567–72. Canberra; Australian Academy of Science and Cambridge; Cambridge University Press.

Stump, E., White, A.J.R. & Borge, S.G. (1986). Reconstruction of Australia and Antarctica: evidence from granites and recent mapping. *Earth and Planetary Science Letters*, **79(3/4)**, 348–60.

Veevers, J.J. (1984). *Phanerozoic Earth History of Australia*, Oxford; Clarendon Press. 418 pp.

Walcott, R.J. (1984). Reconstruction of the New Zealand region for the Neogene. *Palaeogeography, Palaeoclimatology and Palaeoecology*, **46(2)**, 217–31.

Weissel, J.K., Hayes, D.E. & Herron, E.M. (1977). Plate tectonic synthesis: the displacements between Australia, New Zealand and Antarctica since the Late Cretaceous. *Marine Geology*, **25(1/2)**, 231–77.

Willcox, J.B., Symonds, P.A., Hinz, K. & Bennett, D. (1980). Lord Howe Rise, Tasman Sea – preliminary geophysical results and petroleum prospects. *Journal of Australian Geology and Geophysics*, **5(2)**, 225–36.

NB. Some of the data in Anderton *et al.*, (1982) and Hawkes & Mound, (1984) are now published in:

Browne, G.H. & Field, B.D. (1988). A review of Cretaceous-Cenozoic sedimentation and tectonics, east coast, South Island New Zealand. In *Sequences, Stratigraphy, Sedimentology: Surface and Subsurface*, ed. D.P. James & D.A. Leckie, pp. 37–48. Canadian Society of Petroleum Geologists, Memoire 15.

Carter, R.M. (1988). Plate boundary tectonics, global sea-level changes and the development of the eastern South Island continental margin, New Zealand, southwest Pacific. *Marine and Petroleum Geology*, **5(2)**, 89–108.

The crustal blocks of West Antarctica within Gondwana: reconstruction and break-up model

B.C. STOREY

British Antarctic Survey, Natural Environment Research Council, High Cross, Madingley Road, Cambridge, CB3 0ET, UK

Abstract

A heterogeneous stretching model, incorporating simple shear, transcurrent motion and basin development is used to reconstruct West Antarctica within a Middle Jurassic Gondwana. Structural data, and a new interpretation of some of the subice topographic features, suggest no large-scale movement of the individual crustal blocks within West Antarctica. The main movement of these blocks has been accommodated by both stretching of the lithosphere between relatively solid blocks in an analogous way to chocolate-tablet boudinage and by a large component of transcurrent motion. This stretching of the crust has produced discrete zones of thinner crust and a complex array of extensional rifts and pull-apart basins, and was accompanied by widespread magmatic activity. The transcurrent and extensional processes have changed the shape of West Antarctica since break-up; removal of both of these components is used to constrain the original position of the crustal fragments within Gondwana.

Introduction

The position of West Antarctica within Gondwana has been a problem ever since precise Middle Jurassic reconstructions of the supercontinent, based on marine geophysical data, have been attempted. Some reconstructions result in up to 700 km of unacceptable overlap of the Antarctic Peninsula, a Mesozoic magmatic arc, with the continental crust of the Falkland Plateau (Norton & Sclater, 1979) and others have the Antarctic Peninsula lying on the Pacific side of southernmost South America. A solution to this problem has been sought in considering West Antarctica as five distinct microcontinental blocks that moved relative to one another and to the East Antarctic continent during break-up. Different arrangements of the crustal blocks are based on palaeomagnetism, comparative geology and the matching of structural trends (Dalziel & Elliot, 1982).

The model presented here not only considers the above blocks but incorporates both the highly extended crust between them and the kinematic constraints of major shear zones. The model is based on heterogeneous lithospheric stretching, associated simple shear and basin development, and includes a new interpretation of some subice topographic features. There has been little large-scale independent movement of the crustal blocks subsequent to break-up. The main movement has been accommodated by both stretching of the lithosphere between relatively solid blocks in an analogous way to chocolate-tablet boudinage and by a large component of transcurrent motion. This stretching of the crust produced discrete zones of thinner crust and a complex array of extensional rifts and pull-apart basins; it was accompanied by widespread magmatic activity. The transcurrent motion and extension processes have changed the shape of West Antarctica since break-up; removal of both of these components is used to reduce the overlap problem and produce a new configuration of the crustal blocks within West Antarctica. The timing of these movements is not well constrained but was most likely operative, perhaps episodically, since the initiation of break-up at 180 Ma. The occurrence of present day alkali-basalts suggests that some extension may still be taking place. The role of accretion along the western margin of the Antarctic Peninsula is not considered a major component in reducing the overlap of the Antarctic Peninsula with South America.

Crustal blocks

Five crustal blocks within West Antarctica (Fig. 1) are topographically elevated, have a distinctive geological evolution (Dalziel & Elliot, 1982; Storey *et al.*, 1988) and correspond to areas of thicker crust based on gravity anomaly calculations (Groushinsky & Sazhina, 1982). Three of them, Antarctic Peninsula (AP), Thurston Island (TI) and Marie Byrd Land (MBL) formed part of the Mesozoic Pacific margin of the supercontinent. The fourth, the Ellsworth–Whitmore mountains crustal block (EWM), contains part of the Upper Palaeozoic–lower Mesozoic Gondwanide fold belt that affected a Cambrian–Permian sedimentary succession and is believed to be underlain by a Precambrian basement. The fifth, the limits of which are defined geophysically by Storey *et al.* (1988), is unique and is totally formed of Precambrian base-

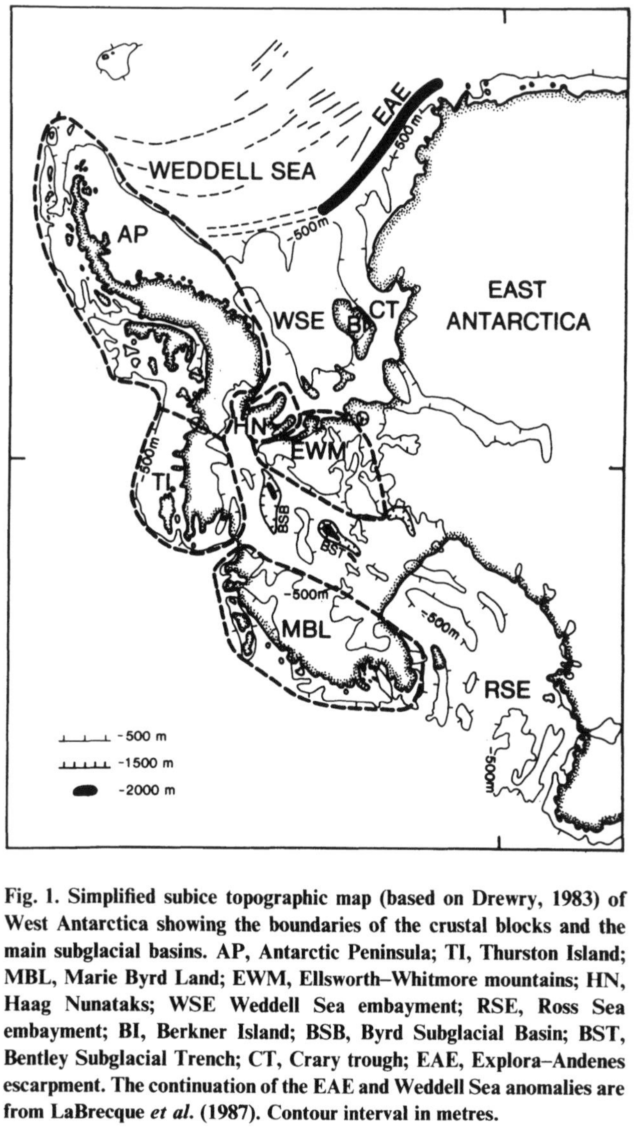

Fig. 1. Simplified subice topographic map (based on Drewry, 1983) of West Antarctica showing the boundaries of the crustal blocks and the main subglacial basins. AP, Antarctic Peninsula; TI, Thurston Island; MBL, Marie Byrd Land; EWM, Ellsworth–Whitmore mountains; HN, Haag Nunataks; WSE Weddell Sea embayment; RSE, Ross Sea embayment; BI, Berkner Island; BSB, Byrd Subglacial Basin; BST, Bentley Subglacial Trench; CT, Crary trough; EAE, Explora–Andenes escarpment. The continuation of the EAE and Weddell Sea anomalies are from LaBrecque *et al.* (1987). Contour interval in metres.

ment that has different geophysical characteristics to the basement beneath the EWM (Garrett, Herrod & Mantripp, 1987).

Although the blocks are separated by crustal discontinuities, some geological correlations can be made between them. The sedimentary rocks of the EWM can be correlated with Cambrian events across the East–West Antarctica boundary in the Thiel Mountains (Storey & Macdonald, 1987), and with the Cambrian–Permian succession of the Pensacola Mountains; *Glossopteris*-bearing Permian strata have now been recorded from both the EWM and AP (Laudon *et al.*, 1985). The Middle Jurassic granitoids of the EWM are contemporaneous with magmatism of the Ferrar Supergroup within the Transantarctic Mountains (TAM) and form part of a widespread within-plate extensional magmatic event associated with break-up that is common to the EWM and East Antarctica (Storey *et al.*, 1988).

Extensional features

Much of West Antarctica is below sea level and the subice topography has a very distinctive horst and graben form and an abundance of rift and rift-like features (Masolov, Kurinin & Grikurov, 1981; Drewry, 1983) (Fig. 1) which are probably unparalleled in any other continent. Some correspond to regions of thinner crust (20–25 km thick) and in some cases contain thick sedimentary successions.

The EWM crustal block for the most part forms a barrier to a continuous channel way across West Antarctica and divides the features into a Weddell Sea and Ross Sea sector. However, there are some deep and narrow grabens on the north and west boundary of the EWM which extend right across the continent. Many of the features are not restricted to West Antarctica and some in the vicinity of the Pensacola Mountains penetrate into East Antarctica. In the Ross Sea sector a major graben, the Wilkes Basin (Steed, 1983), parallels the TAM on the western side.

Sedimentary basins

Thick sedimentary successions are present within both the Ross Sea and Weddell Sea embayments. In the Ross Sea Embayment seismic investigations indicate a series of graben-like troughs infilled with up to 10 km of sediment (Davey, Hinz & Schroeder, 1983). Crustal thicknesses are between 21–30 km and the embayment has been the site of extension and crustal thinning, with associated normal faults and extensional volcanism (Cooper & Davey, 1985; Fitzgerald *et al.*, 1986). Uplift studies in the adjacent TAM suggest most of the extension was during Late Cretaceous and Tertiary times (Fitzgerald *et al.*, 1986). The Weddell Sea embayment is virtually a mirror image on the opposite side of the EWM. Crustal thicknesses are estimated to be 20–25 km, with up to 12–15 km of sedimentary strata on an extended continental basement (Kadmina *et al.*, 1983; Kamenev & Ivanov, 1983) which may be similar to that of Haag Nunataks. There is a complex system of graben-like troughs (e.g. Crary trough) and N–S-trending horsts (e.g. Berkner Island) (Fig. 1).

Although sedimentary basins may be formed either by lithospheric extension and subsidence associated with cooling of the lithosphere, or by loading, it is generally agreed that both the Weddell Sea and Ross Sea embayments formed by extension. The relationship between crustal extension and basin development is complex and may involve either homogeneous stretching of the lithosphere or heterogeneous displacement on lithospheric-scale shear zones. However, an estimate of the amount of extension within the sedimentary basins is important and must be incorporated in any reconstruction of the supercontinent. Data currently available do not permit intricate modelling and both embayments involve complex heterogeneous stretching with localized uplifted areas and faulted basins. Although the amount of crustal thinning is consequently variable throughout the basin, an estimate of the amount of extension for both the Ross and Weddell Sea embayments, by applying a simple pure shear model with an original crustal thickness of 35–40 km and a present day thickness of 25–30 km, is about 25–30%. This is in the order of 250–300 km for each embayment. The N–S orientation of topographic features, the sedimentary basins and the magnetic anomaly associated with the Middle Jurassic gabbro of the Dufek Massif in the Pensacola Mountains (Behrendt *et al.*, 1974) suggest E–W extension within both embayments.

Rifts

Fig. 1 shows the complex system of morphologically-defined grabens within West Antarctica. The most spectacular features occur on the NE side of the Ellsworth Mountains where, from the base of the Rutford Ice Stream to the top of the Ellsworth Mountains, there is 4.5 km of vertical relief. However, the largest area of low relief is present in the centre of West Antarctica where, on the NW side of the EWM, the Byrd Subglacial Basin (BSB) and Bentley Subglacial Trench (BST) are up to 2000 m below sea level (Drewry, 1983). The BSB contains only 0.5 km of sediment, has a magnetic signature and may be floored by volcanic rock (Jankowski, Drewry & Behrendt, 1983).

The origin and significance of these features is speculative. They may either be glacially-scoured land forms, or tectonically-controlled rifts. However, geophysical modelling of the grabens within the HN block suggests they are tectonic features (Garrett *et al.*, 1987) and represent an important episode of extension within West Antarctica. Basins, like the BSB, which contain a magnetic signature, may represent small extensional basins floored by oceanic crust or highly attenuated continental crust.

An analysis of the extensional strain represented by these features is problematical. Their orientation indicates a complex history of extension which may result from a variable incremental strain history. Alternatively, many of the features may not have developed normal to the principal extension direction and may contain a component of strike-slip motion, and some may be controlled by pre-existing basement structures. The BSB in particular is lozenge-shaped (Fig. 1) and is typical of classical rhomb-shaped pull-apart basins which tend to nucleate at releasing fault segments along active strike-slip faults. Pull-apart basins are characterized by local crustal stretching, lithospheric extension, subsidence and alkali volcanism; these are all characteristics of West Antarctica. The Z-shape of BSB and the topographic ridge between the BSB and BST indicate dextral movement within this zone estimated to be about 300 km. The extensional rifts that uplift and extend HN, between the AP and EWM, may also be formed at releasing fault segments of a major fault system that connects the western margin of the WSE and RSE. There is further evidence of dextral strike-slip faulting within the Antarctic Peninsula magmatic arc and fore-arc accretionary complex, including the formation of George VI Sound (Nell & Storey, this volume, p. 443).

Crustal boundaries

For the most part the crustal blocks are bounded by topographic scarps and rift zones that represent zones of crustal thinning. Although commonly associated with a change in magnetic pattern (Drewry, 1983), reflecting magmatism associated with extension, the geology of the crustal blocks is in some cases continuous across the zones of crustal thinning. However, there are two boundaries which may represent important discontinuities in reconstructing the supercontinent: (1) the northern boundary of WSE, and (2) the eastern margin of AP.

Northern boundary of WSE

The WSE is bounded on the northern side by the linear ENE–WSW-trending basement ridge of the Explora–Andenes escarpment, with a wedge of seaward-dipping seismic reflectors on the landward side (Kristoffersen & Haugland, 1986). It is parallel to the earliest recognized magnetic anomalies within the Weddell Sea and is interpreted either as a rifted margin formed during the initial opening phases of the Weddell Sea during the Middle Jurassic, or as a plate boundary with oblique-spreading rather than pure strike-slip faulting (Kristofferson & Haugland, 1986). It is continuous with a strong WNW-striking magnetic anomaly identified by the USAC programme and termed the Orion anomaly (LaBrecque *et al.*, 1987). This continuous feature (Fig. 1) represents the ocean–continent boundary and separates the truly oceanic crust of the Weddell Sea from the extended continental crust beneath the Weddell Sea embayment. It marks a reorganization of the early relative plate motions from a period of E–W extension within the embayment which may be considered as a failed rift, to N–S extension and seafloor spreading within the Weddell Sea. In a reconstruction it forms an important linear structure that has not been substantially disrupted during the subsequent extension and movement of the crustal blocks within West Antarctica.

Eastern margin of the AP

The eastern margin of Palmer Land, AP, forms an important tectonic boundary separating the AP magmatic arc from the WSE. Deformation is concentrated in variably-orientated heterogeneous shear zones incorporating east-directed thrusting, ductile normal shearing and both dextral and sinistral shearing (Meneilly *et al.*, 1987; B.C. Storey, unpublished data). Gabbros, mafic dykes, amygdaloidal basalts and extensional shears may be related to a phase of basin formation and arc extension during the Jurassic. Middle–Upper Jurassic sedimentary rocks of the Latady and Mount Hill formations infilled the extensional basins and were deformed by east-directed folding and thrusting and superimposed dextral shear. Sinistral shears are generally at a high angle to the margin and are most likely conjugate riedal shears. Deformation along this boundary is also indicated on preliminary seismic results (Miller, Lippmann & Kallerhoff, 1984), which show sedimentary rocks within the Weddell Sea to be undisturbed except close to the AP boundary.

Reconstruction and break-up model

In reconstructing the supercontinent Gondwana in the Middle Jurassic there are a number of important considerations concerning West Antarctica.

(1) The crustal blocks do not all have independent geological histories.

(2) The Explora–Andenes escarpment forms a linear feature that has not been disrupted by later crustal block movements. This rules out any relative movement between East and West Antarctica after formation of the escarpment and suggests that

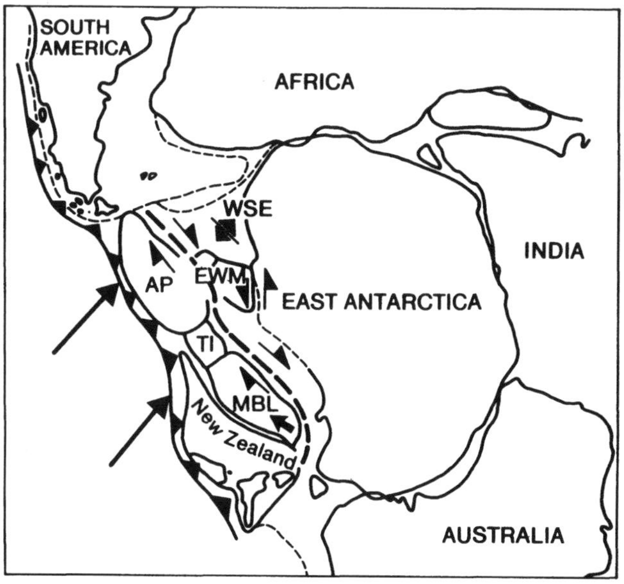

Fig. 2. Reconstruction of West Antarctica within Gondwana showing the extension within the WSE and the transtensional rift. The configuration of the main continents is taken from Lawver & Scotese (1987).

any large displacements of the crustal blocks must predate this feature. Subsequent movements must be restricted to extension and/or block rotations close to their present position.

(3) Some of the subsided fault-bounded regions and sedimentary basins are underlain by thinned continental crust and should be considered in a reconstruction. This suggests the presence of highly extended crustal blocks beneath both the Weddell Sea and Ross Sea embayments. Many reconstructions (reviewed by Dalziel & Elliot, 1982) incorporate a continental block (the Filchner block) in the Weddell Sea embayment region. However the results of this study suggest a 30% estimate of crustal extension must be considered for this block and an analogous block in the Ross Sea embayment, in any reconstruction.

(4) The complex history of extension, as indicated by the geometry of fault scarps throughout West Antarctica, has involved a major period of dextral transcurrent motion with formation of pull-apart basins and a dextral transtensional rift across West Antarctica.

(5) The east coast of the Antarctic Peninsula formed a major dislocation zone during break-up.

A new reconstruction of West Antarctica based on the above considerations is presented in Fig. 2 and the movement vectors of the crustal blocks in Fig. 3. The main components of the break-up model are a major period of E–W extension in the WSE and RSE and a major component of dextral shear along the Pacific margin. The proposed stretching model is considered analogous to chocolate-tablet boundinage and many of the movements of the crustal blocks may be outside of the present limits of palaeomagnetic measurements. The model, however, incorporates limited clockwise rotation of the AP and EWM about a pole close to the present tip of the AP. This agrees with the palaeomagnetic data of Grunow, Kent & Dalziel (1987), E–W spreading in the WSE, and the known magnetic lineations within the Weddell Sea. There may be some relative movement between the AP and EWM, and collision of the EWM block with the southern part of the AP may account for some of the deformation within the Latady and Mount Hill formations on the SE coast of the peninsula.

The strike-slip component of the stretching model incorporates dextral movement of the AP, TI and MBL relative to the EWM and East Antarctica and could account for removal of much of the overlap of the AP with South America in many reconstructions. The AP may consequently have remained pinned to South America prior to opening of the Scotia Sea, and the Weddell Sea may have opened as an intra-Gondwana basin. The dextral component may either have been distributed throughout the AP (Nell & Storey, this volume, p. 443) or contained in a single zone of high strain along the eastern side. The strike-slip model could also account for large *in situ* rotations of the crustal blocks. Although these are not required after 175 Ma by the palaeomagnetic data of Grunow *et al.* (1987), the data of Watts & Brammall (1981) and the discordant structural trends of the EWM relative to East Antarctica suggest rotations may have taken place prior to 175 Ma.

There are, however, few constraints on the age of the different extensional phases within West Antarctica. A major uncertainty surrounds the age of seafloor spreading in the Weddell Sea and the time of formation of the Explora–Andenes escarpment. Anomalies south of the distinctive A33–34 (Late Cretaceous) in the Weddell Sea may have been misidentified as Early Jurassic (LaBrecque & Barker, 1981),

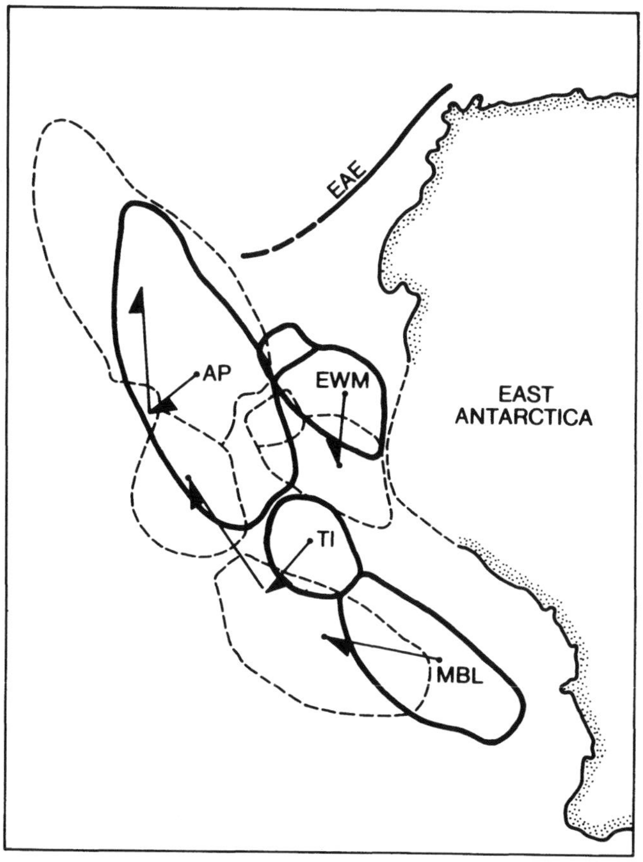

Fig. 3. Proposed movement vectors of the crustal blocks of West Antarctica during break-up.

and the Weddell Sea may be no older than South Atlantic opening (M11 time, 135 Ma) (Barker, Dalziel & Storey, in press). However, the Middle–Late Jurassic age of basin sedimentation along the eastern margin of the AP suggests the period of E–W crustal extension and sedimentation within the WSE may be contemporaneous. This corresponds with the age of the within-plate bimodal magmatic suite of the Ferrar Supergroup and the Ellsworth–Whitmore Granitic Suite (175 Ma), and represents the initial rifting phase of Gondwana. The early separation of Africa and South America from Antarctica identified as M22 (158 Ma) may have been accommodated by this extensional episode within West Antarctica. The subsequent formation of Weddell Sea anomalies and the escarpments (~ 135 Ma) marks a reorganization of the early plate motions from the period of east–west extension which may be considered as a failed rift, to north–south extension and seafloor spreading in the Weddell Sea.

Although Middle Jurassic Ferrar dolerites border the RSE, there is currently no evidence for the early WSE extensional phase in this sector of West Antarctica. The RSE may have formed as a transtensional rift associated with the Late Cretaceous rifting of New Zealand and Australia from Antarctica (Kamp, 1986). The timing of this extensional episode agrees with the fission-track studies of Fitzgerald *et al.* (1986) within the TAM, the uplifted margins bordering the rift system. The extension within the RSE may account for the dextral movement of crustal blocks along the Pacific margin of West Antarctica with the formation of the BSB and BST and the proposed 200–500 km eastward translation of MBL from a position close to the TAM, as predicted by palaeomagnetic data of Grindley & Oliver (1983). The presence of extensional features with marked topographic expression and large alkali-basalt fields of late Cenozoic–Recent age suggests extension may still be taking place within West Antarctica.

A heterogeneous stretching model, incorporating simple shear, transcurrent motion and basin development, is used to reconstruct West Antarctica within a Middle Jurassic Gondwana. The present day configuration of five crustal blocks separated by thinned and extended continental crust formed by: (1) an initial period of rifting and extension in the WSE (175–135 Ma), with clockwise rotation of the AP, TI and EWM accommodating the separation of Africa and South America from Antarctica; (2) change in the plate motions (~ 135 Ma), with opening of the Weddell Sea and South Atlantic; (3) a period of Late Cretaceous–Tertiary extension in the RSE resulting in dextral transcurrent motion of the AP, TI and MBL along the Pacific margin of West Antarctica and the formation of strike-slip basins. Further constraints on this model can only be imposed with more detailed crustal thickness data, palaeomagnetism and fault-zone analysis.

References

Barker, P.F., Dalziel, I.W.D. & Storey, B.C. (in press). Tectonic development of the Scotia Arc region. In *Antarctic Geology*, ed. R. Tingey. Oxford; Oxford University Press.

Behrendt, J.C., Henderson, J.R., Meister, L. & Rambo, W.L. (1974). Geophysical investigations of the Pensacola Mountains and adjacent glacierized areas of Antarctica. *US Geological Survey Professional Paper*, **844**, 28 pp.

Cooper, A.K. & Davey, F.J. (1985). Episodic rifting of Phanerozoic rocks in the Victoria Land Basin, western Ross Sea, Antarctica. *Science*, **229**, 1085–7.

Dalziel, I.W.D. & Elliot, D.H. (1982). West Antarctica: problem child of Gondwanaland. *Tectonics*, **1(1)**, 3–19.

Dalziel, I.W.D., Garrett, S.W., Grunow, A.M., Pankhurst, R.J., Storey, B.C. & Vennum, W.R. (1987). The Ellsworth–Whitmore Mountains crustal block: its role in the tectonic evolution of West Antarctica. In *Gondwana Six: Structure, Tectonics, and Geophysics*, Geophysical Monograph 40, ed. G.D. McKenzie, pp. 173–82. Washington, DC; American Geophysical Union.

Davey, F.J., Hinz, K. & Schroeder, H. (1983). Sedimentary basins of the Ross Sea, Antarctica. In *Antarctic Earth Science*, ed. R.L. Oliver, P.R. James & J.B. Jago, pp. 533–8. Canberra; Australian Academy of Science and Cambridge; Cambridge University Press.

Drewry, D.J. (1983). *Antarctica: Glaciological and Geophysical Folio 2.* Cambridge; Scott Polar Research Institute.

Fitzgerald, P.G., Sandiford, M., Barrett, P.J. & Gleadow, A.J.W. (1986). Asymmetric extension associated with uplift and subsidence in the Transantarctic Mountains and Ross Embayment. *Earth and Planetary Science Letters*, **81(1)**, 67–78.

Garrett, S.W., Herrod, L.D.B. & Mantripp, D.R. (1987). Crustal structure of the area around Haag Nunataks, West Antarctica: new aeromagnetic and bedrock elevation data. In *Gondwana Six: Structure, Tectonics, and Geophysics*, Geophysical Monograph 40, ed. G.D. McKenzie, pp. 109–15. Washington, DC; American Geophysical Union.

Grindley, G.W. & Oliver, P.J. (1983). Palaeomagnetism of Cretaceous volcanic rocks from Marie Byrd Land, Antarctica. In *Antarctic Earth Science*, ed. R.L. Oliver, P.R. James & J.B. Jago, pp. 573–8. Canberra; Australian Academy of Science and Cambridge; Cambridge University Press.

Groushinsky, N.P. & Sazhina, N.B. (1982). Some features of Antarctic crustal structure. In *Antarctic Geoscience*, ed. C. Craddock, pp. 931–6. Madison; University of Wisconsin Press.

Grunow, A.M., Kent, D.V. & Dalziel, I.W.D. (1987). Evolution of the Weddell Sea Basin: new palaeomagnetic constraints. *Earth and Planetary Science Letters*, **86**, 16–26.

Jankowski, E.J., Drewry, D.J. & Behrendt, J.C. (1983). Magnetic studies of upper crustal structure in West Antarctica and the boundary with East Antarctica. In *Antarctic Earth Science*, ed. R.L. Oliver, P.R. James & J.B. Jago, pp. 197–203. Canberra; Australian Academy of Science and Cambridge; Cambridge University Press.

Kadmina, I.N., Kurinin, R.G., Masolov, V.N. & Grikurov, G.E. (1983). Antarctic crustal structure from geophysical evidence: a review. In *Antarctic Earth Science*, ed. R.L. Oliver, P.R. James & J.B. Jago, pp. 498–502. Canberra; Australian Academy of Science and Cambridge; Cambridge University Press.

Kamenev, E.N. & Ivanov, V.L. (1983). Structure and outline of the history of the southern Weddell Sea basin. In *Antarctic Earth Science*, ed. R.L. Oliver, P.R. James & J.B. Jago, pp. 194–7. Canberra; Australian Academy of Science and Cambridge; Cambridge University Press.

Kamp, P.J.J. (1986). Late Cretaceous–Cenozoic tectonic development of the southwest Pacific region. *Tectonophysics*, **121**, 225–51.

Kristoffersen, Y. & Haugland, K. (1986). Geophysical evidence for the East Antarctic plate boundary in the Weddell Sea. *Nature, London*, **322(6079)**, 538–41.

LaBrecque, J. & Barker, P.F. (1981). The age of the Weddell Basin. *Nature, London*, **290(5806)**, 489–92.

La Brecque, J., Brozena, J., Cande, S., Kovacs, L., Raymond, C., Bell, R., Keller, M., Parra, J.C. & Valladares, J. (1987). The USAC program: magnetic anomalies of the western Weddell Basin. In *Abstracts, 5th International Symposium on Antarctic Earth Sciences, Cambridge, August 1987*, pp. 155.

Laudon, T.S., Lidke, D.J., Delevoryas, T. & Gee, C.T. (1985). Sedimentary rocks of the English Coast, eastern Ellsworth Land, Antarctica. *Antarctic Journal of the United States*, **19(5)**, 38–40.

Lawver, L.A. & Scotese, C.R. (1987). A revised reconstruction of Gondwana. In *Gondwana Six: Structure, Tectonics, and Geophysics*, Geophysical Monograph 40, ed. G.D. McKenzie, pp. 17–23. Washington, DC; American Geophysical Union.

Masolov, V.N., Kurinin, R.G. & Grikurov, G.E. (1981). Crustal structures and tectonic significance of Antarctic rift zones. In *Gondwana Five*, ed. M.M. Cresswell & P. Vella, pp. 303–9. New Zealand; University of Wellington.

Meneilly, A.W., Harrison, S.M., Piercy, B.A. & Storey, B.C. (1987). Structural evolution of the magmatic arc in northern Palmer Land, Antarctic Peninsula. In *Gondwana Six: Structure, Tectonics, and Geophysics*, Geophysical Monograph 40, ed. G.D. McKenzie, pp. 209–19. Washington, DC; American Geophysical Union.

Miller, H., Lippmann, E. & Kallerhoff, W. (1984). Marine geophysical work during Antarctic II/4. *Berichte zur Polarforschung*, **19**, 116–29.

Norton, I.O. & Sclater, J.G. (1979). A model for the evolution of the Indian Ocean and the break-up of Gondwanaland. *Journal of Geophysical Research*, **84(B12)**, 6803–30.

Steed, R.H.N. (1983). Structural interpretations of Wilkes Land, Antarctica. In *Antarctic Earth Science*, ed. R.L. Oliver, P.R. James & J.B. Jago, pp. 567–72. Canberra; Australian Academy of Science and Cambridge; Cambridge University Press.

Storey, B.C., Dalziel, I.W.D., Garrett, S.W., Grunow, A.M., Pankhurst, R.J. & Vennum, W.R. (1988). West Antarctica in Gondwanaland: crustal blocks, reconstruction and breakup processes. *Tectonophysics*, **155**, 381–90.

Storey, B.C. & Macdonald, D.I.M. (1987). Sedimentary rocks of the Ellsworth–Thiel mountains ridge and their regional equivalents. *British Antarctic Survey Bulletin*, **76**, 21–49.

Watts, D.R. & Bramall, A.M. (1981). Palaeomagnetic evidence for a displaced terrain in western Antarctica. *Nature, London*, **293(5834)**, 638–42.

Hotlines and Cenozoic volcanism in East Antarctica and eastern Australia

A.G. SMITH[1] & R.A. LIVERMORE[2]

1 Department of Earth Sciences, Downing Street, Cambridge CB2 3EQ, UK
2 British Antarctic Survey, High Cross, Madingley Road, Cambridge CB3 0ET, UK

Abstract

Synthetic hotspot tracks have been constructed for Cenozoic volcanic sites in eastern Australia and East Antarctica using recently revised rotations. For eastern Australia, several tracks reach either the Tasman Sea or Coral Sea margins when those margins were beginning to form. This relationship may be causal. In East Antarctica most tracks cross the Eastern basin of the Ross Sea at about 60–80 Ma, suggesting that the late Cenozoic volcanism in East Antarctica can be plausibly attributed to the migration of a distributed Late Cretaceous–early Cenozoic heat source originally under the Eastern basin.

Introduction

Late Cenozoic volcanism is known from parts of the Transantarctic Mountains (Fig. 1). Some 3000 km to the north lies the southern edge of the Cenozoic volcanic belt of eastern Australia (Fig. 2). Although the two continents were joined together until about 95 Ma ago, or Late Cretaceous time (Mutter *et al.*, 1985; Veevers, 1986), it was only after significant separation had taken place that the Australian volcanism started. Despite their geographical separation we believe that the volcanism in both continents can be understood very simply by placing it in a hotspots reference frame.

Hotspots, hotlines and ocean-floor data

Hotspots are areas of persistent volcanism that seem to be independent of lithospheric plates or their margins. When the effects of plate motions are eliminated, hotspots move relatively slowly with respect to one another (Morgan, 1981). Partly for this reason they are interpreted physically as originating in the lower mantle, forming an 'absolute reference frame' in which to describe plate motions.

Hotspots may affect regions up to 2000 km in diameter. The Iceland hotspot, when initiated, may have induced the break-up of the north-eastern Atlantic, creating a linear zone of intense volcanism at the continental margin at the time of break-up (White & McKenzie, 1989). However, continental volcanism in south-eastern Australia and in East Antarctica took place some tens of millions of years after break-up. Most of this volcanism seems empirically to be related to the position of a linear array of hotspots (a 'hotline') that was originally at the continental margin, rather than to a circular area marking the position of a hotspot in the underlying mantle. The line itself may mark the interaction of such a hotspot with the lithosphere, e.g. it may be a split developed above a hotspot.

Rotations describing the motion of Africa in the hotspots frame (Table 1) have been progressively refined (Duncan, 1981; Morgan, 1981, 1983). It has been debated whether hotspot groups in the Atlantic and Indian oceans move or are stationary relative to one another (Morgan, 1983; Hartnady & le Roex, 1985).

These rotations may be used to predict the tracks of suspected hotspots on other plates by combining them with rotations describing the motions of such plates relative to Africa. Table 1 lists the Euler rotations and sources of the data. New identifications of anomalies 33–34 (75–83 Ma) in the Enderby Basin (Patriat *et al.*, 1985; Royer *et al.*, 1988) suggest rather different East Antarctica–Africa motions to previous models (e.g. Norton & Sclater, 1979). The new interpretation, which must be considered tentative, predicts more erratic motions. Table 1 incorporates rotations for Tertiary and Late Cretaceous anomalies 5–29 from comprehensive studies of the Indian Ocean (Patriat, 1983; Royer *et al.*, 1988). Small additional rotations are applied to correct for the movement of East Africa (Somali plate) with respect to the rest of Africa (Nubian plate).

The revised model for the separation of Australia from Antarctica (Mutter *et al.*, 1985) has been employed (Table 1), involving an early (pre-anomaly 19) phase of very slow spreading.

Continental margins and Cenozoic volcanism in eastern Australia

The north-eastern continental margin of Australia was formed about 62 Ma ago when the Coral Sea started to open

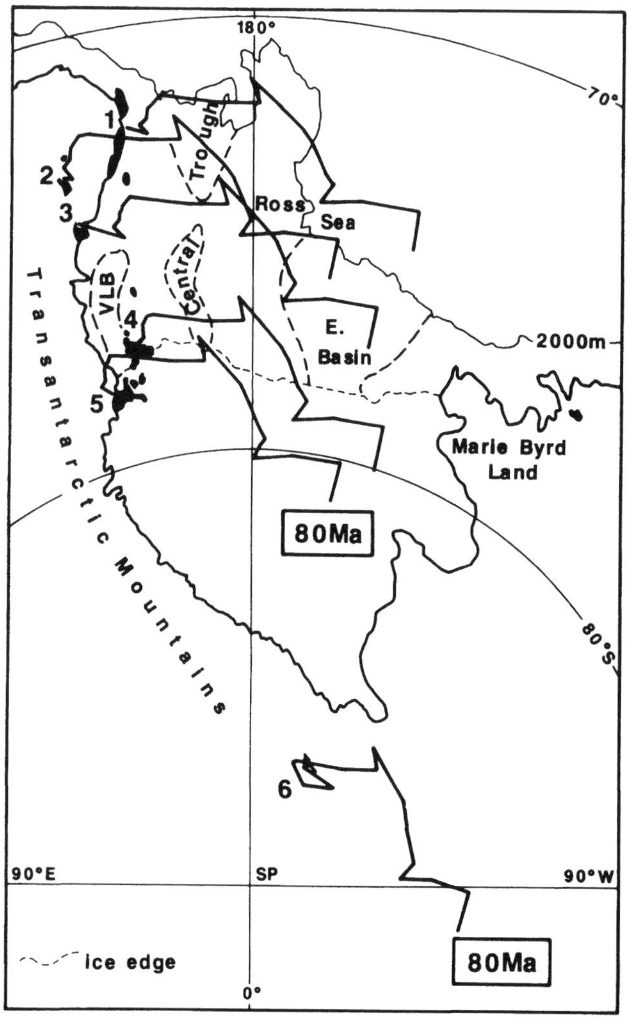

Fig. 1. Physiography and distribution of Cenozoic alkaline volcanic areas (solid black), East Antarctica (modified from Fitzgerald *et al.*, 1986, fig. 1). Lambert equal-area projection centred on South Pole (= SP). Extensional areas are VLB (Victoria Land basin), Central trough and the Eastern basin. Hotspot tracks of Antarctic volcanic areas listed in Table 2, rotated according to parameters of Table 1 back to 80 Ma. The numbers 1–6 refer to volcanoes listed in Table 2 in East Antarctica.

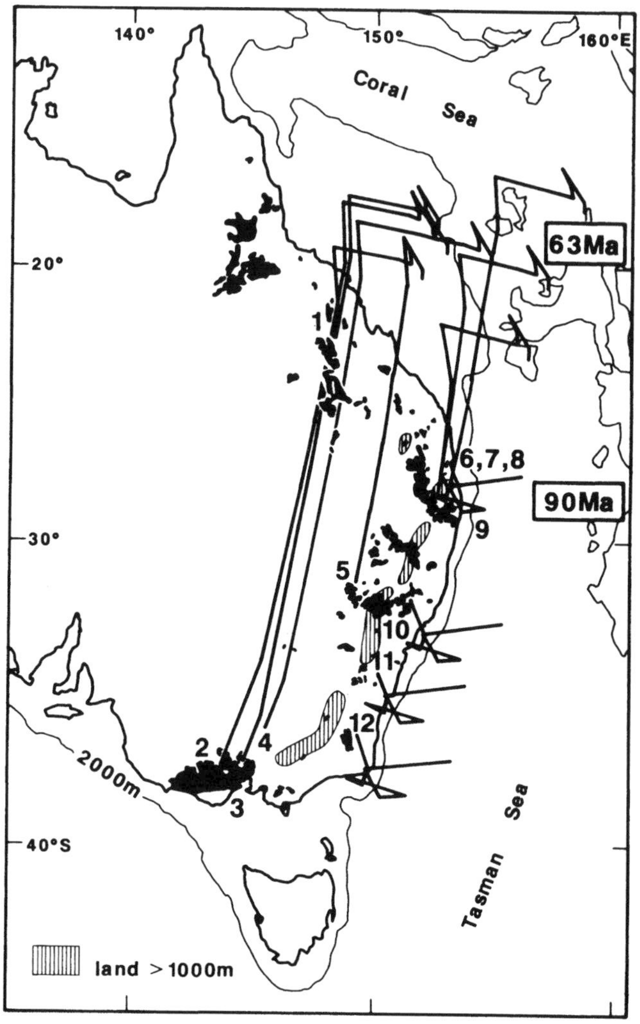

Fig. 2. Physiography of eastern Australia and generalized distribution of Cenozoic volcanic fields (solid black) on mainland Australia. Mercator projection. Hotspot tracks of volcanic areas in eastern Australia listed in Table 2, rotated according to parameters of Table 1. Tracks 1–8 are rotated back to 63 Ma; tracks 9–12 back to 90 Ma. The numbers 1–12 refer to volcanoes listed in Table 2 in eastern Australia.

(Weissel & Watts, 1979). The eastern margin formed when the Tasman Sea opened about 80 Ma ago (Weissel & Hayes, 1977; Shaw, 1978).

In eastern Australia Cenozoic volcanism shows two prominent features: firstly, an earlier basaltic zone of linear lava fields parallel to the Tasman margin (Wellman & McDougall, 1974; Sutherland, 1978) which started to erupt in widely separated localities at about 53–56 Ma (Table 2; Sutherland, 1981); and secondly, a marked southward migration of alkalic central volcanoes in the interval 33 Ma to essentially the present day (Table 2; Sutherland, 1981, 1983).

Sutherland (1981, 1983) showed that when Australian basaltic-volcanic sites in the age range 2–33 Ma were moved back along generalized hotspots tracks, they reached the edge of the Coral Sea at about the time when the Coral Sea was beginning to form. Smith (1982) independently showed a similar relationship for tracks generated in a reference frame fixed to South America, which is similar to the hotspots frame. He also suggested that earlier volcanism at about 55 Ma was the result of Australia migrating over the initial site of Tasman Sea spreading as seen in that frame or a similar frame. By contrast, Pilger (1982) argued that the fit to the hotspots frame was poor and suggested the volcanic traces were due to a migrating stress node.

To look at these relationships more precisely, we assume that volcanism at the sites listed by Sutherland (1981, 1983) started when a hotspot had reached the area. To find out whether the hotspot could also have been part of a hotline that originally lay beneath a newly formed continental margin, we track the mean position of the site back in time along its hotspot track, starting the track at the time of the volcanism as listed in Sutherland (1981, fig. 1; 1983, table 1).

We have tracked the sites of volcanoes associated with the linear lava fields and with the younger central alkalic volcanoes back in time until the tracks reach the present day 2000 m submarine contour, taken here as a reasonable estimate of the position of the continental margin (Fig. 2).

All the earlier basaltic tracks (9–12, Fig. 2) stay close to the site and to the Tasman margin in the interval to 83 Ma, then

Table 1. *Finite rotations used to bring Australia and Antarctica into the hotspots reference frame*

Fragment[a]	Time (Ma)[b]	Euler rotation			Source
		Latitude	Longitude	Angle	
Australia, East Antarctica	19.95	15.9	32.2	− 12.00	1
	37.06	11.9	34.4	− 20.50	2
	42.29	11.5	31.0	− 23.58	3
	44.90	16.4	28.8	− 25.41	4
	90.00	1.5	37.0	− 28.00	5, 6
	100.00	10.3	32.7	− 30.00	2
East Antarctica, Madagascar[c]	9.64	9.1	− 41.6	1.49	7
	19.95	13.4	− 55.1	2.79	7
	42.29	15.3	− 50.4	6.92	7
	44.90	10.2	− 42.8	7.84	7
	48.08	7.9	− 39.6	8.86	7
	52.74	11.3	− 44.9	10.11	7
	57.85	3.8	− 39.7	10.63	8
	63.06	0.6	− 39.2	11.32	8
	64.44	− 0.4	− 39.4	11.59	8
	67.35	1.1	− 41.6	11.84	8
	70.68	− 1.8	− 41.4	13.47	8
	75.29	− 4.0	− 40.3	15.99	9
	82.93	− 1.4	− 40.6	17.84	9
	115.00	− 8.43	− 26.42	42.56	10
Somalia, Arabia	9.64	26.5	21.5	3.86	11
	25.00	26.5	21.5	5.43	12
Arabia, Africa (Nubia)	9.64	36.5	18.0	− 3.08	11
	25.00	36.5	18.0	− 4.34	12
Africa, hotspots	10.00	50.4	323.1	− 2.70	13
	20.00	50.4	323.1	− 5.40	13
	30.00	49.2	324.3	− 8.00	13
	40.00	48.0	322.2	− 10.40	13
	50.00	46.8	318.6	− 12.30	13
	60.00	44.2	311.4	− 14.60	13
	70.00	40.4	303.4	− 17.50	13
	80.00	35.4	301.6	− 21.70	13
	90.00	31.9	308.4	− 24.70	13
	100.00	29.0	313.4	− 27.90	13

[a]First fragment rotates with respect to second (fixed) fragment to produce reconstruction at time given.
[b]Ages of anomalies based on Harland *et al.* (1982).
[c]Madagascar fixed to Somalia during 0–115 Ma.
Sources of rotations as follows: 1, Sclater *et al.* (1981); 2, Weissel, Hayes & Herron (1977); 3, Stock & Molnar (1982); 4, Patriat & Courtillot (1984); 5, Koenig (1980); 6, Mutter *et al.* (1985); 7, Patriat (1983); 8, Royer *et al.* (1988); 9, Patriat *et al.* (1985); 10, Martin & Hartnady (1986); 11, McKenzie, Davies & Molnar (1970); 12, Cochran (1981); 13, Morgan (1983).

move rapidly eastwards into the Tasman Sea. All of the younger alkaline-basalt tracks (1–8, Fig. 2) reach the Coral Sea margin or nearby margins at about 60–70 Ma, though the more easterly of these reach margins whose positions at 60–70 Ma are not clear (Shaw, 1978; Weissel & Watts, 1979).

Continental margins and Cenozoic volcanism in East Antarctica

While extension does exist in East Antarctica, continental extension has not progressed to the point of generating ocean floor (Fitzgerald *et al.*, 1986). The extension itself is distributed over a wide zone giving rise to at least three major basins in the Ross Sea (Victoria Land basin, Central trough and Eastern basin), between the Transantarctic Mountains and Marie Byrd Land (Fig. 1). The whole extended region has been modelled as an asymmetric extension zone (Fitzgerald *et al.*, 1986).

Sporadic sediment outcrops along the Transantarctic Mountains suggest that subsidence of the Victoria Land basin began in Late Cretaceous–early Cenozoic time and that the greatest subsidence has taken place in the Cenozoic (Fitzgerald *et al.*,

Table 2. *Cenozoic volcanic centres in eastern Australia and East Antarctica*

Location	Age (Ma)	Track	Latitude (°S)	Longitude (°E)
Eastern Australia				
Later alkaline-basalt volcanoes (Sutherland, 1983, table 1)				
1. Hillsborough–Buckland	30.3	63	22·7°	148·3°
2. Western Victoria	2.0	63	37·5°	143·8°
3. Macedon	6.1	63	37·4°	144·7°
4. Cobar–Cosgrove	6	63	36·3°	145·6°
5. Nandewar–Orange	15.5	63	31·4°	149·2°
6. Cooroy–Comboyne	22.3	63	28·4°	152·6°
7. Fraser Island–Focal Peak	28	63	28·4°	152·6°
8. Mount Juillerat	16.2	63	28·4°	153·0°
Earlier basaltic volcanoes (Sutherland, 1981, p. 197)				
9. Ipswich	56	90	27·5°	152·7°
10. Mount Barrington	53	90	32·0°	151·5°
11. Mittagong	55	90	34·5°	150·2°
12. Monaro	53	90	36·5°	149·3°
East Antarctica				
Cenozoic volcanoes, mostly alkaline basalt (ages from Armstrong, 1978)				
1. South of Cape Hallet	6	80	72·4°	170·3°
2. Mount Overlord	8	80	73·2°	164·5°
3. Mount Melbourne	0	80	74·3°	164·7°
4. Mount Bird, Ross Island	4	80	77·3°	166·7°
5. McMurdo Sound	14	80	78·3°	163·5°
6. Mount Early	0	80	87	– 155.0°

1986). Late Cretaceous extension is also suggested by the Late Cretaceous age of the oldest sediments in the Great South Basin of New Zealand (Grindley & Davey, 1982), which was only 500 km distant at that time.

Cenozoic volcanism in the Transantarctic Mountains and adjacent areas (Fig. 1) is generally much younger than the eastern Australian volcanism, ranging from present day to about 13 Ma; older K–Ar ages have been interpreted as being unreliable (Armstrong, 1978).

Smith & Drewry (1984) showed that the Cenozoic volcanism in East Antarctica seemed to occur in areas that would have overridden the former position of an extensional margin in the hotspots or South America-fixed reference frames. To apply the hotline model in detail, we need to know the age and position of extensional continental margins in East Antarctica.

Thus we tracked the volcanic sites in East Antarctica back in time to 80 Ma (1–6, Fig. 1). All tracked across the extensional basins and then headed toward the present-day continental margin. Analogy with the younger Australian volcanism suggests that the Eastern basin and other areas at the edge of the Transantarctic Mountains could represent the sites of Late Cretaceous hotlines; they are somewhat more diffuse than those beneath the Coral Sea, as they are now under the present areas of active East Antarctic volcanism.

Discussion

So far we have demonstrated that the difference between the age at which the trace of a volcanic site in the hotspots reference frame reaches the present-day continental margin, and the time when that margin was beginning to form, is small or non-existent. We speculate that volcanism at all Cenozoic volcanic sites discussed here was triggered by thermal anomalies in the asthenosphere which were associated with, and may have been the causes of, continental break-up.

Whether these anomalies reflect persistent hotspot activity in the lower mantle which also affected the asthenosphere throughout the period, or whether they were short-lived lower mantle events associated with break-up that have been decaying ever since, is not clear. In the case of eastern Australia, the presumed Coral Sea hotline must have been a heat source that has persisted for at least 60 m.y. and has affected the asthenosphere for that period. The abundant Recent volcanism in southern Victoria in Australia suggests that the hotspot activity persisted well beyond the period of break-up. Because of the apparently small time lag between volcanism and the arrival of a hotline, it follows that hotlines or their remnants should still lie under the active or Recent volcanic fields of East Antarctica and southern Australia, and should be detectable seismically.

Whether a hotline is also a significant material source for the volcanism above it, or whether it simply contributes additional heat to the lithosphere, with the volcanism being created mostly as a result of partial melting of the lithosphere, is an unanswered question.

Acknowledgements

Part of this work was completed under NERC grant GR3/4405. AGS thanks Victoria University, Wellington, New Zealand, for financial support while visiting the Research School of Earth Sciences. This is Cambridge Earth Sciences contribution 1273.

References

Armstrong, R.L. (1978). K–Ar dating: Late Cenozoic McMurdo Volcanic Group and dry valley glacial history, Victoria Land, Antarctica. *New Zealand Journal of Geology and Geophysics*, **23(6)**, 685–98.

Cochran, J.R. (1981). The Gulf of Aden: structure and evolution of a young ocean basin and continental margin. *Journal of Geophysical Research*, **86**, 263–87.

Duncan, R.A. (1981). Hotspots in the Southern Oceans: an absolute frame of reference for the motion of the Gondwana continents. *Tectonophysics*, **74**, 29–42.

Fitzgerald, P.G., Sandiford, M., Barrett, P.J. & Gleadow, A.J.W. (1986). Asymmetric extension associated with uplift and subsidence in the Transantarctic Mountains and Ross Embayment. *Earth and Planetary Science Letters*, **81**, 67–78.

Grindley, G.W. & Davey, F.J. (1982). The reconstruction of New Zealand, Australia and Antarctica. In *Antarctic Geoscience*, ed. C. Craddock, pp. 15–30. Madison; University of Wisconsin Press.

Harland, W.B., Cox, A.V., Llewellyn, P.G., Pickton, C.A.G., Smith, A.G. & Walters, R. (1982). *A Geologic Time Scale*. Cambridge; Cambridge University Press, 131 pp.

Hartnady, C.J.H. & le Roex, A.P. (1985). Southern Ocean hotspot tracks and the Cenozoic absolute motion of the African, Antarctic and South America plates. *Earth and Planetary Science Letters*, **75(2/3)**, 245–57.

Koenig, M. (1980). Geophysical investigations of the southern continental margin of Australia and Antarctica. PhD thesis, Columbia University, Palisades, New York, 360 pp. (unpublished).

Martin, A.K. & Hartnady, C.J.H. (1986). Plate tectonic development of the southwest Indian Ocean: a revised reconstruction of East Antarctica and Africa. *Journal of Geophysical Research*, **91**, 4767–86.

McKenzie, D.P., Davies, D. & Molnar, P. (1970). Plate tectonics of the Red Sea and East Africa. *Nature, London*, **226**, 243–8.

Morgan, W.J. (1981). Hotspot tracks and the opening of the Atlantic and Indian Oceans. In *The Sea*, vol. 7, ed. C. Emiliani, pp. 443–87. New York; John Wiley & Sons.

Morgan, W.J. (1983). Hotspot tracks and the early rifting of the Atlantic. *Tectonophysics*, **94**, 123–39.

Mutter, J.C., Hegarty, K.A., Cande, S.C. & Weissel, J.K. (1985). Break-up between Australia and Antarctica: a brief review in the light of new data. *Tectonophysics*, **114**, 255–79.

Norton, I.O. & Sclater, J.G. (1979). A model for the evolution of the Indian Ocean and the break-up of Gondwanaland. *Journal of Geophysical Research*, **84**, 6803–30.

Patriat, P. (1983). Evolution du système de dorsale de l'Ocean Indien. Thèse d'Etat, Université de Paris. (unpublished).

Patriat, P. & Courtillot, V. (1984). On the stability of triple junctions and its relation to episodicity in spreading. *Tectonics*, **3**, 317–32.

Patriat, P., Segoufin, J., Goslin, J. & Beuzart, P. (1985). Relative positions of Africa and Antarctica in the Upper Cretaceous: evidence for non-stationary behaviour of fracture zones. *Earth and Planetary Science Letters*, **75(2, 3)**, 204–14.

Pilger, R.H. (1982). The origin of hotspot traces: evidence from Eastern Australia. *Journal of Geophysical Research*, **87**, 1825–34.

Royer, J-Y., Patriat, P., Bergh, H.W. & Scotese, C.R. (1988). Evolution of the Southwest Indian Ridge from the Late Cretaceous (anomaly 34) to the Middle Eocene (anomaly 20). *Tectonophysics*, **155**, 235–60.

Sclater, J.G., Fisher, R.L., Patriat, P., Tapscott, C. & Parsons, B. (1981). Eocene to recent development of the South-west Indian Ridge, a consequence of the evolution of the Indian Ocean Triple Junction. *Geophysical Journal of the Royal Astronomical Society*, **64(3)**, 587–604.

Shaw, R.D. (1978). Sea floor spreading in the Tasman Sea: a Lord Howe Rise–Eastern Australian reconstruction. *Bulletin of the Australian Society of Exploration Geophysicists*, **9(3)**, 75–81.

Smith, A.G. (1982). Later Cenozoic uplift of Africa and SE Australia in a reference frame fixed to South America. *Nature, London*, **296**, 400–4.

Smith, A.G. & Drewry, D.J. (1984). Delayed phase change due to hot asthenosphere causes Transantarctic uplift? *Nature, London*, **309**, 536–8.

Stock, J. & Molnar, P. (1982). Uncertainties in the relative positions of the Australia, Antarctica, Lord Howe, and Pacific Plates since the Late Cretaceous. *Journal of Geophysical Research*, **87(B6)**, 4697–714.

Sutherland, F.K. (1978). Mesozoic–Cainozoic volcanism of Australia. *Tectonophysics*, **48**, 413–27.

Sutherland, F.K. (1981). Migration in relation to possible tectonic and regional controls in eastern Australian volcanism. *Journal of Volcanology and Geothermal Research*, **9**, 181–213.

Sutherland, F.K. (1983). Timing, trace and origin of basaltic migration in eastern Australia. *Nature, London*, **305**, 123–6.

Veevers, J.J. (1986). Breakup of Australia and Antarctica estimated as mid-Cretaceous (95 ± 5 Ma) from magnetic and seismic data at the continental margin. *Earth and Planetary Science Letters*, **77(1)**, 91–9.

Weissel, J.K. & Hayes, D.E. (1977). Evolution of the Tasman Sea reappraised. *Earth and Planetary Sciences Letters*, **36(1)**, 77–84.

Weissel, J.K., Hayes, D.E. & Herron, E.M. (1977). Plate tectonic synthesis: the displacements between Australia, New Zealand and Antarctica since the Late Cretaceous. *Marine Geology*, **25**, 231–77.

Weissel, J.K. & Watts, A.B. (1979). Tectonic evolution of the Coral Sea Basin. *Journal of Geophysical Research*, **84**, 4572–82.

Wellman, P. & McDougall, I. (1974). Cenozoic igneous activity in eastern Australia. *Tectonophysics*, **23**, 1–9.

White, R.S. & McKenzie, D.P. (1989). Magmatism at rift zones: the generation of volcanic continental margins and flood basalts. *Journal of Geophysical Research*, **94**, 7685–730.

Evolution of Cenozoic palaeoenvironments

A review of the Cenozoic stratigraphy and palaeontology of Antarctica

P.N. WEBB

Department of Geology and Mineralogy and Byrd Polar Research Center, The Ohio State University, Columbus, Ohio 43210, USA

Abstract

The distribution and tectonic settings of the major Cenozoic basins have been established at reconnaissance levels using a combination of geological and geophysical data. To date, geological research has emphasized both relatively shallow-water continental shelf and abyssal basin environments. Abyssal sediments include claystones, diatomaceous and calcareous nannoplankton oozes, cherts and chalks. Continental shelf sediments are dominated by biogenic and clastic sediments, including glaciomarine diamictites. Shelf sediments are differentiated into packets of contrasting clastic sediments, often separated by geographically widespread erosional contacts. Sedimentary cycles, representing marine transgressions and regressions, are apparent in many shelf successions and result from regional tectonics (uplift–subsidence), sea-level fluctuation (possibly related to glacial–interglacial events), or some combination of the two. Glaciation in the marine shelf basins of the Ross Sea dates from at least the early Oligocene, but there is evidence from King George Island (South Shetland Islands) for glaciation in the early Palaeogene. Polar abyssal basins first show evidence for glaciation in the Neogene. The continental shelf marine record indicates a complex glacial–interglacial record on Antarctica during the past 36 million years. From the inception of glaciation, marine incursions deep into and across Antarctica are probably associated with periods of ice-sheet recession. The palaeontologic data base has undergone a radical improvement over the past two decades and now contributes to the solution of a wide range of biostratigraphic, palaeontologic, biogeographic, palaeoceanographic and evolutionary studies within and beyond the south polar basins. Diatoms, silicoflagellates, and benthic foraminifera have been used in the development of biostratigraphy. Because of the dynamic character of south polar Cenozoic climate, and the differing regional tectonic histories of individual basins, it is likely that a combination of circum-Antarctic, pan-Antarctic and basin-specific zonations will be necessary to date and correlate sediments and events within and between polar basins and to assess the impact of Antarctic Cenozoic climatic fluctuations on the geological record at hemispheric and global scales.

Introduction

Although Cenozoic investigations were initiated in Antarctica at the turn of the century (Andersson, 1906), comprehensive research efforts commenced less than 30 years ago. This review draws on published data from outcrops and stratigraphic drillholes. Cenozoic rocks crop out in the James Ross Island basin area (Seymour Island, Cockburn Island) and the South Shetland Islands (King George Island) on the Atlantic and Pacific Ocean sides of the Antarctic Peninsula, respectively (Fig. 1). Scattered outcrops are also distributed through the Transantarctic Mountains region (Fig. 1).

Antarctic Cenozoic stratigraphy, palaeontology, basin histories and palaeoclimate studies have been greatly advanced by the completion of 48 drillholes between 1972 and 1988 (Figs 1 & 2). In the Ross Sea and Transantarctic Mountains, drillholes include DSDP 270–273, DVDP 4 & 8–11, MSSTS-1, CIROS-1 & 2 (Figs 1 & 2). Six sites were drilled in the south-east Indian Ocean–south-west Pacific Ocean by DSDP Leg 28 (Sites 265–269, 274), and four sites (DSDP 322–326) drilled in the south-east Pacific Ocean by Leg 35 (Figs 1 & 2). Since early 1987, 25 sites have been drilled at other high latitude locations: ODP Leg 113 (Sites 689–697; Maud Rise, off Queen Maud Land, central Weddell Sea and the South Orkney region); ODP Leg 119 (Sites 736–746; Prydz Bay, Kerguelen Plateau); and ODP Leg 120 (Sites 747–751; Kerguelen Plateau) (Figs 1 & 2).

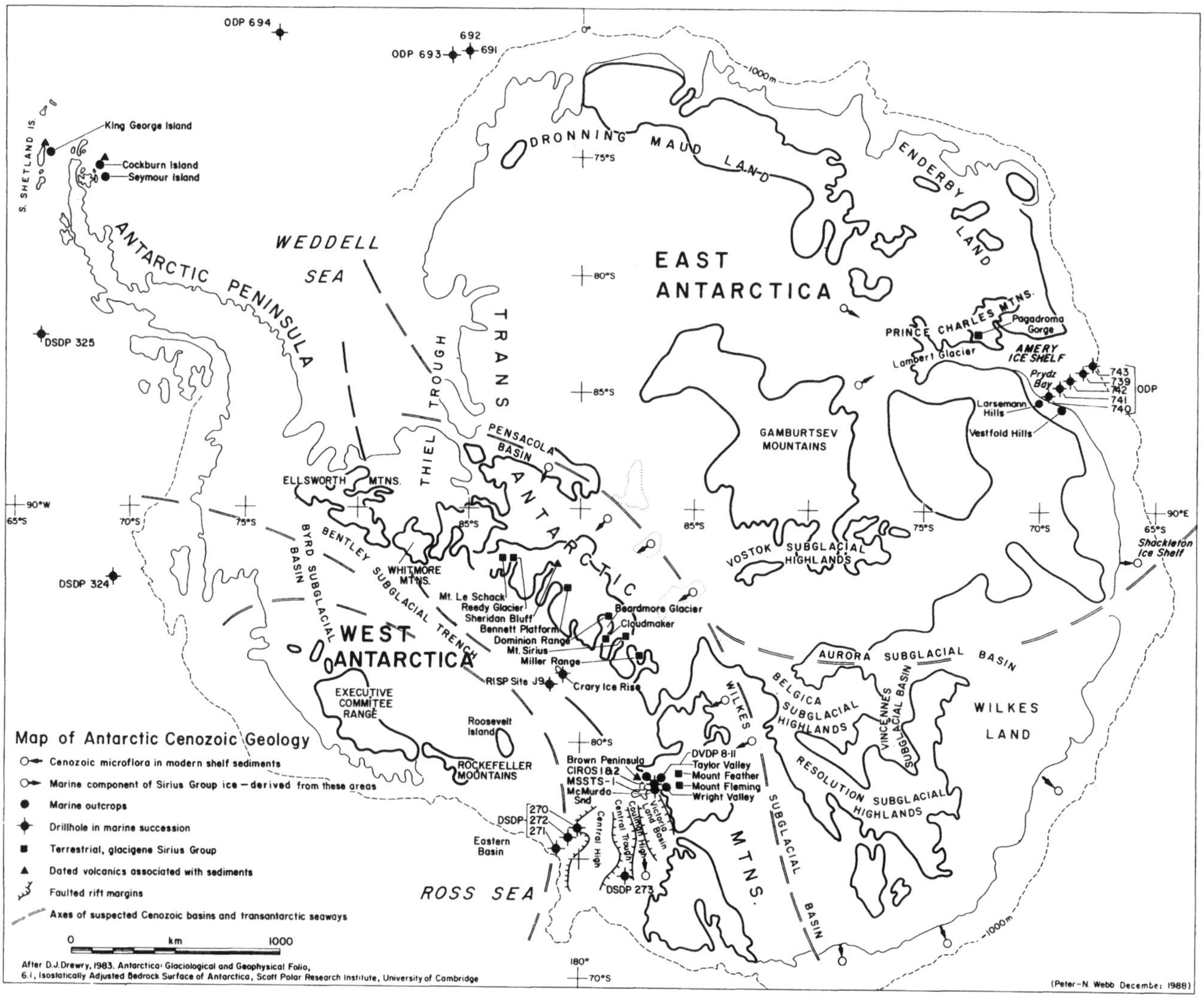

Fig. 1. Map showing location of Cenozoic rocks documented in Antarctica and from the surrounding ocean areas.

Other Southern Hemisphere deep sea drillsites are also shown in Fig. 2.

Ice-transported Cenozoic microfossils recovered at the continental margin point to the existence of major interior basins beneath the present Antarctic ice sheets (Fig. 1).

Structural settings and histories: impacts on Cenozoic basin distribution and development

The influences of structural–tectonic events on basin distribution and development are still poorly understood. Lateral and vertical crustal displacements controlled or influenced the location of basin margins, connections between basins, migration of depocentres, terrestrial palaeotopography and sediment sources, ice nucleation centres and reservoirs, transgressive and regressive sedimentary patterns, terrestrial and marine biotic migration routes and marine connections to the Southern Ocean. Early Cenozoic basin histories were strongly influenced by structural fabrics inherited from Jurassic and Cretaceous break-up and dispersal of the Gondwana continents and associated microcontinents. Some Cenozoic crustal displacements represent a continuation of late Mesozoic activity, whereas others originated during the Cenozoic. Mesozoic and younger tectonic events having an important impact on Cenozoic basin development include: the mid-Cretaceous and later separation of Australia, New Zealand and Antarctica (Laird, 1981; Eittreim, Hampton & Childs, 1985; J.D. Bradshaw, this volume, p. 581), possible Mesozoic extension within the Wilkes Basin (Steed & Drewry, 1982; J.D. Bradshaw, this volume, p. 581), Cretaceous and younger extension and rifting in the Ross Sea (Cooper & Davey, 1985; Cooper, Davey & Hinz this volume, p. 285), Mesozoic transcurrent motion and extension in West Antarctica (Storey, this volume, p. 587), and 5 km of uplift in the Transantarctic Mountains since the Eocene (Fitzgerald *et al.*, 1986).

Sedimentary basins

The distribution of pre-Cenozoic and Cenozoic basins and thicknesses of strata are discussed by Cameron (1981), Behrendt (1983a, b), St John (1986), Cooper & Davey (1985), Cooper *et al.* (this volume, p. 285), and Cooper, Davey & Behrendt (1987), with basin delineation based on topographic, geophysical and geological evidence. Sedimentary basins have

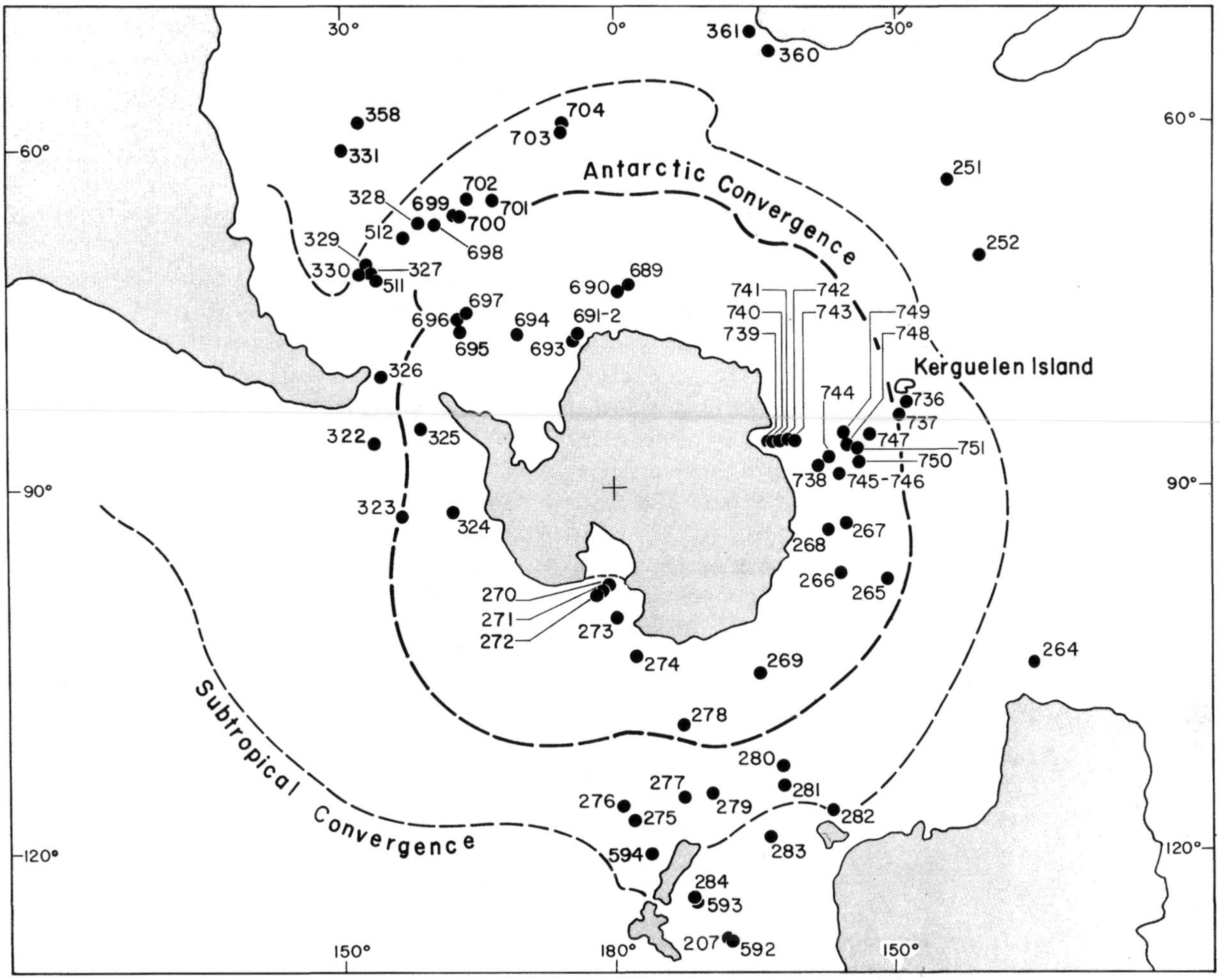

Fig. 2. Map showing location of Cenozoic successions recovered at Deep Sea Drilling Project and Ocean Drilling Program sites in the southern high and temperature latitudes.

also been characterized in terms of tectonic settings; fore-arc and back-arc basins (Gonzaléz Férran, 1985), intra-cratonic basins (Steed & Drewry, 1982; Webb *et al.*, 1986; Cooper, Davey & Hinz, 1988), epi-cratonic basins or embayments (Eittreim *et al.*, 1985; Stagg, 1985; Eittreim & Smith, 1987) and possible failed rifts or aulacogens (Behrendt, 1983a, b; Stagg, 1985).

Stratigraphy

Palaeogene

Parts of the Palaeocene and Eocene are represented on Seymour Island by an 800 m clastic succession (upper part of the López de Bertodano, Sobral and La Meseta formations) (Fig. 1) (Zinsmeister, 1982a; Sadler, 1988). Sadler (1988) emphasized the tabular, lenticular and sedimentary packet nature of this succession, recognized 19 discrete lithologic subunits, and noted that units are characterized by unconformities, fossil lags and glauconite horizons. The presence of sedimentary cycles and erosional episodes of varying temporal extent are clearly implied and a relationship with global sea-level cycles suggested. No pre-Quaternary glacial environments are recognized in the Seymour Island Palaeogene successions.

Birkenmajer (this volume, p. 629) describes a Palaeogene succession from King George Island (Fig. 1) on the Pacific side of the Antarctic Peninsula that is coeval with that at Seymour Island, but has a markedly different character. Here, the succession consists of alternating glaciomarine, volcaniclastic and volcanic rocks. At the base of the succession, glaciomarine sediments are capped by lavas, dated radiometrically at 49 ± 5 Ma and suggesting at least local glaciation (Krakow Glaciation) in the Palaeocene–middle Eocene and possibly in the Late Cretaceous. Higher in the succession the glacial marine Polonez Cove Formation is overlain by the Legru Bay Group, from which radiometric ages of 29.5 ± 2.1 Ma and 25.7 ± 1.3 Ma have been obtained, suggesting a middle or late Oligocene age for the Polonez Cove Glaciation. Lavas associated with the plant-bearing Mount Wawel Formation (interglacial) yielded an uppermost Oligocene age of 24.5 ± 0.5 Ma.

In situ Palaeogene successions of the Ross Sea basins are documented in drillsites DSDP 270, (Hayes *et al.*, 1975), MSSTS-1 (Barrett, 1986; Barrett *et al.*, 1987) and CIROS-1 (Barrett, 1989; Barrett, Hambrey & Robinson, this volume p. 651; Harwood, 1989; Webb, 1989) (Fig. 1). Oligocene sediments are present in all three successions. Recycled microfossils in these sediments point to the existence of Palaeocene and Eocene sediments in the as yet undrilled deeper parts of the

Eastern and Victoria Land basins. CIROS-1 provides the thickest Oligocene sequence (600 m) and consists of well differentiated lithological units, including conglomerate, diamictite, sandstone and mudstone (Barrett *et al.*, this volume, p. 651). It is not apparent whether the multiple glacial events in these early and late Oligocene marine successions represent the fluctuations of local alpine ice from the adjacent Transantarctic Mountains or continental-scale ice masses. The recent recovery of early Oligocene marine sediments with evidence of glacial origins and influences, off Dronning Maud Land and Prydz Bay, points to widespread glaciation during this part of the Palaeogene (Figs 1 & 2).

Deep-water DSDP Sites 267–269 and 274 (south-east Indian–south-west Pacific oceans (Fig. 1), provide successions of Eocene or Oligocene nanno ooze, chalk, and silty claystone (Hayes *et al.*, 1975); in addition a Palaeocene claystone is reported at Site 323 (south-east Pacific Ocean) (Hollister *et al.*, 1976). At none of these deep-water sites is there compelling evidence for glacial marine deposition or glacial conditions on adjacent terrestrial Antarctica in the Palaeogene. At these locations and latitudes glacial phenomena appear in the Miocene (Frakes, 1975; Kemp, Frakes & Hayes, 1975).

Palaeogene successions were recovered during recent Ocean Drilling Program activities in the Weddell Sea (Shipboard Scientific Party, 1987a, b), the Kerguelen Plateau (Shipboard Scientific Party, 1988a, b & c), and off Prydz Bay on the continental margin of East Antarctica (Larsen *et al.*, 1988; Shipboard Scientific Party, 1988a,b) (Figs 1 & 2).

Evidence from the Weddell Sea sector has been interpreted as indicating the existence of relatively warm marine conditions through the Palaeocene and Eocene. On the Maud Rise (Sites 689, 690; ~ 65° S, 0° E) a calcareous–siliceous biogenic sediment transition occurs near the Eocene–Oligocene boundary (37 Ma). This evidence and changes in clay composition were used to interpret strong cooling and/or enhanced aridity in East Antarctica at this time. There is little ice-rafted detritus in the late Oligocene sediments of Maud Rise. Ice-rafted detritus at Site 693 (off Dronning Maud Land) indicates that there were ice accumulations on East Antarctica in the early Oligocene (Figs 1 & 2).

Two of the five sites drilled into the Antarctic continental shelf to the north of Prydz Bay (Figs 1 & 2) penetrated Palaeogene successions. Diatom ages obtained from diamictites at Site 739 indicate that glaciation in this part of Antarctica was in progress during the earliest Oligocene. Older glacial diamictites at Sites 739 and 742 have been assigned a maximum age of late middle Eocene on the basis of magnetostratigraphic and palaeontological data.

Palaeogene successions encountered in the southern Kerguelen Plateau (Fig. 2) include parts of the Palaeocene, Eocene and Oligocene and are dominated by nannofossil chalk and ooze and less abundant chert and porcellanite. The Shipboard Party observed ice-rafted debris in the early Oligocene at Site 747 and noted reports of coeval marine glaciation close to Antarctica off Prydz Bay and Dronning Maud Land.

Recycled Palaeogene diatoms (Harwood, 1986a, this volume, p. 667) and foraminifera (Webb *et al.*, 1984) occur in the terrestrial late Cenozoic Sirius Group of the Transantarctic Mountains (McKelvey *et al.*, this volume, p. 675), suggesting the existence of Palaeogene successions in the intra-cratonic Wilkes and Pensacola basins (Fig. 1).

Neogene

Neogene marine sequences are widely distributed in continental shelf and abyssal settings but rare in outcrop. All Neogene marine sediments have a marine or terrestrial glacial–interglacial origin.

No Miocene successions are known from the Seymour Islands area. At Cockburn Island (Fig. 1), the thin fossil-rich Pecten Conglomerate overlies James Ross Island Volcanic Group rocks, radiometrically dated at 3.65 Ma (Zinsmeister & Webb, 1982; Webb & Andreasen, 1986). At King George Island, the glaciomarine Cape Melville Formation (Cape Melville Glaciation) and volcanic Destruction Bay Formation are cut by dykes radiometrically dated about 20 Ma (early Miocene) (Birkenmajer, this volume, p. 629).

In the Ross Sea Eastern Basin a composite succession provided by DSDP 270 and 272 consists of approximately 0.775 m of lower and middle Miocene silty claystone and silty clay diatomite (Barrett, 1975; Hayes *et al.*, 1975; Leckie & Webb, 1983, 1986; Savage & Ciesielski, 1983). This succession has been subdivided into 10 informal lithological units (Hayes *et al.*, 1975). Multiple glacial events and intervening hiatuses are evident in this succession. Early and middle Miocene sediments are also present in the Central Trough (DSDP 273, Hayes *et al.*, 1975) and the western margins of the Victoria Land Basin (MSSTS-1, Barrett, 1986, Barrett *et al.*, 1987; CIROS-1, Barrett, 1989, Barrett *et al.*, this volume, p. 651). The late Miocene is poorly known in the Ross Sea region. An *in situ* 90 m succession of late Miocene glaciomarine conglomerate, breccia, sandstone, mudstone and diamictite is present in drillhole DVDP 10 and 11, Taylor Valley (Ishman & Webb, 1988) (Fig. 1). Recycled Miocene clasts (including late Miocene sediment) with datable floras and faunas are known in Wright Valley (Brady, 1979; Burckle, Prentice & Denton, 1986), Ross Ice Shelf Project Site J9 (Harwood *et al.*, 1989) and Crary Ice Rise (Scherer *et al.*, 1989). Grounded Pliocene and/or Pleistocene ice sheets appear to have eroded Miocene sediments over wide areas of the Ross Sector, leaving a distinctive unconformity that is well portrayed in offshore seismic traverses.

The Pliocene is represented by the Pecten Gravels of Wright Valley (Webb, 1972, 1974; Burckle *et al.*, 1986), Scallop Hill Formation of southern McMurdo Sound (Speden, 1962; Eggers, 1979; Leckie & Webb, 1979), sediments in DVDP 10 and 11, Taylor Valley (Ishman & Webb, 1988) and CIROS-2 (Barrett *et al.*, 1985) in Ferrar Fjord. Most Pliocene marine sediments are relatively shallow-water bioclastic and/or sandy deposits. Radiometric dates of volcanic material within the Scallop Hill Formation provide an age of 2.62 Ma (late Pliocene) (Webb & Andreasen, 1986).

Pliocene shallow-water coastal marine biofacies are now reported from Cockburn Island, southern McMurdo Sound, the Vestfold Hills (Pickard *et al.*, 1986, 1988) and Larsemann Hills (P.G. Quilty, pers. comm., 1987). D.M. Harwood (pers.

comm., 1987) reports well-preserved Pliocene diatoms in a Sirius Group-like till exposed at Pagadroma Gorge in the northern Prince Charles Mountains, close to the Lambert Glacier region of East Antarctica (McKelvey, 1988) (Fig. 1). These occurrences are taken to indicate that, at many widespread locations around the coastal periphery of Antarctica, there was a significant climatic amelioration, ice-sheet recession and southern penetration of marine waters on one or more occasions during the Pliocene.

The terrestrial 'upper Pliocene–lower Pleistocene' Sirius Group is distributed at high elevation along much of the Transantarctic Mountains (Fig. 1). This unit consists mostly of diamictite, with locally occurring fluvial and lacustrine sediments (Webb *et al.*, 1984; Webb *et al.*, 1986; McKelvey *et al.*, this volume, p. 675). Recycled clasts and matrix sediments within the Sirius Group contain Palaeogene and Neogene siliceous and other microfossils that are thought to be derived from marine successions within the Pensacola and Wilkes basins (Fig. 1). Harwood (1983, 1986a, this volume, p. 667) places a mid–late Pliocene age on the youngest marine diatom floras present in some clasts, suggesting that the terrestrial Sirius Group is possibly latest Pliocene–early Pleistocene in age. *In situ* and near-*in situ* tree stems (Carlquist, 1987) and pollen have been recovered from the Sirius Group of the Dominion Range area of the Transantarctic Mountains, providing additional supporting evidence for a major amelioration of polar glacial climate in the Pliocene–early Pleistocene (Webb & Harwood, 1987). During the Pliocene, major seaways probably extended across large areas of East Antarctica (Webb *et al.*, 1984).

Deep-water DSDP Sites 265–269 and 274 (south-east Indian Ocean–south-west Pacific Ocean) provide Neogene successions that are generally 300 m in thickness and consist of fine-grained sediments dominated by varying proportions of clay, silt, diatoms and calcareous nannoplankton and chalk. Hayes *et al.* (1975) suggested that the upward stratigraphic transition from calcareous to siliceous sediments exhibits pronounced latitudinal diachronism. At the southern sites (267, 268) this transition occurred between the late Oligocene and early Miocene, whereas at the northern site (266) the transition did not occur until the late Miocene. The earliest ice-rafting is evident from the early Miocene in the southernmost site (268) and from the late Miocene in the northernmost site (266). It was noted above that the first appearance of ice-rafting over the Kerguelen Plateau area was during the earliest Oligocene.

Massive and stratified Neogene glacial diamictites, and diamictons, occur at all five sites (739–743) in the prograding continental shelf succession off Prydz Bay (Shipboard Scientific Party, 1988a, b). However, in the southern Kerguelen Plateau (latitudes ~ 54–59° S), Neogene successions at several sites include nannofossil or diatom ooze and subordinate foraminiferal ooze (Shipboard Scientific Party, 1988c).

Neogene sediments are also well developed in the south-east Pacific DSDP Sites 322 and 323 (Bellingshausen Abyssal Plain) and Sites 324 and 325 (Bellingshausen Sea continental rise) (Hollister *et al.*, 1976). Neogene successions in excess of 400, 500 and 700 m are reported from Sites 322, 323 and 325, respectively. Dominant sediment types include sandstone, siltstone, claystone, diatom ooze and diatomite. Ice-rafted debris is scarce at abyssal sites but common at upper continental rise sites (Hollister *et al.*, 1976). The oldest ice-rafted debris is late Miocene at Site 322, probably middle Miocene at Site 323 and early Miocene at Site 325. Hollister *et al.* (1976) suggest that glaciation on adjacent Antarctica was weak in the early Miocene, moderate by the middle Miocene and extensive by the late middle Miocene. This record contrasts markedly with published data from the King George Island area of the south-east Pacific, where multiple glaciation was apparently a feature of the Palaeogene and Neogene (Birkenmajer, this volume, p. 629).

Palaeontology

Shallow-marine lithofacies, particularly in the Antarctic Peninsula Palaeogene, provide a wealth of macrofossils, including vertebrates. Deeper shelf, continental slope and abyssal facies are the domain of rich microfossil assemblages, particularly diatoms, silicoflagellates and benthic foraminifera. Useful regional and site-specific review papers are provided by Zinsmeister (1979, 1982a, b) and Zinsmeister & Camacho (1982), macrofossils; Wrenn & Hart (1988), dinocysts; McCollum (1975), Schrader (1976), Weaver & Gombos (1981) and Harwood (1986a, this volume, p. 667), diatoms; Ciesielski (1975) and Haq & Riley (1976), silicoflagellates; Burns (1975) and Haq (1976), calcareous nannoplankton; Rögl (1976) and Webb (1988), foraminifera.

Biostratigraphy

The relative wealth of macrofossil material recovered from Palaeocene–Eocene and Pliocene–Pleistocene sediments has proven to be of little use biostratigraphically and dating resolution provided by these groups is seldom better than at epoch or subepoch level. The utility of macrofossil groups lies mostly in their application to biogeographic, palaeoceanographic and palaeoecologic studies. Microfossil groups, particularly diatoms, silicoflagellates and benthic foraminifera, are widely distributed both stratigraphically and geographically and occur in a variety of shallow–deep-water facies. Diatoms exhibit varying degrees of similarity to temperate latitude floras, thereby allowing their application to subpolar and polar correlation, and biogeographic and palaeoceanographic studies (Harwood, this volume, p. 667).

In developing microfossil schemes for the south polar basins, we have had to contend with the following phenomena: mixed endemic–cosmopolitan assemblages; diachronous and spatial variations of some taxa, resulting from a continuously evolving ocean-basin geometry; and rapid and gradual climate variations produced as a response to glacial fluctuations in terrestrial Antarctica. We have also had to contend with a paucity of data points (stratigraphic successions), widespread and local hiatuses in the stratigraphic record, incomplete knowledge of true species ranges, differing uses of taxonomy and microfossil reworking.

As more data accumulate from polar basins it has become apparent that many Cenozoic macrofossil and microfossil groups evolved within a southern high latitude circumpolar biogeographic realm. The terms *Paleoaustral*, *Austral* and *Neoaustral* have been utilized in recognizing these Mesozoic and Cenozoic biogeographic elements of polar origin in temperate New Zealand faunas and floras (Fleming, 1979). As knowledge of transantarctic and circumantarctic basins improves, we might expect to see an embellishment of the concept of a distinctive south polar biogeography, with the recognition of subrealm or provincial nomenclature in individual fossil groups. Zinsmeister (1982b) and Zinsmeister & Feldmann (1987) use this approach with the macrofossil Weddellian Province. Because of marked climate fluctuations during the Cenozoic, boundaries of southern biogeographic realms and provinces exhibit continuous geographic change, both within and beyond Antarctica. Faunal and floral reorganizations in the southern high latitudes play a crucial role in the development of polar basin biostratigraphy in many fossil groups. Complex interactions between polar Austral and temperate faunal elements also play a crucial role, and biostratigraphic schemes are unlikely to be explained in terms of simple expansion and contraction of a circumpolar antarctic–subantarctic biogeographic halo. It is against this dynamic biogeographic backdrop that polar biostratigraphy has been and will be developed.

Existing polar microfossil zonations employ standard biostratigraphic procedures; first and last appearance, first and last abundant occurrence, total range, acme range, intermittent range, faunal turnover (origination and extinction of assemblages) etc. Benthic foraminifera and diatom zonations developed in inner continental shelf settings are strongly influenced by local climate and basin events and may not have application beyond a particular basin (Leckie & Webb, 1983, 1986; Harwood, 1986a, b, this volume, p. 667; Webb, 1988). Zonations from deeper shelf and polar abyssal settings, such as those developed for pelagic diatoms and silicoflagellates, are more likely to have a wide circumantarctic utility (Weaver & Gombos, 1981; Gombos & Ciesielski, 1983; Savage and Ciesielski, 1983; Harwood, this volume, p. 667). Two phenomena provide for special difficulties in developing south polar basin biostratigraphy; diachronous faunal–floral biostratigraphic boundaries which result from dynamic fluctuations in Cenozoic marine climate, and extended stratigraphic ranges resulting from microfossil recycling of faunas and floras. The latter phenomenon is an ever-present problem in glaciomarine successions.

Harwood (1986a, b, this volume, p. 667) demonstrated that a Cenozoic diatom biostratigraphy is possible for latitudes higher than 70° S and that such a zonation can be built upon a combination of subantarctic and endemic Antarctic floras. South polar basin diatom biostratigraphy will eventually be measured in terms of several million years per zone. The southern ocean (subantarctic) diatom zonation of Weaver & Gombos (1981) provides for 16 Neogene zones in 24 million years, a resolution of 1.5 m.y./zone. This is about the same level of resolution provided by calcareous nannoplankton and foraminifera in tropical and temperate latitudes.

Planktonic foraminifera of the high-latitude Cenozoic occur in very low diversities and have little biostratigraphic utility (Webb, 1988). Benthic foraminifera are useful, but a high level of biostratigraphic resolution is not anticipated. For example, Webb (1988) recognized eight benthic subdivisions for the late Oligocene (< 30 Ma) to Recent in the Ross Sea basins, a resolution of about 3.5–4 m.y./zone. Many of these broad faunal subdivisions will probably prove to be useful beyond the Ross Sea area. Finer biostratigraphic subdivision, such as the assemblage zones erected by Leckie & Webb (1983, 1986) and Webb, Leckie & Ward (1986), entail shorter spans of time, but because they represent responses to local marine glacial climate, probably have little utility beyond the immediate sedimentary basins.

Concluding remarks

1. Although the location and tectonic setting of the major Cenozoic polar basins can be established in broad terms, precise knowledge of distribution and geometry is poorly constrained. Studies have emphasized two quite different basin settings; relatively complex shallow-water marine envirionments, and abyssal environments. The Victoria Land Basin, in the western Ross Sea, is probably the most thoroughly studied at this time.

2. Although developed under contrasting climatic and structural regimes, Palaeocene–Eocene successions of the inner Weddell Embayment (James Ross Island basin) and Oligocene–Recent successions of the Ross Sea basins provide the most complete stratigraphic coverage for the Cenozoic of Antarctica.

3. Maximum sampled total thickness for continental shelf Cenozoic successions is only about 2350 m: 805 m (early Palaeogene, Seymour Island, Weddell embayment); 600 m (late Palaeogene, Victoria Land Basin, Ross Sea) and 750 m (Neogene, Eastern Basin, Ross Sea).

4. Inner shelf successions consist of a variety of clastic sediments: breccias, conglomerates, sandstones, siltstones, claystones, diatomites and diamictites. Carbonate rocks are not present, although there are rare horizons of carbonate-cemented sediments. Successions of all ages exhibit strong lithological differentiation with widespread intervening disconformities a common occurrence. Sedimentary cycles, representing marine transgressions and regressions, are clearly developed and result from regional tectonics (uplift–subsidence), sea level flutuations (possibly related to glacial–interglacial events), or some combination of the two.

5. Antarctic terrestrial glaciation and its expression in the adjacent marine basins dates from the Oligocene. In some areas, such as the South Shetland Islands and Prydz Bay, it may be an even earlier phenomenon. Multiple glaciation, i.e. alternation of glacial and interglacial events, is evident in the marine continental shelf record and points to dynamic development and decline of continental–terrestrial ice. At different times during the Cenozoic, the Antarctic continent was enveloped by alpine ice, regional ice-caps and pan-Antarctic ice-sheets. During regional and continent-wide ice-recession phases, marine waters penetrated deep into, and probably across, parts of the Antarctic continent. The effects of transcontinental marine circulation (Webb, 1979, 1981; Webb *et al.*,

1984; Harwood, 1986a, b, this volume, p. 667) have not as yet been successfully integrated into hemispheric and global climate and ocean circulation, sea-level fluctuation and sedimentary basin cyclicity–episodicity models.

6. Microfossil zonations are being developed for several south polar basins. Because of the dynamic nature of Cenozoic climate and regional tectonic histories, it seems likely that a combination of circum-Antarctic, pan-Antarctic and basin-specific zonations will be necessary to solve south polar basin problems. An ability to relate and resolve events within and beyond Antarctica and its continental shelf basins is of paramount importance. In this connection, the accurate calibration of polar zonations against magnetic stratigraphy and radiometric scales is essential. This can only be attained by future drilling of onshore and offshore stratigraphic successions.

Acknowledgements

This preparation of this review was supported by grants from the Division of Polar Programs (DPP/NSF 8315553 and 8420622). Alan Cooper and David Harwood reviewed drafts of the manuscript.

References

Andersson, R.J. (1906). On the geology of Graham Land. *Bulletin of the Geological Institution*, Upsala, **7**, 19–71.

Barrett, P.J. (1975). Textural characteristics of Cenozoic pre-glacial and glacial sediments at Site 270, Ross Sea, Antarctica. *Initial Reports of the Deep Sea Drilling Project*, 28, pp. 757–67. Washington, DC; US Government Printer.

Barrett, P.J. (ed.) (1986). *Antarctic Cenozoic history from the MSSTS-1 drillhole, McMurdo Sound.* Wellington; New Zealand DSIR Bulletin, No. 237, 171 pp.

Barrett, P.J. (ed.) (1989). *Antarctic Cenozoic History from the CIROS-1 Drillhole, Mc Murdo Sound.* Wellington; New Zealand DSIR Bulletin, No. 245, 254 pp.

Barrett, P.J., *et al.* (1985). Plio-Pleistocene glacial sequence cored at CIROS-2, Ferrar Fjord, western McMurdo Sound. *New Zealand Antarctic Record*, **6(2)**, 8–19.

Barrett, P.J., Elston, D.P., Harwood, D.M., McKelvey, B.C. & Webb, P.-N. (1987). Mid-Cenozoic record of glaciation and sea-level change on the margin of the Victoria Land Basin, Antarctica. *Geology*, **15**, 634–7.

Behrendt, J.C. (1983a). Geological and geophysical studies relevant to assessment of the petroleum resources of Antarctica. In *Antarctic Earth Science*, ed. R.L. Oliver, P.R. James & J.B. Jago, pp. 423–8. Canberra; Australian Academy of Science and Cambridge; Cambridge University Press.

Behrendt, J.C. (1983b). Petroleum and mineral resources of Antarctica. *United States Geological Survey*, **Circular 909**, 75 pp.

Brady, H.T. (1979). A diatom report on DVDP Cores 3, 4a, 12, 14, 15 and other related surface sections. In *Proceedings of Seminar III on Dry Valley Drilling Project, 1978*, ed. T. Nagata, pp. 165–75. Tokyo; National Institute of Polar Research Memoirs, Special Issue 13.

Burckle, L.H., Prentice, M.L. & Denton, G.H. (1986). Neogene Antarctic glacial history: new evidence for marine diatoms in continental deposits *EOS, Transactions of the American Geophysical Union*, **67(16)**, 295.

Burns, D.A. (1975). Nannofossil biostratigraphy for Antarctic sediments, Leg 28, Deep Sea Drilling Project. *Initial Reports of the Deep Sea Drilling Project*, 28, pp. 589–98. Washington, DC; US Government Printing Office.

Cameron, P.J. (1981). The petroleum potential of Antarctica and its continental margin. *Australian Petroleum Association Journal*, **21**, 99–111.

Carlquist, S. (1987). Upper Pliocene–Lower Pleistocene *Nothofagus* wood from the Transantarctic Mountains. *Aliso*, **11**, 571–83.

Ciesielski, P.F. (1975). Biostratigraphy and paleoecology of Neogene and Oligocene silicoflagellates from cores recovered during Antarctic Leg 28, Deep Sea Drilling Project. *Initial Reports of the Deep Sea Drilling Project*, 28, pp. 625–92. Washington, DC; US Government Printing Office.

Cooper, A.K. & Davey, F.J. (1985). Episodic rifting of Phanerozoic rocks in the Victoria Land Basin, Antarctica. *Science*, **229**, 1085–7.

Cooper, A.K., Davey, F.J. & Behrendt, J.C. (1987). Seismic stratigraphy and structure of the Victoria Land Basin, western Ross Sea, Antarctica. In *The Antarctic Continental Margin: Geology and Geophysics of the Western Ross Sea*, ed. A.K. Cooper & F.J. Davey, pp. 27–65. Houston; Circum-Pacific Council for Energy and Mineral Resources Earth Science Series, 5B.

Cooper, A.K., Davey, F.J., Hintz, K. (1988). Ross Sea – geology, hydrocarbon potential. *Oil & Gas Journal*, **86(45)**, 54–8.

Eggers, A.J. (1979). Scallop Hill Formation, Brown Peninsula, McMurdo Sound, Antarctica. *New Zealand Journal of Geology and Geophysics*, **22(3)**, 353–61.

Eittreim, S.L., Hampton, M.A. & Childs, J.R. (1985). Seismic-reflection signature of Cretaceous continental breakup on the Wilkes Land margin, Antarctica. *Science*, **239**, 1082–4.

Eittreim, S.L. & Smith, G.L. (1987). Seismic sequences and their distribution on the Wilkes Land Margin. In *The Antarctic Continental Margin: Geology and Geophysics of Offshore Wilkes Land*, ed. S.L. Eittreim & M.A. Hampton, pp. 15–43. Houston; Circum-Pacific Council for Energy and Mineral Resources Earth Science Series, 5A.

Fitzgerald, P.G., Sandiford, M., Barrett, P.J. & Gleadow, A.J.W. (1986). Asymmetric extension associated with uplift and subsidence in the Transantarctic Mountains and Ross Embayment. *Earth and Planetary Science Letters*, **81**, 67–78.

Fleming, C.A. (1979). *The Geological History of New Zealand and its Life*. Auckland; Auckland University Press. 141 pp.

Frakes, L.A. (1975). Paleoclimatic significance of some sedimentary components at Site 274. *Initial Reports of Deep Sea Drilling Project*, 28, pp. 785–8. Washington, DC; US Government Printing Office.

Gombos, A.M. & Ciesielski, P.F. (1983). Late Eocene to Early Miocene diatoms from the southwest Atlantic. *Initial Reports of the Deep Sea Drilling Project*, 71, pp. 583–634. Washington, DC; US Government Printing Office.

Gonzalés-Férran O. (1985). Volcanic and tectonic evolution of the northern Antarctic Peninsula–Late Cenozoic to Recent. *Tectonophysics*, **114**, 389–409.

Haq, B.U. (1976). Coccoliths in cores from the Bellingshausen Abyssal Plain and Antarctic Continental Rise (DSDP 35). *Initial Reports of the Deep Sea Drilling Project*, 35, pp. 557–68. Washington, DC; US Government Printing Office.

Haq, B.U. & Riley, A. (1976). Antarctic silicoflagellates from the southeast Pacific, Deep Sea Drilling Project Leg 35. *Initial Reports of the Deep Sea Drilling Project*, 35, pp. 673–92. Washington, DC; US Government Printing Office.

Harwood, D.M. (1983). Diatoms from the Sirius Formation, Transantarctic Mountains. *Antarctic Journal of the United States*, **18(5)**, 98–100.

Harwood, D.M. (1986a). Diatom biostratigraphy and paleoecology with a Cenozoic history of antarctic ice sheets. PhD thesis, Columbus, Ohio State University, 592 pp. (unpublished).

Harwood, D.M. (1986b). Diatoms. In *Antarctic Cenozoic History from the MSSTS-1 Drillhole, McMurdo Sound*, ed. P.J. Barrett, pp. 69–107. New Zealand DSIR Bulletin, No. 237.

Harwood, D.M. (1989). Siliceous microfossils. In *Antarctic Cenozoic History from the CIROS-1 Drillhole, McMurdo Sound*, ed. P.J. Barrett, pp. 67–97. New Zealand DSIR Bulletin, No. 245.

Harwood, D.M. & Scherer R.P. (1989). Diatom biostratigraphy and paleoenvironmental significance of reworked Miocene diatomaceous clasts in sediments from RISP Site J9. *Antarctic Journal of the United States*, **23(5)**, 31–4.

Harwood, D.M., Scherer, R.P., Webb, P.-N. (1989). Multiple Miocene productivity events in West Antarctica as recorded in upper Miocene sediments beneath the Ross Ice Shelf (Site J-9). *Marine Micropaleontology*, **15**, 91–115.

Hayes, D.H., Frakes, L.A., *et al.* (1975). *Initial Reports of the Deep Sea Drilling Project*, 28. Washington, DC; US Government Printing Office. 1017 pp.

Hollister, C.D., Craddock, C. *et al.* (1976). *Initial Reports of the Deep Sea Drilling Project*, 35. Washington, DC; US Government Printing Office. 930 pp.

Ishman, S.E. & Webb, P.N. (1988). Late Neogene benthic foraminifera from the Victoria Land Basin margin, Antarctica: application to glacio-eustatic and tectonics events. *Revue de Paléobiologie*, **Special Volume 2, Part 2**, 523–51.

Kemp, E.M., Frakes, L.A. & Hayes, D.E. (1975). Paleoclimatic significance of diachronous biogenic facies, Leg 28, Deep Sea Drilling Project. *Initial Reports of the Deep Sea Drilling Project*, 28, pp. 909–17. Washington, DC; US Government Printing Office.

Laird, M.G. (1981). The Late Mesozoic fragmentation of the New Zealand segment of Antarctica. In *Gondwana Five*, ed. M.M. Creswell & P. Vella, pp. 311–18, Rotterdam; A.A. Balkema.

Larsen, B., Barron, J.A., Baldauf, J.G. & ODP Leg 119 Scientific Staff (1988). Results from high latitude scientific drilling on the Kerguelen Plateau and Prydz Bay East Antarctica: preliminary results of ODP Leg 119. Geological History of the Polar Oceans: Arctic Versus Antarctic, Nato Advanced Research Workshop, Bremen, Abstract.

Leckie, R.M. & Webb, P.-N. (1979). Scallop Hill Formation and associated Pliocene marine deposits of southern McMurdo Sound. *Antarctic Journal of the United States*, **14(5)**, 54–6.

Leckie, R.M. & Webb, P.N. (1983). Late Oligocene–Early Miocene glacial record of the Ross Sea, Antarctica: evidence from DSDP Site 270. *Geology*, **11**, 578–82.

Leckie, R.M. & Webb, P.N. (1986). Late Paleogene and Early Neogene foraminifera of DSDP Site 270, Ross Sea, Antarctica. *Initial Reports of the Deep Sea Drilling Project*, 90, pp. 1093–142. Washington, DC; US Government Printing Office.

McCollum, D. (1975). Diatom stratigraphy of the Southern Ocean. *Initial Reports of the Deep Sea Drilling Project*, 28, pp. 515–71. Washington, DC; US Government Printing Office.

McKelvey, B.C. (1988). Amery Oasis research testing ice sheet history hypothesis. *Australian National Antarctic Research Expeditions Newsletter*, **Number 53**, 8.

Pickard, J., Adamson, D.A., Harwood, D.M., Miller, G.H., Quilty, P.G. & Dell, R.K. (1986). Early Pliocene marine sediments in the Vestfold Hills, East Antarctica: implications for coastline, ice sheet and climate. *South African Journal of Science*, **82**, 520–1.

Pickard, J., Adamson, D.A., Harwood, D.M., Miller, G.H., Quilty, P.G. & Dell, R.K. (1988). Early Pliocene marine sediments coastline and climate of East Antarctica. *Geology*, **16**, 158–61.

Rögl, F. (1976). Late Cretaceous to Recent foraminifera from the southeast Pacific basin, DSDP Leg 35. *Initial Reports of the Deep Sea Drilling Project*, 35, pp. 539–56. Washington, DC; US Government Printing Office.

Sadler, P.M. (1988). Geometry and stratigraphy of Paleocene units on Seymour Island, northern Antarctic Peninsula. In *Geology and Paleontology of Seymour Island, Antarctica*, ed. R.M. Feldmann & M.O. Woodburne, pp. 303–20. Washington, DC; Geological Society of Amerca Memoir 169.

St John, W. (1986). Antarctica–geology and hydrocarbon potential. In *Future Petroleum Provinces of the World*, ed. M.T. Halbouty, pp. 55–100. Tulsa, Oklahoma; American Association of Petroleum Geologists, Memoir 40.

Savage, M.L. & Ciesielski, P.F. (1983). A revised history of glacial sedimentation in the Ross Sea region. In *Antarctic Earth Science*, ed. R.L. Oliver, P.R. James & J.B. Jago, pp. 555–9. Canberra; Australian Academy of Science and Cambridge; Cambridge University Press.

Scherer, R.P., Harwood, D.M., Ishman, S.E. & Webb, P.-N. (1989). Micropaleontological analysis of sediments from the Crary Ice Rise, Ross Ice Shelf. *Antarctic Journal of the United States*, **23(5)**, 34–6.

Schrader, H.S. (1976). Cenozoic marine planktonic diatom biostratigraphy of the South Pacific Ocean. *Initial Reports of the Deep Sea Drilling Project*, 35, pp. 605–71. Washington, DC; US Government Printing Office, 605–671.

Shipboard Scientific Party, Leg 113 (1987a). Glacial history of Antarctica. *Nature*, **328**, 115–16.

Shipboard Scientific Party, Leg 113 (1987b). Leg 113 explores climatic changes. *Geotimes*, **32(7)**, 12–15.

Shipboard Scientific Party, Leg 119 (1988a). Early glaciation of Antarctica. *Nature*, **333**, 303–4.

Shipboard Scientific Party, Leg 119 (1988b). Leg 119 studies climatic history. *Geotimes*, **33(7)**, 14–16.

Shipboard Scientific Party, Leg 120 (1988c). Leg 120 explores origins and history. *Geotimes*, **33(9)**, 12–16.

Speden, I.G. (1962). Fossiliferous Quaternary marine deposits in the McMurdo Sound region, Antarctica. *New Zealand Journal of Geology and Geophysics*, **5**, 746–77.

Stagg, H.M.J. (1985). The structure and origin of Prydz Bay and Mac-Robertson shelf, East Antarctica. *Tectonophysics*, **114**, 315–40.

Steed, R.H.N. & Drewry, D.J. (1982). Radio echo-sounding investigations of Wilkes Land, Antarctica. In *Antarctic Geoscience*, ed. C. Craddock, pp. 969–75. Madison; University of Wisconsin Press.

Weaver, F.M. & Gombos, A.M. (1981). Southern high latitude diatom biostratigraphy. In *DSDP: A Decade of Progress*, ed. J. Warme, R.G. Douglas & E.L. Winterer, pp. 445–70. Tulsa, Oklahoma; Society of Economic Paleontologists and Mineralogists Special Publication 32.

Webb, P.-N (1972). Wright Fjord, Pliocene marine invasion of an Antarctic dry valley. *Antarctic Journal of the United States*, **7**, 222–32.

Webb, P.-N. (1974). Micropaleontology, paleoecology and correlation of the Pecten gravels, Wright Valley, Antarctica and description of *Trochoelphidiella onyxi* n. sp., n. gen. *Journal of Foraminiferal Research*, **4(3)**, 154–99.

Webb, P.-N. (1979). Paleogeographic evolution of the Ross Sector during the Cenozoic. In *Proceedings of Seminar III on Dry Valley Drilling Project, 1978*, ed. T. Nagata, pp. 206–12. Tokyo; National Institute of Polar Research Memoirs, Special Issue 13.

Webb, P.-N. (1981). Late Mesozoic–Cenozoic geology of the Ross Sector, Antarctica. *Journal of the Royal Society of New Zealand*, **11(4)**, 439–46.

Webb, P.-N. (1988). Upper Oligocene–Holocene foraminifera of the Ross Sea region. *Revue de Paléobiologie*, **Special Volume 2, Part 2**, 589–603.

Webb, P.-N. (1989). Benthic foraminifera. In *Antarctic Cenozoic History from the CIROS-1 Drillhole, McMurdo Sound*, ed. P.J. Barrett, pp. 99–118. New Zealand DSIR Bulletin, No. 245.

Webb, P.-N. & Andreasen, J.E. (1986). K/Ar dating of volcanic material associated with the Pliocene Pecten Conglomerate (Cockburn Island) and Scallop Hill Formation (McMurdo Sound). *Antarctic Journal of the United States*, **21(5)**, 59.

Webb, P.-N. & Harwood, D.M. (1987). Terrestrial flora of the Sirius Formation: its significance for Late Cenozoic glacial history. *Antarctic Journal of the United States*, **22(2)**, 7–11.

Webb, P.-N., Harwood, D.M., McKelvey, B.C. & Stott, L.D. (1984). Cenozoic marine sedimentation and ice-volume variation on the East Antarctic craton. *Geology*, **12**, 287–91.

Webb, P.-N., Leckie, R.M. & Ward, B.L. (1986). Foraminifera (Late Oligocene). In *Antarctic Cenozoic History from the MSSTS-1 drillhole, McMurdo Sound*, ed. P.J. Barrett, pp. 115–25. Wellington; New Zealand Department of Scientific and Industrial Research, Miscellaneous Bulletin 237.

Webb, P.-N., McKelvey, B.C., Harwood, D.M., Mabin, M.C.G., Mercer, J.H. (1986). Sirius Formation of the Beardmore Glacier region. *Antarctic Journal of the United States*, **22(1)**, 8–13.

Wrenn, J.H. & Hart, G.F. (1988). Paleogene dinoflagellate cyst biostratigraphy of Seymour Island, Antarctica. In *Geology and Paleontology of Seymour Island, Antarctica*, ed. R.M. Feldmann & M.O. Woodburne, pp. 321–447. Washington, DC; Geological Society of America Memoir 169.

Zinsmeister, W.J. (1979). Biogeographic significance of Late Cretaceous and Early Tertiary molluscan faunas of Seymour Island (Antarctic Peninsula) to the final break-up of Gondwanaland. In *Historical Biogeography, Plate Tectonics and the Changing Environment*, ed. J. Gray & A. Boucot, pp. 349–55. Corvallis; Oregon State University Press.

Zinsmeister, W.J. (1982a). Review of the Upper Cretaceous–Lower Tertiary sequence on Seymour Island, Antarctica. *Journal of the Geological Society, London*, **139**, 779–85.

Zinsmeister, W.J. (1982b). Late Cretaceous–Early Tertiary molluscan biogeography of the southern circum-Pacific. *Journal of Paleontology*, **56(1)**, 84–102.

Zinsmeister, S.J. & Camacho, H.H. (1982). Late Eocene (to possibly earliest Oligocene) molluscan fauna of the La Meseta Formation of Seymour Island, Antarctica. In *Antarctic Geoscience*, ed. C. Craddock, pp. 299–304. Madison; University of Wisconsin Press.

Zinsmeister, W.J. & Feldmann, R.M. (1987). Antarctica: cradle of evolution of Late Cretaceous–Cenozoic marine faunas of the southern hemisphere. *5th International Symposium on Antarctic Earth Sciences, Cambridge, Abstracts*, 138.

Zinsmeister, W.J. & Webb, P.-N. (1982). Cretaceous–Tertiary geology and paleontology of Cockburn Island. *Antarctic Journal of the United States*, **17(5)**, 41–2.

Foraminiferal biogeography of the Late Cretaceous southern high latitudes

B.T. HUBER

Byrd Polar Research Center and Department of Geology & Mineralogy, The Ohio State University, 125 South Oval Mall, Columbus, Ohio 43210, USA
Present address: Department of Paleobiology, NHB–121, Smithsonian Institution, Washington, D.C. 20560, USA

Abstract

The foraminiferal Austral Province of the Late Cretaceous is defined by the absence of warm-water planktonic and benthonic foraminifera and the presence of distinctive species endemic to the southern high latitudes. Campanian–Maastrichtian planktonic foraminiferal faunas from the Falkland Plateau, northern Antarctic Peninsula and Weddell basin are nearly identical. These assemblages lack thermophilic keeled species, are very low in diversity, and are dominated by simple, globular forms that probably inhabited the uppermost part of the water column. Benthonic foraminiferal assemblages from southern South America, the northern Antarctic Peninsula and New Zealand are quite similar, and indicate that some marine communication must have existed between these regions via trans-Antarctic passages or along the Antarctic margin of the Pacific Ocean. Temporal variations in the northern limits of the Austral Province during the Late Cretaceous are generally difficult to detect due to stratigraphic and geographic gaps in the high-latitude record. At present, the most complete foraminiferal record of palaeoclimatic and palaeoceanographic change for the Cretaceous southern high latitudes comes from the Falkland Plateau; other sites recently drilled in the southern South Atlantic and Weddell basin will provide much new information.

Introduction

Foraminifera have proved to be particularly valuable in palaeoceanographic studies because of their: 1) distinctive patterns of distribution with latitude, depth and local depositional environment; 2) utility in biostratigraphy; and 3) sensitivity to changes in oceanographic conditions. The geographic differentiation of foraminiferal assemblages is the result of a complex interplay of numerous physical and chemical variables. Temporal shifts in the foraminiferal provincial boundaries are predominantly regulated by changes in marine circulation patterns (due to opening or closing of oceanic gateways) and climatic change.

This paper briefly summarizes the distribution of Late Cretaceous foraminifera in the southern high latitudes (poleward of 50° S palaeolatitude), and factors controlling their distribution. Thus, references are mostly limited to Austral studies. Comparison of coeval faunas from the southern margins of the South Atlantic, Pacific and Indian oceans provides valuable insight for reconstruction of palaeoceanographic and palaeoclimatic evolution. However, numerous stratigraphic and geographic gaps in the Cretaceous high-latitude marine record (Figs 1 & 2) still preclude an accurate portrayal of the Late Cretaceous palaeoceanography. Our understanding of this region will considerably advance as new data from Ocean Drilling Program (ODP) Legs 113, 114, 119 and 120 (in the Weddell basin, southern South Atlantic and Kerguelen Plateau) are synthesized.

Southern high-latitude palaeogeography

The Late Cretaceous dispersal history of the remnant Gondwana continents, which included Antarctica, Australia, South America, New Zealand and New Guinea (Fig. 1), strongly affected biotic distributions because of resultant changes in marine circulation. Significant continent motion can be briefly summarized as follows: 1) migration of New Zealand from 80° S at 95 Ma (Oliver *et al.*, 1979) to 62° S by 75 Ma (Grindley *et al.*, 1977); 2) gradual separation of Australia from Antarctica at about 80 Ma (Cande & Mutter, 1982); 3) separation of Antarctica from South America throughout the Late Cretaceous (Lawver, Sclater & Meinke, 1985); and 4) continued widening of the South Atlantic. The southern tip of India was located at about 60° S during the late Albian and migrated northward to about 45° S by the early Maastrichtian (Barron, 1987). Antarctica probably occupied a polar position throughout the period of consideration (Smith, Hurley & Briden, 1981; Barron, 1987). Communication between the

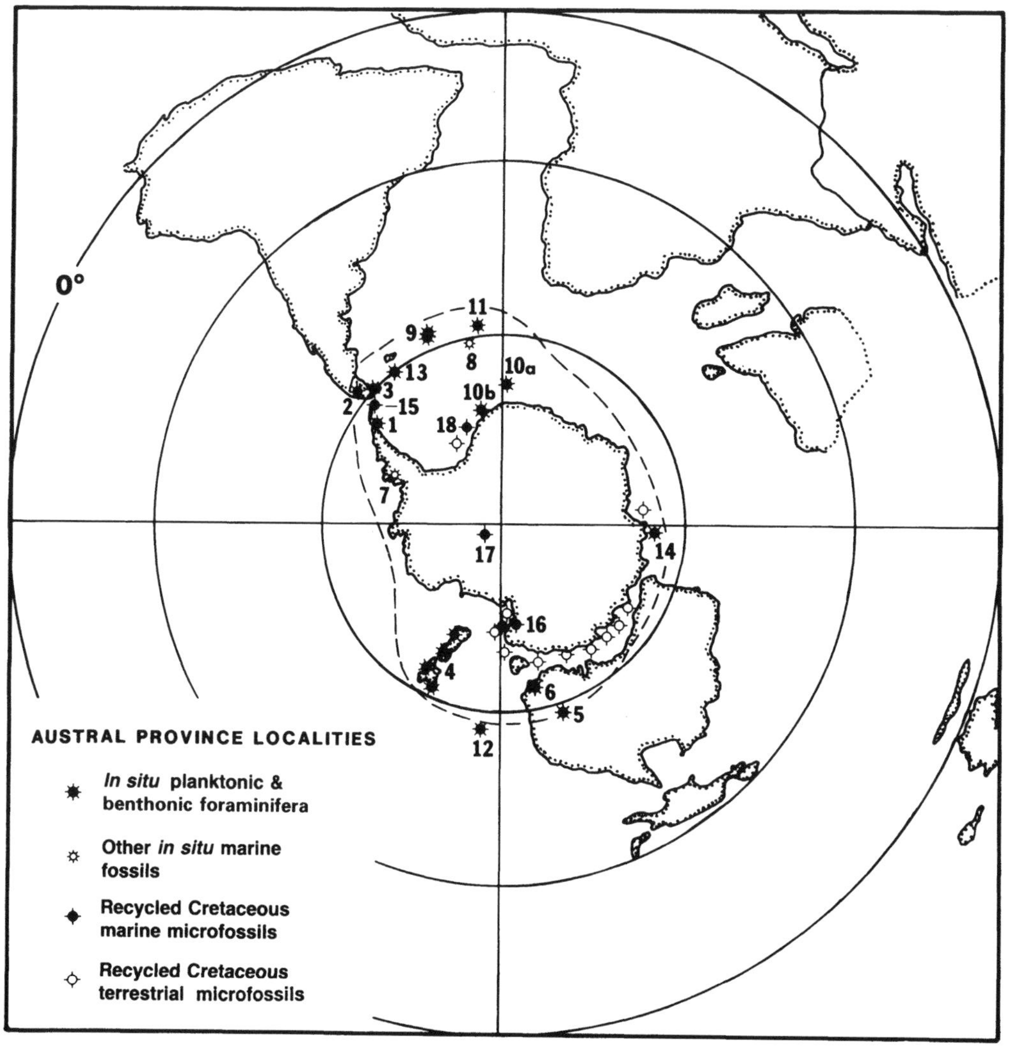

Fig. 1. Palaeogeographical reconstruction map for 80 Ma (after Smith *et al.*, 1981) showing localities of *in situ* marine sediments yielding Cretaceous Austral Province foraminifera (within dashed circle). Other sites shown have yielded reworked Cretaceous marine and terrestrial microfossils. 1, James Ross Island region; 2, southern Chile; 3, southern Argentina; 4, New Zealand; 5, Great Artesian Basin; 6, Otway Basin; 7, Alexander Island; 8, South Georgia; 9, Falkland Plateau (DSDP Holes 327A and 511); 10a, Weddell basin ODP Leg 113 (Holes 689, 690); 10b, Weddell basin, ODP Leg 113 (Holes 692, 693); 11, OPD Leg 114 (Holes 698 and 700); 12, Lord Howe Rise (DSDP Hole 208); 13, Burdwood Bank; 14, Kerguelen Plateau; 15, King George Island; 16, Taylor Valley; 17, Transantarctic Mountains; 18, Weddell basin. See text for references.

North and South Atlantic ocean basins was probably established by the end of the Albian, but gateways for bottom-water circulation were not available until at least the Santonian. Lawver *et al.* (1985) showed that the South Atlantic basin was little more than half its present size by the end of the Cretaceous. The South Pacific basin, on the other hand, was very broad and effectively isolated from the other ocean basins (Zinsmeister, 1982).

The Cretaceous distribution of land and sea within Antarctica is largely conjectural because of a lack of information from the continental interior and surrounding continental margin. Antarctic localities which have yielded Cretaceous marine and non-marine fossils are plotted on Fig. 1 and the stratigraphic ranges for the marine sites are shown in Fig. 2. *In situ* Cretaceous marine sediments are only known from the James Ross Island region, Robertson Island and Alexander Island in the northern Antarctic Peninsula and on King George Island in the South Shetland Islands (Thomson, 1981; Huber, 1988). Recycled marine microfossils of this age have been reported from Taylor Valley and Deep Sea Drilling Project (DSDP) Hole 272 in the Ross Sea region (Webb & Neall,1972; Leckie & Webb, 1985), the Transantarctic Mountains (Webb *et al.*, 1984), Weddell Sea (Truswell & Anderson, 1984) and King George Island (Birkenmajer, Gazdzicki & Wrona, 1983). Cretaceous terrestrial floras were reported in marine sediments from the James Ross Island region and Robertson Island (Duzen, 1908; Askin, 1988), the Weddell Sea (Truswell & Anderson, 1984), the southern margin of the Indian Ocean (Truswell, 1983), the Wilkes Land margin (Domack, Fairchild & Anderson, 1980) and the Ross Sea region (Truswell, 1983; Truswell & Drewry, 1984). These studies suggest a wide geographical distribution of Cretaceous marine and non-marine sediments in subglacial basins of the Antarctic interior. The subglacial topographic map of present

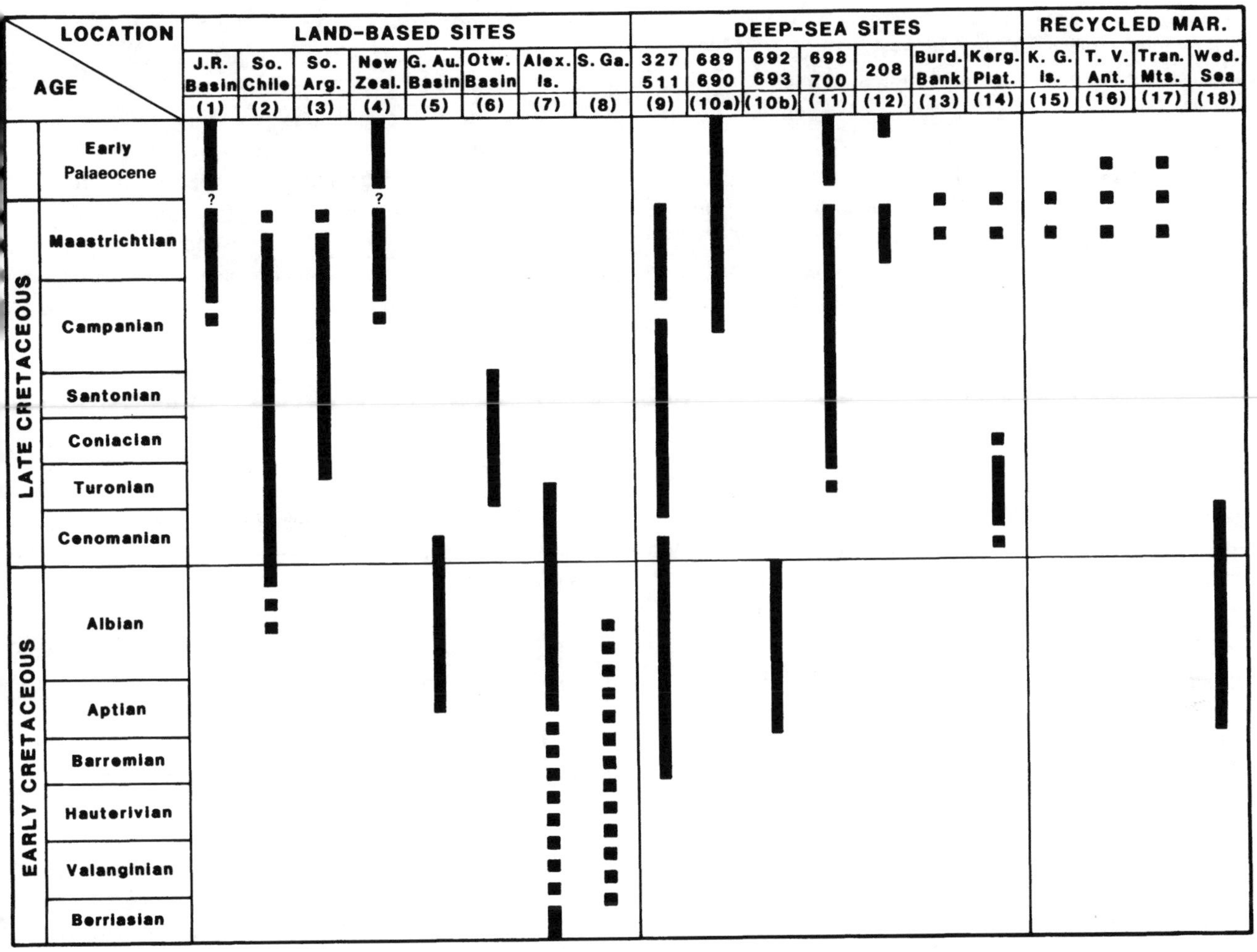

Fig. 2. Stratigraphic ranges of land-based, deep-sea, and recycled marine sediments recovered from within the Austral Province. Numbers in parentheses correspond to site locations of Fig. 1.

day Antarctica (Drewry, 1983), which compensates for isostatic rebound after ice removal, indicates the presence of several large intracratonic basins in both East and West Antarctica. This evidence alone implies that a considerable area of West Antarctica and portions of East Antarctica were below sea level prior to growth of the Antarctic ice-cap.

Provenance studies of sedimentary rocks indicate that the Antarctic Peninsula was an emergent arc during the Late Cretaceous (Dalziel & Elliot, 1982; Macellari, 1988). It is uncertain whether the peninsula was a continuous landmass or a series of islands separated by shallow-marine seaways. Woodburne & Zinsmeister (1984) favour a relatively continuous land connection, extending from southern South America to Australia, to explain the dispersal pattern of Late Cretaceous–early Tertiary marsupials. On the other hand, Zinsmeister (1982), Macellari (1985) and Huber & Webb (1986) show strong affinities among Late Cretaceous shallow-marine faunas from southern South America, northern Antarctic Peninsula, New Zealand and south-east Australia, and indicate that some marine communication must have existed between these regions. Treatment of Antarctica as a wholly emergent continent versus one dissected by seaways leads to considerably different palaeoceanographic and palaeoclimatic interpretations.

Late Cretaceous foraminiferal biogeography

Four major biogeographical divisions can be recognized for the Late Cretaceous based on the diversity and taxonomic character of open-ocean foraminiferal assemblages. These include the tropical Tethyan Province, the subtropical Transitional Province and the cool-temperate Boreal and Austral provinces (Scheibnerova, 1971; Sliter, 1976). The Cretaceous provinces are primarily distinguished using planktonic foraminifera, as shallow-water benthonic assemblages may have been influenced by local environmental conditions and deep-sea benthonic faunas are predominantly cosmopolitan. As in modern foraminiferal distributions, the Cretaceous provincial boundaries are gradational, with few species restricted to a single province.

Planktonic foraminiferal biogeography

The utility of planktonic foraminifera in palaeoceanographic and palaeoclimatic studies has been summarized by Bé (1977) and Lipps (1979). Discussion of some basic principles regarding the functional significance of planktonic foraminiferal test morphology, and their modern vertical and latitudinal distribution in the upper water column, will provide the basis for the uniformitarian assumptions that follow in analysis of the Late Cretaceous high-latitude assemblages.

Studies of Recent planktonic foraminifera show that their distribution roughly parallels latitudinal climatic belts and their taxonomic diversity and evolutionary rates show a substantial poleward reduction (Bé, 1977). Increased seasonality and decreased solar input toward polar regions probably control these trends (Stehli, Douglas & Newell, 1969). Shell morphology varies along latitudinal transects and with depth in the water column. All planktonic species undergo a depth migration during their life cycle, the depth range being different for different groups; it has been suggested that this depth control is maintained by the addition of calcite to the foraminiferal shell during ontogeny (Bé, 1980). Thus, maintenance of neutral bouyancy in a preferred depth habitat is dependent on the equilibrium of the foraminiferal shell density with the ambient water density, the latter being controlled by temperature, salinity and depth. Globulose, thin-walled juveniles, which inhabit the uppermost surface waters, transform their shells during development by chamber thickening and addition of surface ornament, attaining maximum depth just prior to reproduction (Bé, 1980). Small, thin-walled, and globulose ('simple') adult forms inhabit the uppermost surface waters from the equator to the poles, and frequently show a poleward reduction in surface ornament. The adult tests of deeper-dwelling ('complex') forms are generally larger, thicker-walled and more flattened. They are abundant in tropical regions, but are rare in the higher latitudes.

The geographical distribution of fossil planktonic foraminifera was probably influenced by the same factors that affect modern species (Lipps, 1979; Caron & Homewood, 1983). Complex morphotypes, which include keeled and densely ornamented species of the Cretaceous Rotaliporinae and Globotruncanacea, were dominant among low-latitude planktonic assemblages of the warm-water Tethyan Province (Scheibnerova, 1971; Sliter, 1976). On the other hand, the cool-water Austral Province was characterized by the near-absence of complex taxa and dominance of simple planktonic species, such as the Heterohelicinae, Globigerinelloididae and Hedbergellidae (Fig. 3). Comparison of late Campanian–Maastrichtian planktonic assemblages of the Southern Hemisphere shows a decrease in species diversity from a maximum of 60 species in the tropics to < 10 species in the polar regions (Huber, 1988). Furthermore, rugoglobigerine taxa show a poleward reduction in surface ornament among some of the planktonic taxa, suggesting a concomitant shallowing of their depth habitats (Fig. 3).

The northern limit of the Austral Province in the South Atlantic was considered by Krasheninnikov & Basov (1986), on the basis of planktonic foraminiferal assemblages, to pass just north of the Falkland Plateau, well south of the Walvis Ridge, and perhaps south of the African continent (Fig. 1). In the southern Indian Ocean, planktonic foraminifera characteristic of the Austral Province or southern Transitional Province occur only in Cenomanian sediments of the Naturaliste Plateau and central Kerguelen Plateau (Herb & Scheibnerova, 1977). Despite their location at high palaeolatitudes, Santonian and Campanian foraminiferal assemblages from the Carnarvon and Perth basins of Western Australia (Belford, 1960) were probably influenced by warm-water circulation, and were placed by Scheibnerova (1971) in the Transitional Province. Late Campanian–Maastrichtian foraminifera from New Zealand and the Lord Howe Rise are extratropical in character (Webb, 1971, 1973) and are included in the Austral Province.

Several authors have reconstructed the Cretaceous climatic history of the southern high latitudes based on fluctuations in relative abundance of thermophilic keeled taxa, total planktonic species diversity and test preservation. Using data primarily from southern South Atlantic DSDP sites, Krasheninnikov & Basov (1986) recognized three peak warming intervals in the Late Cretaceous: the first persisted from the latest Albian through the early Turonian, the second comprised all of the Santonian, and the maximum warming interval occurred in the mid-Campanian. The intervening periods of cooling included the mid-Turonian through Coniacian and early Campanian, and these were followed by a progressive temperature decline from the late Campanian through the late Maastrichtian.

Herb & Scheibnerova (1977) used changes in the taxonomic content and species diversity of planktonic foraminiferal assemblages from several DSDP sites in the Indian Ocean to recognize the influence of northward migration of the Indian plate. However, they noted that there is insufficient data from the Indian Ocean to define adequately the biogeographic boundary of the Austral Province, and any fluctuations in its position through time.

Oxygen isotope palaeotemperature data from well preserved planktonic foraminifera are needed to test the correlation of temporal variations in shell morphology and climate. Some keeled planktonic species, such as *Abathomphalus mayorensis* and *Globotruncana arca*, apparently adapted to the high-latitude environments as they are abundant in several sites (Huber *et al.*, 1987). The ratio of keeled and non-keeled morphotypes may be a useful indicator of surface-water stratification, but other, perhaps non-environmentally-related, factors may have also controlled their distributions.

The similarity of late Campanian–Maastrichtian planktonic foraminifera from Antarctic and southern South Atlantic sites, including the James Ross Island region, ODP Holes 689B, 690C, 698 and 700 and DSDP Holes 327A and 511, is striking (Huber *et al.*, 1987, Huber, unpublished data). Species of *Globotruncana* are absent from these sites, with the exception of rare occurrences in lower Maastrichtian sediments of the more northerly Hole 327A. Instead, these localities yield species-poor assemblages dominated by long-ranging, simple, globulose species of *Heterohelix*, *Globigerinelloides*, *Hedbergella*, and morphovariants of *Rugoglobigerina*.

Planktonic species restricted to the Austral Province are few. A species originally described from the Falkland Plateau, *Globigerinelloides impensus* Sliter, has recently been recognized at all Leg 113 and 114 sites bearing lowermost Maastrichtian sediments. *Rugotruncana circumnodifer* (Finlay), originally described in New Zealand, has been found on Seymour Island as well as the Maastrichtian sites of ODP Legs 113 and 114. Several new species have been recognized at the Cretaceous Leg 113 and 114 sites, but their distribution was apparently limited to the southern South Atlantic (Huber *et al.*, 1987).

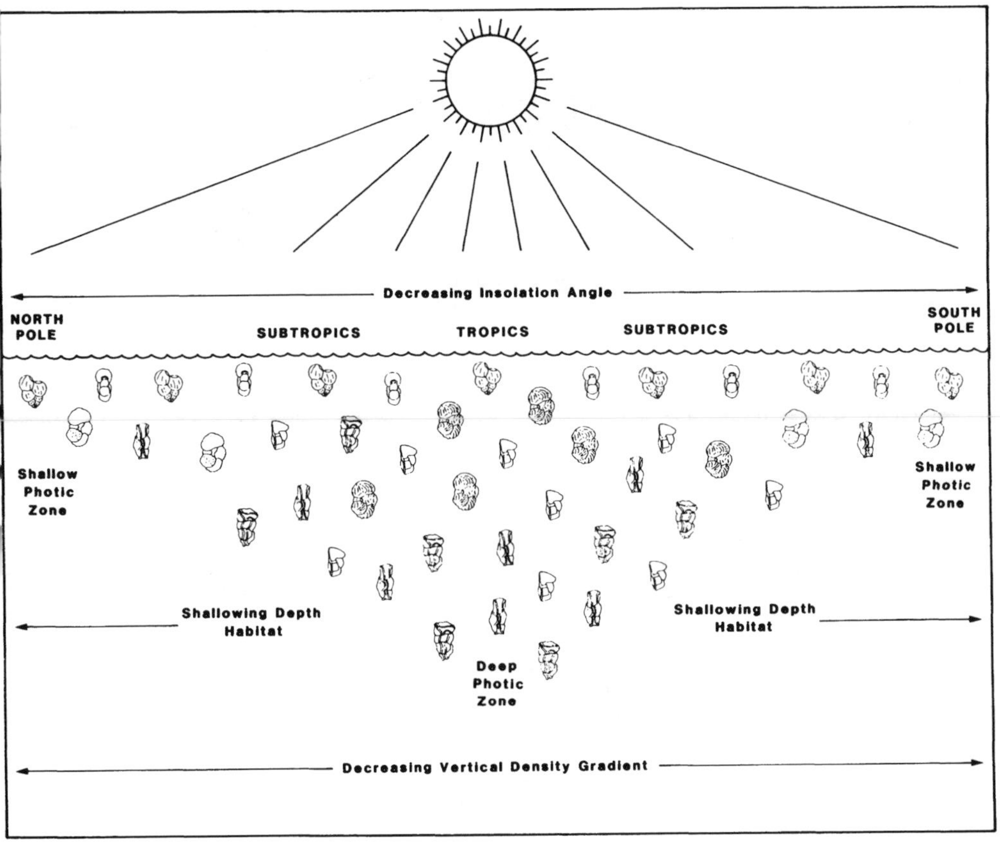

Fig. 3. Diagrammatic portrayal of factors influencing the depth habits of Late Cretaceous planktonic foraminifera. The poleward loss of stenothermal, keeled (deeper-dwelling) morphotypes may have been related to a concomitant shallowing of the photic zone (due to decreasing insolation angle and increased seasonality). Note the poleward reduction in meridional ornament on the figured rugoglobigerine morphotypes.

These will be systematically described in the appropriate ODP proceedings volumes (Huber, unpublished data).

Benthonic foraminiferal biogeography

Shelf assemblages of benthonic foraminifera included in the Austral Province have been characterized by the predominance of agglutinated taxa, low diversity of calcareous benthonic assemblages and absence of larger foraminifera, such as the Orbitoididae and Amphisteginidae (Scheibnerova, 1971; Webb, 1971). Deep-sea benthonic assemblages are not considered useful to identify the high-latitude province since their distribution is more cosmopolitan and they are less influenced by latitudinal and temporal changes in climate.

Several benthonic species were apparently confined to the Austral Province during the late Campanian–Maastrichtian. Huber & Webb (1986) extended the palaeobiogeographic range of *Frondicularia rakauroana* (Finlay) from the southern southwest Pacific (New Zealand and Lord Howe Rise) to Seymour Island (Antarctic Peninsula). The agglutinated species *Dorothia elongata* (Finlay) and *Gaudryina healyi* (Finlay) were also reported from these sites (Huber, 1988), and the additional occurrence of *G. healyi* in Tierra del Fuego was noted by Malumian & Masiuk (1976).

The oldest reported occurrence of the agglutinated genus *Cyclammina* is from late Campanian shelf sediments of the James Ross Island region (Huber, 1988). It also occurs in slightly younger Cretaceous nearshore sediments from the Burdwood Bank (Macfadyen, 1933), New Zealand (Webb, 1971) and in shelf sediments of late Palaeocene age in south-eastern Australia (Ludbrook, 1977). However, cyclamminids did not appear until the latest Palaeocene in lower latitude regions. This not only testifies to the high-latitude origin of this genus, but also reflects its migration from shallow- to deep-water environments, as this taxon is rarely found above bathyal and abyssal depths in modern oceans (Akers, 1954).

The macrofaunal Weddellian Province was recognized by the distribution of Late Cretaceous bivalves, gastropods and ammonites along the southern high-latitude margin of the Pacific Ocean (Zinsmeister, 1982; Macellari, 1985). Except for its restriction to shallow-marine facies, the limits of this province strongly parallel the microfaunal Austral Province.

Conclusions

The foraminiferal Austral Province is characterized by the absence of thermophilic benthonic and planktonic taxa; few species are restricted to it. As in modern biogeographical provinces, the Late Cretaceous provincial boundaries are gradational. More data from high-latitude sites are needed to better characterize temporal variations in the geographical extent of this high-latitude province.

Immigration of thermophilic keeled planktonic foraminifera into the Austral Province (and their relative abundance and diversity) have been used to indicate periods of warming and cooling during the late Cretaceous. Warmer intervals are suggested for the latest Albian–Turonian, Santonian and mid-Campanian, and a progressive cooling trend has been noted from the late Campanian to the late Maastrichtian. These fluctuations compare moderately well with published oxygen isotopic palaeotemperature curves of Douglas & Woodruff (1981) for the Late Cretaceous. More oxygen isotope palaeotemperature analyses are needed for this time interval to validate assumptions such as the importance of temperature in controlling the planktonic foraminiferal distributions.

The record of planktonic foraminiferal palaeobiogeography indicates that latitudinal thermal gradients did exist in the Late Cretaceous, and that the upper water column in tropical regions was more deeply stratified than in the polar oceans. Foraminifera from the northern Antarctic Peninsula and Weddell basin show the most extreme extra-tropical characteristics of Late Cretaceous faunas. This is borne out by the oxygen-isotope palaeotemperature record from Antarctic Peninsula planktonic foraminifera of late Campanian age; calculated surface temperatures of 9–10.5° C (Barrera *et al.*, 1987) are about 10° C cooler than surface-water estimates by Douglas & Savin (1978) from the equatorial Pacific Ocean.

References

Akers, W.H. (1954). Ecologic and stratigraphic significance of the foraminifer *Cyclammina cancellata* Brady. *Journal of Paleontology*, **28(2)**, 132–52.

Askin, R.A. (1988). Campanian to Palaeocene palynological succession of Seymour and adjacent islands, northeastern Antarctic Peninsula. In *Geology and Paleontology of Seymour Island, Antarctica*, ed. R.M. Feldmann & M.O. Woodburne pp. 131–53. Geological Society of America, Memoir, 169.

Barrera, E., Huber, B.T., Savin, S.M. & Webb, P.N. (1987). Antarctic marine temperatures: late Campanian through early Paleocene. *Paleoceanography*, **2(1)**, 21–47.

Barron, E.J. (1987). Global Cretaceous paleogeography–International Geologic Correlation Program Project 191. *Palaeogeography, Palaeoclimatology, Palaeoecology*, **59**, 207–16.

Bé, A.W.H. (1977). An ecological, zoogeographic and taxonomic review of Recent planktonic foraminifera. In *Oceanic Micropaleontology*, ed. A.T.S. Ramsay, pp. 1–100. London; Academic Press.

Bé, A.W.H. (1980). Gametogenic calcification in a spinose planktonic foraminifer *Globigerinoides sacullifer* (Brady). *Marine Micropaleontology*, **5(3)**, 283–310.

Belford, D.J. (1960). Upper Cretaceous Foraminifera from the Toolonga calcilutite and Gingin chalk, Western Australia. *Australia Bureau of Mineral Resources, Geology and Geophysics, Bulletin*, **57**, 1–198.

Birkenmajer, K., Gazdzicki, A. & Wrona, R. (1983). Cretaceous and Tertiary fossils in glaciomarine strata at Cape Melville, Antarctica. *Nature, London*, **303(5912)**, 56–9.

Cande, S.C. & Mutter, J.C. (1982). A revised identification of the oldest seafloor spreading anomalies between Australia and Antarctica. *Earth and Planetary Science Letters*, **58(2)**, 151–60.

Caron, M. & Homewood, P. (1983). Evolution of early planktic foraminifera. *Marine Micropaleontology*, **7**, 453–62.

Dalziel, I.W.D. & Elliot, D.H. (1982). West Antarctica: problem child of Gondwanaland. *Tectonics*, **1(1)**, 3–19.

Domack, E.M., Fairchild, W.W. & Anderson, J.B. (1980). Lower Cretaceous sediment from the East Antarctic continental shelf. *Nature, London*, **287(5783)**, 625–6.

Douglas, R.G. & Savin, S.M. (1978). Oxygen isotopic evidence for the depth stratification of Tertiary and Cretaceous foraminifera. *Palaeogeography, Palaeoclimatology, Palaeoecology*, **3**, 175–96.

Douglas, R.G. & Woodruff, F. (1981). Deep sea benthic foraminifera. In *The Sea*, 7, ed. C. Emiliani, pp. 1233–327. New York; Wiley Interscience.

Drewry, D.J. (1983). *Antarctica: Glaciological and Geophysical Folio*. Cambridge; University of Cambridge.

Duzen, P. (1908). Uber die Tertiare Flora die Seymour-Inseln. *Wissenschaftliche Ergebnisse der Schwedischen Sudpolar-Expedition, 1901–1903*, **3**, 1–27.

Grindley, G.W., Adams, C.J.D., Lumb, J.T. & Waters, W.A. (1977). Paleomagnetism, K–Ar dating and tectonic interpretation of Upper Cretaceous and Cenozoic volcanic rocks of the Chatham Islands, New Zealand. *New Zealand Journal of Geology and Geophysics*, **20(3)**, 425–67.

Herb, R & Scheibnerova, V. (1977). Synopsis of Cretaceous planktonic foraminifera from the Indian Ocean. In *Indian Ocean Geology and Biostratigraphy*, ed. J.R. Heirtyler, H.M. Bolli *et al.*, pp. 399–415. Washington, DC; American Geophysical Union.

Huber, B.T. (1988). Upper Campanian–Paleocene foraminifera from the James Ross Island region (Antarctic Peninsula). In *Geology and Paleontology of Seymour Island, Antarctica*, ed. R.M. Feldmann & M.O. Woodburne pp.163–252. Geological Society of America, Memoir 169.

Huber, B.T. & Webb, P.N. (1986). Distribution of *Frondicularia rakauroana* (Finlay) in the southern high latitudes. *Journal of Foraminiferal Research*, **16(2)**, 135–40.

Huber, B.T., Webb, P.N., Barker, P., Kennett, J.P. & Leg 113 Scientific Party. (1987). Upper Cretaceous planktonic foraminifera from the Weddell Basin, Antarctica (ODP Leg 113). *Geological Society of America, Abstracts with Programs*, **19(4)**, 709.

Kemp, E.M. (1972). Reworked palynomorphs from the West Ice Shelf area, East Antarctica, and their possible geological and paleoclimatological significance. *Marine Geology*, 13, 145–57.

Krasheninnikov, V.A. & Basov, I.A. (1983). Stratigraphy of Cretaceous sediments of the Falkland Plateau based on planktonic foraminifers, Deep Sea Drilling Project, Leg 71. In *Initial Reports of the Deep Sea Drilling Project*, 71, W.J. Ludwig, V.A. Krasheninnikov *et al.*, pp. 789–820; Washington, DC; US Government Printing Office.

Krasheninnikov, V.A. & Basov, I.A. (1986). Late Mesozoic and Cenozoic stratigraphy and geological history of the South Atlantic high latitudes. *Palaeogeography, Palaeoclimatology, Palaeoecology*, **55**, 145–88.

Lawver, L.A., Sclater, J.G. & Meinke, L. (1985). Mesozoic and Cenozoic reconstructions of the South Atlantic. *Tectonophysics*, **114**, 233–54.

Leckie, R.M. & Webb, P.N. (1985). Late Paleogene and early Neogene foraminifers of Deep Sea Drilling Project Site 270, Ross Sea, Antarctica. In *Initial Reports of the Deep Sea Drilling Project*, 90, ed. J.P. Kennett, C.C. von der Borch *et al.*, pp. 1093–142; Washington, DC; US Government Printing Office.

Lipps, J.H. (1979). Ecology and paleocology of planktic foraminifera. In *Society of Economic Paleontologists and Mineralogists, Shortcourse No. 6*, pp. 21–53. Tulsa, Oklahoma; SEPM.

Ludbrook, N.H. (1977). Early Tertiary *Cyclammina* and *Haplophragmoides* (Foraminiferida: Lituolidae) in Southern Australia. *Transactions of the Royal Society of South Australia*, **101(7)**, 165–98.

Macellari, C.E. (1985). Paleobiogeografia y edad de la fauna de *Maorites – Gunnarites* (Amonoidea) de la Antartida y Patagonia. *Ameghiniana*, **21(2–4)**, 223–42.

Macellari, C.E. (1988). Stratigraphy, sedimentology and paleoecology of Late Cretaceous–Paleocene shelf–deltaic sediments of Seymour Island (Antarctic Peninsula). In *Geology and Paleontology of Seymour Island, Antarctica*, ed. R.M. Feldmann & M.O. Woodburne pp. 25–53. Geological Society of America, Memoir 169.

Macfadyen, W.A. (1933). Fossil foraminifera from the Burdwood Bank and their geological significance. *Discovery Reports*, **7**, 1–16.

Malumian, N. & Masiuk, V. (1976). Foraminiferos de la Formacion Cabeza de León (Cretácico superior), Tierra del Fuego, Rep. Argentina. *Asociacion Geológica Argentina, Revista*, **31(3)**, 180–221.

Oliver, P.J., Mumme, T.C., Grindley, G.W. & Vella, P. (1979). Paleomagnetism of the Upper Cretaceous Mt Somers volcanics, Canterbury, New Zealand. *New Zealand Journal of Geology and Geophysics*, **22(2)**, 199–212.

Scheibnerova, V. (1971). Foraminifera and their Mesozoic biogeoprovinces. *Records of the Geological Survey of New South Wales*, **13(3)**, 135–74.

Sliter, W.V. (1976). Cretaceous foraminifera from the southwest Atlantic Ocean, Leg 36, Deep Sea Drilling Project. In *Initial Reports of the Deep Sea Drilling Project*, 36, P.F. Barker, I.W.D. Dalziel *et al.*, pp. 519–73. Washington, DC; US Government Printing Office.

Smith, A.G., Hurley, A.M. & Briden, J.C. (1981). *Phanerozoic paleocontinental World Maps*. Cambridge: Cambridge University Press. 102 pp.

Stehli, F.G., Douglas, R.G. & Newell, N.D. (1969). Generation and maintenance of gradients in taxonomic diversity. *Science*, **164**, 947–9.

Thomson, M.R.A. (1981). Mesozoic ammonite faunas of Antarctica and the break-up of Gondwana. *Proceedings of the Fifth International Gondwana Symposium*, ed. M.M. Cresswell & P. Vella, pp. 269–75. Rotterdam; A.A. Balkema.

Truswell, E.M. (1983). Recycled Cretaceous and Tertiary pollen and spores in Antarctic marine sediments: a catalogue. *Palaeontographica*, **B186**, 121–74.

Truswell, E.M. & Anderson, J.B. (1984). Recycled palynomorphs and the age of sedimentary sequences in the eastern Weddell Sea. *Antarctic Journal of the United States*, **19(5)**, 90–2.

Truswell, E.M. & Drewry, D.J. (1984). Distribution and provenance of recycled palynomorphs in surficial sediments of the Ross Sea, Antarctica. *Marine Geology*, **59**, 187–214.

Webb, P.N. (1971). New Zealand Late Cretaceous (Haumurian) foraminifera and stratigraphy: a summary. *New Zealand Journal of Geology and Geophysics*, **14(4)**, 795–828.

Webb, P.N. (1973). Upper Cretaceous–Paleocene foraminifera from Site 208 (Lord Howe Rise, Tasman Sea), DSDP, Leg 21. In *Initial Reports of the Deep Sea Drilling Project*, 21, R.E. Burns, J. Andrews *et al.*, pp. 541–73. Washington, DC; US Government Printing Office.

Webb, P.N., Harwood, D.M., McKelvey, B.C., Mercer, J.H. & Stott, L.D. (1984). Cenozoic marine sedimentation and ice-volume variation on the East Antarctic craton. *Geology*, **12(5)**, 287–91.

Webb, P.N. & Neall, V.E. (1972). Cretaceous foraminifera in Quaternary deposits from Taylor Valley, Victoria Land. *Antarctic Geology and Geophysics*, ed. R.J. Adie, pp. 653–7. Oslo; Universitetsforlaget.

Woodburne, M.O. & Zinsmeister, W.J. (1984). The first land mammal from Antarctica and its biogeographic implications. *Journal of Paleontology*, **58(4)**, 913–48.

Zinsmeister, W.J. (1982). Late Cretaceous–Early Tertiary molluscan biogeography of the southern circum-Pacific. *Journal of Paleontology*, **56(1)**, 84–102.

First record of dinosaurs in Antarctica (Upper Cretaceous, James Ross Island): palaeogeographical implications

E.B. OLIVERO,[1] Z. GASPARINI,[2] C.A. RINALDI[3] & R. SCASSO[1]

1 Centro de Investigaciones en Recursos Geológicos, Ramirez de Velasco 847, 1414 Buenos Aires, Argentina
2 Museo La Plata, Paseo del Bosque, 1900 La Plata, Argentina
3 Instituto Antártico Argentino, Cerrito 1248, 1010 Buenos Aires, Argentina

Abstract

During the austral summer of 1986, fieldwork on James Ross Island by the Instituto Antártico Argentino resulted in the discovery of the first remains of dinosaurs from the Antarctic continent. These consist of a partial skeleton and bony plates of an armoured ornithischian belonging to the Ankylosauria. The fossil material was found in marine sandy-facies of the Santa Marta Formation (Marambio Group) of Campanian age. The remains were associated with marine invertebrates. At a slightly higher stratigraphic level, marine reptiles related to mosasaurs and plesiosaurs were also found. The occurrence of ankylosaurs on James Ross Island provides important new insight concerning hypotheses of land connections between South America and Antarctica during the Late Cretaceous. An earlier differentiation of the family Ankylosauridae and the distribution of these dinosaurs in Antarctica during the Late Jurassic–Early Cretaceous cannot be completely ruled out. However, a late entrance of northern ankylosaurids into Antarctica, via South America, is considered more likely. Because it was not possible for these ankylosaurs to cross water barriers, their presence indicates that a continuous land connection must have existed between Antarctica and South America for some period of time during the Late Cretaceous.

Introduction

In addition to its severe present climate, a distinctive feature of Antarctica is its complete absence of permanent land vertebrates. Because Antarctica is at present an isolated landmass, encircled by deep oceans, the occurrence of fossil land vertebrates is particularly significant to the concept of continental drift and to our knowledge of former land connections among the southern continents. The finding of Triassic terrestrial reptiles and amphibians (Colbert, 1982), ichnites of large Tertiary ground birds (Covacevich & Rich, 1982) and, particularly, Tertiary marsupials (Woodburne & Zinsmeister, 1984), indicates that Antarctica was faunally an integral part of Gondwana. These data also suggested the probable occurrence of Cretaceous land vertebrates in the Antarctic Peninsula and the possibility of a continental connection between this region and South America.

During the austral summer of 1986, fieldwork by the Instituto Antartico Argentino in James Ross Island (Fig. 1) resulted in the discovery of the first dinosaur from the Antarctic continent. The occurrence of these reptiles provides important new data concerning hypotheses of land connections of the Antarctic Peninsula and supports the idea that a continuous land bridge joined the Antarctic Peninsula and South America during Late Cretaceous times. In this paper we will present a preliminary description of the dinosaur, together with a short explanation of the stratigraphy and geographic location of the fossil locality. A discussion of the palaeogeographic implications of the discovery will also be included.

Description of the fossil material

The dinosaur from James Ross Island is an ankylosaur (Ornithischia) of small size, perhaps a juvenile specimen. Most of the recovered bones are fragmented, but the material is sufficient to warrant identification. The skull of these armoured dinosaurs was protected by thick dermal co-ossifications. In the ankylosaurids the postero-lateral co-ossifications of the skull bear marked lateral projections (Coombs, 1978). Several of these plates are preserved in the Antarctic ankylosaur. A fragment of left dentary is the only preserved part of the mandible. It bears a double row of alveoli and one lingual tooth (Fig. 2*a*). This tooth has the characteristic low, leaf-shaped crown, with apical cusps and lacks a distinct cingulum. The root of the tooth forms a cylindrical stem (Fig. 2*b*).

Other preserved parts of the skeleton consist of vertebrae and ribs, fragments of the limbs, dermal plates and part of the

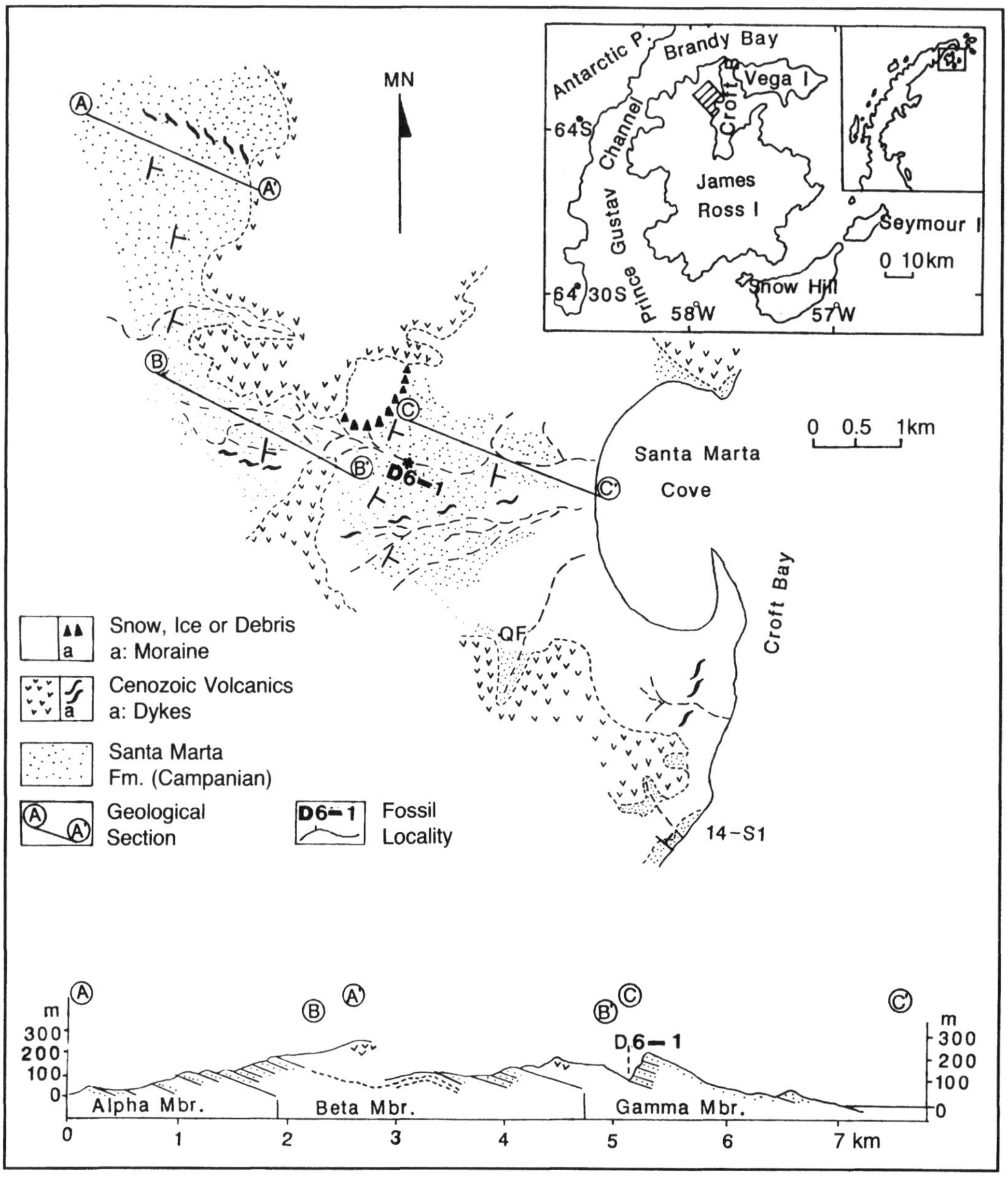

Fig. 1. Location map of the James Ross Island Group and geological sketch of Santa Marta Cove area showing Cretaceous outcrops. The dinosaur site (D6-1) and location of the cross-section of the Santa Marta Formation are also indicated.

tail club. The vertebrae are amphiplatyan and those corresponding to the sacrum are fused. In transverse section the ribs are T-shaped. In dorsal view, the ribs show ossified tendons which are very common in ornithischians.

Several bony plates and dermal ossicles that shielded the body of ankylosaurs were also recovered in the specimen from James Ross Island. Some of these plates are oval, thickened and show a longitudinal crest; others are flat, subcircular and enclosed by smaller polygonal plates. The dermal ossicles are of very small size.

The marked lateral projections of the dermal co-ossifications of the skull (Fig. 2*c*), the tooth without a distinct cingulum (Fig. 2*b*) and the tail club (Fig. 2*d*) allow assignation of this specimen to the family Ankylosauridae (Coombs, 1978; Carpenter & Breithaup, 1986). A more detailed analysis of the anatomy of this dinosaur is given by Gasparini *et al.* (1987).

The material MLP 86-X-28-1 is deposited at Museo La Plata, Division Paleontologia de Vertebrados (La Plata, Argentina).

Stratigraphy

The Cretaceous deposits on James Ross Island comprise two main stratigraphic units: the Gustav Group ((?)Barremian–Santonian) and the Marambio Group ((?)Santonian–Campanian) (Ineson, Crame & Thomson, 1986; Olivero, Scasso & Rinaldi, 1986). The first group crops out along the coast of Prince Gustav Channel and comprises a thick sequence of more than 2000 m of marine conglomerates, breccias, sandstones and mudstones, arranged in a complex succession of coarse- and fine-grained units (Ineson *et al.*, 1986). Most clasts of the conglomerates are composed of low-grade schists, volcanic and granitic rocks, and marine marls and

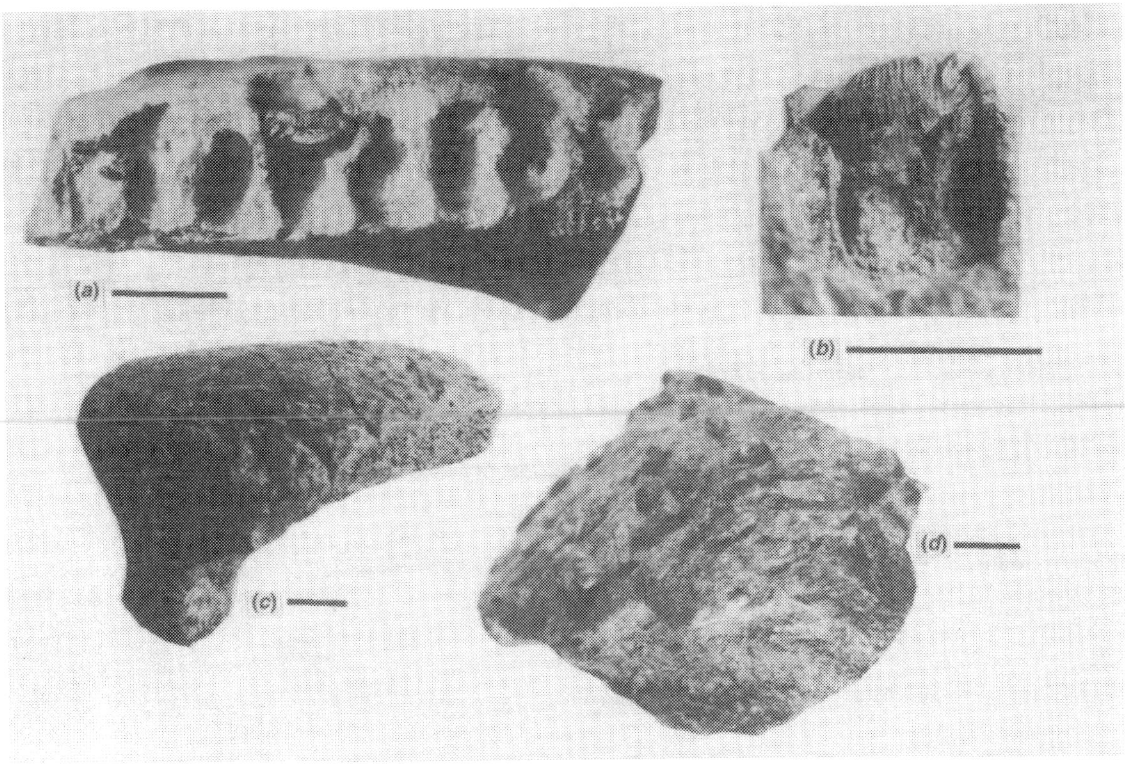

Fig. 2. Specimen MLP 86-X-28-1, Campanian ankylosaurid from Santa Marta Formation, James Ross Island, Antarctica: *a*, left dentary with a lingual tooth; *b*, a lingual tooth without a distinct cingulum; *c*, dermal plate of the skull roof; *d*, part of the tail club. Length of bar is 1 cm.

mudstones of Late Jurassic–Early Cretaceous age (Malagnino *et al.*, 1978; Medina *et al.*, 1982; Ineson *et al.*, 1986). This clearly indicates that the source of the clastic material of the Gustav Group was the Antarctic Peninsula to the west.

The Marambio Group comprises a mainly fine-grained and highly fossiliferous sequence, > 3000 m in thickness, exposed on James Ross, Vega, Cockburn, Snow Hill and Seymour (Marambio) islands. The lower part of the sequence is (?)Santonian–Campanian in age (Olivero *et al.*, 1986). The upper part of the Marambio Group is considered to be Maastrichtian–early Tertiary in age (Macellari, 1986).

The dinosaur remains were found in the basal beds of the Marambio Group in James Ross Island. This section has been differentiated recently as a new stratigraphic unit named the Santa Marta Formation (Olivero *et al.*, 1986). This unit is well exposed in northern James Ross Island, between the head of Brandy Bay and Santa Marta Cove (Fig. 1). It consists of about 1100 m of marine clastic sediments, divided into three members: Alpha, Beta and Gamma. The basal Alpha Member, ~ 480 m in thickness, consists of unconsolidated fine–coarse-grained sandstones and mudstones with occasional layers of conglomerates and coquinas. Marine invertebrate fossils are abundant and diverse only in the middle and the top, with leaves, carbonaceous debris and large trunks occasionally found throughout the Alpha Member.

The Beta Member consists of ~ 350 m of a rhythmic sequence of conglomerates with large clasts up to 40 cm in diameter in a silty or sandy matrix, pebbly sandstones and mudstones. The most common lithologies of clasts in the conglomerates are low-grade schists, volcanic and plutonic rocks, and reworked sandy or silty concretions. Marine fossils, particularly ammonites, are common and more diverse at the base of the member. In the upper part, coquina lenses and beds with trigoniids, baculitids and belemnites are common. Intercalated with these beds are mudstones and sandy siltstones with abundant carbonaceous material and large silicified or carbonized logs up to 1 m in diameter.

The Gamma Member consists of about 280 m of friable fine–medium-grained silty sandstones, frequently with glauconite, and carbonaceous mudstones with occasional layers or lenses of pebbly sandstones or medium conglomerates. Only in the upper two-thirds of this member is the marine invertebrate fauna abundant and diverse, comprising serpulids, corals, gastropods, bivalves, ammonites and echinoids. At locality QF (Fig. 1) a partial skeleton of a plesiosaur and vertebrae of a mosasaur were found associated with this fauna. In the lower third of the Gamma Member, marine invertebrate fossils are rare; no ammonites were recorded, and gastropods and bivalves are scarce and restricted to a few horizons. The ankylosaur skeleton was recovered at locality D6-1 about 90 m above the base of the member (Fig. 1). At this locality the lithology consists of massive strongly bioturbated silty sandstones devoid of glauconite, or with < 5% of this mineral. Trace fossils are abundant with most of them consisting of the ichnogenera *Ophiomorpha*, *Thalassinoides* and *Skolithos*. These sandstones are intercalated with thin layers of carbonaceous mudstones with abundant remains of silicified or carbonized wood. Also associated with the dinosaur skeleton were fish vertebrae and nautilid phragmocones.

Age of Santa Marta Formation

Except for the section with dinosaur bones, the rest of Santa Marta Formation contains a rich assemblage of marine invertebrates. The top of the Alpha Member and the Beta Member bear an abundant and relatively diverse ammonite fauna. On the basis of the presence of species of *Anapachydiscus*, *Eupachydiscus* and *Baculites*, which are common or show strong affinities with species widely distributed in Patagonia, Madagascar, South Africa, Japan and the west coast of North America, Olivero (1984) assigned a Campanian age to this fauna. Recent finds of well preserved nostoceratids and diplomoceratids (consisting of species of *Ainoceras*, *Eubostrychoceras*, *Ryugasella* and *Pseudoxybeloceras*) provide additional support for an Early Campanian age for this section.

Above the stratigraphic interval with the dinosaur bones, marine fossils are again abundant in the upper third of the Gamma Member. Although the ammonite assemblage is less diverse, it is characterized by abundant specimens of *Gunnarites* aff. *antarcticus*.

The evidence derived from ammonite assemblages occurring both below and above the lower third of Gamma Member indicates a Campanian age, most probably Late Campanian, for the beds with dinosaur bones.

Depositional environment

Deposition in slope apron-basin plain and submarine fan settings was proposed for the Gustav Group (Ineson, 1985). It seems likely that most of the Cretaceous sequence of western James Ross Island consists of a thick wedge of marine sediments deposited at the foot of a fault-controlled scarp, with intermittent synsedimentary tectonic activity. At least part of Santa Marta Formation, e.g. the Beta Member, appears to be related to this depositional setting. This is supported by the occurrence of the coarse conglomerates which display a chaotic internal structure suggesting a debris flow type of transport.

Despite the above mentioned features, other data suggest that part of the Santa Marta Formation was deposited in shallow water. This is indicated by the widespread distribution of large fossil tree trunks and plant debris, often occurring within lenses or as thin beds of coal or carbonaceous mudstones. In particular, the uppermost part of the Beta Member is characterized by coquinas and pebbly shell banks composed almost exclusively of trigoniid shells. These are intercalated with fine sandstones and mudstones rich in plant debris. Such beds could represent longshore bars and lagoonal facies, respectively. The succeeding lower section of the Gamma Member, composed of similar fine-grained, strongly bioturbated carbonaceous sediments, could also represent lagoonal and littoral deposits.

All the recovered dinosaur bones belong to a single specimen. Because they were found in a small area of about 2 m × 3 m it seems that the material was not reworked. It was not possible to recover part of the skeleton because the bones were broken into small fragments by the action of ice. These facts, together with the evidence for a shallow-water environment of deposition, indicate that the lower section of Gamma Member has the potential for further dinosaur finds.

An unusual feature of the locality where the ankylosaur bones were discovered is the presence of a relatively large number of nautilid phragmocones, which are assigned tentatively to *Cymatoceras* sp. Nautilid shells are generally very scarce in the Cretaceous and Tertiary deposits of the James Ross Island group and it is interesting to note that the first discovered mammal bones from Antarctica were also associated with a relatively large number of them (Woodburne & Zinsmeister, 1984). This unusually large number of Tertiary nautilids, associated with marsupial bones, was interpreted as the result of the stranding of shells along a beach. It seems likely that the large concentration of nautilid shells at the dinosaur locality could be explained by the same process.

Palaeogeographical implications

The discovery of a Late Cretaceous ankylosaur in Antarctica provides important new data concerning hypotheses of past land connections of the Antarctic Peninsula. In order to explain the occurrence of these reptiles in Antarctica, two main hypotheses can be considered (Gasparini *et al.*, 1987). Firstly, the ankylosaurs were already differentiated at the family level in the Late Jurassic–Early Cretaceous. By this time they were distributed in both Laurasia and Gondwana, and the Antarctic species would be the result of some form of vicariance. The oldest known ankylosaurids are from the Upper Cretaceous (Coniacian) of Asia (cf. Gasparini *et al.*, 1987, and the bibliography therein) and consequently their fossil record does not support this first hypothesis. However, an older non-documented fossil record for this family cannot be completely ruled out.

In the second hypothesis it is considered that Late Cretaceous northern ankylosaurids entered Antarctica via South America; such a derivation is considered to be much more likely. Previous records of ankylosaurs from Patagonia have been questioned and this group of dinosaurs is not yet documented in South America (Bonaparte, 1986). Nevertheless, the family Ankylosauridae is well represented in the Upper Cretaceous of North America and Asia (Coombs, 1978) and other data indicate the possibility of a land vertebrate interchange between North and South America during part of the Late Cretaceous (Bonaparte, 1986). Concerning the problem of how the ankylosaurs were able to get onto the Antarctic Peninsula from South America, a number of independent data strongly suggest a close proximity of these two areas during Late Cretaceous times (Dalziel & Elliot, 1973, 1982; Zinsmeister, 1982, 1987). The finding on Seymour Island of marsupials with South American affinities leaves no doubt about the proximity of these continents during the latest Cretaceous or earliest Tertiary, and strongly suggests a continuous land connection between South America and Antarctica (Woodburne & Zinsmeister, 1984). Such a connection for these continents is also indicated by the distribution of two families of Cretaceous continental turtles (meiolaniids and cheliids) which occur only in Australia and South America (Baez & Gasparini, 1979; Gasparini, De la Fuente & Donadio, 1986).

In previous reconstructions of Gondwana, the area between

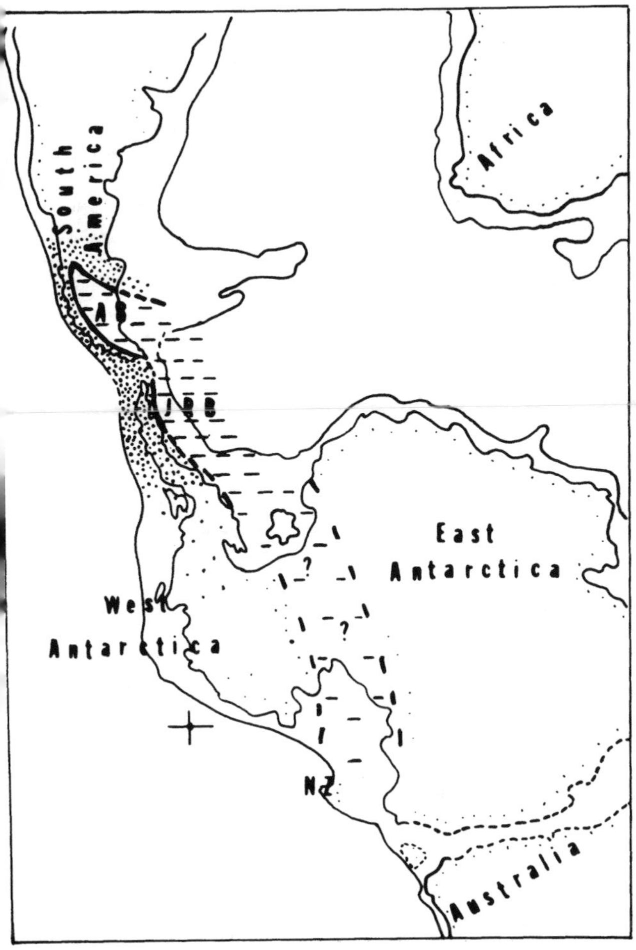

Fig. 3. Palaeogeographic sketch map of the Antarctic Peninsula and southern South America during the Late Cretaceous. Features shown are: inferred continuous land connection (stippled area) between the Antarctic Peninsula and South America; the marine Upper Cretaceous deposits of the Austral Basin (AB) and James Ross basin (JRB); New Zealand (NZ) and the inferred seaway between East and West Antarctica. Data in part from Norton (1982), Riccardi (1987) and Zinsmeister (1987).

Tierra del Fuego and the Antarctic Peninsula is generally depicted as an archipelago (cf. Dalziel & Elliot, 1982). The occurrence of dinosaurs in James Ross Island provides new insight concerning the nature of the land connection between these regions. Because ankylosaurs were large, heavy, armoured dinosaurs, their passive transportation across water barriers (e.g. on natural rafts) seems unlikely. Consequently, if the hypothesis of a late entrance of ankylosaurs into Antarctica is correct, their presence in James Ross Island indicates that a continuous land connection must have existed between Antarctica and South America for some period of time during the Late Cretaceous.

During the Late Cretaceous in the Austral (or Magallanes) Basin of southern Patagonia marine conditions prevailed south of Deseado Massif. Geological data indicate that this basin opened to the Antarctic with elevated terranes situated along the present axis of the Cordillera (Riccardi, 1987). A comparable palaeogeographical setting can be interpreted for the Cretaceous sediments of the Gustav and Marambio groups in Antarctica. Distribution of coarse clastic facies and composition of clasts clearly indicate that this basin was confined to the west by elevated terranes situated along the axis of the Antarctic Peninsula.

In the palaeogeographical scheme of Fig. 3, a continuous land connection, required for the migration of land vertebrates, is shown along the Pacific margin of the Antarctic Peninsula and South America. In this scheme the Late Cretaceous position of the southern Gondwana continents is that of Norton (1982). The location of the seaway between the base of the Antarctic Peninsula and East Antarctica was adopted from Dalziel & Elliot (1982) and Woodburne & Zinsmeister (1984). If the idea of a continuous land connection as explained above is accepted, then this seaway is necessary to justify the affinities between the Late Cretaceous marine invertebrate faunas of Antarctica–Patagonia and New Zealand (Zinsmeister, 1982).

There is independent evidence suggesting a relatively warm climate during the Late Cretaceous, with the presence, particularly in the peninsula region, of an abundant flora (Woodburne & Zinsmeister, 1984; Francis, 1986). The occurrence of trunks, leaves and plant fragments in the Santa Marrta Formation clearly indicates that the land situated to the west of James Ross Island was densely vegetated. According to palaeomagnetic data, the Antarctic Peninsula was situated near its present position by the Late Cretaceous (Codignotto *et al.*, 1978; Dalziel & Elliot, 1982), indicating that this forest was able to grow at the palaeolatitude of James Ross Island (i.e. about 64° S).

Acknowledgements

We would like to thank the Instituto Antártico Argentino, Fuerza Aérea Argentina and the Centro de Investigaciones en Recursos Geológicos (CIRGEO) for providing the means and logistic support. Alejandro López Angriman and Ricardo Roura (University of Buenos Aires) and Mario Buirás and Jorge Amat (Comando Antártico de Ejército) were active collaborators during the fieldwork. We are indebted to Dr William J. Zinsmeister (Purdue University) for the critical reading of part of this study.

The elaboration of this study has been made in part by one of the authors (EBO) at Purdue University (Indiana, USA) during the tenure of a Fellowship of Consejo Nacional de Investigaciones Científicas y Técnicas (Argentina). All the assistance and support provided by the Geosciences Department and Interlibrary Loan Office of Purdue University is greatly appreciated.

References

Baez, A. & Gasparini, Z. (1979). The South American herpetofauna: an evaluation of the fossil record. In *The South American Herpetofauna: its Origin, Evolution and Dispersal*, ed. W. Duellman, pp. 29–54. Lawrence; The University of Kansas, Monograph 7.

Bonaparte, J.F. (1986). History of the terrestrial Cretaceous vertebrates of Gondwana. *IV Congreso Argentino de Paleontologia y Bioestratigrafia, Mendoza. Actas*, 2, 63–95.

Carpenter, K. & Breithaup, B. (1986). Latest Cretaceous occurrence of nodosaurid ankylosaurs (Dinosauria: Ornithischia) in western North America and the gradual extinction of the dinosaurs. *Journal of Vertebrate Paleontology*, **6(3)**, 252–7.

Codignotto, J.O., Llorente, R.A., Mendia, J.E., Olivero, E. & Spikermann, J.P. (1978). Geologia del cabo Spring y de las islas Leopardo, Pinguino y Cesar. *Instituto Antartico Argentino, Contribucion*, **216**, 1–41.

Colbert, E.H. (1982). Mesozoic vertebrates of Antarctica. In *Antarctic Geoscience*, ed. C. Craddock, pp. 619–27. Madison; University of Wisconsin Press.

Coombs, W.P. (1978). The families of the ornithischian dinosaur order Ankylosauria. *Palaeontology*, **21(1)**, 143–70.

Covacevich, V. & Rich, P.V. (1982). New bird ichnites from Fildes Peninsula, King George Island, West Antarctica. In *Antarctic Geoscience*, ed. C. Craddock, pp. 245–54. Madison; University of Wisconsin Press.

Dalziel, I.W.D. & Elliot, D.H. (1973). The Scotia Arc and Antarctic margin. In *The Ocean Basins and Margins: 1. The South Atlantic*, ed. A.E.M. Nairn & F.G. Stehli, pp. 171–246. New York; Plenum Press.

Dalziel, I.W.D. & Elliot, D.H. (1982). West Antarctica: problem child of Gondwanaland. *Tectonics*, **1(1)**, 3–19.

Francis, J.E. (1986). Growth rings in Cretaceous and Tertiary wood from Antarctica and their palaeoclimatic implications. *Palaeontology*, **29(4)**, 665–84.

Gasparini, Z., De la Fuente, M. & Donadio, O. (1986). Los reptiles cenozoicos de la Argentina. Implicancias paleoambientales y evolucion biogeografica. *IV Congreso Argentino de Paleontologia y Bioestratigrafia*, 2, pp. 119–30. Mendoza.

Gasparini, Z., Olivero, E., Scasso, R. & Rinaldi, C. (1987). Un ankylosaurio (Reptilia, Ornithischia) campaniano en el continente antartico. *Anais do X Congreso Brasileiro de Paleontologia*, Rio de Janeiro, 1987, **1**, 131–41.

Ineson, J.R. (1985). Submarine glide blocks from the lower Cretaceous of the Antarctic Peninsula. *Sedimentology*, **32**, 659–70.

Ineson, J.R., Crame, J.A. & Thomson, M.R.A. (1986). Lithostratigraphy of the Cretaceous strata of west James Ross Island, Antarctica. *Cretaceous Research*, **7**, 141–59.

Macellari, C.E. (1986). Late Campanian–Maastrichtian ammonites from Seymour Island, Antarctic Peninsula. *Journal of Paleontology, Memoir* **18**, 1–55.

Malagnino, E.C., Olivero, E.B., Rinaldi, C.A. & Spikermann, J.P. (1978). Aspectos geologicos del borde occidental de la isla James Ross, Antartida. *VII Congreso Geologico Argentino, Neuquén*, Actas, 1, pp. 489–503.

Medina, F.A., Rinaldi, C.A., Del Valle, R.A. & Baldoni, A.M. (1982). Edad de la Formacion Lower Kotick Point en la isla James Ross, Antartida. *Ameghiniana*, **19(3–4)**, 263–72.

Norton, I.O. (1982). Paleomotion between Africa, South America, and Antarctica, and implications for the Antarctic Peninsula. In *Antarctic Geoscience*, ed. C. Craddock, pp. 99–106. Madison; University of Wisconsin Press.

Olivero, E.B. (1984). Nuevos amonites campanianos de la isla James Ross, Antartida. *Ameghiniana*, **21(1)**, 53–84.

Olivero, E.B., Scasso, R.A. & Rinaldi, C.A. (1986). Revision del Grupo Marambio en la isla James Ross, Antartida. *Instituto Antartico Argentino, Contribucion*, **331**, 1–29.

Riccardi, A.C. (1987). Cretaceous paleogeography of southern South America. *Palaeogeography, Palaeoclimatology, Palaeoecology*, **59**, 169–95.

Woodburne, M.O. & Zinsmeister, W.J. (1984). The first land mammal from Antarctica and its biogeographic implications. *Journal of Paleontology*, **58(4)**, 913–48.

Zinsmeister, W.J. (1982). Late Cretaceous–Early Tertiary molluscan biogeography of the southern circum-Pacific. *Journal of Paleontology*, **56(1)**, 84–102.

Zinsmeister, W.J. (1987). Cretaceous paleogeography of Antarctica. *Palaeogeography, Palaeoclimatology, Palaeoecology*, **59**, 197–206.

Palaeoclimatic significance of Cretaceous–early Tertiary fossil forests of the Antarctic Peninsula

J.E. FRANCIS

Department of Geology and Geophysics, University of Adelaide, PO Box 498, Adelaide, South Australia 5001

Abstract

During the Cretaceous and early Tertiary forests grew on the volcanic arc which now forms the Antarctic Peninsula. The wood of the trees was subsequently permineralized and identification of these fossils indicates that the forests were composed of predominantly podocarp and araucarian conifers and some rarer conifer types. Broad-leaved angiosperms became increasingly common during the Late Cretaceous. Fossils of the southern beech (*Nothofagus*) occur for the first time in Campanian strata on Seymour Island and represent the earliest record of *Nothofagus* macrofossils worldwide. These forests were the ancestral stock of the temperate evergreen forests of South America and Australasia today. Growth rings are consistently wide and uniform throughout most of the Antarctic Peninsula Cretaceous and Eocene wood, indicating that the forest environment was very favourable for tree growth. Both the forest composition and the tree-ring patterns are comparable to those of modern cool (or possibly warm) temperate forests such as those in New Zealand or Tasmania. However, there is a radical short-term change in growth ring patterns in upper Maastrichtian and Palaeocene wood from Seymour Island. The rings are extremely narrow indicating that tree growth was very slow. This may have been in response to a marked cooling of the climate during the Cretaceous–Tertiary transition.

Introduction

The fossil floras from the northern Antarctic Peninsula region are important because a fairly continuous record through the Cretaceous and Tertiary sequence is present. These floras document evolution of the Southern Hemisphere vegetation and record the first appearance of the angiosperms in this area, as well as the response of the vegetation to events during the Cretaceous–Tertiary transition.

Fossil wood is an important component of the fossil flora since it occurs in sediments in which foliage has not been preserved. The wood is of value for determining forest composition and for analysis of growth rings, which are a unique record of the terrestrial climate under which the floras grew.

Geological setting

Fossil wood is present in sediments ranging from Early Cretaceous (Barremian) age on Byers Peninsula, Livingston Island, South Shetland Islands (Fig. 1), through to late Eocene–early Oligocene age on Seymour Island. These sediments represent deposition on and around the margins of an emergent volcanic arc. In the earliest Cretaceous the land on the arc was forested and the wood was subsequently buried in non-marine fluvial sediments of the Byers Formation on Livingston Island and the Botany Bay Group on Trinity Peninsula (including the Mount Flora Formation at Hope Bay). Wood from these areas is mostly silicified, the silica having been derived from the volcaniclastic sediments within which the wood is buried.

During the early–mid Cretaceous the eastern margin of the arc was tectonically active and wood was not well preserved within the coarse clastic fan material of the Gustav Group, which now crops out on western James Ross Island (Fig. 1) (Ineson, Crame & Thomson, 1986). By the Late Cretaceous the back-arc basin was progressively filling with muds and silts (Marambio Group on James Ross Island, Lopez de Bertodano Formation on Seymour Island). Wood within these marine sediments is now calcified and occurs within concretions. In the early Tertiary wood was incorporated into nearshore marine sediments of the Cross Valley, Sobral and La Meseta formations, which were deposited from deltas and river systems draining vegetated highlands to the west. These formations now crop out on Seymour Island. Details of the field locations of fossil wood specimens and their ages are given in Francis (1986). Wood specimens are now housed in collections of the British Antarctic Survey, Cambridge and the British Museum (Natural History), London.

During the Cretaceous and Tertiary the Antarctic Peninsula was positioned at fairly high palaeolatitudes. Its exact position

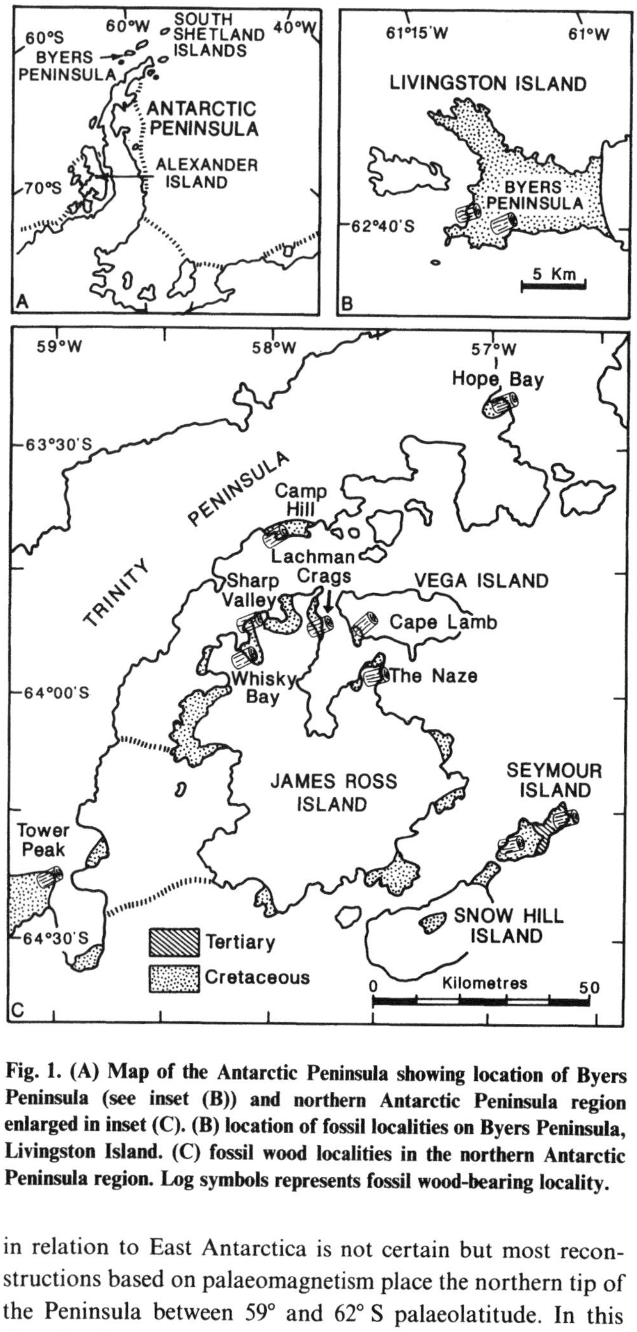

Fig. 1. (A) Map of the Antarctic Peninsula showing location of Byers Peninsula (see inset (B)) and northern Antarctic Peninsula region enlarged in inset (C). (B) location of fossil localities on Byers Peninsula, Livingston Island. (C) fossil wood localities in the northern Antarctic Peninsula region. Log symbols represents fossil wood-bearing locality.

in relation to East Antarctica is not certain but most reconstructions based on palaeomagnetism place the northern tip of the Peninsula between 59° and 62° S palaeolatitude. In this situation the trees would not have been subject to long months of winter darkness. Today the cold Antarctic climate prohibits tree growth, and thus the presence of fossil forests on the Antarctic continent is evidence that the polar climate in these latitudes was much less severe in the Cretaceous–early Tertiary interval.

Forest composition

Wood types

Fossil wood in sediments older than those dated as Campanian is dominantly conifer wood. The predominant type resembles wood of modern Podocarpaceae and is characterized by round, separate bordered pits on the radial walls of the tracheids, mostly uniseriate rays and 1–4 rounded or elliptical cross-field pits on the radial walls of the rays. These woods can be assigned to the fossil form-genera *Mesembrioxylon* Seward or *Podocarpoxylon* Stopes, which by definition include woods comparable to all genera of modern Podocarpaceae. However, preservation of the fossil wood is good enough for further divisions to be distinguished. Most specimens belong to *Podocarpoxylon* Krausel (Fig. 2*a*). In addition, wood of type *Phyllocladoxylon* Krausel is distinguishable as it has large single, simple pores on the radial ray cell walls (Fig. 2*b*). Two specimens have very distinct spiral thickenings on the tracheid walls, which in modern species are characteristic of *Dacrydium cupressinum*. In the Alexander Island fossil forests *Circoporoxylon* is the most common type of fossil wood (Jefferson, 1982). This wood is also similar to that of living podocarps.

Fossil wood with structure resembling that of modern Araucariaceae, *Araucarioxylon* Krausel, is the next most common type present. These woods have distinct biseriate, polygonal tracheid pits in an alternate arrangement (Fig. 2*d*). One specimen from Lachman Crags retains its central pith which consists of parenchyma cells and large, irregular-shaped sclereids (Fig. 2*c*). Some woods contain orange spools of resin in their tracheids, which are reminiscent of those in *Agathoxylon* Krausel.

A rarer type of wood is that with contiguous-bordered pits on tracheid walls, similar to more primitive woods such as *Protocupressinoxylon* Eckhold and other woods associated with *Clasopollis*-bearing plants of the Cheirolepidiaceae.

Angiosperm wood appears first in Campanian sediments on James Ross and Vega islands and subsequently in all overlying formations. In this limited collection 10 angiosperms were identified. The most common type and one of the first to appear is *Nothofagoxylon* Gothan, fossil wood similar to that of southern beech (Figs 2*e* & *f*). The vessels have a ring-porous distribution and tend to be solitary or arranged in groups of 2–4. Rays are predominantly uniseriate and homogeneous, though biseriate and heterogeneous rays do occur. The perforation plates are mostly simple. *Nothofagoxylon* wood was first recorded by Gothan (1908) from Tertiary sediments on Seymour Island (*N. scalariforme*) and has also been described from Tertiary strata of King George Island, South Shetland Islands (*N. antarcticus*) (Torres, 1984). Other angiosperm woods in this collection include three other types, yet to be identified, which are all diffuse porous. One type has very pronounced short, multiseriate rays up to 10 cells wide and long scalariform perforation plates.

Climate interpretation from tree rings

Fossil wood is also useful in that it records in detail within the pattern of the rings the climate under which the trees grew. Growth rings in the wood from the Antarctic Peninsula are well defined and exhibit consistently uniform and fairly wide rings in the Cretaceous and early Tertiary specimens (Fig. 2*g*). Most wood samples have rings with radius of curvature greater than 10 cm, indicating that they came from the outer parts of large trunks or branches and reflect the general climate rather than that of the local forest environment. The average ring width is 2.30 mm for all wood samples, with a

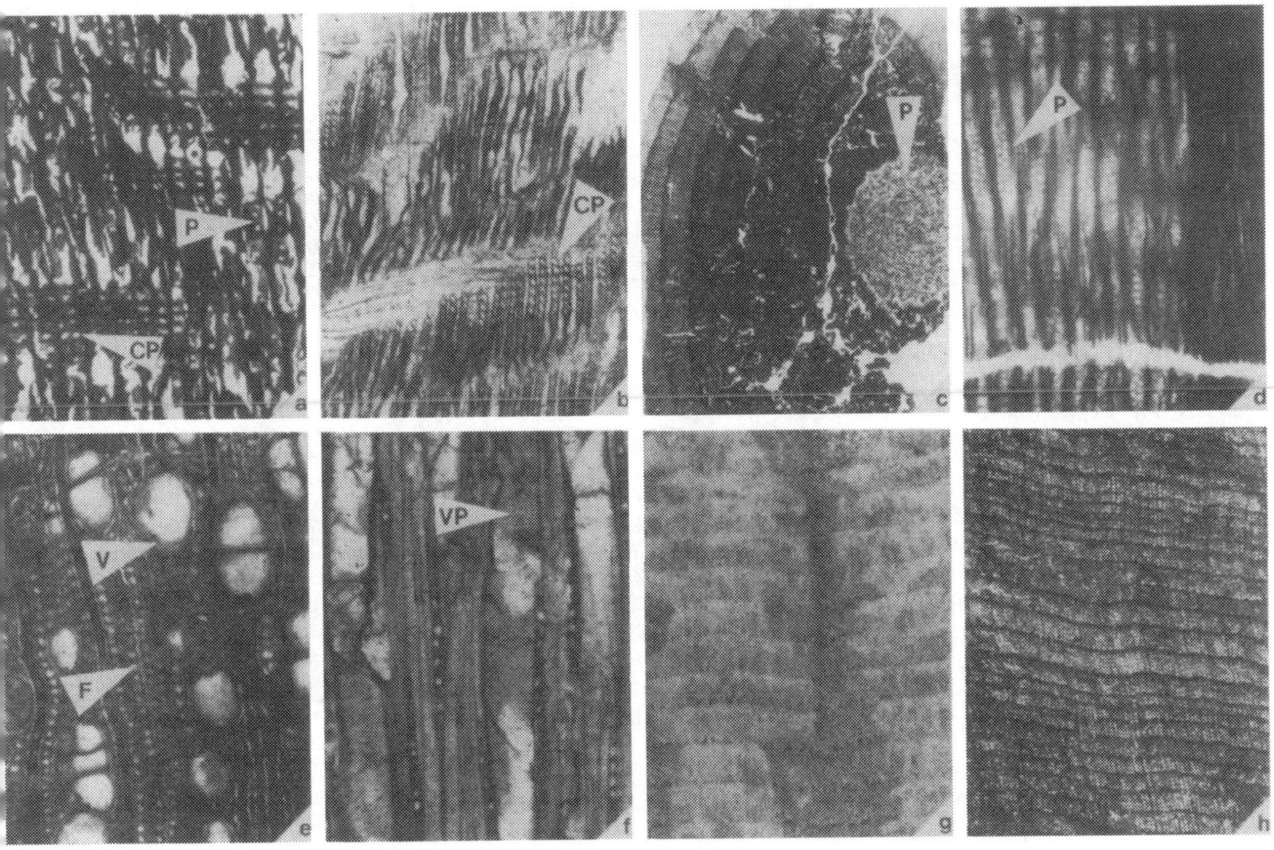

Fig. 2. Fossil wood (numbers refer to BAS collection numbers); (*a*) *Podocarpoxylon*. 4388, Cross Valley Formation, Seymour Island. Note round, separate pits on radial walls of tracheids (P) and up to two cross-field pits on the ray cell walls (CP). Radial section × 170; (*b*) *Phyllocladoxylon*. D.8311.5, Kotick Point Formation, Sharp Valley, James Ross Island. Note large, elliptical cross-field pits (CP). Radial section × 75; (*c*) *Araucarioxylon*. 5060, Marambio Group, The Naze, James Ross Island. Note large central pith (P). Transverse section × 1.5; (*d*) the same specimen. Note alternate arrangement of tracheid pits (P). Radial section × 150; (*e*) *Nothofagoxylon*. D.8321.4, La Meseta Formation, Seymour Island. Note large vessels (V) and thick-walled fibres (F). Transverse section × 55; (*f*) the same specimen. Note opposite arrangement of vessel pitting (VP). Tangential section × 38; (*g*) conifer D.48.5, Mt Flora Formation, Hope Bay. Note the wide (average width 1.48 mm) and very uniform growth rings × 3.5; (*h*) conifer 754, Lopez de Bertodano Formation, Seymour Island. The average ring width is only 0.4 mm, × 20.

range of individual mean ring widths of 0.52–7.50 mm (Francis, 1986). This is reasonably rapid growth but the presence of some trees with average ring widths of > 5 mm shows that at times the environment was extremely favourable for tree growth. More significantly, the ring widths are very uniform throughout each sample. The measure of this variation, the mean sensitivity, yielded a majority of values between 0.1–0.2, indicating that their growth was very consistent from year to year within a uniform environment and was not affected by any local environmental fluctuations.

Individual rings consist of a wide zone of large earlywood cells formed during the growing season, followed abruptly by a narrow zone of small, dense latewood cells. This probably signifies that the growing season was warm and suitable for cell growth but conditions then changed markedly with the onset of the winter season, probably as low light levels or low temperatures retarded cell expansion.

The wide, uniform tree rings and the absence of false or partial rings indicate therefore that the Antarctic Peninsula trees grew in stable forest environments with no obvious external stresses. Such conditions are usually found in forest interiors, in contrast to the environments at the climatically determined limits of tree distribution where low temperatures at high altitude or low rainfall at the lower forest borders produce limiting growing conditions. The very variable ring widths recorded by Jefferson (1982) in wood from the Early Cretaceous forests of Alexander Island may indicate that these more southerly trees were approaching the marginal limits of the Antarctic Peninsula forests.

Earliest record of *Nothofagus* macrofossils

The fossil history of *Nothofagus* is very important since this genus, limited at present to the Southern Hemisphere (excluding Africa), has been regarded as a key to understanding the evolution of life in relation to the break-up of Gondwana. The discovery of *Nothofagoxylon* in Campanian sediments of The Naze, James Ross Island is the oldest recorded occurrence of wood or any other *Nothofagus* macrofossil. Previously the oldest recorded occurrence was of wood and leaves from late Eocene sediments on Seymour Island (Dusen, 1908; Gothan, 1908) and of early Tertiary age from Australia and South America (Tanai, 1986, fig. 12). The pollen, *Nothofagidites*, appears in late Cretaceous sediments in southern South

America (Maastrichtian) and in Australasia. On the Antarctic Peninsula the earliest record of the pollen *Nothofagidites* spp. is from Campanian sediments of Cape Lamb, Vega Island (Askin, 1983). Thus both pollen and wood first appear in the same beds.

Climate deterioration across the Cretaceous–Tertiary boundary

Although most of the Cretaceous and Tertiary trees have wide and uniform growth rings, a radical change in ring characteristics is present in fossil wood from strata adjacent to the Cretaceous–Tertiary boundary. This includes both conifer and angiosperm wood from the Lopez de Bertodano Formation on southern Seymour Island, dated as late Maastrichtian (Huber, Harwood & Webb, 1983) and from the Palaeocene Sobral Formation. Growth rings in seven wood samples of trunk and branch wood are all extremely narrow (Fig. 2*h*). The mean ring width for all these samples is 0.68 mm, ranging from 0.27–1.18 mm for individuals. This represents extremely slow growth. However, the ring widths are still fairly uniform throughout each tree. The mean sensitivity values range from 0.132–0.442, average 0.252; these are again 'complacent' and illustrate that the trees were not affected by a single variable environmental factor. This suggests that the stress was not due to water shortage as false rings are uncommon and such stress usually produces variable rings. The most likely cause may have been low temperatures during the growing season which inhibited growth.

On Seymour Island the Cretaceous–Tertiary boundary appears to be located within an interval of 73 m. One possible position is at a glauconite band approximately 50 m below the base of the Lopez de Bertodano Formation above which Maastrichtian marine molluscs and calcareous microfossils are rare or absent. Alternatively, the boundary may be at the disconformable contact of the Lopez de Bertodano Formation with the overlying Palaeocene Sobral Formation. At this position silicoflagellate assemblages change abruptly (Harwood, 1986) and foraminifera show high extinction rates, inferring a major change in ocean chemistry (Huber, 1986). However, the majority of diatoms show no response to any change and palynomorphs, both marine and terrestrial, exhibit only a gradual change (Askin, 1985).

The climate deterioration reflected in the tree rings does not, however, correspond with these boundary positions. In fact they suggest a longer period of possibly colder climate from the late Maastrichtian to the Palaeocene and therefore tree-ring evidence does not support the theory that a catastrophic event affected the biosphere at the very end of the Cretaceous. A similar sharp climatic deterioration at the end of the Cretaceous was also recorded by Saito & Van Donk (1974) on the basis of oxygen-isotope studies of foraminifera from the south-west Atlantic and the Indian oceans. A cooling of the climate may have affected many aspects of the flora and fauna if the temperatures had fallen below certain threshold levels. By the time of deposition of the Cross Valley Formation in the late Palaeocene the climate had returned to its former warmer levels as shown by rings in the wood from the Cross Valley and La Meseta formations, which are on average > 2 mm in width.

Climate of the Antarctic Peninsula forests

The assemblages of fossil wood indicate that the ancient Antarctic Peninsula forests were composed mainly of podocarp and araucarian conifers, which were subsequently joined by *Nothofagus* and other broad-leaved angiosperms in the Late Cretaceous. This reflects the typical composition of Southern Hemisphere temperate forests today. In addition, fossil foliage in, for example, the Hope Bay and Alexander Island floras indicate that ferns, cycadophytes and bryophytes were also present, probably in the undergrowth. The trees also included ginkgophytes.

These ancient forests covered much of the Gondwanan landmass and consequently remained on the isolated continents after break-up. Temperate evergreen conifer/broad-leaved forests are now found in southern South America (the Valdivian and Magellanic forests), in parts of Australia, Tasmania, New Zealand and in montane New Caledonia and New Guinea, although they were unable to survive in Antarctica when the higher latitudes became colder. These forests live in cool, moist, semi-tropical habitats from sea level to subalpine altitudes (Ovington, 1983). Towards more southerly latitudes species diversity decreases markedly and growth is very slow in the cold but uniform environment with no seasonal extremes, such as in the Magellanic forests of Chile. Rainfall in these areas is very high, often > 2000 mm/y. Growth rings in the fossil wood are wider than those in the southerly Magellanic forests of South America, being more comparable to those in trees from the Tasmanian and New Zealand forests (Francis, 1986) where podocarps and southern beech occur together, particularly in somewhat warmer sites such as lowland areas. Araucarian trees tend to become more dominant in the warmer areas to the north in lower latitudes. The large, entire-margined leaves in the Alexander Island flora, which may not have been able to tolerate freezing temperatures, led Jefferson (1982) to conclude that the climate was probably of warm temperate type. The climate tolerances of the ancient trees may have been somewhat different, particularly before angiosperms appeared and the podocarp and araucarian conifers occupied broader environmental niches. Nevertheless, the remarkable similarity of both composition and growth between the ancient and modern forests suggests that a cool, or possibly warm, temperate climate existed in the past.

Acknowledgements

I am grateful for facilities provided by the British Antarctic Survey, Cambridge, with whom this research was undertaken. Drs B.T. Huber and W.J. Zinsmeister kindly provided samples of wood from Seymour Island. The Trans-Antarctic Association generously donated funds for my attendance at the conference.

References

Askin, R.A. (1983). Campanian palynomorphs from James Ross and Vega Islands, Antarctic Peninsula. *Antarctic Journal of the United States*, **18(5)**, 63–4.

Askin, R.A. (1985). Gradual palynological change at the Cretaceous–Tertiary boundary in Antarctica. Workshop on Cenozoic Geology of the Southern High Latitudes. Sixth Gondwana Symposium. Columbus; Ohio State University. Abstract, 1.

Dusen, P. (1908). Über die Tertiare Flora der Seymour-Insel. *Wissenschaftliche Ergebnisse der Schwedischen Südpolarexpedition*, **3**, 27pp.

Francis, J.E. (1986). Growth rings in Cretaceous and Tertiary wood from Antarctica and their palaeoclimatic implications. *Palaeontology*, **23(4)**, 665–84.

Gothan, W. (1908). Die fossilen Holzer der Seymour-und Snow Hill-Insel. *Wissenschaftliche Ergebnisse der Schwedischen Südpolarexpedition*, **3**, 33.

Harwood, D.M. (1986). Seymour siliceous microfossil biostratigraphy. *Geological Society of America, Abstracts with Programs*, **18(4)**, 292.

Huber, B.T. (1986). Foraminiferal evidence for a terminal Cretaceous oceanic event in Antarctica. *Geological Society of America, Abstracts with Programs*, **18(4)**, 309–10.

Huber, B.T., Harwood, D.M. & Webb, P.N. (1983). Upper Cretaceous microfossil biostratigraphy of Seymour Island, Antarctic Peninsula. *Antarctic Journal of the United States*, **18(5)**, 72–4.

Ineson, J.R., Crame, J.A. & Thomson, M.R.A. (1986). Lithostratigraphy of the Cretaceous strata of west James Ross Island, Antarctica. *Cretaceous Research*, **7(2)**, 141–59.

Jefferson, T.J. (1982). The Early Cretaceous fossil forests of Alexander Island, Antarctica. *Palaeontology*, **25(4)**, 681–708.

Ovington, J.D. (1983). *Temperate Broad-Leaved Evergreen Forests*, Ecosystems of the World, 10. Amsterdam; Elsevier. 241 pp.

Saito, T. & Van Donk, J. (1974). Oxygen and carbon isotope measurements of Late Cretaceous and Early Tertiary foraminifera. *Micropaleontology*, **20(2)**, 152–77.

Tanai, T. (1986). Phytogeographic and phylogenetic history of the genus *Nothofagus* Bl. (Fagaceae) in the Southern Hemisphere. *Journal of the Faculty of Science, Hokkaido University*, Series IV, **21(4)**, 505–82.

Torres, T.G. (1984). Identificación de madera fosil del Terciario de la Isla Rey Jorge, Islas Shetland del Sur, Antártica. In *III Congreso Latinoamericano de Paleontología*, ed. M. del Carmen Perrilliat, pp. 555–65. México; Instituto de Geologia de la UNAM.

Tertiary glaciation in the South Shetland Islands, West Antarctica: evaluation of data

K. BIRKENMAJER

Institute of Geological Sciences, Polish Academy of Sciences, ul. Senacka 3, 31–002 Kraków, Poland

Abstract

Evidence for at least four Tertiary glacial events is provided from King George Island, South Shetland Islands. The oldest (Kraków Glaciation) is documented by fossiliferous glaciomarine sedimentary rocks passing upward into basaltic hyaloclastite capped by basaltic lava with a K–Ar age of ~ 49.4 Ma (Eocene). The next glacial event (Polonez Glaciation) was probably Antarctic-wide and is well documented by lodgement till and the succeeding fossiliferous glaciomarine sedimentary rocks, capped by andesite–dacite lavas yielding a K–Ar age of > 23.6 Ma. Both the Polonez Glaciation deposits and the capping lavas were deeply eroded and redeposited as fluvial and debris-flow clastics during an interglacial-type climatic event (Wesele Interglacial). These clastics are themselves overlain by subaerial, predominantly andesite lavas alternating with lahar-type clastics and cut by narrow buried valleys filled with continental glacial tillite (Legru Bay Group). K–Ar dating of the lavas yielded ages between 29.5 and 25.7 Ma, suggesting a late Oligocene age for the Legru Glaciation. *Nothofagus*-podocarp assemblages, which occur in the upper part of the volcanic–sedimentary Point Hennequin Group (Mount Wawel Formation), dated at 24.5 Ma, may represent another interglacial-type climatic event (Wawel Interglacial). The Melville Glaciation (early Miocene, ~ 22–20 Ma) is well constrained by fossiliferous marine tillite and associated, K–Ar-dated volcanic rocks.

Introduction

The documentation of Tertiary terrestrial tillites and fossiliferous glaciomarine sedimentary rocks, and terrestrial, plant-bearing interglacial deposits on King George Island, South Shetland Islands, is less than a decade old. It began with the discovery of terrestrial and marine tillites associated with the so-called *Pecten* conglomerate (Birkenmajer, 1980), the latter attributed to the Pliocene (Barton, 1965) by comparison with the *Pecten* conglomerate of Cockburn Island (Fig. 1; Andersson, 1906). These deposits on King George Island (Polonez Cove Formation) served as a basis for distinguishing the Polonez Glaciation (Birkenmajer, 1980, 1982a, 1985). K–Ar dating of associated lavas (in Birkenmajer & Gaździcki, 1986), and the stratigraphical evaluation of calcareous nannoplankton (Gaździcka & Gaździcki, 1985; Birkenmajer, 1987) led to a revision of the age of the glacially controlled deposits, from Pliocene to Oligocene.

Interglacial fluvial and debris-flow deposits (Wesele Interglacial) separate the Polonez Cove Formation and its capping lavas (Boy Point Formation) from the Legru Bay Group. The latter group yielded evidence for a local glaciation (Legru Glaciation) of King George Island. Its age, initially correlated with the Plio-Pleistocene boundary (Birkenmajer, 1980), has been revised to late Oligocene by K–Ar dating of andesite lavas associated with terrestrial tillites (Birkenmajer *et al.*, 1986).

Another glaciomarine sequence, termed the Cape Melville Formation (Birkenmajer, 1982b, 1984), has been recognized at Cape Melville and its vicinity. Its early Miocene age was proved by K–Ar dating of associated andesite dykes (Birkenmajer *et al.*, 1985a) and by brachiopod and foraminiferal assemblages (Biernat, Birkenmajer & Popiel-Barczyk, 1985; Birkenmajer & Łuczkowska, 1987).

A mid-Eocene K–Ar date for basalts overlying fossiliferous glaciomarine strata found at Magda Nunatak, King George Bay (Birkenmajer, Paulo & Tokarski, 1985b) provided evidence for the oldest glacial event so far recorded from the South Shetland Islands (Birkenmajer *et al.*, 1986), termed the Kraków Glaciation.

Further K–Ar dating of lavas (Birkenmajer *et al.*, 1983b) associated with plant-bearing strata (Zastawniak, 1981; Zastawniak *et al.*, 1985) indicate that the youngest of these are of latest Oligocene age (~ 24.5 Ma) and may represent an interglacial event between two glacial epochs, the Legru Glaciation (~ 30–25 Ma) and the Melville Glaciation (~ 22–20 Ma).

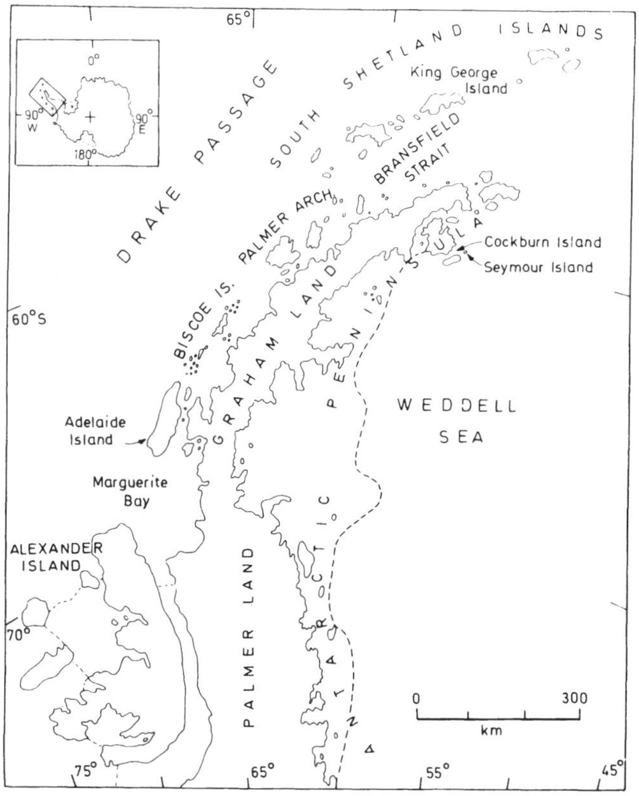

Fig. 1. Sketch maps showing the location of the Antarctic Peninsula and the South Shetland Islands.

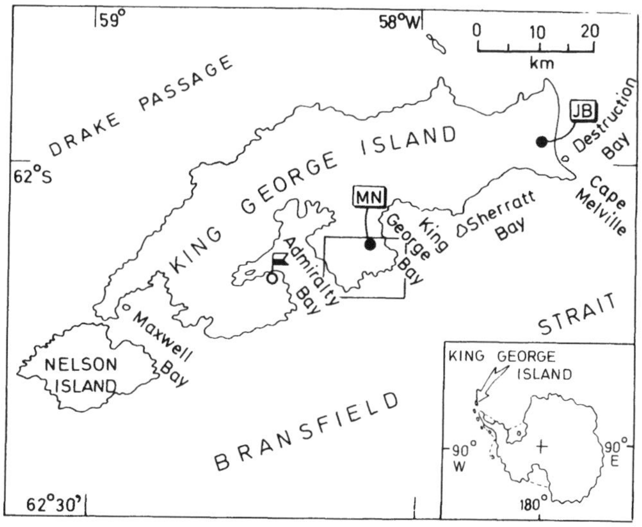

Fig. 2. Key map to show the position of the Polonez Cove–Legru Bay area (box) and the Cape Melville area on King George Island. JB, Jenny Buttress; MN, Magda Nunatak; Arctowski Station (Poland) marked by flag.

Kraków Glaciation

The Kraków Glaciation is represented by an isolated occurrence of fossiliferous glaciomarine clastics with glacially striated, iceberg-rafted dropstones of Antarctic continent provenance at Magda Nunatak (Fig. 2; Birkenmajer *et al.*, 1985b). The glaciomarine deposits (3.6 m thick; bottom not exposed) pass upwards into basaltic hyaloclastite (15–20 m thick), capped by a basaltic lava sheet about 20 m thick. The base of the sequence is unknown.

The glaciomarine deposits yielded Tertiary bivalves and scaphopods of little stratigraphic value (Pugaczewska, 1984), redeposited Cretaceous coccoliths and scarce, poorly preserved Tertiary discoasters (Dudziak, 1984). The matrix of the hyaloclastite overlying the glaciomarine strata yielded Tertiary discoasters: *Discoaster gemmeus* Stradner, *D. lodoensis* Bramlette & Riedel, *Marthasterites tribrachiatus* (Bramlette & Riedel) and *Micrantholithus vesper* Deflandre (determined by Dr J. Dudziak). This assemblage ranges from the uppermost middle Palaeocene through the lower early Eocene. A single K–Ar date of 49.5 ± 5 Ma (mid-Eocene) has been obtained from basaltic lava capping the hyaloclastite (Birkenmajer *et al.*, 1986).

The Kraków Glaciation was probably of minor significance and related to a local glaciation of the highest mountain groups in the Antarctic Peninsula sector. No evidence for glacial climate has so far been detected in apparently continuous (?)middle–(?)lower Oligocene fossiliferous marine strata on Seymour Island (Zinsmeister, 1982). However, there is a break between this marine sequence and the underlying freshwater plant-bearing Palaeocene strata which could correlate with the glacial event recorded from Magda Nunatak. It should be pointed out that lahar-type debris-flow deposits, possibly related to catastrophic melting of local ice-caps on the tops of the highest volcanoes, have been recorded from both Late Cretaceous and early Tertiary volcanic sequences on King George Island (Birkenmajer, 1980, 1982a, 1985).

Polonez Glaciation

The succession of glacially controlled terrestrial and marine deposits of the Polonez Cove Formation, a part of the Chopin Ridge Group, served as a basis for distinguishing the Polonez Glaciation (Birkenmajer, 1980, 1982a). The Polonez Cove Formation rests on basaltic lavas (> 50 m thick) of the Mazurek Point Formation which have yielded a Late Cretaceous K–Ar date of ~ 74 Ma (in Birkenmajer & Gaździcki, 1986). Between the two is an erosional unconformity representing a long stratigraphic hiatus.

The Mazurek Point Formation lavas are cut by deep erosional furrows and depressions infilled with glacifluvial till (Birkenmajer, 1982a). Roches moutonnées and glacial grooves are uncommon, however (Paulo & Tokarski, 1982). Discontinuous basal tillites of the Polonez Cove Formation, the Krakowiak Glacier Member (5–15 m thick), are developed as lodgement till and glacifluvial deposits containing ice-scratched and faceted clasts up to 2 m in diameter. The clasts are mainly of Antarctic continent provenance, although some have a local origin.

There follows a succession of glaciomarine strata resting either upon basal tillites or directly upon basalt. It begins with basaltic conglomerate and breccia containing lenticular *Chlamys* coquinas, designated as the Low Head Member (5–20 m thick). Fine-grained sandstones, siltstones and shales, often with *Penitella* borings (Gaździcki *et al.*, 1982) occur in the middle Siklawa Member (2–10 m thick). Large-scale cross-bedded regressive conglomerates and sandstones (Oberek Cliff Member, 10–40 m thick) terminate the succession. Besides basaltic material, the glaciomarine strata commonly contain iceberg-rafted dropstones (up to 1.5 m in diameter) of Antarctic continent provenance.

The glaciomarine strata of the Polonez Cove Formation contain a rich, mainly shelly fossil fauna (e.g. Gaździcki & Pugaczewska, 1984), characteristic of a very shallow, high-energy, inner-shelf environment. The Tertiary character of this fauna is evident, although no detailed stratigraphic conclusions have been reached due to poor knowledge of contemporaneous faunas in Antarctica and elsewhere (Birkenmajer, 1987). In the calcareous nannofossil assemblage (Gaździcka & Gaździcki, 1985), exclusively Eocene, and most probably recycled, *Chiasmolithus gigas* (Bramlette & Sullivan) occurs together with generally younger taxa, e.g. *C. altus* Bukry & Percival (late Eocene–late Oligocene), *Coronocyclus nitescens* (Kamptner) (Oligocene–Miocene), *Cyclicargolithus floridanus* (Roth & Hay) (early Oligocene–middle Miocene), *Ericsonia muiri* (Black) (late Eocene–early Miocene), *Reticulofenestra bisecta* (Hay, Mohler & Wade) (middle Eocene–late Oligocene), *R. umbilica* (Levin) (middle Eocene–early Oligocene). An Oligocene age for the majority of these forms seems most probable, although some may be recycled (Birkenmajer, 1987).

Andesite–dacite lavas (Boy Point Formation) overlying the Polonez Cove Formation have yielded K–Ar dates of > 23.6 ± 0.3 and > 22.4 Ma (Birkenmajer & Gaździcki, 1986). K–Ar dating of andesite lavas of the Legru Bay Group, which postdate both the Boy Point Formation and the Polonez Cove Formation (29.5 ± 2.1 and 25.7 ± 1.3 Ma) suggests, however, that the Boy Point lavas suffered argon loss due to reheating by the Legru Bay volcanic activity (Birkenmajer *et al.*, 1986). Thus the deposits of the Polonez Glaciation could be of mid- or even early Oligocene age.

Wesele Interglacial

Unfossiliferous coarse-clastic fluvial and debris-flow deposits of the Wesele Cove Formation infill a deep valley cut into the underlying Boy Point Formation lavas and the Polonez Cove Formation. They consist mainly of reworked Boy Point Formation volcanic rocks. These sedimentary rocks are evidence for an interglacial amelioration of climate in the South Shetland Islands following the Polonez Glaciation.

Legru Glaciation

Subaerial, predominantly andesitic lavas alternating with lahar-type debris-flow clastics, cut by narrow buried valleys infilled with continental-type glacial tillites which contain local material from King George Island, have been referred to collectively as the Legru Bay Group (Birkenmajer, 1980, 1982a). Glacial striations on the underlying volcanic rocks, and ice-wedges in the tillites are good evidence for a glacial climatic event, the Legru Glaciation. K–Ar dates of 29.5 ± 2.1 and 25.7 ± 1.3 Ma from the base and the top, respectively, of the 300 m thick lava succession indicate a late Oligocene age for the glaciation (Birkenmajer *et al.*, 1986).

Wawel Interglacial

The youngest Palaeogene floras represented by *Nothofagus*-podocarp assemblages (Zastawniak, 1981; Zastawniak *et al.*, 1985) are recorded from the upper part of the Mount Wawel Formation (> 300 m thick), a higher unit of the Point Hennequin Group (Eocene–Oligocene). Lavas associated with plant-bearing beds have yielded an Oligocene–Miocene boundary K–Ar date of 24.5 ± 0.5 Ma (Birkenmajer *et al.*, 1983b).

Melville Glaciation

The glaciomarine deposits of the Melville Glaciation are represented by the Cape Melville Formation, a part of the Moby Dick Group (Birkenmajer, 1982b, 1984) which is exposed between Cape Melville and Jenny Buttress, in the eastern part of King George Island (Fig. 2). Sedimentary rocks of the Melville Glaciation rest on olivine–augite basalts more than 60 m thick (Sherratt Bay Formation), and with a K–Ar age of > 18 Ma (Birkenmajer *et al.*, 1985a). A stratigraphic hiatus separates the basalts from the overlying marine and glaciomarine strata.

The basalts are overlain successively by the Destruction Bay Formation and the Cape Melville Formation, separated from one another by a slight unconformity. The Destruction Bay Formation (40–100 m thick) is developed best toward the west, and wedges out eastwards. It consists of tuffaceous and psammitic rocks with reworked basaltic material, and displays large-scale cross-bedding. There are horizons rich in marine invertebrates, sometimes with pieces of carbonized wood. There is no indication of glacially controlled sedimentation, and dropstones are missing.

The Cape Melville Formation (about 200 m thick) is represented by glaciomarine deposits: grey to greenish, brownish and black clay–shales and silty shales, with subordinate intercalations of siltstone and fine-grained sandstone, and occasionally thin marl. In the west, the formation rests with local unconformity on the Destruction Bay Formation but in the east it directly overlies the Sherratt Bay basalts. The deposits are rich in iceberg-rafted dropstones (up to 2 m in diameter) of Antarctic continent provenance.

The Destruction Bay Formation has yielded bivalves, gastropods, solitary corals and brachiopods. The brachiopods, represented by several species of the genera *Discinisca*, *Pachymagas*, *Neothyris*, *Rhizothyris*, *Magellania* and (?)*Magella*, are of early Miocene character (Biernat *et al.*, 1985). A *Cibicides–Cibicidoides* foraminiferal assemblage confirms this age (Birkenmajer & Łuczkowska, 1987). Recycled Cretaceous coccoliths (Dudziak, 1984) and belemnites also occur. The sediments were laid down in a shallow-neritic, moderately high-energy marginal shelf–estuarine environment in the east, that deepened to bathyal in the west.

The Cape Melville Formation is very rich in well preserved invertebrate marine fossils (Birkenmajer, 1982b, 1984; Birkenmajer, Gaździcki & Wrona, 1983a): solitary corals, polychaetes, bivalves, gastropods, scaphopods, crabs, etc. The foraminifera comprise two successive early Miocene assemblage zones: (a) *Cibicides–Cibicidoides* zone; and (b) *Cribrostomoides–Cyclammina–Globobulimina* zone (Birkenmajer & Łuczkowska, 1987). There are also recycled Cretaceous coccoliths (Dudziak, 1984) and belemnites. The character of the sediments indicates offshore shelf conditions; the succession of calcareous and arenaceous foraminiferal assemblages suggests bathyal conditions in the lower part of the section, gradually shallowing to depths corresponding to shelf-break and outer shelf, and again deepening to bathyal conditions at the top of the section (Birkenmajer & Łuczkowska, 1987).

The whole Moby Dick Group is crossed by two generations of andesite and basalt dykes. The older dykes have yielded good early Miocene ages of 20.1 ± 0.2 Ma and 19.9 ± 0.3 Ma (Birkenmajer *et al.*, 1985a).

References

Andersson, J.G. (1906). On the geology of Graham Land. *Bulletin of the Geological Institute of Uppsala*, **7(13–14)**, 19–71.

Barton, C.M. (1965). *The Geology of the South Shetland Islands*. III: *The Stratigraphy of King George Island*. London; British Antarctic Survey Scientific Reports, No. 44, 33 pp.

Biernat, G., Birkenmajer, K. & Popiel-Barczyk, E. (1985). Tertiary brachiopods from the Moby Dick Group of King George Island (South Shetland Islands, Antarctica). *Studia Geologica Polonica*, **81**, 109–41.

Birkenmajer, K. (1980). Discovery of Pliocene glaciation on King George Island, South Shetland Islands (West Antarctica). *Bulletin de l'Académie Polonaise des Sciences, Série des Sciences de la Terre*, **27(1–2)**, 59–67.

Birkenmajer, K. (1982a). Pliocene tillite-bearing succession of King George Island (South Shetland Islands, Antarctica). *Studia Geologica Polonica*, **74**, 7–72.

Birkenmajer, K. (1982b). Pre-Quaternary fossiliferous glaciomarine deposits at Cape Melville, King George Island (South Shetland Islands, West Antarctica). *Bulletin de l'Académie Polonaise des Sciences, Série des Sciences de la Terre*, **29(4)**, 331–40.

Birkenmajer, K. (1984). Geology of the Cape Melville area, King George Island (South Shetland Islands, Antarctica): pre-Pliocene glaciomarine deposits and their substratum. *Studia Geologica Polonica*, **79**, 7–36.

Birkenmajer, K. (1985). Onset of Tertiary continental glaciation in the Antarctic Peninsula sector (West Antarctica). *Acta Geologica Polonica*, **35(1–2)**, 1–31.

Birkenmajer, K. (1987). Oligocene–Miocene glaciomarine sequences of King George Island (South Shetland Islands), Antarctica. *Palaeontologia Polonica*, **49**, 9–36.

Birkenmajer, K., Delitala, M.C., Narębski, W., Nicoletti, M. & Petrucciani, C. (1986). Geochronology of Tertiary island-arc volcanics and glacigenic deposits, King George Island, South Shetland Islands (West Antarctica). *Bulletin of the Polish Academy of Sciences, Earth Sciences*, **34(3)**, 257–73.

Birkenmajer, K. & Gaździcki, A. (1986). Oligocene age of the *Pecten* conglomerate on King George Island, West Antarctica. *Bulletin of the Polish Academy of Sciences, Earth Sciences*, **34(2)**, 219–26.

Birkenmajer, K., Gaździcki, A., Kreuzer, H. & Müller, P. (1985a). K–Ar dating of the Melville Glaciation (Early Miocene) in West Antarctica. *Bulletin of the Polish Academy of Sciences, Earth Sciences*, **33(1–2)**, 15–23.

Birkenmajer, K., Gaździcki, A. & Wrona, R. (1983a). Cretaceous and Tertiary fossils in glaciomarine strata at Cape Melville, Antarctica. *Nature, London*, **303(5912)**, 56–9.

Birkenmajer, K. & Łuczkowska, E. (1987). Foraminiferal evidence for a Lower Miocene age of glaciomarine and related strata, Moby Dick Group, King George Island (South Shetland Islands, Antarctica). *Studia Geologica Polonica*, **90**, 81–123.

Birkenmajer, K., Narębski, W., Nicoletti, M. & Petrucciani, C. (1983b). Late Cretaceous through Late Oligocene K–Ar ages of the King George Island Supergroup volcanics, South Shetland Islands (West Antarctica). *Bulletin de l'Académie Polonaise des Sciences, Série des Sciences de la Terre*, **30(3–4)**, 133–43.

Birkenmajer, K., Paulo, A. & Tokarski, A.K. (1985b). Neogene marine tillite at Magda Nunatak, King George Island (South Shetland Islands, Antarctica). *Studia Geologica Polonica*, **81**, 99–107.

Dudziak, J. (1984). Cretaceous calcareous nannoplankton from glaciomarine deposits of the Cape Melville area, King George Island (South Shetland Islands, Antarctica). *Studia Geologica Polonica*, **97**, 37–51.

Gaździcka, E. & Gaździcki, A. (1985). Oligocene coccoliths of the *Pecten* conglomerate, West Antarctica. *Neues Jahrbuch für Geologie und Paläontologie, Monatshefte*, **12**, 727–35.

Gaździcki, A., Gradziński, R., Porębski, S.J. & Wrona, R. (1982). Pholadid *Penitella* borings in glaciomarine sediments (Pliocene) of King George Island, Antarctica. *Neues Jahrbuch für Geologie und Paläontologie, Monatshefte*, **12**, 723–35.

Gaździcki, A. & Pugaczewska, H. (1984). Biota of the '*Pecten* conglomerate' (Polonez Cove Formation, Pliocene) of King George Island (South Shetland Islands, Antarctica). *Studia Geologica Polonica*, **79**, 59–120.

Paulo, A. & Tokarski, A.K. (1982). Geology of the Turrett Point–Three Sisters Point area, King George Island (South Shetland Islands, Antarctica). *Studia Geologica Polonica*, **74**, 81–103.

Pugaczewska, H. (1984). Tertiary Bivalvia and Scaphopoda from glaciomarine deposits at Magda Nunatak, King George Island (South Shetland Islands, Antarctica). *Studia Geologica Polonica*, **79**, 53–8.

Zastawniak, E. (1981). Tertiary leaf flora from Point Hennequin Group of King George Island (South Shetland Islands, Antarctica). *Studia Geologica Polonica*, **72**, 97–108.

Zastawniak, E., Wrona, R., Gaździcki, A. & Birkenmajer, K. (1985). Plant remains from the top part of the Point Hennequin Group (Upper Oligocene), King George Island (South Shetland Islands, Antarctica). *Studia Geologica Polonica*, **81**, 143–64.

Zinsmeister, W.J. (1982). Review of the Upper Cretaceous–Lower Tertiary sequence of Seymour Island, Antarctica. *Journal of the Geological Society, London*, **139(6)**, 779–86.

Late Palaeocene carbonate deposition on the Southwest Indian Ridge

S.W. WISE, JR[1], R.C. TJALSMA[2] & D.H. DAILEY[3]

1 Department of Geology, Florida State University, Tallahassee, Florida 32306, USA
2 Unocal, Science & Technology, PO Box 76, Brea, California 92621, USA
3 PO Box 3808, Tulsa, Oklahoma 74102, USA

Abstract

Islas Orcadas Piston Core 12–14 taken on the southern flank of the south-west Indian Ridge south of Africa recovered upper Palaeocene nannofossil ooze at a present day water depth of 4682 m. Backtracking the core site (presently at 58°26.5′S; 18°14.9′E) demonstrates that carbonate ooze deposition did occur along the more southerly segments of the Southwest Indian Ridge during the early Palaeocene (at which time the core site lay at approximately 56° S in a water depth of about 2700 m). This reinforces speculations that carbonate deposition could have occurred within intracratonic basins of the Antarctic continent during Palaeocene time. The nannofossil content of the core ranges from about 91 to 94%, and the high latitude coccolith assemblage can be assigned a late Palaeocene age corresponding to the high-latitude nannofossil *Heliolithus universus* Zone (which is roughly equivalent to the low-latitude Zones CP5/CP7). Seven species of planktonic foraminifera accompanied by 40 benthic species also indicate a late Palaeocene age (P4/P5). In combination, the calcareous planktonic zonations suggest an age of 59–61 Ma for the material. The 862 cm of nannofossil ooze recovered is overlain by 122 cm of dark yellowish-brown, zeolitic pelagic clay, which contains abundant micromanganese nodules. The carbonate-poor lithology apparently owes its origin to subsidence of the ridge flank sediments below the carbonate compensation depth.

Introduction

Knowledge of the distribution of Palaeogene pelagic carbonate sediments will allow predictions as to whether such sediments may have been deposited in the intracratonic basins of the Antarctic continent itself. An Eocene calcareous nannofossil from the Sirius Formation of the Transantarctic Mountains (Harwood, 1984, fig. 2[9]) suggests that Eocene carbonate sediments were deposited in these basins. Cretaceous and Palaeocene foraminifera have also been reported from Harwood's samples (Webb *et al.*, 1984), but have not yet been documented. As such, the recovery of Palaeocene nannofossil ooze in *Islas Orcadas* Piston Core 12–14 on the southern flank of the Southwest Indian Ridge (at 58° S) (Fig. 1) is of interest and provides a fixed datum point for carbonate deposition in this part of the world. We document here the age and inferred palaeodepth and palaeolatitude of deposition of this sediment, which accumulated on the southernmost portion of the ridge system.

Islas Orcadas (*IO*) Core 12–14 was taken at a water depth of 4682 m at latitude 58°26.5′ S, longitude 18°14.9′ E, where it penetrated a thin wedge of sediment south of a seamount. As

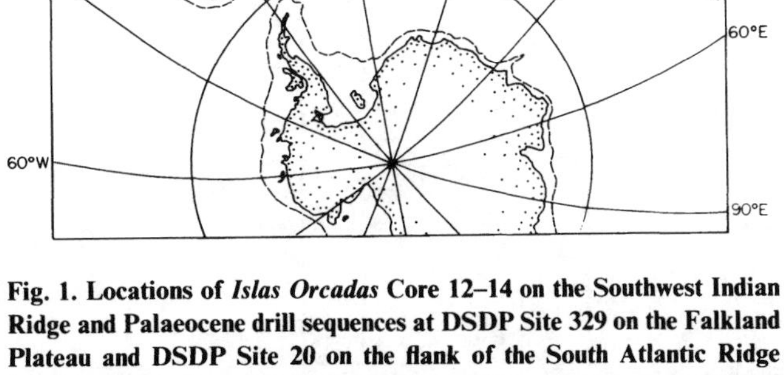

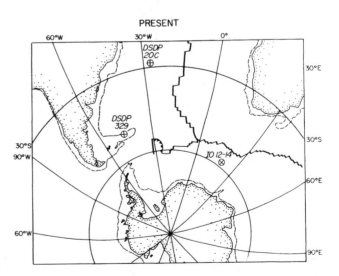

Fig. 1. Locations of *Islas Orcadas* Core 12–14 on the Southwest Indian Ridge and Palaeocene drill sequences at DSDP Site 329 on the Falkland Plateau and DSDP Site 20 on the flank of the South Atlantic Ridge (configuration of present day spreading centres, transform faults and subduction zone are shown as heavy solid lines; from Lawver *et al.*, 1985).

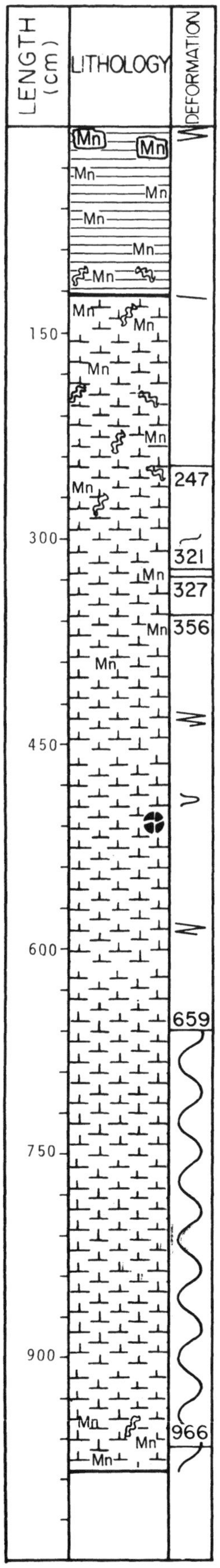

Fig. 2. Lithological column for *Islas Orcadas* Piston Core 12–14 (from Kaharoeddin *et al.* 1979).

IO 12–14 CALCAREOUS NANNOFOSSILS

AGE: LATE PALAEOCENE; ZONE: *Heliolithus universus* (CP5–CP7) — CORE INTERVAL (CM) / SPECIES	807-809	689-691	427-429	298-300	238-240	198-200	178-180	138-140	118-120
Chiasmolithus bidens	V	V	A	A	A	A	A	A	R
Coccolithus pelagicus	C	C	C	C	C	C	C	C	?
'Ericsonia' sp. cf. *subpertusa*	C	C	F				F		
Fasciculithus tympaniformis/involutus	A	A	C	C	C	C	C	C	
Heliolithus sp.					R				
Heliolithus sp. cf. *H. cantabriae*								R	
Heliolithus kleinpellii							R	R	
Heliolithus universus				F	R	F	C	C	
Neochiastozygus concinnus			F					F	
Prinsius bisulcus/martinii	V	V	V	V	V	V	V	V	
Toweius craticulus/tovae	V	V	V	V	V	V	V	V	
Zygodiscus sigmoides	A	C	C	C	C	C	C	C	
Zygodiscus simplex	R		F				R		

Fig. 3. Distribution of calcareous nannofossils in *Islas Orcadas* Core 12–14. Dashed line at 300 cm indicates level below which helioliths are apparently absent.

described by Kaharoeddin *et al.* (1979), the carbonate consists of a very pale orange nannofossil ooze (91–94% nannofossils) with trace amounts of foraminifera (2–3%), diatoms, and insolubles (Fig. 2). Micromanganese nodules are common at certain intervals. At 122 cm the nannofossil ooze changes upwards into a dark yellowish-brown, zeolitic pelagic clay, which also contains abundant micromanganese nodules.

Biostratigraphy

Calcareous nannofossils

The high-latitude calcareous nannofossil assemblage is moderately preserved and dominated by *Chiasmolithus bidens* and species of *Prinsius* and *Toweius* (Fig. 3). The assemblage can be assigned to the upper Palaeocene, high-latitude *Heliolithus universus* Zone of Wise & Wind (1977), based on the presence of the nominative species and the absence of *Discoaster multiradiatus*. This zone corresponds roughly to the low-latitude Zones CP5–CP7 of Okada & Bukry (1980), but in Core 12–14 may equate directly with Zone CP5 due to the absence of any discoasters (such as *D. mohleri*). Discoasters, however, prefer relatively warm waters, and could have been ecologically excluded from these latitudes for the interval represented here. On the other hand, the apparent absence of helioliths below 300 cm (dashed line in Fig. 3) may indicate that the lower portion of the core should be assigned to the *Fasciculithus tympaniformis* Zone (CP4) (but this suggestion would be speculative considering the relatively short section available for study). Of interest is the diversity of the helioliths in the upper portion of the core, some of which approach *Heliolithus kleinpellii* in size and morphology.

Planktonic foraminifera

Dissolution has severely affected the planktonic species and has probably reduced the species diversity as well. The taxonomy is not as straightforward as for equivalent reference material from the Falkland Plateau (Tjalsma, 1977) because several of the diagnostic species are small and are often broken. In particular, most of the *Acarinina* are less abundant and diminutive in size. *Subbotina*, therefore, is the dominant genus throughout the core, while *Acarinina* and *Turborotalia* are far less prominent. *Chiloguembelina* and *Planorotalites* occur only sporadically.

Most common among the species identified are large specimens of the *Subbotina triangularis* group, which show wide morphological variation. *Subbotina varianta*, a small form, is also present. *Acarinina mckannai* is well developed (i.e. closer to normal size) from 718 to 920 cm, whereas less conical morphotypes occur higher in the core; *A. nicoli* occurs throughout the carbonate interval. Other species include *Turborotalia reissi*, *Planorotalites planoconica* and *Chiloguembelina trinitatensis*. The occurrence of *Acarinina mckannai* and *A. nicoli* constrains the biostratigraphical age to the upper Palaeocene Zones P4/P5 of Berggren *et al.* (1985) (see discussion below).

Benthic foraminifera

Quantitative data for benthic foraminifera larger than 150 μm are recorded in Fig. 4. The species are characteristic for the deep-water Palaeocene, therefore the taxonomy is more straightforward than for the planktonics. The assemblages are dominated by *Gavelinella beccariiformis*, which is surprisingly abundant, and to a lesser extent by *Nuttallides truempyi*. Common faunal element includes *Neoponides hillebrandti*, *Bulimina trinitatensis*, *Anomalina praeacuta*, *Oridorsalis* sp., *Aragonia ouezzanensis*, *Cibicidoides* aff. *madrugensis*, *Pullenia coryelli*, *Gyroidinoides globosus* and *Nonion havanense*.

The temporal and spatial distribution of Palaeocene benthics in the South Atlantic was studied by Tjalsma & Lohmann (1983), who noted that Palaeocene deep-water benthics have broad, overlapping bathymetric ranges. As such, it is difficult to assign a specific palaeodepth range to the assemblages of *IO* 12–14, particularly because the core is located in an area far removed from other described South Atlantic Palaeocene localities. Based on the species content only, the *IO* 12–14 assemblages compare best with the upper Palaeocene assemblages of DSDP Hole 20C, located on the west flank of the Mid-Atlantic Ridge with a backtrack depth of 2800–3000 m.

Composite age

Compared with the foraminiferal assemblages from DSDP Site 329 on the Falkland Plateau (Tjalsma, 1977), the *IO* 12–14 fauna is lower in species diversity and lacks the abundance of the various *Acarinina* species (in particular the characteristic *A. tadjikistanensis* group). The assemblages compare best with those encountered in DSDP Core 329–33, the lowermost taken at the site. It contains *A. mckannai/tadjikistanensis* with *Planorotalites* appearing in the middle of the core. As mentioned above, this would equate with planktonic foraminiferal zones P4 and P5.

The nannofossil assemblage of the *IO* Core 12–14 corresponds best with that of the Falkland Plateau DSDP Core 372A-5, Sections 1–4 (4 cm); unfortunately this represents an interval of high dissolution that contains no planktonic foraminifera. As mentioned previously, however, the piston core can be no younger than Zone CP7, which would confine it to the interval of foraminiferal Zone P4. According to the time scale of Berggren *et al.* (1985), the overlap of the foraminiferal (P4/P5) and nannofossil (CP5/CP7) zones dates the core between 59 and 61 Ma (mid-late Palaeocene).

IO 12–14 — BENTHIC FORAMINIFERA

SPECIES / CORE INTERVAL	128-130	228-230	518-520	818-820	916-963
TOTAL NUMBER COUNTED	317	207	244	344	253
PERCENT LISTED	70.9	69.6	74.8	83.2	71.4
Abyssamina quadrata	0.3				
Alabamina creta		0.7	0.8	0.3	
Anomalina praeacuta	6.3	3.0	2.9	2.6	4.0
Aragonia velascoensis	2.8	4.7	3.7	3.2	2.0
Bulimina bradury	0.9	0.3			0.4
Bulimina midwayensis	2.2	1.3	1.2	1.7	
Bulimina trinitatensis	0.6	4.0	8.6	6.1	4.3
Buliminella beaumonti	0.3		0.4	1.2	0.4
Cibicidoides madrugensis	1.6	4.7	1.6	2.9	2.0
Dorothina trochoides		0.3	1.2	0.3	0.8
Gaudryina laevigata					0.4
Gavelinella beccariiformis	15.5	19.2	19.7	27.9	25.3
Gavelinella hyphalus	5.0	1.0		0.3	0.4
Gyroidinoides cf. *quadratus*	0.3	0.7			
Gyroidinoides girardanus	0.6	1.3	0.4	0.3	0.8
Gyroidinoides globosus	0.3	3.7	2.5	2.9	3.2
Gyroidinoides sp. (Tjalsma & Lohmann)	0.6	0.3	1.2	0.6	2.0
Lebticulina cf. *macrodiscus*	2.5	1.0	1.2	0.6	1.2
Lenticulina cf. *turbinatus*	1.3				0.8
Lenticulina velascoensis			0.4		
Lenticulina whitei				0.6	
Neoeponides hillebrandti	7.6	4.4	3.7	5.8	4.7
Nodosaria velascoensis	0.6	0.3			
Nonion havanense	0.9	3.0	3.3	2.9	2.0
Nuttallides truempyi	9.8	5.1	14.3	13.7	4.0
Oridorsalis spp.	1.9	4.0	3.7	3.2	5.5
Osangularia plummerae	0.3	0.7		0.3	
Osangularia velascoensis	0.3	1.3	1.2	1.2	2.0
Pullenia coryelli	4.7	3.0	0.8	1.7	2.4
Pullenia jarvisi	0.3	0.3	0.4	0.6	0.4
Spiroplectammina cf. *jarvisi*		0.7		0.3	0.4
Stilostomella paleocenica	2.2	0.3	0.4		
Tappanina selmensis	0.6			0.3	
Tritaxia paleocenica	0.6	0.3	1.2	1.7	2.0

Fig. 4. Distribution of benthic formaminifera in *Islas Orcadas* Core 12–14. Abundances given as percent of the total assemblage.

Discussion

Although *IO* Core 12–14 did not penetrate to basement, the amount of sediment beneath it, as judged from the seismic

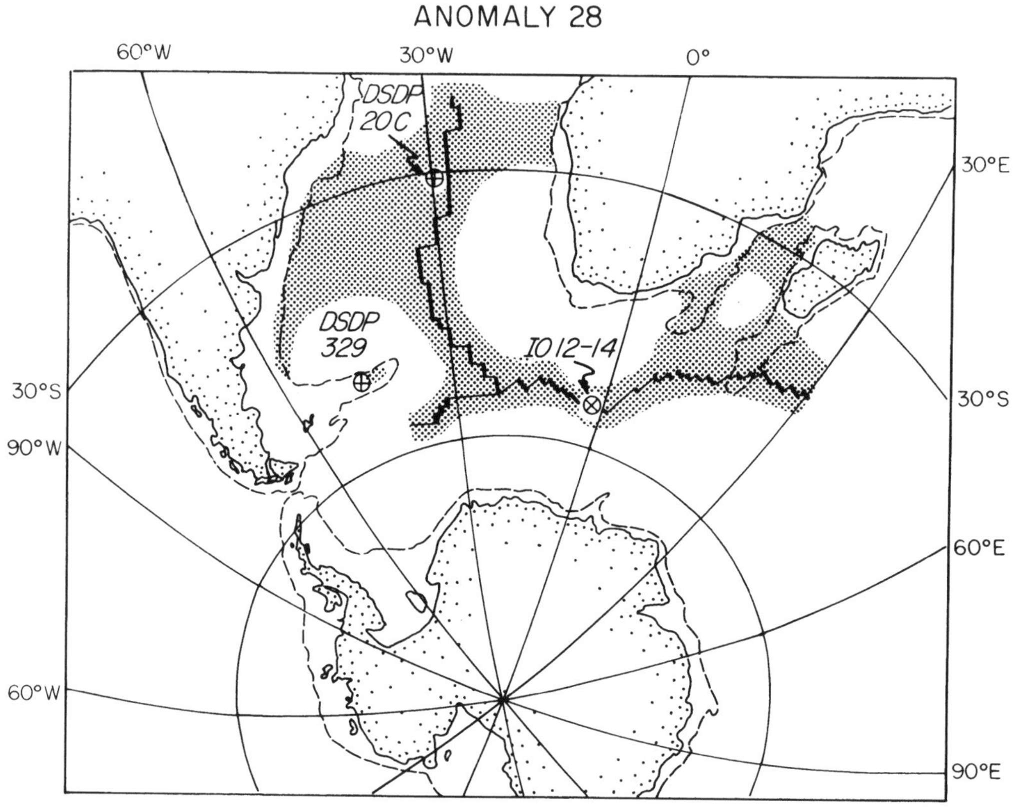

Fig. 5. Reconstruction of South America, Africa and Antarctica for Anomaly 28 time (early Palaeocene); spreading centres and transform faults shown as heavy solid lines (from Lawver *et al.*, 1985). Distribution of carbonate ooze indicated by coarse stippled pattern.

profile (unpublished data), is negligible. Assuming that the underlying sediment is probably Palaeocene in age or not appreciably older, the core locality can be backtracked to the ridge crest on the reconstruction for Anomaly 28 time (early Palaeocene) from Lawver, Sclater & Meinke (1985). This reconstruction (Fig. 5) accords well with the estimated palaeodepth of the benthic foraminiferal assemblage, which is most comparable to that of Hole 20C with a backtrack depth of 2800–3000 m (i.e. slightly below the world average ridge crest depth of 2700 ± 300 m; Sclater, Abbott & Thiede (1977), or of 2518 m for the present day South Atlantic Ridge (Sclater, Anderson & Bell, 1971). As this reconstruction places *IO* 12–14 at or near the ridge crest, it establishes the presence of carbonate ooze (at approximately 56° S) along this southernmost segment of the Southwest Indian Ridge during Palaeocene times. This datum point can be combined with the distribution of Palaeocene carbonate mapped elsewhere in the South Atlantic by McCoy & Zimmerman (1977), and summarized for the Indian Ocean by Sclater & Heirtzler (1977), to produce the regional map for Palaeocene carbonate ooze deposition (> 30% $CaCO_3$) shown in Fig. 5.

The evidence from *IO* 12–14 demonstrates that carbonate deposition was continuous along the ridge system between the South Atlantic and Indian oceans during the Palaeocene. The distribution of Palaeocene carbonate shown in Fig. 5 for the Indian Ocean is similar to that mapped in that region for the Campanian–Maastrichtian and early Eocene by Davies & Kidd (1977). The same is true for the distribution of upper Cretaceous and Eocene carbonate mapped in the South Atlantic by McCoy & Zimmerman (1977). If the evidence cited by Harwood (1984) for the accumulation of Eocene calcareous nannofossils in the intracratonic basins of Antarctica is correct, then it is reasonable to assume that Palaeocene and Campanian–Maastrichtian nannofossils could have been deposited in those same basins as well, assuming adequate communication with the open ocean (see discussion by Toker *et al.*, this volume, p. 639). This seems all the more reasonable considering the relatively deep palaeodepths (2700–3000 m) that nannofossil-rich carbonate oozes were accumulating in the adjacent Southern Ocean, as shown by the present study.

There is a limit, however, on the depth to which carbonate ooze accumulated along the ridge flank during the Palaeocene. This is indicated by the change in lithology near the top of *IO* Core 12–24 from carbonate ooze to non-calcareous clay. This change probably resulted from subsidence of the core site down the ridge flank below the carbonate compensation depth as a consequence of seafloor spreading, an event that apparently took place during late Palaeocene time. Because the core did not penetrate to basement, the palaeodepth at which this transition took place cannot be determined with certainty.

Acknowledgements

Study supported by NSF Grant 84-14268. Mr Wuchang Wei prepared the nannofossil slides and discussions with him and Mr James J. Pospichal were most helpful. Ms Rosemarie

Raymond drafted the figures which were printed by Ms Gabrielle Li.

References

Berggren, W.A., Kent, D.V., Flynn, J.J. & Van Couvering, J.A. (1985). *Cenozoic Geochronology*, **96**, 1407–18.

Davies, T.A. & Kidd, R.B. (1977). Sedimentation in the Indian Ocean through time. In *Indian Ocean Geology and Biostratigraphy*, ed. J.R. Heirtzler, H.M. Bolli, T.A. Davies, J.B. Saunders & J.G. Sclater, pp. 61–85. Washington, DC; American Geophysical Union and Chelsea, Michigan; LithoCrafters, Inc.

Harwood, D.M. (1984), Diatoms from the Sirius Formation, Transantarctic Mountains. *Antarctic Journal of the United States*, **19(5)**, 98–100.

Kaharoeddin, F.A., Eggers, M.R., Graves, R.S., Goldstein, E.H., Hattner, J.G., Jones, S.C. & Ciesielski, P.F. (1979). *ARA ISLAS ORCADAS Cruise 1277 Sediment Descriptions*. Tallahassee, Florida; Sedimentological Research Laboratory Contribution, 47, 108 pp.

Lawver, L.A., Sclater, J.G. & Meinke, L. (1985). Mesozoic and Cenozoic reconstructions of the South Atlantic. *Tectonophysics*, **114**, 233–54.

McCoy, F.W. & Zimmerman, H.B. (1977). A history of sediment lithofacies in the South Atlantic Ocean. In *Initial Reports of the Deep Sea Drilling Project, 39*, ed. P.R. Supko, K. Perch-Nielsen *et al.*, pp. 1047–79. Washington, DC; US Government Printing Office.

Okada, H. & Bukry, D. (1980). Supplementary modification and introduction of code numbers to the 'Low-latitude coccolith biostratigraphic zonation' (Bukry, 1973; 1975). *Marine Micropaleontology*, **5(3)**, 321–5.

Sclater, J.G. & Heirtzler, J.R. (1977). An introduction to deep sea drilling in the Indian Ocean. In *Indian Ocean Geology and Biostratigraphy*, ed. J.R. Heirtzler, H.M. Bolli, T.A. Davies, J.B. Saunders & J.G. Sclater, pp. 1–24. Washington, DC; American Geophysical Union and Chelsea, Michigan; LithoCrafters, Inc.

Sclater, J.G., Abbott, D. & Thiede, J. (1977). Paleobathymetry and sediments of the Indian Ocean. In *Indian Ocean Geology and Biostratigraphy*, ed. J.R. Heirtzler, H.M. Bolli, T.A. Davies, J.B. Saunders & J.G. Sclater, pp. 25–60. Washington, DC; American Geophysical Union and Chelsea, Michigan; LithoCrafters, Inc.

Sclater, J.G., Anderson, R.N. & Bell, M.L. (1971). Elevation of ridges and evolution of the central eastern Pacific. *Journal of Geophysical Research*, **76**, 7888–915.

Tjalsma, R.C. (1977). Cenozoic foraminifera from the South Atlantic, DSDP Leg 36. In *Initial Reports of the Deep Sea Drilling Project*, 36, ed. P.F. Barker, I.W.D. Dalziel *et al.* pp. 493–517. Washington DC; US Government Printing Office.

Tjalsma, R.C. & Lohmann, G.P. (1983). *Paleocene–Eocene Bathyal and Abyssal Benthic Foraminifera from the Atlantic Ocean*. Lawrence, Kansas; Micropaleontology Special Publication, 4, 90 pp.

Webb, P.-N., Harwood, D.M., McKelvey, B.C., Mercer, J.H. & Stott, L.D. (1984). Late Neogene and older Cenozoic microfossils in high elevation deposits of the Transantarctic Mountains: evidence for marine sedimentation and ice volume variation on the East Antarctic craton. *Antarctic Journal of the United States*, **19(5)**, 96–7.

Wise, S.W. & Wind, F.H. (1977). Mesozoic and Cenozoic calcareous nannofossils recovered by DSDP Leg 36 drilling on the Falkland Plateau, southwest Atlantic sector of the Southern Ocean. In *Initial Reports of the Deep Sea Drilling Project, 36*, ed. P.F. Barker, I.W.D. Dalziel *et al.*, pp. 269–491. Washington, DC; US Government Printing Office.

Middle Eocene carbonate-bearing marine sediments from Bruce Bank off northern Antarctic Peninsula

V. TOKER[1], P.F. BARKER[2] & S.W. WISE, JR[3]

1 Faculty of Science, University of Ankara, Ankara, Turkey
2 Department of Geological Sciences, Birmingham University, Birmingham, B15 2TT, UK
3 Department of Geology, Florida State University, Tallahassee, Florida 32306, USA

Abstract

Middle Eocene marine clay containing sparse calcareous microfossils was recovered in *Islas Orcadas* piston core 15–29 from 2707 m of water on the south flank of Bruce Bank, South Scotia Ridge. The diverse and well preserved nannofossil assemblage (about 30 taxa) is assigned to the *Chiasmolithus gigas* Subzone of the *Nannotetrina quadrata* Zone (CP13b). A diverse group of pontosphaerids plus braarudosphaerids and rare *Lithostromation simplex* suggest a continental-margin palaeoenvironment, shallower than the present depth of the site. Benthic foraminifera also indicate a shallower, bathyal palaeodepth of deposition (2000 to about 800 m). We conclude that Bruce Bank probably is a continental fragment of the pre-Oligocene Antarctic–South American isthmus. It subsequently underwent rifting and subsidence in conjunction with seafloor spreading during the opening of Drake Passage. During rifting, the sediment may have been slumped or displaced, as indicated by the presence in the sequence of rare planktonic foraminifera of somewhat younger age. Middle Eocene nannofossil assemblages of the *Nannotetrina quadrata* Zone are present on the Falkland Plateau and as reworked constituents in an Oligocene glaciomarine sequence on King George Island, northern Antarctic Peninsula. A well preserved coccolith specimen of early–middle Eocene age has been reported reworked in tills of the Transantarctic Mountains. These occurrences suggest widespread deposition of lower–middle Eocene carbonate-rich or carbonate-bearing sediments throughout the high latitudes of the Southern Ocean, perhaps even within the cratonic basins of the Antarctic continent itself.

Introduction

Bruce Bank is one of several elevated features along the South Scotia Ridge (Fig. 1) which may represent continental fragments dispersed from the Antarctic–South American isthmus during the opening of Drake Passage. This presumption is supported by recent drilling on the largest of these features, the South Orkney microcontinent (SOM), which yielded shallow-water marine Eocene sediments beneath a break-up unconformity (Leg 113 Shipboard Scientific Party, 1987; Leg 113 Scientific Drilling Party, 1987). We report here Middle Eocene carbonate-bearing sediments sampled on the southern flank of the nearby Bruce Bank, which provide additional information on the palaeotectonic and palaeoceanographic evolution of this region.

The sediments from Bruce Bank were recovered by *Islas Orcadas* (*IO*) Piston Core 15–59 taken in 2707 m of water at latitude 60°33.6′S, longitude 40°13.2′W. As described by Kaharoeddin *et al.* (1980, p. 104), the lithology consists predominantly of an olive-brown clay (65–76%), quartz silt or sand (17–22%), and trace amounts of calcareous nannofossils (1–3%), foraminifera (< 1%), micromanganese nodules, mica, feldspar, volcanic glass, mica and glauconite. Sedimentary clasts up to 15 mm in diameter are common in the upper 127 cm, below which the remainder of the 385 cm long core consists of flow-in, an artifact of the coring process.

Biostratigraphy

Calcareous nannofossils

Although few in number, the calcareous nannofossils are quite well preserved and are assigned to the middle Eocene *Chiasmolithus gigas* Subzone (CP13b) of the *Nannotetrina quadrata* Zone of Okadas & Bukry (1980). This assignment is based primarily on the presence of rare *Chiasmolithus gigas* and rare to few *Nannotetrina fulgens* near the top of the core (Fig. 2). The presence of rare *Reticulofenestra umbilica* in

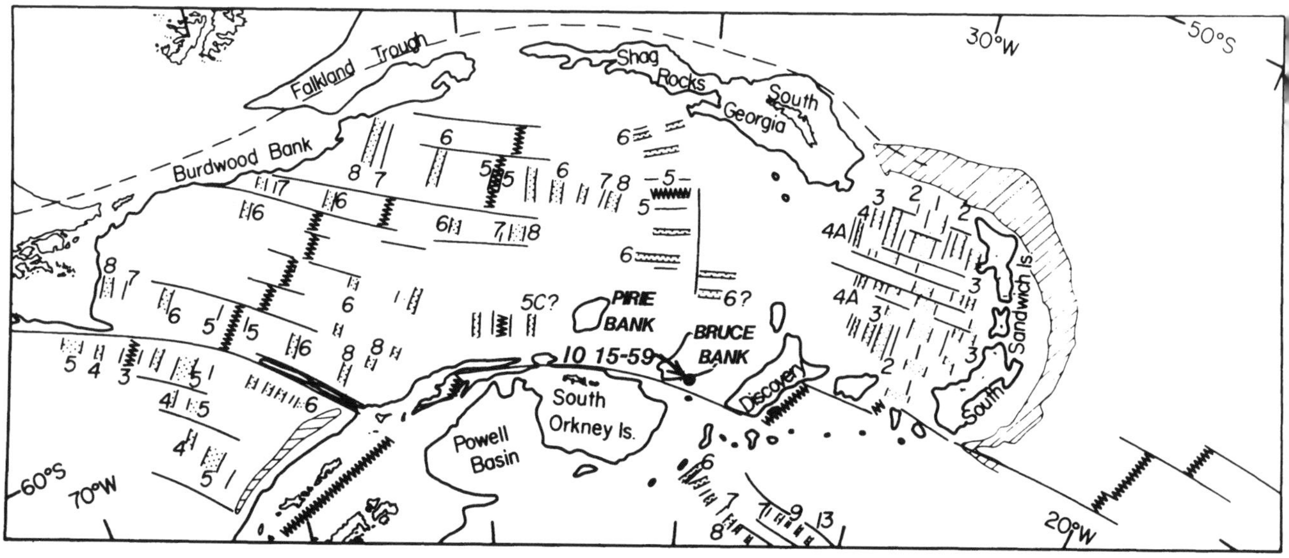

Fig. 1. Location of *Islas Orcadas* Core 15–59 on the southern flank of Bruce Bank, South Scotia Ridge (figure and magnetic anomaly identifications from Barker *et al.*, 1984). Shallow (continental and island-arc) areas defined by 2000-m isobath. Hachured pattern, South Sandwich trench.

IO 15 – 59 CALCAREOUS NANNOFOSSILS

AGE: MIDDLE EOCENE; ZONE (SUBZONE): *Nannotetrina quadrata* / *Chiasmolithus gigas* (CP 13b)													
SPECIES \ CORE INTERVAL (CM)	20–21	30–31	40–41	50–51	60–61	70–71	80–81	90–91	110–111	120–121	126–127	140–141	150–151
Blackites spinosus	C	F	F	F	R	R	F	R	R	R	R	F	
Braarudosphaera bigelowii	F		R	R					R				
Cepekella luminis	R			R	R			R			R	F	
Chiasmolithus expansus	F	F	R	F	F	R	R	R	F	R	F		
Chiasmolithus gigas	R												
Chiasmolithus solitus	F	F	F	C	R	F	F	R	F	F	F	R	R
Coccolithus formosus	R		R		R	R		R				R	
Coccolithus pelagicus	C	F	F	C	F	R	F	F	F	F	F	F	F
Cyclicargolithus pseudogammation	R	R	F	F	F								
"*Cyclococcolithina*" *kingii*	F	F	R	F			R			R			
Discoaster barbadiensis	R	R											
Discoaster bifax	?												
Discoaster distinctus	R	F		R			R			R			
Discoaster wemmelensis	F												
Fasciculithus involutus (RW)	R			R									
Lithostromation simplex				R									
Markalius inversus				F	R	R	R						
Micula decussata (RW)	R												
Nannotetrina fulgens	F	R		R									
Neococcolithes dubius	F	R	R				R	R	R				
Pontosphaera distinctoides	R			R									
Pontosphaera multipora	F		F	F		R	R	F	F	R	F	R	
Pontosphaera plana	F	C	C	C		F	F	F	F	F	R	F	
Pontosphaera puchriporus	F	F		R		F	F	F	C	F	F	F	
Pontosphaera pulcheroides	F	F	F	F					R				
Rhabdosphaera pseudomorionum	R						R						
Reticulofenestra onusta	R			R	R	R	F	F	F	F	F	R	F
Reticulofenestra samodurovii	A	A	A	C	F	C	C	C	C	C	C	C	R
Reticulofenestra umbilica	R		R										
Sphenolithus moriformis			R			R		R					
Watznaueria barnesae (RW)				R									
Zygrablithus bijugatus	F		R	R	R		R	R	F	R	R	R	

Fig. 2. Distribution of calcareous nannofossils in *Islas Orcadas* Core 15–59. RW, reworked taxa from the Cretaceous or Palaeocene.

Sample 20–21 cm could indicate some mixing of younger material into the assemblage, but this is not believed to be the case since small numbers of this taxon have now been observed to overlap the range of *Chiasmolithus gigas* in other mid–high latitude sections (e.g. Applegate & Wise, 1987).

Only two other occurrences of a *Nannotetrina quadrata* assemblage have been reported from the South Atlantic sector of the Southern Ocean. Wise & Mostajo (1983) assigned *Islas Orcadas* Core 16–46 (water depth, 1693 m) from the Maurice Ewing Bank (MEB; eastern Falkland Plateau) to the uppermost portion of this zone, although the zonal marker was not present. The Bruce Bank assemblage differs markedly from that of the MEB core by the presence of *Braarudosphaera bigelowii*, *Lithostromation simplex*, and a diverse group of few to common pontosphaerids. Together these indicate close proximity of the site to a continental margin and possibly a shallower palaeodepth of deposition than for the MEB core.

A second reported nannofossil assemblage assignable to the *Chiasmolithus gigas* Subzone is from King George Island, off the northern Antarctic Peninsula. The nannofossils were found adhering to the tests of planktonic foraminifera recovered from a *Chlamys* coquina (previously known as the '*Pecten* Conglomerate', and now designated the Low Head Member of the Polonez Cove Formation). The unit is considered Oligocene in age and glaciomarine in origin (Birkenmajer & Gaździcki, 1986). Gaździcka & Gaździcki (1985) figured the nannofossil assemblage and dated it as Oligocene based on their identification of *Chiasmolithus altus*, although they illustrate no specimens of that taxon. Due to the presence of *C. gigas*, however, we conclude that the nannofossil assemblage is middle Eocene in age, and that it and the attached foraminifera were reworked together into the younger strata, as suspected by Birkenmajer & Gaździcki, (1986).

Planktonic foraminifera

Islas Orcadas Core 15–59 also contains a few moderately well preserved planktonic foraminifera (Figs 3 & 4) that have been reported from the middle and upper Eocene of New Zealand (Hornibrook, 1965; Jenkins, 1971) and/or the Falkland Plateau (Krasheninnikov & Basov, 1983). The species are all relatively long-ranging forms and no detailed stratigraphic placement is possible. One form, *Globigerina* (= *Subbotina*)

	AGE	EOCENE-OLIGOCENE (?)																
IO 15-59	CORE INTERVAL (CM) / SPECIES	3-5	11-13	21-23	31-33	41-43	51-53	61-63	70-72	81-83	91-93	101-103	111-113	117-119	121-123	171-173	241-243	351-353
PLANKTONIC	*Globigerapsis index*		R			R												
	Globorotalia pseudotopilensis		R						R									
	Globigerina ouachitaensis			R						R								
	Globigerina angiporoides									R	R							
	Globigerina cf. *boweri*										R							
	Globigerina linaperta			R										R				R
	Globigerina cf. *officinalis*						R											
	Globigerina sp.								R			R		R	R			
	Pseudohastigerina micra									R	R	R		R			R	R
BENTHIC FORAMINIFERA	*Anomalinoides semicribrata*						R	R			R	R		R				
	Anomalinoides sp.			R	R	R							R		R	R	R	R
	Bulimina alazanensis		R					R		R		R	R	R	R			
	Cancris sp.												R					
	Cibicidoides sp.			R	R										R			
	Dentalina mucronata											R						
	Gyroidina girardana		R		R			R										
	Gyroidina planulata			R	R			R	R	R	R		R	R				
	Gyroidina sp.											R						R
	Lagena laevis														R			
	Lenticulina sp.							R										
	Melonis sp.			R			R											
	Orthomorphina rohri				R													
	Pleurostomella bellardi			R	R					R								
	Pleurostomella sp.					R			R									
	Robulus sp.		R			R		R		R		R	R	R				
	Spiroplectammina							R	R		R							
	Stilostomella curvatura sp.						R											
	Stilostomella lepidula									R	R	R		R	R			
	Stilostomella sp.						R											
	Trifarina sp.			R						R								

Fig. 3. Distribution of planktonic and benthic foraminifera in *Islas Orcadas* Core 15–59.

angiporoides, however, has been reported as *G.a. minima* only from the upper part of the middle Eocene at Falkland Plateau Site 512 in strata dated as nannofossil Zone CP14a (Krasheninnikov & Basov, 1983; Wise, 1983). The lower range of this species could not be determined at that locality, however, because drilling was terminated at that level.

Benthic foraminifera

The 10 species of benthic foraminifera recognized in the core (Fig. 3) are poorly preserved, long-ranging taxa, although some are common in the middle Eocene section of Site 512 (Basov & Krasheninnikov, 1983). No deep-water arenaceous forms are present, nor are there any miliolids or other indicators of shelf depths. The absence of *Nuttallides truempyi* suggests a palaeodepth of < 2000 m, whereas rare *Cancris* sp. could suggest even shallower depths (E. Thomas, pers. comm., 1987). The palaeo-water depth for the assemblage can best be classified as bathyal, somewhat similar to that of DSDP Site 512, which has a present depth of 1900 m.

Discussion

Islas Orcadas Core 15–59 sampled middle Eocene marine carbonate-bearing clay deposited at bathyal depths, presumably in a marine basin or on the slope of the then extant Antarctic–South American isthmus. Assuming that none of the microfossils have been redeposited, the pristine preservation of

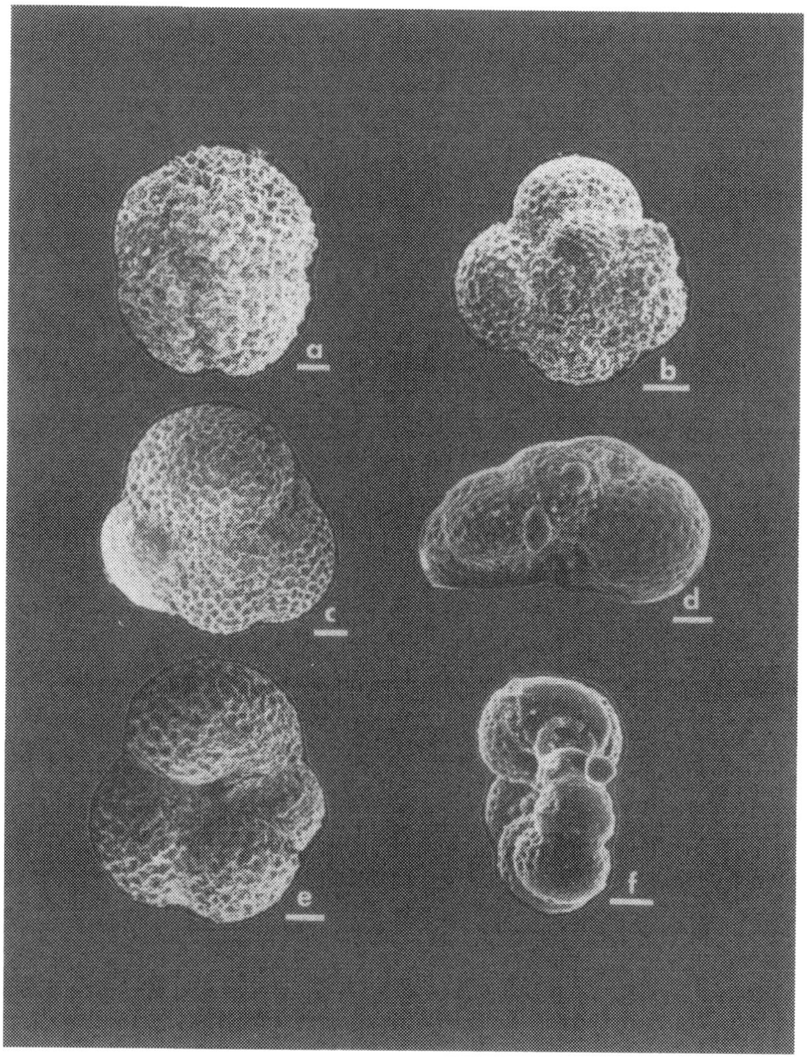

Fig. 4. Planktonic foraminifera from *Islas Orcadas* Core 15–59; scale bars are 20 μm: (a) *Globigerapsis index* (Finlay), umbilical view, Sample 41–43 cm; (b,c,) *Globigerina* sp. cf. *ouachitaensis* Howe & Wallace, b. spiral view, Sample 81–83 cm, c. spiral view, Sample 21–23 cm; (d) *Globigerina* sp., Sample 91–93 cm; (e) *Globigerina angiporoides* Hornibrook, umbilical view, Sample 81–83 cm; (f) *Pseudohastigerina micra* (Cole), apertural view, Sample 81–83 cm.

the nannofossils (as exemplified by taxa such as *Pontosphaera pulchriporus*) and the presence of 'nearshore' indicators (such as *Lithostromation simplex* and *Braarudosphaera bigelowii*) would collectively argue for deposition at water depths considerably shallower than that indicated by the present depth of the site (2707 m). This implies some subsidence of the site since deposition, presumably in conjunction with seafloor spreading following rifting.

Fig. 1 shows the spreading centres, magnetic anomalies and fracture zones identified in the present day Scotia Sea and northern Weddell Sea by Barker, Barber & King (1984). The Scotia Sea has opened through a continuous process of extension, probably entirely back-arc in the broad sense. The oldest identified magnetic anomalies extend back to only 28 Ma, but several small undated extensional basins in the Scotia Sea are probably older. Of the many elevated blocks within and around the Scotia Sea, the nature of only the larger has been established. Among these fragments, the piston core data presented here has revealed the pre-Oligocene age of Bruce Bank. The shallow palaeodepth of the assemblage we have described suggests, but does not establish conclusively, a continental origin.

The spreading history of the Scotia Sea is complicated, but recent calculations of South American–Antarctic plate motion (Barker & Lawver, 1988) provide some constraints. Our reconstruction (Fig. 5) locates Bruce Bank not far from the eastern margin of the South Orkney microcontinent (SOM), where shallow-water Eocene sediments were recently proved by drilling (Leg 113 Shipboard scientific party, 1987). The reconstruction shows a compact arrangement, with no intervening ocean floor, but the continental or other origin of Pirie Bank (P) and other fragments shown has yet to be determined.

Middle Eocene carbonate-rich or carbonate-bearing sediments have now been reported from Bruce Bank (present study), a thick sequence on the Falkland Plateau (67 m minimum, see Wise, 1983; Wise & Mostajo, 1983), and as

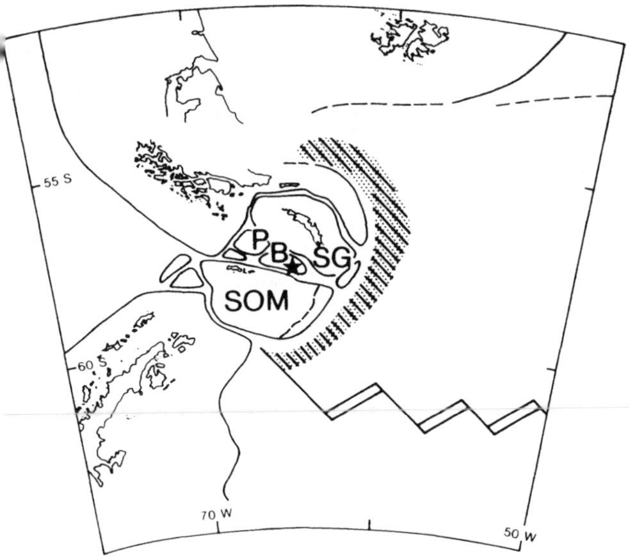

Fig. 5. Middle Eocene reconstruction of the Antarctic–South American isthmus, including Bruce Bank (B, with location of *Islas Orcadas* core 15–59 starred); SG, South Georgia; SOM, the South Orkney microcontinent; and P, Pirie Bank. These fragments were later dispersed during the opening of Drake Passage and the Scotia Sea.

reworked constituents of an Oligocene glaciomarine deposit of the northern Antarctic Peninsula. According to Birkenmajer (1985, fig. 12), the latter material could have been ice-rafted from the interior of the Antarctic continent.

To the above list can probably be added Eocene nannofossils which Harwood (1984) reported to be reworked into Pliocene tills (Sirius Formation) of the Transantarctic Mountains. Harwood (1984, fig. 2[9]) identified one such specimen as *Chiasmolithus bidens* (Palaeocene–lowest Eocene in range), but we assign the specimen to *Chiasmolithus solitus* (lowest Eocene–middle Eocene), a taxon well represented in the middle Eocene sediments of Bruce Bank and especially of the MEB. The widespread occurrence of this species in the Antarctic region and its inclusion in the tills of the Transantarctic Mountains, supports the suggestion of Webb *et al.* (1984) that Eocene carbonate sediments (here suggested to be early–middle Eocene in age) were deposited within the interior seaways (Wilkes–Pensacola basins) of Antarctica prior to the advent of widespread continental glaciation.

The nature and extent of Palaeogene seaways over the Antarctic craton, however, is open to debate. The reconstruction by Webb *et al.* (1984, fig. 1) is based on a map of the present day, predominantly subice bedrock surface of Antarctica (Drewry, 1983, sheet 3). This representation exaggerates the possible extent and depth of the subglacial cratonic basins. The isostatically adjusted bedrock surface (Drewry, 1983, sheet 6) shows subglacial basins of much more modest size and depth. An unknown factor is the extent of glacial erosion of these basins during the Neogene. Some question remains as to whether communication between these basins and the Southern Ocean was sufficient during the Palaeogene to support appreciable numbers of calcareous phytoplankton, or whether the basins were restricted and perhaps anoxic.

Appendix 1

Taxonomy

The following calcaraeous nannofossils are transferred to the genus *Pontosphaera* Lohmann (1902):

Pontosphaera distinctoides (Reinhardt) Toker, Barker & Wise, n. comb.

Basionym: *Discolithus distinctoides* Reinhardt, 1967, p. 212, pl. 3, figs 2, 3 & 6; text–fig. 13.

Pontosphaera pulchriporus (Reinhardt) Toker, Barker & Wise, n. comb.

Basionym: *Discolithus pulchriporus* Rheinhardt, 1967, p. 214, pl. 3, figs 21–23; pl. 7, fig. 3.

Remarks: This rather delicate form is distinguished by perforations on the bridge across the central area. More than two such perforations may be present, as is clearly evident on the holotype.

Acknowledgements

Discussions of benthic foraminiferal palaeobathymetry with Dr Ellen Thomas and planktonic foraminiferal biostratigraphy with Dr Lowell D. Stott were especially helpful. Ms Rosemarie Raymond drafted the fixtures which were printed by Ms Gabrielle Li. Study supported by NSF Grant DPP 84-14268.

References

Applegate, J.L. & Wise, S.W. (1987). Eocene calcareous nannofossils, Deep Sea Drilling Project Site 605, upper continental rise off New Jersey, USA. In *Initial Reports of the Deep Sea Drilling Project*, 93(2), ed. J.E. van Hinte, S.W. Wise *et al.*, pp. 685–8. Washington, DC; US Government Printing Office.

Barker, P.F., Barber, P.L. & King, E.C. (1984). An early Miocene ridge crest–trench collision on the South Scotia Ridge near 36° W. *Tectonophysics*, **102**, 315–32.

Barker, P.F. & Lawver, L.A. (1988). South American–Antarctic plate motion over the past 50 Myr, and the evolution of the South American–Antarctic ridge. *Geophysical Journal*, **94**, 377–86.

Basov, I.A. & Krasheninnikov, V.A. (1983). Benthic foraminifers in Mesozoic and Cenozoic sediments of the southwestern Atlantic as an indicator of paleoenvironment, Deep Sea Drilling Project Leg 71, Falkland Plateau and Argentine Basin. In *Initial Reports of the Deep Sea Drilling Project*, 71(2), ed. W.J. Ludwig, V.A. Krasheninnikov *et al.*, pp. 739–87. Washington, DC; US Government Printing Office.

Birkenmajer, K. (1985). Onset of Tertiary continental glaciation in the Antarctic Peninsula sector (West Antarctica). *Acta Geologica Polonica*, **35(1–2)**, 1–30.

Birkenmajer, K. & Gazdzicki, A. (1986). Oligocene age of the *Pecten* conglomerate on King George Island, West Antarctica. *Bulletin of the Polish Academy of Sciences, Terre*, **34(2)**, 219–26.

Drewry, D.J. (1983). *Antarctica: Glaciological and Geophysical Folio*. Cambridge; Scott Polar Research Institute.

Gaździcka, E. & Gaździcki, A. (1985). Oligocene coccoliths of the *Pecten* conglomerate, West Antarctica. *Neues Jahrbuch für Geologie und Paläeontologie, Monatshefte*, **12**, 727–35.

Harwood, D.M. (1984). Diatoms from the Sirius Formation, Transantarctic Mountains. *Antarctic Journal of the United States*, **19(5)**, 98–100.

Hornibrook, N. de B. (1965). *Globigerina angiporoides* n. sp. from the upper Eocene and lower Oligocene of New Zealand and the status of *Globigerina angipore* Stache, 1965. *New Zealand Journal of Geology and Geophysics*, **8(5)**, 834–8.

Jenkins, D.G. (1971). New Zealand Cenozoic planktonic foraminifera. Wellington; New Zealand Geological Survey Paleontological Bulletin, 42, 278 pp.

Kaharoeddin, F.A., Eggers, M.R., Goldstein, E.H., Graves, R.S., Watkins, D.K., Bergen, J.A. & Jones, S.C. (1980). ARA Islas Orcadas Cruise 1578 Sediment Descriptions. Tallahassee, Florida; Sedimentological Research Laboratory Contribution, 48, 162 pp.

Krasheninnikov, V.A. & Basov, I.A. (1983). Cenozoic planktonic foraminifers of the Falkland Plateau and Argentine Basin, Deep Sea Drilling Project Leg 71. In *Initial Reports of the Deep Sea Drilling Project*, 71(2), ed. W.J. Ludwig, V.A. Krasheninnikov *et al.*, pp. 821–58. Washington, DC; US Government Printing Office.

Leg 113 Scientific Drilling Party (1987). Leg 113 explores climatic changes. *Geotimes*, **32(7)**, 12–15.

Leg 113 Shipboard scientific party (1987). ODP drills in the Weddell Sea. *Nature, London*, **328**, 115–16.

Okada, H. & Bukry, D. (1980). Supplementary modification and introduction of code numbers to the 'Low-latitude coccolith biostratigraphic zonation' (Bukry, 1973; 1975). *Marine Micropaleontology*, **5(3)**, 321–5.

Reinhardt, P. (1967). Zur Taxonomie und Biostratigraphie der Coccolithineen (Coccolithophoriden) aus dem Eozaen Norddeutschlands. *Freiberger Forschungshefte*, **C213**, 201–41.

Webb, P.N., Harwood, D.M., McKelvey, B.C., Mercer, J.H. & Stott, L.D. (1984). Cenozoic marine sedimentation and ice-volume variation on the East Antarctic craton. *Geology*, **12(5)**, 287–91.

Wise, S.W. (1983). Mesozoic and Cenozoic calcareous nannofossils recovered by Deep Sea Drilling Project Leg 71 in the Falkland Plateau Region, southwest Atlantic Ocean. In *Initial Reports of the Deep Sea Drilling Project*, 71(2), ed. W.J. Ludwig, V.A. Krasheninnikov *et al.*, pp. 481–550. Washington, DC; US Government Printing Office.

Wise, S.W. & Mostajo, E.L. (1983). Correlation of Eocene–Oligocene calcareous nannofossil assemblages from piston cores taken near Deep Sea Drilling Sites 511 and 512, southwest Atlantic Ocean. In *Initial Reports of the Deep Sea Drilling Project*, 71(2), ed. W.J. Ludwig, V.A. Krasheninnikov *et al.*, pp. 1171–80. Washington, DC; US Government Printing Office.

Preliminary results of subantarctic South Atlantic Leg 114 of the Ocean Drilling Program (ODP)

P.F. CIESIELSKI[1] & Y. KRISTOFFERSEN[2] & THE SCIENTIFIC PARTY OF OCEAN DRILLING PROGRAM LEG 114

1 Department of Geology, University of Florida, Gainesville, Florida, USA
2 Seismological Observatory, University of Bergen, Bergen, Norway

Abstract

Ocean Drilling Program (ODP) Leg 114 drilled 12 holes at seven sites in the subantarctic South Atlantic during March–May 1987. These sites are located on the Northeast Georgia Rise in the east Georgia Basin (Sites 699 and 700), Islas Orcadas Rise (Site 702), between the Islas Orcadas Rise and Mid-Atlantic Ridge (Site 701) and on the Meteor Rise (Sites 703 and 704). The recovered sediments provide the greatest stratigraphic representation of the Late Cretaceous–Cenozoic ever obtained from the Southern Ocean. Generally well preserved assemblages of all major microfossil groups provide excellent biostratigraphic control. A nearly continuous history of geomagnetic polarity reversals was obtained from the Late Cretaceous–Quaternary, with gaps only in the late Palaeocene–early Eocene and portions of the early–middle Miocene. The bio-magnetostratigraphic framework provided by these sites, and those recovered by Leg 113, will provide the first high-resolution geochronological record of the Late Cretaceous–Cenozoic Southern Ocean. The principal results, as well as the initial tectonic and palaeoenvironmental interpretations of the recently drilled sites, are described here.

Introduction

During late 1986 and early 1987 the Ocean Drilling Program (ODP) conducted its first two drilling cruises to the Southern Ocean; Leg 113 to the Weddell Sea, and Leg 114 to the subantarctic South Atlantic. The latter is the subject of this report. The two cruises represent the first significant effort to obtain lengthy stratigraphical sections from the Southern Ocean in seven years, and together obtained the first detailed Late Cretaceous–Cenozoic sedimentary record from the Antarctic to subantarctic regions of a single sector of the Southern Ocean.

Major objectives in common to these drilling legs are to document the climatic, glacial, and palaeoceanographic history of the region within the context of a more complete and detailed geochronological framework than previously available. An additional goal is to interpret the causes of southern high latitude palaeoenvironmental change. Causes of increased Cenonzoic cooling and glaciation are still poorly known, but probably involved a combination of interrelated effects including tectonics, CO_2 forcing, sea level changes and various feedback mechanisms within the climate system itself.

Scientific objectives specific to Leg 114 are to evaluate: (1) the influence of progressive Antarctic cooling and glacial expansion on the palaeoceanography and biotic evolution of the subantarctic; and (2) the influence of regional tectonics on the glacial and palaeoceanographic history of the Southern Ocean and exchange of Antarctic water masses with the South Atlantic. Among these important regional tectonic events were: the Eocene formation of a deep-water gap between the Islas Orcadas and Meteor rises, which promoted deep water communication between the Antarctic and South Atlantic, and the late Oligocene–early Miocene opening of Drake Passage, which led to the establishment of the Antarctic Circumpolar Current (ACC) (Ciesielski, LaBrecque & Clement, 1987a).

The sedimentary record obtained from Legs 113 and 114 sites (Fig. 1) provides new opportunities to examine high-latitude climate change and oceanographic dynamics. These legs are the first in the Southern Ocean to widely utilize modern non-sediment disturbance, continuous recovery techniques (the Advanced Piston Core (APC) or Extended Core Barrel (XCB)) needed for high-resolution palaeoenvironmental studies. During the six previous Southern Ocean DSDP legs, most holes were rotary-drilled, resulting in drilling disturbance and poor core recovery. Other holes were discontinuously drilled (with washed intervals) to reach basement objectives. Only two sites (594 and 512) were cored with low disturbance techniques (HPC).

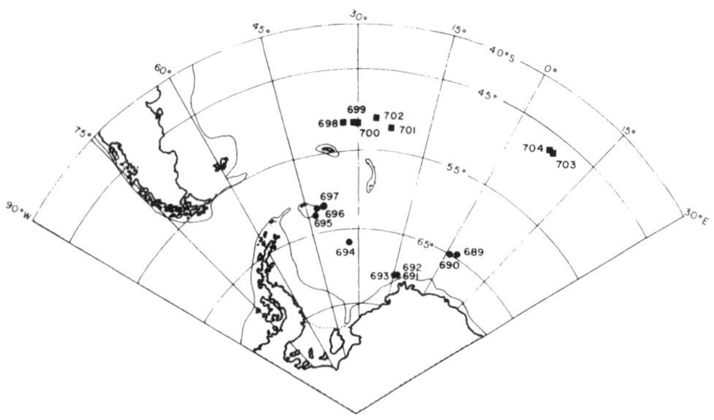

Fig. 1. Location of ODP Leg 113 (circles) and Leg 114 (squares) sites in the South Atlantic sector of the Southern Ocean.

Preliminary results and interpretation

Age and nature of the stratigraphic sequences

A precise geochronological framework has long been the key to interpreting the Cenozoic glacial history of Antarctica and its influence on regional and global oceanography and climate. Previous studies of Antarctic palaeoenvironment based on land and deep-sea sequences have been hindered by the lack of an accurate time scale. This has resulted from the near complete absence of pre-Pliocene palaeomagnetically-dated sequences, poor correlation of Southern Ocean biostratigraphic schemes to standard low-latitude zonations and incomplete stratigraphic representation.

Leg 114 sedimentary sequences provide the most complete stratigraphic representation of a 90 Ma interval of the Late Cretaceous–Quaternary of the subantarctic region of the Southern Ocean (Ciesielski *et al*., 1987b, c). 12 holes were drilled at seven sites along an E–W transect from the Northeast Georgia Rise (NEGR) to the Meteor Rise in water depths ranging from 1807 to 4634 m (Fig. 2). Fig. 3 illustrates the age, thicknesses, and hiatuses present in the seven sites.

Sediments recovered by Leg 114 are almost entirely pelagic types with little terrigenous component (Fig. 2). Late Cretaceous sediments are predominantly nannofossil chalks and limestones. Palaeogene sequences are primarily nannofossil oozes, chalks and limestones displaying progressive diagenesis with increasing age and sub-bottom depth. An exception is the deepest site, Site 701 (4634 m bsl), which accumulated biosiliceous sediments after its middle Eocene subsidence below the carbonate compensation depth (CCD). Calcareous sedimentation was intermittent at six of the seven sites since the earliest Miocene, after which time biosiliceous sedimentation predominated. Only at the shallow and most northerly site, Site 704 (46°52.8′S), did mixed biosiliceous and calcareous sedimentation occur during the Neogene. Abundant discrete volcanic ash layers occur in the Late Cretaceous of Site 700 and throughout the Neogene of Site 701. Chert nodules and stringers occur frequently within Eocene sequences.

One of the greatest successes of Leg 114 was to repeatedly recover lengthy stratigraphic representation of Upper Cretaceous–Cenozoic sequences with rich microfossil assemblages (Figs 2 & 3). Particularly noteworthy was the recovery of thick fossiliferous sections of stratigraphic sequences never before extensively recovered from the Southern Ocean, including the Maastrichtian, early Palaeocene, and early–middle Eocene. The combined palaeomagnetic data from all seven sites provide a near complete history of geomagnetic polarity reversals extending from the Late Cretaceous (Santonian) through the Quaternary, with gaps only in the late Palaeocene–early Miocene. Biostratigraphic studies of Leg 114 sites, when combined with the geomagnetic polarity record, will yield the first detailed geochronological framework for the interpretation of Southern Ocean palaeoenvironments during the last 90 m.y.

Sedimentation in the subantarctic region was relatively continuous during the Late Cretaceous – earliest Miocene, and hiatuses represent short stratigraphic intervals. At six of the seven sites, major hiatuses separate overlying mid – upper Neogene sediments from the mid-Palaeogene to earliest Neogene. Since ~18.5 Ma, all sites, except Site 704, were subjected to numerous erosional and non-depositional episodes. Physical properties measruements suggest a removal of hundreds of metres of sediment at some hiatuses.

Microfossil representation and preservation

Calcareous nannofossils are the dominant microfossil constituent within the Cretaceous, with planktonic and benthic foraminifera the most common minor constituents. Intervals occur with common to abundant calcispherulids (Site 698) and abundant and well preserved radiolarians (Site 700). Late Cretaceous microfossil diversity is generally high, but exhibits signs of diagenesis.

Calcareous microfossils are also the dominant microfossil group in the Palaeogene, although benthic and planktonic foraminifera are abundant in coarse-fraction residues. Above the lowermost Miocene only sporadic low-diversity assemblages of calcareous groups are found, except at the shallow and most northerly Site 704 where they persist through the Neogene.

Diverse assemblages of siliceous microfossils occur throughout portions of the Palaeogene where diagenesis has not resulted in their destruction. Particularly well preserved assemblages occur within the upper Palaeocene, upper Eocene, and Oligocene. The only dominance of siliceous microfossil assemblages within the Palaeogene of Leg 114 sites is within the upper Eocene–Oligocene of deep-water Site 701. In contrast, siliceous microfossils dominate the biogenic fraction of all Neogene sections, except for Site 704 (discussed later).

Tectonic interpretation

The recovered sections give new insight into tectonic and subsidence history, and nature of the NEGR, Islas Orcadas Rise (IOR) and Meteor Rise. Three sites were drilled near the apex (Site 698) and lower eastern flank of the NEGR (Sites 699 and 700), thought to be a fossil arc-massif formed at a convergent plate boundary between the Malvinas plate and the South American plate (LaBrecque & Hayes, 1979). Basement at Site 698 was capped by Campanian or older interbedded

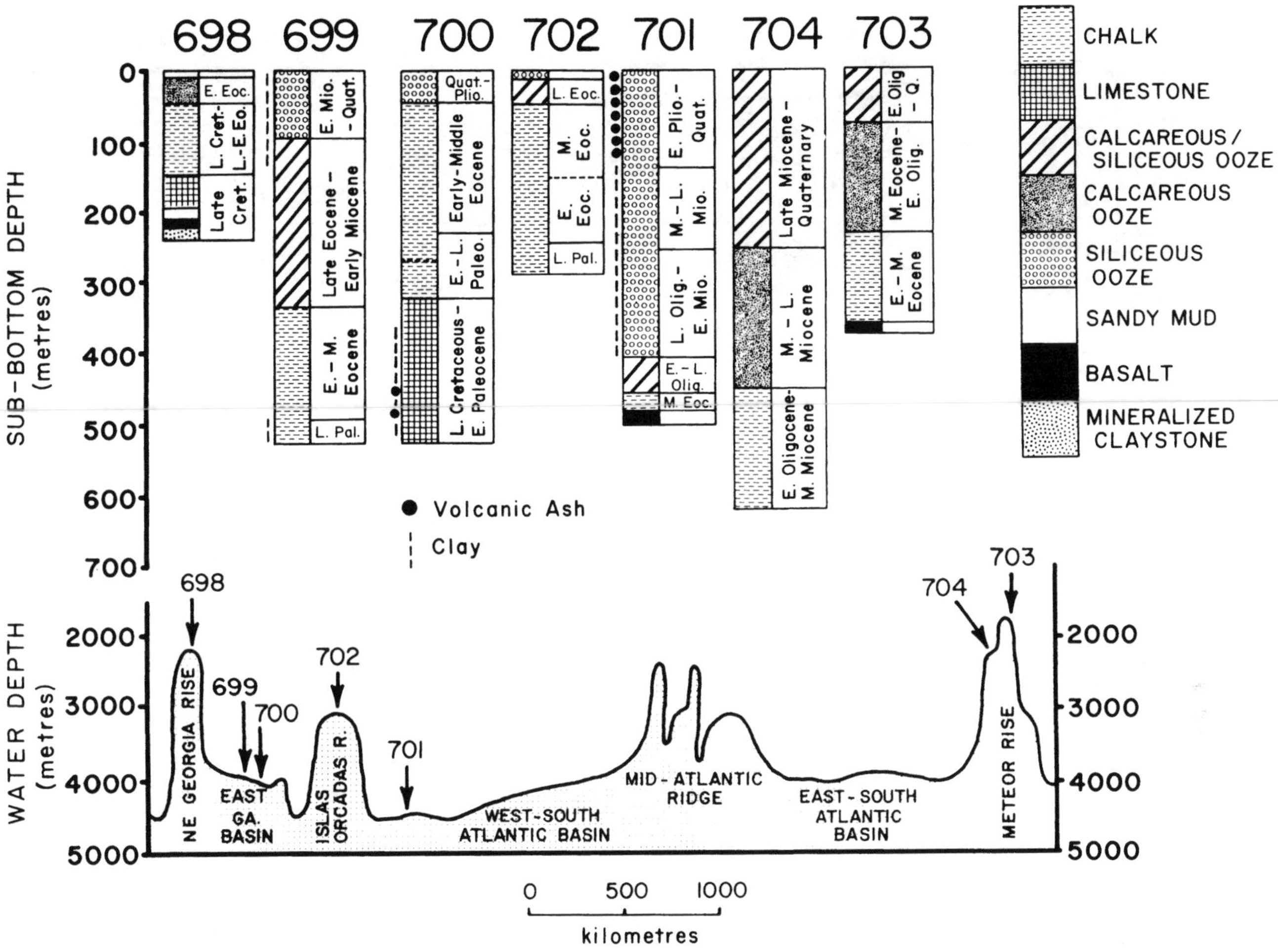

Fig. 2. 12 holes drilled by ODP Leg 114 at seven site locations along a west to east traverse across the subantarctic South Atlantic (Fig. 1). Composite stratigraphic columns (top) reveal the ages, lithologies, and sediment thickness of each site. A schematic cross-section (bottom) shows the site locations along the bathymetry of the transect.

basalt and subaerially-weathered basalt layers. Active regional volcanism occurred during the late Turonian–Santonian; thus, any incipient subduction must be pre-early Campanian. Redeposited rocks in the Oligocene of Site 700, if in place, suggest that a crustal fragment may be part of the structural framework that forms the NEGR.

Major Oligocene faulting occurred on the NEGR involving basement and the overlying sediments between the early Oligocene and Quaternary. In the region of Site 698, the entire sedimentary section was downfaulted into a half-graben with vertical displacement on the major fault of > 1 km. Stratigraphic control and seismic correlations between Sites 699 and 700 clearly demonstrate that corresponding stratigraphic levels and basement were vertically displaced along two major faults with a total throw of ~ 500 m. This later tectonic activity may be related to Scotia Sea opening and interaction with the advancing South Georgia block.

Basal sediments of the IOR (Site 702) indicate that elevation is older than late Palaeocene (> 61 Ma), possibly Late Cretaceous. Early Eocene extension generated numerous small half-grabens over much of the rise, and a major tectonic event formed a N–S-trending horst through the site location between the late Eocene and middle Miocene. By the late Palaeocene most of the rise was still < 2000 m bsl, but subsided to near its present water depth (3083 m bsl) by the mid–late Eocene.

Site 703 was drilled at a depth of 1807 m on the flank of a basement pinnacle of the Meteor Rise, an aseismic ridge extending south-west from the Agulhas Fracture Zone. Having been unsuccessful at obtaining basement on the once conjugate IOR, a major objective at this site was to determine its nature, age, and subsidence history. Weathered (?)early or early middle Eocene porphyritic-plagioclase basalts and basaltic tuffs were encountered which may represent basement or part of a volcanic rubble zone. The oldest *in situ* or reworked microfossils at this site are Eocene, whereas Late Cretaceous microfossils were found on the IOR. Either the IOR is older than the Meteor Rise, or more likely, the Meteor Rise has experienced more recent volcanism which would account for its younger age, shallower basement and more rugged relief. The site was between 600 and 1000 m bsl during the Eocene and subsided to > 1000 m bsl during the Oligocene. Reworked Eocene and Oligocene microfossils indicate the presence of nearby islands which have since subsided below neritic depths.

Site 701 was drilled to basement to the east of the IOR (Fig. 2) on oceanic crust generated by an extension of the Mid-Atlantic Ridge which separated the IOR and Meteor

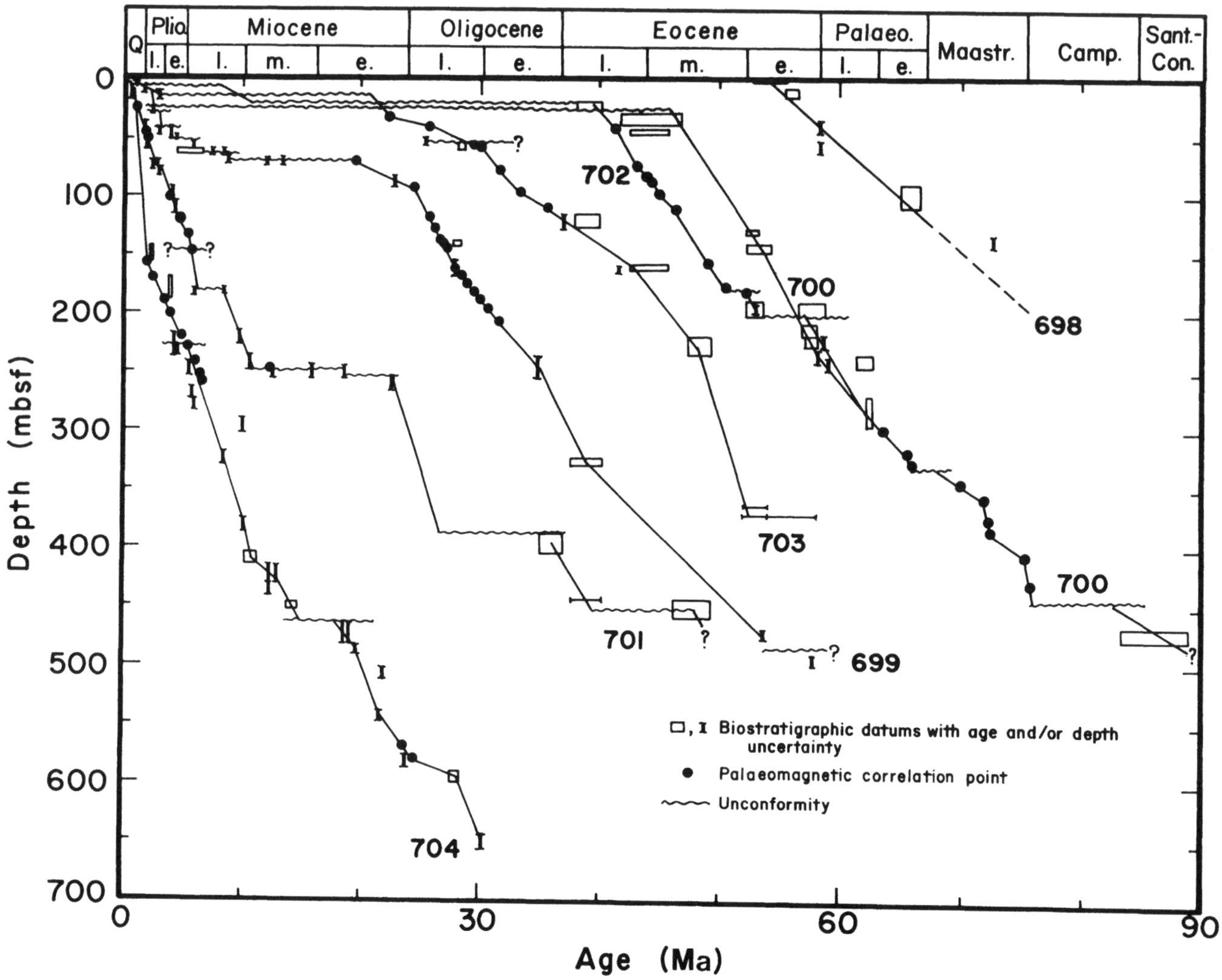

Fig. 3. Sediment ages versus sub-bottom depth for all seven Leg 114 sites. Sites are identified by numbers 698–704.

Rise. Drilling results confirm previous identification of magnetic anomaly patterns, indicating an initiation of Eocene spreading between these features.

Palaeoceanographic interpretation

Planktonic microfossil asssemblages yield a detailed record of the progressive and punctuated cooling of subantarctic surface waters during the Late Cretaceous–Cenozoic. Warmest surface waters in the subantarctic occurred during the Late Cretaceous–middle Eocene, resulting in a relatively high percentage of warm-water taxa such as *Globotruncana*, *Rosita*, *Morozovella*, *Acarinina*, *Discoaster* and *Sphenolithus*. Peak abundances of warm-water planktonic taxa in the Cenozoic occur within the early Eocene (57.8–52 Ma). A mixture of lower-latitude planktonic groups with Austral assemblages occurs in the Late Cretaceous–early Palaeogene. We infer this assemblage mixture to be the result of an influx of shallow and cooler Pacific waters (entering the south-west Atlantic through the unglaciated West Antarctic archipelago) mixing with warmer waters of the southern limb of a larger South Atlantic subtropical gyre.

Surface waters began cooling significantly between 51.0 and 46 Ma, and by the late Eocene (~ 38 Ma) had led to the much cooler water planktonic fauna and flora. This middle Eocene cooling episode begins the progressive and punctuated long-term trend towards subpolar conditions. The middle Eocene cooling coincides in age with the anomaly 18 advent of more rapid seafloor spreading between Antarctica and Australia, which may have caused the establishment of a more vigorous 'shallow circum-Antarctic Current' entering the Atlantic through the West Antarctic seaway. We tentatively infer the cause of the middle Eocene cooling of subantarctic surface waters to be a northward displacement of the large South Atlantic subtropical gyre by the ACC.

Additional episodes of pronounced surface water cooling occurred at or near the middle–late Eocene and early–late Oligocene boundaries, accompanied by the successive disappearance of warm-water species in all major microfossil groups. By the late Oligocene, assemblages were dominated by cosmopolitan and cool-temperate species.

Palaeogene sedimentation in the regions of Leg 114 sites contrasts greatly with that of the Neogene in the same region. Palaeogene sedimentation was more continuous, with the accumulation of calcareous nannofossil ooze above the CCD. Below the CCD (Site 701), biosiliceous sedimentation prevailed with the only significant change in sedimentation being an increase in terrigenous clay deposition during the middle Oligocene. In general, Palaeocene circulation appears to have been weak. Suboxic benthic conditions prevailed for much of the time in response to weak benthic flow and poor mixing. Sites 699 and 700 in the East Georgia Basin reveal no evidence for strong deep circulation through this potential deep-water passage between the Weddell Sea and South Atlantic prior to the Eocene opening of the Islas Orcadas Rise–Meteor Rise gateway.

A brief warming interval occurred at the Oligocene–Miocene boundary followed by the earliest Miocene northward advance of the Polar Frontal Zone (PFZ) and biosiliceous province. This later event pushed the long-term position of the calcareous productivity zone permanently to the north of 51°S. This advance of the PFZ appears to have been intimately related to the opening of the Drake Passage and the establishment of the ACC.

Soon after the eariest Miocene advance of the PFZ, a major increase occurred in the intensity of the Circumpolar Deep Water and Antarctic Bottom Water, producing regional hiatuses. Deep-cutting erosion and non-deposition caused by the full development of the ACC formed hiatuses of different stratigraphic extent. Numerous hiatuses were found within the Neogene of all sites, except Site 704 which remained north of the influence of the ACC throughout most of this period. Multiple erosional and non-depositional episodes resulted in very attenuated Neogene sections at five of the sites (< 70 m thick). At the shallower sites on the NEGR and IOR, erosion cut deeply to expose Eocene sediments (Sites 698, 700, and 702).

The appearance of sand-size and larger ice-rafted detritus appears to be restricted to sediments of late Miocene age or younger, with a significant increase in its abundance in the Pliocene. This finding is consistent with previous studies (Ciesielski, Ledbetter & Ellwood, 1982) which described a rapid middle–late Miocene northward expansion of the zone of ice-rafting around the Antarctic continent. A widespread increase in ice-rafting to the northern Antarctic and subantarctic is attributed to the development of a fully glaciated Antarctic continent with fringing ice shelves. The increased abundance of ice-rafted detritus in the latest Miocene and Pliocene of the south-west Atlantic sector of the Southern Ocean may be related to the establishment of the West Antarctic ice sheet, which did not form until this time according to the recent drilling results of Leg 113 (Barker, Kennett & O'Connell, 1987).

A superb 576 m thick Neogene carbonate–siliceous sequence was recovered from Site 704 on the Meteor Rise. Shipboard studies revealed a strong climate signal characterized by high frequency and amplitude variation in carbonate content, biogenic silica, organic carbon, temperature affinities of microfossils, physical properties and logging parameters. Unlike the other Leg 114 sites which were further to the south in the axis of the ACC, Site 704 was the recipient of almost uninterrupted high accumulation rates of Neogene pelagic sediments. The result was the formation of a remarkable record of the Neogene subantarctic palaeoenvironment.

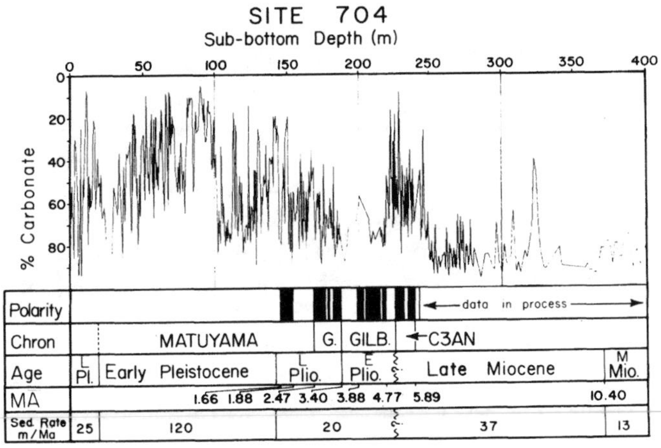

Fig. 4. Fluctuations in calcium carbonate % in the late middle Miocene–Quaternary of Site 704 from the Meteor Rise. Not shown is the older early Oligocene–middle Miocene portion of the record which reveals little variation in carbonate. Also shown are sedimentation rates and palaeomagnetic stratigraphy. Final palaeomagnetic data for the Miocene awaits discrete sample measurements to clarify the reversal pattern for this period of high-frequency reversals.

Changes in the mode and variability of the carbonate record of Site 704 coincide with major palaeoclimatic events. Carbonate content was high, averaging 86% from the early to late Miocene (Fig. 4). The initial significant decrease in carbonate (a 33% drop) occurred in Chron C3AR.6 (~ 6.5–6.37 Ma) and was accompanied by an increase in the amplitude of the carbonate variability. This change is inferred to represent the initial migration of the PFZ to close proximity to Site 704.

Preceding the increase in late Miocene carbonate variability at Site 704 was the deposition of a nearly monospecific ooze of the diatom *Bruniopsis* below the CCD at Site 701. This remarkable finely laminated ooze was deposited during Chron C3AR and also contains abundant silicoflagellates (*Dictyocha*), indicative of extremely warm surface waters. The combined evidence of fine lamination, absence of bioturbation and benthic foraminifera, high organic carbon content, and pyrite strongly suggest an oxygen-poor benthic environment. Monospecific diatom assemblages of similar age were previously noted in the South Atlantic at Sites 520 and 519, where low $\delta^{18}O$ surface-water values were interpreted as warming periods. Comparisons of the Site 704 and 701 upper Miocene records suggest an unusual and perhaps brief period of warm and possibly low-salinity surface waters in the eastern South Atlantic prior to the latest Miocene advance of the PFZ and associated cooler conditions.

Several other changes were noted in the mode and frequency of the Pliocene–Quaternary carbonate record of Site 704, representing fundamental changes in subantarctic surface waters. During the early Gilbert to mid-Gauss (~ 4.8–3.2 Ma), a return to high carbonate values (averaging 77%) with low variability suggests a warm period and Polar Front retreat. During the middle Gauss (3.2–2.9 Ma), carbonate values decreased to 64% and the variability of the signal increased. This event correlates with an advance of the Polar Front and increase in marine oxygen isotope values. These changes preceded the first major influx of ice-rafting to the North Atlantic by 0.5–0.7 m.y. Another increase in the amplitude of the signal followed during the Olduvai subchron (1.66–1.88 Ma), coincident with the Plio-Pleistocene boundary.

The most remarkable sedimentary change at Site 704 occurred during the middle Matuyama Chron and was marked by: (1) decreased carbonate content to as low as 15%; (2) increased biogenic silica accumulation; (3) increased organic carbon deposition; and (4) the onset of strikingly intense colour variations. These cyclic alterations were probably caused by migration of the PFZ over Site 704 in response to glacial–interglacial forcing. Higher silica accumulation occurred during glacial advance of the biosiliceous productivity belt to Site 704, only to be followed by the dominance of carbonate productivity upon the retreat of the front.

In summary, the recently recovered Leg 114 stratigraphic sequences provide a virtually complete record of sedimentation within the subantarctic during the last 90 Ma. Rich microfossil assemblages, and excellent biostratigraphic and magnetostratigraphic control, promise to provide a much more detailed history of the Late Cretaceous–Cenozoic palaeoenvironment of the subantarctic than was previously possible.

Acknowledgements

Support for the preparation of this manuscript was provided by the Joint Oceanographic Institutions Inc. and the National Science Foundation of the United States.

References

Barker, P.F., Kennett, J.P. & O'Connell, S. (1987). *Leg 113 Preliminary Report*, Ocean Drilling Program, Preliminary Reports No. 13. College Station, Texas; Texas A & M University.

Ciesielski, P.F., Kristoffersen, Y. & the ODP Leg 114 Scientific Party. (1987b). Palaeoceanography of the sub-antarctic South Atlantic. *Nature, London*, **328**, 671–2.

Ciesielski, P.F., Kristoffersen, Y. & the ODP Leg 114 Scientific Party. (1987c). Leg 114 finds complete sedimentary record. *Geotimes*, **32(10)**, 23–5.

Ciesielski, P.F., LaBrecque, J.L. & Clement, B. (1987a). *Leg 114 Scientific Prospectus*, Ocean Drilling Program, Scientific Prospectus No. 14. College Station, Texas; Texas A & M University.

Ciesielski, P.F., Ledbetter, M.T. & Ellwood, B.B. (1982). The development of Antarctic glaciation and the Neogene paleoenvironment of the Maurice Ewing Bank. *Marine Geology*, **46**, 1–51.

LaBrecque, J.L. & Hayes, D.E. (1979). Seafloor spreading in the Agulhas Basin. *Earth and Planetary Science Letters*, **45**, 411–28.

Cenozoic glacial and tectonic history from CIROS-1, McMurdo Sound

P.J. BARRETT[1], M.J. HAMBREY[2] & P.R. ROBINSON[3]

1 Antarctic Research Centre, Research School of Earth Sciences, Victoria University of Wellington, Private Bag, Wellington, New Zealand
2 Scott Polar Research Institute, Lensfield Road, Cambridge CB2 1ER, UK
3 TCPL Resources Ltd, 495 Victoria Avenue, Chatswood, N.S.W. 2067, Australia

Abstract

Glacigenic strata of Miocene–early Oligocene age have been cored in the CIROS-1 drillhole 12 km offshore on the western edge of the Victoria Land Basin (77°34′54″S; 164°29′55″E). The core, which was taken continuously from 27 to 702 m below the seafloor, is dominated by muddy sandstone, sandy mudstone and diamictite, with lesser amounts of sandstone and conglomerate. The strata were deposited in a wave-influenced marine deltaic setting with glacial activity much in evidence. The interval from 27 to 366 m sub-seafloor (late Oligocene–early Miocene) was deposited in waters of varying depth and includes a significant proportion of diamictite (40%), interpreted as lodgement and water-lain till. Four periods of glaciation associated with low relative sea level are recorded, and considered to represent episodes of major ice build-up on the continent. The interval below 366 m (early Oligocene) is largely deep-water mudstone with sandstone beds and occasional conglomerate deposited from sediment gravity flows in the lower part. The strata also include scattered subangular–subrounded stones, many of which are faceted and striated, indicating some glaciation on land and ice calving at sea level in this region back to early Oligocene time. The thickness of the Oligocene–early Miocene section and the virtual absence of younger strata suggest this was a significant period of subsidence for the Victoria Land basin and possibly also uplift for the adjacent Transantarctic Mountains.

Introduction

The Antarctic ice sheet is the major potential influence for short period (10^5–10^6 y) changes to world sea level in the Cenozoic, but its early history is yet to be determined. Variations in global ice volume have been inferred from evidence of globally-synchronous onlap and offlap events on continental margins (Miller, Mountain & Tucholke, 1985; Haq, Hardenbol & Vail, 1987), deep-sea unconformities (Barron & Keller, 1982) and oxygen isotopes in deep-sea cores (Shackleton & Kennett, 1975; Shackleton, 1986). Such evidence is indirect, and in some cases capable of quite different interpretations. For example Summerhayes (1985) argued for a regional tectonic interpretation of onlap–offlap curves, and Matthews & Poore (1980) reinterpreted the oxygen isotope data to indicate a significant global ice budget in early Tertiary and even Cretceous time. There was therefore a clear need to seek a more direct approach to determine the extent of Antarctic ice during the Cenozoic.

CIROS-1 was drilled to sample strata close to the present margin of the Antarctic ice sheet in western McMurdo Sound (Fig. 1). The earlier drilling of MSSTS-1 (Barrett *et al.*, 1987) had shown that ice had been more extensive than today for several periods in late Oligocene–Miocene time, and deeper drilling was planned to extend that record back in time. We hoped to see in CIROS-1 variations in climate and relative sea level from palaeontological and sedimentological data. The core was also expected to record the uplift of the adjacent Transantarctic Mountains.

The CIROS-1 drillhole was sited 12 km offshore and 4 km seaward of the zone of normal faulting that marks the western margin of the Victoria Land basin (Cooper & Davey, 1985). The adjacent Transantarctic Mountains locally rise to > 4000 m, and hold back the ice sheet to the west. The seafloor around the site is 200 m deep, and the sea surface is covered with ice 2 m thick for nine months of the year. Sedimentation today takes place in a quiet, relatively sheltered setting with minimal current activity and tides of < 1 m (Barrett, Pyne & Ward, 1983).

Drewry (1981) has cautioned against extrapolating from glacial chronology developed in the McMurdo region because the relationship between local outlet glaciers and the main ice

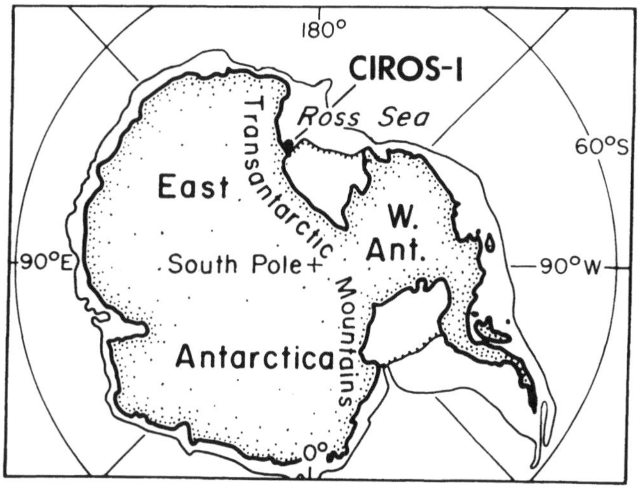

Fig. 1. Antarctica showing the location of CIROS-1.

sheet is complex and non-synchronous. However, the most recent episode of extensive ice in McMurdo Sound did coincide with the last glacial maximum (Denton & Hughes, 1981). If we assume that climatic change in the more distant past affected a large area in the same way, the sedimentary record there should be interpretable in terms of continental rather than just local events.

Seismic surveys show the strata beneath western McMurdo Sound to parallel the seafloor, and dip gently eastward to extend beneath Ross Island (Fig. 2). Thickness of the sequence increases eastward from 1.5 to 4 km (McGinnis *et al.*, 1985). The most recent survey (Fig. 3) shows three distinct sequences: an upper parallel-stratified sequence; a middle lensing sequence with dips reaching 15°; and a lower parallel sequence. CIROS-1 cored through the upper into the middle sequence to a depth of 702 m. Coring was continuous with better than 98% recovery.

Lithology

The cored sequence has been divided lithologically into an upper part (27–366 m), in which diamictite is the most common lithology, and a lower part (366–702 m) dominated by mudstone with relatively little diamictite. The most prominent seismic reflectors in the cored sequence are both in the upper part; they correspond approximately to a lithological change from mudstone to sandstone at 140 m (Reflector Q) and to a mudstone between two thick diamictite beds at 220 m (Reflector R).

The main facies in the core are distinguished on texture (conglomerate, sandstone, mudstone, diamictite). We use the term diamictite to describe a mixture of mud, sand and stones, with the latter forming > 1% of the rock and uniformly spread through it. Robinson *et al.* (1987) provided a detailed core description, a summary which is presented as the lithological log in Fig. 4. The proportions of the main facies types and their characteristics are shown in Table 1.

Development of stratification further distinguishes facies types. Some beds are massive but most show a little stratification. In many places, however, it has been destroyed by burrowing. Even so visible stratification is common and is largely parallel, with only rare instances of cross-bedding, indicating a lack of current activity.

The entire cored sequence appears to be nearly horizontal, for the hole is within 2° of vertical throughout, and horizontal stratification can be found at least every 3 or 4 m. However, the core also reveals a great deal of dipping stratification with every interval of several metres having one or more packets of inclined or folded strata commonly dipping at 20–50°. The wide extent of this deformation and the lack of any clear association with the diamictite beds suggests that most results from gravity-induced soft sediment deformation rather than shearing by grounded ice.

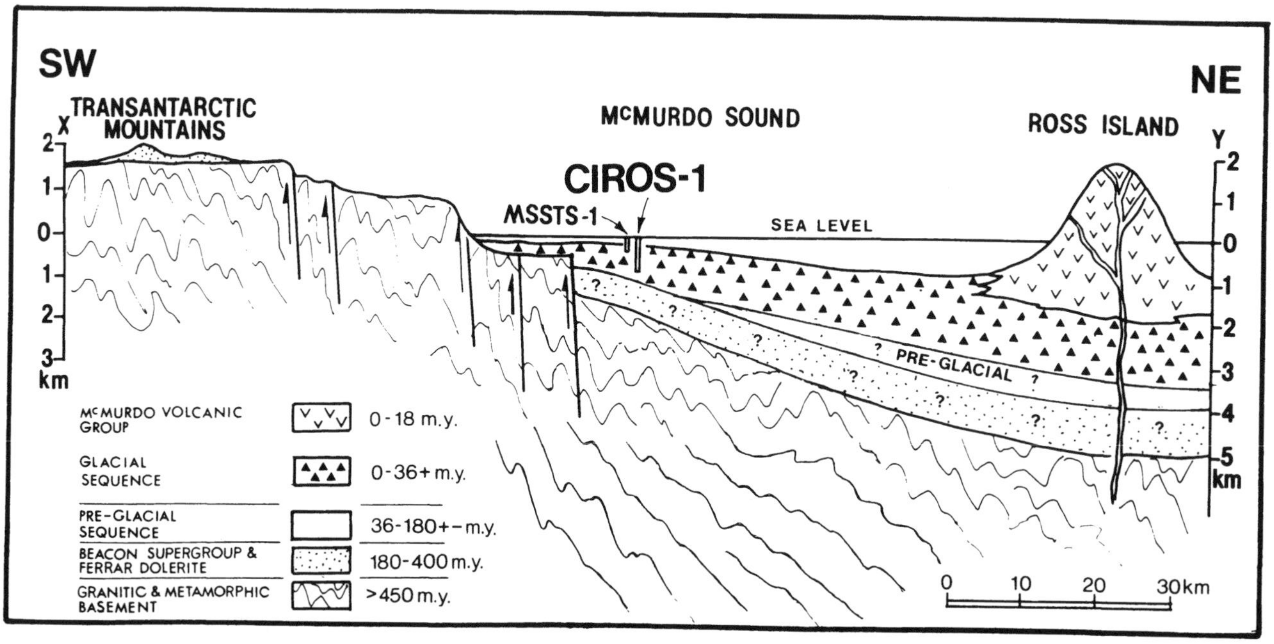

Fig. 2. Geological cross-section of McMurdo Sound showing the location of the CIROS-1 and MSSTS-1 drillholes. The fault system on land has been confirmed by fission–track dating (Gleadow & Fitzgerald, 1987); offshore depth to basement is from McGinnis *et al.* (1985).

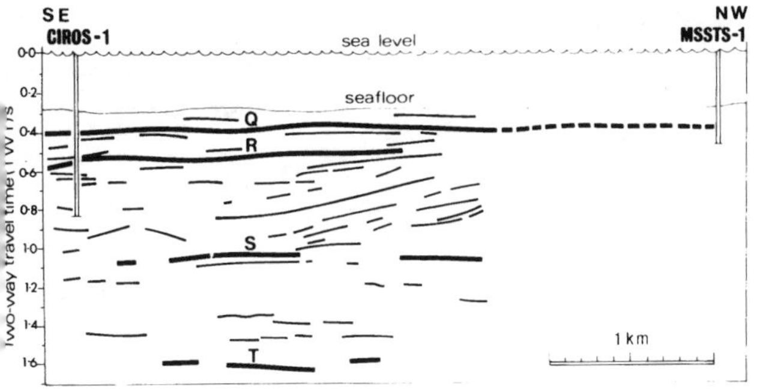

Fig. 3. Interpreted seismic section from CIROS-1 to MSSTS-1 showing the lateral persistence of the prominent reflectors (modified from Davy & Alder, 1989). Lithological correlation with MSSTS-1 is shown by the dashed line.

Clasts are scattered throughout the sequence, and striae and facets indicating basal glacial transport can be found on a significant proportion, typically 3–15%, at every level sampled. The data show clearly that glaciers reached the coast for virtually all of the time represented by the core.

Environment of deposition for the CIROS-1 core is considered to be a wave-influenced delta with sediment supplied mainly from an ancestral Ferrar valley to the west. The setting would be protected from cyclonic storms by the mountains to the west. Predominantly marine deposition is indicated by studies of diatoms and foraminifera. Other fossils in the core include a variety of thin-shelled molluscs, both terrestrial and marine palynomorphs, and the tip of a *Nothofagus* leaf found at 215 m between two thick diamictite beds of late Oligocene age. These will be described in detail in a multi-author volume (Barrett, 1989).

Age

Diatoms from 366 m to the base of the hole are of early Oligocene age, whereas those above 340 m are late Oligocene and early Miocene (D. Harwood, pers. comm.). Foraminifera confirm the early Oligocene age for the lower part of the hole (P.N. Webb, pers. comm.). However, more detailed work through the entire core will be required for dating and correlation beyond the region, the palaeomagnetic stratigraphy and strontium isotopic ages on shell debris are currently under investigation to help with this.

Glacial and sea-level history

The core can be interpreted in terms both of the extent of ice in the vicinity of the drill site and of varying water depth. A facies model relating sediment texture and the abundance and distribution of pebbles in glaciomarine sediments is proposed (Fig. 5A) based on modern sediment distribution in McMurdo Sound (Barrett *et al.*, 1983) and a glaciological model for the release of debris from outlet glaciers and ice shelves proposed by Drewry & Cooper (1981). The temperate fjord models of Powell (1981) and Elverhøi, Lønne & Seland (1983) seems less appropriate because of their dominant meltwater influence. Our model has been applied to the CIROS-1 core to infer the proximity and extent of glacier ice with respect to the drill site

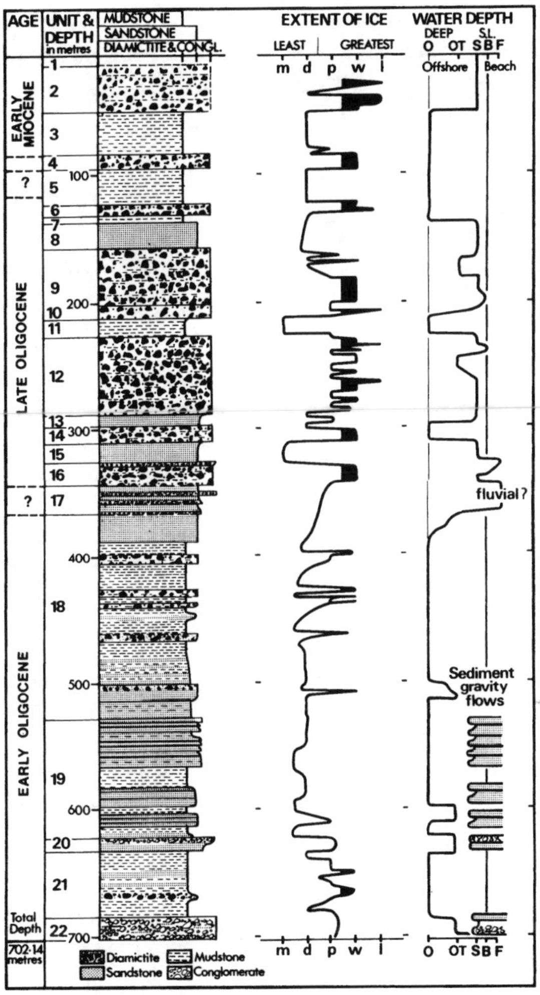

Fig. 4. Lithological log of the CIROS-1 drillhole, showing interpreted variation in extent of ice as recorded at the drillsite and in relative sea level change. (See Fig. 5 and text for derivation of curves).

(Fig. 4). A curve has also been drawn showing variations in relative sea level inferred from sediment texture and structures in the core. The model used for the interpretation (Fig 5B) combines those for a wave-dominated shelf and wave-dominated delta (Elliott, 1986, fig. 7.7 and pp. 129–30, respectively).

The upper part of the core (27–366 m) records several periods with glacier ice more extensive than today and grounding at the drill-site. The first major increase (Unit 16) follows a dramatic shallowing (base of Unit 17), which took place at about the boundary between the early and late Oligocene epochs. This corresponds in time to the largest fall in eustatic sea level in the Cenozoic (Haq *et al.*, 1987), and is consistent with the growth in the continental interior of a large ice sheet that subsequently reached the coast. A further three intervals dominated by diamictite (Units 12, 9 + 10 and 2) also represent periods of low sea level, consistent with major ice build-up; three smaller intervals of diamictite (Units 14, 6 and 4) were deposited during periods of high relative sea level and are presumed to result from local expansion of glacial ice.

These movements of sea level could also be interpreted in terms of tectonic uplift and subsidence, and indeed some

Table 1 *Facies from the CIROS-1 drillhole, with relative abundances, description and interpretation*

Facies	Total thickness and proportion of core	Description	Interpretation
Massive diamictite (Dm)	37.8 (5.7%)	Muddy sandstone or sandy mudstone with matrix-supported clasts (1–20% or rock), of varied size, shape and lithology. Some striated.	Lodgement till or waterlain till.
Weakly stratified diamictite (Dw)	135.5 (20.7%)	As massive diamictite, but with diffuse or wispy stratification.	Waterlain till – proximal glaciomarine.
Massive and weakly stratified sandstone (Sm, Sw)	51.5 (7.7%) 103.9 (15.6%)	Moderately well-sorted – poorly sorted sandstone, often with minor mud and gravel component. Weak stratification, often contorted, irregular, discontinuous.	Nearshore – shore-face with minor ice-rafting in distal glaciomarine setting. Also redeposited in beds 0.1–to 1 m thick below 530 m.
Well-stratified sandstone (Ss)	26.1 (3.9%)	Moderately – well-sorted sandstone with mm-scale stratification.	Shore-face – beach
Massive mudstone (Mm)	15.7 (2.4%)	Poorly sorted, sandy mudstone with dispersed gravel clasts.	Offshore with minor ice-rafting in distal glaciomarine setting.
Weakly stratified mudstone (Mw)	179.9 (27.1%)	As massive mudstone but with weak discontinuous stratification defined by sandier layers	Offshore – deeper near-shore with minor ice-rafting in distal glaciomarine setting.
Well-stratified mudstone (Ms)	19.5 (2.9%)	As massive mudstone, but with discontinuous well defined stratification.	Deeper nearshore with minor ice-rafting in distal glaciomarine setting.
Bioturbated mudstone (Mb)	38.6 (5.8%)	As massive mudstone, but stratification almost totally destroyed by bioturbation.	Offshore–deeper nearshore with minor ice-rafting.
Mudstone Breccia (B)	31.4 (4.7%)	Unstratified – weakly stratified very poorly sorted, sandy mudstone intraformational breccia with up to 70% clasts.	Offshore–deeper nearshore slope deposits with minor ice-rafted component, disrupted by short distance subaqueous debris-flowage.
Rhythmite (R)	4.2 (0.6%)	Graded alternations of poorly sorted muddy sand and sandy mud. regular stratification disrupted by drop-stones.	Turbidity flows derived from subglacial source, with an ice-rafted component.
Conglomerate (C)	20.6 (3.1%)	Unstratified – weakly stratified, poorly sorted, clast–matrix-supported sandy conglomerate.	Fluvioglacial deposits or delta-slope debris flows derived from immediately pro- or subglacial fluvial material.
Brecciated conglomerate at base of hole	2.8 (0.4%)	Poorly sorted, sandy boulder conglomerate, matrix–clast-supported.	Subaqueous fault scarp deposit.

variation in water depth due to tectonism is to be expected. However, we believe the cyclic character, scale and periodicity of the movements are more consistent with glacio-eustatic origin. Clifton, Hunter & Gardner (1988) made a similar but more fully considered case in their interpretation of the Pleistocene Merced Formation of California.

The lower part of the core represents relatively deep-water sedimentation with persistent but limited glacial influence (Fig. 4). The absence of lodgement till suggests that the debris in this part of the core comes from ice calved from local glaciers rather than extensive ice sheet glaciation. Units 19 and 20 include a number of graded sandstone beds from about 5–50 cm thick with sharp, in places loaded, bases and gradational tops. The beds are separated by mudstone and are interpreted as subaqueous sediment gravity slows redepositing sediment on a delta slope. The oldest unit cored (Unit 22) has boulders up to 47 cm across, thought to have come from nearby bedrock, and possibly emplaced in a similar way.

Tectonic history

The considerable thickness of Oligocene and early Miocene strata at CIROS-1, along with the pervasive soft sediment deformation, suggest sedimentation in a subsiding basin close to an actively rising terrane, presumably the Transantarctic Mountains. The lack of late Miocene and Pliocene strata, together with the high seismic velocities of the CIROS-1 and MSSTS-1 core (Davy & Alder, 1989) make it clear that,

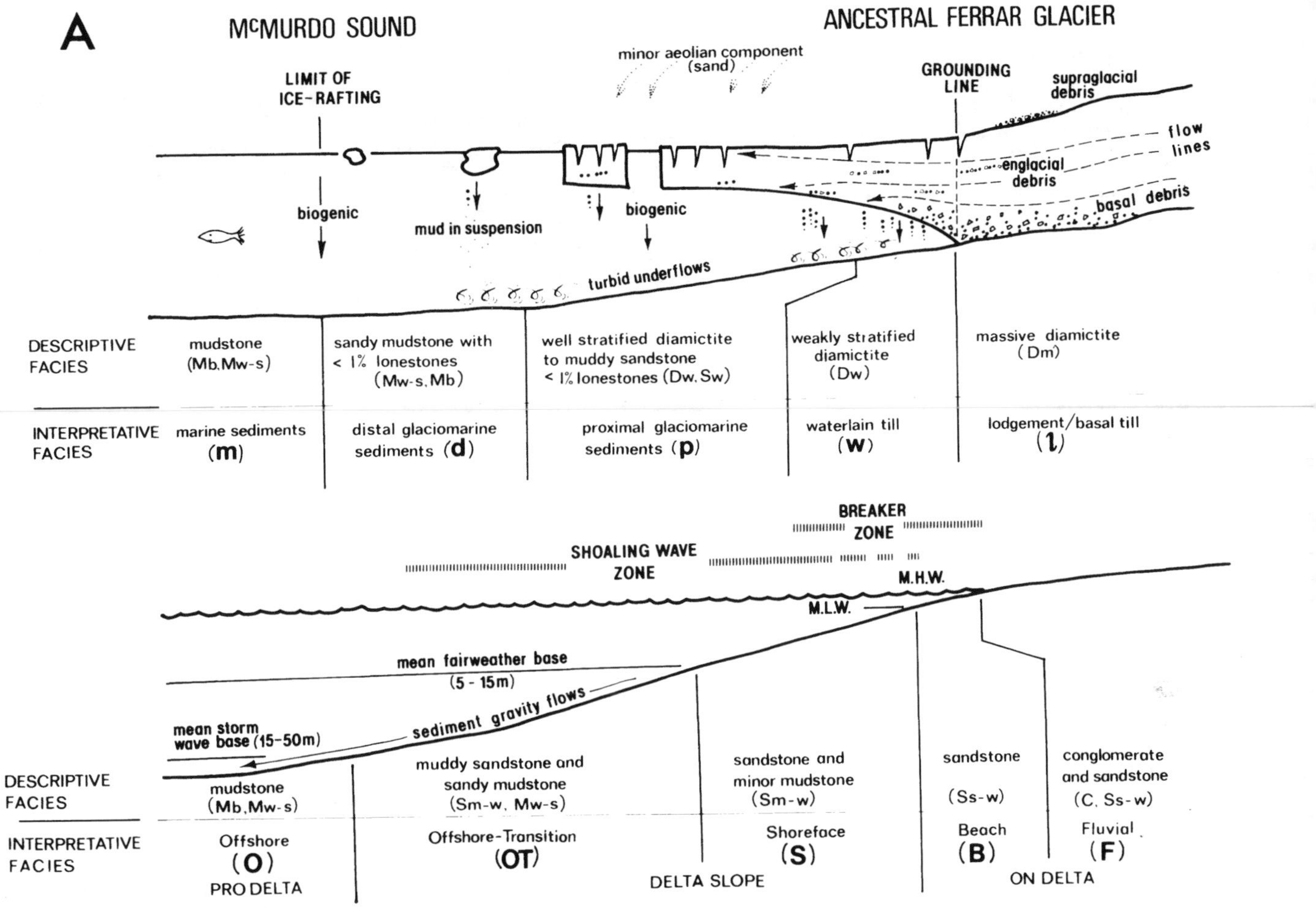

Fig. 5. Sedimentological models used in interpreting the CIROS-1 core. Letters in parentheses are symbols in Fig. 4. (A) Glacial facies representing the spectrum from subglacial deposition (lodgement till) to sedimentation beyond the normal influence of floating ice. (B) Deltaic facies representing the spectrum from fluvial deposition to sedimentation below wave base.

not only has there been little recent sedimentation, but also that a significant thickness of strata, perhaps a kilometre or more, has been subsequently deposited and eroded. This may represent a period of uplift from thermal expansion due to raised isotherms from the extension and crustal thinning implied by the thick Oligocene and early Miocene strata in CIROS-1. Gleadow & Fitzgerald (1987) have already provided sound evidence from apatite fission-track dating for a total of 5 km of uplift in the adjacent mountains over the last 50 m.y. The CIROS-1 record suggests that much of this took place during Oligocene time.

Both clast counts and grain mineralogy show that the main source of the sediment was the Transantarctic Mountains to the west (Fig. 6). The sand fraction is dominated by Early Palaeozoic basement-derived debris, with a small recognizable contribution from the Devonian–Jurassic Beacon sandstone–Ferrar Dolerite cover beds. Intermediate–basic volcanic debris, including glass fragments, persists throughout the core, though it is most common in the upper 100 m, where it can be related to the late Cenozoic McMurdo Volcanic Group. Debris of similar composition has already been described from late Oligocene strata in MSSTS-1 (Gamble, Adams & Barrett, 1986), and the occurrence of similar debris to the bottom of CIROS-1 extends the record of Cenozoic volcanism back still further.

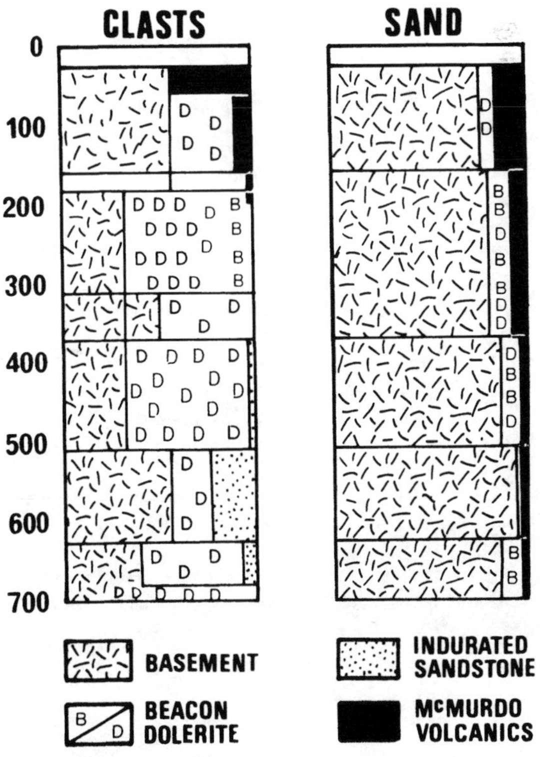

Fig. 6. Average proportions of source rocks contributing to the CIROS-1 core. Data summarized from Barrett (1989).

Conclusions

The CIROS-1 core has provided an excellent stratigraphic record of events close to the Antarctic ice sheet in Oligocene and early Miocene time. Preliminary results indicate an early Oligocene period of deep-water sedimentation with limited glaciation close to sea level followed by a shallowing at about the early–late Oligocene boundary, coinciding with the greatest drop in world sea level in the Cenozoic Era. The late Oligocene–early Miocene strata above indicate much more extensive ice with seven glacial episodes, four of which may be related to build-up of the continental ice sheet and relatively ice-free periods between.

The relatively rapid sedimentation at CIROS-1 records the sinking of the Victoria Land basin and concomitant uplift of the adjacent Transantarctic Mountains through Oligocene and into Miocene time. Since this time a kilometre or more of strata have been eroded, presumably as a consequence of uplift along the margin of the basin.

Acknowledgements

The CIROS project was part of the New Zealand Antarctic Research Programme. It was logistically supported by Antarctic Division, DSIR, and funded by DSIR and the University Grants Committee. We are particularly grateful to the drilling crew, led by Kevin Jenkins, Geophysics Division, DSIR, and the core processing team led by Alex Pyne. We thank Martin Sharp and Julian Dowdeswell for reviewing the paper. MJH thanks the Leverhulme Trust and the Trans-Antarctic Association for financial support. The manuscript was typed by Val Hibbert and the diagram drafted by Ted Hardy.

References

Barrett, P.J., ed. (1989). *Antarctic Cenozoic history from the CIROS-1 drillhole, western McMurdo Sound.* Wellington; DSIR Miscellaneous Bulletin, **245**, 254 pp.

Barrett, P.J., Elston, D.H., Harwood, D.B., McKelvey, B.C. & Webb, P.N. (1987). Middle Cenozoic glaciation and sea level change on the margin of the Victoria Land basin, Antarctica. *Geology*, **15(7)**, 634–7.

Barrett, P.J., Pyne, A.R. & Ward, B.L. (1983). Modern sedimentation in McMurdo Sound, Antarctica. In *Antarctic Earth Science*, eds. R.L. Oliver, P.R. James & J.B. Jago, pp. 550–5. Canberra; Australian Academy of Science and Cambridge; Cambridge University Press.

Barron, J.A. & Keller, G.A. (1982). Widespread Miocene deep-sea hiatuses: coincidence with periods of global cooling. *Geology*, **10(11)**, 577–81.

Clifton, H.E., Hunter, R.E. & Gardner, J.V. (1988). Analysis of eustatic, tectonic and sedimentologic influences on transgressive and regressive cycles in the late Cenozoic Merced Formation, San Francisco, California. In *New Perspectives of basin analysis*, eds C. Paola and K.L. Kleinspehn. New York; Springer Verlag.

Cooper, A.K. & Davey, F.J. (1985). Episodic rifting of Phanerozoic rocks in the Victoria Land basin, western Ross Sea, Antarctica. *Science*, **229(4718)**, 1085–7.

Davy, B. & Alder, G. (1989). Seismic stratigraphy. In *Antarctic Cenozoic history from the CIROS-1 drillhole, western McMurdo Sound*, ed. P.J. Barrett. Wellington; DSIR Miscellaneous Bulletin, **245**, 15–26.

Denton, G.H. & Hughes, T.J. (1981). *The Last Great Ice Sheets*. New York; John Wiley & Sons. 484 pp.

Drewry, D.J. (1981). Pleistocene bimodal response of Antarctic ice. *Nature, London*, **287(5779)**, 214–16.

Drewry, D.J. & Cooper, A.P.R. (1981). Process and models of Antarctic glaciomarine sedimentation. *Annals of Glaciology*, **2**, 117–22.

Elliott, T. (1986). Deltas. Siliciclastic shorelines. In *Sedimentary Environments and Facies*, ed. H.G. Reading, pp. 113–88. Oxford; Blackwell Scientific Publications.

Elverhøi, A., Lønne, Ø. & Seland, R. (1983). Glaciomarine sedimentation in a modern fjord environment, Spitzbergen. *Polar Research*, **1**, 127–49.

Gamble, J.A., Adams, C.J. & Barrett, P.J. (1986). Basaltic clasts from Unit 8. In *Antarctic Cenozoic History from the MSSTS-1 drillhole, western McMurdo Sound*, ed. P.J. Barrett, pp. 145–52. Wellington; DSIR Miscellaneous Bulletin, 237.

Gleadow, A.J.W. & Fitzgerald, P.G. (1987). Uplift history and structure of the Transantarctic Mountains: new evidence from fission-track dating of basement apatites in the Dry Valleys area, southern Victoria Land. *Earth and Planetary Science Letters*, **82**, 1–14.

Haq, B.H., Hardenbol, J. & Vail, P.R. (1987). Chronology of fluctuating sea levels since the Triassic. *Science*, **235(4793)**, 1156–67.

Matthews, R.K. & Poore, R.Z. (1980). Tertiary $\delta^{18}O$ record and glacioeustatic sea-level fluctuations. *Geology*, **8(10)**, 501–4.

McGinnis, L.D., Bowen, R.H., Erickson, J.M., Allred, B.J. & Kreamer, J.L. (1985). East–west Antarctic boundary in McMurdo Sound. *Tectonophysics*, **114**, 341–56.

Miller, K.G., Mountain, G.S. & Tucholke, B.E. (1985). Oligocene glacio-eustasy and erosion on margins of the North Atlantic. *Geology*, **13(1)**, 10–13.

Powell, R.D. (1981). A model for sedimentation by tide water glaciers. *Annals of Glaciology*, **2**, 129–34.

Robinson, P.H., Pyne, A.R., Hambrey, M.J., Hall, K.J. & Barrett, P.J. (1987). *Core Log, Photographs and Grain Size Analyses from the CIROS-1 Drillhole, Western McMurdo Sound, Antarctica*. Wellington; Victoria University of Wellington Antarctic Data Series No. 14. 241 pp.

Shackleton, N.J. (1986). Palaeogene stable isotope events. *Palaeogeography, Palaeoclimatology, Palaeoecology*, **57(1)**, 91–102.

Shackleton, N.J. & Kennett, J.P. (1975). Paleotemperature of the Cenozoic and the initiation of Antarctic glaciation: oxygen and carbon isotope analyses in DSDP sites 277, 279 and 281. In *Initial Reports of the Deep Sea Drilling Project*, 29, ed. J.P. Kennett & R.E. Houtz *et al.*, pp. 743–55. Washington, DC; US Government Printing Office.

Summerhayes, C.P. (1986). Sea level curves based on seismic stratigraphy: their chronostratigraphic significance. *Palaeogeography, Palaeoclimatology, Palaeoecology*, **57(1)**, 27–42.

A subaerial eruptive environment for the Hallett Coast volcanoes

W.C. McINTOSH[1] & J.A. GAMBLE[2]

1 Geoscience Department, New Mexico Institute of Mining and Technology, Socorro, New Mexico 87801, USA
2 Geology Department, Victoria University, Wellington, New Zealand

Abstract

Subaerial volcanic rocks show that the late Miocene–Pliocene (13–2.5 Ma) Hallett Coast shield volcanoes did not erupt and accumulate beneath a continental ice sheet. Subaerial lavas, lava-flow breccias, and subordinate pyroclastic deposits form four large ($> 100\ km^3$), elongate shield complexes. Most of the pyroclastic deposits are subaerial Hawaiian or Strombolian bomb and lapilli units, commonly showing welding or deuteric haematitic oxidation. Associated with the Hawaiian and Strombolian units are less abundant hydrovolcanic (Surtseyan) deposits erupted and deposited subaerially or in shallow water. The deposits are composed of poorly sorted, variably bedded tuffs and lapilli-tuffs, some containing accretionary lapilli. Breccias at the base of most lava flows consist of clast-supported, oxidized, vesicular fragments of holocrystalline lava, indicating subaerial emplacement. Pillows or glassy breccias at the base of a few lava flows probably reflect interaction with limited amounts of ice and snow in a periglacial environment. The predominance of subaerial volcanic rocks shows that a fully expanded Ross Ice Sheet was absent during most of the emplacement of the Hallett Coast volcanoes. However, syn- and post-volcanic advances of a regional ice sheet from the west are recorded along the northern end of Adare Peninsula by a glacial unconformity and erratics.

Introduction

Late Cenozoic alkali volcanic rocks of the McMurdo Volcanic Group are widespread along the western margin of the Ross Sea embayment (Kyle & Cole, 1974). Some units have been interpreted as the products of subglacial eruptions. When combined with radiometric age determinations, these deposits have been used to infer the timing of inception and advances of continental glaciation in Antarctica (Hamilton, 1972; LeMasurier, 1972; Stump, Sheridan & Borg, 1980; Kyle, 1981). The ability to identify confidently volcanic rocks formed in subglacial environments is fundamental to these interpretations.

This paper presents some geological observations from volcanoes in the Hallett volcanic province. The area was visited by the first author and others in January 1983 (LeMasurier *et al.*, 1983) and again in February 1986. The volcanoes of the Hallett province have been interpreted as products of primarily subglacial eruptions (Hamilton, 1972). We believe that these volcanoes were not formed subglacially. This issue has an important bearing on understanding the previous extent of glaciations in the Ross Sea embayment.

Geology of the Hallett Volcanic Province

The Hallett volcanic province primarily consists of four elongate shield volcano complexes, aligned in a north–south direction along the northern Victoria Land coast (Fig. 1). From north to south, these complexes are termed Adare, Hallett, and Daniell peninsulas and Coulman Island. They range from 1770 to 2083 m in elevation and from 30 to 80 km in length. Additional smaller volcanic centres are present on the adjoining mainland and on the Possession Islands (Fig. 1). Throughout much of the Hallett region, the volcanic rocks are well exposed in wave-cut cliffs up to 2000 m high.

Hallett volcanic province rocks show alkalic compositions ranging from basinite to phonolite and trachyte (Hamilton, 1972; Jordan, 1981; P.R. Kyle, unpublished data). Radiometric dates obtained from representative units indicate late Miocene–Pliocene ages (13–2.5 Ma) (Hamilton, 1972; Kreuzer *et al.*, 1981).

Thirteen localities were visited and sampled in this study (Fig. 1) and two stratigraphic sections were measured (Fig. 2). At all localities, the observed volcanic sequences are composed primarily of lava flows, lava-flow breccias, and dikes, accompanied by subordinate amounts (0–15%) of tuffs, lapilli-tuffs, and sedimentary rocks.

Lava flows and lava-flow breccias

The best studied sequence of lavas is the measured stratigraphic section at the south-western edge of Coulman Island (Fig. 2), where lavas and lava-flow breccias comprise

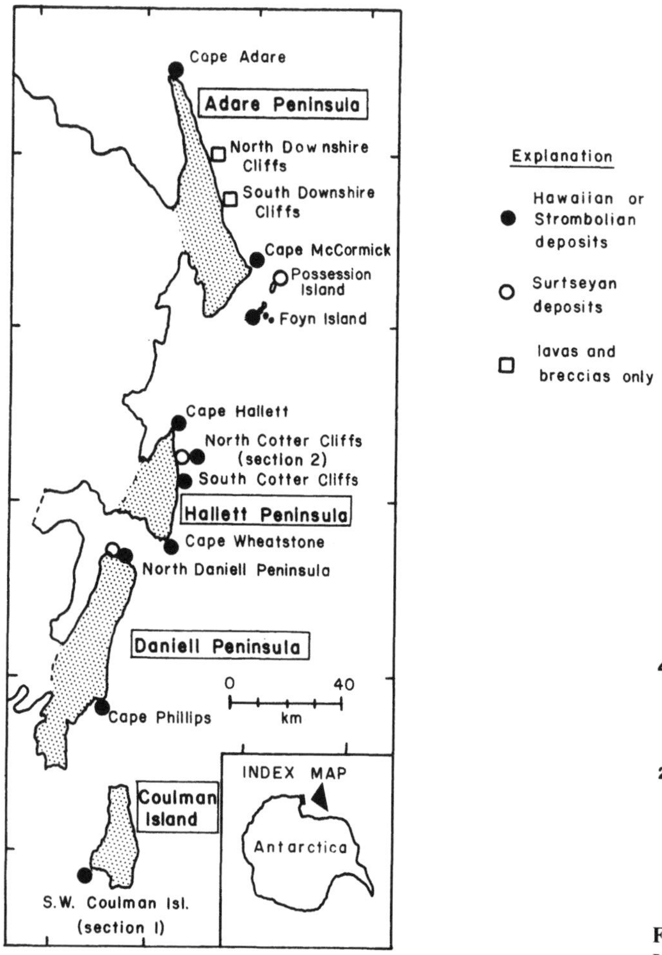

Fig. 1. Map of Hallett volcanic province showing locations studied and types of pyroclastic deposits present. Lavas and associated breccias dominate the volcanic sequences at all 13 localities. Shading denotes the shield volcano complexes.

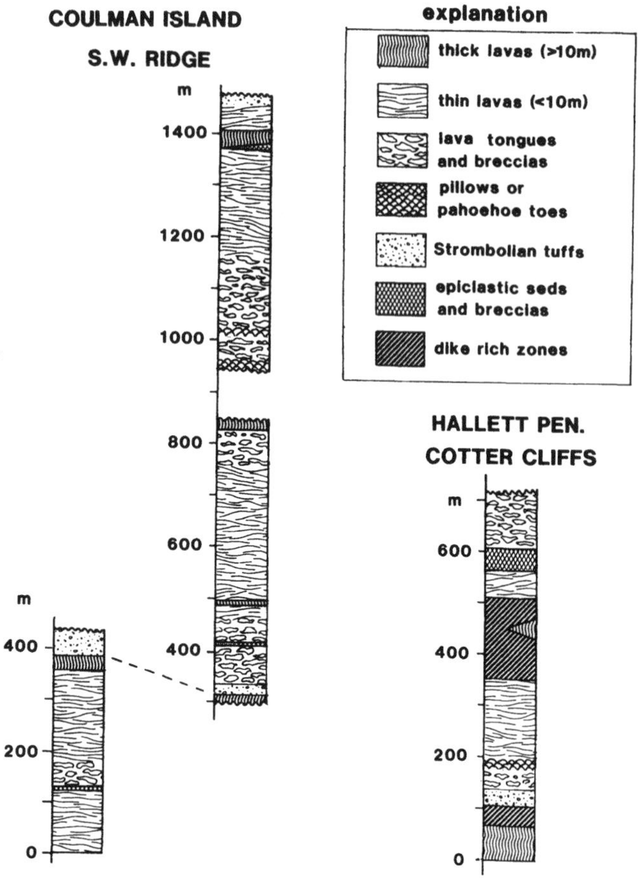

Fig. 2. Measured stratigraphic sections from south-western Coulman Island and Cotter Cliffs on Hallet Peninsula. The left and right portions of the Coulman Island section were correlated across a snow-filled saddle. Exposures extend about 400 m above the top of the Coulman Island measured section and about 800 m above the top of the Cotter Cliffs measured section.

over 98% of the 1385 m thick measured section. The section is dominated by mafic, phenocryst-poor lava flows comprised of thin (0.5–3 m) vesicular tongues which generally lack chilled margins. Associated lava-flow breccias are thin (0.1–1 m) and are composed of clast-supported vesicular fragments of holocrystalline lava commonly showing welding and/or red oxidation.

Less common in this section are intermediate, crystal-rich lavas, typically forming variably vesicular flow tongues, 1–15 m thick, with glassy chilled margins, 0.5–4 cm thick. Most flow tongues are flattened or elongate, although one (near 950 m elevation) shows subspherical pillow-like forms. Breccias at the bases of the flows are 0.5–15 m thick and composed of glassy–holocrystalline, angular–rounded fragments of lava. These breccias are both clast- and matrix-supported and many are zeolitized or palagonitized. Breccias on the upper surfaces of some of the flows show red oxidation.

A measured section at northern Cotter Cliffs on Hallett Peninsula (Fig. 1) also consists predominantly (90%) of lava flows and lava-flow breccias (Fig. 2). The composition and size of flow units appear similar to those in the Coulman Island section, although the section is locally altered by intense dike intrusion. One thick hawaiite flow (180–380 m elevation) is entirely composed of thin (< 1 m) flow tongues which superficially resemble pillows near the base of the unit, but lack associated hyaloclastite breccias and show only thin (< 3 mm) glassy rims.

Sections were not measured at the remaining 11 localities visited (Fig. 1) but, without exception, the observed volcanic sequences were dominated by lavas and lava-flow breccias.

The predominance of lavas in the Hallett Coast volcanic sequences and the nature of the associated lava-flow breccias are consistent with eruption into dominantly subaerial environments. Lava flows associated with red oxidized, matrix-poor, holocrystalline basal breccias certainly formed subaerially. The glassy basal breccias and rare pillows associated with a few lava flows may reflect limited interaction with snow and ice, perhaps in a periglacial environment. However, a subglacial or subaquatic origin for most of the lava flows is ruled out by the rarity of pillows and the confinement of chilled margins to the bases of most flow tongues.

Pyroclastic deposits

Subordinate pyroclastic deposits were observed and sampled at 11 localities (Fig. 1). Some deposits consist of subhorizontal or inclined layers intercalated with lavas or breccias. Other form small, symmetrical cones, typically 0.3 km in diameter, which are deeply buried beneath lava flows. Field observations, coupled with optical and scanning electron microscope (SEM) studies of samples, indicate that these pyroclastic units are primarily Hawaiian or Strombolian deposits accompanied by subordinate hydrovolcanic deposits.

Hawaiian and Strombolian deposits

Subaerially erupted Hawaiian or Strombolian deposits were observed at nine localities (Fig. 1). These deposits range from massive- to thin-bedded and many contain welded spatter and volcanic bombs with associated impact structures. The colours of these deposits vary widely. Many are black with variable reddening characteristic of subaerial oxidation. Others appear pink, grey, or yellow due to zeolitization, hydrothermal alteration, and/or palagonitization.

These deposits are composed primarily of tachylite cinders and sideromelane or tachylite droplet-shaped grains. These droplet grains (Figs 3*a* & *b*) range from 0.1 to 10 mm in diameter and exhibit fluidal shapes, highly vesicular interiors, and smooth, continuous outer skins, resulting from surface tension when fluid lava was fountained into the air (Walker & Croasdale, 1972). The glass droplets in four of the eight deposits show moulded or welded contacts (Figs 3*c*, *d* & *e*), consistent with field observations of welded spatter. Welding and moulding are produced where pyroclastic material accumulates while still plastic and are commonly restricted to subaerial environments (Fisher & Schmincke, 1984). All of the deposits containing droplet grains are well sorted and contain little fine ash. Variable amounts of authigenic calcite, zeolite and silica coat the droplets and fill the voids between grains.

Well-sorted basaltic cinders, glass droplet grains, and welded spatter, commonly accompanied by deuteric reddening, are characteristic products of Hawaiian or Strombolian eruptions, which occur in subaerial environments without significant water entering the vent (Walker & Croasdale, 1972; Sheridan & Wohletz, 1983; Heiken & Wohletz, 1984).

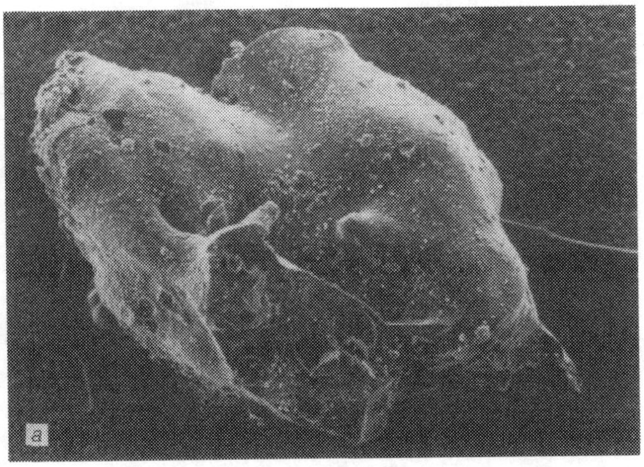

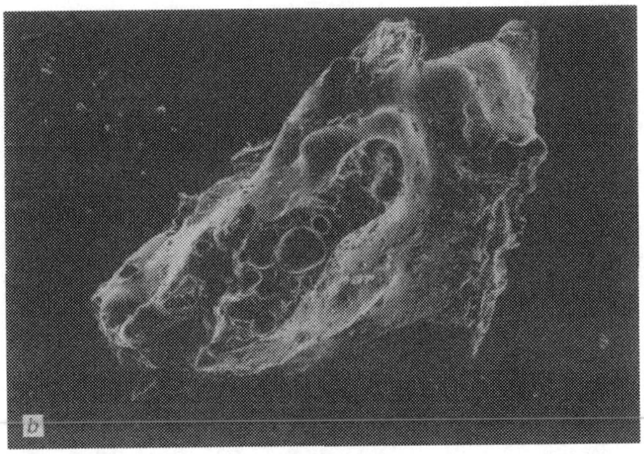

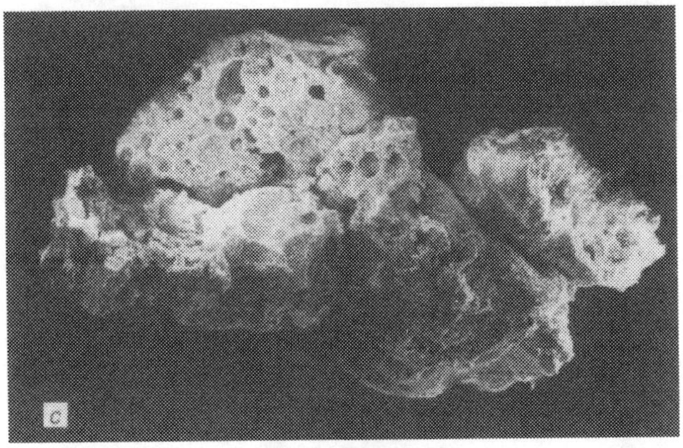

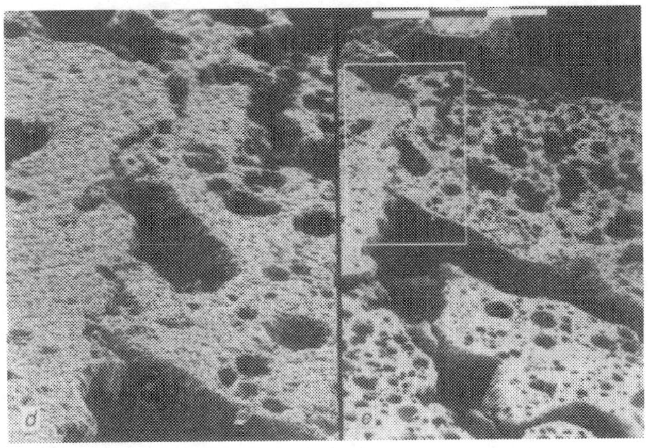

Fig. 3. SEM photographs of representative ash and lapilli particles from Hallett region pyroclastic deposits. (*a* & *b*) Sideromelane droplets from a subaerially erupted Strombolian deposit at Cape Hallett. Each droplet shows a fluidal shape, smooth outer skin, and a vesicular interior (exposed where each droplet was artificially broken). Authigenic calcite and zeolite coat the outer surfaces. Scale bar = 1 mm. (*c*) Four agglutinated tachylite lapilli droplets from a welded spatter deposit on Foyn Island. Authigenic analcime coats the outer surfaces. Scale bar = 1 mm. (*d* & *e*) Detailed views of a welded contact between lapilli from the same deposit as (*c*). (*d*) details the rectangle on (*e*). The surface of the sample was ground flat to reveal the contact. Scale bar = 0.5 mm.

Hydrovolcanic (Surtseyan) deposits

Three of the pyroclastic deposits studied (Fig. 1) are thin-bedded tuffs and lapilli-tuffs, consisting primarily of poorly sorted, fracture-bounded sideromelane fragments and subordinate lithic clasts. The sideromelane fragments range from fine ash to lapilli in size and are typically less vesicular than the glass droplet grains of the Hawaiian and Strombolian deposits. Some of these fragments are well rounded, but most are equant to blocky and bounded by curved fractures which meet to form angular edges.

Fracture-bounded sideromelane clasts are characteristic products of hydrovolcanic eruptions, resulting from quenching, mechanical stress, and steam explosions associated with water–magma interaction (Walker & Croasdale, 1972; Sheridan & Wohletz, 1983; Heiken & Wohletz, 1984). Such eruptions (termed 'Surtseyan', 'hydrovolcanic', or 'phreatomagmatic' by different workers) occur when seawater, lakewater, groundwater, or subglacial meltwater gains access to an erupting basaltic vent.

The three Surtseyan deposits studied from the Hallett volcanic province are products of subaerial or shallow-water eruptive vents. Two of the deposits (north Daniell Peninsula and north Cotter Cliffs; Fig. 1) are directly associated with subaerial Strombolian and Hawaiian deposits. One of these (north Cotter Cliffs; Fig. 1) contains accretionary lapilli, which are characteristic products of ash-laden atmospheric eruption columns (Moore, 1985). The third hydrovolcanic deposit (Possession Island; Fig. 1) contains polished, basaltic beach pebbles, which suggest a littoral eruptive environment analogous to the coral-bearing littoral cones in Hawaii (Hay & Iijima, 1968).

Discussion

As summarized above, the lavas, breccias and pyroclastic deposits observed in the Hallett volcanic province are interpreted to have formed in predominantly subaerial environments. No significant subglacial deposits, such as tillites or thick hyaloclastites, were identified in this study. The distribution of the units studied on the four major shield volcano complexes (Fig. 1) suggests that they are representative of both the dissected interiors and outer carapaces of these volcanic piles. The deposits studied on the Possession Islands may represent either smaller volcanoes or eroded remnants of larger edifices. Altogether, the deposits studied indicate that subaerial conditions prevailed throughout much of the late Miocene–Pliocene eruptive history of the Hallett volcanic province. There is no indication that a continental ice sheet occupied the Ross Sea embayment for extended periods during this time. A number of lines of evidence indicate that the Ross Ice Shelf has grounded and expanded several times previously, to form an ice-sheet extending to the margin of Antarctica's continental shelf (Denton *et al.*, 1975, 1984). Apparently, the Miocene–Pliocene volcanoes of the Hallett volcanic province were erupted prior to or between periods of expanded continental glaciation in the Ross Sea. If long-lived expansions of the Ross Ice shelf did occur during this period, then they apparently did not reach coastal northern Victoria Land.

Several features near the northern tip of Adare Peninsula apparently reflect local advances of a continental ice sheet. Our aerial observations of the cliffs south-east of Cape Adare showed a continuous, undulatory unconformity of probable glacial origin. Glacial tills intercalated with the volcanic sequence have also been reported from these cliffs (Priestley, 1923). Along the north-western edge of Adare Peninsula, palagonitic pillow breccias have been described from elevations of 150–500 m (Jordan, 1981), and glacial erratics derived from highlands to the west mantle the slopes to elevations of 520 m (Hamilton, 1972). These observations suggest that syn- and post-volcanic ice sheets at least 500 m thick have occupied Robertson Bay, to the west of Cape Adare. These ice sheets were apparently locally derived and were not produced by grounding and expansion of the Ross Ice Shelf.

Conclusions

Several lines of evidence indicate that the four major elongate shield volcano complexes of the Hallett volcanic province erupted and accumulated in a predominantly subaerial environment. This evidence includes:

1. The predominance of lavas and associated lava-flow breccias in observed volcanic sequences
2. The non-glassy, clast-supported, deuterically reddened nature of many of these marginal breccias
3. The general paucity of pillows, pillow-breccias, and glassy, matrix-rich hyaloclastites
4. The presence of distinctive Strombolian and Hawaiian pyroclastic deposits showing features characteristic of subaerial eruption and deposition, including glassy cinders and droplet grains, fusiform bombs with impact structures, welded spatter, and deuteric reddening
5. The presence of hydrovolcanic deposits containing accretionary lapilli and beach pebbles, suggesting eruption in a subaerial or shallow-water environment.

The Miocene–Pliocene Hallett Coast volcanoes apparently were erupted prior to or between periods of expanded continental glaciation in the Ross Sea embayment.

Acknowledgements

This work would not have been possible without the help of the officers, pilots, and crew of the USCG *Glacier* and *Polar Star*. We are very grateful to Philip Kyle and Wes LeMasurier for providing the opportunity of Antarctic field work and for reviewing this paper. Nick Banks supplied essential field support on Coulman Island. This work was supported by NSF grants DPP80-20002 and DPP82-18493 to Philip Kyle and DPP80-208336 to Wes LeMasurier.

References

Denton, G.H., Borns, H.W. Jr., Groswald, M.G. & Nichols, R.L. (1975). Glacial history of the Ross Sea. *Antarctic Journal of the United States*, **10**, 160–4.

Denton, G.H., Prentice, M.L., Kellogg, D.E. & Kellogg, T.B. (1984). Late Tertiary history of the Antarctic ice sheet: evidence from the Dry Valleys. *Geology*, **12**, 263–7.

Fisher, R.V. & Schmincke, H.U. (1984). *Pyroclastic Rocks*. Berlin; Springer. 472 pp.

Hamilton, W. (1972). The Hallett volcanic province, Antarctica. *US Geological Survey Professional Paper*, **456-C**, 62 pp.

Hay, R.L. & Iijima, A. (1968). Nature and origin of palagonitic tuffs of the Honolulu Group on Oahu, Hawaii. *Geological Society of America Memoir*, **116**, 331–76.

Heiken, G.H. & Wohletz, K.H. (1984). *Volcanic Ash.* Berkeley; University of California Press.

Jordan, H. (1981). Tectonic observations in the Hallett volcanic province, Antarctica. *Geologisches Jahrbuch*, **B41**, 111–25.

Kreuzer, H., Hohndorf, A., Lenz, H., Vetter, U., Tessensohn, F., Muller, P., Jordan, H., Harre, W. & Besang, C. (1981). K/Ar and Rb/Sr dating of igneous rocks from northern Victoria Land, Antarctica. *Geologisches Jahrbuch*, **B41**, 267–73.

Kyle, P.R. (1981). Glacial history of the McMurdo Sound area as indicated by the distribution and nature of McMurdo Volcanic Group rocks. In *Dry Valley Drilling Project*, Antarctic Research Series, 33, ed. L.D. McGinnis, pp. 403–12. Washington, DC; American Geophysical Union.

Kyle, P.R. & Cole, J. (1974). Structural control of volcanism in the McMurdo Volcanic Group, Antarctica. *Bulletin Volcanologique*, **38(1)**, 16–25.

LeMasurier, W.E. (1972). Volcanic record of Cenozoic glacial history of Marie Byrd Land. In *Antarctic Geology and Geophysics*, ed. R.J. Adie, pp. 251–9. Oslo; Universitetsforlaget.

LeMasurier, W.E., Ellerman, P.J., McIntosh, W.C. & Wright, A.C. (1983). USCGS *Glacier* cruise 1, December 1982–January 1983: reconnaissance of hyaloclastites in the western Ross Sea region. *Antarctic Journal of the United States*, **18(5)**, 60–1.

Moore, J.G. (1985). Structure and eruptive mechanisms at Surtsey Volcano, Iceland. *Geological Magazine*, **122**, 649–61.

Priestley, R.E. (1923). *British (Terra Nova) Antarctic Expedition, 1910–1913. Physiography (Robertson Bay and Terra Nova Regions)*, London; Harrison & Sons. 87 pp.

Sheridan, M.F. & Wohletz, K.H. (1983). Hydrovolcanism: basic considerations and review. *Journal of Volcanology and Geothermal Research*, **17**, 1–29.

Stump, E., Sheridan, M.F. & Borg, S.G. (1980). Early Miocene subglacial basalts, the East Antarctic Ice Sheet, and uplift of the Transantarctic Mountains. *Science*, **207**, 757–8.

Walker, G.P.L. & Croasdale, R. (1972). Characteristics of some basaltic pyroclastics. *Bulletin Volcanologique*, **35**, 305–17.

Origin and age of pectinid-bearing conglomerate (Tertiary) on King George Island, West Antarctica

K. BIRKENMAJER[1], A. GAŹDZICKI[2], R. GRADZIŃSKI[1], H. KREUZER[3], S.J. PORĘBSKI[1] & A.K. TOKARSKI[1]

1 Institute of Geological Sciences, Polish Academy of Sciences, Senacka 3, 31-002 Kraków, Poland
2 Institute of Palaeobiology, Polish Academy of Sciences, Żwirki i Wigury 93, 02-089 Warszawa, Poland
3 Bundesanstalt für Geowissenschaften und Rohstoffe, Stilleweg 2, 3000 Hannover 51, Federal Republic of Germany

Abstract

The pectinid-bearing conglomerate on King George Island, South Shetland Islands, is interpreted as the nearshore deposit of a high-energy wave-dominated coast. K–Ar dating of associated volcanic rocks and calcareous nannoplankton point to an Oligocene age for the conglomerate. It is concluded that the pectinid-bearing sedimentary rocks from King George Island do not correlate with the *Pecten* conglomerate of Cockburn Island, which is regarded as Pliocene in age.

Introduction

A *Pecten*-bearing conglomerate, considered to be an equivalent of the Pliocene *Pecten* conglomerate of Cockburn Island (64°12′S, 56°51′W; Andersson, 1906) was identified on King George Island, South Shetland Islands, by Barton (1965). Birkenmajer (1980, 1982, 1985) recognized the glaciomarine character of the deposit on King George Island and included it as part of his Low Head Member (LHM) of the Polonez Cove Formation, likewise attributed to the Pliocene. Further palaeontological studies of the invertebrate fauna from the LHM generally confirmed its Tertiary age (e.g. Gaździcki & Pugaczewska, 1984).

However, K–Ar dating of the lavas (Boy Point Formation) overlying the Polonez Cove Formation by H. Kreuzer (in Birkenmajer & Gaździcki, 1986) show that the Polonez Cove Formation is older than the Oligocene–Miocene boundary. Further K–Ar dating of the overlying lavas (Boy Point Formation) indicates that a mid-Oligocene age from the Polonez Cove Formation and its pectinid-bearing unit is possible (Birkenmajer *et al.*, 1986). The Oligocene age of the pectinid-bearing LHM is supported by stratigraphical evaluation of associated calcareous nannoplankton (Gaździcka & Gaździcki, 1985).

Stratigraphy

Birkenmajer (1980, 1982) introduced formal lithostratigraphic units for the *Pecten* conglomerate-bearing succession on King George Island (Figs 1 & 2). The base of the succession is formed by Upper Cretaceous subaerial basaltic lavas of the Mazurek Point Formation, more than 100 m thick. An erosional unconformity representing a long hiatus separates the lavas from the overlying Polonez Cove Formation.

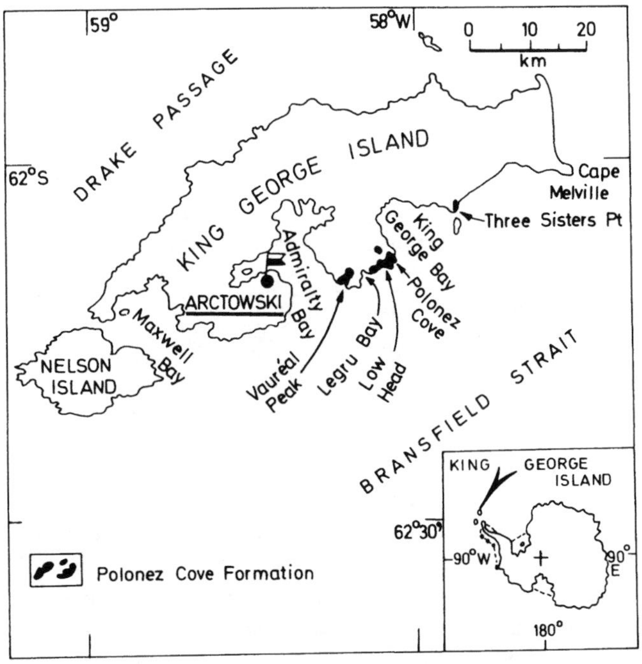

Fig. 1. Sketch map showing the location of Tertiary glacigenic deposits on King George Island (from Birkenmajer, 1985).

The Polonez Cove Formation (5–65 m thick) consists of continental and glaciomarine tillites formed during a glacial epoch called the Polonez Glaciation. The formation has been subdivided into the Krakowiak Glacier Member (diamictites), the Low Head Member (mainly basaltic conglomerates with intercalations of pectinid-bearing coquinas), the Siklawa Member (psammites and pelites), and the Oberek Cliff Member (conglomerates and sandstones) (Fig. 2).

There followed, after an erosional break, andesitic–dacitic lavas of the Boy Point Formation (20–100 m thick), and

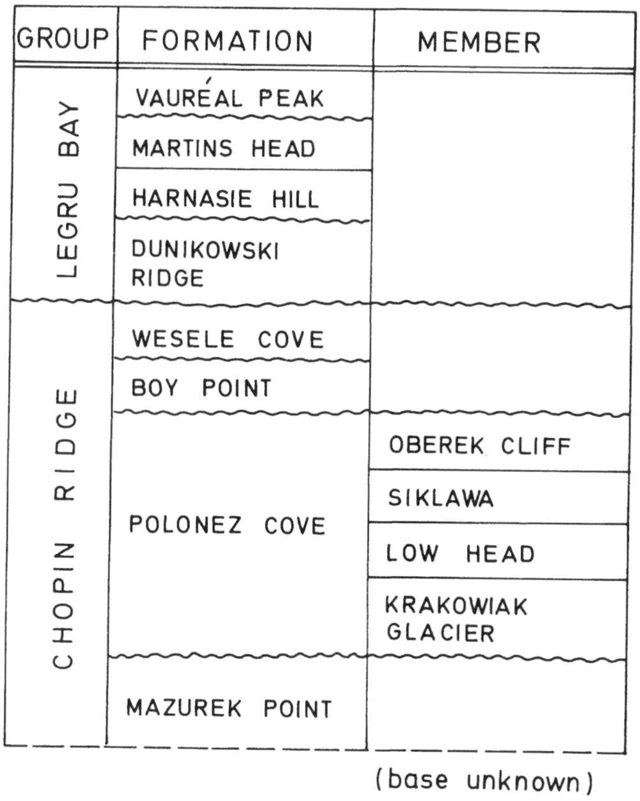
GROUP
FORMATION
MEMBER
LEGRU BAY
VAUREÁL PEAK
MARTINS HEAD
HARNASIE HILL
DUNIKOWSKI RIDGE
CHOPIN RIDGE
WESELE COVE
BOY POINT
POLONEZ COVE
OBEREK CLIFF
SIKLAWA
LOW HEAD
KRAKOWIAK GLACIER
MAZUREK POINT
(base unknown)

Fig. 2. Subdivision of the Chopin Ridge and Legru Bay groups (from Birkenmajer, 1980, 1982).

another separates this formation from the succeeding fluvial and debris-flow conglomerates and breccias of the Wesele Cove Formation (75–100 m thick). The whole succession constitutes the Chopin Ridge Group and is unconformably overlain by a predominantly andesitic, volcanic–volcaniclastic, tillite-bearing complex, the Legru Bay Group (~ 300 m thick).

Sedimentology

The pectinid-bearing sedimentary rocks of the LHM form a fining-upwards unit, 4–20 m thick, which rests on a hydroclastic, olivine–basalt breccia, or on a diamictite assemblage showing evidence of deposition from a grounded continental ice sheet. The unit consists mainly of clast-supported pebble conglomerates, pebbly sandstones and sandstones that are composed of fresh volcanic detritus, chiefly basaltic and basaltic-andesite. Exotic clasts, up to boulder size, are present locally, having been either ice-rafted or reworked from the diamictite substratum (Birkenmajer, 1980, 1982). The sedimentary rocks occur as wedge-shaped to widely channelled beds, 40–120 cm thick, that are massive, flat to low-angle and cross-stratified (Fig. 3*a*). Cross-set dip azimuths show a wide, bimodal distribution. Shelly faunas, dominated by pectinid bivalves (mostly *Chlamys anderssoni* Hennig; Gaździcki, 1984), are fragmented to various degrees, and occur as clusters, lenses and beds within the rudaceous sediment (Fig. 3*b*).

The shell beds (up to 50 cm thick) vary from those consisting of densely packed, mostly horizontally aligned valves (both with concave up and down orientations) to those with a

Fig. 3. (*a*) Bedding style in pectinid-bearing pebbly sandstones. Note the wedge-shaped bed geometries and low-angle planar and trough cross-stratification. (*b*) Close-up view of large-scale cross-stratified conglomerate, showing *Chlamys* concentrates along shallow-dipping pebble laminae.

disrupted framework and shells often in a 'standing-on-end' position. The shelly fauna is dominated by shallow-water biota (Gaździcki & Pugaczewska, 1984), chiefly byssate fissure dwellers and swimmers characteristic of rocky nearshore areas. The pectinid-bearing sedimentary rocks are interpreted as nearshore deposits of a high-energy wave-dominated coast whose configuration was controlled by products of contemporaneous volcanism (Porębski & Gradziński, 1987).

Fossils and age

The fossils described so far from the LHM include: coccoliths, diatoms, chrysomonad cysts, benthonic and planktonic foraminifera, tube worms, bryozoans, brachiopods, bivalves, gastropods, ostracods, crinoids, ophiuroids, and echinoids (Bitner & Pisera, 1984; Gaździcki & Pugaczewska, 1984; Jesionek-Szymańska, 1984; Gaździcka & Gaździcki,

985; Błaszyk, 1987). The bryozoans show very wide age anges: out of 27 taxa described only 13 are reported exclusively rom the Tertiary. Twelve of the 28 bivalve taxa are known rom the Tertiary, and there are two taxa in common with the *Pecten* conglomerate of Cockburn Island, including *Chlamys anderssoni* Hennig. The gastropods are more Tertiary in aspect with eight species out of the 11 taxa determined being exclusively Tertiary; five of these species are present also in the Tertiary ((?)middle–late Eocene–(?)early Oligocene) strata of Seymour Island: *Polinices* cf. *subtenuis* Ihering, *Antarctodarwinella* cf. *nordenskjoeldi* Wilckens, *Trophon* cf. *disparoides* Wilckens and *Cyrtochetus* cf. *bucciniformis* Wilckens (see stratigraphical evaluation by Birkenmajer (1987)).

Eleven coccolith taxa found attached to tests of the planktonic foraminifera *Globigerina* and *Globorotalia* have been determined by Gaździcka & Gaździcki (1985), who correlated the assemblage with the *Chiasmolithus altus* and *Reticulofenestra bisecta* zones, CP-17–CP-19 of the Oligocene. Birkenmajer (1987) raised some doubts as to whether this is a stratigraphically coherent assemblage, and indicated that some coccolith taxa seem to be recycled from older deposits.

K–Ar dating

The basaltic lavas of the Mazurek Point Formation which underlie the Polonez Cove Formation (Fig. 2) yielded a Late Cretaceous age of ~ 74 Ma (H. Kreuzer, in Birkenmajer & Gaździcki, 1986). The andesite–dacite lavas of the Boy Point Formation which overlie the Polonez Cove Formation yielded whole-rock dates of $> 23.6 \pm 0.3$ and $> 22.4 \pm 0.4$ Ma. Another date, 22.3 ± 0.8 Ma, was obtained by H. Kreuzer from a basaltic hydroclastic breccia underlying the pectinid-bearing conglomerate at Low Head (for location, see Tokarski, 1987).

Birkenmajer *et al.* (1986) suggested that the dates from the Boy Point Formation lavas were not the real ages of the lavas but may be a result of reheating and argon loss by the succeeding Legru Bay Group volcanic rocks, dated at 29.5 ± 2.1 and 25.7 ± 1.3 Ma. This conclusion would also apply to the K–Ar date from the basaltic breccia mentioned above.

Conclusion

Notwithstanding the differences in opinion about the age of the Polonez Cove Formation resulting from coccolith dating of the Low Head Member, and K–Ar dating of associated volcanic rocks, an Oligocene (mid-Oligocene?) age for the pectinid-bearing LHM seems to be well founded now. Taking into account that a Pliocene age for the *Pecten* conglomerate from Cockburn Island has been confirmed (Zinsmeister & Webb, 1982), the correlation of this conglomerate with the LHM is less likely, despite the fact that both seem to contain the same characteristic bivalve *Chlamys anderssoni* Hennig.

References

Andersson, J.G. (1906). On the geology of Graham Land. *Bulletin of the Geological Institution of the University of Upsala*, **7(13–14)**, 19–71.

Barton, C.M. (1965). *The Geology of the South Shetland Islands: III. The Stratigraphy of King George Island.* London; British Antarctic Survey Scientific Reports, No. 44, 33 pp.

Birkenmajer, K. (1980). Discovery of Pliocene glaciation on King George Island, South Shetland Islands (West Antarctica). *Bulletin de l'Académie Polonaise des Sciences, Série Sciences de la Terre*, **27(1–2)**, 59–67.

Birkenmajer, K. (1982). Pliocene tillite-bearing succession of King George Island (South Shetland Islands, Antarctica). *Studia Geologica Polonica*, **74**, 7–72.

Birkenmajer, K. (1985). Onset of Tertiary continental glaciation in the Antarctic Peninsula sector (West Antarctica). *Acta Geologica Polonica*, **35(1–2)**, 1–31.

Birkenmajer, K. (1987). Oligocene–Miocene glaciomarine sequences of King George Island (South Shetland Islands), Antarctica. *Palaeontologia Polonica*, **49**, 9–36.

Birkenmajer, K., Delitala, M.C., Narębski, W., Nicoletti, M. & Petrucciani, C. (1986). Geochronology of Tertiary island-arc volcanics and glacigenic deposits, King George Island, South Shetland Islands (West Antarctica). *Bulletin of the Polish Academy of Sciences, Earth Sciences*, **34(3)**, 257–73.

Birkenmajer, K. & Gaździcki, A. (1986). Oligocene age of the *Pecten* conglomerate on King George Island, West Antarctica. *Bulletin of the Polish Academy of Sciences, Earth Sciences*, **34(2)**, 219–22.

Bitner, M.A. & Pisera, A. (1984). Brachiopods from the '*Pecten* conglomerate' (Polonez Cove Formation, Pliocene) of King George Island (South Shetland Islands, Antarctica). *Studia Geologica Polonica*, **79**, 121–4.

Błaszyk, J. (1987). Ostracods from the Oligocene Polonez Cove Formation of King George Island, West Antarctica. *Palaeontologia Polonica*, **49**, 63–81.

Gaździcki, A. (1984). The *Chlamys* coquinas in glaciomarine sediments (Pliocene) of King George Island, West Antarctica. *Facies*, **10**, 145–52.

Gaździcka, E. & Gaździcki, A. (1985). Oligocene coccoliths of the *Pecten* conglomerate, West Antarctica. *Neues Jahrbuch für Geologie und Paläontologie, Monatshefte*, **12**, 727–35.

Gaździcki, A. & Pugaczewska, H. (1984). Biota of the '*Pecten* conglomerate' (Polonez Cove Formation, Pliocene) of King George Island, (South Shetland Islands, Antarctica). *Studia Geologica Polonica*, **79**, 59–120.

Jesionek-Szymańska, W. (1984). Echinoid remains from '*Pecten* conglomerate' (Polonez Cove Formation, Pliocene) of King George Island (South Shetland Islands, Antarctica). *Studia Geologica Polonica*, **79**, 125–30.

Porębski, S.J. & Gradziński, R. (1987). Depositional history of the Polonez Cove Formation (Oligocene), King George Island, West Antarctica: a record of continental glaciation, shallow-marine sedimentation and contemporaneous volcanism. *Studia Geologica Polonica*, **93**, 7–60.

Tokarski, A.K. (1987). Report on geological investigations of King George Island, South Shetland Islands (West Antarctica), in 1986. *Studia Geologica Polonica*, **93**, 123–30.

Zinsmeister, W.J. & Webb, P.N. (1982). Cretaceous–Tertiary geology and paleontology of Cockburn Island. *Antarctic Journal of the United States*, **17(5)**, 41–2.

Cenozoic diatom biogeography in the southern high latitudes: inferred biogeographic barriers and progressive endemism

D.M. HARWOOD

Byrd Polar Research Center and Department of Geology and Mineralogy, The Ohio State University, Columbus, Ohio 43210 USA

Abstract

In the past the paucity of diatom data from marine sequences in Antarctica limited our ability to relate palaeoceanographic events, thought to result from climatic changes in the south polar region, to the physical evidence for these events on the Antarctic continent. Reworked marine diatoms and diatomaceous clasts from the Sirius Group, which provide a unique record of Cenozoic marine sedimentation in East Antarctica, and results from drillholes in McMurdo Sound (MSSTS & CIROS), offer new information to fill this void. In this paper, the composition of fossil diatom assemblages from interior basins of East and West Antarctica and the Southern Ocean are compared in order to help understand the nature, cause and timing of Cenozoic marine communications and/or palaeobiogeographic barriers between these three regions. Available data on diatoms from DSDP cores in the Southern Ocean, from coastal outcrops and drillcores on the Antarctic shelf and margin, and from continental and marine glacial sediments, are reviewed. The history of southern high-latitude diatom palaeobiogeography and the development of modern diatom floras, which are endemic to various water masses in this region, were controlled by the climatic evolution of the Antarctic region.

Introduction

Sufficient data are available to allow speculation on the nature and history of Cenozoic palaeobiogeographic barriers and gradients between the Southern Ocean and Antarctic interior basins. The past 40 m.y. of tectonic, glacial and oceanographic change in the Antarctic region undoubtedly influenced marine diatom palaeobiogeography and biostratigraphy in and around Antarctica. Kennett (1978, 1979) reviewed the development of Southern Ocean planktonic microfossil biogeography. The present discussion focusses on marine diatom biogeography and emphasizes new diatom data from Antarctic interior marine basins, a region not adequately treated in previous biogeographic analyses due to limited data.

Cenozoic evolution of the southern high-latitudes

The late Palaeogene and Neogene witnessed numerous environmental changes that influenced the history of Antarctic diatom biogeography. These include: (1) the development of circum-Antarctic circulation and thermal isolation of Antarctica (Kennett, 1977), leading to the initiation of glaciation by the early Oligocene (Barrett *et al.*, 1987; Barrett, Hambrey & Robinson, this volume, p. 651); Shackleton & Kennett, 1975; Savin, 1977; Matthews & Poore, 1980; Miller & Fairbanks, 1983; Keigwin & Keller, 1984); (2) the growth of several Antarctic ice sheets of varying size, during the late Oligocene, late Miocene and latest Pliocene, which are separated by episodes of ice sheet destruction and marine sedimentation in Antarctic interior basins during the early Miocene and early Pliocene (Harwood, 1983, 1986a, 1987; Webb *et al.*, 1984); (3) the development and northward migration of biogeographic barriers, such as the Polar Front, Antarctic Divergence and Subtropical Convergence (Kennett, 1978, 1979; Ciesielski & Grinstead, 1986), which led to the expansion of the siliceous ooze belt and northward migration of calcareous and ice-rafted sediment facies through the Miocene (Kemp, Frakes & Hayes, 1975); (4) an increase in latitudinal thermal gradient (Savin, 1977), with resultant increase in atmospheric and oceanic circulation, leading to greater upwelling and siliceous biogenic productivity in the Southern Ocean (Brewster, 1980); and (5) uplift of the Transantarctic Mountains, commencing in the middle Eocene (Fitzgerald *et al.*, 1987) with a recent phase of uplift in the Pleistocene or perhaps latest Pliocene (Webb *et al.*, 1986, 1987; McKelvey, Webb, Harwood & Mabin, this volume, p. 675). Cenozoic diatom biogeography and related diatom evolution and migration are considered in the light of these events.

Available diatom data

The analysis attempted here is limited by the paucity of data for comparison, particularly from the Antarctic interior.

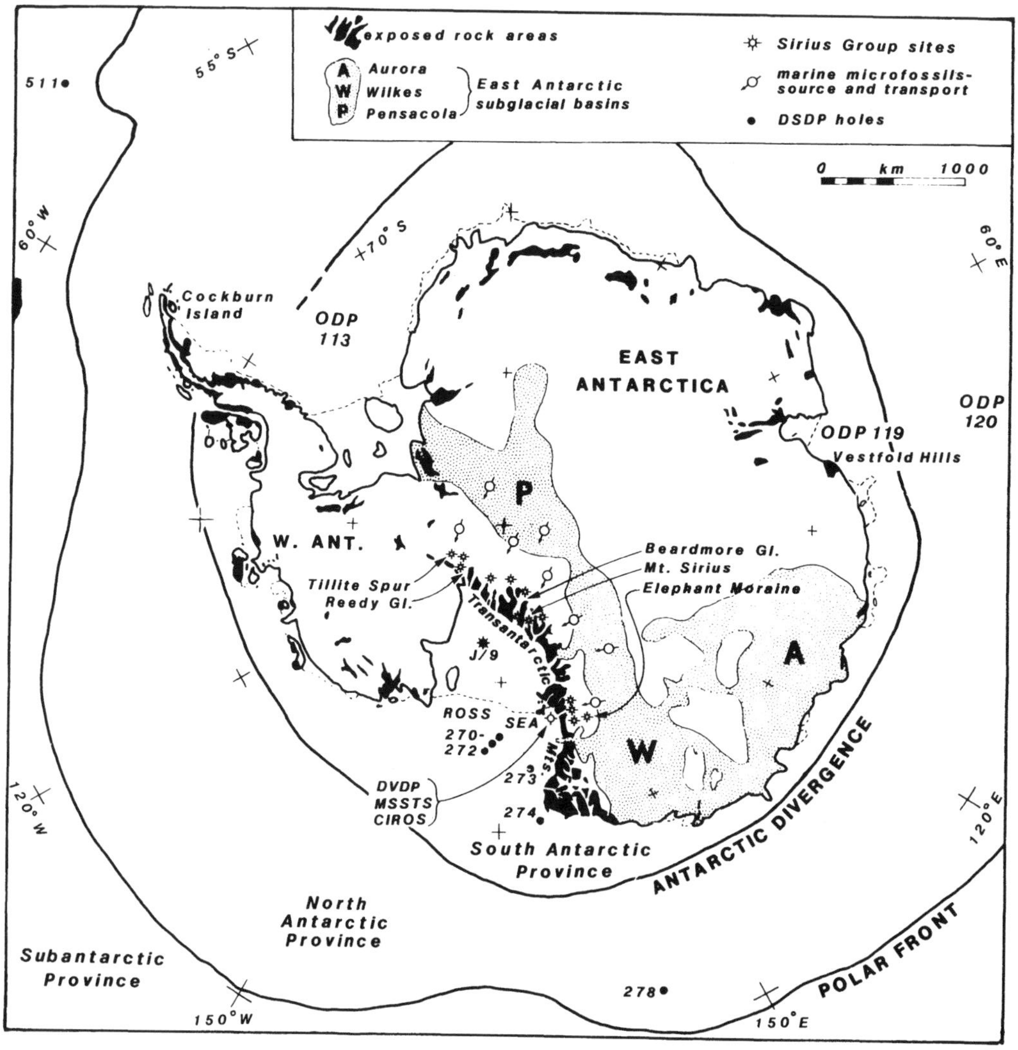

Fig. 1. Map of Antarctica and the Southern Ocean indicating all key localities mentioned in text, including subglacial basins of East and West Antarctica. Likely transport directions of reworked marine microfossils, Sirius Group localities, significant drillholes and modern biogeographic provinces and barriers also shown.

Available data come from a variety of environmental regimes, including *in situ* coastal outcrops, drillholes and piston cores in fjords and on the Antarctic shelf, slope and deep sea (Fig. 1). Diatom assemblages from East Antarctica are known mostly from reworked specimens that were eroded from the Wilkes and Pensacola subglacial basins (Fig. 1), transported by the East Antarctic Ice Sheet and deposited in glacial deposits of the Sirius Group (Harwood, 1983, 1986a; Webb *et al.*, 1984). Although study of these reworked diatom floras presents certain difficulties due to low abundance, multiple assemblage–age mixing, and source regions known only as 'up-glacier', they are the only record of marine sediments from interior basins of East Antarctica. Thus, they provide vital southernmost reference points for palaeobiogeographic studies.

Cenozoic diatom palaeobiogeography of Antarctic waters

Pre-Oligocene

Cretaceous, Palaeocene and Eocene diatom assemblages recovered from the Southern Ocean by the Deep Sea Drilling Project are reviewed in Weaver & Gombos (1981), Fenner (1985) and Harwood (1988). Because pre-Oligocene diatom assemblages are not yet known from Antarctic interior basins, a biogeographic comparison is not attempted here. However, the wide distribution of similar diatom assemblages around the globe suggests minimal provincialism compared to present.

Early Oligocene

Lower Oligocene diatom assemblages from interior basins of East and West Antarctica and the Southern Ocean are of similar floral composition, although West Antarctic assemblages appear to have been more isolated. Of the three regions, the most information is available from the Southern Ocean in DSDP Hole 274 (McCollum, 1975; Fenner, 1984), Hole 280A (Hajós, 1976), Holes 328 and 328A (Gombos, 1977), and Holes 511 and 513A (Gombos & Ciesielski, 1983; Fenner, 1985). In West Antarctica, lower Oligocene assemblages are known from the basal 350 m of the New Zealand CIROS-1 drillhole (Fig. 1) in the western Ross Sea (Barrett *et al.*, this volume, p. 651; Harwood (1989). East Antarctic diatoms floras of this age are known from the Pensacola Basin in reworked sediment clasts, and as isolated diatoms in the Sirius Group (Upper Reedy Glacier; Harwood, 1983, 1986a). Diatom species known from all three regions include: *Coscinodiscus superbus*, *Cotyledon fogedi*, *Hemiaulus characteristicus*, *H. incisus*, *Kisseleviella carina*, *Pseudotriceratium radiosoreticulatum*, *Pyxilla reticulata*, *Triceratium unguiculatum*, *Sphynctolethus pacificus*, *Stephanopyxis eocenica*, *S. hyalomarginata*, and *S. superbus*.

The broad distribution of this assemblage suggests marine communication between these three regions. However, some biogeographic segregation is noted between the Southern Ocean and Antarctic interior basins. The presence of a weakly developed thermal gradient, environmental stress associated with strong seasonality of polar regions, or the shelf environment itself, may have been responsible for: (1) the absence of *Brightwellia*, *Coscinodiscus excavatus* and *Trinacria simulacrum*; (2) the low diversity and low numbers of species of the Asterolampraceae; and (3) the abundance of diatom resting spores in the interior basins of Antarctica, compared to the Southern Ocean (Harwood, 1989).

Assemblages from East Antarctic basins have a greater affinity with Southern Ocean floras than do Ross Sea assemblages; *Sphynctolethus pacificus*, *Coscinodiscus superbus* and *Triceratium unguiculatum* are common in East Antarctic and Southern Ocean assemblages but are rare in the CIROS-1 drillhole in the Ross Sea. This difference may reflect circulation of south-east Indian Ocean waters through the Wilkes and Pensacola basins prior to development of significant deep-water flow across the Tasman Seaway (Kennett, 1977, 1978). Significant compositional differences (involving diatoms listed in the above two paragraphs) between lower Oligocene assemblages in DSDP Hole 274 in the Southern Ocean (Fig. 1) and coeval deep-water assemblages in CIROS-1 suggest that there was little Southern Ocean influence in the western Ross Sea at this time (Harwood, 1989).

Late Oligocene–early Miocene

Diatom assemblages of the Southern Ocean and the Wilkes–Pensacola basins are similar during the late Oligocene–early Miocene. However, coeval Ross Sea assemblages are distinctly different, despite the presence of certain cosmopolitan species common in all three regions. Species of the diatom genus *Rocella* characterize upper Oligocene–lower Miocene assemblages of the Southern Ocean, where they occur in great abundance and in acme horizons in DSDP Hole 278 (Perch-Nielsen, 1976; Schrader, 1976), Hole 516F (Gombos, 1983) and Hole 513A (Gombos & Ciesielski, 1983). Of the sparse assemblages recovered from clasts in the Sirius Group at Mt Sirius (Fig. 1), both *Rocella vigilans* and *R. gelida* are present (Harwood, 1986a), suggesting their common occurrence in the Wilkes–Pensacola basins. In contrast, *Rocella* is very rare in upper Oligocene–lower Miocene Ross Sea sediments from drill cores MSSTS-1 (Harwood, 1986b), CIROS-1 (Harwood, 1989), and absent from lower Miocene diatomaceous sediments from DSDP Hole 270 in the central Ross Sea (Steinhauff *et al.*, 1987). A marine communication between the Wilkes/Pensacola basins and the Southern Ocean and isolation of the Ross Sea during this period is suggested. Diminished communication between the Wilkes–Pensacola basins and the Ross Sea may be due to the barrier of the rising Transantarctic Mountains (Fitzgerald *et al.*, 1987).

The strong affinity of East Antarctic floras with those from the Southern Ocean appears to continue from the early Oligocene. Floral similarity may result from circulation of Southern Ocean waters into the Wilkes and Aurora basins and perhaps across into the Pensacola Basin too (Fig. 1). In contrast, a clockwise gyre may have developed in the Ross Sea, driven at its northern end by strong circum-antarctic circulation, leading to isolated floras in this region of West Antarctica (including the exclusion of *Rocella*).

The inability to apply existing upper Oligocene Southern Ocean diatom zones in drillholes CIROS-1 and MSSTS-1 in the western Ross Sea, due to the rarity or absence of Southern Ocean zonal species, led Harwood (1986b) to propose a zonation for these drillholes. This is based on diatom species with similar stratigraphic ranges in both the Southern Ocean and the Ross Sea. These cosmopolitan southern high-latitude species include: *Azpetta oligocenica*, *Lisitzinia ornata*, *Pseudotriceratium cheneveri*, *Synedra jouseana* and *S. miocenica*. However, the apparent absence of common Southern Ocean and lower-latitude diatom taxa such as *Bogorovia*, *Rossiella*, *Kozloviella*, *Coscinodiscus rhombicus* and *Cestodiscus pulchellus* from the Antarctic interior basins reflects the increasing isolation of the Antarctic region by the developing Antarctic Circumpolar Current and regional cooling. Oxygen isotope studies (Miller & Fairbanks, 1983; Keigwin & Keller, 1984) indicate an ice volume increase in the late Oligocene at ~ 30 Ma. However, apparent marine conditions in the late Oligocene Pensacola Basin, indicated by *Rocella* at Mt Sirius, suggest that initial ice build-up was probably restricted to glaciation in the Transantarctic Mountains and in other high elevation areas of East Antarctica. Alternatively, the marine and glacial history of Pensacola Basin is complex and marked by rapid changes in ice volume (Barrett *et al.*, 1989).

Middle–late Miocene

Information on middle–upper Miocene diatom assemblages from the Antarctic interior is sparse. The few sites known include: (1) reworked diatom assemblages in sediment clasts and matrix sediments from cores taken at the RISP Site J/9 beneath the Ross Ice Shelf (Brady & Martin, 1979; Kellogg & Kellogg, 1986; Harwood, Scherer & Webb, 1989); (2) reworked diatom assemblages from the Wilkes and Pensacola basins in till clasts exposed on the surface of the East Antarctic Ice Sheet at Elephant Moraine and in the Sirius Group at Tillite Spur (Faure & Taylor, 1985; Harwood, 1986a) (Fig. 1); and (3) *in situ* upper Miocene assemblages from the Jason Diamicton in Wright Valley (Burckle, Prentice & Denton, 1986), from drillhole DVDP 11 in the McMurdo Sound area (Brady, 1979; Harwood, 1986a) and from DSDP Hole 272 (McCollum, 1975; Savage & Ciesielski, 1983) in the central Ross Sea. No biogeographic interpretation is attempted here because these assemblages are temporally diverse and the age of reworked material is poorly constrained. However, for the intervals where data are available, the assemblages appear to be widely distributed. High diatom productivity and the absence of ice cover in the Ross Sea and other areas of West Antarctica is suggested by the abundance of reworked clasts of lower Miocene diatomite in the RISP J/9 sediments (Harwood *et al.*, 1989).

Early–mid-Pliocene

Diatom assemblages of this age are of similar composition in the Southern Ocean and Antarctic interior basins. Furthermore, the sequence and ages of Southern Ocean diatom biostratigraphic datums for the lower to mid-Pliocene occur in similar relative stratigraphic order in the CIROS-2 drillhole in McMurdo Sound at 77° S (Harwood, 1986a). These similarities suggest minimal biogeographic segregation between the Southern Ocean and Antarctic interior at this time and broad application of Southern Ocean diatom biostratigraphy within and around Antarctica.

Information on diatom assemblages of this age from West Antarctica comes from drillholes DVDP-10, and DVDP-11 and CIROS-2 (Brady, 1979; Harwood, 1986a) and in glacial erratics in the McMurdo Sound region (Stott *et al.*, 1983). Pliocene diatom floras from the Wilkes and Pensacola basins are widespread in the Transantarctic Mountains, both as reworked specimens and in reworked sediment clasts within the Sirius Group (Harwood, 1983, 1986a, 1986c; Webb *et al.*, 1984, 1987). Known *in situ* diatom-bearing marine deposits in East Antarctica are the early Pliocene diatomaceous sand units at Marine Plain in the Vestfold Hills (Harwood, 1986a; Pickard *et al.*, 1988) and at Casey Station and Larsemann Hills (D.M. Harwood, unpublished data). Studies of Southern Ocean Pliocene diatom floras are reviewed in Weaver & Gombos (1981) and Barron (1985).

Lower Pliocene diatom species common to the Southern Ocean and interior basins of East and West Antarctica include: *Denticulopsis hustedtii*, *D. dimorpha*, *Eucampia antarctica*, *Nitzschia praeinterfrigidaria*, *Thalassiothrix/Thalassionema* and *Thalassiosira torokina*. Not all of these diatoms are restricted to the lower Pliocene. Their occurrence, plus that of other reworked marine microfossils in the Sirius Group, led Webb *et al.* (1984) to suggest an early Pliocene deglaciation of the late Miocene Antarctic Ice Sheet and subsequent marine incursion into interior basins of East, and most likely West, Antarctica. Similar lower Pliocene floras are known from relatively clast-free diatomaceous mudstone units in drillholes CIROS-2, DVDP-10 and 11 on the other side of the Transantarctic Mountains in the western Ross Sea (Harwood, 1986a). The reworked and drillhole diatom data may represent the same Pliocene deglacial–marine event. Warm interglacial conditions and a significant ice-volume reduction are also suggested for the early Pliocene by isotopic and micropalaeontologic studies of numerous DSDP sites in the south Atlantic and south-west Pacific oceans (Hodell & Kennett, 1986; and references therein) and by the Marine Plain deposit (Pickard *et al.*, 1988).

While the above works reflect the initial early Pliocene deglaciation and marine productivity in the interior basins (during what may have been an interval of peak late Neogene warmth), the mid–upper Pliocene assemblage of reworked marine diatoms recovered from numerous Sirius Group deposits (Harwood, 1983, 1985, 1986a) suggest that marine biogenic productivity continued in these basins from the mid–late Pliocene. The younger diatoms, many of which are extant, include: *Actinocyclus actinochilus*, *Coscinodiscus kolbei*, *C. vulnificus*, *Cosmoidiscus insignis*, *Nitzschia kerguelensis*, *Thalassionema nitzschioides*, *T. inura*, *T. lentiginosa*, *T. oliverana* and others. Their age of first appearance constrains the youngest age for marine conditions in the interior basins. All of these diatoms have their first occurrence in the Pliocene or late Miocene, the youngest being *Coscinodiscus vulnificus* and *Nitzschia kerguelensis*, which first appeared between 3.2 and 3.0 Ma (Weaver & Gombos, 1981; Ciesielski, 1983; Harwood, 1986a). Marine conditions, in the absence of ice cover for at least part of the year, must have prevailed until ~ 3.2 Ma. However, a short-lived cooling and ice increase is suggested for the interval 3.9–3.5 Ma by palaeoclimatic data reviewed in Harwood (1986a). Continued late Pliocene warmth and reduced ice volume are supported by the occurrence of *Nothofagus* sp. in the Meyer Desert Formation of the Sirius Group, upper Beardmore Glacier (McKelvey *et al.*, this volume, p. 675).

The wide distribution of the above Pliocene diatom assemblages in interior basins and in the Southern Ocean is in marked contrast to the present and past distinction between Antarctic interior and Southern Ocean floras. Culture studies on *Nitzschia kerguelensis* (Jacques, 1983), and on other common Southern Ocean diatoms (Neori & Holm-Hansen, 1982) present in the Sirius Group, provide temperature limits for Pliocene interior seas; Harwood (1986a, 1986c) suggested a range between 2 and 6 °C and perhaps even as warm as 8 °C.

The lower–mid-Pliocene diatom biostratigraphic record of the southern high latitudes is marked by: (1) the wide distribution of diatom species; (2) an overall increase in diatom diversity; and (3) numerous lowest occurrence datums, marking the appearance of new species which presently dominate the endemic Antarctic floras. This evolutionary radiation may have resulted from species migration into newly opened environmental niches in the rugged glacially-scoured interior coastline following the early Pliocene deglaciation, a situation conducive to vicariant evolution.

Late Pliocene–Pleistocene

The late Pliocene–Pleistocene is characterized by cooling and major ice sheet growth, probably in concert with Northern Hemisphere glaciation, and perhaps influenced by post-Pliocene uplift of the Transantarctic Mountains (Webb *et al.*, 1986; McKelvey *et al.*, this volume, p. 675) and consequent damming of East Antarctic ice drainage, which thickened the ice sheet. The diatom biostratigraphic record is marked by: (1) expulsion of flora from interior basins by ice advance; (2) decline in diatom diversity; and (3) a progression of highest occurrence diatom datums. Ice-related flora became abundant south of the Antarctic Divergence in seasonally ice-covered waters and the modern biogeographic segregation of diatom assemblages developed.

Today, unique diatom assemblages characterize the various biogeographic provinces (Fig. 1) that surround the Antarctic continent (Kozlova, 1966; Hasle, 1969; Abbott, 1974; Fenner, Schrader & Wienigk, 1976; Semina, 1979; Truesdale & Kellogg, 1979; DeFelice & Wise, 1981; Pichon, 1985; Burckle, Jacobs & McLaughlin, 1987). These biogeographic belts include: (1) the South Antarctic flora (south of the Antarctic Divergence and associated with sea ice), dominated by *Nitzschia curta*, *N. cylindrus*, *Thalassiosira antarctica*, *T. gracilis*, *Actinocyclus actinochilus* and *Eucampia antarctica*; (2) the North Antarctic flora (between the Antarctic Divergence and Polar Front region), dominated by *Nitzschia kerguelensis*, *Thalassiosira lentiginosa*, *T. oliverana* and *Azpetia tabularis*; and (3) the flora north of the Polar Front, which include *Thalassionema nitzschioides*, *Hemidiscus cuneiformis*, *Pseudoeunotia doliolus*, *Roperia tesselata* and other species characteristic of tropical and subtropical waters.

Discussion

The following questions should be considered in future work so that we may better understand the processes that affect Cenozoic diatom palaeobiogeography and biostratigraphic dating in the Antarctic region.

To what degree did climate and glacial change control diatom evolution, migration and extinction in the three regions? How do these changes affect application of biostratigraphic zonal schemes?

Is one pan-Antarctic diatom zonation sufficient for warm/-deglacial periods, but are several zonal schemes needed for cold/glacial periods when biogeographic provinciality was better defined?

Is the Antarctic interior environment (fjords, local basins, etc.) more conducive to vicariant evolutionary radiation than the well-mixed Southern Ocean? Are deglacial periods times of diatom radiation in the southern high-latitudes? Are glacial periods times of extinction? Do the numerous first occurrences during the early–mid Pliocene and numerous last occurrences during the latest Pliocene–Pleistocene support this?

How well does the present subglacial geography of East Antarctica (allowing for rebound) reflect the past geographic extent of interior seaways? Has Pleistocene tectonic activity (Webb *et al.*, 1986; McKelvey *et al.*, this volume, p. 675) and Pliocene sediment fill altered the palaeogeography of interior basins?

Much more documentation of Antarctic diatom assemblages, from a variety of environments and latitudes, is needed before we will be able to address the above questions with confidence and reveal the full nature of Cenozoic marine diatom biogeographic evolution in the south polar region. New data need to be collected by future drilling along latitudinal transects from the Southern Ocean onto the Antarctic shelf and into interior basins, utilizing drilling platforms on research ships, ice shelves, ice rises, grounded icebergs, rock platforms and perhaps on the surface of the ice sheet. Until appropriate technology and logistics are available, we must continue to rely on reworked diatom assemblages (and new preparation techniques to enhance their recovery) in order to gain insight into the distribution, composition and age of Cenozoic marine sequences beneath the Antarctic ice sheet.

This paper is meant only as a starting point from which further diatom investigations can test and improve the ideas outlined here. Certainly our understanding of Antarctic diatom palaeobiogeography, biostratigraphy and evolution will improve as more data from better samples become available through recent drilling adjacent to the Antarctic continent by Ocean Drilling Program (ODP) Legs 113 and 119 (Fig. 1), and by continued study of reworked marine microfossils in continental and marine glacial sediments.

Acknowledgements

This paper benefited from comments by L.H. Burckle, B.T. Huber, R.P. Scherer and P.-N. Webb. I thank P.J. Barrett, Victoria University of Wellington and the New Zealand DSIR for supplying sample material from MSSTS and CIROS cores, and the Deep Sea Drilling Project for supplying samples from numerous drillholes. The logistic support of squadron VXE6 in the Transantarctic Mountains is gratefully acknowledged. This research was supported by National Science Foundation grants DPP-8315553 & DPP-8420622 to P.N. Webb and a Geological Society of America Research Award to D.M. Harwood. The paper is Byrd Polar Resarch Center Contribution No. 610.

References

Abbott, W.S. (1974). Temporal and spatial distribution of Pleistocene diatoms from the southeast Indian Ocean. *Nova Hedwigia, Beiheft*, **25**, 291–347.

Barrett, P.J., Elston, D.P., Harwood, D.M., McKelvey, B.C. & Webb, P.-N. (1987). Mid-Cenozoic record of glaciation and sea-level change on the margin of the Victoria Land Basin, Antarctica. *Geology*, **15(7)**, 634–7.

Barrett, P.J., Hambrey, M.J., Harwood, D.M., Pyne, A.R. & Webb, P.–N. (1989). Synthesis. In *Antarctic Cenozoic History from the CIROS-1 Drillhole, McMurdo Sound*, ed. P.J. Barrett, pp. 241–51. Wellington; New Zealand DSIR Bulletin, No. 245.

Barron, J.A. (1985). Miocene to Holocene planktic diatoms. In *Plankton Stratigraphy*, ed. H.M. Bolli, J.B. Saunders & K. Perch-Nielsen, pp. 763–809. Cambridge; Cambridge University Press.

Brady, H.T. (1979). The dating and interpretation of diatom zones in Dry Valley Drilling Project Holes 10 and 11, Taylor Valley, South Victoria Land, Antarctica, Tokyo; *Memoirs of the National Institute of Polar Research*, **13**, 150–63.

Brady, H.T. & Martin, H. (1979). Ross Sea region in the Middle Miocene: a glimpse of the past. *Science*, **203**, 437–8.

Burckle, L.H., Jacobs, S.S. & McLaughlin, R.B. (1987). Late spring diatom distribution between New Zealand and the Ross Sea: correlation with hydrography and bottom sediments. *Micropaleontology*, **33(1)**, 74–81.

Burckle, L.H., Prentice, M.L. & Denton, G.H. (1986). Neogene Antarctic glacial history: new evidence from marine diatoms in continental deposits. *Eos, Transactions of the American Geophysical Union*, **67(16)**, 295.

Brewster, N.A. (1980). Cenozoic biogenic silica sedimentation in the Antarctic Ocean. *Geological Society of American Bulletin*, **Part I, 91**, 337–47.

Ciesielski, P.F. (1983). The Neogene diatom stratigraphy of Deep Sea Drilling Project Leg 71. In *Initial Reports of the Deep Sea*

Drilling Project, 71, ed. W.J. Ludwig & V. Krasheninnikov, pp. 635–65. Washington, DC; US Government Printing Office.

Ciesielski, P.F. & Grinstead, G.P. (1986). Pliocene variations in the position of the Antarctic Convergence in the southwest Atlantic. *Paleoceanography*, **1(2)**, 197–232.

DeFelice, D.R. & Wise, S.W., Jr. (1981). Surface lithologies, biofacies, and diatom diversity patterns as models for delineation of climate change in the southeast Atlantic Ocean. *Marine Micropaleontology*, **6(1)**, 29–70.

Faure, G. & Taylor, K.S. (1985). The geology and origin of the Elephant Moraine on the east antarctic ice sheet. *Antarctic Journal of the United States*, **20(5)**, 111–121.

Fenner, J. (1984). Eocene–Oligocene planktic diatom stratigraphy in the low latitudes and the high southern latitudes. *Micropaleontology*, **30(4)**, 319–42.

Fenner, J. (1985). Late Cretaceous to Oligocene planktic diatoms. In *Plankton Stratigraphy*, ed. H.M. Bolli, J.B. Saunders & K. Perch-Nielsen, pp. 713–62. Cambridge; Cambridge University Press.

Fenner, J., Schrader, H.-J. & Wienigk, H.J. (1976). Diatom phytoplankton studies in the southern Pacific Ocean, composition and correlation to the Antarctic Convergence and its paleoecological significance. In *Initial Reports of the Deep Sea Drilling Project*, 35, ed. C.D. Hollister, C. Craddock *et al.*, pp. 757–813. Washington, DC; US Government Printing Office.

Fitzgerald, P.G., Sandiford, M., Barrett, P.J. & Gleadow, A.J. (1987). Asymmetric extension associated with uplift and subsidence in the Transantarctic Mountains and Ross Embayment. *Earth and Planetary Science Letters*, **81**, 67–78.

Gombos, A.M., Jr. (1977). Paleogene and Neogene diatoms from the Falkland Plateau and Malvinas Outer Basin: Leg 36, Deep Sea Drilling Project. In *Initial Reports of the Deep Sea Drilling Project*, 36, ed. P.F. Barker, I.W.D. Dalziel *et al.*, pp. 575–687. Washington, DC; US Government Printing Office.

Gombos, A.M., Jr. (1983). Survey of diatoms in the upper Oligocene and lower Miocene in Holes 515B and 516F. In *Initial Reports of the Deep Sea Drilling Project*, 72, ed. P.F. Barker, R.L. Carlson, D.A. Johnson *et al.*, pp. 793–804. Washington, DC; US Government Printing Office.

Gombos, A.M., Jr. & Ciesielski, P.F. (1983). Late Eocene to early Miocene diatoms from the southwest Atlantic. In *Initial Reports of the Deep Sea Drilling Project*, 71, ed. W.J. Ludwig & V. Krasheninnikov, pp. 583–634. Washington, DC; US Government Printing Office.

Hajós, M. (1976). Upper Eocene and lower Oligocene Diatomaceae, Archaemonadaceae, and Silicoflagellatae in southwestern Pacific sediments, DSDP Leg 29. In *Initial Reports of the Deep Sea Drilling Project*, 35, ed. C.D. Hollister, C. Craddock *et al.*, pp. 817–83. Washington, DC; US Government Printing Office.

Harwood, D.M. (1983). Diatoms from the Sirius Formation, Transantarctic Mountains. *Antarctic Journal of the United States*, **18(5)**, 98–100.

Harwood, D.M. (1985). Late Neogene climatic fluctuations in the high-southern latitudes: implications of a warm Gauss and deglaciated antarctic continent. *South African Journal of Science*, **81(5)**, 239–41.

Harwood, D.M. (1986a). Diatom biostratigraphy and paleoecology with a Cenozoic history of Antarctic ice sheets. Ph.D. thesis. Ohio State University, Columbus, 592 pp. (unpublished).

Harwood, D.M. (1986b). Diatoms. In *Antarctic Cenozoic History from the MSSTS-1 Drillhole, McMurdo Sound*, ed. P.J. Barrett, pp. 69–109. Wellington; New Zealand DSIR Bulletin, No. 237.

Harwood, D.M. (1986c). Recycled siliceous microfossils from the Sirius Formation. *Antarctic Journal of the United States*, **21(5)**, 101–3.

Harwood, D.M. (1987). Cenozoic Antarctic temperature, ice volume and bottom water events: a comparison of Sirius Formation microfossil data to eustatic and oxygen isotope data. *Geological Society of America, Abstracts with Programs*, **19(6)**, 695.

Harwood, D.M. (1988). Upper Cretaceous and Lower Paleocene diatom and silicoflagellate biostratigraphy of Seymour Island Eastern Antarctic Pensinula. In *Geology and Paleontology of Seymour Island*, ed. R. Feldmann & M.O. Woodburne pp. 55–129. Washington, DC; Geological Society of America Memoir 169.

Harwood, D.M. (1989). Siliceous microfossils. In *Antarctic Cenozoic History from the CIROS-1 Drillhole, McMurdo Sound*, ed. P.J. Barrett, pp. 67–87. Wellington; New Zealand DSIR Bulletin, No. 245.

Harwood, D.M., Greene, D.G. & Webb, P.N. (1989). Multiple Miocene marine productivity events in West Antarctica as recorded in upper Miocene sediments beneath the Ross Ice Shelf (Site J-9). *Marine Micropalaeontology*, **15(1)**.

Hasle, G.R. (1969). An analysis of the phytoplankton of the Pacific Southern Ocean, abundance, composition and distribution during the Brateg expedition, 1947–1948: Hvalradets Skrifter. *Science Results of Marine Biology Research*, **52**, 6–168.

Hodell, D.A. & Kennett, J.P. (1986). Late Miocene–early Pliocene stratigraphy and paleoceanography of the South Atlantic and southwest Pacific Oceans: a synthesis. *Paleoceanography*, **1(2)**, 285–311.

Jacques, G. (1983). Some ecophysiological aspects of Antarctic phytoplankton. *Polar Biology*, **2(1)**, 27–33.

Keigwin, L. & Keller, G. (1984). Middle Oligocene cooling from equatorial Pacific Deep Sea Drilling Project Site 77B. *Geology*, **12(1)**, 16–19.

Kellogg, D.E. & Kellogg, T.B. (1986). Diatom biostratigraphy of sediment cores from beneath the Ross Ice Shelf. *Micropaleontology*, **32(1)**, 74–94.

Kemp, E.M., Frakes, L.A. & Hayes, D.E. (1975). Paleoclimatic significance of diachronous biogenic facies, Leg 28, Deep Sea Drilling Project. In *Initial Reports of the Deep Sea Drilling Project*, 28, ed. D.E. Hayes, L.A. Frakes *et al.*, pp. 908–17. Washington, DC; US Government Printing Office.

Kennett, J.P. (1977). Cenozoic evolution of Antarctic glaciation, the circum-Antarctic current and their impact on global paleoceanography. *Journal of Geophysical Research*, **82**, 3843–60.

Kennett, J.P. (1978). Development of planktonic biogeography in the Southern Ocean during the Cenozoic. *Marine Micropaleontology*, **3**, 301–45.

Kennett, J.P. (1979). Recent zoogeography of Antarctic plankton microfossils. In *Zoogeography and Diversity in Plankton*, ed. J. van der Spoel & A.C. Pierrot-Bults, pp. 328–55. Utrecht; Bunge Scientific Publishers.

Kozlova, O.G. (1966). *Diatoms of the Indian and Pacific Sectors of the Antarctic*. Jerusalem; Israel Program for Scientific Translations, TT66-51154. United States Clearing House for Federal Scientific and Technical Information, (originally published in Russian in 1964), 191 pp.

Matthews, R.K. & Poore, R.Z. (1980). Tertiary $\delta^{18}O$ record and glacioeustatic sea-level fluctuations. *Geology*, **8(10)**, 501–4.

McCollum, D.W. (1975). Diatom stratigraphy of the Southern Ocean. In *Initial Reports of the Deep Sea Drilling Project*, 28, ed. D.E. Hayes, L.A. Frakes *et al.*, pp. 515–71. Washington, DC; US Government Printing Office.

Miller, K.G. & Fairbanks, R.G. (1983). Evidence for Oligocene–middle Miocene abyssal circulation changes in the western North Atlantic. *Nature, London*, **306(5940)**, 250–3.

Neori, A. & Holm-Hansen, O. (1982). Effects of photosynthesis in Antarctic phytoplankton. *Polar Biology*, **1(1)**, 33–8.

Perch-Nielsen, K. (1976). Late Cretaceous to Pleistocene silicoflagellates from the southern southwest Pacific, DSDP, Leg 29. In *Initial Reports of the Deep Sea Drilling Project*, 29, ed. C.D. Hollister, C. Craddock *et al.*, pp. 677–717. Washington, DC; US Government Printing Office.

Pichon, J.J. (1985). Les diatomées traceurs de l'évolution climatique et hydrologique de l'ocean Austral au cours du dernier cycle climatique. Thèse de 3e cycle, Université de Bordeaux I, 279 pp. (unpublished),

Pickard, J., Adamson, D.A., Harwood, D.M., Miller, G.H., Quilty,

P.G. & Dell, R.K. (1988). Early Pliocene marine sediments, coastline and climate of East Antarctica. *Geology*, **16**, 158–61.

Savage, M.L. & Ciesielski, P.F. (1983). A revised history of glacial sedimentation in the Ross Sea region. In *Antarctic Earth Science*, eds. R.L. Oliver, P.R. James & J.B. Jago, pp. 555–9. Canberra; Australian Academy of Science and Cambridge; Cambridge University Press.

Savin, S.M. (1977). The history of the Earth's surface temperature during the past 100 million years. *Annual Review Earth Planetary Science*, **5**, 391–55.

Schrader, H.-J. (1976). Cenozoic marine planktonic diatom biostratigraphy of the southern Pacific Ocean. In *Initial Reports of the Deep Sea Drilling Project*, 35, ed. C.D. Hollister, C. Craddock *et al.*, pp. 605–71. Washington, DC; US Government Printing Office.

Semina, H.J. (1979). The geography of plankton diatoms of the Southern Ocean. *Nova Hedwigia, Beiheft*, **64**, 341–58.

Shackleton, N.J. & Kennett, J.P. (1975). Late Cenozoic oxygen and carbon isotope changes at Deep Sea Drilling Site 284: implications for glacial history of the Northern Hemisphere and Antarctica. In *Initial Reports of the Deep Sea Drilling Project*, 29, ed. J.P. Kennett, R.E. Houtz *et al.*, pp. 801–7. Washington, DC; US Government Printing Office.

Steinhauff, D.M., Renz, M.E., Harwood, D.M. & Webb, P.-N. (1987). Miocene diatom biostratigraphy of Deep Sea Drilling Project Hole 272: stratigraphic relationship to the underlying Miocene of Deep Sea Drilling Project Hole 270, Ross Sea, Antarctica. *Antarctic Journal of the United States*, **22(5)**, 123–5.

Stott, L.D., McKelvey, B.C., Harwood, D.M. & Webb, P.-N. (1983). A revision of the ages of Cenozoic erratics at Mt Discovery and Minna Bluff, McMurdo Sound, Antarctica. *Antarctic Journal of the United States*, **18(5)**, 36–8.

Truesdale, R.S. & Kellogg, T.B. (1979). Ross Sea diatoms: modern assemblage distributions and their relationship to ecologic, oceanographic and sedimentary conditions. *Marine Micropaleontology*, **4**, 13–31.

Weaver, F.M. & Gombos, A.M., Jr. (1981). Southern high-latitude diatom biostratigraphy. In *Deep Sea Drilling Program: A Decade of Progress*, ed. J.E. Warme *et al.*, pp. 445–70. Tulsa; Society of Economic Paleontologists and Mineralogists, Special Publication No. 32.

Webb, P.-N., Harwood, D.M., McKelvey, B.C., Mabin, M.C.G. & Mercer, J.H. (1986). Late Cenozoic tectonic and glacial history of the Transantarctic Mountains. *Antarctic Journal of the United States*, **21(5)**, 99–100.

Webb, P.-N., Harwood, D.M., McKelvey, B.C., Mercer, J.H. & Stott, L.D. (1984). Cenozoic marine sedimentation and ice volume variation on the East Antarctic craton. *Geology*, **12(5)**, 287–91.

Webb, P.-N., McKelvey, B.C., Harwood, D.M., Mabin, M.C.G. & Mercer, J.H. (1987). Sirius Formation of the Beardmore Glacier region. *Antarctic Journal of the United States*, **22(1)**, 8–13.

The Dominion Range Sirius Group: a record of the late Pliocene–early Pleistocene Beardmore Glacier

B.C. McKELVEY[1], P.N. WEBB[2], D.M. HARWOOD[2] & M.C.G. MABIN[2]

1 Department of Geology & Geophysics, University of New England, Armidale, NSW 2351, Australia
2 Byrd Polar Research Center & Department of Geology and Mineralogy, The Ohio State University, Columbus, Ohio 43210, USA

Abstract

On Oliver Platform in the Dominion Range, late Pliocene and/or early Pleistocene Sirius Group strata include the 185 m thick Meyer Desert Formation composed predominantly of semi-lithified fossil-bearing coarse lodgement tillites and minor fluvioglacial and glaciolacustrine deposits. Fossils include *in situ Nothofagus* fragments, palynomorphs, and a variety of Cenozoic marine microfossils reworked from the Pensacola subglacial basin of East Antarctica. Approximately 6 km farther west a 50 m thick section of the younger Mount Mills Formation of the Sirius Group consists of coarse fluvioglacial diamictites, conglomerates and breccias. The Meyer Desert Formation rests disconformably via the glacially cut Dominion Erosion Surface on Mesozoic strata at between 1800 m and 2650 m above sea level. This altitude range reflects original relief on the uplifted erosion surface. The latter is a composite one representing at least two separate phases of cutting separated by uplift. Consequently the mantling Meyer Desert Formation records two (or more) separate periods of deposition, and the sequence has been subsequently uplifted more than 1300 m. The Dominion Range Sirius Group is interpreted as the uplifted late Pliocene and/or early Pleistocene record of Beardmore Glacier, then flowing at an approximately 1300 m lower altitude. There is no reason to believe that it was the product of late Cenozoic glacial overriding of the Transantarctic Mountains and an associated vast volume increase of the East Antarctic Ice Sheet. On the contrary, the Sirius Group strata and the fossil plants indicate a much more temperate setting than at present, with temperatures being in the order of 15–20 °C warmer.

Introduction

The Dominion Range

Separating the head of Beardmore Glacier from the tributary Mill Glacier, the NNW-trending Dominion Range is a partially deglaciated massif covering about 1050 km^2 and rising to 3000 m. The northern and north-eastern part of the range is ice-free and constitutes the 220 km^2 boulder-strewn Meyer Desert. Much of the eastern part of the desert adjacent to Mill Glacier is gently undulating around 1800 m above sea level (a.s.l.). This relative lowland abuts the upfaulted and more rugged western part which includes the flat-topped Oliver Platform. The surface of this feature is stepped and consists of a lower northern bench (~ 1900 m) and a higher southern bench (~ 2200–2500 m) (Fig. 1).

The Dominion Range consists of more than 1200 m of subhorizontal Permian Buckley Formation and Triassic Fremouw Formation strata (i.e. Beacon Supergroup) intruded by sill and sheets of Jurassic Ferrar Dolerite. It is overlain by an irregular composite mantle of late Cenozoic glaciogene deposits (Oliver, 1964; Elliot, Barrett & Mayewski, 1974). The oldest and thickest of these (up to 185 m) is the Sirius Formation (Mercer, 1972; Mayewski, 1975; Mayewski & Goldthwait, 1985) resting on the irregular Dominion Erosion Surface at between 1800 and 2650 m a.s.l. Above 1920 m the top of this unit is now a gently undulating highly weathered ablation surface unmodified by any subsequent glacial activity. Below 1920 m the Sirius Formation is gently incised by the Beardmore Erosion Surface and covered with the Beardmore Glaciation drifts of Mercer (1972). A NNW-trending major fault with a throw of at least 300 m, and here termed the Koski fault, runs the length of Dominion Range through Meyer Desert cutting the Sirius Formation.

Stratigraphy

Along the inland flanks of the Transantarctic Mountains outcrops of the Sirius Formation have been recorded from more than 40 localities, scattered over a distance of 1300 km

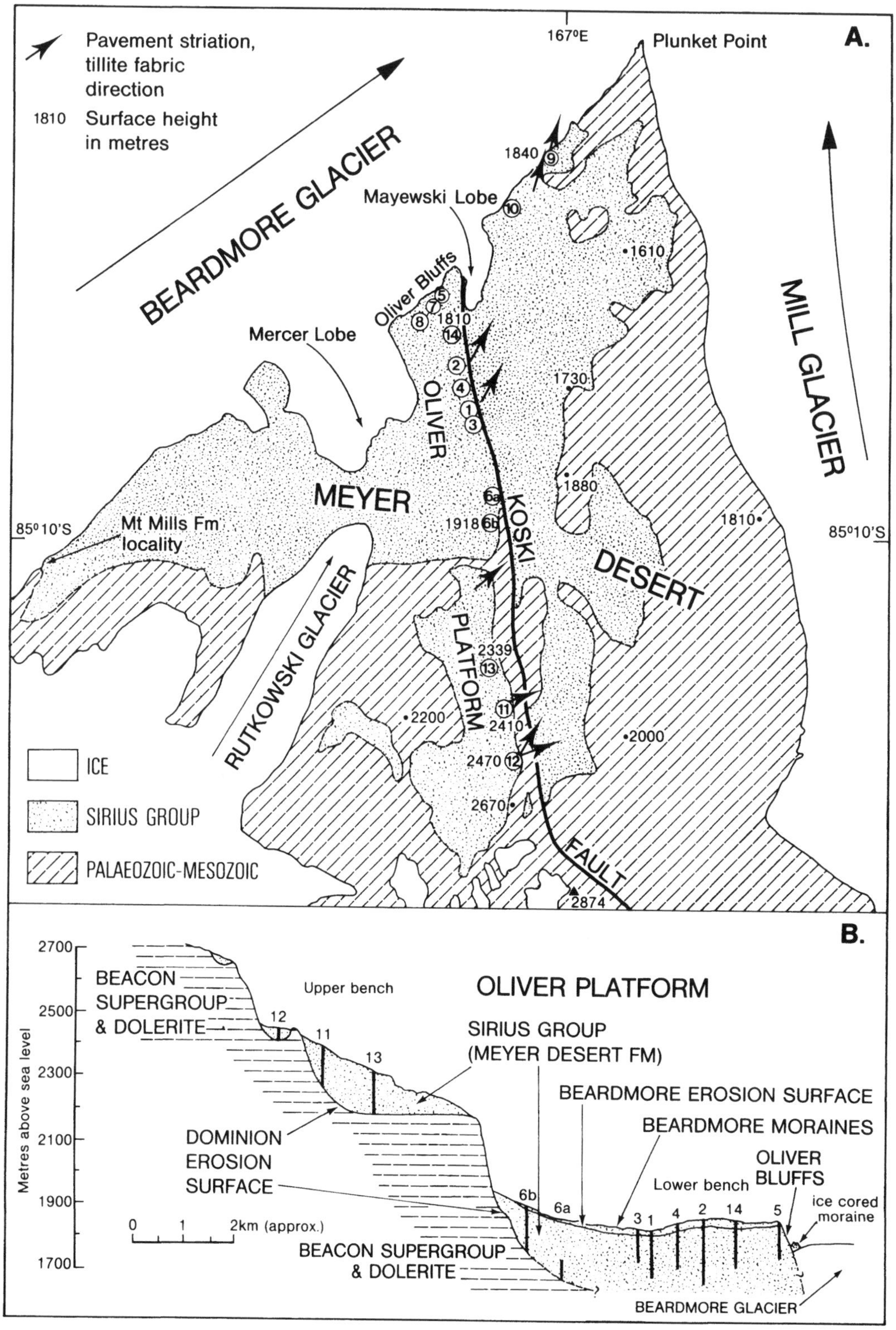

Fig. 1. *A*. Sketch map of the Meyer Desert, northern Dominion Range, showing the Koski fault separating Oliver Platform and the more rugged western part of the desert from the lower undulating portion adjacent to Mill Glacier. The outcrop and probable distribution of the concealed Sirius Group is indicated. Striation directions on the Dominion Erosion Surface and Meyer Desert Formation tillite fabric directions are shown, suggesting that ice flow was similar to today. The Beardmore drifts (Mercer, 1972) are not shown. *B*. This cross-section along Oliver Platform shows a profile of the Dominion Erosion Surface, the approximate position of the Beardmore Erosion Surface, and the position of measured and sampled sections.

Mayewski & Goldthwait, 1985). The majority of these exposures occur at altitudes of between 1750 and 3000 m and most are of diamictite or lodgement tillite. However, other glacigene facies do occur and Mercer (1968) accordingly recognized several informal units at Tillite Spur near the head of Reedy Glacier. Similarly, Mayewski (1975) recognized a highly stratified member as well as the more common massive tillite member at both Bennett Platform (Queen Maud Mountains) and the Dominion Range.

Mercer (1972, p. 432) considered the Sirius Formation 'dates from soon after the establishment of an ice cover on the Transantarctic Mountains, thereby long antedating the Antarctic ice sheet'. He further suggested that, because of faulting observed in the Dominion Range, the glaciers responsible for deposition of the Sirius Formation 'may have formed while uplift of the Transantarctic Mountains was still in progress'. On the other hand Mayewski (1975) envisaged deposition of the Sirius Formation towards the end of a very major continental glaciation (his Queen Maud Glaciation, at > 4.2 Ma) which overrode the Transantarctic Mountains and was responsible for much of their glacial sculpturing. Denton *et al.* (1984) invoked a similar overriding model to account for glacial features in southern Victoria Land. Subsequently Webb *et al.* (1984), in the light of the discovery of Pliocene and older marine microfossils in the Sirius Formation, interpreted the sequence as a record of overwhelming of the Transantarctic Mountains by the East Antarctic Ice Sheet < 3 m.y. ago. The microfossils were reworked by the expanding ice-cap from the preglacial or interglacial Wilkes and Pensacola marine basins located immediately inland of the Transantarctic Mountains. Because the Sirius Formation must be younger than these reworked marine fossils, a late Pliocene or even early Pleistocene age is suggested. Specifically, the Sirius Formation cannot be older than 3.1 Ma (Harwood, 1986a, b).

In this paper we propose that the Sirius Formation be raised in terms of stratigraphic nomenclature to group status. This will allow individual outcrops or geographically related groups of outcrops scattered along the Transantarctic Mountains to be identified as particular formations which in turn can be further subdivided into members. In proposing that different formations of the Sirius Group be recognized we are considerably influenced by the possibility that individual Sirius Group occurrences have different ages.

The Dominion Range Sirius Group

In the Dominion Range the Sirius Group occurs above 1800–2600 m a.s.l. and is composed of the Meyer Desert and Mount Mills formations. The Meyer Desert Formation rests disconformably on the Palaeozoic and Mesozoic Beacon Supergroup strata via the Dominion Erosion Surface (Fig. 1). The base of the Mount Mills Formation was not seen. The Sirius Group is unfolded, but has a slight regional dip (< 5°) to the NE, in the direction of flow of the present Beardmore Glacier. All outcrops show some evidence of slight to moderate soft-sediment deformation. Tillite, conglomerate and breccia clasts of the Sirius Group are overwhelmingly composed of varieties of Ferrar Dolerite. Kirkpatrick Basalt clasts are present, but are much scarcer. Beacon Supergroup clasts are relatively rare but the tillite matrices and interbedded sandstones are predominantly derived from this source. A very few Lower Palaeozoic basement clasts, generally limestones, were observed. All tillite clast imbrication and pavement striation data (Fig. 1) obtained in the Dominion Range, as well as the regional dip, indicate a Sirius Group ice-flow direction similar to that of Beardmore Glacier.

The Sirius Group in the Dominion Range is extensively faulted (Webb *et al.*, 1986). All are normal faults. The largest and oldest, the Koski fault bounding the eastern margin of Oliver Platform (Elliot *et al.*, 1974), has a throw of about 300 m. Most other faults are much smaller with displacements of less than 15 m (see Mercer, 1972, fig. 4). Not all are unequivocal, being identified solely on morphological evidence. Fault scarps are steep and little modified by subsequent erosion: boulder–pedestal development marks the older scarps. Major stratigraphical displacement occurs across only the Koski fault in the Beacon Supergroup 'basement' (Elliot *et al.*, 1974). Several smaller faults, tension fractures, and slump structures cut Beardmore II moraines, but the deposition of these moraines post-dates the Koski fault scarp.

Meyer Desert Formation

Twelve sections of the Meyer Desert Formation, ranging between 32 and 185 m in thickness, were examined and systematically sampled along the northern and eastern faces of Oliver Platform between elevations of approximately 1800 and 2650 m (Fig. 1). Two other sections (86 m and 27 m) were measured immediately east of Mayewski Ice Lobe. Only the sections in the higher (southern) bench of Oliver Platform are complete although they do have weathered (ablated) tops. The other sections are truncated by the Beardmore Erosion Surface.

At the northern end of Oliver Platform, at Oliver Bluffs, a sequence of four (informal) members totals at least 85 m in thickness. Members 1 and 3 are lodgement tillites that enclose the essentially fluvioglacial (including glaciolacustrine) member 2. Member 4 is a more complex unit, reflecting oscillatory and perhaps overall still-stand ice conditions. All strata reflect temperate ice and the frequent presence of abundant meltwater. At several localities *in situ* or nearly *in situ* woody plant fossils (Carlquist, 1987; Webb & Harwood, 1987) have been found at the boundary between members 1 and 2 and members 3 and 4, and from a horizon approximately 4.5 m above the base of member 4. Palynomorphs also occur in the Meyer Desert Formation (Askin & Markgraf, 1986).

Member 1. At least 13 m thick, this member includes three lodgement tillites ranging in thickness from 2 to 6.75 m. These contain clasts up to 2.5 m in diameter, but generally the largest range between only 25 and 50 cm. Where tabular clasts are abundant they are frequently orientated subparallel to bedding. The individual tillites are indistinctly stratified internally and sometimes show crude normal grading. They also contain scattered laterally discontinuous intervals (up to 40 cm) of gently deformed waterlain laminated siltstones,

Fig. 2. Meyer Desert Formation: member 2 conglomeratic fan delta foreset beds and adjacent younger cross-stratified sandstone are overlain disconformably by member 3 lodgement tillite. The relatively steep inclination of the member 2 strata in part results from soft-sediment deformation. (Staff intervals are 25 cm). Locality approximately 250 m north-east of section 7, Oliver Bluffs.

medium–fine sandstone and fine conglomerates. In section 8, a 2.2 m thick pebble breccia and conglomerate interbedded between two tillites thins to a few centimetres over 20 m. The base of member 1 is not known.

Member 2. This highly stratified thin (up to 7.5 m) unit is varied sandstone (often pebbly), siltstone and subordinate conglomerate or breccia, and exhibits considerable rapid facies change (Fig. 2). In marked contrast to members 1 and 3, member 2 contains no lodgement tillites and reflects periglacial or interglacial conditions. In several sections member 2 is only a few centimetres thick, locally 'pinching-out' between enclosing tillite sheets, due to soft sediment deformation or else contemporaneous erosion.

In section 7 a lacustrine sequence ($\sim$ 7.5 m) of laminated sandstones and siltstones interbedded with several thin ($<$ 75 cm) pebbly sandstones or diamictites (see Mercer, 1972, fig. 3 and Webb *et al.*, 1987, fig. 5) passes both laterally and vertically to prograding fan-delta pebble and cobble conglomerates and very coarse sandstones. These latter fluvial rocks are reddish coloured (i.e. oxidized) and have initial dips of up to 5°. Both the lacustrine sandstones and pebbly sandstones show grading and gentle channelling. Strata underlying the pebbly sandstones often exhibit spectacular deformation, including brecciation and convolute lamination. The siltstones frequently contain scattered lonestones, often of coaly material. With the exception of one outcrop of ripple-drift lamination, no cross-bedding or current ripples were observed, the sandstones and pebbly sandstones being essentially resedimented mass movement deposits. No trace fossils of any sort were observed on bedding planes.

Three hundred metres to the southwest (section 8) a wholly fluvial member 2 is only 2 m thick, and consists of laminated sandstones (with occasional lonestones) and thin resedimented pebbly sandstones interbedded with lesser amounts of thin pebble conglomerates and lensoidal cross-bedded coarse sandstones. Individual bed thicknesses range from 5 to 50 cm. Diverse soft sediment deformation varies from simple loading features, to recumbent interstratal folds.

Member 3. At Oliver Bluffs this member ranges in thickness from 9 to 24 m and overlies member 2 with apparent disconformity (see Webb *et al.*, 1987, fig. 4). Erosional relief on the contact is less than 2 m. The base of member 3 is a poorly sorted cobble conglomerate 2–10 m thick. This passes up gradationally to indistinctly stratified lodgement tillite generally similar to member 1. However, large clasts up to 3 m across are common in member 3 and this distinguishes it in the field from members 1 and 4. Similar to member 1, member 3 tillite contains deformed discontinuous lenses of waterlain siltstone, sandstone and pebble conglomerate.

Member 4. This is a slope-forming unit more than 60 m thick at Oliver Bluffs (section 5). It is overlain disconformably (via the Beardmore Erosion Surface) by any of the Beardmore I, II or III moraines of Mercer (1972). Member 4 contrasts with members 1 and 3 in that much of it consists of relatively thin (2 m) diamictite sheets interbedded with either: (i) massive or laminated sandstones (with occasional lonestones); or (ii) thin sequences of interstratified sandstone, siltstones and lensoidal conglomerates. Clasts in member 4 are exclusively doleritic (or of Kirkpatrick Basalt) and generally less than 50 cm in diameter. A few dolerite blocks up to 2.5 m occur. The diamictite matrices are often especially dark, reflecting their Beacon Supergroup coal measure or dark shale provenance. Member 4 overlies member 3 with apparent conformity, although the basal lithologies frequently consist of up to 6 m of poorly sorted breccia or matrix-poor diamictite. At the northern end of Oliver Platform in section 5, a prominent greyish-orange polar soil occurs 52 m above the base of the member.

The younger part of the Meyer Desert Formation exposed on the higher bench of Oliver Platform (e.g. sections 13 & 11, Fig. 1) is similar to member 4 in being noticeably (although indistinctly) stratified. However, there are differences. Numerous weathered (polar soil) intervals individually up to several metres thick and sometimes containing *in situ*, frost-cleaved disintegrated clasts are common, particularly in the uppermost 20–30 m. Within these weathered intervals are thin ($<$ 20 cm) impersistent ablation sequences of lag pebble conglomerate and eolian sandstone laminae. These features indicate the alternation of subaerial exposure with ice cover, and overall a perhaps less intense glacial regime. The stratigraphic relationship of these higher bench sequences of the Meyer Desert Formation to the members exposed at Oliver Bluffs is considered in the Discussion.

Sections 9 (85 m) and 10 (27 m), approximately 2 and 4 km north-east of Oliver Platform, near Plunket Point, also consist of Meyer Desert Formation tillite resting on dolerite basement. Neither section resembles the other. At section 9 the indistincly stratified tillite contains an abundant (~ 50%) dark fine matrix, and overall is relatively fine grained with the larger clasts averaging only 10 cm, although a few of over 1 m were noted. Washed pebble horizons are common. The section 9 sequence is similar to the lower part of The Cloudmaker sequence (see below).

The Meyer Desert Formation at section 10 contrasts with the sequence examined at Oliver Platform. Characterized by rusty red dolerite clasts up to 2.5 m (and quite dissimilar to the underlying dolerite basement), the section consists of poorly sorted, lensoidal basal boulder conglomerates passing up irregularly to fine and coarse diamictites (tillite) that contain lenses of pebbles and cobbles. There is considerable textural variation within outcrops. Initial dips of up to 40° reflect the rugged local relief on the Dominion Erosion Surface.

Approximately 70 km to the north and about 10.5 km south-south-west of The Cloudmaker (84°23′ S, 169°14′ E), an excellent 250 m section of Sirius Group strata contains at least six major stratigraphic units. We consider these rocks to also belong to the Meyer Desert Formation. Most are indistinctly stratified lodgement tillites each usually more than 30 m thick. However, two stratigraphical units 16 m and 27 m thick consist of thin diamictites alternating with lacustrine intervals. Two thin (1 cm) carbonate horizons occur within one of these units about 145.2 m above the section base.

The older tillites, constituting the basal 70 m of section, are remarkably matrix-rich, and resemble glaciomarine deposits; clasts range from only 1 to 20%. The younger tillites contain considerably more clasts and some beds are actually clast-supported diamictites. At approximately 182 m above the section base a 2 m polar soil horizon separates two tillites. Another soil (0.5 m thick) was noted at 152 m, within a tillite unit.

Mount Mills Formation

This coarse diamictite, breccia and conglomerate facies of the Sirius Group is well exposed for over 1.5 km in 50 m high bluffs 6 km west of Oliver Platform (Fig. 3), where it is overlain disconformably via a 2 m soil horizon by Beardmore III moraine. Also present in the formation are numerous beds of very coarse sandstone up to 1.5 m thick. The poorly sorted breccias and conglomerates (i.e. clast-supported diamictites) are almost entirely of doleritic (and Kirkpatrick Basalt) provenance. Discernible in the bluff faces are coarse (fan) foresets showing down-dip grading and channels; weathering surfaces are developed at the tops of some beds. The stratigraphic relationship of the Mount Mills Formation to the Meyer Desert Formation is not yet demonstrated but almost certainly these coarse deposits are younger ablation moraines. At present no other occurrences of this formation are known.

The Dominion Erosion Surface

This disconformity has been surveyed in some detail in the Dominion Range and also at Mount Sirius by Mabin (1986). On the east side of Beardmore Glacier between 2.2 and 3.9 m south-west of Plunket Point (Fig. 1) the disconformity is

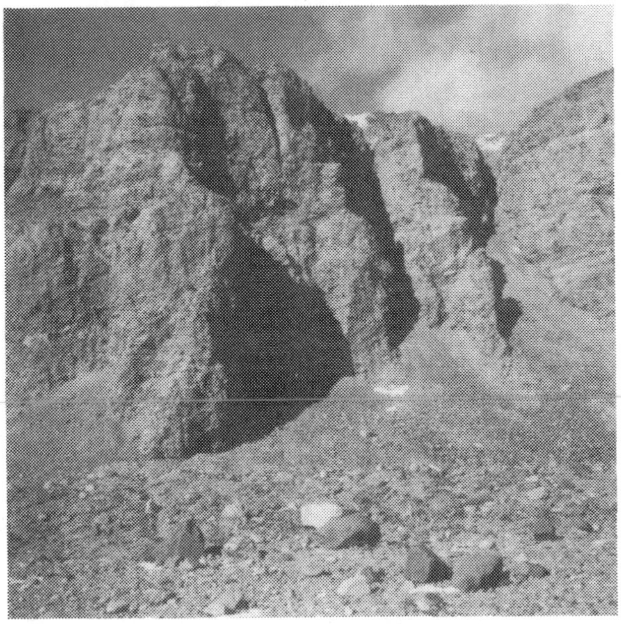

Fig. 3. The Mount Mills Formation exposed 6 km south-west of Oliver Bluffs. The section consists of coarse unfolded fluvioglacial diamictites, conglomerates and interbedded sandstones. Note the inclined stratification (initial dip) and pronounced clast fabrics. Lateral moraine-covered Beardmore Glacier is in the foreground. The height of the exposure is approximately 35 m.

an undulating surface cut on columnar dolerite at about 1750 m a.s.l. and 15–95 m above the glacier. The whaleback undulations vary from 3 to 36 m in height and range from 80 to 300 m apart. Striations on the dolerite trend between 355 and 025°, reflecting diverging and converging of NNE-flowing ice over and around the whalebacks. Other glacial erosion features at this locality include plucking on the downglacier (northern) sides of whalebacks, lunate fractures up to 8 cm across, and small grooves.

In the Koski fault scarp forming the eastern side of Oliver Platform, the Dominion Erosion Surface is exposed intermittently beneath the southern bench between 5.5 km and 11.2 km south of Oliver Bluffs. Cut on Fremouw Formation and on dolerite it rises in irregular steps southwards from 1850 to 2500 m a.s.l. (Fig. 1*B*) corresponding to altitudes of between 50 and 800 m above the present glacier surface at Oliver Bluffs. At four sites numerous striations trend consistently between 025–065°. Nearer Beardmore Glacier the disconformity descends beneath the surface of Meyer Desert to an unknown depth beneath the present glacier level.

Farther down Beardmore Glacier near The Cloudmaker, the Dominion Erosion Surface is a steeply sloping platform 80 m above the present glacier surface and about 1060 m a.s.l. It cuts across nearly vertically dipping basement metasediments and dips 15–20° to the south. In profile it portrays the lower part of a glacial valley or fjord wall, whose floor was near the present glacier surface. Three cross-cutting sets of striations trend at 325, 040 and 080°. The former two sets are compatible with present ice flow in the valley.

About 110 km to the north at the type locality, Mount Sirius (84°08′ S, 163°15′ E), the erosion surface cut in dolerite and elevated 400 m above the surrounding Bowden Nevé exhibits two sets of whaleback undulations; one set up to 20 m deep and 300 m apart, and the other smaller set 1–5 m deep and 20–30 m apart superimposed on these (Mabin, 1986). The erosion surface dips approximately 2° NW and striations at six localities trend between 022 and 105°, with the predominant ice-flow direction being ENE, as reported by Mercer (1972) and McKelvey *et al.* (1984). Also present are crescentic gouges up to 35 cm across, convex in an ENE direction, and at one locality there is a SE-oriented erosional feature (2 m across and 0.5 m deep) indicative of meltwater beneath the ice that carved the surface.

Beneath the Sirius Group at Orr Peak (83°20′ S, 157°48′ E), still farther north in the Miller Range (Webb *et al.*, 1987), the Dominion Erosion Surface is not glacially polished but instead is underlain by *in situ* breccia derived from underlying basement metamorphic rocks that are weathered to a depth of more than 20 cm. The breccia is similar to Mercer's (1968) early Horlick Drift at Tillite Spur in the Reedy glacier area, and interpreted by him there as pre-Sirius strata produced by weathering and solifluction. Mercer (1987) suggested that these breccias and weathering features indicate the 'two sites had not been glaciated during the preceding several million years'. Taking into account the well established latest Miocene glaciation it follows that the breccias, weathering features and overlying erosion surfaces at the two sites may be older than this terminal Miocene event, and owe their preservation to uplift above the Miocene glacier ice level. It is this sort of data that suggests to us the Sirius Group may include strata of significantly different ages.

Discussion

In terms of its facies and geological setting the Meyer Desert Formation is similar to other Sirius Group occurrences along the Transantarctic Mountains, and the model of ice sheet overriding of the Transantarctic Mountains envisaged by Mayewski (1975) and Webb *et al.* (1984) would adequately account for the formation's deposition in the Dominion Range. The one locality of the spectacular Mount Mills Formation does appear (so far) to be unique but it is still compatible with this general model. However, two new types of data from the Dominion Range, as yet unrecorded from other Sirius Group localities, now necessitate reassessment of any glacial overriding model proposed for the Beardmore Glacier region. These are: (1) the *in situ* macroflora; and (2) the widespread post-Sirius faulting.

Mercer (1986) maintains that for a *Nothofagus* sp. flora to have existed at 86° S it would have been similar to the Tundra Magallanica flora (Pisano, 1977) of southernmost South America, and the Sirius Group upon which it grew must have been 500 m (a.s.l.) or lower based on ecological constraints of *Nothofagus* sp. In so arguing he points out that the flora indicates a minimum summer temperature of about 5 °C. If the late Pliocene or Pleistocene altitude of the Meyer Desert Formation at Oliver Bluffs was about 1500 m a.s.l. (calculated using the average 100 m/Ma uplift rates of Smith & Drewry, 1984; Gleadow & Fitzgerald, 1987) then the standard temperature lapse rate of 0.65°/100 m (List, 1949) implies a coastal (sea level) temperature at that time of about 14 °C (similar to that of the present-day Valdivian Rainforest zone of southern Chile between 41 and 43° S). There is absolutely no palaeobotanical evidence of a flora commensurate with such a temperature as this in the Ross embayment, and so we find the suggested altitude of ≤ 500 m an acceptable approximation.

Mercer further points out that a summer temperature of 5 °C at 500 m on the coastal slopes of the Transantarctic Mountains also means ice shelves and floating ice tongues were absent from the Ross embayment, and consequently outlet glaciers calved close to their grounding lines. Drewry (1983) has shown that, in an ice-free Antarctica at isostatic equilibrium, a fjord extended from the Ross embayment deep into the Transantarctic Mountains (to the Dominion Range). Accordingly we envisage the terminus of the ancestral Beardmore Glacier, flanked by flora-covered glacial deposits, calving into the head of Beardmore fjord. Such a setting, particularly in terms of its relatively low altitude, renders acceptable both the interpretation of the coarse Mount Mills Formation rudites as glacial outwash deposits and ablation moraines, and also the glaciolacustrine and fluvioglacial setting of member 2 of the Meyer Desert Formation. Present-day summer surface air temperatures at the mouth of Beardmore Glacier fall in the range of − 10 °C to − 15 °C (Herman & Johnson, 1980, fig. 1.3). Consequently, the interpretation of the Sirius Group's depositional setting presented here necessitates considerable warmth in the late Pliocene or early Pleistocene, continuing from warmer, deglacial conditions in the early Pliocene, when open seas were present on both sides of the Transantarctic Mountains and the East Antarctic ice volume was considerably reduced (Harwood, 1985, 1986a, b).

Consideration of the Oliver Bluffs *Nothofagus* sp. flora in this palaeogeographic scenario necessitates minimum uplift in the order of 1300 m at some time during the last 2.5–3.0 Ma, a rate much greater than the 100 m/Ma average calculated from fission-track data (Gleadow & Fitzgerald, 1987) for much of the Transantarctic Mountains. The occurrence of such a major and young uplift event is perhaps surprising but is certainly supported by the widespread post-Sirius faulting evident in the Dominion Range. Similar faulting has not yet been recorded from elsewhere in the Transantarctic Mountains.

Such uplift carries major implications for estimates of former ice volume based upon moraine elevation data. A very large increase in volume of the East Antarctic Ice Sheet is required by Mayewski & Goldthwait (1985) for their Queen Maud Glaciation to override the 3000–4500 m high crest of the central Transantarctic Mountains. Similarly, the overriding episodes farther north in the Transantarctic Mountains proposed by Denton *et al.* (1984) and the model we proposed (Webb *et al.*, 1984) also involves a vast increase in ice-sheet volume. However, with the identification of ~ 1300 m of post-Sirius uplift in the Beardmore region the requirement there for such a volume increase vanishes. We propose the Dominion Range Sirius Group to have been deposited by an ancestral Beardmore Glacier, flowing at a 1300 m lower altitude down into the head of Beardmore fjord. We envisage this landscape being not markedly different in its drainage pattern to that of the present day. Certainly all ice-flow indicators (both erosion surface striations and clast fabrics) from the Dominion Range, The Cloudmaker and Mount Sirius (McKelvey *et al.*, 1984) are compatible with the present drainage. The lower elevation of the ancestral Beardmore at that time points to the East Antarctic Ice Sheet being considerably smaller than at present.

Geomorphological and stratigraphical considerations point to initial uplift being contemporaneous with sedimentation and predating the faulting in the Dominion Range. We arrive at this conclusion because the base of the Meyer Desert Formation occurs at two levels on Oliver Platform separated by a scarp (Fig. 1*A*). Field mapping of the underlying Beacon Supergroup and dolerite shows this scarp not to be of fault origin, but instead is an exhumed erosion feature on the Dominion Erosion Surface. In section this configuration of the Dominion Erosion Surface along Oliver Platform is that of two terraces separated by the steep (fossil) scarp. Only the sloping back of the lower terrace, where it rises to join this scarp, is so far exposed. The whole profile suggests to us half or one side of a glacial valley-in-valley cross-section.

The upper terrace we see as representing part of the floor and adjacent wall of an ancestral stage of Beardmore Glacier valley. The overlying Meyer Desert Formation, including the small outliers at 2470 m (i.e. section 12) and 2670 m, comprises the oldest Sirius Group strata in the Dominion Range. Subsequent uplift resulted in deeper glacier incision cutting the scarp presently separating the two levels of the Dominion Erosion Surface. This scarp is a fragment of the wall of the younger incised valley form. Its floor (i.e. the lower terrace), although little exposed on Oliver Platform, may be represented by the Dominion Erosion Surface outcrops surveyed by Mabin (1986) near Plunket Point. If this interpretation of the Oliver Platform section is correct, it follows that the Meyer Desert Formation now exposed along the lower bench records a younger depositional event than the strata occupying the upper bench. The small outlier of the Meyer Desert Formation above the upper bench at 2670 m may represent a yet older terrace almost completely destroyed by subsequent erosion. It is possible that the geomorphological disposition of Oliver Bluffs overlooking the present surface of Beardmore Glacier and the similar setting of the Mount Mills Formation indicates a still younger repetition of this valley-in-valley-forming process.

Acknowledgements

This work was supported by National Science Foundation grant DPP 84-20622 and a University of New England research grant. Dr George Denton drew our attention to the exposures at The Cloudmaker. We acknowledge the generous assistance provided by John Splettstoesser, and the VXE-6 aircrews of Beardmore South Camp helicopter detachment. For untiring field assistance we are indebted to Dan Greene, Tim Axelson, Steve Munsell and Chuck McGrosky.

References

Askin, R.A. & Markgraf, V. (1986). Palynomorphs from the Sirius Formation, Dominion Range, Antarctica. *Antarctic Journal of the United States*, **21(5)**, 34–5.

Carlquist, S. (1987). Pliocene *Nothofagus* wood from the Transantarctic Mountains. *Aliso*, **11(4)**, 571–83.

Denton, G.H., Prentice, M.L., Kellogg, D.E. & Kellogg, T.B. (1984). Late Tertiary history of the Antarctic ice sheet: evidence from the Dry Valleys. *Geology*, **12(5)**, 263–7.

Drewry, D.J. (1983). *Antarctica: Glaciological and Geophysical Folio.* Cambridge; Scott Polar Research Institute.

Elliot, D.H., Barrett, P.J. & Mayewski, P.A. (1974). *Reconnaissance Geologic Map of the Plunket Point Quadrangle, Transantarctic Mountains, Antarctica*, US Antarctic Research Program Map, Map A-4. Reston Va.; US Geological Survey.

Gleadow, A.J.W. & Fitzgerald, P.G. (1987). Uplift history and structure of the Transantarctic Mountains: new evidence from fission track dating of basement apatites in the Dry Valleys area, southern Victoria Land. *Earth and Planetary Science Letters*, **82**, 1–14.

Harwood, D.M. (1985). Late Neogene climatic fluctuations in the southern high latitudes: implications of a warm Pliocene and deglaciated Antarctic continent. *South African Journal of Science*, **81(5)**, 239–41.

Harwood, D.M. (1986a). Diatom biostratigraphy and palaeoecology and a Cenozoic history of Antarctic ice sheets. PhD thesis, Ohio State University, Columbus, 604 pp.

Harwood, D.M. (1986b). Recycled siliceous microfossils from the Sirius Formation. *Antarctic Journal of the United States*, **21(5)**, 101–3.

Herman, G.F. & Johnson, W.T. (1980). Arctic and Antarctic climatology of a GLAS general circulation model. *Monthly Weather Review*, **108(12)**, 1974–91.

List, R.J. (1949). *Smithsonian Meteorological Tables*, 6th edn. Washington, DC; Smithsonian Institution Press. 527 pp.

Mabin, M.C.G. (1986). Sirius Formation basal contacts in the Beardmore Glacier region. *Antarctic Journal of the United States*, **21(5)**, 32–3.

Mayewski, P.A. (1975). *Glacial Geology and the late Cenozoic History of the Transantarctic Mountains, Antarctica.* Columbus; Ohio State University, Institute of Polar Studies, Report No. 56, 168 pp.

Mayewski, P.A. & Goldthwait, R.P. (1985). Glacial events in the Transantarctic Mountains: a record of the East Antarctic ice sheet. In *Geology of the Transantarctic Mountains*, Antarctic Research Series 36, ed. M.D. Turner & J.F. Splettstoesser, pp. 275–324. Washington, DC; American Geophysical Union.

McKelvey, B.C., Mercer, J.H., Harwood, D.M. & Stott, L.D. (1984). The Sirius Formation: further considerations. *Antarctic Journal of the United States*, **19(5)**, 42–4.

Mercer, J.H. (1968). Glacial geology of the Reedy Glacier area, Antarctica. *Geological Society of America Bulletin*, **79(4)**, 471–85.

Mercer, J.H. (1972). Some observations on the glacial geology of the Beardmore Glacier area. In *Antarctic Geology and Geophysics*, ed. R.J. Adie, pp. 427–33. Oslo; Universitetsforlaget.

Mercer, J.H. (1986). Southernmost Chile; a modern analog of the southern shores of the Ross embayment during Pliocene warm intervals. *Antarctic Journal of the United States*, **21(5)**, 103–5.

Mercer, J.H. (1987). The Antarctic ice sheet during the late Neogene. In *Palaeoecology of Africa and the Surrounding Islands*, 18, ed. J.A. Coetzee, pp. 21–33. Rotterdam; A.A. Balkema.

Oliver, R.L. (1964). Geological observations of Plunket Point, Beardmore Glacier. In *Antarctic Geology*, ed. R.J. Adie, pp. 248–58. Amsterdam; North-Holland Publishing Company.

Pisano, E. (1977). Fitogeografia de Fuego-Patagonia. I. Comunidades vegetales entre latitudes 52° y 56° S. *Anales del Instituto de la Patagonia*, **8**, 121–250.

Smith, A.G. & Drewry, D.J. (1984). Delayed phase change due to hot

asthenosphere causes Transantarctic uplift. *Nature, London*, **309(5968)**, 536–8.
Webb, P.N. & Harwood, D.M. (1987). Terrestrial flora of the Sirius Formation: its significance for Late Cenozoic glacial history. *Antarctic Journal of the United States*, **22(4)**, 7–11.
Webb, P.N., Harwood, D.M., McKelvey, B.C., Mabin, M.C.G. & Mercer, J.H. (1986). Late Cenozoic tectonic and glacial history of the Transantarctic Mountains. *Antarctic Journal of the United States*, **21(5)**, 99–100.
Webb, P.N., Harwood, D.M., McKelvey, B.C. & Stott, L.D. (1984). Cenozoic marine sedimentation and ice-volume variation on the East Antarctic craton. *Geology*, **12(5)**, 287–91.
Webb, P.N., McKelvey, B.C., Harwood, D.M., Mabin, M.C.G. & Mercer, J.H. (1987). Sirius Formation of the Beardmore Glacier region. *Antarctic Journal of the United States*, **22(1)**, 000–00.

The geology of Marine Plain, Vestfold Hills, East Antarctica

P.G. QUILTY

Australian Antarctic Division, Channel Highway, Kingston, Tasmania 7050, Australia

Abstract

A new genus and species of mid-Pliocene dolphin is being described from diatomites of Marine Plain near Davis Station in the Vestfold Hills. The original discovery consists of an almost complete upper jaw, and skull and other material has since been found. The remains come from a 7–8 m thick marine diatomite covering some 10 km^2, a few metres above sea level. The accompanying macrofauna consists of bivalves and gastropods. Dating was achieved using molluscs, diatoms and racemization of protein, after radiocarbon results showed the sediments to be considerably older than mid-Holocene sediments which are quite widespread in the Vestfold Hills. Vertebrate fossils seem to be common and it is hoped that the area will yield other remains, in particular the seals and birds which lived at the time. The dolphin is the only vertebrate fossil known from Antarctica since the present ecosystem evolved, probably about the time of development of the Antarctic Convergence. Preliminary oxygen isotope studies suggest a palaeotemperature significantly above today's levels, and the sediments are being studied by palynology to determine whether any flora existed at the time.

Introduction

Marine Plain (Fig. 1) has assumed considerable significance since the discovery there by the author of a mid-Pliocene vertebrate fauna including a dolphin and other forms. This paper presents a geological map and discusses the environment of deposition of the fossiliferous sediments.

The locality lies some 10 km south-east of Davis Station in the Vestfold Hills (Fig. 1). Until late 1984 it was believed that scattered patches of thin marine fossiliferous sediments in the vicinity were part of a series of Holocene sediment outcrops widespread in the Vestfold Hills (normally in the radiocarbon age range 5–8 Ka; Adamson & Pickard, 1983). Adamson & Pickard (1983) and Zhang, Zie & Li (1983) recorded three radiocarbon dates in the 28 000–31 000 y range, suggesting even then that the ages quoted were at the limit of those attainable by the radiocarbon technique. Adamson & Pickard (1986), Pickard *et al.* (1986) and Pickard *et al.* (1988) showed, from a variety of techniques (diatom and mollusc stratigraphy, protein racemization), that a mid-Pliocene age (3.5–4.5 Ma) was more likely. Further, the age so obtained is identical with that of the 'Pecten gravels' and correlatives (particularly some Sirius Formation fossils) in the McMurdo Sound–Transantarctic Mountains region (Webb, 1972, 1974; Webb *et al.*, 1984).

The discovery of the fossil dolphin, and subsequent discoveries of other vertebrates by the author in the 1985–86 field season, suggest that the sediments are an important source of vertebrate fossils. They are the only ones known in Antarctica between those of the late Eocene La Meseta Formation (Woodburne & Zinsmeister, 1982, 1983) and the Holocene (for example in the Vestfold Hills area (Adamson & Pickard, 1983, 1986a). The total diversity of the fauna and flora at Marine Plain is yet to be established.

Zhang *et al.* (1983, fig. 2) and Adamson & Pickard (1986a, figs 3 and 6) illustrated a 7 m section through the sequence at Marine Plain. Approximately 10–17 m above sea level, this comprises mid-Pliocene (3.5–4.5 Ma; Adamson & Pickard, 1986a, Pickard *et al.*, 1986) diatomite, overlain disconformably by up to 1.5 m of Holocene glacial deposits. The Pliocene and Holocene units are separated by a sandstone unit (top of Pliocene Section) possibly representing fossil permafrost activity. The Holocene material contains specimens of *Laternula elliptica* (King & Broderip) in living position which have yielded a radiocarbon data of 6490 ± 130 bp (Beta 17523- D.A. Adamson, pers. comm.). This date is uncorrected for reservoir effects. The observations suggest that the young glacial sediments accumulated in a marine environment by deposition at least in part from floating glacial ice (Pickard *et al.*, 1988), rather than being a ground moraine deposit (see below). The former hypothesis goes a long way towards explaining why such a thin veneer of soft Pliocene material could survive an intense glaciation such as the Vestfold Glaciation (latest Pleistocene, 10–25 000 Ka) of Adamson & Pickard (1983, 1986b). The steep banks marginal to lakes are probably retreating due to thermoerosion of permafrost as suggested by Pickard (1986).

The sediments of Marine Plain cover some 10 km^2. Outcrop

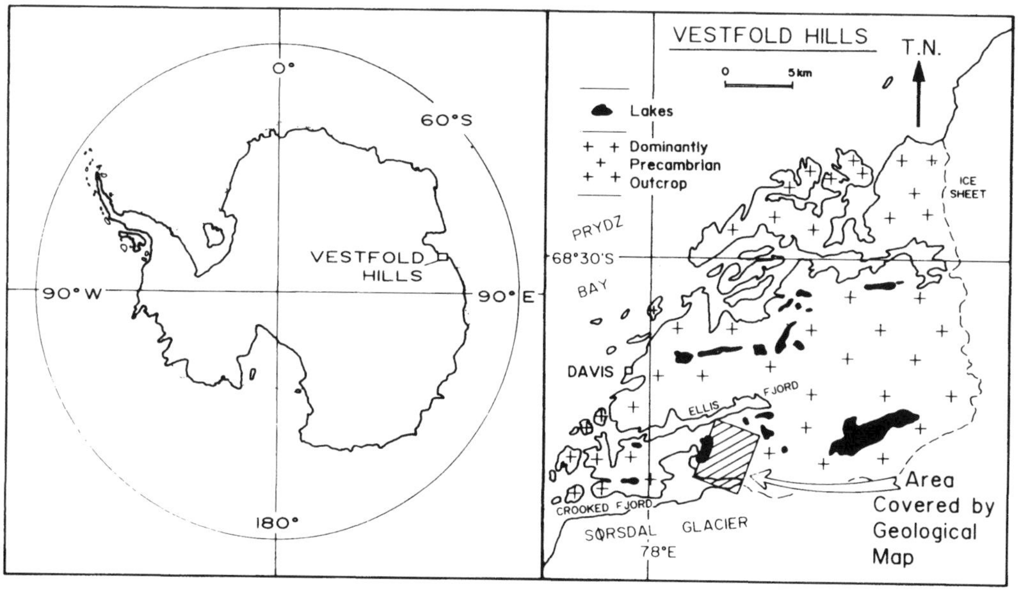

Fig. 1. Location of Marine Plain.

extends north from Iceberg Graveyard and is divided to the north by an outcrop of Precambrian gneisses (Collerson & Sheraton, 1986) into two lobes, perhaps representing original embayments at the time of deposition.

The Pliocene diatomite at Marine Plain appears to be the only such deposit in the Vestfold Hills. Because of its importance and apparent uniqueness, steps have been taken to declare it a Site of Special Scientific Interest under the Antarctic Treaty. A geological map of the area is shown in Fig. 2.

Pliocene diatomite

Pliocene sediments crop out principally in steep banks marginal to small lakes, and all vertebrates located so far have been found in such localities. Although outcrop areas appear to be small on aerial photographs, the veneer of Holocene glacials is so thin that one steps through it regularly when walking over the plains into a 20–30 cm thick dry powder composed mainly of diatoms. This unit produces a great volume of dust during walking and is thus geologically convenient to sample. When disturbed it gives off an odour of sulphur dioxide suggesting that, when fresh, it contained pyrite; this in turn suggests deposition in a reducing environment.

The Pliocene sequence fills the broad, low relief depression that constitutes Marine Plain and the top of the sequence coincides, throughout the area of outcrop, with a marked change in slope and rock type at the base of the Precambrian hills. This base is so clearly marked everywhere that it almost certainly represents an earlier sea level and old coastline. Whether this was the Pliocene or Holocene sea level is unclear, but striated basement just above this level north-west of an unnamed lake in the northern part of the eastern lobe (coordinates 852 857 on the 1:50000 map, Vestfold Hills) suggests that it may have been the Holocene one. It is probably also a useful index to the level from which glacial boulders were dropped onto the shallow Holocene seafloor. Any Pliocene sediments originally occurring above this level would have been stripped off by glaciation.

A mollusc-rich shell bed (Unit 3 of Zhang, 1983) is also a useful marker within the Pliocene in Marine Plain. It can be identified within 2 m of the Holocene debris wherever outcrop permits. *Chlamys tuftsensis* Turner was identified widely, but not north of a low ridge which separates the central northern lake (unnamed) in Fig. 2 from lakes (also unnamed) to the south.

The sequence described by Zhang (1983) is generally applicable throughout Marine Plain in both western and eastern

Table 1. *Oxygen isotope results on Marine Plain molluscs*

	$\delta^{13}C$	$\delta^{18}O$ (PDB)	Calculated temperature (°C)
Modern *Laternula elliptica*, Davis station			
ML-1[a]	1.63	3.43	3.7
ML-2	−0.22	3.63	
ML-3	1.36	3.32	
Holocene *Laternula elliptica*			
H[a]	1.27	2.94	5.3
Pliocene *Chlamys tuftsensis*			
Ct-1[a]	0.22	1.55	10.5

[a]Inhouse sample numbers.
The gas extraction of CT-1 gave a poor yield, and will be repeated shortly. For the moment, the result cannot be considered to be precise. The material is not yet X-rayed to determine whether it is calcite or aragonite. The calculated temperatures are based on the assumption that $\delta^{18}O$ of the sea water = 0‰ (SMOW). This is undoubtedly incorrect because of the addition of freshwater from ice. Thus the calculated temperatures are maxima. It must be emphasized that these are preliminary.

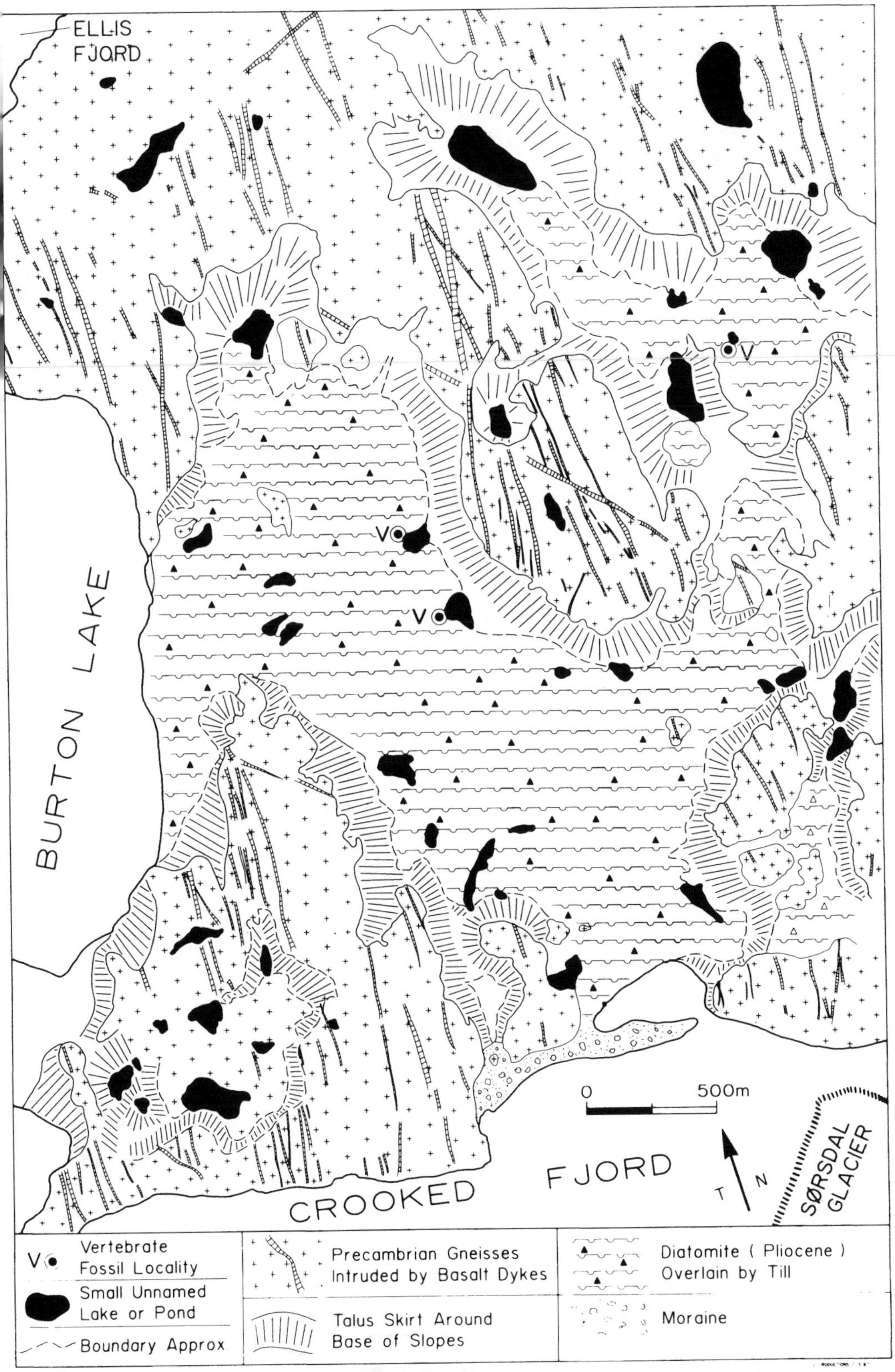

Fig. 2. Geological map of the area, based on three short (2–7 d) field visits and on study of colour aerial photographs of the area. The area covered by the map is contained within the following coordinates on the Second Edition (September 1982) 1:50000 Vestfold Hills map (Australian Division of National Mapping): north-west corner 825 860; north-east corner 860 860; south-east corner 860 820; south-west corner 825 820. This quadrangle is covered by: Run 4, photographs 2–7; Run 5, photographs 11–16 of 26 January 1979 helicopter colour aerial photography flown at 3050 m. Standard photographs were enlarged by some 3:1 and used as a base for field observation and later extrapolation.

areas. The only modification made here to his section is to distinguish the fine diatomaceous dust horizon overlying the sandy fossil permafrost horizon (part of his Unit 4) as a separate unit. During field examination of the Pliocene, no evidence was found of glacial sediments. The upper sandy fossil permafrost contains common small (up to 5 cm) angular boulders, probably of local origin. While these are consistent with very small-scale glaciation, they do not constitute proof of it and the rock texture may have been formed or reconstituted during Holocene permafrost activity.

Pickard (1986) suggested that the environment in the mid-Pliocene may have been much the same as at present. New information indicates that this view may now need to be revised. Preliminary oxygen isotope results from studies on very well preserved molluscs in the Pliocene section give the results shown in Table 1 (A. Chivas, pers. comm.). These results suggest a temperature significantly higher than that of the present day and high enough that some vegetation could be expected, over and above the present moss–lichen assemblages. To test this hypothesis, several samples were analysed palynologically but no spores or pollen of land plants were obtained (however, smooth leiospheres, consistent with a coeval marine source, were recovered; E. Truswell, pers. comm.). Further palynological analysis is planned when the sediment section is sampled carefully from drill cores.

Holocene glacial sediments

The surface of Marine Plain is covered with a veneer of Holocene glacial sediments which consist of unsorted angular to sub-rounded fragments of metamorphic rocks up to approximately 2 m in diameter. They overlie the Pliocene sediments directly and disconformably. The highest sandstone unit of the Pliocene sequence shows permafrost characteristics which may be a result of the Holocene history.

The sediments contain rare bivalves (*Laternula elliptica* (King & Broderip)) in growth position suggesting that deposition occurred in a marine environment and that that environment has not been disturbed since. Thus, at least in some localities, the glacial material fell from floating ice onto the seafloor below, probably a matter of only a few metres.

A prominent characteristic of these sediments on aerial photographs is the presence of 'growth ridges' (Fig. 2) concave to the south and evolving southwards towards Sørsdal Glacier. These 'growth ridges' appear to be stranded ridges of moraine, probably marking successive edges of the glacier or of thin sea ice associated with it. They may thus constitute successive recessional moraine ridges. Such a moraine ridge exists at present on the northern side of Sørsdal Glacier but this is higher and larger than those obvious to the north on aerial photographs. Such moraine ridges could be expected to have disturbed underlying sediment during their formation, although it is possible that the Pliocene sediment acted merely as a basement on which such a ridge accumulated. Minor disturbance may thus have occurred but would not be discernible without excavation beyond the scope of the present study.

The modern situation at Sørsdal Glacier margin probably provides an acceptable model of earlier Holocene conditions in the region. The present form of the glacials is the result of a continuing contraction of the northern margin of Sørsdal Glacier as sea ice or bay ice marginal to the glacier filled its bays with sediment, thus barring its own access.

References

Adamson, D.A. & Pickard, J. (1983). Late Quaternary ice movement across the Vestfold Hills, Antarctica. In *Antarctic Earth Science*, ed. R.L. Oliver, P.R. James & J.B. Jago, pp. 465–9. Canberra; Australian Academy of Science and Cambridge; Cambridge University Press.

Adamson, D.A. & Pickard, J. (1986a). Cenozoic history of the Vestfold Hills. In *Antarctic Oasis*, ed. J. Pickard, pp. 63–97. Sydney; Academic Press.

Adamson, D.A. & Pickard, J. (1986b). Physiography and geomorphology of the Vestfold Hills. In *Antarctic Oasis*, ed. J. Pickard, pp. 99–139. Sydney; Academic Press.

Collerson, K.D. & Sheraton, J.W. (1986). Bedrock geology and crustal evolution of the Vestfold Hills. In *Antarctic Oasis*, ed. J. Pickard, pp. 21–62. Sydney; Academic Press.

Harwood, D.M. (1986). Diatom biostratigraphy and paleoecology with a Cenozoic history of Antarctic ice sheets. PhD thesis, Ohio State University Columbus, 592 pp. (unpublished).

Pickard, J. (1986). The Vestfold Hills: a window on Antarctica. In *Antarctic Oasis*, ed. J. Pickard, pp. 333–51. Sydney; Academic Press.

Pickard, J., Adamson, D.A., Harwood, D.M., Miller, G.H., Quilty, P.G. & Dell, R.K. (1986). Early Pliocene marine sediments in the Vestfold Hills, East Antarctica: implications for coastline, ice sheet and climate. *South African Journal of Science*, **82**, 520–1.

Pickard, J., Adamson, D.A., Harwood, D.M., Miller, G.H., Quilty, P.G. & Dell, R.K. (1988). Late Tertiary marine sediments and coastline of East Antarctica. *Geology*, **16**, 158–61.

Webb, P.-N. (1972). Wright Fjord, Pliocene marine invasion of an Antarctic dry valley. *Antarctic Journal of the United States*, **7**, 227–34.

Webb, P.-N. (1974). Micropaleontology, paleoecology and correlation of the Pecten Gravels, Wright Valley, Antarctica, and description of *Trochoelphidiella onyxi* n. gen., n. sp. *Journal of Foraminiferal Research*, **4**, 185–99.

Webb, P.N., Harwood, D.M., McKelvey, B.C., Mercer, J.H. & Stott, L.D. (1984). Cenozoic marine sedimentation and ice-volume variation on the East Antarctic craton. *Geology*, **12**, 287–91.

Woodburne, M.O. & Zinsmeister, W.J. (1982). Fossil land mammal from Antarctica. *Science*, **218**, 284–5.

Woodburne, M.O. & Zinsmeister, W.J. (1983). A new marsupial from Seymour Island, Antarctic Peninsula. In *Antarctic Earth Science*, ed. R.L. Oliver, P.R. James & J.B. Jago, pp. 320–2. Canberra; Australian Academy of Science and Cambridge; Cambridge University Press.

Zhang Qingsong (1983). Periglacial landforms in the Vestfold Hills, East Antarctica: preliminary observations and measurements. In *Antarctic Earth Science*, ed. R.L. Oliver, P.R. James & J.B. Jago, pp. 478–81. Canberra; Australian Academy of Science and Cambridge; Cambridge University Press.

Zhang Qingsong & Peterson, J.A. (1984). A geomorphology and Late Quaternary geology of the Vestfold Hills, Antarctica. *ANARE Reports*, **133**, 1–84.

Zhang Quinsong, Zie, Y. & Li, Y. (1983). A preliminary study on the evolution of the post-late Pleistocene Vestfold Hills environment, East Antarctica. In *Antarctic Earth Science*, ed. R.L. Oliver, P.R. James & J.B. Jago, pp. 473–7. Canberra; Australian Academy of Science and Cambridge; Cambridge University Press.

Seismic and sedimentological record of glacial events on the Antarctic Peninsula shelf

J.B. ANDERSON, L.R. BARTEK & M.A. THOMAS

Department of Geology and Geophysics, Rice University, Houston, Texas, 77251, USA

Abstract

Seismic-reflection profiles from the continental shelf of the Antarctic Pensinsula reveal a similar sequence of events to other parts of the Antarctic continental shelf (Ross Sea and Weddell Sea). Basal sequences consist of seaward-accreting sediment wedges bounded by widespread unconformities and imply relatively shallow-water deposition. There followed a prolonged period of glacial erosion and deposition, which resulted in the present deep and rugged topography of the shelf. The initial seaward advance of the ice sheet onto the shelf would have occurred as falling sea level exposed the inner shelf, and does not necessarily imply dramatic climatic cooling. Furthermore, once the deep, rugged shelf topography was created, waxing and waning of marine ice sheets became strongly linked to sea level changes and may not have been in harmony with Southern Hemisphere climatic events. Following this episode the shelf was draped by glaciomarine deposits, and deep-sea fans of the Bellinghausen abyssal plain were buried beneath hemipelagic and pelagic sediments. A later ((?)Plio-Pleistocene) episode of ice-sheet expansion on to the shelf resulted in widespread till deposition and renewed deep-sea fan activity.

Introduction

This paper presents a model for the Cenozoic evolution of the Antarctic continental shelf that is based mainly on a seismic stratigraphical analysis of data acquired from the continental shelf of the Antarctic Peninsula (Fig. 1), as well as piston core data. The sequence of events recorded in our data are similar to those inferred from seismic records from the Ross Sea shelf (Hinz & Block, 1984) and Weddell Sea shelf (Haugland, Kristoffersen & Velde, 1985).

The general stratigraphical sequence for the study area is illustrated in Fig. 1. The older section consists of accretionary sequences (S_1) which were subsequently deeply eroded by glacial ice, culminating in an erosional surface that extends to the edge of the continental shelf (U_1); glacial erosion resulted in a foredeepened shelf profile (Fig. 1). The products of glacial erosion fill glacial troughs (glaciomarine deposits) and also include mound-like deposits, presumed to be amalgamated tillites (S_2, Fig. 1). The next episode of shelf development is marked by a shelf-wide draping sequence, presumed to consist of glaciomarine deposits (S_3). This sequence thickens in an onshore direction, into the foredeepened outer shelf basin (Fig. 1).

Piston cores from the shelf penetrated over-compacted diamictons, believed to be tills that were deposited during the most recent glacial advance on to the shelf. A thin veneer of glaciomarine sediments drape these tills, marking the retreat of glacial ice from the shelf and the onset of present marine conditions.

Our seismic stratigraphical interpretations suffer from a lack of age control, and the establishment of the timing of events recorded in these data awaits drilling on the shelf. For now, we propose a model for shelf evolution that is based on similarities in the seismic stratigraphy of the Antarctic Peninsula shelf and that of the Ross Sea shelf, where drilling has been done. Additional age constraints are inferred from DSDP Leg 35 sites 322, 323, 324 and 325 on the adjoining Bellingshausen rise and abyssal floor (Hollister & Craddock, 1976) by attempting to relate sedimentation at these sites to processes that occurred on the shelf.

The model

Stage 1

Initial growth and development of the continental ice sheet lowered sea level enough to expose the shallow shelf, thus allowing fringing parts of the ice sheet to advance (Fig. 2). Conceivably, this was a temperate ice sheet that ablated by melting rather than by iceberg calving. It would have also had a much thinner margin than the present ice sheet, therefore causing only minor isostatic depression of the proglacial shelf. Meltwater streams delivered large quantities of terrigenous sediment to the sea, and thick, extensive accretionary sequences (S_1, Fig. 1) were deposited.

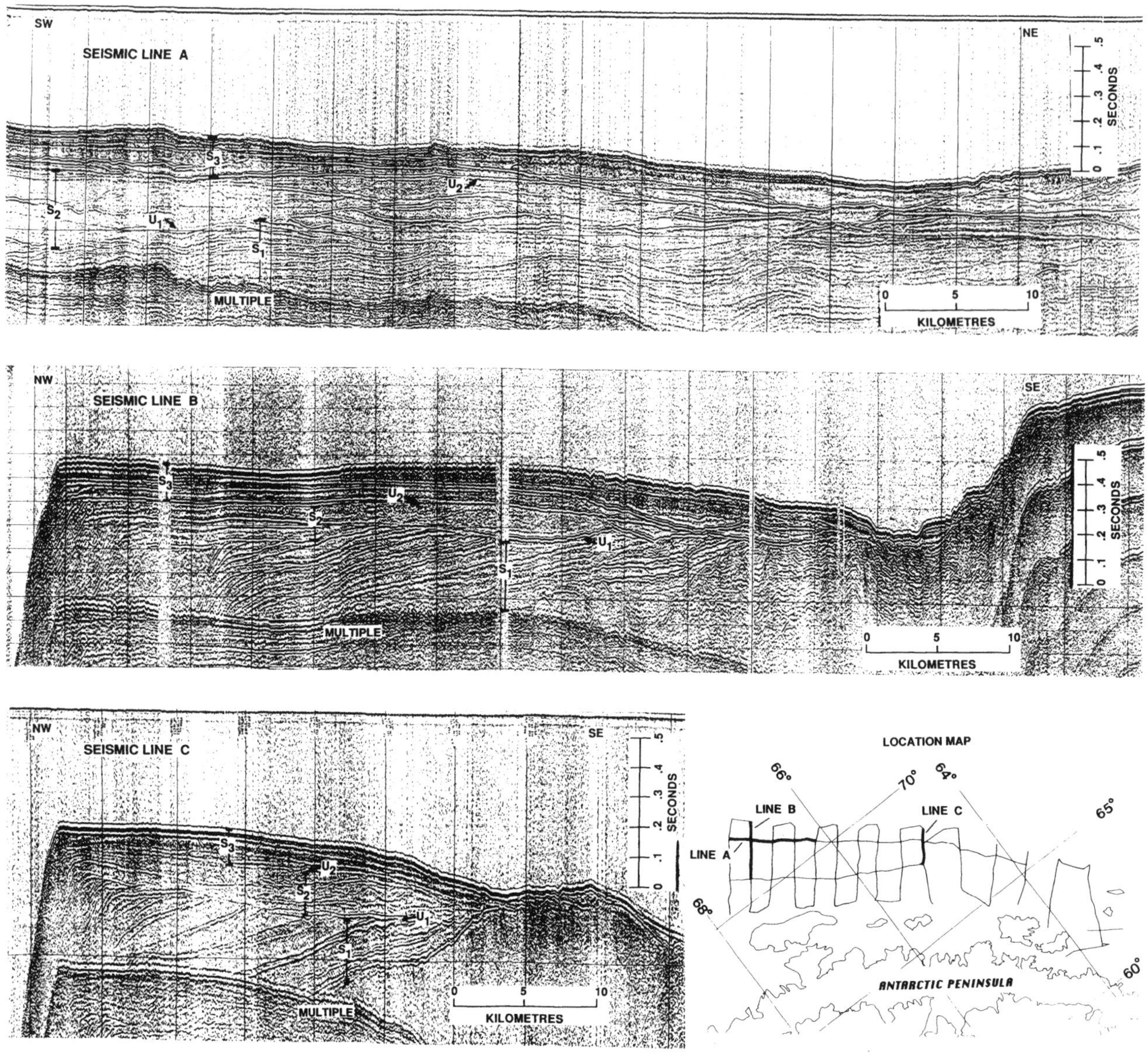

Fig. 1. Seismic-reflection lines from the Antarctic Peninsula continental shelf showing generalized sequence stratigraphy for the region. Inset map shows the locations of lines acquired during the USAP 88 survey.

In the Ross Sea this stage of ice-sheet development may be recorded in the base of DSDP Leg 28 site 270, where marine sands with shallow-water benthic foraminifera of late Oligocene age are buried beneath glaciomarine sediments (late Oligocene–early Miocene age) with benthic foraminifera that reflect sedimentation at water depths well below the reach of waves and wind-driven currents (Leckie & Webb, 1983). An absence of lag deposits above this erosional surface implies glacial erosion.

Stage 2

As sea level continued to fall, the fringes of the ice sheet advanced farther onto the shelf (Fig. 3). Fluctuations in the grounding line of the ice sheet resulted in erosion of glacial troughs and their subsequent filling by glacial and glaciomarine deposits (S_2, Fig. 1). The maximum extent of the ice sheet was situated very near the shelf edge (the seaward extent of U_1, Fig. 1). This was the main period of erosion on this part of the continental shelf.

Glacial ice and associated meltwater streams debouching from the grounded ice front cut deeply into the shelf edge forming large canyons. Turbidity currents flowing down these canyons constructed large fan systems on the deep seafloor. Turbidites near the base of DSDP sites 322 and 325 include well-sorted, quartz sands which imply a fluvial, deltaic or coastal origin for these sands, and a glacial maritime setting very different from that of today; these deposits are of (?)late Oligocene–early to middle Miocene age (Hollister & Craddock, 1976). A gradual increase in the sedimentation of ice-rafted debris occurred at Southern Ocean DSDP sites at the same time (Tucholke *et al.*, 1976), possibly reflecting a decrease in the offshore flux of fine-grained terrigenous sediment as much as, if not more than, an increase in ice rafting. A decrease in the flux of fine-grained sediment to the deep seafloor would be associated with climatic cooling and the change from temperate to polar glacial conditions on the continent.

Seismic records from the Ross Sea continental shelf show a sequence similar to S_2 resting above a prograding sequence

STAGE 1: INITIAL TEMPERATE MARITIME GLACIAL SETTING

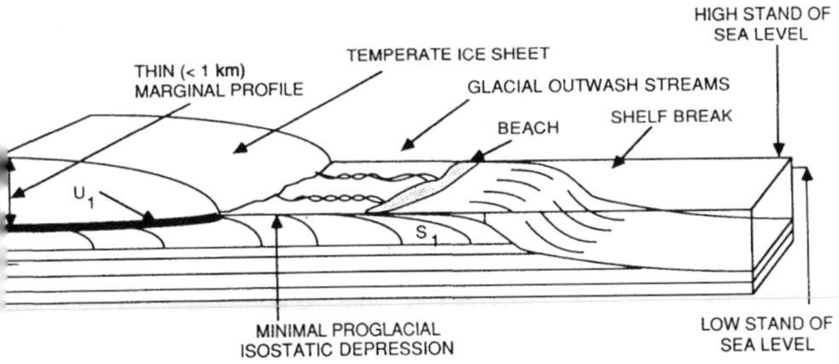

LATE OLIGOCENE ?

-TEMPERATE GLACIAL SETTING WITH GLACIAL OUTWASH STREAMS AND A STRAND PLAIN

U_1 =FIRST GLACIALLY-ERODED (?) UNCONFORMITY

S_1 =SHALLOW-MARINE SHELF SEQUENCE

=QUARTZ-RICH BEACH SANDS

Figs 2–6. Five stages of development of the Antarctic ice sheet. See text for discussion.

STAGE 2: WAXING AND WANING TEMPERATE ICE SHEET

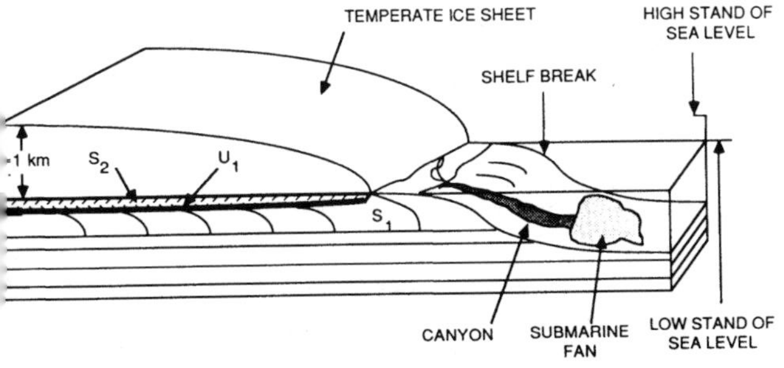

LATE OLIGOCENE TO EARLY / MID-MIOCENE ?

-DEVELOPMENT OF ACTIVE SUBMARINE CANYONS AND FANS
-FANS COMPOSED OF QUARTZ SAND

U_1 =FIRST GLACIALLY-ERODED (?) UNCONFORMITY

S_1 =SHALLOW-MARINE SHELF SEQUENCE

S_2 =SUBGLACIAL AND GLACIAL MARINE SEQUENCE WITH MANY GLACIALLY-ERODED SURFACES

=CANYON FILL (COARSE PEBBLE CONGLOMERATES ETC.)

=QUARTZ-RICH SEDIMENT GRAVITY FLOW DEPOSITS OF FAN (D.S.D.P. LEG 35: SITES 322, 323, AND 325)

=SUBGLACIAL AND GLACIAL MARINE SEDIMENTS

Fig 3

STAGE 3: GLACIAL RETREAT AND DEVELOPMENT OF POLAR ICE SHEET

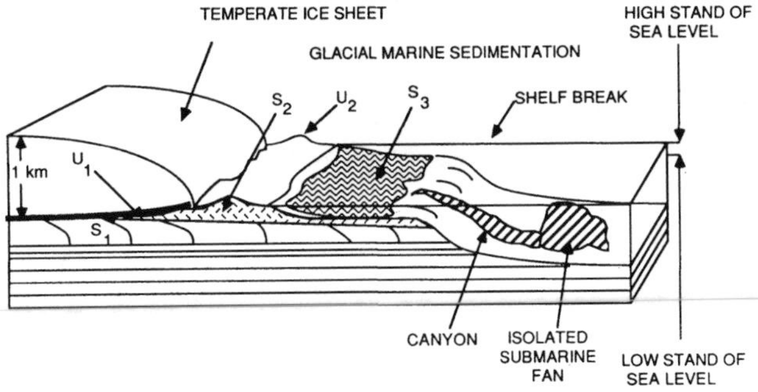

MID–LATE MIOCENE ?

-SEDIMENT SUPPLY TO SUBMARINE CANYONS AND FANS IS CUT OFF, SO THERE IS ONLY PELAGIC AND HEMIPELAGIC SEDIMENTATION IN THE CANYON AND ON THE FAN.

U_1 =THE FIRST GLACIALLY-ERODED (?) UNCONFORMITY

U_2 =LAST GLACIALLY-ERODED SURFACE OF S_2 TIME

S_1 =SHALLOW-MARINE SHELF SEQUENCE

S_2 =GLACIOMARINE SEQUENCE

S_3 =GLACIOMARINE SEQUENCE

=PELAGIC AND HEMIPELAGIC FILL OF CANYONS AND FANS (D.S.D.P. LEG 35, SITES 322, 323, & 325)

=SUBGLACIA AND GLACIAL MARINE SEDIMENTS (S_2)

=GLACIOMARINE SEDIMENTS (S_3)

Fig. 4

STAGE 4: RE-ADVANCE OF ANTARCTIC ICE SHEET

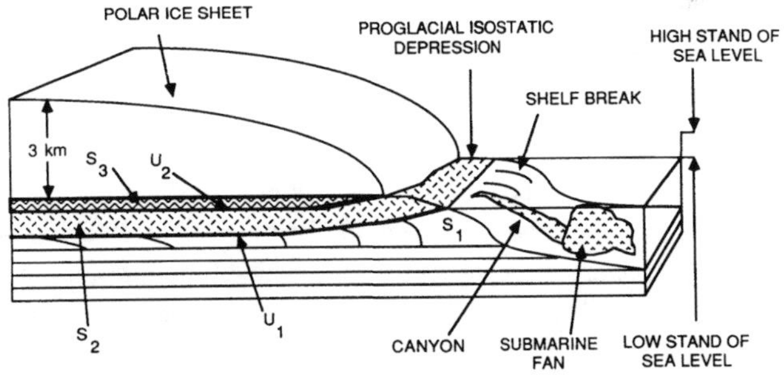

PLIO, – PLEISTOCENE

-CANYONS AND FANS ARE INFILLING WITH POORLY SORTED, LITHIC GRAVELS AND SANDS OF GLACIAL ORIGIN
-TROUGH FORMATION ON SHELF

U_1 =THE FIRST GLACIALLY-ERODED (?) UNCONFORMITY

U_2 =GLACIALLY-ERODED UNCONFORMITY

S_1 =SHALLOW-MARINE SHELF SEQUENCE

S_2 =SUBGLACIAL AND GLACIO-MARINE SEQUENCE

S_3 =GLACIOMARINE SEQUENCE

=SUBGLACIAL AND GLACIO-MARINE SEDIMENTS(S_2)

=POORLY SORTED, LITHIC GRAVELS AND SANDS OF GLACIAL ORIGIN INFILLING CANYONS AND FANS

=GLACIOMARINE SEDIMENTS (S_3)

Fig. 5

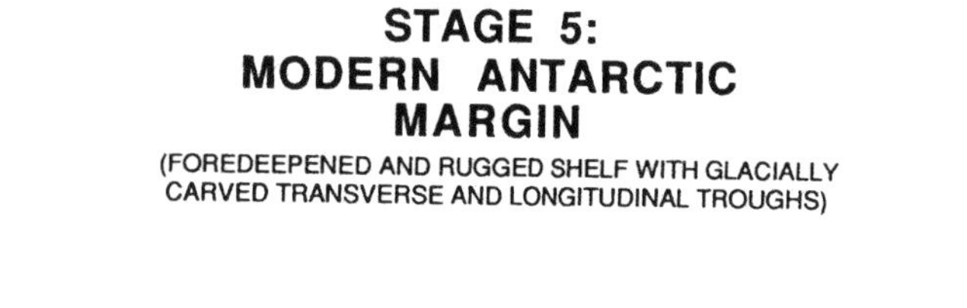

Fig. 6

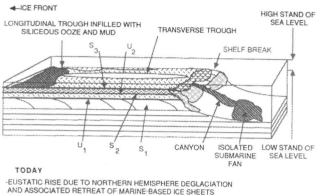

(Hinz & Block, 1984). The upper sequence corresponds approximately to early to middle Miocene deposits drilled during Leg 28. These deposits consist of interbedded pebbly muds, diatomaceous muds, and diamictons, and imply waxing and waning ice sheets.

Stage 3

With continued growth and expansion of the ice sheet (Fig. 4), the combined effects of erosion, isostatic depression, shelf subsidence, and perhaps rising eustatic sea level, brought the base of the ice sheet to a depth at which it could no longer remain grounded. This resulted in decoupling of the ice sheet and retreat to a position on the shelf somewhere near the southern boundary of S_2 (Fig. 1). Thus, glacial retreat from the continental shelf could have occurred without a significant increase in atmospheric temperature. Once the ice sheet detached from the more shallow shelf edge, its retreat on to a deeper inner shelf was probably quite rapid (Thomas & Bentley, 1978).

During Stage 3, sedimentation on the shelf would have been dominated by glaciomarine processes, and sediment bypassing of the shelf would have been minimal (note that S_3 thins appreciably near the shelf edge (Fig. 2)); this would have resulted in the isolation of canyons and associated deep-sea fans. Based on the incomplete recovery at DSDP sites 322, 323, and 325, Stage 3 possibly began at some point during the middle–late Miocene, when turbidites at these sites were draped by pelagic and hemipelagic sediments (Hollister & Craddock, 1976).

Stage 4

Seismic data from the Antarctic Pensinsula region indicate that, since the deposition of S_2 (Fig. 1), glacial erosion has been confined to the inner shelf (Fig. 5). Yet over-compacted diamictons, presumed to be tills, were collected on the outer continental shelf during our 1988 field season. This may imply that glacial ice was coupled to the seafloor in such a manner that deposition, but no erosion, was occurring on the outer shelf.

Evidence that both the East and West Antarctic ice sheets grounded on other parts of the Antarctic continental shelf on one or more occasions during the Plio-Pleistocene comes from piston cores which have penetrated basal tills (subglacial deposits). Basal tills have been recovered from the continental shelves of the Ross Sea, eastern Weddell Sea, and off the George V Coast of East Antarctica (Anderson *et al.*, 1980). The exact age of these tills is poorly constrained.

Piston cores from fan valleys of the Bellinghausen abyssal plain and continental rise have penetrated graded gravels and sands. The main difference between these turbidites and those of the late Oligocene–early Miocene sections of DSDP cores is that the more recent deposits bear clear evidence of their glacial origin, i.e. glacial surface textures on most sand grains and a wide range of particle sizes and compositions (Wright, Anderson & Fisco, 1983). The mineralogical similarity between turbidites and glaciomarine sediments from the adjacent shelves indicates that these fan valleys may have been supplied with sediment directly by subglacial meltwater streams, at times when ice sheets advanced to the shelf edge (Wright *et al.*, 1983). Similar turbidites were cored at DSDP sites 322 and 323 and were dated as Plio-Pleistocene (Hollister & Craddock, 1976). There is also evidence for renewed deep-sea fan activity around other parts of the continental shelf at some time during the Plio-Pleistocene and the indication is that this activity was related to expansion of the ice sheet to the shelf edge in these areas (Wright *et al.*, 1983). The giant Weddell fan is one example (Anderson, Wright & Andrews, 1986), and recent drilling of this feature may help to constrain the timing of this most recent period of fan activity.

Stage 5

Surface sediments of the continental shelf consist of diatomaceous glaciomarine mud which is in sharp contact with relict glacial and glaciomarine sediments. Such a contact is evidence that the most recent retreat of the ice sheet from the shelf occurred very rapidly. The most recent collapse of these marine ice sheets was probably triggered by the late Wisconsin eustatic rise (Thomas & Bentley, 1978). Sediments at the shelf edge consist of poorly sorted lag deposits which reflect impinging deep-sea currents.

Conclusions

Newly acquired seismic data from the continental shelf of the Antarctic Peninsula show several episodes of deposition and erosion which are related to changing glacial conditions on the Antarctic continent. An important component of our model s the fact that the continental shelf was much shallower when he ice sheet first advanced on to it. Given this shallower setting, he initial advance could have occurred under much less severe glacial/climatic conditions than exist at present. Likewise, owering of the shelf through combined glacial erosion and isostasy resulted in a setting suitable only for marine ice sheets. Because marine ice sheets are as sensitive to eustasy as they are to climate, and even more so over short time periods (10^3 y), glacial waxing and waning on the shelf may have been out of phase with climatic changes in the Southern Hemisphere.

References

Anderson, J.B., Wright, R. & Andrews, B. (1986). Weddell Fan and associated abyssal plain, Antarctica: morphology, sediment processes, and factors influencing sediment supply. *Geo-Marine Letters*, **6(2–4)**, 121–9.

Anderson, J.B., Kurtz, D.D., Domack, E.W. & Balshaw, K.M. (1980). Glacial and glacial marine sediments of the Antarctic continental shelf. *Journal of Geology*, **88(4)**, 399–414.

Haugland, K., Kristoffersen, Y. & Velde, A. (1985). Seismic investigations in the Weddell Sea embayment. *Tectonophysics*, **114(4)**, 293–314.

Hinz, K. & Block, M. (1984). Results of geophysical investigations in the Weddell Sea and in the Ross Sea, Antarctica. *Proceedings of the Eleventh World Petroleum Congress (London) 1983*, vol. 2, Geology Exploration Reserves, pp. 79–81. Chichester, New York,; John Wiley & Sons.

Hollister, C.D. & Craddock, C. (1976). Geologic evolution of the South Pacific Basin. In *Initial Reports of the Deep Sea Drilling Project 35*, ed. C.D. Hollister & C. Craddock, pp. 723–43. Washington, DC; US Government Printing Office.

Leckie, R.M. & Webb, P.N. (1983). Late Oligocene–early Miocene glacial record of the Ross Sea, Antarctica: evidence from DSDP Site 270. *Geology*, **11(10)**, 578–82.

Thomas, R.H. & Bentley, C.R. (1978). A model for Holocene retreat of the West Antarctic ice sheet. *Quaternary Research*, **10(2)**, 150–70.

Tucholke, B.E., Hollister, C.D., Weaver, F.M. & Vennum, W.R. (1976). Continental rise and abyssal plain sedimentation in the southeast Pacific Basin-Leg 35 Deep Sea Drilling Project. In *Initial Reports of the Deep Sea Drilling Project 35*, ed. C.D. Hollister & C. Craddock, pp. 359–400. Washington, DC; US Government Printing Office.

Wright, R., Anderson, J.B. & Fisco, P.O. (1983). Distribution and association of sediment gravity flow deposits and glacial/glacial-marine sediments around the continental margin of Antarctica. In *Glacial-Marine Sedimentation*, ed. B.F. Molnia, pp. 233–64. New York; Plenum Press.

Mid-Holocene ice sheet recession from the Wilkes Land continental shelf, East Antarctica

E.W. DOMACK[1], A.J.T. JULL[2], J.B. ANDERSON[3] & T.W. LINICK[2]

1 Geology Department, Hamilton College, Clinton, New York 13323, USA
2 Department of Physics, University of Arizona, Tucson, Arizona 85721, USA
3 Department of Geology & Geophysics, Rice University, Houston, Texas 77251, USA

Abstract

Sediments from the Wilkes Land (140–150° E) continental shelf and slope, East Antarctica, record a history of late Pleistocene–mid Holocene ice sheet retreat and modern, biogenic sedimentation. This chronology is based on five samples of foraminifera ($CaCO_3$) and organic carbon, which were analysed for their ^{14}C content utilizing the University of Arizona tandem accelerator mass-spectrometer. Most of the dates are < 7850 ^{14}C y BP, with one date of 14260 ± 140 y BP from the shelf break. The chronology, stratigraphy and shelf morphology indicate that grounded ice extended out to the shelf edge, but only within the confines of deep-shelf depressions and troughs. Relatively shallow, broad and low relief portions of the shelf apparently escaped glaciation during the most recent glacial maximum. Recession of glacial ice from the shelf was most likely in response to rising relative sea level which initiated the development of floating ice tongues within the ice drainage system. This event took place some time after 9000 y BP and most probably prior to 3000 y BP. Further application of the accelerator method to other sediment cores, which predate the latest ice advance, should help clarify the above hypotheses.

Introduction

Sedimentary deposits on the Antarctic continental margin have great potential for the documentation of late Quaternary glacial fluctuations. However, their utility has been limited because of the difficulty in obtaining accurate age determinations and the uncertainties involved in their sedimentological interpretation. Our increased understanding of Antarctic glaciomarine sedimentation and shelf morphology have greatly reduced the latter problem so that with some confidence one can interpret the depositional processes responsible for shelf sedimentation (Anderson *et al.*, 1983; Barnes, 1987). Accurate age determinations, however, remain a limiting factor. Attempts to utilize diatom biostratigraphy have met with conflicting results, primarily because of the dominance of sediment reworking during glacial expansions onto the shelf, limited resolution and taxonomic discrepancies (Kellogg & Kellogg, 1986).

Conventional ^{14}C methods, as applied to shelf sediments, are generally limited to bulk sampling for organic carbon within post-glacial hemipelagic deposits, often obtained from several cores (Stuiver & Braziunas, 1985). Dates obtained by such sampling have been useful in placing deglacial limits for the Ross Sea at around (prior to) 8000 y BP (Kellogg, Osterman & Stuiver, 1979). Biogenic calcite is generally too sparse within glacial and glaciomarine deposits (diamictons) for conventional dating methods. Though samples of large isolated skeletal fragments (i.e. bryozoans) have been utilized by some workers (Elverhøi, 1981), their occurrence is fortuitous in most glacial and glaciomarine sediments.

The application of accelerator technology to ^{14}C dating represents an advance in attempts to decipher the glacial record of the Antarctic margin. Small samples (< 0.02 g) of biogenic calcite, such as foraminifera, can be dated with resolution which in most cases exceeds the limit for conventional ^{14}C dating (i.e. 30000 y BP). The small sample requirement allows for sampling of precise layers as well as dating of lithologies which have little in the way of datable organic matter. This paper presents the results of preliminary application of the accelerator method to continental shelf sediments of varied lithology which have been recovered from the Wilkes Land margin of East Antarctica (Fig. 1). Discussion regarding the sedimentology and details of age–date interpretation are to be found in other papers (e.g. Domack, 1982 and Domack *et al.*, 1989).

Background

Lithostratigraphy

Poorly sorted terrigenous sediments (diamictons) are widespread on the George V continental shelf (Fig. 1). Such

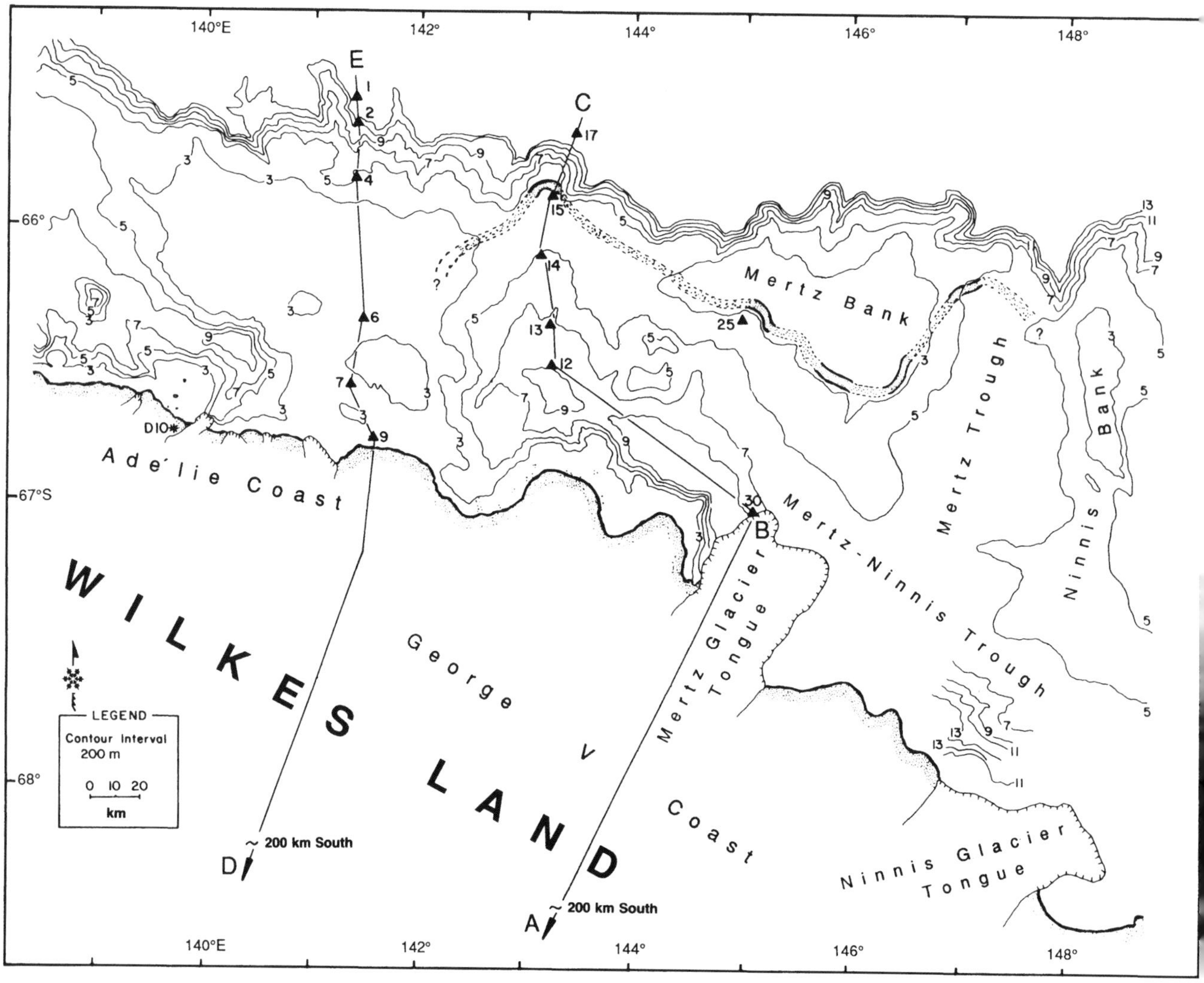

Fig. 1. Bathymetric contours and coastal physiography for the George V and Adélie continental margin of Wilkes Land, East Antarctica. Core locations (solid triangles) along reconstructed and present glacial flow paths are also shown (see Fig. 2). Stippled bars (pecked where inferred) refer to the location of lateral moraines (Barnes, 1987) and related end moraines (see text for discussion); bathymetry is after Chase *et al.* (1987).

deposits have been interpreted as glacial and glaciomarine in origin and thus provide evidence for glacial expansion on to the shelf during the time they were deposited (Anderson *et al.*, 1980). This view if supported by mineralogical and petrological studies of the diamictons, which indicate that significant glacial reworking of submarine outcrops has occurred (Domack, 1987).

Graded sands are a common deposit on the Adélie shelf and contain, in some cases, significant amounts (> 30%) of bioclastic detritus (carbonate). Diamicton and graded sand deposits appear to be, in part, age-equivalent based on observed interbedding of the lithologies in some cores (Domack, 1982). It is likely that density (turbidity) currents played a role in the formation of these sands.

Siliceous mud and ooze (SMO) overlie graded sand and diamicton and clearly postdate glacial recession from the shelf. These sediments are believed to be the product of pelagic and hemipelagic processes under fluctuating (seasonal) sea-ice conditions (Domack, 1988). Marine current activity and iceberg turbation prevents SMO deposition on the shelf in depths < ~ 500 m. Whereas areas shallower than this contain mixtures of bioclastic (carbonate) sandy gravel, SMO and palimpsest clastics, the latter are derived from relict diamicton or graded sand lithologies (Domack, 1980; Dunbar *et al.*, 1985).

Morphology

Early bathymetric data (Vanney & Johnson, 1979) were important in defining the gross morphology of the George V and Adélie (G/A) shelf, which is dominated by a linear coast-parallel trough (Mertz-Ninnis Trough), middle–outer shelf banks (Mertz and Ninnis banks) and transverse depressions (Mertz Trough). Detailed bathymetric observations and side-scan sonar data obtained by the US Geological Survey have revealed the presence of flutings within the Mertz Trough (Barnes, 1987). This set of low-relief parallel ridges trends towards the shelfbreak and is interpreted by Barnes as subglacial in origin. Diamictons are the dominant subsurface deposit in areas which exhibit such features.

Asymmetric submarine ridges have also been observed bordering the landward flank of the Mertz Bank. These features are of significant relief (50 m) and have been interpreted as submarine lateral moraines (Barnes, 1987). Profiles obtained during Operation Deep Freeze 1979 (Anderson *et al.*, 1979) reveal similar asymmetric ridges which occur at the seaward terminus of the Mertz–Ninnis Trough and the Mertz Trough (personal observation). Therefore, it is believed that the moraines mark a series of terminal and lateral ice positions which are defined by the bathymetric trend of major shelf basins and banks (Fig. 1). Glacial diamictons have yet to be recovered in regions seaward of these moraines.

Because significant erosion of pre-glacial lithologies has occurred along the trend of the shelf basins (Domack, 1987), past glacial expansion onto the continental shelf has both been influenced by, and partially amplified the shelf relief. A variety of iceberg gouge marks have also been observed, particularly along the edge of shelf banks, in depths generally < 500 m (Barnes, 1987).

^{14}C dates

Sample selection

Foraminifera were selectively removed from both diamicton and graded sand lithologies over sample intervals of 2–6 cm (Table 1). Foraminifera were scarce within diamictons, particularly basal tills, and only one core (DF-79-15, Fig. 2) proved suitable for sampling. Enough calcite (> 0.02 g) could be obtained for other diamictons only by processing a large volume of sediment over a wide core depth interval. Samples ranged from 0.2 to 0.0558 g and consisted of approximately 200–300 tests per sample; chiefly robust benthic species dominated by *Uvigerina bassensis, Ehrenbergina glabra, Globocassidulina spp.* and *Cassidulnoides spp.* However, diamicton samples contained mixtures of benthic and planktonic forms (Table 1). Dissolution of foraminifera tests was observed only in the samples from core 4 (Fig. 2); individual tests exhibiting such dissolution were not sampled. Two surface samples were selected in order to assess the role of differences in reservoir CO_2 related to glacial melting (sample 9) and upwelling (sample 25) (Domack *et al.*, 1989).

Five samples from the SMO facies were selected from two cores (12 & 13, Fig. 2). These were analysed for ^{14}C from the organic carbon fraction. Prior ^{210}Pb analyses were also conducted on these cores, which established modern sedimentation at rates of 0.3 cm/y (J.K. Cochran, pers. comm., 1987).

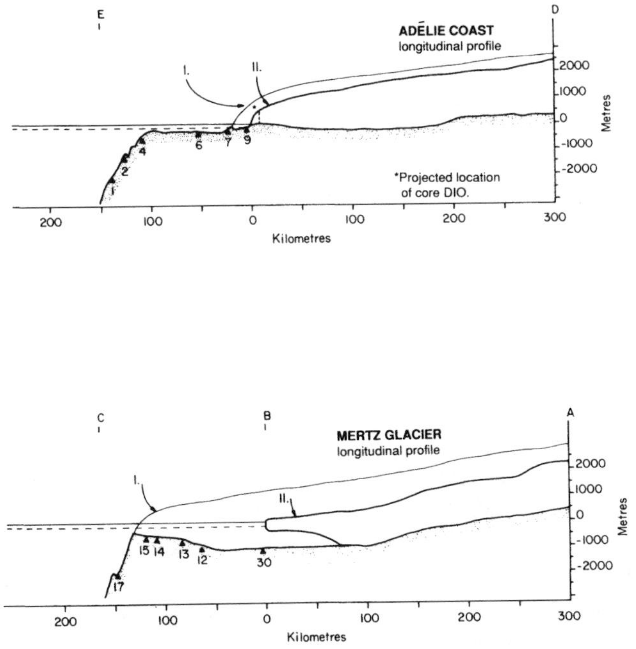

Fig. 2. Glacial and bedrock profiles for the George V (top) and Adélie (bottom) margins (see Fig. 1 for location). Present ice profiles (II) and bedrock elevations are based upon data in Domack (1980), Drewry (1983) and Chase *et al.* (1987). Location of seafloor sediment cores and ice core D10 are also shown. Glacial maximum reconstruction is shown by profile I and present position by profile II. Present and 18000 y BP (pecked) sea levels are also shown. Isostatic depression for maximum reconstruction is not included.

Analysis methods and results

Organic matter and foraminifera calcite were converted into CO_2 and reduced to graphite. Delta ^{13}C corrections were made based on values measured from the graphite. Fractions of modern C, based on the ^{14}C content, were determined at the University of Arizona tandem accelerator mass-spectrometer (TAMS) facility (Linick *et al.*, 1986). Resultant ages ranged from 3230 ± 200 to 14260 ± 140 y BP (Table 1).

Discussion

Significance of ages

Care must be taken in interpreting ^{14}C ages in the Antarctic because both CO_2 reservoir and reworking effects are known to influence strongly resulting ages. The sampling done in this study was designed to evaluate both these effects. Reservoir influences related to glacial melting are believed to exert the greatest control on ^{14}C ages; this is demonstrated by the extrapolated surface (modern) dates of around 5500 y for siliceous mud. ^{210}Pb measurements have established the modern nature of these sediments and reworked (detrital) carbon contributes very little to the total organic fraction (Domack *et al.*, 1989). Since deep-water upwelling in the southern ocean is expected to alter ages by only 400 y or so (Stuiver & Ostlund, 1983; Andree *et al.*, 1985) glacial melt effects are considered to be the main contributors to old surface ages, especially in near-coastal sites (i.e. sample 9, Table 1, Fig. 2).

Stratigraphical interpretations

Adélie Shelf. The oldest sediment appears to be the graded sand obtained in core 4 from the outer shelf–upper slope environment of the Adélie shelf (Fig. 2). Such graded sands are prevalent on this shelf and clearly indicate significant reworking of contained foraminifera. Thus the age of 14260 ± 140 y for sample AA-1406 does not indicate the date of sedimentation at the site of core 4 but the age of foraminifera accumulation (source) somewhere upslope; namely the

Table 1. *List of samples used for ^{14}C dating*

Date number	Core[a]	Samples: Lithology	Depth (cm)	^{14}C age (y BP)	Carbon source[b]	Mass of calcite (g)
AA-1402	15	Diamicton	196–203	6300 ± 120	Benthics & planktonics	0.0206
AA-1403	15	Diamicton	28–33	6780 ± 90	Benthics & planktonics	0.0204
AA-1404	25	Sand	Surface	3230 ± 200	Benthics	0.0200
AA-1405	9	Sand	0–5	5020 ± 180	Benthics	0.0294
AA-1406	4	Graded sand	50–55	14260 ± 140	Benthics	0.0558
AA-1970	12	Siliceous ooze	203–204	6430 ± 80	Organic matter	
AA-1971	12	Siliceous mud	549–550	7350 ± 80	Organic matter	
AA-1972	13	Siliceous ooze	69–70	6230 ± 90	Organic matter	
AA-1973	13	Siliceous ooze	224–225	5800 ± 120	Organic matter	
AA-1974	13	Siliceous ooze	574–575	7850 ± 110	Organic matter	

[a]Deep Freeze (USCGC Glacier) 1979 collection, curated at Florida State University.
[b]Foraminifera calcite or non-carbonate carbon.

outer edge of the Adélie shelf. If glacial ice ever occupied this portion of the shelf, and no ice contact diamictons have yet to be recovered from this area, then it receded some time after 14000 y BP; more likely 10000 y BP if a reservoir correction for sample AA-1406 is applied.

Modern sedimentation on the Adélie shelf is characterized by bioclastic gravels and sands adjacent to the tidewater ice sheet with increasing admixtures of SMO further offshore (Domack, 1988). Thus the change in sedimentation reflected in the down-core to surface deposits is believed to reflect a deepening of waters and a decrease in clastic supply.

George V Shelf. Glaciomarine diamicton at the site of core 15 (Fig. 2) appears to have been deposited beneath an ice shelf (tongue) around 7000–6000 y BP. This is based upon two dates of 6780 ± 90 and 6300 ± 120 y BP (Table 1). If a reasonable reservoir correction of about 4000 y is applied, then the dates suggest that such a deposit is contemporaneous with SMO found further landward (i.e. cores 12 & 13, Fig. 2). Yet 12 kHz reflection data clearly indicate that the diamicton surface lies below the basin muds, in depths of 500 m, and correlates with seafloor exposures of diamicton above 500 m (Domack & Anderson, 1983).

Iceberg turbation is common at depths above 500 m (Barnes, 1987) and may have introduced modern (surface) foraminifera into the diamicton. This unit would then have to be considered as a palimpsest rather than a relict deposit. A date for glacial sedimentation at site 15 cannot be established without knowing the exact ratio of reworked (surface) to relict foraminifera in the sediment; a task difficult to complete given the similarity of surface and down core fauna (personal observation). Thus with the present data the timing of glacial recession is best constrained by post-glacial SMO found within shelf basins and coastward of the moraine system.

The base of SMO within cores 12 and 13 date to between 7350 ± 80 and 7850 ± 110. This places a lower limit for SMO deposition at these sites to around 2500 y when reservoir corrections are applied. Significantly thicker sections of SMO can be found within the Mertz–Ninnis Trough (~28 m; Domack & Anderson, 1983) and within the Mertz Trough (~10 m; Barnes, 1987). A well-established sedimentation rate of 0.31 cm/y can allow for estimates regarding initiation of deposition within the shelf basins (Domack *et al.*, 1989). Thus, post-glacial SMO has been accumulating within the Mertz–Ninnis Trough since 9000 y BP (maximum) and within the Mertz–Ninnis Trough since 3300 y BP (minimum). Therefore, contacts with underlying glacial and glaciomarine diamicton indicate a transition from sub-ice deposition to modern sea-ice dominated settings during mid-Holocene time.

Correlation with other studies

Most late Pleistocene reconstructions for the Antarctic ice sheet make the starting assumption of maximum grounding line extent at around 18000 y BP (e.g. Denton, Hughes & Karlen, 1986). This period of maximum glaciation has been characterized by uniform extent of grounded ice to the edge of the continental shelf (Stuiver *et al.*, 1981) and maximum cold for interior ice drainage systems (Lorius *et al.*, 1979). Retreat from this maximum position to present ice marginal configurations is theorized to have been the result of rising relative sea level and increased melting of fringing ice shelves at around 15000 y BP (Hollin, 1962; Thomas & Bentley, 1978; Johnson & Andrews, 1986). Ice core climatic studies from coastal Adélie Land (D10, Figs 1 & 2) suggest that maximum ice thickness was attained after the period of maximum cold, and thus during a period of regional warming (Raynaud *et al.*, 1979). They further suggested that thinning of about 400 m (250 km inland) occurred relatively rapidly during a 2000–3000 y period some time before 5000 y BP, since which time the glacial system has been in near steady-state equilibrium.

Though the data from the G/A margin are generally consistent with the proposed glacial reconstructions the glacial expansion, as evidenced by ice contact deposits, appears to have been less extensive, being limited to areas of the shelf within the moraine system (Figs 1 & 2). It is further suggested that glacial ice did not cover much of the Adélie shelf region during the latest expansion (Fig. 2). The presence of graded sands on this shelf and their synchroneity with diamictons, found to the east, would imply that an ice margin bordered the eastern edge of the shelf. In fact, this area is characterized by bathymetric 'knolls' (Domack, 1982). Such features on the Mertz Bank were later proved to be part of the submarine moraine system (Barnes, 1987). Glacial retreat from this position removed the source of clastic debris from the eastern and mid-southern Adélie Shelf (Fig. 2). A contemporary rise in sea level then allowed for the burial of the graded sands beneath a veneer of silicous mud and bioclastic carbonate. This event can best be determined as occurring some time after 10000 y BP, during a period of rising relative sea level which persisted (episodically) until 6000 y BP (Carter, Carter & Johnson, 1986).

In contrast, ice occupied the majority of the George V shelf with the exception of the Mertz Bank, which lies seaward of the moraine system (Fig. 1). Glacial recession from the latest maximum position was accompanied by the transition from grounded to floating (ice tongue) conditions and is believed to have been initiated by a change in relative sea level (Fig. 2). This appears to have occurred sometime after 9000 y BP. The fact that the subglacial bed within the Mertz–Ninnis Trough initially slopes toward the interior indicates that even slight rises in sea level may have been sufficient to have caused over 200 km of retreat within the basin. Further, sharp contacts of glacial diamicton and overlying SMO indicate a relatively rapid recessional history. Thomas & Bentley (1978) argued that retreat of marine ice sheets from foredeepened shelves may be quite rapid once the ice sheet decouples from the more shallow shelf-edge. Establishment of floating ice and grounding line retreat within the Mertz–Ninnis glacial complex would have increased the efficiency by which these glaciers drained interior and coastal ice. This would likely have led to a drawdown of Adélie Land ice which drained into the system from the west. The rapid (2000–3000 y) thinning reported by Raynaud *et al.* (1979) from D10 would seem to correlate with the pattern of retreat from the Mertz–Ninnis Trough. Stabilization of the grounding lines may have been due to the reversal in bedrock slope which is associated with the present grounding line positions (Fig. 2). Such changes in bedrock slope are known to control the dynamics of ice shelf (tongue) systems (Thomas, 1979).

Conclusions

Taken together, the data indicate that glacial recession from this portion of the East Antarctic shelf occurred some time during early–mid-Holocene time, some 5000 y later than previously believed. Hypotheses regarding the latest maximum extent of glaciation can be tested further by examination and dating of sediment cores from the upper slope (Fig. 2). Though these cores contain a record which is complicated by sediment gravity flow processes, they also appear to contain a sedimentary sequence which clearly is older than 18000 y BP. Further work on the extent of submarine moraines, particularly along the eastern Adélie shelf, would also provide more conclusive evidence.

Acknowledgements

This work was supported by National Science Foundation grant DPP-86-13565 and a grant from Hamilton College. Gratitude is extended to D. Cassidy at the Florida State University Antarctic Core Facility, for his help in providing samples, as well as M.R.A. Thomson, P. Barnes and an anonymous reviewer for their comments on a draft of this paper.

References

Anderson, J.B., Balshaw, K., Domack, E., Kurtz, D., Milam, R. & Wright, R. (1979). Geological survey of East Antarctic continental margin abroad USCGC *Glacier*. In *Antarctic Journal of the United States*, **14(5)**, 142–4.

Anderson, J.B., Brake, C., Domack, E., Myers, N. & Wright, R. (1983). Development of a polar glacial-marine sedimentation model from Antarctic Quaternary deposits and glaciological information. In *Glacial-Marine Sedimentation*, ed. B.F. Molnia, pp. 233–64. New York; Plenum Press.

Anderson, J.B., Kurtz, D.D., Domack, E.W. & Balshaw, K.M. (1980). Glacial and glacial marine sediments of the Antarctic continental shelf. *Journal of Geology*, **88(4)**, 399–414.

Andree, M., Beer, J., Oeschager, H., Mix, A., Broecker, W., Ragano, N., O'Hara, P., Bonani, G., Hofman, H.J., Mornzoni, E., Nessi, M., Suter, M. & Wolfli, W. (1985). Accelerator radiocarbon ages on foraminifera separated from deep-sea sediments. In *The Carbon Cycle and Atmospheric CO_2: Natural Variations Archean to Present*, Geophysical Monograph 32, ed. E.T. Sundquist & W.S. Broecker, pp. 143–53. Washington, DC; American Geophysical Union.

Barnes, P.W. (1987). Morphologic studies of the Wilkes Land continental shelf, Antarctica – glacial and iceberg effects. In *The Antarctic Continental Margin: Geology and Geophysics of Offshore Wilkes Land*, ed. S.L. Eittreim & M.A. Hampton, pp. 175–94. Houston; Circum-Pacific Council on Energy and Mineral Resources, Earth Science Series, 5A.

Carter, R.M., Carter, L. & Johnson, D.P. (1986). Submergent shorelines in the southwest Pacific: evidence for an episodic post-glacial transgression. *Sedimentology*, **33(5)**, 629–49.

Chase, T.E., Seekins, B.A., Young, J.D. & Eittreim, S.L. (1987). Marine topography of offshore Antarctica. In *The Antarctic Continental Margin: Geology and Geophysics of Offshore Wilkes Land*, ed. S.L. Eittreim & M.A. Hampton, pp. 147–50. Houston; Circum-Pacific Council on Energy and Mineral Resources, Earth Science Series, 5A.

Denton, G.H., Hughes, T.J. & Karlen, W. (1986). Global ice-sheet system interlocked by sea level. *Quaternary Research*, **26(1)**, 3–26.

Domack, E.W. (1980). Glacial marine geology of the George V & Adélie continental shelf, East Antarctica. M.A. thesis, Rice University, Houston, 142 pp. (unpublished).

Domack, E.W. (1982). Sedimentology of glacial and glacial marine deposits on the George V – Adélie continental shelf, East Antarctica. *Boreas*, **11(1)**, 79–97.

Domack, E.W. (1987). Preliminary stratigraphy for a portion of the Wilkes Land continental shelf, Antarctica: evidence from till provenance. In *The Antarctic Continental Margin: Geology and Geophysics of Offshore Wilkes Land*, ed. S.L. Eittreim & M.A. Hampton, pp. 195–203. Houston; Circum-Pacific Council on Energy and Mineral Resources, Earth Science Series, 5A.

Domack, E.W. (1988). Biogenic facies in the Antarctic glacimarine environment: basis for a polar glacimarine summary. *Palaeogeography, Palaeoclimatology, Palaeoecology*, **63**, 357–72.

Domack, E.W. & Anderson, J.B. (1983). Marine geology of the George V continental margin: combined results of Deep Freeze 79 and the 1911–1914 Australasian Expedition. In *Antarctic Earth Science*, ed. R.L. Oliver, P.R. James & J.B. Jago, pp. 402–6. Canberra; Australian Academy of Science and Cambridge; Cambridge University Press.

Domack, E.W., Jull, A.J.T., Anderson, J.B., Linick, T.W. & Williams, C.R. (1989). Application of tandem accelerator mass-spectrometer dating to Late Pleistocene–Holocene sediments of the East Antarctic continental shelf. *Quaternary Research*. **31**, 277–87.

Drewry, D.J. (1983). *Antarctica: Glaciological and Geophysical Folio*. Cambridge; Scott Polar Research Institute.

Dunbar, R.B., Anderson, J.B., Domack, E.W. & Jacobs, S.S. (1985). Oceanographic influences on sedimentation along the Antarctic continental shelf. In *Oceanography of the Antarctic Continental Shelf*, Antarctic Research Series 43, ed. S.S. Jacobs, pp. 291–312. Washington, DC; American Geophysical Union.

Elverhøi, A. (1981). Evidence for a late Wisconsin glaciation of the Weddell Sea. *Nature, London*, **293(5834)**, 641–2.

Hollin, J.T. (1962). On the glacial history of Antarctica. *Journal of Glaciology*, **4**, 173–95.

Johnson, R.G. & Andrews, J.T. (1986). Glacial terminations in the oxygen isotope record of deep sea cores: hypothesis of massive Antarctic ice shelf destruction. *Palaeogeography, Palaeoclimatology, Palaeoecology*, **53(2)**, 107–38.

Kellogg, D.E. & Kellogg, T.B. (1986). Diatom biostratigraphy of sediment cores from beneath the Ross Ice Shelf. *Micropaleontology*, **32**, 74–94.

Kellogg, T.B., Osterman, L.E. & Stuiver, M. (1979). Late Quaternary sedimentology and benthic foraminiferal paleoecology of the Ross Sea, Antarctica. *Journal of Foraminiferal Research*, **9**, 322–35.

Linick, T.W., Jull, A.J.T., Toolien, L.G. & Donahue, D.J. (1986). Operation of the National Science Foundation accelerator facility for radioisotope analysis and results from selected collaborative research projects. *Radiocarbon*, **28(2a)**, 522–33.

Lorius, C., Merlivat, L., Jouzel, J. & Pourchet, M. (1979). A 30,000-year isotope climatic record from Antarctic ice. *Nature, London*, **280(5724)**, 644–8.

Raynaud, D., Lorius, C., Budd, W.F. & Young, N.W. (1979). Ice flow along an IAGP flow line and interpretation of data from an ice core in Terre Adélie, Antarctica. *Journal of Glaciology*, **24(90)**, 103–15.

Stuiver, M. & Braziunas, T.F. (1985). Compilation of isotopic dates from Antarctica. *Radiocarbon*, **27(2a)**, 117–304.

Stuiver, M., Denton, G.H., Hughes, T.J. & Fastook, J.L. (1981). History of the marine ice sheet in West Antarctica during the last glaciation: a working hypothesis. In *The Last Great Ice Sheets*, ed. G.H. Denton & T.J. Hughes, pp. 319–436. New York; John Wiley and Sons.

Stuiver, M. & Ostlund, H.G. (1983). GEOSECS Indian Ocean and Mediterranean radiocarbon. *Radiocarbon*, **25(1)**, 1–28.

Thomas, R.H. (1979). The dynamics of marine ice sheets. *Journal of Glaciology*, **24**, 167–77.

Thomas, R.H. & Bentley, C.R. (1978). A model for Holocene retreat of the West Antarctic ice sheet. *Quaternary Research*, **10(2)**, 150–70.

Vanney, J.R. & Johnson, G.L. (1979). The sea floor morphology seaward of Terre Adélie (Antarctica). *Deutsche hydrographische Zeitschrift*, **32**, 39–88.

Benthic foraminiferal ecology of the Antarctic Peninsula Pacific coast

S.E. ISHMAN

Byrd Polar Research Center and Department of Geology and Mineralogy, The Ohio State University, Columbus, Ohio 43210, USA

Abstract

Foraminiferal analyses of 12 surface sediment samples taken from 12 sites along the west coast of the Antarctic Peninsula demonstrate regional variation in faunal distributions. The three regions investigated, South Shetland Islands, Danco Coast and Marguerite Bay have distinct calcareous benthic-agglutinated foraminiferal distributions. In addition, faunal correlations between these regions demonstrate greater affinities between those faunas from the more restricted South Shetland Islands and Danco Coast bays than those from the Marguerite Bay sites. The faunal distributions observed are related to bottom-water mass variations, which are influenced by: (1) a high-latitudinal thermal gradient influencing climatic patterns and thus localized bottom-water conditions; (2) variation in physiographic configurations of the three regions; and/or (3) production of $CaCO_3$ corrosive bottom waters from adjacent ice shelves in the Marguerite Bay region and little to no corrosive bottom-water production in the South Shetland Islands or Danco Coast regions; or possibly the physico-chemical conditions of the sediments. A greater understanding of modern benthic foraminiferal ecology in the high southern latitudes will lead to more comprehensive palaeoenvironmental interpretations for the Cenozoic.

Introduction

The Cenozoic stratigraphical and palaeontological record of Antarctica is becoming more complete. This was an era of global climatic fluctuation and tectonic change, and for this reason much emphasis has been focussed on the study of Cenozoic Antarctic palaeoenvironments. One approach to palaeoenvironmental studies utilizes fossil benthic marine invertebrates, from which one can make biogeographical and palaeoceanographical interpretations. In order to maximize confidence in this type of approach a comprehensive knowledge of modern benthic marine ecology is necessary. Although there are many classic studies on benthic marine ecology, these are mainly from shelf depths (i.e. < 200 m) of the low-latitude regions (i.e. tropics).

Benthic foraminifera are effective tools as palaeoenvironmental indicators (see Boltovskoy & Wright, 1976) due to their relative abundance (to macrobenthos) in small samples, and extended temporal occurrence. Many studies have discussed foraminiferal ecology of the low latitudes but very few ecological studies exist for high southern latitude foraminifera. Therefore, if palaeoenvironmental interpretations are to be obtained from benthic foraminiferal analyses of Antarctic Cenozoic core material, there must be a comprehensive understanding of modern high southern latitude foraminiferal ecology. The principal objective of this study is to relate the distribution of Recent foraminiferal faunas to modern environmental conditions in the Antarctic Peninsula region. It is hoped that these relationships can be applied down-core to recognize palaeoenvironmental changes that reflect Holocene and older climatic and oceanographic fluctuations and events, as well as penecontemporaneous and post-depositional carbonate solution and agglutinated form degradation.

Previous work

The occurrence of Pleistocene–Recent foraminifera from regions adjacent to the Antarctic coast has been studied by a number of workers. Many of these are illustrated systematic studies. Although such studies provide basic taxonomic information, much more data are needed before palaeoenvironmental interpretations can be made from Holocene and older fossil assemblages. Authors such as Uchio (1960), McKnight (1962), Pflum (1966), Kennett (1966, 1968), Echols (1971), Herb (1971), Anderson (1975a, b) and Quilty (1985) discussed Antarctic foraminifera from surface sediments and the physical and oceanographic factors influencing their distributions.

The relationship between calcareous benthic and agglutinated foraminiferal faunas has been ascribed to depth and the Carbonate Compensation Depth (CCD) in the Antarctic region. Kennett (1966, 1968) concluded that a shallow CCD (550 m) controls the foraminiferal distributions in surface sediments from the Ross Sea. There, calcareous benthic faunas dominate at depths shallower than 550 m, and agglutinated faunas prevail at depths greater than 430 m; the region of overlap contains a mixed fauna. Herb (1971) demonstrated that, within the Bransfield Strait, at depths greater than 200 m, agglutinated faunas are predominant. The distribution of foraminifera from the Scotia Sea to the Antarctic continental shelf (to 500 m depth; Echols, 1971) is associated with a CCD shallowing toward the Antarctic coast.

A second major factor influencing Antarctic foraminiferal distributions is their association with characteristic water masses. Uchio (1960) defined three assemblages (Assemblages I, < 850 m; II, 850–2000 m; and III, > 2000 m) that were coincident with three different water masses: the upper Antarctic Circumpolar, lower Antarctic Circumpolar and Antarctic Bottom waters. Pflum (1966) related the foraminiferal distributions recognized in the region from the Bellingshausen Sea to the Ross Sea to water depth, identifying Shelf Assemblages I and II and a Slope Assemblage. Anderson (1975b), investigating faunal associations in the Weddell Sea, related calcareous benthic and agglutinated foraminiferal distributions to bottom-water masses. He recognized an absence of calcareous benthic taxa where corrosive bottom waters, formed on the shelf and slope of the Weddell Sea, are channelled into the Weddell Sea basin. Quilty (1985) concluded that water mass variation in the Prydz Bay region controlled the lateral distribution of calcareous benthic and agglutinated faunas. He recognized a very abrupt faunal boundary in close proximity to the break between the Amery Depression and the Mac. Robertson Shelf (at approximately 500 m).

Other factors affecting foraminiferal distributions are bottom-current activity and organic carbon content of the surface sediments. McKnight (1962) associated agglutinated faunas with regions of high current activity and high organic carbon concentrations within the sediments. Calcareous benthic faunas, on the other hand, were associated with lower current activity, and showed no relationship to organic carbon content.

Although these studies are comprehensive in their content, several biases have been introduced. Sampling and processing techniques have generated a major bias with respect to the test size distribution of the foraminiferal faunas described. No standardization is apparent among most authors resulting in a bias toward data emphasizing tests in the > 125 μm (3 phi/fine sand) size range. Lack of standardization in size fractions leads to difficulties in comparative studies of circum-Antarctic sites. In addition, Schroder, Scott & Medioli (1987) demonstrated that, by eliminating the smaller benthic foraminifera through processing, a significant portion of the fauna is lost. In the Antarctic faunas this includes a significant portion (> 40%) of the assemblages, both agglutinated and calcareous. This reduces the potential of benthic foraminifera as palaeoenvironmental indicators.

A second factor concerns the differentiation between living and dead foraminifera in surface sediments. Comparisons between the living and dead assemblages provides a measure of reworking or taphonomic loss. This is demonstrated by Ward (1983), where she compares living versus dead assemblages from McMurdo Sound, Antarctica, and recognizes reworking due to downslope transport, and taphonomic loss (i.e. dissolution).

Sample localities

During the Deep Freeze 1986 cruise aboard the USCGC *Glacier*, 121 sediment cores (box cores, piston cores, and grab samples) were collected off the Pacific coast of the Antarctic Peninsula. For this study, 12 samples from box core sites in the South Shetland Islands (Maxwell Bay), Danco Coast region (Charlotte, Andvord and Flandres bays), and Marguerite Bay (Figs 1–4) were examined. Water depths at these sites range from 366 to 1370 m (DF86-80 and DF86-92, respectively). Variability of bottom-water temperature, salinity and dissolved oxygen content between these sites was insignificant. Recovered sediment types range from diatomaceous ooze to diatomaceous mud (the latter containing a significantly greater percentage of terrigenous quartz) (see Anderson *et al.*, 1983; Griffith & Anderson, 1989).

Methodology

Seafloor surface sediments were sampled as box core top samples (5 cm^3 in size). On-board ship these sediment samples were washed through a 63 μm sieve to concentrate the foraminifera, and then treated and stored in a Rose Bengal and alcohol solution to distinguish live from dead foraminifera at the time of collection (see Boltovskoy & Wright, 1976). In the laboratory at Ohio State University these samples were subsampled volumetrically and examined for foraminiferal content while wet. Foraminiferal counts of a minimum of 300 specimens were made systematically from material placed in a petri dish partitioned into a 1 cm-wide grid. Living (stained specimens) and dead forms were tabulated from the total assemblage counts.

Results

General faunal data

A total of 33 genera represented by 64 species were identified from the sediment samples. Both calcareous and agglutinated benthics are present, as well as uncommon occurrences of planktonics. Total species diversities per samples (species numbers) range from 22 to 35. Dominant calcareous benthic species (> 10%) include *Fursenkoina fusiformis* (Williamson), *F. earlandi* (Parr), *F. schreibersiana* (Czjzek), *Bolivina pseudopunctata* Höglund and *B. exilis* Brady. Significant (> 5%) agglutinated species include *Textularia antarctica* (Wiesner), *T. tenuissima* Earland, *T. wiesneri* Earland, *Portatrochammina eltaninae* Echols and *Reophax subdentaliniformis* Parr. Variance between the living assemblages and dead assemblages, as well as their trends, indicates very little reworking or taphonomic alteration of the faunas (Fig. 5). Exceptions are samples DF86-80 (Andvord Bay), and DF86–95 and

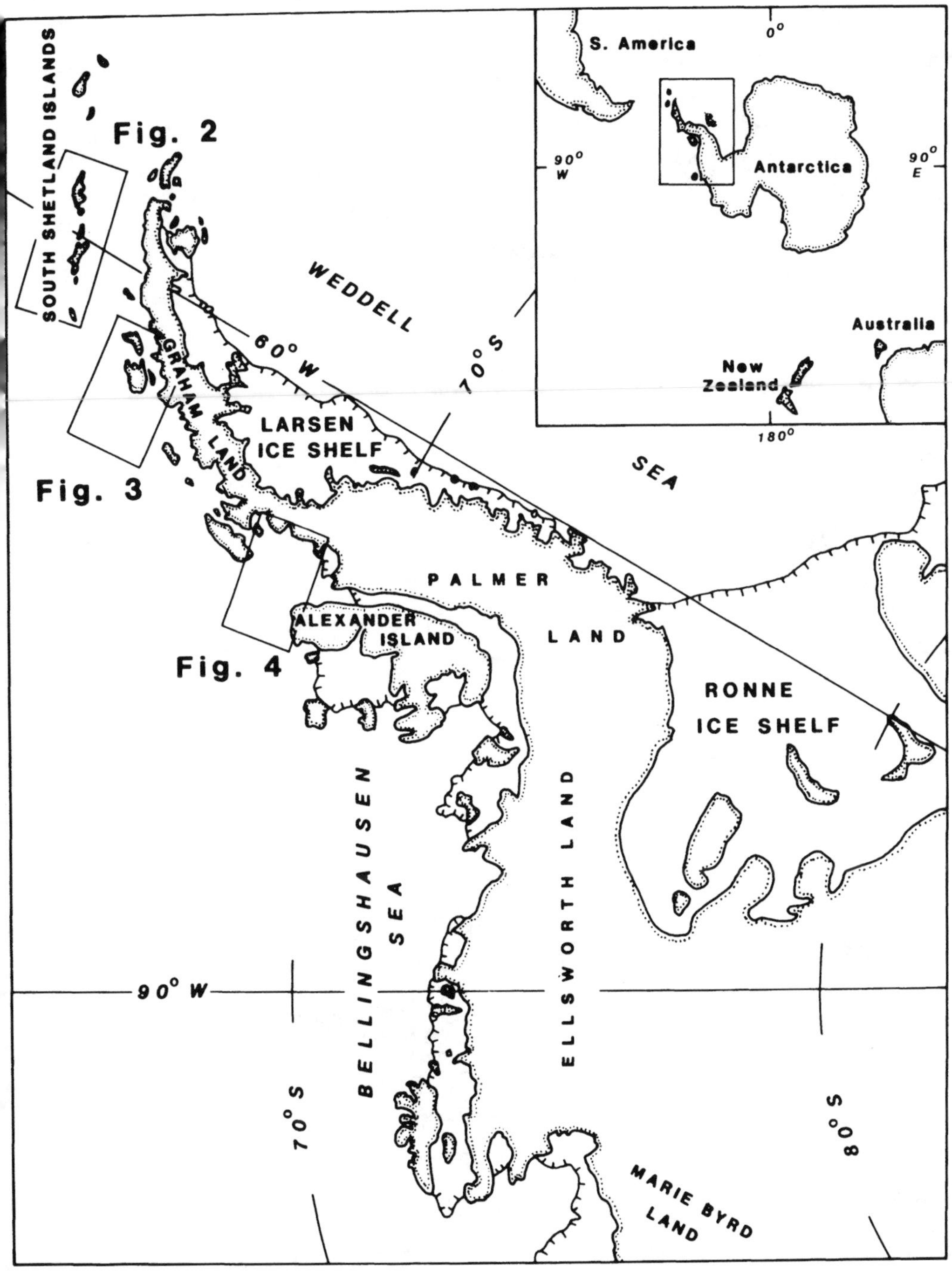

Fig. 1. Locality map showing the Antarctic Peninsula region outlining the three main study areas: Fig. 2, South Shetland Islands; Fig. 3, Danco Coast; Fig. 4, Marguerite Bay.

DF 86–97 (Marguerite Bay). In sample DF86–80 the living calcareous benthic taxa are predominant over the living agglutinated forms; however, in contrast are the dead assemblages where agglutinated taxa prevail over the calcareous benthic taxa. Samples DF86-95 and DF86-97 show a dominance of the living calcareous benthic taxa over the living agglutinated forms. Conversely, the dead agglutinated forms dominate over the calcareous benthic taxa. These relationships suggest taphonomic loss of some of the calcareous benthic forms at these sites for which morphologic observations of the tests indicate post-mortem dissolution (i.e. frosting, loss of final chamber). However, the other sites are taken to represent actual biocoenoses. Consequently, the remaining results are presented as total assemblage data.

Faunal spatial variability

Calcareous benthic: agglutinated foraminifera ratios from these sites display a distinct spatial variation. Samples from the South Shetland Islands and Danco Coast are dominated by a calcareous benthic fauna (Figs 2 & 3), while the Marguerite Bay samples are dominated by agglutinated forms (Fig. 4). Calcareous benthic: agglutinated ratios are 4.55–0.82 from the South Shetland Islands and Danco Coast sites, and

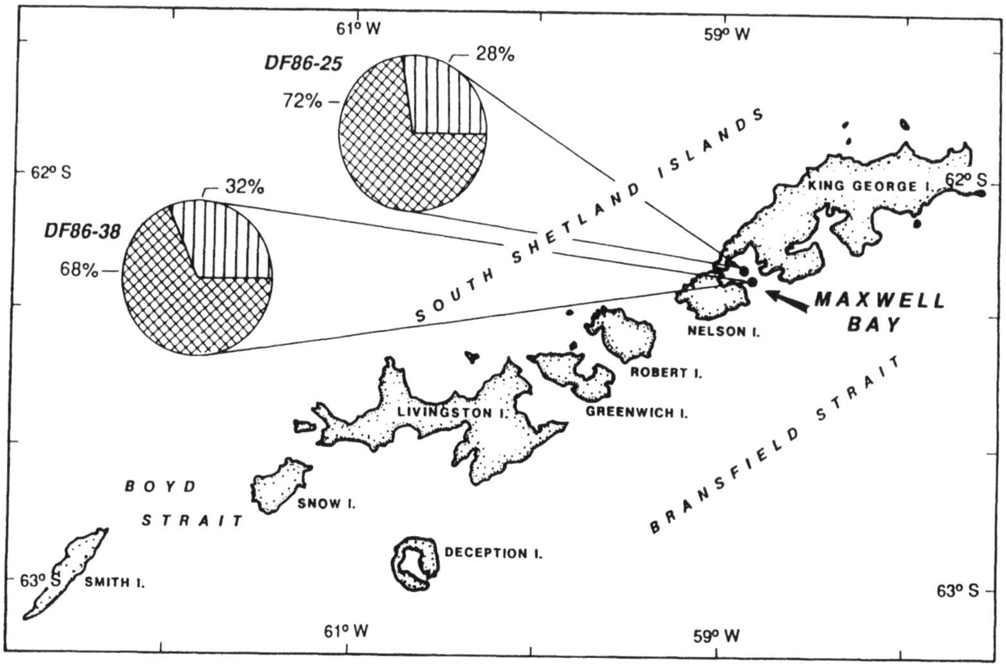

Fig. 2. Map of the South Shetland Islands, Antarctica showing sample localities from Maxwell Bay, King George Island. Pie diagrams illustrate the percentage composition of each asssemblage.

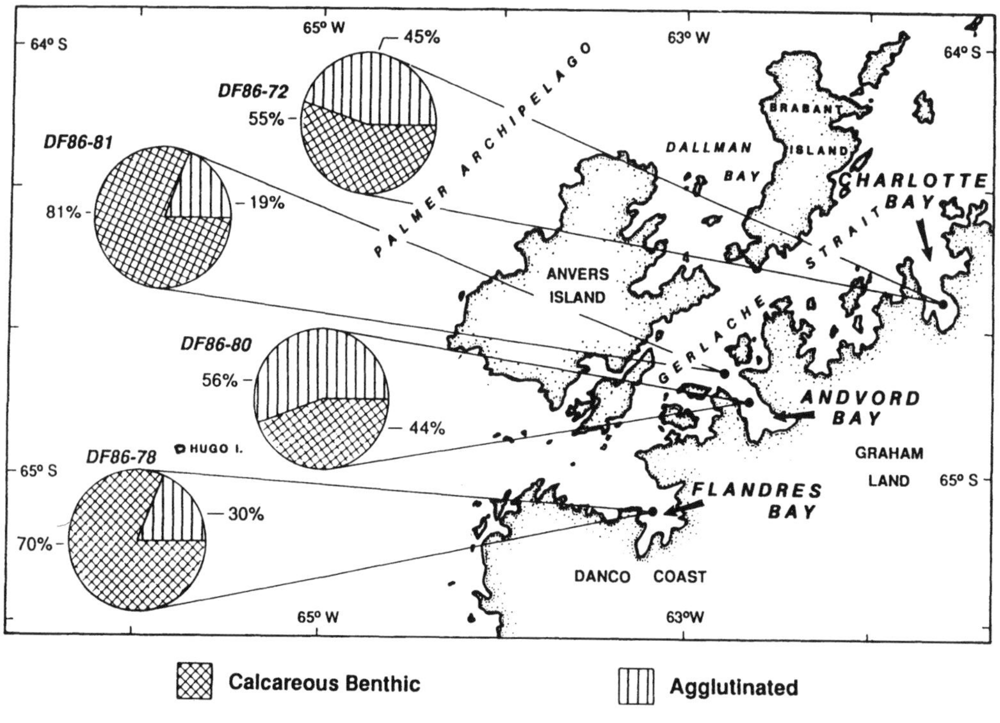

Fig. 3. Map of the Danco Coast region, Antarctica showing sample localities from Charlotte, Andvord and Flandres bays. Pie diagrams illustrate the percentage composition of each asssemblage.

0.56–0.10 from the Marguerite Bay sites. A greater faunal similarity exists between the South Shetland Islands and the Danco Coast assemblages than with the Marguerite Bay faunas. In addition, the inter-region assemblages show very strong affinities to each other.

Discussion

The modern environmental conditions in the Antarctic Pensinsula fjords and bays represent those of an interglacial. However, each region possesses distinct faunal and sedimento-

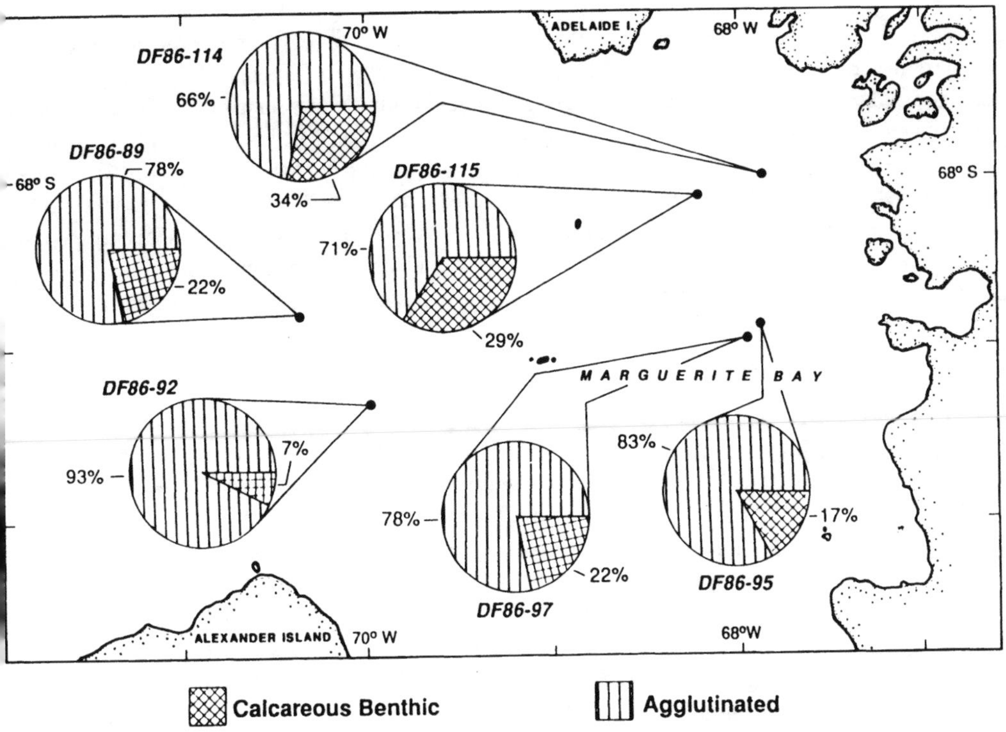

Fig. 4. Map of Marguerite Bay, Antarctica showing sample localities. Pie diagrams illustrate the percentage composition of each asssemblage.

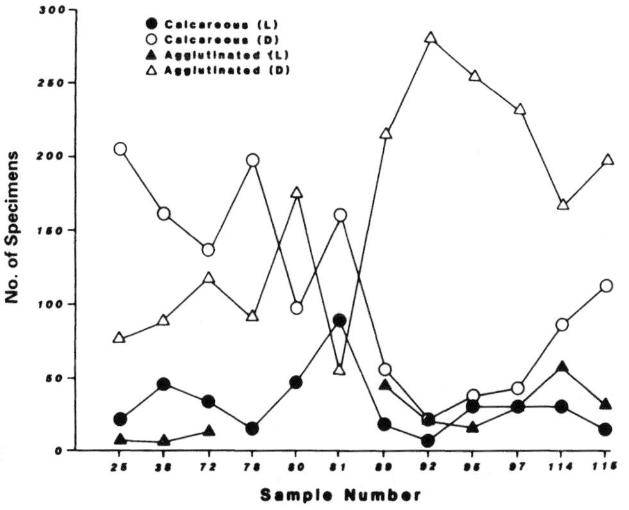

Fig. 5. A comparison of living (L) versus dead (D) foraminiferal assemblages.

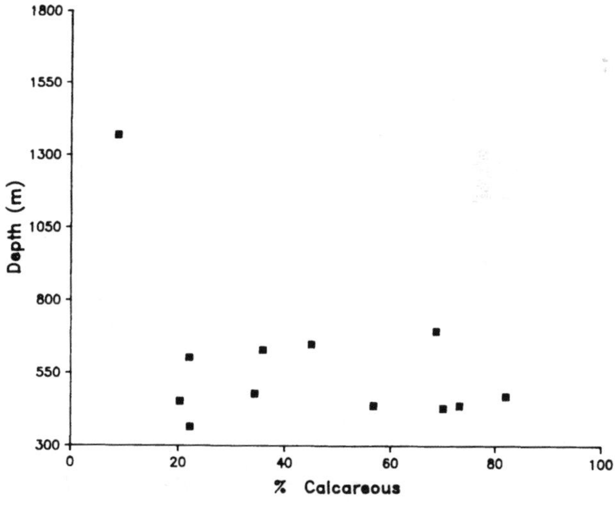

Fig. 6. Plot of total calcareous benthic foraminifera versus water depth. The graph clearly illustrates that there is no correlation between depth and decrease in calcium carbonate (as measured by incorporation into benthic foraminiferal tests).

logical characteristics. The preliminary results display several trends and relationships that lead to the following interpretations. A distinct pattern exists, whereby in the South Shetland Islands and Danco Coast samples the calcareous benthic assemblage predominates, whereas in the Marguerite Bay samples agglutinated forms are dominant. As the sites range in depths from 366 to 1370 m, it is possible that a shallow CCD may be influencing the foraminiferal distributions. However, this suggestion proves invalid when depth is plotted against percentage calcareous benthics (Fig. 6). If a shallow CCD or lysocline were present a linear decrease or well defined break in the percent calcareous benthics with depth would occur. In fact, the data presented demonstrate that there is neither a decrease nor a well defined break in the percent total (L + D) calcareous benthics with depth. With the exception of three faunas, comparisons made between the living and dead assemblages demonstrate little to no taphonomic loss.

The faunal associations demonstrate a high degree of spatial similarity between sites from the same geographical region. In addition, a higher degree of similarity is exhibited between those sites from the smaller, more restricted bays (Maxwell,

Charlotte, Andvord and Flandres bays) than with the faunas from the more open marine Maraguerite Bay (Figs 2–4). Several explanations are offered, the first hinging on the impact of geographical isolation and the influence of local effects. The three sampled areas cover a wide range of latitude (62–69° S; Fig. 1), which accounts for a considerable degree of climatic variability. Climatic responses of bottom- and surface-water conditions (bottom-water rejuvenation and primary productivity, respectively) are in turn important in dictating benthic faunal distributions. Also associated with this factor is the submarine physiographic nature of the three geographical regions. Whereas Maxwell Bay lacks a sill at its mouth, the Danco Coast bays all possess restrictive submarine barriers as well as isolated interbay basins. Marguerite Bay again differs in that it covers a much greater areal extent and contains a very rugged bottom topography (with several deep troughs; Kennedy & Anderson, 1989) and depths exceeding 1300 m. These conditions are significant in influencing bottom-current activity and sedimentation within these regions, and thus benthic marine faunas. A second explanation considers the effects of adjacent terrestrial glacial conditions on these regions. Of the areas discussed, Marguerite Bay is subject to the greatest terrestrial influence of this type. Two major outlet glacial systems, the George VI and Wordie ice shelves, lie in contact with Marguerite Bay. Anderson (1975b) demonstrated that formation and distribution of bottom water is an influential factor in the distribution of calcareous benthic and agglutinated foraminiferal faunas in the Weddell Sea basin. The association of agglutinated foraminiferal faunas with the Marguerite Bay sites suggests that this region also generates bottom water depleted in $CaCO_3$, resulting in their much reduced calcareous benthic populations. In contrast, bottom waters in the restricted bays are not similar to the corrosive bottom waters formed on the continental shelves adjacent to the Antarctic continent, as suggested by the fact that they support a predominantly calcareous benthic population.

Finally, McKnight (1962) related foraminiferal distribution with sediment type and organic carbon. Sedimentation rates in the Bransfield Strait region average 1.8 mm/y (DeMaster *et al.*, 1987) with the highest rates occurring in the bays, up to 3.4 mm/y in Maxwell Bay. Primary productivity (chlorophyll *a* concentrations) reached maximum values of 3.0–12.5 mg/m^3 in the Bransfield Strait and total mass flux in the same region measured 2.9–4.3 g/m^2/d, slightly greater than the SiO_2 flux (at depths < 539 m; Gersonde & Wefer, 1987). This surface productivity is a major contributor to the organic carbon content of the sediment, and potential food source for the benthic community. Organic carbon values for surface sediments from the Antarctic Peninsula bays range from 0.41% (Maxwell Bay) to 1.37% (Andvord Bay). Foraminiferal assemblage distributions are related to the organic carbon content of the sediments. The calcareous benthic faunas are associated with sediments having high organic carbon content. In addition, a direct association was observed between test morphotype, sediment texture and organic carbon content of the surface sediments (Ishman, 1987).

These preliminary observations have clearly demonstrated that benthic foraminiferal distributions in the Antarctic region cannot simply be explained on the premises of water depth and CCD alone. Greater sample coverage, determination of infaunal and epifaunal foraminiferal distribution, as well as the detailed study of the physico-chemical conditions of the bottom waters and sediments will lead to precise faunal-habitat associations. This in turn will provide us with modern analogues on which to base Cenozoic palaeoenvironmental interpretations of the Antarctic region.

Acknowledgements

This work was supported in part by National Science Foundation grants DPP8516908 and DPP8517625 (to Drs John B. Anderson and Peter N. Webb, respectively), and a Sigma Xi Grant-in-Aid of Research. I would also like to acknowledge the officers and crew of the USCGC *Glacier* for their support. In addition, my thanks to Drs J.B. Anderson and D. DeMaster for allowing me to participate in the Deep Freeze '86 cruise. Finally I would like to thank Drs M.R.A. Thomson, J.A. Crame and J.W. Thomson for their review of this manuscript.

References

Anderson, J.B. (1975a). Ecology and distribution of foraminifera in the Weddell Sea of Antarctica. *Micropaleontology*, **21(1)**, 69–96.

Anderson, J.B. (1975b). Factors controlling $CaCO_3$ dissolution in the Weddell Sea from foraminiferal distribution patterns. *Marine Geology*, **19**, 315–32.

Anderson, J.B., Brake, C., Domack, E.W., Myers, N. & Singer, J. (1983). Sedimentary dynamics of the Antarctic continental shelf. In *Antarctic Earth Science*, ed. R.L. Oliver, P.R. James & J.B. Jago, pp. 387–9. Canberra; Australian Academy of Science and Cambridge; Cambridge University Press.

Boltovskoy, E. & Wright, R. (1976). *Recent Foraminifera*. The Hague; Dr. W. Junk b.v. Publishers. 515 pp.

DeMaster, D.J., Nelson, T.M., Nittrouer, C.A. & Harden, S.L. (1987). Biogenic silica and organic carbon accumulation in modern Bransfield Strait sediments. *Antarctic Journal of the United States*, **22(5)**, 108–10.

Echols, R.J. (1971). Distribution of foraminifera in sediments of the Scotia Sea area, Antarctic waters. In *Antarctic Oceanology, I*, Antarctic Research Series, 15, ed. J.L. Reid, pp. 93–168. Washington, DC; American Geophysical Union.

Gersonde, R. & Wefer, G. (1987). Sedimentation of biogenic siliceous particles in Antarctic waters from the Atlantic Sector. *Marine Micropaleontology*, **11(4)**, 311–32.

Griffith, T.W. & Anderson, J.B. (1989). Climatic control of sedimentation in bays and fiords of the northern Antarctic Peninsula. *Marine Geology*, **85**, 181–204.

Herb, R. (1971). Distribution of recent benthonic foraminifera in the Drake Passage. In *Biology of the Antarctic Seas IV*, Antarctic Research Series, 17, ed. G.A. Llano & I.E. Wallen, pp. 251–300. Washington, DC; American Geophysical Union.

Ishman, S.E. (1987). Benthic foraminiferal morphotypes: a strategy for epifaunal and infaunal stabilization. *Geological Society of America, Abstracts with Programs*, **19(7)**, 713.

Kennedy, D.S. & Anderson, J.B. (1989). Glacial-marine sedimentation and quaternary glacial history of Marguerite Bay, Antarctic Pensinsula. *Quaternary Research*, **31(2)**, 255–76.

Kennett, J.P. (1966). Foraminiferal evidence of a shallow calcium carbonate solution boundary, Ross Sea, Antarctica. *Science*, **153**, 191–3.

Kennett, J.P. (1968). The fauna of the Ross Sea: ecology and distribution of foraminifera. *New Zealand Department of Scientific and Industrial Research Bulletin*, **186**, 1–47.

McKnight, W.M., Jr. (1962). The distribution of foraminifera off parts of the Antarctic coast. *Bulletin of American Paleontology*, **44(201)**, 65–158.

Pflum, C.E. (1966). The distribution of foraminifera in the eastern Ross Sea, Amundsen Sea, and the Bellingshausen Sea, Antarctica. *Bulletin of American Paleontology*, **50(226)**, 151–209.

Quilty, P.G. (1985). Distribution of foraminiferids in sediments of Prydz Bay, Antarctica. *Southern Australian Department of Mines and Energy Special Publication*, **5**, 329–40.

Schroder, C.J., Scott, D.B. & Medioli, F.S. (1987). Can smaller benthic foraminifera be ignored in paleoenvironmental analyses? *Journal of Foraminiferal Research*, **17(2)**, 101–5.

Uchio, T. (1960). Benthonic foraminifera of the Antarctic Ocean. *Biological Results of the Japanese Antarctic Research Expedition*, **12**, 1–20.

Ward, B.L. (1983). Benthic foraminifera of McMurdo Sound. In *Antarctic Earth Science*, ed. R.L. Oliver, P.R. James & J.B. Jago, pp. 407–9. Canberra; Australian Academy of Science and Cambridge; Cambridge University Press.

Index

Notes

1. Authors' names are in bold type
2. References to maps, tables, and diagrams are in bold type.
   In some cases there are also textual references on these pages.
3. Extended page references may include figure references.
4. Locality maps are indicated at the beginning of a place name entry.